THIRD EDITION

HEATING AND COOLING OF BUILDINGS

Principles and Practice of Energy Efficient Design

MECHANICAL and AEROSPACE ENGINEERING

Frank Kreith
Series Editor

RECENTLY PUBLISHED TITLES

Air Distribution in Buildings, *Essam E. Khalil*

Alternative Fuels for Transportation, *Edited by Arumugam S. Ramadhas*

Computer Techniques in Vibration, *Edited by Clarence W. de Silva*

Design and Control of Automotive Propulsion Systems, *Zongxuan Sun and Guoming (George) Zhu*

Distributed Generation: The Power Paradigm for the New Millennium, *Edited by Anne-Marie Borbely and Jan F. Kreider*

Elastic Waves in Composite Media and Structures: With Applications to Ultrasonic Nondestructive Evaluation, *Subhendu K. Datta and Arvind H. Shah*

Elastoplasticity Theory, *Vlado A. Lubarda*

Energy Audit of Building Systems: An Engineering Approach, *Moncef Krarti*

Energy Conversion, Second Edition, *Edited by D. Yogi Goswami and Frank Kreith*

Energy Efficiency and Renewable Energy Handbook, Second Edition, *Edited by D. Yogi Goswami and Frank Kreith*

Energy Efficiency in the Urban Environment, *Heba Allah Essam E. Khalil and Essam E. Khalil*

Energy Management and Conservation Handbook, Second Edition, *Edited by Frank Kreith and D. Yogi Goswami*

Essentials of Mechanical Stress Analysis, *Amir Javidinejad*

The Finite Element Method Using MATLAB®, Second Edition, *Young W. Kwon and Hyochoong Bang*

Fluid Power Circuits and Controls: Fundamentals and Applications, *John S. Cundiff*

Fuel Cells: Principles, Design, and Analysis, *Shripad Revankar and Pradip Majumdar*

Fundamentals of Environmental Discharge Modeling, *Lorin R. Davis*

Handbook of Hydrogen Energy, *Edited by S.A. Sherif, D. Yogi Goswami, Elias K. Stefanakos, and Aldo Steinfeld*

Heat Transfer in Single and Multiphase Systems, *Greg F. Naterer*

Heating and Cooling of Buildings: Principles and Practice of Energy Efficient Design Third Edition, *T. Agami Reddy,Jan F. Kreider, Peter S. Curtiss, and Ari Rabl*

Intelligent Transportation Systems: Smart and Green Infrastructure Design, Second Edition, *Sumit Ghosh and Tony S. Lee*

Introduction to Biofuels, *David M. Mousdale*

Introduction to Precision Machine Design and Error Assessment, *Edited by Samir Mekid*

Introductory Finite Element Method, *Chandrakant S. Desai and Tribikram Kundu*

Large Energy Storage Systems Handbook, *Edited by Frank S. Barnes and Jonah G. Levine*

Machine Elements: Life and Design, *Boris M. Klebanov, David M. Barlam, and Frederic E. Nystrom*

Mathematical and Physical Modeling of Materials Processing Operations, *Olusegun Johnson Ilegbusi, Manabu Iguchi, and Walter E. Wahnsiedler*

Mechanics of Composite Materials, *Autar K. Kaw*

Mechanics of Fatigue, *Vladimir V. Bolotin*

Mechanism Design: Enumeration of Kinematic Structures According to Function, *Lung-Wen Tsai*

Mechatronic Systems: Devices, Design, Control, Operation and Monitoring, *Edited by Clarence W. de Silva*

The MEMS Handbook, Second Edition (3 volumes), *Edited by Mohamed Gad-el-Hak*

MEMS: Introduction and Fundamentals

MEMS: Applications

MEMS: Design and Fabrication

Multiphase Flow Handbook, Second Edition, *Edited by Efstathios E. Michaelides, Clayton T. Crowe, and John D. Schwarzkopf*

Nanotechnology: Understanding Small Systems, Third Edition, *Ben Rogers, Jesse Adams, and Sumita Pennathur*

Nuclear Engineering Handbook, *Edited by Kenneth D. Kok*

Optomechatronics: Fusion of Optical and Mechatronic Engineering, *Hyungsuck Cho*

Practical Inverse Analysis in Engineering, *David M. Trujillo and Henry R. Busby*

Pressure Vessels: Design and Practice, *Somnath Chattopadhyay*

Principles of Solid Mechanics, *Rowland Richards, Jr.*

Principles of Sustainable Energy Systems, Second Edition, *Edited by Frank Kreith with Susan Krumdieck, Co-Editor*

Thermodynamics for Engineers, *Kau-Fui Vincent Wong*

Vibration and Shock Handbook, *Edited by Clarence W. de Silva*

Vibration Damping, Control, and Design, *Edited by Clarence W. de Silva*

Viscoelastic Solids, *Roderic S. Lakes*

Weatherization and Energy Efficiency Improvement for Existing Homes: An Engineering Approach, *Moncef Krarti*

THIRD EDITION

HEATING AND COOLING OF BUILDINGS

Principles and Practice of Energy Efficient Design

T. Agami Reddy • Jan F. Kreider
Peter S. Curtiss • Ari Rabl

CRC Press
Taylor & Francis Group
Boca Raton London New York

CRC Press is an imprint of the
Taylor & Francis Group, an **informa** business

CRC Press
Taylor & Francis Group
6000 Broken Sound Parkway NW, Suite 300
Boca Raton, FL 33487-2742

Printed on acid-free paper
Version Date: 20160520

International Standard Book Number-13: 978-1-4398-9989-2 (Hardback)

Visit the Taylor & Francis Web site at
http://www.taylorandfrancis.com

and the CRC Press Web site at
http://www.crcpress.com

Dedicated to Five Generations of Women (TAR):

Krishnamma, Mahalakshmi, Shobha, Agaja, and Maya

Man should not become the "ritual priest of an established knowledge or the slave of the limited intellect" but should use his knowledge and "adaptive skills to work out its truth for the practice of life – otherwise he lives only in the idea."

Sri Aurobindo
The Synthesis of Yoga, circa 1917

Contents

Preface to the Third Edition

To paraphrase from the preface of the second edition (2002): *The art and the science of building systems design evolve continuously as a worldwide cadre of designers, practitioners, and researchers all endeavor to improve the performance of buildings and the comfort and productivity of their occupants.* This holds true even now—perhaps more so due to the recent societal awareness of the impact of building energy use on climate change and sustainability in general. Articles on green, high-performance, and net-zero energy/carbon buildings abound in the popular and semitechnical press. Granted that there is a need for general and multidisciplinary education of building energy science and systems to a broader audience (and there are numerous textbooks that cater to this audience), there is also a need for books that provide the "burden of engineering knowledge."* The third edition, while retaining much of the original material, is thoroughly revised and greatly expanded and completely rearranged. The coverage on transient load analysis and peak load calculation methods and on building mechanical systems has been especially enlarged; current and evolving concepts and systems have been introduced as well.

There are, broadly speaking, three types of technical books: (a) textbooks that emphasize fundamental principles and concepts; (b) specialized books that focus on a narrow topic in their treatment and coverage and can be either very theoretical or very applied; and (c) handbooks for practitioners that provide detailed engineering and technical information through tables, figures, and electronic media. This book would straddle (a) and (c) categories. Building science and basic operating principles of heating, ventilating, and air conditioning (HVAC) equipment are time invariant, but their "fundamental purity" is of limited practical use; on the other hand, the engineering techniques that appear in standards, guidelines, and codes are simplified, somewhat empirical and ad hoc, and are being constantly revised by the professional community (a source of frustration to both student and practitioner).

* If knowledge accumulates as technology progresses, then successive generations of innovators may face an increasing educational burden. Innovators can compensate in their education by seeking narrower expertise, but narrowing expertise will reduce their individual capacities and have negative implications for the organization of innovative activity (Jones 2009).

This textbook is meant to be a bridge providing the scientific principles along with how these concepts are modified into working equations and procedures suitable for actual design practice and analysis. The *art* of building design is something that takes years for a professional to develop, but this needs to be done starting from a scientific base.

This is not a superficial textbook and would need two 15-week semesters to cover it in its entirety. Many engineering schools do devote two semesters, and so this is not an issue. I have been teaching building science and HVAC systems for over 15 years to mostly mechanical and architectural engineers but also to graduate-level architects during which I used earlier editions of this book for about a dozen years. The sequence of the chapters has evolved after many iterations over years of teaching. The recommendation is to cover the first 10 chapters (along with sections of Chapters 23 and 24) in a first course on building loads, supplemented by introducing the students to a building energy simulation software program. The remaining chapters can be covered in one semester with time left over for group projects.

This book appears both in print and as an e-book. Each chapter is written as stand-alone with its abstract, nomenclature, references, and problem set. The second edition of the book had a CD-ROM that came with the book (its capabilities are described in the "Preface" of the second edition). The HCB (Heating and Cooling of Buildings) software has not been revised and is still available for free download at https://www.crcpress.com/Heating-and-Cooling-of-Buildings-Principles-and-Practice-of-Energy-Efficient/Reddy-Kreider-Curtiss-Rabl/9781439899892. The textbook makes references to this software in several places. It is urged that readers make frequent use of this software which enhances the educational value of the book. A solution manual for the end-of-chapter problems is also available along with other instructor resources (PowerPoint slides, pdf version of figures and tables, summary of important concepts of each chapter, etc.). This material is also available on request by qualifying instructors from https://www.crcpress.com/Heating-and-Cooling-of-Buildings-Principles-and-Practice-of-Energy-Efficient/Reddy-Kreider-Curtiss-Rabl/9781439899892.

The problems in this book are arranged by topic. The approximate degree of difficulty is indicated by a parenthetic italic number from 1 to 4 at the end of the problem. Problems are stated most often in IP units; when similar problems are presented in SI units, it is done with approximately equivalent values in parentheses. The IP and SI versions of a problem are not exactly equivalent numerically. Solutions should be organized in the same order as the examples in the text: given, figure or sketch, assumptions, find, lookup values, solution. For some problems, the Heating and Cooling of Buildings software freely available online would be helpful. In some cases, it is advisable to set up the solution as a spreadsheet, so the design variations are easy to evaluate.

T. Agami Reddy
Tempe, Arizona

Reference

Jones, B.F. (2009). The burden of knowledge and the death of the renaissance man: Is innovation getting harder? *Rev. Econ. Stud.*, 76(1), 283–317.

Preface to the Revised Second Edition

The art and the science of building systems design evolve continuously as a worldwide cadre of designers, practitioners, and researchers all endeavor to improve the performance of buildings and the comfort and productivity of their occupants. This book, which presents the technical basis of building mechanical system and lighting systems design, also has evolved due to the efforts of many in the building engineering profession. The authors decided to completely revise the first edition of this book to modernize topic coverage and to include much more reference information in electronic form.

The contents of this book have been expanded to include

- A chapter on economic analysis and optimization
- Heating and cooling load procedures and databases
- More than 200 new homework problems
- New and simplified procedures for ground-coupled heat transfer calculations
- Many updates of fine points in all chapters
- Completely new heating and cooling of buildings software
- All book graphics and appendices supplied on CD-ROM to assist with lecture preparation

One of the most noticeable differences in the second edition of this book is that many of the appendices from the first edition have been moved to the accompanying CD-ROM. This CD-ROM also contains an updated version of the HCB program and links to a website, www.HCBCentral.com. By packaging multimedia along with a textbook, we hope to provide resources to engineering students in a fashion that will eventually be standard practice. There are more than 1000 tables in the electronic appendixes that can be searched by major categories, a table list, or an index of topics. The CD-ROM also directs students to the HCB central website where several hundred links are maintained to help students find manufacturers' and government data, browse in newsgroups, and find any corrections and updates to the text and data tables. The tables are in HTML format, which means that data can be copied and pasted into applications directly from the electronic appendixes. In some cases, the data in the tables and the HCB software may be slightly different from other sources due to the process by which the data have been generated or converted. These differences are noted where appropriate.

We acknowledge the ongoing support of our families while this project was undertaken. Many students and professors shared their ideas with us so that the second edition would be improved in many ways. We particularly thank Professor J. Taylor Beard for his careful reporting of the many suggested changes based on his use of the first edition for many years. The American Society of Heating, Refrigerating, and Air-Conditioning Engineers generously shared the data from their handbook series. Mike Johnston helped to update the solutions manual originally prepared by Wendy Hawthorne. Finally, we are very pleased to have our colleague, Dr. Peter S. Curtiss, PE, join the authorship of this edition.

Jan F. Kreider
Peter S. Curtiss
Ari Rabl
Boulder, Colorado

Acknowledgments to the Third Edition

I express my appreciation to my coauthors, Jan Kreider, Peter Curtiss, and Ari Rabl, and to my editor, Jonathan Plant, for the necessary confidence to entrust me with revising and updating this textbook. Numerous talented and dedicated colleagues contributed in various ways over the several years of my professional career, some by direct association, while others indirectly through their textbooks and papers, both of which were immensely edifying and stimulating to me personally. The list of acknowledgments of such meritorious individuals would be very long indeed, and so I have limited myself to those who either have provided direct valuable suggestions/clarifications or have generously given their time in reviewing certain chapters or portions of this book.

I acknowledge first and foremost Itzhak Maor and Joe Huang for reviewing several parts of this book, suggesting changes, and even providing me with written material to improve preliminary drafts. Other reviewers include Pradeep Bansal, Ian Beausoleil-Morrison, Piotr Domansky, Nasim Karizi, and Max Sherman. I thank the following colleagues for advice, suggestions, and clarifications to my doubts: Chip Barnaby, Fred Bauman, Dan Hahne, Stephen Hancock, Mark Hydeman, David John, Andrew Persily, Chandra Shekhar, Peter Simmonds, Jerry Sipes, and Ian Walker. Ralph Budwig and his students pointed out errors in the previous edition, which is appreciated. I also recognize useful interactions with several graduate students over the years, but I make special mention of Salim Moslehi, Ranojoy Dutta, Mrigesh Roy, Srijan Didwania, and Saurabh Jalori. Finally, I would like to acknowledge one of the coauthor, Peter Curtiss, who developed the HCB software and other electronic resources that came on the CD along with the second edition of this textbook. He has graciously allowed the resources to be available online as free download or free access for everyone to use as they see fit.

Much of the material used in this book is based on ASHRAE handbooks and publications, and I take this occasion to show my appreciation to the excellent work being done by this organization and to its numerous volunteers in painstakingly improving the science and engineering of HVAC&R practice over the last 100 plus years. I also thank Arizona State University for providing me with the institutional environment to undertake this multiyear endeavor.

Finally, I acknowledge the encouragement, patience, and understanding of my wife, Shobha, as well as the tacit support of our children, Agaja and Satyajit, during the several years spent in revising this book. Had it not been for them, this book would have been completed sooner, but then, there is more to life than writing textbooks!

T. Agami Reddy
Tempe, Arizona

Acknowledgments to the Revised Second Edition

A book of this magnitude and scope represents the work of many in addition to the authors. First, we thank our students who gave us the luxury of testing the book in class. Their comments and ideas were essential in finalizing the organization and coverage of this book.

A number of reviewers took considerable time to examine the book closely, finding errors that the authors missed and suggesting improvements in the presentation. The review process began with helpful comments on the original outline by Louis Burmeister, University of Kansas; Thomas Hellman, New Mexico State University; Doug Hittle, Colorado State University; Ronald Howell, University of South Florida; Dennis O'Neal, Texas A&M University; Maurice Wildin, University of New Mexico; and Byron Winn, Colorado State University, and continued with reviews of the draft by John Lloyd, Michigan State University, and Trilochan Singh, Wayne State University. The exceptionally detailed and constructive comments by Professor Wildin on the entire book are especially appreciated. Likewise, Wendy Hawthorne made a careful reading of the text and checked most of the examples, and Bill Shurcliff contributed a thorough review of several chapters. Various colleagues offered comments on parts of this book: Mike Brandemuehl (who also contributed novel end-of-chapter problems in the first half of the book), Manuel Collares-Pereira, Jeff Haberl, J. Y. Kao, John Littler, John Mitchell, Leslie K. Nordford, Mike Riley, Gideon Shavit, and Mike Scofield. Of course, any remaining errors are the sole responsibility of the authors.

Peter Curtiss wrote the software that accompanies this book. The thousands of lines of code that he wrote in producing the final product set a new standard for instructional software used in building systems education. Wendy Hawthorne prepared the solutions manual for the end-of-chapter problems.

Our book called on the work of many others. We are grateful to several people who generously provided material or data: R. Boehm, S. Burek, J. Harris, M. A. Piette, and R. Sullivan. We have tried to give credit to original sources wherever possible and have included complete reference lists in each chapter. Permissions were given liberally, and we thank the numerous original sources for their generosity.

Jan F. Kreider thanks Dottie for her patience during the apparently endless process of writing this book. Her good nature and support are very much appreciated. Kennon Stewart generously provided practical advice from his decades of HVAC design experience. The University of Colorado's College of Engineering provided the facilities used for preparing the book and for some of the experimental results contained within. Denis Clodic, director of the Centre d'Energétique of the Ecole des Mines de Paris, and Flavio Conti at the Joint Research Center of the European Community provided a gracious environment during his sabbatical for the final tasks needed to finish the book.

Ari Rabl thanks several institutions and individuals. The Ecole des Mines offered a propitious setting for the writing, thanks to the encouragement of Jerome Adnot. The Joint Center for Energy Management of the University of Colorado provided support for summer visits. Robert H. Socolow invited him to join the Center for Energy and Environmental Studies of Princeton University and to focus on energy use of buildings. The work with colleagues at Princeton was a uniquely stimulating and productive experience to which this book is a tribute. And, of course, he heartily thanks Brigitte for her moral support.

Jan F. Kreider
Ari Rabl

Acknowledgements to the Revised Second Edition

Authors

T. Agami Reddy is SRP professor of energy and environment at Arizona State University, Phoenix, Arizona, with joint faculty appointments with the Design School and the School of Sustainable Engineering and the Built Environment. During his 30-year career, Professor Reddy has also held faculty and research positions at Drexel University, Philadelphia, Pennsylvania, Texas A&M University, College Station, Texas, and Princeton University, Princeton, New Jersey. He teaches and does research in the areas of sustainable energy systems (green buildings, HVAC&R, solar energy) and building energy data analytics. He is the author of two textbooks and has close to 200 refereed journal and conference papers and several book chapters and technical research reports. He is a licensed mechanical engineer and fellow of both ASME and ASHRAE. He received the ASHRAE Distinguished Service Award in 2008 and was the recipient of the 2014 Yellott Award from the ASME Solar Energy Division.

Jan F. Kreider is a professor emeritus of engineering and founding director of the Joint Center for Energy Management at the University of Colorado, Boulder, Colorado. He received his BSME (magna cum laude) from Case Institute of Technology and his postgraduate degrees from the University of Colorado, Boulder, Colorado. Dr. Kreider is the author of a dozen engineering textbooks and more than 200 technical articles and reports; he has directed more than $15 million in energy-related research during the past decades. He is a fellow of the ASME, a member of ASHRAE and a number of other technical societies, and a winner of ASHRAE's E.K. Campbell Award for excellence in building systems education. He is also the president of a consulting firm specializing in energy systems design and analysis in the building and automotive sectors.

Peter Curtiss has been a practicing engineer for 25 years and currently lives in Boulder, Colorado. He received his BSCE from Princeton University and his postgraduate degrees from the University of Colorado, Boulder, Colorado. He has written more than 40 technical journal articles on subjects ranging from neural network modeling and control in buildings to solar radiation measurements.

He has worked at research institutes in Israel, Portugal, and France as well as in a number of private engineering firms. His current work centers around distributed generation, smart grids, and intelligent power management at the distribution level.

Ari Rabl, until his retirement in 2007, served as senior scientist at the Ecole des Mines de Paris and continued as consultant thereafter. He received his PhD in physics from Berkeley in 1969, and until his move to Paris in 1988, he was research scientist at Argonne National Laboratory, at National Renewable Energy Laboratory, and at Princeton University, Princeton, NJ. He has published about 150 peer-reviewed journal articles in environmental economics, dispersion modeling of pollutants, epidemiology, risk analysis, waste treatment, transportation, and energy policy, as well as 10 patents and 3 books on energy and environment. He received the Frank Kreith Energy Award of the American Society of Mechanical Engineers.

Some Useful Constants

Acceleration due to gravity (standard value in the United States) $g = 32.17\ \text{ft/s}^2$ ($9.81\ \text{m/s}^2$)

Air constants (standard design conditions, sea level) 14.696 psia (101,325 Pa), 70°F (21°C)

Density $\rho = 0.075\ \text{lb}_m/\text{ft}^3$ ($1.20\ \text{kg/m}^3$)

Specific heat $c_p = 0.24\ \text{Btu/(lb}_m\cdot°\text{F)}$ [$1.00\ \text{kJ/(kg}\cdot\text{K)}$]

Sensible load $= \rho c_p \dot{V}(T_o - T_i)$, with $\rho c_p = 1.08$ [Btu/(h·°F)]/(ft³/min)[120(W/K)/(L/s)]

Latent load $= \rho h_{liq\text{-}vap}\dot{V}\Delta W$, with $\rho h_{liq\text{-}vap} = 4840$ (Btu/h)/(ft³/min) [3010W/(L/s)]

Water constants

Density $\rho = 62.32\ \text{lb}_m/\text{ft}^3$ ($998.2\ \text{kg/m}^3$) at 70°F (21°C)

Specific heat $c_p = 1.000\ \text{Btu/(lb}_m\cdot°\text{F)}$[$4.186\ \text{kJ/(kg}\cdot\text{K)}$]

Latent heat of freezing $h_{sol\text{-}liq} = 143.8\ \text{Btu/lb}_m$ (333.8 kJ/kg) at 32°F (0°C)

Latent heat of vaporization $h_{liq\text{-}vap} = 970.3\ \text{Btu/lb}_m$ (2257 kJ/kg) at 212°F (100°C)

Gas constants

Universal $R_u^* = 1545\ \text{ft}\cdot\text{lb}_f/(\text{lb mol}\cdot°\text{R})$ [$8314\ \text{J/(kg mol}\cdot\text{K)}$]

Air $R_a = 53.35\ \text{ft}\cdot\text{lb}_f/(\text{lb}_m\cdot°\text{R})$ [$287\ \text{J/(kg}\cdot\text{K)}$]

Water vapor $R_{H_2O} = 85.78\ \text{ft}\cdot\text{lb}_f/(\text{lb}_m\cdot°\text{R})$ [$462\ \text{J/(kg}\cdot\text{K)}$]

Stefan–Boltzmann constant

$\sigma = 0.1714 \times 10^{-8}\ \text{Btu/(h}\cdot\text{ft}^2\cdot°\text{R}^4)$ [$5.67 \times 10^{-8}\ \text{W/(m}^2\cdot\text{K}^4)$]

Surface heat transfer coefficients (ASHRAE design values)

$h_i = 1.46\ \text{Btu/(h}\cdot\text{ft}^2\cdot°\text{F)}$ [$8.3\ \text{W/(m}^2\cdot\text{K)}$]

$h_o = 4.0\ \text{Btu/(h}\cdot\text{ft}^2\cdot°\text{F)}$ [$22.7\ \text{W/(m}^2\cdot\text{K)}$] at v = 7.5 mph (3.4 m/s) summer

$h_o = 6.0\ \text{Btu/(h}\cdot\text{ft}^2\cdot°\text{F)}$ [$34.0\ \text{W/(m}^2\cdot\text{K)}$] at v = 15 mph (6.7 m/s) winter

Other constants

$g_c = 32.17\ \text{lb}_m\cdot\text{ft/(s}^2\cdot\text{lb}_f)$

Conversion Factors

Acceleration

$1\ m/s^2 = 3.281\ ft/s^2$

$1\ ft/s^2 = 0.3048\ m/s^2$

Area

$1\ m^2 = 10.76\ ft^2$

$1\ km^2 = 0.386\ mi^2$

$1\ ha = {}^*10^4\ m^2 = 2.47\ acre$

$1\ ft^2 = {}^*144\ in.^2 = 0.0929\ m^2$

$1\ mi^2 = 2.590\ km^2$

$1\ acre = 43{,}560\ ft^2 = 4{,}050\ m^2$

Costs

$1\ ¢/kWh = 2.778\ \$/GJ = 2.931\ \$/MBtu$

$1\ \$/m^2 = 0.0929\ \$/ft^2$

$1\ \$/kg = 0.45356\ \$/lb_m$

$1\ \$/L = 3.785\ \$/gal$

$1\ \$/MBtu = 0.948\ \$/GJ = 0.341\ ¢/kWh$

$1\ \$/therm = 9.48\ \$/GJ$

$1\ \$/gal$ (No. 6 fuel oil @ $\eta = 1.0$) $= 6.32\ \$/GJ$

$1\ \$/gal$ (No. 2 fuel oil @$\eta = 1.0$) $= 6.77\ \$/GJ$

Density

$1\ kg/m^3 = 6.2430 \times 10^{-2}\ lb_m/ft^3$

$1\ lb_m/ft^3 = 16.02\ kg/m^3$

Energy or work

$1\ J = {}^*1\left(kg \cdot m^2\right)/s^2 = {}^*10^7\ erg = 0.948 \times 10^{-3}\ Btu$

$1\ kWh = {}^*3.6\ MJ$

$1\ cal = 4.187\ J$

$1\ Btu = 778.16\ ft \cdot lb_f = 1.055\ kJ$

$1\ therm = {}^*10^5\ Btu = 105.5\ MJ$

$1\ quad = {}^*10^{15}\ Btu = 1.055 \times 10^{18}\ J$

$1\ ft \cdot lb_f = 1.3558\ J$

Flow rate (mass)

$1\ kg/s = 2.2046\ lb_m/s$

$1\ kg/s = 132.3\ lb_m/min$

$1\ kg/s = 7937\ lb_m/h$

$1\ lb_m/s = 0.454\ kg/s$

$1\ lb_m/min = 7.56 \times 10^{-3}\ kg/s$

$1\ lb_m/h = 0.1256 \times 10^{-3}\ kg/s$

Flow rate (volume)

$1\ m^3/s = 2119\ cfm$

$1\ m^3/s = 1.585 \times 10^4\ gpm$

$1\ cfm\ (ft^3/min) = 0.4719\ L/s$

$1\ gpm\ (gal/min) = 2.228 \times 10^{-3}\ ft^3/s = 0.0631\ L/s$

Flow rate (volume/area)

$1\ cfm/ft^2 = 5.01\ L/(s \cdot m^2)$

Force

$1\ N = {}^*1\ (kg \cdot m)/s^2 = {}^*10^5\ dyn = 0.2248\ lb_f$

$1\ lb_f = {}^*16\ oz_f = 4.4482\ N$

Heat flux

$1\ W/m^2 = 0.3170\ Btu/(h \cdot ft^2)$

$1\ W/m^2 = 0.0929\ W/ft^2$

$1\ Btu/(h \cdot ft^2) = 3.155\ W/m^2$

$1\ W/ft^2 = 10.76\ W/m^2$

Heat loss coefficient of building

$1\ W/K = 1.895\ Btu/(h \cdot {}^\circ F)$

$1\ Btu/(h \cdot {}^\circ F) = 0.528\ W/K$

Heat transfer coefficient

$1\ W/(m^2 \cdot K) = 0.1761\ Btu/(h \cdot ft^2 \cdot {}^\circ F)$

$1\ Btu/(h \cdot ft^2 \cdot {}^\circ F) = 5.678\ W/(m^2 \cdot K)$

* Exact.

Illuminance

1 lx = *1 lm/m² = 0.0929 fc

1 fc (footcandle) = *1 lm/ft² = 10.76 lx

Length

1 m = 3.281 ft = 39.37 in. = 1.0936 yd

1 km = 0.622 mi

1 cm = *0.01 m = 0.3937 in.

$1\ \text{ft} = *12\ \text{in.} = *\frac{1}{3}\ \text{yd} = *0.3048\ \text{m}$

1 mi = 1.609 km

1 in. = *0.0254 m = *2.54 cm

Mass

1 kg = *10^3 g = 2.205 lb_m

1 ton (metric) = *10^3 kg

1 g = 0.0353 oz_m

1 lb_m = *16 oz_m = *0.45356 kg

1 grain = *1/7000 lb_m = 0.0648 g

1 ton (US long) = *2240 lb_m = 1016 kg

1 ton (US short) = *2000 lb_m = 907 kg

Power

1 W = 3.412 Btu/h

1 kW = 1.341 hp

1 kW = 0.2844 ton refrigeration

1 Btu/h = 0.2931 W

1 hp = 550 (ft · lb_f)/s = 0.7457 kW

1 ton refrigeration = 3.517 kW

1 hp (boiler) = 9.81 kW = 33,475 Btu/h

Pressure

1 Pa = *1 N/m² = *10^{-5} bar = 1.45 × 10^{-4} psi

1 Pa = 4.019 × 10^{-3} inWG

1 atm (std. atmosphere) = 101.325 kPa

1 Pa = 2.088 × 10^{-2} lb_f/ft^2

1 mmHg = 0.01934 psia

1 psi (lb_f · in.²) = 6.894 kPa = 27.7 inWG

1 inWG = 249.1 Pa = 0.0361 psia

1 atm (std. atmosphere) = 14.696 psi

1 inHg = 3.3772 kPa = 0.49115 psia

1 inWG/100 ft = 8.17 Pa/m

Specific enthalpy

1 kJ/kg = 0.4299 Btu/lb_m

1 Btu/lb_m = 2.3266 kJ/kg

Specific heat

1 kg/(kg · K) = 0.2389 Btu/(lb_m · °F)

1 Btu/(lb_m·°F) = 4.1868 kJ/(kg · K)

Temperature

$°C = *(°F - 32) \times \frac{5}{9}$

K = °C + 273.15

°F = *(9/5)°C + 32

°R = °F + 459.67

Thermal conductivity or resistance

1 W/(m · K) = 0.5778 Btu/(h · ft · °F)

1 W/(m · K) = 6.934 Btu · in./(h · ft² · °F)

1 K/W = 0.5275 (°F · h)/Btu

1 Btu/(h · ft · °F) = 1.731 W/(m · K)

1 Btu · in./(h · ft² · °F) = 0.144 W/(m · K)

1 (°F · °h)/Btu = 1.896 K/W

Time

1 year = 8760 h = 3.1536 × 10^7 s

1 day = 86,400 s

Velocity

1 m/s = 196.9 ft/min

1 m/s = 3.281 ft/s

1 km/h = 0.2778 m/s

1 m/s = 2.2371 mi/h

1 rad/s = 9.55 r/min

1 ft/min = *0.00508 m/s

1 ft/s = *0.3048 m/s

1 mi/h = 0.4470 m/s

1 kn = 1.152 mi/h = 0.515 m/s

1 r/min = 0.1047 rad/s

Viscosity (absolute, dynamic)

1 Pa · s = *10 poise (P) = 0.672 $lb_m/(ft \cdot s)$

1 $lb_m/(ft \cdot s)$ = 1.488 Pa · s

Viscosity (kinematic)

1 m^2/s = *10^4 stokes = 10.76 ft^2/s

1 ft^2/s = 0.0903 m^2/s

Volume

1 L = *10^{-3} m^3 = 0.264 gal (US)

1 m^3 = 35.3147 ft^3

1 gal (US) = 3.785 L

1 ft^3 = 0.02832 m^3 = 7.481 gal (US)

1 bbl = 42 gal = 0.159 m^3

Decimal multiples

Factor	SI Prefix	Symbol
10^{18}	exa	E
10^{15}	peta	P
10^{12}	tera	T
10^{9}	giga	G
10^{6}	mega	M
10^{3}	kilo	k
10^{2}	hecto	h
10^{1}	deca	da
10^{-1}	deci	d
10^{-2}	centi	c
10^{-3}	milli	m
10^{-6}	micro	μ
10^{-9}	nano	n
10^{-12}	pico	p

1

Background of the Building Sector and Energy Use Patterns

ABSTRACT This chapter is primarily meant to provide a background and a broad overview of the building sector and the status of energy use in buildings. After a brief historic background, the importance of buildings, both in terms of their asset value and energy consumption, is contextualized by providing pertinent statistics and presenting future trends relevant to the U.S. economy. This is followed by a background on the various categories of end uses in different types of residences and commercial buildings. Next, we present the numerous technical and specialized professionals and companies involved in building design and operation that have evolved over the years. We then discuss some basic concepts on economics (such as payback time and risk). The two types of unit systems, the SI and IP units, and how to perform interconversion calculations are then treated in view of the (unfortunate but realistic) fact that the latter is still widely used in the United States. Finally, we argue on the importance of being able to perform orders of magnitude calculations prior to embarking on a full-fledged analysis. The art of doing such back-of-the-envelope calculations is an essential skill that a scientist or an engineer should acquire and is a cornerstone of intuition. Further, it can often provide a valuable check for the reliability of results from more complicated analyses.

Nomenclature

c_p	Specific heat at constant pressure, kJ/(kg·K) [Btu/(lb$_m$·°F)]
D	Diameter, distance, m (ft)
g	Acceleration due to gravity = 9.8 m/s^2 (32 ft/s^2)
g_c	Conversion factor in IP units, 32.17 lb$_m$·ft/(lb$_f$·s^2)
h	Height, m (ft)
M	Mass, kg (lb$_m$)
m	Mass, kg (lb$_m$)
p	Pressure, Pa (lb$_f$/in.^{2})
T	Temperature, K or °C (°R or °F)
$\dot{V}$	Flow rate, m^3/s or L/s (ft^3/min)
v	Velocity or wind speed, m/s (mph, ft/s)

Greek

ρ	Density, kg/m^3 (lb$_m$/ft^3)
ΔT	Temperature difference, K or °C (°R or °F)

1.1 A Bit of History

The quest for a safe and comfortable environment is older than the human species. Birds build nests; rabbits dig holes. Early human societies succeeded, in some cases, in creating remarkably pleasant accommodations. The cliff dwellings of the Pueblo Indians at Mesa Verde, Colorado, are an eminent example: carved under an overhang to block the summer heat, yet accessible to the warming rays of the winter sun, with the massive heat capacity of the surrounding rocks, they are an early realization of the principles of passive solar architecture.

The ancient Greeks were quite conscious of the benefits to be obtained by good orientation of a building with respect to the sun, and they laid out entire settlements facing south (Butti and Perlin, 1980). Already in classical antiquity, fuel wood was scarce around the Mediterranean (our generation is not the first to be confronted with an energy shortage). The height of comfort in the classical world was achieved in some villas of the Roman Empire: The people who built them pioneered central heating, with a double floor through whose cavity the fumes of a fire were passed. Also in Roman times, the first translucent or transparent window coverings were introduced, which were made of materials such as mica or glass. Thus, it became possible to admit light into a building without letting in wind, rain, or snow.

But it took a long time before buildings reached what we would consider comfortable conditions. When visiting castles built in Europe as late as the sixteenth century and looking at their heating arrangements, one shivers to think how cold winters must have been even for the wealthiest. In an antique store, one of the authors happened upon an old thermometer with a scale where 16°C (60.8°F) was marked as "room temperature." The comfort of heating to 20°C (68°F) or more is fairly recent.

Cooling is more difficult than heating. In the past, the principal method was to ward off the sun, coupled with the use of heavy stonework for thermal inertia—actually quite effective in climates with cool nights. During the Middle Ages, architects of castles like the Alhambra supplemented that approach with skillful use of running water, providing some evaporative cooling. In certain parts of the world, some buildings take advantage of cool breezes that blow regularly from the same direction.

Nights used to be dim because candles and oil lamps were expensive, to say nothing of the quality of the light. In fact before the invention of electric lights, access to daylight was a primary criterion for the design of buildings. Daylight does not penetrate well into the interior; as a rule of thumb, adequate illumination on cloudy days cannot be provided to depths beyond 1.5–2 times the height of the upper window edge. This, together with the need for fresh air before the days of mechanical ventilation, explains the shape of buildings well into the first half of the twentieth century: No room could be far from the perimeter. Total disregard for this constraint did not become practical until the availability of fluorescent lighting (with incandescent lamps, the cost of air conditioning would have been excessive).

Air conditioning is the process of treating air in an internal environment so as to establish and maintain required human comfort standards of temperature, humidity, cleanliness, and air motion. It includes both heating and cooling applications. Effective air conditioning had to await the development of mechanical refrigeration during the first decades of the twentieth century. Routine installation of central air conditioning systems dates from the 1960s. The oil crises of the 1970s stimulated intense research on ways of reducing energy costs. The efficiency of existing technologies is being improved, and "new" technologies such as solar energy and daylighting are being tried. Some approaches are turning out to be more successful than others. This evolution is continuing at full speed, aided by advances in materials and computers. Elucidating these developments is one of the goals of this book.

1.2 Importance of Buildings in the U.S. Economy and Other Countries*

People in modern times spend most of their time in buildings and expend much of their wealth on these buildings. Even without trying for an accurate appraisal of the total real estate in the United States, we can get an idea by looking at the floor area; in 2005, approximately 256×10^9 ft² ($\approx 25 \times 10^9$ m²) in the residential and 80×10^9 ft² ($\approx 8 \times 10^9$ m²) in the commercial sector (EIA, 2005). Construction costs vary a great deal, of course, with the type and quality of building, but very roughly they are in the range of \$125 per square foot (\$1125 per square meter) ±25%. Based on the replacement cost, the total value of the buildings in the United States is on the order of \$42 × 10¹² (\$42 trillion); the GDP, by comparison, was close to \$12 × 10¹² in 2005. More recently, in 2014, the GDP was close to \$16 × 10¹² with the building construction industry accounting for \$1.3 trillion (8.1% of GDP).

Historically, 1.6 million new buildings were constructed in the United States each year, while there are 125.1 million existing buildings. Obviously, one cannot afford to replace the building stock very often. This ensures a certain continuity with the future. However little we may be able to foresee the future of our society, we can expect most of the buildings to last for decades, if not centuries. Rather than replace a building entirely, one may revamp its interior; currently each year some 2% of commercial buildings undergo a major retrofit (Brambley et al., 1988). This implies an awesome responsibility for city planners, architects, and engineers: *do it well or the mistakes will haunt us for a long time.*

The demand for new construction was vigorous during the 1980s. Annual construction rates were around 1.5×10^9 ft² ($\approx 0.14 \times 10^9$ m²) of residential and 1.2×10^9 ft² ($\approx 0.11 \times 10^9$ m²) of commercial floor area; the residential stock was growing at about 1% and the commercial stock at about 2.5% per year. Single-family houses have accounted for about two-thirds of recent residential construction. Forecasting of trends is difficult, especially in a sector that is as cyclical as the construction industry and as sensitive to the state of the economy. The residential market may have seen its peak of growth with the passing of the postwar baby boom, although there may be a continuing push for new and better housing as general living standards improve (a likely trend over the long term). The growing shift from industry to services implies an increasing need for commercial floor space.

Figure 1.1 (from BEDB, 2011) consists of three pie charts: (1) the relative primary energy consumption of the United States to world energy consumption,† (2) the breakdown of the principal end-use sectors of the United States (transportation, industry, commercial, and residential), and (3) the relative contribution of the various primary energy sources relevant to the building sector. It is important to keep the distinction between *primary or source energy* and *site or delivered energy*. The primary or source energy indicates the

* Note that different reports and graphs have slightly different statistics and numbers (uncertainties on the order of 10%).

† China overtook the United States as the largest consumer of energy in 2010.

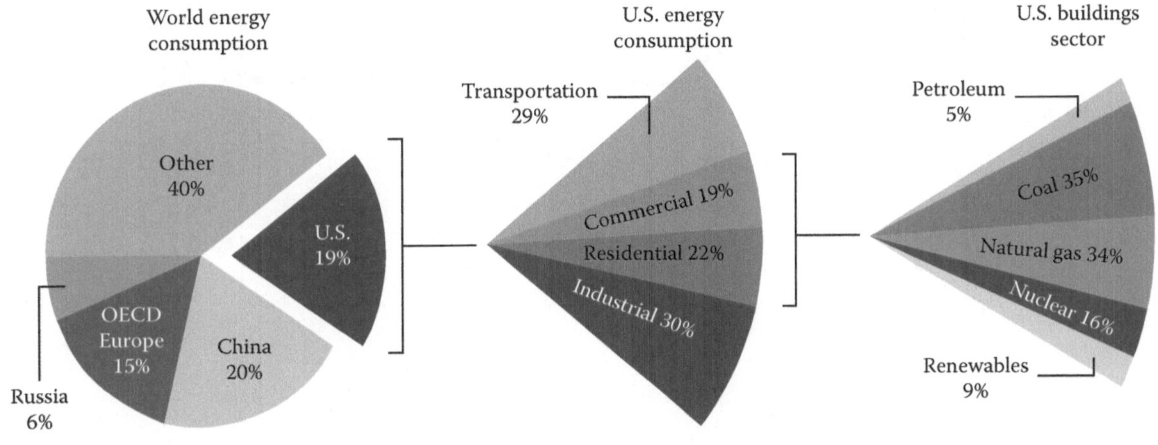

FIGURE 1.1
Primary energy consumption statistics as percentages: worldwide, different end-use sectors in the United States, and sources of primary energy to the building sector. (From BEDB, *Building Energy Data Book*, Energy Efficiency and Renewable Energy Program, U.S. Department of Energy, Washington, DC, 2011, https://ieer.org/wp/wp-content/uploads/2012/03/DOE-2011-Buildings-Energy-DataBook-BEDB.pdf, accessed January 2020.)

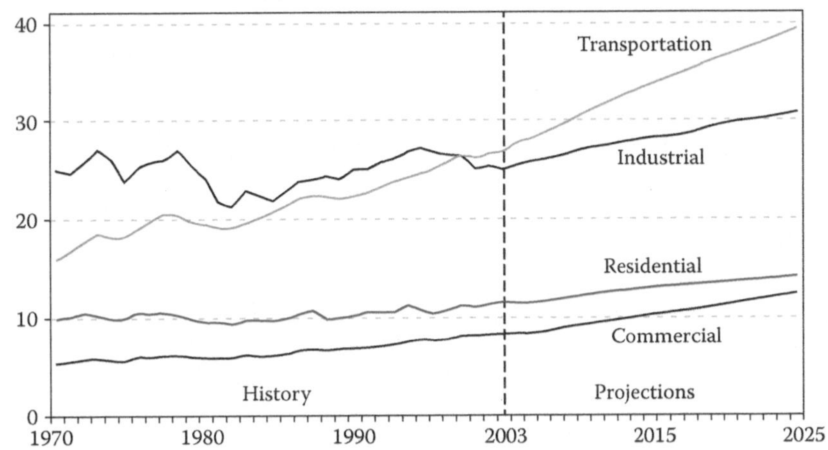

FIGURE 1.2
Historic trends of U.S. delivered (or site) energy, by sector. 1 quad = 1.055×10^{18} J. (From BEDB, *Building Energy Data Book*, Energy Efficiency and Renewable Energy Program, U.S. Department of Energy, Washington, DC, 2011, https://ieer.org/wp/wp-content/uploads/2012/03/DOE-2011-Buildings-Energy-DataBook-BEDB.pdf, accessed January 2020.)

total amount of raw fuel that is required to operate the building and incorporates all transmission, delivery, and production losses. Site or delivered energy, on the other hand, is a measure of the local consumption and is reflected in the utility bills. It does not include energy lost during production, transmission, and distribution to customers. Source energy is much more pertinent than site energy, if the concern is environmental performance evaluation, and is used during energy policy studies. The conversion factor for electricity (which includes production and transmission losses) is taken to be about 0.33 in the United States. Note that this conversion factor depends greatly on the type of power plant generation mix (coal based, hydro, nuclear, renewable, etc.) specific to the region/state/country in question. In terms of site energy, the historic energy use patterns look different since the conversion efficiency from primary energy to electricity is not considered (Figure 1.2). The residential energy consumption is projected to grow at an average of 0.8% per year from 2010 to 2025 that is slightly less than the growth rate for the commercial sector (Kydes, 2008).

Buildings accounted for 41% (40 quads) of the primary energy consumption in the United States in 2010, greater than that attributable to either transportation (29%) or industry (30%). This represented a cost of approximately $400 billion in 2010 dollars. Buildings consumed 74% of the electricity generated in the United States and 34% of the natural gas production. This led to buildings being responsible for 40% of the carbon dioxide emissions in the United States or 7.4% of the total global carbon dioxide emissions (BEDB, 2011). In aggregate, commercial buildings consumed 17.9 quads of primary energy, representing 46.0% of building energy consumption and

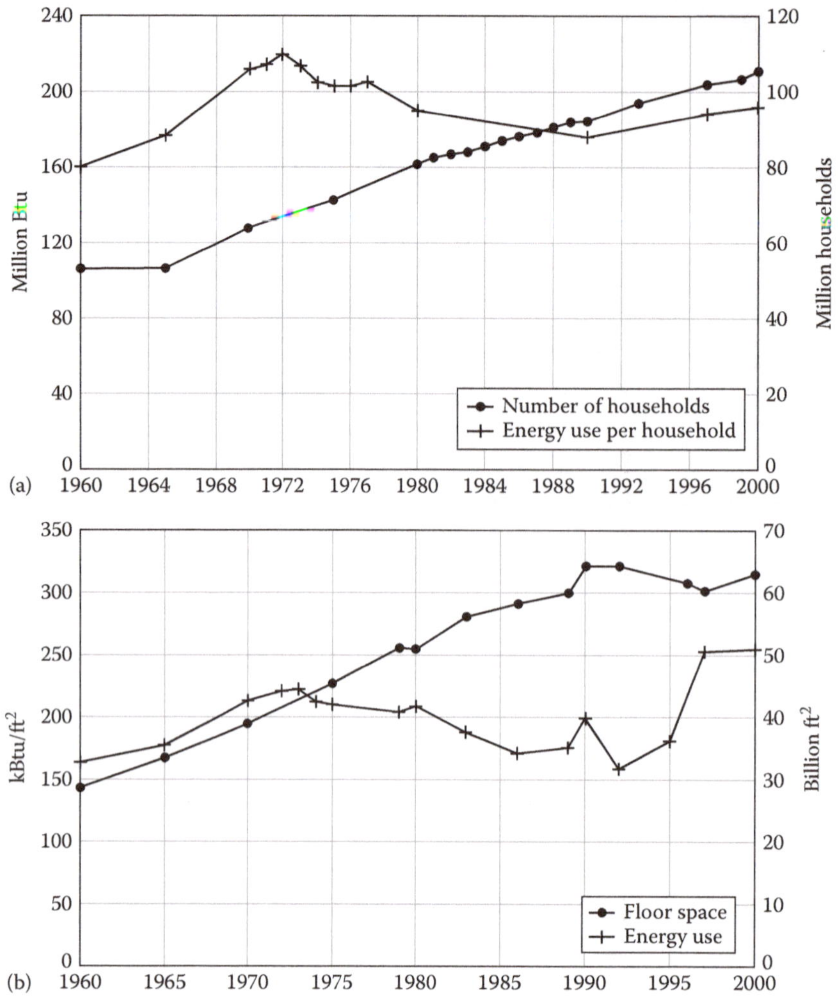

FIGURE 1.3
Building stock and consumption of primary energy in the United States. (a) Number of households and primary energy per household. (b) Commercial floor area and primary energy per floor area. 1 MBtu = 1.055 GJ; 1 kBtu/ft^2 = 11.36 MJ/m^2. Decreases in commercial building area due to redefinition of commercial buildings. (From USDOE, *BTS Core Data Book*, Office of Building Technology, U.S. Department of Energy, Washington, DC, June 1999; USDOC, U.S. statistical abstract, Published annually by the U.S. Department of Commerce, Bureau of the Census, Government Printing Office, Washington, DC, 2001.)

18.9% of the U.S. energy consumption. In comparison, the residential sector consumed 21.0 quads of primary energy, equal to 22.3% of U.S. energy consumption. The total building primary energy consumption in 2009 was about 48% higher than consumption in 1980. Space heating (37%), water heating (12%), space cooling (10%), and lighting (9%) were the dominant end uses in 2010, accounting for 68% of all energy consumed by the buildings sector. The U.S. Department of Energy through its Energy Efficiency and Renewable Energy division has a major program called the Building Technologies Program (BTO) whose mission is "to develop technologies, techniques, and tools for making buildings more energy efficient, productive, and affordable" (http://www1.eere.energy.gov/buildings/printable_versions/index.html).

Several factors have contributed to the growth of energy use in buildings: the population has increased (from 180 million in 1960 to 275 million in 2000 to over 300 million in 2014), the number of households and commercial floor space have increased,* comfort levels have improved, and there are more energy-using devices in buildings. Figure 1.3 provides a more detailed view of this evolution. Frame (a) shows the number of households and the consumption of primary energy per household; frame (b) gives the analogous information for commercial buildings, in terms of floor area. Both follow the same pattern. While the number of households

* The expected growth rates between 2009 and 2035 in U.S. population, households, and commercial floor space are 27%, 31%, and 28%, respectively (BEDB, 2011).

and the commercial floor area have grown steadily, the energy consumption per unit peaked in the early 1970s. Part of that may be due to population shifts to warmer climates, but most of it reflects gains in efficiency in response to the oil shocks.

Economic cycles have a major impact on the building construction industry. In 2005, the Energy Information Administration (EIA) projected that this growth will stagnate due to the recent recession until 2016, after which steady growth is predicted through 2035. Total primary energy consumption in buildings is expected to reach more than 45 quads by 2035, about a 15% increase over 2009 levels. As expected, the number of people employed in architecture and construction has also decreased since 2006. More than 7.9 million people were employed in the two industries then compared to 5.7 million in 2010, a 27% drop, attributed to the recent recession.

It is interesting to consider the importance of cost components involved in a building, from design to operation. Brambley et al. (1988) state that one-time costs (design and construction) represent only about one-fifth of the total life-cycle cost, with the remaining four-fifths being ongoing costs (operation and maintenance [O&M]). Such a comparison is not without ambiguity, involving a choice of weights for present and future expenses (via the discount rate, as discussed in Chapter 23), to say nothing about differences between different types of buildings. We cite this figure merely to point out the dominant role of the cost of O&M. Of the latter, energy represents the lion's share, as can be seen from Figure 1.4, where the ongoing costs, in dollars per unit floor area, are disaggregated according to the categories of administration, cleaning, repair and maintenance, security and grounds, and utilities (i.e., energy, since the cost of water is relatively small). Since 1973, the importance of energy expenditures has increased in both relative and absolute terms. In the residential sector, energy expenditures tend to be smaller, somewhat less than the construction cost, but clearly they are significant here as well. There is an important lesson: *pay attention to energy costs at the design stage.*

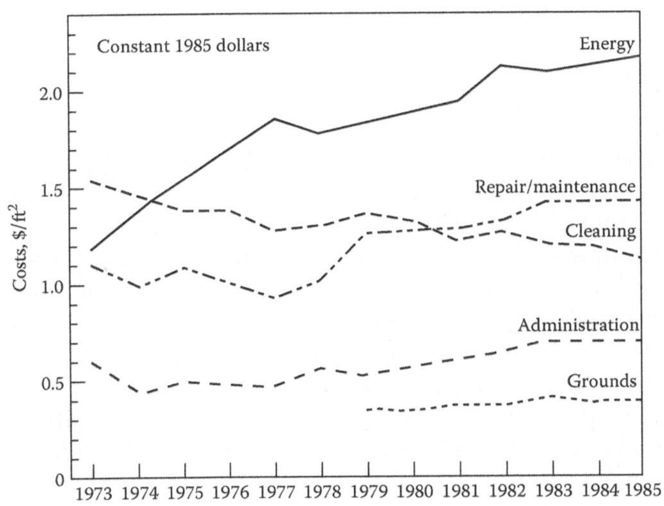

FIGURE 1.4
Average operating costs by component, per unit floor area, for office buildings, corrected for inflation. $\$1/\text{ft}^2 = \$10.76/\text{m}^2$. (From Brambley, M.R. et al., Advanced energy design and operation technologies research: Recommendations for a U.S. Department of energy multiyear program plan, Report PNL-6255, Battelle Pacific Northwest Laboratory, Richland, WA, December 1988.)

1.3 Energy Use Patterns by Building Type and End Use

As described in detail in this book, building energy analysis involves three aspects: the calculation of thermal loads (heat gains in summer and heat losses in winter), the analysis of secondary systems, and the analysis of primary systems that includes energy conversion devices such as chillers and boilers. Secondary systems are energy transfer equipment meant to meet the heating, ventilating, and air conditioning (HVAC) needs of specific zones/spaces within the building. They consist of air and liquid handling equipment, duct and pipe systems, and heating and cooling terminal devices including coils, mixing boxes, and baseboard heating units. It is common in professional practice to use the term HVAC to denote both primary and secondary systems.

1.3.1 Residential Buildings*

There are about 115 million households in the United States, of which 19% were located in the northeast, 24% in the midwest, 35% in the south, and 21% in the west. Over three-quarters of the households (77%) are in urban areas, with 36% in the central city and 41% in suburbia. The remaining households (22%) are in rural areas. The three basic categories of housing type are (1) single-family units (both as detached units and in row houses), (2) multifamily (both low-rise and high-rise apartments), and (3) mobile homes. In 1997, the stock was predominantly single-family units (73%), with apartments accounting for 21% of the total households and 6% for mobile homes. The United States is a nation of homeowners, with 67% of the households owner occupied and the remaining 33% rented. Note that the energy consumption profiles of single-family homes and multifamily homes (apartments) are very different.

* The following statistics are taken from BEDB (2011).

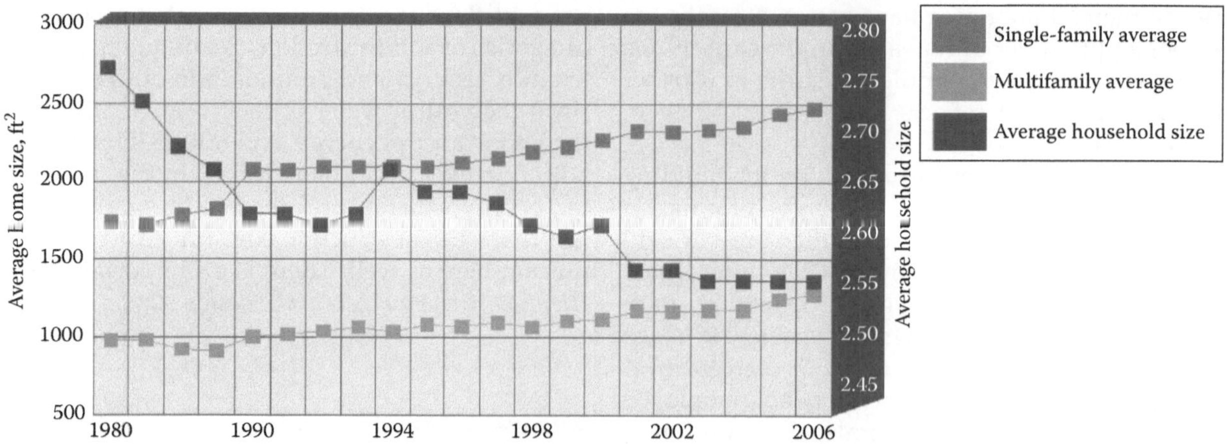

FIGURE 1.5
Historic trends in average U.S. home sizes and number of occupants per home. (From BEDB, *Building Energy Data Book*, Energy Efficiency and Renewable Energy Program, U.S. Department of Energy, Washington, DC, 2011, http://buildingsdatabook.eren.doe.gov/, accessed on January 2015.)

There is a clear trend toward increasing efficiency in residential housing. Homes built between 2000 and 2005 used 44.7 kBtu per heated square foot of heated floor space (or 510 MJ/m^2)—14% less than homes built in the 1980s and 40% less than homes built before 1950. There has also been a trend toward larger home sizes. Specifically, single-family homes built between 2000 and 2005 are 29% larger on average than those built in the 1980s and 38% larger than those built before 1950 (Figure 1.5). Thus, despite better building practices and newer systems, the greater average floor space of new homes has offset their improved efficiency. Concurrently, the number of occupants per home has decreased due to such factors as higher income, smaller families, and deferred marriage.

1.3.2 Commercial Buildings

Commercial buildings are distinguished by type; the commonly adopted classification is shown in Figure 1.6. In 2003, the most recent year for which such data are available, office and retail buildings represented the greatest proportions of commercial floor space (17% and 16%, respectively) and of commercial sector energy consumption (19% and 18%, respectively). Warehouses and storage facilities accounted for 14% of commercial floor space, but had less than half of that of office buildings (92.9 kBtu/ft^2 or 1054 MJ/m^2).

Medical buildings and food sales and service buildings tend to contain energy-intensive end uses, and also tend to be occupied more hours per day and more days per week. Thus, while these buildings represent 8.5% of commercial floor space, they represent close to 19% of commercial primary energy consumption. Statistical records indicate that the energy intensity of commercial buildings has increased very slowly. Between 1980 and 2010, primary energy consumption per square foot increased by 8%. Between 2010 and 2035, EIA actually expects energy intensity to decrease by 6%.

It is also noteworthy that energy intensity varies with square footage (EIA, 1995). For example, Figure 1.7 reveals a clear trend for the eight commercial building size categories. The energy intensity first reduces as the building gets larger due to the decreasing envelope to floor area ratio (conduction losses are proportional to the envelope area), but after 25,000 ft^2 (2,250 m^2), the energy intensity increases since the energy consumed by the secondary systems start becoming important. On an average, secondary systems account for about 30%–40% of the energy used in larger commercial buildings.

1.3.3 End Use

Figure 1.8a and b is pie charts showing the various end uses for residential and commercial buildings in the United States for 2010 (BEDB, 2011). Space heating consumed 45% and 27% of site energy in the residential and commercial sectors, respectively, more than any other end use. Other significant end uses include water heating (18%) and space cooling (9%) for the residential sector. For the commercial sector, the two next largest end uses are lighting (14%) and space cooling (10%). Note that space conditioning and lighting account for almost half of commercial energy consumption, and about 60% for the residential sector.

From Table 1.1, we note that the largest contributor to heating need in residences is due to infiltration followed by conduction through windows and then walls. For commercial buildings, the same three effects appear but in a different order: window conduction, wall conduction, and then infiltration. For cooling loads, solar heat gains through windows is by far the largest contributor,

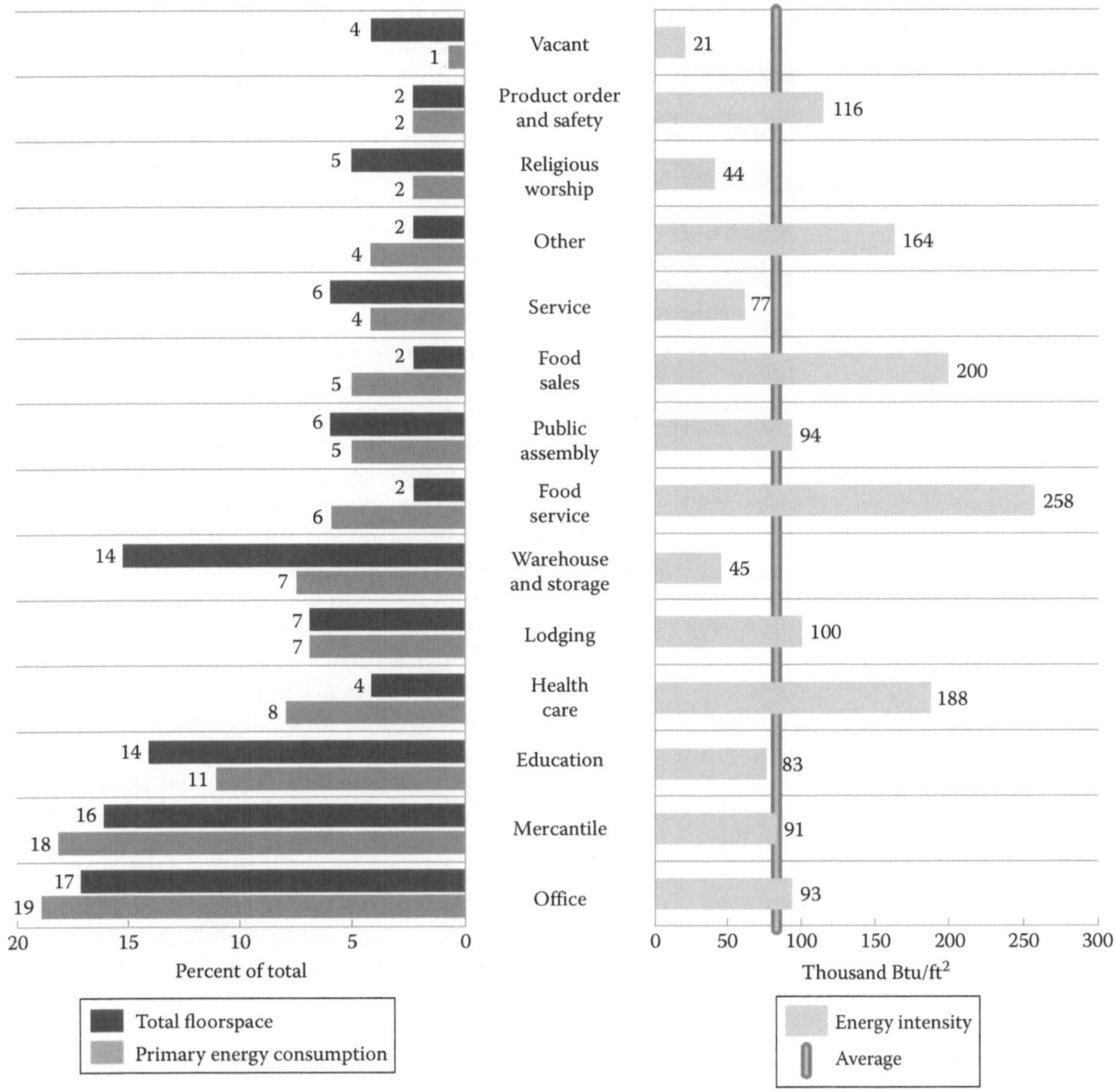

FIGURE 1.6
U.S. commercial building statistics for 2003 (last year where such data were available) by building type: relative percentages of floor space and primary energy consumption and energy intensity (in kBtu/ft^2/year). (From BEDB, *Building Energy Data Book*, Energy Efficiency and Renewable Energy Program, U.S. Department of Energy, Washington, DC, 2011, http://buildingsdatabook.eren.doe.gov/, accessed on January 2015.)

followed by infiltration and heat gains through roofs for residences and by window conduction for commercial buildings. These loads are individually treated in Chapters 4 through 6.

All these considerations indicate a major role for the HVAC design engineer because his or her decisions determine, in large measure, the energy consumption. Yet, lest the reader conclude that energy consumption should be the dominant criterion for the design of commercial buildings, let us add the perspective of the value of the business transacted in the buildings. As an example, suppose an office worker requires 200 ft^2 (≈20 m^2) of gross area and was paid $30,000 per year in 1985 (in exchange for services of roughly equivalent value). Thus, the value of the services, per unit floor area, is $150 per square foot (≈$1500 per square meter) per year, two orders of magnitude larger than the energy cost in Figure 1.4. While this is merely an illustrative example, an even larger value of $300 per square foot per year is cited for office buildings in the United States by Rosenfeld (1990). The value per unit floor area may be even larger, for instance, in retail stores because of the profit to be made on the merchandise. Any reasonable assumptions about salary and density of workers lead to the same basic conclusion: in commercial buildings, the energy costs are tiny in comparison with the value of the business that is transacted. Therefore, we should not risk a reduction in productivity while trying to save

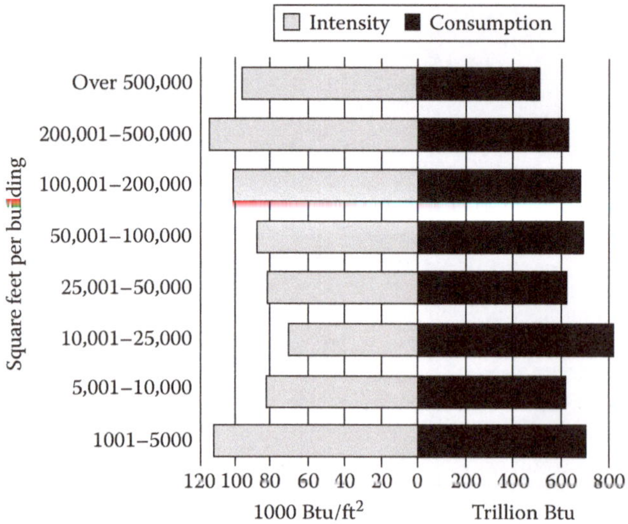

FIGURE 1.7
Energy consumption and usage intensity for eight U.S. commercial building size categories. (From EIA, *Annual Energy Outlook 1995*, U.S. Energy Information Administration, U.S. Department of Energy, Washington, DC, 1995.)

energy in commercial buildings. A drop in productivity as small as 1% could more than wipe out any savings from energy.

As for energy forms used in commercial buildings, their relative contributions are shown in Figure 1.9, from 1960 to 1998. We can see a steady and continuing growth of electricity relative to coal and petroleum. Two phenomena seem to be at play: a shift to electricity because it is more flexible and permits higher end-use efficiency and, as living standards increase, a general trend from dirty to clean energy forms (clean at the site, not necessarily at the source). This evolution is likely to continue, including the growing use of heat pumps.

What may the future bring? Over the long term, the cost of energy and materials tends to decline, thanks to technological progress. This trend is compensated to some extent by the exhaustion of resources and by the increasing costs of environmental protection. Occasionally, the trend is punctuated by shocks, such as the oil crises of the 1970s. The "dash to gas" of recent years increases the risk of a shortage of gas since most of the world's gas supply comes from politically unstable countries.

Concern about air pollution, both outdoor and indoor, has been increasing and is not likely to diminish soon. Since the late 1980s, the risk of global impacts has captured worldwide attention, in particular the greenhouse effect and the depletion of ozone in the upper atmosphere (note that ozone in the upper atmosphere is good for us because it absorbs ultraviolet radiation from the sun, but at the earth's surface it is harmful to plants and to our health). Buildings are implicated in all these problems.

All energy use, other than nuclear or renewable, contributes carbon dioxide (CO_2) to the greenhouse effect. Electricity production from coal or oil entails sizable emissions of NO_x, SO_2, and particulate matter. Natural gas is much cleaner, but the emission of NO_x remains a problem. Chlorofluorocarbons (CFCs), the major cause of ozone depletion, were widely used not only as working fluids in heat pumps and refrigeration equipment, but also in the manufacture of urethane foams for

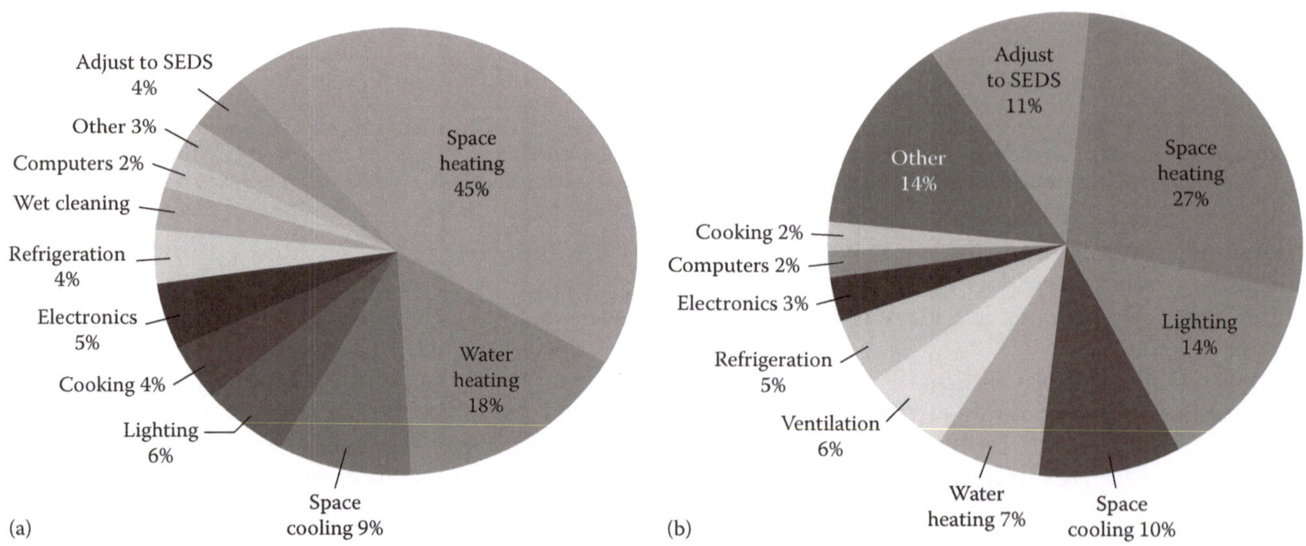

FIGURE 1.8
Pie charts of U.S. energy consumption by end-use (a) residential and (b) commercial buildings. (From BEDB, *Building Energy Data Book*, Energy Efficiency and Renewable Energy Program, U.S. Department of Energy, Washington, DC, 2011, http://buildingsdatabook.eren.doe.gov/, accessed on January 2015.)

TABLE 1.1

Primary Energy Consumption Attributable to Building Envelope Components in 2010 (Quads)

	Residential		Commercial	
Building Component	**Heating**	**Cooling**	**Heating**	**Cooling**
Roofs	1.00	0.49	0.88	0.05
Walls	1.54	0.34	1.48	−0.03
Foundation	1.17	−0.22	0.79	−0.21
Infiltration	2.26	0.59	1.29	−0.15
Windows (conduction)	2.06	0.03	1.60	−0.30
Windows (solar heat gain)	−0.66	1.14	−0.97	1.38

Source: BEDB, *Building Energy Data Book*, Energy Efficiency and Renewable Energy Program, U.S. Department of Energy, Washington, DC, 2011, http://buildings-databook.eren.doe.gov/, accessed on January 2015.

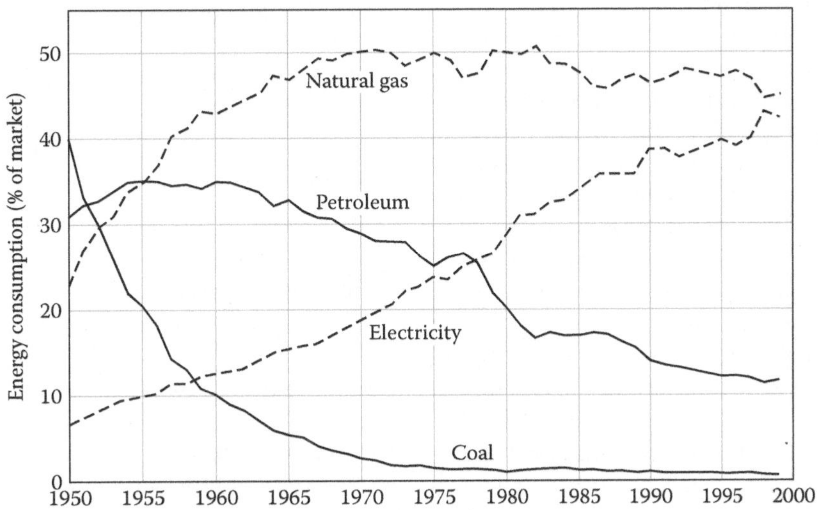

FIGURE 1.9
Relative share of energy sources for commercial buildings in the United States. (From Brambley, M.R. et al., Advanced energy design and operation technologies research: Recommendations for a U.S. Department of energy multiyear program plan, Report PNL-6255, Battelle Pacific Northwest Laboratory, Richland, WA, December 1988.)

insulation and furniture. During the past decade, the Montreal protocol of 1987 has taken effect, limiting the production of CFC11 and CFC12 and eventually banning CFCs altogether, with profound consequences for the design of air conditioning systems. More discussion is provided in Chapter 14.

To emphasize the importance of environmental considerations for the HVAC engineer, we cite from the policy statement defining ASHRAE's concern for the environmental impact of its activities: "…ASHRAE's members will strive to minimize any possible deleterious effect on the indoor and outdoor environment of the systems and components in their responsibility while maximizing the beneficial effects these systems provide, consistent with accepted standards and the practical state of the art…."

1.4 Roles of Building Energy Professionals and HVAC Design Engineers

The building owner usually hires an architectural company to coordinate the overall building planning and design, who in turn hires and supervises engineering firms (civil, mechanical, electrical) and specialized subcontractors (pipe fitters, sheet metal workers, etc.). A typical building life cycle involves various phases; the important ones from the consideration of energy efficiency and use are as follows:

1. *Design phase* (typical duration is 1–2 years) includes designing for energy efficiency and broadly involves (1) the conceptual or schematic

design, where the number of design alternatives involving different types of feasible systems and their associated subsystems are evaluated before a final selection is made, and (2) detailed design and construction documents preparation of the final design selected. Various subactivities are discussed at more length later in this section.

2. *Construction* (typical duration is 1–2 years) that includes (1) civil construction; (2) purchase and installation of the various building mechanical, electrical, and plumbing (MEP) systems; and (3) start-up commissioning to test, adjust, and balance the systems, i.e., assure that various systems are installed properly and are operating as intended.
3. O&M over its active life (typically 50–60 years for the shell and 20–25 years for the MEP systems). The operations of the systems, assuring comfort conditions for occupants while minimizing energy consumption and maintaining and, if necessary, replacing the various MEP systems, are activities performed by the facility staff or by specialized firms hired by the building owner. This aspect has acquired a great deal of importance in view of the increasing importance of energy efficiency and led to the emergence of several professional services companies specializing in tasks such as energy benchmarking and auditing, energy efficiency upgrade identification and assessment of O&M measures, monitoring and verification of energy conservation measures, supervisory control of building energy systems and implementing optimal operation strategies, identifying load control measures to reduce energy costs, performing fault detection and diagnostics of primary equipment as well as continuous commissioning to maintain energy systems at their peak performance levels, and demand response activities during periods when the electricity grid is at its peak condition. There are numerous specialized publications dealing with these issues (see, for example, ASHRAE Applications, 2011; Claridge, 1998; Mull, 2001; Reddy, 2003; Turner, 2006).

The proper design of HVAC systems is subject to several criteria (of which only the first two are relevant to the scope of this book):

1. Standards of comfort—temperature, humidity, air motion, noise, and cleanliness
2. Standards of economic and energetic efficiency—low initial and operating costs and energy consumption subject to the aforementioned criteria
3. Standards of safety—embodied in codes and laws applicable to buildings
4. Desires of the owner as specified initially and on an ongoing basis during the design process at design meetings
5. Effective communication between all design professionals and ongoing documentation of the design process

The goal of rational building design is to provide an environment that is pleasant, comfortable, convenient, and safe—the best one can get for the money one is willing to spend. The design is a collaborative effort between architects and engineers. The domains of primary responsibilities are as follows:

- Architects and civil engineers for building shell and structure
- Mechanical engineers for HVAC equipment, system integration, and controls
- Electrical engineers for wiring, lights, and electric equipment
- Other professionals for acoustics, interior design, security, and safety

Since the oil crises of the 1970s, there has been a growing awareness of the influence of the building shell and the HVAC system on the energy bill. Significant savings can be realized by paying attention to energy consumption during the early design. Thus, considerable feedback between architects and engineers is advisable, for example, to avoid large expanses of glass on south and west facades without control of solar heat gains. This is particularly important in the early steps of the design; otherwise it will be too late to make changes.*

The tasks of the mechanical engineer are to calculate the demands for heating, cooling, and ventilation; to choose the necessary equipment and controls; and to ensure that the components are correctly integrated into the building. Obviously, the capacities of the equipment must be sufficient to maintain comfortable conditions even under extremes of summer heat and winter cold. Comfort depends not only on temperature but also on humidity. Excessive dryness parches the skin and irritates the throat and nose, even to the point of causing nosebleeds in sensitive individuals. The other extreme

* In the past, the load calculations were usually done only after most of the design was already set. Future generations of computers and software are making it easy to update the calculations of loads and costs automatically as the design evolves, thus allowing interactive optimization.

is well known to anyone who has visited England in winter or the southern United States in summer. High humidity interferes with the body's capacity for evaporative temperature regulation; it produces a sensation of clammy dampness in winter and of muggy, "sticky" heat in summer. During the 1980s, concern with indoor air quality added another dimension to the tasks of the design engineer.

The HVAC equipment should cost no more than necessary, and thus its capacity should not be excessive. It follows that the first job of the HVAC engineer is the calculation of *peak loads*. A building must not only function correctly, but also be economical. The annual energy cost is determined by the annual performance. To minimize the true total cost of a building, it is therefore necessary to calculate its annual heating and cooling loads as well, or, what is equivalent, its average loads. Thus, the HVAC design process involves basically the following steps:

- Calculation of peak loads (also referred to as design loads)
- Specification of equipment and system configuration
- Calculation of annual performance (called energy calculations)
- Calculation of costs

The steps are iterative: having determined the total cost of an initial design, one modifies some of its details and recalculates the total cost until one has come sufficiently close to an optimum.

During the past decades, the architectural profession has developed a formal building design procedure consisting of several phases, as outlined in Table 1.2. During each phase, a specific level of detail is developed, and an estimate of project cost based on that design is prepared. The owner of the building reviews the output of each design phase, accepts the results, and authorizes the next phase with a formal, written approval of the current phase. The level of detail increases with each phase.

1. The first phase is *programming*. At this level, initial specifications of the building are made with respect to its volume, usable floor space, shape and orientation, operating schedule, interior temperatures, and ventilation rates.
2. This is followed by the *schematic design phase*. Here, the building loads are estimated, and the components of the HVAC system are specified. The first attempts are made toward optimizing the design.
3. The third phase is *design development*. Loads are calculated with precision, and all technical details are resolved, such as selection of equipment capacities and flow rates. Life-cycle cost (i.e., equipment costs and operating costs, appropriately combined) should be calculated to find the optimum.
4. During the fourth phase, the *construction documents* are prepared. These involve fully written specifications and drawings, including installation diagrams, control procedures, and equipment specifications. A standard format for this purpose has been developed by the Construction Specification Institute. Construction documents are the basis for the bidding phase.
5. The final phase is *contract administration*. The successful bidder is selected, and the building is constructed. As installation is completed, start-up of the equipment begins and a formal acceptance test is done, called commissioning. A user's manual is provided.

TABLE 1.2

First Four Phases of the Building Design Procedure

Design Phase	Professional Services	Drawings or Specifications	Load Calculations	Cost Estimates
Programming	Architect Mechanical engineer	Sketches and models; no specifications	Rules of thumb	Rules of thumb
Schematic design	Architect Mechanical engineer Civil engineer	One-line diagrams; outline specifications	Peak and annual (rough)	Handbook values
Design development	Architect, mechanical engineer, civil engineer, electrical engineer, and other professionals for acoustics, lighting, and landscaping	Intermediate level of detail; drawings and performance specifications	Peak and annual (final computer simulations)	Informal quotes
Construction documents	Same as design development	Final bid packages including detailed specifications and drawings	None	Formal bids

It is important to carry out the work in an orderly and systematic manner. Careful documentation is crucial, to minimize questions and problems that may arise later. It is best to develop good habits from the start; in particular, show job identification, the date, and your initials on each page.

This book may appear more concerned with commercial than residential buildings. There is a reason. Commercial buildings tend to be larger, and their HVAC systems are more complex than those of residential buildings. Furthermore, reliable performance is more important because of the high value of business transacted (the value per unit floor area is much higher than that in the residential sector). Commercial buildings, at least the larger ones, are usually custom designed, while houses and their HVAC systems are more often based on mass production. Thus, the HVAC designer will earn a living mostly with commercial buildings.

1.5 Basic Concepts in Economics of Energy Efficiency

Making a design choice is easy when option A has a lower first cost (purchase and installation) and costs less to operate than an alternative option B. However, usually an item that saves operating costs requires a higher initial investment. The optimal choice depends on the circumstances. Compare, for example, two refrigerators, both providing the same quantity and quality of service. Suppose model A costs \$600 and draws electricity at an average rate of 150 W, while model B costs \$700, but draws only 100 W; the price of electricity is 10¢/kWh. Noting that a year contains 8760 h, we find the difference in annual cost as

$$8760\ \text{h} \times [(0.15 - 0.10)]\ \text{kW}] \times \$0.10/\text{kWh}$$

$$= \$43.80 \text{ annual savings}$$

By paying the extra investment of \$100, one saves \$43.80 each year, for each of the roughly 20 years that a refrigerator can be expected to last. After 100/43.80 = 2.3 years, the extra investment has been paid off, so to speak, and thereafter it yields pure profit year after year. This does not seem to be a bad deal when one compares it to typical interest payments from savings accounts. Is such a comparison appropriate? Yes, although with some complication, because at the end of its lifetime the refrigerator is worn out, whereas the money in a savings account remains.

In any case, it is clear that choosing on the basis of first costs alone is very shortsighted. The cheap refrigerator, model A, would be very expensive in the long run. A rational decision takes all the costs into account that will be incurred over the lifetime of the item in question; in other words, it is based on an analysis of the life cycle cost. One would like to achieve the greatest benefit for the lowest life cycle cost. The building designer is constantly facing decisions of this kind: Is it worth paying the extra first cost for a design or a design element that has lower operating cost? Examples are extra insulation, high-performance glazing, efficient lighting, ducts with reduced pressure drops, and chillers with a high coefficient of performance (COP).

A rigorous analysis of such decisions is fairly involved, and we devote a complete chapter to it (Chapter 23). But for the benefit of readers who will not want to study it, we present at this point a very simple, albeit crude, decision tool. It is called "payback time," defined as the ratio of initial investment to annual savings:

$$\text{Payback time} = \frac{\text{Investment}}{\text{Annual savings}} \tag{1.1}$$

In our example of the refrigerator, we have already calculated the payback time: it is 2.3 years. Payback time fails to take into account a number of effects, such as the difference in value between a dollar that is available today and one that will be available only in the future. But it has the advantage of simplicity. Roughly speaking, at the end of the payback time, the extra investment has paid for itself, and thereafter it brings a profit. The shorter the payback time, the better. For investments without significant risk, one can say as a rule of thumb that the profitability is excellent if the payback time is *less than one-third of the lifetime of the investment*; it is good if the payback time is less than one-half of the lifetime.

Example 1.1: Example of Payback Analysis

Compare two alternatives for the lamp above your desk at home: an incandescent bulb and a compact fluorescent bulb, both giving light of the same quantity and quality. Make the following assumptions.

Given: First cost of incandescent bulb = \$1
First cost of fluorescent bulb = \$15
Power drawn by incandescent bulb = 60 W
Power drawn by fluorescent bulb = 15 W
Lifetime of incandescent bulb = 1000 h
Lifetime of fluorescent bulb = 10,000 h
Light is turned on 2000 h/year
Electricity price paid = \$0.15/kWh

Find: Payback time.

Solution

Investment = difference in initial cost = \$15 − \$1 = \$14. There are savings in electricity and savings in light bulb replacement.

The difference in annual electricity cost = [(60 − 15)W × 2000h × $0.15/kWh/1000(W/kW)] = $13.50. The incandescent bulb has to be replaced twice a year, at a cost of $2 per year. Also, the fluorescent has depreciated $\frac{2000}{10{,}000} \times 15 = \$3$

Therefore, the annual savings with the fluorescent bulb are $13.50 + $2 - $3 = $12.50.

From Equation 1.1, we have payback time = investment/annual savings = $14/$12.50 = 1.12 year.

The lifetime of the fluorescent bulb is 10,000 h/(2,000 h/year) = 5 year.

Comments

1. The payback time is about one-fifth of the lifetime, and the extra investment of $14 for the fluorescent bulb is very profitable.
2. How about risk? In this case, there is no significant technical risk because the technology is well established by now; compact fluorescent bulbs are reliable. But the utilization might be uncertain, depending on the circumstances. For instance, you might decide to buy a better desk lamp and sell the one you have. In that case, the optimal choice depends on the circumstances. The reader can refer to, say Reddy (2011), for a more in-depth discussion on how to perform simple risk analysis.

This example illustrates some general aspects of the investment decisions that will confront the building designer. First, the decision involves a calculation of life-cycle costs, or as a shortcut, an estimate of payback times. Second, it involves an examination of possible risks. Risks can depend on technology (e.g., how reliable is the equipment?) or on utilization (e.g., how will the building be used in the future?). In many cases, such as extra insulation or high-COP chillers, the risk is negligible because the technology is mature and there is no question about the utilization, so the rule of thumb about payback time and profitability can indeed be applied. Since the lifetimes are often on the order of 20 years, payback times of about 10 years may be quite interesting. This stands in sharp contrast to the very short payback times, often less than 2 years that are demanded by many industrial decision makers. The explanation lies in the far greater risks of most industrial investments where it is difficult to predict the future more than a few years ahead. But in more stable situations, longer payback times can prevail; for example, currently solar photovoltaic systems have payback times greater than 15 years (without incentives).

Being concerned with energy in buildings, we mention a further point: the energy used in the process of constructing a building. Fabricating of the steel, glass, concrete, and all the other building materials requires a great deal of energy, and further energy is needed for transport and for construction. This topic, known as *embedded energy*, has attracted much attention. However, we do not believe it to be of immediate concern to the HVAC design engineer as we explain with the following admittedly simplified argument.

In a free market economy, purchase decisions are based on money (and the communist societies have provided a warning example of what happens when one tries to ignore the principles of a free market). Therefore, energy investments must be made on the basis of the monetary value of the energy, not on the basis of its physical quantity. And, of course, one must include all costs, not just the energy.

As for the *embedded energy* in a building, its cost is implicitly contained in the prices paid for the material and for the construction. For example, the charges for delivering a window to a site include the cost of the fuel for the transport vehicle. With costs one can account for energy, but with energy one cannot account for all costs. Embedded energy does not include the cost of labor; therefore, it is not an appropriate criterion for investment decisions.

Of course, an efficient economy also presupposes that social costs are correctly taken into account. As an example of a social cost, suppose you burn a ton of coal without paying for the damage that will be caused by your contribution to acid rain and global warming: You impose a cost on society. The valuation of social costs is a difficult subject. Even more difficult is the implementation of laws, regulations, and taxes that will make people act in a way that is socially optimal. But enormous progress has been made, and continues to be made—that is part of our collective learning process. If policies are formulated correctly, they transmit the correct price signals to everybody, and the resulting decisions are indeed optimal.

It is not the purpose of this book to estimate social costs or to formulate policies. Nor would it be appropriate to try to do so during the design of one building. The designer is not prepared to evaluate the contribution that a building might make to the cost of global warming, but within the context of market prices she or he can and should optimize the design so as to provide the desired conditions of comfort for the lowest life cycle cost. Developing the necessary tools for that job is the purpose of this book.

1.6 Units and Conversions

In almost all countries of the world, engineers use SI units (Système International d'Unités), a standardized and rational system based on metric units. In SI, all units are derived from a few fundamental units: mass, length, time, temperature, amount of substance, and luminous intensity. Prefixes for powers of 10 are added according to the unit conversion table in this book. This table also provides the conversion factors for the old English units, which are still in use in the United States (but no longer in England). These latter are also known as U.S. Customary System or inch-pound (IP) units. In this book, we use the labels "SI" and "IP" in the equation numbers of dimensional equations to indicate the units. The reader can refer to say ASHRAE (1976) for a metric guide relevant to HVAC applications.

In IP, a multiplicity of units obscures the relation between equivalent quantities, and the need for conversion factors lurks at every step. For instance, to remove the heat produced by a 100 kW computer center, an engineer in a country with SI units specifies quite simply a chiller with a capacity of 100 kW; and if it is a gas-fired chiller with COP = 1.1, the engineer knows at once that the gas company will bill for (100/1.1) kWh of fuel for each hour of full-load operation. The engineer's counterpart in the United States has to convert 100 kW first to 28.4 tons of refrigeration to specify the chiller size and then to Btu per hour because gas is metered in therms (1 therm = 10^5 Btu).

The conversion of units could be written in several slightly different ways. We have chosen to multiply each quantity by a conversion factor that is equal to unity and whose units cancel the units we want to eliminate. For instance, to convert 30 in. to feet, multiply by 1 = 1 ft/12 in. and get

$$30\text{ in.} = 30\text{ in.} \times \frac{1\text{ ft}}{12\text{ in.}} = 2.5\text{ ft}$$

This method is systematic and safe.

Note that we include the units with the numbers, for good reason. If the units do not come out correctly, the result must certainly be wrong. Keeping the units, at least up to the point of verifying the correct units for the result, is a convenient trick for avoiding the most common errors to which we are prone in such calculations (forgetting a factor or dividing instead of multiplying).

Example 1.2: Example Illustrating Unit Conversion Calculations

Water of density ρ = 62.44 lb_m/ft^3 (1000 kg/m^3) flows at rate $\dot{V}$ = 10.0 gal/min (0.6308 × 10^{-3} m^3/s) through a pipe of interior diameter D = 1.0 in. (0.0254 m). Find the velocity pressure that is given by the formula

$$p = \frac{\rho v^2}{2} \qquad (1.2)$$

where v = velocity.

Given: ρ = 62.44 lb_m/ft^3 (1000 kg/m^3), D = 1.0 in. (0.0254 m), $\dot{V} = 10.0\text{ gal/min } (0.6308 \times 10^{-3}\text{ m}^3/\text{s})$

Assumptions: Assume for simplicity that all the water flows at the same velocity (actually: the water at the center flows faster than the water near the wall.)

$$v = \frac{\dot{V}}{D^2 \times (\pi/4)} \qquad (1.3)$$

Find: $p = \rho v^2/2$

Solution

In SI units. Here, everything is automatically consistent, and the quantities can be combined directly. From Equation 1.3:

$$v = \frac{0.6308 \times 10^{-3}\text{ m}^3/\text{s}}{(0.0254\text{ m})^2 \times (\pi/4)} = 1.245\text{ m/s}$$

And from Equation 1.2:

$$p = 0.5 \times 1000\text{ kg/m}^3 \times (1.245\text{ m/s})^2$$
$$= 775\text{ kg/(m}\cdot\text{s}^2) = 775\text{ Pa}$$

since the unit of pressure is 1 Pa = 1 N/m^2 = 1 $(kg \cdot m/s^2)/m^2$. This is about 0.8% of atmospheric pressure, the latter being 101 kPa under standard conditions.

In IP units. First, one must convert to consistent units:

$$\dot{V} = 10.0\text{ gal/min} \times \frac{0.00223\text{ ft}^3/\text{s}}{\text{gal/min}} = 0.0223\text{ ft}^3/\text{s}$$

and

$$D = 1.0\text{ in.} = 1.0\text{ in.} \times \frac{1.0\text{ ft}}{12.0\text{ in.}}$$

Hence,

$$v = \frac{0.0223\text{ ft}^3/\text{s}}{(\pi/4) \times (1/12.0)^2\text{ ft}^2} = 4.084\text{ ft/s}$$

$$\text{Pressure } p = 0.5 \times 62.44\text{ lb}_m/\text{ft}^3 \times (4.084\text{ ft/s})^2$$
$$= 520.7\text{ lb}_m/(\text{ft}\cdot\text{s}^2)$$

But this must be converted to IP units for pressure, using the relation between pound-force and pound-mass. This is usually indicated by including the constant

$$g_c = 32.17\ \text{lb}_\text{m}\cdot\text{ft}/(\text{lb}_\text{f}\cdot\text{s}^2)$$

in the denominator of the equation for velocity pressure, as

$$p = \frac{\rho v^2}{2g_c}$$

In fact, g_c is a conversion factor that is equal to unity.

Dividing by g_c, we obtain

$$p = \frac{520.7\ \text{lb}_\text{m}/(\text{ft}\cdot\text{s}^2)}{32.17\ \text{lb}_\text{m}\cdot\text{ft}/(\text{lb}_\text{f}\cdot\text{s}^2)} = 16.2\ \text{lb}_\text{f}/\text{ft}^2$$

There is one more step because HVAC engineers have a predilection for psi (= $\text{lb}_\text{f}/\text{in.}^2$) or for inWG (inches water gauge):

$$p = 16.2\ \text{lb}_\text{f}/\text{ft}^2 \times (1\ \text{ft}/12\ \text{in.})^2 = 0.112\ \text{psi}$$

which in turn is equal to

$$p = 0.112\ \text{psi} \times 27.68\ \text{inWG/psi} = 3.11\ \text{inWG}$$

This is about 0.8% of standard atmospheric pressure, 14.70 psi.

1.7 Orders of Magnitude Calculations

When we start something new, it is a good idea to estimate orders of magnitude before embarking on a full-fledged analysis. The art of doing such back-of-the-envelope calculations is an essential skill of a scientist or an engineer. An understanding of orders of magnitude is a cornerstone of intuition. Often, a decision can be made merely on the basis of a simple estimate. Sometimes, the input is so uncertain as to render the accuracy of detailed calculations meaningless. Besides, a simple calculation, even if of limited accuracy or valid only under restricted circumstances, can often provide a valuable check for the reliability of the result of more complicated analysis (or point out a bug in a computer program).

A related point is the accuracy to which numbers should be stated. Computers can be quite deceptive when printing output to the *n*th significant figure. Presenting more figures than are meaningful is not just illusory but also confusing, making the output more difficult to read. Therefore, *round off a final result to a point consistent with its accuracy*, keeping just an extra figure in case of doubt. But note that the relevant criterion depends on the context. For instance, one might want to test the sensitivity of a result to a small variation in the input, even if the input itself is quite uncertain. Also, during an ongoing calculation one must carry a larger number of decimal points to minimize propagation of errors (fortunately, modern computers are so powerful that we rarely need to worry about rounding before the end).

Here is a simple example to illustrate these points and to develop your intuition about the order of magnitude of energy flows in everyday life.

Example 1.3: Order of Magnitude Calculation

From what height do you have to jump into a bathtub to warm up the water by $\Delta T = 1°\text{C}$ (1.8°F), assuming all the kinetic energy goes into heating the water?

Given: A reasonable estimate of the water in a bathtub would be

M = 100 kg (220.5 lb_m), and your mass might be m = 70 kg (154 lb_m)

Find: Height h such that $mgh = Mc_p\Delta T$ (1.4)

Lookup values: g = 9.80 m/s² (32.17 ft/s²),

$$c_p = 4.186\ \text{kJ}/(\text{kg}\cdot°\text{C})\ [1.0\ \text{Btu}/(\text{lb}_\text{m}\cdot°\text{F})]$$

Solution

In SI units. The required thermal energy is

$$Mc_p\Delta T = 100\ \text{kg} \times 4.186\ \text{kJ}/(\text{kg}\cdot°\text{C}) \times 1°\text{C} = 418.6\ \text{kJ}$$

From Equation 1.4: $h = Mc_p\Delta T/(mg)$ = 418.6 kJ/(70 kg × 9.80 m/s²)

Noting that 1 J = 1 kg × m²/s², one obtains at once h = 610 m.

Since this is only an order-of-magnitude estimate, it is appropriate to state the result as 600 m.

However, if this number is to serve as input for further calculation, the full figures should be kept (e.g., if we want to convert it to IP units for comparison with the result of the following calculation in IP units).

In IP units. The required thermal energy is

$$Mc_p\Delta T = 220.5\ \text{lb}_\text{m} \times 1.0\ \text{Btu}/(\text{lb}_\text{m}\cdot°\text{F}) \times 1.8°\text{F} = 396.9\ \text{Btu}$$

To proceed, one needs to know the equivalence between mechanical and thermal energy in IP units; it is

$$1\ \text{Btu} = 778.16\ \text{ft}\cdot\text{lb}_\text{f} \quad (1.5)$$

The height is found after adding the factor g_c = 32.17 $lb_m \cdot ft/(lb_f \cdot s^2)$:

$$h = \frac{Mc_p \Delta T g_c}{mg}$$

$$= \frac{396.9 \text{ Btu} \times 778.16 \text{ ft} \cdot \text{lb}_f/\text{Btu} \times 32.17 \text{ lb}_m \cdot \text{ft}/(\text{lb}_f \cdot \text{s}^2)}{154 \text{ lb}_m \times 32.17 \text{ ft/s}^2}$$

$$= 2005 \text{ ft} \approx 2000 \text{ ft}$$

Comments

In everyday life, warming up a bath by 1°C does not seem like a big deal. By contrast, climbing 600 m will take an average person around 2 h. In psychological terms, the magnitudes appear very different.

This relation between thermal and mechanical energy quantities of everyday life is quite typical. Thermal energy dominates the energy balance of buildings, while the contribution of gravitational energy is small or negligible. An intermediate role is played by the mechanical energy needed to overcome the friction of transporting fluids around a building. The HVAC designer does have to pay some attention to the energy consumption of pumps and fans. The latter can be a fairly important item because of the large friction of air in typical air distribution systems.

Problems

The main purpose of the problems in Chapter 1 is to develop some familiarity with energy conversions and units, and to acquire a sense of the order of magnitude of energy flows in and around buildings. For several of these problems, you will need to make your own assumptions. State them clearly. Data on properties of materials can be found in the appendices of this book. Approximate degree of difficulty indicated by italic number from 1 to 4.

1.1 You are measuring electricity consumption by various end uses in a building. The sensors you use to make the measurements are accurate to about 1% of the reading. You are able to measure the following items separately and find that

- The telephone system uses 20 W
- The lights use 10.3 kW
- The computers and office equipment use 15.4 kW
- The HVAC system uses 45 kW

If these are the only electrical loads in the building, what is the total electrical consumption? How many figures are significant? *(1)*

1.2 You are an American working with a European customer and must convert all IP units to SI units. What are the equivalent SI units for the following?

- 13.4 MMBtu/h
- 5 tons of refrigeration
- 25 psia
- 120 gal/min
- 0.5 Btu/h/ft²/°F *(1)*

1.3 You see the following numbers in an equipment catalog from a Japanese manufacturer.

- 3.5 dyn/cm²
- 1.3 m³/s
- 23.1 N-m (torque)
- 1.2 kg/m³
- 0.6 kJ/kg/°C

You want to sell them to a U.S. engineer. What are the equivalent IP values? *(1)*

1.4 A building with an interval volume of 12,450 ft³ experiences natural infiltration of 0.333 air change per hour. What is the infiltration rate in cubic meters per second? *(1)*

Helpful hint: 1 year = 8760 h = 3.1536 × 10^7 s (some find it convenient to remember as rule of thumb that 1 year ≈ π × 10^7 s)

1.5 List the major pieces of energy-consuming equipment in your everyday life. For each, estimate the peak power, average power, and annual consumption. *(1)*

1.6 Consider a residential furnace that is rated for a heat output of 50,000 Btu/h. Express the heat output in SI units. How many 100 W light bulbs would produce the same amount of heat? *(1)*

1.7 Convert the horsepower rating of your car (or the car you would like to have) to kilowatts (the unit that is now officially used for cars in Europe). *(1)*

1.8 Estimate the rate of energy transfer during the filling of the tank of a typical car at a gas station (heating value of gasoline ≈ 3.4 × 10^7 J/L). Assume a 40 L gasoline tank and that a fill-up takes 5 min Compare with a typical all-electric residence assuming 1 kW continuous average (without space heating). *(1)*

1.9 Energy prices for fuels are usually quoted in terms of dollars per volume or per weight. Convert the following to cents per kilowatt hour and to dollars per MBtu.

(a) Gasoline at $1.20 per gallon (heating value ≈ 21,000 Btu/lb)

(b) Crude oil at $20 per barrel (heating value ≈ 18,000 Btu/lb)

(c) Coal at $60 per ton (heating value ≈ 12,000 Btu/lb)

(d) Electricity at 10¢/kWh *(2)*

1.10 A typical house in a temperate climate (on the order of 2500 K·days [4500°F·days], e.g., Albuquerque; Louisville, Kentucky; Washington, DC; or Paris) needs about 50–100 GJ of heat per year. Suppose this energy were collected as solar heat in summer at a temperature of 70°C (158°F) and stored. How large a storage tank would be needed if useful heat could be withdrawn as long as the tank is above 30°C (86°F)? Neglect losses from the tank. (Such seasonal storage has been proposed and tested in various places.) *(2)*

1.11 Consider a house with a 1 ton air conditioner, running 500 h/year at full power.

(a) What is the annual thermal energy delivered to the house?

(b) What is the corresponding electricity consumption, if the air conditioner delivers 2.0 J_t of cooling for 1.0 J_e of electricity (COP = 2.0)?

(c) Suppose one stores winter ice in an ice tank for summer cooling. How large a volume is needed if there are no losses from storage? Consider only the latent heat of melting, and take the density of ice as 0.9 ton/m^3. (Such seasonal storage has been proposed and tested in various places.)

(d) How much is the cooling energy of 1 ton (2205 lb_m) of ice worth?

NOTE: The use of residential air conditioning depends on climate and on the construction of the building, but there is a strong behavioral component as well. In midlatitudes of the United States, the seasonal energy consumption for cooling in identical buildings can vary enormously, depending on how the occupants set the thermostat and how they control the heat gains (temperature preferences, utilization of lights and appliances, use of shades, and night ventilation). *(2)*

1.12 A typical personal computer draws 100 W_e, while turned on, almost independent of how it is being used. Estimate the number of hours of utilization per year and the annual energy cost if the price of electricity is 10¢/kWh_e. *(1)*

1.13 The *Reynolds number* Re is a dimensionless group of variables that plays an important role in fluid mechanics. It represents the ratio of inertial and viscous forces in a fluid, and it is defined as

$$\mathrm{Re} = \frac{vD}{\nu}$$

where
v is the velocity
D is the characteristic linear dimension of system
ν is the kinematic viscosity

Evaluate Re for air ($\nu = 15 \times 10^{-6}$ m^2/s) if v = 10 m/s and *D* = 1 m, in both SI and IP units. *(1)*

1.14 The average solar flux incident on a horizontal surface in the United States (averaged over 24 h and 365 days) is about 180 W/m^2 [57.06 Btu/(h·ft^2)]. How much land would be needed to supply the total energy demand of the United States, approximately 80×10^{18} J (76 quad) per year, by means of photovoltaics (efficiency around 10%) and biomass (efficiency around 1%)? Compare with the land area of the United States, 9.16×10^6 km^2 (3.54×10^6 miles^2). *(2)*

1.15 A large coal or nuclear power plant has a peak output on the order of 1 GW_e. Over the year, its actual output is likely to average only 70% of that, due to the need for maintenance and due to an imperfect match between the supply and demand of electricity.

(a) What is the annual energy output, in kWh, in joules, and in Btu?

(b) The year-round average residential electricity use amounts to about 1 kW_e per housing unit in the United States. Assuming 3.2 residents per housing unit, estimate roughly how many power plants are needed to satisfy the residential demand of the New York metropolitan area (population around 20 million).

(c) The heating value of coal is approximately 12,000 Btu/lb, and the conversion efficiency of coal power plants is around 33%. Calculate the quantity of coal needed to satisfy the annual electricity consumption of a housing unit in (*b*). *(2)*

1.16 Suppose an air-to-air heat exchanger (for recovering heat from ventilation air) costs $1200 and reduces the heating bill by $200 per year. What is the payback time? *(1)*

1.17 Consider two choices for the furnace to heat a building: an ordinary gas furnace with efficiency 75% (seasonal average), costing $1000, and a condensing furnace with efficiency 95% (seasonal average), costing \$2000. Suppose the annual heating load is 100 MBtu (105.5 GJ) and the fuel price \$7.00 per MBtu (\$6.64 per GJ). Calculate the payback time for the second furnace relative to the first. Is it a good investment if both furnaces can be expected to last 15 years? *(2)*

1.18 Suppose the annual heating bill of a building is \$1000. How much could you pay for a solar heating system that reduces this bill by one-half if the payback time is to be no longer than 10 years? *(1)*

1.19 Write a 200 word essay on why buildings have been targeted as the most promising sector to reduce energy use? *(2)*

1.20 Go to the Office of Energy Efficiency & Renewable website: energy.gov/eere/buildings/Key-activities energy-efficiency/ and write a 400 word essay summarizing the types of information provided therein. *(2)*

References

ASHRAE (1976). *ASHRAE SI Metric Guide for Heating, Refrigerating, Ventilating and Air-Conditioning.* American Society of Heating, Refrigerating and Air-Conditioning Engineers, Atlanta, GA.

ASHRAE Applications (2011). *Handbook of HVAC Applications.* American Society of Heating, Refrigerating and Air-Conditioning Engineers, Atlanta, GA.

BEDB (2011). *Building Energy Data Book.* Energy Efficiency and Renewable Energy Program, U.S. Department of Energy, Washington, DC, http://buildingsdatabook.eren.doe.gov/, accessed on January 2015.

Brambley, M.R., D.B. Crawley, D.D. Hostetler, R.C. Stratton, M.S. Addison, J.J. Deringer, J.D. Hall, and S.E. Selhowitz (December 1988). Advanced energy design and operation technologies research: Recommendations for a U.S. Department of Energy Multiyear Program Plan, Report PNL-6255, Battelle Pacific Northwest Laboratory, Richland, WA.

Butti, K. and J. Perlin (1980). *A Golden Thread.* Cheshire Books, Palo Alto, CA. A condensed version can be found in chap. 1 of J.F. Kreider and F. Kreith, eds., *Solar Energy Handbook.* McGraw-Hill, New York.

Claridge, D.E., ed. (1998). Special issue on energy conservation and solar buildings. *ASME J. Solar Energy Eng.*, 120(3), American Society of Mechanical Engineers, New York.

EIA (1995). *Annual Energy Outlook 1995.* U.S. Energy Information Administration, U.S. Department of Energy, Washington, DC.

EIA (2005). *Annual Energy Outlook 2005.* U.S. Energy Information Administration, U.S. Department of Energy, Washington, DC, http://www.eia.gov/forecasts/aeo/, accessed on January 2015.

Kydes, A.S. (2008). Outlook for U.S. energy consumption and prices in the midterm, in *Energy Management and Conservation Handbook* (F. Kreith and D.Y. Goswami, eds.). CRC Press, Boca Raton, FL.

Mull, T.E. (2001). *Practical Guide to Energy Management for Facilities Engineers and Plant Managers.* ASME (American Society of Mechanical Engineers) Press, New York.

Reddy, T.A., ed. (2003). Special issue: Emerging trends in building design, diagnostics and operation, *ASME J. Solar Energy Eng.*, 125(3), American Society of Mechanical Engineers, New York.

Reddy, T.A. (2011). *Applied Data Analysis and Modeling for Energy Engineers and Scientists.* Springer, New York.

Rosenfeld, S.I. (1990). Worker productivity: Hidden HVAC cost. *Heat. Piping Air Cond.*, September, 117–119.

Turner, W.C. (2006). *Energy Management Handbook*, 5th ed. The Fairmont Press, Prentice Hall, Lilburn, GA.

USDOC (2001). U.S. statistical abstract. Published annually by the U.S. Department of Commerce, Bureau of the Census, Government Printing Office, Washington, DC.

USDOE (June 1999). *BTS Core Data Book.* Office of Building Technology, U.S. Department of Energy, Washington, DC.

2

Basic Thermal Science

ABSTRACT This chapter serves as a brief review of basic concepts and principles related to thermal science, specifically fluid mechanics, thermodynamics, and heat transfer. These topics are of critical importance in the study of environmental heating and cooling loads, of air-conditioning processes, and of the performance of related equipment and systems. This chapter, given its focus on building-related issues, will also be useful for those students who have already had a first course in engineering thermodynamics and heat transfer. We provide an introduction to several basic thermodynamic properties followed by a review of the ideal gas law and how numerical property values of perfect gases are determined from the tables. Next, we discuss the equations for conservation of mass, momentum, and energy, as typified by the first and second laws of thermodynamics. The laws of heat transfer govern the rate at which heat energy must be supplied to or removed from a building to maintain comfort of occupants or to meet other thermal requirements in buildings and equipment. Finally, we review the key features of the three modes of heat transfer, namely, conduction, convection, and radiation, and illustrate these concepts by way of several examples related to building energy load calculations.

Nomenclature

A	Area, m^2 (ft^2)
c_p	Specific heat at constant pressure, kJ/(kg·K) [Btu/(lb$_m$·°F)]
c_v	Specific heat at constant volume, kJ/(kg·K) [Btu/(lb$_m$·°F)]
D	Diameter, distance, m (ft)
D_h	Hydraulic diameter (Equation 2.58), m (ft)
E	Radiation emissive power, W/m^2 [Btu/(h·ft^2)]
E_b	Blackbody emissive power, W/m^2 [Btu/(h·ft^2)]
$E_{b\lambda}$	Spectral blackbody emissive power, W/(m^2·μ) [Btu/(h·ft^2·μ)]
F	Force, N (lb$_f$)
F	Degrees of freedom
F_{12}	Radiation shape factor between surfaces 1 and 2
F_g	Correction factor for glycol (Equation 2.60)
g	Acceleration due to gravity = 9.8 m/s^2 (32 ft/s^2)
g_c	Conversion factor in IP units, 32.17 lb$_m$·ft/(lb$_f$·s^2)
H	Enthalpy, kJ (Btu)
H	Height, m (ft)
h	Specific enthalpy, kJ/kg (Btu/lb$_m$)
h_{c+r}	Combined heat transfer coefficient for convection and radiation (h_{con} + h_{rad}), W/(m^2·K) [Btu/(h·ft^2·°F)]
h_{con}	Convection heat transfer coefficient, W/(m^2·K) [Btu/(h·ft^2·°F)]
h_i	Indoor surface heat transfer coefficient, W/(m^2·K) [Btu/(h·ft^2·°F)]
h_o	Outdoor surface heat transfer coefficient, W/(m^2·K) [Btu/(h·ft^2·°F)]
h_{rad}	Radiation heat transfer coefficient, W/(m^2·K) [Btu/(h·ft^2·°F)]
k	Thermal conductivity, W/(m·K) [Btu/(h·ft·°F)]
L	Length, m (ft)
l	Length or perimeter, m (ft)
m	Mass, kg (lb$_m$)
$\dot{m}$	Mass flow rate, kg/s (lb$_m$/h)
N	Number of components of mixture
P	Number of phases in mixture
p	Pressure, Pa (lb$_f$/in.^{2})
Q	Energy consumption or heat transfer with subscripts c for cooling and h for heating, kJ (Btu)
$\dot{Q}$	Heat rate or heat flow per unit time with subscripts c for cooling load, h for heating load, lat for latent load, floor for heat flow to floor, etc., W (Btu/h)
$\dot{q}$	Heat flux, W/m^2 [Btu/(h·ft^2)]
R	Thermal resistance, K/W [(°F·h)/Btu] [$\equiv \Delta T/\dot{Q}$]
R	Gas constant, kJ/(kg·K) [ft·lb$_f$/(lb$_m$·°R)]
R_{cd}	Thermal conductive resistance per unit area, (m^2·K)/W [(h·ft^2·°F)/Btu] [$\equiv \Delta T/\dot{q}$]
$\bar{R}_{cd}$	Effective R value averaged over the building element, (m^2·K)/W [(h·ft^2·°F)/Btu]
R^*	Universal gas constant, kJ/(kg·mol·K) [ft·lb$_f$/(lb·mol·°R)]
Re	Reynolds number (Equation 2.21)
r	Radius, m (ft)
S	Conduction shape factor
S	Entropy, kJ/K (Btu/°R)

s	Specific entropy, kJ/(kg·K) [Btu/(lb_m·°R)]
T, t	Temperature, K or °C (°R or °F)
T_n	Temperature of node n, °C (°F)
t	Time, s
U	Overall heat transfer coefficient, W/(m^2·K) [Btu/(h·ft^2·°F)]
U	Internal energy, kJ (Btu)
U_{cd}	Unit thermal conductance (=1/R_{cd}), W/(m^2·K) [Btu/(h·ft^2·°F)]
$\bar{U}$	Effective U value averaged over building element, W/(m^2·K) [Btu/(h·ft^2·°F)]
u	Specific internal energy, kJ/kg (Btu/lb_m)
V	Volume, m^3 (ft^3)
$\dot{V}$	Volumetric flow rate, m^3/s or L/s (ft^3/min)
$\dot{V}_o$	Outdoor airflow rate, L/s (ft^3/min)
v	Velocity or wind speed, m/s (mph, [ft/s])
v	Specific volume, m^3/kg (ft^3/lb_m)
W	Work, kJ (ft·lb_f)
$\dot{W}$	Power, W (ft·lb_f/s)
X, Y, Z	Distance, m (ft)
x, y, z	Distance, m (ft)

Greek

α	Absorptivity
β	Surface tilt angle, degrees
δ	Small increment
δ	Boundary layer thickness, m (ft)
ε	Emissivity
λ	Wavelength
μ	Dynamic or absolute viscosity, Pa·s [lb_m/(ft·s)]
ρ	Density, kg/m^3 (lb_m/ft^3)
ρ	Reflectivity
σ	Stefan–Boltzmann constant, W/(m^2·K^4) [Btu/(h·ft^2·°R^4)]
τ	Shear stress, N/m^2 (lb_f/ft^2)
τ	Transmissivity
υ	Kinematic viscosity, m^2/s (ft^2/s)
ΔT	Temperature difference, K (°F)
Δx	Thickness of layer, m (ft)

Subscripts

abs	Absolute pressure
atm	Local atmospheric pressure
b	Black body
cd	Conduction
comp	Compressor
con	Convection
cond	Condenser
eff	Effective
evap	Evaporator
ext	Exterior
f	Fluid
g	Gauge pressure, glycol
i	Indoor, interior, inner
in	Inlet, flowing in
ins	Insulation
int	Internal
j	Index of surface during radiant heat exchange
max	Maximum
o	Arbitrary reference state, outdoors, out
out	Outlet, flowing out
rad	Radiation
s	Surface
tot	Total
w	Water, water vapor

2.1 Fluid and Thermodynamic Properties

Heating and cooling of buildings involve processes, equipment, and systems that transfer and control heat flows to and from the building interior. A strong understanding of the basic principles studied under thermal sciences is imperative in this regard:

- *Fluid mechanics* is the science dealing with properties of fluids, governing laws and conditions of fluid statics and fluid motion, and with the resistance to flow outside and inside solid surfaces.
- *Thermodynamics* is the science dealing with energy and its transformations and the relationships of the various properties of a substance as it undergoes changes in pressure and temperature.
- *Heat transfer* is the science and art of determining the rate at which heat moves through substances under various externally imposed temperature and/or boundary conditions.

All three are important in the design and analysis of buildings since the performance and efficiency of heating and cooling systems for buildings are determined to a large part by the restrictions that fluid sciences place on HVAC systems and equipment. In this chapter, we review notions of thermal sciences relevant to commercial, institutional, and residential buildings.

A *property* is any attribute or characteristic of matter that can be observed or evaluated quantitatively. The most important ones are as follows: pressure, temperature, specific volume (or its inverse, density), viscosity, specific heat, specific internal energy, specific enthalpy, and specific entropy.

Equilibrium is a condition of balance maintained by an equality of opposing forces. There are different types of equilibrium: thermal, mechanical, and chemical. The study of thermodynamics has to do with determining end states, and not with the dynamics of the process (e.g., how fast the process change occurs). The concept of equilibrium is important as it is only in an equilibrium state that the thermodynamic properties have meaning. We implicitly assume that the system moves from one state of equilibrium to another very slowly as it undergoes a process, a condition called quasi-equilibrium or *quasi-static* process.

1. *Static pressure p* is defined as the stress (force F per unit area A) normal to the surface at any point of a plane in a fluid at rest. In equation form, it is given by

$$p \equiv \lim_{A \to 0} \frac{F}{A} \tag{2.1}$$

The consistent units of pressure are the pascal* in SI and pound-force per square foot in IP. An inconsistent unit of pressure, pound-force per square inch (abbreviated "psi"), is often used in the IP unit system. One psi (1 psi) is equivalent to 6,894 Pa; 1 atm is equivalent to 14.696 psi or 101,325 Pa.

In practice, pressures are often referenced to the local atmospheric pressure. Pressures so expressed are called "gauge pressures." To convert from gauge pressure to absolute pressure, we simply add the local atmospheric pressure to the gauge pressure:

$$p_{abs} = p_g + p_{atm} \tag{2.2}$$

where

p_{abs} is the absolute pressure
p_g is the gauge pressure
p_{atm} is the local atmospheric pressure

In thermodynamic calculations, one must always use the absolute pressure. Gauge pressures can be used whenever pressure differences are involved.

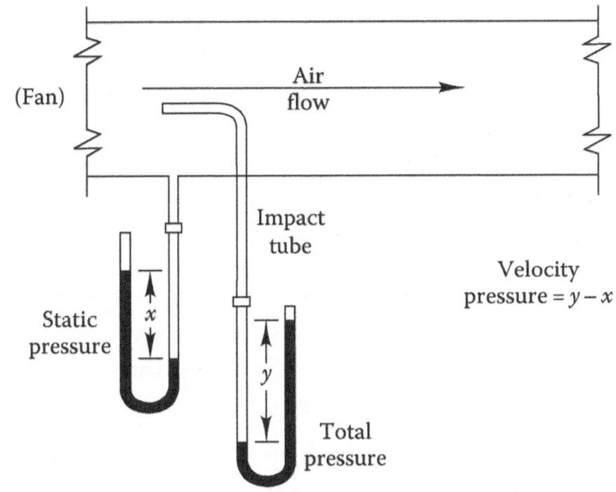

FIGURE 2.1
Pressure measurement arrangements for airflow in a duct.

Figure 2.1 shows how pressure measurements are made in an air duct. A similar approach is used for liquid measurements. The static gauge pressure is the pressurization of the duct air above the local atmospheric pressure. The total pressure is the pressure measured by a tube facing into the flow. The difference between the total and static pressures is due to the velocity of the fluid and is called the "velocity pressure." These concepts are elaborated in Chapter 16.

This measurement technique for air systems leads to another pressure unit commonly used in the HVAC industry. Pressure differences in air systems are often expressed in terms of water column or water gauge (abbreviated WG). In IP units, the unit is inches of water gauge (inWG); in SI units, the water column height can be stipulated in centimeters or millimeters.

2. *Temperature T or t* of a substance indicates its thermal state (i.e., the level of random molecular motion). It is a measure of the potential for heat flow from one substance to another substance in contact with it. Temperature t is a relative scale. Reference points are the freezing point of water (0°C in the Celsius scale and 32°F in the Fahrenheit scale) and the boiling point of water (100°C in the Celsius scale and 212°F in the Fahrenheit scale). Often, the absolute temperature scale T (a state where there is no molecular motion) is more relevant:

SI units: Kelvin scale $T = (t + 273.15)$ K with t in °C (2.3 SI)

IP units: Rankine scale $T = (t + 459.67)$°R with t in °F (2.3 IP)

* To get a feeling for the magnitude of this unit, it may be helpful to recall the famous apocryphal story about the discovery of the law of gravitation and note that an apple weighs about 1 N (newton). When one slices the apple thinly enough to cover 1 m², the pressure is 1 N/m² = 1 Pa.

3. *Density* ρ of a fluid is the mass per unit volume. It is more common to use its reciprocal: the *specific volume v*, the volume occupied per unit mass. The density of air at standard atmospheric pressure and 25°C (77°F) is approximately 1.2 kg/m^3 (0.075 lb_m/ft^3). The density and specific volume of a vapor or gas are affected by both pressure and temperature. Tables are used to determine both; in some cases, they can be calculated from basic thermodynamic property relations. The density of liquids is usually assumed to be a function of temperature only.
4. *Dynamic viscosity* μ is a measure of the molecular resistance to fluid flow, for example, when a solid is to be moved through a fluid or when a fluid is made to flow inside a pipe or duct. This resistance directly determines the nonrecoverable energy under the aforementioned instances. For common fluids such as water and air, the viscosity is the proportionality constant in the expression relating shear stress τ in the fluid to the velocity gradient (dv/dy) in the direction y (see Figure 2.2):

$$\tau = \mu \frac{dv}{dy} \tag{2.4}$$

The common units of viscosity are pound-mass per foot per second (an inconsistent unit derived from the consistent unit pound-force-second per square foot) in IP and pascal-seconds in SI units. Another common viscosity unit is the poise (abbreviated P) from the old cgs system of units (1 Pa·s = 10 P). The viscosity of water at 21.1°C (70°F) is approximately 1 cP, (0.001 Pa·s), or 0.00067 $lb_m/(ft \cdot s)$. The conversion factors among various viscosity units are given in the conversion factor table of this book.

Oil is an example of a highly viscous fluid whereas air has low viscosity. Under certain circumstances such as fluid flow far from any solid surfaces, the effect of viscosity can be neglected. The fluid is then said to be an ideal fluid or an *inviscid fluid*.

Another type of viscosity is often used: the quotient of viscosity and density. This combination of properties is a property itself, which is called the "kinematic viscosity":

$$\upsilon = \frac{\mu}{\rho} \tag{2.5}$$

The viscosity is a strong function of temperature for both air and water. For air, the viscosity is well expressed by a power law proportionality over a limited range of temperatures (for example, 5°C–120°C or 40°F–250°F):

$$\mu \propto T^{0.67} \tag{2.6}$$

in which the proportionality constant will depend on the units used.

In SI units of pascal-seconds, the viscosity of pure water is given by Perry et al. (1999):

$$\mu = \frac{0.1}{2.1482 \times (D' + \sqrt{8078.4 + D'^2}) - 120} \tag{2.7 SI}$$

where

$$D' \equiv (T - 8.435)°C \tag{2.8 SI}$$

The viscosity [$lb_m/(ft \cdot s)$] in IP units is also given as follows:

$$\mu = \frac{1}{32.0 \times (D' + \sqrt{8078.4 + D'^2}) - 1786} \tag{2.7 IP}$$

where

$$D' \equiv (0.556 \times T - 26.21)°F \tag{2.8 IP}$$

A common liquid used in HVAC systems to avoid water from freezing is an aqueous solution of ethylene or propylene glycol.

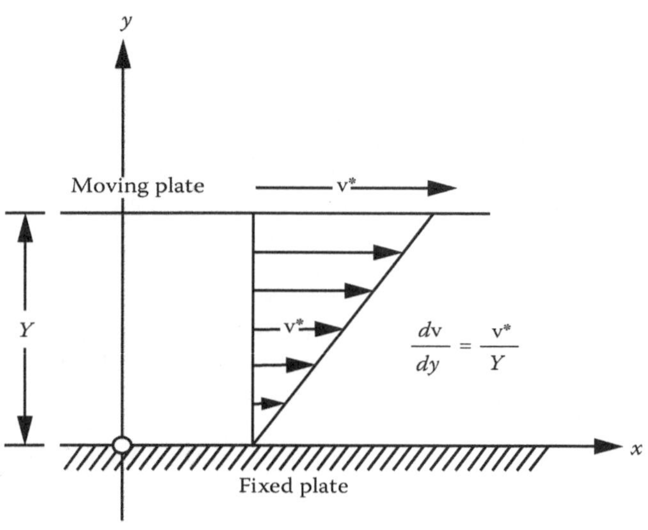

FIGURE 2.2
Velocity profiles and gradients in shear flow.

TABLE 2.1

Properties of 30% (by Weight) Aqueous Solution of Propylene Glycol

Temperature, °F	Viscosity, cP	Density, lb_m/ft^3	Thermal Conductivity, Btu/(h·ft·°F)	Specific Heat, Btu/(lb_m·°F)
20	9.08	64.95	0.242	0.926
40	5.44	64.69	0.249	0.930
60	3.44	64.40	0.255	0.935
80	2.32	64.08	0.261	0.939
100	1.65	63.72	0.266	0.943
120	1.24	63.34	0.270	0.948
140	0.97	62.92	0.274	0.952
160	0.80	62.48	0.278	0.956
180	0.67	62.00	0.281	0.961
200	0.59	61.49	0.283	0.965

Multiply	By	To Obtain
cP	0.001	Pa·s
Lb_m/ft^3	16.01	kg/m^3
Btu/(h·ft·°F)	1.73	W/(m·°C)
Btu/(lb_m·°F)	4.19	kJ/(kg·°C)
cP	0.000672	lb_m/(ft·s)

Source: Born, D.W., *ASHRAE Trans.*, 95(pt. 2), 1989. Copyright ASHRAE, www.ashrae.org.

Glycol solutions are denser and more viscous and provide poorer heat transfer (lower thermal conductivity and lower specific heat) than pure water in the same application. Table 2.1 is an example of some properties of a 30% aqueous solution of propylene glycol that freezes at approximately –15°C (5°F).

5. *Specific heat** c of a substance is the quantity of energy required to raise the temperature of a unit mass by 1 K or 1°R. For gases, one distinguishes between two cases: specific heat at constant pressure c_p and specific heat at constant volume c_v. The former is appropriate in air conditioning because the processes occur at constant pressure. Approximately, c_p = 1.00 kJ/(kg·K) or 0.24 Btu/(lb_m·°F) for dry air, 4.19 kJ/(kg·K) or 1.0 Btu/(lb_m·°F) for liquid water, and 1.86 kJ/(kg·K) or 0.444 Btu/(lb_m·°F) for water vapor.
6. *Specific internal energy* u refers to the energy stored by a substance due to the microscopic motion and/or position of the molecules. This form of energy consists of two parts: the internal kinetic energy due to the velocity of the molecules, and the internal potential energy due to the attractive forces between molecules. Changes in the average velocity of molecules are indicated by temperature changes of the substance. Internal energy is a property that is accurately defined from the first law of thermodynamics as applied to closed systems (Equations 2.26 through 2.28). The units of specific internal energy are in kJ/kg or Btu/lb_m.
7. *Specific enthalpy* h is a very important property defined as $(u + pv)$ where u is the internal energy (itself a property) and $(p \cdot v)$ is the *flow work*, i.e., the work done on the fluid to force it into a control volume. It is thus a measure of the total energy, which includes the energy contained in the system (or that necessary to create the system condition) and the amount of energy required to establish its volume and pressure. Enthalpy is also a property that is accurately defined from the first law of thermodynamics presented in Section 2.5. The units of specific enthalpy are in kJ/kg or Btu/lb_m.
8. *Specific entropy* s is another important property that cannot be directly measured (such as internal energy or enthalpy). It is defined from the second law of thermodynamics presented in Section 2.6. Entropy is a measure of the energy that is not available for work during a thermodynamic process due to the fact that natural processes tend not to be reversible.

* One needs to distinguish between extensive properties and intensive properties. While extensive properties are those, such as volume V (m^3 or ft^3) or enthalpy H (kJ or Btu), that depend on the mass of the substance, intensive properties, such as temperature and pressure, do not. Our notation is to use lower case symbols for intensive properties, i.e., properties per unit mass. Thus, v would denote the specific volume, while u and h the internal energy and enthalpy per unit mass, respectively. The use of the term "specific" (for example, specific heat) is recommended in order to avoid ambiguity.

For example, thermal energy always flows spontaneously as heat from regions of higher temperature to regions of lower temperature. Such processes reduce the state of order of the initial system by homogenization, and therefore entropy is an expression of the degree of disorder or chaos at the microscopic level within the system. The units of specific entropy are kJ/(kg·K) or Btu/(lb_m·°F). The online HCB software contains tables of fluid properties for air, water, glycol solutions, and other liquids used in building HVAC systems. Appendix A includes property tables of dry air (Table A1), liquid water (Table A2) and steam (Tables A3 and A4).

2.2 Determining Property Values

2.2.1 Gibbs Phase Rule

A pure substance is one that is uniform and invariable in chemical composition. Phase is a quantity of matter homogeneous throughout in chemical composition and physical structure. Thus, a pure substance may exist in more than one phase, such as a mixture of liquid water and water vapor (steam).

The *Gibbs phase rule* is a useful rule that specifies the number of independent intensive properties (or degrees of freedom F) needed to completely specify the thermodynamic state of a fluid (liquid or gaseous). It is expressed as $F = 2 + N - P$ where P is the number of phases and N the number of components in the mixture. The thermodynamic state of a single-substance system—e.g., air in a building, steam in a boiler, or refrigerant in an air conditioner—is defined by specifying $F = 2 + 1 - 1 = 2$, i.e., two independent, intensive thermodynamic coordinates or properties. For moist air that is a mixture of dry air and water vapor, $F = 2 + 2 - 1 = 3$, i.e., three independent properties. That this is indeed the case will be borne out in Chapter 13 that deals with psychrometrics.

We have two sources of the properties of materials—either simple equations of state, such as the ideal gas law, or tables computed from complex equations based on careful measurements. In reality, these property equations apply to a narrow range of conditions that fortunately often cover the ranges pertinent to HVAC processes. The use of such formulae simplifies analysis considerably and hence their appeal. In this text, we will be able to use the ideal gas law for dry or moist air and all the gaseous constituents of air. We will use the tabular approach for properties of water and of refrigerants.

2.2.2 Ideal Gas Law

Real gases are often approximated to "ideal" gases because of the resulting simplifications in the analysis. There are two types of ideal gases: semiperfect and perfect. A *semiperfect gas* is one that exactly obeys the following equations:

$$pv = f(T) \quad \text{and} \quad u = f(T) \tag{2.9}$$

where
- p is the absolute pressure
- v is the specific volume
- T is the temperature in absolute units

A *perfect gas*, on the other hand, is one for which the product (pv) is proportional to temperature T. Specifically, the ratio of pressure p times molar volume $\bar{v}$ divided by absolute temperature T is observed to be a constant—the ideal gas constant—as the pressure of the gas is allowed to approach zero. That is,

$$\lim_{p\to 0} \frac{p\bar{v}}{T} = R^* \tag{2.10}$$

in which R^* is the *universal gas constant*. Its value is 8314.41 J/(kg·mol·K) or 1545.35 ft·lb_f/(lb_m·mol·°R). One lb_m·mol is a pound-mole, i.e., the mass of 1 mol of a substance in pound-mass. (Recall by Avogadro's law that 1 g·mol of any substance contains 6.023×10^{23} molecules.) For example, 1 lb_m·mol of nitrogen contains 28 lb_m of nitrogen.

Since we do not use molar volumes in HVAC analysis, we replace Equation 2.10 with the following form using the *gas constant* for a specific gas:

$$pv = RT \tag{2.11}$$

or equivalently,

$$p = \rho RT \tag{2.12}$$

The gas constant R is calculated by dividing R^* by the molecular weight of the gas. Since the molecular weight of air is 28.97, the gas constant for air is R_{air} = 1545.35/28.97 = 53.35 ft·lb_f/(lb_m·°R). In SI units, the gas constant has a value of 287 J/(kg·K). The form of Equation 2.11 is called the "ideal gas law" or "ideal gas equation of state."

Fortunately, the ideal gas law applies at pressures other than close to zero. In fact, the ideal gas law is very accurate for air and its principal constituents up to several atmospheres of pressure and at temperatures involved in building HVAC processes. Therefore, for HVAC engineering calculations involving air, the ideal gas law is entirely satisfactory. Note that neither steam nor refrigerants can be treated as ideal gases. We will treat their properties in Section 12.3.

Ideal gases should not be confused with *pure substances*. A mixture of gases (such as air) is not a pure substance. However, if no change of phase is involved (as in most HVAC processes—see Chapter 13), air can be assumed to be a pure substance for purposes of mathematical analysis.

Example 2.1: Ideal Gas Law

Find the mass of air enclosed in an office room at 20°C (68°F) if the dimensions of the room are 10 m × 10 m × 2.5 m (32.8 ft × 32.8 ft × 8.2 ft). The building is located at sea level.

Given: T = 20°C, sea level

Find: m_{air}

Lookup values: p = 101.325 kPa (standard atmospheric pressure at sea level)

Solution

The ideal gas law applies to this problem. First, we find the density from Equation 2.12:

$$\rho = \frac{p}{RT} = \frac{101{,}325\ \text{Pa}}{287\ \text{J/(kg}\cdot\text{K)} \times (20+273)\ \text{K}} = 1.20\ \text{kg/m}^3$$

The mass can then be deduced as the product of the density and room volume V:

$$m_{air} = \rho V = 1.20\ \text{kg/m}^3 \times (10\ \text{m} \times 10\ \text{m} \times 2.5\ \text{m})$$
$$= 300\ \text{kg}\ (660\ \text{lb}_\text{m})$$

Comments

Relative to the other contents of a typical office, the mass of air is small. As we will see in later chapters on building dynamics, the mass of air can usually be ignored for such calculations.

2.2.3 Tabular Data for Perfect Gases

The specific heats for a semiperfect gas are given by

$$\text{Specific heat at constant volume } c_v = \left(\frac{du}{dT}\right) \tag{2.13}$$

$$\text{Specific heat at constant pressure } c_p = \left(\frac{dh}{dT}\right) \tag{2.14}$$

c_v and c_p are not constant but increase slightly with temperature. On the other hand, the specific heats are constant for perfect gases. Consequently, the properties of perfect gases are easily determined. For example, the specific heat at constant volume c_v and at constant pressure c_p can be used to find changes in the specific internal energy u and the specific enthalpy h.

$$u - u_o = c_v(T - T_o) \tag{2.15}$$

and

$$h - h_o = c_p(T - T_o) \tag{2.16}$$

where the subscripted variables indicate a specified but arbitrary reference state o.

Another useful relationship also exists between c_p and c_v

$$R = (c_p - c_v) \tag{2.17}$$

where R is the gas constant.

The final property needed in second law calculations is the entropy. The specific entropy of an ideal gas (relative to the reference state T_o, v_o, s_o) can be found from

$$s - s_o = c_v \ln\frac{T}{T_o} + R\ln\frac{v}{v_o} \tag{2.18}$$

The variation of the aforementioned properties with temperature is relatively small for the temperature differences mostly encountered in HVAC calculations, and they can be approximated by constant average values. Table 2.2 lists the specific heats and gas constants of a number of common gases. If extended temperature ranges are involved in a computation, tables of properties of air (Keenan and Kaye, 1948) rather than the preceding equations should be used to properly account for the variation of specific heat with temperature. This refinement is rarely needed in building-related calculations.

2.2.4 Mixtures of Perfect Gases

Relationships between various properties of mixtures of perfect gases can be derived quite simply from those of the constituents. Perfect gases do not interact when mixed. Thus, for a given volume V of a mixture of two perfect gases 1 and 2 at the same temperature, the following relationships hold:

$$pV = mRT \tag{2.19}$$

where

$$m = m_1 + m_2 \quad \text{and} \quad R = \frac{m_1R_1 + m_2R_2}{m}$$

TABLE 2.2
Properties of Common Gases

		c_p		c_v		R	
Gas	**Molecular Weight**	**Btu/(lb$_m$·°F)**	**kJ/(kg·°C)**	**Btu/(lb$_m$·°F)**	**kJ/(kg·°C)**	**ft·lb$_f$/(lb$_m$·°R)**	**J/(kg·K)**
Air	28.97	0.240	1.005	0.1715	0.718	53.35	287.1
Hydrogen (H_2)	2.016	3.42	14.32	2.43	10.17	767.0	4127
Helium (He)	4.003	1.25	5.234	0.75	3.14	386.3	2078
Methane (CH_4)	16.04	0.532	2.227	0.403	1.687	96.4	518.7
Water vapor (H_2O)	18.02	0.446	1.867	0.336	1.407	85.6	460.6
Acetylene (C_2H_2)	26.04	0.409	1.712	0.333	1.394	59.4	319.6
Carbon monoxide (CO)	28.01	0.249	1.043	0.178	0.745	55.13	296.6
Nitrogen (N_2)	28.02	0.248	1.038	0.177	0.741	55.12	296.6
Ethane (C_2H_6)	30.07	0.422	1.767	0.357	1.495	51.3	276
Oxygen (O_2)	32.00	0.219	0.917	0.156	0.653	48.24	259.6
Argon (A)	39.94	0.123	0.515	0.074	0.310	38.65	208
Carbon dioxide (CO_2)	44.01	0.202	0.846	0.156	0.653	35.1	188.9
Propane (C_3H_8)	44.09	0.404	1.692	0.360	1.507	35.0	188.3
Isobutane (C_4H_{10})	58.12	0.420	1.758	0.387	1.62	26.6	143.1

Similarly, when two gases are mixed adiabatically with no work being done, the enthalpy and specific heat relationships are given by simple weighting:

$$h = \frac{m_1 h_1 + m_2 h_2}{m} \quad \text{and} \quad c_p = \frac{m_1 c_{p,1} + m_2 c_{p,2}}{m} \tag{2.20}$$

2.3 Types of Flow Regimes

Fluid behavior and subsequent heat transfer coefficients depend on the flow regime, i.e., whether laminar or turbulent. Laminar flow is characterized by fluid particles moving in parallel slices or layers with little or no macroscopic mixing (see Figure 2.3). As the flow velocity increases, the viscous forces holding the particles in check are overwhelmed by the inertia forces, causing a disorder in the particles resulting in fluid particles moving in irregular paths. The velocity at any point, even under steady-state conditions, fluctuates in all directions about a mean value. This type of flow is called turbulent flow that creates greater shear stress in the fluid particles, leading to more irreversibilities and higher pressure losses. However, the convective heat transfer coefficients under turbulent flow are much greater than those under laminar flow. Turbulent flow is mostly encountered in HVAC processes.

When a fluid flows over a thin flat plate of length L, we observe several effects as shown in Figure 2.4. The flow is laminar at the edge of the plate and gradually becomes turbulent along the flow direction. The free-stream velocity v is uniform in the y-direction while the velocity distribution over the plate assumes a profile as shown in Figure 2.4 due to the no-slip condition at the interface. The boundary layer (B.L.) shown is a very thin region of thickness $\delta(x)$ above the plate in

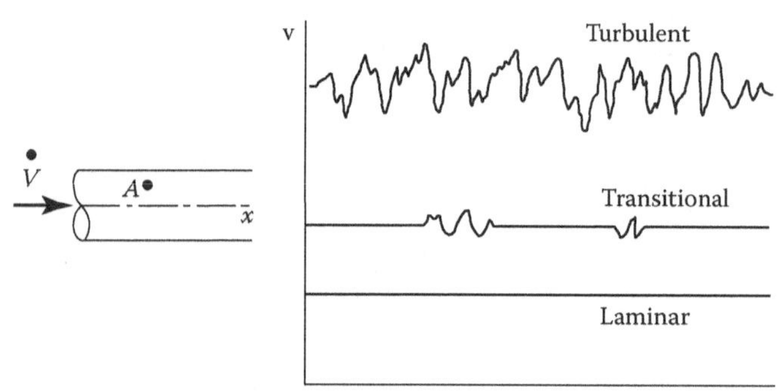

FIGURE 2.3
Different types of flow regimes in a pipe.

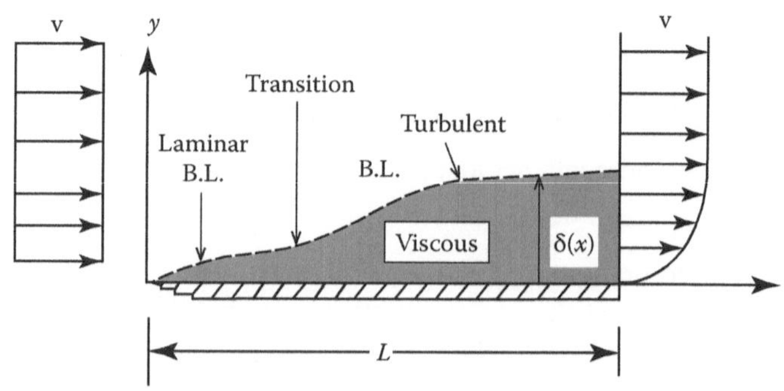

FIGURE 2.4
Fluid boundary layer and flow regimes over a flat plate.

TABLE 2.3

Range of the Reynolds Number for Two Common Geometries

	Laminar	Turbulent	Characteristic Length
Flow inside pipes	<2,100	>4,000	Pipe diameter (round pipe) or hydraulic diameter (noncircular pipe)
Flow over a flat plate	<300,000	>300,000	Distance from leading edge of plate

which the velocity varies from 0 to (0.99v). The viscous shear stresses are assumed to be present only within the boundary layer, and the flow in this region is treated as viscous. The fluid outside the boundary layer is assumed to be inviscid.

The Reynolds number (Re) is a dimensionless number that provides an indication of the type of flow regime. It is defined as the ratio of inertia to viscous forces and is given by

$$\mathrm{Re} = \frac{\rho v D}{\mu} \tag{2.21}$$

where
- ρ is the density of the fluid
- μ is the dynamic viscosity
- v is the fluid average velocity
- D is the characteristic length of the solid

Table 2.3 summarizes flow regimes for two common cases: flow inside pipes or ducts and flow over a solid flat plate. The critical Re number when low transition occurs is specified as a range depending on several parameters, one of which is the smoothness of the solid surface.

2.4 Conservation of Mass and Momentum

The conservation of mass principle simply states that mass of a substance can be neither created nor destroyed in the processes analyzed. Consider a simple flow system, when a fluid stream flows into and out of a control volume. If the mass in the system at time t is $m(t)$, then the mass at time $(t + \delta t)$ is $m(t + \delta t)$. Assuming that during the time increment δt, a small amount δm_{in} enters the system and δm_{out} leaves the system, and the conservation of mass principle states the following:

$$m(t) + \delta m_{in} = m(t + \delta t) + \delta m_{out}$$

which can be simplified as

$$\frac{d\dot{m}}{dt} = \dot{m}_{in} - \dot{m}_{out} \tag{2.22}$$

where $\dot{m} = \delta m/\delta t$.

Under steady-state flow conditions, this simplifies to

$$\frac{d\dot{m}}{dt} = 0 \quad \text{and} \quad \dot{m}_{in} = \dot{m}_{out} \tag{2.23}$$

We can also express this equation in terms of the volumetric flow rate $\dot{V}$:

$$\rho_1 \dot{V}_1 = \rho_2 \dot{V}_2 \tag{2.24}$$

where states 1 and 2 denote different cross sections along the flow direction.

In terms of velocity, Equation 2.24 can be rewritten as

$$\rho_1 v_1 A_1 = \rho_2 v_2 A_2 \tag{2.25}$$

where the areas are represented by A.

Example 2.2: Continuity Equation

Air flows in a duct of 0.03 m^2 (46.5 in.^{2}) cross-sectional area with a velocity of 15.24 m/s (3000 ft/min). This high velocity results in a disturbing noise. If the velocity is to be reduced to 10.1 m/s (2000 ft/min), what should the cross section of the duct be increased to?

Given: $v_1 = 15.24$ m/s, $v_2 = 10.1$ m/s, $A_1 = 0.03$ m^2

Find: A_2

Assumption: The density remains constant; refer to Figure 2.5.

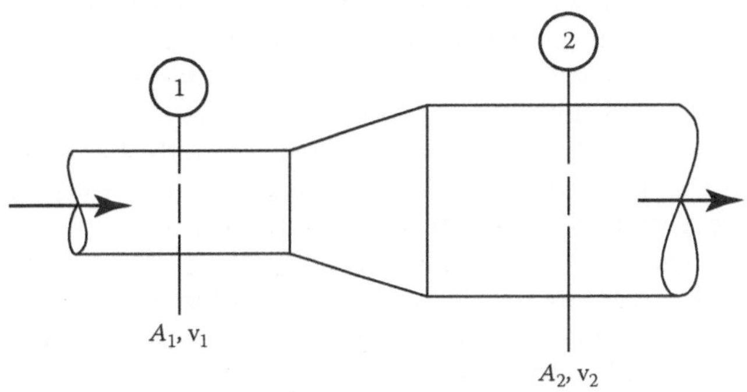

FIGURE 2.5
Steady flow of an incompressible fluid through a duct or a pipe with change in cross-section.

Solution

Using Equation 2.25:

$$A_2 = A_1 \times \left(\frac{v_1}{v_2}\right) = 0.03 \times \frac{15.24}{10.1} = 0.045\ \text{m}^2\ (70.1\ \text{in.}^2)$$

Along with the laws of conservation of mass and conservation of energy (discussed in the next section), momentum is also conserved. The linear momentum of a rigid body is defined as the mass multiplied by its velocity. The momentum remains constant; it can neither be created nor destroyed, but only changed through the action of forces as described by Newton's laws of motion. Momentum considerations are important while dealing with classical mechanics involving bodies or particles in motion that collide. Dealing with momentum is more difficult than dealing with mass and energy because momentum is a vector quantity having both a magnitude and a direction. Angular momentum considerations arise, for example, while studying the design of impellers of pumps and fans.

2.5 First Law of Thermodynamics

Thermodynamics is the study of energy in its various forms and transformations of energy from one form to another. In classical thermodynamics, one considers several forms of energy. Recall that energy is defined as the capacity of producing an effect. It can be stored within the system as mechanical energy (such as potential, kinetic, internal, and rotational) or thermal energy or other forms such as chemical and magnetic. It can also be transferred to or from the system by work or heat transfer. Such interactions cause a *thermodynamic process* to occur associated with a change of the state of the working fluid described by its properties. Heat and work are not properties of a material.

A process can be reversible (or irreversible) depending on whether (or not) the process when made to return to its original state leaves no evidence on both the system and its surroundings. A common example to illustrate irreversibility is that of a piston moving in a cylinder. The friction generated will heat up the cylinder block, while reversing the piston's direction of motion will not negate the heat generated. Also, when processes occur very rapidly, they are usually not reversible, and so quasi-equilibrium is a necessary condition for reversibility in a process. A *cyclic process* is one that consists of two or more processes whereby the initial and final states of the system are identical. Power cycles, for example, are either perfectly cyclic processes or are simplified as such for analysis purposes.

One of the constraints that nature places on processes involving energy conversions is the first law of thermodynamics, commonly called the law of conservation of energy. It is the basis of most of the analysis done in HVAC. We will summarize the two most useful forms of this law: for closed thermodynamic systems, i.e., systems in which there is no mass crossing the boundary, and for open systems in which mass flow does cross the system boundary.

2.5.1 Applied to Closed Systems

A closed thermodynamic system is one in which mass neither enters nor leaves the system during the process under analysis. An example is the combustion of fuel and air confined in a cylinder of an internal combustion engine. The first law for a closed system relates heat added, work produced, and the change of internal energy of the material involved in the process such as a gas enclosed in a cylinder-piston assembly (see Figure 2.6). The closed-system first law is less frequently used in HVAC design than the open-system first law. However, it does occur sufficiently often to warrant review.

There are three heat quantities used in this book. First, the quantity of heat is denoted by Q, which has units of joules (or kilowatt hours) or British thermal units (Btu). Second, the heat rate or power is denoted by $\dot{Q}$, with units of watts or Btu per hour (or foot-pound-force). Third, the heat flux or heat rate per unit area is denoted by $\dot{q}$. Heat flux has units of watts per square meter or Btu per hour per square foot.

In equation form, the first law can be written in three ways. The first is expressed on an extensive basis, i.e., on a total-mass basis:

$$Q - W = \Delta U \tag{2.26}$$

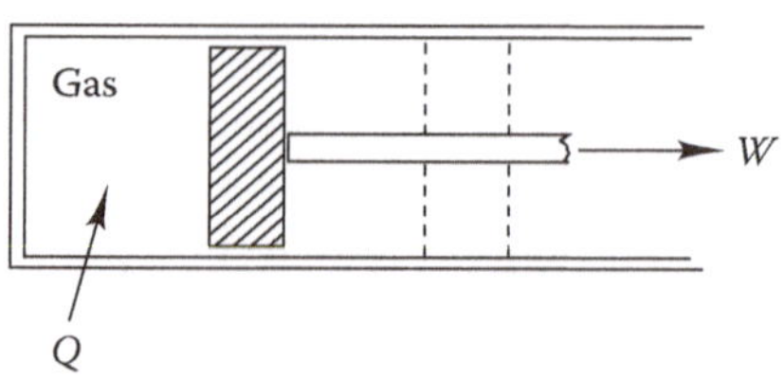

FIGURE 2.6
Schematic diagram of a closed thermodynamic system consisting of gas expanding in a piston–cylinder assembly with work output and heat input.

in which Q is the heat added (taken to be positive when an input to the system). The work W is taken to be positive when it is an output of the system; work has the same units as heat. Finally, ΔU is the change in internal energy of the substance undergoing the process. The internal energy change is positive if the internal energy at the final state is greater than that at the initial state. In Equation 2.26, kinetic and potential energy changes are ignored since they are small and not of interest in building-related processes.

The second form of the first law is written on an intensive (per-unit-mass) basis as follows:

$$q - w = \Delta u \qquad (2.27)$$

In this expression, the symbols denote the intensive analogs of the quantities in Equation 2.26.

It is sometimes useful to write the first law on a time-rate basis as follows:

$$\dot{Q} - \dot{W} = \frac{dU}{dt} \qquad (2.28)$$

The time rate of doing work $\dot{W}$ is called the mechanical power.

Consider Figure 2.7 that depicts the primary components of an air-conditioning system (discussed in detail in Chapter 14). The refrigerant is contained inside the piping, and so this is a closed system if the control volume is drawn so as to encompass the entire system. Further this is a cyclic process, and so Equation 2.28 simplifies to $\dot{Q} - \dot{W} = 0$. Physically this implies that, if the losses/gains of the refrigerant piping are neglected, heat picked up in the evaporator by the refrigerant ($Q_{in,evap}$) plus the work input at the compressor have to be rejected at the condenser ($Q_{out,cond}$). Thus, for this *closed cyclic process*:

$$\dot{Q}_{out,cond} - \dot{Q}_{in,evap} = \dot{W}_{comp} \qquad (2.29)$$

The power consumed by the two fans is external to the system and should not be included in the energy balance of the refrigerant.

2.5.2 Applied to Open Systems

The steady-flow energy equation is the form of the first law most often used in HVAC design and analysis. The applications are broad, including heating and cooling load calculations, chiller design, and heat distribution design among many others. The unsteady form of the first law in which flows, heat rates, and other terms in the energy equation vary with time is encountered only rarely during design, but is needed for transient thermal analyses of buildings. However, in this section we will present only the steady form of the first law.

Figure 2.8 shows the generic open system for which we will write the first law in its three forms—total-mass basis, per-unit-mass basis, and rate basis. The figure shows an input stream and an output stream (of course, more than one input or output stream may be involved) with heat and work transfer

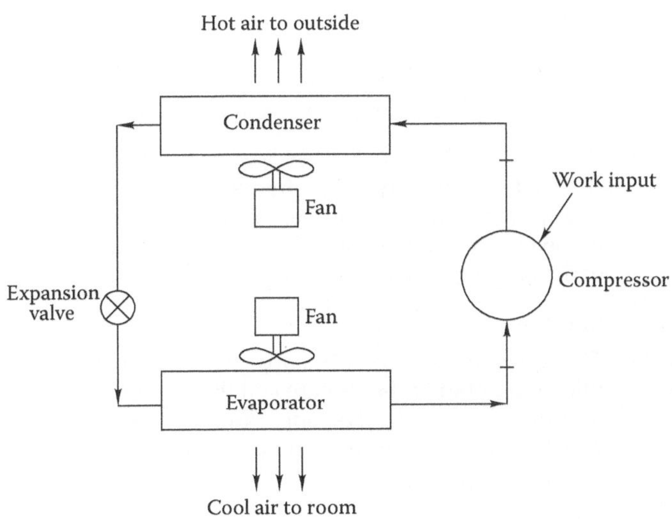

FIGURE 2.7
Schematic diagram of an air-conditioning system using a vapor compression cycle.

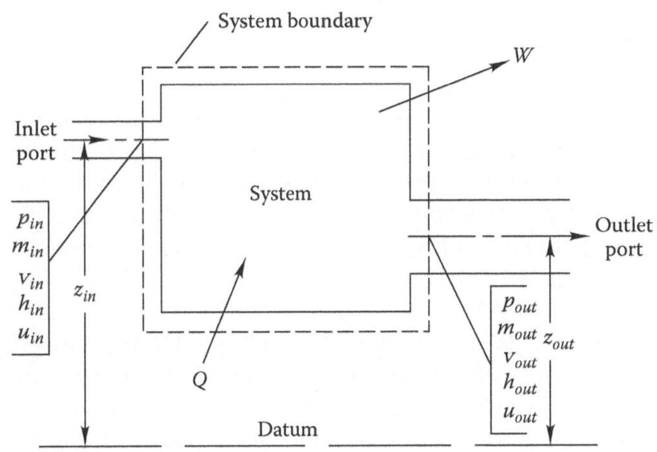

FIGURE 2.8
Schematic diagram of an open thermodynamic system showing work output W and heat input Q along with mass inflow m_{in} and outflow m_{out} across the system boundary.

across the system boundary. On a total-mass (extensive) basis, the open-system first law expression is

$$m_{in}\left(gz_{in}+\frac{v_{in}^2}{2}+h_{in}\right)+Q=m_{out}\left(gz_{out}+\frac{v_{out}^2}{2}+h_{out}\right)+W \quad (2.30)$$

where

- m_{in} and m_{out} are the inlet and outlet mass flow amounts, kg (lb_m)
- z_{in} and z_{out} are the inlet and outlet system port elevations, m (ft)
- v_{in} and v_{out} are the inlet and outlet fluid average velocities, m/s (ft/s)
- h_{in} and h_{out} are the inlet and outlet specific enthalpies, kJ/kg (Btu/lb_m)

The sign convention on heat transfer Q is that heat added to a thermodynamic system is positive; work output by the system W is also positive; this is consistent with signs generally used in the closed-system first law.

The first law on a per-unit-mass (intensive) basis is easily derived from Equation 2.30 by dividing by the mass terms. The steady-state conservation of mass equation ensures that $m_{in} = m_{out}$. Therefore, the first law on an intensive basis (with heat and work transfer terms indicated by lowercase notation q and w):

$$gz_{in}+\frac{v_{in}^2}{2}+h_{in}+q=gz_{out}+\frac{v_{out}^2}{2}+h_{out}+w \quad (2.31)$$

Finally, the rate-basis form of the first law for a steady-state process is

$$\dot{m}_{in}\left(gz_{in}+\frac{v_{in}^2}{2}+h_{in}\right)+\dot{Q}=\dot{m}_{out}\left(gz_{out}+\frac{v_{out}^2}{2}+h_{out}\right)+\dot{W} \quad (2.32)$$

One selects the form to be used in a particular situation based on the type of problem. For example, in analyzing a refrigeration thermodynamic cycle, the mass flow rate may not be known if the required cooling rate is not known. In such a problem, one would use the intensive form, Equation 2.31. On the other hand, if the cooling requirement (kW or Btu/h) of a building is to be found, Equation 2.32 should be used. Note that the closed-system form of the first law can be derived from the open-system form by setting the kinetic, potential, and flow energy terms (that is, pv terms in the enthalpy $h = u + pv$) all to zero. If further, there is no work involved, Equation 2.32 simplifies to an important special case referred to as the "enthalpy equation":

$$\dot{Q}=\dot{m}(h_{out}-h_{in}) \quad (2.33)$$

This equation is used to determine the rate of heat input needed to raise a mass flow rate from an initial state 1 to a final state 2, irrespective of whether the substance is changing phase or not. In case no phase change is involved and the specific heat can be assumed constant over the range of temperature increase, then we get the so-called sensible heat equation:

$$\dot{Q}=\dot{m}c_p(T_{out}-T_{in}) \quad (2.34)$$

The formulation of the first law in IP units presents a particular problem since different units are used for each term. The heat transfer term is most often expressed in Btu or Btu per hour whereas the work term may be in foot-pounds-force. The mass flow term may be in units of pound-mass per hour. The only consistent unit of these three is that used for work. Heat (in Btu) is converted to the consistent unit foot-pounds-force by multiplying by the mechanical equivalent of heat, 778 $(ft \cdot lb_f)$/Btu. Kinetic and potential energy terms involving pound-mass are converted to consistent units by dividing by $g_c = 32.17\ (lb_m \cdot ft)/(s^2 \cdot lb_f)$. In SI, no conversions are needed.

Example 2.3: Cooling of Air Supplied to Room

Consider the room in Example 2.1 that contains 300 kg (660 lb_m) of air. If this air is to be refreshed once each hour and cooled from 29.4°C (85°F) to 12.8°C (55°F), what is the cooling capacity of the cooling coil?

Given: $\dot{m}$, T_{out}, T_{in}

Find: $\dot{Q}$

Lookup value: From Table 2.2, specific heat of air $c_p = 1.00$ kJ/(kg · °C)

Assumptions: Steady-flow process, changes in kinetic and potential energies neglected, no work involved, constant density of air, duct being adiabatic (no heat losses)

Solution

The specified air conditions correspond to the inlet and outlet of the cooling coil that is enclosed in a duct leading to the room. We use Equation 2.34 to find the necessary cooling power needed to cool the air:

$$\dot{Q}=\left(\frac{300\ \text{kg/h}}{3600\ \text{s/h}}\right)\times 1.00\ \text{kJ/(kg}\cdot{}^\circ\text{C)}\times(29.4-12.8)^\circ\text{C}$$

$$=1.38\ \text{kW (4.72 kBtu/h)}$$

2.6 Second Law of Thermodynamics

The second law of thermodynamics places a further restriction on energy conversion processes beyond that imposed by the first law. The second law recognizes that: (1) only certain processes are allowed in an isolated system; for example, thermal energy always flows spontaneously as heat from regions of higher temperature to regions of lower temperature, and (2) some forms of energy are of higher quality than others.

For example, the first law for an isothermal closed system would have us believe that heat and work are equivalent forms of energy in transport across a system boundary. However, the Kelvin–Planck statement of the second law explicitly posits that no process can exist which has as its sole effect the conversion of heat completely to work; hence, work is a higher form of energy than heat. The second law of thermodynamics is of great importance in placing limits on the efficiency of all systems involving thermodynamic cycles such as vapor compression cycles. For example, consider Figure 2.7 showing an air-conditioning system. The second law places limits on the efficiency of the system, i.e., on the ratio ($\dot{Q}_{in,evap}/\dot{W}_{comp}$) of the desired cooling achieved to work input. This efficiency is not constant, however, but depends on the operational temperature levels. Cycles in the framework of the second law are discussed in Chapter 12.

Reducing irreversibility directly improves cycle efficiency. The property called entropy S was described previously in Section 2.1 as a measure of macroscopic molecular chaos. The absolute incremental entropy change dS for a closed system can be determined from

$$dS = \frac{dQ}{T} \tag{2.35}$$

where
dQ is the heat absorbed
T is the absolute temperature

The second law states that when an isolated system undergoes a thermodynamic process, entropy can remain unchanged (reversible process) or can increase (irreversible process), i.e., for a closed system:

$$\Delta S = S_{final} - S_{initial} \geq 0 \tag{2.36}$$

Most real processes are irreversible, and entropy is invariably generated depending on the degree of irreversibility. The second law can also be applied to open systems, and the reader can refer to any thermodynamic textbook (say, Cengal and Boles, 2000) for further discussion.

2.7 Modes of Heat Transfer

Heat transfer occurs across the boundary of a system to another system or surroundings by virtue of a temperature difference between the two systems. The transfer of heat is the principal mechanism by which environmental effects are manifested within buildings. Conduction of heat through a building's skin, transmission of solar radiation through windows, and cooling of occupants by ventilation are all examples of how heat transfer affects the thermal behavior of buildings and their occupants. This chapter is meant to be a review of the heat transfer fundamentals that one needs to calculate heating and cooling requirements of buildings and their HVAC equipment. Standard heat transfer texts listed in the References contain more details on each topic if needed (Cengal, 2002; Holman, 1997; Incropera et al., 2007).

There are three fundamentally different types of heat transfer that we will review generally in this chapter:

1. *Conduction heat transfer*: The result of molecular-level kinetic energy transfers in solids, liquids, and gases. There is a strong correlation between thermal conduction and electrical conduction in solids. Conduction heat flows occur in the direction of decreasing temperature. An example of the flow of heat by conduction is the heat loss through the opaque walls of buildings in winter.
2. *Convection heat transfer*: The result of larger-scale (or bulk) motions of a fluid, either liquid or gas. The higher the velocity of fluid flow, the higher the rate of convection heat transfer, in general. For example, convection heat loss is increased when wind blows over a person's skin compared to that in still air.
3. *Radiation heat transfer*: The transport of energy between two surfaces by electromagnetic waves. The sole requirement for radiation heat transfer to occur is the presence of two surfaces at different temperatures. Radiation must be absorbed by matter to produce internal energy. An example of radiation heat transfer is the energy transported from the sun to earth.

2.8 Conduction Heat Transfer

2.8.1 Fourier's Law of Heat Conduction

Joseph Fourier, in the early nineteenth century, set forth the law of the conduction heat transfer in his fundamental treatise, entitled "The Analytical Theory of Heat." Fourier's law states that the rate of heat transfer by conduction $\dot{Q}$ is proportional to the temperature difference and the heat flow area and is inversely proportional to the distance through which conduction occurs. Fourier's law is a property of matter, not a fundamental law of physics such as the conservation of mass. This law is similar to Ohm's law that governs the flow of electricity in electrical conductors. Recall that current flow is proportional to the voltage difference and inversely proportional to the resistance of the material.

The analogy between heat and current flows is useful for several types of thermal analysis. It is convenient to represent temperatures T as analogous to voltages and heat flows $\dot{Q}$ as analogous to currents. Then, the thermal networks are interpreted by diagrams, where* 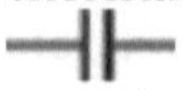 represents a capacitance C and 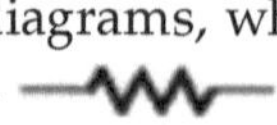 represents a resistance R.

As an equation, Fourier's law in 1-D with constant thermal properties is given by

$$\dot{Q} = -kA\frac{dT}{dx} \tag{2.37}$$

where

k is the thermal conductivity in SI units, W/(m·K), and in IP units of Btu/[h·ft²/(°F/ft)], which converts to Btu/(h·ft·°F)

A is the area through which heat flows

dT/dx is the temperature gradient evaluated at point x where the rate of heat transfer is to be determined

Since heat flows only from a high temperature to a low temperature, a minus sign is needed in Equation 2.37. Note that positive heat transfer is taken to occur in the positive x-coordinate direction. Figure 2.9 shows the temperature profile in a 1-D heat transfer situation with constant thermal conductivity. Heat flows to the right (positive heat flow) if the temperature decreases to the right.

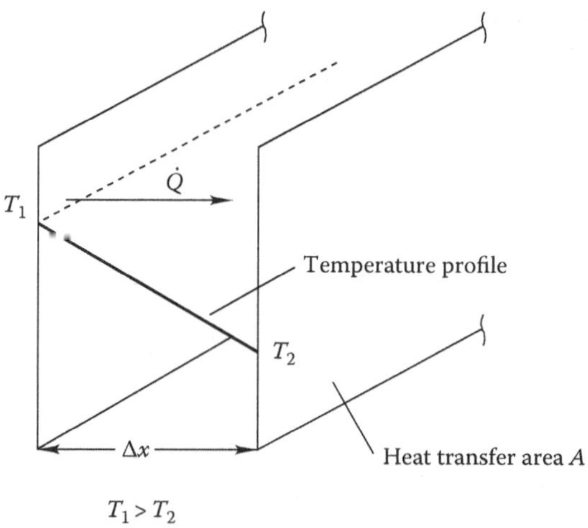

FIGURE 2.9
Heat flow through a plane wall.

2.8.2 Steady-State Conduction in Plane Walls

Fourier's law can be integrated in Cartesian coordinates along a path of constant Δx. For the case shown in Figure 2.9, if the temperature depends only on x, if there are no heat sources in the material, and if the thermal conductivity can be taken as a constant (not a bad assumption for most building materials, as noted here), then the steady-state 1-D form is

$$\dot{Q} = kA\frac{T_1 - T_2}{\Delta x} \tag{2.38}$$

where

k is the thermal conductivity, W/(m·K) or Btu/(h·ft·°F)

T_1 is the higher temperature, K or °F

T_2 is the lower temperature, K or °F

A is the area through which conduction occurs, m² or ft²

Δx is the thickness of material in which conduction occurs, m or ft

This can be expressed slightly differently as

$$\dot{Q} = \frac{T_1 - T_2}{\Delta x/(kA)} \tag{2.39}$$

The denominator is often called the resistance to heat transfer

$$R = \frac{\Delta x}{kA}\ (\text{K/W}) \text{ or } [(\text{h}\cdot{}^\circ\text{F})/\text{Btu}] \tag{2.40}$$

by analogy with electric resistance, which serves as the proportionality constant between voltage and current in Ohm's law. If one takes the temperature difference in the

* Compared to electric circuits, one can note two simplifications: there are no inductances, and the heat stored in a capacitance depends only on a single temperature derivative. Thus, the loose end of each capacitance can be taken as "ground" (zero voltage).

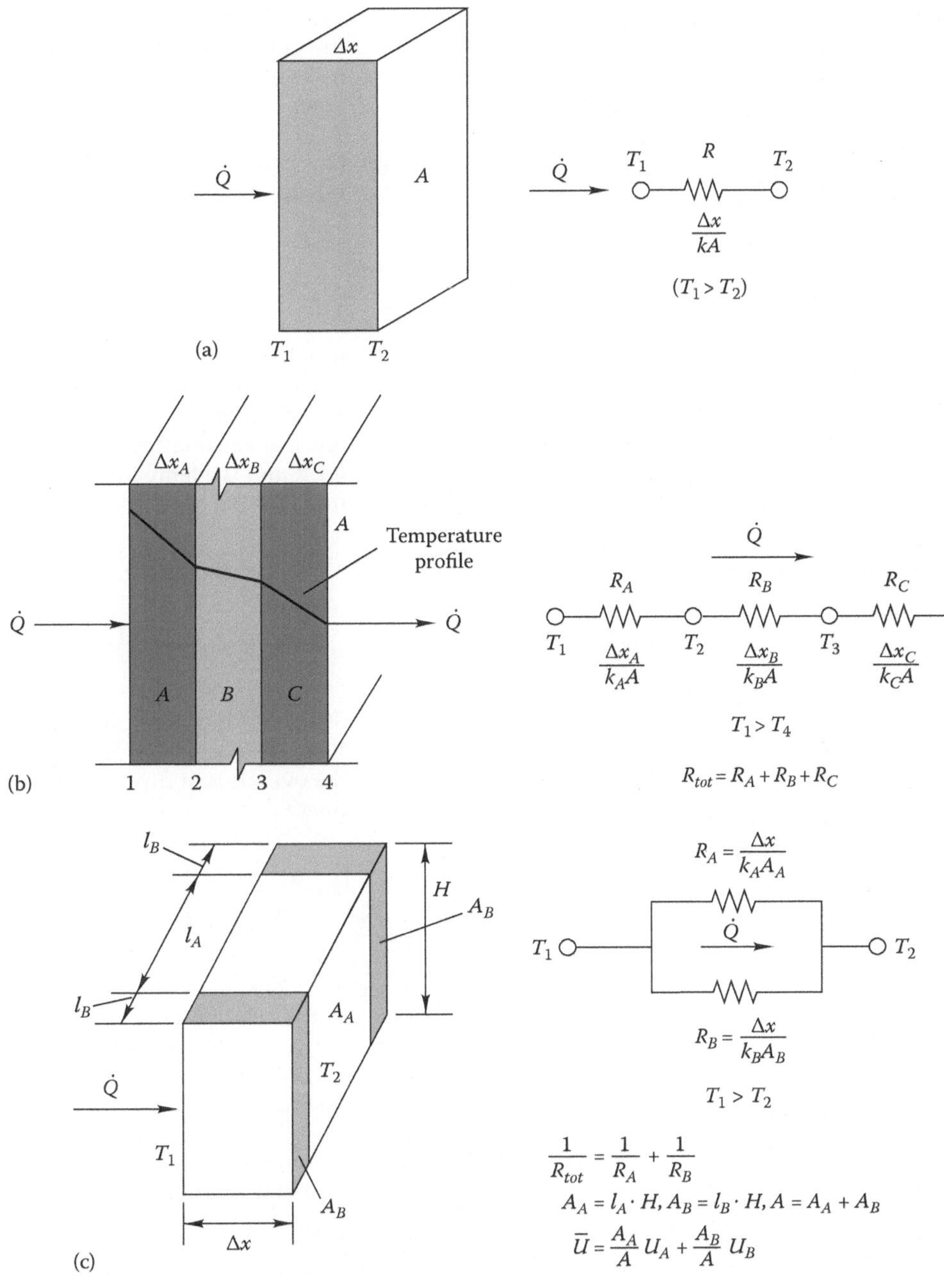

FIGURE 2.10
Electric resistance analog to (a) single-layer heat flow, (b) multiple-layer series-resistance heat flow, and (c) multiple-layer parallel-resistance heat flow.

numerator of Equation 2.39 to be the driving force analogous to voltage and the heat flow to be analogous to current flow, then R is the thermal analog of electric resistance.

The conductive resistance to heat flow per unit area for say, a wall is the commonly used term "R Value":*

$$R_{cd} \equiv AR = \frac{\Delta x}{k}\,[(\text{m}^2\cdot\text{K})/\text{W}] \text{ or } [(\text{h}\cdot\text{ft}^2\cdot{}^\circ\text{F})/\text{Btu}] \quad (2.41)$$

One use of the thermal resistance concept is shown in Figure 2.10. Like electrical resistors, thermal *resistors* can be connected in series and in parallel or in combinations thereof. Figure 2.10b depicts a three-layer wall and shows the equivalent thermal model consisting of three resistors in series,† whereas Figure 2.10c shows a two-part wall, e.g., a residential wall consisting of studs (B) with insulation (A) in between, through which heat

* Some authors use the opposite convention for R and R_{cd}. However, to comply with the electrical analogy, we use the convention noted.

† In most building heat flow analyses, the contact resistance between two composite solid layers is usually neglected.

flows by parallel paths: one through the studs and the other through the insulation.

Another commonly used measure is the unit conductance, or U value. The notation "C-factor" representing steady-state thermal conductance is also used in the building science literature. It is just the inverse of the R value:

$$\text{C-factor} \quad \text{or} \quad U_{cd} \equiv \frac{1}{R_{cd}} \tag{2.42}$$

The use of R_{cd} or U_{cd} to solve a given problem depends on whether the network consists of individual resistances in series or in parallel, as illustrated in the following two examples.

Example 2.4: *R* Value Calculation for a Building Wall: Resistors in Series

The outside wall of a home consists essentially of a 10 cm layer of common brick [$k = 0.68$ W/(m·K)], a 15 cm layer of fiberglass insulation [$k = 0.038$ W/(m·K)], and a 1 cm layer of gypsum board [$k = 0.48$ W/(m·K)].

(a) What is the overall R_{cd} value?
(b) What is the heat flux through this wall if the interior surface temperature is 22°C and the exterior surface is at 5°C?
(c) What are the temperatures at the two interfaces?

Given: $k_1, k_2, k_3, \Delta x_1, \Delta x_2, \Delta x_3$

Figure: See Figure 2.10b.

Assumptions: Steady-state flow, the only heat transfer mode is 1-D conduction, constant k's. Calculation to be done assuming area $A = 1$ m²

Find: Overall R value and $\dot{q}$

Solution

(a) The R values are all given by Equation 2.41. The R value of the outer layer is

$$R_{cd,1} = \frac{0.10\ \text{m}}{0.68\ \text{W/(m}\cdot\text{K)}} = 0.147\ (\text{m}^2\cdot\text{K})/\text{W}$$

The R value of the center layer is

$$R_{cd,2} = \frac{0.15\ \text{m}}{0.038\ \text{W/(m}\cdot\text{K)}} = 3.947\ (\text{m}^2\cdot\text{K})/\text{W}$$

And the R value for the inner gypsum board layer is

$$R_{cd,3} = \frac{0.01\ \text{m}}{0.48\ \text{W/(m}\cdot\text{K)}} = 0.021\ (\text{m}^2\cdot\text{K})/\text{W}$$

The overall R value of the wall is just the sum of the three individual values in keeping with the electric series-resistance model: $R_{cd} = 0.147 + 3.947 + 0.021 = 4.115$ (m²·K)/W or 23.3 (ft²·°F·h)/Btu.

(b) One can find the heat flux by combining Equations 2.39, 2.40 and 2.41, to get

$$\dot{q} = \frac{\Delta T}{R_{cd}}$$

in which $\Delta T = T_{int} - T_{ext} = T_1 - T_4$. Since a unit wall area is used, by definition, R and R_{th} are the same numerically (but not dimensionally). Substituting values for R and the temperature difference, we find that the heat flux is

$$\dot{q} = \frac{(22-5)\ \text{K}}{4.115\ (\text{m}^2\cdot\text{K})/\text{W}}$$

$$= 4.13\ \text{W/m}^2\ [1.31\ \text{Btu/(h}\cdot\text{ft}^2)]$$

(c) Since steady-state flow is assumed, heat fluxes through the interfaces are equal to the overall heat flux deduced under part (b), i.e.,

$$\frac{T_1 - T_4}{R_{cd}} = \frac{T_1 - T_2}{R_{cd,1}} = \frac{T_2 - T_3}{R_{cd,2}} = \frac{T_3 - T_4}{R_{cd,3}} = 4.13\ \text{W/m}^2$$

Since $T_1 = 22$°C and $R_{cd,1} = 0.147$ (m²·K)/W, we can use the relation $(T_1-T_2)/R_{cd,1} = 4.13$ W/m² to deduce $T_2 = 21.39$°C, which can be then used to deduce $T_3 = 5.09$°C in a similar fashion.

The last relationship $(T_3 - T_4)/R_{cd,3} = 4.13$ W/m² can be used as a verification step since one can use the computed value of T_3 to deduce T_4 and see whether it matches the specified value of 5°C.

Comments

The majority (96%) of the R value is due to the insulation. This is the thickest layer of the wall, but more importantly it is also the layer with the lowest thermal conductivity. In the future, one should not routinely add individual R values to find the thermal resistance. This is correct only if the areas of all wall components are the same. Note that in this example we specified surface temperatures, not the interior and exterior air temperatures adjacent to the wall surfaces. This problem must await the discussion of thermal resistance due to convection in Section 2.9.

Finally, observe that the construction of the wall considered here is not practical since there is no structural connection between the outer (brick) and inner (gypsum board) parts. The physical connection must be accounted for by the studs (either metal or wood) whose heat transfer rate is higher than that of the insulation. Example 2.5 considers the framing adjustment that is needed.

Example 2.5: Effect of Studs on Wall Heat Loss: Resistors in Parallel

For structural reasons, the wall described in Example 2.4 must have studs placed every 60 cm (24 in.). The studs are fabricated from wood (conductivity 0.10 W/[m·K]) and are 5 cm wide and 15 cm deep (note that this is slightly larger than the standard nominal 2 in. × 6 in. wood stud used in U.S. construction). Find the R value and heat flux $\dot{q}$, and compare with the results of Example 2.4 to quantify the effect of framing with studs ignoring the brick and gypsum board.

Given: Data from Example 2.4 and stud dimensions

$l_A = 60 - 5 = 55$ cm, $l_B = 5$ cm, $\Delta x = 15$ cm, $k_{stud} = 0.10$ W/(m·K), $k_{ins} = 0.038$ W/(m·K)

Figure: See Figure 2.10c.

Find: Overall R and $\dot{q}$

Solution

From Figure 2.10c, we see that the overall thermal resistance of the parallel network is given by (ignoring the gypsum and brick effects)

$$\frac{1}{R} = \frac{1}{R_{ins}} + \frac{1}{R_{stud}}$$

in which the thermal resistances are (for a unit wall height $H = 1$ m)

$$R_{ins} = \frac{\Delta x}{k_{ins}(H \times l_A)} = \frac{0.15\text{ m}}{0.038\text{ W/(m}\cdot\text{K)} \times (1\text{ m} \times 0.55\text{ m})}$$

$$= 7.18\text{ K/W}$$

$$R_{stud} = \frac{\Delta x}{k_{stud}(H \times l_b)} = \frac{0.15\text{ m}}{0.10\text{ W/(m}\cdot\text{K)} \times (1\text{ m} \times 0.05\text{ m})}$$

$$= 30.0\text{ K/W}$$

The overall thermal resistance is

$$\frac{1}{R} = \frac{1}{7.18} + \frac{1}{30.0} = 0.173\text{ W/K} \quad \text{or} \quad R = 5.79\text{ K/W}$$

Finally, the heat flow through the 60 cm × 1 m wall section is

$$\dot{Q} = \frac{(22-5)\text{ K}}{5.79\text{ K/W}} = 2.93\text{ W}$$

For comparison with Example 2.4, we determine the heat flux $\dot{q}$:

$$\dot{q} = \frac{2.93\text{ W}}{0.60\text{ m} \times 1\text{ m}} = 4.88\text{ W/m}^2$$

The effective R value *averaged over the wall* is

$$\bar{R}_{cd} = \frac{\Delta T}{\dot{q}} = \frac{(22-5)\text{ K}}{4.88\text{ W/m}^2}$$

$$= 3.5\text{ (K}\cdot\text{m}^2\text{)/W [19.8 (h}\cdot\text{ft}^2\cdot{}^\circ\text{F)/Btu]}$$

Comments

The thermal conductivity of wood is about 2.5 times that of insulation. The presence of wood studs has caused the heat flux through this wall to increase by 18% (=4.88/4.13 = 1.18), even though the studs represent only 8% of the wall heat flow area. The stud is a simple example of a *thermal bridge*, a part of the building envelope with higher than average thermal losses. (See Section 2.10.7.) (To be very accurate, recall that we ignored the 4% gypsum and brick effect.)

This example can be simplified by using U values. It is easy to show that the average U_{cd} value of this composite wall is the area-weighted average of the U values of the components. For this example:

$$U_{ins} = \frac{k_{ins}}{\Delta x_{ins}} = \frac{0.038\text{ W/(m}\cdot\text{K)}}{0.15\text{ m}} = 0.0253\text{ W/(m}^2\cdot\text{K)}$$

$$U_{stud} = \frac{k_{stud}}{\Delta x_{stud}} = \frac{0.10\text{ W/(m}\cdot\text{K)}}{0.15\text{ m}} = 0.667\text{ W/(m}^2\cdot\text{K)}$$

The effective U value is then

$$\bar{U}_{cd} = \frac{5\text{ cm}}{60\text{ cm}} \times 0.667 + \frac{55\text{ cm}}{60\text{ cm}} \times 0.253 = 0.288\text{ W/(m}^2\cdot\text{K)}$$

The heat flow through the 60 cm × 1 m wall section is then

$$\dot{Q} = \bar{U}_{cd} \cdot A \cdot \Delta T = 0.288\text{ W/(m}^2\cdot\text{K)}$$
$$\times (0.60\text{ m} \times 1\text{ m}) \times (22-5)\text{ K} = 2.93\text{ W}$$

confirming the previous calculation.

Important rule: *For parallel heat flow paths, the component U values are area averaged; for series heat flow paths, the R values are added* (when the heat flow areas are the same, as is usually the case in buildings).

The two previous examples are typical textbook examples in that (1) steady-state heat conduction is assumed, (2) that construction and installation have been ideal, (3) that there is no contact resistance between solid elements, and (4) that variation of thermal conductivity with temperature is negligible. In practice, the HVAC engineer should exercise caution when calculating building heat losses to account for less than ideal construction. For example, insulation is often compressed or does not completely fill stud cavities in roofs and walls. As a result, the actual thermal resistances are lower than one would calculate by using the aforementioned idealized equations.

2.8.3 Steady-State Conduction in Cylindrical Coordinates

Fourier's law applies to geometries other than plane walls. Of particular interest in buildings is the cylindrical geometry. Heat losses from piping in building HVAC systems are significant parasitic losses that merit attention. The heat transfer through a cylindrical solid such as that in Figure 2.11 is also governed by Fourier's law. When one integrates Equation 2.37 in cylindrical coordinates, using the same assumptions as before, the result is

$$\dot{Q} = \frac{T_1 - T_2}{\ln(r_o/r_i)/(2\pi kL)} \tag{2.43}$$

in which the geometric parameters are shown in Figure 2.11a. The denominator in Equation 2.43 is the thermal resistance for steady, cylindrical conduction.

The concept of resistance enables us to calculate the heat flow through layered, cylindrical walls much as we did for plane walls in Example 2.4. Figure 2.11b shows the model one would use. For example, layer *A* could represent a pipe carrying a fluid, layer *B* could be an insulation layer, and layer C could be a protective jacket over the insulation.

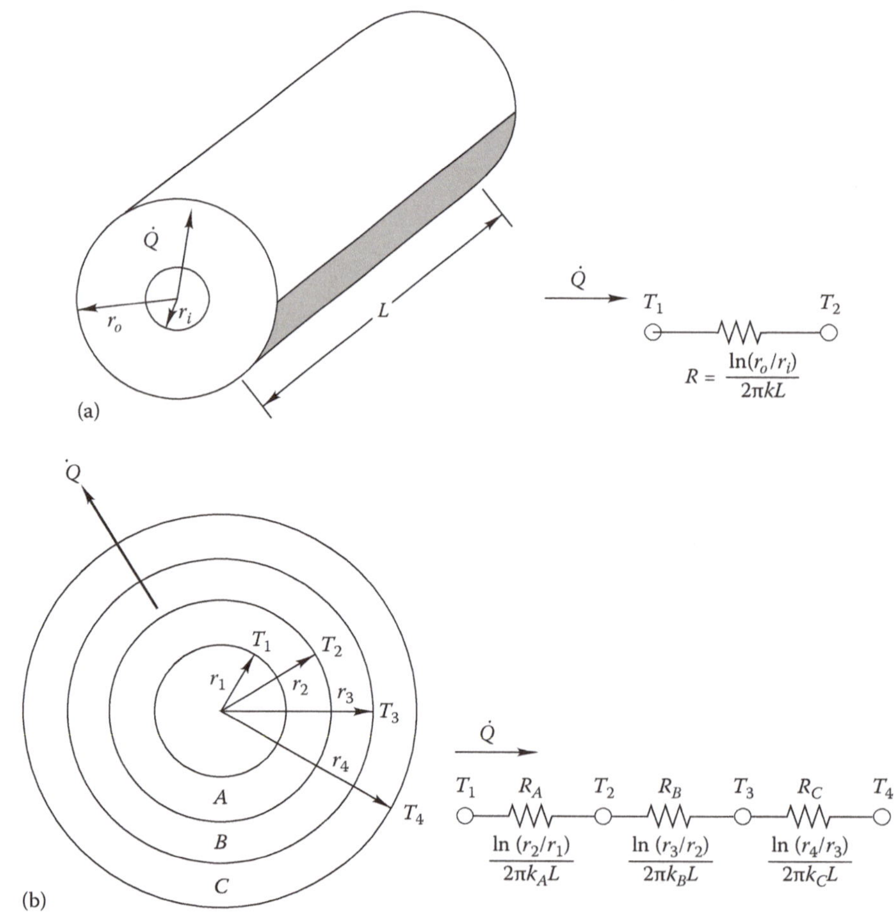

FIGURE 2.11
Conduction in cylindrical coordinates with electric resistance analog circuit. (a) Single-layer conduction. (b) Multiple-layer conduction.

Example 2.6: Heat Loss from Pipe to Room

Steam at 260°C flows through a cast iron pipe of conductivity 80 W/(m·K) with outer diameter 7.5 cm and inner diameter 7.0 cm. The pipe has a 3 cm thick glass wool insulation of conductivity 0.05 W/(m·K). What is the heat loss to the environment per meter length of pipe assuming that the outer layer of insulation has a temperature of 20°C?

Given: $k_{pipe} = 80$ W/(m·K), $k_{ins} = 0.05$ W/(m·K), $T_{pipe} = 260$°C, $T_{ins} = 20$°C, $r_{i,pipe} = 3.5$ cm, $r_{o,pipe} = 3.75$ cm, $r_{ins} = 6.75$ cm

Figure: See Figure 2.11.

Assumptions: There is a steady, 1-D conduction problem with two resistances in series; the pipe surface temperature remains constant over the length being considered.

Solution

We use Equation 2.43 to obtain the heat loss per unit length of pipe:

$$\dot{Q} = \frac{T_{pipe} - T_{ins}}{(\ln(r_{o,pipe}/r_{i,pipe})/2\pi k_{pipe}) + (\ln(r_{ins}/r_{o,pipe})/2\pi k_{ins})}$$

$$= \frac{(260-20)}{(\ln(3.75/3.5)/(2\pi \times 80)) + (\ln(6.75/3.75)/(2\pi \times 0.05))}$$

$$= \frac{240}{(0.0021 + 1.872)} = 128.1 \text{ W/m}$$

Comments

The resistance offered by the iron pipe wall (0.0021 m·°C/W) is negligible compared to that of the insulation and could have been neglected.

2.8.4 Steady-State Conduction in Other Geometries

In building-related heat transfer analyses, there are often situations where the heat conduction is not strictly 1-D. An example is an underground pipe carrying a heated or cooled fluid from a central plant to a building for heating or cooling. There are two methods for treating such cases. The first, which we discuss in this section, uses an expression called the "shape factor" to account for the 2-D effects. The second approach requires a full 2-D analysis and is not covered in detail in this book. However, results from such an analysis are presented in Section 7.3 for several cases having to do with the transfer of heat from the floors and basements of buildings to the ground.

The shape factor approach uses the following basic equation to find steady conduction heat transfer rates:

$$\dot{Q} = kS\Delta T \tag{2.44}$$

where
 S is the shape factor, given in Table 2.4
 k is the thermal conductivity

The following example shows how one uses the shape factor approach to find the loss of heat from a buried pipe carrying a hot fluid.

TABLE 2.4

Conduction Shape Factors

Physical System	Schematic	Shape Factor	Restrictions
Isothermal cylinder of radius r buried in semi-infinite medium having isothermal surface	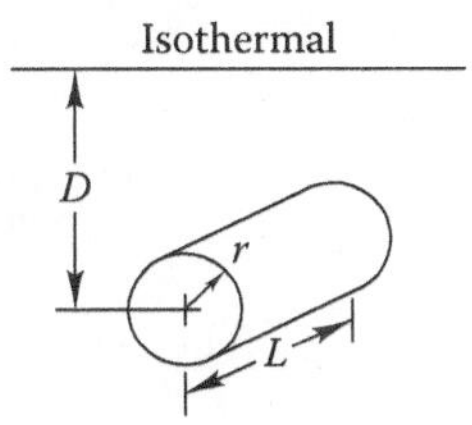	$\frac{2\pi L}{\cosh^{-1}(D/r)}$	$L \gg r$
		$\frac{2\neq L}{\ln(2D/r)}$	$L \gg r, D > 3r$
		$\frac{2\pi L}{\ln(L/r)\{1-\ln[L/2D]/\ln(L/r)\}}$	$D \gg r, L \gg D$
Conduction between two isothermal cylinders buried in infinite medium	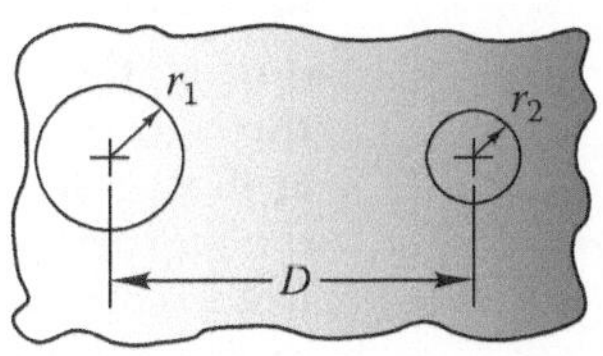	$\frac{2\pi L}{\cosh^{-1}\left(\frac{D^2 - r_1^2 - r_2^2}{2r_1 r_2}\right)}$	$L \gg r_1, r_2\ L \gg D$

Source: Holman, J.P., *Heat Transfer*, 8th ed., McGraw-Hill, New York, 1997. With permission.

Example 2.7: Heat Loss from a Buried Pipe

A pipe with an outer surface temperature of 100°C and a radius of 15 cm is buried 30 cm deep in earth that has a thermal conductivity of 1.7 W/(m·K). If the surface temperature of the earth is 20°C, what is the heat loss for a pipe length of 10 m if the pipe is uninsulated?

Given: k_{earth} = 1.7 W/(m·K), T_{pipe} = 100°C, T_{earth} = 20°C, r_{pipe} = 15 cm, D_{pipe} = 30 cm, L = 10 m

Figure: See Table 2.4, upper panel.

Assumptions: There is steady, 2-D conduction; the thermal resistance between the pipe and earth is negligible. The pipe surface temperature remains constant over the 10 m length being considered.

Find: $\dot{Q}$

Solution

The proper shape factor expression to be used from Table 2.4 depends on the ratio of the pipe length to the pipe radius. From the given data, the first equation in the table applies, i.e.,

$$S = \frac{2\pi L}{\cosh^{-1}(D/r)}$$

The inverse hyperbolic cosine value is

$$\cosh^{-1}\frac{D}{r} = \cosh^{-1}\frac{30}{15} = 1.32$$

from which the shape factor for the 10 m length of pipe is

$$S = \frac{2\pi \times 10}{1.32} = 47.6 \text{ m}$$

and using Equation 2.44, the heat loss is

$$\dot{Q} = kS\Delta T = 1.7 \text{ W/(m}\cdot\text{K)} \times 47.6 \text{ m} \times (100-20)°\text{C}$$

$$= 6,474 \text{ W (22,096 Btu/h)}$$

Comments

Since the pipe is much longer than its diameter, the assumption of 2-D heat flow is appropriate. Note that the value of thermal conductivity to be used in such a calculation depends strongly on the soil moisture content, as discussed in a later section.

Example 2.8: 3-D Heat Losses from a Small Room

Consider Figure 2.12 that shows an exploded view of a small square room of length 2 m and of wall thickness 10 cm. The heat losses occurring through the 4 walls and the ceiling and floor can be easily determined assuming 1-D heat conduction. The heat losses through the 12 edge strips and the 8 corners cannot be treated as 1-D and the shape factor approach is more convenient. The shape factors are given as (Incropera et al., 2007)

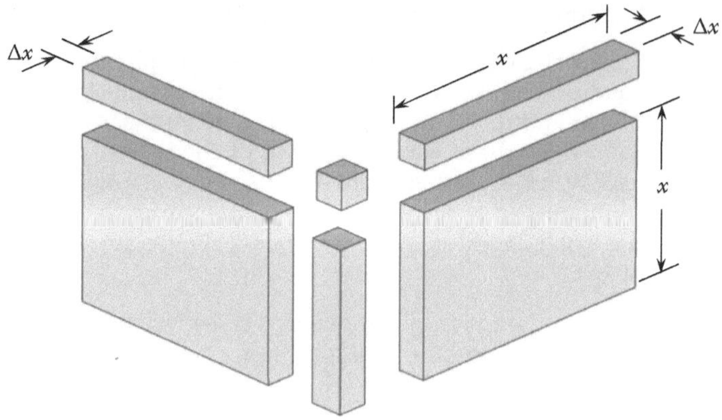

FIGURE 2.12
Dimensions used to estimate 3-D heat losses in a room.

$$S_{wall} = \frac{\text{Area}}{\Delta x}, \quad S_{edge} = 0.54 \times x, \quad S_{corner} = 0.15 \times \Delta x$$

You are asked to evaluate the relative contribution of these edges and corners on the total heat loss.

Given: x = 2 m, Δx = 0.1 m

Figure: See Figure 2.12.

Assumptions: The shape factor approach is used; there are no thermal bridging effects.

Solution

For the walls, ceiling, and floor: shape factor S_{wall} = 6 × (2 × 2/0.1) = 240 m

For the edge strips: shape factor S_{edge} = 12 × (0.54 × 2) = 12.96 m

For the corners: shape factor $S_{corners} = 8 \times (0.15 \times 0.1) = 0.12$ m

The total shape factor S = 240 + 12.96 + 0.12 = 253.08 m

The heat losses are directly proportional to the shape factor (Equation 2.44). Thus, the heat losses through the edges and the corners are only (253.08 − 240)/240 = 0.055% or 5.5% of that of the plane walls. The room selected is a very small one, and if a more realistic sized room was assumed, the effect of the edges and corners would have been even smaller. Hence, neglecting these effects during the calculation of heat losses or gains through building envelope elements introduces at most a few percent error. However, note that the "corner" effect of 2-D heat loss through insulated rectangular ducts carrying conditioned air can be nontrivial since such ducts tend to be relatively small with thick insulation.

2.8.5 Thermal Conductivity of Materials

In previous sections, the thermal conductivity was seen to be the key physical property of materials governing the rate of conduction heat transfer. Table 2.5 illustrates the wide range of conductivities exhibited by common materials. Note the several orders of magnitude difference between gases and metals. Table 2.6 shows values of k in SI and IP units for specific materials often used in buildings and related equipment. More complete tabulations of conductivities, conductances, and resistances of several common building and insulating materials can be found in specialized handbooks such as ASHRAE Fundamentals (2013).

TABLE 2.5

Representative Magnitudes of Thermal Conductivity

Material	Conductivity, Btu/(h·ft·°F)	Conductivity, W/(m·K)
Atmospheric-pressure gases	0.004–0.10	0.007–0.17
Insulating materials	0.02–0.12	0.034–0.21
Nonmetallic liquids	0.05–0.40	0.086–0.69
Nonmetallic solids (brick, stone, concrete)	0.02–1.50	0.034–2.6
Metal alloys	8–70	14–120
Pure metals	30–240	52–410

The thermal conductivity of building materials is often assumed to be a constant, independent of temperature. Figure 2.13 shows why this is an excellent approximation. Over the relatively small range of temperatures encountered in buildings during a year, it is usually adequate to assume that conductivities are constant. If data on the temperature dependence of conductivity are available for a particular material, we suggest that the conductivity at the average temperature be used since the value of k varies almost linearly with temperature.

In the next section on heat transfer by convection, we see that the temperature dependence of conductivity for some liquids and gases is sometimes nonnegligible. In such cases, we can account for this effect in a simple manner, as we shall see shortly.

2.9 Convection Heat Transfer

2.9.1 Defining Equation for h_{con}

When a moving fluid comes in contact with a surface at a different temperature, the resulting heat transfer is called convection. Convection heat transfer is always associated with large-scale or bulk motion of

TABLE 2.6

Values of Thermal Conductivity for Building Materials

Material	k, Btu/(h·ft·°F)	T, °F	k, W/(m·K)	T, °C
Construction materials				
Asphalt	0.43–0.44	68–132	0.74–0.76	20–55
Cement, cinder	0.44	75	0.76	24
Glass, window	0.45	68	0.78	20
Concrete	1.0	68	1.73	20
Marble	1.2–1.7	—	2.08–2.94	—
Balsa	0.032	86	0.055	30
White pine	0.065	86	0.112	30
Oak	0.096	86	0.166	30
Insulating materials				
Glass fiber	0.021	75	0.036	24
Expanded polystyrene	0.017	75	0.029	24
Polyisocyanurate	0.012	75	0.020	24
Gases at atmospheric pressure				
Air	0.0157	100	0.027	38
Helium	0.0977	200	0.169	93
Refrigerant 12	0.0048	32	0.0083	0
	0.0080	212	0.0038	100
Oxygen	0.00790	−190	0.0137	−123
	0.02212	350	0.0383	175

Source: Karlekar, B. and Desmond, R.M., *Engineering Heat Transfer*, West Publishing, St. Paul, MN, 1982. With permission.

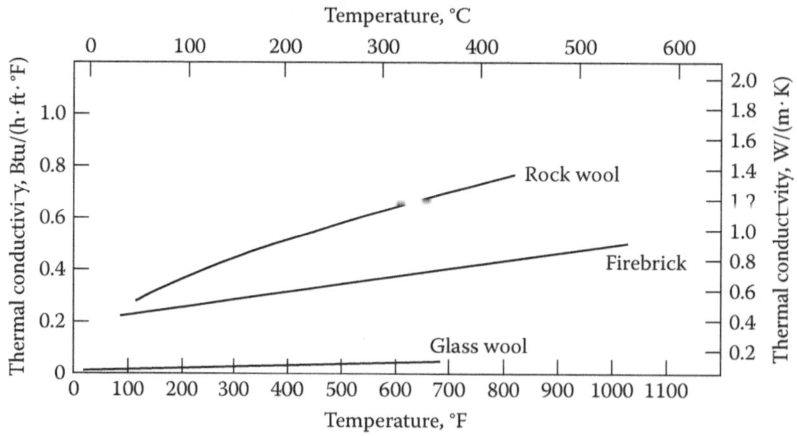

FIGURE 2.13
Thermal conductivity of several materials as a function of temperature.

a fluid—either liquid or gas—over a warmer or cooler surface. In general, the higher the velocity of fluid flow, the higher the rate of convection heat transfer.

Two kinds of convection exist: natural (or free) and forced. *Free convection* results from density differences in the fluid caused by contact with the surface to (or from) which the heat transfer occurs. The gentle circulation of air in a room caused by the presence of a solar-warmed window or wall is a manifestation of free convection. Another example is heat transferred from baseboard systems by natural convection. Buoyancy is the motive force in free convection.

Forced convection occurs when an external force (other than gravity) moves a fluid past a warmer or cooler surface. Usually, the fluid velocities in forced convection are considerably higher than those in free convection, and the rate of heat transfer is generally greater. The improved heat transfer is at the expense of greater consumption of mechanical energy in the forced-flow case. For example, heat transfer from heating and cooling coils is primarily by forced convection since an electric fan drives the air flow. The other commonly encountered phenomena causing forced convection is the wind blowing over the roof, for example.

Convection is a very common mode of heat transfer in buildings, but its detailed analysis is complicated. Newton's law of cooling is to be considered as a simplified approximation useful for both forced and free convection calculations. Simply stated, the rate at which heat is transferred by convection is proportional to the temperature difference and the heat transfer area. The equation expressing Newton's law of cooling is

$$\dot{Q} = h_{con}A(T_s - T_f) = h_{con}A\Delta T \qquad (2.45)$$

where

h_{con} is the convection coefficient, W/(m^2·K), [Btu/(h·ft^2·°F)]
A is the surface area through which convection occurs, m^2 (ft^2)
T_s is the surface temperature, °C (°F)
T_f is the fluid temperature far away from wall, °C (°F)
$\Delta T = T_s - T_f$ is the temperature difference, °C (°F)

Standard heat transfer texts include *dimensionless equations* for finding h_{con}, which are independent of the specific fluid. These typically involve dimensionless numbers such as the Reynolds (forced convection), Grashof (natural convection), Prandtl, and Nusselt numbers (see, e.g., Incropera et al., 2007). A short discussion is provided in Section 12.5 of this book. For this chapter meant to present heat transfer principles more appropriate for analyzing heat flows in building elements, such equations are inconvenient to use since extensive physical property data tables are needed. We will use, instead, *dimensional equations* for air as the working fluid for common situations since such equations save considerable time.

Table 2.7 shows typical values of the convection coefficient. The largest values occur for boiling and condensing water, whereas the lowest values apply for free convection in gases.

2.9.2 Convection Thermal Resistance and *R* Value

The definitions of thermal resistance and the *R* value for conduction have direct analog in convection heat transfer. If we reexpress Equation 2.45 as

$$\dot{Q} = \frac{\Delta T}{R} \qquad (2.46)$$

TABLE 2.7

Magnitude of Convection Coefficients

Arrangement	W/(m²·K)	Btu/(h·ft²·°F)
Air, free convection	6–30	1–5
Superheated steam or air, forced convection	30–300	5–50
Oil, forced convection	60–1,800	10–300
Water, forced convection	300–6,000	50–1,000
Water, boiling	3,000–60,000	500–10,000
Steam, condensing	6,000–120,000	1,000–20,000

The conversion between SI and IP units is 5.678 W/(m²·K) = 1 Btu/(h·ft²·°F).

then the resistance to heat transfer for convection would be

$$R = \frac{1}{h_{con}A} \tag{2.47}$$

The convective thermal resistance value R_{con} is given by

$$R_{con} = \frac{1}{h_{con}} \tag{2.48}$$

The convective R values are used in a later section to simplify heat flow calculations when both conduction and convection modes coexist.

2.9.3 Relevant External Flow Equations

In this section, we summarize convection equations for flows that occur in unconfined geometries. These are often called "external flows." Examples include the airflow over the wall of a building or across a bank of tubes in a heat exchanger.

The equation to be used will depend on the type of flow—laminar or turbulent. In nearly all building-related convection situations where forced convection is involved, the flow will be turbulent (one exception is water flow in pipes for very cold liquid distribution systems). However, free convection can be either laminar or turbulent in buildings. For completeness, we will present expressions for free convection and forced convection under both laminar and turbulent flows.

The following equations are dimensional. In IP units, lengths are in feet, temperatures in degrees Fahrenheit, velocities in feet per second, and convection coefficients in Btu/(h·ft²·°F) unless otherwise noted. In the SI versions of the same equations, lengths are in meters, temperatures in degrees Celsius, velocities in meters per second, and convection coefficients in W/(m²·K). For those properties involving air, physical properties at 70°F (20°C) have been used. Two sets of equations are presented, one for each set of units, identified in the equation number.

1. Free convection in air from a tilted surface is given by the following equations:
   Laminar flow:

$$h_{con} = 1.42\left(\frac{\Delta T \sin\beta}{L}\right)^{1/4} \quad \text{if } L^3 \cdot \Delta T < 1.0 \tag{2.49 SI}$$

$$h_{con} = 0.29\left(\frac{\Delta T \sin\beta}{L}\right)^{1/4} \quad \text{if } L^3 \cdot \Delta T < 63 \tag{2.49 IP}$$

   where
   ΔT is the temperature difference in Equation 2.45
   L is the length of the plate in the direction of the buoyancy-driven flow in meters or feet
   β is the surface tilt angle from the horizontal

   Equation 2.49 applies for surface tilts between 30° and 90°.

   Turbulent flow occurs if the preceding inequalities are reversed. The corresponding equations are as follows:
   Turbulent flow:

$$h_{con} = 1.31(\Delta T \sin\beta)^{1/3} \quad \text{if } L^3 \cdot \Delta T > 1.0 \tag{2.50 SI}$$

$$h_{con} = 0.19(\Delta T \sin\beta)^{1/3} \quad \text{if } L^3 \cdot \Delta T > 63 \tag{2.50 IP}$$

   Equations 2.49 and 2.50 apply to interior wall and window surfaces of buildings and to corresponding exterior surfaces in the absence of wind. Note that for turbulent flows, the physical dimension L does not affect h_{con}.

2. Free convection from horizontal pipes and other horizontal cylinders in air is given by the following equations:
   Laminar flow:

$$h_{con} = 1.32\left(\frac{\Delta T}{D}\right)^{1/4} \quad \text{if } D^3 \cdot \Delta T < 1.0 \tag{2.51 SI}$$

$$h_{con} = 0.27\left(\frac{\Delta T}{D}\right)^{1/4} \quad \text{if } D^3 \cdot \Delta T < 63 \tag{2.51 IP}$$

   in which D is the cylinder's outer diameter (in feet for IP units and meters for SI units).

Turbulent flow:

$$h_{con} = 1.24(\Delta T)^{1/3} \quad \text{if } D^3 \cdot \Delta T > 1.0 \qquad (2.52\ \text{SI})$$

$$h_{con} = 0.18(\Delta T)^{1/3} \quad \text{if } D^3 \cdot \Delta T > 63 \qquad (2.52\ \text{IP})$$

3. For free convection coefficient for warm horizontal surfaces facing up (e.g., flat roofs of buildings warmed by the sun), one uses the following equations:

   Laminar flow:

$$h_{con} = 1.32\left(\frac{\Delta T}{L}\right)^{1/4} \quad \text{if } L^3 \cdot \Delta T < 1.0 \qquad (2.53\ \text{SI})$$

$$h_{con} = 0.27\left(\frac{\Delta T}{L}\right)^{1/4} \quad \text{if } L^3 \cdot \Delta T < 63 \qquad (2.53\ \text{IP})$$

where L is the average length of the sides of the horizontal surface. This expression also applies to cold surfaces facing downward. An example is the inner surface of a plane skylight in the roof of a building in winter.

   Turbulent flow:

$$h_{con} = 1.52(\Delta T)^{1/3} \quad \text{if } L^3 \cdot \Delta T > 1.0 \qquad (2.54\ \text{SI})$$

$$h_{con} = 0.22(\Delta T)^{1/3} \quad \text{if } L^3 \cdot \Delta T > 63 \qquad (2.54\ \text{IP})$$

4. If a warmed surface faces downward, however, the laminar convection coefficient is reduced because of the stable stratification condition. Equation 2.53 can be used, but the dimensional multiplier is changed from 0.27 to 0.12 in IP units. In SI units, the multiplier of 1.32 becomes 0.59. This modified expression also applies for cooled flat surfaces or planes facing upward (e.g., a horizontal skylight's outer surface in summer).

Example 2.9: Free Convection from a Horizontal Hot Roof

Find the free convection coefficient of a 10 m × 6 m horizontal roof of a building along the 10 m side. The roof's surface is heated by the sun to 67°C while the ambient air is at 20°C.

Given: T_s = 67°C, T_f = 20°C and L = 10 m

Find: h_{con}

Solution

Since the flow can be laminar or turbulent, either Equations 2.53 SI or 2.54 SI applies to this case.

We must first determine the type of flow regime, i.e., whether laminar or turbulent flow. With L = 10 m and ΔT = 67 − 20 = 47°C:

$$L^3 \cdot \Delta T = 10^3 \times 47 > 1.0$$

Therefore, the turbulent flow equation, Equation 2.54 SI, applies. Substituting numerical values, we have

$$h_{con} = 1.52 \times (47)^{1/3}$$

$$= 5.498\ \text{W}/(\text{m}^2 \cdot \text{K})\ [0.972\ \text{Btu}/(\text{h} \cdot \text{ft}^2 \cdot {}^\circ\text{F})]$$

Comments

This example illustrates that one should expect small values of h_{con} in free convection. Table 2.7 shows that free convection in air is a relatively ineffective heat transfer mechanism.

5. Forced convection over smooth surfaces or planes does not depend on their orientation.

Laminar flow:

$$h_{con} = 2.0\left(\frac{\text{v}}{L}\right)^{1/2} \quad \text{if } \text{v} \cdot L < 1.4 \qquad (2.55\ \text{SI})$$

$$h_{con} = 0.35\left(\frac{\text{v}}{L}\right)^{1/2} \quad \text{if } \text{v} \cdot L < 15 \qquad (2.55\ \text{IP})$$

Turbulent flow:

$$h_{con} = 6.2\left(\frac{\text{v}^4}{L}\right)^{1/5} \quad \text{if } \text{v} \cdot L > 1.4 \qquad (2.56\ \text{SI})$$

$$h_{con} = 0.54\left(\frac{\text{v}^4}{L}\right)^{1/5} \quad \text{if } \text{v} \cdot L > 15 \qquad (2.56\ \text{IP})$$

where

L is the length of the surface in the direction of flow [m (ft)]

v is the velocity [m/s (ft/s)]

These equations also apply to wind blowing over the roof or wall of a building. The roughness of the exterior finish of a building can affect the heat transfer rate. Figure 2.14 shows the variation of the forced convection air velocity for different types of surfaces. We note that the forced convection coefficient will be twice as large for a rough, stucco wall as for a smooth surface such as glass.

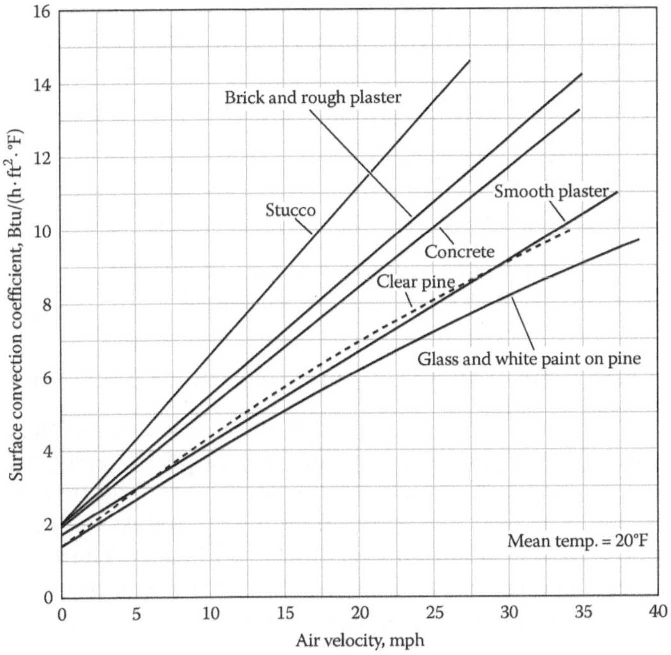

FIGURE 2.14
Surface forced convection coefficient h_{con} for various wall finishes. (For SI values of h_{con} multiply the values in the figure by 5.67.) (From *ASHRAE Handbook of Fundamentals*, American Society of Heating, Refrigerating and Air-Conditioning Engineers, Atlanta, GA, 2005. Copyright ASHRAE, www.ashrae.org.)

Example 2.10: Forced Convection from a Hot Roof

Consider the same 10 m × 6 m flat roof of a building as that assumed in Example 2.9. In this case, wind blows at 10 m/s. As previously, the air temperature is 20°C and the temperature of the roof is 67°C. Calculate the total heat loss when the wind blows along the 10 m long side.

Given: v = 10 m/s, length of surface in direction of flow L = 10 m

Find: h_{con} using Equation 2.56 for forced external flow over plates

Solution

First, we need to determine the flow regime. The condition $v \cdot L = 10\ (\text{m/s}) \times 10\ \text{m} = 100\ \text{m}^2/\text{s} > 1.4$. This corresponds to turbulent flow regime, and Equation 2.56 can be used to calculate the convective heat transfer coefficient:

$$h_{con} = 6.2 \times \left(\frac{10^4}{10}\right)^{1/5} = 24.7\ \text{W}/(\text{m}^2\cdot\text{K})$$

Finally, from Equations 2.46 and 2.47, the total heat loss from the roof is

$$\dot{Q} = h_{con} \cdot A \cdot \Delta T = 24.7\ \text{W}/(\text{m}^2\cdot\text{K}) \times (10 \times 6)\ \text{m}^2 \times (67 - 20)\ \text{K}$$
$$= 69{,}654\ \text{W} = 69.7\ \text{kW}$$

Comment

Note that the convective coefficient is 4.5 times larger than the free convective coefficient determined in Example 2.9. Further, this same example will be solved as Example 12.9 in Chapter 12 using dimensionless equations. The difference between both values is close (within 6%).

2.9.4 Relevant Internal Flow Equations

Flows of fluids confined by boundaries such as the sides of a duct are called internal flows. Since the mechanisms of convection are quite different for such flows vis-à-vis external flows, the expressions for h_{con} are also different. In this section, we present equations for turbulent internal flow in water and air:

1. The forced convection coefficient for fully developed turbulent flow of *air through ducts and pipes* is given by

$$h_{con} = 8.8\left(\frac{v^4}{D_h}\right)^{1/5} \qquad (2.57\ \text{SI})$$

$$h_{con} = 0.5\left(\frac{v^4}{D_h}\right)^{1/5} \qquad (2.57\ \text{IP})$$

in which the hydraulic diameter D_h cm (in.) is defined as four times the ratio of the flow conduit's cross-sectional area divided by the perimeter of the conduit. For the familiar case of a round pipe with diameter D, the hydraulic diameter is

$$D_h = \frac{4(\pi D^2/4)}{\pi D} = D \qquad (2.58)$$

2. Convection coefficient values for fully developed turbulent *water flow through pipes* can be found from

$$h_{con} = 3580 \times (1 + 0.015 \times T)\left(\frac{v^4}{D_h}\right)^{1/5} \qquad (2.59\ \text{SI})$$

In this equation, D_h is in centimeters, and the temperature is in degrees Celsius. Note that all SI dimensional equations in this section assume v to be in meters per second. In IP units:

$$h_{con} = 150 \times (1 + 0.011 \times T)\left(\frac{v^4}{D_h}\right)^{1/5} \qquad (2.59\text{ IP})$$

where D_h is in inches and v in feet per second. The temperature term (in degrees Fahrenheit) accounts for the variation of water viscosity and conductivity with temperature.

Dimensional equations for other fluids involved in building HVAC equipment are given in Mathur (1990).

Example 2.11: Water Flowing Inside a Pipe

Find the internal convective heat transfer coefficient for conditions when water at a temperature of 10°C with a velocity of 2.5 m/s flows through a tube with 8 mm internal diameter. These conditions are common in the water side of the evaporator of a refrigeration system.

Given: v = 2.5 m/s, characteristic length D = 8 mm

Find: h_{con}

Assumption: The flow is turbulent.

Solution

We use Equation 2.59 SI:

$$h_{con} = 3580 \times (1 + 0.015 \times 10\text{ °C})\left(\frac{2.5^4\text{ m/s}}{0.8\text{ cm}}\right)^{1/5}$$

$$= 8960\text{ W/(m}^2 \cdot \text{K)}$$

Comment

This example is identical to Example 12.8 in Chapter 12, which will be solved using dimensionless equations. In that case, we find h_{con} = 7220 W/(m²·K) about a 20% difference.

2.9.5 Tables and Graphs of Convection Coefficients

For preliminary design, approximate values for convection coefficients are adequate. This is particularly true for heating and cooling load calculations. The thermal resistance due to convection is much smaller than that of the insulation in building walls and roofs. Therefore, use of a constant value of h_{con} instead of one depending on the parameters of the previous section is sufficient for the early phases of design.

Components of HVAC systems such as coils and piping that carry water can be exposed to subfreezing temperatures in winter. There are many methods of freeze protection, but one of the most common is to use an antifreeze solution in place of pure water. The standard antifreezes are either ethylene glycol or propylene glycol. Although these substances are well suited to freeze protection, they have lower convection heat transfer values than pure water. Figure 2.15 shows the

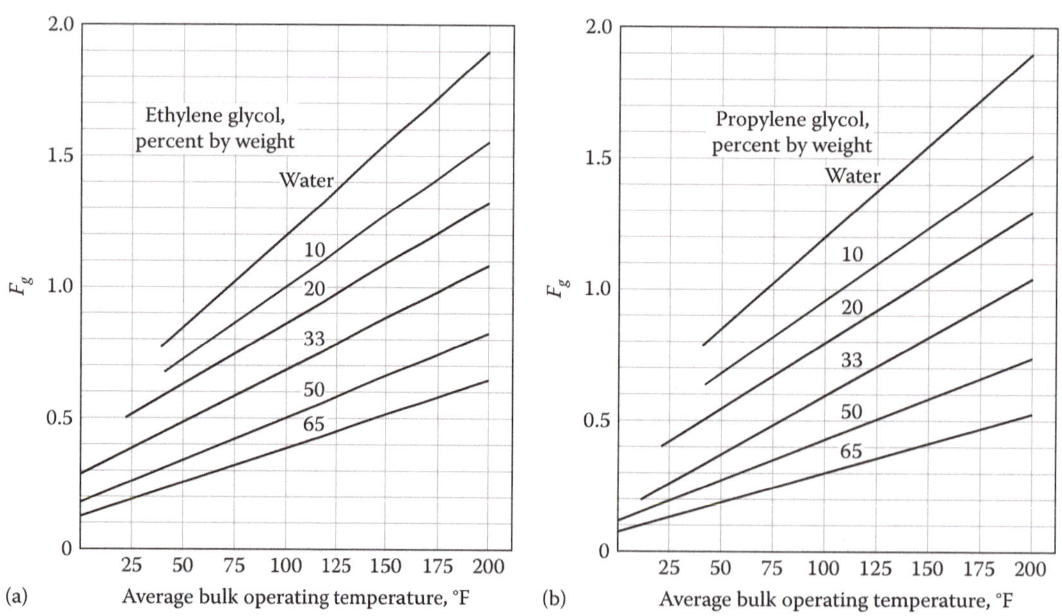

FIGURE 2.15
Glycol correction factor F_g for heat transfer in turbulent flow in pipes: (a) aqueous ethylene glycol solutions and (b) aqueous propylene glycol solutions. (From Nussbaum, O.J., *Heat. Piping Air Cond.*, 62(1), 75, January 1990.)

degradation of h_{con} for various concentrations of antifreeze. To find the convection coefficient with glycol $h_{con,g}$, the value for pure water is multiplied by F_g from the figure (F_g = 1.0 for pure water at 74°F); the temperature effect can also be found from the figure or from Equation 2.59:

$$h_{con,g} = F_g \times h_{con,w} \tag{2.60}$$

Example 2.12: Effect of Glycol on Convection Coefficient

Find the convection heat transfer coefficient for a 50% solution of ethylene glycol at 40°F, flowing at 6 ft/s in a 0.6 in. diameter pipe.

Given: v = 6 ft/s, D = 0.6 in.

Find: $h_{con,g}$

Lookup values: From Figure 2.15 for 50% ethylene glycol, F_g = 0.32.

Solution

Equation 2.59 IP can be used to find the convection coefficient for pure water:

$$h_{con,w} = 150 \times (1 + 0.011 \times 40)\left(\frac{6.0^4}{0.6}\right)^{1/5}$$

$$= 1003 \text{ Btu/(h} \cdot \text{ft}^2 \cdot {}^\circ\text{F)}$$

The glycol coefficient is calculated from Equation 2.60 as follows:

$$h_{con,g} = 0.32 \times 1003 = 321 \text{ Btu/(h} \cdot \text{ft}^2 \cdot {}^\circ\text{F)}$$

Comments

At low temperatures, where they are most needed for freeze protection, glycols have convection coefficients well below those for water. In this example, the coefficient is 68% lower than that for water.

The designer of building retrofits must be aware of this problem if glycol is to be used in a system formerly designed for water. Both heating and cooling capacity can be severely compromised. In addition to reduced heat transfer, glycol solutions also require greater pumping power.

2.9.6 Combined Conduction and Convection

Nearly all heat transfer situations in buildings include more than one mode of heat transfer. For example, heat loss through a wall includes both conduction as treated in Examples 2.4 and 2.5 and convection from the inner and outer surfaces. Figure 2.16a shows the equivalent circuit for this case, and the calculation procedure is illustrated as follows.

Example 2.13: Effect of Convection on Wall R Value

Repeat Example 2.5 for a stud wall to include the effect of inner and outer surface convection coefficients. The inner surface coefficient has been calculated as 0.4 Btu/(h·ft²·°F) (turbulent free convection with a wall-to-room temperature difference of 10°F, from Equation 2.50 IP), and the outer surface coefficient is 3.7 Btu/(h·ft²·°F) (15 mph wind along a 15 ft long wall, from Equation 2.56 IP). Find the overall wall R value.

Given: $h_{con,i}$ or h_i = 0.40 Btu/(h·ft²·°F), $h_{con,o}$ or h_o = 3.7 Btu/(h·ft²·°F)

Figure: See Figure 2.16a.

Find: R_{tot}

Lookup values: From Example 2.5, the R value of the solid wall $R_{cd,wall}$ is 19.8 (h·ft²·°F)/Btu.

Solution

From the equivalent circuit shown in Figure 2.16a, we see that the total wall R value is

$$R_{tot} = \frac{1}{h_i} + R_{cd,wall} + \frac{1}{h_o}$$

Substituting numerical values gives

$$R_{tot} = \frac{1}{0.4} + 19.8 + \frac{1}{3.7}$$

$$= 22.6 \text{ (h} \cdot \text{ft}^2 \cdot {}^\circ\text{F)/Btu } [4.0(\text{m}^2 \cdot \text{K})/\text{W}]$$

Comments

Note that the inner and outer surface air films through which convection occurs account for 13% of the wall's thermal resistance.

For a more accurate calculation, the radiation losses from the walls must be added (treated in the next section). If this is done, the percentage of this wall's resistance due to the surface effects drops to only 4%. The bulk of the insulating effect of a wall occurs in the insulation.

When we have a building element such as a stud wall that consists of wooden studs with insulation in between and is sandwiched between two layers, the heat transfer is strictly 2-D (Figure 2.16b). The heat flow lines would distort in an attempt to preferentially flow through the higher conductivity material, namely, the studs (this is referred to as "thermal bridge"

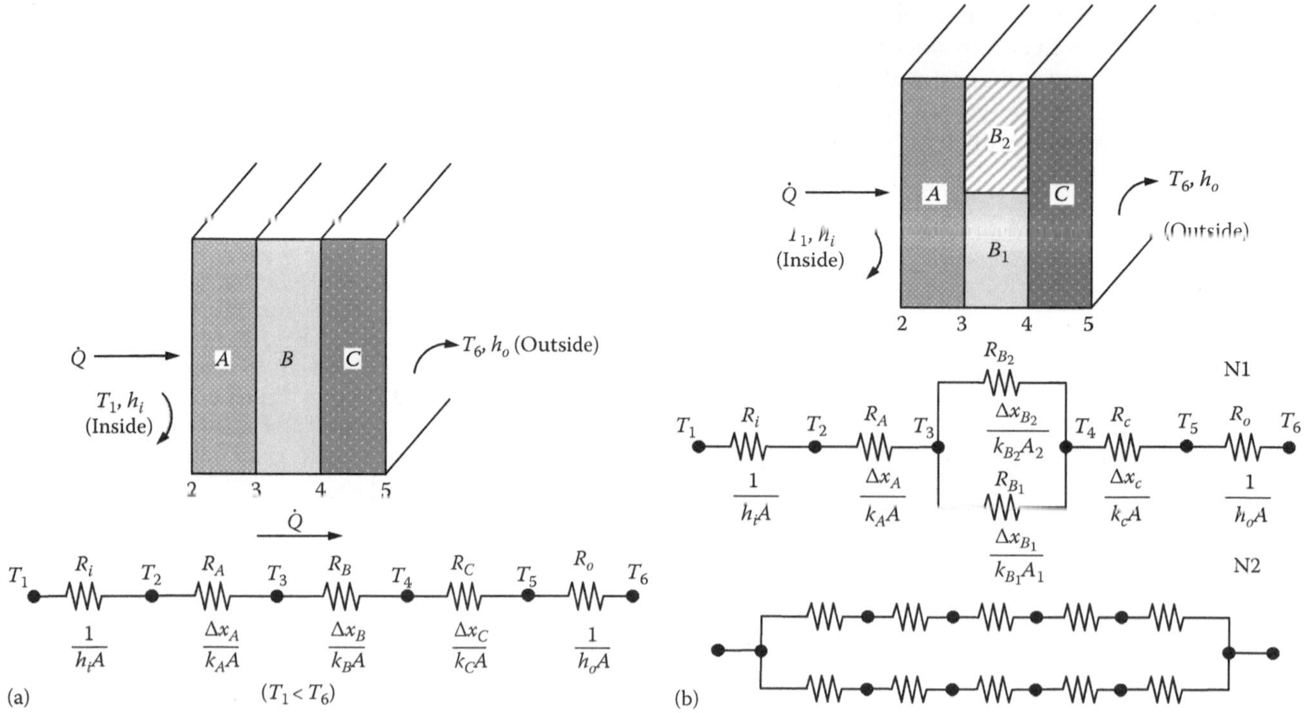

FIGURE 2.16
(a) Sketch and thermal network diagram of a composite solid wall with three elements in series. The inner and outer surface convection coefficients are represented by h_i and h_o. (b) Sketch of a composite wall with parallel elements B_1 and B_2 sandwiched between layers A and C. Such a composite wall can be represented by one of two different thermal network diagrams shown. N1 is the isothermal plane and N2 is the parallel-path representations.

effect—discussed further in Section 2.10.7). Thus, less heat will flow through the insulation (the lower conductivity material) and more through the wooden stud section. Nonetheless, reasonable estimates can be obtained by assuming 1-D flow, i.e., no lateral heat flow in the y-direction. How to best represent the heat flow is a little more subtle since there are two possibilities—networks N1 (the isothermal plane representation) and N2 (the parallel-path method) shown in Figure 2.16b. In the later, the wall is assumed to consist of two entirely separate wall sections. We had solved Example 2.13 assuming, in essence, network N1 since it reduces to that shown in Figure 2.16a. Such a representation would *underestimate* the total wall resistance. However, according to ASHRAE Fundamentals (2013), network N2 will yield acceptable results if all materials involved (the bridge element and all building envelope elements in contact with it) are nonmetals such as wood, drywall, and concrete. On the other hand, if the materials have very different conductivities (nonmetal elements along with structural steel or aluminum frames of windows), it is recommended that the parallel-path network model N1 be adopted. In case of doubt, the analyst should solve the problem following both representations that would provide the limits of the wall resistance value.

Example 2.14: Heat Losses in Parallel

This example illustrates heat flows in parallel occurring from a small building. The walls of a 40 ft × 60 ft office have 8 ft ceilings and the walls contain 2 in. thick doors of total surface area 80 ft² and single-glazed windows of 200 ft² area. What fraction of the total heat loss occurs through the windows.

Given: Thermal conductance of the wall $U_{wall} = 0.08$ Btu/(ft²·h·°F), $U_{window} = 1.13$ Btu/(ft²·h·°F), $A_{door} =$ 80 ft², $A_{window} = 200$ ft^2, conductivity of wooden door $k_{wood} = 0.10$ Btu/(h·ft·°F), thickness $x = 2$ in.

Assumptions: Inside and outside film resistances are 0.68 (h·ft²·°F)/Btu and 0.17 (h·ft²·°F)/Btu respectively.

Solution

The area of the wall is $A_{wall} = [(40 + 60)$ ft × 8 ft] × 2 −80 ft² −200 ft² = 1320 ft².

The R value of the door can be determined as shown in the following table.

Element	Resistance [(h·ft²·°F)/Btu]
Inside film	0.68
Wood	(2 in./12 in./ft)/0.10 = 1.667
Outside film	0.17
Total	2.52

The U-value of the door is given by

$$U_{door} = \frac{1}{R_{door}} = \frac{1}{2.52} = 0.397 \text{ Btu/(h}\cdot\text{ft}^2\cdot{}^\circ\text{F)}$$

The fraction of the total heat loss attributed to the windows is given by

$$\frac{UA_{window}}{[(UA)_{window} + (UA)_{door} + (UA)_{wall}]}$$

$$= \frac{(1.13 \times 200) \text{ Btu/(h}\cdot{}^\circ\text{F)}}{(1.13\times 200 + 0.397 \times 80 + 0.08 \times 1320) \text{ Btu/(h}\cdot{}^\circ\text{F)}}$$

$$= \frac{226}{(226 + 31.76 + 105.6)} = 0.622$$

Thus, 62.2% of the total heat loss is due to the windows. If energy conservation measures are to be implemented, replacing the existing windows with, say a double glazed window, would be an obvious measure to consider.

2.10 Radiation Heat Transfer

Radiation heat transfer is important within conditioned spaces since it is one of the key determinants of both comfort and heat exchange in the zones of a building. Solar radiation is also important in analyzing the thermal behavior of both interiors and exteriors of buildings. It is so important, in fact, that we devote a separate chapter to it. Solar radiation and its effects on buildings are discussed in Chapter 4 and in numerous textbooks such as Goswami et al. (2000). In this section, we review nonsolar radiation heat transfer fundamentals as they apply to buildings.

There are three features of radiation that set it apart from conduction and convection. First, radiation heat transfer occurs by electromagnetic waves with a strong spectral dependence. Second, the microscale details (e.g., color) of surfaces involved in radiation have a first-order effect on the rate of heat transfer. Third, radiation is a highly nonlinear function of temperature.

2.10.1 Thermal Radiation Spectrum and the Stefan–Boltzmann Law

The fundamental physical law governing thermal radiation emission is the Stefan–Boltzmann law. This law states that the heat flux emitted by an ideal radiator called a "blackbody" is proportional to the absolute temperature (units of Kelvin or degrees Rankine) to the fourth power. The proportionality constant denoted by σ is called the Stefan–Boltzmann constant. The Stefan–Boltzmann law is expressed as

$$\dot{q} = \sigma T^4 = E_b \tag{2.61}$$

The term E_b is called the total blackbody emissive power.

The value of the Stefan–Boltzmann constant in IP units is 0.1714×10^{-8} Btu/(h·ft²·°R⁴), and in SI units it is 5.669×10^{-8} W/(m²·K⁴). Recall that absolute temperatures, in degrees Rankine or Kelvin, must be used in radiation calculations.

Since radiation is transmitted by electromagnetic waves, it has a wavelength and frequency. In fact, radiation emitted by a surface has an entire spectrum of wavelengths. The wavelengths of radiation with which one is concerned in buildings range from a fraction of a micrometer to more than 100 μm. The upper curve in Figure 2.17 shows how the spectral emissive power $E_{b\lambda}$ varies with wavelength λ for a surface near room temperature, 530°R (70°F). Note that there is a distinct peak to the distribution.

The Wien displacement law gives the wavelength at which the spectral blackbody emissive power is a maximum:

$$\lambda_{max} T = 5216 \ \mu\text{m}\cdot{}^\circ\text{R} \quad (2898 \ \mu\text{m}\cdot\text{K}) \tag{2.62}$$

In Figure 2.17, the peak calculated from this expression is at 9.84 μm (=5216/530), in agreement with the upper curve on the figure corresponding to $T = 530$°R.

Other than this relation of maximum emission wavelength to temperature, we need not concern ourselves with the details of the spectral distribution. Standard heat transfer texts (such as Incropera et al., 2007) discuss the details.

2.10.2 Gray Surfaces

Real surfaces emit less radiation than ideal "black" ones. The ratio of actual emissive power E to the emissive power of a black surface at the same temperature E_b is called the emissivity. It is defined by

$$\varepsilon = \frac{E}{E_b} \tag{2.63}$$

Although emissivity is a function of wavelength for many materials, a physically realistic simplification is often made by assuming that the emissivity is constant, at least over a limited range of wavelengths. For radiation calculations in buildings, this is an entirely satisfactory assumption. Surfaces for which the emissivity is constant are called "gray surfaces." The lower curve in Figure 2.17 shows the gray body emissive power for a gray surface at 530°R whose emissivity is 0.60.

Table 2.8 lists values for the emissivity of a number of building materials. We shall find such values useful in calculating heating and cooling loads on buildings. As a rule of thumb, this table tells us that the emissivity of most building materials is approximately 0.9 (in the range of typical building temperatures).

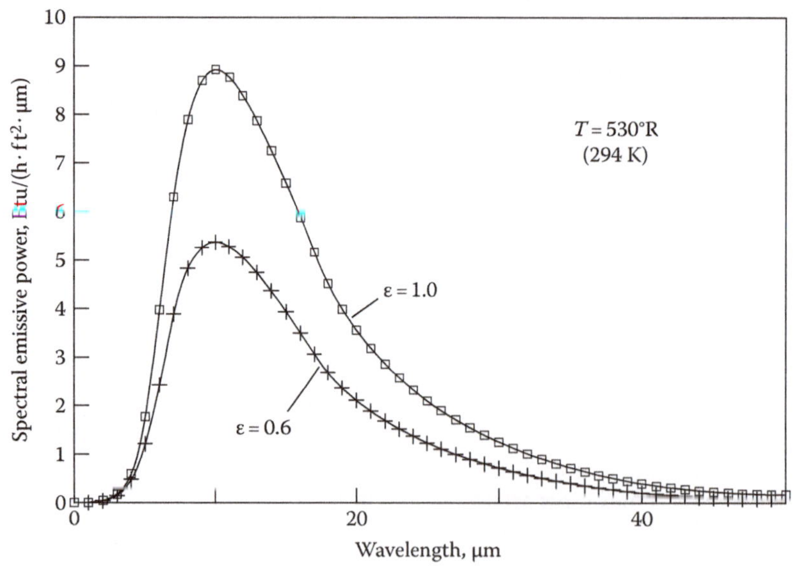

FIGURE 2.17
Example of thermal radiation spectra for black and gray (ε = 0.6 in.) surfaces at room temperature, 70°F (530°R; 294 K).

TABLE 2.8

Emissivities of Some Common Building Materials at Specified Temperatures

Surface	Temperature, °C	Temperature, °F	ε
Aluminum foil			
Bright	40	100	0.05
Brick			
Red, rough	40	100	0.93
Concrete			
Rough	40	100	0.94
Glass			
Smooth	40	100	0.94
Ice			
Smooth	0	32	0.97
Marble			
White	40	100	0.95
Paints			
Black gloss	40	100	0.90
White	40	100	0.89–0.97
Various oil paints	40	100	0.92–0.96
Paper			
White	40	100	0.95
Sandstone	40–250	100–500	0.83–0.90
Snow	−12 to −6	10–20	0.82
Water			
0.1 mm or more thick	40	100	0.96
Wood			
Oak, planed	40	100	0.90
Walnut, sanded	40	100	0.83
Spruce, sanded	40	100	0.82
Beech	40	100	0.94

Source: Sparrow, E.M. and Cess, R.D., *Radiation Heat Transfer*, augmented ed., Hemisphere, New York, 1978. With permission.

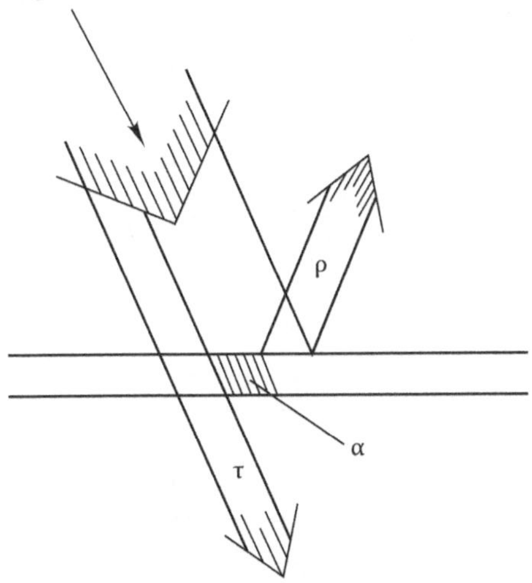

FIGURE 2.18
Schematic representation of transmissivity, absorptivity, and reflectivity. Incoming radiation (=1).

2.10.3 Radiation Properties: Absorptivity, Transmissivity, and Reflectivity

In addition to emissivity, three additional properties of surfaces affect the rate of radiation heat transfer. Figure 2.18 shows how the three properties—absorptivity α, transmissivity τ, and reflectivity ρ—are related. Conservation of energy requires that the sum of these three properties be unity:

$$\alpha + \tau + \rho = 1 \quad (2.64)$$

Therefore, by knowing any two properties, the third can always be found. Although this equation is derived for a single wavelength, it is valid for piecewise gray surfaces if the wavelength range over which the three properties are calculated is the same.

Kirchhoff's identity, stating that absorptivity and emissivity for gray surfaces are equal, is another useful expression:

$$\alpha = \varepsilon \quad (2.65)$$

This expression is also valid for nongray surfaces at a given wavelength.

2.10.4 Shape Factors

Since radiation is transferred by directional beams of radiation, the relative size and location of two (or more) surfaces interchanging radiation are important factors in quantifying the heat transfer rate. The radiation shape factor F_{12} includes all needed geometric information. The shape factor (also called view factor) is the fraction of radiation leaving diffuse surface 1 that is intercepted (but not necessarily absorbed) by surface 2. The shape factor is strictly geometric and does not depend on surface properties such as emissivity or temperature. Figures 2.19 and 2.20 are plots of the shape factors for two parallel surfaces, e.g., between

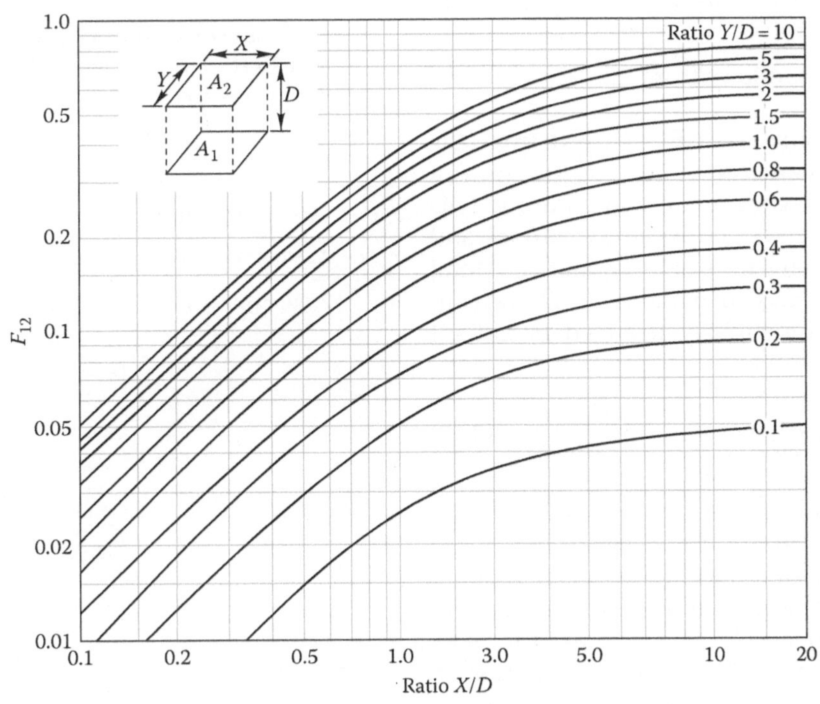

FIGURE 2.19
Shape factor F_{12} for two parallel planes.

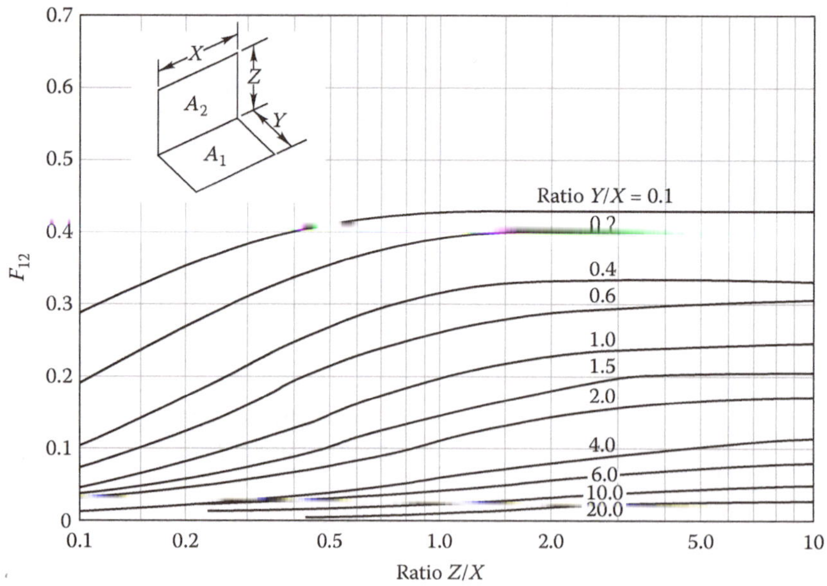

FIGURE 2.20
Shape factor F_{12} for two adjacent orthogonal planes.

the floor and ceiling of a room, and two perpendicular surfaces, e.g., between a floor and an adjacent wall. These charts are plotted from equations given in Siegel and Howell (1981).

Two key relationships exist among shape factors. The first is called the "reciprocity relationship," which is given by

$$A_1F_{12} = A_2F_{21} \quad (2.66)$$

It can be shown that this follows from the second law of thermodynamics.

The second relationship is easily derived from the first law of thermodynamics. It states that the sum of shape factors for a given surface must equal unity. This is easy to see since the radiation leaving a surface and intercepted elsewhere must be exactly equal to the total radiation leaving the surface. In equation form, the conservation of energy for j surfaces involved in radiant transport is given by

$$F_{11} + F_{12} + F_{13} + \cdots + F_{1j} = 1 \quad (2.67)$$

The shape factor F_{11} is nonzero only for concave surfaces, i.e., surfaces that can "see" themselves.

It is important to recall that shape factor algebra described in this section applies only for diffusely emitting and reflecting surfaces. It cannot be used for surfaces such as mirrors that reflect specularly. However, nearly all surfaces in buildings are diffuse.

Equations 2.66 and 2.67, along with a few equations or charts of shape factors, can be used to calculate the shape factors needed for radiation problems within buildings. Example 2.15 illustrates how one calculates shape factors within a room. Example 2.16 shows how these results can be used to find the rate of radiation heat transfer between surfaces in such an enclosure.

Example 2.15: Radiation Shape Factors

One room of a building is 8 ft from floor to ceiling and 16 ft². Find the shape factor from one of the walls to (1) the opposite wall, (2) the adjacent wall, and (3) the ceiling.

Given: $h = 8$ ft, $S = 16$ ft

Figure: See Figure 2.21.

Assumptions: All surfaces are diffuse.

Find: F_{12}, F_{15}, and F_{16}

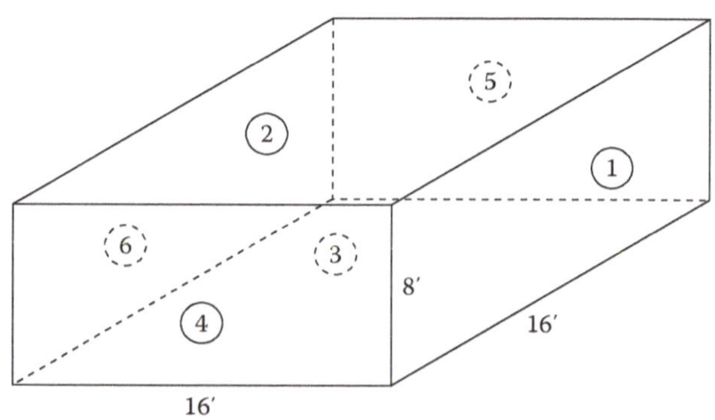

FIGURE 2.21
Isometric sketch of example room.

Solution

To use Figures 2.19 and 2.20, one simply finds the ratios of lengths needed to enter the graphs. First, consider the shape factor between opposite walls. Following the nomenclature of Figure 2.19 is

$$\frac{Y}{D}=\frac{16\text{ ft}}{16\text{ ft}}=1 \quad \text{and} \quad \frac{X}{D}=\frac{8\text{ ft}}{16\text{ ft}}=0.5$$

From Figure 2.19, the shape factor between two opposing walls is read off as

$$F_{16}=0.12$$

For two adjacent walls, the shape factor is found from Figure 2.20. In the nomenclature of this figure,

$$\frac{Y}{X}=\frac{16\text{ ft}}{8\text{ ft}}=2 \quad \text{and} \quad \frac{Z}{X}=\frac{16\text{ ft}}{8\text{ ft}}=2$$

With these two values, one reads from Figure 2.20

$$F_{15}=0.14$$

Finally, for the wall-to-ceiling shape factor, we use Figure 2.20 again with

$$\frac{Y}{X}=\frac{8\text{ ft}}{16\text{ ft}}=0.5 \quad \text{and} \quad \frac{Z}{X}=\frac{16\text{ ft}}{16\text{ ft}}=1$$

The value of this shape factor is

$$F_{12}=0.30$$

Comments

In many shape factor problems, it is not always possible to reach a solution by using shape factor algebra alone. Graphs of shape factors or the equations from which they are derived are often needed. The solution of this problem can be checked by using Equation 2.67. To do this, we sum all the shape factors with initial subscript 1, noting that F_{11} = 0.0. The sum of nonzero shape factors = 0.12 (opposite wall) + 2 × 0.14 (two adjacent walls) + 2 × 0.30 (wall-to-floor and wall-to-ceiling view factors) = 1.00.

2.10.5 Radiative Exchange

With the tools developed in the three preceding sections, we are now able to find the rate of radiation heat transfer between surfaces. In this section, we will treat exchange between two surfaces. A similar method can be used if three or more surfaces are involved. If more than three surfaces are involved, the algebra becomes rather involved and matrix inversion methods are suggested (Siegel and Howell, 1981).

The rate of heat transport between two gray diffuse surfaces that form an enclosure containing a nonabsorbing gas or configured so that *one encloses the other* is given by

General case:

$$\dot{Q}_{12}=\frac{A_1\left(\sigma T_1^4-\sigma T_2^4\right)}{(\rho_1/\varepsilon_1)+(1/F_{12})+(\rho_2 A_1/\varepsilon_2 A_2)} \tag{2.68}$$

We will use this expression in Example 2.16.

This equation applies to a number of cases of special interest such as radiant transport between *two parallel opaque planes spaced closely together*. This condition can occur in buildings between the inner and outer opaque surfaces of an airspace in a wall or in an attic radiant barrier. In these cases, the denominator of Equation 2.68 simplifies since $F_{12} = 1$, $\rho_1 = 1 - \varepsilon_1$, $\rho_2 = 1 - \varepsilon_2$, and $A_1 = A_2 = A$.

Special case:

$$\dot{Q}_{12}=\frac{A_1\left(\sigma T_1^4-\sigma T_2^4\right)}{(1/\varepsilon_1)+(1/\varepsilon_2)-1} \tag{2.69}$$

This result is obtained from Equations 2.64 and 2.65 with $\tau = 0$ to eliminate the reflectivity (since the surfaces are opaque, the transmissivity is zero). An *effective emittance* term is often used for convenience that is defined as

$$\varepsilon_{eff}=\left(\frac{1}{\varepsilon_1}+\frac{1}{\varepsilon_2}-1\right)^{-1} \tag{2.70}$$

A convenient shortcut for calculating radiation heat transfer in buildings is to define a *radiation heat transfer coefficient* h_{rad} as follows:

$$h_{rad}=\frac{\dot{q}_{12}}{T_1-T_2} \tag{2.71}$$

This coefficient can be used just as one uses h_{con} in convection problems. If the two temperatures are close to each other (no more than 50°F or 28°C difference), a good approximation to Equation 2.71 for h_{rad} in two-surface situations given by Equation 2.69 is

$$h_{rad}=4\sigma\varepsilon_{eff}\bar{T}^3 \quad \text{with } \bar{T}\equiv\frac{T_1+T_2}{2} \tag{2.72}$$

The approach is also applicable to Equation 2.68 if the two surfaces are not parallel and of equal area.

The following two examples illustrate how radiation calculations are performed to determine the heat transfer in several building-related situations.

Example 2.16: Room Radiation Heat Transfer

Assume that surface 1 in Example 2.15 is the inner surface of an exterior wall of a poorly insulated building and that its temperature is 60°F on a winter day. All other room surfaces are at 72°F. What is the heat loss from these surfaces to the cool wall? The emissivities of all walls are 0.85.

Given: $\varepsilon_1 = \varepsilon_2 = 0.85$

Wall dimensions as in Example 2.15

$$T_1 = 60°\text{F} = 520°\text{R}$$

$$T_{n\neq 1}(= T_2) = 72°\text{F} = 532°\text{R}$$

Find: $\dot{Q}_{12}$

Figure: Figure 2.21

Solution

Although this problem involves six surfaces, five of the six are at the same temperature T_2 and can, therefore, be treated as one surface as far as radiation transport is concerned. In Equation 2.68, we require values of the shape factor F_{12}, which, by inspection, is unity since all radiation emitted by surface 1 is intercepted by the second surface. Recall that in all radiation calculations, temperatures must be expressed on the absolute temperature scale in degrees Rankine (or Kelvin).

The emissive powers in the numerator are

$$E_{b1} = \sigma T_1^4 = [0.1714 \times 10^{-8}\ \text{Btu/(h}\cdot\text{ft}^2\cdot°\text{R}^4)](520°\text{R})^4$$
$$= 125.3\ \text{Btu/(h}\cdot\text{ft}^2)$$

and

$$E_{b2} = \sigma T_2^4 = 0.1714 \times 10^{-8} \times 532^4 = 137.3\ \text{Btu/(h}\cdot\text{ft}^2)$$

With $F_{12} = 1$,

$$\frac{\rho_1}{\varepsilon_1} + \frac{1}{F_{12}} = \frac{1-\varepsilon_1}{\varepsilon_1} + 1 = \frac{1}{\varepsilon_1}$$

Then, Equation 2.68 takes the following simplified form for this example:

$$\dot{Q}_{12} = \frac{A_1\left(\sigma T_1^4 - \sigma T_2^4\right)}{(1/\varepsilon_1) + (\rho_2 A_1/\varepsilon_2 A_2)} \tag{2.73}$$

The total area of all walls except for 1 (i.e., surface 2) is 896 ft². Substituting numerical values, we have

$$\dot{Q}_{12} = \frac{(16\ \text{ft} \times 8\ \text{ft}) \times (125.3 - 137.3)}{(1/0.85) + (0.15 \times (16 \times 8)/(0.85 \times 896))}$$
$$= -1270\ \text{Btu/h}$$

The sign indicates that heat flows from surface 2 to surface 1.

Comments

The heat loss by radiation is of the same order of magnitude as the convection heat transfer to the wall. A typical convection coefficient for this case is about 0.6 Btu/(h·ft²·°F). If the room air is also 72°F, the heat transport to the cool wall from room air is 0.6 × (16 × 8) × (72 − 60) = 922 Btu/h.

2.10.6 Combined Convection and Radiation

Convection and radiation heat transfer coexist in building heat flows. A common example is the heat transfer from the interior and exterior surfaces of a wall. In such instances, it is more convenient to use tables of combined or effective unit conductances h_{c+r} or unit resistances as shown in Table 2.9 (ASHRAE Fundamentals, 2013). Such tables have been developed based on calculations and experimental tests and include the effect of wall position, direction of heat flow, surface emittance, and whether the air is still (for interior surfaces) or moving (for external surfaces). This table applies for a surface air temperature difference of 10°F and for a surface temperature of 70°F, with heat being transferred from the surface to a virtual blackbody at the same temperature as the ambient air. From Table 2.8, the emittance of a brick wall is 0.93. From Table 2.9, for a vertical interior wall in summer with horizontal flow direction, the average film coefficient is found to be 8.29 W/(m²·°C). For a highly reflective inner surface (emissivity of 0.05), the film coefficient is only 3.4 W/(m²·°C) illustrating the large contribution due to radiation under natural convection conditions. Thus, the surface emissivity can greatly affect the combined radiation and convection surface conductance. The effect of air velocity can also be noted; the value of 8.29 W/(m²·°C) for interior still air increases to 34.0 under a wind velocity of 6.7 m/s for an external surface.

Example 2.17: Radiant Barrier in Residential Wall

A building designer wishes to evaluate the *R* value of a 3.5 in. wide air gap in a wall for its insulating effect. The resistance to heat flow offered by convection is small, so she proposes lining

TABLE 2.9

Surface Unit Conductances[a] and Unit Resistances of Air

		Surface Emittances											
		$\varepsilon = 0.9$				$\varepsilon = 0.2$				$\varepsilon = 0.05$			
		h_{c+r}		R		h_{c+r}		R		h_{c+r}		R	
Position of Surface	**Direction of Heat Flow**	$\frac{Btu}{h \cdot ft^2 °F}$	$\frac{W}{m^2 °C}$	$\frac{h \cdot ft^2 °F}{Btu}$	$\frac{m^2 °C}{W}$	$\frac{Btu}{h \cdot ft^2 °F}$	$\frac{W}{m^2 °C}$	$\frac{h \cdot ft^2 °F}{Btu}$	$\frac{m^2 °C}{W}$	$\frac{Btu}{h \cdot ft^2 °F}$	$\frac{W}{m^2 °C}$	$\frac{h \cdot ft^2 °F}{Btu}$	$\frac{m^2 °C}{W}$
Still air													
Horizontal	Upward	1.63	9.26	0.61	0.11	0.91	5.2	1.10	0.194	0.76	4.3	1.32	0.232
Sloping—45°	Upward	1.60	9.09	0.62	0.11	0.88	5.0	1.14	0.200	0.73	4.1	1.37	0.241
Vertical	Horizontal	1.46	8.29	0.68	0.12	0.74	4.2	1.35	0.238	0.59	3.4	1.70	0.298
Sloping—45°	Downward	1.32	7.50	0.76	0.13	0.60	3.4	1.67	0.294	0.45	2.6	2.22	0.391
Horizontal	Downward	1.08	6.13	0.92	0.16	0.37	2.1	2.70	0.476	0.22	1.3	4.55	0.800
Moving air													
(Any position) Wind velocity of 15 mph or 6.7 m/s (for winter)	Any	6.0	34.0	0.17	0.029								
Wind velocity of 7.5 mph or 3.4 m/s (for summer)	Any	4.0	22.7	0.25	0.044								

Source: Adapted from *ASHRAE Handbook of Fundamentals*, American Society of Heating, Refrigerating and Air-Conditioning Engineers, Atlanta, GA, 2005. Copyright ASHRAE, www.ashrae.org.

[a] Conductances are for surfaces of the stated emittance facing virtual blackbody surroundings at the same temperature as the ambient air. Values are based on a surface air temperature difference of 10°F and for a surface temperature of 70°F.

the cavity's inner and outer surfaces with a highly reflecting aluminum foil film whose emissivity is 5%. Find the R value of this wall cavity, including both radiation and convection effects, if the temperatures of the walls facing the gap are 45°F and 55°F.

Given: $\varepsilon_i = \varepsilon_o = 0.05$

Figure: See Figure 2.22.

Assumptions: The convection coefficient of the air gap is 0.30 Btu/(h · ft² · °F).

The emittance of the cavity surfaces prior to installing the radiant barrier is 1.0. Consider a unit area.

Find: R_{gap}

Solution

As shown in Figure 2.22, this problem involves both series and parallel resistances. The solution to this problem has two parts. We will find the thermal resistance of the wall cavity first without the barrier, and then with the barrier.

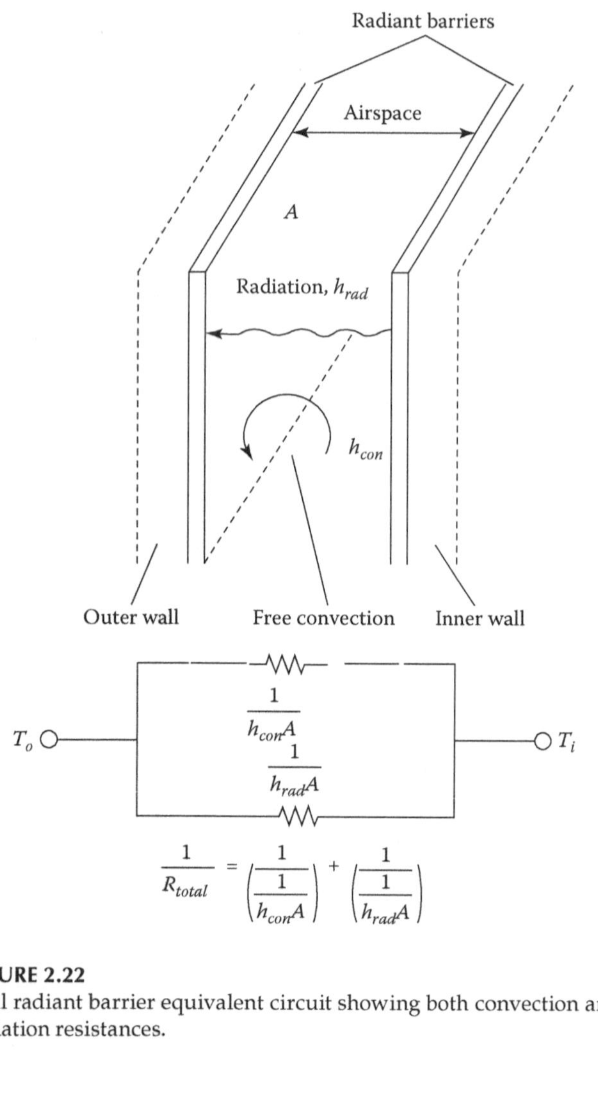

FIGURE 2.22
Wall radiant barrier equivalent circuit showing both convection and radiation resistances.

The radiation coefficient from Equation 2.72 for the standard cavity is

$$h_{rad} = 4 \times (0.1714 \times 10^{-8}) \times \frac{(460+50)^3}{(1/1.0)+(1/1.0)-1}$$

$$= 0.91 \text{ Btu/(h} \cdot \text{ft}^2 \cdot {}^\circ\text{F)}$$

The radiation resistance without radiation shield is

$$R_{rad} = \frac{1}{h_{rad}} = \frac{1}{0.91} = 1.10 \text{ (h} \cdot \text{ft}^2 \cdot {}^\circ\text{F)/Btu}$$

The total convection resistance is

$$R_{con} = \frac{1}{h_{con}} = \frac{1}{0.30} = 3.33 \text{ (h} \cdot \text{ft}^2 \cdot {}^\circ\text{F)/Btu}$$

Finally, these two parallel resistances are added reciprocally to find the total gap resistance

$$\frac{1}{R_{gap}} = \frac{1}{R_{rad}} + \frac{1}{R_{con}} = \frac{1}{1.10} + \frac{1}{3.33} = 1.21$$

from which

$$R_{gap} = 0.83 \text{ (h} \cdot \text{ft}^2 \cdot {}^\circ\text{F)/Btu [0.15 (m}^2 \cdot {}^\circ\text{C)/W]}$$

The calculation is now repeated with the low-emittance radiant shield. The convection coefficient remains the same, but the radiation coefficient is different.

$$h_{rad} = 4 \times (0.1714 \times 10^{-8}) \times \frac{(460+50)^3}{(1/0.05)+(1/0.05)-1}$$

$$= 0.023 \text{ Btu/(h} \cdot \text{ft}^2 \cdot {}^\circ\text{F)}$$

Hence, the resistance $R_{rad} = (1/h_{rad}) = 43.5$ (h · ft² · °F)/Btu

The overall wall cavity resistance with radiant shield is

$$\frac{1}{R_{gap}} = \frac{1}{R_{rad}} + \frac{1}{R_{con}} = \frac{1}{43.5} + \frac{1}{3.33} = 0.323$$

$$R_{gap} = 3.09 \text{ (h} \cdot \text{ft}^2 \cdot {}^\circ\text{F)/Btu [0.54 (m}^2 \cdot \text{K)/W]}$$

Comments

The thermal resistance of this gap is almost 3.7 times that of a standard wall gap. Further, we note that the resistance offered by the radiant barrier is about 13 times that of the convective heat transfer. However, there are practical issues. The low radiant barrier emissivity is difficult to maintain.

Dust accumulation on the surface can reduce emissivity markedly.

Note that the convective coefficient of the air gap was assumed in this example. In reality, one needs to use dimensionless equations to deduce this value more accurately. For heat gains and heat losses, a more convenient approach is followed by practitioners. Extensive tables have been compiled (ASHRAE Fundamentals, 2013), which allow easy determination of the combined convective–radiative thermal resistances for different air space orientations, air space temperatures, and air space gaps. Table 2.10 shows such tabulated values for two air gaps only. Recall that the effective emittance is defined by Equation 2.70, which in this case of both sides lined by radiation shields turns out to be ε_{eff} = $((1/0.05) + (1/0.05)-1)^{-1} = 0.02526$. Also, the mean temperature difference is 50°F and the temperature difference between both surfaces is 10°F. From Table 2.10, for a 3.5 in. air gap with vertical orientation of air space and horizontal heat flow and $\varepsilon_{eff} = 0.03$, the tabular value of the effective combined thermal resistance is found to be slightly greater than 3.63 (as against a value of 3.09 determined in the aforementioned illustrative example). Notice the threefold decrease in the resistance as the emissivity increases from 0.03 to 0.82.

Another common example of a combined convection–radiation situation is the radiant barriers in attics. For example, reflective plastic or aluminum sheets called radiant barriers are often used in residential walls or above residential building ceilings to reduce heat loss by controlling radiation. The radiant barrier can be placed in different locations as shown in Figure 2.23). Placing it on the attic floor insulation has been found to have the most benefit. The roof decking blocks direct solar radiation from entering the house through the ceiling but in the process it heats up and transfers heat by radiation to the top of the ceiling insulation. Because most attics are ventilated, the convective heat gain is relatively small provided the ventilation is adequate. It is estimated that typically radiant barriers can reduce summer ceiling heat gains by 16%–42% that translates to 2%–10% of the air-conditioning costs (Cengal, 2002).

2.10.7 Thermal Bridges

In the context of this book, a thermal bridge is a local area of a building's envelope with relatively lower thermal resistance than those of other portions of the envelope. The wood stud considered in Example 2.5 is an example of a thermal bridge. The presence of the stud caused the wall heat flow to increase by 18%. Similar or greater increases are caused by structural members that penetrate walls (e.g., balcony supports in a high-rise building) or that support walls (e.g., the steel structure in a high rise).

Thermal bridges are unavoidable in conventional building practice, and cause at least two significant difficulties:

1. Heat losses and gains are increased.
2. Lower temperature can cause condensation, leading to moisture problems in winter (material degradation, paint peeling, mold).

Figure 2.24 shows a more complex thermal bridge and its approximation by a thermal network. The construction shown represents a three-layer vertical wall supported by a concrete floor. The concrete floor is the thermal bridge. The thermal network is a combined series–parallel arrangement that can be analyzed by using the rules for series and parallel circuits shown in Figure 2.10. The resistances that are associated with the vertical heat flows include two series heat transfers—through the wall and through a portion of the floor (the subscript B in the figure denotes the thermal bridge area). According to Staelens (1988), this method is accurate to a few percent. Once the total heat flow has been found, the temperatures at each node can also be determined, and one can ascertain whether moisture condensation will occur under winter conditions. In the bridge shown, the critical temperature point where condensation may occur is at the inner corner of the wall at the floor.

2.11 Evaporation and Moisture Transfer

The evaporation of moisture is a primary mechanism for cooling of various surfaces. For example, the evaporation of water from surfaces in evaporative coolers is an important method for cooling buildings in arid parts of the world. In addition, the control of migration of moisture through building materials must be considered in building design to avoid structural damage. Although it is outside the scope of this text to deal with evaporation and moisture transfer in detail, we present the governing equations in the following.

There are two mechanisms of interest. First, the transfer of moisture within building materials is essentially a diffusion process governed by Fick's law (Bird et al., 1960):

$$\dot{m}_w = -DA\frac{\partial C}{\partial x} \tag{2.74}$$

TABLE 2.10
Thermal Resistance of Plane Air Spaces

IP Units													
				Units of Thermal Resistance; °F·ft²·h/Btu									
		Air Space		0.5 in. Air Space					3.5 in. Air Space				
				Effective Emittance ε_{eff}					Effective Emittance ε_{eff}				
Position of Air Space	Direction of Heat Flow	Mean Temp., °F	Temp. Diff., °F	0.03	0.05	0.2	0.5	0.82	0.03	0.05	0.2	0.5	0.82
Horizontal	Up ↑	90	10	2.13	2.03	1.51	0.99	0.73	2.84	2.66	1.83	1.13	0.80
		50	30	1.62	1.57	1.29	0.96	0.75	2.09	2.01	1.58	1.10	0.84
		50	10	2.13	2.05	1.60	1.11	0.84	2.80	2.66	1.95	1.23	0.93
		0	20	1.73	1.70	1.45	1.12	0.91	2.25	2.18	1.79	1.32	1.03
		0	10	2.10	2.04	1.70	1.27	1.00	2.71	2.62	2.07	1.47	1.12
45° slope	Up ↗	90	10	2.44	2.31	1.65	1.06	0.76	3.18	2.96	1.97	1.13	0.82
		50	30	2.06	1.98	1.56	1.10	0.83	2.26	2.17	1.67	1.15	0.86
		50	10	2.55	2.44	1.83	1.22	0.90	3.12	2.95	2.10	1.34	0.96
		0	20	2.20	2.14	1.76	1.30	1.02	2.42	2.35	1.90	1.33	1.06
		0	10	2.63	2.54	2.03	1.44	1.10	2.98	2.87	2.23	1.54	1.16
Vertical	Horizontal →	90	10	2.47	2.34	1.67	1.06	0.77	3.69	3.40	2.15	1.24	0.85
		50	30	2.57	2.46	1.84	1.23	0.90	2.67	2.55	1.89	1.25	0.91
		50	10	2.66	2.54	1.88	1.24	0.91	3.63	3.40	2.32	1.42	1.01
		0	20	2.82	2.72	2.14	1.50	1.13	2.88	2.78	2.17	1.51	1.14
		0	10	2.93	2.82	2.20	1.53	1.15	3.49	3.33	2.50	1.67	1.23
45° slope	Down ↘	90	10	2.48	2.34	1.67	1.06	0.77	4.81	4.33	2.49	1.34	0.90
		50	30	2.64	2.52	1.87	1.24	0.91	3.51	3.30	2.28	1.40	1.00
		50	10	2.67	2.55	1.89	1.25	0.92	4.74	4.36	2.73	1.57	1.08
		0	20	2.91	2.80	2.19	1.52	1.15	3.81	3.63	2.66	1.74	1.27
		0	10	2.94	2.83	2.21	1.53	1.15	4.59	4.32	3.02	1.58	1.34
Horizontal	Down ↓	90	10	2.48	2.34	1.67	1.06	0.77	10.07	8.19	3.41	1.57	1.00
		50	30	2.66	2.54	1.88	1.24	0.91	9.60	8.17	3.86	1.38	1.22
		50	10	2.67	2.55	1.89	1.25	0.92	11.15	9.27	4.09	1.93	1.24
		0	20	2.94	2.83	2.20	1.53	1.15	10.90	9.52	4.87	2.17	1.62
		0	10	2.96	2.85	2.22	1.53	1.16	11.97	10.32	5.08	2.52	1.64

(Continued)

TABLE 2.10 (*Continued*)

Thermal Resistance of Plane Air Spaces

SI Units														
				Units of Thermal Resistance; K·m²/W										
		Air Space		13 mm Air Space					90 mm Air Space					
				Effective Emittance ε_{eff}					Effective Emittance ε_{eff}					
Position of Air Space	Direction of Heat Flow	Mean Temp., °C	Temp. Diff., °C	0.03	0.05	0.2	0.5	0.82	0.03	0.05	0.2	0.5	0.82
Horizontal	Up ↑	32.2	5.6	0.37	0.36	0.27	0.17	0.13	0.50	0.47	0.32	0.20	0.14
		10.0	16.7	0.29	0.28	0.23	0.17	0.13	0.27	0.35	0.28	0.19	0.15
		10.0	5.6	0.37	0.36	0.28	0.20	0.15	0.49	0.47	0.34	0.23	0.16
		−17.8	11.1	0.30	0.30	0.26	0.20	0.16	0.40	0.38	0.32	0.23	0.18
		−17.8	5.6	0.37	0.36	0.30	0.22	0.18	0.48	0.46	0.36	0.26	0.20
45° slope	Up ↗	32.2	5.6	0.43	0.41	0.29	0.19	0.13	0.56	0.52	0.35	0.21	0.14
		10.0	16.7	0.36	0.35	0.27	0.19	0.15	0.40	0.38	0.29	0.20	0.15
		10.0	5.6	0.45	0.43	0.32	0.21	0.16	0.55	0.52	0.37	0.24	0.17
		−17.8	11.1	0.39	0.38	0.31	0.23	0.18	0.43	0.41	0.33	0.24	0.19
		−17.8	5.6	0.46	0.45	0.36	0.25	0.19	0.52	0.51	0.39	0.27	0.20
Vertical	Horizontal →	32.2	5.6	0.43	0.41	0.29	0.19	0.14	0.65	0.60	0.38	0.22	0.15
		10.0	16.7	0.45	0.43	0.32	0.22	0.16	0.47	0.45	0.33	0.22	0.16
		10.0	5.6	0.47	0.45	0.33	0.22	0.16	0.64	0.60	0.41	0.25	0.18
		−17.8	11.1	0.50	0.48	0.38	0.26	0.20	0.51	0.49	0.38	0.27	0.20
		−17.8	5.6	0.52	0.50	0.39	0.27	0.20	0.61	0.59	0.44	0.29	0.22
45° slope	Down ↘	32.2	5.6	0.44	0.41	0.29	0.19	0.14	0.85	0.76	0.44	0.24	0.16
		10.0	16.7	0.46	0.44	0.33	0.22	0.16	0.62	0.58	0.40	0.25	0.18
		10.0	5.6	0.47	0.45	0.33	0.22	0.16	0.83	0.77	0.48	0.28	0.19
		−17.8	11.1	0.51	0.49	0.39	0.27	0.20	0.67	0.64	0.47	0.31	0.22
		−17.8	5.6	0.52	0.50	0.39	0.27	0.20	0.81	0.76	0.53	0.33	0.24
Horizontal	Down ↓	32.2	5.6	0.44	0.41	0.29	0.19	0.14	1.77	1.44	0.60	0.28	0.18
		10.0	16.7	0.47	0.45	0.33	0.22	0.16	1.69	1.44	0.68	0.33	0.21
		10.0	5.6	0.47	0.45	0.33	0.22	0.16	1.96	1.63	0.72	0.34	0.22
		−17.8	11.1	0.52	0.50	0.39	0.27	0.20	1.92	1.68	0.86	0.43	0.29
		−17.8	5.6	0.52	0.50	0.39	0.27	0.20	2.11	1.82	0.89	0.44	0.29

Source: Adapted from ASHRAE, *Handbook of Fundamentals*, American Society of Heating, Refrigerating and Air-Conditioning Engineers, Atlanta, GA, 2013. Copyright ASHRAE, www.ashrae.org.

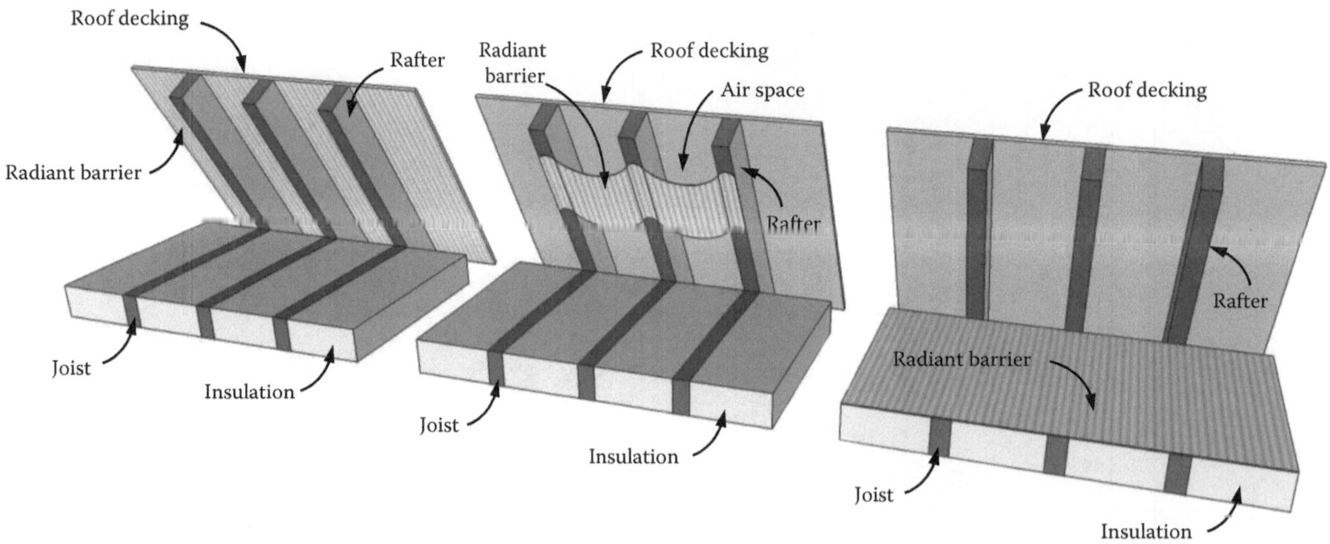

FIGURE 2.23
Radiant barriers in attics can reduce radiation heat gain from roof decking. Three possible locations of placing the radiant film are shown.

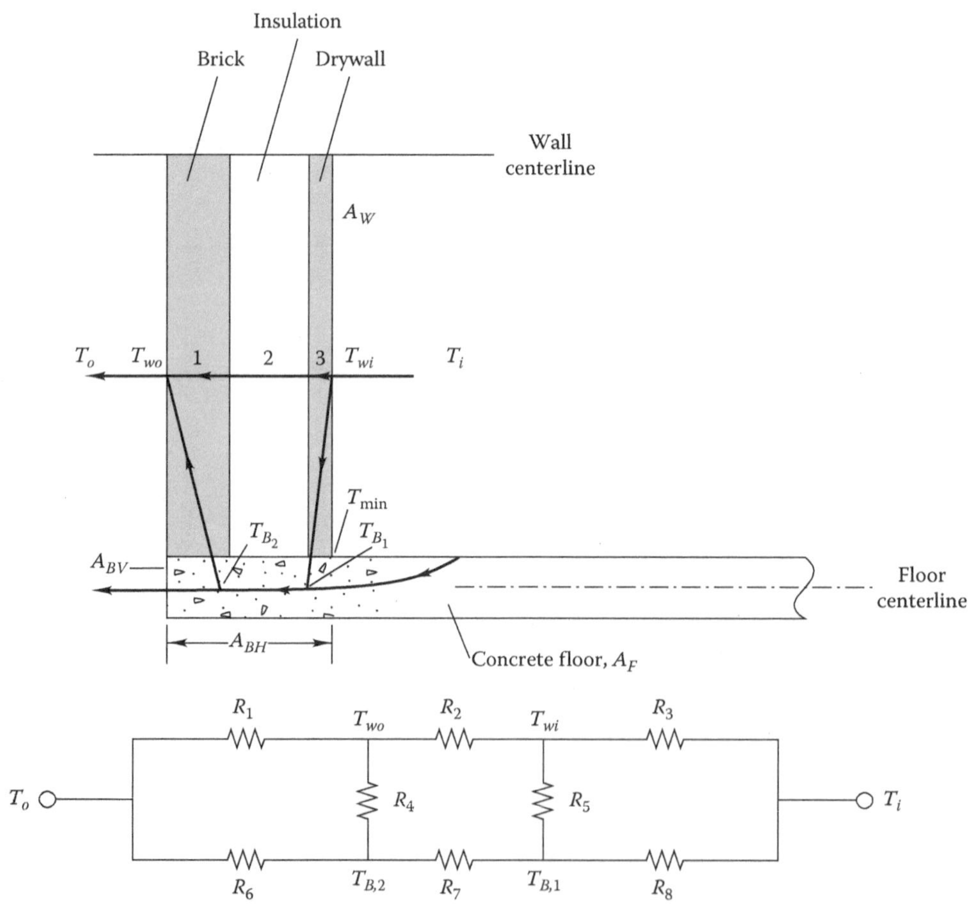

FIGURE 2.24
Example of a thermal bridge (concrete floor) with insulated wall above. The thermal network allows for 2-D calculations of the thermal bridge heat loss. Heavy lines with arrows show various heat flow paths. Each resistance can be found by using basic equations developed earlier for conduction and convection. Resistances R_1, R_6, R_3, and R_8 include surface film effects.

in which D is the diffusivity, A is the area through which mass transfer takes place, and the derivative is the concentration gradient of water vapor in the material in which diffusion is taking place. This equation is analogous to Fourier's law of heat conduction, Equation 2.37; and in cases for which D is constant, the solutions to the conduction equation are solutions to the diffusion equation if the same boundary conditions exist. This expression is used to determine the migration of moisture into and from building contents and structural members.

In winter, the interior of a building is more humid than the exterior. Hence, water vapor will tend to diffuse from the interior to the exterior. If the winter is cold, the migrating moisture may reach a location in the wall where the insulation (or other component of the wall such as a metal stud) is at the dew point (see Chapter 13) and condensation occurs. The condensed water can saturate insulation, rendering it useless as a heat loss barrier. If condensation levels are extreme, wooden structural members can deteriorate or steel ones can rust, compromising the structural integrity of a building.

Moisture migration is often controlled in cold climates by use of a vapor barrier—a material (most often plastic) essentially impervious to moisture. If the vapor barrier is placed on the heated side of building insulation, condensation will be avoided in insulation in winter. The situation in summer is not as clear-cut, but a higher vapor resistance at the building exterior, relative to the interior, is recommended. To be effective, the vapor barrier must be continuous without punctures or tears.

Following the heat–mass transfer analogy for another step, we see that the convective transport of mass is given by

$$\dot{m}_w = h_m A(C_s - C_f) \qquad (2.75)$$

in which h_m is the convective mass transfer coefficient. Its value can be found from convective heat transfer equations by using the proper transformations as described in Bird et al. (1960). C_s and C_f are the surface and fluid concentrations of the species (water vapor in buildings) undergoing the mass transfer process. This equation governs the evaporation of moisture from surfaces. In later chapters, we will use results from this equation to analyze the performance of evaporative coolers and indirectly to find the relation of human comfort to the velocity of air moving over the skin.

2.12 Closure

In this chapter, we have reviewed the key features of heat transfer that apply to buildings. Conduction and convection expressions will be used to find the heating and cooling requirements of buildings in Chapters 7 through 9. A great many equations have been presented in this chapter. Table 2.11 is a summary of the key ones.

Problems

Numbers 1–4 given in parenthesis denote the degree of difficulty.

2.1 Calculate the viscosity of pure water at 80°F and 110°F (27°C and 43°C). *(1)*

2.2 Convert Equation 2.6 to an equality $\mu = kT^{0.67}$ by using the viscosity of air at 50°F, 100°F, and 150°F (10°C, 38°C, and 66°C). How sensitive is the proportionality constant k to the temperature? Does the use of absolute temperature improve the accuracy? *(3)*

2.3 How much heat is added to a fixed volume of 2000 ft^3 (60 m^3) of air to raise it from 60°F (15°C) to 120°F (49°C)? Assume sea level condition. *(1)*

2.4 The 20 m^3 compressed-air storage tank for a building pneumatic control system is at 25°C and 800 kPa. Enough air is allowed to escape to lower the pressure to 600 kPa; the temperature is 20°C. How much air was vented? *(2)*

2.5 If the pressure gauge on a 30 ft^3 (0.85 m^3) oxygen tank reads 30 psig (207 kPa), how much oxygen is in the tank when the temperature is 65°F (18°C) at sea level? *(2)*

2.6 How much air must be added to the tank in Problem 2.5 to raise the pressure at the given temperature to 75 psig (500 kPa)? *(2)*

2.7 How much electric power is needed to pump water from a 100 ft deep (60 m deep) well by using a 69% efficient pump powered by an 88% efficient motor? The friction pressure drop in the 100 ft pipe is 5 ft of water (30 kPa). *(2)*

2.8 Water is pumped through the cooling system of a building at 300 gal/min (19 L/s). If the pressure drop is 18 psia (126 kPa), how much power does the pump exert on the flowing fluid? If the pump has an efficiency rate of 72%, what is the shaft power input to the pump? (*Hint:* When you are using the appropriate form of the first law, assume that water is incompressible.) *(3)*

2.9 Water is pumped through the cooling tower of a cooling plant with a 2 hp (1.5 kW) pump that is 72% efficient. The pressure drop in the piping is 15 psi (105 kPa). What is the water flow rate? Note that the motor imparts 2 hp (1.5 kW) of power to the pump input shaft. *(3)*

TABLE 2.11

Summary of Heat Transfer Equations

Heat Transfer Mode	Equations	Equation Number(s)	Comments
Conduction	$\dot{Q} = \frac{\Delta T}{R}$	(2.39)	Fourier's law
	$R = \frac{\Delta x}{kA}$	(2.40)	Thermal resistance
	$R_{cd} = \frac{\Delta x}{k} = AR$	(2.41)	*R* value
	$U_{cd} \equiv \frac{1}{R_{cd}}$	(2.42)	*U* value
Parallel flow	$\frac{1}{R_{tot}} = \frac{1}{R_1} + \frac{1}{R_2} + \cdots + \frac{1}{R_n}$	Figure 2.10	
Series flow	$R_{tot} = R_1 + R_2 + \cdots + R_n$	Figure 2.10	
Convection	$\dot{Q} = h_{con} A \Delta T$	(2.45)	Definition of h_{con}
	$R = \frac{1}{h_{con} A}$	(2.47)	Thermal resistance
External free convection, horizontal pipes and surfaces, laminar	$h_{con} = \text{const}\left(\frac{\Delta T}{L}\right)^{1/4}$	(2.51) and (2.53)	Dimensional equations
External free convection, horizontal pipes and surfaces, turbulent	$h_{con} = \text{const}(\Delta T)^{1/3}$	(2.52) and (2.54)	Dimensional equations
External forced convection, horizontal pipes and surfaces, laminar	$h_{con} = \text{const}\left(\frac{\text{V}}{L}\right)^{1/2}$	(2.55)	Dimensional equation
External and internal forced convection, horizontal pipes and surfaces, turbulent	$h_{con} = \text{const}\left(\frac{\text{V}^4}{L}\right)^{1/5}$	(2.56), (2.57) and (2.59)	Dimensional equations
Radiation	$\dot{q} = \sigma T^4$	(2.61)	Stefan–Boltzmann equation
	$\dot{Q}_{12} = \frac{A(\sigma T_1^4 - \sigma T_2^4)}{(1/\varepsilon_1) + (1/\varepsilon_2) - 1}$	(2.69)	Parallel surfaces, closely spaced
	$h_{rad} = 4\sigma\varepsilon_{eff} \cdot \bar{T}^3$	(2.72)	Parallel surfaces, closely spaced
	$\dot{Q}_{12} = \frac{A_1(\sigma T_1^4 - \sigma T_2^4)}{(\rho_1/\varepsilon_1) + (1/F_{12}) + \rho_2 A_1/(\varepsilon_2 A_2)}$	(2.68)	Two-surface enclosure

2.10 A building with a floor area of 100,000 ft^2 (10,000 m^2) and inside wall height of 9 ft (3 m) has an air infiltration (i.e., leakage) rate of 0.4 air change per hour. If the outdoor temperature is 0°F (–18°C) and the indoor temperature is 68°F (20°C), how much heat must be provided by the building heating system to warm the cold outside air for two possible building locations—the first at sea level and the second at Leadville, CO [10,000 ft (3,000 m)]? *(3)*

2.11 The ventilation air requirement of a building is 10,000 ft^3/min (4,700 L/s), representing 18% of the total supply airflow to the HVAC system. What is the minimum outdoor temperature at which an economizer can be used without preheating the outdoor air if the building supply air temperature is 55°F (13°C) and the interior temperature of the building is 74°F (23°C)? *(2)*

2.12 Oil is heated to 150°F (65°C) by a steam coil located in the bottom of a horizontal cylindrical tank. If the diameter is 5 ft (1.5 m) and the length 20 ft (6 m), find the heat loss if the tank is insulated with 2 in. (5 cm) of polyisocyanurate and located in a room at 65°F (18°C). Compare the heat loss with that if the tank were insulated with fiberglass of the same thickness. Ignore the resistance to heat transfer offered by the metal tank and by convection at the outer surface of the insulation. *(2)*

2.13 Calculate the thermal resistance of the typical wall shown schematically in Figure P2.13 following the two networks shown in Figure 2.16(b). The inside surface is 0.5 in. (1.2 cm) drywall, and the outside wall surface is 0.44 in. thick hardboard siding (see online HCB software for thermal properties of these materials). The studs, headers, and floor plates in this wall are nominal 2 in. × 6 in. wood studs with an R value of 6.88 (°F·h·ft²)/Btu [1.21 (K·m²)/W] with true finished-wood dimensions of 1.5 × 5.5 in. (3.8 × 13.5 cm). The wall cavities are insulated with fiberglass at $R = 19$ [3.35 (K·m²)/W]. Ignore convection film resistances for now at the inner and outer wall surfaces (together they add less than one unit of R value in IP units). *(3)*

2.14 A roof is constructed as shown in Figure 2.10b. The outer layer is plywood and shingles (total of R = 3 in IP units), and the center layer is 6 in. (15 cm) of fiberglass. The innermost layer is 0.5 in. (1.2 cm) drywall. On a day when the outer surface is 0°F (–18°C) and the indoor surface is 68°F (20°C), where in the wall is the temperature exactly equal to the freezing point of water? Based on this calculation, where should one place a vapor barrier designed to keep water vapor from diffusing into and freezing the insulation? (Since the R value of the wall assembly itself is so much larger than the thermal resistance offered by inner and outer surface air films, you may ignore them for this problem.) *(3)*

2.15 An insulated 24 × 60 in. (60 × 150 cm) rectangular duct carries 12,000 ft³/min (5,700 L) of air at 45°F (7°C) in a low-temperature air-conditioning system at sea level. If the temperature rise in the cold air is not to exceed 1°F in every 100 ft of duct length (2°C in 100 m), what is the R value of insulation needed on the duct? Assume that the combined effect of the convection resistances on the inside of the duct and the outside of the insulation is $R = 1$ in IP units (0.18 SI). This duct is located in an unconditioned roof plenum that reaches 110°F (43°C) on a summer day. *(3)*

2.16 In an energy audit of a building, you find that a 4 in. diameter (10 cm diameter) pipe located in the 80°F (27°C) mechanical room is carrying superheated steam at 400°F (205°C). The pipe is insulated with 2 in. (5 cm) of aluminum-jacketed fiberglass. Find the outer surface temperature of the insulation jacket. Ignore the heat transfer resistance offered by the insulation jacket. Comment on the safety of this installation. *(2)*

2.17 The exterior wall of an old building constructed of face brick and common brick is shown in Figure P2.17. If the outer surface is 10°F (–12°C) and the inner surface is 65°F (18°C), what is the heat flux through the wall? If the air gap is filled with 2.0 in. (3.8 cm) polyisocyanurate board, by what percentage is the heat loss reduced? *(3)*

2.18 The inner and outer walls of a school building are constructed of 4 in. (10 cm) concrete and brick, respectively, connected by 6 in. wide (15 cm wide) grout stiffeners spaced every 4 ft (1.2 m). If the 4 in. deep (10 cm deep) wall cavity is filled

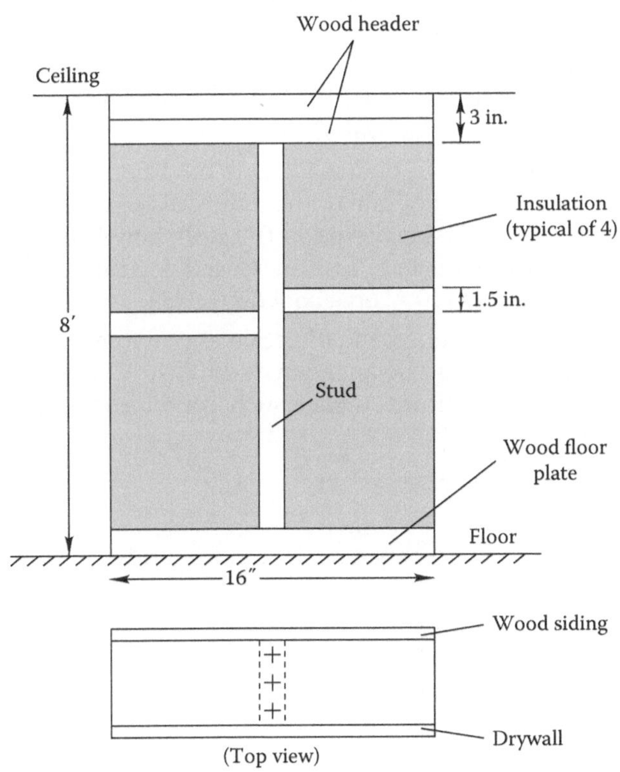

FIGURE P2.13

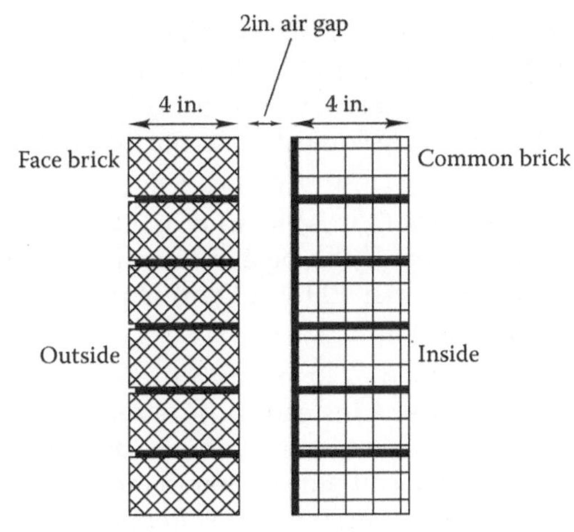

FIGURE P2.17

with extruded polystyrene except for the stiffeners, what is the wall thermal resistance? For convenience, just treat a unit height of wall. Grout is similar to concrete from a thermal viewpoint. *(3)*

2.19 A buried pipe carries 45°F (7°C) chilled water in a 1 miles long (1.6 km long) loop serving a campus cooling system. To save initial cost, a contractor proposes to just bury the 20 in. (50 cm) pipe 2 ft (0.6 m) below the surface, claiming that the ground itself serves as adequate insulation. Find the heat gained by the pipe on a summer day if the ground temperature is 100°F (38°C). Do you recommend insulation? Assume soil conductivity $k = 1.73$ W/(m·K). *(3)*

2.20 If a wall consists of 1 in. (2.5 cm) of plywood exterior siding, 1 in. (2.5 cm) of polyisocyanurate, 3 in. (7.5 cm) of fiberglass, and 0.5 in. (1.2 cm) of gypsum board, find the temperature at the interface between each material, and plot the temperature gradients in each material. In which material is the temperature gradient the steepest? Make your calculations for an outer wall surface temperature of 0°F (−17.3°C) and inner surface temperature of 70°F (21°C). *(3)*

2.21 The difference between the air and the inside wall temperature in a room is 10°F (5.5°C). What is the free convection coefficient if the wall is 8 ft (2.4 m) high? *(1)*

2.22 The surface temperature of an insulated horizontal 12 in. diameter (30 cm diameter) steam pipe varies between 80°F and 150°F (27°C and 66°C) depending on steam usage. What effect does this variation have on the free convection heat transfer coefficient if the pipe is located in a room whose average temperature is 65°F (18°C)? Neglect wall resistance of steam pipe. *(2)*

2.23 On a winter day, a nearly horizontal square skylight in a building is 50°F (10°C). What is the heat loss by free convection from it if its area is 32 ft² (3 m²) and if the room that it illuminates is 72°F (22°C)? *(2)*

2.24 Air blows over the flat roof of a building at 30 mph (14 m/s) and 5°F (−15°C). What is the heat loss from the roof if its area is 320 ft² (30 m²) and its temperature is 20°F (−6°C)? Assume the flat roof to be square. *(2)*

2.25 Chilled water (45°F, 7°C) used to cool a building flows at 6 ft/s (2 m/s) in an 8 in. (20 cm) outside-diameter pipe. If the pipe is uninsulated, what is the heat gain per 100 ft (100 m) of pipe if it is in a 70°F (21°C) room? *(2)*

2.26 Repeat Problem 2.25 if a 33% propylene glycol solution is used instead of pure water. (Glycol is used in some low-temperature cooling systems where chiller coil surface temperatures can drop below the freezing point of water.) (Use $\rho = 64$ lb/ft³ and $c_p = 0.95$ Btu/[lb·°F].) *(3)*

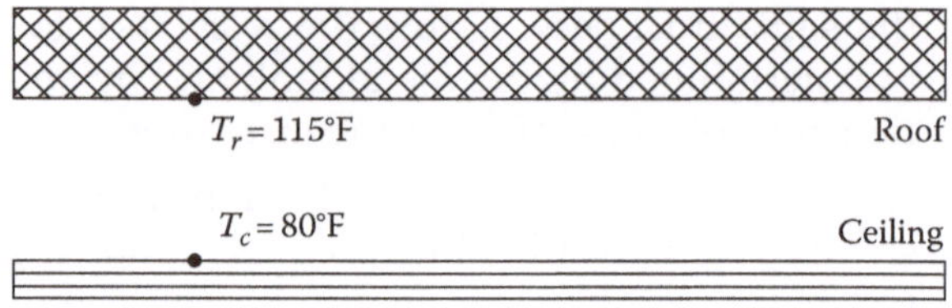

FIGURE P2.27

2.27 Figure P2.27 shows a 10,000 ft² (1,000 m²) square ceiling suspended 2 ft (60 cm) below the roof of a building. The bottom surface of the roof is 115°F (46°C), and the top of the ceiling is 80°F (27°C); both surfaces are gray ($\varepsilon = 0.8$). Find the shape factor between the two surfaces and the radiation heat flux. Comment on the relative magnitude of the convection that also occurs in this situation. *(2)*

2.28 A radiation shield is to be considered for the attic shown in Figure P2.27. It will be located 2 in. (5 cm) above the ceiling and will have an emissivity of 0.05. *Considering only radiation* in the attic, by how much will the shield reduce the heat flow compared to the situation without the shield? *(3)*

2.29 The room shown in Figure P2.29 is heated by a radiant floor maintained at 27°C. The other five surfaces of the space and the room air temperature are all at 20°C. The emissivity of all six surfaces can be taken to be 0.80. Find the total heat transfer from the floor to the other five surfaces by *radiation*. *(2)*

2.30 What is the view factor between the floor and the ceiling in Figure P2.29? Between the floor and the 7 and 5 m wide walls? *(3)*

2.31 If the room in Figure P2.29 is modified by replacing one-half of one of the 5 m walls with a square pane of glass centered in the wall, how is the total floor-to-wall heat loss increased for this room if the glass is 10°C on a cool day? *(3)*

2.32 In dry climates, the effective sky temperature for radiation can be as much as 15°F (10°C) below the air temperature. Under such conditions, find the

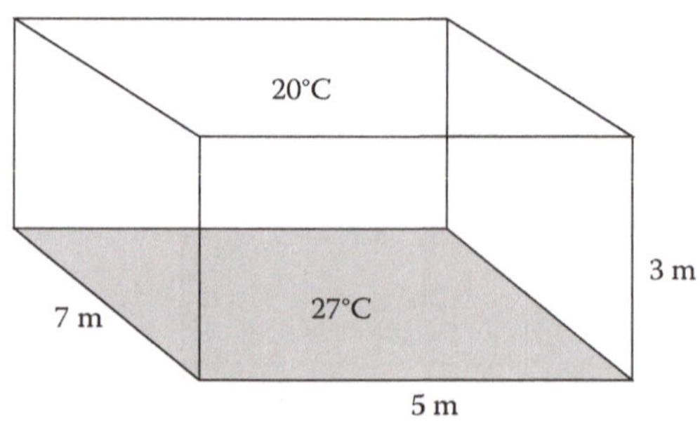

FIGURE P2.29

air temperature at which water could freeze on the top of a cold, sky-facing surface (the surface has no thermal connection to anything else) outdoors if heat is transferred from both sides of the surface by convection with h_{con} = 1.8 Btu/(h·ft²·°F) [10 W/(m²·K)]. *(3)*

2.33 A person stands in the room shown in Figure P2.33. If she can be modeled as a 20 ft² (1.9 m²) cylinder with an emissivity of 0.80, what is the *radiation* heat transfer from her to the room surfaces, all at 70°F (22°C)? Assume the body's effective surface temperature (summer clothing) to be 80°F. *(3)*

2.34 The effective surface temperature of the sun is 5760 K. At what wavelength does the sun emit the most energy? *(1)*

2.35 An incandescent lamp (assumed to be spherical) is housed in a protective plastic sphere for exterior building lighting. If the diameter of the sphere is twice the diameter of the lamp, what is the shape factor between the sphere and the lamp? What are the values of all other shape factors in this system (there are four total)? *(2)*

2.36 A sunspace on a residence is shown in Figure P2.36. Find the shape factors between (a) the floor and the wall, (b) the floor and glass, and (c) the glass and the wall. *(3)*

2.37 What is the heat flux between two gray orthogonal surfaces of equal size having a common edge if one is at 50°F and the other is at 80°F (10°C and 27°C)? The surfaces are 8 ft (2.4 m) square and have emittances of 0.5 and 0.7. *(3)*

2.38 A vertical wall between the residential space and the adjacent garage of a home consists of 2 × 4 in. (5 × 10 cm) framing on 16 in. (40 cm) centers between two 0.5 in. (12 mm) sheets of drywall (gypsum board). The stud space is uninsulated, and the still air temperatures on the two sides of the wall are 72°F and 60°F (22°C and 16°C). Find the resistance of the airspace between the drywall sheets as well as the overall U value of the wall, accounting for the stud thermal bridge. What is the wall heat flux? Solve the problem assuming the two different thermal network configurations shown in Figure 2.16. *(3)*

2.39 Find the wall U value if fiberglass insulation fills the stud space of Problem 2.38. What is the percentage of reduction in wall heat flux due to the addition of this insulation? *(2)*

2.40 Figure P2.40 is a schematic diagram of a long, unventilated attic above a residence. What is the rate of heat transfer through the ceiling for the roof and ceiling properties given in the table

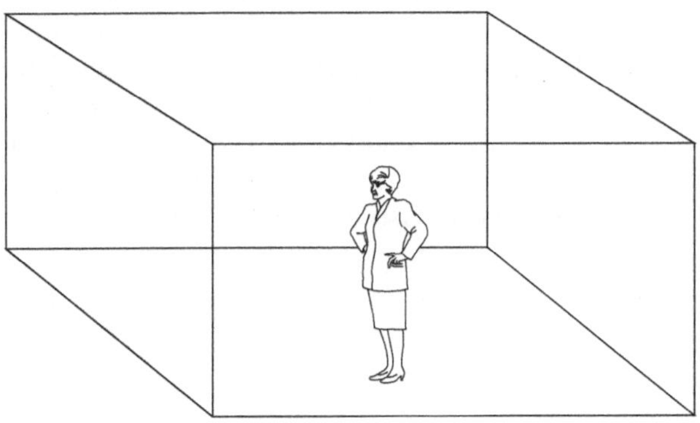

FIGURE P2.33

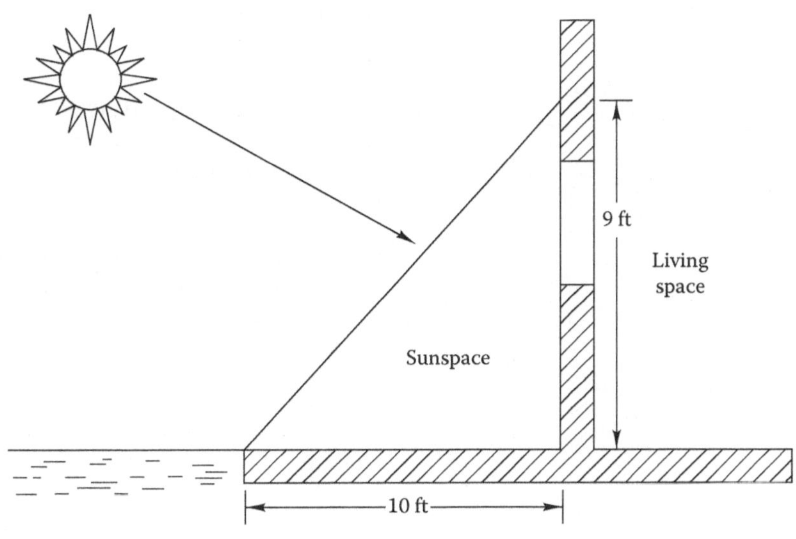

FIGURE P2.36

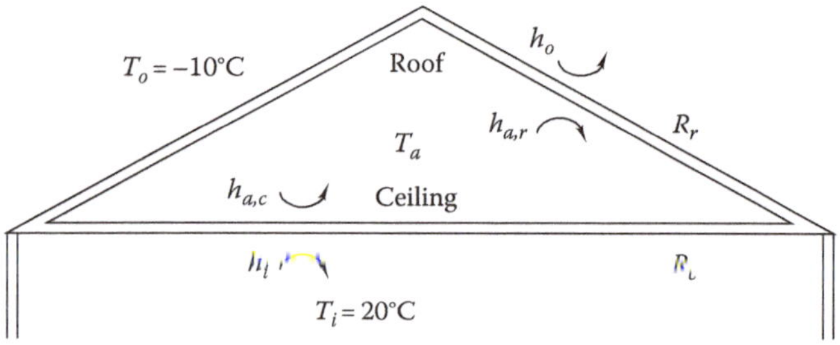

FIGURE P2.40

(the "convection" coefficients include radiation as well)? What is the attic air temperature? *(2)*

	Ceiling	Roof
Area, m²	20	23
R value, (K·m²)/W	5.0	0.5
Convection coefficient, upper, W/(m²·K)	10	25
Convection coefficient, lower, W/(m²·K)	10	8

2.41 A roof assembly is shown schematically in Figure P2.41. Draw the thermal analog circuit for the roof and its thermal bridge (the roof joist). There will be eight total resistances including the convection resistances. Describe each in words and indicate which is likely to be the smallest and which is the largest. *(1)*

2.42 An uninsulated steam condensate return tank is to be insulated with 3 in. (7.5 cm) of fiberglass with thermal conductivity of 0.25 Btu·in./(h·ft²·°F). If the tank is cylindrical with height and radius both equal to 5 ft (1.5 m) and if the tank is at 250°F (121°C) and located in an 80°F (27°C) mechanical room, what will the steam energy savings be per year if the tank is used 8000 h/year? If the steam is valued at $4.00 per million Btu ($3.80 per gigajoule), what is the cost saving due to the installation of insulation? *(3)*

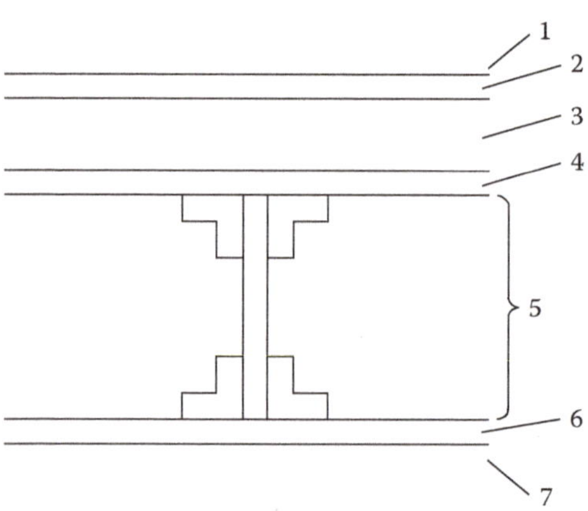

FIGURE P2.41

2.43 Figure 2.24 shows the thermal bridge produced by a concrete floor supporting a wall constructed of brick, insulation, and drywall. The minimum temperature on the building's internal surface is at the corner formed by the inside wall and the floor and is approximately equal to $T_{B,1}$. For a wall constructed of 4 in. (10 cm) brick, 4 in. (10 cm) fiberglass insulation, and 0.5 in. (1.2 cm) drywall and an 8 in. (15 cm) concrete floor, find $T_{B,1}$ for a winter day on which the outdoor temperature is –10°F (–23°C) and the interior is 70°F (21°C). State any assumptions made. *(4)*

2.44 Find the heat transfer from the floor to the *walls and room air* in Problem 2.29 by both convection and radiation. *(4)*

2.45 On a sunny day, 280 Btu/(h·ft²) (880 W/m²) strikes the dark, well-insulated horizontal roof of a building (α = 0.9). There is no wind blowing, so heat is removed only by free convection and radiation from the roof (a negligible amount passes through the roof to the interior). What is the roof temperature under these conditions if the air temperature is 90°F (32°C)? Assume that radiation takes place to the atmosphere, also at 90°F (32°C). *(4)*

2.46 The outside wall of a building absorbs 350 W/m² of solar flux on a 0°C day. If the wall surface temperature is 10°C and the convection coefficient is 35 W/(m²·K), how much heat flows through the wall? *(2)*

References

ASHRAE Fundamentals (2005, 2013). *Handbook of Fundamentals.* American Society of Heating, Refrigerating and Air-Conditioning Engineers, Atlanta, GA.

Bird, R.B., W.E. Stuart, and E.N. Lightfoot (1960). *Transport Phenomena.* Wiley, New York.

Born, D.W. (1989). *ASHRAE Trans.*, 95(pt. 2), ASHRAE.

Cengel, Y.A. (2002). *Heat Transfer,* 2nd ed. McGraw-Hill, New York.

Cengel, Y.A. and M.A. Boles (2000). *Thermodynamics,* 3rd ed. McGraw-Hill, New York.

Goswami, Y., F. Kreith, and J.F. Kreider (2000). *Principles of Solar Engineering.* Taylor & Francis Group, New York.

Holman, J.P. (1997). *Heat Transfer,* 8th ed. McGraw-Hill, New York.

Incropera, F.P., D.P. Dewitt, T.L. Bergman, and A.S. Lavine (2007). *Fundamentals of Heat and Mass Transfer*, 6th ed. Wiley, New York.

Karlekar, B. and R.M. Desmond (1982). *Engineering Heat Transfer*. West Publishing, St. Paul, MN.

Keenan, J.H. and J. Kaye (1948). *Gas Tables*. Wiley, New York.

Mathur, G.D. (1990). Calculating single phase heat transfer coefficients. *Heat. Piping Air Cond.*, 62(3), 103–107.

Nussbaum, O.J. (1990). Using glycol in a closed circuit system. *Heat. Piping Air Cond.*, 62(1), 75–85.

Perry, R.H., D.W. Green, and J.O. Malrnay, eds. (1999). *Perry's Chemical Engineers' Handbook*, 6th ed. McGraw-Hill, New York.

Siegel, R. and J.R. Howell (1981). *Thermal Radiation Heat Transfer*. McGraw-Hill, New York.

Sparrow, E.M. and R.D. Cess (1978). *Radiation Heat Transfer*, augmented ed. Hemisphere, New York.

Staelens, P.G. (1988). Thermal bridges and standardization. *ASHRAE Trans.*, 94(pt. 2), 1793–1801.

3

Human Thermal Comfort and Indoor Air Quality

ABSTRACT This chapter deals with numerous important issues pertinent to designing indoor air environments conducive to human comfort and health. These are largely based on the American Society of Heating, Refrigerating, and Air-Conditioning Engineers (ASHRAE) Comfort Standard 55-2013 and the ASHRAE Ventilation Standard 62.1-2013. Environmental parameters such as temperature, humidity, and wind speed that are critical in maintaining thermal comfort are presented. We introduce the concepts of clothing level insulation (clo) and metabolic rate (met) and discuss their effect on indoor human comfort characterized by the predicted mean vote (PMV) and percentage of people dissatisfied (PPD). The ASHRAE comfort chart is then introduced, and pertinent correlations that allow extrapolations to other environmental conditions are presented and illustrated by way of solved examples. We then describe basic concepts of air quality and commonly found indoor pollutants and their threshold limits. Next, we discuss the role of outdoor ventilation air in controlling the indoor air quality (IAQ) and widely accepted design procedures to determine the necessary outdoor ventilation amounts for different building types and occupancy densities are presented. The effects of short-circuiting of the supply air within the room, the role of filters, and the impact of recycling some of the return air on IAQ are also discussed. Finally, we present a few pertinent studies that deal with the increase in unsolicited complaints (and their financial penalties on operations and maintenance), the reduction in worker productivity, and the increase in sickness and absenteeism if the right indoor environment is not maintained.

Nomenclature

A	Area, m^2 (ft^2)
a, b, c	Coefficients appearing in Equation 3.19
c_p	Specific heat, kJ/(kg·K) [Btu/(lb$_m$·°F)]
clo	Clothing insulation value
C	Concentration, ppm
E_f	Filter efficiency
E_v	Ventilation effectiveness (Equation 3.25)
ET	Effective temperature, °C (°F)
F_r	Flow reduction factor for a variable air volume system
$F_{cl\text{-}n}$	Radiation shape factor from body clothing to surface n of enclosure
$f_{w,sk}$	Fraction of the wetted to the total skin area
h	Height, m (ft)
h_{c+r}	Combined heat transfer coefficient for convection and radiation W/(m^2·K) [Btu/(h·ft^2·°F)]
h_{con}	Convection heat transfer coefficient, W/(m^2·K) [Btu/(h·ft^2·°F)]
h_E	Mass transfer coefficient for water vapor, W·kg$_a$/(m^2·kJ) [lb$_a$/(ft^2·s)]
h_{evap}	Latent heat of evaporation of water, kJ/kg (Btu/lb$_m$)
h_{rad}	Linearized radiation heat transfer coefficient, W/(m^2·K) [Btu/(h·ft^2·°F)]
I_{cl}	Thermal resistance for clothing, clo
I_T	Total thermal resistance between skin and environment air, clo
Le	Lewis number
$\dot{M}$	Metabolic rate per surface area of body, met W/m^2 (Btu/h·ft^2)
$\dot{m}$	Mass rate, kg/s (lb$_m$/h)
N_{pol}	Rate at which contaminant is generated indoor (for CO_2, units are L/s or cfm)
p	Pressure, Pa (lb$_f$/in.^{2})
p_v	Partial vapor pressure of water vapor in air, Pa (lb$_f$/in.^{2})
P_z	Number of people in the ventilation zone
PMV	Predicted mean vote
PPD	Percentage of people dissatisfied
$\dot{Q}$	Heat flow rate, W (Btu/h)
R	Recycle fraction of air
R_a	Outdoor air flow rate per unit occupied area of building, L/(s·m^2) (cfm/ft^2)
R_p	Outdoor air flow rate per person, L/(s·person) (cfm/person)
R_T	Total thermal resistance between body and environment, (m^2·K)/W [(h·ft^2·°F)/Btu]
S	Zone bypass fraction of air supply
T	Temperature, K or °C (°R or °F)
T_a	Dry-bulb air temperature of ambient air, K or °C (°R or °F)
T_{cl}	Temperature of clothing, K or °C (°R or °F)
T_{mrt}	Mean radiant temperature, K or °C (°R or °F)
T_n	Temperature of surface n, K or °C (°R or °F)

T_o Mean monthly outdoor air temperature, K or °C (°R or °F)
T_{op} Operative temperature (Equation 3.11), K or °C (°R or °F)
T_{sk} Temperature of skin, K or °C (°R or °F)
V Volume, m³ (ft³)
$\dot{V}$ Flow rate, m³/s or L/s (ft³/min)
$\dot{V}_o$ Outdoor airflow rate, L/s (ft³/min)
$\dot{V}_e'$ Flow rate at which air is exhausted from building via a central exhaust duct, L/s (ft³/min)
$\dot{V}_e''$ Flow rate at which air is lost from building by exfiltration or local exhaust fans, L/s (ft³/min)
v Velocity or speed, m/s (ft/s)
$\dot{W}$ Rate of work, kJ (Btu/h)
W Humidity ratio of moist air, kg_w/kg_a (lb_w/lb_a)
W^* Saturated humidity ratio of moist air, kg_w/kg_a (lb_w/lb_a)

Greeks

σ Stefan–Boltzmann constant, W/(m²·K⁴) [Btu/(h·ft²·°R⁴)]
Δp Pressure difference, Pa (inWG)

Subscripts

a Indoor or ambient air
bz Breathing zone
cl Clothing and skin
comf Comfort
con Convection
dp Dew point
e Exhaust air
evap Evaporation
i Indoor
lat Latent
o Outdoor
oe Unutilized outdoor air exhausted from building
os Outdoor air to conditioned space
p Person
pol Pollution
r Return
rad Radiation
res Respiration
s Supply
sen Sensible
sk Skin
w Water, wetted
z Zone

3.1 Indoor Environmental Quality

Comfort and health depend on many complex and interrelated phenomena involving objective conditions as well as subjective and psychological perception. However, the HVAC engineer, being neither psychologist nor medical doctor, needs simple objective design criteria that will ensure acceptability by the vast majority of occupants.

Indoor environmental quality (IEQ) is most simply described as characterizing the conditions inside a building. It does not refer to indoor air quality (IAQ) alone, but to the entire environmental quality of a space, which includes thermal, visual, acoustic and aural comfort, access to daylight and views, and occupant control over his/her environment. IEQ has a major impact on occupants' health and also on worker productivity (discussed in Section 3.3.4); and so maintaining proper conditions is of great importance.

A literature survey was conducted by Frontczak and Wargocki (2011) to explore how building indoor environment affects human comfort singly and jointly with other IEQ factors. The paper also examined other factors such as individual characteristics of occupants, building-related factors, occupants' work satisfaction levels, and outdoor climate and seasonal variations. It was found that occupants consistently ranked thermal comfort the highest in most of the studies. Noteworthy is the fact that thermal comfort is defined rather vaguely as "that condition of mind which expresses satisfaction with the thermal environment" (ASHRAE 55, 2013). Despite this ambiguity, much work has been done in this area, and general concepts and design criteria have been established. The scope of this chapter is limited to thermal comfort and IAQ.

3.2 Thermal Comfort

3.2.1 Thermal Balance of the Human Body

The body is a heat engine. It converts chemical energy of the food consumed into both heat to sustain metabolism and to do work. The harder the body exercises in performing work, the greater the need to reject heat in order for the body to maintain thermal balance. The physical basis of comfort lies in the thermal balance of the body. The heat produced by the body's metabolism must be dissipated to the environment; otherwise, the body would overheat. Roughly speaking, if the rate of heat transfer to the environment is higher than the rate of heat production, the body cools down and we feel cold; if the rate is lower, we feel hot. The body self-regulates to a certain extent. For example, when the body is cold, the blood vessels near the skin constrict blood flow and reduce heat loss from the body. When the body is heating up, the reverse occurs coupled with sweating. However, the range of *metabolic regulation conditions* is quite narrow. When exceeded, shivering to increase metabolic rate, or heat exhaustion or heat stroke are familiar results.

Figure 3.1 shows the important environmental factors affecting human comfort. The quantitative effect of these factors on heat losses from the body will be described in this section. Heat loss from the human body is a complex problem in transient heat transfer, involving radiation, convection, conduction and evaporation, along with many other variables, from the wetness of the skin to the composition of the clothing. In this chapter, we discuss only the simplest aspects of the problem that provide a basis for understanding the ASHRAE comfort recommendations that are widely followed worldwide.

The total energy production rate of the body is the sum of the production rates of heat $\dot{Q}$ and of work $\dot{W}$ and can be written in the form

$$\dot{Q} + \dot{W} = \dot{M} \cdot A_{sk} \tag{3.1}$$

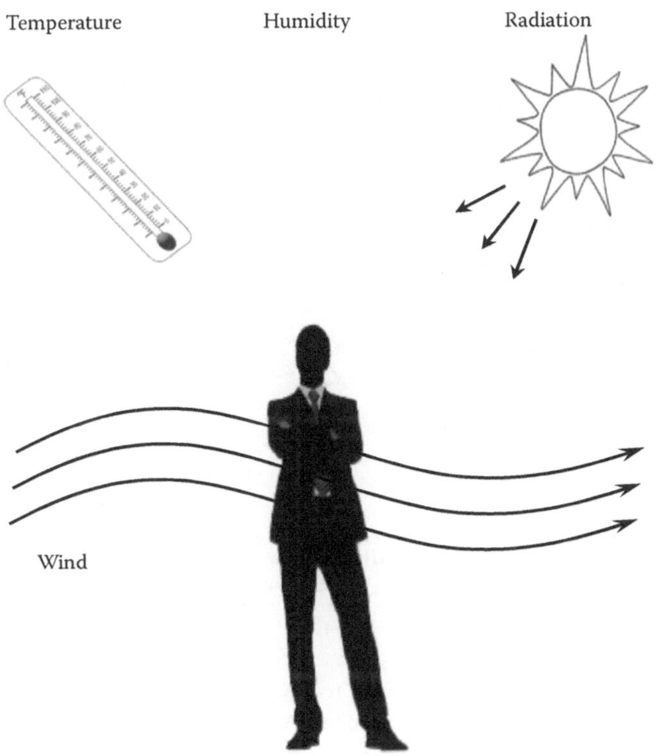

FIGURE 3.1
Factors influencing human comfort.

where
$\dot{Q}$ is the heat production rate
$\dot{W}$ is the rate of work
$\dot{M}$ is the metabolic rate
A_{sk} is the total surface area of skin

The metabolic rate is customarily expressed in units of met (or M), where

$$1\ \dot{M} = 1\ \text{met} = 58.2\ \text{W/m}^2 = 18.4\ \text{Btu/(h}\cdot\text{ft}^2)$$

Metabolic rates for various activities are shown in Table 3.1. Metabolic rate $\dot{M}$ is defined to include the production of work, whereas only the heat production rate matters for thermal comfort. In practice, the difference is entirely negligible compared to overall uncertainties, since rate of work represents at most a few percent of $\dot{M}$ for normal indoor activities. As the area A_{sk} is on the order of 1.5–2 m² (16–22 ft²) for an adult, this implies heat production rates on the order of 100 W (340 Btu/h) for typical indoor activities.

To analyze the dissipation of the heat $\dot{Q}$ to the environment, we neglect storage effects because the temperature of the interior of the body is remarkably constant, 37°C (98.6°F) under normal conditions, maintained by controlling the perspiration rate and the flow of blood to the outer regions of the body. Thus, $\dot{Q}$ can be set as equal to the instantaneous heat flow to the environment. It is convenient to distinguish several major heat transfer modes by writing

$$\dot{Q} = \dot{Q}_{con} + \dot{Q}_{rad} + \dot{Q}_{evap} + \dot{Q}_{res,sen} + \dot{Q}_{res,lat} \tag{3.2}$$

TABLE 3.1
Metabolic Rate $\dot{M}$ for Various Activities

Activity	Met	W/m²	Btu/(h·ft²)
Sleeping	0.7	40	13
Reclining	0.8	45	15
Seated, quiet	1.0	60	18
Standing, relaxed	1.2	70	22
Walking (0.9 m/s, 3.2 km/h, 2.0 mph)	2.0	115	37
Walking (1.8 m/s, 6.8 km/h, 4.2 mph)	3.8	220	70
Office reading, seated	1.0	55	18
Office, walking about	1.7	100	31
House cleaning	2.0–3.4	115–200	37–63
Pick and shovel work	4.0–4.8	235–280	74–88
Dancing, social	2.4–4.4	140–255	44–81
Heavy machine work	4.0	235	74

Source: ASHRAE, *Standard 55-2013: Thermal Environmental Conditions for Human Occupancy*, American Society of Heating, Refrigerating and Air-Conditioning Engineers, Atlanta, GA, 2013. Copyright ASHRAE, www.ashrae.org.

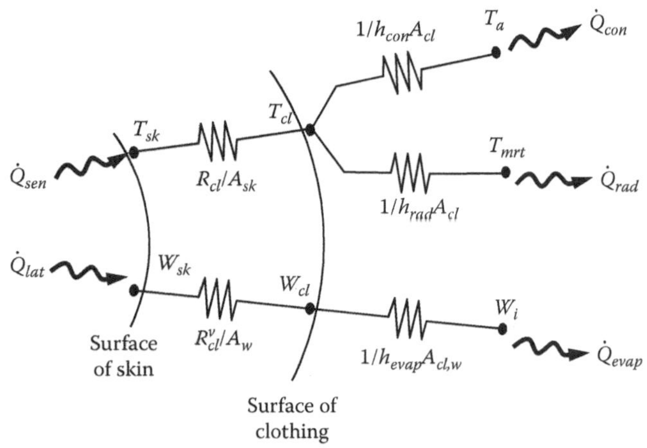

FIGURE 3.2
Thermal network model of sensible and latent heat flows from a human body.

where the first three terms refer to the skin (convection, radiation, and evaporation) and the last two terms to respiration (sensible and latent), as indicated by the subscripts. The respiration heat loss is typically less than 10% of the total heat loss. Figure 3.2 is a simple thermal resistance network representation of the heat loss from a body. How to estimate the various individual heat losses are discussed in the following sections.

3.2.2 Operative Temperature

The convective transfer from the skin of the clothed body can be written as

$$\dot{Q}_{con} = A_{cl} h_{con} (T_{cl} - T_a) \tag{3.3}$$

where

- A_{cl} is the surface area of the clothing and skin in contact with the air
- T_{cl} is the mean temperature of the clothing
- h_{con} is the average convective heat transfer coefficient
- T_a is the dry-bulb temperature of the surrounding or ambient air

Table 3.2 assembles correlations to be used to determine h_{con}, where $\dot{M}$ is the metabolic rate in units of met.

The radiative process is more complicated because different surfaces of the environment (or room) may have different temperatures. For example, during winter, a window exposed to the outside may be at a much lower temperature than that of other surfaces. In summer, a sunlit interior wall may be several degrees warmer that an unlit one. Because these differences are usually small, a linearized radiative heat loss can be assumed without much inaccuracy. To simplify the analysis, it is convenient to define a *mean radiant temperature* (MRT) T_{mrt} of the environment as the temperature of an imaginary isothermal enclosure with which a human body would exchange the same amount of radiation as with the actual environment. Further, the emissivities of various indoor surfaces are close to 0.9, which is sufficiently high that we can assume the surfaces to be black. In that case, the radiative heat loss per unit area of the body is

$$\sigma\left(T_{cl}^4 - T_{mrt}^4\right) = \sigma \sum_n F_{cl\text{-}n}\left(T_{cl}^4 - T_n^4\right) \tag{3.6}$$

where the sum runs over all surfaces with which the body can exchange direct radiation, $F_{cl\text{-}n}$ is the radiation shape factor from the body to the nth surrounding surface, and σ is the Stefan–Boltzmann constant.

Measurements done in an actual room using an instrument called the "Vernon's globe thermometer" allow us to infer MRT (REHVA, 2011). It consists of a hollow sphere 15 cm (6 in.) in diameter painted black with a thermocouple at the center. One measures the globe temperature, the temperature of the surrounding air, and the local air velocity. Subsequently, the MRT can be deduced by setting the convective heat losses/gains to be equal to the radiative heat losses/gains. Since the sum of the shape factors over an enclosure is unity, the T_{cl}^4 terms drop out of Equation 3.6, and T_{mrt} is given by

$$T_{mrt}^4 = \sum_n F_{cl\text{-}n} T_n^4 \tag{3.7}$$

TABLE 3.2
Correlations for Calculating Average Convective Heat Transfer Coefficient

Type of Activity		Correlation	Applicable Range	Units of Velocity	Units of Convective Coefficient	Equation Number
Seated person	IP units	$h_{con} = 0.55$	$0 \le v \le 40$	ft/min	Btu/(h·ft²·°F)	(3.4 IP)
		$h_{con} = 0.061 \times v^{0.6}$	$40 \le v \le 800$	ft/min	Btu/(h·ft²·°F)	
	SI units	$h_{con} = 3.1$	$0 \le v \le 0.2$	m/s	W/(m²·°C)	(3.4 SI)
		$h_{con} = 8.3 \times v^{0.6}$	$0.2 \le v \le 4.0$	m/s	W(/m²·°C)	
Active person in still air	IP units	$h_{con} = (\dot{M} - 0.85)^{0.39}$	$1.1 \le \dot{M} \le 3.0$		Btu/(h·ft²·°F)	(3.5 IP)
	SI units	$h_{con} = 5.7 \times (\dot{M} - 0.85)^{0.39}$	$1.1 \le \dot{M} \le 3.0$		W/m²·°C	(3.5 SI)

Inside buildings, the temperature differences are often small enough that T_{mrt} can be approximated by

$$T_{mrt} \approx \sum_{n} F_{cl\text{-}n} T_n \tag{3.8}$$

For example, if one-half of the surroundings is at 30°C and the other half at 20°C, the difference between these two expressions is only 0.1°C in terms of T_{mrt}. The radiative heat loss of the body can then be written as

$$\dot{Q}_{rad} = A_{cl} h_{rad} (T_{cl} - T_{mrt}) \tag{3.9}$$

where h_{rad} is the radiative heat transfer coefficient. A numerical value of 4.71 W/(m²·°C) [0.83 Btu/(h·ft²·°F)] is suggested for this coefficient for normal nonmetallic clothing.

We define a total heat transfer coefficient

$$h_{c+r} = h_{con} + h_{rad} \tag{3.10}$$

and a so-called operative temperature T_{op}

$$T_{op} = \frac{h_{con} T_a + h_{rad} T_{mrt}}{h_{c+r}} \tag{3.11}$$

Then, neglecting the last three terms in Equation 3.2, we get

$$\dot{Q}_{con} + \dot{Q}_{rad} = A_{cl} h_{c+r} (T_{cl} - T_{op}) \tag{3.12}$$

The radiative and convective heat transfer coefficients are often close to each other, and Equation 3.11 can be approximated such that T_{op} is simply the arithmetic average of T_a and T_{mrt}.

Example 3.1: Mean Radiant and Operative Temperatures

Consider a room 3 m × 3 m × 3 m (≈10 ft × 10 ft × 10 ft); all but one of whose surfaces are at 19°C (66.2°F), while the remaining 3 m × 3 m surface is a window at 10°C (50°F). The room ambient dry-bulb temperature is T_a = 21°C (70°F). Find the MRT and the operative temperature for a person at the center of the room. The air velocity is 0.1 m/s (19.7 ft/min).

Given: Radiation temperatures of surfaces and T_a = 10°C, $\dot{M}$ = 2.6, and v = 0.1 m/s

Find: T_{mrt}, T_{op}

Assumption: h_{rad}/h_{con}

Lookup value: We need the shape factor F from the center of the room to the window. It could be determined from formulas or graphs in books on radiative heat transfer (see Section 2.10). For the determination of MRTs in buildings, it will usually be sufficient to evaluate the solid angle subtended by the surface in question as seen from the center of the body and to approximate the shape factor by this solid angle divided by 4π, the solid angle of a complete enclosure. In the present case, the solid angle is easy to determine because of the symmetry of the room: seen from the center the window fills one-sixth of the total field of view, and therefore, the shape factor F = (1/6) = 0.167.

Solution

The convective heat transfer coefficient h_{con} is determined from Equation 3.4 SI:

$$h_{con} = 3.1\ \text{W/m}^2\cdot°\text{C}$$

In this case, the MRT has only two terms because there are only two different surface temperatures, with corresponding shape factors F and (1 − F). Then, from Equation 3.8,

$$T_{mrt} = F \times 10°\text{C} + (1-F) \times 19°\text{C} = 17.5°\text{C}$$

We use the standard value for the radiative heat transfer coefficient of 4.71 W/(m²·°C)—see Equation 3.9. Thus, from Equation 3.11, the operative temperature is

$$T_{op} = \frac{3.1\ \text{W/(m}^2\cdot°\text{C)} \times 21°\text{C} + 4.71\ \text{W/(m}^2\cdot°\text{C)} \times 17.5°\text{C}}{(3.1 + 4.71)\ \text{W/(m}^2\cdot°\text{C)}}$$

$$= 18.9°\text{C (66°F)}$$

Comment

Assume the operative temperature to be the arithmetic average of the MRT and the space air temperature (correct only when $h_{rad} = h_{con}$) would have yielded

$$T_{op} = \frac{21°\text{C} + 17.5°\text{C}}{2} = 19.25°\text{C (66.6°F)}$$

This is only 0.35°C (0.6°F) different than the more accurate calculation.

3.2.3 Clothing Insulation

The clothing adds thermal resistance from the heat flowing out from the body skin and must be included in the model. The insulating value of clothing is measured in unit of clo, defined as

$$1\ \text{clo} = 0.155\ \text{m}^2\cdot\text{K/W}\ (0.88\ \text{ft}^2\cdot\text{h}\cdot°\text{F/Btu}) \tag{3.13}$$

The unit of clo is based on the insulating value of the typical American man's business suit in 1941. Table 3.3

TABLE 3.3

Typical Clothing Insulation Values for Different Clothing Ensembles

Ensemble	I_{cl} (clo)	I_T (clo)	A_{cl}/A_{sk}
Walking shorts, short-sleeve shirt	0.36	1.02	1.10
Trousers, short-sleeve shirt	0.57	1.20	1.15
Trousers, long-sleeve shirt	0.61	1.21	1.20
Trousers, t-shirt, long-sleeve shirt, suit jacket	0.96	1.54	1.23
Trousers, long-sleeve shirt, long-sleeve sweater	1.01	1.56	1.28
Sweat pants, sweat shirt	0.74	1.35	1.19
Knee-length skirt, short-sleeve shirt, pantyhose, sandals	0.54	1.10	1.26
Knee-length skirt, long-sleeve shirt, full slip pantyhose	0.67	1.22	1.29
Knee-length skirt, long-sleeve shirt, half slip, pantyhose, long-sleeve sweater	1.10	1.59	1.46
Long-sleeve coveralls, t-shirt	0.72	1.30	1.23

I_{cl} is the thermal insulation resistance value for the clothing alone, while I_T is that for the total between skin and ambient air.

1 clo = 0.88 ft²·h·°F/Btu = 0.155 m²·K/W.

gives values of the thermal resistance of various clothing ensembles. The insulation resistances I_{cl} and I_T include convective and radiative effects and are the clo units for the clothing alone and for the total thermal resistance between the skin and the environment (Figure 3.2).

The skin temperature varies with the metabolic rate, but a value of 34.1°C (93.4°F) is typically assumed for seated occupants doing office work. With increasing activity, the metabolic rate increases and the body lowers the skin temperature in order to dissipate it if the environment temperature is kept constant. The following relationship can be used to predict the skin temperature conducive to comfort:

$$T_{sk} = 35.7 - 1.60 \times (\dot{M} - \dot{W})°\text{C} \qquad (3.14\ \text{SI})$$

$$T_{sk} = 96.3 - 2.87 \times (\dot{M} - \dot{W})°\text{F} \qquad (3.14\ \text{IP})$$

where $\dot{M}$ and $\dot{W}$ are the metabolic rate and the work production rate, respectively, in consistent units (W or Btu/h).

The surface area ratio between the total surface area of the clothing and the skin area (A_{cl}/A_{sk}) is also given in Table 3.3. When computing total heat loss, the outside surface area of the clothing (rather than that of the skin) should be used along with I_T. The values listed are valid when MRT is equal to the dry-bulb temperature of the air and when air velocity is less than 0.2 m/s (40 ft/min). The following example illustrates the use of this table.

Example 3.2: Sensible Heat Loss

Consider the same conditions as in Example 3.1. The person is active (met level of 2.6) wearing trousers and a long-sleeve shirt. Calculate the total (convective plus radiative) sensible heat loss.

Given: T_{op} = 18.9°C (66°F)

Find: Q_{sen}

Assumption: Steady-state condition; skin area A_{sk} = 1.8 m² (19.6 ft²) (this is referred to as the DuBois skin surface area) and work rate = 0; $\dot{M}$ = 2.6

Lookup value: From Table 3.3, the total thermal resistance I_T = 1.21 clo. Also, A_{cl}/A_{sk} = 1.20.

Solution

The total thermal resistance is $R_T = I_T \times$ (0.155 m²·°C/W) = 1.21 clo × 0.155 m²·°C/W = 0.188 m²·°C/W

From Equation 3.14, T_{sk} = 35.7 − 1.60 × (2.6 − 0) = 31.54°C

The total external clothing area is then determined:

$$A_{cl} = A_{sk}\left(\frac{A_{cl}}{A_{sk}}\right) = 1.8 \times 1.2 = 2.16\ \text{m}^2$$

Finally, the total sensible heat transfer is given by

$$Q_{sen} = A_{cl} \cdot \frac{(T_{sk} - T_a)}{R_T} = 2.16\ \text{m}^2 \times \frac{(31.54 - 18.9)°\text{C}}{0.188\ \text{m}^2 \cdot °\text{C/W}}$$

$$= 145.2\ \text{W}\ (495\ \text{Btu/h})$$

3.2.4 Humidity Effects

The operative temperature does not include humidity effects. We define a quantity, called "adiabatic equivalent temperature," which is a linear combination of T_{op} and of the vapor pressure of the air, and depends on skin wetness and on clothing permeability. This quantity achieves the goal of combining the three characteristics of the environment (radiation temperature, air dry-bulb temperature, and humidity) into a single temperature index that completely determines the total heat loss from the skin. It is used as the basis of the ASHRAE comfort chart (described in the next section) because any combination of these parameters will cause the same heat loss from the body as would an environment with the same adiabatic equivalent temperature.

Another temperature index, the *effective temperature* ET*, is also used in the analysis of thermal comfort. Like the adiabatic equivalent temperature, it is a linear combination of T_{op} and of the vapor pressure. More strictly, it is the temperature of an isothermal black enclosure with 50% relative humidity where the body surface would experience the same heat loss as in the actual space. The 50% level is chosen as the reference value because it is considered to be the average value over the acceptable range of indoor humidity levels.

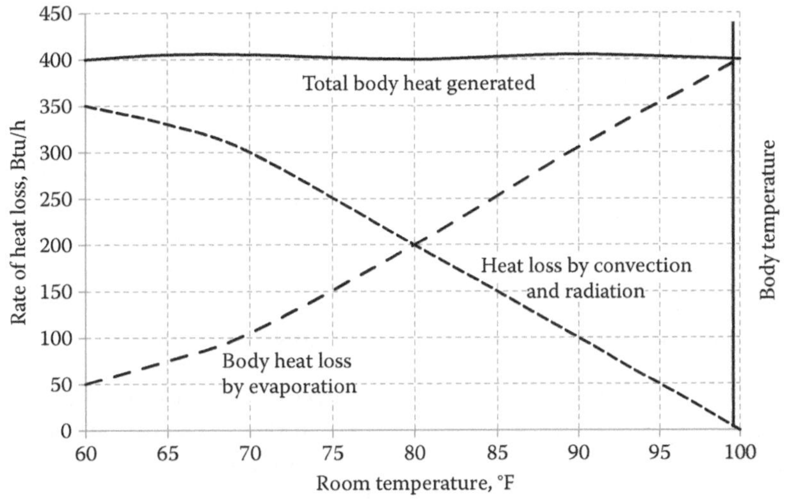

FIGURE 3.3
Variation of heat generated and individual heat losses from a human at rest with variation in ambient temperature.

From Equation 3.2, we note that along with convection and radiation, there are three additional modes of heat transfer. Of these, the evaporative skin loss $\dot{Q}_{evap}$ is an important effect as illustrated in Figure 3.3. As the ambient temperature increases, we notice that the relative contribution of evaporative heat loss increases substantially to compensate for the drop of the convective and radiative heat losses. The evaporative heat loss can be determined by evaluating the mass transfer coefficient for water vapor h_E as

$$h_E = \frac{h_{con}}{\text{Le} \cdot c_{p,a}} \tag{3.15}$$

where
Le is the Lewis number (Le = 0.895 for most applications)
$c_{p,a}$ is the specific heat of ambient air

The evaporative or latent heat loss can then be deduced as (rate of water evaporation)

$$\dot{m}_{evap} = A_{sk} \cdot h_E \cdot f_{w,sk} \cdot (W_{sk}^* - W_a) \tag{3.16}$$

and (latent heat loss)

$$\dot{Q}_{evap} = \dot{m}_{evap} \cdot h_{evap,sk} \tag{3.17}$$

where
$f_{w,sk}$ is the fraction of the wetted skin to the total skin area (which varies from 0.06 for normal skin moisture loss to about 0.5 for comfortable conditions)
W_{sk}^* is the saturated humidity ratio of moist air at the skin temperature
W_a is the humidity ratio of the surrounding air
$h_{evap,sk}$ is the latent heat of evaporation of water at the skin temperature

For simplified calculations, the energy dissipated by sweating for an average person (skin area of 1.8 m^2 or 19.6 ft^2) to maintain comfort can be calculated from

$$\dot{Q}_{evap} = 24.3 \times (\dot{M} - \dot{W} - 1)\ \text{W/m}^2 \tag{3.18 SI}$$

$$\dot{Q}_{evap} = 7.74 \times (\dot{M} - \dot{W} - 1)\ \text{Btu/(h} \cdot \text{ft}^2) \tag{3.18 IP}$$

The following example illustrates the calculation procedure.

Example 3.3: Evaporative Loss

Consider the same conditions as in Examples 3.1 and 3.2 but the dew point temperature* of the space is specified as 12°C (54°F). Calculate the evaporative heat loss assuming a skin wettedness fraction of 0.2.

Given: T_{dp} = 12°C, h_{con} = 3.1 W/(m$^2\cdot$°C), $f_{w,sk}$ = 0.2

Find: $\dot{Q}_{evap}$

Assumption: Steady-state condition; skin area A_{sk} = 1.8 m^2 and skin temperature T_{sk} = 31.54°C; Le = 0.895

Lookup values: The relevant background on psychrometric calculations is given in Chapter 13. We will simply assume the following values as given for the time being:

$$c_{p,a} = 1.007\ \text{kJ/(kg}\cdot\text{°C)},\ W_{sk}^* = 0.030\ \text{kg}_w/\text{kg}_a,$$
$$W_a = 0.008766\ \text{kg}_w/\text{kg}_a,\ h_{evap,sk} = 2427.1\ \text{kJ/kg}$$

* The dew point temperature is the temperature of the air at which water vapor in the air will start condensing out. Relevant concepts are discussed in Chapter 13.

Solution

From Equation 3.15:

$$h_E = \frac{3.1\ \text{W}/(\text{m}^2\cdot{}^\circ\text{C})}{0.895\times 1.007\ \text{kJ}/(\text{kg}_\text{a}\cdot{}^\circ\text{C})}$$

$$= 3.44\times 10^{-3}\ \text{W}\cdot\text{kg}_\text{a}/(\text{m}^2\cdot\text{kJ})$$

From Equation 3.16:

$$\dot{m}_{evap} = 1.8\ \text{m}^2\times 3.44\times 10^{-3}\ \text{W}\cdot\text{kg}_\text{a}/(\text{m}^2\cdot\text{kJ})$$

$$\times 0.2\times(0.030-0.008766)\ \text{kg}_\text{w}/\text{kg}_\text{a}$$

$$= 0.0263\times 10^{-3}\ \text{kg}_\text{w}/\text{s or about } 2.3\ \text{kg/day}$$

And finally, from Equation 3.17:

$$\dot{Q}_{evap} = 0.0263\times 10^{-3}\ \text{kg/s}\times 2427.1\ \text{kJ/kg}\times 1000\ \text{W/kW}$$

$$= 63.8\ \text{W}$$

Note that this is about 43% of the total heat loss of 145.2 W due to convection and radiation (from the previous example).

Comment

The aforementioned result of 2.3 kg/day evaporative loss needs to be compensated by drinking water. This is the basis of the recommendation that a normal person in a normal environment should drink at least 2 L/day.

If we were to use the simplified expression given by Equation 3.18 with work set at zero, we get

$$\dot{Q}_{evap} = 24.3\times(\dot{M}-\dot{W}-1) = 24.3\times(2.6-0-1)$$

$$= 38.9\ \text{W/m}^2$$

With a skin area of 1.8 m², the latent heat loss is about 70 W, which is about 10% of the value of 63.8 W determined by the more accurate method.

3.3 Perception of Comfort

3.3.1 ASHRAE Comfort Chart

The environmental parameters discussed in the previous section are the primary factors used to characterize human comfort. Further, the modeling equations described earlier do not explicitly consider the psychological aspects of thermal comfort that is perceived differently by different humans. An empirical approach relating comfort perceptions with environmental parameters that take into account subjective differences between humans was developed by Fanger (1970). This approach has been modified and complemented by several subsequent field and analytical studies, and forms the basis of the ASHRAE Standard (ASHRAE 55, 2013) as well as that of ISO 7730 (1993).

Thermal comfort is characterized by four physical parameters: air temperature, MRT, air humidity, and air velocity with two additional conditions kept constant, namely activity level and clothing level. Additional criteria for discomfort have also been identified such as local draft, high turbulence, high radiant temperature asymmetry, and unacceptably high vertical air temperature difference (the implications of these effects are discussed later in this section and also in Chapter 20 dealing with air distribution systems).

A thermal sensation index, called the predicted mean vote (PMV), has been proposed to represent occupant acceptability of the indoor environment. This index can be calculated through a complex mathematical correlation of the aforementioned six parameters and relates the imbalance between the actual heat flow from the human body in a given environment and that required for optimum comfort. The PMV index is used to quantify the degree of discomfort and ranges from +3 to −3, with zero indicating neutral or comfort condition. This scale is called the "ASHRAE thermal sensation scale" and is represented as follows:

+3	+2	+1	0	−1	−2	−3
Hot	Warm	Slightly warm	Neutral	Slightly cool	Cool	Cold

Test chamber experiments with actual subjects are finally the most reliable approach, and these were conducted by Fanger (1970). Studies on 1600 college-aged students revealed certain interesting trends between comfort level, temperature, humidity, sex, and length of exposure. An empirical correlation has been subsequently developed:

$$\text{PMV} = a\times T_a + b\times p_v - c \tag{3.19}$$

where the numerical values of the coefficients a, b, and c are given in Table 3.4. The temperature T_a is in °C or °F, and the partial pressure of water vapor in the air (p_v) is in kPa or in psi. In general, a change of 3°C (5.4°F) in temperature or a 3 kPa (0.44 psi) change in water vapor pressure is necessary to change a thermal sensation vote by one category.

Since women prefer slightly higher temperatures than men,* separate correlations are provided for both the sexes and also for a typical combined set of people. Note that duration of occupancy (expressed as "exposure period") is an additional factor affecting PMV as is clear from Table 3.4.

* Some researchers attribute this difference to metabolism rate, while others simply to differences in clothing.

TABLE 3.4

Coefficients for Use in Equation 3.19 to Compute PMV

Exposure Period	Sex	SI Units			IP Units		
		a	*b*	*c*	*a*	*b*	*c*
1.0	Male	0.220	0.233	5.673	0.122	1.61	9.584
	Female	0.272	0.248	7.245	0.151	1.71	12.080
	Combined	0.245	0.248	6.475	0.136	1.71	10.880
3.0	Male	0.212	0.293	5.949	0.118	2.02	9.718
	Female	0.275	0.255	8.622	0.153	1.76	13.511
	Combined	0.243	0.278	6.802	0.135	1.92	11.122

Example 3.4: Difference in PMV due to Exposure Period

Consider the same conditions as in Example 3.3. Calculate the difference in PMV for a combined group of men and women when exposed to 1 and 3 h. Assume MRT is equal to indoor air dry-bulb temperature.

Given: $T_a = 21°C$, $T_{dp} = 12°C$

Find: PMV

Lookup values: The partial pressure corresponding to $T_{dp} = 12°C$ is $p_v = 1.402$ kPa.

Solution

Using Equation 3.19 with the appropriate coefficients from Table 3.4 yields

1 h exposure:

$$\text{PMV} = 0.245 \times 21 + 0.248 \times 1.402 - 6.475$$

$$= -0.98 \quad \text{(slightly cool)}$$

3 h exposure:

$$\text{PMV} = 0.243 \times 21 + 0.278 \times 1.402 - 6.802$$

$$= -1.31 \text{ (slightly cool to cool)}$$

Thus, in this instance, subjects tend to feel cooler as their exposure period is increased.

PMV characterizes the mean thermal sensation or the mean satisfaction level of a group of people exposed to the same indoor environment. Due to individual preferences, one would expect a distribution of votes. An index meant to characterize this variability has been proposed, namely, the percentage of people dissatisfied (PPD). It has been found that PPD is empirically correlated to PMV. When PPD is plotted versus the mean vote of a large group, one typically finds a distribution as shown in Figure 3.4. This graph shows that even under optimal conditions (i.e., mean vote = 0) approximately 5% may be dissatisfied with the thermal environment. A correlation between PPD and PMV has been proposed:

$$\text{PPD} = 100 - 95 \times \exp[-(0.03353 \times \text{PMV}^4 + 0.1297 \times \text{PMV}^2)] \tag{3.20}$$

In view of the fuzziness of the test results, values ranging from PMV = ±0.5 and PPD < 10% are considered acceptable.

The ASHRAE comfort chart in Figure 3.5 indicates the acceptable ranges of operative temperature and humidity for at least 80% of the occupants (10% discomfort allowance for experimental uncertainty and 10% for

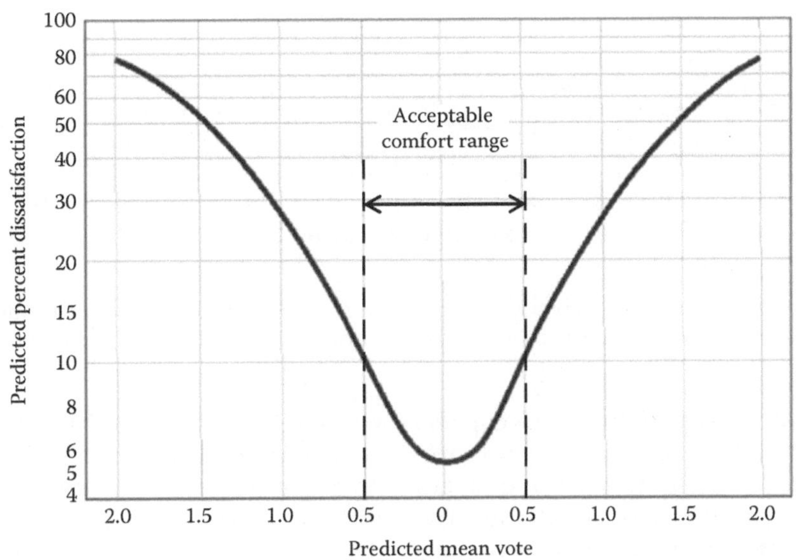

FIGURE 3.4

Percentage of people dissatisfied as a function of mean predicted vote. (From ASHRAE, *Standard 55-2013: Thermal Environmental Conditions for Human Occupancy*, American Society of Heating, Refrigerating and Air-Conditioning Engineers, Atlanta, GA, 2013. Copyright ASHRAE, www.ashrae.org.)

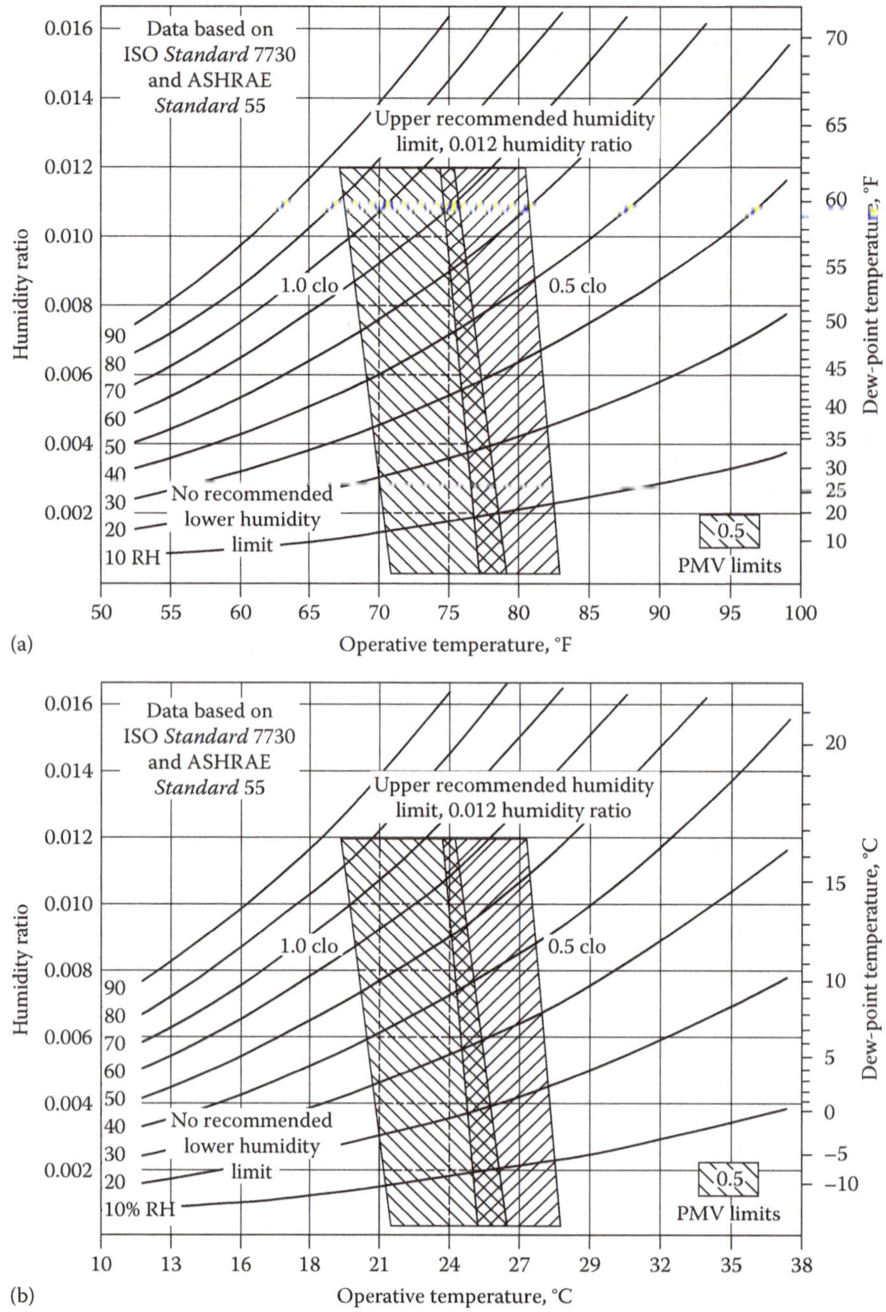

FIGURE 3.5
Acceptable ranges of operative temperature and humidity, for sedentary activity and typical summer and winter clothing. (a) IP Units. (b) SI Units. (From ASHRAE, *Handbook of Fundamentals*, American Society of Heating, Refrigerating and Air-Conditioning Engineers, Atlanta, GA, 2013. Copyright ASHRAE, www.ashrae.org.)

temperature non-uniformity inside room) of the occupants engagedin light sedentary activity, assuming typical summer or winter clothing levels, respectively. These ranges are valid for mean airspeed ≤0.15 m/s (30 ft/min).* The chart is based on steady-state conditions, i.e., for occupants who are in the space for 3 h or more. The comfort chart is similar to the psychrometric chart, but the abscissa is the operative temperature T_{op} rather than the dry-bulb air temperature T_a. The ordinate is the humidity ratio W, and the dew point temperature is indicated on the right-hand scale. The relative humidity lines of the psychrometric chart that have been superimposed in Figure 3.5 are, strictly speaking, correct only if the radiation temperature is equal to the dry-bulb air temperature (in which case $T_{op} = T_a$) (Table 3.5).

* Air velocities up to 50 ft/min go unnoticed by occupants, while velocities >200 ft/min result in drafty conditions. Velocities in the 50–100 ft/min range are pleasant, while 100–200 ft/min conditions are generally pleasant but cause awareness of air movement.

TABLE 3.5

Optimal Operative Temperature and Acceptable Range for Light Sedentary Activity at 50% Relative Humidity and at Mean Airspeed ≤0.15 m/s (30 ft/min)

Season	Typical Clothing	I_{cl} (clo)	Optimum Operative Temperature	Acceptable Range
Winter	Heavy slacks, long-sleeve shirt, and sweater	0.9	22°C 71°F	20°C–23.5°C 68°F–75°F
Summer	Light slacks and short-sleeve shirt	0.5	24.5°C 76°F	23°C–26°C 73°F–79°F
	Minimal	0.05	27°C 81°F	26°C–29°C 79°F–84°F

Source: ASHRAE Standard—55, 2010. *Thermal Environmental Conditions for Human Occupancy*, American Society of Heating, Refrigerating and Air-Conditioning Engineers, Atlanta, GA, 2010. Copyright ASHRAE, www.ashrae.org.

The acceptable values of T_a and W are indicated by the shaded zones. The chart shows different zones for winter and for summer because comfort depends on the insulation value of the clothing: clo = 0.9 during winter and clo = 0.5 during summer. From Figure 3.5, in winter, an effective temperature of 21°C–22°C (70°F–72°F) is optimum for normally clothed people. In summer, about 25°C (76°F) would be optimum. The further one strays from these values, the greater the PPD. The ASHRAE comfort chart specifies steady-state environment zones where at least 80% of occupants ought to be thermally comfortable. This allowance is based on a 10% dissatisfaction criterion in general for the whole body based on PMV–PPD indices as shown in Figure 3.4, plus an additional 10% dissatisfaction percentage due to local thermal discomfort (i.e., range of PMV ±0.5).

Note that these zones are specified in Figure 3.5 in terms of effective temperature ET* (introduced earlier). The left and right boundaries of the shaded zones are lines of constant adiabatic equivalent temperature, hence lines of constant heat loss. They are sloped from upper left to lower right because the evaporative heat loss from the body decreases as the humidity ratio of the air increases. This agrees with the common experience that hot, humid weather is less comfortable than hot, dry weather at the same T_a.

Note that there is no lower relative humidity limits in Figure 3.5. Previous versions of this standard had a lower limit of about 30% relative humidity, but recent versions do not. This change is a result of the contention that there is no research that substantiates the onset of adverse conditions (dry skin, eye irritation, respiratory health) at low humidity levels. A research project is ongoing to review available literature and revisit this rather contentious issue. Higher humidity levels are set to avoid condensation of moisture on building surfaces (which leads to microbial growth) and also to be consistent with the ventilation standard ASHRAE 62.1 (2013). Expectations and recommendations for comfortable temperatures have evolved considerably since the beginning of the century, and only during the last two decades do they seem to have reached general agreement.

Example 3.5: Use of ASHRAE Comfort Chart

Suppose the room of Example 3.1 has a humidity ratio $W_a = 0.005\ kg_w/kg_a$. Do the conditions at the center of the room satisfy the ASHRAE comfort criteria in winter?

Given: T_{op} = 21°C (70°F) and $W_a = 0.005\ kg_w/kg_a$; comfort chart of Figure 3.5

Find: Are these conditions within comfort zone?

Solution

The humidity ratio does satisfy the ASHRAE criterion, even though only barely. At $W_a = 0.005\ kg_w/kg_a$, the minimum T_{op} of the comfort chart is 20.6°C (69°F); the conditions of this example are slightly too cold, but the difference is not significant since the boundaries of the comfort zone are in any case somewhat fuzzy.

Comment

To improve the comfort, one could obviously increase one or several of the temperature components of T_{op}. But an interesting alternative is to raise the humidity ratio W, say to 0.010.

3.3.2 Corrections to ASHRAE Comfort Chart

The ASHRAE comfort recommendations are widely followed worldwide with, of course, appropriate corrections that will now be discussed. For best success, the designer should be aware of these factors. The comfort chart is applicable to an environment with occupants seated at rest or doing light work (met = 1.2), where enclosing surfaces are at the mean temperature equal to the dry-bulb temperature, and with occupants dressed to clo = 0.9 in winter and clo = 0.5 in summer. For different clo values and met values between 1.2 and 3, ASHRAE 55 (2013) recommends that the comfort operative temperature be determined as

$$T_{op,comf} = 27.2 - 5.9 \times \text{clo} - 3.0 \times (1.0 + \text{clo}) \times (\dot{M} - 1.2)°\text{C} \quad (3.21\ \text{SI})$$

$$T_{op,comf} = 81.0 - 10.6 \times \text{clo} - 5.4 \times (1.0 + \text{clo}) \times (\dot{M} - 1.2)°\text{F} \quad (3.21\ \text{IP})$$

where

clo is the clothing insulation
$\dot{M}$ is the metabolic value

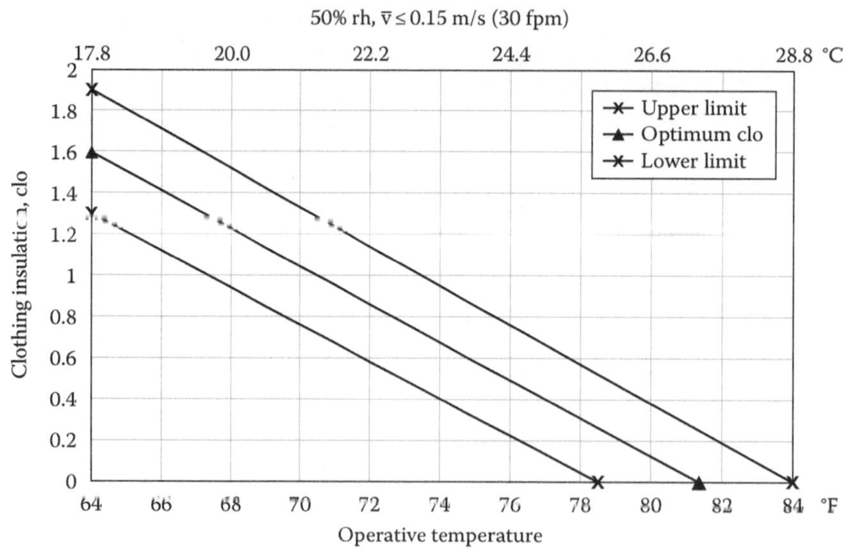

FIGURE 3.6
Clothing insulation necessary to be within the ASHRAE 80% acceptability limits as a function of operative temperature, during light sedentary activity (≤1.2 met). (From ASHRAE, Standard—55, 2004. *Thermal Environmental Conditions for Human Occupancy,* American Society of Heating, Refrigerating and Air-Conditioning Engineers, Atlanta, GA, 2004. Copyright ASHRAE, www.ashrae.org.)

The effect of clothing insulation on the ASHRAE comfort recommendations is shown in Figure 3.6. Note the rather wide uncertainty bands reflective of the large uncertainties inherent in the comfort relations. Further, at lower temperatures the perception of comfort depends also on maintaining sufficiently uniform insulation over the body, in particular the hands and feet. For more than an hour, the minimum operative temperature should not be below 18°C (65°F). The inverse relationship between operative temperature and activity level given by Equation 3.21 is shown in Figure 3.7 for three different clothing levels.

There are several additional physical variables that can affect comfort. Air movement plays a role because the convective heat transfer from the body depends on air velocity. An excess may be perceived as draft, and a lack as stuffiness. The recommendations for airspeed depend on temperature and humidity. As long as the latter is within the comfort zone of Figure 3.5, there is no minimum requirement for airspeed. In hot

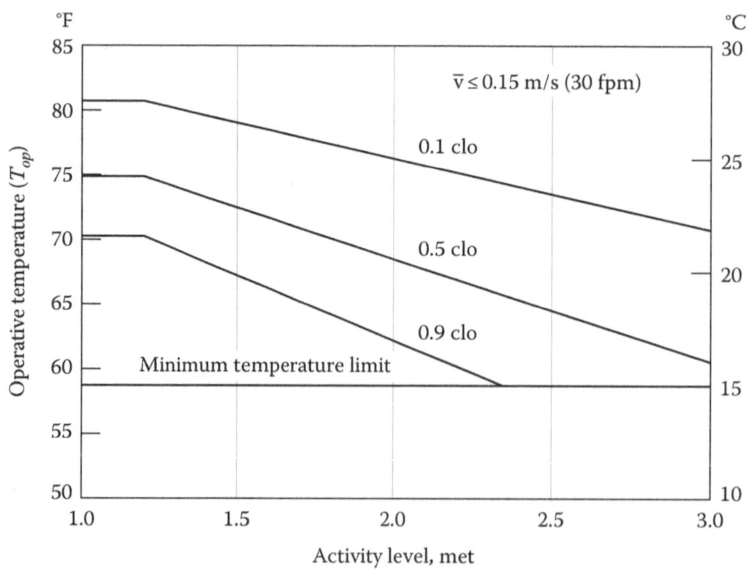

FIGURE 3.7
Recommended operative temperatures for active people as a function of metabolic rate. (From ASHRAE, Standard—55, 2004. *Thermal Environmental Conditions for Human Occupancy,* American Society of Heating, Refrigerating and Air-Conditioning Engineers, Atlanta, GA, 2004. Copyright ASHRAE, www.ashrae.org.)

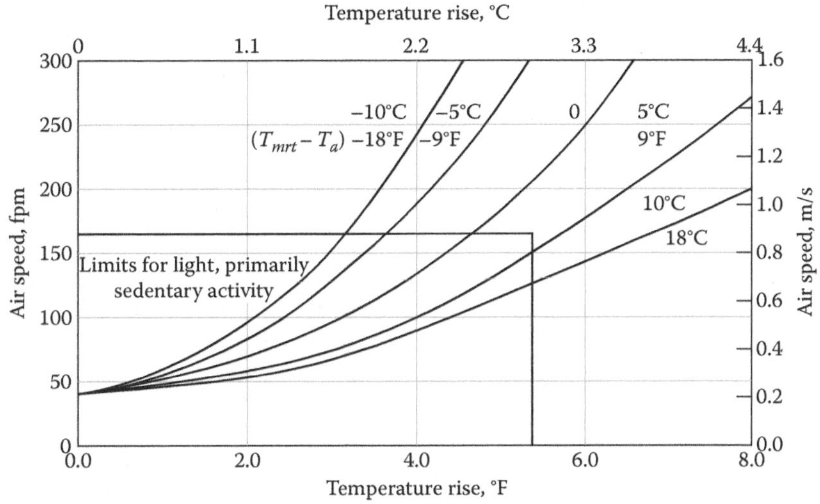

FIGURE 3.8
Airspeed required to increase the air temperature above the summer comfort zone. (From ASHRAE, *Standard—55, 2013: Thermal Environmental Conditions for Human Occupancy*, American Society of Heating, Refrigerating and Air-Conditioning Engineers, Atlanta, GA, 2013. Copyright ASHRAE, www.ashrae.org.)

weather, the upper range of temperatures could be extended if the airspeed is increased. Figure 3.8 indicates how much the air temperature can be increased above the summer comfort zone by increasing the airspeed. However, the acceptability of increased airspeed depends on the control the occupants have; also, one should keep in mind that beyond 160 ft/min (0.8 m/s) loose paper and other light objects may be blown away.

In a cold environment, high air movement would cause draft. Draft is a complex phenomenon that depends, among other factors, on turbulence. Sensitivity to draft is greatest where the skin is exposed at the head and ankles. Figure 3.9 shows the recommended limits for mean airspeed to reduce the risk of draft as a function of air temperature and of turbulence intensity. The latter is defined as the ratio of the standard deviation to the average of the airspeed in the room. In conventionally ventilated spaces, the turbulence intensity is usually in the range of 30%–60%. In rooms with displacement ventilation or without ventilation (see Chapter 20), it may be lower.

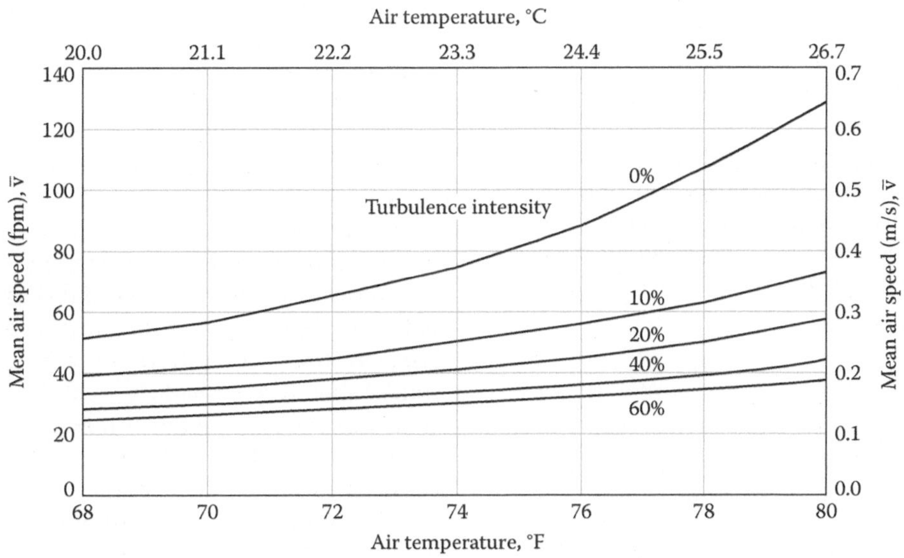

FIGURE 3.9
Recommended limits for mean airspeed to reduce the risk of draft as a function of air temperature and of turbulence intensity. (From ASHRAE, Standard—55, 2004. *Thermal Environmental Conditions for Human Occupancy*, American Society of Heating, Refrigerating and Air-Conditioning Engineers, Atlanta, GA, 2004. Copyright ASHRAE, www.ashrae.org.)

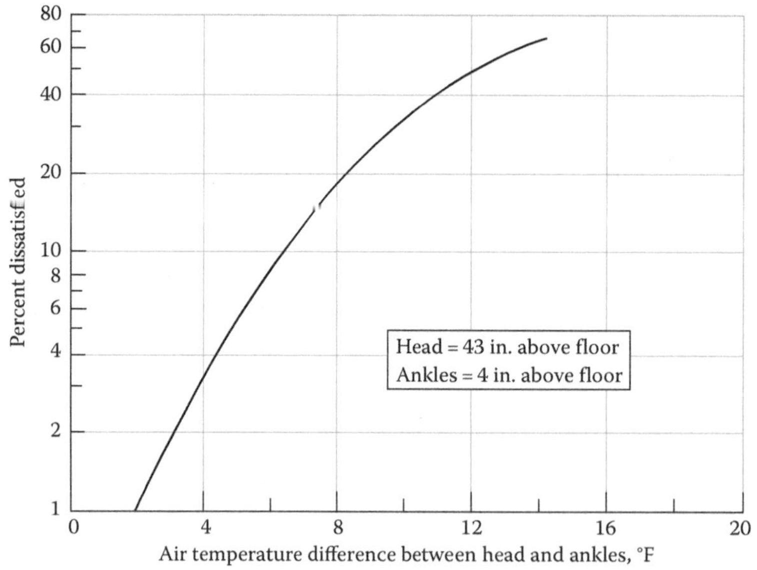

FIGURE 3.10
PPD of seated occupants as a function of air temperature difference between the head and ankles. (From ASHRAE, *Handbook of Fundamentals*, American Society of Heating, Refrigerating and Air-Conditioning Engineers, Atlanta, GA, 2013. Copyright ASHRAE, www.ashrae.org.)

Temperature variations within a room should not be too large. Figure 3.10 shows the PPD as a function of the vertical air temperature difference between head and ankles. For a 5°F (3°C) difference, the PPD is about 7%. It is recommended that vertical temperature difference should not exceed 5°F (3°C) between the levels of 4 in. (0.1 m) and 67 in. (1.7 m) above the floor. Likewise, large temperature drifts should be avoided. It has been found that subjects could tolerate much greater differences if the head were cooler. Specific recommendations and other details can be found in ASHRAE 55 (2013).

Example 3.6: Operative Temperature for Other Conditions

Consider a gymnasium where the metabolic rate of the occupants is 3.0 and the clothing level is 0.3. If the occupants are to experience the same level of comfort as when sedentary, what should be the operative temperature in this space?

Given: $\dot{M} = 3.0$, clo = 0.5

Find: $T_{opt,active}$

Solution

From Equation 3.21:

$$T_{op,active} = 27.2 - 5.9 \times 0.3 - 3.0 \times (1.0 + 0.3) \times (3.0 - 1.2)$$

$$= 18.4°\text{C}\ (65.1°\text{F})$$

Note that the gymnasium need not be kept at this low temperature since typically people do not stay longer than about an hour, while the ASHRAE comfort chart applies to steady-state conditions (occupancy of 3 h), and further people do not exercise vigorously for the entire occupancy period. In order to save energy, an alternative is to increase the temperature by about 1.5°C (2.7°F) and increase the air velocity in the room to about 0.4 m/s (see Figure 3.8).

3.3.3 Adaptive Model

Subsequent research showed that it was extremely difficult for naturally ventilated buildings—in fact, any building lacking extensive use of mechanical systems—to meet these thermal comfort requirements, despite the fact that postoccupancy surveys for these buildings indicated that their occupants were generally satisfied from a thermal comfort standpoint. The predictive model of thermal comfort could not explain findings from many passive buildings. For buildings in which thermal adaptation occurs, research has found improved correlation between occupant comfort and mean outdoor monthly temperatures compared to the existing PPD–PMV model that only considers local effects like air temperature, airspeed, and clothing levels. de Dear and Brager (1998) proposed modifications to thermal comfort standards to account for occupant adaptation in naturally ventilated buildings, later defined by ASHRAE 55 (2013) as "those spaces where the thermal conditions of the space are regulated primarily by the opening and closing of windows by the occupants."

Further, people's perception of comfort is also dependent on local conditions. In countries with warm climates such as Israel, Thailand, Singapore, and the southern parts of China, the neutral or comfort temperature in both residences and commercial buildings was observed to be higher (by 1°C–3°C or 1.8°C–5.5°C) than in nonconditioned buildings as compared to conditioned ones. It was found that in naturally ventilated spaces, larger deviations from the ASHRAE recommendations can be tolerated. People are able to adapt to the thermal environment by means of behavioral adjustments such as (1) changing clothing level, (2) relaxation of expectations, and (3) acclimatization of the conditions to which exposed. This model, referred to as the "adaptive model," allows a wider range of comfort conditions. The relationship between indoor comfort operative temperature $T_{op,comf}$ and mean monthly outdoor air temperature T_o is shown in Figure 3.11 and is given by ASHRAE 55 (2013)

$$T_{op,comf} = 17.8 + 0.31 \times T_o \ (°C) \tag{3.22 SI}$$

$$T_{op,comf} = 54.1 + 0.31 \times T_o \ (°F) \tag{3.22 IP}$$

The figure includes two sets of temperature bands: one for 80% acceptability and another for 90% acceptability when a higher standard of comfort is desired. The lower and upper boundaries for 90% thermal acceptability are approximately −2.2°C (−4.0°F) ≤ $T_{op,comf}$ ≤ + 2.5°C (4.5°F). One should not extrapolate beyond the range of temperatures shown in Figure 3.11. It is suggested that no humidity or airspeed limits need be imposed when using this equation.

3.3.4 Other Considerations

Sometimes, there are objective reasons for different preferences. For example, some dependence of temperature preferences on sex and age has been observed: Women tend to prefer somewhat higher temperatures (as stated earlier), and so do adults over the age of 60. Whether that is due to biological differences or to differences in clothing or activity is of little concern to the building designer because he or she has no influence over either of these factors. It is best to design the building and the HVAC system to allow occupants to attain their preferred conditions.

For instance, the degree of control over one's environment is crucial to understanding the difference between comfort perceptions at home and at work. At home, most people can freely adjust thermostat, open/close windows, and change in lighting levels as they please. Being in control, they are likely to be satisfied, even at times when objective conditions are outside the conventional comfort zone. Owners of a passive solar home are a good example. By and large, people who choose such a home want to live in tune with nature and minimize the consumption of nonrenewable energy. Their willingness to accept indoor temperature fluctuations may appear unbelievable to designers who are accustomed to the stringent temperature tolerances of most office buildings.

At work, by contrast, many people do not have much control over their physical environment: The thermostat settings are determined by others (the boss or coworkers), and ventilation cannot be adjusted independently. This is particularly true with the open office space

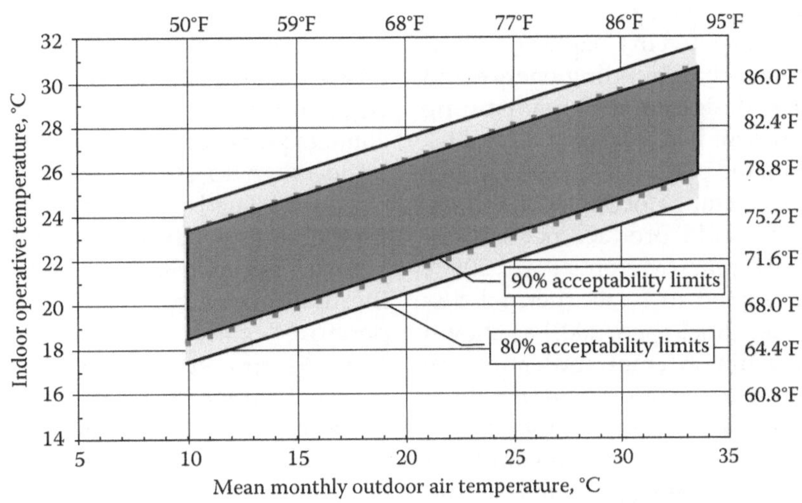

FIGURE 3.11
Variation of acceptable operative temperature range for naturally ventilated buildings. (From ASHRAE, *Standard 55-2013: Thermal Environmental Conditions for Human Occupancy*, American Society of Heating, Refrigerating and Air-Conditioning Engineers, Atlanta, GA, 2013. Copyright ASHRAE, www.ashrae.org.)

layout. Under the same thermal environment, people at work are more likely to be dissatisfied than people at home.

Often, the situation is further complicated by the subtle relationships between objective conditions and subjective perception of comfort. The recent literature highlights the importance of subjective variables such as satisfaction with one's work, aesthetic aspects of one's work environment, the quality of the lighting, colors, views, noise, and odors.

An interesting case in point is the possibility of opening the windows, something that is usually missing in commercial buildings in the United States. Some people feel a certain amount of claustrophobia if they cannot open their windows; they are more likely to find the air stuffy than people who can open their windows—even if the windows actually remain shut and the physical conditions are the same. In fact, people often keep the windows shut anyway, as protection from street noise, dust, or draft. More important than the actual opening is the freedom to choose.

Even though we do not discuss acoustic comfort, we emphasize its importance because the HVAC engineer can make an important contribution by choosing quiet equipment and adding sufficient sound insulation. Otherwise, the hapless occupants may suffer buzzing air vents, humming motors, or droning compressors.

Modeling of occupant comfort has been the subject of numerous studies over the years. For example, studies have looked at comfort in hot and humid locations. Models have been developed to capture comfort under transient metabolic conditions. The interested reader can refer to the published literature; the review papers by Guan et al. (2003) and Frontczak and Wargocki (2011) are recommended.

The thermal environment has a large effect on human performance and learning. As stated in Chapter 1, worker salaries are 100 times greater than building energy and maintenance costs. Hence, designing and maintaining improved indoor environment has a large payoff. Fisk and Rosenfeld (1997) estimated potential annual savings and productivity gains in the range of $29–$168 billion. Reduction in absenteeism would produce net savings of $400 per employee per year (Milton et al., 2000). A preliminary study attempted to relate thermal discomfort to productivity loss (Roelofsen, 2001) as shown in Figure 3.12. This was a first attempt and cannot be considered definitive; however, the figure does illustrate the strong dependence of relative loss of productivity with PPD. Another study analyzed field and simulated results from 11 different sources and generated a trend of relative office worker productivity versus relative temperature (surrogate for comfort deviation). This trend, useful as a general representation of real-world

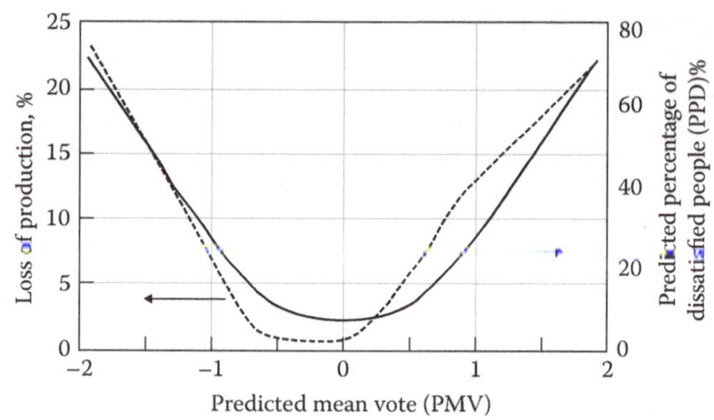

FIGURE 3.12
Illustration of how loss of indoor office occupant productivity closely tracks PPD. (Adapted from Roelofsen, P., The design of the workplace as a strategy for productivity enhancement, in *Proceedings CLIMA 2001*, Napoli, Italy, 2001.)

office work performance, is shown in Figure 3.13. Even though there is scatter, the trend line can be used as a preliminary guide for design, operation, and cost analysis. Note that the plot is fairly shallow for low values of absolute relative temperature difference, but drops off quite sharply after it exceeds about 8°F (4.5°C).

Several studies have been carried out to evaluate the performance of school children as a function of classroom temperature. For 10–12 year olds, the improvement in school exercises improved by 5.7% when temperature was maintained at 22.5°C (72°F) as compared to 26°C (79°F). The relative negative effect of some tasks was as great as 30% when indoor temperature was kept at 20°C (68°F) compared to 27°C–30°C (80°F–86°F). REHVA (2006) provides more in-depth analysis of worker productivity and student learning improvement gains from improved indoor climate.

An allied issue that has acquired great importance in terms of building operation and maintenance (O&M) is unsolicited complaints. When occupants complain, the O&M staff has to rectify the problem at some expense and time. Numerous studies have been undertaken to study this aspect and come up with remedies. The interested reader can refer to ASHRAE Fundamentals (2013) for an introduction to this issue. Figure 3.14 illustrates the trend between the number of thermal related complaints and the mean indoor temperature (again used as a surrogate for comfort). This is based on a study by Federspiel et al. (2003) based on six commercial buildings in three U.S. cities. The *arrival complaints* are those relating to when occupants arrive to work in the morning, while the *operating complaints* are during normal working hours. Occupants are more willing to tolerate large comfort deviations when they come in to work than during normal working hours. For a 100,000 ft^2 building, we note that when

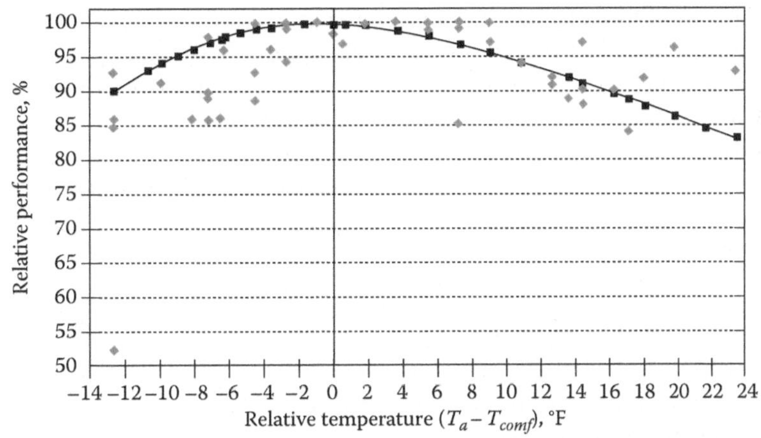

FIGURE 3.13
Relative performance of office worker performance versus deviation from optimal comfort temperature. (From ASHRAE, *Handbook of Fundamentals,* American Society of Heating, Refrigerating and Air-Conditioning Engineers, Atlanta, GA, 2013. Copyright ASHRAE, www.ashrae.org.)

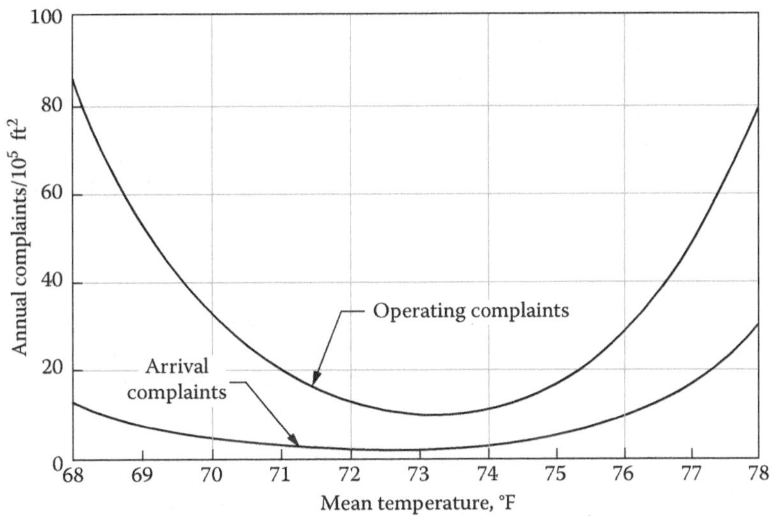

FIGURE 3.14
Variation of rate of unsolicited thermal complaints with mean indoor environment temperature. (From ASHRAE, *Handbook of Fundamentals,* American Society of Heating, Refrigerating and Air-Conditioning Engineers, Atlanta, GA, 2013. Copyright ASHRAE, www.ashrae.org.)

the mean temperature is 3°F (1.5°C) below the optimal of 73°F (22.8°C), one should expect about 30 operating complaints per year.

3.4 Air Quality and Indoor Contaminants

Air quality is essential, not just for comfort but also for health and productivity. Since the 1970s, concern over air quality has risen, in part because buildings have become tighter and ventilation rates have been reduced to conserve energy. Also, there has been recognition of the importance of indoor pollution, both natural (radon) and artificial (e.g., formaldehyde emanating from furnishings and carpets). Sources of indoor air contaminants are building material emissions, equipment emissions, dust, human, animal and insect by-products, toxins and allergens from microbial growth, airborne pathogens, chemicals, and outdoor contaminants drawn into the building.

Sick building has entered our vocabulary. Simple objective parameters such as air exchange rates or formaldehyde concentrations are relatively easy to measure but others are more intangible. The causes of the sick building syndrome are often difficult to pin down completely. They are identified by a spectrum of symptoms such as sensory irritation of eye, sinus congestion, throat dryness and cough, dizziness, skin irritation, fatigue, strange feeling of odor and taste, and other nonspecific hypersensitivity reactions.

Several organizations in the United States such as the Environmental Agency, National Institute for Occupational Safety and Health (NIOSH), and Occupational Safety and Health Administration (OSHA) have identified important outdoor contaminants and prescribed limits in order to avoid undue health-related ailments. Table 3.6 assembles the current standards for some outdoor contaminants.

ASHRAE 62.1 (2013) assembles tables comparing regulations and guidelines pertinent to indoor environments as suggested by several agencies. Table 3.7 is a subset of such tables and includes values of the most important air pollutants or contaminants as recommended by OSHA and NIOSH. Note that IAQ is defined in terms of upper limits for the concentrations of air pollutants. There are several criteria such as threshold limit value, permissible exposure limit, and recommended exposure limit. The interested reader can refer to books such as Heinsohn and Cimbala (2003) for additional information.

TABLE 3.6

Ambient Air Quality Standards for the United States

Contaminant	Level	Averaging Time	Details
Carbon monoxide	35 ppm	1 h	Not to be exceeded more than once a year
Nitrogen dioxide	100 ppb[a]	1 h	98th percentile, averaged over 3 years
Ozone	75 ppb	8 h	Annual 4th highest daily maximum 8 h concentration, averaged over 3 years
Sulfur dioxide	75 ppb	1 h	99th percentile of 1 h daily maximum concentration over 3 years
Particulate $PM_{2.5}$	35 μg/m³	24 h	98th percentile, averaged over 3 years
Particulate PM_{10}	150 μg/m³	24 h	Not to be exceeded more than once a year on average over 3 years
Lead in particle	0.15 μg/m³	3 months	Not to be exceeded

Source: Extracted from ASHRAE, *Standard 62.1-2013: Ventilation for Acceptable Indoor Air Quality*, American Society of Heating, Refrigerating and Air-Conditioning Engineers, Atlanta, GA, 2013. Copyright ASHRAE, www.ashrae.org.

[a] ppb, parts per billion.

TABLE 3.7

Regulations and Guidelines for Indoor Environments (Averaging Times Is 8 h Unless Otherwise Indicated in Square Brackets)

Contaminant	Enforceable and Regulatory (OSHA)	Nonenforceable Guidelines (NIOSH)
Carbon dioxide	5000 ppm	5,000 ppm, 30,000 ppm [15 min]
Carbon monoxide	50 ppm	35 ppm
Formaldehyde	0.75 ppm, 2 ppm [15 min]	0.016 ppm, 0.1 ppm [15 min]
Lead	0.05 mg/m³	0.050 mg/m³
Nitrogen dioxide	5 ppm	1 ppm [15 min]
Ozone	0.1 ppm	0.1 ppm
Particles <2.5 μm	5 mg/m³	—
Total particles	15 mg/m³	—
Sulfur dioxide	5 ppm	2 ppm, 5 ppm [15 min]

Source: Extracted from ASHRAE, *Standard 62.1-2013: Ventilation for Acceptable Indoor Air Quality*, American Society of Heating, Refrigerating and Air-Conditioning Engineers, Atlanta, GA, 2013. Copyright ASHRAE, www.ashrae.org.

The indoor and outdoor concentrations of a contaminant are equal if a building has no sources or sinks. One type of sink is, of course, a filter or cleaning device that removes a contaminant before it enters a building. For some contaminants, absorption by the surfaces in a building can also be a significant sink. A brief description of the most important contaminants is provided in the following:

1. *Carbon dioxide* (CO_2) is exhaled as a by-product of metabolism, making the concentration in buildings higher than that outdoors. It has no harmful effects on the human body, and the concentrations in buildings rarely reach levels high enough to have an appreciable impact on the availability of oxygen needed for breathing. But since elevated concentrations of CO_2 are correlated with human bioeffluents and odors; CO_2 is a convenient proxy indicator for the latter. Standard 62-2010 edition recommends that the indoor concentration of CO_2 be not more than 700 ppm above the outdoor concentration, the latter being in the range of 300–500 ppm.
2. *Carbon monoxide* (CO), the result of incomplete combustion, is toxic because it binds preferentially to hemoglobin, thereby blocking the uptake of oxygen by the blood. Significant quantities can be emitted by motor vehicles,

especially those without properly functioning catalytic converter, and by poorly regulated boilers and furnaces. Obviously, buildings should be designed in such a way that exhaust from such sources cannot enter; e.g., a loading dock must never be placed next to an air intake of the ventilation system.

3. *Sulfur dioxide* (SO_2) is a gas, produced by the combustion of fuels that contain sulfur, in particular, coal and oil. It is an irritant that can cause or aggravate various respiratory problems, both directly and after transformation to sulfates, especially sulfuric acid (the main constituent of acid rain), by chemical reactions in the atmosphere. There are no significant sources of SO_2 in buildings.
4. *Nitrous oxides* (NO_x) designate an unspecified mixture of NO and NO_2. Most high-temperature combustion processes produce some NO that is rapidly oxidized to NO_2 in the atmosphere. Both NO and NO_2 are quite reactive, with complicated chemical reactions that lead to the production of HNO_3 (another constituent of acid rain) and of O_3. Both NO_x and the secondary pollutants created by it are widely believed to cause or aggravate respiratory problems. There are few sources of NO_x in buildings, except open flames, for instance, in gas stoves.
5. *Ozone* (O_3) is a strong oxidant, and exposure to ozone is harmful to most living organisms (however, ozone in the stratosphere is beneficial because it protects us from harmful UV radiation). Ozone air pollution results from the combination of volatile organic compounds (VOCs) and NO_x in the presence of light, often called photochemical smog. Whereas this process is negligible inside buildings, there is some direct production of ozone by copiers, laser printers, and some deodorizers.
6. *VOCs*: Most VOCs have no direct health impacts, but they are undesirable as a precursor of O_3. Among VOCs with direct health impacts, there are formaldehyde and certain aromatic compounds (e.g., benzene) that are carcinogenic. Formaldehyde irritates eyes and mucous membranes and can cause respiratory problems, in addition to being a carcinogen. It is used in the manufacture of many materials for buildings, e.g., carpets, insulation, pressed board, and paper products.
7. *Particulate matter* (PM) often has a subscript indicating the maximum diameter, in micrometers, of the particles that have been counted. Suspended particles in the air, also known as dust, have been increasingly recognized as a health hazard. Dust is a complex mixture of particles of inorganic and organic origin, often contaminated by unhealthy bacteria. There are many different sources, including natural soil, sea salt, pollen, bacteria, and fragments of dead skin. The most harmful particles are emitted by combustion processes, including tobacco smoke. In addition, there are secondary particles, in particular sulfates and nitrates that are created by oxidation of SO_2 and NO_x in the atmosphere. Characteristics of PM are indicated in Figure 3.15, together with cleaning equipment suitable for removing it. Particles larger than 10 μm are of little concern for health because they get filtered by the upper respiratory tract. Air quality regulations are concerned with PM_{10} (inhalable particles) and more recently with $PM_{2.5}$ because the latter are small enough to enter the deepest recesses of the lungs (Figure 3.15).
8. *Radon* is a noble gas, i.e., chemically inert, but it can cause lung cancer because it is radioactive, with a half-life of about 4 days. It is continuously emitted by the decay of naturally occurring radioactive elements in certain geologic formations. If it enters through the foundations into a building, the concentration can reach levels that are considered undesirable. In regions with high levels of radon, it is advisable to reduce the penetration of radon into a building by, e.g., making the foundations sufficiently airtight, or by installing a mechanical air system such as a subslab depressurization fan that sucks the radon from below the slab and vents it outdoors.
9. *Tobacco smoke*: Smoking is a singularly effective means of delivering a host of dangerous contaminants (carbonaceous particles, CO, benzene, nicotine, etc.) into the lungs not only of smokers but also of nonsmokers who share the same space. A pack of cigarettes a day reduces life expectancy by about 8 years, to say nothing of the degraded quality of life due to the associated cancers and cardiopulmonary diseases (Doll et al., 1994). The health impacts of passive smoking are also significant. For the comfort of nonsmokers and for the health of everybody, there is a trend, at least in North America and the EU, to restrict smoking in buildings or confine it to limited areas.

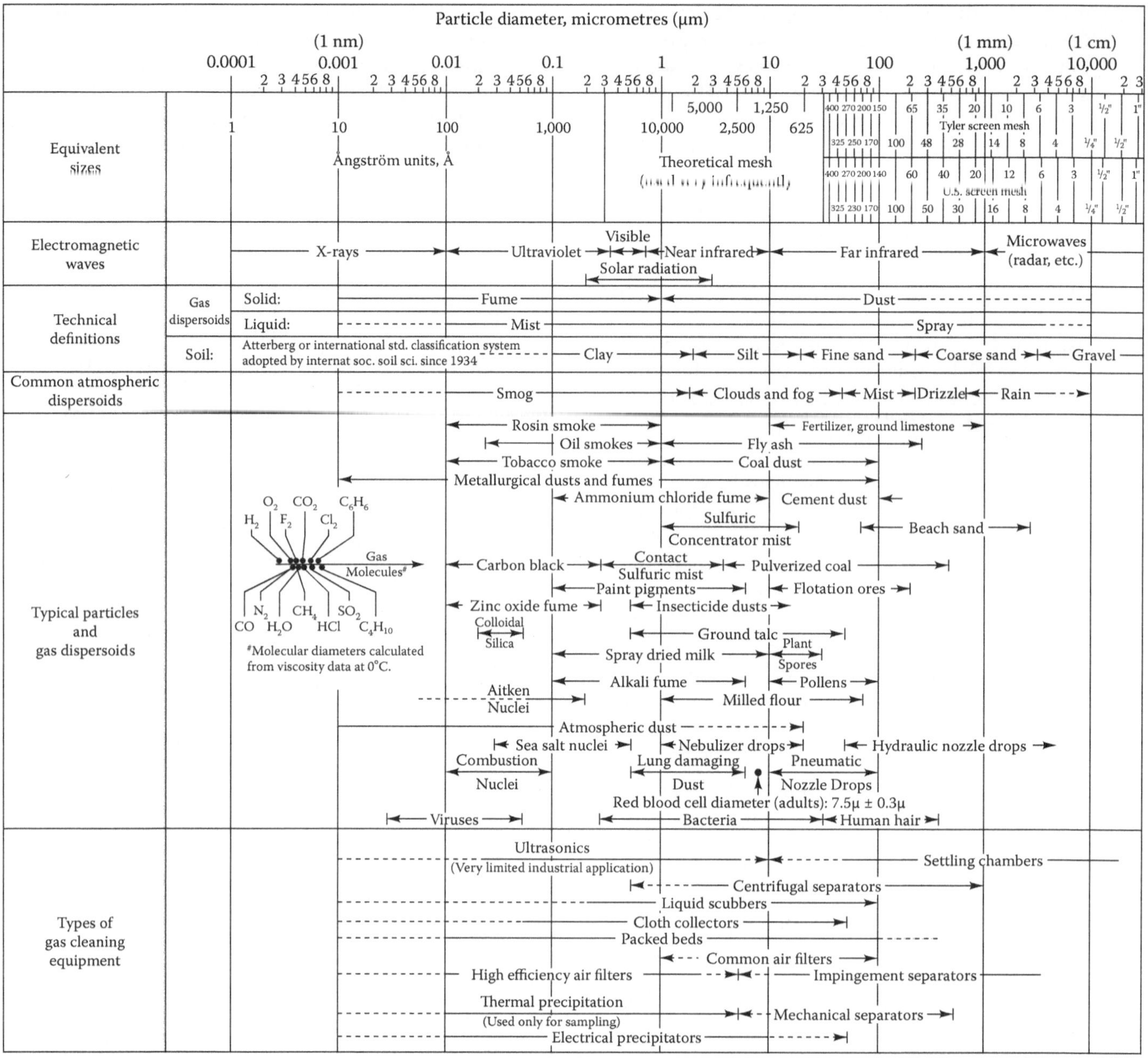

FIGURE 3.15
Characteristics of particulate matter and cleaning equipment suitable for removing it. (From, ASHRAE Standard—62.1, 2004. *Ventilation for Acceptable Indoor Air Quality*. American Society of Heating, Refrigerating and Air-Conditioning Engineers, Atlanta, GA, 2004. Copyright ASHRAE, www.ashrae.org.)

3.5 Control of Indoor Air Quality

3.5.1 General Methods

ASHRAE IAQ (2009) provides a general understanding of the importance of various IAQ issues and describes various practical design, construction, and operation cost-effective technologies suitable to assure and maintain enhanced IAQ in commercial and institutional buildings. Summary guidance is provided in the following discrete categories:

1. Managing the design and construction process that involves selecting proper HVC system types, commissioning, and effective O&M plans

2. Controlling moisture in building assemblies involving measures such as proper material selection and building pressurization
3. Limiting entry of outdoor contaminants by proper location of outdoor intakes and other measures
4. Controlling moisture and contaminants related to mechanical systems
5. Limiting contaminants from indoor sources by appropriate material selection
6. Controlling of contaminants at the source by using appropriate material or local exhaust fans
7. Using more advanced ventilation and filtration methods

Control of contaminants at the source is the preferred method (if feasible) because it avoids the energy consumption for conditioning extra outdoor air. There are many examples. One can choose building materials with minimal outgassing of contaminants. Air from smoking areas should be exhausted directly to the outside rather than being returned to the air distribution system. Local exhaust fans can remove local contaminants before they get mixed with the return air, for instance, from stoves or from rooms where toxic chemicals are stored such as paints, cleaners, or insecticides. However, care should be taken to make sure that the contaminants are actually removed by such exhaust fans: motion of air is not easily controlled by suction alone.

There are several approaches to providing fresh outdoor air in a building. The simplest is to do nothing and rely on natural infiltration through cracks and leaks in the building envelope. In parts of a building where this would not be sufficient, for instance, kitchens and bathrooms, one can enhance the air supply by adding air vents with dampers; for better control one can add exhaust fans. (Air vents are also used in basements and attics to prevent moisture problems.) That is still the customary design for the residential sector in the United States. Unfortunately, with natural infiltration, the air supply will, most of the time, be either too high (causing energy waste) or too low (causing air quality problems).

The best control over air supply can be provided by mechanical ventilation systems; one simply specifies the capacity of the fan(s) and ducts, and one obtains the required outdoor air flow. Mechanical ventilation is the predominant approach for commercial buildings in the United States.

For the design of ventilation systems, it is important to keep in mind the following general rules:

- Avoid air intake near loading docks, furnace vents, and other local sources of air pollution.
- Avoid short-circuiting between exhaust and intake.
- Place exhaust vents or exhaust fans near indoor sources of odor or pollution, e.g., bathrooms, kitchens, and open flames.

3.5.2 Ventilation of Fully Mixed Spaces

ASHRAE 62.1 (2013) deals with ventilation requirements for acceptable IAQ and is meant to ensure that indoor environments are healthy, i.e., they contain contaminants less than the stipulated concentrations. Ventilation with outdoor air is, of course, essential for supplying "fresh air," and usually a well-designed ventilation system, in conjunction with filtering, will be sufficient. But an understanding of a few basic issues is advisable to forestall problems that would be expensive to correct by a retrofit.

The requirement for fresh air in buildings is set by the need to dilute indoor contaminants so that their concentration never exceeds established threshold levels. Building occupants and materials are the main source of pollutants that require control. (The amount of oxygen consumed by human metabolism is very small in typical occupancy densities in buildings and is not the main determinant for fresh air supply.) Contaminants in buildings include radon, formaldehyde, various other organic compounds, particulates, carbon dioxide, tobacco smoke, odors, and nitrogen oxides. Of course, each of these evolves at different rates in different buildings. Since the control of each contaminant in a building is impossible from a practical point of view, carbon dioxide concentration is often taken as an approximate surrogate for all other contaminants at least those related to human activity.

The situation is simple if one can assume the indoor air to be uniformly mixed at all times. In most buildings, there is a fair amount of natural and/or forced convection that mixes the air, and the well-mixed model is often an acceptable approximation. It is also intuitive and simple to analyze. Consider what happens when there is a source of air pollution in a well-mixed building that is ventilated with outdoor air at a rate $\dot{V}_o$. The source can be characterized in terms of the rate N_{pol} at which it adds contaminants to the space. If the outdoor concentration C_o of the contaminant were zero, the indoor concentration C_i would be the ratio $(N_{pol}/\dot{V}_o)$. If C_o is not zero, the term $(N_{pol}/\dot{V}_o)$ represents the increase $(C_i - C_o)$ due to the source. Therefore, the well-mixed model can be written as

$$C_i = C_o + \frac{N_{pol}}{\dot{V}_o} \tag{3.23}$$

The concentrations have units of mass per volume, e.g., $\mu g/m^3$, if N_{pol} is in units of g/s; and the concentrations are dimensionless, e.g., ppm, if N_{pol} is in units of volume per second. Conversely, if source strength N_{pol} and outdoor concentration C_o are given, one can solve this equation for the ventilation rate $\dot{V}_o$ that is necessary to keep the indoor concentration from exceeding a prespecified level C_i.

"Fresh" air can be either outdoor air or treated indoor air with minimal contaminant levels. In some cities, outdoor air is quite contaminated and unsuitable for use in buildings. In such cases, other techniques such as particulate control by mechanical filtration, electronic filtration, air washing to remove gaseous contaminants, or use of activated charcoal are needed to provide fresh air.

Example 3.7: Volumetric Flow Rate for Dilution

The CO_2 production at an activity level of 1.2 met is $N_{CO_2} = 0.005$ L/s. What outdoor airflow $\dot{V}_o$ per occupant is needed at this activity level if the CO_2 concentration is not to exceed $C_i = 1050$ ppm? The outdoor concentration is $C_o = 350$ ppm.

Given: $N_{CO_2} = 0.005\ \mathrm{L/s}$

$C_i = 1050$ ppm
$C_o = 350$ ppm
$C_i - C_o = 700$ ppm

Find: $\dot{V}_o$

Assumptions: Perfect mixing

Solution

Insert N_{CO_2} for N_{pol} in Equation 3.23 and solve for $\dot{V}_o$. Inserting numbers, we find

$$\dot{V}_o = \frac{0.005\ \mathrm{L/s}}{0.00105 - 0.00035} = 7.1\ \mathrm{L/s}\ (15.1\ \mathrm{ft^3/min})$$

Typical ventilation systems in commercial buildings recirculate air at a high rate, the recirculation rate being several times higher than the rate at which fresh air is taken in. Also, the air coming out of the air diffusers has sufficiently high velocities to cause much local mixing. As a result, the air in buildings with mechanical ventilation systems tends to be fairly well mixed. Nonetheless, there may be zones with relatively stagnant air, for instance, between partitions in open-plan offices. In such zones, the local airflow rate is, in effect, lower than the building average, and the resulting local concentration of contaminants is higher. Insufficient mixing can cause local air quality problems.

Before proceeding to guidelines for ventilation, we mention one other flow situation that is easy to analyze. It is called "plug flow" because the fluid is assumed to move like a plug through the space.

Example 3.8: Air Velocity in a Room Assuming Plug Flow

What is the average air velocity if ventilation air is made to flow like a plug from floor to ceiling? Assume that the height of the room is 3.0 m (≈10 ft) and the ventilation rate (total of outdoor air and recirculated air) can be expressed as $\dot{V}/V = 3.0$ air changes per hour.

Given: $h = 3.0$ m and $\dot{V}/V = 3.0\ \mathrm{h^{-1}}$

Find: Velocity v of air

Assumption: Plug flow, from bottom to top

Solution

Consider a room of floor area A. The flow rate $\dot{V}$ is related to velocity v and area A by

$$\dot{V} = \mathrm{v} \times A$$

The volume is

$$V = A \times h$$

Combining the last two equations and solving for v, we find

$$\mathrm{v} = \frac{h\dot{V}}{V} = 3.0\ \mathrm{m} \times 3.0\ \mathrm{h^{-1}} = 9.0\ \mathrm{m/h}$$
$$= 0.0025\ \mathrm{m/s}\ (0.0082\ \mathrm{ft/s})$$

in the direction from bottom to top.

Comments

1. Conservation of mass implies that this is the lowest possible airspeed that achieves the specified ventilation rate.
2. There are ventilation systems (called displacement ventilation discussed in Chapter 20) that achieve this kind of plug flow. They are used in some European buildings (especially in Sweden) and are increasingly used in the United States as well.
3. To achieve well-mixed flow, the velocity must be much larger than this minimum (of course, to maintain comfort it should not exceed the upper limits specified in Section 3.3.1). Data for air diffusers and the resulting air velocities in rooms will be presented in Chapter 20.

3.5.3 ASHRAE Standard 62.1

ASHRAE 62.1 (2013) is a ventilation standard for acceptable IAQ. It specifies minimum ventilation rates and IAQ that minimize the potential for adverse health effects of human occupants. The document is intended for regulatory purposes and is more than a guide to improve IAQ in existing buildings. This standard has been revised several times over the years, and so the latest revision should be consulted by designers.

The standard offers two paths toward compliance: the ventilation rate procedure and the IAQ procedure. The *ventilation rate procedure* is a prescriptive one and is simpler to adopt. It prescribes the minimal outdoor airflow rate that must be delivered to a space for a large number of building types and uses assuming typical contaminant sources. Thus, it is usually chosen for conventional designs. The IAQ path is performance based, which requires a higher level of design expertise. It involves verifying that the air in the building will actually meet the required air quality. It offers the designer much greater flexibility, because any method can be used for treating the air, and the flow of outdoor air can be tailored precisely to the actual requirements.

The prescriptive method is based on the following formula involving number of occupants as well as the area of the zone or space or building:

$$\dot{V}_{bz} = R_p \cdot P_z + R_a \cdot A_z \quad (3.24)$$

where

$\dot{V}_{bz}$ is the outdoor air ventilation rate in the breathing zone of the space (L/s or cfm)
R_p is the outdoor air flow rate per person (see Table 3.8) (L/[s · person] or cfm/person)
R_a is the outdoor air flow rate per unit area (see Table 3.8) (L/[s · m²] or cfm/ft²)
P_z is the number of people in the ventilation zone during typical use
A_z is the zone floor area (m² or ft²)

The relevant numerical values for use in Equation 3.24 are assembled in Table 3.8 for some of the most common applications. The last column in the table indicates the occupancy category by air class. The classes 1 through 4 are assigned by space and the type of HVAC system commonly used. For example, class 1 represents low contamination and low irritation, class 2 is moderate, class 3 is significant, and class 4 is hazardous. The last two categories are not even recommended in the ASHRAE standard except for general manufacturing that is listed as class 3. For sites where the outdoor air is not of sufficient quality, treatment of the outdoor air is also prescribed by ASHRAE 62.1 (2013).

TABLE 3.8
Outdoor Air Requirements for Ventilation for Some Typical Occupancy Categories

	Prescriptive Values				Default Values			
	People Outdoor Rate R_p		Area Outdoor Rate R_a		Occupant Density	Combined Outdoor Air Rate		
Occupancy Category	cfm/person	L/s · person	cfm/ft²	L/s · m²	#/1000 ft² or #/100 m²	cfm/person	L/s · person	Air Class
Daycare (<4 years)	10	5	0.18	0.9	25	17	8.6	2
Classroom (>9 years)	10	5	0.12	0.6	35	13	6.7	1
Lecture classroom	7.5	3.8	0.06	0.3	70	10	5.1	2
Office space	5	2.5	0.06	0.3	5	17	8.5	1
Conference/meeting	5	2.5	0.06	0.3	50	6	3.1	1
Corridors	—	—	0.06	0.3	—	—	—	1
Reception area	5	2.5	0.06	0.3	30	7	3.5	1
Hotel/motel bedroom	5	2.5	0.06	0.3	10	11	5.5	1
Restaurant dining rooms	7.5	3.8	0.18	0.9	70	10	5.1	2
Residential dwellings	5	2.5	0.06	0.3	—	—	—	1
Libraries	5	2.5	0.12	0.6	10	17	8.5	1
Retail sales	7.5	3.8	0.12	0.6	15	16	7.8	2
Supermarket	7.5	3.8	0.06	0.3	8	15	7.6	1
Health club/weight room	20	10	0.06	0.3	10	26	13.0	2
Gambling casinos	7.5	3.8	0.18	0.9	120	9	4.6	1

Source: Extracted from ASHRAE, *Standard 62.1-2013: Ventilation for Acceptable Indoor Air Quality*, American Society of Heating, Refrigerating and Air-Conditioning Engineers, Atlanta, GA, 2013. Copyright ASHRAE, www.ashrae.org.

This table prescribes supply rates of outdoor air required for acceptable indoor air quality. These values have been chosen to dilute human bioeffluents and other contaminants with an adequate margin of safety and to account for health variations among people and varied activity levels.

Example 3.9: Determining Outdoor Ventilation Requirements

What outdoor airflow $\dot{V}_{bz}$ is needed for an office building of 1500 m² floor area if the expected density of occupants is 1 per 15 m²?

Given: 1500 m² floor area and 15 m² per occupant

Find: V_{bz}

Assumptions: The outdoor air has acceptable quality (i.e., it meets the requirements for long-term exposure, Table 3.6).

Lookup values: From Table 3.8, $R_p = 2.5$ L/(s·person) and $R_a = 0.3$ L/(s·m²)

Solution

The expected number of occupants is

$$P_z = \frac{1500}{15} = 100$$

Thus, from Equation 3.24, multiplying this number by the number of occupants, we obtain

$$\dot{V}_{bz} = 2.5\ \mathrm{L/(s \cdot person)} \times 100\ \mathrm{persons} + 0.3\ \mathrm{L/(s \cdot m^2)}$$
$$\times 1500\ \mathrm{m^2} = 700\ \mathrm{L/s}\ (1484\ \mathrm{cfm})$$

Comment

If the occupant density is not specified, the default values in Table 3.8 should be used. In this case, the occupant density would be chosen as of 5 people/100 m² and the required ventilation rate as 8.5 L/person. This would have yielded 637 L/s, which is about 10% lower than the value of 700 L/s obtained in the aforementioned example. Hence, the occupant density is a value that should be selected with care by the designer.

The choice of the CO_2 level in Table 3.8 is, as stated earlier, based on the observation that CO_2 production is closely correlated with metabolic activity and associated body odors. The minimum outdoor air rates in Table 3.8 have been set to avoid stuffiness from body odors, rather than being determined by oxygen demand. In fact, at these outdoor airflow rates, the oxygen content changes only a fraction of a percent as a result of metabolism.

Bioeffluents are emitted by the human body by respiration, metabolic processes, and bacterial decomposition of the skin (Fanger, 2001). They comprise essentially of CO_2 and hundreds of VOCs. Fanger proposed two units to quantify air pollution sources and air pollution perceived by humans:

1. One "olf" is the emission rate of air pollutants from a standard person. It is akin to the metabolic rate for human activity. All other pollution sources are also expressed in terms of this equivalent source strength.
2. One "pol" is the air pollution caused by one standard person (emission rate of one olf) ventilated by 1 L/s of unpolluted air. It is, thus, a measure of the air quality in the space.

Fanger and coworkers developed correlations expressing PPD in terms of the ventilation rate per olf during steady-state conditions. The interested reader can refer to Fanger (2001) for further details.

As a final note, we mention that techniques have been developed from the 1990s where the amount of ventilation air is varied as a function of occupancy. This can result in large energy savings in buildings with variable occupancy (such as classrooms and auditoriums). Sensors that measure CO_2 concentrations in the return air can provide an indirect indication of the number of occupants, and the ventilation rates adjusted accordingly. This technique is called "CO_2-based demand ventilation," and several published papers deal with this subject. The literature review by Emmerich and Persily (1997) is a good starting point for the interested reader.

3.5.4 Air Distribution Efficiency

For the design of ventilation systems, one should pay attention to the fact that in most buildings perfect mixing, as assumed in Equation 3.23, is not achieved. Frequently, a fraction S of the supply air does not reach the spaces of a building where the occupants actually are situated (Figure 3.16). This occupied zone is usually taken to be the space between the floor and a height of approximately 1.80 m (72 in.) and more than 0.6 m (2 ft) from the walls or the air-conditioning equipment. This implies that some of the outdoor air that the ventilation system introduces into the conditioned space bypasses the occupied space and is exhausted to the outside without being used by the occupant; thus, this air cannot be counted as contributing toward meeting the ventilation requirement. To quantify this effect, let us consider the flow rates in a typical air distribution system where a fraction S of the supply air does not reach the occupied zones. This situation is shown in Figure 3.16 with air ducts for pulling in outside air that is then mixed with some amount of recycled air, and then supplied to the space. In overhead fully mixed systems, the amount of supply air has to be large enough that proper mixing and diffusion of the air occurs throughout the space. For this to occur, the amount of air is usually larger than that of the ventilation air. Hence, the common practice of recycling a certain fraction of return air. On the other hand, there is an energy penalty in having to condition outdoor air (as addressed in Chapter 19), and so pulling in just the

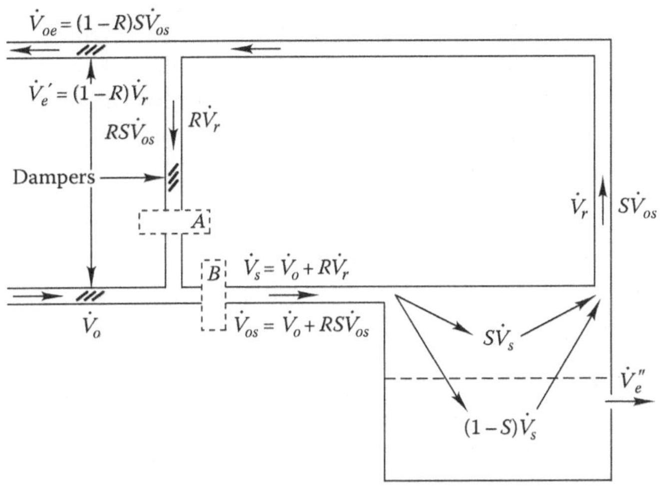

FIGURE 3.16
Flow rates in typical air distribution system (highly schematic). Points *A* and *B* designate possible filter locations.

amount of ventilation air as stipulated by the comfort standard is most conducive to energy efficiency.

The locations of two filters *A* and *B* are also indicated; but their presence will be neglected for the discussion that follows. Their influence is discussed in the next subsection.

The various quantities shown in Figure 3.16 are described as follows:

- $\dot{V}_o$ is the flow rate of outdoor air into building.
- $\dot{V}_s$ is the flow rate of supply air into building.
- $\dot{V}_e'$ is the flow rate at which air is exhausted from building via central exhaust duct.
- $\dot{V}_e''$ is the flow rate at which air is lost from building by exfiltration or local exhaust fans.
- $\dot{V}_r$ is the flow rate of return air from conditioned space.
- $R = 1 - \dot{V}_e'/\dot{V}_r$ = fraction of return air that is recirculated.
- S is the zone bypass fraction of air supply.
- $\dot{V}_{os}$ is the flow rate at which outdoor air is supplied to conditioned space.
- $\dot{V}_{oe}$ is the flow rate at which unutilized outdoor air is exhausted from building.

It is natural to define ventilation effectiveness E_v as the fraction of $\dot{V}_o$ that is really utilized

$$E_v = 1 - \dot{V}_{oe}/\dot{V}_o \tag{3.25}$$

If we assume constant air density, the mass balance for the supply duct can be stated in terms of volumetric flows as

$$\dot{V}_s = \dot{V}_o + R \times \dot{V}_r \tag{3.26}$$

as indicated in Figure 3.16. By analogy to Equation 3.26, $\dot{V}_{os}$ is given as

$$\dot{V}_{os} = \dot{V}_o + R \times S \times \dot{V}_{os} \tag{3.27}$$

since $(S \times \dot{V}_{os})$ is the flow of unutilized outdoor air in the return duct.

Further, $\dot{V}_{oe}$ is given by

$$\dot{V}_{oe} = (1-R) \times S \times \dot{V}_{os} \tag{3.28}$$

Combining Equations 3.25 through 3.28, one readily finds

$$E_v = \frac{1-S}{1-R \times S} \tag{3.29}$$

This is the expression for ventilation effectiveness in terms of the bypass factor *S* and the recycle fraction *R*.

Example 3.10: Ventilation Effectiveness

An office building has been designed to just meet the ASHRAE ventilation standard, but due to the placement of the diffusers, 40% of the supply air does not reach the occupied space. A fraction $R = 0.7$ of the return air is recirculated. If one does not want to change the diffusers, by what factor should one increase the outdoor airflow rate to meet the standard?

Given: $S = 0.4$, $R = 0.7$

Find: Ventilation effectiveness E_v

Solution

From Equation 3.29, the fraction of the outdoor air utilized is $E_v = (1 - 0.4)/(1 - 0.4 \times 0.7) = 0.83$. The outdoor airflow rate would have to be increased by a factor $1/0.83 = 1.2$ to strictly meet the standard.

The earlier discussion pertains to a single zone. In the case of multiple zones served by a single HVAC system, there is another correction to the minimum ventilation air amount that needs to be considered. Multiple spaces may have different ventilation requirements. Further, the thermal loads in these spaces would vary over the year and often by different amounts. Since the single duct supply systems carries the entire supply air amount (ventilation plus recycle), supply air to the individual zones may not contain the required ventilation air amount during different times of the year. In this case, one would have to increase the amount of outdoor air so as to meet the needs of the most critical zone, thereby resulting in a large energy penalty due to overventilation of the other zones.

Hence, ASHRAE 62.1 (2013) contains a calculation procedure where the designer can take "credit" for the unused outdoor air in the return, thereby reducing the degree of overventilation. We will see in Chapter 20 that alternative secondary systems using dedicated outdoor air systems have been developed where this issue is totally avoided.

3.5.5 Filtration Cleaning of Air

A variety of methods are available for cleaning the air in buildings. Most widely used, by far, are filters, and they can be quite effective against dust. The performance of air filters is primarily judged by the following criteria:

- Efficiency (measures ability to remove particles from an airstream)
- Pressure drop across the filter (affects fan power and energy consumption)
- Dust-holding capacity (determines how often the filter should be cleaned or replaced, to avoid excessive flow resistance)

There are three broad categories of air filters:

1. Fibrous media unit filter (to be replaced or cleaned when full)
2. Renewable media filter (where new media are continually introduced into the airstream and old media are removed)
3. Electronic air cleaners (which apply a high electric potential to sweep dust out of the airstream)

Whereas the performance of fibrous media unit filters changes as they fill up with dust, the other two filter categories perform with essentially constant pressure drop and efficiency. To avoid the overloading of unit filters, they should be inspected periodically; or even better, a pressure gauge should be installed to indicate when they are fully loaded.

Roughly speaking, the efficiency of a filter is the percentage of the particles that it removes from an airstream. However, determining the efficiency of a filter is not a simple matter because the efficiency can vary strongly with type and size of particles. This is illustrated in Figure 3.17 where the percentage of particles removed is plotted versus particle size. One sees that smaller particles are more difficult to filter out, which incidentally explains why $PM_{2.5}$ particles penetrate more deeply into the lungs than larger particles.

Several different test methods have been developed, each one appropriate for particular applications. The *weight arrestance* test measures the weight fraction of a standardized dust that the filter removes from the airstream; e.g., if this test yields an efficiency figure of 80%, the filter removes 80% by weight of this type of dust in the airstream. The *dust spot efficiency test* compares the discoloration effect of an airstream with and without the filter. Yet another test is the *DOP* (di-octyl-phthalate) *penetration test* that uses smoke produced by condensation of DOP vapor and measures the smoke concentration upstream and downstream; this latter test is used especially for high-performance filters. Even though the efficiency figures determined by these tests can be completely different (see Table 3.9), they are valuable for ranking and selecting filters for a given application.

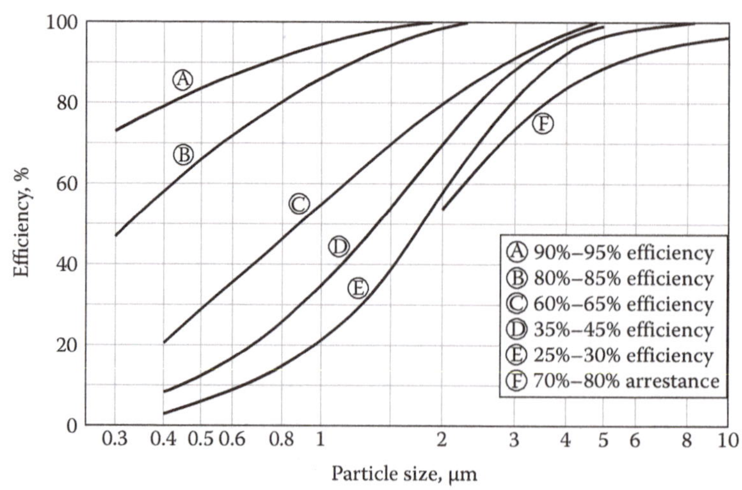

FIGURE 3.17
Efficiency of removing particles versus particle size for several filter types. These curves are approximations based on manufacturer's data. They do not correspond to results of tests recognized by ASHRAE and should not be used for HVAC design. (From ASHRAE, *Handbook of Fundamentals*, American Society of Heating, Refrigerating and Air-Conditioning Engineers, Atlanta, GA, 2013. Copyright ASHRAE, www.ashrae.org.)

TABLE 3.9

Some Characteristics of Filters

Type	Weight Arrestance, %	Dust Spot Efficiency, %	Pressure Drop, Clean, inWG	Pressure Drop, Max., inWG	Comments
Viscous impingement	50–80	<20		0.5	Low cost, good for lint, poor for dust
Dry-type extended surface					
Textile nonwoven media	Up to 80	Up to 60	0.05–0.25	0.5–0.7	Average offices, laboratories, etc.
Fine glass fibers, electret fibers, etc.	Up to 90				Above average offices, laboratories, etc.
HEPA & ULPA	>98	>80	0.5–2.0		Clean rooms, hospitals, etc.

There are a wide variety of filters, and these values are very approximate and should not be used for design; rather, the manufacturer's data should be consulted.

There are a wide variety of filter types. A brief overview is presented in Table 3.9 from ASHRAE Equipment (1988). One type of filter, called viscous impingement filter, is usually a flat panel made of coarse fibers (e.g., glass fibers or metallic wool) with high porosity, coated with a viscous substance to act as adhesive for dust particles. This type of filter has low pressure drop, low cost, high efficiency for lint but low efficiency for dust. Another type is the dry media filter, where fibers are much more closely spaced to form a dense mat, without adhesive on the fibers. Their efficiency can be significantly higher than that of viscous impingement filters.

Where especially clean air is demanded, one uses high-efficiency particulate air (HEPA) or ultralow-penetration air (ULPA) filters, made of submicrometer-size glass fiber paper in an extended surface configuration of deep space folds. They are used in applications such as hospitals and clean rooms. It may be advantageous to use combinations of different filter types, e.g., an inexpensive low-efficiency filter upstream of a more costly high-efficiency filter, thereby extending the life of the latter.

Other methods of air cleaning are used when gases and vapors are to be removed (for which filters are not very effective): (1) washing with water spray (which controls humidity and cleans air at the same time), (2) adsorption (e.g., by charcoal or zeolite), and (3) chemical cleaning (apparatus similar to air washer, but with chemicals instead of plain water).

The filter efficiency required in a given application depends on the quality of the outdoor air, on the rate at which contaminants are produced in the space, on the ratio of the air that is recirculated, and of course on the IAQ that one would like to achieve. The pertinent design equations are generalizations of Equation 3.29, and they can be found in Appendix D of the ventilation standard ASHRAE 62.1 (2013). Here, we show the equations for variable air volume (VAV) systems with constant outdoor airflow (such systems are discussed at length in Chapter 19); they also apply to constant-volume systems (CAV) if the VAV flow reduction factor F_r is set to unity. The equation for filter location *A* in Figure 3.16 is

$$\dot{V}_o = \frac{N_{pol} - E_v \times F_r \times R \times E_f \times C_{bz} \times \dot{V}_r}{E_v(C_{bz} - C_o)} \tag{3.30}$$

and the one for filter location *B* is

$$\dot{V}_o = \frac{N_{pol} - E_v \times F_r \times R \times E_f \times C_{bz} \times \dot{V}_r}{E_v[C_{bz} - (1 - E_f)C_o]} \tag{3.31}$$

where

$\dot{V}_o$ is the flow rate of outdoor air into building
C_{bz} is the concentration of contaminant in breathing zone
C_o is the concentration of contaminant in outdoor air
N_{pol} is the rate at which contaminant is added to space
$\dot{V}_r$ is the flow rate of return air from conditioned space
$R = 1 - \dot{V}_e' / \dot{V}_r$ = fraction of return air that is recirculated
E_v is the ventilation effectiveness
E_f is the filter efficiency
F_r is the flow reduction factor of VAV system (for CAV systems, $F_r = 1$)

These equations apply to single zone systems under steady-state conditions. When the zone outdoor airflow rate $\dot{V}_o$ is specified, these equations allow the breathing zone contaminant concentration C_{bz} to be determined, as illustrated in the following. Methods that account for multiple zones and transient effects are treated in Dolls and Walton (2002).

Example 3.11: Effect of Filters in Air Distribution Ducting

An office building with constant-volume system has a filter in location A with filter efficiency of 70% for environmental tobacco smoke

(ETS) particles. There are 10 occupants, 3 of whom are smokers. Data of NRC (1986) indicate that smoking one cigarette per hour produces between 5 and 10 μg/s of $PM_{2.5}$. Taking the average, one can assume a production rate of ETS of (3 × 7.5) = 22.5 μg/s. The outdoor airflow rate is 20 cfm per person, the supply airflow rate is 50 cfm per person, and the return airflow rate is 40 cfm per person. The ventilation effectiveness is 0.83. Assume that there is no ETS outdoors. What is the concentration of ETS in the breathing zone and how does it compare with the standard of 5 mg/m³ for an 8 h exposure given in Table 3.7?

Given: $\dot{V}_o = 200$ cfm
$\dot{V}_s = 500$ cfm
$\dot{V}_r = 400$ cfm
$E_f = 0.7$
$E_v = 0.83$
N_{pol} = 22.5 μg/s = 1350 μg/min of $PM_{2.5}$ from 3 smokers each smoking 1 cigarette per hour
$C_o = 0$

Find: C_{bz}

Solution

First, one needs the recirculation factor; by solving Equation 3.26,

$$R = \frac{\dot{V}_s - \dot{V}_o}{\dot{V}_r} = \frac{500 - 200}{400} = 0.75$$

Now apply Equation 3.30 with $F_r = 1$ (because it is a constant-volume system), and solve it for C_{bz}, with the result

$$\begin{aligned} C_{bz} &= \frac{N_{pol} + E_v \times C_o \times V_o}{E_v(R \times E_f \times V_r + V_o)} \\ &= \frac{1350\ \mu\text{g/min}}{0.83(0.75 \times 0.7 \times 400\ \text{cfm} + 200\ \text{cfm})} \\ &= 3.97\ \mu\text{g/ft}^3 = 140\ \mu\text{g/m}^3 \text{ of } PM_{2.5} \end{aligned} \tag{3.32}$$

Comment

The details of the formula for the concentration depend on characteristics of the air distribution system. But the key feature is the same for all, namely, that the concentration decreases in inverse proportion with airflow—as expected intuitively.

One of the considerations in designing a filter system is the pressure drop. Obviously, high-pressure drops should be avoided in the interest of energy efficiency. Data for the pressure drop are furnished by the manufacturer, but only at design flow rates. In practice, the flow rates will usually be different, and it is necessary to extrapolate to other flow rates. For this purpose we anticipate a result from Chapter 16, namely, that the pressure drop across an air filter is proportional to the square of the flow rate $\dot{V}$

$$\Delta p = \text{Constant} \times \dot{V}^2 \tag{3.33}$$

This is illustrated in the filter systems design example as follows.

Example 3.12: Pressure Drop in Air Filters

As will be explained in Chapter 19, typical air distribution systems recirculate much of the return air so that the total supply airflow rate is about 3–10 times larger than the outdoor rate. Assume that the outdoor air flow rate for an office building is $\dot{V}_o = 1.0\ \text{m}^3/\text{s}$. Suppose that the total supply flow rate is $\dot{V}_s = 5.0\ \text{m}^3/\text{s}$. One of the filters under consideration comes in units of area 0.6 m × 0.6 m (24 in. × 24 in.) and 0.2 m (8 in.) depth, and the manufacturer's data specify a design flow rate of 0.81 m³/s (1725 cfm) with a pressure drop of 87 Pa (0.35 inWG).

Given: $\dot{V}_s = 5.0\ \text{m}^3/\text{s}$

A_{filter} = 0.6 m × 0.6 m per unit = 0.36 m²

$\dot{V}_{design} = 0.81\ \text{m}^3/\text{s}$ with $\Delta p_{design} = 87$ Pa

Find: Number of units and pressure drop in this application

Solution

The number of filters that needs to be placed in parallel inside the duct is found by dividing the total flow by the design flow per filter unit:

$$\frac{5.0\ \text{m}^3/\text{s}}{0.81\ \text{m}^3/\text{s per unit}} = 6.17 \text{ units}$$

Since we can only install an integral number of units, we must choose either 6 or 7. With 6 units the flow per unit is (5.0 m³/s/6) = 0.833 m³/s. To find the corresponding pressure drop, we apply Equation 3.33, with the result

$$\begin{aligned} \Delta p &= \Delta p_{design}\left(\frac{\dot{V}}{\dot{V}_{design}}\right)^2 = 87\ \text{Pa} \times \left(\frac{0.833\ \text{m}^3/\text{s}}{0.81\ \text{m}^3/\text{s}}\right)^2 \\ &= 92.1\ \text{Pa (0.37 inWG)} \end{aligned}$$

With 7 units in parallel, the corresponding result is 67.7 Pa (0.27 inWG). The pressure drop is significantly lower than before but the duct cross- section has to be increased resulting in higher capital cost.

3.6 Closure

In this chapter, we have focused on the primary concerns of the HVAC engineer that are thermal comfort and air quality, and we have presented the appropriate design criteria. This is an important subject: the trend of the past suggests that people will demand increasingly higher standards of comfort and health. The comfort and health of occupants depend on the amount of clothing as well as the thermal and chemical contents of the building air. Filtration, thermal and humidity control subsystems, and adequate ventilation are essential. The designer has, as one of his or her most important assignments, the assurance of proper comfort and health conditions in occupied spaces in buildings.

ASHRAE 55 (2013) defines comfort limits or ranges for steady-state conditions for a group of people and does not apply for individuals. It also allows deducing a measure of the number of people dissatisfied with their environment when exposed to out-of-range conditions. It is applicable to any building type (commercial and residential) and for new and existing buildings. Providing occupants the ability to control the indoor environment (such as adjustable and personalized conditioned air supply at work table, being able to open/close blinds, and task lighting) was found to enhance human satisfaction with the indoor environment.

The other important standard pertains to ASHRAE 62.1 (2013) dealing with ventilation requirements. This standard has undergone numerous changes over the year reflective of the constantly varying perception and realization of the importance of "fresh" air on occupant health and productivity. Figure 3.18 illustrates the results of one study investigating the relationship between ventilation rate and illness and sick leave. There are several studies that argue the need for ventilation rates higher than those advocated by the current standards; they argue that the resulting benefit to human health outweighs by far the resulting cost of extra energy needed to condition the ventilation air. A good source of further reading is REHVA (2011), which deals with various aspects of both human comfort and IAQ, and contains theoretical aspects, measurements sensors and devices, methodology of how to conduct practical assessments in buildings, as well as a description of numerous case studies.

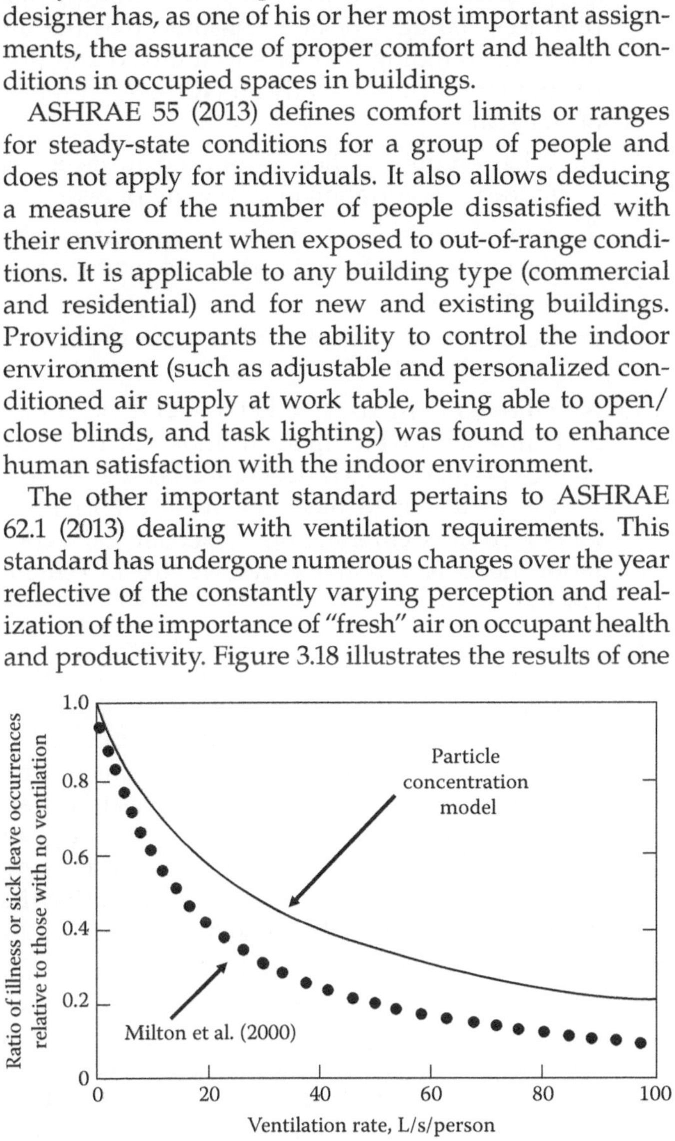

FIGURE 3.18
Predicted trends in illness or sick leave as a function of ventilation rate. (Adapted from REHVA, in P. Wargocki and O. Seppanen, eds., *REHVA Guidebook 6: Indoor Climate and Productivity in Offices. How to Integrate Productivity in Life Cycle Cost Analysis of Building Services*, Federation of European Heating and Air-Conditioning Association, REHVA, Brussels, Belgium, 2006.)

Problems

Numbers 1–4 given in parenthesis denote the degree of difficulty.

3.1 What effects are taken in account by (a) the MRT, (b) the operative temperature, and (c) the adiabatic equivalent temperature? *(1)*

3.2 Consider the ASHRAE comfort chart (Figure 3.5). (a) Why are the temperature limits of the chart inclined rather than parallel to the y axis? (b) Comment on the recommended lower humidity level shown. *(1)*

3.3 Determine whether the following conditions are expected to be comfortable for light office work. Use the ASHRAE Comfort chart.

(a) Summer: air temperature = radiant temperature = 23°C (73.4°F), relative humidity = 60%

(b) Summer: air temperature = 22°C (71.6°F) and radiant temperature = 28°C (82.4°F) (assume that the convective and radiative heat transfer coefficients are equal), relative humidity = 30%

(c) Winter: air temperature = radiant temperature = 20°C (68°F), relative humidity = 30%. *(2)*

3.4 Consider a room with the following conditions with an occupant at the center of room:

- The room dimensions are 2.5 m × 4 m × 5 m (8.2 ft × 13.12 ft × 16.41 ft).
- One side has dimensions 2.5 m × 4 m (8.2 ft × 13.12 ft) and is entirely glazed with interior surface temperature 10°C (50°F).
- The other surfaces are at 20°C (68°F).
- The air is at 22°C (71.6°F) dry-bulb temperature and 30% relative humidity.

(a) What is the MRT?

(b) What is the operative temperature?

(c) Are these conditions within the comfort limits of ASHRAE? *(3)*

3.5 Find the highest permissible T_{op} and W for the summer zone of the ASHRAE comfort chart if the relative humidity is 50%. Assume $T_{op} = T_a$. *(1)*

3.6 Estimate the contribution of thermal storage in the body relative to the steady-state heat loss if the temperature of the body changes by 0.5 K (1°F) over 10 h (typical day–night swing). Assume constant steady-state heat loss corresponding to an activity level of 1 met and a surface area of 1.8 m^2 (19.37 ft^2). Treat the body as isothermal with a heat capacity of 250 kJ/K (130 Btu/°F). *(2)*

3.7 To estimate the effect of hot or cold food on the thermal balance of the body, consider the cooling power of drinking 0.1 L/h (0.0264 gal/h) of cold drinks at 5°C (41°F) (assuming no change in evaporative transfer from the body). Is it significant? *(2)*

3.8 Consider Examples 3.1 and 3.2. You will determine the total sensible heat loss for a sedentary person (met level = 0.9) wearing the same types of clothes. Compare results. *(2)*

3.9 (a) Calculate the 1-h and 3-h PMV and PPD values for the indoor conditions specified by a dry-bulb temperature of 21°C and a dew point temperature of 12°C. Assume clo = 0.9, Met = 0.9 and mixed occupancy. The MRT can be taken to be equal to the room air dry-bulb temperature.

(b) What would be the operative temperature for comfort (use Equation 3.21)? *(2)*

3.10 You are asked to investigate a complaint that an office space being maintained at 76°F and 40% RH in winter is uncomfortably warm. You measure the globe temperature to be 80°F and estimate the mean air velocity to be 40 ft/min. What is your recommendation to correct the situation citing numerical values? *(2)*

3.11 A person feels very comfortable in his house in light clothing when the thermostat is set at 22°C and the MRT = 22°C. During a cold day, the MRT drops to 18°C.

(a) To what value must the indoor dry-bulb air temperature be raised to maintain the same level of comfort?

(b) If the person sits near a window and receives solar radiation, the thermostat can be lowered. What types of factors will influence the amount by which the thermostat can be lowered? *(2)*

3.12 Consider a person standing in a room maintained at 20°C at all times. The inner surfaces of the walls, floors, and ceiling of the house are observed to be at an average temperature of 12°C in winter and 23°C in summer. The exposed surface area, emissivity, and the average outer surface temperature of the person are 1.6 m^2, 0.95, and 32°C, respectively.

Determine the rates of radiation heat transfer between this person and the surrounding surfaces in summer and in winter. *(2)*

3.13 An air handler in a library of area 500 m^2 is to supply a zone that is meant for a maximum of 50 occupants.

(a) Find the required minimum outdoor airflow rate for this zone when fully occupied.

(b) Find the minimum outdoor airflow when totally unoccupied. *(2)*

3.14 Suppose there is a source of NO_x in a building that produces 100 μg/s of NO_x. Assume that the air inside the building is always well mixed and that the outdoor air has already an NO_x concentration of 50 μg/m^3.

(a) What outdoor airflow is needed to meet the indoor concentration of 100 μg/m^3 or less?

(b) If the zone bypass factor is 0.2, what should be the outdoor air flow if there is no recirculation of return air?

(c) Recalculate (b) if 80% of the air is recirculated. *(2)*

3.15 An office building 100 ft × 100 ft × 13 ft (30.48 m × 30.48 m × 3.95 m) has an infiltration rate of 0.6 air change per hour (ACH) during the heating season. Is this sufficient to meet the fresh air requirements of 75 occupants, or is mechanical ventilation necessary if the windows are to remain closed? *(2)*

3.16 A small office building has a volume of about 50,000 ft^3. If the forced ventilation rate if 0.5 ACH and there are 30 occupants who give off 0.01 ft^3 of CO_2 per minute, how does the CO_2 concentration vary throughout the day? Draw a graph from 8 a.m. to 5 p.m. showing the concentration, assuming an ambient concentration of 350 ppm of CO_2 and an initial building concentration of the same amount. Ignore any infiltration effects, and assume that the building air is thoroughly mixed. *(3)*

3.17 A party is held in a house during the winter with all the doors and windows closed. When the guests all leave at midnight, the concentration of CO_2 in the house is 1400 ppm. If the house has a volume of 25,000 ft^3 and an infiltration rate of 0.3 ACH, how long before the CO_2 concentration in the house goes below 500 ppm (assume an ambient concentration of 350 ppm and that no one remains in the house)? You can solve this problem by formulating a first-order differential equation and solve it in closed form, or by discretizing it into hourly time steps and using a spreadsheet. *(3)*

3.18 In Problem 3.17, how long would it take if the doors and windows were opened, increasing the effective ventilation rate to 1 ACH? *(3)*

3.19 The *zone ventilation effectiveness* η_{vent} of a conditioned space is used to quantify the "short-circuiting" of air from the supply diffusers to the return grill. It is defined such that the true amount of supply air delivered to the occupied zone is $(Q_{supply} \cdot \eta_{vent})$. If a clean room requires two air changes per hour and has a zone ventilation effectiveness of 0.8, then what is the required Q_{supply}? *(3)*

3.20 A smoker produces approximately 7.5 μg/s of ETS. What is the minimum required ventilation rate if the goal is to keep the ETS concentration below 50 μg/m³? Assume steady-state conditions and that the air is well mixed in the zone. Note that the production rate is given here in mass per time and the concentration in mass per volume, by contrast to Equation 3.23 and Example 3.7 where the production rate is in volume per time (as appropriate for a gaseous pollutant) and the concentration in ppm. *(2)*

3.21 A smoker producing ETS at a rate of 7.5 μg/s lives in a house of volume V = 500 m³ with an average outdoor airflow rate corresponding to an air exchange rate $\dot{V}_o = 1.0$ ACH. What is the resulting steady-state concentration of ETS if the air is well mixed in the house? Compare with the long-term ETS standard of 50 μg/m³ or less. *(2)*

3.22 You are designing the ventilation system (see Figure P3.22) for a smoking lounge for 10 persons. The outdoor air intake rate $\dot{V}_o$ is 30 L/s/person. Smokers introduce ETS at an average rate of 7.5 μg/s. What is the resulting steady-state indoor air concentration of ETS in μg/m³ if there is no filter and if all 10 occupants are smokers? Neglect any air exchange through the building envelope, and assume perfect mixing in the building. Compare with the long-term ETS standard of 50 μg/m³ or less. *(2)*

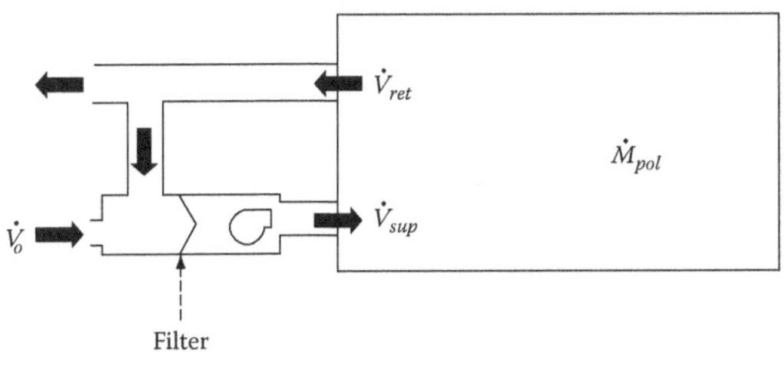

FIGURE P3.22
Typical ventilation system.

3.23 Using the parameters from Problem 3.22, what is the resulting ETS concentration if the filter has a weight arrestance efficiency of 80%? Assume a supply airflow rate $\dot{V}_{sup}$ of four times the outdoor air intake rate. *(4)*

3.24 Consider Example 3.11. Compute the ETS concentration in the breathing zone if the return air flow rate is equal to the supply air flow rate of 50 cfm/person. *(2)*

3.25 Consider Example 3.11 where the filter was located in location *A* of Figure 3.16. Evaluate the ETS concentration in the breathing zone if the filter is moved to location *B*. Which of the two locations results in lower concentrations in the breathing zone? *(2)*

3.26 Do further research and write a 200-word essay on the applicability, limitations, and possible misuse of the adaptive comfort model. *(2)*

References

ASHRAE 55 (2004, 2010, 2013). *Standard 55-2013: Thermal Environmental Conditions for Human Occupancy*. American Society of Heating, Refrigerating and Air-Conditioning Engineers, Atlanta, GA.

ASHRAE 62.1 (2004, 2013). *Standard 62.1-2013: Ventilation for Acceptable Indoor Air Quality*. American Society of Heating, Refrigerating and Air-Conditioning Engineers, Atlanta, GA.

ASHRAE Equipment (1988). *Handbook of Equipment*. American Society of Heating, Refrigerating and Air-Conditioning Engineers, Atlanta, GA.

ASHRAE Fundamentals (2013). *Handbook of Fundamentals*. American Society of Heating, Refrigerating and Air-Conditioning Engineers, Atlanta, GA.

ASHRAE IAQ (2009). *Indoor Air Quality Guide*. American Society of Heating, Refrigerating and Air-Conditioning Engineers, Atlanta, GA.

de Dear, R. and G.S. Brager (1998). Developing an adaptive model of thermal comfort and preference. *ASHRAE Trans.*, 104, 1–18.

Doll, R., R. Peto, K. Wheatley, R. Gray, and I. Sutherland (1994). Mortality in relation to smoking: 40 years' observations on male British Doctors. *Br. Med. J.*, 309, 901–911.

Dolls, W.S. and G.N. Walton (2002). CONTAMW 2.0 User Manual. National Institute of Standards and Technology, Gaithersburg, MD, NSTRI 6921.

Emmerich, S.J. and A.K. Persily (1997). Literature review on CO_2-based demand-controlled ventilation. *ASHRAE Trans.*, 103(2), 229–243.

Fanger, P.O. (1970). *Thermal Comfort: Analysis and Applications in Environmental Engineering*. McGraw-Hill Book Company, New York.

Fanger, P.O. (2001). Chapter 22: Perceived air quality and ventilation requirements, in *Indoor Air Quality Handbook* (J.D. Spengler, J.M. Samet, and J.F. McCarthy, eds.). McGraw-Hill, New York.

Fisk, W.J. and A.H. Rosenfeld (1997). Estimates of improved productivity and health from better indoor environments. *Indoor Air*, 7, 158–172.

Federspiel, C.C., R. Martin, and H. Yan (2003). Thermal comfort models and "Call-out" complaints frequencies, Final Report of ASHRAE Research Project, RP 1129. American Society of Heating, Refrigerating and Air-Conditioning Engineers, Atlanta, GA.

Frontczak, M. and P. Wargocki (2011). Literature survey on how different factors influence human comfort in indoor environments. *Build. Environ.*, 44, 422–937.

Guan, Y., M.H. Hosni, B.W. Jones, and T.P. Gielda (June 2003). Literature review of the advances in thermal comfort modeling. *ASHRAE Trans.*, 109(2), 908–916.

Heinsohn, R.J. and J.M. Cimbala (2003). *Indoor Air Quality Engineering*. Marcel Dekker, New York.

ISO 7730 (1993). *Moderate Thermal Environments—Determination of PMV and PPD Indices and Specification of the Conditions for Thermal Comfort*. International Organization for Standardization, Geneva, Switzerland.

Milton, D., P. Glencross, and M. Walters (2000). Risk of sick-leave associated with outdoor air supply rate, humidification and occupant complaints. *Indoor Air*, 10, 212–221.

NRC (1986). *Environmental Tobacco Smoke: Measuring Exposures and Assessing Health Effects*. Committee on Passive Smoking. Board on Environmental Studies and Toxicology. National Research Council, National Academy Press, Washington, DC. Available from http://www.ulib.org/webRoot/Books/National_Academy_Press_Books/env_tobacco_smoke/0000001.htm. Accessed on February 2001.

REHVA (2006). *REHVA Guidebook 6: Indoor Climate and Productivity in Offices. How to Integrate Productivity in Life Cycle Cost Analysis of Building Services* (P. Wargocki and O. Seppanen, eds.). Federation of European Heating and Air-Conditioning Association, REHVA, Brussels, Belgium.

REHVA (2011). *REHVA Guidebook 14: Indoor Climate Quality Assessment* (S.P. Corgnati and M.G. da Silva, eds.). Federation of European Heating and Air-Conditioning Association, REHVA, Brussels, Belgium.

Roelofsen, P. (2001). The design of the workplace as a strategy for productivity enhancement, in *Proceedings CLIMA 2001*, Napoli, Italy.

4

Solar Radiation

ABSTRACT A basic knowledge of certain key solar radiation concepts is essential to proper building design. Solar radiation contributes greatly to peak summer building loads and also to annual energy consumption. The radiation received by a surface depends on the solar incidence angle and also on local climatic parameters. Accordingly, we begin by presenting solar geometric relationships. This is followed by a discussion of extraterrestrial insolation, which is the basis of most insolation models. Subsequently, we treat terrestrial solar radiation, including the separation of direct and diffuse components needed to calculate the insolation on tilted surfaces. We then describe the most recent American Society of Heating, Refrigerating and Air-Conditioning Engineers (ASHRAE) clear-sky model adopted by the building energy profession (meant for peak-day calculations) along with the isotropic and the ASHRAE anisotropic sky models for computing insolation on tilted surfaces. Finally, we present various statistical correlations between different long-term average solar insolation quantities.

Nomenclature

a, b	Coefficients appearing in Equations 4.41 and 4.42
a', b'	Coefficients appearing in Equation 4.44
ab	Beam air mass exponent
ad	Diffuse air mass exponent
DST	Daylight savings time
E_0	Eccentricity correction factor of the earth's orbit (Equation 4.1)
E_t	Equation of time (Equation 4.5), min
F	Conversion factor, view factor
H_0	Extraterrestrial daily irradiation, MJ/m² (Btu/ft²)
$H_{glo,hor}$	Daily global irradiation at earth's surface, MJ/m² (Btu/ft²)
$H_{glo,vert}$	Daily global irradiation on vertical surface, MJ/m² (Btu/ft²)
I	Irradiance, W/m² [Btu/(h·ft²)]
I_0	Extraterrestrial irradiance, W/m² [Btu/(h·ft²)]
I_{beam}	Beam (direct from solar disk) irradiance, W/m² [Btu/(h·ft²)]
I_{dif}	Diffuse irradiance on horizontal surface, W/m² [Btu/(h·ft²)]
I_{dir}	Direct irradiance, component of beam radiation on surface, W/m² [Btu/(h·ft²)]
$I_{glo,hor}$	Global horizontal irradiance, W/m² [Btu/(h·ft²)]
$I_{glo,T}$	Global irradiance on tilted plane, W/m² [Btu/(h·ft²)]
I_{sc}	Solar constant, W/m² [Btu/(h·ft²)]
K	Optical depth of atmosphere
K_T	Solar clearness index of atmosphere on daily time scale (Equation 4.23)
k	Extinction coefficient of atmosphere
k_T	Solar clearness index of atmosphere on hourly time scale
L	Longitude, degrees
m	Air mass ratio
n	Day of the year counted from January 1 (Table 4.1)
r	Conversion factor given by Equations 4.39 and 4.40
r_n	Sun–earth distance for a specific day n of the year
r_0	Mean sun–earth distance over the year
t	Time
Y	Conversion factor given by Equation 4.35
x	Length, m (ft or in.)
$\overline{x}$	Overbar indicates monthly mean values of quantity x

Greek

α	Altitude angle of sun ($=90° - \theta_s$), degrees
δ	Solar declination, degrees
θ_i	Incidence angle of sun on plane (angle between normal of plane and sun), degrees
θ_p	Zenith angle of plane (tilt from horizontal, upward > 0), degrees
θ_s	Zenith angle of sun, degrees
λ	Latitude, degrees
ρ_g	Reflectivity of ground
τ	Optical depth
τ_{day}	Daylength, hours
ϕ_p	Azimuth of plane (positive for orientations west of south), degrees
ϕ_s	Azimuth of sun, degrees
ω	Hour angle, degrees
ω_{ss}	Sunset hour angle, degrees

Subscripts

civ	Civil
dif	Diffuse
dir	Direct
Gre	Greenwich
glo	Global
grd	Ground
hor	Horizontal
loc	Local
max	Maximum
min	Minimum
norm	Normal
p	Plane
s	Sun
sol	Solar
ss	Sunset
std	Standard
T	Tilted surface, transmission
vert	Vertical

4.1 Introduction

The two primary engineering applications wherein detailed knowledge of solar radiation is essential are solar energy conversion systems (solar thermal or photovoltaic) and building energy systems. While the underlying solar geometry is identical, professionals in both application areas have adopted slightly different methods to estimate solar radiation on planes with arbitrary tilt and orientation as well as how clear-sky (or sunny day) radiation is defined. Given the focus of this book, we shall preferentially present techniques that have been adopted by the building energy community.

Solar radiation is an important component in the energy balance of a building, and one must account for it in a calculation of loads. This is particularly true for perimeter zones and for *peak cooling loads* pertinent to design day calculations (discussed in Chapter 9). In addition to thermal effects, an understanding of solar radiation is necessary for the design of *daylighting* systems (Chapter 22). For peak loads, one needs the characteristics of solar radiation on a short time scale: hourly to daily. Peak heating loads occur during cold winter nights, and in most cases the contribution of solar radiation can be neglected (with the possible exception of buildings with significant thermal storage). For cooling loads, on the other hand, the solar contribution can be very important and should be evaluated with care. Usually, the peak cooling load occurs on sunny days.

To determine the solar contribution to annual building energy consumption, much longer time scales are relevant. Since one is interested in the consumption over the life of a building, the solar radiation data selected for this purpose should cover a whole year and be representative of the average over many years. The fine time resolution of the solar radiation is rarely necessary for this purpose. For hand calculations, one often works in monthly steps, for the building may be sensitive to the month-to-month variations of solar radiation.

Thus, the insolation data needed by the designer are of two types: data that represent sunny days and data that represent long-term averages. If the design is done by means of an hour-by-hour computer simulation, the weather data should be representative of both the long-term average and days with peak loads. Much work has been done to assemble data sets that satisfy these requirements, and now such data are freely available on Internet sites and furnished routinely with the purchase of one of the standard simulation programs. For hand calculations, such an approach would be unmanageable, and simple models are used instead. Such models remain pertinent despite the availability of detailed computer simulation programs; they are, in effect, a concise summary of the essential characteristics of solar radiation and thus are a valuable guide for one's intuition. Being easier to manipulate than large data sets, they are also useful as research tools. Therefore, two solar radiation models are presented, one for clear or sunny days and one for long-term averages.

4.2 Solar Movement and Basic Angles

4.2.1 Earth's Orbit

The earth moves about the sun in an elliptic orbit, completing one revolution per year (=365.25 days). In addition, the earth rotates about its polar axis once per day. The polar axis is inclined at an angle of 23.45° from the normal of the plane of the orbit (called the ecliptic), as shown in Figure 4.1a. These rotations and the inclination between the polar axis and the ecliptic are responsible for the day–night cycle and the seasons. Figure 4.1b shows the position of the earth relative to the sun's rays at the times of the summer and winter solstices, i.e., June 21 and December 21 in the northern hemisphere. Some of the relevant geometric relationships are not intuitively obvious; they can be derived most simply by means of vector algebra. Only the results are shown here; the interested reader can refer to standard texts such as Rabl (1985) for proofs.

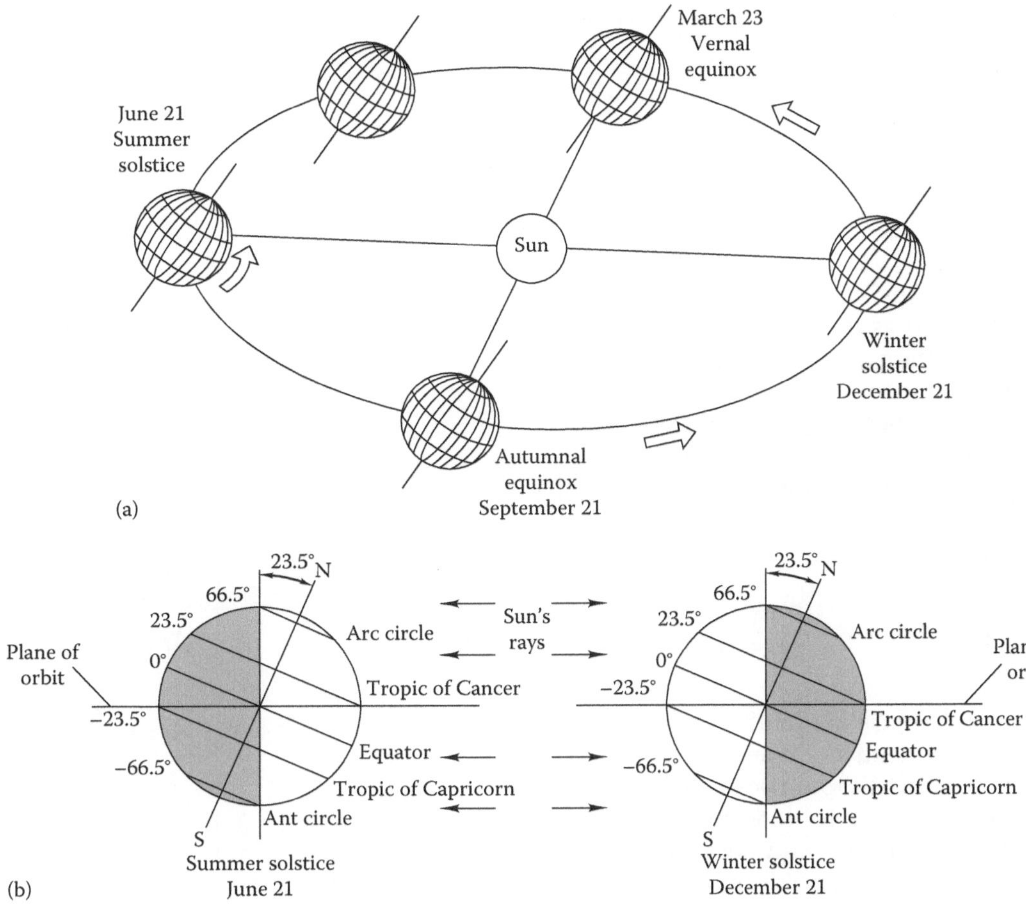

FIGURE 4.1
Geometry of the earth's orbit and inclination of polar axis: (a) entire orbit and (b) enlarged detail, solstices.

The earth orbits about the sun in an approximately circular path with the sun located slightly off center. The earth's mean distance to the sun is about 1.50×10^8 km (9.3×10^7 miles). The annual variation is ±1.7%, the distance being shortest close to December 21 and longest close to June 21. Since the intensity of solar radiation incident upon the top of the atmosphere varies inversely with the square of the earth–sun distance, the earth receives approximately 7% more radiation in December than in June. However, the fact that it is hotter in summer than in winter is due predominantly to the near-normal solar incidence angles during the former time period.

On the winter solstice, all points on the earth's surface above 66.5° north latitude are in total darkness while all regions within 23.5° of the South Pole receive continuous sunlight (Figure 4.1b). At the time of the summer solstice (June 21), the situation is reversed. At the time of the two equinoxes (March 23 and September 21), both poles are equidistant from the sun and all points on the earth's surface have 12 h of daylight and 12 h of darkness.

Because of its tilted axis, the earth's surface has been divided into five climatic zones. The torrid or *tropical zone* includes all locations where the sun is at the zenith (vertically overhead) at least once yearly. The torrid zone extends 23.5° on either side of the equator. The *temperate zones* include all locations where the sun appears above the horizon each day but never reaches the zenith. The temperate zones extend from latitudes 23.5° to 66.5° (north and south). The *frigid zones* include all locations where the sun is below the horizon for at least one full day during the year. These zones extend 23.5° from the poles.

The *eccentricity correction factor* E_0 of the earth's orbit is defined as

$$E_0 \equiv \left(\frac{r_0}{r_n}\right)^2 \tag{4.1}$$

where

r_0 is the mean sun–earth distance

r_n is the actual value for a particular day number n of the year counted from January 1 (i.e., $n = 1$ for January 1, $n = 32$ for February 1, and so on*).

* The online HCB software contains a table listing values of n for each day of the year.

For most engineering applications, a simple correlation is adequate (Iqbal, 1983):

$$E_0 = \left[1 + 0.033 \times \cos\left(\frac{360 \times n}{365.25}\right)\right] \quad (4.2)$$

4.2.2 Solar Declination

For solar energy calculations, it is convenient to use coordinates fixed in the earth as if the sun moved around the earth. Considering the geometry from the point of view of the earth, the sun traverses one circular orbit around the earth on each day. In general, this orbit does not lie in the plane of the equator; rather, the line from sun to earth makes an angle δ relative to the equatorial plane, as shown in Figure 4.2a. This angle is called the solar *declination*, and it is given by*

$$\sin\delta = -\sin 23.45° \times \cos\frac{360° \times (n+10)}{365.25} \quad (4.3)$$

If an analysis requires the use of monthly average days over the year, the recommended values of n and declination δ as shown in Table 4.1 should be used. However, most building analysis is simply done for the 21st of each month due to convenience, and because the two solstices and the two equinoxes occur on or very close to the 21st of the corresponding months. Table 4.1 also indicates the values of n as well as the solar declination for these days.

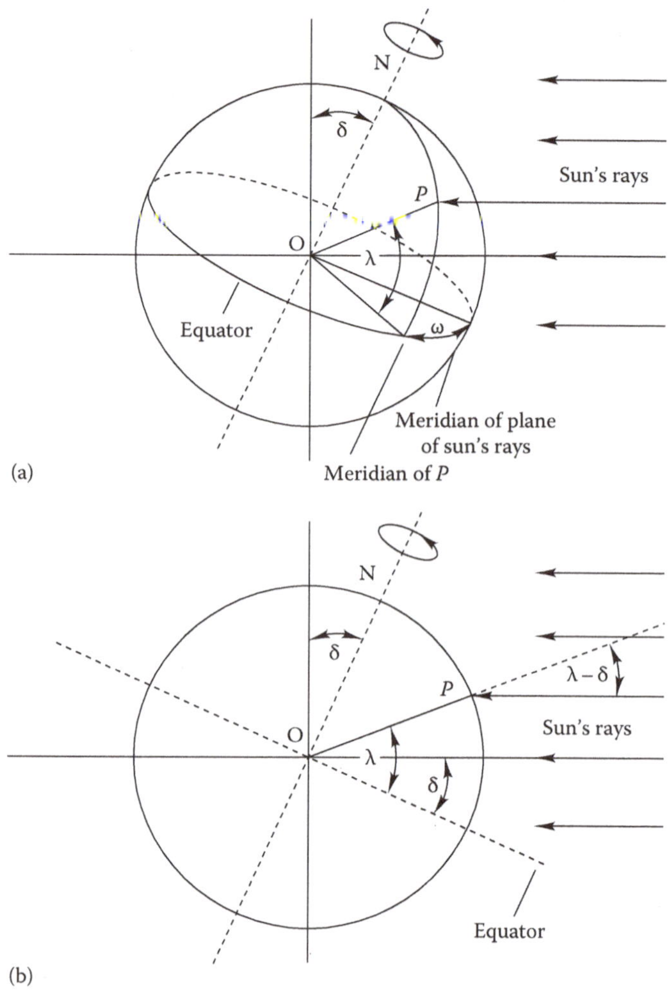

FIGURE 4.2
Latitude λ, hour angle ω, and declination δ. O = center of earth, N = north pole, P = point on earth's surface. (a) 3-D view. (b) Cross-sectional view at solar noon.

4.2.3 Basic Sun–Earth Angles

The equatorial plane is the plane passing through the center of the earth that is perpendicular to the axis of rotation of the earth. Let point P represent a location in the northern hemisphere (see Figure 4.2). The *latitude* λ is defined as the angular distance of the point P north (or south) of the equator. It is the angle between the line OP (point O representing the center of the earth) and the projection of OP on the equatorial plane. By convention, locations in the northern hemisphere are taken to be positive while those in the southern hemisphere are assigned negative values.

The *hour angle* ω is the angle measured in the earth's equatorial plane between the projection of OP and the projection of a line extending from the center of the sun to the center of the earth. The hour angle at a particular location expresses the time of day with respect to solar noon (discussed in Section 4.2.4); thus hour angle is zero at solar noon. One hour of time is represented by (360/24)° = 15° of hour angle. Also, the convention is to take hour angles as positive in the afternoon, and vice versa. Thus, for 11:00, ω = −15°, and for 14:30, ω = +37.5°. The hour angle can be computed from

$$\omega = (t_{sol} - 12\ \text{h}) \times 15° \quad (4.4)$$

where t_{sol} is the solar hour angle.

4.2.4 Solar Time

All solar angle calculations should use solar time. Solar time is based on the apparent motion of the sun as seen from a point on the surface of the earth. *Universal time*, or Greenwich civil time $t_{civ,Gre}$, is the time along the Greenwich meridian (or longitude), adopted as the reference meridian of zero longitude. This is the time

* There are several versions of this correlation in the published literature. All of them give results that are essentially identical.

TABLE 4.1
Solar Declination Values for Average Day and for the 21st Day for Each Month

Month	Date	Day of Year (n)	Declination (Degrees)	Date	Day of Year (n)	Declination (Degrees)
January	17	17	−20.8	21	21	−20.0
February	16	47	−12.8	21	52	−11.1
March	16	75	−2.5	21	80	−0.5
April	15	105	9.1	21	111	11.2
May	15	135	18.5	21	141	19.9
June	11	162	23.0	21	172	23.4
July	17	198	21.1	21	202	20.4
August	16	228	13.3	21	233	11.6
September	15	258	2.3	21	264	0.0
October	15	288	−9.2	21	294	−11.3
November	14	318	−18.6	21	325	−20.2
December	10	344	−23.0	21	355	−23.4

Note: Average day corresponds to the day when the solar declination is closest to the monthly averaged daily values.

quoted in international TV and radio programs. Three quantities are relevant when specifying the time of day at a specified location:

1. *Standard time* t_{std} of the time zone of the specified location is defined by the reference value of the longitude. For instance, in the contiguous United States, the reference meridians for the time zones are 75°W for Eastern, 90°W for Central, 105°W for Mountain, and 120°W for Pacific standard times. Most countries have only one reference meridian; i.e., for India, it is 82.5°E. The standard time is the watch time when daylight savings is not followed.
2. *Local civil time* $t_{civ,loc}$ is the time at the specific location in question. A constant correction is needed, which accounts for the difference in longitude between the reference meridian and the local meridian. Since one full cycle of a day corresponds to 360° longitude, each degree corresponds to (24 h × 60 min)/360° = (1/15) h = 4 min. In most parts of the world, clocks are set to the same time within a time zone covering approximately 15° of longitude (although the boundaries may be quite irregular).
3. *Daylight savings time* (DST) is an artificial change adopted by several countries worldwide. From several decades, most parts of western Europe and North America have instituted an advancement of the clock by 1 h during the summer half of the year (the "spring ahead, fall back" rule). The reason for adopting DST is that it improves the match between human activities and the availability of daylight, thereby resulting in humans consuming less energy. It also results in a reduction in overall building energy consumption (such as lighting electricity use).

Another source of deviation between solar time and local civil time is due to the nature of the earth's orbital motion around the sun. The elliptical orbit induces changes in the orbital angular speed of the earth (according to Kepler's laws). Solar noon is the time when the sun reaches the highest point in the sky; it can differ from noon of local civil time by as much as one-quarter hour. The difference between solar noon and noon of local civil time is called the "equation of time" E_t. It is a function of the time of year (plotted in Figure 4.3) and can be roughly approximated by

$$E_t = 9.87 \times \sin 2B - 7.53 \times \cos B - 1.5 \times \sin B \text{ (min)} \quad (4.5)$$

$$\text{with } B = 360° \times \frac{n-81}{364} \quad \text{for } n\text{th day of year} \quad (4.6)$$

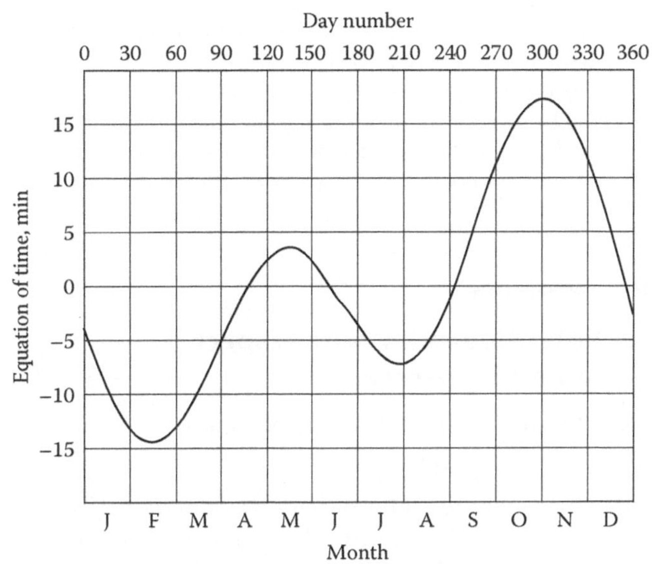

FIGURE 4.3
Equation of time E_t (Equation 4.5) (months are rounded to 30 days).

Solar time t_{sol} is related to standard time t_{std}:

$$\text{In hours } t_{sol} = t_{std} \pm \frac{L_{std} - L_{loc}}{15°/\text{h}} + \frac{E_t}{60\ \text{min/h}} \quad (4.7)$$

$$\text{In minutes } t_{sol} = t_{std} \pm 4' \times (L_{std} - L_{loc}) + E_t \quad (4.8)$$

where L_{std} and L_{loc} designate the longitudes (in degrees) of the time zone and the location, respectively. The plus (+) sign is to be used for locations west of Greenwich and the negative (–) sign for locations east of Greenwich. In regions with daylight saving time, one has to subtract 1 h from daylight savings time to obtain t_{std} during the summer half of the year.

Example 4.1: Solar and Standard Times

At what time is the sun due south in Phoenix, AZ, on July 21? Longitude of Phoenix = L_{loc} = 112°W, and the standard meridian followed is L_{std} = 105°W.

Given: t_{sol} = 12:00

Find: Standard time t_{std}

Lookup values: E_t = −6 min, from Figure 4.3 or Equation 4.5

Solution

Solve Equation 4.8 for standard time:

t_{std} = 12:00 − 4 × [105° − 112°] − (−6min) = 12:34

Thus, the difference between standard time and solar time is only 34 min in this case.

Comments

Arizona does not follow daylight savings because it lies between the Mountain and Pacific reference meridians. During the winter months, it follows the Mountain standard time and in the summer months the Pacific DST. Had daylight savings been adopted in Arizona, solar noon for the aforementioned example would have occurred at 13:34. This 1:5 hr difference between diurnal human activities and solar cycle was deemed too large, and so the state chose not to adopt daylight savings in summer.

4.3 Solar Geometry with respect to Local Observer

4.3.1 Solar Zenith and Azimuth Angles

The movement of the sun with respect to a local observer is more relevant for solar radiation calculations. The sun appears to move across the sky following the path of a circular arc from horizon to horizon. Figure 4.4 shows apparent solar paths for 3 days of the year. The position of the sun at any instant is uniquely characterized by two angles: solar zenith and solar azimuth (these are the standard angles of the spherical coordinate system). These additional angles may be expressed in terms of the three basic angles (latitude, hour angle, and declination).

The incidence angle of the sun on the earth's surface at *P* is the angle between the normal of the surface at *P* and the line from *P* to the sun (Figure 4.2a). The solar zenith angle of the sun θ_s is the incidence angle on a horizontal surface (Figure 4.5). At solar noon, the ray from the sun to *P* is coplanar with the cross section of the earth through *P* and the poles; and so the angles are easy to determine. From Figure 4.2b, one sees immediately that the zenith angle of the sun at noon is equal to the difference between the latitude and the declination. Incidence angles at other times of the day are far less obvious because they involve 3-D geometry. It can be shown that the *zenith angle* θ_s for any latitude λ and any time of day and year is given by

$$\cos\theta_s = \cos\lambda\cos\delta\cos\omega + \sin\lambda\sin\delta \quad (4.9)$$

To indicate the position of the sun uniquely, one needs an additional quantity, the *azimuth* ϕ_s shown in Figure 4.5 (in terms of a rotation about the vertical, ϕ_s is the angle from due south). We can show that it is related to the hour angle ω, declination δ, and zenith angle θ_s by

$$\sin\phi_s = \frac{\cos\delta\sin\omega}{\sin\theta_s} \quad (4.10)$$

It is positive in the afternoon, with ϕ_s pointing west. Instead of the zenith angle, certain professionals (for example, architects) prefer to use the complement, called the "solar altitude angle":

$$\text{Solar altitude angle } \alpha = 90° - \theta_s \quad (4.11)$$

Example 4.2: Calculation of Solar Position

Determine the solar zenith and azimuth angles at 15:00 (3:00 p.m.) solar time in Phoenix, AZ, on June 21 (summer solstice)?

Given: Latitude λ = 33.3°
t_{sol} = 3.00 h
Day of year *n* = 172 (from Table 4.1)

Find: θ_s and ϕ_s

Solution

First, use Equation 4.3 with *n* = 172 to find the declination

$$\sin\delta = -\sin 23.45° \times \cos\frac{360° \times (172 + 10)}{365.25} = 0.398$$

Hence, δ = 23.45°.

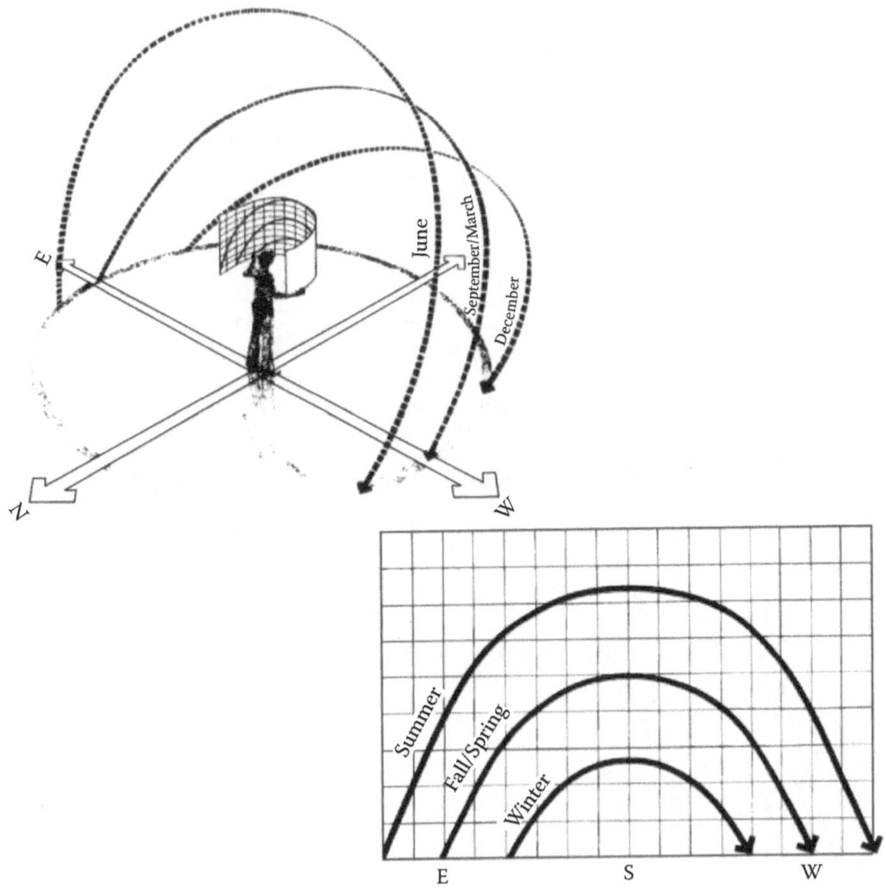

FIGURE 4.4
Solar path diagrams over the day for three different periods of the year with respect to a local observer.

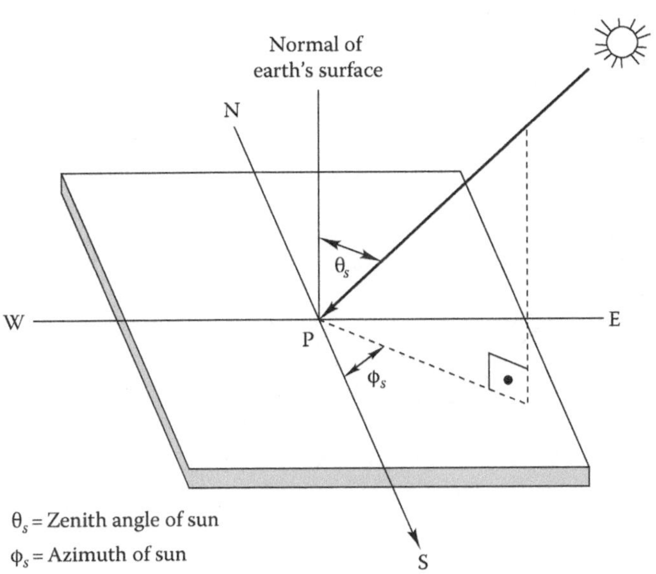

FIGURE 4.5
Zenith angle θ_s and azimuth ϕ_s of sun.

The hour angle $\omega = 45°$ from Equation 4.4.
The zenith angle θ_s is determined from Equation 4.9:

$$\cos\theta_s = \cos 33.3° \times \cos 23.45° \times \cos 45° + \sin 33.3° \times \sin 23.45° = 0.761$$

Hence, $\theta_s = 40.48°$.
The azimuth angle is determined from Equation 4.10:

$$\sin\phi_s = \frac{\cos 23.45° \times \sin 45°}{\sin 40.48°} = 1.0$$

Hence, $\phi_s = 90°$, i.e., the sun is due west.

4.3.2 Sun-Path Diagrams

The position of the sun in the sky is uniquely determined if the solar azimuth angle ϕ_s and the altitude angle α (or the zenith angle θ_s) are known. In the course of the day and year, these solar angles also vary as the sun moves in the sky (see Figure 4.4). A widely used practice to locate the position of the sun in the sky is to draw *sun-path diagrams* (also referred to as solar position plots) for the specific location that are projections of the sun's path on the horizontal plane. Software programs are freely available that can plot such diagrams for any location specified. Usually, two types of sun-path diagrams are used: equidistant projection (Goswami et al., 2000) and cylindrical projection. While the former was used extensively in the past, the cylindrical projection sun-path representation is increasingly used nowadays. Figure 4.6a through c

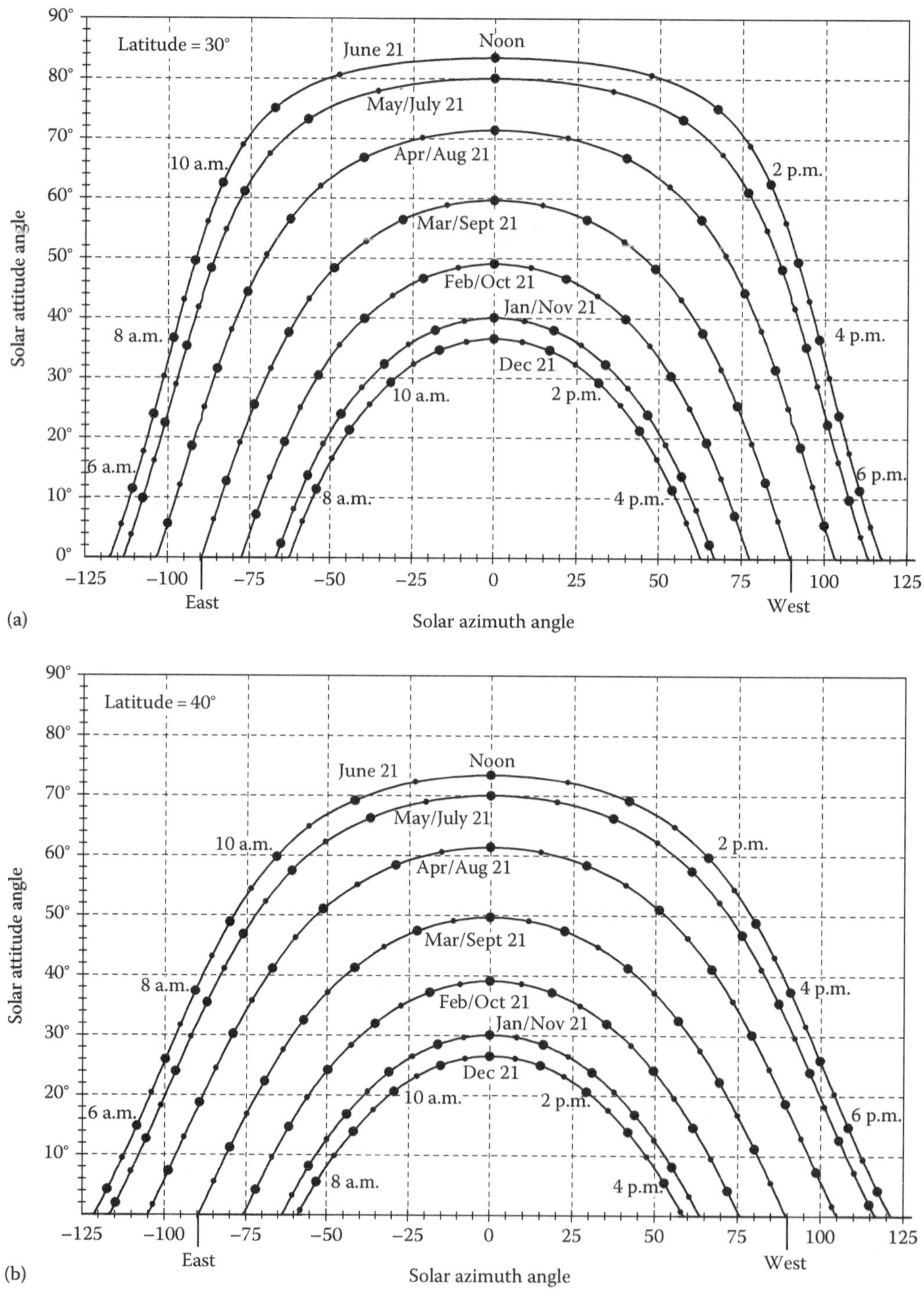

FIGURE 4.6
Cylindrical projection sun-path diagram: solar altitude angle (=90° – θ_s) versus azimuth ϕ_s. Time in legends is solar time. (a) Latitude λ = 30°. (b) Latitude λ = 40°.
(Continued)

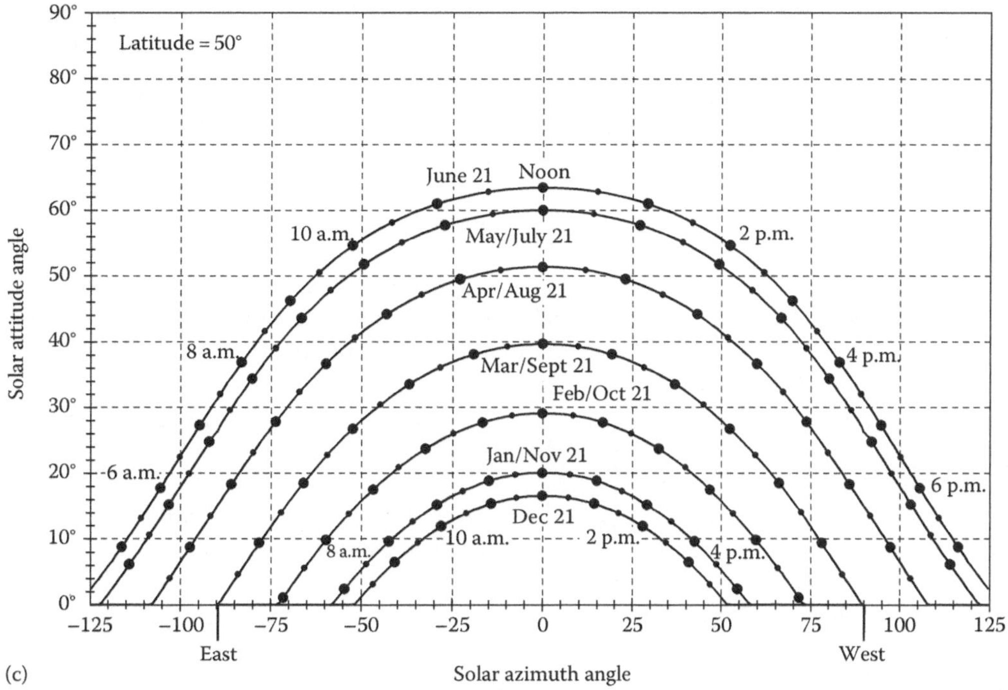

(c)

FIGURE 4.6 ***(Continued)***
Cylindrical projection sun-path diagram: solar altitude angle (=90° − θ_s) versus azimuth ϕ_s. Time in legends is solar time. (c) Latitude λ = 50°.

depicts the sun paths (using cylindrical projection) for latitudes of 30°N, 40°N, and 50°N.

Example 4.3: Using the Sun-Path Diagram

Using the sun-path diagram for 30°N, determine the zenith angle and the azimuth angle for the conditions specified in Example 4.2

Given: Conditions of Example 4.2

Find: θ_s and ϕ_s

Solution

Referring to the June 21st plot (Figure 4.6a) at t_{sol} = 3 p.m., we read off the following approximate values: solar altitude α = 50°, i.e., solar zenith θ_s = 40°, and azimuth angle ϕ_s = 90°W.

Comments

Even though the latitude of Phoenix (=33.3°N) is slightly different than the latitude assumed in this example, we find that these altitude and azimuth angles are very close to those determined in Example 4.2.

4.3.3 Daylength

The daylength, i.e., the time in hours from sunrise to sunset, is easily deduced if the sunset hour angle is known. Sunset occurs when the zenith angle reaches 90°. Setting $\cos\theta_s = 0$ in Equation 4.9, the sunset hour angle ω_{ss} is given by

$$\cos\omega_{ss} = -\tan\lambda \times \tan\delta \quad (4.12)$$

Since the hour angle is specified with respect to solar noon, the daylength τ_{day} can be determined as

$$\tau_{day} = \frac{2\times\omega_{ss}}{15}\ \text{(h)} \quad (4.13)$$

with the multiplier of 2 introduced to also account for the hours from sunrise to noon.

How the sunset time t_{ss} varies with latitude and time of year (specified by declination) is plotted in Figure 4.7. The lengths of day and of night are equal when ω_{ss} is 90°, which happens for all locations when $\delta = 0$. Solving Equation 4.3 for $\delta = 0$, the corresponding days of the year are

$$n = -10 + 90° \times \frac{365.25}{360°} = 82 \quad \text{(March 23)}$$

and

$$n = -10 + 270° \times \frac{365.25}{360°} = 264 \quad \text{(September 21)}$$

These dates are called equinoxes (whose Latin root implies "equal night"), and they mark the official beginning of

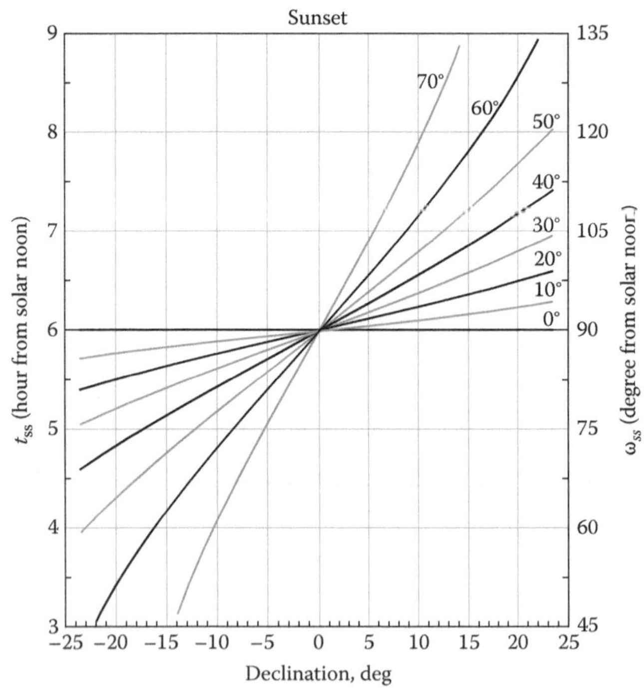

FIGURE 4.7
Sunset time t_{ss} and sunset hour angle ω_{ss} as functions of declination δ, the curves being labeled by latitude λ(0°N, 10°N, ..., 70°N).

spring and of fall. Declination and day length reach their extreme values at the solstices, June 21 and December 21. For latitudes beyond (90° − 23.5°) = 66.5° (the arctic circles), the absolute value of (tan λ tan δ) can exceed unity at certain times of the year, implying that there is no real solution for ω_{ss}. This is the region of midnight sun and of winter days without sunrise.

Example 4.4: Daylength Calculation

Calculate the maximum and minimum daylength over the year for Phoenix, AZ.

Given: Latitude λ = 33.3°

Find: $\tau_{day,\max}$ and $\tau_{day,\min}$, the maximum and minimum daylengths

Solution

The maximum daylength will be on summer solstice, i.e., June 21, when δ = 23.5°.

The minimum daylength will be on winter solstice, i.e., December 21, when δ = −23.5°.

We use Equation 4.12: $\cos\omega_{ss,\max} = -\tan 33.3° \times \tan 23.5° = -0.286$ or $\omega_{ss,\max} = 106.62°$.

Next, we use Equation 4.13: $\tau_{day,\max} = (2 \times 106.62)/15 = 14.22$ h.

Similarly, the minimum daylength is $\tau_{day,\min} = 9.78$ h.

Comments

The maximum and minimum daylengths are symmetric about 12 h. Further, the difference between them increases with higher latitudes.

4.3.4 Incidence Angle on Planes of Arbitrary Tilt and Orientation

For the calculation of incidence angles on arbitrary planes, it is convenient to specify the orientation of the plane in terms of the tilt angle θ_p and azimuth ϕ_p of the surface normal (positive for orientations west of south), as indicated in Figure 4.8. The zenith angle θ_p of a plane is its tilt from the horizontal. In terms of these quantities, the incidence angle θ_i of the sun on the plane (= angle between normal of plane and line to sun) can be written in the form

$$\cos\theta_i = \sin\theta_s \sin\theta_p \cos(\phi_s - \phi_p) + \cos\theta_s \cos\theta_p \quad (4.14)$$

This simplifies considerably for the important case of vertical planes (θ_p = 90°):

$$\cos\theta_i\big|_{vert} = \sin\theta_s \cos(\phi_s - \phi_p) \quad (4.15)$$

Example 4.5: Incidence Angle Calculation

Calculate the incidence angle of the sun on a west-facing vertical wall in Phoenix at 15:00 solar time on June 21 (conditions assumed in Example 4.2).

Given: δ = 23.45°, ω = 45,
From Example 4.2, θ_s = 40.48°, ϕ_s = 90°
For a west-facing wall: θ_p = 90° and ϕ_p = 90°

Find: $\cos\theta_i$

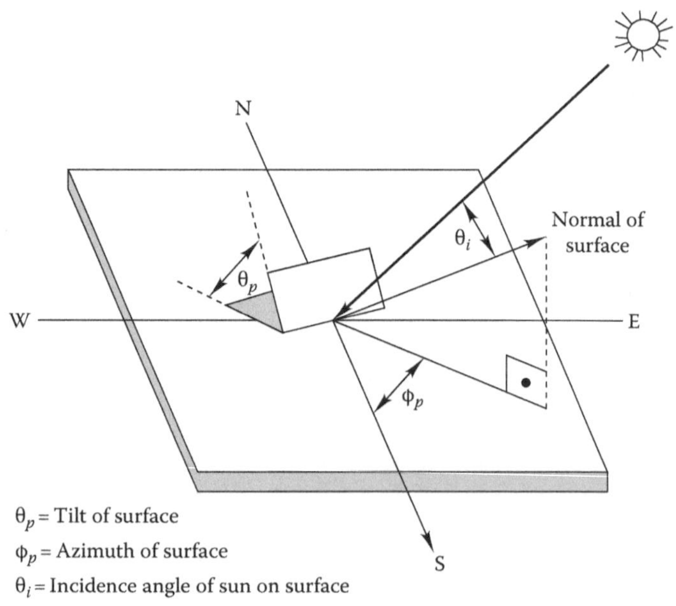

FIGURE 4.8
Zenith angle θ_p and azimuth ϕ_p of a plane and angle of incidence θ_i of sun on this plane.

Solution

From Equation 4.15

$$\cos\theta_i = \sin 40.48° \times \cos(90°-90°) = 0.649$$

Hence, $\theta_i = 49.5°$.

4.3.5 Shading from Far Objects

Rays from the sun to a point of a building are frequently blocked by other buildings, by trees, or by overhangs. Detailed analysis of the geometric relationships can be tedious. Nowadays, computer programs offer welcome relief. But even in the age of computers, it is useful to develop a certain intuition about some general features of shading. Sun-path diagrams (discussed in Section 4.3.2) offer such an intuitive visual representation for ascertaining shading phenomena on buildings and on solar collectors.

Figure 4.9 shows how the sun-path diagram of Figure 4.6 can be used to determine the times of day and year when an object will cast a shade at a point. One superimposes the outline of the horizon, as seen from the point in question, on the sun-path diagram. The point is shaded when the sun path passes below this horizon outline (shown as a heavy dotted line in Figure 4.9).

The practical usefulness of such a representation is illustrated by a simple example. Suppose one wants to build a solar house with a ground-mounted collector at the position of the observer. The times over the year when the collector will be shaded by the tree can be determined from Figure 4.9. There is no shade at all from the middle of February to the end of October. During the winter months, the collector will be shaded between 1 p.m. and 3 p.m.

Example 4.6: Shading Determination

Consider a block similar to that shown in Figure 4.9 whose edges are marked as points *A*, *B*, *C*, and *D*. Points *A*′ and *D*′ are the intersection points of the two sides of the cube onto the north–south

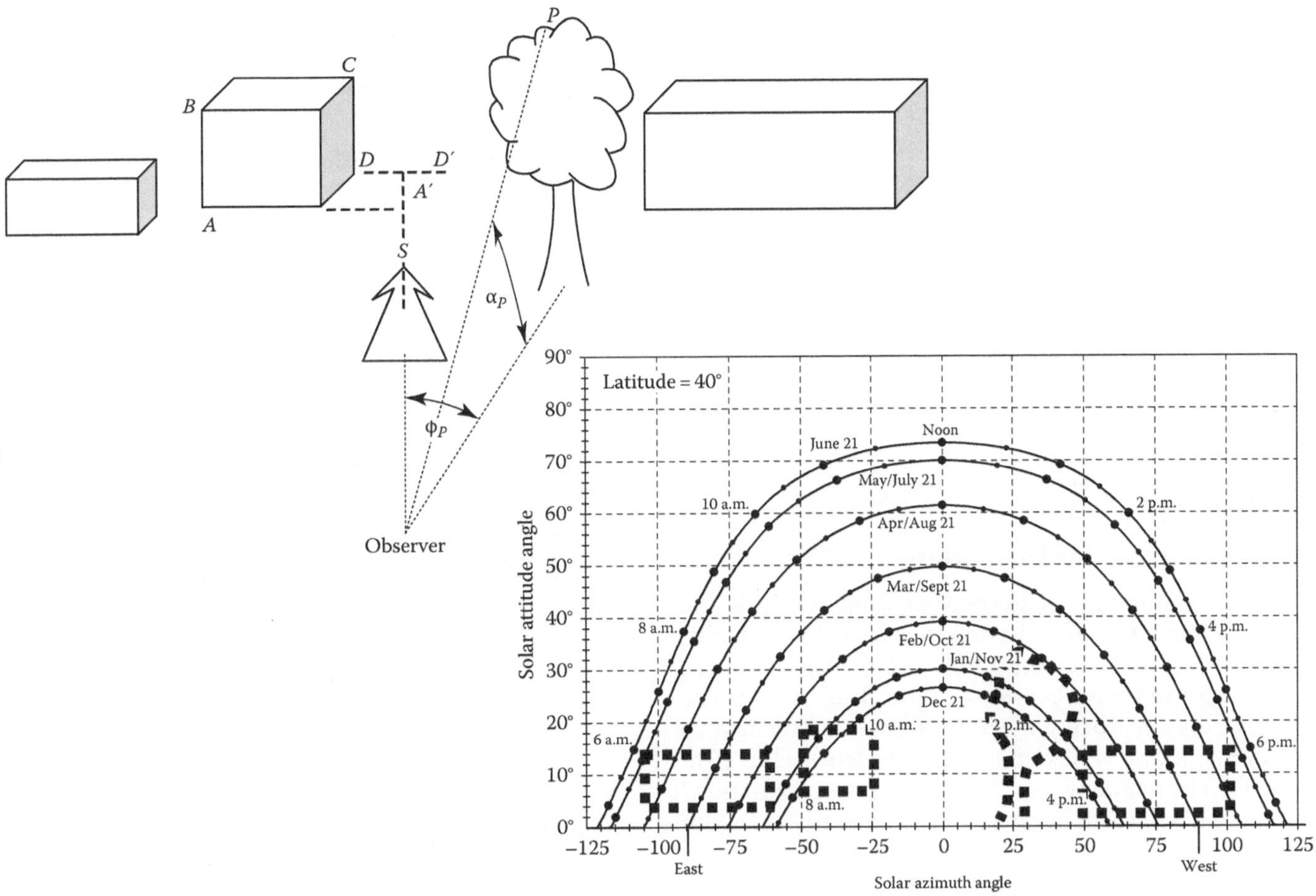

FIGURE 4.9
Outline of horizon superimposed on the sun-path diagram to determine incidence of shading. α_P and ϕ_P are the altitude and azimuth angles of point *P* as seen from the position of the observer.

direction as shown. We wish to determine the shaded area as indicated in Figure 4.9 for the measurements given in the following.

Given: Length OA = 25 m, OD = 24 m, and AB = CD = 10 m
Length OA' = 8.5 m and OD' = 13.5 m

Find: θ_i and ϕ_i

Solution

Let us illustrate the approach with only two points: B and C.

From basic geometry, the altitude angle at point B

$$\tan\alpha_B = \frac{AB}{OA} = \frac{10}{25} \quad \text{or} \quad \alpha_B = 21.8°$$

and the altitude angle at point C

$$\tan\alpha_C = \frac{CD}{OD} = \frac{10}{24} \quad \text{or} \quad \alpha_C = 22.6°$$

The azimuth angle at point B

$$\cos\phi_B = \frac{OA'}{OA} = \frac{8.5}{25} \quad \text{or} \quad \phi_B = 70.1°$$

and the azimuth angle at point C

$$\cos\phi_C = \frac{OD'}{OD} = \frac{13.5}{24} \quad \text{or} \quad \phi_C = 55.8°$$

Comments

These angles allow the two points B and C to be located on Figure 4.9. A similar approach with two other points will provide the necessary shading area specific to the measurements given in this example.

4.4 Extraterrestrial Insolation

A clear terminology is needed to keep track of the various types of solar radiation that need to be distinguished. We will use the term "irradiance" and the symbol I for radiative power in W/m² [Btu/(h·ft²)]. The radiative energy during a certain time interval such as an hour or a day is designated by *irradiation H* in J/m² (Btu/ft²). The vague words "insolation" and "radiation" can serve when distinctions are unnecessary.

4.4.1 Hourly Values Normal to Solar

The *solar constant* I_{sc} is the total solar irradiance, i.e., energy per unit time received on a unit area of surface normal or perpendicular to the solar rays, placed outside the earth's atmosphere at the mean sun–earth distance.

The numerical value of the solar constant has been revised a few times from the time it was first proposed. The current value based on precise measurements done during 1969–1980 is taken to be 1367 W/m² [433.3 Btu/(h·ft²)], which straddles previous values of 1353 W/m² and 1373 W/m². Due to the slight eccentricity of the orbit, the actual value of the extraterrestrial irradiance normal to solar rays $I_{0,norm}$ varies by ±3.3%. A good fit for I_0 is given by the following correlation:

$$I_{0,norm} = E_0 \times 1367 \text{ W/m}^2 \qquad \text{(4.16 SI)}$$

$$I_{0,norm} = E_0 \times 433.3 \text{ Btu/(h}\cdot\text{ft}^2) \qquad \text{(4.16 IP)}$$

where E_0 is the eccentricity correction factor given by Equation 4.2. Since its peak occurs in winter (of the northern hemisphere), the seasonal variations in the northern hemisphere are somewhat smaller than they would be if the earth's orbit were circular.

The extraterrestrial insolation is a useful quantity because many solar radiation models are based on it, e.g., the clear-day model presented in Section 4.6 as well as the generalized correlations presented in Section 4.9.

Example 4.7: Calculation of Extraterrestrial Radiation

Calculate the extraterrestrial irradiance that would be incident on the west-facing vertical wall of Example 4.5 (June 21 at 3 p.m. solar time) if there were no atmosphere.

Given: For June 21, n = 172, θ_i = 49.5°

Find: I_0

Solution

First, we evaluate the eccentricity correction factor from Equation 4.2:

$$E_0 = 1 + 0.033 \times \cos\frac{360° \times 172}{365.25} = 0.9675$$

The extraterrestrial irradiance at normal incidence for the day is given by Equation 4.16:

$$I_{0,norm} = 0.9675 \times 1367 \text{ W/m}^2 = 1322.6 \text{ W/m}^2$$

Accounting for the incidence angle at 3 p.m., the irradiance on the west-facing surface is

$$I_0 = I_{0,norm}\cos 49.5° = 1322.6 \text{ W/m}^2 \times 0.649$$
$$= 858.4 \text{ W/m}^2 \text{ (272.1 Btu/[h}\cdot\text{ft}^2])$$

4.4.2 Daily Values on Horizontal Surface

Also of interest is the total daily extraterrestrial irradiation H_0 on a horizontal surface. It is obtained by integrating the cosine of the solar zenith angle θ_s of Equation 4.9 from sunrise to sunset, multiplied by $I_{0,norm}$. When the sunset hour angle ω_{ss} of Equation 4.12 has a real value, the result is

$$H_0 = \frac{\tau_{day}}{\pi} I_{0,norm} \cos\lambda \cos\delta \left(\sin\omega_{ss} - \frac{\pi\omega_{ss}}{180°} \cos\omega_{ss} \right) \quad (4.17)$$

with τ_{day} = 24 h = 86,400 s. Otherwise, we have the following:

- For days without sun

$$H_0 = 0 \quad \text{when} \tan\lambda \tan\delta < -1 \quad (4.18)$$

- For days with midnight sun

$$H_0 = \frac{\tau_{day}}{\pi} I_0 \sin\lambda \sin\delta \quad \text{when} \tan\lambda \tan\delta > 1 \quad (4.19)$$

H_0 is plotted in Figure 4.10 for the northern hemisphere. If the extraterrestrial radiation I_0 were independent of the day of year n, this plot would be exact for southern latitudes as well, after substituting $\lambda \to -\lambda$ and $n \to n + 183$. The variation of I_0 disturbs this symmetry slightly, by at most ±3% on January 1 and July 1. One feature of this graph may come as a surprise: under the midnight sun, the polar regions would receive more solar radiation per day than the tropics if there were no atmosphere. Of course, the total amount of radiation that reaches the surface during this relatively brief period is quite limited, and the polar regions stay cold.

Example 4.8: Extraterrestrial Radiation on a Horizontal Surface

Calculate the daily total extraterrestrial irradiation on a horizontal surface for June 21 at a latitude equal to that of Phoenix, AZ.

Given: Latitude λ = 33.3° and declination δ = 23.5°
From Example 4.4: $\omega_{ss,\max}$ = 106.62°, $\tau_{day,\max}$ = 14.22 h
From Example 4.7: $I_{0,norm}$ = 1322.6 W/m²
Find: H_0

Solution

Equation 4.17 is used directly:

$$H_0 = \frac{86,400}{\pi} \times 1322.6 \times \cos 33.3° \times \cos 23.5° \times \left(\sin 106.62° - \frac{\pi}{180°} \times 106.62 \times \cos 106.62° \right)$$

$$= 41.57 \times 10^6 \text{ J/m}^2 \quad \text{or} \quad 41.57 \text{ MJ/m}^2$$

This value is consistent with that shown in Figure 4.10.

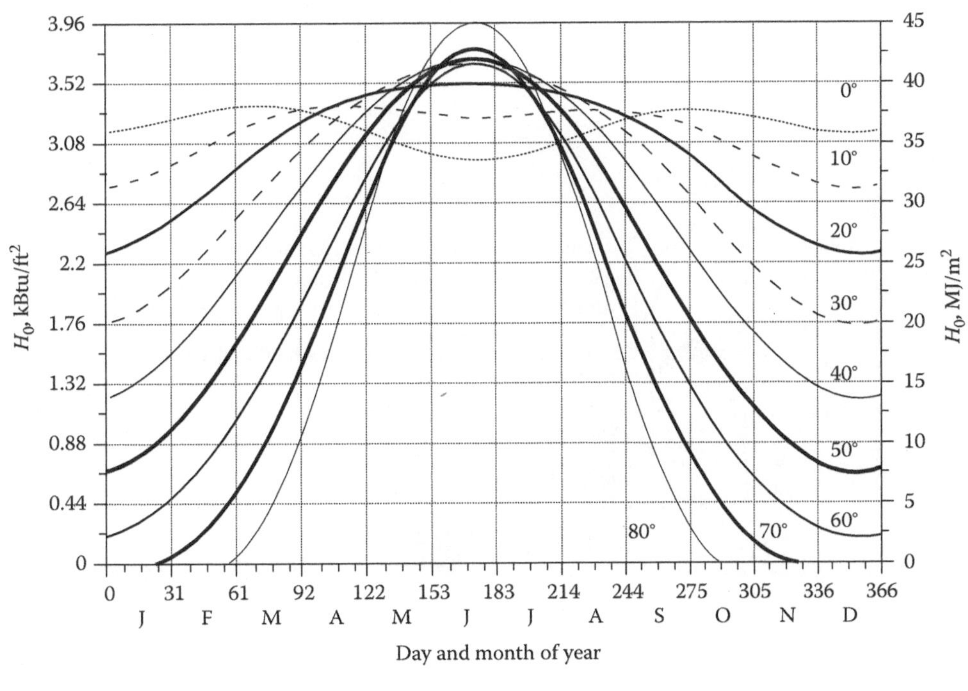

FIGURE 4.10
Extraterrestrial daily horizontal irradiation H_0 as a function of day of year, for latitudes 0°, 10°,..., 80° north.

4.5 Effect of Atmosphere

4.5.1 Air Mass Ratio

The atmosphere both attenuates and scatters the incoming solar radiation. The attenuation or change in irradiance (dI) can be represented by Bouguer law given by

$$dI = \exp(-k \cdot dx) \quad (4.20)$$

where

k is the local atmospheric extinction coefficient

dx is an incremental length of the traversed path

The traversed path of the solar rays depends on the zenith angle (or its complement, the altitude angle) of the sun at the specified day and time as shown in Figure 4.11. If the atmosphere is idealized as a layer of constant density thickness, then the air mass ratio is defined as the actual length traversed by the ray divided by the length it would pass when the sun is at the zenith. For zenith angles between 0° and 70° at sea level, a good approximation is

$$m = \frac{1}{\cos\theta_s} \quad (4.21)$$

For higher zenith angles, a correction due to the earth's curvature needs to be introduced (Iqbal, 1983). Thus, the dimensionless path length ratio $m = 1$ when the sun is directly overhead and $m > 1$ for all other times.

Integrating Equation 4.20 under the assumption of a constant value of k for the entire atmosphere results in

$$I_{norm} = I_{0,norm} \exp(-Km) \quad (4.22)$$

where

$I_{0,norm}$ is the extraterrestrial irradiance normal to solar rays given by Equation 4.16

K is referred to as the "optical depth"

The effects of solar radiation depend not only on the intensity but also on its spectrum. The spectrum varies somewhat with atmospheric conditions. The spectral distribution (intensity as a function of wavelength) is shown in Figure 4.12 for two conditions: air mass zero (i.e., extraterrestrial) and air mass $m = 2$. Notice the effect of molecular absorption due to different atmospheric gases at different wavelengths. A little less than one-half of the energy is in the visible range, 0.4–0.7 μm. Most of the remainder is in the near infrared, up to 2.0 μm. The spectrum becomes important for buildings

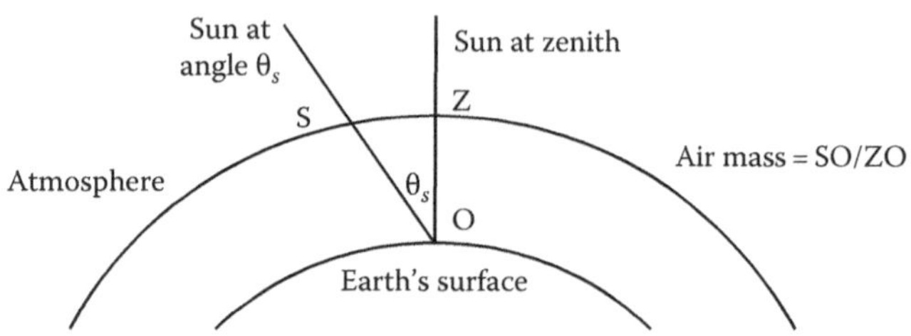

FIGURE 4.11
Concept of air mass.

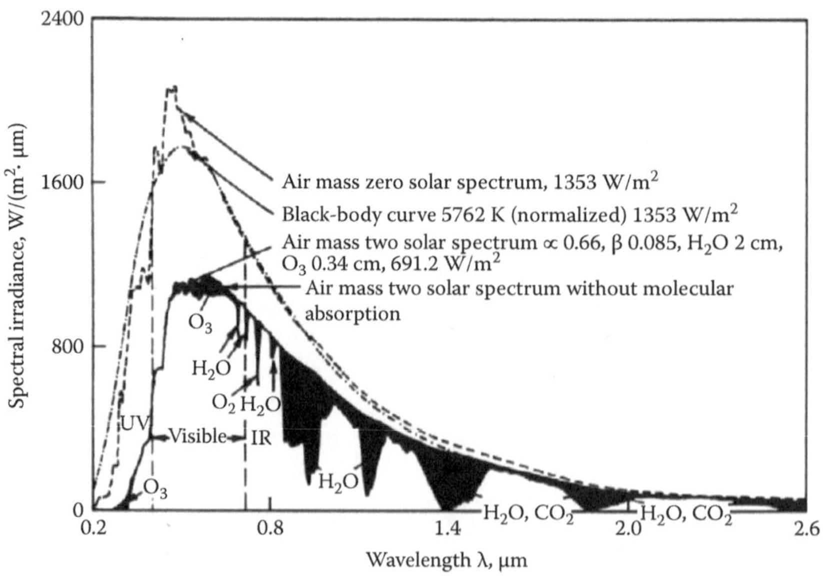

FIGURE 4.12
Solar spectrum for air mass zero (= extraterrestrial) and for air mass two, air mass being defined as 1/cos θ_s. The black portions indicate molecular absorption. (Absolute values of irradiance are based on an old value of solar constant and should be rescaled by 1367/1353.)

when the transmissivity of the window glazing varies with the wavelength (see Section 5.2). Of special interest are glazings that transmit daylight while blocking the infrared portion to minimize cooling loads.

The earlier discussion pertains to beam irradiance only, i.e., the portion of the radiation received directly from the solar disk without any change in direction. The different types of radiation are discussed next.

4.5.2 Components of Solar Radiation

The solar radiation gets attenuated as it passes through the atmosphere (Figure 4.13). The absorption, scattering, and reflection of the solar radiation caused by the air molecules, water vapor, and other particles (such as dust) in the atmosphere as well as radiation exchanges between the ground and clouds result in three types of solar radiation (Figure 4.14). Subscripts are added to distinguish between these types. Radiation normal to the solar disk is called "beam radiation," and is indicated by the subscript "beam," while its projection on any surface is called "direct radiation" and denoted by the subscript "dir." The scattered or *diffuse radiation* emanating from the rest of the sky is designated by the subscript "dif." Even in shade and under overcast conditions, the diffuse radiation may be quite significant while the beam radiation would be essentially zero. The sum of direct and diffuse radiation incident on a surface is usually called "global radiation" (with subscript "glo"), even though the term hemispherical would be more logical since one is referring to radiation incident on one side of a surface only. An additional subscript may indicate the surface on which the radiation is received. For instance, the beam radiation on the horizontal is designated by $I_{dir,hor}$. Instruments to measure global, diffuse, and beam normal radiation are described in numerous solar energy textbooks (for example, Duffie and Beckman, 1991; Goswami et al., 2000; Rabl, 1985).

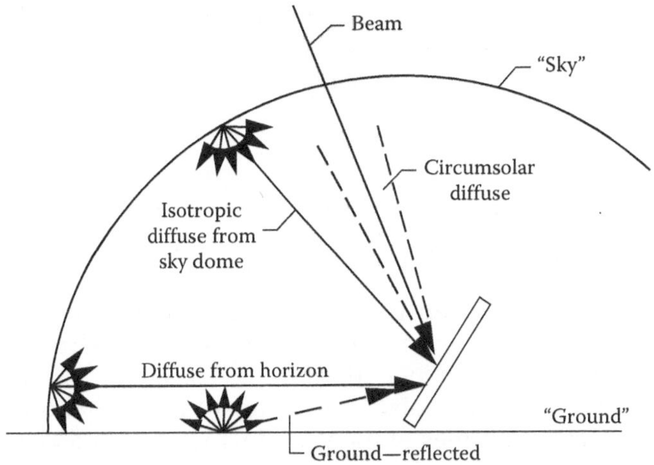

FIGURE 4.14
Different radiation components on a tilted surface.

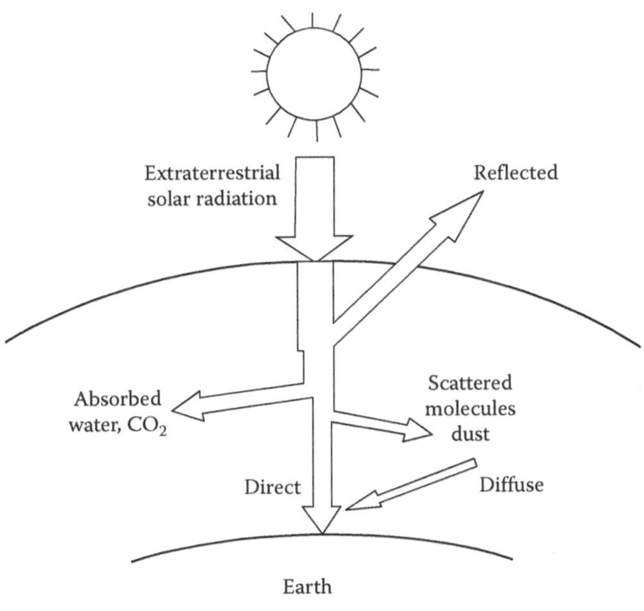

FIGURE 4.13
Attenuation of solar radiation as it passes through the atmosphere.

A very important parameter called the "clearness index" of the atmosphere is used to characterize local climatic conditions with respect to maximum radiation availability, which depends on the geometry of the earth's motion. The *daily* clearness index K_T is defined as

$$K_T = \frac{H_{glo,hor}}{H_{0,hor}} \tag{4.23}$$

where

$H_{glo,hor}$ is the daily global irradiation at the earth's surface on a horizontal surface

$H_{0,hor}$ is the extraterrestrial daily irradiation on the same surface

Thus, the clearness index includes two independent causes for the variability of terrestrial solar radiation: the local atmospheric conditions and the earth's motion that causes H_0 to vary over the year.

On heavily overcast days, K_T may be as low as 0.05–0.1 while on clear days it is around 0.7–0.75 (but not more than 0.8 for a location near sea level). Monthly averages, designated by $\bar{K}_T$, range from 0.3 for very cloudy climates such as upstate New York to 0.75 for the peak of the Sunbelt. Monthly average data for $\bar{K}_T$ are tabulated as also annual clearness indices for numerous locations worldwide (Reddy, 1987). Table 10.1 assembles monthly values for Atlanta, Chicago, Denver, and Phoenix. The online HCB software contains contour maps of monthly $\bar{K}_T$ values for the United States.

4.6 ASHRAE Clear-Sky Irradiance Model

Building peak load calculations used to size equipment are done under well-defined extreme climatic conditions. A critical input is the selection of the beam and diffuse solar irradiation for different hours and different times of the year during so-called clear days. Due to historic reasons, solar energy professionals (see, for example, Rabl, 1985) have adopted calculation methods different from those in building energy. Because of the focus of this book, we present the most current algorithms recommended for building energy calculations following ASHRAE Fundamentals (2013).

The equations for hourly beam (i.e., direct normal) and diffuse irradiance (though it is not strictly correct for diffuse) are patterned along the lines of Bouguer law or more specifically following Equation 4.20. These two components are calculated as

$$I_{beam} = I_{0,norm} \cdot \exp\left(-\tau_{beam} \cdot m^{ab}\right) \quad (4.24)$$

and

$$I_{dif} = I_{0,norm} \cdot \exp\left(-\tau_{dif} \cdot m^{ad}\right) \quad (4.25)$$

where

- I_{beam} is the beam irradiance normal to solar rays
- I_{dif} is the diffuse horizontal irradiance (measured on horizontal surface)
- $I_{0,norm}$ is the extraterrestrial irradiance normal to solar rays
- m is the air mass ratio
- τ_{beam} and τ_{dif} are the beam and diffuse optical depths (dimensionless)
- ab and ad are the beam and diffuse air mass exponents

Values of τ_{beam} and τ_{dif} are location-specific parameters capturing prevalent clear-sky conditions (elevation, precipitable water, aerosols, etc.) on a monthly basis. Tables for numerous locations worldwide have been generated by Thevenard (2009) and Thevenard and Gueymard (2013) for the 21st day of each month. Values for other days of the year should be found by interpolation. Relevant climatic information for about 6430 locations is available on a CD-ROM accompanying the ASHRAE Fundamentals (2013) handbook. A sample of the radiation data needed to determine clear-sky solar irradiance is assembled in Table 4.2 for two locations in the United States (Atlanta, GA, and Phoenix, AZ) with almost identical latitudes.

TABLE 4.2

Monthly Values of Optical Depths for Clear-Sky Solar Beam and Diffuse Irradiance for Two Locations in the United States

	Atlanta, GA				Phoenix, AZ			
Month	τ_{beam}	τ_{dif}	$I_{beam,noon}$[a]	$I_{dif,noon}$[a]	τ_{beam}	τ_{dif}	$I_{beam,noon}$[a]	$I_{dif,noon}$[a]
January	0.334	2.614	885.4	77.5	0.307	2.638	919.6	75.1
February	0.324	2.580	931.9	87.6	0.315	2.557	942.0	89.4
March	0.355	2.474	920.3	104.7	0.338	2.442	937.7	107.9
April	0.383	2.328	902.6	126.4	0.316	2.453	969.2	111.4
May	0.379	2.324	905.7	128.5	0.316	2.412	965.9	117.6
June	0.406	2.270	877.4	135.5	0.322	2.386	955.3	120.7
July	0.440	2.202	846.0	144.5	0.363	2.359	915.3	123.4
August	0.427	2.269	853.4	133.1	0.377	2.366	900.0	120.7
September	0.388	2.428	877.4	109.0	0.351	2.469	915.1	104.4
October	0.358	2.514	883.1	92.9	0.339	2.479	903.0	95.7
November	0.354	2.523	851.8	84.1	0.321	2.554	892.6	80.9
December	0.335	2.618	863.1	73.6	0.305	2.618	900.5	72.6

Source: Extracted from ASHRAE, *Handbook of Fundamentals*, American Society of Heating, Refrigerating and Air-Conditioning Engineers, Atlanta, GA, 2013. Copyright ASHRAE, www.ashrae.org.

The corresponding beam and diffuse radiation values at noon are shown (see Example 4.9 for calculation procedure).

[a] Units of irradiance are in W/m^2. Conversion $1\ W/m^2 = 0.317\ Btu/(h \cdot ft^2)$.

Finally, correlations for the air mass exponents *ab* and *ad* in terms of τ_{beam} and τ_{dif} have been proposed:

$$ab = 1.454 - 0.406 \times \tau_{beam} - 0.268 \times \tau_{dif} + 0.021 \times \tau_{beam} \times \tau_{dif} \quad (4.26)$$

$$ad = 0.507 + 0.205 \times \tau_{beam} - 0.080 \times \tau_{dif} - 0.190 \times \tau_{beam} \times \tau_{dif} \quad (4.27)$$

The use of the aforementioned equations is illustrated by the following example.

Example 4.9: Clear-Sky Calculation

Calculate clear-sky beam normal and diffuse horizontal solar irradiance in Phoenix, AZ, for June 21 at noon solar time.

Given: From Example 4.7, extraterrestrial irradiance at normal incidence for the day is

$$I_{0,norm} = 1322.6 \text{ W/m}^2$$

Latitude $\lambda = 33.3°$, declination $\delta = 23.45°$, and hour angle $\omega = 0°$

Lookup values: From Table 4.2: $\tau_{beam} = 0.322$ and $\tau_{dif} = 2.386$

Find: I_{beam}, I_{dif}

Solution

The zenith angle θ_s is determined from Equation 4.9:

$$\cos\theta_s = \cos 33.3° \times \cos 23.45° \times \cos 0° + \sin 33.3° \times \sin 23.45° = 0.985$$

Hence, $\theta_s = 9.84°$.

The azimuth angle is determined from Equation 4.10:

$$\sin\phi_s = \frac{\cos 23.45° \times \sin 0°}{\sin 9.84°} = 0.0$$

Hence, $\phi_s = 0°$.

From Equation 4.21:

$$m = \frac{1}{\cos\theta_s} = 1.01$$

Then, from Equations 4.26 and 4.27

$ab = 1.454 - 0.406 \times 0.322 - 0.268 \times 2.386 + 0.021 \times 0.322 \times 2.386 = 0.700$

$ad = 0.507 + 0.205 \times 0.322 - 0.080 \times 2.386 - 0.190 \times 0.322 \times 2.386 = 0.236$

and from Equations 4.24 and 4.25

$$I_{beam} = 1322.6 \times \exp(-0.322 \times 1.01^{0.70}) = 955.3 \text{ W/m}^2 \ (302.8 \text{ Btu/[h}\cdot\text{ft}^2])$$

$$I_{dif} = 1322.6 \times \exp(-2.386 \times 1.01^{0.236}) = 120.7 \text{ W/m}^2 \ (38.3 \text{ Btu/[h}\cdot\text{ft}^2])$$

Comments

These irradiance values are consistent with those listed in Table 4.2. This table also assembles similar irradiance values for all months for both Atlanta, GA, and Phoenix, AZ. The same procedure is followed in order to determine beam normal and horizontal diffuse irradiance for hours other than solar noon. The calculation procedure follows the column quantities shown in Table 4.3 for Phoenix, AZ, for the month of June. The irradiance values are symmetric about solar noon, and hence, only values for the forenoon are shown. Figure 4.15 illustrates the diurnal variation of the three different clear-sky radiation components.

The hourly clear-sky values are greater than those assumed for hourly load calculations over the year (namely, TMY data—see Section 4.8). These values are to be used for peak design day load calculations (see Chapter 9) and should not be used for annual building load predictions.

TABLE 4.3

Clear-Sky Calculation Results for Different Hours of the Day for June 21 in Phoenix, AZ

Hour	ω, degrees	θ_s, degrees	ϕ_s, degrees	θ_i, degrees	*m* (—)	I_{beam}, W/m²	I_{dif}, W/m²	$I_{glo,hor}$, W/m²
6	−90	77.38	−109.93	77.38	4.58	519.8	43.4	156.9
7	−75	65.36	−102.86	65.36	2.40	730.2	70.4	374.8
8	−60	53.00	−95.81	53.00	1.66	835.5	89.8	592.6
9	−45	40.48	−87.93	40.48	1.31	895.5	103.8	785.0
10	−30	28.05	−77.29	28.05	1.13	930.7	113.3	934.7
11	−15	16.43	−57.07	16.43	1.04	949.4	118.8	1029.5
12	0	9.84	0.00	9.84	1.01	955.3	120.7	1061.9

Note: The radiation values are symmetrical about solar noon.

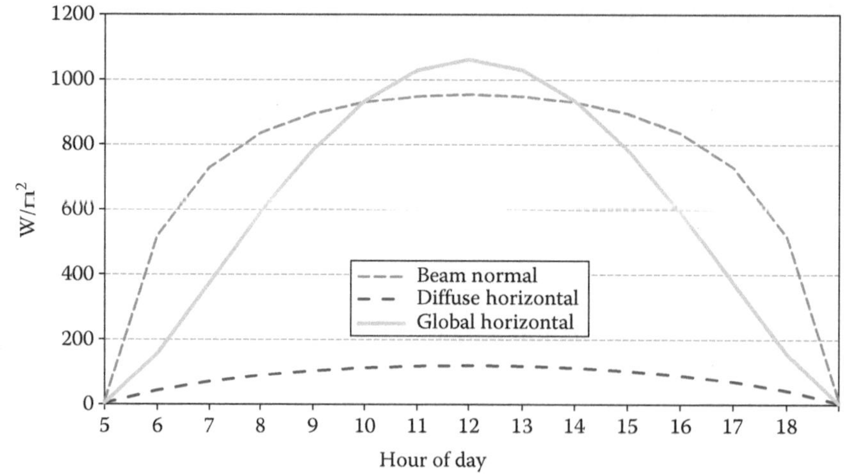

FIGURE 4.15
Diurnal variation in clear-sky radiation values for Phoenix, AZ, during June 21.

4.7 Transposition Models for Tilted and Vertical Surfaces

The previous section described calculation methods to estimate clear-sky beam normal and horizontal diffuse radiation. However, the building designer is more interested in using these values to estimate radiation on surfaces with arbitrary tilt and orientation; algorithms to achieve this are referred to as "transposition models."

Solar radiation on tilted surfaces contains three components: direct radiation, diffuse sky radiation, and radiation reflected from the ground. For diffuse radiation, from sky and from ground, one also needs to consider the angular distribution. The simplest hypothesis is to assume that radiation from the sky is isotropic, i.e., uniform sky dome. More accurate anisotropic models have been developed for solar energy applications (see Duffie and Beckman, 1991) that differ somewhat from anisotropic sky models recommended for building energy applications. The most recent transposition models proposed in ASHRAE Fundamentals (2013) are described in the following. The isotropic sky model that is very simple to use and is still popular among solar professionals though it is somewhat conservative (i.e., tends to under-predict irradiance) is described first.

4.7.1 Isotropic Sky Model

The isotropic sky model is based on the assumption that sky-diffuse and ground-reflected solar components are uniformly distributed over the sky dome (Figure 4.14). If an unshaded flat surface is tilted at an angle θ_p, then the global radiation on a tilted surface $I_{glo,T}$ is the sum of the direct, sky-reflected diffuse, and ground-reflected diffuse components:

$$I_{glo,T} = I_{beam} \cos\theta_i + I_{dif} F_{sky} + I_{glo,hor} \rho_g F_{grd} \tag{4.28}$$

where

$I_{glo,hor}$ is the global irradiance on the horizontal
ρ_g is the reflectivity of the ground (or ground albedo)
F_{sky} and F_{grd} are conversion factors skin to the radiation shape factors; see Section 2.10.4

The first term of Equation 4.28 is due to the direct solar component given by the product of the beam normal irradiance I_{beam} and the cosine of the incidence angle ($\cos\theta_i$). There are two conversion factors:

- For isotropic sky radiation from the sky:

$$F_{sky} = \frac{1+\cos\theta_p}{2} \tag{4.29}$$

- For isotropic radiation reflected from the ground:

$$F_{grd} = \frac{1-\cos\theta_p}{2} \tag{4.30}$$

Since I_{dif} is the diffuse radiation on a horizontal surface, $I_{glo,hor}$ can be determined from

$$I_{glo,hor} = I_{beam} \cos\theta_s + I_{dif} \tag{4.31}$$

Therefore, the global irradiance on the tilted plane is

$$I_{glo,T} = I_{beam}\cos\theta_i + I_{dif}\left(\frac{1+\cos\theta_p}{2}\right) + I_{glo,hor}\cdot\rho_g\left(\frac{1-\cos\theta_p}{2}\right) \tag{4.32}$$

Typical values of the ground albedo or reflectivity from the ground for common surfaces are given in Table 4.4. The reflectivity depends strongly on the condition of the surface. Snow cover, in particular, can greatly increase the reflectivity. Values of 0.2 are often assumed for ground without snow cover and 0.7 with snow cover, unless better information is available. The ground albedo can also vary with the angle of incidence, although it is usually not significant in view of the overall uncertainties, as shown by the data in Kuehn et al. (1998).

For the important case of vertical surfaces, the angle $\theta_p = 90°$ and Equation 4.32 reduces to

$$I_{glo,vert} = I_{beam}\cos\theta_i + \frac{I_{dif}}{2} + \frac{I_{glo,hor}\rho_g}{2} \tag{4.33}$$

Strictly speaking, these equations are correct only when the ground in front of the surface is not shaded. But with partially shaded ground, such as for a west-facing surface in the morning, a correct calculation becomes so complicated that the gain in accuracy has usually not been deemed worth the effort.

TABLE 4.4

Albedo or Reflectivity of Common Exterior Surfaces

Surface	Reflectivity (ρ_g)
Natural surfaces (no vegetation)	
Snow (fresh)	0.75
Soils (clay, loam, etc.)	0.14
Water (relatively large incidence angles)	0.07
Artificial surfaces	
Bituminous and gravel roof	0.13
Blacktop, old	0.10
Building surfaces, dark (red brick, dark paints, etc.)	0.27
Building surfaces, light (light brick, light paints, etc.)	0.60
Concrete, new	0.35
Concrete, old	0.25
Crushed rock surface	0.20
Earth roads	0.04
Vegetation	
Coniferous forest (winter)	0.07
Leaves, dead	0.30
Forests in autumn, ripe field crops, plants	0.26
Grass, dry	0.20
Grass, green	0.26

Sources: From Hunn, B.D. and Calafell, D.O., *Solar Energy*, 19, 87, 1977; Kuehn, T.W. et al., *Thermal Environmental Engineering*, 3rd ed., Prentice-Hall, Englewood Cliffs, NJ, 1998.

Example 4.10: Isotropic Clear Day Calculations

Calculate the clear day irradiance on a south-facing vertical wall for the conditions of Example 4.9 (noon on June 21st) if the reflectivity of the ground is $\rho_g = 0.2$.

Given: $\theta_p = 90°$ for vertical wall and $\phi_p = 0$ for south-facing wall

From Example 4.9, $\phi_s = 0°$, $\theta_s = 9.84°$, $I_{beam} = 955.3$ W/m², and $I_{dif} = 120.7$ W/m²

Find: $I_{glo,vert}$

Solution

From Equation 4.15, $\cos\theta_i|_{vert} = \sin\theta_s\cos(\phi_s - \phi_p)$, which results in $\theta_i|_{vert} = (90° - \theta_s) = 80.16°$

First, find the global horizontal irradiance from Equation 4.31: $I_{glo,hor} = 955.3 \times \cos 9.84° + 120.7 = 1061.9$ W/m². Then, insert into Equation 4.33 for the irradiance on the vertical surface

$$I_{glo,vert} = 955.3\times\cos 80.16° + \frac{120.7}{2} + 1061.9\times\frac{0.2}{2}$$

$$= 163.26 + 60.35 + 106.19$$

$$= 329.8\ \text{W/m}^2\ (104.5\ \text{Btu/[h}\cdot\text{ft}^2])$$

4.7.2 ASHRAE Anisotropic Sky Model

The isotropic sky assumption neglects the observed effects of brightening of the sky around the solar disk (referred to as circumsolar brightening) as well as brightening of the horizon. Hence, while the treatment of beam radiation remains unaltered, the different anisotropic sky models differ from each other and from the isotropic model in how the nonuniform sky effects are treated. The widely used ones are those proposed by Hays and subsequent workers (described in Duffie and Beckman, 1991), by Gueymard (1987) and Perez et al. (1990). Note that these are meant to apply to all conditions, both clear-sky as well as cloudy.

The ASHRAE anisotropic sky model (ASHRAE Fundamentals, 2013) is recommended for building load calculations. It applies only to clear-sky conditions and should not be used for cloudy conditions. The same functional form as that for global radiation on the tilted surface given by Equation 4.32 is assumed with the conversion factor F_{sky} (Equation 4.29) modified as follows (while the conversion factor F_{grd} remains unaltered):

- Sky radiation component:

$$F_{sky} = Y\sin\theta_p + \cos\theta_p \quad \text{if } \theta_p \le 90° \tag{4.34}$$

$$F_{sky} = Y\sin\theta_p \quad \text{if } \theta_p > 90°$$

where

$$Y = \max(0.45, 0.55 + 0.437\cos\theta_i + 0.313\cos^2\theta_i) \quad (4.35)$$

θ_p is the tilt angle of the plane with respect to the horizontal

θ_i is the solar incidence angle on the tilted surface

Note that for vertical surfaces $\theta_p = 90°$ and so Equation 4.35 results in $F_{sky} = Y$. Also, in general, $Y = 0.45$ for $\theta_i > 101°$.

Example 4.11: Anisotropic Sky Calculations

Find the beam, diffuse, and ground-reflected components of clear-sky solar irradiance on a window located in the south wall in Example 4.10.

Given: From Example 4.10, $I_{beam} = 955.3$ W/m², $I_{dif} = 120.7$ W/m², and $I_{glo,hor} = 1061.9$ W/m². Also, the angle of incidence $\theta_i = 80.16°$. The surface tilt is $\theta_p = 90°$, and ground reflectance is assumed to be 0.2.

Solution

From Equation 4.35:

$$Y = \max[0.45, 0.55 + 0.437 \times \cos 80.16° + 0.313 \times \cos^2 80.16°] = \max[0.45, 0.634] = 0.634$$

From Equation 4.34: the conversion factor for diffuse radiation is

$$F_{sky} = 0.634 \times \sin 90° + \cos 90° \quad \text{if } \theta_p \leq 90°$$

which leads to $F_{sky} = 0.634$, while that for ground-reflected radiation is the same as before, $F_{grd} = 0.5$.

Finally, from Equation 4.33,

$$\begin{aligned} I_{glo,vert} &= 955.3 \times \cos 80.16° + 120.7 \times 0.634 + 1061.9 \times \frac{0.2}{2} \\ &= 163.26 + 76.52 + 106.19 \\ &= 346.0 \text{ W/m}^2 \text{ (109.7 Btu/[h} \cdot \text{ft}^2\text{])} \end{aligned}$$

Comments

While the solar irradiance on the window was 329.8 W/m² based on the isotropic sky model, the anisotropic sky model yielded a higher value of 346.0 W/m². The difference is due to the sky-diffuse component that increased from 60.3 to 76.5 W/m², almost a 27% increase!

4.8 Measured Solar Radiation Data Worldwide*

Insolation data have been measured in many locations worldwide (Hulstrom, 1989), the type and quality of data spanning a wide range. Even then, the measured solar radiation data are relatively sparse. Long-term data collection is costly, all the more if one wants hourly rather than daily time resolution or anything beyond global horizontal radiation. Standard weather stations do not measure solar radiation but provide only some indirect estimate such as cloud cover or hours of sunshine. Such information is nonetheless of value because it can be used with various empirical and analytical models to estimate both the total horizontal solar radiation, and subsequently, the beam solar radiation.

Within the past 15 years, there have been tremendous advances in both the technology and spatial/temporal availability of satellite-derived solar radiation. This effort has been driven almost entirely by the solar energy industry for whom accurate assessment and forecast of solar radiation is critical. The ramifications for the building industry are mixed—while satellite-derived solar data are now available down to hourly observations at the 1 km resolution for recent years, these data are available commercially at high cost, although older hourly data sets at lower resolution (United States 10 km resolution 1998–2009, California 1 km resolution 2003–2010) are available as public data (Clean Power Research, 2014). For locations outside the United States, there are no publicly available hourly data sets, although there are detailed maps of average monthly solar radiation (GeoModel, 2014).

For various reasons, measured solar radiation will continue to have many uses, such as for calibrating existing solar models and improving the interpolation between stations where radiation has actually been measured. In the United States, the SOLMET network (NOAA, 1978) had been the major source of measured hourly data, but it was functional only from 1977 through 1980, covered just 26 U.S. stations, and fell into neglect in the early 1980s. Another source of measured solar radiation is the World Radiation Data Centre (WRDC) that was established by the WMO in 1964 and hosted in St. Petersburg Russia, although the website is currently being maintained by NREL (2014). The WRDC was meant to be an archive of measured solar radiation from contributing national meteorological services. The data include hourly solar radiation data from over 40 GAW (Global Atmospheric Watch) stations around the world, plus daily data for about 50 other stations. The temporal coverage is very uneven, from over 20 years down to 1–3 years in many stations.

It should be emphasized that detailed measurements of solar radiation, especially of the direct and diffuse components, are not usually available. For example, contrary to what many people think, the solar radiation contained in the National Solar Radiation Data Base (NSRDB) and the TMY weather files are not measured, but either largely or entirely modeled. In the original 1961–1990 NSRDB and TMY2 weather files, 93% of the solar radiation is modeled and only 7% from

* We thank Joe Huang for updating and enhancing this section.

measurements provided by SOLMET (NREL, 1992). In the 1991–2010 NSRDB update and TMY3, all the solar radiation data are either modeled or based on satellite-derived data (NREL, 2011).

The ASHRAE Handbook CD (ASHRAE Fundamentals, 2013) used two primary sources of weather data in calculating temperature, humidity, and wind design values for 6443 locations: (1) Integrated Surface Dataset data for stations from around the world provided by the National Climatic Data Center located in Nashville, TN, for the period 1986–2010, and (2) the hourly weather records for the period 1986–2010 for 559 Canadian locations. The clear-sky solar radiation is calculated using the ASHRAE Clear-Sky Model 2013 (Thevenard and Gueymard, 2013) for the same 6443 locations, although the model could be applied to any location in the world because the location-specific inputs are based on gridded data base available for the entire world (Gueymard and Thevenard, 2009).

4.9 Statistical Correlation Models

The previous sections dealt with recommended procedures to estimate clear-sky insolation for different months and hours of the days for locations where the necessary model coefficients (such as those shown in Table 4.2) are available. Such information is used for peak load calculations during building design and for HVAC equipment sizing. However, there are also instances when an analyst wishes to generate radiation data for non-clear-sky conditions, say for energy estimation purposes. One could use measured climatic data (as discussed in the previous section) and use say the isotropic sky model or the anisotropic sky models proposed in the solar energy literature to estimate radiation availability on surfaces with arbitrary tilt and orientation. Often, such measured data may not be available, and in order to bridge the gap between what is needed and what has actually been measured, various correlations and models have been developed. Extensive research has been done in the framework of solar energy applications and generalized correlations between the various solar radiation components have been identified.

In this section, we present a method that needs only a minimum of input: the daily clearness index K_T, defined by Equation 4.23. Using only this piece of information, one can calculate the long-term average irradiance on any surface, and the accuracy is sufficient to reproduce the monthly average radiation incident on solar collectors within a few percent (Rabl, 1985; Reddy, 1987). The starting point is thus either $\bar{K}_T$ or equivalently $\bar{H}_{glo,hor}$; the overbar indicates long-term monthly averages.

4.9.1 Monthly Mean Daily Diffuse from Daily Global Insolation

The first step is to estimate the monthly average daily diffuse irradiation on the horizontal surface $\bar{H}_{dif}$ from the correlation

$$\frac{\bar{H}_{dif}}{\bar{H}_{glo,hor}} = 0.775 + 0.347 \times \frac{(\omega_{ss} - 90°)\pi}{180°} - \left[0.505 + 0.261 \times \frac{(\omega_{ss} - 90°)\pi}{180°}\right] \times \cos\frac{360°(\bar{K}_T - 0.9)}{\pi} \tag{4.36}$$

where ω_{ss} is the sunset hour angle given by Equation 4.12. The aforementioned function is plotted in Figure 4.16.

Example 4.12: Monthly Average Diffuse Horizontal Radiation

Find the monthly average daily diffuse irradiation on a horizontal surface for Phoenix, AZ, for the month of June given that $\bar{H}_{glo,hor} = 31.09$ MJ/m².

Given: From Example 4.4, $\omega_{ss} = 106.62°$ for June

From Example 4.8, $\bar{H}_0 = 41.57$ MJ/m²

Find: $\bar{H}_{dif}$

Solution

First, determine the monthly mean clearness index from Equation 4.23:

$$\bar{K}_T = \frac{\bar{H}_{glo,hor}}{\bar{H}_0} = \frac{31.09}{41.57} = 0.748$$

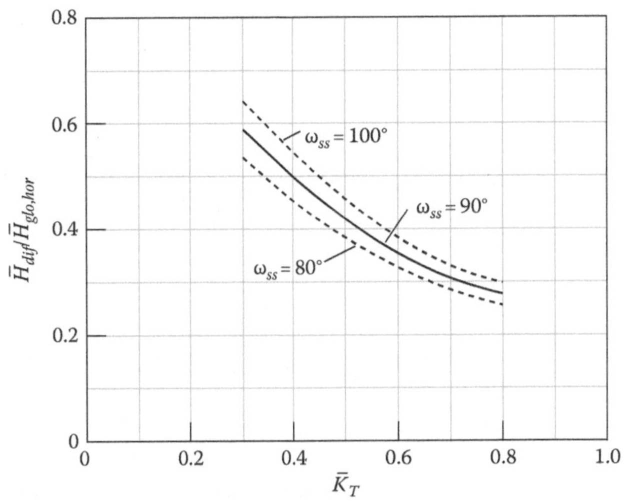

FIGURE 4.16
Correlation of Collares-Pereira and Rabl (1979) for the ratio of long-term averages of daily diffuse and global solar irradiation, as a function of clearness index $\bar{K}_T$ and sunset hour angle ω_{ss}.

Using Equation 4.36:

$$\frac{\bar{H}_{dif}}{\bar{H}_{glo,hor}} = 0.775 + 0.347 \times \frac{(106.62 - 90°)\pi}{180°} - \left[0.505 + 0.261 \times \frac{(106.62 - 90°)\pi}{180°}\right] \times \cos\frac{360°(0.748 - 0.9)}{\pi}$$

$$= 0.8756 - 0.5807 \times 0.954 = 0.3216$$

Finally,

$$\bar{H}_{dif} = 0.3216 \times 31.09 = 10.0 \text{ MJ/m}^2$$

4.9.2 Monthly Mean Hourly from Monthly Mean Daily Insolation

As the second step, one converts the daily irradiation on a horizontal surface to long-term average irradiance at any moment of the day using the correlations

$$\bar{I}_{glo,hor} = r_{glo}(\omega_{ss}, \omega) \times \bar{H}_{glo,hor} \tag{4.37}$$

and

$$\bar{I}_{dif} = r_{dif}(\omega_{ss}, \omega) \times \bar{H}_{dif} \tag{4.38}$$

where ω_{ss} and ω are the hour angles corresponding to sunset time (Equation 4.12 and Figure 4.7) and time of day (Equation 4.4), respectively. The correlation functions are given by

$$r_{dif}(\omega_{ss}, \omega) = \frac{\pi}{\tau_{day}} \times \frac{\cos\omega - \cos\omega_{ss}}{\sin\omega_{ss} - (\pi \times \omega_{ss}/180°) \times \cos\omega_{ss}} \tag{4.39}$$

and

$$r_{glo}(\omega_{ss}, \omega) = (a + b\cos\omega) \times r_{dif}(\omega_{ss}, \omega) \tag{4.40}$$

with

$$\tau_{day} = 24 \text{ h} = 86{,}400 \text{ s}$$

$$a = 0.4090 + 0.5016 \times \sin(\omega_{ss} - 60°) \tag{4.41}$$

and

$$b = 0.6609 - 0.4767 \times \sin(\omega_{ss} - 60°) \tag{4.42}$$

Equations 4.39 and 4.40 are plotted in Figure 4.17a and b. The units are h^{-1} for the left-hand scale and 10^{-6} s^{-1} for the right-hand scale.

The beam normal irradiance can then be determined as

$$\bar{I}_{beam} = \frac{\bar{I}_{glo,hor} - \bar{I}_{dif}}{\cos\theta_s} \tag{4.43}$$

With the isotropy assumption for the diffuse component, one can thus compute the average hourly irradiance on any surface at any time. Integrating from sunrise to sunset, one obtains the daily average irradiation.

Example 4.13: Monthly Mean Hourly Radiation on a Vertical Surface

Using the isotropic sky model, find the long-term average irradiance on the vertical south-facing wall of Example 4.10 (Phoenix, AZ, at noon solar time on June 21) if $\bar{H}_{glo,hor} = 31.09$ MJ/m^2 and $\bar{H}_{dif} = 10.0$ MJ/m^2 (from Example 4.12).

Given: Hour angle $\omega = 0°$, $\theta_i = 80.16°$ and $\theta_s = 9.84°$
From Example 4.4, $\omega_{ss} = 106.62°$ for June 21st

Find: $\bar{I}_{glo,vert}$

Solution

For the conversion from daily to instantaneous insolation, the following quantities are evaluated:
From Equation 4.41

$$a = 0.4090 + 0.5016 \times \sin(106.62 - 60)° = 0.774$$

and from Equation 4.42

$$b = 0.6609 - 0.4767 \times \sin(106.62 - 60)° = 0.314$$

Then, from Equation 4.39

$$r_{dif} = \frac{\pi}{86{,}400} \times \left\{\frac{\cos 0° - \cos 106.62°}{\sin 106.62° - [(\pi \times 106.62)/180°] \times \cos 106.62°}\right\} = 31.34 \times 10^{-6}$$

and from Equation 4.40

$$r_{glo} = (0.774 + 0.314 \times \cos 0°) \times 31.34 \times 10^{-6} = 34.1 \times 10^{-6}$$

Next, the hourly global and diffuse radiation values are determined from Equations 4.37 and 4.38 using the conversion factors r_{glo} and r_{dif} determined earlier:

$$\bar{I}_{glo,hor} = r_{glo}\bar{H}_{glo,hor} = 34.1 \times 10^{-6} \text{ s}^{-1} \times 31.09 \text{ MJ/m}^2 = 1060.2 \text{ W/m}^2$$

$$\bar{I}_{dif} = r_{dif}\bar{H}_{dif} = 31.34 \times 10^{-6} \text{ s}^{-1} \times 10.0 \text{ MJ/m}^2 = 313.4 \text{ W/m}^2$$

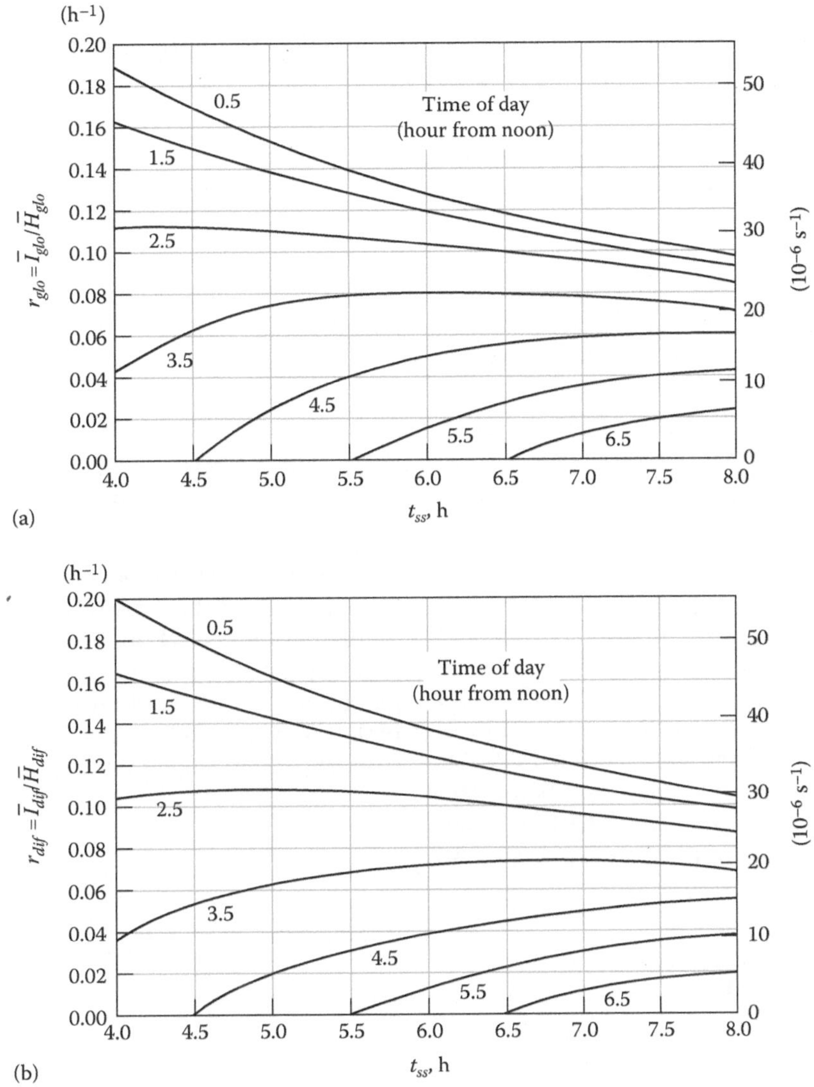

FIGURE 4.17
Correlation between ratio of long-term average instantaneous to daily irradiation on a horizontal surface versus sunset hour t_{ss}: (a) for global insolation and (b) for diffuse insolation.

Then, the beam normal irradiance is found from Equation 4.43:

$$\bar{I}_{beam} = \frac{1060.2 - 313.4}{\cos 9.84°} = 874.5\ \text{W/m}^2$$

Finally, the global irradiance on the vertical south-facing surface is determined from Equation 4.33:

$$\bar{I}_{glo,vert} = 874.5 \times \cos 80.16° + \frac{313.4}{2} + 1060.2 \times \frac{0.2}{2}$$

$$= 149.4 + 156.7 + 106.0 = 412.1\ \text{W/m}^2$$

4.9.3 Monthly Mean Insolation on Vertical Surfaces

Radiation on vertical surfaces is a particularly important quantity since it affects both heat transmission through walls as well as heat loads through windows. Even though the calculations involve fairly complicated combinations of trigonometric functions, it turns out that the results for daily totals can be approximated by a simple correlation, as shown by Potter et al. (1989). Using the data of SERI (1980), they have calculated the monthly average irradiation $\bar{H}_{glo,vert}$ on vertical surfaces of the principal orientations for latitudes from 30° to 45°. They found that the

TABLE 4.5

Coefficients for Equation 4.44: $\bar{H}_{glo,vert} = a'\bar{K}_T + b'$, Btu/(ft^2·day), Proposed by Potter et al. (1989)

	North				East/West				South			
January												
λ =	30°	35°	40°	45°	30°	35°	40°	45°	30°	35°	40°	45°
a′ =	326	217	209	210	1288	1171	1066	927	2783	2925	2861	2651
b′ =	113	122	86	49	−66	−63	−63	−52	−283	−352	−327	258
April/October												
λ =	30°	35°	40°	45°	30°	35°	40°	45°	30°	35°	40°	45°
a′ =	459	425	430	448	1849	1872	1891	1904	1310	1544	1784	2018
b′ =	298	304	285	260	−31	−49	−67	−84	137	85	17	−51
July												
λ =	30°	35°	40°	45°	30°	35°	40°	45°	30°	35°	40°	45°
a′ =	848	817	830	854	1917	2010	2082	2189	599	897	1136	1430
b′ =	251	269	268	266	3	−24	−40	−77	367	291	243	164

relation between $\bar{H}_{glo,vert}$ and $\bar{K}_T$ can be represented by the simple linear model:

$$\bar{H}_{glo,vert} = a'\bar{K}_T + b' \tag{4.44}$$

The coefficients a' and b' depend on the latitude, time of year, and surface orientation, and are listed in Table 4.5. The accuracy is excellent, the average error being less than 1%. The results for January, April/October, and July are shown in Figure 4.18, for north-, east/west-, and south-facing surfaces.

Example 4.14: Monthly Mean Daily Radiation on Vertical Surfaces

Using the Potter et al. correlation given by Equation 4.44, estimate the long-term average irradiation on a west-facing vertical wall in Phoenix, AZ, during July.

Given: Latitude = 33.3°

From Example 4.12, for June $\bar{K}_T = 0.748$ (assume same value for July)

Lookup values: From Table 4.4, by linear interpolation between 30° and 35° latitudes:

$a' = 1982.1$ and $b' = -15.9$

Find: $\bar{H}_{glo,vert}$

Using Equation 4.44

$$\bar{H}_{glo,vert} = 1982.1 \times 0.748 - 15.9$$

$$= 1466.7 \text{ Btu/(ft}^2 \cdot \text{day) (16.65 MJ/m}^2 \cdot \text{day)}$$

4.9.4 Hourly Diffuse from Hourly Global Horizontal Insolation

The previous section presented generalized correlations valid for monthly mean radiation quantities. Generalized correlations have also been developed for specific hourly (as against monthly mean hourly) conversions. A correlation analogous to Equation 4.36 and Figure 4.16 but for hourly time scales is often used. For example, if only the global horizontal irradiance $I_{glo,hor}$ is given, one can estimate the diffuse irradiance I_{dif} on the horizontal surface from the following equation (Erbs et al., 1982):

$$\frac{I_{dif}}{I_{glo,hor}} = \begin{cases} 1.0 - 0.09k_T & \text{for } 0 \le k_T \le 0.22 \\ 0.9511 - 0.1604k_T + 4.388k_T^2 - 16.638k_T^3 + 12.336k_T^4 & \text{for } 0.22 \le k_T \le 0.80 \\ 0.165 & \text{for } 0.80 \le k_T \end{cases} \tag{4.45}$$

where k_T is the hourly clearness index determined as the ratio of terrestrial and extraterrestrial irradiance on a horizontal surface, or

$$k_T = \frac{I_{glo,hor}}{I_{0,norm} \cos\theta_s} \tag{4.46}$$

with θ_s being the solar zenith angle (= incidence angle on the horizontal).

This correlation is plotted in Figure 4.19. The general shape of the curve is consistent with intuition. Low k_T corresponds to overcast skies when all the radiation is diffuse, and so the ratio ($I_{dif}/I_{glo,hor}$) is unity. As k_T increases, the diffuse component, as a fraction of the total, decreases to 0.165 corresponding to the maximum k_T value (around 0.8). Obviously, the exact form of the correlation between diffuse and global insolation depends on the time interval over which the insolation is averaged (hourly, daily, or monthly).

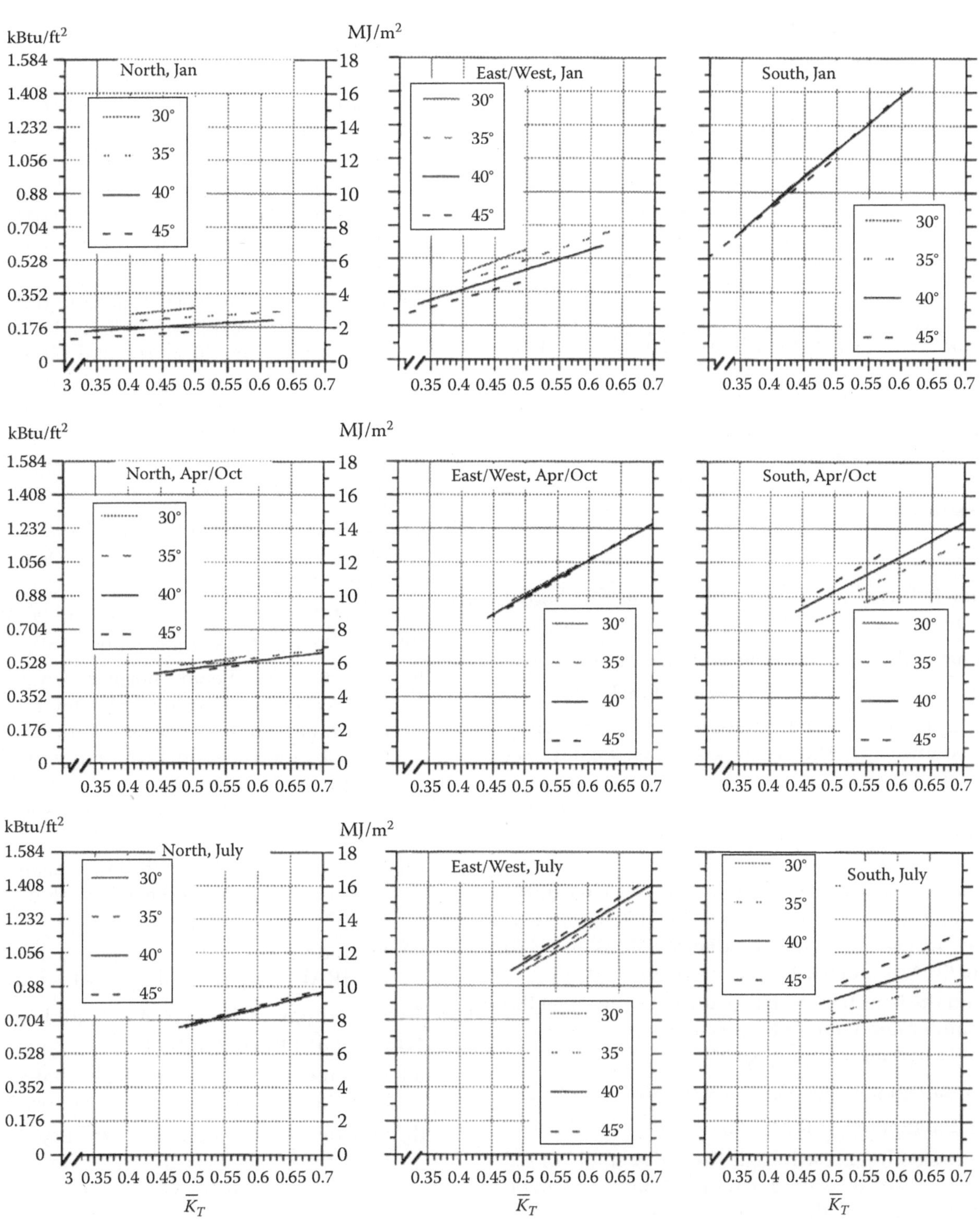

FIGURE 4.18
Monthly average daily global irradiation $\bar{H}_{glo,vert}$ on vertical surfaces in January, April, and July as a function of monthly average clearness index $\bar{K}_T$ and latitude λ (in degrees). (From Potter, R. et al., Simplification of monthly solar radiation calculations for a vertical surface, in *Proceedings of 1989 Conference on American Section, International Solar Energy Society*, Denver, CO, 1989, p. 450.)

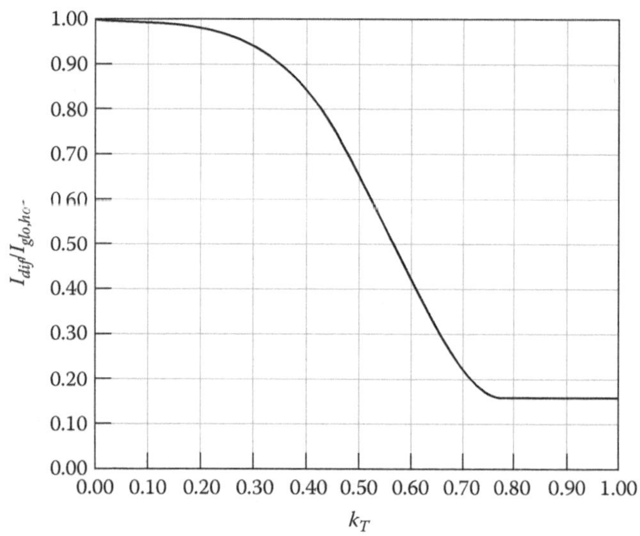

FIGURE 4.19
Correlation Equation 4.44 between the ratio of hourly diffuse irradiance I_{dif}, and global irradiance $I_{glo,\,hor}$ and hourly clearness index k_T.

Problems

Numbers 1–4 given in parenthesis denote the degree of difficulty.

4.1 What is the purpose of the equation of time? *(1)*

4.2 What is the value of the solar constant? *(1)*

4.3 How much does the extraterrestrial normal solar irradiance vary during the course of the day and year? *(2)*

4.4 What approximately is the highest value of the solar irradiance that might be incident on the horizontal surface of a building? *(2)*

4.5 Find the global horizontal irradiance when the direct normal irradiance is 700 W/m^2 [221.9 Btu/(h·ft^2)], the diffuse horizontal irradiance 150 W/m^2 [47.6 Btu/(h·ft^2)], and the angle of incidence is 30°. *(1)*

4.6 Find the length of the day (sunrise to sunset) at the summer and winter solstices in

(a) Honolulu (λ = 21.03°N)

(b) Stockholm, Sweden (λ = 59.35°N)

(c) Singapore (λ = 0°N) *(2)*

4.7 Estimate the azimuth angle (from due south) of the white board in your classroom. What is the angle of incidence of the sun on this surface at solar noon, equinox? *(2)*

4.8 (a) Write an equation for the number of hours per day when direct solar radiation can reach an unshaded fixed surface at arbitrary tilt and zero azimuth.

(b) Evaluate this equation for the case of tilt = latitude at summer solstice and at winter solstice. *(2)*

4.9 Consider a sundial built as a vertical rod of 1 m (3.281 ft) length that casts a shadow on a flat horizontal surface. The location is Princeton, NJ, with approximate latitude 40°N and longitude 75°W.

(a) What time of day (solar time) and time of year is it when the shadow is 0.50 m (1.641 ft) long, pointing due north? How many solutions are there?

(b) What time of day (solar time) and time of year is it when the shadow is 0.50 m (1.641 ft) long, pointing 45° (in the horizontal plane) east of north? (By contrast to part *a*, this requires two equations in two unknowns; but they can be solved in closed form.)

(c) What are the corresponding standard times? *(4)*

4.10 Consider a south-facing unshaded vertical wall at a latitude of 45°N. Assume it to be 1:30 p.m. solar time on January 21.

(a) Find the zenith and azimuth angles of the sun.

(b) Find the incidence angle on the wall.

(c) Suppose the beam normal irradiance is 700 W/m^2 [221.9 Btu/(h·ft^2)] and the diffuse horizontal irradiance is 100 W/m^2 [31.70 Btu/(h·ft^2)]. Find the global irradiance on the wall per unit area if the ground is a diffuse reflector with reflectivity 0.2. Assume isotropic sky model.

(d) How does the answer in part *c* change if the reflectivity is increased to 0.7, a value typical of snow?

(e) How does the answer in part *d* change the ASHRAE anisotropic sky model is used? *(4)*

4.11 Two buildings in a town at a latitude 40°N are arranged like in Figure P4.11.

(a) Find the portion of building *A* that is shaded by building *B* at 11:00 a.m. on August 21.

(b) Superimpose the outline of building *B* as seen from the corner *P* on the sun-path diagram of Figure 4.6. (For simplicity, take only the corners and join them by straight lines, even though, strictly speaking, nonvertical straight lines look curved when viewed in a sun-path diagram.)

(c) Use this diagram to estimate how many hours a window located at the corner *P* can receive direct sunlight on August 21. *(4)*

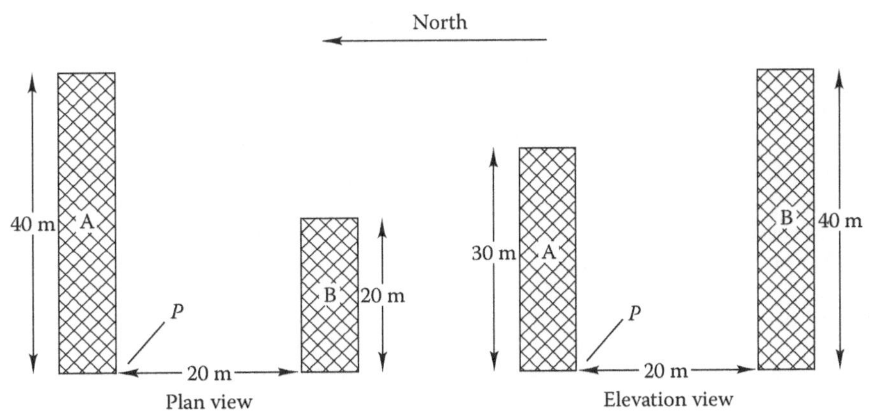

FIGURE P4.11

4.12 You want to measure the height of a telephone pole without climbing it. The latitude is 40°, and the length of the shadow at solar noon on the summer solstice is 5 m. How tall is the telephone pole? *(1)*

4.13 You want to measure the height of a telephone pole without climbing it and without waiting for a solstice. The location is Boulder, Colorado (40.00°N and 105.27°W). You find that the shadow is 10 m at 10 a.m. MDT on May 15. How tall is the telephone pole? *(2)*

4.14 You are designing a sanctuary in Truth or Consequences, New Mexico (33.23°N and 107.27°W), in which you want the sun to illuminate a golden sphere at sunrise on the summer solstice as the sunlight passes through a notch between two rocks. The ground is flat at that location. At what direction (i.e., angle from due east) should the sphere be placed relative to the notch? *(2)*

4.15 At what latitude can you collect the greatest amount of energy on a horizontal surface during the day of June 21 if the sky is clear? At what latitude is the lowest amount collected? *(1)*

4.16 Use the Potter et al. correlation (Equation 4.44) to find the average daily solar radiation on a vertical west-facing window of 10 ft² area in Phoenix, AZ (λ = 33°N), in January ($\bar{K}_T = 0.61$) and June ($\bar{K}_T = 0.76$). *(1)*

4.17 Repeat Examples 4.12 and 4.13 for the month of December given that $\bar{H}_{glo,hor} = 10.57$ MJ/m^2. *(3)*

4.18 You are to calculate clear-sky irradiance at Atlanta, GA, using data provided in Table 4.2.

(a) What are the beam normal and diffuse horizontal irradiance on August 15 at 3 p.m. solar time?

(b) Use the ASHRAE anisotropic sky model to calculate the direct and diffuse radiation for a west-facing vertical wall on August 15 at 3 p.m. solar time. Neglect ground-reflected radiation.

(c) Repeat assuming the isotropic sky model and compare your result with that of (b). *(2)*

4.19 Set up a spreadsheet to calculate and plot, as a function of the time of day, the incidence angle of the sun on the horizontal and vertical surfaces facing the four cardinal directions (akin to Table 4.3). Add comments for documentation, and define names for the key variables (declination, latitude, etc.) to facilitate future reuse and expansion of the spreadsheet. Produce the plots for the 21st day for each of the 12 months, for Atlanta, GA. *(4)*

4.20 Enhance the spreadsheet of Problem 4.19 to reproduce the irradiance values for beam normal and diffuse horizontal at noon for the 21st day of each month according to the ASHRAE clear-sky model (Section 4.6), as shown in Table 4.2. Verify the values shown in Table 4.2 for Atlanta, GA. *(4)*

4.21 Enhance the spreadsheet of Problem 4.20 so as to calculate the global irradiance versus the time of day for the horizontal and vertical surfaces facing the four cardinal directions for each of the 12 months of the year. Compute the values for Atlanta, GA. Assume a ground reflectivity of 0.2. *(4)*

4.22 You will use the generalized correlations from Section 4.9. Compute the monthly mean hourly global radiation on an equator-facing tilted surface in Rome, Italy, for the month of July at 2 p.m. solar time. The tilt angle is 30° and the ground reflectance is 0.2. The following data for Rome are specified: latitude of 41.8°N and clearness index 0.55. Assume the isotropic sky model to apply. *(3)*

References

ASHRAE Fundamentals (2013). *Handbook of Fundamentals.* American Society of Heating, Refrigerating and Air-Conditioning Engineers, Atlanta, GA.

Clean Power Research (2014). www.solaranywhere.com/Public/About.aspx (accessed 2014).

Collares - Pereira, M. and A. Rabl (1979). The average distribution of solar radiation: Correlations between difference and hemispherical and between hourly and daily insolation values. *Solar Energy*, 22, 155–164.

Duffie, J.A. and W.A. Beckman (1991). *Solar Engineering of Thermal Processes*, 2nd ed. Wiley, New York.

Erbs, D.G., S.A. Klein, and J.A. Duffie (1982). Estimation of the diffuse radiation fraction for hourly, daily and monthly-average global radiation. *Solar Energy*, 28, 293.

GeoModel (2014). http://solargis.info/doc/71 (accessed 2014).

Goswami, D.Y., F. Kreith, and J.F. Kreider (2000). *Principles of Solar Engineering*, 2nd ed. Taylor & Francis Group, Philadelphia, PA.

Gueymard, C.A. and D. Thevenard (2009). Monthly average clear-sky broadband irradiance data base for worldwide solar heat gain and building cooling load calculations. *Solar Energy*, 83, 1998–2018.

Gueymard, C.A. (1987). An anisotropic solar irradiance model for tilted surfaces and its comparison with selected engineering algorithms. *Solar Energy*, 38, 367–386 [Erratum, *Solar Energy* 40:175 (1988)].

Hulstrom, R.L., ed. (1989). *Solar Resources.* M.I.T. Press, Cambridge, MA.

Hunn, B.D. and D.O. Calafell (1977). Determination of average ground reflectivity for solar collectors. *Solar Energy*, 19, 87.

Iqbal, M. (1983). *An Introduction to Solar Radiation*. Academic Press, Toronto, Ontario, Canada.

Kuehn, T.W., J.W. Ramsey, and J.L. Threlkeld (1998). *Thermal Environmental Engineering*, 3rd ed. Prentice-Hall, Englewood Cliffs, NJ.

NOAA (1978). SOLMET: Hourly solar radiation—Surface meteorological observations. National Climatic Data Center of the National Oceanic and Atmospheric Administration, Asheville, NC.

NREL (1992). National solar radiation data base User's Manual (1961–1990). National Renewable Energy Laboratory, Golden, CO. Available at http://rredc.nrel.gov/solar/pubs/NSRDB/cover.html. Accessed on October 2014.

NREL (2011). National solar radiation data base, 1991–2010 update: User's manual. Technical Report NREL/TP-581-41364. National Renewable Energy Laboratory, Golden, CO. Available at http://rredc.nrel.gov/solar/old_data/nsrdb/1991-2010/. Accessed on October 2014.

NREL (2014). http://wrdc-mgo.nrel.gov (accessed 2014).

Perez, R., P. Ineichen, R. Seals, J. Michalsky, and R. Stewart (1990). Modeling daylight availability and irradiance components from direct and global irradiance. *Solar Energy*, 44(5), 271–289.

Potter, R., N. Karkamaz, and J.F. Kreider (1989). Simplification of monthly solar radiation calculations for a vertical surface, in *Proceedings of 1989 Conference on American Section, International Solar Energy Society*, Denver, CO, p. 450.

Rabl, A. (1985). *Active Solar Collectors and Their Applications.* Oxford University Press, New York.

Reddy, T.A. (1987). *The Design and Sizing of Active Solar Thermal Systems.* Oxford University Press, Oxford, U.K.

SERI (1980). Insolation data manual. Report SERI/SP-755-789. Solar Energy Research Institute, Golden, CO.

Thevenard, D. (2009). Updating the ASHRAE climatic data for design and standards (RP-1453). ASHRAE Research Project, Final Report. American Society of Heating, Refrigerating and Air-Conditioning Engineers, Atlanta, GA.

Thevenard, D. and C. Gueymard (2013). Updating climatic design data in the ASHRAE 2013 handbook of fundamentals (RP-1613). ASHRAE Research Project, Final Report, American Society of Heating, Refrigerating and Air-Conditioning Engineers, Atlanta, GA.

5

Heat Gains through Windows

ABSTRACT Windows in buildings are major contributors to heat losses/gains impacting both annual energy use and peak design loads. They have relatively low heat resistance to conduction heat flow while being the direct cause of solar heat gains. Despite their poor insulating value, they are essential elements from the standpoints of occupant visual comfort and proper working environment. We first present relevant scientific background on the optical and thermal properties of glazing material. Next, we cover current recommended calculation procedures for determining solar heat gains through glazing material as well as the entire window. This is followed by a treatment of analysis methods, which allow us to ascertain the reduction in solar heat gains due to external overhangs and internal shading devices. Solved examples throughout this chapter illustrate the use of the various calculations procedures as well as provide interpretation of the various optical and thermal values listed in the tables. Finally, we provide a brief discussion of the novel and emerging types of glazing technology. This includes "smart windows" that are glazing materials whose thermal and optical properties can be altered by environmental variables so as to create a more comfortable and energy-efficient work space.

Nomenclature

A	Area, m² (ft²)
A_n^f	Absorptance of glazing layer n (Table 5.1)
a	Window height, m (ft)
b	Window recess, m (ft)
c	Window width, m (ft)
d	Separation between glass panes, mm (in.)
e	Emissivity
F_{shade}	Window shading fraction due to external shading
h_{con}	Convection heat transfer coefficient, W/(m²·K) [Btu/(h·ft²·°F)]
h_i	Indoor surface heat transfer coefficient, W/(m²·K) [Btu/(h·ft²·°F)]
h_o	Outdoor surface heat transfer coefficient, W/(m²·K) [Btu/(h·ft²·°F)]
h_{rad}	Linearized radiative heat transfer coefficient, W/(m²·K) [Btu/(h·ft²·°F)]
h_s	Heat transfer coefficient of space between panes of double glazing, W/(m²·K) [Btu/(h·ft²·°F)]
IAC	Interior solar attenuation coefficient (Equation 5.26)
I	Irradiance, W/m² [Btu/(h·ft²)]
I_{beam}	Beam (direct from solar disk) irradiance, W/m² [Btu/(h·ft²)]
I_{dif}	Diffuse irradiance, W/m² [Btu/(h·ft²)]
I_{dir}	Direct irradiance, component of beam radiation on surface, W/m² [Btu/(h·ft²)]
I_T	Irradiance on tilted plane, W/m² [Btu/(h·ft²)]
k	Thermal conductivity, W/(m·K) [Btu/(h·ft·°F)]
$\dot{Q}$	Heat flow rate, W (Btu/h)
R	Thermal resistance, m²·K/W [(h·m²·°F)/Btu]
R'	Thermal resistance combining external convective resistance and that of glazing, (m²·K)/W [(h·ft²·°F)/Btu]
R^b	Back reflectance of solar radiation of glazing layer (Table 5.1)
R^f	Front reflectance of solar radiation of glazing layer (Table 5.1)
SHGC	Solar heat gain coefficient of window
T	Transmittance for solar radiation (Table 5.1)
T	Temperature, K or °C (°R or °F)
T_{dif}, T_{dir}	Transmittances of glazing for diffuse and direct radiation
T_s	Surface temperature of the glazing, °C (°F)
U	Overall heat transfer coefficient, W/(m²·K) [Btu/(h·ft²·°F)]
x	Distance or x-axis shadow length cast by vertical projection of window, m (ft)
y	Distance or y-axis shadow length cast by horizontal projection of window, m (ft)
z	Distance or window overhang length, m (ft)

Greek

- α Absorptivity, solar altitude angle
- α_i Absorptivity of inner pane
- α_o Absorptivity of outer pane
- γ Profile angle (Equation 5.22)
- δ Solar declination angle
- θ_i Incidence angle of the sun on plane (angle between normal of plane and line to the sun)
- θ_p Zenith angle of plane (tilt from horizontal, up > 0)
- θ_s Zenith angle of the sun
- λ Latitude, wavelength
- ρ Reflectivity
- τ Transmissivity
- ϕ_p Azimuth angle of plane (positive for orientations west of south)
- ϕ_s Azimuth angle of the sun
- ω Solar hour angle
- ΔT_{sol} Temperature rise of the glass from absorption of solar radiation, °C (°F)
- Δx Thickness of glass, m (ft)
- $\Delta\phi$ Solar-wall azimuth angle (Equation 5.21)

Subscripts

- *av* Average
- *con* Convection
- *cond* Conduction
- *cg* Center glass
- *dif* Diffuse
- *dir* Direct
- *eg* Edge of glass
- *f* Frame of window
- *i* Indoor, inner
- *max* Maximum
- *min* Minimum
- *o* Outdoor, outer
- *p* Plane
- *rad* Radiation
- *s* Surface, space
- *sol* Solar
- *tot* Total

5.1 Importance and Design Considerations

Glazing material consists of glass or plastic that transmits a large fraction of the incident solar radiation. Typical building glazing elements are windows, skylights, and glass doors. Windows are provided so that occupants can maintain visual contact with the outside world while benefitting from natural daylight in their working environment. The psychological effect on worker satisfaction and productivity has been demonstrated unequivocally. From the point of view of energy loads, glazing apertures, especially windows, are critical elements of the envelope of a building, providing at once daylight while also admitting solar radiation as heat and causing increased conduction heat losses/gains as compared to opaque walls. With conventional glazing, most of the energy consumption of a building goes, quite literally, out of the windows. In typical houses, roughly one-third of the total heat loss/gain can be attributed to glazing. Also, much of the cooling load is due to solar heat gains through glazing. Finally, significant air infiltration has been found to occur around the edges of windows. On the other hand, windows, if properly designed, could potentially save energy; daylight could reduce electricity consumption of interior lights while solar heating could offset some of the heating energy needs of the building. Clearly, the study of heat transfer through windows merits close attention.

Glazings are selected based on several considerations other than heat gains/losses. Issues to consider are daylighting, privacy, avoidance of glare, condensation control, ultraviolet control, acoustic control, color effects, cold down drafts, and water seepage. These are not addressed in this chapter. Rather, we will limit ourselves to heat gains/losses and energy requirements.

The selection of a window is daunting in view of the numerous commercial products available today. Figure 5.1a provides a sketch of a double-pane window showing common terminology, while photos of a double-pane glass sash section and the frame section of a double sliding window are meant to better illustrate actual window products (Figure 5.1b). We are faced with different combinations of glass material (characterized by optical properties such as emissivity and transmissivity), glazing material, frame material, interpane gas, etc. Further, windows can be installed in one of numerous ways (Figure 5.2), which affect their thermal and air leakage properties. For example, each operable or hinged window perimeter length/section increases infiltration. As for most other commercial products, windows have to be tested and certified by a government-approved agency. A typical example of labeling by the National Fenestration Council is shown in Figure 5.3, which includes four product descriptors related to air leakage, daylighting, solar heat gain, and overall heat loss coefficient. Air leakage is treated in Chapter 6, while a scientific treatment of the last three indices is provided in this Chapter.

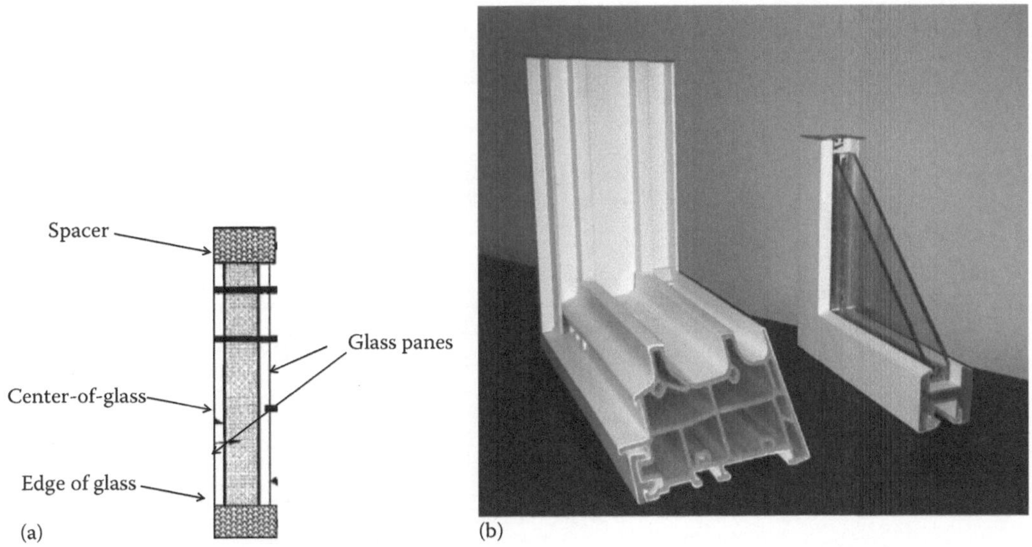

FIGURE 5.1
(a) Sketch of a double-pane window showing common terminology; (b) photos of low-e fixed double-pane glass sash section (on the right) and one with stainless steel spacers and rollers at the bottom (not visible). The I-section provides rigidity and also insulates the glass frame. The frame section for a double sliding window is shown on the left on which the sash slides. The several discrete compartments are meant to provide added rigidity and also increase thermal resistance.

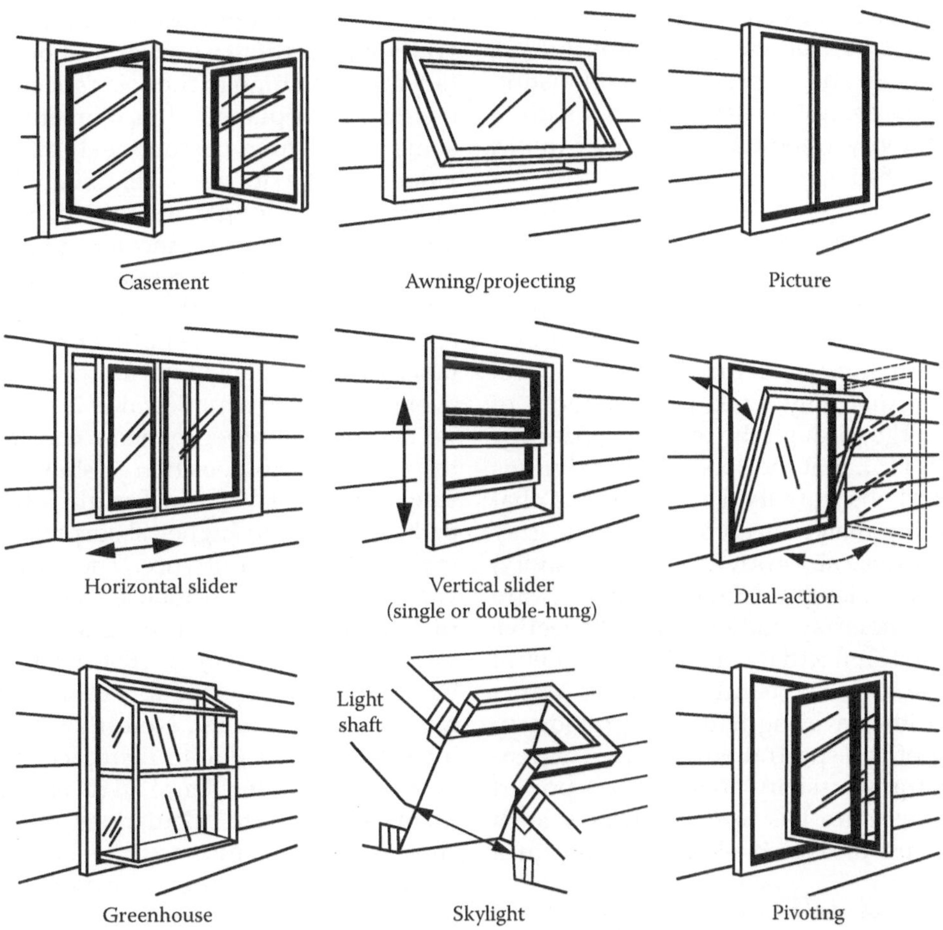

FIGURE 5.2
Types of residential windows. (From ASHRAE, *Handbook of Fundamentals*, American Society of Heating, Refrigerating and Air-Conditioning Engineers, Atlanta, GA, 2013. Copyright ASHRAE, www.ashrae.org.)

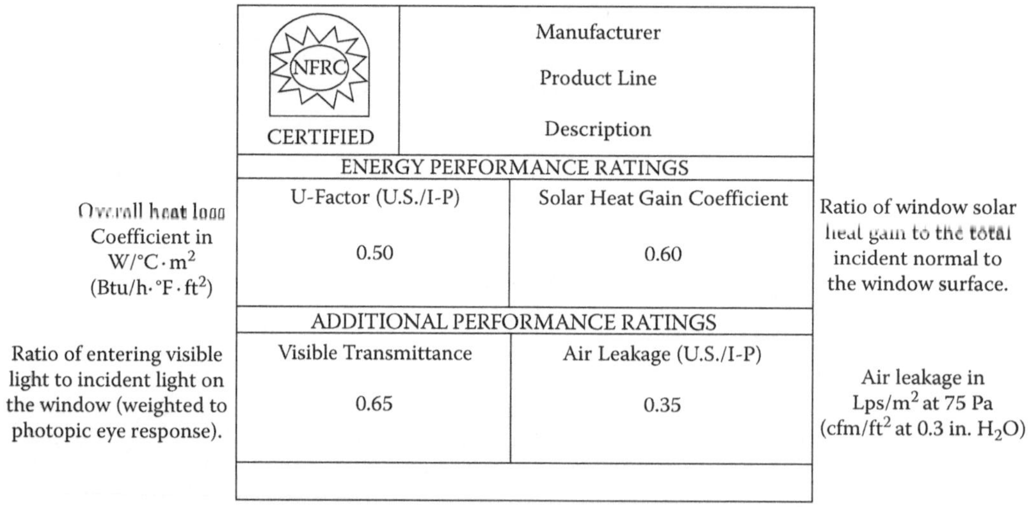

FIGURE 5.3
Example of National Fenestration Rating Council labeling.

5.2 Optical Properties

Of the solar irradiance I_T incident on a single sheet of glazing, a portion (ρI_T) is reflected, a portion (αI_T) is absorbed in the glazing, and a portion (τI_T) is transmitted to the interior. By energy conservation, the sum of reflectivity ρ, absorptivity α, and transmissivity τ equals unity (see Equation 2.64):

$$\rho + \alpha + \tau = 1 \tag{5.1}$$

The radiation transmitted to the interior of the building is assumed to be entirely absorbed, by virtue of the cavity effect, which normally causes the number of reflections to be so large that only a negligible amount of radiation can reemerge through the window opening.

Recall from Chapter 2 that the spectral emissivity is equal to the spectral absorptivity, i.e., $\varepsilon_\lambda = \alpha_\lambda$ and that most opaque building material can be viewed as gray bodies, i.e., bodies whose absorptivity (and emissivity) is independent of wavelength. Glazing materials are perfect examples of nongray bodies, called "selective surfaces." The spectral reflectivity of a glazing material is fairly constant, and so the manner in which transmissivity varies with wavelength would provide an inverse indication of the spectral emissivity. The reason why spectral transmissivity* is usually reported in the published literature has to do with how such optical properties are measured experimentally in a laboratory. Figure 5.4 depicts the spectral transmittance for several types of commonly used single-layer glass under normal solar incidence. The transmittance varies with the amount of trace absorbing materials present. For two types of glass (clear and bronze), the transmittance is fairly high in the solar spectrum of 400–700 nm (i.e., low absorptance), dips momentarily, and increases somewhat. Three coatings exhibit low transmittance values in the infrared region making them almost opaque in the long-wave region. Such glazings let in a large fraction of the incoming solar radiation while absorbing most of the reradiation at long wavelengths. This is referred to as the "greenhouse effect." To a first approximation, wavelength dependence of transmittance over the solar spectrum can be neglected, and an average monochromatic value is often selected.

The optical properties vary considerably with glass thickness and composition as shown in Figure 5.5 for a single glazing layer. Thicker glass has a longer optical path, resulting in higher absorptance and correspondingly lower transmittance. The optical properties also vary with the solar incidence angle θ_i as shown. We note that the optical properties change little till about 45° incidence angle. With increasing θ_i the transmittance tends toward zero due to exponential increase of the reflectivity coefficient. Note that the absorptance first increases with θ_i due to increased optical length, then drops to zero at incidence angle of 90°.

For approximate calculations, the values for monochromatic transmittance, reflectance, and absorptance for various types of glazing systems as shown in Table 5.1 can be used. The solar heat gain coefficient (SHGC) will be discussed in Section 5.4. Note that the sum of the other optical terms does not add to unity. Consider the single glazing system #1a in Table 5.1.

* The distinction between transmittance and transmissivity lies in that the latter applies to unit glazing pane thickness, while the former pertains to the specific glazing system. The same distinction also applies to absorptance vs. absorptivity and to reflectance vs. reflectivity.

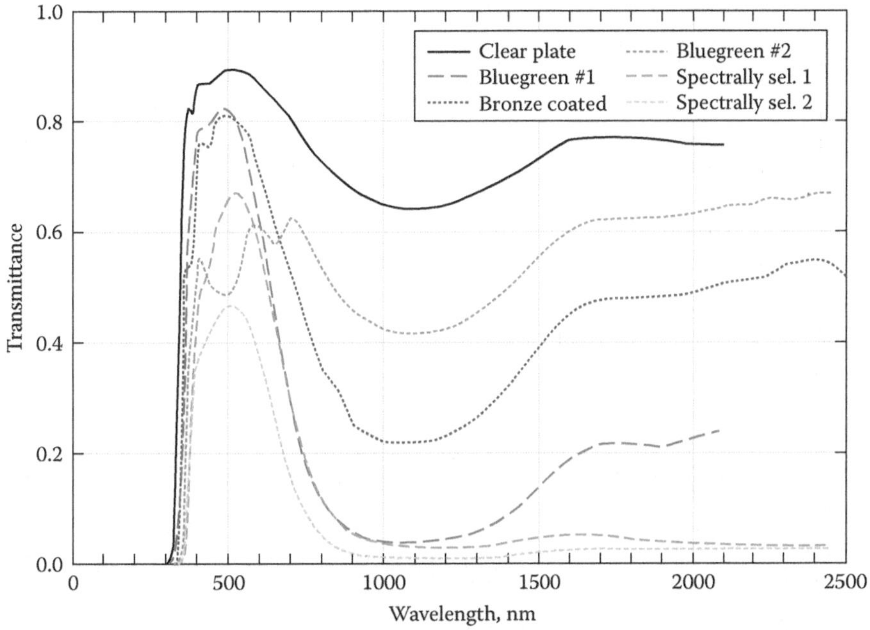

FIGURE 5.4
Spectral transmittance for different types of glass under normal incidence. (From ASHRAE, *Handbook of Fundamentals*, American Society of Heating, Refrigerating and Air-Conditioning Engineers, Atlanta, GA, 2013. Copyright ASHRAE, www.ashrae.org.)

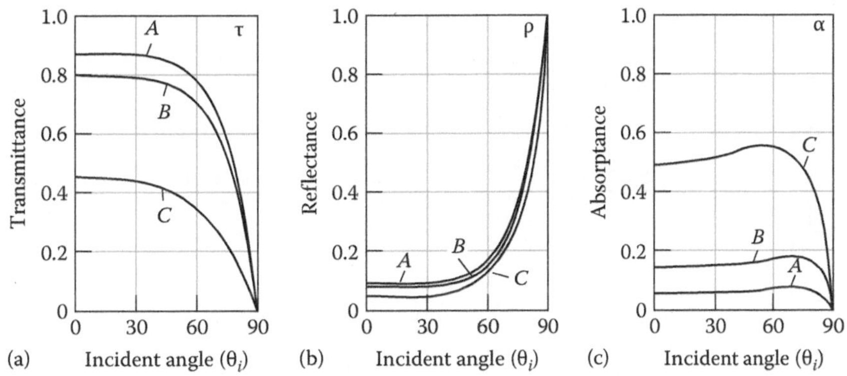

FIGURE 5.5
Variation of solar optical properties as a function of incidence angle θ_i, for three types of glasses: A = DSA (double-strength sheet), B = 6 mm (0.25 in.) clear glass, and C = 6 mm (0.25 in.) gray, bronze, or green tinted heat-absorbing glass. (a) Transmittance. (b) Reflectance. (c) Absorptance. (From ASHRAE, *Handbook of Fundamentals*, American Society of Heating, Refrigerating and Air-Conditioning Engineers, Atlanta, GA, 2013. Copyright ASHRAE, www.ashrae.org.)

In this case, the reflected component of the solar radiation at the back interface R^b of the window pane is essentially absorbed by the window pane and is included in the absorptivity coefficient A_1^f. Thus, at normal incidence, the sum of the transmittance, absorptance, and the front reflectance coefficients add to unit (= 0.83 + 0.09 + 0.08). The manner in which the incident beam solar radiation is distributed between these three quantities is depicted in Figure 5.6 for glazing system #1a using numerical values from Table 5.1. For glazing systems with several glass panes, there are numerous optical and thermal exchanges between the panes, and it is difficult to derive a single absorptance value. That is why individual absorptance values (A^f) of the various layers are shown separately in Table 5.1. For diffuse radiation striking the glazing system, there is no single incidence angle, and so aggregated values are shown (for glazing system #1a, the diffuse transmittance is 0.75, which corresponds to that of beam radiation at 60° incidence angle).

More accurate methods are available to account for the spectral variations in the transmittance and absorptance effects of the glazing material as well as the effect of solar incidence angle that gives rise to an

TABLE 5.1
Optical and Thermal Properties of Different Glazing Window Systems

	Glazing Systems			Center-of-Glazing Properties							Total Window SHGC at Normal Incidence			
				Incidence Angles							Aluminum		Other Frames	
ID	Glass Thickness, in. (mm)			Normal 0.0	40.0	50.0	60.0	70.0	80.0	Diffuse	Operable	Fixed	Operable	Fixed
1a	1/8 (3.2)	*Uncoated single glazing*, CLR	SHGC	0.86	0.84	0.82	0.78	0.67	0.42	0.78	0.75	0.78	0.64	0.75
			T	0.83	0.82	0.80	0.75	0.64	0.39	0.75				
			R^f	0.08	0.08	0.10	0.14	0.25	0.51	0.14				
			R^b	0.08	0.08	0.10	0.14	0.25	0.51	0.14				
			A_1^f	0.09	0.10	0.10	0.11	0.11	0.11	0.10				
5a	1/8 (3.2)	*Uncoated double glazing*, CLR CLR	SHGC	0.76	0.74	0.71	0.64	0.50	0.26	0.66	0.67	0.69	0.56	0.66
			T	0.70	0.68	0.65	0.58	0.44	0.21	0.60				
			R^f	0.13	0.14	0.16	0.23	0.36	0.61	0.21				
			R^b	0.13	0.14	0.16	0.23	0.36	0.61	0.21				
			A_1^f	0.10	0.11	0.11	0.12	0.13	0.13	0.11				
			A_2^f	0.07	0.08	0.08	0.08	0.07	0.05	0.07				
5b	1/4 (6.4)	*Uncoated double glazing*, CLR CLR	SHGC	0.70	0.67	0.64	0.58	0.45	0.23	0.50	0.61	0.63	0.52	0.61
			T	0.61	0.58	0.55	0.48	0.36	0.17	0.51				
			R^f	0.11	0.12	0.15	0.20	0.33	0.57	0.18				
			R^b	0.11	0.12	0.15	0.20	0.33	0.57	0.18				
			A_1^f	0.17	0.18	0.19	0.20	0.21	0.20	0.19				
			A_2^f	0.11	0.12	0.12	0.12	0.10	0.07	0.11				
21a	1/8 (3.2)	*Low-e double glazing, e = 0.1 on surface 2*, LE CLR	SHGC	0.65	0.64	0.62	0.56	0.43	0.23	0.57	0.48	0.50	0.41	0.47
			T	0.59	0.56	0.54	0.48	0.36	0.18	0.50				
			R^f	0.15	0.16	0.18	0.24	0.37	0.61	0.22				
			R^b	0.17	0.18	0.20	0.26	0.38	0.61	0.24				
			A_1^f	0.20	0.21	0.21	0.21	0.20	0.16	0.20				
			A_2^f	0.07	0.07	0.08	0.08	0.07	0.05	0.07				
21c	1/8 (3.2)	*Low-e double glazing, e = 0.1 on surface 3*, CLR LE	SHGC	0.60	0.58	0.56	0.51	0.40	0.22	0.52	0.53	0.55	0.45	0.53
			T	0.48	0.45	0.43	0.37	0.27	0.13	0.40				
			R^f	0.26	0.27	0.28	0.32	0.42	0.62	0.31				
			R^b	0.24	0.24	0.26	0.29	0.38	0.58	0.28				
			A_1^f	0.12	0.13	0.14	0.14	0.15	0.15	0.13				
			A_2^f	0.14	0.15	0.15	0.16	0.16	0.10	0.15				

(Continued)

TABLE 5.1 (*Continued*)

Optical and Thermal Properties of Different Glazing Window Systems

		Glazing Systems		Center-of-Glazing Properties: Incidence Angles							Total Window SHGC at Normal Incidence: Aluminum		Other Frames	
ID	Glass Thickness, in. (mm)			Normal 0.0	40.0	50.0	60.0	70.0	80.0	Diffuse	Operable	Fixed	Operable	Fixed
29a	1/8 (3.2)	*Triple glazing*, CLR CLR CLR	SHGC	0.68	0.65	0.62	0.54	0.39	0.18	0.57	0.60	0.62	0.51	0.59
			T	0.60	0.57	0.53	0.45	0.31	0.12	0.49				
			R^f	0.17	0.18	0.21	0.28	0.42	0.65	0.25				
			R^b	0.17	0.18	0.21	0.28	0.42	0.65	0.25				
			A_1^f	0.10	0.11	0.12	0.13	0.14	0.14	0.12				
			A_2^f	0.08	0.08	0.09	0.09	0.08	0.07	0.08				
			A_3^f	0.06	0.06	0.06	0.06	0.05	0.03	0.06				
29b	1/4 (6.4)	*Triple glazing*, CLR CLR CLR	SHGC	0.61	0.58	0.55	0.48	0.35	0.16	0.51	0.54	0.56	0.46	0.53
			T	0.49	0.45	0.42	0.35	0.24	0.09	0.39				
			R^f	0.14	0.15	0.18	0.24	0.37	0.59	0.22				
			R^b	0.14	0.15	0.18	0.24	0.37	0.59	0.22				
			A_1^f	0.17	0.19	0.20	0.21	0.22	0.21	0.19				
			A_2^f	0.12	0.13	0.13	0.13	0.12	0.08	0.12				
			A_3^f	0.08	0.08	0.08	0.08	0.06	0.03	0.08				
32a	1/8 (3.2)	*Triple glazing e = 0.2 on surface 2*, LE CLR CLR	SHGC	0.60	0.58	0.55	0.48	0.35	0.17	0.51	0.53	0.55	0.45	0.53
			T	0.50	0.47	0.44	0.38	0.26	0.10	0.41				
			R^f	0.17	0.19	0.21	0.27	0.41	0.64	0.25				
			R^b	0.19	0.20	0.22	0.29	0.42	0.63	0.26				
			A_1^f	0.20	0.20	0.20	0.21	0.21	0.17	0.20				
			A_2^f	0.08	0.08	0.08	0.09	0.08	0.07	0.08				
			A_3^f	0.06	0.06	0.06	0.06	0.05	0.03	0.06				
32c	1/8 (3.2)	*Triple glazing e = 0.2 on surface 5*, CLR CLR LE	SHGC	0.62	0.60	0.57	0.49	0.36	0.16	0.52	0.55	0.57	0.46	0.54
			T	0.50	0.47	0.44	0.38	0.26	0.10	0.41				
			R^f	0.19	0.20	0.22	0.29	0.42	0.63	0.26				
			R^b	0.18	0.19	0.21	0.27	0.41	0.64	0.25				
			A_1^f	0.11	0.12	0.13	0.14	0.15	0.15	0.13				
			A_2^f	0.09	0.10	0.10	0.10	0.10	0.08	0.10				
			A_3^f	0.11	0.11	0.11	0.10	0.08	0.04	0.10				

Source: Extracted from ASHRAE, *Handbook of Fundamentals*, American Society of Heating, Refrigerating and Air-Conditioning Engineers, Atlanta, GA, 2013. Copyright ASHRAE, www.ashrae.org.

Clear (CLR), low emissivity (LE), Solar heat gain coefficient (SHGC), solar transmittance (T), front transmittance (R^f), back reflectance (R^b), layer absorptances (A_n^f).

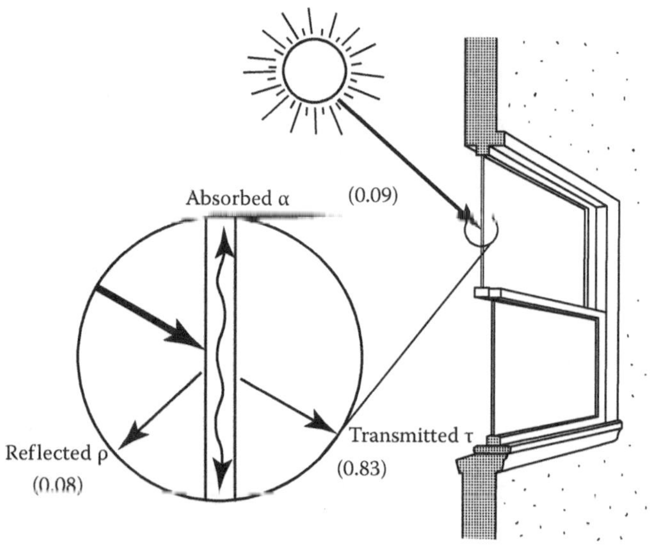

FIGURE 5.6
Distribution of beam radiation falling on glazing product #1a of Table 5.1 at normal solar incidence.

infinite series of multiple internal reflections of the solar rays within the glass pane. The interested reader can refer to solar energy textbooks (for example, ASHRAE Fundamentals, 2013; Goswami et al., 2000; Rabl, 1985).

It is insightful to discuss how an ideal glazing should behave spectrally when coupled to a conditioned space (Figure 5.7). In cold locations, it should let in the solar radiation over the entire solar spectrum (i.e., have high transmittance) while it should act as a good reflector in the long-wave infrared spectrum (i.e., low absorptance and low transmittance) of the radiation emitted by the warm interior surfaces of the building. This is the principle of *low-e coating* on window glasses in cold climates (i.e., low emission over the long-wave spectrum. A less ambiguous term is *high-solar gain low-e glazing system*. On the other hand, in hot locations, the "perfect" glazing material should only let in solar radiation in the visible portion of the solar spectrum (high transmittance over wavelength range of 0.4–0.7 μm) to serve the function of illumination and view to the outside, and block solar radiation from the other spectral bands. It should also block infrared radiation from outdoors to enter the conditioned space. This type of glass is referred to as *low-solar gain low-e glazing system*.

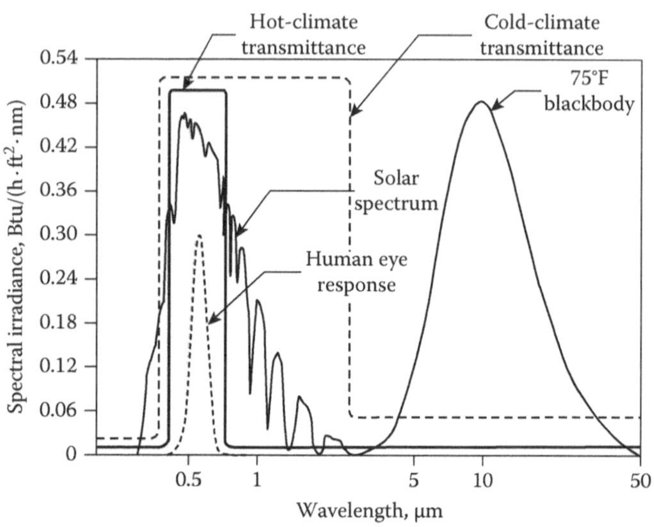

FIGURE 5.7
Conceptual illustration of two spectrally ideal selective glazing surfaces: one for hot climates and one for cold climates. (From ASHRAE, *Handbook of Fundamentals*, American Society of Heating, Refrigerating and Air-Conditioning Engineers, Atlanta, GA, 2013. Copyright ASHRAE, www.ashrae.org.)

5.3 Thermal Properties

5.3.1 Center-Glass U Values

The glazing material such as a glass pane is very thin (typically 5 mm or 0.2 in.). Hence, unlike walls and roofs, the heat capacity of a glazing product is negligible, and conduction heat transfer through windows can be treated as steady state. The energy balance of a single-pane window is shown in Figure 5.8 (assuming 1-D heat flow). Initially, let us leave out the interaction with solar radiation; it will be included in Section 5.4.

For a single-pane glazing, the heat flow from interior to exterior has to pass through three resistances in series: the convective resistance at the interior surface, the conductive resistance of the glass, and the convective resistance at the exterior surface. Therefore, the total resistance R_{total} of the glazing is the sum of the three resistances (as shown in Figure 5.8b), and it is related to the overall heat loss coefficient or U value of the glazing by $UA = 1/R_{total}$, where A is the area of the pane. Canceling the area A, one finds that the U value is given by

$$\frac{1}{U} = \frac{1}{h_i} + \frac{\Delta x}{k} + \frac{1}{h_o} \quad (5.2)$$

where

h_i is the heat transfer coefficient (combined radiation–convection) at the interior surface
h_o is the heat transfer coefficient (combined radiation–convection) at the exterior surface
Δx is the thickness of the glass
k is the conductivity of the glass

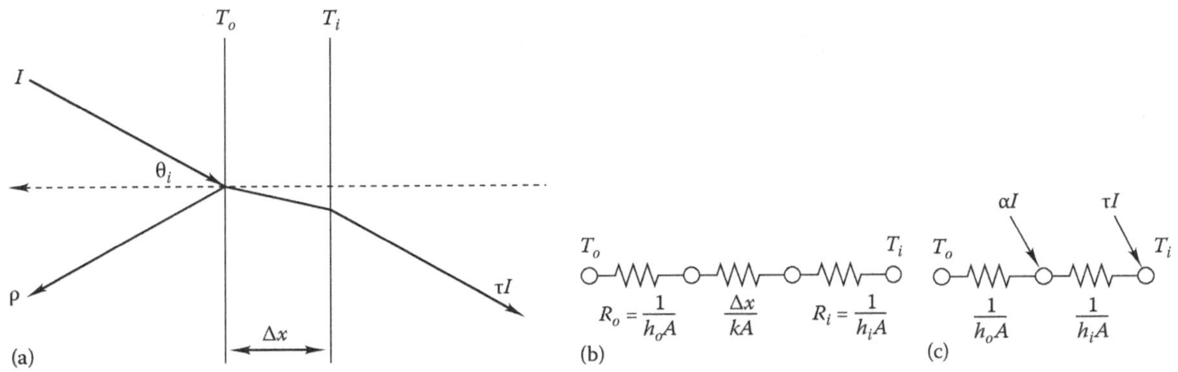

FIGURE 5.8
Energy balance of a single-pane window (neglecting heat capacity of glass). (a) Physical configuration. (b) Thermal network without solar radiation. (c) Thermal network with solar radiation (and neglecting resistance in glass).

The surface heat transfer coefficients can vary with ambient conditions; for highest accuracy, one would have to resort to detailed formulas for the radiative and convective heat transfers (see Section 2.10.6). Recall that in most building applications, one can use lumped parameters h_i and h_o at standard design values—a tremendous simplification. The following ASHRAE Fundamentals (2013) design values combining both convection and radiation effects are often used:

1. For indoor temperature of 21°C (70°F) and a glass emissivity of 0.84 (uncoated glass)

$$h_i = 8.29\ \text{W/(m}^2\cdot\text{K)}\ [1.46\ \text{Btu/(h}\cdot\text{ft}^2\cdot{}^\circ\text{F)}]\quad \text{year-round}$$

2. For a wind velocity of 6.7 m/s or 15 mph and a glass emissivity of 0.84 (uncoated glass)

$$h_o = 28.35\ \text{W/(m}^2\cdot\text{K)}\ [5.0\ \text{Btu/(h}\cdot\text{ft}^2\cdot{}^\circ\text{F)}]$$

The sensitivity of the total U value to variations in h_o is greatest for single glazing; the greater the insulating value of the glazing, the lower does the variability in h_o affect the U value.

Inserting the conductivity of glass $k \approx 0.92$ W/(m · K) [≈ 0.6 Btu/(h · ft · °F)] and $\Delta x = 5$ mm (0.20 in.) as a typical glazing thickness into Equation 5.2 for U, one finds that the resistance of the glass

$$\frac{1}{U} = \frac{1}{8.29} + \frac{0.005}{0.92} + \frac{1}{28.35}$$

$$= 5.5937\ \text{W/(m}^2\cdot\text{K)}\ [0.985\ \text{Btu/(h}\cdot\text{ft}^2\cdot{}^\circ\text{F)}]$$

$$\text{i.e., } R_{total} = 0.1788\ (\text{m}^2\cdot\text{K)/W}\ [1.0\ (\text{h}\cdot\text{ft}^2\cdot{}^\circ\text{F)/Btu}]$$

The glass pane itself only accounts about 2% of the total resistance and so is negligible compared to the surface resistances. Even with acrylic or polycarbonate, whose conductivity is four times lower than that of glass, the contribution of the conductive resistance of the pane is small. Henceforth, we can omit the $(\Delta x/k)$ term from the equations.

The thermal resistance of windows can be enhanced by adding further panes and using low-emissivity films. In fact, the norm nowadays is to use double-pane windows. In that case, the heat transfer mechanism is very similar to the treatment in Section 2.10.6 where convection and radiation heat transfer coexist as in a cavity wall since glazing material behaves like an opaque material (i.e., zero transmittance) for long-wave radiation. In such instances, it is more convenient to use tables of combined or effective unit conductances or unit resistances as listed in Table 2.9.

For a double glazing system and neglecting the resistance of the glass itself (see Figure 2.22), we obtain

$$\frac{1}{U} = \frac{1}{h_i} + \frac{1}{h_s} + \frac{1}{h_o} \tag{5.3}$$

where h_s is the heat transfer coefficient of the space between the panes. The latter is, of course, the sum of the term h_{rad} for radiation and h_{con} for conduction/convection across the airspace:

$$h_s = h_{rad} + h_{con} \tag{5.4}$$

Each additional pane reduces the U value of the glazing by adding another resistance akin to $(1/h_s)$ to the right-hand side of Equation 5.4.

For a double-pane glass, the heat loss from one pane to another due to convection is roughly equal to that due to radiation. Glazing systems with triple-pane glass are

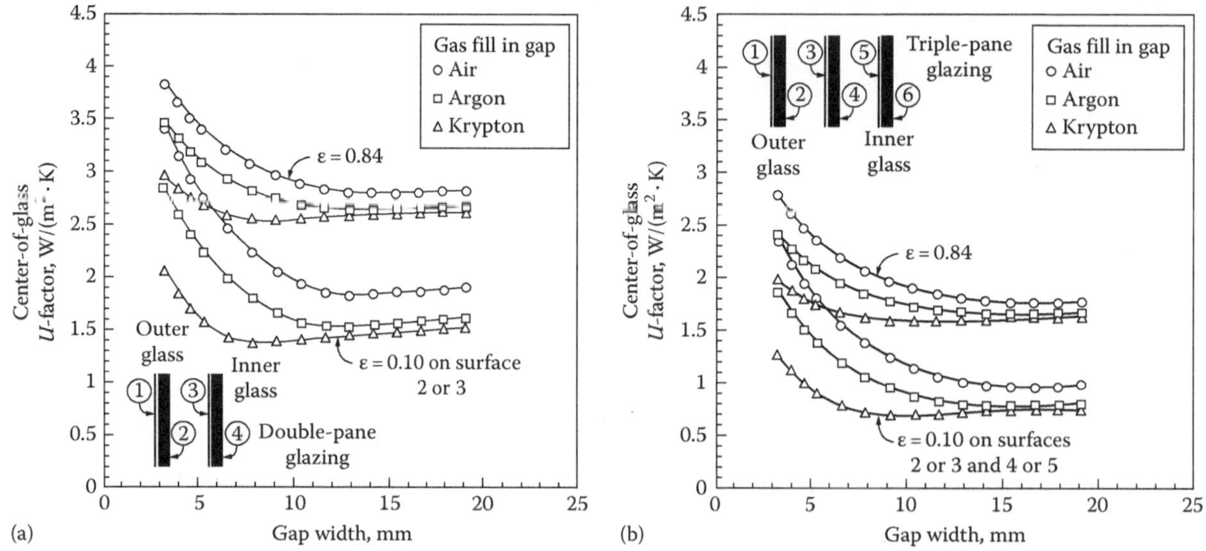

FIGURE 5.9
Variation of center-of-glass U values with gap width for different window emissivities and gas fill materials. (a) Double-pane window. (b) Triple-pane window. (From ASHRAE, *Handbook of Fundamentals*, American Society of Heating, Refrigerating and Air-Conditioning Engineers, Atlanta, GA, 2013. Copyright ASHRAE, www.ashrae.org.)

also available commercially. The heat transfer coefficient between the panes h_s can be minimized by performing the following:

1. *Reducing the convective heat coefficient.* This can be achieved by modifying the gap width between panes and/or using different types of filler gases other than air, which have different physical properties (for example, lower conduction). Common gases used are argon and krypton. The variation of the overall heat loss coefficient U with the separation d between the panes is of interest because the cost of the frame increases with d. When d is small (less than about 8 mm ≈ 0.3 in. for air under normal conditions), there is no convection, and h_{con} varies as k_{air}/d, where k_{air} is the conductivity of air. As d is increased beyond the threshold for convection, h_{con} becomes approximately constant. This effect can be seen in Figure 5.9, where the U value of several window types is plotted as function of the width of the separation between panes. There is no point in making the width larger than about 10 mm (0.40 in.).
2. *Reducing the radiative coefficient* by using low-emissivity films (akin to their use in air gaps within cavity walls treated in Section 2.10.6). These films are usually placed on the inside glass surfaces facing each other across the gap. It is often more cost-effective to simply coat only one surface as shown in Figure 5.9. Note that the U value of such a two-pane fenestration product with ε = 0.10 is reduced by about half as compared to one with no such coating (ε = 0.84). Obviously, using three panes further lowers the U value by another factor of 2 as compared to two-pane glazing, but the cost would also increase and may be unjustified in most cases.

Example 5.1: Effect of Low-Emissivity Films

We will analyze the effect on center-of-glazing overall heat loss coefficient U if the inner sides of a double-glazed window are coated with low-e films. Two cases will be analyzed: coating on both sides (i.e., surfaces 2 and 3 of Figure 5.9) and on one side only (i.e., surface 2 or surface 3 only). The analysis parallels the treatment of analyzing the combined effects of convection and radiation across an air gap; more specifically see Example 2.17.

Given: Emissivity of ordinary glass ε = 0.84, emissivity of a surface coated with low-e film ε = 0.10.

Find: $U_{no\text{-}coat}$, $U_{1\text{-}coat}$, $U_{2\text{-}coat}$

Sketch: Figures 2.22 and 5.9

Assumptions: Neglect resistance of the glass pane, with interpane distance = 0.5 in. (12.7 mm).

The overall convective resistance of the air gap* is $R_{con,2\text{-}3} = 3.5\ \text{h}\cdot\text{ft}^2\cdot{}^\circ\text{F/Btu}$.
The convective resistance of the indoor air film is $R_i = 0.68\ \text{h}\cdot\text{ft}^2\cdot{}^\circ\text{F/Btu}$.

* Section 12.5 presents dimensionless equations that allow this coefficient to be determined.

The convective resistance of the outdoor air film is $R_o = 0.17$ h·ft²·°F/Btu.
The indoor and outdoor air temperatures are 70°F and 30°F, so that $\bar{T} = 50$°F.

Solution

Let us start with the uncoated double-glazed window case. First, we determine the linearized radiative heat transfer coefficient following Equation 2.72:

$$h_{rad,2-3} = \frac{4\sigma\bar{T}^3}{(1/\varepsilon_2)+(1/\varepsilon_3)-1} = \frac{4\times 0.1714\times 10^{-8}\times(50+460)^3}{(1/0.84)+(1/0.84)-1}$$

$$= \frac{0.909}{1.381} = 0.658 \text{ Btu/(h}\cdot\text{ft}^2\cdot{}^\circ\text{F)}$$

This allows computing the interpane heat transfer coefficient (from Equation 5.4):

$$h_s = h_{rad,2-3} + h_{con,2-3} = 0.658 + \frac{1}{3.5} = 0.658 + 0.286$$

$$= 0.944 \text{ Btu/(h}\cdot\text{ft}^2\cdot{}^\circ\text{F)}$$

Note that the radiation contribution to the losses is more than twice than that due to convection. Finally, from Equation 5.2:

$$U = \left(R_i + \frac{1}{h_s} + R_o\right)^{-1} = \left(0.68 + \frac{1}{0.944} + 0.17\right)^{-1}$$

$$= 0.524 \text{ Btu/(h}\cdot\text{ft}^2\cdot{}^\circ\text{F)}$$

In this case, the highest resistance is offered by the air gap between panes. As a result, the temperature gradient across the air gap is the highest. Similar calculation procedures for the other two cases are summarized in the following table:

	$h_{rad,2-3}$ Btu/(h·ft²·°F)	h_s Btu/(h·ft²·°F)	U Btu/(h·ft²·°F)	U W/(m²·°C)
Uncoated	0.658	0.944	0.524	2.973
Coated on one side only (surface 2 or 3)	0.089	0.375	0.284	1.612
Coated on both sides (surfaces 2 and 3)	0.048	0.334	0.260	1.475

The value of 2.97 W/(m²·°C) for the uncoated case is consistent with the value of 2.73 W/(m²·°C) shown in Table 5.3 for double glazing center-of-glass values for 13 mm air gap. For low-e coating of 0.10 on one side, Table 5.3 lists a value of 1.82 W/(m²·°C), which is slightly higher than the value of 1.61 W/(m²·°C) obtained. Part of this difference could be due to the different interior and exterior air temperatures between both cases as well as the values of the convective coefficients assumed.

Comments

The effect of coating both glass pane surfaces reduces the overall heat loss coefficient to half of that of an uncoated one. However, the incremental improvement from coating one side to coating on both sides is small (about 10%), and whether the extra cost of coating a second surface is justified is for the manufacturer to decide.

5.3.2 *U* Values for Entire Windows

The aforementioned formulas do not take into account edge effects. Spacers are used in most glazing products to keep the panes apart while the entire assembly is encased in a frame (see Figure 5.1). In most windows, the conductance of the frame is so different from that of the glazing that it should not be neglected. Furthermore, many multipane window designs use continuous aluminum spacers around the edge of the glazing to keep the panes apart, thus degrading the insulating value. Therefore, ASHRAE Fundamentals (2013) recommends that the overall heat loss coefficient value U_{tot} of a window be calculated as an area-weighted average:

$$U_{av} = \frac{U_{cg}A_{cg} + U_{eg}A_{eg} + U_f A_f}{A_{cg} + A_{eg} + A_f} \tag{5.5}$$

where the subscripts *cg*, *eg*, and *f* refer to the center of the glass, edge of the glass, and frame, respectively. The difference between the center-of-glass and edge-of-glass values can be seen in Figure 5.10 for a variety of spacer materials. Common materials are wood, aluminum, and vinyl. The values for the overall heat transfer coefficients of typical frames are listed in Table 5.2.

The importance of frame and spacer materials has been emphasized by a number of authors. It is particularly true in small windows. Traditionally, aluminum has been the favorite material for the frame and spacer, by virtue of its durability and low cost. Unfortunately, its conductivity is very high. While an aluminum frame does not render single glazing any worse, its use in insulating glazing defeats the purpose. The overall *U* values of a large number of different products are assembled in Table 5.3. This is by no means an exhaustive table and a longer list can be found in ASHRAE Fundamentals (2013). However, given the increasing variety and complexity of fenestration products available today, experienced designers tend to use software programs such as Windows (LBNL, 2011), which has been certified by National Fenestration Rating Council (NFRC) for energy rating of window products.

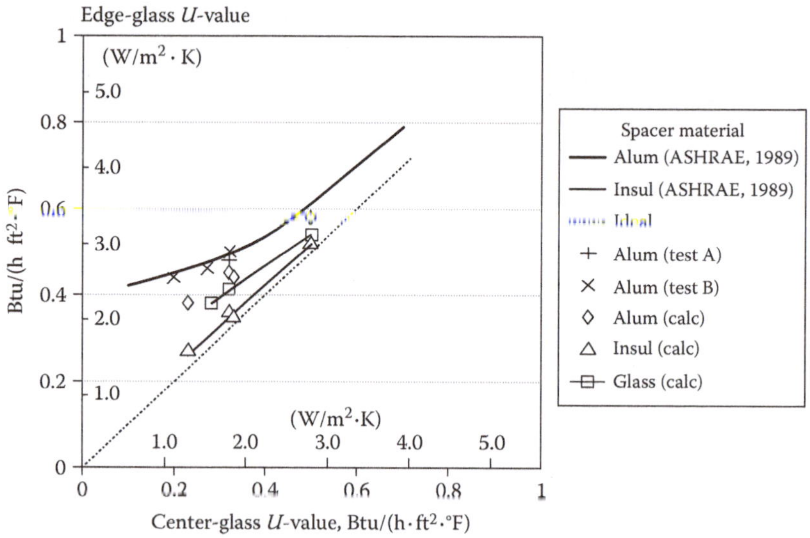

FIGURE 5.10
Center-of-glass and edge-of-glass U values for a variety of spacer materials ("ideal" corresponds to a spacer with the same U value as the glazing). (From McCabe, M.E., Window U-values: Revisions for the 1989 ASHRAE *Handbook of Fundamentals*, *ASHRAE J.*, June, 56, 1989.)

TABLE 5.2
Heat Transfer Coefficients U_f of Typical Frames

Material	Btu/(h·ft²·°F)	W/(m²·K)
Aluminum without thermal break	1.9	10.8
Aluminum with thermal break	1.0	5.68
Wood or vinyl	0.4	2.27

Source: From McCabe, M.E., Window U-values: Revisions for the 1989 ASHRAE *Handbook of Fundamentals, ASHRAE J.*, June, 56, 1989.

Example 5.2: Overall Heat Loss Coefficient of Window

Calculate the U value of a double-glazed air-filled (air gap = 9 mm = 0.35 in.) window of gross area 0.5 m × 1.0 m (1.64 ft × 3.28 ft) with low-emissivity coating of emissivity 0.10 on surface 3, with aluminum spacer and aluminum frame (without thermal break). Frame width = 0.05 m (2 in.) and edge width = spacer width = 0.01 m (0.4 in.).

Given: Center-of-glass U value (double glazed, airspace width 9 mm, $\varepsilon = 0.10$) and edge-of-glass U of for aluminum spacer

Find: U_{av}

Sketch: Figure 5.11

Lookup values: U_{cg} = 1.9 W/(m²·K) [0.335 Btu/(h·ft²·°F)] from Figure 5.9

U_{eg} = 2.8 W/(m²·K) [0.493 Btu/(h·ft²·°F)] from Figure 5.10

U_f = 10.8 W/(m²·K) [1.902 Btu/(h·ft²·°F)] from Table 5.2

Solution

We list the areas and U values in this table:

	Length, m	Width, m	A, m²	U, W/(m²·K)	AU, W/K
Frame	1.00	0.50	1.00 × 0.50 − 0.90 × 0.40 = 0.140	10.800	1.512
Edge of glass	0.90	0.40	0.90 × 0.40 − 0.88 × 0.38 = 0.026	2.800	0.072
Center	0.88	0.38	0.334	1.900	0.635
Sum			**0.500**		**2.219**

Dividing the sum of the UA values by the sum of the A values according to Equation 5.5, we find

$$U_{av} = \frac{2.219\ \text{W/K}}{0.500\ \text{m}^2}$$

$$= 4.44\ \text{W/(m}^2\cdot\text{K)}\ [0.782\ \text{Btu/(h}\cdot\text{ft}^2\cdot{}^\circ\text{F)}]$$

Comments

This is more than twice as large as the U value of the glazing itself. A bad frame has been intentionally selected to demonstrate the dramatic degradation in the performance of a good window.

5.3.3 Surface Temperature of Glazing

Since the perception of comfort depends not only on air temperature but also on radiation temperature, it is of interest to calculate the temperature T_s of the interior glazing surface as a function of outdoor temperature.

TABLE 5.3

U-Factors for Various Fenestration Products with Vertical Installation, W/($m^2 \cdot K$)

	Glass Only		Operable (Including Sliding and Swinging Glass Doors)					Fixed
	Center of Glass	Edge of Glass	Aluminum without Thermal Break	Aluminum with Thermal Break	Reinforced Vinyl/ Aluminum Clad Wood	Wood/ Vinyl	Insulated Fiberglass/ Vinyl	Insulated Fiberglass/ Vinyl
Single glazing								
3 mm glass	5.91	5.91	7.01	6.08	5.27	5.20	4.83	5.40
6 mm acrylic/polycarb	5.00	5.00	6.23	5.35	4.59	4.52	4.18	4.61
3.2 mm acrylic/polycarb	5.45	5.45	6.62	5.72	4.93	4.86	4.51	5.01
Double glazing								
6 mm airspace	3.12	3.63	4.62	3.61	3.24	3.14	2.84	3.04
13 mm airspace	2.73	3.36	4.30	3.31	2.96	2.86	2.58	2.72
6 mm argonspace	2.90	3.48	4.43	3.44	3.08	2.98	2.69	2.86
13 mm argonspace	2.56	3.24	4.16	3.18	2.84	2.74	2.46	2.58
Double glazing, e = 0.60 on surface 2 or 3								
6 mm airspace	2.95	3.52	4.48	3.48	3.12	3.02	2.73	2.90
13 mm airspace	2.50	3.20	4.11	3.14	2.80	2.70	2.42	2.53
6 mm argonspace	2.67	3.32	4.25	3.27	2.92	2.82	2.54	2.67
13 mm argonspace	2.33	3.08	3.98	3.01	2.68	2.58	2.31	2.39
Double glazing, e = 0.40 on surface 2 or 3								
6 mm airspace	2.78	3.40	4.34	3.35	3.00	2.90	2.61	2.77
13 mm airspace	2.27	3.04	3.93	2.96	2.64	2.54	2.27	2.35
6 mm argonspace	2.44	3.16	4.07	3.09	2.76	2.66	2.38	2.49
13 mm argonspace	2.04	2.88	3.75	2.79	2.48	2.38	2.11	2.16
Double glazing, e = 0.20 on surface 2 or 3								
6 mm airspace	2.56	3.24	4.16	3.18	2.84	2.74	2.46	2.58
13 mm airspace	1.99	2.83	3.70	2.75	2.44	2.34	2.07	2.12
6 mm argonspace	2.16	2.96	3.84	2.88	2.56	2.46	2.19	2.26
13 mm argonspace	1.70	2.62	3.47	2.53	2.24	2.14	1.88	1.88
Double glazing, e = 0.10 on surface 2 or 3								
6 mm airspace	2.39	3.12	4.02	3.05	2.72	2.62	2.34	2.44
13 mm airspace	1.82	2.71	3.56	2.62	2.32	2.22	1.96	1.98
6 mm argonspace	1.99	2.83	3.70	2.75	2.44	2.34	2.07	2.12
13 mm argonspace	1.53	2.49	3.33	2.40	2.12	2.02	1.76	1.74
Double glazing, e = 0.05 on surface 2 or 3								
6 mm airspace	2.33	3.08	3.98	3.01	2.68	2.58	2.31	2.39
13 mm airspace	1.70	2.62	3.47	2.53	2.24	2.14	1.88	1.88
6 mm argonspace	1.87	2.75	3.61	2.66	2.36	2.26	2.00	2.02
13 mm argonspace	1.42	2.41	3.24	2.31	2.04	1.94	1.69	1.65

Source: Extracted from ASHRAE, *Handbook of Fundamentals*, American Society of Heating, Refrigerating and Air-Conditioning Engineers, Atlanta, GA, 2013. Copyright ASHRAE, www.ashrae.org.

To convert to Btu/($h \cdot ft^2 \cdot °F$), multiply values shown by 0.1761.

Notes: 1. Based on winter conditions of −18°C outdoor air temperature and 21°C indoor air temperature, with 6.7 m/s outdoor air velocity and zero solar flux. Except for single glazing, small changes in the indoor and outdoor temperatures do not significantly affect overall *U*-factors.

2. Glazing layer surfaces are numbered from outdoor to indoor. All data are based on 3 mm glass, unless otherwise noted. Thermal conductivities are 0.917 W/($m \cdot K$) for glass and 0.19 W/($m \cdot K$) for acrylic and polycarbonate.

3. Standard spacers are metal. Edge-of-glass effects are assumed to extend over the 63.5 mm band around the perimeter.

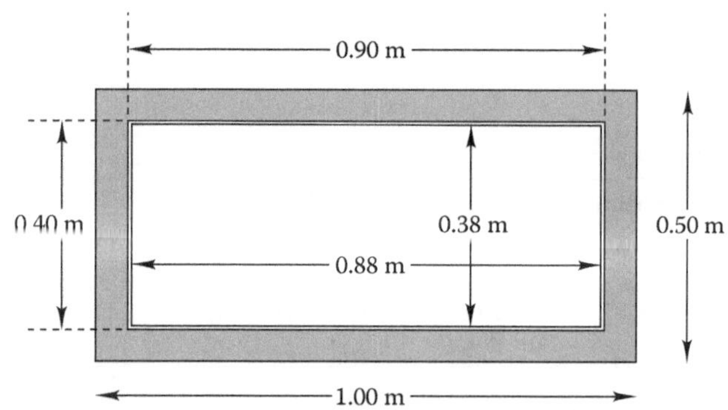

FIGURE 5.11
Dimensions of the frame, edge of glass (spacer), and center of glass for Example 5.2, drawn to scale.

Let us designate by R' the resistance of the glazing without the surface heat transfer coefficient h_i to the indoor air. The total resistance $R = 1/(U_{av})$ of the glazing is, of course, the sum

$$R = R' + \frac{1}{h_i} \tag{5.6}$$

since these two resistances are in series. The heat flow $h_i(T_i - T_s)$ from air to the surface must be equal to the flow from the surface to the outside $(T_s - T_o)/R'$:

$$h_i(T_i - T_s) = \frac{T_s - T_o}{R'} \tag{5.7}$$

Straightforward algebra yields the surface temperature as the following weighted average of indoor and outdoor temperatures:

$$T_s = \left(1 - \frac{U}{h_i}\right)T_i + \frac{U}{h_i}T_o \tag{5.8}$$

This is plotted in Figure 5.12 as function of T_o and for several values of U, assuming T_i = 20°C and h_i = 8.29 W/(m²·K). Note that for an outdoor air temperature of 0°C, the surface temperature T_s for a single-pane glass [U = 6.0 W/(m²·K)] is 5°C, while that for double glazing [U = 3.0 W/(m²·K)], it is 13°C. Since the indoor dew point temperature is around 10°C, there is likely to be water condensation on the single-pane glass when T_o = 0°C.

5.4 Solar Heat Gains

5.4.1 Calculation Procedure

As shown in Figure 5.13, the instantaneous heat flow through the glazing is the sum of

1. Conduction heat gain/loss in the absence of solar radiation
2. Solar radiation striking the glazing, which is transmitted through the window

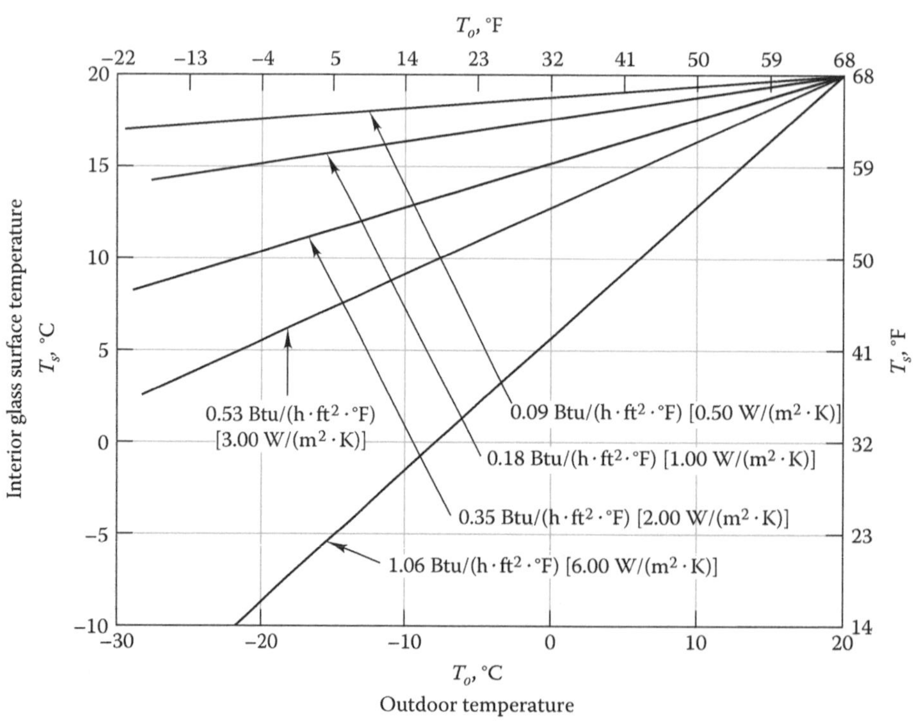

FIGURE 5.12
Interior glass surface temperature versus outdoor temperature, Equation 5.8, for T_i = 20°C and for several U values.

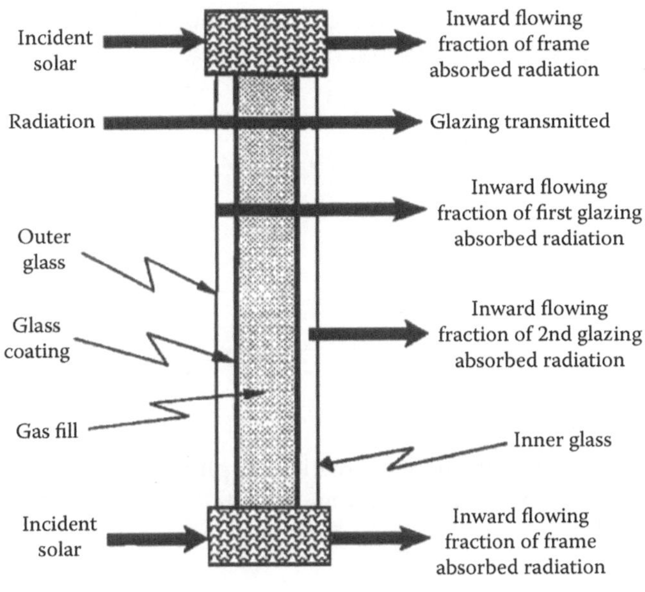

FIGURE 5.13
Components of solar heat gain with a double-pane window. (From ASHRAE, *Handbook of Fundamentals*, American Society of Heating, Refrigerating and Air-Conditioning Engineers, Atlanta, GA, 2013. Copyright ASHRAE, www.ashrae.org.)

3. Portion of the incident solar radiation absorbed by the glazing, which is then transferred inward by convection and radiation (a portion also flows outward, but that is excluded).

Note that, the solar heat gain occurs through both center-of-glazing and window frame, and heat flows from both the panes of the window assembly need to be accounted for. Conduction heat gain/loss has been treated in the previous section. In the absence of solar radiation, the heat gain $\dot{Q}$ of the interior per unit window area A would be only the conductive term corresponding to the U value of the glazing:

$$\frac{\dot{Q}_{cond}}{A} = U(T_o - T_i) \quad \text{if no solar gain} \tag{5.9}$$

The combined effects of (2) and (3) listed earlier are treated together by introducing one term called the "solar heat gain coefficient (SHGC)" defined as

$$\frac{\dot{Q}_{solar,gain}}{A} = \text{SHGC} \cdot I_T \tag{5.10}$$

where I_T is the solar irradiance on the glazing surface. SHGC is the ratio of the solar heat gain through the window and of the solar radiation incident on the window. Thus, its numerical value ranges from 0 to 1. More strictly, the direct, and diffuse (reflected from both the sky and the ground) components of the incident solar radiation have to be separated, and the more accurate operational version of Equation 5.10 is given by

$$\frac{\dot{Q}_{solar,gain}}{A} = \text{SHGC}_{dir} \cdot I_{T,dir} + \text{SHGC}_{dif} \cdot I_{T,dif} \tag{5.11}$$

Note that $I_{T,dir} = I_{beam} \cdot \cos\theta_i$, where θ_i is the solar incidence angle on the window. The values of SHGC for different types of glazing and for different incidence angles for beam and that for diffuse (assumed to be a single averaged value) are given in Table 5.1. More extensive tables can be found in ASHRAE Fundamentals (2013).

Note that the SHGC values in Table 5.1 listed under "total window" apply to the entire window area (glazing plus the frame) while the center-of-glass properties apply to the transparent glazing area only. Certain types of analyses (i.e., when correcting for the effects of external and internal shading devices) require that the SHGC for the center of glass and that of the frame be determined separately since different corrections need to be subsequently applied to each of these components. The corrections are usually minor and often not warranted. The interested reader can refer to McQuiston et al. (2005) for the more involved treatment, while we shall limit ourselves to the simpler approach.

Example 5.3: Calculation of Solar Loads due to Beam and Diffuse Radiation

This example serves to illustrate the use of the data provided in Table 5.1. We will assume that the components of beam, diffuse, and ground-reflected solar radiation on a west-facing vertical wall are specified (see Section 4.7 for calculation procedure), and solar gains through a low-e double-glazed window with fixed Al frame (ID #21a in Table 5.1) are to be calculated.

Given: Solar incidence angle on window $\theta_i = 60°$, beam radiation normal to the solar disk $I_{beam} = 657.6$ W/m², diffuse radiation on window pane $I_{dif,sky} = 102$ W/m², and $I_{dif,ground} = 69.2$ W/m²

Find: SHGC_{dir}, SHGC_{dif}, and $\dot{Q}_{solar,gain}$

Lookup values: From Table 5.1, center-of-glass $\text{COG-SHGC}_{beam} = 0.65$, $\text{COG-SHGC}_{dir,60°} = 0.56$, and $\text{COG-SHGC}_{dif} = 0.57$, and for the entire window, $\text{Window-SHGC}_{beam} = 0.50$

Solution

First, a correction factor for the total window is found as

$$f = \frac{\text{Window-SHGC}_{beam}}{\text{COG-SHGC}_{beam}} = \frac{0.50}{0.65} = 0.769$$

This correction is applied to direct radiation at 60° incidence angle:

$$\text{Window-SHGC}_{dir,60°} = f \times \text{COG-SHGC}_{dir,60°} = 0.769 \times 0.56 = 0.431$$

The same correction is also applied to diffuse radiation:

$$\text{Window-SHGC}_{dif} = f \times \text{COG-SHGC}_{dif} = 0.769 \times 0.57 = 0.438$$

Finally, Equation 5.11 is used to determine the solar heat gains through the window:

$$\frac{\dot{Q}_{solar,gain}}{A} = 0.431 \times 657.6 \times \cos 60° + 0.438 \times (102 + 69.2)$$
$$= 141.7 + 74.98 = 216.7 \text{ W/m}^2$$

5.4.2 SHGC Derivation

Deriving the expression for SHGC for the simple one-pane glazing configuration (Figure 5.8) provides some insight into this coefficient. So as to simplify the derivation, let us overlook the difference between beam and diffuse radiation effects. In the presence of solar radiation, the total heat gain becomes

$$\frac{\dot{Q}}{A} = \frac{\dot{Q}_{cond}}{A} + \frac{\dot{Q}_{solar,gain}}{A} = U(T_o - T_i) + \tau I_T + h_i \Delta T_{sol} \tag{5.12}$$

where the terms on the right represent conduction (without solar), transmitted solar radiation, and extra heat gain due to solar radiation absorbed by the glass, with ΔT_{sol} being the temperature rise of the glass due to absorption of solar radiation.

To render this equation more useful, one eliminates ΔT_{sol} in favor of I_T. Since one can assume linear approximations for all the relevant heat transfer equations, ΔT_{sol} is independent of T_i and T_o. Taking an energy balance of the glass with T_i and T_o as zero and setting the absorbed solar flux αI_T equal to the sum of the heat flows to inside and outside leads to

$$\alpha I_T = h_i\ \Delta T_{sol} + h_o \Delta T_{sol} \tag{5.13}$$

Solving this equation for ΔT_{sol} and replacing it in Equation 5.12 results in

$$\frac{\dot{Q}}{A} = U(T_o - T_i) + \tau I_T + \frac{\alpha I_T h_i}{h_i + h_o} \tag{5.14}$$

The two rightmost terms can be combined as the total solar heat gain $\dot{Q}_{sol}$ through glazing, i.e.,

$$\frac{\dot{Q}}{A} = U(T_o - T_i) + \text{SHGC} \cdot I_T \tag{5.15}$$

where

$$\text{SHGC} = \tau + \alpha \frac{U}{h_o} \quad \text{with } U = \frac{h_i h_o}{h_i + h_o} \tag{5.16}$$

The aforementioned derivation applies to single-pane glasses. For double glazing, an analogous calculation yields

$$\text{SHGC} = \tau + \alpha_o \frac{U}{h_o} + \alpha_i U\left(\frac{1}{h_s} + \frac{1}{h_o}\right) \quad \text{with } \frac{1}{U} = \frac{1}{h_i} + \frac{1}{h_s} + \frac{1}{h_o} \tag{5.17}$$

where

h_s is the heat transfer coefficient of the space between the two panes

α_i and α_o are the absorptivities of the inner and outer panes, respectively

As a check, note that in the limit $1/h_s \to 0$ this formula correctly reproduces the single-pane result with $\alpha_o + \alpha_i = \alpha$.

Example 5.4: Calculation of SHGC from Individual Optical Terms

This example serves to illustrate how the optical values listed in Table 5.1 for different glazing systems can be used to infer the SHGC following Equation 5.17. We will assume the same conditions as those of the low-e double glazing surface with only one surface coated as analyzed in Example 5.1. Further, assume direct solar radiation to be normal to the surface.

Given: Window ID #21a of Table 5.1; from Example 5.1, the overall heat loss coefficient U = 0.284 Btu/(h·ft²·°F); heat transfer coefficient between the two panes h_s = 0.375 Btu/(h·ft²·°F); the convective resistance of the outdoor air film = R_o = 0.17 h·ft²·°F/Btu

Find: SHGC

Lookup values: From Table 5.1 $A_1^f = \alpha_o = 0.20$, $A_2^f = \alpha_i = 0.07$, $T = \tau = 0.59$

Solution

The direct application of Equation 5.17 yields

$$SHGC = \tau + \alpha_o \frac{U}{h_o} + \alpha_i U\left(\frac{1}{h_s} + \frac{1}{h_o}\right) = 0.59 + 0.20 \times \left(\frac{0.284}{(1/0.17)}\right)$$

$$+ 0.07 \times 0.284 \times \left(\frac{1}{0.375} + \frac{1}{(1/0.17)}\right)$$

$$= 0.590 + 0.0097 + 0.0564 = 0.656$$

Comments

This matches closely with the value of 0.65 listed in Table 5.1. Further, this example allows us to ascertain the relative magnitudes of the three elements for the specific glazing type assumed: the portion directly transmitted (0.59/0.656 = 0.900), the portion from the outer glass pane (0.0097/0.656 = 0.015), and the portion from the inner pane (0.0564/0.656 = 0.085). Thus, the solar transmittance τ accounts for about 90% of the total heat load with the inner glass contributing about 8.5% and the outer glass about 1.5%.

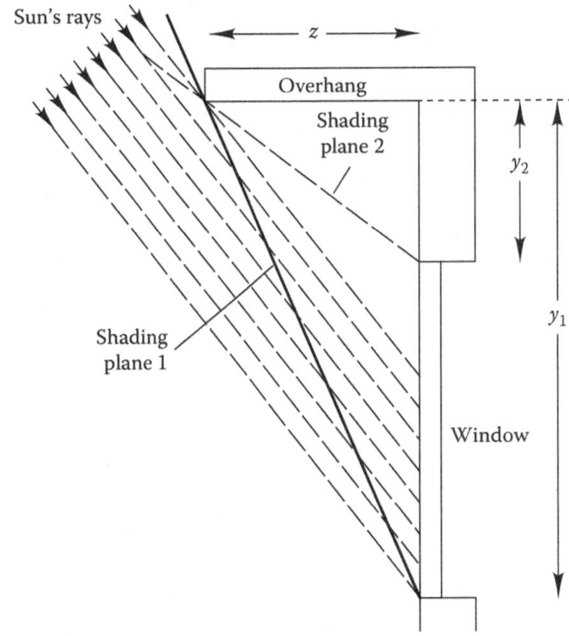

FIGURE 5.14
Shading of a south-facing window from a horizontal overhang.

5.5 External and Internal Shading

5.5.1 Overhangs and Recessed Windows

Sometimes, a designer would like to block direct solar radiation from entering a window during certain times of the day or the year. That may be desirable for two reasons: to reduce the cooling loads and to avoid uncomfortable daylighting in perimeter spaces, resulting in excessive contrast or glare. With fixed devices, it is generally not possible to satisfy all the requirements.

The analysis method discussed in Section 4.3.5 for shading requires a separate horizon outline for each point where shading is to be evaluated. This method is practical when the distance to the shading object is large compared to the dimensions of the surface whose solar exposure is to be studied. But for cases such as an overhang above a window, the dimensions of the shaded surface are not small compared to the shading object, and different points of the window would see different horizon outlines. Though each shading problem is somewhat unique given the variations in terms of location, day of year, window orientation, type of shading device and its dimensions, etc., the general analysis approach is similar.

1. Let us first address the simpler situation so as to provide some insights. *Fixed horizontal overhangs for south-facing windows* are usually very effective, and their analysis is particularly easy for solar noon. Consider a south-facing window with a horizontal overhang of length z as in Figure 5.14. For this case, one can show that the depth y of the shadow attains its extreme value for the day at solar noon (see, e.g., Rabl, 1985). For this instance, we use Equation 4.9 and set hour angle $\omega = 0°$ to obtain

$$\theta_{s,noon} = (\lambda - \delta) \tag{5.18}$$

where λ is the latitude and δ the solar declination angle. Over the year, the minimum and maximum vertical shading lengths due to south-facing overhangs occur at solar noon and are given by (see Figure 5.14)

$$y_{\min} = z\cot(\lambda - \delta) \quad \text{when } \delta < 0 \text{ (winter)} \tag{5.19}$$

and

$$y_{\max} = z\cot(\lambda - \delta) \quad \text{when } \delta > 0 \text{ (summer)} \tag{5.20}$$

Thus, if one chooses z such that the shadow falls on the bottom edge of the window according to Equation 5.20 with a certain declination $\delta > 0$ (summer half of the year), then the window will be completely shaded for all days with declination greater than this value of δ. For passive solar heating, one would like the overhang to allow direct insolation in winter while blocking it in summer. Example 5.5 shows to what extent that is possible.

Example 5.5: Long Horizontal Overhang over a South-Facing Vertical Window

A south-facing window of height $\Delta y = y_1 - y_2 =$ 2 m (6.56 ft) is to receive full sunshine at solar noon until February 15 while being completely shaded beginning May 15 (see Figure 5.14). What is the required length z of the overhang and at what height y_2 should it be above the top of the window? The site is Denver, CO, at latitude 40°N.

Given: λ = 40°, day of the year from January 1, n = 46 for February 15, and n = 135 for May 15

Sketch: Figure 5.14

Find: z

Solution

First, we need the declination for these dates. From Equation 4.3:

$$\delta_{\text{Feb}} = -13.13° \text{ for February 15}$$

$$\text{and} \quad \delta_{\text{May}} = 18.51° \text{ for May 15}$$

Then, we find the position of the shade (= distance below the overhang) from Equations 5.19 and 5.20:

$$y_{\text{Feb}} = z\cot(\lambda - \delta_{\text{Feb}}) = y_2 \quad \text{on February 15}$$

$$y_{\text{May}} = z\cot(\lambda - \delta_{\text{May}}) = y_1 \quad \text{on May 15}$$

The difference $\Delta y = y_{\text{May}} - y_{\text{Feb}}$ should be equal to the height 2 m (6.56 ft) of the window. Combining these equations and solving for z:

$$z = \frac{\Delta y}{\cot(\lambda - \delta_{\text{May}}) - \cot(\lambda - \delta_{\text{Feb}})}$$

$$= \frac{2}{\cot(40 - 18.51)° - \cot(40 + 13.13)°} = 1.18 \text{ m (3.87 ft)}$$

Then, y_{Feb} = 1.12 × cot(40 + 13.13)° = 0.84 m (2.76 ft) = y_2, the height of the overhang above the window. These dimensions are not unreasonable for a house.

2. For more *general cases*, the analysis of shading by nearby objects adopts the *profile angle approach* that allows us to calculate the location of the shadow cast by a point obstacle. Consider the shading of a rectangular window that is set back a distance z into a vertical wall, as shown in Figure 5.15. This case is similar to that of a window with a horizontal overhang just above the upper edge of the window and side vertical fins adjacent to the vertical sides of the window. The isometric view is shown in Figure 5.15. Suppose the position of the sun is given in terms of the azimuth ϕ_s and zenith θ_s, while the azimuth of the wall is ϕ_p (defined, analogous to ϕ_s, as the angle of the surface normal from due south, in terms of a rotation about the vertical; see Figure 4.8). Then, Figure 5.15 depicts the altitude angle $\alpha = 90° - \theta_s$, and the azimuth of the sun relative to the wall, called the "solar-wall azimuth angle," is given by

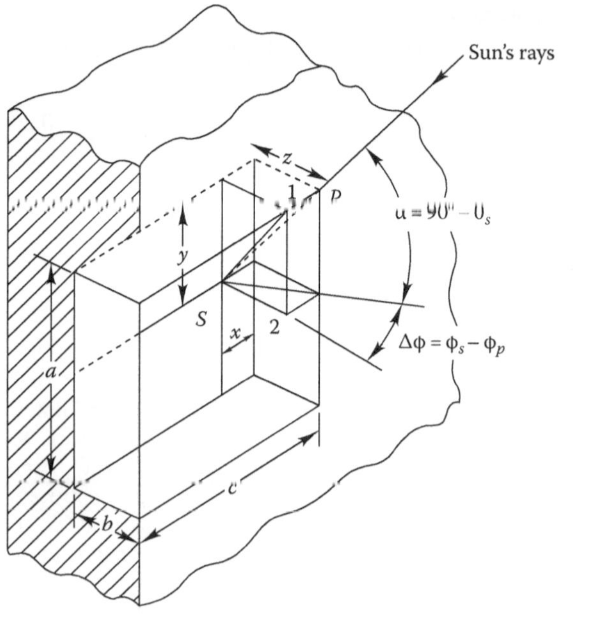

FIGURE 5.15
Coordinates of the shade $S = (x, y, z)$ cast by a point P (in this example, P is the corner of a rectangular window recess).

$$\Delta\phi = \phi_s - \phi_p \tag{5.21}$$

The projected solar altitude angle onto a vertical plane perpendicular to the window is called the "profile angle" γ and is given by

$$\tan\gamma = \frac{\text{length 1-2}}{\text{length } S\text{-2}} = \frac{\tan\alpha}{\cos\Delta\phi} = \frac{\cot\theta_s}{\cos\Delta\phi} \tag{5.22}$$

where points 1 and 2 are shown in Figure 5.15.

Figure 5.16 illustrates how the profile angle varies with solar time for a south-facing surface at 40°N latitude for three different days of the year (the equinox and the two solstices). The envelope defined by these curves represent the annual range of variation of the profile angle. Note that the profile angle is constant for the equinox day and for the hours about solar noon for the other days (suggesting that the vertical height of the shadow cast by the overhang is essentially constant during these 6 h). The profile angle varies greatly for the summer solstice before 9:00 h and after 15:00 h and even exceeds

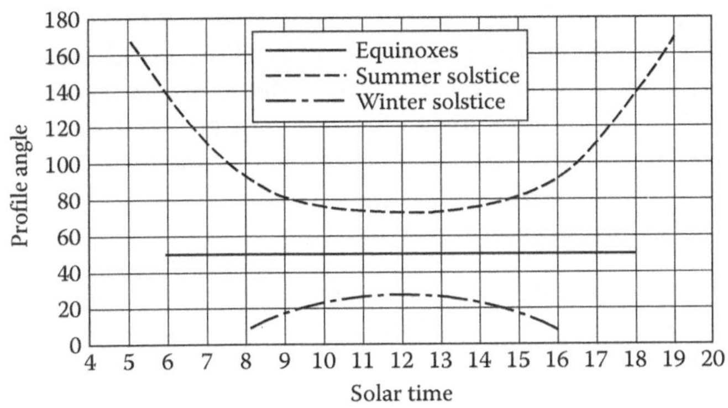

FIGURE 5.16
Variation of profile angle with solar time for south facing surface at 40°N latitude.

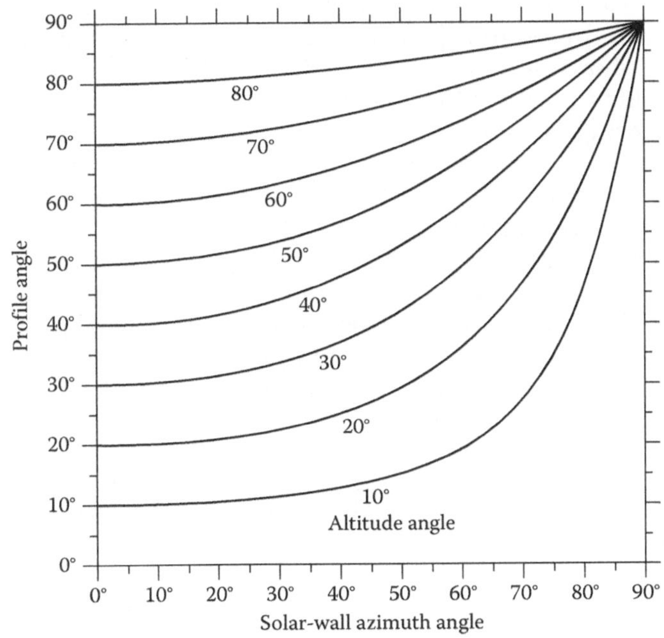

FIGURE 5.17
Variation of profile angle with altitude angle and solar-wall azimuth angle.

90° when the sun is north of east or west. Figure 5.17 is a more general representation of how the profile angle varies with solar altitude and solar-wall azimuth angle (following Equation 5.22). Note that the profile angle is close to the altitude angle varying no more than 10% for up to 40° or so of the solar-wall azimuth angle.

Let x, y, and z be the coordinates of point S, the shadow of the outer corner P of the window (Figure 5.15). These coordinates are relative to P: x is the width of the vertical projection, i.e., shadow cast by the vertical edge, analogously for the height y and the horizontal width z. It is easy to see that x is given by

$$x = z \tan \Delta\phi \tag{5.23}$$

Using vector algebra, one can prove that

$$y = z \cdot \tan\gamma \tag{5.24}$$

Thus, the aforementioned equation allows the vertical projection or shadow length at any point to be determined. From here, the window shading fraction can be deduced.

Example 5.6: Horizontal Overhang over West-Facing Window

Consider the west-facing wall subject to conditions stated in Example 5.3. Assume that it contains a recessed window of height a = 2 m (6.56 ft) and depth b = 0.5 m (1.64 ft). Find the solar gains through the entire window given that the solar zenith and azimuth angles are 40.5° and 90° respectively. Neglect end effects.

Given: Conditions stipulated $\theta_i = 60°$, $I_{beam} = 657.6$ W/m², $I_{dif,sky} = 102$ W/m², and $I_{dif,ground} = 69.2$ W/m²; $SHGC_{dir,60°} = 0.431$ and $SHGC_{dif} = 0.438$; $\theta_s = 40.5°$, $\phi_s = 90°$, $\phi_p = 90°$, $a = 2$ m, and $b = 0.5$ m

Sketch: Figure 5.15

Find: Window shading ratio and solar gains $\dot{Q}_{solar,gains}$

Solution

The profile angle is given by Equation 5.22:

$$\tan\gamma = \frac{\cot 40.5°}{\cos(90° - 90°)} = 1.17, \quad \text{i.e.,}\ \gamma = 49.5°$$

Noting that $z = b = 0.5$ m, we now use Equation 5.24 to calculate length y of the shadow cast by the horizontal edge of the recess:

$$y = z \cdot \tan\gamma = 0.5\ \text{m} \times 1.17 = 0.585\ \text{m}$$

and so

$$\frac{y}{a} = \frac{0.585}{2.00} = 0.293$$

The shadow length of the horizontal projection x is calculated from Equation 5.23:

$$x = 0.5 \times \tan(90° - 90°) = 0$$

This implies that the window is fully illuminated in the frontal direction, i.e., there is no shading due to the vertical side of the window recess. This is not unexpected since the sun-wall azimuth angle is zero. Finally, the total shading fraction due to both projections is simply

$$F_{shade} = \frac{y}{a} = 0.293$$

Thus, 29.3% of the window is shaded.

The sunlit portion of the window receives both beam and diffuse radiation, while the shaded portion only receives diffuse radiation. Then, Equation 5.11 is simply rewritten as

$$\frac{\dot{Q}_{solar,gain}}{A} = \mathrm{SHGC}_{dir} \cdot I_{T,dir}(1 - F_{shade}) + \mathrm{SHGC}_{dif} \cdot I_{T,dif} \quad (5.25)$$

or

$$\frac{\dot{Q}_{solar,gain}}{A} = 0.431 \times 657.6 \times \cos 60° \times (1 - 0.293) + 0.438 \times (102 + 69.2)$$

$$= 100.2 + 74.98 = 175.2\ \mathrm{W/m^2}$$

This is a 20% reduction in the total solar gain as against a window without external shades (see Example 5.3). Note that the treatment of the solar sky diffuse heat gain component is simplified since the window overhang also blocks some of the sky diffuse radiation.

Comments

This example illustrates the fact that overhangs can reduce cooling loads attributed to solar radiation on east- or west-facing windows. In summer, there are long periods when the sun can reach these facades with fairly small angles of incidence. The range of incidence angles is so wide as to make it impossible to block all this radiation, short of eliminating the windows completely. But partial blocking can be achieved by combining horizontal overhangs with vertical shades to the south of each window.

The constraints of passive solar heating cannot be satisfied perfectly during the entire year. The shading effects are necessarily determined by the declination, while the heating and cooling loads reach their extreme values about a month later than the solar declination. The situation is particularly awkward around the equinoxes, yielding either too little heat in spring or too much in the fall. Furthermore, the better one wants to separate the shaded and unshaded phases, the larger the shading device needs to be relative to the window. The problem is much simpler in tropical locations where solar heating is of no interest and the sun is so high that most of it can be blocked from a south- or north-facing window by an overhang of reasonable dimensions.

External shading devices are very effective in reducing solar heat gain since they do not let solar radiation to enter the space; however, they are static devices that cannot be controlled as needed. Another option is to fit exterior sun screens to windows but these will reduce illumination as well. If one wants to achieve better control of solar gains, another option is to vary the transmission of the windows (some technological ways are described in Section 5.6). Adjustable exterior blinds can be unfurled when needed and do allow good control, but they are bulky mechanical devices, are prone to malfunction, and do not last as long as fixed external overhangs which are usually made of concrete.

5.5.2 Internal Shading Devices

Internal shading devices can also be used to reduce heat gains by reflecting back outward some of the incident solar radiation. However, the heat absorbed by these devices will largely be released to the room air. Hence, the effect of such devices is partial to the extent that only a fraction of the heat gain can be negated. Common devices are venetian blinds, roller shades, and draperies. These are used primarily to control glare, provide privacy, for aesthetic reasons, and by using light colored drapes to reflect some of the incoming radiation as shown in Figure 5.18. Further, the effect of venetian blinds placed either outside or inside a double-pane glazing is different, and draperies have to be treated separately. Thermal analysis methods have been developed, but they are rather complex and the interested reader can refer to ASHRAE Fundamentals (2013) and appropriate references.

Briefly, the general analysis approach pertaining to the case of venetian blinds placed outside the glazing panes (the approach also applies to roller shades) is to introduce an interior solar attenuation coefficient (IAC) so that Equation 5.25 for the solar heat gains is expressed as

$$\frac{\dot{Q}_{solar,gain}}{A} = [\mathrm{SHGC}_{dir} \cdot I_{T,dir}(1 - F_{shade}) + \mathrm{SHGC}_{dif} \cdot I_{T,dif}] \times \mathrm{IAC} \quad (5.26)$$

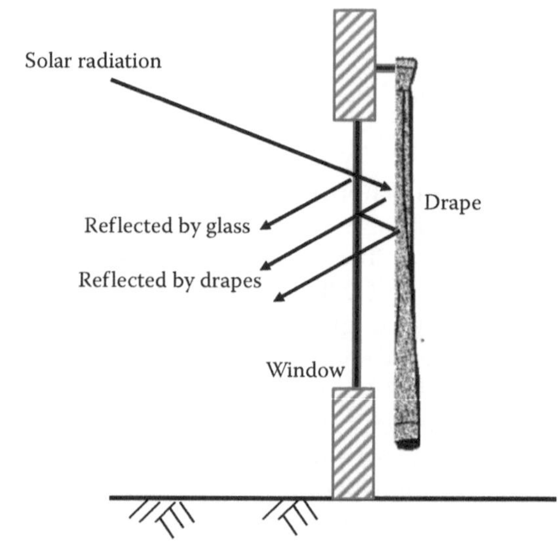

FIGURE 5.18
Draperies reduce heat gain by reflecting back some of the solar radiation.

Strictly speaking, IAC should only be applied to the portion of the solar heat gains through the glazing portion and not to the heat gains through the entire window (which also includes the frame). However, the error introduced is small (a few % only) and the resulting simplification justifies the introduction of such a small error. Values of IAC are assembled in Table 5.4 where we note that these are dependent on both the type of shading device as well as the type of glazing system. For example, consider Example 5.6 where the solar heat gains through a west-facing window with a horizontal overhang were computed. The effect of introducing interior opaque white roller shades (IAC = 0.40 from Table 5.4) would be to reduce the solar heat gains to 40% (i.e., to 175.2 × 0.4 = 70 W/m^2). The combined effect of both the external overhang and the interior shading device has reduced the solar hear gains of the plain window of Example 5.3 by ((216.7-70)/216.7) = 0.677 or 68%.

On a parting note, the SHGC approach combines the transmitted and absorbed solar radiation effects into a single term. However, accurate cooling load calculations require these terms to be treated separately since the transmitted portion of the solar radiation is absorbed by the internal mass of the room and released later in time, while the absorbed effect of the solar radiation is released by convection almost instantaneously. Hence, the aforementioned treatment is a simplified one. Software programs such as Windows (LBNL, 2011) as well as detailed simulation programs such as Energy Plus (2009) treat these terms separately and do not adopt the simplified SHGC approach.

5.6 High-Performance Glazing

5.6.1 Promising Improvements

Windows are a major source of solar heat gains affecting both annual energy use as well as peak design conditions. For the control of solar radiation, the traditional design approaches have been mechanical: overhangs and blinds, fixed or movable. Fixed shading devices lack flexibility for good control of heat gains. Movable devices on the outside of a building are expensive because they must be sturdy enough to withstand wind and weather. For the inside of windows, venetian

TABLE 5.4

Interior Solar Attenuation (IAC) for Single and Double Glazings Shaded by Interior Blinds and Roller Shades

					IAC				
		Glazing Solar Transmittance			Venetian Blinds		Roller Shades		
Glazing System	Nominal Thickness of Each Pane, in.	Outer Pane	Single or Inner Pane	Glazing SHGC	Medium	Light	Opaque Dark	Opaque White	Translucent Light
Single glazing systems									
Clear residential	1/8		0.87–0.80	0.86	0.75	0.68	0.82	0.40	0.45
Clear commercial	1/4 to 1/2		0.80–0.71	0.82					
Clear, pattern	1/8 to 1/2		0.87–0.79						
Tinted	3/16, 7/32		0.74, 0.71						
Above glazings, automated blinds				0.86	0.64	0.59			
Above glazings, tightly closed vertical blinds				0.85	0.30	0.26			
Heat absorbing	1/4		0.46	0.59	0.84	0.78	0.66	0.44	0.47
Reflective coated glass				0.26–0.52	0.83	0.75			
Double glazing systems									
Clear double, residential	1/8	0.87	0.87	0.76	0.71	0.66	0.81	0.40	0.46
Clear double, commercial	1/4	0.80	0.80	0 70					
Heat absorbing double	1/4	0.46	0.80	0.47	0.72	0.66	0.74	0.41	0.55
Reflective double				0.17–0.35	0.90	0.86			
Other glazings (approximate)					0.83	0.77	0.74	0.45	0.52
Range of variation					0.15	0.17	0.16	0.21	0.21

Source: Extracted from ASHRAE, *Handbook of Fundamentals*, American Society of Heating, Refrigerating and Air-Conditioning Engineers, Atlanta, GA, 2013. Copyright ASHRAE, www.ashrae.org.

blinds are classic, but their thermal performance is worse than that of exterior blinds because venetian blinds cause much solar radiation to be absorbed inside the building.

The technology of windows is in rapid evolution, and it is difficult to foresee which products will be the best choice for future buildings. The two most important qualities are optical and thermal ones. The ideal window is a perfect insulator, with variable transmittance for the visible and the infrared portions of the solar spectrum, to permit as much control of daylight and heat gains as possible (see Figure 5.7). Recent years have seen vigorous progress in the development of transparent insulators and of coatings with variable transmittance (discussed in Section 5.6.2), and there is good hope that windows of the future will come quite close to the ideal and be more attractive financially.

There are several methods for reducing the heat loss of windows that involve reducing radiative and/or convective heat losses. Adding further glass panes increases the thermal resistance, but beyond three panes, one faces diminishing returns. Thin plastic films avoid the cost and weight of glass panes. For greatest effect, they should be coated with a heat mirror. Such coatings do not have to interfere with the transmission of solar radiation because the wavelengths of thermal infrared are quite different from those of the infrared and visible portions of the solar spectrum. While uncoated glass or plastic surfaces have an emissivity of 0.84, quite a few durable coatings are now available with emissivities in the range of 0.2–0.4. Heat mirror coatings are a powerful means for reducing the *U* value of windows, as can be seen from Figure 5.9.

Center-of-glass *U* values around 0.7 W/(m^2·K) [0.11 Btu/(h·ft^2·°F)] in a triple-glazed krypton-filled window with two low-emissivity coatings were originally reported by Arasteh et al. (1989). Its overall thickness is 25 mm (1 in.) or less, compatible with the sash and frame requirements of most manufacturers. Such windows are now in commercial production though further work is needed to develop frames and spacers of comparable performance.

To reduce conductive/convective heat losses, one can replace the air between the panes by *inert gases* such as argon or krypton. Figure 5.9 and Table 5.3 show that the improvement over air is appreciable. *Evacuation* would be even better, of course. While the atmospheric pressure would be prohibitive for large unsupported panes, work by Benson et al. (1988) has demonstrated the feasibility of producing flat evacuated windows where the glass panes are separated by tiny glass beads, with about 1 mm (0.039 in.) diameter. The *U* value would be impressive: the center-of-glass value is expected to be 0.35 W/(m^2·K) [0.062 Btu/(h·ft^2·°F)], and with a reasonable frame, the area-averaged value would be around 0.5 W/(m^2·K) [0.088 Btu/(h·ft^2·°F)]. The commercial promise is not yet clear. It will depend on the visual appearance of the glass beads and, of course, on the cost.

Aerogels are another interesting approach. They are basically a glass foam whose structure is smaller than the wavelength of light. Thus, they are quite clear, even at a thickness of several centimeters. Their density is about one-tenth that of glass, while the thermal conductivity is comparable to still air. The conductivity can be cut in half by evacuating the aerogel. Aerogels are quite fragile, and in practical windows they would be sandwiched between glass panes for protection. Fabrication cost and optical clarity need further improvement, but some companies are already close to commercializing the product.

While visual appearance is, of course, a crucial concern for vision windows, there are numerous applications where transparent insulators can be used even if they lack clarity. For many skylights, diffuse glazings are preferable in any case, and for transom windows* they may be acceptable. For flat-plate solar collectors, the specularity of the cover is irrelevant. For all such cases, an interesting option is the use of *plastic honeycombs* made of acrylic or polycarbonate enclosed between flat panes, with cell walls normal to the panes to minimize optical losses. The *U* value is about 1.0 W/(m^2·K) [0.176 Btu/(h·ft^2·°F)] for a thickness of 10 cm (4 in.), while offering a transmissivity for diffuse solar radiation of 0.78. When ordinary glass panes are added on both sides for protection, the transmissivity decreases to 0.60 and the *U* value improves slightly.

Among other possibilities, we might mention windows that use ventilation air to recover heat or to reduce heat losses. For instance, Figure 5.19 shows an *air curtain window* where the heat of the exhaust air circulating through the window panes can partially compensate for the heat loss through the window. As a variation, the flow could be reversed, drawing the supply air through such a window; the heat lost through the inner pane would be brought back into the building. Such windows appear to be expensive, and they are not widely used at present (which in turn prevents the cost reductions achievable by mass production).

The prospects of technology should not make us forget a natural alternative. Deciduous trees or vines provide shade and sunlight in a seasonal pattern well matched to cooling and heating loads. Besides, the beauty of plants has universal appeal. Of course, there are certain expenses for the upkeep, and the control is not perfect. It takes years to grow plants to the desired size—not a negligible problem since buildings should be comfortable right from the start.

* A transom is a transverse horizontal structural beam or bar or a crosspiece separating a door from a window above it.

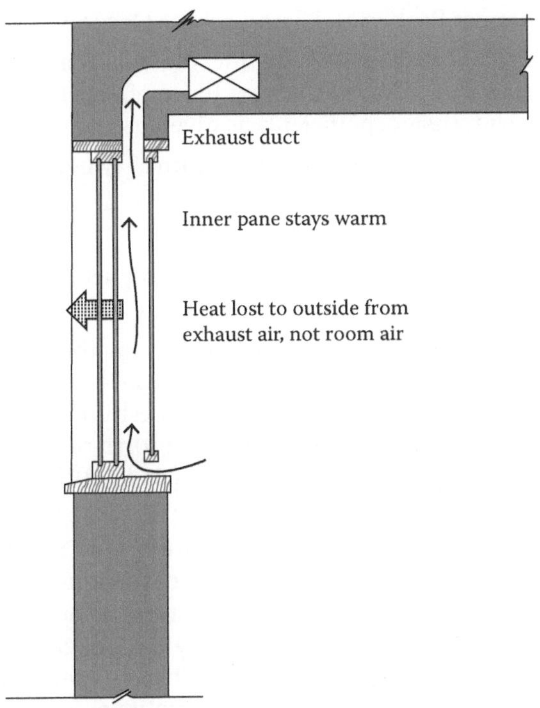

FIGURE 5.19
Air curtain window. (From Nisson, J.D.N. and Dutt, G., *The Superinsulated Home Book*, Wiley, New York, 1985. With permission.)

5.6.2 Smart Windows

Automatic adjustment of mechanical shading devices is possible, but in practice they are problematic in the long run. Far more elegant solutions have evolved. Smart or dynamic or advanced glazing is a new generation technology that alters thermal/optical properties, such as shading coefficients and visible transmittances in response to either an electric charge or an environmental signal. There are essentially four types:

1. *Thermochromic* materials whose transmissivity varies with temperature. As outdoor temperature increases, the visible light transmittance decreases, and vice versa. Gels sandwiched between glass and plastic switch from a clear state when cold to a more diffuse, white, reflective state when hot. In their switched-on state, less visible light is transmitted through the glazing. Though an old concept dating back to the 1870s, certain problems have hindered their widespread commercial development. One of them is the fact that the glazing is prone to chemical leakage around the edges, thereby degrading the optical properties over time.
2. *Photochromic* materials, also an old technological concept, whose transmissivity changes in response to the incident light intensity. In essence, this glass automatically adjusts its visible transmittance according to exterior light exposure. One drawback is that the glazing dims when exposed to winter sun, thereby increasing heating loads. Large sizes of windows are not commercially available.
3. *Liquid crystal display* technology, widely used in wrist watches, is also being developed and modified for use in windows and interior partition. It is primarily meant to provide privacy by diffusing the light and does not reduce the amount of light or solar heat; thus, there are no energy saving benefits. This technology involves a very thin layer of liquid crystals sandwiched between two transparent electrical conductors deposited on heat-treated glass. The conductive coatings are connected to a constant power source. When the power is off, the liquid crystals are in a random and unaligned state that scatters light and causes the glazing to appear translucent. On application of an electric current, the glazing immediately turns hazy clear. Unlike thermochromic glazing technology, liquid crystal requires continuous power supply for the glass to remain clear (24–100 V AC or 0.5 W/ft^2 of glass area).
4. *Electrochromic* (EC) coatings are thin-film coatings applied to a glass or plastic that can change optical and thermal properties of glazing when a small voltage is applied. Rudolph et al. (2009) describe the technology of such devices. The solar transmissivity can be made to vary continuously by applying a voltage that can be tuned and operated either by a preset controller or even manually. Samples have been produced whose transmissivity varies from 0.2 to 0.8. Though EC window prototypes have been installed in a number of buildings in Japan, Europe, and the United States, cost is still the primary drawback to widespread adoption. This technology is the most promising of all "smart" window technologies, and active research and development is ongoing.

Problems

Numbers 1–4 given in parenthesis denote the degree of difficulty.

5.1 What characteristics of a window are relevant for the thermal analysis of a building. *(1)*

5.2 Consider window #21a of Table 5.1 with fixed aluminum frame. What is the instantaneous solar heat gain for direct radiation when the solar irradiance incident on the window is 200 Btu/(h·ft²) (631 W/m²):

(a) Under normal solar incidence

(b) When the solar incidence angle is 50° *(1)*

5.3 Use the Potter et al. correlation (Equation 4.44) to find the average daily solar heat gain through a vertical south-facing window with fixed aluminum frame of 10 ft² area and ID #5b of Table 5.1 in Phoenix, AZ, (λ = 33°N) in January ($\bar{K}_T = 0.61$) and June ($\bar{K}_T = 0.76$).

Assume all radiation to be beam and the effective daily solar incidence angles for beam radiation to be equivalent to those at 10:30 m (solar time) on the 21st day of the month. *(2)*

5.4 A recessed window is located on a vertical wall and oriented due south. Calculate the depth of the recess that will shade the window completely at 10 a.m. solar time on August 23. The window is 1 m wide by 0.5 m high. Perform the calculations for (a) a latitude of 25°N and (b) a latitude of 45°N. *(2)*

5.5 Window type #5b with fixed aluminum frame in Table 5.1 is located on the south-facing vertical wall, that is 2 m tall and 3 m wide. The latitude is 33°N:

(a) Find the global irradiance at noon on June 21 when beam normal irradiance is 900 W/m² [285.3 Btu/(h·ft²)] and the diffuse horizontal irradiance is 100 W/m² [31.70 Btu/(h·ft²)]. Assume isotropic sky model and neglect ground reflectance.

(b) What is the solar heat gain for center of glazing?

(c) A horizontal overhang 0.5 m long is to be designed such that the window is fully shaded on June 21 at noon. How much below the overhang should the window be located?

(d) What is the solar heat gain for center of glazing with such an overhang? *(3)*

5.6 Consider the same window with the horizontal overhang as in Problem 5.5. Assume 3 p.m. solar time when beam normal irradiance is 700 W/m² [221.9 Btu/(h·ft²)] and the diffuse horizontal irradiance is 100 W/m² [31.70 Btu/(h·ft²)]?

(a) What is the solar heat gain including the end effects of the overhang with isotropic sky model?

(b) If venetian blinds of light color are placed behind the window, what would be the reduction in solar heat gain? *(3)*

5.7 Repeat Problems 5.5 and 5.6 for December 21. *(4)*

5.8 Calculate the U value for the following double-glazed windows assuming the temperatures and the heat transfer coefficients as given in Example 5.1:

(a) Ordinary glass with vacuum between the panes

(b) Low-emittance coating with $\varepsilon = 0.05$ on both surfaces facing the gap

(c) Low-emittance coatings as in part (b) but with a vacuum between the panes *(3)*

5.9 Frost can form on a surface when the humidity is sufficiently high and the surface temperature falls below freezing. What is the highest outdoor temperature at which frost can form on the inside of a single-glazed window if it is 60°F indoors? Assume a U value of 1 Btu/(h·ft²·°F). *(2)*

5.10 Windows with air curtains are briefly mentioned in Section 5.6.1. Using resources available on the Internet, describe their operating principles, and give case studies wherein they have been installed in actual buildings and their thermal performance and costs. Set up a thermal network diagram to represent their thermal performance. *(3)*

References

Arasteh, D., S. Selkowitz, and J. Wolfe (1989). The design and testing of a highly insulating glazing system for use with conventional window systems. Report LBL-24903 TA-257 (November). Lawrence Berkeley Laboratory, Berkeley, CA. *J. Solar Energy Eng.*, 111(1), 44–53.

ASHRAE Fundamentals (2009, 2013). *Handbook of Fundamentals*. American Society of Heating, Refrigerating and Air-Conditioning Engineers, Atlanta, GA.

Benson, D.K., T.F. Potter, and C.B. Christensen (1988). Vacuum insulating window R&D: An update, in *Proceedings of American Council for Energy Efficient Economy 1988 Summer Study*, Pacific Grove, CA, Vol. 3, p. 3.21.

Energy Plus (2009). Energy plus building energy simulation software, developed by the National Renewable Energy Laboratory (NREL) for the U.S. Department of Energy, under the Building Technologies program, Washington, DC. http://www.nrel.gov/buildings/energy_analysis.html#energyplus. Accessed on March 2015.

Goswami, D.Y., F. Kreith, and J.F. Kreider (2000). *Principles of Solar Engineering*, 2nd ed. Taylor & Francis Group, Philadelphia, PA.

LBNL (2011). Windows 6.3. Lawrence Berkeley National Laboratory, Berkeley, CA. www.windows.lbl.gov/software/window/6/6.3.36/window_releasenotes.htm. Accessed on June 2014.

McCabe, M.E. (1989). Window *U*-values: Revisions for the 1989 ASHRAE *Handbook of Fundamentals*. *ASHRAE J.*, June, 56.

McQuiston, F.C., J.D. Parker, and J.D. Spitler (2005). *Heating, Ventilating and Air Conditioning*, 6th ed. John Wiley & Sons, New York.

Nisson, J.D.N. (1985). *Windows and Energy-Efficiency: Principles, Practice and Available Products*. Cutter Information Corp., Arlington, MA.

Nisson, J.D.N. and G. Dutt (1985). *The Superinsulated Home Book*. Wiley, New York.

Rabl, A. (1985). *Active Solar Collectors and Their Applications*. Oxford University Press, New York.

Rudolph, S.E., J. Dieckmann, and J. Broderick (2009). Technologies for smart windows. *ASHRAE J.*, July, 104–106. American Society of Heating, Refrigerating and Air-Conditioning Engineers, Atlanta, GA.

6

Infiltration and Natural Ventilation

ABSTRACT Outdoor air infiltration through leakage pathways of the building envelope is an important contributor to building energy loads. The accurate determination of air inflow from a rigorous scientific standpoint is very complex due to temporal and spatial variability of the driving pressure differences across the building envelope and the practical difficulties of determining the location and size of envelope leakage pathways; simplifications are thus required for engineering purposes. We start by illustrating how infiltration affects building loads and by defining certain important terms. An overview of air infiltration in past and current building stock in the United States is then provided. Next, we present calculation methods to estimate the pressure difference across the envelope due to the natural driving forces of wind velocity and stack temperature difference. This is followed by a discussion of a widely used engineering model based on a simplification of the theoretical crack flow equation for determining air infiltration through actual building components and assemblies. We then provide a broad overview of multizone airflow models, followed by two commonly used simplified physical models for single-zone buildings along with illustrative examples. Due to the current interest in adopting hybrid ventilation as a strategy to lower energy use in a building, correlations to estimate natural ventilation caused by deliberate opening of windows and doors are also briefly covered. Finally, we describe different techniques, both in the laboratory and in the field, for measuring air infiltration and interzone flows.

Nomenclature

A	Area, m^2 (ft^2)
a	Exponent for terrain in Equation 6.11
a_s	Stack coefficient of Table 6.3, (L/s)^2/(cm$^4\cdot$K) [(ft^3/min)^2/(in.$^4\cdot$°F)]
a_w	Wind coefficient of Table 6.4, (L/s)^2/[cm$^4\cdot$(m/s)^2], {(ft^3/min)^2/[in.^{4} (mph)^2]}
ACH	Air changes per hour, h^{-1}
C	Flow coefficient in Equation 6.5 m/(s$\cdot$Pa^n) [ft^3/(min$\cdot$inWG^n)]
C	Coefficient in Equation 6.22 for the number of people who pass through the door per hour, [(L/s$\cdot$Pa$^{0.5}$), (ft^3/min)/(inWG)$^{0.5}$]
C	Tracer gas concentration in the indoor space
C_D	Discharge coefficient of opening in Equations 6.3 and 6.30
C_d	Draft coefficient for resistance to airflow between floors
C_p	Pressure coefficient due to wind in Equation 6.8
c_p	Specific heat of air, kJ/(kg$\cdot$K) [Btu/(lb$_m\cdot$°F)]
C_v	Window opening effectiveness in Equation 6.28
ELA	Effective leakage area of Equation 6.25
$F(t)$	Tracer gas injection rate at time t
g	Acceleration due to gravity = 9.81 m/s^2 (32 ft/s^2)
g_c	Conversion factor in IP units, 32.17 lb$_m\cdot$ft/(lb$_f\cdot$s^2)
h	Height m (ft) in Equation 6.11
K	Overall leakage area coefficient in Equations 6.20 and 6.23, m/(s$\cdot$Pa^n) [ft/(min$\cdot$inWG^n)]
k	Coefficient characterizing crack width of operable doors and windows in Equation 6.21
L	Length m (ft)
l	Length of perimeter of component, m (ft)
$\dot{m}$	Mass flow rate, kg/s (lb$_m$/h)
NPL	Neutral pressure line
n	Exponent of airflow equation
p	Pressure, Pa (lb$_f$/in.^{2})
$\dot{Q}$	Heat flow rate, W (Btu/h)
R	Gas constant of air J/(kg$\cdot$K) [ft$\cdot$lb$_f$/(lb$_m\cdot$°R)]
R^2	Coefficient of determination of a regression model
RMSE	Root mean square error
r	Radius, m (ft)
T	Dry-bulb air temperature, K or °C (°R or °F)
T_i	Indoor air temperature, °C (°F)
T_o	Outdoor air temperature, °C (°F)
t	Time, s, min, h
V	Volume, m^3 (ft^3)
$\dot{V}$	Volume flow rate, m^3/s or L/s (ft^3/min)
v	Wind speed, m/s (mph or ft/s)
W	Width
WG	Water gauge
x	Distance, m (ft); quantity defined in Equation 6.18

Greek

δ Boundary layer thickness, m (ft)
θ Wind angle with respect to building
μ Absolute viscosity, (Pa·s) [lb_m/(ft·s)]
ρ Density, kg/m^3 (lb_m/ft^3)
Δh Vertical distance from neutral pressure level, m (ft)
Δp $=p_o - p_i$, pressure difference between outside and inside, Pa (psi, or inWG); general pressure difference
ΔT Indoor–outdoor temperature difference $T_i - T_o$, K (°F)
ΔX Difference in quantity X

Subscripts

i Indoor
inf Infiltration
j Individual leakage sites
f Final
met Meteorological
leak Leakage
o Outdoor
op Opening
p Perimeter
r Reference
t Total
vent Ventilation

6.1 Importance and Basic Definitions

Fresh* or outdoor air in buildings is essential for comfort and health (as discussed in Chapter 3), and the energy for conditioning this air is an important contributor to the total space conditioning load (usually around 10%–20% in larger buildings and as much as 50% for smaller buildings). The designer must pay attention to two competing trends: not enough air (i.e., one risks the sick building syndrome) and too much air (i.e., one wastes energy) (see Example 6.1). The supply of fresh air, or air exchange, is stated as the flow rate $\dot{V}$ of the outdoor air [in m^3/s or L/s (ft^3/min)] that crosses the building boundary and needs to be conditioned. Often, it is convenient to divide it by the volume V of the building and express in units of *air changes per hour* (ACH) = $(\dot{V}/V)$. Even though it is customary to state the airflow as the volumetric rate, the mass flow $\dot{m} = \rho\dot{V}$ is more relevant for building energy calculations. The relation between mass flow and volume flow depends on the density ρ, which varies quite significantly with temperature and pressure.

It is helpful to distinguish three mechanisms that contribute to the total air exchange:

- *Infiltration* is the uncontrolled airflow rate through all the unintentional openings (little cracks and gaps) between different components such as ill-fitting windows or doors in a real building. It is balanced by an equal mass flow called "exfiltration" since mass must be conserved and no change of air mass within the building occurs under steady conditions (see Figure 6.1).
- *Natural ventilation* is the airflow rate induced by deliberate opening of windows or doors. It is a variable quantity depending on prevailing outdoor conditions, and one cannot control it properly.
- *Mechanical or forced ventilation* is the airflow rate intentionally drawn-in by mechanical ventilation such as fans. It can be controlled and varied as necessary to meet indoor IAQ standards (discussed in Section 3.5).

The term "passive ventilation" is also used and applies to both infiltration and natural ventilation. The term "air leakage" is often used interchangeably with infiltration.

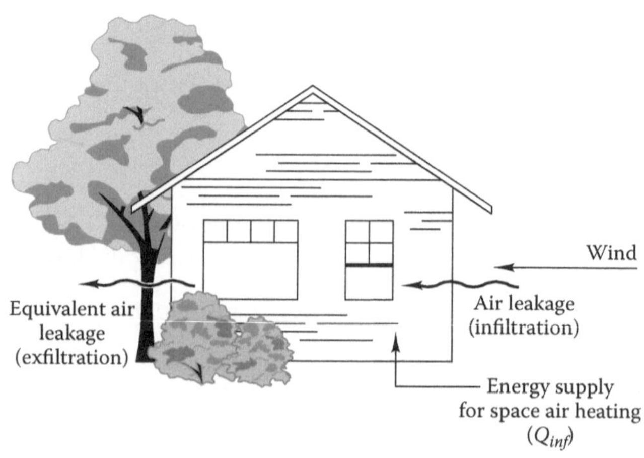

FIGURE 6.1
Sketch showing infiltration through building skin and heat input from building heating system needed to heat infiltrating air.

* Several professionals distinguish between outdoor air and fresh air since the former may not always be "fresh." We have, however, used these terms interchangeably in this book.

Though the terms are related, it is advisable to keep the distinction. Air leakage is a measure of air tightness of the building shell expressed as a leakage area rather than a volume flow rate as in infiltration.

Example 6.1: Energy Implications of Air Infiltration

Suppose that on a windy winter day at sea level, it is determined that air leaks into a residential building at the rate of 1 ACH. In other words, all the warm air in the building is exchanged with outside air once each hour. If the outdoor air temperature is −10°C and the interior temperature is 22°C, how much heat must be provided by the heating system to heat the infiltrating air? The building has 150 m² of floor area and is two stories high with a net interior height of 7 m.

Given: $T_i = 22°C$, $T_o = -10°C$, $V = 150\ m^2 \times 7\ m = 1050\ m^3$, $\dot{V} = 1$ volume/h

Figure: See Figure 6.1.

Assumptions: Steady flow; air is an ideal gas, i.e., the enthalpy depends only on the temperature

Find: $\dot{Q}_{inf}$

Lookup values: Specific heat of air $c_p = 1.01$ kJ/(kg · K). The density of air at 22°C and sea-level barometric pressure is 1.19 kg/m³ (it can be found from the ideal gas law)

Solution

The solution uses the rate form of the open-system first law of thermodynamics. The kinetic and potential energy terms are identical on both sides of the equation if the small openings through which air infiltrates and leaves (exfiltrates) the building are uniformly distributed.

Following the sensible heat equation (Equation 2.34):

$$\dot{Q}_{inf} = \dot{m}c_p(T_o - T_i) \tag{6.1}$$

The mass flow rate can be found from the density and volumetric flow rate:

$$\dot{m} = \rho\dot{V} = \frac{1.19\ \text{kg/m}^3 \times 1050\ \text{m}^3/\text{h}}{3600\ \text{s/h}} = 0.347\ \text{kg/s} \tag{6.2}$$

Finally, the heating to be provided by the heating system is found from Equation 6.1:

$$\dot{Q}_{inf} = 0.347\ \text{kg/s} \times 1.01\ \text{kJ/(kg}\cdot\text{K)} \times [(-10) - 22]°\text{C}$$
$$= -11.2\ \text{kW} \quad \text{(heating load)}$$

Comments

When expressing infiltration in buildings using the air change nomenclature, we need to be clear that the air volume referred to is "inside air." If one air change of outside air were the basis of this method instead, the density of outdoor air would be used in Equation 6.1. For this example, the density of outside air is 12% greater than that of inside air, and the infiltration heat load would also be 12% larger.

6.2 Infiltration Rates across Building Stock

Uncontrolled air exchange or infiltration is highly dependent on natural weather drivers such as wind speed and outdoor temperature. Even with closed windows, it can vary by a factor of 2 or more, being lower in summer than in winter. The variability of air exchange is indicated schematically in Figure 6.2 as the relative frequency of occurrence for three types of houses: a leaky house and a moderately tight house, both with natural infiltration, and a very tight house with mechanical ventilation. The last guarantees adequate supply of fresh air at all times, without the energy waste of conditioning excess air. Techniques for measuring infiltration rates and air exchange rates in residences are described in Section 6.9.

Typical infiltration rates in North American homes vary by an order of magnitude, from a low of 0.2 to over 2 ACH. In the past, not much attention was paid to airtight construction, and older buildings tend to have rather high infiltration rates, in the range of 0.5–2 ACH. With current conventional construction in the United States, one finds lower values, around 0.3–0.7. These values are seasonal averages; instantaneous values vary with wind and the indoor–outdoor temperature difference and can be higher. The results of infiltration rate measurements of two different samples are shown in Figure 6.3. The median seasonal values for a sample of 312 new energy-efficient homes and another of 266 old homes located in different parts of North America are 0.5 ACH and 0.9 ACH, respectively. The measurements were made under unoccupied conditions, and it is said that occupancy influences may increase these values by about 0.10–0.15 ACH (ASHRAE Fundamentals, 2013).

Typical modern U.S. commercial building envelopes are not particularly airtight and result in a significant energy cost. A database of 345 commercial buildings has been compiled by Persily and colleagues at the National Institute of Standards and Technology (NIST). The test results of the infiltration

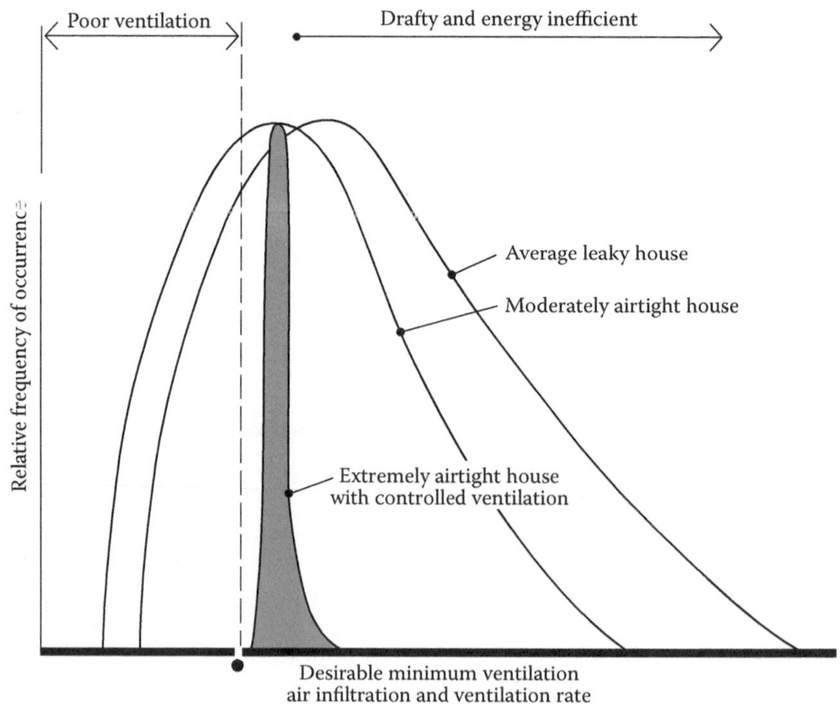

FIGURE 6.2
Variability of air exchange for three types of house plotted as relative frequency of occurrence versus air change rate. (From Nisson, J.D.N. and Dutt, G., *The Superinsulated Home Book*, Wiley, New York, 1985. With permission.)

rates are generally reported in terms of the airflow rate at some reference pressure difference divided by the building volume, floor area, or envelope surface area. NIST results are based on normalizing by above-grade surface area (i.e., five-sided box). The average air leakage at 75 Pa for the NIST buildings was found to be 19.8 $m^3/(h \cdot m^2)$, which is about 20% lower than the average based on 228 buildings reported in 2011 (Emmerich and Persily, 2013). Finally, it was found by analyzing 68 recent buildings built with air barriers that infiltration rates were 66% lower than those without and exhibited much less variability.

When infiltration is insufficient to guarantee adequate indoor air quality, forced ventilation becomes necessary. The required air exchange rate depends, of course, on the density of occupants, i.e., the number of people. In residential buildings, the density is relatively low, and with conventional U.S. construction, infiltration is likely to be adequate. But it is certainly possible to make buildings much tighter than 0.3 ACH of infiltration. In fact, standard practice in Swedish houses is to build them to such high standards that uncontrolled infiltration rates are around 0.2 ACH; consequently, mechanical ventilation supplies just the right amount of needed outdoor air, and an air-to-air heat exchanger minimizes the energy consumption. This is illustrated in Figure 6.4, which shows how ACH varies with temperature difference and wind speed for two classes of buildings. In the United States for buildings with forced ventilation, ASHRAE ventilation standard (ASHRAE 62.1, 2013) applies (as discussed in Section 3.5.3). Good locations to remove indoor air pollution and excessive humidity by mechanical exhaust are the kitchen and bathrooms. In France, building codes require exhaust fans with minimum continuous flow in kitchens and bathrooms of all new residential buildings.

6.3 Basic Flow Equations

6.3.1 Introduction and Types of Airflow Models

To estimate the air exchange rate, the designer has two sources of information: data from similar buildings and mathematical models. The underlying phenomena are very complicated, and a simple comparison with other buildings may not be reliable. However, engineers do use such methods for quick and preliminary estimations, and one commonly used method is described in Section 6.6.2. The modeling approach can be far more precise and depends on the amount of effort the analyst is willing to invest in trying to achieve this accuracy. In essence, there are four different types of working models for determining flow

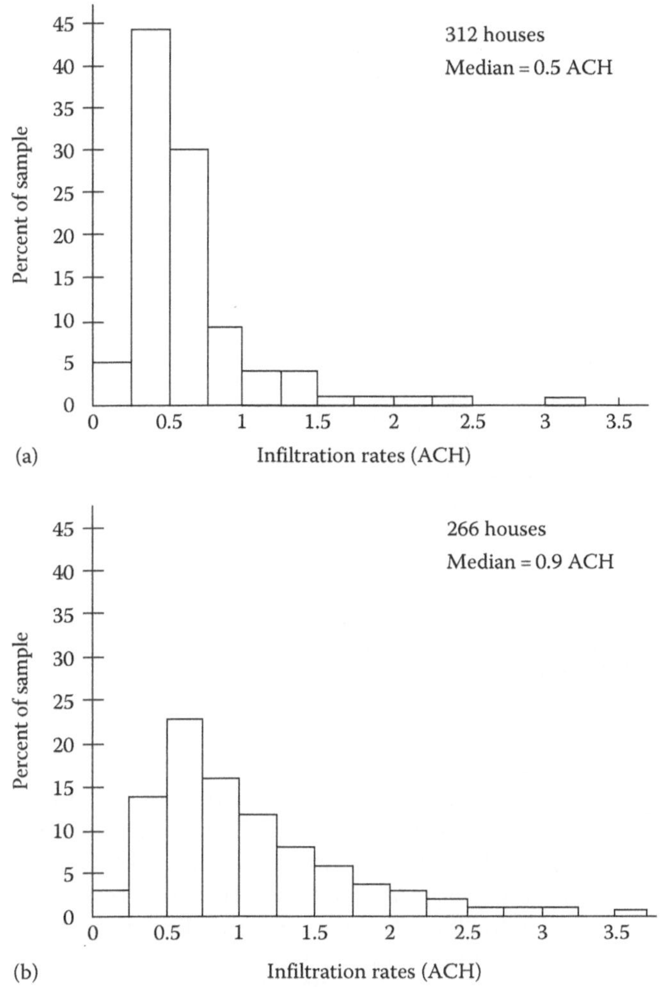

FIGURE 6.3
Histogram of infiltration values: (a) new construction and (b) low-income old housing. (From ASHRAE, *Handbook of Fundamentals*, American Society of Heating, Refrigerating and Air-Conditioning Engineers, Atlanta, GA, 2013. Copyright ASHRAE, www.ashrae.org.)

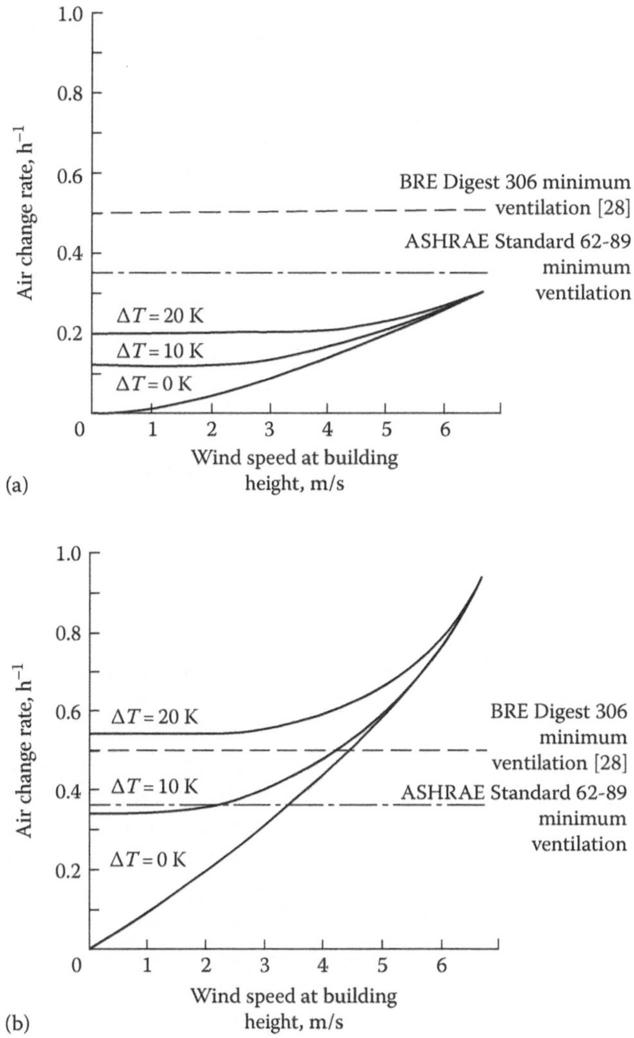

FIGURE 6.4
Air leakage performance of a tight and a moderately tight building: (a) tight building (1.5 ACH at 50 Pa) and (b) moderately tight building (10 ACH at 50 Pa). (From Awbi, H.B., *Ventilation of Buildings*, E &FN Spon, London, U.K., 1991.)

into and within buildings, all of which rely to some extent on the underlying physics of airflow through leakage pathways:

1. *Engineering models* that offer a compromise between scientific rigor and practicality (Section 6.5)
2. *Simplified physical models* for one-zone buildings (described in Section 6.6)
3. *Multizone network models* that are used for predicting infiltration and interzonal flows within buildings necessary for indoor air quality studies (Section 6.7)
4. *Computational fluid mechanics (cfd) models* (though discussed briefly in Section 6.7, they are beyond the scope of this text; the interested reader can refer to Chen and Glicksman, 2001)

Any penetration in the building envelope is a potential air leakage pathway, and leaks to and from unconditioned spaces such as attics and attached garages are also contributors. Typical air leakage locations in houses are shown in Figure 6.5. The air infiltration rate through the building envelope depends on (Awbi, 1991)

1. The size, shape, and distribution of leakage paths
2. The flow characteristics of the leakage paths (laminar or turbulent)
3. The pressure difference across the leakage paths

These pressure imbalances vary both temporally and spatially across the building envelope and even across

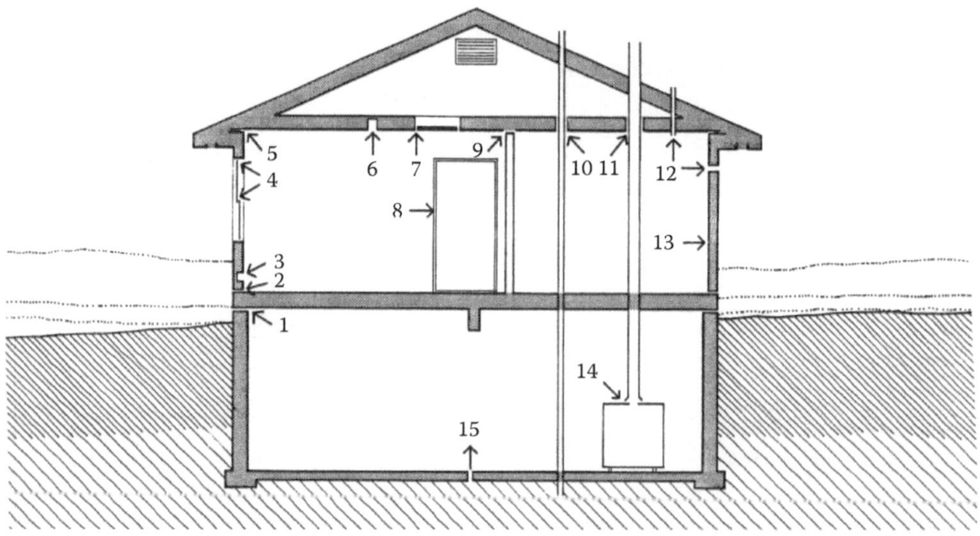

FIGURE 6.5
Typical air leakage sites in a house. (1) joints between joists and foundation, (2) joints between sill and floor, (3) electrical boxes, (4) joints at windows, (5) joints between wall and ceiling, (6) ceiling light fixtures, (7) joints at attic hatch, (8) cracks at doors, (9) joints at interior partitions, (10) plumbing-stack penetration of ceiling, (11) chimney penetration of ceiling, (12) bathroom and kitchen ventilation fans, (13) air-vapor barrier tears, (14) chimney draft air leaks, and (15) floor drain (air enters through drain tile). (From Nisson, J.D.N. and Dutt, G., *The Superinsulated Home Book*, Wiley, New York, 1985. With permission.)

different portions of any single envelope element. The types of building material specificity and data inputs needed are impossible to obtain and are simply overwhelming. Hence, the accurate prediction of infiltration based on purely theoretical grounds is neither practical nor feasible. The engineering approach is to reframe these basic equations in a manner that is practical while trying to maintain as much of the scientific essence as is possible. Thus, some understanding of the flow equations and the factors that affect the pressure distributions across the building envelope is essential.

6.3.2 Crack Flow Equation

Fully turbulent flow occurs in openings of relatively large free area such as a vent or large crack (of 3 mm or larger). The flow rate can be represented by the *standard orifice flow equation*, which can be derived from Bernoulli's equation, which, in turn, is a special case of the first law of thermodynamics applied to open systems (Equation 2.32):

$$\dot{V} = C_D A_{op}\left[\frac{2}{\rho}\Delta p\right]^{1/2} \tag{6.3}$$

where
- $\dot{V}$ is the airflow rate, m³/s (ft³/s)
- C_D is the discharge coefficient of the opening
- A_{op} is the area of opening, m² (ft²)
- $\Delta p = p_o - p_i$ is the pressure difference between outside and inside, Pa (inWG)
- ρ is the air density, kg/m³ (lb$_m$/ft³)

In a sharp-edge orifice flow, the discharge coefficient is almost independent of Reynolds number and has a value of 0.61. For building envelope leaks, values of $C_D = 1$ have been suggested (Sherman and Grimsrud, 1980).

On the other hand, in extremely narrow openings or fine hairline cracks with relatively long flow paths (such as in mortar joints and tight-fitting components), the flow is laminar or viscous. The Couette flow equation for round openings applies:

$$\dot{V} = \frac{\pi r^4}{8\mu L}\Delta p \tag{6.4}$$

where
- r is the radius of opening, m (ft)
- L is the length of flow path, m (ft)
- μ is the absolute viscosity (Pa·s)

In reality, flow through a building crack is neither laminar nor fully turbulent, but some combination of both. Hence, the *crack flow equation* is often used (ASHRAE Fundamentals, 2013):

$$\dot{V} = C\Delta p^n \tag{6.5}$$

where
- C is the flow coefficient that is the function of crack geometry and flow path, often determined experimentally, m/(s·Pa^n) [ft/(min·inWG^n)]
- n is the flow exponent that depends on the nature of flow through the crack

Example 6.2: Leakage through Sharp-Edge Orifice

Determine the airflow rate across a 3 mm (0.12 in.) diameter hole subject to a pressure difference of 10 Pa (1.45 × 10^{-3} psi), which can be treated as a sharp-edge orifice.

Given: Δp = 10 Pa, C_D = 0.61, A_{op} = (3.14/4) × (0.3 × $10^{-2})^2$ = 0.071 × 10^{-4} m²

Lookup values: ρ = 1.20 kg/m³

Find: $\dot{V}$

Solution

From Equation 6.3:

$$\dot{V} = 0.61 \times 0.071 \times 10^{-4} \times \left(\frac{2}{1.2} \times 10\right)^{1/2}$$

$$= 0.177 \times 10^{-4}\ \mathrm{m^3/s} = 0.0177\ \mathrm{L/s}$$

$$= 63.7\ \mathrm{L/h}\ (2.25\ \mathrm{ft^3/h})$$

Comment

A sustained pressure difference of 10 Pa is high for low-rise buildings (though this can be attained for brief periods), but is normal for taller ones. Under such a pressure difference, note that air infiltration rate even through such a small opening is quite large.

6.4 Induced Pressure Differences

The total pressure difference $\Delta p = (p_o - p_i)$ across the building envelope needs to be known in order to determine infiltration rates. This pressure difference is found by superposition as the sum of three pressure terms:

$$\Delta p = \Delta p_{wind} + \Delta p_{stack} + \Delta p_{vent} \quad (6.6)$$

The first is due to wind, the second to the stack effect (like the flow induced in a heated smokestack), and the third to forced ventilation (which may not be a factor in certain cases). We take the pressure differences to be positive when they cause air to flow toward the interior. The flow depends only on the total Δp, not on the individual terms. The relative contribution of the wind, stack, and ventilation terms varies across the envelope, and because of the nonlinear relationship between flow rate and pressure difference, one cannot calculate separate airflows for each of these effects and add them at the end. The total pressure difference needs to be deduced first and then used to determine the total airflow.

6.4.1 Wind Effect

When wind strikes a building, it creates a static pressure distribution on the exterior of the building's envelope. This local wind pressure or velocity pressure is given by Bernoulli's equation:

$$p_{wind} = \frac{\rho}{2}(v^2 - v_f^2)\ \mathrm{Pa} \quad (6.7\ \mathrm{SI})$$

where

v is the wind speed (undisturbed by building), m/s
v_f is the final speed of air at building boundary, m/s
ρ is the air density, kg/m³

In IP units, we have

$$p_{wind} = \frac{\rho}{2g_c}(v^2 - v_f^2)\ \mathrm{lb_f/ft^2} \quad (6.7\ \mathrm{IP})$$

with g_c = 32.17 ($\mathrm{lb_m \cdot ft})/(\mathrm{lb_f \cdot s^2}$), the wind speed being in feet per second and the air density in pound-mass per cubic foot.

Under *standard conditions* of 101.3 kPa (14.7 psi) and 20°C (68°F), the density is

$$\rho = 1.20\ \mathrm{kg/m^3}\ (\rho = 0.075\ \mathrm{lb_m/ft^3})$$

It must be noted that the density of outdoor air can deviate more than 20% above (winter at sea level) or below (summer in the mountains). In IP units, the ratio of ρ to g_c has the value

$$\frac{\rho}{g_c} = 0.00964\ \mathrm{inWG/(mph)^2}$$

under standard conditions if pressure is in inches water gauge and wind speed in miles per hour.

Equation 6.7 applies to winds whose direction is normal to the wall. For other surfaces, very complex pressure distributions are obtained as illustrated by Figure 6.6. Since the final speed v_f is difficult to determine, a convenient shortcut is to use Equation 6.7 with v_f = 0 and multiply it instead by a pressure coefficient C_p:

$$p_{wind} = C_p \frac{\rho}{2} v^2 \quad (6.8)$$

The quantity $[p_{wind}/C_p = (\rho/2)v^2]$ is plotted versus wind speed v in Figure 6.7. The insert in the figure allows the aforementioned quantity to be read off more accurately for lower values of wind velocity. The coefficient C_p is a function of the relative location of the building element with respect to wind direction and also on the

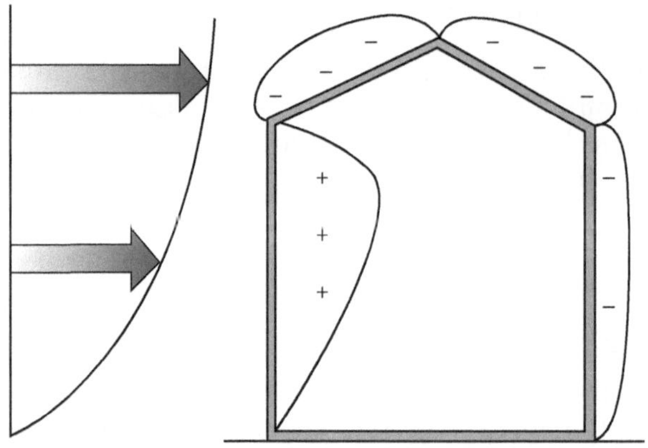

FIGURE 6.6
Wind pressure acting on a building. (From CIBSE AM10, Natural ventilation in non-domestic buildings, The Chartered Institution of Building Services Engineers, London, U.K., 2005.)

roof pitch. Current engineering practice according to ASHRAE literature recommends the use of three types of plots showing the relationship between surface-averaged C_p values and relative wind angle for rectangular buildings:

1. For walls of low-rise buildings (three stories or less) (see Figure 6.8).

   Typical values for average C_p for walls of low-rise buildings are in the range of approximately −0.6 to 0.6, depending on the direction of the wind with respect to the wall (called wind angle θ). An analytical correlation for low-rise buildings is

$$C_p(\theta) = 0.15(\cos^2\theta)^{1/4} + 0.45(\cos\theta) - 0.65(\sin^2\theta)^2 \quad (6.9)$$

2. For roofs of low-rise buildings inclined at less than 20°, $C_p \sim 0.5$.
3. For walls of tall buildings for three different aspect ratios of length L and width W (see Figure 6.9).
4. For roofs of tall buildings for three different aspect ratios of length L and width W (see Figure 6.10).

Numerical values for C_p can be gleaned from these figures where this coefficient is plotted as a function of the angle between the wind and the surface normal. It is always less than unity in absolute terms and can be negative to indicate depressurization of the outdoor surface. The reader can refer to Swami and Chandra (1988) for nonrectangular buildings or to Grosso (1992) for a general model to determine pressure coefficients.

Actually, we are interested in the pressure difference between the interior and exterior of a building. If the interior of an entire floor offers no significant flow resistance, one can find the indoor pressure due

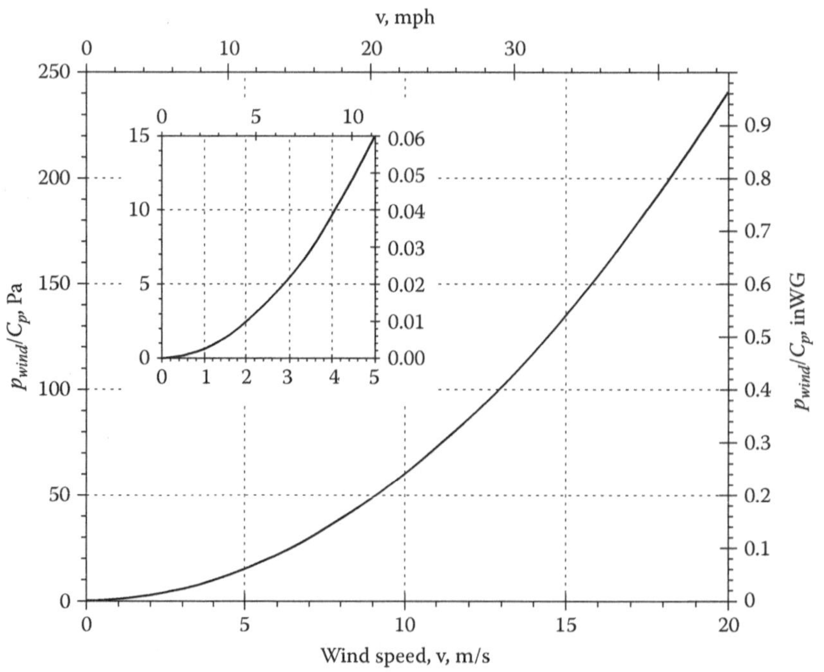

FIGURE 6.7
Wind pressure plotted as $p_{wind}/C_p = (\rho/2)v^2$ versus wind speed v following Equation 6.8.

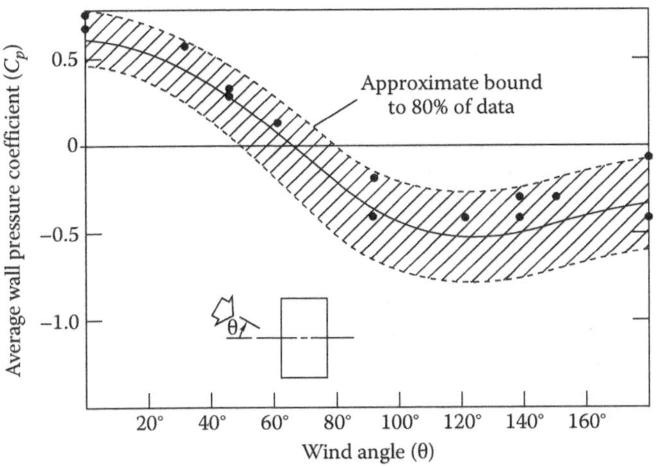

FIGURE 6.8
Typical values of wall- averaged wind pressure coefficient C_p of Equation 6.10 for a rectangular low-rise building as a function of wind direction. (The dots indicate the values from Figure 14.6 of that reference.) (From ASHRAE, *Handbook of Fundamentals*, American Society of Heating, Refrigerating and Air-Conditioning Engineers, Atlanta, GA, 2013. Copyright ASHRAE, www.ashrae.org.)

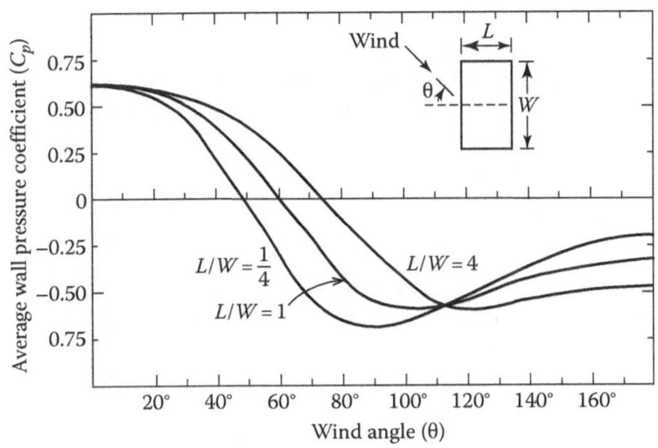

FIGURE 6.9
Wall-averaged wind pressure coefficients for a tall building. (From ASHRAE, *Handbook of Fundamentals*, American Society of Heating, Refrigerating and Air-Conditioning Engineers, Atlanta, GA, 2013. Copyright ASHRAE, www.ashrae.org.)

to wind by averaging the flow coefficient over all orientations of the surrounding wall. Since that average is approximately (−0.2) for a *low-rise building*,* the local pressure difference ($p_o - p_i$) averaged across the wall is, in that case,

$$\Delta p_{wind} = \Delta C_p \frac{\rho}{2} v^2 \tag{6.10}$$

* This can be verified from Figure 6.8 where we find C_p values of 0.6, −0.4, and −0.5 for the windward, leeward, and two perpendicular sides resulting in an average value of (0.6 − 0.4 − 0.5 − 0.5)/4 = −0.2.

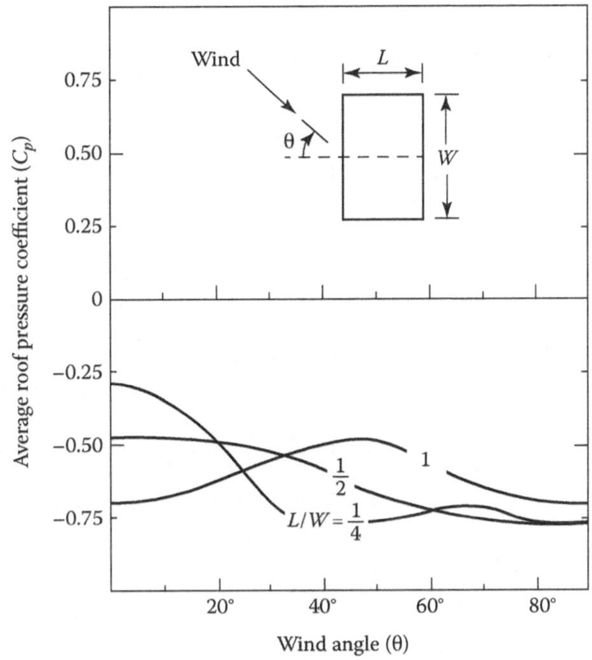

FIGURE 6.10
Average roof wind pressure coefficients for a tall building. (From ASHRAE, *Handbook of Fundamentals*, American Society of Heating, Refrigerating and Air-Conditioning Engineers, Atlanta, GA, 2013. Copyright ASHRAE, www.ashrae.org.)

with $\Delta C_p = C_p - (-0.2)$ being the difference between the local pressure coefficient and the average over all orientations of the building. This approach presumes, obviously, that air leakage is uniformly distributed across all four walls.

A final issue is the determination of the effective wind speed v in Equation 6.8. This value should correspond to the eaves height for a low-rise building and the building height for a high-rise building. Since meteorological data for wind velocity are usually collected at 10 m height and wind speed is strongly modified by terrain and obstacles, being significantly higher far above the ground, a conversion equation based on boundary layer theory is used to correct for this height differential (Spitler, 2010):

$$v = v_{met} \left(\frac{\delta_{met}}{h_{met}} \right)^{a_{met}} \left(\frac{h}{\delta} \right)^{a} \tag{6.11}$$

where

- v and v_{met} are the effective and reference wind speeds, m/s (ft/min)
- δ and δ_{met} are the boundary layer thickness for the local terrain and the boundary layer thickness for the meteorological station, m (ft)
- h and h_{met} are the average height above local obstacles and the height at which wind speed is measured, m (ft)
- a and a_{met} are the exponents for the local terrain and for the meteorological station

TABLE 6.1

Atmospheric Boundary Layer Parameters for Use in Equation 6.11

Terrain Category	Description	Exponent a	Layer Thickness δ, m (ft)
1	Large city centers, in which at least 50% of buildings are higher than 21.3 m (70 ft), over a distance of at least 0.8 km (0.5 mile) or 10 times the height of the structure upwind, whichever is greater	0.33	460 (1509)
2	Urban and suburban areas, wooden areas, or other terrain with numerous closely spaced obstructions having the size of single-family dwellings or larger, over a distance of at least 460 m (1500 ft) or 10 times the height of the structure upwind, whichever is greater	0.22	370 (1214)
3	Open terrain with scattered obstructions having heights generally less than 9.1 m (30 ft), including flat open country typical of meteorological station surroundings	0.14	270 (886)
4	Flat, unobstructed areas exposed to wind flowing over water for at least 1.6 km (1 mile), over a distance of 460 m (1500 ft) or 10 times the height of the structure inland, whichever is greater	0.10	210 (689)

Source: From ASHRAE, *Handbook of Fundamentals*, American Society of Heating, Refrigerating, and Air-Conditioning Engineers, Atlanta, GA, 2013. Copyright ASHRAE, www.ashrae.org.

Values of the relevant parameters for boundary layer thickness and the exponents are given in Table 6.1. Since meteorological stations are usually located in flat open terrain, they often correspond to terrain category 3 in the table so that $a_{met} = 0.14$ and $\delta_{met} = 270$ m. It is recommended that Equation 6.11 not be used when $h < 0$, i.e., when average obstacle height is greater than the building height.

Example 6.3: Wind Pressure on Four Sides of a Tall Building

Find the pressure difference across all four sides on the top floor of a 20-story building located in a large city center. The reference wind velocity is 10 m/s. The building is square and one of the sides is normal to the wind direction. Each floor is 3 m high and the average height of local obstacles around the building is 23.5 m.

Given: $v_{met} = 10$ m/s, building $(L/W) = 1$, building is uniformly leaky on all sides,

h = midheight of 20th floor minus obstacles height = 3 × 20 − 1.5 − 23.5= 35 m.

Meteorological station is located 10 m above the ground and is at terrain category 3 (Table 6.1) with $a_{met} = 0.14$ and $\delta_{met} = 270$ m.

Building corresponds to terrain category 1 with $a = 0.33$ and $\delta = 460$ m.

Lookup values: $\rho = 1.20$ kg/m^3

Find: Δp_{wind} on windward, leeward, and the two sides

Solution

We first use Equation 6.11 to calculate the effective wind velocity on the 20th floor:

$$v = 10 \times \left(\frac{270}{10}\right)^{0.14} \times \left(\frac{35}{460}\right)^{0.33} = 6.8 \text{ m/s}$$

This velocity is also assumed to hold for the roof (strictly speaking, the height should be 35 + 1.5 = 36.5 m not 35 m; but this is a small correction). The following table illustrates the intermediate and final calculations:

Face of Building	Wind Angle	Pressure Coefficient (C_p)	Pressure Coeff. Diff. $\Delta C_p = C_p -$ (Average)	Pressure Diff. Δp_{wind} (Equation 6.10) (Pa)
Windward	0°	(Figure 6.9) 0.62	0.62 − (−0.308) = 0.928	25.7
Leeward	180°	−0.30	−0.30 − (−0.308) = 0.008	0.22
Faces (two sides)	90°	−0.58	−0.58 − (−0.308) = −0.272	−7.55
Roof	0°	(Figure 6.10) −0.70	−0.70 − (−0.308) = −0.392	−10.88
		Average = −0.308		

Comment

The windward face of the building experiences a positive outdoor–indoor pressure difference, which is quite substantial. This would cause large air infiltration. The pressure difference on the leeward side is close to zero, while the two faces and the roof are depressurized, and indoor air would exfiltrate through them.

6.4.2 Stack Effect

The *stack effect* is the result of density differences between air inside and outside the building. In winter, the air inside the building is warmer and hence less dense than the air outside. Therefore, the column of outdoor air would be heavier than that indoors causing a hydrostatic pressure difference. During the cooling season when indoor air is colder than the outside, the effect is reversed. This difference, which is linear with height as shown in Figure 6.11, would cause air to infiltrate from the lower portion of the building. Since the total amount of air inside the building remains the same (due to conservation of mass), the higher portion of the vertical walls would have to exfiltrate. The changeover line is the level of *neutral pressure* that is typically taken to be at the midheight of the building. This presumes that the leaks are uniformly distributed across the building envelope, which is not true in many cases. For high-rise buildings that tend to have more openings (doors, larger windows, etc.) close to the ground level, the neutral pressure line may be lower than the midheight; for residences with chimneys, it could be higher.

The outside–inside pressure difference at a specific location on the wall during winter (when outside air is colder and, hence, denser) is given by

$$\Delta p_{stack} = -(\rho_o - \rho_i) g \Delta h \tag{6.12}$$

where
- ρ_i is the density of air inside building
- ρ_o is the density of air outdoors
- Δh is the vertical distance of specific location minus that of neutral pressure height
- g is the acceleration due to gravity

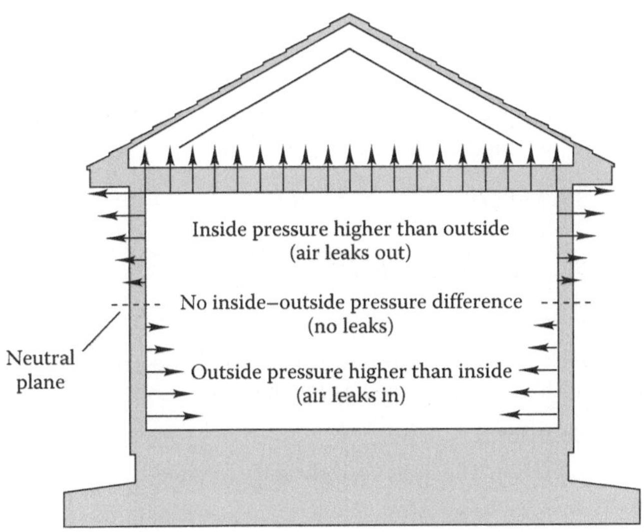

FIGURE 6.11
Air leakage due to stack effect during heating season. (From Nisson, J.D.N. and Dutt, G., *The Superinsulated Home Book*, Wiley, New York, 1985. With permission.)

Note that for wall locations below the neutral pressure line, Δh is negative and so stack pressure is positive and vice versa (as shown in Figure 6.11). Replacing densities by absolute temperatures following the ideal gas law, we get

$$\Delta p_{stack} = -g\Delta h\left(\frac{p_o}{R_o T_o} - \frac{p_i}{R_i T_i}\right) \tag{6.13}$$

Finally, since the outside and inside air pressures are very close compared to the absolute atmospheric pressure and the gas constants for outside and inside air-water vapor mixtures are essentially equal, Equation 6.13 can be recast as

$$\Delta p_{stack} = -g\Delta h\left(\frac{p_i}{RT_i}\right)\left(\frac{T_i}{T_o} - 1\right) = -\rho_i g \Delta h\left(\frac{T_i - T_o}{T_o}\right) \tag{6.14}$$

This equation represents an idealized pressure difference between outdoors and indoors of a building with uniform envelope leakage when there are no internal horizontal separations (as shown in Figure 6.12a). Thus, it applies to areas such as stairwells, atriums, and auditoriums. Another idealized situation corresponds to perfectly isolated or impermeable floors (i.e., when there is no airflow across floors). The resulting pressure difference in such a case is illustrated by Figure 6.12b. Each floor displays its own neutral pressure plane along the external pressure gradient line and acts as if it were independent of adjacent floors.

For real buildings, the airflow resistance of the exterior walls relative to that between floors due to doors and stairways is modeled by introducing a *thermal draft coefficient* C_d and recasting Equation 6.14 as

$$\Delta p_{stack} = -C_d \rho_i g \Delta h\left(\frac{T_i - T_o}{T_o}\right) \tag{6.15 SI}$$

$$\Delta p_{stack} = -C_d \frac{\rho_i g}{g_c} \Delta h\left(\frac{T_i - T_o}{T_o}\right) \tag{6.15 IP}$$

where
- ρ_i is the density of air in building = 1.20 kg/m^3 (0.075 lb$_m$/ft^3)
- Δh is the vertical distance from neutral pressure level, up being positive, m (ft)
- g = 9.81 m/s^2 (32.17 ft/s^2) = acceleration due to gravity [g_c = 32.17 (lb$_m$ · ft)/(lb$_f$ · s^2)]
- T_i and T_o are the indoor and outdoor absolute temperatures, K (°R)
- C_d is the draft coefficient, a dimensionless number

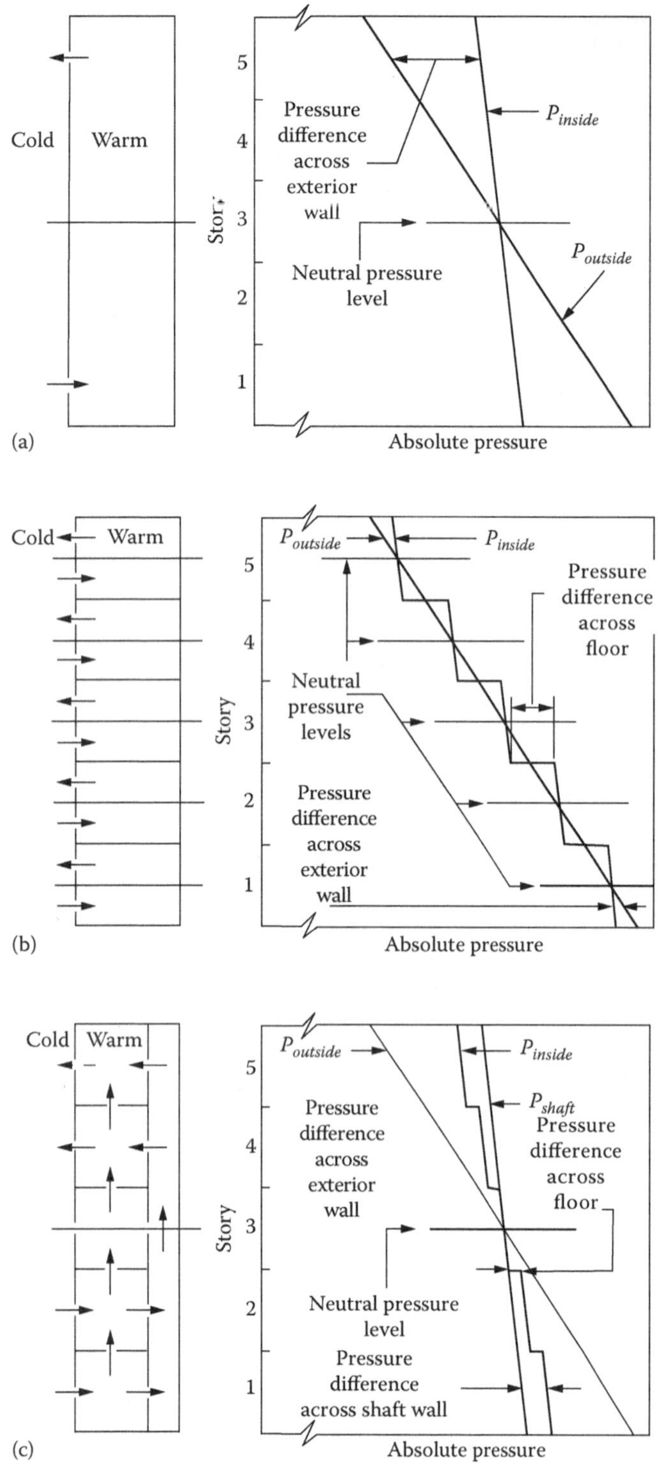

FIGURE 6.12
Pressure profiles due to compartmentalization pressure difference. (a) Building with no internal partition, (b) building with airtight separation of each story, and (c) real building with open shaft. (From ASHRAE, *Handbook of Fundamentals*, American Society of Heating, Refrigerating, and Air-Conditioning Engineers, Atlanta, GA, 2013. Copyright ASHRAE, www.ashrae.org.)

Figure 6.12c illustrates the pressure profiles for a heated building with uniform leakage through the exterior envelope and between floors as well as uniform openings into the vertical shaft at each floor. The thermal draft coefficient ranges from about 0.63 (tight) to 0.82 (loose) for typical modern office buildings and assumes a value of 1.0 if there is no resistance at all (as in an open stairwell).

Equation 6.15 is plotted in Figure 6.13 as a function of $\Delta T = T_i - T_o$ and Δh, assuming indoor air at 24°C (75°F) and $C_d = 1$. Since the relation is linear in Δh, this figure can be read outside the range shown by simply changing the scales of the axes. Briefly, one can say that the stack pressure amounts to

$$\frac{\Delta p_{stack}}{C_d \Delta h \Delta T} = 0.04\ \text{Pa}/(\text{m}\cdot\text{K}) \qquad (6.16\ \text{SI})$$

$$\frac{\Delta p_{stack}}{C_d \Delta h \Delta T} = 0.00014\ \text{lb}_f/(\text{ft}^2\cdot\text{ft}\cdot{}^\circ\text{R}) \qquad (6.16\ \text{IP})$$

The stack effect tends to be relatively small in low-rise buildings, up to about five floors, but in high-rise buildings, it can dominate and should be given close attention.

Example 6.4: Stack Pressure on Top Floor of a Tall Building

Consider the same high-rise 20-story building as in Example 6.3. Calculate the pressure difference due to the stack effect on the 20th floor when the outdoor temperature is −2°C and the indoor at 22°C. Assume a draft coefficient of 0.65 and that the neutral pressure line is 26.5 m from ground level.

Given:

Δh = midheight of 20th floor minus height of the neutral pressure line

$= (3 \times 20 - 1.5) - 26.5 = 32\ \text{m}$

$T_o = -2°\text{C} = 271\ \text{K}$ and $T_i = 22°\text{C} = 295\ \text{K}$, $C_d = 0.65$

Lookup values: $\rho = 1.20\ \text{kg/m}^3$, $g = 9.81\ \text{m/s}^2$

Find: Δp_{stack} (which is assumed to be the same for all four walls as well as for the roof)

Solution

Equation 6.15 SI is used directly to yield

$$\Delta p_{stack} = -0.65 \times 1.2\ \text{kg/m}^3 \times 9.81\ \text{m/s}^2 \times 32\ \text{m} \times \left(\frac{295 - 271}{295}\right) K = -19.92\ \text{Pa}$$

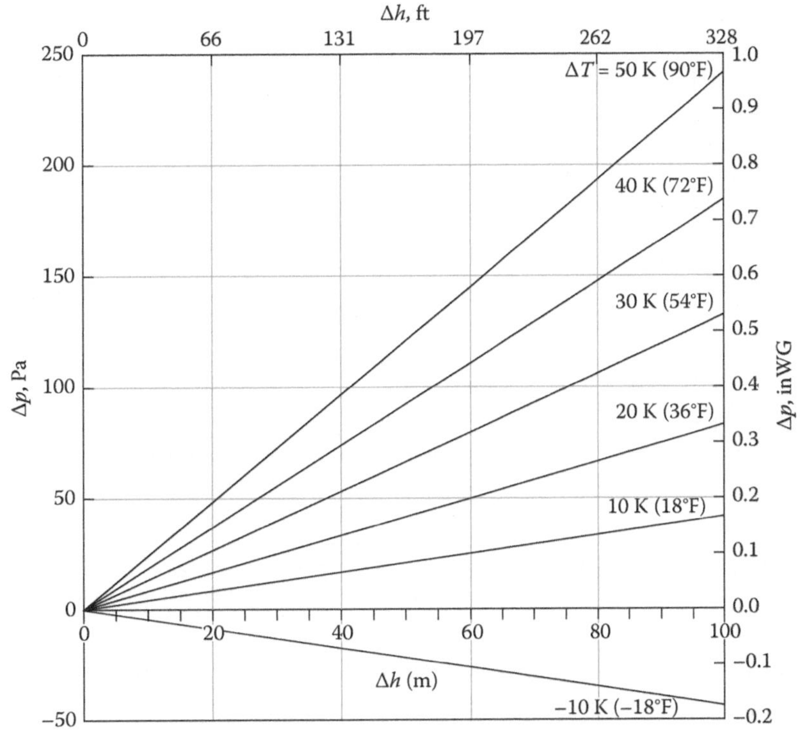

FIGURE 6.13
Variation of pressure difference due to stack effect with vertical distance from neutral pressure line following Equation 6.15 assuming indoor air at 24°C (75°F) and draft coefficient of unity.

Comment

As expected, in winter, floors above the neutral pressure line will have a higher interior pressure than that outdoors. This pressure difference applies to all walls and roof of this top floor space.

Example 6.5: Combining Wind and Stack Effects

Calculate the combined effect of the wind and stack pressure on the top floor of the 20-story building considered in Examples 6.3 and 6.4.

Since the pressures are additive, the calculation is simple to perform using the pressure differences computed previously. A table better illustrates the procedure:

Face of the Building	Wind Pressure Diff. Δp_{wind} (Example 6.3)	Stack Pressure Diff. Δp_{stack} (Example 6.4)	Combined Pressure Diff. Δp (Pa)
Windward	25.75	−19.92	5.83
Leeward	0.22	−19.92	−19.7
Faces (two sides)	−7.55	−19.92	−27.47
Roof	−10.88	−19.92	−30.8

We note that the roof and all walls (except the windward side) experience higher pressures with indoors than outdoors, resulting in substantial exfiltration. The stack effect has enhanced the wind pressure difference on these walls while reducing the outdoor–indoor pressure difference on the windward side. Despite the stack effect, outdoor air will infiltrate from the windward wall. The exfiltration rate will be much higher than the infiltration rate for this floor; the difference is being supplied from the floors underneath.

Figure 6.14 provides a conceptual illustration of the additive behavior of the stack and wind pressures along the height of a building.

6.4.3 Combining Wind, Stack, and Mechanical Ventilation Effects

Finally, certain mechanical systems in buildings such as supply or return air fans and exhaust fans (in vented combustion devices or fume hoods) impact pressure differences across the building envelope. If the intake and exhaust flows are not equal, a pressure difference Δp_{vent} is induced depending on the design and operation of the ventilation system and on the tightness of the building. This would lead to *mechanical ventilation*. In addition, there is some coupling to the wind and stack terms. Thus, the determination of Δp_{vent} may be somewhat difficult. However, as we now show, the situation is simple when Δp_{vent} is larger in magnitude than the wind and stack terms. This is an important case since

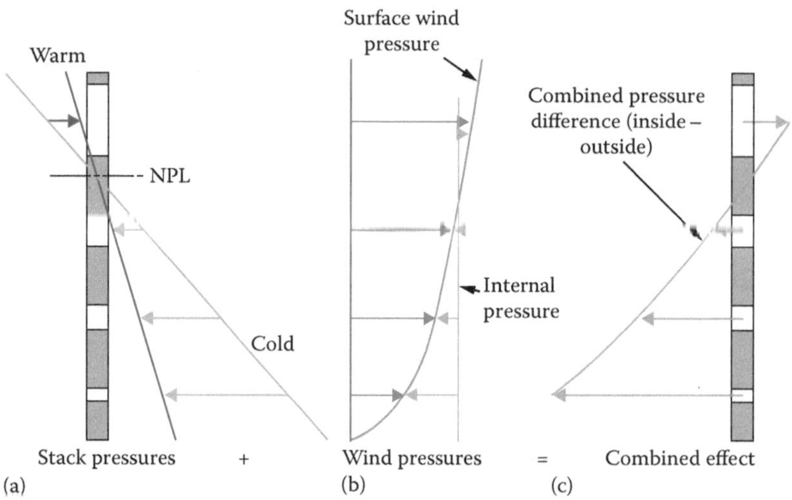

FIGURE 6.14
Superposition of the stack and wind pressures along the height of a building. (a) Stack pressures. (b) Wind pressures. (c) Combined effect. (From CIBSE AM10, Natural Ventilation in non-domestic buildings, The Chartered Institution of Building Services Engineers, London, U.K., 2005.)

many designers aim for slight overpressurization of commercial buildings by making the outdoor air intake larger than the exhaust flow.

Consider how the pressures and flows are related in a fairly tight building where mechanical ventilation maintains overpressure or underpressure Δp_{vent} relative to the outside; for simplicity, we assume it uniform in the entire building. The law of conservation of mass implies that the net airflow provided by the ventilation system equals the net leakage $\dot{V}$ across the envelope, as calculated, according to Equation 6.20 (discussed in the next section), by summing over all leakage sites j of the building envelope:

$$\dot{V} = \sum A_j K_j \Delta p_j^{n_j} \tag{6.17}$$

with

$$\Delta p_j = \Delta p_{wind,j} + \Delta p_{stack,j} + \Delta p_{vent}$$

For simplicity, let us assume one single value n for the exponent. Then, the pressure term on the right-hand side of Equation 6.17 can be rewritten in the form

$$\Delta p_j^n = \Delta p_{vent}^n (1 + x_j)^n$$

with

$$x_j = \frac{\Delta p_{wind,j} + \Delta p_{stack,j}}{\Delta p_{vent}} \tag{6.18}$$

As long as $|x_j| < 1$, the binomial expansion can be used, with the result*

$$\dot{V}_{vent} = \Delta p_{vent}^n \sum A_j K_j \left[1 + n x_j + \frac{n(n-1)}{2} x_j^2 + \cdots\right]$$

The quantity x_j is positive in some parts of the building, negative in others. In fact, if the distribution of cracks is approximately symmetric (top–bottom and windward–leeward) across the entire building, then for each term with positive x_j there will also be one with approximately $(-x_j)$. Thus, the linear terms in the expansion tend to cancel. The higher-order terms are small, beginning with x_j^2 that is multiplied by $[n(n-1)/2]$, a factor that is always less than 1/8 in absolute value since $0 < n < 1$. Thus, the contributions of the x_k-dependent terms are much smaller than that of the leading term. Therefore, *if a building is pressurized to Δp_{vent} by mechanical ventilation and if wind and stack pressures are smaller than Δp_{vent}, then it is indeed a fair approximation to neglect them altogether and write*

$$\dot{V} \approx \Delta p^n \sum A_j K_j \quad \text{with } \Delta p = \Delta p_{vent} \tag{6.19}$$

the sum covering the *entire envelope* of the building. Had we allowed for different exponents n_j in Equation 6.17, Δp with its exponent would remain inside the sum, but the conclusion about the negligibility of stack and wind terms continues to hold.

* The ith power of x_j in this series is multiplied by the binomial coefficient $\binom{n}{i} = \frac{n!}{i!(n-i)!}$.

Example 6.6: Pressure due to Mechanical Ventilation

The mechanical ventilation system of a three-story office building maintains a constant overpressure Δp_{vent} = 25 Pa. Can the wind and stack terms be neglected under the following conditions?

Given: Δh = 0.5 × total building height = 0.5 × (3 × 3 m) = 4.5 m (15 ft)

Δp_{vent} = −25 Pa (−0.1 inWG), T_i = 20°C (68°F), T_o = −10°C (14°F) (winter design condition)

v = 6.7 m/s (15 mph) (typical value assumed for winter design conditions)

Find: Δp_{wind} and Δp_{stack}, and compare with Δp_{vent}

Solution

For the wind term, $|C_P| \lesssim 0.5$ from Figure 6.8. From Figure 6.7 at 6.7 m/s, we find $|\Delta p_{wind}|/C_p$ = 28 Pa; hence $|\Delta p_{wind}| \lesssim 0.5 \times 28$ Pa = 14 Pa.

For the stack term, ΔT = 30 K (54°F). From Figure 6.13, we find $|\Delta p_{stack}|$ = 6 Pa. Hence,

$$\left|\Delta p_{stack} + \Delta p_{wind}\right| = 20 \text{ Pa} < \left|\Delta p_{vent}\right|.$$

During most of the year, v and ΔT are less than the design conditions, and thus the wind and stack terms are smaller than the values calculated here. Their contribution is smaller than *that of the ventilation terms and can, indeed, be neglected.*

The total flow into the building is obtained by summing over all openings j that are overpressurized. In buildings without mechanical ventilation, the pressure differences under natural conditions are positive over part of the building and negative over the rest. If one sums naively over all openings of a building, the result (averaged over momentary fluctuations) is zero, because the quantity of air in a building does not change. The flow into the building must equal the flow out; the former corresponds to positive terms in the sum, the latter to negative terms. What interests us here is the energy needed for conditioning the air that flows into the building. Therefore, the *sum includes only the terms with exterior pressure greater than that indoors, i.e.,* $p_o > p_i$.

To summarize, in many (if not most) buildings *with* mechanical ventilation, one maintains a significant pressure difference between interior and exterior.* If this pressure difference is larger than the pressures induced by wind and temperature, the latter can be neglected and all terms in Equation 6.17 have the same sign.

* Overpressure in the building allows better control and comfort. Underpressure can be maintained with smaller ducts and lower cost, but at the risk of condensation, freezing, and possibly draft.

6.5 Component Models for Air Infiltration

The location, size, and flow characteristics of each crack and opening cannot be identified in practice. Further, different cracks may have different coefficients and exponents, and so it becomes impractical to use the crack equation given by Equation 6.5 as an engineering equation. Openings and fracture cracks in real building envelopes occur randomly around penetrations for services (electricity, gas, and water), around door and window frames, due to gaps in walls/ceiling/floor joints, and from background leakage through porous building components. They also depend on construction type, workmanship during construction, and subsequent weathering and maintenance. One distinguishes between two types of leakage paths: (1) envelope (or background or fabric) leakage and (2) identifiable leakage through components such as operable doors and windows when closed and when opened. It is not possible to determine the value of C for each and every crack, opening, or fracture on the building envelope.

For background leakage, it is better to assume these leaks to be uniformly distributed about the surface area of the building and to specify leakage in terms of surface area of the building envelope element. For identifiable openings, it is convenient to characterize the infiltration in terms of a single equivalent amount of free open area of an orifice that allows the same volume of airflow as the actual component. Such a normalized area is referred to as the "overall leakage coefficient per unit area of the element." This allows Equation 6.5 to be recast as

$$\dot{V} = AK\Delta p^n \tag{6.20}$$

where

- A is the area of building element such as a wall, m² (ft²) (note that sometime a leakage length is used instead for elements such as window sills)
- K is the overall leakage area coefficient of building element, m/(s·Pa^n) [ft/(min·inWG^n)]
- n is the flow exponent, between 0.4 and 1.0 and usually around 0.65 for buildings

To apply Equation 6.20, one needs data for leakage area coefficients and flow coefficients of all the components of a building. Much research has been done to obtain such data, both for components and for complete buildings, e.g., by pressurizing the component or even the whole building (described in Section 6.9). Data for leakage area coefficients K can be found in Table 6.2 for a wide variety of building elements. Note that some elements have very large leakage areas. A door may have as much leakage as that of a 100 m² exterior wall.

TABLE 6.2

Data for Effective Leakage Area Coefficient of Building Components at 4 Pa (0.016 inWG)

Component	Best Estimate	Maximum	Minimum
Sill foundation—wall			
Caulked, in.²/ft of perimeter	0.04	0.06	0.02
Not caulked, in.²/ft of perimeter	0.19	0.19	0.07
Joints between ceiling and walls			
Joints, in.²/ft of wall (only if not taped or plastered and no vapor barrier)	0.07	0.12	0.02
Windows			
Casement			
Weather stripped, in.²/ft² of window	0.011	0.017	0.006
Not weather stripped, in.²/ft² of window	0.023	0.034	0.011
Awning			
Weather stripped, in.²/ft² of window	0.011	0.017	0.006
Not weather stripped, in.²/ft² of window	0.023	0.034	0.011
Single hung			
Weather stripped, in.²/ft² of window	0.032	0.042	0.026
Not weather stripped, in.²/ft² of window	0.063	0.083	0.052
Double hung			
Weather stripped, in.²/ft² of window	0.043	0.063	0.023
Not weather stripped, in.²/ft² of window	0.086	0.126	0.046
Single slider			
Weather stripped, in.²/ft² of window	0.026	0.039	0.013
Not weather stripped, in.²/ft² of window	0.052	0.077	0.026
Double slider			
Weather stripped, in.²/ft² of window	0.037	0.054	0.02
Not weather stripped, in.²/ft² of window	0.074	0.110	0.04
Doors			
Single door			
Weather stripped, in.²/ft² of door	0.114	0.215	0.043
Not weather stripped, in.²/ft² of door	0.157	0.243	0.086
Double door			
Weather stripped, in.²/ft² of door	0.114	0.215	0.043
Not weather stripped, in.²/ft² of door	0.16	0.32	0.1
Access to attic or crawl space			
Weather stripped, in.² per access	2.8	2.8	1.2
Not weather stripped, in.² per access	4.6	4.6	1.6
Wall—window frame			
Wood frame wall			
Caulked, in.²/ft² of window	0.004	0.007	0.004
No caulking, in.²/ft² of window	0.024	0.038	0.022
Masonry wall			
Caulked, in.²/ft² of window	0.019	0.03	0.016
No caulking, in.²/ft² of window	0.093	0.15	0.082
Wall—door frame			
Wood wall			
Caulked, in.²/ft² of door	0.004	0.004	0.001
No caulking, in.²/ft² of door	0.024	0.024	0.009
Masonry wall			
Caulked, in.²/ft² of door	0.0143	0.0143	0.004
No caulking, in.²/ft² of door	0.072	0.072	0.024

(Continued)

TABLE 6.2 (*Continued*)

Data for Effective Leakage Area Coefficient of Building Components at 4 Pa (0.016 inWG)

Component	Best Estimate	Maximum	Minimum
Domestic hot water systems			
Gas water heater (only if in conditioned space), in.^{2}	3.1	3.9	2.325
Electric outlets and light fixtures			
Electric outlets and switches			
Gasketed, in.^{2} per outlet and switch	0	0	0
Not gasketed, in.^{2} per outlet and switch	0.076	0.16	0
Recessed light fixtures, in.^{2} per fixture	1.6	3.10	1.6
Pipe and duct penetrations through envelope			
Pipes			
Caulked or sealed, in.^{2} per pipe	0.155	0.31	0
Not caulked or sealed, in.^{2} per pipe	9.30	1.55	0.31
Ducts			
Sealed or with continuous vapor barrier, in.^{2} per duct	0.25	0.25	0
Unsealed and without vapor barrier, in.^{2} per duct	3.7	3.7	2.2
Fireplace			
Without insert			
Damper closed, in.^{2} per fireplace	10.7	13.0	8.4
Damper open, in.^{2} per fireplace	54.0	59.0	50.0
With insert			
Damper closed, in.^{2} per fireplace	5.6	7.1	4.03
Damper open or absent, in.^{2} per fireplace	10.0	14	6.2
Exhaust fans			
Kitchen fan			
Damper closed, in.^{2} per fan	0.775	1.1	0.47
Damper open, in.^{2} per fan	6.0	6.5	5.6
Bathroom fan			
Damper closed, in.^{2} per fan	1.7	1.9	1.6
Damper open, in.^{2} per fan	3.1	3.4	2.8
Dryer vent			
Damper closed, in.^{2} per vent	0.47	0.9	0
Heating ducts and furnace—forced-air systems			
Ductwork (only if in unconditioned space)			
Joints taped or caulked, in.^{2} per house	11	11	5
Joints not taped or caulked, in.^{2} per house	22	22	11
Furnace (only if in conditioned space)			
Sealed combustion furnace, in.^{2} per furnace	0	0	0
Retention head burner furnace, in.^{2} per furnace	5	6.2	3.1
Retention head plus stack damper, in.^{2} per furnace	3.7	4.6	2.8
Furnace with stack damper, in.^{2} per furnace	4.6	6.2	3.1
Air conditioner			
Wall or window unit, in.^{2} per unit	3.7	5.6	0

Source: ASHRAE, *Handbook of Fundamentals*, American Society of Heating, Refrigerating, and Air-Conditioning Engineers, Atlanta, GA, 2005. Copyright ASHRAE, www.ashrae.org.

For conversion to SI units: 1 in.^{2} = 6.45 cm^2, 1 ft^2 = 0.0929 m^2, and 1 in.^{2}/ft^2 = 69 cm^2/m^2.

Additionally, several correlations have been proposed for specific elements (Spitler, 2010). Here, we treat three cases only: air leakage through curtain walls and operable/revolving doors and windows. The airflows are determined from the following correlations, developed mostly from field tests, where Δp is the total pressure difference at each point of the building, calculated as described earlier. Since the equations are dimensional, all quantities must be used with the specified units. We have added dual scales to the graphs, so Figures 6.15 through 6.18 can be read directly in both systems of units.

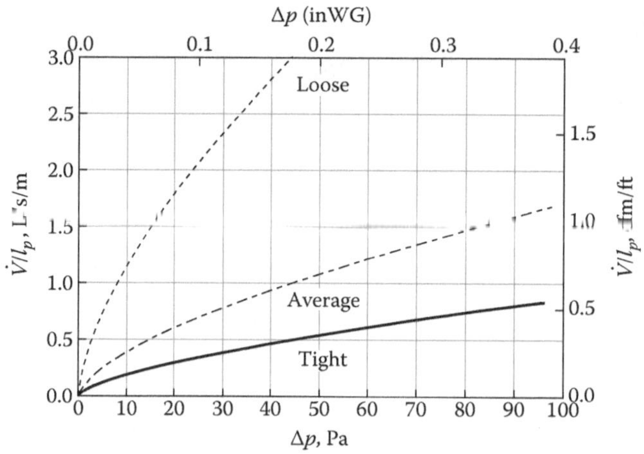

FIGURE 6.15
Window and residential-type door air infiltration $\dot{V}$ per perimeter length l_p. The curves correspond to Equation 6.21, with $n = 0.65$ and coefficient k [(L/s · Pa$^{0.5}$), (ft^3/min · inWG$^{0.65}$)] according to construction type, as shown in the following table:

	Windows		
Coefficient	Wood Double Hung (Locked)	Other Types	Doors (Residential Type)
Tight $k = 1.0$ (IP) = 0.043(SI)	Weather-stripped, small gap width 0.4 mm (1/64 in.)	Weather stripped: wood casement and awning windows, metal casement windows	Very small perimeter gap and perfect-fit weather stripping—often characteristic of new doors
Average $k = 2.0$ (IP) = 0.086 (SI)	Non-weather-stripped, small gap width 0.4 mm (1/64 in.) or weather-stripped, large gap width 2.4 mm (3/32 in.)	All types of sliding windows, weather stripped (if gap width is 0.4 mm, this could be "tight") or non-weather-stripped metal casement windows (if gap width is 2.4 mm, this could be "loose")	Small perimeter gap with stop trim, good fit around door, and weather stripping
Loose $k = 6.0$ (IP) = 0.257 (SI)	Non-weather-stripped, large gap width 2.4 mm (3/32 in.)	Non-weather-stripped vertical and horizontal sliding windows	Large perimeter gap with poor-fitting stop trim and weather stripping or small perimeter gap without weather stripping

1. For *residential-type doors and windows*. The flow per length l_p of perimeter is given by an equation of the form

$$\frac{\dot{V}}{l_p} = k(\Delta p)^{0.65} \quad \text{with } \dot{V}, \text{L/s (ft}^3\text{/min);} \quad l_p, \text{m (ft); } \Delta p, \text{Pa (inWG)} \tag{6.21}$$

Numerical values can be found in Figure 6.15 for windows and residential-type doors for three different construction types (corresponding to different values of k) as a function of pressure difference.

2. *For swinging doors in commercial buildings when closed*: Infiltration rates for commercial-type swinging doors when closed can also be deduced from Equation 6.21 with, however, the exponent set to $n = 0.5$. Figure 6.16 depicts the relationship for four different types of crack widths characterized by the coefficient k. Note that the values of k are greater than for the residential doors.

3. *For commercial-type revolving and swinging doors with traffic*. Obviously, the airflow increases markedly when doors are opened. Figure 6.17 permits an estimate of the airflow through swinging doors, both single bank and vestibule type, as a function of traffic rate. The coefficient C [(L/s · Pa$^{0.5}$), (ft^3/min · inWG$^{0.5}$)] for Figure 6.17a is found from Figure 6.17b for the number of people passing the door per hour. Note that the values of the flow coefficients C are based on a standard-sized door of 1 m × 2.1 m dimensions. The equation for the flow in part (a) is

$$\dot{V} = C(\Delta p)^{0.5} \quad \text{with } \dot{V}, \text{L/s (ft}^3\text{/min);} \quad \Delta p, \text{Pa (inWG)} \tag{6.22}$$

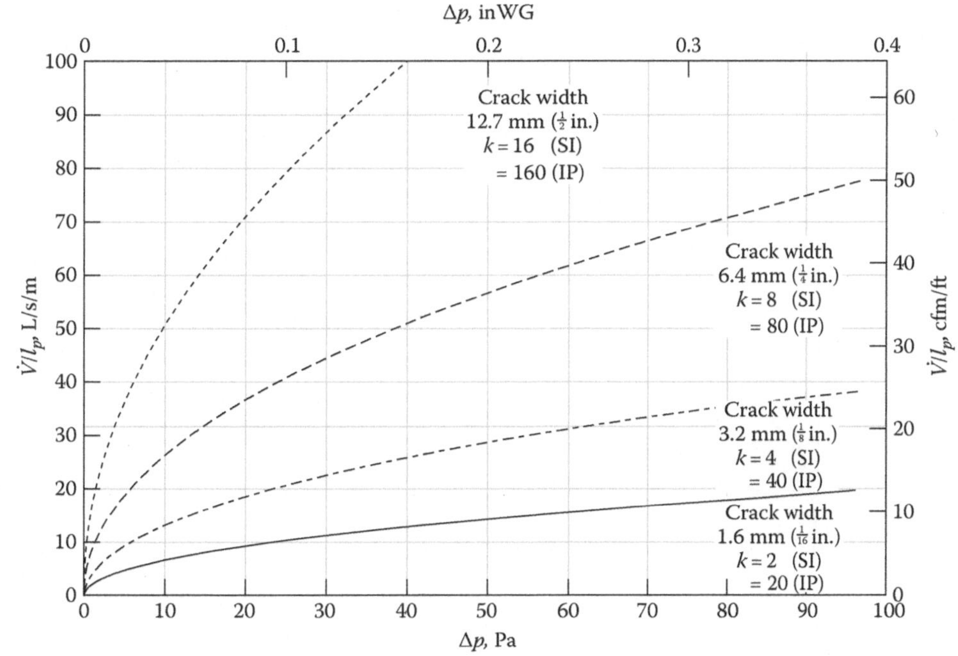

FIGURE 6.16
Infiltration through closed swinging door cracks, $\dot{V}$ per perimeter length l_p. The curves correspond to Equation 6.21, with n = 0.5 and coefficient k [(L/s · Pa$^{0.5}$), (ft^3/min · inWG$^{0.5}$)].

Similar plots have been developed for revolving doors and automatic doors (Spitler, 2010).

4. *Curtain walls for high-rise buildings* (i.e., the non-load-bearing wall construction commonly employed in commercial buildings). The air infiltration flow rate per unit area of *curtain wall* can be determined by the equation

$$\frac{\dot{V}}{A} = K(\Delta p)^{0.65} \text{ with } \dot{V}, \text{L/s (ft}^3\text{/min)};$$
$$A, \text{m}^2(\text{ft}^2); \Delta p \text{ Pa (inWG)} \quad (6.23)$$

It is presented in Figure 6.18 for three construction types, corresponding to the indicated values of the coefficient K [(L/s · m^2 · PA$^{0.65}$) or (ft^3/min · ft^2 · inWG$^{0.65}$)].

Example 6.7: Air Leakage from Single-Story Building

Estimate how much air leaks out of a single-story office building of average construction when it is pressurized to Δp = 50 Pa (0.2 inWG) (assume that wind and stack pressures can be neglected).

Given: One-story building 30 m × 30 m × 2.5 m (98.4 ft × 98.4 ft × 8.2 ft); curtain wall, alternating with floor-to-ceiling windows 3 m × 2.5 m each (9.84 ft × 8.2 ft); windows cover 50% of the entire wall area; four swinging doors 2 m × 1 m (6.56 ft × 3.28 ft), single-bank type, crack width 3.2 mm (1/8 in.); floor area 15 m^2 (161 ft^2) per occupant

Estimate the average traffic over 10 h by assuming occupants make two entries and two exits per day. Assume that 500 visitors enter and leave the building per day. Occupant density is 15 m^2/person.

Assumptions: For simplicity, neglect infiltration through the roof and through cracks at the edges of the roof and the floor.

Find: Air exchange rate $\dot{V}$

Solution

Total area of all four walls = 4 × 30 m × 2.5 m = 300 m^2
Curtain wall area = 300 × 0.5 = 150 m^2
Total number of windows = 150/(3 × 2.5) = 20
Perimeter of windows = 20 × [2 × (3 + 2.5) = 220 m
Perimeter of doors = 4 × [2 × (2 + 1)] = 24 m

$$\text{Number of occupants} = \frac{900 \text{ m}^2}{15 \text{ m}^2} = 60$$

$$\text{Traffic per door} = \frac{(60 \times 4) + (500 \times 2) \text{ visitors}}{4 \text{ door} \times 10 \text{ h}}$$
$$= 31 \text{ passages/h}$$

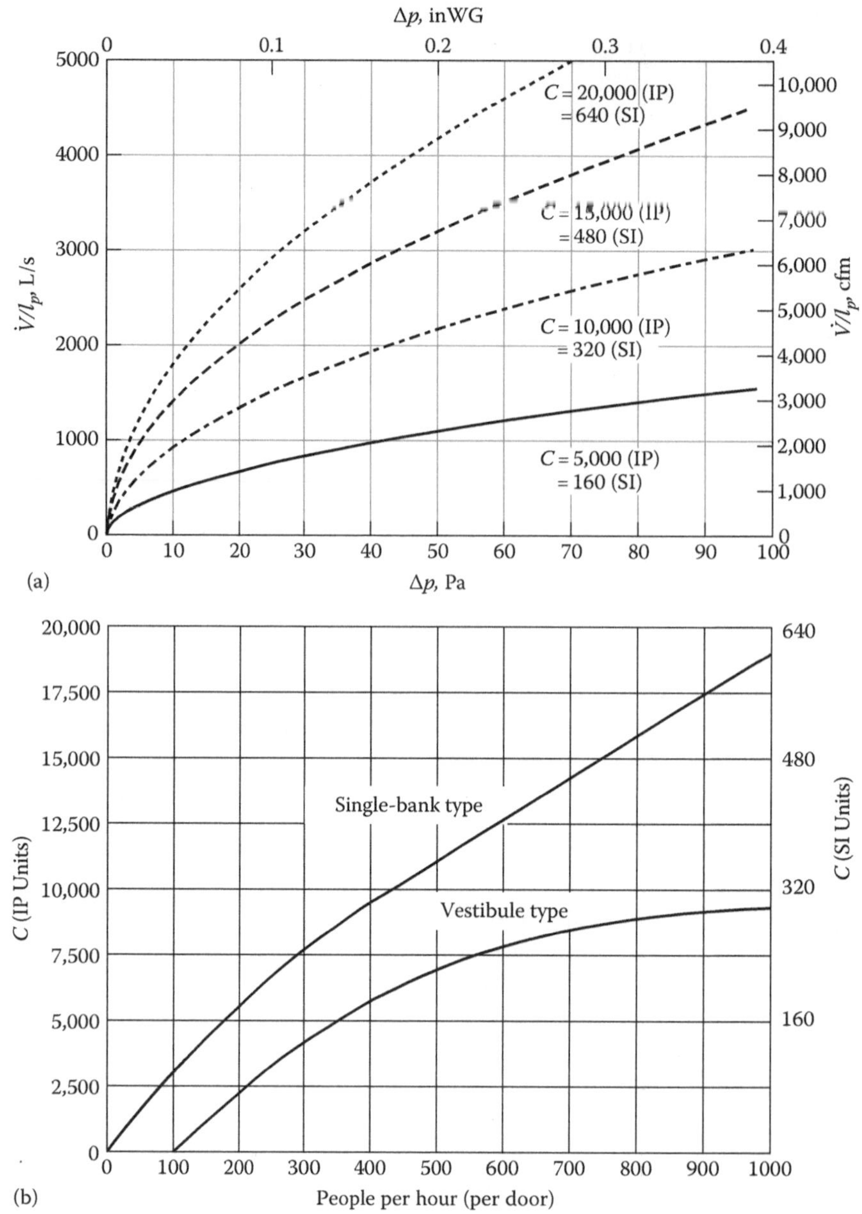

FIGURE 6.17
Infiltration due to door openings as a function of traffic rate: (a) infiltration (with $n = 0.5$) and (b) coefficient C [(L/s·Pa$^{0.5}$), (ft^3/min·inWG$^{0.5}$)].

At $\Delta p = 50$ Pa, we read from Figure 6.15 for "average" construction

$$\frac{\dot{V}}{l_p} = 1.1\ \text{L/(s}\cdot\text{m)} \quad \text{from perimeter of windows}$$

Hence, $\dot{V}_{windows} = 1.1\ (\text{L/s}\cdot\text{m}) \times 220\ \text{m} = 242\ \text{L/s}$. From Figure 6.16, for crack width 3.2mm, $(\dot{V}/l_p) = 28\ \text{L/(s}\cdot\text{m)}$ from perimeter of doors when closed. Hence, $\dot{V}_{doors} = 28\ (\text{L/(s}\cdot\text{m)} \times 24\ \text{m} = 672\ \text{L/s}$. From Figure 6.17b, coefficient C is around 40 in SI units for single bank type of door, and from Equation 6.22 or Figure 6.17a the extra flow through the doors due to traffic, per door, is

$$\dot{V} = 40 \times (50)^{0.5} = 282.8\ \text{L/s/door}$$

Thus, for four doors, $\dot{V}_{doors} = 4 \times 282.8 = 1131.4\ \text{L/s}$. From Figure 6.18, $\dot{V}/A = 1.2\ \text{L/(s}\cdot\text{m}^2)$ from curtain wall; hence,

$$\dot{V}_{curtain\ wall} = 1.2\ (\text{L/s})/\text{m}^2 \times 150\ \text{m}^2 = 180\ \text{L/s}$$

Finally, the total flow is

$$\dot{V} = 242 + 672 + 1131.4 + 180 = 2225.4\ \text{L/s}$$

$$= 2.225\ \text{m}^3/\text{s} \times 3600\ \text{s/h} = 8011\ \text{m}^3/\text{h}\ (4582\ \text{ft}^3/\text{min})$$

Since the building volume is $V = 2250\ \text{m}^3$, this corresponds to $(8011\ \text{m}^3/\text{h}/2250\ \text{m}^3) = 3.56$ ACH.

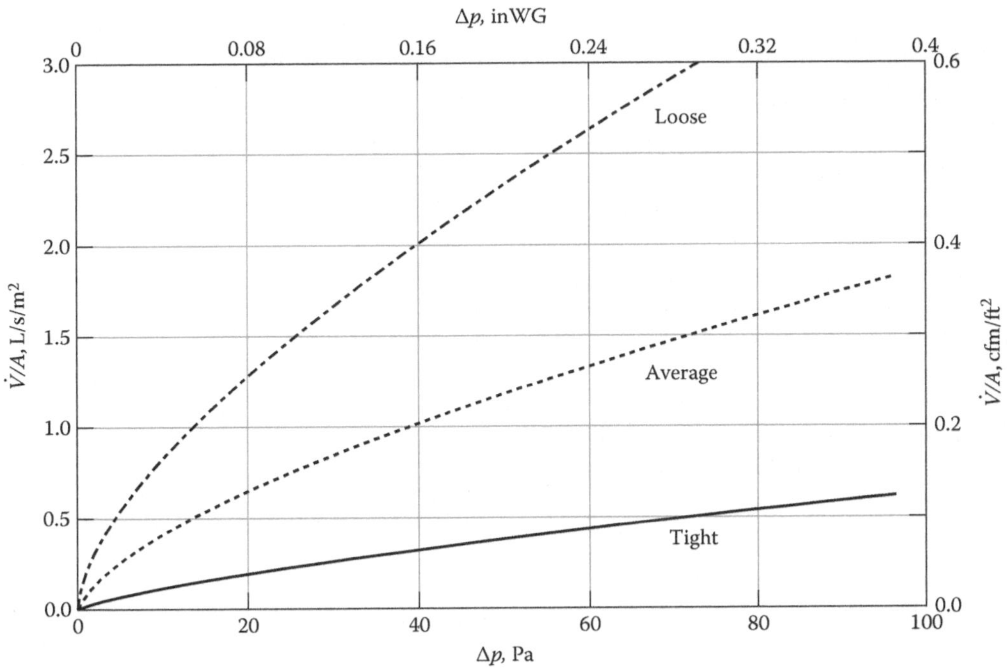

Coefficient (K)			
SI	IP	Description	Construction
0.031	0.22	Tight	Close supervision of workmanship; joints are redone when they appear inadequate
0.093	0.66	Average	Conventional
0.183	1.30	Loose	Poor-quality control or older building where joints have not been redone

FIGURE 6.18
Infiltration per area of curtain wall for one room or one floor. The curves correspond to Equation 6.23, with coefficient K [(L/s·m²·Pa$^{0.65}$), (ft³/min·ft²·inWG$^{0.65}$)] according to construction type.

Note that the dominant contribution comes from the doors, 672 L/s when they are closed and 1100 L/s from traffic. In the interest of energy conservation, tighter doors should be considered and replaced by vestibule-type doors.

Comments

There are 60 permanent occupants and the total floor area of the building is 900 m². According to the ASHRAE Standard 62.1-2013 (ASHRAE Fundamentals, 2013) discussed in Section 3.5.3, one should use the multipliers of 2.5 L/s per person and 0.3 L/s per m² for an office building. Thus, according to the standard, the ventilation flow required is 1.77 m³/s. We note that the calculated value (1.94 m³/s) is about 10% higher than this value, but is within the design safety factor.

There is another problem in applying Equation 6.20 to buildings: The width of an opening can change with pressure. Blower door tests (described in Section 6.9.1) on actual walls have shown that the apparent leakage area can be significantly higher for overpressurization than for underpressurization; external pressure tends to compress the cracks of a building (Lydberg and Honarbakhsh, 1989). Blower door tests do not directly yield the necessary leakage areas. The test data obtained are analyzed in the framework of mathematical models to yield design leakage values. The leakage values shown in Table 6.2 apply for a pressure difference of 4 Pa, a somewhat arbitrary selection as discussed in Section 6.9.1. Thus, Equation 6.20 is an approximation, valid only for a certain range of pressures and flows; different n and K values may have to be used for other ranges. ASHRAE Fundamentals (2013) gives formulas to convert air leakage areas and airflow rates in residences determined under one reference pressure to other operating pressures.

6.6 Simplified Physical Models for Single-Zone Air Infiltration

Simplified models are those where considerable simplifications are made to the basic scientific and engineering principles describing the phenomena; this provides preliminary results at much less computational effort. Only two of the most widely used methods are described in this section, both of which are to be used *for residences or low-rise buildings* whose interior can be treated as a single zone. These are appropriate for buildings with little or no internal resistance to airflow, and thus should not be used to large multizone buildings. Several other models are discussed in Chapter 27 of the ASHRAE Fundamentals Handbook (2013).

6.6.1 Basic Lawrence Berkeley National Laboratory Model

Section 6.5 presented Equation 6.20 that allowed the air infiltration rate through an individual building element to be expressed in terms of a single consolidated "overall leakage coefficient" *K*. This coefficient is determined from a whole building pressurization test for an existing building or can be deduced from Table 6.2 as illustrated in Example 6.8.

The engineering component model approach (Section 6.5) assumes that pressure differences due to weather and mechanical ventilation systems are noninteracting and therefore additive (Equation 6.6). In smaller buildings subject to weather only, it has been found that an alternative empirical formulation is just as good while being more convenient to use with field pressurization tests. The total infiltration flow rate is expressed as the quadrature sum of the flows induced by wind and stack acting alone:

$$\dot{V} = (\dot{V}_{wind}^2 + \dot{V}_{stack}^2)^{1/2} \tag{6.24}$$

This superposition formulation has been evaluated extensively with mathematical modeling and field measurements (Sherman, 1992; Walker and Wilson, 1993). This evaluation supports a model meant for relatively small buildings *without* mechanical ventilation developed earlier by Sherman and Grimsrud (1980) of Lawrence Berkeley National Laboratory (LBNL). This model bypasses Equations 6.17 and 6.20 by correlating the infiltrating airflow directly with the wind speed, temperature difference, and total leakage area. The total leakage area is obtained by adding all the leakage areas of the components, as illustrated in Example 6.7. Once the total leakage area has been found, either by such a calculation or by a pressurization test, the airflow $\dot{V}$ can be estimated by the following LBNL model, as reported in ASHRAE Fundamentals (2013):

$$\dot{V} = A_{leak}\sqrt{a_s\ \Delta T + a_w v^2}\ \text{L/s} \tag{6.25}$$

where

A_{leak} is the effective leakage area (ELA) of building, cm² or in²

a_s is the stack coefficient of Table 6.3, $(\text{L/s})^2/(\text{cm}^4\cdot\text{K})$ $[(\text{ft}^3/\text{min})^2/(\text{in}^4\cdot{}^\circ\text{F})]$

$$\Delta T = |T_i - T_o|\ \text{K}$$

a_w is the wind coefficient of Table 6.4, $(\text{L/s})^2/[\text{cm}^4\cdot(\text{m/s})^2]$ $[(\text{ft}^3/\text{min})^2/(\text{in.}^4\ (\text{mph})^2)]$

v is the wind speed, m/s or mph

The aforementioned model, known as the *basic* LBNL model, has a variety of systematic weaknesses that was investigated by subsequent research (such as the preference pressure, the exponent, the way wind sheltering is done, and superposition). Some of these biases are positive, while some are negative. The ELA values given in Table 6.2 specify a pressure difference of 4 Pa; these ELA values can be used for other conditions as well. This is due in part to the fact that a pressure difference of 4 Pa is an overestimate of the pressures experienced by envelope leaks in actual U.S. residences and partly to the fact that the accuracy of the model is not really good enough to capture such effects (Sherman, 2013). The LBNL model as a whole has been widely adopted despite its limitations and its prediction uncertainty in view of the convenience it provides.

When using Equation 6.25 for load calculations of a new structure, A_{leak} should be taken as equal to the *total* leakage area and not half of this value under the mistaken assumption that half of the leakage would experience infiltration and the other half exfiltration. This equation has been derived to apply to field tests

TABLE 6.3

Stack Coefficient a_s

	Number of Stories		
	One	Two	Three
Stack coefficient a_s, $(\text{ft}^3/\text{min})^2/(\text{in.}^4\cdot{}^\circ\text{F})$	0.0150	0.0299	0.0449
Stack coefficient a_s, $(\text{L/s})^2/(\text{cm}^4\cdot\text{K})$	0.000145	0.000290	0.000435

Source: ASHRAE, *Handbook of Fundamentals*, American Society of Heating, Refrigerating, and Air-Conditioning Engineers, Atlanta, GA, 2013. Copyright ASHRAE, www.ashrae.org.

TABLE 6.4
Wind Coefficient a_w

Shielding Class	Description	Wind Coefficient a_w, (L/s)²/[cm⁴·(m/s)²] Number of Stories One	Two	Three	Wind Coefficient a_w, (ft³/min)²/[in.⁴·(mph)²] Number of Stories One	Two	Three
1	No obstructions or local shielding	0.000319	0.000420	0.000494	0.0119	0.0157	0.0184
2	Light local shielding; few obstructions, a few trees or small shed	0.000246	0.000325	0.000382	0.0092	0.0121	0.0143
3	Moderate local shielding; some obstructions within two house heights, thick hedge, solid fence, or one neighboring house	0.000174	0.000231	0.000271	0.0065	0.0086	0.0101
4	Heavy shielding; obstructions around most of perimeter, buildings, or trees within 10 m in most directions; typical suburban shielding	0.000104	0.000137	0.000161	0.0039	0.0051	0.0060
5	Very heavy shielding; large obstructions surrounding perimeter within two house heights; typical downtown shielding	0.000032	0.000042	0.000049	0.0012	0.0016	0.0018

Source: ASHRAE, *Handbook of Fundamentals*, American Society of Heating, Refrigerating, and Air-Conditioning Engineers, Atlanta, GA, 2013. Copyright ASHRAE, www.ashrae.org.

under house pressurization (or depressurization) where all leaks will experience exfiltration (or infiltration) (Sherman, 2013).

A revised and enhanced model, called "flow coefficient" model, has been proposed, which also applies to simple single-zone houses. The interested reader can refer to Walker and Wilson (1998) for details and to ASHRAE Fundamentals (2013) for an overview. Several other single-zone models have also been developed, for example, the U.K. Building Research Establishment (BRE) model (Warren and Webb, 1980 or Awbi, 1991) and the Canadian AIM-2 model by the University of Alberta (Walker and Wilson, 1998). All these models are very sensitive to the input values provided to them and have been developed on the basis of model parameters deduced from field pressurization tests. The accuracy of the basic LBNL model is said to be ± 25% (Awbi, 1991), while field studies by Persily (1986) found differences of 40% on a long-term basis and up to 100% on individual measurements. The accuracy of the enhanced LBNL model is said to be ± 10% when building parameters are fairly well known. The crux of the problem when using such models for new buildings is that leakage is not solely a material or component property; it is more affected by the system, i.e., construction quality and envelope type. Hence, the best one can do is estimate infiltration based on buildings of similar properties with whatever safety factors deemed appropriate.

Example 6.8: Effective Leakage Area of a Building

Estimate the ELA for a simple wood-frame building. It is built as a rectangular box 12 m × 12 m × 2.5 m (39.4 ft × 39.4 ft × 8.2 ft) with a flat roof, and the walls at the sill are uncaulked and the walls at the roof are not taped or plastered. The windows, which are double hung and not weather stripped, have no caulking and cover 20% of the sides. This basic building will be used for many examples in this book, with further details being specified as needed. Here, we need to spell out characteristics of the airtightness only; they are listed in the following table, which also leads up to the answer.

Given: Single-story wood-frame construction in shape of a rectangular box 12 m × 12 m × 2.5 m, window/wall ratio = 0.20

Figure: Figure 6.19

Solution

Taking numbers from Table 6.2 for each of the components and adding, we obtain the following:

Component	Area, m², or Perimeter, m (Given)	Leakage Area per Area or Perimeter (from Table 6.2)	Leakage Area, cm²
Walls at sill (sill uncaulked)	48 m	0.19 in.²/ft = 4 cm²/m	192
Walls at roof not taped or plastered, no vapor barrier	48 m	0.07 in.²/ft = 1.5 cm²/m	72
Windows: double hung, not weather stripped	24 m²	0.086 in.²/ft² = 6 cm²/m²	144
Window frames: no caulking	24 m²	0.024 in.²/ft² = 1.7 cm²/m²	41
Total			449

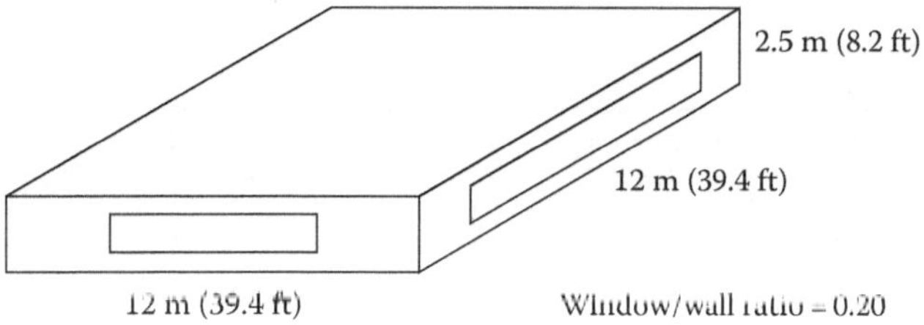

FIGURE 6.19
Sketch of the building used for many of the examples in this chapter.

Comments

This is a shoe-box building, kept as simple as possible with just enough detail to illustrate the principles. For a real building, one would have to include further details such as doors, electric outlets, and fireplace.

Example 6.9: Air Changes of a Building

Find the air exchange rate for the building of Example 6.8 when $\Delta T = T_i - T_o = 20$ K (36°F) and effective design wind speed v = 6.7 m/s (15 mph).

Given: Building of Example 6.8 in suburban neighborhood (heavy shielding)

$$\Delta T = T_i - T_o = 20\ \text{K},\ \text{v} = 6.7\ \text{m/s},\ A_{leak} = 449\ \text{cm}^2$$

Intermediate quantities: Stack coefficient $a_s = 0.000145$ (L/s)²/(cm⁴·K) from Table 6.3; wind coefficient $a_w = 0.000104$ (L/s)²/[cm⁴·(m/s)²] from Table 6.4

Solution

From Equation 6.25, one obtains the infiltration rate

$$\dot{V} = 449\ \text{cm}^2 \times \Big[0.000145(\text{L/s})^2(\text{cm}^{-4}\cdot\text{K}) \times 20\ \text{K} + 0.000104(\text{L/s})^2/[\text{cm}^4\cdot(\text{m/s})^2]\times(6.7\ \text{m/s})^2\Big]^{1/2}$$

$$= 39.1\ \text{L/s} = 140.6\ \text{m}^3/\text{h}$$

Since the volume $V = 360$ m³, the air change rate is 0.39 ACH.

6.6.2 Air Change Method

This method, which is only applicable to residences and small commercial buildings, is even more simplified than the LBNL model. It is based on measurements of actual buildings and requires that the user simply identify a construction class type that best describes the actual building:

- *Loose*: Older buildings or those with poor workmanship showing settlement cracks and poor window or door sealing.
- *Medium*: Buildings constructed using conventional construction practices.
- *Tight*: New buildings constructed with close supervision of workmanship with special precautions taken to prevent infiltration, tight-fitting windows, and doors.

Table 6.5 presents ACH values for residential buildings for different outdoor temperatures under summer and winter conditions, the distinction being that the design wind velocity is different during both seasons (15 mph in winter and 7.5 mph in summer). The indoor temperature is to be taken as 20°C (68°F) in winter and 24°C (75°F) in summer. Notice that several important factors such as surrounding terrain, or even window to wall ratios, are not even considered. It must be used with judgment and common sense. A simple example illustrates this approach.

TABLE 6.5
Empirical Method: Design Overall Infiltration Rates in Air Changes per Hour for Small Buildings

Construction Type	Winter[a] Outdoor Design Temperature, °C (°F)									
	10 (50)	4 (40)	−1 (30)	−7 (20)	−12 (10)	−18 (0)	−23 (−10)	−29 (−20)	−34 (−30)	−40 (−40)
Tight	0.41	0.43	0.45	0.47	0.49	0.51	0.53	0.55	0.57	0.59
Medium	0.69	0.73	0.77	0.81	0.85	0.89	0.93	0.97	1.00	1.05
Loose	1.11	1.15	1.20	1.23	1.27	1.30	1.35	1.40	1.43	1.47
			Summer[b] Outdoor Design Temperature, °C (°F)							
			29 (85)	32 (90)	35 (95)	38 (100)	41 (105)	43 (110)		
Tight			0.33	0.34	0.35	0.36	0.37	0.38		
Medium			0.46	0.48	0.50	0.52	0.54	0.56		
Loose			0.68	0.70	0.72	0.74	0.76	0.78		

Source: ASHRAE, *Handbook of Fundamentals*, American Society of Heating, Refrigerating, and Air-Conditioning Engineers, Atlanta, GA, 2001. Copyright ASHRAE, www.n.org.

[a] Design values for winter are for 6.7 m/s or 24 km/h (15 mph) and indoor temperature of 20°C (68°F).

[b] Design values for summer are for 3.4 m/s or 12 km/h (7.5 mph) and indoor temperature of 24°C (75°F).

Example 6.10: Estimating Air Infiltration Using the ACH Method

Estimate the air infiltration rates in ACH for the same building as that considered in Example 6.9. Assume the construction to be *medium* category since window sills are uncaulked, walls at seam are uncaulked, and walls at roof are not taped or plastered.

Given: Building of Example 6.9, $\Delta T = T_i - T_o = 20$ K. Since the reference $T_i = 20°C$, the outdoor temperature $T_o = 20 - 20 = 0°C$.

Solution

From Table 6.5, the desired estimate of ACH is directly found to be 0.76.

Comment

Note that the calculated ACH value is twice that determined using the simple LBNL model (Example 6.9). If we had assumed the building to be "loose" as one would suppose from the fact that the windows and walls are not properly sealed (as specified in Example 6.9), the ACH would have been 1.2, which is three times the value determined by the LBNL model. This example illustrates to some extent the very large differences in air infiltration rates one could obtain using different methods where one has to "guess" the construction type. This is the reason why editions of the ASHRAE Fundamentals Handbook after 2001 dropped this method altogether.

6.7 Multizone Models

The single-zone models are applicable to relatively small buildings where internal flow resistances are low compared to the envelope leakage or when one is interested in predicting an average concentration of some pollutant over the entire space. For larger buildings or for buildings with multiple thermal zones, the interzonal mass flow interactions become as important as the outdoor infiltration. Moreover, it becomes essential to be able to predict envelope leakage over specific portions of the building that affect the individual zones. Figure 6.20 illustrates these different types of flows for a relatively simple building.

Multizone airflow models (also referred to as network, macro, or compartmental models) discretize the building into a set of nodes connected by links to other nodes, i.e., the actual building is idealized into nodes where each node physically represents the volume of air in a building zone and the links between nodes are mathematical relationships, typically nonlinear such as the crack flow equation (Equation 6.5), between the pressure difference across nodes and the volumetric flow rates across various flow paths. These airflow paths can be windows, doorways, small cracks in building construction, transfer ducts, etc. The fundamental basis of multizone modeling is the "well-mixed" assumption whereby each zone is characterized by a discrete set of state variables, i.e., temperature, pressure, and contaminant concentrations (for IAQ studies). The entire building flow analysis can then be represented by a network where various pressure nodes can be connected either in series or in parallel. The network would consist of equations as shown in the following:

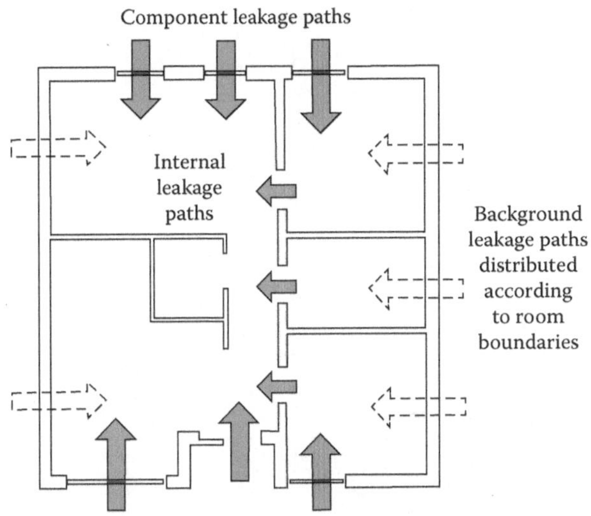

FIGURE 6.20
Illustration of multizone flow paths. (From Liddament, N.W., *Air Infiltration Calculation Techniques—An Applications Guide,* Air Infiltration and Ventilation Center, Bracknell, U.K., 1986.)

- For two leakage paths in series (Figure 6.21a):

$$\Delta p_t = \Delta p_1 + \Delta p_2 \quad \text{and} \quad \dot{V} = C_1 \Delta p_1^{n_1} = C_2 \Delta p_2^{n_2}$$

$$\Rightarrow \Delta p_t = \left(\frac{\dot{V}}{C_1}\right)^{1/n_1} + \left(\frac{\dot{V}}{C_2}\right)^{1/n_2} \tag{6.26}$$

- For two leakage paths in parallel (Figure 6.21b):

$$\dot{V} = \dot{V}_1 + \dot{V}_2 \quad \text{and} \quad \dot{V}_1 = C_1 \cdot \Delta p^{n_1}$$

$$\text{and} \quad \dot{V}_2 = C_2 \cdot \Delta p^{n_2} \Rightarrow \dot{V} = C_1 \cdot \Delta p^{n_1} + C_2 \cdot \Delta p^{n_2} \tag{6.27}$$

Thus, the network can be described by a set of simultaneous equations to which the desired flow boundary conditions are applied. These nonlinear equations are then solved iteratively by determining an internal pressure distribution while imposing that mass balance between infiltrating and exfiltrating air mass be preserved.

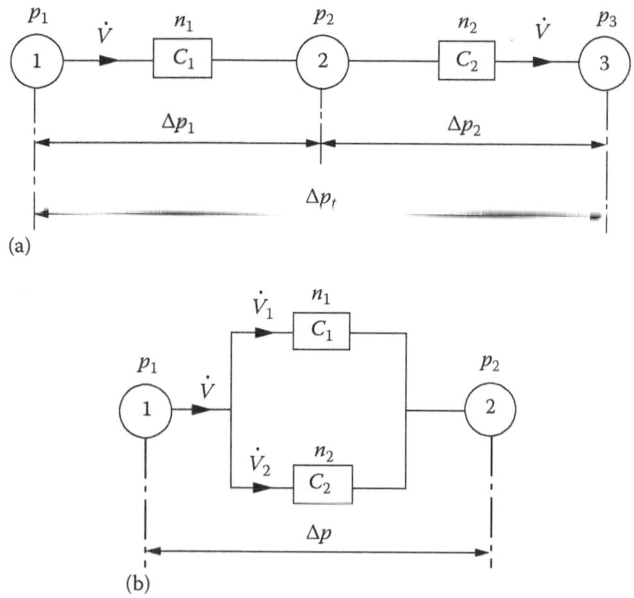

FIGURE 6.21
Flow networks for a multizone building: (a) leakage paths in series and (b) leakage paths in parallel. (From Awbi, H.B., *Ventilation of Buildings*, E &FN Spon, London, U.K., 1991.)

CONTAM is a multizone model software program widely used for indoor air quality and ventilation analysis developed by NIST (NIST, 2008; Walton and Dols, 2005). The network model predicts zone-to-zone airflows based on the pressure-flow characteristics of the path models and pressure differences across the paths. CONTAM has several functions and can be used for a variety of applications. The program is designed to help determine building airflows (through the envelope and between interior zones), contaminant concentrations, and personal exposure. For airflow analysis, the program considers driving forces such as mechanical ventilation system airflows, wind pressures, stack effects, terrain characteristics, and temperature differences. By defining contaminants and specifying the nature of the contaminant source, the program can predict the dispersion of these contaminants through the various airflow paths throughout the building. In calculating concentration levels, the program takes into account how the contaminant may interact with building surfaces (adsorption, desorption, and deposition) and any filtration devices that are present. The primary applications of CONTAM include assessing adequacy of ventilation rates, design and analysis of smoke systems, assessing indoor air quality performance, predicting contaminant dispersion, and estimating personal exposure (Walton and Dols, 2005).

The main difference between multizone and computational fluid dynamics (CFD) models is that the latter involves the numerical solutions of nonlinear partial differential Navier–Stokes equations, continuity equations, and energy conservation equations, whereas multizone models involve the solution of ordinary differential equations (only one independent variable) that can be solved analytically. As stated earlier, multizone models fail to capture spatial variation of variables within a zone, whereas CFD analysis provides this level of detail. However, the increased accuracy comes at the price of computational power and time. CFD programs are expensive, require significant simulation run times, and are limited by their complexity. The successful use of CFD programs and the correct interpretation of the results is dependent on the user's ability to model the boundary conditions and the interior space accurately while having the expertise to interpret the results properly. Multizone models, on the other hand, are much easier to set up and interpret and are adequate in several instances.

6.8 Natural Ventilation Airflow through Large Openings

6.8.1 Background

Natural ventilation is the strategy by which outdoor air is brought into the space by intentional opening of windows, doors, skylights, natural draft ventilators, and similar devices such as wind catchers and other passive devices. Such measures were widely used prior to the advent of mechanical cooling and, even now, common in several developing countries. Modern fully conditioned buildings have tended not to use such passive strategies because of lack of control in flows due to wide variability in weather conditions. However, the drive for energy efficiency has revived interest in natural ventilation. Even the little detriment in thermal comfort is often offset by the flexibility in individualized control provided. Further, studies have shown that those in naturally ventilated buildings tend to suffer less from sick building syndrome and other respiratory diseases than those in fully mechanically ventilated buildings (Sherman and Jump, 2000).

With open windows, the air exchange rate is difficult to predict accurately. It varies with the wind speed and direction as well as window opening area and orientation, and it is highly dependent on the aerodynamics of the building and its surroundings. The designer needs data on ventilation rates with open windows to assess comfort conditions in buildings with operable windows during the transition season between heating and cooling. In Figure 6.22, we present data for natural ventilation that have been measured in a two-story house. Depending on which windows are open and where the ventilation is measured, ACH values vary greatly—from 1 to 20 per h.

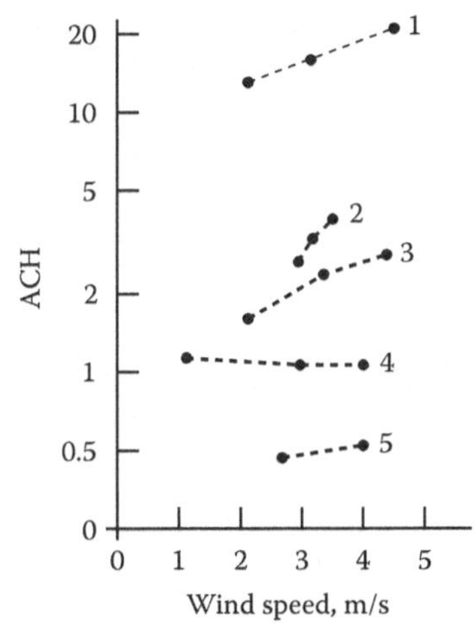

FIGURE 6.22
Measured ventilation rates, as a function of wind speed, in a two-story house with windows open on lower floor. The curves are labeled according to the location of open windows and measuring point as indicated. (From Achard, P. and Gicquel, R., *European Passive Solar Handbook*, Commission of the European Communities, Directorate General XII for Science, Research and Development, Brussels, Belgium, 1986.)

	Open Windows	Measuring Point
1	All upper floor	Upper floor
2	Upper floor windward	Upper floor
3	All upper floor	Lower floor
4	Upper floor leeward	Upper floor
5	None	Whole house

The ventilation rates are influenced by both atmospheric wind and stack effects, and these have to be considered properly because they have different implications depending on the geometry, size, and location of the openings. Intuitively, one would orient the building so that the long side faces the prevailing wind direction, and inlets are located on the high pressure side, i.e., directly into the prevailing wind direction, and outlets where there is depressurization such as in leeward, roof, and sides (see Figure 6.6). To enhance stack ventilation, one would try to space the openings as far apart vertically as possible. Also, it is recommended that the inlet and outlet areas be kept nearly equal since this would lead to the greatest flow per unit area of opening. Several general design guidelines for enhancing natural ventilation are given in ASHRAE Fundamentals (2013) and by Butcher (2007).

6.8.2 Simplified Correlation

The reader can refer to Etheridge and Sandberg (1996) for a fundamental treatment of ventilation flows in large openings. A number of simplified correlations for both two-sided and one-sided opening have been developed based on wind tunnel tests, mathematical modeling, and field measurements, for example, Liddament (1986) and Awbi (1991). Even simpler correlations are given in ASHRAE Fundamentals (2013), which is one of the options available in the EnergyPlus (2009) simulation software. For *ventilation airflow due to wind* $\dot{V}_{wind}$ (in m^3/s or cfm) when inlet and outlet areas of openings are equal:

$$\dot{V}_{wind} = C_v A_{op} \mathrm{v} \tag{6.28 SI}$$

$$\dot{V}_{wind} = 0.88 \times C_v A_{op} \mathrm{v} \tag{6.28 IP}$$

where
- A_{op} is the area of opening, either inlet or outlet (assumed equal) (m^2, ft^2)
- v is the local wind speed (m/s, mph)
- C_v is the dimensionless opening effectiveness, assumed to be in the range of 0.5–0.6 for perpendicular winds and 0.25–0.35 for diagonal winds

For *ventilation airflow due to stack effect* $\dot{V}_{stack}$ (in m^3/s or ft^3/s)

$$\dot{V}_{stack} = C_D A_{op}\left[2g\ \Delta h_{NPL}\left(\frac{|T_i - T_o|}{T_i}\right)\right]^{1/2} \tag{6.29}$$

where
- Δh_{NPL} is the height from midpoint of lower opening to the neutral pressure level (m, ft)
- T_i and T_o are the indoor and outdoor temperatures (K, °R)

It is suggested that the dimensionless discharge coefficient be determined from

$$C_D = 0.40 + 0.0045\ |T_i - T_o| \tag{6.30}$$

This applies to openings where bidirectional flow (see Figure 6.23b) can occur in large openings such as tall doors or loading doors. In case of multiple openings, the flow through the door could be unidirectional (as shown in Figure 6.23a), and a value of $C_D = 0.65$ should be used. Further, when the inlet and outlet openings are not equal, increasing the ratio increases airflow but not in proportion to the added area. The correction curve shown in Figure 6.24 accounts for this behavior, increasing the area ratio by 2, and only enhances flow by about 26%. When openings are not equal, it is suggested that the smaller area be used for A_{op} in Equation 6.29. Finally, the total ventilation rate is calculated as the quadrature sum of both wind and stack effects given by Equations 6.28 and 6.29.

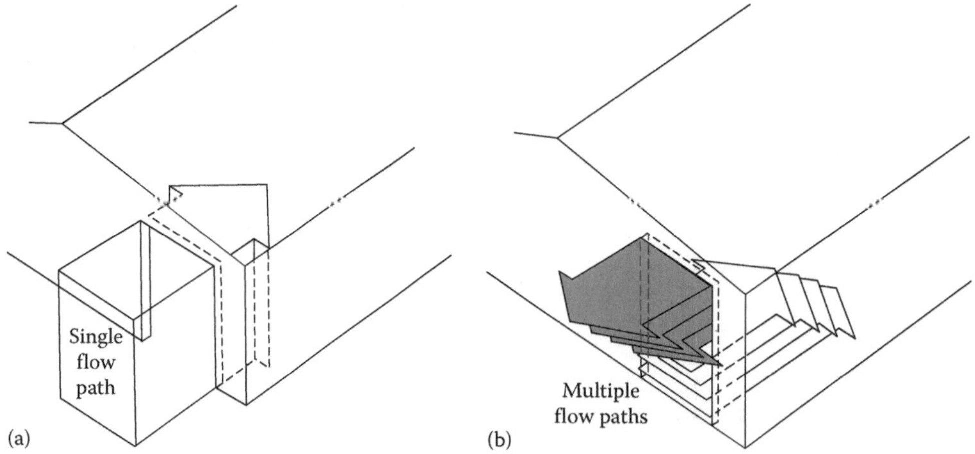

FIGURE 6.23
(a) Unidirectional and (b) bidirectional flows due to stack effect through open doors. (From Liddament, N.W., *Air Infiltration Calculation Techniques—An Applications Guide*, Air Infiltration and Ventilation Center, Bracknell, U.K., 1986.)

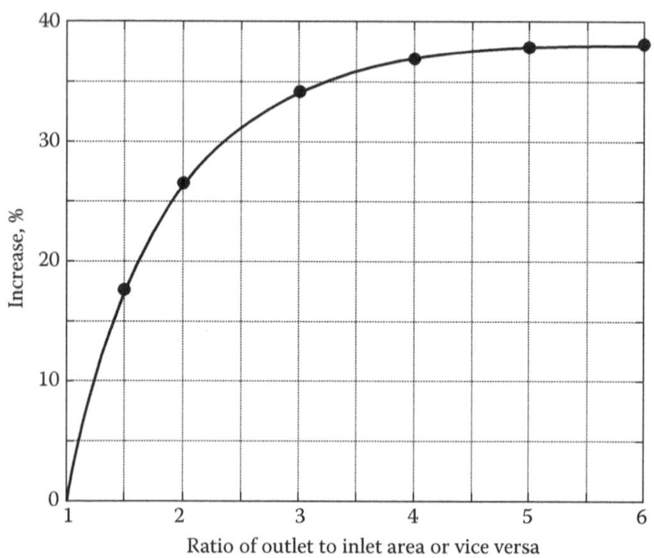

FIGURE 6.24
Increase in flow caused by excess area of one opening over the other. (From ASHRAE, *Handbook of Fundamentals*, American Society of Heating, Refrigerating, and Air-Conditioning Engineers, Atlanta, GA, 2013. Copyright ASHRAE, www.ashrae.org.)

As a final note, it is well known from experience that wind-induced pressure on a building façade is of a turbulent nature with gusts and fluctuations as it flows around a building. This behavior results in pressure fluctuations from the mean value, thereby causing outdoor air to oscillate in and out through the same opening and mixing with the indoor air. The result is akin to a net additional ventilation flow above that predicted by assuming continuous steady-state pressure coefficients. The interested reader can refer to Liddament (1986) or Etheridge and Sandberg (1996) for further discussion. The effect of turbulent fluctuations in the energy use and IAQ of real buildings is uncertain and needs more investigation. Most of the engineering building energy simulation software programs seem to simply ignore this effect.

Example 6.11: Natural Ventilation through Open Windows

Consider the first floor of a house with two windows: one of 1 m² in area and oriented 20° away from the normal of the prevailing wind direction and another of 1.5 m² on the opposite side of the room. If the local wind velocity is 6 m/s, determine the amount of ventilation air flowing into the room when both windows are wide open. Neglect infiltration due to stack effect since the indoor and outdoor temperatures are very close. Assume opening effectiveness to be 0.578.

Given: $C_v = 0.578$

Solution

We use Equation 6.28 to determine the flow rate with both windows assumed to be equal to the smaller one:

$$\dot{V}_{wind} = 0.578 \times 1\ \text{m}^2 \times 6\ \text{m/s} = 3.46\ \text{m}^3/\text{s}$$

The ratio of outlet to inlet window areas = 1.5/1 = 1.5. From Figure 6.24, the percentage increase is about 18%. Thus, the predicted air infiltration is

$$\dot{V}_{wind} = 3.46 \times 1.18 = 4.08\ \text{m}^3/\text{s}$$

This is a large ventilation flow rate and illustrates the major impact that opening of windows has on natural ventilation.

6.8.3 Computer Programs

Other than CONTAM (Walton and Dols, 2005) described earlier, there are several computer programs that can be used to simulate natural ventilation. *Vasari* developed by Autodesk (2013) provides great visualizations of flows in and around the space, but is rather restricted. The only inputs are wind speed and direction; a limitation is that buoyancy is not considered.

Design Builder (2013) is a UK-based software company that allows integrated analysis of building energy, carbon, lighting, and thermal comfort. It is robust software and has a user-friendly interface to EnergyPlus (2009). It considers the effects of both buoyancy and wind driven ventilation. The airflows through each window can be calculated using hourly simulation for a typical day, month, or year. The opening size can be modulated with other controls and schedules that can be set up for each opening. In addition, more in-depth analyses (velocity, temperature, and age of air) can be done at every point in space using the built-in CFD tool.

Virtual Environment, developed by Integrated Environmental Solutions (IES-VE, 2013) in the United Kingdom, is a very robust program. Through two built-in calculation modules, IES-VE can calculate most airflow scenarios: MacroFlo calculates multizone-level air movement for natural ventilation and hourly airflow at openings, while MicroFlo's CFD module yields granular information for building airflow patterns and particulate distributions. Additionally, this program provides a wide range of visualization capabilities, airflow animations, and cross-sectional plots. Through not as visually compelling as Vasari, IES-VE is said to provide sophisticated and trustworthy calculations for natural ventilation and flow rates.

6.9 Measuring Air Infiltration and Interzone Flows

Experimental techniques for measuring air infiltration and interzone flows in buildings are discussed in several ASTM standards and in textbooks (for example, Etheridge and Sandberg, 1996). Only a brief overview is provided here.

6.9.1 Blower Door Tests

The various simplified physical one-zone models have been framed in terms of model parameters that are best determined by house pressurization tests of the actual house. The most common manner of measuring infiltration rates in residences is by artificially pressurizing (or depressurizing) the home using a device called a

FIGURE 6.25
Photo of a blower door test rig. (From CIBSE TM23, Testing buildings for air leakage, CIBSE TM23:2000, The Chartered Institution of Building Services Engineers, London, U.K., 2000.)

"blower door" and measuring the resulting flow rate. This device consists of a door insert with a rubber edge that can provide an airtight seal against the door lamb of one of the doors (usually the main entrance). The blower door has a variable speed fan, an airflow measuring meter, and a pressure difference manometer with two plastic hoses (to allow the inside and outside pressure differential to be measured)—see Figure 6.25. All doors, windows, and fireplace dampers are closed during the testing. The fan speed is first increased incrementally in steps of about 12 Pa so as to depressurize the house, and the pressure difference Δp and airflow rate $\dot{V}$ are measured at each step (more information is provided in ASTM Standard E779-87 (1991)). Proper care is to be taken that the test conditions attain steady-state conditions before recording the results. To obtain accurate data, one needs fairly high pressures, around 50–75 Pa (0.2–0.3 inWG),* which are higher than those encountered during natural conditions in small to medium buildings. Another set of measurements is taken with the airflow direction reversed. The airflow data are then

* In large buildings, it may not be possible to attain such high pressure differences, and maximum values of 25 Pa pressure differences are often used.

regressed against a power law equation of the type given by Equation 6.5, that is, $\dot{V} = C(\Delta p)^{n}$, and the model coefficients are determined as illustrated in Example 6.12. The ELA of the house is subsequently determined as (Sherman and Jump, 2000)

$$\text{ELA} = C\dot{V}_r\left[\frac{\rho}{2\Delta p}\right]^{0.5} \tag{6.31}$$

with reference volume flow rate $\dot{V}_r$ determined from the identified power law equation with $\Delta p = 4$ Pa taken as reference. Note that though the blower door testing and estimation of model parameters by regression is done with pressure differences in the range of 10–50 Pa, extrapolation to the lower values of 4 Pa is required, which may result in inaccurate predictions of $\dot{V}_r$ due to the somewhat imprecise form of the flow–pressure relation assumed (ASHRAE Fundamentals, 2013). Nonetheless, blower doors are widely used devices that not only provide the necessary building-specific leakage data to be used in the various single-zone building models described in Section 6.6, but they also have great diagnostic ability. They help detect hidden air bypasses that would otherwise go uncovered and have been widely used for "house doctoring" activities during retrofitting and weatherization (Nisson and Dutt, 1985). In certain buildings with mechanical ventilation, one could bypass the need for a blower door by using the ventilation system itself provided that a high enough pressure differential can be generated (as in small- to medium-sized buildings).

Blower door testing was not originally intended to provide estimates of air infiltration, but rather to locate and fix air leaks. However, two other metrics directly obtained from blower door tests have been proposed, which are used as indicators of air leakage in a house: CFM50, which is the flow rate in ft^3/min required to maintain a 50 Pa pressure differential, and ACH50, which is the corresponding air change per hour. It has been empirically noticed that such metrics divided by 20 provide a measure of the *seasonal air exchange rate* that is useful for predicting seasonal energy use (Sherman and Jump, 2000).

Example 6.12: Analyzing Blower Door Data

A blower door test is performed on a residence with an internal volume of 480 m^3. The testing is done starting from 10 Pa with increments of 7.5 Pa up to a pressure of 55 Pa and then down again in the same steps for better repeatability. The results are tabulated in the first two columns of Table 6.6. Determine the ELA of this house and the ACH at a pressure difference of 4 Pa.

Given: Blower door data shown in Table 6.6

TABLE 6.6

Blower Door Results of Pressure and Ventilation Flow Rate

Δp, Pa	$\dot{V}$, m^3/h	$\ln(\Delta p)$	$\ln(\dot{V})$
10	0.171	2.30	−1.77
17.5	0.239	2.86	−1.43
25	0.237	3.22	−1.44
32.5	0.343	3.48	−1.07
40	0.390	3.69	−0.94
47.5	0.379	3.86	−0.97
55	0.481	4.01	−0.73
55	0.482	4.01	−0.73
47.5	0.387	3.86	−0.95
40	0.336	3.69	−1.09
32.5	0.347	3.48	−1.06
25	0.300	3.22	−1.20
17.5	0.184	2.86	−1.69
10	0.181	2.30	−1.71

Solution

Equation 6.5 can be converted to a linear model by taking natural logarithms of both the pressure difference and the volume flow rates. The observed data are converted into natural logarithms as shown in Table 6.6. A linear regression is performed, and the following model coefficients (along with their standard errors) are obtained:

$$\ln(\dot{V}) = \underset{(0.170)}{-3.146} + \underset{(0.050)}{0.582} \times \ln(\Delta p) \quad \text{with } R^2 = 0.918 \quad \text{and}$$

$$\text{RMSE} = 0.104$$

The exponential of the intercept term of the linear fit yields the flow coefficient of the crack flow equation (Equation 6.5) resulting in

$$\dot{V} = C\Delta p^{n} = 0.043 \times \Delta p^{0.582} \tag{6.32}$$

Next, we use the aforementioned equation assuming $\Delta p = 4$ Pa, which yields

$$\dot{V}_r = 0.0964 \text{ m}^3/\text{s or } 347 \text{ m}^3/\text{h}$$

Finally, we use Equation 6.31 to determine the ELA:

$$\text{ELA} = 0.043 \times 0.0964 \times \left[\frac{1.2}{2 \times 4}\right]^{0.5}$$

$$= 0.00161 \text{ m}^2 \quad \text{or} \quad 16.1 \text{ cm}^2$$

(Note that ELA is always quoted for a pressure difference of 4 Pa.)

Finally, the ACH = 347/480 = 0.72.

6.9.2 Building Component Tests

Blower door tests yield information about the whole-house leakage that is used by the single-zone models (described in Section 6.6.1). In most part, the ELA information provided in Table 6.2 for individual components is obtained from controlled laboratory tests or from field tests. Though laboratory tests can allow a large number of different specimens to be tested more accurately without weather interfering with the testing, it may yield results significantly different from field conditions because of the way the components are installed at site. Essentially, the test protocol is similar to a blower door test wherein one properly isolates the specimen, creates a preselected pressure difference across the specimen and measures the resulting air infiltration rate (Awbi, 1991). Because of the low flow rates, higher pressure differences of about 200 Pa need to be created. A simple sketch of a test arrangement suitable for field testing is shown in Figure 6.26. The collection chamber is usually inside the building. Slight variants to this basic configuration are usually adopted so as to prevent background leakage, i.e., to prevent air leakage into the collection chamber due to the pressure difference between the chamber and the room.

6.9.3 Tracer Gas Methods

Air exchange rates in buildings can be measured directly using tracer gas techniques. The basic principle is to inject a known amount of gas (which is inert and nonreactive with air), and measure its concentration based on the conservation of mass of the tracer gas. Infiltrating air would tend to dilute the concentration exponentially given by the following differential equation:

$$V\frac{dC(t)}{dt} = F(t) - \dot{V}(t)\cdot C(t) \tag{6.33}$$

where
$C(t)$ is the tracer gas concentration in the space at time t
V is the volume of the space
$F(t)$ is the rate of tracer gas injection
$\dot{V}(t)$ is the airflow rate out of the space at time t

There are several variants of how to conduct the tests. The three most common are as follows:

1. *The decay* or tracer gas dilution method (ASTM E741, 1994): One injects a known amount of tracer gas and measure its dynamic decay in real time, i.e., $C(t)$, from which the airflow rate can be determined by rearranging the following equation and performing a simple regression:

$$C(t) = C_0 e^{-\dot{V}t} \tag{6.34}$$

where C_0 is the initial concentration.

2. *Constant injection* or constant release method: One measures the amount of tracer gas that needs to be injected inside the space for the concentration to remain constant once equilibrium has been attained. The airflow rate is then

$$\dot{V} = \frac{F}{C} \tag{6.35}$$

3. *Constant concentration method:* One uses an automated system to simultaneously measure the tracer gas concentration and to inject the appropriate quantity of tracer gas into the zone to maintain a predetermined constant concentration from which airflow rate is determined as

$$\dot{V}(t) = \left[\frac{F(t)}{C}\right] \tag{6.36}$$

All these methods have their own advantages and disadvantages (see ASHRAE Fundamentals, 2013 or Awbi, 1991). The constant injection method is simplest and is particularly useful for spaces with mechanical ventilation or with high air exchange rates. The decay method is preferred for nonmechanical spaces because

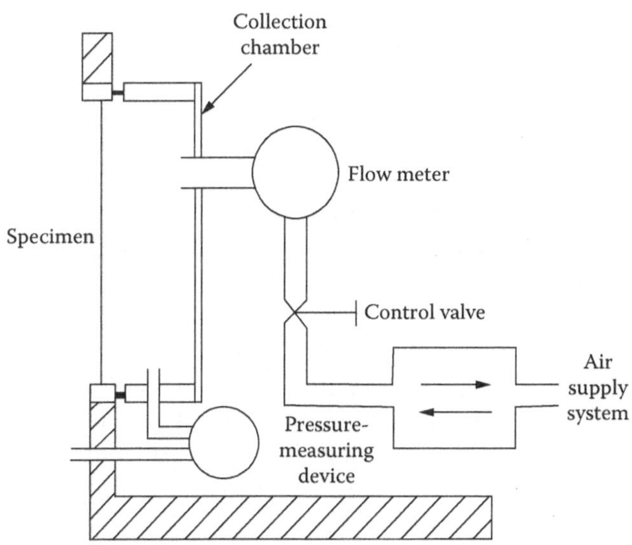

FIGURE 6.26
Airtightness test arrangement of a building component. (From Awbi, H.B., *Ventilation of Buildings*, E &FN Spon, London, U.K., 1991.)

of its lower setup time and simpler instrumentation. However, improper mixing of the tracer gas with the interior air can introduce serious errors.

Tracer gases should be inert, safe, and mix well with air, should not adhere to interior surfaces, should not be present in atmospheric air, and should be easily measurable even at low concentrations. One of the most widely used single tracer gases is sulfur hexafluoride (SF_6), which can be detected at concentrations above 1 part per billion (ppb). The equipment is relatively expensive but allows determination at hourly or even shorter intervals (Grimsrud et al., 1980; Sherman et al., 1980). If several tracer gases are used at the same time, one can infer not only infiltration rates into the space but also interzone flows that are important for IAQ studies. Other common tracer gases are helium, nitrous oxide, methane, and ethane.

Carbon dioxide is interesting as a tracer gas because it is produced by the occupants and can be used for monitoring indoor air quality; as a measure of air exchange, it is uncertain to the extent that the number of occupants and their metabolism are not known. It has also been used as a means of controlling ventilation intake depending on the number of occupants present in the space. Such a strategy, called "demand ventilation," is routinely used as an energy conservation measure especially in highly variable-occupancy spaces (such as classrooms and meeting rooms).

Another low-cost alternative to tracer gas techniques has been developed that uses passive perfluorocarbon sources/emitters and passive samplers/receptors (Dietz et al., 1985). Each source can cover up to several hundred cubic meters of building volume, at a cost of about \$50. Infiltration and interzonal flows can be determined depending on how many emitters and receptors are deployed. However, the method yields only averages over sampling periods of several weeks (in fact, it averages the inverse of the air exchange rate) and is only suitable for studies involving long-term behavior.

6.10 Infiltration Heat Recovery

Infiltration is a major contributor to the energy consumption of buildings, particularly in homes where it often accounts for one-third of the heating and cooling loads. The traditional manner of estimating these loads have been discussed earlier in this chapter (see Section 6.1). The assumption is that there is no thermal interaction between the infiltrating air and the building envelope. Laboratory and field studies have shown that a certain amount of heat recovery does takes place due to the tortuous leakage pathways characteristic of envelope cracks. This is also supported by CFD simulations. This thermal coupling between the infiltration and conduction heat transfer of the building envelope is known as infiltration heat recovery (IHR). Several studies have focused on houses and envelopes that deliberately enhance the IHR such as using dynamic walls and ceilings and empty wall cavities. Such studies have shown energy reduction of the impacts of air infiltration of the order of 5%–20%. In fact, IHR has even been adopted as a design feature in Scandinavia (Brunsell, 1995). However, after a combination of analytical modeling, CFD simulations, laboratory tests, and field evaluations, Walker and Sherman (2003) concluded that IHR is not significant for typical wood-frame house construction with insulated cavities in the United States.

Problems

Numbers 1–4 given in parenthesis denote the degree of difficulty.

6.1 Find the infiltration rate from stack and wind through a curtain wall of area 300 ft² (27.88 m²) if the indoor–outdoor pressure difference is 0.2 inWG (49.8 Pa). *(2)*

6.2 Find the indoor–outdoor pressure difference at the base of a one-story building of height 3 m (9.84 ft) without mechanical ventilation if the pressure coefficient $\Delta C_p = 0.3$, wind speed v = 6.7 m/s (15 mph), and $T_i - T_o = 15$ K (27.0°F). *(2)*

6.3 The reference wind velocity recorded at a meteorological station located in a flat open terrain is 30 mph (13.4 m/s). Find the pressure difference due to wind between the outside and inside of the 30th floor of a multistory building of 50 floors with aspect ratio of 1/4. The wind direction is 30° with respect to one of the longer sides. Assume each floor is 12 ft high (3.5 m) and the height of the local obstacles is 50 ft (15 m). *(3)*

6.4 Consider Example 6.9 where the ACH of a simple building was estimated. Determine the ACH if the walls at both sill and roof are caulked and the windows are weather stripped and caulked. *(2)*

6.5 An undesirable chemical was released accidently in the simple building of Problem 6.4. Using Equation 6.33, estimate the number of hours needed for the concentration to drop (a) by a factor of 2, (b) by a factor of 5, and (c) by a factor of 10. *(2)*

6.6 A one-story building with dimensions 100 ft × 100 ft × 13 ft (30.48 m × 30.48 m × 3.96 m) has an infiltration rate of 0.2 air change per hour during unoccupied periods. During the day, it is occupied by

80 people, each requiring 20 ft³/min (10 L/s) of outdoor air. Find the contribution of the air change to the total heat loss coefficient during unoccupied and occupied periods. Assume that during occupied periods there is only controlled ventilation and no infiltration. U value of roof= 0.24 W/m²·K, U value of opaque portion of walls (all four sides) = 0.40 W/m²·K, U-value of glazing (assume 20% of wall area) = 3.0 W/m²·K *(2)*

6.7 A 10-story office building with floor dimensions 50 ft × 100 ft (15.24 m × 30.48 m) and a height of 130 ft (39.6 m) has curtain walls with windows that are fixed and airtight. The window/wall ratio is 0.5. The draft coefficient for airflow between floors is $C_d = 0.65$. There are two vestibule-type doors on each of the two 100 ft (30.48 m) facades. The traffic rate corresponds to each of the occupants (one per 150 ft² [13.94 m²] of gross floor area) making on average five entrances or exits per 10 h. The indoor and outdoor temperatures are 70°F (21.1°C) and 20°F (−6.7°C), and the wind is incident parallel to the 50 ft (15.24 m) facade at 15 mph (6.7 m/s). Assume that infiltration through the roof is negligible. In other words, the only infiltration occurs through the curtain walls and through the doors.

(a) Compute the pressure differences for each wall due to the stack effect and wind for floors 1, 5, and 10, with due attention to signs (infiltration or exfiltration).

(b) Compute the total infiltration rates for these floors if the ventilation system is balanced for neutral pressure.

(c) Compute the total infiltration rates for these floors if the ventilation system creates an overpressure of 0.1 inWG (24.9 Pa). *(4)*

6.8 There is an odor in your room located on the second floor of a house that you would like to get rid of before your mother visits later this afternoon. The concentration of the odor-causing substance needs to be reduced by a factor of 10 in order to become unnoticeable. Using Figure 6.22, try to estimate the number of hours that the windows should be open before your mother's visit. Assume a wind speed of 3 m/s. *(3)*

6.9 Solve Problem 6.8 using Equation 6.28. The room with dimensions of 5 m × 6 m with a ceiling height of 2.2 m has windows on opposite sides of the room. Each window has an area of 2.5 m². Estimate the number of hours for the odor to reduce by a factor of 10 for wind direction (a) perpendicular to the windows and (b) diagonal wind flow. *(2)*

6.10 Suppose you live in a two-story house in a residential neighborhood that has 10 double-hung 1 m × 1.5 m windows that are not weather stripped. The indoor temperature is constant at 20°C.

(a) Estimate the leakage area from just the windows.

(b) Assuming an average winter wind speed of 2 m/s and an average outdoor temperature of 5°C, what is the infiltration rate?

(c) Estimate the associated heating load in GJ/h with this infiltration assuming 4380 hours/year of heating.

(d) If energy costs are $5/GJ, what is the cost of heating this air?

(e) How much money could you save per year if the windows had weather stripping? *(2)*

6.11 The following table assembles test results of an actual house where blower door tests were performed both before and after weather stripping. Following the procedure described in Example 6.12, determine the two sets of coefficients C and n for tests done before and after the house tightening. Calculate the ELA values for both the sets. *(3)*

Before Weather Stripping		After Weather Stripping	
Δp (Pa)	$\dot{V}$ (m³/h)	Δp (Pa)	$\dot{V}$ (m³/h)
3.0	365.0	2.2	99.2
5.0	445.9	5.5	170.4
5.8	492.7	6.7	185.6
6.7	601.8	8.2	208.5
8.2	699.2	11.6	263.2
9.0	757.5	13.5	283.1
10.0	812.4	15.6	310.2
11.0	854.1	18.2	346.2

References

Achard, P. and R. Gicquel (1986). *European Passive Solar Handbook.* Commission of the European Communities, Directorate General XII for Science, Research and Development, Brussels, Belgium.

ASHRAE 62.1 (2013). *Standard 62.1-2013: Ventilation for Acceptable Indoor Air Quality.* American Society of Heating, Refrigerating and Air-Conditioning Engineers, Atlanta, GA.

ASHRAE Fundamentals (2001, 2013). *Handbook of Fundamentals.* American Society of Heating, Refrigerating, and Air-Conditioning Engineers, Atlanta, GA.

ASTM E741 (1994). E741-83: Standard test method for determining air leakage rate by tracer dilution, *ASTM Book of Standards*, Vol. 04.07. American Society of Testing and Materials, Philadelphia, PA.

ASTM E779 (1991). E779-87: Test method for determining air leakage by fan pressurization, *ASTM Book of Standards*, Vol. 04.07. American Society of Testing and Materials, Philadelphia, PA.

Autodesk (2013). Vasari beta 3 release. San Rafael, CA. http://autodeskvasari.com/. Accessed on May 2014.

Awbi, H.B. (1991). *Ventilation of Buildings*. E &FN Spon, London, U.K.

Brunsell, J. (1995). The indoor air quality and the ventilation performance of four occupied residential buildings with dynamic insulation, in *Proceedings of the 16th AIVC Conference*, Vol. 2, Air Infiltration and Ventilation Center, Coventry, U.K., pp. 471–482.

Butcher, K. (2007). Natural ventilation in non-domestic buildings, reprinted with corrections. The Chartered Institution of Building Services Engineers, London, U.K.

CIBSE AM10 (2005). Natural ventilation in non-domestic buildings. The Chartered Institution of Building Services Engineers, London, U.K.

CIBSE TM23 (2000). Testing buildings for air leakage, CIBSE TM23:2000. The Chartered Institution of Building Services Engineers, London, U.K.

Chen, Q. and L. Glicksman (2001). Chapter 59: Application of computational fluid dynamics for indoor air quality studies, in *Indoor Air Quality Handbook* (J.D. Spengler, J.M. Samet, and J.F. McCarthy, eds.). McGraw-Hill, New York.

DesignBuilder (2013). ver. V3.2. London, U.K. http://www.designbuilder.co.uk/. Accessed on May 2014.

Dietz, R.N., T.W. Ottavio, and C.C. Cappiello (1985). Multizone infiltration measurements in homes and buildings using passive perfluorocarbon tracer method. *ASHRAE Trans.*, 91(pt. 2), 433–444.

Emmerich, S.J. and A.K. Persily (2013). Analysis of the NIST commercial and institutional building envelope leakage database, in *AIVC Airtightness Workshop, Third Tight Vent Workshop on Building and Ductwork Airtightness*, National Institute of Standards and Technology, U.S. Department of Commerce, Washington, DC.

Energy Plus (2009). Energy Plus Building Energy Simulation software, Building Technologies Program. National Renewable Energy Laboratory (NREL), U.S. Department of Energy, Washington, DC. http://www.nrel.gov/buildings/energy_analysis.html#energyplus. Accessed on May 2014.

Etheridge D. and M. Sandberg (1996). *Building Ventilation: Theory and Measurement*. John Wiley & Sons, Chichester, U.K.

Grosso, M. (1992). Wind pressure distribution around buildings: A parametric model. *Energy Build.*, 18(2), 101–131.

Grimsrud, D.T., M.H. Sherman, J.E. Janssen, A.N. Pearman, and D.T. Harrje (1980). An intercomparison of tracer gases used for air infiltration measurements. *ASHRAE Trans.*, 86(pt. 1), 258–267.

IES-VS (2013). Virtual environment. Integrated Environmental Solutions, Glasgow, U.K. http://www.iesve.com/. Accessed on May 2014.

Liddament, N.W. (1986). *Air Infiltration Calculation Techniques—An Applications Guide*. Air Infiltration and Ventilation Center, Bracknell, U.K.

Lydberg, M. and A. Honarbakhsh (1989). Determination of air leakiness of building envelopes using pressurization at low pressures. Document D19:1989. Swedish Council for Building Research, Gävle, Sweden.

Nisson, J.D.N. and G. Dutt (1985). *The Superinsulated Home Book*. Wiley, New York.

NIST (2008). Multizone modeling website. National Institute of Standards and Technology. http://www.bfrl.nist.gov/IAQanalysis/. Accessed on April 2014.

Persily, A.K. (1986). Measurements of air infiltration and airtightness in passive solar homes, in *Measured Air Leakage of Buildings* (H.R. Trechsel and P.L. Lagus, eds.). ASTMSTP 904, American Society of Testing and Materials, Philadelphia, PA, p. 46.

Sherman, M.H. (1992). Superposition in infiltration modeling. *Indoor Air*, 2, 101–114.

Sherman, M.H. (October 2013). Personal communication.

Sherman, M.H. and D.T. Grimsrud (1980). Infiltration-pressurization correlation: Simplified physical modeling. *ASHRAE Trans.*, 86(pt. 2), 778–807.

Sherman, M.H., D.T. Grimsrud, P.E. Condon, and B.V. Smith (1980). Air infiltration measurement techniques, in *Proceedings of First IEA Symposium*, Air Infiltration Centre, London, U.K. Also Lawrence Berkeley Laboratory Report LBL 10705, Berkeley, CA.

Sherman, M.H. and D. Jump (2000). Chapter 6.3: Energy conservation in buildings, in *Handbook of Heating, Ventilation, and Air-Conditioning* (J.F. Kreider, ed.). CRC Press, Boca Raton, FL.

Spitler, J.D. (2010). *Load Calculation Applications Manual*. American Society of Heating, Refrigerating and Air-Conditioning Engineers, Atlanta, GA.

Swami, M.V. and S. Chandra (1988). Correlations for pressure distribution on buildings and calculation of natural ventilation airflow. *ASHRAE Trans.*, 94(1), 243–266.

Walker, I.S. and M.H. Sherman (2003). Heat recovery in building envelopes, in *Proceedings of the 24th AIVC Conference*, Washington, DC. LBNL Report 53484.

Walker, I.S. and D.J. Wilson (1993). Evaluating models for superposition of wind and stack effects in air infiltration. *Build. Environ.*, 28(2), 201–210.

Walker, I.S. and D.J. Wilson (1998). Field validation of equations for stack and wind driven air infiltration calculations. *Int. J. HVAC*&R Res., 4(2), 119–140.

Walton, G.N. and W.S. Dols (2005). CONTAM 2.4 user guide and program documentation. Building Environment Division, Building and Fire Research Laboratory, National Institute of Standards and Technology, Gaithersburg, MD.

Warren, P.R. and B.C. Webb (1980). The relationship between tracer gas and pressurization techniques in dwellings, in *Proceedings of the First IEA Symposium of the Air Infiltration Centre*, London, U.K.

7

Steady-State Heat Flows

ABSTRACT This chapter discusses the various components of heat losses and gains in buildings, which play a major role in *steady-state* load calculations. Loads refer to the thermal heat rates that must be supplied or removed by the HVAC equipment to maintain a space at the desired conditions. These can be broadly divided into three sources: from envelope-related flows, from internal heat sources, and from air infiltration (which was covered in Chapter 6). We start by addressing the prevalent engineering calculation procedures of estimating solar heat gains in buildings through walls, roof, and below-grade surfaces such as basement walls and floor and slab-on-grade floors (heat gains through windows have been treated in Chapter 5). For opaque surfaces on which solar radiation is incident, we introduce the concept of sol–air temperature. This is followed by a treatment of internal heat gains from occupants, lights, and equipment. We then discuss the notion of overall conductive heat loss coefficient and the concept of steady-state thermal balance of a building. How these concepts can be used to analyze unconditioned spaces is also presented. Finally, we address the need for thermal zoning of building spaces, i.e., the separate treatment of different spaces in a building where the loads are too dissimilar to be lumped together for satisfactory indoor thermal comfort.

Nomenclature

A	Area, m² (ft²)
c_p	Specific heat at constant pressure, kJ/(kg·K) [Btu/(lb$_m$·°F)]
c, d	Exponents in Equation 7.12
d	Damping length of the soil (Equation 7.8)
F_{LM}	Motor load factor
F_p	Heat loss coefficient of floor slab per unit length, W/(m·K) [Btu/(h·ft·°F)]
F_{SA}	Special allowance factor for lighting
F_{UL}	Lighting use factor, diversity factor
F_{UM}	Motor use factor
h	Height m(ft)
h_{bw}	Height of basement wall below grade, m (ft)
h_{fg}	Enthalpy of liquid–gas phase change for water, kJ/kg (Btu/lb$_m$)
h_o	Outdoor surface heat transfer coefficient, W/(m²·K) [Btu/(h·ft²·°F)]
I_T	Total solar irradiance on surface, W/m² [Btu/(h·ft²)]
k	Thermal conductivity, W/(m·K) [Btu/(h·ft·°F)]
K_{cond}	Conductive heat transmission coefficient, W/K [(Btu/(h·°F)]
K_{tot}	Total heat transmission coefficient of building, W/K [Btu/(h·°F)]
n	Day of year (n = 1 for January 1)
P	Perimeter, m (ft)
P_L	Total installed lighting electric power, W
P_M	Power rating of electrical motor, W
Q	Energy consumption or heat transfer, kJ (Btu)
$\dot{Q}$	Heat flow or heat loads with subscripts c for cooling load, *h* for heating load, *lat* for latent load, *floor* for heat flow to floor, etc., W (Btu/h)
$\dot{Q}_{gains}$	Heat gains of space, W (Btu/h)
$\dot{q}$	Heat flux, W/m² [Btu/(h·ft²)]
T	Temperature, K or °C (°R or °F)
T_i	Indoor air dry-bulb temperature, °C (°F)
T_o	Outdoor air dry-bulb temperature °C (°F)
T_{os}	Sol–air temperature, °C (°F)
T_s	Surface temperature, °C (°F)
$T_{soil}(z,t_h)$	Soil temperature at depth z at hour t_h of the year, °C (°F)
$\bar{T}_{soil,annual}$	Annual average soil surface temperature, °C (°F)
t_h	Time expressed in hours starting from January 1, hours
U	Overall heat transfer coefficient, W/(m²·K) [Btu/(h·ft²·°F)]
$\dot{V}$	Volumetric flow rate, L/s (ft³/min)
v	Velocity or wind speed, m/s (mph, ft/s)
W	Humidity ratio of air, kg$_w$/kg$_a$ (lb$_w$/lb$_a$)
z	Depth from soil surface, m (ft)
$\bar{x}$	Average or mean value of quantity x

Greek

α	Absorptivity
α_s	Soil diffusivity, m²/s (ft²/h)
η_M	Efficiency of electrical motor
ρ	Density, lb$_m$/ft³ (kg/m³)
θ_p	Zenith angle of plane (tilt from horizontal, upward > 0), degrees

ϕ Phase lag between ambient air and soil surface temperature, days
ΔT_{soil} Annual amplitude of the soil surface temperature, °C (°F)
ΔQ_{sol} Solar heat gain (Equation 7.5), W/m^2 [Btu/(h·ft^2)]
Δq_{ir} Correction to infrared radiation transfer between surface and environment, W/m^2 [Btu/(h·ft^2)]
Δx Thickness of element, m (ft)

Subscripts

air Air infiltration and/or ventilation
av Average
b Basement
bf Basement floor
bw Basement wall
common Common area
cond Conduction
des Design
equip Equipment
g Ground or soil
glaz Glazing
i Indoor
inf Infiltration
j Index for building zone
k Index for element of building envelope or building zone
lat Latent
leak Air leakage
light Lighting
o Outdoor
occ Occupants
N North space
S South space, surface
sen Sensible
sol Solar
stor Storage
tot Total

7.1 Load Calculations

Heating and cooling loads refer to the thermal heat rates that must be supplied to or removed from the interior spaces of a building in order to maintain the desired comfort conditions. That is the demand side of the building, addressed in this chapter and in Chapters 8 and 9. Once the loads have been established, one can proceed to the supply side and determine the sizing and performance of the required air/water distribution systems and the heating and cooling equipment; this is discussed in Chapters 14 through 20.

A load calculation consists of a careful accounting of all the thermal energy terms in a building. One considers all the heat that is generated in the space and heat flows across the envelope; the total energy, including the thermal energy stored in the space, must be conserved according to the first law of thermodynamics. The principal heat flows are indicated in Figure 7.1. While several of these heat flows have been treated in previous chapters, there are some that have not been, namely, conduction heat gains through opaque sunlit walls and roof, heat losses through the floor and basements, and heat gains from internal loads. These are treated in this chapter. Outdoor air, occupants, and certain kinds of equipment contribute both sensible and latent heat loads, and these will have to be distinguished.

While the basic principle is simple, a serious complication can arise from storage of heat in the mass of the building. In practice, this is very important for peak cooling loads, even in lightweight buildings typical in the United States. For peak heating loads, the heat capacity can be neglected unless one would like to consider the effect of thermostat setback during the cold periods and setup during the summer unoccupied periods. For annual energy consumption, the effect of heat capacity depends on the control of the thermostat: it is negligible if the indoor temperature is constant but can be quite significant with thermostat setback or setup. The various modeling approaches that allow consideration of the storage of heat in the building envelope and the interior of the space are discussed in Chapter 8. Methods for calculating design peak loads are addressed in Chapter 9, while annual energy calculations are covered in Chapter 10. The calculation of loads presented here does not take into account the losses in the air/water distribution system. These losses can be quite significant, especially in the case of uninsulated ducts, and

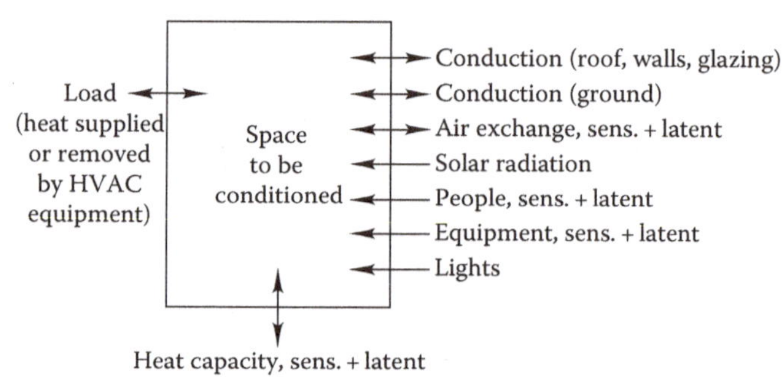

FIGURE 7.1
The heat flow terms in a load calculation.

they should be taken into account in the analysis of the HVAC system (treated in Chapter 19).*

In this chapter, we discuss the various components of heat loss and heat gains prior to the intervention of the effect of thermal storage due to the building mass. Thus, instantaneous gains or losses can be viewed as steady-state heat flows that dictate the load calculations. Most of the following analysis methods follow ASHRAE recommended procedures, while different countries have their own calculation methods. Even in North America, there are other organizations that have developed ANSI-approved procedures along with extensive tables and worksheets to aid the practicing professional. One such example is the Manual J for residential load calculations (ACCA, 2006).

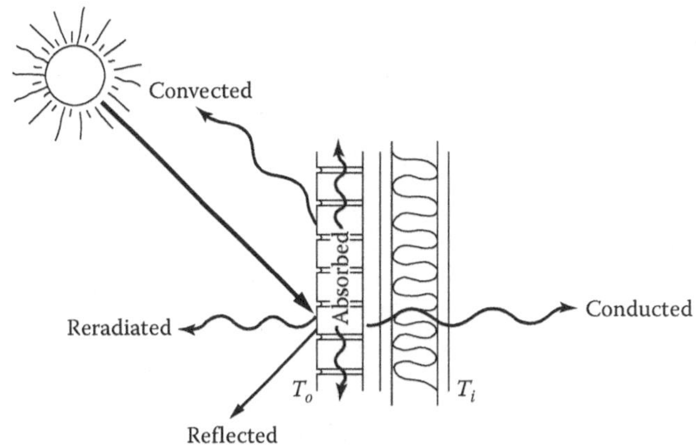

FIGURE 7.2
Heat flows on a sunlit opaque wall.

7.2 Sol–Air Temperature and Instantaneous Conduction Heat Gain

This section develops a concept that is convenient for analyzing solar heat gains due to absorption by opaque exterior surfaces of a building. Recall that if the surrounding environment can be characterized by a single temperature T_o for radiation and for convection, the resulting heat flow per unit area $\dot{q} = \dot{Q}/A$, from environment to surface, is given by

$$\dot{q} = h_o(T_o - T_s) \quad (7.1)$$

where
h_o is the surface heat transfer coefficient (radiation plus convection, as described in Section 2.10.6)
T_s is the surface temperature

If, in addition, the surface has a solar absorptivity α, then the solar radiation absorbed by the wall surface is αI_T. Combining this heat quantity with the conductive flow yields the total heat flow at the surface (see Figure 7.2):

$$\dot{q}_{tot} = h_o(T_o - T_s) + \alpha I_T \quad (7.2)$$

To simplify the load calculations, it is preferable to work with an equation of the form akin to that of Equation 7.1 rather than specify both a temperature and a solar flux. That can be accomplished if one replaces the outdoor air temperature T_o by an equivalent temperature:

$$T_o + \frac{\alpha I_T}{h_o}$$

With such a replacement, Equation 7.1 yields the same result as Equation 7.2.

Not only the solar flux but also infrared exchanges can be treated in this manner—in particular, those with the sky when the radiation temperature of the sky is below the air temperature. Both effects are included in what is called the "sol–air temperature," defined as

$$T_{os} = T_o + \frac{\alpha I_T}{h_o} - \frac{\Delta q_{ir}}{h_o} \quad (7.3)$$

where
h_o is the surface heat transfer coefficient for radiation and convection, W/(m$^2\cdot$K) [Btu/(h$\cdot$ft$^2\cdot$°F)]
I_T is the global solar irradiance on surface, W/m^2 [Btu/(h$\cdot$ft^2)]
Δq_{ir} is the correction to infrared radiation transfer between surface and environment if sky temperature is different from T_o, W/m^2 [Btu/(h$\cdot$ft^2)]

In practice, values of $(\Delta q_{ir}/h_o) \approx 3.9°C \times \cos\theta_p$ ($\approx 7°F \times \cos\theta_p$) where θ_p is the surface tilt angle. Thus, for vertical surfaces $(\Delta q_{ir}/h_o) = 0$, while it is to 3.9°C (7°F) for upward-facing surfaces (the sky overhead is colder than the rest of the environment, typically in the range 5°C–20°C (10°F–40°F). Also, α/h_o varies from 0.052 (m$^2\cdot$K)/W [0.30 (h$\cdot$ft$^2\cdot$°F)/Btu] for dark surfaces to 0.026 (m$^2\cdot$K)/W [0.15 (h$\cdot$ft$^2\cdot$°F)/Btu] for light-colored surfaces.

* For future reference we note that the loads of each zone, as calculated in this chapter, *include* the contribution of the outdoor air change. However, for the analysis of air-based central distribution systems, it is convenient to *exclude* the contribution of ventilation air from the zone loads and to count it instead as load at the air handler. Keeping the load due to ventilation air apart is straightforward because this load is instantaneous. This is illustrated in Chapter 19.

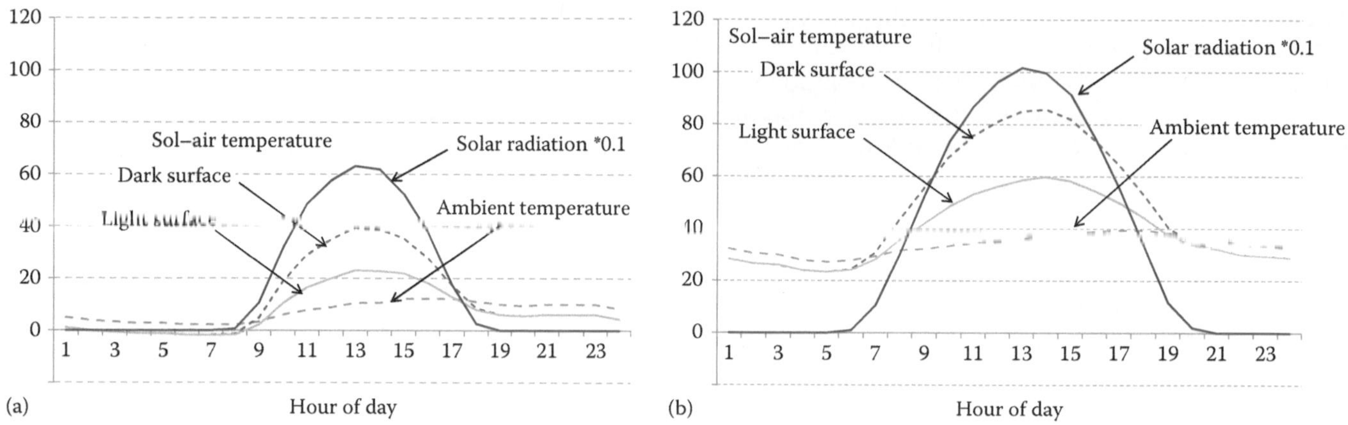

FIGURE 7.3
Diurnal variation of sol–air temperatures (in °C) for a dark- and a light-colored horizontal surface for 2 days in Phoenix, AZ, using TMY3 climatic data. (a) For January 21. (b) For July 21. The corresponding horizontal global radiation (in W/m²) and the ambient dry-bulb temperature (in °C) are also plotted.

In terms of the sol–air temperature, the total heat transfer at the surface is therefore

$$\dot{q}_{tot} = h_o(T_{os} - T_s) \tag{7.4}$$

Thus, a sunlit opaque surface will exchange the same amount of heat with the actual outdoor environment as one with the outdoor temperature equal to the sol–air temperature but under zero solar radiation condition. The sol–air concept is convenient because solar radiation on opaque surfaces will then be automatically included in the resulting load calculations.

In Figure 7.3, we have plotted the diurnal sol–air temperatures on a dark- and a light-colored horizontal surface for 2 days in Phoenix, AZ, using TMY2 climatic data. The corresponding values of hourly global horizontal solar radiation and the ambient dry-bulb temperatures are also shown. These plots illustrate the large differences in sol–air temperature and ambient temperatures during the middle of the day especially in summer, as well as the differences between the dark-colored and the light-colored surfaces. Note that sol–air temperatures exceed 80°C (140°F) in July for the dark surface, an increase of over 40°C (70°F) over the ambient air temperature. The differences are even larger for clear sky conditions (see Section 9.4.1); however, as illustrated in the following example, the magnitude of the increase in heat gains may itself be modest.

As far as the associated heat gains through an opaque wall or roof are concerned, the outdoor temperature appears to be augmented by an amount $\Delta T_{os} = (T_{os} - T_o)$. If the total heat transfer coefficient (including h_o) is U, the extra heat flow is $U\Delta T_{os}$. Keeping only the solar contribution in ΔT_{os}, we find that the solar heat gain $\Delta\dot{Q}_{sol}$ through an opaque surface is

$$\Delta\dot{Q}_{sol} = \frac{U \cdot I_T \cdot \alpha}{h_o} \tag{7.5}$$

Example 7.1: Solar Heat Gain through a Wall

Find the solar heat gain through a wall specified as follows.

Given: U = 0.24 W/(m²·K) [0.042 Btu/(h·ft²·°F)] also called R-23.8

α/h_o = 0.052 (m²·K)/W [0.30 (h·ft²·°F)/Btu]

Solar irradiance on wall I_T = 394.2 W/m² [125 Btu/(h·ft²)] = conditions of Example 4.13

Find: $\Delta\dot{Q}_{sol}$

Solution

From Equation 7.5, we find

$$\Delta\dot{Q}_{sol} = 0.24\ \text{W/(m}^2\cdot\text{K)} \times 394.2\ \text{W/m}^2 \times 0.052\ (\text{m}^2\cdot\text{K})/\text{W}$$

$$= 4.92\ \text{W/m}^2\ [1.56\ \text{Btu/(h}\cdot\text{ft}^2)]$$

Comments

The ratio of solar heat gain to incident solar irradiance is 4.92/394.2 = 0.012, a little more than 1%. Quite generally, solar heat gains *per unit area* through opaque surfaces are one to two orders of magnitude smaller than solar heat gains through glazing.

7.3 Below-Grade Heat Conduction

The transfer of heat between the basement or floor slab of a building and the earth is called "ground coupling" and is essentially a problem of conduction heat transfer. A number of methods for calculating *ground-coupled* heat transfer have been suggested. Krarti (1999) provides a review of many of these methods. Unfortunately, these approaches sometimes disagree greatly in the values of heat transfer rates calculated (Claridge, 1986).

This is particularly the case in residences and single-story buildings with large ratios of floor area to building volume. Ground thermal coupling calculations have probably been the least accurate of any in building thermal analysis. Fortunately, for designers of commercial buildings, this inaccuracy has little overall impact since basement and slab heat losses are a small portion of the total heat loads.

7.3.1 Soil Conductivities

Ground-coupled heat transfer depends on the temperature difference between the indoor air and the soil (treated in Section 7.3.2), the resistance of the building wall or floor material, and the thermal conductivity of the soil. The soil conductivity depends on numerous factors such as the type of soil (soil density and porosity), soil depth, and moisture distribution (which itself is dictated by the soil temperature distribution that varies over the year). Strictly speaking, the below-grade conduction heat transfer problem would have to be framed as a 3-D transient coupled heat and moisture problem—hardly a viable approach for widespread engineering use. Further, soil data for different locations are difficult to acquire. With the intent of simplifying the calculations, Sterling et al. (1993) proposed the concept of *apparent soil thermal conductivity*, which combines the moisture effects into the thermal heat transfer. Table 7.1 lists low and high values for different types of soil. We note wide variation for the thermal conductivity of common soils: between 0.8 and 2.25 W/(m·K) [5.4–15.6 Btu·in./(h·ft²·°F)].

7.3.2 Soil Temperature Profiles

It is necessary to be able to predict soil temperature as a function of time (i.e., hour of the year) and depth z from the soil surface. The soil surface temperature is closely related to the ambient temperature but with a time lag. It can be modeled as

$$T_{soil}(0,t_h) = \overline{T}_{soil,annual} + \Delta T_{soil} \times \sin\left(\frac{2\pi}{8760}t_h - \phi - \frac{\pi}{2}\right) \quad (7.6)$$

TABLE 7.1

Apparent Thermal Conductivity for Selected Soil Types

Soil Type	SI Units, W/(m·K) Low	SI Units, W/(m·K) High	IP Units, Btu·in./(h·ft²·°F) Low	IP Units, Btu·in./(h·ft²·°F) High
Sands	0.78	2.25	5.4	15.6
Silts	1.64	2.25	11.4	15.6
Clays	1.12	1.56	7.8	10.8
Loams	0.95	2.25	6.6	15.6

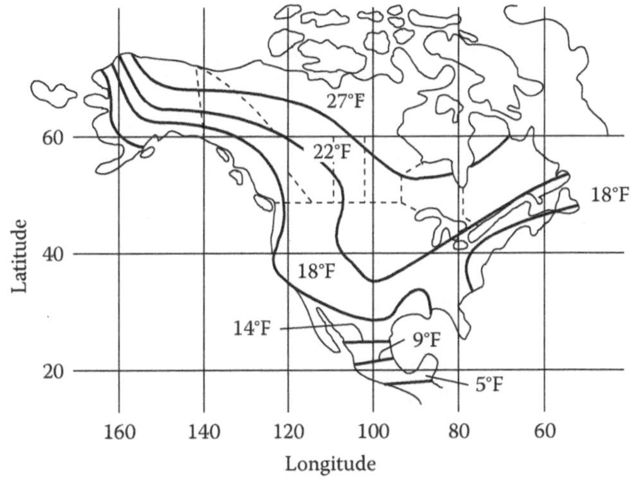

FIGURE 7.4
Map of amplitude of annual soil temperature swings used for ground coupling calculations (°F). For amplitude in SI units, divide by 1.8°F/K. (From ASHRAE, *Handbook of Fundamentals*, American Society of Heating, Refrigerating and Air-Conditioning Engineers, Atlanta, GA, 2013. Copyright ASHRAE, www.ashrae.org.)

where

$\overline{T}_{soil,annual}$ is the annual average ambient temperature (equal to the deep soil average temperature)

ΔT_{soil} is the annual amplitude of the soil surface temperature (equal to that of ambient air temperature), with the ground temperature map for the United States shown in Figure 7.4

ϕ is the phase lag between ambient air and soil surface temperature

t_h is the time expressed in hours starting from January 1

Values of aforementioned quantities for selected U.S. locations are assembled in Table 7.2.

The soil temperature at any depth z can be determined from

$$T_{soil}(z,t_h) = \overline{T}_{soil,annual} + \Delta T_{soil} \times e^{-z/d} \times \sin\left(\frac{2\pi}{8760}t_h - \phi - \frac{z}{d} - \frac{\pi}{2}\right) \quad (7.7)$$

where

d is the damping length given by $\left(\frac{2\alpha_s}{2\pi/8760}\right)^{1/2}$ (7.8)

α_s is the thermal diffusivity of the soil

Note that Equation 7.7 is similar to Equation 7.6 except for the exponential term and the additional lag term. The soil thermal diffusivity α_s is the ratio of the conductivity

TABLE 7.2

Annual Average Soil Temperatures and Amplitudes for Selected U.S. Cities

		SI Units			IP Units		
State	**City**	**Av. Temp., °C**	**Amplitude, °C**	**Phase Angle, radians**	**Av. Temp., °F**	**Amplitude, °F**	**Phase Angle, radians**
Alabama	Auburn	18.3	9.4	0.49	65	17	0.49
Arizona	Tempe	21.1	11.1	0.47	70	20	0.47
California	Davis	18.9	10.6	0.63	66	19	0.63
Colorado	Fort Collins	10.0	13.3	0.54	50	24	0.54
Idaho	Moscow	8.3	10.0	0.73	47	18	0.73
Illinois	Argonne	10.6	12.8	0.70	51	23	0.70
Kansas	Manhattan	12.8	14.4	0.61	55	26	0.61
Kentucky	Lexington	12.8	12.8	0.60	55	23	0.60
Michigan	E. Lansing	10.0	13.3	0.60	50	24	0.60
Minnesota	St. Paul	8.9	13.9	0.65	48	25	0.65
Missouri	Kansas City	12.2	12.2	0.56	54	22	0.56
Montana	Bozeman	6.7	11.7	0.68	44	21	0.68
Nebraska	Lincoln	12.2	15.6	0.52	54	28	0.52
New York	Ithaca	9.4	10.6	0.69	49	19	0.69
Ohio	Columbus	11.7	12.2	0.65	53	22	0.65
Oklahoma	Lake Hefner	17.8	12.8	0.63	64	23	0.63
Oregon	Corvallis	13.3	10.0	0.53	56	18	0.53
South Carolina	Calhoun	17.8	12.2	0.49	64	22	0.49
Tennessee	Jackson	15.6	11.1	0.44	60	20	0.44
Texas	Temple	21.1	11.7	0.58	70	21	0.58
Utah	Salt Lake City	10.6	11.7	0.48	51	21	0.48
Washington	Seattle	11.7	8.3	0.64	53	15	0.64

Source: Krarti, M., *Weatherization and Energy Efficiency Improvements for Existing Homes*, CRC Press, Boca Raton, FL, 2012.

to the product of density and specific heat. All three quantities vary with temperature, and so tables have been assembled that provide values of this quantity for different locations (see, for example, Kusuda and Achenbach, 1965). However, it has been suggested that in the absence of specific data for a given location, an average value of $\alpha_s = 0.56 \times 10^{-6}$ m²/s (0.020 ft²/h) be used.

Example 7.2: Annual Variation of Soil Temperature

Plot the annual variation in the soil temperature for Tempe, AZ, for four different soil depths: at the surface, 1 m depth, 2 m depth, and 3 m depth. Refer to Table 7.2.

Given: $\bar{T}_{soil,annual} = 21.1°C$ and $\Delta T_{soil} = 11.1°C$ and $\phi = 0.47$ radians

Assume: $\alpha_s = 0.56 \times 10^{-6}\ \text{m}^2/\text{s} = 2.016 \times 10^{-3}\ \text{m}^2/\text{h}$

Find: $T_{soil}(z, t_h)$ with $z = 0, 1, 2, 3$ m

Solution

The calculation procedure involves the straightforward use of Equations 7.6 through 7.8.

First, we calculate the damping length $d = 2.371$ following Equation 7.8. Next, Equation 7.7 is repeatedly used. For example, for $z = 2$ m and hour $t_h = 288$, we get

$$T_{soil}(2, 288) = 21.1 + 11.1 \times e^{-(2/2.371)}$$

$$\times \sin\left(\frac{2\pi}{8760}288 - 0.47 - \frac{2}{2.371} - \frac{\pi}{2}\right) = 19.0°C$$

The calculations are easily and conveniently done on a spreadsheet. The annual variations for the three different depths are shown in Figure 7.5. As expected, the surface soil temperature varies the most over the year. The soil temperatures at a depth of 3 m still has an amplitude of about 3.5°C and is shifted by about 3 months as compared to the surface soil temperature. The temperature of the deep soil would reach the annual average value of 21.1°C.

7.3.3 ASHRAE Method for Basement Heat Losses

Heat transfer through below-grade floors and walls of residences has been studied by a number of investigators (Bahnfleth and Pedersen, 1990; Krarti, 1987). According

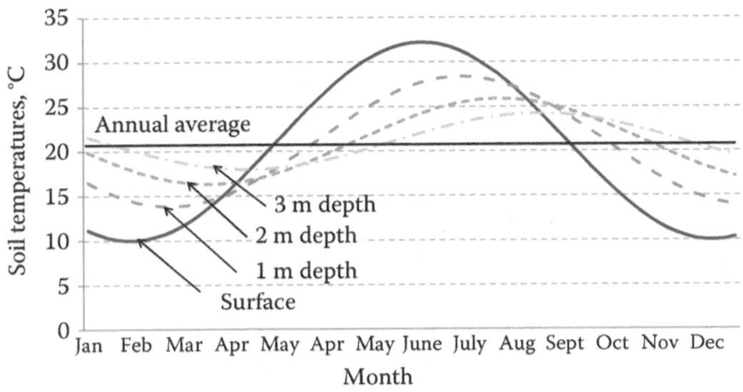

FIGURE 7.5
Annual soil temperatures at different depths for Tempe, AZ.

to Krarti (2012), the percentages of slab-on-grade homes in the United States have increased from 38% in 1991 to over 50% in 2009, residences with basements have declined from a peak of 42% in 1992 to 30% in 2009, and about 20% of all homes built over the last 20 years have crawl spaces. Even though slab and basement losses are often ignored in cooling load calculations for residential commercial buildings (so as to be conservative), it is useful to be cognizant of prevailing professional calculation procedures.

1. Heat transfer from *vertical basement walls* is assumed to be 2-D. Heat flow paths from a basement wall, which when viewed in cross section are circular arcs centered at the intersection of the wall and the earth's surface, as shown in Figure 7.6. The heat loss, finally, occurs to the ambient air. However, this heat loss is quantified using the soil surface temperature which, because of its heat capacity, has a phase shift as discussed in Section 7.3.2.

   The ASHRAE method to determine ground heat losses for hour h of the year counted from January 1 is

$$\dot{Q}_{g,bw}(n) = U_{av,bw} \cdot P_b \cdot h_{bw} \cdot [(T_i(t_h) - T_{soil}(0,t_h))] \quad (7.9)$$

   where
   $U_{av,bw}$ is the average heat transfer coefficient through the basement walls (values assembled in Table 7.3)
   P_b is the building perimeter
   h_{bw} is the height of the basement wall below grade
   T_i is the basement air temperature (which could be conditioned or unconditioned)
   T_{soil} (0, T_h) is the soil or ground surface temperature during hour t_h

   The U value data in Table 7.3 are based on a soil conductivity of 1.38 W/(m·K) [0.8 Btu/(h·ft·°F)]; for other soil conductivity values, use the analytical expressions suggested in ASHRAE Fundamentals (2013). The tables in the online HCB software contain additional data on soil conductivity that are to be used to linearly adjust the values in Table 7.1 for other thermal conductivity k values.

   For peak design day (discussed in Chapter 9), the aforementioned equation can still be used provided T_{soil} (0, t_h) is replaced with the design soil surface temperature $T_{soil,des}$. This latter is determined from $T_{soil,des} = (\bar{T}_{soil,annual} - \Delta T_{soil})$ with $\bar{T}_{soil,annual}$ taken to be the annual average ambient dry-bulb temperature and ΔT_{soil} found from Table 7.2 or Figure 7.4 or following Equation 7.6.

   Note that the aforementioned approach is analogous to the 2-D problem recast into a 1-D conduction equation following the concept of shape factors described in Section 2.8.4.

2. *Basement floor heat losses* are calculated in a similar manner. Table 7.4 gathers the unit heat loss coefficients based on a per-unit basement floor area depending on the shortest width of the basement. To find the basement heat loss, from the table read the unit heat loss and multiply by the floor area and same temperature difference as used for wall losses:

$$\dot{Q}_{g,bf}(t_h) = U_{av,bf} \cdot A_{bf} \cdot [(T_i(t_h) - T_{soil}(0,t_h))] \quad (7.10)$$

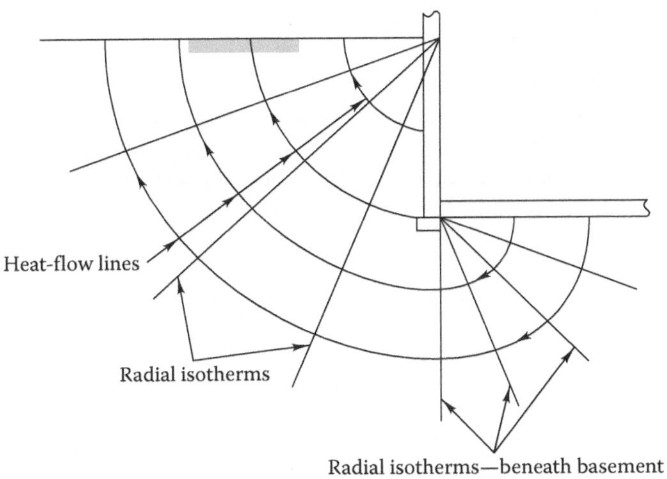

FIGURE 7.6
Radial isotherms for basement heat loss assumed in the ASHRAE method.

TABLE 7.3

Average *U*-Factor for Basement Walls with Uniform Insulation, $U_{av,bw}$ in W/(m²·K) [Btu/(h·ft²·F)]

Depth from Grade, m (ft)	Uninsulated	R-0.88 (R-5)	R-1.76(R-10)	R-2.64 (R-15)
0.3 (1)	2.468 (0.432)	0.769 (0.135)	0.458 (0.080)	0.326 (0.057)
0.6 (2)	1.898 (0.331)	0.689 (0.121)	0.427 (0.075)	0.310 (0.054)
0.9 (3)	1.571 (0.273)	0.628 (0.110)	0.401 (0.070)	0.296 (0.052)
1.2 (4)	1.353 (0.235)	0.579 (0.101)	0.379 (0.066)	0.283 (0.050)
1.5 (5)	1.195 (0.208)	0.539 (0.094)	0.360 (0.063)	0.272 (0.048)
1.8 (6)	1.075 (0.187)	0.505 (0.088)	0.343 (0.060)	0.262 (0.046)
2.1 (7)	0.980 (0.170)	0.476 (0.083)	0.328 (0.057)	0.252 (0.044)
2.4 (8)	0.902 (0.157)	0.450 (0.078)	0.315 (0.055)	0.244 (0.043)

Source: ASHRAE, *Handbook of Fundamentals*, American Society of Heating, Refrigerating and Air-Conditioning Engineers, Atlanta, GA, 2013. Copyright ASHRAE, www.ashrae.org.

The values in the body of the table are heat loss per unit surface area of the basement walls, assuming soil thermal conductivity of 1.38 W/(m·K) [0.8 Btu/(h·ft·°F)]. The *R* values have units of (m²·K)/W [(ft²·h·°F)/Btu].

TABLE 7.4

Average *U*-Factor for Basement Floors (ASHRAE Method), $U_{av,bf}$ in W/(m²·K) [Btu/(h·ft²·h)]

Depth of Floor Below Grade, m (ft)	Shortest Width of Basement, m (ft)			
	6 (20)	7 (24)	8 (28)	9 (32)
0.3 (1 ft)	0.370 (0.064)	0.335 (0.057)	0.307 (0.052)	0.283 (0.047)
0.6 (2)	0.310 (0.054)	0.283 (0.048)	0.261 (0/044)	0.242 (0.040)
0.9 (3)	0.271 (0.047)	0.249 (0.042)	0.230 (0.039)	0.215 (0.036)
1.2 (4)	0.242 (0.042)	0.224 (0.038)	0.208 (0.035)	0.195 (0.033)
1.5 (5)	0.220 (0.038)	0.204 (0.035)	0.190 (0.032)	0.179 (0.030)
1.8 (6)	0.202 (0.035)	0.188 (0.032)	0.176 (0.030)	0.166 (0.028)
2.1 (7)	0.187 (0.032)	0.175 (0.030)	0.164 (0.028)	0.155 (0.026)

Source: ASHRAE, *Handbook of Fundamentals*, American Society of Heating, Refrigerating and Air-Conditioning Engineers, Atlanta, GA, 2013. Copyright ASHRAE, www.ashrae.org.

Soil conductivity is 1.38 W/(m·K) [0.8 Btu/(h·ft·°F)]; the floor is uninsulated.

Example 7.3: Partially Insulated Basement Wall Design Heat Loss

Consider a basement 7 ft deep with dimensions 28 ft × 42 ft (8.5 m × 12.8 m) that is kept at 72°F (22°C). It is insulated to a depth of only 3 ft (0.9 m) with *R*-10 insulation and uninsulated below; find the heat loss through the basement walls and floor when T_{soil} is 50°F (10°C).

Given: Wall depth 7 ft and insulation *R* value 10 (h·ft²·°F)/Btu for the top 3 ft,

basement perimeter P_b = 140 ft, and floor area A_{bf} = 1175 ft²

Assumptions: Steady conduction; building interior temperature T_i = 72°F

Find: $\dot{Q}_g$

Lookup values: For the basement walls, from Table 7.3, $U_{av,bw}$ value for 3 ft depth under the *R*-10 column is 0.07 Btu/(h·ft²·°F). The corresponding values under the uninsulated column for 3 ft and 7 ft are 0.273 Btu/(h·ft²·°F) and 0.170 Btu/(h·ft²·°F).

For the basement floor, from Table 7.4, average *U* value for 7 ft below grade under column 28 ft, $U_{av,bf}$ = 0.028 Btu/(h·ft²·F).

Solution

The total basement unit heat loss is found by summing the average heat losses from the insulated portion and the uninsulated portion of the walls:

$$U_{av,bw} = (3/7)\times 0.07\ \text{Btu/(h}\cdot\text{ft}^2\cdot{}^\circ\text{F)} + (4/7)\times(0.273-0.170)\ \text{Btu/(h}\cdot\text{ft}^2\cdot{}^\circ\text{F)} = 0.089\ \text{Btu/(h}\cdot\text{ft}^2\cdot{}^\circ\text{F)}$$

Then, from Equation 7.9, the basement wall heat loss is

$$\dot{Q}_{bw} = 0.089\ \text{Btu/(h}\cdot\text{ft}^2\cdot{}^\circ\text{F)}\times(140\times 7)\ \text{ft}^2\times(72-50)^\circ\text{F} = 1918.8\ \text{Btu/h}$$

TABLE 7.5

Heat Loss Coefficient F_p of Slab Floor Construction in W/(m·K) [Btu/(h·ft·°F)]

Construction	Insulation	F_p, W/(m·K) [Btu/(h·ft·°F)]
200 mm (8 in.) block wall, brick facing	Uninsulated	1.17 (0.68)
	R-0.95 (R-5.4) from edge to footer	0.86 (0.50)
100 mm (4 in.) block wall, brick facing	Uninsulated	1.45 (0.84)
	R-0.95 (R-5.4) from edge to footer	0.85 (0.49)
Metal stud wall, stucco	Uninsulated	2.07 (1.20)
	R-0.95 (R-5.4) from edge to footer	0.92 (0.53)
Poured concrete wall with heating duct near perimeter[a]	Uninsulated	3.67 (2.12)
	R-0.95 (R-5.4) from edge to footer	1.24 (0.72)

Source: ASHRAE, *Handbook of Fundamentals*, American Society of Heating, Refrigerating and Air-Conditioning Engineers, Atlanta, GA, 2013. Copyright ASHRAE, www.ashrae.org.

Note: The values in the body of the table are heat loss per unit perimeter length, assuming soil thermal conductivity of 1.38 W/(m·K) [0.8 Btu/(h·ft·F)]. The *R* values have units of (m²·K)/W [(ft²·h·°F)/Btu].

[a] Weighted average temperature of heating duct was assumed at 43°C (110°F) during heating season (outdoor air temperature less than 18°C or 65°F).

For the basement floor heat loss, we use Equation 7.10:

$$\dot{Q}_{g,bf} = 0.028\ \text{Btu/(h}\cdot\text{ft}^2\cdot{}^\circ\text{F)} \times (28 \times 42)\text{ft}^2 \times (72 - 50)^\circ\text{F}$$

$$= 724.4\ \text{Btu/h}$$

Finally, the total basement heat loss is 1918.8 + 724.4 = 2643.2 Btu/h

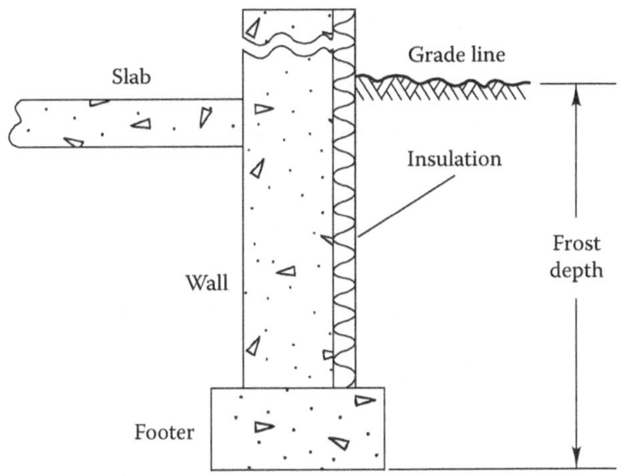

FIGURE 7.7
Schematic diagram of method of insulating slabs and foundations.

7.3.4 ASHRAE Method for Slab on Grade Heat Losses

ASHRAE assumes that heat losses from *slabs on grade* are proportional to the perimeter of the slab, not the area of the slab, as for below-grade slabs. The governing equation based on work in the 1940s at the National Bureau of Standards is as follows:

$$\dot{Q}_{g,slab} = F_p \cdot P_{slab} \cdot (T_i - T_o) \tag{7.11}$$

where the heat loss coefficient F_p is given in Table 7.5, P_{slab} is the perimeter of the slab, and the temperature difference is between the building interior air temperature and exterior air temperature. If slab edge losses are calculated using Equation 7.11, no other slab heat losses need be included.

Note that ground temperatures are not involved since the slab is located close to the grade line. Providing insulation at the edge of a slab will decrease losses significantly. The data in Table 7.5 include this effect for the case of *R*-5.4 insulation applied from the top corner of the slab downward to the frost line. Essentially, the same insulation effect is achieved if one insulates the edge of the slab and beneath the slab, instead of straight down along the foundation, for a distance equal to the frost line depth. Figure 7.7 shows one method of insulating a slab edge. The last row in Table 7.5 has high values of F_p; this indicates that the practice of placing heating ducts in concrete walls is a poor one and should be avoided.

Bahnfleth and Pedersen (1990) have questioned the simple model represented by Equation 7.11. Their work, based on a detailed, transient, 3-D analysis, showed that Equation 7.11 could be in error by up to 50%. Though the time-varying part of the heat loss was well predicted by expressions similar to Equation 7.11, the annual average slab heat loss about which the time variation occurs cannot, however, be expressed accurately in a similar manner. Instead they found that a power law in the slab perimeter-to-area ratio $(P/A)_{slab}$ gave excellent results for the annual mean part of the slab edge heat loss. For the annual part of slab edge loss, these researchers suggest that $(F_p \cdot P_{slab})$ be replaced by a term $c(P/A)^d_{slab}$, where d is reported to be in the range [0.75, 0.90] depending on the ground temperature swing. The improved version of Equation 7.11, yet to be generalized, which is recommended for slab losses at both peak and nonpeak conditions, is

$$\dot{Q}_{g,slab} = \underbrace{c\left(\frac{P}{A}\right)^{d}_{slab} A_{slab}(T_i - T_{soil})}_{\text{Annual average term}}$$

$$\underbrace{+0.13 P_{slab} \Delta T_{soil} \sin\left[(n+\phi)\frac{360}{365}\right] \text{Btu/h}}_{\text{Time-varying term}} \quad (7.12\ \text{IP})$$

where

- the coefficient c has a value of 0.18 in IP units
- n is the day counted from January 1
- ΔT_{soil} is the ground temperature swing or amplitude (read from Table 7.2 or Figure 7.4)
- ϕ is the lag between slab heat loss and day number (about 50 days for the example given by Bahnfleth and Pedersen (1990) for Medford, OR)

7.3.5 True 2-D Methods for Ground Coupling Losses

There are other methods for calculating heat losses from slabs and below-grade basements in residential buildings. Figure 7.8 shows sample results for three annual average conditions in an uninsulated basement following a full 2-D analysis (the problem is symmetric, so only one-half of the basement needs to be shown). In Figure 7.8a, a 2 m deep basement is located in the earth that has a water table of 13°C at 3 m below the surface. Under these summer conditions (21°C earth surface temperature), most heat flow is from the earth's surface to the 13°C water table.

Figure 7.8b shows the same basement under different conditions. Here, the basement is warm; the earth's surface is at an intermediate temperature of 17°C, and the water table remains at 13°C. The isotherms are closely spaced below the basement where most conduction occurs to the water table. In addition, losses occur from the basement wall to the water table. Finally, in Figure 7.8c a coolweather condition exists with the earth's surface at 13°C. In this case, the majority of conductive losses from the wall are to the earth's surface, and the lines of constant flux (normal to the isotherms shown) are approximately circular, as assumed by the ASHRAE method. Hence, the exact 2-D solution supports the use of the ASHRAE method.

The interested reader may refer to Kuehn et al. (1998) who describe a simple closed form solution to the heat transfer problem from basement walls and floors based on research done by Latta and Boileu (1969). The model was found to provide good agreement with detailed computations (Szydlowski and Kuehn, 1981), and so this could potentially be used for load calculations. Chuangchid and Krarti (2000) also developed a simplified design tool in the form of two algebraic equations, one for peak load conditions and the other for annual heat losses, for calculating heat loss from slabs and basements. The equations allow heat losses to be estimated for a wide range of variables such as

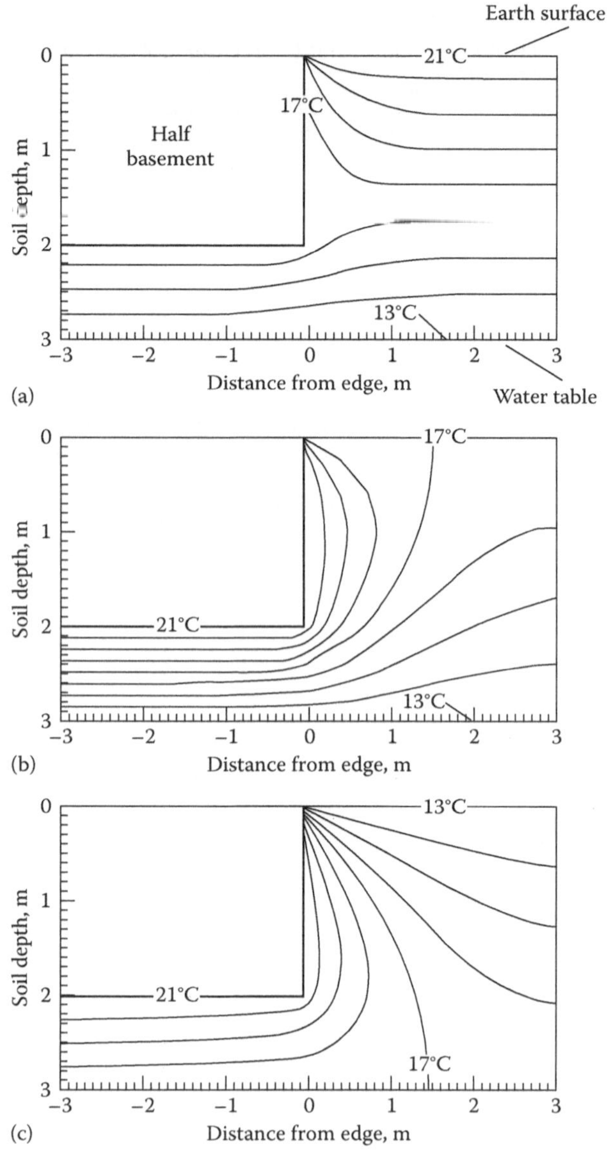

FIGURE 7.8
Isotherms near basement for three cases described in the text, with water table 3m below the surface. and at 13°C. (a) Earth surface temperature of 21°C. (b) Earth surface temperature of 17°C. (c) Earth surface temperature of 13°C. (From Krarti, M., New developments in ground coupling heat transfer, PhD dissertation, University of Colorado, Boulder, CO, 1987.)

foundation size, insulation *R*-values, soil thermal properties, and indoor and outdoor temperatures. A review paper by Zoras (2009) suggests categorizing the various approaches to addressing building earth-contact heat transfer into four groups: analytical/semianalytical, numerical, manual, and design guides. The article then reviews numerous published papers in each of these categories and suggests future developments in the general area.

7.4 Internal Heat Gains

In addition to solar and conductive heat gains, internal heat gains are major contributors especially in low-energy buildings, which affect both cooling and heating loads. Their contribution to summer air conditioning loads is important and needs to be considered carefully. On the other hand, they can partially offset the heating demand of the building or space during winter.

7.4.1 Heat Gains from Occupants

Heat gains from occupants depend on the level of physical activity characterized by the metabolic rate (see Section 3.2). The sensible and latent portions have to be separately considered since they affect the space heat gains differently. The latent load is taken to be an instantaneous cooling load, while the sensible load is partially delayed due to thermal mass of the surrounding space. Further, the convective and radiative portions also need to be treated separately since the convective heat losses are to the space air, while the radiative component affects the surrounding surfaces.

Nominal values of sensible and latent thermal gains for different activity levels are listed in Table 7.6. As the activity level increases, the latent contribution becomes more important (the latent to sensible ratio increasing from 0.5 to 1.7). Also note that the ratio of radiant to sensible portion does not vary much with activity level. The effect of indoor air velocity is a more important factor to the split than is the metabolic rate.

It is instructive to reflect on the origin of this heat gain. The total heat gain must be close to the caloric food intake since most of the energy is dissipated from the body as heat. An average of 100 W corresponds to

$$100\text{ W} = 0.1\,\frac{\text{kJ}}{\text{s}} \times \frac{1\,\text{kcal}}{4.186\text{ kJ}} \times \left(24 \times 3600\,\frac{\text{s}}{\text{day}}\right) = 2064\,\frac{\text{kcal}}{\text{day}}$$

This is indeed a reasonable value compared to the typical food intake (note that the dietician's calorie is really a kilocalorie). The latent heat gain must be equal to the heat of vaporization of the water that is exhaled or transpired. Dividing 30 W (from Table 7.6) by the heat of vaporization of water, we find a water quantity of 30 W/(2450 kJ/kg) = 12.2×10^{-6} kg/s, or about 1.1 kg/24 h. That also appears quite reasonable.

7.4.2 Heat Gains from Lights

Lighting is a major contributor to total energy use in commercial buildings (about 15% on an average—see Figure 1.8b) and, consequently, is a large contributor to the internal loads. There are a large number of lighting systems, the most common being incandescent, fluorescent, gaseous discharge. Recent technological advances in compact fluorescent and light-emitting diodes (LEDs) have resulted in much greater conversion efficiencies; thus, the associated heat gains are much lower than traditional lighting systems. In any case, the available lighting systems have different efficiencies and come in different power ratings. Hence, the heat gains are very specific to the type(s) of lighting used. There are two general ways of estimating instantaneous lighting heat gains, as described here:

1. *Based on installed lighting systems*. The heat gains from all the lighting systems Q_{light} (in W) is estimated as

TABLE 7.6

Nominal Heat Gain Values from Occupants Assuming a Mix of Men and Women

	Total		Sensible		Latent		Radiant to Sensible[a]
Activity	Btu/h	W	Btu/h	W	Btu/h	W	%
Seated at rest	330	95	225	65	105	30	27–60
Seated, light office work	400	115	245	70	155	45	27–60
Standing or walking slowly	450	130	250	75	200	55	38–58
Light physical work	750	220	275	80	475	140	35–49
Heavy physical work	1450	425	580	170	870	255	19–54

[a] The range corresponds to recommended space air velocities (from high to low).

$$\dot{Q}_{light} = P_L \cdot F_{UL} \cdot F_{SA} \tag{7.13}$$

where
- P_L is the total installed lighting wattage estimated by adding the rating of all lamps installed.
- F_{UL} is the lighting use factor (also called "diversity factor"), which is the ratio of the wattage in use to the total installed. It reflects the fact that not all installed lights are in use at the given instant. It is different for different types of buildings; typically, for commercial stores it is unity but is less for office buildings. Further, the use factor can vary with time of day, day of week, and time of year, and so the appropriate value for load estimation needs to be selected with some thought.
- F_{SA} is the special allowance factor that is the actual power consumed by the entire fixture (lamp and ballast) to the rated power consumption. For incandescent lamps, it is unity, while for electronic ballasts, it may be less than unity. For fluorescent fixtures, it can be as high as 2.2, but a value of 1.2 is recommended for general use.

If a space has different types of lighting systems, the individual contributions need to be estimated first and then summed. This needs to be done for each zone separately in case of a multizone building.

2. *Based on space area*. This is a simpler approach especially during the design phase when the final lighting systems have yet to be selected. The approach is simply to assume the maximum allowable lighting power density (LPD) for the space type as stipulated by ASHRAE 90.1 (2010). Some typical values are assembled in Table 7.7. As a rule of thumb, LPDs in office buildings are in the range of around 9.7–13 W/m² (0.9–1.2 W/ft²) for fluorescent lamps and about half of that for LED lighting; for retail stores, an average value is 18 W/m² (1.67 W/ft²).

TABLE 7.7
Lighting Power Density of Some Typical Spaces

Common Space Type	LPD, W/m²	Common Space Type	LPD, W/m²
Office	12	Hospital, emergency	29
Classrooms	15	Convention center, exhibit space	14
Hotel, lobby	12	Restrooms	10
Food preparation	13	Dorms	12
Courtroom	20	Parking garage, garage area	2

Again, it is important to separate the convective and radiative contributions of lighting heat loads, and this aspect depends on type and mount of luminaire and its relative positioning with respect to the space air supply and return registers (discussed in Chapter 9).

7.4.3 Heat Gains from Equipment

Heat gains from equipment such as appliances, motors, computers, and copiers are more subjective, intermittent, and harder to estimate. A conceptual illustration of heat loads generated by a motor-fan assembly is shown in Figure 7.9. The heat gains from equipment operated by motors $\dot{Q}_{equip}$ (in W) can be estimated as

$$\dot{Q}_{equip} = \left(\frac{P_M}{\eta_M}\right) \cdot F_{UM} \cdot F_{LM} \tag{7.14}$$

where
- P_M is the motor power rating
- η_M is the efficiency of motor
- F_{UM} is the motor use factor or fraction of time when operated (it is unity when the motor is operated continuously), which reflects the fact that not all available motors are in use at the given instant
- F_{LM} is the load factor is the fraction of the rated motor delivered under conditions when load is being estimated, which is related to the level of draw down (for example, when computer are idle) and to the part-load efficiency of the equipment

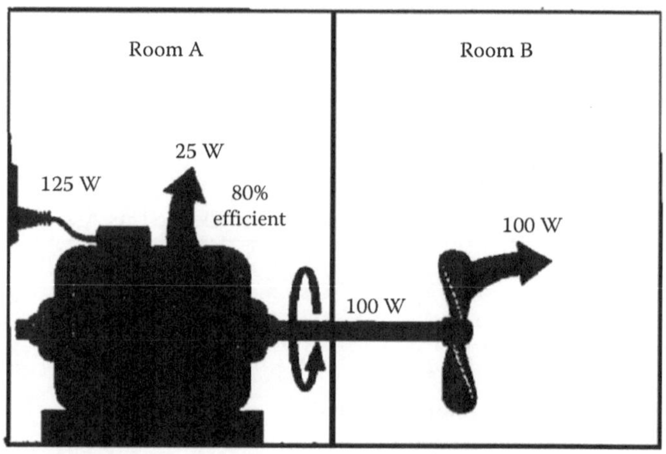

FIGURE 7.9
Heat gains from motor.

For special equipment such as laboratories or kitchens, it is advisable to estimate the heat gains by taking a close look at the inventory of the equipment to be installed, paying attention to the possibility that much of the heat may be drawn directly to the outside by exhaust fans. Such loads should not be considered in the heat gain calculations.

The advent of the computer age has brought a rapid increase in electronic office equipment, and the impact on loads has become quite important, comparable to lighting. The energy consumption for office equipment is uncertain and stochastic: How many of the occupants turn off the computers between uses or keep those running nights and weekends? For lights and for resistive heaters, the nominal power rating (i.e., the rating on the label) is usually close to the power drawn in actual use. But for office equipment, that would be quite misleading; the actual power has been measured to be much lower. Measured diversity factors in office buildings are in the range of 0.37–0.78 with an average of 0.46. Further, many of the equipment may be in standby model for most of the time or may cycle (such as refrigerators). Hence, more discretion is warranted in the use of the various tabular data providing recommendations on thermal gains of typical equipment (such as Table 7.8 for office equipment and Table 7.9 for restaurant equipment).

The interested reader can refer to ASHRAE Fundamentals (2013) or to Hosni et al. (1999) for more detailed information on estimating heat gains from different equipment and to Claridge et al. (2004) for diversity profiles of electricity use in actual office buildings. Typical ranges of average load densities for offices are 0.25–0.75 W/m^2 (2.7–8.0 W/ft^2) based on overall floor area and 1.5–2.0 W/m^2 (16–21.5 W/ft^2) for office cubicles and similar spaces with heavy equipment density.

TABLE 7.8

Typical Heat Gain Rates for Several Kinds of Office Equipment

Equipment	Nameplate Value, W	Heat Gain	
		W	Btu/h
Refrigerator	800	50	171
Desktop computer and monitor	750–1600	70–140	239–478
Laptop computer	50–130	10–35	34–119
Copy machine	1500–2000	500–1000	1706–3412
Laser printer	400–1400	75–130	256–444
Coffee maker			
Sensible	1500	1000	3412
Latent	—	500	1706
Microwave	600	120	409
Vending machine	1000–2000	500–1000	1706–3412

Example 7.4: Total Internal Heat Gains

Consider the building of Example 6.8 that is a rectangular box 12 m × 12 m × 2.5 m (39.4 ft × 39.4 ft × 8.2 ft) (see Figure 6.19). Calculate the total internal gains given the following data.

Given: 10 occupants doing light office work, 3 desktop computers with monitors, 7 laptops, 2 laser printers, 1 photocopier, 1 refrigerator; LPD 12 W/m^2

Find: Total internal heat gains $\dot{Q}_{gains}$

Solution

Such problems are best solved by creating a table using typical values of heat gains from Tables 7.6

TABLE 7.9

Partial List of Recommended Heat Gain for Restaurant Equipment

Appliance Type	Size	Maximum Input Rating		Heat Gain Rate, Btu/h		
				Unhooded		Hooded
		W	Btu/h	Sensible	Latent	Sensible
Barbeque (pressurized), per 5 lb of food capacity	45 lb	470	1,600	550	270	260
Blender, per gallon of capacity	0.25–1.0 gal	1800	6,140	4,060	2080	1980
Cabinet (large hot holding)	16.3–17.3 ft^3	2080	7,100	610	340	290
Cabinet (large hot serving)	37.6–40.5 ft^3	2000	6,820	610	310	280
Cabinet (small hot holding)	3.3–6.5 ft^3	900	3,070	270	140	130
Can opener		170	580	580		0
Coffee brewer	12 cups/2 burners	1660	5,660	3,750	1910	1810
Coffee/hot water holding urn, per gallon of capacity	3.0 gal	460	1,570	580	200	260
Cutter (small)	14 in. bowl	370	1,260	1,260		410
Cutter and mixer (large)	7.5–11.3 gal	3730	12,730	12,730		4060

and 7.8 (since information on diversity factors of various equipment is not specified).

Component	Unit Gain	Total Number/Area	Gain, W
Occupants	95 W	10	950
Desktop computers	110 W	3	330
Laptops	25 W	7	175
Laser printers	100 W	2	200
Photocopier	750 W	1	750
Refrigerator	60 W	1	60
LPD	12 W/m^2	144 m^2	1728
		Total	4193

The internal heat gains are quite large due primarily to lighting and computing equipment, which is typical of modern office buildings. The heat gain due to lighting is 1728 W and that due to all other equipment is 1515 W.

7.5 Treatment of One Zone Spaces

7.5.1 Overall Conductive Heat Transmission Coefficient

The conduction heat flow across the building envelope is usually the largest contributor to the heat loads in a building. As discussed in Chapter 2, heat flows can be assumed to be proportional to the temperature difference when the range of temperatures is sufficiently small; this is usually a good approximation for heat flow across the envelope. Thus, one can calculate the heat flow through each component of the building envelope as the product of its area A, its conductance U, and the difference $(T_i - T_o)$ between the interior and outdoor temperatures. The calculation of U (or its inverse, the Rth value) is described in Chapter 2. Here, we combine the results for the various building components of Figure 7.1 to obtain the total heat flow.

The total conductive heat flow from interior to exterior is

$$\dot{Q}_{cond} = \sum_k U_k A_k (T_i - T_o) \tag{7.15}$$

with the sum running over all parts of the envelope that have a different composition. It is convenient to define a total *conductive heat transmission coefficient* K_{cond}, or *UA* value, as

$$K_{cond} = \sum_k U_k A_k \tag{7.16}$$

so that the conductive heat flow for the typical case of a single interior temperature T_i can be written as

$$\dot{Q}_{cond} = K_{cond}(T_i - T_o) \tag{7.17}$$

In most buildings, the envelope consists of a large number of different parts; the greater the desired accuracy, the greater the amount of detail to be taken into account.

As a simplification, one can consider a few major groups and use effective values for each. The three main groups are glazing, opaque walls, and roof (as stated in Section 7.3, the below-grade heat transfer is usually small except for residences). The reason for distinguishing the wall and the roof lies in the thickness of the insulation: roofs tend to be better insulated because it is easier and less costly to add extra insulation there than in the walls; further, their solar loads are generally higher. If only these three groups are considered, the conductive heat transmission coefficient can be written as

$$K_{cond} = U_{glaz} A_{glaz} + U_{wall} A_{wall} + U_{roof} A_{roof} \tag{7.18}$$

where appropriate effective values are taken for each source. For instance, the value for glazing must be the average over glass and framing, as described in Section 5.3.2. Results for aggregate U values for walls and roofs of typical construction can be found in the online HCB software.

In the energy balance of a building, there is one other term that is proportional to $(T_i - T_o)$. It is the flow of sensible heat [W (Btu/h)] due to air exchange or infiltration:

$$\dot{Q}_{inf,sen} = \rho c_p \dot{V}(T_i - T_o) \tag{7.19}$$

where

ρ is the density of air
c_p is the specific heat of air
$\dot{V}$ is the air exchange rate, m^3/s (ft^3/h)

At standard conditions, 101.3 kPa (14.7 psia) and 20°C (68°F),

$$\rho c_p = 1.2\ \text{kJ/(m}^3\cdot\text{K)}\ [0.018\ \text{Btu/(ft}^3\cdot{}^\circ\text{F)}] \tag{7.20}$$

In IP units, if $\dot{V}$ is in cubic feet per minute, it must be converted to cubic feet per hour by multiplying by 60 (ft^3/h)/(ft^3/min). It is convenient to combine the terms proportional to $(T_i - T_o)$ by defining the total heat transmission coefficient K_{tot} of the building as the sum of conductive and air change terms:

$$K_{tot} = K_{cond} + \rho c_p \dot{V} \tag{7.21}$$

Example 7.5 illustrates the calculation of K_{tot} with typical numbers for a house.

Example 7.5: Total Heat Transmission Coefficient of a Building

Calculate the heat transmission coefficient K_{tot} for the building of Example 6.8 under the following assumptions.

Given: Building of Example 6.8, rectangular box 12 m × 12 m × 2.5 m (39.4 ft × 39.4 ft × 8.2 ft) with flat roof (see Figure 6.19). The insulation is fiberglass with k = 0.06 W/(m·K) [0.10 Btu/(h·ft·°F)], thickness Δx = 0.25 m (10 in.) in the roof, and Δx = 0.15 m (6 in.) in the walls. The windows are double-glazed with

$$U_{glaz} = 3.0\ \mathrm{W/(m^2 \cdot K)[0.53\ Btu/(h \cdot ft^2 \cdot °F)]}$$

and they cover 20% of the sides. The air exchange rate is 0.5 per hour.

Assumptions: For simplicity, treat the opaque surfaces as if they consisted only of the fiberglass, without covering and without the indoor and outdoor surface film coefficients. Neglect thermal bridges due to studs. Assume that all values are independent of temperature and wind.

Find: Heat transmission coefficient K_{tot}

Solution

The calculations are conveniently summarized in this table.

Component	Area A (m²)	Thickness Δx (m)	Conductivity k (W/(m·K))	$U = (k/\Delta x)$ (W/(m²·K))	K_{tot} (W/K)
Roof 12 m × 12 m, flat	144	0.250	0.060	0.24	34.6
Walls (opaque) (height 2.5 m)	96	0.150	0.060	0.40	38.4
Glazing (20% of sides)	24	—	—	3.00	72.0
Air exchanges (0.5/h)	—	—	—	—	60.0
Total					205.0

Thus, the total heat transmission coefficient (conduction plus infiltration) is determined as

$$K_{tot} = 205\ \mathrm{W/K\ [389\ Btu/(h \cdot °F)]}$$

Comments

This example is oversimplified. For greater realism, one can use the U values from the online HCB software. Thermal bridges due to studs in the walls should be included per Example 2.5.

A more refined calculation would take surface heat transfer coefficients into account, as well as details of the construction, using the data from the online HCB software. In practice, such details can take up most of the effort. Example 7.6 shows how the heat loss coefficients vary if the effects of wind on surface heat transfer and infiltration are included.

Example 7.6: Effect of U Value Due to Outdoor Wind Velocity

For the building of Example 7.5, model the infiltration according to the LBNL model of Equation 6.25 and add an outdoor surface heat transfer coefficient (fit to the data in Figure 2.14) on all exterior surfaces:

$$h_o = 10.0 + 3.5\mathrm{v} \quad h_o,\ \mathrm{W/(m^2 \cdot K);\ v,\ m/s} \tag{7.22 SI}$$

$$h_o = 1.8 + 0.19\mathrm{v} \quad h_o,\ \mathrm{Btu/(h \cdot ft^2 \cdot °F);\ v,\ ft/s} \tag{7.22 IP}$$

For simplicity, we neglect h_i for the wall and roof (or assume that it is included in the $(k/\Delta x)$ values of Example 7.5—in any case, it does not vary with v, the variable of interest here).

Given: Leakage area A_{leak} = 449 cm² from Example 6.8; wind and stack coefficients of Example 6.9

Solution

We take the LBNL model of Equation 6.25 for airflow as in Example 6.9 and calculate the U values of the wall and roof according to

$$\frac{1}{U} = \frac{\Delta x}{k} + \frac{1}{h_o}$$

using for $(\Delta x/k)$ the inverse of the values listed for the roof and wall under the $(k/\Delta x)$ column of the table in Example 7.5. For instance, for the walls we have

$$\frac{1}{U} = \frac{1}{0.40} + \frac{1}{10.0 + 3.5\mathrm{v}}\ \mathrm{(m^2 \cdot K)/W}$$

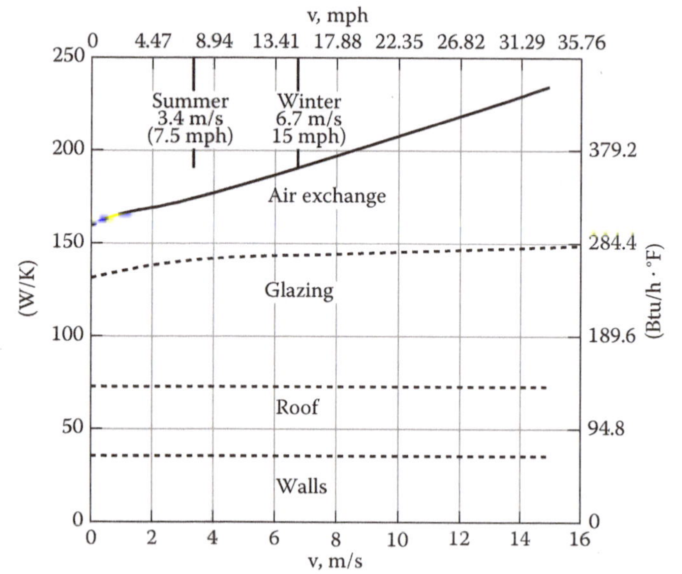

FIGURE 7.10
Variation with wind speed of the heat loss coefficient K_{tot} and its components (indicated by dashed lines) for the house of Example 7.6.

For windows, the U value of 3.0 W/(m²·K) already includes the design surface heat transfer coefficient of h_o = 34 W/(m²·K). Therefore, we calculate the wind dependence of U by first subtracting this contribution and then adding $1/h_o$ at arbitrary wind velocity:

$$\frac{1}{U} = \frac{1}{3.0\ \text{W/(m}^2\cdot\text{K)}} - \frac{1}{34\ \text{W/(m}^2\cdot\text{K)}} + \frac{1}{10.0 + 3.5\text{v}}$$

At h_o = 34 W/(m²·K), we get, as expected, U = 3.0 W/(m²·K).

The result is shown in Figure 7.10. The total heat loss coefficient K_{tot} equals 192 W/K [364 Btu/(h·°F)] at the winter and 176 W/K [334 Btu/(h·°F)] at the summer design values of wind speed. Most of the variation comes from air infiltration. The wall and roof are sufficiently well insulated that h_o has almost no effect.

7.5.2 Heat Balance

Load calculations are straightforward in the static or steady-state limit, i.e., if all inputs are constant. As discussed in Chapter 9, that is usually an acceptable approximation for the calculation of peak heating loads. However, for cooling loads, dynamic effects (i.e., heat storage) must be taken into account because some of the heat gains are absorbed by the mass of the building and do not contribute to the loads until several hours later. Dynamic effects are also important whenever the indoor temperature is allowed to float and are discussed in Chapter 8.

Sometimes, it is appropriate to distinguish several aspects of the load. If the indoor temperature is not constant, the instantaneous load of the space may differ from the rate at which heat is being supplied or removed by the HVAC equipment. The load for the heating or cooling plant is different from the space load if there are significant losses from the distribution system or if part of the air is exhausted to the outside rather than returned to the heating or cooling coil.

The sign convention is that $\dot{Q}$ is positive when there is a heating load and negative when there is a cooling load. In that case, we will add subscripts c and h with the understanding that

$$\dot{Q}_c = -\dot{Q} \quad \text{and} \quad \dot{Q}_h = \dot{Q} \tag{7.23}$$

It is convenient to classify the terms of the steady-state energy balance according to the following groups. The sensible energy terms are

1. Conduction through building envelope other than ground given by Equation 7.17
2. Conduction through floor $\dot{Q}_{floor}$
3. Heat load due to air exchange (infiltration and/or ventilation) given by Equation 7.19
4. Heat gains from solar radiation, from lights, from equipment (appliances, computers, fans, etc.), and from occupants

$$\dot{Q}_{gain} = \dot{Q}_{sol} + \dot{Q}_{light} + \dot{Q}_{equip} + \dot{Q}_{occ} \tag{7.24}$$

Combining the heat loss terms and subtracting the heat gains, one obtains the total *sensible load*

$$\dot{Q}_{sen} = \dot{Q}_{cond} + \dot{Q}_{air} + \dot{Q}_{floor} - \dot{Q}_{gain} \pm \dot{Q}_{stor} \tag{7.25}$$

The term $\dot{Q}_{stor}$ on the right is added to account for storage of heat in the heat capacity of the building (the terms *thermal mass* and *thermal inertia* are also used to designate this effect). A dynamic analysis would include this term; a static analysis neglects it.

The term $\dot{Q}_{floor}$ has been kept as a separate item because it should not be taken as proportional to $(T_i - T_o)$ except in cases like a crawl space, where the floor is in fairly direct contact with outside air. More typical is conduction through massive soil, for which the methods of Section 7.3 are appropriate. In traditional construction, the floor term has usually been small, and often it has been neglected altogether. However, in superinsulated buildings, it can be relatively important.

The total load is the sum of the sensible and the latent loads. The *sensible load* can be expressed in the form

$$\dot{Q}_{sen} = K_{tot}(T_i - T_o) + \dot{Q}_{floor} - \dot{Q}_{gain} \pm \dot{Q}_{stor} \quad (7.26)$$

where K_{tot} is given by Equation 7.21.

The *latent heat gains* are mainly due to air exchange, equipment (such as in the kitchen and bathroom), and occupants. Their sum is

$$\dot{Q}_{lat} = \dot{Q}_{air,\,lat} + \dot{Q}_{lat,\,occ} + \dot{Q}_{lat,\,equip} \quad (7.27)$$

The latent heat gain due to the air exchange is

$$\dot{Q}_{air,\,lat} = \dot{V}\rho h_{fg}(W_o - W_i) \quad (7.28)$$

where

$\dot{V}$ is the volumetric air exchange rate, m^3/s or L/s (ft^3/min)

ρ is the density, kg/m^3 (lb_m/ft^3)

$(\rho \cdot h_{fg})$ = 3010 W/(L/s) [4840 Btu/(h·ft³/min)] at standard conditions

W_i, W_o are the humidity ratios of indoor and outdoor air

During the heating season, the latent gain from air exchange is usually negative (from the sign convention in Equation 7.23) because the outdoor air is relatively dry. A negative $\dot{Q}_{lat}$ implies that the total heating load is greater than the sensible heating load alone—but this is relevant only if there is humidification to maintain the specified humidity ratio W_i. For buildings without humidification, one has no control over W_i, and there is not much point in calculating the latent contribution to the heating load at a fictitious value of W_i.

Example 7.7: Sensible Heating Load

Find the sensible heating load for a house under the following conditions.

Given: Site in Washington, DC

K_{tot} = 186 W/K [353 Btu/(h·°F)]

$\dot{Q}_{gain}$ = 1.0 kW (3.4 kBtu/h)

T_i = 22°C (71.6°F), T_o = −6.3°C (20.7°F)

Assumptions: Steady-state; heat transfer to floor being neglected

Solution

From Equation 7.26

$$\dot{Q}_h = K_{tot}(T_i - T_o) - \dot{Q}_{gain}$$

$$= (186\ \text{W/K})[22 - (-6.3)]\ \text{K} - 1000\ \text{W} = 5264 - 1000\ \text{W}$$

$$= 4.26\ \text{kW}\ (14.55\ \text{kBtu/h})$$

Example 7.8: Latent Heating Load

Estimate the latent load of humidification for the house of Example 7.7 if the air exchange rate is 0.5 per hour and the indoor relative humidity is to be 30% (usually sufficient in winter). What is the uncertainty in the load due to not knowing the humidity of the outdoor air?

Given: Volume of space 360 m³

T_o = −6.3°C (humidity not known)

T_i = 22°C with 30% relative humidity, i.e., indoor air humidity W_i = 0.0048 kg_w/kg_a

The outdoor humidity 0 < W_o < 0.0016 kg_w/kg_a (corresponding to 100% relative humidity)

Air change rate = 0.5 per hour (h^{-1})

Solution

The volumetric airflow rate is

$$\dot{V} = 0.5\ \text{h}^{-1} \times 360\ \text{m}^3 = 180\ \text{m}^3/\text{h} = 50\ \text{L/s}$$

The minimum value of the humidity difference ($W_i - W_o$) is (0.0048 − 0), while the maximum value is (0.0040–0.0016). Thus, the range is 0.0032 < ($W_i - W_o$) < 0.0048. If we assume a mean value of 0.0040, the relative error is (0.0008/0.004) = 0.2. Thus, the uncertainty due to not knowing the outdoor humidity is only 20% at most.

The latent gain at the mean value using Equation 7.28 is

$$\dot{Q}_{infil,lat} = \dot{V}\rho h_{fg}(W_o - W_i) = 50\ \text{L/s} \times 3010\ \text{W/(L/s)}$$

$$\times (-0.004)\ \text{kg}_w/\text{kg}_a = -0.60\ \text{kW}$$

Being negative, it makes a positive contribution to the heating load if one decides to humidify. The sensible load is 4.26 kW from Example 7.7. Thus, the effect of the latent load is quite small in this context because it is only about 14% of the sensible load (= 0.60/4.26) = 0.14.

Example 7.9: Relative Contributions of Various Loads

Consider the building assumed in Examples 7.4 and 7.5, which is a rectangular box 12 m × 12 m × 2.5 m. The individual heat loss coefficients of the various building elements have been determined in Example 7.5, while the internal loads have been determined in Example 7.4. Calculate the contributions of the various heat gains of this building in summer when subject to a temperature difference of 20°C.

Solution

Again, it is most convenient to set up a table of the individual contributions as follows. The contribution of lighting has been listed separately due to its large magnitude.

Component	K_{tot}, W/K	Temperature Difference, °C	Heat Gains, W	Contribution, %
Roof	34.6	20	692	8.3
Walls (opaque)	38.4	20	768	9.3
Glazing	72.0	20	1440	17.4
Air exchange	60.0	20	1200	14.5
Occupants	—	—	950	11.5
Lighting	—	—	1728	20.8
Other equipment	—	—	1515	18.3
Total			8293	100

Note that solar loads through the glazing have been neglected, which would be a large contributor during summer days. In this example, the internal loads (occupants, lights, and equipment) account for over 50% of the total loads. The inefficiency associated with lights and equipment adversely impact building energy use in two ways: the direct cost of electricity and the additional cooling electricity needed to remove their heat gains to the space.

Leaving aside the other equipment contribution, the two components with the largest gains are lighting and glazing. It is no wonder that federal agencies such as the Department of Energy and Environmental Protection Agency in the United States have, during the last three decades, launched numerous R&D as well as energy efficiency deployment programs especially targeted at these two components.

7.5.3 Treatment of Unconditioned Attached Spaces

Buildings also have several unconditioned spaces that are attached thermally to the conditioned space. Examples of these are garages, basements, attics, and crawlspaces. The air temperature within these spaces is to be determined both for load calculations and to ascertain whether it does not reach values either too low or too high to damage the inside furnishings and equipment. The analysis is fairly straightforward and is based on a steady-state thermal balance of the space following Equation 7.26. The approach is illustrated by this simple example.

Example 7.10: Temperature of Unconditioned Attached Space

Consider an unheated garage that has a common wall with the living space as shown in Figure 7.11. Find the air temperature in the garage.

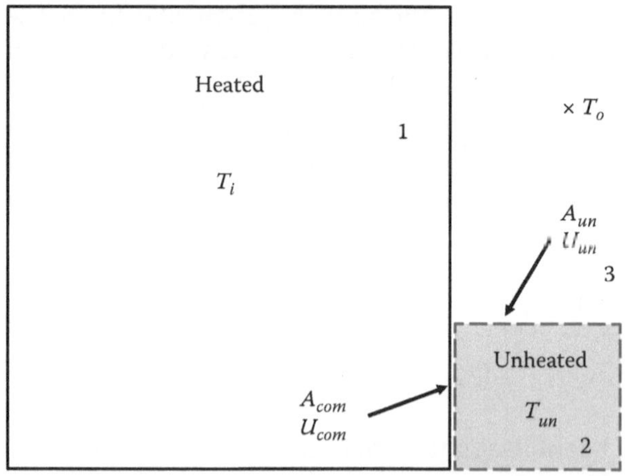

FIGURE 7.11
Sketch for Example 7.11 of an unheated attached garage.

Given: Garage is rectangular with 20 ft × 25 ft sides with a flat roof.

The common wall is 25 ft long and the garage wall height is 8 ft.

The heat loss coefficients are U_{common} = 0.1 Btu/(h·°F·ft²) and $U_{unheated}$ = 0.3 Btu/(h·°F·ft²).

The building interior temperature is T_i = 70°F and the outdoor air temperature is T_o = 30°F.

Figure: Figure 7.11. The interior space, the garage, and the outdoor conditions are represented by states 1, 2, and 3, respectively.

Assumptions: Steady-state conduction; infiltration, solar gains, and heat losses to the ground are neglected

Find: $T_{unheated}$

Solution

Steady-state heat condition implies that heat flow from 1 to 2 should be equal to that from 2 to 3. We now use Equation 7.17 with the aforementioned simplifications to yield the following:

For heat flow from points 1 to 2,

$$\dot{Q}_{1-2} = (UA)_{common}(T_i - T_{unheated})$$

For heat flow from points 2 to 3,

$$\dot{Q}_{2-3} = (UA)_{unheated}(T_{unheated} - T_o)$$

Equating these expressions and solving for the unknown value of $T_{unheated}$ yields

$$T_{unheated} = \frac{(UA)_{common} T_i + (UA)_{unheated} T_o}{(UA)_{common} + (UA)_{unheated}} \quad (7.29)$$

The areas through which heat losses occur are determined as

$$A_{common} = 8 \text{ ft} \times 25 \text{ ft} = 200 \text{ ft}^2$$

$$A_{unheated} = \text{Perimeter} \times \text{height} + A_{roof}$$

$$= (20 + 20 + 25)\text{ ft} \times 8\text{ ft} + 20\text{ ft} \times 25\text{ ft} = 1020\text{ ft}^2$$

Then, from Equation 7.29, we find

$$T_{unheated}$$

$$= [(0.1 \times 200 \times 70) + (0.3 \times 1020 \times 30)] / [(0.1 \times 200) + (0.3 \times 1020)]$$

$$= 32.5°\text{F}$$

Comments

The temperature of the garage is quite close to the outdoor temperature both because it has a much larger area with the outdoors and because of poorer insulation (i.e., higher U value). The heat loss through the garage = 0.1 × 200 × (70 − 32.5) = 759 Btu/h. Compare this to the case when there is no garage, i.e., 0.1 × 200 × (70–30) = 800 Btu/h, a 5% difference. The effect of the garage is to slightly reduce the heat losses from the building.

7.6 Multizoning in Buildings

7.6.1 Need for Zoning

So far, we have considered the interior as a single zone at uniform temperature—a fair approximation for simple houses, for certain buildings without windows (such as warehouses), or for buildings that are dominated by ventilation. But in large or complex buildings, one usually has to calculate the loads separately for a number of different zones. A simple and direct way of determining the number of zones in an existing building is to count the number of thermostats; each zone will have one and only one thermostat controlling its indoor air temperature.

There may be several reasons for zoning. An obvious case is a building where different rooms are maintained at different temperatures, e.g., a house with an attached sunspace. Here, the heat balance equation is written for each zone, in the form of Equation 7.15 but with an additional term

$$\dot{Q}_{j-k} = U_{j-k}A_{j-k}(T_j - T_k) \tag{7.30}$$

for the heat flow between zones j and k.

However, even when the entire building is kept at the same temperature, multizone analysis becomes necessary if the *spatial distribution* of heat gains is too nonuniform. Consider, e.g., a building with large windows on the north and south sides, during a sunny winter day when the gains just balance the total heat loss. Then neither heating nor cooling would be required, according to a one-zone analysis. But how can the heat from the south spaces get to the north spaces of the building?

Heat flow is the product of the heat transfer coefficient and the temperature difference, as in Equation 7.30. Temperature differences between occupied zones are small, usually not more than a few degrees; otherwise, there would be complaints about comfort. The heat transfer coefficients between zones are often not sufficiently large for effective redistribution of heat, especially if there are walls or partitions. This is demonstrated by Example 7.11.

Example 7.11: Interzonal Heat Flows

Consider the loads in adjacent south and north zones of a building on a sunny winter day, assuming for simplicity that the loads are due only to the windows of 10 m² area on each facade. The solar irradiance values are 200 and 10 W/m², respectively. How much heat could flow between these two zones if they are separated by a thin wall with a U value of 4.0 W/(m²·K) and area of 20 m² if the temperature difference is ΔT = 1°C? How does this heat flow compare to the zone loads?

Given: Heat transmission coefficient to outside UA = 3 W/(m²·K)×10 m² = 30 W/K for each of the facades (north and south), outdoor air temperature, T_o = −10°C, solar irradiance on south window is 200 W/m² and that on north window 10 W/m², and indoor temperatures being uniform within each zone, with $T_{i,N}$ = 20°C in the north zone and $T_{i,S}$ = 21°C in the south zone

Solution

The interzone heat transfer coefficient
K_{N-S} = 4.0 W/(m²·K) × 20 m² = 80 W/K

The solar loads are

$$\dot{Q}_{sol,S} = 10\text{ m}^2 \times 200\text{ W/m}^2 = 2000\text{ W on south window}$$

$$\dot{Q}_{sol,N} = 10\text{ m}^2 \times 10\text{ W/m}^2 = 100\text{ W on north window}$$

The zone loads are

$$\text{North zone } \dot{Q}_N = UA(T_i - T_o)_N - \dot{Q}_{sol,N}$$

$$= 30\text{ W/K} \times 30\text{ K} - 100\text{ W}$$

$$= 800\text{ W (heating load)}$$

$$\text{South zone } \dot{Q}_S = UA(T_i - T_o)_S - \dot{Q}_{sol,S}$$

$$= 30\text{ W/K} \times 31\text{ K} - 2000\text{ W}$$

$$= -1070\text{ W (cooling load)}$$

The heat flow between the zones is $\Delta T \times K_{N-S} = 1\ \text{K} \times 80\ \text{W/K} = 80\ \text{W}$. This is small compared to the zone loads, 800 and −1070 W, respectively. Thus, to avoid large temperature differences between both zones, the HVAC system must control the two zones separately.

If there could be perfect heat flow between the zones, the total load would be only 800 − 1070 = −270 W, a small cooling load (which in practice would be reduced to zero by raising T_i a few degrees).

Comments

In this example, the zones are defined by an obvious physical boundary. But even without the partition there would be a problem of unacceptable temperature gradients. Such gradients within a room are more difficult to calculate (and, in this example, we have bypassed the complication by assuming uniformity within each zone). In buildings with air distribution systems, the problem is alleviated to the extent that the system mixes the air between the zones—in effect, enhancing the interzone heat transfer.

We have chosen the numbers in this example somewhat on the extreme side to make a point, but the basic phenomenon is very common: often, the interzone heat transfer is so small that excess heat gains in one zone bring little or no benefit for heating loads in another zone—hence, the thermodynamically perverse fact that many large buildings require simultaneous heating and cooling (described in Section 19.7).

7.6.2 Practical Issues

A zone is a room or a group of rooms in which comfortable conditions can be maintained by a single controlling device. The basic criterion for zoning is the ability to control the comfort conditions; the control is limited by the number of zones one is willing to consider. *To guarantee comfort, the HVAC plant and distribution system must be designed with sufficient capacity to meet the load of each zone.* In choosing the zones for a multizone analysis, the designer should try to match the distribution of heat gains and losses. A common and important division is between interior and perimeter zones, because the interior is not exposed to the changing outdoor environment. Different facades of the perimeter should be considered separately for cooling load calculations, as suggested in Figure 7.12. Corner rooms should be assigned to the facade with which they have the most in common; usually this will be the facade where a corner room has the largest windows. Corner rooms are often the critical rooms in a zone, requiring more heating or cooling (per unit floor area) than single-facade rooms of the same zone.

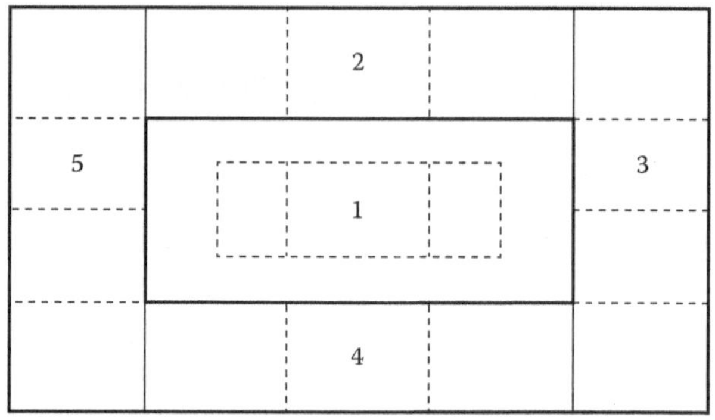

FIGURE 7.12
Example of recommended zoning. Thick lines represent zones, labeled 1 through 5. Dashed lines represent subzones.

Actually, there are different levels to a zoning analysis, corresponding to different levels of the HVAC system. For better comprehension of the discussion here, we anticipate some topics from Chapter 19. In an air system, there are major zones corresponding to each air handler. Within each air handler zone, the air ducts, air outlets, and heating or cooling coils must have sufficient capacity and sufficient controllability to satisfy the loads of each subzone; the design flow rates for each room are scaled according to the design loads of the room. For best comfort (and if cost were no constraint), each zone should have its own air handler and each room its own thermostat. There is a trade-off between equipment cost and achievable comfort, and the best choice depends on the circumstances. If temperature control is critical, one installs separate air handlers for each of the five zones in Figure 7.12 and separate thermostats for each room. To save equipment cost, one often assigns several zones to one air handler and several rooms to one thermostat; but the more divergent the loads, the more problematic the control. For the building of Figure 7.12, a single air handler and five thermostats may be adequate if the distribution of heat gains is fairly uniform and if the envelope is well insulated with good control of solar gains.

Another example is a house whose air distribution system has a single fan (typical of all but the largest houses). Even though there is only one major zone, the detailed design of the distribution system demands some attention to subzones. Within each room, the peak heating capacity should match the peak heat loss. Also, it is advisable to place heat sources close to points with large heat loss, i.e., under windows (unless they are highly insulating).

The choice of zones is not always clear-cut, and the design process may be iterative. Depending on the distribution of gains and losses, one may want to assign several rooms to a zone, one room to a zone, or even several zones to a room (if it is very large). With finer zonal detail, one improves the control of comfort, but at the price of greater calculation effort and higher HVAC system cost. In an open office space, there is no obvious boundary between interior and perimeter; here, a good rule is to make the perimeter zone as deep as the penetration depth of direct solar radiation, typically, a few meters. Spaces connected by open doors, e.g., offices and adjacent hallways, can sometimes be treated as a single zone. Separate zones are advisable for rooms with large computers or energy-intensive equipment. In multistory buildings, one may want to treat the top floor apart from the rest.

The problem of divergent zone loads is one of the prime targets for energy conservation in large buildings (and we will take up the topic again in Chapter 24). The first step is to reduce the loads through the envelope, by improved insulation and control of solar radiation: The smaller the loads, the smaller the differences between the loads. Careful attention must be paid to the choice of its zones. Finally, there is also the possibility of using certain types of HVAC systems that are flexible enough to accommodate zonal heat imbalances such as water loop heat pumps (discussed in Section 14.9).

Problems

Numbers 1–4 given in parenthesis denote the degree of difficulty.

7.1 Find the temperature in the unheated room, shown in Figure P7.1. Some important properties of the space are shown in the table as follows. The unheated room is 8 ft (2.4 m) high. Ignore floor slab losses. *(2)*

Building Component	IP Units: U Value, Btu/(h·ft²·F)	IP Units: Adjacent Air Temp.,°F	SI Units: U Value, W/(m²·K)	SI Units: Adjacent Air Temp.,°C
Exterior wall	0.2	−5	1.13	−21
Ceiling + attic + roof	0.07	−5	0.40	−21
Interior walls	0.06	72	0.34	22

7.2 Repeat Problem 7.1, but include the floor slab losses. Assume that the floor is a slab-on-grade design and consider the full range of loss coefficients given in Table 7.5 (neither the details of wall construction nor the site details are given). Is the ignoring of slab losses in Problem 7.1 justified? *(3)*

7.3 If the house of Example 7.5 has a floor consisting of 4 in. (0.1 m) uninsulated concrete, in direct contact with the ground, how does the total transmission coefficient change? What is the effect of insulating the floor with R-0.95 (R 5.4)? *(2)*

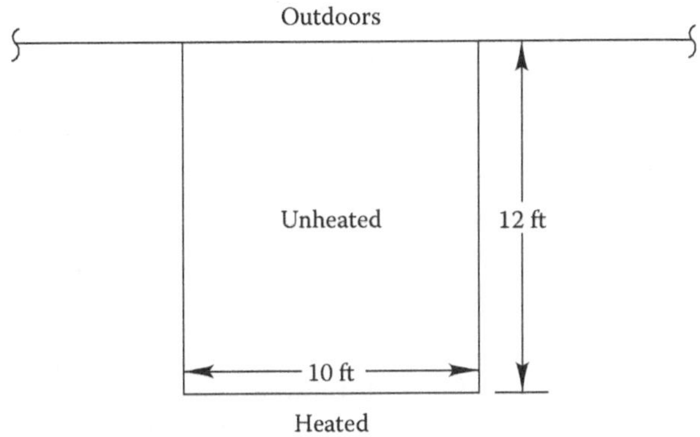

FIGURE P7.1

7.4 Estimate the temperature of the air in an unheated attached garage for these steady-state conditions:

- The dimensions are 5 m × 5 m × 3 m (16.41 ft × 16.41 × 9.85 ft).
- One wall is adjacent to house with U = 0.5 W/(m²·K) [0.088 Btu/(h·ft²·°F)].
- The other walls (including door) and the roof are in contact with the outside and have U = 5 W/(m²·K) [0.88 Btu/(h·ft²·°F)].
- The floor consists of 0.1 m (4 in.) of concrete, and the ground underneath is assumed to have a uniform temperature of 12°C (53.6°F).
- The air exchange rate is 0.1 per hour.
- T_o = −10°C (14°F). *(3)*
- Assume indoor air temperature T_i = 21°C

7.5 The floor of a 100 × 1000 ft (30 × 300 m) warehouse is an uninsulated slab on grade. Find the heat loss from the slab in December and April if the building is located in Denver, CO, using the Bahnfleth and Pederson approach. *(3)*

7.6 A 32 × 100 ft (10 × 30 m), 7 ft deep (2.1 m deep) basement must be maintained at 72°F (22°C) during the winter in Denver, CO. The room above the basement is maintained at 72°F (22°C).

(a) Determine the ground surface temperature at 6 a.m. on January 15.

(b) How much heat is needed if the basement is uninsulated?

(c) If the basement is insulated to R = 10 (R = 1.76)? *(3)*

7.7 Solar radiation of 800 W/m² strikes the flat roof of a room below maintained at 22°C during a day with no wind. The U-value of the roof is

0.25 W/(m^2·°C). Using the concept of sol–air temperature, determine the heat gain through the roof and the roof surface temperature when the outside air temperature is 32°C under the following conditions, assuming h_o = 10 W/(m^2·K):

(a) A dark roof of absorptivity 0.7

(b) A white painted roof of absorptivity 0.25 and long-wave emissivity of 0.6 *(2)*

7.8 Find the sol–air temperature for a vertical surface when the incident solar radiation is 200 Btu/(h·ft^2) (631 W/m^2) and the air temperature 80°F (26.7°C).

(a) Assume α/h_o = 0.15 (h·ft^2·°F)/Btu [0.026 (m^2·K)/W], summer design conditions for light surface.

(b) α/h_o = 0.3 (h·ft^2·°F)/Btu [0.052 (m^2·K)/W], summer design conditions for dark surface. *(2)*

7.9 Find the heat load for a building with total heat loss coefficient K_{tot} = 5000 Btu/(h·°F) (2640 W/K) if T_i = 70°F (21.1°C) and T_o = 0°F (−17.8°C) if one can count on 10 kW of heat gain from equipment that will be left on around the clock. *(2)*

7.10 Find the conductive heat loss coefficient of a one-story building with the following characteristics: dimensions 100 ft × 100 ft × 13 ft (30.48 m × 30.48 m × 3.96 m), steel deck roof with 3.33 in. (0.08 m) insulation (U value of 0.08 Btu/(h·ft^2·°F)), steel siding wall with 4 in. (0.1 m) insulation (U value of 0.066 Btu/(h·ft^2·°F)), window/wall ratio of 0.4, and double-glazed windows with U = 0.50 Btu/(h·ft^2·°F). Neglect heat exchange with the ground. *(2)*

7.11 A wall consists of the following components (order from inside to outside):

0.75 in. (1.8 cm) gypsum plaster, 4.0 in. (0.1 m) glass wool insulation, and 4.0 in. (0.1 m) face brick

(a) Calculate the U value for wind speed of 0 and 15 mph (6.7 m/s).

(b) What is the inside surface temperature of the wall if T_i = 70°F (21.1°C) and T_o = 0°F (−17.8°C)?

(c) Consider qualitatively what would happen if the order of insulation and brick were interchanged. Would the U value change? How would the peak conductive cooling load change? *(3)*

7.12 Wall consists of the following components (order from inside to outside): 0.75 in. of gypsum plaster, 2 in. of foam insulation, and 4 in. of lightweight concrete.

(a) Calculate the U value for wind speeds of 0 and 15 mph.

(b) What is the inside surface temperature of the wall if T_i = 70°F and T_o = 0°F?

(c) Consider qualitatively what would happen if the order of the insulation and concrete were interchanged. Would the U value change? *(3)*

7.13 Two people own a house with a total conductive heat loss coefficient K_{cond} = 120 W/K [227.4 Btu/(h·°F)] and with a ventilation system that is controlled by an occupancy sensor to provide 7.5 L/s (15 ft^3/min) of outdoor air per person for the number of people actually present. They are giving a party for 30 guests. Everyone is so pleased that they are staying until steady-state conditions are established. Assume that the heat gains from other than occupants remain constant at 1 kW (3412 Btu/h).

(a) Estimate the latent and sensible heat gains as well as the outdoor air requirements, before and during the party.

(b) If the outdoor and indoor temperatures are T_o = 0°C (32°F) and T_i = 20°C (68°F), how much does the heating load change relative to steady-state conditions before the party?

(c) Does the latent heat gain have any effect on the heating load if there is no humidity control in the house? *(3)*

7.14 Find the latent cooling load per occupant for sedentary activity if the outdoor airflow rate is 20 ft^3/min per occupant and the outdoor air is 92°F (33.3°C) dry-bulb temperature with 74°F (23.3°C) wet-bulb temperature. Indoor conditions are 70°F (21°C) and 50% relative humidity. *(2)*

7.15 The building considered in Example 7.9 is to be analyzed assuming it to be zoned into an exterior perimeter zone and an interior zone. The exterior zone has a depth of 2 m (6 ft) with 8 occupants, 2 desktop computers, and 5 laptops. The LPD of the exterior zone is 8 W/m^2, while that of the interior zone is 12 W/m^2. Assume that the ventilation air is distributed according to the relative floor areas of the two zones.

(a) Calculate the heat gains/losses of the two zones at different outdoor temperatures ranging from 40°C (104°F) to 0°C (32°F) in decrements of 5°C (9°F) assuming an indoor set point of 21°C (70°F) for heating and 25°C (77°F) for cooling.

(b) Plot these versus outdoor temperature along with the total heat loads of the building.

(c) Calculate and tabulate the contributions of the heat gains and losses due to conduction, ventilation, and internal loads at the two extreme outdoor conditions specified earlier. *(3)*

7.16 A small room contains a horizontal skylight that is 1 m wide by 2.5 m long. The glazing of the skylight is made of single layer of 0.5 mm thick glass

($k = 0.78$ W/m · °C and emissivity of 0.9). The indoor room air temperature is 20°C, while the outdoor air temperature is −8°C and the sky temperature is −30°C. The internal and external convective heat transfer coefficients are given as $h_i = 8.3$ W/m²·K and $h_o = 34$ W/m²·K.

(a) Draw the thermal network diagram.

(b) Calculate the rate of heat loss through the skylight when there is no sunshine.

Assume that the radiation heat transfer from the inside wall surface occurs to the room air temperature, while that of the outside wall occurs to the sky temperature. Neglect resistance offered by the glazing. *(2)*

7.17 You are designing an atrium with single-glazed fenestration, and you are worried about the possibility of the glass cracking when it is heated by the sun and then suddenly hit by cold water from a nearby sprinkler. Estimate the surface temperature of the glazing under the following conditions:

- The outside air temperature is 40°C.
- Solar radiation incident on the glazing is 1000 W/m².
- The inside air temperature is 30°C.
- The U value of the glazing is 6.0 W/(m²·K).
- The solar absorptance of the glazing is 0.5.
- The inside surface convection coefficient is $h_i = 10$ W/(m²·K).
- The outside surface convection coefficient is $h_o = 20$ W/(m²·K). *(3)*

References

ACCA (2006). *Manual J: Residential Load Calculations*, 8th ed., Version 2.00. Approved ANSI Standard, Air Conditioning Contractors of America, Arlington, VA.

ASHRAE 90.1 (2010). *ANSI/ASHRAE/IESNA Standard 90.1-2010: Energy Standard for Buildings Except Low-Rise Residential Buildings*. American Society of Heating, Refrigerating and Air-Conditioning Engineers, Atlanta, GA.

ASHRAE Fundamentals (1997, 2013). *Handbook of Fundamentals*. American Society of Heating, Refrigerating and Air-Conditioning Engineers, Atlanta, GA.

Bahnfleth, W.P. and C.O. Pedersen (1990). A three-dimensional numerical study of slab on grade heat transfer. *ASHRAE Trans.*, 96(pt. 2), 61–72.

Chuangchid, P. and M. Krarti (2000). Parametric analysis and development of a design tool for foundation heat gain for coolers. *ASHRAE Trans.*, 106(pt. 2), 240–250.

Claridge, D. (1986). Building to ground heat transfer, in *Proceedings of American Solar Energy Society Conference*, Boulder, CO, pp. 144–154.

Claridge, D.E., B. Abushakra, J.S. Haberl, and A. Shreshthaputra (2004). Electricity diversity profiles for energy simulation of office buildings (RP-1093). *ASHRAE Trans.*, 110(1), 365–377.

Hosni, M.H., B.W. Jones, and H. Xu (1999). Experimental results for heat gain and radiant/convective split from equipment in buildings. *ASHRAE Trans.*, 105(2), 527–539.

Kusuda, T. and P.R. Achenbach (1965). Earth temperature and thermal diffusivity at selected stations in the United States. *ASHRAE Trans.*, 7(1), 439–447.

Krarti, M. (1987). New developments in ground coupling heat transfer. PhD dissertation, University of Colorado, Boulder, CO.

Krarti, M. (1999). Chapter 6: Building foundation heat transfer, in *Advances in Solar Energy* (Y. Goswami and K. Boer, eds.), CRC Press, Boca Raton, FL, Vol. 13. ASES, pp. 242–308.

Krarti, M. (2012). *Weatherization and Energy Efficiency Improvements for Existing Homes*. CRC Press, Boca Raton, FL.

Kuehn, H.T., J.W. Ramsey, and J.L. Threlkeld (1998). *Thermal Environmental Engineering*, 3rd ed. Prentice Hall, Upper Saddle River, NJ.

Latta, J.K. and G.G. Boileau (1969). Heat losses from home basements. *Canad. Build.*, 19(10), 39–42.

Sterling, R.L., S. Gupta, L.S. Shen, and L.F. Goldberg (1993). Assessment of soil thermal conductivity for use in building design and analysis. Final Research Report, ASHRAE RP-701. American Society of Heating, Refrigerating and Air-Conditioning Engineers, Atlanta, GA.

Szydlowski, R.F. and T.H. Kuehn (1981). Analysis of transient heat loss in earth-sheltered structures. *Underground Space*, 5, 237–246.

Zoras, S. (2009). A review of building earth-contact heat transfer. *Adv. Build. Energy Res.*, 3(1), 289–313.

8

Transient Heat Flow through Building Elements

ABSTRACT In this chapter, we review different mathematical methods to analyze transient heat flows in the building fabric, i.e., elements such as walls and roofs. How these methods are combined to determine building peak or design loads of an entire building or zone is addressed in the next chapter. We start by examining the limitations of a steady-state analysis and the need for dynamic models. Methods for analyzing transient heat conduction through building elements can be classified into continuous and distributed according to how the elements are treated. The partial differential heat equation is an example of the former type of formulation, which can be solved when the initial and boundary conditions are specified. The equations that define the dynamic response of each surface to changes in temperature are then solved incrementally as external temperature conditions change. The distributed methods arrive at a solution by proceeding to discretize the building element into discrete layers. The well-known finite difference approach is one such approach, and this is also described. The time-series formulation is another approach that is more prevalent in building science literature. Both the conduction transfer function (CTF) model and the conduction time-series (CTS) model, fall under this category, and are described. Furthermore, we treat, at some length, the thermal network model, which is a third solution approach. The 1*R*1*C* thermal network is presented and solved as the simplest possible dynamic model. It serves to explain the important concept of the time constant and to estimate the warm-up and cooldown times of a building. We also discuss extensions to more complicated networks. The connection between thermal networks and transfer function models is pointed out, and a solved example illustrates the close agreement in their transient solutions. Finally, the frequency-domain response function method is briefly described.

Nomenclature

A	Area, m^2 (ft^2)
A_n, B_n	Coefficients appearing in Equation 8.4
a_n, b_n	Coefficients appearing in Equations 8.9 and 8.22
b_n, d_n, c_n	Conduction transfer function coefficients appearing in Equation 8.25
CTF	Conduction transfer function model (Equation 8.25)
CTS	Conduction time-series model (Equations 8.30 and 8.31)
C	Capacitor
C	Heat capacity, kJ/K (Btu/°F)
c_n	Conduction time-series coefficients appearing in Equation 8.30
c_p	Specific heat at constant pressure, kJ/(kg·K) [Btu/(lb$_m$·°F)]
$f(t)$	Function of time t
h	Surface heat transfer coefficient combining convection and radiation, W/(m^2·K) [Btu/(h·ft^2·°F)]
K_{tot}	Total heat transmission coefficient of building, W/K [Btu/(h·°F)]
$K(z)$	Transfer function (Equation 8.9)
k	Thermal conductivity, W/(m·K) [Btu/(h·ft·°F)]
L	Length, thickness m (ft)
N	Number of layers used to discretize building element
$\dot{Q}$	Heat rate or heat flow per unit time, W (Btu/h)
$\dot{q}$	Heat flux, W/m^2 [Btu/(h·ft^2)]
R	Resistor
R	Thermal resistance, K/W [(°F·h)/Btu] [$\equiv \Delta T/\dot{Q}$]
$R(t)$	Transient thermal response function
T	Temperature, K or °C (°R or °F)
T_∞	Temperature of air at surface boundary, °C (°F)
t	Time, temporal variable, s or h
U	Overall heat transfer coefficient, W/(m^2·K) [Btu/(h·ft^2·°F)]
$u(t)$	Transient excitation function used in the transfer function method
$V(t)$	Transient excitation function
$y(t)$	Transient response function used in the transfer function method
x	Distance, spatial variable, m (ft)
$\dot{X}$	Time derivative of quantity X
z	Z-transform of a function

Greek

α	Heat diffusivity, m^2/s (ft^2/h)
δ	Time lag
λ	Eigenvalue, decrement factor

ρ Density, kg/m³ (lb_m/ft^3)
τ Time constant, s or h
ϕ_{Bi} Finite difference form of the Biot number (Equation 8.19)
ϕ_{Fo} Finite difference form of the Fourier number (Equation 8.12)
Δt Time interval, time step, s or h
Δx Thickness of building element, m (ft)

Subscripts

con Convection
cond Conduction
des Design
e Envelope
eff Effective
env Envelope
g Glazing
i Interior, inside, indoor
int Internal
m Index for building element node or terms/coefficients in time series
max Maximum
N Number of layers used to discretize building element
n Index for node or terms/coefficients in time series
o Exterior, outside, outdoor
os Sol–air
rad Radiation
r Roof
sol Solar
stor Storage
surf Surface
tot Total
w Wall

8.1 Basic Concepts

8.1.1 Storage Effects and Limits of Static Analysis

In Chapter 2, we presented the basic physical phenomena of heat transfer by conduction, convection, and radiation. Their application to load calculations was discussed in Chapter 7 but was limited to various building heat flows under steady state, or more correctly, quasi-steady-state conditions. A simple static analysis is sufficient for some of the problems the building designer is faced with, but is inadequate in several instances. Building heat loads have to be determined under transient conditions, which arise, for example, due to

1. Thermal mass of the building envelope subject to changes in outdoor ambient air conditions and in solar radiation over the day
2. Transmitted solar radiation, which is first absorbed in the internal mass of the various elements within a room/space (such as slabs, internal walls, and furniture) and released later in time
3. Internal mass interacting with operational schedules causing diurnal fluctuations in internal load (lights, equipment, and occupants)
4. Internal mass interacting with control strategies of the indoor environment (diurnal thermostat setup or setback or variations in supply air volume)

The effect of internal mass storage can be illustrated by considering an extreme example: a building that contains in its interior a large block of solid concrete several meters thick. The conductivity of concrete is relatively low, and diurnal temperature variations do not penetrate deeply into the block; typically, their effects are felt only in the outer layer, to a depth of roughly 0.20 m. Thus, the bulk of the block does not contribute any storage effects on a diurnal time scale. The static heat capacity, defined as the product of the mass and the specific heat, would overestimate the storage potential because it does not take into account the temperature distribution across the block under varying conditions.

To deal with this effect, some people (e.g., Sonderegger, 1977) have used the concept of *effective heat capacity* C_{eff}, defined as the periodic heat flow into and out of a body divided by the temperature swing at the surface. The effective heat capacity depends on the rate of heat transfer and on the frequency. The effective heat capacity is smaller than the static heat capacity, approaching it in the limit of infinite conductivity or infinitely long charging and discharging periods. As a rule of thumb, for diurnal temperature variations, the effective heat capacity of walls, floors, and ceilings is roughly 40%–80% of the static heat capacity, assuming typical construction of buildings in the United States (wood, plaster, or concrete 3–10 cm thick). For items such as furniture that are thin relative to the depth of temperature variations, the effective heat capacity approaches the full static value.

Further complications arise from the fact that the temperatures of different parts of the building are almost never perfectly uniform. As stated earlier, sunlight entering a building is absorbed by the floor, walls, and furniture and raises their temperatures. The air itself does not absorb any appreciable solar radiation and is warmed only indirectly. Thus, the absorbed radiation can cause heat to flow from the building mass to the air even if the air is maintained thermostatically at uniform and constant temperature.

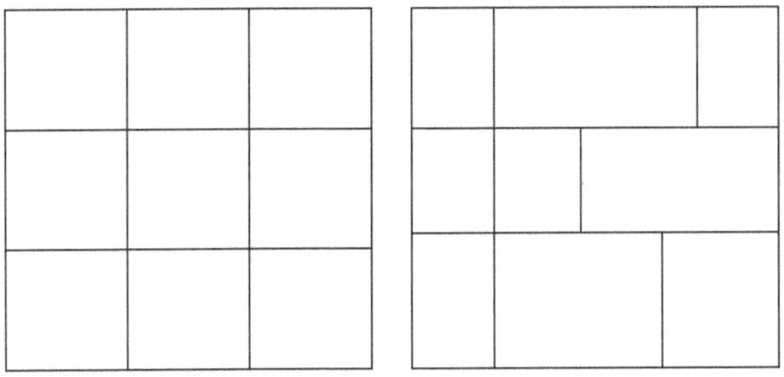

FIGURE 8.1
Variation of schematic floor plan to show that surface area of interior walls is independent of arrangement.

To show that these storage effects can be significant, let us return to the house of Example 7.5 and try to estimate its effective heat capacity. Most of the effective heat capacity lies in the plywood and drywall of the inner surfaces. Glass wool, like most insulation materials, has negligible heat capacity. The outer skin, be it aluminum siding or even brick veneer, does not contribute much because it is only weakly coupled to the interior air. The furniture might be equivalent to one plywood floor. For the interior wall surfaces, let us suppose a floor plan as in Figure 8.1 with nine rectangular rooms (actually, they need not be the same size because the wall surface area depends only on the number of rooms, not on their arrangement). With nine rooms, the number of interior surfaces is 4 for the outer walls and 2 × 4 = 8 for the inner partitions, each contributing 12 m × 2.5 m, for a total of 360 m². Suppose the plywood and the plasterboard are each 1.3 cm thick. Then, the static heat capacity adds up to about 9000 kJ/K, as shown in Table 8.1.

Since these components are fairly thin compared to the penetration depth of diurnal heat pulses, *their effective heat capacity is almost as large as the static one*. We have neglected the wooden support beams in the walls, floor, and roof; their static capacity is about 4500 kJ/K, but they are less directly coupled to the interior, and they might contribute about 2000 kJ/K of dynamic capacity. Thus, the dynamic heat capacity of this type of house is on the order of 11,000 kJ/K. The air accounts for less than 5% and is usually ignored.

A brick or concrete floor would increase the heat capacity significantly. Concrete has a ρc_p value around 1500 kJ/(m³·K). If the concrete floor is 0.1 m thick, its static heat capacity is 144 m² × 0.1 m × 1,500 kJ/(m³·K) = 21,600 kJ/K. Since it is fairly thick, we will assume its effective heat capacity to be around one-half that value. The same house as before with the plywood floor replaced with a concrete one will have an effective heat capacity of about 11,000 − 1,112 + 21,600/2 = 20,688 kJ/K, which is about twice that of the room with plywood floor.

To get an idea of the importance of the heat capacity, suppose that during the course of a summer afternoon the average temperature of the mass increases by 2 K in 5 h, implying a rate of temperature rise of

$$\dot{T} = \frac{2\,\text{K}}{5\,\text{h}} = 0.4\,\text{K/h}$$

This can happen even when the thermostat is set to maintain the air at constant temperature: Under dynamic conditions, air and room mass can have different temperatures. Assuming an effective heat capacity C_{eff} = 11,000 kJ/K, the storage term in the heat balance is

$$\dot{Q}_{stor} = C_{eff}\dot{T} = 11{,}000\,\text{kJ/K} \times 0.4\,\text{K/h} \approx 1.2\,\text{kW}$$

For summer design conditions, the temperature difference $T_i - T_o$ is around 25°C–35°C = −10°C, and the corresponding steady-state conductive heat loss (with K_{tot} = 205 W/K from Example 7.5) is

$$K_{tot}(T_i - T_o) = -2.05\,\text{kW}$$

with a minus sign because it is a heat gain of the building. If there were no storage, the cooling load would be 2.05 kW (for simplicity, we assume that there are no other terms in the thermal balance). But 1.2 kW of this heat gain is soaked up by the storage, and thus the effective

TABLE 8.1
Estimate of Static Heat Capacity for the House of Example 7.5, Not Counting the Wooden Support Beams in Walls, Floor, and Roof

Component	Area A m²	Thickness ΔX m	Volume $V = A\cdot\Delta X$ m³	Specific Heat c_p kJ/(kg·K)	Density ρ kg/m³	ρc_p kJ/(m³·K)	Heat Capacity $C = V\rho c_p$ kJ/K
Drywall, roof	144	0.013	1.87	1.20	800	960.0	1797
Drywall, walls	360	0.013	4.68	1.20	800	960.0	4493
Plywood, floor	144	0.013	1.87	1.09	545	594.1	1112
Furniture (same as floor)							1112
Air	144	2.50	360.0	1.00	1.20	1.20	432
Total (without floor slab)							8946

Material properties from the Online HCB software.

cooling load at this moment is only 2.05–1.2 = 0.85 kW. Under these conditions, the storage effect reduces the cooling load by one-half. The stored heat gain does not appear as load until later: there is a time lag, and this is caused by *thermal inertia* (or simply referred to as *thermal mass*).

This example explains how the heat capacity can cause a time lag in the cooling load and can reduce the peak load. This is particularly important in commercial buildings, where concrete floor slabs are quite common and where a time lag of several hours can shift much of the load past the hours of occupancy, to a time when temperature control is no longer critical.

8.1.2 One-Dimensional Transient Heat Flow Equation

In our present context, a *model* is a mathematical relationship that consists of three components: (1) *driving terms or input variables* (also referred to as perturbations or exciting or forcing functions), (2) *system structure and parameters/properties* that provide the necessary physical description of the systems in terms of physical and material constants; for example, thermal mass, overall heat transfer coefficients, mechanical properties of the elements; and (3) *response function or output variables* that describe system response to the driving terms.

The fundamental equation describing transient heat flow in single homogeneous elements such as slabs, walls, and roofs is well known and can be found in most heat transfer textbooks. Only a brief review is provided here. Most of the analysis methods assume heat flow to be one-dimensional (1-D) occurring only in the direction perpendicular to the exposed surfaces. Consider the wall shown in Figure 8.2 of thickness ΔX and the spatial and temporal dimensions represented by x and t. Hence, the intent is to predict the temperature distribution $T_{wall}(x,t)$ and heat flux distribution $\dot{q}(x,t)$ given a set of initial and boundary conditions. Over the relatively small range of temperatures encountered in buildings, the thermal conductivities can be assumed to be constant. At any section within the wall:

Change of temperature with distance:

$$\frac{\partial T_{wall}(x,t)}{\partial x} = -\frac{1}{k}\dot{q}(x,t) \tag{8.1}$$

Change of heat flux with distance:

$$\frac{\partial \dot{q}(x,t)}{\partial x} = -\rho c_p \frac{\partial T_{wall}(x,t)}{\partial t} \tag{8.2}$$

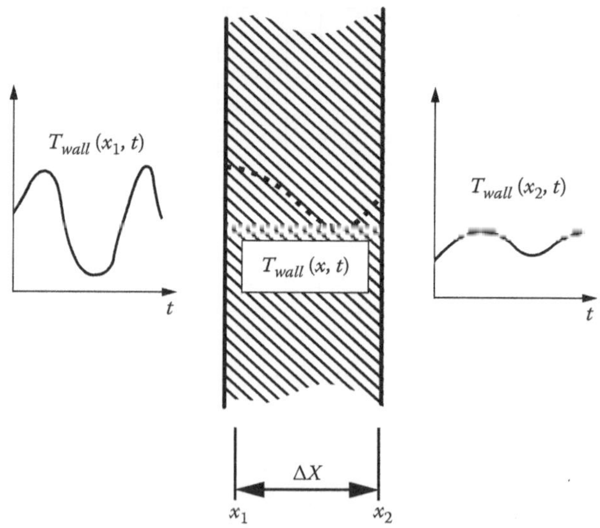

FIGURE 8.2
Illustration of boundary conditions and temperature distribution across a homogeneous wall.

where
- k is the thermal conductivity
- ρ is the density
- c_p is the specific heat at constant pressure

Combining these two equations results in the 1-D Fourier transient heat conduction equation:

$$\frac{\partial T_{wall}(x,t)}{\partial t} = \alpha \frac{\partial^2 T_{wall}(x,t)}{\partial x^2} \tag{8.3}$$

where heat diffusivity $\alpha = k/(\rho c_p)$.

The initial condition specifies the wall temperature distribution at the onset, that is, $T_{wall}(x,0)$, while the two boundary conditions specify $T_{wall}(x_1,t)$ or $\dot{q}(x_1,t)$ and $T_{wall}(x_2,t)$ or $\dot{q}(x_2,t)$. Thus, either the surface temperatures or the surface heat fluxes have to be specified and the solution of this partial differential equation yields the other quantities.

Under steady-state conditions, $(\partial T_{wall}/\partial t) = 0$, which yields $[d^2T_{wall}(x)/dx^2] = 0$ from Equation 8.3. The solution takes the form $T_{wall}(x) = a + bx$, where the two coefficients are determined from the boundary conditions. If the specified temperatures on the two sides of the wall of thickness $\Delta x = L$ are T_1 and T_2, then $a = T_1$ and $b = (T_2 - T_1)/L$. Finally, the steady-state heat flow is given by

$$\dot{q} = -k\left(\frac{dT_{wall}}{dt}\right) = -k \cdot b = -\left(\frac{k}{L}\right)(T_2 - T_1)$$

which is akin to Equation 2.38.

The separation of variables approach (e.g., see Carslaw and Jaeger, 1986) is the general method of solving Equation 8.3. It involves expressing the solution for the temperature as a product of two separate functions, one of space only and the other of time only, which yields two ordinary differential equations. The solution for the temperature distribution across a wall is given as the sum of the steady-state solution plus two infinite series of exponentials:

$$T_{wall}(x,t) = A_0 + B_0 t + \sum_{n=1}^{\infty}[A_n \cos(\lambda_n x) + B_n \sin(\lambda_n x)] \cdot \exp\left(-\lambda_n^2 \cdot \alpha \cdot t\right) \quad (8.4)$$

where
λ is the eigenvalue
the coefficients A_n and B_n are determined from the boundary and initial conditions

The heat flux at the boundaries can then be determined following Equation 8.2.

Note that $n = 0$ yields the steady-state solution. This equation includes a term that represents the amplitude of fluctuations gradually decreasing with depth, a term representing the fact that the peaks decrease with depth, and another to denote that at a given depth x, temperature fluctuates up and down in time. Equation 8.4 for temperature distribution along the wall thickness and over time is general and exact. However, it is quite cumbersome to solve given the infinite Fourier series terms. Fortunately, only a few terms of the infinite series are needed to yield accurate results for building elements subject to periodic excitations.

From a practical viewpoint, it is only the temperature and heat fluxes at the two wall boundaries that are of interest and not the temperature distribution along the wall thickness. A number of solution techniques have been developed for certain special cases that allow their practical application. Some of them are discussed in this chapter.

Also, building elements are usually heterogeneous, and further the temperature of the air films on either side of the wall will be specified and not the wall surface temperatures. Hence, the convective/radiative effects at the wall surfaces have to be included in the framing of the boundary conditions. For the general case, wherein the external side of the wall is subject to the combined effect of solar radiation and ambient air temperature captured by the sol–air temperature T_{os} and when interior air temperature is held constant, we have

Exterior boundary condition:

$$\dot{q}(0,t) = -k\frac{dT_{wall}(0,t)}{dx} = h_o[T_{os}(t) - T_{wall}(0,t)] \quad (8.5)$$

Interior boundary condition:

$$\dot{q}(L,t) = -k\frac{dT_{wall}(L,t)}{dx} = h_i[T_{wall}(L,t) - T_i(t)] \quad (8.6)$$

A conceptual illustration of how the temperature profiles inside a wall change over a day is provided in Figure 8.3 during which the outdoor temperature T_o swings from being higher to lower than the indoor temperature T_i, for an unconditioned space (Santamouris and Asimakopoulos, 1996). Initially, the wall has a uniform

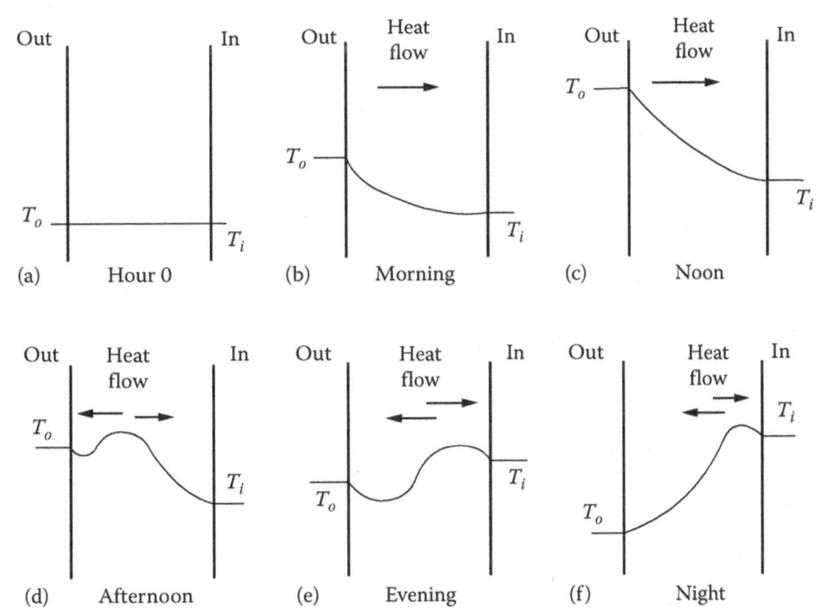

FIGURE 8.3
Diurnal temperature distribution and heat flows through a building element. (From Santamouris, M. and Asimakopoulos, D., *Passive Cooling of Buildings*, James and James (Science Publisher) Ltd., London, U.K., 1996.)

temperature with both the outdoor and indoor air temperatures being equal (insert a). As T_o increases, the outside surface temperature also increases and heat transfer takes place as shown in inserts (b) and (c). In the afternoon, T_o starts to reduce, and the temperature profile within the wall experiences a peak (as shown in insert d), which results in heat flow occurring outward to both the surfaces. With further decreases in T_o, the heat wave propagates through the wall thickness depending on material thermal properties (inserts e and f) with heat flows occurring outward. How the thermal wave propagates across the wall thickness for different types of boundary conditions and different material properties is the transient thermal response one wishes to predict.

8.1.3 Properties of Linear Systems

A system is said to be *linear* if and only if it has the following property: if an input $x_1(t)$ produces an output $y_1(t)$, and if an input $x_2(t)$ produces an output $y_2(t)$, then an input $[c_1x_1(t) + c_2x_2(t)]$ produces an output $[c_1y_1(t) + c_2y_2(t)]$ for all pairs of inputs $x_1(t)$ and $x_2(t)$ and all pairs of real number constants a_1 and a_2. This concept is illustrated in Figure 8.4.

An equivalent concept is the *principle of superposition*, which states that the response of a linear system due to several inputs acting simultaneously is equal to the sum of the responses of each input acting alone. This is an extremely important concept since it allows the response of a complex system to be determined more simply by decomposing the input driving function into simpler terms, solving the equation for each term separately, and then summing the individual responses to obtain the desired aggregated response. A model for this behavior is provided by the convolution theorem

$$f(t) = \int_{-\infty}^{\infty} f_1(\tau) \cdot f_2(t-\tau)d\tau \quad (8.7)$$

where f_1 and f_2 are two functions of time t.

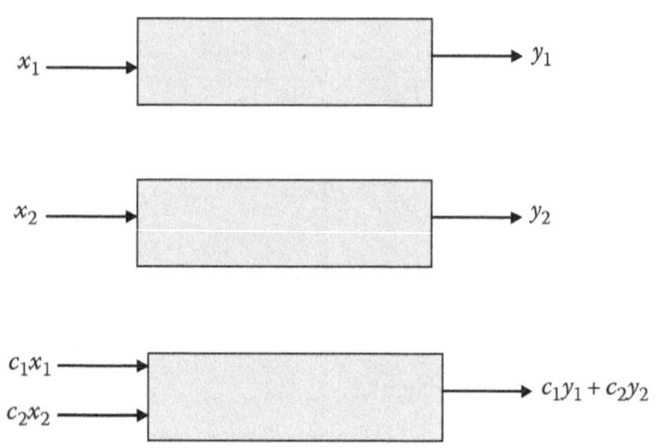

FIGURE 8.4
Principle of superposition of a linear system.

It is more common to use the discretized form that is an infinite sum of the products of the response R of the system τ hours after a unit excitation V is applied to the system

$$f(t) = \sum_{\tau=1}^{\infty} R(\tau) \cdot V(t-\tau) \quad (8.8)$$

For practical implementation, Stephenson and Mitalas (1971) introduced a further transform, called the z-transform, which assumes that the excitation and response functions $V(t)$ and $R(t)$ are sampled at regular time intervals Δ. For example, the output is a train of pulses that can be expressed using the Laplace transform in s-domain as $R(0) + R(\Delta)\cdot e^{-s\cdot\Delta} + R(2\Delta)\cdot e^{-2s\cdot\Delta} + \cdots$. If we substitute $e^{s\cdot\Delta}$ by z, then the output response is $R(0)\cdot z^0 + R(\Delta)\cdot z^{-1} + R(2\Delta)\cdot z^{-2} + \cdots$. The polynomial is the z-transform of the function $R(t)$. The ratio of the excitation and response functions is called the *transfer function* given by

$$K(z) = \frac{R(z)}{V(z)} = \frac{a_0 + a_1z^{-1} + a_2z^{-2} + \cdots}{b_0 + b_1z^{-1} + b_2z^{-2} + \cdots} \quad (8.9)$$

where a and b are the coefficients of the output and input z-transforms. These coefficients are uniquely dependent on the system and not on the boundary or forcing functions. How to calculate these coefficients for actual building elements was also illustrated by Stephenson and Mitalas (1971). Subsequent researchers (e.g., Hittle and Bishop, 1983; Seem et al., 1989) improved and refined the calculation algorithms. The concepts in this section form the basis of the response factor and transfer function methods widely used in North America (discussed further in Section 8.3).

8.1.4 Classes of Analysis Methods

First, one needs to distinguish between *static models* and *dynamic models*. The former lead to steady-state solutions wherein there is no time variation in its input variables, and hence, no change in the output variable as well (as discussed in Chapter 2). Dynamic models, on the other hand, are those that include thermal mass effects and allow transient element or system behavior to be captured with explicit recognition of the time-varying behavior of both output and input variables. The system output or response depends on values of the input variables lagged over time. Thus, the current response, called transient solution, depends on the input/output history of the system. The solution contains time-lagged values of either the forcing functions or the response variable or both. We also distinguish another type of transient solution; a *dynamic-periodic solution* is one when the element or system is subject to a repetitive periodic forcing function (normally a diurnal cycle) and a sufficient number of cycles have elapsed by which time the effect of the assumed initial temperature distribution within the element or system has died out.

One can also distinguish an intermediate case, referred to as *quasi-static solution*. Cases arise when the input variables (such as incident solar radiation on a wall) are constantly changing at a short time scale or high frequency (say, at the minute scale) while it suffices to predict thermal output at say hourly intervals. The dynamic behavior of the wall is poorly predicted at such high-frequency time scales, and so the input variables can be "time-averaged" so as to make them constant during a specific hourly interval. This is akin to introducing a "lowpass filter" for the inputs. Thus, the use of quasi-static models allows one to predict the system output(s) in discrete time-variant steps or intervals during a given day with the system inputs averaged (or summed) during each of the time intervals and fed into the model (as discussed in Chapter 7).

Methods for analyzing transient heat conduction through building elements (homogeneous walls and roofs) may be classified into two broad categories depending on how the elements are considered: (1) *continuous* requiring analytical solutions of elements subject to periodic and step function disturbances or numerical methods (such as finite differences), which are the most flexible but require relatively high computing times; and (2) *distributed*, which require methods such as transfer function (and response factors) and thermal network models, which provide a framework that is general, yet conceptually simple for the modeling of building elements and also of entire spaces and buildings. These methods are discussed later in this chapter.

8.2 Numerical Methods: Finite Differences

Analytical solutions for simple geometries such as solid walls, cylinders, and spheres are given in standard heat transfer textbooks (e.g., Incropera et al., 2007). The excitation functions or boundary conditions are specified as either abrupt step changes or as periodic variations. Time-dependent temperature distributions and instantaneous heat fluxes can be conveniently determined from dimensionless charts. However, these methods are restricted to a relatively small number of geometries and types of boundary conditions.

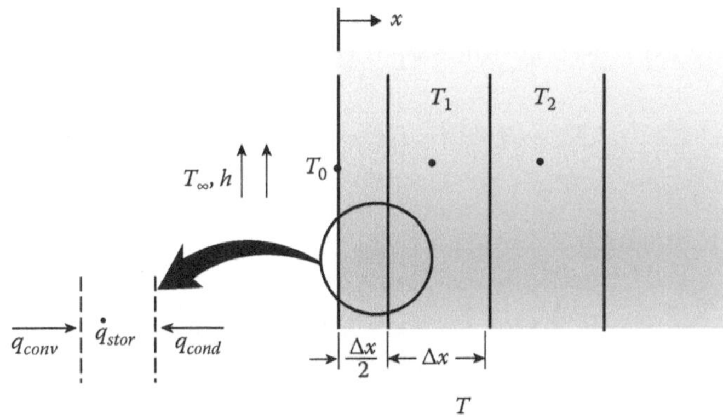

FIGURE 8.5
Discretization of surface and internal nodes of a 1-D conduction problem.

A very flexible and widely used numerical method for solving differential equations with any type of boundary condition is the finite difference approach. It involves discretizing the derivatives of the temperature profile both spatially and temporally. The solid, or in our specific case, the building wall assumed to be a homogeneous element with constant conductivity, is divided into a number of finite layers, as shown in Figure 8.5. Each of the layers or control volume elements is assumed to be at a uniform temperature lumped at its center, that is, all the mass of the wall element is located at the node. The boundary or surface nodes are assumed to be half the control volume of the interior nodes as shown. An energy balance on node m (assuming unit cross-sectional area) over time increment Δt yields

For an interior node m, heat flowing in from node $(m - 1)$ + heat flowing out to $(m + 1)$ = change in heat capacity of node m, or

$$k\frac{(T_{m-1} - T_m)}{\Delta x} + k\frac{(T_{m-1} - T_m)}{\Delta x} = \rho \cdot c_p \cdot \Delta x \frac{\left(T_m^{t+\Delta t} - T_m^t\right)}{\Delta t} \tag{8.10}$$

which simplifies to

$$T_{m-1} - 2 \times T_m + T_{m+1} = \frac{1}{\phi_{Fo}}\left(T_m^{t+\Delta t} - T_m^t\right) \tag{8.11}$$

where

$$\phi_{Fo} = \alpha \cdot \frac{\Delta t}{\Delta x^2} \tag{8.12}$$

and is called the finite difference form of the *Fourier number*.

We now need to solve this equation over time. The simplest is the *explicit finite difference* formulation (also called "forward differencing" scheme) wherein the temperature terms on the left-hand side of Equation 8.11 are expressed at time step t. We then get

$$T_{m-1}^{t} - 2\times T_{m}^{t} + T_{m+1}^{t} = \frac{1}{\phi_{Fo}}\left(T_{m}^{t+\Delta t} - T_{m}^{t}\right) \quad (8.13)$$

from which

$$T_{m}^{t+\Delta t} = \phi_{Fo}\left(T_{m-1}^{t} + T_{m+1}^{t}\right) + (1 - 2\times\phi_{Fo})T_{m}^{t} \quad (8.14)$$

Since the initial conditions $T_{m}^{t=0}$ for all nodes m will be specified, $T_{m}^{t+\Delta t}$ can be determined explicitly for each successive time increment Δt. The smaller the time step, the more accurate the solution. However, computing time increases greatly and so it is advantageous to use as large a time step as possible. Though the explicit finite difference method is easy to use, there is a restriction to the maximum time step that can be selected because of instability issues. The maximum time step is given by the stability criterion, which states that Δt must be selected such that the coefficient associated with the node of interest at the previous time be greater or equal to zero. This implies that for 1-D heat conduction, the term $(1–2 \times \phi_{Fo}) \geq 0$, i.e., $\phi_{Fo} \leq 1/2$ for all interior nodes. This criterion can be alternatively expressed as

$$\Delta t_{\max} \leq \frac{1}{2}\frac{\Delta x^2}{\alpha} \quad (8.15)$$

For example, for a brick wall $\alpha = 0.45 \times 10^{-6}$ m²/s. If a mesh size of 0.01 m is selected, then the simulation time step should be ≤ 111 s = 1.85 min. If a mesh size twice that of the previous value is selected, then the step size can be 4 times greater (≤444 s = 7.4 min), and so on. Usually, it is suggested that a mesh size on the larger side be selected from engineering judgment and simulations performed. Next, the mesh size is reduced by half and simulations redone in order to study differences in the two sets of simulation runs. If the differences are significant, this process is repeated till two successive solutions show little or no difference.

The boundary nodes that are subject to convective and/or radiative heat exchanges have to be analyzed differently than the interior nodes. They are taken to contain half the control volume compared to the interior nodes. The heat balance for such convection nodes, for say node 0, which corresponds to the surface temperature of the wall exposed to an air temperature T_∞:

For an exterior or surface node, at $x = 0$,

$$h_o(T_\infty - T_0) + k\frac{dT}{dx} = \left(\frac{\rho\cdot c_p}{2}\right)\Delta x\frac{dT_0}{dt} \quad (8.16)$$

On forward differencing, we get

$$\left(\frac{\rho\cdot c_p}{2}\right)\Delta x\frac{\left(T_0^{t+1} - T_0^{t}\right)}{\Delta t} = h_o\left(T_\infty^{t} - T_0^{t}\right) + \frac{k}{\Delta x}\left(T_1^{t} - T_0^{t}\right) \quad (8.17)$$

Finally, on rearranging,

$$T_0^{t+1} = \frac{2h_o\Delta t}{\rho\cdot c_p\cdot\Delta x}\left(T_\infty^{t} - T_0^{t}\right) + \frac{2k\Delta t}{\rho\cdot c_p\cdot\Delta x^2}\left(T_1^{t} - T_0^{t}\right) + T_0^{t}$$

$$= 2\phi_{Bi}\phi_{Fo}\left(T_\infty^{t} - T_0^{t}\right) + 2\phi_{Fo}\left(T_1^{t} - T_0^{t}\right) + T_0^{t}$$

$$= 2\phi_{Fo}\left(T_1^{t} + \phi_{Bi}T_\infty^{t}\right) + (1 - 2\phi_{Fo} - 2\phi_{Bi}\phi_{Fo})T_0^{t} \quad (8.18)$$

where the finite difference form of the *Biot number:*

$$\phi_{Bi} = \frac{h_o\Delta x}{k} \quad (8.19)$$

As before, the stability criterion requires that the coefficient for T_0^t be greater than or equal to zero, that is,

$$(1 - 2\phi_{Fo} - 2\phi_{Bi}\phi_{Fo}) \geq 0 \quad \text{i.e., } \phi_{Fo}(1 + \phi_{Bi}) \leq \frac{1}{2}$$

$$\text{i.e., } \Delta t \leq \frac{\Delta x^2}{\alpha}\cdot\frac{1}{2(1 + \phi_{Bi})} \quad (8.20)$$

During an actual simulation, one has to select the more stringent (i.e., smaller) time step of either the interior node or the boundary node. Usually, it is the convective time step of the boundary node given by Equation 8.20 that is more stringent than that of the conductive inner nodes (Equation 8.15).

An alternative scheme that does not suffer from instability issues is the *implicit solution method* (also called "backward differencing" scheme), which is unconditionally stable and any time step can be chosen. In this case, Equation 8.13 is expressed as

$$T_{m-1}^{t+1} - 2\times T_{m}^{t+1} + T_{m+1}^{t+1} = \frac{1}{\phi_{Fo}}\left(T_{m}^{t+\Delta t} - T_{m}^{t}\right) \quad (8.21)$$

Another scheme that provides greater accuracy, called the "central differencing" scheme, is also widely used. In all cases, smaller the time step, the more accurate the solution. The drawback of the backward and central differencing schemes is that they lead to a set of simultaneous algebraic equations of order equal to that of the number of nodes that need to be solved at each time step. The treatment here corresponds to the 1-D problem, but the same approach can be extended to 2-D and

3-D problems. The interested reader can refer to any appropriate heat transfer textbook for more details (say Incropera et al., 2007) or to specialized books dealing with building energy simulation (e.g., Clarke, 1985).

Example 8.1: Transient Response of a Wall Subjected to a Step Change Forcing Function

Consider a concrete wall 20 cm thick of a conditioned space that is subject to an abrupt change in the outdoor air temperature. We shall illustrate the use of the explicit finite difference assuming the wall to be discretized into 5 nodes (nodal spacing is 5 cm).* As shown in Figure 8.5, the nodes are numbered 0, 1, ..., 4 with 0 being the node representing the exterior wall surface and node 4 representing the interior wall surface. Initially, the wall is at 20°C throughout its thickness. The outdoor air temperature is abruptly increased to 35°C. Calculate the time evolution of the temperature profiles inside the wall and determine the heat flow inward. Other information is given next.

Given: Wall properties: $k = 1.73$ W/(m·K), $\alpha = 1.153 \times 10^{-6}$ m²/s, $\Delta x = 5$ cm $= 0.05$ m

$T_i = 20°C$, $T_\infty^{t=0} = 35°C$

Initial temperature profile through wall: $T_m^{t=0} = 20°C$

Convective coefficients: $h_o = 22.7$ W/(m²·K), $h_i = 8.29$ W/(m²·K)

$$\phi_{Bi,o} = \frac{22.7 \times 0.05}{1.73} = 0.6561 \quad \text{and} \quad \phi_{Bi,i} = \frac{8.29 \times 0.05}{1.73} = 0.2396$$

Find: T_m^t for $m = 0, 1, 2, 3, 4$ over time, and $\dot{Q}_{cond}$.

Solution

The simulation time step can be determined.

For the interior nodes (Equation 8.15),

$$\Delta t_{\max} \le \frac{(0.05)^2}{2 \times 1.153 \times 10^{-6}} \le 1084 \text{ s (18 min)}$$

For the boundary node (using the lower Biot value corresponding to the inside surface) (Equation 8.20),

$$\Delta t_{\max} \le \frac{(0.05)^2}{1.153 \times 10^{-6} \times 2 \times (1 + 0.2396)} \le 874.6 \text{ s (14.6 min)}$$

Let us assume a conservative time step of 10 min or 600 s.

The Fourier number is then determined:

$$\phi_{Fo} = \frac{1.153 \times 10^{-6} \times 600}{0.05^2} = 0.27672$$

Using the explicit finite difference formulations, for the first time step of simulation, we get

(a) For the outer wall surface node (Equation 8.18):

$$T_0^1 = 2 \times 0.27672 \times \left(T_1^0 + 0.6561 \times T_\infty^0\right) + (1 - 2 \times 0.27672 - 2 \times 0.27672 \times 0.6561) \times T_0^0$$
$$= 0.55344 \times \left(T_1^0 + 0.6561 \times T_\infty^0\right) + 0.08345 \times T_0^0$$

(b) For the interior nodes $m = 1, 2, 3$ (Equation 8.14):

$$T_1^1 = 0.27672 \times \left(T_0^0 + T_2^0\right) + (1 - 2 \times 0.27672) \times T_1^0$$
$$= 0.27672 \times \left(T_0^0 + T_2^0\right) + 0.44656 \times T_1^0$$

(c) For the inner wall surface node (Equation 8.18):

$$T_4^1 = 2 \times 0.27672 \times \left(T_3^0 + 0.2396 \times T_i\right) + (1 - 2 \times 0.27672 - 2 \times 0.27672 \times 0.2396) \times T_4^0$$
$$= 0.55344 \times \left(T_3^0 + 0.2396 \times T_i\right) + 0.31396 \times T_4^0$$

The first few results along with subsequent intermediate values are assembled in Table 8.2 while the results of the simulations for the first 8 h are shown in Figure 8.6. We note that each of the node temperatures increase exponentially but at varying rates. Even after 8 h, the transients have yet to die out though they are close to the steady-state results shown (left to the reader to compute these values). The outer wall surface has reached 32°C after 8 h, which is lower by a little more than half a degree from its steady-state value. The inner wall node has reached 25.4°C, which is about a degree from its steady-state value. The heat flow inward to the room is also shown. We notice that even after 4 h the heat flowing inward into the conditioned space is lower than its steady-state value by about 16%.

8.3 Time-Series Methods for Conduction Heat Gains

8.3.1 Basis

Two versions of the time-series formulation have been developed for analyzing 1-D transient heat conduction through opaque elements using the properties of linear systems (introduced in Section 8.1.3). One of them is the transfer function model, which is still used in several detailed building energy simulation programs such as

* This is a large time step selected more for educational illustration; in reality, one would select a smaller grid size.

TABLE 8.2

Calculations for the Finite Difference Analysis Method (Example 8.1)

	Outside Air Temperature, °C	Node Temperatures, °C					Heat Flow Inwards, W/m²
Time, min	T_∞	T_0	T_1	T_2	T_3	T_4	$\dot{Q}_{cond}$
0	35.00	20.00	20.00	20.00	20.00	20.00	
10	35.00	25.45	20.00	20.00	20.00	20.00	0.00
20	35.00	25.90	21.51	20.00	20.00	20.00	0.00
30	35.00	26.77	22.31	20.42	20.00	20.00	0.00
40	35.00	27.29	23.02	20.82	20.12	20.00	0.00
—							
60	35.00	28.08	24.08	21.62	20.48	20.18	1.45
120	35.00	29.43	26.05	23.54	21.97	21.29	10.73
240	35.00	30.79	28.18	26.02	24.41	23.37	27.92
360	35.00	31.56	29.38	27.47	25.89	24.64	38.44
480	35.00	32.00	30.09	28.33	26.76	25.39	44.65
Steady state	**35.00**	**32.64**	**31.10**	**29.55**	**28.01**	**26.46**	53.52

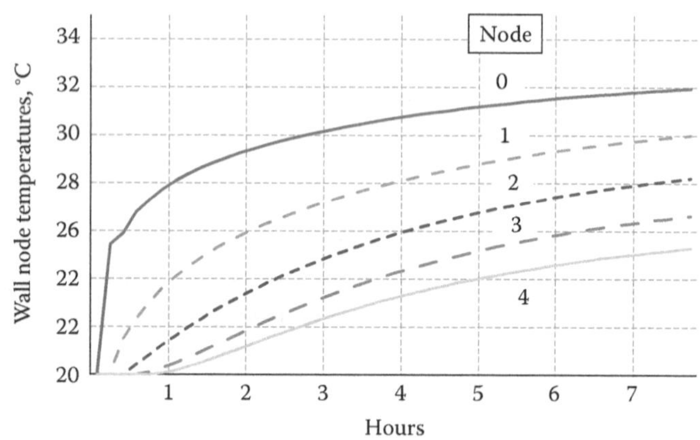

FIGURE 8.6
Transient response of wall nodes when subject to a step change in the forcing function (Example 8.1).

DOE-2 (Birdsall et al., 1990) because of its computational speed and efficient use of computer memory. The load $\dot{Q}$ can be considered to be the *response* of the building or room to the *driving terms* (T_i, T_o, $\dot{Q}_{sol}$, etc.) that act on it. The transfer function method calculates the response of a system by making the following basic assumptions:

1. *Discrete time steps*: All functions of time are represented as series of values at regular time steps (*hourly* in the present case).
2. *Linearity*: The response of a system is a linear function of the driving terms and of the state of the system.
3. *Causality*: The response at time t can depend only on the past, not on the future. In essence, a continuously changing excitation or forcing function is broken up into elementary excitations at discrete time intervals and the system response to each of them are then determined separately (Figure 8.7b). These individual responses are summed to provide the desired system response.

As an example, suppose there is a single driving term $u(t)$ and the response is $y(t)$. To make the expressions more readable, let us indicate the time dependence as a subscript, in the form $y(t) = y_t, u(t) = u_t$, and so on.

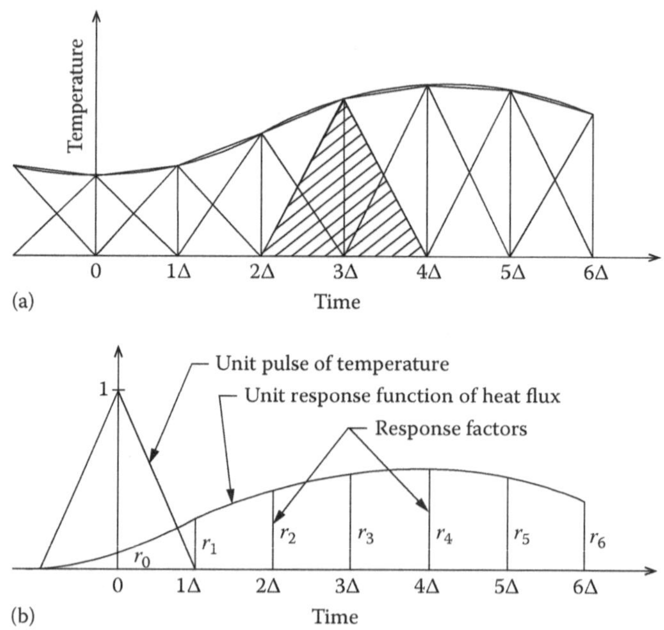

FIGURE 8.7
(a) Linear approximation of an input excitation function by triangular pulses adopted by the thermal response factor (TRF) method; (b) response factors at discrete time steps due to a triangular excitation function on a building element.

Then, according to the transfer function model, the relation between the response and the driving terms is of the form

$$y_t = -(a_1 y_{t-1\Delta t} + a_2 y_{t-2\Delta t} + \cdots + a_n y_{t-n\Delta t}) + (b_0 u_t + b_1 u_{t-1\Delta t} + b_2 u_{t-2\Delta t} + \cdots + b_m u_{t-m\Delta t}) \quad (8.22)$$

where a_1 to a_n and b_0 to b_m are coefficients that characterize the system; they are independent of the driving term or response. Usually, time steps $\Delta t = 1$ h are assumed. Equation 8.22 is obviously linear. It satisfies causality because y_t depends only on the past values of the response ($y_{t-1\Delta t}$ to $y_{t-n\Delta t}$) and on present and past values of the driving terms* (u_t to $u_{t-m\Delta t}$).

The past state of the system enters because of the coefficients a_1 to a_n and b_1 to b_m; this is how thermal inertia is taken into account. The response is instantaneous only if these time-lagged coefficients are zero. The greater their number and magnitude, the greater the weight or influence of the history. The accuracy of the model increases as the number of coefficients is enlarged and as the time step is reduced. For load calculations, hourly time resolution and a handful of coefficients per driving term will suffice. The coefficients are called *transfer function coefficients*.

Incidentally, the relation between u and y could be written in symmetric form as follows:

$$a'_0 y_t + a'_1 y_{t-1\Delta t} + \cdots + a'_n y_{t-n\Delta t} = b'_0 u_t + b'_1 u_{t-1\Delta t} + \cdots + b'_m u_{t-m\Delta t} \quad (8.23)$$

which is equivalent because one can divide both sides of the equation by a'_0. Since the roles of u and y are symmetric, one can use the same model to find, for example, the load (i.e., the heat $\dot{Q}$ to be supplied or removed) as a function of T_i, or T_i as a function of $\dot{Q}$.

Thus, Equation 8.23 can be readily generalized to the case where there are several driving terms. For instance, if the response T_i is determined by two driving terms, heat input $\dot{Q}$, and outdoor temperature T_o, then one can write the transfer function model in the form

$$\begin{aligned} a'_{i,0}T_{i,t} + a'_{i,1}T_{i,t-1\Delta t} + \cdots + a'_{i,n}T_{i,t-n\Delta t} &= a'_{o,0}T_{o,t} + a'_{o,1}T_{o,t-1\Delta t} \\ &+ \cdots + a'_{o,m}T_{o,t-m\Delta t} \\ &+ a'_{Q,0}\dot{Q}_t + a'_{Q,1}\dot{Q}_{t-1\Delta t} \\ &+ a'_{Q,2}\dot{Q}_{t-2\Delta t} \\ &+ \cdots + a'_{Q,r}\dot{Q}_{t-r\Delta t} \end{aligned} \quad (8.24)$$

* A series such as Equation 8.22 is also known as a *time series*.

with the three sets of transfer function coefficients $a'_{i,0}$ to $a'_{i,n}$, $a'_{o,0}$ to $a'_{o,m}$, and $a'_{Q,0}$ to $a'_{Q,r}$. This equation can be considered an algorithm for calculating $T_{i,t}$, hour by hour, given the previous values of T_i and the driving terms T_o and $\dot{Q}$. Likewise, if T_i and T_o were given as driving terms, one could calculate $\dot{Q}$ as response.

8.3.2 CTF Model

The 1-D conductive heat gain (or loss) $\dot{Q}_{cond,t}$ at time t through the roof and walls is calculated according to the conduction transfer function (CTF) model

$$\dot{Q}_{cond,t} = -\sum_{n\geq 1} d_n \dot{Q}_{cond,t-n\Delta t} + A\left(\sum_{n\geq 0} b_n T_{os,t-n\Delta t} - T_i \sum_{n\geq 0} c_n\right) \quad (8.25)$$

where

A is the area of roof or wall, m² (ft²)
Δt is the time step = 1 h
$T_{os,t}$ is the sol–air temperature of outside surface at time t (defined in Equation 7.3 with data in Figure 7.3 and the online HCB software); this is taken as the external thermal boundary condition
b_n, c_n, d_n are the CTF coefficients

The indoor temperature T_i is the indoor thermal boundary condition. Since it is assumed constant, it is multiplied by the sum of the c_n values (and so the individual c_n coefficients are not needed). In general, the initial value $\dot{Q}_{cond,t} = 0$ is not known; its value does not matter if the calculation is repeated over a sufficient number of diurnal cycles (a few days to a week) until the resulting pattern becomes periodic within the desired accuracy (this is illustrated in Example 8.2).

Numerical values of the CTF coefficients for a variety of common building elements have been precalculated with time steps of 1 h, and a partial representative list is provided in Table 8.3: roofs in Table 8.3a and walls in Table 8.3b. These values have been taken from ASHRAE Fundamentals (1997), the last edition where these coefficients have been listed. Thermal properties and code numbers of layers used in roof and wall descriptions are assembled in Table 8.4.† Note that for elements with larger heat capacitance, greater number of lagged values of the heat flows (the d coefficients) and the sol–air temperatures (the b coefficients) are needed. Thus, for steel siding only 2 lag terms in d coefficients and 3 lag terms in b coefficient are needed, while for 200 mm common brick wall, 5 lags are needed for both coefficients.

† These code designations have been modified in subsequent editions of ASHRAE Fundamental Handbooks.

TABLE 8.3

Coefficients of Conduction Transfer Function: SI Units

(Layer Sequence Left to Right = Inside to Outside)		$n = 0$	$n = 1$	$n = 2$	$n = 3$	$n = 4$	$n = 5$	$n = 6$	Σc_n	U	δ	λ
a. Roofs												
Layers E0 A3 B14 E3 E2 A0	b_n	0.00316	0.06827	0.07278	0.00814	0.00007			0.152411	0.314886	2.43	0.94
Steel deck with 125 mm (5 in.) insulation	d_n	1.00000	−0.60061	0.08602	−0.00135							
Layers E0 E1 B15 E4 B7 A0	b_n	0.00002	0.00371	0.01923	0.01361	0.00164	0.00164		0.038233	0.243113	4.85	0.82
Attic roof with 150 mm (6 in.) insulation	d_n	1.00000	−1.34660	0.59384	−0.09295	0.00296	−0.00001					
Layers E0 B22 C12 E3 E2 C12 A0	b_n	0.00336	0.04925	0.03905	0.00213				0.093798	0.784431	5.00	0.56
40 mm (1.67 in.) insulation with 50 mm (2 in.) h.w. concrete	d_n	1.00000	−1.11770	0.23731	−0.00008							
Layers E0 E5 E4 B12 C14 E3 E2 A0	b_n	0.00000	0.00139	0.01234	0.01424	0.00315	0.00013		0.031255	0.325063	6.32	0.60
75 mm (3 in.) insulation w/100 mm (4 in.) l.w. conc.deck and susp. clg.	d_n	1.00000	−1.40605	0.58814	−0.09034	0.00444	−0.00006					
Layers E0 E5 E4 C13 B20 E3 E2 A0	b_n	0.00001	0.00060	0.00197	0.00086	0.00005			0.019817	0.792784	7.54	0.15
150 mm (6 in.) h.w. deck w/20 mm (0.76 in.) insul. and susp. clg.	d_n	1.00000	−1.39181	0.46337	−0.04714	0.00058						
Layers E0 E5 E4 B15 C15 E3 E2 A0	b_n	0.00000	0.00000	0.00002	0.00014	0.00024	0.00011	0.00002	0.002996	0.194878	10.44	0.30
150 mm (6 in.) insul. w/150 mm (6 in.) l.w. conc. deck and susp. clg.	d_n	1.00000	−2.29459	1.93694	−0.75741	0.14252	−0.01251	0.00046				
b. Walls												
Layers E0 A3 B1 B13 A3 A0	b_n	0.04361	0.19862	0.04083	0.00032				0.283372	0.372389	1.30	0.98
Steel siding with 100 mm (4 in.) insulation	d_n	1.00000	−0.24072	0.00168								
Layers E0 E1 B14 A1 A0	b_n	0.00089	0.03097	0.05456	0.01224	0.00029			0.098947	0.314501	3.21	0.91
Frame wall with 13 mm (5 in.) insulation	d_n	1.00000	−0.93389	0.27396	−0.02561	0.00014						
Layers E0 A6 C5 B3 A3 A0	b_n	0.00561	0.04748	0.02052	0.00039				0.074007	0.692064	5.14	0.41
100 mm (4 in.) h.w. concrete with 50 mm (2 in.) insulation	d_n	1.00000	−0.93970	0.04664								
Layers E0 E1 C8 B6 A1 A0	b_n	0.00002	0.00349	0.01641	0.01038	0.00105			0.031356	0.619445	7.11	0.37
200 mm (8 in.) h.w. concrete block with 50 mm (2 in.) insulation	d_n	1.00000	−1.52480	0.67146	−0.09844	0.00239						
Layers E0 A2 C2 B15 A0	b_n	0.00000	0.00003	0.00076	0.00248	0.00170	0.00029	0.00001	0.005274	0.245415	9.36	0.30
Face brick and 100 mm (4 in.) l.w. conc. block with 150 mm (6 in.) insulation	d_n	1.00000	−2.00870	1.37120	−0.37897	0.03962	−0.00165	0.00002				
Layers E0 C9 B6 A6 A0	b_n	0.00000	0.00030	0.00362	0.00561	0.00170	0.00011		0.011343	0.600253	8.97	0.20
200 mm (8 in.) common brick with 50 mm (2 in.) insulation	d_n	1.00000	−1.78160	0.96017	−0.16904	0.00958	−0.00016					
Layers E0 E1 C11 B13 A1 A0	b_n	0.00000	0.00000	0.00010	0.00059	0.00071	0.00023	0.00002	0.001646	0.361789	11.4	0.10
300 mm (12 in.) h.w. concrete with 100 mm (4 in.) insulation	d_n	1.00000	−2.37670	2.04310	−0.79860	0.14868	−0.01231	0.00037				

Source: ASHRAE, *Handbook of Fundamentals*, American Society of Heating, Refrigerating and Air-Conditioning Engineers, Atlanta, GA, 1997. Copyright ASHRAE, www.ashrae.org.

Also shown are the decrement factor λ and time lag δ for periodic heat flow (see Section 8.5.2); U, b_n, and c_n are in W/(m²·°C) (Multiply by 0.1761 to convert to Btu/(h·ft²·°F)); d_n and λ are dimensionless; and δ is in hours.

For definition of code numbers of roofs and wall materials and thermal properties, see Table 8.4.

Though the numerical values of these transfer function coefficients with time lags have no physical meaning, their relative magnitudes provide some insights. For steel siding, the largest value for b_n occurs at $n = 1$, suggesting that the value in sol–air temperature 1 h prior has the largest impact on current heat flow. On the other hand, for the face brick, $n = 3$ has the largest impact. Finally, under periodic heat flow, the time lag δ for such high mass elements is greater and the value of decrement factor λ is smaller.

TABLE 8.4

Thermal Properties and Code Numbers of Layers Used in Roof and Wall Descriptions of Table 8.3: SI Units

Layer ID	Description	Thickness, mm	Conductivity, W/(m·K)	Density, kg/m³	Specific Heat, kJ/(kg·K)	Resistance, (m²·K)/W
A0	Outside surface resistance	—	—	—	—	0.059
A1	25 mm (1 in.) Stucco	25	0.692	1858	0.84	0.037
A2	100 mm (4 in.) face brick	100	1.333	2002	0.92	0.076
A3	Steel siding	2	44.998	7689	0.42	0.000
A6	Finish	13	0.415	1249	1.09	0.031
B1	Air space resistance	—	—	—		0.160
B3	50 mm (2 in.) insulation	51	0.043	32	0.84	1.173
B6	50 mm (2 in.) insulation	51	0.043	91	0.84	1.173
B7	25 mm (1 in.) wood	25	0.121	593	2.51	1.760
B12	75 mm (3 in.) insulation	76	0.043	91	0.84	1.760
B13	100 mm (4 in.) Insulation	100	0.043	91	0.84	2.347
B14	125 mm (5 in.) insulation	125	0.043	91	0.84	2.933
B15	150 mm (6 in.) insulation	150	0.043	91	0.84	3.520
B20	20 mm (0.76 in.) insulation	20	0.043	91	0.84	0.440
B22	42 mm (1.67 in.) insulation	42	0.043	91	0.84	0.968
C2	100 mm (4 in.) low-density concrete block	100	0.381	609	0.84	0.266
C5	100 mm (4 in.) high-density concrete	100	1.731	2243	0.84	0.059
C8	200 mm (8 in.) high-density concrete block	200	1.038	977	0.84	0.196
C9	200 mm (8 in.) common brick	200	0.727	1922	0.84	0.279
C11	300 mm (12 in.) high-density concrete	300	1.731	2243	0.84	0.176
C12	50 mm (2 in.) high-density concrete	50	1.731	2243	0.84	0.029
C13	150 mm (6 in.) high-density concrete	150	1.731	2243	0.84	0.088
C14	100 mm (4 in.) low-density concrete	100	0.173	641	0.84	0.587
E0	Inside surface resistance	—	—	—	—	0.121
E1	20 mm (3/4 in.) plaster or gypsum	20	0.727	1602	0.84	0.026
E2	12 mm (1/2 in.) slag or stone	12	1.436	881	1.67	0.009
E3	10 mm (3/8 in.) felt and membrane	10	0.190	1121	1.67	0.050
E4	Ceiling air space	—	—	—	—	0.176
E5	Acoustic tile	19	0.061	481	0.84	0.314

Source: ASHRAE, *Handbook of Fundamentals*, American Society of Heating, Refrigerating and Air-Conditioning Engineers, Atlanta, GA, 1997. Copyright ASHRAE, www.ashrae.org.

Conversions to IP units: 100 mm = 0.333 ft, 1 W/(m·K) = 0.5778 Btu/(h·ft·°F), 1 kg/m³ = 0.06243 lb_m/ft³, 1 kJ/(kg·K) = 0.2389 Btu/(lb_m·°F), 1 (m²·K)/W = 0.2049°F·ft²·h/Btu

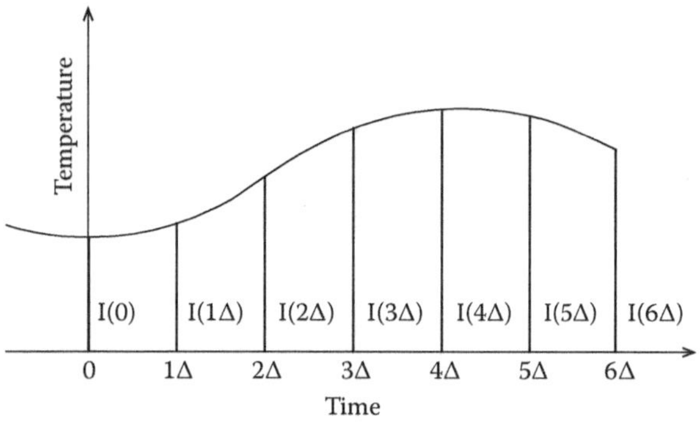

FIGURE 8.8
Linear approximation of input excitation function by a series of discrete pulses adopted by the conduction transfer method.

From a historic perspective, triangular pulses were assumed (Figure 8.7a and b), which led to the thermal response factor (TRF) method wherein the time-lagged values of the response were not included in the model formulation (Mitalas, 1968; Mitalas and Stephenson, 1967; Stephenson and Mitalas, 1967). A very large number of b_n coefficients were then needed for accurate transient response prediction. The CTF, on the other hand, is based on discrete pulses to discretize the excitation function (Figure 8.8), and since it includes time lagged values of the response, a shorter time-series model is adequate.

It is instructive to establish the connection of the transfer function coefficients with the U value. In the steady-state limit, that is, when $\dot{Q}_{cond,t}$, T_{os} and T_i are all constant, Equation 8.25 becomes

$$\dot{Q}_{cond}\sum_{n\geq 0} d_n = A\left(T_{os}\sum_{n\geq 0} b_n - T_i\sum_{n\geq 0} c_n\right) \quad \text{where } d_0 = 1 \qquad (8.26)$$

Since in that limit we also have

$$\dot{Q}_{cond} = AU(T_{os} - T_i) \qquad (8.27)$$

the coefficients of T_{os} and T_i must be equal

$$\sum_{n\geq 0} b_n = \sum_{n\geq 0} c_n \qquad (8.28)$$

and the U value is given by

$$U = \frac{\sum_{n\geq 0} c_n}{\sum_{n\geq 0} d_n} \qquad (8.29)$$

Example 8.2: CTF Analysis for a Wall

Calculate the conductive heat gain per square meter of an exterior vertical wall, of dark color and facing west, for summer design conditions (July 21) if T_i = 25°C.

Given: Exterior wall of 0.10 m (4 in.) concrete with 0.05 m (2 in.) insulation on the outside
T_i = 25°C, Sol–air temperature of Table 8.5.

Find: $\dot{Q}_{cond}$

Lookup values: The transfer function coefficients for walls and roofs are listed in Table 8.3. This wall corresponds to the third entry under the walls section.

$b_0 = 0.00561$	$\Sigma c_n = 0.074007$	$d_0 = 1.0000$
$b_1 = 0.04748$		$d_1 = -0.93970$
$b_2 = 0.02052$		$d_2 = 0.04664$
$b_3 = 0.00039$		$d_3 = 0.0$

The ds are dimensionless, and the bs and cs are in W/(m²·K). All other coefficients are zero, and the U value is 0.693 W/(m²·K).

Solution

Arranging a spreadsheet (Table 8.5) with three long columns (one for t, one for $T_{os,t}$, and one for $\dot{Q}_{cond,t}$, per unit wall area A, according to Equation 8.25, and taking $\dot{Q}_{cond,t} = 0$ for $t > 1$), the results in the following table are obtained (iterative results of successive days are shown in successive columns). Note the three initial rows for t = −2, −1, 0 correspond to the calculations for the previous repetitive day, and so the entries are shown in italics. For the first $(\dot{Q}_{cond,t}/A)$ entry at t = 1, the value is calculated as

$$\frac{\dot{Q}_{cond,1}}{A} = -\frac{d_1\cdot\dot{Q}_{cond,1-1}}{A} - \frac{d_2\cdot\dot{Q}_{cond,1-2}}{A} + b_0\cdot T_{os,1-0} + b_1\cdot T_{os,1-1}$$
$$+ b_2\cdot T_{os,1-2} + b_3\cdot T_{os,1-3} - T_i\sum_{n\geq 0} c_n$$
$$= -(-0.93970)\times 0.00 - 0.04664\times 0.00$$
$$+ 0.00561\times 24.4 + 0.04748\times 25.0$$
$$+ 0.02052\times 26.1 + 0.00039$$
$$\times 27.1 - 25\times 0.074007 = 0.020$$

To initiate the second day calculations, the conductive gains for the last 3 h of the previous day are copied into the first three rows of the column corresponding to the second day, and so on.

We note that at the start of the fifth day the diurnal pattern has stabilized and the dynamic periodic condition has been reached. The diurnal plots of the heat gains for the first day and for the last day are shown in Figure 8.9.

TABLE 8.5

Calculation Results for Example 8.2 Meant to Illustrate the Use of the Conduction Transfer Function Model

t	$T_{os,t}$	$\dot{Q}_{cond,t}/A$	$\dot{Q}_{cond,t+24}/A$	$\dot{Q}_{cond,t+48}/A$	$\dot{Q}_{cond,t+72}/A$	$\dot{Q}_{cond,t+96}/A$
h	°C	W/m²				
−2	*27.2*	*0.00*	*12.659*	*13.396*	*13.438*	*13.440*
−1	*26.1*	*0.00*	*11.426*	*12.080*	*12.117*	*12.119*
0	*25.0*	*0.00*	*10.246*	*10.827*	*10.860*	*10.861*
1	24.4	0.020	9.116	9.631	9.661	9.662
2	24.4	−0.013	8.058	8.515	8.541	8.542
3	23.8	−0.061	7.100	7.506	7.529	7.530
4	23.3	−0.136	6.218	6.577	6.598	6.599
5	23.3	−0.240	5.397	5.716	5.734	5.735
6	25.0	−0.335	4.666	4.949	4.965	4.966
7	27.7	−0.325	4.113	4.364	4.378	4.379
8	30.0	−0.134	3.803	4.026	4.038	4.039
9	32.7	0.225	3.718	3.916	3.927	3.927
10	35.0	0.743	3.842	4.017	4.027	4.028
11	37.7	1.393	4.143	4.299	4.308	4.308
12	40.0	2.170	4.610	4.748	4.756	4.756
13	53.3	3.110	5.274	5.397	5.404	5.404
14	64.4	4.699	6.619	6.728	6.734	6.734
15	72.7	6.995	8.699	8.796	8.801	8.801
16	75.5	9.723	11.234	11.320	11.325	11.325
17	72.2	12.468	13.809	13.885	13.889	13.889
18	58.8	14.749	15.939	16.006	16.010	16.010
19	30.5	15.903	16.959	17.019	17.022	17.022
20	29.4	15.256	16.192	16.245	16.248	16.249
21	28.3	13.949	14.780	14.827	14.830	14.830
22	27.2	12.659	13.396	13.438	13.440	13.440
23	26.1	11.426	12.080	12.117	12.119	12.120
24	25.0	10.246	10.827	10.860	10.861	10.862
Average	**37.95**					**8.965**

Note: Cells in italics under the third column are the initial values for heat gains assumed to start the first iteration. Those in subsequent columns are copied from the simulation results of the last 3 h of the previous day.

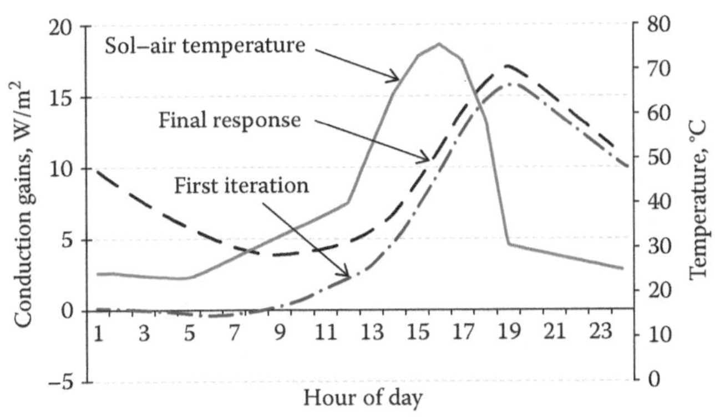

FIGURE 8.9
Results of transient conduction heat gains through the wall determined from the transfer function method (Example 8.2) showing the difference between the first iteration and the final solution.

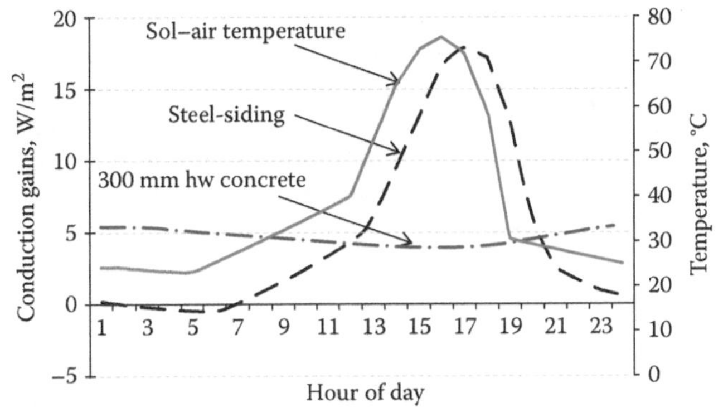

FIGURE 8.10
Plots illustrating the large difference in diurnal heat gain distributions from two different wall constructions when subject to the same boundary sol–air temperature profile.

Comments

The conductive gain reaches its maximum at 19:00, which is 3 h after the peak of the sol–air temperature. As a check, it is strongly recommended that one always verify that the solution satisfies the *steady-state limit*. Multiplying the average temperature difference $T_{os} - T_i = (37.95 - 25)$°C by $U = 0.693$ W/(m²·K), we find 8.974 W/m² while the 24 h average of $\dot{Q}_{cond,t}$ is shown in Table 8.5 as 8.965 W/m²; they are indeed equal within the accuracy of the calculation.

It is insightful to illustrate the differences in time lag and amplitude attenuation or decrement factor in excitation functions with thermal capacity of the building element. Figure 8.10 shows this effect for the light-weight steel-siding wall (wall entry #1 in Table 8.3b) and for the 300 mm hw concrete (wall entry #7), both of which have about the same U-value. As expected, the former has a small time lag of less than 2 h and the amplitude is very slightly lower (from Table 8.3, $\delta = 1.3$ h and $\lambda = 0.98$), whereas the concrete wall has almost damped out the entire excitation function (in this case, $\delta = 11.4$ h and $\lambda = 0.10$).

8.3.3 Conduction Time-Series Model

The conduction time-series (CTS) model for calculating hourly 1-D conductive heat flows through opaque building elements such as roof and wall surfaces is part of the radiant transfer series (RTS) method (ASHRAE Fundamentals, 2013; Spitler, 2010), which is described in Section 9.6. The CTS model is akin to the CTF model that uses CTF coefficients in the sense that both of them use a time-series model involving lagged values and also assume that the indoor space temperature is held constant. Recall that the CTF model uses an iterative process to calculate the conductive heat flow across the opaque elements of the building envelope. Depending on the driving forces and the type of material, this may require several repetitions of each day's values before the iteration converges. The novelty of the CTS model is that it replaces this iteration with a simple summation that does not involve lagged values of the conductive heat gains (in that regard, it is akin to the TRF formulation introduced earlier). In essence, the CTS model is a reformulation of Equation 8.25:

$$Q_{cond,t} = UA \cdot \sum_{j=0}^{23} c_j (T_{os,t-n \cdot \Delta t} - T_i) \tag{8.30}$$

$$\text{Subject to} \sum_{j=0}^{23} c_j = 1.0 \tag{8.31}$$

where

- U is the steady-state overall heat transfer coefficient
- A is the surface area
- c_j is the jth CTS factors that are normalized to unity
- $T_{os,t-n \cdot \Delta t}$ is the lagged sol–air temperature n hours previously
- T_i is the space temperature assumed to be constant

The CTS factors, comprising of 24 coefficients, for various types of walls and roofs can be found from lookup tables similar to those created for the CTF coefficients (Table 8.6). These factors can be viewed as hourly adjustments to the steady-state heat transfer expression that correct for heat absorbed in the wall or roof during previous hours and released during this hour. Figure 8.11 are plots of the CTS factors for hourly lagged values for two wall types: (1) type 27 is a 100 mm light weight concrete wall with board insulation and gyp board; and (2) type 6 is a stud wall with wood siding, sheathing, batt insulation, and 13 mm wood. As expected, the adjustments (i.e., numerical values of the coefficients) for light wall are very large for the first few hours and drop quickly to zero, implying that little heat is stored in such walls. The heavier the wall, the smaller the spikes (more attenuation) and more persistent (longer time delays) are the CTS values. The sum of all the 24 coefficients of each wall must sum to unity as per Equation 8.31. ASHRAE has modified the code numbers of the various layers used till 1997 (see Table 8.4), and the interested reader can refer to ASHRAE Fundamentals (2013) or Spitler (2010) for updated information regarding layer codes and layer properties. The following example illustrates the analysis approach.

TABLE 8.6

Calculation Results for Example 8.3 Meant to Illustrate the Use of the Conduction Time-Series Model

t, h	Cj, %	$T_{os,tr}$, °C	$\frac{\dot{Q}_{cond,t}}{A}$, W/m²
1	1	24.4	6.32
2	10	24.4	4.74
3	20	23.8	3.45
4	18	23.3	2.48
5	14	23.3	1.66
6	10	25.0	0.92
7	7	27.7	0.50
8	5	30.0	0.43
9	4	32.7	0.75
10	3	35.0	1.44
11	2	37.7	2.54
12	2	40.0	3.79
13	1	53.3	5.22
14	1	64.4	7.44
15	1	72.7	11.01
16	1	75.5	15.42
17	0	72.2	19.83
18	0	58.8	23.23
19	0	30.5	24.48
20	0	29.4	22.38
21	0	28.3	18.05
22	0	27.2	14.02
23	0	26.1	10.75
24	0	25	8.27
Average		**37.95**	**8.71**

The wall is a 100 mm concrete light weight wall with board insulation and gypsum board.

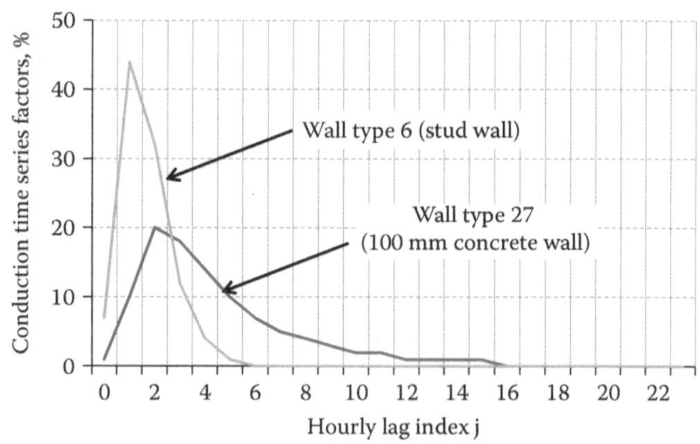

FIGURE 8.11
Conduction time-series (CTS) factors for light and heavy walls.

Example 8.3: CTS Analysis for a Wall

Calculate the conductive heat gain per square meter for a wall type 27 (similar wall construction as that assumed in Example 8.2) and identical orientation. The solar–air values are the same and the interior temperature is kept constant at 25°C.

Given: The sol–air temperatures and CTS values are given in Table 8.6. The U value = 0.673 W/(m²·K), T_i = 25°C.

Find: $\dot{Q}_{cond}$

Solution

We start by creating a spreadsheet (Table 8.6) with three long columns (one for time of day t, one for CTS factors (c_j) and one for $T_{os,t}$). Note that the values of c_j do not correspond to time but with the hourly lag index j from (0–23). The last column assembles the computed values of $\dot{Q}_{cond,t}$, per unit wall area A, according to Equations 8.30 and 8.31. Notice that no iterations are needed. To illustrate the calculations, let us consider hour 22:00. The cell formula is given by

$$\begin{aligned}\frac{\dot{Q}_{cond,1}}{A} &= 0.673\times[0.01\times(27.2-25)+0.1\times(28.3-25)\\&+0.2\times(29.4-25)+0.18\times(30.5-25)+0.14\\&\times(58.8-25)+0.1\times(72.2-25)+0.07\times(75.5-25)\\&+0.05\times(72.7-25)+0.04\times(64.4-25)+0.03\\&\times(53.3-25)+0.02\times(40-25)+0.02\\&\times(37.7-25)+0.01\times(35-25)+0.01\times(32.7-25)\\&+0.01\times(30-25)+0.01\times(27.7-25)+0+\cdots]=14.02\end{aligned}$$

and so on.

Comments

The conductive gain reaches its maximum at 19:00, which is 3 h after the peak of the sol–air temperature. As a check, it is strongly recommended that one always verify that the solution satisfies the *steady-state limit*. Multiplying the average temperature difference $T_{os}-T_i$ = (37.95–25)°C by U = 0.693 W/(m²·K), we find 8.71 W/m² which is identical to the 24 h average of $\dot{Q}_{cond,t}$ shown in Table 8.5 as 8.71 W/m².

8.4 Thermal Networks Models

Thermal network models provide a framework that is general, yet conceptually simple, for the modeling of building elements and also of entire spaces and buildings. They are akin to a simplified finite difference scheme with large time steps and where the slices are based on physical construction. The analogy between heat and electrical current flows has been presented in Section 2.8.1 and various thermal network diagrams have already been previously introduced (e.g., Figures 2.10, 2.11, 2.22, 3.2, and 5.8). In the following sections, we outline a general approach for the dynamic modeling of a building, based on thermal networks. They are intuitive allow a systematic formulation and the solutions of general and complicated problems.

8.4.1 Network Diagrams and Modeling Equations

In this section, we describe how to draw diagrams of thermal networks with heat capacitance and how to obtain the corresponding equations. The thermal network approach approximates a building as being composed of a finite number of parts N, called *nodes*, each of which is assumed to be isothermal. To model heat exchange, the nodes are connected by resistances, thus forming a thermal network. Neighboring nodes are nodes that are directly coupled by conduction, convection, or radiation. The heat flow between neighbors is given by

$$\dot{Q}_{n'-n} = \frac{T_{n'} - T_n}{R_{n'n}} \tag{8.32}$$

where $R_{n'n}$ is the resistance between n' and n. In addition, there may be direct heat input $\dot{Q}_n$ at node n, from heat sources such as solar radiation, lights, or electric resistance heating. Let us designate the heat capacity of node n by C_n and its temperature by T_n. Assuming constant C_n, the rate of change of the heat stored in node n is $C_n\dot{T}_n$,* and by the first law of thermodynamics it must be equal to the total rate of heat input. Thus, the heat balance of node n is a first-order differential equation in T_n:

$$C_n\dot{T}_n = \sum_{n'=1}^{N}\frac{T_{n'} - T_n}{R_{n'n}} + \dot{Q}_n \tag{8.33}$$

As for signs, we note that if $T_{n'} - T_n$ is positive, heat flows from n' to n, making a positive contribution to $\dot{T}_n$. In most cases, a given node can interact directly with only a relatively small number of nodes, and so the number of nonzero terms in this sum is much smaller than N. For example, a homogeneous wall can be modeled as a 1-D network where each node has only two neighbors.

One equation for each node temperature T_n (such as Equation 8.33) form a system of N first-order differential equations with N unknowns. By analogy with electric circuits, it is convenient to represent thermal networks by diagrams. There is a one-to-one correspondence between the diagram and the set of equations of a thermal network. The diagram has the advantage of being much easier to grasp, but the equations are needed for finding the solutions. Once the diagram has been drawn, one can easily write down the equations. There is one first-order differential equation for each node.

* Note that in this section we have used the notation $\dot{T}$ for the first-order time derivative of temperature and $\ddot{T}$ for the second-order time derivative of temperature.

8.4.1.1 Single-Pane Glass

As a simple example, consider a single-pane window between indoor air at T_i and outdoor air at T_o, with surface heat transfer coefficients h_i and h_o (combining radiation and convection). The glass can be approximated as a single isothermal node (Figure 8.12a), with capacitance

$$C = \rho c_p A \Delta x \tag{8.34}$$

where
- ρ is the density
- c_p is the specific heat
- A is the area
- Δx is the thickness

If there is no absorption of radiation in the glass, $\dot{Q}$ is zero and Equation 8.33 for the glass temperature T becomes

$$C\dot{T} = \frac{T_i - T}{R_i} + \frac{T_o - T}{R_o} \tag{8.35}$$

where the resistances are related to the surface heat transfer coefficients by

$$\frac{1}{R_i} = Ah_i \quad \text{and} \quad \frac{1}{R_o} = Ah_o$$

Using combined heat transfer coefficients for convection plus radiation is a simplification that is not appropriate when radiation and convection couple to nodes at different temperatures, for example, radiation to the sky and convection to outdoor air. Figure 8.12b shows how the network has to be modified if a pane of glass faces different radiative and convective temperatures on both sides, for example, sky and outdoor air on one side and indoor air and indoor wall surfaces on the other sides. The corresponding heat balance equation is

$$C\dot{T} = \frac{T_{i,rad} - T}{R_{i,rad}} + \frac{T_{i,con} - T}{R_{i,con}} + \frac{T_{o,rad} - T}{R_{o,rad}} + \frac{T_{o,con} - T}{R_{o,con}} \tag{8.36}$$

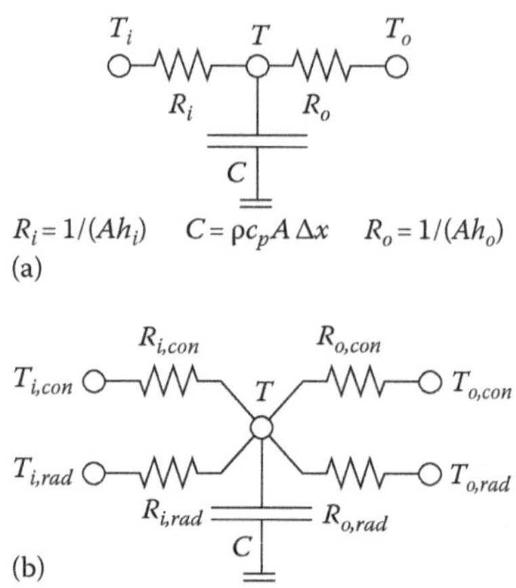

FIGURE 8.12
Thermal network for heat flow through a pane of glass, treating the glass as a single node with heat capacity C and neglecting the resistance of the glass relative to the surface resistances R_i and R_o. (a) With combined heat transfer coefficients (convection plus radiation). (b) With separate heat transfer coefficients for convection and for radiation, coupled to different temperatures.

8.4.1.2 Homogeneous Wall

As a more complicated example, Figure 8.13 shows the thermal network for a homogeneous wall that is approximated by N nodes, where T_i and T_o are the air temperatures on the two sides of the wall and R_i and R_o are the interior and exterior surface resistances. Following the discretization scheme adopted in Section 8.2, the surface or boundary nodes will have half the volume (or thickness) than those of the internal nodes. Then the thickness of the internal nodes is $\Delta x_{int} = L/(N-1)$ and that of a surface node is $\Delta x_{surf} = L/[2(N-1)]$, where L is the thickness of wall. With the same notation as given earlier, the capacitance of the nodes is again given by Equation 8.34. The resistances between nodes are

$$R_{int} = \frac{\Delta x_{int}}{kA} \quad \text{and} \quad R_{surf} = \frac{\Delta x_{surf}}{kA} \tag{8.37}$$

where k is the conductivity {W/(m·K) [Btu/(h·ft·°F)]}. For the heat balance of the internal nodes, we have

$$C_{int}\dot{T}_n = \frac{T_{n-1} - T_n}{R_{int}} + \frac{T_{n+1} - T_n}{R_{int}} \quad \text{for } 2 \le n \le N-1 \tag{8.38}$$

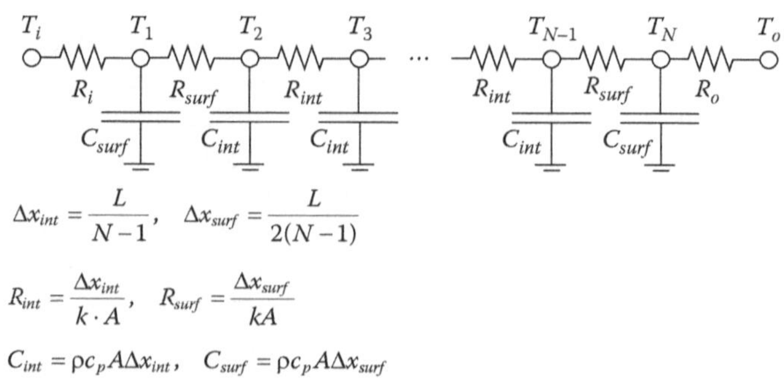

FIGURE 8.13
Thermal network for one-dimensional heat flow through homogeneous slab. Surfaces are in contact with media at temperatures T_i and T_o, the surface heat transfer coefficients being h_i and h_o. Slab has area A and thickness L, and is divided into N layers of thickness.

The equations for the surface or external nodes (Figure 8.13) are

$$C_{surf}\dot{T}_1 = \frac{T_i - T_1}{R_i} + \frac{T_2 - T_1}{R_{surf}} \quad (8.39)$$

and

$$C_{surf}\dot{T}_N = \frac{T_{N-1} - T_N}{R_{surf}} + \frac{T_o - T_N}{R_o} \quad (8.40)$$

This represents a total of N first-order differential equations for the N unknowns T_n, $n = 1 - N$. Sources of heat, for example, solar radiation absorbed at the surfaces, can easily be included. If $\dot{Q}_1$ is absorbed at node 1 and $\dot{Q}_N$ at node N, one adds $\dot{Q}_1$ and $\dot{Q}_N$ to the right-hand side of Equations 8.39 and 8.40, respectively.

8.4.1.3 Heterogeneous Building Elements

Building elements such as roofs and walls are made up of layers of several different substances. Some amount of manipulation is needed to develop a suitable thermal network diagram from which corresponding transient performance models can be developed. The following example illustrates the approach.

Example 8.4: Developing Thermal Network Models of Heterogeneous Layered Elements

Consider the wall type assumed in Example 8.2, which is the third entry in Table 8.3b. The wall consists of four solid layers plus the two air films. These are specified as [E0, A6, C5, B3, A3, A0] the description of which is provided in Table 8.4. As is the norm, the contact resistance between layers is neglected. It is obvious that the 100 mm high-density concrete wall will have almost all the thermal capacity. We wish to develop a suitable thermal network model for this wall assuming a 3-node discretization for the concrete layer.

Solution

Figure 8.14a is a sketch of the wall construction while the initial thermal network representation is provided in Figure 8.14b. Each of the layers has a thermal resistance whose numerical values are given in Table 8.4 and assembled in the following table.

Code	Description of Initial Network Layers	R Value, m²·K/W	C, kJ/K
E0	Inside surface resistance	0.121	—
A6	Finish	0.031	—
C5	100 mm high-density concrete	0.059	188.4
B3	50 mm insulation	1.173	—
A3	Steel siding	0.00	—
A0	Outside surface resistance	0.059	—

The total thermal capacitance of the 100 mm concrete layer listed above is determined from

$$C_{C5} = \rho c_p AL = 2243\ \text{kg/m}^3 \times 0.84\ \text{kJ/(kg}\cdot\text{K)} \times 1\ \text{m}^2 \times 0.1\ \text{m}$$
$$= 188.4\ \text{kJ/K}$$

Since a 3-node representation is required, and recalling that the surface nodes should be half the width of the interior nodes, we deduce that the thermal capacitance of the interior node should

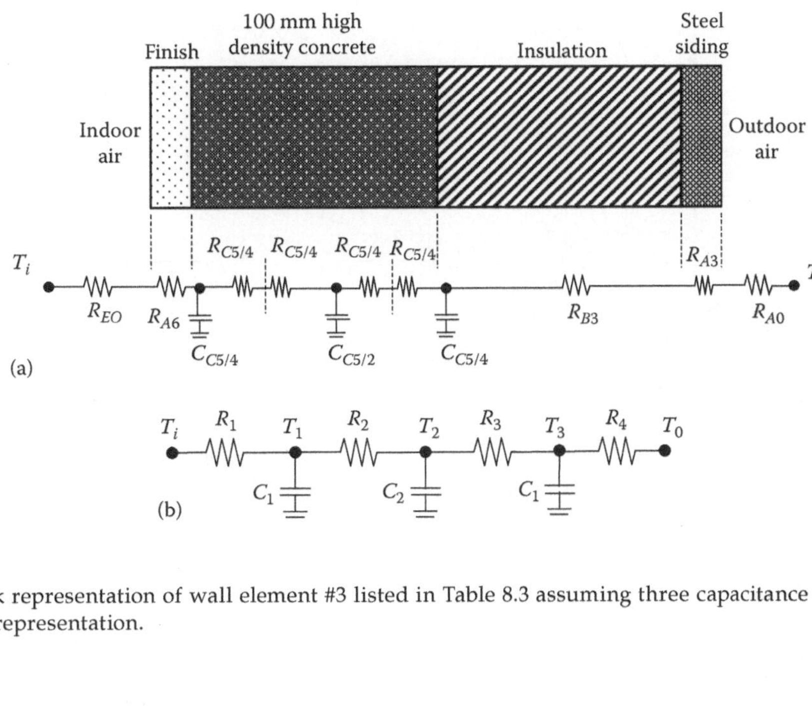

FIGURE 8.14
(a) Initial thermal network representation of wall element #3 listed in Table 8.3 assuming three capacitance nodes for the concrete layer; (b) final thermal network representation.

be = $C_{C5/2}$ = 94.7 kJ/K and that of the surface nodes = $C_{C5/4}$ = 47.35 kJ/K. The final network diagram representation requires that resistors in series be added to obtain a final network as shown in Figure 8.14b. In this case,

$$R_1 = R_{E0} + R_{A6} = 0.152\ \mathrm{m^2 \cdot K/W}$$

$$R_2 = R_3 = R_{C5}/2 = 0.0295\ \mathrm{m^2 \cdot K/W}$$

$$R_4 = R_{B3} + R_{A3} + R_{A0} = 1.232\ \mathrm{m^2 \cdot K/W}$$

The energy balance equations can be finally written down for each of the three nodes. This will result in a set of three coupled first-order differential equations that can be solved using numerical methods as discussed in previous sections.

8.4.1.4 One-Zone Building

There are several possible ways to set up a thermal network to model thermal interactions within a room or space. One such moderately detailed model for a one-zone building is shown in Figure 8.15. It assumes individual thermal networks for the roof, walls, and glazing elements that have different construction properties. These are subject to the temperature difference between the outdoor air temperature T_o and an aggregated indoor envelope surface temperature T_{env}. It has been found that a 3-node thermal network representation for conduction heat transfer of each of the building elements provides a model of the hourly thermal response of actual buildings, which is satisfactory to evaluate different dynamic control methods of the indoor temperature so as to reduce building peak loads (Mitchell and Braun, 2013). Radiation heat transfer interactions then occur to an aggregated interior room mass node (which also includes the floor) at temperature T_r and by convection to the room air at T_i. Solar radiation $\dot{Q}_{sol}$ is shown as being absorbed by the node at T_r. Only a certain fraction f of the internal loads are assumed to strike the interior mass node at T_r (the radiative component) while the rest is the convective component of the internal loads, which appear instantaneously on the room air T_i. Heat is extracted by the HVAC from node T_i, which is assumed to have no additional heat capacity associated with it. An infiltration heat path is also included as shown.

Further features could be added. For instance, in winter the inside surface of a ceiling is likely to be warmer than the surface of a window even if the air temperature in the room is uniform. The resulting radiative exchange between different interior surfaces can be treated by adding radiative conductances between nodes n_{w1}, n_{r1}, and n_{g1} in Figure 8.15. More complex buildings can be modeled similarly, including as necessary several different wall and window types or several zones at different temperatures (e.g., see Clarke, 1985). The number of nodes depends on the desired accuracy. Even three-dimensional effects can be taken into account, for example, for a more accurate treatment of the conduction through the ground. One drawback with the thermal network approach is that it is not unique and several possible networks can be formulated. It is perhaps more suited as an inverse model (see Section 10.6) wherein monitored data of an existing space are used to identify the best model and the associated parameters among several competing network models (Rabl, 1988; Reddy, 1989).

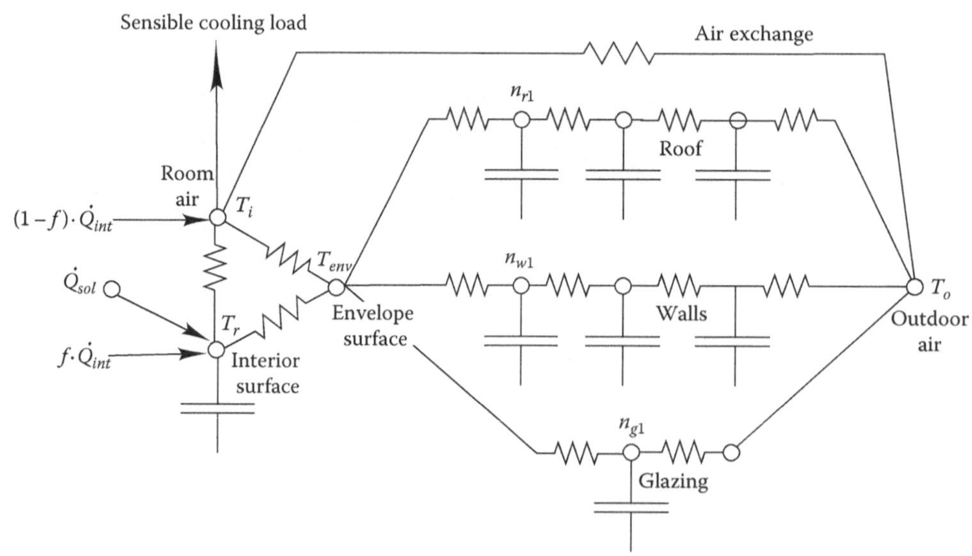

FIGURE 8.15
Thermal network for simple one-zone building whose envelope contains elements with distinct compositions: roof, wall, and double glazing. Infiltration, solar load, and internal loads are also considered.

This approach is suitable for any linear thermal system, that is, one where resistances and capacitances are independent of temperature. That is the case for the envelopes of buildings because the range of temperatures is usually small enough to permit approximation by linearized equations. Some of the parameters may vary with time to account for variable ventilation or variable heat transfer coefficients.

In this manner, any building can be modeled as a thermal network or, equivalently, as a system of coupled first-order differential equations in time. At this point, the formulation is hybrid in the sense that space is treated as a discrete variable, while time is continuous. For a numerical solution, one frequently resorts to discretization of time as well, as explained in Section 8.2.

8.4.2 1R1C Network and Time Constant

To model thermal inertia, a network must contain at least one capacitance. The simplest possibility is the network shown in Figure 8.16, containing one resistance R and one capacitance C, the *1R1C* network. Here, the entire heat capacity of a building is lumped into a single massive node. This model is instructive for explaining the important concept of the *time constant* of a building. Even though the *1R1C* network is quite crude, it gives the designer a simple tool for estimating the warm-up and cooldown times associated with thermostat setback and setback recovery. In fact, most controllers for optimizing the start-up time after setback are based on this network.

The node C is at the indoor temperature T_i, and the resistance R is the inverse of the total heat transmission coefficient K_{tot} of the building

$$R = \frac{1}{K_{tot}} \tag{8.41}$$

Writing down the heat balance equation (8.33) for this network, we find

$$C\dot{T}_i = \frac{T_o - T_i}{R} + \dot{Q} \tag{8.42}$$

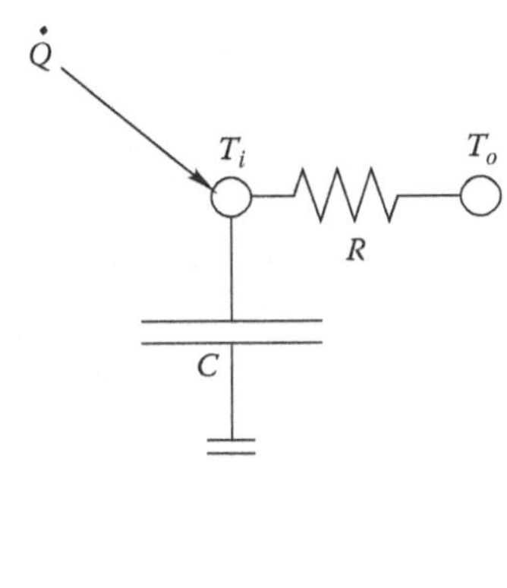

FIGURE 8.16
RC network.

This is a first-order differential equation in the variable T_i. Let us define a quantity τ, called the *time constant*:

$$\tau = RC \tag{8.43}$$

To explain the reason for the name, we rewrite the equation in the form

$$\tau\dot{T}_i + T_i = T_o + R\dot{Q} \tag{8.44}$$

and solve it for the special case when the driving terms T_o and $\dot{Q}$ are constant. In that case, the derivative of the variable

$$T(t) = T_i(t) - T_o - R\dot{Q} \tag{8.45}$$

is equal to $\dot{T}_i$, and so, Equation 8.44 simplifies to

$$\tau\dot{T} + T = 0 \tag{8.46}$$

Its solution is a simple exponential

$$T(t) = T(0)\exp\left(\frac{-t}{\tau}\right) \tag{8.47}$$

where $T(0)$ is the initial value. The time constant sets the time scale for the cooldown; after one time constant, $T(t)$ has decayed to $1/e \approx 0.368$ of its initial value $T(0)$. The longer the time constant of a building, the longer it takes to cool down or warm up. Now, we have a dynamic model, albeit crude, for estimating the effects of thermostat setback and recovery.

Example 8.5: Time Delay in Thermostat Setback

Suppose a building can be described by the *1R1C* network of Figure 8.16. The variables T_i, T_o, and the heat input $\dot{Q}$ are assumed constant until $t = 0$, when the thermostat is setback. How long does it take for T_i to drop from 20°C (68°F) to 15°C (59°F)?

Given: Network of Figure 8.16, with

R = 5.0°C/kW, (implying K_{tot} = 200 W/°C), C = 10 MJ/°C, T_o = 0°C

$\dot{Q}$ as necessary to maintain T_i = 20°C until t = 0 ($\dot{Q} = 0$ for $t > 0$)

Find: Time t at which $T_i(t)$ = 15°C.

Solution

Let $T(t) = T_i(t) - T_o$. For $t > 0$, it satisfies Equation 8.46, and the solution according to Equation 8.47 is

$$T(t) = T(0)\exp\left(\frac{-t}{\tau}\right)$$

with $T(0) = 20°C - 0°C = 20°C$ and time constant

$$\tau = RC = 5.0°C/kW \times 10\ MJ/°C = 5.0 \times 10^4\ s = 13.889\ h$$

We are looking for the value of t at which $T(t) = 15°C - 0°C = 15°C$; hence, we have to find the value of t that solves

$$15°C = 20°C \times \exp\left(\frac{-t}{\tau}\right)$$

The solution is

$$t = -\tau \times \ln\left(\frac{15}{20}\right) = 13.889\ h \times 0.28768 = 4.00\ h$$

Example 8.6: Thermostat Setback Recovery Time

Suppose the heating system of the building in Example 8.5 has been built for design temperatures $T_{i,des} = 20°C$ (68°F) and $T_{o,des} = -20°C$ (−4°F). How long is the setback recovery time for the conditions of Example 8.5?

Given: Conditions of Example 8.5, with peak heat input $\dot{Q}$ designed for $T_i = 20°C$ and $T_o = -20°C$. Heating system has simple on/off control: it runs at full capacity until the desired temperature is reached (this type of control is typical of residential heating systems).

Find: Time t for bringing $T_i(t)$ from 15°C back to 20°C.

Solution

First find the heat input $\dot{Q}$,

$$\dot{Q} = \frac{T_{i,des} - T_{o,des}}{R} = \frac{20°C - (-20°C)}{5.0°C/kW} = 8.0\ kW\ (27.3\ kBtu/h)$$

Take the start of the warmup as $t = 0$. During warmup, the conditions of constant driving terms are again satisfied, and so we can again use the exponential solution (Equation 8.47):

$$T(t) = T(0)\exp\left(\frac{-t}{\tau}\right)$$

If we take

$$T(t) = T_i(t) - T_o - R\dot{Q} \quad \text{with } R\dot{Q} = 5.0°C/kW \times 8.0\ kW = 40°C$$

then the initial value is

$$T(0) = T_i(0) - T_o - R\dot{Q} = 15°C - 0°C - 40°C = -25°C$$

We want to know at what time t we reach $T_i(t) = 20°C$, or

$$T(t) = 20°C - 0°C - 40°C = -20°C$$

Hence, we have to solve the following for time t:

$$-20°C = -25°C \times \exp\left(\frac{-t}{\tau}\right)$$

The solution is

$$t = -\tau \times \ln\left(\frac{-20}{-25}\right) = (-13.889\ h) \times (-0.2231) = 3.10\ h$$

Comments

The warmup time depends on T_o and on $\dot{Q}$. The lower the T_o and the smaller the $\dot{Q}$, the slower the warmup. At $T_o = -10°C$, we would have found

$$T(0) = T_i(0) - T_o - R\dot{Q} = 15°C - (-10°C) - 40°C = -15°C$$

and

$$T(t) = 20°C - (-10°C) - 40°C = -10°C$$

and a warmup time of

$$t = -\tau \times \ln\left(\frac{-10}{-15}\right) = (-13.889\ h) \times (-0.40546) = 5.63\ h$$

How long would the warmup be when T_o equals the design value of −20°C (−4°F)? In that case, we would have to solve the equation with

$$T(0) = T_i(0) - T_o - R\dot{Q} = 15°C - (-20°C) - 40°C = -5°C$$

and

$$T(t) = 20°C - (-20°C) - 40°C = 0°C$$

Now the solution for t diverges—the warmup would take infinitely long. This illustrates our remarks about pickup loads in Section 9.3.

Equation 8.45 can also be solved if the right side is not constant. In that case, the solution is the sum of two terms, one that depends on the initial condition $T_i(t_0)$ and one that depends on the driving term

$$T_i(t) = \exp\left(\frac{t_0 - t}{\tau}\right)T_i(t_0) + \int_{t_0}^{t} dt' \exp\left(\frac{t' - t}{\tau}\right)u(t') \quad (8.48)$$

with the driving term $u(t) = [T_o(t) + R\dot{Q}(t)]/\tau$. That this is indeed a solution can be verified by inserting it into the differential equation.

Unfortunately, the 1*R*1*C* network is not sufficient for all purposes. It treats the building as if all the thermal mass were in the interior, always at the same temperature as the indoor air. In real buildings, the arrangement of masses is much more complicated, as suggested by the network of Figures 8.14 and 8.15. For applications to thermostat control, the *RC* network can give acceptable results if the value of the heat capacity *C* is chosen appropriately. With a judicious choice of *C*, one can make the *RC* network imitate the response of a building to typical changes in heat input (see the discussion of system identification in Section 10.6). This value of *C* is not the static heat capacity, and its determination is not entirely obvious. It can be approximated by the effective diurnal heat capacity, as discussed in Section 8.1.1. Next, we discuss the inclusion of further nodes in a network.

8.4.3 A Network with Two Nodes

To obtain a more realistic model, let us add another resistance and another capacitance, as shown in Figure 8.17 (Rabl, 1988). Compared to the simple 1*R*1*C* network, an extra node with capacitance C_e and temperature T_e has been added between T_i and T_o; it can represent the heat capacity of the envelope while C_i represents the heat capacity of the interior of the building (interior walls and floors, furniture, air). The resistances R_i and R_o on each side of C_e can be numerically different. The total resistance is, of course, the inverse of the heat transmission coefficient

$$R_i + R_o = \frac{1}{K_{tot}} \tag{8.49}$$

It is instructive to solve this case. Now, we have the network equations

$$C_i\dot{T}_i = \frac{T_e - T_i}{R_i} + \dot{Q} \tag{8.50}$$

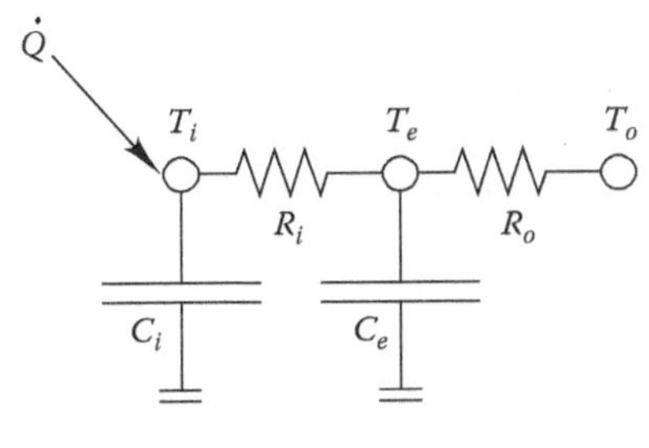

FIGURE 8.17
Thermal network with two resistances and two heat capacities, for Example 8.7.

and

$$C_e\dot{T}_e = \frac{T_i - T_e}{R_i} + \frac{T_o - T_e}{R_o} \tag{8.51}$$

They are not very convenient in this form because they contain the variable T_e that is neither observed nor needed in most applications. It would be preferable to have a model that involves only T_i and the driving terms T_o and $\dot{Q}$. To eliminate T_e, we take the derivative of Equation 8.50 and multiply by R_i:

$$R_iC_i\ddot{T}_i = \dot{T}_e - \dot{T}_i + R_i\ddot{Q}_i \tag{8.52}$$

and on its right-hand side, we insert Equation 8.51 for $\dot{T}_e$. This yields an equation that contains T_e but no derivatives of T_e. Now, we can eliminate T_e by means of Equation 8.50. After rearranging terms, we obtain the desired result

$$\begin{aligned} &T_i + (C_iR_i + C_iR_o + C_eR_o)\dot{T}_i + C_iR_iC_eR_o\ddot{T}_i \\ &= T_o + (R_i + R_o)\dot{Q} + R_iR_oC_e\ddot{Q}_i \end{aligned} \tag{8.53}$$

The price for the elimination of variable T_e has been the introduction of higher derivatives in the other variables T_i and $\dot{Q}$. We now have a second-order differential equation in T_i.

To find out the time constant(s), let us consider once more the special case when the driving terms T_o and $\dot{Q}$ are constant. Then, the variable

$$T(t) = T_i(t) - T_o - (R_i + R_o)\dot{Q} \tag{8.54}$$

satisfies the equation

$$T + (C_iR_i + C_iR_o + C_eR_o)\dot{T} + C_iR_iC_eR_o\ddot{T} = 0 \tag{8.55}$$

Trying an exponential solution

$$T(t) = T_i(0)\exp\left(\frac{-t}{\tau}\right) \tag{8.56}$$

we see that the time constant τ must satisfy

$$\tau^2 - (C_iR_i + C_iR_o + C_eR_o)\tau + C_iR_iC_eR_o = 0 \tag{8.57}$$

This is a quadratic equation, and it has two solutions for τ:

$$\tau_{\pm} = \frac{1}{2}(C_iR_i + C_iR_o + C_eR_o)\left[1 \pm \sqrt{1 - \frac{4C_iR_iC_eR_o}{(C_iR_i + C_iR_o + C_eR_o)^2}}\right] \tag{8.58}$$

Example 8.7: Calculating Time Constants

Find the time constants for the network in Figure 8.17 with the following parameter values. The values of R_o and R_i are chosen to give the same K_{tot} as in Example 8.5, with equal resistance on the inside and on the outside of the capacity C_e of the envelope.

Given: $R_o = 2.5$ K/kW
$R_i = 2.5$ K/kW
$C_e = 5$ kWh/K $= 18.0 \times 10^3$ kJ/K
$C_i = 1$ kWh/K $= 3.6 \times 10^3$ kJ/K

Find: τ_+ and τ_-.

Solution

Inserting the numbers into Equation 8.58, we obtain

$$\tau_\pm = \frac{1}{2}(3.6\times10^3\times2.5+3.6\times10^3\times2.5+18\times10^3\times2.5)$$

$$\times\left[1\pm\sqrt{1-\frac{4\times3.6\times10^3\times2.5\times18\times10^3\times2.5}{(3.6\times10^3\times2.5+3.6\times10^3\times2.5+18\times10^3\times2.5)^2}}\right]$$

$$=\frac{1}{2}\left[63\times10^3\times\left(1\pm\sqrt{1-\frac{1620\times10^6}{(63\times10^3)^2}}\right)\right]=\frac{1}{2}\left[63\times10^3\times(1\pm0.7693)\right]$$

From where

$$\tau_+ = 15.481\ \text{h} \approx 15.5\ \text{h} \quad \tau_- = 2.019\ \text{h} \approx 2.0\ \text{h}$$

Comments

We have tried to choose the numbers so that the models of Examples 8.5 and 8.7 could represent approximately the same building. However, one cannot compare the heat capacities and time constants directly with the single-node model of Example 8.5 because the masses are distributed differently. In the two-node model, the capacity C_e of the envelope is almost twice as large as C_i of the single-node model, but since it is not directly coupled to T_i, its effect on the time constant is reduced. Hence, the value of the larger time constant is comparable to the value 13.89 h found in Example 8.5.

The fact that there are two time constants implies that the general solution of Equation 8.55 is a superposition of two exponentials

$$T_i(t) = A_1 \exp\left(\frac{-t}{\tau_1}\right) + A_2 \exp\left(\frac{-t}{\tau_2}\right) \tag{8.59}$$

The coefficients A_1 and A_2 are determined by the initial conditions, for instance, $T_i(0)$ and $\dot{T}_i(0)$.

This passage from the *RC* network to a network with two nodes illustrates some general phenomena that are encountered in the development of more complicated models. The model with two nodes has two time constants. As a general theorem, one finds that a network with *N* nodes has *N* time constants (usually different from each other). In buildings, there is usually one time constant that is much larger than all the others; this large time constant determines the response of the building to slow changes. When talking about "the time constant" of a building, one refers to this value.

The solution of the 1R1C network involves a single exponential and one initial condition (Equation 8.47 or 8.48). The solution of the two-node network involves two exponentials and two initial conditions (Equation 8.59). This, too, has a natural generalization in the theory of linear differential equations: a network with *N* nodes involves *N* exponentials and *N* initial conditions.

In a complicated network, one will rarely (if ever) be interested in the values of most of the node temperatures. Rather, one would like to know the output (or response) as a function of the input (or driving terms), for example, the indoor air temperature T_i as a function of the outdoor temperature T_o and the heat input $\dot{Q}$. When the uninteresting variables are eliminated, higher derivatives appear in the other variables. If one eliminates all but a single output variable T_i from a network with *N* nodes, one obtains a differential equation of order *N* in T_i.

The procedure we have used in this section for eliminating a variable and for finding time constants and solutions becomes unwieldy when the number of nodes is large. As the number of nodes is increased, matrix algebra provides a simpler and more systematic approach. That is beyond the scope of this book. Instead, we use the two-node network to illustrate how transfer functions can be derived.

8.4.4 Connection with Transient Heat Conduction Equation

Here, we show the relationship between thermal networks and the transient heat conduction equation. Let us consider Equation 8.38 in the limit where the slab thickness Δx goes to zero (and number of layers *N* to infinity, keeping $N\Delta x$ fixed). To simplify, note that *R* and *C* enter only in the combination

$$RC = \frac{\rho c_p}{k}\Delta x^2 \tag{8.60}$$

One can write Equation 8.38 in the form

$$\dot{T}_n = \frac{\alpha}{\Delta x}\left(\frac{T_{n-1}-T_n}{\Delta x}+\frac{T_{n+1}-T_n}{\Delta x}\right) \tag{8.61}$$

where

α is the thermal diffusivity

T_n is the temperature at $x = (n - (1/2))\Delta x$

As $\Delta x \to 0$, the terms on the right-hand side approach the spatial derivatives:

$$\frac{T_{n+1}-T_n}{\Delta x} \to \frac{\partial T}{\partial x} \quad \text{at } x = \left(n+\frac{1}{2}\right)\Delta x$$
$$\frac{T_{n-1}-T_n}{\Delta x} \to -\frac{\partial T}{\partial x} \quad \text{at } x = \left(n-\frac{1}{2}\right)\Delta x \qquad (8.62)$$

The notation with partial derivatives reflects the fact that in this limit T becomes a continuous function of t and of x. The difference of the derivatives at $x = (n + (1/2))\Delta x$ and at $x = (n - (1/2))\Delta x$, divided by Δx, becomes the second derivative. Therefore, Equation 8.38 turns into Equation 8.3, the well-known equation for transient heat conduction in one dimension.

8.4.5 Deducing Transfer Function Coefficients

In this section, we will illustrate how the transfer function coefficients (discussed in Section 8.4.2) can be derived from a network analysis. As an example, let us take the two-node network of the preceding section. It is described by the differential equation in Equation 8.53, which we rewrite here in terms of the time constants as

$$T_i + (\tau_1+\tau_2)\dot{T}_i + \tau_1\tau_2\ddot{T}_i = T_o - (R_i+R_o)\dot{Q} + R_iR_oC_e\ddot{Q}_i \qquad (8.63)$$

since the coefficients of $\dot{T}_i$ and $\ddot{T}_i$ are related to the time constants by

$$\tau_1+\tau_2 = C_iR_i + C_iR_o + C_eR_o \qquad (8.64)$$

and

$$\tau_1\tau_2 = C_iR_iC_eR_o$$

For convenience, we adopt discrete variable notation, and replace t by $(k \cdot \Delta t)$ or $k = (t/\Delta t)$. Using the differencing scheme,

$$x(t) \to x_{k-1}$$
$$\dot{x}(t) \to \frac{x_k - x_{k-2}}{2\Delta t} \qquad (8.65)$$
$$\ddot{x}(t) \to \frac{x_k - 2x_{k-1} + x_{k-2}}{(\Delta t)^2}$$

we obtain

$$a_{i,0}T_{i,k} + a_{i,1}T_{i,k-1} + a_{i,2}T_{i,k-2}$$
$$= a_{o,1}T_{o,k} + a_{Q,0}\dot{Q}_k + a_{Q,1}\dot{Q}_{k-1} + a_{Q,2}\dot{Q}_{k-2} \qquad (8.66)$$

with

$$a_{i,0} = \frac{\tau_1+\tau_2}{2\Delta t} + \frac{\tau_1\tau_2}{(\Delta t)^2}$$

$$a_{i,1} = 1 - \frac{2\tau_1\tau_2}{(\Delta t)^2}$$

$$a_{i,2} = -\frac{\tau_1+\tau_2}{2\Delta t} + \frac{\tau_1\tau_2}{(\Delta t)^2}$$

$$a_{o,1} = 1 \quad (a_{o,0} \text{ and } a_{o,2} \text{ are both } 0)$$

$$a_{Q,0} = \frac{\tau_1\tau_2}{2C_i\Delta t} = -a_{Q,2}$$

$$a_{Q,1} = -(R_i + R_o)$$

Recall from the discussion in Section 8.3 that such a relation between function values at discrete time steps is called a *time series*, and its coefficients are called *transfer function coefficients*. This approach is convenient for numerical calculations on a computer.

Example 8.8: Calculating Transfer Function Coefficients Using RC Networks

Use the transfer functions along with the numbers of Example 8.7 to calculate the time evolution of T_i.

Given: Two-node network with
$R_i = 2.5$ K/kW $= R_o$
$C_i = 1.0$ kWh/K, $C_e = 5.0$ kWh/K
$\tau_1 = 15.481$ h, $\tau_2 = 2.019$ h (from end of Example 8.7)
$T_{o,k} = 0$, $\dot{Q}_k = 0$ for $k \geq 0$
Initial conditions $T_{i,k} = 1$ K for $k \leq 0$

Find: $T_{i,k}$ for $k > 0$.

Solution

The coefficients of the transfer function in Equation 8.66 are easily deduced. For example,

$$a_{i,0} = \frac{\tau_1+\tau_2}{2\cdot\Delta t} + \frac{\tau_1\cdot\tau_2}{(\Delta t)^2} = \frac{15.481+2.019}{2} + \frac{15.481\times 2.019}{1^2} = 40.00$$

The complete set of coefficients are

$$a_{i,0} = 40.00 \quad a_{o,0} = 0.00 \quad a_{Q,0} = 15.63$$
$$a_{i,1} = -61.50 \quad a_{o,1} = 1.00 \quad a_{Q,1} = -5.00$$
$$a_{i,2} = 22.50 \quad a_{o,2} = 0.00 \quad a_{Q,2} = -15.63$$

The a_i and a_o are dimensionless, while the a_Q have dimension of Kelvins per kilowatt. Here, we need only the coefficients a_i, and T_i can be calculated by repeated use of

$$T_{i,k} = \frac{-(a_{i,1}T_{i,k-1} + a_{i,2}T_{i,k-2})}{a_{i,0}}$$

For example,

$$T_{i,1} = \frac{-(a_{i,1}T_{i,0} + a_{i,2}T_{i,-1})}{a_{i,0}}$$

$$= \frac{-(-61.50 \times 1\text{ K} + 22.50 \times 1\text{ K})}{40.00} = 0.98$$

The results listed here show good agreement with the exact solution (Equation 8.59 for $T_i(0) = 0$ and $\dot{T}_i(0) = 0$) in the last column.

k	$T_{i,k-2}$	$T_{i,k-1}$	$T_{i,k}$	$T_{i,exact}$
0	1.00	1.00	1.00	1.00
1	1.00	1.00	0.98	0.99
2	1.00	0.98	0.94	0.95
3	0.98	0.94	0.89	0.91
4	0.94	0.89	0.84	0.87
5	0.89	0.84	0.80	0.82
10	0.66	0.62	0.58	0.60
20	0.35	0.33	0.31	0.32
30	0.18	0.17	0.16	0.17

We have already presented time series in Section 8.3, but without indicating how the transfer function coefficients can be calculated for a given configuration. Here, we have shown how the coefficients can be calculated for a simple thermal network. For networks with a large number of nodes, this method would yield a large number of coefficients, but one can approximate the resulting transfer function by a simpler one with fewer coefficients, such as the ones in Section 8.3.

8.5 Frequency-Domain Methods

8.5.1 Basis

The analysis methods discussed earlier qualify as time-domain methods that use recorded hourly weather data as input and calculate hourly internal conditions as output. Alternatively, frequency-domain methods are more concerned with the cyclical response of the system, exciting it with a known signal (usually a sine wave) and testing for harmonics and resonances within its response. Such methods make use of a number of factors for each material that characterize its dynamic response.

The exciting or driving terms of a building element are periodic in nature. According to the principles of Fourier analysis, any periodic function can be decomposed into a series of sinusoidal functions, called harmonics, whose frequency is a multiple of the first one, also called the fundamental harmonic. By summing up the response to each harmonic, it is possible to obtain the response to the original periodic excitation. The periodicity of the exciting functions is dominated by the diurnal frequency: outdoor temperature, solar radiation, and occupancy all follow a basic 24 h cycle. Any function that repeats itself after 24 h can be represented by a sum of sine and cosine terms with period 2 × 24 h, 3 × 24 h, 4 × 24 h, etc. Therefore, it is appropriate to consider the response of a linear thermal system to sinusoidal driving terms (i.e., sines or cosines of a single frequency). This is helpful for an intuitive understanding of the behavior of real buildings.

Quite generally, if the input of a linear system is sinusoidal, the response is sinusoidal with the same frequency; the amplitude and phase of the response depend on the system parameters. In the case of transient heat flow through a wall, the *time lag* is the phase shift between the heat flux variations on either side of the wall, and the *decrement factor* is the ratio of the amplitude of the output wave to that of the input. This fact can be seen from the solution of the 1*R*1*C* network (Equation 8.48) by inserting sinusoidal driving functions on the right-hand side: the contribution of the initial state, whatever it was, vanishes exponentially with time, and purely sinusoidal response remains, as the reader can verify with some algebra and a table of integrals.

Kuehn et al. (1998) cite previous work done in the 1940s and present closed-form solutions of periodic heat transfer through single and composite homogeneous material when subject to external forcing functions that repeat over consecutive days (the dynamic periodic case). These formulations get progressively more complicated as the number of layers are increased, and therefore, these solutions are generally not used nowadays.

8.5.2 Admittance Method

Another frequency-domain method used in many parts of the world is the admittance method, also referred to as the CIBSE cyclic model (CIBSE Guide A, 2007). It is a dynamic-periodic transient method (implicitly assuming a sequence of identical days for which the external conditions vary on a 24 h cycle) that balances simplicity and accuracy and is mainly used to calculate temperature swings inside buildings. It is particularly useful for those designers who are less experienced in building modeling so that they can gain an understanding of the sensitivity of proposed designs to variations in the basic thermal properties of the construction. It uses a material characteristic known as the admittance of a surface (which is essentially a dynamic *U*-value), as well as thermal lag and decrement factors (described earlier), to define their dynamic response. It is a simplified method suitable for hand calculations and provides estimates of the instantaneous variation of internal conditions about their mean diurnal value. Three pairs of quantities are defined:

1. The thermal *admittance* Y (in units of W/(m$^2\cdot$K)) is the ratio of the heat flow rate entering the interior surface of a wall for each degree of deviation of the *space temperature about its mean* value. For multilayered structures, the admittance is primarily determined by the characteristics of the materials in the layers nearest to the interior surface. For example, the admittance of a heavyweight structure (such as concrete slabs) lined internally with insulation will be close to the value for the insulation alone. However, placing the insulation within the construction or on the outside surface will have little or no effect on the admittance. Y is based on a 24 h sinusoidal cycling of heat flow into the surface and is the reciprocal of the *impedance* (which is the effective resistance of an electric component in an alternating current circuit).
2. The *decrement factor* is the ratio of the rate of periodic heat flow through the structure to the indoor space air for each degree of deviation in *external temperature about its mean* value to the steady-state rate of flow of heat (*U*-value). The decrement factor is normally cited together with the *time lag*, that is, the time shift between the cyclic energy input and the corresponding response of the wall. Thus, the decrement factor provides information about the dampening of the periodic thermal signal passing from outside to inside, whereas the time lag gives the delay between a peak in the outdoor temperature profile and the corresponding peak in the heat flux released to the indoor air.
3. The *surface factor* is the ratio of the variation in heat flow about its mean value readmitted to the space from the surface to the variation of *heat flow about its mean* value absorbed by the surface. It is meant to capture the effects of internal sources of heat including solar gains through windows. The associated time lag is given by the *time factor*. The amplitude of the surface factor decreases and its time lag increases with increasing thermal conductivity, but both are virtually constant with thickness.

The interested reader can refer to CIBSE Guide A (2007) for more details and for tables of these sets of factors for different standard building construction types.

Problems

Numbers 1–4 given in parenthesis denote the degree of difficulty.

8.1 The following three conductive heat load profiles were calculated for south-facing walls in Denver in July. The walls are 4, 8, and 12 in. concrete. Without doing any calculations, identify which wall is which and justify your answer. *(1)*

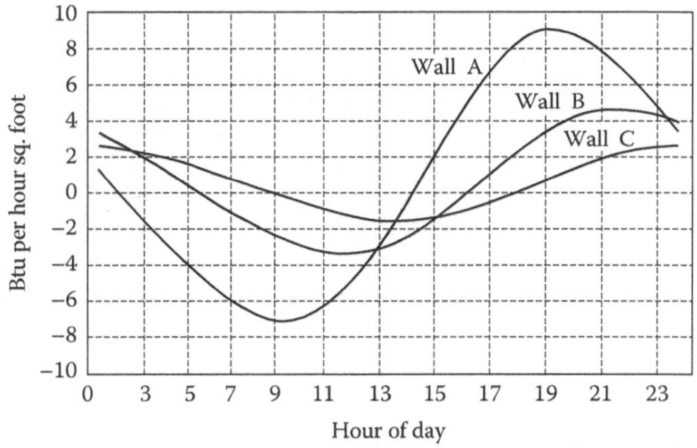

8.2 How much do personal computers increase the indoor temperature in a small office building without air conditioning if they are left on continuously? Assume a floor area of 500 m^2 (5380 ft^2), total heat transmission coefficient K_{tot} = 1200 W/K [2274 Btu/(h·°F)], and heat production per floor area 10 W/m^2 (0.93 W/ft^2) continuous. *(2)*

8.3 Consider the transient response of the 20 cm thick concrete wall that was solved in Example 8.1 using the explicit forward finite difference method. Solve the same problem but by discretizing the wall into 10 nodes. Compare your results with those of the 5-node discretization scheme. *(4)*

8.4 Solve Example 8.1 using the implicit solution method and compare your results with those of the explicit method. *(4)*

8.5 Figure 8.10 illustrates the diurnal periodic transient profiles for two types of walls using data provided in Table 8.3 (first and last entries),

(a) Generate the conduction gain values and compare your results with those shown in the figure.

(b) Instead of an indoor temperature of 25°C, assume a value of 28°C. Compare the resulting conduction heat gain profiles. *(4)*

8.6 You will compare the conductive heat gains through two types of roofs: a light weight (type #1 listed in Table 8.3) and a heavy one (type #6).

There is no solar radiation and the ambient temperature in a sinusoidal manner with a mean of 35°C (95°F) and an amplitude of 10°C (18°F) with a delay of 3 h. Assume an indoor temperature varies of 25°C. Compare the decrement and lag of the heat gain profiles with the corresponding values given in Table 8.3. *(3)*

8.7 What is the time constant of a building with total heat loss coefficient K_{tot} = 30 kW/K [56.85 kBtu/(h·°F)] and effective heat capacity C_{eff} = 2.4 GJ/K (1.26 MBtu/°F)? Estimate how long it takes the indoor temperature to drop from 20°C to 15°C after the heating system is shut off when $T_o = 0$°C. *(2)*

8.8 The air-conditioner of a building is to be switched off during a hot day so as to reduce electricity use during the on-peak period. The building has a time constant of 15 h, and the sum of the internal loads (equipment + lights + people) and the solar loads divided by (UA) = 4°C. Assume $T_{i,b}$ the internal air temperature at the beginning of the float-up to be 18°C. Calculate the number of hours it takes the building to heat up by 4°C, when

(a) Temperature difference between outdoor air temperature and the indoor air temperature at the start of the ramp up is 10°C

(b) When the difference is 20°C *(2)*

8.9 Set up a 3-node thermal network model for a 20 cm thick concrete wall with convective boundaries. Determine its thermal response under conditions specified in Example 8.1. Solve for both internal node temperatures and heat flux inflow to the conditioned space. Compare your results with those of Example 8.1. Discuss causes for differences of the transient response predicted by both methods. *(4)*

8.10 Example 8.4 illustrated the manner in which a 3-node thermal network model can be developed for a heterogeneous layered wall. Assume that there is no solar radiation and the ambient temperature in a sinusoidal manner with a mean of 35°C (95°F) and an amplitude of 10°C (18°F) with a delay of 3 h. Also assume an indoor temperature of 25°C. Determine the diurnal conduction heat gain profiles by two different transient modeling approaches: the thermal network approach and the conduction transfer function approach. Compare and discuss results. *(4)*

8.11 Toward the end of the Roman Empire, the technology of glassmaking had advanced to the point where some villas could be built with a sunspace (called a *solarium*). Make reasonable assumptions about the construction and use a simple thermal network to estimate the temperature in a solarium during a sunny day in Rome when outdoor temperature is given by

$$T_o = \left[1+\cos\left(\pi\frac{t-16\,\text{h}}{12\,\text{h}}\right)\right]\times 4°\text{C}$$

and incident solar irradiation by

$$I(t) = \max\left[0, -0.2+\cos\left(\pi\frac{t-12\,\text{h}}{12\,\text{h}}\right)\right]\times 800\ \text{W/m}^2$$

as a function of time of day t. Plot T_i and T_o. *(4)*

8.12 Suppose a fire breaks out in the dormitory room next to yours. For simplicity, assume the air temperature in your room remains constant at 70°F and that the average air temperature in the adjacent room is 600°F, which is achieved immediately after the fire starts and then remains constant. The wall between the room consists of 8 in. of heavyweight concrete. Unfortunately, you have taped a picture of your sweetheart to this wall. If the paper ignites when it reaches 451°F, how long after the fire starts will the picture catch fire? *(4)*

8.13 The temperature of the return-air plenum above the ceiling in a commercial building is determined by the heat input to this area by lights and the loss to the cooler space below as well as by the flow of air through the plenum. Figure P8.13 is a simplified diagram of such a plenum located above one of the floors in the interior of a building. Derive an equation for the plenum temperature T_p that includes the three effects just listed. Include the ceiling tile R value and area, the return airflow rate $\dot{m}_r$, the lighting fixture top heat loss $\dot{Q}_L$, and the zone temperature. How would your result be modified for the top floor of a building? *(4)*

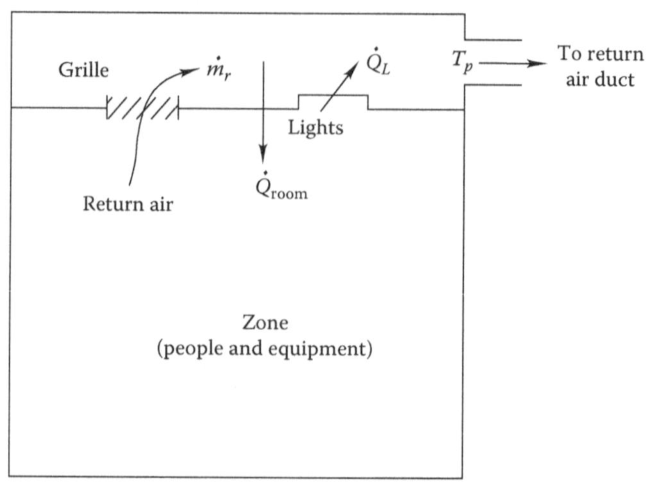

FIGURE P8.13

8.14 The CTS factors for a lightweight stud wall (type 6) are given in the following table. Repeat Example 8.3 using this wall and compare results with the heavy-weight wall. U value is 0.406 W/m²·K *(3)*

Lag, h	CTS, %	Hour	CTS, %	Hour	CTS, %
0	7	8	0	16	0
1	44	9	0	17	0
2	32	10	0	18	0
3	12	11	0	19	0
4	4	12	0	20	0
5	1	13	0	21	0
6	0	14	0	22	0
7	0	15	0	23	0

References

ASHRAE Fundamentals (1997, 2005, 2013). *Handbook of Fundamentals*. American Society of Heating, Refrigerating and Air-Conditioning Engineers, Atlanta, GA.

Birdsall, B., W.F. Buhl, K.L. Ellington, A.E. Erdem, and F.C. Winkelmann (1990). Overview of the DOE2.1 building energy analysis program. Report LBL-19735, rev. 1. Lawrence Berkeley Laboratory, Berkeley, CA.

Carslaw, H.S. and J.C. Jaeger (1986). *Conduction of Heat in Solids*, 2nd ed. Oxford University Press, London, U.K.

CIBSE Guide A (2007). *Environmental Design*, 7th ed. The Chartered Institution of Building Services Engineers, London, U.K.

Clarke, J.A. (1985). *Energy Simulation in Building Design*. Adam Hilger Ltd., Boston, MA.

Hittle, D.C. and R. Bishop (1983). An improved root-finding procedure for use in calculating transient heat flow through multilayered slabs. *Int. J. Heat Mass Transfer*, 26(1), 1685–1694.

Incropera, F.P., D.P. Dewitt, T.L. Bergman, and A.S. Lavine (2007). *Fundamentals of Heat and Mass Transfer*, 6th ed. Wiley, New York.

Kuehn, T.H., J.L. Threlkeld, and J.W. Ramsey (1998). *Thermal Environmental Engineering*, 3rd ed. Prentice-Hall, Englewood Cliffs, NJ.

Mitalas, G.P. (1968). Calculations of transient heat flow through walls and roofs. *ASHRAE Trans.*, 74(pt. 2), 182–188.

Mitalas, G.P. and D.G. Stephenson (1967). Room thermal response factors. *ASHRAE Trans.*, 73(pt. 1), 2.1–2.10.

Mitchell, J.W. and J.E. Braun (2013). *Principles of Heating, Ventilation and Air-Conditioning in Buildings*. John Wiley & Sons, New York.

Rabl, A. (1988). Parameter estimation in buildings: Methods for dynamic analysis of measured energy use. *ASME J. Solar Energy Eng.*, 110, 52–66.

Reddy, T.A. (1989). Application of dynamic building inverse models to three occupied residences monitored non-intrusively, in *Proceedings of the Thermal Performance of Exterior Envelopes of Buildings IV*, ASHRAE/DOE/BTECC/CIBSE, Orlando, FL, 654–671.

Santamouris, M. and D. Asimakopoulos (1996). *Passive Cooling of Buildings*. James and James (Science Publisher) Ltd., London, U.K.

Seem, J.E., S.A. Klein, W.A. Beckman, and J.W. Mitchell (1989). Transfer functions for efficient calculation of multidimensional transient heat transfer. *J. Heat Transfer*, 11, 5–12.

Sonderegger, R.C. (1977). Dynamic models of house heating based on equivalent thermal parameters. PhD thesis, Center for Energy and Environmental Studies Report No. 57, Princeton University, Princeton, NJ.

Spitler, J.D. (2010). *Load Calculation Applications Manual*. American Society of Heating, Refrigerating and Air-Conditioning Engineers, Atlanta, GA.

Stephenson, D.G. and G.P. Mitalas (1967). Cooling load calculations by thermal response factor method. *ASHRAE Trans.*, 73(pt. 2), 2.1–2.11.

Stephenson, D.G. and G.P. Mitalas (1971). Calculation of heat conduction transfer functions for multi-layer slabs. *ASHRAE Trans.*, 77(2), 117–126.

9

Heating and Cooling Design Load Calculations

ABSTRACT The calculation of various driving forces affecting cooling and heating loads of a zone or building along with steady-state and transient heat flow models has been treated in previous chapters. The focus of this chapter is to cover methods that combine all these heat flows and interactions in order to determine design loads on the peak day(s) of the year; annual loads and energy consumption are addressed in the next chapter. The procedure begins with a detailed description of the design climatic conditions. The peak heating load is relatively simple to determine requiring only a few heat transfer phenomena and a steady-state approach. We discuss various issues pertinent to cooling load calculations that require transient heat calculations and involve many more interactive terms to be considered.

We present three different methods to determine design cooling load. The first is the *transfer function method* (TFM) (implemented in the well-known DOE-2 and eQuest simulation programs) that involves three distinct sequential phases: calculation of individual heat gains of three types (conduction, solar, and internal), determination of cooling loads that depend on room and other properties, and the heat extraction rate (applicable when the zone thermostat set point is not kept constant). With the advent of greater computing power, many of the simplifications of TFM can be avoided by going back to the basic heat transfer equations and solving them in a more accurate and less arbitrary manner. We describe the basic principles and assumptions of the *heat balance method* (HBM), and how it is implemented in the EnergyPlus simulation program is discussed. Finally, a third method, namely the *radiant time series* (RTS), is described and illustrated by a solved example. This method, a simplified implementation version of HBM, is well suited for spreadsheet calculations since it is a step-by-step procedure that avoids having to perform iterative calculations like HBM. The RTS calculation procedure is in some ways similar to TFM but is a noniterative modification of the latter and treats internal thermal mass effects differently. Peak or design load calculations have to be done carefully since they directly impact the proper selection and sizing of the heating and cooling equipment.

Nomenclature

A	Area, m² (ft²)
b_n, d_n, c_n	Conduction transfer function coefficients appearing in Equation 8.25
CTF	Conduction transfer function model
CTS	Conduction time series model
c_p	Specific heat at constant pressure, kJ/(kg·K) [Btu/(lb$_m$·°F)]
DB	Dry-bulb temperature, K or °C (°R or °F)
DP	Dew-bulb temperature, K or °C (°R or °F)
DR	Daily range, K or °C (°R or °F)
HBM	Heat balance method for peak load calculation
HR	Humidity ratio, kg$_w$/kg$_a$ (lb$_w$/lb$_a$)
h	Surface heat transfer coefficient combining convection and radiation, W/(m²·K) [Btu/(h·ft²·°F)]
I_T	Global irradiance on tilted plane, W/m² (Btu/(h·ft²))
K_{tot}	Total heat transmission coefficient of building, W/K [Btu/(h·°F)]
MCDB	Mean coincident dry-bulb temperature, K or °C (°R or °F)
MCWB	Mean coincident wet-bulb temperature, K or °C (°R or °F)
p_n, g_n	Coefficients appearing in Equation 9.11
$\dot{Q}$	Heat rate or heat flow per unit time, W (Btu/h)
$\dot{q}$	Heat flux, W/m² [Btu/(h·ft²)]
RoTF	Room transfer function model (Equation 9.7)
RTF	Radiant time factors model (Equation 9.19)
RTS	Radiant time series method for peak load calculation
R	Thermal resistance, K/W [(°F·h)/Btu) [$\equiv \Delta T/\dot{Q}$]
r_j	Fraction of radiant load converted to the space at current time j (Equation 9.19)
SATF	Space air transfer function model (Equation 8.24)
SHGC	Solar heat gain coefficient of window (Equation 5.17)
T	Temperature, K or °C (°R or °F)
T_o	Outdoor dry-bulb temperature, K or °C (°R or °F)

TFM	Transfer function method for peak load calculation
t	Time or Hour
TMY	Typical meteorological year
$\dot{V}$	Volumetric flow rate, m³/s or L/s (ft³/min)
v	Velocity or wind speed, m/s (mi/h, (ft/s)
U	Overall heat transfer coefficient, W/(m²·K) [Btu/(h·ft²·°F)]
v_0, v_1, v_2	Coefficients appearing in Equation 9.7
w_1, w_2	Coefficients appearing in Equation 9.7
WB	Wet-bulb temperature, K or °C (°R or °F)
x	Fraction of daily temperature range (Table 9.3)
WD	Wind direction, degrees
WS	Wind speed, m/s (mph)

Greek

α	Absorptivity
ρ	Density, kg/m³ (lb_m/ft³)
Δq_{ir}	Correction to infrared radiation transfer between surface and environment, W/m² [Btu/(h·ft²)]
$\delta\dot{Q}$	Change in cooling load (Equation 9.10)
δT	Change in thermostat set point temperature (Equation 9.9)
Δt	Time interval, time step, s (h)

Subscripts

av	Average
c	Cooling
cond	Conduction
des	Design
gain	Heat gains
h	Heating
i	Indoor, internal
max	Maximum
min	Minimum
n	Index for transfer coefficients
o	Outdoor, external
os	Sol-air
ref	Reference
sum	Summer
t	Time, time subscript
tot	Total
vent	Ventilation
win	Winter
x	Extraction

9.1 Introduction

The various types of instantaneous heat gains/losses to a space or building have been discussed in previous chapters. They can be broadly classified into (1) conduction heat gains/losses through the building envelope (covered in Chapters 2 and 7), (2) solar heat gains through windows (Chapters 4 and 5), (3) infiltration and ventilation heat gains/losses (Chapter 6), and (4) internal heat gains (Chapter 7). Loads refer to the net thermal heat gains/losses and these have to be met by the HVAC equipment. For maintaining proper indoor conditions conducive to occupant comfort, the HVAC equipment has to be sized properly; sized too large, it would be needlessly expensive in first cost and part-load inefficiencies also penalize operating costs; sized too small, it would provide inadequate comfort during an unacceptably large number of hours over the year. A proper balance is thus necessary; design or peak load calculations have to be performed under prespecified design criteria following a well-accepted calculation methodology.

9.2 Winter and Summer Design Conditions

Loads depend on the weather and on indoor conditions that one wants to maintain. The former is not known in advance. If the HVAC equipment is to guarantee comfort at all times, it must be designed for peak conditions. What are the extremes? For most buildings, it would not be practical to aim for total protection by choosing the most extreme weather on record and adding a safety margin. Such oversizing of the HVAC equipment would be excessive, not just in first cost but also in operating cost. Therefore, a compromise is called for: reduce the cost of the HVAC equipment significantly while accepting the risk of slight discomfort under rare extremes of weather. The greater the extreme, the rarer the occurrence.

To help with the choice of design conditions, weather statistics corresponding to several levels of probability have been determined. They are the conditions that are exceeded at the site in question during a specified percentage of time (i.e., percentiles) and are determined based on historic climatic data over a certain number of years. These annual percentiles, widely accepted by designers, are as follows:

- For heating, 99.6% or 99%
- For cooling, 0.4%, 1%, or 2%

Extensive tables of design conditions for numerous locations worldwide (6443 locations to be specific) have been determined and assembled in the ASHRAE Fundamentals (2013). These represent annual and seasonal (i.e., monthly) combinations of temperature, humidity, solar radiation, and wind speed needed for design calculations. Specifically, the annual values cited are likely to be exceeded over the year (8760 h) at the corresponding percentage probability. Table 9.1 is an extract to illustrate the type of climatic design information provided (this table is specific to the airport at Phoenix, AZ). In order not to overwhelm the reader, we have intentionally left out some information provided in the ASHRAE climatic tables, such as the solar-related monthly atmospheric transmission values (previously shown in Table 4.2), precipitation, and degree-day/degree-hour information (discussed in Section 10.2). Table 9.2 is a more succinct climatic summary, assembling temperature and wind speed data for a few locations in the United States in both SI and IP units.

To see what these statistics imply, consider Washington, DC, where the 1% cooling design value for the dry-bulb temperature is 33.2°C. This would suggest that the ambient temperature is above this value during 0.010 × 8760 h = 87.6 h of the year (which obviously will occur during summer). Since the hottest hours are concentrated during afternoons rather than spread over the entire day, the corresponding number of days can be considerably higher than 1% of the year. Similarly the 99% heating design value for DBT = −6.3°C.

Let us look more closely at the data provided in Table 9.1. We note that for Phoenix, AZ, the hourly dry-bulb (DB) temperature values for the 99.6% and 99% design criteria are 3.7°C and 5.3°C, respectively, and December is the coldest month. For summer conditions, the 0.4%, 1%, and 2% DB design conditions correspond to 43.5°C, 42.4°C, and 41.3°C, respectively, and the hottest month is July. The diurnal DB temperature range for the hottest month is specified as 11.8°C.

Moreover, the tables also show the mean coincident wet-bulb temperatures (MCWB) for cooling conditions, defined as the average wet-bulb temperature during the hours when DB is at the specified levels of 0.4%, 1.0%, and 2%. The MCWB values for Phoenix are 20.9°C, 20.8°C, and 20.7°C, respectively. Also included are the mean coincident wind speed at the 0.4% level (MCWS = 4.2 m/s) and prevailing coincident wind direction (PCWD). For heating conditions (winter), no wet-bulb temperature data are given. Usually, this poses no serious problem because latent loads during the heating season are zero if one does not humidify. If one does humidify, uncertainties in the value of the outdoor humidity have little effect on the latent load because the absolute humidity of outdoor air in winter is very low (see Example 7.8). Nonetheless, dew point (DP) values are listed. Note that values for HR (humidity ratio) and DP (dew point temperature) are also listed. The tables also include design WB temperatures, at the 0.4%, 1.0%, and 2% levels, along with the mean coincident dry-bulb (MCDB) temperatures. This information is needed for analyzing and selecting evaporative equipment such as cooling towers.

The earlier discussion pertains to conditions representative of annual conditions. Seasonally representative conditions are specified by the month-by-month values of DB/MCWB, WB/MCDB, and DR also listed in the lower half of Table 9.1 for four design criteria (0.4%, 2%, 5%, and 10%). Note that the DB and WB values for the first two design criteria are not the same as those listed for the annual values in the upper half of the table. For example, the hottest month for Phoenix is July and the annual DB design values are 43.5°C and 41.3°C corresponding to the 0.4% and 2% probability criteria; these are different than those listed under the July column (the values are 45.4°C and 43.8°C) for the same probability criteria. Hence, design temperatures for annual peak calculations and for individual monthly peak calculations are different.

A certain amount of judgment is needed in the choice of design conditions. For ordinary buildings, it is customary to base the design on the level of 1.0% in summer and 99.0% in winter that will result in smaller equipment sizes and, thus, have a lower initial cost. For critical applications such as hospitals or sensitive industrial processes or for lightweight buildings, one may prefer the more stringent level of 0.4% in summer (and 99.6% in winter). Thermal inertia can help reduce the risk of discomfort: it delays and attenuates the peak loads, as will be explained in the following sections. Therefore, one may move to a less stringent design level for a given application if the building is very massive.*

For peak heating loads, one need not bother with solar radiation because the extremes occur during winter nights. For cooling loads, solar radiation is crucial, and how to calculate clear sky irradiation using the ASHRAE model has been discussed in Section 4.6. For opaque surfaces, the effect of solar radiation is treated by means of the sol–air temperature (Section 7.2); and for glazing, by means of the solar heat gain coefficient SHGC (Section 5.4).

Wind speed is another weather-dependent variable that has a bearing on loads. Traditionally, the ASHRAE Fundamentals (2013) value

$$v_{win} = 15\ \text{mph}\ (6.7\ \text{m/s}) \tag{9.1}$$

* As a guide for the assessment of the relation between persistence of cold weather and thermal inertia, we note that according to studies at several stations, temperatures below the design conditions can persist for up to a week (ASHRAE Fundamentals, 2013).

TABLE 9.1

Design Climatic Data Table for Phoenix, AZ (SI Units)

Phoenix/Sky Harbor, AZ						WMO 722780
Lat. 33.44 °N	**Long. 111.99 °W**	**Elev: 337 m**	**stdP: 97.34**	**Time Zone: -7.00 (NAZ)**	**Period: 1986–2010**	**WBAN: 23183**

Annual Heating and Humidification Design Conditions

Coldest Month	Heating DB		Humidification on DP/MCDB and HR						Coldest Month WS/MCDB				MCWS/PCWD to 99.6% DB	
			99.6%			99%			0.4%		1%			
	99.6%	99%	DP	HR	MCDB	DP	HR	MCDB	WS	MCDB	WS	MCDB	MCWS	PCWD
12	**3.7**	**5.3**	**−15.3**	**1.0**	**15.9**	**−12.8**	**1.3**	**17.8**	**8.4**	**14.1**	**7.4**	**15.1**	**1.6**	**100**

Annual Cooling and Dehumidification, Design Conditions

Hottest month	Hottest Month DB Range	Cooling DB/MCWB						Evaporation WB/MCDB						MCWS/PCWD to 0.4% DB	
		0.4%		1%		2%		0.4%		1%		2%			
		DB	MCWB	DB	MCWB	DB	MCWB	WB	MCDB	WB	MCDB	WB	MCDB	MCWS	PCWD
7	**11.8**	**43.5**	**20.9**	**42.4**	**20.8**	**41.3**	**20.7**	**24.4**	**35.4**	**23.9**	**35.2**	**23.4**	**34.9**	**4.2**	**260**

			Jan	Feb	Mar	Apr	May	Jun	July	Aug	Sep	Oct	Nov	Dec
Monthly design dry-bulb and mean coincident wet-bulb temperatures	0.4%	DB	26.1	28.5	33.6	37.4	41.4	45.0	45.4	44.0	42.2	38.5	32.2	24.7
		MCWB	11.7	13.2	15.2	16.7	18.6	20.3	21.2	21.7	20.8	18.6	15.2	11.6
	2%	DB	23.9	26.4	31.0	35.1	39.2	43.0	43.8	42.6	40.7	36.7	29.6	22.8
		MCWB	11.1	12.4	14.4	16.0	17.7	19.2	21.2	21.8	20.4	18.2	14.5	11.0
	5%	DB	22.0	24.4	28.9	33.2	37.5	41.9	42.6	41.5	39.4	34.8	27.4	21.2
		WCWB	10.6	11.7	13.7	15.1	17.0	19.1	21.3	21.9	20.3	17.5	13.5	10.4
	10%	DB	20.0	22.5	27.0	31.3	36.0	40.6	41.4	40.4	37.9	32.8	25.4	19.2
		MCWB	10.1	11.2	13.0	14.5	16.7	18.9	21.2	22.0	20.0	16.8	12.9	9.7
Monthly design dry-bulb and mean coincident dry-bulb temperatures	0.4%	WB	14.7	15.2	16.4	18.0	19.8	23.0	25.0	25.3	24.5	21.1	17.3	14.3
		MCDB	18.4	23.6	29.2	34.8	36.5	38.8	35.2	37.1	34.5	31.2	24.9	17.9
	2%	WB	13.2	14.1	15.3	16.9	19.0	21.9	24.2	24.5	23.6	20.0	16.1	13.0
		MCDB	18.8	20.9	27.7	33.0	35.2	37.7	35.2	35.9	34.0	31.2	23.9	18.2
	5%	WB	12.2	13.3	14.5	16.0	18.3	20.9	23.8	24.0	22.9	19.0	15.1	12.0
		MCDB	18.6	20.5	26.3	31.2	34.4	37.3	35.0	35.3	33.4	30.7	23.9	17.6
	10%	WB	11.3	12.3	13.6	15.1	17.6	20.0	23.3	23.5	22.2	18.1	14.0	11.1
		MCDB	17.9	19.9	24.6	29.3	33.4	37.2	34.8	35.1	32.8	29.9	23.5	17.3
Mean daily temperature range	5% DB	MDBR	11.4	11.4	12.5	13.2	13.4	13.8	11.8	11.3	12.1	12.5	12.2	11.2
		MCDBR	14.2	14.5	15.5	15.6	15.0	14.9	12.9	12.2	13.2	14.4	14.6	13.6
		MCWBR	6.2	6.1	5.8	5.7	5.0	4.7	3.6	3.3	3.9	5.2	5.8	5.9
	5% WB	MCDBR	10.6	11.6	14.0	14.4	13.8	12.9	11.0	11.0	10.8	11.8	11.9	10.1
		MCWBR	5.1	5.1	5.3	5.6	4.8	4.4	3.1	3.1	3.4	4.6	5.3	5.0

Source: Extracted from ASHRAE, *Handbook of Fundamentals*, American Society of Heating, Refrigerating and Air-Conditioning Engineers, Atlanta, GA, 2013. Copyright ASHRAE, www.ashrae.org.

DB, dry-bulb temperature (°C); DP - dew point temperature (°C); HR, humidity ratio (g/kg); MCDB, mean coincident dry bulb temperature (°C); MCWB, mean coincident wet-bulb temperature (°C); WS, wind speed (m/s); MCWS, mean coincident wind speed (m/s); PCWD,prevailing coincident wind direction; WB, wet bulb temperature (°C); WD, wind direction (0°- North, 90°- East).

TABLE 9.2

Design Weather Data for Selected Locations (SI Units)

		Atlanta, GA	Chicago, IL	Denver, CO	New York	Phoenix, AZ	San Francisco, CA	Seattle, WA	Washington, DC
Latitude		33.64°N	41.99°N	39.83°N	40.66°N	33.42°N	37.62°N	47.46°N	38.87°N
Elevation		313 m/1027 ft	205 m/673 ft	1655 m/5430 ft	7 m/23 ft	339 m/1106 ft	6 m/20 ft	132 m/433 ft	20 m/66 ft
				SI Units					
Heating	Coldest month	1	1	12	1	12	1	12	1
	DBT (°C):								
	99.6%	−5.8	−18.6	−17.5	−10.1	3.7	3.9	−3.8	−8.2
	99%	−3.1	−15.7	−14.1	−7.9	5.3	5.2	−1.3	−6.3
	MCWS (m/s) at 99.6% DBT	5.3	4.9	3.5	7.4	1.6	2.2	4.0	5.4
Cooling	Hottest month	7	7	7	7	7	8	8	7
	Hottest month daily range (°C)	9.4	10.2	15.3	7.4	11.8	8.4	10.4	9.0
	DBT/MCWBT (°C)								
	0.4%	34.4/23.4	33.0/23.5	34.6/15.5	32.1/22.7	43.5/20.9	28.2/17.1	29.6/18.4	34.7/24.3
	1%	33.1/23.3	31.5/22.9	33.3/15.5	30.3/22.1	42.4/20.8	25.6/16.6	27.6/17.6	33.2/23.8
	2%	32.1/23.0	30.0/22.1	31.8/15.4	28.7/21.7	41.3/20.7	23.5/16.2	25.7/17.0	31.9/23.1
	MCWS (m/s) at 0.4% DBT	3.9	5.1	4.1	5.8	4.2	5.7	4.1	4.6
				IP Units					
Heating	Coldest month	1	1	12	1	12	1	12	1
	DBT (°F)								
	99.6%	21.5	−1.5	0.5	13.8	38.7	39.1	25.2	17.3
	99%	26.4	3.7	6.6	17.8	41.6	41.4	29.6	20.7
	MCWS (mph) at 99.6% DBT	11.9	10.9	7.8	16.6	3.7	4.9	9.0	12.0
Cooling	Hottest month	7	7	7	7	7	8	8	7
	Hottest month daily range (°F)	17.0	18.3	27.5	13.3	21.3	15.2	18.7	16.2
	DBT/MCWBT (°F)								
	0.4%	93.9/74.2	91.4/74.3	94.4/60.0	89.8/72.9	110.3/69.6	82.8/62.9	85.3/65.2	94.5/75.7
	1%	91.7/73.9	88.7/73.2	91.7/59.8	86.5/71.8	108.3/69.4	78.1/61.9	81.6/63.7	91.8/74.8
	2%	89.8/73.5	86.0/71.8	89.3/59.6	83.7/71.1	106.4/69.3	74.2/61.2	78.3/62.6	89.5/73.7
	MCWS (m/s) at 0.4% DBT	8.7	11.3	9.3	13.0	9.4	12.8	9.2	10.3

DBT, dry-bulb temperature; MCWBT, mean coincident wet-bulb temperature; MCWS, mean coincident wind speed.

has been recommended for heating loads in winter if there is nothing to imply extreme conditions (such as an exposed hilltop location). For cooling loads, a value one-half as large is recommended:

$$v_{w,sum} = 7.5 \text{ mph } (3.4 \text{ m/s}) \tag{9.2}$$

because wind tends to be less strong in summer than in winter. Of particular interest is the exterior surface heat transfer coefficient (radiation plus convection) h_o for which the recommended design values are

$$h_{o,win} = 6.0 \text{ Btu/(h}\cdot\text{ft}^2\cdot{}^\circ\text{F) [34.0 W/(m}^2\cdot\text{K)]} \tag{9.3}$$

$$h_{o,sum} = 4.0 \text{ Btu/(h}\cdot\text{ft}^2\cdot{}^\circ\text{F) [22.7 W/(m}^2\cdot\text{K)]} \tag{9.4}$$

This coefficient is only one of several components of the calculation of thermal loads, and it enters only through the building heat transmission coefficient K_{tot} defined in Section 7.5.1. The better a building is insulated and tightened, lower is the effect of wind on K_{tot}. With current practice for new construction in the United States, typical wind speed variations may change the heat transmission coefficient by about 10% relative to the value at design conditions. Rather than use such general design values for wind speed, it is recommended that location-specific values be used as far as possible. Subsequently, Equation 7.22 can be used to determine the corresponding surface heat transfer coefficient h_o.

9.3 Design Heating Load Calculation Procedure

Since the coldest weather may occur during periods without solar radiation, it is advisable not to rely on the benefit of solar heat gains when calculating peak heating loads (unless the building contains long-term storage). If the indoor temperature T_i is constant, a static analysis is sufficient, and the calculation of the peak heating load $\dot{Q}_{h,\max}$ is very simple: Find the design heat loss coefficient K_{tot}, multiply by the design temperature difference $(T_i - T_o)$, and subtract the internal heat gains known to occur during the coldest period:

$$\dot{Q}_{h,\max} = K_{tot}(T_i - T_o) - \dot{Q}_{gain} \tag{9.5}$$

In fact, we have demonstrated this procedure already with the house of Example 7.7, which we recall at this point. It is a building located in Washington, DC, with

$$K_{tot} = 186 \text{ W/K [353 Btu/(h}\cdot{}^\circ\text{F)]} \quad \text{and}$$

$$\dot{Q}_{gain} = 1.0 \text{ kW (3.4 kBtu/h)}$$

We had selected the 99% design condition:

$$T_o = -6.3^\circ\text{C } (20.7^\circ\text{F}) \quad \text{and} \quad T_i = 22^\circ\text{C } (71.6^\circ\text{F})$$

resulting in a design heat load of

$$\dot{Q}_{h,\max} = 4.26 \text{ kW (14.55 kBtu/h)}$$

Example 7.8 has shown that the latent load of humidification (if a humidifier is used) would increase the load by about 14%.

The traditional steady-state calculation of peak heating loads was well justified for buildings with constant thermostat set point. But thermostat setback has become common practice for energy conservation. Thermostat setback can have a sizable impact on peak heating loads because setback recovery occurs during the early morning hours on top of the peak heat loss. What would happen if the thermostat were set back at night? For a rough indication, recall from Section 8.1.1 that this house (of lightweight construction typical in the United States) would require heat input at the rate of 1.2 kW if its temperature were to be increased by 2°C in 5 h. For setback recovery after winter nights, one might want rates that are several times faster, say, 4°C in 2.5 h. Assuming that the heat input is proportional to the warm-up rate,* the extra load for setback recovery (also known as the *pickup load*) would be 4 × 1.2 kW = 4.8 kW, comparable to the static design heat load. In this case, the capacity of the heating system would have to be doubled relative to the case without setback.

In a given situation, the required extra capacity depends on the amount of setback $T_i - T_o$, the acceptable recovery time, and building construction. For reasonable accuracy, a dynamic analysis is recommended. Optimizing the capacity of the heating system involves a trade-off between energy savings and capacity savings, with due attention to part-load efficiency. As a general rule for residences, ASHRAE Fundamentals

* Actually, at faster rates the effective heat capacity is smaller (the heat pulse takes longer than 1 h to penetrate the entire mass), and so the real increment for setback recovery is less. We do not know how much without doing a detailed dynamic analysis, but in any case this discussion confirms the warning about the incompatibility of static analysis and setback recovery.

(2013) recommends oversizing by about 40% for a night setback of 10°F (5.6°C) to be increased to 60% oversizing if there is additional setback during the day. In any case, some flexibility can be provided by adapting the operation of the building: If the capacity turns out insufficient, the occupant can reduce the depth and duration of the setback during the coldest periods. In commercial buildings with mechanical ventilation, the demand for extra capacity during setback recovery is reduced if the outdoor air intake is lowered during unoccupied periods (see Section 3.5.3 where ASHRAE Standard 62.1, 2013 is discussed).

9.4 Subtleties with Cooling Load Calculations

9.4.1 Calculation of Design-Day Sol–Air Temperatures

The concept of sol–air temperature and how to determine their numerical values were discussed in Section 7.2. Recall that this concept is convenient since it allows the determination of the combined effects of outdoor temperature and absorbed solar irradiation on conduction heat gains through opaque surfaces. The discussion on sol–air temperature and the generation of hourly solar-air temperatures in Phoenix for January 21 and July 21 (Figure 7.3) may have incorrectly suggested to the reader that design day hourly sol–air temperature values could be simply determined using typical meteorological day (TMY) data of hourly ambient temperature and solar radiation for the hottest month. The problem with such an approach is that TMY data are representative of *mean climatic conditions* over several years at that location and would thereby underrepresent peak day behavior. This would result in undersizing equipment that would lead to occupant complaints. The TMY data are appropriate for annual energy calculations but not for peak day design calculations.

Design heating load calculations require a single steady-state analysis as discussed in the previous section. Design cooling loads require a dynamic calculation (due to reasons discussed in Chapter 8 and in the next subsection) and may require, in addition to the peak annual load calculation, the determination of peak loads for individual months. This is because peak cooling loads are greatly affected by solar radiation and by the specific size and orientation of windows, and so the cooling loads for several months have to be determined in order to identify the annual maximum.

We shall describe how the hourly sol–air temperatures for the annual design day are determined, and a similar approach can be followed for individual monthly calculations:

1. Select for the hottest month listed, the monthly clear sky atmospheric optical depth coefficients specified in the design tables for the specific location (such as Table 4.2).
2. Compute the clear sky hourly beam normal and diffuse horizontal values following the ASHRAE model, namely Equations 4.24 and 4.25, and illustrated in Example 4.9.
3. Use the ASHRAE anisotropic sky model to determine the global radiation on surfaces with different orientations and tilts (see Section 4.7.2 and Example 4.11).
4. Compute the hourly ambient dry-bulb temperature for the design day (to be discussed in the following).
5. Finally, determine the hourly sol–air temperatures on the various surfaces following Section 7.2.

Step (4) requires the determination of the hourly ambient dry-bulb values (T_o) during the design day of the selected month. The climatic tables list the dry-bulb range (DR) for the hottest month along with the design dry-bulb temperature value ($T_{o,\max}$). The diurnal hourly values can then be generated from

$$T_o(h) = T_{o,\max} - \mathrm{DR} \times x \tag{9.6}$$

where x is the fraction of the daily temperature range given in Table 9.3. The resulting diurnal profile is close to sinusoidal.

In Figure 9.1, we have plotted, for two different climatic locations (Atlanta and Phoenix), the diurnal peak day sol–air temperatures T_{os} for horizontal and vertical surfaces facing the four cardinal directions based on the 1% ASHRAE summer design conditions. These sets

TABLE 9.3

Fraction (x) of Daily Temperature Range Used for Peak Day Calculation

Time, h	Fraction	Time, h	Fraction	Time, h	Fraction
1	0.88	9	0.55	17	0.14
2	0.92	10	0.38	18	0.24
3	0.95	11	0.23	19	0.39
4	0.98	12	0.13	20	0.50
5	1.00	13	0.05	21	0.59
6	0.98	14	0.00	22	0.68
7	0.91	15	0.00	23	0.75
8	0.74	16	0.06	24	0.82

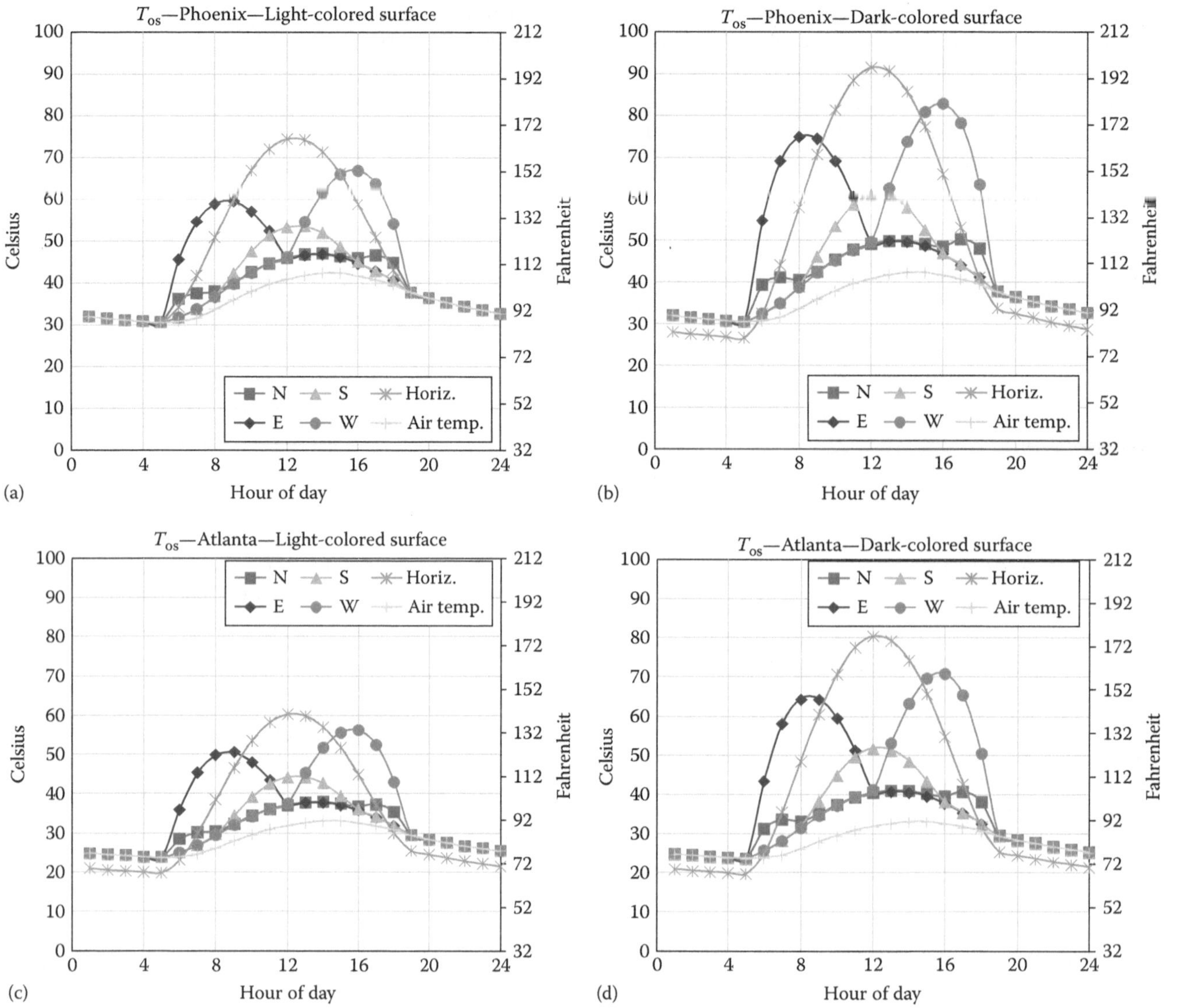

FIGURE 9.1
Diurnal sol–air temperatures for various surface orientations compared to ambient dry-bulb profile during the peak hottest day. (a) Phoenix, light-colored surface; (b) Phoenix, dark-colored surface, (c) Atlanta, light-colored surface, and (d) Atlanta, dark-colored surface.

of plots are meant to illustrate the large differences between the ambient T_o and the sol–air T_{os} temperatures for all surfaces (the exception is the north-facing surface that only receives diffuse radiation). They also highlight the effects of light and dark surfaces with α/h_o = 0.18 and 0.30 (h·ft²·°F)/Btu [0.032 and 0.052 (m²·K)/W] respectively (see Section 7.2). The sol–air temperatures at other values of α/h_o can be obtained from these plots by scaling the difference between the curves for T_o and T_{os} for the surface in question (for the horizontal this is not quite correct because of Δq_{ir}). As expected, T_{os} values are greater for Phoenix, which is both sunnier and hotter than Atlanta (there is about a 12°C difference in horizontal surface values of T_{os} at noon). The high T_{os} values for horizontal surfaces several hours before and after noon are to be noted. Also, at noon, there is about 20°C difference between lighter and darker surfaces for both locations reinforcing the use of lighter surfaces for roofs in order to minimize envelope heat gains.

9.4.2 Need to Consider Storage Effects

Heat capacity tends to be more important for proper determination of cooling than for heating loads, for a number of reasons. Summer heat flows are more peaked than those in winter. Peak heating loads correspond to times without sun, and the diurnal variation of $T_i - T_o$ is small compared to its maximum in most climates. By contrast, for peak cooling loads, the diurnal variation of

$T_i - T_0$ is comparable to its maximum, and solar gains are crucial. Also, in climates with cold winters, heating loads are larger than cooling loads, and the storage terms, for typical temperature excursions, are relatively less important in winter than in summer. Finally, the initial and operating cost implications of oversizing cooling equipment, should proper consideration not be given to storage and transient effects, is much more severe than oversizing heating equipment.

Storage effects for latent loads are difficult to analyze (see, e.g., Fairey and Kerestecioglu, 1985; Kerestecioglu and Gu, 1990), and most of the current computer programs for building simulation, such as DOE2.1 (Birdsall et al., 1990), do not account for moisture exchange with the building mass. In practice, this neglect of moisture storage is usually not a serious problem. Precise humidity control is not very important in most buildings. Where it is important, e.g., in hospitals, temperature and humidity are maintained constant around the clock. When the air is at constant conditions, the moisture in the materials does not change much and storage effects can be neglected; but in buildings with intermittent operation, these effects can be large, as shown by Fairey and Kerestecioglu (1985) and by Wong and Wang (1990).

Because of thermal inertia, it is advisable to distinguish between several heat flow rates. The *heat gain* is the rate at which heat is transferred to (say, via transient heat conduction through the walls or roofs discussed in Section 8.3) or generated in the space. The *cooling load* is the rate at which the cooling equipment would have to remove thermal energy from the air in the space in order to maintain constant temperature and humidity. The difference in diurnal variations of both the gain and the cooling load is conceptually illustrated in Figure 9.2. Finally, the *heat extraction rate* is the rate at which the cooling equipment actually removes thermal energy from the space when the indoor air temperature is varied during the day.*

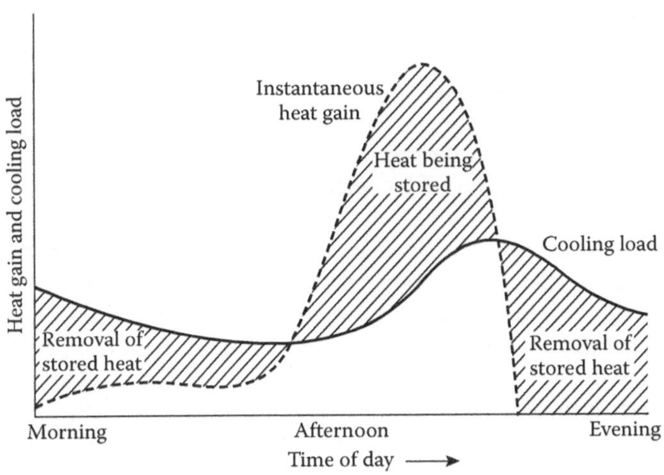

FIGURE 9.2
Plots to illustrate differences in diurnal variation of heat gains and cooling loads.

Conductive heat gains and radiative heat gains are not directly absorbed by the indoor air; a certain portion of these flows is first absorbed by the interior thermal mass (Figure 9.3) thereby increasing its temperature relative to the indoor air. Only gradually are they transferred to the air. Thus, their contribution to the cooling load is delayed, and there is a difference between heat gain and cooling load. Averaged over time, these rates are, of course, equal, by virtue of the first law of thermodynamics. The heat extraction rate is equal to the cooling load only if the temperature of the indoor air is constant (as assumed in this section). Otherwise, the heat flow to and from the building mass causes the heat extraction rate to differ from the cooling load (this feature is also analyzed later in this chapter).

The following *recommendations for the importance of dynamic effects* emerge from the earlier discussion:

1. They can significantly reduce the peak cooling loads, with or without thermostat setup.
2. They can be neglected for peak heating loads, except if thermostat setback recovery is to be applied (in that case a pick-up allowance is simply provided).
3. They play a major role in the determination of peak cooling loads.
4. For the calculation of annual consumption, they can have an appreciable effect if the indoor temperature is not kept constant.
5. Storage of *latent* heat is neglected for most applications.

To preserve much of the simplicity of the static approach in a method for peak cooling loads, ASHRAE had developed the CLTD/SCL/CLF method (ASHRAE Fundamentals, 1989) that modifies the terms of a static calculation to account for thermal inertia. This method is rather empirical with limited accuracy; even then it has been widely used by smaller design firms and HVAC energy service companies during the last 25–30 years since it is amenable to spreadsheet calculations. However, the onset of building energy simulation programs and an ever-growing pool of analysts knowledgeable in the use of such programs has gradually made this method obsolete. The previous two

* In later chapters on HVAC systems, we will encounter yet another rate, the *coil load*; it is the rate at which the cooling coil removes heat from the air, and it can be different from the heat extraction rate due to losses in the distribution system.

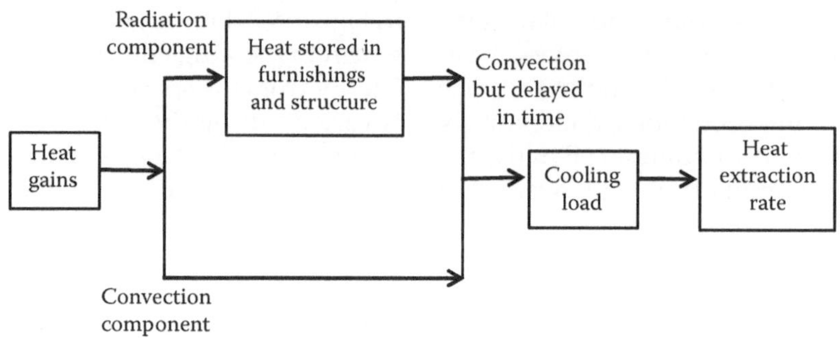

FIGURE 9.3
Schematic illustrating relationships between heat gain, cooling load, and heat extraction rate.

editions of this textbook described this method at some length, but it has been dropped in the current version. A description of this method along with an illustrative program is provided in the online HCB software.

9.4.3 Extensions to Multiple Zones

The calculation of peak *heating* loads and capacities can often be done without considering separate perimeter zones because peak heating loads occur when there is no sun; with uniform internal gains, the corresponding thermal balance is uniform around the perimeter. But while the calculation can be carried out for a single zone, the operation requires multiple zones: The heating system must allow separate control of different facades to compensate for the variability of solar gains during the day. For *cooling* loads, a multizone analysis is essential, even at the calculation stage, because the loads occur when the sun is shining.

As discussed in the previous section, peak cooling loads require a dynamic analysis whereas peak heating loads can be estimated quite well by static models (at least in the absence of thermostat setback). Compared to heating loads, the calculation of cooling loads of large buildings is thus doubly complicated: It requires multiple zones and dynamic analysis if one wants reasonable accuracy.

A related issue is the coincidence between peak loads of different zones. To determine the capacity of the central plant, one needs to know the peak load of the totality of zones served by the plant. This is usually less than the simple sum of the individual peak loads because of noncoincidence. The term *diversity* is used to designate the ratio of the actual system peak to the sum of the individual peak loads. In practice, one often finds a diversity around 0.6–0.8 for large buildings or groups of buildings (e.g., university campuses); for better estimates at the design stage, computer simulations are recommended.

9.5 Transfer Function Method for Cooling Load Calculations

9.5.1 Basis of the Method

The conduction transfer function (CTF) model was introduced in Section 8.3.2 as a popular method to analyze transient conduction heat flows through building elements (such as walls and roofs). In essence, the conductive heat gains $\dot{Q}_{cond}$ can be considered to be the *response* of the building element to the *driving terms* (T_i, T_{os}) that act on it and time-lagged values of $\dot{Q}_{cond}$. The response is formulated as a linear time series such as

$$\dot{Q}_{cond,t} = -\sum_{n\geq 1} d_n \dot{Q}_{cond,t-n\Delta t} + A\left(\sum_{n\geq 0} b_n T_{os,t-n\Delta t} - T_i \sum_{n\geq 0} c_n\right) \tag{8.25}$$

where

- A is the area of roof or wall, m² (ft²)
- t is the time subscript
- Δt is the time step = 1 h
- $T_{os,t}$ is the sol–air temperature of outside surface at time t
- T_i is the internal temperature kept constant
- b_n, c_n, d_n are the CTF coefficients

This equation can be considered an algorithm for calculating $\dot{Q}_{cond}$, hour by hour, given the previous values of $\dot{Q}_{cond}$, T_i and the driving terms T_{os}. Once the necessary numerical values of the transfer function coefficients have been obtained (such as in Table 8.3), the calculation of peak loads is simple enough for a spreadsheet. One specifies the driving terms for the peak day and iterates an equation akin to Equation 8.25 until the result converges to a steady daily pattern. These CTF coefficients are also included in the online HCB software supplied with this book.* In the remainder of this section, we discuss how this CTF model has been extended and combined with other sources of heat gain to form the ASHRAE transfer function method (TFM) (ASHRAE Fundamentals, 1997). A simplified version of TFM is available as part of the online HCB software. The method involves three steps:

1. Calculation of the conductive heat gain (or loss) for each distinct component of the envelope using Equation 8.25 and illustrated by Example 8.2
2. Calculation of the load of the room at constant temperature, based on this conductive heat gain (or loss), as well as any other heat sources in the room
3. Calculation of the heat extraction (or addition) rate for the cooling (or heating) device and thermostat set points of the room

Steps 2 and 3 will be elaborated in the following.

9.5.2 Cooling Load at Constant Temperature

The following are the broad types of heat gains:

1. Conduction through exterior envelop elements (walls, roof, windows, etc.) and solar radiation absorbed by interior shade
2. Solar transmittivity through glass without interior shade and the radiative component of heat from occupants and equipment
3. Lights (note that the coefficients depend on the arrangement of the lighting fixture and the ventilation system)
4. Convective gains (from air exchange, occupants, equipment)

The calculation of the conductive heat gain (or loss) as described in Section 8.3 is to be repeated for each element of the room envelope that has a distinct composition. The relation between these conductive gains and the total load depends on the construction of the entire room. For example, a concrete floor can store a significant fraction of the heat radiated by lights or by a warm ceiling, thus postponing its contribution to the cooling load of the room.

For each heat gain component $\dot{Q}_{gain}$, the corresponding cooling load $\dot{Q}_c$ (or reduction of the heating load) at constant T_i is calculated by using another set of coefficients, the coefficients υ_n and w_n, of the *room transfer function* (RoTF):

$$\dot{Q}_{c,t} = \upsilon_0\dot{Q}_{gain,t} + \upsilon_1\dot{Q}_{gain,t-\Delta t} + \upsilon_2\dot{Q}_{gain,t-2\Delta t} + \cdots - w_1\dot{Q}_{c,t-\Delta t} - w_2\dot{Q}_{c,t-2\Delta t} - \cdots \quad (9.7)$$

with the subscript t indicating time, as before. The coefficient w_0 of $\dot{Q}_{c,t}$ is not shown because it is set equal to unity. The RoTF coefficients for a variety of room construction types are listed in Tables 9.4 and 9.5. In these tables, all coefficients with index 2 or higher are zero. Since w_0 is unity, only a single coefficient w_1 is shown. Again, it is instructive to take the steady-state limit and check the consistency with the first law of thermodynamics. It requires that the sum of the v_n values equal the sum of the w_n values:

$$\sum_{n\geq 0} \upsilon_n = \sum_{n\geq 0} w_n \quad \text{where } w_0 = 1 \quad (9.8)$$

The reader can verify that the entries of Tables 9.4 and 9.5 do indeed satisfy this condition.

Equation 9.7 has to be applied separately to each of the heat gain types in Table 9.5, and the resulting cooling load components $\dot{Q}_{c,t}$ are added to obtain the total cooling load of the room at time t.

Example 9.1: Cooling Load Calculation for a Room

Find the cooling load for a room kept at 25°C that has the following three types of heat gain:

1. Conduction through 10 m² of the exterior west-facing wall of Example 8.2.
2. 300 W of electric lights that are turned on from 9:00 to 18:00 (hour ending); the light fixtures are vented.
3. Air exchange (ventilation plus infiltration) at a constant rate of 5.208 L/s (corresponds to 0.5 air change per hour for a room floor area of 15 m² and height of 2.5 m)—the air exchange can consequently be assumed to be "medium."

In addition, we will assume that the envelope construction type is medium, 150 mm (6 in.) concrete floor, while the room contains "ordinary" furniture. This room could be a windowless office area surrounded on all but one side by rooms at the same temperature. In practice, the ventilation rate may be greatly reduced when the office is unoccupied, but here we assume constant air exchange for simplicity.

* Some of the numbers may be slightly different because these coefficients have undergone small modifications over the years.

TABLE 9.4

w_1 Coefficient of the Room Transfer Function (RoTF) (w_0 = 1.0 and Higher Terms Are Zero)

	Room Envelope Construction[b]				
	50 mm (2 in.) Wood Floor	75 mm (3 in.) Concrete Floor	150 mm (6 in.) Concrete Floor	200 mm (8 in.) Concrete Floor	300 mm (12 in.) Concrete Floor
	Specific Mass per Unit Floor Area, kg/m² (lb$_m$/ft²)				
Room Air[a] Circulation and S/R Type	50 (10)	200 (40)	370 (75)	590 (120)	780 (160)
Low	−0.88	−0.92	−0.95	−0.97	−0.98
Medium	−0.84	−0.90	−0.94	−0.96	−0.97
High	−0.81	−0.88	−0.93	−0.95	−0.97
Very high	−0.77	−0.85	−0.92	−0.95	−0.97
	−0.73	−0.83	−0.91	−0.94	−0.96

Source: ASHRAE, *Handbook of Fundamentals*, American Society of Heating, Refrigerating and Air-Conditioning Engineers, Atlanta, GA, 1997. Copyright ASHRAE, www.ashrae.org.

[a] Circulation rate

Low: Minimum required to cope with cooling load from lights and occupants in interior zone. Supply through floor, wall, or ceiling diffuser. Ceiling space not used for return air, and h = 2.27 W/(m²·K) [0.4 Btu/(h·ft²·°F)], where h = inside surface convection coefficient used in calculation of w_1 value.

Medium: Supplied through floor, wall, or ceiling diffuser. Ceiling space not used for return air, and h = 3.41 W/(m²·K) [0.6 Btu/(h·ft²·°F).

High: Room air circulation induced by primary air of induction unit, or by room fan and coil unit. Ceiling space used for return air, and h = 4.54 W/(m²·K) [0.8 Btu/(h·ft²·°F).

Very high: Used to minimize temperature gradients in a room. Ceiling space used for return air, and h = 6.81 W/(m²·K) [2 Btu/(h·ft²·°F)].

[b] Floor covered with carpet and rubber pad; for a bare floor or if covered with floor tile, take next w_1 value down the column.

Given: A = 10 m² exterior wall with U = 0.693 W/(m²·K). The values of $\dot{Q}_{cond}$ are listed under the second column of Table 9.6, 300 W lights from 9:00 to 18:00, air exchange at a constant rate of $\dot{V}$ = 5.208 L/s; hence

$$\dot{V}\rho c_p = 6.25 \text{ W/K}$$

envelope construction type medium, 150 mm (6 in.) concrete floor, room air circulation type medium, ordinary furnishings, and vented light fixtures.

Find: $\dot{Q}_c$ for T_i = 25°C

Lookup values: The room transfer coefficients are as follows:

(i) For cooling load, from Table 9.4, w_1 = −0.94; this is the same for all heat gain types.
(ii) For wall heat gains, from Table 9.5, υ_0 = 0.681, $\upsilon_1 = 1 + w_1 - \upsilon_0 = -0.621$.
(iii) For lights, $\upsilon_0 = 0.65$, $\upsilon_1 = 1 + w_1 = -0.59$.

Solution

We have set up a spreadsheet with three sets of columns: (a) for wall conduction, (b) for lights, and (c) infiltration. For each of these, the cooling load is calculated according to Equation 9.7 with the appropriate coefficients. Not knowing the initial conditions, we repeat the calculation until the convergence is acceptable. The last column shows the total cooling load.

(a) *Conductive* component $\dot{Q}_{c,t}$ of cooling load at time t. The second column is the heat gain $\dot{Q}_{cond}$ from Table 8.5 of Example 8.2 multiplied by A = 10 m². Under the third column, for the first $\dot{Q}_{c,t}$ entry at t = 0, we assume $\dot{Q}_{c,1-1} = 0$ as the initial value and find

$$\dot{Q}_{c,1} = \upsilon_0\dot{Q}_{cond,1} + \upsilon_1\dot{Q}_{cond,1-1} - w_1\dot{Q}_{c,1-1}$$
$$= 0.681 \times 96.62 + (-0.621) \times 108.61 - (-0.94) \times 0.0$$
$$= -1.65$$

and so on. Hourly values for four complete iterations are shown.

(b) Cooling load $\dot{Q}_{c,t}$ due to heat gain $\dot{Q}_{lights}$ from *lights*. For the first $\dot{Q}_{c,t}$ entry at t = 9, the detail of the calculation is

$$\dot{Q}_{c,9} = \upsilon_0\dot{Q}_{lights,9} + \upsilon_1\dot{Q}_{lights,9-1} - w_1\dot{Q}_{c,9-1}$$
$$= 0.65 \times 300 + (-0.59) \times 0 - (-0.94) \times 0.0 = 195$$

and so on. Hourly values for three complete iterations are shown.

TABLE 9.5

v Coefficients of the Room Transfer Function (RoTF)[a] (Only v_0 and v_1 are Nonzero)

Heat Gain Component	Room Envelope Construction[b]	Dimensionless v_0	Dimensionless v_1
Solar heat gain through glass[c] with no interior shade; radiant heat from equipment and people	Light	0.224	$1 + w_1 - v_0$
	Medium	0.197	$1 + w_1 - v_0$
	Heavy	0.187	$1 + w_1 - v_0$
Conduction heat gain through exterior walls, roofs, partitions, doors, windows with blinds, or drapes	Light	0.703	$1 + w_1 - v_0$
	Medium	0.681	$1 + w_1 - v_0$
	Heavy	0.676	$1 + w_1 - v_0$
Convective heat generated by equipment and people, and from ventilation and infiltration air	Light	1.000	0.0
	Medium	1.000	0.0
	Heavy	1.000	0.0

Heat Gain from Lights[d]

Furnishings	Air Supply and Return	Type of Light Fixture	v_0	v_1
Heavyweight simple furnishings, no carpet	Low rate; supply and return below ceiling ($\dot{V} \le 0.5$)[e]	Recessed, not vented	0.450	$1 + w_1 - v_0$
Ordinary furnishings, no carpet	Medium to high rate, supply and return below or ceiling ($\dot{V} \ge 0.5$)	Recessed not vented	0.550	$1 + w_1 - v_0$
Ordinary furnishings, with or without carpet on floor	Medium to high rate, or induction unit or fan and coil, supply and return below, or through ceiling, return air plenum ($\dot{V} \ge 0.5$)	Vented	0.650	$1 + w_1 - v_0$
Any type of furniture, with or without carpet	Ducted returns through light fixtures	Vented or free-hanging in air-stream with ducted returns	0.750	$1 + w_1 - v_0$

Source: ASHRAE, *Handbook of Fundamentals*, American Society of Heating, Refrigerating and Air-Conditioning Engineers, Atlanta, GA, 1997. Copyright ASHRAE, www.ashrae.org.

[a] The transfer functions in this table were calculated by procedures outlined in Mitalas and Stephenson (1967) and are acceptable for cases where all heat gain energy eventually appears as cooling load. The computer program used was developed at the National Research Council of Canada, Division of Building Research.

[b] The construction designations denote the following:

Light construction: Such as frame exterior wall, 50 mm (2 in.) concrete floor slab, approximately 146 kg of material/m² (30 lb/ft²) of floor area

Medium construction: such as 100 mm (4 in.) concrete exterior wall, 100 mm (4 in.) concrete floor slab, approximately 341 kg of building material/m² (70,341 lb/ft²) of floor area

Heavy construction: Such as 150 mm (6 in.) concrete exterior wall, 150 mm (6 in.) concrete floor slab, approximately 635 kg of building material/m² (130 lb/ft²) of floor area

[c] The coefficients of the transfer function that relate room cooling load to solar heat gain through glass depend on where the solar energy is absorbed. If the window is shaded by an inside blind or curtain, most of the solar energy is absorbed by the shade and is transferred to the room by convection and long-wave radiation in about the same proportion as the heat gain through walls and roofs; thus the same transfer coefficients apply.

[d] If room supply air is exhausted through the space above the ceiling and lights are recessed, such air removes some heat from the lights that would otherwise have entered the room. This removed light heat is still a load on the cooling plant if the air is recirculated, even though it is not a part of the room heat gain as such. The percent of heat gain appearing in the room depends on the type of lighting fixture, its mounting, and the exhaust airflow.

[e] $\dot{V}$ is room air supply rate in (ft³/min)/ft² of floor area.

(c) *Ventilation*. Since ventilative heat gain is instantaneous, its transfer function coefficients are $\upsilon_0 = 1$, $\upsilon_1 = 0 = \upsilon_2$, and $w_1 = 0$; hence Equation 9.7 simplifies to the steady-state model:

$$\dot{Q}_{c,vent,t} = (T_{o,t} - T_{i,t})\dot{V}_t \rho c_p$$

(for simplicity we consider only sensible load)

Note that outdoor air temperature T_o values are used, not the sol–air temperature values.

For the first $\dot{Q}_{c,vent}$ entry at $t = 1$, the detail of the calculation is

$$\dot{Q}_{c,vent} = (24.4°\text{C} - 25.0°\text{C}) \times 6.25\ \text{W/K} = -3.75\ \text{W}$$

and so on.

The last column of Table 9.6 shows the *total* load $\dot{Q}_{c,tot}$ = conduction + lights + ventilation = sum of loads from parts (a, b, and c). Figure 9.4 plots the diurnal variation of the total cooling load as well as the separate conduction and lights contributions. Hence, the late afternoon peakiness in the

TABLE 9.6

Calculations for Example 9.1

		(a)					(b)					(c)		
Time (t) h	$\dot{Q}_{cond}$ W	$\dot{Q}_{c,t}$ W	$\dot{Q}_{c,t+24}$ W	$\dot{Q}_{c,t+48}$ W	$\dot{Q}_{c,t+72}$ W	$\dot{Q}_{c,t+96}$ W	$\dot{Q}_{lights}$	$\dot{Q}_{c,t}$	$\dot{Q}_{c,t+24}$	$\dot{Q}_{c,t+48}$	$\dot{Q}_{c,t+72}$	T_o	$\dot{Q}_{c,vent}$	$\dot{Q}_{c,tot}$
−2	*134.40*	*0.00*	*95.77*	*116.90*	*121.69*	*122.7744*								
−1	*121.19*	*0.00*	*89.09*	*108.96*	*113.46*	*114.4769*	*0*	*0.00*	*32.90*	*40.36*	*42.05*			
0	*108.61*	*0.00*	*82.45*	*101.13*	*105.36*	*106.3132*	*0*	*0.00*	*30.93*	*37.94*	*39.52*			
1	96.62	−1.65	75.85	93.41	97.38	98.28	0	0.00	29.07	35.66	37.15	24.4	−3.75	131.69
2	85.42	−3.38	69.47	85.97	89.71	90.56	0	0.00	27.33	33.52	34.92	24.4	−3.75	121.73
3	75.30	−4.95	63.53	79.05	82.56	83.35	0	0.00	25.69	31.51	32.83	23.8	−7.50	108.68
4	65.99	−6.47	57.90	72.48	75.78	76.53	0	0.00	24.15	29.62	30.86	23.3	−10.63	96.76
5	57.35	−8.01	52.50	66.21	69.31	70.02	0	0.00	22.70	27.84	29.01	23.3	−10.63	88.40
6	49.66	−9.32	47.56	60.44	63.36	64.02	0	0.00	21.34	26.17	27.27	23.3	−10.63	80.66
7	43.79	−9.78	43.68	55.80	58.54	59.16	0	0.00	20.06	24.60	25.63	23.8	−7.50	77.29
8	40.39	−8.88	41.38	52.76	55.34	55.92	0	0.00	18.85	23.12	24.09	25	0.00	80.02
9	39.27	−6.69	40.56	51.26	53.68	54.23	300	195.00	212.72	216.74	217.65	26.6	10.00	281.88
10	40.28	−3.24	41.16	51.22	53.50	54.02	300	201.30	217.96	221.73	222.59	28.3	20.63	297.23
11	43.08	1.27	43.02	52.47	54.62	55.10	300	207.22	222.88	226.43	227.23	30.5	34.38	316.71
12	47.56	6.83	46.07	54.96	56.97	57.43	300	212.79	227.51	230.84	231.60	32.2	45.00	334.03
13	54.04	13.69	50.58	58.93	60.82	61.25	300	218.02	231.86	234.99	235.70	33.8	55.00	351.95
14	67.34	25.17	59.84	67.70	69.47	69.88	300	222.94	235.95	238.89	239.56	34.4	58.75	368.19
15	88.01	41.78	74.37	81.75	83.42	83.80	300	227.56	239.79	242.56	243.19	35	62.50	389.49
16	113.25	61.74	92.37	99.31	100.88	101.24	300	231.91	243.40	246.01	246.60	34.4	58.75	406.59
17	138.89	82.29	111.09	117.61	119.09	119.42	300	236.00	246.80	249.25	249.80	33.8	55.00	424.22
18	160.10	100.13	127.20	133.33	134.72	135.04	0	44.84	54.99	57.29	57.81	32.7	48.13	240.97
19	170.22	110.62	136.07	141.83	143.14	143.43	0	42.15	51.69	53.85	54.34	30.5	34.38	232.15
20	162.49	108.93	132.85	138.27	139.49	139.77	0	39.62	48.59	50.62	51.08	29.4	27.50	218.35
21	148.30	102.48	124.96	130.06	131.21	131.47	0	37.24	45.67	47.58	48.02	28.3	20.63	200.11
22	134.40	95.77	116.90	121.69	122.77	123.02	0	35.01	42.93	44.73	45.14	27.2	13.75	181.91
23	121.20	89.09	108.96	113.46	114.48	114.71	0	32.90	40.36	42.05	42.43	26.1	6.88	164.01
24	108.62	82.45	101.13	105.36	106.31	106.53	0	30.93	37.94	39.52	39.88	25	0	146.41
Average	89.65	35.83	77.46	86.89	89.02	89.51	112.50	92.31	107.93	111.46	112.27		20.70	222.48

The values in the first three rows in italics are meant to initiate the time - lagged calculation.
The three values in column 2 are the same as those for hours 22, 23 and 24 of the column 2.
The three values in column 3 are simply assumed to be zero.
The three values in column 4 are the same as those for hours 22, 23 and 24 of column 3, and so on.

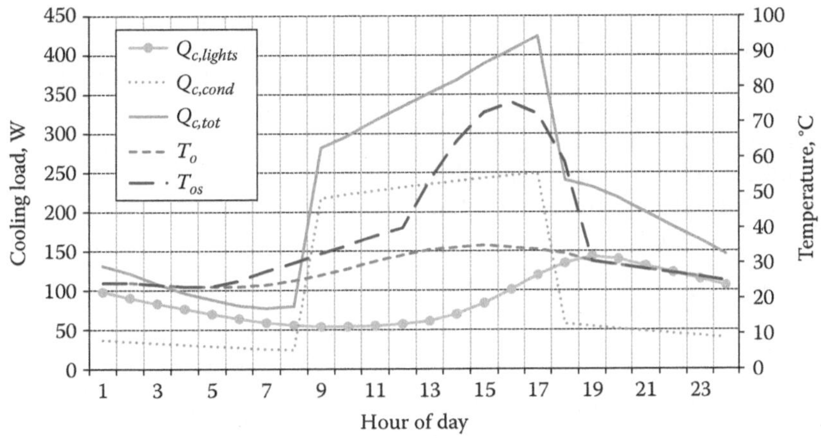

FIGURE 9.4
Diurnal variation of the total peak cooling load along with those due to conduction and lights (for the west-facing room of Example 9.1). The corresponding outdoor temperature and sol–air temperatures for this peak summer day are also shown.

sol–air temperature profile due to the west-facing wall along with the lights being on till 18:00 have resulted in a large cooling load late in the afternoon.

Comments

The peak load is 424 W at 17:00. To check the consistency of the results, note that the average of $\dot{Q}_{c,tot}$ in the last column of part (a) is 89.51 W, in acceptable agreement with $UA(T_{os} - T_i)_{av} = 10\ \text{m}^2 \times 0.693\ \text{W/(m}^2\cdot{}^\circ\text{C)} \times (37.95\text{–}25)^\circ\text{C} = 89.65$ W (with $T_{os,av}$ from Example 8.2).

Since the lights are on 9 h/day, their average power is 300 W × (9/24) = 112.5 W, in good agreement with the 24 h average of the $\dot{Q}_{c,t+72}$ column of part (b).

The sum of the averages for conduction, lights, and ventilation (last column of part c)

$$89.51 + 112.27 + 20.70 = 222.48\ \text{W}$$

It is interesting to compare these results with a steady-state calculation found by simply adding the instantaneous heat gains. The instantaneous contribution from the lights is 300 W, constant from 9:00 to 17:00. From Table 8.5, we find that the sol–air temperature for a dark, west-facing wall reaches its peak

$$T_{os,\max} = 75.5^\circ\text{C} \quad \text{at } 16:00$$

and the corresponding conduction is

$$UA(T_{os} - T_i)_{\max} = 10\ \text{m}^2 \times 0.693\ \text{W/(m}^2\cdot{}^\circ\text{C)} \times (75.5 - 25)^\circ\text{C} = 350\ \text{W}$$

The ventilation heat gain, listed in part (c) of Table 9.6, is seen to reach its peak of 62.5 W at 15:00, and at 16:00 it is 58.75 W. Therefore, the peak load of the steady-state calculation occurs between 15:00 and 16:00. Noting that at 15:00 the value of the total conductive term is 389.5 W, we see that from 15:00 to 16:00 the decrease in ventilation load is small compared to the increase in conduction. Therefore, we can say that the steady-state peak is very close to 16:00, with 350 W for conduction, 58.75 W for ventilation, and 300 W for lights, giving a total of 709 W. The real cooling load peak from the transient analysis (= 424 W) reaches only 60% of this steady-state value, and it occurs about 1.5 h later, at 17:00.

While specific numbers vary a great deal with the circumstances, the general pattern is common to all peak cooling loads: *Thermal inertia attenuates and delays the peak contributions of individual load components.* The total peak is usually less than the result of a steady-state calculation, although it could be more if the time delays act in the sense of making the loads coincide. Daily average loads, by contrast to peak loads, can be determined by a static calculation, if the average indoor temperature is known; that follows from the first law of thermodynamics. But if the thermostat allows floating temperatures, the indoor temperature is, in general, not known without a dynamic analysis.

With the transfer function equations described so far, we can calculate peak cooling loads when the indoor temperature T_i is constant. We now address the generalization to variable T_i.

9.5.3 Variable Indoor Temperature and Heat Extraction Rate

The indoor temperature T_i may vary, not only because of variable thermostat set points but also because of limitations of the HVAC equipment (capacity, throttling range, imperfect control). The extension to variable T_i requires one additional transfer function equation called *space air transfer functions* (SATFs).

Recall that the behavior of a room can also be described by a relation such as Equation 8.24 that links the output (room temperature T_i) to all the relevant input variables (outdoor temperature T_o, heat input or extraction by the HVAC system $\dot{Q}$, solar heat gains, etc.)

$$\begin{aligned} & a_{i,0}T_{i,t} + a_{i,1}T_{i,t-1\Delta t} + \cdots + a_{i,n}T_{i,t-n\Delta t} \\ & \quad = a_{o,0}T_{o,t} + a_{o,1}T_{o,t-1\Delta t} + \cdots + a_{o,m}T_{o,t-m\Delta t} \\ & \qquad + a_{Q,0}\dot{Q}_t + a_{Q,1}\dot{Q}_{t-1\Delta t} + a_{Q,2}\dot{Q}_{t-2\Delta t} + \cdots + a_{Q,r}\dot{Q}_{t-r\Delta t} \end{aligned} \tag{8.24}$$

A separate set of transfer function coefficients is needed for each input variable with different time delay characteristics; here, we have indicated only T_o and $\dot{Q}$ explicitly. Now, consider two different control modes, mode 1 with the constant value $T_{i,ref}$ assumed in Section 9.5.2 and mode 2 with arbitrary T_i, all input being the same except for $\dot{Q}$. Let

$$\delta T_i = T_{i,ref} - T_i \tag{9.9}$$

and

$$\delta \dot{Q} = \dot{Q}_{ref} - \dot{Q} \tag{9.10}$$

designate the differences in T_i and $\dot{Q}$ between these two control modes. Taking the difference between Equation 8.24 for mode 1 and for mode 2, we see that all variables other than δT_i and $\delta \dot{Q}$ drop out. The transfer function between δT_i and $\delta \dot{Q}$ is called the *space air transfer function* (SATF), and following ASHRAE practice, its coefficients are designated by p_n (= $a_{Q,n}$) and g_n (= $a_{i,n}$):

$$\sum_{n \geq 0} p_n \delta \dot{Q}_{t-n\Delta t} = \sum_{n \geq 0} g_{n,t} \delta T_{i,t-n\Delta t} \tag{9.11}$$

A subscript t has been added to g_n to allow the transfer function to vary with time if the air exchange rate varies. Numerical values as per ASHRAE can be obtained from Table 9.7. While p_n is listed directly, g_n is given in terms of g_n^* from which g_n is calculated according to

$$\begin{aligned} g_{0,t} &= g_0^* A + p_0 K_{tot,t} \\ &- g_1^* A + p_1 K_{tot,t-\Delta t} \\ &= g_2^* A \end{aligned} \tag{9.12}$$

where

A is the floor area

$K_{tot,t}$ in W/K [Btu/(h · °F)] is the total heat transmission coefficient of the room

The latter is the sum of conductive and air change terms according to Equation 7.21:

$$K_{tot,t} = K_{cond} + \rho c_p \dot{V}_t$$

and a subscript t for time dependence has been added to allow for the possibility of variable air change. Of course, K_{cond} is the sum of the conductance-area products for the envelope of the room.

To verify the consistency of these coefficients with the first law of thermodynamics, let us take the steady-state limit where $\delta \dot{Q}$, δT_i, and $K_{tot,t}$ are constant and can be pulled outside the sum. Replacing the g_n by Equation 9.12, we find

$$\delta \dot{Q} \sum p_n = \delta T_i \left(A \sum g_n^* + K_{tot} \sum p_n \right) = \text{Steady-state limit} \tag{9.13}$$

A look at the numerical values of g_n^* in Table 9.7 shows that their algebraic sum is zero. Thus, the equation reduces to

$$\delta \dot{Q} = K_{tot} \times \delta T_i \tag{9.14}$$

as it should. Since K_{tot} is positive, $\delta \dot{Q}$ and δT_i have the same sign; this means that $\delta \dot{Q}$ is positive for positive heat

TABLE 9.7

Normalized Coefficients of Space Air Transfer Function (SATF)

	p_0	p_1	g_0^*	g_1^*	g_2^*	g_0^*	g_1^*	g_2^*
Room Envelope Construction	**Dimensionless**			**Btu/(h · ft² · °F)**			**W/(m² · K)**	
Light	1.00	−0.82	1.68	−1.73	0.05	9.54	−9.82	0.28
Medium	1.00	−0.87	1.81	−1.89	0.08	10.28	−10.73	0.45
Heavy	1.00	−0.93	1.85	−1.95	0.10	10.50	−11.07	0.57

Source: ASHRAE, *Handbook of Fundamentals*, American Society of Heating, Refrigerating and Air-Conditioning Engineers, Atlanta, GA, 1997. Copyright ASHRAE, www.ashrae.org.

input to the room. If we want to state cooling loads $\dot{Q}_c$ as positive quantities, we should therefore take $\dot{Q}_c = -\dot{Q}$ and $\delta\dot{Q}_c = -\delta\dot{Q}$. In particular, if we call the cooling load at temperature T_i the *heat extraction rate* $\dot{Q}_x$ (because that is the rate at which the HVAC equipment must extract heat to obtain the temperature T_i), we can write Equation 9.11 in the form

$$\sum_{n\geq 0} p_n(\dot{Q}_{c,ref} - \dot{Q}_{x,t-n\,\Delta t}) = \sum_{n\geq 0} g_{n,t}(T_{i,t-n\Delta t} - T_{i,ref}) \quad (9.15)$$

where $\dot{Q}_{c,ref}$ is the cooling load at the constant temperature $T_{i,ref}$.

Using Equation 9.11, one can calculate $\delta\dot{Q}$ for any δT_i, or δT_i for any $\delta\dot{Q}$. It can also be used for a mixed regime where δT_i is specified for certain hours and $\delta\dot{Q}$ for others. The calculation proceeds from 1 h to the next, solving for δT_i or $\delta\dot{Q}$ as appropriate. The daily cycle is iterated until the result converges to a stable pattern.

Example 9.2: Calculation of Heat Extraction Rate

Find T_i and $\dot{Q}_c$ for the room of Example 9.1 if the thermostat is set at $T_{i,ref}$ = 25°C from 8:00 until 18:00 (hour ending) and if the air conditioner is turned off from 18:00 until 8:00. The air conditioner has sufficient capacity to satisfy the peak load.

Given: Room of Example 9.1, with control schedule thermostat set at 25°C from 8:00 to 18:00, free-float otherwise.

$A_{floor} = 15$ m²
$K_{cond} = UA = 6.93$ W/K from Example 8.2
$\dot{V}\rho c_p$ = 6.25 W/K from Example 9.1 and still constant
$\dot{Q}_{c,ref} = \dot{Q}_{c,tot}$ in the last column of Table 9.6c, which is the output of Example 9.1

Find: T_i, $\dot{Q}_x$

Assumptions: Same as in Example 9.1, but T_i = 25° C only from 8:00 until 18:00, while it floats freely during the other hours.

Lookup values: Looking up Table 9.7 for medium construction, we find the following coefficients for the RoTF, where g_n is obtained from g_n^* by using Equation 9.12 with A_{floor} = 15 m² and $K_{tot} = K_{cond} + \dot{V}\rho c_p = 6.93 + 6.25 = 13.18$ W/K. Thus,

$$g_{0,t} = 10.28 \times 15 + 1 \times 13.18 = 167.38$$

$$g_{1,t} = -10.73 \times 15 - 0.87 \times 13.18 = -172.42$$

$$g_{2,t} = 0.45 \times 15 = 6.75$$

Assembling the results in a table, we find:

n	p_n	g_n^*	g_n
0	1.0	10.28	167.38
1	−0.87	−10.73	−172.42
2	0	0.45	6.75

Solution

We start by setting up a spreadsheet with rows for hour of the day starting from the onset of the cooling period, i.e., 8 a.m. Four columns are created as shown in Table 9.8 with time t, the cooling load $\dot{Q}_{c,t}$ from the last column of Table 9.7, $\delta T_{i,t} = T_{i,t} - T_{i,ref,t}$ and $\delta\dot{Q}_{c,t} = \dot{Q}_{x,t} - \dot{Q}_{c,ref,t}$.

(i) From t = 8 to 17 (as well as for t < 0), we set $\delta T_i = 0$ and solve for $\delta\dot{Q}_c$ according to Equation 9.11. For the first $\delta\dot{Q}_{c,t}$ entry at t = 8, the detail of the calculation under the fourth column is

$$\delta\dot{Q}_1 = -p_1\delta\dot{Q}_{1-1} + g_{n,0}\delta T_{i,1-0} + g_{n,1}\delta T_{i,1-1} + g_{n,2}\delta T_{i,1-2}$$
$$= -(-0.87) \times 77.29 + 167.38 \times 0.0 + (-172.56) \times 0.0$$
$$+ 6.75 \times 0.0 = 67.24$$

(ii) For the remaining hours, we set $\delta\dot{Q}_c = \dot{Q}_{c,t}$ and solve for δT_i according to Equation 9.11. For the 18:00 entry under the third column,

$$\delta T_{i,1-0} = (\delta\dot{Q}_1 + p_1\delta\dot{Q}_{1-1} - g_{n,1}\delta T_{i,1-1} - g_{n,2}\delta T_{i,1-2})/g_{n,0}$$
$$= [240.97 + (-0.87) \times 19.20 - (-172.42) \times 0.0$$
$$- 6.75 \times 0.0]/167.38 = 1.34$$

(iii) The cycle is repeated for 4 days, and the results are printed side by side. The last columns show the actual values of temperature $T_i = T_{i,ref} + \delta T_i$ and the heat extraction rate $\dot{Q}_{c,x} = \dot{Q}_{c,ref} - \delta\dot{Q}_c$. The repetitive day results have not quite converged, and an additional day iteration is probably warranted.

The results are plotted in Figure 9.5a and b and compared with the constant set point case of Example 9.1.

Comments

1. To check the consistency with the steady-state limit, evaluate the conduction and ventilation loads at the new average of T_i of 26.4°C. The load from

TABLE 9.8
Calculations for Example 9.2

Time (*t*)	$\dot{Q}_{c,tot}$	δT_1	$\delta\dot{Q}_c$	δT_1	$\delta\dot{Q}_c$	δT_1	$\delta\dot{Q}_c$	δT_1	$\delta\dot{Q}_c$	δT_1	$\delta\dot{Q}_c$	T_1	$\dot{Q}_{c,x}$
h	W	°C	W	°C	W	°C	W	°C	W	°C	W	°C	W
		t		*t* + 24		*t* + 48 h		*t* + 72 h		*t* + 96 h			
8	80.02	0	67.24	0	−226.83	0	−297.10	0	314.27	0	−318.47	25	308.48
9	281.68	0	58.50	0	−185.37	0	−243.64	0	−257.88	0	−261.36	25	543.23
10	297.23	0	50.90	0	−161.27	0	−211.97	0	−224.35	0	−227.38	25	524.6
11	316.71	0	44.28	0	−140.30	0	−184.41	0	−195.19	0	−197.10	25	514.53
12	334.03	0	38.52	0	−122.06	0	−160.44	0	−169.81	0	−172.10	25	506.13
13	351.95	0	33.51	0	−106.20	0	−139.58	0	−147.74	0	−149.73	25	501.68
14	368.19	0	29.16	0	−92.39	0	−121.43	0	−128.53	0	−130.27	25	498.45
15	389.29	0	25.37	0	−80.38	0	−105.65	0	−111.82	0	−113.33	25	502.82
16	406.59	0	22.09	0	−69.93	0	−91.91	0	−97.29	0	−98.60	25	505.18
17	424.22	0	19.20	0	−60.84	0	−79.97	0	−84.64	0	−85.78	25	510.00
18	240.97	1.34	240.97	1.76	240.97	1.86	240.97	1.88	240.97	1.89	240.97	26.89	0
19	232.15	1.51	232.15	1.94	232.15	2.05	232.15	2.07	232.15	2.08	232.15	27.08	0
20	218.35	1.60	218.35	2.03	218.35	2.13	218.35	2.16	218.35	2.16	218.35	27.16	0
21	200.11	1.65	200.11	2.07	200.11	2.17	200.11	2.20	200.11	2.20	200.11	27.20	0
22	181.91	1.68	181.91	2.10	181.91	2.20	181.91	2.22	181.91	2.23	181.91	27.23	0
23	164.01	1.70	164.01	2.11	164.01	2.21	164.01	2.24	164.01	2.24	164.01	27.24	0
24	146.21	1.71	146.41	2.11	146.41	2.21	146.41	2.24	146.41	2.24	146.41	27.24	0
1	131.69	1.72	131.69	2.12	131.69	2.21	131.69	2.24	131.69	2.24	131.69	27.24	0
2	121.73	1.74	121.73	2.14	121.73	2.24	121.73	2.26	121.73	2.26	121.73	27.26	0
3	108.68	1.74	108.68	2.14	108.68	2.23	108.68	2.25	108.68	2.26	108.68	27.26	0
4	96.76	1.74	96.76	2.13	96.76	2.22	96.76	2.24	96.76	2.25	96.76	27.25	0
5	88.40	1.74	88.40	2.13	88.40	2.22	88.40	2.24	88.40	2.25	88.40	27.25	
6	80.66	1.75	80.66	2.13	80.66	2.22	80.66	2.24	80.66	2.25	80.66	27.25	0
7	77.29	1.77	77.29	2.15	77.29	2.24	77.29	2.26	77.29	2.27	77.29	27.27	0
Average	222.28		103.25		35.15		18.88		4.90		13.93	26.28	208.55

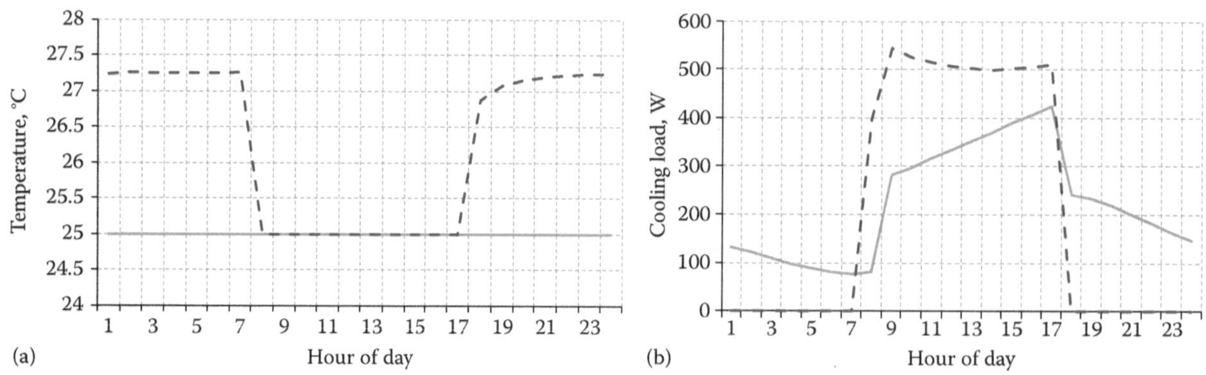

FIGURE 9.5
(a) Indoor temperature and (b) cooling load for room of Example 9.2. Solid line = constant set point at 25°C; dashed line = set point at 25°C from 8:00 to 18:00, floating the rest of the time.

lights (on for 9 h) remains, of course, unaffected. The loads at average conditions are

$$\text{Conduction} = UA(T_{os} - T_i)_{av}$$
$$= 10\ \text{m}^2 \times 0.693\ \text{W/(m}^2 \cdot {}^\circ\text{C)}$$
$$\times (37.95 - 26.28)^\circ\text{C} = 80.87\ \text{W}$$

$$\text{Ventilation} = \rho c_p (T_o - T_i)_{av} = 6.250 \times (28.31 - 26.28)^\circ\text{C}$$
$$= 12.69\ \text{W}$$

$$\text{Lights} = (300 \times 9)/24 = 112.25\ \text{W}$$

Their sum is 205.81 W, in very good agreement with the average 208.6 W of $\dot{Q}_x$.

2. Compared to the fixed thermostat set point in Example 9.1, the setup reduces the average load from 222.5 to 208.6 W, a saving of over 6%.
3. Savings from thermostat setup depend on the relation between indoor and outdoor temperatures and on the relative importance of conduction, ventilation, and internal gains. In this example, internal gains dominate and setup recovery increases the peak load (a situation that is unlikely when skin loads dominate).
4. In climates with cool, dry nights, significant *savings* may be achieved *by increasing the ventilation at night* (in this example, we have assumed constant ventilation; its load contribution becomes negative at night, but the magnitude is relatively small). If outdoor air at night is cool but moist, absorption of humidity in the building may increase the latent load during the following day.

Example 9.2 assumed that the cooling capacity of the cooling system was adequate to meet the extra cooling rate. Capacity limitations of the HVAC equipment can also be included. At each hour with thermostatic control, we can check whether the actual load (at T_i = thermostat set point) exceeds the capacity. If it does, one solves for T_i instead, setting $\dot{Q}_x$ equal to the capacity for this hour.

Likewise, one can account for the throttling range of a control system that modulates the heat extraction rate according to the control law shown in Figure 9.6. Stated as an equation, this means that the heat extraction rate $\dot{Q}_x$ is determined by the room temperature T_i according to

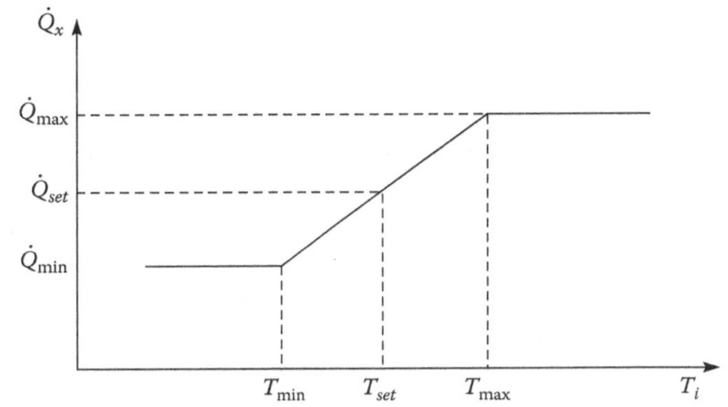

FIGURE 9.6
Control law of Equation 9.16 for heat extraction rate $\dot{Q}_x$ (solid line) as a function of room temperature T_i.

$$\dot{Q}_{x,t} = \begin{cases} \dot{Q}_{max} & \text{for } T_{i,t} > T_{max} \\ \dot{Q}_{set} + \dot{Q}'(T_{i,t} - T_{set}) & \text{for } T_{min} < T_{i,t} < T_{max} \\ \dot{Q}_{min} & \text{for } T_{i,t} < T_{min} \end{cases} \quad (9.16)$$

where

$$\dot{Q}' = \frac{\dot{Q}_{max} - \dot{Q}_{min}}{T_{max} - T_{min}} \quad (9.17)$$

and T_{set} is the thermostat set point; we have added the subscript t to indicate that this equation applies instantaneously at each hour t. At each new hour t, Equations 9.15 and 9.16 can be considered as a system of two equations for two unknowns: $\dot{Q}_{x,t}$ and $T_{i,t}$. After finding the solution, one repeats the process for the next hour, and so on.

The calculation procedure of the TFM approach is summarized in Figure 9.7. It involves calculating the three different types of heat gains, then determining their individual cooling loads (with constant indoor set point temperature), adding them to find the total cooling load, and finally, finding the heat extraction rate under a variable thermostat set point diurnal schedule. There are three sets of transfer coefficients:

1. CTFs that depend on construction type of walls and roofs.
2. RoTFs that depend on room envelope and construction, type of lighting fixtures, and type of air circulation in the room.
3. SATFs that depend on room envelope construction and type.

Note that several iterations over the 24 h of the peak day are needed at different stages before periodic steady patterns are attained.

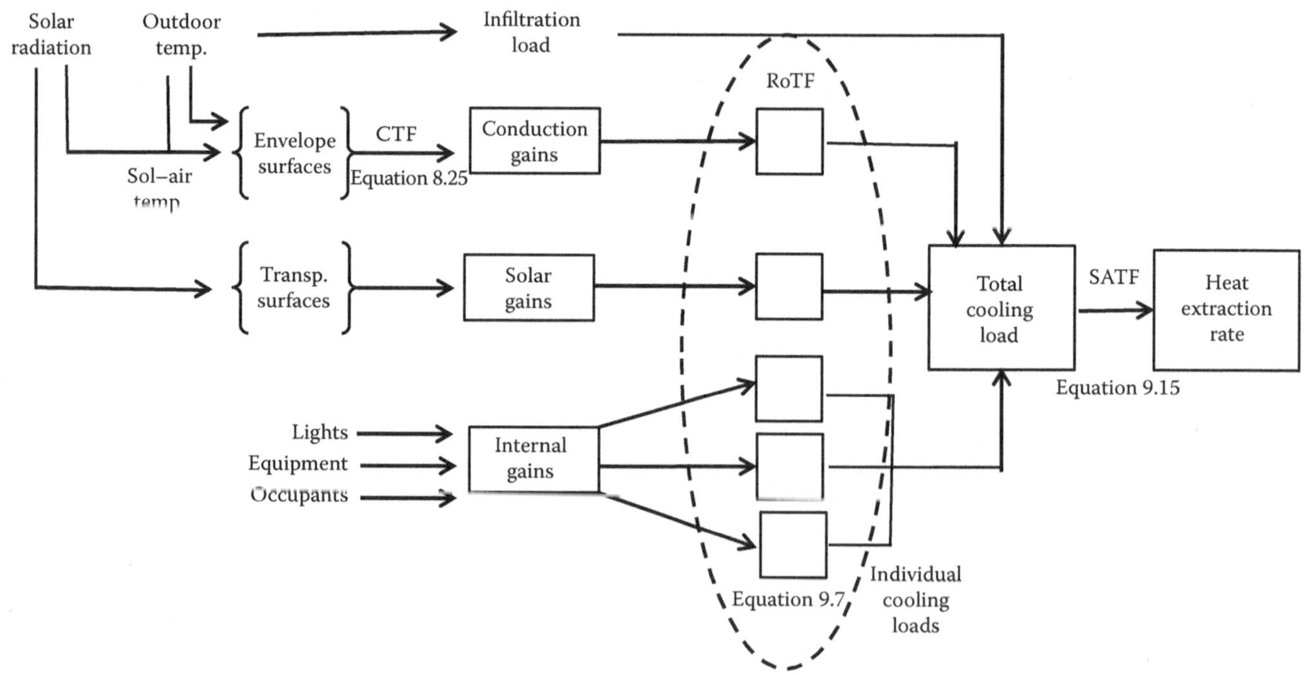

FIGURE 9.7
Schematic of the transfer function method (TFM) involving three sets of hourly transfer function coefficients (CTF, RoTF, and SATF). These are used during separate iterative calculations repeated over 24 h till a periodic steady daily pattern is reached. Only a single envelope surface and a single transparent surface are shown.

9.6 Heat Balance Method

The load calculation methods described earlier, namely TFM (which appeared last in the 1997 edition of ASHRAE Fundamentals), and the CLTD/SCL/CLF method (which appeared last in the 1989 edition of ASHRAE Fundamentals), are transformation-based procedures. The fundamental heat transfer interactions are modified in certain ways to speed up the simulations when applied to buildings with numerous zones. However, this process involves injecting not only a certain degree of arbitrariness but also obscuring certain heat transfer phenomena and interactions. Such simplified methods were developed over 40 years back when computing speed was a major factor. The advent of faster and more powerful computers has eliminated such restrictions, and allows more transparent and rigorous algorithms to be developed. The most widely used in North America is the heat balance method (HBM)* that is implemented in the EnergyPlus simulation program (2009). A basic description of HBM is provided as follows; more details can be found in ASHRAE Fundamentals (2013) and Spitler (2010).

* The term "heat balance" is used rather than "energy balance" to indicate that sensible loads only are involved. Latent loads are dealt with separately and added to the sensible loads while determining equipment loads.

Figure 9.8 is a schematic of the heat transfer processes and interactions of HBM as applied to a single zone. These have to be replicated for each zone in a multiple-zone building. Three distinct heat balances are involved on three elements:

1. Outside surface(s) of the zone element(s)
2. Inside surface(s) of the zone element(s)
3. Zone air volume

The various heat transfer processes affecting elements (1) and (2) as shown in Figure 9.8 have to be considered for each surface element enclosing the zone, and this is done using their primordial forms. The sol–air temperature concept (Section 7.2) is not used since heat balances (elements (1) and (2)) consider the absorbed radiation and the conduction effects separately. For transparent surfaces, the SHGC concept (Section 5.4.2) is not used. The solar radiation gains through glazing include the transmitted component and the absorbed portion on each glass layer (that is why HBM requires the input of the optical properties of each glass layer). The transmitted component will appear in the heat balance for each internal surface that receives solar radiation either directly or indirectly through internal reflections. The portion of the solar radiation absorbed in the outer glass layer will appear in heat balance (1), and that absorbed in the inner glass layer will appear in heat balance (2).

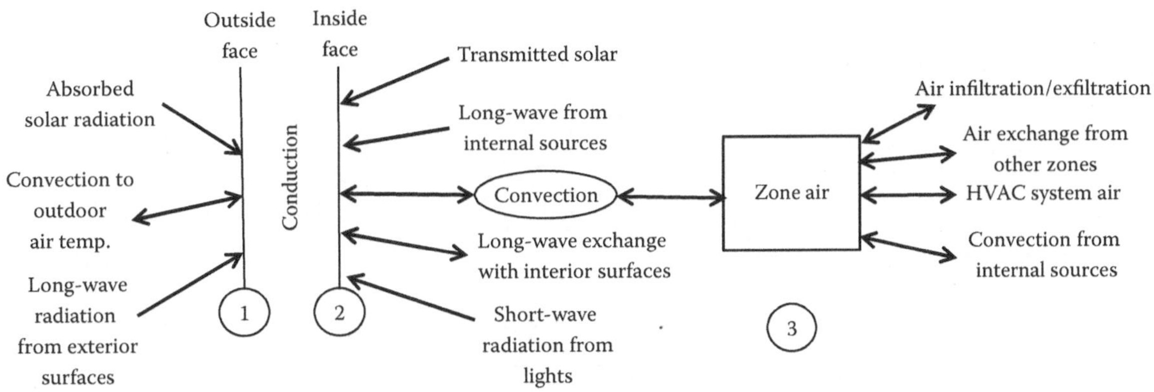

FIGURE 9.8
Schematic of the heat balance method (HBM) showing various heat transfer interactions as applied to a single zone and a single opaque surface. Three distinct heat balances are involved shown as (1), (2), and (3).

Further, HBM considers *directional* optical properties, i.e., as a function of the angles of incidence on each transparent surface due to solar as well as from internal surface reflections. Furnishings, partitions, and other interior elements inside the zone are also explicitly considered in the convective and radiative exchanges between the surfaces and the zone air. These are treated in the same manner as those for the opaque element.

Certain assumptions are made regarding the inside surfaces of the room (walls, roof, windows, floor, etc.):

1. Uniform in surface temperature
2. Uniform long-wave and short-wave irradiation
3. Diffuse radiating surfaces
4. 1-D transient heat conduction through the opaque element

The overall algorithm allows some flexibility. For example, the wall conduction process can be modeled by any one of the methods discussed in Chapter 8. The two solution approaches implemented in EnergyPlus software are the numerical finite difference and the conduction transfer function methods. Because the three heat balances require temperature and heat fluxes to be known at each stage, the numerous heat transfer equations have to be solved simultaneously for each zone. EnergyPlus software adopts an iterative numerical approach within each time-step of the peak day simulation until convergence is achieved (generally, 4–6 iterations).

9.7 Radiant Time-Series Method

In the ever-present quest for more accurate load prediction methods, new procedures are being developed all the time. One such procedure is the ASHRAE radiant time-series (RTS) method that appeared first in the 2001 edition of the ASHRAE Fundamentals Handbook. It was developed as a spreadsheet version of HBM (thus, an alternative* to the CLTD/SCL/CLF method). The RTS method exhibits similarities in some ways to TFM (Spitler and Fisher, 1999) and in other ways to HBM. These similarities and principal differences will become apparent in the discussion that follows. For additional details, refer to Spitler et al. (1997), ASHRAE Fundamentals (2013), and Spitler (2010).

The RTS method is a simplified version of HBM. It is based on a "reduced" heat balance procedure that offers insights into the building heat transfer flows without loss in computing accuracy while being implementable on a spreadsheet program. It does not constrain the heat transfer processes in terms of imposing heat balances as does HBM. RTS is said to be rigorous in that it uses the basic governing heat transfer equations, while being sequential, thereby eliminating the need for iterative calculations needed in HBM. Further, it directly provides information (as does TFM) on the contribution of the different individual building components/elements that constitute the total cooling load. This is important to the designer who can then use judgment to make informed decisions on the selection of materials and element sizes (such as windows) that would reduce peak loads while meeting prestipulated architectural and other constraints. Such insight is not provided by HBM.

Figure 9.9 provides a schematic of the computational steps involved in the RTS method. Notice that the heat

* ASHRAE is careful to state that the new methods (HBM and RTS methods) should not be viewed as superseding or invalidating or discrediting the older CLTD/SCL/CLF and TFM methods. The reasons stated for introducing the former are to provide improved accuracy, better transparency in the analysis, and reduced dependence on purely subjective inputs. A comparison of the diurnal cooling load profiles predicted by the different peak-day design methods can be found in ASHRAE Fundamentals (2001)—Chapter 29.

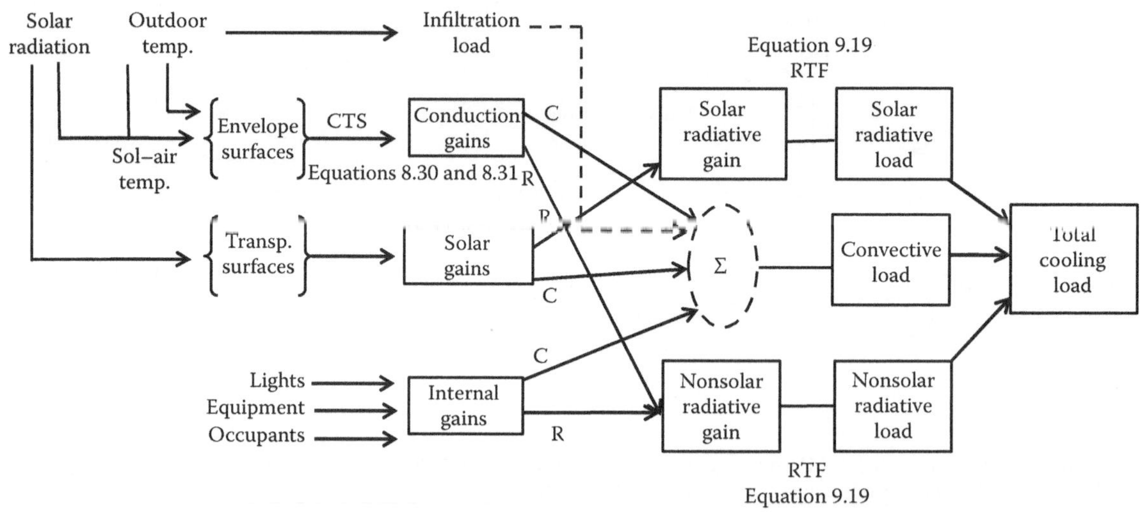

FIGURE 9.9
Schematic of the radiant time series (RTS) method wherein each of the heat gains are split into convective (C) and radiative (R) components. Three sets of 24 hourly coefficients are used: one set of conduction time series (CTS) and two sets of radiant time factors (RTFs)—one for solar gains and one for nonsolar gains.

gains are treated in the same fashion as does TFM; the sol–air temperature for opaque surfaces and the SHGC for transparent surfaces (see Figure 9.7). There are some important differences, however. Instead of using the conduction transfer function (CTF) model as does TFM for building envelope conduction gains, the conduction time series (CTS) model is adopted. The primary advantage (and rationale) of the CTS model (described and illustrated in Section 8.3.3) is that the iterative computations over the peak day for the hourly conductive heat flows are avoided. The transfer function coefficients have been replaced by a set of 24 response factors that allow the hourly heat gains over the peak day to be determined by a simple summation (see Example 8.3).

A second difference between TFM and RTS method is the manner in which each heat gain is split into radiative and convective components using predetermined fractions. Table 9.9 shows recommended values for the radiative fraction; the convective fraction is simply

TABLE 9.9
Typical Radiative Fraction of Building Heat Gains[a]

Heat Gain Type	Typical Radiative Fraction
Conductive heat gain: walls and floors	0.46
roofs	0.60
windows: SHGC > 0.5	0.33
SHGC < 0.5	0.46
Solar heat gain through fenestration: without interior shading	1.0
with interior shading: SHGC > 0.5	0.33
SHGC < 0.5	0.46
Suspended fluorescent lighting, unvented	0.5–0.57
Recessed fluorescent luminaire lighting: without lens	0.48–0.68
with lens, vented to supply and return air	0.61–0.73
Compact fluorescent luminaire	0.95–1.0
Office equipment, with/without fan	0.1/0.3
Occupants—typical office conditions	0.6

Source: ASHRAE, *Handbook of Fundamentals*, American Society of Heating, Refrigerating and Air-Conditioning Engineers, Atlanta, GA, 1997. Copyright ASHRAE, www.ashrae.org.

[a] Sum of radiative and convective fractions add to unity.

$$\text{Load convective fraction} = 1 - \text{load radiative fraction} \quad (9.18)$$

The convective fraction immediately becomes a cooling load on the building HVAC system. The radiative portion is absorbed by the building materials, furniture, etc., and is convected into the space as a time-lagged and attenuated cooling load. This is done by introducing 24 hourly values of a series called radiant time factors (RTF). Two sets of RTF must be determined for each zone: one for the transmitted solar heat gain and one for the radiation from internal sources and building envelope surfaces (called *nonsolar*). The difference between the two is that the former is assumed to be absorbed by the floor only while the latter is assumed to be evenly distributed throughout the space. The radiant loads are converted to hourly cooling loads through the use of radiant time factors. Similar to the response factors, the time factors estimate the cooling load based on past and present heat gains:

$$\dot{Q}_{c,t} = \sum_{j=0}^{23} r_j \dot{Q}_{gain,t-j\Delta t} \quad (9.19)$$

where

r_o is the fraction of the load convected to the space at the current time

r_1 is the fraction at the previous hour, and so forth

This step replaces the zone transfer function of the TF method.

The RTF are determined through a zone heat balance model as described by Spitler (2010). Recall that the numerical values of the hourly CTS coefficients reflect the fraction of an earlier heat gain at the opaque element (wall or roof) that appears as an indoor heat gain during the current hour. By definition, the 24 hourly fractional values should sum to unity (or 100 if in percentages). RTF coefficients are conceptually similar to the CTS coefficients but relate to the radiant heat gains and depend on zone/room construction. The hourly values of the CTS and RTF coefficients capture the time-delay effects of one construction versus another. Numerical values in percentages can be found in tables such as Table 9.10 for a few wall types and zone construction categories. The TFM and RTS methods have to be performed at hourly time steps during the peak day while HBM does not contain predetermined values and allows time steps shorter than one hour to be selected as desired. Finally, the RTS method applies only to the case when the indoor temperature is kept constant throughout the peak day.

A summary of the calculation steps of the RTS method is provided in Table 9.11. The following example will illustrate the overall computational approach.

Example 9.3: Calculation of Cooling Loads Using the RTS Method

Consider the room assumed in Example 9.1 with the following modifications:

(a) The analysis is to be done on July 21 at Phoenix, AZ.
(b) The 10 m^2 exterior wall is now south-facing and is of heavy construction (type 27 of Table 9.10) and dark in color, assuming all other walls of the room are internal facing.
(c) There is a 4 m^2 window of type 21a (low e double glazing) of Table 5.1. There are no external overhangs nor internal shading.
(d) As in Example 9.1, the indoors is kept at a constant temperature of T_i = 25°C and has the same lighting power and schedule (assume compact fluorescent) and air exchange rate due to infiltration. The room is of heavy construction with no carpet.

Given: A_{wall} = 10 m^2 exterior wall, A_{window} = 4 m^2 window, 300 W lights from 9:00 to 18:00, air exchange at a constant rate of $\dot{V}$ = 5.208 L/s; hence,

$$\dot{V}\rho c_p = 6.25 \text{ W/K}$$

Find: $\dot{Q}_c$ for T_i = 25°C

Lookup values: U_{wall} = 0.673 W/(m$^2\cdot$K) from Table 9.10. U_{window} = 1.613 W/(m$^2\cdot$K) from Example 5.4 and Table 5.1.

Other values are stated during the computation steps.

Solution

We have set up a spreadsheet with one row for each hour of the peak day and several columns numbered 1–22 as shown in Table 9.12. There are in essence six subtables wherein we calculate (a) driving forces, steps 1–5; (b) wall conduction heat gains, steps 6–9; (b) window gains, steps 10–12; (c) heat gains from lighting, step 13; (d) ventilation load, step 14; (e) zone solar and nonsolar radiative components, steps 15–16; and (f) nonsolar/solar radiative and convective components and total cooling load, steps 17–22. Calculations for each of these columns are described as follows.

1. Hour.
2. Fraction (x) of daily temperature range (Table 9.3).
3. Hourly outdoor dry-bulb temperature T_o using Equation 9.6. The 1% summer design dry-bulb temperature for Phoenix = 42.4°C and the daily range DR = 11.8°C (Table 9.1).

TABLE 9.10

CTS and RTS Coefficients (in Percentages) for Representative Wall, Roof, and Zone Construction without Carpet with 50% Glass

	CTS				Nonsolar RTS		Solar RTS	
Lag	**Wall Type 6**	**Wall Type 27**	**Roof Type 4**	**Roof Type 14**	**Light-50**	**Heavy-50**	**Light-50**	**Heavy-50**
0	7	1	1	1	44	25	45	
1	44	10	17	2	19	9	19	12
2	32	20	31	8	11	6	11	7
3	12	18	24	11	7	5	7	5
4	4	14	14	11	5	5	5	4
5	1	10	7	10	3	4	3	4
6	0	7	4	9	3	4	3	3
7	0	5	2	7	2	4	2	3
8	0	4	0	6	1	3	1	3
9	0	3	0	5	1	3	1	3
10	0	2	0	5	1	3	1	3
11	0	2	0	4	1	3	1	3
12	0	1	0	3	1	3	1	2
13	0	1	0	3	1	3	0	2
14	0	1	0	3	0	2	0	2
15	0	1	0	2	0	2	0	2
16	0	0	0	2	0	2	0	2
17	0	0	0	2	0	2	0	2
18	0	0	0	1	0	2	0	2
19	0	0	0	1	0	2	0	2
20	0	0	0	1	0	2	0	2
21	0	0	0	1	0	2	0	2
22	0	0	0	1	0	2	0	2
23	0	0	0	1	0	2	0	1

Number	Type	Description	U-Value, W/m^2·K	Total R	Mass, kg/m^2	Thermal Capacity, kJ/m^2·K
Wall 6	Stud wall	Wood siding, sheathing, batt insulation, 13 mm wood	0.406	2.5	25.6	32.7
Wall 27	Precast conc.	100 mm LW concrete, board insulation, gyp board	0.673	1.5	143.9	124.7
Roof 4	Sloped frame	Asphalt shingles, wood sheath, batt ins, gyp board	0.235	4.2	34.7	47.0
Roof 14	Concrete	Membrane, sheath, insul board, 100 mm LW conc	0.304	3.3	149.2	134.9

Const.	Exterior Wall	Roof/Ceiling	Partition	Floor	Furnishings
Light	Steel siding, 50 mm insulation, air space, 19 mm gyp	100 mm LW conc., air space, acoustic tile	19 mm gyp, air space, 19 mm gyp	Acoustic tile, ceiling air space, 100 mm LW conc	25 mm wood at 50% floor area
Heavy	100 mm face brick, 50 mm insul., 200 mm HW conc., 19 mm gyp	200 mm LW conc., air space, acoustic tile	19 mm gyp, 200 mm HW conc. block, air space, 19 mm gyp	Acoustic tile, ceiling air space, 200 mm LW conc.	25 mm wood at 50% floor area

Source: ASHRAE, *Handbook of Fundamentals*, American Society of Heating, Refrigerating and Air-Conditioning Engineers, Atlanta, GA, 2013.

TABLE 9.11

Summary of the Radiant Time Series (RTS) Method

Calculation steps for each hour of the peak day (24 h total)

(0) Sol–air temperature for each building envelope element
- Hourly ambient temperature.
- Hourly solar irradiation on each element.
- Calculate sol–air temperatures.

(1) For each opaque envelope element (wall, roof, etc.)
- Use conduction time series (CTS) to determine hourly transient conduction gains (lagged).
- Split into convective/radiative components (for walls 54%/46%, for roofs 40%/60%).

(2) For each transparent surface (window, glazing)
- Calculate conductive heat gain.
- Calculate solar gain using the SHGC model.
- Split into convective/radiative components: without interior shading: 0%/100% with interior shading: see Table 9.12

(3) For each source of internal heat source, calculate heat gains:
- Occupants: sensible (convective/radiative components: 40%/60%)
- Latent (convective: 100%)
- Lighting: sensible (see Table 9.12)
- Equipment: sensible (see Table 9.12)

(4) Infiltration and ventilation heat gains (100% convective)
Repeat from step (0).

(5) For each hour, sum individual convective cooling load components from steps (1 to 4).

(6) For each hour, sum radiative components for nonsolar loads from steps (1 and 3).
Use nonsolar radiant time factors (RTF) to determine hourly radiation heat gains (lagged).

(7) For each hour, sum radiative components of solar loads from step (2).
Use solar radiant time factors (RTF) to determine hourly radiation heat gains (lagged).

(8) For each hour, sum convective and radiative portions to determine total cooling load.

4. Global solar irradiation (I_T) on a south-facing vertical surface in July (hottest month). The calculational algorithms are explained in Section 4.7 and illustrated in Example 4.11. These values have been copied from a separate spreadsheet.
5. Sol–air temperature T_{os} on south-facing vertical wall following Equation 7.3 and Example 7.1.
6. Hourly CTS values for wall type 27 (Table 9.10).
7. Hourly conduction heat gains $Q_{cond,t}$ through A_{wall} are calculated from Equations 8.30 and 8.31 as illustrated in Example 8.3.
8. Hourly convective gains through wall assuming a split of 54% (Table 9.9).
9. Hourly radiative gains through wall assuming a split of 46%.
10. Solar heat gain coefficients (SHGC) for the type 21a window. The calculation procedure requires treating the beam and diffuse solar components differently and also correcting beam SHGC for incidence angle effect as illustrated in Example 5.4. As a simplification, we assume entire radiation to be beam. The incidence angles have been determined in a separate spreadsheet for each hour and we list the corresponding SHGC values computed.
11. Solar heat gains due to transmitted and absorbed components = A_{window} × SHGC × I_T.
12. Solar heat gains due to window conduction = $A_{window} \times U_{window} \times (T_o - T_i)$.
13. Heat gains from lights. From Table 9.9, we assume radiative split = 1.0 for compact fluorescent lighting.
14. Infiltration heat loads = $\dot{V}\rho c_p \times (T_o - T_i)$ assumed instantaneous.
15. Sum all nonsolar radiative gains = columns (9 + 13).
16. Sum solar radiative gains = column 11.
17. Hourly nonsolar RTS factors for heavy zone construction with 50% glazing and no carpet (Table 9.10).
18. Hourly solar RTS factors for heavy zone construction with 50% glazing and no carpet (Table 9.10).
19. Nonsolar radiative cooling load component. An approach similar to step 7 but using Equation 9.19 is adopted using entries from columns 15 and 17.
20. Solar radiative cooling load component. An approach similar to step 7 but using Equation 9.19 is adopted using entries from columns 16 and 18.
21. Sum all convective loads = columns (8 + 12 + 14).
22. Total cooling load = columns (19 + 20 +21).

Figure 9.10 plots the hourly values of the CTS and RTF series for the wall type 27 and heavy zone construction assumed in the solved example. Notice that the two radiant factors are fairly close until lags of 10 h after which they deviate a little (about 2%). The CTS curve peaks at 3 h lag. Figure 9.11 shows the total cooling load during the peak day along with the three individual contributions of the convective and radiant loads. The total cooling load peaks between 14:00 and 15:00 h. The important contributor is the solar radiative component that is fairly symmetrical about noon for this south-facing space. The convective component is slightly higher than the nonsolar radiative components. We leave it up to the reader to perform a steady-state limit consistency check like that shown in Example 9.2.

TABLE 9.12

Calculations for Example 9.3

1	2	3	4	5	6	7	8	9	10	11	12
		DBT	Solar	Sol–Air		Wall Conduction Gains				Window	
Hour	Fraction	T_o	I_T	T_{os}	CTS	Total	Conv	Rad	SHGC	Trans	Cond
	—	°C	W/m²	°C	%	W	W	W	—	W	W
1	0.88	32.02	0.00	32.02	1	84.41	45.58	38.83	0	0	45.27
2	0.92	31.54	0.00	31.54	10	75.76	40.91	34.85	0	0	42.22
3	0.95	31.19	0.00	31.19	20	67.97	36.70	31.27	0	0	39.94
4	0.98	30.84	0.00	30.84	18	61.25	33.08	28.18	0	0	37.65
5	1	30.60	0.00	30.60	14	55.45	29.94	25.51	0	0	36.13
6	0.98	30.84	31.17	32.46	10	50.67	27.36	23.31	0.23	28.68	37.65
7	0.91	31.66	68.06	35.20	7	48.34	26.10	22.23	0.5	136.13	42.98
8	0.74	33.67	103.80	39.07	5	50.24	27.13	23.11	0.6	249.13	55.93
9	0.55	35.91	197.52	46.18	4	57.27	30.92	26.34	0.64	505.66	70.39
10	0.38	37.92	298.45	53.44	3	71.42	38.57	32.85	0.65	775.97	83.33
11	0.23	39.69	364.85	58.66	2	93.84	50.67	43.17	0.65	948.61	94.75
12	0.13	40.87	387.95	61.04	2	121.60	65.66	55.94	0.65	1008.66	102.37
13	0.05	41.81	364.85	60.78	1	149.75	80.86	68.88	0.65	948.61	108.46
14	0.00	42.40	298.45	57.92	1	173.45	93.66	79.79	0.65	775.97	112.26
15	0.00	42.40	197.52	52.67	1	189.04	102.08	86.96	0.64	505.66	112.26
16	0.06	41.69	103.80	47.09	1	194.22	104.88	89.34	0.6	249.13	107.70
17	0.14	40.75	68.06	44.29	0	189.23	102.18	87.04	0.5	136.13	101.61
18	0.24	39.57	31.17	41.19	0	177.61	95.91	81.70	0.23	28.68	93.99
19	0.39	37.80	0.00	37.80	0	163.19	88.12	75.07	0	0	82.57
20	0.50	36.50	0.00	36.50	0	147.21	79.50	67.72	0	0	74.20
21	0.59	35.44	0.00	35.44	0	131.26	70.88	60.38	0	0	67.35
22	0.68	34.38	0.00	34.38	0	117.03	63.20	53.83	0	0	60.49
23	0.75	33.55	0.00	33.55	0	104.65	56.51	48.14	0	0	55.16
24	0.82	32.72	0.00	32.72	0	93.91	50.71	43.20	0	0	49.84

	13	14	15	16	17	18	19	20	21	22
	Heat Gains		Sum Radiative		Nonsolar	Solar	Nonsolar	Solar	Sum	Total
Hour	Lights	Infil	Nonsolar	Solar	RTS	RTS	Radiative	Radiative	Convec	Loads
	W	W	W	W	%	%	W	W	W	W
1	0	43.85	38.83	0.00	25	27	123.03	142.90	134.70	400.62
2	0	40.90	34.85	0.00	9	12	118.07	135.14	124.03	377.24
3	0	38.69	31.27	0.00	6	7	113.11	130.08	115.33	358.52
4	0	36.48	28.18	0.00	5	5	108.20	127.59	107.21	343.00
5	0	35.00	25.51	0.00	5	4	103.37	125.94	101.07	330.38
6	0	36.48	23.31	28.68	4	4	98.67	131.75	101.49	331.91
7	0	41.64	22.23	136.13	4	3	94.31	160.35	110.72	365.38
8	0	54.18	23.11	249.13	4	3	93.55	198.21	137.23	428.99
9	300	68.19	326.34	505.66	3	3	162.59	277.17	169.50	609.27
10	300	80.73	332.85	775.97	3	3	184.74	378.13	202.63	765.49
11	300	91.79	343.17	948.61	3	3	199.24	466.66	237.21	903.11
12	300	99.16	355.94	1008.66	3	3	211.90	525.44	267.19	1004.54
13	300	105.06	368.88	948.61	3	2	225.21	543.65	294.39	1063.24
14	300	108.75	379.79	775.97	3	2	235.50	518.40	314.68	1068.58
15	300	108.75	386.96	505.66	2	2	245.17	448.84	323.10	1017.10
16	300	104.33	389.34	249.13	2	2	253.71	360.78	316.90	931.39
17	300	98.43	387.04	136.13	2	2	257.89	298.75	302.21	858.85

(Continued)

TABLE 9.12 (*Continued*)
Calculations for Example 9.3

	13	14	15	16	17	18	19	20	21	22
	Heat Gains		Sum Radiative		Nonsolar	Solar	Nonsolar	Solar	Sum	Total
Hour	Lights	Infil	Nonsolar	Solar	RTS	RTS	Radiative	Radiative	Convec	Loads
	W	W	W	W	%	%	W	W	W	W
18	0	91.05	81.70	28.68	2	2	191.91	245.08	280.95	717.94
19	0	79.99	75.07	0.00	2	2	173.11	213.09	250.68	636.88
20	0	71.88	67.72	0.00	2	2	162.69	196.19	225.57	584.45
21	0	65.24	60.38	0.00	2	2	154.87	184.14	203.46	542.47
22	0	58.60	53.83	0.00	2	2	146.91	173.60	182.29	502.80
23	0	53.44	48.14	0.00	2	2	138.93	162.76	165.11	466.80
24	0	48.28	43.20	0.00	2	1	130.97	152.38	148.82	432.18

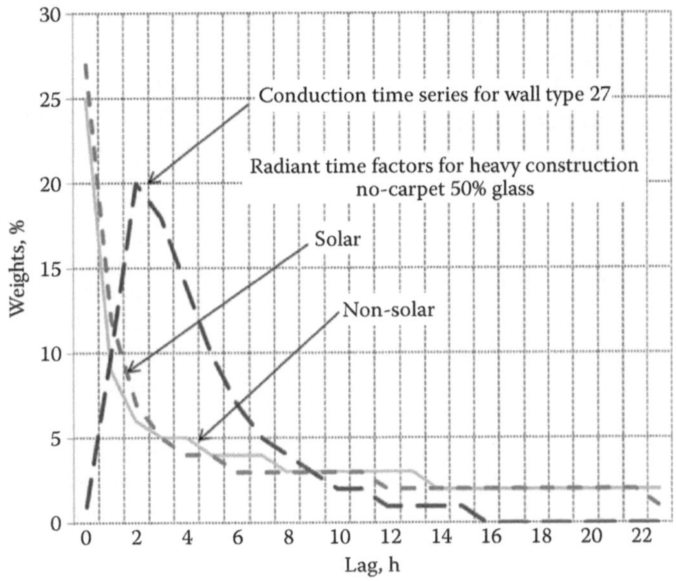

FIGURE 9.10
Conduction time series and the radiant time factors used in Example 9.3.

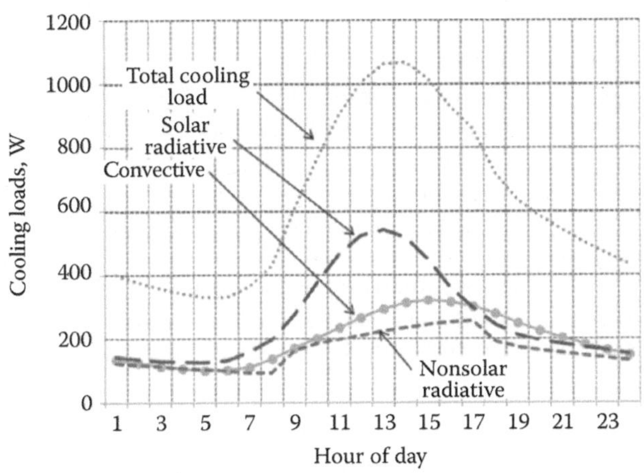

FIGURE 9.11
Peak day cooling loads and their components for Example 9.3.

Comments

It is urged that the student develop spreadsheets as described in this chapter to perform and automate these various calculations. This gives familiarity with the various equations and allows some insights to be gained. For more professional work involving load calculations, ASHRAE had developed spreadsheet versions of the RTS method that are available on a CD accompanying Spitler (2010). The interested reader can refer to this text for a detailed solved example illustrating the use of the RTS method to a two-story 2,800 m^2 (30,000 ft^2) office building.

Problems

Numbers 1–4 given in parenthesis denote the degree of difficulty.

9.1 Find the heat load for the following building. The design outdoor temperature is T_o = 0°F (−17.8°C) at night and until 8:00 a.m. At night the building is unoccupied, at indoor temperature T_i = 50°F (10°C) and without internal gains, and the total heat loss coefficient K_{tot} is 30 kBtu/(h·°F) (15.8 kW/K). Starting at 8:00, the building is occupied by 500 occupants, with 250 kW (853 kBtu/h) of gains from lights and equipment, and T_i = 70°F; the ventilation system is also turned on at 8:00, increasing the total heat loss coefficient K_{tot} to 45 kBtu/(h·°F) (23.7 kW/K). Assume steady-state conditions; i.e., neglect pickup loads. Does the peak occur during occupied or unoccupied periods? *(2)*

9.2 Consider a single-family, detached, single-story house located in suburban Long Island. The location has a design outdoor temperature of 13°F, and the indoor temperature is typically maintained at 70°F during the heating season. The house is built on a slab-on-grade and has no garage. The house has the following characteristics:

General	Length	56 ft
	Width	32 ft
	Wall height	8 ft
	Perimeter	176 ft
	Floor area	1,792 ft²
	Gross wall area	1,408 ft²
	Net wall area	1,114 ft²
	Window area (openings)	252 ft²
	Door area (openings)	42 ft²
	Roof area	2,060 ft²
	Gross volume (living area)	14,336 ft³
Foundation	Slab-on-grade with 8 in. foundation walls and vertical board insulation on outside of foundation wall	
	R value of insulation	5.4 (h·ft²·°F)/Btu
Construction net *U* values	Walls	0.084 Btu/(h·ft²·°F)
	Windows	0.420 Btu/(h·ft²·°F)
	Doors	0.350 Btu/(h·ft²·°F)
	Roof/ceiling (based on ceiling area)	0.030 Btu/(h·ft²·°F)

Using the information given, perform the following calculations to determine the design heating load for the house:

(a) What is the design heat loss due to transmission heat transfer through the combined above-grade, building envelope (walls, windows, door, and ceiling)?

(b) What is the design heat loss due to heat transfer through the slab-on-grade floor?

(c) The infiltration leakage area has been determined to be 91.6 in.² (590 cm²). If the design winter wind velocity is 15 mph (6.7 m/s), what is the volumetric flow rate of infiltration air?

(d) What is the design heat load for the house? Ignore the effects of any moisture requirements, and make the conservative assumption that there are no heat gains due to solar, lights, equipment, or people. Assume that the air density is 0.075 lb/ft³. (Note that 1 ft³/h = 0.00787 L/s.)

(e) If this house were located in Phoenix, AZ, what would be the design outdoor temperature used to calculate the design heating load? *(4)*

9.3 Consider a single-family detached residence. It is a single-story building with an attached heated garage. (Assume that the garage is insulated the same and maintained at the same level of comfort as the house living space.) The house has the following characteristics.

General	Gross floor area (including garage)	2972 ft²
	Living gross floor area	2516 ft²
	Perimeter (including garage)	360 ft
	Wall height	8 ft
	Gross wall area	2880 ft²
	Window area (openings)	332 ft²
	Door area (openings)	70 ft²
	Garage door area	120 ft²
	Roof area (including garage)	3420 ft²
Wall construction	Wood siding, bevel, 0.5 in. × 8 in. lapped	
	Plywood sheathing, 0.5 in.	
	Framing, nominal 2 × 4, 16 in. center, with fiberglass batt insulation	
	Gypsum, 0.5 in.	
Roof construction	Wood shingles, cedar shake	
	Felt building paper	
	Plywood sheathing, 0.5 in.	
	Framing, nominal 2 × 6 trusses, 24 in. centers	
	23° roof pitch	
Ceiling construction	Cellulose insulation, blown in, 8 in.	
	Framing, nominal 2 × 6 trusses, 24 in. centers	
	Gypsum, 0.5 in.	
Window construction	Insulating glass, double, 0.5 in. airspace, wood sash and frame	
	Loosely drawn interior drapery	
Door construction	Solid-core wood doors, 1.75 in., no glazing, no storm doors	
Garage door construction	U = 0.55 Btu/(h·ft²·°F)	

(a) Calculate overall *UA* values for each element of the building envelope, i.e., wall, roof, window, door.

(b) What is the heating load contribution due to conduction if the outdoor temperature is constant at 5°F and the indoor temperature is maintained at 70°F?

(c) How much would the conduction portion of the building heat loss be reduced by replacing the 0.5 in. plywood sheathing in the walls with 0.75 in. styrofoam (expanded, extruded polystyrene)?

(d) If you needed to reduce conduction through the envelope by 20%, how would you do it? *(4)*

9.4 This problem uses the house description from Problem 9.3. This house is to be built in suburban Denver. It will be built on a 0.3 acre lot, which is large compared to most suburban densities, and will have several small trees and shrubs near the house. It is estimated that the house will have an air change rate of 0.5 air change per hour under typical winter conditions of 10 mph wind speed and an outdoor temperature of 30°F. (Assume the house is maintained at 70°F.)

(a) Estimate the effective infiltration leakage area of the house using the LBNL model.

(b) Calculate the heating load due to infiltration under design winter conditions of 15 mph wind speed and 2°F outdoor temperature. (2)

9.5 This problem uses the house description from Problem 9.4. The house has a heated full basement under all but the garage. (The garage measures 22.5 ft × 19.5 ft.) The basement foundation wall is insulated on the outside with R4 insulation. The wall is 7.5 ft high with 1.5 ft above grade. The garage is built on a slab-on-grade floor. The slab foundation wall is also insulated below grade with the same insulation. Assuming that the average winter air temperature is 25°F and that the soil conductivity is 0.8 Btu/(h·ft·°F), estimate the design heat loss from the basement and garage slab. Clearly state your assumptions. *(2)*

9.6 Use the transfer function method to solve this problem. The installed lighting power in a building with A_{floor} = 10,000 ft² (929 m²) is 1.8 W/ft² (19.4 W/m²). The construction type is heavy. Find the contribution to the cooling load at 17:00 if the lights are left on from 9:00 to 17:00. *(2)*

9.7 Set up a spreadsheet for the transfer function method to solve Examples 9.1 through 9.2. For the outdoor temperature T_o, assume sinusoidal variation with average 75°F (23.9°C) and peak 89°F (31.7°C) (5% design conditions for Denver). Examine how much the peak cooling load and the daily cooling energy could be reduced if one ventilates the building with 1, 5, and 20 air changes per hour whenever T_o is below T_i. *(4)*

9.8 Use the transfer function method to determine the peak cooling load of the room specified in Examples 9.1 and 9.2 assuming a heavy wall construction (entry #6 in Table 8.3). Use the dry-bulb temperature, sol–air temperatures, solar radiation, and SHGC values from Table 9.12. *(4)*

9.9 To the room described in Problem 9.7, a south-facing window of 4 m² area of type 21a (low double e glazing of Table 5.1) is added. There are no overhangs nor internal shading. Determine the peak cooling load of the room. *(8)*

9.10 Use the transfer function method to estimate how much the heat from personal computers increases the indoor temperature in a small office building without air conditioning, if they are left on 8 h/day. Does the answer depend on the temperature that the building would have in the absence of computers? Assume

- Floor area of 500 m² (5380 ft²)
- Total heat transmission coefficient K_{tot} = 1200 W/K [2274 Btu/(h·°F)]
- Heat production per floor area 10 W/m² (107.6 W/ft²) while turned on
- Construction-type medium *(3)*

9.11 Use the transfer function method to evaluate two types of roofs: roof type #1 and #6 from Table 8.3. Compare the cooling loads for June in Phoenix, AZ, for these types of roofs when they are painted white (absorptivity of 0.1) and dark (absorptivity of 0.9). Use the global horizontal irradiation values from Table 4.3 and the DBT value from the third column of Table 9.12. Assume room specified in Examples 9.1 and 9.2. *(4)*

9.12 Use the RTS method to solve this problem. Consider the design of a small office building with the following specifications:

- Site Phoenix, AZ
- T_i constant 24 h/day at 78°F (25.6°C) in summer
- Size 30 ft × 30 ft × 8 ft (9.14 m × 9.14 m × 2.44 m), orientation due south
- Open floor plan to allow treating it as one zone (assume light construction)
- Equal windows on all facades, with window/wall ratio = 0.25
- 0.5 air change per hour from 8:00 a.m. to 6:00 p.m., 0.2 air change per hour the rest of the time *(4)*
- Double-pane glass window type #5b of Table 5.1

- Lights and office equipment 3.0 W/ft^2 (32.28 W/m^2) from 8:00 a.m. to 6:00 p.m., 0.5 W/ft^2 (5.38 W/m^2) the rest of the time
- Occupants 0.01 per square foot (0.108 per square meter) from 8:00 a.m. to 6:00 p.m.
- Neglect heat exchange with the ground

You will select any one roof and one wall type from Table 9.10. Make use of the spreadsheets for solar irradiation calculation you have developed to solve Problems 4.19 through 4.21.

(a) Calculate the peak and the daily total cooling load. Determine the relative contributions of each type of load.

(b) Try to reduce the peak cooling load by selecting a different roof type.

(c) Try to reduce the peak cooling load further by selecting a different type of wall.

(d) Try to reduce the peak cooling load still further by means of exterior shading devices that keep direct solar radiation from all the windows. *(4)*

9.13 Repeat the analysis of Problem 9.12 for Atlanta, GA. *(4)*

References

ASHRAE Fundamentals (1989, 1997, 2001, 2013). *Handbook of Fundamentals*. American Society of Heating, Refrigerating and Air-Conditioning Engineers, Atlanta, GA.

ASHRAE 62.1 (2013). *Standard 6.1-2013: Ventilation for Acceptable Indoor Air Quality*. American Society of Heating, Refrigerating and Air-Conditioning Engineers, Atlanta, GA.

Birdsall, B., W.F. Buhl, K.L. Ellington, A.E. Erdem, and F.C. Winkelmann (1990). Overview of the DOE2.1 building energy analysis program. Report LBL-19735, rev. 1. Lawrence Berkeley Laboratory, Berkeley, CA.

EnergyPlus (2009). Energy Plus Building Energy Simulation software, Building Technologies Program. National Renewable Energy Laboratory (NREL) for the U.S. Department of Energy, Washington, DC. http://www.nrel.gov/buildings/energy_analysis.html#energyplus. Accessed May 2014.

Fairey, P.W. and A.A. Kerestecioglu (1985). Dynamic modeling of combined thermal and moisture transport in buildings: Effects on cooling loads and space conditions. *ASHRAE Trans.*, 91(pt. 2A), 461.

Kerestecioglu, A.A. and L. Gu (1990). Theoretical and computational investigation of simultaneous heat and moisture transfer in buildings: 'Evaporation and Condensation' theory. *ASHRAE Trans.*, 96(pt. 1), 455–464.

Mitalas, G.P. and D.G. Stephenson (1967). Room thermal response factors. *ASHRAE Trans.*, 73(pt. 1), 2.1–2.10.

Spitler, J.D. (2010). *Load Calculation Applications Manual*. American Society of Heating, Refrigerating and Air-Conditioning Engineers, Atlanta, GA.

Spitler, J.D. and D.E. Fisher (1999). On the relationship between the radiant time series and transfer function methods for design cooling load calculations. *Int. J. HVAC&R Res.*, 5(2), 125–138.

Spitler, J.D., D.E. Fisher, and C.O. Pedersen (1997). The radiant time series cooling load calculation procedure. *ASHRAE Trans.*, 103(2), 503–515.

Wong, S.P.W. and S.K. Wang (1990). Fundamentals of simultaneous heat and moisture transfer between the building envelope and the conditioned space air. *ASHRAE Trans.*, 96(pt. 2), 73–83.

10

Simplified Annual Energy Estimation Methods and Inverse Modeling

ABSTRACT In this chapter, we present useful and widely used simplified analysis methods that allow annual (or seasonal or monthly) primary energy (either heating or cooling) needs of a building to be determined without having to resort to a full-fledged hour-by-hour simulation over the year. The degree-day method relies on a single-weather datum to characterize the severity of the weather of the location in question. The annual (or seasonal or monthly) energy consumption is then directly proportional to the corresponding degree-day value. It is applicable to residential and light-commercial buildings where energy use is primarily driven by building loads and that normally have simple HVAC equipment whose efficiency can be characterized by a single constant value. An extension of the basic degree-day method is the bin method that gets its name from the way the weather data are assembled. Here, the weather is broken up in outdoor dry-bulb temperature bins and the cumulative number of hours in each bin is determined for the location in question from long-term climatic records. Then, calculations for each bin are done from which the annual or seasonal energy use can be estimated. These methods are widely used for preliminary sensitivity analyses of design options and also by engineering service companies contemplating energy efficiency upgrades in existing buildings. Hence, advantages as well as limitations of these methods are described in this chapter along with numerous solved examples to enhance comprehension. In the last section, we introduce the concept of inverse modeling and provide an overview of single-variate and multivariate steady-state models along with dynamic and hybrid modeling approaches. This modeling approach is being increasingly used when the building already exists and one wishes to improve its operational energy efficiency by making use of its actual energy consumption monitored at different time scales: monthly, daily, hourly, and subhourly. Inverse models are useful for verifying energy efficiency measures that have been implemented, for energy management, for fault detection, and for supervisory control, to name a few applications. We close with a brief description and summary of the principal methods.

Nomenclature

A	Diurnal temperature swing (Equation 10.28)
$a,b,c,\ldots,g$	Regression coefficients in the variable base degree-day and the multivariate-linear change-point inverse models
b_0, b_1, b_2	Regression coefficients in the monthly mean temperature inverse models
C	Cost, $
$\text{CDD}(T_{bal})$	Cooling degree-days for base T_{bal}, K·days (°F·days)
CFD	Cumulative frequency distribution
COP	Coefficient of performance of a vapor cooling system
c_p	Specific heat of air, kJ/(kg·K) [Btu/(lb_m·°F)]
E	Electricity use, kWh
$F(x)$	Gaussian cumulative frequency distribution of x
$f(x)$	Gaussian probability density function of x
$\text{HDD}(T_{bal})$	Heating degree-days for base T_{bal}, K·days (°F·days)
I	Indicator variable used in inverse modeling
K_{cond}	Conductive heat transmission coefficient, W/K [(Btu/(h·°F)]
$\bar{K}_T$	Monthly average clearness index of location
K_{tot}	Total heat transmission coefficient of building, W/K [Btu/(h·°F)]
k	Thermal conductivity, W/m·K [Btu/(h·ft·°F)]
N	Number of days
p_{fuel}	Price of fuel or electricity, cents/kWh or $/ GJ
R^2	Coefficient of determination of a regression model
RMSE	Root mean square error of a regression model
Q	Energy use with subscripts c for cooling, h for heating, kJ (Btu)
$\dot{Q}$	Heat flow with subscripts c for cooling load, h for heating load, W (Btu/h)
$\dot{Q}_{gain}$	Heat gains of building, W (Btu/h)
$\dot{Q}_{nonsol}$	Heat gains of building excluding solar, W (Btu/h)

$\dot{Q}_{sol}$	Solar heat gains of building, W (Btu/h)
T	Temperature, K or °C (°R or °F)
T_{bal}	Balance-point temperature of building, °C (°F)
T_{base}	Base temperature identical to the balance-point temperature (term used in the context of climatic data), °C (°F)
T_i	Indoor air dry-bulb temperature, °C (°F)
T_{max}	Temperature above which air conditioner is turned on, °C (°F)
T_o	Outdoor-air dry-bulb temperature °C (°F)
$\bar{T}_{o,m}$	Monthly mean outdoor-air dry-bulb temperature, °C (°F)
$\bar{T}_{o,m}(t)$	Monthly mean outdoor-air dry-bulb temperature at hour t of day, °C (°F)
$\bar{T}_{o,yr}$	Annual average outdoor-air dry-bulb temperature, °C (°F)
t	Time of day, hour of day, hour
U	Overall heat transfer coefficient, W/(m²·K) [Btu/(h·ft²·°F)]
$\dot{V}$	Volume flow rate, m³/s or L/s (ft³/min)
x	Regressor variable(s) in a regression model
$\bar{x}$	Over bar indicates monthly (or annual) mean values of quantity x
y	Response variable in a regression model
Z_b	Variable defined by Equations 10.21 and 10.22

Greek

ε	Effectiveness of heat recovery device
η	Efficiency of heating or cooling equipment
θ	Normalized temperature variable (Equation 10.19)
σ_{yr}	Annual standard deviation of monthly average outdoor temperatures, K (°F)
σ_m	Standard deviation of daily outdoor temperatures for each month, K (°F)
ρ	Density, lb_m/ft^3 (kg/m³)
Δx	Change in the quantity x
Δx	Thickness of insulation, cm (in.)

Subscripts

bin	Bin
bldg	Building
c	Cooling
cond	Conduction
cp	Change point
dp	Dew point
h	Heating, hour
int	Internal
m	Month
max	Maximum
meas	Measured
nonsol	Nonsolar
sol	Solar
surf	Surface
yr	Year

10.1 General Approaches

The previous three chapters dealt with steady-state and dynamic methods to estimate the various elements of a design load calculation. These are relevant to analysis involving hourly time scales, either for an individual hour for peak design heating load calculations or for a whole day for peak design cooling calculations. Along with being able to determine the peak loads so as to size equipment, analytic methods are also needed to estimate seasonal and/or annual energy use. One could of course use the same load calculation procedures and run them through an entire year's worth of climatic and building operational inputs. With the advent of computer simulation programs, this is what is generally adopted in design offices. However, there are several instances when a quicker method is preferable, for example, during the preliminary stages of an energy audit when different retrofits are being evaluated to improve the energy efficiency of the building. Such methods fall under the classification of *annual energy estimation methods.*

The yearly energy consumption Q_{yr} is needed to evaluate the operating cost—an essential step if one wants to come close to an optimal design in the sense of minimizing the life-cycle cost of a building. Consumption Q_{yr} is the time integral of the instantaneous consumption over the heating (or cooling season); the instantaneous consumption is the instantaneous heating load $\dot{Q}_h$ divided by the instantaneous efficiency η_h of the heating (or cooling) equipment. This can be expressed in either the integral form or as a hourly summation given by

$$Q_{h,yr} = \int \frac{\dot{Q}_h(t)}{\eta_h(t)}\,dt \quad \text{or} \quad = \sum^{yr} \frac{\dot{Q}_h(h)}{\eta_h(h)} \qquad (10.1)$$

Several methods are available, depending on the complexity of the case and the amount of detail one wishes to take into account. This chapter will discuss two most prevalent types: degree-day methods and bin methods.

Degree-day methods are the simplest and are a subclass of analysis methods referred to as *single measure* methods (ASHRAE Fundamentals, 2013). They are appropriate if the building is operated in the same manner (for example, the indoor thermostat is set at a constant value throughout the year) and if the efficiency of the HVAC equipment can be considered constant for all the hours of operation over the year or the season. For situations where efficiency or conditions of utilization vary with outdoor temperature, one can calculate the consumption for different values or bins of the outdoor temperature and multiply it by the corresponding number of hours over the year or season; this approach is used in the *bin methods* (see Section 10.4). As a starting point for these steady-state methods, one needs the value of outdoor dry-bulb air temperature T_o below which heating becomes necessary (or above which cooling is needed); that is called the "balance-point temperature" (discussed in Section 10.2). The severity of a climate can be characterized concisely in terms of degree-days.

When the indoor temperature is allowed to fluctuate, one leaves the domain of simple steady-state models. Various correction terms have been developed to permit an approximate treatment of some dynamic effects with steady-state methods. For greatest accuracy, the use of full-fledged dynamic models is recommended, a choice that is becoming ever more natural with the advance of computer technology. We do not describe dynamic correction terms for steady-state methods because we believe that interest in them is waning in favor of dynamic simulation programs.

But even in an age when computers can calculate the energy consumption of a building at the touch of a key, the concepts of degree-day and balance-point temperature remain valuable tools for intuition and certain types of evaluations involving HVAC system variants. Many engineering firms offering energy audit, energy retrofit, and HVAC upgrade services tend to develop in-house programs (often using the ubiquitous spreadsheet) based on degree-day and bin methods.

10.2 Degree-Day Method

10.2.1 Balance-Point Temperature

Consider a small building in a cold location maintained at an indoor air temperature T_i with "free" heat gains $\dot{Q}_{gain}$ (from the sun, occupants, lights, etc.). The *balance-point temperature* T_{bal} of this building is defined as that value of the outdoor dry-bulb temperature T_o where, for the specified value of T_i, the total heat loss is equal to the free heat gain

$$K_{tot}(T_i - T_{bal}) = \dot{Q}_{gain} \tag{10.2}$$

where K_{tot} [W/K or Btu/(h·°F)] is the total heat loss coefficient of the building including ventilation/infiltration effects. If the peak winter design load of the building is known, it can be conveniently determined from design temperatures:

$$K_{tot} = \frac{\dot{Q}_{design}}{T_{i,design} - T_{o,design}} \tag{10.3}$$

Note that this simplified method of estimating K_{tot} should not be used with peak summer design load because the effect of including peak solar gains can bias the estimate.

Finally, the balance-point temperature is

$$T_{bal} = T_i - \frac{\dot{Q}_{gain}}{K_{tot}} \tag{10.4}$$

In words, T_{bal} is the outdoor dry-bulb temperature above which cooling is required and below which heating is required (provided, of course, that T_i, Q_{gain}, and K_{tot} remain constant during the season/year).

10.2.2 Heating Degree-Day Method

Heating is needed only when T_o drops below T_{bal}. The rate of energy consumption of the heating system is

$$\begin{aligned} \dot{Q}_h &= \frac{K_{tot}}{\eta_h}[T_{bal} - T_o(t)] \quad \text{when } T_o < T_{bal} \\ &= 0 \quad \text{otherwise} \end{aligned} \tag{10.5}$$

where η_h is the efficiency of the heating system (also designated by AFUE, for annual fuel-use efficiency, details, and modeling expressions to be discussed in Section 15.4). For now, we make the simplifying assumption that η_h, T_{bal}, and K_{tot} are constant. Then, the annual heating consumption can be written as an integral:

$$Q_{h,yr} = \frac{K_{tot}}{\eta_h}\int [T_{bal} - T_o(t)]^+ dt \tag{10.6}$$

where the plus sign superscript on the bracket indicates that only positive values are to be counted. This integral of the temperature difference is a convenient summary of the effect of outdoor temperatures on a building. In practice, it is approximated by a sum of averages over short-time intervals (daily or hourly), and the result is called "degree-days" or "degree-hours."

If daily average values of outdoor temperature T_o are used for evaluating the integral, one obtains the degree-days for heating $HDD(T_{bal})$ as

$$HDD(T_{bal}) = 1\text{ day} \times \sum_{\text{days}} (T_{bal} - T_o)^+ \qquad (10.7)$$

with dimensions of K·days (°F·days). It is a function of T_{bal}, reflecting the roles of T_i, heat gain, and the loss coefficient. Note that T_{bal} is also called the "base" of the degree-days, a term often used by those analyzing climatic data.

In summary, one has to estimate the balance-point temperature specific to the building and then determine the degree-days specific to the location in question based on that temperature value, after which Equation 10.7 can be used. Hence, this method is referred to as the "variable-base degree-day" (VBDD) in contrast to the two earlier variants called the (original) degree-day and the modified degree-day. These variants are very unreliable for engineering calculations, but are still used for policy type of analyses.

The question arises whether shorter time intervals should be used for calculating degree-hours. With a summation over hourly values of T_o, one would obtain the degree-hours in the form

$$HDD(T_{bal}) = \frac{1\text{ day}}{24\text{ h}} \times 1\text{ h} \times \sum_{\text{hours}} (T_{bal} - T_o)^+ \qquad (10.8)$$

Because of the nature of short-term fluctuations of T_o, the magnitude of the HDD increases somewhat as the averaging interval is reduced. One could argue that the daily interval is more appropriate because the time constant (see Section 8.4.2) of most buildings is closer to a day than to an hour, and roughly speaking, the thermal inertia of a building averages the effect of fluctuations over the time constant. In any case, most degree-day tabulations do not indicate the details of the calculation (which explains some of the discrepancies between different sources).

Degree-days or degree-hours for a balance-point temperature of 18°C in Europe or 65°F (18.3°C) in the United States have been widely tabulated (see Table 10.1), based on the observation that T_{bal} has represented average conditions in typical buildings (that the appropriate value has been changing with improving insulation is a point that we discuss next). This numerical value is assumed whenever T_{bal} is not indicated explicitly. A contour map of heating degree-days for the United States is shown in

TABLE 10.1
Heating Degree-Days and Other Pertinent Climatic Indices for Selected Locations

	Jan	Feb	Mar	Apr	May	June	July	Aug	Sept	Oct	Nov	Dec	Annual
Atlanta, GA													
HDD (18.3°C)	393	315	252	108	41	12	8	9	23	99	239	375	1874
HDD (65°F)	707	567	454	194	74	22	14	16	41	178	430	675	3373
$\bar{K}_T$	0.43	0.46	0.49	0.53	0.53	0.52	0.51	0.52	0.50	0.53	0.50	0.43	0.50
$\bar{T}_o$ (°C/°F)	6/43	7/45	11/52	16/61	21/70	24/75	26/79	25/77	22/72	17/63	11/52	6/43	16/61
Chicago, IL													
HDD (18.3°C)	700	586	486	264	127	32	17	20	64	189	411	627	3523
HDD (65°F)	1260	1055	875	475	229	58	31	36	115	340	740	1129	6341
$\bar{K}_T$	0.41	0.44	0.47	0.49	0.51	0.55	0.54	0.54	0.53	0.51	0.42	0.36	0.48
$\bar{T}_o$ (°C/°F)	−4/25	−3/27	3/37	10/50	16/61	21/70	24/75	23/73	19/66	13/55	5/41	−2/28	10/51
Denver, CO													
HDD (18.3°C)	609	505	488	308	182	80	37	45	109	251	433	563	3610
HDD (65°F)	1096	909	878	554	328	144	67	81	196	452	779	1013	6498
$\bar{K}_T$	0.62	0.62	0.63	0.62	0.61	0.64	0.63	0.63	0.65	0.65	0.61	0.60	0.63
$\bar{T}_o$ (°C/°F)	−1/30	0/32	3/37	9/48	14/57	19/66	23/73	22/72	17/63	11/52	4/39	0/32	10/50
Phoenix, AZ													
HDD (18.3°C)	251	175	134	52	14	2	0	1	2	28	128	233	1020
HDD (65°F)	452	315	241	94	25	4	0	2	4	50	230	419	1836
$\bar{K}_T$	0.60	0.65	0.68	0.74	0.76	0.75	0.69	0.69	0.71	0.69	0.64	0.59	0.68
$\bar{T}_o$ (°C/°F)	11/52	13/55	15/59	20/68	25/77	29/84	33/91	32/90	29/84	22/72	15/59	11/52	21/70

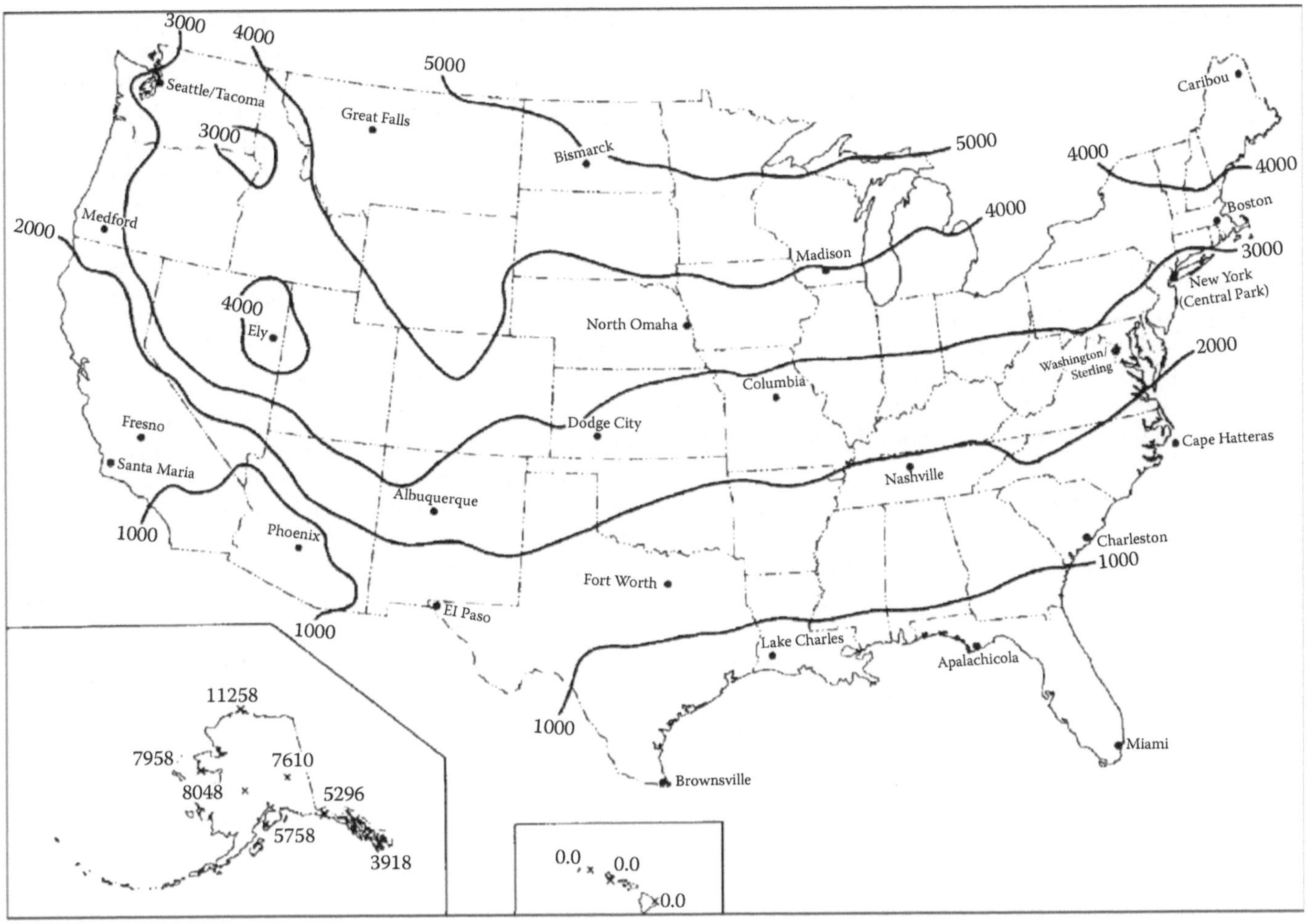

FIGURE 10.1
Annual heating degree-days (K·days) for the United States, for base of 65°F (18.3°C). Inserts show Alaska and Hawaii. To convert to °F·days, multiply by 9/5. (From SERI, Solar radiation energy resource atlas of the United States, Report SERI/SP-642-1037, Solar Energy Research Institute, Golden, CO, 1981.)

Figure 10.1. The extension of degree-day data to different bases is discussed in the next section.

In terms of degree-days, the annual heating consumption is

$$Q_{h,yr} = \frac{K_{tot}}{\eta_h} \text{HDD}(T_{bal}) \tag{10.9}$$

Example 10.1: Annual Heating Energy

Find the annual heating bill for a house in New York City under the following conditions.

Given: Heat transmission coefficient $K_{tot} = 205$ W/K [388.5 Btu/(h·°F)]

Heat gain $\dot{Q}_{gain} = 569$ W (1941 Btu/h)
Indoor temperature $T_i = 21.1$°C (70°F)
Efficiency of heating system $\eta_h = 0.75$
Fuel price \$8/GJ (\$8.44/MBtu)

Assumptions: K_{tot}, $\dot{Q}_{gain}$, T_i, and η_h are constant.

Lookup values: HDD = 2800 K·days (5040°F·days) from Figure 10.1 or Figure 10.3.

Solution

First, we must evaluate the balance-point temperature (Equation 10.4):

$$T_{bal} = T_i - \frac{\dot{Q}_{gain}}{K_{tot}} = 21.1°\text{C} - \frac{569 \text{ W}}{205 \text{ W/K}} = 18.3°\text{C}$$

Coincidentally, T_{bal} happens to have the standard value, and we can proceed directly with the value HDD = 2800 K·days. Inserting this into Equation 10.9 (remember the conversion of units from days to seconds), we obtain the annual heating energy

$$\begin{aligned} Q_{h,yr} &= \frac{K_{tot}\text{HDD}}{\eta_h} \\ &= 205 \text{ W/K} \times 2800 \text{ K·days} \times 24 \text{ h/day} \times \frac{3600 \text{ s/h}}{0.75} \\ &= \frac{49.58 \text{ GJ}}{0.75} = 66.11 \text{ GJ (62.67 MBtu)} \end{aligned}$$

Finally, the annual cost is 66.11 GJ × \$8/GJ = \$529/year.

10.2.3 Cooling Degree-Day Method

Cooling degree-days can be defined in a manner analogous to Equations 10.7 and 10.8 based on either daily or hourly values. If daily values are assumed, then

$$\text{CDD}(T_{bal}) = 1\ \text{day} \times \sum_{\text{days}} (T_o - T_{bal})^+ \qquad (10.10)$$

The definition of the balance-point temperature of a building is the same for cooling as for heating. However, the numerical values of degree-days for cooling (CDD) are generally different from HDD because T_i, Q_{gain}, and K_{tot} can be different. Note that annual heating and cooling degree-days, if based on the same T_{bal}, are related by

$$\text{HDD}(T_{bal}) - \text{CDD}(T_{bal}) = 365\ \text{days} \times (T_{bal} - \overline{T}_{o,yr}) \qquad (10.11)$$

where $\overline{T}_{o,yr}$ is the annual average of T_o. This follows from the fact that the terms of Equation 10.7 are nonzero if and only if the terms of Equation 10.11 vanish, and vice versa. Standard cooling degree-day data are widely available for a base of 18.3°C or 65°F; for example, Figure 10.2 shows the data for the United States.

Sometimes, it is desirable to have heating and cooling degree-days for time periods shorter than a year, and listings of monthly values are indeed quite common. The relation of Equation 10.11 continues to hold, of course, with the appropriate number of days on the right side. Monthly data for a few locations are assembled in Table 10.1, while a longer list can be found in the online HCB software.

How the annual heating degree-days vary with T_{bal} for a particular site, New York, is shown in Figure 10.3a. This plot has been generated by evaluating Equations 10.7 and 10.8 with data for the number of hours per year during which T_o is within 5°F temperature intervals centered at 72, 67, 62, ..., −8°F (from the weather data tables on the online HCB software). The data for the number of hours in each interval (or bin) are included as labels

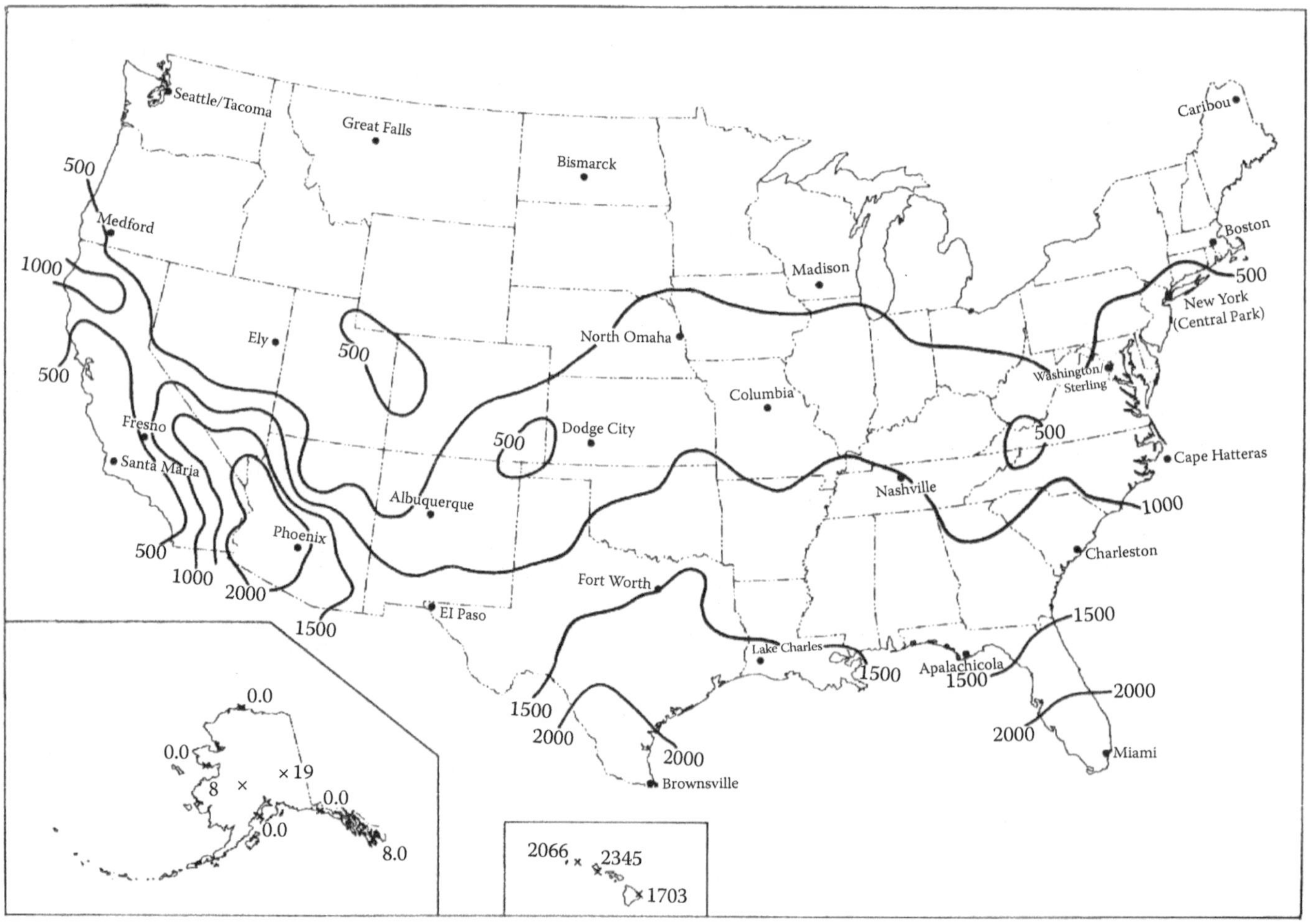

FIGURE 10.2
Annual cooling degree-days (K·days) for the United States, for base of 65°F (18.3°C). Inserts show Alaska and Hawaii. To convert to °F·days, multiply by 9/5. (From SERI, Solar radiation energy resource atlas of the United States, Report SERI/SP-642-1037, Solar Energy Research Institute, Golden, CO, 1981.)

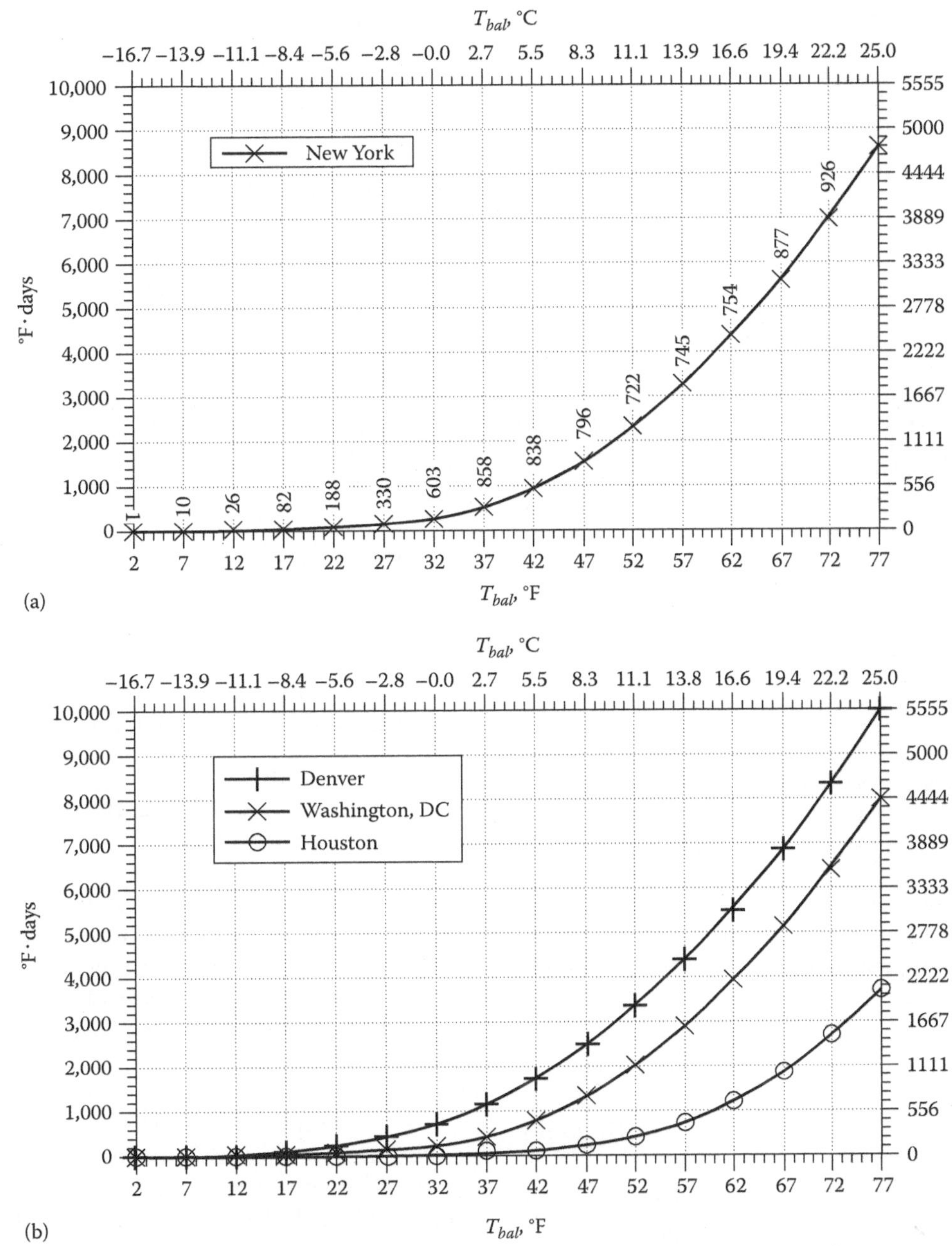

FIGURE 10.3
Annual heating degree-days HDD (T_{bal}) as function of T_{bal}. Based on the weather data tables on the online HCB software. (a) For New York, with labels h/year when T_o is in 5°F (2.7 K) bin centered at the point (interpreting x axis as T_o values). (b) For Denver, Houston, and Washington, DC, degree-days only.

in this plot. Analogous curves, without these labels, are shown in Figure 10.3b for three further sites: Houston; Washington, DC; and Denver.

A graphical illustration of how CDD is calculated is provided by Figure 10.4 using TMY2 data for Phoenix, AZ. Outdoor average daily temperatures T_o for the months of March (a mild month) and July (a very hot month) are plotted along with the 65°F line as shown. The hatched areas are indicative of the CDD (65°F) values for the 2 months. For the month of March, there are days when $T_o < 65$°F, and such days do not contribute to the CDD value, i.e., no cooling is necessary and the air conditioner does not operate. During the month of July, we note that $T_o > 65$°F for all the days, and so the air conditioner has to run continuously. In such cases, Equation 10.10 simplifies to

$$\mathrm{CDD}(T_{bal}) = N_m \times (\bar{T}_{o,m} - T_{bal}) \tag{10.12}$$

where N_m is the number of days in the month.

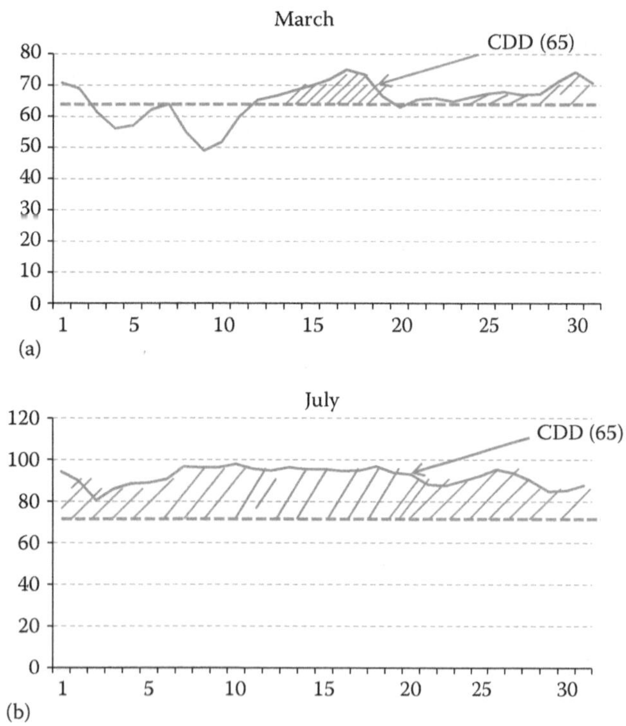

FIGURE 10.4
Graphical illustration of how CDD (65°F) is determined from climatic records. The plots are for 2 months in Phoenix, AZ, based on TMY2 data. (a) March. (b) July.

Empirical expressions to deduce the annual HDD to any arbitrary base temperatures when HDD(65°F) is known have also been proposed; for example, Randolph and Masters (2008):

$$\text{HDD}(T_{bal}) = \text{HDD}(65°\text{F}) - [0.021 \times \text{HDD}(65°\text{F}) + 114] \times (65 - T_{bal}) \quad (10.13\ \text{IP})$$

Such generalized correlations have has not been fully vetted, and should be used with some caution.

Example 10.2: Annual Cooling Energy

Estimate the annual cooling degree-days in New York for a base of T_{bal} = 23.1°C (73.6°F) and for a base of T_{bal} = 18.1°C (64.7°F).

Given: HDD(T_{bal}) can be deduced from Figure 10.3a.*

Find: CDD(T_{bal})

Lookup values: $\overline{T}_{o,yr}$ = 12.5°C, from the weather data tables on the online HCB software.

Solution

Reading Figure 10.3a, we find

$$\text{HDD}(23.1°\text{C}) \approx 4080\ \text{K} \cdot \text{days}\ (7344°\text{F} \cdot \text{days})$$

From Equation 10.11 we have

$$\text{CDD}(23.1°\text{C}) = \text{HDD}(23.1°\text{C}) - 365\ \text{days} \times [(23.1 - 12.5)\ \text{K}] \approx 208\ \text{K} \cdot \text{days}\ (375°\text{F} \cdot \text{days})$$

Likewise for 18.1°C,

$$\text{HDD}(18.1°\text{C}) \approx 2778\ \text{K} \cdot \text{days}\ (5000°\text{F} \cdot \text{days})$$

and

$$\text{CDD}(18.1°\text{C}) = \text{HDD}(18.1°\text{C}) - 365\ \text{days} \times [(18.1 - 12.5)\ \text{K}] \approx 734\ \text{K} \cdot \text{days}\ (1321°\text{F} \cdot \text{days})$$

The annual cooling energy use could be determined by an equation analogous to Equation 10.9:

$$Q_{c,yr} = \frac{K_{tot}}{\eta_c} \text{CDD}(T_{bal}) \quad (10.14)$$

However, there are some practical reasons why such an estimation could give misleading results. The equation strictly applies for a building whose K_{tot} value remains unchanged throughout the season. That assumption is acceptable during the heating season when windows are kept shut and the air exchange rate is fairly constant. But during the cooling season, one can get rid of heat gains and postpone the onset of cooling by opening the windows in residences or increasing the ventilation rate in large commercial buildings (in buildings with mechanical ventilation, this is called the "economizer mode," discussed in Section 19.6.2). The air conditioner is needed only when the outdoor temperature exceeds the threshold $T_{\max}$. This threshold is given by an equation analogous to Equation 10.3, where the closed-window heat transmission coefficient K_{tot} is replaced by its value $K_{\max}$ for open windows:

$$T_{\max} = T_i - \frac{\dot{Q}_{gain}}{K_{\max}} \quad (10.15)$$

Now, $K_{\max}$ can vary considerably with wind speed, but to simplify the discussion, let us assume a constant value. The resulting cooling load is shown schematically in Figure 10.5 as a function of T_o. The solid line is the load with open windows or increased ventilation;

* One could also have used the generalized Equation 10.13 IP but using data from the actual location is more accurate.

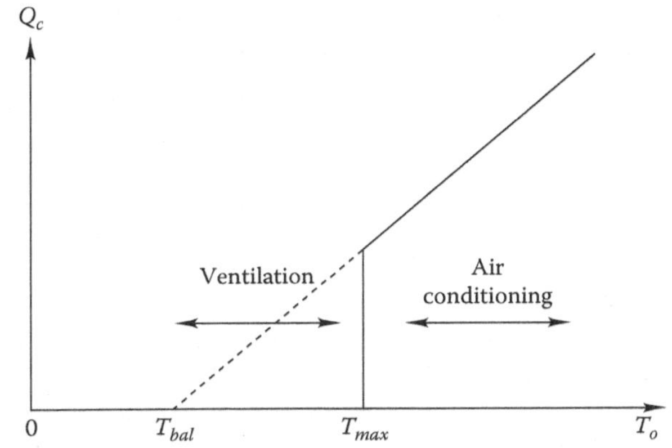

FIGURE 10.5
Cooling load as a function of outdoor temperature T_o. Below T_{max}, cooling can be avoided by using ventilation.

the dashed line is the load if K_{tot} is kept constant. The annual cooling load for this mode is the sum of the cooling load during the days when $T_o > T_{max}$ with windows closed plus that needed to cool down the building from T_{max} to T_{bal} during these same days:

$$Q_{c,yr} = K_{tot}[\text{CDD}(T_{max}) + (T_{max} - T_{bal}) \times N_{max}] \quad (10.16)$$

where CDD(T_{max}) is the cooling degree-days for base T_{max} and N_{max} is the number of days during the season when T_o rises above T_{max}. This model of air conditioning is very schematic, of course. In practice, heat gains and ventilation rates are variable, and, more importantly, the erratic occupant behavior in utilizing the windows and air conditioner has an important effect on cooling energy use.

Example 10.3: Effect of Window Opening on Cooling Energy

Estimate the annual cooling load, with and without a window opening, for a house in New York under the following conditions:

Given: Heat transmission coefficient K_{tot} = 205 W/K [388.5 Btu/(h·°F)] with closed windows at 0.5 air change/h

K_{cond} = 145 W/K due to conduction alone

Heat gains $\dot{Q}_{gain}$ = 1200 W (4094 Btu/h)

T_i = 24.0°C (75.2°F)

Average air change rate with open windows = 10 air changes/h (a reasonable value in view of Figure 6.22)

Find: $Q_{c,yr}$

Assumptions: K_{tot}, K_{max}, $\dot{Q}_{gain}$, and T_i are constant.

Solution

The balance-point temperature for cooling is (from Equation 10.4)

$$T_{bal} = T_i - \frac{\dot{Q}_{gain}}{K_{tot}} = 24.0°\text{C} - \frac{1200 \text{ W}}{205 \text{ W/K}} = 18.1°\text{C}$$

The corresponding number of cooling degree-days has already been found in Example 10.2:

$$\text{CDD}(T_{bal}) = 734 \text{ K} \cdot \text{days}$$

If windows were always *closed*, the annual cooling load would be

$$Q_{c,yr} = K_{tot} \times \text{CDD}(T_{bal})$$
$$= 205 \text{ W/K} \times 734 \text{ Kdays} \times 24 \text{ h/day} \times 3600 \text{ s/h}$$
$$= 13.0 \text{ GJ}$$

If windows are *opened for* ventilative cooling, we first have to find K_{max}. Noting that at 0.5 air change/h, the contribution of the air change is

$$K_{tot} - K_{cond} = 205 - 145 \text{ W/K} = 60.0 \text{ W/K}$$

Hence, at 10 air changes/h this term must be multiplied by 10/0.5, giving a value

$$K_{max} = K_{cond} + 60.0 \text{ W/K} \times \frac{10}{0.5} = 145 + 1200 \text{ W/K}$$
$$= 1345 \text{ W/K}$$

Next, we find T_{max} from Equation 10.15:

$$T_{max} = T_i - \frac{\dot{Q}_{gain}}{K_{max}} = 24.0°\text{C} - \frac{1200 \text{ W}}{1345 \text{ W/K}} = 23.1°\text{C (73.6°F)}$$

When T_o is between T_{bal} = 18.1°C and T_{max} = 23.1°C, the building can be maintained at comfortable conditions by opening the windows.

The required number of cooling degree-days was calculated in Example 10.2 as

$$\text{CDD}(23.1°\text{C}) = 208 \text{ K} \cdot \text{days}$$

To find the number of days N_{max} when T_o is above T_{max}, we take the data for the number of hours in each temperature bin, as shown by the labels in Figure 10.3a. Adding these numbers, we find that there are 7756 h when T_o is below 24.2°C (75.5°F). Since the total number of hours in a year is 8760 h, we deduce that (8760–7756 h)= 1004 h are *above* 24.2°C. Around this temperature range, the distribution of hours per bin is quite uniform at 926 h/5°F, a fact that allows us to interpolate the number of hours above 73.6°F as 1004 h + 926 h × [(75.5 – 73.6)°F]/5°F = 1355 h. Hence, N_{max} = 56.5 days.

Finally, we can insert these values into Equation 10.16 to obtain the cooling load as

$$Q_{c,yr} = 205 \text{ W/K} \times [208 \text{ K} \cdot \text{days} + (23.1 - 18.1)°\text{C} \times 56.5 \text{ days}]$$
$$\times 24 \text{ h/day} \times 3600 \text{ s/h}$$
$$= 8.7 \text{ GJ}$$

Comments

The difference between the loads with and without a window opening is quite large, on the order of 50% in this case (very dependent on K_{tot}, K_{max}, $\dot{Q}_{max}$, T_i, and the climate). This is one of several reasons why actual residential cooling loads are highly sensitive to occupant behavior.

A basic assumption of the degree-day method, namely, that T_{bal} is constant throughout the season, is, of course, not well satisfied in practice. Solar gains are zero at night, and internal gains tend to be highest during the evening. The pattern for a typical house is shown in Figure 10.6. As long as T_o stays always below T_{bal}, the variations average out without changing the consumption. But for the situation in Figure 10.6, when T_o rises above T_{bal} from shortly after 10:00 a.m. to 10:00 p.m., the consequences for the energy consumption depend on the thermal inertia and on the control of the HVAC system. If this building had low inertia and temperature control were critical, heating would be needed at night and cooling during the day. In practice, this effect is reduced by thermal inertia and by the deadband of the thermostat that allows T_i to float somewhat.

The closer T_o is to T_{bal}, the greater the uncertainty (this is clear from Figure 10.4). If the occupants keep the windows closed during mild weather, T_i will rise above the set point. If they open the windows, the potential benefit of the heat gain is reduced. In either case, the true values of T_{bal} become quite uncertain. Therefore, *the degree-day method is unreliable for estimating the consumption during mild weather*, and so is any steady-state method, for that matter.

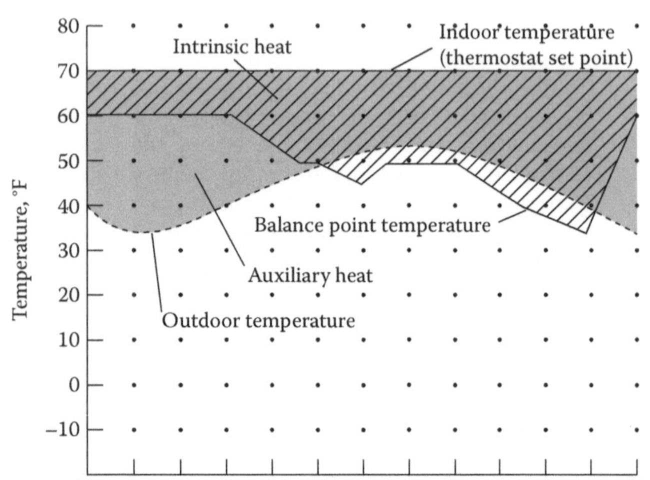

FIGURE 10.6
Variation of balance-point temperature and internal gains for a typical house. (From Nisson, J.D.N. and Dutt, G., *The Superinsulated Home Book*, Wiley, New York, 1985. With permission.)

In fact, the consumption becomes very sensitive to occupant behavior and cannot be predicted with certainty.

Despite these problems, the degree-day method can give remarkably accurate results for the annual heating energy of single-zone buildings (Claridge et al., 1987). There are several reasons. Typical buildings have time constants that are roughly on the order of 1 day, and their thermal inertia averages in effect over the diurnal variations, especially if T_i is allowed to float. Furthermore, the uncertainty of the energy consumption in mild weather is small in absolute terms; hence, a relatively large error here does not carry much weight in the total for the season.

An important point to note about annual energy calculations, based on the steady-state methods described in this section, is that the heat gains must be the average values of the period in question, *not* the peak values. In particular, the solar radiation should be based on the averages (see Section 4.9), *not* the peak value (Section 4.6).

10.3 Models for Estimating Degree-Days under Different Base Temperatures

The calculation of Q_h from the degree-days HDD(T_{bal}) in Example 10.1 was easy since T_{bal} happened to have the conventional value. That is usually not the case. The average value of T_{bal} varies widely from one building to another because of widely differing personal preferences for the settings of thermostat and of thermostat setback and because of different building characteristics. In response to the fuel crises of the 1970s, heat transmission coefficients have been reduced, and thermostat setback has become a common practice. At the same time, the energy use by appliances has increased. These trends reduce T_{bal}, and evidence for this has been well documented (for example, Fels and Goldberg, 1986). Hence, degree-days with a base of 65°F or 18°C must be employed with caution. For determining degree-day values at other base temperatures (referred to as variable base degree-days or VBDD), a number of correlations have been proposed, which are described in the following section.

10.3.1 Erbs et al. Method

Several authors have proposed formulas for estimating VBDD relative to an arbitrary base temperature, T_{base}, when daily outdoor temperature data are not available for evaluating the integral in Equation 10.6. The basic idea is to assume a typical probability distribution of temperature data, characterized by its average $\bar{T}_o$ and by its standard deviation σ. Erbs et al. (1983) have developed a model that needs as input only the averages $\overline{T}_{o,m}$ for each month of the year. The standard deviations σ_m for each month are then estimated from the following correlation:

$$\sigma_m = 1.45 - 0.0290\overline{T}_{o,m} + 0.0664\sigma_{yr}$$

(dimensional equation, T and σ in °C) (10.17 SI)

where σ_{yr} is the standard deviation of the monthly mean temperatures given by

$$\sigma_{yr} = \sqrt{\frac{1}{12}\sum_{m=1}^{12}(\overline{T}_{o,m} - \overline{T}_{o,yr})^2} \quad (10.18)$$

about the annual average $\overline{T}_{o,yr}$.

To obtain a simple expression for the degree-days, a normalized dimensionless temperature variable θ is defined as

$$\theta = \frac{T_{bal} - \overline{T}_{o,m}}{\sigma_m\sqrt{N_m}} \quad (10.19)$$

where N_m is the number of days in the month. While temperature distributions can be quite different from month to month and location to location, most of this variability can be accounted for by the average and the standard deviation of T_o. Being centered around $\overline{T}_{o,m}$ and scaled by σ_m, the quantity θ largely eliminates these effects. In terms of θ, the monthly heating degree-days for any location are very well approximated by the following distribution:

$$\text{HDD}(T_{bal}) = \sigma_m N_m^{3/2}\left\{\frac{\theta}{2} + \frac{\ln[\exp(-a\theta) + \exp(a\theta)]}{2a}\right\} \quad (10.20)$$

with $a = 1.698$

Erbs et al. (1983) have verified, for nine locations spanning most climatic zones of the United States, that the annual heating degree-days can be estimated with a maximum error of 175 K·days if one uses this equation for each month. For cooling degree-days, they found that the largest error is 150 K·days. Such errors are quite acceptable, representing less than 5% of the total.

10.3.2 Schoenau and Kehrig Method

Schoenau and Kehrig (1990) have proposed a slight modification to the Erbs et al. (1983) approach, which is said to be much more accurate (as reported by Thevenard, 2011). Hence, the former has replaced the latter as the one recommended for building studies (ASHRAE Fundamentals, 2013). The improved accuracy comes at the expense of additional data requirements. While the Erbs et al. correlation only needed monthly average ambient temperature to deduce the HDD for any base, the Schoenau and Kehrig approach requires, in addition, the monthly standard deviation σ_m of daily average temperatures for each month. Subsequently, a normalized value similar to θ (Equation 10.19) is defined as follows for each month:

$$\text{For heating } Z_b = \frac{T_{bal} - \overline{T}_{o,m}}{\sigma_m} \quad (10.21)$$

$$\text{For cooling } Z_b = \frac{\overline{T}_{o,m} - T_{bal}}{\sigma_m} \quad (10.22)$$

It is then suggested that both monthly HDD and CDD be determined from

$$\text{DD}(T_{bal}) = N_m \cdot \sigma_m[Z_b \cdot F(Z_b) + f(Z_b)] \quad (10.23)$$

where

N_m is the number of days in the month

f and F are the probability density function and the cumulative distribution function, respectively, of the Gaussian distribution given by

$$f(Z) = \frac{1}{\sqrt{2\pi}}\exp\left(-\frac{Z^2}{2}\right) \quad \text{and} \quad F(Z) = \int_{-\infty}^{Z} f(z)\cdot dz \quad (10.24)$$

Both these functions are easily determined since they are built-in functions in many scientific calculators and spreadsheet programs. Thevenard (2011) has suggested that in case the monthly standard deviation values needed for the Schoenau and Kehrig approach are not available, then the Erbs et al. correlation given by Equation 10.17 SI can be used to estimate these values. The prediction accuracy is then poorer and is said to fall midway between the Schoenau and Kehrig approach and the Erbs et al. approach.

There are instances when the Erbs et al. and the Schoenau and Kehrig approaches are not needed for estimating monthly energy use. Recall from Section 10.2.3 that if the base temperature is significantly higher (about 10°C or so) than the monthly average temperature, it is unlikely that any day of the month will have a temperature value lower than the base temperature. Then the expression of Equation 10.23 simplifies to Equation 10.12. This is clear from Figure 10.4b for Phoenix, AZ, during July when the outdoor temperature is always higher than 65°F. An analogous argument holds for heating degree-day estimation as well.

Example 10.4: Monthly Heating Degree-Day Estimation

Find the heating degree-days for New York using the model of Erbs et al. (1983).

Given: Monthly averages of T_o from the online HCB software as reproduced in the second column.

Find: HDD(T_{bal}) for T_{bal} = 18.3°C (65°F)

Solution

The results have been calculated with a spreadsheet and are reproduced as follows. The second column lists the values of monthly average outdoor temperature, and N_m is the number of days in the month from which $\bar{T}_{o,yr}$ and σ_{yr} can be deduced (as shown). Intermediate quantities are shown in the fourth and fifth columns. The sixth column shows the results, and for comparison the seventh column shows the actual values calculated by the online HCB software (direct evaluation of Equation 10.7 using bin data).

			σ_m	θ	HDD (T_{bal})	HDD (T_{bal})
Month	$\bar{T}_{o,m}$	N_m	**Equation 10.17**	**Equation 10.19**	**Equation 10.20**	**HCB Software**
Jan	0.1	31	2.031	1.61	565	565
Feb	0.8	28	2.01	1.645	490	492
Mar	5.1	31	1.886	1.257	411	412
Apr	11.2	30	1.709	0.759	219	215
May	16.8	31	1.546	0.174	81	76
Jun	22	30	1.396	−0.484	12	0
Jul	24.8	31	1.314	−0.888	3	0
Aug	23.8	31	1.343	−0.735	5	0
Sep	20.2	30	1.448	−0.240	26	16
Oct	14.8	31	1.604	0.392	128	116
Nov	8.6	30	1.784	0.993	294	293
Dec	1.9	31	1.978	1.489	509	508
$\bar{T}_{o,yr}$	12.51			**Sum:**	**2742**	**2693**
σ_{yr}	8.789					

Comments

The agreement is excellent, except for the summer months where the true values are zero.

It was stated earlier that when $(T_{bal} - \bar{T}_{o,m}) > 10$°C (approximately), then HDD $= N_m(T_{bal} - \bar{T}_o)$ is a very good approximation. We find this to be the case for January when (18.3 − 0.1) × 31 = 564.2, which is almost identical to the HDD(Jan) value of 565 K·days shown. However, for May, we expect this approximation to fail, which it does since (18.3 − 16.8) × 31 = 46.5. This value is much lower than the HDD(May) value of 81 K·days shown in the table.

10.3.3 Parametric Analysis

The annual energy estimation methods are well suited for performing parametric studies during the preliminary design phase. In *parametric studies*, one varies one parameter at a time and displays the effect on loads (peak and average) and on costs. Good graphic display of the results is important so the designer can see in which direction to pursue the search. It is instructive to try to identify, as far as possible, the contributions of individual elements of the design, e.g., the heat loss due to air exchange. This will be illustrated in Example 10.5.

To illustrate the method of parametric studies, let us take a simple building like the one of Example 6.8 and ask how the energy consumption could be reduced. Starting from an initial design, called "reference," we add various features to reduce the energy consumption. Since the ultimate goal is to find the design that is the most cost-effective, we also consider the incremental cost of each additional energy conservation feature. As a crude indicator of cost-effectiveness, we calculate the payback time; a more rigorous economic analysis would evaluate the life-cycle savings or the rate of return, as explained in Chapter 23 on engineering economics.

Also, to highlight the essential features of a parametric study, we have simplified the example as much as possible. The building is a rectangular box, as shown in Figure 6.19. We assume steady-state conditions, with all building parameters constant, in particular the indoor temperature, the air change rate, and the heat gains. We neglect heat exchange with the ground. We take the heating system efficiency constant at η = 1.0, as appropriate for electric resistance heating without distribution losses. And we consider only a small number of design variations of the envelope.

Example 10.5: Degree-Day Method for Analyzing Design Alternatives

Consider several envelope design variations for a house of boxlike shape and wood-frame construction. For each variation, calculate the annual heating energy as well as the payback time relative to the reference design.

Given: House of Figure 6.19
Rectangular box 12 m × 12 m × 2.5 m (39.4 ft × 39.4 ft × 8.2 ft)
Area of roof = 144 m²
Area of opaque walls = 96 m²
Area of glazing = 24 m², which is 20% of the wall area
Heat gains $\dot{Q}_{gain}$ = 1 kW (3412 Btu/h) constant year-round
Volume = 360 m³

Outdoor air at 0.5 air change per hour, hence

$$\dot{V}_\rho C_p = 0.5 \times 360 \text{ m}^3/\text{h} \times 1.2 \text{ kJ}/(\text{m}^3 \cdot \text{K}) = 60 \text{ W/K}$$

(see Equation 7.20)

T_i = 20°C (68°F) constant year-round
Heating system efficiency η = 1.0
Energy price p_{fuel} = \$19.44/GJ = 7¢/kWh (\$20.50/MBtu), a typical value for electric heating on the east coast of the United States

Site: New York City

Reference design (design 1): U values of opaque surfaces equivalent to 0.10 m (4 in.) of fiberglass for walls and 0.15 m (6 in.) for roof; conductivity of fiberglass k = 0.060 W/(m·K) [0.10 Btu/(h·ft·°F)]; double-glazed windows with U = 3.0 W/(m²·K) [0.53 Btu/(h·ft²·°F)].

Variations to be considered: Add successively the following energy conservation features:

- Design 2: Increase roof insulation thickness from 0.15 to 0.25 m; the incremental cost per volume of fiberglass is \$18/m³.
- Design 3: The heat recovery from exhaust ventilation with heat exchanger effectiveness ε = 0.75; the incremental cost is \$1300.
- Design 4: Reduce U value of windows from 3.0 to 1.5 W/(m²·K) (low emissivity, triple glazed); the incremental cost per area is \$140/m².
- Design 5: Increase wall insulation thickness from 0.1 to 0.15 m; the incremental cost per volume of fiberglass is \$18/m³, plus the incremental cost of changes in wall construction (0.15 m studs instead of 0.10 m) amounting to \$4.00/m² of the wall area.

Assumptions: All parameters are constant; use the VBDD method. Calculate the U values of the wall and roof by considering only the fiberglass insulation, without film coefficients and support structure. There is no heat exchange with the ground.

Lookup values: Degree-day data from Figure 10.3

Solution

The problem has been solved with a spreadsheet, and the essential points are summarized in the tables as follows and in Figure 10.7. We explain the details for the case of design 2.

The U values of the walls and roof are calculated as $U = k/\Delta x$, where Δx = thickness. For example, the U value of the roof is

- For the reference design: 0.060/0.15 W/(m²·K) = 0.40 W/(m²·K)
- For the improved design 2: 0.060/0.25 W/(m²·K) = 0.24 W/(m²·K)

Hence, the difference in U values is ΔU = 0.16 W/(m²·K). The difference in the contribution to K_{tot} is ΔK_{tot} = 144 m² × 0.16 W/(m²·K) = 23.04 W/K, and the resulting K_{tot} for Design 2 is 224.2 W/K.

The balance-point temperature is calculated according to Equation 10.4 as

$$T_{bal} = T_i - \frac{\dot{Q}_{gain}}{K_{tot}} = 20°\text{C} - \frac{1000 \text{ W}}{224.2 \text{ W/K}} = 15.54°\text{C}$$

The corresponding degree-days are DD(T_{bal}) = 2139 K·days (read from the IP scales of Figure 10.3 and converted to SI).

From Equation 10.9, the annual heating energy with η = 1.0 is found (after converting to consistent units):

$$Q_{h,yr} = 224.2 \text{ W/K} \times 2139 \text{ K} \cdot \text{days} = 41.42 \text{ GJ}$$

Taking the difference from the consumption for the reference design, we find annual energy savings of ΔQ = 6.04 GJ; they are worth $\Delta Q p_{fuel}$ = 6.04 GJ × \$19.44/GJ = \$117.40.

For the associated cost, one can assume that only the additional insulation material needs to be taken into account because neither additional structural support nor labor is necessary in new construction. Thus, the incremental cost of the roof insulation is ΔC = 144 m² × 0.10 m × \$18/m³ = \$259.

Dividing the investment ΔC by the annual savings, we obtain the *payback time* as

$$\frac{\$259}{\$117.40/\text{year}} = 2.2 \text{ year}$$

This means that the extra cost of the roof insulation will have paid for itself after 2.2 years and is worth the investment. The lifetime of the features is on the order of 20 years or longer. As a rule of thumb, if the lifetime is several times longer than the payback time and if there is negligible risk, the investment is very good for payback times shorter than 5 years and fairly good for payback times shorter than 10 years.

When several energy conservation features are under consideration, payback times could be evaluated for each feature separately or for any combination of features. For simplicity, we show only the cumulative cost (as the features are added on top of each other) and the payback time for the corresponding combinations.

It is instructive to plot, as in Figure 10.7, the components of K_{tot} and the heating energy Q_h; thus, one can clearly see which loss terms are the most important ones.

		Change		Contribution to K_{tot}				
Design	ΔU, W/(m²·K)	ΔK_{tot}, W/K	ΔC, $	Roof	Walls	Glazing	Air	K_{tot}, W/K
1. Reference				57.6	57.6	72.0	60.0	247.2
2. Addition of roof insulation	0.16	23.04	259	34.6	57.6	72.0	60.0	224.2
3. Addition of HX		45.00	1300	34.6	57.6	72.0	15.0	179.2
4. Low-U glazing	1.50	36.00	3360	34.6	57.6	36.0	15.0	143.2
5. Addition of wall insulation	0.20	19.20	470	34.6	38.4	36.0	15.0	124.0

T_{bal}, °C	$DD(T_{bal})$, K·days	Q, GJ	ΔQ, GJ	ΔQp_{fuel}, $	Cumulative ΔC	Payback Time, year
15.95	2222	47.46	0.00	0.00	0	
15.54	2139	41.42	6.04	117.40	259	2.2
14.42	1917	29.67	17.79	345.98	1559	4.5
13.01	1639	20.27	27.19	528.71	4919	9.3
11.93	1444	15.47	31.99	622.07	5390	8.7

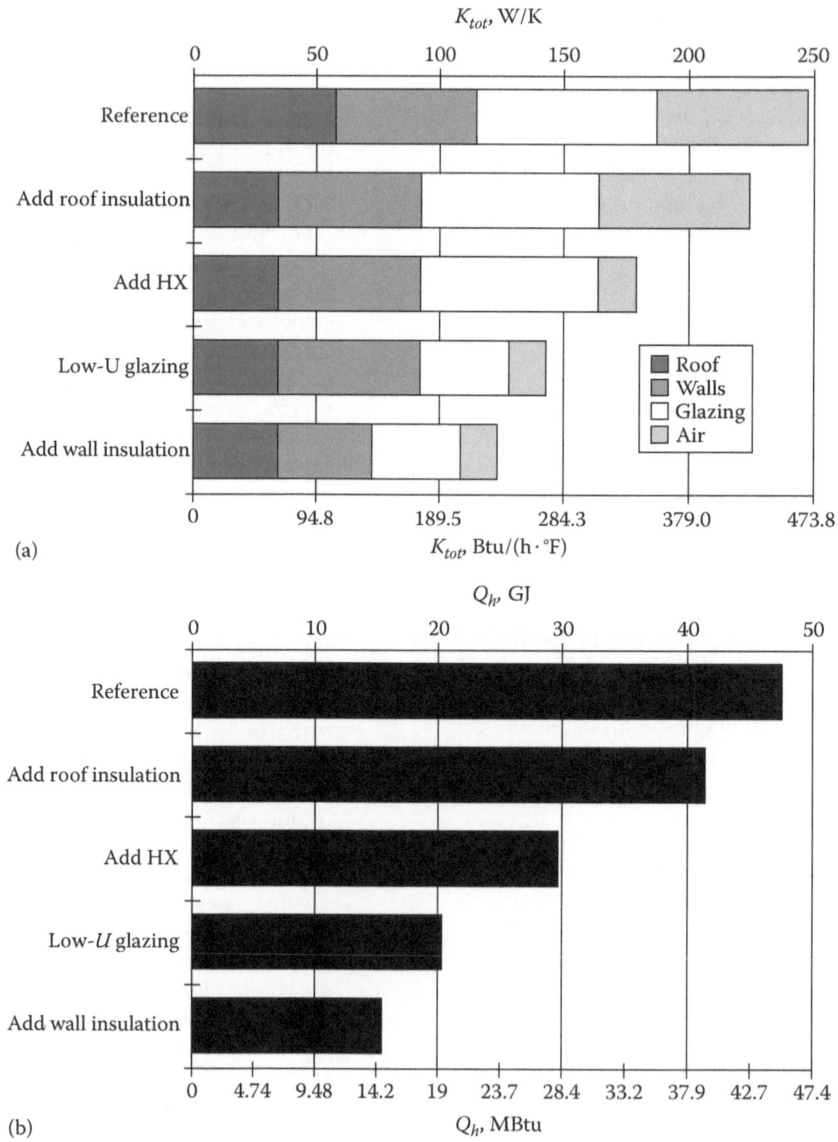

FIGURE 10.7
Parametric study of Example 10.5. (a) Contributions to total heat loss coefficient K_{tot} and (b) annual heating energy Q_h.

Comments

1. The reference design is representative of conventional construction during the 1980s. The features considered here do not require any unusual technologies, yet they reduce the consumption by a factor of 3.
2. In a real application, one should not limit oneself to envelope modifications. Other heating systems should be considered, in particular, a heat pump or a condensing furnace.
3. Considering only cumulative design modifications, one cannot identify and rank the cost-effectiveness of each individual feature. For instance, the reversal of payback times between designs 4 and 5 implies that extra wall insulation is more cost-effective than the low-U glazing, at least with the earlier cost assumptions. Also, it may turn out that several features are cost-effective by themselves, but not in combination (the load remaining after adding one feature may be too small to justify the other).
4. There is a certain risk of being misunderstood or quoted out of context when one states specific numbers for payback times, since the results are highly variable with the assumptions. The cost of energy-saving features can vary quite a bit from one supplier to another. In some regions of the United States and in much of Canada, the number of degree-days is twice as large as in New York City; the corresponding savings would be twice as large and the payback time one-half as long. On the other hand, natural gas might be available at less than one-half the energy price assumed here, implying payback times twice as long.

In this example, it is easy to identify the heat loss contributions of individual design elements because the annual consumption is the product of degree-days and K_{tot} and the contributions of the individual design elements to K_{tot} are explicit. When the conditions are not suitable for a steady-state analysis, the identification of individual contributions may be less obvious, especially if the computer simulation program does not list the contributions to K_{tot}. In such a case, it is advisable to begin by considering an upper and a lower bound of a parameter. For instance, setting the U value of the walls equal to zero, one obtains an upper limit on the savings achievable by wall insulation. The process of parametric search is all the more useful if it can be carried out interactively: one sees immediately in which sense to continue.

10.4 Bin Method

10.4.1 Basic Approach

There are many applications where the degree-day method should not be used, even when corrected for variable base, when the heat loss coefficient K_{tot} and the efficiency η of the HVAC system cannot be treated as constant, or when the balance-point temperature varies diurnally or seasonally. Further, the effect of solar loads is not accounted for properly, thereby biasing the results. The efficiency of a heat pump, e.g., varies strongly with outdoor temperature (discussed in Section 14.5). Also, the efficiency of the HVAC equipment may be affected indirectly by T_o when the efficiency varies with the load, a common situation for boilers and chillers. Furthermore, in most commercial buildings, the occupancy has a very pronounced pattern, which affects the heat gains, the indoor temperature, and the ventilation rate.

In such cases, a steady-state calculation can yield good results for the annual energy consumption if different temperature intervals and time periods are evaluated separately. This approach is called the "bin method," because the consumption is calculated for several values of the outdoor temperature T_o and multiplied by the number of hours N_{bin} in the temperature interval (or bin) centered on that temperature:

$$Q_{bin} = N_{bin} \times \frac{K_{tot}}{\eta} \times (T_{bal} - T_o)^+ \qquad (10.25)$$

The plus subscript on the parentheses indicates that only positive values are to be counted; no heating is needed when T_o is above T_{bal}. This equation is evaluated for each bin, and the total consumption is the sum of the Q_{bin} values over all bins.

In the United States, the necessary data, called "bin data," have been generated in temperature intervals of 5°F (2.8°C) (ASHRAE, 1984) and widely available. Hourly bin data for a few U.S. locations have been assembled in Table 10.2 for temperatures below 72°F (22.2°C). As an illustration, the histogram of the bin data for Denver, CO, is shown in Figure 10.8. Notice that the bin data distribution for this location (but not in general) is near Gaussian except for the portion near

TABLE 10.2
Hourly Bin Data for Selected Locations in the United States for Heating Calculations

Temp. Bin		Atlanta, GA	Chicago, IL	Denver, CO	New York, NY	Phoenix, AZ	San Francisco, CA	Seattle, WA	Washington, DC
72°F	22.2°C	1185	762	549	926	762	285	258	960
67°F	19.4°C	926	769	684	877	776	665	448	766
62°F	16.7°C	823	650	783	751	767	1264	750	740
57°F	13.9°C	784	592	731	745	769	2341	1272	673
52°F	11.1°C	735	569	678	722	659	2341	1462	690
47°F	8.3°C	676	543	704	796	540	1153	1445	684
42°F	5.6°C	598	591	692	838	391	449	1408	790
37°F	2.8°C	468	800	717	858	182	99	914	744
32°F	0°C	271	822	721	603	57	10	427	542
27°F	−2.8°C	112	551	553	330	8		104	254
22°F	−5.6°C	44	335	359	188			39	138
17°F	−8.3°C	19	196	216	2			20	54
12°F	−11.1°C	8	117	119	26			3	17
7°F	−13.9°C	2	85	78	10				2
2°F	−16.7°C		59	36	1				
−3°F	−19.4°C		25	22					
−8°F	−22.2°C		12	6					
−11°F	−25°C		3	1					

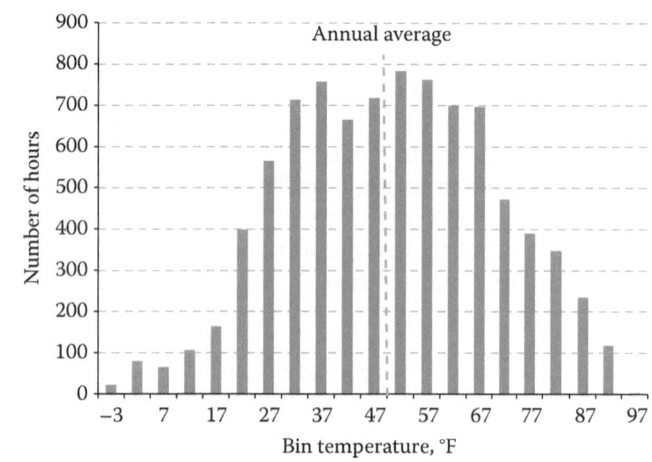

FIGURE 10.8
Bin data for Denver, CO.

the peak (the annual average temperature is 50.1°F). For enhanced flexibility, the data may also be generated for each month and for several periods of the day. Erbs et al. (1983) have proposed such an empirical method along the same lines as the approach described in Section 10.3.1, and this is described in Section 10.4.2.

For the data included with the online HCB software for this book, the day is divided into six intervals of 4 h each; e.g., during the time from midnight to 4:00 a.m. the average February in New York has 24 h when the temperature is between 29°F and 34.5°F (See table on next page).

Example 10.6: Bin Method to Determine Annual Heating Needs of an Office Building

Find the heating energy consumption of a simple commercial building in New York in February, for the following conditions:

Given: Occupied 7 days per week, 8:00 to 18:00; unoccupied otherwise

$$K_{tot} = 2.0\ \text{kW/K}\ [3.8\ \text{kBtu/(h}\cdot{}^\circ\text{F)}]\quad \text{occupied}$$

$$= 1.0\ \text{kW/K}\ [1.9\ \text{kBtu/(h}\cdot{}^\circ\text{F)}]$$

unoccupied (reduced because ventilation system shut off)

$$\dot{Q}_{gain} = 20.0\ \text{kW}\ (68.2\ \text{kBtu/h})\quad \text{occupied}$$

$$= 1.0\ \text{kW}\ (3.4\ \text{kBtu/h})\quad \text{unoccupied}$$

$$T_i = 20.0^\circ\text{C} = \text{constant}$$

Efficiency of heating system η = 1.0. Bin data of online HCB software, reproduced on next page.

Find: Q_h

Assumptions: At steady state

Solution

First, the balance-point temperatures are calculated:

$$T_{bal} = 20°C - \frac{20\ \text{kW}}{2.0\ \text{kW/K}} = 10°C \quad \text{occupied}$$

$$T_{bal} = 20°C - \frac{1\ \text{kW}}{1.0\ \text{kW/K}} = 19°C \quad \text{unoccupied}$$

The numbers of hours (No.) for different outdoor temperature bins and for different hourly segments of the days for New York are given in the spreadsheet below. Then, for each bin (temperature and time of day), the heating load Q_h is calculated according to Equation 10.25 as shown in the spreadsheet. For instance, the bin of 5.6°C from 9:00 to 12:00 contains 25 h, and so we have Q_h = 2.0 kW/K × (10°C − 5.6°C) × 25 h = 220 kWh, taking the K_{tot} and T_{bal} values for occupied periods weighted according to the number of hours of each type. The corresponding value in the table is shown as 222.2 kWh; the difference is due to the rounding in converting the bin temperature of 42°F into 5.6°C (rather than as 5.56°C, which would have yielded the value shown in the following table).

The sums of the rows and columns are shown at the right and the bottom; the grand total heating load comes to

$$12{,}082.4\ \text{kWh} = 12{,}082.4\ \text{kWh} \times \frac{3.6}{1{,}000}\ \text{GJ/kWh}$$

$$= 43.5\ \text{GJ}\ (41.2\ \text{MBtu})$$

Comments

To keep the example simple, we have assumed the same schedule for each day of the week. More realistic schedules (such as weekdays/weekends) can easily be accommodated by this method: one takes for each bin the average over all relevant conditions, weighted by the corresponding fraction of time. For example, if the building were unoccupied on weekends, the entries for the 9:00 to 12:00 hourly period for T_b= 5.6°C bin would be calculated as

$$Q_h = [2.0\ \text{kW/K} \times (10°C - 5.6°C) \times 25\ \text{h} \times (5/7)] + [1.0\ \text{kW/K} \times (19°C - 5.6°C) \times 25\ \text{h} \times (2/7)] = 252.85\ \text{kWh}.$$

Thus, the heating energy requirement increases because of the absence of "free" heat during the weekends.

Bin	1–4 h		5–8 h		9–12 h		13–16 h		17–20 h		21–24 h		
T_{bin}	No.	Q_h	No.	Q_h	No.	Q_h	No.	Q_h	No.	Q_h	No.	Q_h	Sum (kWh)
52°F/11.1°C	0	0.0	0	0.0	0	0.0	5	0.0	1	3.9	0	0.0	**3.9**
47°F/8.3°C	0	0.0	1	10.7	7	23.3	16	53.3	12	84.0	1	10.7	**182.0**
42°F/5.6°C	10	134.4	13	174.8	25	**222.2**	30	266.7	21	234.5	15	201.7	**1,234.3**
37°F/2.8°C	33	535.3	32	519.1	27	390.0	22	317.8	31	475.3	33	535.3	**2,772.9**
32°F/0°C	24	456.0	16	304.0	20	400.0	25	500.0	30	585.0	34	646.0	**2,891.0**
27°F/−2.8°C	13	283.1	18	392.0	12	306.7	3	76.7	9	213.0	12	261.3	**1,532.8**
22°F/−5.6°C	17	417.4	11	270.1	16	497.8	9	280.0	0	0.0	9	221.0	**1,686.3**
17°F/−8.3°C	11	300.7	17	464.7	2	73.3	2	73.3	8	256.0	4	109.3	**1,277.3**
12°F/−11.1°C	3	90.3	0	0.0	3	126.7	0	0.0	0	0.0	4	120.4	**337.4**
7°F/−13.9°C	1	32.9	4	131.6	0	0.0	0	0.0	0	0.0	0	0.0	**164.4**
Sum (kWh)		**2250.2**		**2266.9**		**2040.0**		**1567.8**		**1851.8**		**2105.8**	**12,082.4**

We will take up the bin method again in Sections 14.7, 15.4, and 19.5 to evaluate the annual performance of vapor compression equipment, boilers, and secondary all-air systems, respectively.

10.4.2 Model for Generating Bin Data

The bin method requires data on the frequency distribution of hourly outdoor temperature at the location. In case such hourly temperature records are not available, one could use a generalized correlation proposed by Erbs et al. (1983), which predicts the monthly cumulative frequency distribution (CFD) in terms of the normalized dimensionless temperature variable θ given by Equation 10.19. The cumulative fraction of the number of hours in the month below a balance or base temperature T_{bal} is given by

$$\mathrm{CFD}(T_{bal}) = [1 + \exp(-3.396 \times \theta)]^{-1} \qquad (10.26)$$

The number of hours during a given month for a specified bin can then be deduced as the difference between the CFD values corresponding to the higher and lower temperatures of the bin multiplied by the number of hours in that month. Repeating the calculations for the 12 months would yield the desired number of hours for that specific bin over the year. Thus, the approach only requires the knowledge of the average temperatures $\bar{T}_{o,m}$ for each of the 12 months in a location. Note the similarity between the generalized approach to determine VBDD (Section 10.3.1) and this approach. This feature is not surprising since the VBDD is simply the area under the CFD for a specified base temperature. This correlation for CFD has been determined based on nine locations only, and hence it has to be used with caution.

Example 10.7: Generating Annual Bin Data for New York City

Generate bin data for New York City in February and compare the distribution with that given in Example 10.6. Use the monthly mean temperature values from Example 10.4.

Lookup values from Example 10.4: $\bar{T}_{o,m}$ = 0.8°C and σ_m =2.01

Find: N_{bin}

Solution

The calculation results are assembled in the following table. The normalized dimensionless temperature variable θ is determined for the high and low temperature values for each of the bins following Equation 10.19. For the lowest temperature bin corresponding to a midpoint of −13.9°C, the lower temperature value would be (−13.9 − 2.8/2 = −15.3), and its highest value would be (−13.9 + 2.8/2 = −12.5°C), and so on as shown. The values of the corresponding normalized temperatures are

$$\theta(-15.3°\mathrm{C}) = \frac{-15.3 - 0.8}{2.01\sqrt{28}} = -1.514$$

$$\text{and} \quad \theta(-12.5°\mathrm{C}) = \frac{-12.5 - 0.8}{2.01\sqrt{28}} = -1.250$$

Using Equation 10.26 results in CFD (−15.3°C) = 0.0058 and CFD (−12.5°C) = 0.0141, which leads to N_{bin} (−13.9°C) = (0.0141 − 0.0058) × 24 h × 28 days = 9.48 − 3.91= 5.6 h.

Repeating the calculation for each of the 12 months and then adding the hourly occurrences would yield the annual bin data distribution.

Low Temp. Value of Bin (°C)	θ Equation 10.19	CFD Equation 10.26	# of Hours	Bin Temperature (°C)	Number of Hours	
					Calculated	Actual
−15.3	−1.514	0.0058	3.91	−13.9	5.6	5
−12.5	−1.250	0.0141	9.48	−11.1	13.2	10
−9.7	−0.987	0.0338	22.72	−8.3	30.2	44
−6.9	−0.724	0.0788	52.96	−5.6	63.3	62
−4.1	−0.461	0.1730	116.25	−2.8	111.1	67
−1.3	−0.197	0.3384	227.39	0	146.0	149
1.5	0.066	0.5556	373.39	2.8	133.0	178
4.3	0.329	0.7535	506.37	5.6	86.3	114
7.1	0.592	0.8820	592.71	8.3	44.4	37
9.9	0.856	0.9481	637.14	—	—	—

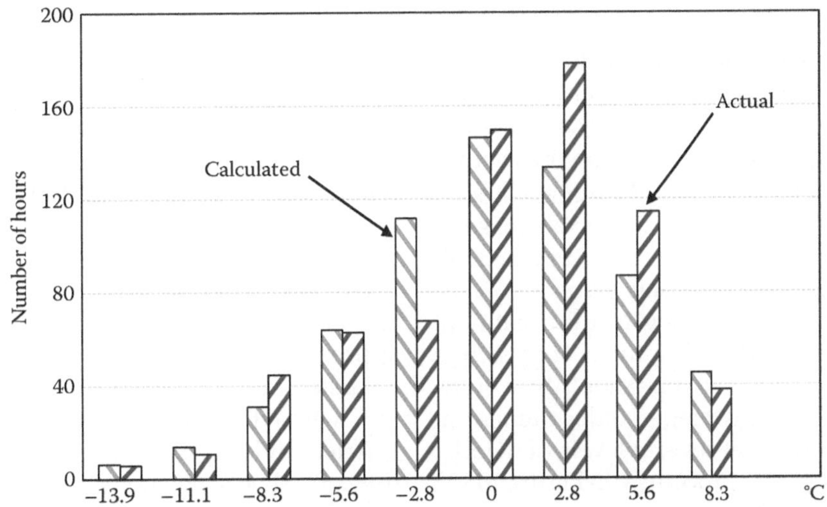

FIGURE 10.9
Comparison of actual versus empirically generated bin data for New York City for February following Example 10.7.

The results are plotted in Figure 10.9 against the actual values drawn from Example 10.7. We note that both the histograms have the same overall shape though the actual numerical values are not that close for some bins. This supports the recommendation that such correlations are to be used with some caution.

10.4.3 Diurnal Temperature Distribution

For some applications, one needs the temperature distribution during the day, e.g., to account for the effect of occupancy schedules on the consumption of commercial buildings. Erbs et al. (1983) offer a formula for estimating the average temperature $\bar{T}_{o,m}(t)$ for any hour t of a month:

$$\frac{\bar{T}_{o,m}(t)-\bar{T}_{o,m}}{A} = 0.4632\cos(t*-3.805) + 0.0984\cos(2t*-0.360)+0.0168\cos(3t*-0.822) + 0.0138\cos(4t*-3.513) \quad (10.27)$$

where

$\bar{T}_{o,m}$ is the monthly average temperature

$$t* = \frac{2\pi\times(t-1\text{ h})}{24\text{ h}}$$

($t = 1$ h corresponding to 1:00 a.m. and so on)

A is the diurnal temperature swing (peak to peak), correlated via

$$A = 25.8\bar{K}_T - 5.21 \quad \text{(dimensional equation, in °C)} \quad (10.28\text{ SI})$$

$$A = 1.8\times(25.8\bar{K}_T - 5.21) \quad \text{(dimensional equation, in °F)} \quad (10.28\text{ IP})$$

with the monthly average solar clearness index $\bar{K}_T$ (defined in Equation 4.23 as the ratio of terrestrial and extraterrestrial solar radiation, with data on the online HCB software). Note that this equation applies to monthly average distributions and should not be used for design-day calculations for which purpose Table 9.3 should be used.

Example 10.8: Outdoor-Air Temperature Distribution for Specific Hour of Day

Calculate the monthly average outdoor temperature for Phoenix at 3 p.m. during the month of July.

Lookup values: From Table 10.1, $\bar{K}_T = 0.69$ and $\bar{T}_o = 33°C$

Find: $\bar{T}_{o,m}(t)$ for t = 3 p.m. = 15 h

Solution

From Equation 10.28 SI, the diurnal amplitude swing is $A = 25.8 \times 0.69 – 5.21 = 12.59°C$.

Next, we calculate

$$t* = \frac{2\pi\times(t-1\text{ h})}{24\text{ h}} = \frac{2\pi\times(15-1)}{24} = 3.665$$

Then from Equation 10.27,

$$\frac{\bar{T}_{o,m}(t)-\bar{T}_{o,m}}{A} = 0.5245$$

Finally,

$\bar{T}_{o,m}(3\text{ p.m.}) = 0.5245\times 12.59 + 33 = 39.6°C$

10.4.4 Inclusion of Solar Gains

There are various refinements that can be included in a bin calculation. Of special interest is the seasonal variation of solar gains. If, as in Example 10.6, a separate calculation is done for each month, $\dot{Q}_{gain}$ could be based on the average solar heat gain of the month. The diurnal variation of solar gains can be accounted for by calculating the average solar gain for each of the hourly time periods of the bin method.

If such a detailed calculation of solar gains is considered too tedious or if great accuracy in the treatment of solar gains is not required, a simplified approach is to assume a linear correlation between monthly average solar heat gains and monthly average outdoor temperature. Here, too, one can work at several levels of detail.

We begin by illustrating this procedure in Figure 10.10a for the simplest case: a building with constant operating conditions every hour of the day and every day of the year (i.e., T_i, K_{tot}, and nonsolar heat gains $\dot{Q}_{nonsol}$ are all constant). In this case, it suffices to calculate the daily average heating load $\dot{Q}_1$ for the peak winter day and the daily cooling load $-\dot{Q}_2$ for the peak summer day:

$$\dot{Q}_1 = K_{tot}(T_i - T_{o,1}) - \dot{Q}_{nonsol} - \dot{Q}_{sol,1} \tag{10.29}$$

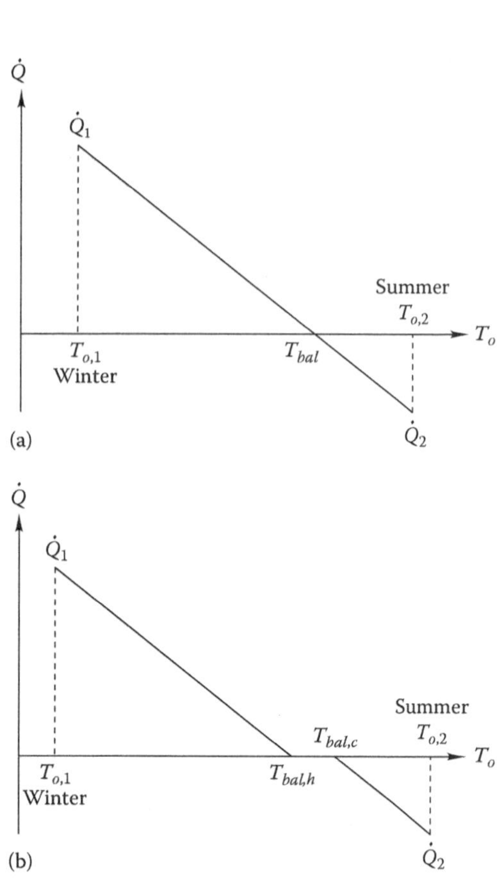

FIGURE 10.10
Load as function of outdoor temperature T_o (ventilative cooling is not shown). (a) Simplest case, with T_i constant year-round. (b) Higher set point in summer than in winter.

and

$$\dot{Q}_2 = K_{tot}(T_i - T_{o,2}) - \dot{Q}_{nonsol} - \dot{Q}_{sol,2} \tag{10.30}$$

where $T_{o,1}$ and $T_{o,2}$ are the corresponding values of the outdoor temperature T_o and $\dot{Q}_{sol,1}$ and $\dot{Q}_{sol,2}$ are the corresponding daily average solar gains. These values are then plotted versus the corresponding values $T_{o,1}$ and $T_{o,2}$ of the outdoor temperature T_o. Connecting these two points by a straight line, one obtains a simple linear approximation for the daily load as a function of T_o. The balance-point temperature of Equation 10.4 is the value of T_o where this line crosses the T_o axis.*

After dividing by the equipment efficiency, we find the annual heating energy by summing over all temperature bins below the balance point, each bin weighted by the corresponding number of days. A similar approach can be done for cooling above the balance-point temperature, although ventilative cooling should be taken into account according to Figure 10.5.

Next, we show in Figure 10.10b how this procedure is modified to take into account the common practice of setting the thermostat a few degrees higher in summer than in winter. Evaluating $\dot{Q}_2$ for the summer value of T_i, one finds that the cooling part of the line is shifted to the right. As a further refinement, one can carry out a separate calculation of the solar gain for the transition season, when T_o is close to the values of T_{bal} found earlier. Then, the corrected values of the points $T_{bal,h}$ and $T_{bal,c}$ where the lines intersect the T_o axis are calculated with the midseason solar gain. Note that the slopes for heating and for cooling can be different.

Finally, the plots shown in Figure 10.10b can be developed separately for different time periods of the day and for different occupancy conditions, say, 12:00 to 16:00 Saturdays. By using the bin data for the corresponding periods, the calculation can thus take the operating schedules of commercial buildings into account.

10.4.5 Inclusion of Dynamic Effects

So far, we have discussed only steady-state methods for energy consumption. They are correct when the indoor temperature T_i is constant. But how about cases where T_i varies? Savings from thermostat setback in winter and setup in summer can be appreciable. For example, ASHRAE Fundamentals (2013) cites typical savings in the range of 7%–13% of the annual heating energy if

* An alternative method for calculating solar gains in the bin method was proposed by Vadon et al. (1991), who developed correlations that directly yield the solar gains coincident with each bin of T_o.

the thermostat is set back by 10°F (5.6°C) for 8 h each winter night; of course, the precise numbers in a given case are highly dependent on climate and on building construction.

If the indoor temperature is set back during unoccupied periods, the bin method becomes less reliable because the exact temperature profile during setback is not known. One can obtain an approximate answer for the heating energy by assuming that the temperature will instantaneously drop to the setback value at the start and jump back to the normal occupancy value at the end of the unoccupied period; with that assumption the bin calculation of Example 10.6 can be used (with the appropriate values of T_i during occupied and unoccupied periods). The error of this approximation can be significant for mild climates or deep thermostat setbacks.

An analogous calculation can be carried out for the cooling season. Of course, the heat input, in particular the solar radiation, should correspond to average—not peak—conditions. Thermostat setup is more problematic for the calculation of cooling energy than for heating energy because the temperature profile during the setup period is poorly known (quite often, the building does not even reach the upper temperature limit, if any, at which the cooling equipment turns on again).

Example 10.9: Effect of Thermostat Setback on Heating Energy

Consider a house in New York with $K_{tot} = 200$ W/K, internal heat gains of 1000 W, and heating system efficiency of 80%. Assume negligible heat capacity. You are asked to calculate the energy savings (in GJ) in heating energy between keeping the thermostat fixed at 21°C throughout the year and lowering it to 16°C for 12 h of the day (of course, during winter periods at nighttime).

Given: K_{tot}, efficiency of heating system $\eta = 0.8$

$$\dot{Q}_{gain} = 1000 \text{ W}$$

$$T_i = 21.0°\text{C daytime and } T_i = 16.0°\text{C nighttime}$$

Find: $T_{bal,day}$ *and* $T_{bal,night}$, HDD($T_{bal,day}$), and HDD ($T_{bal,night}$)

Use Equation 10.9 to calculate $Q_{h,yr}$ with and without night setback.

Solution

We use Equation 10.4 to find the balance points.
During daytime,

$$T_{bal} = T_i - \frac{\dot{Q}_{gain}}{K_{tot}} = 21°\text{C} - \frac{1000 \text{ W}}{200 \text{ W/K}} = 16°\text{C}$$

During nighttime,

$$T_{bal} = 16°\text{C} - \frac{1000 \text{ W}}{200 \text{ W/K}} = 11°\text{C}$$

From Figure 10.3, HDD (16°C) = 2335 K · days/year and HDD (11°C)= 1275 K · days/year
Without night setback:

$$Q_{h,yr} = \frac{K_{tot}\text{HDD}}{\eta_h}$$

$$= 200 \text{ W/K} \times 2335 \text{ K} \cdot \text{days} \times 24 \text{ h/day} \times \frac{3600 \text{ s/h}}{0.8}$$

$$= 50.4 \times 10^9 \text{ J/year}$$

With night setback:

$$Q_{h,yr} = 200 \text{ W/K} \times \frac{(2335 + 1275) \text{ K} \cdot \text{days}}{2}$$

$$\times 24 \text{ h/day} \times \frac{3600 \text{ s/h}}{0.8}$$

$$= 39.0 \times 10^9 \text{ J/year}$$

Comments

This is a 23% reduction in annual heating energy use. However, this calculation is approximate and will tend to overpredict the savings due to neglecting the building thermal mass.

Besides variable thermostat settings, there is another effect that can make T_i vary: during the milder parts of the heating season, the instantaneous balance-point temperature can rise above T_o for part of the day, as already described in Figure 10.6. Assuming the windows are kept closed, much of the excessive heat gain may be stored in the mass of the building, thus reducing the heating load when T_o finally drops below T_{bal}. Such storage implies an increase of T_i above the thermostat set point, even if the latter is constant. This effect is particularly important in buildings with large solar gains and high inertia. Obviously, an uncorrected steady-state method should not be used for passive solar buildings.

If T_i at each hour were known, one could calculate the corresponding values of T_{bal} and proceed with the bin method as before. Unfortunately, the evolution of T_i is not known in general; it depends on the thermal inertia of the building and on the temperature difference ($T_i - T_o$). By examining some typical temperature profiles, one can derive certain correlations for correction terms. Several shorthand methods have been developed that include correction terms for variable T_i in steady-state methods. As an example, we mention the *solar load ratio method* (see Section 24.3.3) for passive solar buildings; it modifies the degree-day result by a correction term that has been derived from a large number of dynamic computer simulations (Balcomb et al., 1982).

Of course, the introduction of correction terms tends to spoil the simplicity of the steady-state approach, and with the evolution of computer technology, one may reach a point where a dynamic calculation will be easier than the old shorthand methods with correction terms.

The principles of a dynamic calculation of energy consumption are the same as for a dynamic calculation of peak loads. One simply repeats the calculation, time step by time step, for the entire year, including the efficiency of the HVAC equipment as appropriate. An hourly time step is commonly used. For an annual energy calculation, one uses hourly weather data for each day of the year instead of a sequence of identical days. This procedure is realized in the DOE2.1 computer simulation program (Birdsall et al., 1990), which is essentially based on the transfer function method described in Sections 8.3 and 9.5.

The weather data must be representative of long-term averages. One can save some calculational steps by using less than 365 days for a year, scaling up the results at the end; of course, the days for the calculation must be chosen carefully to be representative. For example, for a number of European sites, a set of so-called short reference year data is available that replaces the year by 8 weeks.

10.5 Advantages and Limitations

Table 10.3 assembles some of the important advantages and limitations of the degree-day and bin methods. Despite their limitations, these methods are widely used in energy policy-related studies, and also by engineering service companies for preliminary evaluations of HVAC systems design alternatives in existing buildings where equipment performance is dependent on their part-load performance, which in turn is dependent on the ambient conditions; thus, it behooves the student to be knowledgeable with the bin method.

10.6 Inverse Modeling

10.6.1 Background

At the design stage, one needs to calculate the performance of a building, based on its detailed description (say, gleaned from blueprints). Building energy *simulation models* are mechanistic (i.e., based on a mathematical formulation of the physical behavior) and deterministic (i.e., there is no randomness in the inputs or outputs). The simulation predicts hourly energy use during the entire year from which monthly total energy use and peak use along with utility rates provide an estimate of the operating costs of the building. This modeling approach is generally useful for evaluating different design options before the actual system is built. This is an instance of what is sometimes called the "forward problem," by contrast to the "inverse problem" where one relies on monitored performance data in order to deduce important characteristics of the building envelope and of the HVAC system. The latter is actually an example of a very general subject, arising in fields as diverse as engineering and economics and also known as *system identification*.

Inverse statistical modeling is the process of identifying a predictive model structure and estimates of the model parameters from measured system data. It helps one achieve a better understanding of the system dynamics by combining the basic physics of the system (via the selection of proper functional forms of the

TABLE 10.3

Summary of Advantages and Limitations of Degree-Day and Bin Methods

	Advantages	Limitations
Degree-day method	• Very simple to use • Appropriate for preliminary analysis involving building loads • Widely used especially for heating energy estimation • Strictly applicable to single-zone buildings	• Allows little flexibility to capture time-dependent changes in building load and in HVAC equipment operation • Cannot capture part-load efficiency changes or cycling losses in equipment • Can be applicable under steady-state assumption though some dynamic cases can be analyzed
Bin method	• More accurate than degree-day method • Appropriate for preliminary analysis involving building loads • Can capture changes in load and HVAC equipment operation during day and weekday/weekend • Can capture equipment efficiency changes to some extent • Can be used for multizone buildings	• Requires more climatic data information than bin methods • Solar loads treated somewhat empirically • Requires more information about change in equipment efficiency with load • Can be applicable under steady-state assumption though some dynamic cases can be analyzed

regression equation) with statistical methods for identifying model parameters. Inverse models can be static (or steady-state) or dynamic models; the distinction is easily made depending on whether the model contains time-lagged variables (either in the regressor or the response variables). The steady-state inverse models are unable to capture to dynamic effects (such as the thermal mass effects of the building) and may not perform well for buildings that exhibit such behavior or for short-time steps (such as 15 min or 1 h periods).

Some of the applications in existing buildings for which the inverse approach has been shown to be useful are as follows (Reddy, 2011):

1. *Commissioning tests*: To evaluate whether a component or a system is installed and commissioned properly.
2. *Comparison with design intent*: To compare actual consumption with design predictions. In case of discrepancies, are they due to anomalous weather, to unintended building operation, to improper operation, or to other causes?
3. *Demand side management*: To predict reduction in electrical costs if certain operational changes are made, which alter the building loads during on-peak hours of the day, for example, increasing thermostat settings and reducing ventilation rates and/or indoor lighting levels.
4. *Operation and maintenance*: To predict energy savings from planned retrofits to building shell and changes to air handler operation, or from changes in the various control settings, or from equipment replacements (say, replacing an old chiller with a new and more energy-efficient one).
5. *Monitoring and verification* (M&V): To verify, once energy conservation measures are implemented, that the actual savings are consistent with the anticipated savings and due to the retrofit and not due to other causes, e.g., the weather or changes in building occupancy.
6. *Automated fault detection, diagnosis, and evaluation*: To automatically detect faults in HVAC equipment that reduce operating life and/or increase energy use and to determine the financial implications of this degradation.
7. *Optimal supervisory operation**: To characterize HVAC equipment (such as chillers, boilers, fans, and pumps) in their installed state and optimize their control and operation.

For several of the earlier applications, a baselining methodology is crucial to verify savings from energy conservation programs. For M&V specifically, the intent is to develop baseline energy consumption prediction models using inverse methods from preretrofit consumption data and then to use these models to predict the energy consumption during postretrofit. The difference between the measured and predicted postretrofit data thus gives an estimate of the savings achieved. The baseline models are developed using monthly, daily, or hourly energy consumption as the response variable versus certain parameters of interest as the regressor variables. There are many issues related to the development of a formal baselining methodology at the whole building level (see, for example, ASHRAE 14, 2002). Some of these issues along with the related studies are presented in the following sections.

Steady-state inverse models can also be used by energy management and control systems to predict energy use (Kreider and Haberl, 1994). Hourly or daily comparisons of measured energy use against predicted energy use can be used to determine whether systems are being left on unnecessarily or are in need of maintenance. Combinations of predicted energy use and a knowledge-based system have been shown to be capable of indicating above-normal energy use and diagnosing the possible cause of the malfunction if sufficient historical information about malfunction signatures has been previously gathered (Haberl and Claridge, 1987).

A model for the inverse problem can be quite different from the conventional forward models. It should contain only a small number of adjustable parameters, because the information content of the data is rather limited, being collected under fairly repetitive conditions and subject to errors. Thus, the model itself can be quite simple, although a fairly sophisticated procedure is often needed to ensure reliable determination of the adjustable model parameters. Developing an inverse model involves the following steps:

1. Choose a model (functional form) making sure that it captures the crucial physical features of the situation.
2. Select one (or more influential) variable as regressor and one dependent or response variable y_{meas}.
3. Determine the parameters of the model by regression, say, minimizing the squared differences between y_{model} and y_{data}, summed over all data points (usually this will be done by linear least squares, using commercially available regression software).
4. Test the model against another set of data (this is called cross validation).

* This aspect is also an important consideration at the design stage.

5. As much as possible, express the results in terms of quantities with direct physical interpretation, such as heat transmission coefficient, admittances, and time constants, to determine whether they are realistic.

There is no unique procedure that will guarantee the best results. The main difficulty lies in the choice of variables to be included and the functional form of the regression model. Interactive regression software is advisable (i.e., programs that allow the user to change the number of variables without having to reread the data). There are several guidelines for the selection of the most pertinent variables (see any book dealing with regression analysis, say, Reddy, 2011): (1) the coefficient of determination R^2 should be close to unity, (2) the standard error of the regression or the root mean square (RMSE) should be small, and (3) the standard error of the parameters should be relatively small (based on a Statistical t-test). With respect to inverse modeling, although R^2 is a proper metric to use when the primary objective is to evaluate different models and determine the best one when fitting a given data set, the RMSE becomes more relevant when the objective changes to evaluating the actual energy savings determined from these models (Reddy and Claridge, 2000).

10.6.2 Single-Variate Models for Monthly Data

One can distinguish between two types of models: steady state and dynamic. *Static or steady-state models* are those that do not consider such effects such as thermal mass or capacitance that cause short-term temperature transients and will not contain time-lagged regressor or response variables. Generally, these models are appropriate for monthly, daily, and hourly data and are often used for baseline model development. *Dynamic models* capture effects such as building warm-up or cooldown periods and peak loads and are appropriate for building load control, fault detection, and equipment control. Such methods are not discussed in this book, and the interested reader can refer to technical papers in handbooks (say, Kreider, 2001) or in special journal publications (such as Claridge, 1998, and Reddy, 2003).

Energy signature models are the simplest example of inverse modeling (Hammarsten, 1987). The name refers to the behavior of the energy consumption as a function of outdoor conditions, as shown in a plot like Figure 10.10. The most influential variable is the outdoor dry-bulb temperature, and a model that involves this variable only qualifies as a single-variate model. Such a model is most appropriate with monthly energy consumption data that are conveniently available from utility bills. Two mathematical formulations to model monthly building energy use are most widely used, and they are described in this section. Classical linear functions are usually not appropriate because of the presence of functional discontinuities, called "change points" due to the presence of control mechanisms—thermostats in residences, HVAC operating and control schedules, and economizer cycles in commercial buildings.

10.6.2.1 Variable-Base Degree-Day Model

The VBDD method discussed in Section 10.3 is the basis of the functional form for this type of inverse model. It is most suitable for shell-dominated buildings such as residences and small commercial buildings wherein energy use is not strongly influenced by the HVAC equipment. Let us consider the monthly electrical energy use of the building *under cooling mode*. Equation 10.14 can be expressed as

$$E_{c,m} = \frac{K_{tot}}{\text{COP}} \cdot \text{CDD}_m(T_{bal}) \tag{10.31}$$

where
- COP is the coefficient of performance of the air conditioner (AC)
- $E_{c,m}$ is the monthly electricity consumed by the AC

The whole-house monthly electricity consumption $E_{bldg,m}$ is the sum of the baseload (i.e., non–AC related) and $E_{c,m}$, and can be expressed as

$$E_{bldg,m} = a + b \times CDD_m(T_{bal}) \tag{10.32}$$

where the model parameters can be interpreted as
- a is the baseload monthly electricity use
- $b = (K_{tot}/\text{COP})$

This is referred to as a three-parameter (3-P) VBDD model (the parameters of which include a, b, and T_{bal}). Note that this is similar to a linear spline model with one of the segments being horizontal or constant. Obviously, the monthly degree-days should be calculated based on the actual billing periods and not on the calendar months.

Another often encountered behavior is the four-parameter (4-P) VBDD model for an all-electric home that is a spline model with two nonconstant linear segments:

$$E_{bldg,m} = a + b \times \text{CDD}_m(T_{bal}) + c \times \text{HDD}_m(T_{bal}) \tag{10.33}$$

The 3-P model based on the VBDD concept is also called the Princeton Scorekeeping Method (PRISM) (Fels, 1986). It was originally developed for modeling heating energy in residences. Adding parameters does not usually guarantee an improvement of the results, even if the standard error or R^2 appears better. Residential cooling is often quite erratic for a meaningful determination of the corresponding PRISM parameters. PRISM can also be used for commercial buildings, but attention should be paid to the role of weekends, variable ventilation rates (economizer), and simultaneous heating and cooling. It must be cautioned that the interpretation of the parameters may not be straightforward (Fels and Goldberg, 1986; Rabl et al., 1986).

Example 10.10: Modeling of Utility Bill Data Using the VBDD Method

The following example (Sonderegger, 1998) presents the results of a utility bill analysis. Assume that values of the utility bills over an entire year have been collected. To obtain the model coefficients through regression, the utility bills must be normalized by the length of the time interval between utility bill read dates. This is equivalent to expressing all utility bills, degree-days, and other independent variables, by their daily averages.

Figure 10.11 illustrates how well a regression fit captures measured baseline energy use in a hospital building. Cooling degree-days are found to be a significant variable, with the best fit corresponding to a base temperature of 54°F (12.2°C). Except for two data points (shown as unfilled squares), all other points lie along the best fit straight line. There are good reasons why individual utility bills may be unsuitable to develop a baseline and should be excluded from the regression. For example, a bill may be atypically high because of a one-time equipment malfunction that was subsequently repaired. Another common reason is due to short-lived changes in occupancy (occupants are away or have guests for several days). However, it is often tempting to look for reasons to exclude bills that fall far from "the line" and not question those that are close to it. For example, bills for periods containing vacations or production shutdowns may look anomalously low, but excluding them from the regression would result in a chronic overestimate of the future baseline during the same period. Such instances are circumstance specific and certain amount of judgment is then warranted.

10.6.2.2 Monthly Mean Temperature Model

Instead of using the VBDD concept, another approach called MMT model has become very popular because of its simplicity and convenience. It involves using the monthly mean temperature $\bar{T}_{o,m}$ as the regressor variable for modeling monthly mean energy use. As before, the $\bar{T}_{o,m}$ should correspond to the utility billing period and not merely to the calendar month. The functional notations used for the various types of single-variate inverse models are tabulated in Table 10.4 and plotted in Figure 10.12.

Figure 10.12a shows a simple one-parameter, or constant, model. Figure 10.12b applies to a steady-state two-parameter (2-P) model for cooling only (CO) energy use where b_0 is the y-axis intercept and b_1 is the slope of the regression line. The 2-P model represents cases when cooling (or heating) is always required. Figure 10.12c shows a 3-P heating only (HO) change-point model. This model is typical of natural gas energy use in a single-family residence that uses gas for space heating and domestic water heating. In this case, b_0 represents the baseline energy use at the change point, and b_1 is the slope of the regression

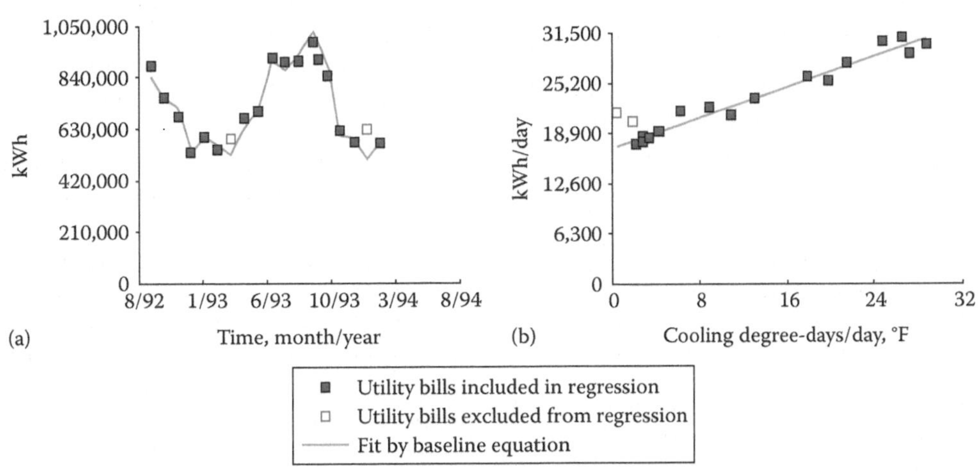

FIGURE 10.11
Variable-base degree-day model identification using electricity utility bills at a hospital. Note that some utility bills have been excluded from the regression due to a degree-day threshold. (a) Left graph shows bills versus time, while (b) right graph shows bills versus cooling degree-days. (From Sonderegger, R.C., *ASHRAE Trans.*, 104(2), 859, 1998.)

TABLE 10.4

Single-Variate Monthly Mean Temperature (MMT) Models Widely Applied to Utility Bill Data

Model Type	Form	Examples
One-parameter or constant (1-P)	$E_m = b_0$	Non-weather-sensitive energy consumption
Two-parameter (2-P)	$E_m = b_0 + b_1 \times \bar{T}_{o,m}$	Consumption varies linearly with outdoor temperature
Three-parameter change point (3-P)	Cooling $E_m = b_0 + b_1 \times (\bar{T}_{o,m} - T_{cp})^+$ Heating $E_m = b_0 + b_1 \times (T_{cp} - \bar{T}_{o,m})^+$	Seasonal weather-sensitive energy use in smaller buildings with flat baseline (electricity in summer for cooling, fuel use in winter)
Four-parameter change point (4-P)	$E_m = b_0 + b_1 \times (T_{cp} - \bar{T}_{o,m})^+ + b_2 \times (\bar{T}_{o,m} - T_{cp})^+$	Energy use in larger commercial buildings, applicable to both heating and cooling
Five-parameter change point (5-P)	$E_m = b_0 + b_1 \times (T_{cp,1} - \bar{T}_{o,m})^+ + b_2 \times (\bar{T}_{o,m} - T_{cp,2})^+$	Heating and cooling supplied by same energy source resulting in two change points

Note: The notation $()^+$ denotes that the term within the brackets should be set to zero if it is negative.

line for values of ambient temperature less than the change point, T_{cp} while b_2 is the slope of the regression line for values of ambient temperature greater than the change point T_{cp}. Figure 10.12d applies to a 3-P CO model for cooling energy use and Table 10.4 provides the appropriate analytic expression. Figure 10.12e and f illustrates 4-P models for heating and cooling, respectively. The model parameter b_0 represents the baseline energy exactly at the change point T_{cp}, and b_1 and b_2 are the lower and upper region regression slopes for ambient air temperature below and above the change point. Finally, Figure 10.12g shows a five-parameter (5-P) model that is useful for modeling buildings that are electrically heated and cooled. The 5-P model has two change points and would be appropriate for building, which are operated with different thermostat set points during winter and summer.

Example 10.11: Modeling of Utility Bill Data Using the MMT Method

Electricity utility bills of an actual residence in Houston, TX, have been normalized by building floor area and the number of days in the month. These are assembled in Table 10.5 along with the corresponding month and monthly mean outdoor temperature values for Houston (the first three columns of the table). The intent is to identify a 4-P MMT model from these data (the functional form is shown in Table 10.4).

The scatter plot of the data along with some trend lines drawn visually (Figure 10.13) suggests that the change point is in the range 17°C–21°C. Let us perform the calculation assuming a value of 17°C. Based on this assumption, the last two columns of the table have been generated to correspond to the two regressor variables of the 4-P model. A linear multiple regression yields

$$E_m = 0.2050 - 0.00589 \times (T_{cp} - \bar{T}_{o,m})^+ + 0.01496 \times (\bar{T}_{o,m} - T_{cp})^+ \quad (10.34)$$

with $R^2 = 0.996$ and RMSE = 0.0055 with all three parameters being statistically significant.

The reader can repeat this analysis assuming a different value for the change point (say T_{cp} = 18°C) in order to study the sensitivity of the model to the choice of the change-point value. The regression identifies three parameters with the fourth parameter being the change point T_{cp} (which needs to be determined by trial and error) which yields the lowest RMSE.

10.6.2.3 Discussion

The energy use regression models are not of the standard type of spline models because the change-point parameter (either T_{bal} for VBDD model or T_{cp} in the MMT model) is one of the parameters to be determined. Thus, though these models look linear, they are, strictly speaking, nonlinear models. Corresponding software developed using these methods rely on a search method to perform the regression and thereby determine the best fit model parameter values (Kissock et al., 2001).

The advantage of these steady-state inverse models is that their use can be easily automated and applied to large numbers of buildings where monthly utility billing data and average daily temperatures for the billing period are available (Marchio and Rabl, 1991). Change-point regression models work best with heating data from buildings with systems that have little or no part-load nonlinearities (i.e., systems that become less efficient as they begin to cycle on–off with part loads). In general, change-point regression models do not predict cooling loads as well because outdoor humidity has a large influence on latent loads on the cooling coil. Other factors that decrease the accuracy of change-point models include solar effects, thermal lags, and on–off HVAC schedules. Hence, 4-P models exhibit a better statistical fit than 3-P models in buildings with continuous, year-round cooling or heating (i.e., grocery stores and office buildings with high internal loads). However, every

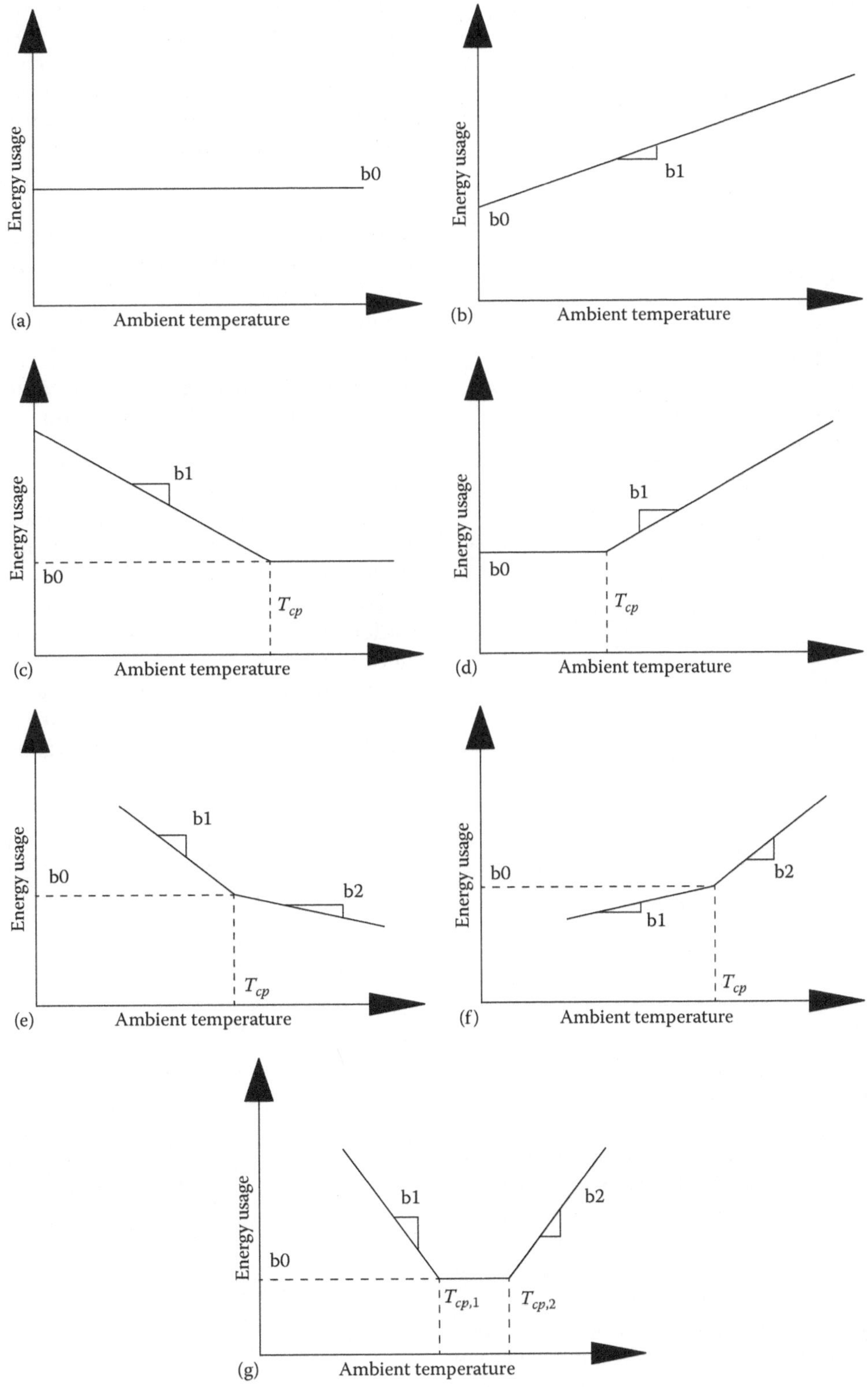

FIGURE 10.12
Steady-state, single-variate monthly mean temperature models appropriate for modeling energy use in residential and commercial buildings. See Table 10.4 for functional forms and nomenclature. (a) 1-P model. (b) 2-P model for cooling. (c) 3-P model for heating. (d) 3-P model for cooling. (e) 4-P model for heating. (f) 4-P model for cooling. (g) 5-P model.

TABLE 10.5

Utility Bill Data and Analysis Results for Example 10.11 Assuming a Change-Point Temperature of 17°C

Month	$\bar{T}_{o,m}$, °C	E_m, kWh/m²/day	$(T_{cp} - \bar{T}_{o,m})^+$, °C	$(\bar{T}_{o,m} - T_{cp})^+$, °C
Jan	11	0.1669	6	0
Feb	13	0.1866	4	0
Mar	16	0.1988	1	0
Apr	21	0.2575	0	4
May	24	0.3152	0	7
June	27	0.3518	0	10
July	29	0.3898	0	12
Aug	29	0.3872	0	12
Sept	26	0.3315	0	9
Oct	22	0.2789	0	5
Nov	16	0.2051	1	0
Dec	13	0.1790	4	0

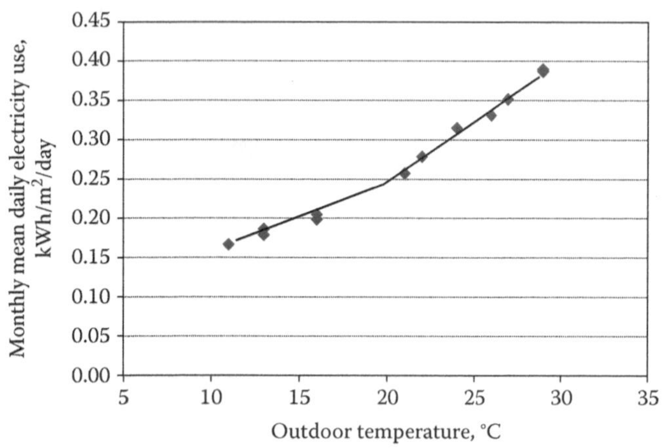

FIGURE 10.13
Simplified illustration of a 4-P cooling change-point model of building electric use with outdoor temperature.

model should be checked to ensure that the regression is not falsely indicating an unreasonable relationship.

A major advantage of using a steady-state inverse model to evaluate the effectiveness of energy conservation retrofits lies in its ability to factor out year-to-year weather variations by using a normalized annual consumption (Fels, 1986). Basically, the annual energy conservation savings can be calculated by comparing the difference obtained by multiplying the preretrofit and postretrofit parameters by the weather conditions for the average year. Typically, 10–20 years of average daily weather data from a nearby weather service site are used to calculate 365 days of average weather conditions, which are then used to calculate the average preretrofit and postretrofit conditions.

When the data are reasonably good and the basic assumptions of the model are close to actual conditions, the VBDD approach has been found to work very well (in the sense of yielding good correlation coefficients, R^2 above 0.8, and small standard errors, a few percent). That is likely to be the case in residential buildings heated by one of the commercial energy sources (i.e., no significant use of firewood). In that case, the normalized consumption can be determined with a standard error of 1%–3%. But the standard error of the individual parameters tends to be much larger (10% or more) because of compensating errors. That is a general phenomenon in system identification: for the quantity chosen as the dependent variable for the regression, the error is small, while the parameters of the model can remain quite uncertain.

10.6.3 Single- and Multivariate Models for Daily and Hourly Data

Steady-state single-variate inverse models using daily data have also been applied with some success to certain types of buildings, such as shell-dominated ones (Kissock et al., 1998). In such a case, the VBDD and MMT models, described earlier for utility bill data analysis, become identical in their functional form. Single-variate models can also be applied to daily data to compensate for differences such as weekday and weekend use by separating the data accordingly and identifying models for each period separately. They can also be used to identify energy savings resulting from energy efficiency improvements in buildings in case they are strongly affected by one variable only. This is illustrated in Figure 10.14 for a large university building in Texas that underwent a major retrofit to the air handling system (from a constant air volume to a variable air volume system). Notice the

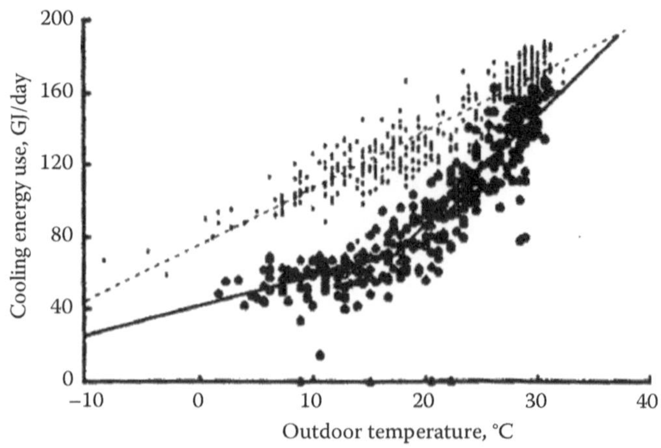

FIGURE 10.14
Scatter plot showing how daily cooling energy use varies with outdoor temperature for a large university building in Texas (From Kissock, J.K. et al., *ASME J. Solar Energy Eng.*, 120(3), 168, August 1998). The faint dots correspond to initial energy use, while the dark circles correspond to energy use after energy efficiency improvements were made to the building.

close to a 2-P behavior prior to the retrofit and the pronounced 4-P behavior after the retrofit.

Daily and hourly data sets tend to be larger and can support models with more number of regressor variables. Commercial buildings, in general, have higher internal heat generation with simultaneous heating and cooling energy use due to multiple zones and are strongly influenced by HVAC system type and control strategy. This makes energy use in commercial buildings less strongly influenced by ambient temperature alone. Broadly speaking, there are two types of steady-state multivariate model groups:

1. *Multivariate-linear change-point (MLCP) regression models* where the set of data observations are treated without retaining the time-series nature of the data (Katipamula et al., 1998)
2. *Fourier series models* that retain the time-series nature of the building energy use data and capture the diurnal and seasonal cycles according to which buildings are operated using trigonometric functions (Dhar et al., 1998)

The goal of modeling energy use by the multivariate approach is to characterize building energy use with a few readily available and reliable input variables. The MLCP models are a logical extension to single-variate models provided that the choice of the variables to be included and their functional forms are based on the engineering principles on which HVAC systems and other systems in commercial buildings operate. For M&V applications wherein a baseline model is being identified, the model should contain variables not affected by the retrofit and likely to change (for example, climatic variables) from preretrofit to postretrofit periods. Other less obvious variables such as changes in operating hours, in baseload, and in occupancy levels should be included in the model if these are not part of the efficiency measures but are likely to vary during the postretrofit period.

Several standard statistical indices exist for evaluating the goodness of fit of the model and the degree of influence that each of the independent variables exerts on the response variable (for example, see Reddy, 2011). Although energy use is, in fact, dependent on several variables, there are strong practical incentives for identifying the simplest model that results in acceptable accuracy. Multivariable models require more metering and are unusable even if one of the variables becomes unavailable. In addition, some of the regressor variables may be correlated. This condition, called "multicollinearity," can result in large uncertainty in the estimates of the regression coefficients and can also lead to poorer model-prediction accuracy compared to a model where the regressors are not linearly correlated. The functional basis of air-side heating and cooling use in various HVAC system types has been addressed by Reddy et al. (1995) and subsequently applied to monitored data in commercial buildings (Katipamula et al., 1998). Because none of the quadratic and cross-product terms of the engineering equations are usually picked up by the multivariate models, one is often left with models for energy use that are strictly linear.

In addition to outdoor dry-bulb temperature, internal electric equipment and lighting load E_{int}, solar loads Q_{sol}, and latent effects via the outdoor dew-point temperature T_{dp} are candidate regressor variables. In commercial buildings, a major portion of the latent load is due to fresh air ventilation. However, this load appears only when the outdoor-air dew-point temperature exceeds the cooling-coil temperature. Hence, the term $(T_{dp} - T_{surf})^+$ (where the + sign indicates that the term is to be set to zero if negative and T_{surf} is the mean surface temperature of the cooling coil [typically about 11°C–13°C or 51.5°F–55.5°F]) is a more realistic descriptor of the latent loads than is T_{dp} alone. The use of $(T_{dp} - T_{surf})^+$ as a regressor in the model is a simplification that seems to yield good accuracy. Therefore, multivariate inverse models often adopt the following structure:

$$E = a + b \times T_o + c \times I + d \times I \times T_o + e \times T_{dp}^{+} + f \times Q_{sol} + g \times E_{int} \quad (10.35)$$

where the indicator variable I is introduced to handle the change in slope of the energy use E (which could be the whole building electricity use or the building thermal load) due to T_o. The variable I is set equal to 1 for T_o values to the right of the change point (i.e., for high T_o range) and set equal to 0 for low T_o values. As with the single-variate segmented models (i.e., 3-P and 4-P models), the regression involves a search method in order to determine the change point that minimizes the total sum of squares of residuals.

The aforementioned MLCP inverse model form has been found to be very accurate for daily time scales and slightly less so for hourly time scales. This is because changes in the way the building is operated during the day and the night lead to different relative effects of the various regressors on energy use, which cannot be accurately modeled by one single hourly model. Breaking up the energy use data in hourly bins corresponding to each hour of the day and then identifying 24 individual hourly models lead to significantly greater accuracy (Katipamula et al., 1995).

It is best to use data sets for inverse model identification, which encompasses the entire range of variation of the climatic variation such as a whole year. The accuracy of inverse models identified from data sets of less than a year has been investigated by several researchers during the last 20 years, and some practical guidelines have emerged. The interested reader can refer to Singh et al. (2014) for a summary of the research findings to date and typical suggestions.

10.6.4 Dynamic Models

Dynamic inverse models are usually used with hourly or subhourly data in cases (1) where the thermal mass of a building is sufficiently significant to delay the heat gains or losses or (2) when simulations of equipment are performed at short-time steps (needed for control applications). Dynamic models traditionally required the solution of a set of differential equations. The disadvantages of dynamic inverse models include their complexity and the need for more detailed measurements to tune the model. Unlike steady-state inverse models, dynamic inverse models usually require a high degree of user interaction and knowledge of the building or system being modeled.

Several residential-energy studies using dynamic inverse models based on parameter estimation approaches have been reported, most of them involving intrusive data gathering. Rabl (1988) has classified the various types of dynamic inverse models used for whole building energy use and drawn attention to the common underlying features of these models. There are essentially four different types of model formulations: thermal network models, time-series models involving autoregressive moving-average (ARMA) models, differential equation models, and modal models. All these qualify as parameter estimation approaches. Table 10.6 lists several pertinent studies in each category.

Another important application of inverse modeling is the development of shortcuts to speed up large computer programs. Much time can be saved if, instead of recalculating a certain process in full detail at each step, one can approximate it by simpler expressions. An example is the use of transfer functions instead of the original transient heat conduction equations. In a similar spirit, the behavior of a large network can often be approximated by an equivalent and much simpler network. For instance, simulating a house with a network model containing some one hundred nodes, Neveu et al. (1986) found that the results for the heating loads could be reproduced within a fraction of a percent by a two-node approximation.

10.6.5 Calibrated Simulation Models

A hybrid approach, called "calibrated simulation modeling," which uses a forward model involving a detailed

TABLE 10.6

Classification of Methods for Thermal Analysis of Buildings

Method	Forward	Inverse	Hybrid	Comments
Steady-state methods				
Simple linear regression (Kissock et al., 1998)		X		One dependent parameter, one independent parameter
Multiple linear regression (Katipamula et al., 1998)		X	X	One dependent parameter, multiple independent parameters
VBDD method	X			Variable reference temperatures
ASHRAE bin method and inverse bin method	X	X	X	Hours in temperature bin multiplied by load for that bin
Single-variate change-point models: 3-P (CO, HO), 4-P, and 5-P (HC)		X	X	Uses daily or monthly utility billing data and average period temperatures
Modified bin method	X		X	Modified bin method with cooling load factors
Fourier series (Dhar et al., 1998)		X		Uses hour and/or day as additional regressors in the model
Dynamic methods				
Thermal network (Sonderegger, 1977)	X	X	X	Equivalent thermal parameters (forward and inverse modes)
Thermal network (Rabl, 1988; Reddy, 1989)		X		Inverse mode only
Response factors (Stephenson and Mitalas, 1967)	X			Tabulated or as used in simulation programs
ARMA model (Jalori and Reddy, 2015; Rabl, 1988)		X		Multiple-input ARMA model
PSTAR (Subbarao, 1986, 1988)	X	X	X	Combination of ARMA and Fourier series, including loads in time domain
Modal analysis (Bacot et al., 1984)	X	X	X	Building described by diagonalized differential equation using nodes
Differential equation (Rabl, 1988)		X		Analytical linear differential equation
Computer simulation (DOE2, EnergyPlus)	X		X	Hourly simulation programs with system models
Computer emulation (EnergyPlus, TRNSYS)	X		X	Subhourly simulation programs
Artificial neural networks (Kreider and Haberl, 1994; Kreider and Wang, 1992)		X	X	Connectionist models

CO- Cooling only; HO- heating only ; HC- heating cooling.

simulation model in conjunction with elements of system identification, has also attracted a great deal of attention. The advantage of this approach is that it would have the potential to satisfy many of the applications for which inverse models are suitable for (as listed earlier). The calibrated simulation approach typically involves (1) going back to the blueprints of the building and of the HVAC system and repeating the analysis performed at the design stage while using actual building schedules and operating modes and (2) performing a calibration or tuning of the simulation model (i.e., varying the inputs in some fashion) since actual performance is unlikely to match observed performance. This process is, however, tedious and much effort has been invested by the building professional community in this regard with mixed success (Reddy, 2006).

Although at first thought this might appear to be a simple process, there are several practical difficulties in achieving a calibrated simulation, including the measurement and adaptation of weather data for use by the simulation programs, the choice of methods used to calibrate the model, and the choice of methods used to measure the required input parameters for the simulation (i,e., the weight of the building, infiltration coefficients, and shading coefficients). In the scientific sense, truly *calibrated* models have only been achieved in a very few applications since they require a very large number of input parameters, a high degree of expertise, and enormous amounts of computing time, patience, and financial resources—much more than most practical applications would allow. The calibrated simulation approach has been the focus of several research projects and technical papers in the last decade, and slow but steady progress is being made in this area (ASHRAE 14, 2002; Reddy et al., 2007).

10.6.6 Summary of Methods

Table 10.6 provides a classification of different methods of analyzing building energy use. Change-point models with outdoor temperature as the only regressor and multiple linear regression are most widely used during inverse analysis. Table 10.7 assembles various criteria of the different modeling approaches that are useful for selecting an inverse model:

1. Usage of the model (diagnostics D, energy savings calculations ES, design DE, and control C)
2. Degree of difficulty in understanding and applying the model
3. Time scale for data used by the model (hourly H, daily D, monthly M, and subhourly S)
4. Calculation time
5. Input variables used by the models (temperature *T*, humidity *H*, solar *S*, wind *W*, time *t*, thermal mass tm)

Space does not permit a more detailed treatment of each modeling approach; but based on the guidance in the two tables, the advanced student can make a selection and apply it to measured building data sets. The interested reader is referred to ASHRAE Fundamentals (2013) for more details of the methods and to numerous handbooks and reports that deal with such issues.

TABLE 10.7

Different Selection Criteria of Common Inverse Model Approaches

	Usage[a]	Difficulty	Time Scale[b]	Calc. Time	Variables[c]	Accuracy
Simple linear regression	ES	Simple	D, M	Very fast	*T*	Low
Multiple linear regression	D, ES	Moderate	D, M	Fast	*T, H, S, W, t*	Medium
ASHRAE bin method and inverse bin method	ES	Moderate	H	Fast	*T*	Medium
Change-point models	D, ES	Moderate	H, D, M	Fast	*T*	Medium
ASHRAE TC 4.7 modified bin method	ES, DE	Moderate	H	Medium	*T, S, tm*	Medium
Thermal network	D, ES, C	Complex	S, H	Fast	*T, S, tm*	High
Fourier series analysis	D, ES	Moderate	S, H	Fast	*T, H, S, W, t*	Medium
ARMA model	D, ES, C	Complex	S, H	Medium	*T, H, S, W, t, tm*	High
Modal analysis	D, ES, C	Complex	S, H	Medium	*T, H, S, W, t, tm*	High
Differential equation	D, ES, C	Very complex	S, H	Fast	*T, H, S, W, t, tm*	High
Computer simulation (component based)	D, ES, C, DE	Very complex	S, H	Slow	*T, H, S, W, t, tm*	Medium
Computer simulation (fixed schematic)	D, ES, DE	Very complex	H	Slow	*T, H, S, W, t, tm*	Medium
Computer emulation	D, C	Very complex	S, H	Very slow	*T, H, S, W, t, tm*	High
Artificial neural networks	D, ES, C	Complex	S, H	Fast	*T, H, S, W, t, tm*	High

[a] Usage shown includes diagnostics D, energy savings calculations ES, design DE, and control C.
[b] Time scales shown are hourly (H), daily (D), monthly (M), and subhourly (S).
[c] Variables include temperature *T*, humidity *H*, solar *S*, wind *W*, time *t*, and thermal mass *tm*.

To put things in perspective, the historical development of inverse building energy analysis methods followed three broad lines (Jalori and Reddy, 2015):

1. *Self-help analysis methods*: The first wave of tool development included rather limited and general-purpose building energy analysis methods. These focused on analyzing monthly utility bills (along with limited on-site measurements) in order to perform general evaluations, identify broad energy conservation measures, and then verify savings after they have been installed.
2. *Customized tools and services*: The next trend was the emergence of energy consulting and services companies who enhanced and customized the previous methods and extended their usefulness using daily and/or hourly (or subhourly) data. Energy service companies also started offering shared savings projects to larger clients. Generally, only large industries or commercial enterprises could afford these services; small and medium individual buildings did not benefit greatly from this trend.
3. *Big data analytics*: The current trend in the industry is to leverage the recent evolution in various fields such as data mining and information extraction algorithms along with widespread availability of smart meter data, database storage, and management systems. The advanced capability involves remote auditing, monitoring and verification of large number of buildings at relatively low cost as well as load analysis, fault detection and identification of operational energy saving opportunities. Data analysis techniques include methods beyond traditional statistics and data analysis and involved data mining, machine learning, and artificial intelligence techniques. Cloud-based solutions are also being offered by numerous companies that are attractive to the mass market.

Problems

Numbers 1–4 given in parenthesis denote the degree of difficulty.

10.1 For a house with total heat loss coefficient K_{tot} = 200 W/K [379 Btu/(h·°F)], calculate the magnitude of the internal heat gains for the following conditions:

- There are four occupants.
- The average heat gain from lights and equipment is 600 W (2047 Btu/h).
- The average (24 h) solar flux through the windows is 25 W/m² [7.93 Btu/(h·ft²)].

Find the balance-point temperature if the indoor temperature is 21°C (69.8°F). *(2)*

10.2 Find the annual energy consumption of a house with total heat loss coefficient K_{tot} = 700 Btu/(h·°F) (369.6 W/K) in a climate with 7000°F·days (3889 K·days). Assume that the balance-point temperature is 65°F and the heating system efficiency is 80%. *(1)*

10.3 Consider a house with total heat loss coefficient K_{tot} = 200 W/K [379 Btu/(h·°F)], indoor temperature 21°C (69.8°F), average heat gains 1200 W (4094 Btu/h), and heating system efficiency of 80%. Calculate the annual energy consumption for Washington, DC, using (a) the VBDD method with the data of Figure 10.3 and (b) the bin data on the online HCB software. *(2)*

10.4 A commercial building in Boston has the following specifications:

- T_i constant 24 h/day at 70°F (21.1°C) in winter
- Size 30 ft × 30 ft × 8 ft (9.14 m × 9.14 m × 2.44 m), orientation due south
- Open floor plan
- Equal windows on all facades, with window/wall ratio = 0.25
- Double-pane glass with U = 0.5 Btu/(h·ft²) [2.84 W/(m²·K)] and SHGC = 0.82
- 0.5 air change/h from 8:00 a.m. to 6:00 p.m.; 0.2 air change/h the rest of the time
- Lights and office equipment 3.0 W/ft² (32.28 W/m²) from 8:00 a.m. to 6:00 p.m.; 0.5 W/ft² (5.38 W/m²) the rest of the time
- Occupant per ft² (0.108 per m²) from 8:00 a.m. to 6:00 p.m.
- Heating system efficiency 80%
- Coupling to the ground being neglected

Calculate

(a) The peak heating load

(b) The balance-point temperatures corresponding to daytime and nighttime conditions

(c) The heating energy for January (use bin data of the online HCB software) *(4)*

10.5 A house in Phoenix, AZ, has a cooling load (including piping loss) of 30,000 Btu/h at outdoor design conditions (115°F) with an indoor temperature of 78°F. What is the electricity consumption of an AC system with an SEER of 10 Btu/W-h if electricity costs 10¢/kWh? Assume the annual degree-days at base 65°F for Phoenix for heating is 923°F·day and that the house has a balance-point temperature of 68°F. The annual mean temperature of Phoenix is 75°F.

NOTE: SEER—seasonal energy efficiency ratio is the cooling provided (in Btu/h) per unit electricity input (in W). (2)

10.6 Use the following steps to estimate the annual heating energy of a building you know (e.g., the building where you live; it is preferable to choose a building that does not use air conditioning during the heating season):

(a) Estimate the areas of roof, walls, and windows.

(b) Estimate the *U* values by assuming typical construction.

(c) Estimate the air change rate.

(d) Estimate the balance-point temperature.

(e) Use the VBDD method to obtain the annual heating energy.

(f) Compare with utility bills if available. Which assumptions or parameters would you change to get agreement? (*Note:* The agreement of overall results does not guarantee correctness of a model—there could be compensating errors.) *(4)*

10.7 Use Figure 10.3 and the bin data on the online HCB software to find the cooling degree-days for a balance-point temperature of 72°F (22.2°C) in Washington, DC. *(1)*

10.8 Use the Erbs correlation (Section 10.3.1) to estimate the monthly heating degree-days for Phoenix, AZ. The monthly mean outdoor temperature values required for the Erbs correlation are assembled in the table that follows along with the HDD (18.3°C) values. Compare the HDD values computed with those shown in the table. *(2)*

10.9 Use the Erbs correlation (Section 10.4.2) to generate the hourly bin data for Phoenix, AZ, for the entire year. You will use the monthly mean outdoor temperature values given in the table of Problem 10.9 to compute the monthly distributions and then add them to compute the annual values (see Example 10.7). Compare your results with those listed in Table 10.2. *(4)*

10.10 To speed up an hourly simulation program, someone proposes replacing the weather tape (with 365 days) by 2 × 3 typical days, arguing that three seasons (winter, spring/fall, and summer), with one clear and one cloudy day each, suffice to represent all the essential features. Discuss by considering the following points:

(a) What features of the weather are needed for peak loads and those for annual energy?

(b) If a single day can represent the summer (winter) design conditions, is it acceptable to calculate the summer (winter) peak load by running the simulation for just 24 h, or does the procedure have to be modified?

(c) Would your answer for (b) change for buildings that are unoccupied during weekends? *(2)*

10.11 In energy signature models, why is it (scientifically) better to correlate the consumption with degree-days than with T_o? *Hint:* Consider a period where one-half of the days have T_o above and one-half below the balance-point temperature T_{bal}. *(1)*

10.12 Suppose the house of Example 7.5 (Figure 6.19) has a forced-air heating system with some air ducts placed directly along the outside walls. The total duct surface is 20 m², and the portion in contact with the outside wall is 4 m²; the surface heat transfer coefficient inside the duct is 20 W/m². The hot air is distributed at 60°C when the furnace is running; when the furnace is off, assume for simplicity that the duct is at T_i = 20°C. Thus, on average the temperature rise above T_i of the duct is proportional to the load. Assume design outdoor temperature of −10°C.

(a) How much do these duct losses increase the peak heating load, in absolute and in relative terms?

	Monthly Av. Temp., °C	HDD, °C·days/year		Monthly Av. Temp., °C	HDD, °C·days/year		Monthly Av. Temp., °C	HDD, °C·days/year
Jan	11	251	May	25	14	Sept	29	2
Feb	13	175	June	29	2	Oct	22	28
Mar	15	134	July	33	0	Nov	15	128
Apr	20	52	Aug	32	1	Dec	11	233

(b) How much does the annual energy consumption for heating increase, in absolute and in relative terms?

(c) How would these answers change if the duct were insulated with 2.5 cm of glass wool?

(d) What is the payback time of adding this glass wool insulation if the house is heated with natural gas at $5/GJ, the furnace efficiency is 90%, and the glass wool costs $25? (4)

10.13 Suppose the house of Example 7.5 (Figure 6.19) has a hydronic heating system with radiators placed along the outside walls. The total wall surface facing the radiators is 10 m². The surface heat transfer coefficient from radiator to wall is 10 W/m². The difference between radiator temperature and T_i is proportional to the load, the peak radiator temperature being 80°C.

(a) Calculate how much the temperature of the wall next to the radiators increases.

(b) How much does this effect increase the peak heating load, in absolute and in relative terms?

(c) How much does the annual energy consumption for heating increase, in absolute and in relative terms?

(d) How do these numbers change if 5 cm of fiberglass insulation is placed on these portions of the wall?

(e) What is the payback time of adding this fiberglass insulation if the house is heated with natural gas at $5/GJ, the furnace efficiency is 90%, and the glass wool costs $25? *(4)*

10.14 Consider a house located in New York with K_{tot} = 500 Btu/(h·°F) and T_i = 70°F, assumed constant during the entire heating season.

(a) What average internal heat gain due to solar, lights, people, and equipment would be required to give a balance temperature of 65°F?

(b) Using the degree-day data of Figure 10.3, what is the annual heating energy consumption if the house is heated with a furnace having a constant efficiency of 82%?

(c) What would be the annual heating energy consumption if the house were moved to a location with the bin data in the following table? *(3)*

Temperature, °F	N_{bin}, h
52.5	32
57.5	180
62.5	300
67.5	620

10.15 Consider the building from Example 10.1 and Figure 6.19. Assume that the internal temperature remains constant through the year and that the house is slab-on-grade construction with uniform horizontal insulation under the slab. The insulation is 2.5 cm of polyisocyanurate [conductivity of 0.02 W/(m·K)].

(a) What are the annual mean and peak heat gain/loss from this slab?

(b) How does the annual heating bill change if you take into account the ground-coupled heat loss? *(4)*

10.16 Example 10.5 illustrates how the degree-day method can be used to evaluate the cost-effectiveness of different energy efficiency strategies for the building envelope.

(a) Reproduce the calculations for the payback analysis for all the options clearly showing all the steps involved.

(b) Calculate the energy savings and the payback time, relative to the reference design, for each of the four design variations by itself. *(3)*

10.17 Calculate the annual heat lost per square meter (foot) of roof area as a function of the insulation thickness Δx. Even though, strictly speaking, the balance-point temperature changes with insulation, assume for simplicity that the number of degree-days remains constant at 2000 K·days (3600°F·days). Using the cost data of Example 10.5, perform the following:

(a) Calculate the annual energy and the cost of the additional insulation.

(b) Plot the energy savings and the insulation cost versus insulation thickness from Δx = 0.05 to 0.5 m (0.164 to 1.64 ft).

(c) At what thickness is the payback time (relative to a thickness of 0.05 m) equal to 10 years? *(2)*

10.18 Consider a house in Washington, DC, with a heat loss coefficient K_{tot} = 389 Btu/(h·°F), a constant indoor temperature of 68°F, and average internal gains of 3400 Btu/h. Compare the annual energy cost of heating this house with the following three options:

(a) A conventional natural gas furnace (efficiency of 75%)

(b) A condensing natural gas furnace (efficiency of 93%)

(c) A heat pump with these characteristics:

$$\begin{aligned} \text{COP} &= 2.6 \quad 50°\text{F} < T_o \\ &= 2\,2 \quad 30°\text{F} < T_o \le 50°\text{F} \\ &= 1.7 \quad 15°\text{F} < T_o \le 30°\text{F} \\ &= 1.0 \quad T_o \le 15°\text{F} \end{aligned}$$

Bin temperature data for Washington can be found in Table 10.2. Natural gas costs \$6/GJ and electricity costs 8¢/kWh. *(3)*

10.19 The occupant of the house in Problem 10.18 installs a thermostat that automatically reduces the set point temperature from 68°F to 60°F during the hours of 8 p.m. through 8 a.m. What is the simple payback time if this thermostat costs \$100 and the house uses the conventional gas furnace? Neglect any transient behavior of the heating system or any thermal storage effects in the house. *(2)*

10.20 Evaluate the contribution of outdoor air to the annual heating energy of a house for two design options: for natural infiltration and mechanical ventilation. Make the following assumptions. The house has volume $V = 360\ m^3$ (12,708 ft^3) and conductive heat loss coefficient $K_{cond} = 150$ W/K [284.3 Btu/(h · °F)]. With mechanical ventilation, there is a constant outdoor airflow of 10 L/s (20 ft^2/min) per occupant, for an occupancy of 3. For natural infiltration, use the LBL model of Section 6.6.1 with the coefficients for a two-story building at a site with shielding class 3; and choose the leakage area such that the infiltration rate is 3 × 10 L/s when the indoor–outdoor temperature difference $T_i - T_o = 5$ K (9°F) and the wind speed = 5 m/s (11.5 mph). Use the bin data for Chicago (Table 10.2), and assume a constant wind speed of 5 m/s (11.5 mph) for simplicity. How does the heating energy reduction compare with the energy for the fan if the fan draws 100 W continuously? *(3)*

References

ASHRAE (1984). *Bin Weather Data* (RP 385, L. Degelman). American Society of Heating, Refrigerating and Air-Conditioning Engineers, Atlanta, GA.

ASHRAE 14 (2002). *ASHRAE Guideline 14-2002. Measurement of Energy and Demand Savings.* American Society of Heating, Refrigeration and Air-Conditioning Engineers, Atlanta, GA.

ASHRAE Fundamentals (2013). *Handbook of Fundamentals.* American Society of Heating, Refrigerating and Air-Conditioning Engineers, Atlanta, GA.

Bacot, P., A. Neveu, and J. Sicard (1984). Analyse Modale Des Phenomenes Thermiques en Regime Variable Dans le Batiment. *Revue Generale de Thermique*, 267, 189.

Balcomb, J.D., R.W. Jones, R.D. McFarland, and W.O. Wray (1982). Expanding the SLR method. *Passive Solar.*, 1(2), 67–90.

Birdsall, B., W.F. Buhl, K.L. Ellington, A.E. Erdem, and F.C. Winkelmann (1990). Overview of the DOE2.1 building energy analysis program. Report LBL-19735, rev. 1. Lawrence Berkeley Laboratory, Berkeley, CA.

Claridge, D.E. (August 1998). Special Issue Editor, Conservation and solar buildings—Methods for analysis of measured energy data in commercial buildings. *ASME J. Solar Energy Eng.*, 120(3), 149–229.

Claridge, D.E., M. Krarti, and M. Bida (1987). A validation study of variable-base degree-day cooling calculations. *ASHRAE Trans.*, 93(2), 90–104.

Dhar, A., T.A. Reddy, and D.E. Claridge (August 1998). Modeling hourly energy use in commercial buildings with Fourier Series functional forms. *ASME J. Solar Energy Eng.*, 120(3), 217–223.

Erbs, D.G., S.A. Klein, and W.A. Beckman (June 1983). Estimation of degree-days and ambient temperature bin data from monthly-average temperatures. *ASHRAE J.*, 25(6), 60–65.

Fels, M.F., ed. (1986). Special double issue, Measuring energy savings: The scorekeeping approach. *Energy Build.*, 9(1 and 2), 180.

Fels, M.F. and M. Goldberg (1986). Refraction of PRISM results into components of saved energy. *Energy Build.*, 9, 169–180.

Haberl, J.S. and D.E. Claridge (1987). An expert system for building energy consumption analysis: Prototype results. *ASHRAE Trans.*, 93(1), 979–998.

Hammarsten, S. (1987). A critical appraisal of energy signature models. *Appl. Energy*, 26, 97–110.

Jalori, S. and T.A. Reddy (June 2015). A unified inverse modeling framework for whole building energy interval data: Daily, hourly baseline modeling and short-term load forecasting. *ASHRAE Trans.*, 121 (part 2), 156–169.

Katipamula, S., T.A. Reddy, and D.E. Claridge (June 1995). Effect of time resolution on statistical modeling of cooling energy use in large commercial buildings. *ASHRAE Trans.*, 101(2), 3894–3895.

Katipamula, S., T.A. Reddy, and D.E. Claridge (August 1998). Multivariate regression modeling. *ASME J. Solar Energy Eng.*, 120(3), 177–184.

Kissock, J.K., J.S. Haberl, and D.E. Claridge (July 2001). Inverse modeling toolkit: User's guide, ASHRAE final research report for RP-1050. American Society of Heating, Refrigerating and Air-Conditioning Engineers, Atlanta, GA.

Kissock, J.K., T.A. Reddy, and D.E. Claridge (1998). Ambient temperature regression analysis for estimating retrofit savings in commercial buildings. *ASME J. Solar Energy Eng.*, 120(3), 168–176, August.

Kreider, J. (2001). *Handbook of HVAC.* CRC Press, Boca Raton, FL.

Kreider, J.F. and J. Haberl (1994). Predicting hourly building energy usage: The great predictor shootout—Overview and discussion of results. *ASHRAE Trans.*, 100(2), 1104–1118.

Kreider, J.F. and X.A. Wang (1992). Improved artificial neural networks for commercial building energy use prediction. *ASME J. Solar Energy Eng.*, 92, 361–366.

LBL (1982). *DOE2 Engineers Manual.* Lawrence Berkeley Laboratory report number LBL-11353 (LA-8520-M, DE83004575). National Technical Information Services, Springfield, VA.

Marchio, D. and A. Rabl (1991). Energy-efficient gas-heated housing in France: Predicted and observed performance. *Energy Build.*, 17, 131–139.

Neveu, A., P. Bacot, and R. Regas (1986). Modèles d'évolution thermique des bâtiments: Conditions pratiques d'identification. *Revue Générale de Thermique*, 296, 413.

Nisson, J.D.N. and G. Dutt (1985). *The Superinsulated Home Book*. Wiley, New York.

Rabl, A. (1988). Parameter estimation in buildings: Methods for dynamic analysis of measured energy use. *ASME J. Solar Energy Eng.*, 110, 52–66.

Rabl, A., L.K. Norford, and G.V. Spadaro (1986). Steady state models for analysis of commercial building energy data, In *American Council for Energy Efficient Economy Summer Study*, Santa Cruz, CA, August 1986.

Randolph, J. and G.M. Masters (2008). *Energy for Sustainability*. Island Press, Washington, DC.

Reddy, T.A. (1989). Application of dynamic building inverse models to three occupied residences monitored non-intrusively, in *Proceedings of the Thermal Performance of Exterior Envelopes of Buildings IV*, ASHRAE/DOE/BTECC/CIBSE. Orlando, FL, 654–671.

Reddy, T.A. (2003). Special Issue Editor, Emerging trends in commercial building design, diagnosis and operation. *ASME J. Solar Energy Eng.*, 235(11), 235–371. August Issue.

Reddy, T.A. (January 2006). Literature review on calibration of building energy simulation programs: Uses, problems, procedures, uncertainty and tools. *ASHRAE Trans.*, 112(1), 226–240.

Reddy, T.A. (2011). *Applied Data Analysis and Modeling for Energy Engineers and Scientists*. Springer, New York.

Reddy, T.A. and D.E. Claridge (January 2000). Uncertainty of measured energy savings from statistical baseline models. *HVAC&R Res. J.*, 6(1), 3–20.

Reddy, T.A., S. Katipamula, J.K. Kissock, and D.E. Claridge (February 1995). The functional basis of steady-state thermal energy use in air-side HVAC equipment. *ASME J. Solar Energy Eng.*, 117, 31–39, 226–240.

Reddy, T.A., I. Maor, and C. Ponjapornpon (March 2007). Calibrating detailed building energy simulation programs with measured data—Part I: General methodology. *HVAC&R Res. J.*, 13(2), 221–241.

SERI (1981). Solar radiation energy resource atlas of the United States. Report SERI/SP-642-1037. Solar Energy Research Institute, Golden, CO.

Schoenau, G.J. and R.A. Kehrig (1990). A method for calculating degree-days to any base temperature. *Energy Build.*, 14, 299–302.

Singh, V., T.A. Reddy, and B. Abushakra (January 2014). Predicting annual energy use in buildings using short-term monitoring: The dry bulb temperature analysis (DBTA) method. *ASHRAE Trans.*, 120(1), 397–405.

Sonderegger, R.C. (1977). Dynamic models of house heating based on equivalent thermal parameters. PhD thesis, Center for Energy and Environmental Studies Report No. 57, Princeton University, Princeton, NJ.

Sonderegger, R.C. (1998). Baseline equation for utility bill analysis using both weather and non-weather related variables. *ASHRAE Trans.*, 104(2), 859–870.

Stephenson, D.G. and G.P. Mitalas (1967). Cooling load calculations by thermal response factor method. *ASHRAE Trans.*, 73. (Pt. 2), 2.1–2.11.

Subbarao, K. (1986). Thermal parameters for single and multi-zone buildings and their determination from performance data. SERI Report SERI/TR-253-2617. Solar Energy Research Institute, Golden, CO.

Subbarao, K. (1988). PSTAR—Primary and Secondary Terms Analysis and Renormalization: A unified approach to building energy simulations and short-term monitoring. Report SERI/TR-254-3175. Solar Energy Research Institute, Golden, CO.

Thevenard, D. (2011). Methods for estimating heating and cooling degree-days to any base temperature. *ASHRAE Trans.*, 117 (1), 884–891.

Vadon, M., J.F. Kreider, and L.K. Norford (1991). Improvement of the solar calculations in the modified bin method. *ASHRAE Trans.*, 97(2), 204–211.

11

Description of Typical Building HVAC Systems and Components

ABSTRACT This chapter provides a general background into building energy systems with photos and sketches to illustrate the variety of HVAC systems and equipment. The *anatomy* of a building's HVAC system consists of four main parts: (1) primary systems or central plant, (2) distribution system, (3) terminal devices, and (4) controls. Recall that conditioning a space involves cooling or heating and providing adequate conditioned ventilation air to the space. Secondary systems are those that distribute the cooling (or heating) thermal energy produced by the primary systems, namely the chillers (or boilers), to the building spaces needing to be conditioned. This can be done using either water or air as the working fluid, and so it is common to recognize three generic types of secondary systems, namely, all-water, all-air, and air-water systems. Alternatively, HVAC systems can also be categorized into three groups: unitary, centralized, and district. The heating and cooling energy is then transferred to the room air by distribution systems. An important, and often neglected, aspect is the control system that maintains proper functioning of the various components of the HVAC system. We describe each of these system categories at length along with discussion and photographs or sketches of a few pertinent components. The material in this chapter would be useful to a reader with limited practical knowledge of these systems since it would allow him/her to visualize the equipment or system while it is being discussed both qualitatively and in an analytical framework in the remaining chapters of this book.

11.1 Primary and Secondary HVAC Systems

Heating and cooling devices (for small buildings) and equipment/systems (for larger buildings) are those that produce the heating or the cooling necessary to condition the outdoor ventilation air and to meet the building loads. They are called *primary systems* or "plant" by some building analysts and HVAC system modelers. These devices are usually the major source of energy consumption in the form of electricity and/or natural gas (or fuel oil or even coal). Perhaps the best known example of a heating device is the *boiler* (Figure 11.1) and that of a cooling system is the vapor compression *cooling system* (Figure 11.2). In the gas boiler, the heat generated by burning natural gas goes to convert water to steam that is then sent to the building. A valve modulates gas flow depending on the pressure sensor reading, while the flue gases escape out of a flue stack with an adjustable damper. In the case of the chiller, the heat drawn from the space is picked up by the refrigerant in the evaporator at a lower temperature and rejected to the outdoor air in the condenser at a higher temperature. The compressor draws electric power in order to accomplish this heat transfer.

The heating or cooling thus generated has to be distributed to the specific rooms and zones in the building. This energy distribution is done by *secondary systems*, also referred to as "systems." Depending on the working fluid, usually either water (or steam) or air streams, which pick up the heat/coolth from the primary devices and distribute it to the building, these systems are called hydronic or air systems respectively. The energy is then transferred either directly to the room air by terminal devices such as radiators or fan coil units or transferred to an air stream, which is then supplied to the room via *terminal boxes* and *room diffusers* (see Figure 11.3).

The terminal boxes are located in the space above the suspended ceiling in a zone. The air then travels from them through ductwork (flexible or rigid) to the diffusers. There are many different variations of terminal boxes used in commercial buildings. The space above the suspended ceiling can be used as a *return air plenum* because air from the zone can be made to pass through this space before it returns to the air distribution system. In some cases, the return grill is connected directly to the return air duct—this is called a *ducted return*. It is much easier to control the amount of return air in a room with ducted return; these kinds of systems are common in laboratories and chemical rooms that use variable air volume systems. However, a nice feature of using an open return plenum is that the return grills can be placed where they are needed, such as over a heat source. This is much more difficult on a ducted return, since rigid ducts are usually used.

Figure 11.4 is a conceptual sketch showing the various heat transfer loops needed to provide cooling to large buildings using an air delivery system. We note

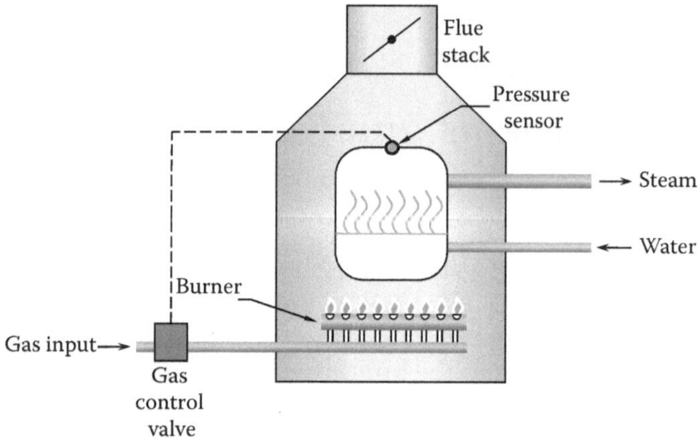

FIGURE 11.1
Basic boiler schematic.

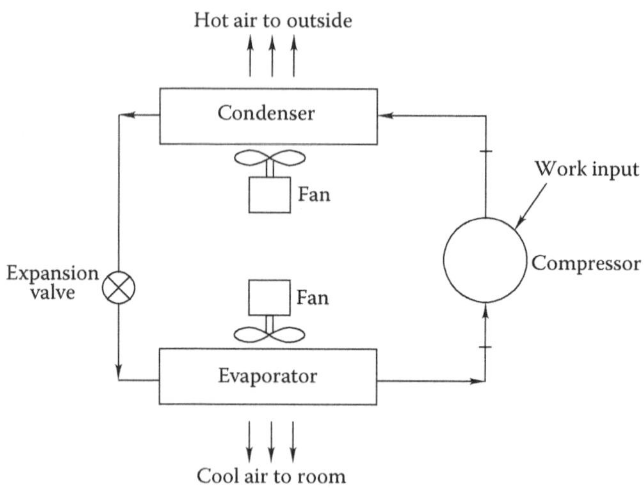

FIGURE 11.2
Components of a vapor compression cycle.

that in order to remove the heat gains from the conditioned space, there may be two secondary system loops and three primary system loops as shown. The cooling generated in the refrigeration loop is used to produce chilled water, which is then sent to cooling coils inside an air duct used to cool the supply air stream to the space. Heat has to be dissipated in the condenser of the refrigeration loop. This is achieved by first transferring the heat to cooling water, which then transfers it to the ambient air in a cooling tower. For buildings with smaller loads, some of these loops are not needed. For example, the refrigeration heat can be transferred directly to the ambient air, as in air-cooled condensers. The chilled water loop can be dispensed with by having the supply air stream pick up the coolth from the evaporator coil directly. This reduces the first cost at the expense of some loss in overall thermal efficiency.

Secondary systems essentially consist of

1. Distribution system: ducts and pipes to carry the working fluids
2. Equipment to move the working fluid such as fans and pumps
3. Devices such as heating and cooling coils to transfer heat between fluid and air to be supplied to a room.
4. Terminal devices to control, deliver, and distribute heating or cooling air to various points inside a room such as radiators or fan-coil units, or room diffusers
5. Automatic control devices such as pressure and temperature sensors, valves, dampers, thermostats, etc.

In large buildings, the energy consumed by the pumps and the fans to achieve this can be substantial, up to 40%

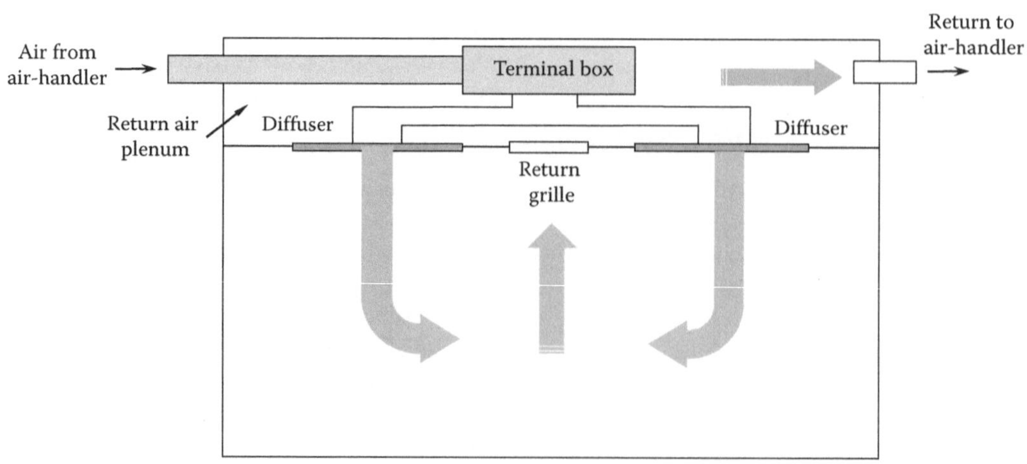

FIGURE 11.3
Schematic of zone terminal box and room air flows.

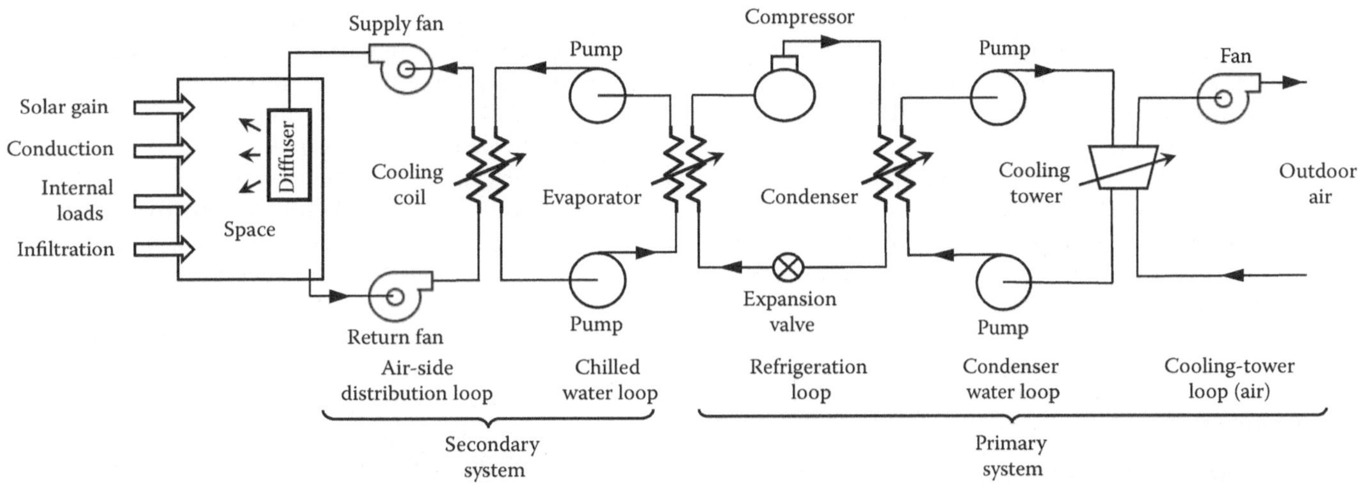

FIGURE 11.4
Complete sketch of an all-air secondary system with a water-cooled primary system showing some of the important components. Note that the five heat transfer loops needed to remove the heat gains to the space. As the space loads get smaller, some of these loops can be combined or eliminated.

or more of the total building energy use. Usually, however, the secondary energy is much less, in the range of 5%–20% in small to medium buildings.

This chapter will provide a brief background description of some of the numerous systems and devices that have evolved over the last century pointing to subsequent chapters in this book, which treat specific components and systems. These chapters provide details on how to perform a mathematical analysis of such components and systems. This involves representing each and every component by a mathematical function using fundamental heat transfer principles, manufacturer rated performance, and curve-fit procedures to account for part-load performance. These component models are then assembled together to represent the behavior of the entire system. Solving this set of equations under specified climatic and load input conditions along with selected operating conditions imposed constraints is referred to as *system simulation*.

11.2 Types of Secondary Systems

There are a large variety of secondary HVAC system configurations, but generally, they are grouped into three generic types based on the manner in which cooling/heating energy from the primary equipment is distributed to the space:

1. *All-air systems*, wherein both the sensible and latent heat loads of the space are met by an appropriately conditioned *single* supply air stream to each individual space. This air stream also delivers the required ventilation air flow rate. These systems are discussed in Chapter 19, and since an understanding of psychrometrics is essential to their analysis and design, Chapter 13 is devoted to this subject matter.
2. *All-water or hydronic systems*, wherein the space cooling (or heating) function is performed by chilled water (or hot water) circulated from a central cooling plant (or boiler plant) to heat exchangers or terminal units located in, or immediately adjacent to, the conditioned space. Heat is transferred to the room air by either natural convection or forced convection via fans or blowers. Only the sensible heat loads are met by these systems. These systems are addressed in Chapter 18.
3. *Air-water and hybrid systems* are those where the functions of meeting space loads and drawing in and conditioning the required ventilation air flow are treated by two separate HVAC systems. The thrust to develop such alternative air delivery systems is largely driven by the need to improve energy efficiency while enhancing IAQ. In such systems, chilled (or hot) water terminal units or heat pumps can be used to meet the sensible loads of the space, while the space latent load and ventilation requirements are met by a supply of conditioned air, referred to as primary air. Recent system variants have evolved that rely on radiant panels to provide space cooling and heating to the occupant in addition to the traditional all-water and

all-air systems. A radiant panel is defined as a temperature-controlled surface (less than 300°F or 150°C) where 50% or more of the design heat transfer occurs by radiation (ASHRAE Systems, 2012). Recently, a new system configuration called the dedicated outdoor air system (DOAS) has become widely popular. Many of the current innovations in HVAC&R system design are occurring within this generic system type, which are discussed in Chapter 20.

11.3 Broad Classification of HVAC Systems

An alternative classification scheme is based on size, construction, and operating characteristics of HVAC systems. We distinguish three broad categories:

1. *Unitary or packaged systems*, wherein all the basic components needed for air conditioning have been factory-assembled into one or two pieces of equipment, transported to the site, and hooked up with a minimum of on-site labor. Split systems, which consist of separate components, also fall in this category since several components (such as fans, filters, heating source, heating and cooling coils, controls, condenser, etc.) come pre-packaged together from the factory. These systems, also referred to as decentralized systems, are meant to meet the loads of small to mid-sized buildings of almost all building types where initial costs and simplified system installation are important issues. Though these systems have lower first costs, they usually last only 8–15 years, are less efficient, and require more maintenance than centralized systems. Often, there is no clear separation between secondary and primary systems. Mostly air is used as the heat transfer medium. A few such systems are described in Section 11.4.
2. *Centralized or built-up systems*, wherein the various components are transported individually and assembled at site by the contractor. These systems are meant to meet the loads of medium and large buildings. While air or water can both be selected as the heat transfer medium for small and medium buildings, water is generally superior in the case of large systems. Centralized systems are used in almost all building types, but particularly in large buildings (ASHRAE Systems, 2012). These HVAC system types are described in Section 11.5.
3. *Campus or district systems* are meant to meet loads of several buildings in an urban neighborhood or in a campus such as universities or mixed use manufacturing facilities. These primary systems will be located in a dedicated building or facility called the physical plant. The heating and cooling is then distributed to the various individual buildings via chilled water or steam loops (which can run into several miles), which are well-insulated and buried underground. The heat and/or coolth are transferred via building heat exchangers to secondary systems within each building. Some handbooks include district systems into the centralized category (e.g., ASHRAE Systems, 2012). A brief overview is provided in Section 11.6.

11.4 Unitary Systems

The most obvious type of unitary systems is the *single unit system*. The various individual components of such a unitary system come mostly pre-packaged with the contractor needing minimal effort at site for hookup. This is obvious for the case of the totally encased *room air-conditioner* shown in Figure 11.5. All the components are contained within a single box, which only needs to be positioned into a wall or window opening. Other types are through-the-wall room HVAC systems, air-cooled heat pumps that provide heating or cooling as needed (see Section 14.5). Unitary systems are generally found in buildings three-story or less such as retail spaces, small offices, classrooms, motels, etc. Their capacities range from 5 to 460 kW (1.5 to 130 tons), but are available in pre-established capacity increments (O'Neal and Bryant, 2001). Some designers combine them with central HVAC systems, wherein unitary systems are used for the building perimeter zones.

All-air split systems are more prevalent in newer housing and smaller commercial buildings. The typical residential HVAC system is the split AC to cool the air and a furnace to heat air (Figure 11.6). A sketch of a gas fired furnace along with several components is shown in Figure 11.7. The heat transfer is achieved via coils inside the common duct, which then distributes air to all the rooms of the house. The layout is similar to the rooftop unit (described later) except that here the condenser coil and the compressor of the refrigeration system are located outdoors (either on the roof or beside the house on the ground) with refrigerant piping connection to the evaporator coil (called a direct expansion or DX coil) located inside the air duct of the building (see Figure 11.8a). Split air source heat pump units are

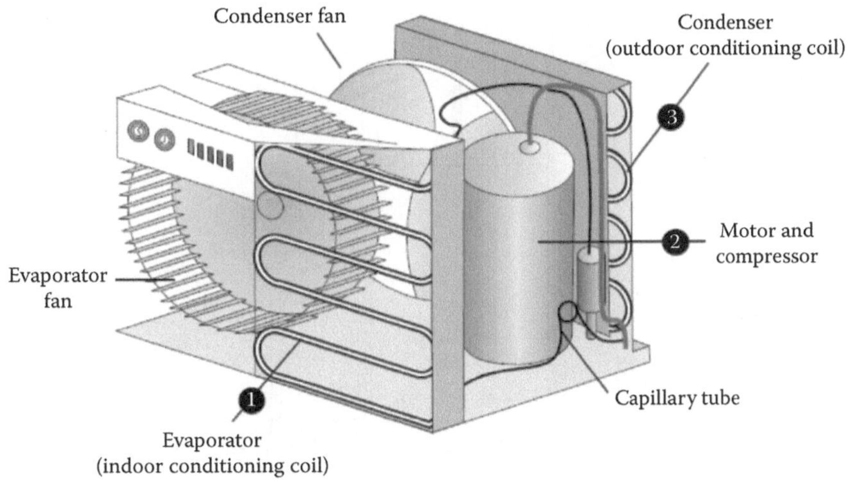

FIGURE 11.5
Room air-conditioner. (From www.electrical-knowhow.com, accessed on September 2014.)

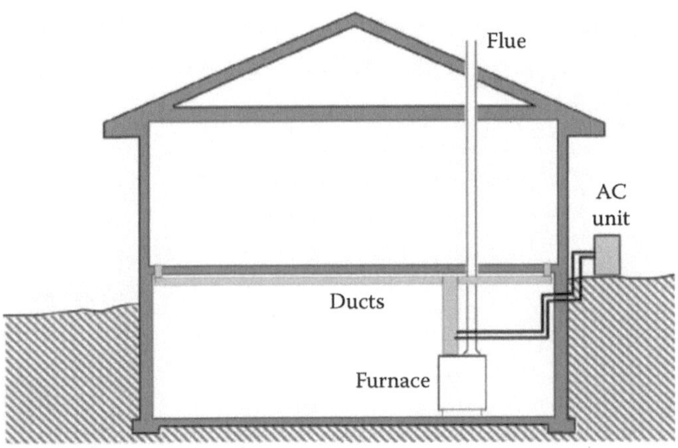

FIGURE 11.6
Typical residential HVAC system components—split AC and furnace to heat air.

also commercially available. A picture of such packaged units can be seen on top of multi-family city buildings (Figure 11.8b). In the last decade or so, multiple-unit variable refrigerant flow (VRF) systems have also been developed; these are extensively used in the far-east while their market share in the United States is slowly increasing. The reader can refer to ASHRAE Systems (2012) for more details about these systems, including advantages and limitations.

A special and widely used unitary system is the *rooftop unit* or *packaged unit*. This kind of equipment is very common on large retail buildings, warehouses, and small commercial buildings. The rooftop unit is a self-contained air-handling unit (AHU) that uses direct expansion (DX) cooling and gas heating and usually provides air directly to the zone being conditioned (Figure 11.9). Note that the air-cooled condenser fans (there are six of them shown) can be controlled individually depending on the cooling load on the chiller. Figure 11.10 shows a schematic of a typical rooftop unit showing the different sections devoted to moving the air and to providing cooling through the DX coil. Heating of the air stream is done by the flue gases from burning natural gas supplied to the heating coils. The return room air stream mixes with a certain amount of outdoor air, and the mixed air is conditioned before sending it back to the space below. The room supply fan and the condenser fan assure the proper functioning of the two air streams. "Large box stores" and supermarkets may use two dozen such units to provide air conditioning to different sections of the building. Because of their low cost, such rooftop units are widely used in low-rise office buildings (typically less than three to four stories).

The thermal performance and specifications of unitary and split systems meant for heating and cooling applications are discussed in Chapter 19. Advantages and disadvantages of packaged unitary systems are listed in Table 11.1.

11.5 Centralized Systems

Centralized systems are those whose individual components are assembled and built up by the contractor at site. Historically, *hydronic systems* (or all-water systems) were widely used in residences and smaller commercial buildings. Figure 11.11 is an isometric view of a basic two bedroom house where the heating is supplied by hot water from a boiler located in the basement of the house. The piping is laid out to run through all the rooms of the house with terminal units positioned under windows to compensate for the greater heat losses occurring at

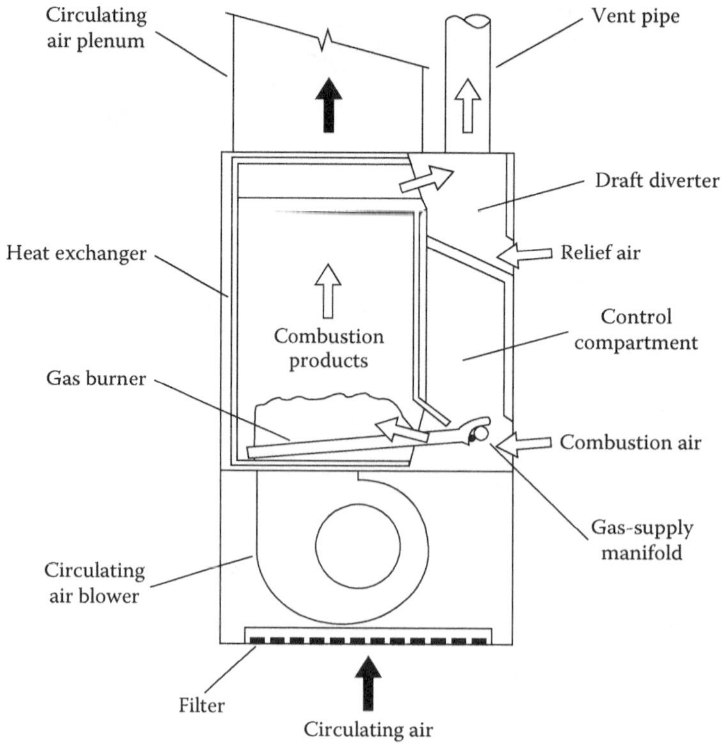

FIGURE 11.7
Examples of furnaces for residential space heating: vertical.

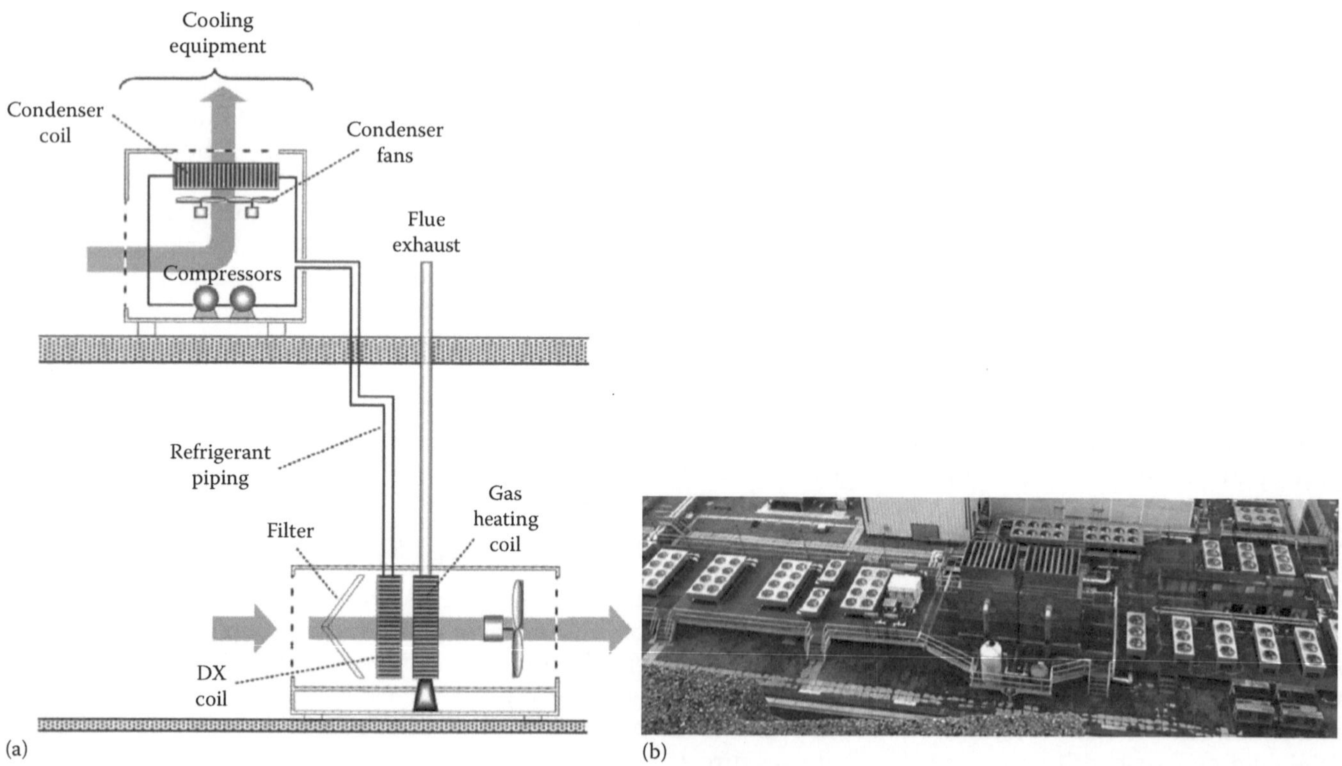

FIGURE 11.8
(a) Split central air system for residential and small commercial buildings. (From Curtiss, P. and Breth, N., *HVAC Instant Answers*, McGraw-Hill, New York, 2002.) (b) Photograph showing the numerous air-cooled condensers of various sizes (only two water cooling towers can be seen in the forefront), which take up most of the roof space of tall apartment buildings in cities.

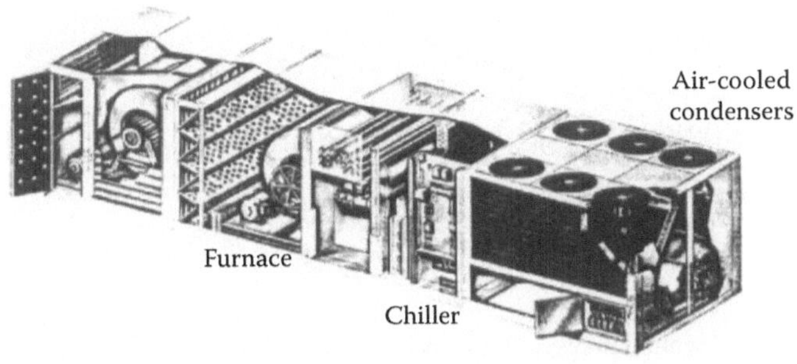

FIGURE 11.9
Assembled view of a packaged roof top unit. (From Pita, E.G., *Air Conditioning Principles and Systems*, 4th ed., Prentice Hall, Upper Saddle River, NJ, 2002.)

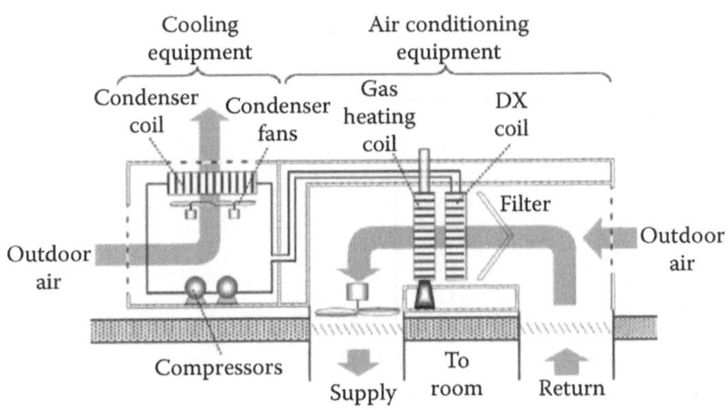

FIGURE 11.10
Sketch of a roof top unit showing individual components and air flows. (From Curtiss, P. and Breth, N., *HVAC Instant Answers*, McGraw-Hill, New York, 2002.)

TABLE 11.1
Advantages and Disadvantages of Packaged Equipment

Disadvantages	Advantages
Limited performance choices because of fixed component sizing	Individual room control is allowed
Generally not good for close humidity control	Cooling/heating available at any time for any space
Space control is on-off causing temperature swings	Individual ventilation can be included in unit
Relatively short life	Capacities certified by manufacturer
Energy use higher than central systems because of fixed capacity increments and tendency to oversize	Equipment in unoccupied spaces can be easily turned off
Full use of economizer cycle usually not possible	Operation is usually very simple
Limited air distribution control	Requires less floor space than central systems
Sound level inside room may be objectionable	Low first costs
Poor aesthetics	Flexibility in locating equipment so as to reduce duct lengths
Air filtering option is rather limited	Simple installation, no need for trained personnel
May be difficult to remove water condensate	
Maintenance can be an issue	

Source: Adapted from O'Neal, D.L. and Bryant, J.A., Chapter 4.2: Air conditioning systems, in J.F. Kreider, ed., *Handbook of Heating, Ventilation and Air Conditioning*, CRC Press, Boca Raton, FL, 2001.

those locations. Terminal units come in a variety of designs and shapes, and the HVAC designer needs to select the right configuration and the lengths of the various individual units depending on the operating water temperatures and the room loads. The old-fashioned *cast-iron radiator* shown in Figure 11.12a is now obsolete and replaced with units that transfer heat to the room air more efficiently. One example is the *fan coil unit* shown in Figure 11.12b. There are also several variants of the piping distribution network, such as series-connected, parallel-connected, and series-parallel. In recent years, the radiant underfloor heating systems that are sized and laid out for individual rooms have become quite popular especially for residences (especially in Europe). Analysis and design aspects of hydronic systems are discussed in Chapter 18.

All-air HVAC systems for larger buildings have many more components (see Figure 11.13). A mechanical room in the basement or occupying part of a floor in taller buildings would contain one or more boilers, one or more chillers, and even several central air-handling units. The heat absorbed by the refrigerant during the cooling process must be rejected at the condenser to an environmental sink, which is usually the ambient air. This can be done directly to the air, or indirectly, first to water and then to the ambient air. The former requires an additional loop and a *cooling tower* as shown in Figures 11.13 and 11.14. In larger buildings, the associated system cost is still lower than that of an air-cooled condenser due to enhanced heat transfer rates from refrigerant to cooling water. Several allied equipment and components are shown, and these are discussed in Chapter 17. For smaller building loads, it is more economical to select *air-cooled condensers* (Figure 11.15) similar to those in a rooftop unit (Figure 11.9). Figure 11.8b is a photograph depicting the numerous air cooled condensers that occupy much of the roof space in tall apartment buildings in cities. Two water-cooled condensers can be seen in the forefront of the picture.

The air-conditioning and distribution is also similar to the roof-top unit but much more involved. Most units draw in some outside air and mix it with return air from

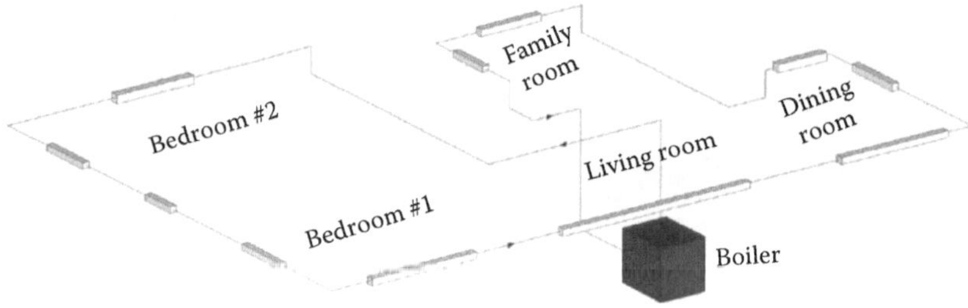

FIGURE 11.11
Layout of a hydronic system with terminal heaters in each room. A series piping layout is shown though other types of layouts (such as parallel or series-parallel) are also common.

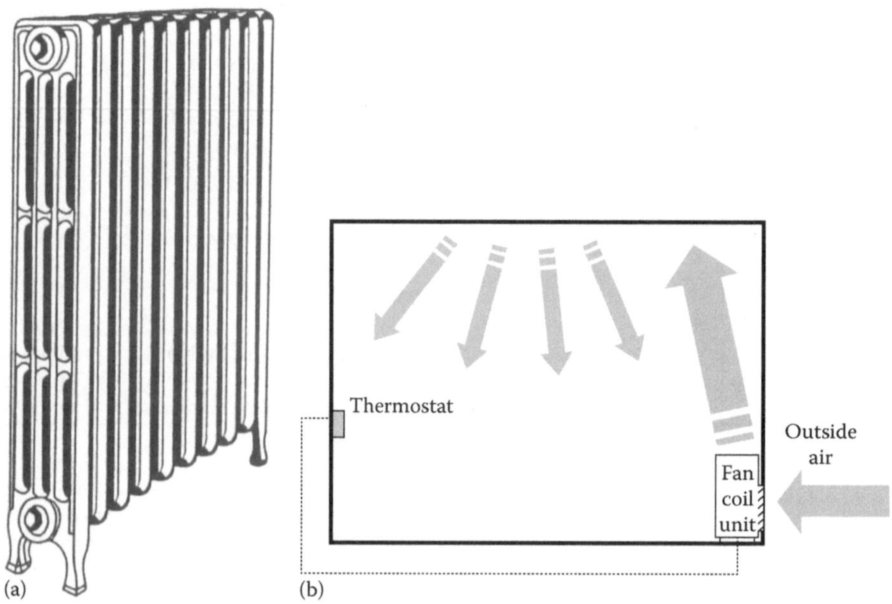

FIGURE 11.12
Two types of hydronic terminal units: (a) old-fashioned cast-iron radiator and (b) fan coil unit.

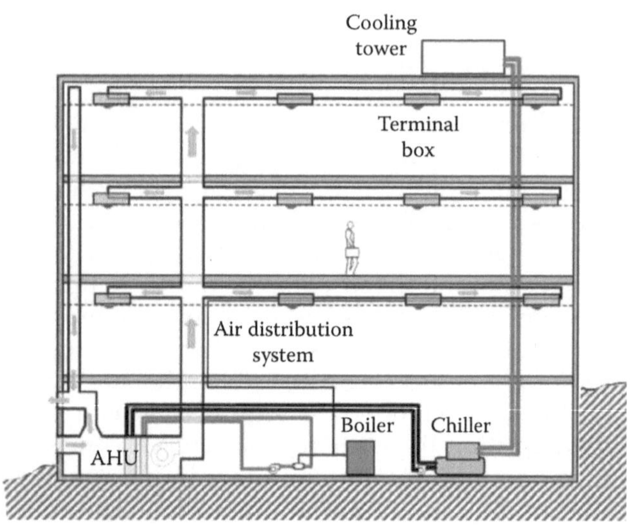

FIGURE 11.13
Typical all-air HVAC system components for a commercial building. (From Curtiss, P. and Breth, N., *HVAC Instant Answers*, McGraw-Hill, New York, 2002.)

the building so as to reduce energy needs to condition the outside air. The mixed air is conditioned by the cooling coil and/or the heating coil that are supplied with chilled water or hot water from the chiller and the boiler respectively (as shown in Figure 11.16). These coils also come in numerous designs and variations (one common design is shown in Figure 11.17), and these need to be selected and sized carefully especially the cooling coils since they have to both cool and dehumidify the air as needed. Coils and other components are treated in Chapter 17.

Often, the outdoor to return air ratio has to be modulated, and this is done by motorized dampers or by pneumatic actuators to vary the opening angle of the outdoor air damper blades (Figure 11.18). Some air-handlers, however, use 100% outside air (e.g., in some hospitals). Energy efficiency of such buildings as well as in more traditional ones can be greatly improved by using heat recovery units, of which there are several generic types such as run-around coils, plate heat exchangers, heat

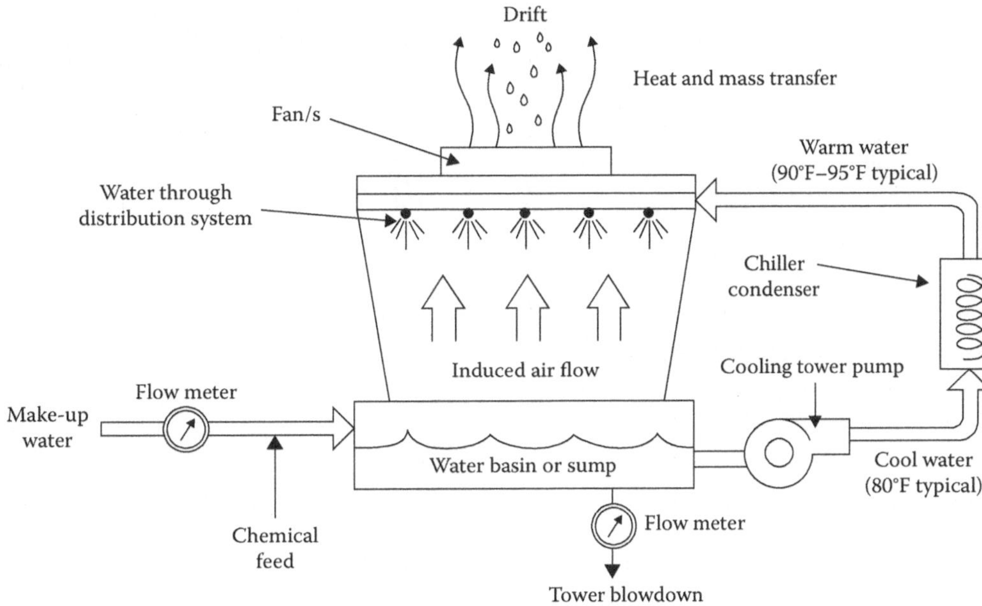

FIGURE 11.14
Mechanical draft cooling tower. (From O'Neal, D.L. and Bryant, J.A., Chapter 4.2: Air conditioning systems, in J.F. Kreider, ed., *Handbook of Heating, Ventilation and Air Conditioning*, CRC Press, Boca Raton, FL, 2001.)

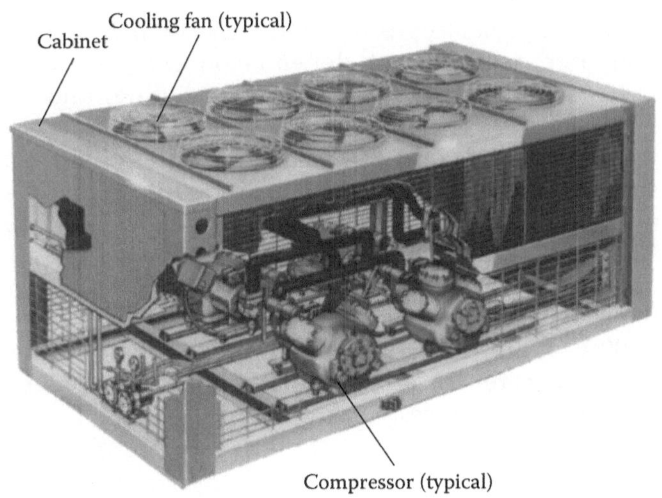

FIGURE 11.15
Air-cooled condenser.

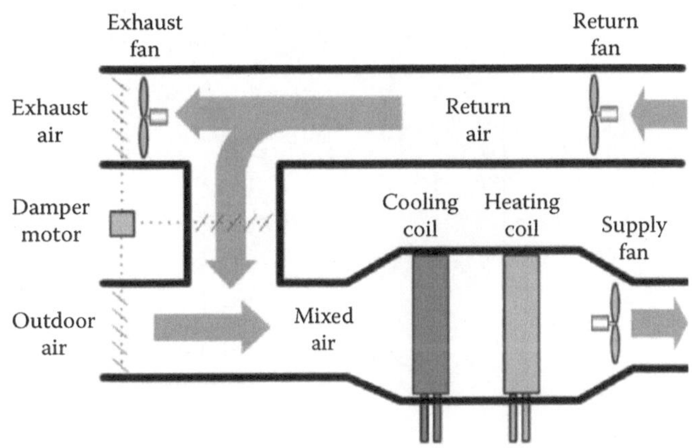

FIGURE 11.16
Single-duct air-handling unit. (From Curtiss, P. and Breth, N., *HVAC Instant Answers*, McGraw-Hill, New York, 2002.)

pipes, and rotary devices. One such device, the *rotary energy exchanger*, which allows both heat and moisture transfer across the building supply and exit airstreams, is shown in Figure 11.19. These devices are described in Section 19.6.

The air distribution to various portions of the building could be done either by single duct or dual duct systems. Further, these systems could be operated under *constant air volume* (CAV) or *variable air volume* (VAV) operation. While the fan (or fans) runs at full output at all times for the former, VAV systems vary the airflow rate to provide only as much air as is necessary to

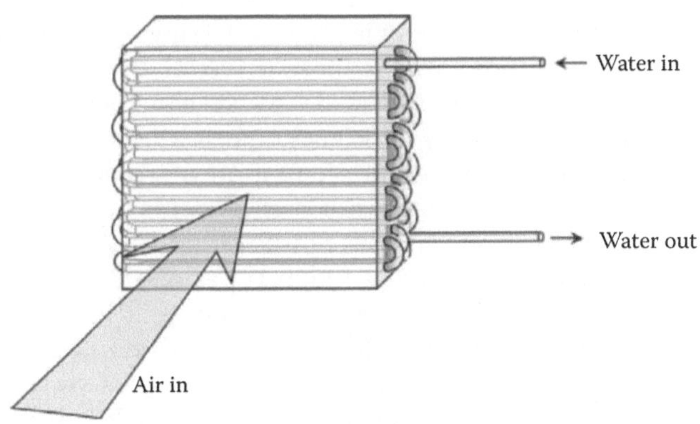

FIGURE 11.17
Water-to-air cross-flow heat exchanger.

FIGURE 11.18
Pneumatic actuator on tight-seal outdoor air dampers.

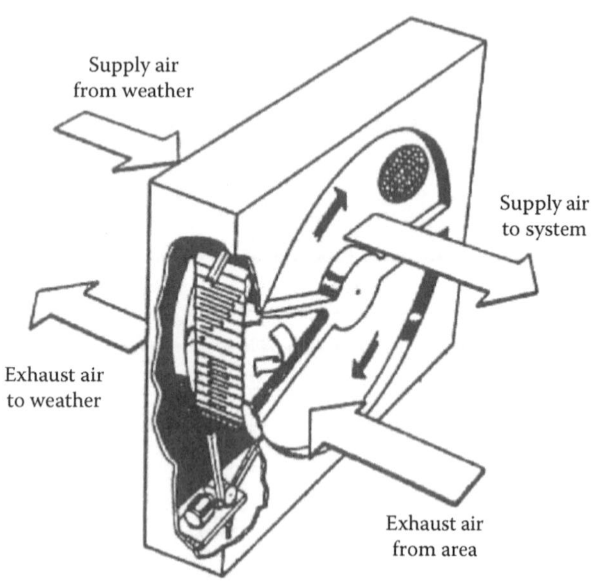

FIGURE 11.19
Heat recovery device. (From ASHRAE Systems, *HVAC System and Equipment Handbook*, American Society of Heating, Refrigerating and Air-Conditioning Engineers, Atlanta, GA, 2012.)

FIGURE 11.20
Large air-handler unit showing access door. (From Curtiss, P. and Breth, N., *HVAC Instant Answers*, McGraw-Hill, New York, 2002.)

the building. The airflow is modified using inlet vanes, outlet dampers, or variable speed drives on the fan. VAV systems have largely replaced CAV systems because of their superior energy efficiency, while the basic VAV system has also undergone mutation in recent years. A description and analysis of such systems and subsequent modifications such as fan-powered VAV systems are covered in Chapter 19. Other types of air distribution methods that are more energy efficient have also evolved in the last two decades or so. Underfloor air distribution and displacement ventilation are two such methods, and these are described in Section 20.5. In the ongoing effort to reduce energy use while improving indoor air quality, several hybrid designs have evolved in recent years, and these are also discussed in Sections 20.7 and 20.9.

As with other types of AHUs, the final configuration of the equipment depends on the amount of space available, the allowable roof loading, and the distance between the potential rooftop locations and the conditioned zone. The size of an air-handling unit varies greatly depending on the amount of air passing through the unit. Those that handle less than 5000 CFM (2400 L/s) are no more than large fan-coil units, often with barely enough room to change the filters. Medium-size units (up to about 20,000 CFM or 9,600 L/s) are the size of cars. Some AHUs are even larger (Figure 11.20) with full-size doors for access during maintenance.

11.6 District Systems

District heating and cooling systems are centralized systems where the heating and cooling are generated in one facility and distributed to several buildings. These have been in existence for decades and service the needs of downtown business districts of several large cities in the United States, state and federal governmental building clusters, as well as institutional and university campuses. Such systems are also widespread in Europe. The basic criterion for such systems to be cost-effective is that the cooling or heating load density (cooling load per unit area) should be relatively high. There are several

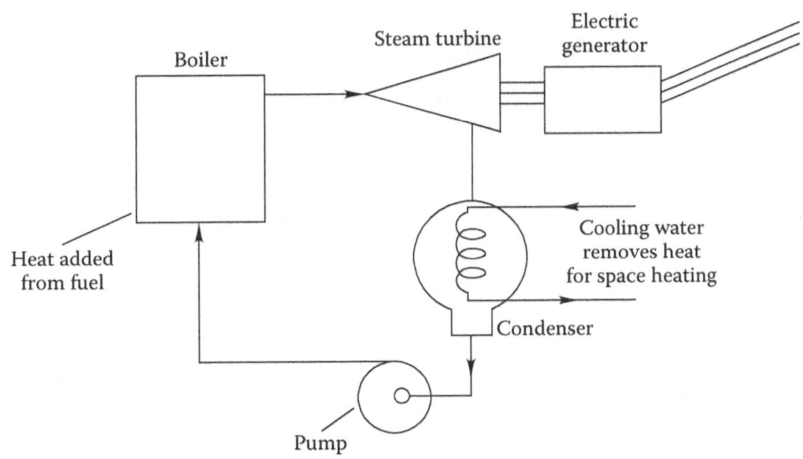

FIGURE 11.21
Schematic diagram of a cogeneration steam power plant for simultaneously producing electric power and heating needs of a building.

technologies that can improve the cost-effectiveness of such systems; for example, incorporating a thermal energy storage or a cogeneration power plant (see Figure 11.21). These two technologies are discussed in Sections 18.9 and 15.6, respectively.

A typical chilled-water distribution system consists of a *primary loop* and a *secondary loop* as shown in Figure 11.22. The primary loop is often used to obtain rough control of the water temperature and the valve to the secondary loop is used to maintain more precise temperature control of the distributed water. The boiler or chiller provides heating or cooling to the primary loop, and also controls its temperature. The three-way valve allows some of the water in the primary loop to enter the secondary loop. The valve modulates flow to maintain a pre-set temperature in the secondary loop. A *nulling loop* (sometimes called a *bridle* or *decoupler*) is provided so that if the three-way valve is completely closed there is still circulation in the secondary loop. The figure shows two loads, but in practice there can be dozens of loads. The loads are the heating or cooling coil requirements in the AHU or any other equipment that uses the hot or cold water.

The primary loop setpoint temperature is usually maintained in the range of 35°F–45°F (1.7–7.2°C), about 5°F–10°F (2.7–5.5°C) below the desired setpoint temperature range (40°F–50°F [4.4–10°C]) in the secondary loop. As the cooling load increases, the temperature of the return water in the secondary loop also increases. The temperature sensor records this as a rise in the secondary supply water temperature, and the temperature controller opens the three-way valve a little to allow more primary loop water into the secondary loop. As the three-way valve allows more and more of the secondary return water into the primary loop, the chiller stages up until the capacity of the chiller is met.

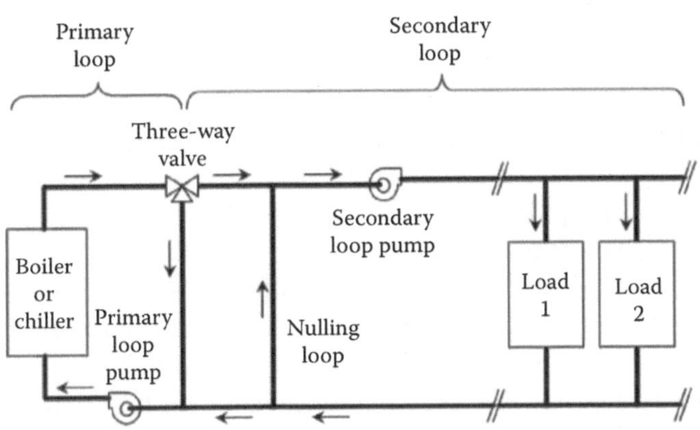

FIGURE 11.22
Basic water distribution loop.

FIGURE 11.23
Cut-away of a large fire tube boiler.

FIGURE 11.24
Photograph of a cooling plant with three large centrfugal chillers. The insulated piping are those carrying chilled water while the bare piping carry water to and back from the cooling towers. (From www.energy.iastate.edu, accessed on September 2014.)

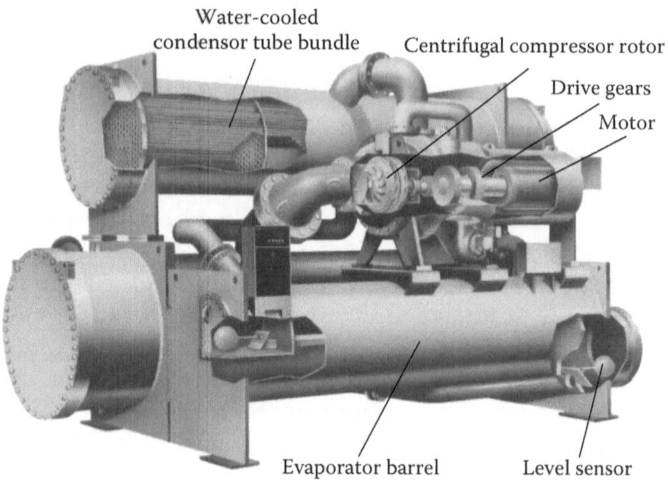

FIGURE 11.25
Centrifugal chiller cut away.

Boilers in such systems are very large, and resemble *industrial boilers* (Figure 11.23). Several boilers are usually installed with one or several coming online or offloaded as the building loads vary in time, i.e., diurnally, weekly, and annually. The same is true of the large multiple chillers, which form the cooling plant, as shown in Figure 11.24. The various components in the photograph can be identified from the cutaway depicted in Figure 11.25. The numerous pumps and distribution loops for these chillers supply chilled water to the primary loop (Figure 11.26), and also to the numerous cooling water towers (Figure 11.27). The cooling towers are usually sized to match individual chiller capacity, and so as many cooling towers are installed as the number of chillers. For ease in maintenance and ordering of spare parts, the norm is to select chillers, cooling towers, boilers, and pumps of the same size.

FIGURE 11.26
Photo of a large cooling plant with pumps and piping needed to circulate the chilled water to the building and the condenser water to the cooling towers. (From www.energy.iastate.edu, accessed on September 2014.)

FIGURE 11.27
Induced draft cooling tower system (four separate towers are shown).

Problems

Numbers 1–4 given in parenthesis denote the degree of difficulty.

11.1 What is the difference between primary and secondary systems? *(1)*

11.2 What is the difference between all-water and air-water secondary systems? *(1)*

11.3 Describe the purpose and functioning of the four main components of a vapor compression system. *(1)*

11.4 Figure 11.4 shows the various heat exchanger loops of a complete HVAC&R system. Redraw this system for the case of a DX cooling coil and air-cooled condenser. *(1)*

11.5 Discuss the applicability, advantages, and disadvantages of air-cooled and water-cooled cooling systems. *(2)*

11.6 What is an AHU? Describe the various components of an AHU. *(1)*

11.7 Discuss why all-air systems are so popular in North America. *(2)*

11.8 Describe the purpose and functioning of a cogeneration system. *(1)*

11.9 What type of energy storage is most common in buildings? What is its purpose? *(1)*

11.10 Discuss differences in HVAC systems used in small versus large buildings. *(2)*

References

ASHRAE Systems (2012). *HVAC System and Equipment Handbook*. American Society of Heating, Refrigerating and Air-Conditioning Engineers, Atlanta, GA.

Curtiss, P. and N. Breth (2002). *HVAC Instant Answers*. McGraw-Hill, New York.

O'Neal, D.L. and J.A. Bryant (2001). Chapter 4.2: Air conditioning systems, in *Handbook of Heating, Ventilation and Air Conditioning* (J.F. Kreider, ed.). CRC Press, Boca Raton, FL.

Pita, E.G. (2002). *Air Conditioning Principles and Systems*, 4th ed. Prentice Hall, Upper Saddle River, NJ.

12

Thermal Principles Relevant to Equipment and Systems

ABSTRACT The important thermodynamic properties and a basic statement of the two laws of thermodynamics were presented in Chapter 2. That material was adequate for the first 10 chapters of this book dealing with building loads. However, analysis of HVAC equipment and systems requires more in-depth understanding of thermodynamic and heat transfer principles. This chapter reviews some of the relevant topics that will aid in better appreciation of the remainder of the equipment-related chapters of this book. Specifically, we start by presenting different types of thermodynamic processes along with applications of the first law to closed and open processes (relevant to boilers, turbines, etc.). Next, we introduce the second law as applied to ideal power cycles, namely, the power, refrigeration, and heat pump cycles. This is followed by a treatment of phase diagrams and tabular data for determining state properties of pure substances, and an introduction to the characteristics of homogeneous binary mixtures. We then cover important dimensionless numbers (such as Reynolds, Nusselt, Prandtl, and Rayleigh) and appropriate heat transfer correlations meant to be applied under different flow regimes and common geometries. Finally, we present a basic introduction to the thermal performance of the ubiquitous heat exchanger, and discuss two general analysis and design approaches, the LMTD method and the effectiveness-NTU method.

Nomenclature

A	Area, m² (ft²)
COP	Coefficient of performance of a cooling device
$\dot{C}$	Heat capacitance of fluid stream, W/K [Btu/(h·°F)]
c_p	Specific heat at constant pressure, kJ/(kg·K) [Btu/(lb$_m$·°F)]
c_v	Specific heat at constant volume, kJ/(kg·K) [Btu/(lb$_m$·°F)]
D	Diameter, m (ft)
F	Heat exchanger correction factor (Equation 12.33)
Gr	Grashof number (Equation 12.17)
g	Acceleration due to gravity
H	Enthalpy, kJ (Btu)
H	Height, m
h	Specific enthalpy, kJ/kg (Btu/lb$_m$)
h	Heat transfer coefficient, W/(m²·K) [Btu/(h·ft²·°F)]
h_{con}	Convection heat transfer coefficient, W/(m²·K) [Btu/(h·ft²·°F)]
k	Thermal conductivity W/(m·K) [Btu/(h·ft·°F)]
LMTD	Logarithmic mean temperature difference (Equation 12.31), K (°F)
L	Length or characteristic length, m (ft)
m	Mass, kg (lb$_m$)
m	Exponent in Equation 12.19
$\dot{m}$	Mass flow rate, kg/s (lb$_m$/h)
Nu	Nusselt number (Equation 12.16)
NTU	Number of transfer units (Equation 12.37)
n	Polytropic process exponent
n	Exponent in Equation 12.19
P	Correction factor defined in Figure 12.14
PF	Performance factor of a heat pump
Pr	Prandtl number (Equation 12.15)
p	Pressure, Pa (lb$_f$/in.²)
Q	Energy consumption or heat transfer with subscripts c for cooling and h for heating, kJ (Btu)
$\dot{Q}$	Heat flow with subscripts c for cooling load, h for heating load, *lat* for latent load, *floor* for heat flow to floor, etc., W (Btu/h)
R	Thermal resistance, K/W [(°F·h)/Btu) [$\equiv \Delta T/\dot{Q}$]
R	Correction factor defined in Figure 12.14
Ra	Rayleigh number (Equation 12.18)
Re	Reynolds number (Equation 12.14)
r	Radius, m (ft)
S	Entropy, kJ/K (Btu/°R)
s	Specific entropy, kJ/(kg·K) [Btu/(lb$_m$·°R)]
T	Temperature, K or °C (°R or °F)
U	Overall heat transfer coefficient, W/(m²·K) [Btu/(h·ft²·°F)]
U	Internal energy, kJ (Btu)
u	Specific internal energy, kJ/kg (Btu/lb$_m$)
V	Volume, m³ (ft³)
V	Velocity or wind speed, m/s (mi/h, (ft/s)
v	Specific volume, m³/kg (ft³/lb$_m$)
W	Work, kJ (ft·lb$_f$)
$\dot{W}$	Power, W (ft·lb$_f$/s)
x	Quality of boiling fluid d (Equation 12.7)
x	Concentration of a binary mixture (Equation 12.13)

Greeks

α	Thermal diffusivity
β	Coefficient of volume expansion = $(1/T)$ for ideal gases (such as air)
γ	Adiabatic pressure exponent
ε	Heat exchanger effectiveness
η	Efficiency
μ	Dynamic or absolute viscosity, Pa·s [$lb_m/(ft \cdot s)$]
ρ	Density, kg/m^3 (lb_m/ft^3)
υ	Kinematic viscosity, m^2/s (ft^2/s)
Δx	Change or difference in quantity x

Subscripts

A, B	constituents of a homogeneous binary mixture
a	Ambient
c	Cold, cooling
comb	Combined
con	Convection
final	Final
h	Hot or high
hp	Heat pump
i	Internal or inlet or flowing in
ini	Initial
isen	Isentropic
isot	Isothermal
l	Low
liq	Liquid
max	Maximum
min	Minimum
o	Outlet or external or flowing out
sat	Saturated
tot	Total
vap	Vapor

12.1 First Law: Heat and Work Interactions

12.1.1 Recap

Sections 2.1 through 2.6 reviewed the basic fluid and thermodynamic concepts and principles such as important fluid properties, the concept of quasi-static processes, the ideal gas law, determining perfect gas property data from tables, formulations of the first law of thermodynamics as applied to closed and open systems, and a description of the second law.

Pure substances are those that are uniform and invariable in chemical composition. Two common examples are (1) atmospheric air that is a mixture of several gases, and (2) a single substance in more than one phase (mixture of water and water vapor, or a mixture of refrigerant consisting of liquid and vapor phases). The Gibbs phase rule specifies the number of independent properties needed to completely specify the state. The thermodynamic state of a pure single-substance fluid (gas or liquid) is characterized by its properties, and a thermodynamic process is one which consists of a series of quasi-static equilibrium states which change in a specified manner. A thermodynamic process is often accompanied by heat and work interactions between the system and the surroundings; with the sign convention that heat added to the system is positive, as is work done by the system (a historic legacy from the heat engine era).

For a thermodynamic cycle, the first law applied to closed systems (see Equation 2.32) is expressed in terms of cyclic integrals of heat input (Q) and work output (W):

$$\oint \delta Q = \oint \delta W \tag{12.1}$$

For *open systems* when the potential energies are negligible compared to the heat and work quantities, the steady-state form (when inflow fluid mass flow rate equals that of outflow) of the first law is simply given by (this follows from Equation 2.32)

$$\dot{m}\left(\frac{\mathrm{v}_i^2}{2} - \frac{\mathrm{v}_o^2}{2} + h_i - h_o\right) + \dot{Q} = +\dot{W} \tag{12.2}$$

where
- $\dot{m}$ is the fluid mass flow rate
- v is the fluid velocity and the fluid-specific enthalpy $h = u + pv$ where u is the internal energy and p and v are the fluid pressure and specific volume, respectively

12.1.2 Important Steady-State Processes

A process occurring inside a closed system is called a nonflow process. We can identify the following *nonflow cases* as summarized in Table 12.1 (see any thermodynamics book, for example, Cengel and Boles, 2000). Let a gas, assumed to be ideal, be contained in the piston-cylinder with states 1 and 2 representing the initial and final states. The general expansion process is shown in Figure 12.1 using pressure–volume coordinates. An infinitesimal change in volume dV results in work output represented by the shaded area dA. The work delivered by the system is then given by the total area under the curve as shown in Figure 12.1. The expressions for work delivered under different types of processes are given in Table 12.1. Since the thermodynamic process occurs slowly, the kinetic and potential energies are negligible, and so is the internal friction.

The general case corresponds to the polytropic process given by pV^n = constant. The exponent n is called the polytropic coefficient. During an isometric process no work is delivered. For a constant pressure process, $n = 0$, while for a constant temperature process, $n = 1$. Enthalpy remains constant in an isenthalpic process, such as in the expansion valve of a refrigeration cycle. An adiabatic process is one where the gas is expanded without adding or removing heat. If, in addition, the process occurs without any friction loss, the entropy remains constant, and this is called an isentropic process. For an adiabatic process, $n = \gamma = (c_p/c_v)$, i.e., the ratio of the two specific heats. Its numerical value is 1.66 for ideal monoatomic gases while $\gamma(air) = 1.4$ (air is predominantly a diatomic gas).

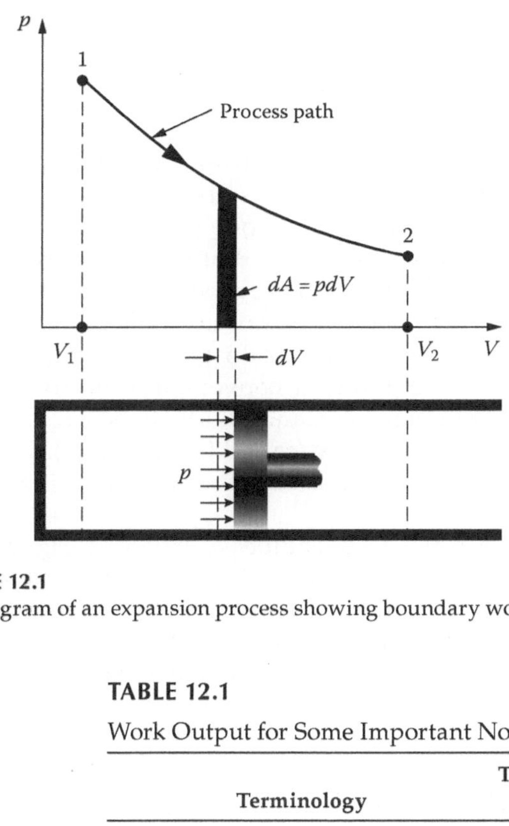

FIGURE 12.1
p–V diagram of an expansion process showing boundary work.

Example 12.1: Work Needed for Isothermal and Adiabatic Compression

Consider a piston-cylinder system containing 0.5 m³ of air at 100 kPa and 20°C, which is to be compressed to 0.1 m³ volume. Determine the work needed when the compression is isothermal and when it is isentropic (reversible adiabatic).

Given: $\gamma(air) = 1.4$, $p_1 = 100$ kPa, $V_1 = 0.5$ m³ and $V_2 = 0.1$ m³

Figure: See Figure 12.1.

Assumptions: Compression process is in quasi-equilibrium, ideal gas assumption for air.

Find: W_{isot} and W_{isen}

Solution

From Table 12.1, for isothermal compression,

$$W_{isot} = p_1V_1 \ln\left(\frac{V_2}{V_1}\right)$$

$$= 100 \text{ kPa} \times 0.5 \text{ m}^3 \times \ln\left(\frac{0.1}{0.5}\right)\frac{1 \text{ kJ}}{1 \text{ kPa}\cdot\text{m}^3} = -80.47 \text{ kJ}$$

TABLE 12.1
Work Output for Some Important Nonflow Processes for Ideal Gases

	Terminology	Type of Reversible Process	Exponent n	Work Output
1	Polytropic	General	—	$W = \int p\,dV = \frac{(p_2V_2 - p_1V_1)}{1-n}$
2	Isometric	Constant volume	—	$W = 0$
3	Isobaric	Constant pressure	$n = 0$	$W = p(V_2 - V_1)$
4	Isothermal	Constant temperature	$n = 1$	$W = p_1V_1 \ln\left(\frac{V_2}{V_1}\right)$
5	Isenthalpic	Constant enthalpy	—	$W = 0$
6	Isentropic (reversible adiabatic)	Constant entropy	$n = \gamma$	$W = \frac{p_1V_1^{\gamma}(V_2^{1-\gamma} - V_1^{1-\gamma})}{(1-\gamma)}$

The final and initial states are represented by subscripts 2 and 1, respectively.

For isentropic compression,

$$W_{isen} = \frac{p_1 V_1^{\gamma}(V_2^{1-\gamma} - V_1^{1-\gamma})}{(1-\gamma)}$$

$$= \frac{100\ \text{kPa} \times (0.5\ \text{m}^3)^{1.4} \times (0.1^{1-1.4} - 0.5^{1-1.4})}{(1-1.4)} \frac{1\ \text{kJ}}{1\ \text{kPa m}^3}$$

$$= -113.0\ \text{kJ}$$

Comments

The work is negative implying that work has to be done on the system. The work needed for the isentropic compression is greater than that needed for isothermal compression (in this example, by about 40%). This increase in work is significant, and impacts many HVAC components such as refrigerant compressors or air compressors. Hence, when the compression ratio is large, it is common design practice to use two stages of compression (i.e., a low-pressure compressor and a high-pressure compressor) with intercooling between the two stages.

Flow processes are those that occur in open systems when the fluid crosses the control volume, and the velocities may be significant enough to warrant their inclusion. In such cases, Equation 12.2 can be used. Note that the manner in which the control volume is drawn dictates whether the component or system is to be treated as an open or closed system. For example, even though the air-conditioning system of Figure 11.2 is a closed cycle, the compressor element consisting of the inlet and outlet piping and the component itself could be viewed as an open system since there is fluid flow into and out of the control volume.

Some important processes as they relate to HVAC equipment are shown in Table 12.2, where 1 and 2 designate the inlet and outlet state points.

12.2 Second Law Applied to Ideal Carnot Cycles

In the analysis of building energy systems, the second law of thermodynamics is most important for placing limits on the efficiency of all systems involving thermodynamic cycles. These cycles include heat pump and refrigeration cycles, which are discussed in this chapter. We will first discuss the *idealized cycles* assuming quasi-static or thermodynamic equilibrium to exist at all times (i.e., the processes proceed very slowly so that all parts of the process are reversible). This is not true of course in practice, and real processes and systems are treated in Chapter 14 for cooling and heat pump cycles. However, the reversible paradigm offers considerable insight into important thermal cycles.

12.2.1 Power Cycle

A French scientist, Sadi Carnot, was the first person to appreciate the limitations placed in converting heat energy to work, which arose in the historic context of heat engines. He found that while all work can be converted into heat, the reverse does not hold; thus, all forms of energy are not created equal. This led to two developments: the formulation of an expression for the maximum conversion efficiency (called the Carnot efficiency), and an idealized cycle that can achieve this efficiency (called Carnot cycle). These are independent of any actual thermodynamic working fluid.

Consider the heat engine sketched in Figure 12.2a that receives a certain amount of heat Q_h at a high-temperature source T_h and converts a portion of the energy into work W_{net} while discharging the rest to a low-temperature source at temperature T_l. Then, the

TABLE 12.2

Flow Processes in Important HVAC Equipment

	Equipment	Process	Heat Input/Work Input	Comments
1	Boiler (water) and evaporators (refrigeration)	Heat input causes fluid to change from liquid to vapor.	$\dot{Q} = \dot{m}(h_2 - h_1)$	KE and PE effects negligible.
2	Condensers (water or refrigeration)	Heat rejection causes fluid to change from vapor to liquid.	$\dot{Q} = -\dot{m}(h_2 - h_1)$	"
3	Cooling towers	Heat rejection causes water to cool down.	$\dot{Q} = -\dot{m}(h_2 - h_1)$	"
4	Compressors in refrigeration	Work needed to compress vapor to a high pressure.	$\dot{W} = -\dot{m}(h_1 - h_2$	Irreversibility due to fluid friction cannot be generally ignored; PE negligible.
5	Turbine or reciprocating engine	Work delivered due to expansion of fluid from high pressure and temperature.	$\dot{W} = \dot{m}\left(\frac{v_1^2}{2} - \frac{v_2^2}{2} + h_1 - h_2\right)$	"

PE, potential energy; KE, kinetic energy.

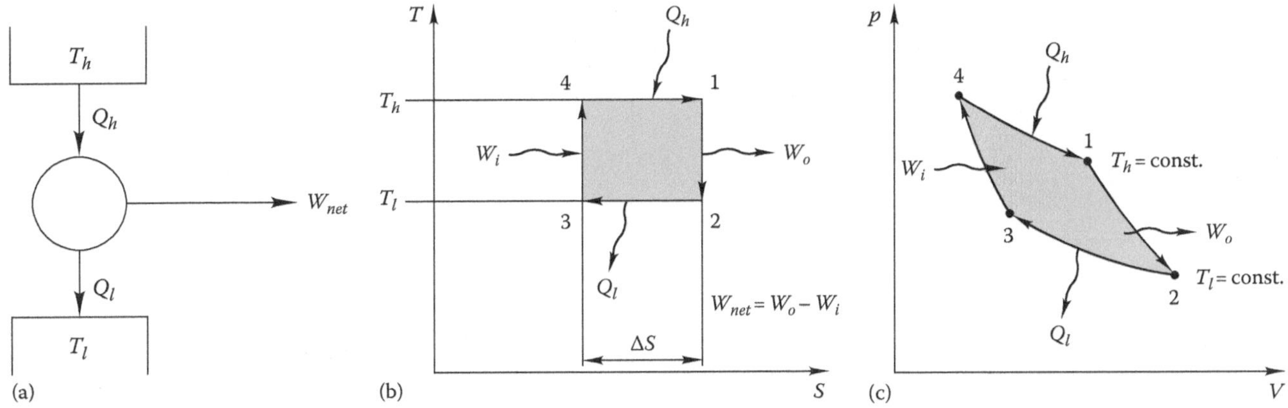

FIGURE 12.2
(a–c) Schematic diagram of Carnot heat engine with cycle plotted on T–S and p–V coordinates.

Carnot cycle efficiency (refer to any standard thermodynamics text) is given by

$$\eta_{Carnot} \equiv \frac{W_{net}}{Q_h} = \frac{T_h - T_l}{T_h} \tag{12.3}$$

where the temperatures are in absolute units. This expression places an upper limit on the efficiency at which a heat engine can convert heat to work when the source and sink temperatures are specified. Though it is an unattainable ideal, it is used as a standard of comparison for actual conversion cycles.

The Carnot heat engine cycle consists of four reversible processes as shown in two different sets of property coordinates: the temperature–entropy (*T–S*) and the pressure–volume (*p–V*) (see Figure 12.2b and c). Beginning at the upper left-hand corner and moving clockwise, the following applies:

1. *Reversible isothermal expansion*: Heat (Q_h) is added to the working fluid at the source temperature T_h while the gas expands and produces work with the process occurring isothermally.
2. *Reversible adiabatic expansion*: Heat input is discontinued and the adiabatic expansion is allowed to continue thereby producing work (W_o).
3. *Reversible isothermal compression*: Heat (Q_l) is rejected to a low-temperature reservoir T_l while providing work to compress the gas so that the process occurs isothermally.
4. *Reversible adiabatic compression*: Work (W_i) is consumed in compressing the fluid adiabatically to its initial condition.

The shaded area on the *T–S* diagram in Figure 12.2b and that on the *p–V* diagram in Figure 12.2c represent the net work output of the cycle.

12.2.2 Refrigeration Cycle

A Carnot refrigerator is one that is operated in the opposite direction to the Carnot heat engine as shown in Figure 12.3. The work input to the cycle W_{net} has the net effect of removing heat Q_l from the low-temperature reservoir at T_l and rejecting it to the high-temperature reservoir at T_h. Note that from the first law, the heat rejection $Q_h = Q_t + W_{net}$. For example, in summer, heat can be extracted from the indoors, "pumped up a temperature gradient," and rejected to the hotter outdoor air. Since this process cannot occur naturally, a thermodynamic cycle is required with some amount of external work needed to operate it. The amount of work needed is shown by the shaded area on the *T–S* diagram in Figure 12.3b and also by that on the *p–V* diagram in Figure 12.3c.

If one applies the first law to each process in the refrigeration cycle shown in Figure 12.3, the efficiency of the process is easily determined. Surprisingly, the efficiency is found to be greater than 1. In keeping with common engineering practice, the widespread convention is to reserve the term *efficiency* for a ratio of effect to cause less than 1 and when energy input and output are of the same form of energy (such as in a boiler for example). As a result, for refrigeration and cooling cycles, the term *coefficient of performance* (*COP*) is used as the figure of merit. The Carnot cooling* COP is

$$\text{COP}_{Carnot,c} \equiv \frac{Q_l}{W_{net}} = \frac{T_l}{T_h - T_l} \tag{12.4}$$

* A dimensional number called the *energy efficiency ratio* (*EER*) is often used in the United States instead of the dimensionless COP. See Section 14.6 for details.

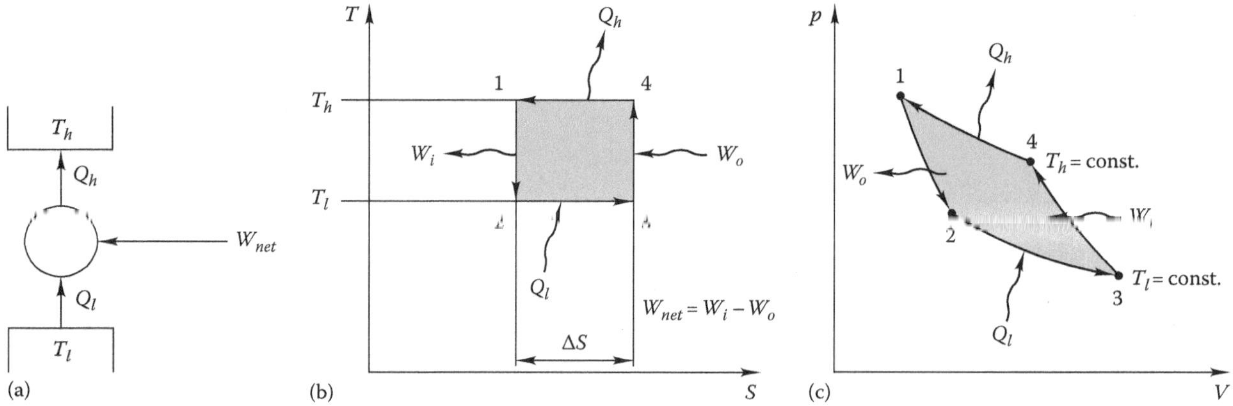

FIGURE 12.3
(a–c) Schematic diagram of Carnot refrigerator with cycle plotted on *T–S* and *p–V* coordinates.

Again, the temperatures must be in absolute units. Inspection of Equation 12.4 suggests that $COP_{Carnot,c}$ can be enhanced by increasing the low temperature T_l and/or reducing the high temperature T_h. However, these cannot be controlled at will but are imposed by the circumstances. For example, the highest value of T_l is dictated by the temperature at which the cold space or product needs to be maintained, while the lowest value of T_h is dictated by the sink temperature, i.e., the temperature to which heat is to be rejected (often, the ambient dry-bulb temperature). Real refrigeration cycles and chiller systems are discussed in Sections 14.1, 14.2 and 14.4.

12.2.3 Heat Pump Cycles

The heat pump is a machine that uses the same equipment as a refrigerator and has a cycle that looks exactly like the cooling refrigeration cycle. Instead of cooling being the desired effect, it is intended to supply heat Q_h to the higher temperature indoors (T_h) by pumping a certain amount of heat Q_l from a low temperature T_l. Thus, the cycle shown in Figure 12.3 applies to the heat pump cycle as well, but the definition of the COP differs because now the heat added to the building (the high-temperature reservoir) is the desired effect. Therefore, the numerator in Equation 12.4 is replaced with Q_h, and the COP expression becomes

$$COP_{Carnot,h} = PF_{Carnot,hp} \equiv \frac{Q_h}{W_{net}} = \frac{Q_l + W_{net}}{W_{net}}$$
$$= COP_{Carnot,c} + 1 = \frac{T_h}{T_h - T_l} \qquad (12.5)$$

The convention is not to use COP to express the efficiency of a heat pump, but to use the term *performance factor* (PF). This convention is not universally followed, however, and using COP is also acceptable. Again, the conditions that maximize PF_{hp} are to maintain T_l as high as possible and T_h as low as possible. Real heat pumps are discussed in Section 14.5.

Example 12.2: Carnot Heat Pump

If a Carnot heat pump operates between an outdoor temperature of −10°C and a building interior temperature of 22°C, what is the COP?

Given: $T_l = 273 + (-10) = 263$ K, $T_h = 273 + 22 = 295$ K

Figure: See Figure 12.3.

Assumptions: Reversible, Carnot cycle

Find: $PF_{Carnot,hp}$

Solution

The last part of Equation 12.5 gives the solution directly:

$$PF_{Carnot,hp} = \frac{295}{295 - 263} = 9.22$$

Comments

If we were able to operate a heat pump of this efficiency each kW of electricity used to operate the heat pump (W_{net} in Figure 12.3) would result in 9.22 kWh of heat being provided to the building's interior. Of course, practical heating equipment cannot achieve PFs of this level for reasons we shall discuss in greater detail in Chapter 14.

12.2.4 Combined Carnot Heat Engine and Refrigerator

The refrigeration and heat pump cycles discussed earlier require work to be done on the system. This work, often in the form of electricity, has to be generated in a power plant. Let us consider the ideal combined cycle

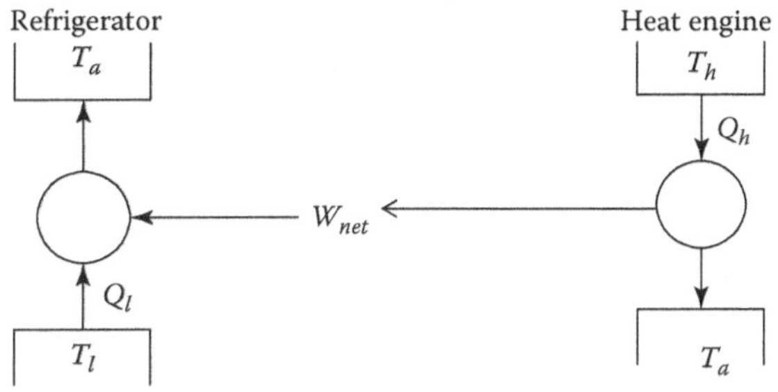

FIGURE 12.4
Schematic diagram of combined Carnot heat engine and refrigerator.

shown in Figure 12.4 where the work needed (W_{net}) to operate the ideal Carnot refrigeration cycle between high and low temperatures T_a and T_l is generated by a Carnot power cycle operating between high and low temperatures T_h and T_a. The combined effect of both cycles is that we now have a system that takes Q_h as heat input at T_h and produces cooling Q_l at temperature T_l. Note that the power cycle and the refrigeration cycles both reject heat to T_a, which could be the ambient sink temperature. The COP of the combined cycle is then

$$\mathrm{COP}_{Carnot,comb} \equiv \frac{Q_l}{Q_h} = \frac{W_{net}}{Q_h} \cdot \frac{Q_l}{W_{net}}$$

$$= \frac{T_h - T_a}{T_h} \frac{T_l}{T_a - T_l} = \frac{T_l}{T_h}\left(\frac{T_h - T_a}{T_a - T_l}\right) \tag{12.6}$$

It is clear that in order to enhance $\mathrm{COP}_{Carnot,comb}$, T_h and T_l have to be kept as high as possible while T_a has to be kept as low as possible. The earlier expression is relevant for policy studies involving proper accounting of energy conversion from a societal (or source energy) perspective as against from the building owner (or site energy) point of view. From a societal point of view, Q_h is associated with the primary energy burnt at the electric power plant that produces a cooling effect Q_l at the building level.

Example 12.3: Ideal Combined Cycle

We will assume the same operating conditions as in Example 12.2 with the high source temperature of the power cycle being 457°C. Compare the COP of the ideal refrigeration cycle and that of the combined power-refrigeration cycle.

Given: $T_l = 273 + (-10) = 263$ K, $T_a = 273 + 22 = 295$ K, $T_h = 273 + 457 = 730$ K

Figure: See Figure 12.4.
Assumptions: Reversible Carnot cycle

Find: $\mathrm{COP}_{Carnot,c}$ and $\mathrm{COP}_{Carnot,comb}$

Solution

Directly using Equation 12.4 yields

$$\mathrm{COP}_{Carnot,c} = \frac{T_l}{T_a - T_l} = \frac{263\ \mathrm{K}}{(295 - 263)\ \mathrm{K}} = 8.22$$

Also, from Equation 12.6,

$$\mathrm{COP}_{Carnot,comb} = \frac{T_l}{T_h}\left(\frac{T_h - T_a}{T_a - T_l}\right) = \frac{263\ (730 - 295)\ \mathrm{K}}{730\ (295 - 263)\ \mathrm{K}} = 4.90$$

Comments

The ideal COP of the combined cycle is much lower (about 60%) than that of the vapor compression cycle. This example serves to illustrate the large differences to be expected in energy efficiency studies that deal with either site energy or source energy. The second law efficiency is a better indicator in both instances, but especially so for the latter.

12.3 Pure Substances

Heating and cooling processes that do not result in a phase change are called sensible heating or cooling processes. However, heat addition or removal often results in substances changing phase, i.e., passing from liquid to vapor and vice versa. Understanding the behavior of substances such as steam (or refrigerants) during their transition between liquid and vapor phases is very important in air-conditioning and refrigeration systems as well as in humid air processes (discussed in Chapter 13). The ideal gas law cannot be used during such cases, and one must use property tables or charts specific to the substance.

12.3.1 Phase Diagrams of Water Vapor and Steam

As a pure substance, say liquid water, is heated at constant pressure, it passes from the liquid region (called subcooled) to the liquid–vapor region (called wet region or mixed phase) to the vapor region (called superheated). Figure 12.5 shows a pressure–volume diagram (p–v diagram) for water including the superheated region where liquid–vapor (dynamic) equilibrium conditions can exist. Two lines of constant temperature T_1 and T_2 are shown. The shape of the dome reflects the common experience that water boils at a higher temperature as the pressure is increased.

According to the Gibbs phase rule (see Section 2.2.1), two independent properties are needed to specify a pure substance (such as water or a refrigerant) in the

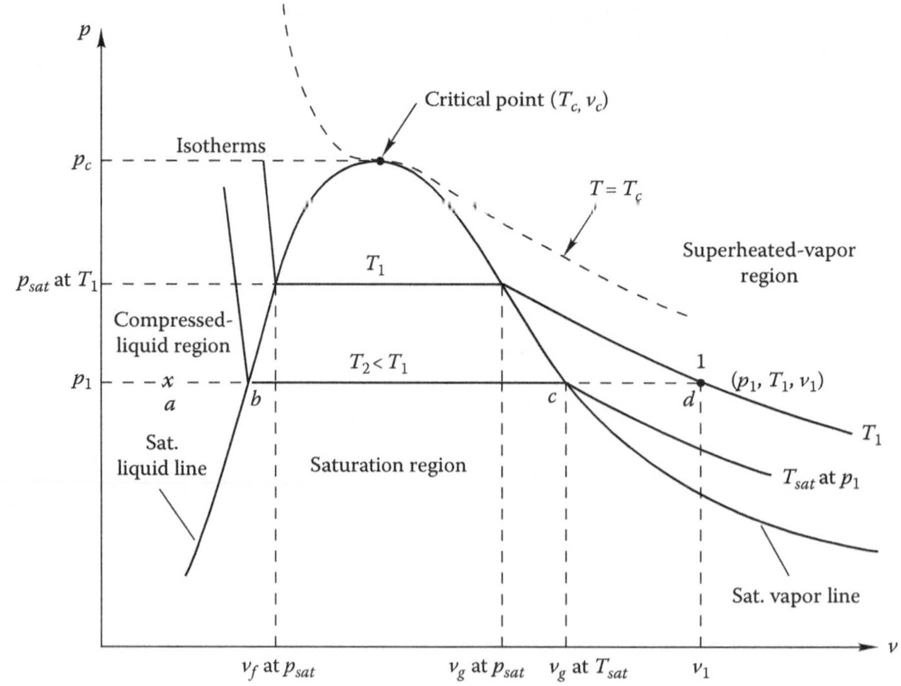

FIGURE 12.5
Pressure–volume diagram for water.

subcooled or superheated regions where only one phase exists, while one property is needed during the mixed phase region.

The constant pressure heating process is shown by points *a–b–c–d*. The temperature T_2 during process *b–c* remains constant and we note the large change in specific volume $v_{vap} \gg v_{liq}$. The key feature of this diagram is the shape of the saturation curve and the regions defined by it. To the left of the critical point and the saturation curve, only compressed liquid water exists; to the right of the saturation curve, only superheated vapor is present. In the liquid region, properties are primarily a function of temperature; pressure has a secondary effect. Conditions precisely on the saturation curve are called *saturated-liquid conditions* (100% liquid) to the left of the critical point and *saturated-vapor* (100% vapor) to the right. The peak of the dome is the critical point 374.1°C (705.4°F) and 22.09 MPa (3206.2 psia) above which a clear distinction between liquid and vapor phases cannot be made.

Beneath the saturation curve, "wet steam" or mixed phase conditions exist. Wet steam is a mixture of saturated liquid and water vapor. Since pressure and temperature are not independent properties beneath the saturation curve, a separate thermodynamic condition called the *quality* x is defined as

$$x = \frac{m_{vap}}{m_{tot}} \tag{12.7}$$

Since the mixture contains only liquid and vapor, the mass that is not vapor must be liquid. Hence,

$$\frac{m_{liq}}{m_{tot}} = 1 - x \tag{12.8}$$

The quality is used frequently to find thermodynamic properties such as internal energy, specific volume, enthalpy, and entropy for wet steam mixtures. For example, one would use the quality to find the specific volume of wet steam as follows. By definition, the specific volume of steam is the volume of liquid water plus the volume of water vapor divided by the total mass. From the definition,

$$v = \frac{V}{m_{tot}} = \frac{m_{liq}v_{liq} + m_{vap}v_{vap}}{m_{tot}} \tag{12.9}$$

in which the subscript *liq* refers to the liquid phase and *vap* to the vapor (or gaseous) phase.

Equation 12.9 can also be written as:

$$v = v_{liq}(1 - x) + xv_{vap} \tag{12.10}$$

If we define the phase-change specific volume as

$$v_{liq-vap} = (v_{vap} - v_{liq})$$

then Equation 12.10 can be reexpressed as

$$v = v_{liq} + x v_{liq-vap} \tag{12.11}$$

Mass-weighted equations analogous to Equations 12.10 and 12.11 can be derived for other properties such as internal energy, enthalpy, and entropy for the mixed phase region. These will be used in later chapters for the analysis and design of chillers, boilers, and heat pumps.

12.3.2 Determining Property Values

Steam, because of its high latent heat, is widely used to provide heating to air. In the majority of conditions, it cannot be treated as an ideal gas. In fact, there is no convenient equation for finding the properties of steam. As a result, one must resort to the steam tables for properties of wet and superheated steam.

Complete steam tables need an entire book (for example, Keenan et al., 1978). However, since we are interested in a relatively restricted portion of the full range of properties, simplified tables are adequate. Properties of liquid and saturated water vapor are given in Table A3 in the Appendix based on saturated pressure (first column) or the corresponding temperature (second column). The specific volume of saturated liquid and vapor, specific enthalpy, and specific entropy at the saturated liquid condition and at the saturated vapor condition are listed in separate columns as shown. Separate tables are also available to determine properties of superheated vapor (an excerpt is shown in Table A4 as an illustration). Here, two properties are needed to specify the state, and so several mini-tables for different pressures are assembled together for different temperature values. The specific volume, internal energy, enthalpy, and entropy values for different combinations can be read off from these mini-tables. Linear interpolation is often adequate for determining property values falling between these discrete values of pressure and temperature.

Example 12.4: Wet Steam Properties

Find the temperature, specific volume, and enthalpy of wet steam if the quality is 10% (very wet steam). The pressure is 150 kPa (0.15 MPa).

Given: $x = 0.10$, $p_{tot} = 150$ kPa

Assumptions: Thermodynamic equilibrium exists.

Find: T, v, h

Lookup values: Use Table A3 for the properties of saturated liquid and water vapor at 150 kPa: $T = 111.4°C$, $v_{liq} = 0.001\ m^3/kg$, $v_{vap} = 1.159\ m^3/kg$, $h_{liq} = 467.11$ kJ/kg, $h_{vap} = 2693.6$ kJ/kg

Solution

Use Equation 12.11 to find the specific volume and enthalpy. Finding the temperature is just a matter of looking up the saturation temperature corresponding to the stated pressure, i.e., $T = 111.4°C$ (this has already been completed during the lookup value determination).

For the specific volume, we have

$$\begin{aligned} v &= v_{liq} + x v_{liq-vap} \\ &= 0.001 + 0.10 \times (1.159 - 0.001) = 0.117\ \text{m}^3/\text{kg} \end{aligned}$$

Likewise, for the enthalpy,

$$\begin{aligned} h &= h_{liq} + x h_{liq-vap} \\ &= 467.11 + 0.1 \times (2693.6 - 467.1) = 689.8\ \text{kJ/kg} \end{aligned}$$

When heat engines using water as the working fluid are analyzed, the *temperature–entropy* (*T–S*) diagram is most convenient since isentropic processes will appear as vertical lines on such a plot. Such a *T–S* plot for water is depicted by Figure 12.6. The shape of the saturation curve is similar to that in the *p–V* diagram. The corresponding regions—superheated, saturated, wet, and compressed liquid—are located similarly to those regions in the *p–V* diagram.

Example 12.5: Using Superheat Tables

Water at the rate of 50 kg/s is to be heated, vaporized, and superheated in a pressurized boiler from 20°C to 150°C at a pressure of 0.3 MPa. If the boiler efficiency is 90%, determine the amount of heat input to the boiler?

Given: $T_i = 20°C$ (subcooled), $T_o = 150°C$ (superheated), $\eta = 0.9$

Assumptions: The enthalpy of subcooled liquid is equal to that of the saturated liquid at the same temperature.

Lookup values: From Table A4 ($p = 0.3$ MPa and $T = 150°C$), enthalpy $h_o = 2761.0$ kJ/kg and from Table A3 (saturated steam table for $T = 20°C$), $h_i = 80$ kJ/kg (interpolated).

Find: Q_{in}

Solution

The solution is found directly from the enthalpy equation for boilers given in Table 12.2:

$$\begin{aligned} \dot{Q}_{in} &= \dot{m}(h_o - h_i)/\eta = 50\ \text{kg/s} \times (2761.0 - 80)\ \text{kJ/kg}/0.9 \\ &= 148.9\ \text{kW} \end{aligned}$$

For computer calculations, several empirical correlations exist with varying degrees of accuracy

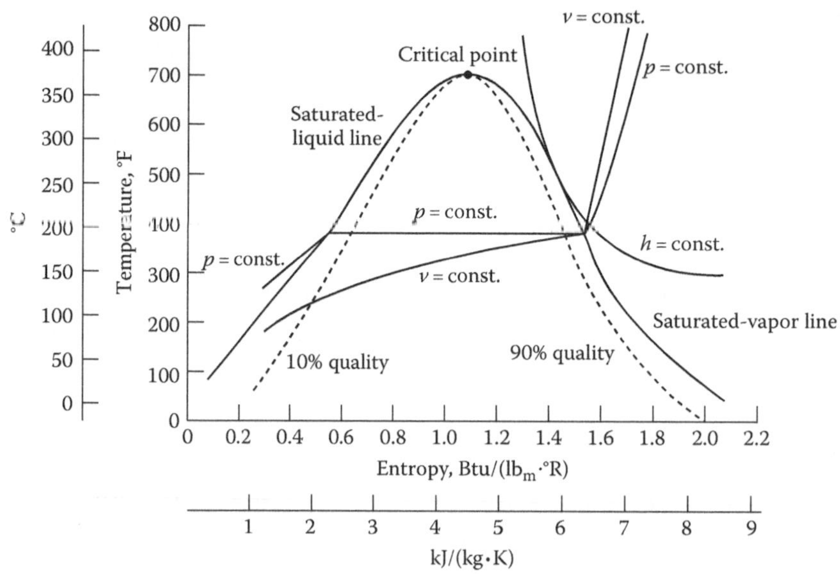

FIGURE 12.6
Temperature–entropy diagram for water.

and associated complexity. One widely used correlation used to determine the water vapor saturation pressure over liquid vapor for the temperature range of 0°C–200°C (32°F–392°F) is given by (ASHRAE Fundamentals, 2013)

$$\ln p_{sat} = a/T + b + c \cdot T + d \cdot T^2 + e \cdot T^3 + f \cdot \ln T \quad (12.12)$$

where
T is in absolute units (K or °R)
p_{sat} is in Pa or in psia

The numerical values of the coefficients are given in Table 12.3.

12.3.3 Refrigerants

Refrigerants are the working fluids in air conditioners, chillers, and refrigerators. The majority are chlorinated and fluorinated hydrocarbons (CFCs), although a number of other organic and inorganic compounds can be used. With the known deleterious effects of some conventional CFCs on the integrity of the earth's stratospheric ozone, the use of these compounds is being significantly curtailed from the last two decades. The two most widely used, harmful CFCs are refrigerant 11 (R11) and refrigerant 12 (R12). Both R11 and R12 are being phased out in favor of R123 and R134a, respectively. In this section, we will use R134a as the example refrigerant, although the reader is cautioned that there are still many systems that use R22 and will continue to do so as long as the supply does not get too expensive. Refrigerants are discussed at more length in Section 14.11.

Refrigerants are not ideal gases under the conditions present in cooling equipment. As a result, we cannot use the ideal gas approach to find thermodynamic state coordinates, and tables have to be used instead. Saturated and superheat thermodynamic properties of R22 and R134a are given in Tables A5 through A8. These tables contain the same sort of data as the previous sample steam tables.

A useful design tool for refrigeration cycle analysis is a pressure–enthalpy (p–h) diagram. In Sections 14.1 and 14.2, the reason for this will become apparent. The basic advantage of the p–h diagram is that three of the four steps in a basic refrigeration cycle can be plotted as straight lines. Of course, thermodynamic properties can be read directly from the chart instead of looking them up in a table. Figure 12.7 is the p–h diagram for R134a. The online HCB software contains thermodynamic property tables and charts (both p–h and T–S) for refrigerants 11, 12, 13, 14, 22, 114, 134a, 500, and 502. The reader is encouraged to try out the software to quickly compare the p–h and T–S diagrams for several refrigerants.

TABLE 12.3
Values of Coefficients Appearing in Equation 12.12

Coefficient	SI Units	IP Units
a	-5.8002206×10^{3}	-1.0440397×10^{4}
b	1.3914993	-1.1294650×10^{1}
c	$-4.8640239 \times 10^{-2}$	$-2.7022355 \times 10^{-2}$
d	4.1764768×10^{-5}	1.2890360×10^{-5}
e	$-1.4452093 \times 10^{-8}$	$-2.4780681 \times 10^{-9}$
f	6.5459673	6.5459673

Example 12.6: R134a Properties

Find the enthalpy change in a process in which 38 lb_m of R134a at 80% quality is heated to 300°F in a constant-pressure process at 200 psia ($lb_f/in.^2$ absolute).

Given: $p_{ini} = 200$ psia, $x_{ini} = 0.80$

Figure: Figure 12.7, the *p–h* diagram for R134a

Assumptions: Thermodynamic equilibrium exists.

Find: The enthalpy change ΔH

Solution

The solution to this problem is most quickly done by using the *p–h* diagram since the process is just a horizontal line at 200 psia with the left endpoint at the 80% quality-line intersection. The right end of the process line is at the intersection of the 200 psia line and the 300°F line. The enthalpy at the left endpoint can be approximated by first finding the enthalpies at saturated liquid (55 Btu/lb_m) and saturated vapor (120 Btu/lb_m) and then interpolating. The enthalpy at the right endpoint can be read directly from the *p–h* diagram.

Thus, the left and right endpoints are

$$h_1 = 55 + (120 - 55) \times 0.8 = 107 \text{ Btu/lb}_m$$

and

$$h_2 = 165.0 \text{ Btu/lb}_m$$

The enthalpy change is the product of the enthalpy change per unit mass and the mass:

$$\Delta H = 38 \text{ lb}_m \times (165 - 107) \text{ Btu/lb}_m = 2204 \text{ Btu}$$

Comments

For a more accurate calculation, the reader may use the saturated table (Table A6) and the superheated R134a tables (Table A8).

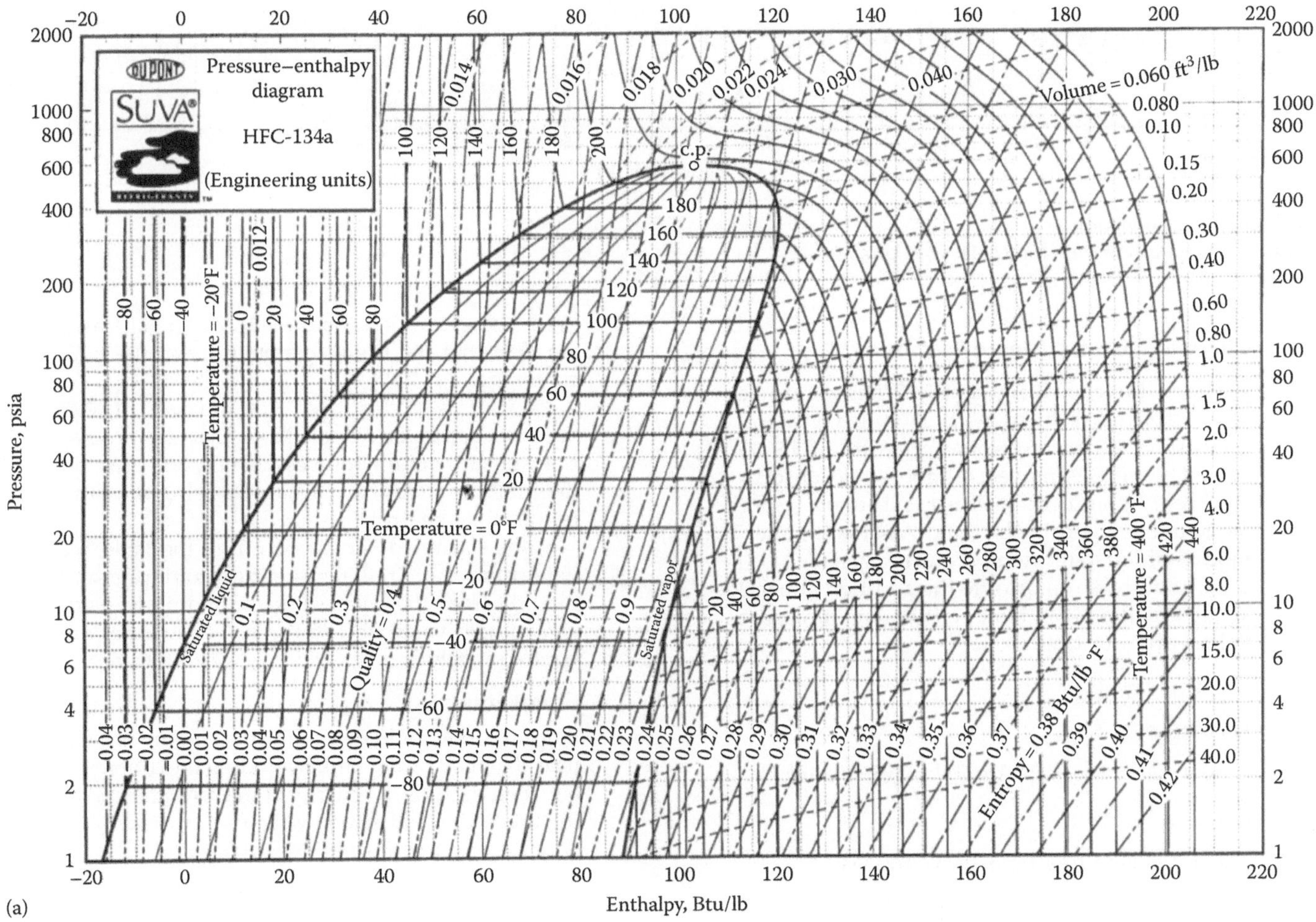

(a)

FIGURE 12.7
Pressure–enthalpy diagram for R134a. (a) IP units.

(Continued)

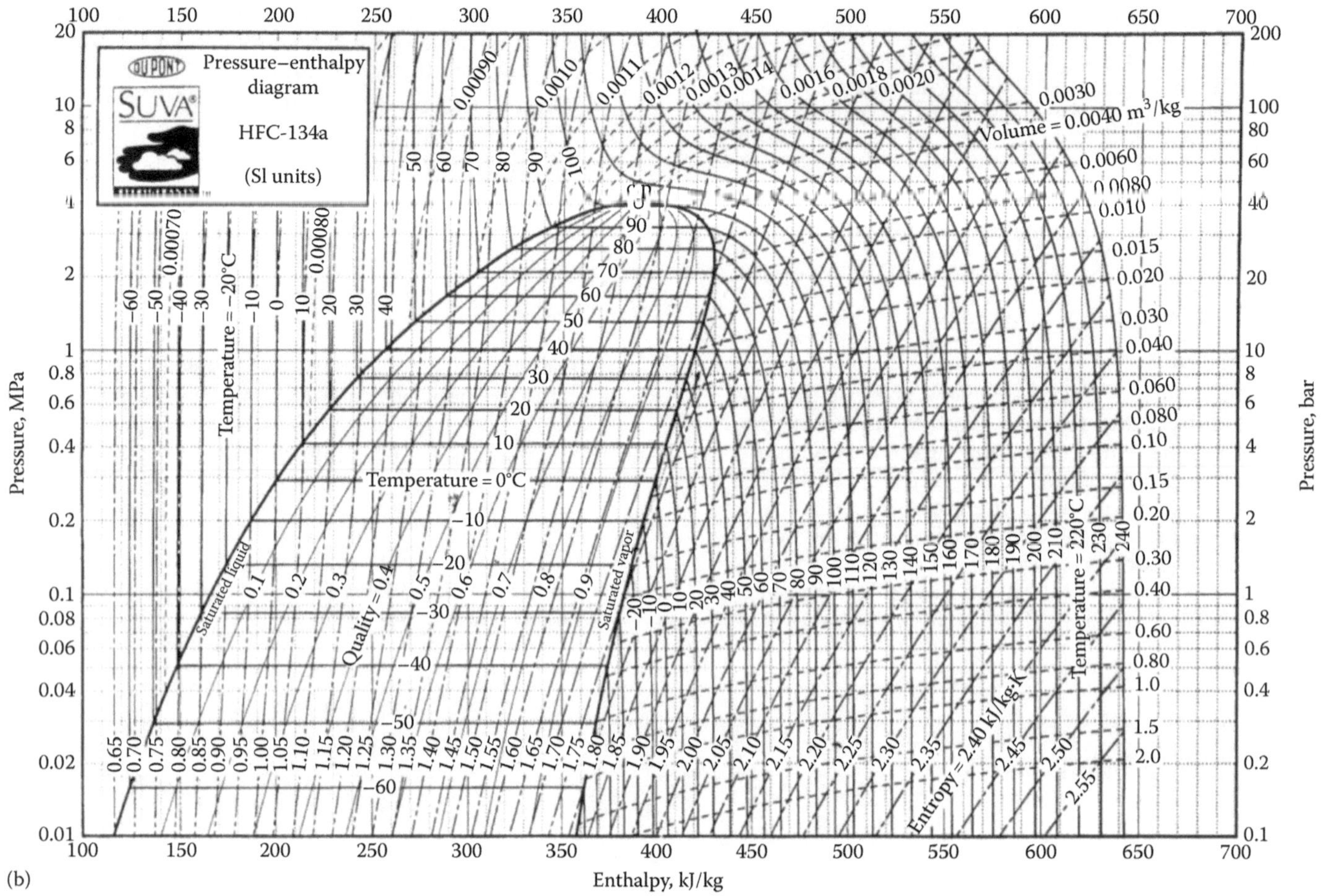

FIGURE 12.7 (*Continued*)
Pressure–enthalpy diagram for R134a. (b) SI units.

Example 12.7: First Law of Thermodynamics: Closed System

A closed tank of refrigerant R22 has a volume of 700 ft³ and is filled with 2800 lb_m of a wet mixture of R22 at an initial pressure of 136.19 psia. The tank is moved outdoors and absorbs sufficient solar heat to cause saturated vapor alone to exist within the tank. What is the thermodynamic state (temperature, specific volume, and pressure) prior to the addition of heat? What are the final temperature and pressure values after heating?

Given: V, m, p_{ini}

Find: $v_{ini}, p_{final}, T_{final}$

Lookup values: The R22 table (Table A5) is used.

Solution

The initial (and final) specific volume is

$$v = \frac{700\ \text{ft}^3}{2800\ \text{lb}_\text{m}} = 0.25\ \text{ft}^3/\text{lb}_\text{m}$$

From the saturation table for R22 at 136.19 psia, we read that the initial temperature of the tank contents is $T_i = 70°F$.

As heat is added, the wet mixture of R22 becomes saturated vapor. Note that the process is a constant-volume process; hence, one can find the final thermodynamic state by using the saturation tables again.

In the saturated vapor column of the table, we determine the final temperature and pressure directly from the entries corresponding to a *saturated vapor* value of 0.25 ft³/lb_m. The values are approximately $T_{final} = 102°F$, $p_{final} = 216.45$ psia.

Comments

If needed, one could determine the amount of solar heat absorbed by the tank by finding the difference between the internal energy at the start and end of the process, as indicated by the first law (the work done is zero since the volume is constant):

$$Q = \Delta U$$

The interested reader can verify that the heat absorbed is approximately 79 kBtu.

12.4 Homogeneous Binary Mixtures

Binary mixtures are formed by combining two pure substances. They can be of two types:

1. A *homogeneous mixture* is one that is uniform in composition and cannot be separated into its constituent parts by purely mechanical means such as settling or centrifuging. The various properties such as density, pressure, and temperature are uniform throughout the mixture. A common example is dry air and water vapor mixture.
2. A *heterogeneous mixture* is one whose composition is nonuniform (an example is fog) and whose constituent parts can be separated by ordinary mechanical means.

Homogenous binary mixtures involving liquid and vapor phases arise in absorption refrigeration systems (Section 14.3); and so some of their basic characteristics are presented here. Such mixtures are also involved in industrial and food processing industries where say a substance A is to be separated from another substance B by distillation.

The thermodynamic state of a binary mixture comprising of two substances A and B can be determined if three independent properties are specified. Usually, the mass fraction x_A (or x_B) along with p and T are used where

$$x_A = m_A/(m_A + m_B) \tag{12.13}$$

A temperature–concentration diagram of a binary mixture is shown in Figure 12.8a for a given pressure. How the diagrams are displaced with pressure is illustrated in Figure 12.8b. Binary mixtures, contrary to pure substances, do not have a single boiling or condensation points. The temperature where the two curves on the left axis meet corresponds to the boiling point of pure substance B at the corresponding pressure, while that on the right axis to the boiling point of substance A. Obviously, substance A is more volatile of the two since it has the lower boiling temperature. We can distinguish three regions. The upper region corresponds to superheated vapor, the region between the two curves to the liquid–vapor region where both vapor and liquid coexist, and the low region corresponds to the subcooled liquid state. The separation lines are often referred to as the dew-point line (or condensing line) and the bubble-point line (or the boiling line) as indicated.

Consider a subcooled mixture of concentration $x_{B,1}$ specified by point 1 (see Figure 12.9) that is being heated up at constant pressure. The concentration of the liquid mixture remains unaltered till the boiling temperature T_2 is reached. Any further heat addition results in a mixture of liquid and vapor (such as state point 2′). The mass concentration of substance B in the liquid

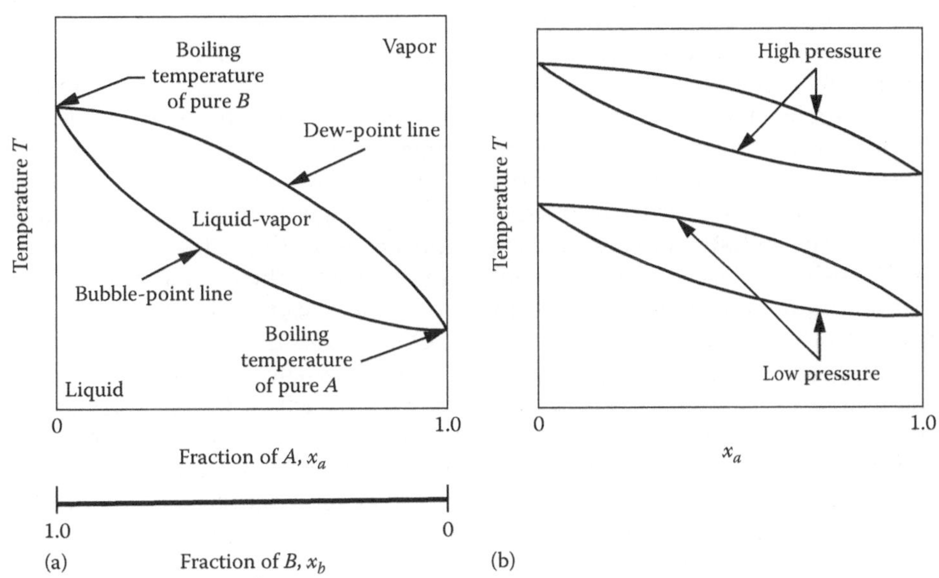

FIGURE 12.8
Temperature–concentration diagram of a homogenous binary mixture (a) at a constant pressure and (b) for two different pressures.

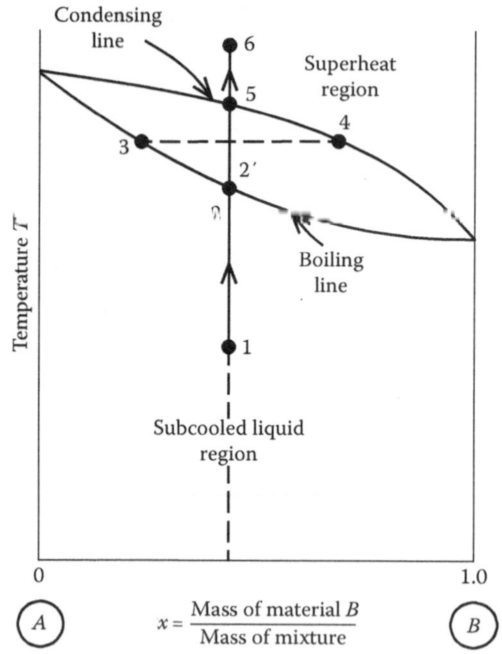

FIGURE 12.9
Evaporation process for a homogeneous binary mixture at a constant pressure.

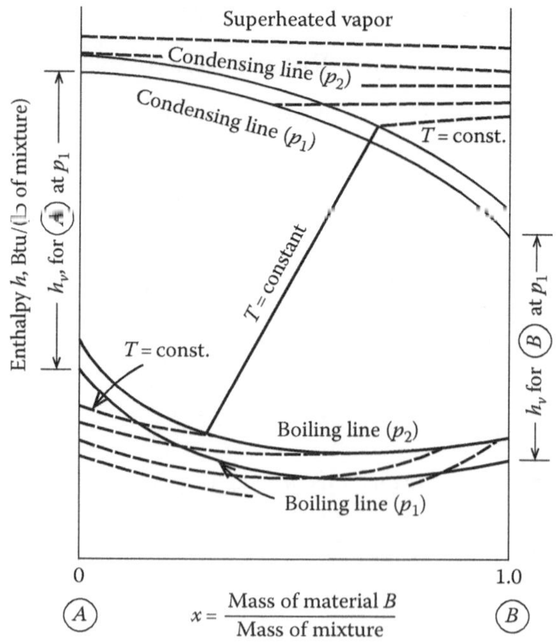

FIGURE 12.10
Enthalpy–concentration diagram for a homogeneous binary mixture.

would be $x_{B,3}$ while that in the vapor phase would be $x_{B,4}$ (as shown by the dotted line). As more heat is added, the remaining liquid mixture would gradually vaporize. What is important to note is that the relative mass concentrations of substances A and B in both liquid or vapor phases will be constantly changing with temperature during the boiling process 2–5. When the temperature of the mixture reaches T_5, the mass concentration $x_{B,3}$ of substance B (now entirely vapor) is back to its initial value $x_{B,1}$. Similar considerations also arise during the condensation process of a binary mixture.

Analysis of absorption systems requires determining energy flows associated with state changes. The enthalpy–concentration diagrams (shown in Figure 12.10) are very useful for this purpose. We note that the condensing and boiling lines at a given pressure are now separated by the enthalpy of vaporization of substances A and B as shown. Several lines of constant T are shown in the liquid and vapor regions for different pressures while only one line corresponding to pressure p_1 is shown in the saturation region so as not to clutter up the diagram. Such charts are useful when analyzing various processes of binary mixtures, for example, throttling, heating, and cooling of two binary streams with and without heat input or with heat removal. The basic principles described in this section are meant to serve as the needed background to analyzing absorption refrigeration systems discussed in Section 14.3. For further reading, there are several textbooks, for example, Kuehn et al. (1998).

12.5 Convective Heat Transfer Correlations

Convection heat transfer was described in Section 2.9.1 and is expressed by Equation 2.45. Order of magnitude values of the convective heat transfer coefficient were given in Table 2.7. It was stated that this is more a definition of the convective heat transfer coefficient than a phenomenological law (unlike say Fourier's law of heat conduction). Extensive experiments and associated analysis have led to the development of a number of correlations applicable to different situations (flow inside or over a pipe, flow over a plate, flow over a series of pipes, etc.) as a function of flow velocity, fluid properties, flow regime, and flow geometries. Applied heat transfer is more concerned with performing proper experiments, and representing these heat transfer coefficients in succinct correlations for practitioners to use.

One could use dimensional equations specific to the fluid as described in Section 2.9. A more common approach is to use relevant correlations involving dimensionless parameters, i.e., sets of parameter groups that do not have units and that have a clear physical representation. One such number called the Reynolds number has been described in Chapter 2 and given also by Equation 12.14. The advantage of using dimensionless numbers is that they are independent of the fluid used; so the same correlations would hold for example, to water or to air. The most

TABLE 12.4

Commonly Used Dimensionless Numbers

Name	Definition	Nomenclature	Description	Equation Number
Reynolds number	$\text{Re} = \frac{\rho v D}{\mu}$	ρ—Density of the fluid μ—Dynamic viscosity v—Fluid average velocity *D*—Characteristic length	Ratio of inertia to viscous forces—used to characterize flow regime	12.14
Prandtl number	$\text{Pr} = \frac{\mu c_p}{k}$	c_p—Specific heat at constant pressure *k*—Conductivity	Relative measure of the ratio of momentum to energy transport	12.15
Nusselt number	$\text{Nu} = \frac{h_{con} D}{k}$	h_{con}—Convective coefficient	Ratio of convective to conduction heat transfer	12.16
Grashof number	$\text{Gr} = \frac{g\beta(\Delta T)L^3}{\upsilon^2}$	*g*—Acceleration due to gravity	Ratio of buoyancy to viscous forces	12.17
Rayleigh number	$\text{Ra} = \frac{g\beta(\Delta T)L^3}{\upsilon\alpha}$	β—Coefficient of volume expansion = (1/*T*) for ideal gases (such as air) Δ*T*—Temperature difference between fluid and solid *L*—Characteristic length υ—Kinematic viscosity α—Thermal diffusivity	Ratio of buoyancy to viscous forces	12.18

important parameter sets, called *dimensionless numbers* are assembled in Table 12.4 along with their operational definitions.

Correlations in the form of power laws have been developed for different flow configurations. Generally, they are expressed in the following forms:

$$\text{Forced convection}: \text{Nu} = \text{CRe}^m\text{Pr}^n \quad (12.19)$$

$$\text{Natural convection}: \text{Nu} = \text{CRa}^m\text{Pr}^n$$

where the coefficient C and the exponents m and n are specific to the flow regime, and the geometry.

A short list of some of the frequently used correlations for common geometries is assembled in Table 12.5. These are some of the simpler formulations, with more complex and more accurate correlations available in the published literature. The reader can refer to standard textbooks such as Cengel (2002) or Holman (1997) or Incropera et al. (2007) or in reference books such as the ASHRAE Fundamentals (2013).

The following three examples illustrate the use of these types of correlations.

TABLE 12.5

Convective Heat Transfer Correlations for Some Commonly Encountered Flow Regimes and Geometries

Type of Convection	Geometry	Flow Regime	Correlation	Equation Number
Forced internal flow	Circular tubes	Turbulent Re > 2500	$\text{Nu}_D = 0.023\text{Re}_D^{0.8}\text{Pr}^n$ $n = 0.4$ heating $n = 0.3$ cooling	12.20
Forced external flow	Plates	Laminar Re < 3 × 10⁵	$\text{Nu}_L = 0.664\text{Re}_L^{1/2}\text{Pr}^{1/3}$	12.21
		Turbulent Re > 5 × 10⁵	$\text{Nu}_L = 0.037\text{Re}_L^{0.8}\text{Pr}^{1/3}$	12.22
Natural external flow	Vertical single plate	Laminar $\text{Ra}_L < 10^9$	$\text{Nu}_L = 0.59\text{Ra}_L^{1/4}$	12.23
		Turbulent $\text{Ra}_L > 10^9$	$\text{Nu}_L = 0.1\text{Ra}_L^{1/3}$	12.24
Natural external flow	Horizontal cylinder	Laminar $\text{Ra}_D < 10^7$	$\text{Nu}_D = 0.48\text{Ra}_D^{1/3}$	12.25
		Turbulent $\text{Ra}_D > 10^7$	$\text{Nu}_D = 0.125\text{Ra}_D^{1/3}$	12.26
Natural internal flow	Vertical parallel plates of height *H* and spacing *L*	$\text{Ra}_L < 10^3$ $10^7 > \text{Ra}_L > 10^4$ $\text{Ra}_L > 10^7$	$\text{Nu}_L = 1$ $\text{Nu}_L = 0.42\text{Ra}_L^{1/4}\text{Pr}^{0.012}\left(\frac{L}{H}\right)^{0.3}$ $\text{Nu}_L = 0.046\text{Ra}_L^{1/3}$	12.27

L, flow length; *D*, diameter.

Example 12.8: Water Flow Inside Tubes*

The internal convective heat transfer coefficient is to be computed for conditions when water at a temperature of 10°C with a velocity of 2.5 m/s flows through a tube with 8 mm internal diameter. These conditions are common in the water side of the evaporator of a refrigeration system.

Given: v = 2.5 m/s, characteristic length D = 8 mm

Find: h_{con} using Equation 12.20 for cooling condition

Lookup values: From property tables of water, ρ = 1000 kg/m^3, μ = 0.00131 Pa·s, c_p = 4190 J/(kg·K), k = 0.573 W/(m·K)

Solution

First, the Reynolds number is determined from Equation 12.14 as

$$\mathrm{Re} = \frac{\rho VD}{\mu} = \frac{1000\ \mathrm{kg/m^3} \times 2.5\ \mathrm{m/s} \times 0.008\ \mathrm{m}}{0.00131\ \mathrm{Pa \cdot s}}$$
$$= 15,267$$

The flow is clearly turbulent.

Next, the Prandtl number is computed from Equation 12.15 as follows:

$$\mathrm{Pr} = \frac{4190\ \mathrm{J/kg} \times 0.00131\ \mathrm{Pa \cdot s}}{0.573\ \mathrm{W/(m \cdot K)}} = 9.6$$

We then use Equation 12.20 with n = 0.3 to find

$$\mathrm{Nu}_D = 0.023 \times 15{,}267^{0.8} \times 9.6^{0.3} = 100.8$$

from which the convective heat transfer coefficient is deduced from Equation 12.16 as

$$h_{con} = \frac{0.573\ \mathrm{W/(m \cdot K)} \times 100.8}{0.008\ \mathrm{m}} = 7220\ \mathrm{W/(m^2 \cdot K)}$$

Comments

This example is identical to Example 2.11 that was solved using simplified dimensional equations. In that case, h_{con} = 5171 W/(m^2 K), which is about a 25% difference from the value found using the more accurate dimensionless equation.

Most heat exchangers are designed such that turbulent conditions prevail so as to enhance heat transfer rates and reduce heat exchanger surface areas. Further, it has been found that entrance effects in heat exchangers increase heat transfer rates by 10%–15% as compared to using such generalized expressions that are meant for fully developed flow inside long tubes.

A treatment akin to the aforementioned can also be used to analyze internal heat transfer in air ducts, for example, to estimate heat losses/gains to a cold/hot attic.

Example 12.9: Forced Airflow on Roof of Building†

Consider a 10 m × 6 m flat roof of a building at sea level over which wind blows at 10 m/s. The air temperature is 20°C while the temperature of the roof is 67°C. Calculate the convective heat transfer coefficient when the wind blows along the 10 m long side.

Given: v = 10 m/s, characteristic length D = 10 m

Find: h_{con} using Equation 12.22 for forced external flow over plates

Lookup values: From property tables of air (Table A1) at a mean temperature of (20°C + 74°C)/2 = 47°C: ρ = 1.110 kg/m^3, μ = 1.94 × 10^{-5} Pa·s, c_p = 1007 J/(kg·K), conductivity k = 0.0275 W/(m·K)

Solution

First, the Reynolds number is determined from Equation 12.14 as

$$\mathrm{Re} = \frac{\rho \mathrm{v} D}{\mu} = \frac{1.11\ \mathrm{kg/m^3} \times 10\ \mathrm{m/s} \times 10\mathrm{m}}{1.94 \times 10^{-5}\ \mathrm{Pa \cdot s}} = 5.72 \times 10^6$$

The flow is clearly turbulent.

Next, the Prandtl number is computed from Equation 12.15 as

$$\mathrm{Pr} = \frac{1007\ \mathrm{J/kg} \times 1.94 \times 10^{-5}\ \mathrm{Pa \cdot s}}{0.0275\ \mathrm{W/(m \cdot K)}} = 0.71$$

We then use Equation 12.22 and find

$$\mathrm{Nu}_L = 0.037 \times (5.72 \times 10^6)^{0.8} \times 0.71^{1/3} = 8405$$

from which the convective heat transfer coefficient is deduced from Equation 12.16 as

$$h_{con} = \frac{0.0275\ \mathrm{W/(m \cdot K)} \times 8405}{10\ \mathrm{m}} = 23.11\ \mathrm{W/(m^2 \cdot K)}$$

Comments

This value is very close (to within 6%) of that found in Example 2.10 using the simplified dimensional equation.

* Same inputs as Example 2.11 that was solved using a dimensional correlation.

† Same inputs as Example 2.10 solved using a dimensional correlation.

If the wind were to blow along the shorter roof length of 6 m, a similar procedure would yield a value of h_{con} = 22.13 W/(m²·K), a 4.5% reduction. Though this is not a large difference, it can become important when the roof aspect ratio were to increase, i.e., become more rectangular.

Example 12.10: Natural Convective Flow Inside an Air Space

Consider double-glazed window plates with spacing of 12.7 mm and height of 0.6 m kept at temperatures of 17°C and 0°C. Calculate the natural convective heat transfer coefficient.

Given: Spacing L = 0.0127 m, height H = 0.60 m

Find: h_{con} using Equation 12.25 for natural internal flow over plates

Lookup values: From property tables of air (Table A1) at a mean temperature of (0°C + 17°C)/2 = 8.5°C: $\rho = 1.271$ kg/m³, $\mu = 1.75 \times 10^{-5}$ Pa·s, c_p = 1004 J/(kg·K), k = 0.0246 W/(m·K),

$$\beta = \frac{1}{(273.15+8.5)} = 0.00355\ \mathrm{K}^{-1}$$

$$\upsilon = 1.40\times10^{-5}\ \mathrm{m^2/s},\ \alpha = 1.95\times10^{-5}\ \mathrm{m^2/s}\ \text{and Pr} = 0.717$$

Solution

First, the Rayleigh number is determined from Equation 12.18 as

$$\begin{aligned}\mathrm{Ra} &= \frac{g\beta(\Delta T)L^3}{\upsilon\alpha}\\ &= \frac{9.81\ \mathrm{m/s^2}\times0.00355\ \mathrm{K^{-1}}\times(17°\mathrm{C}-0°\mathrm{C})\times0.0127^3\ \mathrm{m^3}}{1.40\times10^{-5}\ \mathrm{m^2/s}\times1.95\times10^{-5}\ \mathrm{m^2/s}}\\ &= 4442\end{aligned}$$

Next, we will determine the Nusselt number using Equation 12.27:

$$\begin{aligned}\mathrm{Nu}_L &= 0.42\ \mathrm{Ra}_L{}^{1/4}\mathrm{Pr}^{0.012}\left(\frac{L}{H}\right)^{0.3}\\ &= 0.42\ \times4442^{1/4}\times0.717^{0.012}\times\left(\frac{0.0127}{0.6}\right)^{0.3} = 1.083\end{aligned}$$

from which the convective heat transfer coefficient is deduced from Equation 12.16 as

$$h_{con} = \frac{0.0246\ \mathrm{W/(m\cdot K)}\times1.083}{0.0127\ \mathrm{m}} = 2.10\ \mathrm{W/(m^2\cdot K)}$$

Comments

The Nusselt number is close to unity, suggesting that the interpane heat transfer (neglecting the radiative component) is mostly by conduction. The spacing is small enough to have largely suppressed the onset of a convection loop.

The interpane convective resistance value was specified as an input during the simplified treatment for double glazed windows in Example 5.1. A value of 3.5 h·ft²·°F/Btu was selected, which is equal to a convective heat transfer coefficient = (1/3.5) = 0.286 Btu/(h·ft²·°F) = 1.62 W/(m²·K), a difference of about 22%. This is partly due to the slightly different assumptions made in the operating conditions for both examples.

12.6 Heat Exchangers

12.6.1 Background

Heat exchangers are devices in which two fluid streams exchange thermal energy by conduction/convection. They are perhaps the most ubiquitous of all thermal equipment and can be found in most HVAC systems and in industrial and chemical processes as well. During the heat exchange process, one fluid is heated and the other is cooled. The fluids may be gases, liquids, or vapors (either condensing or boiling). For example, in buildings, heat exchangers are used:

- To heat or cool air in heating or cooling coils in an air-handler unit through which hot water (or steam) or cold water (or refrigerant) is circulated
- To generate hot water or steam in a boiler by heat transfer from combustion gases
- To reject heat in a condenser of a chiller from the refrigerant to the ambient air
- To transfer heat from the circulating water in the evaporator of a chiller to the boiling refrigerant

In a broad sense, there are two types of heat exchangers:

1. *Indirect*: Heat exchangers where the two fluid streams are physically separated from each other by a solid surface. Examples are heating or cooling coils inside air ducts, baseboard heaters in rooms, evaporators and condensers in refrigeration systems, heat exchangers in boilers, furnaces, and hot water heaters.
2. *Direct*: Heat exchangers where the two streams come in contact with each other; a common example is cooling towers where the condenser water is evaporatively cooled by the air stream.

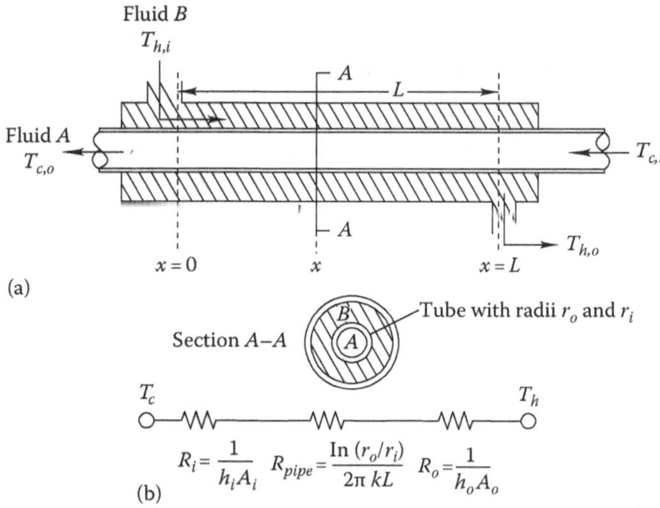

FIGURE 12.11
Counterflow shell-and-tube heat exchanger: (a) schematic diagram showing fluid temperatures and (b) equivalent thermal circuit.

The simplest indirect heat exchanger design consists of two concentric pipes. If the fluids flow in the same direction, the heat exchanger design is called *parallel flow*; if the flow direction of one stream is changed, the exchanger is of *counterflow* design (as shown in Figure 12.11a). A third type is the *cross-flow* where the two fluid streams enter the heat exchanger perpendicular to each other. Heat transfer in many types of heat exchangers can be enhanced by the use of fins (which enhances external convective coefficient) or by fluting the inside of pipes to enhance internal heat transfer coefficient. Such topics can be found in advanced handbooks such as Roshenow et al. (1998).

The rate of heat transfer occurring in a heat exchanger is much larger than the heat losses to the ambient. In that case, a simple energy balance holds for the heat transfer rate $\dot{Q}$:

$$\dot{Q} = \dot{C}_c\left(T_{c,o} - T_{c,i}\right) = \dot{C}_h\left(T_{h,i} - T_{h,o}\right) \qquad (12.28)$$

where
- heat capacitance rates are denoted by $\dot{C} = \dot{m} \times c_p$, with $\dot{m}$ being the mass flow rate and
- c_p, the specific heat at constant pressure

Heat transfer from one fluid stream to the other in a heat exchanger involves two convection processes, those between each fluid and the surface between the streams. In addition, heat is conducted through the common tube surface. The overall heat transfer conductance U_oA_o combines both modes of heat transfer by using the idea of thermal resistance that is shown in the equivalent circuit in Figure 12.11b. The overall conductance for this series heat transfer process is the reciprocal of the sum of the three resistances, namely, R_i the internal convective resistance, the tube wall resistance R_{pipe} and the external convective resistance R_o:

$$U_oA_o = \frac{1}{R_i + R_{pipe} + R_o} \qquad (12.29)$$

in which the resistances are given by the respective convection and conduction equations and are all based on the tube external area. To account for the buildup of surface films of fouling substances, a fouling resistance is sometimes added to the denominator of Equation 12.29. Values can be found in standard heat transfer texts. Generally, the internal fouling resistance is of the order of 15% of the internal convective heat transfer resistance.

There are generally two methods of analyzing heat exchanger thermal performance depending on how the problem is specified. They yield exactly the same result but one is preferable over the other under different situations. One of them is the logarithmic mean temperature difference (LMTD), which is described in the next section.

12.6.2 LMTD Approach

Consider an indirect counter flow heat exchanger. The temperature difference between the two streams is not constant and varies with the flow length as shown in Figure 12.12. The heat rate can be expressed as

$$\dot{Q} = \text{UA} \times (\text{LMTD}) \qquad (12.30)$$

where
- $\dot{Q}$ is the heat rate, W (Btu/h)
- UA is the product of overall U value and heat transfer area A of active heat exchanger surface (see Figure 12.11b), W/K (Btu/(h·°F))
- LMTD is the log mean temperature difference for pure parallel or counterflow configurations, K (°F)

The LMTD is the logarithmic average temperature difference between the two fluid streams. In simple heat exchangers without the change of phase, it is given by

$$\text{LMTD} = \frac{\Delta T_1 - \Delta T_2}{\ln(\Delta T_1/\Delta T_2)} \qquad (12.31)$$

where ΔT_1 and ΔT_2 are the temperature differences between the two streams at the inlet and the outlet of the heat exchanger (see Figure 12.12). For a counterflow heat exchanger, we have

$$\Delta T_1 = T_{h,i} - T_{c,o} \quad \text{and} \quad \Delta T_2 = T_{h,o} - T_{c,i} \qquad (12.32)$$

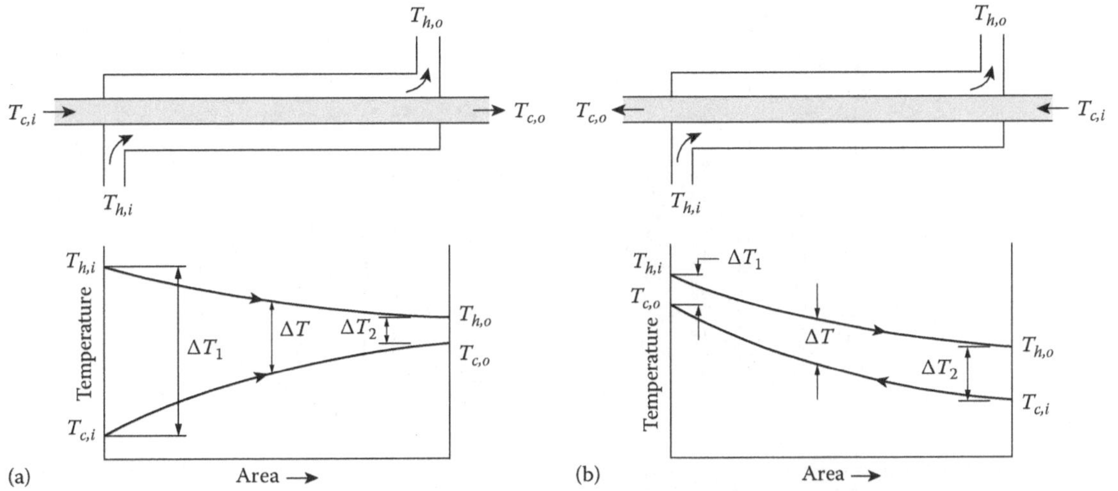

FIGURE 12.12
Temperature profiles along the length of two basic heat exchanger configurations. (a) Parallel flow heat exchanger and (b) counterflow heat exchanger.

Practical heat exchangers are much more sophisticated than the simple design considered earlier. Figure 12.13 is a sketch of the commonly encountered shell-and-tube heat exchanger for liquid heat transfer. While one of the fluids circulates through the tubes, the other fluid on the shell side is made to zigzag along the length of the heat exchanger perpendicular to the tube banks by means of a series of baffles as shown. Such directional changes in the fluid flow enhance heat transfer. There are numerous variants to this basic design; one of them is to have numerous tube passes as against the single pass shown. The types of heat exchangers shown in Figure 12.14a are basically a single-pass cross-flow heat exchanger with one of the fluids unmixed (the one inside the tubes) and the other mixed (the one on the shell side). Generally, manufacturers plot the thermal performance of their heat exchangers in terms of a correction factor F applied to Equation 12.30 as

$$\dot{Q} = F \times \text{UA} \times (\text{LMTD}) \tag{12.33}$$

This factor can be read from charts such as that shown in Figure 12.14b for a single-pass cross-flow heat exchanger with the hot fluid mixed and the cold fluid unmixed. A series of curves are plotted in terms of two dimensionless parameters P (ratio of actual temperature

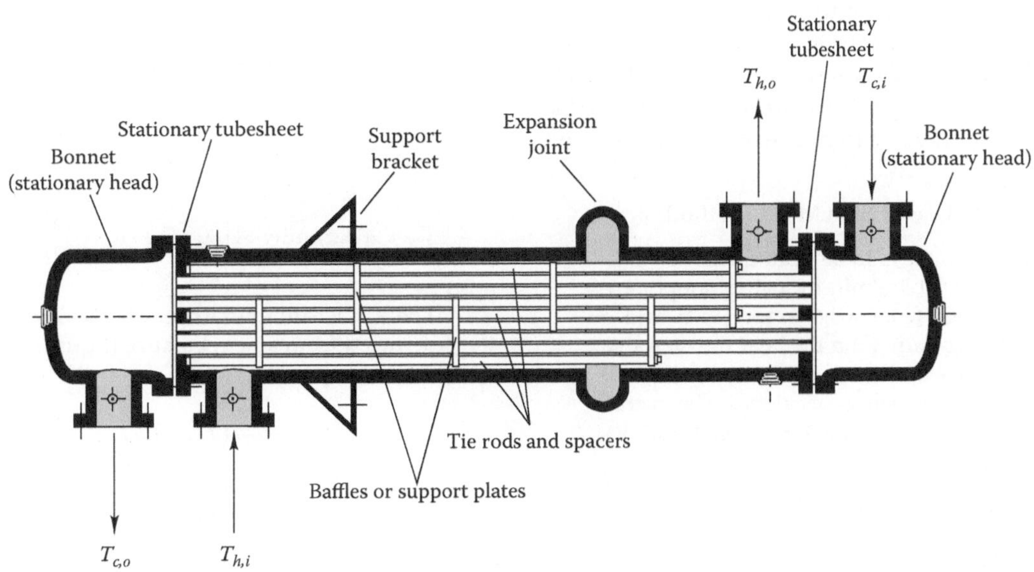

FIGURE 12.13
Typical shell-and-tube heat exchanger showing piping connections, baffles, tube sheets, and supports.

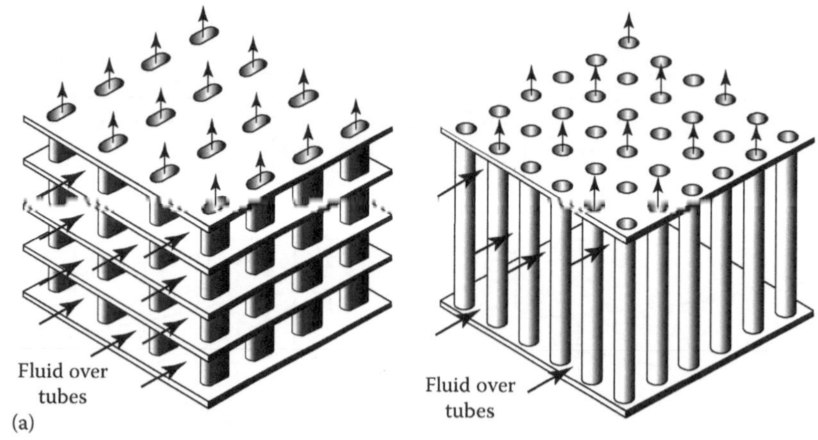

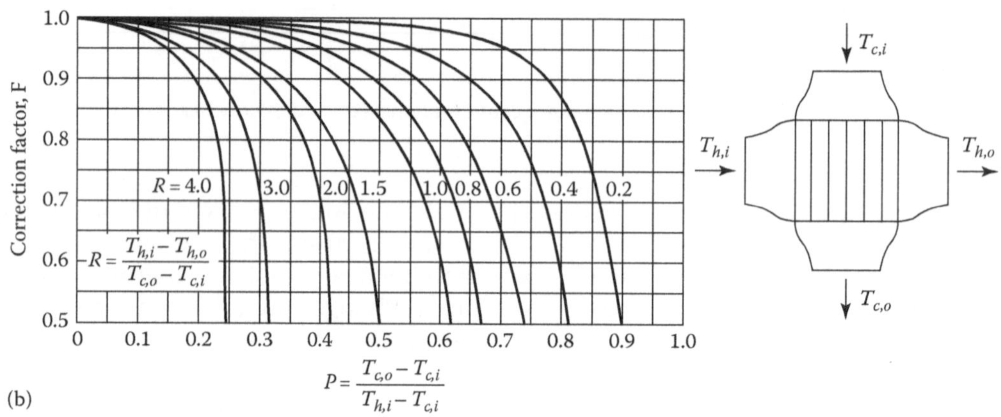

FIGURE 12.14
Cross-flow heat exchanger with one of the fluids mixed: (a) Two common configurations and (b) correction factor F (of Equation 12.32) for single-pass cross-flow heat exchangers with the hot fluid mixed and the cold fluid unmixed.

difference of cold fluid and the max temperature difference) and R (the ratio of the temperature difference of the hot fluid to that of the cold fluid)—the mathematical expressions are shown in Figure 12.14. More extensive figures can be found in a number of sources including Turton et al. (1986) and Bowman and Turton (1990). The factor F is embodied in the software used by manufacturers to assist designers with heat exchanger selection.

Example 12.11: Use of the LMTD Method

Potable service water is to be heated from 20°C to 74°C at a rate of 70 kg/min by using nonpotable pressurized water from a boiler at 110°C and a flow rate of 90 kg/min. Find the heat transfer area needed if a cross-flow heat exchanger with one fluid mixed design is being considered. The overall U value for this design is specified as 320 W/(m²·K).

Given: $T_{ci} = 20°C$, $T_{c,o} = 74°C$ $T_{h,i} = 110°C$, $U_o = 320$ W/(m²·K), $\dot{m}_c = 70$ kg/min, $\dot{m}_h = 90$ kg/min

Figure: See Figure 12.14

Assumptions: Jacket losses are negligible.

Find: LMTD, F and A_o

Lookup values: $c_p = 4180$ J/(kg·K)

Solution

The cold-side capacitance rate is

$$\dot{C}_c = 70\ \text{kg/min} \times 4180\ \text{J/kg}\cdot\text{K} \times \frac{1\ \text{min}}{60\ \text{s}} = 4877\ \text{W/K}$$

and the hot-side capacitance rate is

$$\dot{C}_h = 90\ \text{kg/min} \times 4180\ \text{J/kg}\cdot\text{K} \times \frac{1\ \text{min}}{60\ \text{s}} = 6270\ \text{W/K}$$

The outlet temperature of the hot stream is deduced from a heat balance (Equation 12.28):

$$T_{h,o} = T_{h,i} - \frac{\dot{C}_c}{\dot{C}_h}(T_{c,o} - T_{c,i})$$

$$= 110° - \frac{4877}{6270} \times (74°\text{C} - 20°\text{C}) = 68°\text{C}$$

Next, from Equation 12.32, $\Delta T_1 = T_{h,i} - T_{c,o} = 110°C - 74°C = 36°C$ and $\Delta T_2 = T_{h,o} - T_{c,i} = 68°C - 20°C = 48°C$

This allows the LMTD to be determined from Equation 12.31:

$$\text{LMTD} = \frac{\Delta T_1 - \Delta T_2}{\ln(\Delta T_1/\Delta T_2)} = \frac{36°\text{C} - 48°\text{C}}{\ln(36/48)} = 41.7°\text{C}$$

The two dimensionless parameters P and R are determined from the expressions given in Figure 12.14:

$$P = \frac{T_{c,o} - T_{c,i}}{T_{h,i} - T_{c,i}} = \frac{74 - 20}{110 - 20} = 0.6$$

and

$$R = \frac{T_{h,i} - T_{h,o}}{T_{c,o} - T_{c,i}} = \frac{110 - 68}{74 - 20} = 0.778$$

The correction factor F can be simply from read off from Figure 12.14 as $F = 0.88$.

Finally, Equation 12.30 is rearranged for the required heat transfer area to be determined:

$$A_o = \frac{\dot{C}_c(T_{c,o} - T_{c,i})}{F \times \text{U} \times (\text{LMTD})}$$

$$= \frac{4877\ \text{W/K} \times (74 - 20)°\text{C}}{0.88 \times 320\ \text{W/(m}^2\cdot\text{K)} \times 41.7°\text{C}} = 22.4\ \text{m}^2$$

12.6.3 Effectiveness-NTU Approach

The other widely used analysis method for predicting the heat transfer rate in heat exchangers is referred to as the ε-NTU, or *effectiveness number-of-transfer-units* approach. The reader is referred to Kays and London (1964) for details.

The *effectiveness* is defined as the ratio of actual to maximum possible heat transfer rates:

$$\text{Effectiveness } \varepsilon \equiv \frac{\text{Actual heat transfer rate}}{\text{Maximum possible heat transfer rate}}$$

$$= \frac{\dot{Q}}{\dot{Q}_{\max}} = \frac{(T_{h,i} - T_{h,o})}{(T_{h,i} - T_{c,i})} \times \frac{\dot{C}_h}{\dot{C}_{\min}} \qquad (12.34)$$

The maximum possible rate of heat transfer is limited to the product of the maximum temperature difference existing across the heat exchanger and the minimum fluid capacitance rate $\dot{C}_{\min} = \min(\dot{C}_h, \dot{C}_c)$. If one solves Equation 12.34 for the actual heat rate, one obtains

$$\dot{Q} = \varepsilon \dot{C}_{\min}(T_{h,i} - T_{c,i}) \qquad (12.35)$$

in which the temperatures are shown in Figure 12.11.

The ε-NTU approach uses an additional equation for effectiveness as a function of fluid capacitance rates and exchanger U value to complete the right-hand side of Equation 12.35. This performance equation applies for all heat exchangers, but the expression for the effectiveness differs with the heat exchanger configuration and design.

For example, the effectiveness of single-pass *parallel-flow* shell-and-tube heat exchangers is given by

$$\varepsilon = \frac{1 - \exp\left[-\text{NTU}\left(1 + \dot{C}_{\min}/\dot{C}_{\max}\right)\right]}{1 + \dot{C}_{\min}/\dot{C}_{\max}} \qquad (12.36)$$

where

$\dot{C}_{\max}$ is the larger of the two capacitance rates

$\dot{C}_{\min}$ is the smaller

NTU is the *number of transfer units* defined as

$$\text{NTU} \equiv \frac{\text{U}_o A_o}{\dot{C}_{\min}} \qquad (12.37)$$

A special case arises when one of the fluids is changing phase, i.e., its temperature remains constant. This occurs, for example, in refrigeration systems discussed in Section 14.4. In this case, $\left(\dot{C}_{\min}/\dot{C}_{\max}\right) = 0$, resulting in

$$\varepsilon = 1 - \exp(-\text{NTU}) \qquad (12.38)$$

Figure 12.15 is a plot of parallel-flow effectiveness versus NTU for different ratios of the capacitance rates.

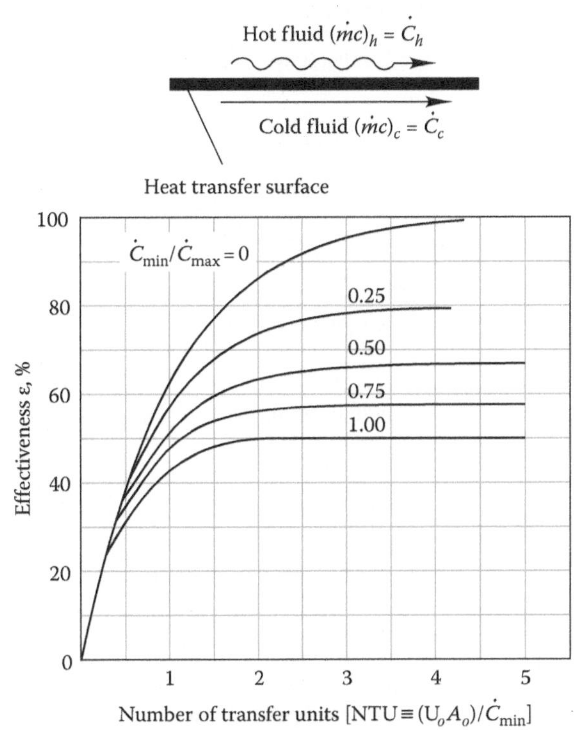

FIGURE 12.15
Parallel-flow heat exchanger effectiveness as a function of NTU.

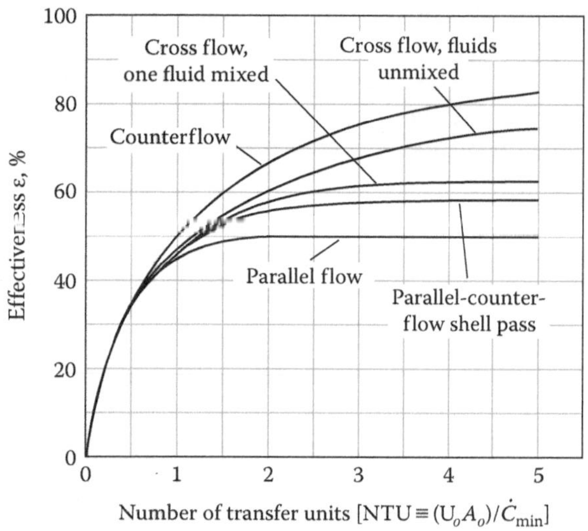

FIGURE 12.16
Comparison of effectiveness of several heat exchanger designs for equal hot- and cold-side capacitance rates, $\dot{C}_{min} = \dot{C}_{max}$.

Generally, effectiveness increases with NTU but there is a strong diminishing-returns effect. For NTU > 3.0, there is little improvement in effectiveness with increasing NTUs. For a matched-capacitance situation (when the capacitance rates of both fluid streams are equal, i.e., $\dot{C}_{max} = \dot{C}_{min}$), the effectiveness is limited to 0.5 in parallel-flow exchangers.

How the effectiveness varies with NTU for different types of flow configurations is plotted in Figure 12.16 for the matched-capacitance case. Notice that counterflow effectiveness (uppermost curve) is significantly greater than that for parallel flow (lowest curve). Counterflow heat exchangers have larger effectiveness values than parallel-flow devices for given capacitance rates. This is a result of the fact that a larger average temperature difference exists between the two streams over a larger part of the heat exchanger than in the parallel-flow case.

Table 12.6 assembles effectiveness expressions for some important flow configurations as a function of

TABLE 12.6

Heat Exchanger Effectiveness Relations, $N = \text{NTU} = \dfrac{U_o A_o}{\dot{C}_{min}}$, $C = \dfrac{\dot{C}_{min}}{\dot{C}_{max}}$

Flow Geometry	Relation
Double pipe	
Parallel flow	$\varepsilon = \dfrac{1-\exp[-N(1+C)]}{1+C}$
Counter flow	$\varepsilon = \dfrac{1-\exp[-N(1-C)]}{1-C\exp[-N(1-C)]}$ for $C<1$
	$\varepsilon = \dfrac{N}{1+N}$ for $C=1$
Cross-flow	
Both fluids mixed	$\varepsilon = 1-\exp\left\{\dfrac{1}{Cn}[\exp(-NCn)-1]\right\}$ where $n = N^{-0.22}$
Both fluids mixed	$\varepsilon = N\left[\dfrac{N}{1-\exp(-N)}+\dfrac{NC}{1-\exp(-NC)}-1\right]^{-1}$
$\dot{C}_{max}$ mixed, $\dot{C}_{min}$ unmixed	$\varepsilon = \dfrac{1}{C}\{1-\exp[-C+C\exp(-N)]\}$
$\dot{C}_{max}$ unmixed, $\dot{C}_{min}$ mixed	$\varepsilon = 1-\exp\left\{-\dfrac{1}{C}[1-\exp(-NC)]\right\}$
Shell and tube	
One shell pass; two, four, and six tube passes	$\varepsilon = 2\left[1+C+\sqrt{1+C^2}\,\dfrac{1+\exp\left(-N\sqrt{1+C^2}\right)}{1-\exp\left(-N\sqrt{1+C^2}\right)}\right]^{-1}$

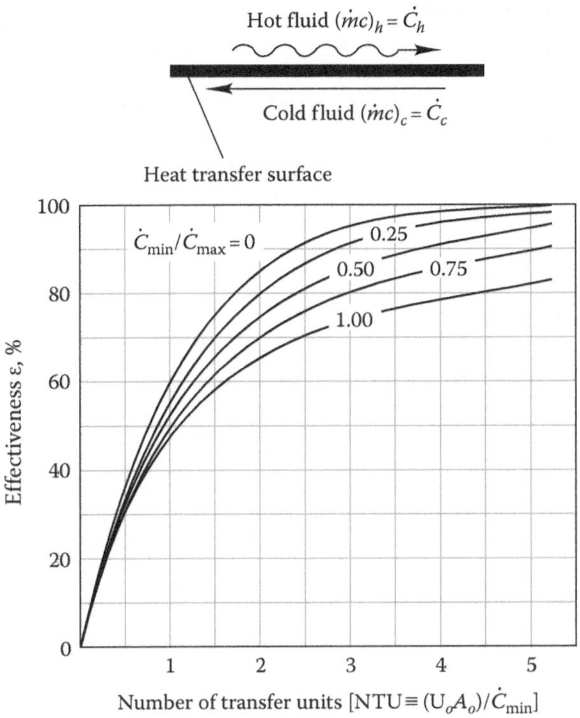

FIGURE 12.17
Counterflow heat exchanger effectiveness as a function of NTU.

NTU and factor C. The effectiveness for the parallel heat exchanger is given by the first expression in Table 12.6. The same parameters appear in this equation as in Equation 12.36, namely, NTU and the two capacitance rates. Figure 12.17 is a plot of counterflow heat exchanger effectiveness. Notice that effectiveness values are well above those shown in Figure 12.15 for parallel-flow exchangers at identical capacitance and NTU values.

Example 12.12 illustrates the use of the effectiveness approach to calculate a number of thermal performance characteristics of a heat exchanger used to heat water in a building.

Example 12.12: Use of the Effectiveness-NTU Method

The same example as the previous one (Example 12.11) will be analyzed under a slightly modified specifications. Here, the heat exchanger area is known, but the exit temperature of the hot fluid is not. Also, we will simply assume a single-pass counterflow heat exchanger. The problem calls for determining the heat transfer rate and the exit temperatures of both streams.

Given: $T_{c,i} = 20°C$, $T_{h,i} = 110°C$, $U_o = 320$ W/(m^2·K), $A_o = 20$ m^2, $\dot{m}_c = 70$ kg/min, $\dot{m}_h = 90$ kg/min

Figure: See Figure 12.11.

Assumptions: Jacket losses are negligible.

Find: ε, $T_{c,o}$, $T_{h,o}$

Lookup values: $c_p = 4180$ J/(kg·K)

Solution

As previously, cold-side capacitance rate is $\dot{C}_c = 4877$ W/K and the hot-side capacitance rate is $\dot{C}_h = 6270$ W/K.

The minimum capacitance is for the cold stream. The capacitance ratio is

$$\frac{\dot{C}_{min}}{\dot{C}_{max}} = \frac{4877}{6270} = 0.778$$

The NTU value is found from the following definition:

$$NTU = \frac{U_oA_o}{\dot{C}_{min}} = \frac{320 \times 20}{4877} = 1.31$$

Reading from Figure 12.17 (alternatively the second equation in Table 12.6 can be used), we find that the effectiveness is

$$\varepsilon = 0.60$$

It is now a simple matter to find first the heat rate and then the outlet temperatures of each stream. From Equation 12.35,

$$\dot{Q} = \varepsilon\dot{C}_{min}\left(T_{h,i} - T_{c,i}\right)$$
$$= 0.60 \times 4877 \text{ W/K} \times (110°C - 20°C) = 263.4 \text{ kW}$$

The heat rate for either stream is also given by the product of the mass flow, specific heat, and stream temperature rise:

$$\dot{Q} = \dot{m}c_p\left(T_i - T_o\right)$$

One now solves for the outlet temperatures. For the hot stream,

$$T_{h,o} = T_{h,i} - \frac{\dot{Q}}{\dot{C}_{max}}$$
$$= 110 - \frac{263,400}{6,270} = 68.0°C$$

and for the cold stream,

$$T_{c,o} = 20 + \frac{263,400}{4,877} = 74.0°C$$

Comments

For improved accuracy, one could use the equation for effectiveness listed in Table 12.6. However, for design purposes, the curves are satisfactory. Alternatively, manufacturers' data can be consulted for effectiveness values.

Table 12.6 gives a summary of effectiveness equations versus NTU and capacitance ratio for a number of heat exchangers commonly used on buildings except for plate heat exchangers. For these devices, it is suggested that manufacturers' data be consulted since they are not true counter-flow, parallel-, or cross-flow heat exchangers and are constructed in many different variations.

12.6.4 Comparison of Methods

Examples 12.11 and 12.12 illustrated the use of the LMTD and the NTU methods for a similar heat exchanger analysis problem. Both the methods can be used for design as well as for performance prediction—they will give identical results. However, the NTU method is more straightforward for cases in which outlet temperature(s) are not known. Such situations can also be solved by the LMTD method. But that would be an iterative process, and so less convenient. In the former case, effectiveness calculations made at one temperature would apply for a relatively wide range of temperatures (if the flows remain fixed) since relevant water properties do not change much with temperature. Hence, it is a convenient approach for predicting performance of an actual operating heat exchanger subject to inlet temperature fluctuations. This is the reason why it is sometimes referred to as a "feed-forward" method. The NTU method is the more accepted method for heat exchanger design and operation.

The LMTD approach is more convenient for design instances when the entering and leaving fluid temperatures of both fluid streams are known along with the mass flow rate of one of the streams. Then, one can deduce the LMTD, determine the correction factor F specific to the generic heat exchanger configuration at hand, and then compute the needed heat exchange area. This is the approach used by manufacturers for suggesting to the client the specific heat exchanger model necessary to meet the design loads.

Problems

Numbers 1–4 given in parenthesis denote the degree of difficulty.

12.1 Complete the following table for water. *(2)*

T, °C	p, kPa	v, m³/kg	Phase
60		3.25	
	175		Saturated vapor
300	300		
100	10		
		0.001097	Saturated liquid
1000	10		

12.2 Complete the following table for water. *(2)*

T, °F	p, psia	v, ft³/lb$_m$	Phase
170		30.0	
	60		Saturated vapor
400	20		
240	5.0		
		0.017209	Saturated liquid
220	14.7		

12.3 Complete the following table for R22. *(2)*

T, °F	p, psia	v, ft³/lb$_m$	Phase
		0.6561	Saturated vapor
45	90.791		
10		0.5	
	83.280		Saturated liquid

12.4 Complete the following table for R134a. *(2)*

T, °F	p, psia	v, ft³/lb$_m$	Phase
		2.0528	Saturated vapor
150	50.0		
−10		0.090	
	1.00		Saturated liquid

12.5 A rigid tank with a volume of 5 m³ contains 10 kg of saturated liquid–vapor mixture at 75°C. Heat is added to the water until all liquid is completely evaporated. What is the temperature? *(1)*

12.6 Write a computer program to fit saturated water *vapor* enthalpy data as a function of temperature between 40°F and 125°F (0°C and 50°C). Use a third-degree polynomial. *(3)*

12.7 Write a computer program or use a spreadsheet to fit saturated R134a *vapor* enthalpy data as a function of temperature between −20°F and 40°F (–30°C and 5°C). Use a third-degree polynomial. *(2)*

12.8 Find the enthalpy, specific volume, and temperature of 70% and 90% quality steam at 40 psia (280 kPa). *(1)*

12.9 Water is heated from 50°F to 140°F (10°C to 60°C) by mixing saturated steam at 25 psia (175 kPa) with it. If hot water is required at 200 gal/min (12 L/s), what is the steam rate in pounds per hour (kilograms per second)? *(2)*

12.10 Refrigerant 134a is condensed from a saturated vapor to saturated liquid at 100°F by using water entering at 60°F. If the refrigerant flow rate is 500 lb/h and the water flow rate is 1500 lb/h, what are the heat rejection rate and the water outlet temperature? *(2)*

12.11 Refrigerant 22 is throttled from a saturated liquid at 136.2 psia (950 kPa) to 55 psia (380 kPa). What are the temperature drop, quality, and specific volume at the end of the throttling process? *(2)*

12.12 Refrigerant 134a is throttled from a saturated liquid at 70°F (21°C) to a saturated liquid–vapor mixture at 36.1 psia (250 kPa). What are the temperature drop, quality, and specific volume at the end of the throttling process? *(2)*

12.13 Air is heated in a duct with a 30 kW electric resistance coil. If the sea-level airflow is 300 m^3/min of 15°C air, what is the heating unit outlet temperature? We know that 300 W is lost from the surface of the heating coil housing. *(2)*

12.14 Air at 90°F (27°C) and 14.7 psia (101.3 kPa) is heated by passing over a steam coil. If steam enters the coil at 30 psia (210 kPa), 400°F (205°C), and 30 lb/min (0.23 kg/s) and leaves at 25 psia (175 kPa) and 200°F (94°C), what is the air mass flow rate that produces a leaving air temperature of 140°F (60°C)? *(2)*

12.15 Standard atmospheric air enters the evaporator of a residential air conditioner at 350 ft^3/min (118 L/s) and 85°F (29°C). Refrigerant 22 at 69.6 psia (480 kPa) and 30% quality enters the coil at 3.5 lb/min (0.027 kg/s) and leaves as a saturated vapor. Calculate the exit air temperature from the evaporator and the rate of heat transfer from the air if the effect of water vapor in the air can be ignored due to low humidity. *(2)*

12.16 Standard atmospheric air enters the cooling coil of an air handling unit in a small commercial building at 10,000 ft^3/min (4,700 L/s) and 83°F (28°C). Refrigerant 22 at 76 psia (500 kPa) and 35% quality enters the coil at 100 lb/min (0.75 kg/s) and leaves as a saturated vapor. Calculate the exit air temperature from the coil and the rate of heat transfer from the air if the effect of water vapor in the air can be ignored due to low humidity. *(2)*

12.17 A polytropic process law is given by $pv^n = k$, where k is constant. Derive an equation for the work done during a closed polytropic process beginning at (p_1, v_1) and ending at (p_2, v_2). *(2)*

12.18 Refrigerant 134a at 140°F (60°C) and 125 psia (840 kPa) enters the condenser of a chiller at 800 lb/h (0.10 kg/s) and exits as a saturated liquid at the same pressure. If the cooling water enters at 85°F (30°C) and leaves at 104°F (40°C), what is the mass flow rate of the cooling water? Assume that water is a compressed liquid at 30 psia (205 kPa). *(2)*

12.19 An air conditioner removes heat from a residence at 1200 kJ/min while drawing power at the rate of 10 kW. What is the COP of this air conditioner, and how much heat is rejected to the outdoors? *(1)*

12.20 The cooling system on a small office building maintained at 20°C removes heat from the building to offset summer heat gains. The heat gain from outside is 60 kW, and gains from people and computers within the office amount to 20 kW. If the COP of the cooling system is 2.4, what is the electric power consumption of the cooler? *(1)*

12.21 A homeowner is considering replacing the electric resistance heating system in his residence with a heat pump. During one winter month, the resistance heaters (electric baseboards) consumed 2000 kWh, priced at \$0.09/kWh. If a heat pump could operate during the same period with a COP of 2.4, what would be the savings in utility bills for that month? *(2)*

12.22 A tightly sealed building is to be warmed from 41°F (5°C) to 73°F (23°C) with a heat pump having a COP of 3.0. If the power input to the heat pump is 17,750 Btu/h (5.2 kW), how long will it take to warm the interior? Assume that the heat capacity of the building interior and its furnishings is equivalent to 1300 lb (600 kg) of wood [specific heat 0.5 Btu/(lb · °F), 2.1 kJ/(kg · K)]. *(2)*

12.23 An air-conditioning system on a house maintains the interior at 68°F (20°C) when the outside temperature is 95°F (35°C). The compressor motor is rated at 4.5 hp (3.4 kW). What is the maximum rate at which heat can be removed from the building interior? *(1)*

12.24 What is the minimum power input required to maintain the interior of a building at 72°F (22°C) if heat gains from the 92°F (33°C) exterior are 8000 Btu/min (140 kW) and internal gains from people and lights are 2000 Btu/min (36 kW)? *(1)*

12.25 If a building loses heat at the rate of 17,000 Btu/(h · °F) (9,000 W/K) temperature difference from indoors (72°F, 22°C) to outdoors, what is the lowest outdoor temperature at which a Carnot heat pump can maintain the interior temperature of the building? This heat pump has a nominal power input, when it is running, of 80 kW. (Note that the heat pump controls the indoor temperature by cycling on and off; but when it is on, it always draws 80 kW.) *(2)*

12.26 One of the factors that degrades the actual COP of heat pumps relative to the Carnot maximum is the need to transfer heat between heat sources and sinks and the Carnot cycle working fluid itself. Suppose that a building heat pump with low- and high-temperature reservoirs at 40°F (4.5°C) (outdoor air) and 72°F (22°C) (indoor air temperature) delivers 30,000 Btu/h (8.8 kW) of heat to the building interior. If the temperature differences needed to transfer heat from these two reservoirs to the Carnot fluid are both 10°F (5.5 K), what is the COP of this heat pump, accounting for heat transfer penalties? Compare it to the ideal Carnot COP without this penalty. *(2)*

12.27 Generalize the results of Problem 12.26, and plot the COP of this heat pump for a range of outdoor temperatures between 0°F and 60°F (–18°C and 16°C). The heat transfer resistance between the two reservoirs and the heat pump working fluid is independent of the operating temperature. On the same graph, plot the ideal Carnot heat pump COP for comparative purposes. *(3)*

12.28 A double-pipe counterflow heat exchanger transfers heat between two water streams. What is the overall heat transfer conductance U_oA_o if 200 gal/min (13 L/s) is heated in the tubes from 50°F to 100°F (10°C to 38°C) with 200 gal/min (13 L/s) of inlet water on the shell side at 115°F (46°C)? What is the effectiveness? *(2)*

12.29 Repeat Problem 12.28, except use saturated steam at 230°F (110°C) as the heat source. Sufficient steam is supplied to avoid subcooling below 230°F (110°C). *Hint:* The flow rate of the condensing steam does not matter. *(2)*

12.30 Repeat Problem 12.28 (with the same two fluid inlet temperatures and same volumetric flow rates), but assume that the fluid being heated is 33% propylene glycol. Account for the differences in specific heat and density as well as the difference in the heat transfer coefficient for glycol. Assume that 90% of the resistance to heat transfer in the heat exchanger of Problem 12.28 is attributed to convection, equally split between the two water-to-tube surfaces. *(3)*

12.31 Suppose that the heat exchanger in Problem 12.28 is connected by mistake as a parallel-flow heat exchanger instead of a counterflow exchanger. What will the effectiveness be if the value of U_oA_o is unaffected by the change? *(2)*

12.32 A double-pipe counterflow heat exchanger transfers heat between two water streams. What is the overall heat transfer conductance U_oA_o if 300 gal/min (19 L/s) is heated in the tubes from 50°F to 100°F (10°C to 38°C) with 500 gal/min (31 L/s) of inlet water on the shell side at 115°F (46°C)? What are the effectiveness and LMTD? *(2)*

12.33 Repeat Problem 12.32, except use saturated steam at 220°F (105°C) as the heat source. The outlet of the steam side of the heat exchanger is saturated water at 220°F (105°C). *(2)*

12.34 Repeat Problem 12.32, but assume that the fluid being heated is 50% propylene glycol. Account for both the differences in specific heat and density and the difference in heat transfer coefficient for the glycol. Assume that 90% of the resistance to heat transfer in the heat exchanger of Problem 12.32 is attributed to convection, equally split between the two water-to-tube surfaces. *(3)*

12.35 A counterflow heat exchanger has an entering hot-water stream at 12 kg/s and 60°C and a cold stream at 15 kg/s and 8°C. What is the effectiveness if the cold-water stream leaves at 42°C? What is the overall heat transfer conductance U_oA_o? *(2)*

12.36 If the heat transfer surface in Problem 12.35 becomes fouled by a substance that decreases the overall heat transfer conductance by 18% (a common event if water treatment is not continuously applied), what are the effectiveness and heat transfer rate under the same entering-water conditions? By how much must the hot water temperature be increased to produce the same heat transfer rate as with the clean heat exchanger? *(3)*

12.37 Suppose that the heat exchanger in Problem 12.35 is connected by mistake as a parallel-flow heat exchanger instead of a counterflow exchanger. What will the effectiveness be if the value of U_oA_o is unaffected by the change? *(2)*

12.38 Figure P12.38 is a schematic diagram of a heat recovery system, consisting of a cross-flow/counterflow heat exchanger and filters, on a

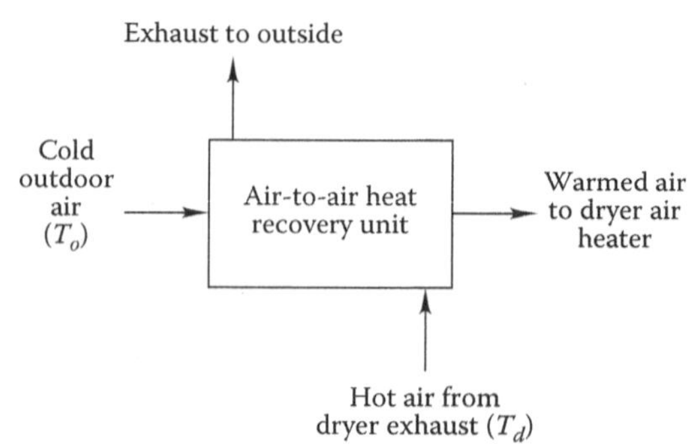

FIGURE P12.38

commercial laundry that operates 24 h/day, 7 days/week. Hot, moist air leaves the large clothes dryer at 10,000 *standard* ft^3/min (4,700 L/s) and 220°F (105°C). If the effectiveness of the heat recovery unit is 0.75, what is the annual energy savings produced by the heat recovery unit? Assume an average outdoor air temperature of 50°F *(2)*

12.39 Calculate the part-load heat transfer curve for a tube-in-tube, water-based counterflow heat exchanger in which flow is varied from 10% to 100% of design on the hot-water side. Refer to Figure P12.39. Flow is constant on the cold side. Include the effect of water velocity on the heat transfer coefficient on the hot side using equations from Chapter 12. For the purposes of this problem, assume a simple heat exchanger consisting of a 30 ft (10 m) length of nominal $\frac{3}{4}$ in. (19 mm) copper pipe inserted in the center of a larger pipe so that the flow areas of the hot and cold streams are the same. At full flow, the fluids on both sides of the heat exchanger flow at 6 ft/s (2 m/s), and the heat transfer coefficients on both sides are the same at full flow. The hot-water inlet temperature is 120°F, and the LMTD is 15°F (8°C) at full flow. Ignore the heat transfer resistance of the common pipe wall and the effect of fluid temperature on water properties. Express your results as a graph of the percent of full-flow heat transfer rate on the ordinate versus percent of hot fluid flow on the abscissa. If the results for this problem are typical of counterflow heat exchanges in general, what are the key findings regarding effectiveness ε? *(4)*

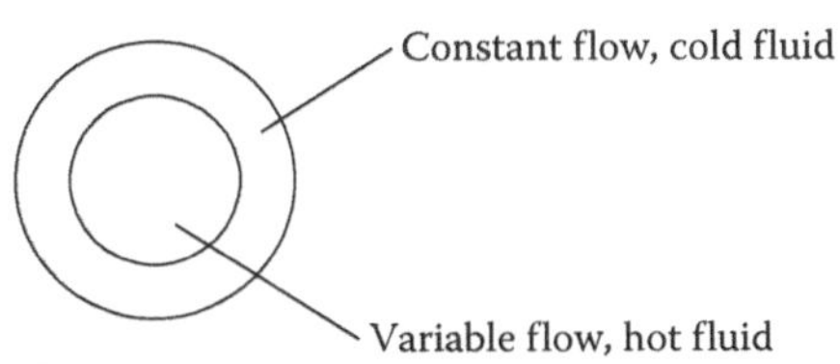

FIGURE P12.39

References

ASHRAE Fundamentals (2013). *Handbook of Fundamentals*. American Society of Heating, Refrigerating and Air-Conditioning Engineers, Atlanta, GA.

Bowman, J. and R. Turton (1990). Quick design and evaluation: Heat exchangers. *Chem. Eng.*, 97(July), 92–99.

Cengel, Y.A. (2002). *Heat Transfer*, 2nd ed. McGraw-Hill, New York.

Cengel, Y.A. and M.A. Boles (2000). *Thermodynamics*, 3rd ed. McGraw-Hill, New York.

Holman, J.P. (1997). *Heat Transfer*, 8th ed. McGraw-Hill, New York.

Incropera, F.P., D.P. Dewitt, T.L. Bergman, and A.S. Lavine (2007). *Fundamentals of Heat and Mass Transfer*, 6th ed. Wiley, New York.

Kays, W.M. and A.L. London (1964). *Compact Heat Exchangers*. McGraw-Hill, New York.

Keenan, J.H., F.G. Keyes, P.G. Hill, and J.G. Moore (1978). *Steam Tables*. Wiley, New York.

Kuehn, T.H., J.L. Threlkeld, and J.W. Ramsey (1998). *Thermal Environmental Engineering*, 3rd ed. Prentice-Hall, Englewood Cliffs, NJ.

Roshenow, W.M., J.P. Hartnett, and Y. Cho, eds. (1998). *Handbook of Heat Transfer*. McGraw-Hill, New York.

Turton R., D. Ferguson, and O. Levenspiel (1986). Charts for the performance and design of heat exchangers. *Chem. Eng.*, 93(August), 81–88.

13

Psychrometric Properties and Processes

ABSTRACT This chapter starts by discussing the composition of atmospheric air followed by a description of the important moist-air thermodynamic properties. Next, we present tabulated data and analytical equations to determine psychrometric properties of moist air common in air-conditioning processes. Important processes involving mixing of streams, cooling/heating, and humidification/dehumidification are then discussed in detail. All the analytical and design information in this chapter is based on the application of the first law of thermodynamics and the conservation of mass along with state properties of air and water vapor. Finally, we discuss the construction and use of the psychrometric chart, an essential tool to HVAC designers. A sound mastery of the material in this chapter is essential for proper comprehension and analysis of how secondary HVAC systems are designed and operated. Students who may have been previously exposed to psychrometrics during their thermodynamics course are urged to review this chapter.

Nomenclature

a, b	Coefficients appearing in Equation 13.2
ADP	Apparatus dewpoint temperature, K or °C (°F)
BP	Bypass factor in Equation 13.48
COP	Coefficient of performance
c_p	Specific heat at constant pressure, kJ/(kg·K) [Btu/(lb$_m$·°F)]
H	Height, elevation, m (ft)
h	Specific enthalpy, kJ/kg [Btu/lb$_m$]
h_{vap}	Specific enthalpy of saturated water vapor kJ/kg [Btu/lb$_m$]
h_{liq^*}	Specific enthalpy of saturated liquid at the thermodynamic wet-bulb temperature, kJ/kg [Btu/lb$_m$]
K	Parameter given by Equation 13.20
M	Molecular weight
m	Mass, kg (lb$_m$)
$\dot{m}$	Mass flow rate, kg/s (lb$_m$/h)
p	Pressure, partial pressure, Pa (lb$_f$/in.^{2})
p_a	Partial pressure of dry air, Pa (lb$_f$/in.^{2})
p_{tot}	Total pressure, Pa (lb$_f$/in.^{2})
p_{vap}	Partial pressure of water vapor, Pa (lb$_f$/in.^{2})
$\dot{Q}$	Energy use or load, W (Btu/h)
$\dot{Q}_{lat}$	Latent cooling load, W (Btu/h)
$\dot{Q}_{sen}$	Sensible cooling load, W (Btu/h)
$\dot{Q}_{tot,c}$	Total cooling load, W (Btu/h)
$\dot{Q}_h$	Heating load, W (Btu/h)
R	Gas constant, J/(kg·K) [ft·lb$_f$/(lb$_m$·°R)]
RH	Relative humidity
s	Specific entropy, kJ/(kg·K) [Btu/(lb$_m$·°R)]
SHR	Specific heat ratio (Equation 13.47)
T	Temperature, K or °C (°R or °F)
T_{db}	Dry point temperature, K or °C (°R or °F)
T_{dp}	Dew-bulb temperature, K or °C (°R or °F)
T_{wb}	Wet-bulb temperature, K or °C (°R or °F)
V	Volume, m^3 (ft^3)
v	Specific volume, m^3/kg, (ft^3/lb$_m$)
W	Humidity ratio, kg$_w$/kg$_a$ (lb$_w$/lb$_a$)

Greek

ρ	Density, kg/m^3 (lb$_m$/ft^3)
ϕ	Relative humidity

Subscripts

a	Air, dry air
c	Critical, cooling
cc	Cooling coil
d	Dry
$d\text{-}s$	Difference between saturated and dry property values of air mixture
lat	Latent
liq	Liquid
m	Air–water mixture, mixed air condition
ref	Reference
sat	Saturated
sen	Sensible
tot	Total
vap	Vapor, vaporization
w	Water

13.1 Definition and Importance of Psychrometrics

Psychrometry is the science of measurement of moisture content in a mixture of air and water vapor. Of more practical import, psychrometrics deals both with the determination of moist-air properties knowing the state of the mixture and with the analysis of the mass and energy exchanges required to accomplish various relevant processes. A thorough understanding of psychrometry is essential to analyze the various processes related to air conditioning in buildings.

Atmospheric air is never totally dry; it always contains varying degrees of water vapor. Just like relatively small amounts of carbon drastically impact the physical properties of iron alloys, small amounts of moisture in the indoor air have a major and direct effect on the comfort of building occupants. As discussed previously in Chapter 3, comfort and health require that the moisture content of air be controlled to within a relatively narrow range. In hot and humid climates, moisture must be removed by using a dehumidification process that is often a part of the air cooling system itself. In dry, cold weather, humidity is added to the air by humidifiers, which are often a part of the heating system. Since the latent heat of water is large (about 2450 kJ/kg or 1050 Btu/lb_m at room temperature and pressure), either the removal or the addition of moisture to air in a building can involve significant amounts of energy. Moisture control in buildings is also critical to limit moisture collection in or on building components, such as insulation, window glass, and structural members, which could cause physical damage as well as lead to health ailments to occupants.

13.2 Composition and Pressure of Atmospheric Air

Atmospheric air is a mixture of several gases, water vapor, and numerous pollutants (the last two vary considerably from location to location). The composition of dry air is relatively constant and varies slightly with time, location, and altitude. The standard composition of dry air has been specified by the International Joint Committee on Psychrometric Data in 1949 as shown in Table 13.1. ASHRAE Fundamentals (2013) gives the following definition of the U.S. Standard Atmosphere:

TABLE 13.1
Primary Constituents of Atmospheric Air with All Water Vapor and Contaminants Removed

Constituent	Molecular Mass	Volume Percentage
Nitrogen	28.016	78.09
Oxygen	32.000	20.95
Argon	39.944	0.93
Carbon dioxide	44.010	0.03

1. Acceleration due to gravity is constant at 9.807 m/s^2 (32.174 ft/s^2).
2. Temperature at sea level is 15°C or 288.1 K (59.0°F).
3. Pressure at sea level is 101.039 kPa (29.921 in. of mercury or 14.969 lb/in.^{2} absolute [psia]).
4. The atmosphere consists of dry air that behaves as a perfect gas.*

The atmospheric pressure should be known in most calculations involving buildings. The density of air and hence, the barometric pressure both vary with altitude. The pressure at different altitudes is given in standard gas tables. One widely used correlation to calculate the total atmospheric pressure p_{tot} (in kPa) for up to 11,000 m altitude is given by the following correlation (ASHRAE Fundamentals, 2013):

$$p_{tot} = 101.325 \times [1 - 2.25577 \times 10^{-5} \times H]^{5.2559} \quad (13.1\ \text{SI})$$

where
H is the altitude above sea level in m
p_{tot} is in kPa

The atmospheric air density can then be found from Equation 2.12, i.e., from $\rho = p/RT$.

In IP units, the equivalent expression for up to 36,000 ft altitude is

$$p_{tot} = 14.696 \times [1 - 6.8754 \times 10^{-6} \times H]^{5.2559} \quad (13.1\ \text{IP})$$

where
H (ft) is the altitude above sea level
p_{tot} is in psia

* A perfect gas is an ideal gas whose specific heat can be assumed constant at different temperatures (see Section 2.2.3).

TABLE 13.2

Numerical Values of Parameters Appearing in Equation 13.2

	$H \le 1200$ m (4000 ft)		$H > 1220$ m (4000 ft)	
Constant	**SI Units**	**IP Units**	**SI Units**	**IP Units**
a	101.325	29.92	99.436	29.42
b	−0.01153	−0.001025	−0.010	−0.0009

Alternatively, for altitudes up to 18,300 m (60,000 ft), the following simpler equation can be used:

$$p_{tot} = a + b \cdot H \tag{13.2}$$

where
Constants a and b are given in Table 13.2
H is the elevation above sea level
p is the pressure in kPa or in. Hg

In HVAC applications, the mixture of various constituents that compose dry air is considered to be a single gas. The molecular mass of dry air can be assumed to be 28.965. The gas constant for air is $R_a = 287$ J/(kg · K) or 53.35 ft · $lb_f/(lb_m \cdot °R)$.

Example 13.1: Atmospheric Pressure

Calculate the atmospheric pressure at Phoenix, AZ, whose altitude is 1120 ft (341.4 m).

Given: $H = 1120$ ft.

Find: p_{tot}.

Solution

Using Equation 13.1 IP, we find

$$p_{tot} = 14.696 \times [1 - 6.8754 \times 10^{-6} \times 1120]^{5.2559}$$

$$= 14.696 \times 0.9602 = 14.11 \text{ psia}$$

If, instead, we were to use Equation 13.2, we get

$$p_{tot} = 29.92 - 0.001025 \times 1120 = 28.57 \text{ in. Hg}$$

$$= 14.0 \text{ psia } (96.5 \text{ kPa})$$

In this case, the difference between both correlations is about 0.7%.

13.3 Psychrometric Properties of Moist Air

13.3.1 Ideal Gas Assumption

Ideal gases were already discussed in Section 2.2.2. Air–water vapor mixtures are assumed to behave as ideal gases with constant specific heat (referred to as *perfect gases*) despite the fact that water vapor is essentially a superheated mixture. Even under saturated-air conditions, the assumption is valid for water vapor since moist air is a very dilute mixture of water vapor and dry air, i.e., the mass concentration of water vapor to that of dry air is extremely small (in the range of 0.005–0.015 for normal HVAC applications). The ideal gas assumption is said to involve errors of less than 1% in building heating and cooling calculations (Kuehn et al., 1998).

The amount of water vapor contained in the air may vary from zero (totally dry) to a maximum determined by the temperature and pressure of the mixture. Moist air up to about 3 atm can be assumed to obey the perfect gas law. Assuming dry air to consist of one gas only, the total pressure p_{tot} of moist air given by the Gibbs–Dalton Law for a mixture of perfect gases is equal to the individual contributions of dry air and water vapor:

$$p_{tot} = p_a + p_{vap} \tag{13.3}$$

where
p_a is the partial pressure of dry air
p_{vap} is the partial pressure of water vapor

It is because $p_{vap} \ll p_a$ that we can implicitly assume water vapor to also follow the ideal gas law.

13.3.2 Important Moist-Air Properties and Tables

There are a number of important properties relevant to air and water vapor mixtures. Before discussing the thermodynamics of moist-air processes, we shall define their most important properties and indicate where each is used.

1. The *dry-bulb temperature* T_{db}, which is the most important property, is defined as the temperature of a moist-air mixture measured by a perfectly dry sensor, such as a thermocouple or thermometer.
2. The *partial pressure* p_{vap} of water vapor in a homogeneous air–water vapor mixture is the pressure exerted by water vapor alone.
3. The *relative humidity* (RH) ϕ is the ratio of the partial pressure of water vapor p_{vap} to the saturation pressure of water vapor p_{sat} at the existing dry-bulb temperature. The saturation pressure is the pressure exerted by saturated water vapor at T_{db}; it can be read from steam tables such as Table A3 or Keenan et al. (1978). The RH is the most common measure of moisture content—it is also equal to the ratio of moisture contained

in an air–water vapor mixture relative to that of a saturated mixture at the same temperature (Equation 13.8).*

4. The *humidity ratio* W is the ratio of the mass of water vapor to the mass of *dry* air in a moist-air mixture. The humidity ratio is used to calculate the enthalpy of moist air. It has units of lb_w/lb_a or kg_w/kg_a, where the subscript "a" denotes dry air and "w" for water or water vapor.
5. The *dew point temperature* T_{dp} is the temperature at which a moist-air mixture at a given humidity ratio becomes saturated; i.e., moisture begins to condense from the air, as it is cooled at constant pressure. The dew point is used in design calculations to find the temperature at which condensation must be accounted for in equipment design.
6. The *wet-bulb temperature* T_{wb} is the temperature of a thermometer with a wetted wick over which air flows at a specified velocity. The wet-bulb temperature is an intermediate quantity used in many psychrometric calculations. Designers use it as a quick indication of the moisture content of air.
7. *Specific volume* v is the volume of the mixture (say in m^3 or in ft^3) per unit mass of dry air. It is the reciprocal of the mass density of air. The specific volume is a quantity that is useful in the estimation of air mass flow rates from specified volumetric flow rates, in fan power requirements and in air distribution duct sizing.
8. *Specific enthalpy* h of moist air is equal to the sum of dry air enthalpy and that of water vapor. It is a measure of the total (sensible plus latent) energy contained in the moist air. This property is used extensively while analyzing energy flows in most psychrometric processes.

There are other lesser-used moist-air properties, two of which are (1) *absolute humidity* or water vapor density that is the ratio of the mass of water vapor to the total volume of the sample and (2) *specific humidity* that is the ratio of the mass of water vapor to the total mass of moist air in the sample. Note that specific volume and specific enthalpy are defined per mass of *dry* air and not per mass of moist air (as is specific humidity). This convention of using mass of dry air to define properties dispels the confusion created when mass transfer takes place (i.e., when water vapor is either added or removed from the air).

* Although the relative humidity is strictly defined as a ratio of mass fractions, the ideal gas assumption enables us to express the relative humidity more conveniently for working purposes as the ratio of pressures.

Properties of moist air can be determined: (1) from the psychrometric chart (described in Section 13.5) (this approach is the most convenient for hand calculations and for understanding the entire air-conditioning process; however, it is rather approximate and not suitable for detailed design or analysis calculations) or (2) from using a combination of equations based on ideal gas law assumptions and correlations describing property values at saturated conditions (assembled in tables such as Table 13.3). The methodology, discussed in Section 13.4, is best suited for detailed analysis involving computer software.

13.4 Analytical Approach to Determining Moist-Air Properties

13.4.1 Saturated Moist-Air Tables

Thermodynamic properties of moist air can be determined from Table 13.3, which is an excerpt of the complete moist-air table (available in the online HCB software). This table contains air properties at 1 atm of pressure. It includes the following data from left to right: The second column is the saturated humidity ratio W_{sat} followed by columns of specific volume v, enthalpy h, and entropy s, each at three conditions—(1) dry air (denoted by *"a"*), (2) difference between saturated air and dry air (denoted by *"d-s"*), and (3) saturated moist air (denoted by *"sat"*). The reference state for enthalpy and entropy is 0°C in SI units (and 0°F for IP units) and 1 atm pressure in Table 13.3. The last column is the saturated vapor pressure while the two columns next to it are the enthalpy and entropy at saturated conditions per unit mass of the dry air mixture.

For moist air that is neither saturated nor perfectly dry, the following equations are used to find the enthalpy and specific volume. Other properties are found similarly:

$$h = h_a + \frac{W}{W_{sat}} h_{d-s} \tag{13.4}$$

$$v = v_a + \frac{W}{W_{sat}} v_{d-s} \tag{13.5}$$

The ideal gas law can be used to find expressions relating the RH and the humidity ratio to standard thermodynamic properties such as specific volume and pressure. To do this, consider the process 1–2 shown in Figure 13.1 (the T–s diagram for water) in which superheated water vapor at T_{db} (point 1) is cooled at constant pressure p_w to a saturated-vapor state at the dew point T_{dp} (point 2). The mixture is assumed to be homogeneous

TABLE 13.3

Thermodynamic Properties of Moist Air at Standard Barometric Pressure (14.696 psia)

Temp., °F	Saturated Humidity Ratio W_{sat}, lb_w/lb_a	Volume, ft^3/lb_m Dry Air v_a	$v_{d\text{-}s}$	v_{sat}	Enthalpy, Btu/lb_m Dry Air h_a	$h_{d\text{-}s}$	h_{sat}	Entropy, $Btu/(lb_m$ Dry Air·°R) s_a	$s_{d\text{-}s}$	s_{sat}	Enthalpy[a] h_{liq}, Btu/lb_m	Entropy[a] s_{liq}, $Btu/(lb_m \cdot °R)$	Vapor Pressure[b] p_{sat}, in.Hg
10	0.0013158	11.832	0.025	11.857	2.402	1.402	3.804	0.00517	0.00315	0.00832	−154.13	−0.3141	0.062901
12	0.0014544	11.883	0.028	11.910	2.882	1.550	4.433	0.00619	0.00347	0.00966	−153.17	−0.3120	0.069511
14	0.0016062	11.933	0.031	11.964	3.363	1.714	5.077	0.00721	0.00381	0.01102	−152.20	−0.3100	0.076751
16	0.0017724	11.984	0.034	12.018	3.843	1.892	5.736	0.00822	0.00419	0.01241	−151.22	−0.3079	0.084673
18	0.0019543	12.035	0.038	12.072	4.324	2.088	6.412	0.00923	0.00460	0.01383	−150.25	−0.3059	0.093334
20	0.0021531	12.085	0.042	12.127	4.804	2.303	7.107	0.01023	0.00505	0.01528	−149.27	−0.3038	0.102798
22	0.0023703	12.136	0.046	12.182	5.285	2.537	7.822	0.01123	0.00554	0.01677	−148.28	−0.3018	0.113130
24	0.0026073	12.186	0.051	12.237	5.765	2.793	8.558	0.01223	0.00607	0.01830	−147.30	−0.2997	0.124396
26	0.0028660	12.237	0.056	12.293	6.246	3.073	9.318	0.01322	0.00665	0.01987	−146.30	−0.2977	0.136684
28	0.0031480	12.287	0.062	12.349	6.726	3.378	10.104	0.01420	0.00728	0.02148	−145.31	−0.2956	0.150066
30	0.0034552	12.338	0.068	12.406	7.206	3.711	10.917	0.01519	0.00796	0.02315	−144.31	−0.2936	0.164631
—	—												
55	0.009233	12.970	0.192	13.162	13.213	10.016	23.229	0.02715	0.02048	0.04763	23.11	0.0459	0.43592
56	0.009580	12.995	0.200	13.195	13.453	10.397	23.850	0.02762	0.02122	0.04884	24.11	0.0478	0.45205
57	0.009938	13.021	0.207	13.228	13.694	10.790	24.484	0.02808	0.02198	0.05006	25.11	0.0497	0.46870
58	0.010309	13.046	0.216	13.262	13.934	11.197	25.131	0.02855	0.02277	0.05132	26.11	0.0517	0.48589
59	0.010692	13.071	0.224	13.295	14.174	11.618	25.792	0.02901	0.02358	0.05259	27.11	0.0536	0.50363
60	0.011087	13.096	0.233	13.329	14.415	12.052	26.467	0.02947	0.02442	0.05389	28.11	0.0555	0.52193
61	0.011496	13.122	0.242	13.364	14.655	12.502	27.157	0.02994	0.02528	0.05522	29.12	0.0575	0.54082
62	0.011919	13.147	0.251	13.398	14.895	12.966	27.862	0.03040	0.02617	0.05657	30.11	0.0594	0.56032
63	0.012355	13.172	0.261	13.433	15.135	13.446	28.582	0.03086	0.02709	0.05795	31.11	0.0613	0.58041
64	0.012805	13.198	0.271	13.468	15.376	13.942	29.318	0.03132	0.02804	0.05936	32.11	0.0632	0.60113
65	0.013270	13.223	0.281	13.504	15.616	14.454	30.071	0.03178	0.02902	0.06080	33.11	0.0651	0.62252
66	0.013750	13.248	0.292	13.540	15.856	14.983	30.840	0.03223	0.03003	0.06226	34.11	0.0670	0.64454
67	0.014246	13.273	0.303	13.577	16.097	15.530	31.626	0.03269	0.03107	0.06376	35.11	0.0689	0.66725
68	0.014758	13.299	0.315	13.613	16.337	16.094	32.431	0.03315	0.03214	0.06529	36.11	0.0708	0.69065
69	0.015286	13.324	0.326	13.650	16.577	16.677	33.254	0.03360	0.03325	0.06685	37.11	0.0727	0.71479
70	0.015832	13.349	0.339	13.688	16.818	17.279	34.097	0.03406	0.03438	0.06844	38.11	0.0746	0.73966
71	0.016395	13.375	0.351	13.726	17.058	17.901	34.959	0.03451	0.03556	0.07007	39.11	0.0765	0.76528
72	0.016976	13.400	0.365	13.764	17.299	18.543	35.841	0.03496	0.03677	0.07173	40.11	0.0783	0.79167
73	0.017575	13.425	0.378	13.803	17.539	19.204	36.743	0.03541	0.03801	0.07343	41.11	0.0802	0.81882
74	0.018194	13.450	0.392	13.843	17.779	19.889	37.668	0.03586	0.03930	0.07516	42.11	0.0821	0.84684
75	0.018833	13.476	0.407	13.882	18.020	20.595	38.615	0.03631	0.04062	0.07694	43.11	0.0840	0.87567

Source: ASHRAE, *Handbook—Fundamentals*, ASHRAE, Atlanta, GA, 2013. Copyright ASHRAE, www.ashrae.org.

[a] Saturated conditions and per unit mass of moist air mixture.

[b] Multiply inHg by 0.49115 to convert to psia.

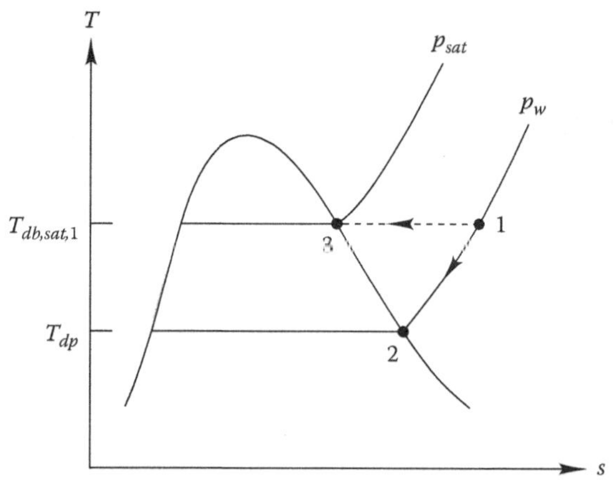

FIGURE 13.1
The T–s diagram for steam showing dry-bulb and dew point temperatures and saturation line.

and at thermodynamic equilibrium. In buildings, moist-air mixtures are usually between 0°C and 60°C (32°F and 140°F); therefore, the ideal gas laws will give accurate results for building design purposes. Figure 13.1 also shows the saturation pressure p_{sat} located on the saturation curve at point 3, the temperature corresponding to the dry-bulb temperature of the moist air at condition 1.

Example 13.2: Specific Enthalpy

Determine the specific enthalpy of moist air at a dry-bulb temperature of 56°F (13.3°C) and humidity ratio of 0.005 lb_w/lb_a.

Given: T_{db} = 56°F and $W = 0.005\ lb_w/lb_a$.

Find: h.

Solution

Lookup values from Table 13.3 at T_{db} = 56°F are

$$W_{sat} = 0.009580\,\text{lb}_\text{w}/\text{lb}_\text{a},\quad h_a = 13.453\,\text{Btu}/\text{lb}_\text{a},$$

$$h_{d-s} = 10.397\,\text{Btu}/\text{lb}_\text{a}$$

Inserting these values in Equation 13.4

$$h = 13.453\,\text{Btu}/\text{lb}_\text{a} + \left(\frac{0.005}{0.009580}\right) \times 10.397\,\text{Btu}/\text{lb}_\text{a}$$

$$= 18.88\,\text{Btu}/\text{lb}_\text{a}\ (5.53\,\text{W})$$

13.4.2 Humidity Ratio from Relative Humidity and Dry-Bulb Temperature

Let us begin with the working expression for RH in terms of the partial pressures as described in Section 13.3.2:

$$\phi = \frac{p_{vap}}{p_{sat}} \tag{13.6}$$

The ideal gas law can be used to replace this pressure ratio by a ratio of specific volumes, and since $T_{db} = T_{sat}$

$$\phi = \frac{RT_{db}/v_{vap}}{RT_{db}/v_{sat}} = \frac{v_{sat}}{v_{vap}} \tag{13.7}$$

If we consider equal volumes V of unsaturated and saturated mixtures (at the same temperature), the ratio of the masses of water vapor in the mixtures can be found from the definition of specific volume (= V/m). Using this substitution in Equation 13.7, we have

$$\phi = \frac{m_{vap}}{m_{sat}} \tag{13.8}$$

Like the RH, the humidity ratio can be expressed in a more useful fashion by using the ideal gas law to replace the mass fraction with a ratio of pressures. The saturation pressure of water vapor can be calculated from Equation 12.12 or it can also be looked up in the steam tables in the Table A3.

The ideal gas law written on an extensive basis with the gas constant R and molecular weight M is

$$p_{tot}V = m\left(\frac{R}{M}\right)T \tag{13.9}$$

Equation 13.9 can be solved for the mass. For the masses of air and water vapor, we have

$$m_a = \frac{p_a V M_a}{RT} \tag{13.10}$$

and

$$m_w = \frac{p_{vap} V M_w}{RT} \tag{13.11}$$

The ratio of m_w to m_a is the humidity ratio W by definition; therefore,

$$W \equiv \frac{m_w}{m_a} = \frac{p_{vap}M_w}{p_a M_a} \tag{13.12}$$

We now substitute the numerical values of the two molecular weights—28.96 for air and 18.02 for water—to get

$$W = \frac{p_{vap}}{p_a}\left(\frac{18.02}{28.96}\right) = 0.622 \times \frac{p_{vap}}{p_a} \tag{13.13}$$

Since the total of the partial pressure of water vapor and dry air is the atmospheric pressure p_{tot}, by Dalton's law, we can replace p_a in Equation 13.13 as follows:

$$W = 0.622 \times \left(\frac{p_{vap}}{p_{tot} - p_{vap}} \right) \tag{13.14}$$

Because p_{vap} is much less than p_{tot}, the humidity ratio and partial pressure are related approximately linearly. We will use this observation later in construction of the psychometric part. However, since the atmospheric pressure depends on the location, we must use the proper elevation of the location for which the psychrometric calculations are being done. For altitude effects, see the tables in the online HCB software.

An alternate form of Equation 13.14 using the RH (Equation 13.6) is sometimes useful:

$$W = 0.622 \times \left(\frac{\phi p_{sat}}{p_{tot} - \phi p_{sat}} \right) \tag{13.15}$$

The dry-air basis for the preceding quantities p_a and m_a is used since the composition of dry air is essentially invariant, whereas the composition of moist air obviously varies with the moisture content. The use of the dry-air basis leads to less involved calculations than a moist-air basis, although the latter could be used equally well from a theoretical point of view.

Example 13.3: Humidity Ratio

Calculate the humidity ratio of a typical air-conditioned indoor space specified by a dry-bulb temperature of 75°F (23.9°C) and 50% RH.

Given: $T_{db} = 75$°F and $\phi = 0.50$.

Find: W.

Solution

Lookup value from steam tables (Table A3): the saturated water pressure at 75°F is $p_{sat} = 0.435$ psia. Next, we use Equation 13.15 to yield the desired answer:

$$W = 0.622 \times \left(\frac{0.5 \times 0.435}{14.696 - 0.5 \times 0.435} \right) = 0.0093 \text{ lb}_w/\text{lb}_a$$

Comments

Alternatively, we could determine p_{sat} from Equation 12.12 with numerical values of the coefficients given in Table 12.3 (with temperature $T = 75 + 460 = 535$°R):

$$\ln p_{sat} = -1.0440397 \times 10^4/T - 1.1294650 \times 10$$
$$-2.7022355 \times 10^{-2} T + 1.2890360 \times 10^{-5} T^2$$
$$-2.4780681 \times 10^{-9} T^3 + 6.5459673 \cdot \ln T = -0.8328$$

from which $p = 0.4348$ psia. The result is identical to the value read off the steam tables.

13.4.3 Adiabatic Saturation and Thermodynamic Wet-Bulb Temperatures

In principle, either the RH or the humidity ratio could be used to specify the moisture content of atmospheric air. The problem was that, historically, neither could be measured directly with inexpensive and durable instruments (nowadays, semiconductor devices are available, but their reliability in the field is still questionable). Hence, an indirect method using a straightforward temperature measurement was widely adopted. We will first discuss the heat and mass transfer of an ideal, though impractical, measurement system and then move on to include two practical modifications.

Consider the apparatus shown in Figure 13.2 in which moist air at temperature T_{db1} enters an adiabatic enclosure (i.e., no heat transfer occurs *across* the surface bounding the apparatus). Since the air is not fully saturated, a certain amount of water would evaporate into the air whose latent heat is supplied by the air. Hence, the moist-air dry-bulb temperature decreases, while its humidity ratio increases. The entering moist air becomes increasingly saturated as a result. Since heat is removed from the airstream to accomplish the evaporation, the outlet temperature T_{db2} is lower than the inlet temperature. The temperature reached by air when it passes through a spray of water (where intimate contact between water and air is best achieved) is such that thermal and vapor pressure equilibrium is established between the air and the water; this is called the *adiabatic saturation temperature*.

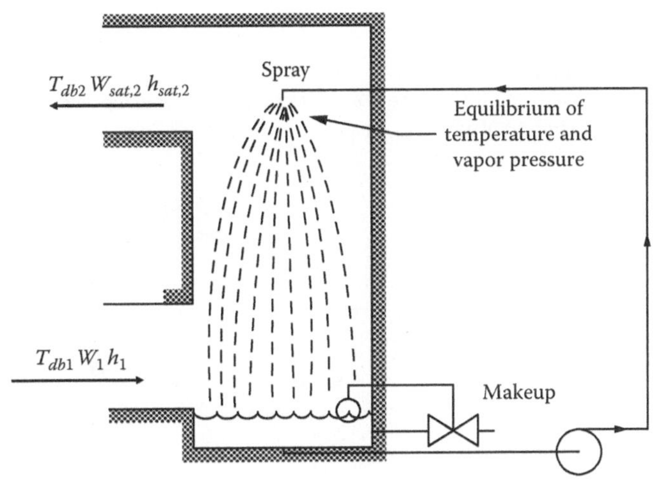

FIGURE 13.2
Schematic diagram showing adiabatic saturation process equipment.

A small amount of makeup water to compensate for the evaporated water is supplied at a temperature equal to that of the sump. The thermodynamic *wet-bulb temperature* is the temperature of the sump water in an adiabatic saturator. An energy balance on the adiabatic saturator shown in Figure 13.2 yields

$$h_{sat,2} - h_1 = (W_{sat,2} - W_1) \cdot h_{liq}^* \qquad (13.16)$$

where

h_{liq}^* is the enthalpy of saturated liquid at the thermodynamic wet-bulb temperature

$h_{sat,2}$ and $W_{sat,2}$ are the enthalpy and humidity ratio of moist air at saturated conditions

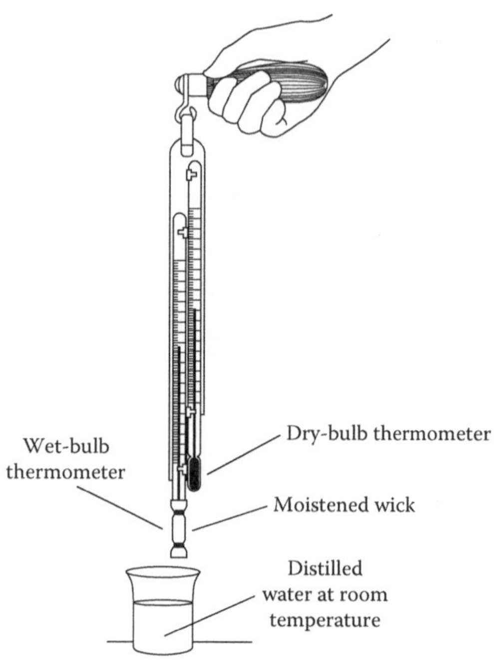

FIGURE 13.3
Sling psychrometer device for conveniently measuring wet- and dry-bulb temperatures.

These three quantities are sole functions of temperature T_{db2}. Since h_1 is a function of T_{db1} and W_1, the sole unknown quantity W_1 can be inferred iteratively. Because of the latent heat term, a process that occurs along a constant wet-bulb temperature will not be strictly isenthalpic (i.e., not a constant enthalpy process).

The adiabatic saturator is not a practical device for measuring the wet-bulb temperature. Instead, a hand-held sling psychrometer is used as a standard measuring equipment in many laboratories and HVAC field engineering services (Figure 13.3). The apparatus consists of two thermometers, one measuring the dry-bulb temperature and the other, which has a wetted wick covering the bulb, measuring the wet-bulb temperature. The instrument has a handle that allows the thermometers to be rotated so as to simulate air movement adequate for proper heat transfer between the bulb and the ambient air. As air flows over the moistened wick surrounding the "wet bulb," water evaporates, extracting heat from the thermometer and lowering its temperature. At equilibrium, the thermometer bulb will be at a temperature at which the rate of evaporation times the heat of vaporization just balances the heat transferred to the bulb. The wet-bulb temperature is below the dry-bulb temperature because heat must be transferred to the wet bulb in order to evaporate moisture. Though the wet-bulb temperature is not the same as the adiabatic saturation temperature, the differences are small. A detailed discussion of these differences can be found in Kuehn et al. (1998).

An alternative to the sling psychrometer, more suitable for automated and continuous measurement, is an apparatus consisting of a small duct through which air flows over a stationary wet-bulb thermometer (shown in Figure 13.4). The suggested velocity of air over the wet bulb is 2.5–5 m/s (7.5–15 ft/s). The wick over the bulb is made of fine cotton, which must be kept clean. Periodic replacement is needed depending on the water mineral content and air cleanliness. The time interval between services can be increased if distilled water is used. Nowadays, electronic devices are used to measure humidity levels in air and instruments described above are mostly of historic interest.

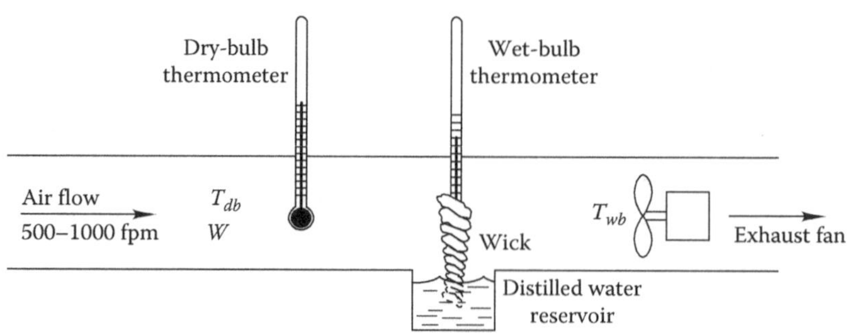

FIGURE 13.4
Psychrometry apparatus for measuring wet- and dry-bulb temperatures.

13.4.4 Relative Humidity from Wet-Bulb and Dry-Bulb Temperatures

The devices described earlier provide a close measure of the thermodynamic wet-bulb temperature. To find the RH directly from the wet-bulb temperature (and the concurrently measured dry-bulb temperature), the following approach by Pallady (1984, 1989) is recommended since it is more accurate than the expression given by ASHRAE Fundamentals (2013).* The following algorithm is reported to be accurate to 0.3% on average (not accounting for errors in temperature measurement) and is based on the van der Waals equation. It uses the critical pressure p_c and temperature T_c of water to find the vapor pressure and humidity, as shown in the following. The RH is calculated from this alternative working form of Equation 13.6 in which the numerator p_{vap} has been replaced by $p_{sat}(T_{wb}) - p'$:

$$\phi = \frac{p_{sat}(T_{wb}) - p'}{p_{sat}(T_{db})} \tag{13.17}$$

where

- $p_{sat}(T_{wb})$ is the saturation pressure at wet-bulb temperature T_{wb} (temperature units to be used with equations in this section are *absolute*—K or °R)
- $p_{sat}(T_{db})$ is the saturation pressure at dry-bulb temperature T_{db}, K or °R
- p' is the partial pressure of water vapor due to *depression of wet-bulb temperature* below the dry-bulb temperature (it does not express a physical process result but in just an intermediate quantity

Further, p' is given by

$$p' = p_{tot}\left(\frac{T_{db} - T_{wb}}{1514}\right)\left(1 + \frac{T_{wb} - 273.2}{873}\right) \tag{13.18 SI}$$

or, in IP units,

$$p' = p_{tot}\left(\frac{T_{db} - T_{wb}}{2725}\right)\left(1 + \frac{T_{wb} - 492}{1571}\right) \tag{13.18 IP}$$

The two saturation partial pressures in Equation 13.17 are found at dry- and wet-bulb temperatures from

$$p_{sat}(T) = p_c[10^{K(1-T_c/T)}] \tag{13.19}$$

where $p_{sat}(T)$ is the saturation pressure of water at temperature T. This expression for saturation pressure is to be applied at both the dry- and wet-bulb *absolute* temperatures. The critical pressure of water p_c is 22.1 MPa (3226 psia), and the critical temperature T_c is 647.30 K (1165.67°R).

The parameter K depends on the temperature and is given by

$$K = 4.39553 - 6.2442\left(\frac{T}{1000}\right) + 9.953\left(\frac{T}{1000}\right)^2 - 5.151\left(\frac{T}{1000}\right)^3 \tag{13.20 SI}$$

in which T is the dry- or wet-bulb temperature in Kelvins. In IP units, this dimensional expression is

$$K = 4.39553 - 3.469\left(\frac{T}{1000}\right) + 3.072\left(\frac{T}{1000}\right)^2 - 0.8833\left(\frac{T}{1000}\right)^3 \tag{13.20 IP}$$

in which T is the dry- or wet-bulb temperature in degrees Rankine.

In this method, the partial pressure of water vapor p_{vap} needed to compute ϕ from Equation 13.17 is found indirectly from the difference between the saturation partial pressure *at the wet-bulb temperature* $p_{sat}(T_{wb})$ (note that this is different from p_{sat} used in the previous section) and the partial pressure of water vapor p' due to the *depression of the wet-bulb temperature* below the dry-bulb temperature. These two partial pressures are not used directly in design but are convenient for finding the RH.

To summarize the calculation procedure:

- Find both $p_{sat}(T_{db})$ and $p_{sat}(T_{wb})$ from Equation 13.19 using the known wet- and dry-bulb temperatures.
- Next, determine p' from Equation 13.18.
- Finally, calculate the RH, using Equation 13.17.

The following example illustrates the method.

Example 13.4: Calculation of Relative Humidity from Dry- and Wet-Bulb Temperatures

Find the RH and the humidity ratio of moist air at 204°F (95.6°C) if the wet-bulb temperature is measured at 190°F (87.8°C) by a sling psychrometer. The atmospheric pressure is that of standard atmosphere, 14.696 psia (101 kPa).

---

* Mathur (1989) reports the accuracy of the ASHRAE equation to be 1.2%–1.5% between 50°F and 150°F (35°C and 90°C).

Given: T_{db} = 204°F = 664°R, T_{wb} = 190°F = 650°R, and p_{tot} = 14.696 psia.

Find: ϕ, *W*.

Solution

The approach described in the previous paragraph will be used. From Equation 13.20, the *K* values at the dry- and wet-bulb temperatures are found to be

$$K(T_{db}) = 3.18814 \quad \text{and} \quad K(T_{wb}) = 3.19611$$

The values of $p_{sat}(T_{db})$ and $p_{sat}(T_{wb})$ are then determined from Equation 13.19:

$$p_{sat}(T_{db}) = 3226 \times [10^{3.19611 \times (1-1165.67/664)}] = 12.58 \text{ psia}$$

$$p_{sat}(T_{wb}) = 3226 \times [10^{3.19611 \times (1-1165.67/650)}] = 9.40 \text{ psia}$$

Next, Equation 13.18 is used to compute p':

$$p' = 14.696 \text{ psia} \times \left(\frac{664-650}{2725}\right) \times \left(1 + \frac{650-492}{1571}\right)$$
$$= 0.0831 \text{ psia}$$

Finally, the RH is found from Equation 13.17:

$$\phi = \frac{9.40 - 0.0831}{12.58} = 0.74$$

while the humidity ratio is readily deduced from Equation 13.15, where p_{sat} was evaluated at T_{db}:

$$W = 0.622 \times \left(\frac{\phi \times p_{sat}}{p_{tot} - \phi \times p_{sat}}\right) = 0.622 \times \left(\frac{0.74 \times 12.58}{14.696 - 0.74 \times 12.58}\right)$$
$$= 1.08 \text{ lb}_w/\text{lb}_a$$

Comments

Observe that the capacity of hot air to absorb water vapor is significant. Both the RH and the humidity ratio can also be found by using the online HCB software.*

13.4.5 Dew Point Temperature from Partial Water Vapor Pressure

The dew point temperature is shown on the saturation curve in Figure 13.1. It is the temperature at and below which moist air is saturated, i.e., contains the maximum possible amount of water vapor. The humidity ratio of air remains constant as it is cooled until the dew point is reached. Below the dew point, condensation occurs and the humidity ratio decreases.

The dew point is important in building design since condensation must be avoided within building elements such as walls or on HVAC system components such as chilled-water pipes. Of course, condensation is desirable in air-conditioning systems since it is the mechanism by which humidity is removed from humid outdoor air before introducing such air into the interior spaces of the buildings for ventilation purposes.

The dew point temperature can be found from p_{vap} by using the following expression (for temperatures above the freezing point and between 0°C and 93°C) (ASHRAE, 2013):

$$T_{dp} = 6.54 + 14.526\alpha + 0.7389\alpha^2 + 0.0949\alpha^3 + 0.4569 p_{vap}^{0.1984} \,°\text{C} \tag{13.21 SI}$$

in which $\alpha = \ln p_{vap}$ and p_{vap} has units of kilopascals (kPa).

In IP units, the expression for the dew point temperature (in the range 32°F–200°F) is

$$T_{dp} = 100.45 + 33.193\alpha + 2.319\alpha^2 + 0.1707\alpha^3 + 1.2063 p_{vap}^{0.1984} \,°\text{F} \tag{13.21 IP}$$

in which p_{vap} has units of pounds per square inch absolute (psia).

A convenient rule of thumb states that at 70% RH a difference of 5.5°C (10°F) between the dew point and dry-bulb temperature exists. A difference of 11°C (20°F) between dew point and dry-bulb temperature corresponds roughly to a RH of 50%. These and similar rules for other dew point–dry-bulb differences apply to within a few percent accuracy in the restricted range of dry-bulb temperatures involved in building design.

Example 13.5: Dew Point Temperature

A metal water cup is filled with water in a room at standard atmospheric pressure whose dry-bulb temperature is 20°C (68°F) and RH of 50%. Crushed ice is added incrementally to the water until condensation appears on the outside of the water cup. What will be the water temperature when this occurs?

Given: T_{db} = 20°C and ϕ = 50%.

Find: T_{dp}.

Solution

From Equation 13.20 with *T* = (20 + 273.15) K = 293.15 K, we get p_{sat} = 2317 Pa.

Next, we calculate $p_{vap} = p_{sat} \times \phi$ = 2317 × 0.5 = 1158.5 Pa or 1.1585 kPa.

* The psychrometrics section of the HCB software can be used to solve nearly all the examples in this chapter along with many of the end-of-chapter problems.

Finally, substituting this value in Equation 13.21 SI with $\alpha = \ln(1.1585)$, we get

$$T_{dp} = 6.54 + 14.526 \times \alpha + 0.7389 \times \alpha^2 + 0.0949 \times \alpha^3 + 0.4569 \times p_{vap}^{0.1984} = 9.3°\text{C } (48.7°\text{F})$$

13.4.6 Moist-Air Enthalpy and Specific Heat

The enthalpy of moist air is the sum of the enthalpy of the dry air and that of the water vapor contained in the mixture. On an intensive basis (i.e., per unit mass of dry air), the specific enthalpy h of the mixture is

$$h = h_a + Wh_{vap} \tag{13.22}$$

where h_{vap} is the specific enthalpy of water vapor.

Because of the low partial pressure of water vapor in moist air, the perfect gas approximation can be invoked that provides a convenient way of expressing specific heat and enthalpies as functions of temperature only. Most steady-state analysis of HVAC processes only requires the determination of enthalpy changes, that is, Δh. As a result, the reference temperature on which enthalpy values are tabulated is immaterial. For convenience, therefore, the standard base temperature for moist-air properties has been taken to be 0°F or 0°C, depending upon which units are being used. In terms of temperature, we express the enthalpy of moist air as

$$h = c_{p,a}T_{db} + W\left(h_{vap,ref} + c_{p,vap}T_{db}\right) \tag{13.23}$$

where
- h is the enthalpy of moist air, kJ/kg_a (Btu/lb_a)
- $c_{p,a}$ is the specific heat of dry air, $kJ/(kg_a \cdot °C)$ [$Btu/(lb_a \cdot °F)$]
- $c_{p,vap}$ is the specific heat of superheated water vapor, $kJ/(kg_w \cdot °C)$ [$Btu/(lb_w \cdot °F)$]
- T_{db} is the dry-bulb temperature, °C (°F)
- $h_{vap,ref}$ is the specific enthalpy of saturated water vapor at reference temperature, kJ/kg_w (Btu/lb_w)
- W is the humidity ratio, kg_w/kg_a (lb_w/lb_a)

The water vapor enthalpy includes the energy required to heat liquid water to the dew point temperature plus the latent heat at the dew point temperature plus the energy required to superheat the water vapor. Rather than having to use the dew point temperature that makes it awkward, a convenient simplification is made to Equation 13.22 where the vaporization is assumed to occur at the reference temperature and then superheated to T_{db}. Substituting numerical values for the two specific heats and for the extrapolated value of enthalpy of saturated water vapor (at 0°F or 0°C), we have the following pair of dimensional equations:

$$h = 1.0 \times T_{db} + W\left(2501.3 + 1.86 \times T_{db}\right) \text{ kJ/kg}_a \tag{13.23 SI}$$

where
- T_{db} is in °C
- $h_{liq\text{-}vap,ref}$ = 2501.3 kJ/kg_w is the enthalpy of saturated water vapor at 0°C

$$h = 0.240 \times T_{db} + W\left(1061.2 + 0.444 \times T_{db}\right) \text{ Btu/lb}_a \tag{13.23 IP}$$

where
- T_{db} is in °F
- $h_{liq\text{-}vap,ref}$ = 1061.2 Btu/lb_w is the enthalpy of saturated water vapor at 0°F

These equations are somewhat approximate because of the perfect gas assumption that implies constant c_p values. For greater accuracy, the tables of moist-air properties discussed in the previous section can be directly used as illustrated in Example 13.2.

Example 13.6: Determination of Moist-Air Properties

Calculate values of humidity ratio, specific volume, and enthalpy for air at dry-bulb temperature T_{db} = 60°F (15.6°C) and RH ϕ = 80%.

Given: T_{db} = 60°F, ϕ_1 = 0.8, and gas constant R_a = 53.352 $ft \cdot lb_f/(lb_m \cdot °R)$.

Assumptions: The atmospheric pressure is the standard atmosphere, 14.696 psia.

Find: W, v, and h.

Solution

Lookup value from the steam tables (Table A3): saturated vapor pressure at 60°F is p_{sat} = 0.2563 psia.

The vapor pressure of the moist air is $p_{vap} = p_{sat} \cdot \phi$ = 0.2563 psia × 0.8 = 0.2050 psia.

From Equation 13.15, the humidity ratio is

$$W = 0.622 \times \left(\frac{p_{vap}}{p_{tot} - p_{vap}}\right) = 0.622 \times \left(\frac{0.205}{14.696 - 0.205}\right) = 0.0088 \text{ lb}_w/\text{lb}_a$$

and the density

$$\frac{1}{v} = \rho = \frac{p_a}{R_a T_{db}} = \frac{(14.696 - 0.205) \text{ lb}_f/\text{in.}^2 \times 144 \text{ in.}^2/\text{ft}^2}{53.352 \text{ ft} \cdot \text{lb}_f/(\text{lb}_a \cdot °\text{R}) \times (60+459.67)°\text{R}} = 0.07526 \text{ lb}_a/\text{ft}^3$$

Thus, specific volume is $v = (1/0.07526) = 13.287\ \text{ft}^3/\text{lb}_a$ ($0.830\ \text{m}^3/\text{kg}_a$).

Finally, enthalpy is determined from Equation 13.23:

$$
\begin{aligned}
h &= 0.240 \times T_{db} + W(1061.2 + 0.444 \times T_{db}) \\
&= 0.240\ \text{Btu}/(\text{lb}_a \cdot {}^\circ\text{F}) \times 60^\circ\text{F} + 0.0088\ \text{lb}_w/\text{lb}_a \\
&\quad \times (1061.2\ \text{Btu}/\text{lb}_w + 0.444\ \text{Btu}/(\text{lb}_w \cdot {}^\circ\text{F}) \times 60^\circ\text{F}) \\
&= 24.0\ \text{Btu}/\text{lb}_a\ (55.8\ \text{kJ}/\text{kg}_a)
\end{aligned}
$$

Often, we need to distinguish between the sensible and latent components of the enthalpy quantity. A simplified approach widely adopted by practicing HVAC engineers is to express enthalpy as

$$h = h_{sen} + h_{lat} = c_{p,m}(T_{db} - T_{ref}) + W \cdot h_{liq-vap,ref} \quad (13.24)$$

where the heat of vaporization is $2501.3\ \text{kJ/kg}_w$ (corresponding to 0°C) in SI units and $1075\ \text{Btu/lb}_w$ in IP units (corresponding to a reference temperature of 32°F).

The humid air specific heat at constant pressure $c_{p,m}$ can be expressed as

$$
\begin{aligned}
c_{p,m} &= c_{p,a} + W \cdot c_{p,vap} \\
&= 1.0\,\text{kJ}/(\text{kg}_a \cdot {}^\circ\text{C}) + W\,\text{kg}_w/\text{kg}_a \times 1.86\,\text{kJ}/(\text{kg}_w \cdot {}^\circ\text{C}) \\
&\quad \text{in SI units } (\text{kJ}/\text{kg}_a \cdot {}^\circ\text{C}) \\
&= 0.24\,\text{Btu}/(\text{lb}_a \cdot {}^\circ\text{F}) + W\,\text{lb}_w/\text{lb}_a \times 0.444\,\text{Btu}/(\text{lb}_w \cdot {}^\circ\text{F}) \\
&\quad \text{in IP units } (\text{Btu}/\text{lb}_a \cdot {}^\circ\text{F}) \qquad (13.25)
\end{aligned}
$$

Note that since $h_{liq-vap}$ depends on the temperature at which water evaporates into steam and not at a specific base temperature as assumed, these simplified expressions tend to introduce small errors into how the total energy is split into the sensible and latent components.

13.5 Psychrometric Chart

The thermodynamic state of an air–vapor mixture is fully determined if three independent intensive properties are specified. Since we can assume for most of the HVAC processes being studied that the total atmospheric pressure does not change, a chart known as the psychrometric chart, applicable to a specific value of total pressure (commonly the standard atmospheric pressure), is used. The psychrometric chart not only provides a quick means of determining values of moist-air properties but is also very useful in solving numerous process problems with moist air while allowing quick visualization of the process path. Hence, for better comprehension, we describe the manner in which it is generated along with the description of the pertinent moist-air properties.

The psychrometric chart is an xy plot with dry-bulb temperature T_{db} as the abscissa and the humidity ratio W as the ordinate.* Since these are two independent thermodynamic variables, all other properties of moist air can be expressed as functions of them at a given atmospheric pressure. On the standard psychrometric chart, isolines of the following moist-air properties are plotted: (1) RH, (2) wet-bulb temperature, (3) vapor pressure, (4) specific volume, and (5) enthalpy. The presence of so many lines on the chart may appear confusing to the beginner, but this is soon dispelled by a little familiarity in usage.

Figure 13.5a through e shows how lines of constant properties listed appear on the psychrometric chart. Figure 13.6 is the psychrometric chart in IP units, while Figure 13.7 corresponds to air–water mixtures represented in SI units. The charts shown are for the "normal" temperature range. The expressions developed earlier depend on the local atmospheric pressure, which is assumed to be 1 atm if the elevation is not much different from sea level. However, at high altitude, the values of thermodynamic variables can be affected considerably by the lower local barometric pressure. As a result, high-altitude psychrometric charts must be used. Charts for several elevations are included in the online HCB software. Additional charts are available from ASHRAE for both low and high temperature ranges and for altitudes between sea level and 2260 m (7500 ft).

In the psychrometric chart, the RH lines curve upward from the lower left to the upper right of the chart. The saturation curve (100% RH) can be plotted from data in the steam tables at the desired atmospheric pressure. Since saturated-steam properties are given in steam tables as a function of pressure, one uses Equations 13.19 and 13.20 along with Equation 13.14 to find the psychrometric chart ordinate, the humidity ratio, corresponding to the selected vapor pressure p_{vap}. Lines of constant RH less than 100% are plotted by linearly scaling vertical distances between the abscissa and the saturation curve by their respective values of RH. The RH values are written along the curves. The increment between lines is 10% in the two charts. On the 100% humidity line, the wet-bulb, dry-bulb, and dew point temperatures are identical by their definitions.

* Close inspection of the psychrometric charts shows that the coordinates are not quite orthogonal. The dry-bulb lines are nearly vertical but slope slightly to the left.

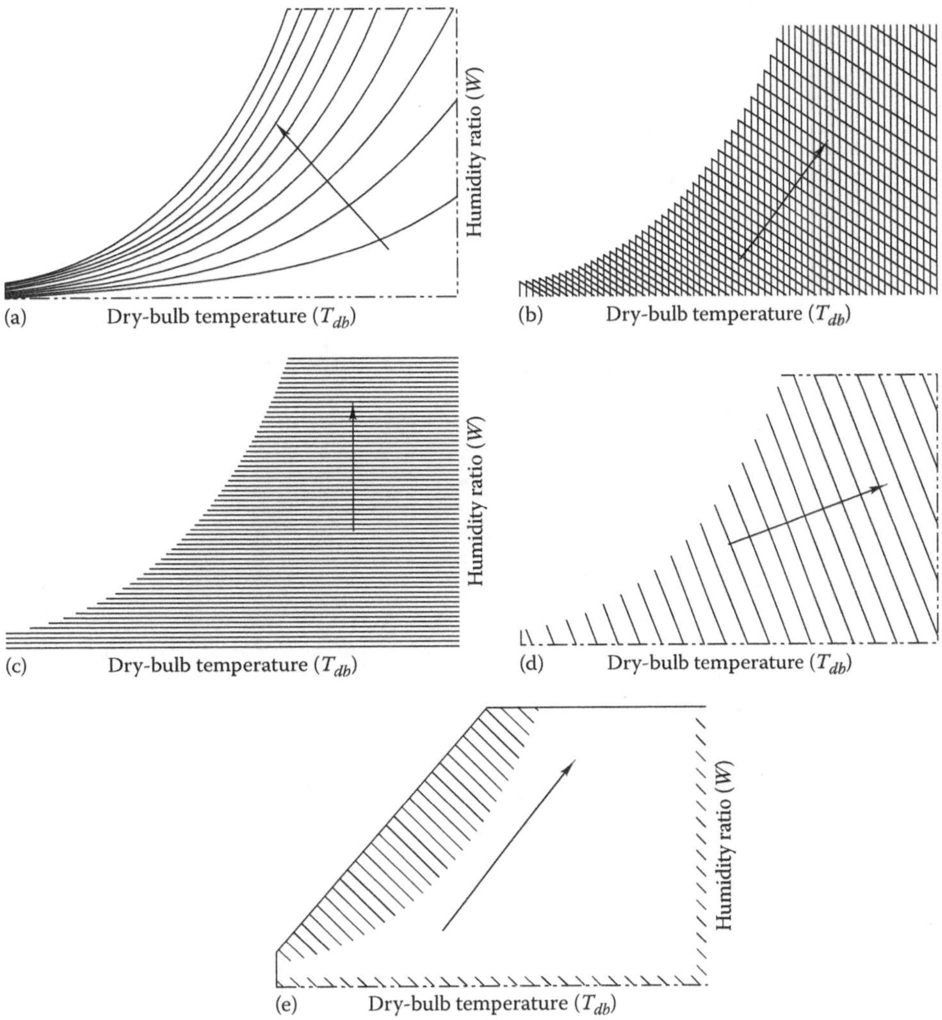

FIGURE 13.5
Skeleton psychrometric chart showing lines of (a) constant relative humidity, (b) constant wet- and dry-bulb temperatures, (c) constant humidity ratio, (d) constant specific volume, and (e) constant enthalpy.

Wet-bulb temperature lines slope downward from left to right in the chart. By solving Equations 13.17 through 13.20 for dry-bulb temperatures as a function of the RH for selected values of the wet-bulb temperature, one can plot lines of constant wet-bulb temperature, as shown in the figure; values are written along the lines. The increment between labeled lines in the two charts is 5°C (5°F). The T_{wb} lines are straight due to a fortuitous combination of heat and mass transfer properties for air and water vapor (Stoecker and Jones, 1982).

Enthalpy lines also slope downward from left to right in the chart. Values are noted at the left and right ends of the lines. The increment between marked lines in the two figures is 10 kJ/kg$_a$ (5 Btu/lb$_a$). Enthalpy lines and wet-bulb temperature lines are very nearly parallel, but lines of constant wet-bulb temperature are slightly steeper than lines of constant enthalpy. Therefore, it is best to use the auxiliary enthalpy scales (1) along the left diagonal border of the chart and (2) along the bottom and right side of the chart along with a straightedge for accurate readings. Enthalpy lines are plotted by selecting a value of enthalpy, inserting it into Equation 13.23, and finding the temperature T_{db} for several values of the humidity ratio. The results, when plotted, provide the enthalpy lines on the psychrometric chart. For more accurate values, one can use Table 13.3 of moist-air properties as described earlier.

Specific-volume lines slope steeply downward from left to right of the chart at an angle of about 20° from the vertical. Lines of constant specific volume are found by solving Equation 13.10 for the pressure $p_a = (p_{tot} - p_{vap})$ as a function of the temperature and specific volume. Then, for a given specific volume, one finds $(T - p_{vap})$ pairs that are plotted. The specific-volume values are written along the lines. The increment between marked lines is 0.02 m^3/kg$_a$ (0.5 ft^3/lb$_a$) in the two charts.

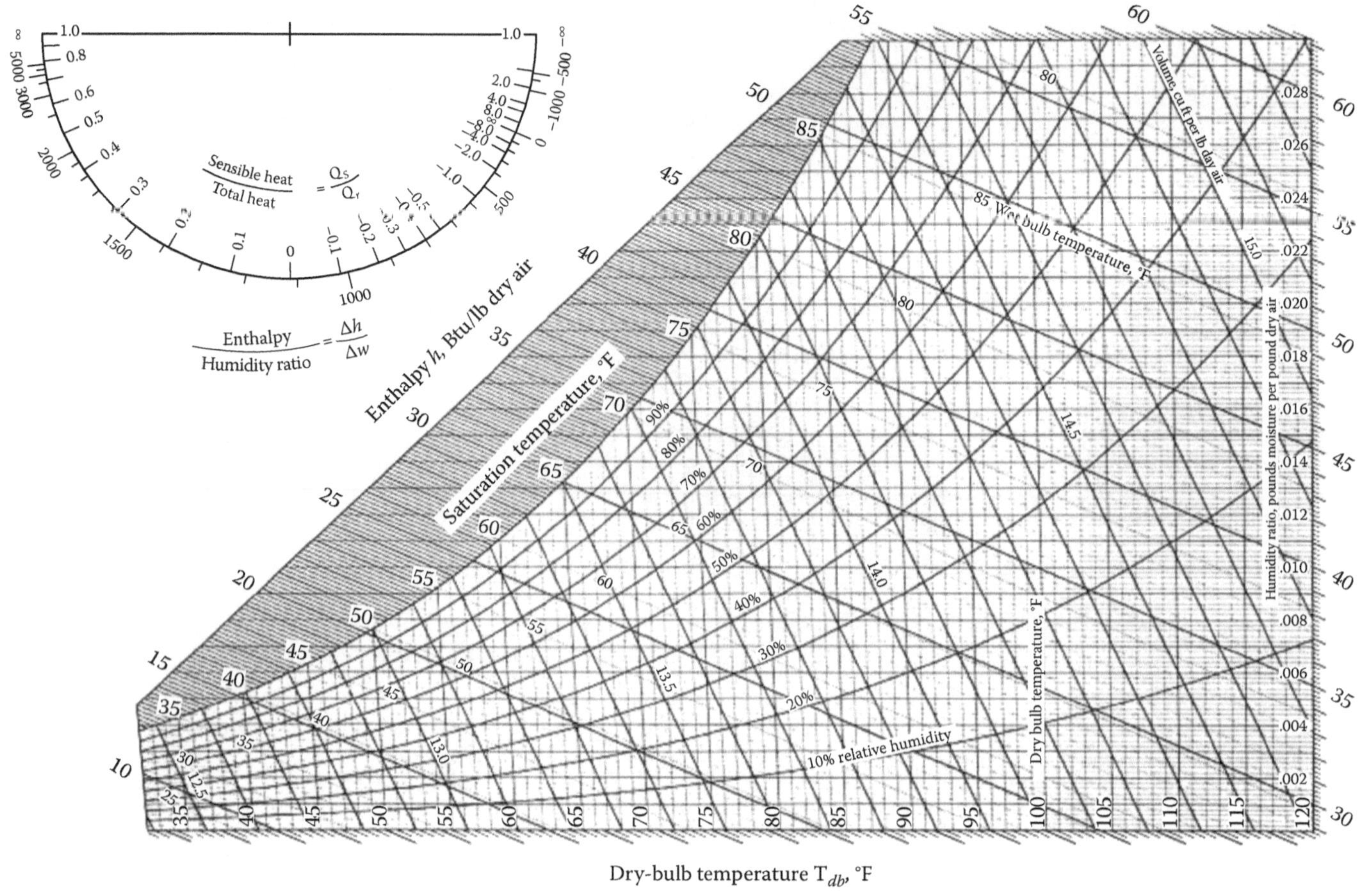

FIGURE 13.6
Psychrometric chart for sea level in IP units. (From ASHRAE, *Handbook of Fundamentals*, ASHRAE, Atlanta, GA, 2013. Copyright ASHRAE, www.ashrae.org.)

Vapor pressure p_{vap} lines are horizontal and parallel to lines of constant humidity ratio. Values are found by solving Equation 13.14 for the vapor pressure as a function of the humidity ratio. In some psychrometric charts (though not in Figures 13.6 and 13.7), these values are written along an auxiliary ordinate that is nearly linear. Likewise, dew point temperature lines are horizontal and parallel to lines of constant humidity ratio. Values are found from Equation 13.21 and are optionally written along an auxiliary, nonlinear ordinate. However, the dew point is just as easily read along the saturation curve itself. Details of the geometric construction of the psychrometric chart can also be found in Kuehn et al. (1998).

Example 13.7: Use of the Psychrometric Chart

Find the five listed properties for atmospheric air in a building maintained at 70°F (21°C). The wet-bulb temperature is measured to be 60°F (15.5°C), a typical interior condition in winter in heating climates.

Given: T_{db} = 70°F and T_{wb} = 60°F.

Figure: See Figure 13.6.

Assumptions: The atmospheric pressure is the standard atmosphere, 14.696 psia.

Find: ϕ, h, v, T_{dp}, and p_{vap}.

Solution

This problem can be solved easily by using the IP psychrometric chart. Note that only one of the two independent variables (T_{db}) used to plot the chart is known, but two different *independent* properties are known. Hence, the problem can be solved.

First, find the intersection of the 70°F dry-bulb vertical line and the 60°F wet-bulb line. This intersection is approximately at the 57% RH point.

At the same intersection point, the following values can be read by interpolation. The humidity ratio is 0.0088 lb_w/lb_a. The enthalpy is read from the scales at the upper left and bottom of the chart and determined as 26.5 Btu/lb_a. By interpolation the specific volume is 13.55 ft^3/lb_a. A better approach than this rough interpolation is to calculate the specific volume from the partial

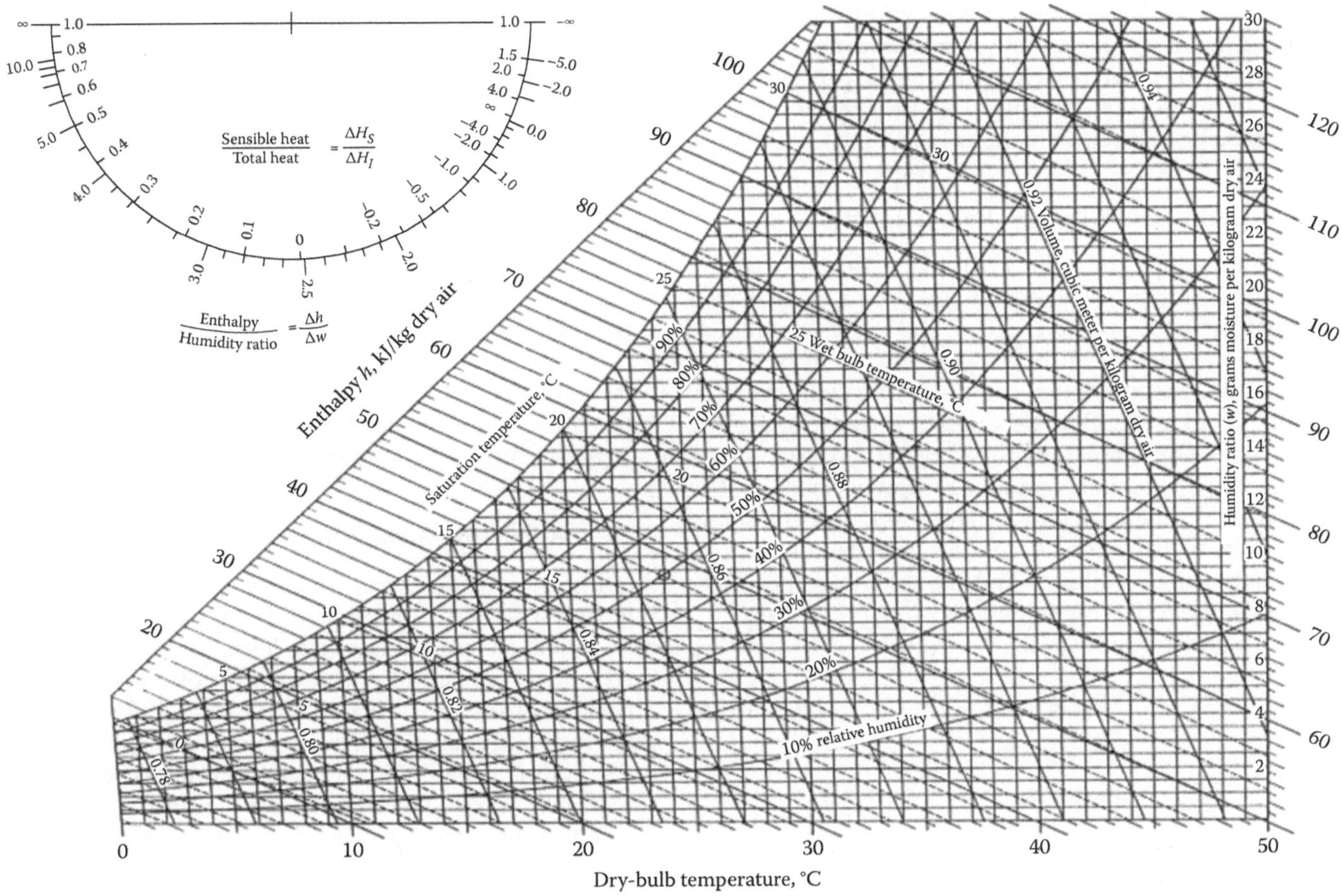

FIGURE 13.7
Psychrometric chart for sea level in SI units. (From ASHRAE, *Handbook of Fundamentals*, ASHRAE, Atlanta, GA, 2013. Copyright ASHRAE, www.ashrae.org.)

pressure p_{vap} and the dry-bulb temperature by using the ideal gas law:

$$v = \frac{R_a T_{db}}{p_{tot} - p_{vap}}$$

The partial pressure of water vapor p_{vap} cannot be found from the chart but is easy to find from Equation 13.14 since the humidity ratio W is already known:

$$p_{vap} = \frac{p_{tot}}{1 + 0.622/W} = \frac{14.696 \text{ psia}}{1 + 0.622/0.0088} = 0.21 \text{ psia}$$

An alternate method of finding p_{vap} is to use the basic equation for the RH (Equation 13.6)

$$p_{vap} = \phi \cdot p_{sat} = 0.57 \times 0.3633 \text{ psia} = 0.21 \text{ psia}$$

in which the saturation pressure has been determined from the second column of Table 12.3.

The value of p_{vap} can be used in the aforementioned ideal gas law to find a more accurate value of the specific volume. The dew point is found at the intersection of the humidity ratio line (0.0088) and the saturation curve. The dew point is 54°F.

Comments

The psychrometric chart can be read to about a few percent accuracy. For better accuracy, the online HCB software, tables, or psychrometric equations can be used.

Although the availability of computer programs allow for more accurate and faster determination of the moist-air properties (as well as the conversion of different system of units and arbitrary atmospheric pressures), the psychrometric chart is still used extensively by HVAC professionals in several aspects of design and analysis.

13.6 Basic Psychrometric Processes

Surprisingly, most HVAC systems using air as the heat transfer fluid involve only a few basic thermodynamic processes; a thorough understanding of which is sufficient for us to perform heating/cooling and moisture exchange analysis of any air system; however, complex it may appear. The five basic psychrometric processes discussed in this section are critical for proper appreciation of Chapter 19 dealing with all-air HVAC systems.

13.6.1 Underlying Conservation Equations

Consider a duct containing a device through which moist air is flowing. The device could be a cooling or a heating coil, and/or a humidifier. The analysis of moist-air processes flowing through such a device is based on the laws of conservation of mass and of energy at states (1) and (2) of Figure 13.8. Although in actual practice the properties of the moist air may not be uniform across the duct cross section (especially downstream of the device), such phenomena are neglected, and the focus is on bulk or fully mixed conditions. Further, assuming steady-state conditions and the duct to be perfectly insulated, the following equations apply:

Mass balance of dry air is

$$\dot{m}_{a1} = \dot{m}_{a2} \tag{13.26}$$

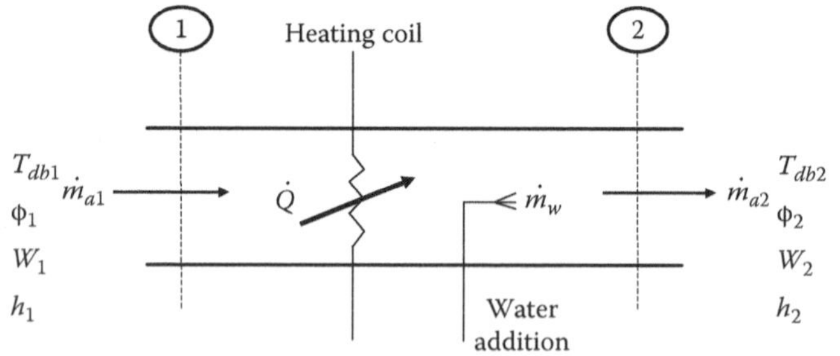

FIGURE 13.8
Portion of duct wherein air is conditioned from state 1 to state 2.

Mass balance on water vapor is

$$\dot{m}_{a1}W_1 + \dot{m}_w = \dot{m}_{a2}W_2 \tag{13.27}$$

Heat balance is

$$\dot{m}_{a1}h_1 + \dot{m}_w h_w + \dot{Q} = \dot{m}_{a2}h_2 \tag{13.28}$$

where

$\dot{Q}$ is the rate of heat added to the airstream (in W or Btu/h)
$\dot{m}_a$ is the mass flow rate of dry air (in kg_a/s or lb_a/h)
$\dot{m}_w$ is the rate of water (or water vapor) added to the airstream (in kg_w/s or lb_w/h)

13.6.2 Adiabatic Mixing of Airstreams

An important process in buildings is the adiabatic mixing of airstreams, i.e., mixing without addition or extraction of heat or without any heat loses/gains through the duct walls. For example, outdoor air is often mixed with a portion of the air returned from a conditioned space upstream of a cooling coil prior to delivery to the building's occupied space. This mixing is undertaken because the airflow should be larger than the minimum airflow needed to meet the space loads so as to ensure proper air motion and comfort in enclosed spaces. Whenever two moist airstreams are mixed, both energy and mass of each species—dry air and water vapor—must be conserved. By using these principles, it is straightforward to find the mixed exit condition and to plot the entire mixing process on a psychrometric chart.

Figure 13.9a depicts the mixing of two airstreams of different flow rates, temperatures, and moisture contents. Using the nomenclature of this figure, one can

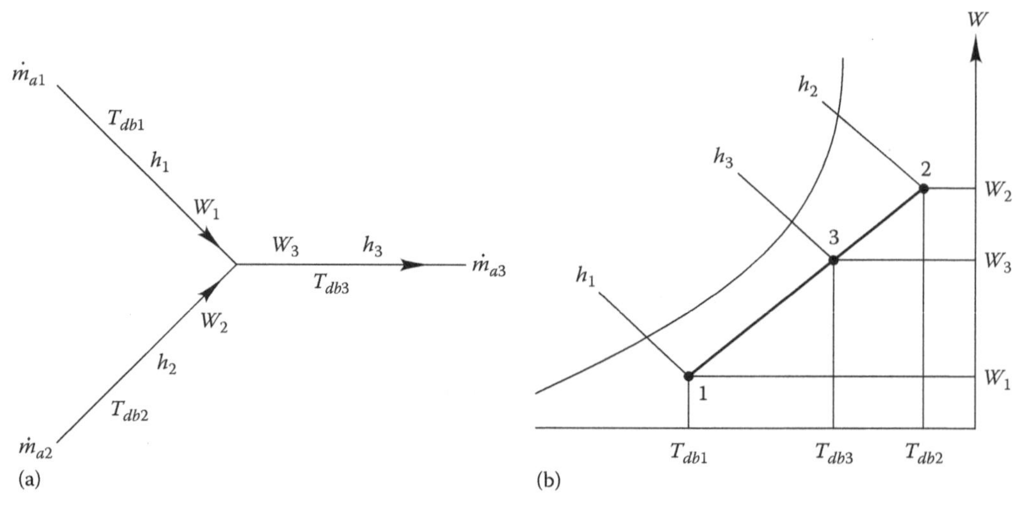

FIGURE 13.9
Adiabatic mixing: (a) schematic diagram of flow streams and (b) plot of process on psychrometric chart.

write the conservation of energy and conservation of mass (for both dry air and water), respectively, as follows:

Mass balance of dry air is

$$\dot{m}_{a1} + \dot{m}_{a2} = \dot{m}_{a3} \quad (13.29)$$

Mass balance of water vapor is

$$\dot{m}_{a1}W_1 + \dot{m}_{a2}W_2 = \dot{m}_{a3}W_3 \quad (13.30)$$

Heat balance is

$$\dot{m}_{a1}h_1 + \dot{m}_{a2}h_2 = \dot{m}_{a3}h_3 \quad (13.31)$$

The outlet airflow can be eliminated in Equations 13.30 and 13.31 by using Equation 13.29 to establish the following ratios:

$$\frac{\dot{m}_{a1}}{\dot{m}_{a2}} = \frac{h_2 - h_3}{h_3 - h_1} = \frac{W_2 - W_3}{W_3 - W_1} \quad (13.32)$$

If we consider the enthalpy and humidity ratio coordinates on the psychrometric chart in Figure 13.9b, this expression represents a straight line on the chart as a result of the relationships between similar triangles. The mixed conditions correspond to a point between the two inlet conditions at distances from these conditions inversely proportional to the ratios of mass flow (e.g., line segment 2–3 is inversely proportional to airflow $\dot{m}_{a2}$ at point 2). The expressions for mixed air humidity ratio and enthalpy are

$$W_3 = \frac{\dot{m}_{a1}W_1 + \dot{m}_{a2}W_2}{\dot{m}_{a1} + \dot{m}_{a2}} \quad (13.33)$$

$$h_3 = \frac{\dot{m}_{a1}h_1 + \dot{m}_{a2}h_2}{\dot{m}_{a1} + \dot{m}_{a2}} \quad (13.34)$$

It is more convenient to use temperatures rather than enthalpies during hand calculations. A simplifying approach widely used by practicing HVAC professionals is to adopt the mixed-air specific heat concept introduced in Equation 13.25. The exit conditions can also be solved analytically to yield

$$T_{db3} = \frac{\dot{m}_{a1}c_{p,m1}T_{db1} + \dot{m}_{a2}c_{p,m2}T_{db2}}{\dot{m}_{a1}c_{p,m1} + \dot{m}_{a2}c_{p,m2}} \quad (13.35)$$

A further simplification (also widely used) is to assume that the specific heats of the three air-mixture conditions are equal; in that case, the specific heat terms can be dropped from Equation 13.35. Since enthalpy is a linear function of the dry-bulb temperature according to Equation 13.35, the temperature of the mixture is also proportionately located as shown along the abscissa. Thus, these three moist-air properties (humidity ratio, enthalpy, and dry-bulb temperature) are just the weighted averages of the inlet properties. Note that other properties such as the outlet wet-bulb temperature, the RH, and the specific volume must be found from psychrometric calculations; they are not necessarily the weighted average of the inlet values.

If the inlet stream-specific volumes are not too different, one can divide Equation 13.34 by the average specific volume of the two inlet streams to replace the mass flow rate terms by corresponding volumetric flow rate terms (units of L/s or ft³/min are typical). This form of the equation is useful for preliminary design since volumetric flow rates are most often specified in HVAC analysis.

Example 13.8: Adiabatic Mixing of Airstreams

A stream of 5,000 cfm (2,359 L/s) outdoor air at 40°F (4.4°C) dry-bulb temperature and 35°F (1.7°C) wet-bulb temperature is adiabatically mixed with 15,000 cfm (7,078 L/s) of recirculated air at 75°F (23.9°C) dry-bulb and 50% RH. Find the dry-bulb temperature and wet-bulb temperature of the resulting mixture.

Given: $\dot{V}_1 = 5000$ cfm, $T_{db1} = 40°F$, $T_{wb1} = 35°F$, $T_{db2} = 75°F$, and $\phi_2 = 50\%$.

Figure: See Figure 13.9.

Assumptions: The atmospheric pressure is the standard atmosphere, 14.696 psia.

Find: T_{db3} and T_{wb3}.

Solution

Lookup values using psychrometric chart are as follows:

$$W_1 = 0.00315\ \text{lb}_w/\text{lb}_a,\quad v_1 = 12.65\ \text{ft}^3/\text{lb}_a,\quad h_1 = 13.2\ \text{Btu/lb}_a$$

$$W_2 = 0.00915\ \text{lb}_w/\text{lb}_a,\quad v_2 = 13.68\ \text{ft}^3/\text{lb}_a,\quad h_2 = 28.32\ \text{Btu/lb}_a$$

Mass flow rate of stream 1 is

$$\dot{m}_{a1} = \frac{\dot{V}_1}{v_1} = \frac{5000\ \text{ft}^3/\text{min}}{12.65\ \text{ft}^3/\text{lb}_a} = 395.3\ \text{lb}_a/\text{min}$$

Mass flow rate of stream 2 is

$$\dot{m}_{a2} = \frac{\dot{V}_2}{v_2} = \frac{15{,}000\ \text{ft}^3/\text{min}}{13.68\ \text{ft}^3/\text{lb}_a} = 1096.5\ \text{lb}_a/\text{min}$$

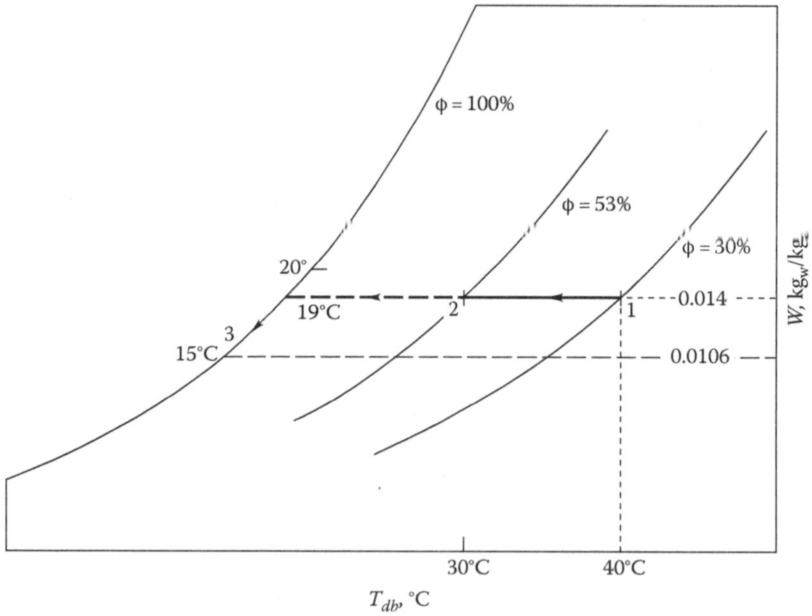

FIGURE 13.10
Cooling processes plotted on psychrometric chart: sensible cooling only (1–2) (Example 13.10).

Humidity ratio of mixed air is

$$W_3 = \frac{(\dot{m}_{a1} \cdot W_1 + \dot{m}_{a2} \cdot W_2)}{(\dot{m}_{a1} + \dot{m}_{a2})}$$

$$= \frac{395.3 \times 0.00315 + 1096.5 \times 0.00915}{1491.8} = 0.00756 \text{ lb}_w/\text{lb}_a$$

Similarly, the enthalpy of the mixture can be computed: $h_3 = 24.3$ Btu/lb$_a$.

Finally, the dry-bulb temperature and the wet-bulb temperature are found to be

$$T_{db3} = 66°\text{F (18.9°C)} \quad \text{and} \quad T_{wb3} = 56.5°\text{F (13.6°C)}$$

13.6.3 Sensible Heating and Cooling

A process is called *sensible* (either heating or cooling) when it involves a change in dry-bulb temperature only (i.e., the moisture level specified by the humidity ratio is unchanged during the process). This could apply to either heating (an increase in T_{db}) or to cooling (a decrease in T_{db}). In such a case, $\dot{m}_w = 0$, and $W_1 = W_2$, and Equation 13.28 reduces to

$$\dot{Q}_{sen} = \dot{m}_a(h_2 - h_1) \tag{13.36}$$

where $\dot{m}_a = \dot{m}_{a1} = \dot{m}_{a2}$.

A simplified expression useful for hand calculations is to use the specific heat equation:

$$\dot{Q}_{sen} = \dot{m}_a \cdot c_{p,m}(T_{db2} - T_{db1}) \tag{13.37}$$

where the mixed-air specific heat $c_{p,m}$ is given by Equation 13.25.

The process of sensible heating or cooling is represented by a straight line on the psychrometric chart as shown in Figure 13.10. Such a process occurs when moist air flows across a heating coil or when it flows over a cooling coil provided condensation does not occur. This is illustrated in the following two examples.

Example 13.9: Sensible Heating

Moist air enters a steam-heating coil at 40°F (4.4°C) dry-bulb temperature and 36°F (2.2°C) wet-bulb temperature at a rate of 2000 ft^3/min (943.8 L/s). The air leaves the coil at a dry-bulb temperature of 90°F (32.2°C). Determine

(a) The heat transfer rate that occurs at the coil
(b) The amount of steam needed if saturated steam at 220°F (100°C) enters the coil and leaves as a condensate at 220°F

Given: $\dot{V}_1 = 2000$ cfm, $T_{db1} = 40°$F, $T_{wb1} = 36°$F, and $T_{db2} = 90°$F.

Figure: See Figure 13.11, process 1–2.

Find: $\dot{Q}_h$.

Solution

Assumptions: Thermodynamic equilibrium exists. The local atmospheric pressure is 1 atm, or 14.696 psia.

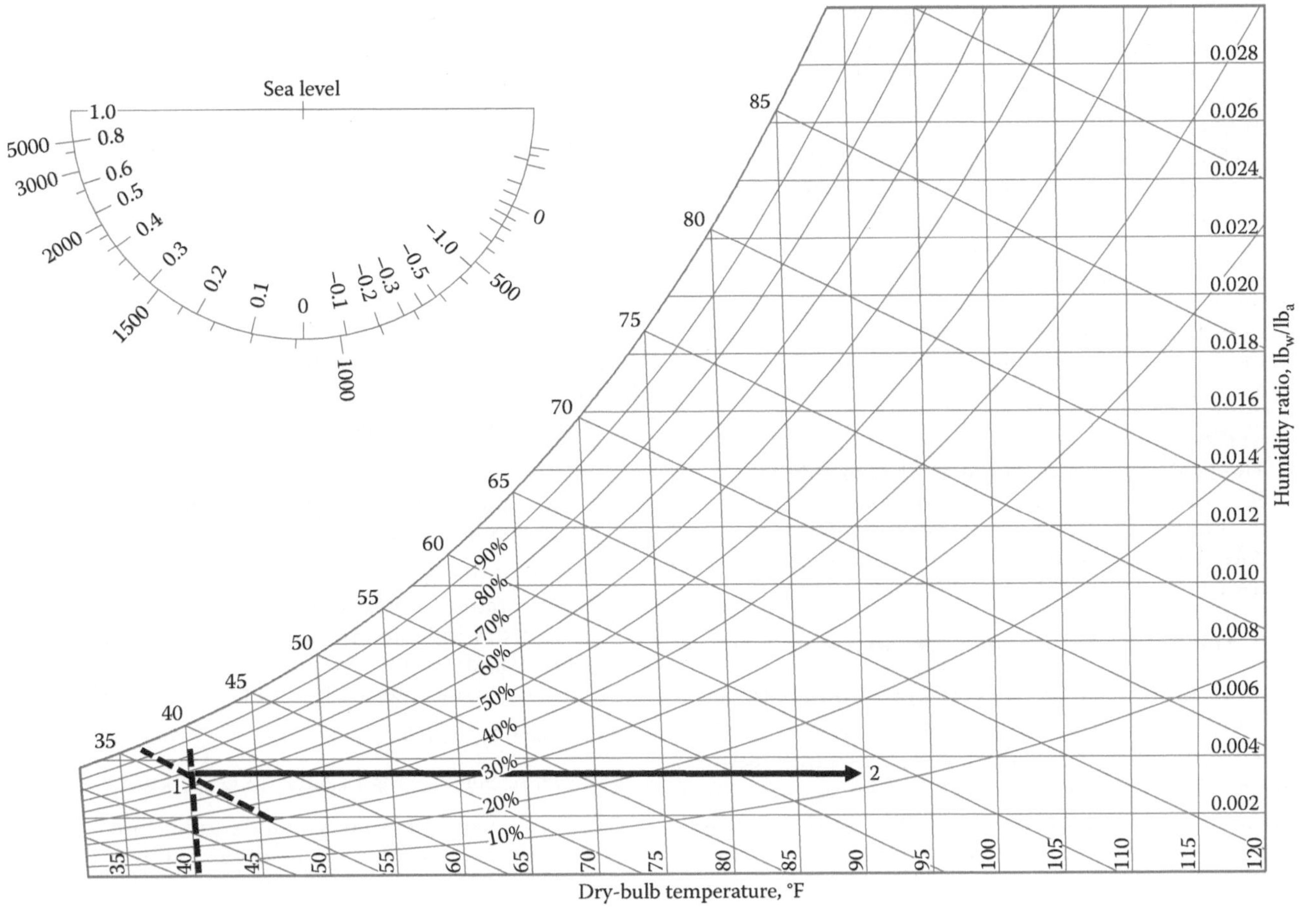

FIGURE 13.11
Sensible heating process (Example 13.9).

Lookup values: Let states 1 and 2 represent the entering and exiting airstream conditions, respectively. From the psychrometric chart, v_1 = 13.66 ft^3/lb$_a$, h_1 = 13.47 Btu/lb$_a$, and h_2 = 26.5 Btu/lb$_a$. From the steam tables, h_3 = 1153.4 Btu/lb$_w$, and h_4 = 168.1 Btu/lb$_w$.

The mass flow rate of moist air is

$$\dot{m}_a = \frac{\dot{V}_1}{v_1} = \frac{2000\ \text{ft}^3/\text{min}}{12.66\ \text{ft}^3/\text{lb}_\text{a}} = 160\ \text{lb}_\text{a}/\text{min} = 9479\ \text{lb}_\text{a}/\text{h}$$

Finally, from Equation 13.36,

$\dot{Q}_h = 9{,}479\,\text{lb}_\text{a}/\text{h} \times (26.5 - 13.47)\,\text{Btu}/\text{lb}_\text{a} =$ $123{,}511\,\text{Btu}/\text{h}\ (36.2\,\text{kW})$.

The amount of steam required is easily determined from an energy balance on the heating coil:

$$\dot{m}_{steam} = \dot{m}_a \frac{(h_2 - h_1)}{(h_3 - h_4)} = 9479 \times \left(\frac{26.5 - 13.47}{1153.4 - 168.1}\right)$$
$$= 125.3\ \text{lb}_\text{w}/\text{h}\ (56.5\ \text{kg}_\text{w}/\text{h})$$

Example 13.10: Sensible Cooling Process on Psychrometric Chart

Moist air is cooled from 40°C (104°F) and 30% RH to 30°C (86°F). Does moisture condense? What are the values of the RH and the humidity ratio at the process endpoint?

Given: T_{db1} = 40°C, ϕ_1 = 0.30, and T_{db2} = 30°C.

Figure: See Figure 13.10, process 1–2.

Find: W_2 and ϕ_2.

Solution

First, we must determine whether moisture condenses. In terms of psychrometric variables, we need to determine whether the final temperature is above or below the dew point. By referring to the SI psychrometric chart (Figure 13.7), we observe that the dew point temperature T_{dp} = 19°C. Therefore, moisture does not condense, and this is a *sensible cooling* process with no latent heat removal. The problem can be solved using the properties of moist air as plotted in Figure 13.10 shown by points 1–2.

Second, we note that the *humidity ratio* will not change since no moisture condensation has occurred. Therefore, the humidity ratio at the end of the process will be identical to that at the start, 0.014 kg_w/kg_a.

Third, the *relative humidity will* change since the relative ability of the cooler air to hold moisture is less than that of the warm air at the start of the process. Since this process is a constant humidity-ratio process, it is plotted as a horizontal line on the psychrometric chart. One moves to the left from the initial point until the intersection with the T_{db} = 30°C vertical line. At this point, we read a *relative humidity value* of 53%.

Comments

The heat removed from the air during this process can be read from the psychrometric chart by finding the enthalpy at the two endpoints. Alternatively, the online HCB software can be used to provide more accurate results.

13.6.4 Adiabatic Saturation and Evaporative Cooling

Section 13.4.3 described the concept of adiabatic saturation and its relation to the thermodynamic wet-bulb temperature. For practical purposes, this temperature is essentially the same as the wet-bulb temperature measured by a sling psychrometer. In this section, we will analyze adiabatic saturation and its relation to the evaporative cooling process (ASHRAE Fundamentals, 2013).

When the evaporation process does not reach the thermodynamic wet-bulb temperature, Equation 13.16 can be modified to

$$h_1 + (W_2 - W_1)h_{w2} = h_2 \tag{13.38}$$

where h_{w2} is the enthalpy of liquid water at T_{db2}.

By definition of an adiabatic process, water is supplied to the basin at T_{db2} (see Figure 13.2) with enthalpy h_{w2} to maintain the water level. The outlet temperature T_{db2} corresponding to a condition with T_{db1} at the inlet is the wet-bulb temperature of the environmental inlet air. We first use Equation 13.24 to find the enthalpy of moist air recalling that the specific heat of water $c_{p,w}$ is 1.0 Btu/(lb · °F) and that the reference temperature for water is taken by convention to be 32°F:

$$h_{w2} = c_{p,w}(T_2 - 32) = T_2 \text{ Btu/lb}_a \quad (13.39 \text{ SI})$$

Next, we use the equation for liquid-water enthalpy to solve for the humidity ratio W_1. Starting with Equation 13.38 to solve for W_1, we then use Equation 13.23 IP for enthalpy values and replace $\dot{m}_w = \dot{m}_a(W_2 - W_1)$ to find

$$W_1 = \frac{(1093 - 0.556T_{wb2})W_2 - 0.240(T_{db1} - T_{wb2})}{1093 + 0.444T_{db1} - T_{wb2}} \quad (13.40 \text{ IP})$$

Note that W_1 is nearly a linear function of the dry-bulb temperature since 1093 ≫ $0.444T_{db1}$ for a given wet-bulb temperature T_{wb2}. This explains why constant-wet-bulb-temperature lines look straight on the psychrometric chart.

In SI units, the expression for the inlet humidity ratio is

$$W_1 = \frac{(2501 - 2.381T_{wb2})W_2 - 1.0(T_{db1} - T_{wb2})}{2501 + 1.805T_{db1} - 4.186T_{wb2}} \quad (13.40 \text{ SI})$$

Example 13.11 illustrates how the humidity ratio can be found by analyzing data from an adiabatic saturator (essentially the same as a psychrometer).

Example 13.11: Adiabatic Saturation Process on Psychrometric Chart

If an adiabatic saturator (an idealization of an evaporative cooler) operates at 1 atm with an inlet dry-bulb temperature of 82°F (28°C) and an exit dry-bulb temperature of 65°F (18.5°C), what are the entering humidity ratio and RH?

Given: T_{db1} and T_{db2}.

Figure: See Figures 13.2 and 13.12 (process 1–2).

Assumptions: Thermodynamic equilibrium exists. The local atmospheric pressure is 1 atm, or 14.696 psia.

Find: W_1 and ϕ_1.

Lookup values: At 65°F, the saturation pressure is 0.306 psia from Table A3.

Solution

As indicated in Figure 13.12, this process is a constant wet-bulb temperature process, not a

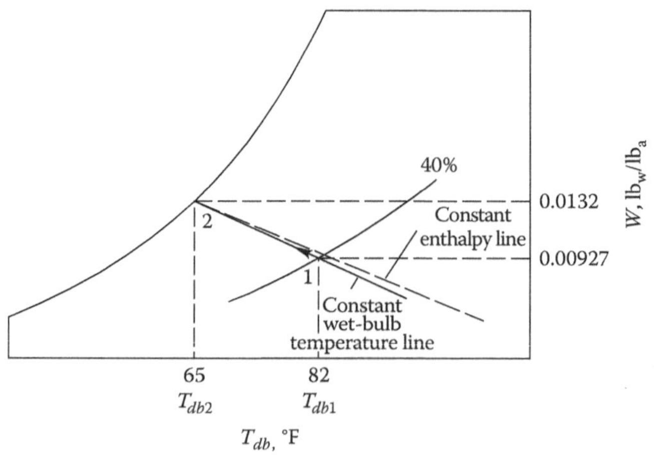

FIGURE 13.12
Adiabatic saturation process plot. The difference between the enthalpy and wet-bulb lines has been exaggerated. (The numerical values apply to Example 13.11.)

constant-enthalpy process because water supplied to maintain the sump water level carries enthalpy with it. From Equation 13.14, the humidity ratio at the exit can be found to be

$$W_2 = 0.622 \times \left(\frac{0.306}{14.696 - 0.306} \right) = 0.0132 \text{ lb}_w/\text{lb}_a$$

Equation 13.40, IP can now be used to find the inlet humidity ratio:

$$W_1 = \frac{(1093 - 0.556 \times 65) \times 0.0132 - 0.240 \times (82 - 65)}{1093 + 0.444 \times 82 - 65}$$

$$= 0.00927 \text{ lb}_w/\text{lb}_a$$

To find the inlet RH, Equation 13.14 can be solved for the water vapor pressure:

$$W_1 = 0.00927 = 0.622 \left(\frac{p_{vap,1}}{14.696 - p_{vap,1}} \right)$$

from which $p_{vap,1} = 0.216$ psia.

At 82°F, the saturation pressure is 0.542 psia from Table A3. Finally, from the definition of RH, we have

$$\phi_1 = \frac{p_{vap,1}}{p_{sat,1}} \times 100\% = \frac{0.216}{0.542} \times 100 = 40\%$$

Comments

It is a common human experience to have an evaporating liquid cool one's skin. Adiabatic saturation is the basic process involved in evaporative cooling of airstreams. It provides an economical method for reducing the dry-bulb temperature of air in climates with low ambient air humidity. Evaporative cooling is essentially an *incomplete adiabatic saturation process*. In actual practice, the exit airstream is often not entirely saturated or at the thermodynamic wet-bulb temperature. This same phenomenon is used in mechanical equipment for buildings in several ways, for example, space cooling can be achieved in dry climates by passing air over or through a wet medium or spray. Sensible heat is removed from the airstream by evaporating some of the liquid. Evaporative cooling equipment is discussed in Section 20.8. Cooling towers (discussed in Section 17.7) are used to reject heat from chillers using the same principle.

We can do a simple psychrometric calculation at this point, however, to illustrate how the evaporative process operates.

Example 13.12: Residential Evaporative Cooler

A residence in the Sonora Desert of Arizona is equipped with an evaporative cooler with an airflow rate of 3000 ft³/min (1416 L/s) sized by the rule of thumb of 2 ft³/min per square foot of residence area. If the outdoor conditions in summer are 110°F (43.5°C) and 20% RH, what is the water flow rate if 80% of the available dry-bulb temperature depression is achieved?

Given: $\dot{V}_1 = 3000$ cfm, $T_{db1} = 110$°F, and $\phi_1 = 0.2$.

Figure: See Figure 13.13.

Assumptions: The site is at sea level. (The Sonora Desert is about 300–400 m above sea level, but to avoid the complications of using the basic psychrometric equations, we will assume a sea-level elevation and use the sea-level psychrometric chart.)

Find: $\dot{m}_w$.

Lookup values: $W_1 = 0.011$ lb$_w$/lb$_a$, $T_{wb2} = 75.5$°F (the wet-bulb temperature is the minimum temperature achievable in this process).

Solution

The exit dry-bulb temperature is first determined given that 80% of available dry-bulb temperature depression is achieved:

$$0.8 = \frac{110 - T_{db2}}{110 - 75.5} \quad \text{or} \quad T_{db2} = 82.4°\text{F}$$

To find the exit humidity ratio, one moves along the 75.5°F wet-bulb line until the 82.4°F dry-bulb temperature is reached. At this point, we read the exit humidity ratio as $W_2 = 0.0176$ lb$_w$/lb$_a$.

From the chart, we also read the inlet-specific volume as 14.6 ft³/lb$_a$.

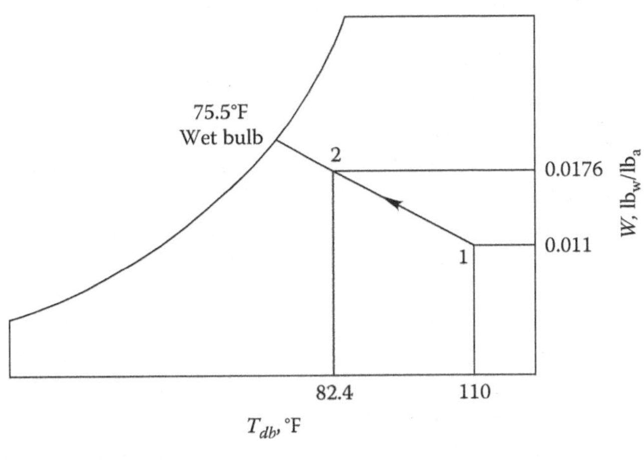

FIGURE 13.13
Evaporative cooler process plot (Example 13.12).

Then, the needed water supply rate is

$$\dot{m}_w = \dot{m}_a W = \frac{\dot{V}\Delta W}{v}$$

$$= 3000\,\text{ft}^3/\text{min} \times \frac{(0.0176 - 0.011)\ \text{lb}_w/\text{lb}_a}{14.6\ \text{ft}^3/\text{lb}_a}$$

$$= 1.36\ \text{lb}_w/\text{min}$$

Comments

This flow rate, which amounts to about 9.5 gal/h (0.01 L/s), is not excessive for a desert climate when compared to the savings in electric energy needed for mechanical cooling. For example, if water costs $2/1000 gal (typical of an arid western state in the United States), the water consumed costs less than 2¢/h. The electricity to operate the fan in the evaporative cooler would cost another 2–3¢/h. On the other hand, an electric cooling system would cost about four to eight times as much, depending on the weather and load as well as the electric system's efficiency.

13.6.5 Cooling and Dehumidification

Summertime occupant comfort in buildings in warm climates requires the removal of heat from the interior of buildings. Since occupants and outdoor ventilation air also introduce humidity into buildings, comfort requires control of humidity levels as well. Air cooling and dehumidification are usually accomplished in the same piece of equipment, called the *air conditioner* in small buildings or the *air handler cooling coil* in large buildings. To achieve space cooling, one must supply air to the conditioned space at a dry-bulb temperature sufficiently below the space temperature to remove sensible heat gains. To achieve dehumidification, this same air must also be supplied at sufficiently low moisture content to be able to remove moisture gains in the space. Figure 13.14a and b depict the general cooling–dehumidification process 1–2 both schematically and on a psychrometric chart. This change of air conditions is most often achieved in a single piece of equipment—the cooling coil—installed in the air supply duct. A cold liquid circulates through the coil at a temperature sufficiently low to remove needed amounts of latent and sensible heat. If the coil surface temperature is below the dew point at inlet condition 1, moisture is condensed to liquid water. The liquid is drained from the coil as shown.

Note that the final condition of air leaving the cooling coil is not on the saturated curve (though this would be an idealized process). A simplified visual explanation as to why this occurs is as follows. The cooling coil is made up of several rows of coils with complex fin geometries with chilled water supplied in a combination of parallel and series flow configurations (see Figure 11.17). The actual process path of the air as it flows over the numerous rows of the cooling coil depends on the type of surface design, on surface temperature distribution both longitudinally and across the coil face, and also on flow conditions (velocity, flow regime). Some of the air molecules can be viewed as coming in direct and intimate contact with the cold coil surface and are thereby cooled and dehumidified down to point 3 (see Figure 13.14b), while some of the air entirely bypasses the cooling coil surface and essentially leaves at state 1. The mixture of both these airstreams results in a bulk condition represented by state point 2 that lies on the straight line joining the two end points 1 and 3 (as discussed in Section 13.6.2). Some textbooks show this process as a curved line for a more realistic representation, but we have adopted the simpler straight-line representation.

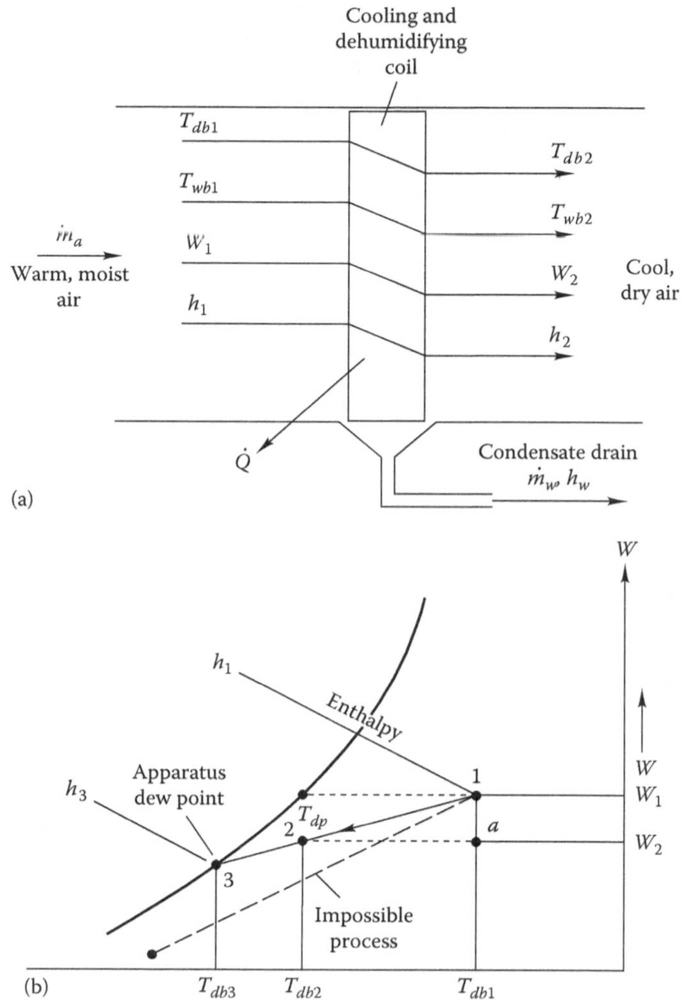

FIGURE 13.14
Cooling and dehumidification process plot: (a) equipment schematic and air properties and (b) process plotted on psychrometric chart. Note that process 1–4 is impossible since the process line does not intersect the saturation curve; there is no apparatus dew point.

Example 13.13: Latent and Sensible Cooling Process on Psychrometric Chart

Consider an ideal process where moist air is cooled from 40°C (104°F) and 30% RH to 15°C (50°F). Does moisture condense? What are the values of the RH and the humidity ratio at the process endpoint?

Given: $T_{db1} = 40°C$, $\phi_1 = 0.30$, and $T_{db2} = 15°C$.

Figure: See Figure 13.15, process 1–3.

Find: W_3 and ϕ_3.

Solution

The inlet conditions are identical to those in Example 13.10 while the air is cooled to a lower temperature. The dew point is known to be 19°C for the moist air at the beginning of the process. Therefore, moisture condenses since the outlet temperature (= 15°C), for this case is below this dew point. This is a process where both sensible and latent cooling occur.

Inspection of process 1–3 in Figure 13.15 shows that the *humidity ratio changes* since moisture condensation has occurred. The saturated air at the exit temperature of 15°C has a humidity ratio of 0.0106 kg_w/kg_a. Finally, the *relative humidity changes* to the saturated value, i.e., RH of 100%, as shown in Figure 13.15.

This process can be viewed as a two-part process: (1) constant-humidity-ratio cooling until saturation and (2) constant-relative-humidity cooling with condensation of moisture. Figure 13.15 shows the process as it is displayed within the HCB software. Analysis of such problems is illustrated by the following example. However, some pertinent equations need to be introduced first.

The heat and mass transfer across the cooling coil can be expressed in terms of the initial and final states.

The rate of water condensation is

$$\dot{m}_w = \dot{m}_a(W_1 - W_2) \tag{13.41}$$

The rate of total heat transfer is

$$\dot{Q}_{tot,c} = \dot{m}_a\left[(h_1 - h_2) - h_{liq}\left(W_1 - W_2\right)\right] \tag{13.42}$$

in which h_{liq} is the enthalpy of liquid water at the temperature at which it leaves the cooling coil.

Though the moisture that condenses out of the airstream does so at various temperatures ranging from the initial dew point to its final saturation temperature, an average value is often assumed.

The aforementioned equation gives the total rate of heat transfer from the moist air. The last term is usually small compared to the other term and is often neglected (this is illustrated in Example 13.14).

Cooling and dehumidification process involves both sensible and latent heat transfer. The sensible heat transfer $\dot{Q}_{sen}$ results in a decrease in dry-bulb temperature, while the latent heat transfer $\dot{Q}_{lat}$ is associated with the decrease in humidity ratio. These quantities can be estimated as follows. Let point "*a*" be the intersection point between the constant dry-bulb temperature line from point 1 and the constant humidity ratio line at point 2 (see Figure 13.14). Then, rate of sensible heat transfer is

$$\dot{Q}_{sen} = \dot{m}_a(h_a - h_2) \tag{13.43}$$

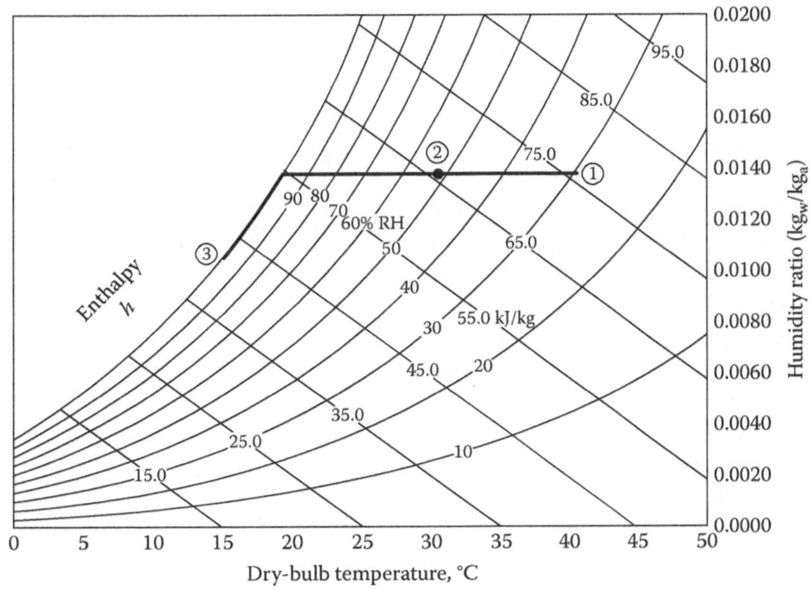

FIGURE 13.15
Sensible and latent cooling processes as calculated with the online HCB software (Example 13.3).

and the rate of latent heat transfer is

$$\dot{Q}_{lat} = \dot{m}_a(h_1 - h_a) \tag{13.44}$$

Analogous to a sensible heating and cooling process, simplified expressions are also often used for hand calculations (see Equations 13.23 and 13.24):

The rate of sensible heat transfer is

$$\dot{Q}_{sen} = \dot{m}_a c_{p,m}(T_{db1} - T_{db2}) \tag{13.45}$$

The rate of latent heat transfer is

$$\dot{Q}_{lat} = \dot{m}_a h_{vap}(W_1 - W_2) \tag{13.46}$$

where

$c_{p,m}$ is determined from Equation 13.25
and $h_{vap} = h_{liq\text{-}vap,ref}$ as in Equation 13.24

It is customary to characterize the relative contributions of sensible and latent heat transfer rates by the *coil sensible heat factor* (SHF_{cc}) defined as follows:

$$SHF_{cc} = \frac{\dot{Q}_{sen}}{\dot{Q}_{sen} + \dot{Q}_{lat}} \tag{13.47}$$

The SHF_{cc} of the coil is an important operational parameter specified by the manufacturer, and the HVAC system designer has to select a cooling coil with the right numerical value to assure satisfactory operation of the HVAC system. The SHF_{cc} should not be confused with the room SHF_{space} that relates to the ratio of the sensible to the total loads of the room or space.

Another important parameter is the *apparatus dew point* (*ADP*) temperature (also referred to as the coil dew point temperature), which is given by T_{db3} (Figure 13.14b). This temperature value represents an average cooling coil surface temperature that would produce the same conditioning as the actual process. The ADP value in turn impacts the operation of the chiller system that has to produce chilled water at the desired temperature. Note that line 1–4 in Figure 13.14b represents an impossible process since there is no possible ADP, i.e., no surface sufficiently cold enough for moisture to condense. Although such processes can be drawn on the chart, they cannot occur in practice, and so remedial design and operational modifications need to be made (this aspect is discussed more fully in Section 19.2.4).

Another important parameter is the *coil bypass (BP) factor*, which is the fraction of the airstream that passes through the cooling coil that is not being conditioned at all (recall the simplified explanation given earlier of what happens to the airstream while it passes through the cooling coil). This coil bypass factor BP_{cc} is given by (Figure 13.14)

$$BP_{cc} = \frac{\text{Air mass flow rate which bypasses coil surface}}{\text{Total air mass flow rate}} = \frac{\text{Segment } 2-3}{\text{Segment } 1-3} = \frac{T_{db2} - T_{db3}}{T_{db1} - T_{db3}} \tag{13.48}$$

Sometimes, the *coil contact factor* $CF_{cc} = (1 - BP_{cc})$ is used. One would think that manufacturers ought to design coils with zero bypass factors since it would represent better cooling coil performance. The reason for not doing so is that there is a trade-off between higher contact factor and resulting pressure drop across the coil which the circulating fan must overcome. Thus, coils with higher contact factors tend to require higher fan electric power. Typical ranges for the BP factor are 0.3–0.5 for coils in residences and 0.1–0.2 for those in commercial buildings.

Example 13.14: Cooling of Moist Air

Outdoor air at sea level enters an air conditioner's cooling coil at 2000 ft³/min (944 L/s). The inlet dry-bulb condition is 100°F (38°C) and the wet-bulb condition is 75°F (24°C). The air is cooled to a dry-bulb temperature of 55°F (13°C) and 90% RH. Assume that the liquid water leaves the coil at 50°F.

(a) Find the total heat transfer rate as well as the sensible and latent cooling loads on the coil.
(b) Determine the sensible heat factor, apparatus dew point, and the bypass factor of the cooling coil.

Given: $\dot{V}_1 = 2000$ cfm, $T_{db1} = 100°F$, $T_{wb1} = 75°F$, $T_{db2} = 55°F$, $\phi_2 = 0.9$, and $T_{cc,liq} = 50°F$.

Figure: See Figure 13.16.

Find: $\dot{Q}_{tot,c}$, $\dot{Q}_{sen}$, $\dot{Q}_{lat}$, SHF_{cc}, ADP, BP_{cc}.

Lookup values: The enthalpy of liquid water at 50°F is 18.06 Btu/lb_w from Table A3. From the sea-level psychrometric chart, we determine the following inlet and outlet values in IP units:

State	W, lb_w/lb_a	h, Btu/lb_a	v, ft^3/lb_a
1	0.013	38.4	14.4
2	0.0083	22.2	—

Solution

(a) First, we will find the mass flow rate:

$$\dot{m}_a = \frac{2000\ ft^3/min \times 60\ min/h}{14.4\ ft^3/lb_a} = 8333\ lb_a/h$$

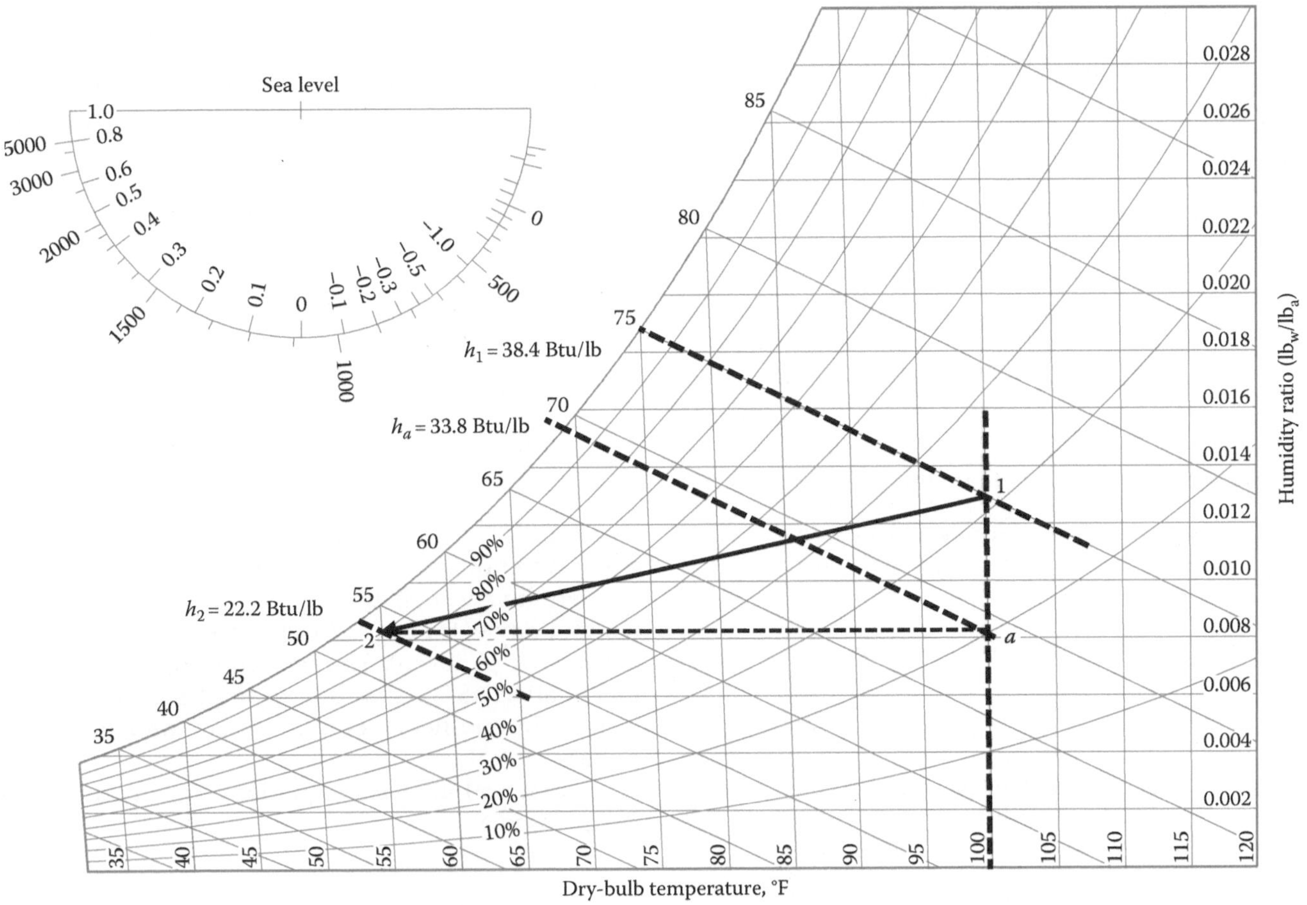

FIGURE 13.16
Cooling of moist air (Example 13.14).

Then, using Equation 13.42, the heat rate can be found since all values on the right-hand side are known:

$$\begin{aligned}\dot{Q}_{tot} &= \dot{m}_a\left[(h_1 - h_2) - h_{liq}(W_1 - W_2)\right]\\ &= 8,333\ \text{lb}_a/\text{h} \times [(38.4 - 22.2)\ \text{Btu/lb}_a\\ &\quad - 18.06\ \text{Btu/lb}_w \times (0.013 - 0.0083)\ \text{lb}_w/\text{lb}_a]\\ &= (134,995 - 707.3)\ \text{Btu/h} = 134.3\ \text{kBtu/h}\ (39.3\ \text{kW})\\ &= 11.2\ \text{tons of cooling}\end{aligned}$$

In order to determine the contributions of the sensible and latent coil loads, we locate point "*a*" as the intersection on the psychrometric chart of a horizontal line drawn from point 2 and the vertical line drawn from point 1. The enthalpy is found to be $h_a = 33.8$ Btu/lb_a.

From Equation 13.43,

$$\begin{aligned}\dot{Q}_{sen} &= 8,333\ \text{lb}_a/\text{h} \times (33.8 - 22.2)\ \text{Btu/lb}_a\\ &= 96,663\ \text{Btu/h} = 8.06\ \text{tons}\end{aligned}$$

and from Equation 13.44,

$$\begin{aligned}\dot{Q}_{lat} &= 8,333\ \text{lb}_a/\text{h} \times (38.4 - 33.8)\ \text{Btu/lb}_a\\ &= 38,332\ \text{Btu/h} = 3.19\ \text{tons}\end{aligned}$$

Finally, the total cooling load on the coil is

$$\dot{Q}_{tot,c} = \dot{Q}_{sen} + \dot{Q}_{lat} = 134,995\ \text{Btu/h} = 11.25\ \text{tons}$$

It is instructive to compare these values with those from the simplified approach often adopted for hand calculations. The average specific heat of the air water mixture is (Equation 13.25)

$$\begin{aligned}c_{p,m} &= c_{p,a} + W \cdot c_{p,vap} = 0.24 + \frac{0.444 \times (0.013 + 0.0083)}{2}\\ &= 0.245\ \text{Btu/(lb}_a \cdot {}^\circ\text{F)}\end{aligned}$$

Then, from Equation 13.45,

$$\dot{Q}_{sen} = 8,333 \text{ lb}_a/\text{h} \times 0.245 \text{ Btu/lb}_a \cdot °\text{F} \times (100-55) \text{ Btu/lb}_a = 91,871 \text{ Btu/h} = 7.66 \text{ tons}$$

And from Equation 13.46,

$$\dot{Q}_{lat} = 8,333 \text{ lb}_a/\text{h} \times 1075 \text{ Btu/lb}_{liq} \times (0.013 - 0.0083) \text{ lb}_w/\text{lb}_a = 42,102 \text{ Btu/h} = 3.51 \text{ tons}$$

Finally,

$$\dot{Q}_{tot,c} = 91,871 + 42,102 = 133,973 \text{ Btu/h} = 11.16 \text{ tons}$$

Note that the differences in the sensible, latent, and total loads by both methods are −5%, +10%, and less than 1%, respectively.

(b) The cooling coil sensible heat factor is

$$\text{SHF}_{cc} = \frac{\dot{Q}_{sen}}{\dot{Q}_{sen} + \dot{Q}_{lat}} = \frac{8.06}{11.25} = 0.716.$$

We draw a straight line between points 1 and 2 extending till the saturated line (not shown in Figure 13.16). The corresponding dry-bulb temperature T_3 is the ADP, and this is read off as ADP = T_{db3} = 50°F.

Finally, from Equation 13.48, we find

$$\text{BP}_{cc} = \frac{T_{db2} - T_{db3}}{T_{db1} - T_{db3}} = \frac{55-50}{100-50} = 0.1.$$

Comments

This cooling effect must be provided by the chiller. Note that the amount of energy leaving the coil in the form of liquid condensate for this problem (707 Btu/h as determined earlier) is very small, about 0.5% of the total coil load.

The quantity of water removed is significant. Using Equation 13.41, we find that 8333 × (0.013 − 0.0083) = 39.2 lb_w/h (4.7 gal/h or 17.8 L/h) is condensed out. The designer must include a method of routing this condensed water to a drain for proper disposal.

13.6.6 Heating and Humidification

A generalization of the adiabatic saturation process discussed earlier is a process in which heat is added during evaporation. This method is used to simultaneously humidify and heat the air in buildings in winter, when low-humidity conditions could affect health. The humidification is typically in the form of water spray or water vapor (steam). If no moisture is added, only *sensible* heating has been done, but if moisture and heat are added, both sensible and *latent* heating, respectively, have been accomplished. The general results in this section also apply to the special case of only sensible heating when no moisture is added.

The mass and energy conservation equations can be written as (see Figure 13.17a) follows:

The rate of water evaporation is

$$\dot{m}_w = \dot{m}_a(W_2 - W_1) \tag{13.49}$$

The rate of heat transfer is

$$\dot{Q}_h = \dot{m}_a[(h_2 - h_1) - (W_2 - W_1)h_{liq,2}] \tag{13.50}$$

where h_{liq} is the enthalpy of the humidification medium, either water or steam as appropriate.

Alternatively, moisture can be added to the airstream without any extra heat addition (i.e., the heating coil is inactive or removed). The heat input term ($\dot{Q}_h$) is zero in Equation 13.50. Practical examples of this process include humidifiers, cooling towers, and evaporative coolers.

For practical analysis, Equations 13.49 and 13.50 are reexpressed as

$$h_1 + \frac{\dot{Q}_h}{\dot{m}_a} + (W_2 - W_1)h_{liq,2} = h_2 \tag{13.51}$$

where

$\dot{Q}_h$ is the rate of heat addition to the moist air
$\dot{m}_a$ is the dry-air mass flow rate

Makeup water is supplied with enthalpy $h_{liq,2}$ at T_{db2}.

If we were designing heating and humidification systems, a graphical construct in the psychrometric chart can be useful. Figure 13.17b shows the warming and humidification process when using steam. The definition of the humidity ratio and the conservation of mass can be used to relate water and mass flow rates. From this, the ratio of the enthalpy change to the humidity ratio change in Equation 13.51 can be found by using the conservation of mass of water $\dot{m}_w = \dot{m}_a(W_2 - W_1)$. Thus,

$$\frac{\Delta h}{\Delta W} = \frac{\dot{Q}_h + \dot{m}_w h_{liq,2}}{\dot{m}_w} \tag{13.52}$$

The ratio (enthalpy/humidity ratio) is independent of the airflow rate, and we can construct a line with this slope on the psychrometric chart. The constructed line will have a slope related to the space heating load and the moisture added to the airstream by the steam.

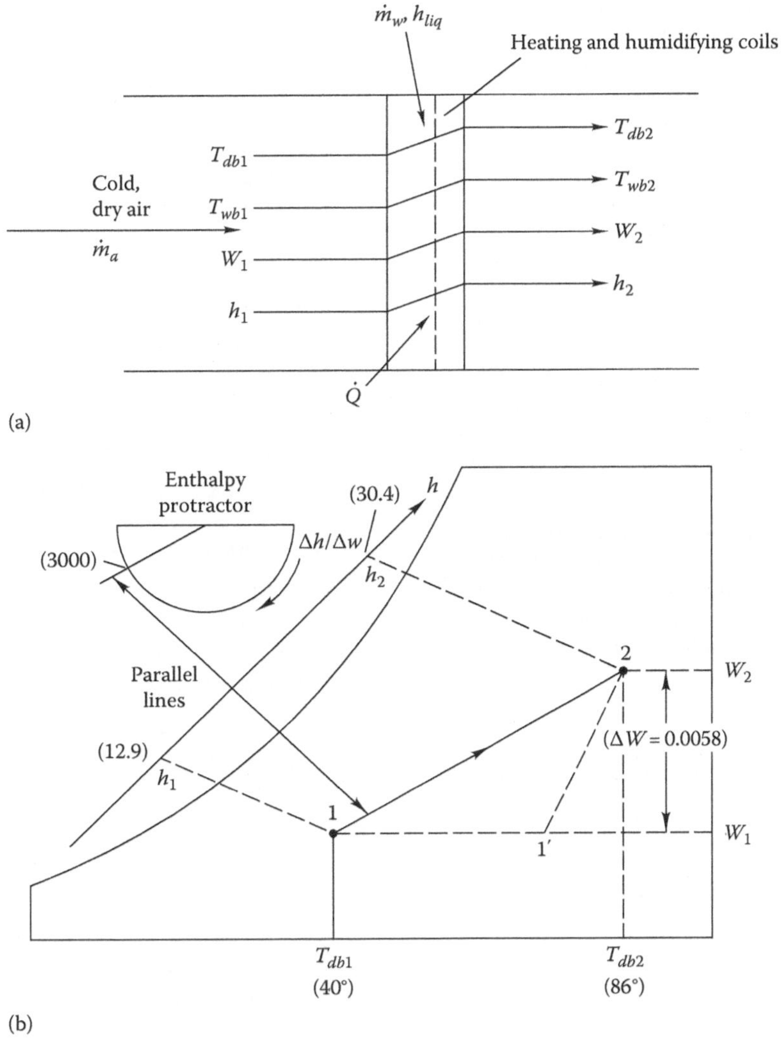

FIGURE 13.17
Heating and humidification process plot: (a) equipment schematic and air properties and (b) process plotted on psychrometric chart. (The numerical values in parentheses apply to Example 13.15.)

The principal utility of this ratio is that for a desired outlet (or inlet) state, the conditions of the inlet (or outlet) point must lie on a straight line with slope given by Equation 13.52. Often, psychrometric charts include a protractor-like construction in the upper left-hand corner that enables the user to establish the slope and then pass a straightedge through the known inlet point in order to find the outlet condition (see Figure 13.6).

Figure 13.17b shows how the protractor is used for the air warming and humidification process 1–2. In actual practice, air is first warmed and then humidified in separate equipment, as shown by the two-step process 1–1′ and 1′–2. Of course, if this graphical approach is not sufficiently accurate, the expression for enthalpy of moist air developed earlier or look up tables can be used.

Example 13.15: Warming and Humidification of Cold, Dry Air

Outdoor air at 40°F (4.5°C) and 60% humidity is heated and humidified by saturated steam at 230°F (110°C). The airflow rate is 15,000 ft³/min (7,080 L/s), and heat is added to the air at the rate of 765,000 Btu/h (224 kW) while it absorbs 415 lb_m/h (0.0524 kg/s) of steam. What are the dry- and wet-bulb temperatures at the exit of this heater/humidifier if it is located at sea level?

Given: $T_{db1} = 40°F$, $\phi_1 = 60\%$, $\dot{V}_a = 15{,}000\ ft^3/min = 900{,}000\ ft^3/h$, $\dot{m}_w = 415\ lb_w/h$, and $\dot{Q}_h = 765{,}000$ Btu/h.

Figure: See Figure 13.17b.

Find: T_{db2} and T_{wb2}.

Lookup values: Steam enthalpy at 230°F from Table A3 by interpolation, $h_{vap} = 1157$ Btu/lb$_w$ = $h_{liq,2}$.

The inlet condition moist-air enthalpy is read from the psychrometric chart (Figure 13.6) as h_1 = 12.9 Btu/lb$_a$. Also, Specific volume $v_1 = 12.65$ ft^3/lb$_a$.

Solution

This problem can be solved using the graphical approach with the enthalpy protractor in Figure 13.6. The enthalpy/humidity ratio needed to use the graphical construction is found from Equation 13.52:

$$\frac{\Delta h}{\Delta W} = \frac{765{,}000 \text{ Btu/h} + 415 \text{ lb}_w/\text{h} \times 1{,}157 \text{ Btu/lb}_w}{415 \text{ lb}_w/\text{h}}$$

$$= 3{,}000 \text{ Btu/lb}_a$$

This value is located on the enthalpy protractor, and a line is constructed in the chart with the same slope passing through the known process start point 1. To find the termination point 2 of the process, find the exit enthalpy as follows. First, find the humidity ratio increase ΔW:

$$\Delta W = \frac{\dot{m}_w}{(\dot{V}_a / v_1)} = \frac{415 \text{ lb}_w/\text{h}}{(900{,}000 \text{ ft}^3/\text{h}/12.65 \text{ ft}^3/\text{lb}_a)}$$

$$= 0.0058 \text{ lb}_w/\text{lb}_a$$

The end-state enthalpy is then found as follows:

$$h_2 = h_1 + \frac{\Delta h}{\Delta W}\Delta W = 12.9 + 3000 \times 0.0058 = 30.4 \text{ Btu/lb}_a$$

The outlet enthalpy is used to terminate the construction line at point 2, as shown in Figure 13.17. At point, we can read from the chart the following values of the outlet wet- and dry-bulb temperatures: T_{wb} = 66°F (18.9°C) and T_{db} = 86°F (30°C).

Comments

This example demonstrates how the protractor construction can save considerable time. If the chart had not been used, then complicated, iterative solutions involving a number of equations in Section 13.4 would have been needed.

Example 13.16: Heating and Humidification

Outdoor winter air enters a heating and humidification system at a rate of 900 cfm (425 L/s) and at 41°F (5°C) dry-bulb and 35°F (1.7°C) wet-bulb temperatures. The air is to be heated to 88°F (31°C) and humidified to a humidity ratio of 0.008 lb$_w$/lb$_a$. Determine the amount of sensible heating that needs to be provided by the heating coil and the amount of saturated steam at 212°F (100°C) that needs to be provided for humidification.

Given: T_{db1} = 41°F, T_{wb1} = 35°F, T_{db3} = 88°F, W_3 = 0.008 lb$_w$/lb$_a$, and $\dot{V}_a$ = 900 ft^3/min.

Figure: See Figure 13.18.

Find: $\dot{Q}_h$ and $\dot{m}_w$.

Lookup values: Saturated steam enthalpy at 212°F from Table A3 in h_{vap} = 1150.5 Btu/lb$_w$ (which is equal to $h_{liq,2}$). From psychrometric chart, h_1 = 13 Btu/lb$_a$, v_1 = 12.67 ft^3/lb$_a$, W_1 = 0.003 lb$_w$/lb$_a$, and h_3 = 30 Btu/lb$_a$.

Solution

The mass flow rate of dry air is

$$\dot{m}_a = \frac{\dot{V}_a}{v_1} = \frac{900 \text{ ft}^3/\text{min}}{12.67 \text{ ft}^3/\text{lb}_a} = 71.03 \text{ lb}_a/\text{min} = 4262 \text{ lb}_a/\text{h}$$

The mass flow rate of steam is

$$\dot{m}_w = \dot{m}_a(W_3 - W_1) = 4262 \text{ lb}_a/\text{h} \times (0.008 - 0.003) \text{ lb}_w/\text{lb}_a$$

$$= 21.31 \text{ lb}_w/\text{h} \ (9.6 \text{ kg/h})$$

An energy balance on the humidifier only is used to determine the air enthalpy at state 2:

$$h_2 = h_3 - \left(\frac{\dot{m}_w}{\dot{m}_a}\right) \times h_{vap}$$

$$= 30 \text{ Btu/lb}_a - \left(\frac{21.31}{4262}\right) \times 1150.5 \text{ Btu/lb}_a$$

$$= 24.25 \text{ Btu/lb}_a$$

From the psychrometric chart, this corresponds approximately to an exit condition of T_{db2} = 87°F, i.e., the humidification occurred essentially at constant T_{db}.

Finally, the heating coil energy is easily found as

$$\dot{Q}_h = \dot{m}_a(h_2 - h_1) = 4{,}262 \text{ lb}_a/\text{h} \times (24.25 - 13) \text{ Btu/lb}_a$$

$$= 47{,}936.8 \text{ Btu/h} \ (14.05 \text{ kW})$$

13.7 Closure

This chapter has provided a discussion along with relevant equations by which one can analyze various psychrometric processes that occur in HVAC systems. A proper understanding of this material is critical for proper comprehension of the all-air system addressed in Chapter 19. Table 13.4 summarizes the important equations presented in this chapter, while Figure 13.19 depicts different possible psychrometric processes.

The direction and slope of the conditioning line between incoming air at T_{db1} and exiting air depends on the enthalpy of the moisture added. For process (1) of

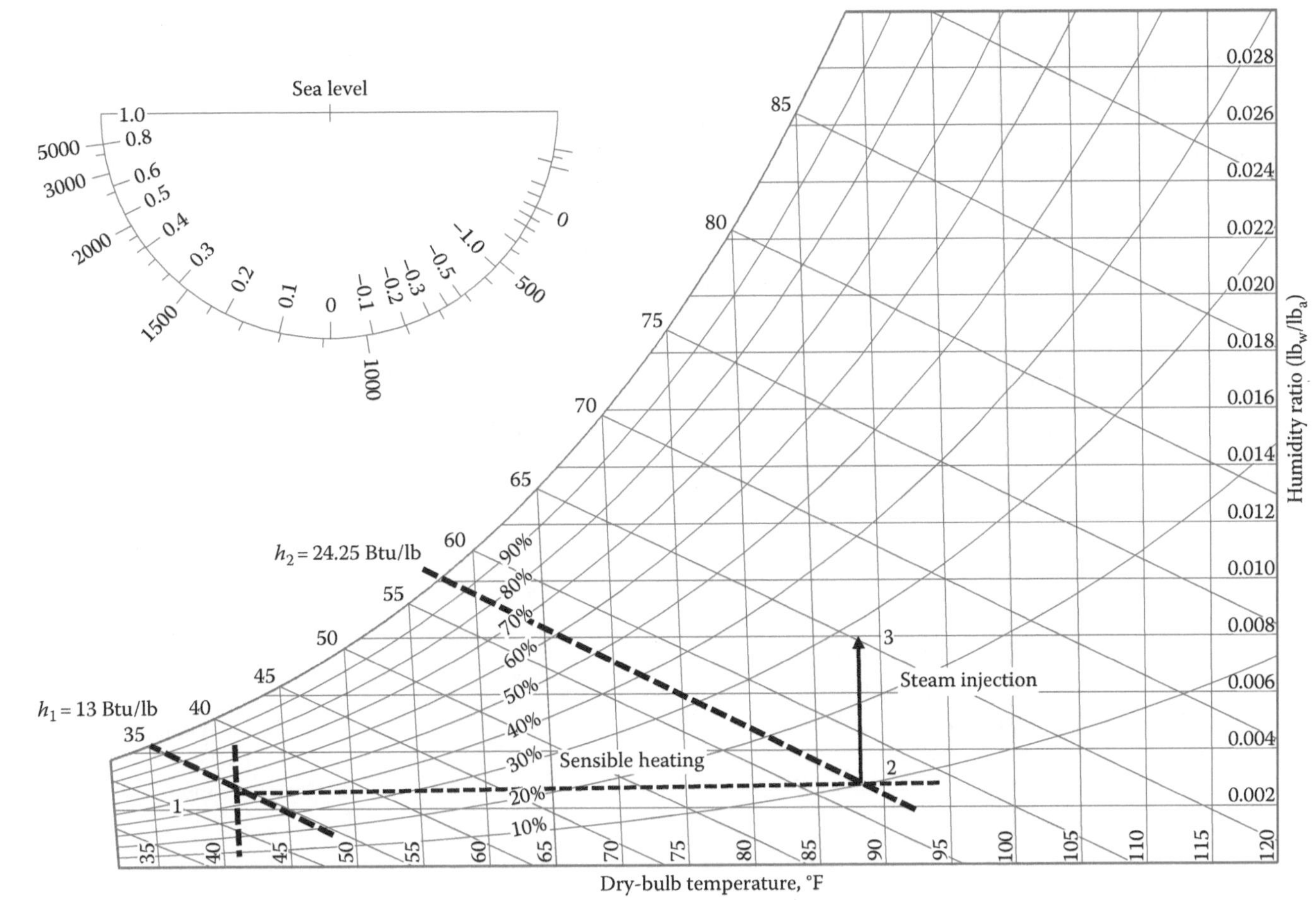

FIGURE 13.18
Heating and humidification process (Example 13.17).

Figure 13.19 to occur, the enthalpy of steam added must be equal to that of saturated water vapor ($h_{liq} = h_{vap}$) at the air dry-bulb temperature. The enthalpy and wet-bulb temperature of the air will increase, but the dry-bulb temperature of the air will remain constant.

For process (8) to occur, the humidifying spray must be water at the wet-bulb temperature of the air. The exiting airstream will have the same wet-bulb temperature as the entering airstream. The dry-bulb temperature of the air will decrease. The leaving enthalpy of the air will be close to the entering enthalpy because the constant enthalpy lines on the psychrometric chart are approximately parallel to the constant wet-bulb lines.

If the enthalpy of the steam added is greater than the enthalpy of saturated water vapor at the air dry-bulb temperature ($h_{liq} > h_{vap}$), then the airstream will be sensibly heated during humidification (process (2) in Figure 13.19). In contrast, if the steam enthalpy is lower than the enthalpy of saturated water vapor at the entering air dry-bulb temperature, the air will be sensibly cooled during the humidification process.

Also shown in Figure 13.19 are processes involving dehumidification. While this usually entails passing air over a cooling coil and cooling it below its dew point temperature, another less energy-intensive approach is to use chemical dehumidification or adsorbent cooling (process 4). Silica gel is a commonly used material that has the property that it can adsorb or strip away the water vapor from the air. This is accomplished practically by either passing air through a desiccant wheel coated with silica gel or through desiccant spays. Desiccant cooling is discussed in Section 20.9.

In this chapter, the basics of psychrometrics and a number of applications have been developed. The key assumption made was that air–water vapor mixtures behave as ideal gases. We have also illustrated how the psychrometric chart is a convenient tool for plotting processes involving heating, cooling, mixing, humidifying, and dehumidifying of moist airstreams. When accuracy better than that possible with a chart is needed, the equations presented in Section 13.4 should be used.

TABLE 13.4
Equation Summary

Topic	Equations	Equation Number	Notes
Psychrometric properties	$h = h_a + \frac{W}{W_{sat}} h_{d-s}$	13.4	Tabular air properties interpolation
	$\phi = \frac{p_{vap}}{p_{sat}}$	13.6	Relative humidity
	$W = 0.622 \frac{p_{vap}}{p_{tot} - p_{vap}}$	13.14	Humidity ratio
	$\phi = \frac{p_{sat}(T_{wb}) - p'}{p_{sat}(T_{db})}$	13.17	ϕ from T_{dp} and T_{wb}
	$p_{sat}(T) = p_c[10^{K(1-T_c/T)}]$	13.19	Auxiliary equation used with (Equation 13.18) for ϕ from T_{db} and T_{wb}
	$T_{dp} = 6.54 + 14.526\alpha + 0.7389\alpha^2 + 0.0949\alpha^3 + 0.4569 p_{vap}^{0.1984}$	13.21 SI	Dew point temperature
	$h = c_{p,a} T_{db} + W(h_{liq-vap,ref} + c_{p,vap} T_{db})$	13.23	Moist-air enthalpy
Psychrometric processes	$\frac{\dot{m}_{a1}}{\dot{m}_{a2}} = \frac{h_2 - h_3}{h_3 - h_1} = \frac{W_2 - W_3}{W_3 - W_1}$	13.32	Adiabatic mixing
	$W_1 = \frac{(2501 - 2.381 T_{wb2}) W_2 - 1.0 (T_{db1} - T_{wb2})}{2501 + 1.86 T_{db1} - 4.186 T_{wb2}}$	13.40 SI	Adiabatic saturation
	$\dot{Q}_{tot,c} = \dot{m}_a \left[(h_1 - h_2) - h_{liq} (W_1 - W_2) \right]$	13.42	Energy to cool and dehumidify air
	$\text{SHF}_{cc} \equiv \frac{\dot{Q}_{sens}}{\dot{Q}_{sens} + \dot{Q}_{lat}}$	13.47	Coil sensible heat ratio
	$\frac{\Delta h}{\Delta W} = \frac{\dot{Q}_h + \dot{m}_w h_{liq,2}}{\dot{m}_w}$	13.52	Warming and humidifying

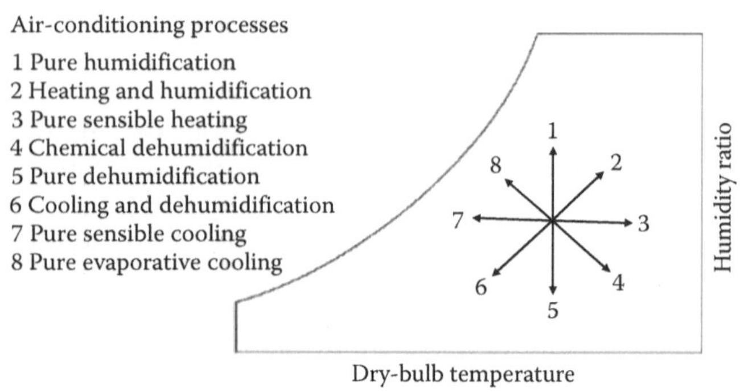

FIGURE 13.19
Different psychrometric process.

Problems

Numbers 1–4 given in parenthesis denote the degree of difficulty.

13.1 Find the density of air at the elevation of Boulder, Colorado [5,500 ft (1,677 m)], and at the summit of Pike's Peak (14,110 ft [4,302 m]) for an air temperature of 60°F (15°C). *(1)*

13.2 Calculate the humidity ratio, enthalpy, and specific volume of saturated air at 14.696 psia (101.3 kPa), using the ideal gas law and table of moist-air properties at 20°F (−6.5°C) and at 70°F (21°C). *(1)*

13.3 The air in a room at sea level* is 68°F (20°C) and 50% relative humidity (RH). Will moisture condense on a window whose surface is at 45°F (7°C)? If the room is 15 ft (4.5 m) square and 8 ft (2.5 m) high, how much water is contained in the room? *(1)*

13.4 For a site where the atmospheric pressure is 13.5 psia (93 kPa), find the relative humidity, humidity ratio, dew point, and enthalpy for a condition where the dry-bulb temperature is 95°F (35°C) and the wet-bulb temperature is 60°F (15.5°C). *(2)*

13.5 For a site at sea level, find the relative humidity, humidity ratio, dew point, and enthalpy for a condition where the dry-bulb temperature is 100°F (38°C) and the wet-bulb temperature is 55°F (13°C). *(2)*

13.6 At 5000 ft (1500 m) where the atmospheric pressure is 12.2 psia (84 kPa), find the relative humidity,

* In the problems, the term *sea level* means that the atmospheric pressure is 14.696 psia (101.325 kPa). The effect of altitude on density can be found from Equation 13.1 or from the appropriate table in the online HCB software.

humidity ratio, dew point, and enthalpy for a condition where the dry-bulb temperature is 90°F (32°C) and the wet-bulb temperature is 55°F (13°C). *(2)*

13.7 Write a computer program or use a spreadsheet to derive an empirical third-order polynomial equation for the enthalpy of saturated sea-level air as a function of the dry-bulb temperature in the range 10°F–130°F (−12°C to 55°C). *(3)*

13.8 Calculate the dew point temperature for a sea-level site where the dry- and wet-bulb temperatures are 75°F and 55°F (24°C and 13°C), respectively. *(2)*

13.9 A chilled-water line carries chilled water at 45°F (7°C) through a room at 70°F (21°C) and 60% relative humidity (at sea level). How much fiberglass insulation is needed on the pipe to avoid condensation? *(3)*

13.10 During very cold weather at a 5000 ft high (1500 m high) mountain site in Colorado, the interior surface of a single-glazed window can reach 40°F (4°C). If the room is at 68°F (20°C), what is the maximum relative humidity that can exist in the room without condensations occurring on the window? *(2)*

13.11 If the relative humidity at a sea-level site is 50% for a dry-bulb temperature of 80°F (26.5°C), calculate the moist-air enthalpy. *(2)*

13.12 Use the tables of moist-air properties to find the enthalpy and specific volume for air at 70°F (21°C) if the humidity ratio is 0.008. *(1)*

13.13 Use the tables of moist-air properties to find the enthalpy and specific volume for air at 125°F (51°C) if the humidity ratio is 0.040. *(1)*

13.14 A glass of ice water at 40°F (4.5°C) condenses moisture on its exterior outdoors on an 85°F (29°C) day. What is the minimum relative humidity needed for the condensation to occur if the location is at 5000 ft (1500 m)? *(2)*

13.15 Complete the following table using the psychrometric chart or tables and equations for moist-air properties: *(2)*

p_{atm}	T_{db}	W	h	ϕ	T_{wb}	T_{dp}
14.696 psia	95°F			50%		
29.92 inHg	70°F				50°F	
101.325 kPa		0.0070	55 kJ/kg			
101.325 kPa	20°C	0.010				
12 psia	55°F					45°F
101.325 kPa	30°C				30°C	
101.325 kPa	−10°C			90%		

13.16 Solve Problem 13.15 using the online HCB software. *(2)*

13.17 Calculate and plot the dew point and enthalpy of air at 70°F (21°C) dry-bulb temperature for relative humidities of 10%, 30%, 50%, 70%, and 90%. *(3)*

13.18 Calculate the humidity ratio, enthalpy, and specific volume for saturated air at 30°F (−1°C) and 80°F (27°C) at sea level. *(2)*

13.19 Calculate the humidity ratio, enthalpy, and specific volume for saturated air at 40°F (4°C) and 80°F (27°C) at a 5000 ft (1500 m) elevation. *(2)*

13.20 Complete the following table, using either the psychrometric chart or the online HCB software:

Point	p_{atm}, psia	T_{db}, °F	W	h	ϕ	T_{wb}	T_{dp}
A	14.696	95			45%		
B	14.696	70				56°F	

What is the sensible heat ratio for the cooling process *AB*? *(2)*

13.21 Complete the following table, using either the psychrometric chart or the online HCB software:

Point	p_{atm}, kPa	T_{db}, °C	W	h	ϕ	T_{wb}	T_{dp}
A	101.325	35			50%		
B	101.325	20				14°C	

What is the sensible heat ratio for the cooling process *AB*? *(2)*

13.22 An automotive air conditioner cools 250 ft³/min (120 L/s) of 95°F (35°C) and 35% relative humidity sea-level air to saturation at 45°F (7°C). How much moisture must be drained from the evaporator per hour? If the COP of this air conditioner is 2.2 and the air-conditioning compressor belt drive has an efficiency of 75%, how much power is extracted from the automobile engine to operate the air conditioner? *(3)*

13.23 Describe the steps in the algorithm used to calculate the humidity ratio from measured wet- and dry-bulb temperatures. *(2)*

13.24 Construct the relative humidity curves on a psychrometric chart for 2500 ft altitude in IP units, using the graphics routines in a commercial spreadsheet software package. Check your solution by comparison with Figure 13.6 and the electronic appendices. Optionally, you could add enthalpy and wet-bulb temperature lines. *(4)*

13.25 Solve Example 13.4 for a 5000 ft (1500 m) altitude. *(3)*

13.26 Air at 100°F (38°C) and 30% relative humidity is cooled to 70°F (21°C). How much moisture condenses per pound (kilogram) of air? *(2)*

13.27 Air at 100°F (38°C) and 30% relative humidity is cooled to 55°F (13°C). How much moisture condenses per pound (kilogram) of air? *(1)*

13.28 One hundred ft³/min (47 L/s) of air is to be humidified by saturated steam at 212°F (100°C). Sea-level air enters the steam humidifier at 55°F (13°C) dry-bulb and 39°F (4°C) wet-bulb temperatures. What is the steam flow rate if the air is humidified to 80% relative humidity? What is the temperature of the humidified air? *(2)*

13.29 Air enters a steam humidifier at 50°F (10°C) dry-bulb and 35°F (2°C) wet-bulb temperatures at a flow rate of 1000 ft³/min (470 L/s). The air is humidified by saturated steam at 230°F (110°C). What is the steam flow rate if the air is humidified to 90% relative humidity? What is the temperature of the humidified air? The location of the humidifier is at 5000 ft (1500 m) altitude. *(2)*

13.30 Suppose that an HVAC design engineer used the sea-level psychrometric chart instead of the 5000 ft chart to solve Problem 13.29. What would the error in the steam flow rate be? *(3)*

13.31 On a winter day, outdoor air at 10°F (–12°C) and 70% relative humidity is heated to 70°F (21°C) in a residence. If the occupants require that the indoor humidity be 60% for comfort reasons, how much moisture per unit of dry air must be added to the outdoor air if the atmospheric pressure is 14.696 psia (101.325 kPa)? *(2)*

13.32 Air at 34°F dry-bulb temperature and 33°F wet-bulb temperature must be heated to 68°F, the interior temperature of a residence at sea level. The airflow rate is 800 ft³/min. What is the heat rate? How much of it is due to the presence of moisture? *(2)*

13.33 Humidity is added to the outdoor air in Problem 13.31 by evaporating water sprayed into the airstream. How much energy must be added to the air to increase the humidity level to the required 60%? *(2)*

13.34 Moist air at 70°F (21°C) dry-bulb temperature and 45°F (7°C) wet-bulb temperature is humidified to a final dew point temperature of 55°F (13°C) by addition of saturated 230°F (110°C) steam. If the airflow rate is 200 lb/min (1.5 kg/s), what is the required steam flow rate? What is the final dry-bulb temperature? *(2)*

13.35 Outdoor air at 40°F (4.5°C) and 60% humidity is heated and humidified by steam at 230°F (110°C). The airflow rate is 30,000 ft³/min (14,000 L/s), and heat is added to the air at the rate of 1,500,000 Btu/h (440 kW) while it absorbs 800 lb_m/h (0.10 kg/s) of steam. What are the dry- and wet-bulb temperatures at the exit of this heater/humidifier if it is located at sea level? *(3)*

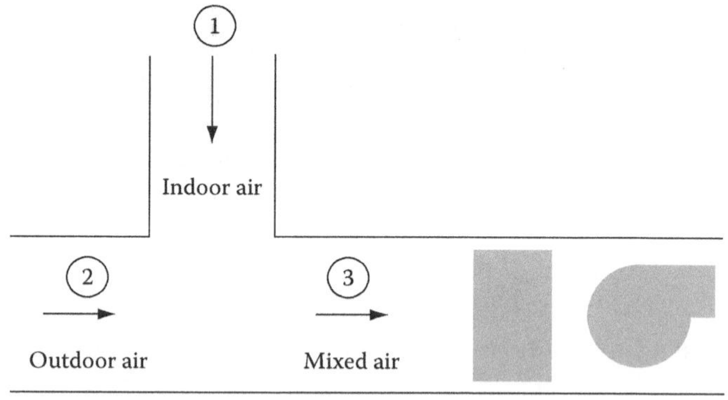

FIGURE P.13.36

13.36 Figure P.13.36 shows the typical arrangement for mixing two airstreams at the air handler of a commercial building HVAC system at sea level. Outdoor air at 35°C dry-bulb temperature and 25°C wet-bulb temperature mixes with 25°C and 50% relative humidity return indoor air from the building in the mass flow ratio of 1:5. What are the enthalpy, relative humidity, humidity ratio, and dry-bulb temperature of the mixed air? If the fan flow rate is 12 kg/s, how much energy and moisture must be removed from the mixed airstream to provide 25°C, 50% relative humidity supply air to the building? *(3)*

13.37 Cold air at 10°C dry-bulb temperature and 5°C wet-bulb temperature is mixed with warm, moist air at 25°C dry-bulb temperature and 20°C wet-bulb temperature in a ratio of 1:2, respectively. Use the sea-level psychrometric chart to find the mixed-air condition: dry-bulb and wet-bulb temperatures, relative humidity, humidity ratio, and specific volume. *(2)*

13.38 Repeat Example 13.14, using moist-air property tables instead of the psychrometric chart. *(3)*

13.39 An evaporative cooler is able to cool air by 85% of the difference between the entering air dry- and wet-bulb temperatures. If inlet air is at 100°F (38°C) and 25% relative humidity, what is the outlet condition (T_{db}, ϕ, and W)? How much water is evaporated if the airflow is 10,000 ft³/min (4,700 L/s) at 3,000 ft (1,000 m) elevation (p_{atm} = 13.2 psia [91 kPa])? *(3)*

13.40 Repeat Example 13.12, using the moist-air property tables and psychrometric equations rather than the psychrometric chart. How close is the agreement between the two? *(3)*

13.41 Because the 2000 ft³/min (950 L/s) exhaust fan in a chemistry laboratory has no on/off switch, it operates continuously. How much energy is wasted because this fan operates needlessly for 14 h/day for the 150-day cooling season if the interior air condition is 70°F and 50% relative humidity? On average, the outdoor condition at this sea-level site is 80°F (26°C) and 50% relative humidity. The air conditioner's COP is 2.4. If electric power costs 8¢/kWh, what is the value of this wasted energy? *(3)*

13.42 At a site at 5000 ft (1500 m), air is cooled from 80°F (27°C) dry-bulb temperature and 75°F (24°C) wet-bulb temperature to saturation at the 55°F (13°C) outlet of a cooling coil. How much water is removed? How much latent heat is removed, and how much sensible heat? What is the SHR? *(3)*

13.43 At a sea-level site, air is cooled from 80°F (27°C) dry-bulb temperature and 75°F (24°C) wet-bulb temperature to saturation at the 55°F (13°C) outlet of a cooling coil. How much water is removed? How much latent heat is removed, and how much sensible heat? What is the SHR? *(3)*

13.44 An office is occupied by 30 persons who each produce 200 Btu/h (58 W) of sensible heat and 0.25 lb/h (0.1 kg/h) of moisture. The office is to be maintained at 72°F (22°C) and 50% relative humidity. Conditioned air is supplied at 60°F (15°C) to meet the sensible and latent loads. What is the SHR? To meet the loads, what must the humidity ratio and mass flow rate of the supply air be? The office is located at sea level. *(3)*

13.45 A high-school classroom is occupied by 20 students who each produce 180 Btu/h (53 W) of sensible heat and 0.20 lb/h (0.09 kg/h) of moisture. The class is to be maintained at 68°F (20°C) and 55% relative humidity. Conditioned air is supplied at 60°F (15°C) to meet the sensible and latent loads. What is the SHR? To meet the loads, what must the humidity ratio and mass flow rate of the supply air be? The office is located at 5000 ft (1500 m) altitude. *(3)*

13.46 Find the cooling coil load for the following entering and required leaving air conditions for an airflow of 10,000 ft³/min (4,700 L/s) at sea level: *(3)*

	Inlet State		Outlet State	
Case	**Dry Bulb, °F**	**Wet Bulb, °F**	**Dry Bulb, °F**	**Relative Humidity, %**
1	80	67	50	90
2	80	60	50	90
3	45	44	50	90

13.47 Complete the following table of cooling coil performance that compares the loads at various coil outlet temperature conditions: *(3)*

Point	$T_{db,in}$, °F	W_{in}	$T_{db,out}$, °F	$T_{dp,out}$, °F	W_{out}	Latent Load, Btu/h	Sensible, Load, Btu/h	SHR
1	78	0.010386	65	58.2				
2	78	0.010386	63	57.5				
3	78	0.010386	61	56.3				
4	78	0.010386	59	55.0				
5	78	0.010386	57	53.8				
6	78	0.010386	55	52.0				

13.48 What is the inlet humidity ratio to an ideal evaporative cooler (i.e., one that cools inlet air to the inlet air wet-bulb temperature) that produces saturated outlet air at 63°F (17°C) from 85°F (29°C) dry-bulb temperature inlet air? The cooler is at sea level. *(1)*

13.49 Work Problem 13.48 for a site where the atmospheric pressure is 12.8 psia (88 kPa). *(1)*

13.50 An adiabatic saturator operating at sea level with entering air of 80°F (26°C) has a leaving air temperature of 65°F (18°C). Compute the entering-air humidity ratio and relative humidity. *(2)*

13.51 Two airstreams, both at 5000 cfm, are well mixed inside an air-handling unit. One airstream is at 80°F and 80% relative humidity, and the other is at 50°F and 80% relative humidity. What are the resulting mixed airstream temperature and relative humidity? *(1)*

13.52 Which sample of moist air has the higher density: 50°F at 10% relative humidity or 50°F at 90% relative humidity? *(1)*

13.53 Which sample of moist air has the higher density: 50°F at 50% relative humidity or 80°F at 50% relative humidity? *(1)*

13.54 An economizer mode attempts to mix outside air and building return air to minimize the amount of energy needed to condition the resulting mixed airstream to match the desired supply air conditions. Suppose the conditions are outside air at 90°F and 40% relative humidity, the return air at 80°F and 70% relative humidity, and the supply air set point of 55°F at 80% relative humidity. Should the economizer control use mostly outside air or mostly building return air? *(2)*

13.55 A very large building contains approximately 1 million ft³ of air at 70°F and 60% relative humidity. Assuming a ventilation rate of 0.5 air change per hour and ambient air design conditions of 92°F dry-bulb and 76°F wet-bulb temperature, how much water is removed each hour from the outdoor air entering the building? *(2)*

13.56 You exhale air at about 80°F and 50% relative humidity. What outdoor air conditions must be met before you start to see your breath? *(1)*

13.57 Air leaves a cooling coil at 55°F and a humidity ratio of 0.008 lb water/lb air. Does this supply air meet the requirement of a minimum of 85% relative humidity? This air then passes through a fan that heats up the airstream by 2°F before being supplied to the building. Is the 85% relative humidity requirement met? *(2)*

13.58 Air at 90°F and 50% relative humidity is cooled to 70°F. How much moisture condenses out of the air? *(1)*

13.59 Air at 90°F and 50% relative humidity is cooled to 60°F. How much moisture condenses out of the air? *(1)*

13.60 Data from a coastal weather station record a daytime high temperature of 90°F and a relative humidity of 37%. At night, the temperature drops to 60°F and the relative humidity reaches 100%. How much has the humidity ratio varied throughout the day? *(1)*

13.61 Most relative humidity sensors have an accuracy of about ±3%. If an airstream is measured to have a temperature of 55°F and a relative humidity of 90%, assuming the stated error of the humidity sensor, what is the possible range of air enthalpy for this airstream? *(2)*

13.62 A cooling coil at sea level removes 100,000 Btu/h from an airstream. The airflows into the coil at 10,000 cfm. What is the leaving air temperature from the coil if the entering air is at 80°F and 50% relative humidity? *(2)*

13.63 In Problem 13.62, what is the leaving air temperature if the coil is located in Leadville, Colorado, at 10,000 ft above sea level? *(2)*

References

ASHRAE Fundamentals (1989, 2013). *Handbook of Fundamentals*. American Society of Heating, Refrigerating and Air-Conditioning Engineers, Atlanta, GA.

Keenan, J.H., F.G. Keyes, P.G. Hill, and J.G. Moore (1978). *Steam Tables*. Wiley, New York.

Kuehn, T.H., J.L. Threlkeld, and J.W. Ramsey (1998). *Thermal Environmental Engineering*, 3rd ed. Prentice-Hall, Englewood Cliffs, NJ.

Mathur, G.D. (1989). Predicting wet vapor saturation pressure. *Heat. Piping Air Cond.*, 61, 103–104.

Pallady, P.H. (1984). Evaluating moist air properties. *Chem. Eng.*, 91, 117–118, October.

Pallady, P.H. (1989). Computing relative humidity quickly. *Chem. Eng.*, 96, 255–257, November.

Stoecker, W.F. and J.W. Jones (1982). *Refrigeration and Air Conditioning*. McGraw-Hill, New York.

14

Chillers and Heat Pump Cycles and Systems

ABSTRACT In Chapter 12, we covered the Carnot cycle based on the second law of thermodynamics as applied to power, refrigeration, and heat pump cycles. This chapter extends this treatment to the description and analysis of practical vapor compression refrigeration and heat pump cycles, and then to actual cooling systems with heat exchangers both at the evaporator and the condenser. In 2013, vapor compression systems accounted for 99% of the cooling needs and 11% of heating needs in U.S. residential and commercial buildings. We also discuss absorption cycles whereby cooling can be achieved, not by work input, but by heat input to the system. The attraction of heat pumps is that they can deliver more thermal power than they consume electrically during an appreciable portion of the heating season. In moderate climates requiring both heating and cooling, the heat pump can also be operated as an air conditioner during summer, thereby avoiding the additional cost of a separate air conditioning system. The design methodology to size heat pumps and the associated auxiliary heater is presented. We then discuss the differences between air, water and ground source heat pumps, and their advantages and disadvantages. How cooling equipment performance is affected by off-rating conditions and degraded by part-load operation is described, and several energy analysis methods widely used to model this phenomenon are presented and illustrated with several solved examples. Though not attainable in practice, the concept of theoretical efficiency limits of cooling equipment specified by the second law of thermodynamics can provide useful insights. We derive such expressions for ideal and actual heat pumps and illustrate their use via simple examples. The last section includes an overview of refrigerants, their method of designation, and desirable thermodynamic and physical properties, along with international efforts to minimize their adverse effect on climate change and human health.

Nomenclature

A,B,C,D	Weights appearing in Equation 14.26
a,b,c	Model coefficients appearing in Equations 14.32 and 14.33
AC	Absorption cooling
$\dot{C}$	Heat capacitance of fluid stream, W/K [Btu/(h·°F)]
DBT	Dry-bulb temperature °E (°F)
CAP_FT	Chiller capacity at full load, W (Btu/h)
COP	Coefficient of performance of a cooling device (Equation 14.4)
CR	Compression ratio (Equation 14.12)
C_d	Degradation coefficient of unitary equipment due to cycling
c_p	Specific heat at constant pressure, kJ/(kg·K) [Btu/(lb$_m$·°F)]
DX	Direct Expansion
EER	Energy efficiency ratio, Btu/(W·h)
EIR_FPLR	Energy input ratio at part load
EIR_FT	Energy input ratio at full load
GWP	Global warming potential of a refrigerant
h	Specific enthalpy, kJ/kg (Btu/lb$_m$)
HX	Heat exchanger
IPLV	Integrated part-load value of a chiller
K_{tot}	Overall heat loss/gain coefficient for a house or space, W/K [Btu/(h·°F)]
kW/ton	Chiller electric consumption in kW divided by cooling capacity in tons
LCCP	Life cycle climate performance
$\dot{m}$	Mass flow rate, kg/s (lb$_m$/h)
NTU	Number of transfer units, defined by Equation 12.36
ODP	Ozone depletion potential of a refrigerant
PF	Performance factor of a heat pump
PLF	Part-load factor, defined by Equation 14.28
PLR	Part-load ratio, defined by Equation 14.27
p	Pressure, Pa (lb$_f$/in.^{2})
Q	Energy consumption or heat transfer with subscripts c for cooling and h for heating, KJ(Btu)
$\dot{Q}$	Heat flow rate with subscripts c for cooling load, h for heating load, lat for latent load, sen for sensible load, etc., W (Btu/h)
SEER	Seasonal energy efficiency ratio, Btu/(W·h)
SPF	Seasonal performance factor of a heat pump
s	Specific entropy, kJ/(kg·K) [Btu/(lb$_m$·°R)]
T	Temperature, K or °C (°R or °F)
U	Overall heat transfer coefficient W/(m^2.K) [Btu/(h.ft^2.°F)]
$\dot{V}$	Volume flow rate, m^3/s or L/s (ft^3/min)
VC	Vapor compression
v	Specific volume, m^3/kg (ft^3/lb$_m$)
$\dot{W}$	Power, W (ft·lb$_f$/s or HP)

WBT	Wet-bulb temperature, °C (°F)
X	Mass fraction or concentration of binary solution
1,2,3,4	State points of refrigerant for standard VC cycle
1′,2′,3′,4′	State points of refrigerant for modified VC cycle
1*,2*,3*,4*	State points of Carnot cycle
x^*	Value of x at rated conditions

Greek

ε	Heat exchanger effectiveness
η	Efficiency
η_R	Refrigerating efficiency (Equation 14.11)
ρ	Density, kg/m³ (lb$_m$/ft³)

Subscripts

abs	Absorber
aux	Auxiliary
b	Balance point
bin	Outdoor temperature bin
c	Cooling
cd	Condenser water
ch	Chiller or evaporator water
comp	Compressor
cond	Condenser
cool	Cooling load
e	Electricity
evap	Evaporator
gen	Generator
HX	Heat exchanger
h	Heating, high
hp	Heat pump
i	Indoor
in	Inlet, input
isen	Isentropic
l	Low
mod	Modified VC cycle
o	Outdoor
out	Outlet, output
Rank	Rankine
r	Refrigerant
refr	Refrigeration effect
sink	Where heat is rejected
source	From where heat is extracted
std	Standard
ss	Steady-state
sup	Supply
sys	system
tot	Total
yr	Year

14.1 Standard Vapor Compression Cycle

14.1.1 Description

The ideal power and refrigeration cycles discussed in Section 12.2 cannot be realized in practice because of thermodynamic, thermal and practical limitations on components and materials. Pragmatic considerations require that the Carnot cycle itself be modified slightly, and thermodynamic irreversibilities in various components such as heat exchangers and compressors be considered. Further, the working fluid of an actual refrigeration cycle also affects the cycle efficiency. As a result, the ideal Carnot cycles discussed in Section 12.2 have to be altered appropriately as discussed in this section.

For practical purposes, the Carnot heat engine is replaced by a slightly modified cycle, the standard *Rankine power cycle,* which is plotted on a T–s diagram in Figure 14.1. Though the similarities to Figure 12.2 are apparent, there

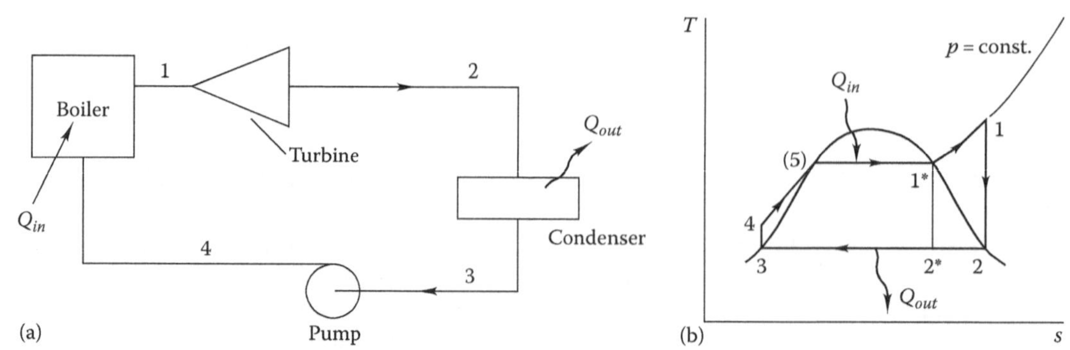

FIGURE 14.1
Schematic diagram of the Rankine power cycle: (a) arrangement of mechanical components; (b) cycle on T–s coordinates showing wet expansion.

are, however, significant differences. The fluid expansion (work producing process done in the turbine) if done in the wet condition, i.e., from state point 1* to 2* would erode and destroy the turbine blades. Hence, expansion in the wet steam region is avoided by superheating the saturated dry steam from point 1* to point 1 so that the expansion can occur entirely in the dry steam region (process 1–2). Further, steam expansion in the turbine is never isentropic and is another source of inefficiency. Also, the compression of a liquid requires much less work than compressing wet steam. Hence, the steam is cooled down till it reaches the saturated liquid stage (point 3) before it is compressed in a pump and introduced into the boiler. Note, the two heat exchange processes shown (2–3: vapor condensation, and 4–1: water evaporation and superheat). The heat addition process 4–1 does not occur isothermally, another departure from the ideal Carnot cycle. The effect of such unavoidable practical modifications to the theoretical Carnot cycle is to reduce the Carnot efficiency given by Equation 12.3.

Similarly, practical reasons also warrant that the Carnot refrigeration cycle be modified as shown in Figure 14.2. This is called the *standard vapor compression (VC) refrigeration cycle.* Again, so as not to compromise compressor life, the compression is performed entirely in the superheated region (process 1–2) as compared to process 1*–2* for the Carnot cycle. In addition, the standard VC cycle differs in the fluid expansion step. No attempt is made to extract any work in this process since including an additional piece of equipment such as a turbine is not economical in view of the small amount of work one can recover. Instead, a simple throttling device is used. Since throttling is an isenthalpic process, it appears as a vertical line (process 3–4) on the p–h diagram in Figure 14.2c. In fact, all four processes appear as straight lines on the p–h diagram; therefore, these thermodynamic coordinates are more convenient than the T–s diagram for analyzing VC refrigeration cycles (Figure 14.2b). The refrigeration industry works almost exclusively using the p–h diagram that was introduced earlier in Section 12.3.3. The relationships of the various properties of interest, namely the pressure, temperature, enthalpy, entropy, and specific volume can be gleaned from the p–h diagram for refrigerant R-134a (Figure 12.7).

14.1.2 Analysis Equations

The standard *VC refrigeration cycle* shown in Figure 14.2 consists of the following processes:

1. *Process 1–2*: Dry compression of saturated vapor to high pressure assumed to be isentropic.
2. *Process 2–3*: Condensing the hot vapor to saturated liquid by heat rejection to the environment assumed to occur at constant pressure.

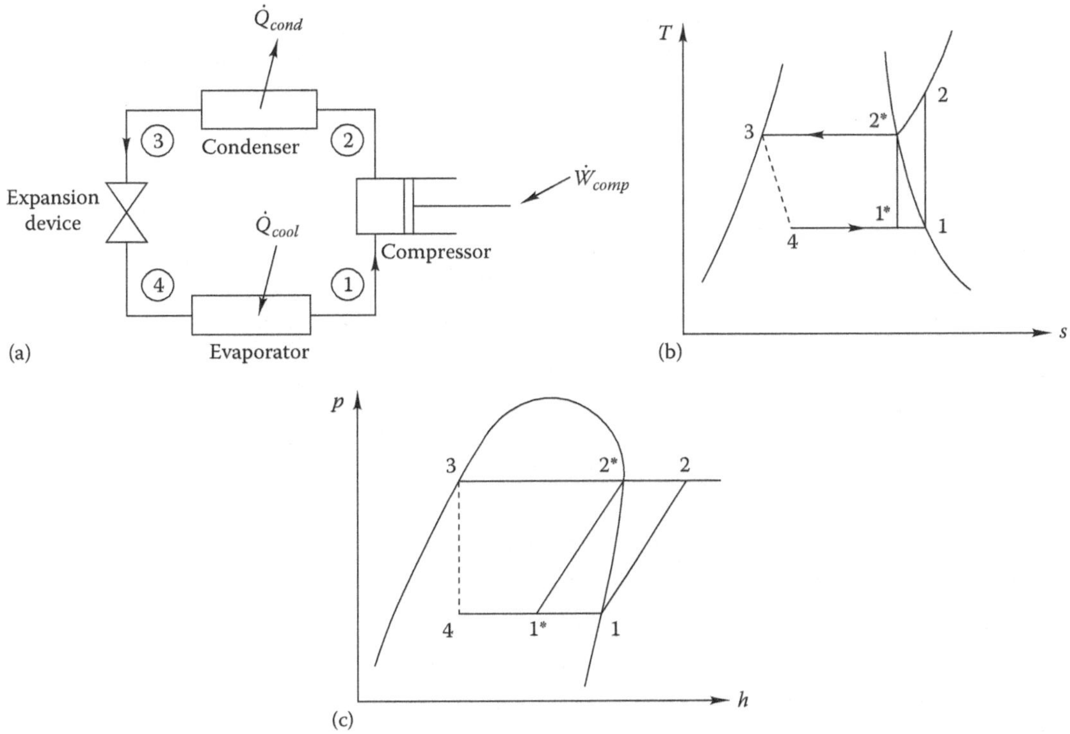

FIGURE 14.2
Schematic diagram of the standard VC refrigeration cycle: (a) arrangement of mechanical components, (b) cycle on T–s coordinates showing wet compression, and (c) cycle on p–h coordinates showing wet compression.

3. *Process 3–4*: Isenthalpic expansion or throttling of the high-pressure liquid to low-pressure wet mixture. The process is shown as a dotted line since it is not in equilibrium as is assumed for most thermodynamic processes.
4. *Process 4–1*: Heat addition to the cold fluid by evaporation to saturated vapor assumed to occur at constant pressure.

Point 1 in the process represents the inlet to the compressor. Compressor power $\dot{W}_{comp}$ raises the pressure and temperature of the working fluid to a level (point 2) at which heat $\dot{Q}_{cond}$ can be readily rejected to the environment. The heat rejection step results in a phase change of the working fluid from superheated vapor to saturated liquid. The liquid is throttled from the high-side pressure to the low-side pressure in an isenthalpic process. At the end of this irreversible process, the refrigerant is a saturated mixture of liquid and vapor at a temperature selected to enable effective transfer of heat from the space to be cooled. This heat input $\dot{Q}_{cool}$ causes the wet vapor refrigerant to evaporate completely, resulting in a state of saturated vapor at the compressor inlet. The various quantities of interest are shown in Table 14.1.

In IP units, the rate of refrigeration is often expressed as *tons of cooling*. This unit of cooling effect is equivalent to 12,000 Btu/h (3.5 kW). It was originally defined on the basis of melting of 1 ton (2,000 lb_m) of ice (heat of fusion 144 Btu/lb_m) per day (24 h), giving a cooling rate equal to 12,000 Btu/h (3,515 W).

Example 14.1: Standard VC Cycle

A refrigeration cycle uses refrigerant 134a (also denoted by R-134a) and operates between −14°C and 30°C. The cooling capacity is 5.0 kW. Find the compression ratio, power input, COP, kW/ton, and the refrigerating efficiency of this cycle.

Given: T_{cond} = 30°C, T_{evap} = −14°C, $\dot{Q}_{cool}$ = 5.0 kW

Figure: See Figure 14.2.

Assumptions: All processes are steady. The cycle follows the standard VC cycle.

Find: CR, $\dot{W}_{comp}$, COP_{std}, kW/ton, and η_R

Lookup values: The relevant thermodynamic properties for R-134a can be determined from Table A6. Italicized values are specified in the problem, the single value shown bolded is read from the thermodynamic chart or from Table A8, and the remainder are determined from the property tables. Values in parentheses are not needed for the solution but are included as reference data.

Solution

Part of the solution is embodied in the property values assembled in Table 14.2.

(i) The saturated pressure at a temperature T_{evap} = −14°C is the low-side pressure equal to 0.1707 MPa (24.8 psia) and the high-side pressure at T_{cond} = 30°C is 0.7701 MPa (113.3 psia).
(ii) Next, we begin at point 1 where the enthalpy and entropy are first found from the property table.

TABLE 14.1

Quantities of Interest While Analyzing Standard VC Refrigeration Cycles

	Quantity	Given By	Units	Equation Number
1	Work of compression	$W_{comp} = (h_2 - h_1)$	kJ/kg or Btu/lb_m	(14.1)
2	Refrigerating effect	$Q_{refr} = (h_1 - h_4)$	kJ/kg or Btu/lb_m	(14.2)
3	Heat rejection	$Q_{cond} = (h_2 - h_3)$	kJ/kg or Btu/lb_m	(14.3)
4	Coefficient of performance	$COP_{std} = (h_1 - h_4)/(h_2 - h_1)$	—	(14.4)
5	Cooling capacity[a]	$\dot{Q}_{cool} = \dot{m}_r(h_1 - h_4)$	kW or Btu/h or tons	(14.5)
6	Refrigerant flow rate	$\dot{m}_r = \dot{Q}_{cool}/(h_1 - h_4)$	kg/s or lb_m/min	(14.6)
7	Compressor power	$\dot{W}_{comp} = \dot{m}_r(h_2 - h_1)$	kW or HP	(14.7)
8	Heat rejection rate	$\dot{Q}_{cond} = \dot{m}_r(h_2 - h_3)$	kW or Btu/h	(14.8)
9	Volume flow rate of refrigerant per kW of refrigeration	$\dot{V}_r = \dot{m}_r \cdot v_1$	$m^3/(s \cdot kW)$	(14.9)
10	kW/ton[b]	$(\dot{W}_{comp}/\dot{Q}_{cool}) = 3.516/COP$	kW/ton	(14.10)
11	Refrigerating efficiency	$\eta_R = (COP_{std}/COP_{Carnot})$	—	(14.11)
12	Compression ratio	$CR = (p_{cond}/p_{evap})$	—	(14.12)

Refer to Figure 14.2c for the state point definitions. The symbol h denotes the specific enthalpy.

[a] 1 ton of cooling = 3.517 kW = 12,000 Btu/h.

[b] This is a commonly used dimensional measure of performance, proportional to the inverse of COP.

TABLE 14.2

R-134a Properties for Example 14.1 Extracted from Table A6 or Read from the *p–h* Diagram (Figure 12.7) and Shown Bolded

Point	*T*, °C	*p*, MPa	*h*, kJ/kg	*s*, kJ/(kg·K)
1	*−14°*	0.1707	390.33	(1.7367)
2	**37°**	0.7701	**421.50**	(1.7367)
3	*30°*	0.7701	241.65	(1.1432)
4	*−14°*	0.1707	241.65	—

(iii) We then use the fact that entropies at points 1 and 2 are equal. So, using the entropy value from point 1 and the known high-side pressure, properties at 2 can be found from the superheat Table A8. Alternatively, one could simply use the *p–h* diagram (Figure 12.7), and find the intersection of a constant entropy line drawn from point 1 and the constant pressure horizontal line; this is point 2.

(iv) Since point 3 represents saturated liquid at p_2, the properties at 3 can be read from the saturated-liquid column of the property table.

(v) Finally, the conditions at 4 can be found since the pressure and enthalpy are both known (i.e., $h_3 = h_4$). (The entropy values at points 3 and 4 are not needed for the cycle calculation and need not be determined unless one is concerned about the irreversibility inherent in the throttling process 3–4.)

The compression ratio is given by Equation 14.12:

$$CR = \frac{0.7707}{0.1701} = 4.53$$

The refrigerant mass flow is (Equation 14.6)

$$\dot{m}_r = \frac{\dot{Q}_{cool}}{(h_1 - h_4)} = \frac{5.0\text{ kW}}{(390.33 - 241.65)\text{ kJ/kg}} = 0.0336\text{ kg/s}$$

The compressor power input is (Equation 14.7)

$$\dot{W}_{comp} = 0.0336\text{ kg/s} \times (421.5 - 390.24)\text{ kJ/kg} = 1.05\text{ kW}$$

The COP of the standard VC cycle is thus

$$COP_{std} = \frac{\dot{Q}_{cool}}{\dot{W}_{comp}} = \frac{5.00}{1.05} = 4.76$$

The Carnot refrigeration cycle efficiency is (Equation 12.4)

$$COP_{Carnot} = \frac{(-14 + 273)}{[30 - (-14)]} = 5.886$$

From which the refrigerating efficiency is (Equation 14.11)

$$\eta_R = \frac{COP_{std}}{COP_{Carnot}} = \frac{4.76}{5.886} = 0.809$$

Following Equation 14.10, the kW/ton of the machine = 3.516/4.76 = 0.739 kW/ton.

Comments

The careful reader will observe that the throttling process 3–4 represents an irreversible process in the ideal vapor compression cycle as indicated by the entropy increase. This irreversibility is forced on us by the difficulties of trying to build a wet-mixture turbine that could survive in a refrigeration machine. However, if such a turbine could be made,* some of the power input to the compression process could be avoided since it would be provided by this hypothetical turbine.

Though using the *p–h* diagram is convenient to analyze refrigeration cycles, the state properties cannot be determined accurately. Even small differences can result in large changes in certain quantities, such as the COP. Hence, for any sort of accuracy, it is urged that points 1 and 3 be read off the saturated tables while state 2 be determined from the superheat tables (such as Tables A7 and A8 for R-22 and R-132a respectively). Alternatively, software programs (there are several available) such as the online HCB software accompanying this textbook should be used for determining thermodynamic properties to achieve the needed accuracy. You are encouraged to try the software for the problems in this chapter and for generally analyzing such vapor compression cycles and systems.

14.2 Modified and Actual VC Cycles

14.2.1 Modified VC Cycle

The standard VC cycle still has some practical limitations, and further idealizes the thermodynamic processes in some components. In this section, we will describe a few of the important modifications. We will then consider

* A few small-scale chiller manufacturers have indeed developed such turbines.

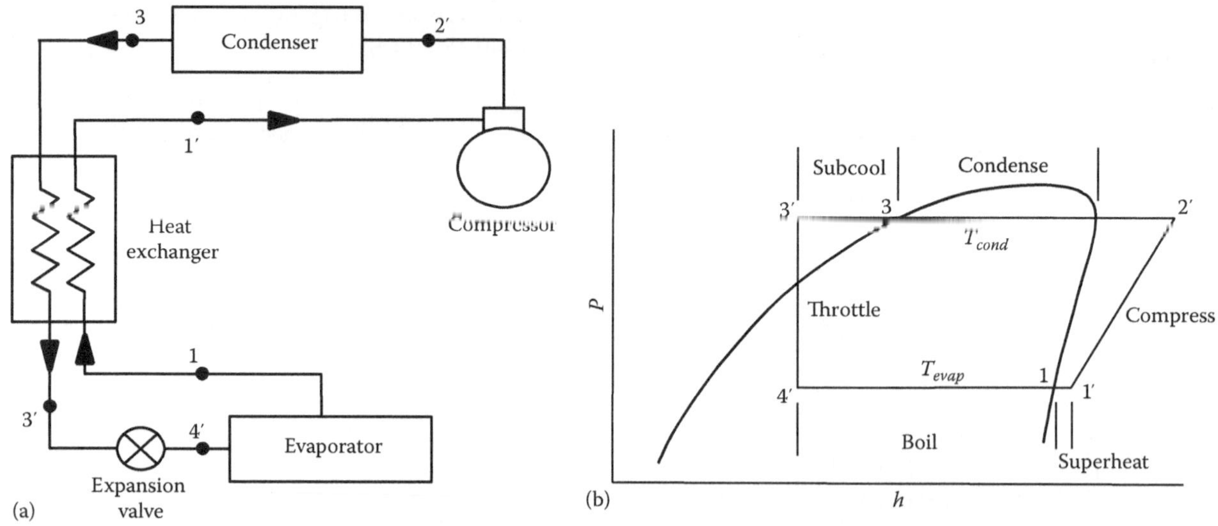

FIGURE 14.3
Modified VC cycle with a heat exchanger meant to superheat vapor leaving the compressor and subcool condenser liquid: (a) system diagram; (b) cycle on *p–h* diagram.

the single-stage cycle in Section 14.2.2 and briefly introduce two-stage cycles in Section 14.2.3.

A departure from the standard cycle has to do with the refrigerant leaving the evaporator and entering the compressor. The state of the refrigerant was assumed to be dry saturated vapor. This is impractical in reality since it is impossible to assure that all the refrigerant leaves at this state. Thermodynamic cycles are based on quasi-equilibrium processes, and so state 1 would represent an *average* condition that can be taken to be the dry saturated vapor. Actual cycles operating in finite time will consist of a mixture of superheated and slightly wet saturated vapor. The trace refrigerant liquid will still be injurious to the compressor. The pragmatic solution to avoid droplets is to superheat the vapor leaving the evaporator. The liquid refrigerant at state 3 leaving the condenser is hotter than the dry saturated vapor at condition 1, and so simply introducing a liquid-to-suction heat exchanger (as shown in Figure 14.3a) can provide the necessary superheating (point 1′ in Figure 14.3b) without having to resort to an external source of heating. In the process, the liquid leaving the condenser is subcooled to state 3′ which is further advantageous. The subcooled liquid will reduce bubble formation during the expansion process and result in smoother refrigerant flow through the expansion device. Assuming adiabatic operation of the heat exchanger, the following heat balance will hold:

$$(h_3 - h_3') = (h_1' - h_1) \quad (14.13)$$

Although the compression process is assumed to be isentropic (as sketched in Figure 14.2), actual compressors are not isentropic due to internal friction and heat losses, and due to the small pressure differential between the fluid at the entrance and exit of the compressor that are respectively needed to make the vapor flow into and then leave the compressor. The actual compressor outlet condition has a higher entropy (shown as point 2″ in Figure 14.4), and thus moves further into the superheat region, i.e., to the right of point 2′ by an amount depending on the compressor efficiency. Compressor isentropic efficiency is the ratio of the theoretical enthalpy rise between points 2′ and 1′ to the actual enthalpy rise in the physical equipment:

$$\eta_{isen} = \frac{h_2' - h_1'}{h_2'' - h_1'} \quad (14.14)$$

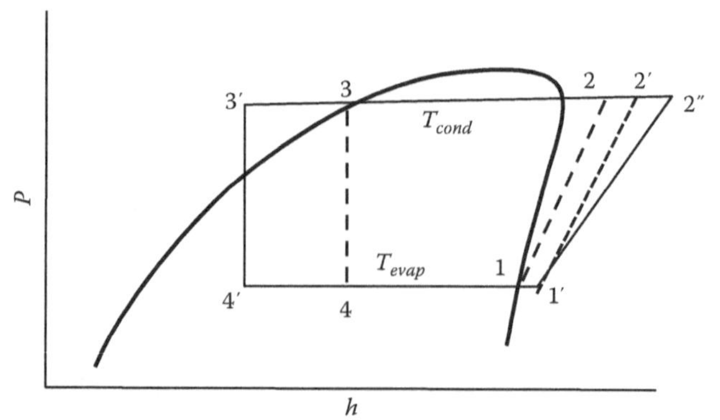

FIGURE 14.4
p–h diagram showing the various state points for the three cycles: the standard VC cycle (1–2–3–4), the modified VC cycle (1′–2′–3′–4′), and cycle with non-isentropic compression (1′–2″–3′–4′).

Note that the state point 4′ entering the evaporator is pushed further toward the saturated liquid line, and one obvious inference is that the refrigerating effect will be increased, and so the COP of the cycle should improve. The latter is not necessarily true since superheating the refrigerant vapor pushes point 1′ further into the superheat regions where the lines of constant entropy flare out. This results in additional work needed for compression. Further, the specific volume of the refrigerant increases with degree of superheat, which results in the compressor drawing in less mass flow rate of refrigerant, thereby reducing cooling capacity. Thus, there may in fact be a thermodynamic penalty. Example 14.2 will illustrate this statement.

In most chiller designs, the superheat heat exchanger is placed outside the cooling air stream, and so the cooling capacity of the chiller unit is given by $(h_1 - h_3')$. This is the general analysis approach followed in most books and manuals. However, there are a few chiller equipment designs where the refrigerant superheat process also contributes to the cooling process, and in such cases cooling capacity is given by $(h_1' - h_3')$.

Example 14.2: Modified Vapor Compression Cycle

Consider the same example as Example 14.1 where an R-134a cycle operates between −14°C and 30°C with a cooling capacity of 5.0 kW. The compressor isentropic efficiency is given as 0.8, while the vapor is superheated by 6°C. Find the power input, COP, kW/ton, and the refrigerating efficiency of this cycle, and compare these against the results of Example 14.1.

Given: $T_{cond} = 30°C$, $T_{evap} = -14°C$, $\dot{Q}_{cool} = 5.0$ kW, $\eta_{isen} = 0.8$, $T_1' - T_1 = 6°C$

Figure: See Figure 14.4.

Assumptions: All processes are steady. The cycle follows the modified VC cycle.

Find: $\dot{W}_{comp}$, COP_{mod}, kW/ton, and η_R

Lookup values: Thermodynamic properties are read from the R-134a saturation tables (Table A6) or from the online HCB software. As in Example 14.1, the low-side pressure is 0.1707 MPa and the high-side pressure is 0.7701 MPa.

Solution

We start with the fact that the vapor at state 1′ is superheated by 6°C. Hence, for a temperature of (−14 + 6 =) −8°C and pressure of 0.1707 MPa, the vapor enthalpy $h_1' = 395.24$ kJ/kg. With the condition that the process is isentropic from 1′ to 2′, i.e., $s_1' = s_2'$, we can determine the enthalpy of point 2′. We determine $h_2' = 427.5$ kJ/kg.

We can deduce the enthalpy of the subcooled fluid leaving the condenser (Equation 14.13)

$$h_3' = h_3 - (h_1' - h_1) = 241.65 - (395.24 - 390.33)$$
$$= 236.74 \text{ kJ/kg}$$

Next, we compute the refrigerant mass flow from Equation 14.6 with $h_4' = h_3'$

$$\dot{m}_r = \frac{\dot{Q}_{cool}}{(h_1 - h_4')} = \frac{5.0 \text{ kW}}{(390.33 - 236.74) \text{ kJ/kg}}$$
$$= 0.0326 \text{ kg/s} \quad \text{(earlier it was 0.0336)}$$

We rearrange Equation 14.14 to get the enthalpy of the vapor leaving the compressor:

$$h_2'' = h_1' + \frac{(h_2' - h_1')}{\eta_{isen}} = 395.24 + \frac{(427.50 - 395.24)}{0.8}$$
$$= 435.565 \text{ kJ/kg}$$

This allows the compressor power input to be determined (from Equation 14.7)

$$\dot{W}_{comp} = 0.0326 \text{ kg/s} \times (435.565 - 395.24) \text{ kJ/kg}$$
$$= 1.315 \text{ kW} \quad \text{(earlier it was 1.05)}$$

The COP of the modified VC cycle:

$$COP_{mod} = \frac{5.0}{1.315} = 3.8 \quad \text{(earlier it was 4.76)}$$

From Equation 14.10, the kW/ton of the machine = 3.516/3.8 = 0.925 kW/ton (earlier it was 0.739)

The Carnot efficiency remains unaltered and is the same as that of Example 14.1, i.e., COP_{Carnot} = 5.886. The refrigerating efficiency from Equation 14.11 is finally determined:

$$\eta_R = \frac{COP_{mod}}{COP_{Carnot}} = \frac{3.8}{5.886} = 0.646 \quad \text{(earlier it was 0.809)}$$

Comments

Due to the isentropic efficiency of the compressor, the COP and the kW/ton have become worse than those for the standard VC cycle of Example 14.1. In fact, multiplying the COP found in Example 4.1 by η_R =0.8 (the value assumed in Example 14.2) yields a COP for the modified cycle of 3.81. The further difference in COP between 3.81 and 3.40 found earlier is because of the superheat effect.

14.2.2 Actual Single-Stage Cycle

The modified VC cycle described earlier differs from the actual cycle in other ways and requires detailed specifications of the actual system piping; and so are overlooked in most textbooks. These include:

1. *Compressor*—friction losses, heat losses, or heat gains
2. *Condenser*—friction losses, subcooling to ensure pure liquid at throttling valve inlet
3. *Evaporator*—friction losses, superheating to ensure pure vapor at compressor inlet
4. *System losses*—pressure drops, heat gains, and heat losses in all liquid and vapor refrigerant lines; compressor shaft friction

These thermodynamic and fluid flow irreversibilities of the mechanical components of the system distort the standard VC cycle shown in Figure 14.2 to the actual cycle shown in Figure 14.5. Each of these irreversibilities causes inefficiencies; the one exception is the possible *cooling of refrigerant in the compressor*. If cooled, the refrigerant specific volume will be reduced, and so will the corresponding power input. For this reason, one finds cooling fins around the compression chamber in small vapor compression units.

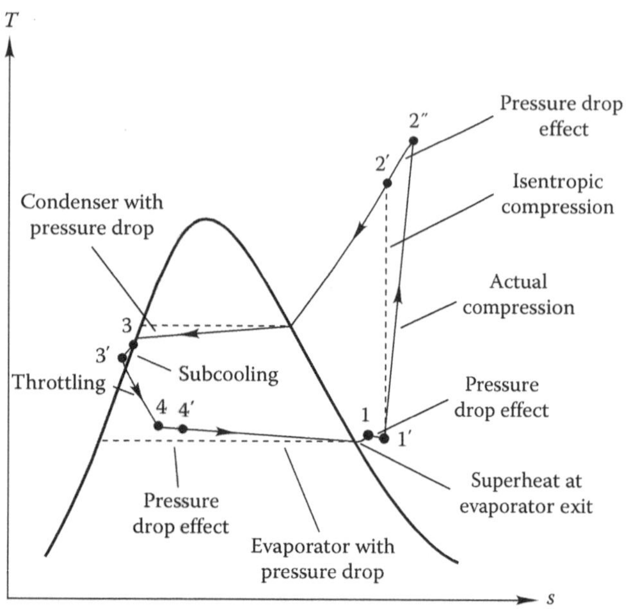

FIGURE 14.5
The actual VC cycle on a *T–s* diagram showing friction losses, compressor inefficiencies, enthalpy changes, superheating, and subcooling in a real cycle. Line segments such as 1–1′ or 2″–2′ account for piping losses, and sloped evaporator and condenser paths include pressure drop effects.

14.2.3 Multistage Cycles

There are other VC cycles used in actual systems. The common ones are multistage cycles and cascade cycles. In *multistage cycles*, the compression is achieved in stages with several compressors in series. If the pressure ratio is high, the refrigerant temperature during compression increases greatly and more work is required as a result. Cooling the refrigerant during each stage reduces the total power requirement. Instead of using an external cooling source, some of the evaporating refrigerant from the stage above is used for refrigerant intercooling. Such cycles are used in industrial applications when refrigerants like ammonia are used. One can also have cycles with more than one evaporator. Such systems are used in process applications that require two evaporating temperatures, say an evaporating temperature of 5°C for air conditioning a building and an evaporating temperature of −10°C for food preservation. Instead of two refrigeration systems, one system could cater to both applications. Another type of VC cycle is the *cascade cycle* where two or more single-stage cycles are used in series. The evaporator of the higher stage cycle is coupled to the condenser of the stage immediately below it and so on. Cascade cycles are used extensively for cryogenic applications. These cycles are not discussed here because they are rarely, if at all, used for building cooling applications. The interested reader can refer to standard textbooks like Stoecker and Jones (1982), Kuehn et al. (1998), or Mitchell and Braun (2013).

14.3 Absorption Cooling

Absorption cycles are those where the cooling can be achieved, not by work input, but by heat input to the system. The ideal COP of the absorption cycle is much lower than that of the VC cycle and this is also reflected in real machines. However, it is somewhat misleading to compare the COPs directly since the absorption system is powered by thermal energy that is lower form of energy than the electricity needed to operate the VC cycle. If we assume a typical electric power plant generation and transmission efficiency of 32%, then the overall thermal to cooling efficiency of the VC cycle (Example 14.2) will be equal to (3.40 × 0.32 = 1.09), which is much closer to that of the absorption system.

14.3.1 Absorption Cycle

Absorption cooling (AC) refrigeration differs from VC cooling only in the method of compressing the refrigerant. The compressor in VC is replaced with a small liquid pump and heat addition. The absorption and desorption of the refrigerant from the pumped liquid are achieved by heat transfer alone. Therefore, the use of electric power is minimal with absorption equipment. This approach has considerable merit when a source of heat between 100°C and 200°C (200°F–400°F) is available at competitive cost. In this section, we discuss the ideal single-stage AC cycle. Nearly all modern absorption machines use two- or three-stage cycles. In multiple-stage equipment, generators at different temperatures, heated by different sources, are used to improve the COP.

Figure 14.6 is a schematic diagram of the absorption cycle that involves a homogeneous mixture of two substances (see Section 12.4 for a basic discussion). In the AC cycle, refrigerant pressurization is achieved by dissolving the refrigerant in a liquid absorbent, pressurizing the solution by using a liquid pump that requires very little work input, and, finally, driving the low-boiling refrigerant from the high-pressure, refrigerant-rich absorbent in the generator by the addition of heat. The refrigerant must be more volatile than the absorbent so that separation from the absorbent occurs easily in the generator. After the refrigerant is released at high pressure, the weak absorbent returns to the absorber where refrigerant is again reabsorbed into the absorbent. The absorption of refrigerant is an exothermic reaction releasing heat. Hence heat has to be removed from the AC equipment in two places—the condenser (as with VC equipment) and the absorber.

The effective performance of the AC depends on the two substances that comprise the refrigerant–absorbent pair. The pair must have a high degree of stability for long-term operations, be nontoxic and nonflammable and must not cause long-term environmental harm. The two most common refrigerant–absorbent pairs are water–lithium bromide (LiBr–H_2O) and ammonia–water (NH_3–H_2O). In the former, the refrigerant is low-pressure water vapor; in the latter, it is ammonia. Pure LiBr is a solid that forms a homogeneous solution when mixed with water. An outstanding feature of this binary pair is that LiBr is nonvolatile. Only the water vapor is driven off in the generator, and so no rectification column (meant to remove traces of the absorbent from the refrigerant vapor) is required. (This is not the case for NH_3–H_2O mixtures where traces of water vapor will be present along with the ammonia vapor leaving the generator.) Thus, LiBr–H_2O systems are simpler since a rectifying column is not required and can operate at lower generator temperatures, which are achievable by burning waste biomass or from hot water from solar flat-plate collectors. The condition of the state points to the left of the dotted line of Figure 14.6 is similar to those of the standard VC cycle. State point 3 is superheated water vapor, while those of points 4 and 5 are saturated liquid and saturated vapor respectively.

However, LiBr has two drawbacks, given the use of water as the refrigerant: The evaporator temperature cannot be much below 40°F (5°C), resulting in very low system pressures, and care must be taken to operate the generator at temperatures sufficiently high to avoid crystallization of LiBr salts.

Performance of the AC system is measured by its COP. However, the principal input to this process is not shaft power, but heat. Hence, the COP is given by

$$COP_{AC} = \frac{\dot{Q}_{evap}}{\dot{Q}_{gen}} \tag{14.15}$$

where

$\dot{Q}_{evap}$ is the heat removal rate at evaporator, Btu/h (kW)

$\dot{Q}_{gen}$ is the heat supply rate to generator, Btu/h (kW)

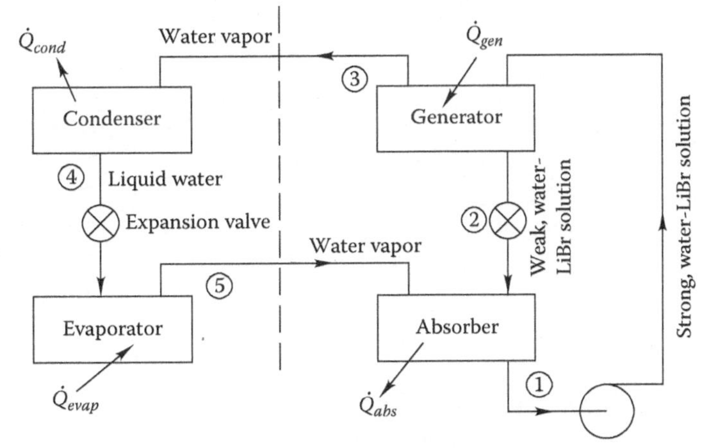

FIGURE 14.6
Schematic diagram of simple LiBr absorption cycle. To the left of the vertical dashed line, the system components are similar to those of the VC system.

The COP for LiBr systems ranges from 0.5 to slightly above 1.2 for the most sophisticated, steam-fired systems. Ammonia AC systems are older, and predominantly used in industrial refrigeration applications rather than for air-conditioning. They have the advantage that the evaporator operates at pressures above atmospheric and that the evaporator can operate at temperatures below those possible with the LiBr system (below 0°C). However, extra components are needed, as noted earlier, and due to the higher operating pressures, heavier construction is necessary. The toxicity of ammonia is also a concern. The COPs of both of these common absorption systems are about the same.

14.3.2 Analysis of Lithium Bromide Cycle

The analysis of the AC cycle is somewhat different from that for the VC cycle, which uses a pure substance. Instead, a homogeneous mixture of absorbent and refrigerant is the working fluid. The concentration of the refrigerant is expressed as the *mass fraction X* [lb_m LiBr/lb_m mixture (kg LiBr/kg mixture)]. This is one of the thermodynamic state variables used to conduct cycle analyses for this equipment. A brief introduction was provided in Section 12.4. The refrigerant is the pure substance, which has the lower boiling point.

Figure 14.7 is an example of the charts one can use to find the enthalpy of the mixture as a

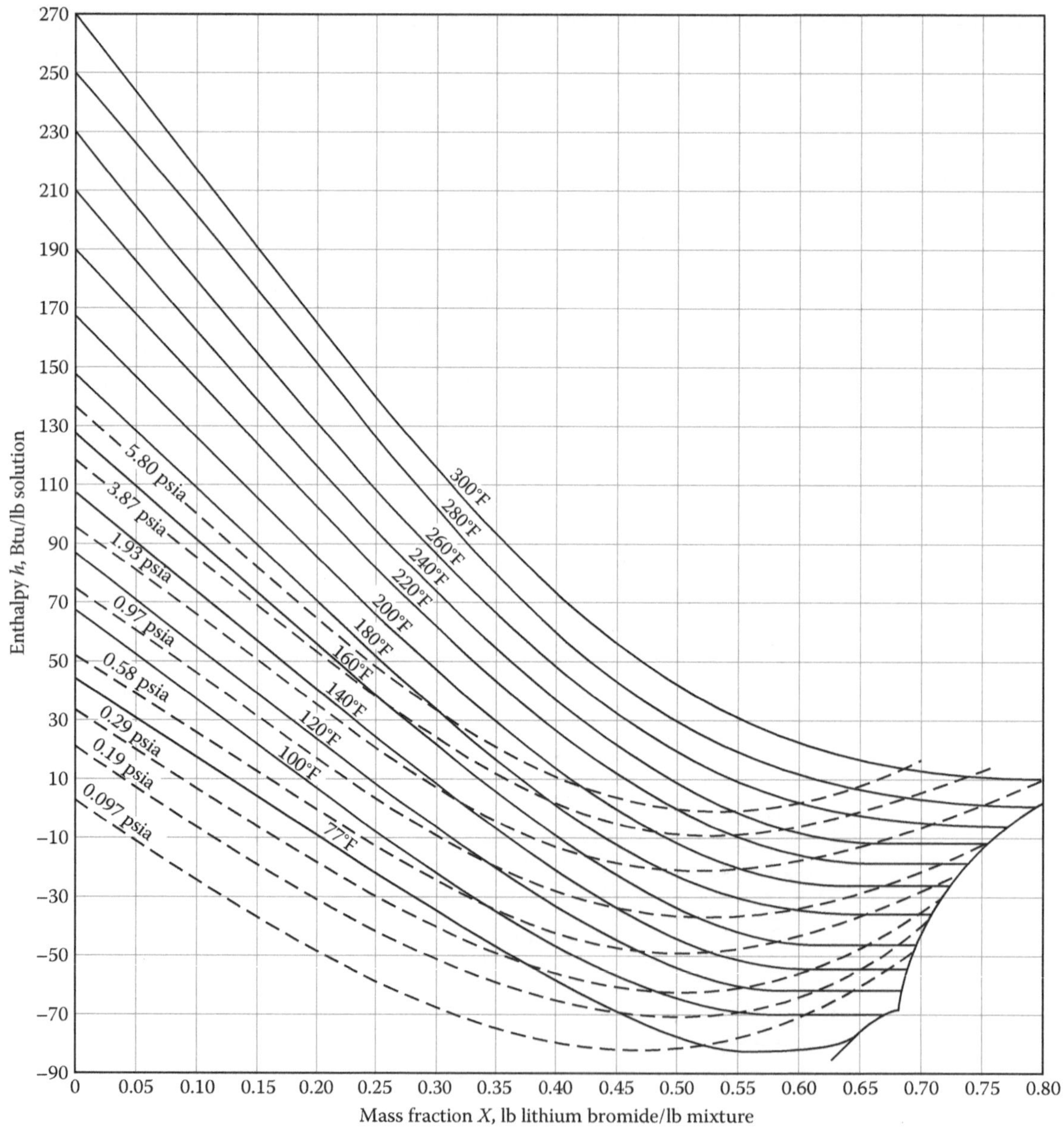

FIGURE 14.7
The LiBr enthalpy–concentration diagram. 1 psia = 6.89 kPa; 1 Btu/lb = 2.32 kJ/kg. (Courtesy of Institute of Gas Technology, Chicago, IL.)

function of concentration and pressure (or temperature). Plotted along the abscissa is the mass fraction X or concentration of LiBr in solution. The ordinate is the enthalpy h of the solution at the value of either temperature or pressure read from the curves in the figure. The chart applies for saturation and subcooled conditions.

To solve problems involving the LiBr cycle, one often knows a pressure and temperature condition but not the mass fraction or enthalpy. In those cases, find the intersection of the known pressure and temperature lines in the figure, and then read the corresponding values of enthalpy (needed for energy balance calculations) and concentration (needed for mass balance calculations) from the axes of Figure 14.7. Example 14.3 illustrates the method.

Example 14.3: Lithium Bromide Absorption Cycle

A lithium bromide chiller operates between a condensing temperature of 104°F (40°C) and an evaporator temperature of 50°F (10°C). Heat is added to the generator at 212°F (100°C) and removed from the absorber at 86°F (30°C). If the pump flow rate is 4800 lb_m/h (0.605 kg/s), what are the COP and the heat rates at each component of the cycle: generator, absorber, condenser, and evaporator?

Given: $T_{gen} = 212°F$, $T_{con} = 104°F$, $T_{evap} = 50°F$, $T_{abs} = 86°F$ (italicized in Table 14.3) $\dot{m}_{pump} = 4800$ lb/h (0.61 kg/s)

Figure: See Figure 14.6.

Assumptions: All components operate at equilibrium. State points 4–5 are at liquid and vapor saturated conditions respectively. Ignore all pressure and heat losses in piping.

Find: $\dot{Q}_{cond}$, $\dot{Q}_{evap}$, $\dot{Q}_{abs}$, $\dot{Q}_{gen}$, COP

Solution

The solution consists of, first, finding thermodynamic properties at all state points 1–5. Second, we use mass and energy balance equations to find the heat and mass flow terms.

TABLE 14.3

Thermodynamic Properties for Example 14.3

State	Temp., °F	p, psia	X	h, Btu/lb_m
1	*86*	0.178	0.50	−71[a]
2	*212*	1.070	0.67	−24[a]
3 (superheated vapor)	*212*	1.070	0.0	1150.5[b]
4 (saturated liquid)	*104*	1.070	0.0	72[b]
5 (saturated vapor)	*50*	0.178	0.0	1083.3[b]

[a] Read from Figure 14.7.
[b] Read from steam tables.

The property table is completed as follows:

(a) First, the known temperatures in the second column are entered. The given temperatures within each component are assumed to be the temperatures of fluid streams *exiting* that component.
(b) Second, pressures are entered in the table. We refer to the saturated steam tables (Table A3) to determine the condenser and evaporator pressures p_3 and p_5 since the corresponding temperatures are specified. Consequently, the pressures at all state points are now known. Note that pressures in the LiBr cycle are sub-atmospheric and small.
(c) Next, the LiBr mass fractions at points 1 and 2 can be deduced from Figure 14.7 as the intersection of the known pressure and temperature lines (the online HCB software can be used for more accurate determination). Of course, for states 3, 4, and 5 consisting of pure water vapor, the mass fraction is zero.
(d) Finally, from Figure 14.7 (or from the online HCB software), the enthalpies at points 1 and 2 are deduced. The enthalpies at points 3, 4, and 5 are easily determined from the steam superheat and saturated tables (Tables A3 and A4).

At this point, a mass balance is written on the generator, and then each heat flow term can be found. The total mass balance and the LiBr mass balance written on the generator are

$$\dot{m}_2 + \dot{m}_3 = \dot{m}_1 = 4800\ \text{lb}_\text{m}/\text{h} \tag{14.16}$$

$$\dot{m}_1 X_1 = \dot{m}_2 X_2 \tag{14.17}$$

From these expressions, the two unknown mass flow rates can be found:

$$\dot{m}_2 = \dot{m}_1\left(\frac{X_1}{X_2}\right) = 4800\left(\frac{0.50}{0.67}\right) = 3582\ \text{lb}_\text{m}/\text{h} \tag{14.18}$$

$$\dot{m}_3 = \dot{m}_1 - \dot{m}_2 = 4800 - 3582 = 1218\ \text{lb/h} = \dot{m}_4 = \dot{m}_5 \tag{14.19}$$

The four heat rate terms can now be found from energy balances on the four components of the AC cycle. The arrows in Figure 14.6 show the direction of the heat flow:

$$\dot{Q}_{gen} = \dot{m}_3 h_3 + \dot{m}_2 h_2 - \dot{m}_1 h_1 = 1218 \times 1150.5 + 3582 \times (-24) - 4800 \times (-71) = 1656\ \text{kBtu/h} \tag{14.20}$$

$$\dot{Q}_{cond} = \dot{m}_3(h_3 - h_4) = 1218 \times (1150.5 - 72) = 1314 \text{ kBtu/h} \quad (14.21)$$

$$\dot{Q}_{abs} = \dot{m}_2 h_2 + \dot{m}_5 h_5 - \dot{m}_1 h_1 = 3582 \times (-24) + 1218 \times (1083.3) - 4800 \times (-71) = 1574 \text{ kBtu/h} \quad (14.22)$$

$$\dot{Q}_{evap} = \dot{m}_5(h_5 - h_4) = 1218 \times (1083.3 - 72) = 1232 \text{ kBtu/h}$$

As a check, by the first law of thermodynamics, since pumping power is negligible,

$$\dot{Q}_{evap} = -\dot{Q}_{gen} + \dot{Q}_{abs} + \dot{Q}_{cond} = -1656 + 1574 + 1314 = 1232 \text{ kBtu/h}$$

Finally, the COP, defined as the ratio of cooling produced in the evaporator to the heat added in the generator, is

$$\text{COP} = \frac{\dot{Q}_{evap}}{\dot{Q}_{gen}} = \frac{1232 \text{ kBtu/h}}{1656 \text{ kBtu/h}} = 0.74$$

Comments

The solution to AC cycle problems involves the use of two sets of thermodynamic tables or charts—those for pure water and those for LiBr. Although the reference states for various LiBr charts found in the literature may vary from those in Figure 14.7, the energy flows will be the same. Figure 14.7 has the advantage that all thermodynamic variables are included in a single LiBr chart rather than two charts, as in other references.

Although the COP of an absorption cycle is much lower than that for a VC cycle, remember that the heat input to a power plant that produces the electricity to power VC machines is about three times the power plant's electric output. Therefore, the ratio of cooling effect to *thermal* energy input (to the power source) is not much different for VC and AC cycles.

This section has described the basic AC system. A number of refinements are often used to improve performance. For example, a heat exchanger is often used across the LiBr streams entering and leaving the generator. By extracting heat from the generator exit stream and adding it to the inlet stream, the amount of external heat required for the generator can be reduced. Higher generator temperatures can improve the COP within limits. Two-stage systems with interstage heat exchange can have better efficiency than single-stage systems. These refinements are discussed in ASHRAE (2012), ASHRAE Fundamentals (2013), and in textbooks such as Kuehn et al. (1998).

14.4 Chiller Systems

14.4.1 System Analysis

A *chiller* is an assembly of equipment used to produce chilled water for cooling of spaces within buildings. In addition to the evaporator, condenser, expansion device, and compressor, heat exchangers between the heat rejection subsystem and cooling distribution subsystem are also included in the chiller assembly. Figure 14.8 is a conceptual sketch of a liquid chiller distinguishing the vapor compression cycle and the chiller system and showing the individual components.

Section 14.2 dealt with refrigeration cycles, where we limited ourselves to analyzing the refrigerant

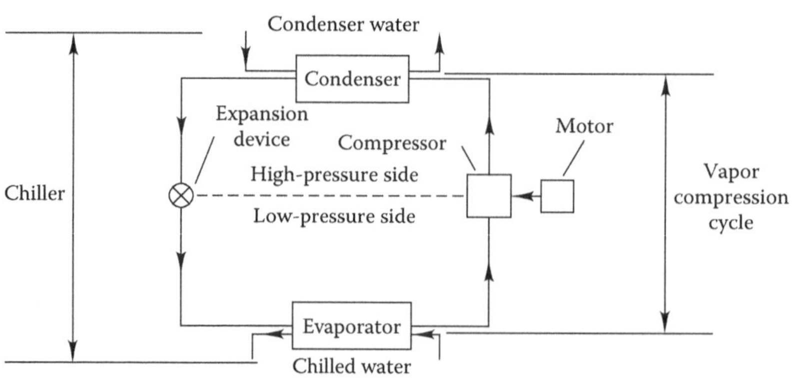

FIGURE 14.8
Schematic diagram showing essential components of basic liquid chiller and relation to vapor compression cycle.

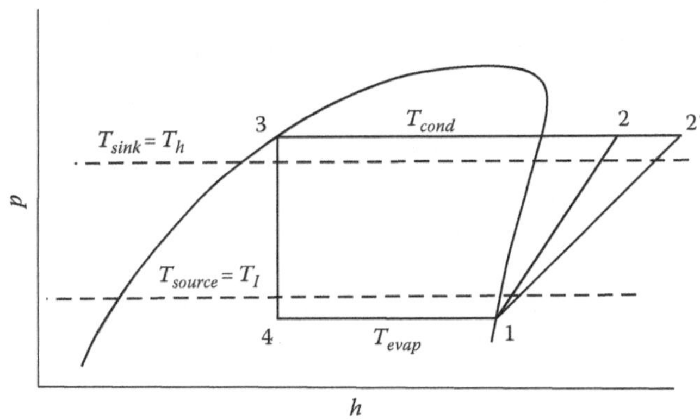

FIGURE 14.9
The standard VC refrigeration cycle drawn on a p–h diagram along with the sink and source temperatures.

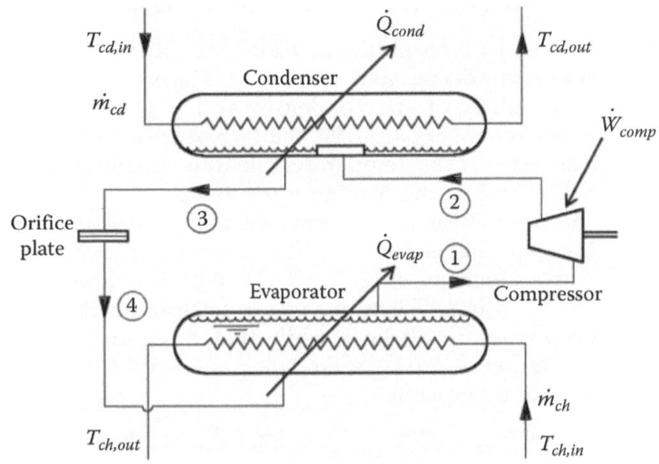

FIGURE 14.10
Sketch of a water-cooled refrigeration system with flooded evaporator and condenser showing the two coolant water loops.

properties in the piping and components when the high and low refrigerant condensing and boiling temperatures were specified. This will not be the case while analyzing an actual situation that calls for determining the high and low refrigerant temperatures given the actual heat source and sink temperatures (also referred to as coolant temperatures). Since actual heat exchangers need a certain temperature difference in order for them to operate, the refrigerant temperature in the condenser will be higher than $T_{sink} = T_h$ while the refrigerant would boil in the evaporator at a temperature lower than $T_{source} = T_l$ (see Figure 14.9). The wider refrigerant temperature levels will reduce the cycle COP as is obvious from the expression for COP_{Carnot} given by Equation 12.4. A more effective heat exchanger will tend to reduce these temperature differences; unfortunately, this would usually require a larger surface area of heat exchange and hence, higher costs. Typically, the temperature of air or water used to cool the condensing refrigerant will be 10°F–15°F (5.5°C–8.3°C) lower while the air or water to be cooled in the evaporator will be 10°F–15°F (5.5°C–8.3°C) warmer than the evaporating refrigerant (Mitchell and Braun, 2013).

The superheating and the subcooling modifications discussed earlier apply mostly to small- to medium-capacity refrigeration systems. For large systems, the evaporators and condensers are designed differently and operated under flooded conditions, i.e., the refrigerant is on the shell side and the boiling and condensation occur as in pool (see Figure 14.10). The conditions of points 1 and 3 are well represented by saturated vapor and saturated liquid respectively (as was assumed for the standard VC cycle).

Analysis of heat exchangers was treated in Section 12.5. The expressions for effectiveness of different types of heat exchanger configurations were presented. For the case when one of the fluids is either condensing or evaporating, i.e., its temperature remains constant, Equation 12.35 applies:

$$\varepsilon = 1 - \exp(-\text{NTU}) \tag{14.23}$$

where NTU is the number of heat transfer units (defined by Equation 12.37).

In addition to the equations assembled in Table 14.1, we need to include those for heat exchanger performance. For the condenser, this can be represented in terms of either the coolant inlet or outlet temperatures $T_{cd,in}$ or $T_{cd,out}$:

$$\begin{aligned}\dot{Q}_{cond} &= (\dot{m}_{cd} \cdot c_p)\varepsilon_{cond}(T_{cond} - T_{cd,in})\\ &= \frac{(\dot{m}_{cd} \cdot c_p)\varepsilon_{cond}}{(1-\varepsilon_{cond})}(T_{cond} - T_{cd,out})\end{aligned} \tag{14.24}$$

Similarly, at the evaporator end

$$\begin{aligned}\dot{Q}_{evap} &= (\dot{m}_{ch} \cdot c_p)\varepsilon_{evap}(T_{ch,in} - T_{evap})\\ &= \frac{(\dot{m}_{ch} \cdot c_p)\varepsilon_{evap}}{(1-\varepsilon_{evap})}(T_{ch,out} - T_{evap})\end{aligned} \tag{14.25}$$

Practically, most chillers have to be modulated in step with variations in the building cooling load requirements. This is done by a thermostatic valve that maintains the fluid temperature leaving the evaporator $T_{ch,out}$ at a pre-set level (discussed further in Section 17.3). This is the reason why the expressions are framed in terms of either the inlet or the outlet conditions depending on the manner in which the problem is stated. The following example illustrates the procedure for analyzing the entire chiller.

Example 14.4: Analysis of Chiller Systems

Consider a large chiller using R-134a with water-cooled condenser and evaporator. The effectiveness values of the condenser and evaporator heat exchangers are 0.90 and 0.75, respectively. The water inlet temperature to the condenser is 25°C while the leaving water temperature in the evaporator is controlled at 2°C. The cooling capacity is 500 kW. The compressor isentropic efficiency is given as 0.9, with no vapor superheating. The water flow rate through the evaporator is 8.0 kg/s while that through the condenser is 12.0 kg/s. Find the power input and COP of this system.

Given: $T_{cd,in}$ = 25°C, $T_{ch,out}$ = 2°C, $\dot{Q}_{cool}$ = 500 kW, $\dot{m}_{cd}$ = 12.0 kg/s, $\dot{m}_{ch}$ = 8.0 kg/s, η_{isen} = 0.90, ε_{evap} = 0.75, ε_{cond} = 0.9

Figure: See Figure 14.10.

Assumptions: All processes are steady state. The cycle follows the standard VC cycle while considering the fact that the compression is not isentropic. Specific heat of water c_p = 4.186 kJ/kg.

Find: $\dot{W}_{comp}$, COP

Solution

This problem cannot be solved sequentially like in the two previous examples. One could adopt an iterative approach or better still use an equation solver software program. There are different ways by which an iterative procedure can be followed. One approach is to assume an *initial guess value* for the refrigerant condensing temperature $T_3 = T_{cond}$ =40°C.

(i) Rewrite Equation 14.25 in terms of the evaporator temperature:

$$T_{evap} \equiv T_1 = T_{ch,out} - \dot{Q}_{evap}\frac{(1-\varepsilon_{evap})}{(\dot{m}_{ch}\cdot c_p)\varepsilon_{evap}}$$

$$= 2 - 500 \times \frac{(1-0.75)}{(8\times 4.186\times 0.75)} = -3.0°\text{C}$$

(ii) The entropy at state 1 is found to be s_1 = 1.7255 kJ/kg, which is also equal to s_2. This allows the enthalpies at states 1 and 2 to be found as h_1 = 396.84 kJ/kg and h_2 = 425.04 kJ/kg.

(iii) At state 3, the refrigerant is saturated liquid with h_3 = 256.41 kJ/kg = h_4.

(iv) The refrigerant mass flow rate:

$$\dot{m}_r = \frac{\dot{Q}_{cool}}{(h_1 - h_4)} = \frac{500 \text{ kW}}{(396.84 - 256.41)\text{ kJ/kg}}$$

$$= 3.56 \text{ kg/s}$$

(v) As in Example 14.2, we use Equation 14.14 to get the enthalpy of the vapor leaving the compressor:

$$h_2' = h_1 - \frac{(h_2 - h_1)}{\eta_{isen}} = 396.84 + \frac{(425.04 - 396.84)}{0.9}$$

$$= 428.17 \text{ kJ/kg}$$

(vi) The compressor power input (from Equation 14.7)

$$\dot{W}_{comp} = 3.56 \text{ kg/s} \times (428.17 - 396.84) \text{ kJ/kg}$$

$$= 111.55 \text{ kW}$$

(vii) The heat rejection rate at the condenser can be found very simply from an energy balance (this will give identical results as using Equation 14.8)

$$\dot{Q}_{cond} = \dot{Q}_{cool} + \dot{W}_{comp} = 500 + 111.55 = 611.55 \text{ kW}$$

(viii) We can now use Equation 14.24 to verify our initial guess of T_3 = 40°C

$$T_{cond} = T_{cd,in} + \frac{\dot{Q}_{cond}}{(\dot{m}_{cd}\cdot c_p)\varepsilon_{cond}}$$

$$= 25 + \frac{611.55}{(12\times 4.186\times 0.9)} = 38.5°\text{C}$$

(ix) The difference between the aforementioned value and the initial guess value is small; a second iteration would likely provide the necessary convergence. If the present difference is acceptable, the chiller actual COP is easily determined:

$$\text{COP} = \frac{500 \text{ kW}}{111.55 \text{ kW}} = 4.48$$

This system was a water-cooled chiller. Air-cooled chillers that are smaller in capacity have a number of differences in equipment and operation. They will, almost invariably, be operated such that there is vapor superheat, and the condenser and evaporator will not be of the flooded type. Further, the air to be cooled will be passed directly over the evaporator with the refrigerant boiling inside the tubes. This is called a direct expansion or DX evaporator (and is widely used in split systems in residential and small commercial buildings—see Chapter 11). On the condenser side, the outdoor air can be passed directly through the heat exchanger where the refrigerant will be condensing as in small split domestic units, or a liquid condenser

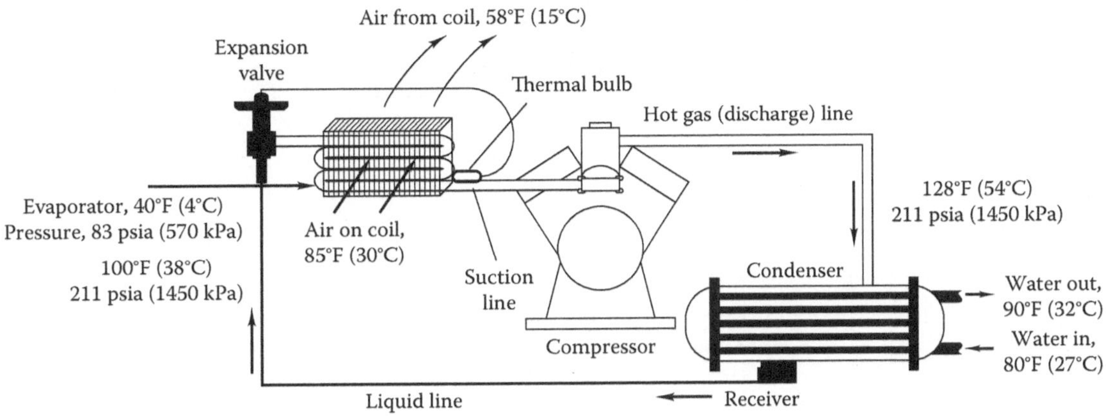

FIGURE 14.11
Vapor compression cycle equipment with typical R-22 operating temperatures and pressures. A direct expansion (DX) evaporator is used. The state points are based on a compressor efficiency of 85% and an evaporator outlet superheat of 9°F (5°C). Pressure drops are not noted.

could also be included in larger buildings. Typical operating conditions are shown in Figure 14.11 for a R-22 DX system with a water-cooled condenser. Air-cooled chillers tend to have lower COPs than water-cooled chillers and are generally more convenient to install and operate since they do not need a water supply or cooling towers. Their lower initial costs make them suitable for small to medium cooling load applications.

14.4.2 Chiller Performance Maps

The performance of chillers varies with operating conditions such as ambient air temperature or cooling water set temperature. The rated conditions pertain to a set of chiller operating conditions specified by equipment standards (this aspect is described in detail in Section 14.6), which usually are reflective of the equipment operating under peak load conditions. Under off-rated operating conditions, the thermodynamic cycle is modified with respect to that under design conditions and the various quantities shown in Table 14.1 are affected. This issue is also addressed in Section 14.7.

Manufacturers provide either tables or figures, called *chiller performance maps*, of how their equipment performs under off-rated operating conditions. An illustrative performance map is shown in Figure 14.12 which depicts how the cooling capacity and input electric power vary with evaporating temperature for different condensing temperatures. Note that this figure is based on the refrigerant operating conditions, and not on the air or water-coolant temperatures. This figure is for a system with R-22 four-cylinder hermetic compressor with air-cooled DX and an air-cooled condenser, but the plots look similar for other types of chiller systems as well. Cooling capacity increases with increasing evaporating temperature and decreasing condensing temperature, while power input increases as both these temperatures increase. Using Figure 14.12, for an evaporating temperature of 45°F (7.2°C) and a condensing temperature of 118°F (47.8°C), the power is 13.5 kW and the capacity is 160,000 Btu/h (46.9 kW) yielding a COP of about 3.5 or 1.0 kW/ton (typical value for a small unit).

These trends can be gleaned from Figures 14.13 and 14.14, which show how the standard VC cycle changes with variation in the evaporator and the condenser temperatures, respectively. As the evaporator temperature increases with fixed condenser temperature (Figure 14.13), compression work $(h_2 - h_1)$ is markedly less than that of $(h_{2a} - h_{1a})$, and so the refrigerant flow rate will increase. The net effect is an increase in power input. At the same time, the cooling effect is increased a little due to the shape of the saturation dome (from 4a-1a to 4-1); and so the effect of the increased refrigerant flow rate is to *decrease* the cooling capacity.

On the other hand, in Figure 14.14, increasing condensing temperature will reduce the refrigerant flow because the compressor has to pump the refrigerant through a higher pressure ratio. The cooling effect is also reduced (from 4–1 to 4a–1) and with the reduced refrigerant flow, the cooling capacity will decrease. Note that even with the reduced refrigerant flow rate, the power input increases due to the higher pressure ratio. The COP behavior is akin to that suggested by the Carnot cycle: COP increases as the pressure ratio decreases.

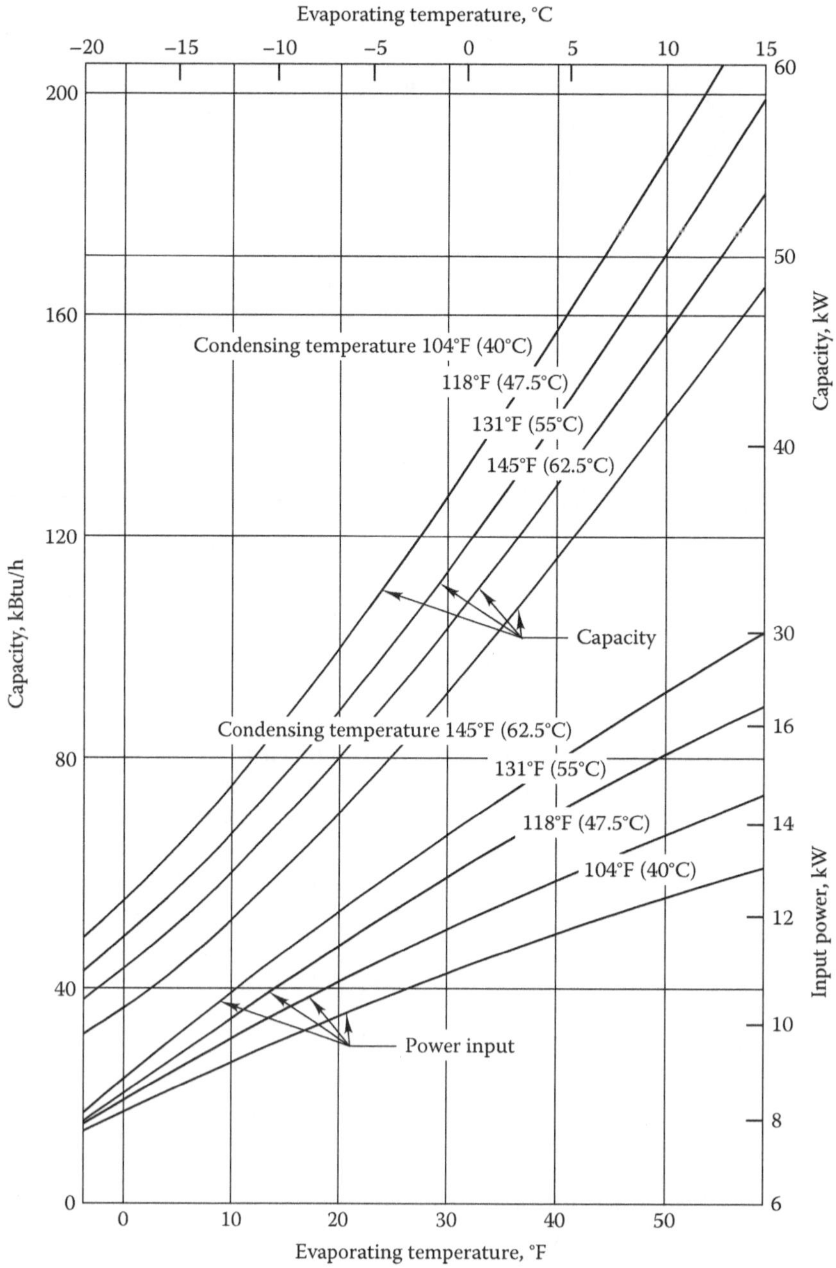

FIGURE 14.12
Example of "performance map" for reciprocating chiller; based on R-22, 10°F subcooling, 20°F superheat, and 1725 rpm compressor speed. (From ASHRAE, *Handbook of Systems and Equipment*, American Society of Heating, Refrigerating and Air-conditioning Engineers, Atlanta, GA, 2012. Copyright ASHRAE, www.ashrae.org.)

Tables 14.4 and 14.5 are samples of performance tables provided by manufacturers for an air-cooled and for a water-cooled chiller respectively. These correspond to specific evaporator and condenser heat exchanger sizes matched with the compressor in question (similar tables also provide information of system performance with other heat exchanger–compressor pairs). Notice in Table 14.4 that the operating conditions are specified in terms of a suction temperature, which is the evaporator temperature, i.e., that of the refrigerant boiling in the DX evaporator. For a water-cooled chiller, the performance is expressed in terms of the two water flow temperatures. The condenser entering water temperature is dictated by the outdoor wet-bulb temperature (typically 5°F–15°F differential), but this quantity does not explicitly appear in Table 14.5. The capacity of the water-cooled chiller is typically larger while its operating temperatures are typically lower as compared to those of an air-cooled chiller.

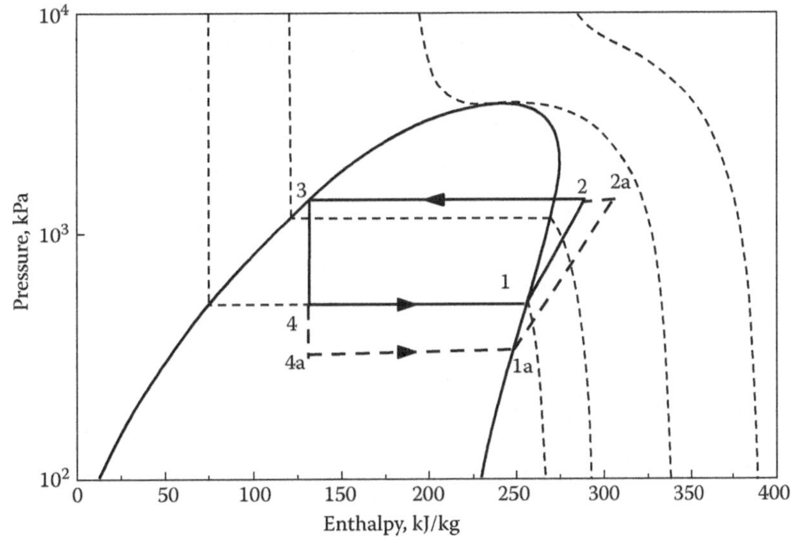

FIGURE 14.13
Effect of changing evaporator temperature on the VC cycle.

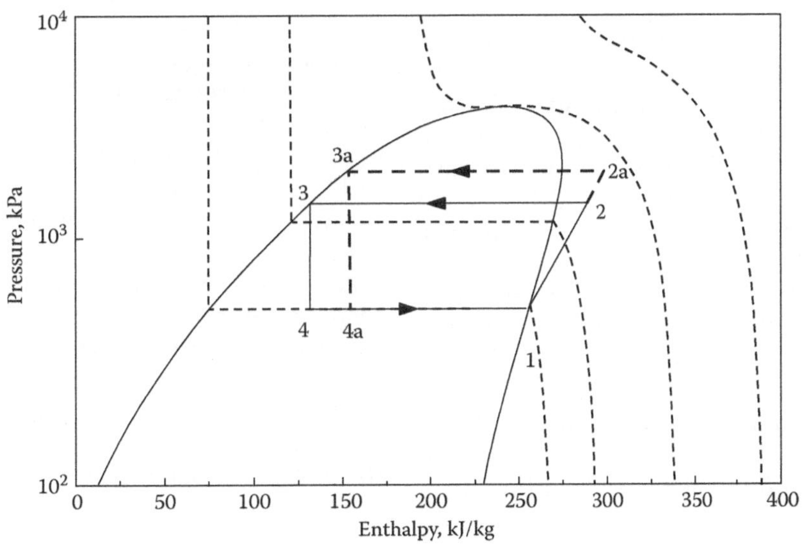

FIGURE 14.14
Effect of changing condenser temperature on the VC cycle.

TABLE 14.4
Sample of How Manufacturers Present Data on the Performance of an Air-Cooled DX Chiller Unit

	Ambient Dry-Bulb Temperature (T_o)							
	90°F		95°F		100°F		105°F	
Suction Temp., °F	**Cooling Capacity, tons**	**Electric Input, kW**	**Cooling Capacity, tons**	**Electric Input, kW**	**Cooling Capacity, tons**	**Electric Input, kW**	**Cooling Capacity, tons**	**Electric Input, kW**
30	10.9	12.6	10.6	12.8	10.2	13.0	9.7	13.4
35	12.1	13.2	11.8	13.7	11.4	13.9	11.1	14.2
40	13.3	14.1	12.8	14.2	12.4	14.6	12.0	14.9
45	**14.5**	**14.8**	14.5	15.2	13.8	15.4	13.3	16.0

AHRI rated conditions are shown bolded.

TABLE 14.5

Sample of How Manufacturers Present Data on the Performance of an Water-Cooled Chiller Unit

	Temperature of Water Entering Condenser ($T_{cd,in}$)							
	80°F		85°F		90°F		95°F	
Chilled Water Leaving Temp. ($T_{ch,out}$) °F	Cooling Capacity, tons	Electric Input, kW	Cooling Capacity, tons	Electric Input, kW	Cooling Capacity, tons	Electric Input, kW	Cooling Capacity, tons	Electric Input, kW
42	23.0	20.3	22.9	20.9	22.3	21.4	21.6	22.0
44	24.4	20.5	**23.7**	**21.0**	23.0	21.7	22.4	22.3
45	24.8	20.6	24.1	21.2	23.4	21.8	22.8	22.4
46	25.2	20.7	24.5	21.5	23.8	21.9	23.1	22.6
48	26.0	20.9	25.3	21.7	24.7	22.2	24.0	22.9
50	26.8	21.1	26.1	21.9	25.5	22.5	24.8	23.2

AHRI rated conditions are shown bolded.

14.5 Air Source Heat Pumps

14.5.1 General Description

For climates with moderate heating and cooling needs, heat pumps offer an attractive energy-efficient alternative to furnaces and boilers. Heat pump cycles are thermodynamically identical to vapor compression cycles consisting of one compressor and two heat exchanger coils, and have been introduced in Section 12.2.3. Refrigeration and cooling systems are used to produce a cooling effect as their output by extracting heat from a low-temperature source and rejecting it to a high-temperature sink. However, the heat rejected to the environment by cooling systems can be of value when heat is needed for the application. A *heat pump* extracts heat from environmental air or other medium-temperature sources (such as a nearby surface water pond, the ground, groundwater, or building heat recovery systems), raises its temperature sufficiently to be of value in meeting space heating (or other applications), and delivers it to the space. During summer, the same unit provides cooling, thereby avoiding the additional cost of a separate air conditioning system. This is done by using a reversing valve meant to reverse the direction of refrigerant flow as shown in Figure 14.15 along with minor system modifications (additional expansion valve, two check valves, and piping). Figure 14.16 shows an air-source heat pump for a residence where space heating is done with outdoor air as the heat source.

An ideal Carnot heat pump would also appear as a rectangle in the *T–s* diagram (recall Figure 12.3). The coefficient of performance (COP, i.e., output divided by input) alternatively referred to as the performance factor (PF) of a Carnot heat pump was given by Equation 12.5 and was shown to be inversely proportional to the difference between the high- and low-temperature

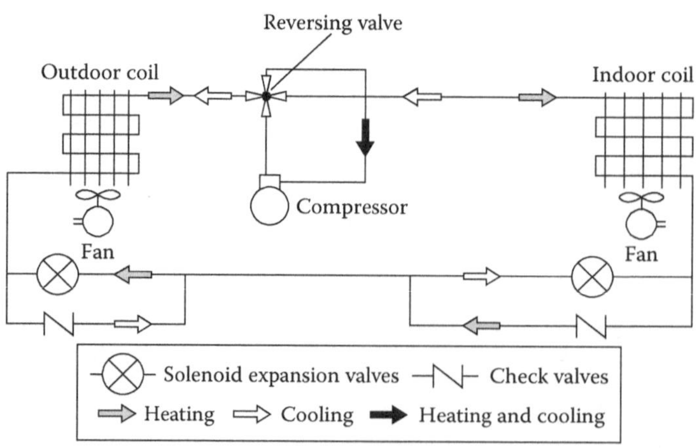

FIGURE 14.15
Air-to-air heat pump diagram. A reciprocating compressor is used. This design allows operation as a heat pump or an air conditioner by reversing the refrigerant flow.

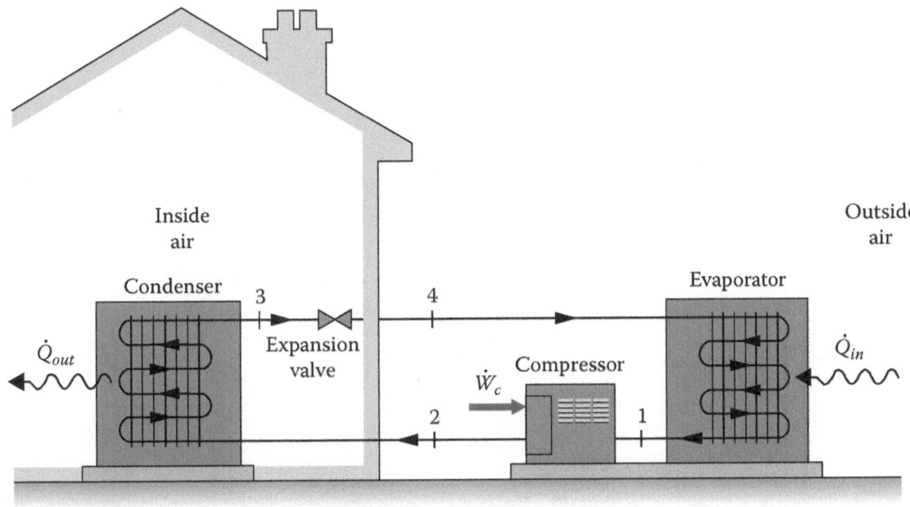

FIGURE 14.16
Sketch of an air source heat pump for a residential application.

reservoirs. The same result applies generally to heat pumps using actual fluids. For refrigeration cycles supplying cooling to a space, it is the condenser coolant fluid temperature (coupled to the sink or outdoor air temperature) which varies over the year. For heat pumps, it is the condenser that is coupled to the space and remains fairly constant year-round, while the low-side evaporator temperature tracks the varying outdoor temperature. As a result, the *capacity and COP or PF of air source heat pumps are strong functions of outdoor temperature*. This feature of heat pumps must be accounted for by the designer since heat pump capacity diminishes as the space heating load on it increases. Heat pumps can be supplemented by fuel heat or electric resistance heating, depending on the cost of each as discussed in Section 14.5.3.

Heat pump efficiencies can be greatly improved if water and ground are used as the heat extraction source. Such heat pumps are not subject to outdoor temperature variations and the narrower temperature difference between the low and high sides as compared to air-source heat pumps improves their COP or PF. These types of systems are discussed in Sections 14.8 and 14.9.

14.5.2 Operational Considerations

Heat pumps are available in sizes ranging from small residential units (10 kW) to large central systems (up to 15 MW) for commercial buildings. Large systems produce heated water at temperatures up to 105°C (220°F), which can be used for process heat applications. Central systems can use both environmental and internal building heat sources. In many practical circumstances, the heat gains in the core zones of a commercial building could satisfy the perimeter heat losses in winter. A heat pump can be used to efficiently condition both types of zones simultaneously (elaborated in Section 14.9).

The *outdoor and indoor heat exchangers* use forced convection on the air side to produce adequate heat transfer coefficients. In the outdoor exchanger, the temperature difference between the boiling refrigerant and the air is between 6°C and 14°C (10°F and 25°F). If the heat source is internal building heat, water is used to transport heat to the heat pump evaporator, and smaller temperature differences can be used.

There are several factors that reduce the efficiency of air source heat pumps:

1. A persistent problem is the *accumulation of frost* on the outdoor coil at coil surface temperatures just above the freezing point. The problem is most severe for humid climates (relative humidity around 60%), specifically between temperatures from 4°C to –7°C (40°F to 20°F). The frosting reduces evaporator airflow, which reduces heat transfer, thereby decreasing cooling capacity and COP. Figure 14.17a illustrates this behavior for a field tested heat pump.
2. *Reverse-cycle defrosting* can be accomplished by briefly operating the heat pump as an air conditioner (by reversing the flow of refrigerant) and turning off the outdoor fan. Hot refrigerant flowing through the outside melts the accumulated frost. This energy penalty must be accounted for in calculating the COP of heat pumps. Defrost control can be initiated by time clock or, better still, by a sensor measuring either the refrigerant condition (temperature or pressure) or ideally by monitoring the air pressure drop across the coil (see Figure 14.17b).

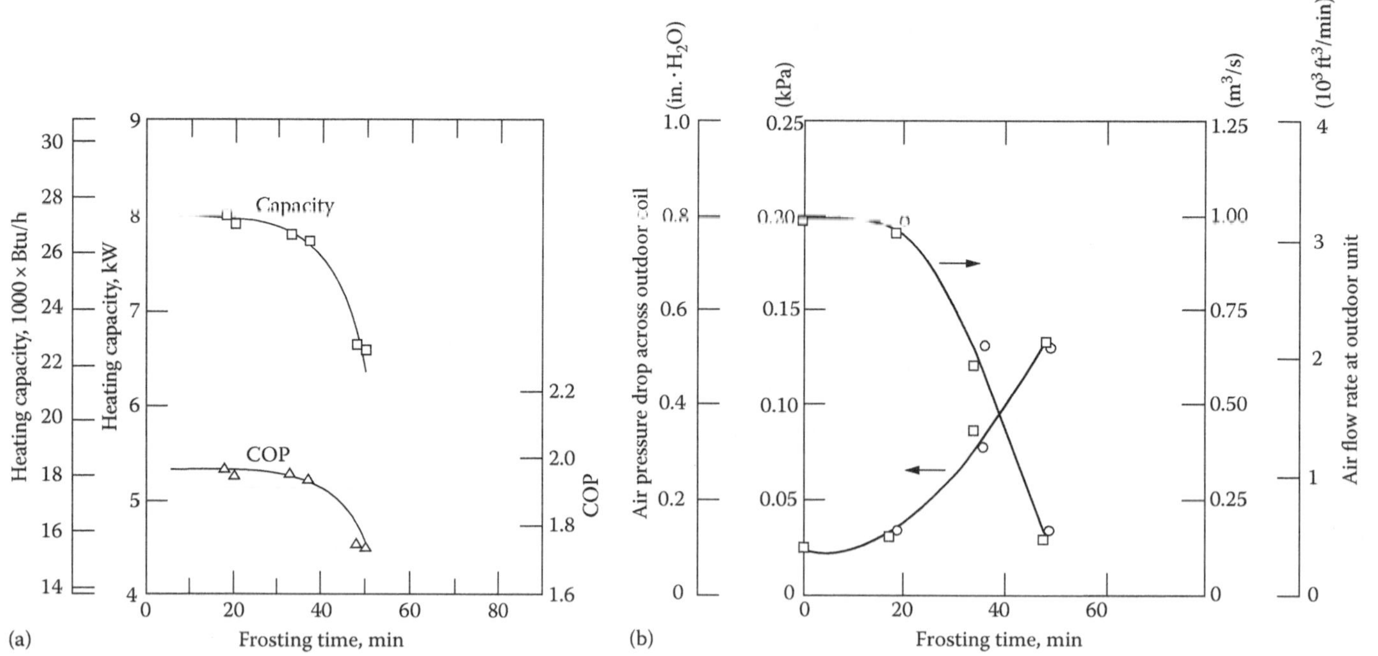

FIGURE 14.17
Performance degradation of air source heat pumps with frosting time: (a) effect on heating capacity, (b) effect on pressure drop and air flow rate over coil. (From Domingorena, A.A. and Ball, S.J., Performance evaluation of a selected three-ton air-to-air heat pump in the heating mode, Report ORNL/CON-34, Oak Ridge National Laboratory, Oak Ridge, TN, 1980.)

3. *Cycling losses* occur because the unit has too much capacity under part load. Some heat pump units take 10 min or more to reach steady state operating during which time the unit delivers less heating than under steady state. To avoid part-load penalties in moderate weather, a variable-speed compressor drive is often used.
4. *Improper refrigerant charge* is a major issue that plagues a large fraction of smaller units that have been operating for a few years. Refrigerant leaks out of the system over time, which severely degrades performance.

The realities of heat pump performance in the field as discussed earlier reduce the capacity of actual systems from the Carnot ideal. Figure 14.18 shows ideal Carnot COP values as a function of source temperature (say, the ambient air temperature) for a high-side temperature of 21°C (70°F). The intermediate curve shows performance for a Carnot heat pump with actual (i.e., finite temperature difference) heat exchangers. Finally, the performance of an actual heat pump is shown in the lower curve. Included in the lower curve are the effects of heat exchanger losses, use of real fluids, compressor inefficiencies, and pressure drops. Thus, the COP of actual machines is much lower (about 50%) than that for an ideal Carnot cycle with heat exchanger penalties. This is discussed further in Section 14.10.

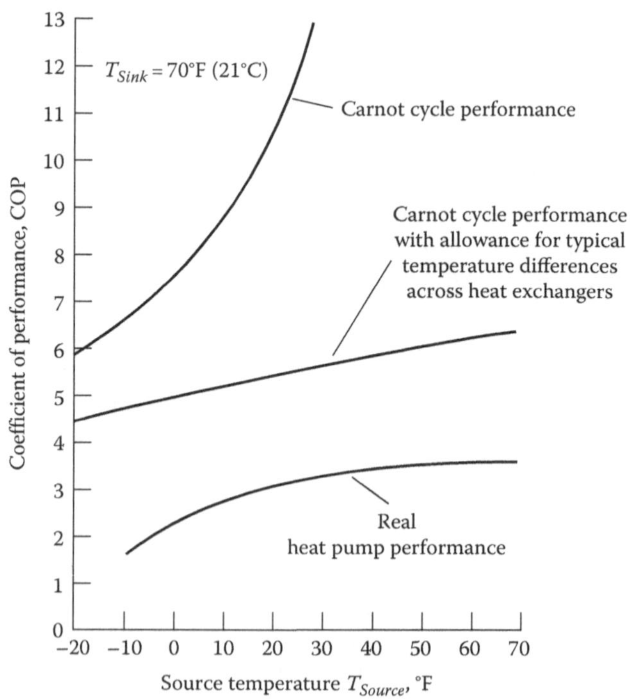

FIGURE 14.18
COP of ideal Carnot heat pump cycle, Carnot cycle (with heat exchanger penalty), and real heat pumps.

14.5.3 Heat Pump Manufacturer Performance Data

The strong dependence of heat output of an air source heat pump with ambient temperature is depicted in Figure 14.19 for four different units of different capacities. Both heating capacity and power draw increase with increasing ambient air temperature. The heat pump heating capacity is thus not well matched with

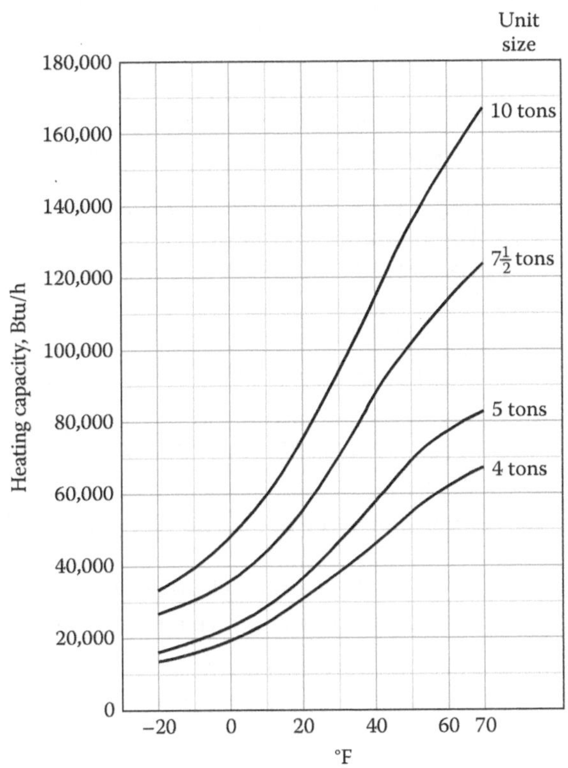

FIGURE 14.19
Heating performance curves as a function of ambient temperature for four different nominal heat pump units.

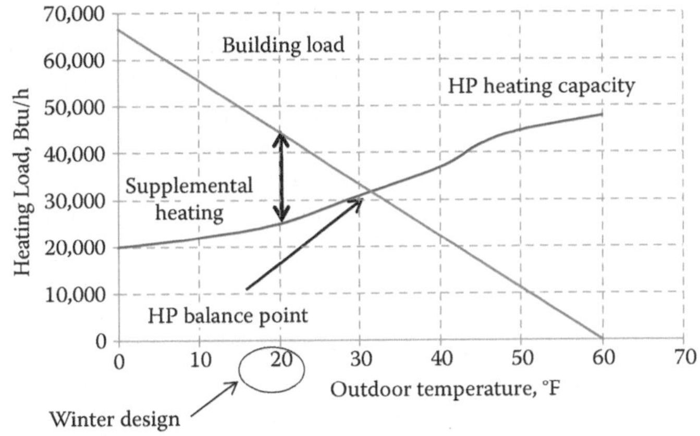

FIGURE 14.20
Graph of building heat load, heat pump capacity, and auxiliary heat quantity as a function of outdoor air temperature for the air source residential heat pump of Example 14.5. Note that the heat pump balance point has no relation to the space heating and cooling load balance-point temperature.

the space heating loads that decrease as the outdoor temperature increases. Figure 14.20 depicts the conflicting characteristics of heat pumps and building loads in the heating season. As the ambient temperature drops, building loads increase but heating capacity drops. The point at which the two curves intersect is called the *heat pump balance point*. To the left, auxiliary heat is needed; to the right, the heat pump must be modulated since it has excess capacity. Typically, residential heat pumps are selected so that their balance points are between 35°F and 45°F (1.7°C–7.2°C).

Table 14.6 illustrates the type of information often provided by unitary heat pump manufacturers. The left side of the table corresponds to the performance of the heat pump under cooling mode, while the right

TABLE 14.6
Sample Manufacturer Performance Data for a Unitary Residential Split Heat Pump Unit

Cooling Mode						Heating Mode				
Outdoor Dry-Bulb, °F	Total Cooling, Btu/h	Sensible Cooling, Btu/h	Electric Power, W	Suction Pressure, psig	Discharge Pressure, psig	Outdoor Dry-Bulb, °F	Heating Capacity, Btu/h	Electric Power, W	Suction Pressure, psig	Discharge Pressure, psig
115	36,000	25,200	6620	81	370	60	48,000	5400	65	302
105	39,000	27,300	6320	78	335	50	44,900	5100	55	260
95[a]	42,000	29,400	6000	75	330	**45**[a]	42,000	4900	50	250
85	44,000	30,800	5650	71	265	40	37,000	4100	45	238
75	46,000	32,200	5220	66	225	30	31,000	3600	37	216
						20[a]	25,000	3200	30	200
						10	22,000	3000	22	182
						0	20,000	2800	17	170

The tests have been conducted with indoor conditions: 80°F dry-bulb and 67°F wet-bulb for cooling, and 70°F dry-bulb for heating.
[a] Corresponds to rated conditions (see Section 14.6).

side of the table to that under heating mode. Under different outdoor dry-bulb conditions, the steady-state cooling capacities (total and sensible), as well as the electric power draw by the compressor are specified. The refrigerant working pressures in the evaporator (suction) and the condenser (discharge) are also indicated. The values of outdoor temperature (shown in bold) correspond to rated performance, an aspect discussed in Section 14.6. A heat pump unit is first selected such that its cooling capacity meets the building design cooling load. The size of the auxiliary heater is then determined as the deficit between the building design heating load and the heat pump capacity at that outdoor temperature. The following example illustrates the approach of how tables such as Table 14.6 are used for design purposes.

Example 14.5: Heat Pump Selection Procedure

A residence has a total heat transmission coefficient K_{tot} = 1110 Btu/(h·°F). The summer and winter outdoor design conditions are 105°F and 20°F respectively. The summer and winter balance points of the house are 70°F and 60°F respectively. The summer cooling load has a sensible heat factor of 0.7. Determine

(a) Whether the heat pump with characteristics shown in Table 14.6 is appropriate
(b) The heat pump balance point temperature
(c) The size of the auxiliary electric heater in Watts
(d) The approximate cooling COP and heating PF values under the summer and winter design conditions

Given: K_{tot} = 1110 Btu/(h·°F), $T_{b,summer}$ = 70°F and $T_{b,winter}$ = 60°F, $T_{o,design\ summer}$ = 105°F, $T_{o,design\ winter}$ = 20°F

Figure: See Figure 14.20.

Solution

(a) First, we calculate the cooling load on the house at the design summer condition:

$$\dot{Q}_{cool,design} = K_{tot}(T_o - T_b)_{summer}$$

$$= 1100 \text{ Btu/(h}\cdot\text{°F)} \times (105-70)\text{°F} = 38{,}850 \text{ Btu/h}$$

The design cooling capacity of the heat pump at the summer design condition of 105°F is 38,850 Btu/h, which is slightly lower than the heat pump total cooling capacity of 39,000 Btu/h at the same outdoor temperature value. Also, the sensible factor of the HP is = (27,300/39,000) = 0.7, which matches the building sensible heat factor. Thus, we conclude that this heat pump is appropriate.

(b) From Figure 14.20, the heat pump balance point temperature is close to 32°F
(c) The size of the auxiliary heater is the difference between the building heating load and the HP capacity at the winter design condition of 20°F.
The heating load on the house at the design winter condition:

$$\dot{Q}_{heat,design} = K_{tot}(T_b - T_o)_{winter} = 1100 \text{ Btu/(h}\cdot\text{°F)}$$

$$\times (60-20)\text{°F} = 44{,}400 \text{ Btu/h}$$

From the heat pump performance table, the heating capacity at 20°F is only 25,000 Btu/h.
Hence, the auxiliary electric heater size = 44,400 − 25,000 = 19,400 Btu/h = 5.69 kW.
(d) The cooling COP of the heat pump unit =

$$COP_{cool} = \frac{\text{Total cooling load at 105°F}}{\text{Electric power drawn}}$$

$$= \frac{39{,}000 \text{ Btu/h}}{3.412 \text{ Btu/(h}\cdot\text{W)} \times 6{,}320 \text{ W}} = 1.81$$

The heating COP or PF of the heat pump unit =

$$PF_{HP\ unit} = \frac{\text{Total heating load at 20°F}}{\text{Electric power drawn by heat pump}}$$

$$= \frac{25{,}000 \text{ Btu/h}}{3.412 \text{ Btu/(h}\cdot\text{W)} \times 3{,}200 \text{ W}} = 2.29$$

Comments

The performance factor of the entire heat pump system under design heating load will be much lower since it should include the electric power draw of the heat pump as well as the backup electric heat. In this case:

$$PF_{HP\ system} = \frac{25{,}000 \text{ Btu/h}}{3.412 \text{ Btu/(h}\cdot\text{W)} \times (3{,}200+5{,}690) \text{ W}} = 0.82$$

14.5.4 Control and Other Considerations

Controls for heat pumps are more complex than those for fuel-fired systems since outdoor conditions, coil frosting, and building heat load must all be considered. In addition, to avoid excessive electric demand charges, the controller must avoid *coincident* operation of resistance heat and the compressor at full capacity (attempting to meet a large load on a cold day).

Heat pump efficiency is greater if lower delivery temperatures can be used. To produce adequate space heat in such conditions, a larger coil may be needed in the airstream. However, if the coil is sized for the cooling load, it will nearly always have adequate capacity for heating. In such a case, adequate space heat can be provided at relatively low air temperatures of 35°C–43°C (95°F–110°F). Further, supplemental heat should always be added downstream of the heat pump condenser. This ensures that the condenser will operate at as low a temperature as possible, thereby improving the COP.

Recovery from night thermostat setback must be carefully thought out by the designer if an air source heat pump is used. A step increase in the thermostat setpoint on a cold winter morning will inevitably cause the auxiliary heat source to come on. If this heat source is electricity, high electric demand charges may result, and the possible economic advantage of the heat pump will be reduced. One approach to avoid activation of the electric resistance heat elements is to use a longer warm-up period with gradually increasing thermostat setpoint. A smart controller could control the setup time based on known heat pump performance characteristics and the outdoor temperature. Alternatively, fuel could be used as the auxiliary heat source. During building warm-up, all outside air dampers remain closed, as is common practice for any commercial building heating system.

This section has discussed practical aspects of heat pumps with an illustrative application to space heating. Heat pumps have other applications in buildings:

- *Operation in buildings with simultaneous heating and cooling needs*: A system of this type is discussed in Section 14.9.
- *Heat pump water heater*: This system heats water by operating a heat pump using indoor air as the source and the water to be heated as the sink. These systems are small and primarily used for residential applications.
- *Ventilation air heat recovery*: Since all buildings with mechanical ventilation exhaust some warmed air in winter, a heat pump can be operated between this exhaust and the fresh air supply. The COP for this application is constant.

14.5.5 Electric Resistance Heating

Electricity can be used as the heat source in both furnaces and boilers. They will have efficiencies close to unity, and in that sense, are less attractive than heat pumps whose PF > 1. Electric units are available in the full range of sizes from small residential furnaces (5–15 kW) to large boilers for commercial buildings (200 kW to 20 MW). Electric units have four attractive features: (1) relatively lower initial cost, (2) efficiency near 100%, (3) near-zero part-load penalty, and (4) flue gas vents not necessary.

The high cost of electricity (both energy and demand charges) diminishes the apparent advantage of electric boilers and furnaces, however. Nevertheless, they continue to be installed where first cost is a prime concern. The prudent designer should consider the overwhelming life cycle costs of electric systems. Electric boiler and furnace sizing follows the methods outlined in Chapter 15 for fuel-fired systems. In many cases, the thermodynamic and economic penalties of pure resistance heating can be reduced by using electric heat pumps.

Environmental concerns must also be taken into account in considering electric heating. Low conversion and transmission efficiencies (relative to direct combustion of fuels for water heating) result in relatively higher CO_2 emissions. Also, SO_2 emissions from coal power plants are an environmental concern.

14.6 Rating Standards

Although the COP of a vapor compression unit is the natural, dimensionless performance index that arises in the thermodynamic analysis of heat pumps and air conditioners, the HVAC industry sometimes uses a dimensional performance measure more in line with the performance units shown in Tables 14.4 and 14.5 (tons and kW). The *energy efficiency ratio* (EER) is the ratio of cooling capacity (Btu per hour) to the electric input rate (Watts). Thus, EER has the units of Btu per Watt per hour. The dimensionless COP = EER/3.413 Btu/(W·h).

14.6.1 AHRI Standards

The performance of most HVAC equipment is affected by the conditions under which they operate. The Air-Conditioning, Heating and Refrigeration Institute (AHRI) is the trade organization representing North American air-conditioning and refrigeration manufacturers that sets and publishes standard on the operating conditions and the experimental procedure needed to rate equipment. The intent of these standards is to provide a rational means of comparing the performance of HVAC equipment from different manufacturers under repeatable and controlled test conditions. The standards are intended for the guidance of the industry, including manufacturers, engineers, installers, contractors, and users. They establish definitions, requirements for testing and rating, minimum data for requirements for published ratings, marking and nameplate data, and conformance conditions for water-chilling packages using the vapor compression cycle. AHRI also certifies performance of equipment through independent tests.

TABLE 14.7
AHRI Operating Testing Conditions for Rating Unitary Equipment

Type of Equipment	Outdoor Air Temperature, °F	Indoor Entering Air Temperature	Rating Standards
Room AC	95	80°F DBT/67°F WBT	Assoc. of Home Appliance Manufacturers (AHAM, 1992) and ANSI/ASHRAE Standard 58-1999 (ASHRAE, 1999)
PTAC/PTHP (cooling)	95	80°F DBT/67°F WBT	ANSI/AHRI Standard 310/380 (2004)
PTAC/PTHP (heating)	47	70°F DBT	ANSI/AHRI Standard 310/380 (2004)
Split AC and HP (cooling)	82	80°F DBT/67°F WBT	AHRI Standard 210/240 (2008)
Split AC and HP (heating)	47	65°F DBT	AHRI Standard 210/240 (2008)

DBT, dry-bulb temperature; WBT, wet-bulb temperature.

Unitary equipment for cooling consists of all components needed for air conditioning, factory-assembled into one or two pieces of equipment. The attraction of unitary systems is that they require little design or field-assembly work. Standard units up to 50 tons (175 kW) are available from many manufacturers. Unitary commercial and residential cooling systems have standard rating methods in the United States that prescribe testing and calculation procedures. The most common, single rating number is the seasonal energy efficiency ratio (SEER). It is essentially a seasonally averaged EER. However, the ratio is not dimensionless, being expressed as the ratio of cooling effect (Btu/h) to the electric power input (kW). Table 14.7 assembles the operating test conditions for unitary chillers, and unitary heat pumps based on vapor compression cycles. Other standards apply to heat-operated equipment (absorption chillers and heat pumps).

The EER rating is a steady-state efficiency measure at some standard or rated defined operating conditions. For most unitary equipment, these conditions, from Table 14.7, are 80°F dry-bulb and 67°F wet-bulb for indoor conditions, and 95°F dry-bulb and 75°F wet-bulb for outdoor conditions). However, the chiller unit or system operates under part-load during the entire season depending on the climate zones in which the unit has been installed. In 1978, the U.S. Congress mandated the requirement that units be labeled using the SEER (Seasonal EER) index that takes into consideration how certain variables affecting the performance of the unit vary in that climatic zone. The current minimum federal SEER is 13 though commercial units are available with SEER values as high as 23. Since the SEER reflects performance in conditions milder than those for peak or rated conditions, SEER values for an air conditioner are generally higher than the EER rating values by 2–3 points.

AHRI standard 550/590 (AHRI, 2011) applies to water-cooled chillers. It sets procedures for determining chiller performance under standard and application (non-standard) rating conditions, at full- and part-load capacity, and the method for calculating the capacity tolerance the chiller must meet to comply with its stated rating. Published AHRI ratings for all water-chilling packages must include the mandatory Standard Rating with 29.4°C (85°F) condenser water flowing at a rate of 0.054 L/s per kW (3.0 gpm per ton) and a fouling factor of 0.000044 $m^2 \cdot °C/W$ (0.00025 $h \cdot ft^2 \cdot °F/Btu$), and 6.7°C (44°F) chilled water flowing at a rate of 0.043 L/s per kW (2.4 gpm per ton) with a fouling factor of 0.000018 $m^2 \cdot °C/W$ (0.0001 $h \cdot ft^2 \cdot °F/Btu$). Recommendations for application ratings that are intended for part-load conditions are also provided. These include measurements under varying conditions, within the operating range of the equipment, as follows: condenser water from 18.3°C (65°F) to 40.6°C (105°F) (at ≤2.78°C (5°F) increments) and chilled water from 4.44°C (40°F) to 8.89°C (48°F) (at ≤1.1°C (2°F) increments).

14.6.2 IPLV Rating

Part-load AHRI ratings are calculated using the IPLV (integrated part-load value) to determine the part-load energy efficiency at 100%, 75%, 50%, and 25% load points. To determine COP and EER for water-cooled chillers, the following relation is used:

$$\text{IPLV} = 0.01 \times A + 0.42 \times B + 0.45 \times C + 0.12 \times D \quad (14.26)$$

where

A = COP or EER at 100% and $T_{cd,in}$ = 29.4°C (85°F)
B = COP or EER at 75% and $T_{cd,in}$ = 23.9°C (75°F)
C = COP or EER at 50% and $T_{cd,in}$ = 18.3°C (65°F)
D = COP or EER at 25% and $T_{cd,in}$ = 18.3°C (65°F)

and chiller water outlet temperature $T_{ch,out}$ is kept constant at 6.7°C (44°F).

The IPLV rating reduces the weight under rated full-load performance and includes part-load performance. These IPLV weights have been loosely deduced from building load distributions identified from simulations of typical buildings in 129 U.S. cities. Since condensing temperature is lower during off-rated milder operating conditions, the IPLV rating is higher than that under rated conditions. AHRI standard also recommends that in case the chiller is to be operated over a narrower range of load variation than that assumed in Equation 14.26, then the four data points should be selected uniformly over the operating range of chiller loading.

It must be noted that the IPLV metric was derived to provide a representation of the average part-load efficiency for a single chiller only. It is not meant to be an efficiency standard (although this is a common misinterpretation). Also, it should not be used for multiple chiller installations that form the great majority (over 90%) of all central chiller plant installations. A good discussion of the advantages and limitations of IPMV is provided by Geister and Thompson (2009).

Example 14.6: IPLV Calculation

Determine the IPLV for a water-cooled chiller whose COP values have been determined as follows for the four test conditions specified by the AHRI 550/590 standard.

Given: COP (*A*) = 6.2, COP(*B*) = 8.2, COP(*C*) = 12.0, COP(*D*) = 12.0

Find: COP(IPLV)

Solution

Using Equation 14.26 results in

$$\mathrm{COP(IPLV)} = (0.01 \times 6.2) + (0.42 \times 8.2) + (0.45 \times 12) + (0.12 \times 12) = 10.35$$

Note that the COP under rated conditions corresponding to 100% load condition, i.e., point *A* is COP* = 6.2. Clearly, and as stated earlier, COP(IPLV) > COP*.

14.7 Part-Load Performance

Unitary equipment, which include chillers, heat pumps, boilers, pumps, and fans, rarely operates at its design rated capacity. Even under operating conditions other than the rated conditions, load requirements are often lower than the corresponding steady-state output of the equipment (recall the performance of the heat pump above its balance point temperature). The equipment is then said to operate under part-load operation. Part-load operation of such equipment degrades steady-state efficiency, and this effect needs to be modeled.

14.7.1 Unitary Chillers: Degradation Coefficient Method

Reduced cooling capacity can be accomplished by reducing the power input to the compressor. The two most common methods of capacity control for unitary equipment are: (1) on/off and (2) step. The former is discussed in this section, while the later type of capacity control involves stepping the compressor or motor, i.e., operating it at different speeds. For example, boiler and cooling tower fan speeds can be varied, thereby increasing/reducing air flow rates in discrete steps. Medium and large chillers have more sophisticated control involving capacity modulation, and how to model such behavior is covered in the next two subsections.

The simplest control is to cycle equipment on–off when their capacity exceeds the load requirements. This cycling degrades performance compared to steady operation, i.e., the thermal efficiency of the equipment becomes progressively poorer as cycling increases. This performance degradation is often modeled by the part-load factor (PLF)–part-load ratio (PLR) method. The PLR of a chiller is simply the ratio of the actual cooling load $\dot{Q}_{cool}$ to be met to the cooling capacity $\dot{Q}_{cool,ss}$ at full-load or steady-state operation under the same operating condition. Thus, it represents the theoretical run-time fraction, and is given by

$$\mathrm{PLR} = \min\left\{1, \frac{\dot{Q}_{cool}}{\dot{Q}_{cool,ss}}\right\} \tag{14.27}$$

The part-load factor (PLF), on the other hand, is the ratio of the COP under cycling to that under full load, and is given by

$$\mathrm{PLF} = \frac{\mathrm{COP}}{\mathrm{COP}_{ss}} = \frac{\dot{Q}_{cool}}{\dot{Q}_{cool,ss}} \frac{\dot{W}_{comp,ss}}{\dot{W}_{comp}} \tag{14.28}$$

The compressor power under part-load operation is easily found as

$$\dot{W}_{comp} = \dot{W}_{comp,ss} \times \left(\frac{\mathrm{PLR}}{\mathrm{PLF}}\right) \tag{14.29}$$

The relationship between PLR and PLF is shown in Figure 14.21. The relationship is fairly linear up to about PLR values of about 0.4, and so the following model form is often assumed:

$$\mathrm{PLF} = 1 - C_d \times (1 - \mathrm{PLR}) \tag{14.30}$$

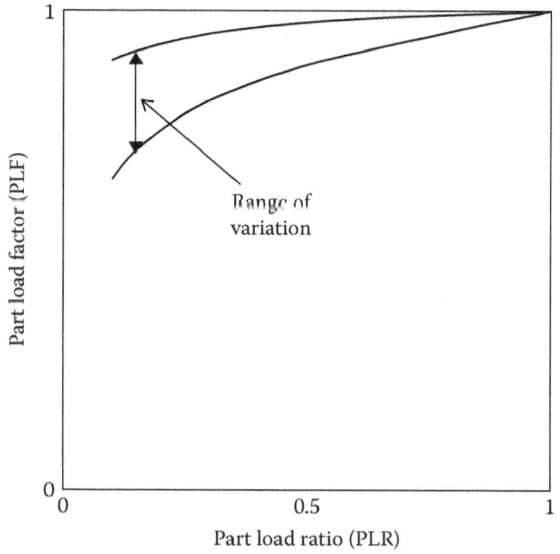

FIGURE 14.21
Modeling part load of unitary equipment using the PLR–PLF method. The relationship is close to linear for values of PLR until about 0.4.

where C_d is the degradation coefficient supplied by the manufacturer. Its value is usually in the range 0.15–0.25. If a value has not been provided, it is suggested that a conservative value of 0.25 be chosen.

Example 14.7: Predicting Part-Load Behavior Using the Degradation Factor

We will use the results of Example 14.4 to illustrate the use of the degradation factor approach to predicting part-load performance of a unitary chiller. The same chiller is operated under part load where the cooling load to be met is only 350 kW. What will be the corresponding electric draw of the compressor? The manufacturer has not specified a value for the degradation factor.

Given: $\dot{Q}_{cool}$ = 350 kW, and from Example 14.4, $\dot{W}_{comp,ss}$ = 111.55 kW and $\dot{Q}_{cool,ss}$ = 500 kW.

Assume: C_d = 0.25

Find: $\dot{W}_{comp}$ and COP

Solution

From Equation 14.27: PLR = (350/500) = 0.7
From Equation 14.30: PLF = 1–0.25 × (1–0.7) = 0.925
Finally, from Equation 14.29:

$$\dot{W}_{comp} = 111.55 \times \left(\frac{0.7}{0.925}\right) = 84.42 \text{ kW}$$

which results in a part-load COP = (350/84.42) = 4.15
The steady-state COP_{ss} = (500/111.55) = 4.48; thus, cycling has reduced the COP by 7.4%.

For heat pumps, the amount of auxiliary heat needed and the type (electricity, natural gas, oil, or other) must be determined by an economic analysis and fuel availability (heat pumps are often used when fossil fuels are unavailable). This requires that an analysis be performed considering the combined effect of ambient source temperature on system output and efficiency as well as the part-load performance. The COP of an air source heat pump is greatly dependent on environmental conditions. Hence, predicting seasonal or annual energy use entails using off-rated performance data supplied by the equipment manufacturer, modeling part-load equipment performance degradation following the simplified PLF–PLR method, and weighting the results with bin weather data for the location in question (see Section 10.4 where bin data were first introduced in the context of building load calculations). Example 14.8 illustrates the approach, and also serves to provide a sense of the magnitude of part-load operation effect for a residential-scale air source heat.

Example 14.8: Seasonal Heat Pump Performance Calculated by the Bin Method

A residence in a heating climate has a total heat transmission coefficient K_{tot} = 650 Btu/(h °F) (343 W/K). We need to evaluate an air source heat pump with a capacity of 39,900 Btu/h (11.7 kW) at 47°F (8.3°C) (standard rating point in the United States). Find the heating season electric energy usage, seasonal COP [often called the *seasonal performance* factor (SPF)], and energy savings relative to electric resistance heating. Use the bin data and heat pump performance data given in Table 14.8. Assume a degradation coefficient of 0.25. The house heating base temperature is 65°F (18.3°C), accounting for internal gains.

Given: K_{tot} = 650 Btu/(h·°F), bin data in Table 14.8

Assume: C_d =0.25

Find: SPF, Q_{yr} (with and without heat pump)

Solution

The solution will use weather and performance data from Table 14.8 whose columns are

1. Center point of temperature bin T_{bin}
2. Heating demand $\dot{Q} = K_{tot} \times (65°F - T_{bin})$
3. Heat pump heating capacity at the bin temperature, supplied by the manufacturer
4. Heat pump electric power input at the bin temperature, supplied by the manufacturer
5. COP_{ss} determined as the ratio of columns (3) and (4), this includes defrost penalty
6. Part-load ratio (PLR) using Equation 14.27
7. Part-load factor (PLF) using Equation 14.30.

TABLE 14.8

Heat Pump and Building Load Data (Example 14.8)

1	2	3	4	5	6	7	8	9	10	11	12
Bin Temp., °F	Heating Load, Btu/h	HP[a] Capacity, Btu/h	HP[a] Electric, W	HP COP_{ss}, —	PLR, —	PLF, —	HP COP, —	HP Output, Btu/h	HP Electric, Btu/h	Auxiliary Electric, W	Heating System COP, —
62	1,950	27,100	3600	2.206	0.072	0.768	1.694	1,950	1,151	0	1.69
57	5,200	26,400	3500	2.211	0.197	0.799	1.767	5,200	2,943	0	1.77
52	8,450	25,500	3400	2.198	0.331	0.833	1.831	8,450	4,616	0	1.83
47	11,700	24,300	3300	2.158	0.481	0.870	1.878	11,700	6,229	0	1.88
42	14,950	22,400	3200	2.052	0.667	0.917	1.881	14,950	7,948	0	1.88
37[b]	18,200	20,400	3100	1.929	0.892	0.973	1.877	18,200	9,698	0	1.88
32[b]	21,450	18,300	3000	1.788	1.000	1.000	1.788	18,300	10,236	3,150	1.60
27	24,700	16,400	2900	1.657	1.000	1.000	1.657	16,400	9,895	8,300	1.36
22	27,950	14,600	2800	1.528	1.000	1.000	1.528	14,600	9,554	13,350	1.22
17	31,200	13,000	2700	1.411	1.000	1.000	1.411	13,000	9,212	18,200	1.14
12	34,450	11,700	2600	1.319	1.000	1.000	1.319	11,700	8,871	22,750	1.09
7	37,700	10,600	2500	1.243	1.000	1.000	1.243	10,600	8,530	27,100	1.06
2	40,950	9,500	2400	1.160	1.000	1.000	1.160	9,500	8,189	31,450	1.03
−3	44,200	8,600	2300	1.096	1.000	1.000	1.096	8,600	7,848	35,600	1.02
−8	47,450	7,700	2200	1.026	1.000	1.000	1.026	7,700	7,506	39,750	1.00

[a] Supplied by manufacturer.
[b] Heat pump balance point is between these two temperature values.

8. Heat pump COP under part-load operation when unit cycles = (PLF) × COP_{ss}
9. Heat pump output under part load:
   - Above the heat pump balance point, taken to be the heating demand $\dot{Q}$ (column 2)
   - Below the heat pump balance point, it is equal to the heat pump capacity (given from manufacturer's data—column 3)
10. Heat pump electric input under part load (column 9/column 8)
11. Auxiliary power; the positive difference, if any, between heating demand $\dot{Q}$ (column 2) and heat pump output under part load (column 9)
12. Heating *system* COP under part load given by $\dot{Q}$ plus the auxiliary power divided by the sum of auxiliary power and heat pump input (column 2/[column 10 + column 11]).

The energy calculations simply involve multiplying the loads by the number of hours for each temperature bin. These calculations are summarized in Table 14.9. We will work through the calculations for the 22°F bin in detail to clarify the process. The first two columns of this table are the bin weather data. The third column is the heating energy by bin. For this bin, the bin energy is

$$Q_{22} = \frac{27{,}950\ \text{Btu/h} \times 359\ \text{h}}{1{,}000{,}000} = 10.03\ \text{MBtu}$$

Since 22°F is below the balance point, the heat pump capacity is less than the load, as shown in Table 14.8. The heat pump output is

$$Q_{out,22} = \frac{14{,}600\ \text{Btu/h} \times 359\ \text{h}}{1{,}000{,}000} = 5.24\ \text{MBtu}$$

From Table 14.8, the COP at this temperature is 1.528. Therefore, the electricity input during this temperature bin is

$$Q_{in,22} = \frac{Q_{out,22}}{\text{COP}} = \frac{5.24\ \text{MBtu}}{1.528} = 3.43\ \text{MBtu}$$

Because the heat pump cannot meet the load, some auxiliary heat is needed

$$Q_{aux,22} = 10.03 - 5.24 = 4.79\ \text{MBtu}$$

Finally, the total electric input is the sum of the heat pump and supplemental electricity requirements

$$Q_{tot,22} = 3.43 + 4.79 = 8.22\ \text{MBtu}$$

The rest of Table 14.9 is completed in this manner. The bottom line in the table contains energy totals. With the heat pump, the total electricity requirement is 66.1 MBtu/year (19,360 kWh/year). If pure resistance heating was used, the total electricity requirement would be 98.36 MBtu/year (28,820 kWh/year). The variations

TABLE 14.9
Heat Pump Energy Calculations (Example 14.8)

1	2	3	4	5	6	7
Bin Temp., °F	Bin Time, h	Heating Energy Required, MBtu	HP Thermal Output, MBtu	HP Electric Input, MBtu	Auxiliary Electric Input, MBtu	Total Electric Input, MBtu
62	783	1.53	1.53	0.90	0.00	0.90
57	731	3.80	3.80	2.15	0.00	2.15
52	678	5.73	5.73	3.13	0.00	3.13
47	704	8.24	8.24	4.39	0.00	4.39
42	692	10.35	10.35	5.50	0.00	5.50
37[a]	717	13.05	13.05	6.95	0.00	6.95
32[a]	721	15.47	13.19	7.38	2.27	9.65
27	553	13.66	9.07	5.47	4.59	10.06
22	359	10.03	5.24	3.43	4.79	8.22
17	216	6.74	2.81	1.99	3.93	5.92
12	119	4.10	1.39	1.06	2.71	3.76
7	78	2.94	0.83	0.67	2.11	2.78
2	36	1.47	0.34	0.29	1.13	1.43
−3	22	0.97	0.19	0.17	0.78	0.96
−8	6	0.28	0.05	0.05	0.24	0.28
Total	**6415**	**98.36**	**75.8**	**43.5**	**22.56**	**66.1**

[a] Heat pump balance point is between these two temperature values.

in the energy drawn by the heat pump system, which includes the heat pump and the electric auxiliary heat, are shown in Figure 14.22 for different bin temperatures.

The SPF for the heat pump is the seasonal output divided by the seasonal input to the heat pump:

$$SPF_{hp} = \frac{Q_{out,yr}}{Q_{in,yr}} = \frac{75.8 \text{ MBtu}}{43.5 \text{ MBtu}} = 1.74$$

The SPF for the heating system is the seasonal heat load plus the auxiliary heater electric input divided by the seasonal input to the heat pump and the auxiliary heater:

$$SPF_{sys} = \frac{Q_{out,yr}}{Q_{in,yr} + Q_{in,aux,yr}} = \frac{(75.8 + 22.56) \text{ MBtu}}{(43.5 + 22.56) \text{ MBtu}} = 1.49$$

Comments

This example has shown how the part-load characteristics of air source heat pumps are influenced by ambient temperature, and that this characteristic of the equipment must be considered in annual energy calculations. The advantage of a constant-temperature heat source is apparent from this example. If groundwater or building exhaust air (both

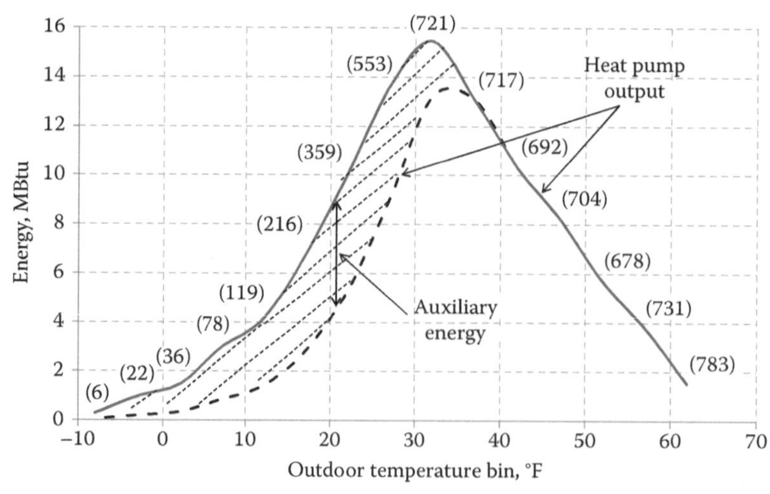

FIGURE 14.22
Heat pump energy use by the bin method, Example 14.8. The numbers at each bin temperature indicate the number of hours of occurrence at that bin.

essentially at constant temperature) were used as the heat source rather than outdoor air, one can avoid the dropoff in capacity that occurs in the air source device just when heat is most needed.

Rather than using the PLR–PLF method, an alternative rather simplified approach to account for cycling effects that reduce actual heating capacity of the heat pump has been proposed (ASHRAE Fundamentals, 1989):

$$\text{Actual capacity} = 0.75\,(\text{nominal capacity}) + 0.25\,(\text{bin load}) \qquad (14.31)$$

This applies to operating conditions higher than the balance point temperature; below it, no adjustment is needed. The online HCB software can be used to solve heat pump problems of this type, and includes the defrost penalty on capacity. You are encouraged to try it out for this problem and to study the effect of heat pump size on the seasonal COP.

14.7.2 Medium to Large Chillers: Simplified Part-Load Model

Medium to large chillers have controls that allow modulation, i.e., continuous capacity variation over a certain range of PLR values. Power drawn by vapor compression refrigeration equipment depends on the cooling load demanded by the building (or heating in case of a heat pump) but also on the operating temperature levels, usually the condenser water inlet temperature and the evaporator leaving temperature (called the chiller setpoint temperature). If we neglect the effect of the two temperatures assuming them to be closely controlled at some preassigned values to the chiller load, then the chiller power input can be deduced from the rated condition and the PLR defined by Equation 14.27:

$$\dot{W}_{comp} = \frac{\dot{Q}^{*}_{cool}}{\text{COP}^{*}}[a + b(\text{PLR}) + c(\text{PLR})^2] \quad 0.1 < \text{PLR} \le 1.0 \qquad (14.32)$$

where

$\dot{W}_{comp}$ is the compressor power drawn at part load

$\dot{Q}^{*}_{cool}$ is the cooling capacity at rated condition

COP* is the COP at rated condition

a,*b*,*c* are the chiller-specific part-load coefficients

Absorption cycle chiller part-load performance can be calculated in a similar manner. For example, the steam rate $\dot{Q}_{in}$ input at the generator at part load is given by (LBL, 1982)

TABLE 14.10

Part-Load Coefficients for Generic Vapor Compression Chillers

Chiller Type	*a*	*b*	*c*
Hermetic compressor	0.160	0.316	0.519
Reciprocating compressor	0.023	1.429	−0.471
Centrifugal compressor	0.049	0.545	0.389

Source: From LBL, *DOE-2 Engineers Manual*, Lawrence Berkeley Laboratory Report, Berkeley, CA, 1982.

$$\dot{Q}_{in} = \frac{\dot{Q}_{in,full}}{\text{PLR}}[a + b(\text{PLR}) + c(\text{PLR}^2)] \quad 0.1 < \text{PLR} \le 1.0 \qquad (14.33)$$

The part-load coefficients *a*,*b*,*c* must be determined from manufacturers' data, or generic coefficients can be used as given in Table 14.10. Note that during actual operation the condenser temperature drops with building load as both are associated with reduced outdoor temperature. Hence, there is a loose relationship between both the cooling tower return temperature and the chiller load (i.e., the building load), and this must be reflected properly by the numerical values of these coefficients. The online HCB software contains PLR coefficients for six specific chillers from three different manufacturers. Although these averaged values are used by building designers, it is more accurate to use the method described in Section 14.7.3.

There is a consistency check that the PLR coefficients must meet since, at PLR = 1, the input power must match the peak rating. This requires that

$$a + b + c = 1 \qquad (14.34)$$

For example, for the hermetic compressor, this requirement is met nearly exactly:

$$0.160 + 0.316 + 0.519 = 0.995 \approx 1.0$$

Finally, we point out that the PLR equation may not be accurate for PLR < 0.1. Chillers do not often operate at such low capacities, so this does not present a major problem. At such low PLR values, the chiller controller would shut off the unit because of lubrication problems and other issues.

If part-load equation coefficients are known, the seasonally averaged COP can be found if the cooling-season load profile is known (as illustrated by Example 14.8). In addition, the COP at part load can be found as illustrated in the following example.

TABLE 14.11
Chiller Part-Load COP

PLR	0.10	0.20	0.30	0.40	0.50	0.60	0.70	0.80	0.90	1.00
COP	2.03	3.28	3.98	4.33	4.47	4.47	4.41	4.30	4.16	4.00

Example 14.9: Simplified Part-Load Performance Model

Find the part-load COP of a hermetic chiller, using the data in Table 14.10, if the full-load rated COP* = 4.0. Plot the results in the range of PLR = [0.1, 1.0].

Given: $a = 0.160$, $b = 0.316$, $c = 0.519$

Find: $COP_{part\text{-}load}$

Solution

The chiller COP at part load $COP_{part\text{-}load}$ is given by

$$COP_{part\text{-}load} = \frac{\dot{Q}_{cool}}{\dot{W}_{comp}} \quad (14.35)$$

The denominator of Equation 14.35 can be found from Equation 14.32 at given levels of cooling load $\dot{Q}_{cool}$. Specifically, the expression for $COP_{part\text{-}load}$ is found to be:

$$COP_{part\text{-}load} = \frac{COP^* \times PLR}{a + b \times PLR + c \times PLR^2} \quad (14.36)$$

Table 14.11 assembles the results, and Figure 14.23 shows the COP as a function of PLR.

Comments

The performance of this chiller is relatively uniform in the PLR range between 0.3 and 1.0 in contrast to boilers and furnaces described in Chapter 15 where efficiency drops strongly for PLR < 1. Note that the COP at part load appears to be larger than at full load. This is an artifact of using PLR coefficients based on constant evaporator and condenser temperatures. The reader should be aware that the part-load data used in this example are not universally applicable to this generic chiller type. Manufacturer's data must be used for reliable assessments of part-load performance.

When several chillers of different sizes are available in a building, part-load information can be used to determine which chillers should be operating (Austin, 1991). Proper sequencing of chillers to minimize energy use is the job of the control system, discussed in Chapter 21.

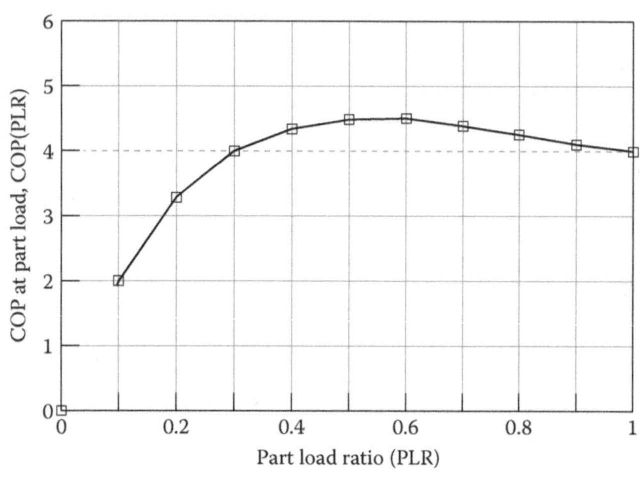

FIGURE 14.23
Chiller COP at part load (Example 14.9).

14.7.3 Medium to Large Chillers: DOE Model

For accurate building energy simulation models, the simplified part-load performance models discussed in the previous section are inadequate, and so more accurate modeling approaches have been developed. Such algorithms have been embedded in most of the detailed simulation programs developed in North America. Since they were originally part of the DOE-2 simulation software (LBL, 1982), this model structure is commonly referred to as the DOE model. The DOE model approach is fairly involved, and consists of several submodels as described in the following:

1. Obtain, from the equipment manufacturer, the performance map covering the full range of operating conditions. It is mandatory to acquire the cooling capacity $\dot{Q}^*_{cool}$ and power draw $\dot{W}^*_{comp}$ under rated conditions.
2. Identify correction models for available full-load performance under off-design or arbitrary conditions. Since the chillers will be usually water-cooled, a second-order polynomial model is fit to the chiller map of a quantity called the CAP_FT (the capacity at full load) $\equiv (\dot{Q}_{cool,ss}/\dot{Q}^*_{cool})$, which is the ratio of the available steady-state cooling capacity under different condenser water inlet temperature and evaporator water outlet temperature to the cooling capacity under rated conditions.
3. Compute a quantity called the energy input ratio (EIR_FT) $\equiv (\dot{W}_{comp,ss}/\dot{Q}_{cool,ss})/(\dot{W}^*_{comp}/\dot{Q}^*_{cool})$, which is

the ratio of the full-load steady-state *unit power* under various operating conditions (specified in the performance map) to that under rated conditions. Then, identify a second-order polynomial model of EIR_FT versus with the same two water temperature values.

4. Under any part-load condition, the operating capacity will be known from the building load requirements. So, the PLR defined by Equation 14.27 can be determined. The corresponding PLF values defined by Equation 14.28 can be predicted from PLR based on predeveloped models for different generic chiller types using a first-order polynomial model. Note that a new term is introduced, namely EIR_FPLR $\equiv (\dot{W}_{comp}/\dot{W}_{comp,ss})$ since, instead of using COP, the industry norm is to use kW/ton.
5. The power drawn by the vapor compression equipment under arbitrary operating conditions and under part-load conditions can finally be determined as

$$\dot{W}_{comp} = \dot{W}^*_{comp} \times (\text{CAP\_FT} \times \text{EIR\_FT} \times \text{EIR\_FPLR}) \quad (14.37)$$

Table 14.12 assembles the various polynomial model coefficients pertinent to a generic centrifugal chiller wherein the rated size and performance can be changed arbitrarily. There are 15 coefficients in total, and so at least 30 or more performance points on the performance map are needed if this approach were to be tailored to a specific chiller make and model. As in the previous section, it is more accurate to identify polynomial model coefficients using actual data from chiller manufacturers than this generic approach. Such polynomial models are called black-box models since little or no thermodynamic or heat transfer considerations are used in developing the functional forms. There has been an ongoing thrust to develop physics-based models that typically require far fewer coefficients and tend to be more robust. One such example is the modeling approach proposed by Gordon and Ng (2000) for vapor compression and absorption chillers and heat pumps. The interested reader can refer to Reddy (2011) for a treatment of regression and other statistical techniques applied to chillers and other HVAC equipment data.

Example 14.10: DOE-2 Model for Part-Load Performance

Find the electric power drawn by a water-cooled chiller operating at 50% capacity under a condenser water inlet temperature of 80°F and chilled water setpoint temperature of 46°F. Use the data in Table 14.12. The rated kW/ton is 0.6 kW/ton and the full-load rated cooling capacity is 1.2 MBtu/h (100 tons).

Given: rated condition: kW/ton* = 0.6 kW/ton and $\dot{Q}^*_{cool}$ =1.2 MBtu/h, $T_{ch,out}$ = 46°F and $T_{cd,in}$ = 80°F.

Find: $\dot{W}_{comp}$

Solution

First, we calculate the power at rated conditions: $\dot{W}^*_{comp} = 100 \text{ tons} \times 0.6 \text{ kW/ton} = 60 \text{ kW}$

As described in step (b) earlier, we use the model coefficients of Table 14.12, and find the following model:

$$\begin{aligned}\text{CAP\_FT} = &-0.29861976 + 0.02996076 \times T_{ch,out} \\ &- 0.00080125 \times T^2_{ch,out} + 0.01736268 \times T_{cd,in} \\ &- 0.00032606 \times T^2_{cd,in} + 0.00063139 \times T_{ch,out} \times T_{cd,in}\end{aligned} \quad (14.38)$$

TABLE 14.12

Polynomial Model Regression Coefficients for the DOE-2 Model for Water-Cooled Generic Centrifugal Chillers

CAP_FT		EIR_FT		EIR_FPLR	
Terms	Coefficients	Terms	Coefficients	Terms	Coefficients
Intercept	−0.29861976	Intercept	0.51777196	Intercept	0.17149273
$T_{ch,out}$	0.02996076	$T_{ch,out}$	−0.00400363	$PLR_{part\text{-}load}$	0.58820208
$T^2_{ch,out}$	−0.00080125	$T^2_{ch,out}$	0.00002028	$PLR^2_{part\text{-}load}$	0.23737257
$T_{cd,in}$	0.01736268	$T_{cd,in}$	0.00698793		
$T^2_{cd,in}$	−0.00032606	$T^2_{cd,in}$	0.00008290		
$T_{ch,out}\ T_{cd,in}$	0.00063139	$T_{ch,out}\ T_{cd,in}$	−0.00015467		

Source: From LBL, *DOE-2 Engineers Manual*, Lawrence Berkeley Laboratory Report, Berkeley, CA, 1982.
All temperatures are in °F.

Replacing numerical values for $T_{ch,out}$ and $T_{cd,in}$ in Equation 14.38 yields CAP_FT = 1.00987.

Similarly, we find EIR_FT = 0.89693.

Next, using PLR =0.5, we can compute EIR_FPLR = 0.52494.

One can directly conclude that though the cooling load has reduced by half from the off-design conditions, the power draw due to part-load cycling operation did not reduce as much since its value is now 0.52494, i.e., about a 5% penalty compared to the value of PLR = 0.5.

Finally, we use Equation 14.37 to find the electric power draw:

$$\dot{W}_{comp} = 60\text{ kW} \times 1.00987 \times 0.89693 \times 0.52494$$

$$= 60\text{ kW} \times 0.4755 = 0.0951\text{ MBtu/h} = 28.5\text{ kW}$$

Finally, it would be instructive to illustrate the regression model building process itself in case an analyst has performance data of a particular chiller.

Example 14.11: Identifying a Regression Model from Chiller Performance Map Data

We will use the water-cooled chiller performance data in Table 14.5 to identify regression models. The data in the table are rearranged so as to be suitable for regression (either a spreadsheet program or a software package can be used). One can evaluate different functional forms of the models but the norm is to simply use black-box linear models as described earlier. One could adopt stepwise linear regression in case of higher orders, and such techniques can be found in most statistical regression textbooks, for example, Reddy (2011).

Solution

The cooling capacity and the compressor power at full load are regressed against the two fluid temperatures whose variations represent the usual operating range of this chiller. The results of the regression models are summarized in Table 14.13.

The equations of the fitted model with coefficient of determination R^2 and standard error (SE):

$$\dot{Q}_{cool} = 16.247 - 0.128 \times T_{ch,out} + 0.4155 \times T_{cd,in}$$

with Adjusted $R^2 = 99.2\%$ and SE = 0.118 tons

and

$$\dot{W}_{comp} = 4.7969 + 0.124 \times T_{ch,out} + 0.1306 \times T_{cd,in}$$

with Adjusted $R^2 = 99.1\%$ and SE = 0.074 kW

Note that first-order linear models are excellent with very high R^2 values, and all terms being statistically significant at the 95% confidence level as indicated by the Student's t-value and the associated probability values. A convenient visual manner of evaluating how well these models have captured the data is to generate x–y plots as shown in Figure 14.24.

Comments

The reader may find it rather surprising that first-order linear models have captured the performance map data so well, while the DOE model advocates a second-order polynomial. The data in Tables 14.4 and 14.5 were extracted from actual manufacturer catalogs with anonymity being maintained by not citing the manufacturer or the equipment model number. We strongly suspect that the data are not actual measured data but synthetic data generated from simulation models developed by the manufacturers themselves from in-house testing. This is a commonly adopted procedure, and some amount of discretion is warranted especially in cases where this could compromise the results of an analysis. Note that the DOE models described earlier and the associated model coefficients are based on actual experimental tests and are not based on synthetically generated data.

14.8 Ground Source Heat Pumps

The performance of heat pumps is highly sensitive to the operating conditions. The efficiency of most air-source heat pumps drops dramatically at low temperatures, generally making them unsuitable for cold climates. The ground offers a convenient alternative to ambient air as a source of heat at temperatures closer to the space to be heated or cooled. The deep soil temperature year-round is fairly constant and close to the annual outdoor

TABLE 14.13

Regression Results for Example 14.11

	Model for Cooling Capacity, tons				Model for Power Input, kW			
Parameter	Estimate	Standard Error	t-Statistic	P-Value	Estimate	Standard Error	t-Statistic	P-Value
Constant	16.247	0.565953	28.7081	0.0000	4.7969	0.355661	13.487	0.0000
$T_{ch,out}$	−0.128	0.004299	−29.7765	0.0000	0.124	0.002701	45.902	0.0000
$T_{cd,in}$	0.4155	0.0092115	45.1078	0.0000	0.1306	0.0057889	22.563	0.0000

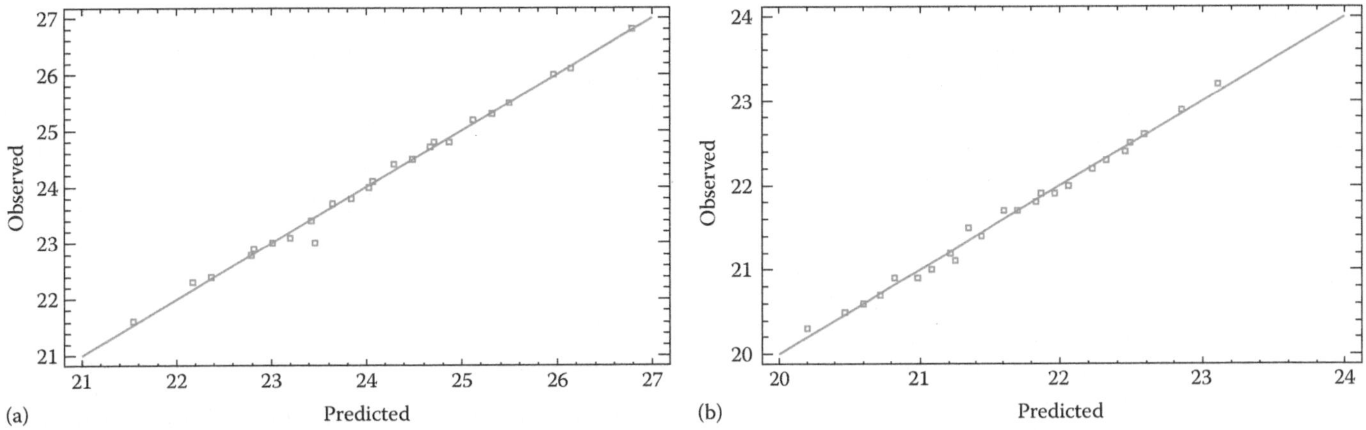

FIGURE 14.24
Goodness of fit of the linear first-order regression models for water-cooled chiller using data from Table 14.4 (Example 14.11). (a) Cooling capacity model. (b) Power input model

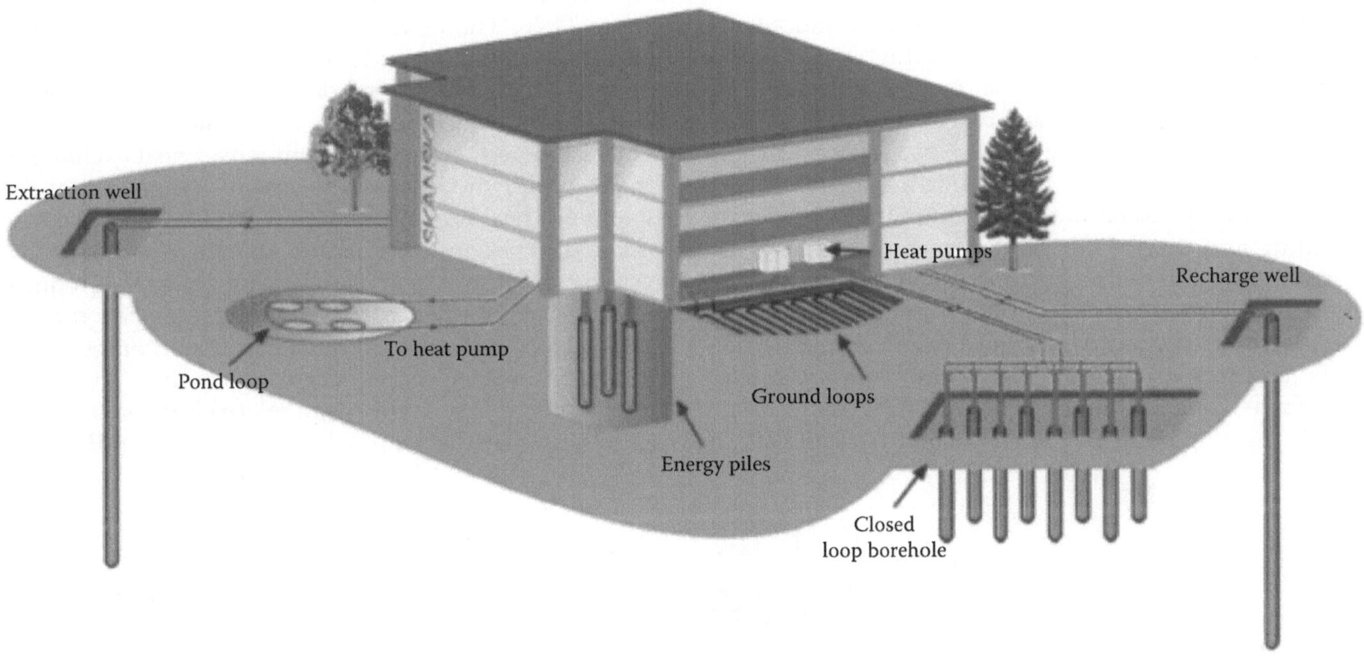

FIGURE 14.25
Different ground-source heat pump configurations. (From Geothermal website, www.alibaba.com.)

air temperatures. The large fluctuations in ambient air temperatures during the year that impose a severe penalty on cycle efficiency can thereby be greatly reduced. Ground-source heat pumps* is a broad term that applies to a variety of systems using heat from the ground soil or heat from surface or underground water (Kavanaugh and Rafferty, 1997).

Several ground source heat pump designs are shown in Figure 14.25. There are essentially two subcategories: water source and ground coupled heat pumps. Both of them involve modifying the outdoor heat exchanger coil so that the DX refrigerant coil now transfers heat to water rather than to air. Often, plate and frame heat exchanger designs are adopted whose improved effectiveness will result in smaller heat exchanger areas as compared to air source heat pumps. In *water source heat pumps*, this water can be piped to exchange heat to a close-by surface water body (such as lakes, ponds, or rivers) or even to underground lakes and rivers. Water temperatures over the day and over the year fluctuate less than ambient air, and this would improve VC cycle efficiencies.

* These are also referred to, erroneously in our opinion, as geothermal heat pumps. Geothermal energy, as traditionally understood, is heat from deep within the earth's crust at temperatures and pressures high enough to operate heat engines.

However, most residences and commercial buildings usually do not have such water bodies in close proximity, and so much attention has been focused on *ground coupled heat pump* systems that are able to provide heating and cooling to buildings in a very energy efficient manner. There has been a steep increase in the number of installations of such systems in the past 15–20 years (an average annual worldwide rate of increase of 10%), and the industry has reached a certain maturity. The primary drawbacks are the cost associated with digging and laying the ground heat exchangers. Several designs have evolved in terms of ground heat exchangers, three common ones being depicted in Figure 14.25. One could deploy the piping vertically or horizontally in trenches at least 4 ft deep (which reduces digging costs but requires more surface land and more soil temperature fluctuations) with either rigid piping or slinky coils (which can reduce ground area needed).

One important variable that dictates economics is the thermal conductivity of the surrounding soil. Another factor that has a critical impact on the long-term efficiency of such systems is the extent to which the temperature gradient around the ground heat exchanger gets modified over the years. If the building is located in a cooling dominated climate, the heat extracted from the building and dissipated into the ground increases the surrounding soil temperature, dries out the soil thereby reducing thermal conductivity, and gradually degrades heat pump efficiency. Similar heat pump efficiency degradation can also occur in cold climates that are heating dominated due to the ground becoming increasingly cooler. Ground coupled heat pumps have been used in residences as well as in medium and large commercial buildings such as schools. The interested reader can refer to Kavanaugh and Rafferty (1997) for details into the various types of ground source heat pumps, their engineering design methods, description of various auxiliary equipment such as pumps and piping, factors that degrade performance, their economics, and the associated regulatory aspects.

14.9 Decentralized Water Loop Heat Pumps

Rather than using large centralized systems, each room or zone of a building is equipped with an individual heat pump unit that is then connected together with other unitary heat pumps in a water loop as shown in Figure 14.26. The heat pumps in such a decentralized configuration can provide both heating and cooling as needed. Under heating mode, the water loop becomes the source of heat input at the evaporator with the zone heat exchanger acting as the condenser heat exchanger. Under cooling mode, the zone heat exchanger acts as the evaporator heat exchanger with water loop acting as the sink at the condenser. A bigger advantage is that such a system can remove heat from certain zones that require cooling and pump it to other zones that may require heating (such as from external zones in winter requiring heating to interior spaces requiring cooling). This type of heat interchange maintains a relatively constant water temperature in the loop, which reduces overall power

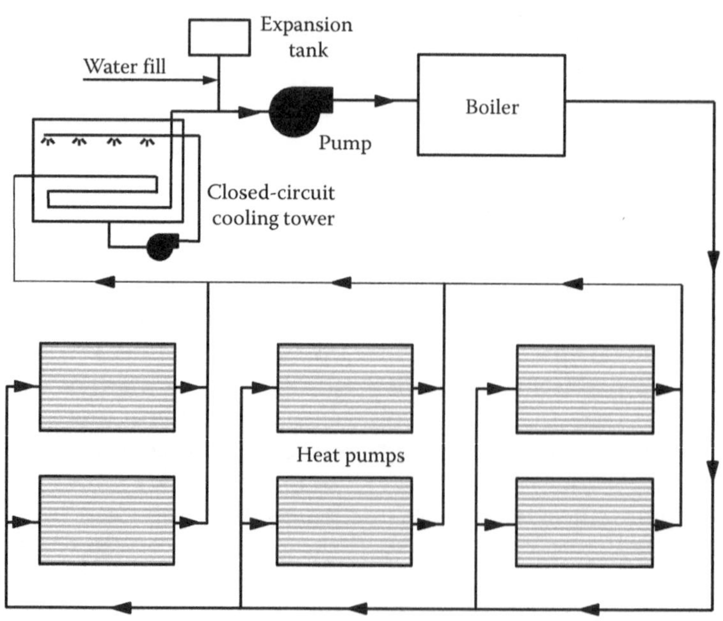

FIGURE 14.26
Decentralized water loop systems with several unitary heat pumps.

input, thereby enhancing overall system COP. Under the "balanced" condition of net cooling being equal to net heating, no external cooling or heating source is needed and the electric draw of the heat pumps is the only energy consumption of the system. More usually, some source of heating or cooling is necessary.

As stated earlier, heat pumps will be sized to meet the peak cooling conditions. So, in milder climates, the only source of cooling is often a cooling tower, which is activated when the water loop temperature reaches, say 35°C (90°F) or so. Under heating mode, a separate boiler will be activated when the loop water temperature reaches 15°C (60°F) or so (Figure 14.26). Since the water source heat pumps operate under a much reduced temperature difference as compared to air source heat pumps, their efficiency will be higher, which makes them very advantageous. One could also use a solar thermal collector to heat up a water storage, which can further reduce the conventional fuel requirements of the boiler. The heat pump units and their space heat exchangers come in different shapes and sizes and can be adapted to ceiling spaces, small equipment rooms, or as room consoles. They are used extensively in multi-family buildings, dormitories, certain types of motels, and even in certain commercial buildings.

Decentralized heat pump systems with multiple units (or for that matter all heat pump systems) only meet heating and cooling loads of the space. They do not, as such, have provision for supplying outdoor ventilation air as required for human comfort. Separate ventilation or primary systems can be integrated nicely with unitary heat pump systems, and these are discussed in Chapter 20. Because there are so many individual units, their maintenance costs are higher than centralized systems. Table 14.14 summarizes advantages and disadvantages of air and water source heat pumps in terms of both thermal performance and practical issues.

14.10 Theoretical Performance Indices for Heating and Cooling

The concept of efficiency is of fundamental importance (especially when evaluating the annual efficiency) for any equipment whose part-load efficiency is not constant. The solved examples in this chapter are meant to illustrate how a typical cycle or system analysis is to be performed. However, it is useful for scientific understanding to acquire a broader perspective of efficiency by seeking to define theoretical limits for heating and cooling applications involving vapor compression equipment.

At first sight, such a question may not seem meaningful. After all, in theory, one could just keep adding insulation and heat recovery until the heat load vanishes; such a limit is not very informative. However, the question assumes a different hue if we take the heat loss coefficient and the heat gains of a building as specified starting points, and then ask ourselves how much energy is needed if we have at our disposal a source of high thermodynamic quality such as electricity. The corresponding thermodynamic limit for the heating system efficiency is derived in the following subsections, beginning with a simple idealized situation and then adding processes that are needed to accomplish the necessary heat transport into an actual building. This leads to a definition of second-law efficiency. The equations are also relevant for cooling.

14.10.1 Limit for Ideal Heat Pumps

Suppose that a building has a heating load $\dot{Q}_h$ and electricity is available. We could use an electric resistance heater, but a heat pump would be more efficient. It could extract heat from the outdoor air at T_o and deliver heat at T_i, consuming work $\dot{W}_h$ (mechanical or electrical), as

TABLE 14.14

Advantages and Disadvantages of Air and Water Source Heat Pumps

Type	Advantages	Disadvantages
Air source	Indoor distribution permits air conditioning and humidity control	Defrost required
	Outdoor air source readily available	Low capacity at cold outdoor temperature
	Simple installation	Lower efficiency because of large evaporator $\Delta T \approx 30°F$
	Least expensive	Indoor air distribution temperature must be high for comfort reasons
	Established commercial technology	Reliability at low temperature is only fair, due to frosting effects
		Must keep evaporator clear of leaves, dirt, etc.
Water source	Multiple family and commercial installations as central system	Needs water source at useful temperature
	In commercial installations, good coupling to cooling towers	Efficiency penalty due to space heat exchanger ΔT
	No refrigerant reversal needed; reverse water flow instead	

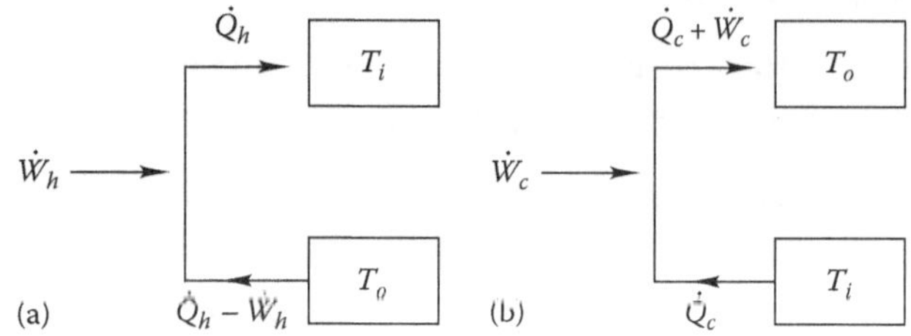

FIGURE 14.27
Thermodynamics of heat pump (a) for heating load $\dot{Q}_h$ and (b) for cooling load $\dot{Q}_c$.

indicated in Figure 14.27a. A heat pump is a heat engine running backward. From Equation 12.3, for an ideal Carnot heat engine between a hot reservoir at T_i and a cold reservoir at T_o, the work output $\dot{W}_h$ and the heat input $\dot{Q}_h$ are related by

$$\dot{W}_h = \dot{Q}_h \frac{T_i - T_o}{T_i} \tag{14.39}$$

An ideal Carnot heat pump is simply an ideal Carnot heat engine run in reverse. $\dot{W}_h$ and $\dot{Q}_h$ change signs, but Equation 14.39 remains unchanged. Inserting typical temperatures, for example, $T_i = 20°C$ and $T_o = 0°C$, we find

$$\dot{W}_h = \dot{Q}_h \frac{20\ \text{K}}{(273+20)\ \text{K}} = 0.068\dot{Q}_h$$

This means that an ideal heat pump can deliver 1/0.068 = 14.7 kWh_t of heat for each 1 kWh_e of electricity for these conditions (we have added subscripts t and e to distinguish thermal and electric energy forms).

Based on considerations such as these, some energy analysts have introduced the concept of the *second-law efficiency* of a process. It is defined as the ratio of the efficiency of the process to the efficiency of an ideal reversible process that satisfies the same demand (i.e., a process whose efficiency is as high as permitted by the second law of thermodynamics). For this example of T_i = 20°C and T_o = 0°C, it is very easy to infer the second-law efficiency of electric resistance heating. For 1 kWh_e of electric energy, the resistance heater delivers 1.0 kWh_t of heat compared to 14.7 kWh_t with an ideal heat pump. Hence, the second-law efficiency of resistance heating is only 1/14.7 = 6.8%, even though the ordinary (i.e., first-law efficiency) is 100%.

Next, consider the case where $T_o > T_i$ and the building has a cooling load $\dot{Q}_c$. The corresponding energy flows are indicated in Figure 14.27b. For a Carnot process, the load $\dot{Q}_c$ and the work input $\dot{W}_c$ are related by

$$\dot{W}_c = (\dot{Q}_c + \dot{W}_c)\frac{T_o - T_i}{T_o}$$

or

$$\dot{W}_c = \dot{Q}_c \frac{T_o - T_i}{T_i} \tag{14.40}$$

Recall from Section 12.2.2 that the ratio $\dot{Q}/\dot{W}$ of delivered thermal energy to work input is the *coefficient of performance* (COP). The COP of an ideal heat pump under heating mode is

$$\text{COP}_{Carnot,h} = \frac{T_i}{T_i - T_o} \tag{14.41}$$

Note that this expression is identical to Equation 12.5 with different terminology. For this example of T_i = 20°C and T_o = 0°C, we have a heating COP = 14.7.

The COP of an ideal air conditioner is

$$\text{COP}_{Carnot,c} = \frac{T_i}{T_o - T_i} \tag{14.42}$$

For example, if T_o = 35°C and T_i = 25°C, an ideal air conditioner has a COP of (273 + 25)/10 = 29.8; it delivers 29.8 kWh_t of cooling for each 1 kWh_e of electric input.

14.10.2 Finite Airflow Rates in Building

Since a Carnot heat pump does not use air as the working fluid, the heat $\dot{Q}_h$ must somehow be transferred to the air in the building; likewise, the heat drawn from the outdoor air must be transferred to the heat pump. That involves two heat exchanges. As a first step, let us consider an air heating system that supplies hot air at flow rate $\dot{V}$ and temperature T_{sup}. The heat supplied by this airstream is

$$\dot{Q}_h = \dot{V}\rho c_p (T_{sup} - T_i) \tag{14.43}$$

and it must be equal to the heating load $\dot{Q}_h = K_{tot}(T_i - T_o)$ if this heating system is to do its job.

This implies the relation

$$T_{sup} - T_i = (T_i - T_o)\frac{K_{tot}}{\dot{V}\rho c_p} \tag{14.44}$$

The supply air temperature T_{sup} would be equal to T_i only in the limit of infinite flow rate. In practice, typical values of the supply air are 30°C–40°C (86°F–104°F) for heating and 12°C–16°C (53.6°F–60.8°F) for cooling. To find the resulting COP of the heat pump, we replace T_i by T_{sup} in Equation 14.41. The corresponding results for cooling can be calculated in an analogous manner.

Example 14.12: Heat Pump Carnot COP with Finite Air Flow Rate

Find the COP of a heat pump for a house with a typical value $K_{tot} = 250$ W/K [474 Btu/(h·°F)] if T_i = 20°C (68°F), T_o = 0°C (32°F), and the airflow rate of the heating system is $\dot{V} = 0.42$ m³/s.

Given: T_i, T_o, K_{tot}, $\dot{V}$

Find: COP
Lookup values: $\rho C_p = 1.2$ kJ/(m³·K)

Solution

The quantity $\dot{m}c_p = \dot{V}\rho C_p = 0.42\ \text{m}^3/\text{s} \times 1.2\ \text{kJ/(m}^3\cdot\text{K)}$ = 0.5 kW/K. Inserting this into Equation 14.44, the supply temperature to the house is

$$T_{sup} = 20°\text{C} + (20°\text{C} - 0°\text{C})\frac{250\ \text{W/K}}{0.5\ \text{kW/K}} = 30°\text{C}$$

$$= 303.15\ \text{K}\ (86°\text{F})$$

The corresponding COP is

$$\text{COP} = \frac{T_{sup}}{T_{sup} - T_o} = \frac{303.15}{303.15 - 273.15} = 10.11$$

which is much less than the value of 14.7 with infinite flow rate.

14.10.3 ΔT of Heat Exchangers

Further approaching realism, we shall incorporate two heat exchangers to the heat pump, one between the outside air and the cold side (= evaporator) and one between the hot side (= condenser) and the indoor air (Figure 14.28). The working fluid is isothermal in the evaporator and in the condenser, since we are assuming a Carnot cycle for now (a Rankine cycle will be considered in the next subsection). In terms of the heat exchanger effectiveness ε, the heat flow is given by

$$\dot{Q}_h = \varepsilon_i \dot{C}_i (T_{cond} - T_{sup}) \quad (14.45)$$

where
- ε_i is the heat exchanger effectiveness on indoor air side
- $\dot{C}_i = (\dot{m}c_p)_i$ = heat capacity of indoor air flow rate
- T_{sup} is the supply air temperature to the building
- T_{cond} is the temperature of refrigerant in condenser, which is higher than T_{sup} of Equation 14.44 because ε_i is less than unity

For the outdoor side, one has an analogous equation

$$\dot{Q}_h - \dot{W}_h = \varepsilon_0 \dot{C}_0 (T_o - T_{evap}) \quad (14.46)$$

where $\dot{C}_0$ = heat capacity of outdoor air flow rate.

The heat exchanger effectiveness for these equations can be obtained from Table 12.6. Still assuming the heat pump to be a Carnot heat pump, we can calculate the COP of this system by solving the last two equations for T_{cond} and T_{evap} and by using these temperatures instead of T_i and T_o in Equation 14.41

$$\text{COP} = \frac{T_{cond}}{T_{cond} - T_{evap}} \quad (14.47)$$

While T_{cond} can be obtained directly from Equation 14.45, we first have to eliminate $\dot{W}_h$ from Equation 14.46 before we can solve for T_{evap}. This can be done by setting

$$\dot{W}_h = \dot{Q}_h\left(\frac{T_{cond} - T_{evap}}{T_{cond}}\right)$$

according to Equation 14.47. Inserting this into Equation 14.46 and rearranging results in

$$T_{evap} = \frac{\varepsilon_o \dot{C}_o}{\varepsilon_o \dot{C}_o + \varepsilon_i \dot{C}_i (1 - T_{sup}/T_o)} T_o \quad (14.48)$$

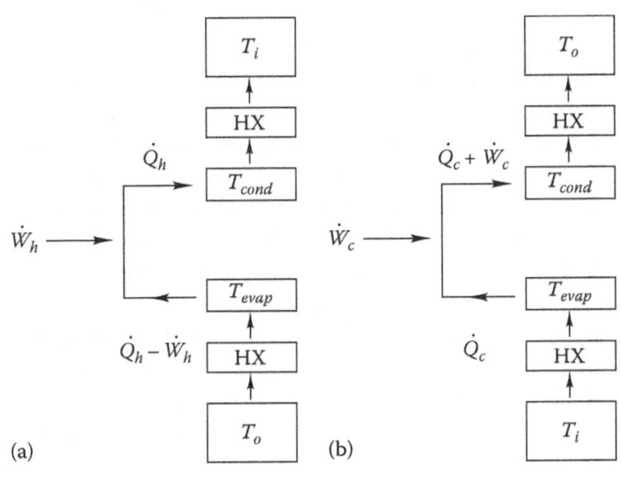

FIGURE 14.28
Thermodynamics of heat pump with heat exchangers (HX) between heat pump and air: (a) for heating; (b) for cooling.

Example 14.13: Heat Pump Carnot COP with Heat Exchangers

Find the COP of a Carnot heat pump for the conditions of Example 14.12 if $\dot{C}_i = \dot{C}_o = 0.5$ kW/K, and if the heat exchangers are of identical counterflow design, with areas $A_{HXi} = A_{HXo} = 4$ m², and heat transfer coefficients $U_{HXi} = U_{HXo} = 200$ W/(m²·K).

Given: $T_{sup} = 30°C$, $T_o = 0°C$,
$(K_{tot})_{HXi} = (K_{tot})_{HXo} = 0.80$ kW/K
$\dot{C}_i = \dot{C}_o = 0.5$ kW/K on indoor and outdoor sides

Find: COP

Lookup values: Heat exchanger effectiveness from $C_{min}/C_{max} = 0$ curve of Figure 12.17 for NTU = $(K_{tot})_{HX}/\dot{C} = 1.6$ is $\varepsilon = 0.80 = \varepsilon_i = \varepsilon_o$.

Solution

The heating load of Example 14.12 is

$$\dot{Q}_h = (20°C - 0°C)(250 \text{ W/K}) = 5 \text{ kW}$$

Solving Equation 14.45 for T_{cond}, we obtain

$$T_{cond} = T_{sup} + \frac{\dot{Q}_h}{\varepsilon_i \dot{C}_i} = 30°C + \frac{5 \text{ kW}}{0.80 \times 0.5 \text{ kW/K}}$$

$$= 42.50°C = 273.15 + 42.50 = 315.65 \text{ K}$$

From Equation 14.48, noting that $\varepsilon_i \dot{C}_i = \varepsilon_o \dot{C}_o$ and that the temperature ratio must be evaluated in unit of Kelvins, we obtain

$$T_{evap} = \left(\frac{1}{1 + 1 - T_{sup}/T_{cond}}\right) T_o = \left(\frac{1}{1 + 1 - 303.15/315.65}\right) \times T_o$$

$$= 0.962 \times T_o = 262.75 \text{ K} = -10.40°C$$

Finally, the COP is

$$\text{COP} = \frac{T_{cond}}{T_{cond} - T_{evap}} = \frac{315.65 \text{ K}}{315.65 \text{ K} - 262.75 \text{ K}} = 5.97$$

Comments

This is only about 40% of the value 14.7 with infinite flow rates and without heat exchanger penalties.

14.10.4 Actual Heat Pumps

Practical heat pumps and air conditioners (of the compression type) are based on the Rankine cycle and were discussed in the early sections of this chapter. The COP of an ideal Rankine heat pump with phase-change temperatures T_{cond} and T_{evap} can be written as a fraction f_{Rank} of the COP of the Carnot heat pump with upper and lower temperatures T_{cond} and T_{evap}:

$$\text{COP}_{Rank,ideal} = f_{Rank}\left(\frac{T_{cond}}{T_{cond} - T_{evap}}\right) \tag{14.49}$$

The factor f_{Rank} depends on the working fluid, but does not vary strongly over a typical range of operating temperatures. For heat pumps and air conditioners with Freon, f_{Rank} is in the range of 0.7–0.8.

The COP of an actual Rankine engine is further reduced by a factor f_{loss} because of various losses. The latter are dominated by friction in the compressor, so much so that one can approximate f_{loss} by the compressor efficiency. Typical values are also in the range of 0.7–0.8. Since this loss factor is relatively insensitive to temperature, one can approximate the efficiency of an actual Rankine cycle as

$$\text{COP}_{Rank,real} = f_{Rank} f_{loss}\left(\frac{T_{cond}}{T_{cond} - T_{evap}}\right) \tag{14.50}$$

Comparison with results in Figure 14.18 shows that this approximation is good to about 10% over the range of typical operating temperatures. Since typical values of $f_{Rank} f_{loss}$ are around 0.5, we can extract the rule of thumb that *the COP of actual heat pumps and air conditioners is about one-half of the Carnot COP based on the actual temperatures of the evaporator and condenser.* Thus for the conditions of Example 14.13, the COP of an actual heat pump can be expected to be around 3.0. If the areas or heat transfer coefficients of the heat exchangers are smaller than in Example 14.13, the COP would be even lower.

14.11 Refrigerants

14.11.1 Background

Refrigerants are the working fluids that remove heat from the cold space and reject heat to a hotter sink. Primary refrigerants are used in vapor compression cycles, while secondary refrigerants are used for transportation at low temperatures. For example, 100 years back, meat and other products were transported by locomotives across the United States in wagons packed with ice. Secondary refrigerants (also referred to as "heat transfer fluids") are used in some building cooling equipment, but they are more widely used for certain specialized energy applications (for example, antifreeze liquids). We shall limit our discussion to the former class of refrigerants that are much more appropriate for building cooling systems.

Obviously, the choice of suitable refrigerants is influenced by how their thermodynamic properties impact the energy efficient operation of the vapor compression system (covered in Section 14.11.3).

In addition, the selection is dictated by other considerations; the most critical ones are safety, toxicity, stability, reliability, material compatibility (discussed in Section 14.11.4), and in the last couple of decades, by its environmental impact (Section 14.11.5). Issues such as ozone depleting potential (ODP), global warming potential (GWP), and life cycle climate performance (LCCP) have become very important considerations. This has resulted in international efforts to phase out certain well-established refrigerants, and also such efforts are the drivers to develop more benign ones.

14.11.2 Classification and Method of Designation

There is no single set of optimum characteristics for refrigerants (especially for thermodynamic properties), and so tradeoffs are to be made among desirable characteristics. Thus, a variety of refrigerants having a range of properties have been developed to meet the requirements of various applications. Many of these refrigerants are mixtures of different substances whose composition and saturation temperature do not vary when boiled and have properties different from each constituent. These are called *azeotropes* and behave like simple pure substances. These refrigerants were the ones assumed in the previous sections of this chapter. On the other hand, there are other mixtures, called *zeotropic* blends, which change composition when boiled at constant pressure, and so can be separated into their components by distillation. These are used extensively in absorption systems, and also in VC systems (for example, the zeotrope R-410a is a replacement to R-22).

The broad classes of refrigerants are listed in Table 14.15 along with other descriptors:

1. *Inorganic refrigerants* were the only natural refrigerants available until the 1930s. They are usually not manufactured chemically but need to be refined. An exception is ammonia, which is still widely used in industrial refrigeration, while water as a refrigerant, due to its high freezing temperature, finds application in some absorption systems.
2. *Halocarbon refrigerants* are those that contain one or more of the three halogens: chlorine, fluorine, and bromine. They have been developed in the 1930s and currently have the largest market share. There are several companies that produce these refrigerants popularly called Freon, though strictly speaking it is a trade name given by the du Pont company.
3. *Hydrocarbon refrigerants*, used extensively in cars, are blends of flammable substances such as propane and butane. They have been in use from several decades prior to the advent of halocarbon refrigerants. They are said to be nontoxic and nonozone depleting, but their production has been banned since 1995 due to safety concerns.
4. *Organic refrigerants* are those that contain one or more of carbon, hydrogen, and nitrogen atoms. They are also naturally occurring.

TABLE 14.15

Classification of Common Refrigerants

Refrigerant Class	Numerical Designation	Chemical Name	Chemical Formula	Safety[a]	ODP[a]	GWP[a] (100 Year)
Inorganic	717	Ammonia	NH_3	B2	0	<1
	744	Carbon dioxide	CO_2	A1	0	1
	718	Water	H_2O	A1	0	<1
Halocarbon	11	Trichloromonofluoromethane	CCl_3F	A1	1	4,600
	12	Dichlorodifluoromethane	CCl_2F_2	A1	0.82	10,600
	22	Monochlorodifluoromethane	$CHClF_2$	A1	0.034	1,700
	134a	Tetrafluoroethane	CH_2FCF_3	A1	0	~1,400
	502[b]	—	$CClF_2CF_3$	A1	0.221	4,500
Hydrocarbon	50	Methane	CH_4	A3	0	23
	170	Ethane	C_2H_6	A3	0	~20
	290	Propane	C_3H_8	A3	0	~20
Organic	600	Butane	C_4H_{10}	A3	0	~20
	630	Methylamine	CH_3NH_2	—	0	—

ODP is the ozone depletion potential and GWP is a global warming potential indictor relative to carbon dioxide.

[a] Courtesy of Calm and Hourahan (2001).

[b] Azeotrope of R-22 and R-115. Chemical name is Hydrogenated Chlorofluorocarbon/Chlorofluorocarbon Blend.

Rather than use chemical formulas for the refrigerants, the norm is to use a generic numbering scheme defined by Standard 34 (ANSI/ASHRAE, 2013). All refrigerant numbering is preceded by the capital letter "R" for refrigerant followed by a numerical designation. The numbering basis of identifying both halocarbons and hydrocarbons are identical. It is based on the number of fluorine atoms (right digit of number), number of hydrogen atoms plus 1 (center digit), and number of carbon atoms minus 1 (left digit, unless zero, which is suppressed). For example, the chemical formula for dichlorodi-fluoromethane is CCl_2F_2. Therefore, the first digit is zero (suppressed); the center digit is 1 since there are no hydrogen atoms. The right digit is 2 since there are two fluorine atoms. Hence, the refrigerant number is R-12.

Inorganic compounds are numbered by adding 700 to their molecular weight, a rather arbitrary methodology. Therefore, ammonia (molecular weight = 17) is numbered R-717. Azeotropic mixtures of refrigerants use a serial numbering system beginning with 500 are not related to the chemical composition. As an example, a mixture of R-22 and R-115 on a mass basis of 48.8% and 51.2% respectively is designated by R-502. Zeotropic blends are also arbitrarily designated by the 400 series followed by the weight proportion of the components. For example, a mixture of 92% R-502 (azeotrope of R-22 and R-115) with 8% propane (R-290) will be designated by R-290/22/115(8/45/47). Miscellaneous organic refrigerants are designated by the series 600.

Finally, ASHRAE has added a composition-designated prefix. Refrigerant numbers are preceded with letters designating the element present. For example, B for bromine, C for chlorine, F for fluorine, and again C for carbon. Thus, R-22, which is a hydrochlorofluorocarbon, is designated by HCFC-22, and R-502s by HCFC/CFC-502. Another common group is hydrofluorocarbons or HFCs, which contains no chlorine; examples are R-125, R-134a,* and R-143a. A new class of refrigerants called HFO (hydrofluoro olefins) has also been developed, which has a low global warming potential rating and is used in automotive air-conditioning.

14.11.3 Thermodynamic Properties

Thermodynamic properties for several common refrigerants are given in charts and tables and have been treated in Section 12.3.3 while discussing pure substances. A complete set of properties is included in the online HCB software. Desirable characteristics are

1. High latent enthalpy of vaporization, i.e., a large refrigeration effect per unit mass
2. High critical pressure, which results in reduced compressor power
3. Positive evaporating pressures will prevent atmospheric air leaking into refrigerant piping during operation
4. Low but positive condensing pressures will allow lightweight equipment and piping to be used
5. Low temperature of vapor leaving compressor

Table 14.16 assembles some of these important characteristics for a few common refrigerants when used in a standard VC cycle operating between the same two evaporation and condensation temperatures of −15°C and 30°C respectively. Even for the ideal cycle the differences can be quite revealing. The pressures show a range significant enough to affect the thickness of condenser and evaporator tubing walls. The evaporator and condenser pressures must be close to the atmospheric pressure (in order to avoid thick piping and increased leaks through joints), with a slight over-pressurization being preferable so as to avoid air leaking into the refrigerant piping. Note that the R-114 evaporator is sub-atmospheric. The lower the compression ratio, the better in terms of piping selection and load on the compressor. Another practical consideration relating to proper mechanical operation is the temperature of the refrigerant leaving the compressor (state point T_2). Higher the value, more critical is the need to assure proper lubrication so as to avoid premature compressor failure; inadvertent scalding to maintenance technicians is another factor.

Flow rates, and therefore compressor sizes, vary by as much as 40%. The COP_{std} values for these five refrigerants vary by only 10% at the stipulated operating temperatures T_{evap} = −15°C and T_{cond} = 30°C. Notice that R-717 (ammonia) has a much greater refrigerating effect, lower flow rate, and higher COP. This is why, despite its adverse toxicity effects and high compressor exit temperature, it is widely used especially in industrial and refrigeration systems.

The refrigerating efficiencies η_R range from 0.77 to 0.84 (Table 14.16) which is consistent with the value of 0.809 found in Example 14.1. Figure 14.29 depicts the variation of η_R for three refrigerants for a constant condensation temperature. Note that the variation of η_R with decreasing evaporator temperature is very close to linear, a behavior that would be useful to model part-load efficiencies of vapor compression systems.

* R-134a contains hydrogen, fluorine, and carbon. There are four fluorine atoms per molecule are and three of them are bonded to one carbon atom and the fourth to a second carbon atom. R-134, on the other hand, is a structure where the fluorine atoms are bonded symmetrically, two to each carbon atom. As a result, the thermodynamic properties are quite different.

TABLE 14.16

Comparison of Different Refrigerants in terms of Performance between the Operating Temperatures of T_{evap}= –15°C and T_{cond} = 30°C

Refrigerant	Evaporation Pressure, MPa	Condensation Pressure, MPa	Comp. Ratio, —	Compressor Exit Temp., °C	Refrigerating Effect, kJ/kg	Refrigerant Flow Rate, kg/s·kW	COP_{std}	Refrigerating Efficiency, —
R-12	0.183	0.745	4.07	38	116.58	0.00858	4.69	0.818
R-22	0.296	1.192	4.03	53	162.46	0.00616	4.75	0.829
R-114	0.047	0.252	5.41	30	99.19	0.01008	4.44	0.774
R-134a	0.160	0.770	4.81	43	150.71	0.00664	4.42	0.771
R-502	0.349	1.319	3.78	37	104.39	0.00958	4.43	0.773
R-717	0.236	1.164	4.94	98	1102.23	0.00091	4.84	0.844

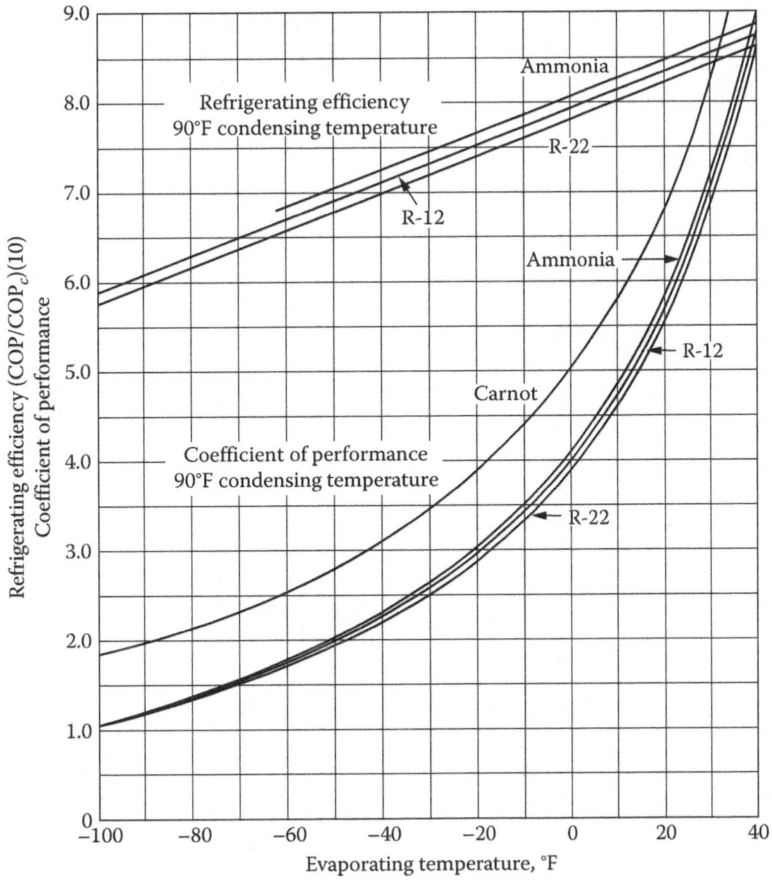

FIGURE 14.29
Variation of COP and refrigeration efficiency with operating conditions for theoretical standard VC cycle for three common refrigerants. (From ASHRAE *Handbook of Fundamentals* (2013). American Society of Heating, Refrigerating and Air-conditioning Engineers, Atlanta, GA.)

14.11.4 Physical and Chemical Characteristics and Health Safety

The ASHRAE Fundamentals (2013) lists dozens of refrigerants and their thermodynamic properties. Only a few of these are practical for building systems due to toxicity restrictions, high cost, flammability, inappropriate thermodynamic or transport properties, or environmental concerns. Desirable traits are

1. Good thermodynamic properties such as density and specific heat
2. Good transport characteristics, i.e., properties conducive to high heat transfer such as thermal conductivity and viscosity
3. Noncorrosive in the presence of water and inert in terms of material used in the piping and components

4. High dielectric strength of vapor (a factor in hermetically sealed compressors where vapor comes in contact with motor windings)
5. Low oil solubility (to maintain lubrication and prevent oil logging in the evaporator)
6. Low water solubility (so as to avoid freezing the expansion device)
7. Inert (nonreacting with piping materials)
8. Stable over time

Health safety is a major concern and much work has been done in this regard. Primarily, this aspect relates to *toxicity*, i.e., the extent to which the refrigerant is harmful to humans due to inadvertent exposure, and *flammability*, i.e., its risk in catching fire and/or supporting combustion when mixed with air. Standard 34 (ANSI/ASHRAE, 2013) classifies the toxicity level of a refrigerant into two groups, A and B, with the latter denoting "Evidence of toxicity identified" and Level A as "Toxicity not identified." The flammability classification is numeric, with 1 denoting no flame propagation, class 2 representing lower flammability limit (LFL), and class 3 being highly flammable. A subclass 2L has also been introduced in order to expedite the adoption of HFO refrigerants. Note from Table 14.15 that hydrocarbons and butane fall under class 3 while ammonia, which is known to be highly toxic, is classified as B.

14.11.5 Environmental Concerns

During the early 1990s, 90% of all refrigerants used consisted of R-11, R-12, R-22, and R-502 (an azeotropic mixture of R-22 and R-115). These chlorofluorocarbons

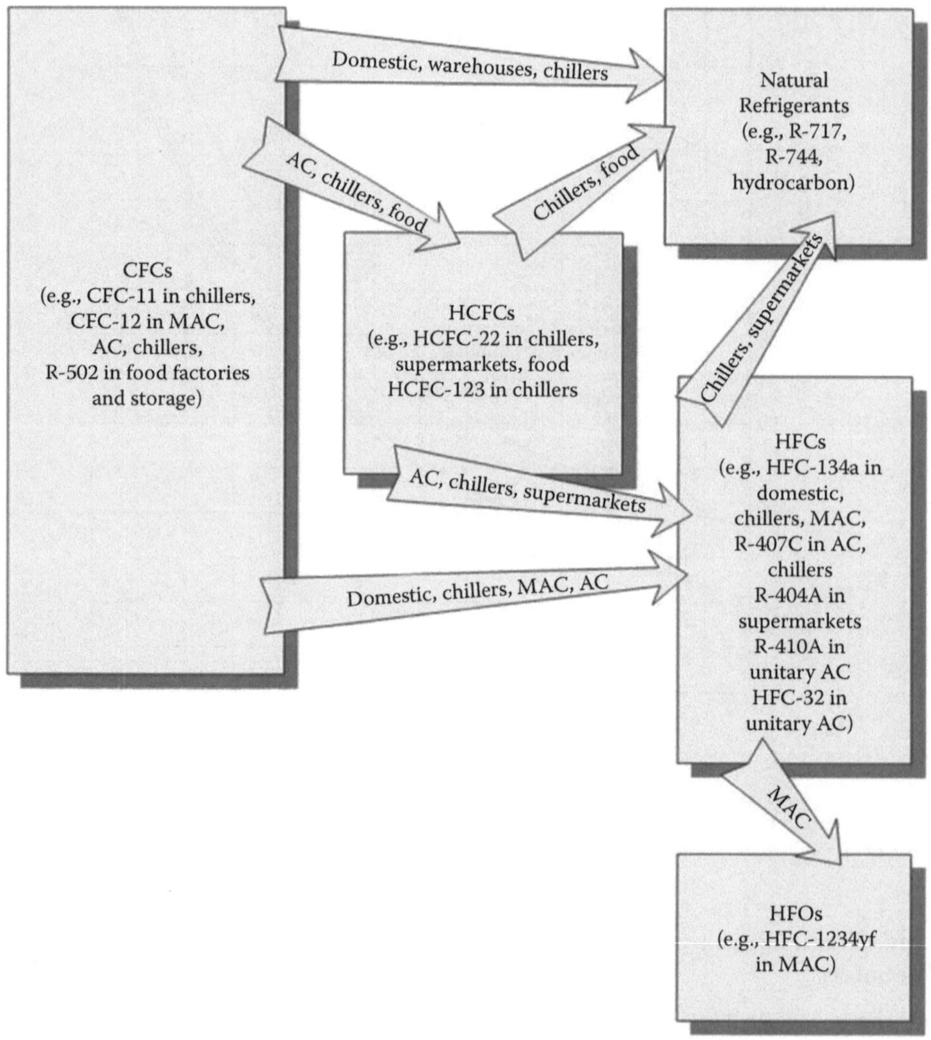

FIGURE 14.30
Map of refrigerant evolution showing routes being taken in the phase-out of CFCs and HCFCs. Refrigerants in the future are likely to be the natural refrigerants, HFCs, unsaturated HFCs (also known as HFOs), and possibly blends of these refrigerants. AC - air-conditioners, MAC - Mobille air-conditioners. (From ASHRAE position document on refrigerants and their responsible use, American Society of Heating, Refrigerating and Air-Conditioning Engineers, Atlanta, GA, 2012. Copyright ASHRAE, www.ashrae.org.)

(CFCs)* are chemically stable for terrestrial applications, inexpensive, and nontoxic compounds. However, they have been found to destroy the stratospheric ozone layer, which protects us from ultraviolet (UV) solar radiation. Increased UV flux levels at the earth's surface can lead to such harmful effects as increased skin cancer and other health-related tissue damage. The ozone depletion potential (ODP) is a normalized index of the ozone destruction potential with the value of 1.0 assigned to R-11. The ODP values of some of the common refrigerants are also shown in Table 14.15.

International efforts to address the ODP risk was initiated by several countries in the early 1980s culminating in the 1987 Montreal Protocol, which limited the production of certain CFCs to 50% of their 1998 levels and, by 1992, to freeze their production at their 1986 levels. This was amended later by the Copenhagen Protocol in 1994, which stopped the production of these CFCs by 1995. The refrigerants of primary concerns are the CFCs, which include R-11, R-12, and R-502; these were the most destructive to the ozone layer and have been the first to be phased out. Regulations are in place to phase out HCFCs; for example, R-123 is now being used to temporarily replace R-11 but is causing efficiency reduction and certain operational problems, and is likely to be phased out in 10 years. On the other hand, R-134a appears to be a satisfactory substitute for R-12. The HFC compounds are the least harmful to the ozone layer since they do not contain any chlorine, but their use is also being restricted due to global warming potential. Some of these "drop-in" replacements, in turn, are likely to be replaced by more benign compounds in the future such as natural refrigerants as shown in Figure 14.30. Some of the compounds proposed for the first round of replacements have problems, including toxicity, higher cost, reduced equipment capacity, and incompatibility with existing compressor seal designs and lubricating oils. The refrigerant replacement process is moving at different speeds in different countries. In some countries, CFCs are already prohibited, in others their use is in decline as these countries move gradually to complete prohibition.

The major environmental concern currently is not ozone, but rather the impact of refrigerants on climate change. Released refrigerants in the lower atmosphere trap the infrared radiation from the earth, thereby contributing to the greenhouse effect. Another normalized index, called the global warming potential (GWP), is now used to characterize this effect with that of carbon dioxide over a time duration of 100 years taken as the reference value of unity. Notice the very large ODP values for the halocarbon family of compounds (Table 14.15).

* Refrigerant 22 is an HCFC (hydrogenated CFC) in which the carbon atoms are not fully populated with halogens. It is less harmful in the upper atmosphere than true CFCs.

Problems

Numbers 1–4 given in parenthesis denote the degree of difficulty.

14.1 A refrigerator uses refrigerant 134a and operates between a low-side pressure of 43 psia (0.30 MPa) and a high-side pressure of 180 psia (1.2 MPa). The refrigerant mass flow is 1000 lb/h (0.126 kg/s). Find the cooling produced, power input, and COP of this machine. *(2)*

14.2 Repeat Problem 14.1 for a real cycle in which the compressor efficiency is 85% and 9°F (5°C) superheat is used at the compressor inlet. Find the cooling produced, power input, and COP of this machine. Assume that superheat at compressor inlet is provided by a HX coupled to the condenser exit. *(2)*

14.3 The throttling step of the Rankine refrigeration cycle represents a loss of potential power generation because a turbine could be inserted in place of the throttling valve to produce additional shaft power that could increase the cycle refrigeration effect. Setting aside the practical problems with a turbine for a wet vapor, how much work could be produced by an isentropic turbine for the cycle in Problem 14.1? What would the COP be if the power so produced were used for additional power input for vapor compression? *(2)*

14.4 An ideal chiller uses refrigerant 22 and operates between a low-side pressure of 76 psia (0.52 MPa) and a high-side pressure of 300 psia (2.1 MPa). The chiller capacity is 100 tons (350 kW). Find the refrigerant flow rate, power input, and COP of this machine, as well as the evaporator and condensing temperatures. *(2)*

14.5 Work Problem 14.4 with a real compressor efficiency of 85% and 9°F (5°C) of superheat at the compressor inlet. You can check your solution with the HCB software. *(2)*

14.6 An ideal chiller uses refrigerant 22 and operates between a low-side pressure of 65 psia (0.45 MPa) and a high-side pressure of 250 psia (1.7 MPa). The chiller capacity is 200 tons (700 kW). Find the refrigerant flow rate, power input, and COP of this machine, as well as the evaporator and condensing temperatures. *(2)*

14.7 Work Problem 14.6 for the same chiller capacity with a real compressor efficiency of 85% and 9°F

(5°C) of superheat at the compressor inlet. You can check your solution with the online HCB software. *(2)*

14.8 An ideal chiller uses refrigerant 134a and operates between a low-side pressure of 36 psia (0.25 MPa) and a high-side pressure of 200 psia (1.4 MPa). The chiller capacity is 500 tons (1750 kW). Find the refrigerant flow rate, power input, and COP of this machine, as well as the evaporator and condensing temperatures. *(2)*

14.9 Work Problem 14.8 with a real compressor efficiency of 85% and 9°F (5°C) of superheat at the compressor inlet. You can check your solution with the online HCB software. *(2)*

14.10 An ideal R134a chiller operates with a condensing temperature of 125°F (52°C) and an evaporator temperature of 35°F (2°C). If the cooling rate is 20 tons (70 kW), what are the refrigerant flow rate, condenser pressure, evaporator pressure, and COP? Compare the COP with that of a Carnot refrigeration cycle operating between the same condensing and evaporating temperatures. Discuss the reason(s) for the difference. *(3)*

14.11 What is the power input [kW/ton (kW/kW)] required for an ideal R22 chiller operating between a condensing temperature of 130°F (54°C) and an evaporating pressure of 55 psia (380 kPa)? *(2)*

14.12 What is the power input [kW/ton (kW/kW)] required for an R22 chiller operating between a condensing temperature of 130°F (54°C) and an evaporating pressure of 55 psia (380 kPa) if the compressor efficiency is 85% and if 9°F (5°C) of superheat exists at the compressor inlet? You can use the HCB software to check your solution. *(2)*

14.13 Derive an equation for the maximum COP of an absorption refrigeration cycle with the same components used for the Carnot refrigeration cycle, namely, isothermal reservoirs (four) and reversible, isentropic processes. Ignore pump work. *(2)*

14.14 Using the approach of Example 14.3, evaluate the performance of a LiBr absorption cycle with the following temperatures: condensing temperature, 100°F (38°C); evaporating temperature, 44°F (7°C); generator temperature, 210°F (99°C); and absorber temperature, 93°F (34°C). Find the COP and cooling rate if the pump flow rate is 100 lb/min (0.76 kg/s). *(3)*

14.15 A Carnot heat pump is operated between a constant indoor temperature (heat sink) at 70°F (21°C) and a varying outdoor temperature (heat source). Ignoring the temperature difference required for heat transfer from the source and sink, use a spreadsheet to determine the seasonal COP of this ideal device in Nashville, Tennessee, on a building with a heat loss coefficient of 500 Btu/(h·°F) (265 W/K) if the building balance-point temperature is 64°F (18°C). The capacity of this ideal heat pump is 16,000 Btu/h (4.7 kW) at 32°F (0°C). Whenever it is on, the heat pump power input is at its 32°F level. The bin data can be obtained from the online HCB software. *(4)*

14.16 Work Problem 14.15 with an ideal Carnot heat pump, but assume that a 10°F (5.5°C) temperature difference is required for heat transfer at both the heat source and the sink. *(4)*

14.17 Plot the COPs of two chillers whose part-load characteristics are given in the online HCB software chiller part-load ratio table: the 925-ton (0.625 kW/ton) unit by manufacturer A and the 900-ton (0.624 kW/ton) unit by manufacturer B. Plot the COP as a function of PLR in the interval (0.10, 1.0). Although the full-load COP of each is almost the same, the part-load performance is not. Comment on the part-load characteristics and which chiller is likely to use less energy during a cooling season. *(3)*

14.18 Plot the COP of an absorption chiller whose part-load characteristic is given by Equation 14.33 with $a = 0.11$, $b = 0.36$, and $c = 0.53$ with a full-load COP = 1.2. Plot the COP as a function of PLR in the interval (0.10, 1.0). Comment on your plot relative to operation at part load. *(2)*

14.19 What is the power input to the chiller in Example 14.9 for a part-load ratio of 0.64 if the peak chiller capacity is 50 tons (175 kW)? *(2)*

14.20 A commercial building has a peak cooling load of 875 tons (3080 kW) at 95°F (35°C) and zero cooling load at 55°F (13°C). Assume that the cooling load is a linear function of the outdoor dry-bulb temperature. A single, large centrifugal chiller with a capacity of 925 tons (3250 kW) (5% safety factor) is to be evaluated. It is the unit made by manufacturer A (0.692 kW/ton) in the online HCB software of the chiller PLR table. What are the annual energy consumption and the annually averaged COP for a building in Chicago equipped with this chiller? *(3)*

14.21 The building in Problem 14.20 can be cooled either by the specified chiller or by three chillers, each capable of meeting one-third of the load (use three 300-ton chillers by manufacturer A). Since the large chiller has more capacity than needed most of the time, it may use more energy than the small chillers operating more closely to their full capacity. Rework Problem 14.20 using the three-chiller strategy. For simplicity, assume that the lead chiller (the first one to be operated) operates up to full capacity before the second is activated; the same strategy is used to operate the third chiller relative to the second. *(4)*

14.22 A commercial building has a peak cooling load of 740 tons (3080 kW) at 100°F (38°C) and zero cooling load at 50°F (10°C). Assume that the cooling load is a linear function of the outdoor dry-bulb temperature. A single, large chiller with a capacity of 925 tons (3250 kW) (25% safety factor) is to be evaluated. It is the unit made by manufacturer A (0.692 kW/ton) in the online HCB software chiller PLR table. What are the annual energy consumption and the annually averaged COP in a building in Albuquerque, New Mexico, equipped with this chiller? Study the effects of the 25% oversizing of this chiller on annual energy consumption by using a chiller exactly matching the load [740 tons (3080 kW)] but having the same part-load characteristics as the 925-ton (3250 kW) chiller. *(4)*

14.23 Work Problem 14.20 with a 925tons (3250 kW) absorption chiller described in Problem 14.18 instead of with a vapor compression chiller. *(3)*

14.24 The building in Problem 14.23 can be cooled either by the single specified absorption chiller or by two absorption chillers—one capable of meeting one-third of the load and the other capable of meeting two-thirds. Rework Problem 14.23, using the two-chiller strategy. For simplicity, assume that the lead chiller (the first one to be operated) operates up to full capacity before the second is activated. Depending on the load in a particular bin, either chiller may be the lead chiller. (In actual practice, one would use a more sophisticated control approach to further minimize energy consumption.) *(4)*

14.25 Figure 14.19 shows the capacity of several residential heat pumps. If the heat loss coefficient of a duplex residence is 1700 Btu/(h·°F) (900 W/K) with a heating balance-point temperature of 66°F, (a) What are the heat pump balance points for each of the four heat pumps shown? (b) If the design outdoor air temperature is 0 °F, what capacity auxiliary heaters should be installed for each heat pump? *(2)*

14.26 Figure 14.19 shows the capacity of several residential heat pumps. Each can meet the heating requirements of a building with a heat pump balance point in the range between 20°F and 40°F (–7°C and 5°C). Describe how you would select one unit from the four shown to minimize the homeowner's cost, including both the first cost for the heat pump (increases with size) and the ongoing operating costs for electric power and auxiliary resistance heat (this last cost decreases with increasing heat pump size). *(3)*

14.27 (a) Table P14.27 lists the capacities (including fans and defrost effects) of several sizes of heat pump. If a residence has a heat loss coefficient of 700 Btu/(h·°F) (365 W/K) and a balance-point temperature of 68°F (20°C), what will be the SPF of the model C heat pump in Boise, Idaho? What is the overall heating system SPF, including resistance heat? *(2)*

(b) Repeat (a) with the next-larger- and next-smaller-capacity (at 32°F) heat pumps given in the table. Comment on the results and how this influences the selection of a heat pump. *(2)*

TABLE P14.27

Performance of different heat pump models[a]

		Outdoor Dry-Bulb Temperature																		
Model		**72°**	**67°**	**62°**	**57°**	**52°**	**47°**	**42°**	**37°**	**32°**	**27°**	**22°**	**17°**	**12°**	**7°**	**2°**	**−3°**	**−8°**	**−13°**	**−18°**
A	kBtu/h	27.6	27.6	27.1	26.4	25.5	24.3	22.4	20.4	18.3	16.4	14.6	13.0	11.7	10.6	9.5	8.6	7.7	7.1	6.6
	kW	3.7	3.6	3.6	3.5	3.4	3.3	3.2	3.1	3.0	2.9	2.8	2.7	2.6	2.5	2.4	2.3	2.2	2.1	2.0
B	kBtu/h	35.6	35.4	35.0	34.1	32.9	31.4	29.2	27.0	24.6	22.5	20.6	18.5	16.6	14.8	13.2	11.7	10.3	9.2	8.2
	kW	4.3	4.3	4.2	4.1	3.9	3.8	3.7	3.6	3.5	3.4	3.3	3.2	3.1	3.0	2.9	2.8	2.7	2.5	2.4
C	kBtu/h	34.4	34.2	33.8	33.0	31.8	30.3	28.2	25.9	23.7	21.5	19.6	17.6	15.8	14.0	12.6	11.1	9.8	8.9	7.9
	kW	4.2	4.1	4.0	3.9	3.8	3.7	3.6	3.5	3.4	3.3	3.2	3.1	3.0	2.9	2.8	2.7	2.5	2.4	2.3
D	kBtu/h	45.2	45.0	44.5	43.4	41.8	39.9	37.0	33.8	30.7	27.8	25.1	22.4	19.9	17.6	15.4	13.5	11.7	10.3	9.0
	kW	5.1	5.0	4.9	4.7	4.6	4.4	4.3	4.1	4.0	3.9	3.7	3.6	3.4	3.3	3.2	3.0	2.9	2.7	2.6
E	kBtu/h	44.1	43.9	43.3	42.3	40.7	38.8	36.0	32.8	29.7	26.8	24.1	21.5	19.1	16.9	15.0	13.1	11.5	10.2	9.0
	kW	5.1	5.0	4.9	4.7	4.6	4.4	4.3	4.1	4.0	3.9	3.7	3.6	3.4	3.3	3.2	3.0	2.9	2.7	2.6
F	kBtu/h	50.4	50.2	49.6	48.4	46.6	44.4	41.3	37.9	34.5	31.4	28.5	25.6	22.9	20.5	18.3	16.2	14.4	13.0	11.7
	kW	6.2	6.0	5.9	5.7	5.6	5.4	5.2	5.1	4.9	4.7	4.6	4.4	4.3	4.1	4.0	3.8	3.6	3.5	3.3

[a] kBtu/h are the heating capacities and kW is the compressor power.

TABLE P14.28

Performance of a heat pump for different indoor and outdoor temperatures

	Heating Capacity, kBtu/h, at Indicated Indoor Dry-Bulb Temp.				Compressor Power, kW, at Indicated Indoor Dry-Bulb Temp.			
Outdoor Temp.	**60**	**70**	**75**	**80**	**60**	**70**	**75**	**80**
−20	10.4	9.3	8.8	8.2	1.8	1.9	1.9	2.0
−10	11.9	10.5	9.8	9.1	2.1	2.2	2.2	2.3
0	15.0	13.4	12.7	11.9	2.3	2.5	2.5	2.6
10	19.4	17.7	16.9	16.0	2.6	2.7	2.8	2.9
20	24.8	23.0	22.1	21.2	2.9	3.0	3.1	3.2
30	30.7	28.8	27.9	27.0	3.1	3.3	3.3	3.4
40	36.9	35.0	34.0	33.0	3.3	3.5	3.6	3.7
45	39.6	38.0	37.2	36.4	3.4	3.6	3.7	3.8
50	41.6	40.0	39.2	38.3	3.5	3.7	3.8	3.9
60	44.6	43.0	42.1	41.3	3.7	3.9	4.0	4.1
70	45.9	44.1	43.3	42.4	3.8	4.1	4.2	4.3
80	44.8	43.0	42.1	41.3	4.0	4.2	4.3	4.4

Correction factors for other airflows (value at 1200 ft^3/min times correction factor = value at new airflow)

Airflow	1050	1200	1350
Heating capacity	0.980	1.000	1.020
Compressor, kW	1.025	1.000	0.975

14.28 Table P14.28 shows how the *indoor* temperature of a building affects the performance of a heat pump. For the building described in Problem 14.27, find the annual electric energy use (including resistance heat) at two indoor temperatures—70°F and 75°F (21°C and 24°C)—in Denver, CO. Use the online HCB software if you prefer. You will need to make several key assumptions. State them all. *(3)*

14.29 Define a building in the online HCB software such that it has a heat loss coefficient of 800 Btu/(h·°F) (420 W/K) with a balance temperature of approximately 64.5°F (18.1°C) if the thermostat setting is 68°F (20°C). Consider heat pump models D and F in Table P14.27. What is the balance point for each heat pump applied to this building? *(2)*

14.30 Define a building in the online HCB software such that it has a heat loss coefficient of K_{tot} = 800 Btu/(h·°F) (422 W/K); with a specified thermostat setting of 68°F (20°C) it is to have a balance temperature of approximately 64.5°F (18°C). Consider heat pump models D and F in Table P14.27. If this building is located in Boston, find the total electric consumption and the overall heating system SPF using the online HCB software. On the basis of these calculations, which heat pump would you recommend based on energy consumption? (Of course, the final selection must include the effect of the costs of the two heat pumps.) *(3)*

14.31 If the building conductance in Problem 14.30 were reduced 18% by improving the thermal integrity of the envelope, what would the cost savings in electricity be if the model D heat pump were used? The cost of electricity is 10¢/kWh including demand charges. *(2)*

14.32 Repeat Problem 14.27(a) for a *ground source heat pump* for which the heat source temperature is constant year-round at 48°F (9°C). Note that this problem is considerably simpler than Problem 14.27 because the COP is constant. *(2)*

References

AHAM (1992). *ANSI/AHAM Standard RA C-1-1992, Room Air Conditioners*. Association of Home Appliance Manufacturers, Washington, DC.

AHRI (2004). *ANSI/AHRI Standard 310/380: Standard for Packaged Terminal Air-Conditioning and Heat Pumps*. Air Conditioning, Heating and Refrigeration Institute, Arlington, VA.

AHRI (2008). *ANSI/AHRI Standard 210/240: Performance Rating of Unitary Air-Conditioning and Air-Source Heat Pump Equipment*. Air Conditioning, Heating and Refrigeration Institute, Arlington, VA.

AHRI (2011). *AHRI Standard 550/590: Standard for Performance Rating of Water Chilling and Heat Pump Water-Heating Packages using the Vapor Compression Cycle*. Air Conditioning, Heating and Refrigeration Institute, Arlington, VA.

ANSI/ASHRAE (2013). *ASHRAE Standard 34-92, Number Designation and Safety Classification of Refrigerants.* American Society of Heating, Refrigerating and Air-Conditioning Engineers, Atlanta, GA.

ASHRAE (1999). *ANSI/ASHRAE Standard 58-1999. Method of Testing for Rating Room Air Conditioners and Packaged Terminal Air Conditioner Heating Capacity.* American Society of Heating, Refrigerating and Air-Conditioning Engineers, Atlanta, GA.

ASHRAE (2012). ASHRAE position document on refrigerants and their responsible use. American Society of Heating, Refrigerating and Air-Conditioning Engineers, Atlanta, GA.

ASHRAE Fundamentals (1989/2013). *Handbook of Fundamentals.* American Society of Heating, Refrigerating and Air-Conditioning Engineers, Atlanta, GA.

ASHRAE Systems (2012). *Handbook of Systems and Equipment.* American Society of Heating, Refrigerating and Air-Conditioning Engineers, Atlanta, GA.

Austin, S.B. (1991). Optimum chiller loading. *ASHRAE J.*, 33, 40–43. American Society of Heating, Refrigerating and Air-Conditioning Engineers, Atlanta, GA.

Calm, J.M. and G.C. Hourahan (2001). Refrigerant data summary. *Eng. Syst.* 18(11), 74–88, November.

Domingorena, A.A. and S.J. Ball (1980). Performance evaluation of a selected three-ton air-to-air heat pump in the heating mode. Report ORNL/CON-34. Oak Ridge National Laboratory, Oak Ridge, TN.

Geister, W.R. and M. Thompson (2009). A closer look at chiller ratings. *ASHRAE J.*, 51(12), 22–32, December.

Gordon, J.M. and K.C. Ng (2000). *Cool Thermodynamics.* Cambridge International Science Publishing, Cambridge, U.K.

Kavanaugh, S.P. and K. Rafferty (1997). *Ground-Source Heat Pumps.* American Society of Heating, Refrigerating and Air-Conditioning Engineers, Atlanta, GA.

Kuehn, T.H., J.L. Threlkeld, and J.W. Ramsey (1998). *Thermal Environmental Engineering,* 3rd ed. Prentice-Hall, Englewood Cliffs, NJ.

LBL (1982). *DOE-2 Engineers Manual.* Lawrence Berkeley Laboratory Report, Berkeley, CA.

Mitchell J.W. and J.E. Braun (2013). *Principles of Heating, Ventilation and Air-Conditioning in Buildings.* John Wiley & Sons, New York.

Reddy, T.A. (2011). *Applied Data Analysis and Modeling for Energy Engineers and Scientists.* Springer, New York.

Stoecker, W.F. and J.W. Jones (1982). *Refrigeration and Air Conditioning.* McGraw-Hill, New York.

15

Combustion Heating Equipment and Systems

ABSTRACT This chapter describes fossil fuel–fired furnaces and boilers—devices that convert the chemical energy in fuels to heat or produce thermal energy for meeting building heat loads. We start by reviewing the basic principle of stoichiometric combustion and the need for excess air intake and then describing different types of fuels and their properties. Next, notions pertinent to combustion efficiency and operational efficiency as well as the impact of part-load operation are discussed. We then describe different types of furnaces and boilers along with their design, selection, and operational and safety controls. How the efficiency of the boiler can be tracked by monitoring the flue gas characteristics is also addressed. We end by providing a broad overview of combined heat and power (CHP) systems that are a more efficient and proven alternative to simultaneously meet electricity, heating, and cooling loads especially in large buildings and campus type of environments.

Nomenclature

A, B, C	Model coefficients appearing in Equation 15.7
AFUE	Annual fuel utilization efficiency
c_p	Specific heat at constant pressure, kJ/(kg·K) [Btu/(lb$_m$·°F)]
E	Electric power, W (ft·lb$_f$/s or HP)
EUF	Energy utilization factor, (Equation 15.19)
HHV	Higher heating value, MJ/kg or MJ/m^3 (Btu/lb$_m$ or Btu/gal or Btu/ft^3)
EUF$_{VW}$	Value-weighted EUF, (Equation 15.20)
FESR	Fuel energy savings ratio, (Equation 15.23)
f	Weight factor between heat and work, $0.33 < f < 0.5$
$f_{excess\ air}$	Excess air fraction, (Equation 15.2)
h	Specific enthalpy, kJ/kg (Btu/lb$_m$)
LHV	Lower heating value, MJ/kg or MJ/m^3 (Btu/lb$_m$ or Btu/gal or Btu/ft^3)
$\dot{m}$	Mass flow rate, kg/s (lb$_m$/h)
N	Number of temperature bins
n	Number of hours, h
P	Motive power, W
p	Pressure, Pa (lb$_f$/in.^{2})
PLR	Part-load ratio, (Equation 15.6)
Q	Energy consumption or heat transfer, J (Btu)
$\dot{Q}$	Heat load or flow with subscripts c for cooling load and h for heating load W (Btu/h)
T	Temperature, K or °C (°R or °F)
t	Time, s

Greek

η	Efficiency
λ	Thermal power ratio
Δx	Change in quantity x

Subscripts

boil	Boiler
blowdown	Boiler blowdown
CHP	Combined heat and power
comb	Combustion
elec	Electricity
full	Full load
H	Heating equipment
h	High
HR	Heat recovery
G	Electric generator
gen	Generated or produced
in	Input
j	Suffix for temperature bin
l	Low
leak	Leakage, loss
m	Middle, mid
out	Output
P	Prime mover
r	Rejected
SHP	Separate heat and power
ss	Steady state
u	Useful
yr	Annual

15.1 Principles of Combustion

15.1.1 Introduction

The primary sources of heat for building heating systems are fossil fuels, natural gas, and various grades of fuel oil and coal. Electricity is used under certain circumstances for heat in commercial buildings (such as electric strip heating or from heat pumps discussed in Chapter 14), although the economic penalties for so doing are significant. Solar power can be converted to heat for applications in commercial buildings including perimeter zone heating and service water heating. The direct use of sunlight for lighting is described in Chapter 22.

Fossil fuel–fired *furnaces and boilers*—devices that convert the chemical energy in fuels to heat—have been introduced in Chapter 11. This chapter will discuss them in more detail. Furnaces heat airstreams that are used in turn for heating the interior of buildings. Forced-air heating systems supplied with heat by furnaces are the most common type of residential heating system in the United States. Boilers are pressure vessels used to transfer heat, produced by burning a fuel, to a fluid, usually water (see Figures 11.1 and 11.23). The water may be converted into steam or simply heated up sensibly to a higher temperature. The key distinction between furnaces and boilers is that air is heated in the former and water is heated in the latter. It is beyond the scope of this book to describe in detail the design of boilers and furnaces or how they convert chemical energy to heat. Rather we provide the basic information needed by HVAC designers for these two classes of equipment. Since boilers and furnaces operate at elevated temperatures (and pressures for boilers), they are hazardous devices. As a result, a body of standards has been developed to ensure the safe operation of this equipment (refer to ASME standards on boilers and pressure vessels). We briefly describe this equipment-related aspect as well.

15.1.2 Types of Fuels and Their Properties

The fuels used for producing heat in boilers and furnaces can be classified as either gaseous, liquid, or solid (such as coal). Regardless of the state, the fuel must be in a gaseous or vapor form for the combustion to occur and should also be in intimate contact with the oxygen from the air. This requires that liquid fuels be atomized and solid fuels be finely ground prior to being fed into the combustion chamber.

Most fuels used are hydrocarbon fuels that consist of carbon, hydrogen, sulfur, and other trace elements. During complete combustion, these elements get converted into CO_2, H_2O, and SO_2, respectively, while releasing heat. Natural gas is the most commonly used gaseous fuel. Its exact composition varies with its source. It is a mixture of methane (55%–98%), higher hydrocarbons (primarily ethane), and noncombustible gases. Natural gas is extensively used in buildings because of its convenience (piping directly to the boiler or furnace) and its clean-burning qualities (contains very little pollutants after it is refined). Liquid fuels are mostly extracted from crude petroleum, and the most common form is fuel oil (graded from 1 to 6 as the oil gets heavier). Fuel oil no. 1 is kerosene and used in small space heaters. Fuel oil no. 2 is the most often used in furnaces and boilers in residences and small commercial buildings. Higher-grade oils are cheaper but need special equipment for combustion. Fuel oils have to be stored in tanks and do leave a hard-to-remove residue after they are burnt. Solid fuels are wood, coal, coke, and other fuels including refuse-derived fuels; however, they find limited applications in modern buildings.

The *heating value, or heat of vaporization, or calorific value* of a fuel is the quantity of heat generated or released during the combustion process. It is a constant value for a given type of fuel composition. Combustion produces water vapor as a by-product, and whether or not we include the energy contained in this water vapor has led us to distinguish between two types of heating values: higher and lower. HHV is the total amount of energy released per unit of the fuel. Lower heating value (LLV) or net heating value is the energy released minus the latent heat of vaporization and is thus a measure of the energy available for the actual heat transfer process. HHV can be determined from the LLV if the enthalpy of vaporization is known. These values are often determined experimentally in a bomb calorimeter.

Heating values are expressed in MJ/kg (or Btu/lb_m) for solid fuels, in MJ/m^3 (or Btu/gal) for liquids, and in MJ/m^3 (or Btu/ft^3) for gaseous fuels. Depending on the specific industry, the standard reference temperature and pressure differ, usually 15°C, 20°C, or 25°C and 103.3 kPa (60°F, 68°F, and 77°F and 14.7 psia). Heating values of selected fuels are given in Table 15.1. The given values of specific volume (or density) allow the HHV or LHV expressed on a mass basis to be converted into volume basis. For example, the HHV for methane in MJ/m^3 = 55.5 × 0.679 = 37.7 MJ/m^3. For problems involving mixtures of several gases, the mass fractions of each of the constituent gases are first determined knowing the volumetric (or molar) composition and the molecular weights and then multiplied by the corresponding heating values. The following example illustrates this procedure.

TABLE 15.1

Heating Values of Some Common Gaseous Fuels

Substance	Molecular Symbol	Molecular Weight	Higher HV, Btu/lb_m	Lower HV, Btu/lb_m	Specific Volume, ft^3/lb_m[a]	Higher HV, MJ/kg	Lower HV, MJ/kg	Density, kg/m^3[a]
Carbon (to CO_2)	C	12	14,093	14,093	—	32.788	32.780	—
Hydrogen	H_2	2	61,095	51,623	188.0	142.107	118.680	0.085
Carbon monoxide	CO	28	4,347	4,347	13.5	10.111	10.111	1.187
Methane	CH_4	16	23,875	21,495	23.6	55.533	49.997	0.679
Ethane	C_2H_6	30	22,323	20,418	12.5	51.922	47.492	1.28
Propane	C_3H_8	44	21,669	19,937	8.36	50.402	46.373	1.92

All values corrected to 60°F (16°C) and sea level.

[a] At 32°F (0°C) and sea level.

Example 15.1: Determining Heating Values of Mixtures

A sample of natural gas has the following composition by volume: 95% CH_4, 4% C_2H_6, and 1% CO_2. Calculate the HHV of this fuel mixture per mass.

Find: HHV

Assumptions: HHV for CO_2 is zero, ideal gas assumption.

Lookup values: Molecular weight of CH_4 = 16, C_2H_6 = 30, and CO_2 = 44

From Table 15.1, HHV (CH_4) = 55.533 MJ/kg and HHV(C_2H_6) = 51.922 MJ/kg

Solution

For ideal gases, the volume fractions are equivalent to the mole fractions.

The molecular weight of the fuel is 0.95 × 16 + 0.04 × 30 + 0.01 × 44 = 16.84.

The mass fraction of CH_4 is $\frac{0.95\times 16}{16.84} = 0.903$ and that of C_2H_6 $\frac{0.04\times 30}{16.84} = 0.0713$.

Finally, $HHV_{mixture} = 0.903 \times 55.533 + 0.0713 \times 51.922 = 53.85$ MJ/kg.

15.1.3 Combustion Analysis

Combustion is the process by which chemically bound energy is released in an exothermic reaction of carbon and hydrogen to produce heat, carbon dioxide, and water vapor. Combustion analysis involves using the basic chemical reaction equation and the known composition of air to determine the composition of flue gases. The minimum amount of air needed to completely combust the fuel is known as the *stoichiometric* or theoretical air. In this case, there should be no free oxygen present, i.e., all the oxygen in the air supply will be attached with the combustion products. The chemical reaction for stoichiometric combustion of methane is as follows:

$$CH_4 + 2O_2 \rightarrow CO_2 + 2H_2O \quad (15.1)$$

Recalling that the molecular weights are approximately

- Hydrogen (H_2), 2
- Methane (CH_4), 16
- Oxygen (O_2), 32
- Carbon dioxide (CO_2), 44
- Water (H_2O), 18

we can easily deduce that 4.0 lb_m of oxygen per pound of methane is required for complete combustion. Recall that air is 23.15% oxygen by weight or 20.95% by volume. Thus, theoretically, we require (1/0.2315) × 4 = 17.3 lb_m (or kg) of air per lb_m (or kg). It is easy to show that on a volumetric basis (recall Avogadro's law, which states that 1 mol of any gas at the same temperature and pressure occupies the same volume), the equivalent requirements are 2.0 ft^3 of oxygen per cubic foot of methane for complete combustion. This oxygen requirement is equivalent to (1/0.2095) × 2 = 9.54 ft^3 of air per cubic foot of methane. Thus, the air–fuel ratio for complete combustion of methane is 9.54 ft^3 air/ft^3 of fuel by volume or 17.3 lb_m air/lb_m of fuel by mass. Since natural gas is a mixture of several substances, tables have been generated to simplify combustion analysis problems. Table 15.2 assembles pertinent data for the stoichiometric air–fuel ratios for some commonly used fuels along with details of the flue products.

A rule of thumb to check the preceding calculation has been proposed: 0.9 ft^3 of air is required for 100 Btu of fuel heating value (about 0.25 m^3 of air per 1 MJ of heating value). For example, the heating value of natural gas is about 1000 Btu/ft^3, requiring 9 ft^3 of air according to the above rule. This compares well with the value of 9.54 ft^3 previously calculated.

Example 15.2: Stoichiometric Combustion of a Gaseous Fuel

Determine the stoichiometric air–fuel ratio by volume for combusting natural gas with the following composition by volume: 85% methane

TABLE 15.2

Stoichiometric Air–Fuel Ratio and Flue Products for Some Common Gaseous Fuels

		By Volume: m^3/m^3 of Fuel (ft^3/ft^3)						By Mass: kg/kg of Fuel (lb/lb)					
		Combustion			Flue Products			Combustion			Flue Products		
No.	Substance	O_2	N_2	Air	CO_2	H_2	N_2	O_2	N_2	Air	CO_2	H_2	N_2
1	Carbon	31.0	3.76	4.76	1.0	—	3.76	2.66	8.86	11.53	3.66	—	8.86
2	Hydrogen	0.5	1.88	2.38	—	1.0	1.88	7.94	26.41	34.34	—	8.94	26.41
3	Carbon monoxide	0.5	1.88	2.38	1.0	—	1.88	0.57	1.90	2.47	1.57	—	1.90
4	Methane	2.0	7.53	9.53	1.0	2.0	7.53	3.99	13.28	17.27	2.74	2.25	13.28
5	Ethane	3.5	13.18	16.68	2.0	3.0	13.18	3.73	12.39	16.12	2.93	1.80	12.39
6	Propane	5.0	18.82	23.82	3.0	4.0	18.82	3.63	12.07	15.70	2.99	1.63	12.07

(CH_4), 6.2% ethane (C_2H_6), 6.0% hydrogen (H_2), 0.4% carbon monoxide (CO), and 2.4% nitrogen (N_2).

Find: Air–fuel ratio

Assumptions: Complete combustion, combustion gases being ideal gases

Lookup values: From Table 15.2

Solution

The air required for stoichiometric combustion is determined by simple weighting:

Volume of air = 0.85 × 9.53 + 0.062 × 16.68 + 0.06 × 2.38 + 0.004 × 2.38 + 0.0 × 0.024 = 9.29 m^3/m^3

Example 15.3: Finding Precombustion Composition of a Fuel

The inverse problem entails finding the precombustion composition of the fuel knowing the composition of the flue gas. The flue gas composition of a hydrocarbon fuel on a volume basis was found to be 14.1% CO_2, 3.1% O_2, 0.3% CO, 0.4% H_2, and 82.1% N_2. Balance the chemical equation and determine the air–fuel ratio by volume.

Find: The air–fuel ratio

Assumptions: Complete combustion, combustion gases being ideal gases

Lookup values: From Table 15.2

Solution

We introduce unknown coefficients x_1, x_2, and y to setup a preliminary chemical reaction:

$$C_{x_1}H_{x_2} + y \times O_2 + 3.76 \times y \times N_2$$

$$\rightarrow 14.1CO_2 + 3.1O_2 + 0.3CO + 0.4H_2 + 82.1N_2 + z \times H_2O$$

(Note that the volume ratio of nitrogen to oxygen in atmospheric air is (79%/21%) = 3.76.)

These coefficients can be determined from mole balances of different elements:

- Carbon balance: $x_1 = 14.1 + 0.3 = 14.4$.
- N_2 balance: $y = 82.1/3.76 = 21.835$.
- O_2 balance: $y = 21.835 = 14.1 + 3.1 + 0.3/2 + z/2$. Thus, $z = 8.97$.
- H_2 balance: $x_2 = 2 \times 0.4 + 2 \times z$. Thus, $x_2 = 18.74$.

The chemical reaction can now be balanced:

$$C_{14.4}H_{18.74} + 21.835O_2 + 82.1N_2$$

$$\rightarrow 14.1CO_2 + 3.1O_2 + 0.3CO + 0.4H_2 + 82.1N_2 + 8.97H_2O$$

The air–fuel ratio can then be determined:

Air–fuel ratio by volume = (21.835 mol O_2 + 82.1 mol of N_2)/mole of fuel = 103.9 m^3 of air per m^3 of fuel

15.1.4 Excess Air

Combustion air is often provided in excess of the theoretical minimum or stoichiometric amount to guarantee complete combustion. Incomplete combustion poses safety concerns since it yields highly toxic carbon monoxide (CO) in the flue gas. This is because oxygen is more strongly attracted to hydrogen than it is to carbon. Incomplete combustion is to be avoided not only as energy waste but also as air pollution. The amount of excess air involved in combustion is usually expressed as the *excess air fraction* $f_{excess\ air}$:

$$f_{excess\ air} = \frac{\text{Air supplied} - \text{Stoichiometric air}}{\text{Stoichiometric air}} \qquad (15.2)$$

In combustion calculations for gaseous fuels, the air amounts in Equation 15.2 are usually expressed on a volumetric basis, whereas for all other fuels, a mass basis is used.

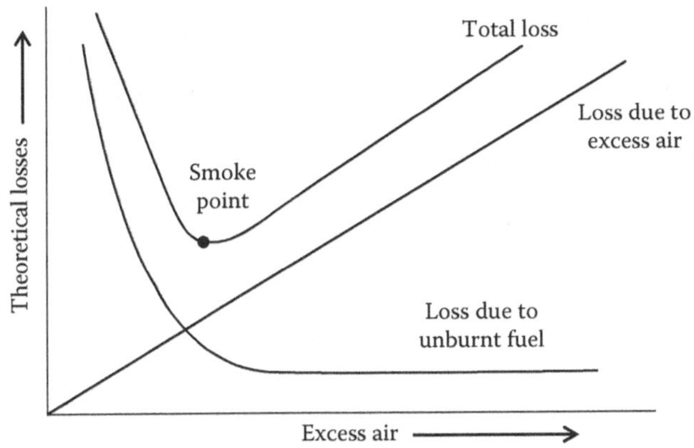

FIGURE 15.1
Variation of boiler flue gas losses with excess airflow.

The amount of excess air provided is critical to the efficiency of a combustion process. Excessive air both reduces combustion temperature (reducing the heat transfer rate to the working fluid) and results in excessive heat loss through the flue gases (Figure 15.1). Insufficient excess air results in incomplete combustion and loss of chemical energy in the flue gases. The amount of excess air provided varies with the fuel and with the design of the boiler (or furnace). Recommendations of the manufacturer should be followed. The optimum excess air fraction is usually between 10% and 50%. Unfortunately, it is common practice to provide 50%–100% excess air that significantly reduces boiler efficiency (Mull, 2001). How the excess air fraction can be used as a way of monitoring the combustion efficiency of boilers and furnaces while in operation is discussed in Section 15.5.

15.2 Furnaces

15.2.1 Classification and Description

Modern furnaces directly provide heat to the building supply airstream. This is done by combusting the fuel supplied by a gas-supply manifold into a combuster or firebox, drawing in ambient air, mixing the two streams, and igniting them. The heat in the combustion products is then transferred by forced convection to the circulating airstream via a heat exchanger. This airstream is then supplied to the space (see Figure 11.6). The exhaust gases are cooled during the heat exchange process but remain hot and buoyant enough to exit the flue pipe. The primary components are the heat exchanger, fuel burner, air blower, controls, and the insulated cabinet or firebox with the vent pipe. There are numerous designs to achieve this; four residential classifications based on airflow type are discussed here.

The *upflow* furnace shown previously in Figure 11.7 has a blower located below the firebox heat exchanger with heated air exiting the unit at the top. Return air from the heated space enters this furnace type at the bottom. The upflow design is used in full-size mechanical rooms where sufficient floor-to-ceiling space exists for the connecting ductwork. This is the most common form of residential furnace.

Downflow furnaces (Figure 15.2a) work in reverse: air flows downward as it is heated by passing over the heat exchanger. This design is used in residences without basements or in upstairs mechanical spaces in two-story buildings. *Horizontal* furnaces of the type shown in Figure 15.2b use a horizontal airflow path with the air mover located beside the heat exchanger. This design is especially useful in applications where vertical space is limited, e.g., in attics or crawl spaces of residences.

A combination of upflow and horizontal furnaces is available and is named the *basement* (lowboy) furnace (Figure 15.2c). With the blower located beside the firebox, air enters the top of the furnace, is heated, and exits from the top. This design is useful in applications where headroom is restricted.

The combustion side of the heat exchanger in gas furnaces can be at either atmospheric pressure (the most common design for small furnaces) or superatmospheric pressures produced by combustion air blowers. The latter are of two kinds: forced-draft (blower upstream of combustion chamber) or induced-draft (blower downstream of combustion chamber) furnaces that have better control of parasitic heat losses through the stack. As a result, efficiencies are higher for such *power combustion furnaces*.

In addition to natural gas, liquefied propane gas (LPG) and fuel oil can be used as energy sources for furnaces. LPG furnaces are very similar to natural gas furnaces. The only differences between the two arise due to the difference in energy content (1000 Btu/ft^3 for natural gas and 2500 Btu/ft^3 for propane) and supply pressure to the burner. Gas furnaces can be adapted for LPG use, and vice versa in many cases. Fuel oil burner systems differ from gas burner systems owing to the need to atomize oil before combustion. The remainder of the furnace is not much different from a gas furnace except that heavier construction is often used.

Other furnaces for special applications are also available. These include (1) unducted space heaters located within the space to be heated and relying on natural convection for heat transfer to the space, (2) wall furnaces attached to walls and requiring very little space, and (3) direct-fired unit heaters used for direct space heating in commercial and industrial applications. Unit heaters are available in sizes from 25,000 to 320,000 Btu/h (7 to 94 kW), and these are classified as light commercial. Output capacities of large commercial furnaces may range from 150,000 Btu/h (44 kW) to over 2 million Btu/h (600 kW). Such equipment is constructed with material with more structural strength and also includes more sophisticated controls.

On commercial buildings, one often finds furnaces incorporated into *package units* (or "rooftop units") consisting of air conditioners and gas furnaces (or electric resistance coils)—see Figure 11.9. Typical sizes of these units range from 5 to 50 tons of cooling (18 to 175 kW) with a matched capacity to 50% oversized furnace. Smaller units are designed to be used for a single zone in either the heating-only or the cooling-only mode. Larger units above 15 tons (53 kW) can operate simultaneously in heating and cooling modes to condition several zones. In the heating mode, these commercial-size package units operate with an air temperature rise of about 85°F (47°C).

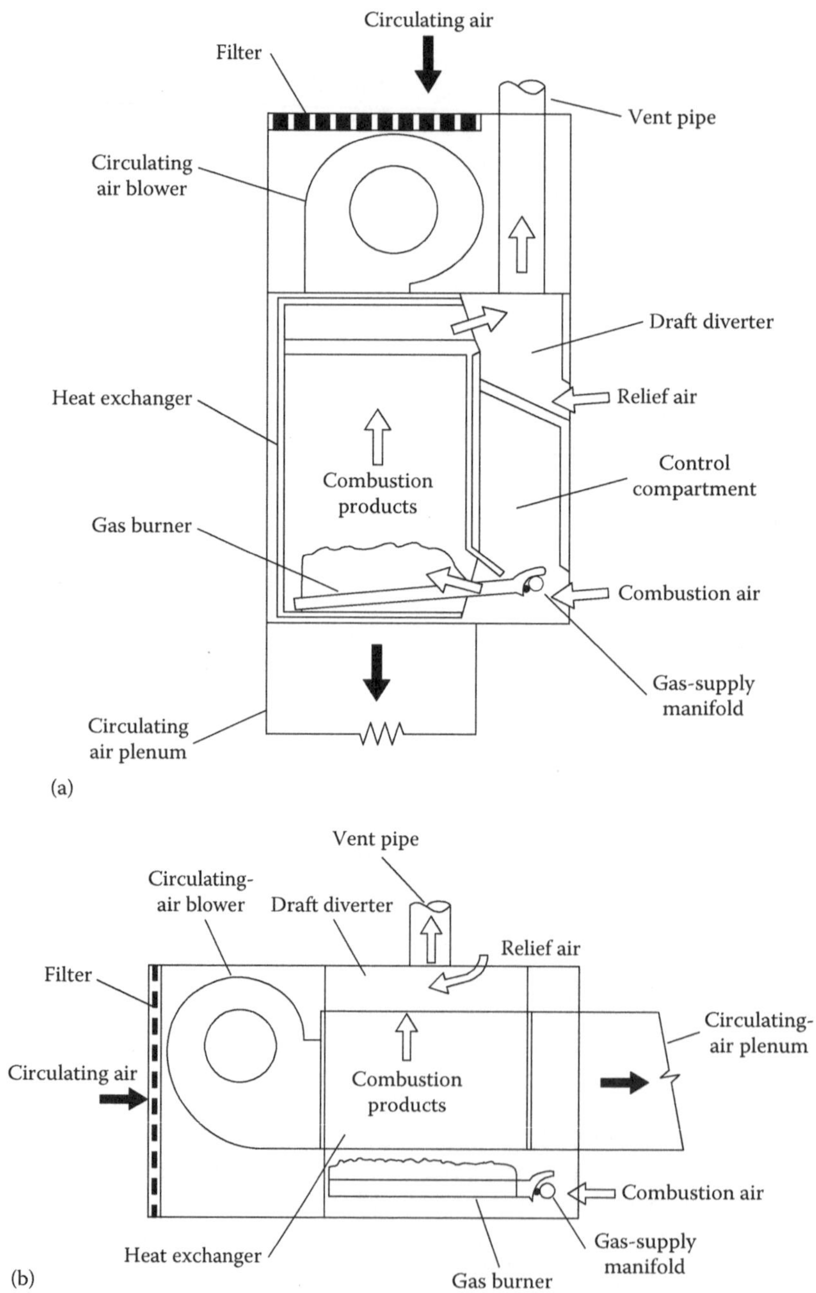

FIGURE 15.2
Examples of furnaces for residential space heating: (a) downflow and (b) horizontal. *(Continued)*

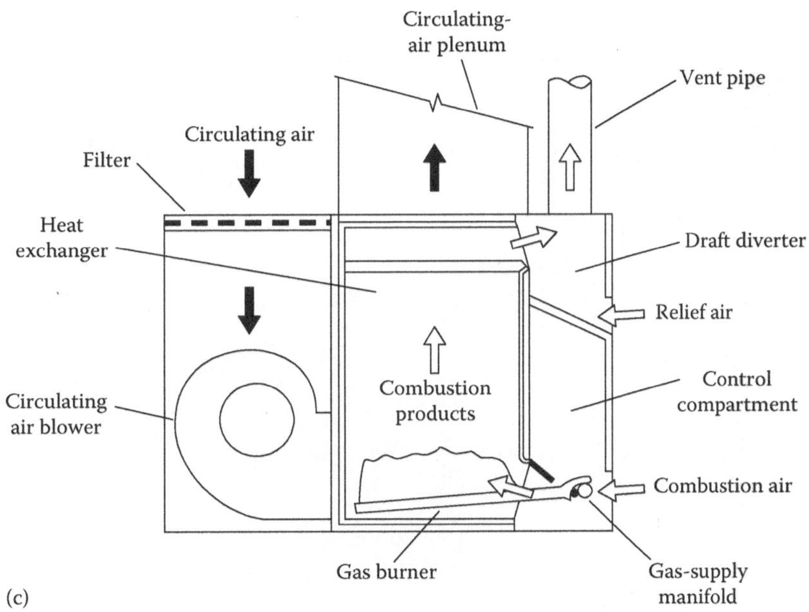

FIGURE 15.2 ***(Continued)***
Examples of furnaces for residential space heating: (c) lowboy.

15.2.2 Furnace Design and Selection

Selection of a furnace is straightforward once the fuel source and design heat loads of the space are determined. The following factors must be accounted for in furnace sizing and type selection:

- *Design heat loss* of area to be heated, in Btu per hour or kilowatts
- Morning *recovery capacity from night setback*
- Constant *internal gains or waste heat recovery* that reduces the needed heat rating of a furnace
- *Humidification load*
- *Fan and housing size* sufficient to accommodate air conditioning system, if any
- *Duct heat losses* if heat so lost is external to the heated space
- *Available space* for furnace location

Residential furnaces are available in sizes ranging from 35,000 to 175,000 Btu/h (10 to 51 kW). Commercial sizes range up to 1,000,000 Btu/h (300 kW).

15.2.3 Furnace Controls

In essence, there are two types of controls (Pita, 2002).

1. *Safety controls* that shut down the furnace if certain safety limits are exceeded. For example, if the combustion temperature gets dangerously high, such as 200°F (93°C) (which could happen if the building supply air were to stop due to blocked airflow caused by dirty filters or due supply fan malfunction), the gas supply would be shut off. Another type of safety control involves switching off the gas supply if the flame were to die out unexpectedly.
2. *Operating controls* that assure proper functioning of the unit. These regulate the burner as the heating demand changes and also include a sequence of steps for proper start-up or shutdown of the unit. For example, a typical start-up sequence would be as follows: (a) the space thermostat closes a circuit, (b) the air circulating fan starts, (c) the pressure switch checks whether the air circulation is adequate, (d) the combustion fan starts, (e) that this fan is operating properly is checked, (f) the timed purge cycle (typically 30–60 s) exhausts combustion gases that may remain in the furnace from the previous cycle, (g) the main gas valve opens, (h) the spark ignition circuit is activated, and (i) safety controls check for proper operation. Manual (as against automatic) controls to either start up or shut off the unit are also incorporated.

Figure 15.3 is a sketch of a residential gas-fired boiler showing the various components as well as the location of the controls, flame sensor, and pressure switches. Gas burners can be of two types depending on how the air and the fuel are fed into the combustion chamber. Atmospheric gas burners (shown in Figure 15.4) include

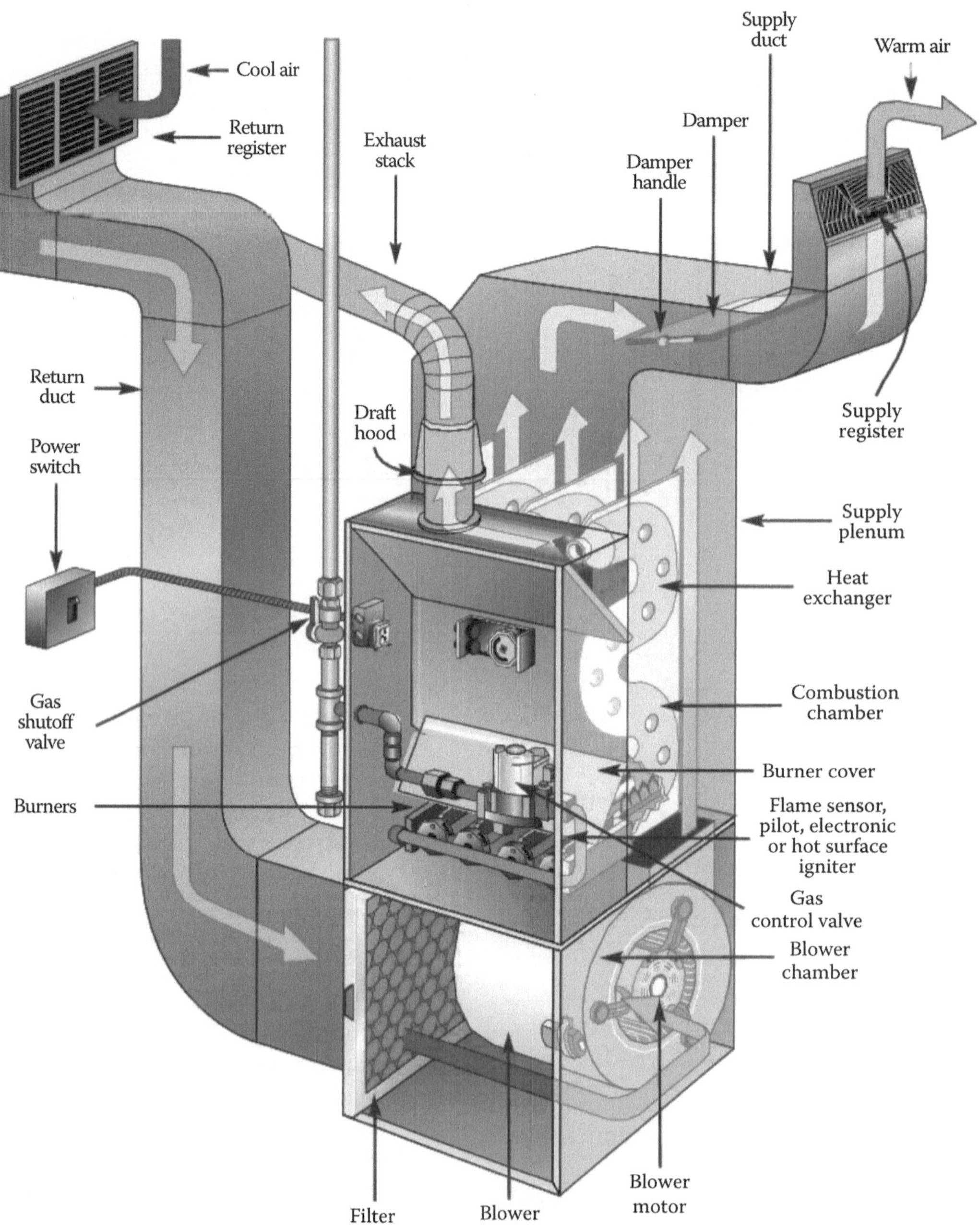

FIGURE 15.3
Sketch of a gas-fired furnace showing various components. (Downloaded from www.mybutterfly.com, accesed on June 2014.)

a venturi tube that creates a depressurization and draws primary air that mixes with the gas supply. As the load of the space varies, the gas supply is modulated, which in turn varies the primary air supplied. The remainder of the air is drawn by natural draft as secondary airflow. The other type of gas burner is the power gas burner that uses fans to deliver the air and create turbulence to enhance air and gas mixing.

There are essentially three approaches to achieve fuel ignition. The *standing pilot method* involves a continuously burning small gas flame that ignites the primary fuel supply as needed. This was historically the most common method; however, because it is wasteful, this design has been mostly banned on new equipment. The *intermittent pilot ignition* involves creating a spark when there is a call for heating. The pilot flame then lights the main fuel supply. Both the pilot and the main flame are switched off when heating is no longer called for. The third and the most modern approach is *direct spark ignition*. Here, there is no pilot, and a spark ignites the fuel directly once it is fed into the combustion chamber.

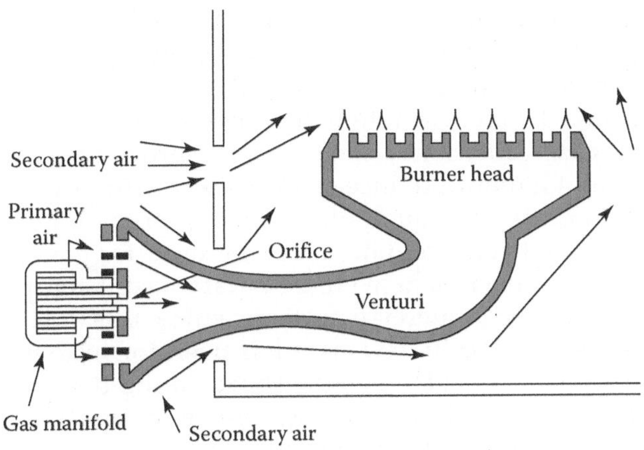

FIGURE 15.4
Cross section of an atmospheric gas burner.

Furnaces (and boilers) need to maintain proper draft, i.e., a pressure differential large enough to allow the inflowing air and gas to overcome friction in the combustion chamber and the vent pipe. This draft can be created naturally by the buoyancy of the hot flue gases in residential furnaces or by a small fan for larger furnaces. In residential furnaces, the required draft is typically about 0.002 inWG (0.5 Pa). Draft should be kept constant for a given fuel firing rate. Too low a draft will result in incomplete combustion, too high in thermal inefficiency. Changes in outdoor conditions, such as outdoor temperature and wind, have an effect on this draft and so some means of control is required. This can be achieved quite simply by using a draft hood as shown in Figure 15.5a. A momentary increase in flue pressure will increase the air intake from the room through the hood and not affect the air drawn in to the combustion chamber. A downdraft can be canceled by the draft hood diverting the air into the space rather than into the combustion chamber. For oil-fired equipment and some power gas-fired units, a damper is used to regulate the draft (Figure 15.5b). Closing or opening the damper can counter momentary fluctuations in the draft pressure.

15.2.4 Furnace Efficiency and Energy Calculations

Efficiency of a device on the basis of the first law of thermodynamics is defined as the useful energy delivered by the energy input. For boilers and furnaces, three different efficiency indices are usually pertinent: (1) combustion efficiency, (2) overall steady-state efficiency, and (3) seasonal or annual efficiency.

The *combustion efficiency* η_{comb} is a measure of how completely the burner is able to burn the fuel. It is defined as the energy contained in the fuel supplied less that carried away by the flue losses, all divided by the former:

$$\eta_{comb} = \frac{\left(\dot{m}_{fuel} h_{fuel} - \dot{m}_{flue} h_{flue}\right)}{\dot{m}_{fuel} h_{fuel}} \tag{15.3}$$

in which the subscripts identify the fuel input and flue gas exhaust mass flow rates $\dot{m}$ and enthalpies h. Gas flows are usually expressed in cubic feet per hour

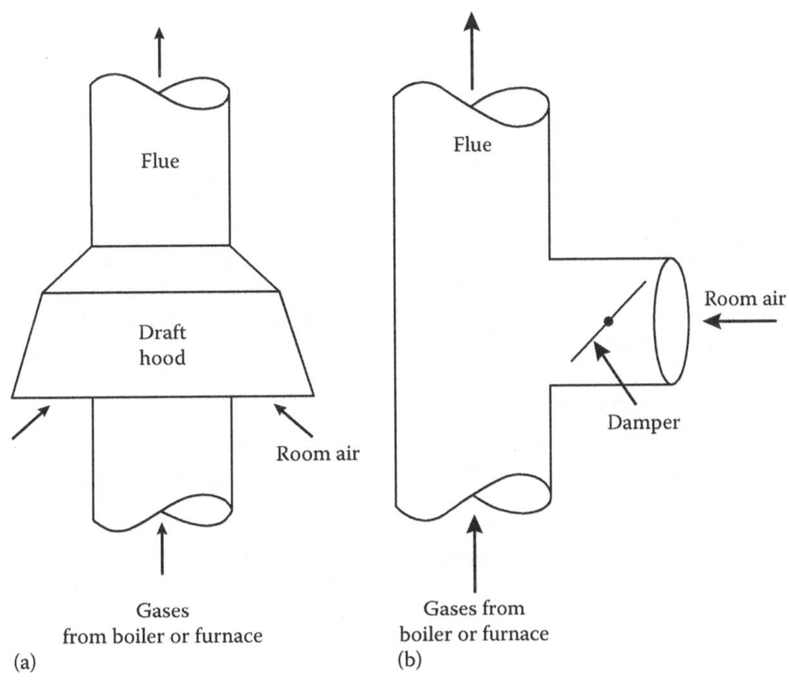

FIGURE 15.5
Control devices for combustion flue gases: (a) draft hood and (b) damper valve.

(liters per second). To find the mass flow rate, one must know the density, which in turn depends on the gas main pressure. The ideal gas law can be used for such calculations. Efficiency values are specified by the manufacturer at a single value of fuel input rate. Of course, one could use either greater or lower flow rates, but the design flow rate is that at which the manufacturer's efficiency value applies. At higher fuel input rates, the furnace could overheat with hazardous results. Industry standards dictate the temperature limits allowable in furnaces, thereby limiting the fuel input rates.

The *overall steady-state furnace efficiency* $\eta_{furnace,ss}$ (or instantaneous furnace efficiency) is a better measure of the entire furnace's thermal conversion efficiency. It is defined in terms of net useful heat, i.e., accounts for heat losses by convection and radiation through the jacket of the furnace as well as heat carried away by the flue gases. It is somewhat dependent on the specific installation and would be lower than the combustion efficiency.

This instantaneous efficiency is of limited value in selecting furnaces owing to the fact that furnaces often operate in a cyclic, part-load mode where instantaneous efficiency may be lower than that at peak operating conditions. Part-load efficiency is low since cycling causes inefficient combustion, cyclic heating and cooling of the furnace heat exchanger mass, and thermal cycling of the distribution ductwork. A more useful performance index is the *annual fuel utilization efficiency* (AFUE), which also accounts for other loss mechanisms over a season. These include stack losses (sensible and latent), cycling losses, infiltration, and pilot losses. A standard (ANSI/ASHRAE 103, 2007) is used for determining the AFUE for residential furnaces.

Table 15.3 shows typical values of AFUE for residential furnaces (ASHRAE Systems, 2012). The table shows that efficiency improvements can be achieved by eliminating standing pilots, by using a forced-draft design or by condensing the products of combustion to recover latent heat normally lost to the flue gases. Efficiency can also be improved by using a vent damper to reduce stack losses during furnace-off periods. Although this table is prepared from residential furnace data, it can be used for commercial-size furnaces as well. Few data have been published for commercial systems since this has not been mandated by law, as it has been for residential furnaces. The AFUE has the shortcoming that a specific usage pattern and equipment characteristics are assumed. In the next section, we discuss a more accurate method for finding annual performance of heat-producing primary systems.

TABLE 15.3

Typical Values of AFUE for Furnaces

Type of Gas Furnace	AFUE, %
1. Atmospheric with standing pilot	64.5
2. Atmospheric with intermittent ignition	69.0
3. Atmospheric with intermittent ignition and automatic vent damper	78.0
4. Same basic furnace as type 2, except with power vent	78.0
5. Same as type 4 but with improved heat transfer	81.5
6. Direct vent with standing pilot, preheat	66.0
7. Direct vent, power vent, and intermittent ignition	78.0
8. Power burner (forced draft)	75.0
9. Condensing	92.5
Type of Oil Furnace	**AFUE, %**
1. Standard	71.0
2. Same as type 1 with improved heat transfer	76.0
3. Same as type 2 with automatic vent damper	83.0
4. Condensing	91.0

The AFUE can be used to find the annual energy consumption directly from its definition. The fuel consumption during an average year $Q_{fuel,yr}$ is given by

$$Q_{fuel,yr} = \frac{Q_{yr}}{\text{AFUE}} \text{ MBtu/year(GJ/year)} \qquad (15.4)$$

where Q_{yr} is the annual heat load. By using this approach, it is a simple matter to find the energy savings one might expect, on average, by investing in a more efficient furnace.

Example 15.4: Energy Savings from Using a Condensing Furnace

A small commercial building is heated by an old atmospheric-type gas furnace. The owner proposes to install a new pulse-type (condensing) furnace. If the annual heat load Q_{yr} on the warehouse is 200 GJ, what energy savings will the new furnace produce?

Given: Q_{yr}, furnace type

Assumptions: AFUE is an adequate measure of seasonal performance, and furnace efficiency does not degrade with time.

Find: $\Delta Q_{fuel} = Q_{fuel,old} - Q_{fuel,new}$

Lookup values: AFUEs from Table 15.3

$$\text{AFUE}_{old} = 0.645, \quad \text{AFUE}_{new} = 0.925$$

Solution

The energy savings are easily deduced from Equation 15.4 as

$$\Delta Q_{fuel} = Q_{yr}\left(\frac{1}{\text{AFUE}_{old}} - \frac{1}{\text{AFUE}_{new}}\right)$$

Substituting the tabulated values for AFUE, we have

$$\Delta Q_{fuel} = 200 \times \left(\frac{1}{0.645} - \frac{1}{0.925} \right) = 93.9 \text{ GJ/year}$$

Comments

The saving of energy from using the modern furnace is substantial, almost equivalent to 50% of the annual heating load. For a more precise calculation for a given furnace, one could use the bin method, described in Chapter 10. This approach requires more detailed part-load data from a manufacturer but will give a more accurate result. We present such an example in Section 15.4 relevant to boilers.

In addition to energy consumption, the designer must be concerned with myriad other factors in furnace selection. These include as follows:

- Air-side temperature rise; duct design and airflow rate affected
- Airflow rate; duct design affected
- Control operation (for example, will night or unoccupied day–night setback be used? Is fan control by thermal switch or time-delay relay?)
- Safety issues (combustion gas control, fire hazards, high-temperature limit switch), introduced briefly in Section 15.2.3.

Chapters 16 and 20 will deal with other aspects of furnace-based heating systems design including duct sizing and layout, as well as air distribution within rooms, respectively.

15.3 Boilers

15.3.1 Classification and Description

A boiler is a device made from copper, steel, or cast iron to transfer heat from a combustion chamber (or electric resistance coil) to water in the liquid phase, vapor phase, or both. Boilers are classified both by the fuel used and by the operating pressure. Fuels include gas, fuel oils, wood, coal, refuse-derived fuels, and electricity. In this section, we focus on fossil fuel–fired boilers.

Boilers produce either hot water or steam at various pressures. Although water does not literally boil in hot water "boilers," they are called boilers nevertheless. Steam is an exceptionally effective heat transport fluid due to its very large heat of vaporization.

Boilers for buildings are classified as follows:

1. *Low-pressure boilers*: Steam boilers with operating pressures below 15 psig (100 kPa) and hot water boilers with pressures below 150 psig (1000 kPa). Temperatures are limited to 250°F (120°C).
2. *High-pressure boilers*: Steam boilers with operating pressures above 15 psig (100 kPa) and hot water boilers with pressures above 150 psig (1000 kPa). Temperatures are above 250°F (120°C).

Heat rates for steam boilers are often expressed in pound-mass of steam produced per hour (or kilowatts). The heating value of steam for these purposes is rounded to 1000 Btu/lb_m. Steam boilers are available at heat rates of 50–50,000 lb_m/h of steam (15–15,000 kW). This overlaps the upper range of furnace sizes noted in Section 15.2. Steam produced by boilers is used in buildings for space heating, water heating, and absorption cooling. Water boilers are available in the same range of sizes as steam boilers: 50–50,000 kBtu/h (15–15,000 kW). Hot water is used in buildings for space and water heating.

Since the energy contained in steam and hot water within and flowing through boilers is very large, an extensive codification of regulations has evolved to ensure safe operation. In the United States, the ASME Boiler and Pressure Vessel Code (ASME BPVC, 2013) governs construction of boilers. For example, the code sets the limits of temperature and pressure on low-pressure water and steam boilers listed earlier.

Large boilers are constructed from steel or cast iron. Cast-iron boilers are modular and consist of several identical heat transfer sections bolted and gasketed together to meet the required output rating. Steel boilers are not modular but are constructed by welding various components together into one assembly. Heat transfer occurs across tubes containing either the fire or the water to be heated. The former are called "fire-tube boilers," and the latter are "water-tube boilers." Either material of construction can result in equally efficient designs. Small, light boilers of moderate capacity are sometimes needed for use in buildings. For these applications, the designer should consider the use of copper boilers.

Figure 15.6 shows a cross section of a steam boiler of the type used in buildings, and Figure 15.7 is a cutaway photograph of a steam boiler.

15.3.2 Boiler Design and Selection

Boilers are specified based on a few key criteria. In this section, we list these but do not discuss the internal

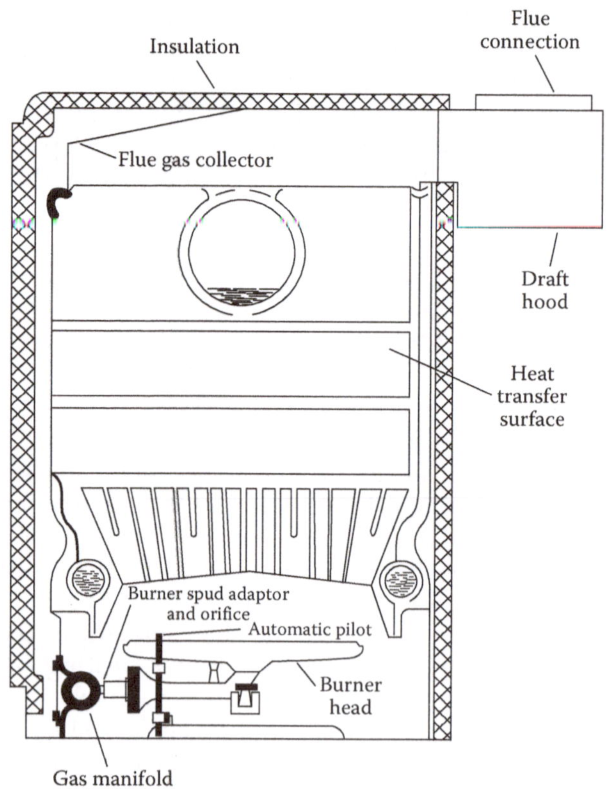

FIGURE 15.6
Cross-sectional drawing of boiler showing burner, heat exchanger, and flue connection.

design of boilers and their construction. Boiler selection is based on the following criteria:

- *Boiler fuel*: Type, energy content, and heating value including altitude effects if gas fired (no effect for coal or fuel oil boilers)
- *Required heat output*: Net output rating, kBtu/h (kW)
- *Operating pressure and working fluid*
- *Efficiency and part-load characteristics*
- *Other*: Space needs, control system, combustion air requirements, safety requirements, and ASME code applicability

The boiler heat output required for a building is determined by summing the *maximum heating requirement* of all zones or loads serviced by the boiler during peak demand for steam or hot water and adding to that (1) parasitic losses including piping losses and (2) initial loop fluid warm-up. As described in Chapter 9, simply adding all the *peak heating unit capacities* of all the zones in a building can result in an oversized boiler since not all zones require peak heating simultaneously. The ratio of the sum of all zone loads under peak conditions to the total heating capacity installed in a building is called the "diversity." Additional boiler capacity may be needed to recover from night setback in massive buildings.

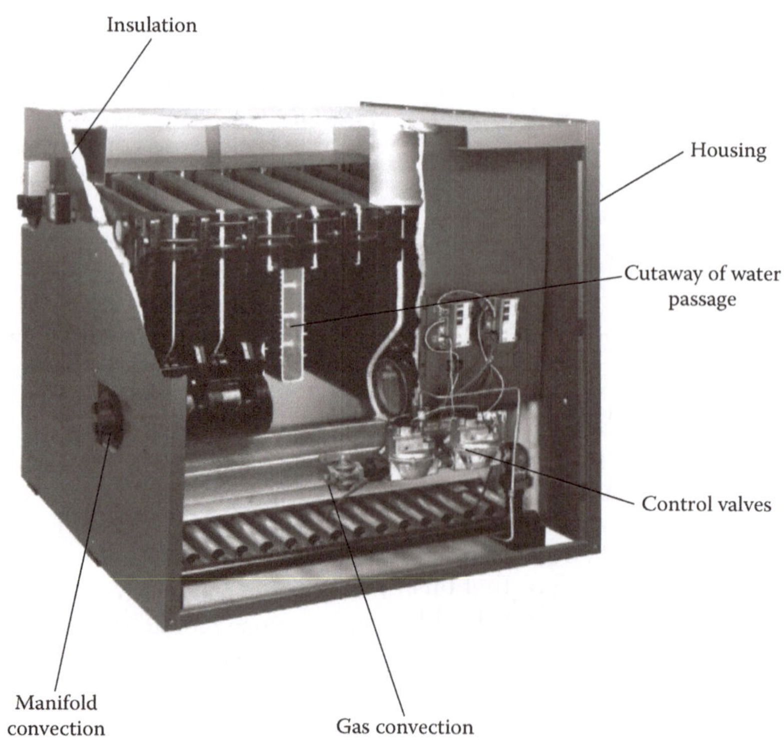

FIGURE 15.7
Cutaway photograph of boiler. (Courtesy of Weil-McLain, Inc., Michigan City, IN.)

This transient load is called the "pickup" load and must be accounted for in both boiler and terminal heating unit sizing. Chapter 9 contained a detailed description of pickup loads along with an example of estimating the needed extra capacity.

Just as for furnaces, *the overall* steady-state boiler efficiency $\eta_{boiler,ss}$ is defined as the gross output (energy in the steam generated) divided by the heat contained in the natural gas burnt under continuous firing conditions. Thus, this figure of merit quantifies the fraction of the input chemical energy that is transferred to the working fluid under steady-state continuous operation. The jacket losses are included in this definition, which is to some extent installation specific. Common range is 70%–90%. The ambient air temperature in the boiler room has an effect on boiler efficiency (the higher the ambient air, the better the efficiency since some of the sensible heating of the combustion air is avoided). Typically, boiler efficiencies are based on an ambient air temperature of 80°F or 26.5°C (Mull, 2001).

Alternatively, boilers can be tested in laboratories and rated in accordance with standards issued by the Hydronics Institute (formerly IBR, the Institute of Boiler and Radiator Manufacturers, and the SBI, the Steel Boiler Institute), the American Gas Association (AGA), and other industry groups. In addition to cast-iron boiler ratings, IBR ratings are industry standards for baseboard heaters and finned-tube radiation. The ratings of the Hydronics Institute apply to steel boilers. (IBR and SBI are trademarks of the Hydronics Institute.) SBI and IBR ratings apply to oil- and coal-fired boilers, while gas boilers are rated by the AGA.

Boilers are often sized by their *sea-level input* fuel ratings. Of course, this rating must be multiplied by the applicable efficiency to determine the gross output of the boiler. In addition, if a gas boiler is not to be located at sea level, the effect of altitude must be accounted for in the rating. Some boiler designs use a forced-draft burner to force additional combustion air into the firebox, to offset part of the effect of altitude. Also, enriched or pressurized gas may be provided at high altitude so that the heating value per unit volume is the same as that at sea level. If no accommodation to altitude is made, the output of a gas boiler drops by approximately 4%/1000 ft (13%/1 km) of altitude above sea level. Therefore, a gas boiler located in Denver, Colorado (5000 ft, 1500 m), will have a capacity of only 80% of its sea-level rating.

Table 15.4 shows the type of data provided by manufacturers for the selection of boilers for a specific project. Reading across the table, the fuel input needs are first tabulated for the 13 boiler models listed. Column 5 is the sea-level boiler output at the maximum design heat rate. Columns 6–8 convert the heat rate to steam (as ft^3/h and in Btu/h) and hot water production rates. Column 10 expresses the heat rate in still a different way by using units of boiler horsepower (= 33,475 Btu/h or 9.8 kW). Columns 11–14 provide information needed for designing the combustion air supply system and the chimney.

A convenient rule of thumb can be used to check boiler selection in heating climates in the United States: The input rating (e.g., columns 2 and 3 of Table 15.4) when converted to Btu per hour expressed on the basis of *per heated square foot* of building is usually in the range of one-third to one-fifth of the design temperature difference (difference between indoor and outdoor winter design temperatures). For example, if the design temperature difference for a 100,000 ft^2 building is 80°F, the boiler input is expected to range between 1,600 kBtu/h [= (80/5) × 100,000 ft^2] and 2700 kBtu/h [= (80/3) × 100,000 ft^2]. The difference between the two depends on the energy efficiency of the building envelope and its infiltration controls. Boiler efficiency also has an effect on this design check.

Economic criteria including initial cost and life cycle operating cost must be considered.* Although high-efficiency boilers may cost more initially, it is very often worthwhile to make the investment when evaluated over the life cycle. However, in many cases, first cost is the primary determinant of selection. In these cases, the HVAC engineer must point out to the building owner or architect that the building lifetime penalties of using inexpensive, but inefficient heating equipment are considerable, many times the initial cost difference.

The designer is advised to avoid the customary tendency to oversize furnaces. An oversized furnace operates at a lower efficiency than a properly sized one owing to the penalties of part-load operation. If a correct heat load calculation is done (with proper attention to the recognized uncertainty in building infiltration losses), only a small safety factor should be needed, say, 10%. The safety factor is applied to account for heat load calculation uncertainties and possible future, the modest changes in building load due to usage changes. Oversizing of furnaces also incurs other penalties including excessive duct size and cost along with poorer control of comfort due to larger temperature swings in the heated space.

15.3.3 Boiler Controls

Just as for furnaces, boilers also have safety limit controls as well as operational controls. For example, if the water level falls below a safe level, a sensor will shut off the burner. Other safety controls have to do with high or low fuel pressure, high or low fuel oil temperature, and flame failure. A critical safety device is a relief valve in case steam pressure builds up to a critical level.

* The techniques used to make the final selection appear in Chapter 23 on economics.

TABLE 15.4

Example of Manufacturer's Boiler Capacity Table

Boiler Unit Number, Steam, or Water, °F	IBR[a] Burner Capacity			Gross IBR Output, Btu/h	Net IBR Ratings			Net Heat Transfer Area, ft² H_2O	Boiler hp	Net Firebox Volume, ft³	Stack Gas Volume, ft³/min	Positive Pressure in Firebox, inWG	IBR Chimney Size Vent Dia., in.
	Light Oil, gal/h	Gas, kBtu/h	Min. Gas Press. Req'd., inWG		Steam, ft²/h	Steam, Btu/h	Water, Btu/h						
(1)	(2)	(3)	(4)	(5)	(6)	(7)	(8)	(9)	(10)	(11)	(12)	(13)	(14)
486	6.30	882	5.5	720,000	2,250	540,100	626,100	4,175	21.5	11.02	395	0.34	10
586	8.25	1155	7.0	940,000	2,940	705,200	817,400	5,450	28.1	14.45	517	0.35	10
686	10.20	1428	5.5	1,160,000	3,625	870,200	1,008,700	6,725	34.6	18.08	640	0.35	10
786	12.15	1701	6.0	1,380,000	4,355	1,044,700	1,200,000	8,000	41.2	21.61	762	0.36	12
886	14.10	1974	5.0	1,600,000	5,115	1,227,900	1,391,300	9,275	49.6	25.14	884	0.37	12
986	16.05	2247	6.0	1,820,000	5,875	1,409,800	1,582,600	10,550	54.3	28.67	1006	0.38	14
1086	18.00	2520	6.5	2,040,000	6,600	1,583,900	1,773,900	11,825	60.9	32.20	1128	0.39	14
1186	19.95	2793	7.0	2,260,000	7,310	1,754,700	1,965,200	13,100	67.5	35.73	1251	0.40	14
1286	21.95	3073	7.0	2,480,000	8,025	1,925,500	2,156,500	14,375	74.1	39.26	1376	0.41	14
1386	23.90	3346	6.5	2,700,000	8,735	2,096,300	2,347,800	15,650	80.6	42.79	1498	0.42	14
1486	25.90	3626	7.5	2,920,000	9,445	2,267,100	2,539,100	16,925	87.2	46.32	1623	0.43	16
1586	27.85	3899	7.5	3,140,000	10,160	2,437,900	2,730,400	18,200	93.8	49.85	1746	0.44	16
1686	29.75	4165	8.5	3,350,000	10,835	2,600,900	2,913,000	19,420	100.1	53.38	1865	0.45	16

Source: Courtesy of Weil-McLain, Inc., Michigan City, IN.
Note: 1 bhp = 33,475 Btu/h – 9.8 kW.
[a] IBR, Institute for Boiler and Radiator Manufacturers.

Additional accessories include (1) a gauge glass for an operator to verify the water level in the boiler, (2) pressure gauges and thermometers at the boiler inlet and outlet, and (3) sensors that measure CO_2 and other flue gas elements.

Proper control of boilers in response to varying outdoor conditions can improve efficiency and occupant comfort. A standard feature of boiler controls is the *boiler reset* system. Since full boiler capacity is needed only at peak heating conditions, better comfort control results if capacity is reduced with increasing outdoor temperature. Capacity reduction of zone hot water heating is easy to accomplish by simply reducing the water temperature supplied by the boiler. For example, the reset schedule might specify boiler water at 210°F at an outdoor temperature of –20°F and at 50°F outdoors a water temperature of 140°F. This schedule is called a one-to-one schedule because the boiler output drops by 1.0°F for every degree rise in outdoor temperature.

15.3.4 Boiler Efficiency

Boilers also have three efficiency indices similar to furnaces (see Section 15.2.4). The use of Equation 15.3 requires accurate measurement of the flow rates of the fuel and the flue gas (discussed further in Section 15.5.1). These measurements are difficult to make (uncertainties can be as large as 50%), and so a simpler indirect equation is often used to determine *boiler combustion efficiency*:

$$\eta_{boiler} = \frac{\text{HHV} - \text{Heat losses}}{\text{HHV}} \tag{15.5}$$

It can be determined experimentally to ascertain the condition of the fuel combustion equipment, including burners, heat transfer surfaces, and combustion air supply equipment.

The heat loss terms include five sources: (1) sensible heat loss in flue gases, (2) latent heat loss in flue gases due to combustion of hydrogen, (3) heat loss in water in combustion air, (4) heat loss due to incomplete combustion of carbon, and (5) heat loss from unburned carbon in ash (coal and fuel oil).

The largest source of heat loss is due to sensible flue gas heat loss attributed to the temperature of the exiting flue gases and to the excess air fraction (since this excess air has to be heated to the exit flue gas temperature). These exiting temperatures range from 400°F to 600°F (220°C to 330°C) depending on the type of fuel. The various sources of heat losses can be estimated using information provided by the Boiler Efficiency Institute (Dyer and Maples, 1991). They range from 75% to 86% for most noncondensing boilers, while condensing boilers range from 88% to 95% efficiency (since these recover some of the heat contained in the water of the combustion air). Note that contrary to most equipment (such as chillers and pumps), boilers do not operate the most efficiently at full load; their best operating range is the 65%–75% range of full-load capacity.

As noted earlier, efficiency under specific test conditions has very limited usefulness in calculating the annual energy consumption of a boiler because of significant drop-off of efficiency under part-load conditions. For small boilers (up to 300 kBtu/h [90 kW]), the Department of Energy has set forth a method for finding the AFUE (defined in Section 15.2.4 on furnaces). The annual energy consumption must be known in order to perform economic analyses for optimal boiler selection.

15.3.5 Factors Affecting Boiler Efficiency

There are several primary sources of inefficiency that are discussed here (Mull, 2001). The first is related with the various causes of incomplete combustion: (1) improper fuel and air missing in the combustion chamber causing local fuel-rich and fuel-lean pockets, (2) insufficient air supply, (3) insufficient residence time of the reactant that leads to incomplete combustion reaction, (4) quenching of combustion reactions due to flame impingement on a cold surface, and (5) the flame temperature being too low that slows down combustion reactions. These causes can degrade combustion efficiency and further cause air pollution and hazardous gases such as carbon monoxide. Another source of inefficiency is tube fouling. This is caused by deposits on the heat exchanger tubes: on the fire side by deposits of ash, soot, and mineral and on the water side by crystalline deposits (silica, magnesium, calcium, etc.). This additional resistance results in higher flue gas temperatures (a rule of thumb is that boiler efficiency drops by 1% for every 4°F (2°C) increase in stack temperature). Water samples are taken periodically to monitor this buildup, and tube cleaning has to be done at regular intervals.

Another reason for boiler inefficiency is cycling loss due to part-load operation, which is discussed in the next section. A third source is the need for boiler blowdowns, i.e., purging the steam/hot water in the boiler and replacing it with makeup water. This is necessary to maintain low concentrations of dissolved and suspended solids in the boiler water and to remove any sludge, both of which can reduce the effectiveness of the heat exchange. Ways of minimizing this loss by proper water treatment, reducing other types of losses (such as jacket losses and steam leaks), and using steam traps (see Section 18.5.2), flue gas economizer, preheating air supply, and fuel additives to improve combustion are discussed by Mull (2001).

15.4 Seasonal Energy Calculations

15.4.1 Part-Load Models

For larger boilers, data specific to a manufacturer and an application must be used to determine annual consumption. Efficiencies of fossil fuel boilers vary with heat rate depending on the internal design. If the boiler has only one or two firing rates, the continuous range of heat inputs needed to meet a varying heating load is achieved by cycling the boiler on and off. However, as the load decreases, efficiency decreases since the boiler spends progressively more time in transient warm-up and cooldown modes, during which relatively less heat is delivered to the load. At maximum load, the boiler cycles very little, and efficiency can be expected to be near the rated efficiency of the boiler. Part-load effects can reduce average efficiency to less than one-half of the peak efficiency. Of course, for an oversized boiler, the average efficiency is well below the peak efficiency since it operates at part load for the entire heating season. This operating cost penalty persists for the life of a building long after the designer who oversized the system has forgotten the error.

The approach to modeling part-load performance parallels the one used to model chillers; see Section 14.7. The *part-load ratio* (PLR) is a quantity between 0 and 1, which is a measure of the theoretical run-time fraction based on thermal load defined as

$$\text{PLR} \equiv \frac{\dot{Q}_{out}}{\dot{Q}_{out,full}} \tag{15.6}$$

where

$\dot{Q}_{out}$ is the boiler heat output at part load, Btu/h (kW)
$\dot{Q}_{out,full}$ is the rated heat output at full load, Btu/h (kW)

It is not practical to calculate from basic principles the manner in which boiler *input* varies with the value of PLR since the processes to be modeled are very complex and nonlinear. The approach used for boilers (and other heat-producing equipment) involves using test data to calculate the boiler input needed to produce an output $\dot{Q}_{out}$. If efficiency were constant and if there were no standby losses, the function relating input to output would merely be a constant, namely, the efficiency. For real equipment, the relationship is more complex. A common function used to relate input to output (i.e., to PLR) is a simple polynomial (at least for a boiler) such as*

$$\frac{\dot{Q}_{in}}{\dot{Q}_{in,full}} = A + B(\text{PLR}) + C(\text{PLR})^2 + \cdots \tag{15.7}$$

where

$\dot{Q}_{in}$ is the fuel input (or input energy) required to meet the part-load level corresponding to PLR
$\dot{Q}_{in,full}$ is the fuel input at rated full load on the boiler

One could also have modeled part-load effects using a similar approach wherein a polynomial model is used not in terms of input thermal energy but in terms of boiler efficiency (i.e., actual by rated boiler efficiency). The reason why this approach is not common is due to the fact that higher-order models are warranted in such a case that makes it less robust as a modeling approach. Finally, for small boilers (or furnaces), one could also adopt the simpler to use PLF–PLR method by defining a equipment degradation coefficient (see Section 14.7.1 as applied to vapor compression equipment).

The first term of Equation 15.7 represents standby losses, e.g., those resulting from a standing pilot light in a gas boiler. Since the part-load characteristic is not far from linear for most boilers, a quadratic or cubic expression is sufficient for annual energy calculations. Part-load data are not as readily available as standard peak ratings. If available, the data may often be in tabular form. The designer will need to make a quick regression of the data to find model coefficients *A*, *B*, and *C* using commonly available spreadsheet or statistical software (see Example 14.11), and the model can then be used for annual energy calculations, as described in the following section.

In the remainder of this section, we examine a particularly simple application of a boiler—building space heating—to illustrate the importance of part-load effects. The *annual* energy input $Q_{in,yr}$ of a space heating boiler can be calculated from the basic equation

$$Q_{in,yr} = \int_{yr} \frac{\dot{Q}_{out}(t)}{\eta_{boiler}(t)}\,dt \tag{15.8}$$

where

η_{boiler} is the boiler efficiency—a function of time since the load on the boiler varies with time
$\dot{Q}_{out}(t)$ is the *boiler heat output*, which varies with time as well

The argument of the integral is just the instantaneous, time-varying energy input to the boiler. However, since the needed output—not the input—is usually known as a result of building load calculations, the form in Equation 15.8 is the most convenient to adopt. The time dependence in this expression is determined, in turn, by the temporal variation of load on the boiler as imposed by the HVAC system in response to climatic, occupant, and other time-varying loads.

* This simple polynomial form is used for illustration purposes in this book, recognizing that more complex forms such as those used in ASHRAE Standard 90.1 (2010) may be more accurate.

15.4.2 Bin Method Calculation

A simple case is a boiler used solely for space heating. As described in Chapter 7, the heating load is determined to first order by the difference between indoor and outdoor temperatures, all characteristics of the building's load and use remaining fixed. Therefore, the heat rate in Equation 15.8 is determined by the outdoor temperature if the interior temperature remains constant. In this very simple case, one could replace the integral in Equation 15.8 with a sum and utilize the bin approach as follows:

$$Q_{in,yr} = \sum_{j=1}^{N} \frac{\dot{Q}_{out}(T_j) \cdot n_j(T_j)}{\eta_{boiler}(T_j)} \qquad (15.9)$$

where

$\eta_{boiler}(T_j)$ is the efficiency of boiler in a given ambient temperature bin j; the efficiency depends strongly, but indirectly, on ambient temperature T_j since the load, which determines PLR, depends on temperature

$\dot{Q}_{out}(T_j)$ is the boiler load (i.e., building heat load) that depends on ambient temperature

$n_j(T_j)$ is the number of hours in temperature bin j for which the values of efficiency and heat input apply

This expression assumes that the sequence of hours during the heating season is of no consequence, i.e., the thermal mass effects of the space are negligible. Example 15.5 illustrates how the bin weather data described in Chapter 10 can be used to properly account for part-load efficiency of a boiler used for space heating.

Example 15.5: Annual Energy Consumption of a Gas Boiler

A gas boiler is used to supply space heat to a building. The load varies linearly with ambient temperature, as shown in Table 15.5. If the efficiency of the boiler is 80% at peak-rated conditions, find the seasonal average efficiency, annual energy input, and annual energy output, using the data in the table. The boiler input at rated conditions is 8750 kBtu/h corresponding to –12.5°F, the temperature bin at which the load is 7000 kBtu/h.

This boiler is turned off in temperature bins higher than 57.5°F, roughly corresponding to the limit of the heating season; therefore, the standby losses above this temperature are zero.

The values of the coefficients in the part-load characteristic, Equation 15.7, are specified as

$$A = 0.1 \quad B = 1.6 \quad C = -0.7$$

Given:

- Bin data in the first two columns of Table 15.5 (the online HCB software contains bin data files)
- Load data in the third column of Table 15.5 (note the linearity of load with ambient temperature)
- Part-load characteristic equation

$$\frac{\dot{Q}_{in}}{\dot{Q}_{in,full}} = 0.1 + 1.6 \times \text{PLR} - 0.7 \times \text{PLR}^2 \qquad (15.10)$$

Assumptions: The bin approach is sufficiently accurate for this problem.

TABLE 15.5

Summary of Solution for Example 15.5, Boiler Energy Analysis

Calculating Annual Boiler Energy Use							
Bin Range, °F	Bin Size, h	Heating Load, kBtu/h	PLR	$\dot{Q}_i$, kBtu/h	Boiler Effic.	Fuel Used, MBtu	Net Output, MBtu
55–60	762	0	0.00	875	0.000	667	0
50–55	783	500	0.07	1844	0.271	1,444	391
45–50	716	1000	0.14	2750	0.364	1,969	716
40–45	665	1500	0.21	3594	0.417	2,390	997
35–40	758	2000	0.29	4375	0.457	3,316	1,516
30–35	713	2500	0.36	5094	0.491	3,632	1,782
25–30	565	3000	0.43	5750	0.522	3,249	1,695
20–25	399	3500	0.50	6344	0.552	2,531	1,396
15–20	164	4000	0.57	6875	0.582	1,127	656
10–15	106	4500	0.64	7344	0.613	778	477
5–10	65	5000	0.71	7750	0.645	504	325
0–5	80	5500	0.79	8094	0.680	647	440
–5 to 0	22	6000	0.86	8375	0.716	184	132
Annual						22,439	10,525

Find: $\bar{\eta}_{boiler}$, $Q_{in,yr}$, $Q_{out,yr}$

Lookup values: Table 15.5

Solution

Equation 15.9 is expressed in terms of the efficiency at part load. It can be easily rearranged in terms of the part-load energy input:

$$Q_{in,yr} = \sum_{j=1}^{j=N} \dot{Q}_{in}(T_j)n_j(T_j)$$

The resulting calculations appear in the right half of Table 15.5. We will work through in detail the calculations for the 37.5°F bin to see how the table is completed. The load at this bin is 2000 kBtu/h.* The value of PLR is the ratio of this load to the peak load 7000 kBtu/h:

$$\text{PLR} = \frac{2000 \text{ kBtu/h}}{7000 \text{ kBtu/h}} = 0.286$$

Equation 15.10 is now used to find the energy input. This expression is the dimensionless form of the argument of the summation in Equation 15.10:

$$\frac{\dot{Q}_{in}}{\dot{Q}_{in,full}} = 0.1 + 1.6 \times 0.286 - 0.7 \times 0.286^2 = 0.50$$

Hence, the heat input is

$$\dot{Q}_{in} = 0.50 \times \dot{Q}_{in,full} = 0.50 \times 8750 \text{ kBtu/h} = 4375 \text{ kBtu/h}$$

This is the value entered in the fifth column of the table. The sixth column is the boiler efficiency in each bin, defined as the ratio of the load (2000 kBtu/h) to the heat input just calculated (4375 kBtu/h)

$$\eta_{boiler,37.5} = \frac{2000 \text{ kBtu/h}}{4375 \text{ kBtu/h}} = 0.457$$

The part-load effect is immediately obvious—the efficiency is much less than the rating at full load of 0.80.

Finally, the fuel consumed is the product of the fuel input rate (4375 kBtu/h) and the number of hours in the 37.5°F bin (758 h):

$$Q_{i,37.5} = 4375 \text{ kBtu/h} \times 758 \text{ h} = 3316 \text{ MBtu}$$

The heat produced by the boiler in the 758 h corresponding to the 37.5°F bin is as follows:

$$Q_{out,37.5} = 2000 \text{ kBtu/h} \times 758 \text{ h} = 1516 \text{ MBtu}$$

These last two numbers are the entries in the rightmost two columns in Table 15.5.

Finally, to find the total annual energy used by the boiler, one sums the "fuel used" column of the table to find that 22,439 MBtu is used to meet the annual load of 10,525 MBtu. The ratio of these two numbers is the overall annual boiler efficiency: 47%. This value is 41% less than the peak efficiency of 80%. Clearly, part-load effects play a major role during annual energy calculations.

Comments

Calculations of this type are well suited to spreadsheet solutions. This problem was particularly simple since the building load was linear with the ambient temperature. If the boiler had been used to supply heat for an absorption chiller whose load depended on solar flux and dry- and wet-bulb temperatures, the same general approach could have been used but would have been more complex.

One method of avoiding the poor efficiency of this system is to use two (or more) smaller boilers, the combined capacity† of which totals the needed 7000 kBtu/h. Properly chosen, the smaller boilers will operate more nearly at full load for more of the time, resulting in higher seasonal efficiency. Multiple-boiler systems also offer standby security; if one boiler should fail, the other could carry at least part of the load. A single-boiler system would entirely fail to meet the load. However, two (or more) smaller boilers cost more than one large boiler with the same total capacity.

The final decision must be made based on economics, giving proper account to the increased reliability of a system composed of several smaller boilers. Constraints are imposed on such decisions by initial budget, fuel type, owner, and architect decisions and available space.

15.5 Improving and Monitoring Thermal Performance

15.5.1 Direct Method of Efficiency Determination

As stated earlier, it is often impractical to determine efficiency directly because of the difficulty in measuring the flow rates and other terms that need to be included in the calculation. Nevertheless, we shall illustrate this approach for the sake of completeness. According to the Boiler Efficiency Institute (Dyer and Maples, 1991), the

* kBtu = 1000 Btu; MBtu = 1000 kBtu.

† The careful reader will note that the boiler design point of −12.5°F does not occur in the bin data since bin data are based on a typical year, whereas the design temperature extreme will not occur in an average year. The boiler "oversizing" due to this effect is also part of the cause for reduced annual boiler efficiency.

efficiency of a steam boiler η_{boil} can be found from field measurements by

$$\eta_{boiler} = \frac{\dot{Q}_{steam}}{\dot{m}_{fuel} \times \text{HHV}} \quad (15.11)$$

where

$\dot{Q}_{steam}$ is the steam output rate, Btu/h (kW)
$\dot{m}_{fuel}$ is the fuel supply rate, lb_m/h (kg/s)
HHV is the HHV of fuel, Btu/lb_m (kJ/kg)

An illustrative example is presented here in order to demonstrate this approach.

Example 15.6: Direct Efficiency Measurement

A boiler generates 1000 lb_m /h of superheated steam at temperature = 500°F and pressure = 250 psia. The following information is given about its performance:

(a) Natural gas fuel use = 35,000 ft^3/day.
(b) Condensate is saturated liquid at 100°F.
(c) The boiler blower and the condensate pump consume 12 hp.
(d) About 5% of the steam leaks out from the system.
(e) There is a 5 min boiler blowdown each day using 10% of the steam produced during that time.

Assumptions: HHV for natural gas = 1000 Btu/ft^3

Find: η_{boiler}

Lookup values: Enthalpy $h_{superheat}$ = 1265 Btu/lb_m and $h_{condensate}$ = 68 Btu/lb_m

Solution

Useful energy contained in the steam delivered per day to the load:

$$\dot{m}_{steam} = \left(\dot{m}_{steam,gen} - \dot{m}_{steam,blowdown} - \dot{m}_{steam,leak}\right)$$
$$= 1000\ \text{lb}_\text{m}/\text{h} \times [24\ \text{h/day} - 0.1 \times (5/60) - 24 \times 0.05]$$
$$= 22{,}792\ \text{lb}_\text{m}/\ \text{day}$$

Energy contained in the fuel =

$$Q_{fuel} = m_{fuel} \times \text{HHV} = 35{,}000\ \text{ft}^3/\text{day} \times 1{,}000\ \text{Btu/ft}^3$$
$$= 35 \times 10^6\ \text{Btu/day}$$

Energy consumed by the boiler blower and pump per day =

$$Q_{elec} = 12\ \text{hp} \times 0.746\ \text{kW/hp} \times 24\ \text{h/day} \times 3412\ \text{Btu/kWh}$$

$$Q_{elec} = 0.733 \times 10^6\ \text{Btu/day}$$

Finally, the steady-state efficiency of the boiler over a day

$$\eta_{boiler} = \frac{\dot{m}_{day,steam}(h_{superheat} - h_{condensate})}{Q_{fuel} + Q_{elec}}$$
$$= \frac{22{,}792\ \text{lb}_\text{m}/\text{day} \times (1{,}265 - 68)\ \text{Btu/lb}_\text{m}}{(35 \times 10^6 + 0.733 \times 10^6)\ \text{Btu/day}}$$
$$= 0.763 \text{ or } 76.3\%$$

Comments

If the steam losses attributed to leakage and blowdown were totally eliminated, it will be found that the boiler efficiency would increase to 80.4%, about 4% percentage points or a 5% increase.

15.5.2 Indirect Method: Flue Gas Analysis

The indirect combustion method assumes that the total flue-input energy is either transferred to the working fluid or lost with the flue gas due to the various sources listed in Section 15.3.4. The combustion of fuel in a boiler is a chemical reaction and as such is governed by the laws of stoichiometry. In this section, we discuss the combustion of natural gas (for our purposes, it is assumed to be 100% methane) in boilers as an example of fuel burning for heat production. We also outline how the flue gas from a boiler can be analyzed to ascertain the efficiency of the combustion process. Continuous monitoring of flue gases by a building's energy management system can result in early identification of boiler combustion problems. In a new building, one should test a boiler to determine its efficiency as installed and to compare output to that specified by the designer.

Flue gas analysis is a method of determining the amount of excess air in a combustion process. This information can be used to find an approximate value of boiler efficiency. Periodic regular analysis can provide a trend of boiler efficiency with time, indicating possible problems with the burner or combustion equipment in a boiler or furnace. Flue gas analysis is often expressed as the volumetric fraction of flue gases—oxygen, nitrogen, and carbon monoxide. If these three values are known, the excess air (%) when natural gas (assuming 100% methane) is burnt can be found from (ASHRAE Fundamentals, 2013)

$$f_{excess\ air} = \frac{O_2 - 0.5 \times CO}{0.264 \times N_2 - (O_2 - 0.5 \times CO)} \quad (15.12)$$

in which the chemical symbols represent the volume fractions (units of percent) in the flue gas analysis.

The previous discussion relates to the combustion of methane and the rate at which air is to be supplied for proper combustion. Of course, many other fuels are used to fire boilers. Table 15.6 contains data that can be used to

TABLE 15.6

Combustion Efficiency as a Function of Two Easy to Monitor Flue Gas Variables

Natural Gas	% CO_2	Difference between Flue Gas and Room Temperature, °F										
		300	350	400	450	500	550	600	650	700	750	800
	4.0	25.1	27.7	30.4	33.1	35.8	38.3	40.9	43.5	46.2	48.8	
	4.5	23.6	25.9	28.3	30.7	33.0	35.4	37.8	40.1	42.6	44.8	47.2
Fuel analysis	5.0	22.2	24.4	26.8	28.7	30.9	33.0	35.7	37.3	39.7	41.8	43.8
1120 Btu/ft³	5.5	21.2	23.4	25.2	27.3	29.2	31.3	33.2	35.3	37.3	39.2	41.0
% by volume	6.0	20.4	22.3	24.1	25.8	27.8	29.6	31.5	33.3	35.2	36.8	38.8
CH_4 79.9	6.5	19.8	21.4	23.2	24.8	26.5	28.3	30.0	31.7	33.5	34.6	36.8
C_2H_6 17.3	7.0	19.1	20.7	22.3	23.9	25.5	27.1	28.8	30.4	32.0	33.8	35.3
CO_2 0.3	7.5	18.5	20.0	21.5	23.0	24.6	26.1	27.7	29.1	30.8	32.2	33.8
N_2 2.5	8.0	18.0	19.5	20.9	22.4	23.8	25.2	26.7	28.1	29.5	31.0	32.4
	8.5	17.6	19.0	20.4	21.7	23.1	24.5	25.8	27.1	28.6	29.9	31.3
	9.0	17.2	18.5	19.9	20.1	22.5	23.8	25.2	26.4	27.8	29.0	30.3
	9.5	16.9	18.1	19.5	20.7	21.9	23.1	24.4	25.6	26.9	28.2	29.4
	10.0	16.6	17.8	19.0	20.2	21.4	22.6	23.8	25.0	26.2	27.4	28.6

quickly estimate the excess air from a flue gas analysis of natural gas made up of different constituents as shown. The concentrationss of CO_2 and O_2 are the two key constituents measured while performing a flue gas analysis. Figure 15.8 graphically depicts how the two concentrations vary with excess air fraction for different types of fuels. These plots are generic since natural gas, oil and coal are fuels whose composition varies with location.

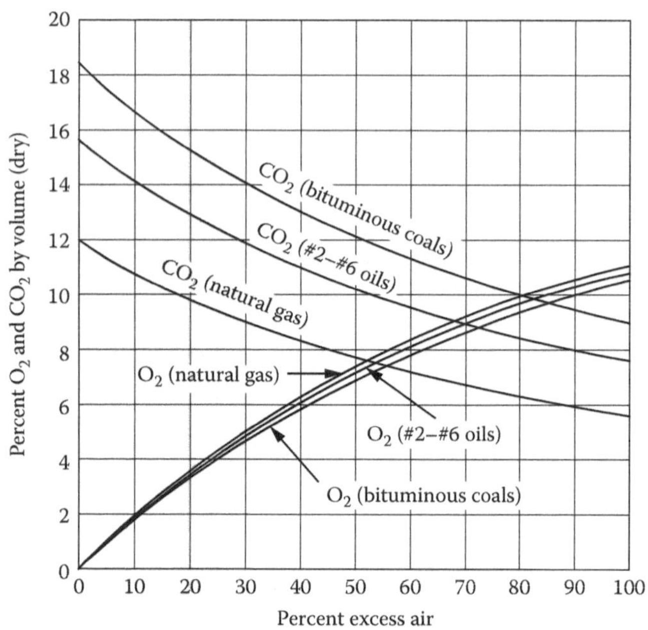

FIGURE 15.8
How excess air affects the composition of CO_2 and O_2 in the flue gas of different types of fuel. (From Mull, T.E., *Practical Guide to Energy Management for Facilities Engineers and Plant Managers*, ASME Press, American Society of Mechanical Engineers, New York, 2001; Courtesy of American Society of Mechanical Engineers, New York.)

Example 15.7: Flue Gas Analysis

The volumetric analysis of flue gas from combustion of methane in a gas boiler is measured to be 10.5% carbon dioxide, 3.2% oxygen, 86.3% nitrogen, and 0% carbon monoxide. The boiler manufacturer recommended a value of 20% excess air. Is the actual fraction consistent with the recommendation?

Given: Flue gas composition tabulated earlier

Find: $f_{excess\ air}$

Solution

Equation 15.12 is used as follows:

$$f_{excess\ air} = \frac{3.2\% - 0.5 \times 0\%}{0.264 \times 86.3\% - [3.2\% - (0.5 \times 0\%)]} = 0.163$$

The excess air is 16.3%, which is less than the recommended value of 20%. So, the boiler is not operating optimally, and adjusting the blower fan is warranted.

If we had used the plot for O_2 in Figure 15.6 corresponding to natural gas, we would have found a value close to 16%, while had we used the plot for CO_2, we would have found about 14%. So such plots are meant for approximate evaluations.

Along with a figure such as Figure 15.8, boiler manufacturers also provide tables that can be conveniently used to deduce boiler combustion efficiency (following the definition of Equation 15.5) from two simple measurements of the flue gas, namely, the flue gas temperature and the excess air percentage (which can be fairly well determined from the CO_2 percentage). Table 15.6 is a typical example of how these two quantities affect the boiler fractional heat loss for a specific type of natural gas composition. For example, if for this

TABLE 15.7

Stoichiometric and Excess Air Values of CO_2 for Combustion of Common Fossil Fuels

Type of Fuel	Theoretical or Maximum CO_2, %	CO_2 at Given Excess Air Values		
		20%	40%	60%
Gaseous fuels				
Natural gas	12.1	9.9	8.4	7.3
Propane gas (commercial)	13.9	11.4	9.6	8.4
Butane gas (commercial)	14.1	11.6	9.8	8.5
Mixed gas (natural and carbureted water gas)	11.2	12.5	10.5	9.1
Carbureted water gas	17.2	14.2	12.1	10.6
Coke oven gas	11.2	9.2	7.8	6.8
Liquid fuels				
No. 1 and No. 2 fuel oil	15.0	12.3	10.5	9.1
No. 6 fuel oil	16.5	13.6	11.6	10.1
Solid fuels				
Bituminous coal	18.2	15.1	12.9	11.3
Anthracite	20.2	16.8	14.4	12.6
Coke	21.0	17.5	15.0	13.0

Source: ASHRAE, *Handbook of Fundamentals*, American Society of Heating, Refrigerating and Air-Conditioning Engineers, Atlanta, GA, 2013. Copyright ASHRAE, www.ashrae.org.

boiler, the flue gas composition of CO_2 concentration was 10%, while the difference between the flue gas temperature and the boiler room was 500°F; then, we note that the fractional boiler loss is 21.4%, i.e., the boiler has a combustion efficiency of (100 – 21.4 = 88.6%). Inspection of this table clearly reveals that combustion efficiency increases with decreasing flue gas temperature and increasing CO_2 concentration. The CO_2 concentrations and stoichiometric and excess airflow rates for a number of different fuels are assembled in Table 15.7.

Coal and fuel oil contain carbon and hydrogen along with sulfur, the combustion of all of which produces heat. However, sulfur oxide formed during combustion is a corrosive acid if dissolved in liquid water. To avoid corrosion of boilers and stacks, liquid water must be avoided anywhere in a boiler by maintaining sufficiently high stack temperatures to prevent condensation. Stainless-steel stacks and fireboxes provide an alternative solution, since they are not subject to corrosion; but they are very costly. In addition, sulfur oxides are one of the sources of acid rain. Therefore, these emissions must be carefully controlled.

15.6 Combined Heating and Power Systems

15.6.1 Introduction

Power generation systems create large amounts of heat in the process of converting fuel to electricity. More than two-thirds of the energy content of the input fuel converted to heat is wasted in many older central generating plants. As an alternative, an end user with significant thermal and power needs can generate both thermal and electric energy in a single CHP system located at or near its facility. Figure 15.9 gives a rather idealized conceptual example of the efficiency difference between separate and CHP. A typical CHP system converts 80 out of 100 units of input fuel to useful energy—30 to electricity and 50 to heat. By contrast, traditional separated heat and power components require 163 units of energy to provide the same amount of heat and power. Thus, with today's technologies, CHP systems can cut fuel use nearly 40% (Roop and Kaarsberg, 1999).

15.6.2 Available Technologies

Large-scale CHP systems have been widely used for several decades in such industries as pulp and paper, chemical manufacturing, and petroleum refining. Their use in commercial buildings is more recent, and such systems requires more careful design and operation due to the high diurnal and seasonal variability of electrical and thermal loads in commercial buildings.

Commercially, available CHP technologies for distributed generation include diesel engines, natural gas engines, steam turbines, gas turbines, microturbines, and phosphoric acid fuel cells. Table 15.8 summarizes the characteristics of commercial CHP prime movers. The table shows the wide range in CHP capacity—from 1 kW Stirling engine CHP systems to 250 MW gas turbines.

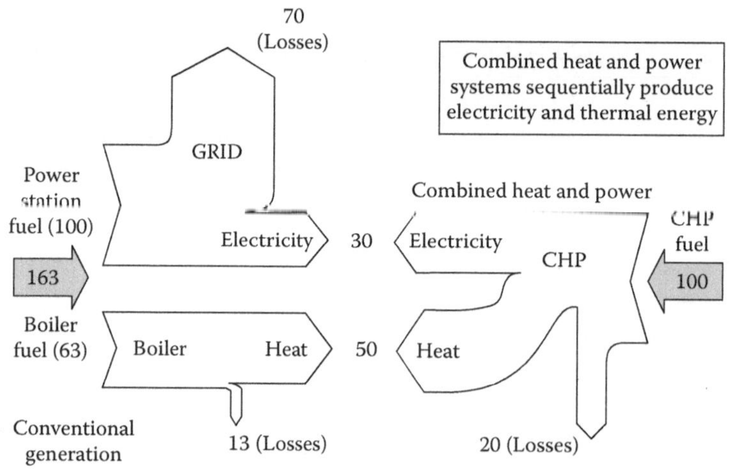

FIGURE 15.9
Illustrative example of energy flows in cogenerated vs. separate power and heat production technologies.

TABLE 15.8
Comparison of CHP Technologies

	Diesel Engine	Natural Gas Engine	Gas Turbine	Microturbine	Fuel Cells	Stirling Engine
Electric efficiency (LHV)	30%–50%	25%–45%	25%–40% (simple) 40%–60% (combined)	20%–30%	40%–70%	25–40%
Part load	Best	OK	Poor	Poor	N/A	OK
Size, MW	0.05–5	0.05–5	3–200	0.025–0.25	0.2–2	0.001–0.1
CHP installed cost, $/kW	800–1500	800–1500	700–900	500–1300	>3000	>1000
Start-up time	10 s	10 s	10 min–1 h	60 s	3–48 h	60 s
Fuel pressure, psi	<5	1–45	120–500	40–100	0.5–45	N/A
Fuels	Diesel, residual oil	Natural gas, biogas, propane	Natural gas, biogas, propane, distillate oil	Natural gas, biogas, propane, distillate oil	H_2, natural gas, propane	All
Uses for heat recovery	Hot water, LP steam, district heating	Hot water, LP steam, district heating	Heat, hot water, LP–HP steam hot water, district heating	Heat, hot water, LP steam	Hot water, LP steam	Direct heat, hot water, LP steam
CHP output, Btu/kWh	3400	1000–5000	3,400–12,000	4,000–15,000	500–3700	3000–6000
Usable temp. for CHP, °F	180–900	300–500	500–1,100	400–650	140–700	500–1000

Source: Onsite Sycom Energy Corporation, Market Assessment of CHP in the State of California, Report to the California Energy Commission, Carlsbad, CA, September 1999 (except for Stirling engine data).

CHP can operate on topping or bottoming cycles. In *topping cycles*, energy from the fuel generates shaft or electric power first, and thermal energy from the existing stream is recovered for other applications such as process heating and thermally activated cooling. This is sketched in Figure 11.21 where some of the electricity can operate electric vapor compression chillers, while the recovered waste heat can be used in absorption chillers. Note that there will still be a certain amount of heat wastage since the heat recovery device is never perfect. A rule of thumb is that 50%–60% of the waste heat from the gas turbine can be recovered. In a *bottoming cycle*, shaft or electric power is generated later from thermal energy left over meeting the demands of a higher-temperature thermal load. Typically, for larger loads (say, exceeding 5 MW), a *combined cycle* is also used, where the thermal output from the gas turbine prime mover is recovered and used either directly as heat, to generate additional shaft power (or electricity) using a steam turbine operating at a lower temperature.

15.6.3 Definitions of Different Thermodynamic Efficiency Metrics

Cogeneration or CHP systems, thus, capture the waste heat energy from electric generation to meet a wide variety of thermal needs, including hot water, steam, and process heating or cooling. The efficiency of the various equipment that comprise CHP systems (and most engineering equipment, for that matter) is better determined from manufacturer's catalogs or from actual testing. However, it is important to be conceptually clear on the manner in which the CHP industry defines overall system efficiencies since these can be defined in several ways. A somewhat ad hoc mix of first and second law efficiencies has been adopted by practitioners of this technology.

15.6.3.1 First Law Efficiency of Individual Equipment

Power efficiency of the prime mover (usually an internal combustion engine or a turbine, either gas or steam) characterizes its conversion efficiency from the heat contained in the combusted fuel to motive power or net useful power:

$$\eta_P = \frac{\text{Net power (useful)}}{\text{Rate of heat contained in combusted fuel}} = \frac{P}{\dot{m}_{fuel} \cdot HHV} \tag{15.13}$$

The efficiency of the boiler or heater, designated in this section as η_H, is given by Equation 15.11.

The *efficiency of the electric generator* is given as

$$\eta_G = \frac{\text{Net electric power}}{\text{Net motive power}} = \frac{E}{P} \tag{15.14}$$

Often, the electric generator is coupled to the prime mover, and the manufacturer will quote a combined thermal to electric efficiency that is referred to as "overall power efficiency":

$$\eta_E = \eta_P \cdot \eta_G \tag{15.15}$$

15.6.3.2 Ideal System Efficiencies

Though ideal Carnot efficiencies are never achieved in practice, they serve as benchmarks or reference efficiencies against which actual system efficiencies can be compared (see Sections 12.2 and 14.1). Consider the schematic in Figure 15.10a that shows a simple power plant cycle involving heat supply $\dot{Q}_h$ and heat rejection $\dot{Q}_l$ at constant source and sink temperatures, respectively (designated as T_h and T_l with temperatures in absolute units). Then, the ideal thermal efficiency of such a simple power plant is given by Equation 12.3 and expressed as follows:

$$\eta_P = \frac{P}{\dot{Q}_h} = \frac{T_h - T_l}{T_h} \tag{15.16}$$

Figure 15.10b depicts a combined cycle power plant where the heat rejected from the higher prime mover is used as a source of heat input to the lower prime mover.

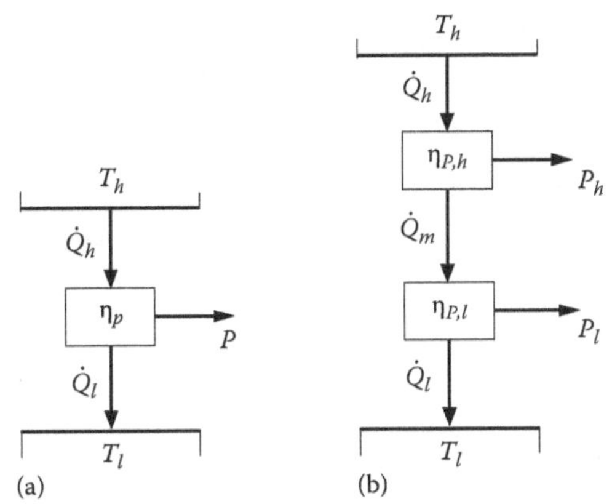

FIGURE 15.10
Schematics for deriving expressions for ideal Carnot efficiencies of (a) simple and (b) combined power cycles accepting heat from one high temperature and rejecting heating at a low temperature.

The expression for ideal *efficiency for a combined cycle power plant* in such a case is as follows:

$$\eta_P = \frac{P_h + P_l}{\dot{Q}_h} = \frac{\eta_{P,h} \cdot \dot{Q}_h + \eta_{P,l} \cdot \dot{Q}_m}{\dot{Q}_h} = \frac{\eta_{P,h} \cdot \dot{Q}_h + \eta_{P,l} \cdot (1 - \eta_{P,h}) \cdot \dot{Q}_h}{\dot{Q}_h} = \eta_{P,h} + \eta_{P,l} \cdot (1 - \eta_{P,h}) \tag{15.17}$$

where $\dot{Q}_m$ is the heat transfer between the high and low prime movers (see Figure 15.10), which is assumed to occur with no heat loss between the two cycles.

15.6.3.3 System Efficiencies

Consider Figure 15.11a that depicts a separate heat and power (SHP) plant comprised of a separate prime mover and a heating system. The heat content of the total fuel consumed per unit time ($\dot{Q}_{in,SHP}$) is the sum of that consumed by the heating equipment ($\dot{Q}_{in,H}$) and that consumed by the prime mover ($\dot{Q}_{in,P}$). The *overall efficiency of a SHP plant* is then

$$\eta_{SHP} \equiv \frac{E + \dot{Q}_u}{\dot{Q}_{in,P} + \dot{Q}_{in,H}} = \frac{E + \dot{Q}_u}{\dfrac{E}{\eta_P \cdot \eta_G} + \dfrac{\dot{Q}_u}{\eta_H}} \tag{15.18}$$

Consider the case when a CHP plant is operated such that the exhaust heat *exactly* meets the thermal load required $\dot{Q}_u$, while the prime mover generates electric power (see Figure 15.11b). The criterion of performance for the total CHP system efficiency is referred to as the "energy utilization factor" (*EUF*):

$$\text{EUF} \equiv \frac{E + \dot{Q}_u}{\dot{Q}_{in,CHP}} = \eta_{P,CHP} \cdot \eta_G (1 + \lambda) \tag{15.19}$$

where λ is the useful thermal to power ratio $= (\dot{Q}_u / E)$.

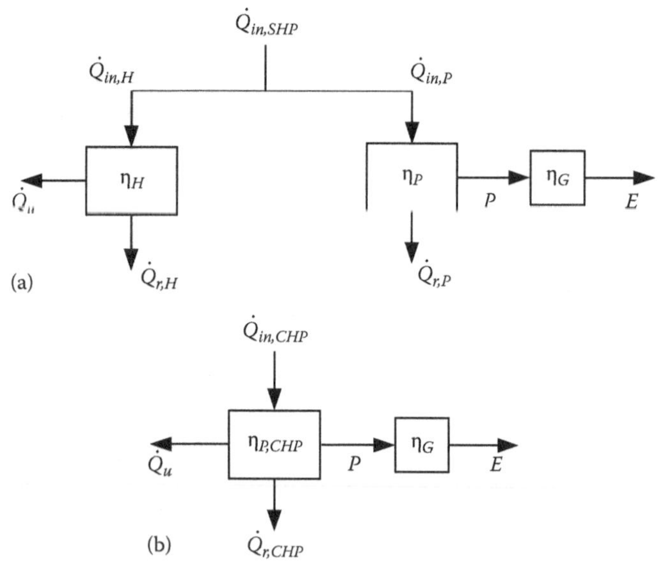

FIGURE 15.11
Schematics for deriving expressions for first law efficiencies of (a) SHP and (b) CHP systems.

A limitation of EUF is that both work and heat output are given equal values, which is contrary to the basic tenet of the second law of thermodynamics, which places more value on work as compared to heat. A *value-weighted EUF* (EUF_{VW}) is commonly used in practice:

$$\text{EUF}_{VW} \equiv \frac{E + f \cdot \dot{Q}_u}{\dot{Q}_{in,CHP}} = \eta_{P,CHP} \cdot \eta_G \cdot (1 + f \cdot \lambda) \quad (15.20)$$

where f is the weighting factor between heat and work. Typical range of f values proposed in the literature seems to be $0.33 < f < 0.5$, where the lower value of 0.33 is suggested by Horlock (1997) and the higher value has been adopted as the standard by the Federal Energy Regulatory Commission. Thus,

$$\text{EUF}_{VW}(\text{FERC}) = \frac{E + (\dot{Q}_u/2)}{\dot{Q}_{in,CHP}} \quad (15.21)$$

This discussion assumed that the thermal load and electric power needs of the application *exactly matched* those provided by the CHP. In a practical situation, and especially so in a CHP plant for a building(s), the instantaneous thermal and power loads to be met will usually vary greatly both diurnally and over the year. Hence, the CHP plant would have to be operated such that E and Q_u quantities can be varied (over a certain range of values) as required. Further, there may be a deficit in either electric load or thermal load outputs that will have to be met by purchases from the electric grid of from a separate boiler or heating system.* In such situations, overall CHP efficiency may not be the primary thermodynamic criterion of performance to consider. A more useful criterion with direct bearing to economic assessment of CHP plants is the one that characterizes the fuel savings ($\Delta Q_{in,save}$) from the CHP system compared to a SHP system:

$$\Delta Q_{in,save} = \sum^{hours} \left(\dot{Q}_{in,SHP} - \dot{Q}_{in,CHP} \right) \quad (15.22)$$

We can then define another performance measure, namely, the "fuel energy savings ratio" (*FESR*), which is the ratio of the fuel savings to the fuel required in the space heating and power plant:

$$\text{FESR} = \frac{\Delta Q_{in,save}}{Q_{in,SHP}} \quad (15.23)$$

These notions and equations are illustrated by the following example.

Example 15.8: Evaluating Benefits of CHP

The heat and power loads of an industrial site are 25 and 15 MW, respectively. These loads are currently being met by a boiler of efficiency 0.75 and a conventional power plant of efficiency 0.40. An analyst is evaluating the benefits of installing a CHP plant with a power efficiency of 0.25 and a heat recovery efficiency of 37.5% implying that the remaining 37.5% of the heat input is not recovered, i.e., wasted). Determine the following:

(a) The resulting energy savings and the FESR if the CHP is operated to meet the power loads
(b) The energy savings and the FESR if the CHP is operated to meet the heat loads
(c) The EUF in both the two cases assuming a weight factor between heat and work to be 0.5

Given:

$$Q_u = 25 \text{ MW}, \quad E = 15 \text{ MW}, \quad \eta_H = 0.75,$$
$$\eta_P \cdot \eta_G = 0.40, \quad \eta_{CHP,P} \cdot \eta_G = 0.25,$$
$$\eta_{CHP,HR} = 0.375, \quad f = 0.5$$

Find: $\Delta Q_{in,save}$, FESR, EUF
See Figure 15.12.

* One could also operate CHP plants such that excess electricity generated can be sold back to the electric grid and waste heat recovered in excess of the thermal load required is dumped. Usually, the latter situation is very rare since CHP operators tend to avoid such a strategy.

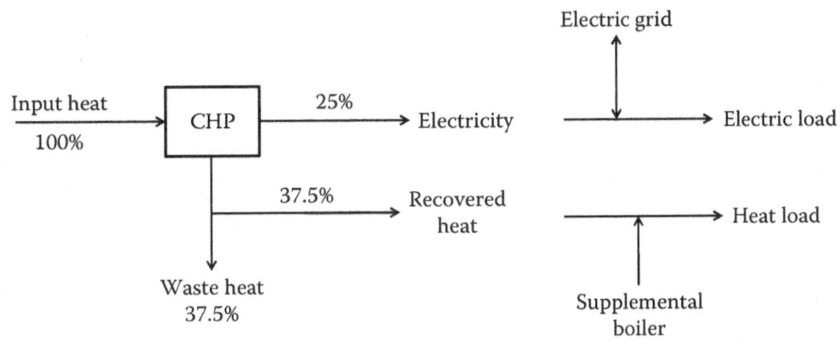

FIGURE 15.12
Schematic for the CHP system assumed in Example 15.8.

Solution

For a SHP plant, the total energy supply

$$\dot{Q}_{in,SHP} = \frac{\dot{Q}_u}{\eta_H} + \frac{E}{\eta_P \cdot \eta_G} = \frac{25}{0.75} + \frac{15}{0.4} = 70.83 \text{ MW}$$

(a) For a CHP plant operated to meet the power loads, the thermal energy supply should be

$$\dot{Q}_{in,CHP} = \frac{E}{\eta_{CHP,P} \cdot \eta_G} = \frac{15}{0.25} = 60 \text{ MW}$$

Thus, the CHP system will need 60 MW of thermal energy to meet the power load of 15 MW. The remaining (60 – 15) = 45 MW are available to meet the thermal load of 25 MW.

Since the heat recovery efficiency is 0.375 (Figure 15.12), the recovered heat that can be used to meet the thermal load is 60 × 0.375 = 22.5 MW. Thus, (25 – 22.5) = 2.5 MW, has to be met by the boiler. The boiler heat input needed is

$$\dot{Q}_{in,boiler} = 2.5/0.75 = 3.33 \text{ MW}$$

The energy savings if a CHP plant is installed is

$$\Delta Q_{in,save} = 70.83 - (60 + 3.33) = 7.5 \text{ MW}$$

FESR is given by Equation 15.23:

$$\text{FESR} = \frac{\Delta Q_{in,save}}{Q_{in,SHP}} = \frac{7.5}{70.83} = 0.106$$

(b) For a CHP plant operated to meet the thermal loads, the thermal energy supply should be

$$\dot{Q}_{in,CHP} = \frac{\dot{Q}_h}{\eta_{HR} \times (1 - \eta_{CHP} \cdot \eta_G)} = \frac{25}{0.5 \times (1 - 0.25)} = 66.67 \text{ MW}$$

Thus, the CHP system will need 66.67 MW of thermal energy to meet the thermal load of 25 MW. The prime mover will generate $E = \dot{Q}_{in,CHP} \times (\eta_{CHP} \cdot \eta_G) = 66.67 \times 0.25 = 16.67$ MW. Thus, a power excess of (16.67 – 15) = 1.67 MW is generated, which can be sold to the electric utility if such a sell back arrangement is in place.

The energy savings for the CHP plant is

$$\Delta Q_{in,save} = 70.83 - 66.67 = 4.16 \text{ MW}$$

FESR is given by Equation 15.23:

$$\text{FESR} = \frac{\Delta Q_{in,save}}{Q_{in,SHP}} = \frac{4.16}{70.83} = 0.059$$

(c) The FESR value calculated above does not combine electrical and thermal savings into a single number. Also as stated earlier, FESR does not consider the higher value of electric power as compared to thermal energy. This value-weighted term is given by $f = 0.5$, meaning that thermal energy is only half as valuable as power. This consideration leads to adopting Equation 15.21:

When CHP is sized to meet power load,

$$\text{EUF}_{VW} = \frac{E + f \cdot \dot{Q}_u}{\dot{Q}_{i,nCHP}} = \frac{15 + 25/2}{(60 + 3.33)} = 0.434$$

Similarly, when CHP is sized to meet thermal load,

$$EUF_{VW} = \frac{16.67 + 25/2}{66.67} = 0.438$$

Comments

The aforementioned example illustrates the performance characteristics of two different operating modes of a CHP. Note that both of them have $EUF_{VW} > \eta_P = 0.40$, indicating that overall CHP system is more energy efficient even though the CHP prime mover has a much lower conversion efficiency (assumed as 0.25) than that of the conventional power plant. Finally, the relative prices of thermal energy and electricity will determine which of the two modes is more cost-effective.

15.6.4 Designing CHP Systems for Buildings

CHP systems can provide cost savings as well as substantial emissions reductions for industrial, institutional, and commercial users. Selecting the right CHP technology for a specific application depends on many factors, including the amount of power needed, the duty cycle, space constraints, thermal needs, emission regulations, fuel availability, utility prices, and interconnection issues (Borbely and Kreider, 2001). Designing a technically and economically feasible CHP system for a specific building application requires detailed engineering and site data. Engineering information should include electric and thermal load profiles, capacity factor, fuel type, and performance characteristics of the prime mover (for a case study, see Maor and Reddy [2009]). Additional criteria can also be considered, for example, site-specific criteria such as maximum noise levels and footprint.

Turner (2006) distinguishes between three different types of CHP design options; this is important since this would impact the manner in which the CHP system would be subsequently operated and controlled during regular service. We have expanded the list to the following five CHP design options:

1. *Sized for isolated operation* where the site is standalone, i.e., the site does not have grid power, and hence, all the thermal and electric needs have to be met by the CHP system. Excess standby capacity for scheduled and unscheduled maintenance as well as momentary demand spikes and energy creep issues must also be considered.
2. *Sized by electric baseload* where the CHP is sized such that it meets the minimum electric billing demand (which can be gleaned from historic utility bills). Supplemental power is purchased from the electric grid, while any thermal energy shortfalls have to be met by a separate heating system.
3. *Sized by thermal baseload* where the CHP is sized so that most of the thermal energy is met with heat recovered from the prime mover, with any excess electric power sold to the electric grid and any shortfalls met by supplemental grid power.
4. *Sized for intermediate loads* where some amount of thermal load and some amount of electric load are met by the CHP plant. This is probably the most common design option since in actual reality the final CHP design and equipment sizing will depend on location-specific economics and issues such as energy security and reliability. Economic issues would involve considering not only the cost of thermal and electric energies but also operation and maintenance costs of the equipment as well as environmental costs.
5. *Sized for peaking loads* where the CHP system is specifically designed to curtail electric demand by utility peak shaving and thereby save on demand charges.

Successful on-site CHP projects require careful evaluation of the feasibility of the CHP system to the site under investigation. The process of performing the feasibility study for a CHP is a phased process starting from a preliminary screening study and ending with a detailed and comprehensive study. A feasibility study essentially consists of

1. Obtaining information on existing or proposed facilities including electrical, heating, and cooling load data
2. Developing technically feasible solutions to effectively and efficiently meet the facility's load requirements
3. Conducting an economic analysis that involves calculating estimated energy usage and cost, preparing budget cost estimates, and calculating life cycle costs to determine the recommended plant size and configuration.

State-of-the-art CHP installations with proper thermal/electric balance can have design efficiencies of 80% and will still result in significant overall energy savings. On-site use of CHP systems also reduces transmission and distribution system line losses to zero from typical central unit line losses of 4%–7%. The reader interested in a more in-depth understanding of CHP systems has numerous resources available to him. There are several formal textbooks that present the thermodynamics in a scientific manner (for example, Horlock, 1997; Khatchenko, 1998), as well as those meant more for the practicing engineer (for example, Borbely and Kreider, 2000; Orlando, 1996; Petchers, 2003), and also resource guides and market scoping studies (Aspen, 2000; MAC, 2005; Onsite, 2000).

Problems

Numbers 1–4 given in parenthesis denote the degree of difficulty.

15.1 A residence in Chicago uses a type 3 (Table 15.3) gas furnace for space heating with a thermostat setting of 72°F (22°C). During the immediately past winter with 6500°F-days (3600°C-days), 100 MBtu (105 GJ) of gas energy was used. If the thermostat had been set at 68°F (20°C), what would the gas consumption have been? *(2)*

15.2 The owner of a residence in Boston decides to replace her failed, old, standard oil-fired furnace with a new condensing type. How many gallons of oil could be saved if the annual heat load on this residence is 120 MBtu (127 GJ) on average? The heating value of 1 gal of oil is 140,000 Btu (39,000 kJ/L). If the cost of oil is \$1.05/gal (27¢/L), what is the annual cost saving due to the furnace upgrade, and what is the payback period if the more efficient furnace costs \$800 more than would replacement with the original design? *(2)*

15.3 Work Problem 15.2 for Denver, Colorado, where the annual load on the building is 125 MBtu (130 GJ). *(2)*

15.4 Flue gas analysis of a boiler gives the following results: CO_2, 9%; O_2, 2%; N_2, 87%; and CO, 2%. What is the excess air fraction? Would you suggest any adjustments to the boiler? If so, what is needed? *(2)*

15.5 Flue gas analysis of a boiler gives the following results: CO_2, 11%; O_2, 5.0%; N_2, 83.5%; and CO, 0.5%. What is the excess air fraction? Would you suggest any adjustments to the boiler? If so, what is needed? *(2)*

15.6 Equation 15.1 is the chemical equation for the complete combustion of methane, the principal constituent of natural gas. How many pounds (kilograms) of (a) oxygen and (b) standard air are needed to completely burn 10 lb (4.5 kg) of methane? *(2)*

15.7 Derive the result stated following Equation 15.1 that 8.7 ft^3 of standard air is required to combust 1 ft^3 of methane at sea level. *(5)*

15.8 If the heating value of natural gas is 1000 Btu/ft^3 (37 MJ/m^3) at standard conditions, what is the heating value at 4000 ft (1200 m)? *(1)*

15.9 Using a rule of thumb, estimate *for preliminary purposes only* what the *input* heat rating for a boiler would be in Chicago for a well-insulated, 200,000 ft^2 (20,000 m^2) warehouse. Check the calculation, assuming that the warehouse is square in plan, 10 ft (3 m) high, flat roofed, and insulated to R 20 (R 3.5) and has an air exchange rate of 0.7 per hour. Ignore floor slab losses because the perimeter is small relative to the building surface area, and use a reasonable boiler efficiency. What combination of boilers (specify the model numbers) would you select from Table 15.4 for this building if the owner requested that at least two boilers be installed, to ensure some building heat even if one boiler were to fail? Assume that the warehouse is kept at 60°F and the winter design temperature is 0°F. *(4)*

15.10 Work Problem 15.9 for Denver, Colorado, accounting for elevation effects on both boiler capacity and heat load. *(4)*

15.11 A large boiler is supplied with 90,000 ft^3/h (2550 m^3/h) of natural gas with a HHV of 1,000 Btu/ft^3 (37 MJ/m^3) and 40% excess air. If combustion is complete and the stack gases leave the boiler at 300°F (105°C), how much superheated steam at 20 psia and 300°F (140 kPa and 150°C) is produced? What is the boiler efficiency? Water is supplied to the boiler at 200°F (94°C), and boiler thermal losses (other than stack gas losses) are 5% of the heat rate. *(4)*

15.12 Describe in words the function of boiler controls. Why is their proper operation so important in an energy-conserving building? *(2)*

15.13 A boiler has a peak efficiency of 90% at the peak load of 5 MBtu/h (1465 kW). Figure P15.13 shows the part-load characteristic. If the actual load is 45% of the peak load, what is the efficiency? If the load is 80% of the peak? *(1)*

15.14 The part-load characteristics of a boiler are given in the table that follows. Find the coefficients in the part-load equation, Equation 15.7, assuming a linear characteristic (that is, *C* and higher-order

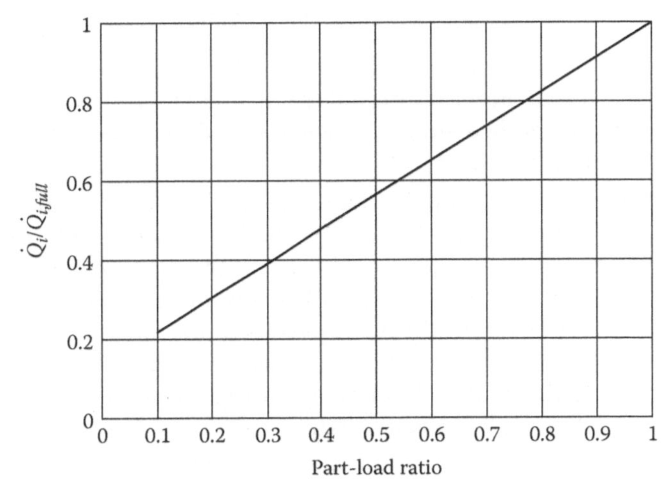

FIGURE P15.13

coefficients are zero). Plot the efficiency of the boiler as a function of PLR. *(2)*

PLR	0.2	0.4	0.6	0.8	1.0
$\dot{Q}_{i,full}$	0.31	0.44	0.63	0.81	1.00

15.15 A hospital in Denver, Colorado, requires heating to 72°F (22°C) during the entire heating season. The heat load at –2.5°F (–19°C) is 10.0 MBtu/h (2950 kW), and the balance-point temperature is estimated to be 62°F (17°C). The boiler capacity (corrected for altitude) is 11.0 MBtu/h (3240 kW) with a full-load efficiency of 82%. Either write a computer program or use spreadsheet software to solve this problem using the bin data for Denver. The part-load equation for the boiler is

$$\frac{\dot{Q}_i}{\dot{Q}_{i,full}} = 0.10 + 0.90(\text{PLR})$$

What are the annual energy consumption, the annually averaged efficiency, and the annual PLR of the boiler? *(4)*

15.16 A college dormitory in Pittsburgh requires heating to 69°F (20°C) during the entire heating season. The heat load at 1°F (–17°C) is 4.0 MBtu/h (1180 kW), and the balance-point temperature is estimated to be 62°F (17°C). The boiler capacity is 4.6 MBtu/h (1400 kW) with a full-load efficiency of 84%. Either write a computer program or use a spreadsheet with the bin data from the online HCB software to solve this problem. The part-load equation for the boiler is

$$\frac{\dot{Q}_i}{\dot{Q}_{i,full}} = 0.12 + 0.88(\text{PLR})$$

What are the annual energy consumption, the annually averaged efficiency, and the annual PLR of the boiler? *(4)*

15.17 The building in Example 15.5 can be heated either by the boiler specified in the example or by three boilers, each of one-third the capacity but with the same part-load characteristics. Since the boiler in the example has more capacity than needed most of the time, less energy may be used with smaller boilers operating more closely to their full capacity. Add the columns needed to the spreadsheet in Table 15.5 to study this option. For simplicity, assume that the lead boiler (the first one to be operated) operates up to full capacity before the second is activated; the same strategy is used to operate the second boiler relative to the third. *(4)*

15.18 The building in Example 15.5 can be heated either by the boiler specified in the example or by two boilers, one with one-third the example boiler capacity and one with two-thirds the capacity. Both have the same part-load characteristics as the boiler in Example 15.5. Since the boiler in the example has more capacity than needed most of the time, less energy may be used with smaller boilers operating more closely to their full capacity. Add the columns needed to the spreadsheet in Table 15.5 to study this option. For simplicity, assume that the smaller boiler operates up to full capacity before the second is activated. *(4)*

15.19 As the energy manager of a building, you want to establish an energy balance on a boiler to periodically check its efficiency. The heat input is easy since you know the amount of natural gas consumed. The boiler produces z lb of steam per hour with 10% new feedwater (10% of the steam is lost due to poor steam trap performance and consumption for air humidification) supplied at 55°F (13°C) and 90% condensate return at y°F. On average, 5% of the steam is exhausted for the purpose of solids control. Describe how you would do an energy balance on the steam side of the boiler and calculate the boiler efficiency. *(4)*

15.20 An energy audit of a large building discovers that 20 billion Btu (20 billion kJ) of energy is being lost per year due to steam leaks. If the steam is saturated at 130 psia (900 kPa), how much water is lost due to the steam leaks? If water costs $1.00 per 1000 gal ($0.25 per 1000 L), what is the value of the lost water? How does this cost compare to the cost of lost steam, valued at $3.00 per 1000 lb ($6.50 per 1000 kg)? *(2)*

15.21 Many university campuses in the United States are installing large cogeneration facilities to reduce their utility bills. Consider a hypothetical, unrealistically simple case where a university has a constant electric load of 10 MW year-round. The thermal load follows the *load duration curve* as shown in Figure P15.21a. Assume the price of electricity is constant at $0.035/kWh year-round and the price of gas is $0.45/therm year-round. Suppose gas turbine cogeneration systems are available in any size and that of the energy contained in the gas, 30%

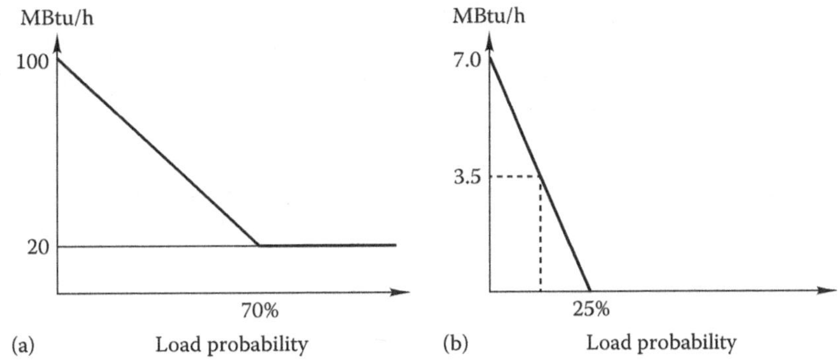

FIGURE P15.21
(a) University campus and (b) an office building. The line represents the fraction of time (*X* axis) that the load is at or above the corresponding value on the *Y* axis. For example, for the load duration curve on the right, the dotted line shows that 12.5% of the time the load is above 3.5 MBtu/h. Also note that these examples are highly simplified and that in practice the curves are less regular than shown here.

is converted to electricity and 45% is available as usable heat (with 25% as waste heat). What are the energy savings if (a) the system is sized to meet the electrical load (10 MW), (b) the system is sized to meet the peak thermal load (100 MBtu/h), and (c) the system is sized to meet the base thermal load (20 MBtu/h)? Assume boiler efficiency of 80% *(4)*

15.22 Consider an office building that uses a 500 kW generator to produce power on site and reduce the overall peak demand. Recovered heat from the generator is used to satisfy the thermal load. The thermal load duration curve for this building is shown in Figure 15.21b. Assume the electrical and thermal efficiencies of the generator are 30% and 50%, respectively, and that electricity costs 2¢/kWh with a demand charge of $20 per peak kilowatt and gas costs $6.50/MBtu. If the building electrical load is 200 kW during unoccupied hours and 1500 kW during occupied hours, do most of the savings come from reduced electricity use or from reduced gas consumption used for heating? Assume that thermal load is needed during occupied periods only and that CHP is operated only during occupied periods. *(5)*

References

ANSI/ASHRAE Standard 103 (2007). *Standard 103-2007: Method of Testing for Annual Fuel Utilization Efficiency of Residential Central Furnaces and Boilers*. American Society of Heating, Refrigerating and Air-Conditioning Engineers, Atlanta, GA.

ASHRAE 90.1 (2010). *Standard 90.1-2010: Energy Standard for Buildings Except Low-Rise Residential Buildings*. American Society of Heating, Refrigerating and Air-Conditioning Engineers, Atlanta, GA.

ASHRAE Fundamentals (2013). *Handbook of Fundamentals*. American Society of Heating, Refrigerating and Air-Conditioning Engineers, Atlanta, GA.

ASHRAE Systems (2012). *Handbook of HVAC Systems and Equipment*. American Society of Heating, Refrigerating and Air-Conditioning Engineers, Atlanta, GA.

ASME BPVC (2013). *Boiler and Pressure Vessel Code*. American Society of Mechanical Engineers, New York.

Aspen (March 2000). Combined heat and power: A federal manager's resource guide. Prepared by Aspen Systems Corporation for the US Department of Energy's Federal Energy Management Program, Rockville, MD.

Borbely, A.M. and J.F. Kreider (2001). *Distributed Generation: The Power Paradigm for the New Millennium*. CRC Press, Boca Raton, FL.

Dyer, D.F. and G. Maples (1991). *Boiler Efficiency Improvement*, 5th ed. Boiler Efficiency Institute, Auburn University, Auburn, AL.

Horlock, J.H. (1997). *Cogeneration-Combined Heat and Power (CHP): Thermodynamics and Economics*. Krieger Publishing Company, Malabar, FL.

Khartchenko, N. (1998). *Advanced Energy Systems*. Taylor & Francis, Washington, DC.

MAC (2005). Combined heat and power (CHP) resource guide. Prepared by Midwest CHP Applications Center, University of Chicago and Avalon Consulting Co., Chicago, IL.

Maor, I. and T.A. Reddy (December 2009). Chapter 21: Case study: Optimal sizing using computer simulations—New school, in *Sustainable On-Site CHP Systems: Design, Construction, Operations* (M. Meckler and L. Hyman, eds.). McGraw Hill, New York, pp. 345–354.

Mull, T.E. (2001). *Practical Guide to Energy Management for Facilities Engineers and Plant Managers*. ASME Press, American Society of Mechanical Engineers, New York.

Onsite (January 2000). The market and technical potential for combined heat and power in the commercial/institutional sector. Prepared by Onsite Sycom Energy Corporation for the US Department of Energy's Energy Information Agency, Carlsbad, CA.

Onsite Sycom Energy Corporation (September 1999). Market assessment of CHP in the State of California. Report to the California Energy Commission, Carlsbad, CA

Orlando, J. (1996). *Cogeneration Design Guide.* American Society of Heating, Refrigerating and Air-Conditioning Engineers, Atlanta, GA.

Petchers, N. (2003). *Combined Heating, Cooling and Power Handbook: Technologies and Applications.* Fairmont Press Inc./Marcel Dekker Inc., Lilburn, GA.

Pita, E.G. (2002). *Air Conditioning Principles and Systems*, 4th ed. Prentice Hall, Upper Saddle River, NJ.

Roop, J.M. and T.M. Kaarsberg (1999). Combined heat and power: A closer look, in *Proceedings of the 21st National Industrial Energy Technology Conference,* Houston, TX, May 12, 1999.

Turner, W.C. (2006). *Energy Management Handbook*, 5th ed. The Fairmont Press/Prentice Hall, Lilburn, GA.

16

Pumps, Fans, and System Interactions

ABSTRACT We start by first reviewing the principles that govern incompressible fluid flow in ducts and pipes. This is followed by a description of methods and relevant data to determine pressure losses in straight pipes and ducts, and due to fittings to be found in piping and ducting networks. We then address prime movers that include pumps and fans along with pertinent equations and graphical ways of representing their performance. Fan laws and pump affinity laws are a set of relations that predict the effect on fan or pump performance (pressure and power) when such quantities as temperature and pressure of fluid, operating speed, and fan or pump size area change; these concepts are presented. Next, we discuss issues to be considered when integrating prime movers into complete piping or ducting networks. We then describe different ways of fan control to vary airflow, namely, by dampers (used at each zone), inlet guide vanes (mounted on fans), and variable speed motors. Issues in duct design are discussed, and various methods of sizing ducts are presented with illustrative examples. Finally, we describe different ways by which fluid flow can be measured.

Nomenclature

A	Area, m² (ft²)
A, B	Calibration constants (Equation 16.53)
C	Coefficient applicable to venture and orifice flow meters (Equations 16.48 and 16.51)
C_f	Dimensionless duct fitting coefficient (Equation 16.23)
D	Diameter, m (ft)
D_h	Hydraulic diameter (Equation 16.7), m (ft)
E	Factor defined by Equation 16.49
f	Friction factor
g	Acceleration due to gravity = 9.81 m/s² (32.2 ft/s²)
g_c	Conversion factor in IP units, 32.17 $lb_m \cdot ft/(lb_f \cdot s^2)$
H	Head of pressure produced by fan or pump, m (ft)
H	Height of rectangular duct, m (in.)
h	Specific enthalpy, kJ/kg (Btu/lb_m)
h	"Head" referring to pressure (and pressure difference) in terms of the physical height of a fluid, m, mm (ft)
$h_{friction}$	Pressure drop due to friction expressed as head, m, mm (ft)
i	Current draw of hot-wire anemometer, A
K	Dimensionless proportionality coefficient (Equation 16.4)
K_f	Dimensionless pipe fittings coefficient (Equation 16.15)
L	Length, m (ft)
$\dot{m}$	Mass flow rate, kg/s (lb_m/h)
N	Fan or pump speed, rev/min or rpm
N_s	Specific speed of pump (Equation 16.29)
P	Perimeter, m (ft)
p	Pressure, Pa ($lb_f/in.^2$)
PLR	Part load ratio (Equation 16.43)
$\dot{Q}$	Rate of heat addition, W (Btu/h)
Re	Reynolds number
r	Radius, m (ft)
T	Temperature, K or °C (°R or °F)
u	Specific internal energy, kJ/kg (Btu/lb_m)
u_o	Tangential velocity at outer tip of fan, m/s (ft/min)
$\dot{V}$	Flow rate, m³/s or L/s (ft³/min)
v	Velocity or wind speed, m/s (mi/h, (ft/s)
v_o	Velocity of fluid leaving blade of fan, m/s (ft/min)
v_{to}	Tangential velocity of fluid leaving blade of fan, m/s (ft/min)
W	Width, m (ft)
$\dot{W}$	Power, W (Btu/h)
z	Elevation, m (ft)

Greek

β	Throat to pipe diameter ratio of fluid flow measuring instrument (Equations 16.49 and 16.51)
ε	Roughness factor
η	Efficiency
θ	Angle
μ	Dynamic or absolute viscosity, Pa·s $[lb_m/(ft \cdot s)]$
υ	Kinematic viscosity, m²/s (ft²/s)
ρ	Density, kg/m³ (lb_m/ft^3)
Δp	Pressure drop, Pa ($lb_f/in.^2$)

Subscripts

a	Air
act	Actual
eq	Equivalent
f	Fitting
H	Hot wire anemometer
h	Hydraulic
i	Inflow, inlet
l	Length
o	Outflow, outlet
q	Heat
SEF	System effect factor
shaft	Shaft work
std	Standard
sys	System
tot	Total

16.1 Modified Equation of Motion

The energy needed for heating or cooling buildings is supplied by liquids or/and gases that transfer heating/cooling from the primary equipment to the various rooms and zones of a building. The most common liquids are water (and aqueous solutions) and refrigerants, whereas the most common gas is atmospheric air. The equation representing the first law of thermodynamics for an open system was discussed in Section 2.5.2 and given by Equations 2.30 through 2.32. The same equation also governs fluid motion in pipes and ducts in buildings. It is most useful when expressed on a rate basis in steady state as

$$\dot{Q} - \dot{W}_{shaft} = \dot{m}_o\left(gz_o + \frac{v_o^2}{2} + h_o\right) - \dot{m}_i\left(gz_i + \frac{v_i^2}{2} + h_i\right) \quad (16.1)$$

where
- $\dot{Q}$ is the rate of heat addition
- $\dot{m}$ is the mass flow rate
- v is the average velocity
- z is the elevation above datum
- $\dot{W}_{shaft}$ is the shaft power
- h is the enthalpy
- o, i are the outflow and inflow points

In steady state, the conservation of mass law requires that the two mass flow rates be the same. If the heat flow and shaft work terms are zero in Equation 16.1, the resulting expression is the Bernoulli equation for incompressible isothermal flow:

$$gz_o + \frac{v_o^2}{2} + \frac{p_o}{\rho} = gz_i + \frac{v_i^2}{2} + \frac{p_i}{\rho} \quad (16.2\ \text{SI})$$

Customarily, the units used for density in IP applications of the Bernoulli equation are pound-mass per cubic foot. For dimensional consistency, the first and second terms on both sides of Equation 16.2 must be divided by the conversion factor g_c [32.2 $lb_m \cdot ft/(lb_f \cdot s^2)$]. No correction of this type is needed with SI units.

HVAC designers often express the pressure terms in Equation 16.1 in terms of "head," i.e., the physical height of a column of fluid (usually water) in a manometer such as that shown in Figure 2.1. Equation 16.1 is converted to the head basis by dividing it by the mass flow rate and acceleration of gravity g:

$$h_q - h_{shaft} = \left(z_o + \frac{v_o^2}{2g} + \frac{u_o}{g} + \frac{p_o}{\rho_o g}\right) - \left(z_i + \frac{v_i^2}{2g} + \frac{u_i}{g} + \frac{p_i}{\rho_i g}\right) \quad (16.3)$$

in which u represents the internal energy and the head terms h_q and h_{shaft} correspond to the heat and work terms in Equation 16.1; do not confuse the head terms with the enthalpy, both of which customarily use the same symbol h.

The Bernoulli equation (Equation 16.2) can be generalized to the case of isothermal flow induced by a pump through piping with friction by adding a friction term to the right-hand side. Such an equation is commonly referred to as the *extended Bernoulli equation* and is most often used in HVAC fluid flow design:

$$z_i + \frac{v_i^2}{2g} + \frac{p_i}{\rho g} + H_{pump} = z_o + \frac{v_o^2}{2g} + \frac{p_o}{\rho g} + h_{friction} \quad (16.4\ \text{SI})$$

where H_{pump} is the increase in pressure head induced by the pump.

The frictional head loss $h_{friction}$ is usually expressed as a multiple of the velocity head $v^2/(2g)$ as follows:

$$h_{friction} \equiv \frac{\Delta p_{friction}}{\rho g} = K\frac{v^2}{2g} \quad (16.5)$$

The dimensionless proportionality constant K depends on the flow situation, type of fluid, and size and shape of the fluid conduit.

The velocity v used in the friction term depends on the reference velocity for K. It is not necessarily either v_i or v_o. How to compute friction losses in pipes and ducts is discussed later.

Example 16.1: Open System

A pump has to deliver water from the first floor of a building to the roof at an elevation of 50 m. The absolute pressure at the pump entrance is 0.3 atmosphere and the discharge is at atmospheric pressure. The velocity of the water at the pump entrance is 1.0 m/s and that at the exit is 2.0 m/s. The friction in the pipe is 6 m of water. What is the required pump exit pressure?

Given: v_1 = 1.0 m/s, v_2 = 2.0 m/s, $z_0 - z_i$ = 50 m, $h_{friction}$ = 6 m

Assumptions: Isothermal flow

Find: H_{pump}

Lookup values: Acceleration due to gravity g = 9.81 m/s², Density of water ρ = 998.2 kg/m³

Solution

We convert the pressure difference specified in atmospheres into head in water gauge (WG) as

$$\frac{\Delta p}{\rho g} = \frac{(1-0.3)\ \text{atm} \times 101{,}325\ \text{Pa/atm}}{998.2\ \text{kg/m}^3 \times 9.81\ \text{m/s}^2} = 7.24\ \text{mWG}$$

We can now use Equation 16.5:

$$\begin{aligned} H_{pump} &= z_o - z_i + \frac{v_o^2 - v_i^2}{2g} + \frac{p_o - p_i}{\rho g} + h_{friction} \\ &= (50-0) + \frac{2^2 - 1^2}{2 \times 9.81} + 7.24 + 6 \\ &= 63.29\ \text{mWG or } 6.12\ \text{atm} \end{aligned}$$

Thus, the pump has to increase the pressure of the fluid by 6.12 atm for the aforementioned system to perform as stipulated.

16.2 Pressure Losses in Liquid and Air Systems

Flow of fluids through conduits (ducts and pipes) is subject to parasitic losses due to friction and bends/changes in cross section. In this section, methods for finding the pressure drop in pipes and ducts are summarized. Since the flow in HVAC systems is turbulent,* only turbulent pressure drop equations will be preferentially discussed.

16.2.1 Pressure Losses in Straight Piping

The pressure drop in a straight pipe is given by the D'arcy–Weisbach equation:

$$\Delta p_{friction} = f\left(\frac{L}{D_h}\right)\left(\rho\frac{v^2}{2}\right) \tag{16.6}$$

where

$\Delta p_{friction}$ is the pipe pressure drop
L is the pipe length
v is the average fluid velocity
D_h is the *hydraulic diameter*, defined as four times the ratio of flow area A (i.e., conduit cross section) to wetted perimeter P (e.g., the circumference of fully filled pipe or duct):

$$D_h \equiv \frac{4A}{P} \tag{16.7}$$

Note that the proportionality constant K in Equation 16.4 is related to the friction factor f by $K = f \times (L/D_h)$ in Equation 16.6 for straight pipe.

The friction factor f is found from an independent equation that involves fluid and pipe properties along with the average velocity v. A number of empirical expressions for f have been proposed. A simple, explicit equation is (Perry et al., 1999)

$$f = \frac{1.325}{\left\{\ln\left[\varepsilon/(3.7D_h) + 5.74/\text{Re}^{0.9}\right]\right\}^2} \tag{16.8}$$

where

ε/D_h is the roughness-to-diameter ratio, shown in Figure 16.1 for common types of pipe
Re is the *Reynolds number*, defined as $\text{Re} \equiv v \cdot D_h/\nu$
ν is the kinematic viscosity from Equation 2.5

Equation 16.8 is as accurate as the more commonly used implicit Colebrook equation for f given by

$$\frac{1}{\sqrt{f}} = -0.87 \times \ln\left(\frac{\varepsilon}{3.7 \times D_h} + \frac{2.52}{\text{Re} \times \sqrt{f}}\right) \tag{16.9}$$

* So-called low-temperature cooling systems may involve laminar flow at low loads. See Kreider (1985) for laminar-flow friction factor expressions.

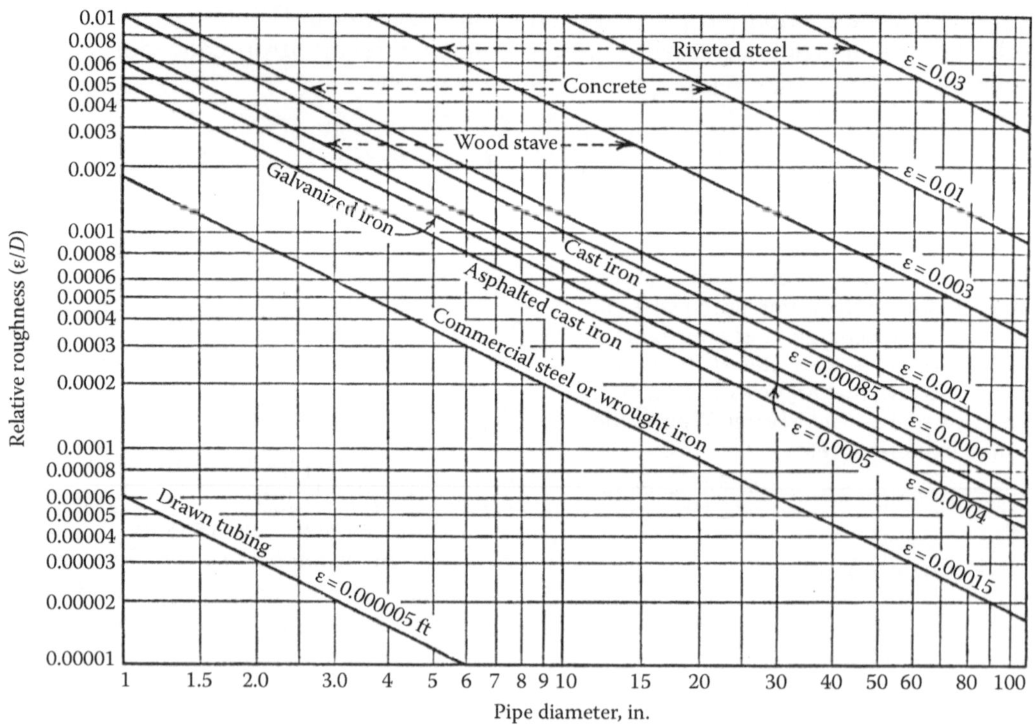

FIGURE 16.1
Pipe internal roughness ratio (ε/D) versus pipe diameter D (in.). The absolute roughness ε is shown in feet. (From Tuve, G., *Mechanical Engineering Experimentation*, McGraw-Hill, New York, 1961. With permission.)

Figure 16.2 is a plot of the friction factor as a function of the Reynolds number with the pipe roughness ratio as a parameter. For rapid calculations during preliminary design, it is adequate for estimating the friction factor. As shown on this chart, called the *Moody diagram* (Moody, 1944), turbulent flow exists for Reynolds numbers above about 3000. Also note that the friction factor is independent of the Reynolds number for sufficiently large values of Re for rough pipe. This is the area of the Moody diagram to the right of the dashed line.

Another method of estimating the pipe pressure drop is to use a dimensional chart such as that given in Figure 16.3a and b. Pressure drop (in feet of fluid flowing per 100 ft of pipe length) is plotted versus flow (in gallons per minute) with actual pipe diameter (in inches) as a parameter. Lines sloping downward from the left are used to find the fluid velocity. Figure 16.3a and b applies for pure water in a steel pipe that has been in place for several years. For new copper or smooth plastic pipe, the pressure drops read from this chart should be multiplied by the constant 0.62. Glycols are often used for freeze protection in HVAC system piping exposed to cold air in heating coils or elsewhere. Figure 16.3c is a plot of the correction that must be applied to values read from Figure 16.3a and b when ethylene glycol is used for freeze protection. In addition to increased pressure drops, glycol solutions have poorer heat transfer properties than pure water, as discussed in Section 2.1. Proper attention also needs to be given to the corrosive nature of glycols (Born, 1989).

Example 16.2: Using the D'arcy–Weisbach Equation

Find the flow given the pressure drop and pipe size. Water at 38° C flows through 400 mm cast-iron pipe with a pressure drop of 225 Pa/m. What is the volumetric flow rate?

Given: $\Delta p/L$ = 225 Pa/m (typical design rule of thumb), D = 400 mm

Fluid: Water at 38°C

Assumptions: Fully developed, turbulent flow

Find: $\dot{V}$

Lookup values: Kinematic viscosity at 38°C: $\nu = 0.6831 \times 10^{-6}\ \text{m}^2/\text{s}$

Pipe roughness ε = 0.00085 ft (from Figure 16.1) from which the roughness ratio is

$$\frac{\varepsilon}{D} = \frac{0.00085\ \text{ft} \times 304.8\ \text{mm/ft}}{400\ \text{mm}} = 0.00064$$

Solution

Equation 16.8 cannot be used to solve this problem directly since neither the friction factor f

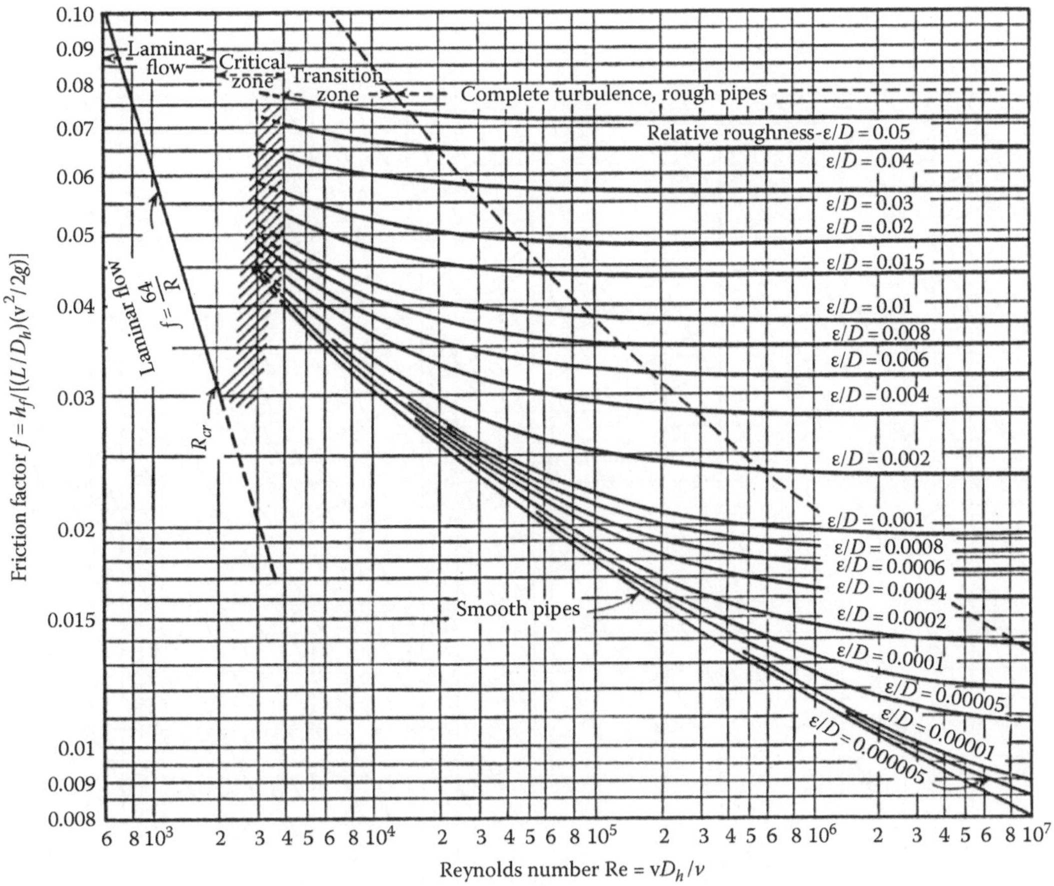

FIGURE 16.2
The Moody diagram with friction factor as a function of Reynolds number and relative roughness as a parameter. The hydraulic diameter is D_h. (From Tuve, G., *Mechanical Engineering Experimentation*, McGraw-Hill, New York, 1961. With permission.)

nor the velocity v in the Reynolds number is known. However, a rearrangement of the D'arcy–Weisbach (Equation 16.6) in terms of volumetric flow rate with $\Delta p = \rho g h_f$ and $\dot{V} = A\text{v}$ results in

$$\frac{h_{friction}}{L} = \frac{f(L/D)(\text{v}^2/2)}{gL} = f\frac{\dot{V}^2/(2D)}{g[(\pi/4)D^2]^2} \qquad (16.10)$$

Taking the square root of Equation 16.10 and solving for $f^{-1/2}$, we have

$$1/\sqrt{f} = \frac{\dot{V}\sqrt{8}}{\pi\sqrt{g(h_{friction}/L)D^5}} \qquad (16.11)$$

The Reynolds number can also be expressed as a function of volumetric flow rate as follows:

$$\text{Re} \equiv \frac{\text{v}D}{\upsilon} = \frac{\dot{V}}{(\pi/4)D^2} \times \frac{D}{\upsilon} = \frac{4\dot{V}}{\pi D\upsilon} \qquad (16.12)$$

When Equations 16.11 and 16.12 are inserted into Equation 16.9, we get

$$\dot{V} = -0.966\sqrt{gD^5(h_{friction}/L)} \times \ln\left[\frac{\varepsilon}{3.7D} + \frac{1.782\upsilon D}{\sqrt{gD^5(h_{friction}/L)}}\right] \qquad (16.13)$$

For this example, the quantity under the radical has the value [recall that $h_{friction} = \Delta p/(\rho g)$]

$$\sqrt{gD^5\left(\frac{h_{friction}}{L}\right)} = \left(9.81 \times 0.4^5 \times \frac{225}{9.81 \times 993}\right)^{1/2} = 0.0481\ \text{m}^3/\text{s} \qquad (16.14)$$

From Equation 16.13, the volumetric flow rate is

$$\dot{V} = -0.966 \times 0.0481 \times \left[\ln\left(\frac{0.00064}{3.7} + \frac{1.782 \times 0.6831 \times 10^{-6} \times 0.4}{0.0481}\right)\right]$$

$$= 0.400\ \text{m}^3/\text{s}\ (400\ \text{L/s},\ 6350\ \text{gal/min})$$

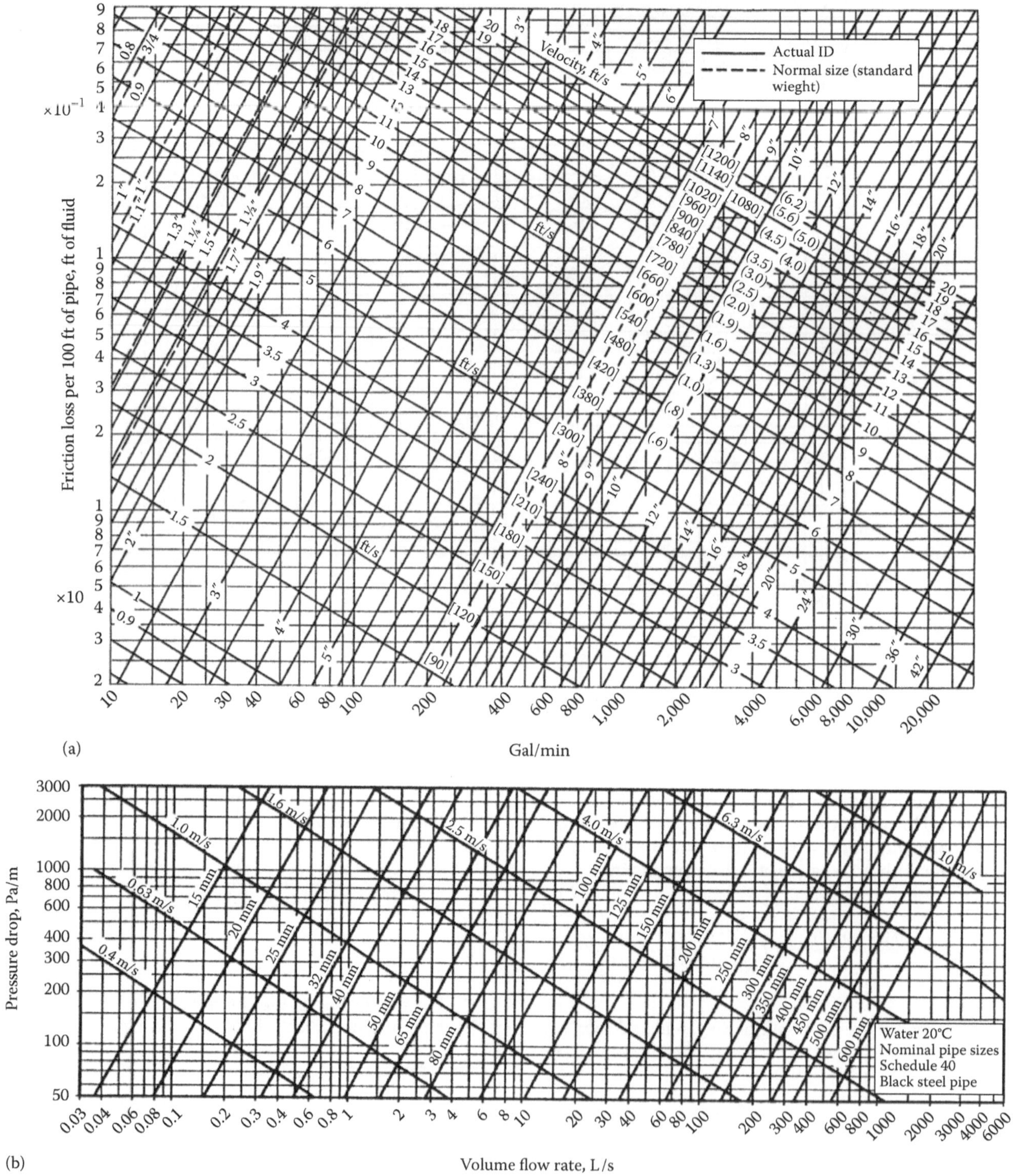

FIGURE 16.3
Pipe friction chart for seasoned steel pipe in (a) IP units in the range of 10–30,000 gal/min with friction factor ranging from 0.2 to 100 ftWG and (b) SI units. *(Continued)*

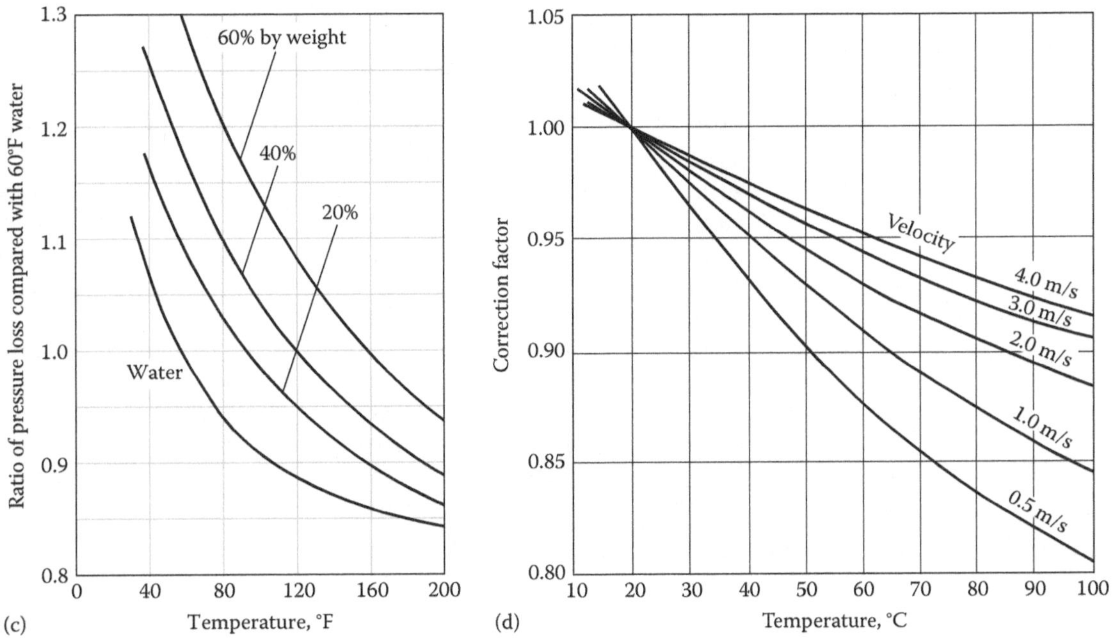

FIGURE 16.3 (*Continued*)
(c) Correction factor for ethylene glycol flowing in steel pipe. Multiplier is applied to values read from the pipe friction chart. (From Nussbaum, O.J., *Heat. Piping Air Cond.*, 62, 75, January 1990.) (d) Pressure drop correction factor for different water temperatures (to be applied to Figure 16.3b).

Comments

Designers often use rules of thumb such as 3–6 ftWG (0.9–1.8 mWG) of pressure drop per 100 ft (30 m) of pipe length (or Pa/m) for preliminary design. This example illustrates how to accurately find the flow in such a case. Of course, a rough solution can be estimated from Figure 16.3b, which gives a flow of 400 L/s that is consistent with the earlier example.

Note from Figure 16.3b that the pipe friction chart applies to a fluid temperature of 20°C. Since density is an important property that impacts pressure drop, the earlier example had to use fluid property values at the specified 38°C. A simpler approach is to use correction charts, as shown in Figure 16.3d, that allow corrections for temperature at different flow velocities. For example, the flow velocity in the earlier example is 3.18 m/s, and the correction factor for 38°C is close to 0.975. Thus, the actual pressure drop will be lower by this amount than that read off from Figure 16.2b.

Pipe sizes are stated in terms of the *nominal* diameter, which is roughly related to the actual inside diameter of the pipe. Table 16.1 is an excerpt from the detailed table of steel pipe properties available in the online HCB software. For a standard pipe (schedule 40), a nominal 1 in. pipe is seen to have an outside diameter of 1.315 in. and an inside diameter of 1.049 in. with a resulting wall thickness of 0.133 in. An extrastrong pipe (schedule 80) of the same nominal size has the same outside diameter but a smaller inside diameter since its wall thickness is greater. Plastic pipe is manufactured in dimensions based on standard steel dimensions but with greater wall thickness for the same operating pressure.

Nominal copper piping sizes are all 1/8 in. less than the physical outside diameter. Table 16.2 assembles sizes available in type L copper pipe. Type L is the standard weight of drawn copper piping for pressurized piping applications. The online HCB software contains additional data on type K piping (a thicker, soft tempered tubing suitable for field bending) and type M tubing (a thinner-walled material that cannot be bent and is intended for lower-pressure use).

The selection of *pipe size* involves a trade-off between noise, material cost, and pumping power. We note from Example 16.2 that the pressure drop (and pumping power) decreases with the fifth power of the pipe diameter, whereas pipe cost increases roughly linearly with the diameter. One manufacturer recommends the sizes of piping for specific flow rates shown in Table 16.3. According to Waller (1990), noise is produced by rapid flow transitions; in straight pipe, water without entrained air produces little noise at velocities below 30 ft/s (10 m/s). Erosion (in copper piping) is another consideration that limits water velocities to no more than 6 ft/s (2 m/s). ASHRAE Fundamentals (2013) recommends that velocities be limited to 5 ft/s (1.7 m/s) in plastic pipe.

TABLE 16.1

Standard Dimensions of Selected Small Steel Pipes

Nominal Pipe Size, in.	Nominal Metric Size, mm	Schedule No.	Outside Diameter, in.	Inside Diameter, in.	Wall Thickness, in.	Flow Area, in.^{2}
½	13	40 (S)	0.840	0.622	0.109	0.3040
		80 (X)	0.840	0.546	0.147	0.2340
¾	19	40 (S)	1.050	0.824	0.113	0.5330
		80 (X)	1.050	0.742	0.154	0.4330
1	25	40 (S)	1.315	1.049	0.133	0.8640
		80 (X)	1.315	0.957	0.179	0.7190
1¼	32	40 (S)	1.660	1.380	0.140	1.495
		80 (X)	1.660	1.278	0.191	1.283
1½	38	40 (S)	1.900	1.610	0.145	2.036
		80 (X)	1.900	1.500	0.200	1.767
2	50	40 (S)	2.375	2.067	0.154	3.355
		80 (X)	2.375	1.939	0.218	2.953

S, standard pipe; X, extrastrong pipe. 1 in. = 2.540 cm; 1 in.^{2} = 6.452 cm^2. See also the online HCB software.

TABLE 16.2

Standard Dimensions of Copper Pipe

Classification	Nominal Tube Size, in.	Outside Diameter, in.	Stubbs Gauge	Wall Thickness, in.	Inside Diameter, in.	Flow Area, in.^{2}
Type L	⅜	½	19	0.035	0.430	0.146
	½	⅝	—	0.040	0.545	0.233
	¾	⅞	—	0.045	0.785	0.484
	1	1⅛	—	0.050	1.025	0.825
	1¼	1⅜	—	0.055	1.265	1.256
	1½	1⅝	—	0.060	1.505	1.78
	2	2⅛	—	0.070	1.985	3.094
	2½	2⅝	—	0.080	2.465	4.77
	3	3⅛	—	0.090	2.945	6.812
	3½	3⅝	—	0.100	3.425	9.213

1 in. = 2.540 cm; 1 in.^{2} = 6.452 cm^2. See also the online HCB software.

16.2.2 Pressure Losses in Pipe Fittings

Straight lengths of pipe are connected by elbows and tees and interrupted by valves to control flow. They must also include ancillary equipment such as filters. Pipe fittings such as valves, elbows, tees, and other equipment have pressure drops (referred to as *minor losses*) that are of the same order of magnitude as pressure losses through a straight pipe. As noted earlier, the pressure drop through fittings is given by Equation 16.4. We denote the pressure loss coefficient for fittings by K_f. Therefore, the fitting pressure loss is

$$\Delta p_{friction} = K_f \rho \frac{v^2}{2} \tag{16.15}$$

Or in terms of head of flowing fluid, the pressure loss is

$$h_{friction} = K_f \frac{v^2}{2g} \tag{16.16}$$

The total pressure loss in a single loop of uniformly sized pipe consisting of lengths of straight pipe and fittings is

$$h_{friction,tot} = \frac{v^2}{2g}\left(\frac{fL}{D} + \sum_{fittings} K_f\right) \tag{16.17}$$

TABLE 16.3

Recommended Pipe Sizes, Flows, and Velocities

	Steel Pipe				Copper Tubing			
Nominal Pipe or Tubing Sizes, in.	Minimum gal/min	Velocity, ft/s	Maximum gal/min	Velocity, ft/s	Minimum gal/min	Velocity, ft/s	Maximum gal/min	Velocity, ft/s
½	—	—	1.8	1.9	—	—	1.5	2.1
¾	1.8	1.1	4.0	2.4	1.5	1.0	3.5	2.3
1	4.0	1.5	7.2	2.7	3.5	1.4	7.5	3.0
1¼	7.2	1.6	16	3.5	7	1.9	13	3.5
1½	14	2.2	23	3.7	12	2.1	20	3.6
2	23	2.3	45	4.6	20	2.1	40	4.1
2½	40	2.7	70	4.7	40	2.7	75	5.1
3	70	3.0	120	5.2	65	3.0	110	5.2
3½	100	3.2	170	5.4	90	3.1	150	5.2
4	140	3.5	230	5.8	130	3.5	210	5.6
5	230	3.7	400	6.4	—	—	—	—
6	350	3.8	610	6.6	—	—	—	—
8	600	3.8	1200	7.6	—	—	—	—
10	1000	4.1	1800	7.4	—	—	—	—
12	1500	4.3	2800	8.1	—	—	—	—

Source: Courtesy of Taco, Inc., Cranston, RI.

Note that the pipe length L includes the length of both the straight pipe and the fittings. The K_f factors account for the *excess* pressure drop above that which would occur if the physical length of the fitting were replaced by straight pipe of the same length. Figure 16.4 shows sketches of common fittings used in HVAC systems. The K_f value for many fittings is a strong function of pipe size. This information is contained in Figure 16.5. The left column lists the fitting type, and the left scale is the internal pipe diameter in inches. The right scale shows the K_f values. Each large dot in the center of the figure has a number associated with it corresponding to an entry in the list to the left. To find the value of K_f for a specific fitting, pass a straightedge through the pipe diameter and the dot corresponding to the fitting. The extension of the straightedge to the right scale gives the value of K_f. Fittings for which K_f is independent of the pipe size have their coded dots located on the K_f axis in Figure 16.5. If a size change occurs in a fitting (e.g., items 46–55 in Figure 16.5), the velocity to be used in Equations 16.16 and 16.17 is the velocity in the smaller-diameter connection.

Example 16.3: Pumped-Fluid Pipe Circuit

Figure 16.6 shows a simple circuit for pumping pure ethylene glycol from an outdoor holding tank to an indoor tank at the same elevation. Plastic pipe is used; its actual inside diameter is 5.05 in. (nominal 5 in. pipe), and the roughness height is 0.00002 ft. If the flow rate is 500 gal/min on a day when the fluid temperature is 77°F, what is the pump shaft input power requirement if the pump efficiency is 67%?

Given: D = 5.05 in. = 0.42 ft, ϵ = 0.00002 ft, $\dot{V}$ = 500 gal/min = 1.12 ft³/s, $\eta_{pump} = 0.67$

Figure: See Figure 16.6.

Find: $\dot{W}_{pump}$

Lookup values: Glycol viscosity $\mu = 109 \times 10^{-4}$ $lb_m/(ft \cdot s)$
Glycol density $\rho = 68.47$ lb_m/ft^3
Both are from the online HCB software.

Solution

The first step is to determine the K_f values for each fitting by using Figure 16.5. For the square-edged exit from the tank (fitting 23), K_f = 0.5. Likewise, for the pipe exit to the receiving tank, K_f = 1.0 (one velocity head is lost). The flanged gate valve (assumed to be fully open) has K_f = 0.04 according to the nomograph (fitting 35). Finally, the flanged swing check valve has K_f = 2.0 (fitting 32). The sum of the K_f values is 3.54.

In order to determine friction drop in the 350 ft of pipe, several steps are needed. The mean velocity is

$$v = \frac{\dot{V}}{A} = \frac{1.12\ \text{ft}^3/\text{s}}{(\pi/4) \times (5.05/12)^2\ \text{ft}^2} = 8.0\ \text{ft/s}$$

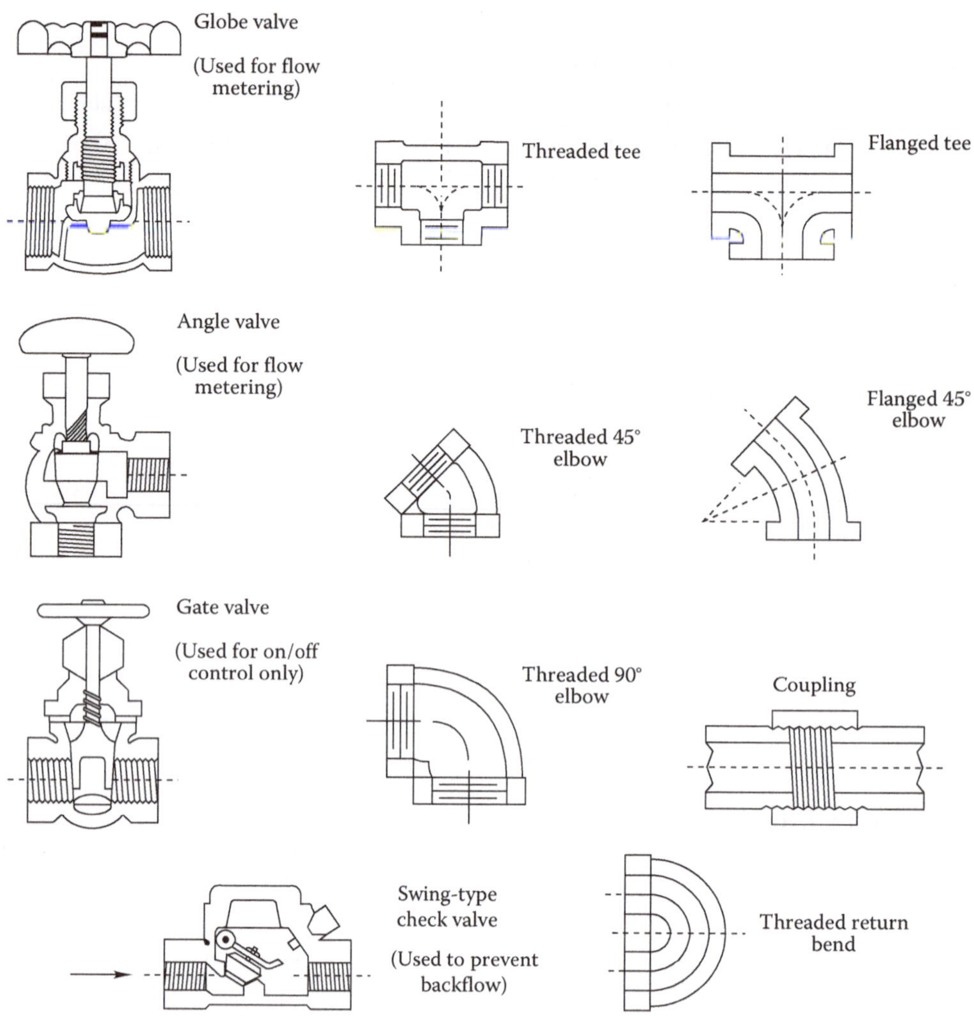

FIGURE 16.4
Typical pipe fittings used in HVAC systems.

Next, the Reynolds number is calculated:

$$\mathrm{Re} = \frac{8.0\ \mathrm{ft/s} \times (5.05/12)\ \mathrm{ft} \times 68.47\ \mathrm{lb_m/ft^3}}{109 \times 10^{-4}\ \mathrm{lb_m/(ft \cdot s)}} = 21{,}150$$

Since the Reynolds number is above the laminar limit (~3000), the flow is turbulent. Equation 16.8 is used to find the friction factor *f*:

$$f = \frac{1.325}{\{\ln[0.00002/(3.7 \times 0.42) + 5.74/21{,}150^{0.9}]\}^2} = 0.0256$$

The reader can confirm this value approximately by referring to Figure 16.2. The pump head is found from Equation 16.16:

$$h_{friction,tot} = \frac{8.0^2}{2g}\left[0.0256 \times \left(\frac{350}{0.42}\right) + 3.54\right] = 24.8\ \mathrm{ft}$$

Finally, the power imparted to the fluid is found from

$$\dot{W}_{pump} = \dot{V}\Delta p = \dot{V}\rho g h_{friction,tot}$$

$$= 1.12\ \mathrm{ft^3/s} \times \left[\frac{68.47\ \mathrm{lb_m/ft^3}}{32.2\ \mathrm{lb_m \cdot ft/(lb_f \cdot s^2)}} \times 32.2\ \mathrm{ft/s^2} \times 24.8\ \mathrm{ft}\right]$$

$$= 1902\ \mathrm{ft \cdot lb_f/s} = 3.46\ \mathrm{hp} \qquad (16.18)$$

Since the pump is 67% efficient, the pump shaft input power is

$$\dot{W}_{shaft} = \frac{3.46}{0.67} = 5.2\ \mathrm{hp}$$

Comments

Where IP units are used, pumps are rated in terms of gallons per minute of flow and feet of head. The solution of this problem is given in *feet of glycol*, not feet of water, the usual fluid used to specify the pump pressure rise.

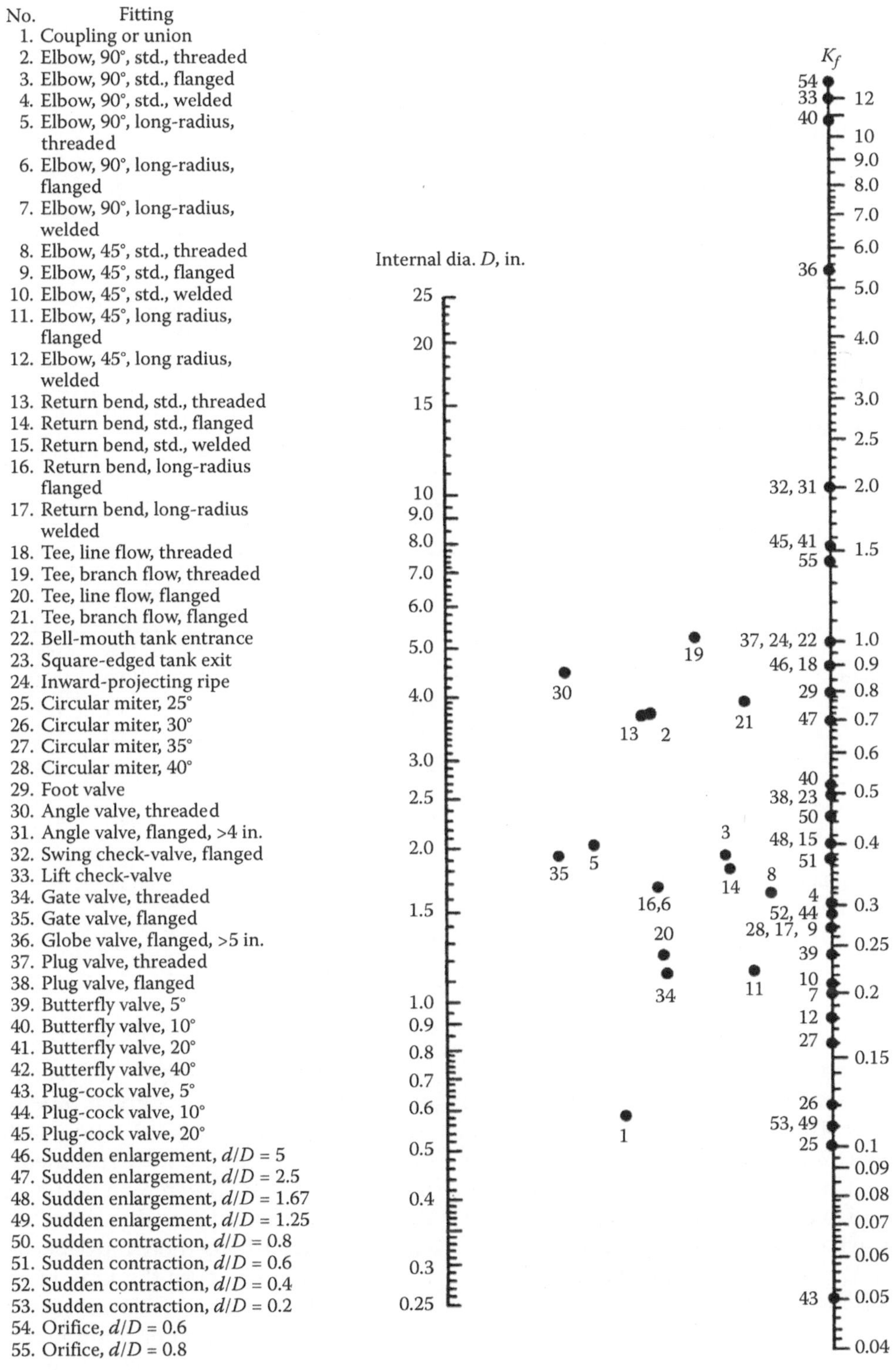

FIGURE 16.5
Nomograph for pipe fitting friction factors. (From Rao, K.V.K., *Chem. Eng.*, 89, 1982.)

During preliminary design, piping layouts are not drawn in sufficient detail to allow a proper calculation of fitting pressure drops. At this stage of design, it is appropriate to estimate fitting losses as 50% of the straight-pipe loss (which can be determined during early design). To this must be added the pressure drops of control valves and equipment such as coils, boilers, and chillers.

Pressure losses of fittings are sometimes presented in terms of the *equivalent length* of straight pipe that would have the same pressure loss as the fitting (see Table 16.4).

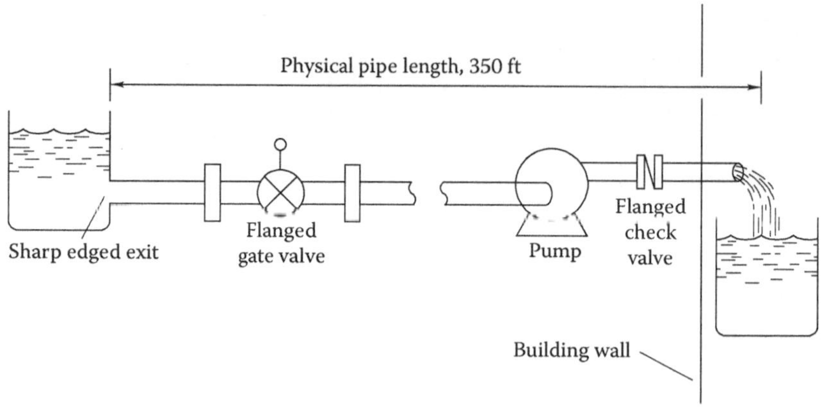

FIGURE 16.6
Piping diagram for Example 5.2.

TABLE 16.4
Equivalent Feet of Pipe for Fittings and Valves

	Nominal Pipe Size, in.												
	½	¾	1	1¼	1½	2	2½	3	4	5	6	8	10
45° elbow	0.8	0.9	1.3	1.7	2.2	2.8	3.3	4.0	5.5	6.6	8.0	11.0	13.2
90° elbow standard	1.6	2.0	2.6	3.3	4.3	5.5	6.5	8.0	11.0	13.0	16.0	22.0	26.0
90° elbow long	1.0	1.4	1.7	2.3	2.7	3.5	4.2	5.2	7.0	8.4	10.4	14.0	16.8
Gate valve open	0.7	0.9	1.0	1.5	1.8	2.3	2.8	3.2	4.5	6.0	7.0	9.0	12.0
Globe valve open	17	22	27	36	43	55	67	82	110	134	164	220	268
Angle valve	7	9	12	15	18	24							
Tee-side flow	3	4	5	7	9	12	14	17	22	28	34	44	56
Swing check valve	6	8	10	14	16	20	25	30	40	50	60	80	100
Tee-straight throughflow	1.6	2.0	2.6	3.3	4.3	5.5	6.5	8.0	11.0	13.0	16.0	22.0	26.0
Radiator angle valve	3	6	8	10	13								
Diverting tee		20	14	11	12	14	14	14					
Flow check valve		27	42	60	63	83	104	125	126				

A simpler and more convenient approach of determining pressure losses in fittings and valves is to use tables such as the abbreviated one shown in Table 16.4. The equivalent length of straight pipe for the particular fitting is directly determined for the corresponding nominal pipe diameter. Thus, for a open gate valve fitted to a 4 in. pipe, the equivalent length is 4.5 ft of the straight pipe. If a fitting is to be replaced by an equivalent length L_{eq} of pipe, then the following equality must hold, according to Equations 16.6 and 16.16:

$$f\left(\frac{L_{eq}}{D}\right)\left(\rho\frac{v^2}{2}\right) = K_f\rho\frac{v^2}{2} \qquad (16.19)$$

From this, we can solve for the equivalent length:

$$L_{eq} = \frac{K_f}{f}D \qquad (16.20)$$

This relationship shows the fundamental shortcoming with the equivalent-length approach. Even though K_f and D are constant for a given pipe fitting under various flow conditions, the friction factor is *not*, unless one is operating in the fully turbulent region where f is independent of Re (the region to the right of the dashed line in Figure 16.2). Since this is not always the case, the equivalent-length method must be used with caution. Later, when we discuss the flow of air in ducts where the Reynolds number is lower, the equivalent-length method is an even poorer approximation. The equivalent-length method is *not recommended* for use in HVAC design involving air ducts.

The pressure drop for *steam flow* in steel pipe can be found from Figure 16.7 in a two-step process. Figure 16.7a is used to find the pressure drop and velocity for atmospheric-pressure steam. Then, Figure 16.7b is used to correct the velocity for the actual steam pressure. A correction for the pressure drop is not made. In Figure 16.7a, an example is shown by the dashed line.

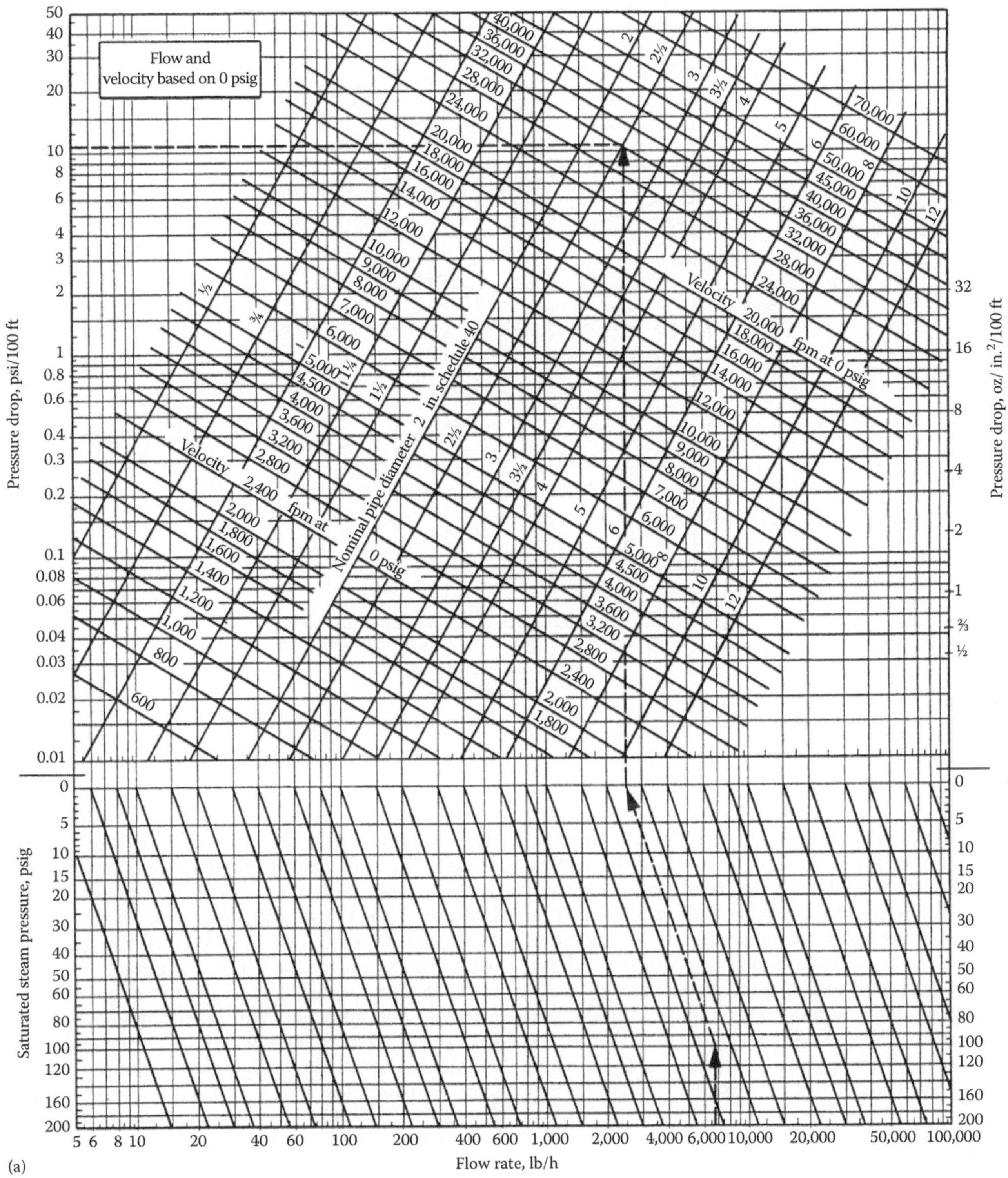

(a)

FIGURE 16.7
(a) Steam flow friction charts. After you find the values for 0-psig steam, use the second part of the figure to correct for the actual steam pressure. Dashed lines represent the text example. (From ASHRAE, *Handbook of Fundamentals*, American Society of Heating, Refrigerating and Air-Conditioning Engineers, Atlanta, GA, 2001. Copyright ASHRAE, www.ashrae.org.) *(Continued)*

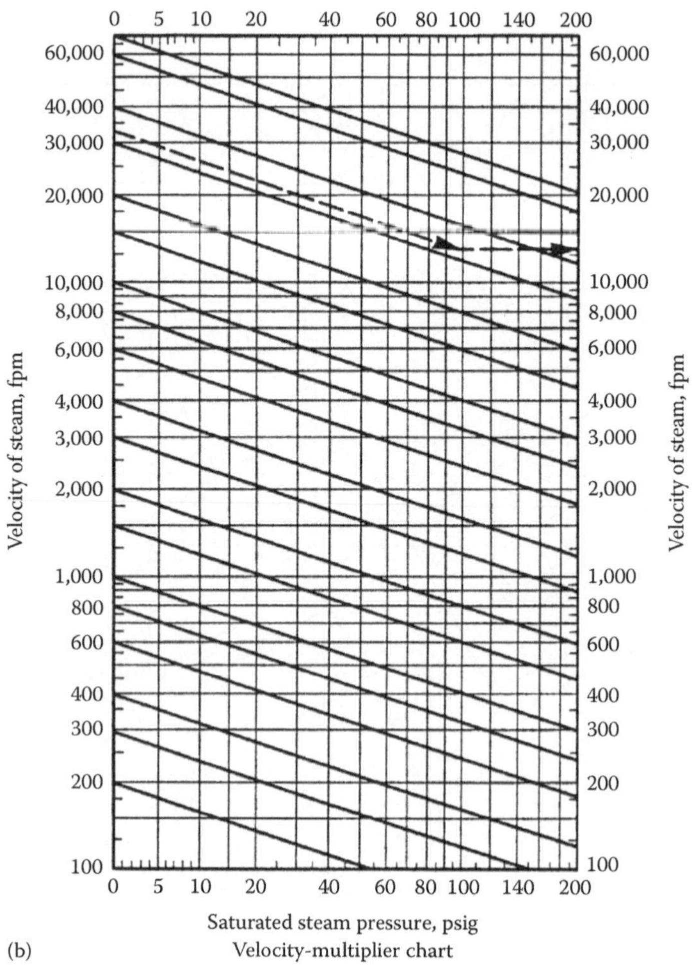

(b)

FIGURE 16.7 ***(Continued)***
(b) Chart used to adjust 0 psig velocities from Figure 16.7a.

Steam at 100 psig and 6700 lb_m/h is flowing through a pipe with a pressure drop of 11 psi/100 ft. The intersection of the specified pressure drop line and the vertical steam flow line (about 2,500 lb_m/h at 0 psig) indicates that a 2.5 in. line should be used and that a 32,000 ft/min velocity of 0 psig steam would result. The fact of entering Figure 16.7b at 32,000 ft/min and dropping down and to the right indicates that the actual velocity of 100 psig steam would be about 13,000 ft/min. Typical steam velocities should not exceed 15,000 ft/min. Additional information on steam piping is contained in Grimm and Rosaler (1990) and ASHRAE Fundamentals (2013).

16.2.3 Pressure Losses in Ducts and Duct Fittings

The dimensionless equations governing the pressure loss in airflow confined in circular or rectangular ducts are the same as those that apply for the flow of water, Equations 16.6 and 16.8. Likewise, the pressure drop through duct fittings is given by an expression similar to Equation 16.15.

In dimensional form, however, one cannot use the pipe friction chart in Figure 16.3 for airflow. Instead Figure 16.8, prepared for standard air (ASHRAE Fundamentals, 2013), must be used. This particular chart applies for round, galvanized-steel ducts with joints every 48 in. (equivalent to an average roughness $\epsilon = 0.0003$ ft). The chart may be used without correction (to ±5% accuracy) for altitudes up to 1500 ft, for temperatures between 40°F and 100°F, for other duct materials with medium smoothness, and duct pressurizations of ±20 inWG. For materials or environments outside this range, the friction factor expression, Equation 16.8, must be used. Note that common fiberglass ductwork is not classified as medium smooth, and Figure 16.8 should not be used for it. Wright (1945) describes the construction of the duct friction chart. The norm in IP units is to express pressure drop in ducts in inches of water gauge per 100 ft of straight duct and that for pipes in feet of water head per 100 ft of straight pipe.

Since ducts consume a relatively large building volume compared to pipes, it is sometimes necessary to use rectangular ducts to fit within confined spaces between floors or between the ceiling and the floor above in commercial buildings. Rectangular ducts of the same cross-sectional area as round ducts have a higher frictional loss. However, Figure 16.8a can be used for rectangular ducts if an *equivalent diameter for friction* is used instead. The equivalent diameter for friction can be calculated from

$$D_{eq} = 1.30 \times \frac{(WH)^{0.625}}{(W+H)^{0.25}} \tag{16.21}$$

where

D_{eq} is the diameter of circular duct with volumetric flow rate and friction drop equivalent to specified rectangular duct, m (in.)

W is the width of rectangular duct, m (in.)

H is the height of rectangular duct, m (in.)

Note that a circular duct with this diameter will *not* have the same average velocity as the rectangular duct. Rectangular ducts should be as square as possible. Aspect ratios of 4:1 or less are best (if space permits), and ratios greater than 8:1 should be avoided. An additional complication present in airflows but not in water flows is the variation of the pressure drop with density. According to Equation 16.6, the pressure drop at a given velocity (not a given mass flow rate) is proportional to the density. Therefore, if one knows the pressure drop at a given density (condition 1), the pressure drop at a different density (condition 2) is given by

$$h_{friction,2} = h_{friction,1} \times \frac{\rho_2}{\rho_1} \tag{16.22}$$

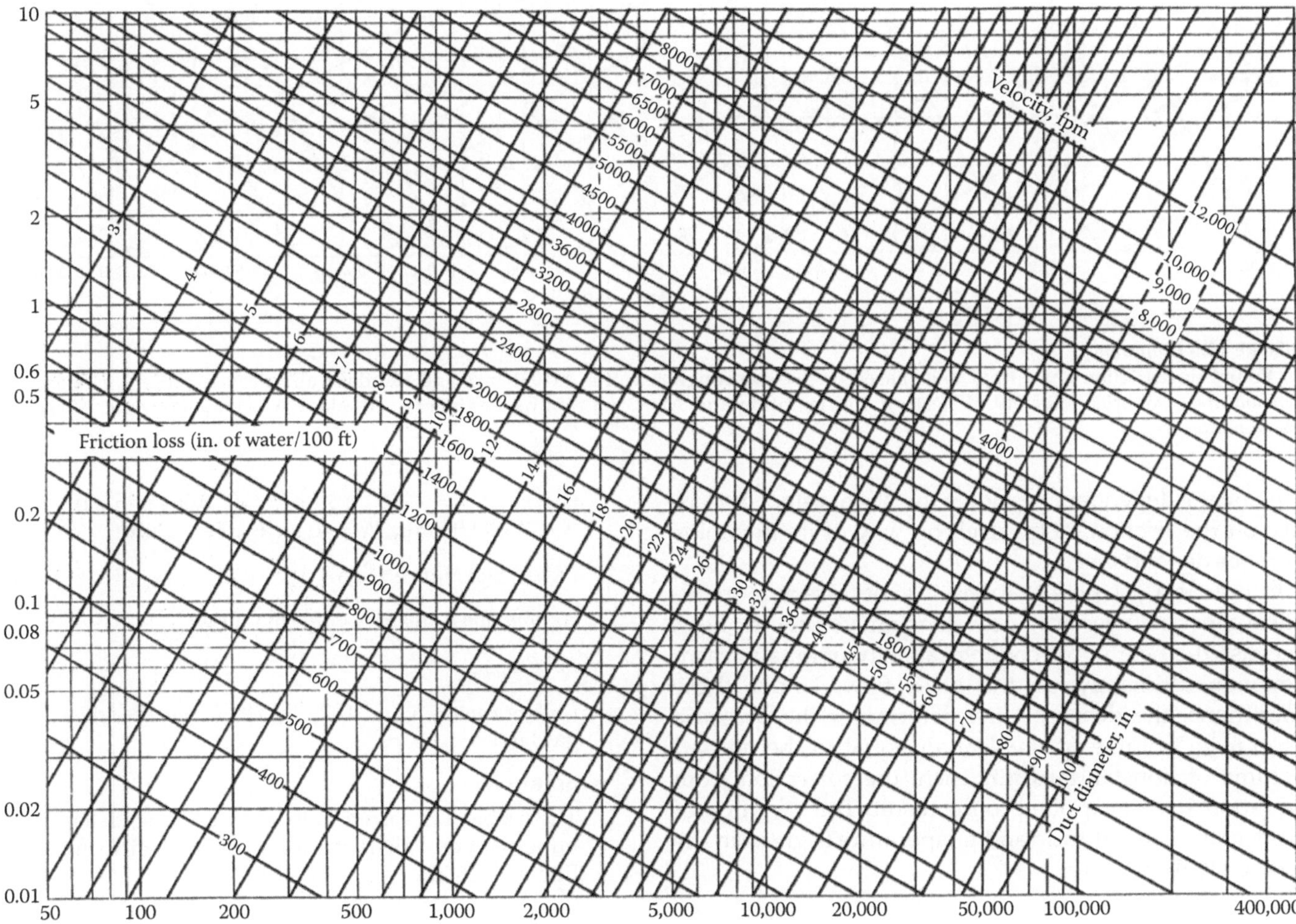

(a)

FIGURE 16.8
(a) Round duct friction chart in IP units based on average roughness of ε = 0.0003 ft and standard air. (From ASHRAE, *Handbook of Fundamentals*, American Society of Heating, Refrigerating and Air-Conditioning Engineers, Atlanta, GA, 2001. Copyright ASHRAE, www.ashrae.org.)

(Continued)

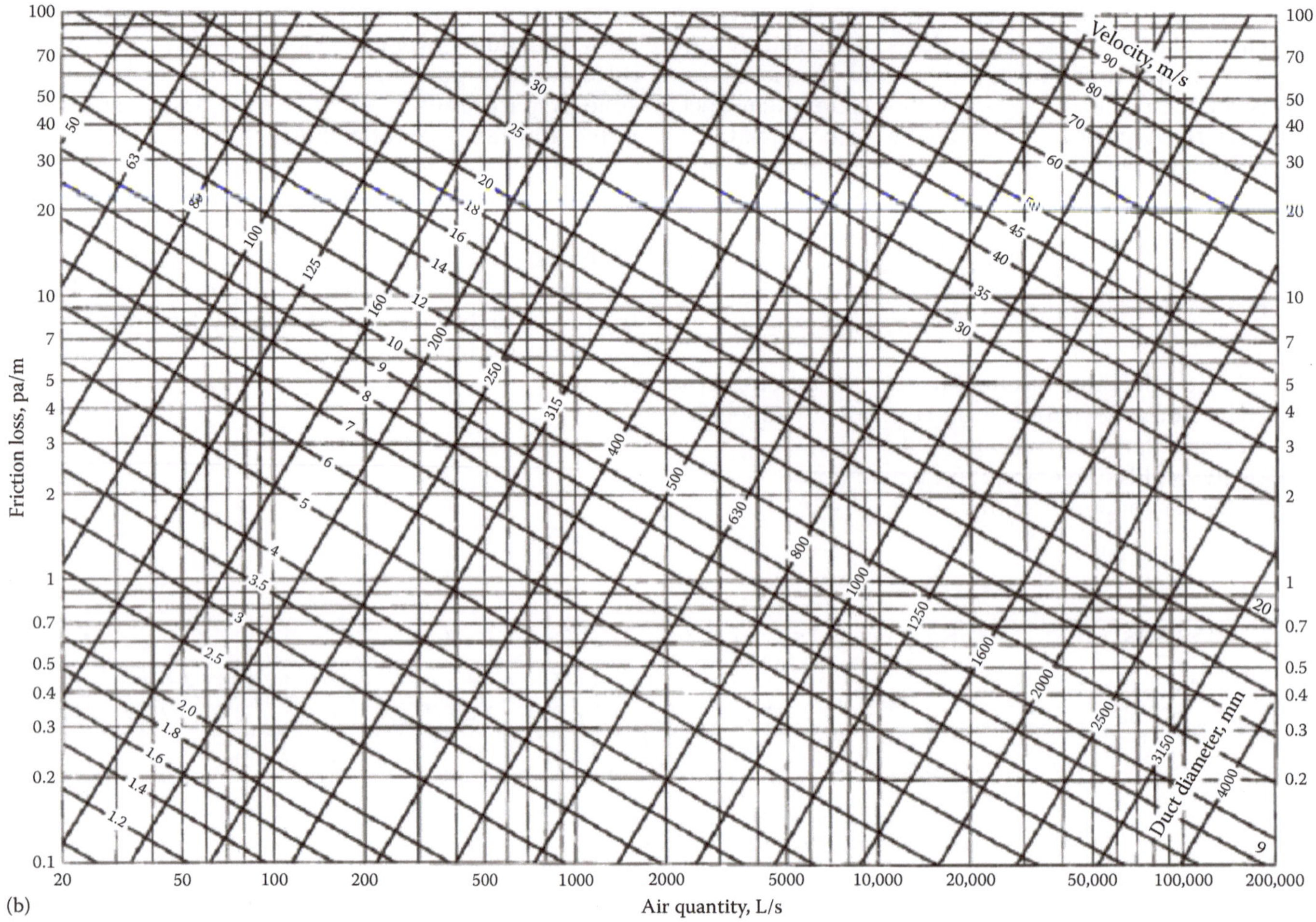

(b)

FIGURE 16.8 *(Continued)*
(b) Round duct friction chart in SI units based on average roughness of ε = 0.091 mm and standard air. (From ASHRAE, *Handbook of Fundamentals*, American Society of Heating, Refrigerating and Air-Conditioning Engineers, Atlanta, GA, 2001. Copyright ASHRAE, www.ashrae.org.)

This linear dependence ignores the slight effect of density on the friction factor itself by way of its Reynolds number dependence.

In practical design situations, it is necessary to maintain a constant mass flow, not a constant volumetric flow, because the heating or cooling capacity of an airstream depends ultimately on the mass flow. Under these conditions, the designer first determines the pressure drop for a volume of *standard* air equal to the needed volume of nonstandard air. Then, the correction from Equation 16.21 is applied to the result. The variation of air density with altitude and temperature is given in the online HCB software.

Example 16.4: Duct Friction

What is the pressure drop in a 250 ft rectangular duct of dimensions 20 in. × 40 in. for an actual airflow volume of 14,000 ft^3/min at an altitude where the density is 0.064 lb_m/ft^3?

Given: L = 250 ft, ρ_{act} = 0.064 lb_m/ft^3, $\dot{V} = 14{,}000$ ft^3/min, W = 40 in. and H = 20 in.

Figure: See Figure 16.8a.

Assumptions: Steady, isothermal flow

Find: $h_{friction}$

Lookup values: ρ_{std} = 0.075 lb_m/ft^3 (from the ideal gas law)

Solution

The problem is solved in three steps. First, determine the equivalent standard round duct using Equation 16.21:

$$D_{eq} = 1.30 \times \frac{(40 \times 20)^{0.625}}{(40 + 20)^{0.25}} = 30.46 \text{ in.}$$

Next, the friction loss for the actual volume of air is read from Figure 16.8a. From the figure at a duct size of 30.5 in., we read 0.25 inWG/100 ft. Therefore,

$$h_{friction,std} = \frac{0.25\ \text{inWG}}{100\ \text{ft}} \times 250\ \text{ft} = 0.63\ \text{inWG}$$

Finally, the density correction is applied according to Equation 16.22:

$$h_{friction,act} = 0.63\ \text{inWG} \times \left(\frac{0.064}{0.075}\right) = 0.54\ \text{inWG}$$

Comments

If the mass flow, not the volumetric flow, were specified at nonstandard conditions, there would be two effects. The velocity in the duct would vary inversely as the air density, and the pressure drop would vary linearly with the density. The net effect on the pressure drop is calculated by the aforementioned two-step procedure if the duct friction chart is used.

Pressure losses in duct fittings are given by the general expression:

$$\Delta p_{friction} = C_f \rho \frac{v^2}{2} \tag{16.23}$$

where C_f is the dimensionless fitting (or local loss) coefficient; it is the airflow analog of K_f used in pipe fitting calculations. (As described in the previous section, the equivalent-length approach is not appropriate for duct pressure drop calculations.) The analogous expression in terms of head is as follows:

$$h_f = C_f \frac{v^2}{2g} \tag{16.24}$$

Table 16.5 is an example of a fitting loss tabulation. This table applies for smooth elbows in round ducts with fully developed entering flow. Depending on the elbow radius r, the C_f value can be read from the second line of the table. This value is denoted by $C_{f,90}$ since it applies for a 90° elbow. For other elbow angles between 0° and 180°, a correction $C_{f,\theta}$ is needed. It is read from the fourth line of Table 16.5. The online HCB software contains selected values of loss coefficients for many round duct fittings. The fittings listed include: (1) round converging duct entries; (2) smooth elbows; (3) three-, four-, and five-piece elbows; (4) rectangular elbows; (5) elbows with turning vanes; (6) transitions; (7) converging tees; and (8) diverging tees.

The C_f values for rectangular duct and combined rectangular-round ducts can be found from extensive tables in ASHRAE Fundamentals (2013) or Idelchik et al. (1986). Example 16.5 illustrates the use of Table 16.5 for a single fitting.

Example 16.5: Pressure Loss in an Elbow

What is the pressure drop in an 8 in. diameter smooth 90° elbow through which 250 ft³/min of standard air is flowing? The radius of the elbow is 8 in.

Given: $\dot{V} = 250$ ft³/min, $r/D = 8/8 = 1.0$

Table: See Table 16.5.

Find: $\Delta p_{friction}$

Lookup values: From Table 16.5, $C_f = 0.22$, $\rho = 0.075\ \text{lb}_m/\text{ft}^3$

Solution

The average air velocity is

$$v = \frac{\dot{V}}{A} = \frac{250\ \text{ft}^3/\text{min}}{(\pi/4)\left(\frac{8}{12}\right)^2 \text{ft}^2} = 716\ \text{ft/min}$$

The pressure drop is found from Equation 16.23:

$$\Delta p_{friction} = 0.22 \times \frac{0.075\ \text{lb}_m/\text{ft}^3}{32.2\ \text{lb}_m \cdot \text{ft}/(\text{lb}_f \cdot \text{s}^2)} \times \frac{(716/60)^2\ \text{ft}^2/\text{s}^2}{2}$$

$$= 0.036\ \text{lb}_f/\text{ft}^2$$

The aforementioned unit for pressure drop, denoted by $\Delta p_{friction}$, is not used in practice. Instead, the pressure drop converted to head (or inches of water gauge) is as follows:

$$h_{friction} = 0.036\ \text{lb}_f/\text{ft}^2 \times \frac{1}{144}\ \text{ft}^2/\text{in.}^2 \times 27.7\ \text{inWG}/(\text{lb}_f/\text{in.}^2)$$

$$= 0.0069\ \text{inWG}$$

TABLE 16.5

Friction Coefficients for Smooth-Radius Duct Elbows

Coefficients for 90° elbows											
r/D	0.5	0.75	1.0	1.5	2.0	2.5					
$C_{f,90}$	0.71	0.33	0.22	0.15	0.13	0.12					
Angle correction factors K_θ											
θ	0	20	30	45	60	75	90	110	130	150	180
$C_{f,\theta}$	0	0.31	0.45	0.60	0.78	0.90	1.00	1.13	1.20	1.28	1.40

Comments

The pressure loss at this typical flow is small. Elbows in round ducts with moderate turning radii do not have excessive losses.

The lengthy unit conversion in this example can be avoided by using a *dimensional equation* for *standard air*. In IP units, if velocity is in feet per minute, the pressure drop in head in a fitting is given by

$$h_{friction} = C_f\left(\frac{v}{4005}\right)^2 \text{ inWG} \qquad (16.25 \text{ IP})$$

In SI units with the velocity in meters per second, the fitting pressure loss in pascals for standard air is

$$\Delta p_{friction} = C_f\left(\frac{v}{1.29}\right)^2 \text{ Pa} \qquad (16.25 \text{ SI})$$

These equations can be used for nonstandard air at the same velocity by multiplying the calculated pressure loss for standard air by the ratio of the actual air density to that of standard air.

16.2.4 System Characteristic Curves

Let the "system" be assumed to be the entire piping or ducting network used to transport liquids or air. The pressure drop in the fitting and straight pipe given by Equation 16.17 can be recast into the *system characteristic* for a piping circuit:

$$\Delta p_{tot} = \left[f\left(\frac{L}{D}\right) + \sum K_f\right]\left(\frac{1}{2}\rho v^2\right) \qquad (16.26)$$

where
f is the pipe friction factor
v is the velocity
K_f is the resistance coefficient of the fitting

Some designers use the equivalent-length method instead of K_f. However, equivalent lengths are valid for only one (usually unspecified or unknown) velocity. Therefore, the results may be less accurate than if one uses the K_f approach.

Since piping systems operate within a relatively narrow flow range, the values of f and K_f are approximately constant, with the result that the system characteristic for a pipe loop can be expressed as

$$\Delta p_{tot} = K_{sys}\dot{V}^2 \qquad (16.27)$$

where $\dot{V}$ is the volumetric flow rate.

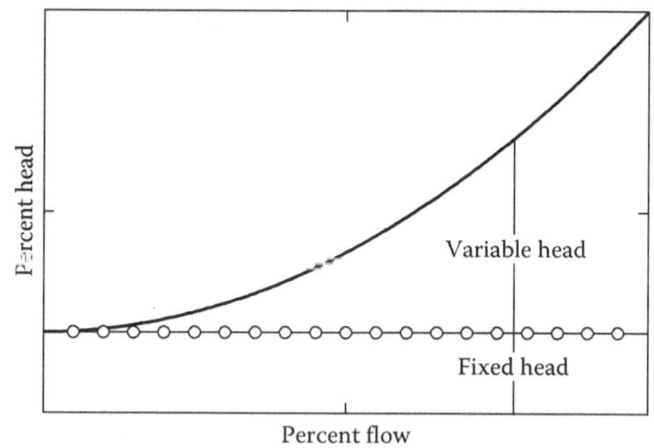

FIGURE 16.9
Characteristic curve of a basic piping system.

The head versus flow characteristics of a piping system is shown in Figure 16.9. The fixed head corresponds to the static lift pressure at zero flow and is given by the difference in elevation between the inlet and outlets for an open system, while for a closed water system, it is equal to zero (for air systems, the fixed head can be taken as zero because of the relatively low density of air). The variable head due to friction is represented by Equation 16.27 as a quadratic curve. This form of representation of piping (and ducting) system behavior is particularly useful since pump (and fan) curves are also expressed in terms of Δp and $\dot{V}$ as discussed in the next section.

16.3 Prime Movers

The pressure losses that occur in liquid and air systems are balanced by pressure increases in pumps and fans, respectively. In this section, we give a description of these prime movers and some basic concepts. The application of these components to entire systems is described in greater detail in Chapters 18 and 19.

16.3.1 Fans

The most common method for increasing pressure in HVAC fluid loops is by using centrifugal fans and pumps. A photograph of a fan is shown in Figure 16.10. Within the housing of centrifugal machinery, the shaft torque is converted to a force that propels fluid by increasing the pressure forces acting on it. Figure 16.11 shows a single blade of the type used in either a fan or a pump. The blade rotates counterclockwise (in the *forward-curved* design) with outer tip tangential velocity u_o. The fluid leaving the blading has a velocity v_o with

FIGURE 16.10
Photograph of a typical centrifugal fan used for building exhaust.

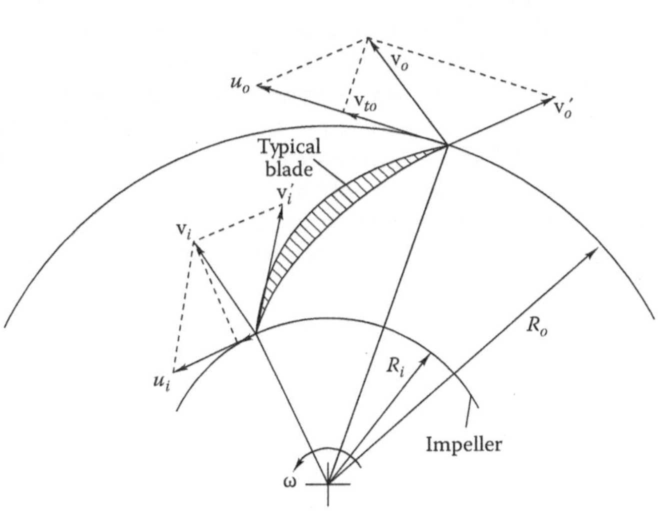

FIGURE 16.11
Velocity diagram for centrifugal machinery. Here v′ is the fluid velocity relative to the blade, *u* is the blade tip velocity, and v is the absolute fluid velocity.

a useful tangential component v_{to}. As shown in Kreider (1985), the theoretical head produced by the centrifugal device is given by

$$H_{ideal} = \frac{u_o v_{to}}{g} \tag{16.28}$$

The actual performance of pumps and fans must account for losses and nonideal flow within the blading. This is provided by manufacturers as plots containing additional data such as efficiency, input power, and other design information.

Manufacturers present data for real machinery in terms of a head-capacity curve, as shown in Figure 16.12. The data are measured experimentally by using standard testing procedures. Their principal application is to select the *operating point* of a system that is the point at which the system pressure loss (given by Equation 16.17 for liquid or air systems) is just balanced by the pressure developed by the pump or fan. If Equation 16.17 is plotted on Figure 16.12, it is approximately a parabola* passing through the origin. The operating point is the intersection of the two curves, as shown in the figure. How the efficiency and power input vary with flow rate can also be evaluated.

16.3.2 Pumps

There are different types of pumps (refer to handbooks such as Rishel et al., 2006 for a detailed description), but centrifugal pumps are most commonly used in HVAC applications. Sectional view of such a pump is shown in Figure 16.13. The primary components are the impellers or rotating vanes, the guide vanes that provide rotation to the flow before entering the impeller, and the casing that directs the discharge flow out of the pump through the volute perpendicular to the shaft. The water enters the impeller at low velocity and low pressure, it is then thrown outwards by the impellers by centrifugal forces, and consequently leaves the casing at high velocity and pressure.

The relationships between head, capacity, power, and efficiency of a pump are traditionally represented graphically by performance curves. A typical pump curve is shown in Figure 16.14. In addition to the flow versus pressure characteristic with which we are now familiar, other data are given. The dashed lines are lines of constant horsepower whereas the solid lines labeled with percentages are constant-efficiency lines. Of course, one selects a pump so that the required flow/pressure rise condition occurs near the maximum efficiency point. For example, suppose the pump of Figure 16.14 is needed to provide 750 gal/min (47 L/s) at 85 ft (26 m) of head. From the figure, it is seen that a 10 in. (25 cm) diameter impeller can meet this condition with a power input of 20 hp (15 kW). The efficiency of the pump is shown to be about 81%.

Below the pump capacity, power, and efficiency curves, a curve is shown to find the *net positive suction head* (NPSH). This design parameter is the value of static pressure (at the pump inlet flange) that the manufacturer specifies to avoid cavitation—the damaging flash of hot liquid to vapor (or the release of soluble

* The curve is not exactly parabolic since f and K_f can vary weakly with v.

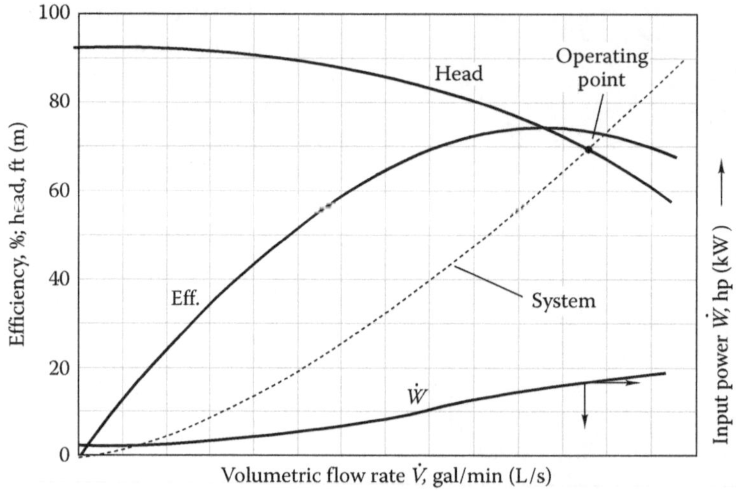

FIGURE 16.12
Typical pump or fan performance curve (solid curve). System curve is approximately parabolic (dashed). Intersection point is the system operating point.

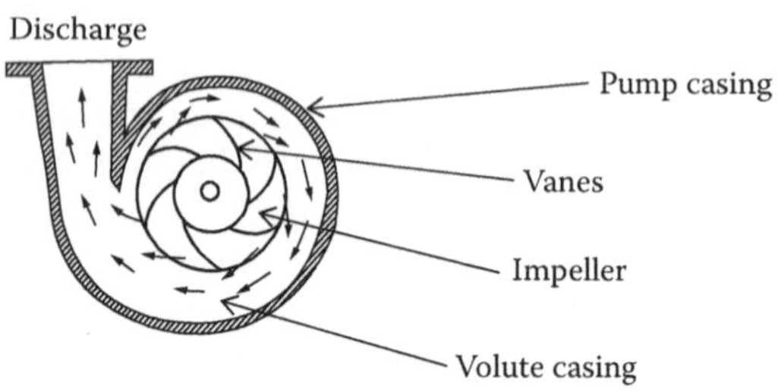

FIGURE 16.13
Cross section of a typical centrifugal pump.

gases from the pumped liquid) and its recondensation in pump blading. Cavitation is likely to occur at the pump inlet due to the local acceleration of fluid that occurs at the pump impeller entry. Cavitation must be avoided in any proper piping system design. The designer must guarantee that the *available NPSH*, called the *NPSHA*, is greater than the *required NPSH*, called the *NPSHR*, read from the pump curve. The NPSHR increases approximately as the square of the pump flow rate. The design pump NPSH establishes

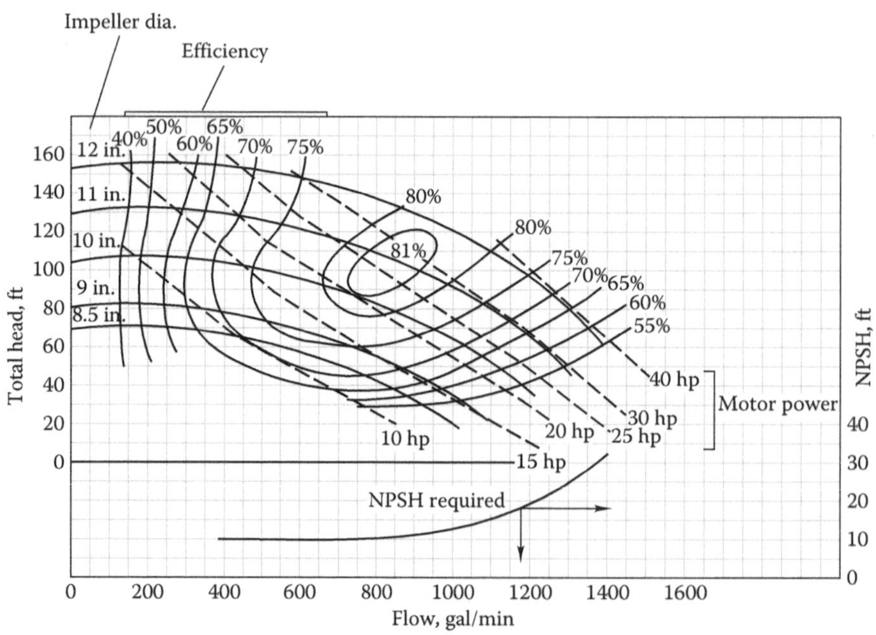

FIGURE 16.14
Example pump curve showing constant-efficiency lines (labeled in %), NPSHR, and input power (labeled in hp units) for five impeller sizes (labeled in inches).

the reference pressure above which the entire piping system operates.

Adequate NPSH is not often a problem with closed systems but must be carefully considered with pumps in open systems, such as cooling tower sump pumps where the amount of static head at the pump inlet is limited. To meet the NPSHR criterion for a cooling tower, the pump's inlet must be located a distance below the tower sump equal to the NPSHR plus any pipe friction losses between the sump outlet and the pump inlet. In open loops, the design is not much different from that for closed systems, except that system pressurization must rely on gravity. The major practical effect is that pumps must be positioned so that they do not lose their prime or have inadequate inlet NPSH.

The concept of *specific speed* N_s of a pump is useful for designers since it allows a convenient relationship between pump head and pump capacity:

$$N_s = \frac{N \cdot \dot{V}}{H^{3/4}} \quad (16.29\ \text{IP})$$

where
- N is the pump speed in rev/min or rpm
- $\dot{V}$ is the pump flow rate in gal/min
- H is the pump head in ft

For example, for a pump rotating at 1760 rpm with a capacity of 2000 gal/min and 100 ft head, the specific speed = 2489. Most HVAC pumps have N_s values in the range of 1000–5000, with exceptions being some large cooling tower pumps that may be as high as 8000 (Rishel et al., 2006). High specific speed pumps are those meant for low head and high-capacity applications and are built with shorter and wider vanes (and vice versa).

16.3.3 Expressions for Power

The power required to produce a fluid pressure rise across a pump or fan can be found from the first law of thermodynamics given by Equation 16.1. Kinetic and potential energy difference terms are small, if not zero; heat transfer is small; and internal energy changes are negligible. The remaining terms in the first law result in

$$\dot{W}_{fluid} = \dot{m} \times \left(\frac{p_i - p_o}{\rho} \right) \quad (16.30)$$

where
- $\dot{W}_{fluid}$ is the power input to fluid
- $\dot{m}$ is the mass flow rate
- $p_i - p_o$ is the inlet to outlet pressure rise

Alternatively, the shaft power can be expressed in terms of volumetric flow and pressure rise by using the equivalence $\dot{m} = \rho\dot{V}$:

$$\dot{W}_{fluid} = \dot{V} \times (p_i - p_o) \quad (16.31)$$

The power given by these expressions is seen to be negative, indicating that work is done on the fluid. The shaft power input is greater than the power input to the fluid because of turbulence and other inefficiencies within pumping machinery. The ratio of the fluid power to the shaft power is the *pump efficiency* η_p:

$$\eta_p = \frac{\dot{W}_{fluid}}{\dot{W}_{shaft}} \quad (16.32)$$

The fan efficiency is defined in a similar manner.

Sometimes, it is useful to express the shaft power input in dimensional equations that employ the units used by practitioners. For *pumps*, a dimensional equation is sometimes useful for designers using the inconsistent IP units of gallons per minute and feet of head. The motor horsepower for pumps can be found from the dimensional equations:

$$\dot{W}_{shaft} = \frac{\dot{V}\ \text{gal/min} \times H\ \text{ft}}{3960 \times \eta_{pump}}\ (\text{hp}) \quad (16.33\ \text{IP})$$

or

$$\dot{W}_{shaft} = \frac{\dot{V}\ \text{m}^3/\text{h} \times H\ \text{m}}{369 \times \eta_{pump}} = \frac{\dot{V}\ \text{L/s} \times H\ \text{m}}{102 \times \eta_{pump}}\ (\text{kW}) \quad (16.33\ \text{SI})$$

In both equations, the head H is in feet (meters) of water. For other fluids, the shaft work expressions should be multiplied by the actual pumped-fluid specific gravity.

To illustrate the use of this equation, consider the example earlier of a pump having to move 750 gal/min (47 L/s) through an 85 ft (26 m) of head operating point noted earlier and the efficiency of 81% read from Figure 16.14:

$$\dot{W} = \frac{750\ \text{gal/min} \times 85\ \text{ft}}{3960 \times 0.81} = 19.9\ \text{hp}$$

The value read from the curve matches the calculated power very closely.

The dimensional equations for *fans* are similar, but the constants in the denominator differ since the customary units used for fan calculations differ from those used with pumps. The shaft power for fans:

$$\dot{W}_{shaft} = \frac{\dot{V}\ \text{ft}^3/\text{min} \times H\ \text{inWG}}{6356 \times \eta_{fan}}\ (\text{hp}) \qquad (16.34\ \text{IP})$$

or

$$\dot{W}_{shaft} = \frac{\dot{V}\ \text{m}^3/\text{s} \times H\ \text{cmWG}}{10.06 \times \eta_{fan}} = \frac{\dot{V}\ \text{L/s} \times \Delta p\ \text{kPa}}{985.5 \times \eta_{fan}}\ (\text{kW}) \qquad (16.34\ \text{SI})$$

The shaft power is also referred to as the brake horsepower. In order to determine the actual electrical draw by the fan or pump, one needs to include the efficiency of the electric motor.

16.3.4 Fan Laws

Since flow requirements in HVAC systems vary considerably in response to building loads that depend on the weather, time of day, and season of the year, fans and pumps may need to function at a number of different operating points, determined, e.g., by different fan or pump rotational speeds. As the speed, head, and flow rate vary, the power requirements also vary. The relationships among these key variables are called the *fan laws* for fans and the *affinity laws* for pumps. Table 16.6 summarizes the six fan laws (derived in standard fluid mechanics texts). The only requirement for the fan laws to apply is that the *efficiency remain constant* and that *geometric similarity in blading and housing be retained* as sizes change. This table gives the relationships among the following parameters that can vary during the operation of a fan or of one fan geometrically similar to another fan:

- Rotational speed N
- Volumetric flow rate $\dot{V}$
- Shaft power $\dot{W}$
- Pressure rise h
- Air density ρ
- Rotor diameter D

For example, the first line of the table shows that an increase in the speed of a fan of fixed size with constant air density increases the flow rate linearly, the head quadratically, and the power cubically.

Example 16.6: Fan Laws: Fan Speed Increase

If a fan delivers 20,000 ft³/min of standard air at a static pressure of 3.0 inWG at a fan speed of 500 r/min and a power input of 18 hp, what speed, input power, and static pressure are required if the flow rate is increased to 28,000 ft³/min by increasing the rotational speed?

Given: $\dot{V}_1 = 20{,}000$ ft³/min, $N_1 = 500$ r/min, $H_1 = 3.0$ inWG, $\dot{W}_1 = 18$ hp, $\dot{V}_2 = 28{,}000$ ft³/min

TABLE 16.6

Summary of Fan Laws

	Rotational Speed (N)	Flow Rate ($\dot{V}$)	Fan Power ($\dot{W}$)	Produced Head (H)	Air Density (ρ)	Rotor Diameter (D)
Speed change	Varies	$\dot{V}_2 = \dot{V}_1 \frac{N_2}{N_1}$	$\dot{W}_2 = \dot{W}_1 \left(\frac{N_2}{N_1}\right)^3$	$H_2 = H_1 \left(\frac{N_2}{N_1}\right)^2$	Fixed	Fixed
Diameter change (fixed tip speed)[a]	$N_2 = N_1 \frac{D_1}{D_2}$	$\dot{V}_2 = \dot{V}_1 \left(\frac{D_2}{D_1}\right)^2$	$\dot{W}_2 = \dot{W}_1 \left(\frac{D_2}{D_1}\right)^2$	Fixed	Fixed	Varies
Diameter change[a]	Fixed	$\dot{V}_2 = \dot{V}_1 \left(\frac{D_2}{D_1}\right)^3$	$\dot{W}_2 = \dot{W}_1 \left(\frac{D_2}{D_1}\right)^5$	$H_2 = H_1 \left(\frac{D_2}{D_1}\right)^2$	Fixed	Varies
Density change	Fixed	Fixed	$\dot{W}_2 = \dot{W}_1 \frac{\rho_2}{\rho_1}$	$H_2 = H_1 \frac{\rho_2}{\rho_1}$	Varies	Fixed
Density change	$N_2 = N_1 \left(\frac{\rho_1}{\rho_2}\right)^{1/2}$	$\dot{V}_2 = \dot{V}_1 \left(\frac{\rho_1}{\rho_2}\right)^{1/2}$	$\dot{W}_2 = \dot{W}_1 \left(\frac{\rho_1}{\rho_2}\right)^{1/2}$	Fixed	Varies	Fixed
Density change (fixed mass flow)	$N_2 = N_1 \frac{\rho_1}{\rho_2}$	$\dot{V}_2 = \dot{V}_1 \frac{\rho_1}{\rho_2}$	$\dot{W}_2 = \dot{W}_1 \left(\frac{\rho_1}{\rho_2}\right)^2$	$H_2 = H_1 \frac{\rho_1}{\rho_2}$	Varies	Fixed

[a] At the same rating point, e.g., at the maximum efficiency point.

Table: See Table 16.6.

Assumptions: The system to which the fan is connected is unchanged, and the fan efficiency is unchanged as well.

Find: N_2, H_2, $\dot{W}_2$

Solution

The first line of Table 16.6 is used, since it applies to speed changes with fixed air density and fan size. The flow rate, input power, and produced head are found by using the equations in the third, fourth, and fifth columns of the table. To find the head and power, the new fan speed must be known. It is found by solving the flow speed equation for rotational speed since the flow at both conditions is known:

$$N_2 = N_1 \frac{\dot{V}_2}{\dot{V}_1} = 500\ \text{r/min} \times \left(\frac{28{,}000\ \text{ft}^3/\text{min}}{20{,}000\ \text{ft}^3/\text{min}} \right)$$

$$= 700\ \text{r/min} \quad (16.35)$$

The new head is now found from the fifth column of the table:

$$H_2 = H_1 \left(\frac{N_2}{N_1} \right)^2 = 3.0\ \text{inWG} \times \left(\frac{700}{500} \right)^2 = 5.88\ \text{inWG} \quad (16.36)$$

Finally, the power input is found from the equation in the fourth column of Table 16.6:

$$\dot{W}_2 = \dot{W}_1 \left(\frac{N_2}{N_1} \right)^3 = 18\ \text{hp} \left(\frac{700}{500} \right)^3 = 49.4\ \text{hp} \quad (16.37)$$

Comments

The fan laws are among the most important tools for the air system designer. It is important to understand their applications while keeping in mind the limitations described earlier.

The movement of air through HVAC systems in buildings is caused by pressure forces produced by fans. In the past, the energy consumption of fans has often been overlooked, yet it can be very significant. It represents about 30%–40% of the electricity used in commercial buildings (Coad, 1989). Fan electricity consumption can easily be larger than the consumption of the chiller because fans operate for many more hours per year, albeit at lower power rates, than chillers do. It is, therefore, very important that every feasible means be used to control fan power and to design duct systems that are energy efficient. These aspects are described later. The designer must also specify the most efficient motors that are economically feasible for fans (and pumps) used in secondary HVAC systems (Greenberg et al., 1988).

It is worthwhile to reemphasize that fans are *volumetric devices* causing the flow of a given volume of air under specified conditions. However, at high altitudes, the mass flow produced by fans is lower because the density is lower, even though the volumetric flow remains the same at the same fan speed. One can imagine two identical air distribution systems operating with identical fan speeds at two different altitudes. The system at lower altitude will have a flow rate depending on the fan capacity, duct system, and air density. The system at high altitude will have the same volumetric rate since the fan speed is the same, but the pressure drop will be proportionally less by the same amount as the density is less (since the system pressure drop depends nearly linearly on the density). Since the horsepower is the product of the pressure rise and flow rate, it, too, is reduced by the density ratio. Figure 16.15 shows the performance of a fan and duct system at sea level and at an altitude where the density is one-half that at sea level. The reduced *mass* flow at high altitude results in reduced coil performance. The online HCB software contains tabulated data for air density at various altitudes.

16.3.5 Pump Affinity Laws

The performance of pumps operating at fixed efficiency is also governed by a set of very useful rules known as the *pump affinity laws*. Table 16.7 summarizes the affinity laws relating pump flow, speed, and pressure rise at various operating conditions. The affinity laws are essentially the same as the fan laws in Table 16.6, except that the number of equations is smaller since liquids are essentially incompressible and density effects can be neglected. System and pump curves have the same interpretation as for air systems. Their point of intersection determines the operating point for pumped systems.

16.4 System and Prime Mover Interactions

16.4.1 Relationship between Pressure and Velocity in a Duct or Pipe

In Section 2.1, we had introduced the concepts of static, velocity, and total pressures of airflow inside a duct and illustrated these using Figure 2.1. These notions also apply to flow in pipes. The total pressure is the

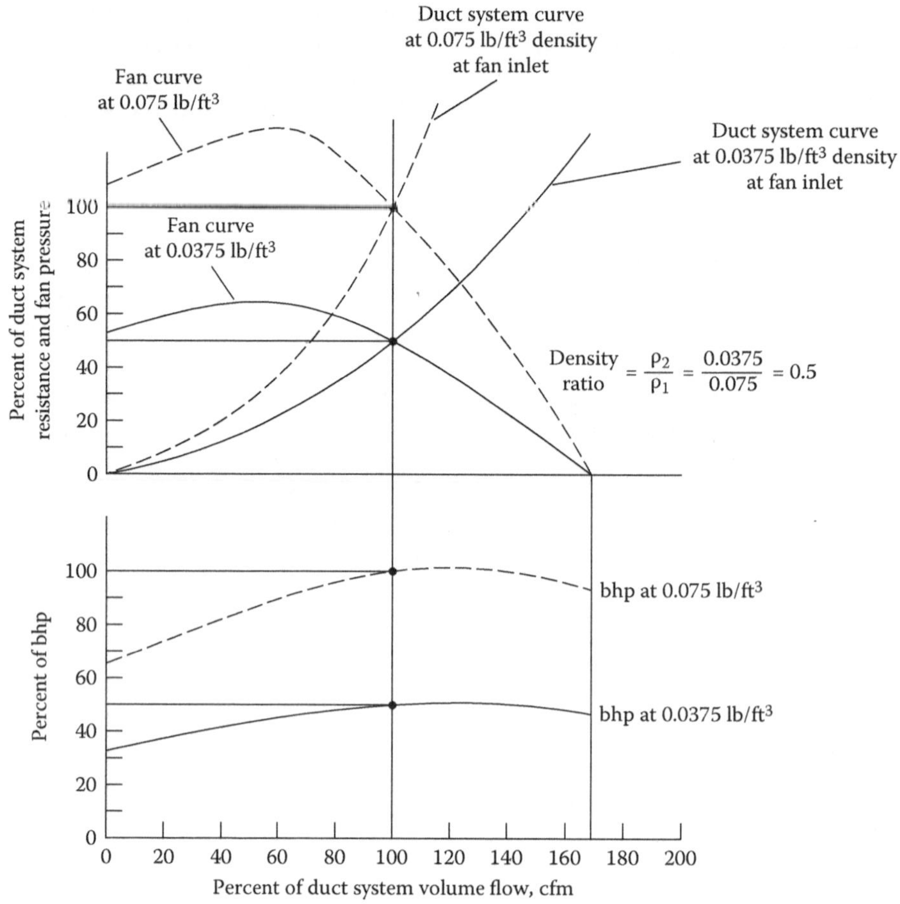

FIGURE 16.15
Effect of altitude on duct resistance and fan power. The low-density case is an extreme corresponding to the density of process air at 10,000 ft altitude and 250°F. (The bhp is the brake horsepower input to the fan.)

TABLE 16.7
Summary of Pump Affinity Laws

	Flow Rate ($\dot{V}$)	Pump Power ($\dot{W}$)	Produced Head (H)
Speed change (fixed rotor size)	$\dot{V}_2 = \dot{V}_1 \frac{N_2}{N_1}$	$\dot{W}_2 = \dot{W}_1 \left(\frac{N_2}{N_1}\right)^3$	$H_2 = H_1 \left(\frac{N_2}{N_1}\right)^2$
Diameter change (fixed rotor speed)[a]	$\dot{V}_2 = \dot{V}_1 \left(\frac{D_2}{D_1}\right)^3$	$\dot{W}_2 = \dot{W}_1 \left(\frac{D_2}{D_1}\right)^5$	$H_2 = H_1 \left(\frac{D_2}{D_1}\right)^2$
Density change	Fixed	$\dot{W}_2 = \dot{W}_1 \frac{\rho_2}{\rho_1}$	Fixed

[a] At the same rating point, e.g., at the maximum efficiency point.

most relevant pressure to consider since it has direct bearing on the fan pressure requirement. It will always reduce along the direction of fluid flow due to friction (as is obvious from Equation 16.5), but changes in velocity at one duct section (due to changes in duct or pipe cross section) can increase or reduce local velocity head and thereby change static pressure accordingly. Thus, one should not confuse *total and static pressures* (the latter is the pressure that the duct walls have to bear and should be thick enough to prevent bulging). This is illustrated in Figure 16.16a, where the change in the cross section causes the fluid velocity to decrease, thereby increasing the static pressure even though the total pressure keeps decreasing with distance. Note that the entrance and exit pressure losses have also been included. The plot of the pressure head versus pipe or duct distance is known as the *pressure gradient diagram*. This is very useful for designing piping and ducting systems and is also used for diagnosing problems associated with improper flow in existing fluid systems (see ASHRAE Systems, 2012 or other references). Figure 16.16b shows the pressure profile for a simple ducting and fan system meant for a small commercial or residential building. The fan increases the total pressure so as to overcome the pressure drop in various duct elements and also provide enough velocity pressure at the diffuser exits to deliver the required airflow volumes to the various rooms.

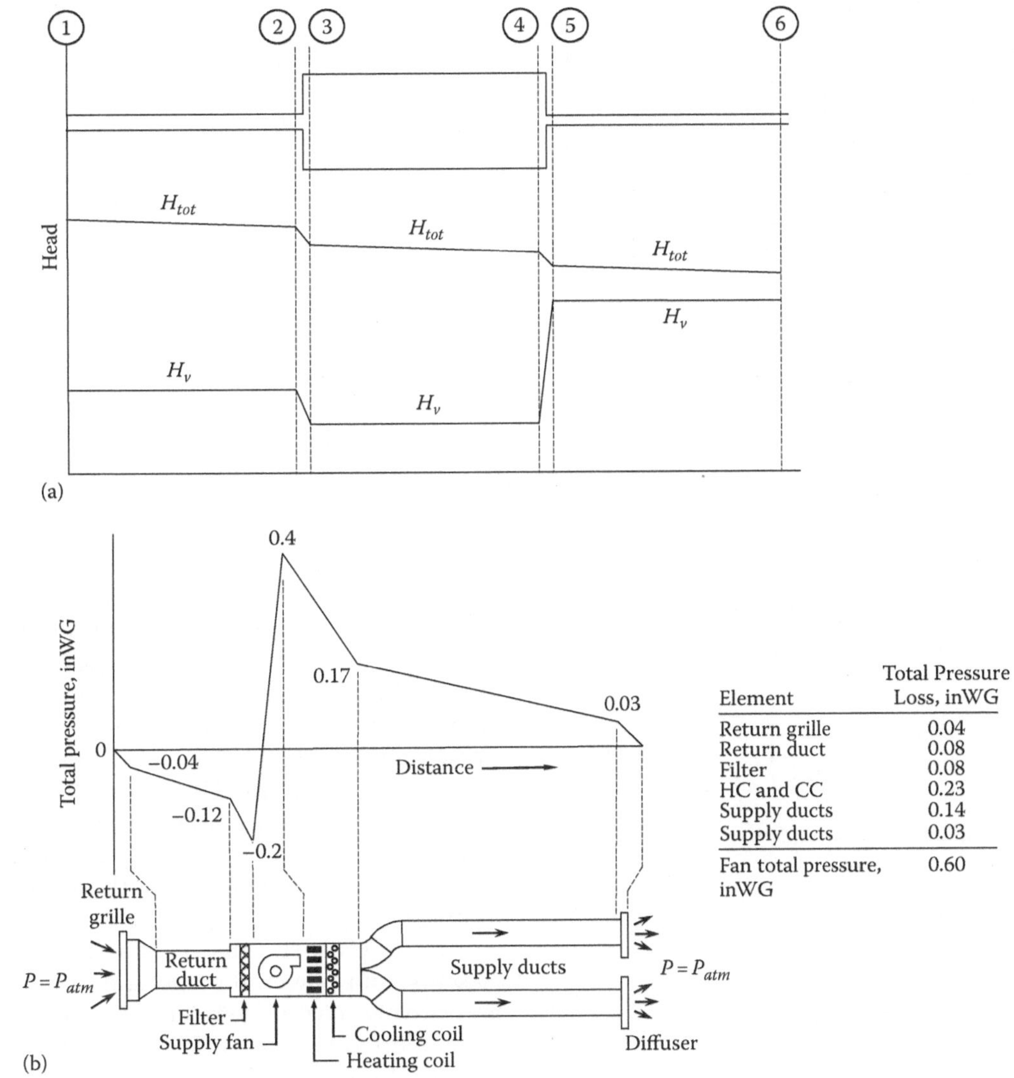

Element	Total Pressure Loss, inWG
Return grille	0.04
Return duct	0.08
Filter	0.08
HC and CC	0.23
Supply ducts	0.14
Supply ducts	0.03
Fan total pressure, inWG	0.60

FIGURE 16.16
(a) Schematic of a pressure gradient diagram showing how the total and velocity pressures vary along the length of a pipe or duct. (b) Example of pressure profile for a small commercial building with a supply fan.

16.4.2 Combined Prime Mover and System Behavior

The analytical method of expressing the system curve is particularly convenient for finding the operating point when a fan or pump is connected in a system. The analytical fan (or pump) curve is readily found from tables of fan (or pump) data provided by manufacturers using regression methods. Section 14.7 described algorithms for modeling part-load performance of chillers under varying operating conditions that have been embedded in most of the detailed simulation programs developed in North America (for example, the DOE-2 simulation software, LBNL, 1982). Similarly, part-load models using polynomial functions for variable speed fans (and pumps) are also available, and attempts have been made to improve these models for more realistic design purposes. The interested reader can refer to Stein and Hydeman (2004) who evaluated different fan models suitable for calibration against actual manufacturer data and suggested the use of a characteristic fan model.

Example 16.7: Fan Operating Point

A curve fit of data from a fan manufacturer shows that the equation for the *fan curve* at 1100 r/min (115 rad/s) is given by the dimensional equation (pressure is in inches water gauge and flow is in cubic feet per minute):

$$\Delta p_{fan} = 4 - 0.2\left[\left(\frac{\dot{V}}{1000}\right)^2 - \frac{\dot{V}}{1000}\right] \qquad (16.38)$$

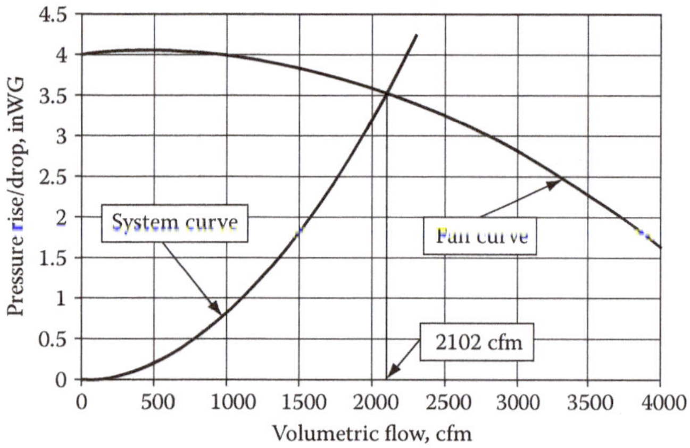

FIGURE 16.17
Fan and system curves for Example 16.7.

With pressure referring to the static pressure (see Equation 16.40), find the operating point if the *system curve*, including the system effect factor, is given by the following dimensional equation:

$$\Delta p_{sys} = 0.8 \times \left(\frac{\dot{V}}{1000}\right)^2 \quad (16.39)$$

If the fan efficiency is 61%, what is the required motor size?

Given: Fan curve, system curve, and $\eta_{fan} = 0.61$

Figure: See Figure 16.17.

Find: Operating point (p and $\dot{V}$), $\dot{W}$

Solution

The operating point for this fan system occurs where the system pressure drop and flow exactly match the available fan pressure rise at the same flow. To find the solution, one can either plot the two curves, as shown in Figure 16.17, or solve analytically by equating the fan curve and the system curve. If the latter approach is used, we get

$$0.8 \times \left(\frac{\dot{V}}{1000}\right)^2 = 4 - 0.2 \times \left[\left(\frac{\dot{V}}{1000}\right)^2 - \frac{\dot{V}}{1000}\right]$$

The quadratic formula is used to find the flow rate at the operating point:

$$\dot{V} = 2102 \text{ ft}^3/\text{min } (992 \text{ L/s})$$

The system pressure at this flow rate can be found from Equation 16.39:

$$\Delta p_{sys} = 0.8 \times \left(\frac{2102}{1000}\right)^2 = 3.54 \text{ inWG}$$

The ideal power from the basic equation is (Equation 16.31)

$$\dot{W}_{ideal} = \dot{V} \times \Delta P$$

$$= 2{,}102 \text{ ft}^3/\text{min} \times \frac{3.54 \text{ in.} \times 5.2 \text{ lb}_f/(\text{ft}^2 \cdot \text{in.})}{33{,}000 \text{ ft} \cdot \text{lb}_f/(\text{min} \cdot \text{hp})}$$

$$= 1.17 \text{ hp } (0.87 \text{ kW})$$

The actual power is the ideal power divided by the efficiency:

$$\dot{W}_{act} = \frac{\dot{W}_{ideal}}{\eta_{fan}} = \frac{1.17}{0.61} = 1.92 \text{ hp}$$

The system curve and fan curve must intersect to the right of the maximum pressure point to ensure stable operation, as shown in Figure 16.17. The physical reason for instability is that there is always a tendency for air from the high-pressure side of the fan (outlet) to flow toward the low-pressure side (inlet). To the right of the maximum pressure point, this presents no problem since the momentum in the air ensures throughflow. However, at flow conditions to the left of the pressure maximum, backflow can occur in a cyclic fashion. This can cause large intermittent forces on the fan blading with subsequent physical damage.

Given the constraints on maximum efficiency, flow stability, and power, the operating range of a given fan is relatively narrow compared to the entire span of the usual plotted fan curve. This restriction is further complicated by the fact that a nominally constant-volume system does not actually operate at constant volume, since filters become dirty or cooling coils have varying amounts of condensed moisture. The careful designer will evaluate the expected range of system curves to ensure that all possible operating points lie to the right of the pressure maximum and not far from the efficiency optimum. Fortunately, we have a wide selection of fans from which to choose, and so the seemingly large number of criteria can be met.

16.4.3 Elements in Parallel and in Series

A piping or ducting network can contain several elements in series or in parallel. An equivalent composite system characteristic can be easily generated using the analogy to electric systems: resistances in series add up, and resistances in parallel add up as inverses. Figure 16.18a illustrates the case of flow in series, and so the equivalent curve shown by "ac" is simply generated by adding the head or pressure drop at different values of flow rates. For networks

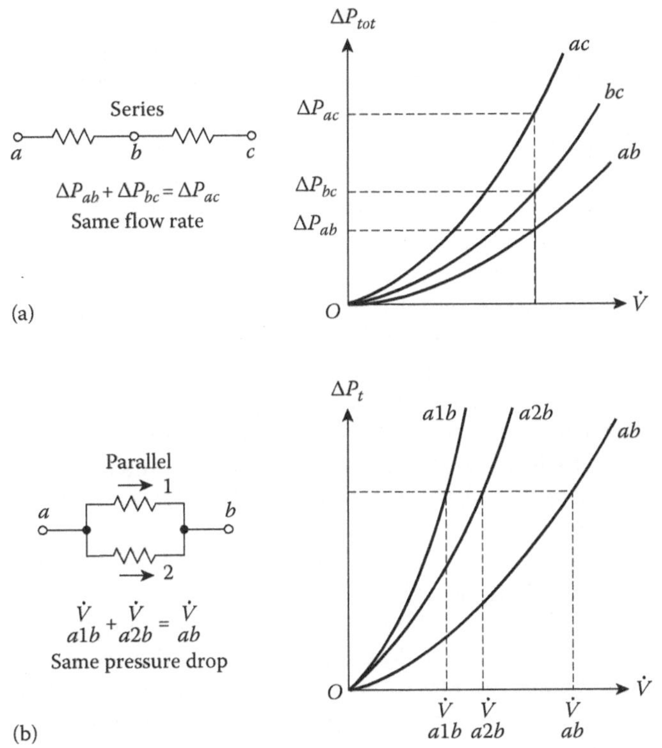

FIGURE 16.18
Equivalent characteristic curves for piping and ducting networks. (a) Series flow arrangement and (b) parallel flow arrangement.

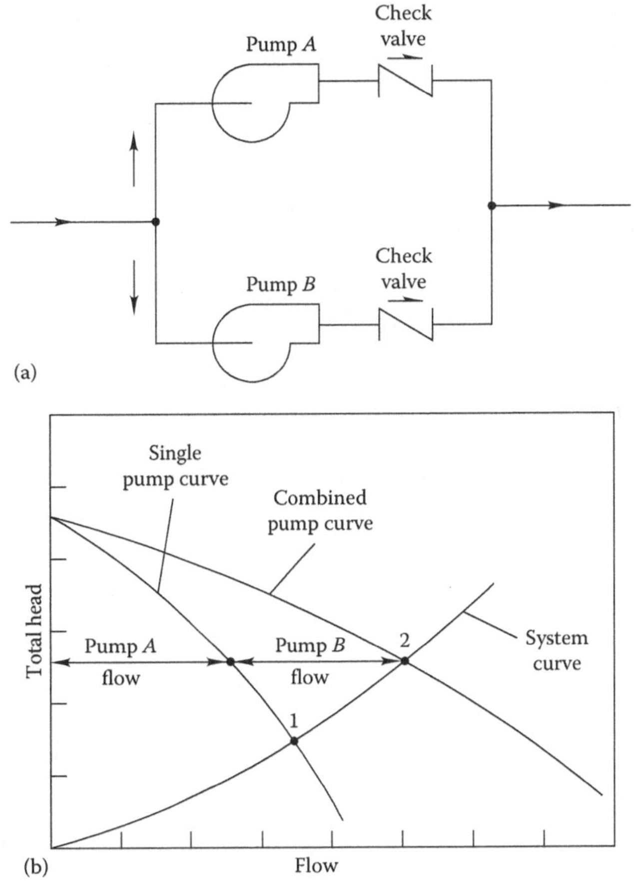

FIGURE 16.19
Piping system with two equal size pumps (or fans) connected in parallel (a) Sketch, (b) pump and system curves and operating points.

in parallel, the equivalent system characteristic is deduced by adding the volume flow rates at different values of heads or pressure drops (as shown in Figure 16.18b).

Systems for large buildings are more complex than the simple systems shown to this point in this section. They often contain several pumps, control valves, and various pieces of terminal heating or cooling equipment. The flow through multiple loops is relatively more complex as a result. For example, large systems will often require standby capacity if a pump should fail. One method of accomplishing this is to use two identical, *parallel-connected* pumps, each with the capacity to provide one-half the design flow. Figure 16.19 shows the arrangement to be used and the resulting pump curves. The curve of parallel-connected identical pumps is just the curve of a single pump except that the flow rate is doubled. The system curve intersects the dual-pump curve at operating point 2 (denoting two pumps) with corresponding flow $\dot{V}_2$.

If one pump should fail, the operating point is determined by the system curve intersection with the single-pump curve at point 1 (the check valve ensures that reverse flow through the inactive pump does not occur). It is seen that the flow rate is not one-half of that which exists for two active pumps. In fact, the flow for a single pump in this situation is about two-thirds of the flow of both pumps taken together. For most HVAC systems, this flow rate is adequate temporarily for most conditions, since the full capacity is needed only at peak-load conditions.

Series pumping arrangements are sometimes used when high-pressure drops must be overcome without the use of special positive-displacement pumps. If identical pumps are used in series, the combined, effective pump curve is that of a single pump except that the developed head at any flow is double that of a single pump. The series arrangement is less common than the parallel one in HVAC systems.

In essence, there are three general methods of controlling flow in piping systems as the load varies. The variable speed motor control (similar to variable speed and discussed in Section 16.5.2) is the most energy efficient but has the highest first cost. Designing the system with multiple pumps and switching one or more as needed can also provide flow modulation. Such an arrangement has a distinct advantage in terms of maintenance as compared to one large pump. However, such a flow control is not very precise especially when large variations in water fluid are needed. A less energy-efficient, but cheaper alternative suitable for smaller systems is to use valve flow control (this is also discussed in Section 16.5.2 and Chapter 21).

16.5 Types of Fans and Their Control

16.5.1 Fixed-Speed Operation

Manufacturers present "fan curves," i.e., curves showing the relationship between pressure developed by the fan and the volume flow rate under fixed-speed conditions, as discussed in Section 16.3.1. Figure 16.20 illustrates how the shape of these curves vary for three common types of centrifugal fans distinguished by type of blading arrangement. Plots of how the power and the efficiency vary with volume flow rate are also shown.

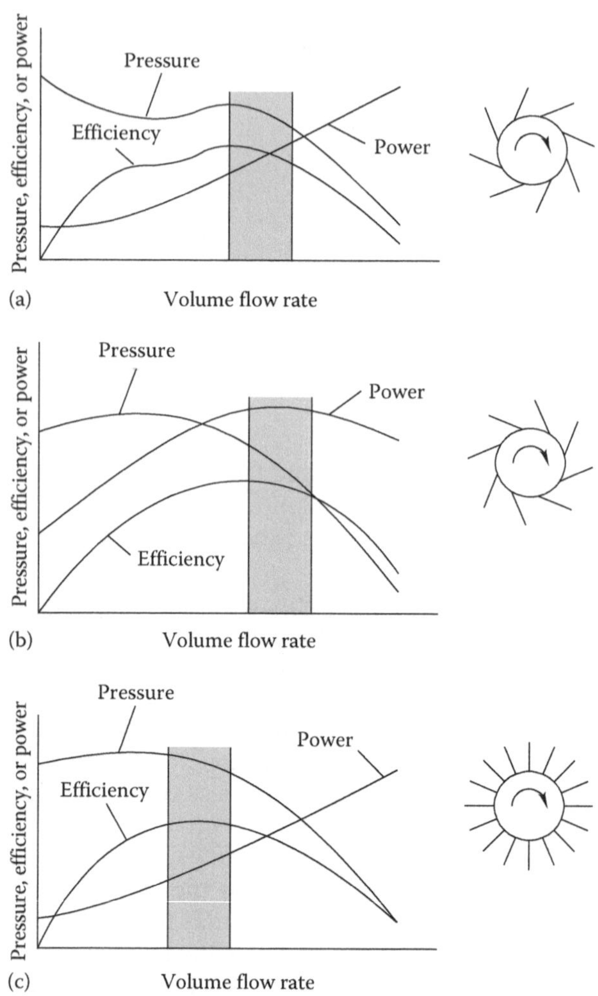

FIGURE 16.20
(a) Forward-curved, (b) backward-curved, and (c) radial fan blading arrangements and performance curves. The shaded region is the suggested operating range near peak efficiency, it is to the right of the peak pressure point to ensure stability.

Forward-curved blading results in relatively lower efficiency. Consequently, this design is used for low-speed, low pressure rise cases. Because of the dip in the pressure curve, this fan must always be operated to the right of the maximum pressure point for stability. It is also important to note that the power curve continues to rise with the volume. If this fan is operated with inadequate load (during system construction or later for service with ducts disconnected), the motor can be overloaded since the input power increases continuously with the flow. The low operating speed is an advantage, however, since stresses in the rotor will be smaller and bearing life longer.

Backward-curved blading results in the best efficiency of any centrifugal HVAC fan design. The noise level is relatively low, and the power curve is well behaved and stable over a broad flow range. This is the most common design in HVAC systems, even though the fan speed must be higher for a given flow rate than for other designs.

Radial blading designs lie between the two previous approaches. However, the power curve is not self-limiting. The efficiency is about the same as for forward-curved designs, while the speed is lower than that for backward-curved fans. The noise level is moderate.

Many hundreds of fans are available for a given flow rate and pressure rise application. If several fans of similar cost are available for a project, the one having the highest efficiency at the operating point should be selected. This operating point (referred to as the maximum efficiency point) is where the fan curve and the duct system curve intersect.

The *static pressure* plotted in fan curves such as those in Figures 16.15 and 16.20 has a special meaning. Fans are tested under specific conditions including a static pressure tap at the fan outlet and a smooth transitional fitting (not connected to a duct system) at the fan inlet. The fan static pressure rise, which is plotted on fan curves, is defined as the outlet *static* pressure minus the inlet *total* pressure since the fan outlet static pressure is referenced to the *test room pressure* in which the fan is housed, not to an inlet duct. In equation form,

$$\Delta p_{static} = p_{static,o} - p_{total,i} \tag{16.40}$$

Recall this special definition when using this pressure rise to match calculated system static pressure drop. Also, if as-tested performance is to be realized in the field, the inlet condition must be the same as the test—uniform flow with no prerotation.

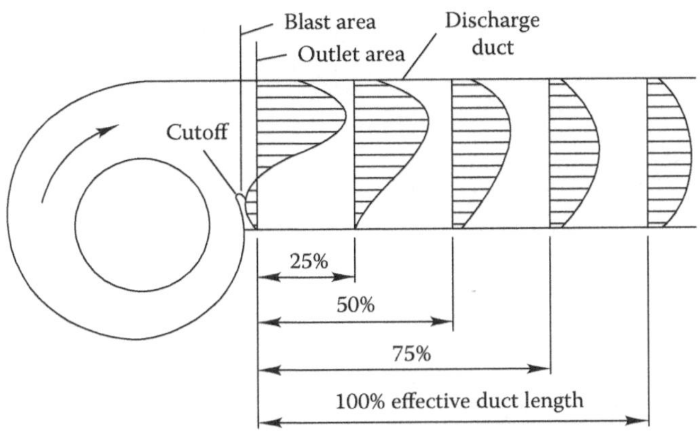

FIGURE 16.21
Adjustment of fan outlet velocity profiles with distance from fan outlet connection.

Finally, for proper fan operation, the fan outlet duct should be a straight section of sufficient length that the nonuniform flow at the fan outlet becomes approximately uniform. As a rule of thumb, the designer should allow at least 2.5 duct diameters of downstream straight duct to achieve uniform flow for air velocities less than 2500 ft/min (13 m/s), as shown in Figure 16.21. For each additional 1000 ft/min above 2500, an additional duct diameter in the outlet duct length must be added. This duct length required to achieve fully developed flow is called the *effective duct length*. If space restrictions in the mechanical room limit duct sizes to less than this value, the *additional pressure drop* must be accounted for, as described in AMCA (1973), where several hundred outlet (and inlet) conditions are documented and their extra pressure drops are listed.

The designer must be aware that parasitic pressure losses due to inadequate straight ducts at the fan itself can result in pressure drops of the same order of magnitude as total system friction losses and that those losses will cost the owner extra operating energy for the life of the system. One method of quantifying the effect of less-than-ideal duct connections to fans make use of the *system effect factor* (SEF). The system effect is based on an expression similar to that for duct pressure drops:

$$\Delta p_{SEF} = K_{SEF}\left(\frac{1}{2}\rho v^2\right) \tag{16.41}$$

The value of K_{SEF} depends on the fan inlet and outlet duct arrangement in a complicated way. The online HCB software contains the data needed to find K_{SEF} for some common fan connection geometries by using the idea of effective duct length described earlier. For those not included there, the reader is referred to ASHRAE Systems (2012).

16.5.2 Variable-Volume Operation

One of the most common air systems installed in commercial buildings involves varying the supply airflow rates (within limits) to meet varying space loads. The variation in flow is of interest since the power used to move air varies as the cube of the flow. For a factor-of-2 reduction in flow, the power needed and energy consumption theoretically decrease by a factor of 8 *if* efficiency remains unchanged (from the fan laws). This thermal analysis of variable-air-volume HVAC systems will be discussed in detail in Chapter 19. In this section, we present and compare three methods of providing variable airflow.

The first method of flow control introduces restrictions (e.g., dampers) in the airflow circuit. By thus increasing the resistance in a flow loop, the system curve becomes steeper and the flow lower, as shown in the left system curve in Figure 16.22a. In this figure, two operating points are shown (1 and 2) for two system resistance curves with different damper settings. The flow rate is reduced from $\dot{V}_1$ to $\dot{V}_2$. The amount of reliable control achievable by this method is limited, given the constraints on operating range discussed earlier. The dampers most often used with this approach are *fan outlet dampers* to throttle the flow. However, the pressure drop across the dampers represents wasted fan power. This method is inexpensive from an equipment viewpoint but costly operationally.

Adjustable *inlet vanes* can be used to control flow more efficiently. These vanes impart a spin to the air prior to its entry into the fan blading. The result is a modified fan curve as shown by the solid lines in Figure 16.22b. This approach is more efficient than the outlet damper approach since fan power is not wasted across dampers, but rather is partly used to produce angular momentum at the fan inlet. The dashed curves show how the fan power decreases with flow, with the result that electricity savings can be achieved due to the use of inlet vanes.

The most efficient method for fan flow control uses *variable-speed** fan motors, but it has the highest cost. The result of speed changes is to change the fan curve, which now intersects the system curve at progressively lower flow rates as the fan speed slows. On the undampered (right) system curve, reducing fan speed reduces flow from point 1 to point 2. Thus, flow reduction from $\dot{V}_1$ to $\dot{V}_2$ can be achieved without any unwanted pressure drop. In practice, one would not use both dampers and a variable-speed drive to control airflow.

* The term *variable frequency* is sometimes used interchangeably with *variable speed* to describe this method of flow control because speed reduction of the most common electric motors is accomplished by reducing the frequency of the motor power supply.

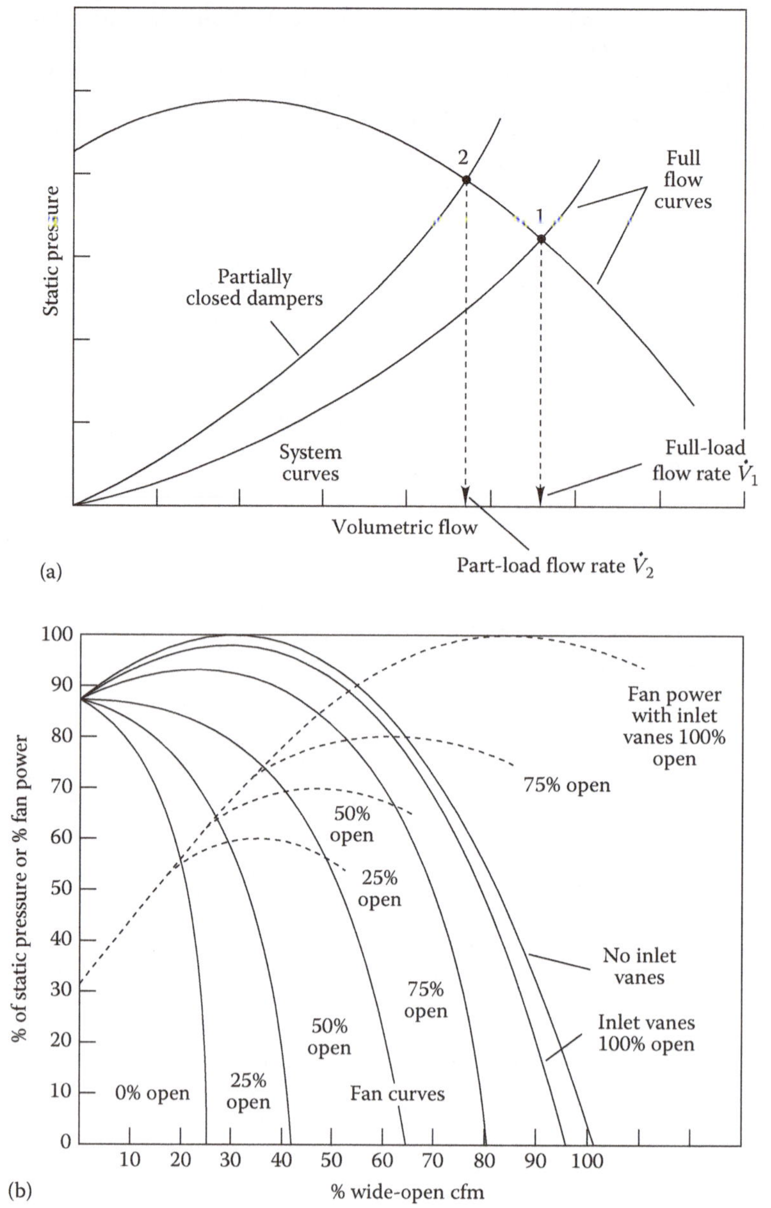

FIGURE 16.22
Variable-volume fan and system curves for (a) outlet damper flow control method and (b) inlet vane control method.

One measure of the energy savings that can be expected from the three flow control methods is the part-load power consumption, expressed as a function of the part-load ratio (in this case defined as the fraction of design or rated fan flow). A second- or third-order polynomial is appropriate according to Brandemuehl et al. (1993).

For variable-speed drives, the suggested expression is*

$$\frac{\dot{W}}{\dot{W}_{rated}} = 0.00153 + 0.0052 \times \text{PLR} + 1.1086 \times \text{PLR}^2 - 0.1164 \times \text{PLR}^3 \quad (16.42)$$

where

$\dot{W}$ is the power input at part load, hp (kW)

$\dot{W}_{rated}$ is the power input at rated design flow, hp (kW)

$\text{PLR} = \frac{\dot{V}}{\dot{V}_{rated}}$ is the ratio of flow at part load to design rated flow (16.43)

* The fan laws would suggest a part-load curve involving only a cubic term. However, not all assumptions involved in deriving the fan laws, e.g., no flow separation or constant efficiency at any speed, necessarily apply to actual fans in actual systems.

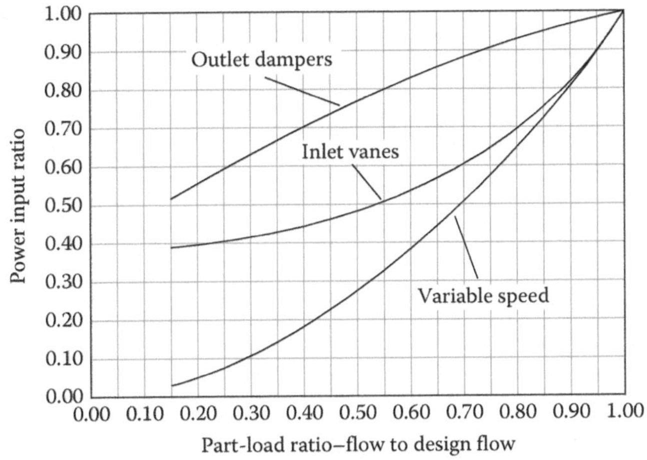

FIGURE 16.23
Part-load fan characteristics for outlet damper, inlet vane, and variable-speed control methods. (From DOE2.1, Several manuals for the DOE2.1 software—2.1D building description language summary (DE-890-17726) and 2.1A engineers' manual (DE-830-04575), The National Technical Information Service, Springfield, VA, 1981.)

A similar expression applies to *outlet dampers*:

$$\frac{\dot{W}}{\dot{W}_{rated}} = 0.371 + 0.973 \times \mathrm{PLR} - 0.342 \times \mathrm{PLR}^2 \quad (16.44)$$

and for *inlet vane control*:

$$\frac{\dot{W}}{\dot{W}_{rated}} = 0.351 + 0.308 \times \mathrm{PLR} - 0.541 \times \mathrm{PLR}^2 + 0.872 \times \mathrm{PLR}^3 \quad (16.45)$$

Figure 16.23 compares the power inputs of the three methods of fan flow control as a function of PLR. It is clear that the variable-speed approach has the largest savings over a wider operating range. The equations given earlier are not universal, only typical. If part-load data from manufacturers become available for flow control methods, they should be used instead of the preceding expressions. The data that do exist are consistent; in order of increasing efficiency, the three flow control systems are dampers, inlet vanes, and variable-speed drives.

Example 16.8: Comparison of Different Flow Control Methods

This example is meant to compare the relative power usage reductions achieved by the different flow control methods, namely, control of fan outlet dampers, adjustable inlet guide vanes, and fan speed control. Consider an airflow system in which the flow needs to be reduced to 60% of its full-load capacity. We will calculate the resulting full-load or rated reduction fraction in power by these control methods.

Given: PLR = 0.6

Find: $\frac{\dot{W}}{\dot{W}_{rated}}$ for all three control methods

Solution

For variable-speed drives (Equation 16.42):

$$\frac{\dot{W}}{\dot{W}_{rated}} = 0.00153 + 0.0052 \times 0.6 + 1.1086 \times 0.6^2 - 0.1164 \times 0.6^3 = 0.379$$

For outlet dampers (Equation 16.44):

$$\frac{\dot{W}}{\dot{W}_{rated}} = 0.371 + 0.973 \times 0.6 - 0.342 \times 0.6^2 = 0.832$$

and for inlet vane control (Equation 16.45)

$$\frac{\dot{W}}{\dot{W}_{rated}} = 0.351 + 0.308 \times 0.6 - 0.541 \times 0.6^2 + 0.872 \times 0.6^3 = 0.529$$

Hence, the inlet guide vane control consumes about 40% more (= 0.529/0.379 = 1.4) in electric power draw compared to variable speed drive, while this ratio is even greater when compared to outlet damper control (= 0.832/0.379 = 2.2).

Comments

To conclude, it is of interest to compare the power draw with that predicted by the ideal fan laws. The power reduces by the cube of the flow ratio. Thus, ideally, the power ratio should be = $(0.6)^3$ = 0.216. By that token, even the variable speed drive control draws about 57% more than this ideal value.

To conclude, it is instructive to discuss what really happens to the characteristic curves and the operating points under different flow control methods. Variable-speed drives are often used for pumps to reduce operating costs. Figure 16.24 shows how the original pump curve 1 is affected by reducing the speed (curve 2). This is the more energy efficient way to reduce flow rate. Note that decreasing the prime mover speed is analogous to using a smaller impeller size in the volute. For example, let us assume that the current operating point is given by point *A*. If we wish to reduce volumetric flow from $\dot{V}_1$ to $\dot{V}_2$, we have two options:

1. Use a throttle valve and close it partially to create an additional system pressure drop that modifies the piping system curve counterclockwise as shown in Figure 16.24. The new operating point is now *B*. However, the total head is higher than needed, and so the throttle valve is partially closed so as to induce an

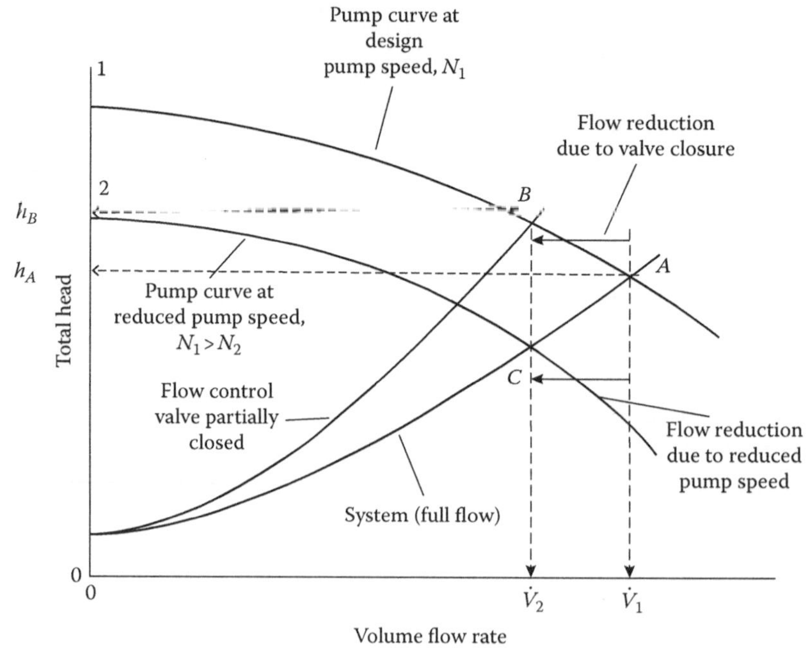

FIGURE 16.24
Flow control can be achieved either by partial closure of control valves or by variable speed drives.

extra pressure drop given by $(h_B - h_C)$. This is essentially an eddy loss that results in waste heat generation. The required pumping power is close to what it was originally.

2. Vary the impeller speed such that the characteristic curve 1 is lowered to curve 2. The new operating point is given by C since the original piping system curve is not affected. Note that the head at point C is lower, but that will not affect system performance. The required pumping power is now greatly reduced since power reduces with the cube of the pump speed (if efficiency remains constant), as is the case with prime movers. The variable speed drive mechanism is most energy efficient; it is, however, more expensive than other control methods in first cost and hence is generally economical for larger systems where the energy savings during operation are larger.

16.6 Duct Design Methods

16.6.1 Background

The design of duct systems for hot and cold air distribution builds upon fundamental material presented earlier in this chapter. Several design methods are described in this section, and the two most common are discussed in detail with examples. The objective of duct design is to deliver the amount of air (at the proper condition) needed to meet the loads in each zone of a building. Duct design is constrained by many factors. One of the most important—available space—is often beyond the purview of the HVAC engineer. Other constraints include the need to (1) meet loads in a variety of zones, (2) meet economic criteria, (3) minimize operating energy subject to previous constraint, and (4) control noise levels.

The sequence of events in duct design is as follows. A preliminary system is laid out on a set of preliminary drawings including all structural members. Once a layout has been made, ducts are sized based on needed air quantities in each zone and for each terminal device. Pressure drop calculations are made at this point, and a fan is selected. The next iteration of the design will need to account for potential flow imbalances in the original design, duct runs of excessive pressure drop, and noise problems. After one or more iterations to accommodate these criteria, a set of final design drawings is prepared. At least one cost estimate is necessary as part of the design process, when the design is deemed sufficiently complete. Computer-aided design (CAD) of duct systems is commonplace in large and medium-size design offices; it replaces the formerly tedious, manual iterations needed for duct design. In addition, various design options are easy to compare by using CAD approaches.

16.6.2 Different Classes and Types of Methods

Duct designs are divided into two generic classes. *Low-velocity systems* have velocities below about 2500 ft/min (13 m/s) and pressure drops of less than 0.15 in./100 ft (1 Pa/m), whereas *high-velocity systems* have velocities up to 4500 ft/min (23 m/s) and pressure drops per unit lengths of less than 0.7 in./100 ft (4.7 Pa/m). Figure 16.25 shows the recommended velocity and friction rates for the two regimes. High-velocity systems can use smaller ducts if space is a problem, but fan power levels are higher. They are usually to be found in large commercial buildings where thermal loads are high requiring large airflow rates. Low-velocity systems are used if fan operating costs are to be lower and if adequate building space for larger ducts exists. High-velocity systems require special ductwork (round or oval spiral ducts) to control leaks and properly confine the high-pressure air so as to reduce noise. Flow balancing is an important issue, and duct design and ductwork have to be designed carefully.

There are five methods of duct design of which the first three are meant for low velocity and the last two for high velocity systems (Table 16.8):

1. The simplest is the *constant-velocity* method whereby ducts are sized so that the velocity is constant everywhere in all the branches of the system. Of course, flow imbalance can result, and special measures may be needed to achieve balance. Constant velocities are often needed in industrial exhaust systems where particulates must remain entrained until the air is exhausted from a facility. The designer's judgment is needed to achieve a good design.
2. A second method called the *balanced-capacity or pressure method* progressively changes velocity so

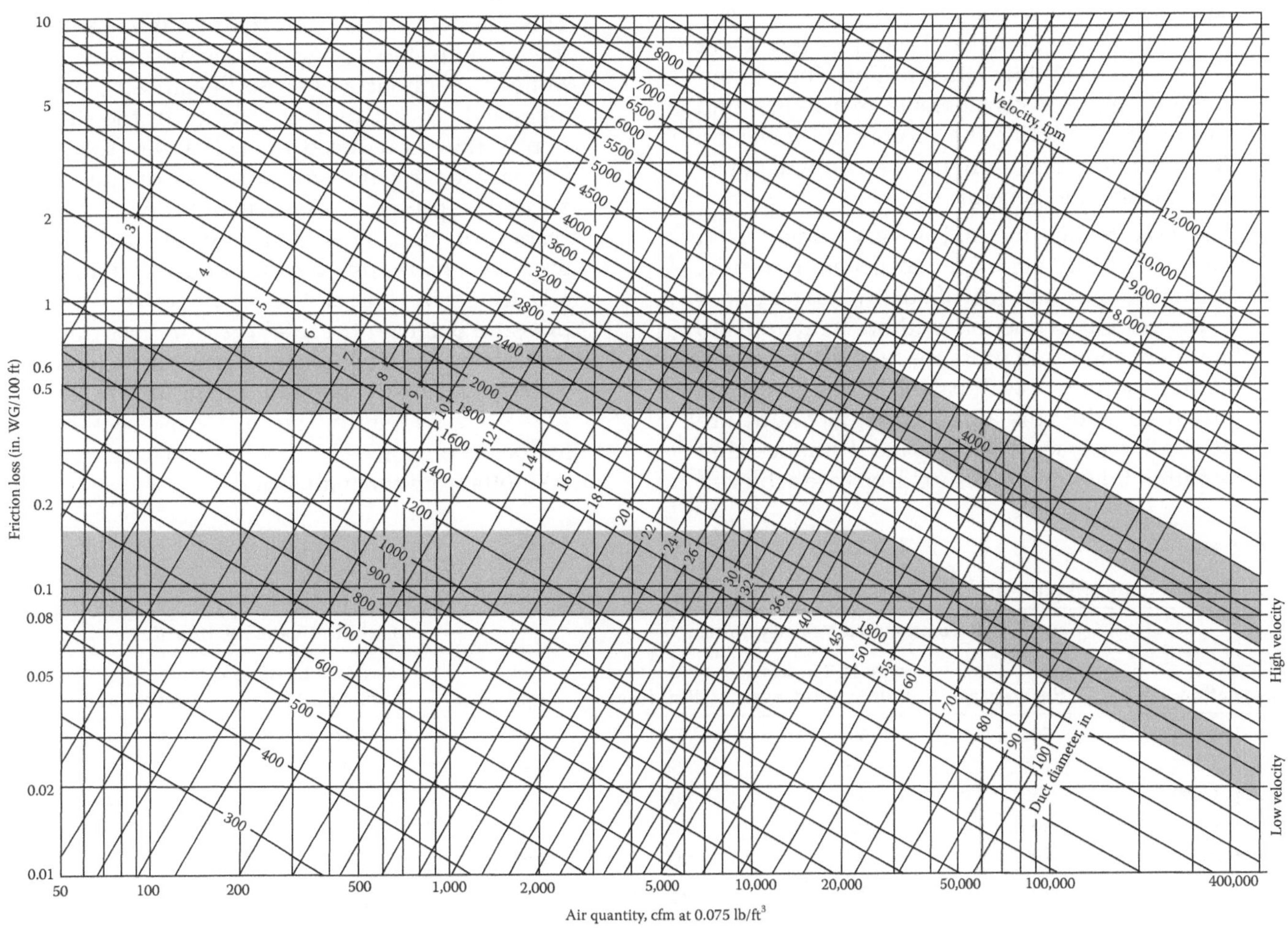

FIGURE 16.25
Recommended operating ranges for low- and high-velocity air systems; pressure drop versus flow rate. (From ASHRAE, *Handbook—HVAC Systems and Equipment*, American Society of Heating, Refrigerating and Air-Conditioning Engineers, Atlanta, GA, 2012. Copyright ASHRAE, www.ashrae.org.)

TABLE 16.8

Different Types of Duct Sizing Methods

Method	Classification	Basis of Duct Sizing
Equal velocity	Low velocity	For simple systems, equal velocity in duct branches (lot of judgment needed)
Equal friction	Low velocity	Equal friction loss per unit duct length for all branches
Balanced capacity	Low velocity	Equal pressure drops from fan to outlets of all branches
Static regain	High velocity	For large installations, same static pressure along duct length
T- method	High velocity	Based on life cycle costing (LCC) optimization using dynamic programming

that the pressure drop in each branch is the same, thereby guaranteeing system flow balance. This method is suggested for use in systems such as the variable-air-volume system, where precise flow balance is not necessary (because other means of flow control are used in such systems). A similar method is the *velocity reduction* method that progressively reduces air velocity in ducts with distance from the fan. Table 16.9 lists recommended velocities for various parts of ventilation systems. It is seen that the highest velocity occurs at the fan outlet, and progressively lower speeds are used downstream. This duct design method requires considerable experience and judgment in selecting the appropriate velocities. A system designed with this method will not be self-balancing in general.

3. The *equal-friction* method is very widely used for low velocity systems, and generally results in smaller sizes and cost of ducting as compared to the velocity reduction method. It attempts to maintain the same pressure gradient throughout the system. A well-balanced design can be produced with this approach if duct runs are of similar capacity. *Well-balanced* implies that required flows and pressure drops are achieved without resort to brute-force methods of flow control such as excessive dampering. If the system has various duct lengths, this method will give a different solution from the balanced capacity method since the pressure gradient will differ in accordance with duct length differences. An illustrative example of this method is provided in Example 16.9.
4. Very large high-velocity ducting systems are usually sized by the *static regain* method. This method is based on the requirement that the system static pressure remain about the same throughout a system. Specifically, ducts are sized so that the air velocity in the direction of flow is reduced in such a way that the increase (or "regain") in static pressure just balances the pressure losses in the downstream section of duct. This is accomplished by progressively increasing the duct cross section; ideally, the conversion of velocity pressure to static pressure balances the total pressure loss due to friction in the duct section. For practical reasons, instead of constantly changing the duct size along its length, the duct cross section is increased only at certain points (such as branch take-offs from the main duct).

TABLE 16.9

Recommended and Maximum Duct Velocities

	Recommended Velocity, ft/min			Maximum Velocity, ft/min		
Designation	Residences	Schools, Theaters, Public Buildings	Industrial Buildings	Residences	Schools, Theaters, Public Buildings	Industrial Buildings
Outside air intakes[a]	500	500	500	800	900	1200
Filters[a]	250	300	350	300	350	350
Heating coils[a]	450	500	600	500	600	700
Air washers	500	500	500	500	500	500
Suction connections	700	800	1000	1000	1400	1400
Fan outlets	1000–1600	1300–2000	1600–2400	1500–2000	1700–2800	1700–2800
Main ducts	700–900	1000–1300	1200–1800	800–1200	1100–1600	1300–2200
Branch ducts	600	600–900	800–1000	700–1000	800–1300	1000–1800
Branch risers	500	600–700	800	650–800	800–1200	1000–1600

Source: Courtesy of Reynolds Metal Company, Richmond VA.

[a] The velocities are for the total face area, not the net free area. Other velocities are for the net free area. Divide ft/min by 197 to convert to m/s.

Consider Example 2.2. The diverging duct with cross-sectional areas A_1 = 46.5 in.^{2} and A_2 = 70.16 in.^{2} caused the velocity to decrease from 3000 to 2000 ft/min. Recall that the sea-level velocity pressure for standard air is given by $p_v = (1/2)\rho v^2$. In IP units, the difference in velocity pressure at two locations is (Equation 16.25 IP)

$$p_{v1} - p_{v2} = \left(\frac{3000}{4005}\right)^2 - \left(\frac{2000}{4005}\right)^2$$

$$= 0.311 \text{ inWG} \qquad (16.46 \text{ IP})$$

This pressure difference is the static pressure regain if there is no friction and other dynamic losses. Typically, designers assume a recovery factor, in the range of 0.7–0.9 to account for such duct pressure losses. If a value of 0.8 is assumed, it would imply that instead of 0.311 inWG increase in static pressure from section 1 to section 2 of the duct, the actual increase will be (0.311 × 0.8 =) 0.249 inWG.

A nearly uniform static pressure throughout a duct system is desirable since it simplifies terminal box and other distribution equipment selection and allows for easier system balancing. The disadvantages are that very low velocities and large duct sizes at the end of long runs may be needed, and the calculations are iterative and quite complex. This method is best implemented by using commercially available computer software since the iterative calculations are tedious and time-consuming to perform by hand. An illustrative example is given in Example 16.10.

5. The design methods described earlier require several years of experience and are subject to individual preferences. Optimization methods of duct sizing have been developed wherein the objective function can be framed as minimizing life cycle cost (LCC) that includes first costs (material plus labor) as well as operating costs such as fan power consumption. Optimization methods based on genetic algorithms have also been developed. The interested reader can refer to Tsal et al. (1988) and to Elite (2009).

In summary, it is recommended that the equal-friction method be used for intermediate-size systems and for low-pressure branch ducts in large systems. The static regain method is often used in large buildings with high-pressure systems or for the main distribution ducts in medium-size systems. For small buildings with simple systems, the velocity reduction method is adequate. Constant-velocity designs are usually restricted to industrial systems.

16.6.3 Equal Friction Method

Example 16.9: Equal-Friction Duct Design

Use a *design pressure drop* of 0.1 inWG/100 ft to design the simple duct system shown in Figure 16.26. The fitting pressure drops are taken from tables from the online HCB software (similar to Table 16.5). Use round ductwork in standard 1 in. increments.* The pressure loss at each branch outlet grille is equivalent to 20 ft of duct according to the grille manufacturer. The elbows at *G* and *E* are full-radius elbows (r/D = 1.0). The tees are of conical design, and the location is at sea level.

Although they are inconsistent, use the units of the problem statement in the solution since they are those used by duct designers in the United States.

Given: $\dot{V}$ (per figure), $\frac{\Delta p}{L} = 0.1$ inWG/100 ft

Figure: See Figure 16.26.

Assumptions: Ignore pressure losses due to duct size transitions. The location is at sea level.

Find: Branch pressure drops Δp

Lookup values: From the online HCB software, we can determine pressure drop coefficients C_f for tees and elbows. For example, C_f = 0.22 for an r/D = 1.0 elbow. Tabulating the tee coefficients will need to await duct velocities calculated. The values of C_f for all fittings are shown in the ninth column of Table 16.10.

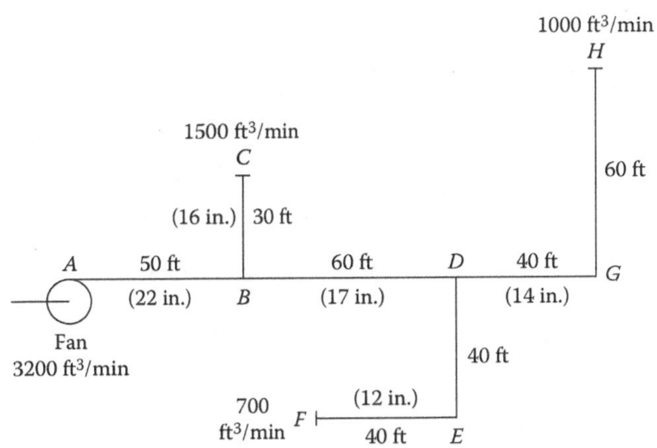

FIGURE 16.26
Duct layout for Example 16.9.

* If round ducts cannot be fitted above ceilings in a building, their equivalent in rectangular ducts can be found from Equation 16.21. Different C_f values will apply and can be found in the online HCB software.

TABLE 16.10

Results of Example 16.9: Duct Design

Section	Length, ft	$\dot{V}$, ft³/min	$\Delta p/L$, inWG/100 ft	Duct Loss, inWG	Dia., in.	Velocity, ft/min	p_v, inWG	C_f	Fitting Loss, inWG	Total Loss, inWG
A–B	50	3200	0.10	0.05	22	1210	0.091	—	—	0.050
B–C	30	1500	0.10	0.03	16	1070	0.071	0.48	0.034	0.064
C	20	1500	0.10	0.02	—	1070	0.071	—	—	0.020
B–D	60	1700	0.10	0.06	17	1080	0.073	0.011	0.001	0.061
D–G	40	1000	0.10	0.04	14	935	0.055	0.013	0.001	0.041
G–H	60	1000	0.10	0.06	14	935	0.055	0.22	0.012	0.072
H	20	1000	0.10	0.02	—	935	0.055	—	—	0.020
D–E	40	700	0.10	0.04	12	890	0.049	0.51	0.025	0.065
E–F	40	700	0.10	0.04	12	890	0.049	0.22	0.011	0.051
F	20	700	0.10	0.02	890	—	0.049	—	—	0.020

Solution

The results of the solution are tabulated in Table 16.10. The table is completed from left to right as follows:

(a) First the duct length and volumetric flow values are entered from information shown in Figure 16.26.

(b) The next column (pressure drop per 100 ft) is specified in the problem statement. It is constant everywhere since we are using the equal-friction approach.

(c) Multiplying the duct length by this factor gives the values in the fifth column of Table 16.10.

(d) From Figure 16.8a, the duct size is read from the known volumetric flow rate and design pressure drop. These values are entered in the sixth column of the table. For example, a 3200 ft³/min flow at 0.1 inWG/100 ft pressure drop occurs if a 22 in. diameter duct is used.

(e) From the volumetric flow and the diameter, it is easy to find the air velocity in each branch.

(f) From the velocity, the sea-level velocity pressure for standard air is found: $p_v = (1/2)\rho v^2 = \left(\frac{v}{4005}\right)^2$

(g) From the online HCB software using fitting loss coefficient tables, C_f can now be found, since the velocities are known. For the straight-through loss in the tees at *B* and *D*, one needs the velocity ratios. At *B* the ratio is 1080/1210 = 0.89, and at *D* the ratio is 935/1080 = 0.89. From these ratios the respective C_f coefficients are found to be 0.011 and 0.013, from the table labeled "main" in the "wye, diverging" table. These C_f values are entered into the ninth column of Table 16.10.

(h) The branch line loss coefficients are found in a similar manner. At *B*, the velocity ratio is 1070/1210 = 0.88, and at *D* it is 890/1080 = 0.82. From the online HCB software tables for "tee, diverging, round, conical branch" entries, the C_f coefficients are interpolated as 0.48 and 0.51, respectively. At this point, the fitting pressure losses can be calculated from the basic equation:

$$\Delta p_{fit} = C_f\left(\frac{1}{2}\rho v^2\right) = C_f \times p_v \qquad (16.47)$$

(i) The final column of the table is the sum of straight duct and fitting pressure loss (column 5 + column 10).

For the three branches, the pressure drops are

$$\Delta p_{ABDGH} = 0.050 + 0.061 + 0.041 + 0.072 + 0.020 = 0.244 \text{ in.}$$

$$\Delta p_{ABC} = 0.050 + 0.064 + 0.020 = 0.134 \text{ in.}$$

$$\Delta p_{ABDEF} = 0.050 + 0.061 + 0.065 + 0.051 + 0.020 = 0.247 \text{ in.}$$

(j) The maximum loss is 0.247 inWG (61 Pa). This is the value that would be used for fan selection (along with other pressure drops including coils, the system effect factor, and filters; the velocity pressure p_v, dissipated in the conditioned room, must also be added to the total for fan sizing).

Comments

The two longer branches in this example have essentially the same pressure drop and are therefore self-balancing. The shorter duct with only one-half the pressure drop of the other branches will require a *balancing damper* to provide approximately another 0.113 inWG pressure drop. This will result in a system in which the pressure drop in each branch is balanced, i.e., essentially the same. In practice, all ducts in an air distribution system will have balancing dampers. Even when fully open, these dampers will have an associated pressure drop. Note that we have ignored the duct size transitions that occur at points *B* and *D* in both downstream branches. These four resistances are small. Abrupt transitions are to be avoided due to excessive pressure loss.

Return air systems are designed by using the same technique as in this example. The only difference lies in the pressure coefficient for tees because the flow converges in return ducts while it diverges in supply ducts. The online HCB software contains data for both cases. A separate return fan is often used in the return air system as discussed in Chapter 19.

16.6.4 Static Regain Method

Example 16.10: Static Regain Duct Design

Using the same duct layout as in Example 16.9, use the static regain method to determine the duct diameter for section *B–D*. Assume a static regain factor of 0.8, and take the duct diameter for section *A–B* to be 22 in. (as in previous example).

Given: Duct layout, lengths, and flows as in Example 16.9.

Figure: See Figure 16.26 and Table 16.10.

Assumptions: Ignore pressure losses due to duct size transitions. Assume round ducts.

Find: Duct diameters at points *B*, *D*

Solution

1. The velocity through the main duct section *A–B* is 1210 ft/min.
2. The static pressure regain at point B due to duct diameter increase should compensate for the dynamic losses in the downstream section *B–D*. Assume a velocity of 800 ft/min. The volumetric flow through section *B–D* is 1700 cfm and the length is 60 ft. From Figure 16.8 the pressure drop per 100 ft = 0.045 inWG/100 ft. The friction loss through section *B–D* = 0.045 × (60/100) = 0.027 inWG. Using Equation 16.46, along with the regain factor,

$$\text{Pressure regain at } B = 0.8 \times \left[\left(\frac{1210}{4005}\right)^2 - \left(\frac{800}{4005}\right)^2\right]$$

$$= 0.0411 \text{ inWG}$$

   The regain is too large compared to the target of 0.027 inWG.
3. A second iteration assuming velocity of 900 ft/min would result in a pressure drop per 100 ft = 0.06 inWG/100 ft, and the pressure drop through section *B–D* = 0.06 × (60/100) = 0.027 inWG. The pressure regain at *B* is now 0.03 inWG, which is acceptable.
4. The corresponding round duct size for section *B–D* is close to 18 in. diameter.

Although the majority of this section dealt with airflow design of ducts, their thermal design is also important. Since conditioned air from the central plant is often at a temperature quite different from that of the space in which ducts are located, heat losses or gains must be controlled by the use of duct insulation. Warm air ducts in unheated spaces (e.g., crawl spaces in residences) and cool air ducts in uncooled spaces (residential attics, commercial building zone plenums) are prime candidates for added insulation. The heat loss or gain from ducts must be accounted for in secondary system design. Insulation can be applied either to the duct exterior or can be used as the duct liner, where it also serves to absorb sound. Relevant equations to determine heat losses/gains of air flowing through ducts, and thereby deduce the insulation needed to limit heat losses, are given in Sections 2.8 and 2.9 (also see Problem 2.15).

Typically, 1–2 in. (2.5–5 cm) thicknesses of rigid duct insulation provide the needed insulation. ASHRAE Standard 90.1 (2007) recommends that the *R* value (IP units) of duct insulation be chosen according to $R = \Delta T_{duct}/15$, where ΔT_{duct} is the air temperature difference from inside to outside of the duct; this *R* value is for the required insulation only and does not include inner and outer surface convection resistances. Ducts conducting cold air may require external vapor barriers (with sealed seams) to avoid condensation and insulation degradation. Most building codes also require that duct insulation meet standard fire hazard requirements.

The duct designer will specify other features besides its size. Duct leakage is a significant problem (unsealed metal ducts can lose up to 15% of the airflow at 1 inWG duct pressure, whereas sealed ducts will lose less than 1%); therefore, sealed ductwork must be included in the specifications. The thickness gauge of duct metalwork must be determined; it depends on air pressure and duct dimensions. Duct supports, turning vanes, layout to avoid building structural members, and other fabrication details are the responsibility of the design engineer. During and after construction, the design engineer should also inspect the system to ensure conformance with the design documents.

16.7 Fluid Flow Measurement

Flow measurement in liquid streams and airstreams in buildings is needed for control and energy management purposes. It is not possible to measure velocity directly, unless expensive Doppler-type devices are used. Instead, we typically rely on equipment that produces a pressure drop that can be converted to velocity by an equation specific to the equipment. Several devices are in common use—the venturi, orifice plate, and Pitot tube. The first and second produce an average flow value whereas the third produces a point measurement.

16.7.1 Venturi Meter

Figure 16.27 shows a venturi meter with two pressure taps. The *average throat velocity* is given by

$$\mathrm{v} = CE\sqrt{\frac{2\Delta p}{\rho}} \tag{16.48}$$

where E depends on β, the throat-to-pipe-diameter ratio:

$$E = (1-\beta^4)^{-1/2} \tag{16.49}$$

The venturi coefficient C corrects for differences in velocity between the uniform streamline flow (assumed in the Bernoulli equation from which Equation 16.48 is derived) and the actual flow through a venturi including viscous effects. For the universal venturi tube,

$$C = 0.9797 \tag{16.50}$$

When IP units of density (lb_m/ft^3) are used in Equation 16.48, the density value must be divided by g_c [32.2 $(lb_m \cdot ft)/(s^2 \cdot lb_f)$] for dimensional consistency.

16.7.2 Orifice Plate

Figure 16.28 shows an orifice plate used to produce a pressure difference related to the fluid velocity in the pipe (or duct) in which it is inserted. Equation 16.48 is used to find the average velocity, but the orifice coefficient for the pressure tap locations shown in the figure is given by

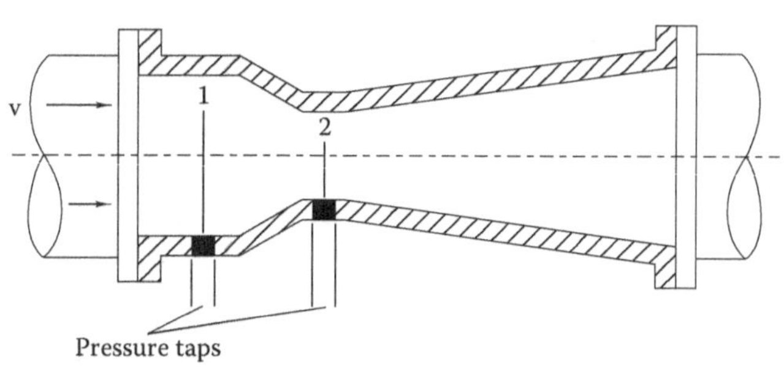

FIGURE 16.27
Cross section of venturi meter.

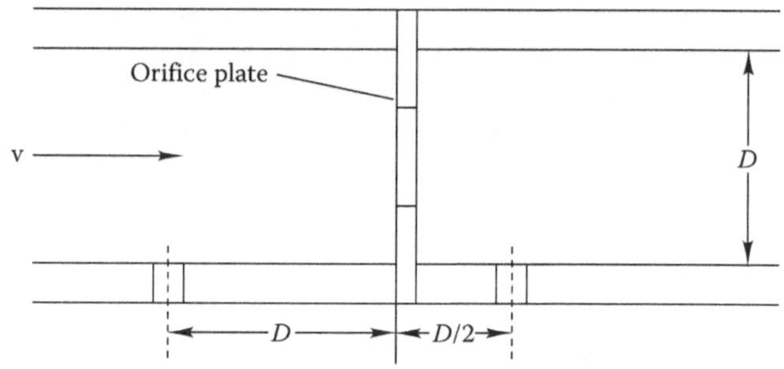

FIGURE 16.28
Orifice plate cross section with D and $D/2$ pressure taps.

$$C = 0.5959 + 0.0312\beta^{2.1} - 0.184\beta^8 + 0.039\frac{\beta^4}{1-\beta^4} - 0.0158\beta^3 + 91.71\frac{\beta^{2.5}}{\mathrm{Re}_{D_{pipe}}^{0.75}} \tag{16.51}$$

The Reynolds number is based on the pipe diameter. Since the velocity v is not known a priori, all terms but the last are used to estimate of v. This value is then used to find the Reynolds number in the final term, and the final value of v is calculated. A different expression for C is recommended if different pressure tap locations are used (Kreider, 1985; Miller, 1983).

Orifice plates cause a significant *parasitic pressure drop* in any fluid stream in which they are located, but they have a low first cost. Venturi meters are more expensive initially but have much smaller pressure penalties than orifice plates. Life-cycle power costs must be considered in the selection of flow measurement equipment that are to be permanently installed.

Incidentally, Equation 16.48 also applies for small orifices that occur in the envelopes of buildings (see Section 6.3 and Equation 6.3). Of course, in this instance one is not measuring flows in ducts, but rather attempting to find air leakage rates given the pressure difference between the interior and exterior of a building. Instead of the square root, however, building leakage appears to depend on the pressure difference raised to an exponent between 0.60 and 0.65.

16.7.3 Pitot Tube

The velocity at a *point* rather than average velocities are sometimes needed in fluid streams in buildings for sensor calibration or equipment control. The Pitot tube

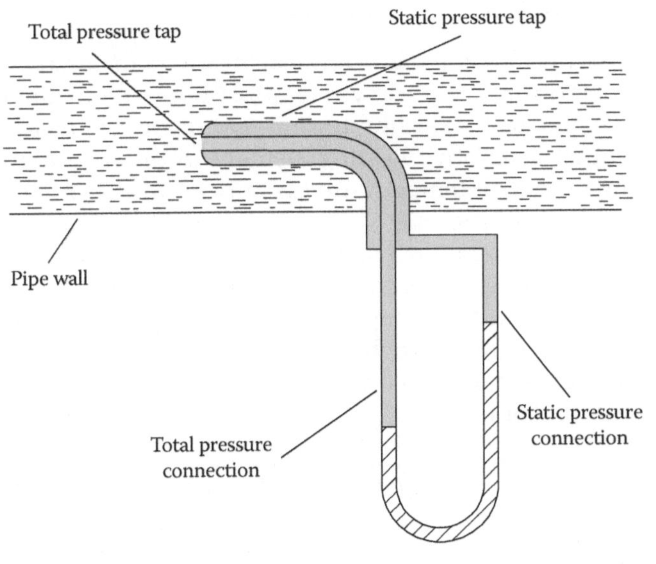

FIGURE 16.29
Pitot tube. When the densities of the flowing fluid and the manometer fluid are different, the pressure exerted by the column of flowing fluid (crosshatched) must be accounted for in the Pitot equation. However, this is not required when a water column is used to measure airflow since the density of air is much less than the manometer fluid, namely water.

shown in Figure 16.29 is used for this purpose. The local velocity is given by

$$v = \sqrt{\frac{2\Delta p}{\rho}} \tag{16.52}$$

Some manufacturers provide a multiplying coefficient similar to C in Equation 16.48. However, if none is provided, Equation 16.51 is used directly without adjustment. Pitot tubes can also be used to find the average velocity in a duct by averaging several point velocity measurements.

Example 16.11: Pitot Tube

A Pitot tube in an airstream registers a pressure reading of 2.1 inWG. What is the velocity of this airstream?

Given: $\Delta p = 2.1\ \text{inWG} \times [(144\ \text{lb}_f/\text{ft}^2)/27.7\ \text{inWG}] = 10.9\ \text{lb}_f/\text{ft}^2$

Figure: See Figure 16.29.

Assumptions: Standard air in steady flow; $\rho = 0.075\ \text{lb}_m/\text{ft}^3$

Find: v

Solution

Equation 16.52 can be directly used to find the velocity:

$$v = \sqrt{\frac{2 \times 10.9\ \text{lb}_f/\text{ft}^2}{(0.075\ \text{lb}_m/\text{ft}^3)/[32.2\ (\text{lb}_m \cdot \text{ft})/(\text{lb}_f \cdot \text{s}^2)]}} = 97\ \text{ft/s}$$

16.7.4 Turbine Flow Meters

Turbine flow meters use a different approach to measure average velocity (or, equivalently, volumetric flow) in pipes. Figure 16.30 shows how a turbine rotor is caused to rotate by the passage of fluid. Turbines are capable of 1% accuracy over long periods if properly maintained and if the measured fluid is kept clean by continuous filtration. Although expensive, one turbine meter in each fluid loop—e.g., the building's chilled-water loop—can serve as the measurement standard against

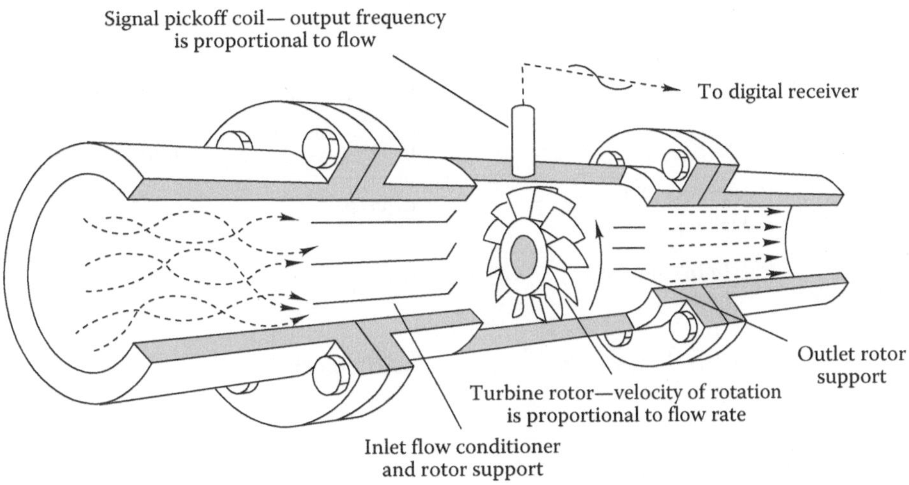

FIGURE 16.30
Cutaway drawing of turbine flow meter.

which other flow devices can be periodically calibrated. Venturi meters are a good compromise between the permanent pressure penalty of orifice plates and the initial cost of turbine meters when only average velocity (or, equivalently, the volumetric flow) is needed. Benedict (1984) discusses details of fluid flow measurement.

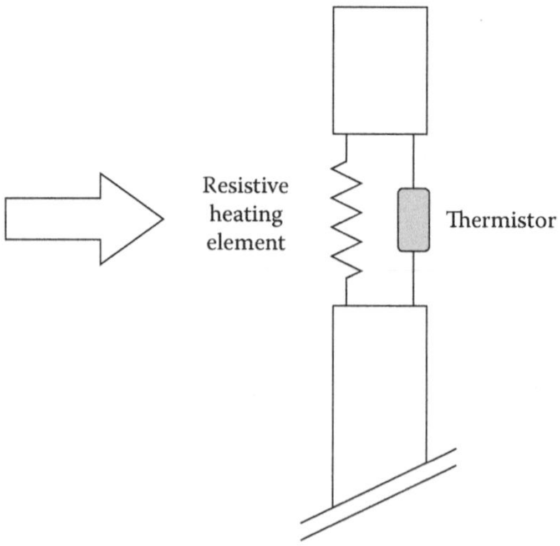

FIGURE 16.31
Hot-wire anemometer.

16.7.5 Hot-Wire Anemometers

Hot-wire anemometers rely on heating an airstream to determine the airflow rate. Most hot-wire anemometers operate by varying the amount of current passing through a resistive heating element in order to maintain a constant temperature (usually around 200°F) at a downstream thermistor. See Figure 16.31. Other types keep the current constant and measure the change of downstream temperature. Since the response of the thermistor will also depend on the air temperature, it is necessary to measure this value as well.

The current draw of a constant temperature hot-wire anemometer is

$$i = \sqrt{(A + B\sqrt{\dot{m}})(T_H - T_a)} \tag{16.53}$$

where
- A and B are the calibration constants
- $\dot{m}$ is the mass flow rate
- T_H and T_a are the hot-wire and air temperatures, respectively

Note that this equation identifies the mass flow rate, not the volumetric flow rate. The mass flow rate is, of

TABLE 16.11
Equation Summary

Topic	Equation	Equation Number	Notes
Pressure drop	$\Delta p_f = f\left(\frac{L}{D_h}\right)\left(\rho\frac{v^2}{2}\right)$	16.6	D'arcy–Weisbach equation
	$f = \frac{1.325}{\left\{\ln\left[\varepsilon/3.7D + 5.74/\mathrm{Re}^{0.9}\right]\right\}^2}$	16.8	Pipe turbulent friction factor
	$h_{friction} = K_f\frac{v^2}{2g}$	16.16	Pipe fitting pressure drop
	$h_{friction} = C_f\frac{v^2}{2g}$	16.24	Duct fitting pressure drop
Pumps and fans	$\dot{W}_{fluid} = \dot{V}(p_i - p_o)$	16.31	Pump or fan power input to fluid
	$\dot{W}_{shaft}(\mathrm{hp}) = \frac{\dot{V}\ \mathrm{gal/min} \times H\ \mathrm{ft}}{3960 \times \eta_{pump}}$	16.33 IP	Dimensional pump input power equation
	$\dot{W}_{shaft}(\mathrm{hp}) = \frac{\dot{V}\ \mathrm{ft^3/min} \times H\ \mathrm{inWG}}{6356 \times \eta_{fan}}$	16.34 IP	Dimensional fan input power equation
	$N_2 = N_1\frac{\dot{V}_2}{\dot{V}_1}, H_2 = H_1\left(\frac{N_2}{N_1}\right)^2, \dot{W}_2 = \dot{W}_1\left(\frac{N_2}{N_1}\right)^3$	16.35 through 16.37	Fan laws
Flow measurement	$v = \sqrt{\frac{2\Delta p}{\rho}}$	16.52	Pitot tube
	$v = CE\sqrt{\frac{2\Delta p}{\rho}}$	16.48	Venturi meter and orifice plate

course, independent of density and of thermal conductivity. The density is used only to find the resulting air velocity in the vicinity of the sensor.

16.8 Closure

Much of the scientific material presented in this chapter can be found in textbooks on fluid mechanics. For additional details, the reader is referred to textbooks such as Kreider (1985), White (1998), or Streeter et al. (1997). This chapter also discussed practical issues and design considerations useful to the HVAC designer. Table 16.11 summarizes the important equations developed in this chapter having to do with the fluid mechanics of HVAC systems.

Problems

Numbers 1–4 given in parenthesis denote the degree of difficulty.

16.1 What is the velocity pressure in standard air flowing at 200 ft/s (6 m/s)? Express the answer in inches water gauge (pascals). *(1)*

16.2 What is the velocity pressure in standard air flowing at 200 ft/min (1.0 m/s)? Express the answer in inches water gauge (pascals). *(1)*

16.3 Figure P16.3 shows a tank drain. At what point does the water stream strike the ground as a function of z_1 and z_2? Ignore friction effects. *(2)*

16.4 What are the pressure differences $p_2 - p_1$ and $p_3 - p_1$ in the tee shown in Figure P16.4 if $D_2 = D_1/2 = 2D_3/3 = 4$ in. (10 cm)? The inlet water flow rate of 3 ft³/s (0.085 m³/s) is equally split between the two exit streams. Ignore friction losses. *(2)*

16.5 Water flows at 6 ft/s (2 m/s) through a straight 500 ft (150 m) pipe up a grade to a tank such that the pipe outlet is 30 ft (10 m) above the inlet where the pump outlet pressure is 150 psig (1 MPa). What is the pressure just prior to the pipe exit if friction losses are ignored for this 2 in. diameter (5 cm diameter) pipe? How much power is provided to the fluid by the pumping system if the pump inlet pressure is 30 psig? *(2)*

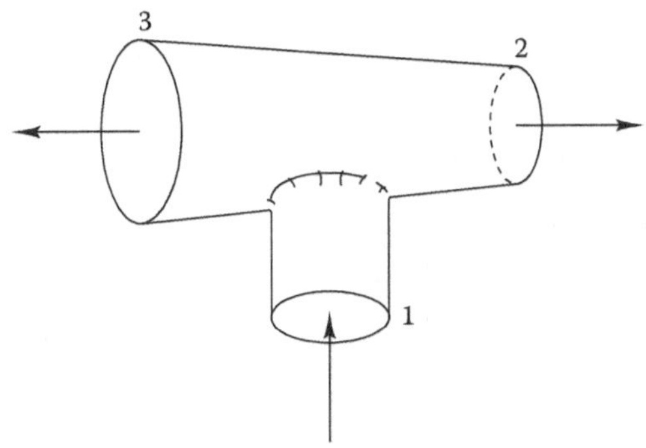

FIGURE P16.4

16.6 What is the pressure p_2 in Figure P16.6 if $y_1 = 60$ cm, $y_2 = 1$ m, $y_3 = 20$ m, and $\dot{V} = 0.006$ m³/s? Neglect the friction losses in the 3 cm pipe used throughout. The nozzle diameter is 1 cm. *(1)*

16.7 How much shaft work is required to pump 10 gal/min (6 L/s) of octane from one tank into another 50 ft (15 m) above? Friction losses are 6% of the static head, and the pipe size is uniform. *(2)*

16.8 Calculate the velocity head in water flowing at 6 ft/s (2 m/s) in feet of water gauge (pascals). *(1)*

16.9 Calculate the friction factor, using Equation 16.8, in schedule 40, nominal, 4 in. commercial steel pipe in which 60°F (15°C) water flows at 7 ft/s (2 m/s). What would the friction factor be if the fluid were 40% ethylene glycol? *(2)*

16.10 Calculate the friction factor, using Equations 16.8 and 16.9, in schedule 40, nominal, 2 in. commercial steel pipe in which 180°F (82°C) water flows at 5 ft/s (1.5 m/s). *(2)*

16.11 Water at 100°F (38°C) flows through nominal 5 in. diameter, schedule 40 commercial steel pipe with

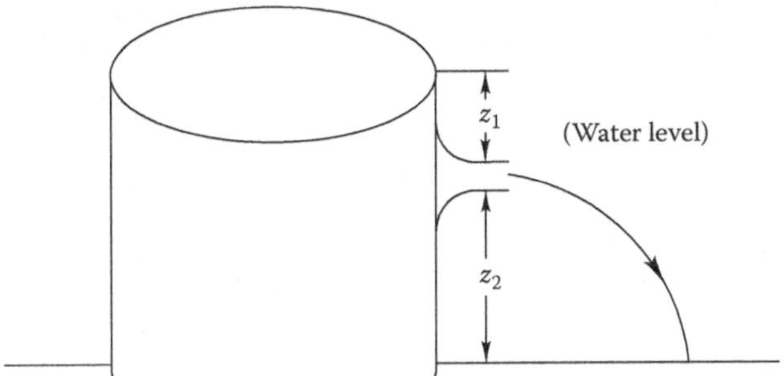

FIGURE P16.3

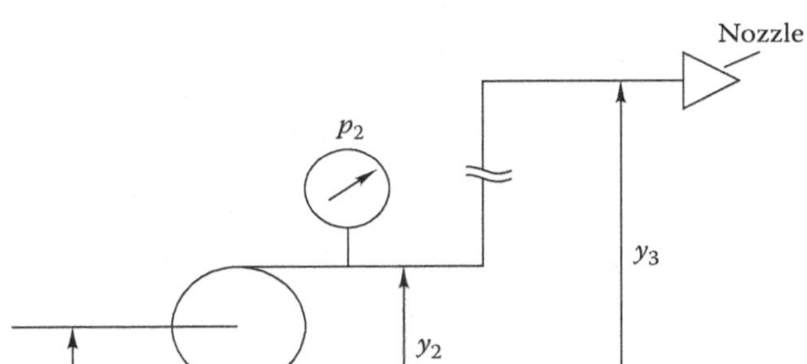

FIGURE P16.6

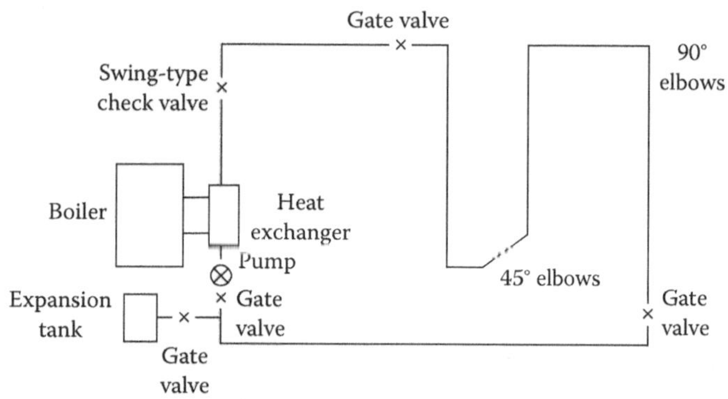

FIGURE P16.13

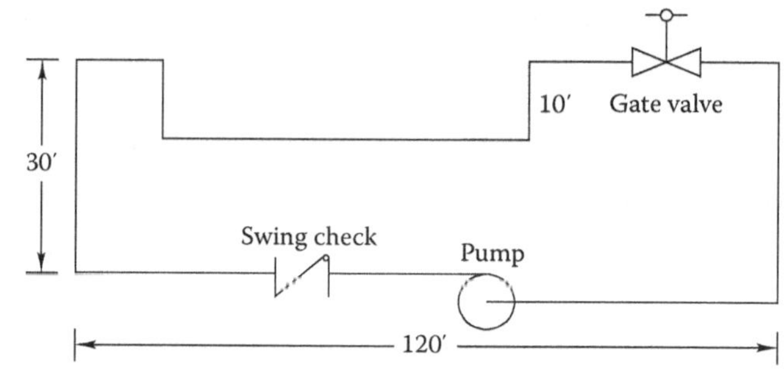

FIGURE P16.15

a typically used pressure drop of 4 ftWG/100 ft (4 m/100 m) of pipe. What is the volumetric flow rate? *(2)*

16.12 Chilled water at 55°F (13°C) flows through nominal 5 in. diameter, schedule 40 commercial steel pipe with a pressure drop of 5 ftWG/100 ft (5 m/100 m) of pipe. What is the volumetric flow rate? *(2)*

16.13 The schematic diagram in Figure P16.13 represents the hot water distribution system in a residential baseboard heating system. The physical length of the pipe is 300 ft (90 m) of nominal $\frac{3}{4}$ in. type L copper pipe (baseboard units are essentially straight lengths of pipe and are included in the 300 ft). The fluid flow rate is 4 gal/min (0.25 L/s) of 180°F (82°C) water. The boiler heat exchanger has the same pressure drop as 30 ft (10 m) of straight pipe; the valves are threaded, and the elbows are soldered. What is the pressure drop that the pump must overcome if all valves are wide open? *(4)*

16.14 What would the pump motor energy savings be in Problem 16.13 if nominal 1 in. type L copper pipe and fittings were used instead of $\frac{3}{4}$ in. and if the heating system operated 3500 h/year? The combined pump motor efficiency is 0.70. *(4)*

16.15 Figure P16.15 is the schematic diagram of a water piping loop. What is the pressure drop if the flow rate is 200 gal/min through 4.5 in. nominal schedule 40 steel pipe if all fittings are threaded? What is the water velocity in the pipe? *(2)*

16.16 Figure P16.15 is the schematic diagram of a water piping loop. What is the pressure drop if the flow rate is 10 L/s through 100 mm nominal schedule 40 steel pipe? What is the water velocity in the pipe? *(2)*

16.17 A chilled-water loop carrying 200 gal/min (13 L/s) consists of 200 ft (60 m) of straight nominal 3 in. schedule 40 steel pipe with one branch tee, one line flow tee, eight standard 90° elbows, and four gate valves. All fittings are flanged. Additional pressure drop occurs in the chiller (equivalent to 20 ft of water) and in the air handler cooling coil (equivalent to 30 ft of water). Determine and comment on the water velocity and pressure drop of the straight pipe. *(3)*

16.18 Work Problem 16.17 with nominal 3.5 in. schedule 40 steel pipe and compare to the same nominal size of type L copper pipe. *(3)*

16.19 What is the equivalent diameter (for pressure drop) of a duct 20 in. × 48 in. (50 cm × 120 cm)? *(1)*

16.20 If 0.25 inWG (60 Pa) is available to overcome friction in 300 ft (90 m) of 25 in. diameter (60 cm diameter) ductwork, what is the volumetric flow rate at sea level?* What is the air velocity? *(2)*

16.21 If 0.25 inWG (60 Pa) is available to overcome friction in 300 ft (90 m) of 20 in. diameter (50 cm diameter) ductwork, what is the volumetric flow rate at 5000 ft (1500 m) elevation? What is the air velocity? *(2)*

16.22 The velocity in a duct is 1800 ft/min (9 m/s). If 5000 ft³/min (2530 L/s) is to flow in this duct, what are the diameter and pressure drop in 150 ft (45 m). *(1)*

16.23 The pressure drop in the duct system of an HVAC system carrying 5000 ft³/min (2360 L/s) of 55°F (13°C) air is 2.0 inWG (500 Pa) at sea level. If the HVAC system is duplicated at an altitude of 5000 ft (1500 m), what will the pressure drop be if the volumetric flow rate remains the same? *(1)*

16.24 Repeat Problem 16.23 except that the mass flow must remain constant, not the volumetric flow rate. *(2)*

16.25 If the sea-level system in Problem 16.23 carries air at 120°F (49°C) rather than at 55°F (13°C), what will the pressure drop be for the same volumetric flow? *(1)*

* In the problems, *sea level* means that the atmospheric pressure is 14.696 psia (101.325 kPa). The effect of altitude on density can be found from Equation 13.1 or 13.2.

16.26 Standard air [ρ = 0.075 lb/ft^3 (1.2 kg/m^3)] flowing through a heating coil (turned off) and downstream ductwork has a pressure drop of 1.0 inWG (250 Pa); the coil pressure drop is 0.3 inWG. If sufficient heat is added to the air to decrease the density to 0.062 lb/ft^3 (1.0 kg/m^3), what is the system pressure drop? What is the design point for the fan for this system? *(2)*

16.27 A duct 4 ft × 4 ft carries 25,000 ft^3/min. What is the pressure drop per 100 ft? *(1)*

16.28 An air distribution system carrying 10,000 ft^3/min (4,700 L/s) consists of 750 ft (230 m) of 24 in. diameter (60 cm diameter) ductwork, six $r/D = 2$, 90°, smooth elbows, one filter [0.2 inWG (50 Pa)], and a full-diameter butterfly damper. Find the system pressure drop if the damper is wide open ($\theta = 0°$) at sea level. *(2)*

16.29 Consider the air system described in Problem 16.28. As the butterfly damper is closed in 10° increments from fully open to 60°, how does the system pressure drop increase, assuming that sufficient fan capacity exists to force the required air amount through the system for all damper positions? Plot the system pressure drop versus damper position. *(4)*

16.30 Work Problem 16.28 at an elevation of 6000 ft (1800 m). *(2)*

16.31 Figure P16.31 shows an air distribution system schematically. Find the pressure drop between the fan outlet and point *B* if the duct velocity is 1800 ft/min (9 m/s) everywhere. Ignore, for simplicity, the pressure drop for (1) the tees (straight throughflow) and (2) duct-size changes. The diffuser pressure drop is 0.1 inWG (25 Pa). *(4)*

16.32 Rework Problem 16.31 if the duct velocity is reduced to 1500 ft/min (7.6 m/s). *(4)*

16.33 What is the shaft power input to a 68% efficient fan that moves 5000 ft^3/min (2400 L/s) against a 2.0 inWG (500 Pa) pressure drop at sea level? What would the power input be at 5000 ft (1500 m) altitude? *(1)*

16.34 What is the shaft power input to an 85% efficient pump that moves 1000 gal/min (64 L/s) against a 3 psi (21 kPa) pressure drop? What is the electric power consumption if an 87% efficient motor is used to drive the pump? *(1)*

16.35 A duct system is designed to carry 9000 ft^3/min (4200 L/s) with a pressure drop of 0.80 inWG (200 Pa). If the duct is connected to the oversized fan whose fan curve is shown by the down- and right-sloping solid line in Figure P16.35, what will be the flow rate and pressure drop in the system? Curve *B* in the figure is the duct resistance curve for this case. *(2)*

16.36 The flow rate determined in Problem 16.35 is approximately 2000 ft^3/min larger than the design. If a flow damper is inserted in the system, how much pressure drop must a damper produce to reduce the flow to the design level of 9000 ft^3/min (4200 L/s)? *(2)*

16.37 The flow rate determined in Problem 16.35 is approximately 2000 ft^3/min larger than the design. If the fan speed is reduced to provide flow equal to the design level of 9000 ft^3/min (4200 L/s), what will the new fan speed be? Curve *A* in the figure is the fan curve for the reduced-speed case. *(2)*

16.38 The system curve shown in Figure P16.35 by a solid line has a system operating point of 12,500 ft^3/min (5,900 L/s) and 0.97 inWG (240 Pa). If a damper that has a 0.4 inWG (100 Pa) pressure drop is inserted in the system, what will the new flow rate be? *(2)*

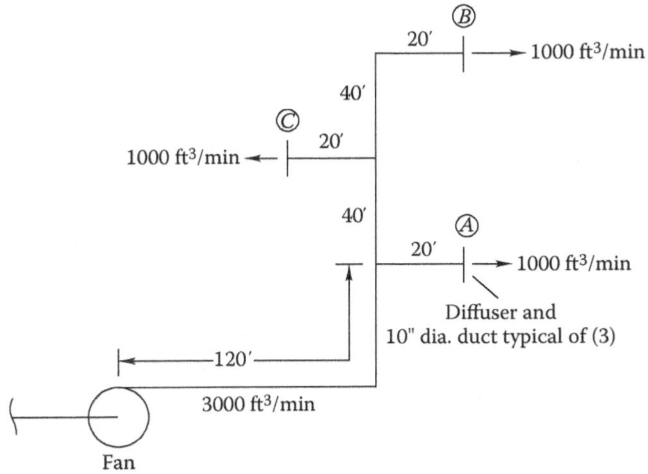

FIGURE P16.31

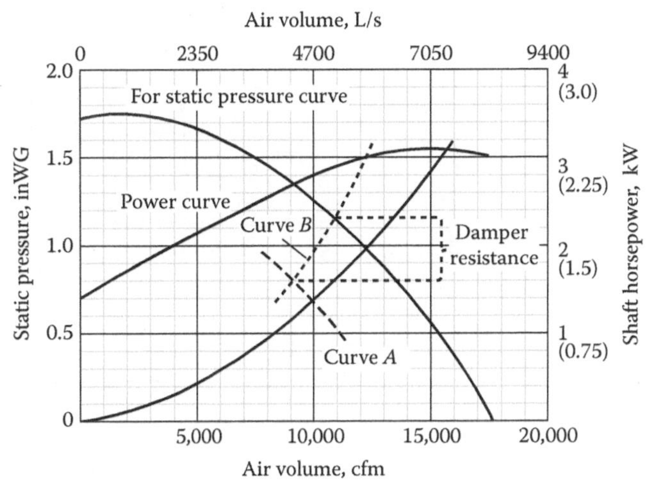

FIGURE P16.35

16.39 The system curve shown in Figure P16.35 by a solid line has a system operating point of 12,500 ft³/min (5,900 L/s) and 0.97 inWG (240 Pa). If the fan speed is reduced from 575 to 475 r/min, what will the system flow rate and fan power be? State any assumptions. *(2)*

16.40 The system curve shown in Figure P16.35 by a solid line has a system operating point of 12,500 ft³/min (5,900 L/s) and 0.97 inWG (240 Pa). If the fan speed is increased from 575 to 675 r/min, what will the system flow rate and fan power be? State any assumptions. *(2)*

16.41 Plot the fan efficiency curve for the fan whose characteristics are shown in Figure P16.35. *(3)*

16.42 If the fan whose sea-level characteristics are shown in Figure P16.35 is moved to a 5000 ft (1500 m) altitude, how will the system, fan, and power curves change? *(2)*

16.43 A chilled-water system requires a 100 gal/min (63 L/s) pump with 40 ftWG (115 kPa) of head. Select the most appropriate pump from the pump curves in Figure P16.43. State the efficiency, motor power input, impeller size, and NPSH requirement for your selection. Note that the pump may not be a precise match to the specified flow condition. *(2)*

16.44 Suppose that you selected a 1750 r/min pump with a 7.0 in. impeller in Problem 16.43. What will the actual flow be? Plot the system and pump curves to solve this problem. Since the flow will be higher than that specified in Problem 16.43, a flow control valve in series with the pump is partly closed. What will the efficiency and pump power be if the control valve is closed sufficiently to produce the desired 100 gal/min (6.3 L/s)? Plot the system curve for the partially closed valve situation also. Is there a more energy-efficient way to achieve design point operation than by use of this valve? *(3)*

16.45 One more energy-efficient approach to controlling the flow in Problem 16.44 is to slow the pump down with a variable-speed drive until the flow is at the desired level of 100 gal/min. Plot the new pump curve and the original system curve with the control valve wide open. Use the pump affinity laws to find the new pump curve and the power input. What is the pump efficiency? *(3)*

16.46 Describe analytically the effect of pipe size on pump power consumption. If the cost of the pipe depends on the amount of material in the pipe (recall that larger pipes have thicker walls), how would you go about selecting pipe sizes based on the joint minimization of pipe material cost and pump initial and operating costs? The costs of pump and motor assemblies vary according to the 0.6 power of their size. *(4)*

16.47 The wind speed is measured by a Pitot tube facing upwind. If the reading on a manometer connected to the tube is 1.23 in. (3.1 cm) gauge oil (specific gravity of 0.826), what is the wind speed in feet per second (meters per second) at sea level and at 2500 ft (750 m) above sea level? *(2)*

16.48 The velocity of water at the centerline of a pipe is measured by a Pitot tube facing upstream. If the reading on a manometer connected to the tube is 6 in. (15 cm) of water, what is the water velocity in feet per second (meters per second)? *(2)*

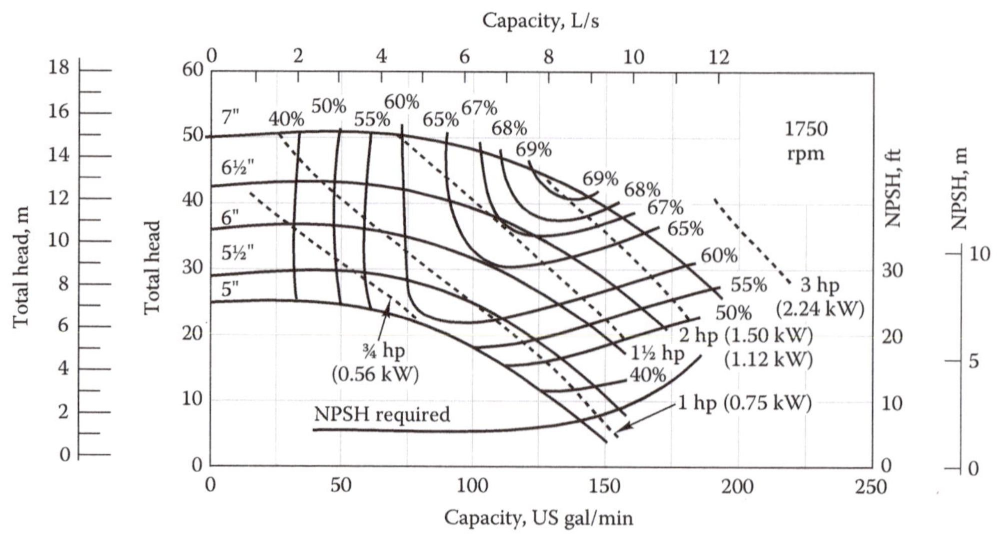

FIGURE P16.43

16.49 Size a venturi meter through which water flows to produce a pressure differential of 6.0 psia (41 kPa). The water flows in an 8 in. (20 cm) pipe at 6 ft/s (2 m/s). What is the throat diameter of a universal venturi tube for this application? *(2)*

16.50 Water at 40°C flows through a 120 mm diameter pipe. It is proposed to measure the flow with a 30 mm, sharp-edged orifice with *D* and *D*/2 taps. If the measured pressure difference is 30 kPa, what is the flow rate? *(2)*

16.51 Select the appropriate orifice to measure 50°F (10°C) water flowing at 700 lb/s (320 kg/s) in an 18 in. diameter (45 cm diameter) pipe. The value of Δp must not exceed 80 inWG (20 kPa). An iterative solution will be required. *(3)*

16.52 Compare the pressure loss coefficient *C* for flow into the two input connections of a converging, equal-main-area, round tee for which the branch flows are equal. What is the effect of the branch flow area ratio (between 0.3 and 1.0) on this result? *(2)*

16.53 Which of three designs of equal-area 90° elbows (consider round or square) has the smallest pressure drop; which is the largest? Consider three types: smooth-radius, round; three-, four-, and five-piece round; and mitered rectangular elbows. Discuss only the case for an elbow radius/duct diameter ratio of 1.0. *(2)*

16.54 Identical round duct transitions with an area ratio of 2:1 are used in the supply (transition is to a smaller duct size as air is diverted to branches) and return ducting (transition is to a larger duct size) in a building air supply and return loop. In which application is the pressure drop through the transition greater if it is fabricated with an angle of 15°? *(2)*

16.55 Identical tees are used in identical converging and diverging flows. In which application is the *main* branch pressure drop larger? *(2)*

16.56 One branch of a duct system has one less three-piece, $r/D = 1$, round 90° elbow than the other. To balance the flow, a butterfly damper (whose diameter is one-half the duct's diameter) is used in the former duct. At what angle should the butterfly damper be positioned to balance the flow? *(2)*

16.57 Rework Example 16.53 at 4000 ft (1220 m) altitude with the same $\Delta p/L$ value. Note that the only entries that need to be changed in Table 16.10 are those that are affected by the velocity pressure. *(2)*

16.58 Rework Example 16.53 for the greatest pressure drop branch, but only include 15°, round duct transitions at each point of duct diameter change. How significant is the addition of the transitions to the fan rating? *(2)*

16.59 Figure P16.59 is the schematic diagram for an air distribution system at sea level that is to be analyzed by the equal-friction method [$\Delta p/L$ = 0.25 inWG/100 ft (2.0 Pa/m)]. The ductwork is round; and smooth-radius, round, $r/D = 1$ elbows are used. What pressure rise must the fan develop to deliver air in this system to branch *ABCD* if the duct outlets are of negligible pressure drop? The diverging tees at *B* and *C* are round, conical branch designs; all size transitions are 15° converging designs. *(4)*

16.60 The pressure drop in duct *CE* in Figure P16.59 is to be made the same as in branch duct *CD* (refer to Problem 16.59 for the design static drop to be used in *CD*; point *C* is upstream of the tee) to balance the system. The key difference between the two branches, other than the flow rate, is that the *CD* flow is straight through the tee and the *CE* flow is into the branch line. The duct lengths are about the same. What should the diameter of branch *CE* be to produce the same pressure drop as in branch *CD*? *(3)*

16.61 Figure P16.61 is the schematic diagram for an air distribution system at sea level that is to be analyzed by the equal-friction method [$\Delta p/L$ = 0.15 inWG/100 ft (1.2 Pa/m)]. The ductwork is round; and smooth-radius, round, $r/D = 1$ elbows are used. What pressure rise must the fan develop to deliver air in this system to air outlet 5 if the duct outlets have a pressure drop equivalent to 25 ft (8 m) of straight duct? The diverging tees at *B*, *C*, *D*, and *E* are round, conical branch designs; all size transitions are 15° converging designs, and the balancing damper downstream of tee *E* is a wide-open butterfly type. *(4)*

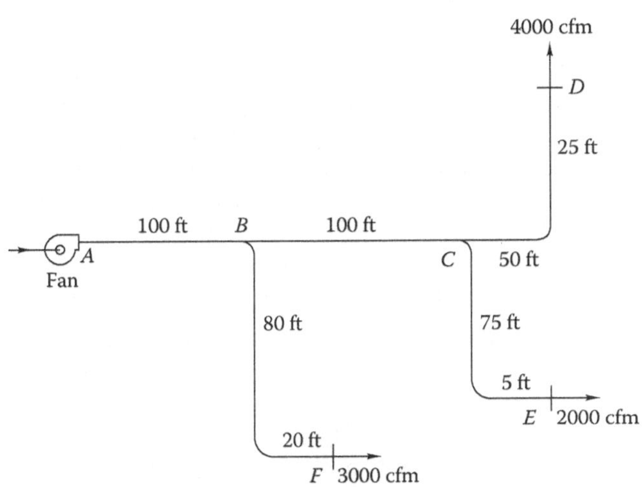

FIGURE P16.59

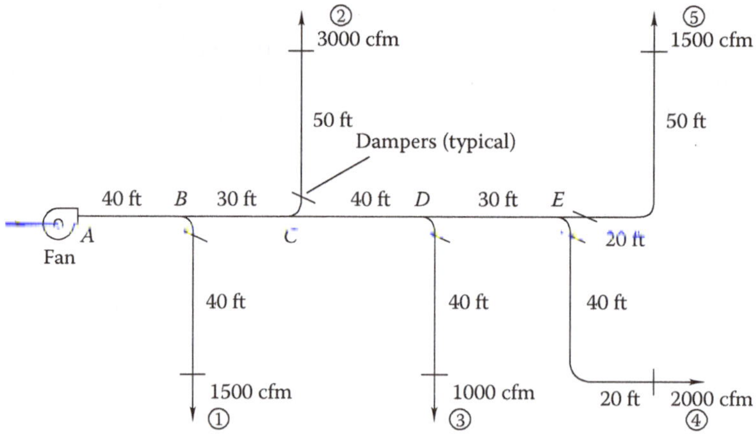

FIGURE P16.61

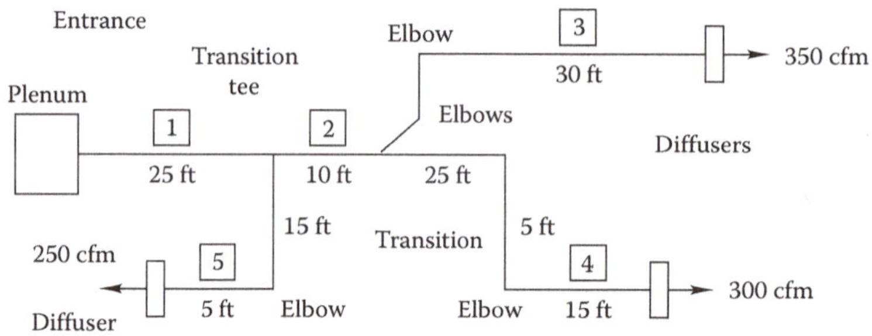

FIGURE P16.62

16.62 Select the five round duct sizes for the residential heating system shown in Figure P16.62. The main supply duct velocity should not exceed 1000 ft/min (5 m/s), and the branch duct air velocity should be no more than 600 ft/min (3 ft/s). *(2)*

16.63 Rework Problem 16.62, assuming that rectangular ducts no deeper than 8 in. (20 cm) can be accommodated between the floor joists of this residence. *(2)*

16.64 Write a spreadsheet template or computer program for determining the pressure drop in the round ductwork system of Problem 16.61 using the equal-friction method. It should be general enough to accommodate various values of design $\Delta p/L$, airflow rates, air density, number of tees (main pressure drop), duct lengths, number of elbows, and outlet diffuser pressure drops. *(4)*

16.65 A 15 kW fan will be equipped with either a VSD or inlet vanes to control flow. If the application requires that the fan operate the following number of hours at the indicated part-load levels, what is the kWh savings due to use of the VSD? If the VSD controller costs \$700 more than adjustable inlet vanes, what is the payback period for the VSD approach, assuming that electric power costs 8 ¢/kWh? *(3)*

PLR	0.20	0.40	0.60	0.80	1.00
Hours	300	700	900	250	50

16.66 Work Problem 16.65, except compare the VSD approach with outlet dampers for flow control. *(2)*

16.67 The flow through a pump is 500 gal/min (30 L/s) against a head of 20 ftWG (60 kPa). What is the electric power required if the motor efficiency is 91% and the pump efficiency is 72%? *(1)*

16.68 Figure P16.68 is a pump curve for one manufacturer's series of 6 in. (15.2 cm) to 9.75 in. (24.8 cm) pumps. The 100% system curve intersects the 7.5 in. (19.3 cm) rotor diameter curve at 100 gal/min (6.3 L/s). At this full-load operating point, what are the values of the pump power, efficiency, and NPSH? What are the same parameters for the system curve in which flow is reduced by 50% by using a flow control valve? *(2)*

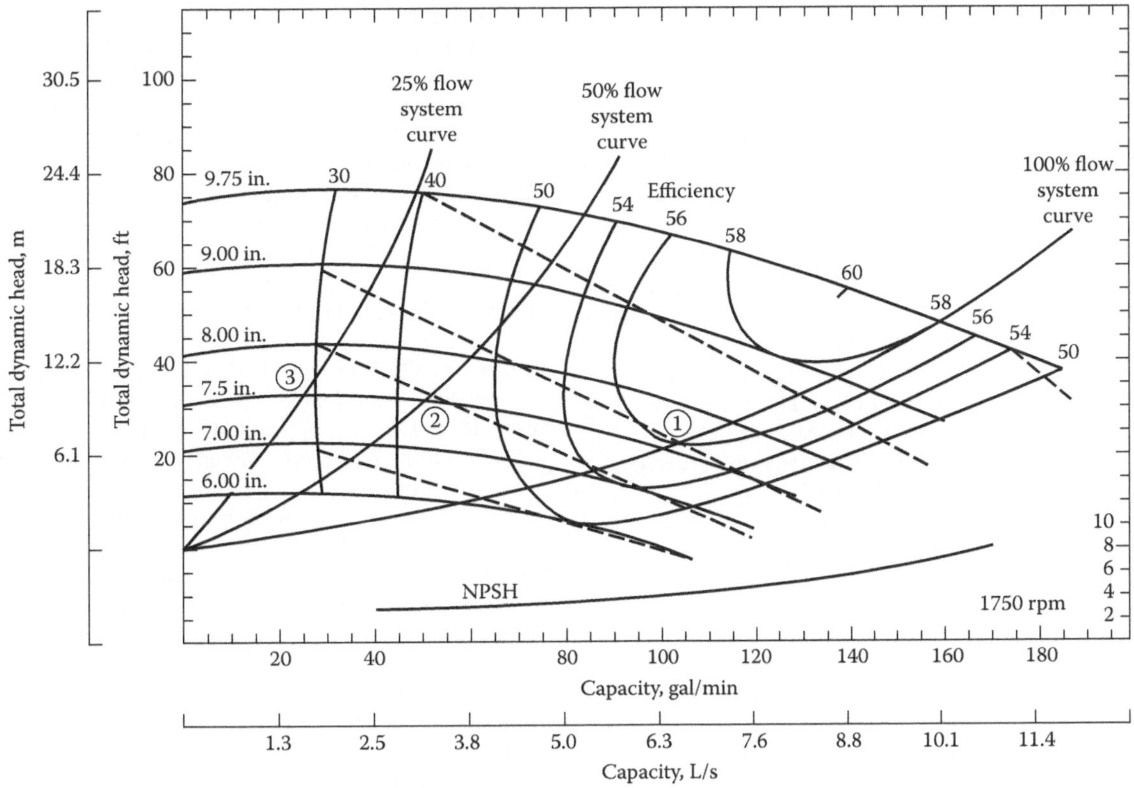

FIGURE P16.68

16.69 Flow is reduced to 50% of the full flow of 90 gal/min (5.7 L/s) using the 7.5 in. (19.3 cm) pump in Figure P16.68 by partially closing a flow control valve. Rather than use this energy-inefficient approach, an engineer proposes to reduce the pump rotational speed from the nominal value shown in the figure. What will be the new pump speed as well as the new values of pump power and efficiency? How will the NPSH value be affected? *(2)*

16.70 What is the available net positive suction head (NPSHA) for the system shown in Figure P16.70? Particular care is needed in the design of open, cooling tower pump systems. *(2)*

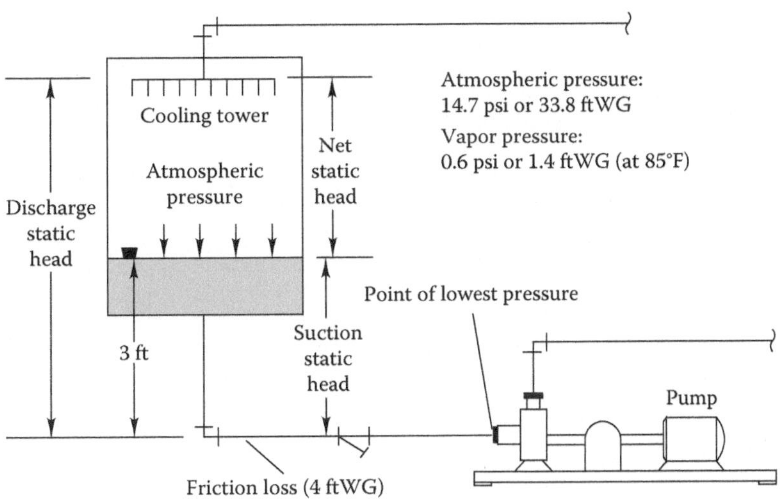

FIGURE P16.70

References

AMCA (1973). *Fans and Systems*. Publication 201-73, Air Movement and Control Association, Inc., Arlington Heights, IL.

ASHRAE 90.1 (2007). *ANSI/ASHRAE/IESNA Standard 90.1-2010: Energy Standard for Buildings Except Low-Rise Residential Buildings*. American Society of Heating, Refrigerating and Air-Conditioning Engineers, Atlanta, GA.

ASHRAE Fundamentals (2001/2013). *Handbook of Fundamentals*. American Society of Heating, Refrigerating and Air-Conditioning Engineers, Atlanta, GA.

ASHRAE Systems (2012). *HVAC Systems and Equipment Handbook*. American Society of Heating, Refrigerating and Air-Conditioning Engineers, Atlanta, GA.

Benedict, R.P. (1984). *Fundamentals of Temperature, Pressure and Flow Measurements*. Wiley, New York.

Born, D.W. (1989). Inhibited glycols for corrosion and freeze protection in water-based heating and cooling systems. *ASHRAE Trans.*, 95(2), 969–975.

Brandemuehl, M.J., S. Gabel, and I. Andresen (1993). HVAC 2 toolkit: Algorithms and subroutines for secondary HVAC system energy calculations. American Society of Heating, Refrigerating and Air-Conditioning Engineers, Atlanta, GA.

Coad, W.J. (1989). The air system in perspective. *Heat. Piping Air Cond.*, 61, 124, 128.

DOE2.1 Software (1981). 2.1D building description language summary (DE-890-17726) and 2.1A engineers' manual (DE-830-04575). The National Technical Information Service, Springfield, VA.

Elite (2009). Ductsize- HVAC duct sizing and analysis software following the static regain and equal friction methods. www.elitesoft.com. Accessed on March 2014.

Greenberg, S., J.P. Harris, H. Akbari, and A. DeAlmeida (March 1988). Technology assessment: Adjustable-speed motors and motor drives. Report LBL-25080. Lawrence Berkeley Laboratory, Berkeley, CA.

Grimm, N.R. and R.C. Rosaler (1990). *Handbook of HVAC Design*. McGraw-Hill, New York.

Idelchik, I.E., G.R. Malyavskaya, and E. Fried (1986). *Handbook of Hydraulic Resistance*. Hemisphere, New York.

Kreider, J.F. (1985). *Principles of Fluid Mechanics*. Allyn and Bacon, Boston, MA.

LBNL (1982). DOE-2 engineers manual. Lawrence Berkeley National Laboratory Report, Berkeley, CA.

Miller, R.W. (1983). *Flow Measurement Engineering Handbook*. McGraw-Hill, New York.

Moody, L.F. (1944). Friction factors for pipe flow. *ASME Trans.*, 66, 671.

Nussbaum, O.J. (January 1990). Using glycol in a closed circuit system. *Heat. Piping Air Cond.*, 62, 75–85.

Perry, R.H., D.W. Green, and J.O. Malrnay, eds. (1999). *Perry's Chemical Engineers' Handbook*, 6th ed. McGraw-Hill, New York.

Rao, K.V.K. (1982). Nomograph for pipe fitting friction factors. *Chem. Eng.*, 89.

Rishel, J.B., T.H. Durkin, and B.L. Kincaid (2006). *HVAC Pump Handbook*, 2nd ed. McGraw-Hill, New York.

Stein, J. and M.M. Hydeman (2004). Development and testing of the characteristic curve fan model. *ASHRAE Trans.*, 110(2), 347–356.

Streeter, V.L., E.B. Wylie, and D. Bedford (1997). *Fluid Mechanics*. McGraw-Hill, New York.

Tsal, R.J., H.F. Behls, and R. Mangle (1988). T-method duct design, part I: Optimization theory, part II: Calculation procedure and economic analysis. *ASHRAE Trans.*, 94(2), 90–111.

Tuve, G. (1961). *Mechanical Engineering Experimentation*. McGraw-Hill, New York.

Waller, B. (October 1990). Piping from the beginning. *Heat. Piping Air Cond.*, 62, 51–71.

White, F. (1998). *Fluid Mechanics*, 4th ed. McGraw-Hill, New York.

Wright, D.K. (1945). New friction factor for round ducts. *ASHVE J.*, 51, 303–316.

17

Cooling System Equipment

ABSTRACT Chapter 12 covered the methods commonly used to analyze heat exchanger performance, while Chapter 14 provided a description and analysis equations relevant to practical vapor compression refrigeration and heat pump cycles and systems. In this chapter, we cover the major components and equipment of cooling systems. The various important generic types of compressors, namely, reciprocating, rotary, scroll, screw, and centrifugal, are first described and an intercomparison made. Next, we describe the different types and operating principles of expansion devices, evaporators, and condensers and their practical applications are discussed. We then address the performance of heating coils followed by a theoretical treatment of cooling coils, under both dry and wet operations. Finally, a description of cooling tower operation follows, with an in-depth theoretical analysis along with an illustrative example. Elements over which the HVAC designer has little control (e.g., the internal design of a compressor or expansion device) are not discussed in detail in this chapter.

Nomenclature

A	Area, m² (ft²)
COP	Coefficient of performance of a cooling device
c_p	Specific heat at constant pressure, kJ/(kg·K) [Btu/(lb$_m$·°F)]
F	Heat exchanger correction factor defined by Equation 12.33
g_c	Conversion factor in IP units, 32.17 lb$_m$·ft/(lb$_f$·s²)
h	Specific enthalpy, kJ/kg (Btu/lb$_m$)
h	Convective heat transfer coefficient, W/K (Btu/h·°F)
$h_{a,sat,i}$	Enthalpy of saturated air at the tower water inlet temperature kJ/kg(Btu/lb$_m$)
LMTD	Logarithmic mean temperature difference, defined by Equation 12.30, K (°F)
m	Power law exponent (Equation 17.18)
$\dot{m}$	Mass flow rate, kg/s (lb$_m$/h)
$\dot{m}_a$	Mass flow rate of air on dry-air basis, kg/s (lb$_m$/h)
NTU	Number of transfer units, defined by Equation 12.37
n	Polytropic process exponent
p	Pressure, Pa (lb$_f$/in.²)
$\dot{Q}$	Heat flow with subscripts c for cooling load, h for heating load, W (Btu/h)
R	Thermal resistance, K/W[(°F·h)/Btu) [$\equiv\Delta T/\dot{Q}$]
R	Capacity ratio, defined by Equation 17.16
r	Radius, m (ft)
r_o	Rotor outer radius, m (ft)
s	Specific entropy, kJ/(kg·K) [Btu/(lb$_m$·°R)]
T	Temperature, K or °C (°R or °F)
T	Shaft torque, N·m (lb$_f$·ft)
T_{db}	Dry-bulb temperature of air, °C (°F)
T_{wb}	Wet-bulb temperature of air, °C (°F)
T_w	Water temperature, °C (°F)
U	Overall heat transfer coefficient, W/(m²·K) [Btu/(h·ft²·°F)]
$(\tilde{U}A)_{wet}$	Enthalpy-based overall heat transfer coefficient, kg/s (lb$_m$/h)
v_{to}	Fluid exit tangential velocity, m/s (ft/s)
V	Volume, m³ (ft³)
$\dot{V}$	Volume flow rate, m³/s or L/s (ft³/min)
v	Specific volume, m³/kg (ft³/lb$_m$)
W	Work, kJ (ft·lb$_f$)
W	Absolute humidity ratio of air, kg$_w$/kg$_a$ (lb$_w$/lb$_a$)

Greek

ε	Heat exchanger effectiveness
η	Efficiency
ω	Angular velocity, s⁻¹
Δx	Change or difference in quantity x

Subscripts

a	Air
c	Cold
cl	Clearance (volume)
$comp$	Compressor
db	Dry bulb
des	Design
f	Fin (on tubes)
i	Entering, in, inlet, inner, internal
$isen$	Isentropic
$liq\text{-}vap$	Liquid to vapor

m	Metal, mean
o	Leaving, out, outlet, outer, external
ref	Refrigerant
s	Supply or makeup
sat	Saturated
swept	Swept by piston in a cylinder
test	Test condition
tower	Cooling tower
vol	Volume
vap	Vapor
w	Water, water vapor
wb	Wet bulb

17.1 Introduction

In essence, a *chiller* is an assembly of equipment used to produce chilled water or cold air for cooling of spaces within buildings. Chapter 14 described various vapor compression cycles and systems. The primary components of a vapor compression cooling system are the compressor, the evaporator, the condenser, and the expansion device. Additional heat exchanger loops such as cooling tower loops to enhance the effectiveness of heat rejection are important subsystems of larger cooling plants. These subsystems along with their components are studied in this chapter. Heat transfer from the chilled water (or hot water/steam) and the air supply to the rooms is done in cooling (or heating) coils. Though they are not, strictly speaking, part of the cooling plant, these coils are also important components of the cooling distribution system and are also discussed in this chapter.

17.2 Compressors

17.2.1 Categories of Chillers

The compressor is the heart of the vapor compression system. It sucks refrigerant vapor from the evaporator, compresses it to a high pressure, and delivers the superheated vapor to the condenser. There are two broad categories of compressors in vapor compression systems:

1. *Positive displacement* compressors where the increase in the refrigerant vapor pressure is achieved by reducing the volume in the compression chamber through the application of external work. The four common types of designs are as follows: reciprocating, rotary, screw, and scroll.
2. *Dynamic* compressors where the kinetic energy of the refrigerant vapor is first increased by a rotating element, which is then converted into a pressure increase. The most widely used type is the centrifugal compressor.

These compressor types differ in their physical construction and exhibit different characteristics in terms of efficiency under different design conditions, part-load efficiency behavior, total cooling capacity, sound, and cost. It is up to the designer to evaluate these characteristics in terms of the needs specific to the application and select the most appropriate type.

17.2.2 Generic Compressor Power

Prior to discussing the most common types of compressors, it is instructive to study the generic compressor from a thermodynamic point of view. Recall that processes involving refrigerants cannot be accurately represented by the ideal gas law, and so property tables have to be used.

The compression process in not isentropic but polytropic, i.e., can be expressed by pv^n = constant, where n is the polytropic exponent. The polytropic work input for the open-cycle compression cycle, shown in Figure 17.1, is given by

$$W_{polytropic} = \int_i^0 v\,dp = \frac{np_iv_i}{n-1}\left[\left(\frac{p_o}{p_i}\right)^{(n-1)/n} - 1\right] \qquad (17.1)$$

where

$W_{polytropic}$ is the polytropic work input, kJ/kg (ft lb_f/lb_m)
n is the polytropic exponent
p_i, p_o are the compressor inlet and outlet pressures, Pa (lb_f/ft^2)
v_i is the fluid inlet specific volume, m^3/kg (ft^3/lb_m)

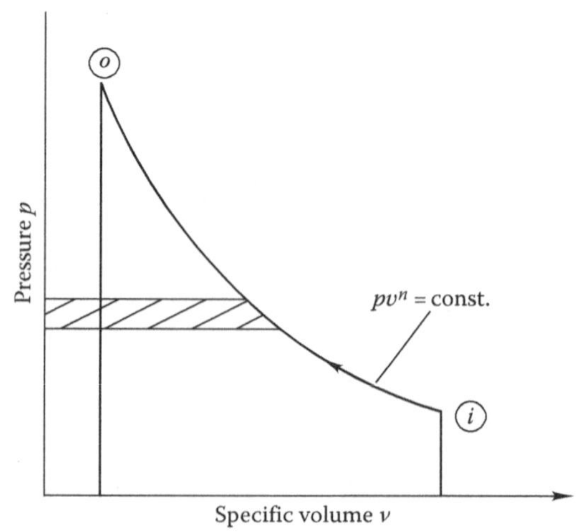

FIGURE 17.1
Polytropic compression process p–v diagram used for chiller compressor work calculation.

This form of the polytropic work expression is the most appropriate compared to several similar expressions available since the inlet and outlet pressures are customarily specified in compressor calculations. Further, we consider a *theoretical* polytropic exponent for refrigerants useful to screen refrigerants. Its value would be between 1 (isothermal process for which the ideal gas law applies) and the isentropic process exponent (ratio of the specific heats - see Section 2.1); [R-12 (1.0911), R-22 (1.1056), R-123 (1.0716), and R-134A (1.0706)]. The real or *actual* compressor exponent n would depend on compressor type and design variations (typical range, 1.1–1.3).

The *isenropic efficiency* of a compressor defined as the ideal (i.e., isentropic) compression work divided by that under polytropic compression is convenient during analysis. The computational procedure is illustrated in the following example.

Example 17.1: Vapor Compressor Analysis

A compressor using R-22 operates between an inlet condition of 90.8 psia (626 kPa) and a discharge pressure of 290 psia (2000 kPa). Refrigerant enters the compressor as saturated vapor. The polytropic exponent for the actual process is 1.2. Find the isentropic compressor efficiency.

Given: $p_i = 90.8$ psia $= 13{,}075.2$ lb_f/ft^2, $p_o = 290$ psia, $n = 1.2$

Figure: See Figure 17.1.

Assumptions: The polytropic model is sufficiently accurate, refrigerant at inlet is saturated vapor.

Find: $\eta_{comp,isen}$

Lookup values: From appendix Table A5-IP at saturated pressure of 90.8 psia, the specific volume $v_i = 0.6029$ ft^3/lb_m and enthalpy $h_i = 108.6$ Btu/lb_m.

Solution

The actual work input is found from Equation 17.1:

$$W_{polytropic} = 1.2 \times 13{,}075.2\ \text{lb}_f/\text{ft} \times \left(\frac{0.6029\ \text{ft}^3/\text{lb}_m}{1.2-1}\right)\left[\left(\frac{290}{90.8}\right)^{(1.2-1)/1.2} - 1\right]$$

$$= 10{,}100\ (\text{ft}\cdot\text{lb}_f)/\text{lb}_m$$

The ideal work is based on an isentropic compression process. The entropies at the start and end of this process are equal, i.e.,

$$s_i = 0.2191\ \text{Btu}/(\text{lb}_m\cdot{}^\circ\text{R})$$

From the superheat table (Table A7), the ideal outlet enthalpy (corresponding to an exit temperature of about 158°F) is determined as:

$$h_o = 120.9\ \text{Btu/lb}_m$$

The ideal (isentropic) compressor work is

$$W_{ideal} = h_O - h_i = (120.9 - 108.6)\ \text{Btu/lb}_m \times 778\ (\text{ft}\cdot\text{lb}_f)/\text{Btu}$$

$$= 9569.4\ (\text{ft}\cdot\text{lb}_f)/\text{lb}_m$$

Finally, the compressor efficiency is found as the ratio of ideal to actual work:

$$\eta_{comp,isen} = \frac{9569.4}{10{,}100} = 0.947$$

Comments

The *isentropic* efficiency is meant to characterize the efficiency of the compressor *machinery*. It includes other losses such as flow losses in centrifugal compressor blading or in reciprocating compressor valves, driveshaft losses, and piping pressure losses. It is best to use chiller manufacturer performance data in order to infer its numerical value.

17.2.3 Reciprocating Compressors

The most common refrigerant compressor is the reciprocating compressor. It is used in cooling systems ranging from small residential units on the order of a few hundreds of Btu per hour (tenths of a kilowatt) output to large units used in commercial buildings rated at hundreds of tons (hundreds of kilowatts) of cooling. These compressors have several pistons that move back and forth in the cylinders (akin to those in an automobile engine). Figure 17.2a shows the schematic arrangement used in each cylinder. The inlet refrigerant stream enters through a mechanical valve during the suction stroke. At the end of the stroke, the valve closes and the compression begins. Near top dead center, the outlet valve opens, allowing the compressed refrigerant to enter the outlet manifold. Small pressure drops occur across each valve, as indicated by the dashed horizontal lines in the figure. Note that the abscissa of Figure 17.2b is *extensive* volume. The volume changes shown are due to the piston's movement. The specific volume varies with mass flow into and from the cylinder volume.

Figure 17.2b is idealized to the extent that the intake (1–2) and exhaust (3–4) processes to the compressor are shown to be constant pressure. Process 4–1 is an expansion of the compressed refrigerant remaining in the *clearance volume* $V_{cl} = V_4$ after the exhaust valve closes. This refrigerant must be expanded to a sufficiently low pressure only after which refrigerant will flow through the inlet valve. The expanded refrigerant is then mixed with inlet refrigerant at a different thermodynamic condition. Process 2–3 is the polytropic compression

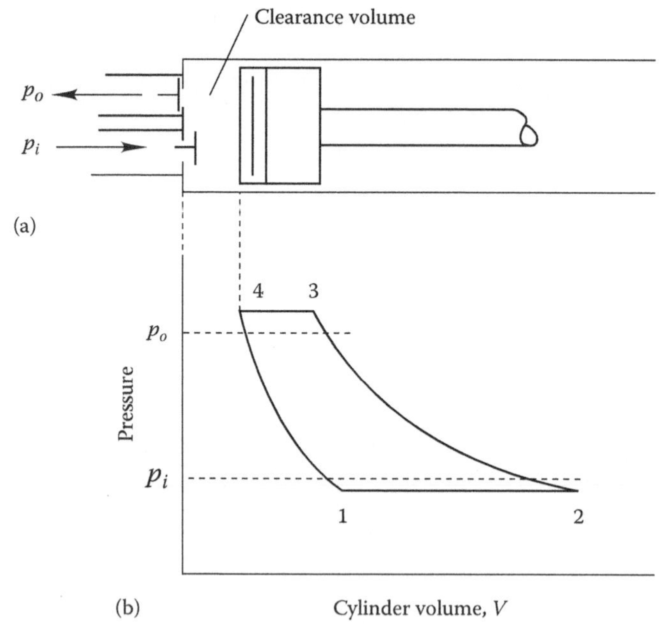

FIGURE 17.2
Reciprocating chiller: (a) piston-and-valve arrangement and (b) the corresponding p–V diagram.

described in the earlier section. Hot, compressed gases may undergo heat transfer during the exhaust process 3–4.

There is yet another source of ineffectiveness of the compressor and that has to do with the small amount of high-pressure refrigerant vapor left in the clearance volume of the compression chamber (shown as point 4 in Figure 17.2). A certain volume of the cylinder (shown as point 1 in Figure 17.2) is now needed for this residual vapor to reexpand, and only the rest (volume between points 2 and 1) is available to suck in fresh refrigerant vapor. A common measure to quantify this effect is the *volumetric efficiency*, defined as the ratio of actual mass of vapor compressed to that which would be compressed if the entire cylinder displacement volume were filled with refrigerant at the inlet condition:

$$\eta_{vol} = \dot{m}_{ref} \frac{v_i}{\dot{V}_{swept}} \tag{17.2}$$

where
η_{vol} is the volumetric efficiency
$\dot{m}_{ref}$ is the refrigerant mass flow, kg/s (lb_m/h)
v_i is the inlet refrigerant specific volume, m^3/kg (ft^3/lb_m)
$\dot{V}_{swept}$ is the piston displacement or cylinder swept volume per unit time, m^3/s (ft^3/h)

This theoretical volumetric efficiency depends on size of the clearance volume, on cylinder volume displacement, and on inlet and exhaust conditions. The refrigerant used also has an effect on the efficiency value. For a well-designed compressor matched with its refrigerant, theoretical volumetric efficiencies will exceed 90%. Actual volumetric efficiencies including piston ring and valve leakage and thermal effects can be 10%–15% lower, depending on the pressure ratio and mechanical details. The *volumetric efficiency* is a measure of the effectiveness of a compressor as a refrigerant "mover," while the *isentropic compressor efficiency* is a measure of the thermodynamic efficacy of the compression process itself. The part-load performance of reciprocating chillers at various speeds is better than that for some other designs, such as centrifugal compressors, since the effect of rotational speed on the various sources of inefficiency is relatively smaller.

The various efficiencies just described can be combined to find the COP of an actual vapor compression cycle by using the following expression:

$$\text{COP} = \frac{\dot{Q}_c}{[(\Delta h)_{s=\text{const}} / \eta_{comp}](\eta_{vol}\dot{V}_{swept}/v_i)} \tag{17.3}$$

where Q_c is the cooling capacity or load and other terms have been described previously.

Actual compressor capacity data are generally presented as compressor maps discussed previously in Section 14.4.2 (Figure 14.12). The lower family of curves represents the power input for R-22 machines at various evaporator and condensing temperatures. The upper curves represent the cooling capacity produced at the same set of conditions applying to the lower family of curves. The ratio of the two ordinates is the COP (taking proper account of units for the IP calculation). Power input increases with increasing evaporator and condensing temperatures.

An idea of the physical dimensions of the compressor would be useful. The catalog of an R-22 reciprocating compressor (four cylinders, hermetic) operating at 29 r/s, a condensing temperature of 40°C, and an evaporating temperature of −4°C shows a refrigeration capacity of 115 kW. At this operating point, the motor (whose efficiency is 90%) draws 34 kW. The bore of the cylinders is 87 mm and the piston stroke is 70 mm. The performance is based on 8°C subcooling, the actual volumetric efficiency is 77.4%, and the compressor efficiency is 71%.

Compressors can be classified as single-stage or multistage and by type of motor drive (electrical or mechanical), by type of control, and by drive enclosure (ASHRAE Systems, 2012). In terms of the last classification, two configurations of reciprocating compressors are the most common:

1. *Hermetic* compressors consist of a motor and compressor contained in a single gastight housing unit using a common driveshaft. The motor is exposed to the refrigerant; therefore, heat produced by motor (and compressor) inefficiencies is added to the circulating refrigerant, increasing heat rejection requirements. However, the motors themselves in hermetic compressors may be more efficient due to better cooling. Such compressor designs are relatively small in capacity and commonly used for residential and small commercial applications such as air conditioners, refrigerators, and freezers. Hermetic compressors typically operate 10–20 years without needing maintenance. Chillers of this type using CFCs may be difficult to retrofit with new refrigerants that have less adverse ozone-layer depletion and global warming impacts.
2. *Open* compressors use a separate external motor to drive the compressor. The drive can be by direct coupling, belts, or gears. In addition to the electric motors, the drive could be steam turbines, or gasoline/diesel engines. These are easier to repair in the field than are hermetic compressors. However, seals are needed on the connecting shaft to prevent refrigerant leaks from the compressor casing. Open compressors are able to incorporate more cylinders and higher outputs than hermetic units.

Control of the capacity of reciprocating compressors can be achieved by on/off control or by varying the motor drive speed (single speed or variable speed) or by unloading the compressor that involves holding the intake valves open on part of the cylinders of a multicylinder device so as to create a situation where there is no refrigerant flow. Cylinders are unloaded, depending on the compressor design, in such a way that rotational imbalance is not induced.

17.2.4 Rotary Compressor

Rotary compressors due to their low noise and vibration levels are used in applications from home refrigerators, to automotive air conditioning, to low-temperature refrigeration such as freezing plants. Sizes range from 1/6th to 3 tons of cooling. Two different designs are common: rotating vane and rolling piston. Figure 17.3 shows cross-sectional drawings of a rotating-vane compressor with two-vane (the more common design) and four-vane designs. There are no suction or discharge valves, while the rotor center is off-set compared to that of the housing. A series of blades slide inside slots as the rotor turns. Near the inlet port the gap between the rotor and housing is large. As the rotor turns, refrigerant trapped between each pair of blades, the rotor, and the housing is

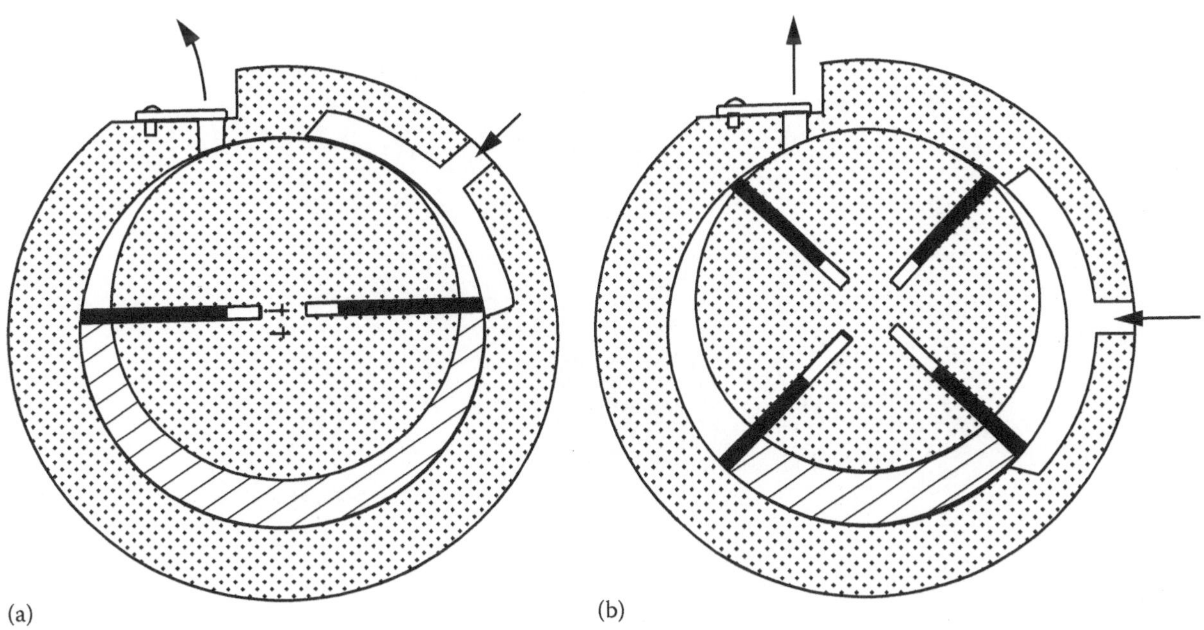

FIGURE 17.3
Multiple-vane rotary compressors: (a) two-vane compressor and (b) four-vane compressor.

compressed because the rotor-to-housing gap decreases as the discharge port is approached. The blades basically move in translation relative to the rotor, while the rotor itself is driven by the compressor drive system.

Small rotary machines have the advantage that they can be started with inexpensive motors that ordinarily do not have high starting torque abilities. These machines are inherently balanced so that rotational balance is not a concern. Rotary compressors are also relatively light and wear well. They have high volumetric efficiencies because of their small clearance volume and low reexpansion loss. Compression ratios are limited to about 7:1 due to the excessive leakage and high blade and driveshaft stresses that occur at higher pressures. Oil lubricates the blades and bearings as well as cools the compressor via an external heat exchanger on larger units. Part-load performance is good down to 20% part-load ratio, but their overall efficiencies are rather low.

17.2.5 Scroll Compressors

Scroll compressors are positive displacement devices that are quite widely used for medium-sized air conditioners including residential coolers and heat pumps as well as some automotive applications. First patented in the early 1900s, commercial models appeared only in the 1980s because of the tight machining tolerances needed. Sizes range from 1.5 to 10 tons (5 to 35 kW). Their improved efficiency over earlier designs is the prime reason why their market share is likely to increase in the future within the small to medium systems category. Figure 17.4 shows a cross-sectional view of a scroll compressor at several positions.

The scroll itself is an involute spiral confined on its sides by flat planes. There are pairs of scrolls, one fixed and the other moving. The moving scroll is driven in eccentric translation by the crankshaft. The moving scroll does not rotate but orbits in a circle around an axis that is off-center with respect to the fixed scroll. The seal between the moving and fixed scrolls is achieved by precision machining and seals at the involute tips.

Compression is accomplished as follows. Gas enters the scroll on the outer circumference. The meshed involutes form crescent-shaped pockets that reduce in volume as the moving scroll translates. The trapped gas is compressed as the crescents reduce in size from the outer periphery inward, as shown in subsequent positions in the figure. At the limit of inward movement, the discharge port is uncovered, and the high-pressure gas is discharged. Thus, the refrigerant vapor is continuously compressed and pushed from the outside toward the exhaust port at the center. The scroll compressor has few moving parts, no suction, or discharge valves and has low noise and vibration. Because it has no clearance volume, its efficiency is relatively high. Off-peak performance is better than that of reciprocating compressors if deviation from the design pressure ratio is not too large. At standard rating conditions, the COP of air conditioners using scroll compressors is about 2.9–3.2 (ASHRAE Fundamentals, 2013).

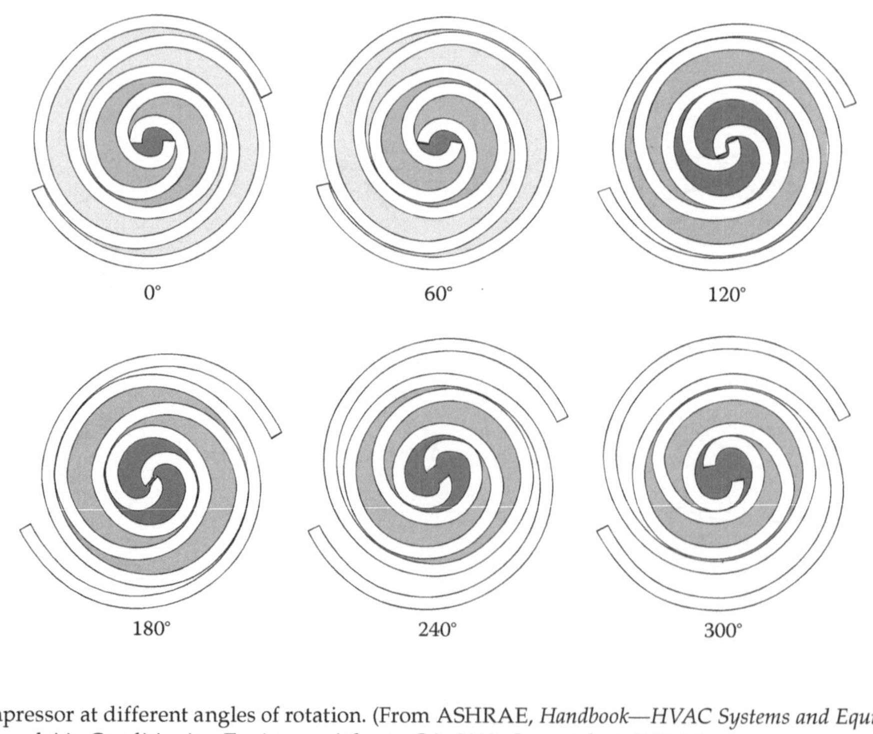

FIGURE 17.4
Operation of a scroll compressor at different angles of rotation. (From ASHRAE, *Handbook—HVAC Systems and Equipment*, American Society of Heating, Refrigerating and Air-Conditioning Engineers, Atlanta, GA, 2012. Copyright ASHRAE, www.ashrae.org.)

17.2.6 Screw Compressors

Screw compressors are intermediate in size between small reciprocating machinery and large centrifugal units. They were developed in the 1930s and deployed from the 1960s onward. Typical sizes are on the order of a few hundreds to a few thousands of cooling tons (kilowatts) of output. In this range, they are more compact than reciprocating equipment of the same capacity.

Figure 17.5 shows the double-screw design (single-screw design is also available) with its two mating counterrotating rotors and housing. The compression effect is accomplished by differing rotation rates of the two rotors. In the compressor shown, the male rotor has four lobes, while the female rotor has six. Therefore, the female rotor rotates at two-thirds the rate of the male rotor. Refrigerant is moved axially, entering at one end of the compressor at the top and leaving the other end at the bottom. A typical speed for the male rotor is 3600 rpm.

Compression occurs as follows. Two lobes facing the inlet manifold separate, forming a volume of increasing size into which refrigerant flows. As the rotor continues to turn, two volumes of gas are trapped between the rotors and the casing. When two lobes remesh, the volume of gas shrinks and the refrigerant is compressed. Finally, the trapped compressed volume moves to the discharge port, where it is squeezed out axially and radially. A seal is maintained between the two rotors by an oil film.

The efficiency of screw compressors is fairly high at some optimal pressure ratio but drop off quickly with changing pressure ratios. However, the isentropic efficiency drop can be greatly reduced by varying the compressor volume ratio (Figure 17.6). Performance penalties are not negligible at small part-load ratios even though this chiller design can operate down to 10% of rated capacity.

17.2.7 Centrifugal Compressors

Centrifugal compressors, the most common type of dynamic compressor, are used when larger capacities than those offered by reciprocating compressors are needed. These compressors can produce between 200 kW and 10 MW of cooling output. The design concept is at least 100 years old and is similar to that of a centrifugal pump. Fluid to be compressed enters at the center of the rotor and flows to the periphery of the rotor under the centrifugal force produced by backward-curved blading in the rotor. The angular momentum imparted to the refrigerant is converted to kinetic energy at the tips of the impeller, which is then converted into a pressure increase inside the diffusers between the rotating vanes and the outer casing. Several stages of compression are common in one machine. Overall, compression efficiencies of 70%–80% are normal. Refer to Figure 11.25 for a cutaway drawing of a large centrifugal chiller.

Figure 17.7 shows the blades of a one-stage centrifugal compressor rotating counterclockwise. The angular rotation and the tangential direction of the vapor when leaving the rotor blades are also shown. The shaft work input is the product of the torque and rotational speed:

$$\dot{W}_i = T \times \omega = \dot{m}_{ref} \times (\mathrm{v}_{to} r_o) \times \omega \tag{17.4}$$

where
- $\dot{W}_i$ is the shaft power input, W (ft·lb$_f$/s)
- T is the shaft torque, N·m (lb$_f$ ft)
- ω is the angular velocity, s^{-1}
- $\dot{m}_{ref}$ is the refrigerant flow rate, kg/s (lb$_m$/s)
- v_{to} is the fluid exit tangential velocity, m/s (ft/s)
- r_o is the rotor outer radius, m (ft)

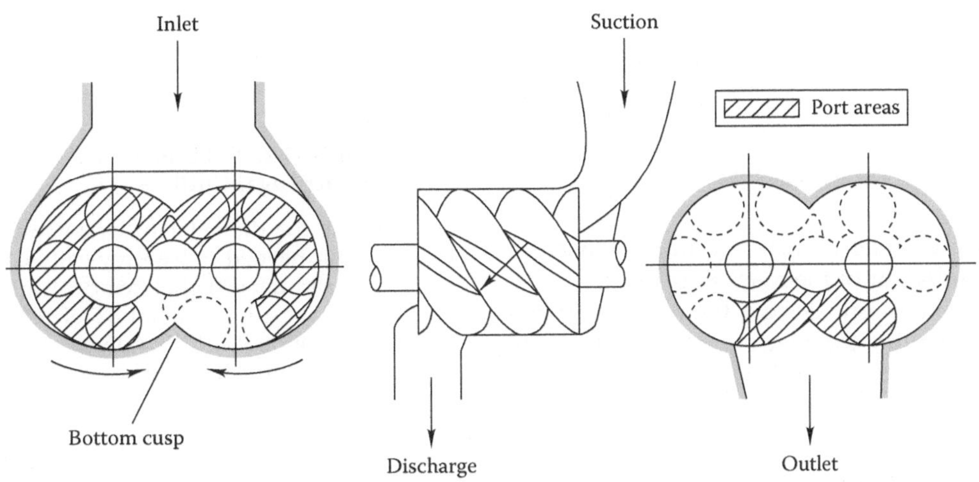

FIGURE 17.5
Dual-screw compressor. (From ASHRAE, *Handbook—HVAC Systems and Equipment*, American Society of Heating, Refrigerating and Air-Conditioning Engineers, Atlanta, GA, 2012. Copyright ASHRAE, www.ashrae.org.)

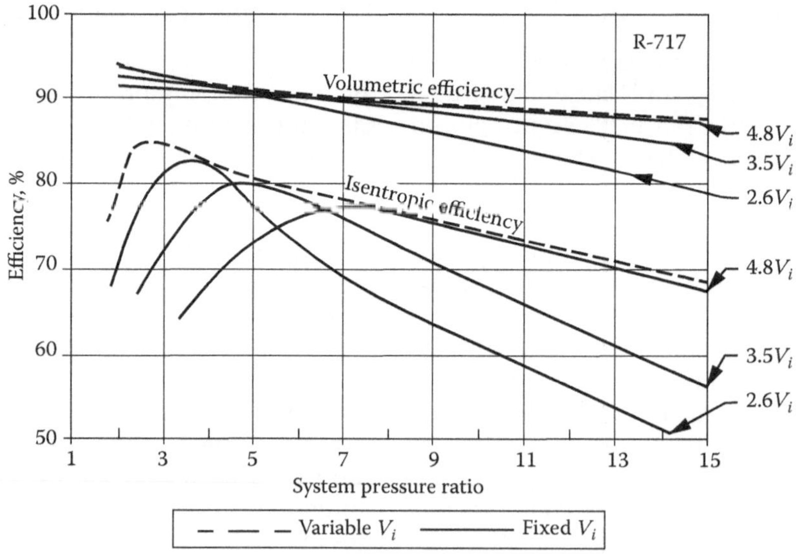

FIGURE 17.6
Typical screw compressor efficiency curves for R-717. V_i is the compressor volume ratio which is varied by a slide valve that changes the size and shape of the discharge port. (From ASHRAE, *Handbook—HVAC Systems and Equipment*, American Society of Heating, Refrigerating and Air-Conditioning Engineers, Atlanta, GA, 2012. Copyright ASHRAE, www.ashrae.org.)

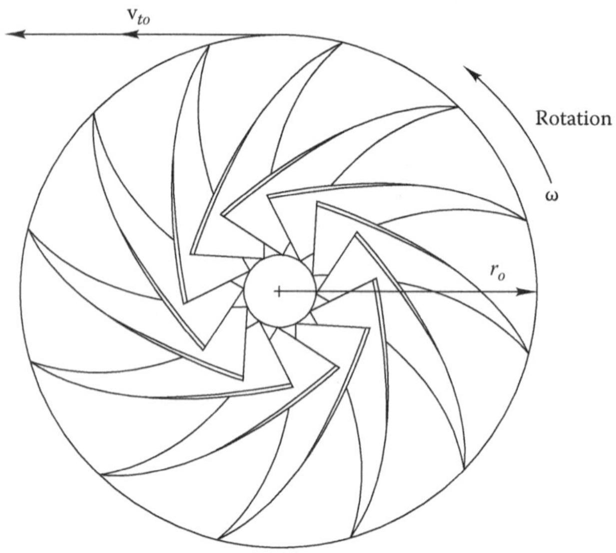

FIGURE 17.7
Centrifugal chiller rotor velocity diagram.

(For IP units, the conversion constant g_c must be used in Equation 17.4 as a divisor of the mass flow rate.)

With proper aerodynamic design, the refrigerant exit tangential velocity from the blading is little different from the rotor tip velocity. Therefore,

$$v_{to} = \omega \times r_o \tag{17.5}$$

Using this result in Equation 17.4, we have

$$\dot{W}_i = \dot{m}_{ref} \times v_{to}^2 \tag{17.6}$$

By the first law of thermodynamics, one can also find the required compressor input power. All terms drop out of Equation 2.32 except the enthalpy terms with the result that

$$\dot{W}_i = \dot{m}_{ref}(h_o - h_i) \tag{17.7}$$

Equating the two expressions provides the following result for velocity needed to produce a given enthalpy increase. The required enthalpy increase is known from a cycle analysis such as that performed in Section 14.2.2. This result is approximate due to idealizations made earlier:

$$v_{to} = \sqrt{h_o - h_i} \tag{17.8}$$

Example 17.2: Rotational Speed for Centrifugal Chillers

What is the diameter of a centrifugal compressor rotor needed to isentropically compress R-22 from a saturated suction temperature of 5°C (41°F) to a condensing temperature of 50°C (122°F) and a pressure of 1.94 MPa (145 psia)? The compressor rotates at 1800 r/min (188 rad/s) at design conditions.

Given: $\omega = 1800$ r/min × 2π rad/r × 1 min/60 s = 188 s^{-1}
$T_i = 5°C$, $T_o = 50°C$, $p_o = 1.94$ MPa

Assumptions: The idealizations of the previous section apply.

Find: r_o

Lookup values: From Table A5: h_i = 406.6 kJ/kg (saturated) and from Table A7, h_o = 435 kJ/kg (superheated)

Solution

Equation 17.8 is used to find the tangential velocity. From the known angular velocity, the rotor diameter can be found. The tangential velocity is

$$v_{to} = [(435.0 - 406.6)\ \text{kJ/kg} \times 1000\ \text{J/kJ}]^{1/2} = 168.5\ \text{m/s}$$

The rotor outer radius is then

$$r_o = \frac{v_{to}}{\omega} = \frac{168.5\ \text{m/s}}{188\ \text{s}^{-1}} = 0.90\ \text{m (3.0 ft)}$$

Comments

This result is refrigerant specific, but for the common CFCs, there is not much difference. For example, for R-12, the rotor radius for the aforementioned example would be 0.76 m. However, for ammonia operating between the same two conditions, the enthalpy change would have been 52 kJ/kg and the rotor radius about 1.20 m. This is a relatively large rotor size to be economically manufactured. Of course, the rotor size can be reduced if the rotor is spun faster. Centrifugal forces increase with the square of angular velocity and linearly with the rotor radius.

The mass flow rate of refrigerant through centrifugal compressors is set by the *depth* of the impeller, i.e., the dimension into the plane of Figure 17.7.

A number of practical considerations affect centrifugal chiller performance. Part-load *control* can be accomplished either by prerotation vanes at the inlet or by variable-speed drive. The former is relatively effective at moderate reductions of capacity but is a source of inefficiency at low capacity, where the vanes act as little more than flow obstructions. Variable-speed drives are more effective at maintaining good efficiency at reduced capacity. The load of a centrifugal compressor can be reduced to as little as 10% of the design load.

Surging is the term applied to unstable flow in centrifugal compressors, which is characterized by loud objectionable noise and wide fluctuations of load on the motor and compressor, a phenomenon that also can happen in centrifugal pumps and fans. It results from flow separation at reduced flow rate, i.e., at reduced capacity. Operation in this range is to be strictly avoided to reduce equipment damage. Automatic chiller controls prevent operation in this area.

17.2.8 Intercomparison

Compressor types can be compared in different ways. Table 17.1 provides such a comparison for some of the important attributes. In the low range, rotary and scroll machines compete with reciprocating machines with screw compressors finding a niche in the higher range between reciprocating and centrifugal machines. The two most common types still remain the reciprocating and the centrifugal compressors. Figure 17.8 suggests that a decrease in refrigeration capacity from 240 to 100 tons is experienced by the centrifugal compressor when the evaporator temperature reduces by only 10°F (5.5°C), while for a reciprocating machine, the required change is close to 30°F (16°C). This implies that the centrifugal compressor will maintain a near constant evaporator temperature even when the refrigeration load changes greatly, a distinct advantage in certain applications.

Figure 17.9a shows how cooling capacity and power drawn at constant speed vary with condenser temperatures. The condenser temperature has a much large effect on the cooling capacity of a centrifugal compressor than on that of a reciprocating machine. While the former cannot function beyond a condenser temperature of about 106°F, the reciprocating machine will keep producing a cooling effect even when the condenser temperature keeps increasing. This underlines the importance of having a cooling tower to lower the condenser temperature for centrifugal chillers, while air-cooled condensers could be compatible with

TABLE 17.1

Comparison of Different Chiller Compressor Types

Chiller Type	Capacity Range, kW	Operating Cost	Full-Load COP	Part-Load Efficiency	Part-Load Limit, %	Compression Ratios Up To
Reciprocating	2–600	High	2.5–4.0	Very good	30	10:1
Rotary vane	0.5–10	Low	3.0–4.0	Very good	20	7:1
Scroll	5–200	Medium	2.9–3.2	Good	50	10:1
Twin screw	150–4,000	High	5.0–7.0	High	5	15:1
Centrifugal	400–10,000	High	5.0–7.0	Poor	10	3:1

Note: Numbers shown are approximate and each technology is generally more attractive at different capacity ranges.

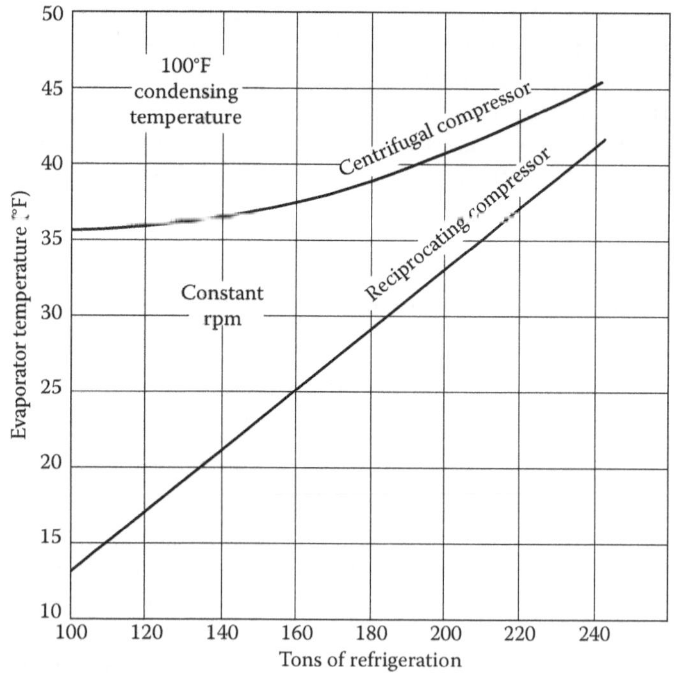

FIGURE 17.8
Comparison of centrifugal and reciprocating compressors: effect of evaporator temperature on refrigeration capacity. (From Dossat, R.J., *Principles of Refrigeration*, Prentice-Hall, Upper Saddle River, NJ, 1997.)

reciprocating compressors. Figure 17.9b allows a comparison of the power requirements of both compressor types. While the power reduces with increasing condenser temperature (and falloff in capacity) for the centrifugal compressor, the reciprocating machine shows a small increase in power even as the cooling capacity decreases.

Finally, Figure 17.10a and b shows the refrigerating capacity and power characteristics of both machines with speed changes. The centrifugal machine exhibits a much greater sensitivity to speed changes (a drop of 50% in capacity with only a 12% speed reduction) than does the reciprocating machine.

17.3 Expansion Devices

Expansion devices in refrigeration systems serve two primary purposes:

1. Create a pressure drop between high and low sides of the system.
2. Control refrigerant flow rate to the evaporator.

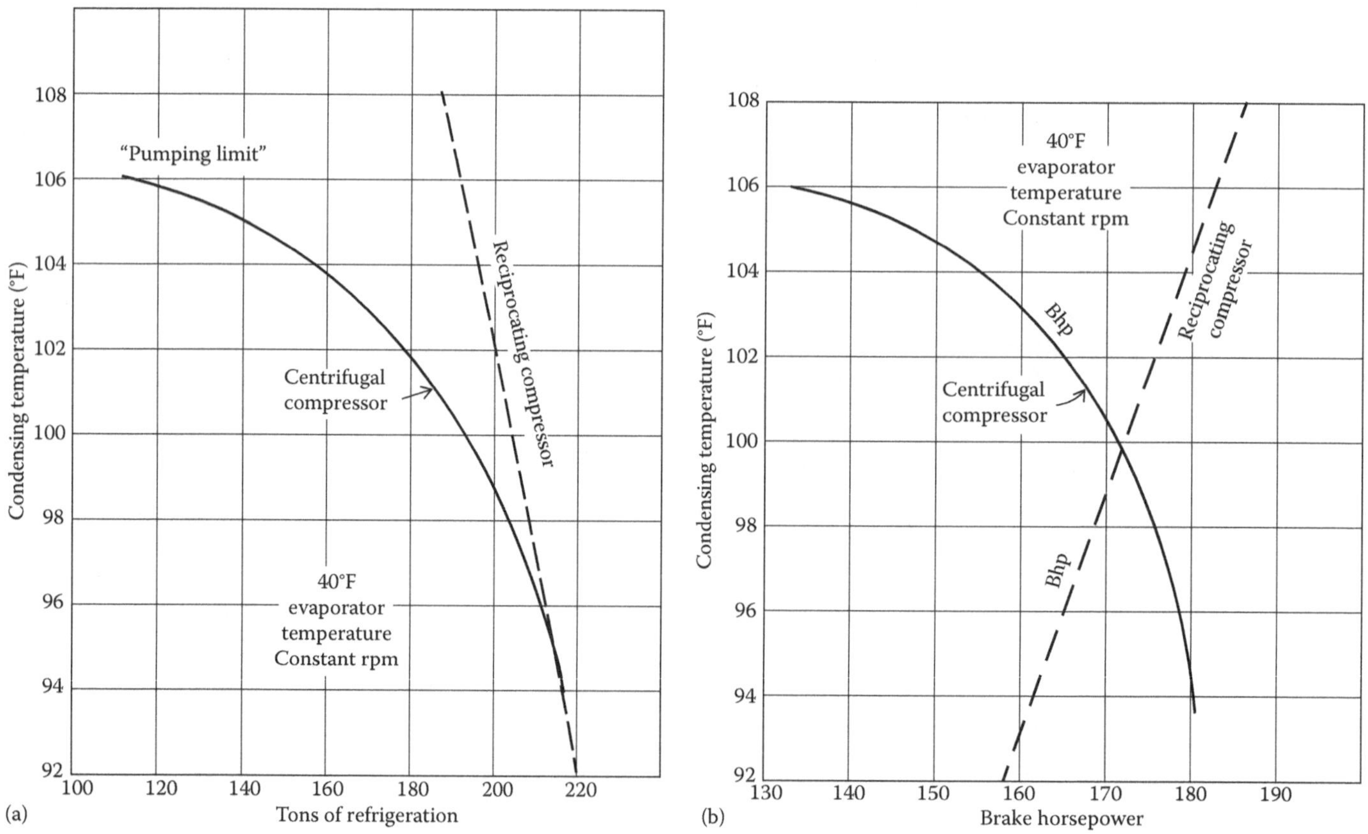

FIGURE 17.9
(a and b) Comparison of centrifugal and reciprocating compressors: effect of condensing temperature on refrigeration capacity and brake horsepower. (From Dossat, R.J., *Principles of Refrigeration*, Prentice-Hall, Upper Saddle River, NJ, 1997.)

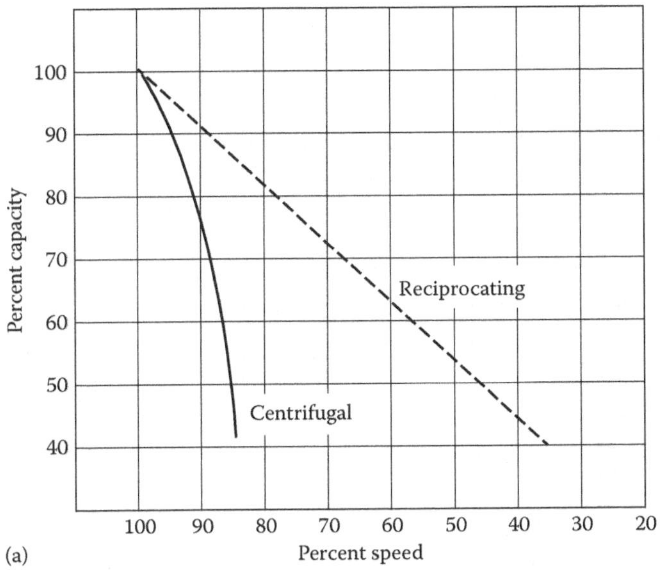

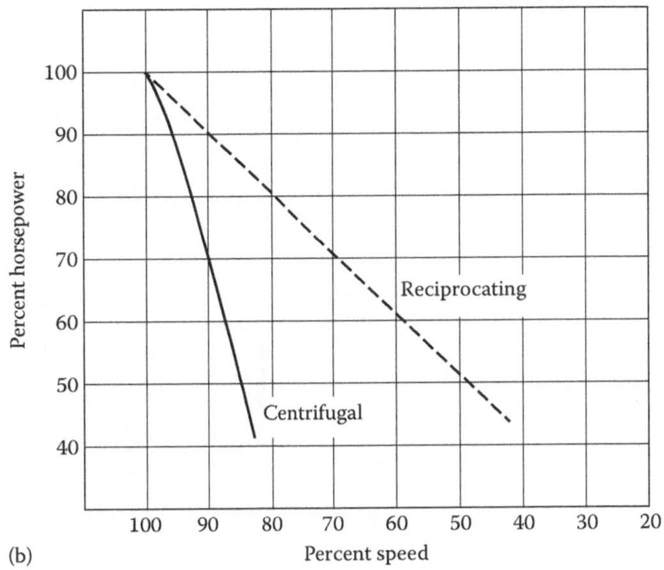

FIGURE 17.10
Comparison of centrifugal and reciprocating compressors: effect of speed on (a) refrigeration capacity and (b) brake horsepower. The evaporator and condensing temperatures are held constant. (From Dossat, R.J., *Principles of Refrigeration*, Prentice-Hall, Upper Saddle River, NJ, 1997.)

Three types of expansion devices are commonly used: (1) capillary tubes, (2) short-tube orifices, and (3) thermostatic expansion valves (TXVs), electronic expansion valves, and float valves. In this section, we will briefly describe capillary tubes and TXVs.

Capillary tubes (Figure 17.11) are small-bore, long tubes used in small cooling systems up to a few tons (few kilowatts) in size. Tubes 1–2 mm in diameter and up to a few meters long are used to produce the high- to low-side pressure drop. As liquid refrigerant passes through the tube, the pressure drops due to friction; the reduced pressure causes the liquid to boil. As liquid is converted to vapor, the velocity increases; this acceleration causes additional pressure drop. The tube is essentially a passive device and works best under one set of operating conditions; it cannot accommodate a large range of load and system pressures.

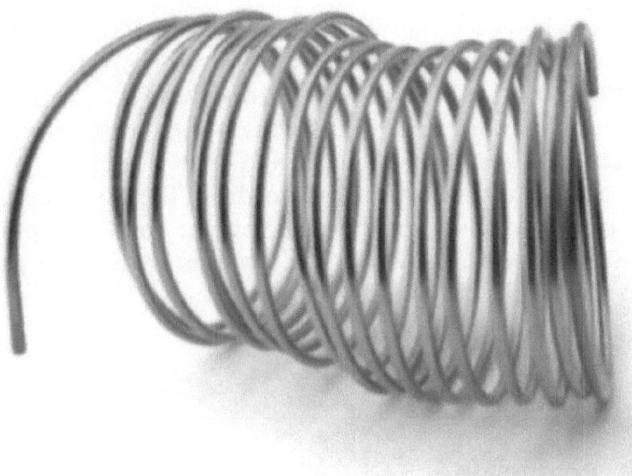

FIGURE 17.11
Capillary tube.

From basic fluid mechanics, it is clear that a capillary tube will allow higher refrigerant flow under higher pressure differences (unless the flow becomes choked). High head pressures can be caused by high condenser temperatures (as a result of hot weather) or due to a large liquid inventory that has the effect of reducing the useful condensing heat transfer area of the tubes. However, if the load is unchanged as high-side pressure increases, some accommodation between the constant refrigerant enthalpy rates is needed for constant load and the increased flow produced by the increasing system pressure difference. The remaining degree of freedom is the evaporator temperature and corresponding suction pressure. The suction pressure will adjust itself within limits by rising to reduce system flow.

There is also the situation of reduced load under cooler condensing conditions. In this case, the liquid refrigerant inventory in the evaporator will increase to the limit permitted by the size of the refrigerant charge. Of course, the charge must be selected so that all refrigerant can be held in the evaporator without any liquid entering the compressor inlet, a destructive event that must be avoided. Capillaries are sized by a measure of

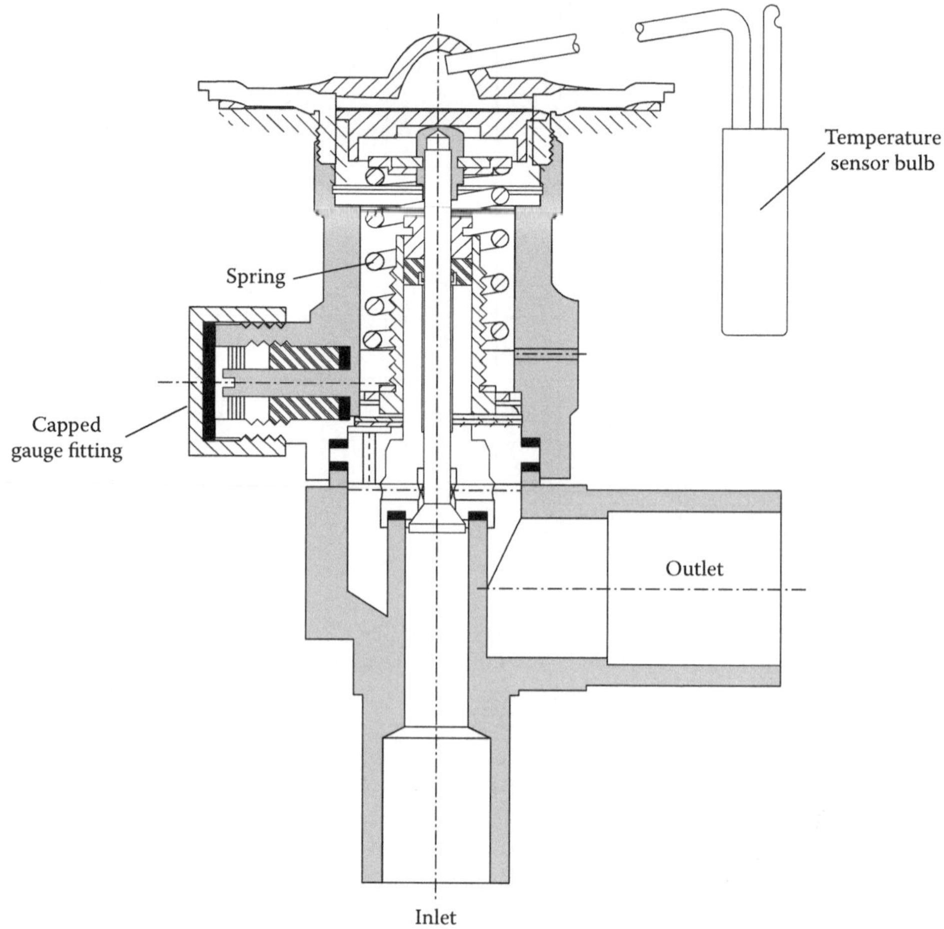

FIGURE 17.12
Detailed drawing of a thermostatic expansion valve.

trial and error for one condition, and deviation there from will reduce system efficiency. Stoecker and Jones (1982) describe the details of capillary selection.

Capillaries have the advantage of being inexpensive and passive. They have the disadvantages of a relatively narrow operating range, susceptibility to clogging by small particles or to crimping of the tube during installation, and the requirement of proper charging within rather narrow limits. Instead of a capillary system, a fixed orifice can be used to produce the needed pressure drop. Its behavior is similar to that of a capillary, but it is less likely to clog because of its larger bore.

Figure 17.12 shows the various components of a TXV used as the expansion device. It is the most popular method of controlling refrigerant flow and producing the needed pressure drop in medium-capacity systems. The operation is based on maintaining a constant degree of superheat of the vapor leaving the evaporator. Figure 17.13 illustrates the working more clearly. The sensor bulb is filled with a small amount of the refrigerant and is in close thermal contact with the suction line. A higher degree of superheat temperature at the evaporator exit implies that more refrigerant flow to the evaporator is needed to meet the cooling load (and vice versa). The superheat temperature increases the pressure of fluid in the bulb, and once it is high enough to force the diaphragm downward by overcoming the spring force plus the evaporator pressure, refrigerant from the condenser will flow through the seat opening to the evaporator coil. The spring selected determines the amount of superheat needed to open the valve. If the evaporator has a large pressure drop, the TXV design must be modified, but the basic operation is the same. TXV is especially suitable for systems with wide and frequent load fluctuations.

17.4 Evaporators and Condensers

The designer of HVAC systems during preliminary design will often first calculate the required heat rate of an exchanger based, e.g., on heating or cooling load

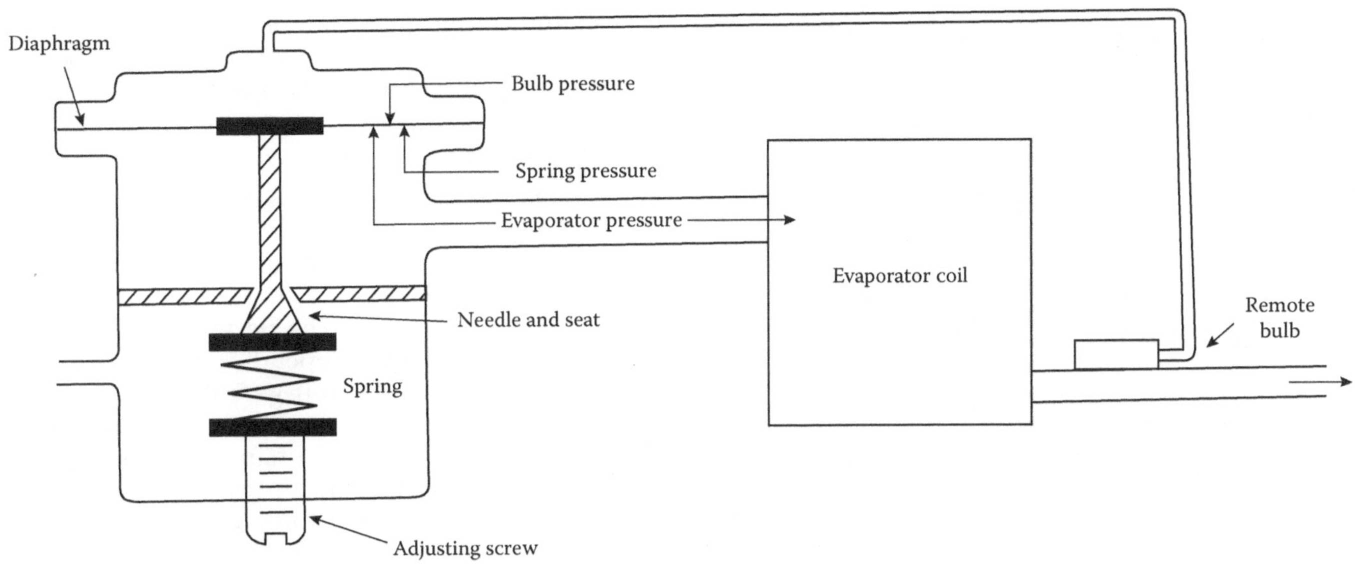

FIGURE 17.13
Sketch slowing the location, operation, and control of a thermostatic expansion valve.

calculations, known heat rejection rates from chillers, or domestic hot water loads. Next, typical fluid temperature drops are specified based on experience; a typical water-side temperature drop in an air heating coil in a duct is 20°F (10°C). Finally, required flows are calculated from the known heat rate and the specified temperature drop. These data are entered into the heat exchanger specification section of the building's contract documents.

During final design, other details must be considered including heat exchanger pressure drop, effects of fouling on performance, avoidance of tube vibration in shell-and-tube exchangers, water quality and potential scaling problems, physical location and mounting, service access, piping arrangements, insulation needs (if any), materials, and flow control. Many of these items can be handled with the assistance of manufacturers, who often have computer programs to assist the designer with final selection; manufacturers also provide all physical dimensions needed for piping layout. A list of potential heat exchanger problems that should be avoided at final design is given in Yokell (1990).

In Chapter 12, we discussed two methods (LMTD method and the effectiveness method) to analyze the thermal performance of common heat exchangers that transfer heat without phase change (i.e., without condensation or evaporation of water vapor). These methods also apply to analysis of evaporators and condensers. In this and the next two sections, we describe the different types of heat exchanger equipment and present some of the basic modeling approaches. Only a broad overview is provided, and the interested reader can refer to more detailed texts such as Dossat (1997) or ASHRAE Systems (2012).

17.4.1 Evaporators

Evaporators are heat exchangers in which a volatile fluid (or refrigerant) is boiled or vaporized so as to remove heat from either a space or to cool a fluid (a liquid, usually water, or air). One way of classifying evaporators is according to the method of liquid feed: (1) dry expansion or DX, where the amount of refrigerant fed into the evaporator is limited by the TXV or capillary tube (as discussed in Section 17.3), (2) flooded type (discussed in Section 17.4.1), and (3) liquid overfeed wherein the amount of refrigerant circulated is in excess of what can be evaporated; such designs are commonly employed in multiple evaporator systems and are not relevant for building applications.

Another way of distinguishing evaporators is by application:

1. *Natural convection air evaporators* with direct expansion (or DX) are used typically in household refrigerators, display cases, and large storage rooms. Such applications require low velocities and minimum dehydration of the products. The air circulation is by natural convection, which is influenced by the shape, size, and location of the evaporator and that of the product to be refrigerated. The freezer section of the refrigerator uses the standard serpentine plate evaporator shown in Figure 17.14. Bare-pipe evaporator designs are also common, and sometimes baffles are used in certain types of applications to enhance air circulation.
2. *Forced convection air evaporators* are used in commercial refrigeration and also to cool

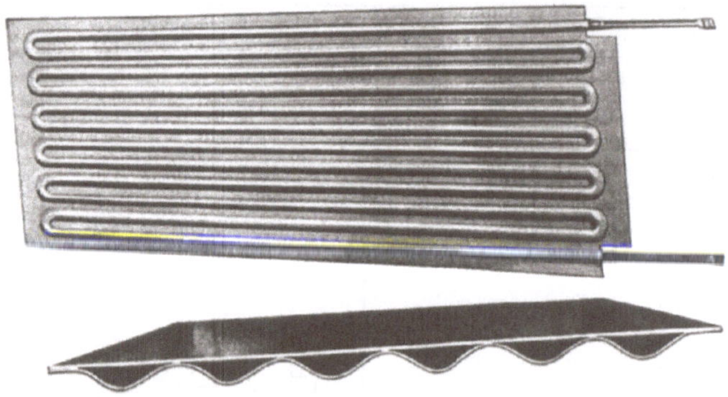

FIGURE 17.14
Typical serpentine plate evaporator used in freezer section of a household refrigerator.

building spaces using fan coil units (described in Section 18.3.2). These are essentially bare-tube or finned-tube coils enclosed in a metal housing through which air is circulated by one or more fans. A ceiling mounted unit is shown in Figure 17.15, while designs are available to mount them on the wall or on the floor. Since the refrigerant boils inside the tubes, these would qualify as DX coils as well.

FIGURE 17.15
Photograph of a typical forced convection fan coil unit with a DX evaporator supplying cool air to a space.

3. *Water chillers or coolers* generally cool water. They are of two types:
   a. DX chilled water evaporators that are shell-and-tube designs (see Figure 12.13) where the water to be chilled is circulated on the shell side and the refrigerant boils inside the tubes. The performance of the chillers greatly depends on proper distribution of the refrigerant through the numerous tubes and also incorporates devices meant to enhance internal heat transfer. Often, this type of coolers are used with smaller tonnage chillers using positive displacement compressors such as reciprocating, rotary, scroll, and screw machines.
   b. Flooded shell-and-tube evaporators where the water to be chilled is passed through tubes and the refrigerant boils outside the tubes on the shell side (Figure 17.16). As the refrigerant boils, a mist that consists of vapor and liquid forms over the tube bundle. Eliminators or fins are used to separate the vapor and liquid. This type of evaporators is best suited for large tonnage chillers with rotary or centrifugal compressors.

Consider Equation 12.29, which is the equation for the overall conductance of a heat exchanger. Three resistances come into play: the internal convective resistance (R_i), the tube wall resistance (R_{tube}), and the external convective resistance (R_o). Of the three factors,

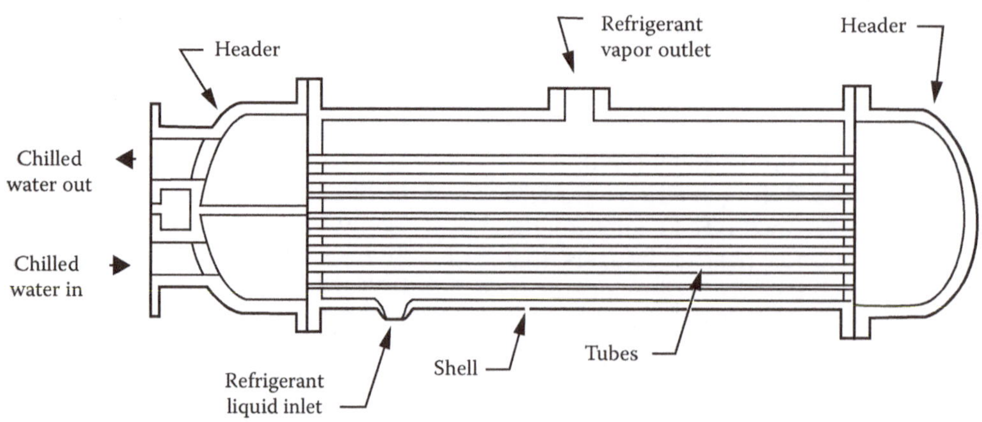

FIGURE 17.16
Flooded evaporator. (From ASHRAE, *Handbook—HVAC Systems and Equipment*, American Society of Heating, Refrigerating and Air-Conditioning Engineers, Atlanta, GA, 2012. Copyright ASHRAE, www.ashrae.org.)

R_{tube} is generally the least influential. Convective heat transfer coefficients are greater with water than with gases or vapors.

Flooded evaporators, because they are filled with liquid on the outside of the tubes, have higher convective coefficient values than dry-expansion type. Also, heat transfer both inside and outside the tube can be enhanced by increasing turbulence, which is achieved by several devices such as internal fluting or fining (such as twisted aluminum spines in the center of the tube). In liquid cooling applications, where the liquid is in contact with the outside of the tubes, bare-tube evaporators are adequate and fining is unnecessary.

In air-cooled evaporators where the air flows outside the tubes, the capacity of the evaporator is limited by the resistance of the outside surface heat transfer coefficient. In such cases, fins are used to enhance this heat transfer coefficient (Figure 17.17). For example, the thermal performance of DX evaporators (usually shell-and-finned-tube heat exchangers) can be enhanced by using fins. The detailed heat transfer analysis of fins can be found in any textbook, for example, that of Cengel (2002). Generally, DX machines have lower COP values than do water-based chillers because they tend to have air-cooled condensers and further their compressors are less efficient.

In general, the evaporator heat rate decreases with decreasing chilled water or airflow rate at a given inlet temperature. The capacity increases with reduced evaporator temperature or with higher incoming air or water temperatures. These trends are apparent from the prior discussion of heat exchangers in Chapter 12. Details of evaporator design are not discussed in this book and can be found in more specialized texts (such as Dossat, 1997).

17.4.2 Condensers

All cooling equipment must reject heat to a heat sink equal to the sum of the heat removed from the cooled building spaces and the compressor work (or the absorber heat input for an absorption cooling system). There are two standard methods for rejecting heat: air cooled, wherein the heat is directly rejected to the ambient air, and water cooled, wherein the heat is first rejected to cooling water and then to the ambient air in cooling towers (discussed in Section 17.7).

An *air-cooled condenser* cools refrigerant directly by forcing outdoor air to flow over the outer surface of tubes in which the condensing refrigerant is circulated. In this section, we describe both forms of heat rejection to outdoor air from central plant cooling systems.

Small chillers (less than 200 tons or 700 kW) often use air-cooled condensers rather than water-cooled condensers and cooling towers. Air-cooled condensers are simply cross-flow heat exchangers with refrigerant flowing and condensing inside of finned tubes. Fans force air over the exterior surface of the tubes and the fins. These condensers have modest maintenance requirements but may result in higher energy consumption and shorter compressor life, because condensing pressures are relatively high

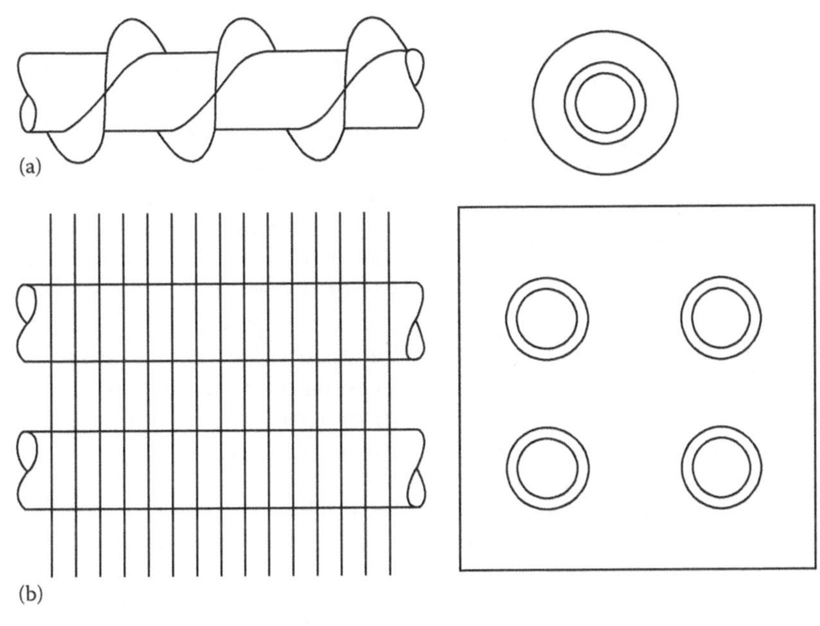

FIGURE 17.17
Two types of fins on tubes to enhance heat transfer: (a) circular plate fins and (b) bar fins.

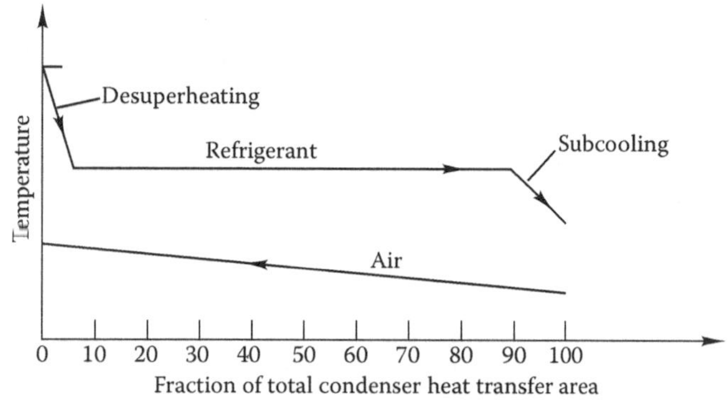

FIGURE 17.18
Refrigerant and air temperature profiles within an air-cooled condenser.

on summer days given typical air-to-refrigerant temperature differences.

Although much of the heat transfer involves isothermal phase change, some initial desuperheating and final liquid subcooling are usual. Figure 17.18 shows an example of how the three heat exchange processes occur, and Figure 11.15 is a photograph of an air-cooled condenser. The ε-NTU method described in Section 12.6.3 can be used to calculate the performance of an air-cooled condenser of this type by dividing it into three heat exchange sections—two sensible and one latent.

The design of air-cooled condensers requires the specification of the following:

1. *Heat rejection rate.* Determined by chiller or air conditioner peak heat discharge rate.
2. *Airflow rate.* A balance between excessive pressure drop for high flow rates and high initial cost for large heat transfer surface. Usual values are 600–1200 ft^3/min/ton (80–160 L/s/kW). The fan power consumption is typically 0.1–0.2 hp/ton (20–40 W/kW).
3. *Temperature difference* (refrigerant to entering air). Affected by decision in item 2 and heat rate from item 1. Typical values are 15°F–40°F (8°C–22°C).
4. *Noise.* Large air-cooled condensers can be noisy. They should be located so that noise produced is not a nuisance to building occupants.
5. *Unobstructed airflow.* A supply of outdoor air is needed for proper operation. Short-circuiting of warm air leaving the condenser back to the condenser inlet must be avoided. The area near the condenser air inlet must be kept clean so that the coils do not become blocked off or fouled.

In addition to problems associated with high-temperature operation, air-cooled condensers require special controls at low outdoor temperatures to avoid cooling condensed refrigerant to such an extent that it will not expand completely through the expansion valve. The fan speed can be controlled to avoid excessive subcooling, or the condenser fan can be cycled on and off. Dampers to control condenser airflow have also been used, but not entirely successfully.

High altitude causes a reduction in air-cooled condenser capacity since the mass flow rate is reduced (because of lower air density), even though the fan produces the same volumetric flow. At 5000 ft (1500 m) elevation, air-cooled condenser capacity is reduced by 10%.

The performance of air-cooled condensers can be improved if the air-side surface is kept wet with purified water. Evaporation from the condenser will enhance performance markedly because the driving potential for a coil cooled by evaporation is the wet-bulb, not the dry-bulb, temperature. Since the wet-bulb temperature is often 15°F–25°F (8°C–14°C) below the dry-bulb temperature, *evaporative condensers* will operate at temperatures substantially lower than those of air-cooled equipment and somewhat lower than those of water-cooled condensers employing a cooling tower. Low condenser temperatures result in lower compressor power needs and longer compressor life. Figure 17.19 is a sketch of an evaporative condenser, also called an indirect-contact condenser.

The key consideration in the design of evaporative condensers is the water composition. If minerals are not controlled, they will accumulate on the condenser surface and foul it, reducing heat transferability in relatively short order. Biological growth can also foul the surface. The cost of water must also be considered in arid climates.

*Water-cooled condenser*s are used in large commercial air conditioning units (typically over 100 tons) and in central plants. They are more energy efficient than air-cooled condensers because the larger heat transfer coefficients of water as compared to air result in lower refrigerant temperatures (10°F–15°F or 5°C–8°C), thereby allowing the compressor to operate at lower head pressures. However, they are more expensive and require more routine maintenance than air-cooled condensers. They are constructed in three designs:

1. Shell-and-tube designs, which are the most popular and have been discussed previously (see Figure 12.13).
2. Shell-and-coil designs, which are used in smaller units. They are smaller in size and

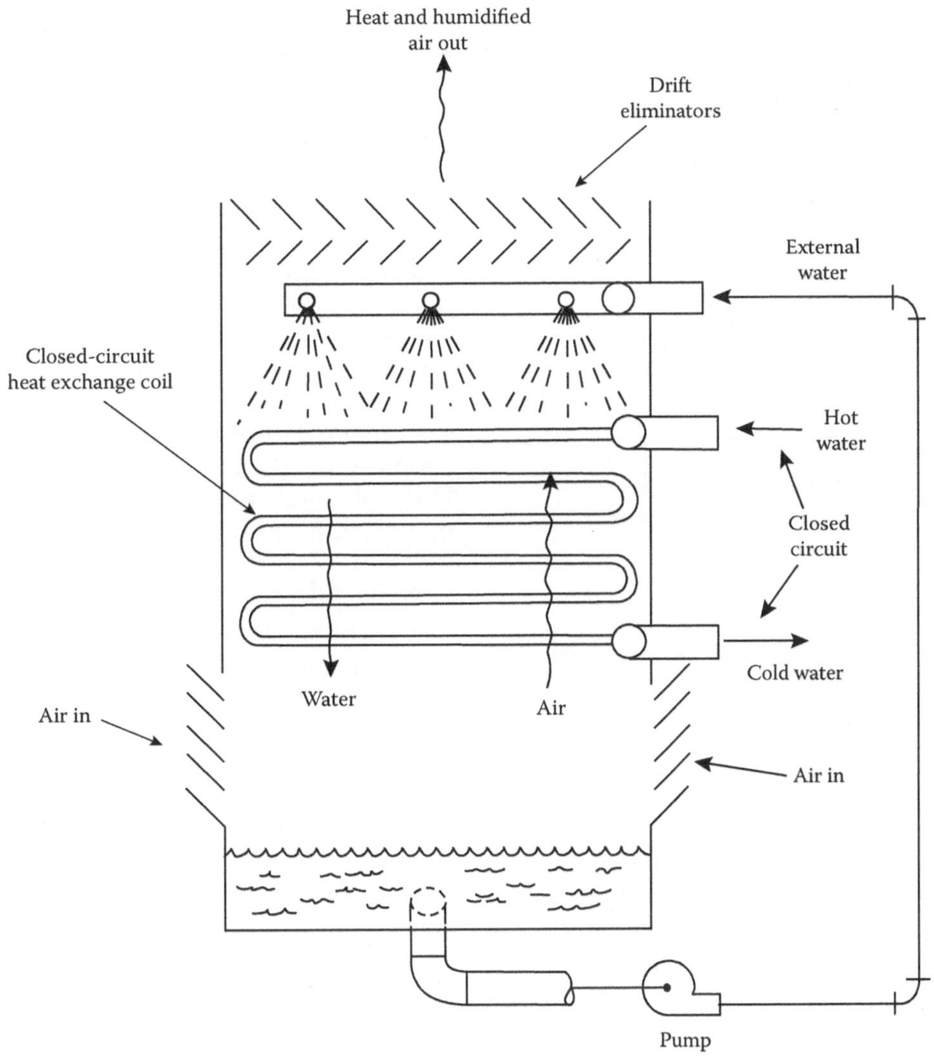

FIGURE 17.19
Indirect-contact evaporative condenser. (From Marley, *Cooling Tower Fundamentals*. Marley Cooling Tower Co., Mission, KS, 1983.)

cheaper since the tubes are not straight like in shell-and-tube designs but are in a coil.

3. Tube-within-a-tube designs, where two concentric tubes are used. Such designs are generally used for small cooling systems.

The interested reader can refer to more advanced texts such as Stoecker and Jones (1982), Dossat (1997), or ASHRAE Systems (2012).

17.5 Heating Air Coils

From Section 12.6.2, we recall that practical heat exchangers with complex flow geometries can be modeled following Equation 12.33 reproduced:

$$\dot{Q} = F \times (UA) \times (\text{LMTD}) \quad (12.33)$$

where

UA is the overall heat transfer coefficient multiplied by the coil area

LMTD is the logarithmic mean temperature difference given by Equation 12.31

F is a correction factor that can be read from charts in a number of sources, including Turton et al. (1986) and Bowman and Turton (1990)

Of course, the calculation of the factor *F* is embodied in the software used by manufacturers to assist designers with heat exchanger selection.

A cross-flow air to liquid heat exchanger is shown in Figure 17.20. Manufacturers also provide performance data in extensive charts. Table 17.2 shows an

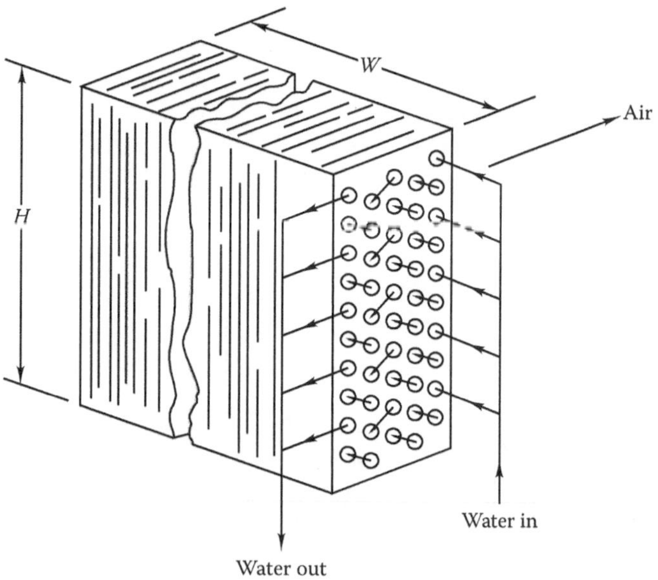

FIGURE 17.20
Cross-flow air heating coil showing 10 tube passes per row. Heat exchangers with at least this number of tube passes can be well modeled as counterflow heat exchangers.

example of such data. The body of the table shows heat transfer rates (at sea level) for a range of airflow rates and water flow rates. The heat rates are for a given difference between the entering air dry-bulb ($T_{db,i}$) temperature and the entering water temperature ($T_{w,i}$), as shown in Section 12.6 in the discussion of heat exchanger effectiveness.

The last row at the bottom of Table 17.2 includes an adjustment factor for other temperature differences. The numbers in this table are just the ratio of the temperature difference shown to the rating-point temperature difference of 145°F. The assumption is made that the heat exchanger effectiveness is independent of the operating temperature—a good assumption for air and water. (Note that Table 17.2 is valid only at sea level; for high altitudes, the heat rates shown must be multiplied by the ratio of local air density to standard air density.) Table 17.2 also shows the water-side pressure drop at various flow rates. If other liquids are used, the pressure drop will differ—glycols are often used for freeze protection in coils; pressure drops with glycol are larger than those for water. Hot-water coils are usually designed in the following operating ranges:

- *Air-side velocity*: 3–25 ft/s (1–8 m/s)
- *Hot-water temperatures*: 120°F–250°F (50°C–120°C)
- *Water velocities*: 0.5–8 ft/s (0.2–2.5 m/s)

When steam is used in heating coils, it is usually low-pressure steam at 2–10 psig (14–70 kPa). Water-side temperature drops are in the range of 10°F–20°F (5°C–10°C) for both heating and cooling coils.

Coils and shell-and-tube heat exchangers rely on forced convection for effective heat transfer. Free convection in air can also be effective for direct space heating as discussed in Section 18.3. Figure 18.6 shows an example of a baseboard unit operated with hot water. High water temperatures (215°F is a typical

TABLE 17.2
Example of Manufacturer's Two-Row Heating Coil Capacity Table

Water Flow Rate, gal/min	Water Pressure Drop, ft	Capacity, kBtu/h											
		Air Flow Rates, ft³/min											
		200	250	300	350	400	450	500	550	600	650		
0.3	<0.1	12.6	13.4	13.9	14.4	14.7	15	15.2	15.4	15.5	15.7		
0.5	0.1	17.1	18.8	20.2	21.3	22.3	23	23.7	24.3	24.8	25.3		
0.8	0.2	19.3	21.6	23.5	25.1	26.4	27.6	28.6	29.5	30.4	31.1		
1	0.3	20.6	23.3	25.5	27.4	29	30.5	31.8	33	34	35		
1.5	0.5	22	25.1	27.8	30.1	32.2	34	35.6	37.1	38.5	40		
2	0.9	22.8	26.2	29.1	31.7	34	36	37.9	40	41.2	42.6		
2.5	1.3	23.3	26.8	30	32.7	35.1	37.3	39.4	41.2	43	44.5		
3	2.6	23.6	27.3	30.5	33.4	35.9	38.3	40.4	42.4	44.2	45.9		
($T_{w,i}-T_{db,i}$)	160°	150°	140°	130°	120°	110°	100°	90°	80°	70°	60°	50°	40°
Factor	1.10	1.03	0.97	0.90	0.83	0.76	0.69	0.62	0.55	0.48	0.41	0.34	0.28

Source: Courtesy of Anemostat, Inc., Scranton, PA.
These data are based on an ($T_{w,i}-T_{db,i}$) value of 145°F. For other ($T_{w,i}-T_{db,i}$) values, use the correction factor listed on the last row of the table. $T_{w,i}$ is the entering water or stream temperature and $T_{db,i}$ is the entering dry-bulb temperature. The data applies to sea level operation.

rating point for convectors) induce free convection in the air surrounding the pipes containing the heating water. To enhance heat transfer, the pipes are usually fitted with air-side fins (which increase the area of heat transfer), which are thermally bonded to the pipes. Still better performance can be achieved by using a small fan to force air over the fin tubes in baseboard heaters.

The principles of coil selection and sizing used for heating and cooling coils and for shell-and-tube heat exchangers are also used for other HVAC equipment, such as air-cooled condensers, water tank immersion heaters using steam, chiller condensers and evaporators, and air-to-air heat exchangers for heat recovery in ventilation air. Example 17.3 shows how a coil is selected for space heating.

Example 17.3: Heating Coil Selection

The heat load for a space to be heated by a forced-air heating coil is 20,000 Btu/h (5,860 W) in a building at an altitude where the air density is 80% that at sea level. The space is maintained at 70°F (21°C), and hot water is supplied from the heating plant at 180°F (82°C). Select a coil from Table 17.2 for this application. For proper air distribution in the space, the airflow rate must be at least 500 ft³/min (236 L/s).

Given: $\dot{Q} = 20{,}000$ Btu/h, $T_{db,i} = 70$°F, $T_{w,i} = 180$°F, $\dot{V}_a = 500\ \text{ft}^3/\text{min}(236\ \text{L/s})$

Find: Coil water flow

Lookup values: See Table 17.2.

Solution

The heat rate must be adjusted by two factors before the table can be used. First, since the temperature difference is 110°F, not the standard 145°F used for the tabular values, the heat rates in the table must be multiplied by the factor 0.76, read from the bottom of Table 17.2. Second, the heat rate is 20% less than that at sea level due to the high altitude of the site. Hence, the value of heat rate to be read from the table is

$$\dot{Q}_{table} = \frac{20{,}000\ \text{Btu/h}}{0.76 \times 0.80} = 32{,}900\ \text{Btu/h}$$

In the table under the 500 ft³/min column, one finds a heat rate of 35.6 kBtu/h corresponding to a water flow rate of 1.5 gal/min. A higher airflow rate, say 550 ft³/min, could be used. For this airflow, one finds a heat rate of 33 MBtu/h at a lower water flow rate of 1 gal/min. Which operating value is selected from the table ultimately depends on the airflow or water flow available at the location of this terminal heating unit and the operating costs.

Comments

When using tables for coil selection, the designer should always pick a value equal to or larger than the heat load. This provides a small capacity margin if the coil specified should fall below specifications. The water-side pressure drop shown in the second column of Table 17.2 is needed for piping systems design. Recall that the larger the product of pressure drop and flow, the larger the pump operating energy needed. The same rule applies for fan energy use; if the 550 ft³/min coil is used, the fan energy will increase by the cube of the flow rate (see the discussion of fan laws in Section 16.3.4).

17.6 Wet Cooling Air Coils

Heat transfer processes in the previous section involved only sensible heat exchanges. For heating coils, this is always the case; for cooling coils, only if the dew point of entering air is lower than the dew point of the surface of the coil is this true. However, the surface of cooling coils is often below the dew point of the air passing over them during the cooling season in moderately humid and humid climates. In this section, we discuss briefly how coil calculations are done if the surface is wet. A third case also exists as a partially wet and partially dry coil. This intermediate case is handled by treating the coil as two heat exchangers—one wet and the other dry—with a common boundary (ASHRAE Systems, 2012).

The driving force for combined heat and mass transfer at the surface of a wet cooling coil is the air enthalpy rather than the air temperature. For example, the local heat transfer from a wet surface is given by (Kuehn et al., 1998)

$$\dot{Q}_{wet} = (\tilde{U}A)_{wet}(h_a - h_{a,sat,m}) \tag{17.9}$$

where

$\dot{Q}_{wet}$ is the heat transfer rate from cooling coil liquid via wetted surface to airstream, Btu/lb$_m$ (kW)

$(\tilde{U}A)_{wet}$ is the *enthalpy-based* overall heat transfer coefficient, lb$_m$/h (kg/s) (note that the units differ from other $\tilde{U}$A quantities used elsewhere in this book)

h_a is the enthalpy of air flowing over coil, Btu/lb$_m$ (kJ/kg)

$h_{a,sat,m}$ is the saturated air enthalpy at *mean water temperature* inside coil tubes, Btu/lb$_m$ (kJ/kg)

The overall enthalpy-based heat transfer coefficient (between coil fluid and airstream) is calculated from

$$(\tilde{U}A)_{wet} = \frac{A_{wet}}{b(R_i + R_m) + c_{p,a}(R_f + R_o)} \quad (17.10)$$

where

A_{wet} is the external wet surface area of coil

R_i is the internal water-side thermal resistance, (h ft$^2\cdot$°F)/Btu [(m$^2\cdot$K)/W]

R_m is the metal tube conductive resistance

b is the slope of approximate, locally linear equation $h_{sat} = a + bT_{sat}$ relating saturated air enthalpy to saturation temperature T_{sat}, Btu/(lb$_m\cdot$°F) [kJ/(kg$\cdot$K)]

$c_{p,a}$ is the specific heat of air

$R_f = \frac{1-\eta_o}{\eta_o h_o}$ = thermal resistance of fins in cooling coil

$\eta_o = 1 - \frac{A_{fin}}{A_{fin} + A_{pipe}}(1-\eta_f)$, in which η_f is standard fin efficiency (refer to any heat transfer textbook, say Cengel, 2002)

$R_o = \frac{1}{h_o}$ = thermal resistance at outer surface of the tube

This $\tilde{U}A$ value can be used in a *log mean enthalpy* expression analogous to Equation 12.30 to find the total heat rate. The enthalpy differences in this expression are the difference between air enthalpy and *saturated* air enthalpy (evaluated at the *cooling water* temperatures) at the inlet and outlet of the coil, as shown in Figure 17.21.

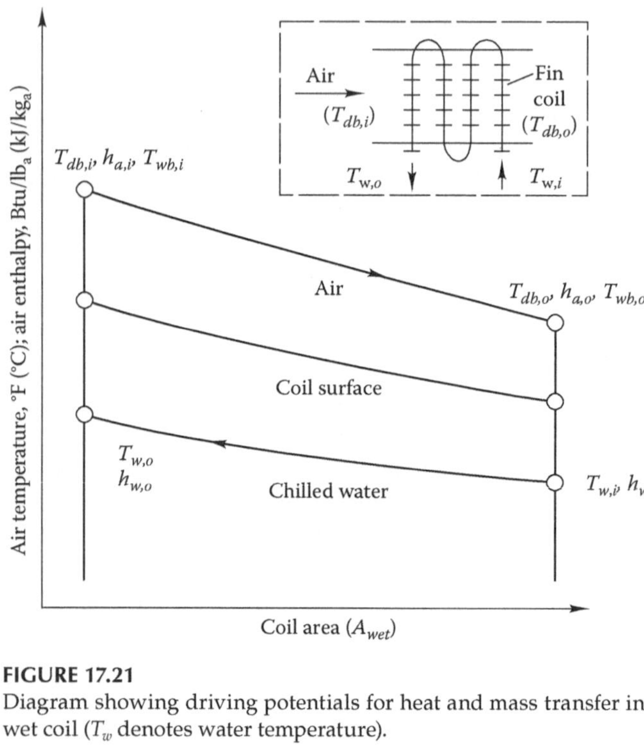

FIGURE 17.21
Diagram showing driving potentials for heat and mass transfer in a wet coil (T_w denotes water temperature).

Alternatively, the ε−NTU method developed in Section 12.6.3 can be used to find the heat transfer rate. An expression similar to Equation 12.37 is used except that the $\tilde{U}A$ value in NTU is replaced by $\tilde{U}A$ above. The overall driving force (i.e., temperature difference) is replaced by the enthalpy difference between coil inlet air and that of saturated air at the coil water inlet temperature. Finally, the capacitance ratio (assuming that the airstream is the lower-capacitance stream) is replaced by $\dot{m}_a/(\dot{m}_w c_{p,w}/b)$. Example 17.4 indicates how this method can be applied to a wet coil.

Example 17.4: Performance of a Wet Cooling Coil

A testing and balancing contractor tests a coil to determine if its performance is as specified. The coil specified by the designer had the following characteristics (refer to Figure 17.21 for nomenclature) at the design point:

1. *Entering air*: $T_{db,i}$ = 95°F (35°C), $T_{wb,i}$ = 75°F (23.9°C)
2. *Leaving air*: $T_{db,o}$ = 66°F (18.9°C), $T_{wb,o}$ = 61°F (16.1°C)
3. *Entering coil water temperature*: $T_{w,i}$ = 55°F (12.7°C)
4. *Airflow rate*: 20,000 ft^3/min (9,440 L/s)

During the coil test, the contractor measures the following data. Since test conditions rarely are the same as design conditions, the operating condition is different. Use these measured data to determine whether the coil has the specified cooling capacity.

- *Entering air*: $T_{db,i}$ = 90°F (32.2°C), $T_{wb,i}$ = 73°F (22.8°C)
- *Leaving air*: $T_{db,o}$ = 63°F (17.2°C), $T_{wb,o}$ = 59°F (15.0°C)
- *Entering coil water temperature*: $T_{w,i}$ = 52°F (11.1°C)
- *Airflow rate*: 20,000 ft^3/min (9,440 L/s)

The air side is the minimum-capacitance side of the coil, and the airflow during the test is the same as that at design conditions.

Given: Design and measured data as given earlier

Figure: See Figure 17.21.

Assumptions: The coil is tested at sea level. Its surface is wet—confirmed by plotting the process on a psychrometric chart.

Find: Cooling capacity of coil

Lookup values: Enthalpies are read from the psychrometric chart as follows:

$$h_{a,i,des} = 38.7 \text{ Btu/lb}_m$$

$$h_{a,o,des} = 27.2 \text{ Btu/lb}_m$$

$$h_{a,i,test} = 36.8 \text{ Btu/lb}_m$$

$$h_{a,o,test} = 25.8 \text{ Btu/lb}_m$$

$$h_{w,i,des} = 23.2 \text{ Btu/lb}_m (\text{saturated at } 55°\text{F})$$

$$h_{w,i,test} = 21.5 \text{ Btu/lb}_m (\text{saturated at } 52°\text{F})$$

Solution

The way to compare the performance of a coil to the design specifications is to calculate the effectiveness for both cases. For the *design* conditions, the enthalpy form of Equation 12.34 is

$$\varepsilon_{des} = \frac{\dot{Q}_{des}}{\dot{m}_a(h_{a,i} - h_{w,i})} \quad (17.11)$$

$$\dot{m}_a = 20{,}000 \text{ ft}^3/\text{min} \times 0.075 \text{ lb}_m/\text{ft}^3 \times 60 \text{ min/h}$$
$$= 90{,}000 \text{ lb}_m/\text{h}$$

$$\dot{Q}_{des} = \dot{m}_a \Delta h_a = 90{,}000 \text{ lb}_m/\text{h}$$
$$\times (38.7 \text{ Btu/lb}_m - 27.2 \text{ Btu/lb}_m)$$
$$= 1{,}035 \text{ kBtu/h}$$

The design effectiveness is

$$\varepsilon_{des} = \frac{1{,}035{,}000 \text{ Btu/h}}{90{,}000 \text{ lb}_m/\text{h} \times (38.7 - 23.2) \text{ Btu/lb}_m} = 0.742$$

Under *test* conditions, the same calculations give these results:

$$\dot{Q}_{test} = \dot{m}_a \Delta h_a = 90{,}000 \text{ lb}_m/\text{h} \times (36.8 - 25.8) \text{ Btu/lb}_m$$
$$= 990 \text{ kBtu/h}$$

The calculated effectiveness is

$$\varepsilon_{test} = \frac{990{,}000 \text{ Btu/h}}{90{,}000 \text{ lb}_m/\text{h} \times (36.8 - 21.5) \text{ Btu/lb}_m} = 0.719$$

The measured effectiveness is 97% of that specified by the designer, a small difference within the tolerance of the measurement errors.

Comments

This type of calculation is necessary whenever results from a test and balance contractor are being evaluated. Testing will almost never be done at design conditions for either heating or cooling coils. The concept of heat exchanger effectiveness provides a convenient method for using the results at test conditions to determine coil performance at the design condition.

Cooling coil design and selection are largely dictated by the dehumidification requirements. The air and water velocities are in the same ranges as those for heating coils (stated earlier). The coolant water temperature is in the range of 35°F–45°F (1.5°C–7°C). The entering air dry-bulb temperatures are typically around 70°F–80°F (21°C–27°C) and 30%–50% relative humidity (RH), while those for leaving air are typically 50°F–60°F (10°C–15.5°C) in temperature and 80%–100% RH (Mitchell and Braun, 2013).

The number of rows of the coil, the face velocity of the air, and the refrigerant temperature influence the cooling capacity and the amount of moisture removal, as shown in an illustrative catalog data of Table 17.3

TABLE 17.3

Example of Manufacturer's DX Cooling Coil Performance Data

		Face Velocity = 2 m/s		Face Velocity = 3.0 m/s	
Refrigerant Temperature, °C	Rows of Tubes	Final Dry-Bulb Temperature, °C	Final Wet-Bulb Temperature, °C	Final Dry-Bulb Temperature, °C	Final Wet-Bulb Temperature, °C
1.7	2	17.0	16.2	18.6	17.3
	3	14.7	14.1	16.3	15.6
	4	12.6	12.3	14.6	14.0
	6	9.8	9.6	11.7	11.4
7.2	2	19.6	17.9	21.1	18.7
	3	17.5	16.5	18.9	17.5
	4	16.1	15.3	17.4	16.4
	6	13.9	13.4	15.4	14.7

These data are with air entering at $T_{db,i} = 31°C$ and $T_{wb,i} = 21.7°C$.

(nowadays, coil selection software is more commonly used). Coils with 4–6 rows are commonly selected for building air cooling applications. We note the following features that are useful to keep in mind during coil selection:

1. Coil capacity can also be increased by increasing the number of rows (Figure 17.20). In such a case, the exiting airstream gets colder (lower dry-bulb temperature T_{db}) and dryer (lower wet-bulb temperature T_{wb}) and approaches the saturated condition of the water inlet temperature. However, the incremental cooling and dehumidifying capacity reduces with each successive row since the temperature and enthalpy differences between airstream and local coil surface keeps reducing as the air passes over the bank of rows. Further, increasing the number of rows increases air pressure drop, thereby needing higher fan power.
2. Increasing the air velocity (and thereby increasing the volumetric airflow rate) enhances coil capacity but the leaving air is less cool and dry. An additional fan power penalty is incurred.
3. Increasing the refrigerant temperature reduces both the cooling and the dehumidification capability but improves the chiller COP. So, a global optimization needs to be undertaken.

Many of the aforementioned features are illustrated by the results assembled in Table 17.4. A cooling coil of one square meter face area is assumed, and the total cooling capacity and the amount of water removed have been computed for the different operating conditions pertinent to Table 17.3. We note that both the cooling capacity and the water removal rate increase with the number of rows. While the incremental cooling capacity reduces, that of the amount of water removed actually increases. For example, for a face velocity of 2 m/s and refrigerant temperature of 1.7°C, the first two cooling coils have a cooling capacity of 42.7 kW, while the next two rows only have (68.3–42.7) = 25.6 kW. This trend is consistent for other conditions as well. On the other hand, the first two coils remove 11.53 kg/h of moisture, while the next two remove (32.56–11.53) = 21.03 kg/h, i.e., subsequent cooling rows remove more humidity. This is consistent with our expectation since the air condition as it flows over successive coils gets closer to saturation. The effect of increasing the face velocity increases the total cooling capacity but results in less water being removed from the airstream. The effect of increasing the refrigerant temperature reduces both the cooling capacity and the amount of water removal.

17.7 Cooling Towers

A cooling tower cools condenser water by evaporation i.e., by bringing it in direct contact with an outdoor air stream. Moderate-to-large chillers (larger than 150–200 tons or 525–700 kW) use water as a condenser cooling medium. The lower condensing temperatures achieved with water cause the chiller to operate more efficiently than if an air-cooled condenser were used. A cooling tower in its most basic and simplest form is shown in Figure 17.22. Water exits the spray manifold, falls under the influence of gravity, and entrains atmospheric air with it. While in direct contact with the air, some of the liquid water is evaporated. The heat of evaporation is extracted from the

TABLE 17.4

Cooling Capacities and Water Removal Rates for a 1 m^2 Coil Surface for the DX Coil Specified in Table 17.3

Refrigerant Temperature, °C	Rows of Tubes	Face Velocity = 2 m/s: Total Cooling Capacity, kW	Face Velocity = 2 m/s: Amount of Water Removed, kg/h	Face Velocity = 3.0 m/s: Total Cooling Capacity, kW	Face Velocity = 3.0 m/s: Amount of Water Removed, kg/h
1.7	2	42.7	11.53	52.2	8.85
	3	57.0	23.75	70.4	22.55
	4	68.3	32.56	86.4	36.49
	6	84.0	45.17	110.6	55.68
7.2	2	30.4	3.08	36.4	0.44
	3	40.6	10.27	50.0	7.28
	4	49.0	17.27	62.0	16.40
	6	61.5	27.36	79.5	30.86

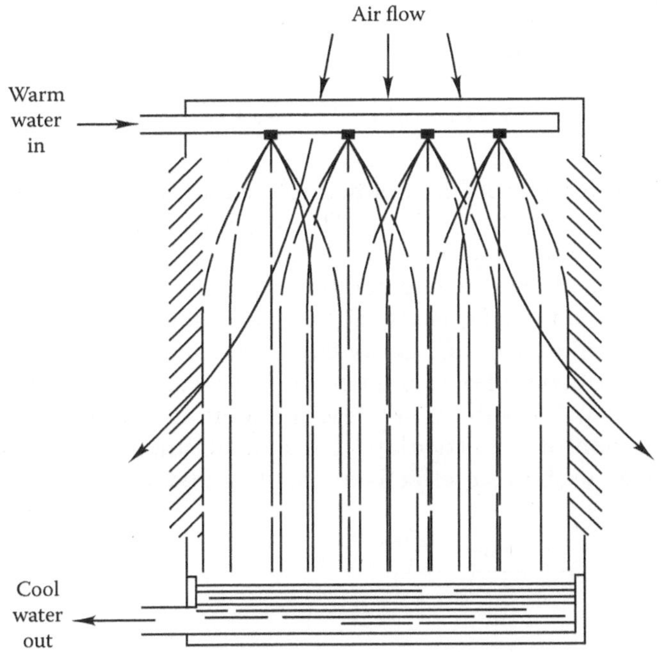

FIGURE 17.22
Basic atmospheric cooling tower.

water stream, thereby cooling it. Cooled water collects at the bottom of the tower, where it is removed and pumped to the chiller for condenser cooling. The tower shown in Figure 17.22 is called an "atmospheric tower" since it uses entrained atmospheric-pressure air for liquid cooling. This simple design is relatively ineffective for large heat rejection rates and requires considerable energy input to create the water spray at the inlet. It is rarely used.

Mechanical draft towers use fans or blowers to cause air to counterflow upward against the down-flowing water droplets. Figure 17.23a shows a *forced-draft* cooling tower. In this equipment, air is forced to flow against the water droplet stream by use of an external centrifugal fan. One problem with this device is the relatively close proximity of the humid air exiting the tower and the high-velocity fan inlet. This can cause short-circuiting, whereby humid air is drawn back through the cooling tower, and reduced performance will result. Cooling towers can be also used for "tower cooling" of buildings in winter to reduce energy consumption (compared to the former practice of operating chillers in winter). However, in subfreezing conditions, the fan near the base of a forced-draft tower can freeze or collect ice on the fan blading, thereby unbalancing it. This tower design is being used less in northern locations for this reason.

Induced-draft towers use a fan to draw air through the tower fill, as shown in Figure 17.23b. This approach solves both the short-circuiting and the freezing problems of the forced-draft tower. The former is avoided since exit velocities are several times higher for this design than for forced-draft towers. The exit air plume will, therefore, be less likely to mix with the air at the fan inlet. Fan freezing is avoided since the fan is located in the warm exit airstream.

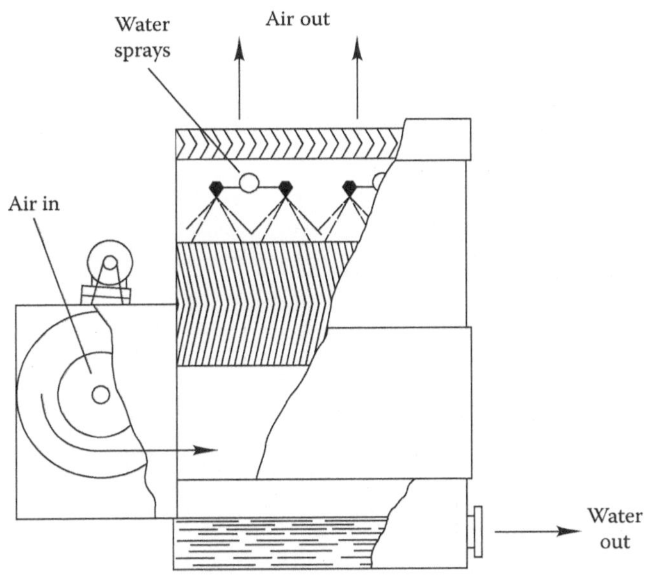

(a)

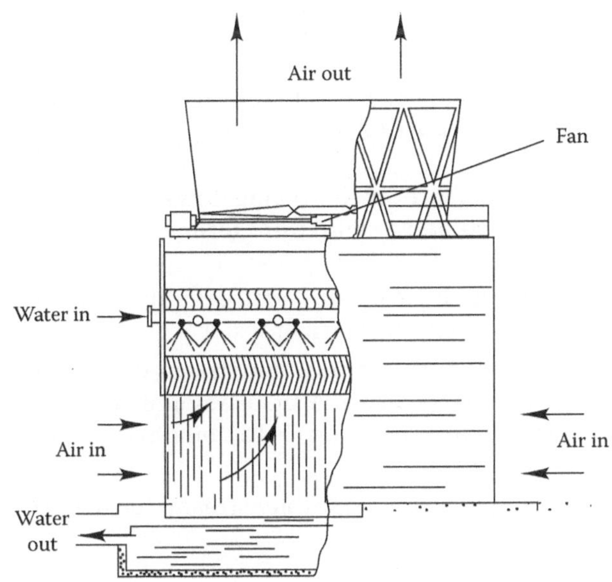

(b)

FIGURE 17.23
(a) Forced-draft counterflow cooling tower. (b) Induced-draft counterflow cooling tower.

Tower design requires specification of several quantities related to the maximum capacity of the tower:

1. *Design entering air condition*: This is usually selected at the 1% wet-bulb design temperature (see Section 9.2) (Grimm and Rosaler, 1990).
2. *Heat rejection rate* is set by peak condenser load of chiller at design cooling load conditions. Tower size increases linearly with load.
3. *Water flow rate* is determined by chiller condenser specifications.

   Manufacturers (Marley, 1983) suggest that water flow not be modulated at part load because proper flow distribution is difficult to maintain and freezing could occur in stagnant flow areas in winter. The fan power in a cooling tower is much larger than the pumping power; therefore, adjustment of the water flow rate does not reduce power consumption much anyway. Typical flow rates are 1.5–3 gal/min/ton (0.025–0.05 L/s/kW of cooling).
4. *Range* is the change in the water temperature across the tower ($T_{w,i} - T_{w,o}$). This is determined from items 2 and 3. Tower size decreases with increasing range subject to constraints on condenser water temperature. The range is usually selected between 10°F and 20°F (6°F–11°C).
5. *Approach* is the difference between the water temperature leaving the cooling tower and the entering air wet-bulb temperature ($T_{w,o} - T_{wb,i}$). Most cooling towers can achieve an approach as low as 7°F–9°F (4°C–5°C) though generally the values are higher. The temperature of water leaving the cooling tower is equal to that entering the condenser. The lower this value, the better will be the COP of the chiller.

An energy balance on a cooling tower involves two principal terms and a small third term:

1. Enthalpy increase in airstream due to heat transfer from the warmer water
2. Enthalpy decrease in water stream due to heat transfer to cooler air. A small amount of water is lost due to evaporation (1–4%), but this mass can be neglected
3. Enthalpy decrease in water stream due to evaporation of water

In equation form, the energy balance must be expressed in differential form since the mass flow rate of *liquid water* changes with the location in the tower. The local energy balance at any point in the tower, as shown in Figure 17.24a, is (ignoring heat transfer through the tower casing and

$$\dot{m}_a \cdot dh_a = -\dot{m}_w \cdot dh_w + \dot{m}_a \cdot dW \cdot h_{liq\text{-}vap} \qquad (17.12)$$

where all terms are defined relative to the differential volume in Figure 17.24a. This expression is solved by numerical integration as shown in ASHRAE Systems (2012) after ignoring the second term on the right-hand side of Equation 17.12. This cumbersome, iterative approach can be avoided by using the method described by Lowe and Christie (1961) and Braun et al. (1989a). We will briefly describe their approach in this section. It is particularly convenient since it uses the heat exchanger effectiveness idea first described in

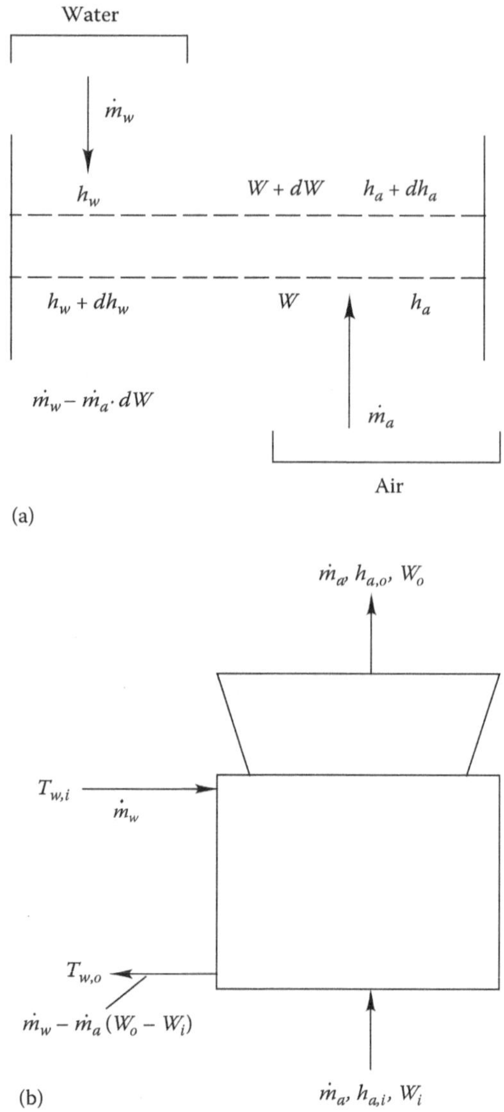

FIGURE 17.24
(a) Cooling tower differential volume energy and mass balance terms. (b) Cooling tower overall mass and energy balance terms.

Section 12.5.3 and avoids the complication of a differential equation.

An overall energy balance on the tower shown in Figure 17.24b is

$$\dot{m}_a(h_{a,o} - h_{a,i}) = \dot{m}_w(h_{w,i} - h_{w,o}) + \dot{m}_a(W_0 - W_i)h_{liq\text{-}vap} \quad (17.13)$$

where

$\dot{m}_a$ is the airflow through tower on dry-air basis, lb_m/h (kg/s)

$h_{a,o}$ and $h_{a,i}$ are the exit and inlet moist air specific enthalpies, Btu/lb_m (kJ/kg)

$\dot{m}_w$ is the tower water mass flow rate, lb_m/h (kg/s)

$h_{w,i}$ and $h_{w,o}$ are the tower water inlet and exit specific enthalpies, Btu/lb_m (kJ/kg)

W_o and W_i are the tower outlet and inlet airstream absolute humidity ratios, lb_w/lb_a (kg_w/kg_a)

$h_{liq\text{-}vap}$ is the specific enthalpy of vaporization, Btu/lb_m (kJ/kg)

According to Section 12.5.3, the effectiveness of a heat exchanger is the ratio of the actual heat transfer rate to the maximum rate permitted by the second law of thermodynamics. In the context of a cooling tower, the maximum heat transfer rate is limited by the enthalpy difference between inlet air and outlet air *if it were saturated at the temperature of the warm inlet water* (this is the upper bound of enthalpy theoretically achievable by the airstream in a tower). The tower heat transfer effectiveness ε_{tower} is then

$$\varepsilon_{tower} = \frac{\dot{Q}}{\dot{m}_a(h_{a,sat,i} - h_{a,i})} \quad (17.14)$$

in which $h_{a,sat,i}$ is the enthalpy of saturated air at the cooling tower inlet water temperature. The tower inlet water is at essentially the same temperature as the chiller condenser outlet water stream.

By analogy with a counterflow heat exchanger, the tower effectiveness is also given by

$$\varepsilon_{tower} = \frac{1 - \exp[-NTU \times (1 - R)]}{1 - R \times \exp[-NTU \times (1 - R)]} \quad (17.15)$$

The *capacity ratio* R is defined as

$$R \equiv \frac{\dot{m}_a c_{p,a,sat}}{\dot{m}_w c_{p,w}} \quad (17.16)$$

where

$c_{p,w}$ is the specific heat of liquid water

$c_{p,a,sat}$ is the effective specific heat of *saturated air*, given by

$$c_{p,a,sat} = \frac{h_{a,sat,i} - h_{a,sat,o}}{T_{w,i} - T_{w,o}} \quad (17.17)$$

where

$h_{a,sat,i}$ is the enthalpy of saturated air at the tower water inlet (warm) temperature

$h_{a,sat,o}$ is the enthalpy of saturated air at the tower water outlet (cool) temperature

The specific heat of saturated air depends on temperature. Hence, it must be evaluated at the known tower operating conditions at the chiller design point. The number of transfer units (NTU) must be calculated from tower manufacturer data. Braun et al. (1989a) found that a simple power law was adequate for this purpose. They suggest

$$\text{NTU} = a\left(\frac{\dot{m}_w}{\dot{m}_a}\right)^m \quad (17.18)$$

in which a and m are dimensionless constants found from curve-fitting catalog data. The value of a is between 1.0 and 3.0 for towers, and m ranges between 0.2 and 0.6.

When the effectiveness has been determined, the tower air exit conditions and heat rate can be found. For example, the exit air enthalpy is

$$h_{a,o} = h_{a,i} + \varepsilon_{tower}(h_{a,sat,i} - h_{a,i}) \quad (17.19)$$

If water evaporative loss is ignored, the tower water outlet temperature is found from

$$T_{w,o} = T_{w,i} - \frac{\dot{m}_a(h_{a,o} - h_{a,i})}{\dot{m}_w c_{p,w}} \quad (17.20)$$

The value of $c_{p,a,sat}$ is a weak function of tower outlet temperature, so one or two iterations of Equations 17.15 through 17.20 may be needed for precise calculations. The basis for the first estimate of $c_{p,a,sat}$ could be the entering air wet-bulb temperature.

The attractiveness of the preceding method is that it is closed form and can be used for a wide range of operating conditions with good accuracy. Such calculations are needed when annual energy consumption for building cooling systems must be found. All that is required for a tower heat rejection rate determination is the air inlet condition (e.g., wet- and dry-bulb temperatures) and the tower water inlet temperature, which is known from the chiller condenser outlet. An alternate closed-form solution to the cooling tower problem is given by Whillier (1967) and is used in an example in Braun et al. (1987).

Example 17.5: Cooling Tower Design and Part-Load Operation

Use the effectiveness method to evaluate a cooling tower at design conditions and to check its operation at off-peak conditions. At peak load, a 525 ton (1845 kW) chiller consumes 0.67 kW/ton (this is equivalent to a COP of 5.25).* Using the NTU relation given here, find the required airflow rate at full load and several additional points of lower heat rejection rate. Additional details of the load, climate, and heat rejection systems design are given as follows.

Given: At design conditions:

$$T_{w,i} = 105°\text{F},\ T_{w,o} = 85°\text{F},\ T_{wb,i} = 78°\text{F},\ T_{db,i} = 91°\text{F}$$

$\text{NTU} = 2 \times (\dot{m}_w / \dot{m}_a)^{0.3}$ (from manufacturer's data for a specific tower, this is not a generally applicable equation; each tower is different)

Figure: See Figure 17.24.

Assumptions: In accordance with good cooling tower design, the water flow rate remains constant to ensure proper water flow distribution. Only airflow is varied to change the tower capacity. The tower is at sea level.

Find: $\dot{m}_a$ at full and part load

Lookup values: $h_{a,i}$ = 41.2 Btu/lb$_m$ (Figure 13.6)
$h_{a,sat,i}$ at 105°F = 81.4 Btu/lb$_m$, $h_{a,sat,o}$ at 85°F = 49.5 Btu/lb$_m$, $c_{p,w}$ = 1.0 Btu/(lb$_m \cdot$°F)

Solution

First, the tower heat rejection is found. It is the sum of the heat removed from the space and the compressor power:

$$\dot{Q}_{tower} = 525 \text{ tons} \times (12{,}000 + 0.67 \times 3413) \text{ Btu}/(\text{h} \cdot \text{ton}) = 7{,}500 \text{ kBtu/h}$$

The tower water flow rate can be found since the tower water range is known (105°F − 85°F = 20°F):

$$\dot{m}_w = \frac{7{,}500{,}000 \text{ Btu/h}}{1 \text{ Btu}/(\text{lb} \cdot °\text{F}) \times 20°\text{F}} = 375{,}000 \text{ lb/h}$$

Equations 17.14 through 17.18 are now used to find the peak-load performance of the tower. Since the airflow rate appears implicitly in these equations, an iterative solution is needed. We set up the equations, substituting all known numerical values, and then we present the results of the iterative solution. The iteration is started with an estimate of the tower effectiveness, which is refined at each iteration until convergence is reached.

The effective specific heat of *saturated air* $c_{p,a,sat}$ is given by Equation 17.17:

$$c_{p,a,sat} = \frac{81.4 - 49.5}{105 - 85} = 1.59 \text{ Btu}/(\text{lb}_\text{m} \cdot °\text{F})$$

The coupled equations governing tower performance are solved in the following order. First, Equation 17.14 is used to find the first estimate of the tower airflow rate. If we use an initial estimate of effectiveness halfway between the thermodynamic limits of 0.0 and 1.0,

$$\varepsilon_{tower} = 0.500 \tag{17.21}$$

Then, the airflow rate by rearranging Equation 17.14 is

$$\dot{m}_a = \frac{7{,}500{,}000}{0.500 \times (81.4 - 41.2)} = 373{,}000 \text{ lb/h}$$

Second, the capacitance ratio R can be found from Equation 17.16:

$$R = \frac{373{,}000 \times 1.59}{375{,}000 \times 1} = 1.58$$

Third, the NTU value is found from Equation 17.18 as follows:

$$\text{NTU} = 2 \times \left(\frac{375{,}000}{373{,}000}\right)^{0.3} = 2.00$$

Finally, the *new value* of the tower effectiveness can be found by inserting these values of R and NTU into Equation 17.15. If it is sufficiently close to that given in Equation 17.21, the iteration is complete. If the difference between the two values is less than the acceptable tolerance, say, ±0.1%, the iteration is at an end. If not, the iteration is repeated, beginning from Equation 17.21 and using the new value of the effectiveness found:

$$\varepsilon_{tower} = \frac{1 - \exp[-2.00 \times (1 - 1.58)]}{1 - 1.58 \times \exp[-2.00 \times (1 - 1.58)]} = 0.542$$

This value of ε_{tower} is too far from the estimate to be considered accurate, so the iteration is repeated. The final results for this problem after several iterations† are

$$NTU = 2.131, \quad R = 1.288,$$

$$\dot{m}_a = 303{,}700 \text{ lb}_\text{m}/\text{h}(38.35 \text{ kg/s}), \text{ and } \varepsilon_{tower} = 0.614$$

* This power input is to the compressor only; it does not include pumps, controls, or the cooling tower fans.

† This *simple iteration* method converges, but it converges slowly. Better methods that converge more quickly include the Newton–Raphson method. Computer software packages, such as Mathematica, exist specifically for this purpose.

TABLE 17.5

Cooling Tower Part-Load Performance for Example 17.5

Airflow, lb_m/h	R	NTU	ε_{tower}	$\dot{Q}_{tower}$, Btu/h	$h_{a,o}$, Btu/lb_m	$T_{w,o}$, °F	$h_{a,sat,o}$, Btu/lb_m	$c_{p,a,sat}$, Btu/($lb_m \cdot$°F)
350,000	1.48	2.04	0.56	7,944,742	63.90	83.81	47.91	1.59
303,700[a]	1.29	2.13	0.61	7,500,000	65.90	85.00	49.30	1.61
300,000	1.28	2.14	0.62	7,426,315	65.95	85.20	49.53	1.61
250,000	1.07	2.26	0.68	6,790,050	68.36	86.89	51.63	1.64
200,000	0.88	2.42	0.74	5,931,386	70.86	89.18	54.66	1.69
150,000	0.68	2.63	0.81	4,860,225	73.60	92.04	58.74	1.75
100,000	0.47	2.97	0.88	3,534,868	76.55	95.57	64.26	1.82
50,000	0.24	3.66	0.95	1,913,422	79.47	99.90	71.73	1.90

[a] Design point, i.e., heat rejection is 7,500,000 Btu/h.

The fan airflow rate is the key result of the calculation. The designer will specify this along with the cooling tower heat rate, range, and approach.

At part load, the tower performance can be found as the aforementioned by using a new value of the airflow rate and finding the corresponding heat rejection rate. At reduced heat rejection rates at part load, the tower exit water temperature is higher than the design point of 85°F; this has a small effect on $c_{p,a,sat}$. Table 17.5 summarizes the performance of this tower at flow rates ranging from slightly above the design flow rate to 50,000 lb_m/h, representing about 15% of full load.

Comments

The effectiveness increases as the airflow decreases at part load because the residence time of air within the tower is longer. However, the heat rate decreases with airflow since Equation 17.16 shows that the heat rate decreases linearly with decreasing flow, whereas the effectiveness increases with decreasing flow but at a less than linear rate. Problem 17.5 requires the reader to calculate the entries in Table 17.5.

The cooling tower capacity is controlled by adjusting the fan speed or less efficiently by bypassing either air or water around the tower at part load to control the water outlet temperature. Internal tower water flow itself should not be modulated to vary capacity since distribution within the cells and the fill of the tower depend on proper water flow rates. In multiple-cell cooling towers with variable-fan-speed capability, the following control strategy should be used: Operate as many cells as needed to meet heat rejection requirements with all fans on the lowest-speed setting. When capacity is added, increment the lowest-speed (with off [i.e., the natural-draft mode] being the "lowest" of the low-speed settings) fan first (Braun et al., 1989b). All fans should be on at low speed before any fan is operated at a higher speed as the load is increased. Equal water flow distribution among tower cells leads to lowest energy consumption.

Cooling towers can be used in the absence of any mechanical cooling operation to cool buildings in swing seasons and in winter (Murphy, 1991). They are commonly referred to as water economizers (in contrast to air-side economizers discussed in Section 19.6.2). Figure 17.25 shows a simple method of implementing *tower cooling*. When weather conditions are appropriate, the cooling tower operates with the chiller off and provides cold water to the chilled water distribution system by bypassing the chiller, as shown by the solid lines in the figure (dashed lines show the usual mechanical cooling configuration). A serious problem that must be addressed with this approach is the mixing

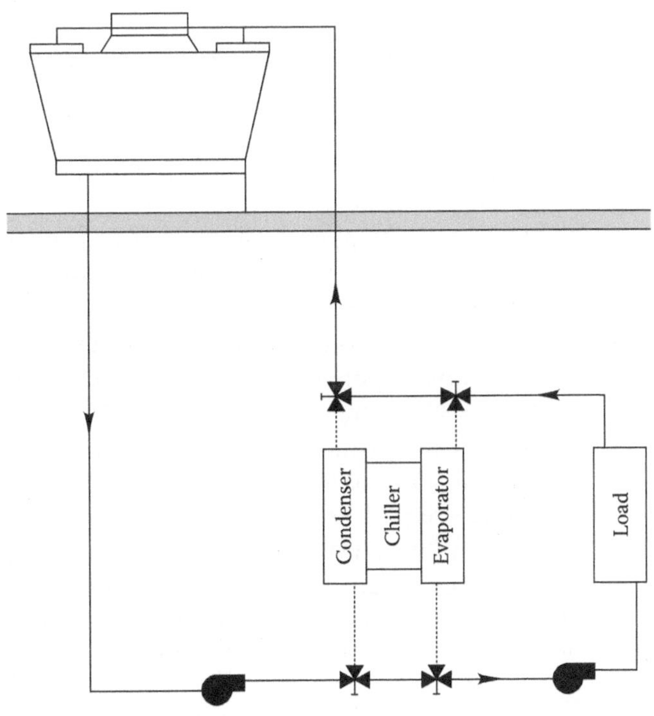

FIGURE 17.25
Schematic diagram of tower cooling system piped in parallel with mechanical chiller.

of cooling tower water with treated chilled water loop fluid. Filtration of the former can help alleviate contamination and heat exchanger surface fouling in the cooling coils. Alternatively and much more commonly, a heat exchanger can be placed between the tower loop and the chiller water loop.* A very effective heat exchanger such as a plate heat exchanger can transfer the cooling effect from the tower loop to the chilled water loop with little temperature penalty. This system is the *indirect* tower cooling design.

Episodes of *Legionella pneumophila* (Legionnaires' disease) have been associated with cooling towers on buildings. Prudent preventive maintenance to avoid this disease includes keeping cooling tower sumps clean, using chemical inhibitors, circulating water daily (even if tower is not required for cooling purposes), and avoiding contamination of cooling water from other streams in a building. The interested reader can refer to Barbaree (1991) for more discussion.

Cooling tower maintenance is very important. Formerly, various organic and inorganic compounds using heavy metals were used, and they performed very well. It is not possible to use these materials any more, but ozone produced on site and new organic materials can help control pH, scaling, corrosion, and biological growth.

Problems

Numbers 1–4 given in parenthesis denote the degree of difficulty.

17.1 (a) The R22 compressor in Problem 14.6 has a polytropic exponent of 1.35. What is the compressor efficiency?

(b) What is the diameter of the centrifugal chiller compressor rotor for the R22 chiller? Assume a rotor speed of 1800 r/min. *(2)*

17.2 (a) The R134a compressor in Problem 14.8 has a polytropic exponent of 1.32. What is the compressor efficiency?

(b) What is the diameter of the centrifugal chiller compressor rotor for the R134a chiller? Assume a rotor speed of 3600 r/min. *(2)*

17.3 A 300 ton (1050 kW) chiller is operated with 4 gal/min of condensing water per ton (0.072 L/s per kW) of refrigeration. Under design conditions of 95°F dry-bulb and 75°F wet-bulb (35°C and 24°C) temperatures, the cooling tower is designed to cool condensing water from 100°F to 85°F (38°C to 29°C). The air under these conditions leaves the tower at 92°F (33°C) and 90% RH. What is the required air mass flow rate? How much water is consumed? What are the tower range and the inlet air to water temperature difference (called the *approach*)? *(2)*

17.4 A 1000 ton (3400 kW) chiller is operated with 3.5 gal/min of condensing water per ton (0.063 L/s per kW) of refrigeration. Under design conditions of 93°F dry-bulb and 71°F wet-bulb (34°C and 21°C) temperatures, the cooling tower is designed to cool condensing water from 100°F to 83°F (38°C to 28°C). The air under these conditions leaves the tower at 95°F (35°C) and 85% RH. What is the required air mass flow rate? How much water is consumed? *(2)*

17.5 Construct the part-load curve for the cooling tower in Example 17.5 for a range of airflow rates between 50,000 and 350,000 lb_m/h (6 and 45 kg/s). Plot the heat rejection rate and effectiveness versus air mass flow, assuming that the water flow rate remains constant. Check your solution with Table 17.5.† *(4)*

17.6 Rework Example 17.5 for a tower range of 16°F (9°C) instead of the value used in the example. The tower heat load and water flow rate remain the same. *(4)*

17.7 Suppose that a more effective cooling tower is used instead of the tower specified in Example 17.5. Specifically, the tower constant is 2.7 instead of 2.0. How does this affect the needed airflow for the same heat rejection rate? *(2)*

17.8 What is the effectiveness of a cooling tower that rejects 600,000 Btu/h (180 kW) under design condition of 105°F (41°C) entering water temperature with 92°F (33°C) and 40% RH entering air if the airflow rate is 300 lb/min (2.2 kg/s)? *(2)*

17.9 What is the effectiveness of a cooling tower that rejects 150 kW under design conditions of 39°C entering water temperature with 32°C and 28% RH entering air if the airflow rate is 2.2 kg/s? *(2)*

17.10 The condensing temperature of the chiller in Problem 14.10 is reduced by 12°F (6.6°C) due to the use of an evaporative condenser. What is the COP of the chiller with the addition of the evaporative feature? Is the change significant? The refrigerant mass flow rate is the same as for Problem 14.10. *(3)*

* Since the quality of water passing through the tower side of the heat exchanger is not tightly controlled, heat exchanger fouling may occur. The HVAC designer must leave sufficient space to disassemble and clean this heat exchanger periodically.

† Problems 17.5 and 17.6 require the enthalpy of saturated air as a function of temperature. In IP units, an equation relating the two is $h_{a,sat} = 98.92 - 2.362T + 0.02092T^2$, where the temperature is in degrees Fahrenheit and enthalpy is in Btu/lb. Use this equation only within the temperature range of these problems.

17.11 An HVAC engineer selects the design condition of return air from building zones during the cooling season to be 75°F (24°C) and 50% RH. The coil load calculation reveals that the sensible heat ratio is 0.6. Comment on this design in the context of achievable apparatus dew point (ADP). *(2)*

17.12 An HVAC engineer selects the design condition of return air from building zones during the cooling season to be 75°F (24°C) and 50% RH. The coil load calculation reveals that the sensible heat ratio is 0.7. Comment on this design in the context of achievable ADP. *(2)*

17.13 In Chapter 13, we observed that certain latent plus sensible air cooling processes that can be drawn on a psychrometric chart cannot actually be achieved because no *ADP* exists. A coil entering air condition that illustrates this is 70°F (21°C) and 50% RH with SHR = 0.5 (relatively high latent load). If you were faced with such a return air condition, describe how you could make system modifications that would allow an ADP to be achieved and the desired coil outlet condition to be reached. *(2)*

References

ASHRAE Fundamentals (2013). *Handbook of Fundamentals.* American Society of Heating, Refrigerating and Air-Conditioning Engineers, Atlanta, GA.

ASHRAE Systems (2012). *Handbook of Systems and Equipment.* American Society of Heating, Refrigerating and Air-Conditioning Engineers, Atlanta, GA.

Barbaree, J. (1991). Controlling Legionella in cooling towers. *ASHRAE J.*, 33, 38–42.

Bowman, J. and R. Turton (1990). Quick design and evaluation: Heat exchangers. *Chem. Eng.*, 97(July), 92–99.

Braun, J.E., S.A. Klein, and J.W. Mitchell (1989a). Effectiveness models for cooling towers and cooling coils. *ASHRAE Trans.*, 95(2), 164–174.

Braun, J.E., J.W. Mitchell, S.A. Klein, and W.A. Beckman (1987). Performance and control characteristics of a large cooling system. *ASHRAE Trans.*, 93(1), 1830–1852.

Braun, J.E., J.W. Mitchell, S.A. Klein, and W.A. Beckman (1989b). Applications of optimal control to chilled water systems without storage. *ASHRAE Trans.*, 95(1), 663–675.

Cengel, Y.A. (2002). *Heat Transfer,* 2nd ed. McGraw-Hill, New York.

Dossat, R.J. (1997). *Principles of Refrigeration.* Prentice-Hall, Upper Saddle River, NJ.

Grimm, N.R. and R.C. Rosaler (1990). *Handbook of HVAC Design.* McGraw-Hill, New York.

Kuehn, T.H., J.L. Threlkeld, and J.W. Ramsey (1998). *Thermal Environmental Engineering,* 3rd ed. Prentice-Hall, Upper Saddle River, NJ.

Lowe, H.J. and D.G. Christie (1961). Heat transfer and pressure drop data on cooling tower packings and model studies of the resistance of natural draft towers to airflow. *ASME Heat Transfer Proc.*, 113, 933–950.

Marley (1983). *Cooling Tower Fundamentals.* Marley Cooling Tower Co., Mission, KS.

Mitchell, J.W. and J.E. Braun (2013). *Principles of Heating, Ventilation and Air-Conditioning in Buildings.* John Wiley & Sons, New York.

Murphy, D. (1991). Cooling towers used for free cooling. *ASHRAE J.*, 33(6), 16–22.

Stoecker, W.F. and J.W. Jones (1982). *Refrigeration and Air Conditioning.* McGraw-Hill, New York.

Turton, R., D. Ferguson, and O. Levenspiel (1986). Charts for the performance and design of heat exchangers. *Chem. Eng.*, 93(August), 81–88.

Yokell, S. (1990). A Working Guide to Shell and Tube Heat Exchangers, MC Graw-Hill, New York.

Whillier, A. (1967). A fresh look at the calculation of performance of cooling towers. *ASHRAE Trans.*, 82(1), 269–282.

18

Hydronic Distribution Equipment and Systems

ABSTRACT This chapter starts by classifying hydronic systems and then describing various common piping layouts (series, one-pipe, two-pipe, and four-pipe layouts). Next, we address different types of the traditional terminal units, namely, convectors, cabinet heaters, and fan-coil units, which transfer heat from the circulating fluid to the room air primarily by convection. This is followed by a description of different types of low-temperature radiant panels for heating and cooling (ceiling, floor, and wall mounted) where the primary mode of heat transfer is by radiation. Heat analysis equations along with guidelines for good design practice are provided with solved examples. We then discuss auxiliary equipment such as service hot water and steam equipment along with expansion tanks that are a critical component in closed systems. We present the basic functioning of two-way and three-way valves to modulate flow through coils and other heat transfer equipment. Subsequently, piping networks meant for multiple-chiller plants providing cooling to several buildings in a campus are described along with ways to control them under variable load conditions. Finally, we provide a brief overview of the benefits, types, design approaches, and technologies relevant to both ice and chilled water cool storage systems.

Nomenclature

A	Area, m² (ft²)
AUST	Area-weighted average temperature of uncontrolled indoor surfaces, °C (°F)
CC	Cooling coils
c_p	Specific heat at constant pressure, kJ/(kg·K) [Btu/(lb_m·°F)]
CTES	Cool thermal energy storage
F	View factor
f_{useful}	Usable storage capacity of hot water storage tank
g	Acceleration due to gravity = 9.81 m/s² (32 ft/s²)
h	Head referring to pressure, m or cm (ft)
L	Length, m (ft)
$\dot{m}$	Mass flow rate kg/s (lb_m/h)
p	Pressure, Pa (lb_f/in.²)
p_a	Atmospheric pressure, Pa (lb_f/in.²)
$\dot{Q}$	Heat flow with subscripts c for cooling load, *h* for heating load, W (Btu/h)
$\dot{q}$	Heat flux, W/m² [Btu/(h·ft²)]
T	Temperature or dry-bulb temperature, K or °C (°R or °F)
V	Volume, m³ or L (ft³)
$\dot{V}$	Flow rate, m³/s or L/s (ft³/min)
v	Specific volume, m³/kg (ft³/lb_m)

Greek

α	Linear coefficient of expansion, °C⁻¹ (°F⁻¹)
ε	Emissivity, effectiveness
ρ	Density, kg/m³ (lb_m/ft³)
ϕ	Relative humidity
σ	Stefan–Boltzmann constant, W/(m²·K⁴) [Btu/(h·ft²·°R⁴)]
Δh	Loss of head or pressure drop, m (ft)
ΔT	Temperature difference, °C (°F)
Δt	Time duration, h or s

Subscripts

av	Average
c	Cooling
ch	Chiller
con	Convection
dis	Discharge
db	Dry-bulb
dp	Dew point
eff	Effective
exp	Expansion tank
h	Heating
in	Inlet
max	Maximum
out	Outlet
rad	Radiation
r	Recovery
rp	Radiant panel
set	Set point

sys System
tank Storage tank
tot Total
tu Terminal or hydronic unit
uhs Unheated surface
v Valve
w Winter

18.1 Hydronic System Classification

In Chapter 11 (Section 11.2), the three generic types of secondary systems were described. Hydronic (also called all-water) systems were historically the first type to evolve, followed by all-air systems (discussed in Chapter 19), and then by air–water systems (discussed in Chapter 20). In recent years, new secondary systems designs have been developed, which modify air–water systems into hybrid systems so as to improve energy efficiency and enhance indoor air quality (also addressed in Chapter 20). Hydronic systems are those wherein the space cooling (or heating) function is performed by chilled water (or hot water) circulated from a central cooling plant (or boiler plant) to heat exchangers or terminal units located in, or immediately adjacent to, the conditioned space. Heat is finally transferred to the room air either by natural convection or forced convection via fans or blowers. Water is an excellent medium to transport heat. It has high specific heat, which is nontoxic, inexpensive, and readily available. It is an energy-efficient transport medium for heat at varying temperature levels by altering the pressure as well as for *coolth* (with addition of antifreeze such as glycol if temperatures lower than freezing are required).

In all newer buildings, the historic gravity-fed systems have been replaced with forced circulation systems using pumps. Hydronic or hot water heating systems can be either recirculating (or closed loop) or once through (open loop). The latter systems will typically have inlet suction pressures close to the atmospheric pressure, while the discharge pressure is typically higher depending on the application. Typical examples are cooling towers, air washers, and spay humidifiers. Some amount of water will be lost during operation, thereby requiring makeup water supply. Closed-loop systems are normally operated at pressures greater than atmospheric, and makeup water is needed only periodically. Only when chemical levels exceed certain thresholds is there need to flush out the water and replace it with fresh water. Most of the terminal units described in Section 18.3 and the loop that supplies water to cooling and heating coils of all-air systems have to be configured as closed-loop systems. Expansion tanks are critical to compensate for changes in volume due to density variations related to temperature of the circulating water, and these are described in Section 18.5.3.

Hydronic systems are usually classified by temperature (refer to, for example, Mull, 2001):

- *Low-temperature hot water systems* are the most widely used for residential and smaller commercial and institutional buildings whose loads are less than about 1.5 MW (5×10^6 Btu/h). These systems are generally used for space heating loads and domestic hot water. The maximum temperature is less than 120°C (250°F), while the maximum pressure is not to exceed 1100 kPa (160 psia); more generally, the working pressure is in the 200 kPa (30 psia) range. Steam-to-water or water-to-water heat exchangers are often used. The system temperature drop (supply to return) is in the range of 6°C–24°C (10°F–40°F).
- *Medium-temperature hot water systems* are commonly used for space heating in commercial and institutional buildings as well as in industrial applications with process loads. The loads range from 1.45 to 1.75 MW ($5–6 \times 10^6$ Btu/h). The design supply temperatures are in the range of 120°C–175°C (250°F–350°F), with pressure ratings for boilers and piping of about 1030 kPa (150 psia).
- *High-temperature hot water systems* are generally limited to campus-type district heating applications with supply temperatures in the range of 175°C–230°C (350°F–400°F). The system temperature drop can be up to 55°C (100°F). The pressure rating is about 300–350 psia.

Obviously, different categories require different boilers and associated equipment. Increasing the supply water temperature will require heavier and thicker equipment and pipe walls. It also increases the danger of burns and scalding. Present hot water systems design practice for residences is to assume a system temperature drop of 10°C (20°F). This corresponds to a heat carrying capacity of about 45 kW/L flow (10,000 Btu/h per gpm flow). Chilled water supply temperatures are in the range of 4°C–10°C (40°F–50°F) with system temperature drops of 2.5°C–8°C (5°F–°15°F).

Hydronic systems are well suited for perimeter spaces with seasonal cooling and heating requirements. Because of their limited space requirements, all-water systems are well suited for retrofit applications. They take up little to no space in the central machine room and do not need ducts. They can provide individual room control and can be coupled with heat recovery and solar heating systems. All-water fan-coil systems are common in churches and other monumental or historic buildings. However, they are not well suited for interior spaces and for spaces requiring close control of humidity and requiring proper ventilation air. Their maintenance is relatively high and repair has to be performed in occupied spaces.

18.2 Types of Hydronic Distribution Circuits

The simplest and lowest cost hydronic system is one where all the terminal units are connected in series (Figure 18.1). A designer may consider such a system for applications such as residences and small commercial/institutional buildings. The disadvantages of a *series hydronic system* are obvious: (1) In case one unit needs maintenance, the entire system has to be shut down, (2) flow rate to each unit has to be the same, and (3) the water temperature gradually decreases such that downstream terminal units are exposed to lower temperature (during heating) requiring that such units be generally larger (if heating loads in each space are close to each other).

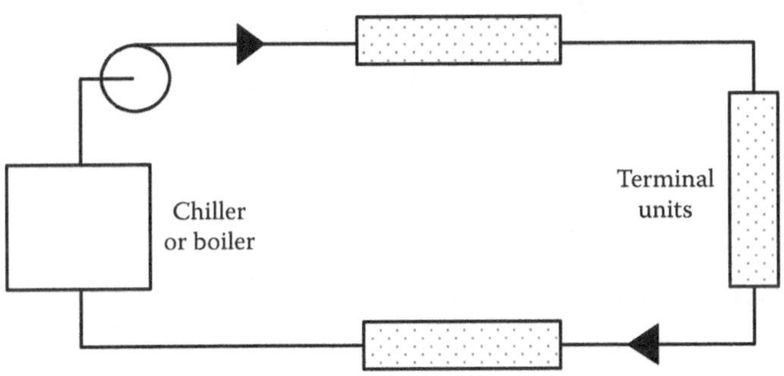

FIGURE 18.1
Sketch of a series-loop hydronic system.

Another hydronic system layout, also meant for small buildings because of its low cost, is the *one-pipe monoflow system* (Figure 18.2). Here, maintenance to individual units does not require shutting down the entire system, and further, different flow rates to terminal units can be specified. This system, however, requires the use of special flow diverting tee fittings (as shown in Figure 18.2b). These not only increase first cost, but the pressure drops across such fittings result in higher pumping power of the circulating pump. This piping layout still suffers from the drawback that the supply water temperature to the terminal heaters during heating (and vice versa for cooling) gradually decreases depending on their downstream location.

A further improvement, and a system layout in wide use, is the *two-pipe system* (Figure 18.3). All the previously stated drawbacks are now overcome at the expense of a slight increase in cost. This layout offers great flexibility to the designer in terms of being able to specify different flow rates through different terminal units in case of different heating (or cooling) loads in different rooms/spaces in the building. Two designs are the most common: the *direct-return* or the "out-and-back" layout (Figure 18.3) and the *reverse-return* layout (Figure 18.4). The latter is generally superior from a hydraulic viewpoint since the pipe runs for water flow through each unit are much closer, and so the system will tend to be self-balancing. The drawback of reverse return is that the piping length needed is usually longer than the direct return, thereby increasing first cost of the system. As a general guideline, an imbalance of 10–15 ft of head (0.3–0.45 m) is deemed acceptable (Lorsch, 1993).

All the aforementioned configurations allow either hot water or cold water to be circulated at any time, but not simultaneously. In case the building needs heating in certain spaces and cooling in others (as is the norm in larger commercial/institutional buildings), the two-pipe

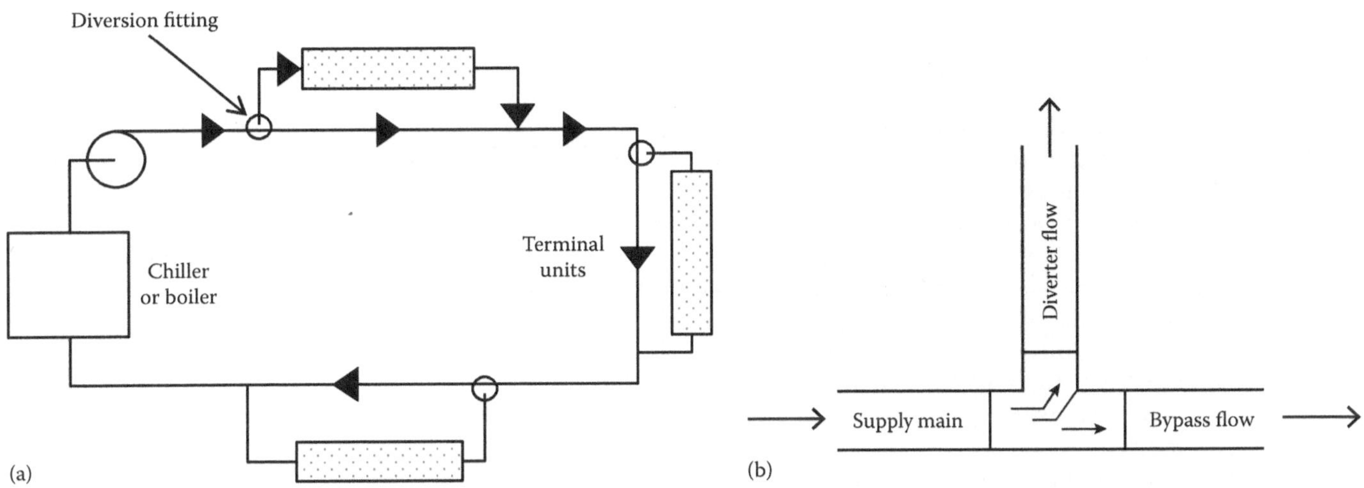

FIGURE 18.2
(a) Sketch of a one-pipe monoflow system. (b) Flow diverting tee fitting.

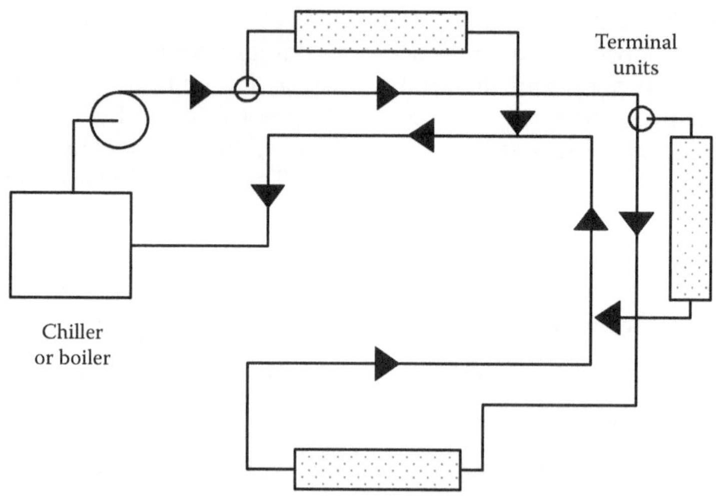

FIGURE 18.3
Direct-return two-pipe hydronic system.

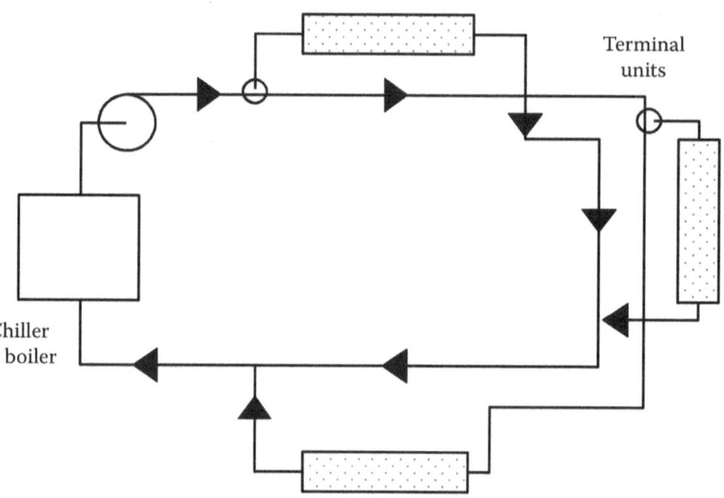

FIGURE 18.4
Reverse-return two-pipe hydronic system.

system cannot be used. In such cases where simultaneous heating and cooling loads have to be met, the *four-pipe system* (Figure 18.5) is clearly superior (either direct or reverse return). Despite its higher first costs, this piping configuration is widely used in most modern buildings. Three-pipe systems that can provide simultaneous heating and cooling have also been developed, but most building codes have banned this configuration. Valves start to leak over time, thereby leading to unneeded mixing of hot and cold water streams resulting in a large energy penalty.

In large and very tall buildings, a common hydronic configuration is to adopt combination layouts. This reduces piping length and greatly reduces initial and

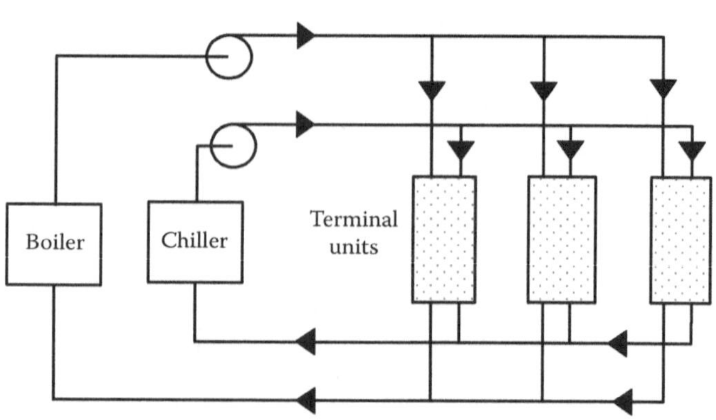

FIGURE 18.5
Sketch of a four-pipe direct-return hydronic system.

operating costs and helps in maintenance. Essentially, one can use multiple monoflow two-pipe systems for different floors of a building.

18.3 Traditional Terminal Units

Terminal units are essentially heat transfer devices that allow the transfer of cooling (or heating) from the circulating water to the space or room air. This heat transfer is almost entirely achieved by convection, with radiation being a very small component in certain types of designs. Terminal units are often located along exposed walls and under windows (if possible) to prevent cold downdrafts. The design of these devices differs depending on whether cooling or heating is to be achieved and also on the magnitude of heat flux to be transferred. Typically, residential applications require lower heat rates to be transferred as against commercial or nonresidential applications. Some of the principal types of terminal devices used traditionally are presented in this section, while some more recent designs are discussed in the next section. Refer to publications such as ASHRAE Systems (2012), Pita (2002), Mull (2001), or Bobenhausen (1994) for more detailed treatment.

18.3.1 For Heating Applications

The most prevalent type of heater is the *baseboard or fin-tube units* (Figure 18.6). They are used in both residential and commercial spaces (requiring relatively high heat inputs), differing not in their design but in their size and construction. As shown in the figure, heat is delivered by hot water flowing through a copper or steel piping (1/2–3/4 in. or 1.5–2.2 cm for residences and 3/4–2 in. or 2.2–5 cm for commercial units) that contains closely spaced fins to enhance heat transfer. These fins are made of thin gauge aluminum or steel plates. Typically, terminal units can range from 2–3 to 25–30 ft in length (0.6–0.9 to 7.5–9 m) and contain 25–60 fins/ft length or 80–200 fins/m). This arrangement is enclosed in metal grills to (1) prevent scalding or burn by accidental contact and (2) control the direction of heated air outflow. Units in commercial spaces can have more than one row of tubes. They sometimes come with damper covers that can be used to vary the amount of air induced by natural convection into the unit and thereby control the amount of heating to be released to the space. The terminology "baseboard" stems from the fact that they are normally located on the floor and against the base of the wall. Rating of these units is discussed in Section 18.3.3.

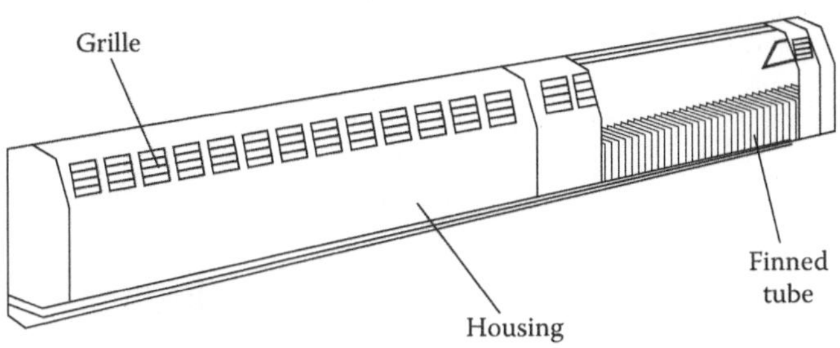

FIGURE 18.6
Baseboard convector. (Courtesy of Sterling Radiator, Westfield, MA.)

Some of the others types of heating terminal units are described here:

1. *Cast-iron radiators* (Figure 11.12) are historically old designs, made of hollow sections of cast iron or steel, which evolved from devices that used steam for heating rather than hot water as the working fluid. Except for replacement parts in older installations, they have been largely replaced by convector radiators.
2. *Convectors* (Figure 18.7) are the modern-day version of cast-iron radiators. Space air is induced through the inlet grille by buoyancy, heats up as it flows through the heating element (or finned tube), and flows out through the outer grille.

FIGURE 18.7
Photograph of a wall-mounted convector.

The entire assembly is enclosed by a front cover, which can be opened during maintenance. Such units can also be tiered by more than one row of hot water finned pipes that can be wall hanging or floor mounted either recessed or free standing.

3. *Cabinet unit heaters* (Figure 18.8) are those that enhance heat transfer rates by using a fan to force air through the heating element (which improves the heat transfer coefficient). As shown in the figures, they can be floor mounted or wall mounted and essentially contain the same components. Unlike the terminal units relying on natural convection, these units have an air filter (the fan can overcome this pressure drop), which improves the room air quality. These units are often used in vestibules.
4. *Space heaters* are very similar to cabinet unit heaters but are used in industrial and commercial spaces, such as in semioccupied spaces (warehouses and parking garages) and large box retail buildings. In such applications, noise and lack of elegance can be tolerated if this results in higher energy efficiency and lower cost. These units mostly use a powerful propeller fan along with adjustable louvers to control air direction. Some space heater models are also designed for operation with steam.

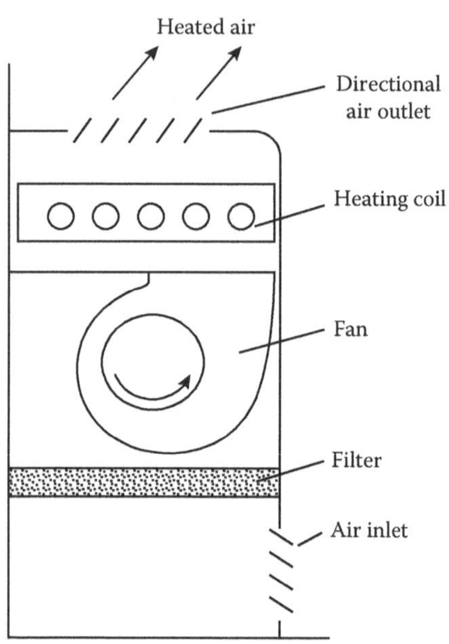

FIGURE 18.8
Floor-mounted forced airflow cabinet unit heater.

18.3.2 For Heating/Cooling Applications

Fan-coil units are large capacity terminal devices that can provide heating (such as cabinet unit heaters) as well as cooling to the space. As shown in Figure 18.9, this is simply achieved by circulating either hot or cold water through the coil, whose heat or coolth is transferred to air that is forced by the fan to flow through the finned coils prior to discharge to the room. Practical HVAC systems designs favor the use of two separate sets of coils, one for heating and one for cooling. At any given time, only one set of coils is active, but separating the function allows for coil designs better suited for either heating or cooling. These units can be mounted in various horizontal and vertical arrangements. It is essential to provide an insulated drain pan beneath the coil to collect and drain out the water condensed from the air.

While an outside air vent (or damper) is shown in the figure, this is optional and, even when provided, may not guarantee the required ventilation air at all times without the danger of overventilation, thereby reducing energy efficiency.

The fan operation results in some amount of noise that can be muted but not eliminated. This may be an issue in certain applications such as classrooms or lecture halls. Models are commercially available where the fan is eliminated and airflow is enhanced by induction. Certain amount of airflow (called primary air) is sent via a separate duct to the terminal unit that is made to flow through small nozzles, thereby creating a low-pressure zone in the plenum chamber. This depressurization then *induces* room air to flow through the heating/cooling coils, a phenomenon akin to that produced by a fan. However, the induced depressurization is rather

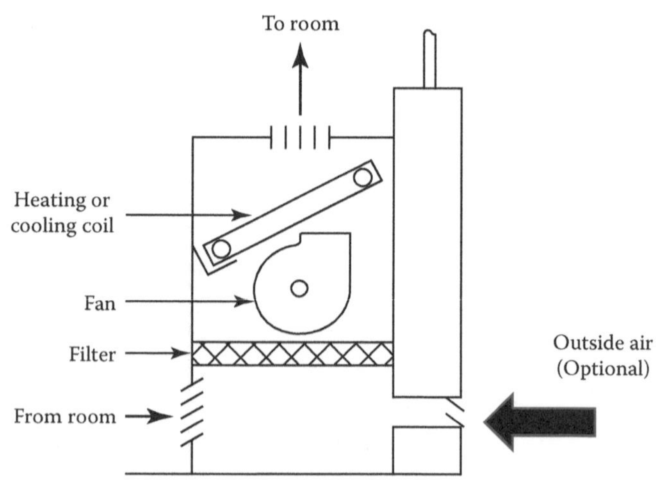

FIGURE 18.9
Fan-coil unit.

low, and therefore, the resulting airflow is much lower than what could be achieved with a fan.

A major advantage of these units is that a thermostat can be used to control water flow rate, thereby achieving a preset temperature set point within the space. These units are widely used in VAV systems and are discussed at more length in Section 19.3.

18.3.3 Terminal Unit Manufacturer Ratings

Manufacturers publish equipment catalogs such as that shown in Table 18.1 based on standard testing procedures established by various authorities such as the Hydronic Institute. Factors that affect capacity are (1) tube size, (2) water flow rate, (3) average water temperature in the radiator, and (4) room air temperature, generally taken to be 65°F (18.3°C).

Recall the sensible heat balance equation given by

$$\dot{Q}_h = \rho_w \cdot \dot{V}_w \cdot c_{p,w} \cdot \Delta T \qquad (18.1)$$

where

ρ_w is the density of water
$\dot{V}_w$ is the volumetric water flow rate
$c_{p,w}$ is the specific heat of water at constant pressure
ΔT is the water temperature drop across the terminal unit

In IP units and assuming typical values of density of 8.33 lb_m/gal and specific heat of 1.0 Btu/(lb_m °F), a simplified sensible heat equation for water is often used:

$$\dot{Q}_h = 500 \times \dot{V}_w \times \Delta T \qquad (18.2\ \text{IP})$$

where the heat rate is in Btu/h, the temperature difference is in °F, and $\dot{V}_w$ is in gal/min.

In SI units, the expression simplifies further since the density and specific heat values are close to unity:

$$\dot{Q}_h = \dot{V}_w \times \Delta T \qquad (18.2\ \text{SI})$$

with the heat rate in kW, the volume flow rate in L/s, and the temperature difference in °C.

The sensible heat equation directs provides an estimate of the *heat carrying capacity* of the hydronic system.

It would be instructive to compare water and air as heat transport mediums. Consider a system where 1 ton of cooling (12,000 Btu/h or 3.5 kW) is to be transported. If we assume a 20°F (11.1°C) difference between room supply and room thermostat temperatures, then the all-air system would require about 545 cfm (257 L/s) of air. On the other hand, for an all-water system with a typical 5°F (2.8°C) temperature difference, Equation 18.2 would predict a water flow rate of about 4.8 gpm (0.3 L/s) or about 0.6 cfm. Next, assuming typical pressure drop of 3 inWG (7.6 cm-WG) for an all-air system with a 0.7 efficiency fan, the required fan power is close to 0.37 HP (0.28 kW). Contrast this to a water pump with an efficiency of 0.7 and having to overcome a head loss of 75 ft (23 m); the power required turns out to be about 0.13 HP (0.1 kW). This is a factor of 3.8, or close to a fourfold reduction! Hence, moving heat or coolth around in pipes requires much less pumping power and also needs much less physical space. However, air systems are typically less expensive and can provide ventilation air along with free cooling during certain times of the year (see Section 19.6.2).

Example 18.1: Sizing Terminal Units

The design heating load of a space is 37,500 Btu/h (11 kW). A low-temperature hydronic system is to be designed with a maximum temperature supply of 220°F (100°C) and a temperature drop not to exceed 25°F (13.9°C). Find the water flow rate

TABLE 18.1

Sample of a Manufacturer Rating Catalog for Typical Baseboard Radiator with Copper Tubing and Aluminum Fins Assuming Air Entering at 65°F

Nominal Tube Size, in.	Flow Rate, gpm	Velocity ft/s	Hot Water Ratings, Btu/h/ft Length at Following Average Water Temperatures, °F: 170	180	190	200	210	220	230	240
	1	0.6	510	580	640	710	770	840	910	970
	2	1.2	520	590	650	730	790	860	930	990
3/4	3	1.8	530	600	670	740	800	870	950	1010
	4	2.4	540	610	680	750	810	890	960	1030
	1	1.2	550	620	680	750	820	880	950	1020
	2	2.4	560	630	700	770	840	900	970	1040
1/2	3	3.6	570	640	710	780	850	920	990	1060
	4	4.8	580	660	720	790	870	930	1000	1080

and the length of the terminal unit specified by Table 18.1.

Given: $\dot{Q}_h = 37{,}500$ Btu/h, $\Delta T_{tu} = 25°F$

Assumptions: Steady flow is assumed.

Find: $\dot{V}_w, L_{tu}$

Solution

First, we calculate the flow rate. Rearranging Equation 18.2 yields

$$\dot{V}_w = \frac{\dot{Q}_h}{500 \times \Delta T} = \frac{37{,}500 \text{ Btu/h}}{500 \times 25°\text{F}} = 3 \text{ gal/min}$$

Inspection of Table 18.1 reveals that one could consider either the ¾ in. pipe or the ½ in. pipe. This is where recommendations based on heuristic good design practice have to be followed. ASHRAE Systems (2012) recommends that water velocities in pipes not exceed 4 ft/s for pipes 2 in. or smaller to avoid noise. If we follow this recommendation and keep in view that smaller pipes are less expensive, we would select the ½ in. pipe that has a water velocity of 3.6 ft/s.

The mean water temperature in the terminal unit is, to a close approximation, simply the arithmetic average of the inlet and outlet water temperatures, i.e., $T_{av} = T_{in} - \frac{\Delta T}{2} = 220 - \frac{25}{2} = 207.5°F$. From Table 18.1, the heat output per unit length under these conditions is found by linear interpolation to be 832.5 Btu/ft. Hence, the length of the terminal unit needed is L_{tu} = (37,500/832.5) = 45 ft (13.5 m).

Comments

It is unlikely that one would select one single baseboard heater this long. Rather, this length will be distributed along the perimeter of the space or room to enhance more uniform room heating and hence provide better occupant comfort. Ideally, the baseboard units will be placed under windows that are sources of large heat loss, and each of the units will be analyzed and sized as individual units following the procedure illustrated earlier.

18.4 Low-Temperature Radiant Panels

The distinction between terminal units discussed earlier and radiant panels is that heat transfer to the room air from the former occurs largely by convection, while radiation is the primary mode of heat transfer in the latter. Radiant panels for heating and cooling were developed in the 1950s–1960s. There has been renewed interest and revival in recent years because of their esthetics, their nonintrusiveness, and especially their use as a solution to the more stringent building energy codes being enacted. In this section, we present some basic principles of low-temperature space heating. More details can be found in specialized publications such as ASHRAE Systems (2012). The discussion of radiant panels addressed here is meant for low-temperature space heating and cooling, as against high-temperature radiant heating applications that use either electricity or natural gas to produce a high-temperature source from which radiation can be directed for localized heating.

18.4.1 Description

Space heating systems in many parts of the world use warmed floors and/or ceilings. Although this system is less common in the United States as compared to Europe, the good comfort and quiet operation provided by such design options make them worth considering for some applications. Radiant systems are well suited to operation with heat pump, solar, and other low-temperature systems. There are three generic types of radiant cooling/heating systems: (1) panel systems, (2) core-cooled ceiling or floor pipes, and (3) water grids using capillary tubes.

Radiant panels can be placed on ceilings or on walls to provide both heating and cooling, while floor-mounted systems are better suited for heating. They can be used as stand-alone units but are often used in conjunction with air–water systems (discussed in Chapter 20). The term "radiant" is somewhat of a misnomer since up to 30% (for ceiling mounted) and 50% (for floor embedded) of the heat transferred from radiant panels could be actually by convection. In fact, the industry's nomenclature is quite specific: a radiant panel is defined as a temperature-controlled surface (less than 300°F or 150°C and generally around 120°F or 50°C) where 50% or more of the design heat transfer occurs by radiation (ASHRAE Systems, 2012).

Low-temperature radiant panels are temperature-controlled surfaces that can provide heating and sensible cooling to the occupants by radiation directly, and indirectly via convection, and exchange to the indoor space. They have fast response times and are suitable for rooms with quick changing loads. Figure 18.10a shows a simple radiant panel construction for a ceiling surface mount for a room with a plenum. Hot or cold water is circulated through the serpentine (or parallel) tubing, often made of copper, which is thermally bonded to the sheet that is made of either aluminum or copper (Figure 18.10b). A blanket of insulation is placed over the top to limit upward heat loss to the unheated plenum above. The panels usually cover only a portion of the ceiling and can be deployed so as to be esthetically pleasing. Another common design variant is a *ceiling-suspended metal panel*. The

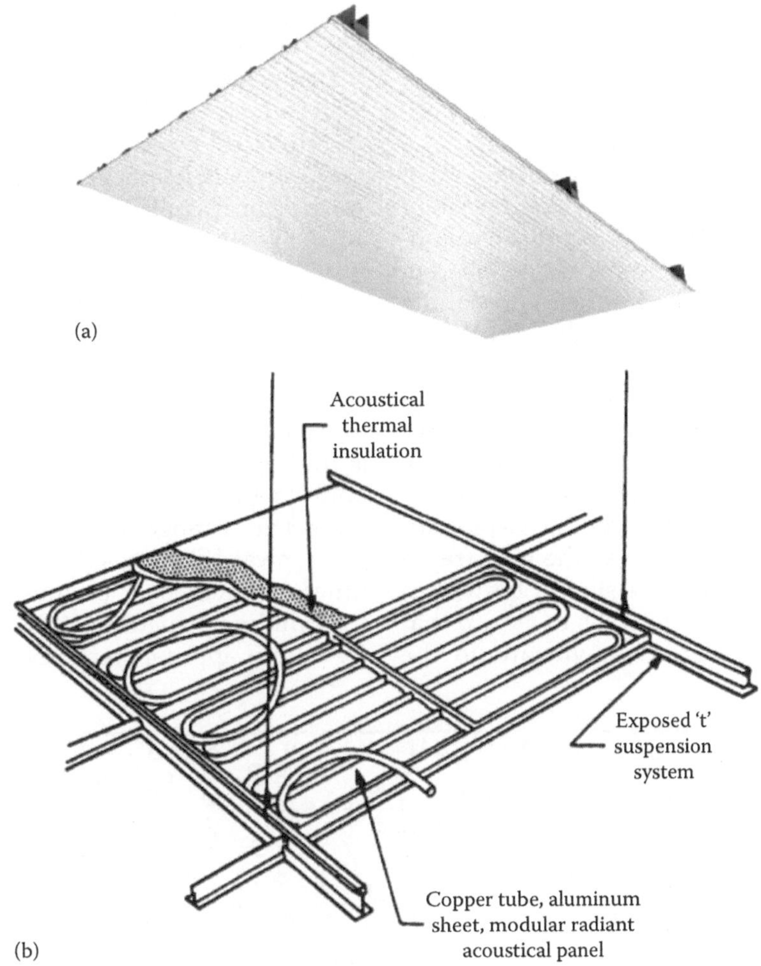

FIGURE 18.10
Ceiling-mounted radiant panels. (a) Front view of panel and (b) tube layout. (Image courtesy of Price Industries, Suwanee, GA.)

two aforementioned designs are suitable as retrofit or renovation options. Radiant panels get less effective with ceiling height; in such cases, panels located on walls are better design options.

An alternative to traditional radiant panels has also been developed, namely, *chilled sails*. The sails are somewhat similar to linear radiant panels without insulation and consist of a serpentine tube through which chilled water is passed with the tube thermally bonded with slats. They are mostly suspended from the ceiling, either open to the room or behind a perforated ceiling (they can also be installed along a wall). They allow air to pass through openings between sails (Figure 18.11), thereby increasing the capacity of the units (roughly double that of radiant panels). The chilled sails cool the surrounding space air by creating a natural convective flow with the warmer space air that augments radiation heat transfer from the cooler sail surface temperature. Thus, they combine the radiant cooling effects of standard radiant panels with a convective component. Sails can also be used for

FIGURE 18.11
Chilled sails. (Image courtesy of Price Industries, Suwanee, GA.)

heating, but they are mostly using for cooling applications. Typical applications include offices, meeting rooms, theaters, lobbies/foyers, waiting areas, and studios where architectural appeal is an important consideration (Price, 2011).

Note that radiant panels for cooling can only meet the room sensible loads and are not suitable for latent load removal. In such cases, they have to be used in conjunction with hybrid air–water systems as described in Chapter 20. Generally, cold ceiling panels can handle loads in the range of 30–70 Btu/h·ft^2 (95–220 W/m^2) with up to 50% of the ceiling space utilized for cooling. They can respond faster with load changes, but they cost more and need colder water temperatures of about 56°F–59°F (13.3°C–15°C). Another attractive design for new construction is to use *core-cooled thermo active slabs*, wherein bare concrete ceiling slabs with water coils embedded within serve as the heat transfer panel (Figure 18.12a). The coils should be within 1 in. (2.5 cm) of the lower surface so as to assure that more cooling will be transferred downward to the space below. Insulating the floor of the space above will improve overall performance.

Thermoactive cooling slabs need larger surface area than radiant panels since their capacity is lower (about 25 Btu/[h·ft^2] or 80 W/m^2). They are less expensive and need moderate temperatures (64°F–75°F or 18°C–24°C). In general, since chilled water temperatures are higher than all-air systems, chillers can operate at higher temperatures resulting in better efficiency and reduction in energy costs.

Radiant floors are generally best for supplying heat to the space. They are embedded in the concrete slab as shown in Figure 18.12b. The concrete provides heat capacity to the room that reduces peak loads and shifts them to later in the evening. A variant of embedded design is the "topping slab" design. It features a separate slab built on the structural slab separated by an insulation layer. This arrangement directs most of the heat to flow only upward. This has the advantages of faster response times, and, moreover, permits design independence from the building structure. Yet another variant well suited for retrofit applications is the "over-floor" design where

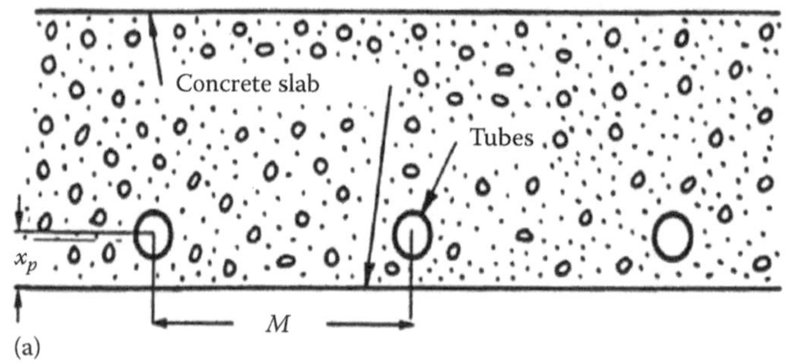

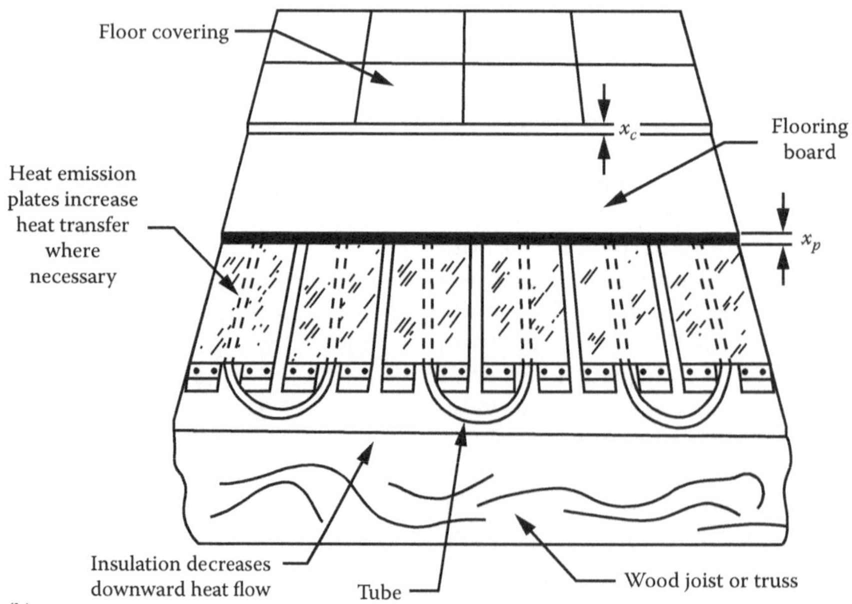

FIGURE 18.12
Common embedded pipe and tube systems. (a) Coils in structural concrete ceiling slabs and (b) coils on structural floor slab.

the piping can be embedded in a separate wooden flooring placed above the structural slab. One could also supply cooling, but this could cause discomfort due to cold feet and is less effective. However, for high ceiling and wide spaces (such as retail and airports) or for low-rise buildings, they may be the only all-water system option. The use of carpets, furniture, and other items can hinder proper operation of such systems.

A type of system whose operation is similar to core-cooled hydronic systems is the *ventilated slab system* that uses air as the heat transfer medium (Zmeureanu and Fazio, 1988). Here, air channels are formed within precast slabs either in the ceiling or the floor depending on whether cooling or heating is required. The cooled or heated air then heats up the slab, which then cools or heats the space by convection and radiation heat transfer, akin to core-cooled hydronic systems. Used predominantly in northern Europe, Australia, and the Middle East, they capitalize on the thermal heat capacity of the slabs to reduce energy use and peak cooling loads.

Yet another type of radiant system is the *capillary tube network*. Figure 18.13 is a photograph of such a pipe network of radiant cooling coils on the ceiling of a building under construction. The various manifolds covering the ceiling uniformly provide an even surface temperature. The tubes can be embedded in plaster or gypsum board or even mounted on ceiling panels.

Control of radiant heating systems has proven to be a challenge in the past due to the large time constant of these systems. Both under- and overheating of the space are common problems. If the outdoor temperature drops rapidly, these systems will have difficulty responding quickly. Conversely, after a morning warm-up followed by high solar gains on a sunny winter day, the radiant system may overshoot. Systems with pipes embedded in concrete slabs are especially sluggish. The current generation of "smart" controls should help improve the comfort control of these systems.

18.4.2 Performance Equations

The radiation heat output per unit area or heat flux $\dot{q}_{rad}$ of radiant panels that is absorbed by an unheated surface is given by the Stefan–Boltzmann equation, discussed in Section 2.10.5:

$$\dot{q}_{rad} = \varepsilon_{eff} \cdot F_{rp-uhs} \cdot \sigma[(T_{rp} + 460)^4 - (T_{uhs} + 460)^4] \quad (18.3\ \text{IP})$$

where

T_{rp} is the heating/cooling panel surface temperature in °F

T_{uhs} is the area-weighted temperature in °F of all unheated indoor surfaces of walls, ceilings, floors, windows, etc., but excluding active panel surface (this is akin to the mean radiant temperature concept discussed in Chapter 3); a designated terminology is used in the literature, namely, area-weighted unheated surface temperature (AUST)

$\varepsilon_{eff} = (1/\varepsilon_{rp} + 1/\varepsilon_{uhs} - 1)^{-1}$ = effective emittance of space, where *rp* and *uhs* refer to the heated panel and the unheated surfaces, respectively; ε_{eff} is approximately 0.87 for the most indoor spaces

F_{rp-uhs} is the view factor between panel heating/cooling surface and unheated surfaces; its value is 1.0 in present case

σ is the Stefan–Boltzmann constant

FIGURE 18.13
Photograph of a capillary tube network of radiant ceiling cooling coils.

The resulting expression widely used for low-temperature radiant heating and cooling panels is (ASHRAE Systems, 2012)

$$\dot{q}_{rad} = 0.15\times10^{-8}[(T_{rp}+460)^4 - (\text{AUST}+460)^4] \quad (18.4\ \text{IP})$$

where T_{rp} and AUST are in °F with the sign convention that positive values imply heating and negative values represent cooling.

The aforementioned expression is graphically plotted in Figure 18.14a for heating and in Figure 18.14b for cooling. For interior spaces in a building, the value of AUST can be taken to be the indoor air dry-bulb temperature. However, for exterior spaces with windows and external heat gains, the AUST value may be significantly higher and needs to be determined for the specific instance. From Figure 18.14a, we read that the radiation heat flux is about 50 Btu/h · ft² (158 W/m²) for AUST = 70°F (21.1°C) and an effective mean panel surface of 110°F (43.3°C).

The convection from the heating surface can be found by using the standard free-convection expressions akin to those given in Section 2.9.3. The exact expressions for the convective heat flux $\dot{q}_{con}$ proposed by ASHRAE Systems (2012) are given in Table 18.2 and plotted in Figure 18.15. Note that the indoor space dry-bulb temperature T is now used and the expressions are slightly nonlinear because of changes in the effective convective coefficient. The heat fluxes vary considerably with panel type; floor heating and ceiling cooling is the largest and ceiling heating the lowest (as one would expect from buoyancy considerations). Note that leaving cold strips induces better convection, and that is why the associated curve in Figure 18.15 is higher than that of an all-heated ceiling panel. For a temperature difference of 40°F (22.2°C), the floor-heating equation suggests a convective heat transfer flux of about 40 Btu/h · ft² (126 W/m²), which is less than that for radiant heat transfer determined previously. Hence, this meets the criterion that a radiant panel is one whose heat transfer by radiation should be 50% or more.

18.4.3 Design Suggestions

Radiant panel systems need to be designed with some care but generally are similar to the approach used for terminal units described in the previous section. The designer's job is to determine the panel area needed, its operating temperature, the heating liquid flow rate, and construction details. The panel size is

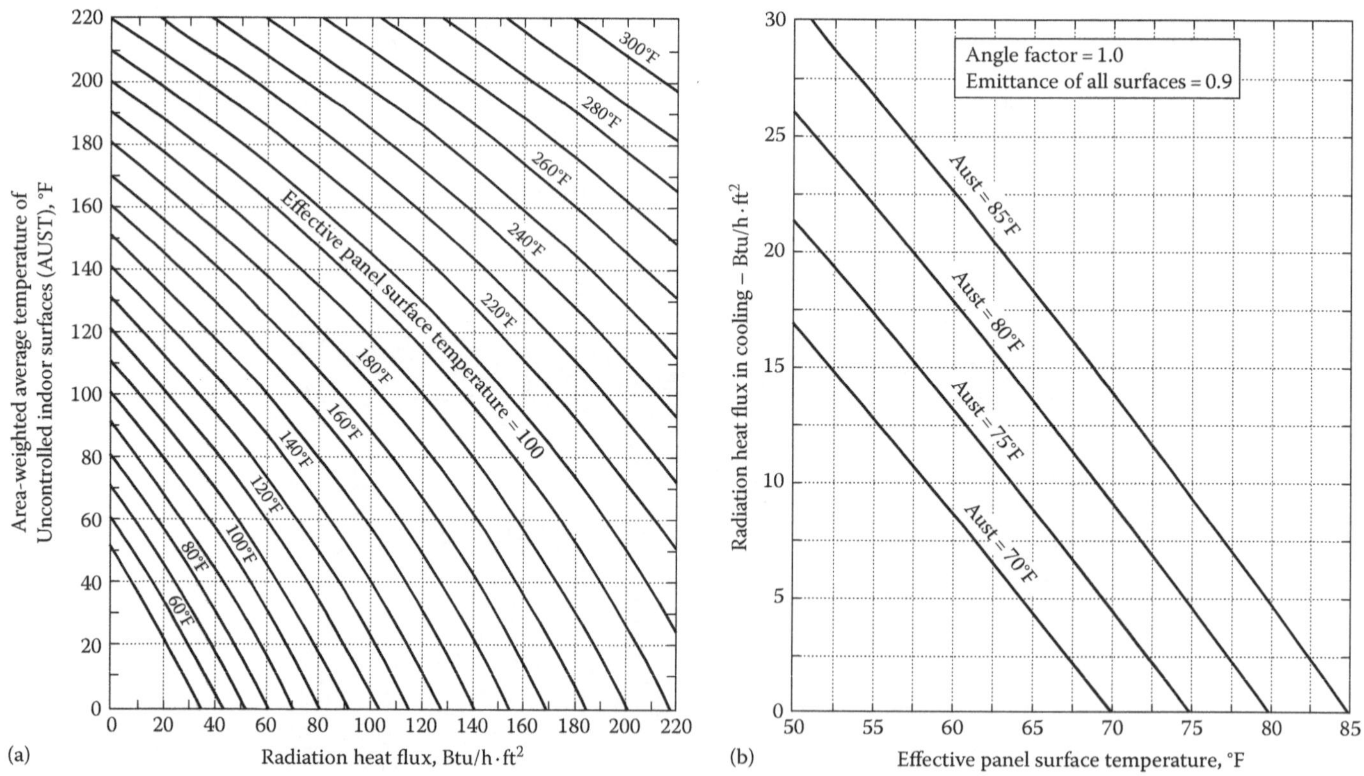

FIGURE 18.14
Radiation heat flux plots. (a) Heated ceiling, floor, or wall panel surfaces and (b) cooled ceiling or wall panel surface. (From *ASHRAE Handbook—HVAC Systems and Equipment*, American Society of Heating, Refrigerating, and Air-Conditioning Engineers, Atlanta, GA, 2012. Copyright ASHRAE, www.ashrae.org.)

TABLE 18.2

Natural Convection Heat Flux Equations for Radiant Panels

Type of Panel	Proposed Equation in IP Units	Equation Number
All-heated ceiling surface	$\dot{q}_{con} = 0.02 \times (T_{rp} - T)^{0.25}(T_{rp} - T)$	(18.5)
Heated ceiling with cold unheated strips	$\dot{q}_{con} = 0.13 \times (T_{rp} - T)^{0.25}(T_{rp} - T)$	(18.6)
Heated floor or cooled ceiling surface	$\dot{q}_{con} = 0.31 \times \lvert(T_{rp} - T)\rvert^{0.31}(T_{rp} - T)$	(18.7)
Heated or cooled wall panel surface	$\dot{q}_{con} = 0.26 \times \lvert(T_{rp} - T)\rvert^{0.32}(T_{rp} - T)$	(18.8)

Source: ASHRAE, *Handbook—HVAC Systems and Equipment*, American Society of Heating, Refrigerating and Air-Conditioning Engineers, Atlanta, GA, 2012. Copyright ASHRAE, www.ashrae.org.

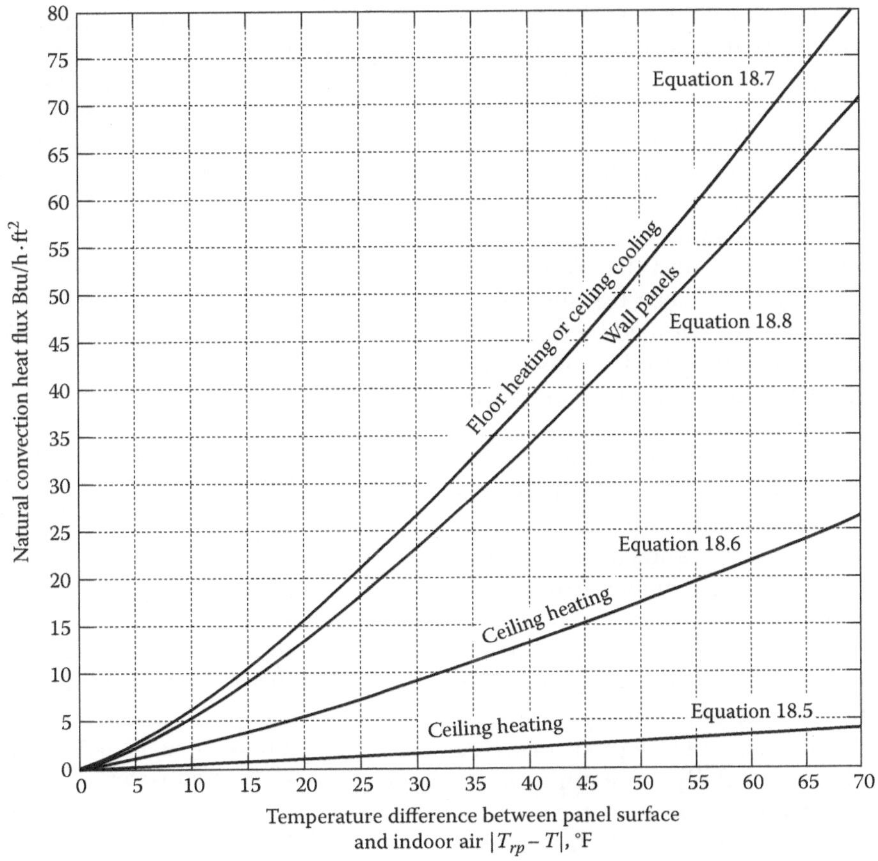

FIGURE 18.15
Heat transfer due to natural convection at floor, ceiling, and wall panel surfaces. (From *ASHRAE Handbook—HVAC Systems and Equipment*, American Society of Heating, Refrigerating, and Air-Conditioning Engineers, Atlanta, GA, 2012. Copyright ASHRAE, www. shrae.org.)

determined based on standard heat load calculations for the space. Proper account should be made of any losses from the back of the radiant panels to unheated spaces. Heating panel temperatures should not exceed 85°F (29.5°C) for floors and 115°F (46°C) for ceilings. Hot water temperatures are typically 120°F (49°C) for floors and up to 155°F (69°C) for ceilings. It is typical to design hydronic systems for heating panels with a water temperature drop of 20°F (11°C).

Panels can be piped in a series configuration if pipe runs are not excessively long (the final panels in a long series run will not perform up to specifications due to low fluid temperatures). Long series loops also have excessively high-pressure drops. If large areas are to be heated, a combination of series and parallel connections can be used. Manufacturer's advice should be sought regarding the number of panels that can be connected in series without performance penalties.

Heating panel design involves the following process:

1. Determine the room heat load.
2. Decide on the location of panels (roof or floor).
3. Find panel heat flux, including both radiative and convective contributions at 80°F (27°C) for floor panels and 110°F (43°C) for ceiling panels.
4. Divide heat load by heat flux to find needed panel area.
5. If panel area exceeds available floor or ceiling area, raise panel temperature (not exceeding temperatures noted earlier) and repeat steps 3 and 4.
6. If the panel area is still insufficient, consider both floor and ceiling panels to improve thermal quality of room insulation.

If radiant floors are to be built during building construction rather than used as prefabricated panels in ceilings, the following guidelines can be used: Tubing spacing for a system of the type shown in Figure 18.12b should be between 6 and 12 in. (15 and 30 cm). The tubing diameter generally ranges between 0.5 and 1.0 in. (1.2 and 2.5 cm). Flow rates are determined by the rate of heat loss from the panel, which in turn depends on the surface temperature, and hence the fluid temperature. This step in the design is iterative.

For cooling, one has to avoid condensation on the cooling panels. This requires that the panel temperature be above the dew point temperature of the room air. One also needs to know the allowable temperature increase of the cold water as it flows through the cooling panel (a typical value is around 5°F or 2.8°C). The exact values of these temperatures will depend on the thermal resistances of the heat flow paths that are dictated by the construction details of the panels, and though these can be calculated from first principles (see ASHRAE Systems, 2012), it is usually more appropriate and simpler to use manufacturer-specified recommendations for selecting such design values. In fact, rather than use the fundamental equations for radiation and convective heat transfer presented in the previous section, it may be preferable in practical design situations to simply use manufacturer-specified performance data similar to that in Table 18.1.

Example 18.2: Design Example for a Radiant Cooling Panel

Consider an interior space of a building with a ceiling area of 200 ft² (18.6 m²) that needs to be cooled with radiant panel on the ceiling. The total panel area has to be determined subject to the following specification.

Given: Space heat load $\dot{Q}_{tot}$ = 4000 Btu/h (1.17 kW)

Indoor space condition:

Dry-bulb temperature $T_{db,space}$ = 75°F (23.9°C)
Relative humidity $\phi_{space} = 48\%$

Manufacturer recommends: Panel surface should be 5°F (2.8°C) above room dew point.

Find: Required ceiling panel area A_{rp}

Assumption: AUST is equal to the space dry-bulb temperature.

Solution

First, we determine the dew point temperature of the space: T_{dp} = 55°F.

The average panel temperature is $T_{rp,av}$ = 55°F + 5°F = 60°F.

The heat transfer due to radiation can be found from Figure 18.14b or using Equation 18.4 IP. For AUST = 75°F, we find $\dot{q}_{rad} = 13\ \text{Btu/(h}\cdot\text{ft}^2)$.

The heat transfer due to convection can be found from either Figure 18.15 or Equation 18.7. With a temperature difference $|T_{rp} - T_{db,space}|$ = |60°F − 75°F| = 15°F, we find $\dot{q}_{con} = 11\ \text{Btu/(h}\cdot\text{ft}^2)$.

The required ceiling panel area is $A_{rp} = \dfrac{4000\ \text{Btu/h}}{(13+11)\ \text{Btu/(h}\cdot\text{ft}^2)} = 167\ \text{ft}^2 (15.5\ \text{m}^2)$.

Comments

The required ceiling panel area is less than the ceiling area available, and so this size of radiant panel can be accommodated. One would also perform a calculation for winter conditions when heating is needed, and determine whether this area would be adequate or not.

The careful designer will also determine and evaluate the cold water supply and return temperatures. For this, one needs to

1. Know the suggested temperature increase of the cold water as it flows through the panel (a typical value is 5°F or 2.8°C)
2. Consider a small temperature allowance (typically 1°F or 0.5°C) to allow for normal drift in temperature controls for water and air systems and for any room relative humidity changes
3. Know the manufacturer-suggested temperature differential between the average panel temperature and the average water temperature (say, 3°F or 1.7°C)

With the aforementioned values, the average cold water temperature is (60°F − 3°F − 1°F) = 56°F (13.3°C), while the supply water temperature is about (56°F − 5/2°F) = 53.5°F (12°C). Considering temperature gains of a few degrees in typical chilled pipes, the chiller can be operated at 50°F or so (about 10°C), which is higher than the set point for all-air systems; thus resulting in improved chiller COP. This is one of the energy efficiency benefits of low-temperature cooling panels as stated earlier.

18.5 Auxiliary Heating Equipment

18.5.1 Service Hot Water

Heated water is used in buildings for various purposes including basins, showers, and sinks or for custodial service or specialty services such as kitchens in restaurants and the like. In this section, we give an overview of service (or *domestic*) water heating methods for buildings. For details, refer to ASHRAE Applications (2011).

Water is heated either by equipment that is part of the space heating system (i.e., the boiler) or by a stand-alone water heater. The stand-alone equipment is similar to a small boiler except that water chemistry must be accounted for by using anodic protection for the tank and by water softening in geographic areas where hardness can cause scale (lime) deposits in the water heater tank. Equipment can also be classified by the fuel used: gas fired, oil fired, electric, and indirect (where the heating medium is steam or hot water generated from another boiler).

Two types of systems are used for water heating: *instantaneous* (*tankless or minimal storage*) and *storage*. The former heats water on demand as it passes through the heater, which uses either steam or hot water. Output temperatures can vary with this system type unless a control valve is used on the heated water side (it should not be used on the heat supply) of the water heater. Instantaneous water heaters are best suited to relatively uniform loads. They avoid the cost and heat losses of the storage tank but require larger and more expensive heating elements. A hot water with storage can be categorized as an instantaneous type when its heating rate divided by the storage capacity is less than 4000 Btu/(h·gal) or 300 W/L.

Storage-type systems are used to accommodate varying loads or loads where large peak demands make it impractical to use instantaneous systems. Water in the storage tank can be heated by an immersion steam coil, or by direct firing, or by an external heat exchanger. In sizing this system, the designer must account for standby losses from the tank jacket and connected hot water piping. For any steam-based system, cold supply water can be preheated by using the steam condensate.

To size the equipment, two items must be known: (1) hourly peak demand for the year, gal/h (L/h), and (2) daily consumption, gal/day (L/day). Of course, the volumetric usage rates must be converted to energy terms by multiplying them with the specific heat and water temperature rise:

$$\dot{Q}_h = \dot{m}_w \cdot c_{p,w} \cdot (T_{set} - T_{supply}) \qquad (18.9)$$

where

$\dot{Q}_h$ is the heat rate, on daily or hourly basis, Btu/day or Btu/h (kWh/day or W)
$\dot{m}_w$ is the water mass flow rate, on daily or hourly basis, calculated from the volumetric flow, lb_m/d (kg/d)
T_{set} is the required hot water supply temperature or set-point temperature, °F (°C)
T_{supply} is the temperature of water from the supply mains, °F (°C)
$c_{p,w}$ is the specific heat of water, Btu/(lb_m·°F) (kJ/kg·°C)

Table 18.3 summarizes average and maximum daily and the maximum hourly water demands for various types of buildings. Table 18.4 lists representative water heater temperatures for several end uses ranging from 95°F to 180°F (35°C–82°C). When using the lower settings in the table, the designer must be aware of the potential for *Legionella pneumophila* (Legionnaires' disease). This microbe has been traced to infestations of showerheads; it is able to grow in water maintained at 115°F (46°C). This problem can be limited by using domestic water temperatures near 140°F (60°C). For residences, a rule of thumb is 15–20 gal/day/person (57–76 L/day/person), while the peak draw in a typical family is 10–12 gal/h/person (38–45 L/h/person).

Figure 18.16 shows a typical piping arrangement for a hot water boiler along with accessories. Cold makeup water from the mains supply is introduced at the bottom of the tank so as to maintain thermal stratification. There is, however, a certain amount of mixing that occurs, and thus, not all the thermal energy of the hot water can be usefully extracted at the desired temperature set point. A dilution fraction is introduced to account for the *usable* storage capacity. Hence, hot water can be supplied from a storage-type system at the maximum rate of (assuming jacket losses to be small) given by

$$\dot{V}_w = \dot{V}_r + \frac{f_{useful} V_{tank}}{\Delta t} \qquad (18.10)$$

where

$\dot{V}_w$ is the volumetric hot water supply rate, gal/h (L/s)
$\dot{V}_r$ is the water heater recovery rate, gal/h (L/s)
f_{useful} is the usable storage capacity of hot water system, (typical range 0.60–0.80)
V_{tank} is the tank volume, gal (L)
Δt is the duration of peak demand, h (s)

TABLE 18.3

Hot Water Demands and Use for Various Types of Buildings

Type of Building[a]	Maximum Hour	Maximum Day	Average Day
Men's dormitories	3.8 gal (14.4 L)/student	22.0 gal (83.4 L)/student	13.1 gal (49.7 L)/student
Women's dormitories	5.0 gal (19 L)/student	26.5 gal (100.4 L)/student	12.3 gal (46.6 L)/student
Motels: No. of units[b]			
20 or less	6.0 gal (22.7 L)/unit	35.0 gal (132.6 L)/unit	20.0 gal (75.8 L)/unit
60	5.0 gal (19.7 L)/unit	25.0 gal (94.8 L)/unit	14.0 gal (53.1 L)/unit
100 or more	4.0 gal (15.2 L)/unit	15.0 gal (56.8 L)/unit	10.0 gal (37.9 L)/unit
Nursing homes	4.5 gal (17.1 L)/bed	30.0 (113.7 L)/bed	18.4 gal (69.7 L)/bed
Office buildings	0.4 gal (1.5 L)/person	2.0 gal (7.6 L)/person	1.0 gal (3.8 L)/person
Food service establishments			
Type A: full-meal restaurants and cafeterias	1.5 gal (5.7 L)/max meals/h	11.0 gal (41.7 L)/max meals/h	2.4 gal (9.1 L)/average meals/day[c]
Type B: drive-ins, grills, luncheonettes, sandwich, and snack shops	0.7 gal (2.6 L)/max meals/h	6.0 gal (22.7 L)/max meals/h	0.7 gal (2.6 L) average meals/day[c]
Apartment houses: number of apartments			
20 or less	12.0 gal (45.5 L)/apt.	80.0 gal (303.2 L)/apt.	42.0 gal (159.2 L)/apt.
50	10.0 gal (37.9 L)/apt.	73.0 gal (276.7 L)/apt.	40.0 gal (151.6 L)/apt.
75	8.5 gal (32.2 L)/apt.	66.0 gal (250 L)/apt.	38.0 gal (144 L)/apt.
100	7.0 gal (26.5 L)/apt.	60.0 gal (227.4 L)/apt.	37.0 gal (140.2 L)/apt.
200 or more	5.0 gal (19 L)/apt.	50.0 gal (195 L)/apt.	35.0 gal (132.7 L)/apt.
Elementary schools	0.6 gal (2.3 L)/student	1.5 gal (5.7 L)/student	0.6 gal (2.3 L)/student[b]
Junior and senior high schools	1.0 gal (3.8 L)/student	3.6 gal (13.6 L)/student	1.8 gal (6.8 L)/student[b]

Source: ASHRAE, *Handbook—HVAC Systems and Equipment*, American Society of Heating, Refrigerating and Air-Conditioning Engineers, Atlanta, GA, 2012. Copyright ASHRAE, www.ashrae.org.

[a] The average usage of a U.S. residence is 60 gal/day (227 L/h) with a peak usage of 6 gal/h (22.7 L/h) (ASHRAE Applications, 2011).

[b] Interpolate for intermediate values.

[c] Per day of operation. Temperature basis 140°F (60°C).

TABLE 18.4

Representative Hot Water Use Temperatures

Use	Temperature °F	Temperature °C
Lavatory		
Handwashing	105	40
Shaving	115	45
Showers and tubs	110	43
Therapeutic baths	95	35
Commercial and institutional laundry	180	82
Residential dishwashing and laundry	140	60
Surgical scrubbing	110	43

Note: Table values are water use temperatures, not necessarily water heater set points.

Several techniques for sizing hot water system heating rate and storage size are described in various texts, for example, Mull (2001) or ASHRAE Applications (2011). An often adopted simplified sizing method relies on using Table 18.3 and figures such as Figure 18.17 (curves for only two typical applications are shown). These typical plots are based on actual metering but may not be representative of all buildings within the same building type. The recovery rate or capacity is the amount of hot water, in say gal/h, which can be continuously produced by the hot water system at the set-point temperature. Any combination of storage size and recovery rate falling on the proper curve would satisfy the requirements of the specific type of building. It is clear that there is a trade-off between useable storage capacity and recovery capacity. Increasing the storage size allows the designer to specify a smaller size of water heater, and vice versa. Using the minimum recovery rate and maximum storage capacity results in the smallest hot water capacity able to meet the building requirements.

Let us assume that there are 200 occupants in an office building, and the storage tank configuration suggests a dilution fraction of 0.7, i.e., only 70% of the storage tank volume can be extracted at the desired useful temperature with the other 30% being degraded due to mixing of the hot water tank with the incoming cold water from the mains. From Figure 18.17a, the minimum recovery rate is 0.1 gal/(h·person), while the corresponding usable storage capacity is 1.6 gal/person. Thus, $0.1 \times 200 = 20$ gal/h recovery is needed, while the corresponding storage capacity assuming a dilution factor of 0.7 is equal to

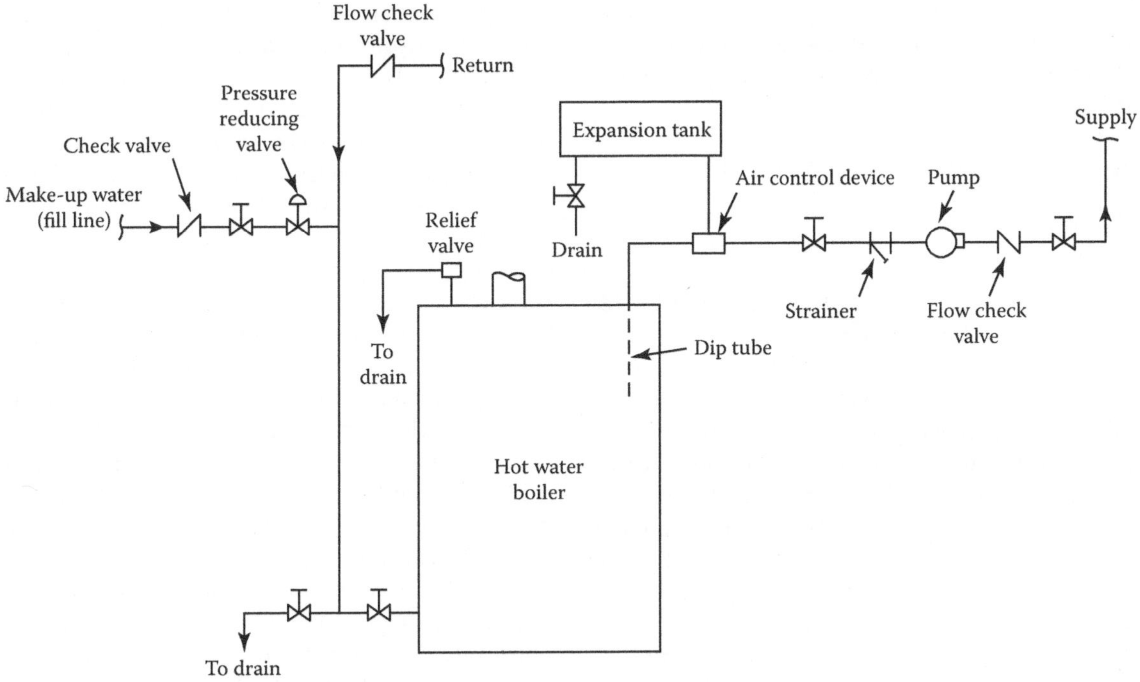

FIGURE 18.16
Piping arrangement and accessories for a hot water heater in a small building.

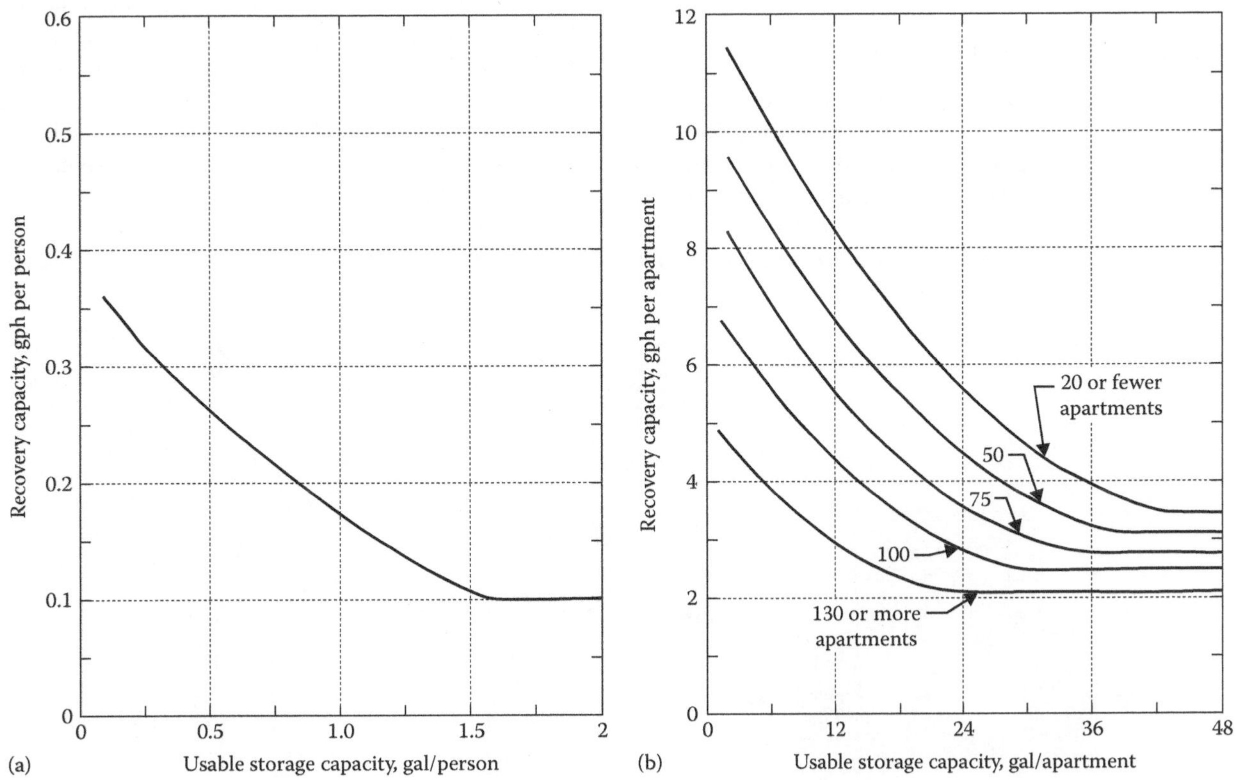

FIGURE 18.17
Hot water recovery capacity versus usable storage capacity. (a) Office buildings and (b) apartments. (From, *ASHRAE Handbook—HVAC Systems and Equipment*, American Society of Heating, Refrigerating and Air-Conditioning Engineers, Atlanta, GA, 2011. Copyright ASHRAE, www.ashrae.org.)

(1.6 × 200)/0.7 = 457 gal. If the hot water set-point temperature is 110°F and the cold water source temperature is specified as 50°F, then the water heater output energy rate should be

$$Q_{max} = 166.6\ \text{lb}_m/\text{h} \times 1.0\ \text{Btu}/(\text{lb}_m \cdot °\text{F}) \times (110°\text{F} - 50°\text{F})$$
$$= 9996\ \text{Btu/h}$$

where the water mass flow rate is determined as (20 gal/h × 8.33 lb_m/gal) = 166.6 lb_m/h.

A water heater of this heating capacity (with a small safety allowance) can then be selected for this building. If the designer chooses to be cautious and adopt a higher value (but less than the maximum hourly draw of 0.4 gal/h/person listed in Table 18.3 for office buildings), then obviously, a water heater with a larger energy output rating will have to be selected. We leave it to the reader to infer the corresponding heat output rate and the storage capacity of the water heater.

More accurate design procedures for commercial buildings involve counting the number of different fixtures in each of the domestic hot water loops, referring to lookup tables to determine individual fixture flows/requirements, and then using semiempirical equations to determine the peak and daily draws of the water loops.

18.5.2 Steam Systems and Equipment

Steam systems have additional components needed to provide safety or adequate control in building thermal systems. Unless applications specifically require superheated steam, most steam systems use saturated steam. Due to heat losses from piping, some of this steam can condense back to water. This entrained condensate must be separated. In this section, we provide an overview of the most important components, including steam traps and relief valves.

Steam traps are used to separate both steam condensate and noncondensable gases from live steam in steam piping systems and at steam equipment. Steam traps isolate or confine steam in heating coils, e.g., while releasing condensate to be revaporized in the boiler. The challenge in trap selection is to ensure that the condensate and gases are removed promptly and with little to no loss of live steam. For example, if condensate is not removed from a heating coil, it will become waterlogged and will have a much reduced heating capacity. We briefly describe the most common types of traps and identify ones that should be used in HVAC applications.

Thermodynamic traps are simple and inexpensive. The most common type, called the "disk trap," is shown in Figure 18.18a. This type of trap operates on kinetic energy changes as condensate flows through and flashes into steam within the trap. Steam flashed (i.e., converted from hot liquid to vapor) from hot condensate above the disk holds the trap closed until the disk is cooled by cooler condensate. Steam line pressure then pushes the disk open. It remains open until all cool condensate has been expelled and until hot condensate accumulates and flashes again; only then does the valve close. The disk action is accelerated by the flow of condensate beneath the disk; the high velocities produce a low-pressure area there in accordance with Bernoulli's equation, and the disk slams shut. These traps are rugged and make a characteristic clicking sound, facilitating operational checking. They can stick open if a particle lodges in the seat. This design has relatively high operating cost due to its live steam loss.

Thermostatic traps use the temperature difference between steam and condensate to control condensate flow.

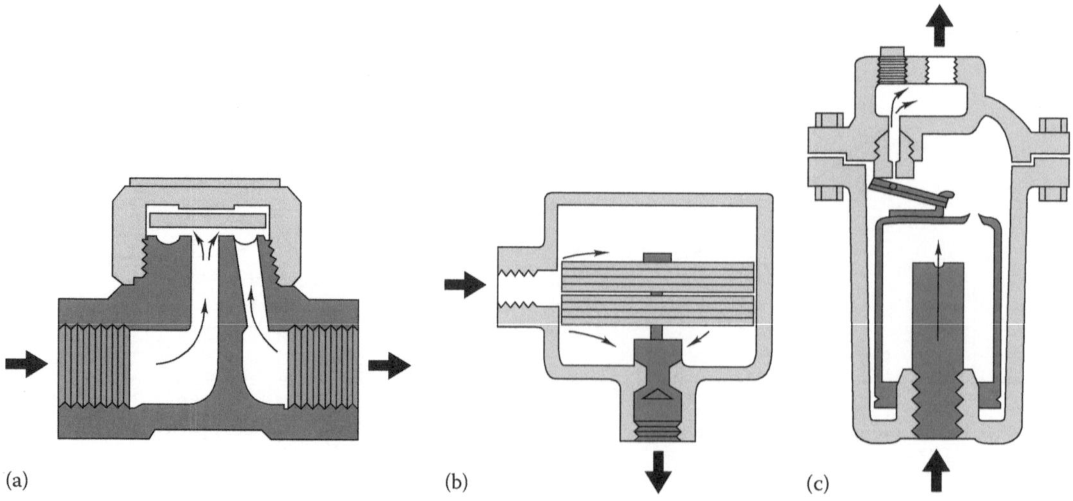

FIGURE 18.18
Steam traps. (a) Thermodynamic disk trap, (b) thermostatic trap, and (c) mechanical, inverted-bucket trap.

One type of thermostatic trap is shown in Figure 18.18b. The bimetal unit within the housing opens the valve as condensate cools, thereby allowing condensate to exit from the trap. Significant subcooling of the condensate is needed to open the valve, and operation can be slow. Other, more complex designs have more rapid response and reduced need for subcooling. The bimetal element can be replaced with a bellows filled with an alcohol–water mixture, permitting closer tracking of release setting as the steam temperature changes. A trap that has a temperature/pressure characteristic greater than the temperature/pressure (T/p) of saturated steam (see saturation curve in Figure 12.5 or saturated-steam data in Tables A3) will lose live steam, whereas a trap with a T/p characteristic lying below the steam curve will build up condensate. The ideal trap has an opening T/p characteristic identical with the T/p curve of saturated steam.

Mechanical traps operate on the density difference between condensate and live steam to displace a float. Figure 18.18c shows one type of mechanical trap, called the "inverted-bucket trap," which uses an open, upside-down bucket with a small orifice. Steam flowing with the condensate (which fills the housing outside of the bucket) fills the inverted bucket and causes it to float, since the confined steam is less dense than the liquid water surrounding the bucket. Steam bleeds through the small hole in the bucket and condenses within the trap housing. As the bucket fills with condensate, it becomes heavier, eventually sinks, and opens the valve. Steam pressure forces condensate from the trap. The design of this trap continuously vents noncondensable gases, although the capacity for noncondensable gas flow (mostly air) rejection is limited by the size of the small hole in the top of the bucket. This hole is limited in size by the need to control parasitic steam loss through the same hole. Dirt can block the hole causing the trap to malfunction. The trap must be mounted vertically. An inverted-bucket trap will have significantly smaller parasitic live steam losses than the thermodynamic disk trap.

Steam traps are used to drain condensate from steam headers and from equipment where condensing steam releases its heat to another fluid. Steam piping is sloped so that condensate flows to a collecting point where it is relieved by the trap. At equipment condensate collection points, the trap is placed below the equipment where the condensate drains by gravity. Figure 18.19 shows a typical trap application for both purposes. The left trap drains the header, and the right trap drains the condensate produced in the heating coil.

Selection of traps requires knowledge of the condensate rejection rate (lb_m/h, kg/s) and the suitability of various trap designs to the application. Table 18.5 summarizes the applications of the three types of traps discussed earlier (Haas, 1990).

The operating penalties for malfunctioning steam traps (clogged, dirty, or corroded) can dwarf the cost of a trap because expensive heat energy is lost if live steam leaks out from malfunctioning traps. One of the first things to inspect in an energy audit of a new or existing

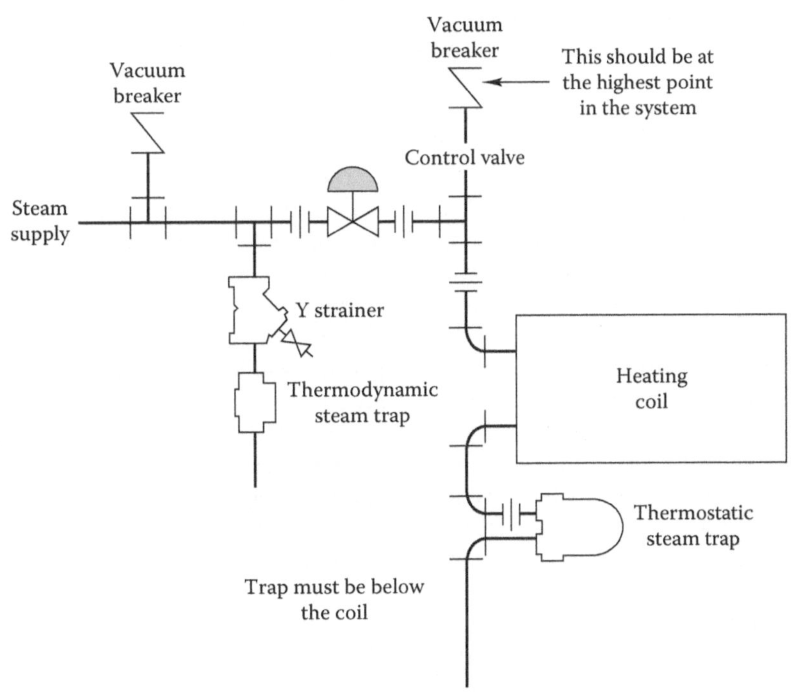

FIGURE 18.19
Piping arrangement for heating coil steam trap application.

TABLE 18.5

Operating Characteristics of Steam Traps

System Needs	Thermodynamic	Float Thermostatic	Inverted Bucket
Maximum pressure, psig	1,740	465	2,755
Maximum capacity, lb_m/h	5,250	100,000	20,500
Discharge temperature, °F	Hot	Hot	Hot
		(Close to saturated-steam temperature)	
Discharge	On/off	Continuous	On/off
Air venting	Good	Excellent	Fair
Dirt handling	Fair	Good	Good
Freeze resistance	Good	Poor	Poor
Superheat	Excellent	Poor	Fair
Water hammer	Excellent	Fair	Excellent
Varying load	Good	Excellent	Good
Change in psi	Good	Excellent	Fair
Back pressure	Maximum 80%	Good	Good
Usual failure	Open	Closed/air vent open	Open

Source: From Haas, J.H., *Chem. Eng.*, 97(January), 151, 1990.

steam system is the condition of the traps. For example, if steam is produced in a gas-fired boiler of typical efficiency, a 0.25 in. (0.64 cm) leak in a steam trap will lose about $2000 worth of steam in a year. (The cost of gas in this example is $3.00/1000 ft^3 [$0.11/1 m^3], the usual units used by utilities; this converts to approximately $3.00/1 MBtu or $2.84/1 GJ.)

A pressure *relief valve* is needed to control possible overpressure in boilers for safety reasons. Valves are specified by their ability to pass a given amount of steam or hot water at the boiler outlet condition. This dump rate can be specified in units of either mass per time or energy flow per time. Pressure relief valves must be used wherever heat can be added to a confined volume of water. Water could become confined in the piping of an HVAC system, e.g., if automatic control valves failed while closed or if isolation valves were improperly closed by a system operator. Not only boilers must be protected, but also heat exchangers and water pipe lengths that are heated externally by steam tracing or solar heat. The volume expansion characteristics of water can produce tremendous pressures if heat is added to confined water. For example, water warmed by only 30°F (17°C) will increase in pressure by 1100 psi (7600 kPa)! The method for sizing boiler relief valves is outlined in Wong (1989). The discharge from boiler relief valves must be piped to a drain or other location where injury from live steam will be impossible. Expansion of fluid in piping is also accommodated by *expansion tanks;* they are described in the next subsection.

There are numerous other components used in steam systems such as pressure reducing and regulating valves, temperature control valves, desuperheaters, steam heat exchangers, steam and condensate meters, and condensate return systems. Details of such equipment can be found in specialized handbooks such as ASHRAE Systems (2012) or Mull (2001).

18.5.3 Expansion Tanks

An expansion tank is an essential component for *closed* systems; its thermal function is to allow water to expand into a designated airspace to avoid high stresses that would otherwise occur in a piping system if no such air space for water expansion existed. Unconfined water expands almost linearly with temperature in the range of temperatures encountered in HVAC systems. For example, water expands about 2% between 40°F and 150°F (5°C and 65°C); it expands a little more than 3% between 40°F and 200°F (5°C and 94°C). Expansion tanks are needed in all closed systems, but not in systems *open* to the atmosphere.

Figure 18.20 shows an expansion tank and air separator connected near pump suction piping not far from the pump; this establishes a constant-pressure reference point in the system. For a hot water boiler, the expansion pump should be located at the boiler outlet or air vent with the pump located just downstream. The expansion tank pressure is set to equal at least the required net positive suction head (NPSHR; see Section 16.3.2) plus all friction losses between the expansion tank fitting and the pump inlet, as shown in Figure 18.20. Adequate NPSH is not often a problem with closed systems but must be carefully considered with pumps in open systems, such as cooling tower sump pumps. It is important to note that no matter how large the water piping system volume, only one expansion tank should be provided in the hydronic system.

Often, air (or nitrogen) charge is added to the tank after the system is filled at its operating pressure. There are two types of configurations: *closed tank with an open air–water*

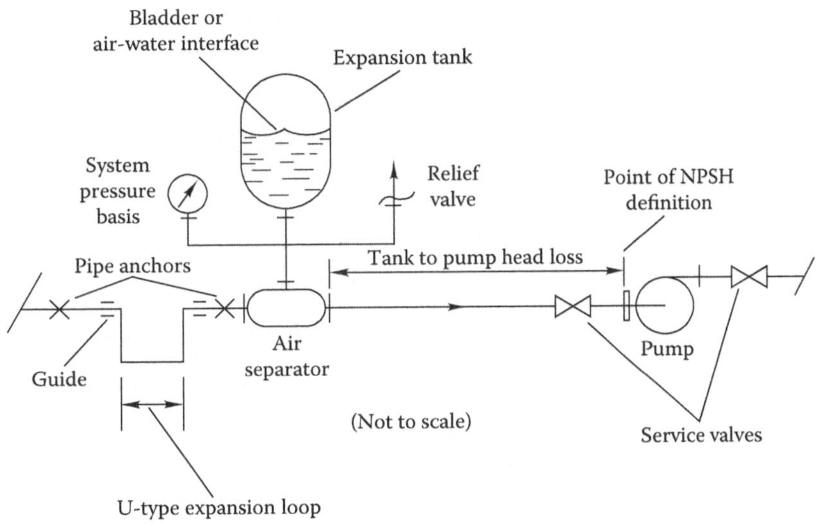

FIGURE 18.20
Expansion tank connection upstream of pump.

interface or one with a *flexible rubber diaphragm* (or bladder) at the interface that avoids dissolving the gas charge and the system water. Further, it is assumed that the tank is isothermal and does not track system operating temperature; this is accomplished by using an uninsulated tank thermally isolated from the active pipe loop by a small-diameter pipe. Expansion tanks with diaphragms can be sized using the following expression:

$$V_{exp} = \frac{V_{sys}\left[\left(\frac{v_2}{v_1} - 1\right) - 3\alpha\,\Delta T\right]}{\left(1 - \frac{p_1}{p_2}\right)} \tag{18.11}$$

where
- V_{exp} is the needed working volume of tank
- V_{sys} is the volume of water in entire system
- v_1 and v_2 are the specific volumes of water at minimum and maximum design system temperatures, respectively (see the following discussion of p_1 and p_2)
- α is the linear coefficient of thermal expansion for piping material; 0.000065/°F (0.000117/K) for steel and 0.000093/°F (0.000167/K) for copper
- ΔT is the difference between maximum and minimum piping system operating temperatures
- p_1 is the tank gas pressure at system initial fill temperature (for hot water systems) or operating temperature (for chilled water systems)
- p_2 is the maximum acceptable tank gas pressure corresponding to condition for v_2

Another method of pressurizing an expansion tank without a bladder is to simply fill the system. Air trapped in the tank will serve as an air cushion for the system. Expansion tanks of this type are larger than the bladder tank for the same service since the denominator in Equation 18.11 is replaced, for these tanks, by the quantity $(p_a/p_1 - p_a/p_2)$, in which p_a is the local atmospheric pressure. Even though this type of tank will be larger, it is preferred for hot water systems since air that comes out of solution at the air separator will collect at the top of the tank. Air is continuously introduced into hot water systems along with boiler makeup water. For chilled water systems, a bladder-type design is used since cold water retains air in solution.

The air separator shown in Figure 18.20* is a special fitting where low velocities and a vortex are produced to separate entrained air from system water. The air then rises to the expansion tank, where it collects and can be periodically released if necessary. The amount of air in a water loop is limited; if there are no water leaks requiring continuous makeup water, the amount of air in the tank will stabilize.

Chilled water piping design is less complex since the conditions under which chilled water is used are less critical. For example, systems are often closed and operated at conditions where the NPSH needs of pumps are easy to meet. The temperature swings of chilled water piping are often less than those for hot water piping since the pipes are insulated (with moisture-proof insulation) and not exposed to the sun. Hence, the expansion tanks can

* Good HVAC piping practice requires that each component be able to be isolated by service valves for replacement or service, hence the valves shown at the inlet and outlet of the pump in Figure 18.16. In this chapter we do not always show the isolation valves in system diagrams, but the designer must include them. Likewise, equipment such as chillers and boilers must be equipped with drain valves for service purposes.

be smaller than those in hot water systems. It is important to recognize, however, that chilled water systems can become relatively warm when the chiller is not operating. Fluid expansion under these conditions must be considered in the expansion tank sizing (see discussion of Equation 18.11).

In addition to fluid expansion, the designer must account for expansion (and contraction) of piping itself due to temperature changes in either hot or chilled water systems. The forces produced by pipe expansion can become extremely large. Figure 18.20 shows one method of accommodating pipe expansion by using an elastically deformable expansion loop. The size of expansion loops depends on the distance between expansion loops, pipe diameter, expected temperature change, and loop design—U bend (as shown in the figure), L bend, or Z bend. The online HCB software contains nomographs for sizing expansion loops in steel and copper piping networks.

Example 18.3: Sizing an Expansion Tank

Size a diaphragm expansion tank for a large hot water system of volume 7.6 m^3 (268 ft^3). The highest point in the system is 12 m (39.4 ft) above the expansion tank. The system is initially filled with water at 20°C (68°F). The operating temperature is 90°C (194°F). The maximum pressure in the system is to be 250 kPa gauge (35.2 psig), and the location is at sea level.

Given: $V_{sys} = 7.6\ \text{m}^3, p_2 = 101 + 250 = 351\ \text{k Pa (absolute)}$

Lookup values:

$$v_1(20°\text{C}) = 1.0017 \times 10^{-3}\ \text{m}^3/\text{kg},$$

$$v_2(90°\text{C}) = 1.0361 \times 10^{-3}\ \text{m}^3/\text{kg}, \quad \alpha = 0.117 \times 10^{-2}\ \text{K}^{-1}$$

Assumption: Steel tank

Find: V_{exp}

Solution

First, calculate pressure p_2. The additional pressure due to the 12 m water column after cold water is filled:

$$p_1 = \frac{g \times h}{v_1} = \frac{9.81\ \text{m/s}^2 \times 12\ \text{m}}{1.0017 \times 10^{-3}} = 117.5\ \text{kPa (gauge)}$$

$$= 117.5 + 101 = 218.5\ \text{kPa (absolute)}$$

Inserting appropriate values into Equation 18.11,

$$V_{exp} = \frac{7.6 \times \left[\left(\frac{1.036 \times 10^{-3}}{1.0017 \times 10^{-3}} - 1\right) - 3 \times 0.117 \times 10^{-3} \times (90 - 20)\right]}{\left(1 - \frac{218.5}{351}\right)}$$

$$= 0.195\ \text{m}^3 = 195\ \text{L} (51.5\ \text{gal})$$

Comments

Notice that the size of the expansion tank needed, though small, compared to the total system volume (it is about 2.5% of the system volume in this example) is essential; otherwise, the piping would leak and even rupture.

18.6 Piping Systems Design

18.6.1 Design Considerations

Various issues pertinent to hot water and steam systems have been discussed in the previous section. Much of the information presented on this topic also applies to chilled water systems design, but the latter are generally less involved. The objective of pipe design is to deliver the amount of water* at the proper temperatures needed to meet the loads in each zone of a building. As with duct design (discussed in Section 16.6), there are a number of constraints to piping design. They include available space, the need to meet loads in a wide variety of zones under all load conditions, economic criteria, and minimal operating costs. The piping designer's task is to determine the needed water flow in a building including the primary heating and cooling systems as well as the water-based secondary system, if any. All-air HVAC systems may not need a secondary liquid system.

After the first flow estimation step based on zone loads, a tentative primary and secondary piping system is laid out on a set of preliminary drawings. Once a layout has been made, pipes are sized based on needed heating or cooling rates at each central plant component and in each zone for each terminal device. The online HCB software contains a table that can be used to assist with quick sizing of water, steam, condensate, and natural gas pipe sizing. Pressure drop calculations can be made at this point, and a pump is selected. A cost estimate is made. The next design refinement accounts for potential flow imbalances in the original design. Ultimately, a set of final design drawings is prepared.

Piping is manufactured from various materials including steel, copper, polyvinyl chloride (PVC), and stainless steel. For hot water and steam, black steel piping is used; for chilled water (and refrigerants such as R-22), either steel or hard copper is used. Cold water and condensate drainage use either metal or PVC piping. Chemically, reactive liquids require stainless-steel or glass-lined steel piping.

Piping systems consist of pumps, filters, valves, piping runs, pipe fittings, tanks, heating and cooling generation

* We will use the term "water" generically to include antifreeze solutions as well.

and transfer devices, and specialty items depending on the application. The velocity through piping systems is limited by criteria related to either by noise (due to air entrained in the water) or by pipe erosion (due to air bubbles and fine sediments in the water). Flow separation at bends and cavitation at changes in cross section are additional reasons for noise. An extreme example is the "water hammer" caused by a valve being closed abruptly resulting in the momentum of the water being converted to series of pressure waves. In piping runs in general (for heating and cooling fluids), velocities should be in the range of 4–10 ft/s (1.2–3 m/s) for pipe diameters less than 2 in. (5 cm). A typical design value of 4 ft/s (1.2 m/s) is often used to control noise near occupied spaces. Even if noise is not a problem, velocities in copper piping should be limited to 7 ft/s (2 m/s) to avoid erosion; for steel pipes 2 in. (5 cm) or less, velocity should be less than 15 ft/s (5 m/s). For large diameter pipes, a preliminary design option for pressure drop is 4 ft of water head for 100 ft (0.4 kPa/m). The final design may deviate from this value depending on an economic analysis that considers pipe cost, pumping power, and insulation prices.

Piping systems are usually configured such that they have several branches in parallel. In such cases, valves have to be used to *balance* the system. Each valve is throttled to a position that creates an additional pressure drop across the whole branch, thereby resulting in the specified or required flow rate. The pressure losses described in Section 16.2 are valid only for fully open valves and do not provide any information on partially open valves. The longest circuit in the system is selected, the corresponding pressure drop calculated with valves fully open, and this determines the pressure drop in the other circuits as well.

18.6.2 Design Methodology

The general design procedure for hydronic piping systems (which may contain several parallel branches) can be summarized as follows:

1. Select the type of terminal units best suited for application.
2. Choose the type of piping arrangement best suited for application.
3. Sketch the piping system and terminal units—select the shortest piping length to the largest load.
4. Assume (guess) the trial value(s) of water temperature drop(s) across terminal units.
5. Calculate the corresponding water flow rates.
6. Check whether flow rates and flow velocities are within manufacturer recommended range (should meet the design criteria: *open loop*, 5–10 ft/s, and closed loop 4 ft/s if the diameter is <2.0 in.).
7. If not satisfactory, go back to step 4 and assume another value. Iterate if necessary.
8. Calculate the pressure drop across various sections of the piping circuit.
9. Determine the largest pressure drop and select the appropriate pump.
10. Balance the circuit using valves as necessary.
11. Calculate the water temperature change through each unit, based on the capacity of each unit.
12. Select an appropriate supply temperature of water.
13. Determine the water temperature entering and leaving each terminal unit.
14. Select terminal units from manufacturer catalogs.

The following sample example will illustrate part of the design procedure, specifically steps 4 through 7 for a common type of closed-loop system.

Example 18.4: Determining Pressure Drops of Closed-Loop Circuits

Consider the system layout with pipe lengths, pipe diameters, and water flow rates as shown in Figure 18.21. This may be the cooling system of a large building with a single chiller supplying chilled water to two heat exchanger coils CC_1 and CC_2. There are three globe valves (V_1, V_2, and V_3) and one gate valve (V_4) as shown. The pressure drop across the chiller is given as 10 psi (68.9 kPa) and that across the cooling coils as 4 psi (27.5 kPa). Other auxiliary equipment have been neglected in order to keep this example simple. The appropriate pump head has to be determined for this system to operate as specified.

Given: Table 16.4 to determine equivalent pressure drops of straight pipe.

Assumptions: Steady flow is assumed and pressure drop at pump suction and discharge are not considered.

Find: Δh_{pump}

Solution

Since this is a closed loop, we are not concerned with height differentials. It is simpler to organize the calculations in tabular form as shown in Table 18.6. Let us assume that the pipe diameters are 3 in. for the section 6–1–2–3 since it carries 200 gpm and the remaining piping to be 2 in. The velocities from Figure 16.3 are 3.25 and 2 ft/s, respectively, which are acceptable design values (one could also have selected slightly smaller pipe

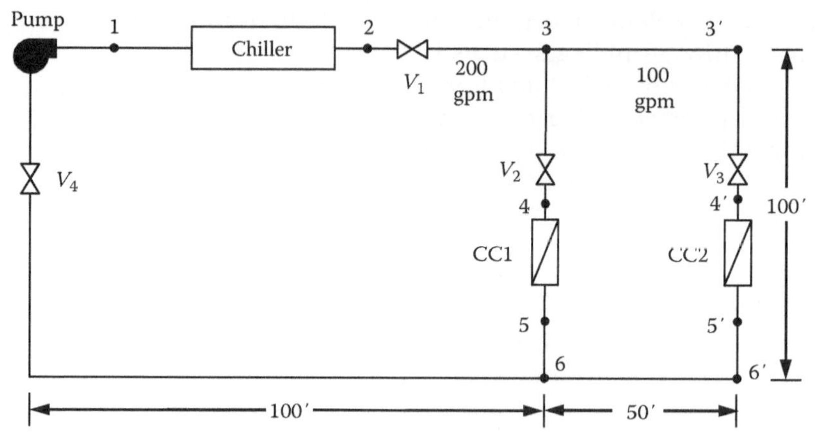

FIGURE 18.21
Sketch of the hydronic system analyzed in Example 18.4.

TABLE 18.6
Pressure Drop Calculations for Example 18.4

Section	Flow, gpm	Component	Pipe Diameter, in.	Velocity, ft/s	Effective Pipe Length, ft	Head Loss, ft/100 ft	Total Head, ft
		Straight pipe			300		
		Gate valve V_4 (open)			3.2		
6–1–2–3	200	Globe valve V_1 (open)	3	3.25	82	1.4	5.6
		90° elbows standard (2 nos.)			8 × 2		
				Total	401.2		
		Chiller (10 psi)					21.9
						Total	27.5
		Straight pipe			100		
3–6	100	Globe valve V_2 (open)	2	2.0	55	0.7	1.25
		Side flow tees (2 nos.)			12 × 2		
				Total	179		
		CC1 (4 psi)					8.76
						Total	10.0
		Straight pipe			200		
		Globe valve V_3 (open)			55		
3–3′–6′–6	100	Straight flow tees (2 nos.)	2	2.0	5.5 × 2	0.7	1.94
		90° elbows standard (2 nos.)			5.5 × 2		
				Total	277		
		CC2 (4 psi)					8.76
						Total	10.7

diameters). The corresponding pressure drops in straight length pipes in feet of water head across 100 ft of straight pipe are also read off from Figure 16.3 and inserted in the table. The straight length equivalent values for the fittings have been read off from Table 16.4 and inserted in the table. For example, the equivalent length for the gate valve V_4 in the 3 in. diameter pipe is 3.2 ft. The pressure drop across the chiller and the cooling coils are easily converted in head and have also been tabulated.

Note that the calculations have been done with the valves fully open. As a result, the pressure drops across the parallel branches 3–6 and 3–3′–6′–6 are not equal. There is an imbalance of 1.4 ft of water head, and this has to be compensated for by partially closing valve V_2 so as to create an additional pressure drop, thereby allowing the required flow rates to be realized. This aspect is further discussed in the next section.

To determine the total pressure drop, the pressure drop across only the larger of the two parallel paths has to be selected, namely, 3–3′–6′–6. Thus, the total pressure head that the pump has to overcome is (27.5 + 10.7 =) 28.2 ft of the water head (8.8 m of water head) with a flow capacity

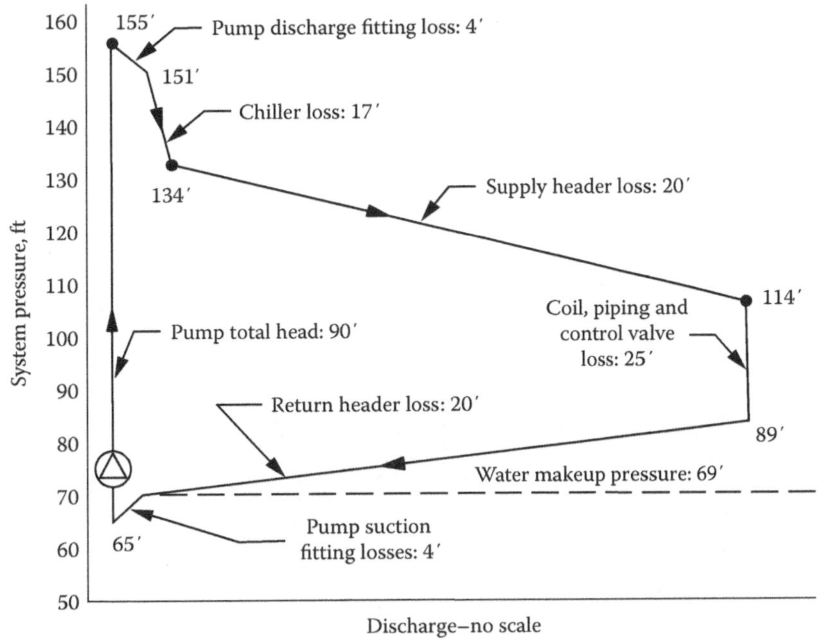

FIGURE 18.22
Example of a hydraulic pressure gradient diagram for a piping layout.

of 300 gpm (18.9 L/s). The designer can then consult the pump curves of several models and manufacturers and select an appropriate pump.

The concept of pressure gradient as a means of evaluating the hydraulic losses along the length of the piping system has been presented earlier in Section 16.4.1 (and by Figure 16.16) for air ducting systems. Such plots are also often generated for piping systems to provide insights into whether the design needs modification. For example, if the pressure drop in the piping compared to the other components is high, the designer may wish to evaluate larger pipe diameters. Such a hydraulic gradient diagram is shown in Figure 18.22 for a piping system.

18.7 Modulating Valves and Capacity Control

Example 18.4 illustrated how valves could be used to balance a hydronic system. An allied aspect is flow control. During operation, it is necessary to control the flow of water to heating/cooling coils in response to load variations. There are two control methods. The simple on–off option is only realistic for simple system with only one zone and involves (1) using the room thermostat to cycle the primary equipment (boiler or air conditioner) and (2) a water temperature sensor or aquastat placed at the supply or return of the load. If the actual value differs from a predetermined set point, the pump can be either cycled with the load device as is common for residential systems or left running as is usually done in commercial hot water or chilled water systems.

For more complex buildings with multiple terminal units, modulating valves are first used to control water flow to the individual units, and the source system is then controlled so as to adequately meet the capacity requirements of the numerous units. Control valves vary the amount of water flowing through the terminal unit; there are two common types as illustrated in Figure 18.23. The two-way valve (Figure 18.23a) is varied from full-open to full-closed position so as to modulate the flow rate and maintain a preset temperature difference across the terminal unit or coil. Hence, the flow rate in the branch is varied. In the three-way mixing valve (Figure 18.23b), the flow rate to the branch remains constant, but part of the water flow is diverted by the balance valve so as to maintain the same preset temperature difference across the terminal unit. However, since the flow through and the bypassed water flows mix downstream of the load, the temperature drop across the entire branch changes with the load. Thus, a two-way valve is a constant ΔT but variable-flow device, while a three-way mixing valve is a constant flow but variable ΔT device.

In terms of overall load control of the primary device (say a chiller), the performances of two-way and the three-way valves are not identical. The basic operational difference is that when say the loads decrease,

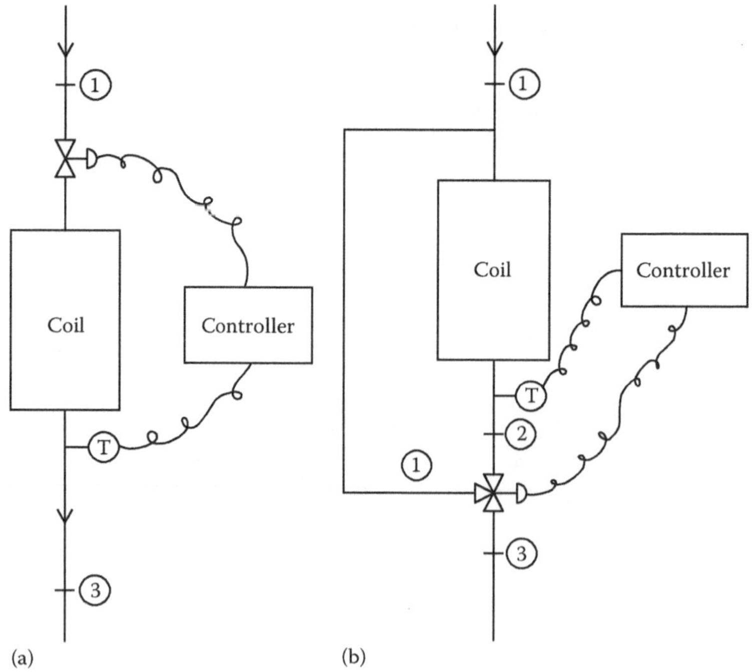

FIGURE 18.23
Flow control through coils. (a) Two-way valve and (b) three-way valve.

the water flows in a system controlled by two-way valve has to be reduced, thereby causing the primary equipment (chiller or boiler) to operate under a variable load condition. This is not the case for a hydronic system with three-way valves, which allows a constant flow to be maintained across the primary equipment. Three-way valves have been more popular as a result even though they are more expensive. Two-way valves are cheaper and are more energy efficient when combined with staged or variable-flow pumping in large systems (as discussed in the next section). Details of the flow characteristics of different types of valves are presented in Chapter 21 that deals with HVAC control systems.

A valve can be modeled as a variable resistance in a fluid flow stream. A term called "valve authority" is defined as $[\Delta p_v/(\Delta p_v + \Delta p_{sys}]$, where Δp_v is pressure drop across the valve when fully open and Δp_{sys} that across branch circuit without the valve (Brandemuehl et al., 1993). It is a measure of the extent to which valve operation can affect the flow rate. The authority is a useful design criterion when specifying a control valve. For a side branch of a hydronic circuit with a coil, a typical design value for valve authority is 0.5. That implies that if we neglect losses due to straight pipe and fittings, the valve has approximately the same pressure drop as that of the coil under full load conditions. More detailed discussion is provided in Section 21.4.2.

18.8 Large Cooling Systems

18.8.1 Distribution Piping Configurations

Different types of hydronic distribution configurations were discussed previously in Section 18.2. In essence, the same types of configurations also apply, but the fact that campuses and large buildings use multiple-pump stations with numerous terminal units requires special considerations. A single loop may heat or cool many zones on several floors of a tall building. The layout of a piping loop to minimize installation and operating costs involves several trade-offs. For example, consider the two-pipe system discussed earlier. Figure 18.24 shows the two generic designs of systems used for connecting multiple cooling coils to a central plant (this discussion applies to heating systems as well). Recall that the system (Figure 18.24a) uses less piping than the reverse return but has major flow imbalance problems since the flow resistance through branch 1 is much less than that through branch 4 with its greater length of piping. A design which specifies additional control valves (as shown in Figure 18.24a) uniquely to respond to load and are also meant to balance flow should be avoided.

The reverse-return approach, shown schematically in Figure 18.24b, has approximately equal pipe lengths for each cooling coil. As a result, the flow is balanced automatically. In fact, if the pressure drop through the

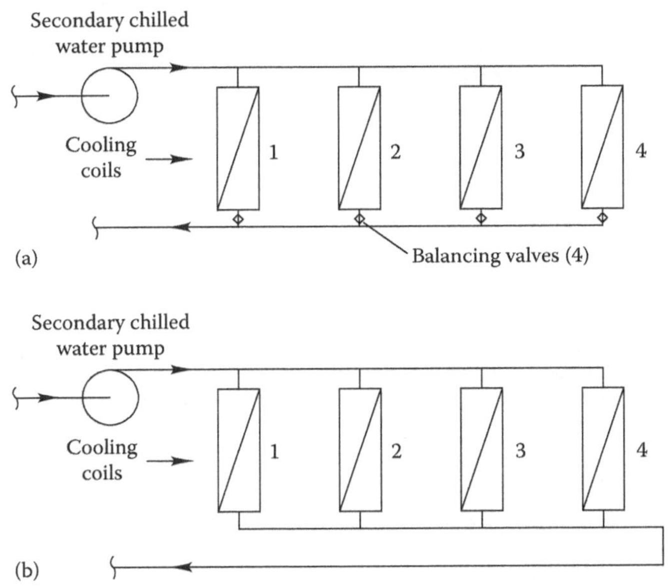

FIGURE 18.24
Piping configurations for multiple cooling coils. (a) Direct return and (b) reverse return.

header is one-tenth that through the branch lines, flow will balance to within 3% in each branch. Balancing valves can often be eliminated, but an additional length of pipe is needed on the longer-return pipe run. Each cooling coil can be controlled more properly by this approach. The earlier discussion is illustrated by Figure 18.25, which shows the pressure profiles for the two system configurations. Though the total head on the central plant is unchanged, the pressure drop across each of the individual loads is much more balanced in the reverse-return configuration. The large differences in the pressure at the three parallel branches in the straight return piping layout have to be equalized by using balancing valves, which is a control requirement hard to maintain over time.

In complex systems, there may be conflicting flow requirements in various parts of a circuit. For example, a chiller may be designed to operate at a 10°F (5.5°C) temperature change in the evaporator, but the designer may specify a 20°F (11°C) temperature drop across the cooling coils to save pumping power. Or to save pumping costs, a variable-flow distribution system may be used on

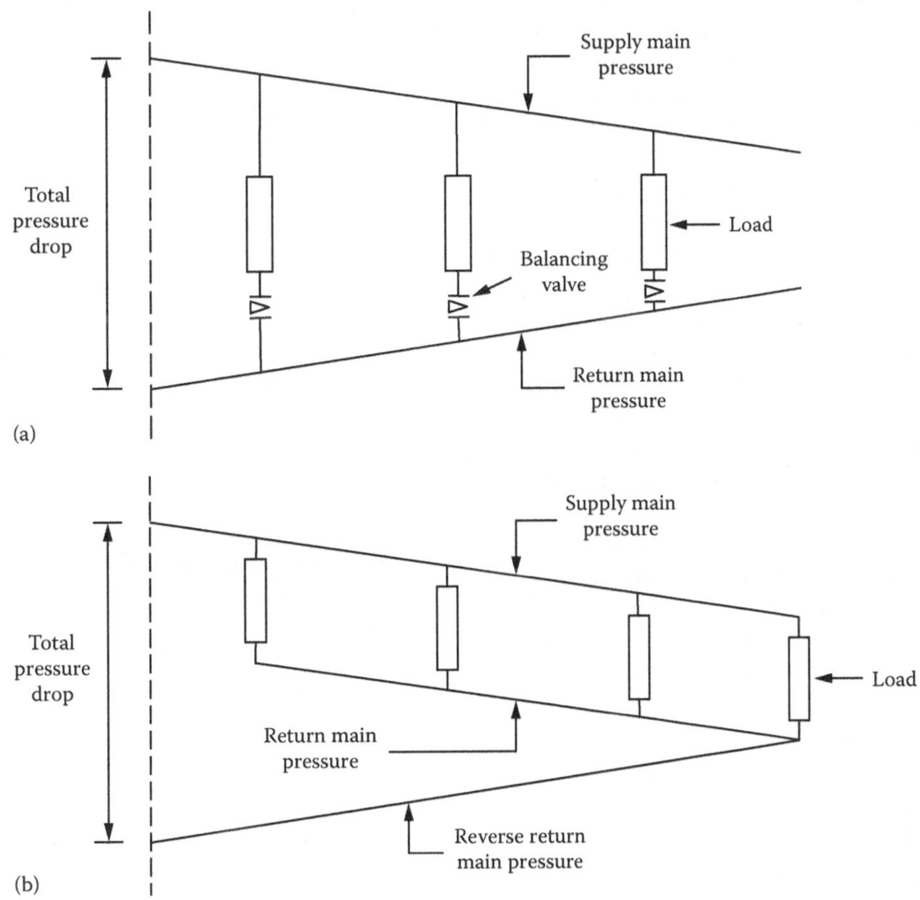

FIGURE 18.25
Pressure profiles for two-pipe systems. (a) Direct return and (b) reverse return.

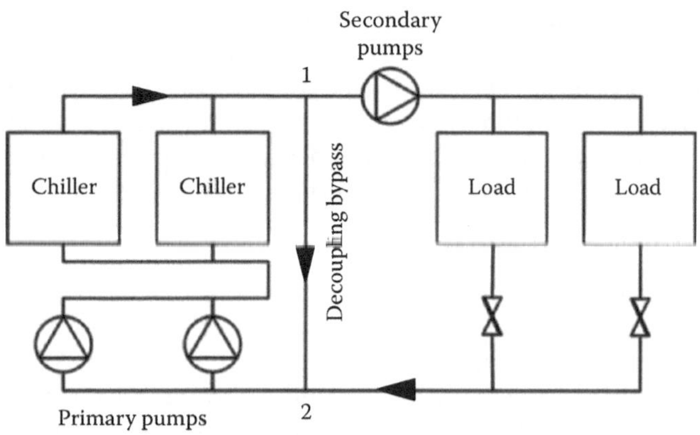

FIGURE 18.26
Variable-flow chilled water system with common pipe to decouple the primary and secondary loops. Common pipe is designed to be short and of large diameter so that pressure drop is very small relative to system pressure drop.

the building side (either by multiple-pump staging or by variable-speed pumping), while the chillers require constant flow. One method of overcoming the control difficulties caused by variable relationships between chiller(s) and building loads in such cases is to use a *primary and secondary* loop design, as shown in Figure 18.26.

The basic concept of primary–secondary systems design is to have a relatively short common section of pipe between the two systems that hydraulically decouples the two loops. The primary circuit flow is determined by *central plant* flow specifications, and the secondary circuit flow is determined by requirements of the *secondary* distribution system. Note that the use of two-way valves allows the secondary flow rate to be varied continuously as the total load varies. Thus, the common practice is to use variable-speed pumps in the secondary loop, so as to save pumping power under low load conditions. If the common-pipe section is short and of appropriate large diameter, its pressure drop will be very small and the flow in one loop will be independent of flow in the other. The flow in the common pipe can be in either direction depending on the relative magnitudes of the primary and secondary loop flows. The direction can change during a day depending on loads or on operating system on and off times. The primary flow rate is normally larger than the secondary flow rate. Supply chilled water temperatures are close to 45°F (6.7°C), while return water temperatures are 55°F–64°F (12.8°C–17.8°C).

Example 18.5: Primary and Secondary Loop Flows

The flow through a chiller is 500 gal/min (32 L/s). At a given time of day, the load on the secondary system is 625,000 Btu/h (183 kW) with a temperature rise of 10°F (5.5°C). Find the secondary loop flow and the temperature drop across the chiller–evaporator under these operating conditions.

Given: $\dot{V}_{ch} = 500\ \text{gal/min}$

$$\dot{Q}_{ch} = 625{,}000\ \text{Btu/h}$$

$$\Delta T_{coils} = 10°\text{F}$$

Figure: See Figure 18.26.

Assumptions: Steady flow is assumed, and heat transfer from the surface of primary and secondary piping is ignored.

Find: $\dot{V}_{coils}$, $\varnothing T_{ch}$

Lookup values: $\rho_w = 8.33\ \text{lb}_\text{m}/\text{gal}$, $c_{p,w} = 1\ \text{Btu/(lb}_\text{m}\cdot°\text{F)}$

Solution

The temperature drop across the chiller–evaporator can be found from the sensible heat balance equation:

$$\dot{Q}_{ch} = \dot{V}_{ch} \times \rho_w \times c_{p,w} \times \Delta T_{ch} \tag{18.12}$$

from which the chiller water temperature drop is

$$\Delta T_{ch} = \frac{625{,}000\ \text{Btu/h}}{500\ \text{gal/min} \times 8.33\ \text{lb}_\text{m}/\text{gal} \times 60\ \text{min/h} \times 1\ \text{Btu/(lb}_\text{m}\cdot°\text{F)}}$$
$$= 2.5°\text{F}$$

The conservation of energy, namely, $\dot{Q}_{ch} = \dot{Q}_{coils}$ allows us to determine the flow rate through the coils of the load (the density of water in primary and secondary loops is the same):

$$\dot{V}_{coils} = \dot{V}_{ch} \times \frac{\Delta T_{ch}}{\Delta T_{coils}}$$

$$= 500\ \text{gal/min} \times \frac{2.5°\text{F}}{10°\text{F}} = 125\ \text{gal/min} \tag{18.13}$$

By the conservation of mass at either tee joining the two loops, the flow in the common pipe is found to be

$$\dot{V}_{common} = 500 - 125 = 375\ \text{gal/min (23.7 L/s)}$$

The direction of flow is from tee 1 to tee 2 as shown in Figure 18.26.

Comments

As load decreases in a secondary cooling loop, so must either the flow or the fluid temperature rise. The two methods of control commonly used (two- or three-way valves controlled by a coil output or zone temperature sensor) were discussed in the previous section.

18.8.2 Rules for Chiller System Operation and Control

Taking into account the part-load characteristics of chillers, cooling towers, fans, and pumps, Braun et al. (1989) have developed a set of rules for minimizing the energy consumption of chiller plants in commercial buildings. In a case study, the authors showed that chiller energy consumption could be reduced from 26% to 43% compared to conventional fixed-speed equipment, if these rules are followed. The designer will use this information in preparing the chiller control specification.

1. If the cooling tower is constructed of several cells with variable-speed drives, operate all cells at the same speed.
2. If the cooling tower is constructed of several cells with multiple-speed fans, increment the lowest-speed fans first when additional tower capacity is needed.
3. Variable-speed condenser water pumps should be controlled with their associated chillers to give peak pump efficiency. If chillers have more than one fixed-speed pump, all pumps should be controlled so that they all operate at the same speed (assuming the common practice that the several pumps are all the same model). As described earlier, *great care must be used if cooling tower water flow is to be modulated*, since flow imbalance and freezing may occur.
4. Multiple-chiller plants should be operated so that they all have the same chilled water set points and so that condenser and evaporator flows are proportional to each chiller's cooling capacity. There is not much difference in energy consumption for multiple-chiller plants operated in series or parallel.
5. Parallel air handlers should all operate with the same air set points.
6. Optimal sequencing of multiple chillers depends on the specific characteristics of the chillers under part-load operation, as described in Section 14.7. No general rules exist.

It was possible to find this relatively compact and simple set of rules because all the key characteristics of an *optimally operating chiller plant*—chilled water temperature, supply air temperature, cooling tower airflow rate, and cooling tower water flow rate (if allowed to vary)—are related nearly linearly to cooling load and wet-bulb temperature. Control of building relative humidity (rarely done except for critical applications) imposes an additional constraint on the chiller plant. Under typical commercial building conditions, humidity control will always result in more energy consumption compared to the case when it is not controlled.

There are numerous technical articles written by very experienced consulting engineers on the issue of optimized design and control of chilled water plants (see, for example, Taylor 2012). The interested reader can also refer to specialized publications such as the self-directed learning book by ASHRAE-SDL (2013).

18.9 Cool Thermal Energy Storage

18.9.1 Purpose and Benefits

Thermal energy storage (TES) systems are of two types: hot water and cool storage (using either chilled water or ice storage). We shall only deal with the latter type in this section. The reader interested in heat storage can refer to the ASHRAE Systems (2012) for additional information. Generally, cool TES (CTES) is not meant to be an energy conservation option at the facility, but a cost-saving measure. The electric rate structure, i.e., the manner in which electric utilities charge customers for the electricity consumed, involves a flat customer charge for the electric connection, an energy use charge (related to the energy in kWh consumed during a month), and a demand charge based on the peak or maximum use of the facility during the month (often, based on the maximum kWh during a 15 or 30 min time period expressed in kW).* There are literally hundreds of variants of these basic charges since many of the three thousand or so electric utilities in the United States offer a variety of rate structures for different types of customers (typically, residential, small commercial, large commercial, industrial, etc.).

In essence, different utilities have different: (1) *energy use charges* that may vary with time of day and by season and also by the magnitude of the energy use (called block or tiered rates) and (2) *demand charges* that may vary by season and also by the magnitude of the demand use during that month (block rates) and/or in conjunction with a percentage of the previous year demand (ratchet rates). Usually, residential and small commercial customers are not subject to demand charges. The reason for this complexity is because of the large diurnal, weekday-to-weekend and seasonal load variations in the different buildings that also result in large fluctuations in the electric load to be met by these utilities. In order to avoid demand shortages, a utility either have to increase its own power generation capacity or enter in purchase agreements with other utilities

* Refer to Section 23.2.4 for more discussion.

usually at a high premium. These extra costs and the risk-related monetary costs (such as having to negotiate for electricity purchase over a future period in case it is needed) are then passed on to the customer base. In the majority of cases, the electric load demand is high during a period of several hours from late morning to late afternoon and low during the rest of the day. These on-peak and off-peak periods have different energy use and demand rates so as to incentivize customers to alter their load profiles to better suit the generation capability of the electric utility.

A representative example of an electric rate structure applicable to a large university campus in the southwest of the United States is as follows:

- On-peak time is from 11 a.m. to 9 p.m. during weekdays.
- On-peak rates are \$0.05/kWh and \$18/kW demand.
- Off-peak rates are \$0.03/kWh and \$3/kW demand.

Under such circumstances, it would benefit customers to try to modify their diurnal load profiles. Professional services are offered by several specialized engineering firms under the name of *demand side management* or load management. Such measures generally fall under two categories:

1. *Peak shaving* where, for example, nonessential equipment are switched off or installing demand limiting switches to other equipment or replacing old equipment or equipment shifting, i.e., say replacing a vapor compression chiller by an absorption chiller or even installing a combined heat and power system.
2. *Peak shifting* by precooling the mass of the thermal building (precooling the building during off-peak hours and setting up the thermostat in a staged manner during the on-peak hours). This strategy is being implemented in some buildings, but the more historic and widespread approach is to use an external cool storage system.

A traditional cooling plant will involve one or more traditional vapor compression chillers that will be able to meet the peak design load. Introducing a CTES system allows shifting some or all of the on-peak electric use of the vapor compression chillers to the off-peak period, thereby reducing their on-peak usage and demand. Ideally, the CTES is depleted at the end of the on-peak period of the design day, and the same chillers are used to recharge the CTES during the off-peak period such that the process can be repeated the next day. Thus, the chillers are operated at higher capacity (than the instantaneous off-peak building cooling demand) and/or longer hours, thereby incurring an increased electric consumption during off-peak hours. Many electric utilities promote CTES by offering financial incentives since the displaced electric demand reduces their diurnal peak generation electric demand while increasing electricity sales; these outcomes are advantageous for their operating generation costs and can result in deferring installation of additional generation capacity. It is worthwhile to point out a common claim made, namely, that cooling plants with CTES will actually consume less electricity. The basis of this claim is that operating the chillers during off-peak hours when the outdoor temperatures are lower than those during the on-peak period would lead to increased chiller COP, and the associated electricity savings would offset the CTES related heat losses. This claim has been found to be often not borne out by operational results in actual systems; the current thinking is that this benefit is rather circumstance specific and cannot be claimed in most instances.

18.9.2 Storage Mediums

There are essentially three types of storage mediums (Dorgan and Elleson, 1993):

1. *Chilled water* where the cooling capacity is due to sensible heat. This is the simplest method of storage, and it is easy to interface/retrofit with cooling plant piping networks. A typical set point for storage tank charging is 40°F (5°C) with a return temperature from the load of 60°F (15°C), conditions easily supplied by conventional chillers. Thus, for a 20°F (11°C) temperature difference, a storage capacity of 1 ton-h (12.66 MJ) can be provided by 10.7 ft^3 (0.34 m^3) of water. The economics favor large sizes.* For example, a typical water TES system may have a volume of 1.3 million gallons (4,900 m^3) with a storage capacity of over 14,000 ton-h (177,240 MJ) capacity with a peak output of 2,300 tons (8,000 kW) and allow an electric demand shift of 1.7 MW.

   Commercial installations with tank sizes three to four times larger have been installed. It is thermally advantageous to maintain a temperature gradient along the height of the tank. Because of

* An alternative design configuration is to use multiple tanks where better stratification can be maintained by operating the tanks separately. Further, the tanks can also be buried underground so as to be unobtrusive.

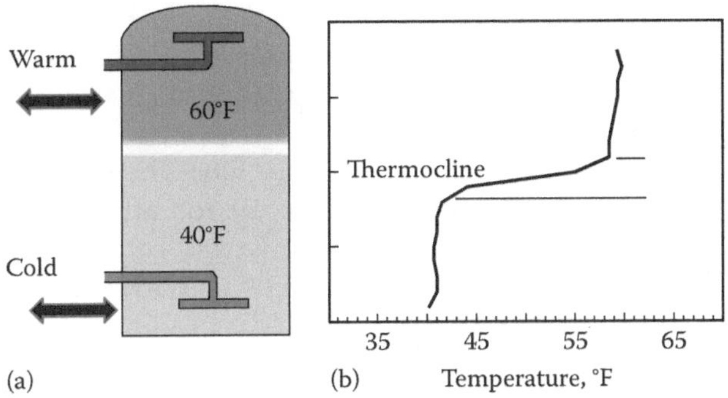

FIGURE 18.27
Naturally stratified chilled water storage tank with diffusers and typical stratification temperature profile.

buoyancy, cold water around 40°F (4.5°C) will collect at the bottom of the tank and warmer water at 60°F (15.5°C) will rise to the top of the tank. This thermal stratification tends to occur naturally and remain as such if the cold and warm water streams are fed by properly designed diffusers to reduce flow velocities (and thereby minimize mixing) as shown in Figure 18.27. Three zones tend to occur: two mixed spaces at either end of the tank and a stable layer of about 3.3 ft (1 m) in height containing much of the vertical temperature gradient, called the "thermocline." During charging, the chilled water from the chillers will be fed into the tank from the bottom and the warm water will exit from the top diffuser. This will gradually raise the thermocline layer, and thereby increase the usable cooling capacity. During discharge, the cold water is drawn from the bottom of the tank and delivered to the building load, while the warm return water is fed from the top. This would result in the thermocline moving downward.

One can define *storage efficiency* as the ratio of the amount of energy removed from storage to the amount put in during one charge/discharge cycle. It is, thus, a measure which largely accounts for the thermal heat losses from the tank walls. However, a more meaningful metric is the ratio of the capacity that can be discharged from the storage tank to its theoretical capacity, which is called the *figure of merit*. This measure includes the effects of erosion (i.e., broadening) of the thermocline due to the thermal conduction between the hot and cold layers and also of the effect of mixing in the bottom zone as the storage discharges, which causes the water discharge temperature to rise above a preset usable threshold (say, 50°F or 28°C). Typically, the figure of merit is in the range of 85%–95% for well-designed systems.

2. *Ice storage* where the cooling capacity is provided by the latent heat of fusion of ice. A refrigerant or a secondary fluid (such as glycol) is cooled to 15°F–26°F (−9°C to −3°C) by special chillers that are then used to charge the ice tank. Typical storage capacity in actual systems ranges from 2.4 to 3.3 ft^3/ton-h (or 6.3 to 8.5 L/MJ) even though the ideal ratio of 20°F (11.1°C) sensible heat capacity to latent heat is about seven times (recall that the latent heat of ice is 144 Btu/lb$_m$ or 334 kJ/kg). Rather than one large tank, the needed storage capacity is generally achieved by coupling several small tanks of 100–200 ton-h (1200–2500 MJ). One common design, called the ice-on-coil, which involves building up a concentric layer of ice on the piping (plastic piping is mostly used in the recent designs) as chilled water is circulated through the pipes, is illustrated in Figure 18.28. Several ice storage designs have been commercially developed, and some of these are described in Section 18.9.4.
3. *Eutectic salts* that are mixtures of salt hydrates and water with a solidifying temperature of about 47°F or 8°C (which varies with the composition of the salt). This allows the benefit of the latent heat of fusion to be exploited. Their storage capability is lower than ice storage, requiring about 6 ft^3/ton·h (16 MJ). In a CTES system, these eutectic salts are placed inside plastic containers that

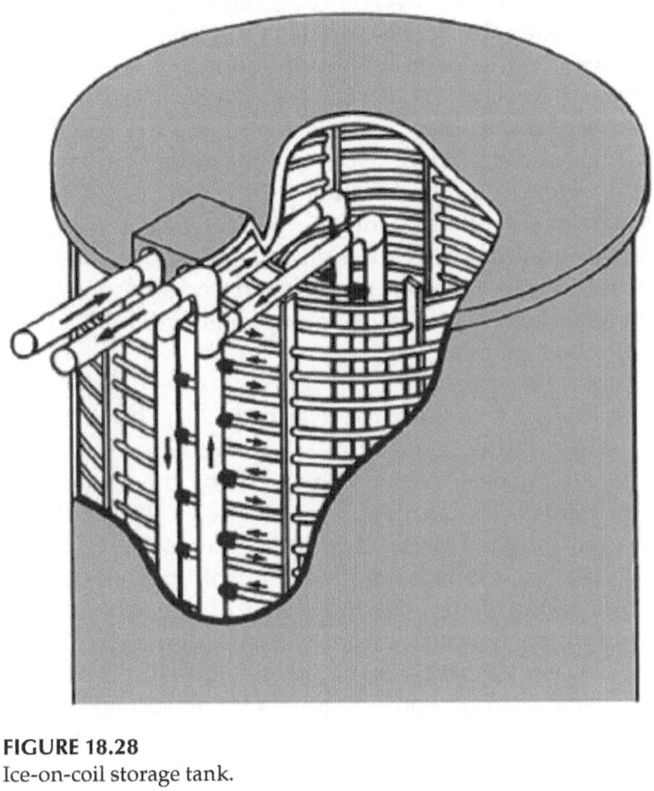

FIGURE 18.28
Ice-on-coil storage tank.

are immersed within an insulated chilled water tank. The charging is achieved bypassing chilled water between these containers that freezes the salts. During discharging, the warm water is circulated between these containers, thereby cooling the water. Hence, traditional chillers can be used and the interface with existing hydronic piping networks is convenient. The number of such systems actually installed to date is relatively small.

18.9.3 Design and Control Considerations

The design of CTES systems depends on how they are to be operated. Not only can they be operated on a diurnal 24 h cycle, but in certain large buildings (such as hotels), their operation could be on a weekly cycle. The CTES under weekly cycle is charged during the weekend when the electric rates are low and gradually discharged over the other days. This reduces initial costs since the chillers and cooling equipment such as cooling towers and pumps can be downsized. A CTES can even replace a chiller in a multi-chiller installation and can provide emergency cooling capability in critical building applications (such as hospitals or data centers). CTES can also be sized to cater to buildings that experience large thermal loads for only for a few hours of the week (such as convention centers or churches that are only fully occupied). Another extreme example is the use of CTES in closed sports stadiums where the cooling loads for an afternoon are met by large CTES systems that have been gradually charged over several days during off-peak hours. Generally, CTES systems are more viable when serving cooling loads of short duration.

Let us consider the common instance where a CTES is to be sized to meet a diurnal load and experiences a 24 h charge/discharge cycle. A typical design day can be selected to analyze and size the system (a detailed design and economic analysis will involve a complete annual simulation with numerous secondary considerations—see, for example, Dorgan and Elleson, 1993). There are usually three design options with one having two subcategories:

1. *Option 1: Full storage* (Figure 18.29a). The strategy is to shift the entire on-peak cooling load to the off-peak period. Thus, during the on-peak period, the chillers are switched off and the entire building cooling load is met by the CTES. Such a system is simple to control. Typically, the cooling system operates at full capacity during all nonpeak hours of the design day. This strategy results in the largest shift in on-peak demand but would require a larger storage and refrigerant equipment size. Rarely is this option the best design choice.
2. *Option 2: Partial storage, load leveling* (Figure 18.29b). In this case, only part of the on-peak load is shifted to the off-peak period in such a manner that the electric draw of the chillers is constant during the entire day. The cooling plant operates at full capacity for all 24 h of the design day. This design would require the minimum refrigeration capacity with moderate storage size. This strategy is particularly attractive for applications where the peak cooling load is much higher than the average load.
3. *Option 3: Partial storage, demand limiting* (Figure 18.29c). This is a middle ground strategy between the previous two design options. Here, the cooling equipment is operated at a reduced capacity during the on-peak period in such a manner that the total 24 h chiller(s) electricity cost is minimum and/or the demand is kept below a certain level. This option is usually the most cost optimal. However, the CTES should be discharged in a controlled manner that requires not only the ability to predict thermal and electric demand over the day but also a monitoring system for proper operation.

Example 18.6: Idealized Comparison of Different CTES Sizing Options

Some of the basic concepts presented earlier will be illustrated by means of an example. Consider a building whose cooling load in tons is shown in the second column of Figure 18.30. We will assume that there are no heat losses from the piping and the CTES tank and that the figure of merit of the storage tank is 100%. The on-peak period is taken to be from 11:00 to 18:00 h ending (i.e., for 8 h). The chiller sizing and hourly cooling output as well as the state of charge of the CTES under different design options are summarized on an hour-by-hour basis in Figure 18.30.

(a) For the no-CTES plant option, the maximum capacity of the chiller should be 220 tons of cooling, which is the peak cooling load of the building at hour 15:00 (second column).

(b) For the case of a full storage design option, the chiller is to be operated at constant cooling output of 168 tons only during the off-peak period and entirely switched off during on-peak periods. The required chiller size is 168 tons, which is equal to 2688 tons (the daily total building cooling load) divided by 16 (the number of hours of chiller operation). The calculations are done such that at the end of the peak period (18:00) the storage is

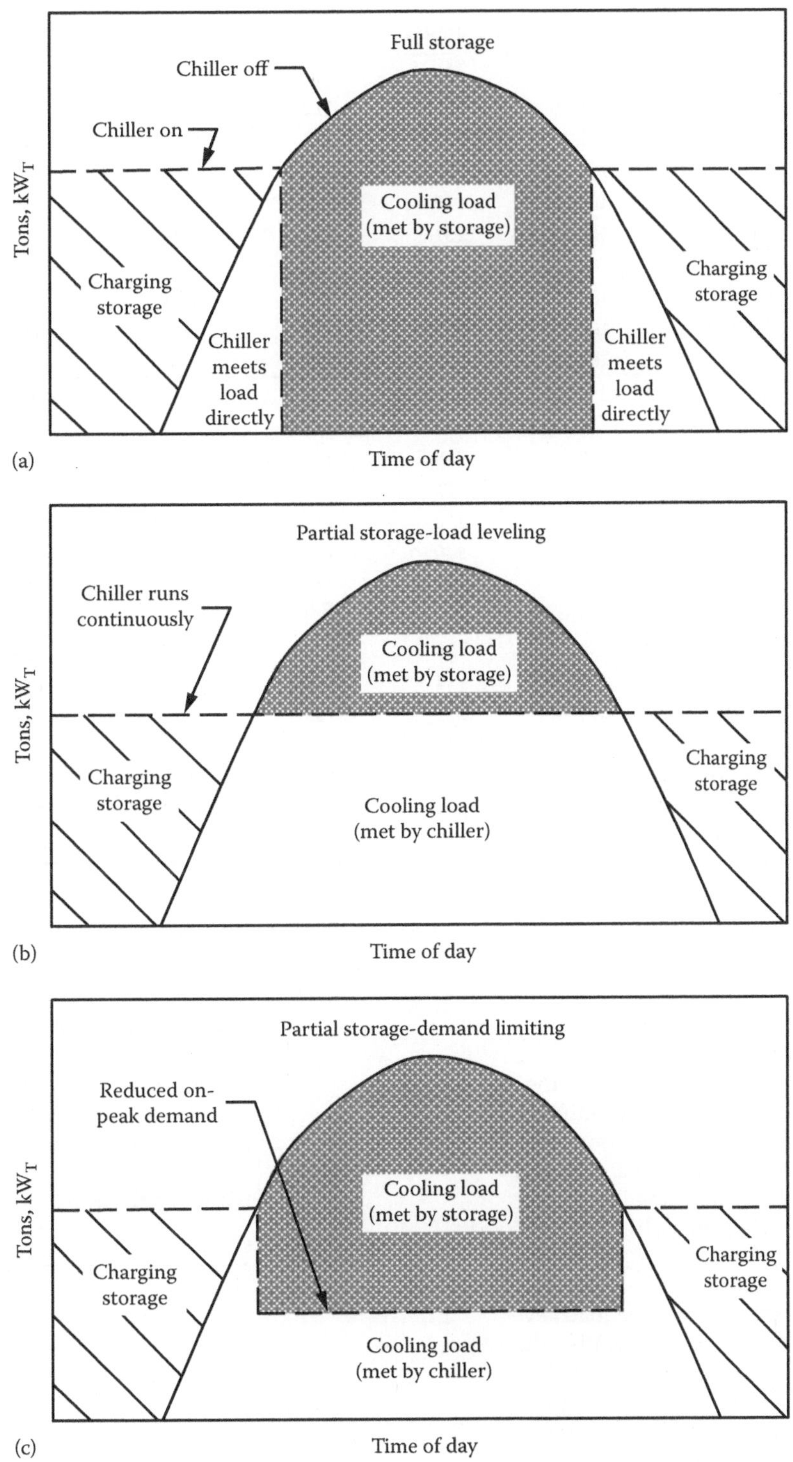

FIGURE 18.29
The three basic design options. (a) Full storage; (b) partial storage, load leveling; and (c) partial storage, demand limiting.

		Full storage		Partial storage-load leveling		Partial storage-demand limiting	
	Cooling load	Chiller	CTES charge	Chiller	CTES charge	Chiller	CTES charge
Hour ending	Tons	Tons	Ton-hours	Tons	Ton-hours	Tons	Ton-hours
1	50	168	689	112	297	138	479
2	53	168	804	112	356	138	564
3	50	168	922	112	418	138	652
4	52	168	1038	112	478	138	738
5	58	168	1148	112	532	138	818
6	65	168	1251	112	579	138	891
7	72	168	1347	112	619	138	957
8	102	168	1413	112	**629**	138	993
9	119	168	1462	112	622	138	1012
10	128	168	**1502**	112	606	138	**1022**
11	150	0	1352	112	568	60	932
12	166	0	1186	112	514	60	826
13	198	0	988	112	428	60	688
14	216	0	772	112	324	60	532
15	**220**	0	552	112	216	60	372
16	212	0	340	112	116	60	220
17	181	0	159	112	47	60	99
18	159	0	0	112	0	60	0
19	110	168	58	112	2	138	28
20	92	168	134	112	22	138	74
21	71	168	231	112	63	138	141
22	58	168	341	112	117	138	221
23	55	168	454	112	174	138	304
24	51	168	571	112	235	138	391
Total	2688	2688		2688		2688	

FIGURE 18.30
Chiller and CTES sizing analysis results for Example 18.6. The on-peak hours and cooling loads are shown boxed. Numbers in bold represent maximum values over 24 hours.

totally depleted. The storage inventory increases cumulatively as the difference between the chiller output of 168 tons and the corresponding hourly building cooling load (for example, for hour 19:00, the CTES charge = 0 + 168 − 110 = 58 and so on). We note that a smaller chiller than case (a) is adequate, while the storage capacity required is 1502 ton·h (full charge occurs at hour 10:00)

(c) For the case of partial storage load leveling, the chiller size is simply the total diurnal building cooling load of 2688 tons divided by 24, which equals 112 tons. The hourly CTES inventory is determined in a manner similar to that described in (b). The CTES capacity required is only 629 ton·h, while the full charge occurs at hour 8:00.

(d) For the case of partial storage demand limiting, we have assumed that the on-peak chiller load should be limited to 60 tons (somewhat arbitrarily). From Figure 18.30, we note that a chiller size of 138 tons and a CTES capacity of 1022 ton·h would be adequate.

The pertinent results are summarized here.

	No CTES	Full Storage	Partial-Load Leveling	Partial-Demand Limiting
Chiller plant size, tons	220	168	112	138
CTES capacity, ton·h	—	1502	629	1022

Of course such a design analysis should be done for all days of the year and adopt a life cycle costing approach (see Section 23.2). For the aforementioned example, the on-peak and off-peak electric consumption and demand would be determined for each month based on the chiller COP and part-load performance. If the on-peak and off-peak utility rates were provided, one could calculate the annual cost savings for each of these three design choices (strictly speaking, for the partial storage demand limiting case, one should optimize for the chiller design capacity and not simply assume an on-peak value of 60 tons as selected). The optimal design selection will also be dictated by the chiller costs, the cost of the CTES, and several factors that influence the initial costs and operating costs of the system (for example, pumping power). This example is simply meant to illustrate an idealized thermal analysis.

The manner in which the chillers and the CTES system are configured and operated during periods when the loads are below the design loads greatly impacts the electric energy consumed by the cooling plant. The operation has to be properly optimized, and the interested reader can refer to specialized research publications, for example, Henze (2003) for further details.

The arrangement of the chillers and the CTES can be classified as either series or parallel. Figure 18.31 shows the two different *series configurations*, both of which provide the ability to operate the chiller and CTES system individually or in unison at different levels of contribution. An advantage with this layout is that it does not require a change in flow path during

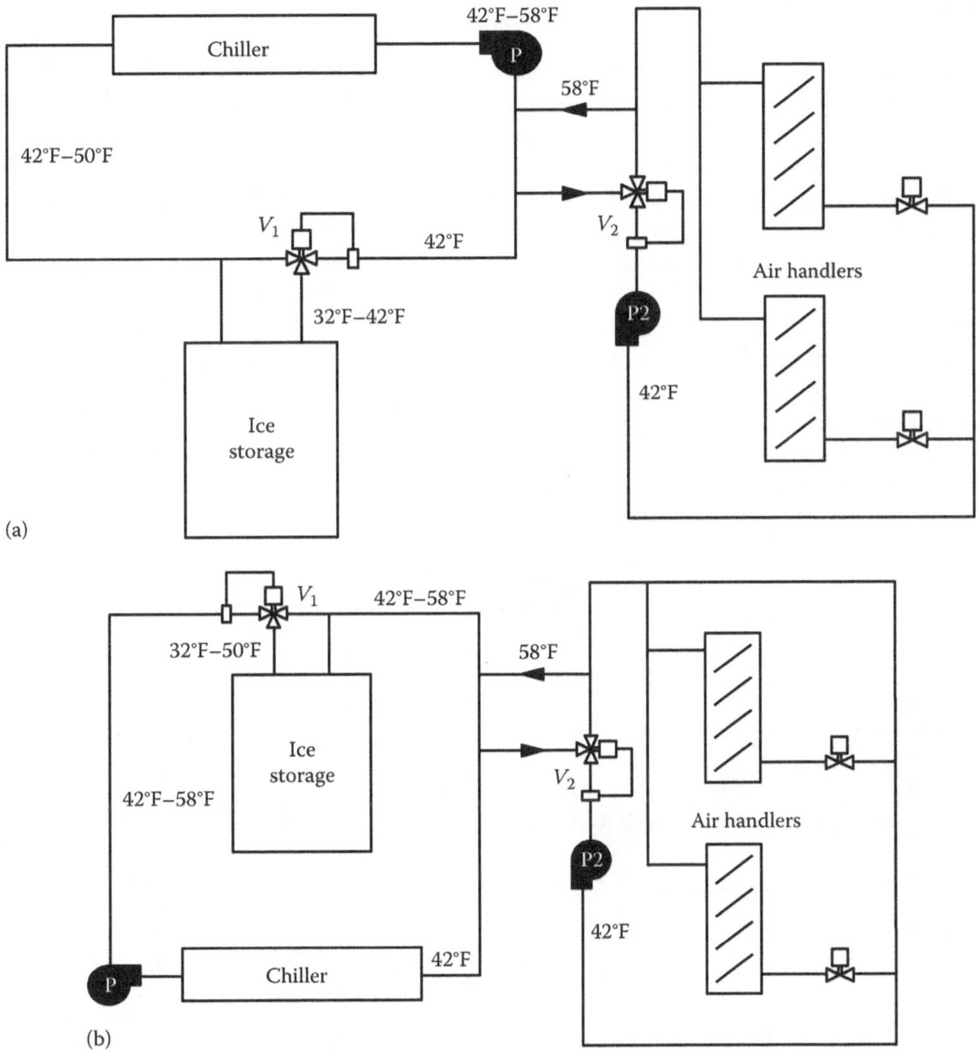

FIGURE 18.31
The two series flow cooling system configurations. (a) Chiller priority and (b) storage priority. (From Lorsch, H.G., *Air-Conditioning Systems Design Manual*, American Society of Heating, Refrigerating and Air-Conditioning Engineers, Atlanta, GA, 1993.)

charging and discharging. The first figure (Figure 18.31a) depicts the chiller upstream of the CTES and is referred to as a "chiller-priority configuration." This configuration allows a flexible operating control strategy; the chiller meets as much of the load as possible and the three-way control valve allows flow through the CTES to provide supplemental cooling in case the load exceeds the capacity of the chiller. Since the chiller sees warm fluid returning from the building load, its operation would be more efficient (higher COP). On the other hand, the usable portion of the total nominal storage capacity is reduced because of the lower storage temperature. This strategy is used in instances where the cost of stored energy is higher than the cost of direct cooling or when the utility savings through on-peak demand reduction is high. The control is simple to implement and is thus commonly used in most chilled water systems and in ice storage systems as well.

Figure 18.31b shows the chiller downstream configuration, also referred to as "storage priority," which aims at depleting as much of the cooling from storage as possible prior to requiring supplemental assistance from the chiller. While the chillers would tend to operate less efficiently, a higher usable storage capacity can be achieved. However, the control sequence tends to be more complex.

There are several variants of the parallel flow configuration, one of which is shown in Figure 18.32. This is more suitable for retrofit applications that require a fixed temperature difference. The relative contribution of the chiller and the storage varies during the discharge cycle. Since both of them have the same entering water temperature, the control is easily achieved by setting a fixed leaving temperature. A disadvantage is that there are many more interconnections between the chiller and the CTES to accommodate the changes in flow paths between charging and discharging cycles (ASHRAE Systems, 2012).

18.9.4 Ice Storage Designs

A number of different ice storage technologies have been developed commercially. The most common ones are as follows (Dorgan and Elleson, 1993):

1. *Ice harvesting* produces high continuous discharge rates by forming sheet ice on the evaporator surface of the chiller which is periodically released into a water-filled storage tank.
2. *Encapsulated ice* wherein water inside submerged containers freeze and thaw as cold or warm coolant is circulated through the tank holding the containers.
3. *Ice-on-coil designs* where ice is formed on submerged pipes or tubes (often made of plastic) through which refrigerant or secondary fluid (glycol-water) is circulated during charging. Typically, the thickness of the ice around the tubes is about 1.5 in. or 3.5 cm. Discharging can be done in one of two ways (see Figure 18.33). The external melt discharge occurs by circulating brine (say 25% ethylene glycol and 75% water) through the space that surround the ice pipes, and the melting of the ice occurs from the

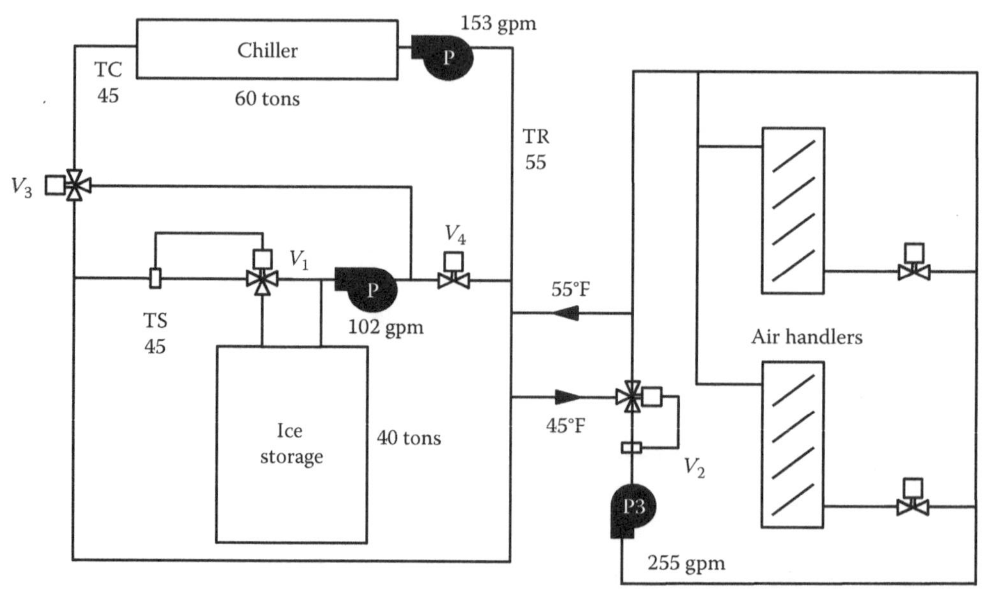

FIGURE 18.32
A commonly used parallel flow chiller and storage system configuration.

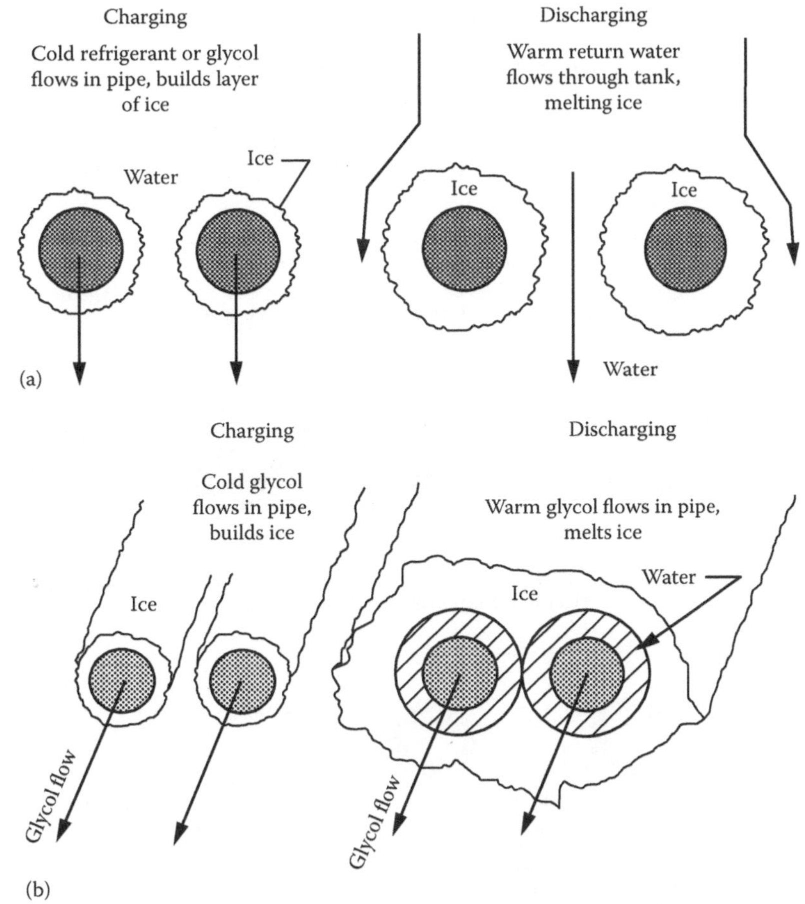

FIGURE 18.33
Ice on coil storage configurations. (a) External melt and (b) Internal melt. (From Dorgan, C.E. and Elleson, J.S., *Design Guide for Cool Thermal Storage*, American Society of Heating, Refrigerating and Air-Conditioning Engineers, Atlanta, GA, 1993.)

outside down to the tubes. The discharge for the internal melt design involves passing the brine inside the tubes and letting the ice melt from the outer tube annulus outward. For energy-efficient operation, all ice should be melted each day so that the charging cycle for the next day starts with bare pipes. Both these variants are suitable for use with the three systems design options described earlier. However, storage priority is usually recommended for external melt systems, while chiller priority is more suitable for internal melt systems.

The accurate modeling of charging and discharging processes is fairly complex. During charging, the water outside the tubes has to be first cooled sensibly down to the freezing point. In that case, the thermal resistance network is represented by Figure 18.34a. There are three resistances in series: the internal fluid to wall convective resistance, the resistance of the tube wall, and the external convective resistance of the water around the pipes. Once the freezing point is reached, ice starts to build up outside the tubes and heralds in an unconstrained growth phase. An additional conductive resistance of the ice layer has now to be included in the thermal network (see Figure 18.34b). This is a resistance that increases as the ice layer gets thicker. Once the ice layer formed around adjacent tubes touch one another, a third mode of charging, namely constrained charging then takes over (Figure 18.34c). When operating an external melt ice-on-coil system, one should limit the buildup of ice around the pipes and minimize bridging of ice between neighboring tubes so as not to penalize the discharging process. When internal melt systems are discharged, the ice layer around the tubes starts melting outward, i.e., an annulus of freezing water is formed first between the tubes and the outer ice annulus, which then gradually increases in diameter. The thermal resistance offered by this growth is rather difficult to model accurately. Typically, the tubes in an internal melt system occupy 10% of the tank volume, while another 10% is left empty to allow for expansion of water when it freezes. The ice-on-coil internal

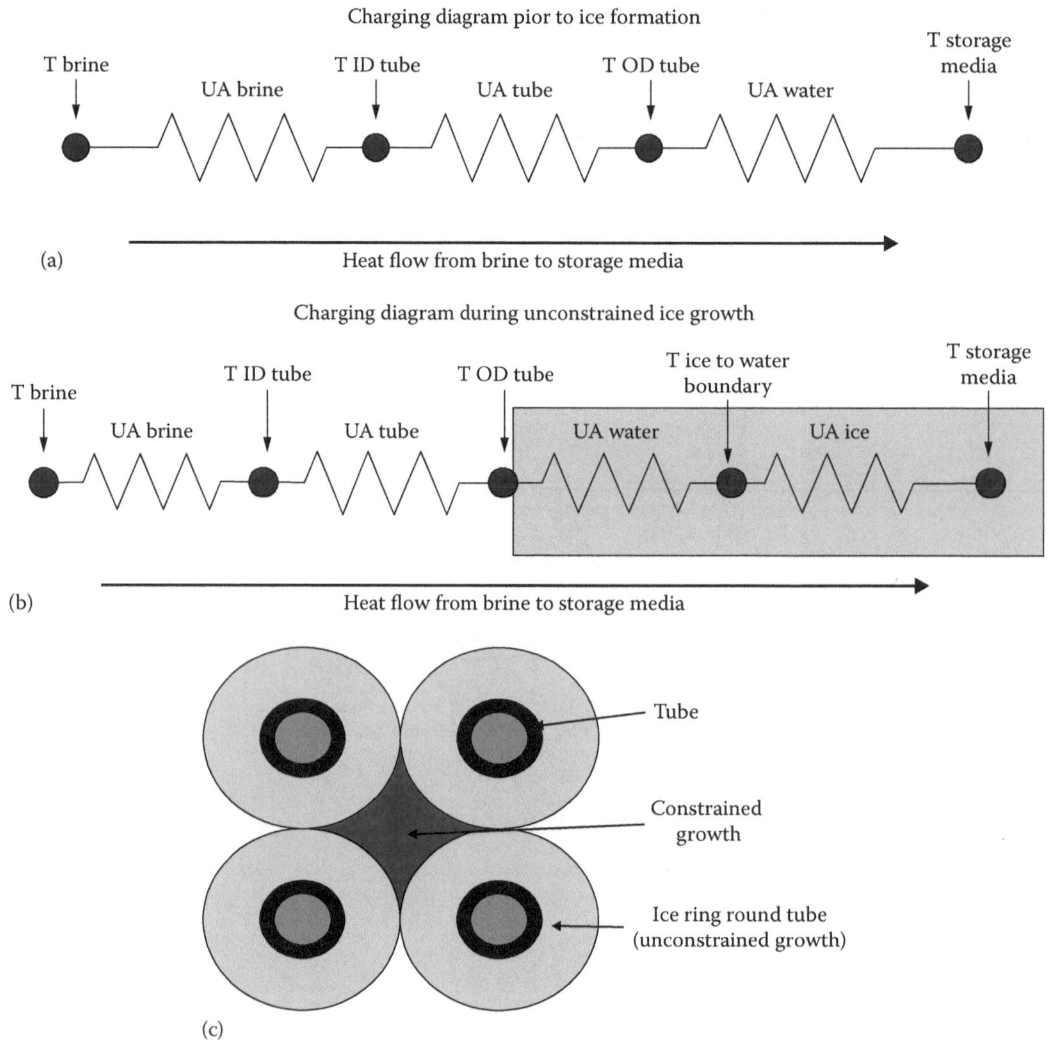

FIGURE 18.34
Different stages of charging an ice-on-coil system: (a) first stage where ice has yet to be formed, (b) second stage of charging where ice forms around the coils and grows in an unconstrained manner, and (c) constrained growth where the ice layer is formed around adjacent tubes that touch one another and only the water in between the tubes is gradually frozen.

melt design is perhaps the most prevalent among ice storage systems installed in recent years.

The cooling rate and the outlet temperature of an internal melt ice storage system vary with time (or more accurately as the CTES is gradually discharged), and this behavior is shown in Figure 18.35. The thermal cooling draw will be high during the initial stages and gradually decrease to zero as the cooling energy stored in the tank gets depleted. The outlet water temperature will be low initially and gradually rise to a value close to the inlet water temperature. These patterns also depend on the flow rate of the secondary fluid being circulated through the tubes.

There are several technical papers that have developed detailed models for ice-on-coil systems. These are rather cumbersome to use for practical design and sizing purposes. An elegant modeling approach is that

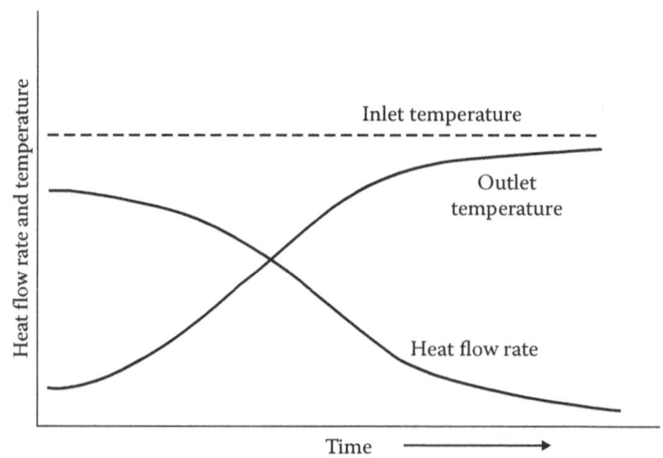

FIGURE 18.35
Outlet temperature and heat flow from an ice storage tank.

proposed by Jekel et al. (1993) who applied the concept of heat exchanger effectiveness (see Section 12.6.3) to model the time-dependent thermal behavior of ice storage tanks. Recall that the effectiveness is the ratio of the actual to the maximum heat transfer rates. In this case, the effectiveness ε_{tank} is defined by

$$\varepsilon_{tank} = \frac{(T_{tank,in} - T_{tank,out})}{(T_{tank,in} - T_{freezing})} \tag{18.14}$$

where

$T_{tank,in}$ and $T_{tank,out}$ are the inlet and outlet temperatures of the circulating water

$T_{freezing}$ is the freezing water temperature (can be assumed to be 32°F or 0°C)

As discussed earlier, the time-dependent variation of the thermal resistances that dictate the charging/discharging rate as well as the outlet secondary fluid temperatures can be captured by performance curves of effectiveness versus fraction of cooling capacity discharged from the CTES (see Figure 18.36). The effectiveness is close to unity during the initial stages of the discharge cycle (the outlet temperature depicted in Figure 18.35 is low, while the rate of discharge is at its maximum). Gradually, the effectiveness decreases and reaches zero when the storage is fully discharged. Further, the effectiveness varies with circulating fluid flow rate (as it does for traditional heat exchangers also), and so a series of curves are necessary to model the effect of different fluid flow rates. These characterization curves have to be determined for the specific design of the ice storage system being designed and can be generated either from manufacturer data, or from field tests, or from detailed modeling. This approach is, however, approximate and the reader can refer to Kuehn et al. (1998) for a discussion on the errors likely to be introduced.

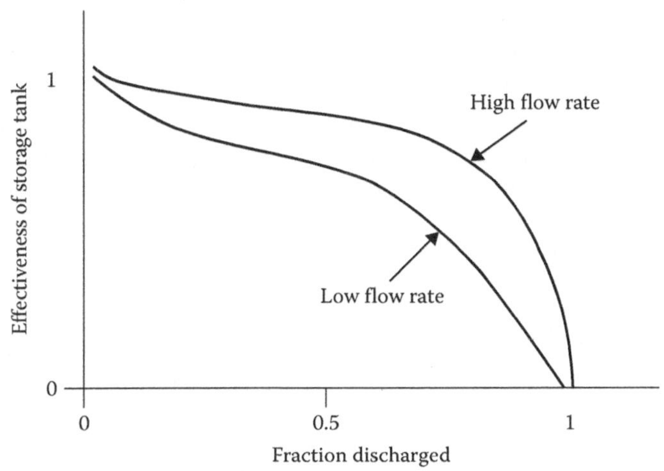

FIGURE 18.36
Characteristic curves of an ice storage tank: effectiveness versus fraction of storage capacity discharged with high and low fluid flow rates.

It is important to point out that the modeling and analysis of CTES systems involves not only determining the design capacity or size of the CTES for the design option chosen (full or partial storage) but also making sure that it can provide the required rate of discharge at any given time during the operation. The latter issue is especially critical during the end of the peak period when the CTES is close to deletion. In fact, it has been found that the design capacity of the CTES systems is more often dictated by the required discharge rate than the total integrated load shift requirement. The modeling approach proposed by Jekel et al. is also convenient for this purpose (refer to Mitchell and Braun, 2013 for an in-depth discussion). This issue is illustrated by the following example.

Example 18.7: Verifying whether CTES Can Meet the Required Discharge Draw

Consider Figure 18.30, which lists the hour-by-hour draw (in ton·h) of the CTES for three different design options for specified design day building cooling loads. Let us consider the partial storage demand limiting option, which requires a CTES storage size of 1022 ton·h. The rate at which the CTES has to be discharged can be determined from the last column. Let us analyze this situation for hour 16:00. Assume that the coolant flow rate is 70,000 lb_m/h and that the entering temperature is 60°F.

(a) The desired rate of discharge is 372 – 220 = 152 ton·h.
(b) The average energy remaining in storage during the hour is (372 + 220)/2 = 296 ton·h.
(c) The fraction of CTES discharged is [1 – (296/1022)] = 0.71.
(d) The corresponding effectiveness value is determined from the characteristic curve of the specific CTES system. Say, it is found to be $\varepsilon_{tank} = 0.68$.
(e) The corresponding coolant leaving temperature can be determined by rearranging Equation 18.14 as

$$T_{tank,out} = T_{tank,in} - \varepsilon_{tank}(T_{tank,in} - T_{freezing})$$
$$= 60 - 0.68 \times (60 - 32) = 41.0°\text{F}$$

(f) The actual rate of discharge can then be determined:

$$\dot{Q}_{tank,dis} = \dot{m}_w \times c_{p,w}(T_{tank,in} - T_{tank,out})$$
$$= 70{,}000\,\text{lb}_m/\text{hr} \times 1\ \text{Btu/(lb}_m\text{°F)} \times (60 - 41)\text{°F}$$
$$= 1{,}330{,}000\ \text{Btu} = 110.8\ \text{ton}\cdot\text{h}$$

This is much lower than the desired discharge rate of 152 ton·h. Hence, the CTES tank size has to be increased to circumvent this shortage. Strictly speaking, the earlier calculation procedure for sizing the CTES has to be done for both charging and discharging cycles in order to verify that the building cooling rates can be satisfied during all hours of the design day.

Problems

Numbers 1–4 given in parenthesis denote the degree of difficulty.

18.1 One hundred gallons (380 L) of water is confined in a length of nominal 4 in. (10 cm) steel pipe. If the pressure in the pipe is maintained essentially constant by an expansion tank while the water warms from 100°F to 200°F (38°C to 93°C), how much water will move into the expansion tank? *(2)*

18.2 Five hundred gallons (1900 L) of water is confined in a completely filled steel tank, the pressure in which is maintained essentially constant by an expansion tank. If the water is heated from 50°F to 250°F (10°C to 121°C), how much water will move into the expansion tank? *(2)*

18.3 If the pipe in Problem 18.1 were to expand freely, what would the length change be? Design a U-type expansion loop for the pipe in Problem 18.1 so that this expansion can be accommodated without any length change between the fixed anchors attached to the ends of the pipe. *(2)*

18.4 Work Problem 18.3 for copper pipe with anchors placed 50 ft (15 m) apart. *(2)*

18.5 Size an expansion tank for a residential heating system at sea level containing 150 gal (550 L) of water if the water temperature increases from 50°F to 230°F (10°C to 110°C). The system maximum pressure is 60 psig (420 kPa), and the initial expansion tank nitrogen charge pressure is 15 psig (105 kPa). *(2)*

18.6 Repeat Problem 18.5, but assume that the expansion tank was filled with atmospheric air prior to charging. No special gas charge was added to the tank. *(2)*

18.7 What are the peak (Btu/h, kW) and daily (Btu/day, MJ/day) water heating loads for a full-service restaurant that serves 40 meals/h (during the 3 h lunch peak) and 280 meals/day (daily total)? The water source temperature is 48°F (9°C). How large should the storage tank be to meet the peak load if the recovery rate is 40 gal/h (150 L/h)? State any assumptions needed. *(3)*

18.8 What are the peak (Btu/h, kW) and daily (Btu/day, MJ/day) water heating loads for a 150-unit apartment in which the 2 h peak occurs in the morning? The water source temperature is 50°F (10°C). How large should the storage tank be to meet the peak load if the recovery rate is 900 gal/h (3500 L/h)? State any assumptions needed. *(3)*

18.9 Consider Example 18.6. The example illustrated the approach to analyze the thermal-based sizing of three cool storage options. You will now supplement this by a cost analysis. Assume a conventional chiller unitary initial cost of \$600/ton and that of an ice making chiller to be \$750/ton and that of the cool storage to be \$30/ton·h. The utility rates are specified as \$0.05/kWh for off-peak periods and \$0.10/kWh for on-peak ones, while the demand charge is \$25/kW during on-peak periods only. Also assume a chiller efficiency of 0.8 kW/ton. Neglect heat losses and other considerations.

(a) With the aforementioned data, reanalyze the optimal demand limiting sizing option and determine the optimal on-peak chiller load based on a month consisting of 20 identical week days only. Recall that the demand charges are to be applied only once during a month, while the energy charges relate to the total electricity consumed over the month.

(b) Evaluate the various options in terms of cost assuming a 20-year equipment life and neglect time value of money. Again assume the year to consist of 240 identical days. *(4)*

18.10 It was pointed out in Section 18.9.4 that an important consideration while sizing ice storage systems was to make sure that the desired rate of discharge was being met for all hours during the design day. The effectiveness versus fraction discharged for an ice storage system is given by Figure 18.36. Analyze the three storage options shown in Table 18.6 to determine whether the required rate of discharge is being met during the on-peak hours.

Assume the set-point temperature of the ice storage tank is 40°F, while the inlet fluid temperature is 55°F and the design flow rate is 3 gpm/ton. *(4)*

References

ASHRAE Applications (2011). *Handbook of HVAC Applications*. American Society of Heating, Refrigerating and Air-Conditioning Engineers, Atlanta, GA.

ASHRAE Systems (2012). *Handbook of Systems and Equipment*. American Society of Heating, Refrigerating and Air-Conditioning Engineers, Atlanta, GA.

ASHRAE-SDL (2013). *Fundamentals of Design and Control of Central Chilled Water Plants*. Self-directed Learning Series, American Society of Heating, Refrigerating and Air-Conditioning Engineers, Atlanta, GA.

Bobenhausen, W. (1994). *Simplified Design of HVAC Systems*. John Wiley & Sons, New York.

Brandemuehl, M.J., S. Gabel, and I. Andresen (1993). *HVAC 2 Toolkit: Algorithms and Subroutines for Secondary HVAC System Energy Calculations*. American Society of Heating, Refrigerating and Air-Conditioning Engineers, Atlanta, GA.

Braun, J.E., J.W. Mitchell, S.A. Klein, and W.A. Beckman (1989). Applications of optimal control to chilled water systems without storage. *ASHRAE Trans.*, 95(1), 663–675.

Dorgan, C.E. and J.S. Elleson (1993). *Design Guide for Cool Thermal Storage*. American Society of Heating, Refrigerating and Air-Conditioning Engineers, Atlanta, GA.

Haas, J.H. (January 1990). Steam traps—Key to process heating. *Chem. Eng.*, 97, 151–156.

Henze, G.P. (2003). An overview of optimal control for central cooling plants with ice thermal energy storage. *ASME J. Solar Energy Eng.*, 125(3), 302–309; Reddy, T.A., ed. *Emerging Trends in Building Design, Diagnostics and Operation*. American Society of Mechanical Engineers, New York.

Jekel, T.B., J.W. Mitchell, and S.A. Klein (1993). Modeling ice-storage tanks. *ASHRAE Trans.*, 99(1), 1016–1024.

Kuehn, T.H., J.L. Threlkeld, and J.W. Ramsey (1998). *Thermal Environmental Engineering*, 3rd ed. Prentice-Hall, Upper Saddle River, NJ.

Lorsch, H.G. (1993). *Air-Conditioning Systems Design Manual*. American Society of Heating, Refrigerating and Air-Conditioning Engineers, Atlanta, GA.

Mitchell, J.W. and J.E. Braun (2013). *Principles of Heating, Ventilation and Air-Conditioning in Buildings*. John Wiley & Sons, New York.

Mull, T.E. (2001). *Practical Guide to Energy Management for Facilities Engineers and Plant Managers*. ASME Press, American Society of Mechanical Engineers, New York.

Pita, E.G. (2002). *Air Conditioning Principles and Systems*, 4th ed. Prentice Hall, Upper Saddle River, NJ.

Price (2011). *Engineer's HVAC Handbook: A Comprehensive Guide to HVAC Fundamentals*. Price Industries Limited, Winnipeg, Manitoba, Canada.

Taylor, S.T. (June 2012). Optimizing design and control of chilled water plants. *ASHRAE J.*, 56–74.

Wong, W.Y. (May 1989). Safer relief valve sizing. *Chem. Eng.*, 96, 137–140.

Zmeureanu, R. and P. Fazio (1988). Thermal performance of a hollow core concrete floor system for passive cooling. *Build. Environ.*, 23(3), 243–252.

19

All-Air Systems

ABSTRACT This chapter deals with all-air systems, which are the most popular secondary system type in the United States. We first describe the working of single-duct single-zone HVAC systems along with how to size and analyze such systems for cooling and heating design conditions and how to draw the process diagrams on a psychrometric chart. Subsequently, we discuss system layout, working principles, reasons for evolution, advantages, and areas of applicability of the following generic all-air systems for single-zone and multizone buildings: single-duct constant air volume (CAV) and variable air volume (VAV) systems, dual-duct CAV systems, and fan-powered VAV systems. The operating conditions and the reasons why simultaneous heating and cooling (called coil bucking) occur in CAV and VAV systems are discussed. We then use solved examples to analyze part-load performance of the various all-air systems and illustrate the use of the bin method to estimate annual heating, cooling, and fan energy of secondary systems coupled with primary systems (such as chillers and boilers). Next, the energy efficiency benefits of using economizer cycles and energy recovery devices as well as the implementation of different operating and control strategies (such as deck reset, discriminating temperature controls, and critical zone reset) are described. Finally, we discuss two efficiency metrics that quantify energy penalties due to coil bucking.

Nomenclature

COP	Coefficient of performance
c	Specific heat at constant pressure, kJ/(kg·K) [Btu/(lb$_a$·°F)]
E	Electricity use, kWh
EDE	Energy delivery efficiency (Equation 19.32)
HDD	Heating degree days, °C-days (°F-days)
h	Specific enthalpy, kJ/kg ([Btu/lb$_w$)
h_{vap}	Latent heat of vaporization of water, kJ/kg [Btu/lb$_m$)
K_{wt}	Total heat transmission coefficient of building, W/K [Btu/(h·°F)]
$\dot{m}$	Mass flow rate, kg/s (lb$_m$/h)
PLR	Part-load ratio
p	Pressure, partial pressure, Pa (lb$_f$/in.^{2})
Q	Energy Consumption, J (Btu)
$\dot{Q}$	Energy use or load, W (Btu/h)
$\dot{Q}_{cc}$	Load on the cooling coil, W (Btu/h)
$\dot{Q}_{hc}$	Load on the heating coil, W (Btu/h)
SHR	Sensible heat ratio (Equation 19.1)
T	Temperature, K or °C (°R or °F)
T_{bal}	Balance point temperature of space or building, K or °C (°R or °F)
T_{db}	Dry-bulb temperature, K or °C (°R or °F)
T_{wb}	Wet-bulb temperature, K or °C (°R or °F)
v	Specific volume, m^3/kg, (ft^3/lb$_m$)
$\dot{V}$	Volumetric flow rate, L/s (ft^3/h)
W	Humidity ratio, kg$_w$/kg$_a$ (lb$_w$/lb$_a$)
$\dot{W}$	Mechanical power, W (hp)
WG	Water gauge, a measure of pressure drop
$\dot{x}$	Time rate of quantity x
$\bar{x}$	Average value of quantity x

Greek

ε	Effectiveness
ρ	Density, kg/m^3 (lb$_m$/ft^3)
φ	Relative humidity
η	Efficiency
Δ	Change

Subscripts

A, B	Zones *A* and *B* of building
a	Air, dry air
ann	Annual
bal	Balance
bldg	Building
boil	Boiler
cc	Cooling coil
ch	Chiller
cond	Conduction
cool	Cooling
gain	Internal heat gain
heat	Heating

hc	Heating coil
in	Inlet
fan	Fan or blower
lat	Latent
liq	Liquid
m	Mixed air condition, mass
max	Maximum
min	Minimum
multiz	Multizone
o	Outdoor air condition
out	Outlet
ph	Preheat coil
r	Return, recycled air condition
space	Space or room or zone
supply	Supply air to space
sen	Sensible
tot	Total
w	Water
1-z	One zone

19.1 Basic Principles

19.1.1 Introduction

Traditionally, one distinguishes between three major types of secondary systems: all-water or hydronic systems (treated in Chapter 18), all-air, and air–water systems. However, the last decade has seen the emergence of a number of novel system variants and hybrid systems (presented in Chapter 20), and so some of the previously clear boundaries have become blurred. Nonetheless, a basic treatment of all-air systems is essential in order to understand these recent developments. All-air systems, which are very widely used in the United States, have been around from the last 50 years with important system improvements gradually occurring over the years. There are a large number of all-air system configurations coupled with different operational strategies to fit the myriad conditions of indoor requirements for human comfort (Sun, 1994) in applications such as buildings (of which there are numerous different types), process applications, and also for special environments requiring close control of temperature and humidity (clean rooms, computer rooms, server farms, hospital operating rooms, ...). An all-air system provides complete sensible and latent cooling, preheating, and humidification capacity to the building supply air, and usually no additional cooling or dehumidification equipment is required at the zone (however, some amount of heating may be needed).

Air-conditioning a space (i.e., either heating or cooling) is achieved by convection heat transfer rather than by radiation in the majority of all-air systems. The indoor air is conditioned using, in most cases, cold/hot streams of water generated from primary equipment (such as chillers and boilers) located in the building itself or piped from a central physical plant. The earliest methods for cooling a building used natural or mechanical ventilation with outdoor air. The ventilation needs of a building in terms of occupant comfort, and indoor environmental quality (IEQ) considerations were discussed in Section 3.5.

Building loads vary greatly during a single day, as shown in Chapter 9. It is a considerable challenge for the HVAC designer to conceive of and complete a systems design that will provide comfort during not only the design day but also the entire year under widely varying climatic and occupancy conditions that typify modern building use. The energetic or economic measure of the effectiveness of the HVAC system depends almost solely on operational efficiency at part-load conditions since the system is operated about 99% of the year at less than peak-load conditions. In this chapter, we will see that peak-load conditions are used to initially size equipment but that the much larger, long-term cost of HVAC systems depends on their operation at off-design conditions. For that reason, the issues of part-load system operation and annual energy estimation are addressed in this chapter.

This chapter is divided into several sections. The first section describes the various individual components and the functioning of the basic single-duct all-air system and distinguishes between CAV and VAV operation. Sections 19.2 and 19.3 discuss operation, design, sizing, and part-load energy consumption calculations of CAV and VAV systems, respectively, supplying cooling and heating needs to single-zone buildings. Section 19.4 deals with similar aspects but for multiple zone buildings. Section 19.5 discusses the procedure and presents the manner in which annual energy use can be estimated. An example compares the annual energy use of different all-air systems providing air conditioning to a building with two zones. Section 19.6 discusses various control and operational practices by which energy efficiency in such all-air systems can be enhanced, along with pertinent models for analyzing their effect on the basic HVAC all-air systems. Finally, the last section describes two indices that can be used to characterize and analyze the energy efficiency of multizone HVAC systems in a building either during design or during operation.

19.1.2 Basic One-Zone AC System

Systems of the type shown in Figure 19.1 are commonly used for supplying air for both heating and cooling purposes in buildings. It consists of three sets of coils: the preheat coil (PH), the cooling coil (CC), and the reheat coil (RC). The simplest HVAC system will be adequate to illustrate the basic concepts, while several

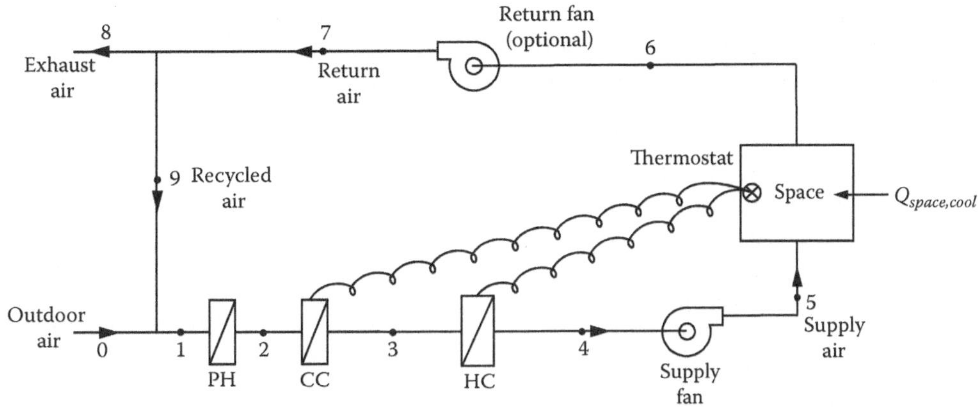

FIGURE 19.1
Schematic diagram of the basic single-zone HVAC system with some of the major energy equipment. The important psychrometric state points (0–9) and the return, supply, exhaust, and ventilation flows are also shown.

enhancements are discussed later in this chapter. When the system in Figure 19.1 is used for cooling, the air typically leaves the cooling coil (connected to the primary system chiller) at about 12.8°C (55°F) and 90% relative humidity in conventional systems. Ordinarily, some dehumidification is also achieved in the cooling coil. It would be the best if the preheat and heating coils are not active during the cooling season. On the other hand, when heating is required, the air delivery temperature for heating is often in the range 32°C–40°C (90°F–105°F) in conventional systems. During heating, it would be the best if the cooling coil is inactive in such simple systems.

Two fans are generally specified. The supply air fan draws air through the coils and causes it to flow into the space to be conditioned. According to the conservation of mass, the amount of air supplied to the space must also be removed from it, except for differences that may be caused by infiltration or exfiltration. The return air duct and fan serve to remove air from the zone. A return air fan is required in larger system to avoid large overpressurization of the supply ducts and of the space. It is optional for smaller systems in order to reduce initial costs. The supply fan shown downstream of the coils in Figure 19.1 (*draw-through arrangement*) could be placed upstream of the coils in the *blow-through configuration*. In draw-through units, cool air from the cooling coil is heated by flow work produced by the fan (and further heated by the fan motor if it is the airstream), thereby requiring that the cooling coil subcool the air to provide expected room supply air. Blow-through systems avoid this problem by having the cooling coils remove fan and motor heat directly, so that subcooling is not needed. However, this design is more expensive, requires more floor area, and can have problems with cooling coil condensate entrainment in the airstream. Draw-through units are the most common design option.

Under conditions of significant heating or cooling loads, considerable energy and cost are expended to condition room air. Instead of discarding this airstream after a single pass through the space, a portion called the *return air*, is often reused or recycled via the return air duct with a corresponding amount drawn in from outdoors. This recycling results in significant energy efficiency, and therefore most of the all-air HVAC systems incorporate such an arrangement.

Ventilation is defined as the amount of outdoor air supplied to a space by mechanical or natural means. It serves two purposes: addition or removal of heat and/or humidity from occupied spaces, and supply of fresh air to meet occupant health requirements. If the room is perfectly tight, the amount of room air that must be discarded or exhausted is equal to the ventilation airflow rate that is drawn into the system. The control of the recycled and outdoor air intake amounts is achieved by the joint action of a set of three inter-linked dampers (Figure 19.2). The exhaust and intake dampers operate in the same

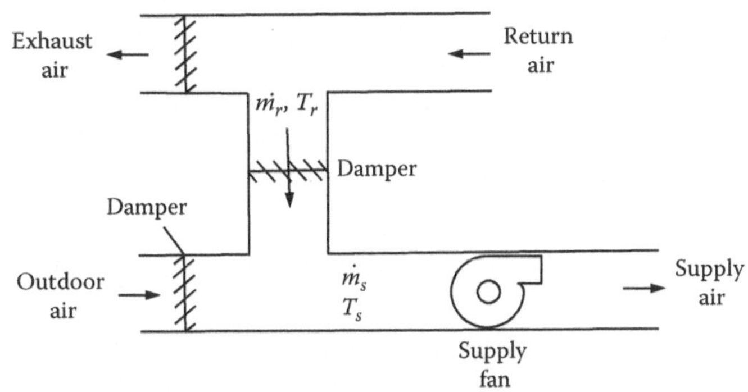

FIGURE 19.2
The three sets of dampers and their locations that are meant to vary the return and exhaust airflows along with the outdoor air ventilation amount.

sense, whereas the damper in the return air duct is reverse acting. The outdoor air damper can consist of two portions, one fixed so as to always allow the minimum amount of outdoor air to be drawn in (often an injection fan is used), and another where the damper blades can be varied so as to modulate the outdoor air inflow. Meeting indoor comfort conditions described in Chapter 3 will often require conditioning this air by heating/cooling it or humidifying/dehumidifying it or both.

During intermediate seasons when neither extreme heating nor extreme cooling is needed, it is possible to use more outdoor air than the minimum required for health conditions. This type of operation is called an "economizer cycle" and would result in an energy saving (discussed in Section 19.6).

The zone load depends on both internal loads and loads driven by external conditions such as solar gain and ambient temperature that affect both the transmission load and the ventilation load. The total cooling coil load $\dot{Q}_{cc,tot}$ for the space shown in Figure 19.1 is given on the *air side* of the coil by

$$\dot{Q}_{cc,tot} = \dot{m}_a(h_{cc,in} - h_{cc,out}) \tag{19.1}$$

where

h denotes the airstream enthalpies

the subscripts *in* and *out* correspond to inlet and outlet cooling coil conditions, respectively

The coil inlet enthalpy is determined by the outside air fraction at the air mixing station upstream of the coils and by both the return and outdoor air enthalpies:

$$h_{cc,in} = \frac{\dot{m}_r h_r + \dot{m}_o h_o}{\dot{m}_r + \dot{m}_o} \tag{19.2}$$

where $\dot{m}_r$ and $\dot{m}_0$ are the mass flow rates of the recycle and outdoor air. Note that the return air conditions are identical to those of the space (except for differences in the air mass flow rates).

The cooling load on the coil can also be equated to the temperature rise on the *liquid side* (connected to the chilled water loop) by

$$\dot{Q}_{cc,tot} = \dot{m}_{liq} c_{liq}(T_{liq,out} - T_{liq,in}) \tag{19.3}$$

The system airflow rates in the single-zone system are determined by the heating and cooling loads. The cooling load often requires the greater flow rate. By default, since only one fan with a fixed flow rate is usually available, the heating season airflow is, therefore, the same as the cooling season airflow.

The required volumetric airflow to the space

$$\dot{V}_{a,cool} = \frac{\dot{Q}_{space,tot}}{\rho_a(h_{space} - h_{cc,out})} \tag{19.4}$$

where

$\dot{Q}_{space,tot}$ is the *total* (sensible plus latent) cooling load of the space, W (Btu/h)

h_{space} is the enthalpy of the air leaving the space, kJ/kg$_a$ (Btu/lb$_a$)

$h_{cc,out}$ is the enthalpy of air leaving the cooling coil, kJ/kg$_a$ (Btu/lb$_a$)

The enthalpy of moist air is found from equations developed in Chapter 13 or from the online HCB software. There is some ambiguity as to when to use which equation. It is most appropriate to use the enthalpy equation during design or peak-day analysis, which is meant to determine the sizing of the various components. During part-load operation, HVAC control is achieved by the space thermostat that is based on the space dry-bulb temperature, and so the sensible heat equation is more appropriate.

The building air-conditioning process during summer operation (with the heating coils inactive) consists of three steps (refer to Figure 19.1):

1. Room air at condition 6 is mixed with outdoor air required for ventilation at condition 0 to result in air at state 1. For this discussion, the fan temperature rise is ignored. Thus, states (4, 5) are identical and so are states (6–9) in their properties but not in their airflow rates. Note that the mixed-air condition 1 is found by using the rule for mixing of airstreams discussed in Section 13.6.2.
2. This mixed air (state 1) enters the cooling coil at state 2 (states 1 and 2 are identical since the preheat coil is inactive), where it is cooled and dehumidified to condition 3.
3. The cooled dehumidified air is supplied at state 5 to the space.

Figure 19.1 shows the process schematically, and Figure 19.3 is a plot of the entire process on the psychrometric chart. In winter, the process operates similarly with heat addition instead of extraction from the mixed airstream. Because outdoor air in winter is very low in humidity ratio, some mechanism of adding water to the airstream is required. Such systems are discussed in Section 19.2.3.

The *sensible heat ratio* (*SHR*), or sensible heat factor, is used to analyze air-conditioning processes. It is defined

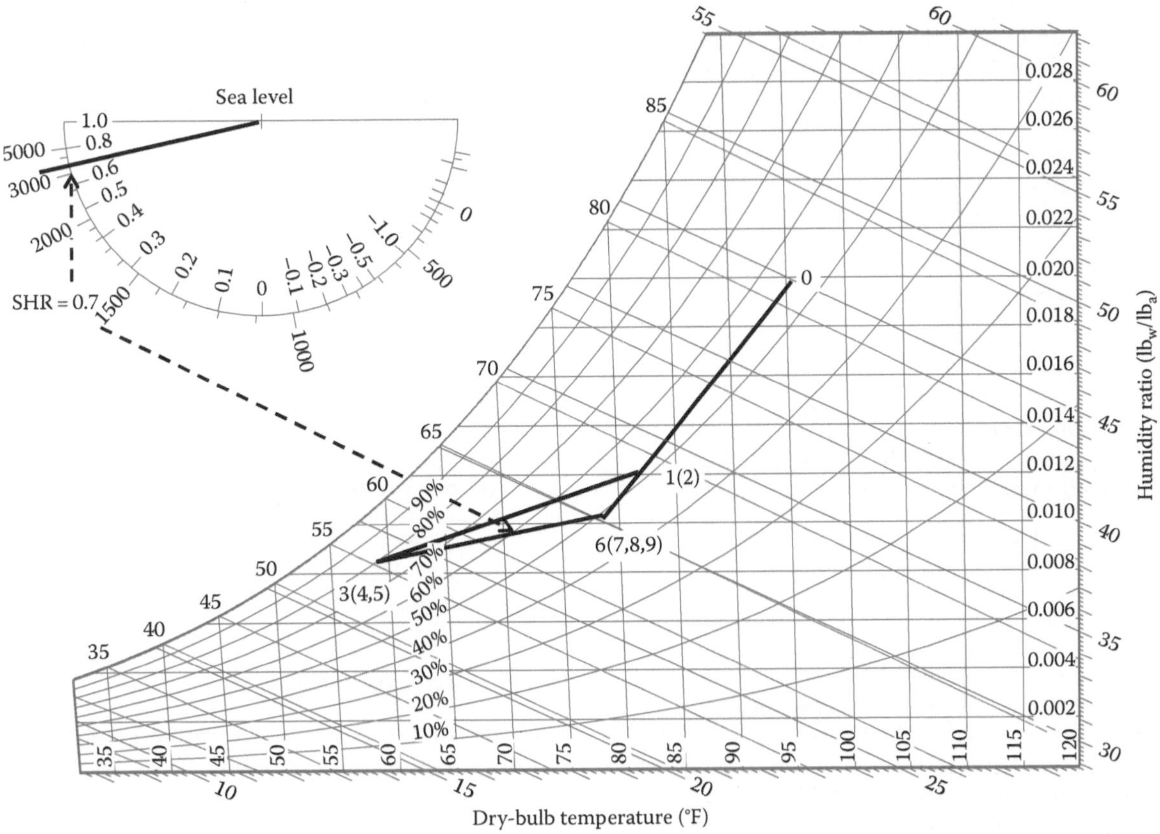

FIGURE 19.3
The air-conditioning process diagram on a psychrometric chart. The state points correspond to those in Figure 19.1, while the numerical values apply to Example 19.1.

as the ratio of the sensible heat to the total (latent + sensible) heat removed:

$$\mathrm{SHR} \equiv \frac{\dot{Q}_{sen}}{\dot{Q}_{sen} + \dot{Q}_{lat}} \tag{19.5}$$

The protractor included with the psychrometric chart has an extra scale useful for cooling system calculations. The *inner* scale is the SHR* scale. Its use is illustrated by Example 19.1.

Example 19.1: Building Air-Conditioning

A space is to be maintained at 78°F (25.5°C) dry-bulb temperature and 50% RH. The total cooling load (heat to be removed from the space to maintain comfort) is 120,000 Btu/h (35 kW) of which 70% is sensible heat. Ventilation air at 1000 ft³/min (472 L/s) is required on the peak day when the outdoor conditions are 95°F (35°C) and 55% RH. The room supply air temperature is 20°F (11.1°C) below that of the room set point temperature (a typical value for commercial buildings at peak operating conditions). What are the space air supply flow rate and the cooling coil rating (total cooling capacity and SHR)?

Figure: See Figures 19.1 and 19.3.

Assumptions: The location is at sea level. The supply air temperature to the space is 58°F. The duct heat transfer and the fan air temperature rise are ignored for simplicity.

Given: $\mathrm{SHR}_{space} = 0.7$, $\dot{Q}_{space,\,cool} = 120{,}000$ Btu/h, $\dot{V}_0 = 1{,}000\ \mathrm{ft^3/min} = 60{,}000\ \mathrm{ft^3/h}$

$T_{db,6} = 78$°F, $T_{db,o} = 95$°F, $\phi_o = 55\%$, $T_{db,5} = 58$°F (outdoor and indoor air conditions)

Find: $\dot{m}_a$, $\dot{Q}_{cc,tot}$, SHR_{cc}

Lookup values: Specific volume $v_o = 14.4\ \mathrm{ft^3/lb_a}$, $W_o = 0.0197\ \mathrm{lb_w/lb_a}$, $W_6 = 0.0103\ \mathrm{lb_w/lb_a}$

Solution

The psychrometric chart (Figure 13.6) will be used to illustrate the graphical solution approach.

1. *Locate specified points* 0 and 6 on the psychometric chart.
2. *Locate supply air point 5.* The conditioned space inlet and outlet conditions lie on

* There are two SHR factors: one that applies to the space and one to the cooling coil. This distinction and its cause are discussed in Example 19.1.

line 3–6 in Figure 19.3. Condition 6 (78°F, 50% humidity) is already known from the problem statement. The slope of line 3–6 is determined from the inner scale of the protractor as shown for SHR = 0.7. The location of point 3 is the intersection of the line 3–6 and a vertical line drawn from $T_{db,3}$ = 58°F. This corresponds to a RH value of 80%.

3. *Find the space supply airflow rate.* The first law for this process is

$$\dot{Q}_{space,cool} = \dot{m}_a\left(h_6 - h_5\right) \tag{19.6}$$

The enthalpies can be read from Figure 13.6: $h_5 = h_3 = 23$ Btu/lb$_a$, $h_6 = 30$ Btu/lb$_a$ The air mass flow rate is thus

$$\dot{m}_a = \frac{120{,}000 \text{ Btu/h}}{(30-23) \text{ Btu/lb}_a}$$

$$= 17{,}140 \text{ lb}_a\text{/h } (3770 \text{ ft}^3\text{/min})$$

4. *Locate mixed air point 1.* The end points 0 and 9(6) are known, and so the inverse relation between mass flows and line segment lengths (Equation 13.32) is used to locate point 1 (see Section 13.6.2). The outdoor air mass flow is

$$\dot{m}_{a,0} = \frac{\dot{V}_0}{v_0} = \frac{60{,}000 \text{ ft}^3\text{/h}}{14.4 \text{ ft}^3\text{/lb}_a} = 4{,}170 \text{ lb}_a\text{/h}$$

The ratio of air mass flows is (4,170/17,140) = 0.243. Therefore, point 1 is located 24.3% of the distance from point 6 along line segment 6–0. The properties at point 1 can be read from the psychrometric chart. They are T_{db} = 82°F, T_{wb} = 69°F, and moist air enthalpy h_a = 33.4 Btu/lb$_a$. Alternatively, we can use the weighted average rule to calculate the mixed air humidity ratio:

W_1 = [0.243 × 0.0197 lb$_w$/lb$_a$ + (1 − 0.243) × 0.0103 lb$_w$/lb$_a$] = 0.01216 lb$_w$/lb$_a$

Similarly, assuming constant-specific heats, the mixed air dry-bulb temperature is as follows:

$T_{db,1}$ = [0.243 × 95°F + (1 − 0.243) × 78°F)] = 82°F (consistent with the chart reading). This allows point 1 to be fixed on the psychrometric chart.

5. *Determine cooling coil load.* Line 1–3 can now be constructed by connecting points 1 and 3. The slope of the resulting line is transposed to the protractor, and the *coil's sensible heat ratio* SHR$_{cc}$ can be read off as 0.55. The corresponding relative humidity is 80%, and the humidity ratio is 0.0084 lb$_w$/lb$_a$. Finally, the coil heat removal rate (or "coil cooling load") can be found.

$$\dot{Q}_{cc,tot} = \dot{m}_a\left(h_1 - h_3\right) = 17{,}140 \text{ lb}_a\text{/h} \times (33.4-23) \text{ Btu/lb}_a$$

$$= 178{,}300 \text{ Btu/h} \quad \text{or} \quad 14.9 \text{ tons } (52.2 \text{ kW})$$

Comments

Peak-condition coil information (air and water flow rates, SHR, and heat rate) is supplied to manufacturers for coil selection, costing, and fabrication purposes. Note that the cooling coil load and zone cooling loads are quite different because the coil must remove zone heat and humidity gains as well as the latent and sensible loads imposed by ventilation (or outdoor) air. Hence, in most cases, both the SHR quantities will be different, i.e., SHR$_{Room}$ ≠ SHR$_{cc}$.

19.1.3 Airstream Heating due to Fans

Relevant material on prime movers such as fans and pumps was presented in Chapter 16. The concept of fan efficiency was introduced. The motor power input to the fan (called the brake HP) increases the static pressure of the circulating air, while the rest is the inefficiency of the fan that results in waste heat being generated and goes to heat-up the circulating fluid. In determining the capacity of cooling coils, it is necessary to account for the *temperature* rise that occurs along with the pressure rise across a fan, because it can impose an additional cooling requirement. From the first law, we can write (replacing the enthalpy equation with the sensible heat equation)

$$\frac{\dot{W}_{fluid}}{\eta_{fan}} = \dot{m}_a \cdot c_a \cdot \Delta T_{db} \tag{19.7}$$

where
$\dot{W}_{fluid}$ is the mechanical power input to the fluid
η_{fan} is the fan efficiency

Combining this equation with Equation 16.30 results in

$$\Delta T_{db} = \frac{\Delta p}{\eta_{fan} c_a \rho_a} \tag{19.8}$$

which simplifies to

$$\Delta T_{db}(°\text{C}) = 0.807 \times \frac{\Delta p(\text{kPa})}{\eta_{fan}} \tag{19.9 SI}$$

or

$$\Delta T_{db}(°F) = 0.363 \times \frac{\Delta p(\text{inWG})}{\eta_{fan}} \quad (19.9\ \text{IP})$$

For a pressure rise in a fan of 250 Pa (1.0 inWG), the temperature rise with a 100% efficient fan is 0.2°C (0.36°F). For a 70% efficient fan, the temperature rise is 0.29°C (0.52°F). In commercial building HVAC systems, a pressure rise of several inches water gauge is common. The resulting temperature rise can be 1.0–2.0°C (2–3°F) and *must be considered* when cooling coils and central cooling plants are sized. Note that "fan heat" occurs at the location of the air supply fan; therefore, if the fan is *downstream* of a cooling coil, the cooling load appears as a zone load whereas for a fan located *upstream* of the coil, the load appears directly as an additional load on the coil. Additional fan heat is also produced by return fans.

Equation 19.8 applies when the HVAC fan motors are located outside the airstream. In case the *motor is located inside the airstream,* the air temperature rise is enhanced due to motor inefficiencies:

$$\Delta T_{db} = \frac{\Delta p}{\eta_{fan}\eta_{motor}c_a\rho_a} \quad (19.10)$$

where η_{motor} is the motor efficiency.

A similar thermal effect occurs with pressure increases in water flow through pumps. However, the density and specific heat of water are much larger than those for air, and according to the Equation 19.10, the temperature rise will be small. For example, the temperature rise in water with a pressure increase across a typical pump operating at pressure differences of 3 atm (100 ft of head) is 0.028°C (0.05°F). The dissipation of pressure energy by viscosity in piping also produces heat. For this example, the temperature rise is 0.07°C (0.13°F). Hence, the total temperature increase is 0.1°C (0.18°F)—about a 1% effect ignored by most designers. A similar fluid heating in insulated ducts does not occur since adiabatic flow with friction in a compressible fluid such as air results in a minute temperature drop (<0.005°C or 0.01°F) within the ducting system.

Example 19.2: Effect of Fan Heat

We will rework Example 19.1 but include the effect of airstream heating of the supply fan. The supply fan pressure drop is 3 inWG (750 Pa). The efficiency of the supply fan is 70%, while the fan motor efficiency is 84%—both of them are located within the airstream.

Figure: See Figures 19.1 and 19.4.

Assumptions: The location is at sea level. The supply air condition to the space is the same as that in Example 19.1 (58°F and 0.0084 lb_w/lb_a).

The duct heat gains/losses are ignored.

Given: $\Delta p_{fan} = 3$ inWG, $\eta_{fan} = 0.7$, $\eta_{motor} = 0.84$

$$\text{SHR}_{space} = 0.70, \quad \dot{Q}_{space,cool} = 120{,}000\ \text{Btu/h},$$
$$\dot{V}_0 = 1{,}000\ \text{ft}^3/\text{min} = 60{,}000\ \text{ft}^3/\text{h}$$

$T_{db,5}$ = 58°F, outdoor and indoor air conditions are identical to those in Example 19.1.

Find: $\dot{m}_a$, $\dot{Q}_{cc,tot}$, SHR_{cc}

Solution

1. *Locate specified points* 0 and 6 on psychrometric chart. Note that the state points 3 and 4 are no longer the same as in Figure 19.3 because of the supply fan reheat. The supply airflow rate remains unchanged since T_5 is the same. Consequently, point 1 is also unchanged.
2. *Locate point 3*. Using Equation 19.9 IP but including the motor efficiency (since the motor is located in the airstream), we can calculate the supply fan reheat:

$$T_{db,5} - T_{db,3} = 0.363 \times \left(\frac{3\ \text{in WG}}{0.7 \times 0.84}\right) = 1.85°F$$

The cooling coil set point temperature has to be lower to compensate for the airstream heating Thus, point 3 is specified given $T_{db,3}$ = 58 − 1.85 = 56.15°F and humidity ratio $W_3 = W_5 = 0.0084\ lb_w/lb_a$. Then, the enthalpy at point 3 is determined as $h_3 = 22.6\ \text{Btu}/lb_a$.

3. *Determine cooling coil load.*

$$\dot{Q}_{cc,tot} = \dot{m}_a(h_1 - h_3) = 17{,}140\ \text{lb}_a/\text{h} \times (33.4 - 22.6)\ \text{Btu/lb}_a$$
$$= 185{,}112\ \text{Btu/h} = 15.4\ \text{tons}$$

Comments

The cooling coil load has increased by 3.4% from the previous case. This example illustrates the fact that the airstream heating due to the fan is small but not negligible and needs to be considered in accurate design and analysis calculations.

19.1.4 Fundamental Operational Difference between CAV and VAV Systems

Example 19.1 illustrated the graphical approach to analyzing a single-duct single-zone HVAC system at peak conditions. This allows sizing the various

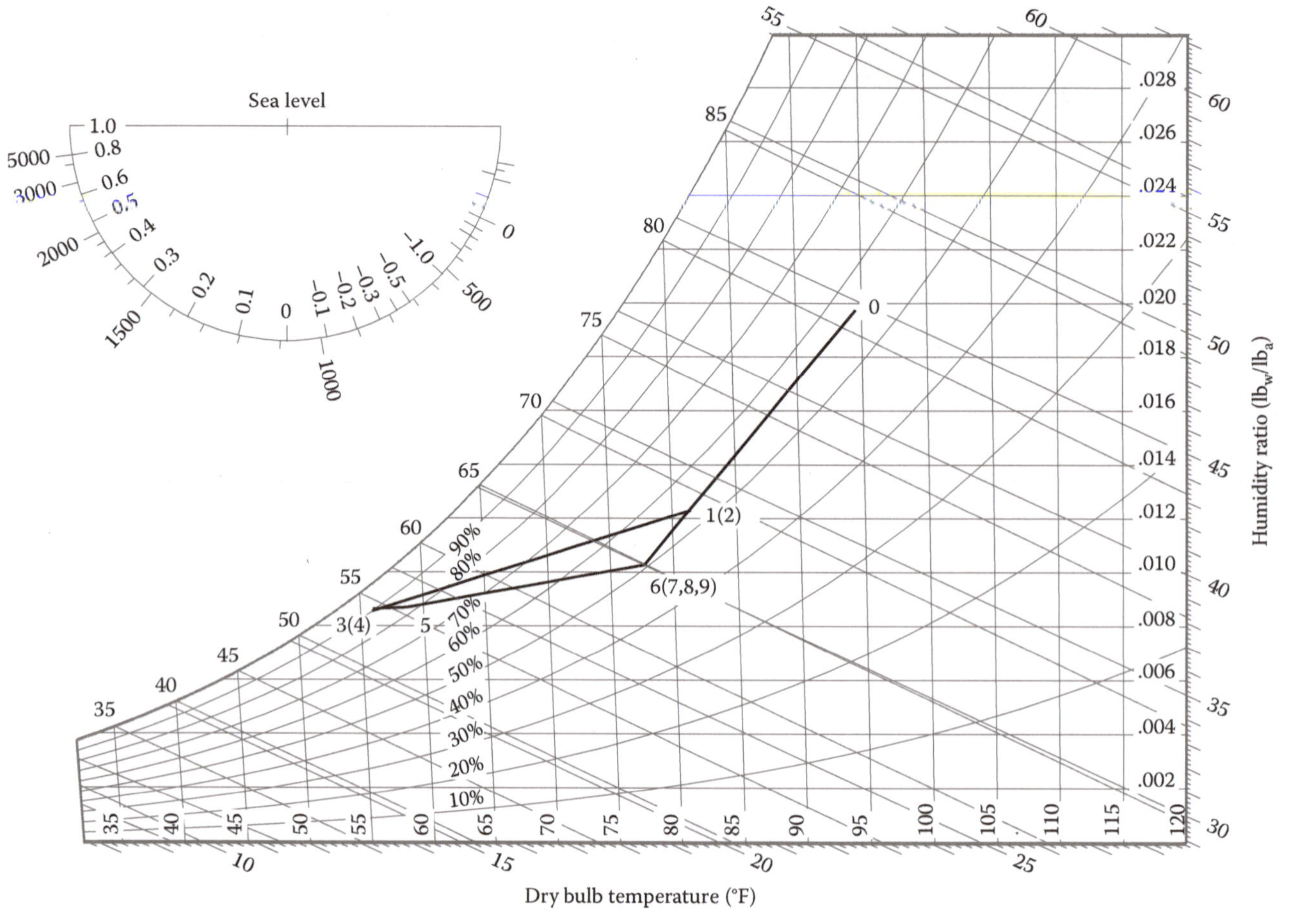

FIGURE 19.4
Same as Figure 19.3 but including the effect of supply fan reheat. Numerical values apply to Example 19.2.

components (such as cooling/heating coils, fan, duct). Under part-load conditions, the cooling (and heating) space loads $\dot{Q}_{space,tot}$ are lower than the peak values, and so certain operational changes are warranted in order to ensure satisfactory operation of the HVAC system. The control is achieved via the space thermostat that, however, only senses dry-bulb temperature and maintains a preselected value of $T_{db,space}$. The design equation given by Equation 19.1 has to be recast in terms of temperatures to meet the space sensible heat load:

$$\dot{Q}_{space,sen} = \dot{m}_a \cdot c_a \cdot \left(T_{db,space} - T_{db,supply}\right) \tag{19.11}$$

The space dry-bulb temperature can be maintained at its preselected set point temperature by one of the two different ways of control:

1. By holding the system airflow rate $\dot{m}_a$ constant and letting the space thermostat modulate the supply air temperature $T_{db,supply}$ (basis of the constant air volume or CAV* system)
2. By modulating the space supply airflow rate $\dot{m}_a$ while holding the supply air inlet temperature constant (basis of the variable air volume or VAV system discussed in Section 19.3)

The CAV system operation is called the *fixed-volume* (or variable temperature) method. Ideally, heating and cooling coils should not operate simultaneously to avoid energy waste. When both coils operate at the same time, they are said to "fight" or "buck" each other. If, at any given time, only the heating or cooling coil is permitted to operate by the control system, the fixed-volume single-zone system can be quite energy efficient. Unfortunately, for CAV operation, this is only achieved at peak design conditions. Coil bucking always occurs during part-load operation (Example 19.5) and is a major source of energy

* The commonly used acronym for constant air volume systems is CV. However, we have intentionally used the acronym CAV so as to be consistent with that widely used for variable air-volume systems, viz., VAV.

inefficiency in CAV systems. It can be considerably reduced if CAV systems are replaced by VAV systems—discussed in Section 19.3.

19.2 Single-Zone Single-Duct CAV Systems

19.2.1 Description

In this section, we discuss the design and analysis of CAV systems, also referred to as fixed-volumetric flow-rate air systems, for heating and cooling a single zone. The basic operation of such systems has been previously described in Section 19.1.2. The design aspect involves analyzing the HVAC system under peak conditions (both summer and winter) so that equipment selected are able to satisfy the required airflow rate and meet the peak cooling and heating space loads. A number of system variants have evolved over the years in order to reduce coil bucking, and some of these are also described. The analysis under part-load conditions is an important aspect in the determination of annual energy requirements of the HVAC system and is also treated in Section 19.2.4.

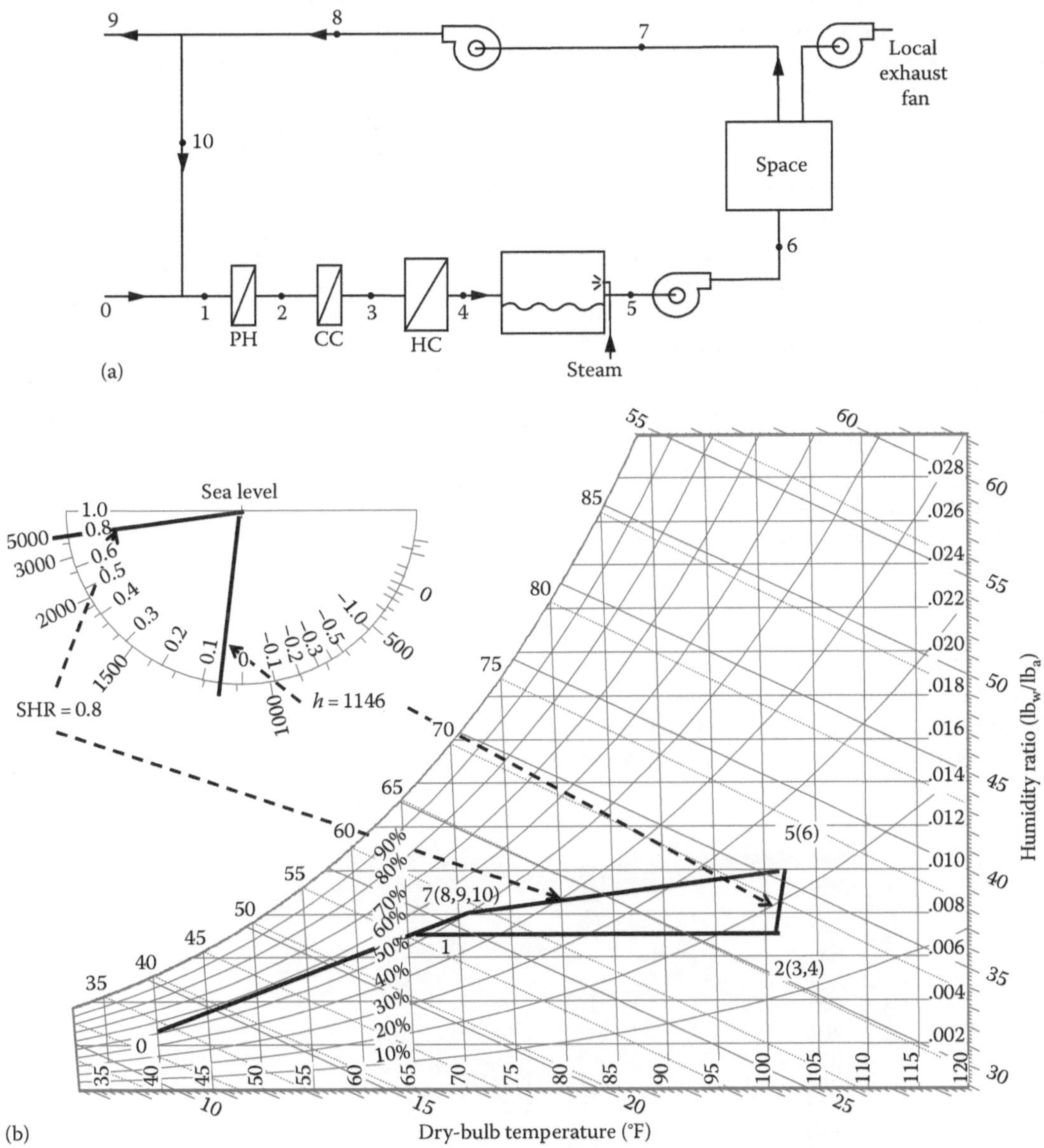

FIGURE 19.5
(a) Schematic diagram of the complete single-duct single-zone constant air volume system with ducted return. The preheater and the steam humidifier would operate during winter, while the cooling coil and the reheat coil would be active when cooling loads are to be met. (b) Psychrometric process diagram with state points as per Figure 19.4.

Figure 19.5a shows an air-based system for heating or cooling a single zone. Mixed air at point 1 flows through the filter (not shown), preheat coil, cooling coil, heating coil, steam humidifier, supply fan and then to the conditioned space. In smaller buildings, this secondary system equipment along with a return/outside air mixing chamber and flow control dampers is customarily housed in one custom-built assembly, called the *air-handling unit* (AHU) (Figure 19.6). The air handler is usually located near the primary plant (i.e., chiller and boiler) to offer convenient piping connections to the chilled and hot water coils. Insulated ducts carry conditioned air to the single zone in this system.

Several optional components are shown in the single-zone system of Figure 19.5a. The preheat coil is located just before the cooling coil and is a critical component especially in very cold locations. Its functions are: (1) to prevent mixed air temperature at point 1 reaching such low values (say, 55°F or less) under very cold outdoor conditions that the cooling coils may freeze up and incur physical damage, (2) to prevent water being sprayed in the air washer from freezing, and (3) to provide some flexibility in the air relative humidity so that one can control the amount of water being sprayed into the airstream under varying outdoor conditions.

The second option, a humidifier, is needed to avoid excessive air dryness during the heating season but does not operate in the cooling season. The humidifier is located downstream of all coils and filters to avoid liquid moisture collection on them if water droplets from the humidifier do not completely evaporate within the air-handling unit. The humidification load is added to the zone sensible heat load, while sizing the heating coil since the heating coil must provide the energy to evaporate moisture for humidification. Otherwise, the evaporating water would cool the heated air below design conditions. Either steam or evaporative-media humidifiers are the most common sources for humidity in commercial buildings.

The third optional component is an exhaust fan that may be needed for local exhaust from a fume hood, stove, or toilet. In many cases, the return fan, can be avoided in single-zone systems if air is relieved from the space in a lowpressure-drop return system. If large outdoor air fractions are used, however, a return fan will be needed to avoid over-pressuring the space in the single zone.

19.2.2 Cooling Mode: Peak Design Conditions

Example 19.1 illustrated the fact that the CAV system for a one-zone space can be designed so as to only require cooling energy (without any heating) under peak-load summer conditions. This is not always possible. One important instance where heating is needed even under peak-load conditions arises when the peak space latent load fraction is high such as in hot and humid locations. This solved example will discuss why this occurs and illustrate how to analyze such cases.

Example 19.3: Peak Design for High Latent Load Spaces

We shall use the same design specifications as that of the CAV system analyzed under peak-load condition in Example 19.1 with one major difference. Instead of the space SHR being equal to 0.7, we are specified a value of 0.5, i.e., the sensible and latent loads are equal. The cooling coil leaving air temperature is as before, at 58°F and 80% RH. We will determine the supply air mass flow rate, and cooling coil and reheat coil loads in this case.

Figure: See Figures 19.5a and 19.7.

Assumptions: The location is at sea level. The duct heat transfer and the fan air temperature rise are ignored for simplicity.

Given: $\text{SHR}_{space} = 0.50$, $\dot{Q}_{space,tot} = 120{,}000\ \text{Btu/h}$, $\dot{Q}_{space,sen} = \dot{Q}_{space,lat} = 60{,}000\ \text{Btu/h}$

Outdoorairconditions:$T_{db,o} = 95°\text{F}$, $\phi_o = 0.55$, $\dot{V}_0 =$ $= 1000\ \text{ft}^3/\text{m}$

Cooling coil leaving air conditions: $T_{db,3} =$ 58°F and $\phi_3 = 0.8$.

FIGURE 19.6
Cutaway drawing of a packaged air-handler unit.

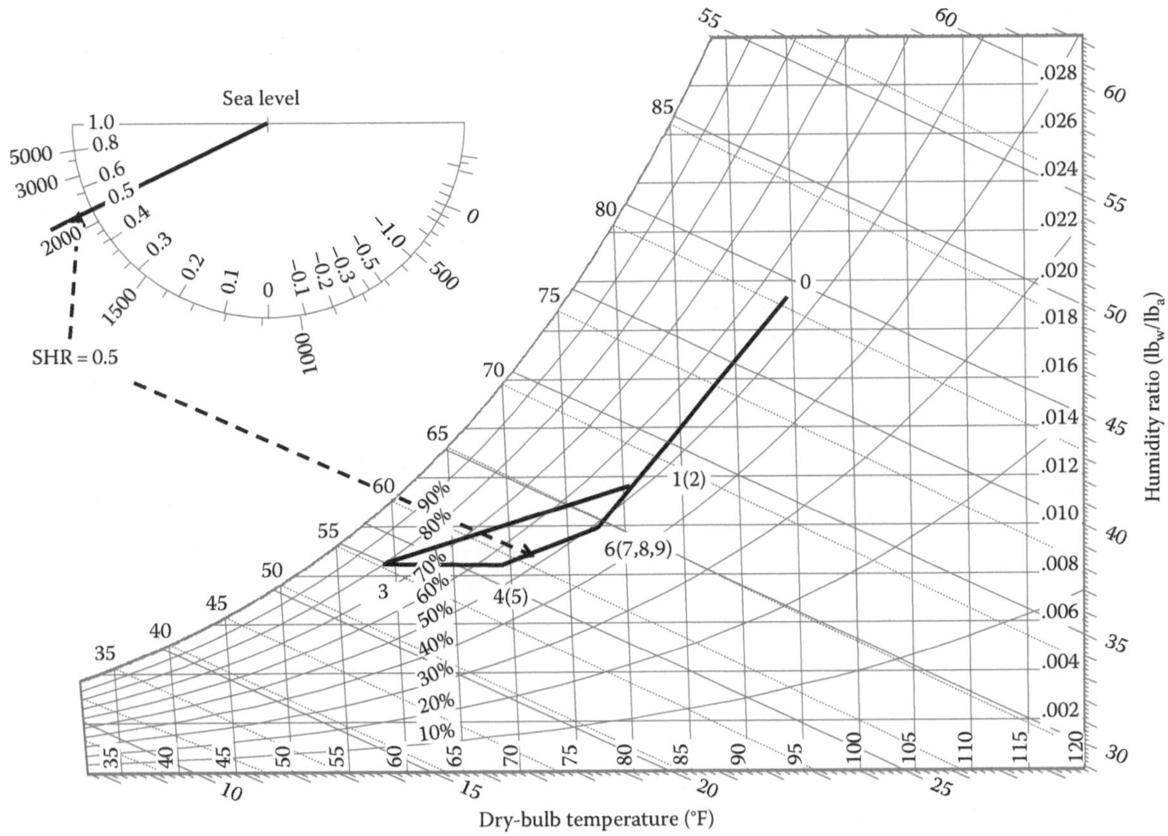

FIGURE 19.7
Process diagram for single-duct constant air volume system to air-condition a single space with high latent loads. The state points correspond to Figure 19.1, while the numerical values apply to Example 19.3. Note that the supply fan reheat is neglected.

Space condition: $T_{db,7} = 78°F$ and $\phi_7 = 0.5$.

Find: $\dot{m}_a$, $\dot{Q}_{cc,tot}$, $\dot{Q}_{cc,sen}$, and $\dot{Q}_{hc,tot}$

Lookup values: Specific volume $v_0 = 14.4$ ft³/lb$_m$, humidity ratio $W_0 = 0.0197$ lb$_w$/lb$_a$, $h_0 = 44.52$ Btu/lb$_a$, $W_3 = 0.0082$ lb$_w$/lb$_a$, $h_3 = 22.85$ Btu/lb$_a$, $W_7 = 0.0103$ lb$_w$/lb$_a$, $h_7 = 30$ Btu/lb$_a$.

Solution

The solution via the graphical procedure is adopted.

1. *Locate specified points* 0, 3, and 7 on the psychrometric chart (see Figure 19.7), which are specified from the problem statement.
2. *Locate supply air point 5.*
   Previously, the slope of line 5–7 was determined from the inner scale of the protractor. The location of point 6 is the intersection of the line drawn from point 7 with a slope equal to 0.5 and the horizontal line drawn from point 3. This is found to be $T_{db,5} = 69.1°F$ and $h_5 = 25.5$ Btu/lb$_a$ and $W_5 = 0.0082$ lb$_w$/lb$_a$.
3. *Determine space supply airflow rate.*
   The enthalpy balance equation (Equation 19.1) is rearranged to yield

$$\dot{m}_a = \frac{\dot{Q}_{space,cool}}{h_7 - h_5}$$

$$= \frac{120{,}000 \text{ Btu/h}}{(30 - 25.5) \text{ Btu/lb}_a} = 26{,}667 \text{ lb}_a\text{/h}$$

4. *Determine mixed air condition point 1.*
   The outdoor air mass flow is the same as before $\dot{m}_{a,0} = 4170$ lb$_a$/h.
   The ratio of air mass flows is (4,170/26,667) = 0.156. Therefore, point 1 is located 15.6% of the distance from point 7 along line segment 7–0. The properties at point 1 can be read from the psychrometric chart or determined by calculation. The dry-bulb temperature could be calculated by using the weighted average rule, e.g., the mixed-air temperature $T_{db,1} = [0.156 \times 95°F + (1 - 0.156) \times 78°F)] = 80.65°F$. Similarly, we find and the humidity ratio $W_1 = 0.0118$ lb$_w$/lb$_a$ and enthalpy $h_1 = 32.31$ Btu/lb$_a$.
5. *Determine cooling coil load.*

$$\dot{Q}_{cc,tot} = \dot{m}_a (h_1 - h_3) = 26{,}087 \text{ lb}_a\text{/h} \times (32.31 - 22.85)$$

$$\text{Btu/lb}_a = 246{,}783 \text{ Btu/h (20.6 tons)}$$

6. *Determine reheat coil load.*

$$\dot{Q}_{hc} = \dot{m}_a(h_4 - h_3)$$
$$= 26{,}087\ \text{lb}_a/\text{h} \times (25.5 - 22.85)\ \text{Btu/lb}_a$$
$$= 69{,}131\ \text{Btu/h}$$

Comments

Note the substantial increase in supply airflow rate (from 17,140 to 26,667 lb_a/h and the substantial additional reheat). The primary reason for the need of reheat is that the SHR_{space} line is so steep that it does not intersect the saturation curve, and so no feasible apparatus dew point temperature value can be deduced. The practical connotation is that it is not possible to simply circulate chilled water through the cooling coil and achieve the desired cooling and dehumidification process in one step. Thus, if a CAV system is used, one is forced to provide reheating even at peak load, i.e., under design conditions. This is a major energy penalty in hot and humid locations in several coastal locations worldwide.

19.2.3 Heating Mode: Peak Design Conditions

Under winter conditions, outdoor air not only is cold but is much drier. So HVAC systems meant to condition spaces during winter need to both heat and humidify the supply airstream. Several system configurations are possible. Two of the most common ones will be discussed in this section.

In very cold locations, a common design option is to use steam for humidification as shown in Figure 19.5a. The process diagram is shown in Figure 19.5b. The preheater (PH) (line 1–2) and steam injection (line 4–5) are shown as active, while the heating coil (HC) is shown as inactive. Though steam injection simultaneously humidifies and heats the airstream, having an additional heating coil will provide the needed flexibility in control under diverse part-load conditions during which the relative amounts of sensible and latent heats can vary.

The airflow required for single-zone *heating* is

$$\dot{V}_{a,heat} = \frac{\dot{Q}_{space,heat}}{\rho_a c_a (T_{db,supply} - T_{db,space})} \qquad (19.12)$$

where

$\dot{V}_{a,heat}$ is the heating airflow rate, m^3/s (ft^3/h)
$\dot{Q}_{space,heat}$ is the heating load, W (Btu/h)
$T_{db,supply}$ is the supply air temperature leaving the steam injector

The fan size selection for a single-zone fixed-volume system is based on the larger of the two flow rates calculated previously:

$$\dot{V}_a = \max\left(\dot{V}_{a,cool}, \dot{V}_{a,heat}\right) \qquad (19.13)$$

This flow rate is assumed to be constant throughout the year for the CAV system.

Example 19.4: Winter Peak Design

Consider the CAV system analyzed under peak cooling load condition in Example 19.1. The same space is to be maintained at 72°F (25.5°C) and 50% RH. The total heating load of the space is 150,000 Btu/h (44 kW) with an SHR of 0.8. The outdoor design condition is 40°F (4.4°C) dry bulb and 40% RH. The same amount of ventilation air (1000 ft^3/min or 472 L/s) is needed. Design the necessary heating and humidification equipment assuming saturated steam at 200°F (93°C) is available and the supply airflow rate is kept at the summer design peak value of 17,140 lb_a/h (2.2 kg/s). The supply air dry-bulb temperature should not exceed 105°F (40.5°C).

Figure: See Figure 19.5a and b.

Assumptions: The location is at sea level. The duct heat transfer and the fan air temperature rise are ignored for simplicity.

Given: $\dot{m}_a = 17{,}140\ \text{lb}_a/\text{h}$, $\text{SHR}_{space} = 0.80$, $\dot{Q}_{space,tot} = 150{,}000\ \text{Btu/hs}$

Outdoor air conditions: $T_{db,o} = 40°\text{F}$, $\phi_o = 0.40$, $\dot{V}_0 = 1000\ \text{ft}^3/\text{min}$

Space conditions: $T_7 = 72°\text{F}$ and $\phi_7 = 0.5$, $T_{db,6} < 105°\text{F}$

Find: $\dot{m}_{steam}$, $\dot{Q}_{ph}$, and $\dot{Q}_{hc}$

Lookup values: Specific volume v_0 = 12.65 ft^3/lb_a, humidity ratio W_0 = 0.0021 lb_w/lb_a, W_7 = 0.0084 lb_w/lb_a, h_7 = 26.4 Btu/lb_a, specific heat of air c_a = 0.24 $\text{Btu/lb}_a \cdot °\text{F}$, and enthalpy of steam h_{steam} = 1146 Btu/lb_w.

Solution

We will illustrate the use of the graphical procedure as previously.

1. *Locate specified points* 0 and 7 on the psychrometric chart (see Figure 19.5b), which are specified from the problem statement.
2. *Determine outdoor air mass flow rate.*
   The outdoor air mass flow is given by

$$\dot{m}_{a,0} = \frac{\dot{V}_0}{v_0}$$
$$= \frac{1000\ \text{ft}^3/\text{min} \times 60\ \text{min/h}}{12.65\ \text{ft}^3/\text{lb}_a} = 4743\ \text{lb}_a/\text{h}$$

Note that this mass flow rate is over 13% higher than the previous case and is attributed to the difference in the specific volumes of outdoor air between summer and winter conditions.

3. *Locate mixed air condition point 1.*
   The inverse relation between mass flows and line segment lengths (Equation 13.32) is used to locate 1. The ratio of air mass flows is (4,743/17,140) = 0.277. The dry-bulb temperature and humidity ratios can be calculated from the chart or just as simply using the weighted average rule:
   a. *The mixed-air temperature*: $T_{db,1}$ = [0.277 × 40°F + (1 − 0.277) × 72°F] = 63.15°F
   b. *The humidity ratio*: W_1 = [0.277 × 0.0021 lb_w/lb_a + (1 − 0.277) × 0.0084 lb_w/lb_a] = 0.00665 lb_w/lb_a.

   The corresponding moist-air enthalpy is read from the psychrometric chart, h_1 = 22.4 Btu/lb_a.
4. *Determine supply air condition point 5.*
   As previously discussed, the slope of line 5–7 is equal to the inner scale of the protractor corresponding to SHR_{space} = 0.80. We determine enthalpy h_5 of the specified supply air mass flow rate as follows:

$$h_5 = h_7 + \frac{\dot{Q}_{space,heat}}{\dot{m}_a} = 26.4\ \text{Btu/lb}_a + \frac{150{,}000\ \text{Btu/h}}{17{,}140\ \text{lb}_a/\text{h}}$$

$$= 35.15\ \text{Btu/lb}_a$$

   The intersection of the enthalpy line corresponding to 35.15 Btu/lb_a and the line 6–7 yields a point whose corresponding dry-bulb value is $T_{db,5}$ = 102°F (this is acceptable since it is lower than the stipulated maximum value of 105°F). The corresponding humidity ratio W_5 = 0.0096 lb_w/lb_a.
5. *Determine the amount of steam needed.*
   This is easily calculated since the entering and leaving humidity ratio of the airstreams are known:

$$\dot{m}_{steam} = \dot{m}_a(W_5 - W_4)$$

$$= 17{,}140\ \text{lb}_a/\text{h} \times (0.0096 - 0.00665)\ \text{lb}_w/\text{lb}_a$$

$$= 50.56\ \text{lb}_w/\text{h}$$

6. *Compute sensible heating needed.*
   We can use different combinations of preheat and reheat amounts to attain the desired effect. One option is to assume that all heating is provided by the preheater. This is then a simple heating and humidification problem as treated in Section 13.6.6. We draw a line from point 6 parallel to the outer scale of the protractor corresponding to 1146 (since enthalpy of steam h_{steam} = 1146 Btu/lb_w) that intersects the horizontal line from point 1 at point 2. This is determined to be $T_{db,2}$ = 100.8°F and h_2 = 31.6 Btu/lb_a.

   Finally, the sensible heat provided by the heating coil is found:

$$\dot{Q}_{hc} = \dot{m}_a(h_2 - h_1) = 17{,}140\ \text{lb}_a/\text{h} \times (31.6 - 22.4)\ \text{Btu/lb}_a$$

$$= 157{,}688\ \text{Btu/h}$$

Figure 19.8a depicts another system configuration with a preheat coil (PC), an air washer where liquid water is sprayed into the airstream, and a reheat coil (HC). Note that such a configuration avoids the need to specify a steam generator (as is needed for the system shown in Figure 19.5a). The process diagram of such a system is drawn on the psychrometric chart in Figure 19.8b. The air washer provides the necessary humidification to the airstream to meet the requirements of both conditioning the dry outdoor ventilation air and that of the space. Because of the sensible heat reduction associated with the evaporative cooling process (line 2–3), both the preheater (line 1–2) and the heater (line 3–4) have to supply large amounts of heat.

19.2.4 Part-Load Operating Conditions

The previous examples illustrated the analyses procedures under peak or design conditions that allow the overall HVAC system to be configured and the various individual equipment sized for selection purposes. We need to analyze system performance in terms of the cooling and heating energy consumption of the CAV system for a one-zone space under part-load conditions. As stated earlier, a CAV system under part-load operation holds the system airflow rate constant and lets the space thermostat modulate the supply air temperature. The analysis procedure is best described by way of the following example. While Section 19.1.2 illustrated the use of the graphical approach to solve problems, we shall perform the same analysis using quantitative methods, while relying on the psychrometric chart to determine state properties and to get a sense of the overall system functioning.

Example 19.5: Analysis of CAV Systems under Part-Load Operation

Consider the CAV system that was analyzed under peak-load condition in Example 19.1. This example will serve to illustrate how the CAV system needs to be operated under part-load condition so as to maintain satisfactory indoor space

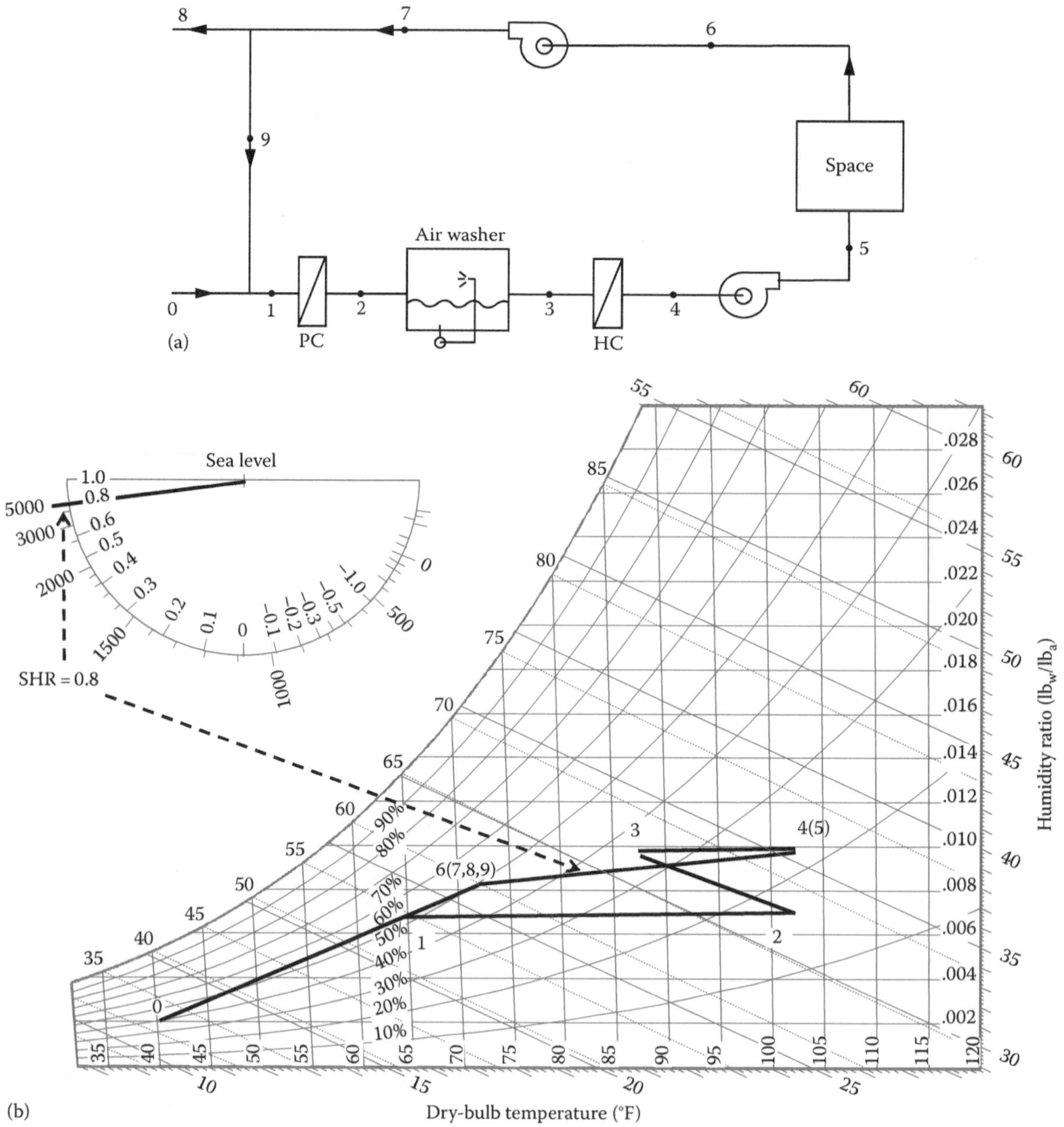

FIGURE 19.8
(a) Schematic of the winter heating and humidification system with injection of liquid water. The cooling coil is not shown. (b) Psychrometric process diagram with state points.

conditions. We will determine the cooling coil loads as well as the reheat energy required in this case. We assume the same zone and cooling coil outlet temperatures and the amount of ventilation air as stipulated in Example 19.1. The outdoor air conditions are 80°F and 60% RH, while the space cooling load is 70,000 Btu/h with $SHR_{space} = 0.7$.

Figure: See Figures 19.1 and 19.4.*

Assumptions: The location is at sea level. The supply air mass flow rate = 17,140 lb$_a$/h. The duct heat transfer and the fan air temperature rise are ignored for simplicity. The preheat and steam humidifier are inactive.

Given: $SHR_{space} = 0.70$, $\dot{Q}_{space,tot} = 70{,}000$ Btu/h, $\dot{Q}_{space,sen} = 70{,}000 \times 0.7 = 49{,}000$ Btu/h

Outdoor conditions: $T_{db,o} = 80°F$, $\phi_o = 0.6$, $\dot{V}_0 = \dot{V}_0 = 1000$ ft^3/min

Cooling coil conditions: $T_{db,3} = 58°F$ and $\phi_3 = 0.8$.

Space condition: $T_{db,6} = 78°F$ and supply air mass flow rate = 17,140 lb$_a$/h

Find: $\dot{Q}_{cc,tot}$, $\dot{Q}_{cc,sen}$, and $\dot{Q}_{hc,tot}$

* Numerical values shown in the figure do not correspond to this example.

Lookup values: Specific volume v_0 = 13.9 ft³/lb$_m$, humidity ratio W_0 = 0.01325 lb$_w$/lb$_a$, W_3 = 0.0082 lb$_w$/lb$_a$, specific heat of air c_a = 0.24 Btu/(lb$_a$ · °F), and latent heat of vaporization h_v = 1075 Btu/lb$_w$.

Solution

1. *Calculate outdoor air mass flow rate.*

$$\dot{m}_{a,0} = \frac{\dot{V}_0}{v_0} = \frac{1000\ \text{ft}^3/\text{min} \times 60\ \text{min/h}}{13.9\ \text{ft}^3/\text{lb}_\text{a}} = 4320\ \text{lb}_\text{a}/\text{h}$$

2. *Calculate supply air temperature* point 5 using sensible heat balance since the room thermostatic control is based on dry-bulb temperature.
   Using sensible heat balance equation:

$$\dot{Q}_{space,sen} = \dot{m}_a \cdot c_a \cdot (T_{db,6} - T_{db,5})$$

$$T_{db,5} = T_{db,6} - \frac{\dot{Q}_{space,sen}}{\dot{m}_a \times c_a}$$

$$= 78°\text{F} - \frac{49{,}000\ \text{Btu/h}}{17{,}140\ \text{lb}_\text{a}/\text{h} \times 0.24\ \text{Btu}/(\text{lb}_\text{a} \cdot °\text{F})}$$

$$= 66.1°\text{F}$$

3. *Verify indoor comfort.*
   Calculate humidity ratio of air leaving room using a latent heat balance

$$\dot{Q}_{space,lat} = \dot{m}_a \cdot h_v \cdot (W_6 - W_5)$$

or

$$W_6 = W_5 + \frac{\dot{Q}_{space,lat}}{\dot{m}_a \times h_{vap}} = 0.0082\ \text{lb}_\text{w}/\text{lb}_\text{a}$$

$$+ \frac{70{,}000\ \text{Btu/h} \times (1 - 0.7)}{17{,}140\ \text{lb}_\text{a}/\text{h} \times 1{,}075\ \text{Btu/lb}_\text{w}} = 0.00934\ \text{lb}_\text{w}/\text{lb}_\text{a}$$

   This corresponds to an indoor relative humidity of ϕ_o = 45%, which along with a dry-bulb temperature of 78°F is satisfactory for indoor human comfort.
4. *Calculate mixed-air condition point 1.*
   A common assumption that simplifies the analysis is to assume a constant-specific heat of moist air c_a. The energy balance equation can then be expressed as

$$\dot{m}_{a,1} T_{db,1} = (\dot{m}_{a,1} - \dot{m}_o) T_{db,7} + \dot{m}_o T_{db,o}$$

or

$$17{,}140\ \text{lb}_\text{a}/\text{h} \times T_{db,1} = (17{,}140 - 4{,}320)\ \text{lb}_\text{a}/\text{h} \times 78°\text{F}$$
$$+ 4{,}320\ \text{lb}_\text{a}/\text{h} \times 80°\text{F}$$

   resulting in $T_{db,1}$ = 78.5°F.
   Similarly, the humidity is determined as W_1 = 0.0103 lb$_w$/lb$_a$.
5. *Determine cooling coil loads (process 1–3).*
   Using the simplified expressions for sensible and latent loads, we have
   a. *Sensible load (Equation 13.45)*

$$\dot{Q}_{cc,sen} = 17{,}140\ \text{lb}_\text{a}/\text{h} \times 0.24\ \text{Btu}/(\text{lb}_\text{a} \cdot °\text{F}) \times (78.5 - 58)°\text{F}$$
$$= 84{,}329\ \text{Btu/h} = 7.03\ \text{tons}$$

   b. *Latent load (Equation 13.46)*

$$\dot{Q}_{cc,lat} = 17{,}140\ \text{lb}_\text{a}/\text{h} \times 1075\ \text{Btu/lb}_\text{w}$$
$$\times (0.0103 - 0.0082)\ \text{lb}_\text{w}/\text{lb}_\text{a}$$
$$= 38{,}931\ \text{Btu/h} = 3.24\ \text{tons}$$

   c. *Total load*

$$\dot{Q}_{cc,tot} = 84{,}329 + 38{,}931 = 123{,}260\ \text{Btu/h} = 10.27\ \text{tons}$$

6. *Determine reheat coil load.*
   Using the sensible heat equation, reheat load

$$\dot{Q}_{rh,tot} = 17{,}140\ \text{lb/h} \times 0.24\ \text{Btu}/(\text{lb}_\text{a} \cdot °\text{F}) \times (66.1 - 58)°\text{F}$$
$$= 33{,}320\ \text{Btu/h}$$

Comments

The reheat is an energy penalty in that the single airstream supply to the space has to be overcooled in order to provide the dehumidification necessary for occupant comfort and then reheated in order to negate this overcooling. This is a good illustration of the concept of coil bucking stated earlier.

19.3 Single-Zone Single-Duct VAV Systems

19.3.1 Description of the Basic System

A more widely used method for reducing the energy penalties associated with CAV systems is to reduce the supply airflow at part-load conditions. Such systems, called *variable air volume* (VAV) systems (briefly introduced in Section 19.1.4), have become almost standard for all but small-sized buildings in the United States, and for certain special-purpose rooms (such as rooms in the building interior, conference rooms). Although this design is more energy efficient than CAV systems, it can have problems of its own that can test the ingenuity of

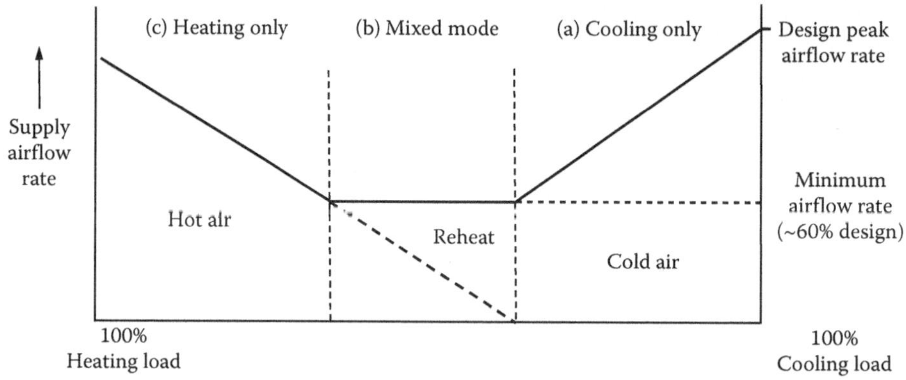

FIGURE 19.9
Modulation of room supply airflow for a single-zone variable air volume system as space loads vary over the year. Note the three modes of operation denoted by (a), (b), and (c).

the HVAC design engineer. Several variants of VAV systems have evolved over time, which are also described in subsequent sections.

In essence, VAV systems modulate system peak airflow from full-load levels whenever loads are less than peak loads. Since flow is reduced, energy transfer at the air handler coils is reduced, and fan power is also reduced markedly. The reduction in energy transfer is essentially proportional to the airflow reduction. One would also expect the fan power reduction to be proportional to the cube of the airflow reduction, but fan power is not reduced as much because of the need to maintain duct pressure (this is described later). The VAV system is the only commonly used design that reduces energy consumption significantly as the load is reduced. The vast majority of these systems are designed for cooling. (The conditions that might lead us to adopt a VAV heating-only system are rarely encountered in commercial buildings.) Some automotive air-conditioning systems are stepped-VAV designs where the flow can be reduced in discrete steps by the blower speed selector switch.

The HVAC system layout for a typical one-zone VAV system is identical to that of CAV system with a preheat coil, a cooling coil, and a reheat coil just before the zone supply inlet. Under peak cooling (or heating) conditions, the VAV system operates identically to a fixed-volume system with the air handler operating at maximum flow and maximum cooling coil capacity. The heating (or cooling) coils are inactive. Hence, the capacity of the cooling coil (or heating coil) and the fans of a VAV system are identical to those of a CAV system.

19.3.2 Different Operating Modes

It is under part-load operation that the distinctions between CAV and VAV systems emerge. We can distinguish between the three operating modes of the basic VAV system (Figure 19.9). As the cooling load reduces, the system airflow has to be correspondingly reduced with the load. This is achieved by the zone thermostat acting on the zonal VAV damper *and* the fan speed controller modulating the supply fan so as to maintain a preset duct pressure (measured by a pressure sensor inside the air duct). Three fan volume control methods are commonly used, and these are discussed in Section 16.5. The fundamental VAV system is a cooling-only system (*mode a*) that modulates *system* airflow in response to cooling loads as sensed by a dry-bulb temperature thermostat. The air supply is at a fixed temperature, viz., equal to the cooling coil leaving air temperature (if fan heating is neglected). Since the VAV system is primarily a cooling system, it is best used where cooling is required during the majority of the year. Buildings with significant internal gains and those located in warm climates are good candidates.

In order to ensure proper mixing of the cold supply air once it enters the space, the supply air volume cannot be reduced beyond a certain value, typically 60% of the maximum peak value. Under such cases, the VAV system is essentially operated as a CAV system but with the supply airflow set at its minimum allowable value. This operation is shown as *mode b* in Figure 19.9. There will be some coil bucking under this mode, i.e., the air cooled by the cooling coil has to be reheated. As the heating load increases, the supply airflow rate can be ramped up again with the cooling coil being inactive and heating provided by the preheat and reheat cooling coils. This is *mode c* or the heating-only VAV system operation that occurs much less frequently than the other two modes because of the rather high internal loads common to most modern buildings. In certain instances when mode b occurs rarely during the year, a common HVAC design practice is to use electric strip heaters around the zone perimeter rather than using a terminal reheat system because of the resulting reduction in cost and maintenance.

The need to supply ventilation air and to meet perimeter (or top floor) heating loads in winter complicates the cooling-only VAV system. Heat can be added by a coil in the VAV box or by zone reheat using electric strip heating or perimeter baseboard or fan coil units as shown in Figure 19.10a. Some heating can be done from heat sources within a zone, such as lights and computers, but this cannot be relied on to warm up exposed zones after periods of no occupancy in winter. Hence, local heat for warm-up, at the least, is required. There may also be a need for heat in internal rooms that are not continuously occupied since the dampers on VAV boxes cannot close entirely, particularly if there is a physical minimum damper closure to ensure adequate ventilation airflow. Without internal gains and without heat, such intermittently occupied spaces can drop in temperature. Heating is also needed if humidity control is a requirement in a zone. For laboratory spaces that need to maintain a slight overpressurization, a venturi valve is required to control the make-up air (as shown in the photo of Figure 19.10b).

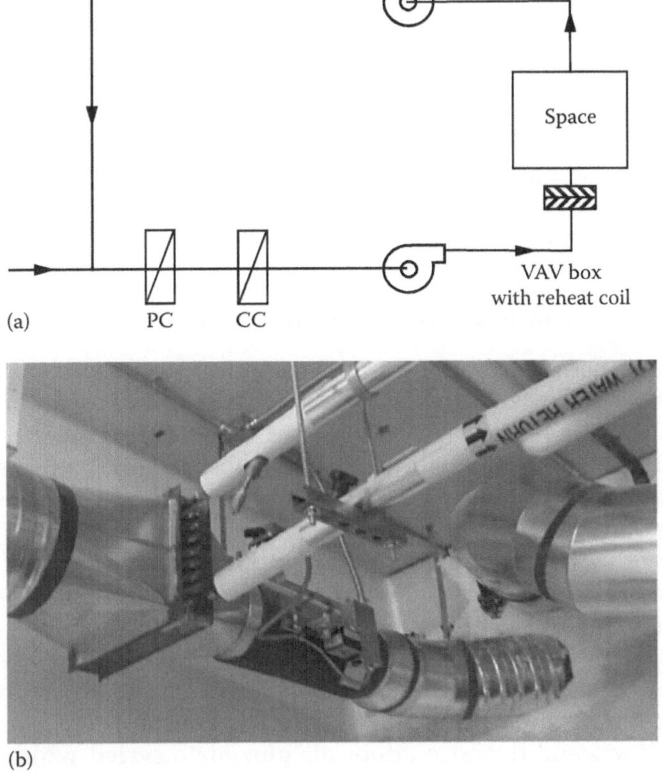

FIGURE 19.10
Terminal reheat configuration for a variable air volume (VAV) system. (a) Sketch of a single-zone ducted return system for a small laboratory. (b) Photograph showing VAV duct, venturi valve for pressure control of make-up air, and heating coil. (Courtesy of Arizona State University, Tempe, AZ.)

Example 19.6: VAV Analysis under Part-Load Operation

Consider Example 19.5 where a CAV system was assumed. We will solve this problem assuming VAV system operation.

Figure: Figures 19.1 and 19.3.*

Assumptions: The location is at sea level. The duct heat transfer and the fan air temperature rise are ignored for simplicity. The preheat coil is inactive.

Given: $\dot{Q}_{space,tot} = 70{,}000\,\text{Btu/h}, \dot{Q}_{space,sen} = 49{,}000\,\text{Btu/h}$

Outdoor conditions: $T_o = 80°\text{F}$, $\phi_o = 0.6$, $\dot{V}_0 = \dot{V}_0 = 1000\ \text{ft}^3/\text{min}$, or $\dot{m}_0 = 4320\ \text{lb}_a/\text{h}$

Cooling coil conditions: $T_{db,3} = 58°\text{F}$ and $\phi_3 = 0.8$.

Space condition: $T_{db,6} = 78°\text{F}$.

Find: $\dot{Q}_{cc,tot}$, $\dot{Q}_{cc,sen}$, and $\dot{Q}_{hc,tot}$

Lookup values: $W_0 = 0.01325\ \text{lb}_w/\text{lb}_a$, $W_3 = 0.0082\ \text{lb}_w/\text{lb}_a$, specific heat of air $c_a = 0.24\ \text{Btu/(lb}_a\cdot°\text{F)}$, and latent heat of vaporization $h_{vap} = 1075\ \text{Btu/lb}_w$.

Solution

1. *Calculate supply air mass flow rate* with supply air temperature assumed to be equal to that of the cooling coil set temperature. From a sensible heat balance equation (Equation 19.11),

$$\dot{m}_a = \frac{\dot{Q}_{space,sen}}{(T_{db,6} - T_{db,5}) \times c_a} = \frac{49{,}000\ \text{Btu/h}}{(78-58)°\text{F} \times 0.24\ \text{Btu/(lb}_a\cdot°\text{F)}}$$

$$= 10{,}208.3\ \text{lb}_a/\text{h}$$

2. *Verify indoor comfort*. Calculate humidity ratio of air leaving room from a latent heat balance:

$$W_6 = W_5 + \frac{\dot{Q}_{space,lat}}{\dot{m}_a \times h_{vap}} = 0.0082\ \text{lb}_w/\text{lb}_a$$

$$+ \frac{70{,}000\ \text{Btu/h} \times (1-0.7)}{10{,}208.3\ \text{lb}_a/\text{h} \times 1075\ \text{Btu/lb}_w} = 0.0101\ \text{lb}_w/\text{lb}_a$$

This condition corresponds to an indoor relative humidity of $\phi_o = 49\%$, which along with a dry-bulb temperature of 75°F is within the indoor human comfort range.

3. *Calculate mixed-air condition.*

$$\dot{m}_{a,1} \times T_{db,1} = (\dot{m}_{a,1} - \dot{m}_o) \times T_{db,7} + \dot{m}_o \times T_{db,o}$$

* Numerical values shown in the figure do not correspond to this example.

or

$$10{,}208.3\ \text{lb}_a/\text{h} \times T_{db,1} = (10{,}208.3 - 4{,}320)\ \text{lb}_a/\text{h} \times 78°\text{F} + 4{,}320\ \text{lb}_a/\text{h} \times 80°\text{F}$$

resulting in $T_{db,1}$ = 78.85°F.
Similarly, the humidity ratio is determined as $W_1 = 0.0114\ \text{lb}_w/\text{lb}_a$

4. *Determine cooling coil loads.*
Using the simplified expressions from Section 13.6.5, we have
   a. *Sensible load (Equation 13.45)*

$$\dot{Q}_{cc,sen} = 10{,}208.3\ \text{lb}_a/\text{h} \times 0.24\ \text{Btu}/(\text{lb}_a \cdot °\text{F}) \times (78.85 - 58)°\text{F}$$
$$= 51{,}082\ \text{Btu/h} = 4.26\ \text{tons (down from 7.03 tons in Example 19.5)}$$

   b. *Latent load (Equation 13.46)*

$$\dot{Q}_{cc,lat} = 10{,}208.3\ \text{lb}_a/\text{h} \times 1{,}075\ \text{Btu}/\text{lb}_w \times (0.0114 - 0.0082)\ \text{lb}_w/\text{lb}_a$$
$$= 35{,}116\ \text{Btu/h}$$
$$= 2.93\ \text{tons (down from 3.24 tons)}$$

   c. *Total load*

$$\dot{Q}_{cc,tot} = 51{,}082 + 35{,}116 = 86{,}198\ \text{Btu/h}$$
$$= 7.18\ \text{tons (down from 10.27 tons)}$$

5. *Calculate reheat coil load.*
In this case, reheat coil load is zero.

Comments

In the aforementioned example, the reduced airflow has eliminated the need for reheat but has somewhat compromised occupant comfort due to the slightly elevated humidity in the space (RH increased from 45% to 49%). The flow reduction of (10,208.3/17,140) = 0.595 is right at the lower end of the norm allowed. Had it been lower, the proper operating strategy would be to adopt *mode b* shown in Figure 19.9. This would have entailed a small amount of reheat and increased cooling energy use as well.

19.4 All-Air Systems for Multiple Zones

19.4.1 Basic Single-Duct Systems

The single-zone single-duct CAV system is the simplest secondary system that one can imagine. However, multiple-zone systems are the norm in commercial buildings because of the diversity of loads and relatively stringent comfort requirements. One could imagine adding zones to the single-zone system shown in Figure 19.1, but control would be poor unless energy-inefficient terminal reheat coils at the entrance to the individual spaces are provided. Their use is curtailed by energy codes except for special circumstances where accurate zone temperature or humidity control is needed.

The CAV system for multiple zones operates with a fixed-volume flow rate and fixed supply fan outlet temperature. The air supplied to each zone is at a constant temperature, set by the main cooling coil controller. Typically, a set point of 13°C (55°F) is selected to meet the design or peak sensible and latent cooling loads. Variation of loads within a zone or among zones below the peak is accounted for by adding heat to the air at each zone with a reheat coil (electrical, steam, or hot water). The source of energy waste is obvious—cooled air is reheated prior to release to the zone(s). To minimize energy waste, the cold air temperature should be reset to the highest possible temperature that will just meet the cooling load. Section 19.5 discusses this aspect and illustrates the calculation procedures via a solved example.

Simple single-duct VAV systems are much more appropriate than CAV systems for meeting multizone requirements, but they also may require some amount of terminal reheat using reheat coils as shown in Figure 19.10. If heating is needed (usually in a perimeter zone), reheat can also be supplied by a separate subsystem such as baseboard radiation or electric strip heating. Section 19.5 also discusses this aspect and illustrates the calculation procedures via a solved example.

19.4.2 Fan-Powered Single-Duct VAV Systems

Because distribution of air within a zone depends on the velocity of air entering a zone (discussed in Section 20.3), the reduced airflow needed to meet off-peak cooling loads can drastically affect zone airflow patterns and, hence, comfort. As a result, the simple VAV system is often modified to provide constant airflow in a *zone* with varying conditioned air input to each zone. The conditioned air supplied to a zone from the HVAC system air handler is called *primary air*. Constant room airflow with varying primary airflow is accomplished by mixing room air (sometimes called *secondary air*) with primary air within a VAV box. The constant total airflow (conditioned outdoor air plus air recycled within the zone by the VAV box) is typically 4–6 air changes/h (or 0.6–1.0 ft³/[min·ft²]) (Holness, 1990) in commercial buildings, irrespective of the primary air supply volume.

The first method for combining primary and secondary air is illustrated in Figure 19.11. In this *induction* method, primary air entrains secondary air in a

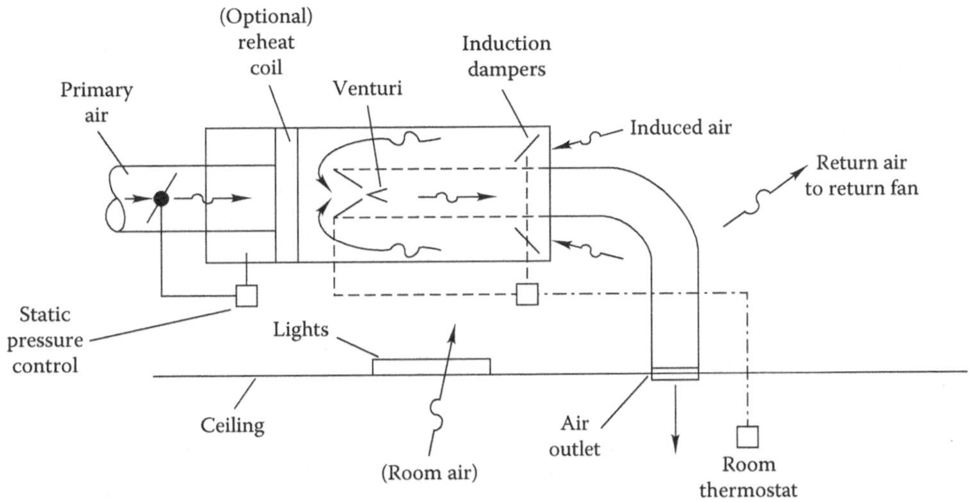

FIGURE 19.11
Schematic diagram of induction variable air volume box.

specially designed VAV terminal box located in the ceiling plenum* of a zone. Return air from zones normally passes through grilles located in ceiling tiles into the plenum, from which it flows to a return fan. An adequate supply of return air is therefore available in the plenum for mixing with primary air. Since the room air volume flow rate remains essentially constant in induction systems due to the action of the dampers shown, satisfactory room air circulation is maintained as cooling loads vary. Effective induction systems require increased duct pressure to accomplish entrainment compared to the basic VAV system without induction. Therefore, operating costs are higher.

Another method to guarantee good room air distribution is to use *fan-powered* VAV mixing boxes. In this method, a small fan is used to mix primary air (the amount controlled by the primary air damper that is controlled in turn by the room thermostat) and secondary air and to ensure proper distribution throughout the area served by a VAV box. The mixed air thus produced has a varying temperature but a constant airflow as the load varies. Fan-powered units are either of *parallel* or *series* design. Series systems are constant-volume flow systems that use a continuously operating fan located in the mixed airstream to provide essentially constant zone airflow. Constant flow is needed in some large zones not only during part-load cooling conditions but also during heating conditions to avoid stratification. Figure 19.12 is a schematic diagram of a two-zone space being conditioned by a series fan-powered VAV terminal. Note that the secondary flows of each zone can be controlled separately. Figure 19.13a shows closeup sketches for fan-powered series and parallel VAV boxes, while Figure 19.13c is a cutaway of a parallel box.

The parallel approach uses an intermittently operating fan. Parallel boxes rely on primary cooling airflow to adequately distribute zone air during most of the cooling season. However, when cooling needs are low or during the heating season, the fan operates in parallel boxes to provide air distribution. Figure 19.14 shows primary and room air characteristics of both types of fan-powered VAV boxes. The key distinction is that series boxes produce constant room airflow during either heating or cooling modes, whereas the parallel design does not. The operation of the parallel box fan is controlled by the room thermostat. The interested reader can refer to the ASHRAE literature wherein several papers have been published on series and parallel fan-powered terminal units (see, for example, Davies et al., 2009 who developed performance models and validating them against careful laboratory measurements).

The control of VAV systems involving a number of devices (thermostat, valves) is discussed in detail in Chapter 21. However, it is important to understand the general features of the VAV system control in order to design the secondary system properly. Control of these systems is complex and has been one of the sources of difficulty in some installed systems. First, one must ensure adequate flow at the zone most remote from the air handler. This is traditionally accomplished by

* Exhausting room air can be achieved either by a ducted return or a plenum return arrangement. The former is similar to a supply air duct layout with return grills located in the ceiling or walls that send air through branches to a main return air trunk. Though easy to clean and maintain while providing greater fire protection, disadvantages are that they increase floor height and need higher fan power. The plenum return design uses no branches, but the room air is sent through return grills into the space created by a false ceiling. The plenum is sealed from the sides and the air is channeled to an AHU through masonry ducts with little pressure drop since the whole ceiling space acts as a return path.

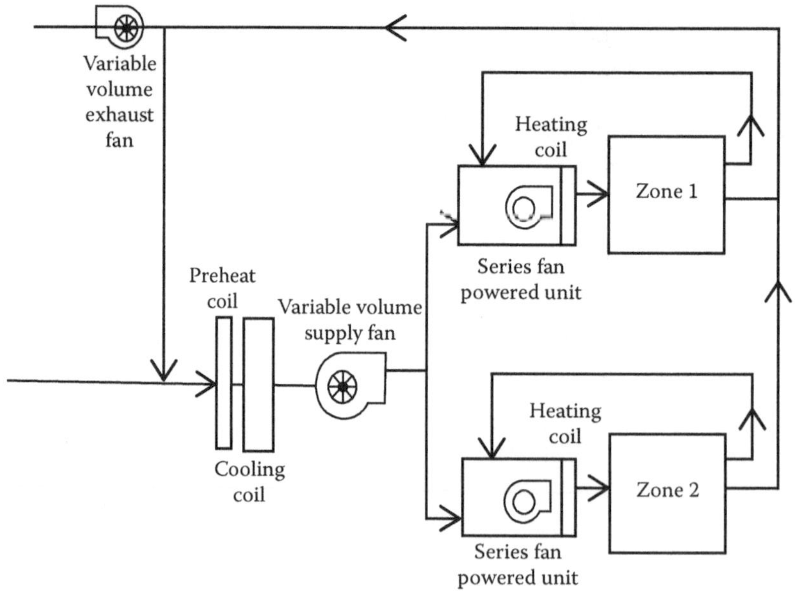

FIGURE 19.12
Sketch of a two zone space conditioned by a series fan-powered variable air volume system.

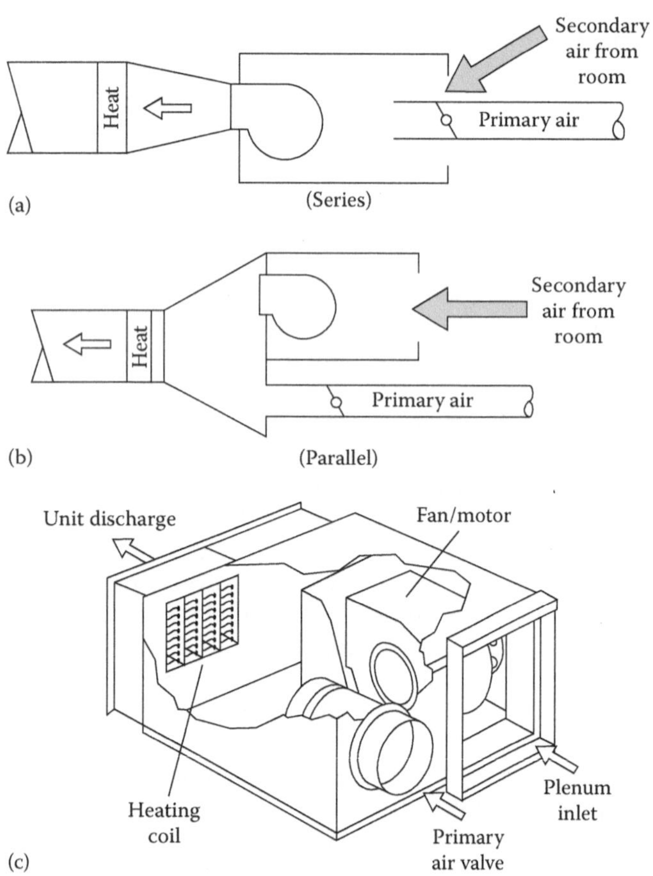

FIGURE 19.13
(a and b) Schematic diagrams of series and parallel fan-powered variable air volume (VAV) box. (c) Cutaway drawing of parallel VAV box. (Courtesy of Titus, Inc., Plymouth, IN.)

controlling the supply fan speed with a pressure signal measured near the end of the duct. The actual airflow through the zones is controlled by each thermostat's control of the damper position (see Figure 19.15a). Because of the need to maintain static pressure at the end of a duct run, the fan power savings that one might estimate by using the fan laws are not achievable (Englander and Norford, 1988).* However, if the VAV system supply fan is controlled instead in such a way that one box is fully open at all times, then greater fan power savings can be realized because the extra fan power used to pressurize the end of a duct system is no longer needed.

Because of exhaust and exfiltration from zones, the return flow will be less than the supply flow. Return fans are variable speed and can be controlled on a flow basis if the supply and exhaust flows are known. For cost reasons, these flows are not always measured. An alternative method of controlling the return fan (and exhaust dampers) is by building pressure control.†

* One also needs to consider safety. For example, if a variable-speed fan drive were to fail at full speed in a system with normally closed (NC) dampers, duct seams could be ruptured. A duct maximum pressure sensor, interlocked with the fan drive, could protect the system. Also, NC boxes open more slowly than the time taken for the fan to reach full speed; this also requires design attention such as start-up fan speed controller.

† Building pressure control is important for infiltration management. A negative building (with exhaust exceeding outdoor air supply) will be less expensive to construct since the air-handling equipment and ducts are smaller. However, uncontrolled infiltration may cause discomfort and local freezing of pipes if they are near winter air leaks. A neutral to slightly positive building pressure with exhaust about equal to or less than supply will greatly reduce infiltration problems.

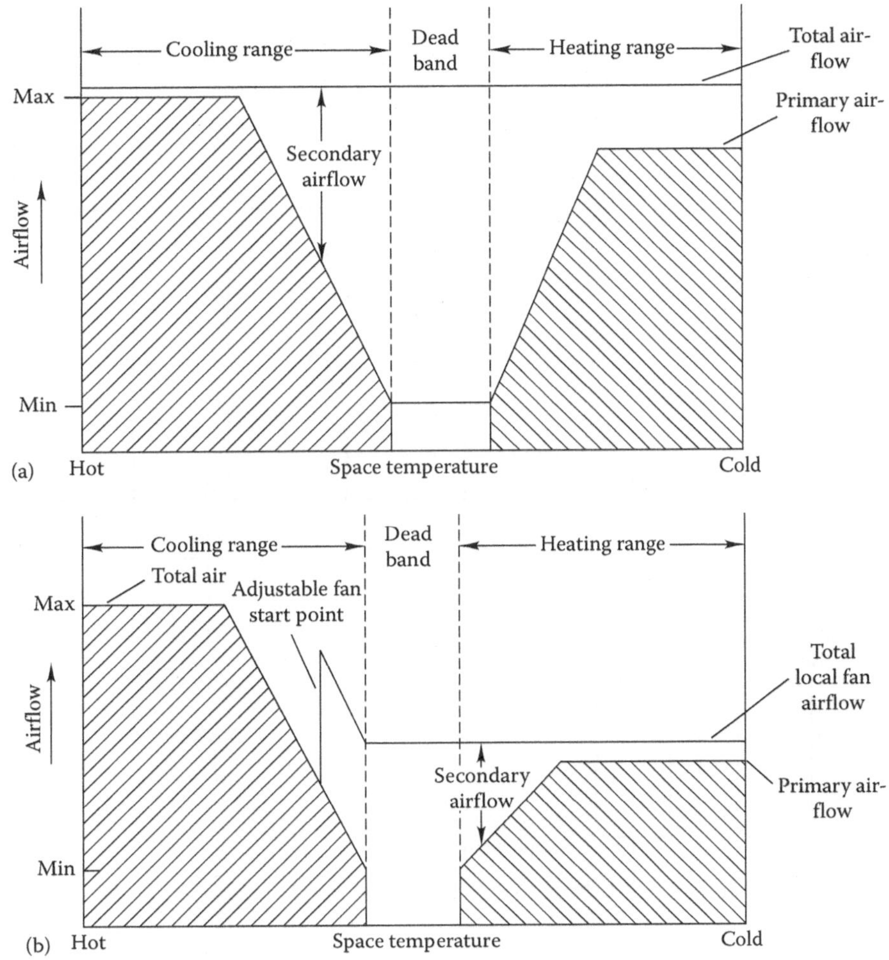

FIGURE 19.14
(a) Flow characteristics for series fan-powered variable air volume (VAV) boxes. The filled areas represent primary airflow. The total airflow is constant. (b) Flow characteristics for parallel fan-powered VAV boxes. Filled areas represent primary airflow. Total box airflow is higher during cooling than heating. Fan operates only at zone temperatures below the adjustable set point.

This approach can be difficult since measuring a representative building pressure is sometimes problematic. A variable-volume relief fan is sometimes used in VAV systems. It is located in the exhaust duct (shown in the upper left-hand corner of Figure 19.15a). It operates only when the amount of outside air exceeds the minimum (i.e., during economizer cycle operation discussed in Section 19.6.2) and is controlled by the building static pressure sensor or by airflow measurements in the outside and exhaust air ducts.

Another important flow characteristic of VAV boxes is the dependence of flow on *supply duct pressure*. Duct pressure at one point is controlled by the fan speed controller. However, pressure can vary at other points along the duct as more or less air is required by various zones. It is desirable to have the primary flow to all VAV boxes be independent of local supply duct pressure, to ensure stable system operation. Mixing boxes designed to operate in this way are called *pressure-independent* VAV boxes. Pressure independence is provided by local microprocessors or by mechanical devices responding to duct pressure and zone air velocity. The designer must verify that claims of pressure independence are in fact borne out by independent test data. VAV boxes that are sensitive to supply duct pressure are called *pressure dependent;* the air valve position in these boxes is controlled by the zone thermostat.

The engineering of VAV systems requires *very careful attention to design*. First, as for all systems, the zone loads must be determined accurately to minimize part-load penalties. If calculated loads are higher than actual loads, the VAV system will always operate at part load, and potential savings will not be realized due to part-load inefficiencies; this penalty applies even more strongly to CAV systems. On the other hand, if zone loads are undercalculated, the VAV box will remain open more than expected, thereby starving other zones in the system. A particular problem exists with zones

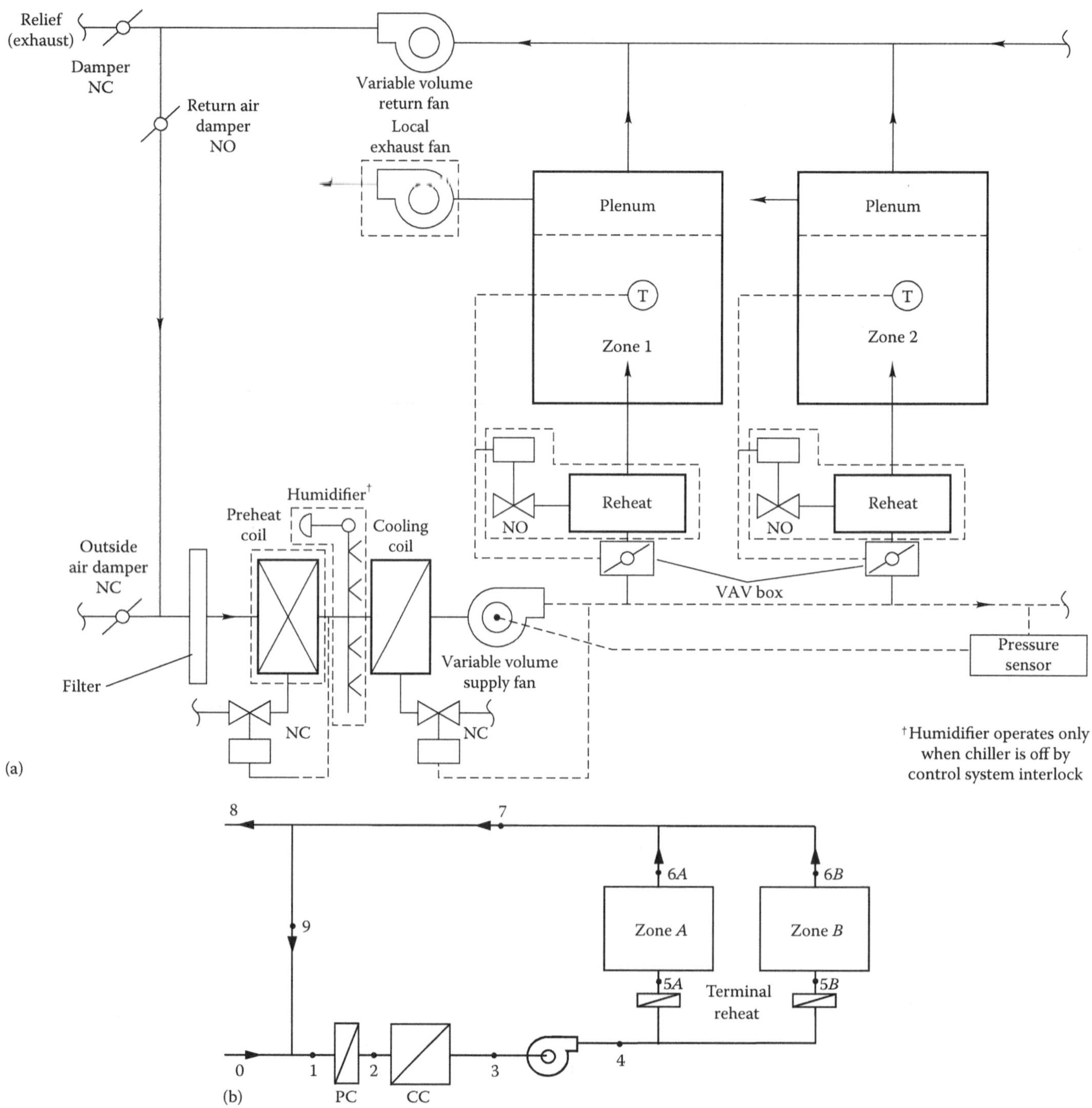

FIGURE 19.15
(a) Plenum return single-duct variable air volume system for a two-zone building. (b) Simplified sketch with state points for system analysis.

having high latent loads controlled by dry-bulb temperature thermostats. The dry-bulb temperature thermostat may close off the VAV box flow without meeting the latent load.

Flow balancing is also essential if adequate air is to be supplied to all zones. Avery (1986) and Guntermann (1986) describe a number of other VAV system problems that can be avoided by careful design. Table 19.1 assembles some of the *advantages* of properly designed VAV systems over other all-air systems. ASHRAE 62.1 (2013) discusses additional advantages and design precautions for VAV systems.

TABLE 19.1

Advantages of Properly Designed Variable Air Volume Systems

1. Fan power and central plant energy savings can be significant when VAV systems are applied correctly to appropriate buildings. Cumulative savings of operating costs are usually much greater than the initial cost of installing VAV boxes at the zones.
2. System flexibility is large. Future load changes are easy to accommodate.
3. Balancing of VAV systems is readily achieved if adequate fan capacity exists. If loads exceed design estimates, poor flow balance will probably occur. (The velocity reduction method described in Section 16.6 is adequate for sizing most VAV systems, since flows vary greatly from peak to minimum load.)
4. Noise levels are lower most of the time than for fixed-volume systems since maximum flow occurs infrequently (only under maximum load conditions).

19.4.3 Dual-Duct and Multizone Systems

The *dual-duct* CAV system shown in Figure 19.16a has a single supply and return fan but two sets of ducts—one for cold air and the other for hot air. The heating and cooling coils are energized by the central plant as in the single-zone system. The two sets of ducts terminate at a mixing box at each zone. The relative amounts of hot and cold airstreams supplied to each zone are controlled by reverse-acting dampers operated by the thermostat in each zone. The airflow is maintained constant to each zone by the action of these dampers. The dual-duct system has several advantages, while the larger number of *compelling disadvantages* has resulted in few such systems being built today (see Table 19.2).

The *system* airflow in a dual-duct system is most often determined by the full cooling airflow requirement (see Example 19.8 in this chapter). Therefore, an equation similar to Equation 19.1 applies except on a system-wide basis. The total system mass flow rate of air required by a cooling system under peak cooling conditions is

$$\dot{m}_{a,cc,peak} = \frac{\dot{Q}_{space,tot}}{(h_{cc,in} - h_{cc,out})} \tag{19.14}$$

in which the coil cooling load is taken to be the sum of sensible and latent loads for the entire collection of zones served by one air handler.

Under part-load conditions, the dual-duct CAV system air handler flow is the same as under full-load conditions but is split between the hot and cold "decks." The conservation of mass requires that

$$\dot{m}_a = \dot{m}_{a,cc} + \dot{m}_{a,hc} \tag{19.15}$$

To find the cold and hot deck flows individually, the loads on the zones A and B must be known. Recalling that the zone thermostats are the controlling elements, zone sensible heat loads determine the zonal flows originating from the cooling coil ($\dot{m}_{a,cc,A}$ and $\dot{m}_{a,cc,B}$) and those from heating coil ($\dot{m}_{a,hc,A}$ and $\dot{m}_{a,hc,B}$)—see Figure 19.16b.

Energy balances assuming constant-specific heat of air lead to

$$\dot{m}_{a,cc,A}T_{db,3} + \dot{m}_{a,hc,A}T_{db,4} = \dot{m}_{a,A}T_{db,5A}$$

and

$$\dot{m}_{a,cc,B}T_{db,3} + \dot{m}_{a,hc,B}T_{db,4} = \dot{m}_{a,B}T_{db,5B} \tag{19.16}$$

while mass balance considerations result in

$$\dot{m}_{a,cc,A} + \dot{m}_{a,hc,A} = \dot{m}_{a,A} \quad \text{and} \quad \dot{m}_{a,cc,B} + \dot{m}_{a,hc,B} = \dot{m}_{a,B} \tag{19.17}$$

Simple algebraic manipulation of these two equations provides the necessary working relations:

$$\dot{m}_{a,cc,A} = \dot{m}_{a,A}\frac{(T_{db,4} - T_{db,5A})}{(T_{db,4} - T_{db,3})} \quad \text{and} \quad \dot{m}_{a,hc,A} = \dot{m}_{a,A} - \dot{m}_{a,cc,A}$$

and

$$\dot{m}_{a,cc,B} = \dot{m}_{a,B}\frac{(T_{db,4} - T_{db,5B})}{(T_{db,4} - T_{db,3})} \quad \text{and} \quad \dot{m}_{a,hc,B} = \dot{m}_{a,B} - \dot{m}_{a,cc,B} \tag{19.18}$$

Example below illustrates how to use the equations for a single-zone.

Example 19.7: Dual-Duct System Part-Load Operation

A dual-duct conditioned single zone has a peak heating load of 7.50 kW (25,600 Btu/h) and a peak sensible cooling load of 6.50 kW (22,200 Btu/h). The zone temperature is to be held at 25°C (77°F) (also assumed for simplicity to be the return air temperature).

The two duct supply temperatures at the zone are 40°C (104°F) and 14°C (57°F). If the cooling load is 3.50 kW (11,950 Btu/h) at a part-load point, what are the heating and cooling rates at the zone?

Given: Hot and cold deck and room temperatures:

$$T_{db,4} = 40°\text{C}, T_{db,3} = 14°\text{C}, T_{zone} = 25°\text{C}$$

Design peak loads: $\dot{Q}_{\max,cool} = 6.5$ kW, $\dot{Q}_{\max,heat} = 7.5$ kW

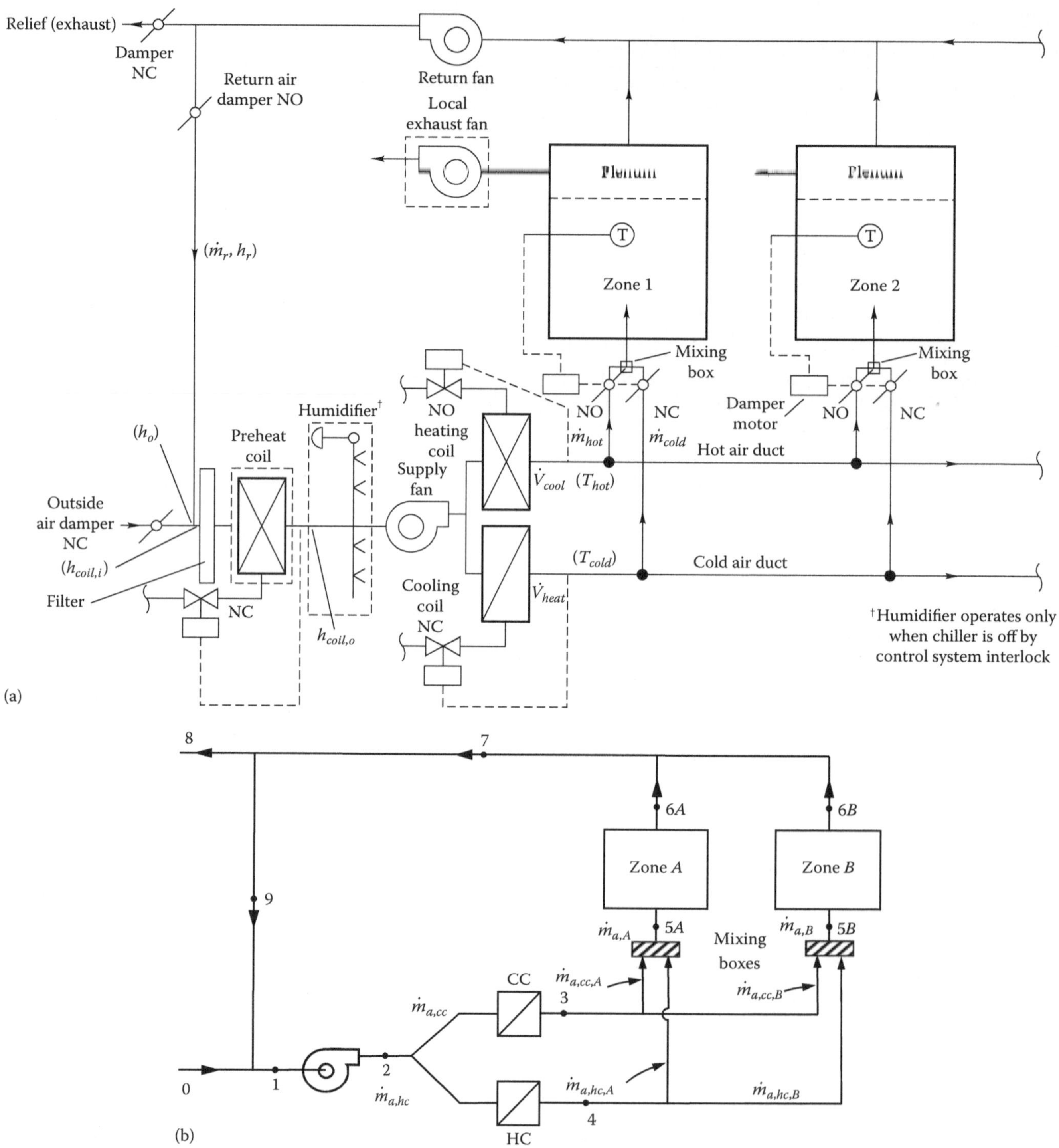

FIGURE 19.16
(a) Dual-duct constant air volume system for a two-zone space. (b) Simplified sketch with state points for system analysis.

Part load: $\dot{Q}_{part,cool} = 3.5$ kW

Figure: Figure 19.16b (consider only one of the zones).

Assumptions: Ignore the air temperature rise across the fan and the heat losses and gains in the ductwork. Consider only sensible loads for simplicity.

Find: $\dot{Q}_{cc}$, $\dot{Q}_{hc}$

Solution

The solution involves two steps. First, the airflow at peak load is determined. Since the dual-duct system is a constant-volume system, this same flow will apply to part-load conditions. Second, solve for the heating and cooling rates at part load.

TABLE 19.2

Attributes of Dual-Duct HVAC Systems

Advantages	Disadvantages
1. All space conditioning needs are taken care of at the central plant—no HVAC primary heating or cooling equipment is located in the zones. 2. Since warm air is available year-round, temperature control of lightly loaded zones is good during the cooling season. (If heat were not available, leaks in cold duct dampers could cause lightly loaded zones to be too cold.) Control response to load changes is rapid. 3. Duct sizing calculations are not as critical since the dampers at each zone absorb any pressure imbalance caused by design inaccuracy or part-load operation. 4. If humidity control is needed, this system is a good candidate.	1. Ducts tend to leak. 2. Zone damper leakage requires oversizing both the heating plant (to account for cold air damper leakage during peak heating) and the cooling plant (to account for warm air damper leakage during peak cooling). 3. The space required for two full-sized ducts and the requirement that each zone have two duct connections make duct design more difficult. Duct velocities and pressures are higher because two ducts must fit into the space of one. 4. Overall, system control stability requires that the zone mixing boxes be capable of constant volume control. A separate volume control damper is sometimes needed at each zone's mixing box. 5. The system is *not energy-efficient* since reheating of cold air by warmed air and the reverse occur. 6. Initial cost is higher than that for single-duct systems. 7. An economizer should not be applied to the basic dual-duct system because the heating coil inlet air will be at the cold deck set point, not at the mixed-air temperature. Heating energy requirements will be much higher and can exceed the cooling energy savings. Hence, the benefits of an economizer are not available to dual-duct systems.

The airflow rate needed for peak heating is found from Equation 19.12

$$\dot{m}_{a,hc,peak} = \frac{7.5\ \text{kW}}{(40-25)°\text{C} \times 1.0\ \text{kJ/(kg}\cdot\text{K)}} = 0.50\ \text{kg/s}$$

and for peak cooling, the airflow rate is

$$\dot{m}_{a,cc,peak} = \frac{6.5\ \text{kW}}{(25-14)°\text{C} \times 1.0\ \text{kJ/(kg}\cdot\text{K)}} = 0.59\ \text{kg/s}$$

The higher flow rate (0.59 kg/s) will be used for both heating and cooling since the constant-volume dual-duct system uses a single constant-volume fan.

The supply air temperature entering the zone at part load (3.5 kW cooling) is found from a sensible heat balance

$$\dot{Q}_{part,cool} = 3.5\ \text{kW} = 0.59\ \text{kg/s} \times 1.0\ \text{kJ/(kg}\cdot\text{K)} \times (25 - T_5)°\text{C}$$

from which the supply air temperature is found to be $T_5 = 19.1°C$.

At the part-load condition, we use Equation 19.18

$$\dot{m}_{a,cc} = 0.59\ \text{kg/s} \times \frac{(40-19.1)°\text{C}}{(40-14)°\text{C}} = 0.475\ \text{kg/s}$$

$$\text{and} \quad \dot{m}_{a,hc} = 0.59 - 0.475 = 0.115\ \text{kg/s}$$

Finally, the cooling and heating coil loads are

$$\dot{Q}_{cc} = 0.475\ \text{kg/s} \times 1.0\ \text{kJ/(kg}\cdot\text{K)} \times (25-14)°\text{C}$$
$$= 5.23\ \text{kW (17,850 Btu/h)}$$

$$\dot{Q}_{hc} = 0.115\ \text{kg/s} \times 1.0\ \text{kJ / (kg}\cdot\text{K)} \times (40-25)°\text{C}$$
$$= 1.73\ \text{kW (5900 Btu/h)}$$

The difference between these two coil loads is the 3.5 kW cooling rate, as required by the problem statement.

Comments

This example illustrates the key problem with dual-duct systems from an energy viewpoint. The heating energy in this example is not needed at all. However, it is present due to the system's basic configuration requiring constant airflow. To compound the problem, not 3.50 kW but 5.23 kW of cooling is needed, since the unwanted 1.7 kW of heat must be extracted by the cooling system in addition to the zone's load. The energy waste is obvious.

The fundamental reason that this waste occurs is that heating is needed to warm a larger-than-necessary volume of air to maintain zone temperatures at part load. The cold deck temperature cannot be adjusted upward since 14°C is needed for proper dehumidification. However, if the air volume could be reduced at part load, the energy waste could be avoided. This is the motivation for dual-duct VAV systems that, though uncommon, do exist in several large buildings in the United States (often existing dual-duct CAV systems are retrofitted to dual-duct VAV systems for enhancing energy efficiency).

One method of simplifying the dual-duct system and reducing cost is to mix hot and cold airstreams at the central air handler instead of at each zone and to use only one duct for distribution to the zones. This *multi-zone* system is shown in Figure 19.17a. Zone thermostat controls for all conditioned spaces are centrally located

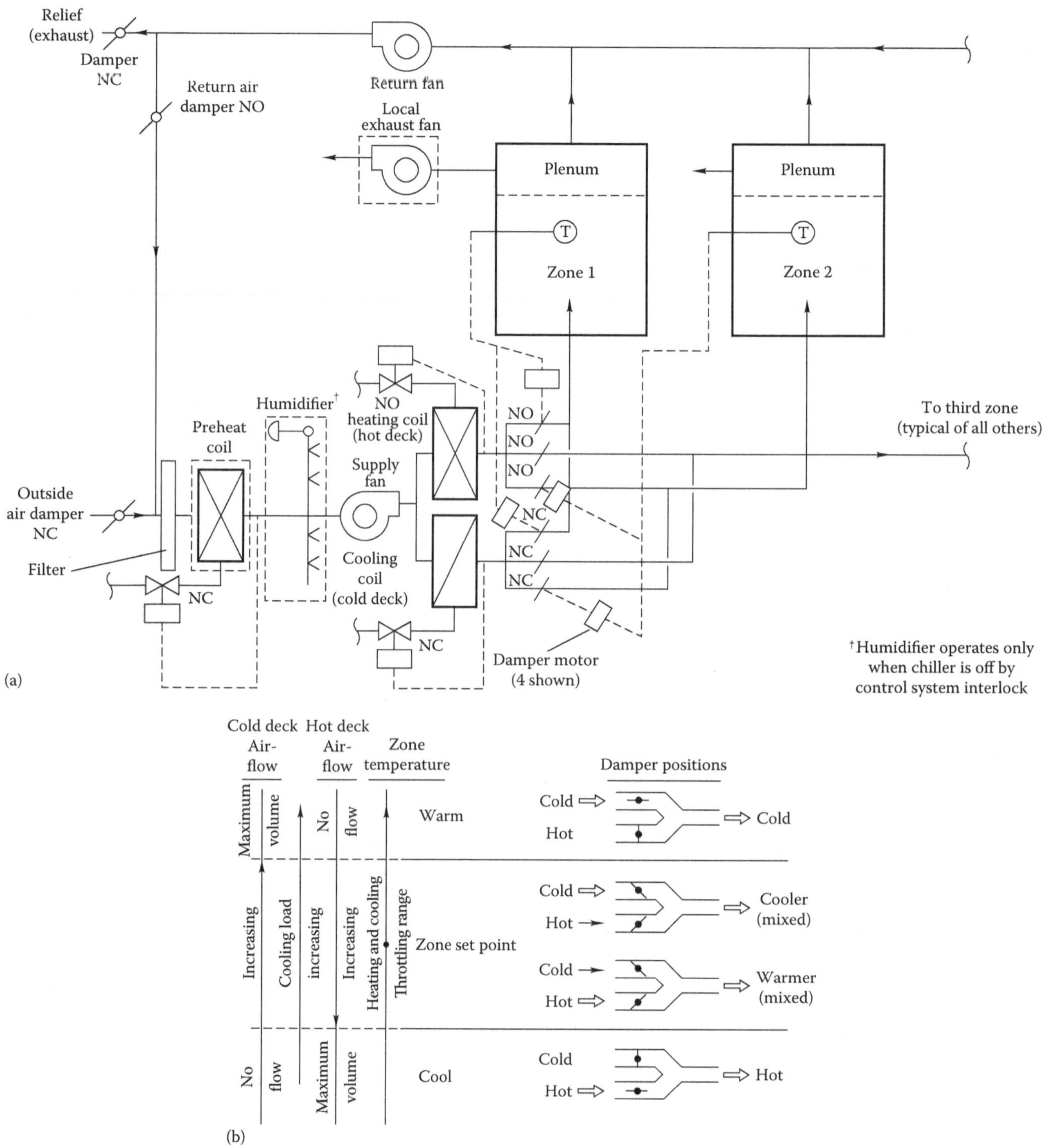

FIGURE 19.17
(a) Multizone HVAC system for a two-zone space. (b) Multizone system damper operation under various heating and cooling load conditions.

at the main air handler unit (see Figure 19.17b) rather than at the zone mixing boxes located over the occupied spaces, as in the dual-duct system. Zone control precision is moderately good with the multizone system.

Damper leakage can be a problem since leakage affects the entire hot or cold deck airflow. Leakages of 10% are common, although 3%–5% dampers are available. Leakage forces the designer to oversize cooling and heating plants by the leakage amount. In hot, humid climates, multizone systems should not be used, since hot deck damper leakage bypasses humid air and can increase the humidity of zone air to unacceptable levels. Other practical problems include the need for running many small ducts—one to each zone—and problems with properly mixing hot and cold airstreams among various zone ducts at part load. The energy penalties of dual-duct systems also apply to multizone systems.

The multizone system is less flexible than the dual-duct system since at most 12–14 zones can be served by packaged multizone units. A small zone connected to a multizone unit with a number of large zones with changing loads can have unpredictable airflow and comfort control. Finally, it is more difficult to add a new zone to a multizone unit than to a dual-duct system. As with dual-duct systems, few multizone systems are being built nowadays.

19.5 Design Sizing and Energy Analysis

19.5.1 Sizing under Design Conditions

Selection of secondary equipment is driven by peak loads imposed by the building on the primary and secondary systems. Once the *peak* loads are calculated (see Chapter 9), the secondary system can be sized. The operation, sizing, and analysis under part load of the commonly used secondary HVAC systems, namely, single-duct CAV (Section 19.2.4), VAV systems (Section 19.3.2), and dual-duct CAV systems (Section 19.4.3), were discussed in the framework of *single-zone systems*. The extension to multiple zones follows the same general principles and is covered in this section. The approach used to size secondary equipment including coils, fans, and ducts is illustrated only for the VAV system since it is similar for dual-duct and other systems as well.

Steps in sizing the secondary system components in a single-duct VAV system serving multiple zones (refer to Figure 19.15b) are as follows:

0. Determine the peak heating and cooling (sensible and latent) loads for each zone and allow for pickup loads and safety factors as appropriate (Chapter 9).
1. Establish minimum ventilation airflow rate (Table 3.8) and cooling coil leaving humidity ratio (related to cooling coil effectiveness and hence to coil cost) during the cooling season. The outdoor air loads are included in steps 8 and 10.

Cooling Design Day

2. Compute each zone's peak supply airflow rate based on a preselected value of the zonal supply air temperature difference—the difference between the zone air temperature and the zone cooling air supply temperature. This temperature difference is the designer's decision, usually 20°F–22°F (11°C–12°C). Sum all zone flow rates to find the total system airflow. This is the *fan size. Duct sizing* can now be completed by using the techniques described in Section 16.6.
3. Calculate fan power.
4. Determine the cooling air supply temperature at the zones by subtracting the supply air temperature difference from the zone temperature.
5. Calculate the required cooling coil leaving dry-bulb temperature to meet the criterion of step 4, accounting for duct heat gains and the air temperature rise through the fan.
6. Find each zone's return air temperature, accounting for return duct heat gains.
7. Calculate the flow-weighted average return air humidity ratio and temperature, including return fan temperature rise (if any). Verify that occupant comfort conditions are met.
8. Compute the mixed-air (return and outside airstreams) temperature and humidity ratio. This is the cooling coil inlet air condition.
9. Find the cooling coil sensible and latent loads. This is the *cooling coil rating* (accounting for piping heat gains). This is also the chiller size when safety factors and pickup loads are added.

Heating Design Day

10. Compute the preheat coil load. The preheat coil heats outdoor air from the design winter temperature to the coil leaving temperature, determined in step 4. This step sizes the *preheat coil.*
11. Compute the zone heating requirement. It is the sum of the zone heat load from step 1 and the heat required to warm zone supply air from the supply air temperature (see step 4) to the zone dry-bulb temperature. Account for any supply duct heat losses. This step sizes the zone *reheat coil and/or zone baseboard* heating units.

The heat load from steps 10 and 11 along with piping losses, pickup loads, and safety factors determines the central system *boiler size*. Often a diversity factor is applied to the same zone loads, reflecting the fact that the peak loads on all zones are not coincident.

Example 19.8 shows how a secondary system is sized for a small, two-zone commercial building for which the loads are known. Note that the calculation sequence described earlier is followed till step 10 after which it deviates a little since one of the zones calls for cooling even during the peak winter design day.

Example 19.8: Single-Duct VAV Secondary System Sizing (Design or Peak Conditions)

A two-zone building is to be equipped with a VAV system with preheat and reheat under the load conditions noted as follows. Size the supply fan, cooling coil, preheat coil, and reheat coil (or baseboard heating) for this building. Note that these will also apply to a CAV system under peak design conditions.

Given: The following data are known from the load calculation and by the designer's specification (these loads exclude the contribution from outdoor air):

	Zone *A* (Exterior)	Zone *B* (Interior)
Sensible peak summer design cooling load	$\dot{Q}_{sen,A} = 224,844$ Btu/h	$\dot{Q}_{sen,B} = 103,308$ Btu/h
Latent peak summer design cooling load	$\dot{Q}_{lat,A} = 56,000$ Btu/h	$\dot{Q}_{lat,B} = 20,000$ Btu/h
Sensible peak winter design load	$\dot{Q}_{heat,A} = 143,000$ Btu/h	$\dot{Q}_{cool,B} = 49,092$ Btu/h
Zone temperature T_{zone}	75°F	75°F

Other:

(a) Summer design dry- and wet-bulb temperatures: 97°F and 76°F (resulting in 0.0144 lb_w/lb_a humidity ratio at peak conditions).
(b) Winter design temperature: 7°F and air assumed to be totally dry.
(c) Supply system pressure drop at full airflow: 3.0 inWG.
(d) Fan efficiency: 60%.
(e) Supply fan air temperature rise: 1°F.
(f) Return air fan air temperature* rise: none.
(g) Ventilation airflow rate: 2400 ft³/min.
(h) Cooling coil outlet humidity ratio (at typical 85% RH): 0.0077 lb_w/lb_a.
(i) Zone primary supply air temperature difference ΔT_{supply}: 20°F.
(j) The maximum temperature to which space supply air can be heated is 105°F.

Figure: See Figure 19.15b.

Assumptions:

(i) Ignore factors not included in the list given earlier, such as duct heat losses and gains.
(ii) The location is assumed to be at sea level.
(iii) Rather than use enthalpy balances on the zones, the flows will be designed based on sensible heat loads (this is meant to simplify the problem).
(iv) Peak loads are coincident; no diversity adjustment is used.
(v) During peak heating, latent loads on the space are negligible.
(vi) During winter, room humidifiers are used to maintain the necessary humidity levels.
(vii) The supply airflow to the zones cannot be reduced to less than 60% by mass of the full-load design value by mass.

Lookup value: Outdoor-specific volume = v_0 = 14.35 ft^3/lb_a, c_a = 0.24 Btu/($lb_a \cdot$°F), and h_v = 1075 Btu/lb_w.

Find: Coil loads (preheat, cooling, and terminal reheat) and fan size.

Solution

The solution will follow the numbering system used in the preceding discussion.

1. Calculate ventilation airflow rate.

$$\dot{m}_{a,o} = \frac{\dot{V}_a}{v_0} = \frac{2400 \text{ cfm}}{14.35 \text{ ft}^3/\text{lb}_a} = 167.2 \text{ lb}_a/\text{min}$$

Cooling Design Day ($T_{db,0}$ = 97°F)

2. Determine air mass flow rates (Equation 19.11).

$$\dot{m}_{a,A} = \frac{224,844 \text{ Btu/h}}{0.24 \text{ Btu/(lb}\cdot°\text{F)} \times 60 \text{ min/h} \times 20°\text{F}} = 780.7 \text{ lb}_a/\text{min}$$

$$\dot{m}_{a,B} = \frac{103,308 \text{ Btu/h}}{0.24 \text{ Btu/(lb}\cdot°\text{F)} \times 60 \text{ min/h} \times 20°\text{F}} = 358.7 \text{ lb}_a/\text{min}$$

Thus, the total airflow is $\dot{m}_a = 1,139.4$ lb_a/min or 14,983 ft³/min(based on cooling coil leaving condition, specific volume = 13.15 ft^3/lb_a).

* The return fan heat is often less than the supply fan heat because airflow is less (local exhausts remove air locally), and return ducting via ceiling plenums has smaller pressure drop than found in supply ducts.

3. The system pressure drop is 3.0 inWG. Therefore, the fan power is given by Equation 16.34 IP:

$$\dot{W} = \frac{\dot{V}_a \cdot \Delta P}{6,356 \times \eta_{fan}} = \frac{14,983\ \text{ft}^3/\text{min} \times 3.00\ \text{inWG}}{6,356 \times 0.6} = 11.8\ \text{hp}$$

Note that this is the fan brake hp and does not include the electric motor efficiency

4. Determine cooling air supply temperature.

$$T_{supply} = T_{zone} - \Delta T_{db,supply} = 75°\text{F} - 20°\text{F} = 55°\text{F}$$

5. Determine cooling coil leaving temperature.

$$T_{coil,out} = T_3 = T_{supply} - \Delta T_{db,fan} - \Delta T_{db,duct}$$
$$= 55°\text{F} - 1°\text{F} - 0°\text{F} = 54°\text{F}$$

6. Determine zone return air condition. According to the problem statement, there are no return duct heat gains; therefore, the return dry-bulb temperature for each zone is 75°F.
7. Compute average return air humidity and temperature. The return air average moisture content under steady state is the given zone supply air humidity ratio (0.0077 $\text{lb}_\text{w}/\text{lb}_\text{a}$) increased by the total latent load:

$$\bar{W}_6 = W_3 + \frac{Q_{lat}}{\dot{m}_a \cdot h_{vap}} = 0.0077\ \text{lb}_\text{w}/\text{lb}_\text{a}$$
$$+ \frac{(56,000 + 20,000)\ \text{Btu/h}}{1,139.4\ \text{lb/min} \times 60\ \text{min/h} \times 1,075\ \text{Btu/lb}}$$
$$= 0.00873\ \text{lb}_\text{w}/\text{lb}_\text{a}$$

The average return air temperature is $T_7 = T_9 = 75°\text{F}$. This condition corresponds to an average space $RH_6 = 48\%$ that meets human comfort criteria.

8. Compute mixed-air condition. To find the mixed-air temperature entering the cooling coil, an energy balance is performed at point 1 where the return and outside air ducts connect. Since the ventilation airflow rate is 167.2 lb_a/min, the mixed-air temperature under design conditions is* as follows:

$$T_m = T_9\left(1 - \frac{\dot{m}_0}{\dot{m}_a}\right) + T_0\frac{\dot{m}_0}{\dot{m}_a} = 75.0°\text{F} \times \left(1 - \frac{167.2}{1139.4}\right) + 97°\text{F} \times \left(\frac{167.2}{1139.4}\right)$$
$$= 78.2°\text{F}$$

Likewise, the mixed-air humidity ratio is found by a water vapor mass balance:

$$W_m = 0.00873\ \text{lb}_\text{w}/\text{lb}_\text{a} \times \left(1 - \frac{167.2}{1139.4}\right) + 0.0144\ \text{lb}_\text{w}/\text{lb}_\text{a}$$
$$\times \left(\frac{167.2}{1139.4}\right) = 0.0096\ \text{lb}_\text{w}/\text{lb}_\text{a}$$

9. Compute cooling coil loads (preheat coil is inactive in summer). The coil sensible load is

$$\dot{Q}_{cc,sen} = \dot{m}_a c\ \Delta T_{coil} = 1139.4\ \text{lb}_\text{a}/\text{min} \times 60\ \text{min/h} \times 0.24\ \text{Btu/(lb} \cdot °\text{F)}$$
$$\times (78.2 - 54)°\text{F}$$
$$= 397,529\ \text{Btu/h or } 33.13\ \text{tons}$$

The latent load is found from:

$$\dot{Q}_{cc,lat} = m_a \cdot h_{vap} \cdot \Delta W_{cc} = 1,139.4\ \text{lb}_\text{a}/\text{min} \times 60\ \text{min/h} \times 1075\ \text{Btu/lb}$$
$$\times (0.00957 - 0.0077)\ \text{lb}_\text{w}/\text{lb}_\text{a}$$
$$= 137,428\ \text{Btu/h} = 11.45\ \text{tons}$$

The total coil load or coil cooling rate is

$$\dot{Q}_{cc,tot} = \dot{Q}_{cc,sen} + \dot{Q}_{cc,lat} = 534,632\ \text{Btu/h} = 44.55\ \text{tons}$$

Heating Design Day ($T_{db,0} = 7°\text{F}$)

10. Determine preheat coil load. The peak load on the preheat coil is determined based on the load needed to heat the cold outdoor ventilation air on the winter design day to the nominal coil outlet temperature (54°F in this example). Therefore, the preheat (subscript "ph") coil size is given by (assuming that outdoor air intake is the same as under peak summer condition)

$$\dot{Q}_{ph} = \dot{m}_a c_p (T_3 - T_0) = 167.2\ \text{lb}_\text{a}/\text{min} \times 60\ \text{min/h}$$
$$\times 0.24\ \text{Btu/(lb}_\text{a} \cdot °\text{F)} \times (54 - 7)°\text{F} = 113,161\ \text{Btu/h}$$

It must, however, be noted that the preheat coil will be activated if $T_1 < 54°\text{F}$ and will raise the supply airstream temperature to 54°F.

11. Determine airflow rates from Equation 19.11
    a. Zone *A* requires heating: assume maximum allowable supply air temperature of 105°F:

$$\dot{m}_{a,A} = \frac{143,000\ \text{Btu/h}}{0.24\ \text{Btu/(lb} \cdot °\text{F)} \times 60\ \text{min/h} \times (105 - 75)°\text{F}}$$
$$= 331.0\ \text{lb}_\text{a}/\text{min}$$

* Although mass flow rates should theoretically be used, engineers often use volumetric flow rates instead during preliminary analysis, since the density of all airstreams is about the same and the inaccuracy introduced is generally small.

Recall the stipulation that minimum airflow rate should not be less than 60% of the peak, or $\dot{m}_{a,\min} = 0.6 \times 780.7 = 468.4\ \text{lb}_a/\text{min}$. So, we set $\dot{m}_{a,A} = 468.4\ \text{lb}_a/\text{min}$.

b. Zone *B* requires cooling: assume minimum coil set point temperature of 54°F:

$$\dot{m}_{a,B} = \max\left\{\begin{matrix}(0.6 \times 358.7), \\ \left[\dfrac{49{,}092\ \text{Btu/h}}{0.24\ \text{Btu/(lb}\cdot{}^\circ\text{F)} \times 60\ \text{min/h} \times (75-55)^\circ\text{F}}\right]\end{matrix}\right\}$$

$$= \max(215.2, 170.4) = 215.2\ \text{lb}_a/\text{min}$$

Note that even for zone *B*, the stipulated minimum is reached.

Total airflow rate is 468.4 + 215.2 = 683.7 lb_a/min (rounded).

12. Determine supply air temperatures to the zones based on sensible heat balances.

$$T_5 = T_6 \pm \frac{\dot{Q}_{space}}{\dot{m}_a c_p}$$

Zone A: heating

$$T_{db,5A} = 75^\circ\text{F} + \frac{143{,}000\ \text{Btu/h}}{0.24\ \text{Btu/(lb}_a\cdot{}^\circ\text{F)} \times 60\ \text{min/h} \times 468.4\ \text{lb}_a/\text{min}}$$

$$= 96.2^\circ\text{F}$$

Zone B: cooling

$$T_{db,5B} = 75^\circ\text{F} - \frac{49{,}092\ \text{Btu/h}}{0.24\ \text{Btu/(lb}_a\cdot{}^\circ\text{F)} \times 60\ \text{min/h} \times 215.2\ \text{lb}_a/\text{min}}$$

$$= 59.2^\circ\text{F}$$

13. Determine mixed-air temperature.

$$T_2 = 75.0^\circ\text{F} \times \left(1 - \frac{167.2}{683.7}\right) + 7^\circ\text{F} \times \left(\frac{167.2}{683.7}\right) = 58.4^\circ\text{F}$$

Likewise, the mixed-air humidity ratio can be found assuming room air to be at the cooling coil leaving humidity ratio $W_1 = W_3 = 0.0077\ \text{lb}_w/\text{lb}_a$. This is not needed, however, for our calculations, since there will not be any latent load on the coil during the winter design condition.

14. Compute cooling coil load. The needed supply air temperature to Zone *B* is 59.2°F (from step 12) and 1°F temperature rise by the fan, which should be taken into account. Hence, the coil leaving temperature should be 58.2°F.

$$\dot{Q}_{cc,sen} = 683.7\ \text{lb}_a/\text{min} \times 60\ \text{min/h} \times 0.24\ \text{Btu/(lb}_a\cdot{}^\circ\text{F)}$$

$$\times (58.4 - 58.2)^\circ\text{F} = 2060\ \text{Btu/h or } 0.172\ \text{tons}$$

(in practice such a small cooling load would not justify switching on the chiller)

15. Compute reheating (assuming the chiller to be activated).

Zone A:

$$\dot{Q}_{hc,A} = 468.4\ \text{lb}_a/\text{min} \times 60\ \text{min/h} \times 0.24\ \text{Btu/(lb}_a\cdot{}^\circ\text{F)}$$

$$\times (96.2 - 59.2)^\circ\text{F} = 249{,}846\ \text{Btu/h}$$

Zone B:

$$\dot{Q}_{hc,B} = 215.2\ \text{lb}_a/\text{min} \times 60\ \text{min/h} \times 0.24\ \text{Btu/(lb}_a\cdot{}^\circ\text{F)}$$

$$\times (59.2 - 59.2)^\circ\text{F} = 0\ \text{Btu/h}$$

The total peak heating requirement is the sum of the preheat and reheat amounts:

$$\dot{Q}_{heat,tot} = 113{,}161 + 249{,}846 + 0 = 363{,}007\ \text{Btu/h}$$

This heat rate must be increased by pickup loads from night setback, safety factors, and piping losses to size the coils and the boiler.

Comments

Since this is a simple two-zone building, no diversity was assumed in the cooling or heating loads. Although diversity is rarely taken into account in heating loads unless internal or solar gains coincide with peak heating conditions, some diversity can be assumed in cooling loads when the building consists of a larger number of zones. With two zones, it is safer not to assume any diversity.

This example shows that the secondary systems have a considerable effect on the sizing of the primary heating and cooling plants. In any building more complex than a single-family residence, the secondary system loads are usually greater (and in many instances, *much greater*) than space zone loads. This effect must be considered in sizing the central plant. In this example at design conditions, the *zone cooling load* is about 330 kBtu/h, while the *coil cooling load* is 535 kBtu/h. The difference is due to the outdoor air that is conditioned at the central air handler coil.

In this example, we have used a given value of fan temperature rise. The actual rise can be computed as discussed in Section 19.1.3. Although the return fan has not been sized in this example, it can be done exactly as it was done with the supply fan. The duct pressure drop and return fan efficiency must be known to complete the calculation.

19.5.2 Performance under Part-Load Conditions

This section illustrates the calculation procedure for determining coil loads during part-load operation for each of the three all-air systems considered.

Example 19.9: Comparison of Energy Use of Different Secondary Systems under Part Load

(a) *Single-Duct VAV System* (Figure 19.15b)

Consider the same building with two zones as in Example 19.8 for which a VAV system was sized. Analyze this system under part-load operation under the following specifications:

	Zone *A* (Exterior)	Zone *B* (Interior)
Sensible cooling load	$\dot{Q}_{sen,A} = 143,100$ Btu/h	$\dot{Q}_{sen,B} = 91,260$ Btu/h
Latent cooling load	$\dot{Q}_{lat,A} = 36,000$ Btu/h	$\dot{Q}_{lat,B} = 20,000$ Btu/h
Zone temperature	$T_6 = 75°F$	$T_6 = 75°F$

Given:

1. Outdoor air conditions: $T_{db,o}$ = 77°F and $W_o = 0.0126$ lb$_w$/lb$_a$.
2. Ventilationairflowrate$\dot{m}_{a,o} = 167.2$ lb$_a$/min (same as the peak summer design flow rate).
3. The design flow rates to each zone: $\dot{m}_{a,A} = 780.7$ lb$_a$/min and $\dot{m}_{a,B} = 358.7$ lb$_a$/min.
4. Cold deck temperature $T_3 = 54°F$ and 85% RH (resulting in $W_3 = 0.0077$ lb$_w$/lb$_a$).
5. The supply airflow to the zones cannot be reduced to less than 60% of the full-load design value by mass.
6. Fan efficiency is 60%.

Assumptions:

1. The latent load in the interior zone is unchanged with outdoor temperature.
2. Ignore factors not included in the list given earlier, such as duct heat losses and gains.
3. The location is assumed to be at sea level.
4. Peak loads are coincident; no diversity adjustment is used.
5. Supply fan air temperature rise is 1°F, and assume the same 3 inWG pressure drop in the duct.
6. The preheat coil is inactive.

Find: Coil loads and fan power

Lookup values: Specific volume v_0 = 13.95 ft³/lb$_a$ and v_3= 13.13 ft³/lb$_a$

(i) Outdoor airflow rate is specified: $\dot{m}_{a,0}$ = 167.2 lb$_a$/min

(ii) Calculate zone supply airflows assuming supply temperatures to be equal to cold deck temperature T_3 = 54°F plus fan reheat of 1°F:

$$\dot{m}_{a,A} = \max\left\{\begin{matrix}(780.7 \text{ lb}_a/\text{min} \times 0.6), \\ \left[\dfrac{143,100 \text{ Btu/h}}{0.24 \text{ Btu/(lb}_a \cdot °\text{F}) \times 60 \text{ min/h} \times (75-55)°\text{F}}\right]\end{matrix}\right\}$$

$$= \max\{468.4, 496.9\} = 496.9 \text{ lb}_a/\text{min}$$

Note that minimum flow condition has not been reached.

Similarly,

$$\dot{m}_{a,B} = \max\left\{\begin{matrix}(358.7 \text{ lb}_a/\text{min} \times 0.6), \\ \left[\dfrac{91,260 \text{ Btu/h}}{0.24 \text{ Btu/(lb}_a \cdot °\text{F}) \times 60 \text{ min/h} \times (75-55)°\text{F}}\right]\end{matrix}\right\}$$

$$= \max\{215.2, 316.9\} = 316.9 \text{ lb}_a/\text{min}$$

i.e., minimum flow condition has not been reached.

Total supply airflow is

$$\dot{m}_{a,2} = 496.9 + 316.9 = 813.8 \text{ lb}_a/\text{min}$$

or 10,685 ft³/min (assuming $v_3 = 13.13$ ft³/min)

(iii) Determine zone supply air dry-bulb temperatures:
Since the minimum flow conditions have not been reached, we have

$$T_{supply,A} = 55°\text{F and } T_{supply,B} = 55°\text{F}$$

(iv) Determine fan brake horse power from Equation 16.34 (actually the static pressure will drop a little, but this is neglected):

$$\dot{W} = \frac{10,685 \text{ ft}^3/\text{min} \times 3.00 \text{ inWG}}{6,356 \times 0.6} = 8.41 \text{ hp}$$

Note that this does not include fan motor efficiency

(v) Check whether average space humidity levels are acceptable:

$$\bar{W}_6 = 0.0077 \text{ lb}_w/\text{lb}_a + \frac{(36,000+20,000)\text{Btu/h}}{813.8 \text{ lb}_a/\text{min} \times 1075 \text{ Btu/lb}_w \times 60 \text{ min/h}}$$

$$= 0.00877 \text{ lb}_w/\text{lb}_a$$

which for a space at 75°F dry-bulb temperature corresponds to RH_5 = 48% (acceptable).

(vi) Determine mixed-air condition:
Dry-bulb temperature:

$$T_1 = \left(\frac{167.2}{813.8}\right)\times 77°\text{F} + \left(1-\frac{167.2}{813.8}\right)\times 75°\text{F} = 75.4°\text{F}$$

Humidity ratio:

$$W_1 = \left(\frac{167.2}{813.8}\right)\times 0.0126\ \text{lb}_w/\text{lb}_a + \left(1-\frac{167.2}{813.8}\right)\times 0.00877\ \text{lb}_w/\text{lb}_a$$
$$= 0.0096\ \text{lb}_w/\text{lb}_a$$

(vii) Calculate cooling coil loads:

$$\dot{Q}_{cc,sen} = 813.8\ \text{lb}_a/\text{min}\times 60\ \text{min/h}\times 0.24\ \text{Btu/(lb}_a\cdot°\text{F)}$$
$$\times(75.4-54)°\text{F} = 250{,}893\ \text{Btu/h} = 20.9\ \text{tons}$$

$$\dot{Q}_{cc,lat} = 813.8\ \text{lb}_a/\text{min}\times 60\ \text{min/h}\times 1075\ \text{Btu/(lb}_w)$$
$$\times(0.0096-0.0077)\ \text{lb}_w/\text{lb}_a = 97{,}337\ \text{Btu/h} = 8.11\ \text{tons}$$

Total cooling coil load $Q_{cc,tot}$ = 348,231 Btu/h or 29.0 tons

(viii) Calculate reheat at terminals for Zones *A* and *B*:

$$\dot{Q}_{hc,A} = 0 \quad \text{and} \quad \dot{Q}_{hc,B} = 0$$

(b) *Single-Duct CAV System* (Figure 19.15b)

Consider the same building with two zones as for the single-duct VAV system analyzed earlier. Compute the cooling and heating loads if the secondary system used to condition the two zones is a single-duct CAV system.

0. Same outdoor airflow rate $\dot{m}_{a,0}$ = 167.2 lb_a/min. Also, total supply air mass flow rate $\dot{m}_a = 1139\ \text{lb}_a/\text{min}$ while the design flow rates to each zone:

$$\dot{m}_{a,A} = 780.7\ \text{lb}_a/\text{min} \quad \text{and} \quad \dot{m}_{a,B} = 358.7\ \text{lb}_a/\text{min}$$

1. Determine zone supply air dry-bulb temperatures assuming the airflows to individual zones to be equal to those determined under peak conditions (Example 19.8):

The supply temperature of zone *A* is

$$T_{db,A} = 75°\text{F} - \frac{143{,}100\ \text{Btu/h}}{780.7\ \text{lb}_a/\text{h}\times 0.24\ \text{Btu/(lb}_a\cdot°\text{F)}\times 60\ \text{min/h}}$$
$$= 62.3°\text{F}$$

and that of zone *B* is

$$T_{db,B} = 75°\text{F} - \frac{91{,}260\ \text{Btu/h}}{358.7\ \text{lb}_a/\text{h}\times 0.24\ \text{Btu/(lb}_a\cdot°\text{F)}\times 60\ \text{min/h}}$$
$$= 57.3°\text{F}$$

2. Fan power is 12 hp (same as under peak load since flow is unchanged).
3. Check whether average space humidity levels are acceptable:

$$\bar{W}_6 = 0.0077\ \text{lb}_w/\text{lb}_a + \frac{(36{,}000+20{,}000)\ \text{Btu/h}}{1{,}139.4\ \text{lb}_a/\text{h}\times 1{,}075\ \text{Btu/lb}_w\times 60\ \text{min/h}}$$
$$= 0.0085\ \text{lb}_w/\text{lb}_a$$

which corresponds to RH_5 = 45% (acceptable)

4. Determine mixed-air condition:
Dry-bulb temperature:

$$T_{db,1} = \left(\frac{167.2}{1139.4}\right)\times 77°\text{F} + \left(1-\frac{167.2}{1139.4}\right)\times 75°\text{F} = 75.3°\text{F}$$

Humidity ratio:

$$W_1 = \left(\frac{167.2}{1139.4}\right)\times 0.0126\ \text{lb}_w/\text{lb}_a + \left(1-\frac{167.2}{1139.4}\right)$$
$$\times 0.0085\ \text{lb}_w/\text{lb}_a = 0.0091\ \text{lb}_w/\text{lb}_a$$

5. Calculate cooling coil loads:

$$\dot{Q}_{cc,sen} = 1139.4\ \text{lb}_a/\text{min}\times 60\ \text{min/h}\times 0.24\ \text{Btu/(lb}_a\cdot°\text{F)}$$
$$\times(75.3-54)°\text{F} = 349{,}477\ \text{Btu/h} = 29.12\ \text{tons}$$

$$\dot{Q}_{cc,lat} = 1{,}139.4\ \text{lb}_a/\text{min}\times 60\ \text{min/h}\times 1{,}075\ \text{Btu/lb}_w$$
$$\times(0.0091-0.0077)\ \text{lb}_w/\text{lb}_a = 102{,}888\ \text{Btu/h}$$
$$= 8.57\ \text{tons}$$

Total cooling coil load $Q_{cc,tot}$ = 37.7 tons.

6. Calculate reheat at terminals for zones *A* and *B*:

$$\dot{Q}_{hc,A} = 780.7\ \text{lb}_a/\text{min}\times 60\ \text{min/h}\times 0.24\ \text{Btu/(lb}_a\cdot°\text{F)}$$
$$\times(62.3-55)°\text{F} = 82{,}067\ \text{Btu/h}$$

$$\dot{Q}_{hc,B} = 358.7\ \text{lb}_a/\text{min}\times 60\ \text{min/h}\times 0.24\ \text{Btu/(lb}_a\cdot°\text{F)}$$
$$\times(57.3-55)°\text{F} = 11{,}880\ \text{Btu/h}$$

Total load on reheat coils:

$$\dot{Q}_{hc,tot} = \dot{Q}_{hc,A} + \dot{Q}_{hc,B} = 11{,}880\ \text{Btu/h}$$

(c) *Dual-Duct CAV System* (Figure 19.16b)

Consider the same building with two zones as for the single-duct CAV system analyzed earlier. Compute the cooling and heating loads if the secondary system used to condition the two zones is a dual-duct CAV system with the cold deck set at the same cooling coil conditions as previously. Because of the location of the fan, the fan heats up the mixed airstream (T_2 = 75.3°F + 1°F = 76.3°F), while, in order to be consistent, the cold deck is set at $T_{db,3}$ = 55°F. The hot deck temperature is taken to be $T_{db,4}$ = 105°F to be consistent with the peak design procedure described in Example 19.8.

Steps 1–5 are identical to the single-duct CAV system case.

6. Compute the airflow rates using Equation 19.18:

$$\dot{m}_{a,cc,A} = 780.7\ \text{lb}_a/\text{min} \times \frac{(105-62.3)°\text{F}}{(105-55)°\text{F}} = 667.7\ \text{lb}_a/\text{min}$$

$$\text{and}\quad \dot{m}_{a,hc,A} = 780.7 - 666.7 = 114.0\ \text{lb}_a/\text{min}$$

$$\dot{m}_{a,cc,B} = 358.7\ \text{lb}_a/\text{min} \times \frac{(105-57.3)°\text{F}}{(105-55)°\text{F}} = 342.2\ \text{lb}_a/\text{min}$$

$$\text{and}\quad \dot{m}_{a,hc,B} = 358.7 - 342.2 = 16.5\ \text{lb}_a/\text{min}$$

7. Determine the airflow over the cooling and heating coils:

$$\dot{m}_{a,cc} = 666.7 + 342.2 = 1008.9\ \text{lb}_a/\text{min}$$

$$\text{and}\quad \dot{m}_{a,hc} = 130.5\ \text{lb}_a/\text{min}$$

8. Calculate cooling coil load:

$$\dot{Q}_{cc,sen} = 1{,}008.9\ \text{lb}_a/\text{min} \times 60\ \text{min/h} \times 0.24\ \text{Btu}/(\text{lb}_a \cdot °\text{F}) \times (76.3-55)°\text{F} = 322{,}343\ \text{Btu/h} = 26.86\ \text{tons}$$

$$\dot{Q}_{cc,lat} = 1{,}008.9\ \text{lb}_a/\text{min} \times 60\ \text{min/h} \times 1{,}075\ \text{Btu/lb}_w \times (0.0091-0.0077)\ \text{lb}_w/\text{lb}_a = 91{,}104\ \text{Btu/h} = 7.6\ \text{tons}$$

Total cooling coil load $Q_{cc,tot}$ = 34.46 tons.

9. Calculate heating coil load:

$$\dot{Q}_{hc} = 130.5\ \text{lb}_a/\text{min} \times 60\ \text{min/h} \times 0.24\ \text{Btu}/(\text{lb}_a \cdot °\text{F}) \times (105-76.3)°\text{F} = 53{,}726\ \text{Btu/h}$$

Comments

The secondary system is meant to meet the zone loads plus the energy needed to condition the ventilation outdoor air to the space condition. We will assume that the ideal indoor space condition is 75°F and 50% RH (at which condition, the humidity ratio is 0.0093 lb_w/lb_a). Hence, the "ideal" load is given by

$$\begin{aligned}\dot{Q}_{cc,ideal} &= (\text{Sensible} + \text{latent loads})\ \text{for Zone } A \text{ and } B \\ &\quad + (\text{Sensible} + \text{latent loads})\ \text{for ventilation air} \\ &= (143{,}100 + 91{,}260 + 36{,}000 + 20{,}000)\ \text{Btu/h} \\ &\quad + 167.2\ \text{kg/min} \times 60\ \text{min/h} \\ &\quad \times [0.24\ \text{Btu}/(\text{lb}_a \cdot °\text{F}) \times (77-75)°\text{F} + 1{,}075\ \text{Btu/lb}_w \\ &\quad \times (0.0126 - 0.0093)\ \text{lb}_w/\text{lb}_a] \\ &= 330{,}763\ \text{Btu/h} = 27.6\ \text{tons}\end{aligned}$$

The following table assembles the summary results of all three all-air systems under the part-load operation specified in Example 19.9.

	Ideal Loads	Single-Duct CAV	Single-Duct VAV	Dual-Duct CAV
Cooling (tons)	27.6	37.7	29.0	34.5
Heating (Btu/h)	—	93,947	0	53,726
Fan power (hp)	—	12.0	8.41	12.0
Thermal efficiency	—	0.61	0.95	0.71

We note that the VAV system outperforms the other two systems and is very close to the ideal loads. It is the best choice since it requires the least cooling energy, no heating energy, and lower fan electricity use. If we define efficiency in terms of the ideal loads, then for the VAV system, it is (27.6/29) = 95%, while that for the single-duct CAV = (27.6 tons × 12,000 Btu/h·ton)/(37.7 tons × 12,000 Btu/h·ton + 93,947 Btu/h) = 0.61. The two CAV systems are similar thermodynamically, and the differences in energy use determined are partly due to the manner in which we selected the hot deck temperature, due to the way we adjusted for the supply fan being upstream, and also due to slightly different resulting humidity levels in the space that affect the latent loads. The single-duct CAV dehumidifies the entire airstream, while, in the dual-duct system, it is only the portion of the airstream flowing over the cooling coil that is dehumidified. This has an impact on humidity levels inside the two zones.

The aforementioned examples will be used to illustrate an important issue with single-duct multizone VAV system part-load operation. The relative outdoor air fractions to each of the spaces vary in accordance with the relative zone loads, and this would lead to inadequate outdoor air amount supply to some of the spaces. For example, under design conditions (Example 19.8), the fraction of supply airflow to Zone *A* is (780.7/1139.4 = 0.685). Since, the outdoor air and return airstreams are mixed together, this will also be the fraction of the total outdoor air that flows into Zone *A*. Under part-load operation, even though the same amount of outdoor air is drawn in Example 19.9, the fraction is (496.9/813.8 = 0.611). Hence, zone *A* is underventilated, while Zone *B* is overventilated. In order to compensate for such variations, ASHRAE 62.1 (2013) stipulates a higher outdoor air amount resulting in energy penalties. A type of secondary system that avoids such penalties is described in Section 20.7.2.

19.5.3 Energy Analysis Using the Bin Method

The peak condition sizing procedure described in Section 19.5.1 determines the size of the secondary system equipment once the system configuration has been decided upon (a VAV system was assumed). However, it does not assist with assessing the relative economic and energy efficiency benefits of different systems. Section 19.5.2 illustrates the relative efficiencies of a single-duct CAV system, a single-duct VAV system, and a dual-duct CAV system at one part-load condition. In this section, we describe how the year-long performance of an HVAC system can be calculated if the loads are known.

The "bin" weather analysis presented in Section 10.4 can be used to find the annual energy consumption of a VAV system. The following example illustrates this approach.

Example 19.10: VAV System Calculations Based on the Bin Approach

The same two-zone building of Example 19.9 is assumed, and its annual cooling and heating energy is to be determined using the bin method.

Given: The zone envelope loads along with the corresponding binned dry-bulb temperature data (for Andrews Air Force Base, Maryland) are given in Table 19.3. Chiller and boiler part-load coefficients are included in the following part-load equations.

The same zone conditions, outdoor ventilation requirements, cooling coil outlet condition, and fan reheat are assumed. The following equipment has been selected based on the peak summer design condition with appropriate safety factors (see Example 19.8).

Fan size: 16,000 ft^3/min (7,550 L/s), equipped with outlet dampers for flow control.

Fan motor power: 14 hp (10.4 kW)—note that this includes the electric motor efficiency.

Chiller output capacity: 600,000 Btu/h (50 tons, 176 kW).

Chiller full-load COP: 3.9.

Boiler capacity: 500,000 Btu/h (176 kW).

Boiler full-load efficiency: 0.80.

Minimum airflow: Zone A, 468 lb$_a$/min; Zone B, 215.2 lb$_a$/min.

Figure: See Figure 19.15b.

Assumptions: The following simplifying assumptions are made:

(i) Ducts are assumed to be adiabatic.
(ii) The system is located at sea level.
(iii) The building is conditioned 24 h/day.
(iv) Cooling tower and pump energy calculations will not be done.

Part-load models:

1. The fan power at part load with outlet damper control is given by Equation 16.44 as

$$\dot{W}_{fan} = 14\text{ hp} \times \left(0.371 + 0.973 \times PLR_{fan} - 0.342 \times PLR_{fan}^2\right)$$

where
14 hp is the rated fan motor power
PLR is the part-load ratio defined by Equation 16.43

2. Chiller power is found from Equation 14.32 with part-load coefficients specified by the manufacturer for the selected chiller:

$$\dot{W}_{ch} = \frac{600{,}000\text{ Btu/h}}{3.9} \times \left(0.179 + 0.739 \times PLR_{ch} + 0.082 \times PLR_{ch}^2\right)$$

where full-load COP is 3.9 and the design rated chiller capacity is 600,000 Btu/h and PLR is defined by Equation 14.27.

3. The boiler input heat rate is given by (Equation 15.7)

$$\dot{Q}_{boil} = \frac{500{,}000\text{ Btu/h}}{0.80} \times \left(0.0 + 1.368 \times PLR_{boil} - 0.368 \times PLR_{boil}^2\right)$$

where
500,000 Btu/h is the rated boiler capacity
0.80 is the boiler full-load efficiency

Find: Chiller, fan, and boiler annual energy consumption.

Solution

Problems of this type lend themselves conveniently to spreadsheet solution. The solution to this problem is shown as a spreadsheet by Table 19.3. We will work through the entries in *the non-peak part-load 77°F bin* in detail (since both heating and cooling needs have been worked out in Example 19.9) and we shall generally describe the balance of the table entries. Note that two parts of the primary plant are also included in the table to

TABLE 19.3

Variable Air Volume System Bin Calculation Summary

	Zone 1 (Perim.)	Zone 2 (Core)					Part-Load Quadratic Constants *A*	*B*	*C*
T_{zone}	75°F	75°F							
Min. zone	468.4 lb_a/h	215.2 lb_a/h		Fan size	16,000 ft³/min	Fan	0.371	0.973	−0.342
$T_{cc,o}$	54°F	54°F		Fan power	14 hp				
ΔT_{fan}	1°F	1°F		Chiller size	600,000 Btu/h	Chiller	0.179	0.739	0.082
Outside air airflow	2,400 total ft³/min		Basic data	Chiller full-load COP	3.9				
Supply air temp.	55°F	55°F	$\rho_a = 0.075$ lb_a/ft^3	Boiler size	500,000 Btu/h	Boiler	0	1.368	−0.368
Return air temp.	75°F	75°F	$c_a = 0.24$ Btu/(lb_a·°F)	Boiler, full-load eff.	0.8				

			Zone B					Zone A				
Bin, °F	W_0, lb/lb	Hours	Sens. Load, Btu/h	Latent Load, Btu/h	Zone Flow, lb/min	T_{5B}, °F	$Q_{hc,B}$, Btu/h	Sens. Load, Btu/h	Latent Load, Btu/h	Zone Flow, lb/min	T_{5A}, °F	$Q_{hc,A}$, Btu/h
97	*0.014*	*6*	*103,308*	*20,000*	358.7	55.0	0	*224,844*	*56,000*	780.7	55.0	0
92	*0.014*	*72*	*100,296*	*20,000*	348.3	55.0	0	*204,408*	*51,000*	709.8	55.0	0
87	*0.013*	*243*	*97,284*	*20,000*	337.8	55.0	0	*183,972*	*46,000*	638.8	55.0	0
82	*0.013*	*428*	*94,272*	*20,000*	327.3	55.0	0	*163,536*	*41,000*	567.8	55.0	0
77	*0.013*	*631*	*91,260*	*20,000*	316.9	55.0	0	*143,100*	*36,000*	496.9	55.0	0
72	*0.012*	*925*	*88,248*	*20,000*	306.4	55.0	0	*122,664*	*31,000*	468.4	56.8	12,242
67	*0.010*	*858*	*85,236*	*20,000*	296.0	55.0	0	*102,228*	*26,000*	468.4	59.8	32,678
62	*0.008*	*755*	*82,224*	*20,000*	285.5	55.0	0	*81,792*	*21,000*	468.4	62.9	53,114
57	*0.007*	*688*	*79,212*	*20,000*	275.0	55.0	0	*61,356*	*16,000*	468.4	65.9	73,550
52	*0.005*	*686*	*76,200*	*20,000*	264.6	55.0	0	*40,920*	*11,000*	468.4	68.9	93,986
47	*0.004*	*665*	*73,188*	*20,000*	254.1	55.0	0	*20,484*	*6,000*	468.4	72.0	114,422
42	*0.003*	*734*	*70,176*	*20,000*	243.7	55.0	0	*48*	*1,000*	468.4	75.0	134,858
37	*0*	*708*	*67,164*	*20,000*	233.2	55.0	0	*−20,388*	*0*	468.4	78.0	155,294
32	*0*	*621*	*64,152*	*20,000*	222.8	55.0	0	*−40,824*	*0*	468.4	81.1	175,730
27	*0*	*362*	*61,140*	*20,000*	215.2	55.3	0	*−61,260*	*0*	468.4	84.1	196,166
22	*0*	*212*	*58,128*	*20,000*	215.2	56.2	0	*−81,696*	*0*	468.4	87.1	208,208
17	*0*	*101*	*55,116*	*20,000*	215.2	57.2	0	*−102,132*	*0*	468.4	90.1	222,089
12	*0*	*51*	*52,104*	*20,000*	215.2	58.2	0	*−122,568*	*0*	468.4	93.2	235,969
7	*0*	*14*	*49,092*	*20,000*	215.2	59.2	0	*−143,000*	*0*	468.4	96.2	249,846
	Total System											

(Continued)

TABLE 19.3 (*Continued*)

Variable Air Volume System Bin Calculation Summary

Bin, °F	TotFlow, lb/min	IntLatent, Btu/h	W_b, lb/lb	T_1, °F	W_1, lb/lb	Q_{ph}, Btu/h	T_3, °F	$Q_{cc,sens}$, Btu/h	$Q_{cc,lat}$, Btu/h	$Q_{cc,tot}$, Btu/h	$Q_{hc,tot}$, Btu/h
97	1,139.4	76,000	0.00873	78.2	0.0096	0	54.0	397,529	137,103	534,632	0
92	1,058.0	71,000	0.00874	77.7	0.0096	0	54.0	360,870	126,643	487,513	0
87	976.6	66,000	0.00875	77.1	0.0095	0	54.0	324,211	116,171	440,382	0
82	895.2	61,000	0.00876	76.3	0.0095	0	54.0	287,552	106,764	394,316	0
77	813.8	56,000	0.00877	75.4	0.0096	0	54.0	250,893	97,337	348,231	0
72	774.8	51,000	0.00872	74.4	0.0095	0	54.0	227,089	89,603	316,692	12,242
67	764.4	46,000	0.00863	73.3	0.0089	0	54.0	211,888	60,742	272,630	32,678
62	753.9	41,000	0.00854	72.1	0.0085	0	54.0	196,687	37,300	233,987	53,114
57	743.5	36,000	0.00845	71.0	0.0080	0	54.0	181,486	15,610	197,096	73,550
52	733.0	31,000	0.00836	69.8	0.0076	0	54.0	166,285	0	166,285	93,986
47	722.6	26,000	0.00826	68.5	0.0074	0	54.0	151,084	0	151,084	114,422
42	712.1	21,000	0.00816	67.3	0.0071	0	54.0	135,883	0	135,883	134,858
37	701.6	20,000	0.00814	65.9	0.0062	0	54.0	120,682	0	120,682	155,294
32	691.2	20,000	0.00815	64.6	0.0062	0	54.0	105,481	0	105,481	175,730
27	683.7	20,000	0.00815	63.3	0.0062	0	54.0	91,167	0	91,167	196,166
22	683.7	20,000	0.00815	62.0	0.0062	0	55.2	66,878	0	66,878	208,208
17	683.7	20,000	0.00815	60.8	0.0062	0	56.2	45,272	0	45,272	222,089
12	683.7	20,000	0.00815	59.6	0.0062	0	57.2	23,666	0	23,666	235,969
7	683.7	20,000	0.00815	58.4	0.0062	0	58.2	2,060	0	2,060	249,846

Bin, °F	Fan PLR	Fanpower, kW	Fan Energy, kWh	Chiller PLR	Chiller Power, kW	Chiller Energy, kWh	Boiler PLF	Boiler Power, Btu/h	Boiler Energy, kBtu
97	0.936	10.26	62	0.89	40.70	244	0.00	0	0
92	0.870	10.01	721	0.81	37.59	2,706	0.00	0	0
87	0.803	9.73	2,364	0.73	34.52	8,389	0.00	0	0
82	0.736	9.42	4,031	0.66	31.57	13,511	0.00	0	0
77	0.669	9.07	5,725	0.58	28.66	18,083	0.00	0	0
72	0.637	8.90	8,230	0.53	26.69	24,689	0.02	20,797	19,237
67	0.628	8.85	7,593	0.45	23.98	20,572	0.07	54,898	47,102
62	0.620	8.80	6,644	0.39	21.63	16,330	0.11	88,230	66,614
57	0.611	8.75	6,020	0.33	19.42	13,359	0.15	120,794	83,106
52	0.602	8.70	5,968	0.28	17.59	12,067	0.19	152,590	104,677
47	0.594	8.65	5,752	0.25	16.70	11,103	0.23	183,617	122,105
42	0.585	8.60	6,311	0.23	15.81	11,603	0.27	213,876	156,985
37	0.577	8.55	6,051	0.20	14.92	10,566	0.31	243,366	172,303
32	0.568	8.49	5,275	0.18	14.04	8,721	0.35	272,088	168,967
27	0.562	8.46	3,061	0.15	13.22	4,786	0.39	300,042	108,615
22	0.562	8.46	1,793	0.11	11.83	2,508	0.42	316,154	67,025
17	0.562	8.46	854	0.08	10.61	1,071	0.44	334,394	33,774
12	0.562	8.46	431	0.04	9.39	479	0.47	352,281	17,966
7	0.562	8.46	118	0.00	8.19	115	0.50	369,807	5,177
			77,006			**180,903**			**1,173,654**

Note: The values in italics are inputs to the analysis table.

the right. The solution has three parts—fan calculations, cooling system calculations, and heating system calculations.

(a) *Fan Energy Calculations*
From Example 19.9, the total airflow is $\dot{V}_{tot} = 813.8\ \text{lb}_a/\text{min} \times 13.13\ \text{ft}^3/\text{lb}_a = 10{,}685\ \text{ft}^3/\text{min}$.

The fan part-load ratio is $\text{PLR}_{fan} = \dfrac{10{,}685}{16{,}000} = 0.669$.

The fan power $\dot{W}_{fan}$ is given by model (1):

$$\dot{W}_{fan} = 14\ \text{hp} \times (0.371 + 0.973 \times 0.669 - 0.342 \times 0.669^2)$$
$$= 12.16\ \text{hp} \times 0.746\ \text{kW/hp} = 9.07\ \text{kW}$$

In the 77°F bin, there are 631 h annually: therefore, the annual fan energy is

$$E_{fan} = 9.07\ \text{kW} \times 631\ \text{h} = 5725\ \text{kWh}$$

(b) *Cooling System Calculations*
From Example 19.9, the total cooling load is $\dot{Q}_{cc,tot} = 348{,}231\ \text{Btu/h}$.

The chiller part-load ratio is $PLR_{ch} = \dfrac{348{,}231}{600{,}000} = 0.58$.

The chiller power is given by model (2):

$$\dot{W}_{ch} = \frac{600{,}000}{3.9} \times (0.179 + 0.739 \times 0.58 + 0.082 \times 0.58^2)$$
$$= 97{,}692\ \text{Btu/h} \times 0.2931\ \text{W}\cdot\text{h/Btu} = 28{,}633\ \text{W} = 28.6\ \text{kW}$$

In the 77°F bin, there are is 631 h annually; therefore, the annual chiller energy is

$$E_{ch} = 28.6\ \text{kW} \times 631\ \text{h} = 18{,}083\ \text{kWh}$$

(c) *Heating System Calculations*
From Example 19.9, the total heating load (which is actually the reheat in Zone *B*) is 0 Btu/h. Since the boiler capacity is 500,000 Btu/h, the boiler part-load ratio is

$$\text{PLR}_{boil} = \frac{0}{500{,}000} = 0.0$$

The boiler input heat rate for this bin is given by (model 3)

$$\dot{Q}_{boil} = \frac{500{,}000}{0.80} \times (0.0 + 1.368 \times 0.0 - 0.368 \times 0.0^2) = 0\ \text{Btu/h}$$

The boiler energy use is the heat rate multiplied by 631 h in this bin:

$$Q_{boil} = 0\ \text{kBtu/h} \times 631\ \text{h} = 0\ \text{kBtu}$$

The reader can verify these values in Table 19.3. Similarly, the calculations are done for each bin. It is urged that the reader generate these values in a spreadsheet and compare them against those shown in Table 19.3. Note that the minimum supply air stipulation is met for Zones *A* and *B* for outdoor temperatures of 72°F and 27°F, respectively. Hence, reheating is needed for Zone *A* from the 72°F bin downward. However, there is no need to reheat the supply air for Zone *B* at temperatures below 27°F since the cooling coil set point (T_3) can be reset to the supply air supply temperature minus the 1°F fan reheat ($T_{5,B}$). The annual values of total fan electric energy use, chiller electricity use, and boiler heating energy requirements for this example are shown along the bottom row of Table 19.3. As a check, the total number of bin hours must equal 8760.

Figure 19.18a depicts the linear variation of the sensible and latent loads for both zones with outdoor temperature. The coil loads, shown in Figure 19.18b however, show a bisegmented almost linear behavior, a very common behavior in buildings with more than one zone.* The hinge points of 75°F and 55°F for the cooling coil and the reheat coil, respectively, are referred to as change points in the monitoring and verification (M&V) professional literature dealing with building energy data analytics. These can be associated with the set point temperatures of the cooling coil, and that of the zones (as well as other set points not considered in this simple example). The interested reader can refer to Reddy et al. (1995) for further details. The energy consumption at different bins for the chiller electricity and the boiler thermal energy are plotted in Figure 19.18c. The energy usage is higher at the mid-temperature range due to larger number of hours and to the fact that coil bucking is occurring (i.e., simultaneous heating and cooling).

Comments

This example forms the basis of a number of homework problems that will complete the energy consumption picture. These will include cooling tower and pump bin calculations. Once you have set up the spreadsheet, it is a simple matter to consider more efficient means of airflow control such as variable-speed drives and inlet vane control. Various chiller and boiler types with different COPs are easy to study by changing the appropriate values in the input data list.

The bin method approach is computationally efficient and is widely used by the HVAC engineering community for certain types of analysis. However, the methodology does not permit the accurate calculation

* Refer to Section 10.6 dealing with inverse modeling of building energy use.

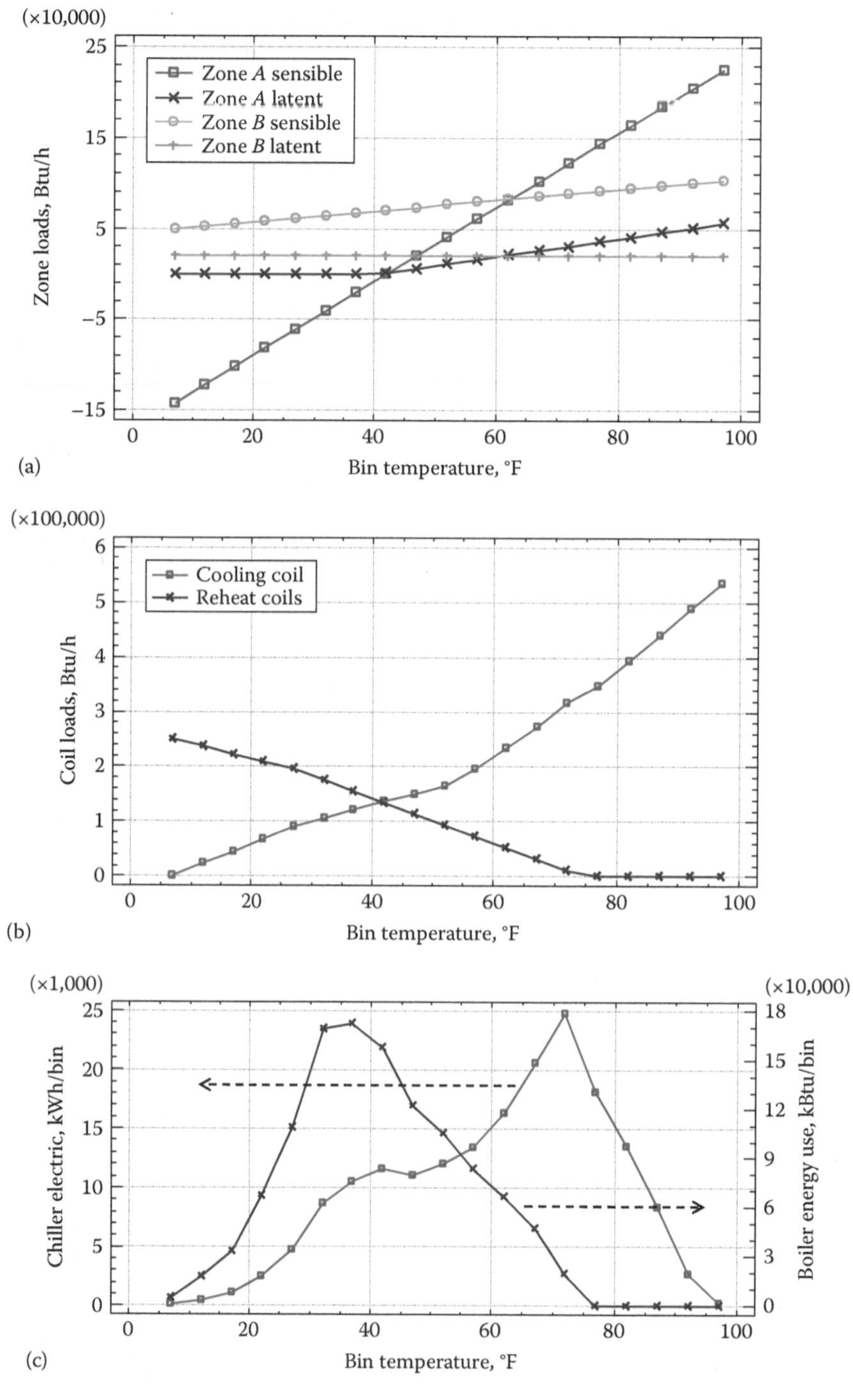

FIGURE 19.18
(a) Variation of zone loads with bin temperature for Example 19.10. (b) Variation of coil loads with bin temperature. Notice that the load line experiences a change in slope at the hinge points (for the cooling coil at around 55°F, while for the heating coil it is around 75°F). These hinge points are called change points in the building energy data analytics literature dealing with developing inverse models based on monitored data (see Sections 10.2 and 10.3). (c) Variation of the chiller electricity use and the boiler thermal energy over the year. Note the wide temperature range over which both heating and cooling are needed in the building.

of transient loads and time-of-day performance indices such as peak electric demand. The bin approach is also not able to assess the performance of any HVAC system including either hot or cold storage subsystems. However, in many cases, the bin approach can give an indication of the *comparative* performance of various systems on a given building. Numerous commercial versions of bin-based building analysis computer programs have been developed over the years.

If the shortcomings of the bin approach preclude its use in a given problem, an hourly chronological simulation scheme has to be used. Here, the preceding series of calculations are repeated on an hour-by-hour basis for each of the 8760 h of a typical year. The computational run time needed is considerably longer, but the analysis flexibility is commensurately greater as well. Simulation software with tens of thousands of lines of code allow us to analyze the detailed performance of many different types of HVAC systems in a wide variety of building types and shapes, climates, and operational schedules. Many standard, publicly available packages are available such as Energy Plus and DOE2.1, while numerous proprietary softwares have also been developed by HVAC equipment companies and engineering firms.

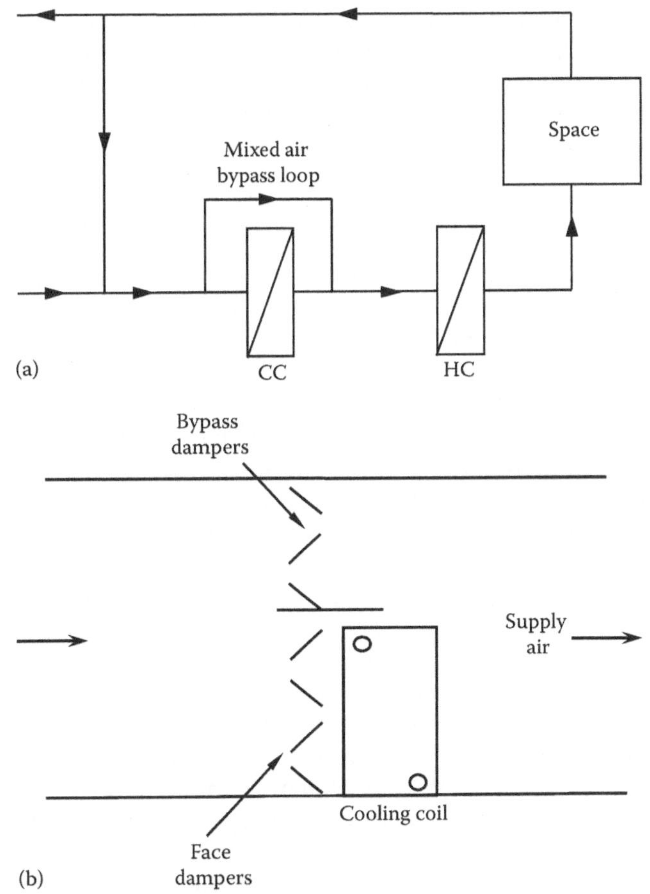

FIGURE 19.19
(a) HVAC system modification involving a mixed air bypass loop. (b) Face and bypass damper arrangement.

19.6 Energy Efficiency Design and Operation Practices

19.6.1 HVAC System Variants

The various fixed- or constant-volume systems described in the preceding sections have significant inefficiencies and energy waste at part load as demonstrated by the solved examples. The annual fan electricity use in the air handlers of a CAV system is large since peak airflow rates are used even at part-load conditions. A number of new secondary systems designs have evolved over the past decade or two, and some of the important ones are described in Chapter 20 involving air–water systems and novel hybrid systems.

Cooling coil bypass arrangement is a traditional approach meant to reduce energy use of terminal reheat systems at part-load conditions in humid locations. Two system variants will be briefly described that minimize or eliminate the need for reheat. Figure 19.19a shows an all-air HVAC system where part of the mixed air is made to bypass the cooling coil (this is achieved by a face-and-bypass damper arrangement—see Figure 19.19b), while Figure 19.20 shows another arrangement where part of the recycle airstream bypasses the cooling coil. Both systems had similar thermal performance when the bypass fraction is properly optimized for each system (Reddy, 1994). The heating energy is reduced several times as compared to the traditional CAV system needs, and the cooling coil thermal energy use is also lower. Such systems are more advantageous under high outdoor air recycle fractions. The primary problem with such systems is the loss in dehumidification capability of the

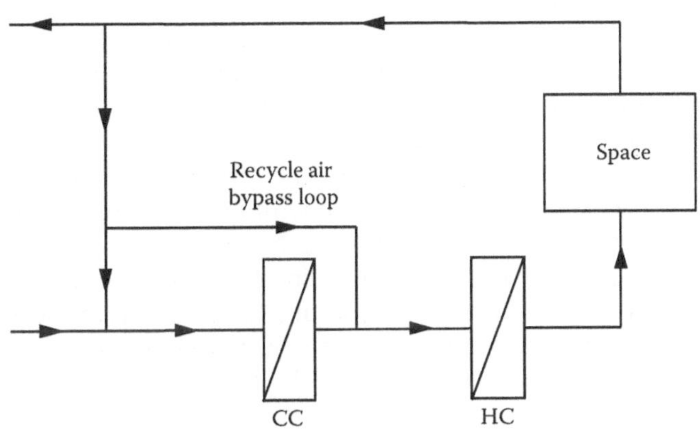

FIGURE 19.20
HVAC system modification involving a recycle air bypass loop.

cooling coil that adversely affects the indoor space relative humidity, and this may be an issue under extreme operating conditions. Hence, the current HVAC design practice recommendation is not to use such systems since better all-air systems design choices are available.

19.6.2 Air-Side Economizer Cycle

The interior spaces of commercial buildings often have large heat gains due to lighting, equipment (such as computers), and occupants. Therefore, even in cold and mild weather, many commercial buildings may require cooling. Cooling that uses outdoor air without operating a cooling coil is called the *air-side economizer cycle.* The economizer is not a cycle in the thermodynamic sense, but the name is in common use.

Whenever the outdoor dry-bulb temperature is below the space set point temperature (typically 75°F or 24°C) as encountered during spring and fall, then outdoor air can be used in the amount required for cooling in the dry-bulb *temperature economizer* cycle. However, the economizer approach can be used under a broader set of conditions than just those meeting the dry-bulb temperature criterion. In fact, some cooling effect (sensible plus latent) can theoretically be achieved any time that the enthalpy of outdoor air is less than that of the required indoor air condition. This leads to the more general case of the *enthalpy economizer.* To achieve proper interior wet-bulb temperature conditions, supplementary mechanical cooling may be needed. However, the amount of mechanical cooling will be less than that needed if the economizer approach was not used at all.

Figure 19.21 shows the portions of the psychrometric chart where the two economizer cycles could be used if a building is to be maintained at 75°F (24°C) and 50% relative humidity with a 55°F (13°C) cooling coil outlet temperature. To the left of the 75°F (24°C) dry-bulb temperature line, the dry-bulb temperature economizer will use outdoor air for cooling. Note that *area D* above the room condition enthalpy line has higher enthalpy than room air and will result in a cooling penalty (due to higher latent loads) if a dry-bulb temperature economizer is used. To the left of 75°F (24°C) line and below the room enthalpy line (*regions A and C* combined), the enthalpy economizer will introduce additional outdoor air. *Region C* is ignored by the dry-bulb temperature economizer. Although the enthalpy here is lower than the room enthalpy, cooling it to the cooling coil dry-bulb set point temperature of 55°F (13°C) results in an energy penalty; therefore, *area C* is of marginal cooling value. Of course, in the remainder of the weather domain covering the balance of the chart to the right (*region B*), mechanical cooling is needed and minimal outdoor air is used just to satisfy ventilation needs.

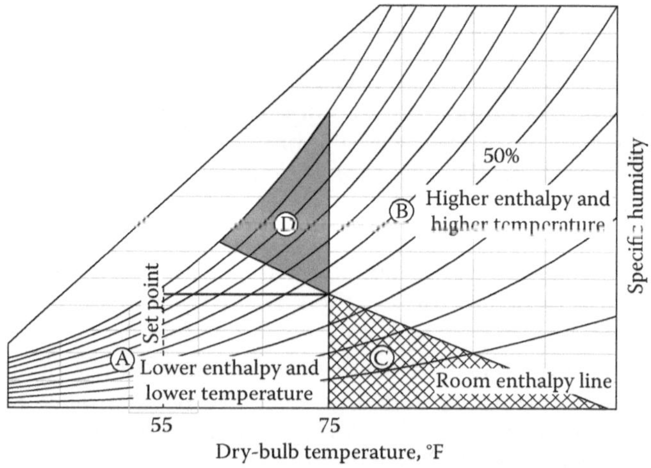

FIGURE 19.21
Enthalpy and temperature economizer operating ranges. The constant-enthalpy line represents the room enthalpy criterion.

Though the enthalpy-based economizer is theoretically more effective, their reliability has been a major source of practical concern that has restricted their widespread adoption. Reliable, durable, and inexpensive enthalpy sensors do not exist, and so activation of enthalpy-based economizers involves using measurements of relative humidity and dry-bulb temperature. Unfortunately, a humidity sensor meeting the listed criteria does not exist either. Hence, most designers settle for dry-bulb temperature economizers. Spitler et al. (1987) have shown by means of computer simulations that most of the benefits of an enthalpy economizer can be achieved by a temperature economizer partly because cooling coils are controlled based on the dry-bulb outlet temperature. Economizers of either type are more effective in moderate, dry climates than in hot, humid ones.

The use of the economizer cycle is very effective in northern latitudes, while it is seldom used in lower latitudes where the low outdoor air conditions are met so rarely that the extra expense in duct system modifications cannot be economically justified. Economizers can be applied to many HVAC systems (but not all) that have been discussed in this chapter. This introductory section applies to all such systems. The outdoor airflow has to be modulated properly; otherwise, the benefit of reduced cooling energy can be compromised. The following example illustrates the control strategy.

Example 19.11: Building Process—The Economizer

Consider a CAV system operated with a constant supply flow rate $\dot{m}_{a,supply}$ at supply temperature $T_{db,supply}$ using a temperature controlled economizer cycle.

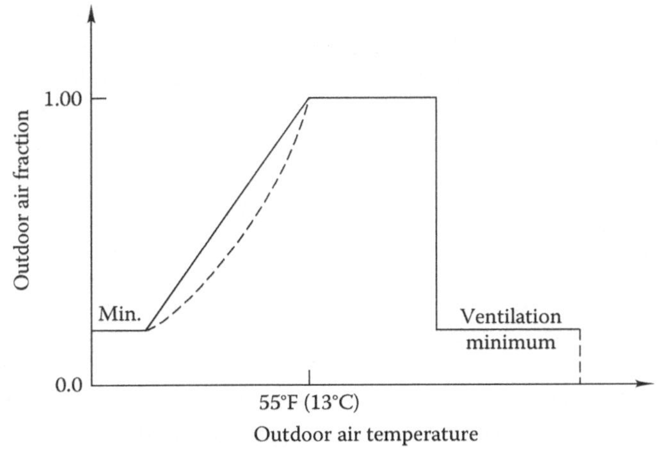

FIGURE 19.22
Outside airflow characteristic for fixed-volume system with economizer.

(a) Derive the equation relating the amount of outdoor air $m_{a,o}$ needed as a function of the outdoor air temperature $T_{db,o}$.
(b) Derive the equation for the outside temperature above which the economizer is inadequate for cooling.
(c) If the return temperature from a zone is 25°C (77°F) and if at least 25% outdoor air* is required at all times to meet ventilation requirements, below which outdoor temperature will the outdoor inlet damper be at its minimum setting? The design cooling system air supply temperature is to be 13°C (55°F).

Given: Constant $T_{db,supply}$, $\dot{m}_{a,supply}$.

Figure: See Figure 19.22.

Assumption: Steady state exists and a temperature economizer is used.

Find: Expression for $\dot{m}_{a,o}$ and $T_{db,o}$ at the limiting condition called the *economizer cutoff point*.

Solution

(a) A sensible heat balance at the mixing point yields

$$\dot{m}_{a,r}c_aT_{db,r} + \dot{m}_{a,o}c_aT_{db,o} = \dot{m}_{supply}c_aT_{db,supply} \quad (19.19)$$

while the conservation of mass equation is given by

$$\dot{m}_{a,r} + \dot{m}_{a,o} = \dot{m}_{a,supply} \quad (19.20)$$

These two equations can be solved for the outside airflow rate needed to maintain a given supply temperature with given outdoor and return air temperatures. Assuming the specific heat values to be constant (a minor approximation),

$$\dot{m}_{a,o} = \dot{m}_{a,supply} \times \left(\frac{T_{db,r} - T_{db,supply}}{T_{db,r} - T_{db,o}}\right) \quad (19.21)$$

(b) This equation can be rearranged to yield the expression for the outdoor temperature capable of meeting design values of supply flow rate and temperature:

$$T_{db,o} = \left(1 - \frac{\dot{m}_{a,supply}}{\dot{m}_{a,o}}\right) \times T_{db,r} + \frac{\dot{m}_{a,supply}}{\dot{m}_{a,o}} \times T_{db,supply} \quad (19.22)$$

There are two limits to economizer performance. At the upper limit of economizer use (at the point where mechanical cooling is just about to be turned on), there is no return air (all air from the building is exhausted), and the supply temperature is essentially the same as the outdoor air. The air supply mass flow is as follows:

$$\dot{m}_{a,o} = \dot{m}_{a,supply} = \dot{m}_{a,max} \quad (19.23)$$

At the other extreme, some outdoor air is needed during occupied periods whether space cooling is needed or not. This minimum flow $\dot{m}_{a,\min}$ can be deduced from ASHRAE ventilation standard (ASHRAE 62.1, 2013). We can use Equations 19.22 and 19.23 to find the minimum outdoor temperature below which the economizer should be switched off. Below this temperature, heat needs to be added to meet the supply air temperature requirement.

$$T_{db,o,\min} = \left(1 - \frac{\dot{m}_{a,supply}}{\dot{m}_{a,\min}}\right) \times T_{db,r} + \frac{\dot{m}_{a,supply}}{\dot{m}_{a,\min}} \times T_{db,supply} \quad (19.24)$$

At outdoor temperatures less than this value, the minimum airflow required by ventilation $\dot{m}_{\min}$ is used, and heat is added. The aforementioned equations are

* In the problem statement, the basis of the 25% outside requirement was not stated—whether on a mass or a volume basis. In the United States, it is customary to specify airflow on a volumetric basis, whereas the first law is properly used only on a mass basis. The difference between a volumetric basis and a mass basis would be the presence of density terms in the aforementioned equations for the mass basis. Since air density depends on absolute temperature, and the temperatures involved in HVAC systems are not widely different on an absolute basis, the densities of all airstreams are not much different. This argument is the basis for the customary use of volumetric flow percentages by HVAC designers.

conceptually illustrated by Figure 19.22 in terms of an outdoor air fraction.* During hot weather, minimum outside or ventilation air is used. As the outdoor temperature drops to the point where the economizer can begin to save cooling energy, maximum outdoor air is used since it is less costly to condition the outdoor air than to cool and dehumidify warm, humid indoor air at the same dry-bulb temperature.† At still cooler outdoor temperatures below the nominal building cold air supply temperature (approximately 55°F or 13°C), the amount of outdoor air needed for cooling is gradually reduced to the minimum. During cold weather, only the ventilation minimum outdoor air is used. Although this usage schedule of outdoor air has been discussed in the context of a single zone, it applies to large multiple-zone buildings as well. Note that the supply fan shown in the figure causes a modest temperature rise (1°F–2°F) in the air due to flow work and fan inefficiencies.

(c) The minimum outdoor temperature is determined from Equation 19.24

$$T_{db,o,\min} = \left(1 - \frac{1}{0.25}\right) \times 25°\text{C} + \left(\frac{1}{0.25}\right) \times 13°\text{C}$$

$$= -23°\text{C}\ (-9.4°\text{F})$$

Comments

This outdoor temperature is well below the freezing point of water; hence, the designer must guarantee that this very cold air is properly mixed with warm return air. Cold air and warm air do not mix easily. A special air handler component called an *air blender* is needed, or else antifreeze must be used in the coils. Antifreeze is expensive, needs periodic replacement, and has poorer heat transfer characteristics than water. Whatever approach is used, it is the designer's responsibility to avoid coil freezing by specification of controls, outside air dampers, coil fluids, and use of redundant safety measures.

19.6.3 Exhaust Air Energy Recovery Devices

Exhaust air energy recovery systems are being increasingly used to lower HVAC cooling and heating energy needs. The leaving conditioned exhaust air can be used to precondition the entering outside air. In winter, the entering outdoor air will be warmed; in summer, it will be cooled. There are several types of such devices, e.g., run-around coils, heat wheels (Figure 19.23a), heat exchangers, and heat pipes. The run-around system comprises of two air-to-liquid heat exchangers as shown in Figure 19.23a,b. It is very flexible in installation with no cross-contamination between airstreams, but it is the least efficient. If the two airstream ducts are in close proximity, heat wheels, heat pipes, and plate heat exchangers are preferred for energy recovery purposes.

Heat wheels, also known as rotary air-to-air energy exchangers, are made up of a revolving disk filled with an exchange medium suitable for sensible and total heat transfer. Commonly used media are aluminum, copper, and stainless steel for sensible heat exchange, while desiccant-treated materials (discussed in Section 20.9) are used for total heat recovery. As the wheel rotates (typically 1–2 times/min), heat (and moisture) is picked up from the hotter (and more humid) airstream and released to the colder (and less moist) one.

The performance of both types of devices is quantified by an *effectiveness* ε (assumed to be constant during operation even when the temperatures and humidities vary during the year but not the flow rates), which represents the actual energy exchange to the maximum potential (Section 12.6.3). Since both the airstreams are equal, the expression for a sensible heat exchanger effectiveness simplifies to

$$\varepsilon = \frac{T_{db,0,in} - T_{db,0,out}}{T_{db,0,in} - T_{db,r,in}} \tag{19.25}$$

where

$T_{db,0,in}$ and $T_{db,0,out}$ are the inlet and outlet dry-bulb temperatures of the ventilation (or outdoor) air through the heat exchanger

$T_{db,r,in}$ is the inlet temperature of the room exhaust air

Typical effectiveness values are 50%–80% for sensible heat exchange and 55%–85% for total heat exchange. Air-side pressure drops are in the range of 0.4–0.7 inWG across the wheel which the fan has to overcome. There is also some amount of cross-leakage between the two airstreams within the wheel (1%–10%) that degrades performance.

Example 19.12: Energy Benefit of Exhaust Air Energy Recovery

The following example will illustrate the thermal benefits of an energy recovery unit assuming a building with 100% outdoor air. The exhaust air temperature is always kept constant at 75°F year-round. The building supply temperature is assumed to be 55°F and 100% saturated

* The sloped line is straight if dry-bulb temperature control is used. If enthalpy control were used instead, the line would be concave to the left as shown.

† The outdoor condition where increased outdoor air should be introduced instead of using return air varies from one location to the next. The designer must make this evaluation on a case-by-case basis.

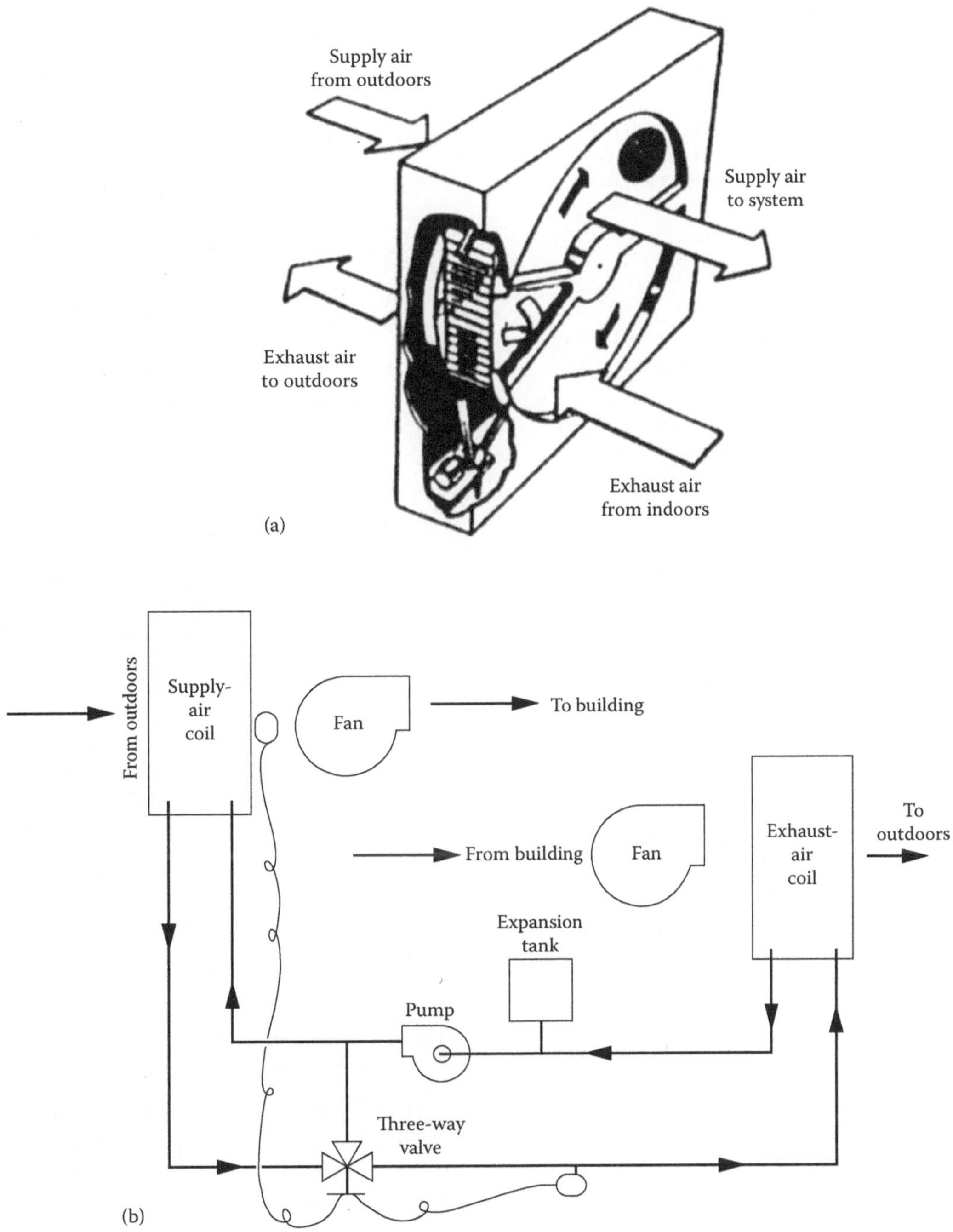

FIGURE 19.23
Airstream heat recovery devices. (a) Run-around coil is a sensible device. (b) Enthalpy wheel is a total (sensible plus latent) device. (From, *ASHRAE Handbook of Fundamentals*, ASHRAE, Atlanta, GA, 2013. Copyright ASHRAE, www.ashrae.org.)

during summer and 90°F during winter operation. The ventilation and exhaust airstreams are both equal to 10,000 cfm.

(a) Determine the capacity of the cooling and heating coils without energy recovery under the following peak or design outdoor air conditions:
  (i) *Winter*: 40°F and 40% RH
  (ii) *Summer*: 98°F and 40% RH
(b) Determine the capacity of the cooling and heating coils when an energy recovery unit of effectiveness of 0.6 is included.

Given: $T_{db,0,in}$, $T_{db,r,in}$, ε, and $\dot{V}_{supply}$.

Figure: See Figure 19.24.

Assumption: CAV operation with mass flow rate determined from summer conditions and remains constant.

Find: $Q_{cc,tot}$ with and without energy recovery during peak summer, and Q_{hc} with and without energy recovery during peak winter conditions.

Solution

Summer conditions: For $T_{db,0,in}$ = 98°F and ϕ_o = 40%, specific volume v_0 = 14.415 ft³/lb$_a$, and humidity ratio W_0 = 0.0156 lb$_w$/lb$_a$ and h_0 = 40.8 Btu/lb$_a$.

Thus, mass flow rate of ventilation/space supply air is

$$m_{a,o} = \frac{10{,}000 \text{ cfm} \times 60 \text{ min/h}}{14.415 \text{ ft}^3/\text{lb}_a} = 41{,}623.3 \text{ lb}_a/\text{h}$$

(a1) Summer: Without energy recovery system
Enthalpy of air leaving the cooling coil: For $T_{db,cc}$ = 55°F and $\phi_{a,cc}$ = 100%, h_2 = 23.6 Btu/lb$_a$.
Total load on the cooling coil is

$$Q_{cc,tot} = m_{a,o} \times (h_0 - h_2) = 41{,}623.3 \text{ lb}_a/\text{h} \times (40.8 - 23.6) \text{ Btu/lb}_a$$
$$= 716{,}724 \text{ Btu/h} = 60 \text{ tons}$$

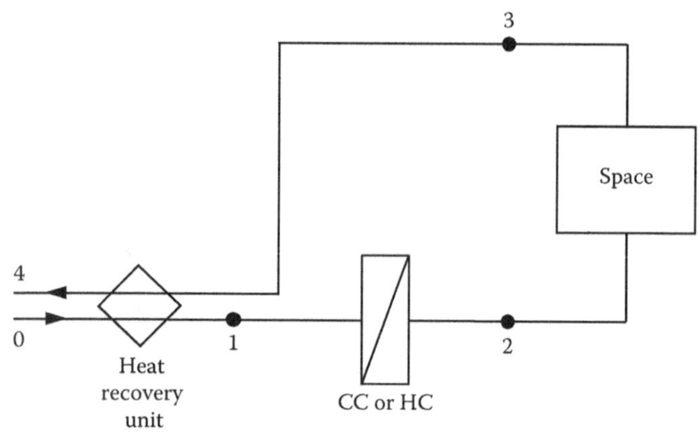

FIGURE 19.24
Sketch of the HVAC system with energy recovery unit for Example 19.12.

(b1) *Summer*: *With* energy recovery system
From Equation 19.25,

$$T_{db,1} = T_{db,0} - \varepsilon \times (T_{db,0} - T_{db,3}) = 98°\text{F} - 0.6 \times (98 - 75)°\text{F}$$
$$= 84.2°\text{F}$$

At T_{db} = 84.2°F and W_o = 0.0156 lb$_w$/lb$_a$, enthalpy h_1 = 38.6 Btu/lb$_a$. Therefore, total load on the cooling coil is

$$Q_{cc,tot} = m_{a,o}(h_1 - h_2) = 41{,}623.3 \text{ lb}_a/\text{h} \times (38.6 - 23.6) \text{ Btu/lb}_a$$
$$= 624{,}349 \text{ Btu/h} = 52.0 \text{ tons}$$

or a reduction of 13%.

(a2) *Winter conditions*: For $T_{db,in}$ = 40°F and ϕ_o = 40%, we find h_0 = 13 Btu/lb$_a$.

Winter: *Without* energy recovery system
Enthalpy of air leaving the heating coil (sensible heating from $T_{db,hc}$ = 40°F and ϕ_o = 40% to $T_{db,2}$ = 90°F) h_2 = 24 Btu/lb$_a$.
Total load on the heating coil is

$$Q_{hc,tot} = m_o(h_0 - h_2) = 41{,}623.3 \text{ lb}_a/\text{h} \times (24 - 13) \text{ Btu/lb}_a$$
$$= 521{,}170 \text{ Btu/h}$$

(b2) *Winter*: *With* energy recovery system
From Equation 19.25,

$$T_{db,1} = T_{db,0} + \varepsilon \times (T_{db,0} - T_{db,3}) = 40°\text{F} + 0.6 \times (75 - 40)°\text{F}$$
$$= 61°\text{F}$$

The corresponding enthalpy h_1 = 17.0 Btu/lb$_a$. Therefore, total load on the heating coil is

$$Q_{hc,tot} = m_o(h_1 - h_2) = 41{,}623.3 \text{ lb}_a/\text{h} \times (24 - 17) \text{ Btu/lb}_a$$
$$= 291{,}363 \text{ Btu/h}$$

or a reduction of 44%.

19.6.4 Ways to Improve Energy Efficiency under Part-Load Conditions

There are a number of ways by which energy use can be reduced by optimizing operation of the HVAC system during part-load operation or by system retrofits. Some of these are described:

1. *Cold deck reset*: CAV terminal reheat systems rely on heating coils placed above the space to raise the supply air temperature so as to avoid overcooling. This is wasteful in energy, and control methods are used to reduce this reheat. In single-zone spaces, the set point of the air leaving the cooling coil (referred to as the supply air deck temperature) can be reset,

i.e., gradually modified based on some feedback signal such as the outdoor temperature value. During summer periods, one would leave the deck temperature at its base condition of, say, 55°F (13°C), which is suited for some maximum summer design condition. During milder weather when space loads are lower, one does not need such a low deck temperature and this value can be gradually increased. Such a control strategy, called cold deck reset, would reduce the cooling energy requirements while also reducing the reheat energy. Though humidity removal capability is somewhat compromised, it is usually not a major issue if implemented properly on a case-by-case basis. For dual-duct systems, one could also reset the hot deck temperature, with the set temperature being higher during cold outdoor temperature conditions, and vice versa.

2. *Discriminating temperature control*: In case of multiple zones, a similar reset scheme for CAV systems, called discriminating temperature control, is often adopted. In this case, the approach involves controlling the supply air temperature to respond to feedback from the zone thermostats. The supply air temperature will be set to be only as low as that required for the zone with the greatest cooling load. This control strategy can also be applied to dual-deck systems, i.e., the hot deck and cold deck temperatures need be only as high and as low as the zones needing the greatest heating and cooling.
3. *Critical zone reset*: For VAV systems, a control strategy called critical zone reset is adopted wherein the VAV terminals for each zone are monitored to identify the "worst" zone, i.e., the zone requiring the highest airflow. The supply fan is unloaded accordingly to keep the VAV terminal serving this zone fully open. This lowers duct pressure and the supply fan energy consumption is reduced. In addition, the amount of chilled water needed to maintain the supply air set point temperature is also reduced.
4. *Alternative all-air system configurations*: Energy conservation retrofits to dual-duct CAV systems are usually cost-effective. The simplest measure is to retrofit it to a dual-duct VAV system as discussed earlier. Still another refinement to the dual-duct system is the *dual-fan dual-duct system*. This more complex system consumes less energy since only the amount of cold or warm air needed to meet zone loads is supplied. Cold air and warm air are not mixed as in the basic dual-duct or multizone approach. Both of the separate hot and cold streams are variable-volume in this design. Yet another energy-efficient system alternative is to modify the ducting and add a third deck (or duct) that allows supply air to each zone to bypass the heating and cooling coils completely. Such a *3-deck system* eliminates the mixing of the hot and cold stream altogether. The interested reader can refer to Sun (1994) for further description.

19.7 Energy Penalties due to Mixing of Hot and Cold Streams

The ideal HVAC system is intended to meet the space loads and the ventilation loads only. A major cause of inefficiency in multizone buildings is the simultaneous heating and cooling of different zones. This can be viewed as a transmission or distribution or energy delivery inefficiency. Two different ways to quantify this penalty have been proposed and are discussed in this section.

19.7.1 Multizone Efficiency Index

Consider a HVAC system in question with the performance of an "ideal" one-zone design where perfect mixing of the air ensures that all heat gains are utilized to offset heating loads before resorting to the economizer or chiller. All systems except the secondary HVAC system are kept the same in this comparison, in particular, the building envelope and the schedules for occupancy, thermostat, and minimum outdoor air requirement. A *multizone efficiency index* $\eta_{multiz,heat}$ for heating is defined as the ratio of the annual heating energy $Q_{ann,heat,1-z}$ for the ideal one-zone system divided by the annual heating energy $Q_{ann,heat}$ for the system in question, with analogous quantities for cooling:

$$\eta_{multiz,heat} = \frac{Q_{ann,heat,1\text{-}z}}{Q_{ann,heat}} \quad \text{and} \quad \eta_{multiz,cool} = \frac{Q_{ann,cool,1\text{-}z}}{Q_{ann,cool}} \tag{19.26}$$

The highest possible value of this index is unity. The smaller this index, the greater the overconsumption due to simultaneous heating and cooling. The following example illustrates the calculation and use of this index during economizer operation of a multizone building.

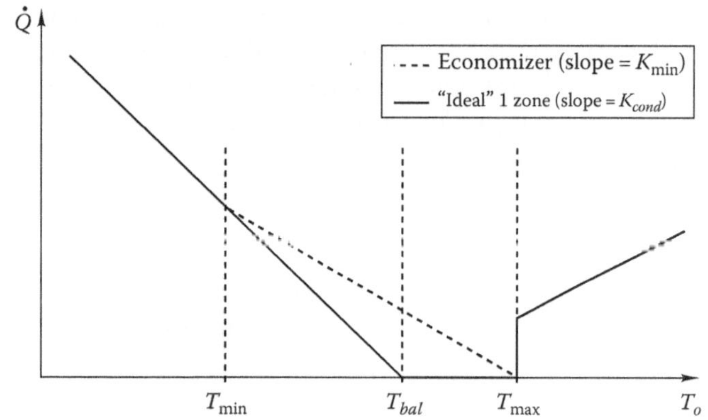

FIGURE 19.25
Energy use for heating and cooling of multizone building with economizer, Q versus $T_{db,O}$ for Example 19.13.

Example 19.13: Determination of Multizone Index for a Building

Consider a building* that can be characterized by the following quantities and assumptions:

Conductive heat loss coefficient K_{cond} = 15 kW/K.

Total heat loss coefficient $K_{tot} = K_{min} = 15 + 10 = 25$ kW/K during cold weather (with economizer off).

$K_{tot} = K_{max} = 15 + 25 = 40$ kW/K during mild weather (with economizer on).

To allow simple steady-state analysis, assume the building is occupied and conditioned around the clock, with internal heat gains $\dot{Q}_{gain}$ = 300 kW and $T_{db,space}$ = 20°C.

Assumptions: There is a single air handler, the economizer control is designed to maintain constant supply air temperature, and the return air temperature is equal to $T_{db,space}$.

Then, the economizer equations of Section 19.6.2 are applicable, and the graph of $\dot{Q}$ versus T_o has a slope of K_{cond} when $T_{min} < T_o < T_{max}$ (economizer regime). When $T_o < T_{min}$, the economizer is turned off and the slope is K_{min}. This is shown in Figure 19.25, where two consumption patterns are indicated, as labeled: economizer (between T_{min} and T_{max}, excess heat gains are vented by economizer) and "ideal" one zone (assuming perfect mixing within building).

Find: $\eta_{multiz,heat}$ of Equation 19.26.

Lookup values: Annual heating degree-hours† for several bases (temperatures are determined from bin data of HCB software), HDD(12.5°C) = 36,000 K · h, HDD(8°C) = 20,440 K · h, and HDD(1.25°C) = 5,840 K · h.

Solution

The economizer is turned off when T_o is below T_{min}:

$$T_{min} = \frac{K_{min}T_{bal} - K_{cond}T_{max}}{K_{min} - K_{cond}} \tag{19.27}$$

The annual heating consumption $\dot{Q}_{ann,heat,1\text{-}z}$ is easy to find because it corresponds to the straight line for $T_o < T_{bal}$. Hence, it is given by

$$\dot{Q}_{ann,heat,1\text{-}z} = K_{min} \times \text{HDD}(T_{bal}) \tag{19.28}$$

where HDD(T_{bal}) is the annual heating degree-hours for base T_{bal} and

$$T_{bal} = T_i - \frac{\dot{Q}_{gain}}{K_{min}} \tag{19.29}$$

The annual heating consumption $\dot{Q}_{ann,heat}$ with the economizer mode is a little more complicated because it corresponds to the solid line for $T_o < T_{min}$ and to the dashed line for $T_{min} < T_o < T_{max}$. It is given by

$$\dot{Q}_{ann,heat} = K_{cond} \times \text{HDD}(T_{max}) + (K_{min} - K_{cond}) \times \text{HDD}(T_{min}) \tag{19.30}$$

The first term in this equation would be the answer if the consumption could follow the dashed line even below T_{min}; the second term adds the correction below T_{min}.

Inserting numerical values, we find

$$T_{bal} = 20 - \frac{300}{25} = 8°\text{C} \quad \text{for which}$$

$$\text{HDD}(8°\text{C}) = 20,440 \text{ K} \cdot \text{h}$$

and

$$T_{max} = 20 - \frac{300}{40} = 12.5°\text{C} \quad \text{for which}$$

$$\text{HDD}(12.5°\text{C}) = 36,000 \text{ K} \cdot \text{h}$$

The economizer is turned off when T_o is below T_{min}:

$$T_{min} = \frac{25 \times 8 - 15 \times 12.5}{25 - 15} = \frac{200 - 187.5}{10} = 1.25°\text{C}$$

* These numbers are representative of a 12,000 m² office building in New Jersey which one of the authors have monitored in detail.

† See Sections 10.2 and 10.3 for a discussion of the degree-day calculation method.

The corresponding degree-hours are HDD(1.25°C) = 5840 K·h.
The heating loads are

$$\dot{Q}_{ann,heat,1-z} = 25\ \text{kW/K} \times 20{,}440\ \text{K}\cdot\text{h} = 511\ \text{MWh}$$

$$\dot{Q}_{ann,heat} = 15\ \text{kW/K} \times 36{,}000\ \text{K}\cdot\text{h} + (25-15)\ \text{kW/K} \times 5{,}840\ \text{K}\cdot\text{h} = 598\ \text{MWh}$$

The multizone index for the economizer mode is

$$\eta_{multiz,heat} = \frac{\dot{Q}_{ann,heat,1-z}}{\dot{Q}_{ann,heat}} = \frac{511}{598} = 0.85$$

The overconsumption during the heating season is 598 − 511 = 87 MWh, or 87/598 = 15%.

Comments

There is overconsumption due to multiple zones even though the chiller does not run at all. In effect, the economizer is a form of simultaneous heating and cooling because it increases the airflow for the entire building even though some zones require heating. This penalty could be avoided by means of heat recovery and thermal storage, and the overconsumption index quantifies the potential savings. A simple solution that avoids at least some of this overconsumption is to install two separate air handlers, one for the interior zone and one for the exterior zone.

19.7.2 Energy Delivery Efficiency

Another approach called the energy delivery efficiency (EDE) has been proposed by Reddy et al. (1994) meant to rate the energy delivery of multizone systems considering simultaneous heating and cooling penalties in actual buildings. Taking the control volume to include both the HVAC system and the building, and viewing internal loads such as lighting to be generated inside the control volume, a heat balance neglecting transient effects associated with thermal mass yields

$$Q_{bldg} = Q_{cool} - Q_{heat} \tag{19.31}$$

where

Q_{bldg} is the net building heat gains or net cooling load
Q_{cool} is the measured whole-building cooling thermal energy supplied by the cooling coil
Q_{heat} is the measured whole-building thermal heating energy supplied by the heating coil

The value ($Q_{cool} - Q_{heat}$) can be viewed as the amount of energy needed to condition the space that would be required had no mixing of cold and hot airstreams taken place. This amount is, thus, a sort of absolute thermodynamic minimum. In reality, the building consumes total thermal energy amounting to (Q_{cool} + Q_{heat}). Consequently, the EDE_{Ideal} that rates the amount of simultaneous heating and cooling is defined as

$$\text{EDE}(1-\text{zone}) = \frac{\text{Thermodynamic minimum energy use}}{\text{Actual energy use}} = \left|\frac{(Q_{cool} - Q_{heat})}{(Q_{cool} + Q_{heat})}\right| \tag{19.32}$$

with absolute values being taken in order not to have negative values of efficiency when the building requires more heating energy than cooling energy.

However, medium to large buildings tend to be multizoned, and so a more realistic baseline minimum energy use would be that needed to condition a two-zone approximation of the building. The rationale for a two-zone treatment is that a two-zone treatment with one exterior or perimeter zone and one interior or core zone is a good compromise between accuracy and simplification. Steady-state expressions for the *minimum* heating and cooling energy use of a two-zone building in terms of building envelope parameters, ventilation rates, and specified internal loads for different outdoor temperatures have been derived that provide a basis of comparing actual energy use in a building to this minimum (Reddy et al., 1995). The approach has been primarily used to perform diagnostics in an actual building wherein measurements of the heating and cooling energy consumption are available.

The diagnostic ability of this approach is illustrated for an actual building in Texas that underwent retro-commissioning of the air-side deck reset schedule (Figure 19.26). Frames (a) and (b) depict how area-normalized chilled water and hot water energy use, respectively, vary with outdoor temperature $T_{db,o}$ during both the pre- and post-retrofit periods. Frames (c) and (d) indicate how actual diurnal EDE indices compare with $\text{EDE}_{\text{Ideal,2-zone}}$ using physical parameters derived from monitored data (see Reddy et al. 1994 for details). The improvement in EDE due to the retrofit is very clearly illustrated, and further, one notes that the HVAC system is operating in a close to ideal manner after the retrofit since the data points scatter quite closely around the ideal 2-zone EDE line (frame d).

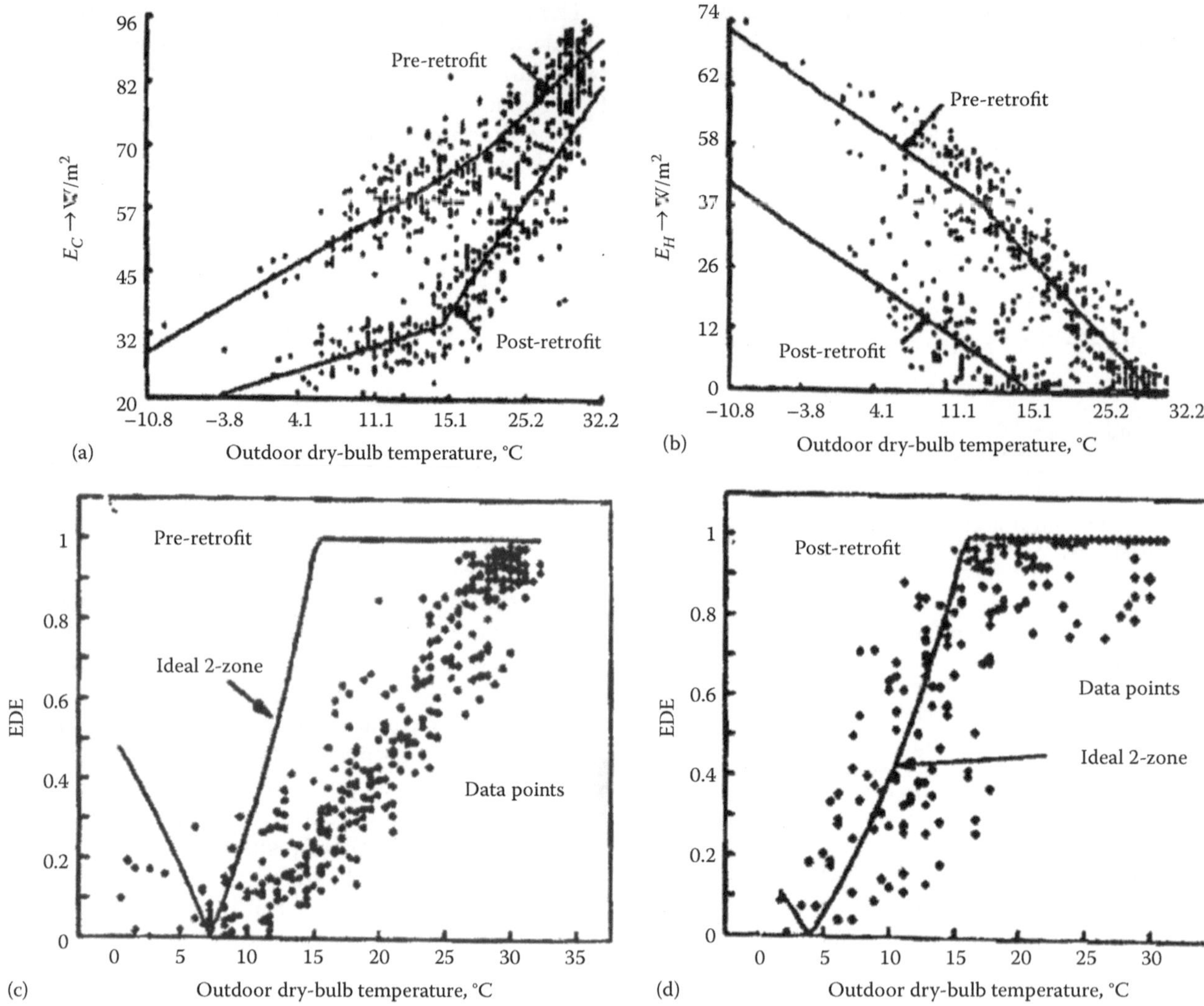

FIGURE 19.26
Illustration of the energy delivery efficiency (EDE) index as a diagnostic concept for an actual commercial building that underwent an air-side retrofit. (a) Area normalized daily cooling energy consumption. (b) Area normalized daily heating energy consumption. (c) Values of daily EDE index prior to the retrofit. (d) Values of daily EDE index after the retrofit. (From Reddy et al., 1995).

19.8 Closure

A summary of the advantages and disadvantages of all-air systems is provided in Table 19.4 (ASHRAE, 1993). There are numerous useful references for those wishing more fundamental and design expertise. A publication by Knebel (1983) presents a brief background of existing energy analysis procedures and provides a mathematical foundation for simplified simulation of various all-air HVAC systems. Much of the simulation methodology covered in this chapter has origins from this text. ASHRAE (1993) is a design manual that covers all air-conditioning systems. Written by a team of professional experts, this manual contains useful design tips meant for entry-level engineers and also for engineering students involved in senior design or capstone projects. It is meant to bridge the gap between theoretical knowledge acquired in schools and actual design practice. The book by Sun (1994) focuses exclusively on central air-handling systems and provides detailed practical information and system sketches on numerous types of energy-efficient systems and subsystems (such as economizer cycles, heat recovery systems), some of which were discussed in this chapter. It is a quick and easy to use guide to selecting and analyzing air-handling systems for a wide variety of building types and locations.

TABLE 19.4

Advantages and Disadvantages of All-Air Systems

Advantages	Disadvantages
Best choice for close control of zone temperature and humidity.	May limit the extent to which energy use can be reduced in low or zero energy buildings.
Major equipment is centrally located allowing for nonintrusive maintenance.	Duct space requirements add to building height.
Providing heating and cooling in different zones simultaneously is easily achieved.	Air balancing is difficult.
Can handle space churn to some extent.	May be more expensive than other secondary systems in first cost and operating cost.
Seasonal changeover is easily achieved by HVAC controls.	Noise in fan operation may be a problem in certain types of spaces.
Well suited for air-side economizer, heat recovery, and large outside air requirements.	May not be satisfactory for perimeter spaces in cold locations.
No drain pipe or power wiring in occupied areas.	Difficult to correct indoor conditions in individual rooms if improperly designed initially.
Less property damage if air ducts leak.	May lead to shoddy construction since air leakage from ducts causes no visible damage and so may be overlooked.

Problems

Numbers 1–4 given in parenthesis denote the degree of difficulty.

19.1 A CAV system with a separate furnace and humidifier maintains a single-zone space at 75°F (23.8°C) and 50% RH under a winter condition of 35°F (1.7°C) dry-bulb temperature and 80% RH. The total space heat load is 225,000 Btu/h (65.9 kW) and the latent load is 56,250 Btu/h (14.5 kW). The outdoor air quantity needed for ventilation is 1000 cfm. The supply air to the space is at 120°F (48.9°C). Determine (a) the quantity of air supply to the room, (b) humidity ratio of supply air, (c) the heating required at the furnace, and (d) the amount of steam needed. Assume that saturated steam at 200°F (93.3°C) is used with an enthalpy of 1146 Btu/lb. *(2)*

19.2 Consider the dual duct system of Example 19.7. You will investigate what happens to the cooling and heating loads when the deck temperatures are modified: hot deck to 120°F (48.9°C) and cold deck to 55°F (12.8°C). Comments on the effect of this change. *(2)*

19.3 In a VAV system, find the range of primary airflow requirements for a 75°F (24°C) and 40% relative humidity zone whose load (entirely sensible) varies from 2000 to 5000 Btu/h (600 to 1500 W). Air is supplied to the zone at 56°F (13°C). *(1)*

19.4 In a VAV system, find the range of primary airflow requirement for a 75°F (24°C) zone whose load of 2000–5000 Btu/h (600–1500 W) has an SHR that varies from 0.7 to 0.9. Air is supplied to the zone at 56°F (14°C) and 85% relative humidity, and the control is by dry-bulb temperature. *(1)*

19.5 Determine the air handler fan flow rate for cooling a small commercial building consisting of four zones, the peak load of each of which is 30,000 Btu/h (9 kW) including ventilation air. Coil conditions are 84°F (29°C) and 38% relative humidity inlet and 53°F (12°C) and 90% relative humidity outlet. The diversity among the four zones is 80%. What is the size of the main duct and each of the four branch ducts, if all are square in cross section, that you would suggest for this building? *(3)*

19.6 Size the preheat coil for a VAV system with 100,000 actual ft^3/min (47,000 actual L/s) system flow if the outside air fraction is 19% in Billings, Mont. *(2)*

19.7 Complete a VAV systems design, following the process used in Example 19.8, for the following two-zone building:

	Zone 1 (Exterior)	Zone 2 (Interior)
Sensible peak cooling load	$\dot{Q}_{sen,1} = 150{,}000$ Btu/h	$\dot{Q}_{sen,2} = 250{,}000$ Btu/h
Latent peak cooling load	$\dot{Q}_{lat,1} = 40{,}000$ Btu/h	$\dot{Q}_{lat,2} = 60{,}000$ Btu/h
Heating peak load	$\dot{Q}_{heat,1} = 300{,}000$ Btu/h	$\dot{Q}_{heat,2} = 70{,}000$ Btu/h
Zone temperature T_{zone}	75°F	75°F

Other: Design conditions are for Denver (altitude 5280 ft [1610 m]). The supply system pressure drop at full airflow is 3.5 inWG (28 Pa), and the fan efficiency is 63%. Take the supply fan air temperature rise to be 1.5°F (0.8°C) at peak conditions, and the return fan air temperature rise is 1.0°F (0.6°C). The required fresh air ventilation flow rate is 3200 ft^3/min (1500 L/s), and the cooling coil outlet humidity is 90%. Use a zone supply air temperature difference ΔT_{supply} of 20°F (11°C). *(4)*

19.8 A constant-volume reheat system is needed to accurately control conditions in a building designed to the specifications of Problem 19.7. Size the cooling, reheat, and preheat coils and other equipment for this system to be located in New York City. *(4)*

19.9 Freeze protection of a reheat coil in either constant- or variable-volume systems can be accomplished by the use of antifreeze, but at the expense of reduced heat transfer, higher pumping power, and increased maintenance costs. However, another approach is to locate it in the return duct downstream of the building exhaust, where freezing air does not exist. The shortcomings of antifreeze are thereby avoided. Discuss this approach. Can you identify any difficulties? *(2)*

19.10 Work Example 19.10 with the bin dry-bulb temperature data listed in the table for Denver, CO. Assume RH = 40% for all bins. *(4)*

Bin, °F	97	92	87	82	77	72	67	62	57	52	47	42
Hours	7	71	174	291	384	494	618	794	776	739	729	752
Bin, °F	37	32	27	22	17	12	7	2	−3	−8	−13	
Hours	724	704	555	394	243	137	84	54	22	13	9	

19.11 Use the calculated airflow rates in Example 19.10 to assess the performance of a VSD instead of the outlet dampers used in the example. What are the savings in fan power that would be achieved if this more efficient method of volume control were used? *(2)*

19.12 The boiler system in Example 19.10 is oversized by 100% and operates mostly under part load. What would be the annual boiler energy input if we wish to evaluate two alternative scenarios (a) only one boiler with half of the total capacity is installed, i.e., with boiler size of 250,000 Btu/h, (b) two identical boilers of 125,000 Btu/h each were installed? *(2)*

19.13 If the chiller in Example 19.10 were oversized by 20% to 60 tons (210 kW), what would the annual energy penalty be compared to the original design? *(2)*

19.14 If the air condition to be maintained in a building is 72°F and 50% relative humidity and the outdoor humidity ratio remains constant at 0.009 lb_w/lb_a, evaluate the performance of two methods of economizer control: dry-bulb temperature and enthalpy. The outdoor dry-bulb temperature for the day in question is given by $T_a = 58 - 12 \cos(15t)$, in which t is the time measured in military time and T_a is in degrees Fahrenheit. The design air delivery temperature is 55°F in the dry-bulb temperature control approach; above 55°F the outside air damper is fully open, and below 55°F, it is modulated as needed down to its 20% flow minimum. The enthalpy controller operates the same as a dry-bulb temperature controller below 55°F, whereas above 55°F, the outside air damper remains fully open whenever the outside air enthalpy is below the room enthalpy; otherwise it is at its minimum position. What is the percentage of savings in cooling energy during this day? You may wish to specify a numerical value of total flow rate to solve the problem, although it is not necessary. Comment on the feasibility of enthalpy control for this situation. *(4)*

19.15 Occasionally it is necessary to control both dry-bulb temperature and humidity in a zone. This can be accomplished by adjusting the apparatus dew point (ADP) and supply airflow rate so that both latent and sensible loads are met. Find the ADP and coil airflow rate to control a zone to 75°F (24°C) and 40% relative humidity if the sensible load is 300,000 Btu/h (88 kW), the latent load is 30,000 Btu/h (8.8 kW), and the design outdoor conditions are 95°F (35°C) dry-bulb temperature and 40% relative humidity. Assume cold air leaves the cooling coil fully saturated. *(4)*

19.16 An economizer must deliver 10,000 ft³/min (4,700 L/s) at 56°F (13.5°C) dry-bulb temperature. The return air from the building is at 75°F (24°C) and 50% relative humidity, and the outside air condition is 35°F (2°C) and 40% relative humidity. How much outside air is needed, and what is the mixed-air humidity ratio? *(3)*

19.17 An economizer must deliver 50,000 ft³/min (23,500 L/s) at 55°F (13°C) dry-bulb temperature. The return air from the building is at 72°F (22°C) and 45% relative humidity, and the outside air condition is 32°F (0°F) and 50% relative humidity. How much outside air is needed, and what is the mixed-air humidity ratio? *(3)*

19.18 Make a dimensional plot of the economizer outside airflow characteristic similar to Figure 19.22 for a 20,000 ft³/min (9,500 L/s) fixed-volume system with a minimum flow set point of 18%, assuming dry-bulb temperature control. The supply air condition is required to be 55°F (13°C), and the room air return condition is 77°F (25°C) and 50% relative humidity. For simplicity, ignore fan heat. At 10°F (−12°C) the outside air setting is the 18% minimum. How is the high-temperature return to minimum flow determined? *(3)*

19.19 Repeat Problem 19.18 with enthalpy control. To identify the point at which one will switch to and from the minimum airflow, you will need to make an assumption regarding outside air conditions of temperature and humidity. *(3)*

19.20 Some buildings use *night purging*, where cool night air is used to cool the building mass in the evening to reduce cooling energy used the following day. However, bringing in the cool night air can also cause the building and its contents to absorb a lot of moisture that can actually increase the latent load. If the building return air is a constant 78°F at 70% relative humidity without night purging and 76°F at 77% relative humidity, does night purging make sense? Explain your answer. *(3)*

References

ASHRAE (1993). *Air-Conditioning Systems Design Manual* (H. Lorsch, ed.). American Society of Heating, Refrigerating and Air-Conditioning Engineers, Atlanta, GA.

ASHRAE *Fundamentals* (2013). *ASHRAE Handbook of Fundamentals.* ASHRAE, Atlanta, GA.

ASHRAE 62.1 (2013). *Standard 62-2013: Ventilation for Acceptable Indoor Air Quality.* American Society of Heating, Refrigerating and Air-Conditioning Engineers, Atlanta, GA.

Avery, G. (1986). VAV—Designing and controlling an outside air economizer. *ASHRAE J.*, December, 12, 26–29; Avery, G. (1989). The myth of pressure-independent VAV terminals. *ASHRAE J.*, 31, 28–30.

Davis, M., J. Bryant, and D.L. O'Neal (2009). Modeling the performance of single-duct VAV systems that use fan powered terminal units. *ASHRAE Trans.*, 115(1), 307–313.

DOE2.1 (1982). Several manuals for the DOE2.1 software are available from the National Technical Information Service, Springfield, VA. (2.1D Building Description Language Summary [DE-890–17726] and the 2.1A Engineers' Manual [DE-830-04575]).

Energy Plus (2009). Energy Plus Building Energy Simulation software, developed by the National Renewable Energy Laboratory (NREL) for the U.S. Department of Energy, under the Building Technologies program, Washington, DC. http://www.nrel.gov/buildings/ energy_analysis.html#energyplus.

Englander, S.K. and L.K. Norford (1988). Fan energy savings: Analysis of a variable-speed drive retrofit, in *Proceedings of the 1988 ACEEE Summer Study on Energy Efficiency in Buildings*, American Council for an Energy Efficient Economy, Washington, DC.

Guntermann, A.E. (August 1986). VAV system enhancements. *Heat Piping Air Cond.*, 58(8), 67–78.

Holness, G.V.R. (February 1990). Human comfort and IAQ. *Heat Piping Air Cond.*, 62, 43–52.

Knebel, D. (1983). *Simplified Energy Analysis Using the Modified Bin Method.* American Society of Heating, Refrigerating and Air-Conditioning Engineers, Atlanta, GA.

Reddy, T.A. (1994). Enhancing thermal energy efficiency of terminal reheat HVAC systems by coil-bypass, in *Solar Engineering 1994* (D.E. Klett, R.E. Hogan and T. Tanaka, eds.). ASME, New York; *Proceedings of the ASME/JSME/JSES International Solar Energy Conference*, San Francisco, CA, March 1994, pp. 475–482.

Reddy, T.A., S. Katipamula, J.K. Kissock, and D.E. Claridge (February 1995). The functional basis of steady-state thermal energy use in air-side HVAC equipment. *ASME J. Solar Energy Eng.*, 117, 31–39.

Reddy, T.A., J.K. Kissock, S. Katipamula, and D.E. Claridge (May 1994). An energy delivery efficiency index to evaluate simultaneous heating and cooling effects in large commercial buildings. *ASME J. Solar Energy Eng.*, 116, 79–87.

Spitler, J.D., D.C. Hittle, D.L. Johnson, and C.O. Pederson (1987). A comparative study of the performance of temperature-based and enthalpy-based economy cycles. *ASHRAE Trans.*, 93(2), 13–22.

Sun, T.Y. (1994). *Air Handling System Design.* McGraw-Hill, New York.

20

Room Air Distribution and Hybrid Secondary Systems

ABSTRACT This chapter covers various aspects of room air distribution as well as recent developments in hybrid components and systems. We first discuss different types of the basic air–water systems to set the stage for subsequent material. Next, we provide an overview of air distribution jets and airflow patterns, resulting from the combined effects of forced flow and natural buoyancy effects inside rooms. This is then followed by a discussion of different design issues and components relevant to fully mixed air distribution systems. Next, we describe other air distribution systems, such as the underfloor and the ventilation displacements methods We then discuss ceiling-mounted chilled beams, a device gaining popularity, which are hydronic devices that provide sensible cooling to the space by convection. Hybrid secondary systems are being increasingly used due to the flexibility they provide along with energy and indoor air quality advantages over traditional-air systems. One such system, namely, the dedicated outdoor air system (DOAS) is presented. Finally, we treat two other low-energy cooling techniques, namely, evaporative cooling and desiccant cooling technologies and systems. Evaporative processes using water can reduce energy consumption in many ways, including reducing cooling loads, meeting part of the sensible load, or improving the performance of air-cooled condensers on chiller plants. Desiccant systems pre-dry the supply or ventilation airstream, thereby reducing the load on the mechanical cooling systems and the electrical demand. The thermal energy needed to regenerate the desiccant can be obtained from low-temperature low-grade energy sources.

Nomenclature

A	Area, m^2 (ft^2)
A_{eff}	Net flow area or outlet area of jet, m^2 (ft^2)
A_o	Area of jet flow at source, m^2 (ft^2)
ACH	Air changes per hour, h^{-1}
ADPI	Air distribution performance index
B_{cb}	Width of chilled beam, mm (in.)
$\dot{C}$	Heat capacity, kJ/K (Btu/°F)
CL	Cooling load, kW (Btu/h)
COP	Coefficient of performance
DOAS	Dedicated outdoor air systems
HDS	Hybrid desiccant systems
c_p	Specific heat at constant pressure, kJ/(kg·K) [Btu/(lb$_m$·°F)]
D	Diameter, m (ft)
EDT	Effective draft temperature (Equation 20.3)
f_o	Free area of bottom perforated plate of chilled beam, %
H_{cb}	Height of chilled beam, mm (in.)
HDD	Heating degree days, °C-days (°F-days)
h_v	Heat of vaporization of water, kJ/kg [Btu/lb$_m$]
K	Dimensionless constant (Equation 20.2)
L	Characteristic length, m (ft)
$\dot{m}$	Mass flow rate, kg/s (lbm/h)
NC	Noise criteria
PF	Performance factor (Equation 20.8)
p	Pressure, Pa (lb$_f$/in.^{2})
p_S	Static pressure drop, Pa (lb$_f$/in.^{2})
p_T	Total pressure drop, Pa (lb$_f$/in.^{2})
p_v	Partial pressure of water vapor load, Pa (lb$_f$/in.^{2})
$\dot{Q}$	Energy use or load with subscripts c for cooling, h for heating, W (Btu/h)
$\dot{Q}_{cc}$	Load on the cooling coil, W (Btu/h)
RPF	Return plenum fraction (Equation 20.5)
r	Radius, distance from jet source, m (ft)
SHR	Sensible heat ratio
SPF	Supply plenum fraction (Equation 20.5)
T	Throw of jet, m (ft)
T	Temperature, K or °C (°R or °F)
TDV	Thermal displacement ventilation
UFAD	Underfloor air distribution
v	Specific volume, m^3/kg, (ft^3/lb$_m$)
v	Velocity, m/s (ft/min)
$\dot{V}$	Volumetric flow rate, L/s (ft^3/h)
W	Humidity ratio, kg$_w$/kg$_a$ (lb$_w$/lb$_a$)
UCLR	Underfloor cooling load ratio (Equation 20.4)
WG	Water gauge, a measure of pressure drop
Z_{cb}	Slab clearance between ceiling and upper edge of chilled beam, mm (in.)
ZF	Zone fraction (Equation 20.5)
x	Distance, m (ft)
$\dot{x}$	Time rate of quantity x
$\overline{x}$	Average of quantity x

Greek

ε Effectiveness
ρ Density, kg/m³ (lb_m/ft^3)
ϕ Relative humidity
Δx Change in quantity x

Subscripts

a	Air, dry air
bldg	Building
cc	Cooling coil
cb	Chilled beam
cool	Cooling
d	Desiccant
db	Dry bulb
dp	Dew point
evap	Direct evaporative
ew	Enthalpy heat wheel
hx	Heat exchanger, indirect evaporative
in	Inlet
lat	Latent
max	Maximum
min	Minimum
o	Outdoor air condition
out	Outlet
p	Primary air
reg	Regeneration
space	Space or room or zone
sen	Sensible
sw	Sensible heat wheel
tot	Total
w	Water
wb	Wet bulb

20.1 Introduction

Air distribution involves four components: (1) fans to move the air, (2) ducts to direct the air to the terminal unit, (3) terminal units to control the amount and temperature of room supply air, and (4) diffusers to deliver the air properly to the entire space. The air supply to the rooms or zones is done by overhead or underfloor distribution ducts, and how to size such fans and ducts has been covered in Section 16.6. In this chapter, we shall address components related to (4) such as mixing devices and room diffusers.

Recall that *secondary systems* are those that meet the sensible and latent loads of various zones of a building, while also conditioning the ventilation air drawn into the building to meet occupant comfort. The first type of system to evolve was the all-water system (covered in Chapter 18), wherein only the zone loads were met by hot/chilled water supplied directly to the zones, and with outdoor intake and conditioning being perfunctory or none at all. All-air systems (covered in Chapter 19) were introduced next about 50 years back, and we have covered pertinent aspects related to advantages and thermal analysis of different types of all-air systems for one zone and multiple zone buildings: peak sizing, part-load performance, annual energy prediction methods, and ways to improve energy efficiency.

The third generic type of secondary system is the air–water system, wherein the functions of meeting space loads and drawing in and conditioning the required ventilation airflow are met by two different HVAC systems. This system will be discussed in this chapter along with other types of hybrid systems that have been evolving in recent years. The thrust to develop such alternative air delivery systems is largely driven by the need to improve energy efficiency in HVAC systems and to meet the various targets set forth by professional societies such as ASHRAE (for example, ASHRAE 189 (2009) or by federal agencies (all new federal buildings built after 2030 should be net-zero energy). These initiatives are discussed in Chapter 24.

The hybrid systems have not been in use long enough for their energy calculations relevant for design and year-round operation to have been fully vetted with actual performance evaluations (Int-Hout, 2012). This is in sharp contrast to overhead systems (traditional or fan powered) whose calculations are well established. Though hybrid system configuration and design are to some extent "new territory," these systems will become increasingly mainstream, and so this chapter discusses the operational principles of these systems and presents their basic calculation procedures with examples.

20.2 Basic Air–Water Systems

Air is ultimately the principal medium for heating or cooling spaces in all buildings except those that are conditioned by radiant systems. In Section 11.3, systems that provide warmed or cooled air from the central plant were introduced. In this section, we consider systems that use heated or cooled water from a central primary system to heat or cool zones. The energy transfer from liquid to air in *air–water systems* occurs at terminal devices in each zone. Temperature control is achieved by controlling water or air temperatures (or both). Humidity control is often a by-product of cooling, but humidification during the heating season is generally not employed in commercial buildings.

As stated in Section 18.1, the interest in using water as a heat transport medium is that it has a significantly higher volumetric heat capacity (density times specific heat) than air, by a factor of 3500 difference, in fact. As a result, less water need be moved, conduits can be much smaller, and pumping power is much reduced. If a central air system is used, only the needed ventilation air must be provided, thereby reducing duct sizes significantly. Recall that in typical commercial office buildings, for example, ventilation is only about 15%–30% of the total airflow. This *primary* air is provided by a constant-volume central system. In some smaller buildings, outdoor air is not provided by the central system at all—it is introduced into the zone directly through outside openings. Such an arrangement referred to as "trickle ventilation" is shown in Figure 20.1. If wind and stack effects are important for the building, this approach can be unreliable in terms of maintaining required ventilation air supply and zone pressurization. To achieve proper interior air distribution, zone air is recirculated along with the supplied fresh air. Return air systems can be eliminated in many cases since the modest amount of outdoor air supplied to a zone offsets exfiltration, provides needed zone pressurization, or is balanced with local exhaust flows.

All-water systems (discussed in Chapter 18) have several inherent disadvantages that cannot be avoided with either design or operation. First, maintenance (e.g., changing filters, repairing piping) must be done in the zone space housing the fan coil unit instead of in the equipment room as in air-based systems. Second, if significant latent cooling loads are present, a means is needed for disposing of water condensate from zone latent loads at each unit ventilator. These condensate disposal systems must be cleaned periodically. In addition, filtration at local fan coil units is problematic since filters are generally small (and often inefficient), requiring frequent replacement. As noted earlier, ventilation can be questionable unless provided by a separate central ventilation system. Humidity control is unreliable in many cases without special measures.

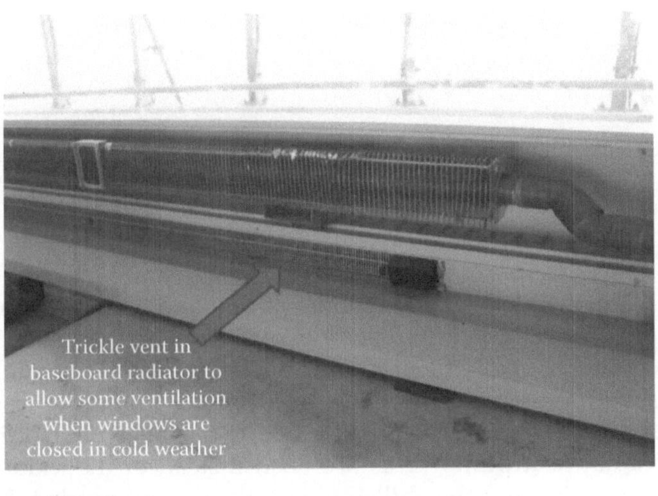

FIGURE 20.1
Photo of a room with trickle ventilation along with baseboard heating. (Image courtesy of Erik Kolderup.)

Figure 20.2a shows the simplest air–water system that is a variant to the hydronic system, but here, there is an explicit path for ventilation air to be ducted into the spaces. This *two-pipe system* uses an air-to-water heat exchanger for both heating and cooling the zone. The coil is housed in a *fan coil* (or unit ventilator) system located in a zone. The two-pipe system is provided with either hot or cold water for heating or cooling depending on the load and season of the year. The change from heating to cooling is made at the central plant for the entire building. Control is achieved by a zone thermostat that controls coil water flow.

The two-pipe system has inherent practical difficulties in some climates. For example, on a sunny fall or spring day, a zone with southern exposure may need cooling, while a zone on the north side may need heating. Simultaneous heating and cooling is not possible. If a single pump is used, as shown in Figure 20.2a, only one flow stream is available at all times. Coil or central plant characteristics may dictate different heating and cooling flows for proper performance. During very cold weather, outdoor air supply dampers at each unit ventilator will be closed (if a separate ventilation air system is not used), limiting the ventilation air supply. Coil freezing must be prevented under low-temperature conditions. Two-pipe systems are not considered practical or energy efficient in climates where frequent heating-to-cooling changeovers are needed.

One method of solving some of the problems of the two-pipe system is to use the somewhat more complex *four-pipe system*. The configuration of the unit ventilator for a four-pipe system is shown in Figure 20.2b. With the four-pipe system, heating and cooling are both available for different zones that may need each at the same time. In addition, since two pumps are used, the proper flow for each loop (and its primary system) can be maintained. Of course, the cost is higher, but control is much better. Figure 20.3 provides a cross-sectional diagram of the four-pipe unit with details of the various individual components.

In both two-pipe and four-pipe systems, central plant controls may adjust water supply temperatures based on outdoor weather. For example, on a mild day, chilled

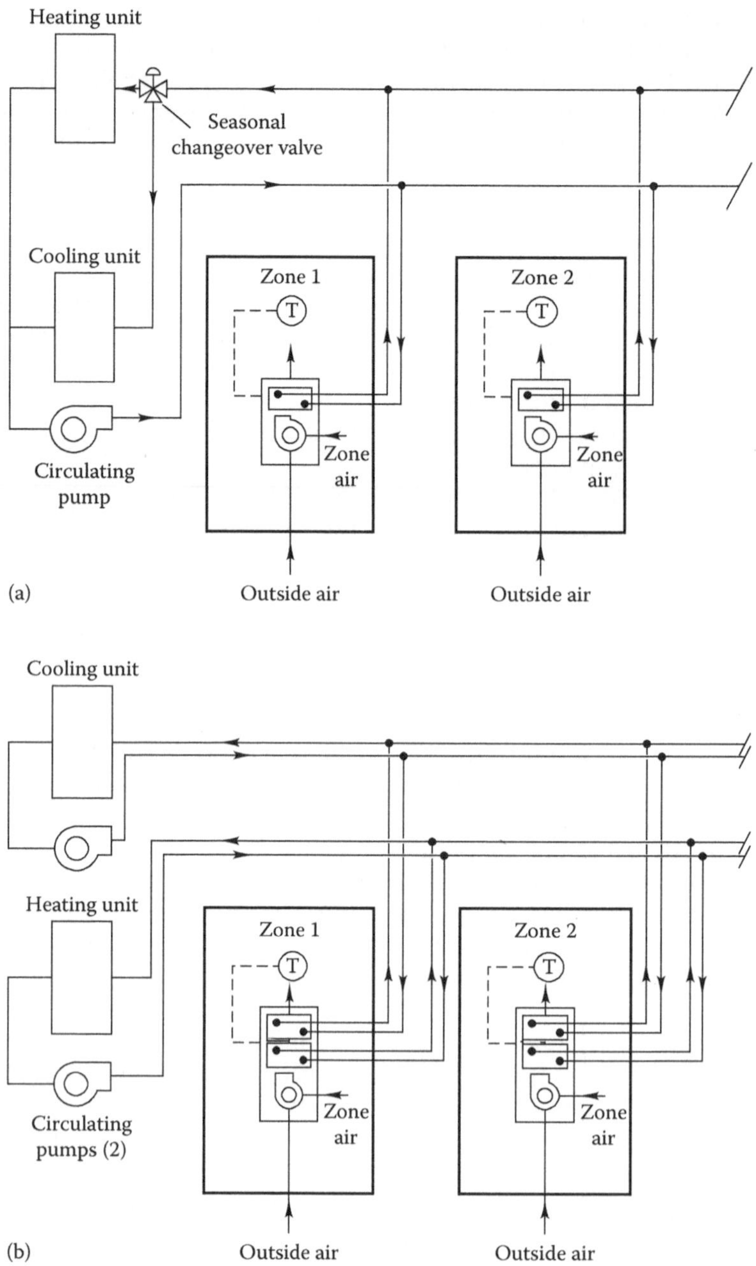

FIGURE 20.2
(a) Two-pipe air–water system with local outside air supply and fan coil terminal units. A single circulation pump is used. Outdoor air can be supplied locally (by unit ventilators) or centrally via a constant-volume system. (b) Four-pipe air–water system with local outside air supply and fan coil terminal units. Separate chilled and hot water pumps are used.

water can be warmer than on a hot day and still meet the load. Energy savings can be significant if water temperatures are *reset* to match the load conditions in both air–water and all-air systems. Advantages and disadvantages of air–water systems are summarized in Table 20.1.

Central air–water systems are also commonly used. One such example is illustrated in Figure 20.4 where individual induction units interconnected by a closed water loop meet the sensible and latent loads of the individual rooms, while a *separate* primary ventilation system supplies the code-required outdoor airflow rate. Heating and cooling to the induction units and to the ventilation air are provided by the boilers and chilled water plant as shown. A variant of this system layout is one where individual heat pump units are used instead of the induction units, and the chilled water plant is not needed. In such a case, the primary system has its own conditioning system (say, a roof top unit and heat recovery wheels as

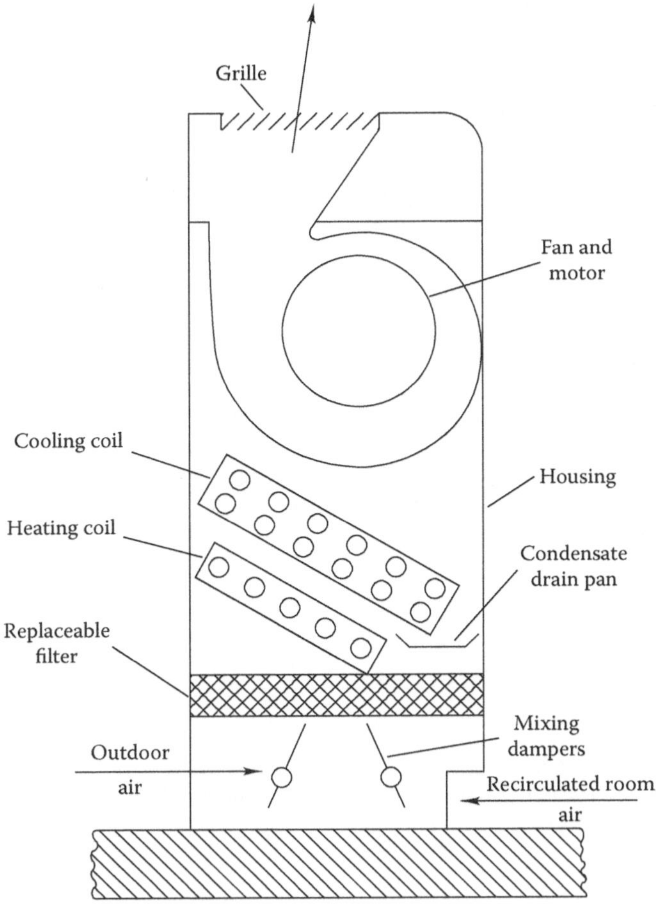

FIGURE 20.3
Cross-sectional diagram of four-pipe fan coil unit showing outside air supply and two separate coils.

TABLE 20.1
Advantages and Disadvantages of Air–Water Systems

Advantages	Disadvantages
Ideally, combines the best features of all-water and all-air systems	If designed badly, the worst features of all-water and all-air systems can result.
Low transport energy needed while meeting ventilation comfort requirements	Requires more maintenance because of numerous individual zone units.
Air duct size reduced	Controls tend to be more complicated.
Central air handler size reduced	Not suitable for high exhaust air requirements.
Good for spaces with highly variable loads	Severe damage if pipes leak.
Easy to shut off water supply in unoccupied spaces	Cannot shut off primary air to unoccupied spaces.
Good for exterior spaces	Humidity control may be difficult.
Can simultaneously heat and cool different spaces	
Can be combined with ventilation heat recovery devices (heat wheels)	
Can be combined with water loop heat pumps	

appropriate) that conditions the ventilation air to that of the space thermostat set point temperature. The individual heat pumps are sized to meet the design cooling loads of the rooms they serve, and because of the closed water loop connecting all heat pumps, the energy efficiency of the entire system is improved. During hot summer days (say, when outdoor air dry-bulb temperature $T_{db,o} > 90°F$ or 32°C), the cooling tower is operated to provide evaporative cooling (free water cooling economizer). When the outdoor temperature drops, the heat pumps can start providing heat to the rooms, which is supplemented by boiler heat as shown. Because of the complexity in controls, many such "heat pump" systems are designed only as cooling devices, and when $T_{db,o}$ drops to below, say, 40°F or 4.5°C, the boiler starts supplying the required heat. Such systems are the most suitable for exterior spaces of office buildings not requiring close humidity control and for buildings with variable vacancy such as dorms and hotel rooms where the occupants can switch off the heat pump on leaving the room. The heat pump with a separate ventilation system was the first attempt to decouple the room loads and that associated with the

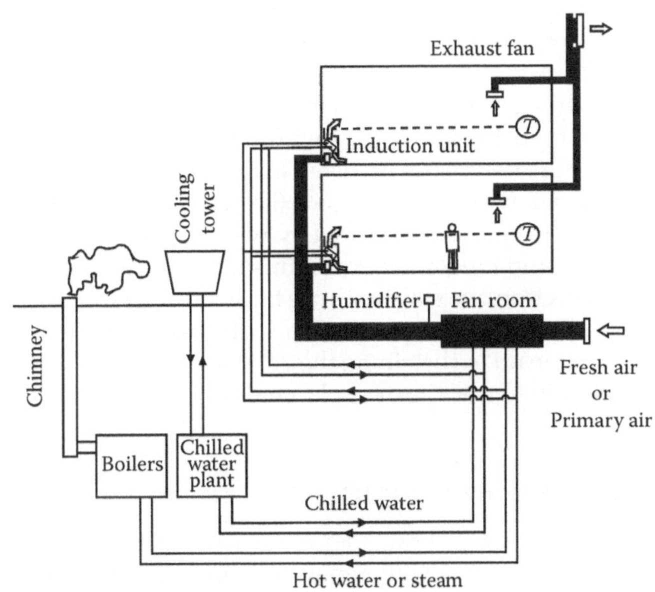

FIGURE 20.4
Central closed-loop individual water heat pump units with a central ventilation system (not shown).

ventilation air. Numerous variants have evolved based on this basic concept, and such hybrid systems are described in Section 20.7. The area of research, development, and demonstration of such hybrid systems is one that is being ardently pursued by the HVAC community.

20.3 Air Distribution in Rooms

20.3.1 Basic Considerations

After having been delivered to a room by the duct system, conditioned air must be effectively distributed in the space to ensure the comfort of its occupants. The designer would want to create conditions conducive to proper room temperature and humidity, air motion, and air quality in the room, if not throughout the entire room, at least in the occupied areas. Air stagnation pockets that tend to have large temperature gradients, drafts and gusts, and noise due to air diffusion (or dispersion) should be avoided.

Common devices used for air distribution are *diffusers* that are outlets through which air is supplied to the space (further discussed in Section 20.4.2), *grilles* (or coverings for diffusers or for openings for return air) consisting of a frame enclosing a set of either vertical or horizontal vanes or both, and *registers* that is a grille with an adjustable damper for volume flow rate control and for changing direction of air entering the room. Proper design of room air distribution systems involves (1) selecting type of diffuser and return air grills, (2) determining their proper placement that would depend on room geometry as well as whether heating or cooling is required, (3) sizing the devices depending on the heating and cooling capacities needed, (4) ascertaining that the associated noise is lower than predetermined limits, and (5) studying their interaction with other heating and cooling distribution devices within the room.

There are two issues involved in the selection of diffusers: the behavior of free air jets (i.e., when the incoming air not affected or obstructed by walls, ceiling, or other surfaces) and the interaction of this jet with the room air. In this section, we start with the basic principles of these two issues and then proceed to a discussion of devices that are meant to distribute the air throughout the room, and finally to their correct placement.

20.3.2 Behavior of Air Jets

Figure 20.5 shows how a jet of air issuing from a duct outlet on a vertical wall behaves upon entering a room. The upper diagram shows that an *isothermal jet* (i.e., its temperature is the same as that of the room air) slowly diffuses into the surrounding still air. There are no buoyancy effects, and so the centerline of the jet remains horizontal. The *throw* is the distance which the jet travels before the velocity reduces to a predefined value (usually, 50 ft/min (0.25 m/s), 100, 150, or 200 ft/min as per ASHRAE 70 (2011). Note that the jet of primary airstream diverges (average of about 22° angle) due to entrainment of room air as a result of friction (air molecule momentum transfer) that gradually reduces the jet velocity while increasing the jet diameter. The latter phenomenon is referred to as the *spread*.

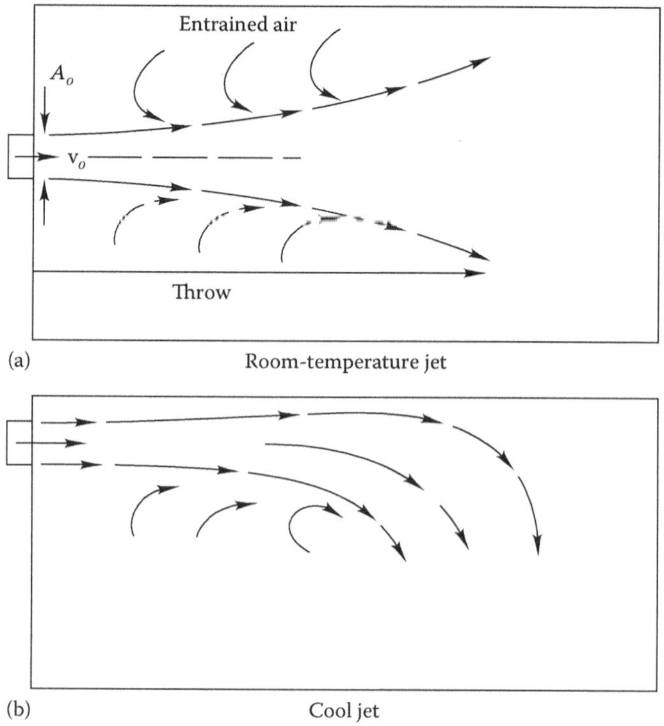

FIGURE 20.5
Air diffusion patterns from a vertical outlet. (a) Isothermal and (b) cool jet.

Traditionally (see, for example, ASHRAE Fundamentals, 2013), the full length of air jets can be divided into four zones:

Zone 1: A short core zone closest to the outlet face, wherein the velocity (and temperature for a nonisothermal jet) remains essentially constant.

Zone 2: A transition zone whose length is dependent on the type of outlet, its aspect ratio, and initial flow turbulence.

Zone 3: The zone of greatest importance in terms of room air distribution where fully turbulent flow is established, with the length being from 25 to 100 diameters of the outlet.

Zone 4: The zone where the jet velocity (and temperature) degrades precipitously and loses noticeable form. A lower threshold of 50 ft/min (0.25 m/s) is usually assumed since people tend to sense air movement above that value.

If the jet is round, the velocity v at any point in the jet at standard conditions in air is given by (Schlichting 1979)

$$v = \frac{7.41 \times v_o \sqrt{A_o}}{x \times (1 + 57.5 \times r^2/x^2)^2} \quad (20.1)$$

where
v_o is the velocity at jet source
A_o is the jet flow area at source
x is the distance from jet source
r is the distance from jet centerline

This equation suggests that, in essence, jet centerline velocity decreases inversely with distance from the source. One can define a jet radius in several ways, e.g., it could be the distance from the centerline at which the velocity is 1% of the centerline velocity. The jet radius r increases with distance from the source.

Example 20.1: Velocity of an Air Jet

Air at standard conditions issues from a 15 cm diameter (6 in. diameter) circular opening with a uniform velocity of 3 m/s (9.8 ft/s). What is the centerline velocity 3 m (9.8 ft) from the jet source?

Given: Diameter $D_o = 0.15$ m, $v_o = 3$ m/s.
Figure: See Figure 20.5.

Find: $v(x = 3, r = 0)$.

Solution

Equation 20.1 can be used to solve the problem by setting $r = 0$ m. The jet source area is

$$A_o = \frac{\pi}{4} D^2 = 0.7854 \times (0.15)^2 = 0.0177 \text{ m}^2$$

The jet centerline velocity is

$$v = \frac{7.41 \times 3.0 \text{ m/s} \times \sqrt{0.0177} \text{ m}^2}{3.0 \text{ m}}$$

$$= 0.985 \text{ m/s (3.2 ft/s or 192 ft/min)}$$

Comments

Since conditioned air jets are usually either warmer or cooler than room air, they will not have a horizontal centerline. They will either rise if warm or drop if cool. The *drop* is the vertical distance between the lower edge of a horizontal jet from when it leaves the diffuser till it reaches the end of its throw. Figure 20.5b illustrates the behavior of a *nonisothermal* cooler air jet. This example shows that jets at typical duct air velocities will penetrate too far into a room for proper air circulation. Cool air should not have a velocity of more than 45–50 ft/min (0.25 m/s) in occupied rooms, e.g., for good comfort. Therefore, special fittings called diffusers are fitted to the ends of ducts to better distribute airflow in rooms.

Figure 20.5b also illustrates another phenomenon (called the *Coanda effect*) when a jet is projected close to (less than about a foot or 0.3 m) and roughly parallel to a room surface (either ceiling or vertical wall). The entrainment is limited to one side only, and the jet tends to "stick" to the surface (in this figure the ceiling) because of the negative pressure created between the jet and the surface. Though the velocity is lower as compared to the free jet case, the throw of the jet is increased, while the drop for horizontal jets is reduced. This phenomenon can be used to advantage by the designer while selecting diffuser sizes and their placement since the supply air under the ceiling can be made to spread out further than otherwise, thereby having to specify fewer diffusers for the space.

20.3.3 Classification of Air Diffusion Methods

Typical occupied spaces can be viewed as consisting of an occupied zone and an unoccupied zone. The former is where occupants normally reside. For a person's head, or specifically the nose, height is taken to be 67 in. (1.7 m) for standing occupants; this could be altered depending on type of space occupancy. Higher spaces and the volume that includes a perimeter strip of 3.3 ft (1 m) from any exterior wall or windows plus another strip of 1 ft (0.3 m) of any internal wall is assumed to be unoccupied. Room air distribution strategies are classified into three types (Figure 20.6):

1. *Fully mixed systems*: This was the type previously considered in Chapters 3 and 19 where the whole room air is at essentially fully mixed conditions. There is little or no thermal or pollutant stratification vertically, and this is achieved by supplying large amounts of conditioned air into the room either from overhead, or sometimes by underfloor, diffusers. The conditioning of the space is achieved by diluting space air with the supply air.
2. *Partially mixed systems*: This arises when the occupied space is fully mixed and maintained at a condition distinctly different from the unoccupied zone. The unoccupied volume is usually stratified into three zones whose relative lengths may vary (see Figure 20.6). Most underfloor air distribution (UFAD) systems are examples of this type, and these are discussed in Section 20.5.1.
3. *Fully stratified systems*: The space is fully stratified vertically, i.e., there is a distinct thermal gradient that arises naturally by distributing the air into the room using a technique called thermal displacement ventilation (TDV), discussed in Section 20.5.3.

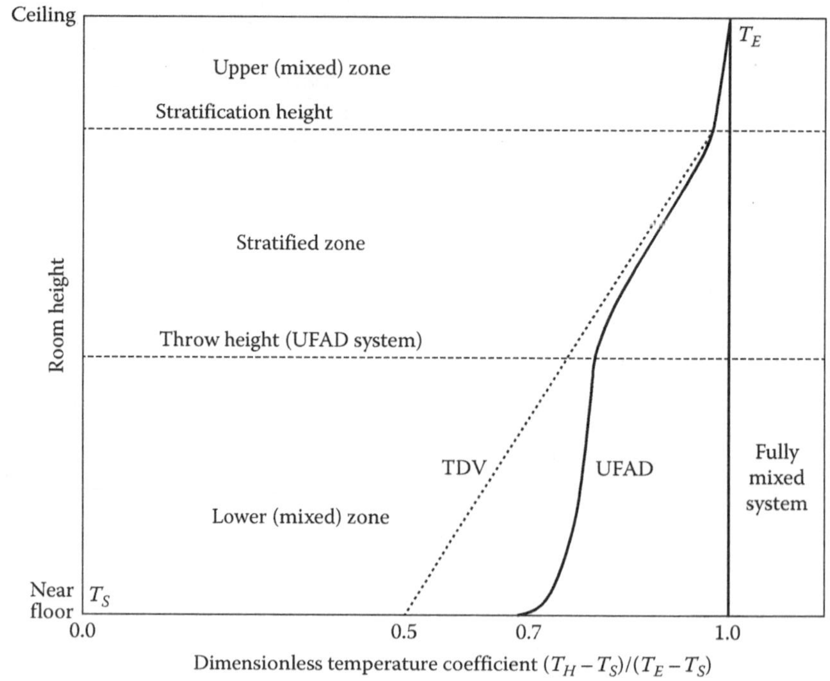

FIGURE 20.6
Vertical temperature stratification profiles in a room caused by different room air distribution methods. T_H is the temperature at height H in the room, TDV, thermal displacement ventilation; UFAD, under floor air distribution system. (From *ASHRAE Handbook of Fundamentals*, ASHRAE, Atlanta, GA, 2013. Copyright ASHRAE, www.ashrae.org.)

These conditions have to be created and maintained by proper design selection of the amount of air to be distributed, its temperature, type and location of diffusers, and their throw characteristics. The spatial distributions of thermal loads of the space are also important factors for the proper design of such distribution systems. Such aspects are discussed in the next two sections.

20.4 Fully Mixed Room Distribution Systems

20.4.1 Qualitative Flow Patterns

Proper room air distribution is an interplay of three types of influences. The first is how the heated/cooled jet pumped into the room by the supply fan through the diffusers interacts with the room air (or in case of natural ventilation, the opening of windows and doors). The second is the interaction of the buoyancy or natural convection effects caused by temperature differences between room air and room surface temperatures (such as heat losses through windows in winter causing the window pane to be cooler than other room surfaces). An additional significant influence is that due to furniture and partitions in rooms, but because of its circumstance-specific nature, this is not considered in the following text.

Some designers use general guidelines for first deciding on diffuser locations based on buoyancy considerations. For example, in locations requiring more heating than cooling (say, heating degree days [HDD] > 2000°F-days or 1100°C-days), floor or sill diffusers located under windows is the better choice. For locations requiring more cooling than heating (say, HDD < 2000°F-days), ceiling distribution is preferred. Supply airflow rates in excess of 3 cfm/ft^2 or 5 L/(m^2·s) (values in typical office buildings are at most 1.0–1.5 cfm/ft^2 or 5–7.5 L/(m^2·s) that translates usually to 5–15 ACH) require special care in their air distribution systems design. For that matter, clean rooms or computer server rooms may require up to 8–10 cfm/ft^2 (40–50 L/(m^2·s)).

A qualitative understanding of the temperature distributions throughout the room under the normal types of air distribution methods would provide further insights. Figure 20.7 illustrates the manner in which air is introduced into a room from a floor perimeter diffuser fan across the room with a thermostat set at temperature "T." The contour for the air distribution is drawn for air velocity of 50 ft/min (0.25 m/s), which is the acceptable value for human comfort. Because of the Coanda effect coupled with the buoyancy force, the hot air gets to be well distributed across the room volume going up one vertical wall, spreading across the ceiling and down the opposite wall. There is, however, thermal stratification

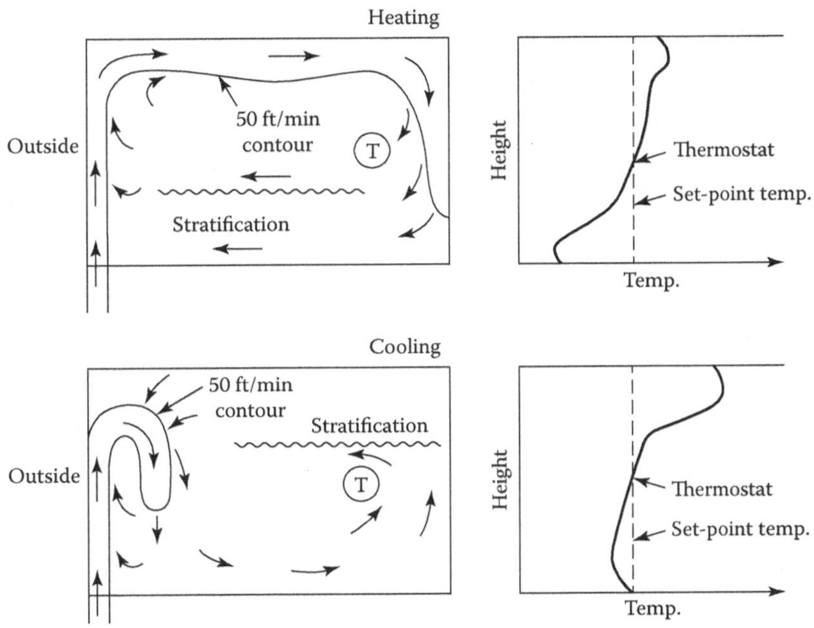

FIGURE 20.7
Airflow patterns and vertical thermal stratification for a floor-perimeter diffuser under heating and cooling modes. (From, *ASHRAE Handbook of Fundamentals*, ASHRAE, Atlanta, GA, 2005, www.ashrae.org.)

close to the floor (temperature is lower) even though the upper half of the room is fairly uniform thermally.

The patterns are very different under cooling mode. Buoyancy works to disrupt the Coanda effect, and so the plume drops down before it reaches the ceiling. This results in a thermal stratification at the upper end of the space, but since this zone is above the occupied space, there may be little or no adverse effect for occupant comfort. Note that these patterns are qualitative and if the air velocity were to be increased/decreased, the patterns would alter greatly. Further, the location of the return grills (not considered in this figure) can also alter these patterns; however, this location is usually not influential since often it is located in the stagnant portion of the space. A careful designer could select the diffuser throw and the placement of the return grills to overcome much of the thermal stratification depicted in the figure.

Ceiling diffusers are more common in commercial spaces since the ductwork can be hidden behind the false ceiling nor blocked by the layout of the furniture and partitions. Figure 20.8 illustrates the airflow patterns created from a ceiling diffuser under cooling and heating modes. Under cooling mode, the Coanda effect helps to better spread the jet across the ceiling, and the cold air then drops due to buoyancy forces. The room is well mixed, and there is no stratification in the occupied zone of the room. However, under heating mode, the buoyancy force is so dominant that the supply air does not mix well with the room air, and there is a very pronounced thermal stratification. This would be unacceptable for occupant comfort, and so a separate perimeter heating system may have to be specified.

The previous discussion of flow patterns is rather simplistic and meant to impart only a basic understanding of the forces at play. Prediction of actual air patterns in rooms is very complex for which computational fluid dynamic (CFD) models would be necessary. These patterns also depend on the type and construction details of the diffusers, and these are discussed next.

20.4.2 Types of Diffusers

As discussed earlier, diffusers spread air from the terminal box into the zone. Most diffusers are designed to spread the air out across the top of the ceiling so that cool zone air can slowly drop down into the zone without creating drafts. The big challenge is to locate the diffuser so that it effectively distributes air in the zone without short-circuiting air back into the return air grill, thereby avoiding the creation of drafts. Sometimes, a baseboard is required not for heating but simply to keep air traveling up the walls and windows.

Diffusers are mostly mounted in the ceiling plane and are connected to the distribution duct system above the ceiling. They can be either square or round and can also be situated behind light fixtures as shown in Figure 20.9. Further, diffusers can have vanes that may or may not be adjustable (Figure 20.10) to accommodate changes in supply airflow rate as in variable air volume (VAV) systems (these are called *active diffusers*). Also, diffusers

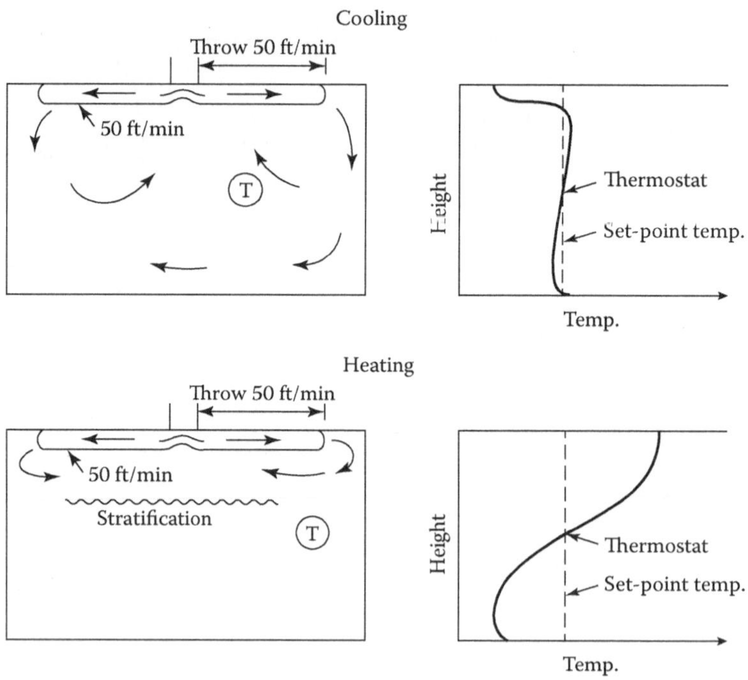

FIGURE 20.8
Airflow patterns and vertical thermal stratification for a ceiling diffuser under heating and cooling modes. (From *ASHRAE Handbook of Fundamentals*, ASHRAE, Atlanta, GA, 2005, www.ashrae.org.)

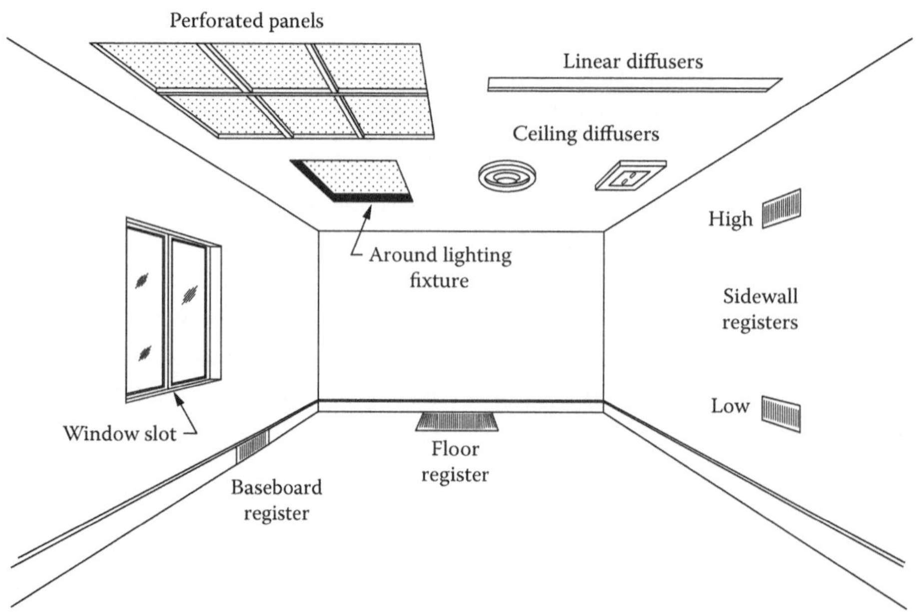

FIGURE 20.9
Common types of registers and diffusers.

can have asymmetric throws and with different angles of throw. The location of the air diffusion devices in the room plays a key role in maintaining thermal comfort. The throw, spread, and drop are affected by the diffuser design and the relative location of the room walls. The selection of air diffusers and grilles relies in large part on manufacturer's data, since the design of these components, and hence their performance, varies greatly from one model to another. ASHRAE Fundamentals (2013) suggests a classification of diffusers into five groups (Groups A–E) depending on where they are mounted (ceiling or floor), the direction of throw (horizontal or vertical) and

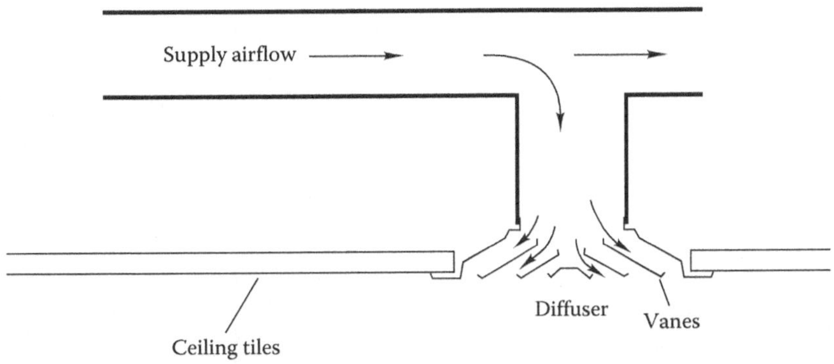

FIGURE 20.10
Cross-sectional sketch of ceiling diffuser.

type of jet (nonspreading and spreading). Such details are beyond this textbook, and the interested reader can refer to the aforementioned reference for further details.

For rooms with ceilings higher than the nominal 10 ft (3 m), special diffuser designs are needed. This is particularly critical when one is attempting to heat tall spaces such as auditoriums and gymnasiums from duct systems located near the ceiling.

20.4.3 Noise and Pressure Drop

The selection of diffusers is greatly determined by the noise they make as air passes through them and the dampers. The noise has no distinguishable frequency, and its loudness can interfere with speech. ASHRAE (2013) discusses the development of a single number scale called the *noise criteria* (NC) value depending on the type of activity within the space that quantifies the maximum allowable air noise level for that space. For a quiet room, the NC value should be low (15–30), for offices (35–40), while for a factory higher values are acceptable (45–60).

Manufacturers provide catalog tables where sound, pressure, and throw values have been measured according to an experimental test standard, for example, ASHRAE 70 (2011). The standard is meant to apply for ideal inlet conditions and so does not include the effect of flex duct, neck mounted dampers, close coupling, or 90° elbows leading to the diffusers themselves (John, 2012). An ASHRAE research project evaluated such effects and found that these had significant impact on diffuser performance. Design guidelines in the form of correction tables for NC and pressure drop were developed, which used in conjunction with diffuser manufacturer catalogs would yield a more realistic prediction of how these devices perform in the field. This research found that a damper can have a significant impact on total pressure through a diffuser (the pressure drop can be several times greater) and that the damper attached to a diffuser should not be used to balance the supply airflow; if such fine-tuning is needed, a separate inlet damper should be provided at the terminal box.

Some amount of noise is desirable both psychologically and thermally. Occupants feel reassured that conditioned air is being supplied to the room when they hear the noise from the distribution systems and tend to complain less. An oversized diffuser will be so quiet that almost no noise is heard. Also, the mixing in the room may be poorer in a VAV system: draft during maximum airflows and dumping during minimum airflow operation.

The NC value of a diffuser is determined by measuring sound pressure levels at varying frequency bands and then weighting them by the octave of the sound emitted. These values are also supplied in manufacturer's data such as those shown in Figure 20.11, for a round diffuser that is based on an NC rating of 35. The throw of the diffuser must be related to zone size to ensure that no dead airspaces exist. The terminal velocity V_T* is the average air velocity at the throw distance T (i.e., the half-width of the room, as shown). The figure is used by entering it with the design value of air volumetric flow (known from load calculations) and throw (based on room size). At the intersection of the throw and flow lines, one drops down vertically to find the size (i.e., throat diameter D_s). From the size, the designer reads the diffuser pressure drop, neck size, and diffuser outlet velocity. Room air velocities of 50 ft/min (0.25 m/s) are desirable for sedentary occupants (ASHRAE Fundamentals, 2013). The three values of room velocities V_R along with corresponding terminal velocities V_T, shown in the left center of the figure, are typical for offices.

In the absence of manufacturer's data, or during preliminary design, the throw can be estimated from the dimensional equation

$$T = K\frac{\dot{V}}{\sqrt{A_{eff}}} \tag{20.2}$$

where

the throw T is in feet
the net flow area or outlet area A_{eff} is in square feet
the flow rate $\dot{V}$ is in cubic feet per minute

* Upper case V has been used for velocity only in this section to be consistent with Figure 20.11.

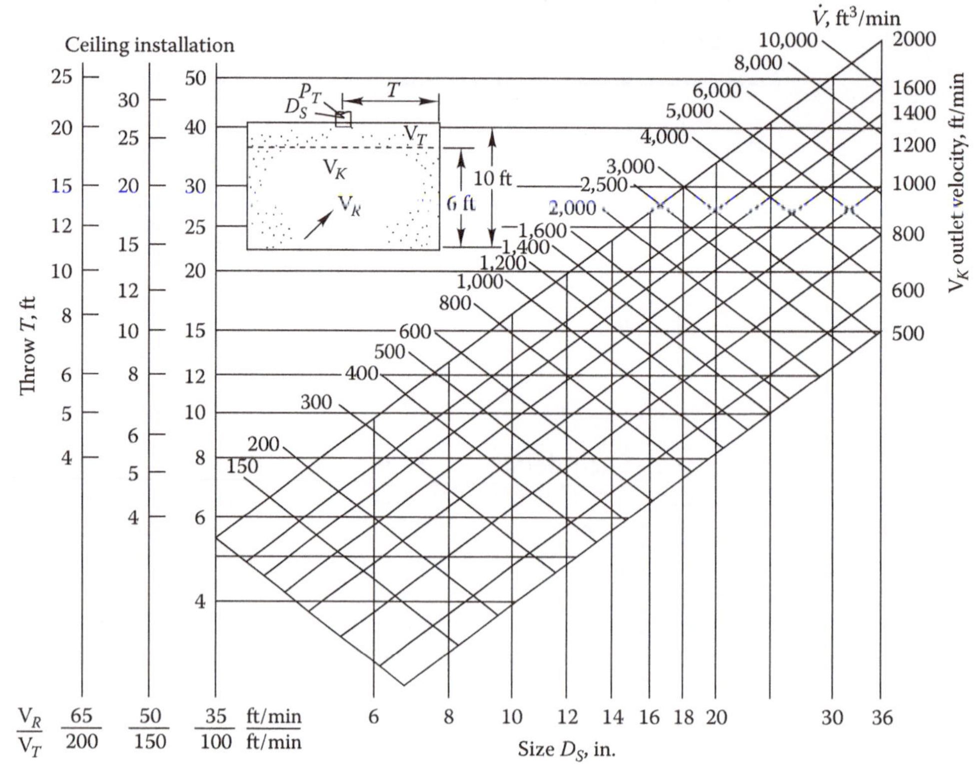

Dimensions

Listed size D_S, in.	6	8	10	12	14	16	18	20	24	30	36
A_K outlet area	0.16	0.28	0.44	0.66	0.91	1.2	1.5	1.9	2.8	4.3	6.2
A_N neck area	0.20	0.35	0.55	0.79	1.1	1.4	1.8	2.2	3.1	4.9	7.1

Pressure drop table

V_K outlet velocity	500	600	700	800	900	1000	1200	1400	1600	1800	2000
P_T w/#4 damper	0.02	0.03	0.04	0.05	0.07	0.09	0.12	0.17	0.22	0.28	0.34
P_S w/#4 damper	0.01	0.01	0.02	0.02	0.03	0.04	0.06	0.08	0.11	0.14	0.17
P_T w/o #4 damper	0.02	0.02	0.03	0.04	0.05	0.06	0.09	0.12	0.16	0.20	0.25
P_S w/o #4 damper	<0.01	<0.01	0.01	0.01	0.02	0.02	0.03	0.04	0.05	0.07	0.08

Pressure accuracy is ±0.01 in or 10%, whichever is greater.
When diffusers are used on exposed duct, multiply the throw (T) by 0.07.

Symbols: V_T – Terminal velocity (ft/min) T – Throw (ft) P_T – Total pressure (inWG)
V_R – Room velocity (ft/min) A_K – Outlet area (ft²) P_S – Static pressure (inWG)
V_K – Outlet velocity (ft/min) A_N – Neck area (ft²)

FIGURE 20.11
Example of air diffuser performance data table for a noise criteria level of 35. Symbols are defined in the figure. (Image courtesy of Allied Thermal Corporation, New York.)

The dimensional constant K varies approximately linearly between 0.012 and 0.0075 between terminal velocities of 100 and 200 ft/min, respectively.

Example 20.2: Selection of a Diffuser

Select the proper round ceiling diffuser with the following requirements: supply air volume of 1500 ft³/min with a throw of 10 ft. The noise should not exceed NC of 35. Determine outlet velocity and the corresponding total and static pressure drop assuming a diffuser with damper control is selected.

Given: $\dot{V} = 1500$ ft³/min and $T = 10$ ft.

Figure: See Figure 20.11.

Find: D_S, p_T, p_S, V_K.

Assumptions: Room velocity $V_R = 50$ ft/min.

Solution

The problem is solved graphically using Figure 20.11. The horizontal line $V_R = 50$ ft/min and the line $\dot{V} = 1500\ \text{ft}^3/\text{min}$ intersect at a point that corresponds to about $D_S = 18$ in. The outlet velocity is read from the figure as $V_K = 1000$ ft/min. From the upper portion of the table, the corresponding outlet area $A_K = 1.5\ \text{ft}^2$ and neck area $A_N = 1.8\ \text{m}^2$. The lower portion of the table corresponding to this value of V_K yields $p_T = 0.09$ in and $p_S = 0.04$ in.

Comments

We can compare the A_K value determined previously with that predicted by Equation 20.2. If we assume $K = 0.01$, $T = 10$ ft, and $\dot{V} = 1500\ \text{ft}^3/\text{min}$, we get an outlet area $A_{eff} = 1.5\ \text{ft}^2$, which is consistent with the value of A_K determined in Example 20.2.

20.4.4 EDT, ADPI, and AER

Irrespective of how the supply air is distributed to the room, there is bound to be some amount of thermal stratification as well as stagnation pockets. This would affect occupant comfort as shown in Figure 20.12, which also depends on whether it is at ankle region or the neck region of the occupant. For an air velocity of 50 ft/min (0.25 m/s), a 3°F (1.7°C) reduction at the ankle region would cause 10% of the occupants to express discomfort, while the same amount at the neck region causes 40% of the occupants to be dissatisfied. Higher dry-bulb air temperature (T_{db}) and higher air velocity (v) can offset each other's effect to a small degree as discussed in Chapter 3, which can also be gleaned from Figure 20.12. At any location x in the room, an empirical correlation proposed by Ryberg and Norback (1949) can be used to determine an effective draft temperature (EDT):

$$\text{EDT} = (T_{db,x} - \bar{T}_{db}) - M(\text{v}_x - \bar{\text{v}}) \tag{20.3}$$

where the overbars represent average room values. The empirical constant $M = 7.0°\text{C}\cdot\text{s/m}$ ($0.07°\text{F}\cdot\text{min/ft}$) when dry-bulb temperatures are in °C (°F) and velocities in m/s (ft/min).

Note that the previous equation suggests that a 1°C change in temperature is equivalent to 1/7 = 0.142 m/s (or 1°F to a 15 ft/min) change in velocity. It has been found empirically that a majority of people involved in sedentary activities are comfortable when the value of EDT lies between −1.7°C and 1.1°C (−3°F and 2°F) when the air velocity is less than 0.36 m/s (70 ft/min). These conditions serve as the basis for the *air distribution performance index* (ADPI), which is defined as the percentage of locations in the occupied zone of a room, which meets the previous criteria. The ADPI is primarily used to evaluate overall room comfort under cooling mode, though it can also be applied for heating. The highest value it can assume is 100 (since it is a percentage measure) and is not affected by relative humidity or radiant temperature of the room. Heavy heating or cooling loads tend to lower the ADPI value.

ADPI values are used to select diffuser selection and placement in a rational manner. These values are

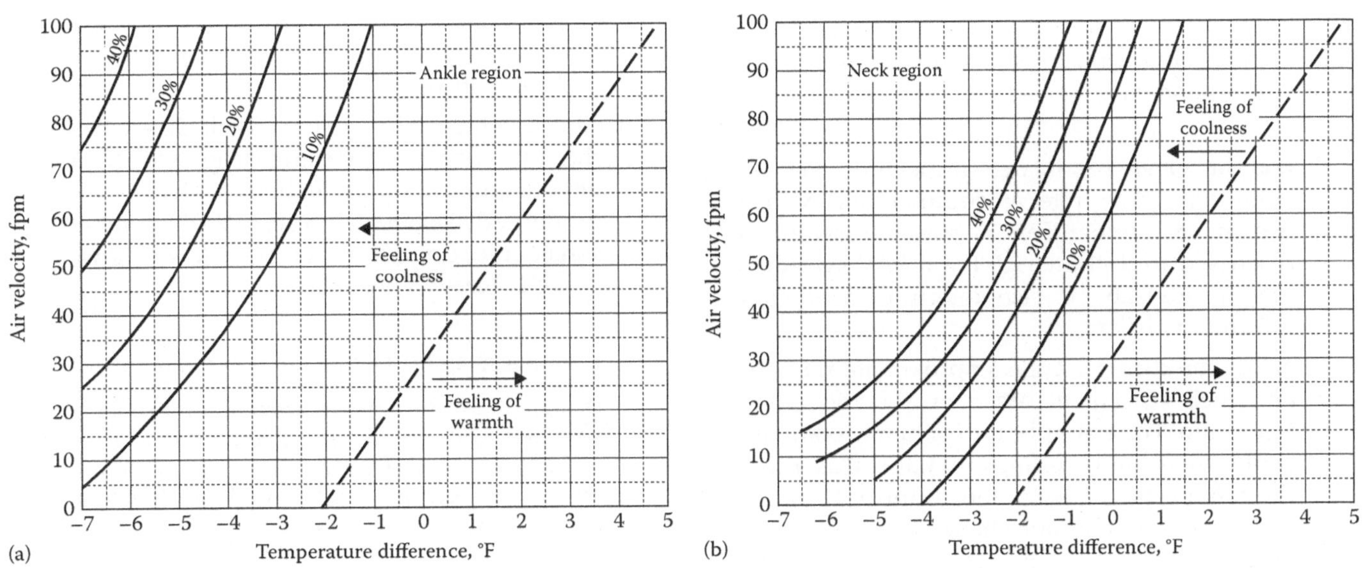

FIGURE 20.12
Effect of room air velocity on occupant comfort for (a) ankle and (b) neck regions. (From *ASHRAE Handbook of Fundamentals*, ASHRAE, Atlanta, GA, 2005, www.ashrae.org.)

experimentally determined and quoted by diffuser manufacturers, but these ADPI values are strictly valid for empty rooms with cold air supply, and so some amount of judgment is finally needed during design specification.

Another index relevant to how the diffuser interacts with the room is the effective *air exchange rate* (AER). AER is similar to air changes per hour (ACH), which is a measure of the number of times a volume equivalent to that of the room is supplied by the room distribution system per hour. It was mentioned earlier that ACH values of 5–15 were common in all-air systems. However, ACH does not usually include the recirculated amount (in, say, a fan-powered VAV system) or the induced room air in the space due to the incoming jet. The induction ratios in typical diffusers (high sidewall grill, ceiling round, ceiling slots) are typically in the range of 4:1 to 10:1. Even if we assume the lower value of 4 for a VAV system supplying 1 cfm/ft^2 in a room with a 9 ft ceiling height, the effective AER = $[(4 \times 1)/9] \times 60 = 27$, quite a high exchange rate. The concept of effective AER is especially pertinent in hybrid distribution systems where only the ventilation air volume is being supplied by the distribution system (discussed in Section 20.7).

20.4.5 Air Distribution Systems Design Procedure

The design process does not lead to a unique solution, and so numerous good design solutions are possible. These feasible solutions must all meet certain performance criteria, from which personal preferences of the designer would dictate the final selection. The following section outlines a general procedure for designing air distribution systems based on manufacturer tables. These tables are meant to illustrate the design process and are by no means comprehensive. Specialized publications or software are used by practicing professionals.

1. For the given room, determine the room air volumetric flow rate at summer and winter design conditions. This is straightforward as discussed in Section 19.2. For a VAV system, it may be advisable to also determine acceptable minimum airflow.
2. Select the diffuser type, approximate number, and their placement based on the room configuration. For example, for a square room, one could use a single ceiling circular diffuser or if the room is large, use of four uniformly placed diffusers is an obvious choice. For a rectangular room of aspect ratio close to 2, one would assume two locations, and so on. Equation 20.2 can be used for a first estimate of the diffuser throw from which an initial estimate of the number of diffusers can be made. The airflow rate to be handled by each diffuser would be the total room supply airflow rate divided by the number of diffusers. Table 20.2 is an abbreviated table providing initial guidance on selection of diffuser type.
3. Determine the characteristic length L depending on type of diffuser and their characteristics from Table 20.2.
4. Determine the value of diffuser throw $T_{0.25}$ (or T_{50}*) that maximizes ADPI for each diffuser using manufacturer catalog tables such as Table 20.3. Often the optimum value cannot be met, and so the last two columns in the table give the acceptable range of values for the throw ratio that will yield an ADPI value greater than the value listed in the last but one column.
5. Select the appropriate diffuser for each location from the manufacturers' catalogs based on the flow rate and throw determined earlier. Table 20.4 is applicable for round diffusers, while similar tables for other diffuser types can also be found in the literature.
6. Check using tables such as Table 20.4 whether the corresponding pressure drops and NC values are satisfactory under both peak design conditions. Also evaluate these, if VAV operation is considered, under minimum airflow conditions as well. Iterate if necessary.

TABLE 20.2

Common Diffuser Types and Their Characteristic Lengths

Diffuser Type	Approximate Air Loading per Floor Space, L/(s·m^2) [cfm/ft^2]	Characteristic Length L
High sidewall grille	0.6–1.2 [3–6]	Distance to wall perpendicular to jet
Circular ceiling diffuser	0.9–5.0 [5–25]	Distance to closest wall or midway to nearest ceiling diffuser
Sill grille	0.8–2.0 [4–10]	Length of room in direction of jet

Example 20.3: Air Distribution Systems Design

Design an air distribution system for a room of size 25 ft × 50 ft with a sensible cooling loading of 22 Btu/(h·ft^2) of floor space and a 20°F difference between room and supply air temperatures. Assume that circular ceiling diffusers are to be used.

* $T_{0.25}$ (T_{50}) designate the throw for which the jet velocity reaches 0.25 m/s (50 ft /min).

TABLE 20.3

Typical Diffuser Selection Guidelines

Terminal Device	Room Load, W/m² (Btu/h·ft²)	$T_{0.25}/L$ (T_{50}/L) Max. ADPI[a]	Maximum ADPI	For ADPI Greater Than	Range of $T_{0.25}/L$
High side-wall grilles	250 (80)	1.8	68	—	—
	190 (60)	1.8	72	70	1.5–2.2
	125 (40)	1.6	78	70	1.2–2.3
	65 (20)	1.5	85	80	1.0–1.9
Circular ceiling grilles	250 (80)	0.8	76	70	0.7–1.3
	190 (60)	0.8	83	80	0.7–1.2
	125 (40)	0.8	88	80	0.5–1.5
	65 (20)	0.8	93	90	0.7–1.3
Sill grille, straight vanes	250 (80)	1.7	61	60	1.5–1.7
	190 (60)	1.7	72	70	1.4–1.7
	125 (40)	1.3	86	80	1.2–1.8
	65 (20)	0.9	95	90	0.8–1.3

[a] $T_{0.25}$, throw to 0.25 m/s velocity, T_{50}, throw to 50 ft/min velocity.

TABLE 20.4

Typical Circular Ceiling Diffuser Catalog Data

Size	Neck Velocity, ft/min	500	600	700	800	900	1000
8″	Total pressure, inWG	0.052	0.075	0.101	0.130	0.166	**0.205**
	Flow rate, cfm	175	210	245	280	315	**350**
	Throw, ft	4	5	6	7	8	**9**
	NC	15	21	26	31	34	**37**
10″	Total pressure, inWG	0.043	0.062	0.084	0.108	0.138	**0.170**
	Flow rate, cfm	270	330	380	435	490	**545**
	Throw, ft	5	6	7	8	9	**10**
	NC	11	17	21	26	30	**33**
12″	Total pressure, inWG	0.042	0.060	0.081	0.105	0.134	**0.166**
	Flow rate, cfm	390	470	550	630	705	**785**
	Throw, ft	6	7	8	10	11	**12**
	NC	11	17	22	26	30	**33**
14″	Total pressure, inWG	0.061	0.087	0.118	0.152	0.194	**0.240**
	Flow rate, cfm	530	635	745	850	955	**1060**
	Throw, ft	8	9	11	12	14	**15**
	NC	18	23	28	32	36	**40**

Given: Room dimensions and thermal loading.

Figure: See Tables 20.2 through 20.4.

Find: Diffuser size and neck velocity.

Solution

We follow the design procedure described earlier:

1. The supply airflow rate per unit floor area is determined from a simple heat balance equation

$$\dot{V} = 22\ \text{Btu}/(\text{h}\cdot\text{ft}^2)/[0.075\,\text{lb}_\text{m}/\text{ft}^3 \times 0.24\ \text{Btu}/(\text{lb}_\text{m}\cdot{}^\circ\text{F}) \times 20^\circ\text{F} \times 60\ \text{min/h}] = 1.02\ \text{cfm/ft}^2$$

2. Since the room length is twice its width, it is logical to assume two rows of diffusers located in the center of each of the two 25 ft × 25 ft squares. The supply diffusers will be on the ceiling with the return grille somewhere close to the floor on the walls so as to create good mixing.
3. From Table 20.2, the characteristic length $L = 25/2 = 12.5$ ft.
4. From the third column of Table 20.3, the optimum value for T_{50}/L for circular diffusers is 0.8, irrespective of the room loading.
5. The throw to 50 ft/min for maximum ADPI should be = 12.5 × 0.8 = 10.0 ft. For the room whose loading is 20 Btu/(h·ft²), the maximum ADPI from Table 20.3 is about 93.

6. The airflow rate through each of the two diffuser = (25 ft × 25 ft) × 1.02 cfm/ft² = 636.55 cfm. Since a throw of 10 ft is desired, we find from Table 20.4 that we could use either a 12 in. size with a neck velocity 800 ft/min and NC = 26, or a 14 in. size with a neck velocity of 600 ft/min with NC = 23. The NC values of both these selections are acceptable for a quiet office setting. The pressure drops for the two sizes are 0.105 and 0.087 inWG, and the designer could opt to go with the one with the smaller pressure drop even though it is likely to cost more.

20.5 Other Types of Room Air Distribution Methods

The fully mixed air delivery system is an excellent system to meet the thermal needs of the space. However, it is not the best method for removing airborne pollutants and odors and gases from furnishings and even occupants in an energy-efficient manner. Mixing in the room is essentially caused by the room inlet velocity of the air jet generated by the static pressure increase in the fan. Two air delivery systems that use buoyancy to some extent to cause room air circulation are discussed. They are generally superior in terms of contaminant removal and occupant comfort. Since they need to move only the amount of air needed to meet the minimum ventilation codes of the space (typically 2.5–3 ACH as compared to 5–15 ACH for traditional fully mixed HVAC distribution systems) and further can operate with higher supply air temperatures during cooling mode, they tend to be more energy efficient overall while being much quieter during operation.

20.5.1 UFAD

UFAD systems are becoming increasingly common in America and Europe, especially for cooling the space. Very simply, the conditioned air is supplied vertically into the occupied space from an open airway created between a structural concrete slab and the underside of a raised floor that is slightly pressurized. Thus, there are no hard connections to the diffusers as in ducted overhead distribution systems.* Depending on the throw of the floor diffusers from the underfloor plenum, the room could be made to behave either as a fully mixed space or as a partially stratified space (see Figure 20.6). Moreover, the occupants are able to modify the air supply from the diffuser closest to them so as to tailor the local space to their needs. UFAD systems are variants of the same basic type of equipment used by traditional overhead fully mixed distribution systems, but with somewhat different considerations that are discussed in the following text.

Figure 20.13 shows a typical office room with raised floor creating an airway for the cool air to flow. Typical underfloor plenum heights are 30–45 cm (12–18 in.) with floor tiles mounted on metal pedestals. The plenums are maintained at a pressure of about 1.2–2.5 mmWG (0.05–0.1 inWG) above that of the space, and so the air leakage between the seams of the tiles and the room, as well as leakage between the tiles and the perimeter joint of the room, must be minimized by proper sealing and caulking during construction. Improper interior wall sealing has been found to cause a lot of the supply air to bypass the space altogether, and this has been identified as a major drawback in many actual UFAD installations. The air is diffused into the room so that comfort conditions are maintained in the occupied zone, while the unoccupied zone can be allowed to assume higher temperatures, thereby resulting in some HVAC energy saving. Some typical air temperature values are shown in Figure 20.13. The air is supplied to the space at 60°F–65°F (15.5°C–18.3°C) that is higher than the 55°F (12.8°C) or so supply air temperature typical in fully mixed overhead distribution systems. The return grilles are located at ceiling level or at least above the occupied zone.

Four of the common types of diffusers used in UFAD systems are shown in Figure 20.14:

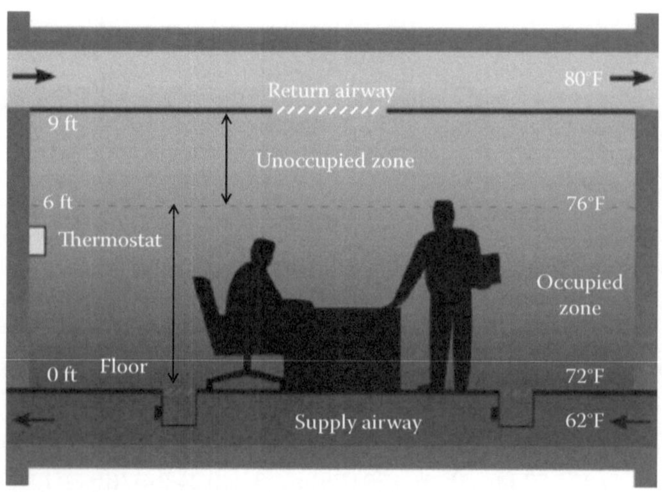

FIGURE 20.13
A typical underfloor air distribution space with representative air temperatures and a phantom line separating the occupied and unoccupied zones. (From Filler, M., *ASHRAE J.*, 46(10), 39, October 2004.)

* There are some specialized circumstances where hard connections (or ducted plenum) designs are adopted.

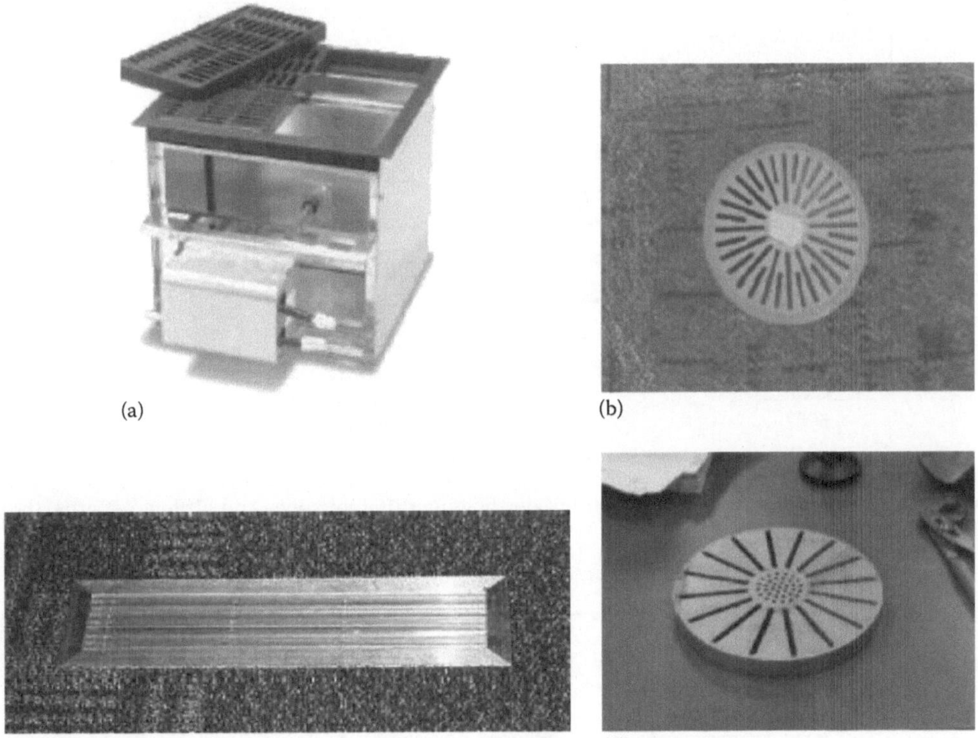

(a) (b) (c) (d)

FIGURE 20.14
Four different types of underfloor air distribution diffusers. (a) Variable area, (b) swirl, (c) linear bar grille, and (d) swirl, horizontal discharge (HD). (Image courtesy of Price Industries, Suwanee, GA.)

1. Variable area square directional diffuser that automatically vary the diffuser opening ratio
2. Perimeter linear bar grills
3. Interior swirl diffusers with vertical throw
4. Horizontal discharge swirl diffusers

Figure 20.15 illustrates airflow in a space supplied by a vertical swirl diffuser. Often, the UFAD design involves one diffuser per occupant whose flow can be adjusted by the occupant as needed. The stratification is affected by the air velocity. If airflow is too great, then there is little stratification, and much of the energy-saving benefits of UFAD are lost. If airflow is too low, the resulting stratification will create a greater temperature difference between the ankle and the head that may be unacceptable if it exceeds 4°F–5°F (2.2°C–2.7°C). Thus, the design intent is to maintain this temperature differential by suitably adjusting the supply air velocity.

Typically under cooling mode, supply air velocities are less than 150 fpm (0.7 m/s), while providing a swirl (as shown in Figure 20.15) that enhances mixing due to the higher local velocity created. However, a clear space, also called the *adjacent zone*, of about 2 ft (0.6 m) has to be provided around each diffuser since the high-velocity discharge will cause discomfort due to draft to any occupant in that zone. The phantom line shown as X_{50} is the *throw height* at which the bulk air velocity reaches the occupant comfort velocity of 50 fpm (0.25 m/s). The other fictitious line shown as SH (*stratification height*) is the boundary between the high-velocity fully mixed zone and the low-velocity thermally stratified zone where the air movement is primarily caused by buoyancy. This temperature profile is illustrated in Figure 20.6 with clear distinction between the high-velocity mixed air space and a low-velocity stratified space. Figures 20.13 and 20.15 show the return airway at the upper end of the space that is fully mixed consisting of the warm contaminated air plumes rising from the stratified spaces. As throw and mixing are reduced, UFAD systems tend to approach the operation of displacement ventilation systems addressed in the Section 20.5.3. As the vertical throw is increased, the SH and the throw height are eliminated, and the air distribution resembles that of a forced air system with uniform mixing. Maximum vertical throws of diffusers should be less than 2–3 ft (0.6–0.9 m) from the ceiling so as not to eliminate the stratified upper zone.

The effect of discharge velocity on the temperature stratification for swirl diffusers is illustrated in Figure 20.16 (Webster et al., 2002). The profiles shown were measured

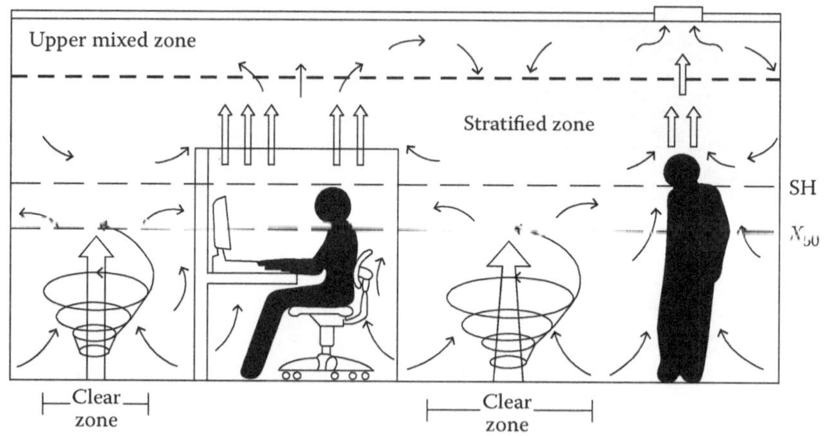

FIGURE 20.15
Sketch of a partially mixed room with under floor air distribution showing different zones and airflow patterns. (From Webster, T. et al., *ASHRAE J.*, 44(5), 28, May 2002.)

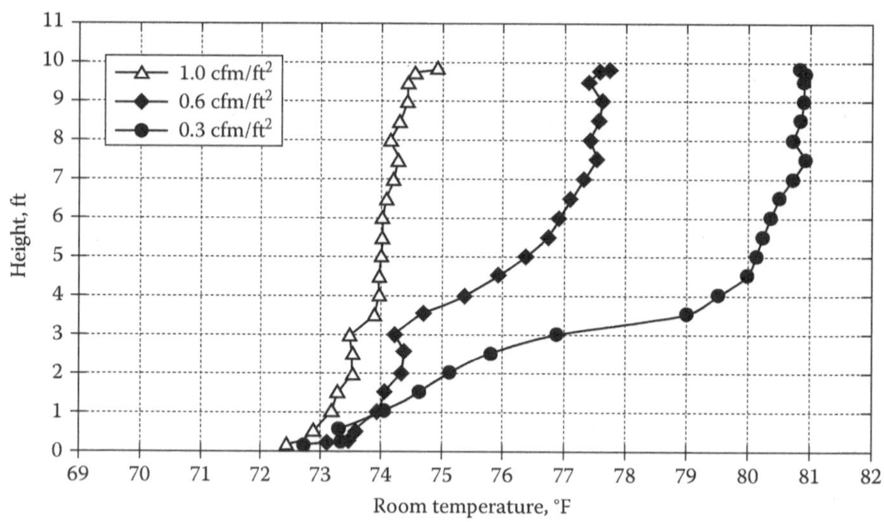

FIGURE 20.16
Effect of room airflow on thermal stratification using swirl diffusers. (From Webster, T. et al., *ASHRAE J.*, 44(5), 28, May 2002.)

in a room with a total heat load of 5.2 W/ft² (56 W/m²) and a supply temperature of 64°F (18°C) and room airflows of 1.0, 0.6, and 0.3 cfm/ft² (5.2, 3.0, and 1.5 L/s·m²). As expected, the thermal gradient is reduced in proportion to air velocity. For the highest flow rate of 1 cfm/ft², the temperature difference between the head and the foot is about 1.3°F (0.72°C), representing an overventilated space. For the lowest airflow rate of 0.3 cfm/ft², the difference is close to 7°F (3.9°C), exceeding the limit of a 5°F (2.8°C) differential advocated for occupant comfort as per ASHRAE 55 (2013). The intermediate flow rate of 0.6 cfm/ft² (3.0 L/s·m²) results in less than a 4°F (2.2°C) differential that would be acceptable.

UFAD systems tend to be more appropriate for cooling interior spaces than perimeter ones. Controlling perimeter space loads is more challenging because heating and cooling loads generally vary over the year. Often hybrid systems (discussed in Section 20.7) involving two different types of distribution systems are used such as UFAD and radiant cooling panels or chilled beams.

UFAD systems are best suited for general open office areas, conference rooms, exhibit spaces, clean rooms, hospitals, churches, libraries, computer rooms, and public lobbies. Areas such as laboratories, child care centers, and storage rooms are well suited for UFAD systems. The construction costs of UFAD systems are higher, while the raised floor itself is an additional cost. UFAD systems require higher airflow rates than traditional all-air systems since the supply air temperatures are higher; that is why these systems are often used in hybrid configuration (Section 20.7). The benefits other than reduced energy costs are due to higher supply temperature, more hours of economizer operation (Section 19.6.2), and lower fan power (since duct pressure drops are lower). Further, these systems offer the occupants better control of the space, and also these systems can be varied easily to accommodate changes in space layout due to remodeling or churn. Finally, whether a UFAD system is more cost-effective than a traditional overhead fully mixed system depends on the specific application and the preference of the owner. The interested reader can refer to numerous technical articles on UFAD such as thermal stratification results from test chambers (Webster et al., 2002), lessons learnt from actual installations (Daly, 2002; Montgomery, 2009), and best practice considerations during design and commissioning (Filler, 2004; Montanya et al., 2009). ASHRAE has also published a comprehensive design guide (ASHRAE UFAD, 2013).

20.5.2 UFAD Design Calculation Procedure

The thermal energy requirements for both UFAD and TDV systems are different from those of fully mixed systems. The net heat zone loads need not be met since only the occupied zone needs to be maintained at the desired occupant conditions. The temperature above the occupied space can be higher, and also some of the heat loads attributed to the nonoccupied zone need not be met since they are removed by the exhausting airstream. For example, in UFAD systems, convective loads nearer the ceiling (especially lighting) reduce total required space air supply and fan power. Further, the raised floor of UFAD results in heat gains not present in fully mixed systems (Figure 20.17). Current research results indicate that on average 20%–40% of the total room cooling load is transferred to the supply plenum, and so one needs to distinguish between the three spaces in an UFAD (the supply plenum, the zone, and the return plenum). The dynamic load calculation procedures described in Chapters 8 and 9 are no longer directly applicable since such room-specific vertical spatial disaggregation of loads is not well captured by traditional load calculation procedures. CFD calculation methods can be adopted, but they are not widely used as yet for routine design calculations.

Extensive simulations and experimental research led to a simplified UFAD design tool meant for both peak and energy cooling load predictions and is described in this section (Bauman et al., 2007, 2010; Schiavon et al., 2010). Here, the space is divided into three spaces: the supply plenum, the occupied zone, and the return plenum space. The approach was to use a detailed simulation program to generate peak design cooling loads in each of these three spaces ($CL_{UFAD\text{-}SP}$, $CL_{UFAD\text{-}Z}$, and $CL_{UFAD\text{-}RP}$) in a wide variety of climates and building types subject to UFAD. These peak loads were then used to develop correction factors that could be applied to traditional dynamic peak load calculations assuming fully mixed spaces. As depicted in Figure 20.18, the factor UCLR (underfloor cooling load ratio) is

$$UCLR = CL_{UFAD}/CL_{OH} \tag{20.4}$$

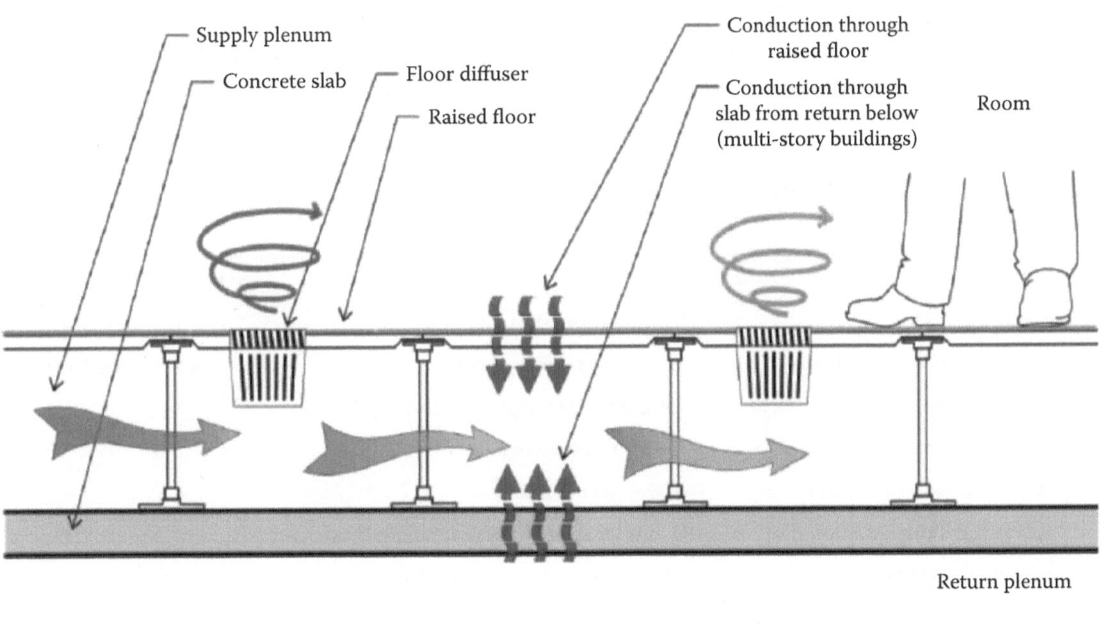

FIGURE 20.17
A typical underfloor air distribution space showing the supply plenum and the various heat gains from the space as well as the return air plenum from the floor underneath. (Image courtesy of Price Industries, Suwanee, GA.)

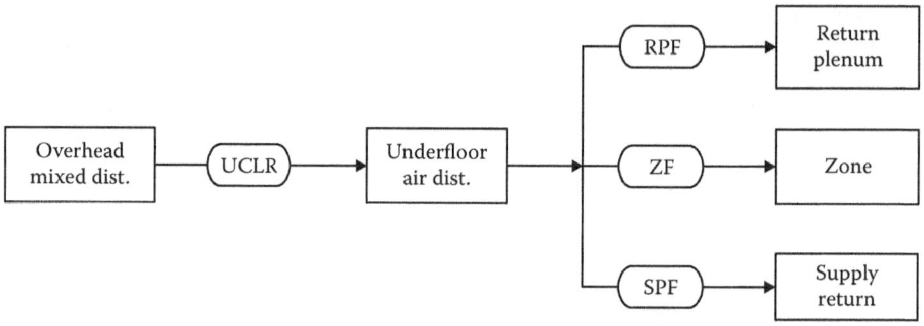

FIGURE 20.18
Flow diagram of underfloor air distribution design tool showing corrective terms to be applied to an overhead fully mixed air distribution system.

where CL_{UFAD} and CL_{OH} are the peak cooling loads for the UFAD system and the overhead distribution systems, respectively.

Subsequently, the UCLR is then further separated into three additional factors (which due to the normalization constraint need to add to unity) defined as

$$\text{The supply plenum fraction (SPF)} = CL_{UFAD\text{-}SP}/CL_{UFAD}$$
$$\text{The zone fraction}(ZF) = CL_{UFAD\text{-}Z}/CL_{UFAD}$$
$$\text{The return plenum fraction (RPF)} = CL_{UFAD\text{-}RP}/CL_{UFAD} \tag{20.5}$$

The results were analyzed, and it was found that (1) floor level and zone location were significant factors; (2) others such as structure type, supply air temperature, and internal heat gain have a small but not insignificant influence; and (3) factors such as window-to-wall ratio, plenum configuration, presence of carpet, and climate do not have a significant effect. Finally, empirical polynomial expressions were developed as

$$\begin{aligned} UCLR &= a_0 + a_1 + a_2 \\ (SPF)^{1/2} &= b_0 + b_1 + b_2 + b_3 \\ RPF &= c_1 \\ ZF &= 1 - SPF - RPF \end{aligned} \tag{20.6}$$

The numerical values for the empirical coefficients are assembled in Table 20.5.

Table 20.6 assembles the mean and the standard deviation of these factors found from the extensive simulations performed by Schiavon et al. (2010). Except for RPF, the standard deviation is much smaller than the mean, thereby suggesting that the mean values are meaningful and can be used with some confidence. Note that UCLR has a median value of 1.19 implying that UFAD systems have a 19% higher peak load than overhead distribution. However, the zone fraction ZF is only 63% due to the stratification effect as discussed earlier. Note that the peak cooling loads of UFAD are generally higher than the fully mixed ones primarily because of low thermal

TABLE 20.5
Regression Coefficients for the Underfloor Air Distribution Factors in Equation 20.6

UCLR	SPF	RPF
$a_0 = 0.9528$	$b_0 = 0.6179$	$c_1 = 0.01$ for ground floor
$a_1 = 0.0$ for floor level	$b_1 = 0.0$ for interior zone	0.01 for middle floor
$a_2 = 0.1572$ for in-between floors	−0.2095 for perimeter zone	0.30 for top floor
0.2379 for top floor	$b_2 = 0$ for ground level	
$a_3 = 0$ for north orientation	0.1242 for in-between floors	
0.1739 for east orientation	−0.0896 for top floor	
0.0999 for south orientation	$b_3 = 0$ if interior zone or on first floor	
0.1349 for west orientation	0.0396 if perimeter zone and in-between floor	
0.0802 for interior zone	0.1642 if perimeter zone and on top floor	

Source: Schiavon, S. et al., Development of a simplified cooling load design tool for underfloor air distribution systems, Final Report to CEC PIER Program, July 2010, http://escholarship.org/uc/item/70f4n03z.

TABLE 20.6

Pertinent Statistics of the Numerous Simulations Performed for Underfloor Air Distribution Systems

Parameter	Mean	Std. Dev.	Median	Minimum	Maximum
UCLR	1.18	0.13	1.19	0.82	1.55
SPF	0.28	0.11	0.27	0.07	0.72
ZF	0.62	0.16	0.63	0.25	0.93
RPF	0.10	0.15	0.02	0.00	0.51

Source: Schiavon, S. et al., Development of a simplified cooling load design tool for underfloor air distribution systems, Final Report to CEC PIER Program, July 2010, http://escholarship.org/uc/item/70f4n03z.

storage capacity since the raised floor panels in UFAD systems have lighter mass. Also, heat losses to the cooler underfloor plenum increase the thermal loads.

These equations should be used with caution since they are generalized expressions and the methodology has not been fully vetted as yet. The primary reason for describing this procedure is its elegant simplicity in how the loads assignment is done, while maintaining some of the rigor needed in analyzing UFAD systems. The earlier simplified design methods were largely meant for meeting cooling loads in interior zones, while this approach is more general. The development of simplified design methods for UFAD systems is currently an area of active research interest.

Example 20.4: Analysis of an UFAD Systems

Analyze the thermal benefits of using an UFAD as compared to the traditional fully mixed air distribution system for the following office room: windowless room with a west facing wall situated on the first floor and surrounded on all sides by rooms at the same temperature (except on the west side of course).

Given: From Table 20.5, we find

$$a_0 = 0.9528, \quad a_1 = 0.0, \quad a_2 = 0.1572,$$

$$b_0 = 0.6179, \quad b_1 = -0.295, \quad b_2 = 0, \quad b_3 = 0$$

$$\text{RPF} = 0.01$$

Solution

From Equation 20.6, we find

$$\text{UCLR} = a_0 + a_1 + a_2 = 1.11$$

$$\text{SPF} = (b_0 + b_1 + b_2 + b_3)^2 = 0.1043$$

$$\text{RPF} = 0.01$$

$$\text{ZF} = 1 - \text{SPF} - \text{RPF} = 0.8857$$

Thus, the UFAD has a peak that is 11% higher (since UCLR = 1.11) than that of a traditional fully mixed overhead distribution system. The zone fraction, on the other hand, is about 12% lower since ZF = 0.886. These results are consistent with the results of Table 20.6 and the discussion previously.

20.5.3 Displacement Ventilation Distribution

Originally developed for industrial space cooling applications, TDV systems are being widely used in commercial buildings (especially in Europe) as well with high-ceiling rooms (greater than about 2.3 m or 10 ft). They are good candidates for classroom, auditoriums, conference rooms, restaurants, gymnasiums, lobbies, airport terminals, and industrial spaces. Here, the mixing between the room air and the supply air is minimized as much as possible, with buoyancy being the primary driver for room airflow. Hence, TDV can only be used for cooling applications (during summer or for year-round cooling as is often needed in interior spaces). Further, TDV is most suitable when pollution sources and heat sources are in close proximity and the floor is relatively free of obstructions. If the space requires heating, a second HVAC system has to be used.

Ventilation air is supplied close to the floor at very low velocities of 0.15–0.25 m/s (30–50 fpm or less) at relatively high supply temperatures of 18°C–20°C (65°F–68°F) for space temperatures at 23°C–24°C. The cool air tends to hug the floor and forms a pool of cold air because of its negative buoyancy. At specific locations in the room where the cool air comes in contact with heat sources, it replaces the air around the heat source (therefore, it has a high air exchange efficiency*) and rises toward the ceiling. The rising air produces a vertical airflow pattern near each occupancy, called a thermal plume (Figure 20.19). Note that many of the air pollutants generated within the space are associated with thermal loads as well (such as body odor from occupants, electric equipment outgassing). Warm contaminated air accumulates in a fully mixed space close to the ceiling (this upper zone is outside the occupied zone) and is exhausted near the ceiling. Return air temperatures are around 80°F–85°F (26.5°C–29.5°C). Thus, the TDV system is essentially a one pass (or plug flow) system with buoyancy-assisted mechanical ventilation where the overall floor-to-ceiling airflow pattern is created by natural buoyancy, while the removal of the warm contaminated air is by mechanical exhaust. The low velocity is achieved by passing the supply air through large

* Air exchange efficiency is an indicator of indoor air quality. It is defined as the ratio of the shortest possible time needed to replace the air in the room and the average time of air exchange (which is the reciprocal of ACH). For a fully missed air distribution system, the air exchange efficiency is 0.5.

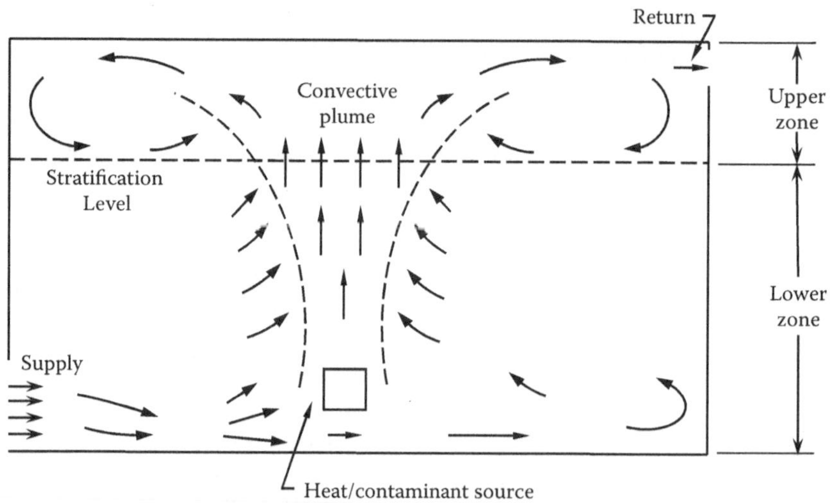

FIGURE 20.19
Schematic of a displacement ventilation system showing the two zones and the creation of the thermal plume around the heat source.

FIGURE 20.20
Typical displacement ventilation diffusers. (Image courtesy of Price Industries, Suwanee, GA.)

opening diffusers usually configured as large-area free-standing floor pedestals (with 360° or quarter round patterns) or low side-wall flat panels that can be integrated into the wall space (Figure 20.20).

The supply air accelerates, while it falls toward the floor due to gravity (this zone is called the "near zone"), and then the velocity decreases as it spreads across the room (see Figure 20.21). The *adjacent zone* is defined as the area near the ankle level where the local velocity exceeds 50 fpm (0.25 m/s) and causes occupant discomfort. The highest velocity is near the floor and is much lower above a height of 8 in. (0.2 m) from the floor. Seated occupants should not be placed in the near zone to avoid discomfort due to draft. In an ideal displacement ventilation system, all the occupants are exposed to clean cool ventilation air (Figure 20.22). Once contaminated and heated, the air rises upward in a plume. However, the air velocities of the plumes are very low, and so even small air currents set up by occupant movement inside the room or door opening/closing can cause the plume to be disrupted and cause undesirable mixing.

The vertical stratification should be maintained properly. The design room temperature is often defined with respect to a reference height of 1.1 m (3.6 ft) from the floor that is typically the neck-to-head height of a seated person. The temperature differential should be less than 3°C (5.4°F) when the room is maintained at 19°C–21°C (34°F–38°F) during winter or 21°C–22°C (38°F–40°F) during summer.

Similar to UFAD systems, displacement ventilation only meets sensible loads. Space latent loads (along with conditioning the outdoor air) must be met by the supply of airstream conditioned in a separate air handler unit. Displacement ventilation systems aim to minimize mixing of supply air with room air, instead maintaining conditions in the occupied zone as close as possible to that of the conditioned supply air, leading to an improved ventilation effectiveness (see Section 3.5.4). ASHRAE 62.1 (2013) allows a reduction in the required minimum ventilation rate of 1.2, resulting in a 16% reduction in the outdoor air intake. Note that lights and other heat-generating equipment are usually located above the occupied space, and so the supply airstream does not have to meet these loads. As a result, the energy needed for conditioning the supply airstream is much less than that required for the fully mixed system This is a large energy saving in terms of avoided energy to condition this outdoor air as well as having to meet only some

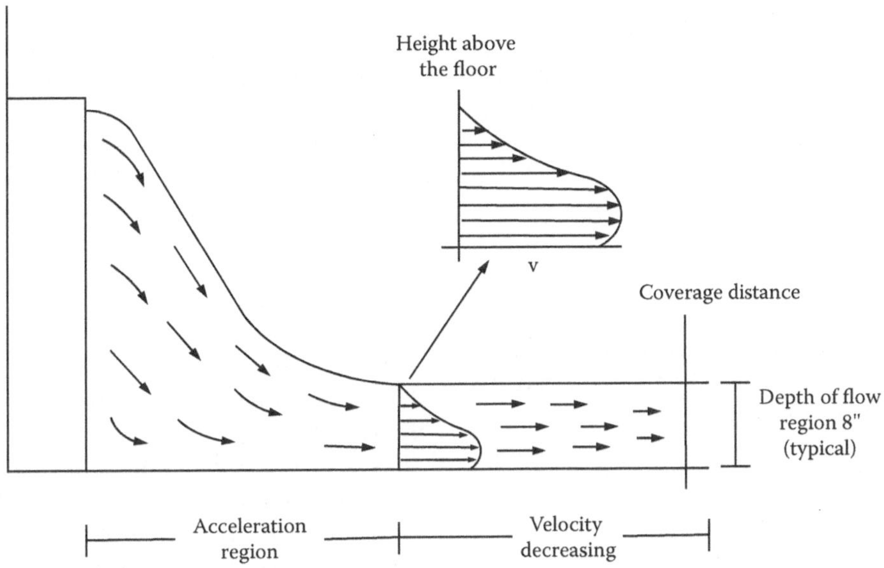

FIGURE 20.21
Idealized diffuser airflow pattern.

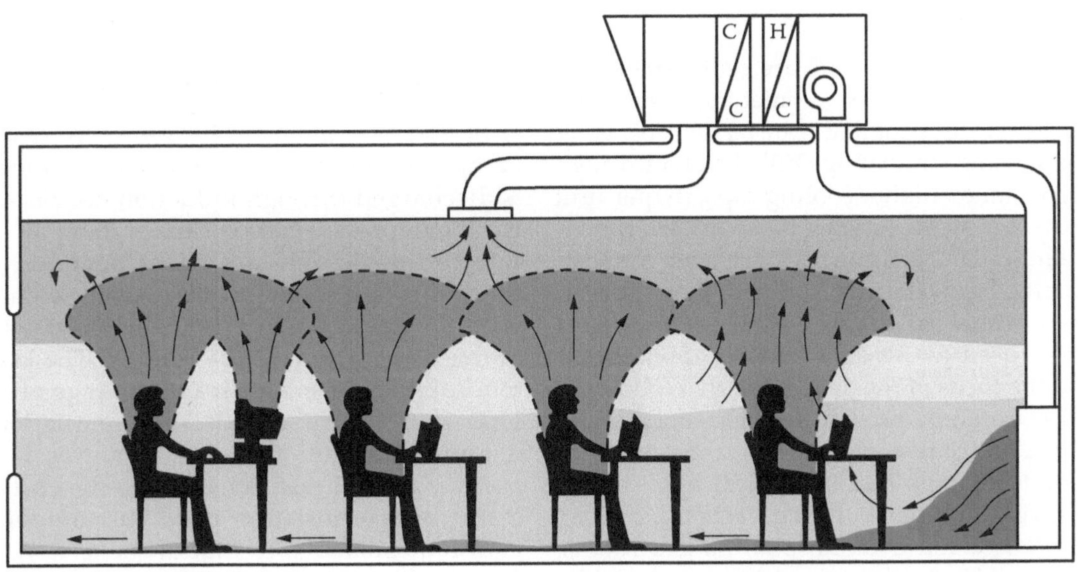

FIGURE 20.22
Sketch of plumes generated in a space with several seated occupants. (Image courtesy of Price Industries, Suwanee, GA.)

of the space heat loads. An important challenge to the designer for both systems is how to heat up the cold airstream leaving the cooling coil at 55°F (13°C) to 65°F (18°C) without wasting energy by reheat.

There are several advantages of displacement ventilation air distribution systems (Liddament, 2001):

- Smaller ventilation systems compared to conventional all-air systems
- Energy-saving benefits similar to UFAD (lower fan energy, higher supply temperature, more economizer hours over year for suitable locations)
- Improved air quality, and also such systems can be used to achieve better smoke control
- Low noise (good for classrooms)
- Can be used in conjunction with DOAS (see Section 20.7.2) where required minimum ventilation provided at all times

Some of the disadvantages are as follows:

- The floor space available for occupants is reduced since diffusers cannot be obstructed, and occupants cannot be seated near diffusers nor in the adjacent zone.
- Precise temperature and airflow control is needed to establish and maintain operating conditions.
- The air pollution in the upstream portion can become very concentrated and, in case they do fall back into the occupied zone, can cause severe occupant discomfort.
- Limited heating and cooling capacity (meant for low room loads, about 30 W/m^2), and so an additional heating/cooling system may be needed.

20.6 Chilled Beams

Active and passive chilled beams are ceiling-mounted room air circulation devices (sometimes they are mounted on side walls also) that transfer sensible cooling to the space using circulating chilled water through coils. They have much higher cooling capacity per unit area than radiant systems (greater by about 4–6 times). Strictly speaking, such beams can also be used for heating by circulating hot water, and so often, these devices are referred to simply as "beams." However, they tend to be primarily used for sensible cooling applications, which will be the focus of the discussion that follows.

These devices can only provide sensible cooling, and so dehumidification/latent cooling has to be supplied by some complementary means. Typical applications are interior spaces in office buildings, schools, hotels/dorms, laboratories, and even hospital patient rooms. Areas where humidity loads are variable such as lobby areas and egress spaces are unsuitable.

There are two types of beams (passive and active), and the differences between both are described briefly in this section. The interested reader can refer to specialized publications such as Price (2011) for more detailed discussion.

20.6.1 Passive Chilled Beams

Passive chilled beams (Figure 20.23a) are inverted finned-tube linear slot diffusers with cold water at elevated temperatures to avoid condensation (around 60°F or 15.5°C) flowing inside the tubes; they are often used in conjunction with underfloor and displacement ventilation systems (discussed in Section 20.5). The air between the fins of the coil is cooled, drops into the space below, and creates a buoyancy-driven flow. This buoyancy is maintained as long as chilled water is circulated through the coil. The downward plume from the passive beams hits the floor and spreads out with a filling-box regime forming a layer of cool air that expands upward from the floor. The beam plume interacts with the buoyancy of a displacement ventilation system, and the combined system is effectively a mixed system. Care should be exercised during design that the return air grilles should not be too close to the chilled beams since this can cause reverse flows through beams.

Passive chilled beams have no supply air, and so a separate air system is required for ventilation (outdoor air) and space dehumidification. There are two common designs: *exposed passive beams* where the beam is installed below the ceiling and *recessed passive beams* where they are part of the ceiling. Passive chilled beams are not suitable for use in rooms that are to be overpressurized (say, to avoid outdoor air infiltration).

20.6.2 Active Chilled Beams

Active beams have been in use for over two decades, being first introduced in Europe. As the primary air is discharged through induction nozzles inside a cavity, the depressurization causes room air to be drawn into the device via a grill or opening, which then flows through a low-pressure drop water coil, mixes with the primary supply air, and is discharged back to the room (Figure 20.24a and b). The HVAC air system supplies primary air in the range of 57°F–65°F or 14°C–18°C (ASHRAE, 2011) for ventilation and space dehumidification. The *air-side fraction* is the ratio of *primary* airflow rate delivered to the chilled beam (to meet room ventilation code and outdoor air latent load requirements) to that of the *total room air supply* from the beams to the indoor space. For office spaces with typical CAV systems, the air-side fraction is generally around 0.15 (0.15 cfm/ft^2 for ventilation air and 1 cfm/ft^2 for supply air from the beams to the room), while for schools, it is generally around 0.33 (0.5 cfm/ft^2 for ventilation air and 1.5 cfm/ft^2 for supply air). The ratio between primary air and total room supply air from the beams typically varies from 1:5 to 1:3. Chilled beams are best for spaces with low air-side fractions.

It is recommended that active chilled beams be designed to operate at minimum primary airflow rates. The beam output is controlled by either an on–off valve or a mixing valve that regulates the water temperature in the coil. This valve closes when the thermostat set

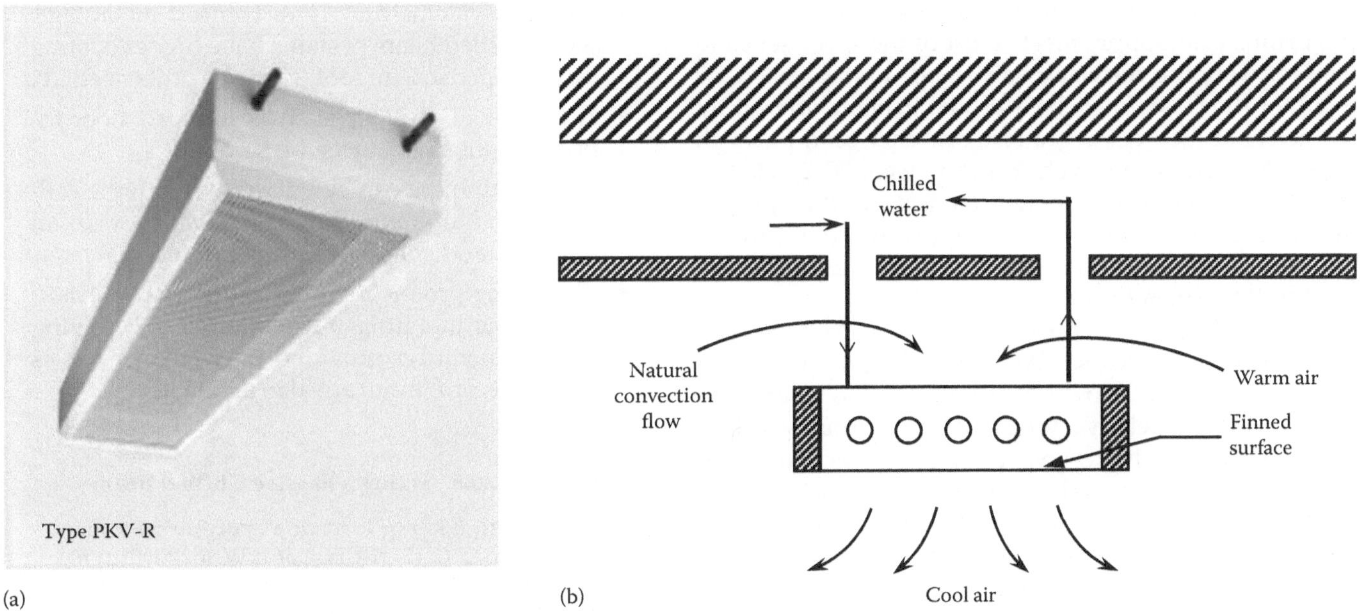

FIGURE 20.23
Ceiling-mounted passive chilled beam. (a) Picture of a unit. (Image courtesy of Price Industries, Suwanee, GA.) (b) Airflow patterns with interior components.

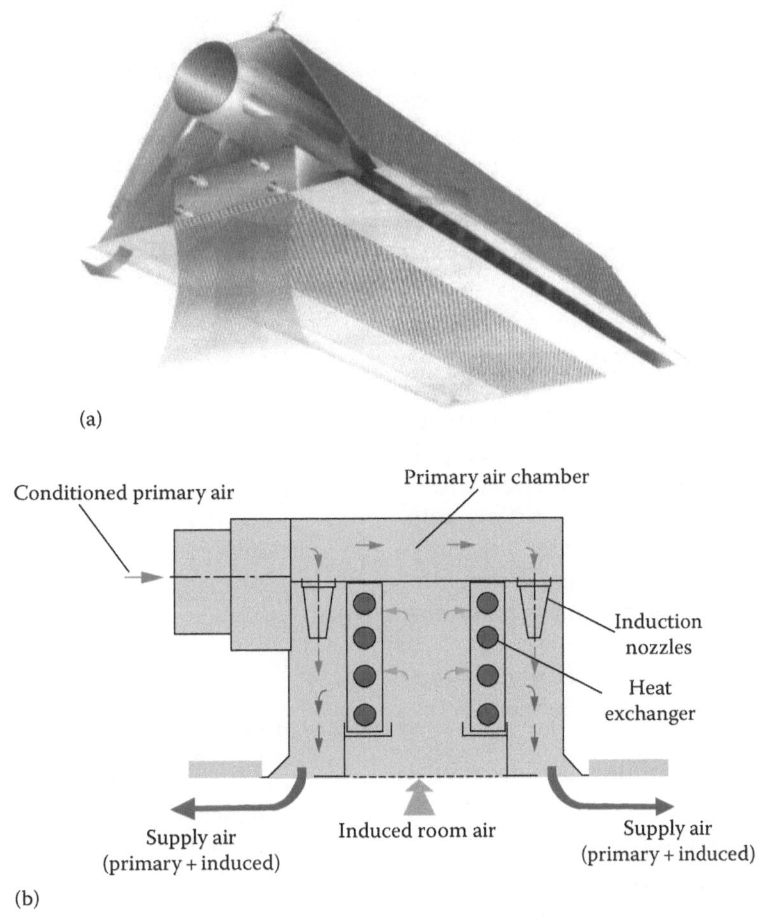

FIGURE 20.24
Ceiling-mounted active chilled beams. (a) Picture of a unit. (Image courtesy of Price Industries, Suwanee, GA.) (b) Airflow patterns with interior components.

point is reached. When the HVAC system is oversized, the primary air supply meets most of the space sensible loads as well, and the water valve tends to stay closed for much of the time. In such cases, the benefits offered by the active chilled beam are lost, and they tend to act as expensive diffusers (Livchak and Lowell, 2012). Hence, proper estimate of the cooling loads as well as proper design of both the primary air supply and the diffuser is essential to derive the full operational and indoor air quality (IAQ) benefits of such systems.

Current designs of chilled beams tend to have constant primary airflows. VAV beam designs are being introduced in the market that will be able to cater to spaces with widely varying cooling output. Chilled beams are relatively expensive, and so, fewer units with high capacities are usually specified. The risk of such a design is that the air supply can be compromised since overthrow can cause uncomfortable conditions at the midpoints between diffusers, and excessive spacing can also lead to poor air distribution.

Beams operate based primarily on convection with radiation being a secondary mode of heat exchange (unlike radiant panel devices). The general feeling among experienced designers is that HVAC systems using beams provide better thermal comfort with lower first and operating energy costs than traditional HVAC systems. However, there is still some controversy regarding this issue; for example, a recent study seems to suggest that traditional fan-powered VAV systems perform better when evaluated for a hospital in California (Stein and Taylor, 2013).

The discharge temperature from chilled beams can be varied by controlling the flow rate of cool water to the coil. The inlet pressures are typically lower than 0.2–1 inWG (50–250 Pa). The temperature increase through the coil is 3.5°F–7°F (2°C–4°C). Active chilled beams have cooling capacities up to 40 Btu/(h·ft^2) or 120 W/m^2. The heating and cooling capacities depends on several factors: difference between mean beam temperature and room temperature, primary airflow rate, water flow rate, air-side pressure drop, water-side pressure drop, control strategy, style of beam, beam discharge airflow patterns, space humidity ratio, and noise considerations.

There are some similarities between chilled beams and radiant cooling panels (discussed in Section 18.4):

- Both provide only sensible cooling at zone level via chilled water.
- Both cannot provide dehumidification.
- Both need controls to avoid condensation. This is achieved by keeping the chilled water temperature above the dew point temperature.
- Radiant systems may have comfort advantage, while chilled beam systems have higher cooling capacity per square foot (so fewer units needed).
- A benefit of chilled beam is reduced floor-to-floor height since ducts are smaller.
- Both equipment can be integrated with the ceiling tiles that along with low noise provide an aesthetic and comfortable environment.
- The sizing procedure is similar to that of traditional terminal units as in Example 18.1 relying on manufacturer-provided performance tables (software programs are also available).

Example 20.5: Sizing a Passive Chilled Beam

The design cooling load of a space to be maintained at 25°C (77°F) is 2.46 kW (8200 Btu/h). Determine the dimensions of a passive chilled beam (length, width, height, and slab clearance height) whose specifications are provided by the manufacturer and summarized in Figure 20.25 for a particular set of design and operating conditions. Assume the temperature difference between room air and chilled mean temperatures is 10°C (18°F) and free area of the bottom perforated cover plate of the chilled beam is 50%.

Given: $\dot{Q}_c = 2.46$ kW, $\Delta T_{cb} = 10$°C, $f_o = 0.5$,
$T_{db,space} = 25$°C.

Find: $B_{cb}, H_{cb}, Z_{cb}, L_{cb}, \dot{V}_w$.

Solution

The cooling capacity shown in Figure 20.25 is per unit length of the chilled beam. There are different combinations possible, and we will select one such combination. Let us assume B_{cb} = 460 mm, Z_{cb} = 230 mm, H_{cb} = 300 mm, and f_o = 0.5. From Figure 20.25, we find that such a chilled beam will provide 345 W/m. Thus, the length of the chilled beam needed is:

$$L_{cb} = \frac{2.46 \text{ kW}}{0.345 \text{ kW/m}} = 6.96 \text{ or } 7.0 \text{ m}.$$

Depending on the dimensions of the room and the placement of the occupants, we can select one or more rows of chilled beams, to be suspended from the ceiling and connected in parallel, whose total length is 7 m. Say, we select two parallel rows. We will assume that each of these chilled beam panels will be supplied with chilled water that will have a temperature drop of 2°C (a typical design choice). Then, the inlet and outlet water

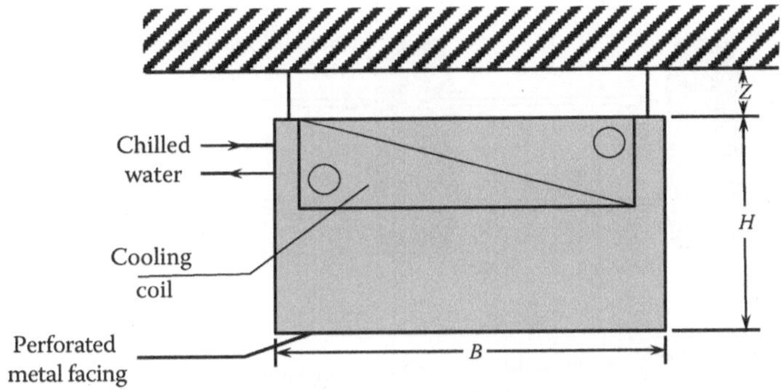

		B = 180 mm, Z = 90 mm				B = 320 mm, Z = 160 mm				B = 460 mm, Z = 230 mm				B = 600 mm, Z = 300mm			
f_o --->		20%	34%	50%	100%	20%	34%	50%	100%	20%	34%	50%	100%	20%	34%	50%	100%
H, mm	110	66	76	81	85	129	152	162	169	193	229	243	253	257	304	324	338
	200	80	94	100	105	160	188	200	209	240	283	300	314	319	376	399	416
	300	92	109	115	121	184	217	230	240	277	335	345	361	368	433	459	480

Cooling capacity per unit length (W/m) for a 10 K temperature difference.

FIGURE 20.25
Example of a specification sheet of a passive chilled beam manufacturer for (Z/B) ~ 0.5. Terminology: f_o free area of bottom perforated cover plate (%), B width (mm), H height (mm), and Z suspension height from lower edge of ceiling slab to upper edge of chilled beam (mm).

temperatures to the chilled beam are $T_{cb,in} = 14°C$ and $T_{cb,out} = 16°C$. The designer will evaluate whether this is a reasonable temperature above the dew point of the room so as to avoid condensation. Note that the lower the supply temperature to the chilled beam, the greater the cooling capacity; however, this would reduce chiller coefficient of performance (COP).

Finally, the required water flow rate through each of the two chilled beam units is provided by the sensible heat equation:

$$\dot{V}_w = \frac{\dot{Q}_c}{\rho_w \cdot c_{p,w} \cdot \Delta T}$$

$$= \frac{(2.46/2)\ \text{kW}}{998.2\ \text{kg/m}^3 \times 4.186\ \text{kJ/(kg}\cdot°\text{C)} \times 2°\text{C}}$$

$$= 0.000147\ \text{m}^3/\text{s} = 0.147\ \text{L/s} = 530\ \text{L/h}$$

Comments

The height dimension H_{cb} of the chilled beam unit has an important effect on the amount of room air drawn through the coil. The taller the height, the greater will be the induced airflow. The slab gap dimension Z_{cb} is a measure of the restriction placed on the air path, and the greater the gap, the higher the cooling capacity of the beam (the increase is asymptotic though). The free area at the bottom perforated cover plate (variable f_o) provides a means of controlling the downward air velocity and spreading out the airflow so as to avoid cold draft to occupants.

Nowadays, chilled beam manufacturers provide software programs rather than tables to help in designing and selecting chilled beams depending on the application requirements. These programs allow the designer to evaluate the effect of varying different parameters prior to making the final selection. More details of the sizing software capability can be found in specialized handbooks such as Price (2011).

20.7 Hybrid Secondary Systems

20.7.1 Basic Principle

As mentioned in Section 20.1, hybrid systems (sometime referred to as *parallel systems*) are those that decouple the space heating and cooling loads and the conditioning of the ventilation air by using separate subsystems: (1) to meet the heating and cooling loads of the space and (2) to condition the ventilation air. Numerous variants have been developed and tested in the last two decades, while new system configurations are still being researched into. In this section, we will only describe one widely acclaimed hybrid system, called dedicated outdoor air system (DOAS), which is often used with UFAD systems supplemented by radiant panels or chilled beams (described in Sections 18.4 and 20.6 respectively). Other cooling subsystems, such as evaporative cooling and desiccant systems, can also be used with DOAS, and these are covered later on in

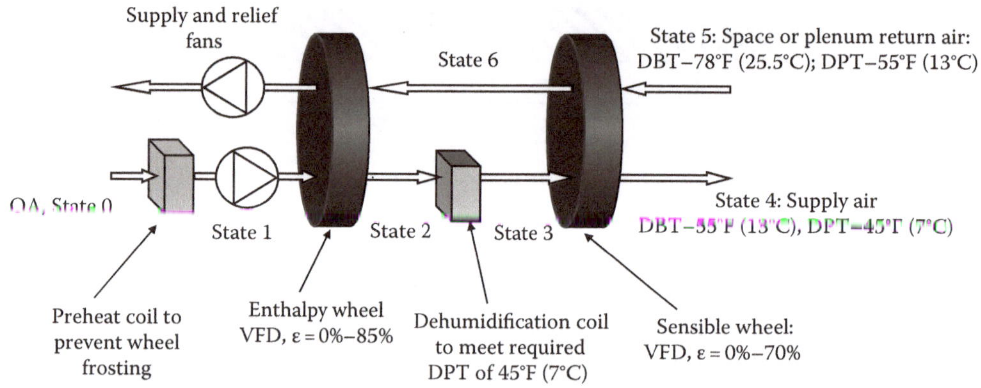

FIGURE 20.26
General arrangement of the dedicated outdoor air system with state points. (From Mumma, S.A., Dedicated OA systems, comments/letters: IAQ applications, *Winter*, 20–22, 2002.) DPT- Dew point temperature; DBT- Dry-bulb temperature; VFD - Variable frequency drive.

this chapter. The primary reasons why hybrid systems are so attractive are because of not only the associated improved air quality and the energy efficiency but also the flexibility they provide to the designer in combining different types of components and subsystems most suitable for the building and its location, while specifying subsystems that he or she has a preference for.

20.7.2 DOAS

VAV systems suffer from two other possible drawbacks. One is that for a multizone building, outdoor air in excess of the minimum required amount has to be drawn in since the outdoor air and return airstreams are mixed together, and the amount of air supply to individual zones is based on thermal space loads; hence, the relative outdoor air fraction to each of the spaces varies as the zone loads vary over the course of the year.* In order to compensate for such variations, ASHRAE 62.1 (2013) stipulates a higher outdoor air amount that has energy penalties associated with conditioning it. No such excess air is required for DOAS. DOAS, or some earlier variant of it, has been proposed in the early 1980s. The HVAC industry was slow to adopt the system until the last 10 years or so. DOAS are said to be the most reliable system for introducing proper amounts of outdoor air for multiple zones during the full operating range of the system. DOAS units are 100% outdoor air units that directly supply this air to the various zones as required. Thus, unlike VAV systems in multizone spaces, DOAS ensures 100% outdoor air requirements to all spaces at all times, thereby enhancing indoor air quality. This is a compelling advantage for assurance verification in a court of law. Further, microbial growth is inhibited since the ventilation air provides all the latent load.

The second drawback is that VAV systems may compromise the indoor relative humidity when the relative sensible to latent fractions vary over the year. DOAS separate the loads, by introducing the outdoor air at *lower* humidity ratio than the desired space humidity ratio in order to allow the zone HVAC unit to handle *only* the space sensible cooling load. This approach allows the individual space units to handle only the space cooling and heating loads.

DOAS are a new paradigm in HVAC systems (Mumma, 2001b). Though flexible in terms of component selection, the following requirements have to be met in order to *qualify* as a DOAS:

1. Design the outdoor air-conditioning system to be distinct from the space conditioning system; this ensures proper ventilation to all occupied spaces.
2. Ensure an energy-efficient means of conditioning the outdoor air and the latent load of the space that minimizes reheat energy.
3. Adopt cost-effective ways of providing dehumidification to the outdoor airstream by using appropriate heat recovery equipment (such as desiccant wheels).
4. Select proper systems for meeting space sensible heat loads. There are several workable designs possible depending on specific circumstances and the personal preferences of the designer. Radiant panels and chilled beams with UFAD are good design choices.

There are several ways by which various components can be assembled to form a DOAS. One general DOAS arrangement is shown in Figure 20.26.

* This aspect has been illustrated for the dual-zone VAV system in Example 19.9.

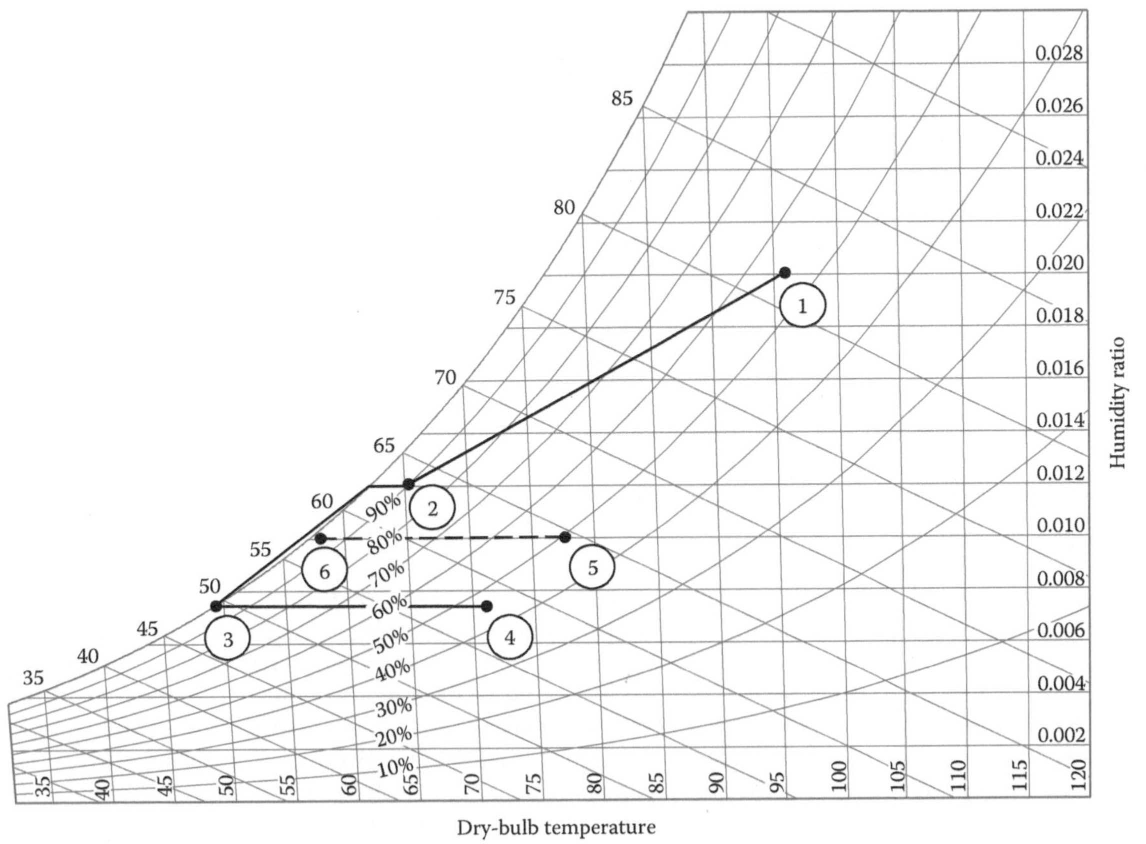

FIGURE 20.27
Dedicated outdoor air system processes on the psychrometric chart. The numerical values correspond to Example 20.6.

Outdoor air at state 0 is cooled and dehumidified by the enthalpy wheel (effectiveness can reach maximum values of 85%) whose honeycomb structures are filled with passive desiccant material—shown as process 1–2 on the psychrometric chart in Figure 20.27. One could also use a sensible heat exchanger (plate or even a heat pipe), but an enthalpy wheel is more energy efficient. So as not to damage the desiccant material, a preheater is provided during winter operation. The supply air then passes over a dehumidification coil to cool as necessary but more importantly to dehumidify the air to a state that can meet the latent loads of the space as well. Typically, if the space dew point is to be maintained at 50°F–55°F (10°C–13°C), the dew point of this air has to be around 42°F–48°F (6°C–9°C), which is less than that of a typical CAV or a VAV system. In case the outdoor air is dry enough, the cooling coil need not dehumidify the air. The general recommendation is that a DX coil not be used (Mumma, 2001b); a chilled water coil is preferred because of the additional control it can provide. However, recent advances in DX systems designs (variable speed compressors) allow a DX system to be used under certain circumstances. Finally, another sensible wheel (shown as one with a variable frequency drive [VFD] with an effectiveness of around 70%) is used to heat the supply air as necessary since the deep cooling in the coil (in order to lower the dew point temperature) may have resulted in excessive cooling. Hence, a VFD is advisable. The reheating is done by heat exchange between the supply and return airstreams, but if a source of cheap or free sensible heat is available, this wheel can be eliminated. One could also use a heat pipe to transfer the heat between the two airstreams. Significant energy savings are realized since generally 20%–30% less outdoor air must be conditioned as compared to a VAV system for multizone spaces (ASHRAE Applications, 2015).

The DOAS unit discussed earlier only conditions the OA and meets the latent loads of the space. The balance of the sensible loads of the spaces must be met by a separate cooling system. One such system is shown in Figure 20.28. Here, the same chiller or a separate chiller supplies the cold water for the radiant panels placed in the room. There are numerous other

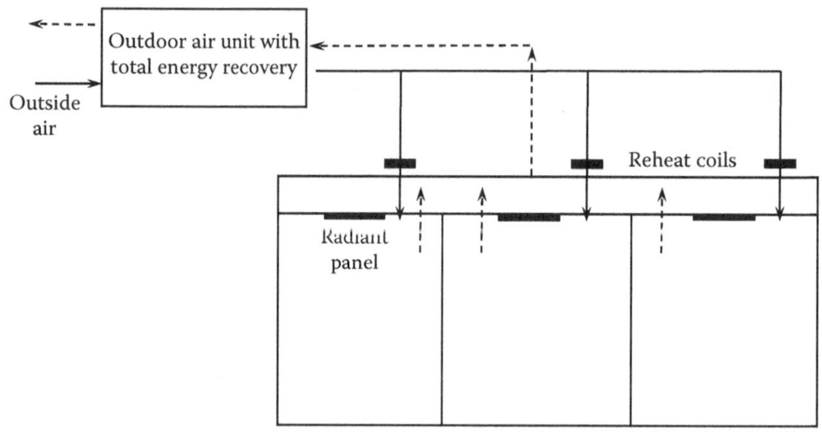

FIGURE 20.28
Dedicated outdoor air system with parallel radiant cooling system.

choices, for example, a parallel VAV system without airside economizer (for simplicity), packaged unitary heat pump system, fan coil units, ceiling radiant panels, chilled beams (with high aspiration rates—as high as 1:36—available commercially so as to meet the desired ADPI), and evaporative cooling and desiccant cooling systems. Further, one could use any of the three types of room air distribution systems: fully mixed, UFAD, or displacement ventilation.

A USDOE (2002) report ranked radiant ceiling cooling, enthalpy heat recovery, and DOAS as the three most energy-saving technologies among 55 of the most promising ones considered (Figure 20.29). In fact, DOAS and radiant panels are shown to have simple paybacks of less than a year (probably a stretch). The combination of all three technologies, however, seems a logical design preference for the next generation of secondary systems. Of some interest is the recent article by Stein and Taylor (2013) who presented results of a head-to-head competition where several mechanical engineering companies participated in the design of three different secondary systems for the same building. The comparisons between variable air volume with reheat (VAVR), a DOAS with active chilled beams, and a hybrid of the two revealed that the VAVR was the clear winner with lower first cost as well as lower energy costs. Though these results are somewhat specific to the building (medical center for graduate studies) of 5,249 m^2 (56,500 ft^2) and to the location (Davis, California), it does illustrate the fact that the cost effectiveness of VAVR vs. DOAS-based systems are somewhat subjective, and should be

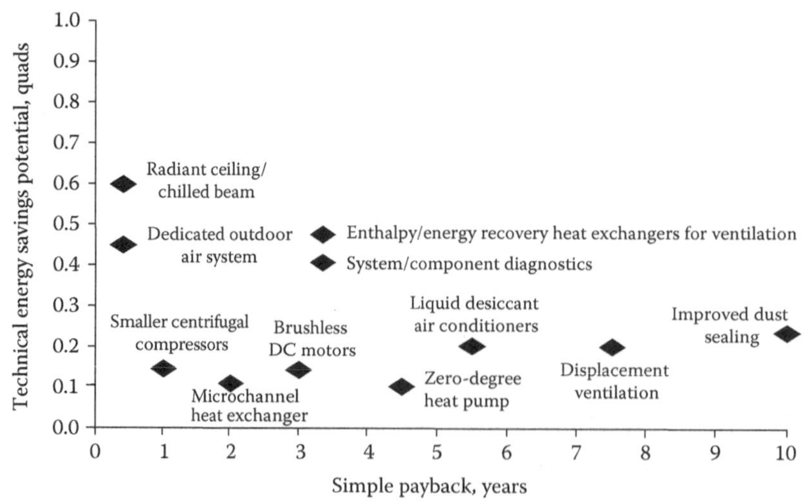

FIGURE 20.29
Simple payback estimates of promising HVAC technologies. (Redrawn from USDOE, *Energy Consumption Characteristics of Commercial Building HVAC*, USDOE, Washington, DC, 2002.)

carefully analyzed during the design phase. On the other hand, Sastry and Rumsey (2014) presented the results of a 2-year side-by-side monitoring program of a large office building in the hot and dry location of Hyderabad, India, where they found that a radiant cooling system with DOAS consumed 34% less energy than a VAV system. This difference could be partially attributed to the lower outdoor air code requirements in India as compared to the United States. In any case, further analytical and field research is warranted before specific recommendations emerge.

Example 20.6: DOAS Design

This example serves to illustrate the design procedure for a DOAS unit when part of a DOAS with two wheels, a cooling coil and a preheat coil (Figure 20.26). During a typical design process involving a DOAS unit, the designer specifies the desired *leaving conditions* typically (ASHRAE Applications, 2015):

- *Cooling (summer) conditions*: 60°F–70°F (15°C–18°C) and 40–60 gr_w/lb_a* (0.006–0.009 kg_w/kg_a) humidity ratio
- *Heating (winter) conditions*: 65°F–68°F (18°C–20°C) and 30–40 gr_w/lb_a (0.004–0.006 kg_w/kg_a) humidity ratio

For this example, we will assume a DOAS providing outdoor air to three classrooms requiring a total ventilation air of 1200 ft³/min (566 L/s). The total cooling load (heat to be removed from the space to maintain comfort) is 80,000 Btu/h (35 kW) of which 80% is sensible heat. This corresponds to the peak summer design day when the entering outdoor air conditions are 95°F (35°C) and 55% RH. The same airflow rate of 1200 ft³/min (566 L/s) is available for energy recovery by exhausting it through the DOAS unit. The desired space leaving air conditions are specified as 78°F (25.6°C) and 50% RH. The effectiveness value of the enthalpy wheel is 0.8 (includes both sensible heat and latent heat transfer) and that of the sensible wheel is 0.7 (sensible heat transfer only).

Figure: See Figures 20.26 and 20.27.

Assumptions: The location is at sea level. The duct heat transfer and the fan air temperature rise are ignored for simplicity. The air leaves the cooling coil as saturated air. The preheat coil is inactive.

Given: $SHR_{space} = 0.8$, $\dot{Q}_{space,cool} = 80{,}000$ Btu/h, $\dot{V}_0 = 1{,}200\ \text{ft}^3/\text{min} = 72{,}000\ \text{ft}^3/\text{h}$, and $T_{db,o} = 95°F$, $\phi_o = 55\%$ (outdoor air condition), $T_{db,5} = 78°F$, $\phi_5 = 50\%$ (return air condition), $\varepsilon_{ew} = 0.8$, $\varepsilon_{sw} = 0.7$.

Find: $\dot{Q}_{cc}$.

Lookup values: Specific volume $v_o = 14.4\ \text{ft}^3/\text{lb}_a$, $W_o = 0.0197\ \text{lb}_w/\text{lb}_a$, $W_5 = 0.0103\ \text{lb}_w/\text{lb}_a$.

Solution

1. *Locate specified points 0 (or 1) and 5* on the psychometric chart.
2. *Determine condition of air entering the sensible wheel—state 3.* We will use the latent heat balance to determine the humidity ratio of the air entering the space:

$$W_3 = W_4 = W_5 - \frac{\dot{Q}_{space,lat}}{(\dot{V}_0/v_0)\times h_v}$$

$$= 0.0103\ \text{lb}_w/\text{lb}_a - \frac{80{,}000\ \text{Btu/h}\times(1-0.8)}{(72{,}000\ \text{ft}^3/\text{h}/14.4\ \text{ft}^3/\text{lb}_a)\times 1{,}075\ \text{Btu/lb}_w}$$

$$= 0.0073\ \text{lb}_w/\text{lb}_a$$

In order to achieve the desired supply humidity ratio of 0.0073 lb_w/lb_a, the cooling coil needs to remove moisture, which requires cooling the air to 50°F (assuming saturated condition). This is point 3.

3. *Determine condition of supply air—state 4.* The effectiveness equation on the sensible heat wheel can be used to determine the supply air temperature:

$$\varepsilon_{sw} = \frac{T_{db,4} - T_{db,3}}{T_{db,5} - T_{db,3}}$$

or

$$T_{db,4} = T_{db,3} + \varepsilon_{sw}(T_{db,5} - T_{db,3})$$

$$= 50°F + 0.7\times(78-50)°F = 69.6°F$$

4. *Determine condition of air leaving the sensible wheel—state 6.* Since the two airstream flow rates across the sensible heat wheel are equal, an energy balance yields $T_{db,6} = T_{db,5} - \varepsilon_{sw}(T_{db,5} - T_{db,3})$, or $T_{db,6} = 78 - 0.7 \times (78 - 50) = 58.4°F$.
5. *Determine condition of air leaving the enthalpy wheel—state 2.* Rather than use enthalpies, we will simply determine the temperature and humidity separately:

$$\varepsilon_{ew} = \frac{T_{db,0} - T_{db,2}}{T_{db,0} - T_{db,6}}$$

or

$$T_{db,2} = T_{db,0} - \varepsilon_{ew}(T_{db,0} - T_{db,6})$$

$$= 95°F - 0.8\times(95-58.4)°F = 65.7°F$$

* HVAC industry often uses the unit of Grains (gr); 7000 gr = 1 lb.

and

$$\varepsilon_{ew} = \frac{W_0 - W_2}{W_0 - W_6}$$

or

$$W_2 = W_0 - \varepsilon_{ew}(W_0 - W_6)$$
$$= 0.0197 - 0.8 \times (0.0197 - 0.0103)$$
$$= 0.0122\ \text{lb}_w/\text{lb}_a$$

6. *Determine cooling coil load.* The coil sensible load is

$$\dot{Q}_{cc,sen} = \dot{m}_a \cdot c_{p,a} \cdot \Delta T_{coil}$$
$$= (72{,}000/14.4)\ \text{lb}_a/\text{h} \times 0.24\ \text{Btu}/(\text{lb}_a \cdot {}^\circ\text{F}) \times (65.7 - 50)^\circ\text{F}$$
$$= 18{,}840\ \text{Btu/h or } 1.57\ \text{tons}$$

The cooling coil latent load is

$$\dot{Q}_{cc,lat} = m_a \cdot h_v \cdot \Delta W_{cc}$$
$$= (72{,}000/14.4)\ \text{lb}_a/\text{h} \times 1075\ \text{Btu/lb}_w \times (0.0122 - 0.0073)\ \text{lb}_w/\text{lb}_a$$
$$= 26{,}337.5\ \text{Btu/h} = 2.19\ \text{tons}$$

Thus, the total coil load or coil cooling rate is

$$\dot{Q}_{cc,tot} = \dot{Q}_{cc,sen} + \dot{Q}_{cc,lat} = 45{,}177\ \text{Btu/h} = 3.76\ \text{tons}$$

Comments

In addition to the DOAS cooling coil, the designer has to select a sensible cooling system to remove the sensible loads of the space. Several options are available: chilled beams or radiant cooling panels or DX systems such as mini split units (VRF or variable refrigerant flow units), or chilled water-based fan coils units.

20.8 Evaporative Cooling Equipment

Another method for producing a cooling effect uses the ability of low-humidity air to evaporate water in an adiabatic process, with the result that the dry-bulb temperature of the air is lowered. This process is called the *evaporative cooling process*. It is restricted to drier areas of the world using outdoor air or to systems using desiccants to dry moist outdoor air. Many buildings in desert climates can have all their sensible cooling loads satisfied by evaporative equipment that uses significantly less power than vapor compression equipment. A drawback is that water supply should be readily available.

20.8.1 Background

The evaporation process described in Section 13.6.4 can be applied to cooling of airstreams if a source of sufficiently dry air is available. This very dry air can be humidified by direct contact with water to produce sensible cooling. In this section, we give an overview of the operation of commercial evaporative cooling equipment. These systems have become increasingly attractive because (1) refrigerants are not involved, (2) energy consumption can be reduced in several regions of the United States, and (3) ventilation required for human health is easily provided (during the cooling season).

Another evaporative process for reducing mechanical cooling needs in buildings (particularly buildings with large roof areas in dry climates) is the roof spray system, in which a thin, evaporating layer is maintained on flat-roof buildings. For example, evaporation from the roof can reduce that cooling load by 3–4 Btu/(h·ft^2) (9–13 W/m^2), amounting to a 40%–50% reduction in peak roof cooling loads in a 95°F (dry bulb)/60°F (wet bulb) (35°C/15.5°C) climate. Water flow to the roof sprays is controlled by roof temperature, so standing water is avoided.

20.8.2 Direct Evaporative Coolers

Figure 13.13 depicts the evaporative cooling process on the psychrometric chart. The essential components of a small, side-draft *direct* evaporative cooler are shown in Figure 20.30a. The evaporative medium (one such pad is shown behind the fan) is constructed of specially fabricated paper, fiberglass, or aspen wood fibers. The pads are kept saturated by the small pump that raises water from the sump, through a distribution system, to the top of the pads on a continuous basis. The water then falls through the pads back to the sump. Evaporation occurs adiabatically to air drawn through the pads by the centrifugal blower. Systems of this basic design are available in flow ranges between 2,000 and 20,000 ft^3/min (1 and 10 m^3/s). Water treatment (softening, filtration) is important if pad lifetime is to be maximized. The face velocity should be limited to 5 ft/s (1.5 m/s) to avoid entraining excessive water in the airstream, from which it could be carried into the building being cooled. In a draw-through system as shown, the cooled air also cools the motor, improving its efficiency, and lengthening its life.

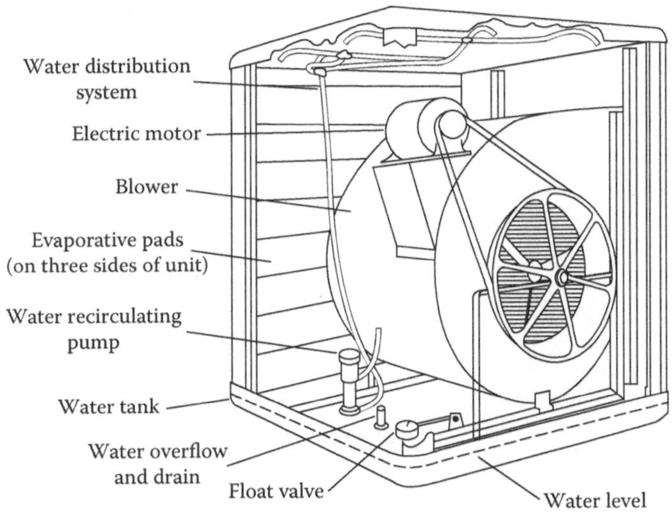

(a)

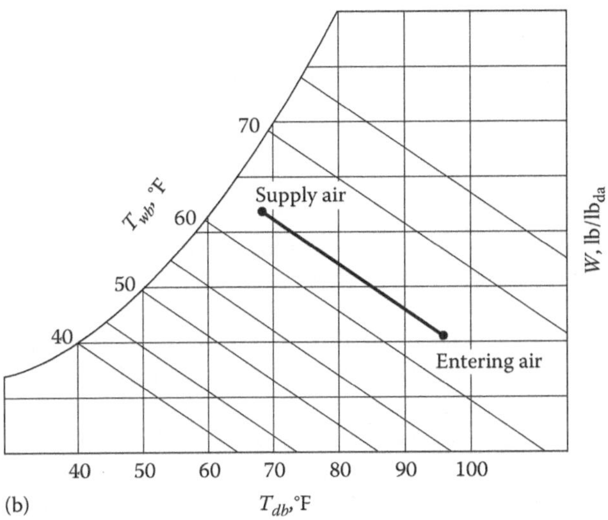

(b)

FIGURE 20.30
(a) Direct evaporative cooler showing key components, including wetted pads, blower, motor, and water supply system. (b) Typical direct evaporative cooling process plotted on the psychrometric chart.

Larger direct evaporative coolers use different media with greater thicknesses. However, the same fundamental approach is used. Pads on larger units can be made from rigid cellulose or fiberglass (Figure 20.31) and may be up to 1.5 ft (45 cm) thick. Coolers using rigid media can produce cooled airflows up to 200,000 ft^3/min (94 m^3/s).

The temperature of evaporatively cooled air cannot be below the wet-bulb temperature of the entering airstream (Figure 20.30b). In humid climates, this places a limit on the amount of sensible cooling that can be realistically expected from evaporation processes. The *effectiveness* ε of evaporative coolers (sometimes called

FIGURE 20.31
Sample of evaporative cooling media. (Courtesy of Munters, Inc., Kista, Sweden.)

the *saturation efficiency*) is defined as the dry-bulb temperature depression divided by the difference between the entering dry- and wet-bulb temperatures:

$$\varepsilon_{evap} = \frac{T_{db,in} - T_{db,out}}{T_{db,in} - T_{wb,in}} \tag{20.7}$$

The effectiveness of practical direct systems ranges from 80% to 90%. Of course, evaporative coolers provide no latent cooling since the relative humidity of the direct cooler outlet air is higher than that of the inlet air.

Heat and mass transfer coefficients for evaporative cooler media have been measured and correlated by Dowdy et al. (1986) and by Liesen and Pederson (1991). The heat and mass transfer coefficients were found to depend on the pad thickness and on the Reynolds and Prandtl numbers. Figure 20.32 shows example of effectiveness data for various thicknesses of pad and various airflow rates (at sea level).* For thick pads, effectiveness approaching 100% can be achieved, but the pressure drop and associated fan power are quite high. Inlet air temperature does not have a strong effect on effectiveness.

Example 20.7: Direct Evaporative Cooler Exit Air Condition

What is the relative humidity of air leaving a direct evaporative cooler in the Arizona desert where outdoor conditions are 105°F (40.6°C) dry-bulb and 65°F (18.3°C) wet-bulb temperatures? The cooler effectiveness is 80%.

* Problem 20.22 gives an empirical equation for a common evaporative cooler medium. It can be used instead of reading data from Figure 20.32.

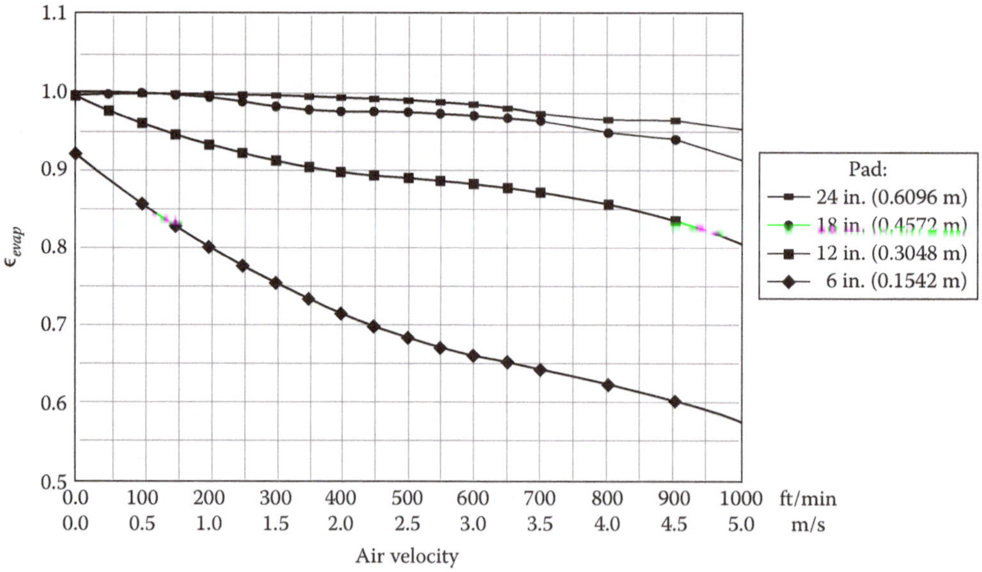

FIGURE 20.32
Example of evaporative cooler effectiveness for various pad thicknesses and airflow rates. (From *ASHRAE Handbook of Fundamentals*, ASHRAE, Atlanta, GA, 2005, www.ashrae.org.)

Given: $T_{db,in} = 105°F$, $T_{wb,in} = 65°F$.

Find: Relative humidity ϕ.

Solution

Equation 20.7 can be solved for the outlet dry-bulb temperature:

$$T_{db,out} = T_{db,in} - \varepsilon_{evap}(T_{db,in} - T_{wb,out})$$

$$= 105°F - 0.80 \times (105 - 65)°F = 73°F$$

The exit wet-bulb temperature is known since evaporative cooling is a constant wet-bulb process, i.e., $T_{wb,out} = T_{wb,in} = 65°F$.

From the psychrometric chart for sea level (Figure 13.6), we find

$$\phi = 66\%$$

Comments

Although the humidity is not excessive, the dry-bulb temperature is not sufficiently low to accomplish much sensible cooling to comfort conditions. The space could be held at 82°F–85°F (28°C–29°C) with this source of supply air. Long-term evaporative effectiveness of 80% relies on proper cooler maintenance. Although 80% is easily achievable with well-designed, new units, it may be difficult to achieve this over three years or more unless pads are replaced or well maintained.

Figure 20.33 shows summer wet-bulb temperature data at the 5% frequency level for the United States (wet-bulb temperatures shown will not be exceeded more than 5% of the cooling season—June through September). The rule of thumb is that evaporative cooling should be considered if the 5% wet-bulb temperature is less than 75°F (24°C). According to this rule, the map shows that all the western United States, the north central states, and the northeastern states are good candidates for evaporative cooling.

In addition to space cooling, direct evaporative precoolers, in suitable climates, can be used to reduce the dry-bulb temperature of outdoor air used to cool conventional air-cooled condensers. This results in lower condenser temperatures, which in turn reduces compressor head pressure and power needs, while increasing capacity. These improve the chiller COP value. The modest energy input to an evaporative cooler upstream of an air-cooled condenser essentially consists of extra fan power needed to overcome the additional pressure drop offered by the evaporative medium. This usually amounts to 0.2 inWG (0.5 cmWG). The improvement in compressor efficiency will nearly always outweigh the small extra electric power needs of the direct cooler stage. Costs of water and its treatment must also be considered before a final decision is made. Example 13.12 illustrated how such analyses can be done.

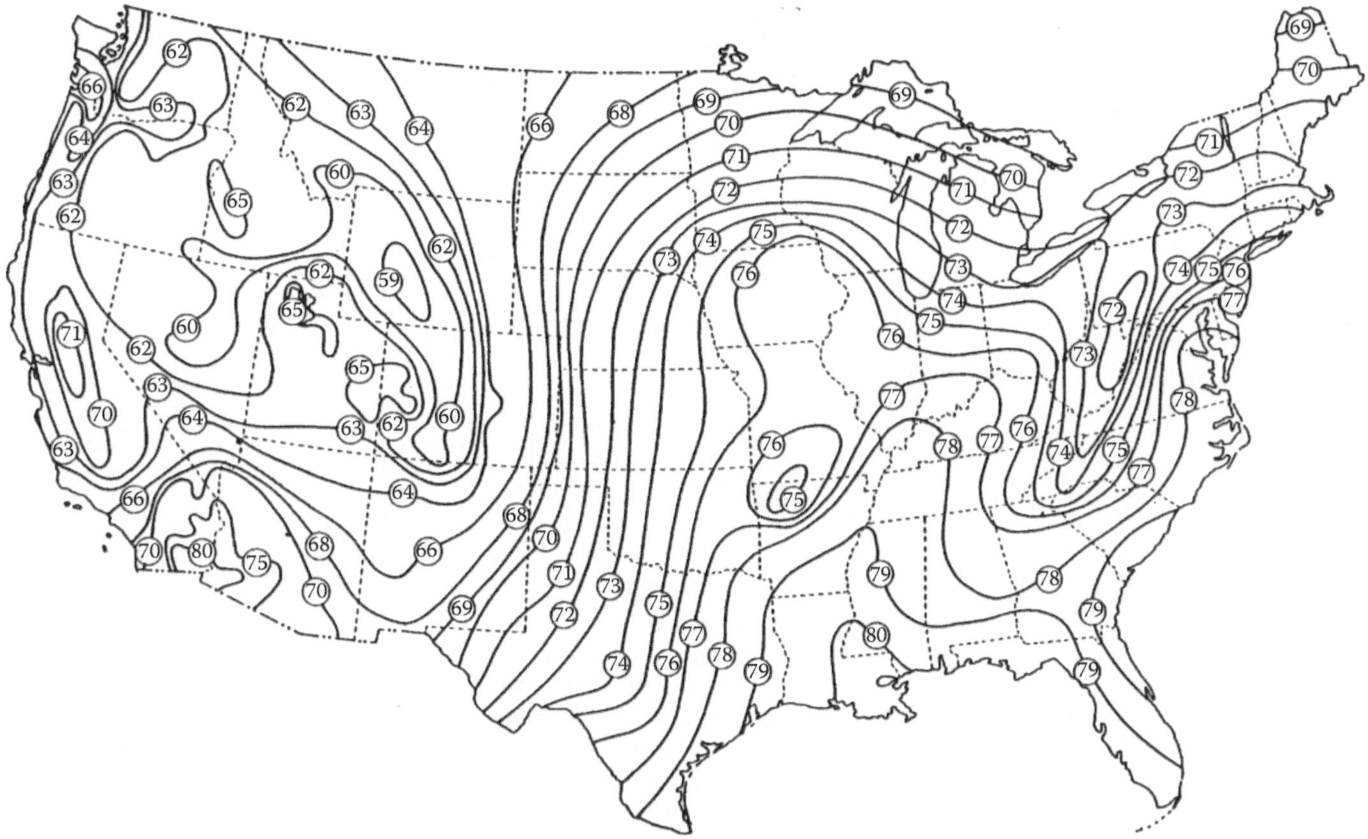

FIGURE 20.33
Map of lines of constant summer wet-bulb temperature in the United States at the 5% frequency level. (Courtesy of SPX Cooling Technologies, Inc., Overland Park, KS.)

20.8.3 Indirect Evaporative Coolers

Indirect evaporative coolers have two outdoor airstreams with the primary airstream cooled sensibly without humidifying it (Figure 20.34). The secondary airstream is evaporatively cooled by using a direct evaporative cooler. Some of the sensible cooling effect so produced is transferred to the primary stream by means of a heat exchanger; this cooled air is then supplied to the conditioned space. The primary stream may consist of some return air from the building plus the outside ventilation air as shown in the figure. Unlike the adiabatic process, in a direct cooler, the indirect cooling of the primary airstream is not an adiabatic process, since no moisture is added. A true cooling effect is produced without the trade-off of added humidity in the building. Indirect coolers can be used by themselves or with mechanical cooling (or with a direct evaporative cooler, as described in Section 20.8.3) depending on the climate.

The exit dry-bulb temperature from an indirect cooler cannot be less than the entering wet-bulb temperature of the entering secondary airstream. Therefore, the performance of an indirect cooler (including the direct cooler) is rated by a factor similar to the effectiveness, called the *performance factor* (PF). It is the ratio of the primary air dry-bulb temperature drop to the difference between the entering air dry-bulb and wet-bulb temperatures.

$$\mathrm{PF} = \frac{T_{db,o} - T_{p,out}}{T_{db,o} - T_{wb,o}} \tag{20.8}$$

where the subscripts *o* and *out* denote outdoor air and indirect cooler outlet conditions, respectively, as indicated in Figure 20.34. Typical values of PF range from 60% to 80%.

The PF can be expressed in a more convenient working equation involving the familiar direct cooler effectiveness ε_{evap} and primary-to-secondary heat exchanger effectiveness ε_{hx} as follows. The indirect cooler heat exchanger effectiveness is (for the case of no building return air, i.e., $T_{db,o} = T_{p,in}$)

$$\varepsilon_{hx} = \frac{\dot{C}_{bldg}(T_{p,in} - T_{p,out})}{\dot{C}_{\min}(T_{p,in} - T'_{db,o})} \tag{20.9}$$

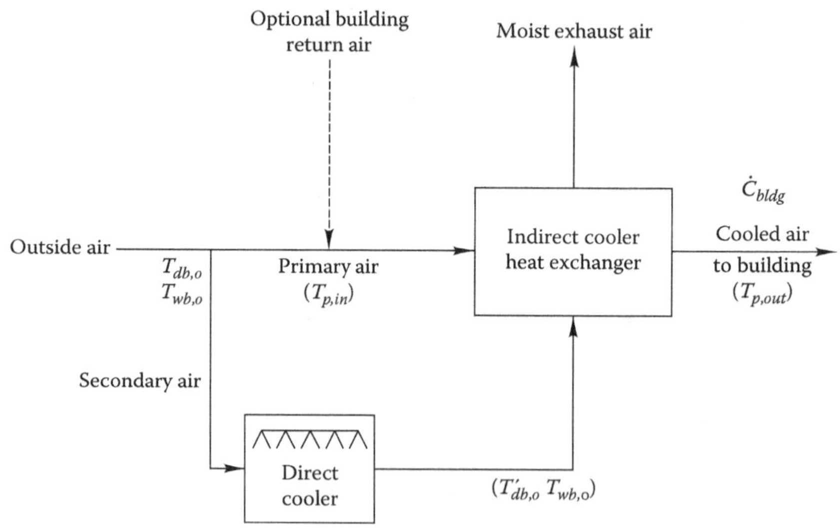

FIGURE 20.34
Indirect evaporative cooler system diagram. Return air from building is optional.

where

$\dot{C}_{bldg}$ is the capacitance rate for building supply air, Btu/(h·°F) (W/K)

$\dot{C}_{min}$ is the minimum capacitance rate in the heat exchanger

$T'_{db,o}$ is the dry-bulb temperature of air leaving the direct cooler

Combining Equations 20.7 through 20.9 gives

$$\text{PF} = \varepsilon_{hx}\varepsilon_{evap}\frac{\dot{C}_{min}}{\dot{C}_{bldg}} \quad (20.10)$$

Finally, the cooling rate of an indirect cooler is

$$\dot{Q}_{cool} = \text{PF} \times \dot{C}_{bldg} \times (T_{db,o} - T_{wb,o}) \quad (20.11)$$

Indirect coolers can also be used as air precoolers for chiller coils if the local outdoor air is not of sufficiently low enthalpy to carry the entire cooling load by direct evaporative cooling. This approach is beneficial since chiller power input can be reduced during the entire cooling season. In addition, a smaller chiller can be purchased, initially reducing capital investment. Indirect systems can provide more of the annual cooling load of buildings than one might think, even in humid climates. For example, Supple (1982) reports that 30% of the annual cooling load in Chicago on a typical building can be provided by this means. An evaporative cooler, even in a humid climate, could be used in the return airstream to remove heat produced by lighting and return fans.

Indirect cooling systems can use exhaust building air (relatively dry and cool) instead of outdoor air in humid climates. Such systems can be as effective as indirect systems in dry climates. There are limitations due to the air quantities involved, but this system should be considered instead of the usual "heat" recovery system, which retrieves only (a fraction of) the sensible cooling effect in building exhaust air. The latent cooling effect is lost when only sensible heat exchange is used between intake and exhaust air.

Direct and indirect evaporative coolers can be used in series to produce additional air cooling. Primary air first passes through the indirect system within which the dry-bulb and wet-bulb temperatures are reduced (at constant humidity ratio), as shown on the psychrometric chart in Figure 20.35. At the exit of the indirect stage, the cooled air passes through a direct evaporative cooler in which the dry-bulb temperature drops further as the humidity ratio increases in a constant-wet-bulb temperature process. In dry climates, COP values up to 15 can be achieved with *indirect–direct* systems. Recall that the COP is defined as the air cooling produced divided by the electric input, primarily to the air handler. Figure 20.36 shows an indirect–direct system used to assist mechanical cooling. When outdoor conditions are appropriate, total building cooling can be accomplished without mechanical cooling; the cooling coil is bypassed in these cases. Both direct and indirect stages are bypassed when the economizer cycle can carry the cooling load in winter.

Other enhancements to evaporative coolers can further improve performance. For example, the water used to wet the pads can be cooled by a small chiller. Colder water will produce both lower air temperatures

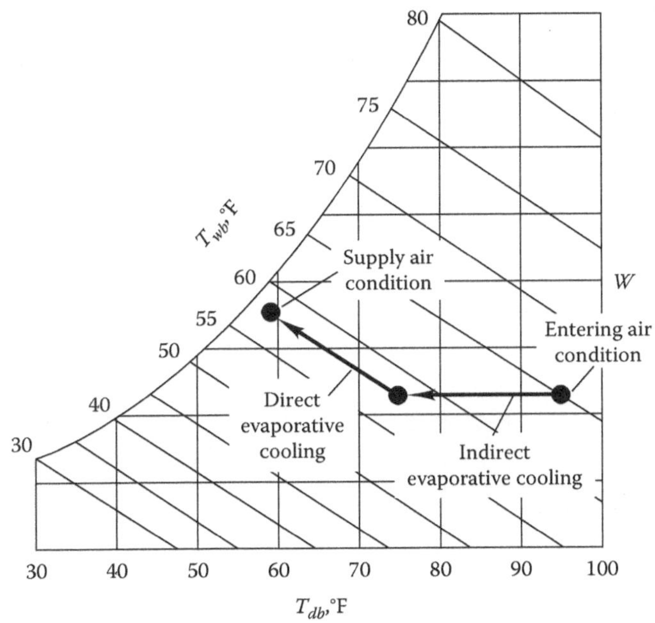

FIGURE 20.35
Indirect and direct evaporative cooling processes plotted on a psychrometric chart. Indirect process is at constant humidity ratio, whereas direct process is at constant enthalpy, i.e., nearly constant wet-bulb temperature.

and lower humidities in the cooler outlet stream. This simple step can reduce cooling costs by one-third in favorable climates.

Control of indirect–direct systems follows this sequence: On the first call for cooling, outdoor air is introduced directly into the building if the air temperature is low enough to offset cooling loads (this is the economizer cycle described in Section 19.6.2). With increasing cooling load, the *indirect* stage is activated and operates until still more cooling is needed. Finally, the *direct* stage is operated to reduce the dry-bulb temperature still more. This is the limit of evaporative cooling that can be provided. At this point, mechanical cooling as shown in Figure 20.36 is activated. During the heating season, the direct stage can be operated if needed to humidify building air; proper attention must be paid to freeze protection in control systems design for this winter use of evaporative cooling equipment.*

20.8.4 Design Considerations

The sizing of evaporative coolers can be accomplished by various methods ranging from rules of thumb regarding volumetric flow per unit floor area (for residences) to performance specifications at the design cooling temperature and the coincident peak amount of latent and sensible heating needed. A typical rule for residences (Watt, 1986) states that between 15 and 30 ACH should be used depending on the local difference between the dry-bulb and wet-bulb temperatures. This method assumes that evaporatively cooled air should be supplied at a temperature 8°F (4.5°C) below the room-temperature set point. The required flow rate is then the sensible load divided by the product of the density, specific heat, and temperature rise:

$$\dot{V}_a = \frac{\dot{Q}_{cool}}{\rho_a \times c_{p,a} \times \Delta T} \tag{20.12}$$

where $\Delta T = 8°F$ (4.4°C).

More careful sizing is needed for commercial buildings. Often the evaporative cooler sizing is based not on meeting the peak load (as is the case for mechanical cooling) but on reducing operating costs by an optimal amount. In such a design exercise, the cooling effect produced by several sizes of coolers is computed and compared to the load. When capacity is sufficient to meet the load, the mechanical chiller is not operated, thereby saving electricity charges. It is rarely economical (except in arid locations) to invest in an evaporative cooler sufficiently large to meet the design cooling load. A full-capacity mechanical chiller is assumed to be installed, and a moderately sized evaporative cooler acts as an energy saver when load and climatic conditions so allow. Example 20.8 illustrates how a designer would go about finding the seasonal cooling effect produced by an evaporative cooler.

Example 20.8: Use of Bin Method for Evaporative Cooler Performance Calculations

Find the total sensible cooling produced by an indirect evaporative cooler with performance factor *PF* = 70% during the cooling season at the site for which the weather data are given. The data are for June, July, and August in Denver, CO.† The outdoor air supply rate is 50,000 ft³/min (23,600 L/s) as might be appropriate for a small commercial building of 25,000 ft² (2,320 m²).

Given: $\dot{V}_a = 50{,}000$ ft³/min, PF = 0.70, and outdoor air density = 0.060 lb_m/ft^3.

Bin dry-bulb (and mean coincident wet bulb) temperature data are shown in Table 20.7 (data aggregated into large bins in the interest of economy of calculations in this example).

Figure: See Figure 20.34.

Find: Cooling effect produced for the months of June, July, and August.

* A heat exchange loop from building exhaust air to the outside air supply duct can reduce the pad-freezing potential.

† The online HCB software also contains dry- and wet-bulb temperature bin data (based on the TMY database). The data are somewhat different for some sites from those used in this example, which are taken from *Engineering Weather Data* (AFM 88-29, 1978) published by the Government Printing Office.

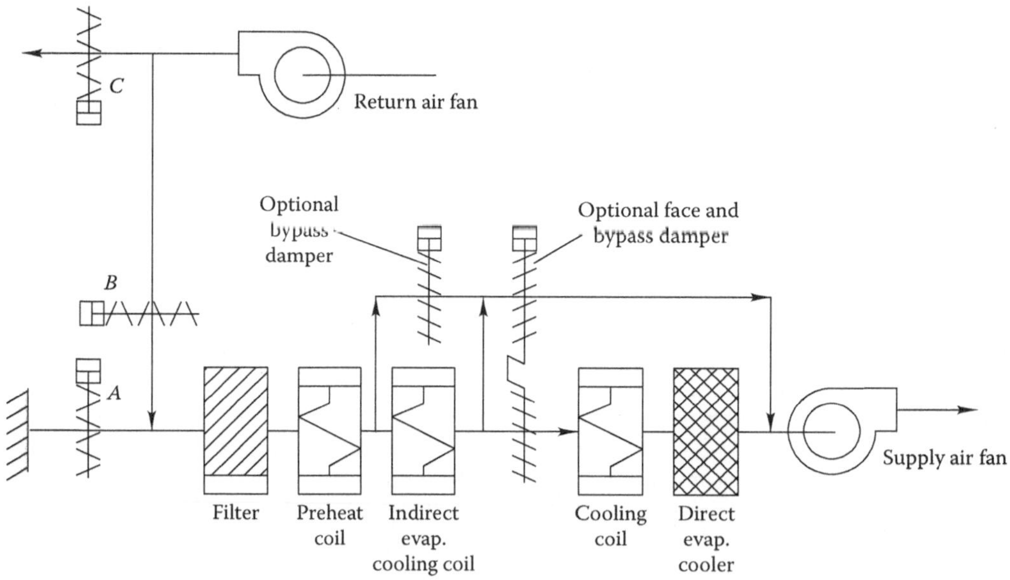

FIGURE 20.36
Evaporatively assisted mechanical cooling system with bypass arrangements to permit 100% evaporative operation or economizer cycle operation.

TABLE 20.7
Indirect Evaporative Cooler Performance in Denver

Bin, °F	Hours per Bin	$T_{db,o}$, °F	$T_{wb,o}$, °F	$\dot{Q}_{cool}$, kBtu/h	Q_{bin}, Mbtu	$T_{p,out}$, °F
90–99	78	95	60	1058	82.6	70
80–89	371	85	59	786	291.7	67
70–79	541	75	58	514	278.1	63
60–69	750	65	55	302	226.8	58
				Total	879.2	

Solution

It is the custom in evaporative cooler calculations to define the indices of performance on a temperature basis, not on an enthalpy basis, as is done for the evaporative process in cooling towers. This has been reflected in the derivation of Equations 20.8 through 20.11. We use this approach in this solution. Since the value of PF is given, only Equation 20.11 is used to solve this problem. The solution is summarized in Table 20.7.

For the 90°F–99°F bin, the calculations are given as follows. The cooling rate is found from Equation 20.11 (the building airstream is the minimum capacitance rate stream in this example):

$$\dot{Q}_{cool} = 0.7 \times 50{,}000\ \text{ft}^3/\text{min} \times 0.060\ \text{lb/ft}^3$$

$$\times 0.24\ \text{Btu/lb}\cdot{}^\circ\text{F} \times 60\ \text{min/h} \times \frac{(95-60)^\circ\text{F}}{1{,}000}$$

$$= 1{,}058\ \text{kBtu/h}$$

The total cooling effect is the product of the cooling rate and the number of bin hours:

$$Q_{bin} = n_{bin}\dot{Q}_{cool} = \frac{78\ \text{h} \times 1058\ \text{kBtu/h}}{1000} = 82.6\ \text{MBtu}$$

The total cooling effect produced in the three-month period is 879.2 MBtu (927.6 GJ) or 73,200 ton·h.

By using a typical rule of thumb for evaporative coolers of 0.2 kW/ton of electric input at peak conditions (the 90°F–99°F bin), the COP of this unit is as follows:

$$\text{COP}_{peak} = \frac{12{,}000\ \text{Btu/h}}{0.2\,\text{kW} \times 3{,}413\ \text{Btu/kWh}} = 17.6$$

At off-peak conditions, less cooling is produced since the difference between the wet- and dry-bulb temperatures is smaller. However, the fan power is the same. Therefore, the COP is lower in the low-temperature bins.

Another performance characteristic also needs to be checked. The temperature supplied by the indirect cooler must be found to determine whether additional mechanical cooling is needed for space conditioning. Of course, the most critical period is the warmest part of the cooling season. The direct cooler outlet temperature can be found from Equation 20.11 and from the first law:

$$\dot{Q}_{cool} = \text{PF} \times \dot{C}_{bldg} \times (T_{db,o} - T_{wb,o}) = \dot{C}_{bldg} \times (T_{db,o} - T_{p,out})$$

Solving the right-hand expression for the outlet temperature gives

$$T_{p,out} = (1-\text{PF}) \times T_{db,o} + \text{PF} \times T_{wb,o}$$

The rightmost column in Table 20.7 shows the result for the four temperature bins of this example. The outlet temperature in the 85°F and 95°F bins is too warm to be used directly in a space. However, the cooling produced can offset some of the mechanical cooling needed if the direct cooler is used in a system of the type shown in Figure 20.36. In the cooler bins (75°F and 65°F), the evaporatively cooled air could be used directly to cool a space.

Comments

Note that the bin data were not given for temperatures below 60°F. For temperatures below this, evaporative cooling is not used—outdoor air is used directly in an economizer cycle if cooling is required. Of course, most of the hours below 60°F in June through August occur at night, when cooling may not be needed at all, but precooling could be accomplished during nighttime.

This example is the first step in selecting an evaporative cooler to reduce mechanical cooler operating time and to save on operating costs. The next step would compare evaporative cooler capacity to the load to determine when chiller operation can be avoided. Finally, chiller power savings are calculated by using the known chiller part-load characteristics. Included in this comparative assessment must be the costs of water and its demineralization, moving relatively larger volumes of air, and larger duct requirements for evaporative systems. In 1991 dollars, evaporative cooling systems cost about $0.35/ft^3/min of airflow.

There is an additional, little-known economic benefit to using evaporative coolers. Since the water is evaporated and not returned to the sewer system, many municipal water districts permit submetering of evaporative cooler water and do not make the usual sewer tax assessment on the submetered usage! The sewer cost is about 50% of the water bill in the United States for water returned to the sewer by whatever means.

If an evaporative cooler is to be used as the *only cooling source* for a building, one must select the proper *design wet-bulb* temperature. Cooling loads are often determined by using the 1.0% design condition described in Chapter 9. For example, the online HCB software gives the 1.0% values of dry-bulb and mean coincident wet-bulb temperatures as 90°F and 59°F (32°C and 15°C) for Denver, Colorado, a location well suited for evaporative cooling. However, the 1% *wet-bulb condition* is quite different at 63°F (17°C). This difference represents roughly a 4°F (2.2°C) increase in evaporative cooler outlet, a major reduction in performance compared to what would be expected if the 59°F (15°C) value had been used. Experienced designers design evaporative coolers with at least the 1% wet-bulb temperature value and the 1% dry-bulb design temperature. More conservative design would use the 0.4% not 1.0% conditions. This sizing should be checked against two other points: (1) maximum wet-bulb and mean coincident dry-bulb temperatures and (2) maximum dry-bulb and mean coincident wet-bulb temperatures.

According to Watt (1986), *Legionella pneumophila* (Legionnaires' disease) will not multiply in basins of evaporative coolers. The primary reason is that the temperature of this water is always very cool (even when the cooler is turned off in direct sun) and rarely stagnant during the cooling season. This bacterium breeds at 120°F–140°F (49°C–60°C) in stagnant water, according to Watt. However, Barbaree (1991) noted that growth can occur at as low as 77°F (25°C). Control can be achieved by using biocides, pH adjustment, and periodic cooler cleaning. The same preventive maintenance is recommended for cooling towers. Alternatively, the amount of standing water in the sump can be minimized by using a sump moisture sensor to control water supplied to the evaporative media.

20.9 Desiccant Cooling Systems

20.9.1 Operational Principles

In a traditional HVAC system, the cooling and dehumidification of the supply airstream is achieved by the same piece of equipment, i.e., the cooling coil. This coupling requires that even in cases when the sensible load is low, the airstream has to be cooled down to a low temperature simply in order to achieve the needed dehumidification so as to meet the latent loads. Reheating is then needed, resulting in energy wastage.

The active desiccant dehumidification process is close to the adiabatic process line but proceeds in a direction opposite to that of evaporative cooling when drawn on the psychrometric chart. (In an actual process, there is a small deviation due to the residual heat transferred from the regeneration). Moisture is removed from the air by bringing it in contact with a substance having a high affinity for extracting water called a *desiccant*. The phenomenon is akin to a sponge soaking up water. The extraction of water results in the latent heat of vaporization being released, which heats up the air.

The driving force for the process is the difference in the partial water vapor pressures in air ($p_{v,a}$) and in the desiccant ($p_{v,d}$). We note from the series of curves in Figure 20.37 that the water vapor pressure in the desiccant increases fairly linearly with temperature, while it asymptotes with desiccant moisture content. When $p_{v,a} > p_{v,d}$, water vapor is transferred from the air

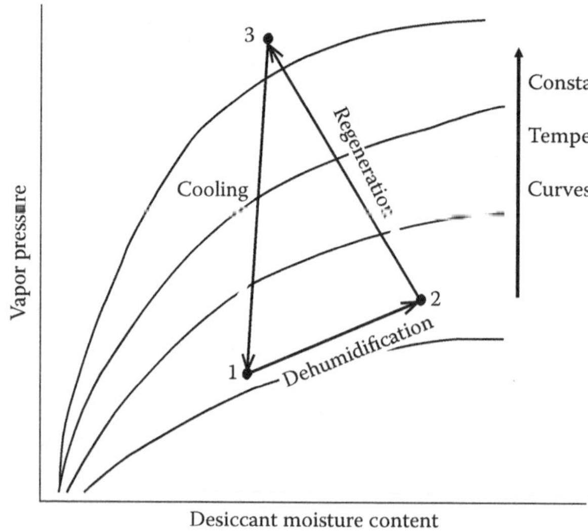

FIGURE 20.37
The desiccant cooling and reaction processes on a pressure versus moisture content state diagram.

to the desiccant initially at state 1 until equilibrium is reached at which state the desiccant is saturated with water (state 2 in Figure 20.37). This process is accompanied by heating of the airstream. In order to regenerate the desiccant for reuse, the desiccant has to be heated (so some source of thermal energy must be available), which increases the water vapor pressure on its surface. If air with lower vapor pressure is brought in contact with this desiccant, the moisture passes from the desiccant to the air. This process is called *regenerating* (also called *reactivation of*) the desiccant bed. Finally, the desiccant has to be cooled back to the initial state (process 3–1 in Figure 20.37). The reader can refer to such texts as Hodges et al. (1998) for further reading.

20.9.2 Description of Systems

Figure 20.38 shows a sketch of the basic system in which the outdoor or the recirculated moist air to be dried is passed through the desiccant bed and leaves it drier but warmer. The regeneration airstream is heated in order to regenerate the desiccant device after which the moist air is exhausted out. The dehumidified supply airstream has to be cooled, while the regeneration air has to be heated in order to regenerate the air. Several different variants of complete air systems with desiccant cooling have been developed, which we will refer to as hybrid desiccant systems (HDS). In the final count, the amount of energy saved depends primarily on the ability of the HDS to shift part of latent cooling load to a low-grade (or waste) source of thermal heat and to eliminate reheat (plus the extra fan power to move the air through the heat recovery devices as illustrated later). Low-grade heat sources at around 180°F–250°F (80°C–120°C) are usually: electric resistance heat, solar water heating, steam from boilers, or natural gas burners (the last is very common in commercial building applications). Currently, there are desiccant-based systems utilizing vapor compression equipment where the heat rejected from the condenser is used for regeneration.

There are several ways by which the process of continuous dehumidification of an airstream can be achieved: liquid-spray towers, solid packed tower, rotating horizontal beds, and rotating wheels that are impregnated with desiccant material. The last type is

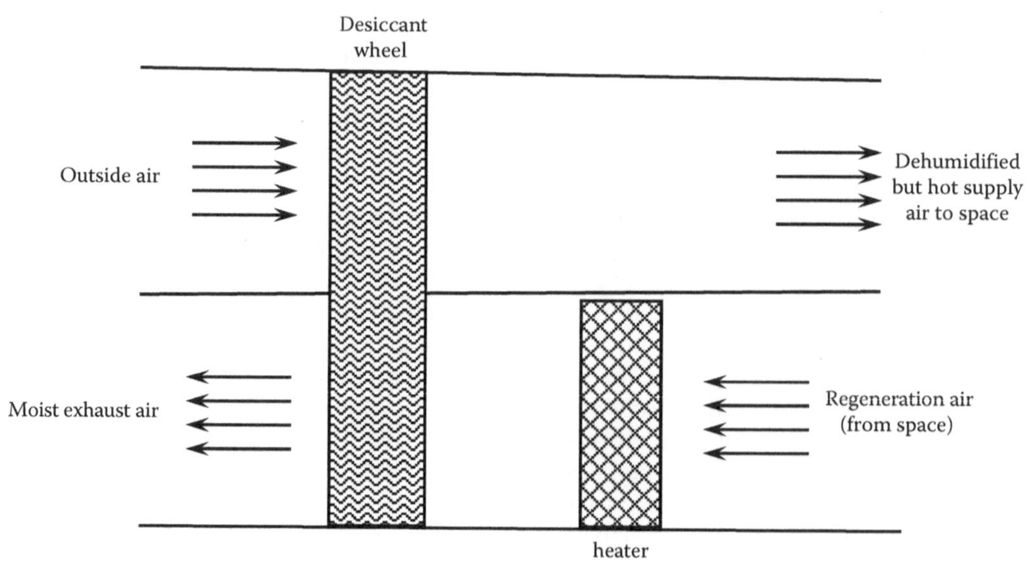

FIGURE 20.38
The basic desiccant and regeneration diagram with process (space supply) and return airstreams.

very common and will be assumed for the discussion that follows. Enthalpy recovery wheels and desiccant wheels look similar, but are designed with different considerations in mind. While both the heat and the mass transfer rates are important in the former, the desiccant material in the latter is meant to primarily absorb moisture. The substrate, which does not absorb moisture, is a honeycomb-like matrix that is impregnated on both sides with desiccant material, about 25 micron thickness. The mass fraction of the desiccant material is relatively low, roughly 15%–30%. Desiccant wheels are typically made to rotate at about 20–30 rotations per hour in order to achieve the desirable heat and mass transfer performance. Note that such desiccant systems become increasingly expensive as more components are added. The need to operate them at different outdoor conditions during the year also results in complex control systems. For example, when the outdoor conditions are relative dry (say, humidity ratios < 0.008 kg_w/kg_a), then the desiccant wheel need not be operated at all.

One of the several variants of a HDS is shown in Figure 20.39a with associated state points depicted on the psychrometric chart in Figure 20.39b. Several pieces of equipment are incorporated to enhance energy efficiency. There are two separate airstreams: the *process airstream*, i.e., the air that needs to be conditioned and supplied to the space, and the *reactivation or regeneration airstream*. The outside air at state point 1 is dried and heated by the desiccant bed to point 2. It is then sensibly cooled to point 3 at the sensible heat exchanger by the return and cooler air from the space. Further sensible cooling of the airstream to point 4 is provided by mechanical cooling at the cooling coil (sometimes called a "post-cooling coil") after which the air is supplied to the space. One could conceptually use the room return air (point 5) heated at the sensible to regenerate the desiccant wheel by further heating the air from point 6 to point 7 and then exhausted outdoors (point 8). However, it is more common not to use this space return air for regenerating the desiccant wheel but to use to a separate outdoor airstream for this purpose, as discussed in Section 20.9.2.

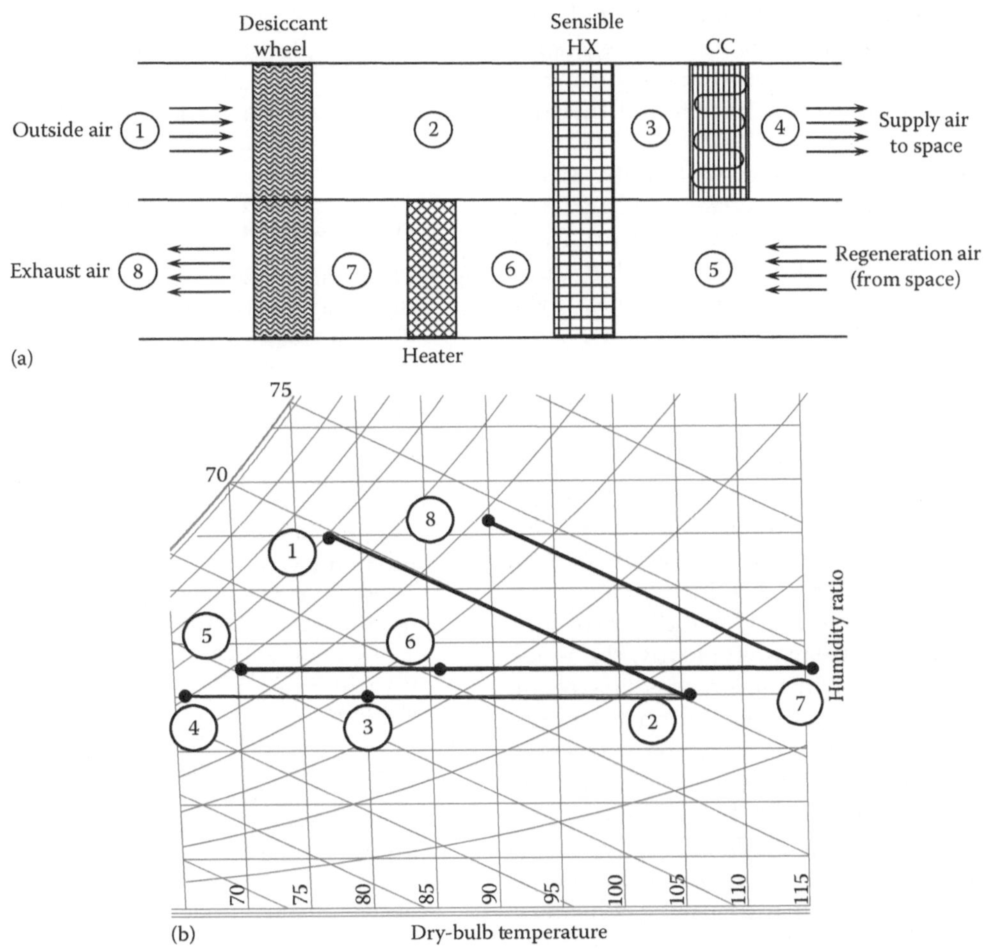

FIGURE 20.39
The more energy-efficient version of Figure 20.38. (a) The additional desiccant wheel and the cooling coil and (b) the two airstream state points on a psychrometric chart.

FIGURE 20.40
Active desiccant-based dehumidification system. (Courtesy of Munters, DryCool, Baltimore, MD.)

One variant to this system is to replace the heat exchanger by a second desiccant wheel (FEMP, 1997). An artist's rendering of such a complete desiccant unit is shown in Figure 20.40. Alternative designs are to replace mechanical cooling by an evaporative cooler or use solar heating to heat up the air needed for regeneration. A number of researchers have studied this cycle, or innovative variations of it, and have found thermal COPs in the range of 0.5–2.58.

Desiccant systems can be used with both traditional fully mixed air delivery systems as well as with DOAS type systems where only the ventilation airstream needs to be conditioned. Typically, fully mixed air delivery systems are used for applications where active dehumidification is required along with lower dew point temperatures than those that can be provided from conventional vapor compression–based systems. These applications can be ice rinks, special pharmaceutical applications, hospitals, supermarkets, etc. Finally, it must be pointed out that the normal design practice is not to use the return air from the space to reactivate the desiccant wheel, but to use a separate outdoor airstream. In such a case, a common operational practice is to have different airflow rates for the two airstreams. Typically, the reactivation airstream is less than the process airstream (by a factor of 1:3) for higher regeneration temperature, and so the desiccant wheel is configured such that only 25% of the wheel surface is exposed to the reactivation airstream, while 75% is exposed to the process airstream to be dehumidified. This typically occurs for systems with higher regeneration temperatures commonly achieved by using direct gas heaters or steam coils (around 220°F–260°F or 100°C–125°C). For instances with lower regeneration temperature (around 180°F or 82°C), the process and the regeneration airflow can be almost equal, and 50% of the wheel is exposed to the process airstream and 50% to the regeneration airstream.

Variables affecting performance of a desiccant wheel are (FEMP, 1997) (1) process air moisture, (2) process air temperature, (3) process air velocity, (4) reactivation air temperature, (5) reactivation air moisture, (6) reactivation air velocity, (7) amount of desiccant exposed to airstream, and (8) desiccant adsorption characteristics.

Example 20.9: Desiccant Systems Design with DOAS

This example, which is similar to Example 20.6, serves to illustrate the design procedure for a desiccant-based DOAS as shown in Figure 20.40 with two wheels (desiccant and sensible) and a post-cooling coil. The sensible effectiveness value of the sensible wheel is 0.7. The desiccant wheel adsorbent is silica gel, designed for high regeneration temperature utilizing a direct fired gas burner. Air for regeneration is introduced directly from the outdoor, and the building exhaust air is used only with the sensible heat recovery wheel. As explained in Example 20.6, the designer specifies the desired DOAS leaving conditions.

The desiccant-based DOAS unit provides outdoor air to 12 classrooms requiring ventilation air at a flow rate of 4800 ft^3/min (2265 L/s). This is for the peak summer design day when the entering outdoor air conditions are 95°F (35°C) and 40% RH. The same airflow rate of 4800 ft^3/min (2265 L/s) is available for energy recovery by exhausting it through the sensible wheel of the DOAS. The regeneration air for the desiccant wheel is introduced directly from the outdoor. The desired DOAS unit leaving conditions are 70°F (21°C) and 0.0085 lb_w/lb_a (0.0085 kg_w/kg_a), while the space is to be maintained at 78°F (25.6°C) and 40% RH.

Figure: See Figure 20.41.

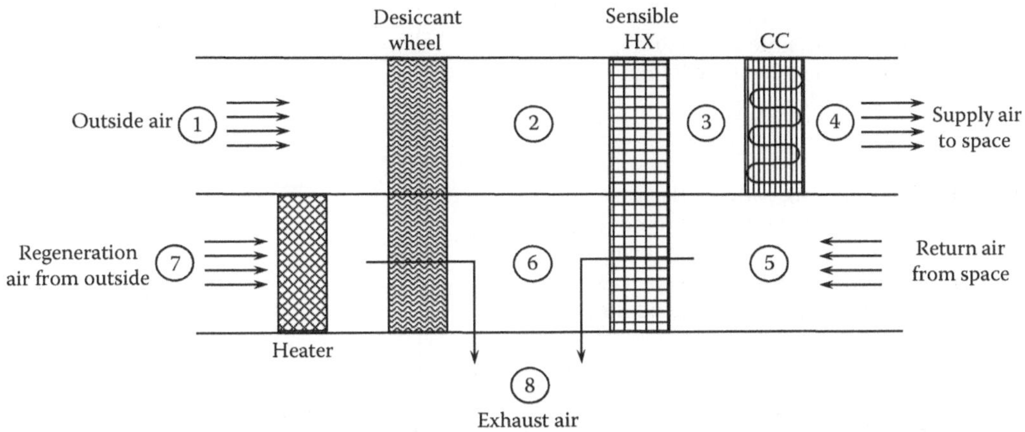

FIGURE 20.41
The desiccant and regeneration process diagram for Example 20.9.

Assumptions: The location is at sea level. The duct heat transfer and the fan air temperature rise are ignored for simplicity. The outdoor air and the exhaust airflow rates from the building through the sensible heat exchanger are equal. However, the regeneration airflow will be 1/3 of the desiccant process airflow rate, i.e., (4800/3) = 1600 cfm.

Given: $\dot{V}_0 = 4{,}800 \text{ ft}^3/\text{min} = 288{,}000 \text{ ft}^3/\text{h}$.

$$T_{db,5} = 78°\text{F},\ \phi_5 = 40\%\ \text{(indoor condition)},$$

$$T_{db,o} = 95°\text{F},\ \phi_o = 55\%\ \text{(outdoor condition)},$$

$$\varepsilon_{sw} = 0.7$$

Find: $\dot{Q}_{cc}$, $\dot{Q}_{reg}$.

Lookup values: Specific volume v_o = 14.3 ft³/lb$_a$, W_o = 0.0143 lb$_w$/lb$_a$.

Solution

1. *Locate specified points 1 and 5* on the psychometric chart.
2. *Select appropriate desiccant wheel model.*

   We have used a software (NovelAire, 2009) to select the desiccant wheel model that can handle 4800 and 1600 cfm for the process and regeneration airstreams, respectively. The regeneration temperature is specified as 260°F with 25% of the desiccant wheel to be used for regeneration.

3. *Determine condition of air leaving desiccant wheel—point 2.*

   The software predicted that the desiccant wheel can reduce the humidity ratio from 0.0143 to 0.00825 lb$_w$/lb$_a$. The software also predicts a moisture crossover of at least 0.00028 lb$_w$/lb$_a$ between the exhaust and the process airstream. Thus, W_2 = 0.00825 + 0.00028 = 0.000853 lb$_w$/lb$_a$. The outlet dry-bulb temperature of the process airstream is predicted as $T_{db,2}$ = 138.4°F.

4. *Determine temperature of air leaving sensible heat wheel—point 3.*

   From the effectiveness equation, we can determine $T_{db,3}$:

$$T_{db,3} = T_{db,2} - \varepsilon_{sw}(T_{db,2} - T_{db,5}) = 138.4 - 0.7 \times (138.4 - 78) = 96.1°\text{F}$$

5. *Determine cooling coil load.*

   The post-cooling coil sensible load to cool the air from point 3 to point 4:

$$Q_{cc,sen} = m_a \cdot c_{p,a}\,\Delta T_{coil} = \left(\frac{288{,}000 \text{ ft}^3/\text{h}}{14.3 \text{ ft}^3/\text{lb}_a}\right) \times 0.24 \text{ Btu}/(\text{lb}_a \cdot °\text{F}) \times (96.1 - 70)°\text{F}$$

$$= 126{,}156 \text{ Btu/h or } 10.5 \text{ ton.}$$

   The latent load on the coil is zero since all the dehumidification has been done by the desiccant wheel, and only sensible cooling is needed by the post-cooling coil. The total cooling coil load is thus

$$\dot{Q}_{cc,tot} = \dot{Q}_{cc,sen} + \dot{Q}_{cc,lat} = 126{,}156 + 0 \text{ Btu/h} = 10.5 \text{ tons}$$

6. *Determine thermal energy needed for regeneration.*

   The regeneration energy load is predicted by desiccant selection software as

$$\dot{Q}_{reg} = 284{,}835 \text{ Btu/h}$$

This quantity could have been equally calculated from the sensible heat equation: amount required to heat 1600 cfm of air from 95°F to 260°F.

Comments

In addition to the DOAS cooling coil, the designer has to select a sensible cooling system to remove the sensible loads of the space. A chilled beam or a radiant cooling panel or DX systems such as mini split units (VRF) or chilled water–based fan coils units could be used.

20.9.3 Desiccant Materials

There are two kinds of desiccants:

1. *Absorbents* undergo a physical/chemical change when moisture is absorbed. Liquids such as lithium chloride, lithium bromide, and ethylene glycol are common examples.
2. *Adsorbents* where no physical change occurs, and the retention of moisture is mainly a surface phenomenon. Commonly used substances in this category are solids such as silica gel, zeolites, activated aluminas (oxides and hydrides of aluminum), and molecular sieves (or synthetic zeolites). New materials such as titanium silicate and synthetic polymers are also being developed.

Figure 20.42 shows the equilibrium absorption capacity of several substances. Note that the molecular sieve has the highest capacity up to 30% relative humidity, while silica gel is better between 30% and 75%—the typical humidity range for buildings. Liquid desiccants offer a number of advantages over solid desiccants. The ability to pump a liquid desiccant makes it possible to use solar energy for regeneration more efficiently. It also allows several small dehumidifiers to be connected to a single regeneration unit. Since a liquid desiccant does not require simultaneous regeneration, the liquid may be stored for later regeneration when solar heat is available. A major disadvantage is that the vapor pressure of the desiccant itself may be enough to cause some desiccant vapors to mix with the air. This disadvantage, however, may be overcome by proper choice of the desiccant material.

In commercial applications, well-maintained desiccant systems should last about 10,000–100,000 h (10–15 years) before they have to be replaced (ASHRAE Fundamentals, 2013). Solid desiccants are less chemically reactive and more sensitive to clogging (which reduces effective surface area) and to hygrothermal stress due to rapid expansion and contraction with temperature.

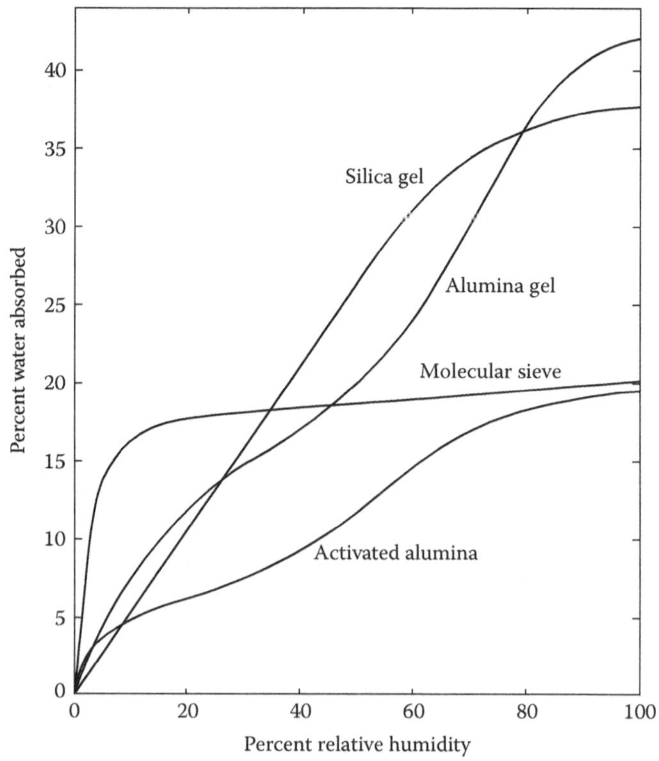

FIGURE 20.42
The water absorption characteristics of different desiccants versus relative humidity of the air.

20.9.4 Miscellaneous Issues

Desiccant systems are used in applications where the prime consideration is usually other than purely the energy efficiency standpoint. In certain applications, there is no alternative to such systems; for example, in processes requiring very low dew point temperatures (less than 30°F or 0°C) or in buildings with very high latent loads such as ice rinks and supermarkets. Desiccant-based systems, and especially HDS, allow reductions in the size of conventional systems. This reduces cost but also decreases electrical demand costs. Hence, HDS have been used in conjunction with DOAS in such traditional building types as hotels, schools, and dormitories. In general, cooling coils are a better choice for spaces with dew point temperatures greater than 40°F (4°C), while active desiccants are preferable for lower dew point temperatures. An exception to this rule is when a source of cheap heat (such as waste heat from an engine-driven chiller or a process) is available to reactivate the desiccant.

Hybrid systems allow independent control of both temperature and humidity and improve the air quality. Microbial growth in ducts and condensate drains is reduced especially in humid climates. Kovak and Heimannn (1997) state that active desiccants are not the best method for dehumidification in DOAS in spite of evidence that desiccants have a beneficial microbial sanitizing influence.

In summary, desiccant systems are best suited to handle (1) all-air systems with 100% outdoor air such as in restaurants or certain laboratories or (2) air–water hybrid systems where only the ventilation air is to be conditioned. They have an economic advantage in terms of operating costs, when the latent load fractions are high and when the regeneration of the desiccant can be done using a relatively cheap source of heat. Initial costs of desiccant material as well as the several auxiliary equipments need are the primary hindrance to their widespread adoption.

Problems

Numbers 1–4 given in parenthesis denote the degree of difficulty.

20.1 Select a diffuser from Figure 20.11 to distribute 2000 ft³/min (950 L/s) of air at a low velocity of 35 ft/min (0.2 m/s) in a 40 ft (7 m) square room. What are the recommended diffuser throat diameter and its static pressure drop if the diffuser is not equipped with a damper? *(2)*

20.2 Consider Example 20.3 where the diffuser size for a room was determined.

(a) How would the design change if the same room had a sensible load of 35 Btu/(h·ft²)? You have the flexibility to alter, by a little amount, the temperature difference between room and supply airstreams.

(b) Qualitatively discuss the differences in design if high sidewall grills are to be used. *(3)*

20.3 Summarize advantages and limitations of underfloor and displacement ventilation systems compared to traditional fully mixed systems. *(2)*

20.4 Summarize the UFAD design calculation procedure (Section 20.5.2). Discuss the conceptual approach along with advantages and limitations of the methodology. *(2)*

20.5 Consider the UFAD analysis procedure illustrated in Example 20.4. Analyze all 16 combinations of (1) ground and top floors, (2) perimeter zone and interior zone, and (3) four directions. Tabulate results and discuss general trends observed. *(2)*

20.6 Example 20.5 illustrated the sizing procedure for a passive chilled beam. You will analyze the same situation with a cooling load in the space of 5.0 kW. *(2)*

20.7 You are to design a chilled beam and DOAS hybrid system for a room whose peak loads are specified in Example 19.1. Sketch the complete systems with all components, and analyze the loads on the various equipment. Make realistic assumptions needed. *(3)*

20.8 Example 19.5 illustrated the performance of the secondary system under part-load performance. Perform a similar analysis using the chilled beam and DOAS hybrid systems designed in Problem 20.7. *(3)*

20.9 Research the DOAS concept and, from the general literature, identify different types of hybrid systems that have been proposed and/or built. Discuss benefits, limitations, and areas of application. *(2)*

20.10 What is the exit air condition (temperature and humidity) from a direct evaporative cooler at sea level with inlet air at 100°F (38°C) dry-bulb temperature and wet-bulb temperature of 60°F (16°C) if the effectiveness is 85%? *(1)*

20.11 What is the exit air condition (temperature and humidity) from a direct evaporative cooler at 5000 ft (1500 m) altitude with inlet air at 100°F (38°C) dry-bulb temperature and wet-bulb temperature of 60°F (16°C) if the effectiveness is 85%? *(1)*

20.12 What is the sensible cooling produced by a direct evaporative cooler with 9000 standard ft³/min (4230 L/s) of inlet air at 95°F (35°C) dry-bulb temperature and wet-bulb temperature of 59°F (15°C) if the effectiveness is 84%? *(1)*

20.13 What is the water consumption in the evaporative cooler in Problem 20.12? *(1)*

20.14 What is the exit air temperature from an indirect evaporative cooler at sea level if the inlet air is 93°F (34°C) and 60°F (16°C) if the PF is 65%? *(1)*

20.15 A classroom is to be sensibly cooled with a direct evaporative cooler (87% effectiveness) when the outside air condition is 95°F dry-bulb and 67°F wet-bulb (38°C and 19°C) temperatures. The room temperature is 78°F (25°C), and the sensible cooling load is 62,000 Btu/h (18 kW). What airflow is required to cool the space? *(2)*

20.16 List the 10 best states in the United States for considering evaporative cooling based on historical summer wet-bulb temperatures. *(2)*

20.17 Describe how one could calculate the reduction in flat-roof cooling load during a clear summer day that could be achieved by the use of roof sprays in a dry climate. Include the use of sol–air temperature, the psychrometric chart, and local weather data along with the roof properties. *(2)*

20.18 Repeat Example 20.8 for a direct evaporative cooler that has an effectiveness of 87%. Compare the results with the indirect cooler considered in Example 20.8. *(2)*

20.19 The following table summarizes the summer bin data for a location in the western United States (assume sea-level elevation). Find the sensible cooling energy provided by an 85% effective, direct evaporative cooler with an airflow rate of 3000 ft^3/min. The peak building sensible cooling load is 36,000 Btu/h at 95°F and decreases linearly with dry-bulb temperature to zero at 70°F. Note that the capacity of the cooler increases with decreasing outdoor temperature while the load decreases; therefore, the cooler will cycle on and off at temperatures below 95°F. Summarize your solution by completing the last three lines of the table. What is the seasonal COP if the fan motor is rated at 0.25 hp? *(3)*

	Temperature Bin		
	1	2	3
Dry-bulb range, °F	91–100	81–90	71–80
Coincident wet-bulb temperature, °F	62	57	54
Bin hours	120	420	490
Cooling load, Btu/h			
Available sensible cooling rate, Btu/h			
Bin total sensible cooling, Btu			

20.20 The following table summarizes the summer bin data for a location in the western United States (assume sea-level elevation). Find the sensible cooling provided by an 85% effective, direct evaporative cooler with an airflow rate of 1300 L/s. The peak building load is 10 kW at 38°C and decreases linearly with the dry-bulb temperature to zero at 21°C. Note that the capacity of the cooler increases with decreasing outdoor temperature, while the load decreases; therefore, the cooler will cycle on and off at temperatures below 38°C. Summarize your solution by completing the last three lines of the table. What is the seasonal COP if the fan motor is rated at 180 W? *(3)*

	Temperature Bin		
	1	2	3
Dry-bulb range, °C	33–38	27–32	22–26
Coincident wet-bulb temperature, °C	17	14	12
Bin hours	120	420	490
Cooling load, kW			
Available sensible cooling rate, kW			
Bin total sensible cooling, kWh			

20.21 Repeat Example 20.8 for Chicago, a more humid site for which the relevant weather data are summarized as follows. *(3)*

Bin, °F	Hours per Bin	Wet-bulb temperature, °F
90–99	51	76
80–89	380	71
70–79	794	66
60–69	716	61

20.22 The effectiveness of one of the most common evaporative media used in large evaporative coolers for commercial buildings can be calculated from

$$\varepsilon_{evap} = 1 - e^{-4.76t/v^{0.33}}$$

where
v is the air velocity in feet per second
t is the media pad thickness in feet

Plot the effectiveness for a 6 in. thick pad for a range of air velocities between 2 and 20 ft/s. Make a similar plot for a fixed velocity of 8 ft/s for pad thicknesses varying between 2 and 12 in. In view of these results (assuming that the pressure drop through the pad increases linearly with thickness), discuss the trade-offs involved in selecting the optimal pad thickness, including the effects of fan power and cooling produced. *(3)*

20.23 Consider the chiller of Problem 14.6 to be coupled to an evaporative cooler as shown in Figure 20.36. If outdoor air at 95°F (38°C) dry-bulb and 65°F (18°C) wet-bulb temperatures passes through an 85% effective, direct evaporative cooler prior to entering the air-cooled condenser, what improvements in chiller performance will result? Assume, for every degree of condenser inlet air temperature drop, that the condenser temperature also drops one degree. Find the change in capacity and COP resulting from the use of the cooler. *(4)*

20.24 Write an essay of about 400 words describing some of the advantages and drawbacks of desiccant systems. Describe their historic evolution, application areas, maturity in terms of technology, cost, and current market penetration. What types of hybrid systems could benefit from this technology? *(2)*

20.25 Solve this problem following steps described in Example 20.9 using design conditions for the high humidity space specified in Example 19.3. To simplify the problem assume that there is no steam leakage in the desiccant wheel and that it can lower the airstream to the desired humidity level. Make appropriate assumptions. *(3)*

References

ASHRAE 55 (2013). *Standard 55-2013: Thermal Environmental Conditions for Human Occupancy*. American Society of Heating, Refrigerating and Air-Conditioning Engineers, Atlanta, GA.

ASHRAE 62.1 (2013). *Standard 6.1-2013: Ventilation for Acceptable Indoor Air Quality*. American Society of Heating, Refrigerating and Air-Conditioning Engineers, Atlanta, GA.

ASHRAE 70 (2011). *Method of Testing for Rating the Performance of Air Outlets and Inlets*. ANSI/ASHRAE Standard 70-2006 (RA 2011), American Society of Heating, Refrigerating and Air-Conditioning Engineers, Atlanta, GA.

ASHRAE 189.1 (2009). *Standard for the Design of High-Performance Green Buildings Except Low-Rise Residential Buildings*. American Society of Heating, Refrigerating and Air-Conditioning Engineers, Atlanta, GA.

ASHRAE Applications (2015). *HVAC Applications*. American Society of Heating, Refrigerating and Air-Conditioning Engineers, Atlanta, GA.

ASHRAE Fundamentals (2005, 2013). *Handbook of Fundamentals*. American Society of Heating, Refrigerating and Air-Conditioning Engineers, Atlanta, GA.

ASHRAE UFAD (2013). *Underfloor Air Distribution (UFAD) Design Guide*. American Society of Heating, Refrigerating and Air-Conditioning Engineers, Atlanta, GA.

Barbaree, J. (1991). Controlling *Legionella* in cooling towers. *ASHRAE J.*, 33, 38–42.

Bauman, F., S. Schiavon, T. Webster, and K.H. Lee (September 2010). Cooling load design tool for UFAD. *ASHRAE J.*, 52(9), 62–71.

Bauman, F., T. Webster, and C. Benedek (October 2007). Cooling airflow design calculations for UFAD. *ASHRAE J.*, 49(10), 36–44.

Daly, A. (May 2002). Underfloor air distribution: Lessons learnt. *ASHRAE J.*, 44(6), 21–24.

Dowdy, J.A., R.L. Reid, and E.T. Handy (1986). Experimental determination of heat- and mass-transfer coefficients in Aspen pads. *ASHRAE Trans.*, 92(2), 60–70.

FEMP (April 1997). *Two-Wheel Desiccant Dehumidification System*. Federal Energy Management Program, U.S. Department of Energy, Washington, DC.

Filler, M. (October 2004). Best practices for underfloor air systems. *ASHRAE J.*, 46(10), 39–44.

Hodges, B.K., J.W. Stevens, and A. Jalazadeh (1997). Desiccant dehumidification curriculum module for engineering/technology HVAC courses. Department of Mechanical Engineering, Mississippi State University, Starkville, MS.

Int-Hout, D. (December 2012). Methods for effective room air distribution—Part two. *ASHRAE J.*, 54(12), 38–41.

John, D.A. (April 2012). Effects of typical inlet conditions on air outlet conditions. *ASHRAE J.*, 54(4), 16–22.

Kovak, B. and P.R. Heimann (April 1997). The sanitizing effects of desiccant-based cooling. *ASHRAE J*, 39(4), 60–64.

Liddament, M.W. (2001). Chapter 13: Ventilation strategies, in *Indoor Air Quality Handbook* (J.D. Spengler, J.M. Samet, and J.F. McCarthy, eds.). McGraw-Hill, New York.

Liesen, R.J. and C.O. Pederson (1991). Development and demonstration of an evaporative cooler simulation model for the BLAST energy analysis computer program. *ASHRAE Trans.*, 97(2), 866–873.

Livchak, A. and C. Lowell (April 2012). D'ont turn active beams into expensive diffusers. *ASHRAE J.*, 54(4), 52–60.

Montanya, E.C., D. Keith, and J. Love (July 2009). Integrated design and UFAD. *ASHRAE J.*, 51(7), 30–40.

Montgomery, R.D. (June 2009). UFAD commissioning for air force base. *ASHRAE J.*, 51(6), 38–45.

Mumma, S.A. (2001a). Dedicated OA systems, comments/letters: IAQ applications *Winter*, 20–22.

Mumma, S.A. (2001b). Overview of integrating dedicated outdoor air systems with parallel terminal systems. *ASHRAE Trans.*, 107(1), 545–552.

NovelAire (2009). Desiccant wheel software ver. 3.20. NovelAire Technologies, Baton Rouge, LA.

Price (2011). *Engineer's HVAC Handbook: A Comprehensive Guide to HVAC Fundamentals*. Price Industries Limited, Winnipeg, Manitoba, Canada.

Rydberg, J. and P. Norback (1949). Air distribution and draft, ASHVE Research Report No. 1362. *ASHVE Trans.*, 5, 225.

Sastry, G. and P. Rumsey (May 2014). VAV vs Radiant: Side-by-Side Comparison, *ASHRAE Journal*, 56(5), 16–24.

Schiavon, S., F. Bauman, K.H. Lee, and T. Webster (July 2010). Development of a simplified cooling load design tool for underfloor air distribution systems. Final Report to CEC PIER Program, University of California Berkeley, CA. http://escholarship.org/uc/item/70f4n03z.

Schlichting, H. (1979). *Boundary Layer Theory*, 7th ed. McGraw-Hill, New York.

Stein, J. and S.T. Taylor (May 2013). VAV reheat versus active chilled beams and DOAS. *ASHRAE J.*, 55(5), 18–32.

Supple, R. (1982). Evaporative cooling for comfort. *ASHRAE J.*, 24(8), 42.

USDOE (2002). *Energy Consumption Characteristics of Commercial Building HVAC*. USDOE, Washington, DC.

Watt, J.R. (1986). *Evaporative Air Conditioning Handbook*. Chapman & Hall, New York.

Webster, T., F. Bauman, and J. Reese (May 2002). Underfloor air distribution: Thermal stratification. *ASHRAE J.*, 44(5), 28–36.

21

HVAC Control Systems

ABSTRACT HVAC controls have been likened to the nervous system of the human body which receives and sends signals from and to the various organs to enable them to perform their functions properly. One of the major causes of energy inefficiency in actual buildings is due to controls not performing as intended. Though this is often an issue with building operation and maintenance, the building designer should also be familiar with HVAC controls in order to be able to perform his/her job satisfactorily. This chapter deals with several aspects relevant to building HVAC control systems. The introductory review includes a brief discussion of local and supervisory control, the elements of a control system, and the role of the control designer. We then describe the four different modes of feedback control after which a discussion of the major types of control hardware (pneumatic, electronic, and DDC) is presented. Subsequently, various control components are discussed followed by examples of important HVAC control systems and control sequences relevant to large cooling systems. This chapter, though primarily meant to cover local control issues, also touches upon building automation systems (BAS) and energy management systems (EMS) or energy management or control systems (EMCS). We also briefly describe data exchange/communications protocols such as BACnet. This chapter concludes by providing basic mathematical concepts of local control theory related to modeling, control stability assessment, and control system simulation.

Nomenclature

A	Valve authority (Equation 21.17)
a, b, c	Coefficients appearing in Equation 21.22
C_v	Dimensional flow coefficient of valve (Equation 21.18)
$E(s)$	Laplace transform of error e
e	Error
$F(s)$	Laplace transform of the forcing function
$f(t)$	Forcing function in time domain
$G(s)$	Combined transfer function of controller and process (Equation 21.33)
I	Current, amp.
K	Steady-state heat loss coefficient, (Equation 21.30) W/K (Btu/(h·°F))
K	Valve size constant (Equation 21.12)
K_d	Derivative controller constant
K_i	Integral controller constant
K_p	Proportional controller constant
k	Temperature coefficient of resistor, K^{-1} ($°F^{-1}$)
k	Valve proportionality constant (Equation 21.11)
L	Time lag or delay, s
L_{stat}	Time lag of thermostat, s
M	Magnitude of step change
$\dot{m}$	Mass flow rate, kg/s (lb_m/h)
NC	Normally closed
NO	Normally open
O	Response
p	Pressure, Pa ($lb_f/in.^2$)
$\dot{Q}$	Heat rate or heat flow per unit time, W (Btu/h)
R	Resistance, ohm
R_o	Resistance at reference temperature, Ω
s	Complex variable
T	Temperature, °R or °F (K or °C)
TF	Transfer function (Equation 21.24)
t	Time, s or h
V	Voltage, controller output
V_o	Constant value of controller output with no error
$\dot{V}$	Volumetric flow rate, m^3/s or L/s (ft^3/min)
$V(s)$	Laplace transform of controller output
v	Specific volume, m^3/kg (ft^3/lb_m)
$Y(s)$	Laplace transform of output function
$y(t)$	Independent variable function of time
z	Valve stem position

Greek

ε	Coil effectiveness
τ	Time constant
Δp	Pressure drop, Pa (lb_f/in^2)
ΔV_{max}	Throttling range, °C (°F)

Subscripts

max	Maximum
min	Minimum
s	Sensor

set Set point
stat Thermostat
sys System
v Valve

21.1 Introductory Concepts

21.1.1 Need for Control

In the previous chapters, we assumed that indoor environmental conditions are maintained at the desired setpoints by the HVAC system sized for that purpose. We discussed how to size systems and made calculations of energy consumption at peak and off-peak conditions. HVAC system controls are the information link between varying energy demands on a building's primary and secondary systems and the (usually) approximately uniform demands for indoor environmental conditions. Without an adequately designed and properly functioning control system, even the most expensive, most thoroughly designed HVAC system will not operate as expected. It simply will not control indoor conditions to provide comfort.

The HVAC designer must design a control system that

- Sustains a comfortable building interior environment
- Maintains an acceptable indoor air quality
- Is as simple and inexpensive as possible and yet meets HVAC system operation criteria reliably for the system lifetime
- Results in energy efficient HVAC system operation under all conditions

According to Coad (1990), the "majority" of HVAC control systems installed in the past 60 years in buildings (in the United States) are, if not maintained, not "performing as intended." A considerable challenge is presented by this state of affairs to the HVAC system designer. One reason for poor control operation is inadequate design or unclear assignment of responsibility for control system design. In this chapter, devoted entirely to controls, we present the rudiments of control design from the point of view of the HVAC system designer. There are numerous references available to the interested reader on the subject of HVAC control; some of them are ASHRAE Fundamentals (2013), Haines (1987), Honeywell (1988), Letherman (1981), Levenhagen and Spethmann (1993), Horan (1997), and Shadpour (2001).

21.1.2 Local versus Supervisory Control

There are two types of control systems: local loop and supervisory. *Local loop control* involves maintaining the various subsystems or pieces of equipment at their desired operating conditions. For example, a simple residential HVAC system can involve a small chiller and fans to move the air over the DX coil and into the room. The thermostat maintains a preset temperature by switching the fans and the chiller on/off. This is an example of a single-loop control. On the other hand, a large building with several zones and with multiple boilers, chillers, and cooling towers (see Figure 11.4) has several control loops. Since the equipment operate primarily under part-load conditions, (1) the various pieces of equipment need to be sequenced (or scheduled, i.e., shut down or brought online) and then (2) each piece of equipment brought online needs to be operated such that the overall energy consumption (electricity and gas) of the physical plant is minimized. Such a strategy, called *supervisory control*, requires global optimization, but this offers the advantage of considerable energy savings (ASHRAE Applications, 2015; Braun et al., 1989). Section 18.8.2 discusses some issues and proposes control sequences relevant to large cooling systems. Much of this chapter deals with local control issues, with Section 21.5.4 addressing supervisory control rather cursorily. Section 21.6 introduces the basic concepts of networking of various building systems so as to provide holistic control, automation, and energy management. Data exchange/communications protocols such as BACnet are also briefly described.

21.1.3 Elements of a Control System

A control system essentially has three elements:

1. *Sensors*, which allow the determination of the existing conditions of the controlled variable. The most common ones in HVAC are thermostats (which measure temperature), humidity sensors, flow sensors, and pressure sensors. There are different types of sensors for the same variable to be controlled; for example, flow sensors can be vane, paddle, ultrasonic, etc. The common types of sensors are described in Section 21.4.1, while Table 21.1 assembles some of the important sensor operating characteristics.
2. *Controllers*, which compare the existing controlled variable condition to what is desired and then transmit a signal to a device to correct for any difference. Some controllers can be integral with the sensors such as the house

TABLE 21.1
Sensor Operating Characteristics

Characteristic	Description	Example/Discussion
Full-scale range	Operating range of sensor	Input range: −10°F to 200°F Output range: 0–5 V
Signal span	Difference between maximum and minimum values of the sensor operating range	Input span = 200°F − (−10°F) = 210°F Output span = 5 − 0 = 5 V
Accuracy	Maximum difference between actual value of measured variable and value indicated by sensor divided by span	Let measured value = 72°F and actual value = 75°F, then accuracy = absolute (72 − 75)/210 = 0.014 or 1.4%
Sensitivity	Change in output signal in response to change in input signal	Sensitivity = output span by input span = (5/210) = 0.0238 V/°F
Resolution	Smallest change in measured variable required to produce a detectable change in sensor's output signal	Too little a resolution is not always desirable since noise in the signal can be recorded
Time constant	Lag time or time taken for a sensor to reach 63% of the total change in response to a step change in the input signal	Concept similar to that appearing in transient heat conduction (Section 8.4) and also discussed in Section 21.7.2

TABLE 21.2
Control Terminology

Terminology	Description of Device
Direct action	Increase in input signal results in an increase in output signal
Reverse action	Increase in input signal results in a decrease in output signal
Normally open	In case of power loss or control signal failure, the fail-safe position of the damper or valve is to be fully open
Normally closed	In case of power loss or control signal failure, the fail-safe position of the damper or valve is to be fully closed
Proportional band	The throttling range of the controlled variable over which the controller output goes from fully opened to fully closed
Error signal	The difference between the setpoint and the actual control point
Dead band	Range of the controlled variable over which the output of the controller does not change

thermostat. The common types of controllers are pneumatic, electric, electronic, and microprocessor, and these are discussed in Section 21.3. Important control terminology is provided in Table 21.2.

3. *Actuators,* which adjust component/device performance to achieve the desired controlled variable condition by acting on the command from the controller. For example, actuators turn equipment on/off or adjust fluid flow by modulating valves or dampers. Actuators are the interface between the control system and the mechanical/electrical control equipment.

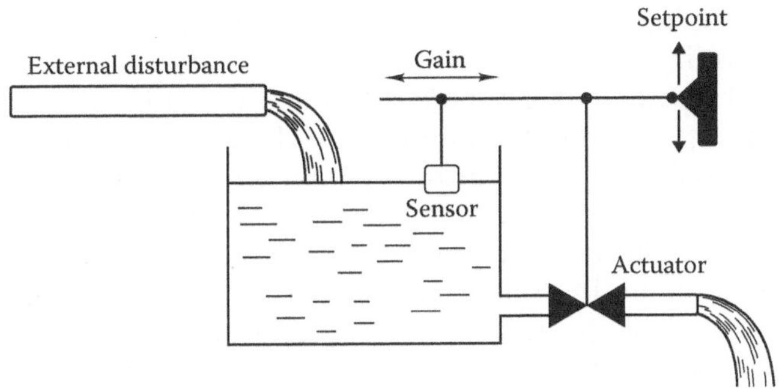

FIGURE 21.1
Simple water-level controller. The setpoint is the full water level; the error is the difference between the full level and the actual level.

Figure 21.1 shows a familiar control problem. It is necessary to maintain the water level in the tank under varying inlet flow rates. The float operates a valve, controlling flow from the tank. This simple system includes all the elements of a control system:

- *Sensor* is the float that reads the water level.
- *Controlled system characteristic* is the water flow, often termed the *controlled variable.*
- *Controller* is the linkage connecting float to valve stem.
- *Actuator* (*controlled device*) is the internal valve stem mechanism that sets the valve (the final control element) position.

This system is called a *closed-loop* or *feedback* system because the sensor (float) is directly affected by the action of the controlled device (valve). In an *open-loop* system, there is no sensor to measure the controlled variable. An example would be a method of controlling the valve based on an external parameter such as the time of day that may have an indirect relation to water consumption from the tank. There are four common methods of control, of which Figure 21.1 shows but one. In the next section, we describe each with relation to an HVAC system example.

21.1.4 Duties of the Control Designer

To achieve proper control based on the control system design, the HVAC system itself must be constructed and calibrated according to the mechanical system drawings. These must include properly sized primary and secondary systems. In addition, air stratification must be avoided, proper provision for control sensors is required, freeze protection is necessary in cold climates, and proper attention must be paid to minimizing energy consumption subject to reliable operation and occupant comfort. Proper maintenance is essential as well.

The principal, ultimate controlled variable in buildings is the zone temperature (and, to a lesser extent, air quality in some buildings). Of course, the control of zone temperature involves many other types of control within the primary and secondary HVAC systems, including boiler and chiller control, pump and fan control, liquid and airflow control, humidity control, and auxiliary system control (e.g., thermal storage control). This chapter discusses only *automatic control* of these subsystems. Honeywell (1988) defines an *automatic control system* as "a system that reacts to a change or imbalance in the variable it controls by adjusting other variables to restore the system to the desired balance."

The HVAC engineer identifies the control sequences at an advanced stage of the design process. At that point, the HVAC system (both primary and secondary) has been conceptualized, HVAC sizing has been done at least in a preliminary way, and zoning has been completed. The HVAC engineer's or control designer's key task is to crystallize every detail of the control system's logic under all operating conditions. The products of the control design exercise are as follows:

1. Control specification
2. Control drawings
3. Control system equipment lists (schedules)—sensors, controllers, actuators, control equipment (valves, dampers, etc.), software

21.2 Modes of Feedback Control

The four common modes of relating the error (difference between desired setpoint and sensed value of controlled variable; see Figure 21.1) to the corrective action to be taken by the controller are as follows: (1) two-position (on/off), (2) proportional, (3) integral, and (4) derivative.

The last three are usually used in a variety of combinations with one another. Figure 21.2a shows a steam coil used to heat air in a duct. The simple control system shown includes an air temperature sensor, a controller that compares the sensed temperature to the setpoint, a steam valve controlled by the controller, and the coil itself. We will use this example system as a point of reference when discussing the various control system types in the succeeding text. Figure 21.2b is the *control diagram* corresponding to the physical system shown in Figure 21.2a. The various terms and notations are discussed in Section 21.7.

21.2.1 Two-Position

Two-position control applies to an actuator that is either fully open or fully closed. In Figure 21.2a, the valve is a two-position valve if two-position control is used. The position of the valve is determined by the value of the coil outlet temperature. Figure 21.3 depicts two-position control of the valve as follows: If the air temperature drops below 95°F (35°C), the valve opens and remains open until the air temperature reaches 100°F (38°C). The differential is usually adjustable, as is the temperature setting itself. Two-position control is the least expensive method of automatic control and is suitable for the control of HVAC systems with large time constants. Examples include residential space and water heating systems. Systems that are fast reacting should not be controlled by using this approach since inaccurate control may result.

21.2.2 Proportional (P)

Proportional control corrects the controlled variable in proportion to the difference between the controlled variable and the setpoint. For example, a proportional controller would make a 10% increase in the coil heat output rate in Figure 21.2 if a 10% decrease in the coil outlet air temperature were sensed. The proportionality constant between the error and the controller output is called the *gain* K_p. A proportional controller can be modeled as follows:

$$V = V_o + K_p e \tag{21.1}$$

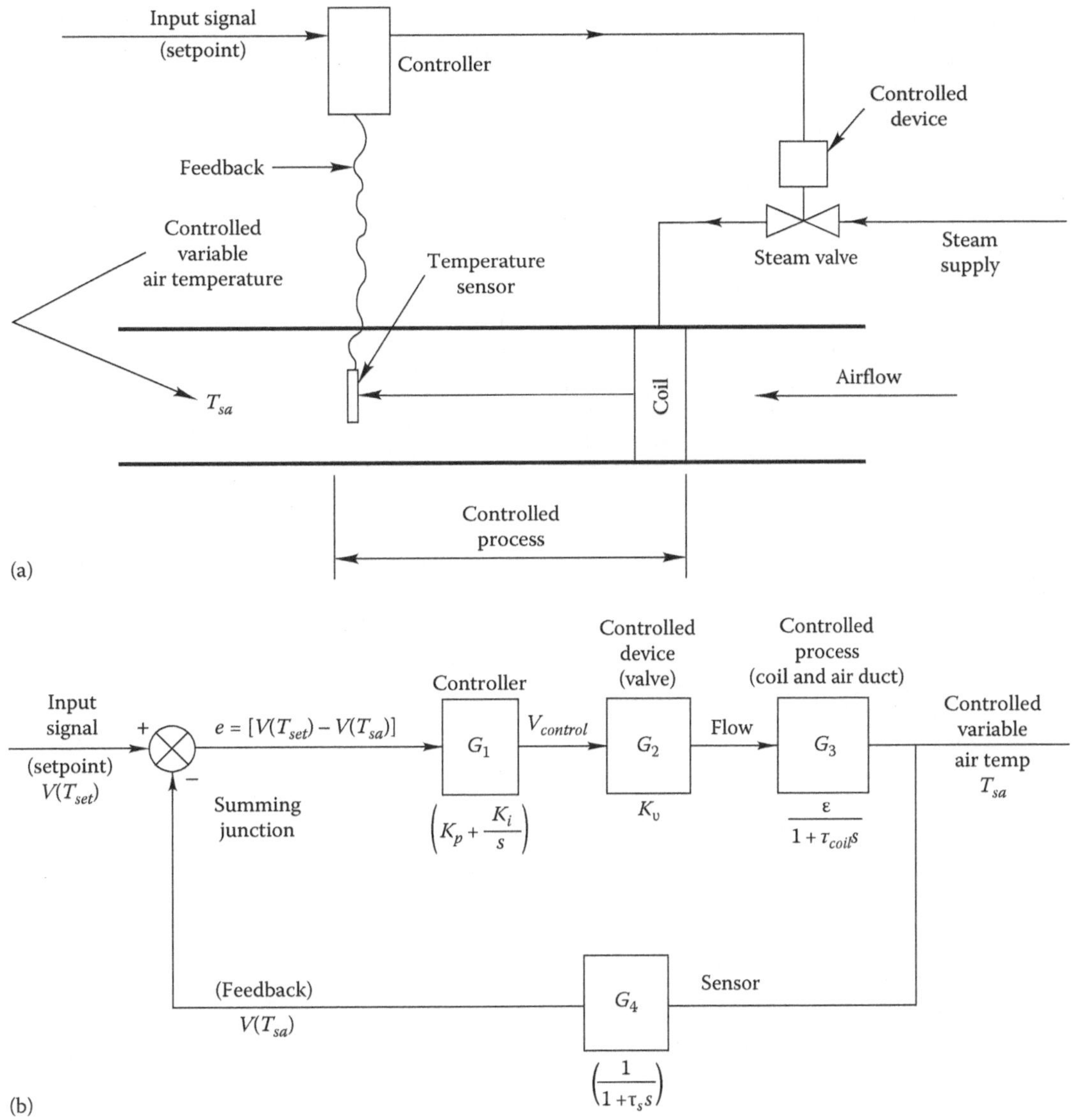

FIGURE 21.2
(a) Simple heating coil control system showing the process (coil and short duct length), controller, controlled device (valve and its actuator), and sensor. The setpoint entered externally is the desired coil outlet temperature. (b) Equivalent control diagram for heating coil. The *G*'s represent functions relating the input to the output of each module (the expressions in parentheses are examples of the *G*'s discussed in Section 21.7 on Laplace transforms). Voltages *V* represent both temperatures (setpoint and coil outlet) and the controller output to the valve in electronic control systems.

where
- V is the controller output; symbol V is used since in electronic controls the controller output is often a voltage
- V_o is the constant value of controller output when no error exists at control range midpoint
- e is the error

In the case of the steam coil, the *error* e is the difference between the sensed air temperature T_{sensed} and the needed air temperature (i.e., the setpoint T_{set}):

$$e = T_{set} - T_{sensed} \tag{21.2}$$

A progressively lower coil air outlet temperature results in a greater error e, and, consequently, the control action would lead to a greater steam flow rate.

The *throttling range* ΔV_{max} is the total change in the controlled variable that is required to cause the actuator or controlled device to move between its limits. For example, if the nominal temperature of a zone is 72°F (22°C) and the heating controller throttling range is 6°F (3°C), then the heating control undergoes its full travel between a zone temperature of 69°F and 75°F (21°C and 24°C). This control, whose characteristic is shown in Figure 21.4, is *reverse-acting*; i.e., as the temperature

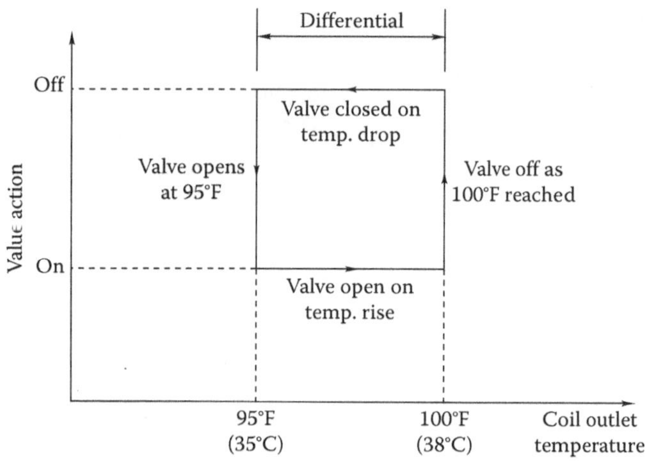

FIGURE 21.3
Two-position (on/off) control characteristic for the steam valve in Figure 21.2a.

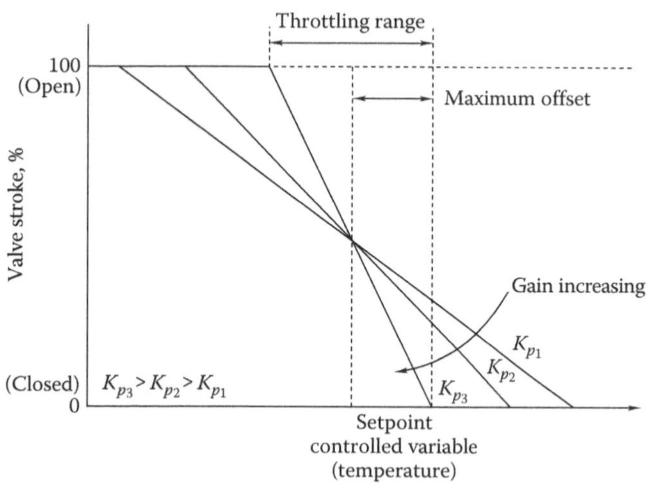

FIGURE 21.4
Proportional control characteristic showing various throttling ranges and the corresponding proportional gains K_p. This characteristic is typical of a heating coil temperature controller.

(controlled variable) increases, the heat output of the coil decreases (the heating valve position decreases).

Another way of expressing the control characteristic is by its gain. The gain is inversely proportional to the throttling range, as shown in Figure 21.4. The narrower the throttling range (i.e., the steeper the characteristic), the greater the numerical value of the slope of the line and the greater the gain. Beyond the throttling range, the system is out of control. In actual hardware, one can set the setpoint and either the gain or the throttling range (most common), but not both of the latter two since one determines the other. Proportional control by itself is not capable of reducing the error in heat output (except at the midpoint) to zero, since an error is needed to produce the capacity required for meeting a load, as we will see in Example 21.1. This unavoidable value of the error in proportional systems is called the *offset*. From Figure 21.4, it is easy to see that the offset is larger for systems with smaller gains. As we will see later, there is a limit to which one can increase the gain to reduce offset, because high gains can produce control instability, i.e., unpredictable and unsteady control.

Example 21.1: Proportional Gain Calculation

If the steam heating coil in Figure 21.2a has a heat output that varies from 0 to 20 kW as the outlet air temperature varies from 35°C to 45°C in an industrial process (with fixed input temperature), what is the coil gain and what is the throttling range? Find an equation relating the heat rate at any sensed air temperature to the maximum rate in terms of the gain and setpoint.

Given: $\dot{Q}_{max} = 20$ kW, $\dot{Q}_{min} = 0$ kW, $T_{max} = 45$°C, $T_{min} = 35$°C, $T_{set} = 40$°C (coil output at actuator midtravel)

Figure: See Figure 21.2a.

Assumption: Steady-state operation

Find: K_p, ΔV_{max}

Solution

The throttling range is the range of the controlled variable (air temperature) over which the controlled system (heating coil) exhibits its full-capacity range. The temperature varies from 35°C to 45°C; therefore the throttling range is

$$\Delta V_{max} = 45 - 35 = 10°\text{C} \tag{21.3}$$

The proportional gain is the ratio of the controlled system (coil) output to the throttling range. For this example, the gain is

$$K_p = \frac{\dot{Q}_{max} - \dot{Q}_{min}}{\Delta V_{max}} = \frac{(20-0)\text{ kW}}{10\text{ K}} = 2.0\text{ kW/K} \tag{21.4}$$

The controller characteristic can be found by inspecting Figure 21.4. It is assumed that the average air temperature T_{set} (40°C) occurs at the average heat rate (10 kW). The equation of the sloped lines shown in Figure 21.4 is

$$\dot{Q} = K_p(T_{set} - T_{sensed}) + \frac{\dot{Q}_{max}}{2} = K_p e + \frac{\dot{Q}_{max}}{2} \tag{21.5}$$

The quantity $T_{set} - T_{sensed}$ is the *error e* and is the signal to the control system that a valve-opening change is needed to meet the desired setpoint.

Inserting the numerical values, we have

$$\begin{aligned}\dot{Q} &= (2.0\text{ kW/K})(40 - T_{sensed}) + 10\text{ kW}\\ &= (2.0\text{ kW/K})(T_{set} - T_{sensed}) + 10\text{ kW}\end{aligned} \tag{21.6}$$

Comments

In an actual steam coil control system, it is the steam valve that is controlled directly in order to indirectly control the heat rate of the coil. This is typical of many HVAC system controls in that the desired control action is achieved indirectly by controlling another variable, which in turn accomplishes the desired result. That is why the controller and controlled device are often shown separately, as in Figure 21.2b.

This example illustrates in a simple system why proportional control always requires an error signal with the result that there is always an *offset* (i.e., the desired setpoint and the actual temperature are always different except at one point when $\dot{Q} = \dot{Q}_{max}/2$) with this control mode. Equation 21.5 shows that for any coil heat output other than $\dot{Q}_{max}/2$, an error must exist. This error is the offset and is smaller for higher coil gains.

Proportional control is used with stable, slow systems that permit the use of a narrow throttling range and resulting small offset. Fast-acting systems need wide throttling ranges to avoid instability and large offset results. In Section 21.7.3, we will examine the quantitative criteria for determining the stability of proportional control systems.

21.2.3 Proportional Integral

Integral control is often added to proportional control to eliminate the offset inherent in proportional-only control. The result, *proportional plus integral control*, is identified by the acronym *PI*. Initially, the corrective action produced by a PI controller is the same as for a proportional-only controller. After the initial period, a further adjustment due to the integral term reduces the offset to zero. The rate at which this occurs depends on the time scale of the integration. In equation form, the PI controller is modeled by

$$V = V_o + K_p e + K_i \int e\,dt \tag{21.7}$$

An alternate form is sometimes used:

$$V = V_o + K_p\left(e + K_i' \int e\,dt\right) \quad \text{where } K_i' = \frac{K_i}{K_p}$$

where K_i is the integral gain constant. It has units of reciprocal time and is the number of times that the integral term is calculated per unit time. This is also known as the *reset rate*; *reset control* is an older term used by some to identify integral control.

The integral term in Equation 21.7 has the effect of adding a correction to the output signal V as long as the error term exists. The continuous offset produced by the proportional-only controller can thereby be reduced to zero because of the integral term. The time scale $K_i' = K_p/K_i$ of the integral term is on the order of 1–60 min. PI control is used for almost all applications in which the sensor is near the process and for fast-acting systems for which accurate control is needed. Examples include mixed-air controls, duct static pressure controls, and coil controls. Because the offset is eventually eliminated with PI control, the throttling range can be set rather wide, to improve stability under a wider range of conditions than that provided by proportional control alone. We will learn in Section 21.7 that integral control can cause stability problems under certain conditions; therefore, it has mixed benefits.

21.2.4 Proportional Integrative Differential

Derivative control is used to speed up the action of PI control. When derivative control is added to PI control, the result is called *proportional integrative differential (PID) control*. The derivative term added to Equation 21.7 generates a correction signal proportional to the time rate of the change of error. This term has little effect on a steady proportional system with uniform offset (time derivative is zero) but initially, after a system disturbance, produces a larger correction more rapidly. Equation 21.8 includes the derivative term in the mathematical model of the PID controller:

$$V = V_o + K_p e + K_i \int e\,dt + K_d \frac{de}{dt} \tag{21.8}$$

or, alternatively,

$$V = V_o + K_p\left(e + K_i' \int e\,dt + K_d' \frac{de}{dt}\right) \quad \text{where } K_d' = \frac{K_d}{K_p}$$

where K_d is the derivative gain constant. The time scale K_d/K_p of the derivative term is typically in the range of 0.2–15 min. Since HVAC systems do not often require rapid control response, the use of PID or PD control is less common than the use of PI control. However, it is used, for example, whenever there is a long delay between the process (e.g., a heating coil) and the sensor (e.g., a zone's thermostat). Since a derivative is involved, any noise in the error (i.e., sensor) signal must be avoided to maintain good control. One application in buildings where PID control has been effective is in duct static pressure control, a fast-acting subsystem that could otherwise be unstable.

Figure 21.5 compares the reaction of the three systems discussed above to a step change in load on a coil. The proportional system never achieves the setpoint. However, since the integral term generates a correction proportional to the area under the error curve, it slowly reduces the error to zero, but the maximum error is greater than for the PID approach, which takes action faster to reduce the error than does the PI system alone. The PD system (not shown) never achieves zero error since the zero derivative of the constant offset produces no control action.

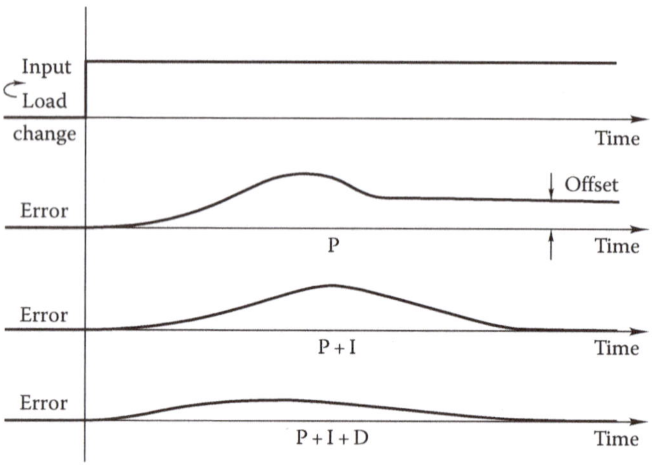

FIGURE 21.5
Performance comparison of P, PI, and PID controllers when subjected to a uniform input step change.

21.3 Basic Control Hardware

In this section, we describe the various physical components needed to achieve the actions required by the control strategies discussed in the previous section. Since there are two fundamentally different control approaches—pneumatic and electronic—the following material is so divided. Sensors, controllers, and actuators for principal HVAC applications are described. The design of these components is discussed in Section 21.3, while Section 21.5 describes the design of several complete control systems.

21.3.1 Pneumatic

The first widely adopted automatic control systems used compressed air as the operating medium. In the early 1990s, about 50% of the sales of one large controls company was pneumatic equipment. This has, however, changed greatly; pneumatic controls are almost never used in new construction, and during old building renovation are being replaced by DDC controls. Pneumatic controls deliver compressed air (in the range of 15–25 psig [100–170 kPa gauge] in the United States) to the actuators via pneumatic tubing (often 1/4th in. diameter or 6 mm). They are fast acting, but are used in stand-alone mode and cannot be networked. Though they have been largely superseded, we shall give an overview of how these devices operate.

Since temperature is most often the parameter to be controlled, the most common pneumatic *sensor* is the temperature sensor. Figure 21.6 shows one method of sensing temperature and producing a control signal

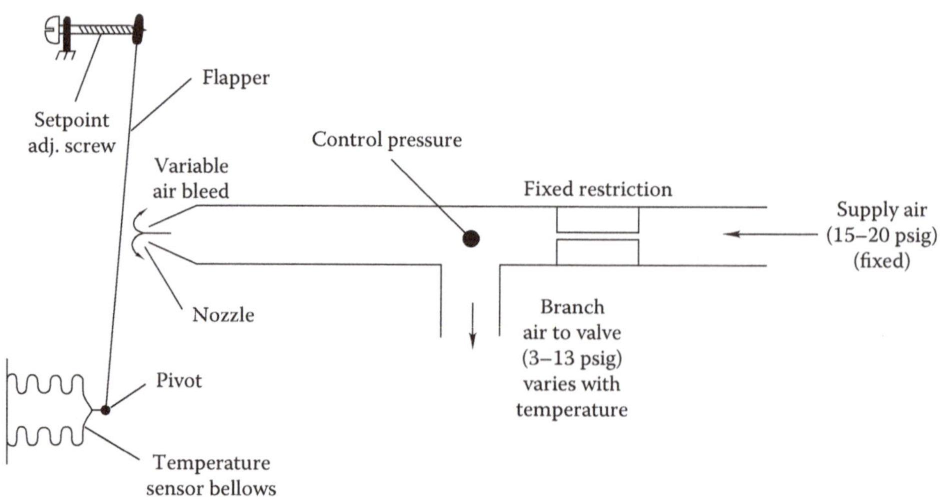

FIGURE 21.6
Drawing of pneumatic thermostat showing adjustment screw used to change temperature setting.

from it. Main supply air is delivered by a compressor to the zone thermostat. An amount of this air is bled from the nozzle depending on the position of the flapper and the size of the restrictor (diameters on the order of thousandths of an inch [hundredths of millimeters] are typical). The pressure in the branch line that controls the valve ranges between 3 and 13 psig (20 and 90 kPa) typically. In simple systems, this pressure from a thermostat could operate an actuator such as a control valve for a room heating unit. In this case, the thermostat is both the sensor and the controller—a rather common configuration. The consumption of air in this sensor is quite small.

Many other temperature sensor approaches can be used. For example, the bellows shown in Figure 21.6 can be eliminated, and the flapper can be made of a bimetal strip. As the temperature changes, the bimetal strip changes curvature, opening or closing the flapper-nozzle gap. Another approach uses a remote bulb filled with either liquid or vapor that pushes a rod (or a bellows) against the flapper to control the pressure signal. This device is useful if the sensing element must be located where direct measurement of temperature by a metal strip or bellows is not possible, such as in a water stream or high-velocity ductwork. The bulb and connecting capillary size may vary considerably by application. Bulbs can be up to several feet long in freeze control sensors, and capillaries up to 30 ft long are available.

Pressure sensors may use either bellows or diaphragms to control the branch line pressure. For example, the motion of a diaphragm may replace that of the flapper in Figure 21.6 to control the bleed rate. A bellows similar to that shown in the figure may be internally pressurized to produce a displacement that can control the air bleed rate. A bellows produces significantly greater displacements than a single diaphragm.

Humidity sensors in pneumatic systems are made from materials that change size with the moisture content. Nylon or other synthetic hygroscopic fibers that change size significantly (i.e., 1%–2%) with humidity are commonly used. Since the dimensional change is relatively small, mechanical amplification of the displacement is used. The materials that exhibit the desired property include nylon, hair, and cotton fibers. Since the properties of hair vary with age, the more stable material—nylon—is preferred and most widely used (Letherman, 1981). Humidity sensors for electronic systems are quite different and are discussed in Section 21.3.2.

An *actuator* converts pneumatic energy to motion—either linear or rotary. It creates a change in the controlled variable by operating control devices such as dampers or valves. Figure 21.7 shows a pneumatically operated control valve. The valve opening is controlled by the pressure in the diaphragm acting against the spring. The spring is essentially a linear device. Therefore, the motion of the valve stem is essentially linear with air

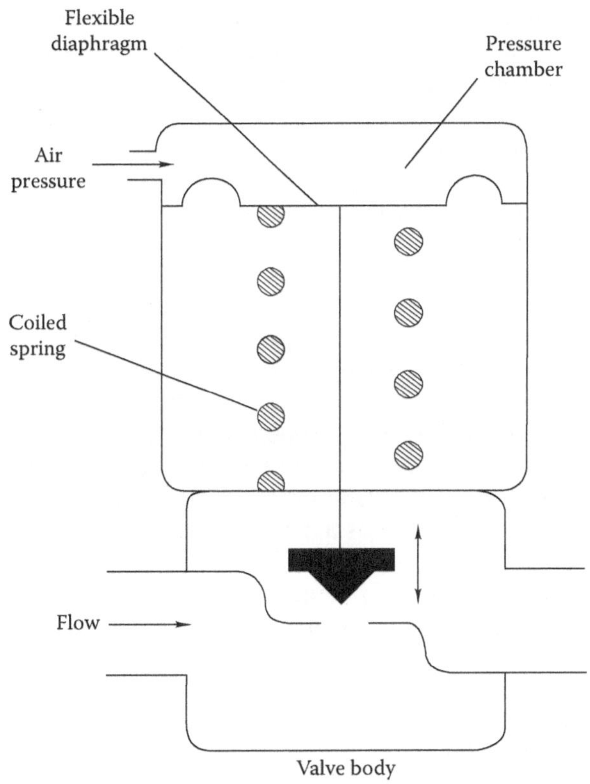

FIGURE 21.7
Pneumatic control valve showing counterforce spring and valve body. Increasing pressure closes the valve.

pressure. However, this does not necessarily produce a linear effect on flow, as discussed in Section 21.4. Figure 21.8 shows a pneumatic damper actuator. Linear actuator motion is converted to rotary damper motion by the simple mechanism shown. The details of control valve and damper design are included in Section 21.4.

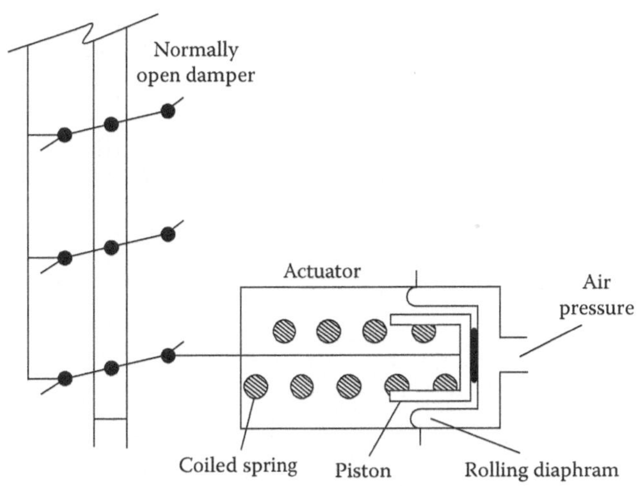

FIGURE 21.8
Pneumatic damper actuator. Increasing pressure closes the parallel-blade damper.

Pneumatic *controllers* produce a branch line (see Figure 21.6) pressure that is appropriate to produce the needed control action for reaching the setpoint. They are manufactured by a number of control firms for specific purposes. Classifications of controllers include the sign of the output (direct or reverse acting) produced by an error, by the control action (proportional, PI, or two-position), or by the number of inputs or outputs. Figure 21.9 shows the essential elements of a dual-input, single-output controller. The two inputs could be the heating system supply temperature and outdoor temperature used to control the output water temperature setting of a boiler in a building heating system. This is essentially a boiler *temperature reset* system that reduces heating water temperature with increasing ambient temperature for better system control and reduced energy use. (Note that *reset* has a different meaning than when used with an integral controller.)

The air supply for pneumatic systems must produce very clean, oil-free, dry air. A compressor producing 80–100 psig (5.5–6.8 atm) is typical. Compressed air is stored in a tank for use as needed, avoiding continuous operation of the compressor. The air system should be oversized by 50%–100% of estimated, nominal consumption. The air is then dried to avoid moisture freezing in cold control lines in air handling units and elsewhere. Dried air should have a dew point of –30°F (–34°C) or less in severe heating climates. In deep-cooling climates, the lowest temperature to which compressed air lines are exposed may be the building cold air supply. Next, the air is filtered to remove water droplets, oil (from the compressor), and any dirt. Finally, the air pressure is reduced in a pressure regulator to the control system operating pressure of approximately 18 psig (124 kPa gauge). Control air piping uses either copper or nylon (in accessible locations).

21.3.2 Electric and Electronic

Electric and electronic controls experienced a market share increase from the 1980s, but are being replaced by DDC in recent years. They are stand-alone controllers with no network capability. They are typically powered by a 24 VAC power source, which is usually a transformer mounted on the terminal unit. Signals between the thermostat and the controller are electronic/analog. Their precise control, flexibility, compatibility with microcomputers, and reliability are advantages in commercial buildings, but they are more costly per control loop. With the continuous decrease in microprocessor cost and associated increase in capability, the cost penalty, if any, is expected to virtually disappear, especially when it is calculated on a per-function basis. In this section, we survey the sensors, actuators, and controllers used in modern electronic control systems for buildings.

Figure 21.10 shows a simple *analog* electronic control system used to control the temperature of a hot-wire (or hot-film) anemometer for airflow measurement. Since the heat loss from the heated wire (a few ten-thousandths of an inch in diameter) is a function of the velocity, the velocity can be found by measuring the I^2R loss in the hot wire if the wire temperature remains constant. The resistance, and hence temperature, will remain constant if its value is the same as that of R_3, a precision isothermal reference resistor. The controller produces a

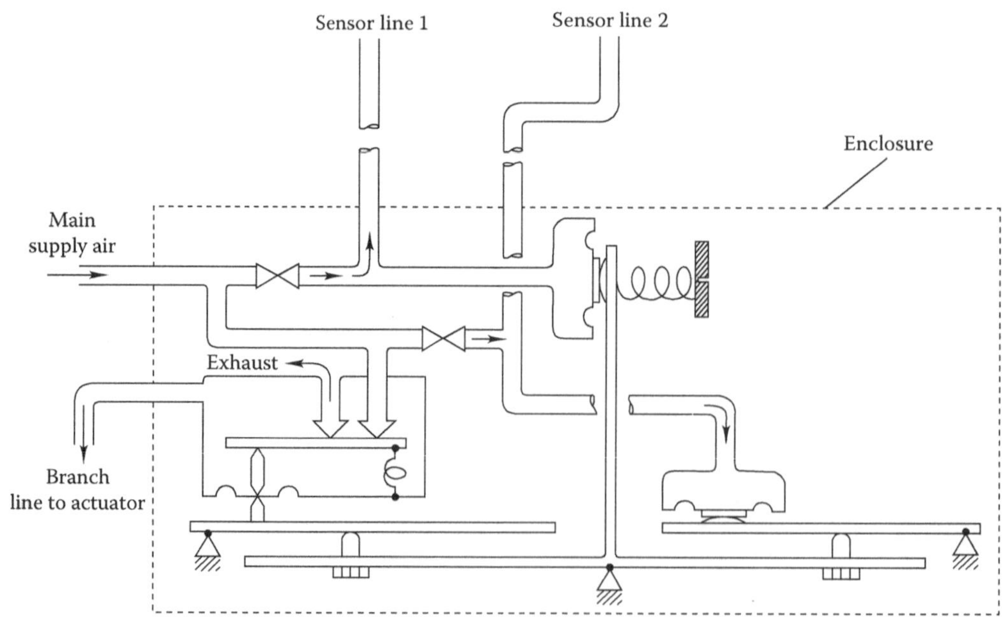

FIGURE 21.9
Example of pneumatic controller with two inputs (sensors 1 and 2) and one control signal output (branch line to actuator).

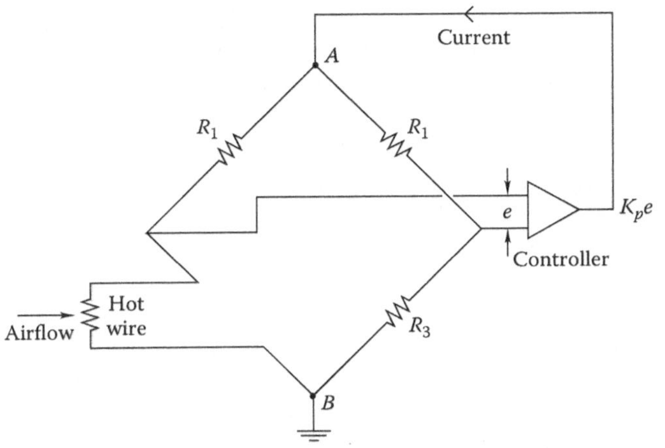

FIGURE 21.10
Hot-wire anemometer analog controller. The voltage between points *A* and *B* is related to the airflow value over the hot wire.

current output proportional to the difference in voltage *e* between the two branches of the Wheatstone bridge. An increase in current results from an increase in airflow over and heat transfer from the heated wire. After calibration, the voltage drop across the hot wire (or across the vertical diagonal of the bridge) is a direct and accurate indication of the airflow over the wire.

The hot-wire system embodies all features of an electronic control system. The sensor component is the differential voltage measurement *e* across the bridge. The proportional controller and actuator is the voltage-to-current device shown in the figure. The controlled variable is the wire resistance that indirectly accomplishes the desired control goal—a constant-temperature hot-wire anemometer.

21.3.3 Direct Digital Controllers

Direct digital controllers (DDC) are the most popular and flexible option for most modern buildings. They enhance the previous analog-only electronic system with digital computer features. The term *digital* refers to the use of digital computers in these systems. Modern DDC systems use analog sensors (converted to digital signals within a computer) along with digital computer programs to control HVAC systems. DDCs are powered by a 24 VAC power source. The output of this microprocessor-based system can be used to control electronic, electric, or pneumatic actuators or a combination of them. DDC systems have the advantage of reliability and flexibility compared to other types of control hardware. For example, it is easier to accurately set control constants in computer software than to make adjustments at a controller panel with a screwdriver. DDC systems offer the option of supervisory control and HVAC diagnostic knowledge-based systems since the sensor data used for control are very similar to those used in an Energy Management System (EMS) . Pneumatic systems do not offer this ability. Figure 21.11 shows a schematic diagram of a direct digital controller. The entire control system must include sensors and actuators not shown in this controller-only drawing. As discussed in Section 21.6.1, DDC controls can be networked into a building automation system that can support numerous building automation and energy conservation tasks such as night setback, fire smoke control, occupancy monitoring, energy monitoring, and remote troubleshooting (Price, 2011).

Temperature measurements for DDC applications are made by three principal methods:

1. Thermocouples
2. Resistance temperature detectors (RTDs)
3. Thermistors

Each has its advantages for particular applications. Thermocouples consist of two dissimilar metals chosen to produce a measurable voltage at the temperature of interest. The voltage output is low (millivolts) but is a well-established function of the junction temperature. Except

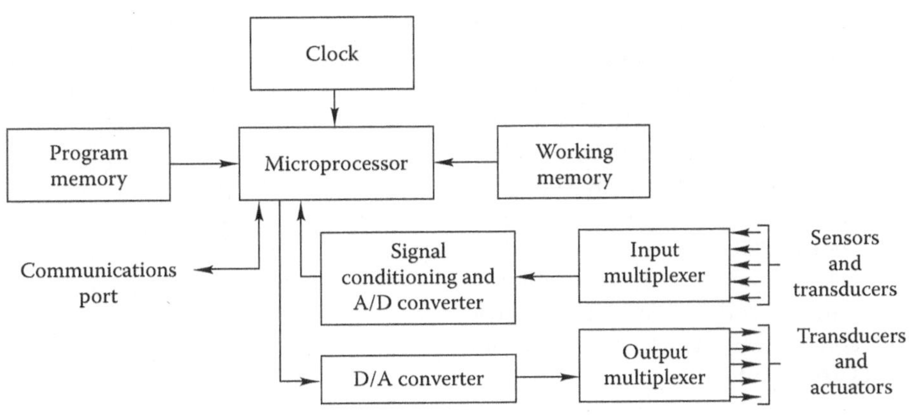

FIGURE 21.11
Block diagram of a direct digital controller.

for flame temperature measurements, thermocouples produce voltages too small to be useful in most HVAC applications (e.g., a type-J thermocouple produces only 5.3 mV at 100°C).

RTDs use small, responsive sensing sections constructed from metals whose resistance-temperature characteristic is well established and reproducible. To first order,

$$R = R_o(1 + kT) \tag{21.9}$$

where

R is the resistance, Ω
R_o is the resistance at reference temperature (0°C), Ω
k is the temperature coefficient of resistance, °C^{-1}
T is the RTD temperature, °C

This equation is easy to invert to find the temperature as a function of resistance. Although complex higher-order expressions exist, their use is not needed for HVAC applications.

Two common materials for RTDs are platinum and Balco (a 40% nickel, 60% iron alloy). Nominal values of k, respectively, are 3.85×10^{-3} and 4.1×10^{-3} K^{-1}.

Resistance is measured indirectly by measurements of voltage and current. Therefore, the controller must supply current to the RTD. The current can cause self-heating and consequent errors. These are avoidable by using higher-resistance RTDs. In addition, lead wire resistance can cause lack of accuracy for the class of platinum RTDs whose nominal resistance is only 100 Ω because the lead resistance of 1–2 Ω is not negligible by comparison with that of the sensor itself.

Thermistors are semiconductors with the property that resistance is a strong but nonlinear function of temperature, given approximately by

$$R = Ae^{B/T} \tag{21.10}$$

where A is related to the nominal value of resistance at the reference temperature and is on the order of 0.06 to 0.07 for a nominal 10 K sensor (at 20°C). The exponential coefficient B (a weak function of temperature) is on the order of 5400°R–7200°R (3000–4000 K). The nonlinearity inherent in thermistors can be reduced by connecting a properly selected fixed resistor in parallel with it. The resulting linearity is desirable from a control system design viewpoint. Thermistors can have problems with long-term drift and aging; the designer and control manufacturer should consult on the most switchable thermistor design for HVAC applications.

Humidity measurements are needed for control of enthalpy economizers, as discussed in Chapter 19. Humidity may also need to be controlled in special environments such as clean rooms, hospitals, and spaces that house computers. The relative humidity, dew point, and humidity ratio are all indicators of the moisture content of air. The psychrometer described in Chapter 13 or the sensors described in the previous section (for pneumatic system applications) can be used with DDC systems but are impractical. Instead, an electrical, capacitance-based approach is preferred. The high dielectric constant of water absorbed into a polymer causes a significant change in capacitance. This is used along with a local electronic circuit to produce a linear voltage signal with humidity. If not saturated by excessive exposure to very high humidities, these devices produce reproducible signals without excessive hysteresis (Huang, 1991). Response times are on the order of seconds if the local air velocity over the sensor is above a few feet per second. Another solid-state device for humidity measurement uses the variation in resistance of a thin film of lithium chloride. The resistance change is significant but also depends on the temperature. The most common sensor type used presently employs the capacitance approach. Figure 21.12 is a photograph of a commercial humidity and temperature sensor.

Pressure measurements are made by electronic devices that depend on a change of resistance or capacitance with imposed pressure. Figure 21.13 shows a cross-sectional drawing of each. In the resistance type, stretching of the membrane lengthens the resistive element, (a strain gauge), thereby increasing resistance. This resistor is an element in a Wheatstone bridge; the resulting bridge

FIGURE 21.12
Photograph of humidity and temperature sensor. (Courtesy of Phys-Chem Scientific Corporation, New York.)

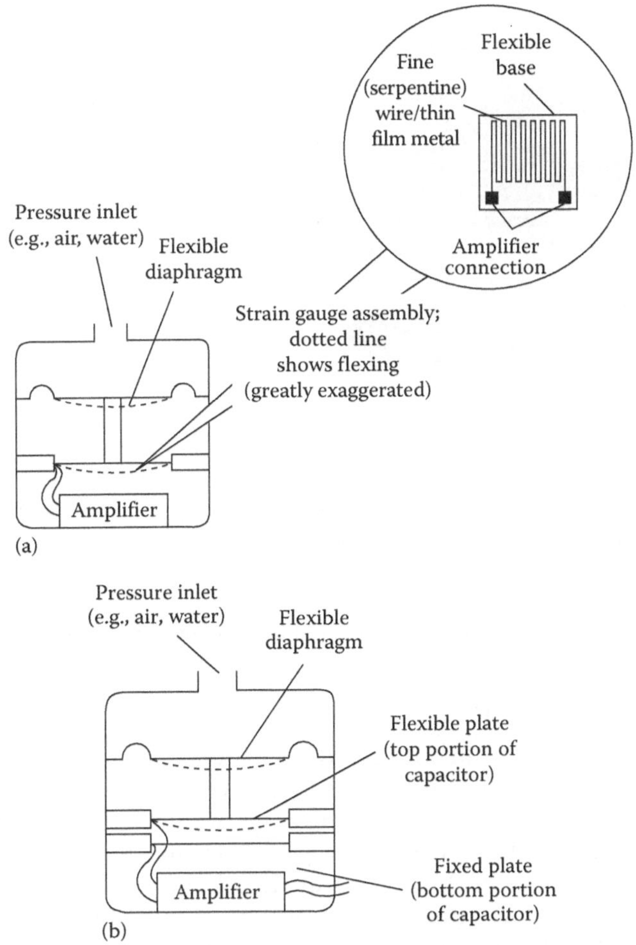

FIGURE 21.13
(a) Resistance- and (b) capacitance-type pressure sensors.

voltage imbalance is linearly related to the imposed pressure. The less common capacitive type unit has a capacitance between a fixed and a flexible metal plate that decreases with pressure. The capacitance change is amplified by a local amplifier that produces an output signal proportional to the pressure.

Flow measurement or indication is also needed in DDC systems. Pitot tubes (or arrays of tubes), hot films, thermistors, and other flow measurement devices described in Section 16.7 can be used to measure either airflow or liquid flow in secondary HVAC systems. Airflow information is important for proper control of building pressure, VAV system control, and outside air supply. Water flow rates are needed for chiller and boiler control and for secondary-system liquid-loop control. In some cases, the quantitative value of flow is not needed, only a knowledge that flow exists. Sensors for this are called *flow switches*. They are electromechanical switches that change from open to closed (or vice versa) upon the existence of flow. Similar switches are used in some control system designs to sense the damper position (open or closed).

Temperature, humidity, and pressure *transmitters* are often used in HVAC systems. They amplify signals produced by the basic devices described in the preceding paragraphs, and they produce a standardized electric signal over a standard range, thereby permitting standardization of this aspect of DDC systems. The standard ranges are as follows: current: 4–20 mA (dc) and voltage: 0–10 V (dc). Although the majority of transmitters produce such signals, the noted values are not universally used. Stand-alone transmitters cost more than those integrated with the controller.

Figure 21.11 shows the elements of a direct digital controller. The heart of the controller is the microprocessor that can be programmed in either a standard or a system-specific language. Control algorithms (linear or not), sensor calibrations, output signal shaping, and historical data archiving can all be programmed as the user requires. A number of firms have constructed controllers on standard personal-computer platforms. It is beyond the scope of this book to describe the details of programming HVAC controllers, since each manufacturer uses a different approach. The essence of any DDC system, however, is the same as shown in the figure. Honeywell (1988) discusses DDC systems and their programming in greater detail.

Actuators and positioners for electronic control systems include

- *Motors*—operate valves, dampers
- *Variable-speed controls*—pump, fan, chiller drives
- *Relays and motor starters*—operate other mechanical or electric equipment (pumps, fans, chillers, compressors), electric heating equipment
- *Transducers*—convert, e.g., electric signal to pneumatic (EP transducer)
- *Visual displays*—not actuators in the usual sense but used to inform system operator of control and HVAC system function

We discuss several of these in the context of design in Section 21.5 Figure 21.14 shows a cutaway of a DDC system actuator—an electric motor used for damper control. The motor is controlled by a signal from the direct digital controller. As an option, it can be equipped with a rotary potentiometer to feed back the damper position to the controller microprocessor. Pneumatic actuators with EP (electric to pneumatic) transducers are often used since they are more powerful and less costly than electrical actuators.

The power supply to a DDC system and critical actuators must be maintained in case of a power outage. This is accomplished by use of an *uninterruptible power supply* (UPS) that is able to operate the DDC central processor and important peripherals for a period until power

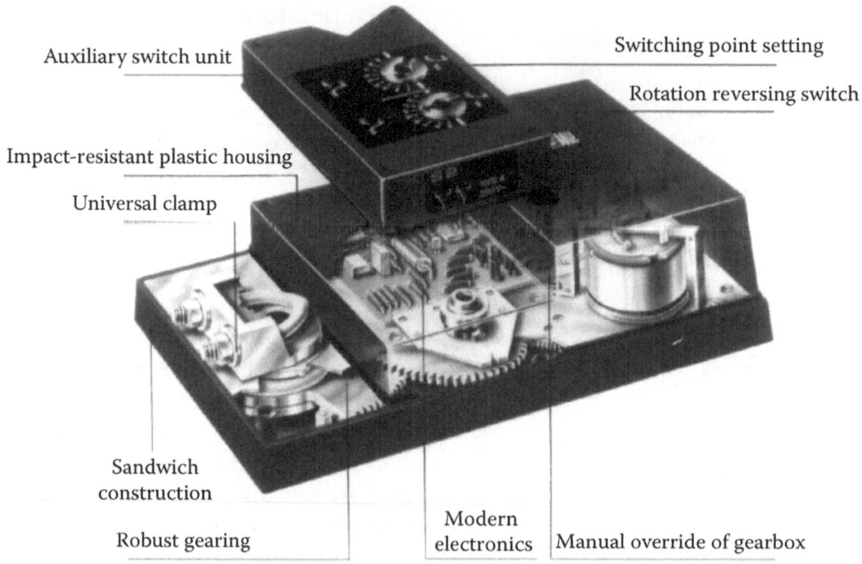

FIGURE 21.14
Electric damper motor. Cutaway showing motor, gear drive, shaft attachment, and circuit board. (Courtesy of Belimo Company, New York.)

is restored. The length of the UPS capacity duration must be decided by the designer in view of local power quality and the time needed to conduct an orderly shutdown of the HVAC system.

Pneumatic and DDC systems have both advantages and disadvantages. Pneumatic systems have the advantage of lower cost, inherently modulating actuators and sensors, more powerful actuators, explosion-proof components, and diagnostic simplicity. Disadvantages include the need for a compressor producing clean and dry air, the cost of air piping, and the need for regular component calibration. DDC systems can be very precise (limited by sensor and actuator accuracy), can accommodate complex control algorithms and scheduling, easily accept changes to control constants, produce data usable by EMSs, and allow central control of a group of buildings. Present disadvantages of DDC systems include cost (this penalty is decreasing, as discussed earlier, but wires and power supplies cost more than pneumatic piping) and training needs of maintenance personnel who may be more familiar with pneumatic controls.

21.4 Basic Control System Design Considerations

This section discusses selected topics in HVAC control system design, including control system zoning, valve and damper selection, and control logic diagrams.

21.4.1 Sensors

The ultimate purpose of an HVAC control system is to provide adequate ventilation while controlling the zone temperature (and secondarily air motion and humidity) to conditions that ensure maximum comfort and productivity of the occupants. In other chapters of this book, we have discussed topics bearing on this issue, including heating and cooling loads, zone air supply, and zone air distribution. From a controls viewpoint, the HVAC system is assumed to be able to provide comfort conditions if controlled properly.

The thermal criteria for zone selection are discussed in Section 7.6. Basically, a zone is a portion of a building whose loads differ in magnitude and time sufficiently from other areas so that separate portions of the secondary HVAC system and control system are needed to maintain comfort. Typical zoning is done based on *exposure*—four cardinal directions and a core zone with little outside exposure—and *schedule*. Corner rooms with double exposure and top-floor spaces require special attention. In very tall buildings, upper floors may have sufficiently different exposure (wind and solar) that different zoning may be needed. Internal gain distributions within the core zone may dictate separate zones within the core. Rooms with large electrical loads (printing, computing, manufacturing machinery) will be controlled separately.

Having specified the zones, the designer must select the location of the thermostat (and other sensors, if used). According to Madsen (1989), the best placement for a thermostat is in a central location within a zone rather than at the typical wall locations used. It was found that zone temperature control was better (smaller temperature swings) and that energy consumption was reduced for a central location rather than for a wall location. Both field and laboratory tests supported this conclusion. Electronic sensors can be easily located anywhere; pneumatic thermostats are less easy to locate anywhere but on a wall.

If a central location is not possible, the usual wall location must be used. The thermostat should be located so that it senses the average temperature. Therefore, cold outside walls, areas exposed to direct sunlight, and parts of the zone directly in the airstream of either heating or cooling equipment should be avoided. A nominal mounting height that places the thermostat within the occupied level of a zone should also be used; height should be about eye level.

Thermostat signals are either passed to the central controller or used locally to control the amount and temperature of conditioned air or coil water introduced into a zone. The air is conditioned either locally (e.g., by unit ventilator or baseboard heater) or centrally (e.g., by the heating and cooling coils in the central air handler). In either case, a flow control actuator is controlled by the thermostat signal. In addition, airflow itself may be controlled in response to zone information in variable-air-volume (VAV) systems. Except for variable-speed drives used in variable-volume air or liquid systems, flow is controlled by valves or dampers. The design selection of valves and dampers is discussed next.

21.4.2 Steam and Liquid Flow Control

The flow-through valve, such as that shown in Figure 21.7, is controlled by the valve stem position that determines the flow area. The variable flow resistance offered by valves depends on their design. The flow characteristic may or may not be linear with position. Figure 21.15 shows common flow characteristics. Note that the plotted characteristics apply only for *constant valve pressure drop*. The characteristics shown are idealizations of actual valves. Commercially available valves will resemble but not necessarily exactly match the curves shown.

The *linear* valve has a proportional relation between the volumetric flow $\dot{V}$ and the valve stem position z:

$$\dot{V} = kz \tag{21.11}$$

where k is the proportionality constant. The flow in *equal-percentage valves* increases by the same fractional amount for each increment of opening. In other words, if the valve is opened from 20% to 30% of full travel, the flow will increase by the same percentage as if the travel had increased from 80% to 90% of its full travel.

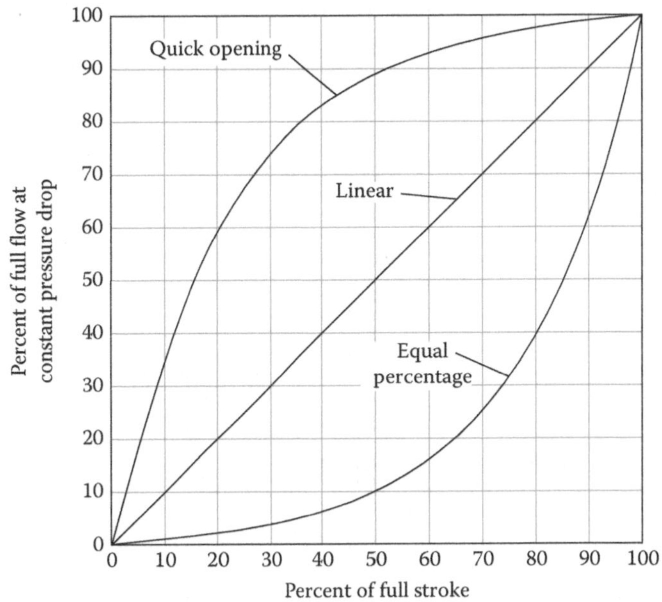

FIGURE 21.15
Quick-opening, linear, and equal-percentage valve characteristics.

However, the absolute volumetric flow increase for the latter case is much greater than for the former. The equal-percentage valve flow characteristic is given by

$$\dot{V} = Ke^{kz} \tag{21.12}$$

where K is the valve size constant. Quick-opening valves do not provide good flow control but are used when rapid action is required with little stem movement for on/off control.

Example 21.2: Equal-Percentage Valve

A valve at 30% of travel has a flow of 4 gal/min. If the valve opens another 10% and the flow increases by 50% to 6 gal/min, what are the constants in Equation 21.12? What will the flow be at 50% of full travel?

Figure: See Figure 21.15.

Assumptions: Pressure drop across the valve remains constant.

Find: k, K, $\dot{V}_{50}$

Solution

Equation 21.12 can be evaluated at the two flow conditions. If the results are divided by each other, we have

$$\frac{\dot{V}_2}{\dot{V}_1} = \frac{6}{4} = e^{k(z_2 - z_1)} = e^{k(0.4-0.3)} \tag{21.13}$$

In this expression, the travel z is expressed as the dimensionless *fraction* of the total travel and is dimensionless. Solving this equation for k gives

$$k = 4.05 \text{ (no units)}$$

From the known flow at 30% travel, we can find the second constant K:

$$K = \frac{4\ \text{gal/min}}{e^{4.05 \times 0.3}} = 1.19\ \text{gal/min} \tag{21.14}$$

Finally, the flow is given by

$$\dot{V} = 1.19e^{4.05z} \tag{21.15}$$

At 50% travel, the volumetric flow rate can be found from this expression:

$$\dot{V}_{50} = 1.19e^{4.05 \times 0.5} = 9.0\ \text{gal/min} \tag{21.16}$$

Comments

This result can be checked since the valve is an equal-percentage valve. At 50% travel, the valve has moved 10% beyond its 40% setting at which the flow was 6 gal/min. Another 10% stem movement will result in another 50% flow increase from 6 to 9 gal/min, confirming the solution.

The plotted characteristics of all three valve types assume constant pressure drop across the valve. However, in an actual system, the pressure drop across a valve will not remain constant. But, for the valve to maintain its control characteristics, the pressure drop across it must be the majority of the entire loop pressure drop. A valve designed to have a full-open pressure drop equal to that of the balance of the loop, will allow good flow control. This introduces the concept of *valve authority*, defined as the valve pressure drop as a fraction of total system pressure drop:

$$A \equiv \frac{\Delta p_{v,open}}{\Delta p_{v,open} + \Delta p_{sys}} \tag{21.17}$$

For proper control, the full-open valve authority should be at least 0.50. If the authority is 0.5 or more, control valves will have installed characteristics not much different from those shown in Figure 21.15. If not, the valve characteristic will be distorted at low flow since the majority of the system pressure drop will be dissipated across the valve.

Valves are further classified by the number of connections or ports. Figure 21.16 shows sections of typical *two-way* and *three-way* valves. Two-port valves control flow through coils or other HVAC equipment by varying the valve flow resistance as a result of flow area changes. As shown, the flow must oppose the closing of the valve. If not, at near closure the valve would slam shut or oscillate, both of which cause excessive wear and noise. The three-way valve shown in the figure is configured in the *diverting* mode. That is, one stream is split into two

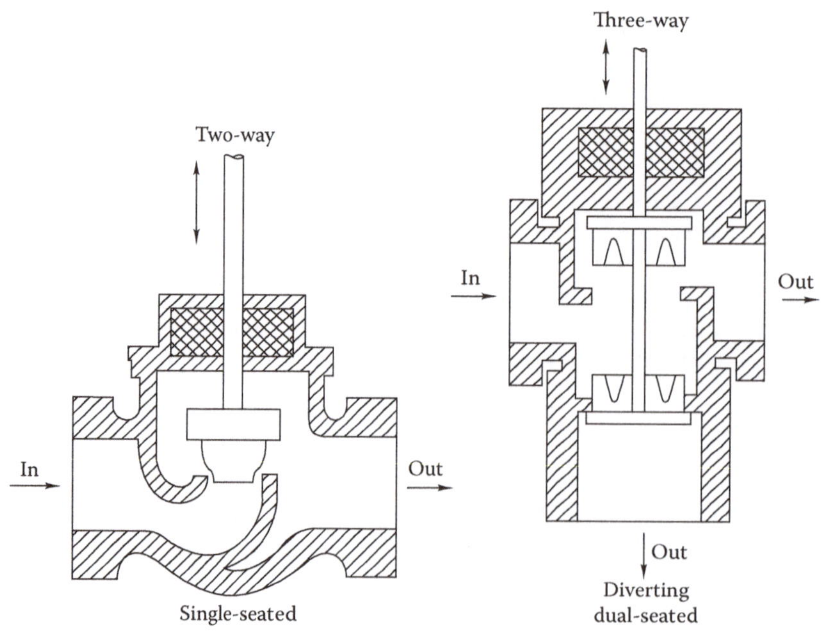

FIGURE 21.16
Cross-sectional drawings of direct-acting, single-seated, two-way valve and dual-seated, three-way, diverting valve.

depending on the valve opening. The three-way valve shown is double-seated (single-seated three-way valves are also available); therefore, it is easier to close than a single-seated valve, but tight shutoff is not possible.

Three-way valves can also be used as *mixing* valves. In this application, two streams enter the valve and one leaves. Mixing and diverting valves *cannot be used interchangeably* since their internal design is different to ensure that they can each seat properly. Particular attention must be paid by the installer to be sure that connections are made properly; arrows cast in the valve body show the proper flow direction. Figure 21.17 shows an example of a three-way valve for mixing and one for diverting applications.

The *valve capacity* is denoted in the industry by the dimensional flow coefficient C_v, defined by

$$\dot{V}\,(\text{gal/min}) = C_v\sqrt{\Delta p}\,(\text{psi}) \tag{21.18 IP}$$

where C_v is specified as the flow rate of 60°F water that will pass through the fully open valve if a pressure difference of 1.0 psi is imposed across the valve. If SI units (liters per second and megapascals) are used, the numerical value of C_v is 24% larger than that in IP units. Once the designer has determined the value of C_v, manufacturers' tables can be consulted to select a valve for the known pipe size. If a fluid other than water is to be controlled, the C_v value found from Equation 21.18 should be multiplied by the square root of the fluid's specific gravity.

Steam valves are sized by using a similar dimensional expression:

$$\dot{m}\,(\text{lb/h}) = 63.5C_v\sqrt{\frac{\Delta p\,(\text{psi})}{v\,(\text{ft}^3/\text{lb})}} \tag{21.19 IP}$$

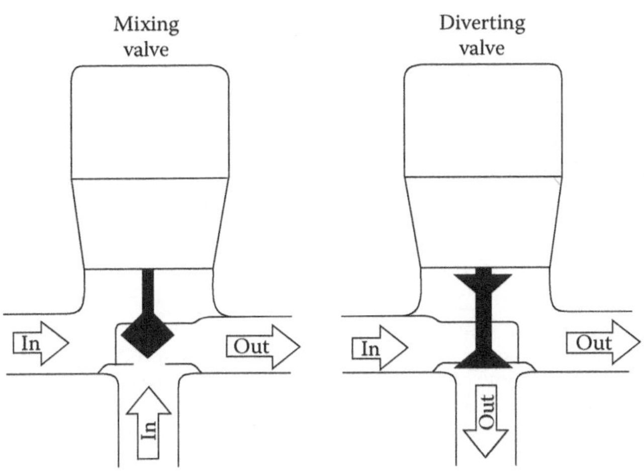

FIGURE 21.17
Three-way mixing and diverting valves. Note the significant difference in internal construction. Mixing valves are more commonly used.

in which v is the steam specific volume and $\dot{m}$ is the mass flow rate. If the steam is highly superheated, multiply C_v found from Equation 21.19 by 1.07 for every 100°F of superheat. For wet steam, multiply C_v by the square root of the steam quality. Honeywell (1988) recommends that the pressure drop across the valve for use in the equation be 80% of the difference between the steam supply and return pressures (subject to the sonic flow limitation discussed in the following). Tables in the appendices on the online HCB software can be used for preliminary selection of control valves for either steam or water.

For a specific application, the type of valve (linear or not) selected should have a controlled system that is as linear as possible. Control valves are commonly used to control the heat transfer rate in coils. For a linear system, the *combined characteristic of the actuator, valve, and coil should be linear.* This will require entirely different valves, e.g., for hot water and steam control, as we shall see. This difference is the reason for using linear and equal-percentage values.

Figure 21.18 shows the part-load performance of a hot water coil used for air heating; at 10% of full flow, the heat rate is 50% of its peak value. From Section 12.6, we note that the heat rate in a cross-flow heat exchanger increases roughly in an exponential fashion with the flow rate, a highly nonlinear characteristic. This heating coil nonlinearity results from the longer water residence time in a coil at reduced flow and the relatively large temperature difference between air being heated and the water heating it.

However, if we were to control the flow through this heating coil by an equal-percentage valve (positive exponential increase of flow with valve position), the combined valve-plus-coil characteristic would be roughly linear. Referring to Figure 21.18, we see that 50% of stem travel corresponds to 10% flow. The third graph in the figure is the combined characteristic. This near-linear subsystem is much easier to control than if a linear valve were used with the highly nonlinear coil. Hence the rule: Use equal-percentage valves for heating coil control.

Linear two-port valves are to be used for steam flow control to coils, since the transfer of heat by steam condensation is a linear constant-temperature process—the more steam supplied, the greater the heat rate, in exact proportion. Note that this is a completely different coil flow characteristic from that for hot water coils. However, steam is a compressible fluid, and the sonic velocity sets the flow limit for a given valve opening when the pressure drop across the valve is more than 60% of the steam supply line absolute pressure. As a result, the pressure drop to be used in Equation 21.19 is the *smaller* of (1) 50% of the absolute steam line pressure upstream of the valve and (2) 80% of the difference between the steam supply and return line pressures.

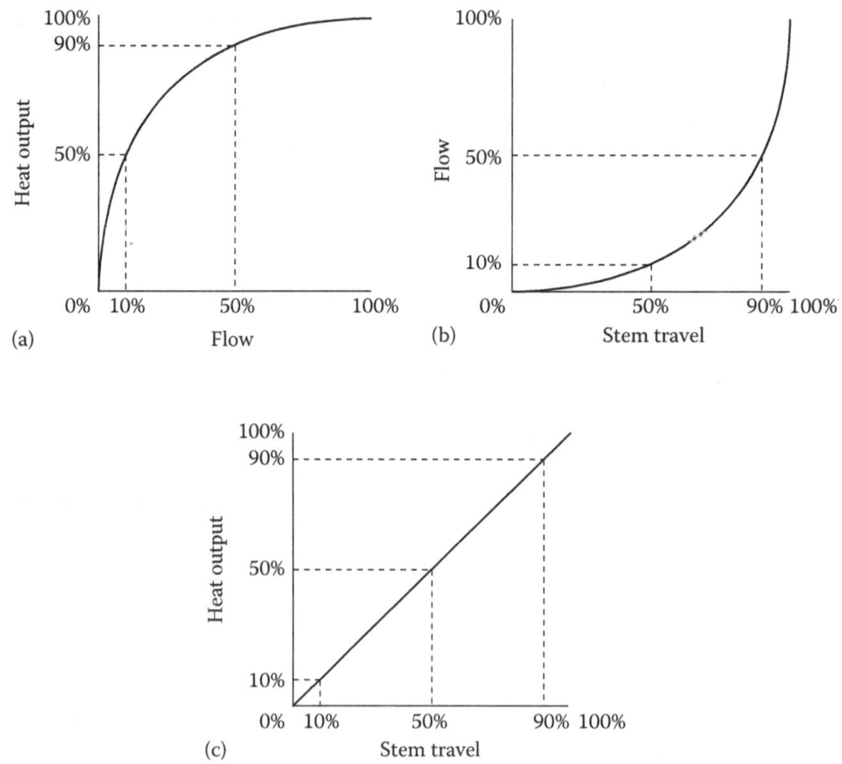

FIGURE 21.18
(a) Heating coil, (b) equal-percentage valve, and (c) combined coil-plus-valve linear characteristic.

The 80% rule gives good valve modulation in the subsonic flow regime (Honeywell, 1988).

Chilled water control valves should also be linear since the performance of chilled water coils (smaller air-water temperature difference than in hot water coils) is more similar to that of steam coils than to that of hot water coils.

Two- or three-way valves can be used to control flow at part load through heating and cooling coils, as shown in Figure 21.19. The control valve can be controlled from the coil outlet water or air temperature. Two- or three-way valves achieve the same local result at the coil when used for part-load control. However, the designer must consider effects on the balance of the secondary system (pump size and power, flow balance) when selecting the valve type.

In essence, the two-way valve flow control method results in variable flow (tracking variable loads) with constant coil water temperature change, whereas the three-way valve approach results in roughly constant secondary loop flow rate but smaller coil water temperature change (beyond the local coil loop itself). Since some chillers and boilers require that the flow remain within a rather narrow range, the energy and cost savings that could accrue due to the two-way valve, variable-volume system are difficult to achieve in small systems unless the two-pump, primary/secondary loop approach described in Section 18.8.1 is employed. If this dual-loop approach is not used, the three-way valve method is required to maintain constant boiler or chiller flow. In large systems, a primary/secondary design with two-way valves is preferred.

The location of the three-way valve at a coil must also be considered by the designer. Figure 21.19b shows the valve used downstream of the coil in a mixing bypass mode. If a balancing valve is installed in the bypass line and is set to have the same pressure drop as the coil, the local coil loop will have the same pressure drop for both full and zero coil flow. However, at the valve mid-flow position, the overall flow resistance is less, since two parallel paths are involved, and the total loop flow increases to 25% more than that at either extreme.

Alternatively, the three-way valve can also be used in a diverting mode, as shown in Figure 21.19c. In this arrangement, essentially the same considerations apply as for the mixing arrangement discussed before.* However, if a circulator (small pump) is inserted as

* A little known disadvantage of the three-way valve control has to do with the *conduction* of heat from a closed valve to a coil. For example, the constant flow of hot water through two ports of a *closed* three-way heating coil control valve keeps the valve body hot. Conduction from the closed hot valve mounted close to a coil can cause sufficient air heating to actually decrease the expected cooling rate of a downstream cooling coil during the cooling season. Three-way valves have a second practical problem: Installers often connect three-way valves incorrectly, given the choice of three pipe connections and three pipes to be connected. Both these problems can be avoided by using two-way valves.

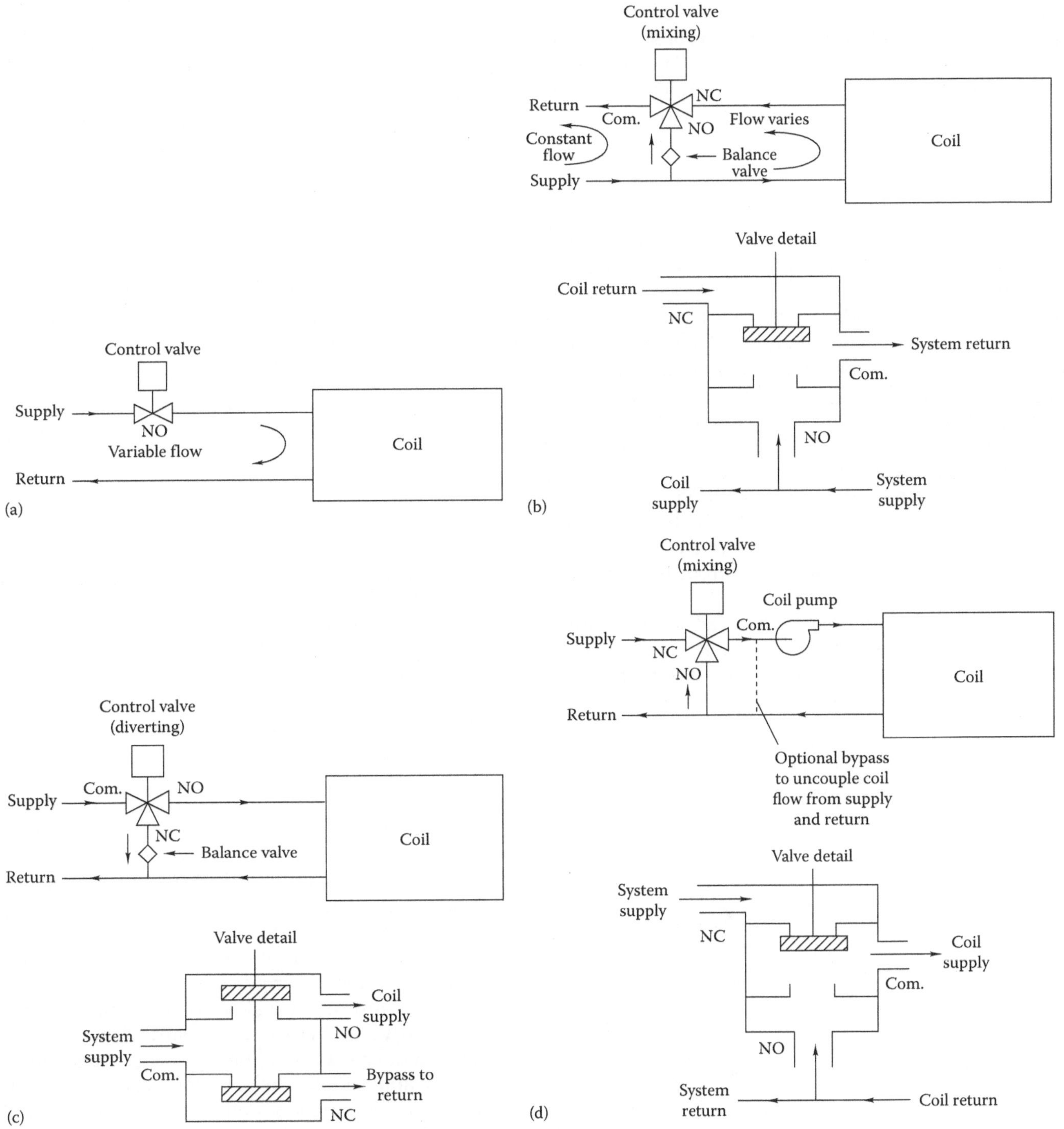

FIGURE 21.19 Various control valve piping arrangements: (a) two-way valve, (b) three-way mixing valve, (c) three-way diverting valve, and (d) pumped coil with three-way mixing valve.

shown in Figure 21.19d, the direction of flow in the branch line changes and a mixing valve is used. Pumped coils are used because control is improved. With constant coil flow, the highly nonlinear coil characteristic shown in Figure 21.18 is reduced, since the residence time of hot water in the coil is constant, independent of the load. However, this arrangement has the same effect on the external secondary loop as a two-way valve. As load is decreased, flow into the local coil loop also decreases. Therefore, the uniform secondary loop flow normally associated with three-way valves is not present unless the optional bypass is used.

For HVAC systems requiring precise control, high-quality control valves are required. The best controllers and valves are of "industrial quality." The additional cost for these valves compared to conventional building hardware results in more accurate control and longer lifetime.

21.4.3 Airflow Control

Dampers are used to control airflow in secondary HVAC air systems in buildings. As discussed in Section 16.5, variable airflow rates are also produced by variable fan speeds. In this section, we discuss the characteristics of dampers used for flow control in systems where constant-speed fans are involved. Figure 21.20 shows cross sections of the two common types of dampers used in commercial buildings. Parallel-blade dampers use blades that all rotate in the same direction. They are most often applied to two-position locations—either open or closed. Their use for flow control is not recommended. The blade rotation changes the airflow direction, a characteristic that can be useful when airstreams at different temperatures are to be effectively blended.

Opposed-blade dampers have adjacent counter-rotating blades. The airflow direction is not changed with this design, but pressure drops are higher than for parallel blading. Opposed-blade dampers are preferred for flow control. Figure 21.21 shows the flow characteristics of these dampers to be closer to the desired linear behavior. The parameter α on the curves is the ratio of system pressure drop to fully open damper pressure drop.

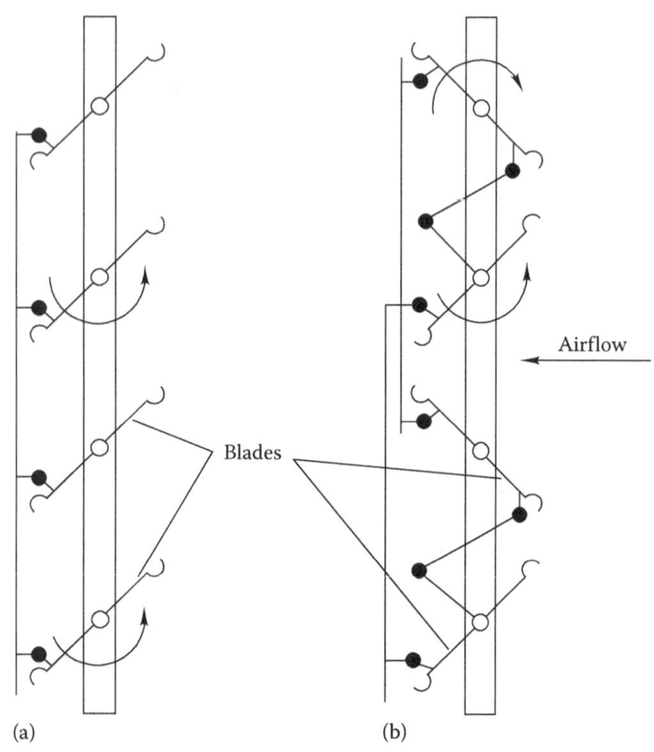

FIGURE 21.20
Diagram of (a) parallel- and (b) opposed-blade dampers.

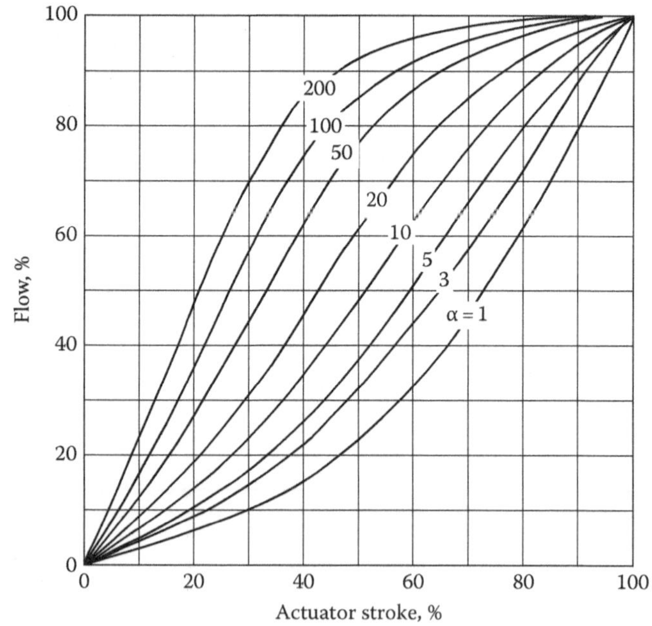

FIGURE 21.21
Flow characteristics of opposed-blade dampers. The parameter α is the ratio of system resistance (not including the damper) to damper resistance. The ideal linear damper characteristic is achieved if this ratio is about 10 for opposed-blade dampers.

Damper leakage is always a concern of the designer in critical locations such as at outdoor air intakes. Extra expense should be expected for low-leakage dampers in such applications. Damper leakage is expressed in units of leakage volumetric flow per unit damper area (e.g., cubic feet per minute per square foot) at a specified pressure difference. The manufacturer's literature should be consulted. A typical damper may leak 40 $ft^3/(min \cdot ft^2)$ at 1.0 inWG, while a low-leakage damper may leak only 10 $ft^3/(min \cdot ft^2)$ at the same pressure difference. The designer will specify acceptable leakage rates depending on the application. The actuator torque, air velocity, materials of construction, and maximum pressure difference (damper fully closed) must also be considered by the HVAC engineer.

A common application of dampers controlling the flow of outside air uses two sets in a *face and bypass* configuration, as shown in Figure 21.22. The preheat coil is used for preheating cold outdoor air upstream of the system heating coil, as described in Section 19.1. For full heating, all air is passed through the coil, and the bypass dampers are closed. If no heating is needed in mild weather, the coil is bypassed (for minimum flow resistance and fan power cost, flow through fully open

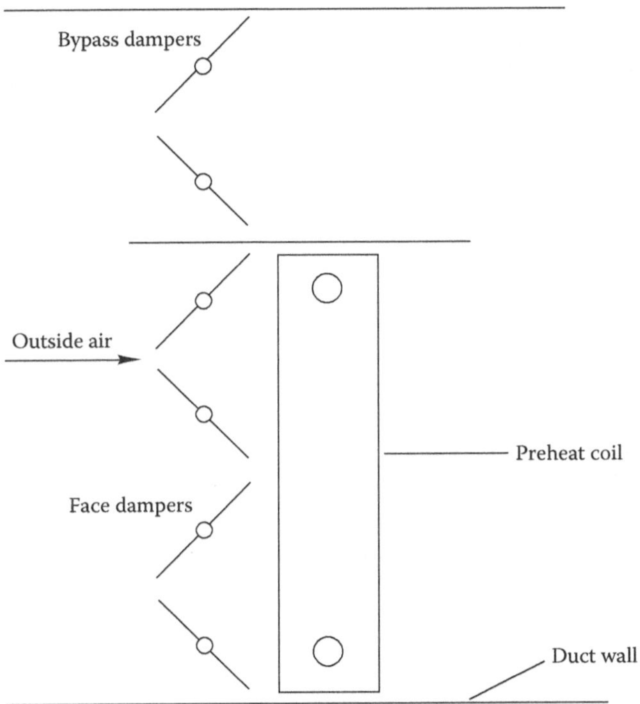

FIGURE 21.22
Face and bypass dampers used for preheat coil control.

face and bypass dampers can be used if the preheat coil water flow is shut off). Between these extremes, flow is split between the two paths. The face and bypass dampers are sized so that the pressure drop in the full-bypass mode (damper pressure drop only) and the full-heating mode (coil plus damper pressure drop) is the same.

21.5 Examples of HVAC System Control Systems

Several widely used control configurations for specific tasks are described in this section. These have been selected from the hundreds of control system configurations that have been used for buildings. The goal of this section is to illustrate how control components described previously are assembled into systems and what design considerations are involved. For a complete overview of HVAC control system configurations, the reader is referred to ASHRAE (2012), Grimm and Rosaler (1990), and Honeywell (1988). The illustrative systems in this section are drawn in part from the last reference.

In this section, we will discuss seven control systems in common use. Each system will be described by using a schematic diagram, and its operation and key features will be discussed in the accompanying text.

21.5.1 Outside Air Control

Figure 21.23 shows a system for controlling outside and exhaust air from a central air handling unit equipped for economizer cooling when available. In this and the following diagrams, the following symbols are used:

C	Cooling coil
DA	Discharge air (supply air from fan)
DX	Direct-expansion coil
EA	Exhaust air
H	Heating coil
LT	Low-temperature limit sensor or switch; must sense lowest temperature in air volume being controlled
M	Motor or actuator (for damper or valve), variable-speed drive
MA	Mixed air
NC	Normally closed
NO	Normally open
OA	Outside air
PI	Proportional plus integral controller
R	Relay
RA	Return air
S	Switch
SP	Static pressure sensor used in VAV systems
T	Temperature sensor; must be located to read average temperature representative of volume being controlled

This system is able to provide the minimum outside air during occupied periods, to use outdoor air for cooling when appropriate by means of a temperature-based economizer cycle, and to operate fans and dampers under all conditions. The numbering system used in the figure indicates the sequence of events as the air-handling system begins operation after an off period.

1. The fan control system turns on when the fan is turned on. This may be initiated by a clock signal or a low-temperature space condition. The OA dampers are at the minimum position before the fan starts.
2. The space temperature signal determines if the space is above or below the setpoint. If above, the economizer feature will be activated and will control the outdoor and mixed-air dampers. If below, the outside air damper is set to its minimum position.
3. The mixed-air PI controller controls both sets of dampers (OA/RA and EA) to provide the desired mixed-air temperature.
4. When the outdoor temperature rises above the cutoff point for economizer operation, the outdoor air damper is returned to its minimum setting.
5. Switch S is used to set the minimum setting on outside and exhaust air dampers manually.

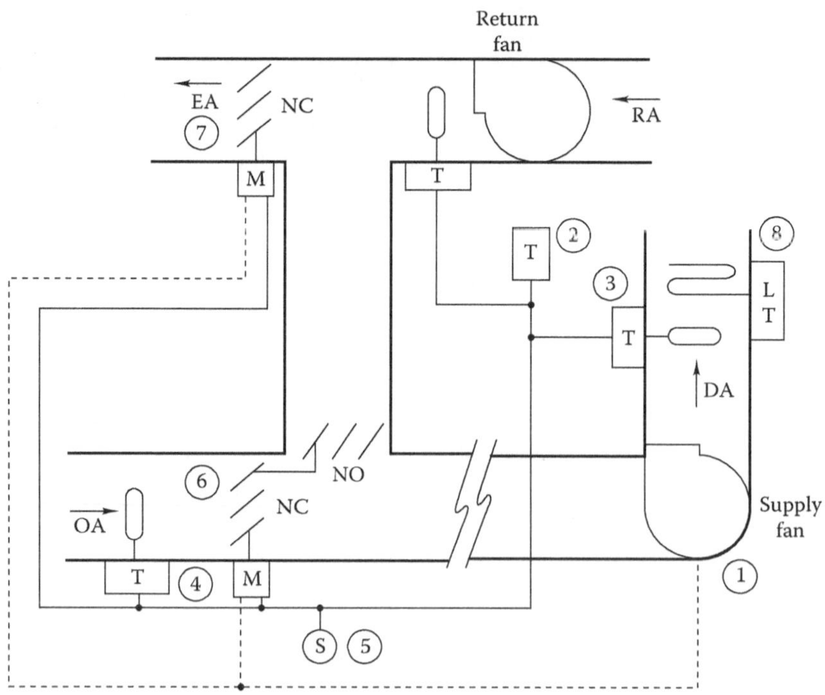

FIGURE 21.23
Outside air control system with economizer capability.

This is ordinarily done only once during building commissioning and flow testing.

6. When the supply fan is off, the outdoor air damper returns to its NC position and the return air damper returns to its NO position.
7. When the supply fan is off, the exhaust damper also returns to its NC position.
8. Low temperature sensed in the duct will initiate a freeze protect cycle. This may be as simple as turning on the supply fan to circulate warmer room air. Of course, the OA and EA dampers remain tightly closed during this operation.

21.5.2 Heating Control

If the minimum air setting is large in the preceding system, the amount of outdoor air admitted in cold climates may require preheating, as discussed earlier and in Section 19.1. Figure 21.24 shows a preheat system using face and bypass dampers. (A similar arrangement is used for direct-expansion [DX] cooling coils.) The equipment shown is installed upstream of the fan in Figure 21.23. This system operates as follows:

1. The preheat subsystem control is activated when the supply fan is turned on.
2. The preheat PI controller senses temperature leaving the preheat section. It operates the face and bypass dampers to control the exit air temperatures between 45°F and 50°F.
3. The outdoor air sensor and associated controller control the water valve at the preheat coil. The valve may be either a modulating valve (better control) or an on/off valve (less costly).
4. The low-temperature (LT) sensors activate coil freeze protection measures, including closing dampers and turning off the supply fan.

Note that the preheat coil (and all coils in this section) is connected so that the hot water (or steam) flows counter to the direction of airflow. In Section 12.6, it was pointed out that counterflow provides a higher heating rate for a given coil than parallel flow does. Mixing of heated and cold bypass air must occur upstream of the control sensors. Stratification can be reduced by using sheet-metal *air blenders* or by propeller fans in the ducting. The preheat coil should be located in the bottom of the duct. Steam preheat coils must have adequately sized traps and vacuum breakers to avoid condensate buildup that could lead to coil freezing at light loads.

The face and bypass damper approach enables air to be heated to the required system supply temperature without endangering the heating coil. (If a coil were as large as the duct—no bypass area—it could freeze when the hot water control valve cycles open and close to maintain discharge temperature.) The designer should consider pumping the preheat coil as shown in Figure 21.19d to maintain water velocity above the 3 ft/s needed to avoid freezing. If glycol is used in the system, the pump is not necessary, but heat transfer will be reduced.

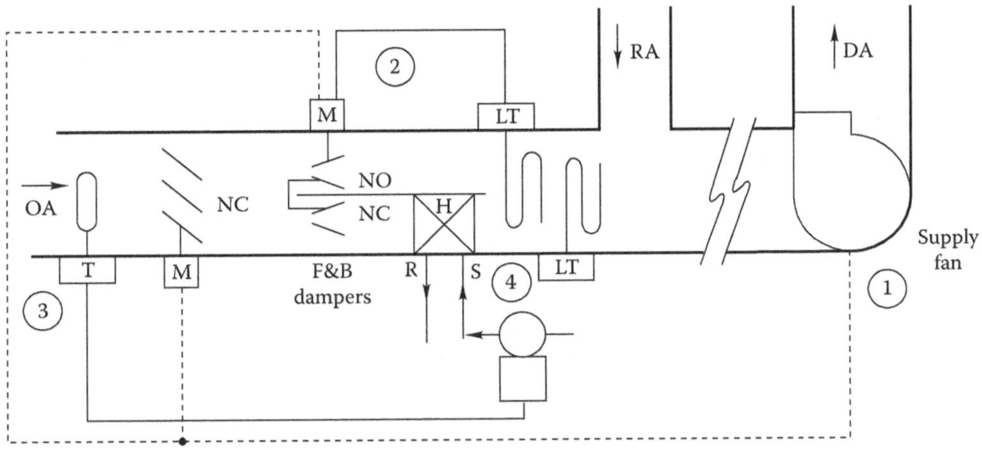

FIGURE 21.24
Preheat control system. Counterflow of air and hot water in the preheat coil results in the highest heat transfer rate.

During winter, heat must be added to the mixed airstream to heat the outside-air portion of mixed air to an acceptable discharge temperature. Figure 21.25 shows a common heating subsystem controller used with central air handlers. (It is assumed that the mixed-air temperature is above freezing by action of the preheat coil, if needed.) This system has an added feature of adjusting the coil discharge temperature based on the ambient temperature since the amount of heat needed decreases with increasing outside temperature. This feature, called the coil discharge *reset*, provides better control and can reduce energy consumption. The system operates as follows:

1. During operation, the discharge air sensor and PI controller control the hot water valve.
2. The outside-air sensor and controller increases the *setpoint* of the discharge air PI controller as the ambient temperature drops.
3. Under sensed low-temperature conditions, freeze protection measures are initiated as discussed earlier.

Reheat at zones in VAV or other systems uses a system similar to that just discussed. However, the boiler water temperature is reset, and no freeze protection is normally included. The air temperature sensor is the zone thermostat for VAV reheat, not a duct temperature sensor.

21.5.3 Cooling Control

Figure 21.26 shows the components in a cooling coil control system for a single-zone system. Control is similar to that for the heating coil discussed earlier, except that the zone thermostat (not a duct temperature sensor) controls the coil. If the system were a central system serving several zones, a duct sensor would be used. Chilled water supplied to the coil partially bypasses and partially flows through the coil, depending on the coil load. The use of three- and two-way valves for coil control was discussed in detail earlier. The valve-NC connection is used as shown so that valve failure (control power outage or loss of air supply) will not block secondary loop flow.

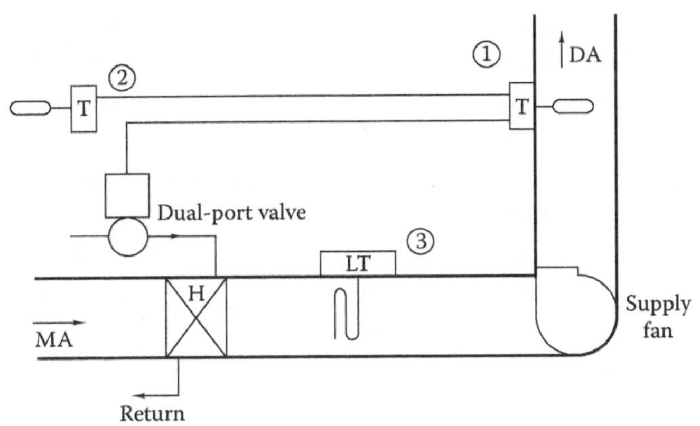

FIGURE 21.25
Heating coil control subsystem using two-way valve and optional reset.

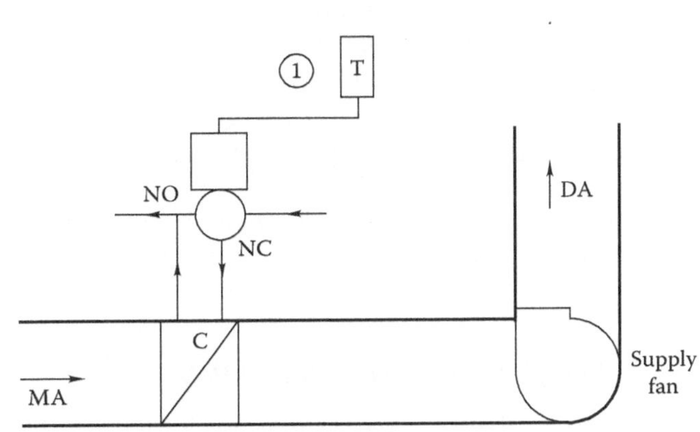

FIGURE 21.26
Cooling coil control subsystem using three-way diverting valve.

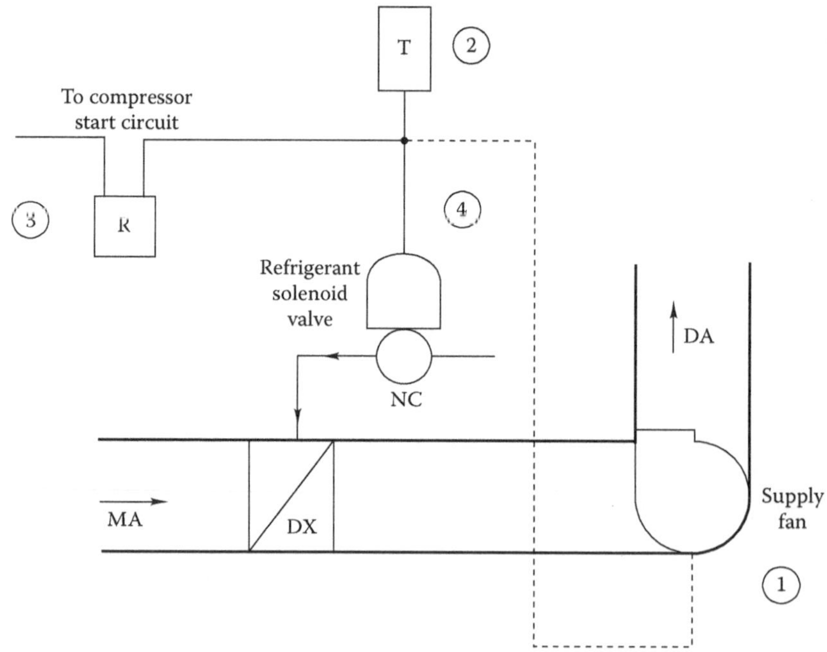

FIGURE 21.27
DX cooling coil control subsystem (on/off control).

Figure 21.27 shows another common cooling coil control system. In this case, the coil is a DX coil, and the controlled medium is refrigerant flow. DX coils are used when precise temperature control is not required. Here since the coil outlet temperature drop is large whenever refrigerant is released into the coil because refrigerant flow is not modulated; it is most commonly either on or off. The control system sequence is as follows:

1. The coil control system is energized when the supply fan is turned on.
2. The zone thermostat opens the two-position refrigerant valve for temperatures above the setpoint and closes it in the opposite condition.
3. At the same time, the compressor is energized or de-energized. The compressor has its own internal controls for oil control and pump-down.
4. When the supply fan is off, the refrigerant solenoid valve returns to its NC position and the compressor relay to its NO position.

At light loads, bypass rates are high, and ice may build up on coils. Therefore, control is poor at light loads with this system.

21.5.4 Complete Systems: CAV and VAV

The preceding five example systems are actually control subsystems that must be integrated into a single control system for the HVAC system's primary and secondary systems. In the remainder of this section, we briefly describe two complete HVAC control systems widely used in commercial buildings. The first is a fixed-volume or constant air volume (CAV) system, and the second is a variable air volume (VAV) system.

Figure 21.28 shows a CAV, central system air-handling system equipped with supply and return fans, heating and cooling coils, and an economizer for a single-zone application. If the system were to be used for multiple zones, the zone thermostat shown would be replaced by a discharge air temperature sensor. This fixed-volume system operates as follows:

1. When the fan is energized, the control system is activated.
2. The minimum air quantity is set (usually only once during commissioning, as described earlier).
3. The OA temperature sensor supplies a signal to the damper controller.
4. The RA temperature sensor supplies a signal to the damper controller.
5. The damper controller positions the dampers to use outdoor or return air, depending on which is cooler.
6. The mixed-air low-temperature controller controls the OA dampers to keep excessively low-temperature air from entering the coils. If a preheat system were included, this sensor would control it.

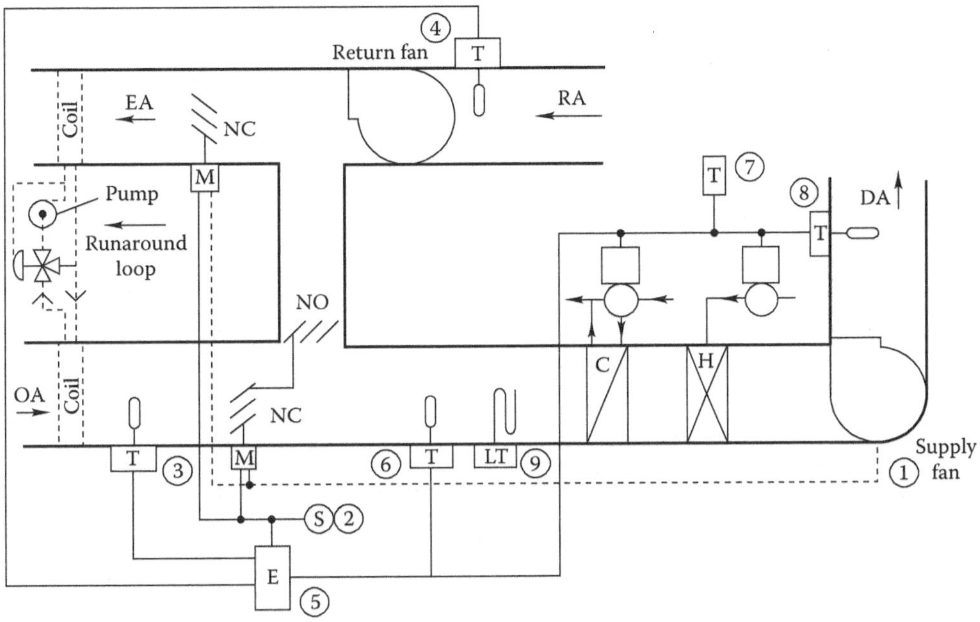

FIGURE 21.28
Control for a complete CAV system. Optional runaround heat recovery system is indicated on left with dashed lines.

7. Optionally, the space sensor could reset the coil discharge air PI controller.
8. The discharge air controller controls the
   a. Heating coil valve.
   b. Outdoor air damper.
   c. Exhaust air damper.
   d. Return air damper.
   e. Cooling coil valve after economizer cycle upper limit is reached.
9. The low-temperature sensor initiates freeze protection measures, as described previously.

A method for reclaiming either heating or cooling energy is shown by dashed lines in Figure 21.28. This "runaround" system extracts energy from exhaust air and uses it to precondition outside air. For example, the heating season exhaust air may be at 75°F, while outdoor air is at 10°F. The upper coil in the figure extracts heat from the 75°F exhaust and transfers it through the lower coil to the 10°F intake air. To avoid icing of the air intake coil, the three-way valve controls this coil's liquid inlet temperature to a temperature above freezing. In heating climates, the liquid loop should also be freeze-protected with a glycol solution. Heat reclaim systems of this type can also be effective in the cooling season, when outdoor temperatures are well above the indoor temperature.

A VAV system has additional control features, including a motor speed (or inlet vane) control and a duct static pressure control. Figure 21.29 shows a VAV system serving both perimeter and interior zones. It is assumed that the core zones always require cooling during the occupied period. The system shown has a number of options and does not include every feature present in all VAV systems. However, it is representative of VAV design practice. The sequence of operation is as follows:

1. When the fan is energized, the control system is activated. Prior to activation, during unoccupied periods, the perimeter zone baseboard heating is under the control of room thermostats.
2. Return and supply fan interlocks are used to prevent pressure imbalances in the supply air ductwork.
3. The mixed-air sensor controls the outdoor air dampers (and/or preheat coil, not shown) to provide the proper coil air inlet temperature. The dampers will be at their minimum position at about 40°F typically.
4. The damper's minimum position controls the minimum outdoor airflow.
5. As the upper limit for economizer operation is reached, the OA dampers are returned to their minimum position.
6. The return air temperature is used to control the morning warm-up cycle after night setback. (This option is available only if night setback is used.)
7. The OA damper is not permitted to open during morning warm-up by action of the relay shown.

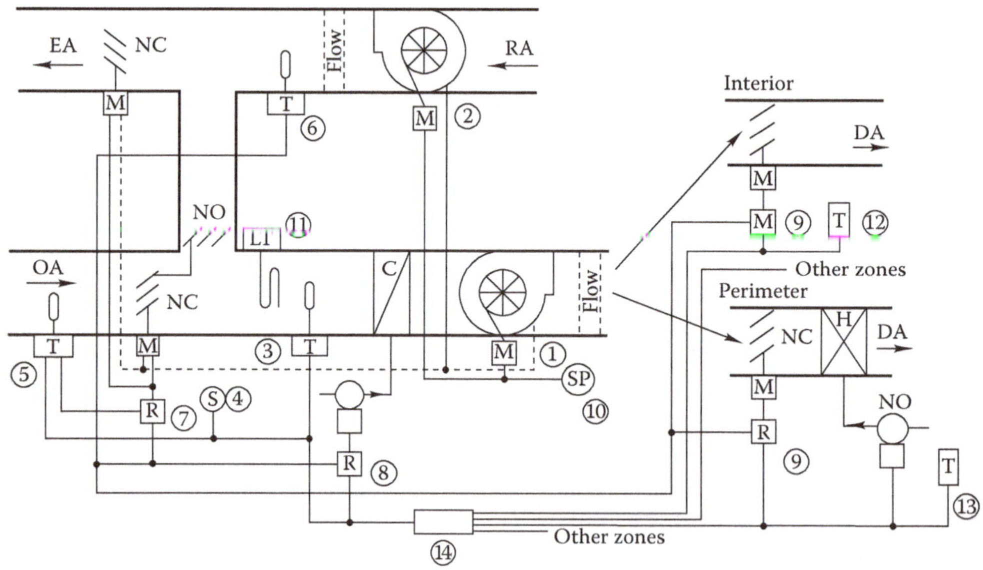

FIGURE 21.29
Control for complete VAV system. Optional supply and return flow stations are shown with dashed lines.

8. Likewise, the cooling coil valve is de-energized (NC) during morning warm-up.
9. All VAV box dampers are moved full open during morning warm-up by action of the relay override. This minimizes warm-up time because the boxes will almost always be in the closed position. Perimeter zone coils and baseboard units are under the control of the local thermostat.
10. During operating periods, the PI static pressure controller controls both supply and return fan speeds (or inlet vane positions) to maintain approximately 1.0 inWG of static pressure at the pressure sensor location (or optionally to maintain building pressure). An additional pressure sensor (not shown) at the supply fan outlet will shut down the fan if fire dampers or other dampers should close completely and block airflow. This sensor overrides the duct static pressure sensor shown. (Alternatively, the supply fan can be controlled by duct pressure and the return fan by building pressure.)
11. The low-temperature sensor initiates freeze protection measures.
12. At each zone, room thermostats control VAV boxes (and fans, if present); as the zone temperature rises, the boxes open more.
13. At each perimeter zone, room thermostats close VAV dampers to their minimum settings and activate zone heat (coil and/or perimeter baseboard) as the zone temperature falls.
14. The controller, using temperature information for all zones (or at least for enough zones to represent the characteristics of all zones), modulates OA dampers (during economizer operation) and the cooling control valve (above economizer cycle cutoff) to provide air sufficiently cooled to meet the load of the warmest zone.

The duct static pressure controller is critical to proper operation of VAV systems. As stated in Section 19.6.4, if proper duct pressurization at the end of long runs is not present, flow starvation can occur at VAV terminals located far from the air handler. The static pressure controller must be of PI design since a proportional-only controller would permit the duct pressure to drift upward as cooling loads drop due to the unavoidable offset in P-type controllers. In addition, the control system should position inlet vanes closed during fan shutdown to avoid overloading on restart.

Return fan control is best achieved in VAV systems by an actual flow measurement in supply and return ducts, as shown by dashed lines in the figure. The return airflow rate is the supply rate less local exhausts (fume hoods, toilets, etc.) and exfiltration needed to pressurize the building. Then, the duct pressure sensor is used only for controlling maximum duct pressure.

VAV boxes are controlled locally, assuming that adequate duct static pressure exists in the supply duct and that supply air is at an adequate temperature to meet the load (this is the function of the controller described in item 14 above). Figure 21.30 shows a *local* control system used with a series-type fan-powered VAV box. This particular system delivers a constant flow rate to the zone,

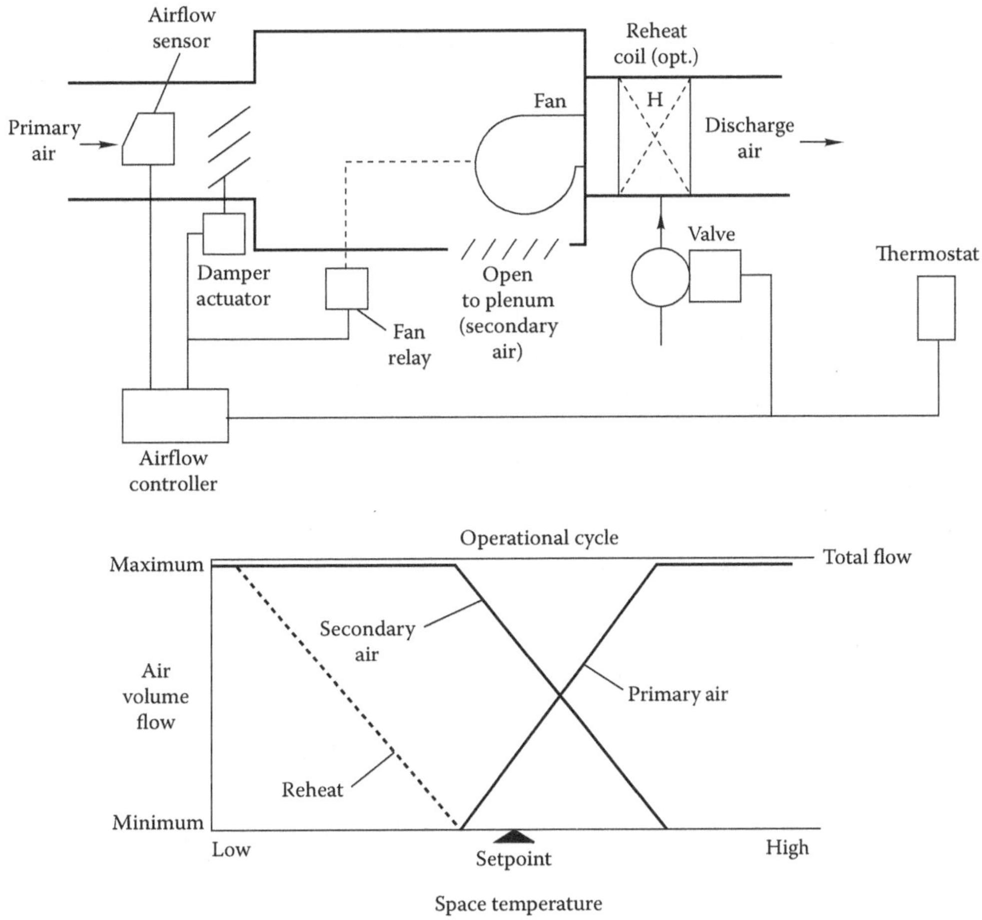

FIGURE 21.30
Series-type, fan-powered VAV box control subsystem and primary flow characteristic. The total box flow is constant at the level identified as maximum in the figure. The difference between primary and total airflow is the secondary air recirculated through the return air grille. Optional reheat coil requires airflow shown by the dashed line.

to ensure proper zone air distribution, by action of the airflow controller. Primary air varies with the cooling load, as shown in the lower part of the figure and as discussed in Section 19.4. Optional reheat is provided by the coil shown.

21.5.5 Other Systems

Fire and smoke control is important for human safety in large buildings. The design of smoke control systems is dictated by national codes. The principal concept is to eliminate smoke from the zones where it is present while keeping adjacent zones pressurized to avoid smoke infiltration. Some components of space conditioning systems, e.g., fans, can be used for smoke control, but HVAC systems are generally not smoke control systems by design.

Electric systems are primarily the responsibility of the electrical engineer or a design team. However, HVAC engineers must ensure that the electrical design accommodates the HVAC control system. Interfaces between the two occur where the HVAC controls activate motors on fans or chiller compressors, pumps, electric boilers, or other electric equipment.

In addition to electrical specifications, the HVAC engineer often conveys electrical control logic by using a *ladder diagram*. An example is shown in Figure 21.31 for the control of the supply and return fans in a central system. The electric control system is shown at the bottom and operates on low voltage (24 or 48 V ac) from the control transformer shown. The supply fan is started manually by closing the "start" switch. This activates the motor starter coil labeled 1M, thereby closing the three contacts labeled 1M in the supply fan circuit. The fourth 1M contact (in parallel with the start switch) holds the starter closed after the start button is released.

The manual-off-auto switch is typical and allows both automatic and manual operation of the return fan. When it is on the manual position, the fan starts. In the auto position, the fans will operate only when the adjacent 3M contacts are closed. Either of these actions activates the relay coil 2M, which in turn closes the three 2M contacts

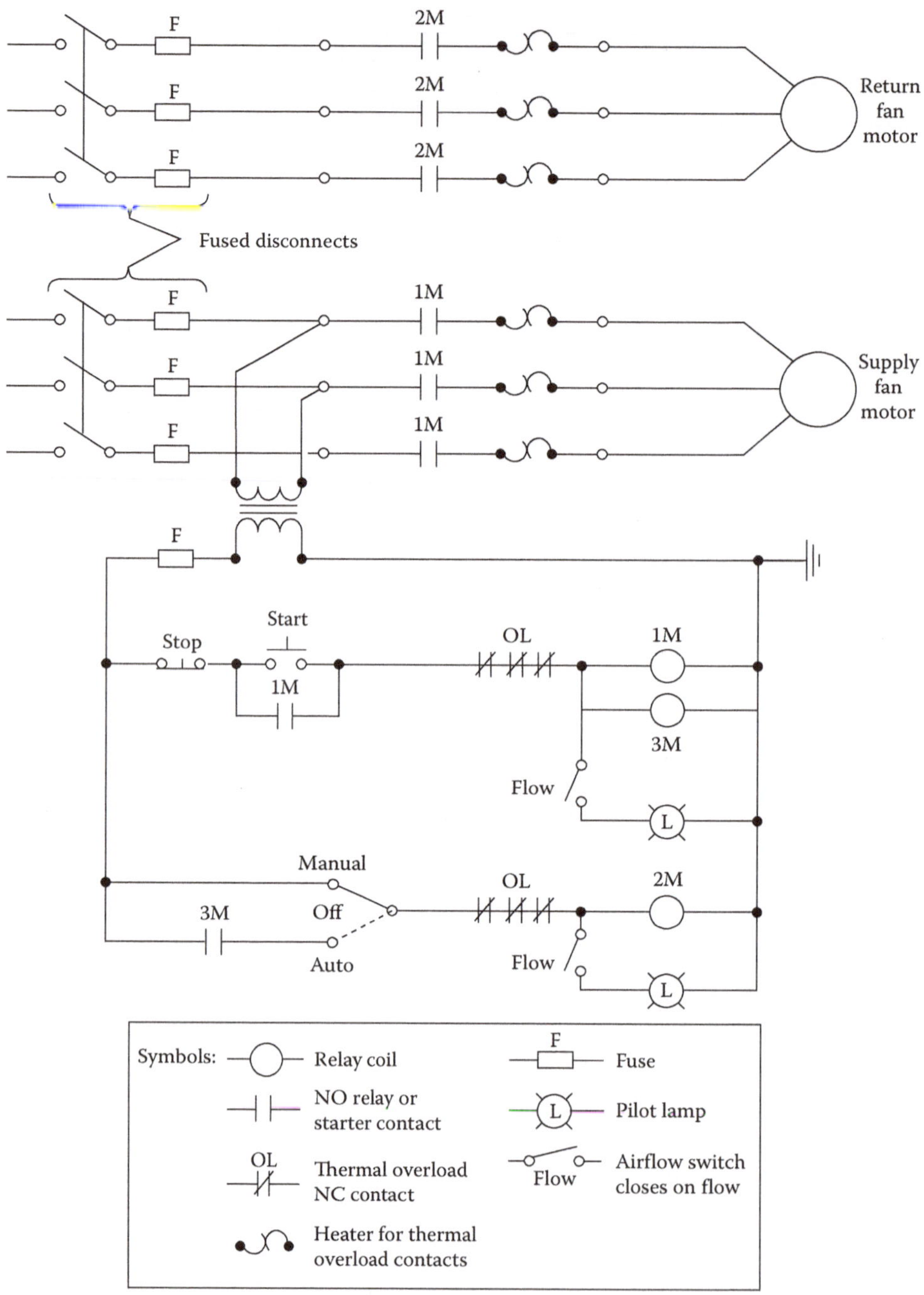

FIGURE 21.31
Ladder diagram for supply and return fan control. Manual-off-auto switch permits manual or automatic control of the return fan.

in the return fan motor starter. When either fan produces actual airflow, a flow switch is closed in the ducting, thereby completing the circuit to the pilot lamps (L). The fan motors are protected by fuses and thermal overload heaters. If motor current draw is excessive, the heaters (OL) shown in the figure produce sufficient heat to open the normally closed thermal overload contacts.

This example of a ladder diagram is primarily illustrative and is not typical of an actual design. In a fully automatic system, both fans would be controlled by 3M contacts actuated by the HVAC control system. In a fully manual system, the return fan would be activated by a fifth 1M contact, not by the 3M automatic control system.

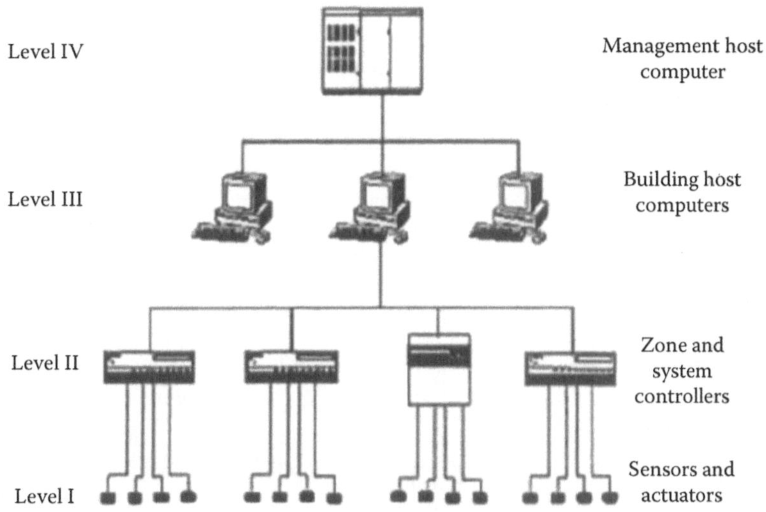

FIGURE 21.32
Example of a hierarchical network of a BAC system. (From Swan, W., *Eng. Syst.*, 13(7), 24, July 1996, http://www.bacnet.org/Bibliography/ES-7-96/ES-7-96.htm.)

21.6 Building Automation

21.6.1 Networking Architectures

HAVC system control at a campus level involves four levels of hierarchy (Figure 21.32). Level I is the *unit level control* where a controller is needed for each individual HVAC component. Level II is the *system level control*, which comprises of several unit-level controls that are tied together by communication networks, and thus provide control to individual zones and related systems. Examples are air-handlers and chilled water loops. Level III is the *building level control* where several system level controls are tied together and decisions (such as supervisory control) can be made by the building manager to minimize energy use throughout that building. Dedicated chillers or boilers are some of the pertinent equipment to be controlled, but lighting and fire/safety alarming can also be combined. Finally, level IV (if appropriate) is at the *whole campus level*, commonly referred to as wide area network (WAN) where numerous individual buildings are controlled in unison by the facility manager. There are network architectures other than hierarchical that are in common use. Figure 21.33 shows two such architectures for HVAV, lighting and fire/safety linked in a ring and a star configuration. The advantages and limitations of each architecture can be found in specialized books such as McGowan (1991). Figure 21.34 conceptualizes a network that integrates several engineering functions to reduce energy use and O&M (operation and maintenance) costs in a building.

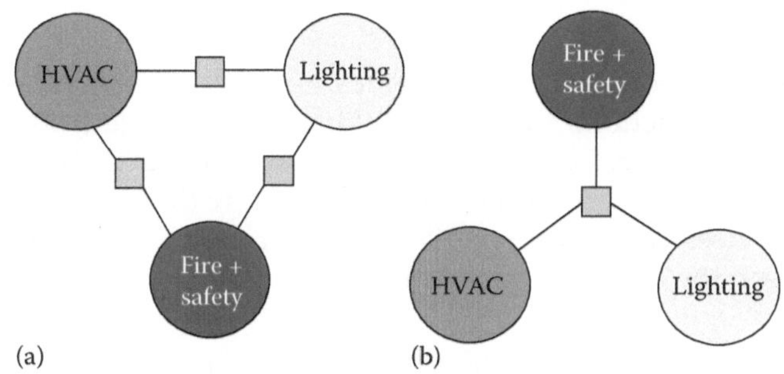

FIGURE 21.33
Integration of BAC, fire/safety, and lighting networks: (a) Ring architecture and (b) star architecture.

The increasing adoption of DDC controls and integration of various computer networks is leading to so-called *intelligent buildings*. An intelligent building gives the occupants access to the full range of information and communications options and allows the building owner to manage the resources and costs in an efficient manner. The ultimate goal of intelligent building design is to integrate all of the information and control functions done by the occupants within the building and by the manager of the building into one single computerized system. These functions include

- Control of building systems
- Energy efficiency
- Life safety systems
- Telecommunications systems
- Workplace automation

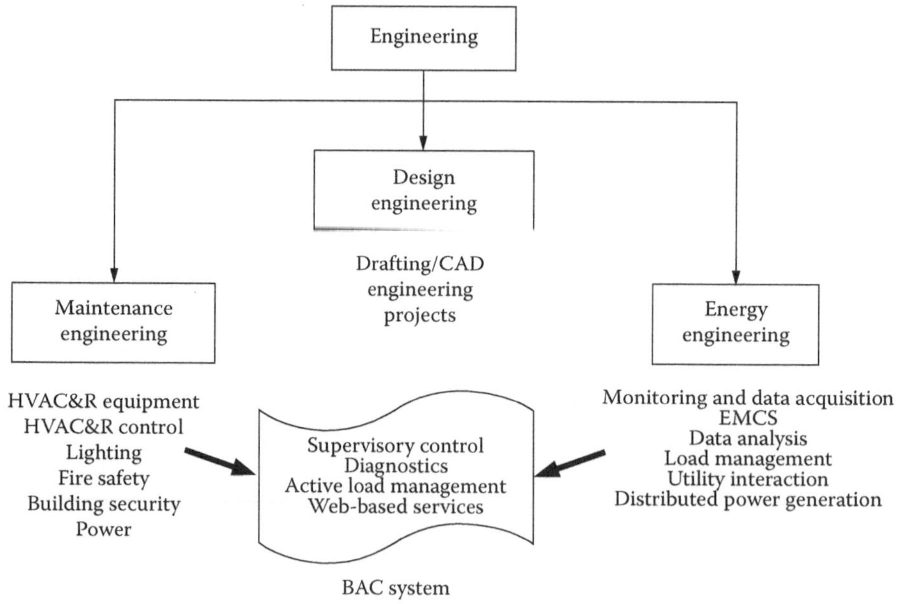

FIGURE 21.34
Current trend in integrating various engineering services.

Furthermore, using a wide-area network (WAN), all of these functions in many buildings can be controlled from a central source. Extending this concept to transportation, energy generation, and area-wide emergency services can enable the "intelligent city" that has long been the dream of urban planners.

21.6.2 BACnet Data Exchange Protocol

BACnet™ (Building Automation and Control Network) is a standard communication protocol developed by the American Society of Heating, Refrigeration and Air Conditioning Engineers (ASHRAE), which has been adapted by the American National Standards Institute (ANSI) and also by the International Standards Organization (ISO) as ISO-16484-5. BACnet (ANSI/ASHRAE Standard 135, 2012) is used in the control of HVAC (heating, ventilation, and air conditioning), building security, lighting, fire, access control, waste management, and other control equipment in buildings. Protocols are rules that need to be followed, which allow handling not only different types of signals and controllers but also the detection of bad data and facilitate quick data transfer. The main reason for the creation of BACnet was to enable building owners to use control equipment from different manufacturers without having to worry about interoperability.

Before the introduction of BACnet, most HVAC equipment manufacturers had their own proprietary communications protocols. The typical user complaint was that if additions or alterations were to be made to existing HVAC DDC systems, the user was forced to select products offered by the original system's manufacturer, even if they were not the most suitable or cost effective. The BACnet protocol solves the interoperability problem by requiring manufacturers of control equipment to include one or several of the "Objects" and their properties, so that other control equipment can communicate with them. In addition, BACnet also endorsed existing open system interconnection (OSI) network protocols and architecture, so that BACnet networks could communicate using existing and reliable LAN (local area networks) and higher layer protocols. This process was designed to lead to cost-effective and reliable network operation (since one could match and seamlessly operate different vendor products), and the results from the field supported this premise.

BACnet uses an "object-oriented" method for organizing the data contained in a particular device or a controller. Since these devices will be connected to other BACnet devices, organizing the data structure within the devices in a standardized object-oriented manner creates a seamless foundation to the BACnet network structure or architecture. All BACnet-compatible instruments and sensors must contain some of the standard BACnet specified objects and their properties. Devices may contain one or many of the BACnet specified objects, while these objects may also contain any number of the BACnet specified properties. However, devices do not need to contain all of the objects and properties specified; only those that are relevant to the particular sensor type need to be included. Manufacturers are also free to add their own proprietary objects and properties that can only communicate with other equipment controllers

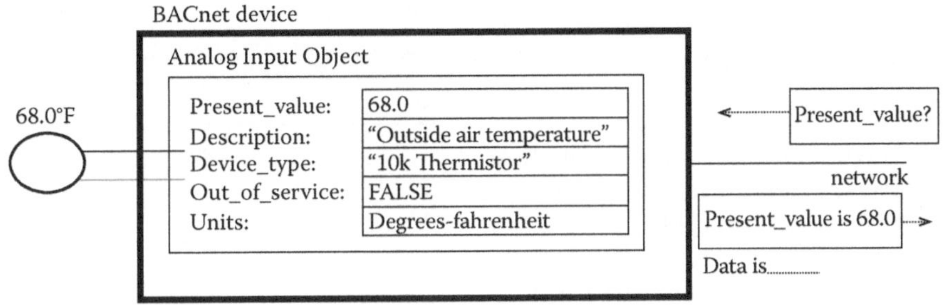

FIGURE 21.35
An analog Input Object in BACnet protocol.

made by the same manufacturer, as long as they do not interfere with the operation of the BACnet network.

Objects, such as *Analog Input* (Figure 21.35) and *Analog Output*, contain many properties, such as *Object_Identifier, Present_Value* and *Units*, which operators or other BACnet controllers may use to monitor or control the devices. For example, a temperature sensor must include at the very minimum the *Analog Input* and *Analog Output objects*. An Analog Output of this device could have properties called *Present_Value, Description, Device_Type, Out_Of_Service*, and *Units*. These properties will each have values, such as the current reading of the temperature in °F, which other sensors in the BACnet network may be able to access, monitor, and even control. *Properties* such as *Present_Value* are constantly updated and could change several hundred times a day, while other properties such as *Device_Type* are set by the manufacturer and need never be changed.

Though BACnet network protocol was developed based on the OSI model, BACnet developers decided to choose only the services that are essential for building automation and control, and implemented them on the BACnet model. Hence, only four out of the seven layers were adopted: application, network, data link, and physical layers. Several articles have been published over the years describing existing; new features of the BACnet standard (Swan, 1996; Bushby and Newman, 2002). Figure 21.36 is a conceptual sketch of various zone controllers tied in a bus architecture using a BACnet router for two-way communication over the Internet. Wireless handheld control devices are becoming popular as well.

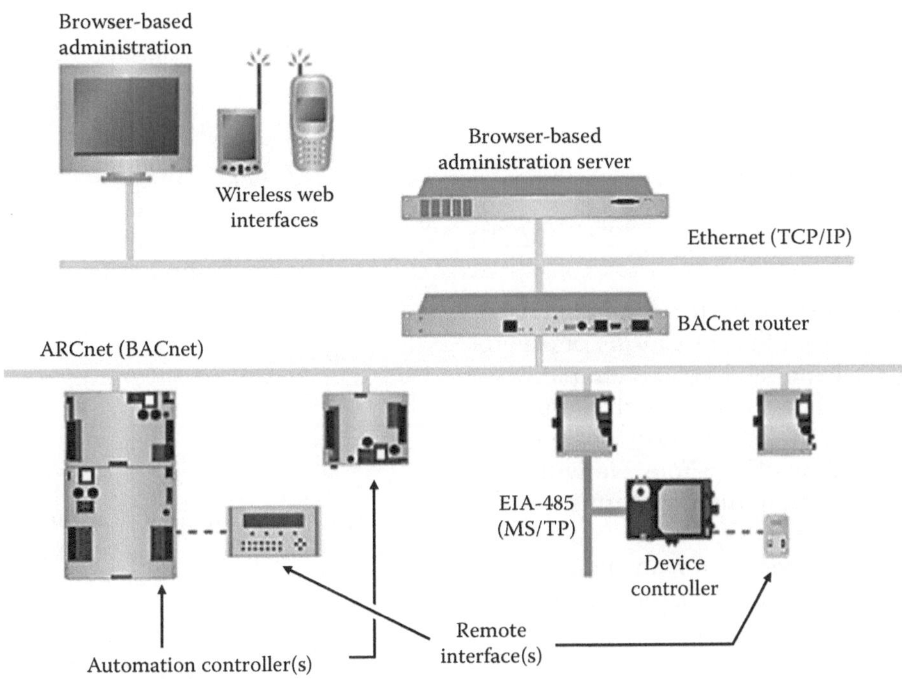

FIGURE 21.36
Sketch of a network architecture of a BAC system communicating over the Internet.

21.7 Topics in Advanced Control System Design

The previous sections have described control system configurations and hardware design but not how one selects the type of control—P, PI, or PID—to be used or what the control constants K_p, K_i, and K_d should be. In this section, we will discuss some of the considerations that lead to selection of control algorithms. We will find Laplace transforms very useful for this purpose and will review this tool first. In this section, we will discuss primarily linear systems or systems that can be considered linear over a portion of their operating range.

21.7.1 Laplace Transforms

The Laplace transform is an integral transform that can be used to convert differential equations to algebraic equations. Solving the algebraic equation and inverting the result provide the solution to the differential equation. In HVAC control design, we are interested in solving the differential equations that model HVAC controls and the process being controlled. Even if the equations are not solved explicitly, the nature of the transform indicates how the control system will behave.

By definition, the Laplace transform $F(s)$ of a function of time $f(t)$ is found from

$$F(s) = \int_0^\infty e^{-st} f(t)\,dt \qquad (21.20)$$

in which s is the complex variable

$$s = \sigma + j\omega \qquad (21.21)$$

Standard mathematics textbooks, for example, Wylie and Barrett (1982), show that the transforms of several common functions are expressed by the simple polynomials (Table 21.3). For HVAC system control analysis, we are most interested in the system response without reference to initial conditions. Therefore, we will drop the initial-condition terms from the transforms used in this section.

If a linear second-order system for an independent variable $y(t)$ has a forcing function $f(t)$, the transform of the equation with coefficients a, b, and c is

$$(as^2 + bs + c)Y(s) = F(s) \qquad (21.22)$$

TABLE 21.3

Summary of Laplace Transforms

$F(s)$	$f(t)(t > 0)$	Notes
1	$\delta(t)$	Unit impulse
$\exp(-Ts)$	$\delta(t - T)$	Delayed impulse
$\frac{1}{s+a}$	$\exp(-at)$	Exponential decay
$\frac{1}{(s+a)^n}$	$\frac{t^{n-1}\exp(-at)}{(n-1)}$	$n = 1, 2, 3, \ldots$
$\frac{1}{s}$	$u(t)$ or 1	Unit step
$\frac{1}{s^2}$	t	Unit ramp
$\frac{1}{s^n}$	$\frac{t^{n-1}}{(n-1)}$	$0! = 1, n = 1, 2, 3, \ldots$
$\frac{\omega}{s^2+\omega^2}$	$\sin \omega t$	Sine
$\frac{s}{s^2+\omega^2}$	$\cos \omega t$	Cosine
$\frac{1}{(s+a)(s+b)}$	$\frac{e^{-at} - e^{-bt}}{b-a}$	$a \neq b$
$\frac{s}{(s+a)(s+b)}$	$\frac{ae^{-at} - be^{-bt}}{a-b}$	$a \neq b$
$\frac{1}{s(s+a)}$	$\frac{1-e^{-at}}{a}$	Exponential decay
$\frac{1}{s^2+2\zeta\omega_n s+\omega_n^2}$	$\frac{1}{\omega_d} e^{-\zeta\omega_n t} \times \sin \omega_d t \left(\omega_d = \omega_n \sqrt{1-\zeta^2}\right)$	$0 < \zeta < 1$
$\frac{s}{(s+a)^2}$	$(1 - al)e^{-at}$	
$\frac{a^2}{(s+a)^2 s}$	$1 - e^{-at}(1 + at)$	
$sF(s) - f(t = 0)$	$\frac{df(t)}{dt}$	Derivative
$\frac{F(s)}{s} + \frac{\int t(t)dt \vert_{t=0}}{s}$	$\int f(t)\,dt$	Integral

Solving for $Y(s)$ in the s domain is simply a matter of dividing both sides of the equation by the polynomial. The result is

$$Y(s) = \frac{F(s)}{as^2 + bs + c} \qquad (21.23)$$

The *transfer function* TF is the ratio of output to the forcing function. In this example,

$$\mathrm{TF} = \frac{Y(s)}{F(s)} \quad (21.24)$$

To find the solution in the time domain, one uses results from the first and second columns of Table 21.3. If $F(s)$ in Equation 21.23 were a constant, the solution would be exponential or sines and cosines or an exponential–sinusoidal combination.

Example 21.3: Laplace Transform

The Laplace transform of a differential equation is

$$Y(s) = \frac{s+3}{s(s+1)^2} \quad (21.25)$$

What is the inverse $y(t)$?

Given: $Y(s)$

Find: $y(t)$ if $y(0) = 0$

Solution

First, the given function must be rearranged so that the results in Table 21.3 can be used to invert the result. It is easy to show that Equation 21.25 can be factored as follows:

$$Y(s) = \frac{3}{s} - \frac{3}{s+1} - \frac{2}{(s+1)^2} \quad (21.26)$$

Referring to Table 21.3, we see that each term type appears in the first column, and we can write the inverse function by inspection. The solution is

$$y(t) = 3 - 3e^{-t} - 2te^{-t} \quad (21.27)$$

Comments

The solution is a constant with two exponential decay terms. For large times, note that $y(t) = 3$. By a theorem in Laplace transforms, it is known that this limit is the same as a related limit in the s domain:

$$\lim_{x\to\infty(t)} y = \lim_{s\to\infty} sY(s) \quad (21.28)$$

From Equation 21.28, we see that the long time limit can be found without inverting the Laplace transform.
A complementary theorem also relates the initial value of a function to its transform:

$$\lim_{x\to 0(t)} y = \lim_{s\to\infty} sY(s) \quad (21.29)$$

21.7.2 Laplace Transforms for HVAC Equipment

A basic feedback controller can be represented by the schematic diagram in Figure 21.37. The error e is processed by a controller that produces a signal V controlling the process. For a room during the heating season, the controlled variable is room temperature and the controlling system is the thermostat sensor and its controller. The controlled process is the heat input into the room from the heating terminal device and the heat loss from the room.

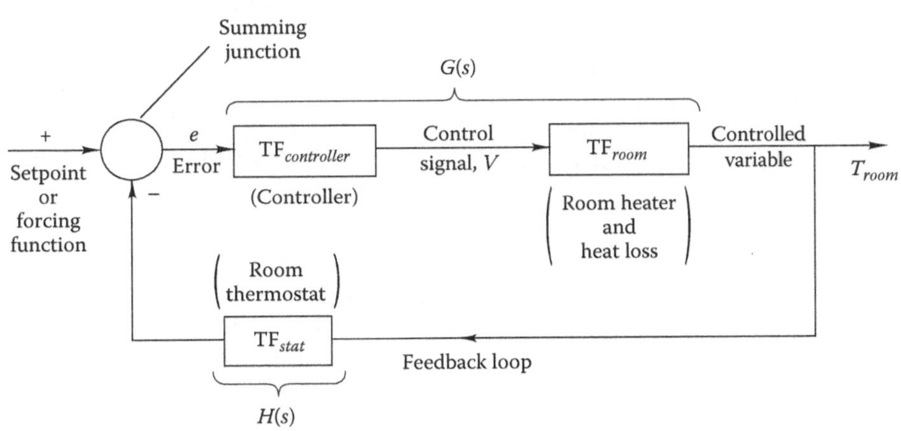

FIGURE 21.37
Simple feedback control loop for room-temperature control. Error e is processed by the controller, which produces a control signal to the controlled process (the room heating equipment and the room heat loss characteristic). The thermostat may have its own dynamics characterized by its transfer function TF_{stat}.

To write an equation for each block in the diagram, the appropriate Laplace transforms must be known. According to Letherman (1981), the characteristics of some common building components can be represented by the expressions given in the following. For the room itself, the transfer function TF is

$$\mathrm{TF}_{room} = \frac{Ke^{-sL_{room}}}{1+\tau_{room}s} \tag{21.30}$$

where

TF_{room} is the room transfer function (the "process")
K = steady-state loss coefficient
L_{room} is the time delay (see Table 21.3, line 2)
τ_{room} is the room time constant, the ratio of lumped room thermal capacitance to loss coefficient

The transfer function of a PID controller is

$$\mathrm{TF}_{controller} = \frac{V(s)}{E(s)} = K_p + \frac{K_i}{s} + K_d s \tag{21.31}$$

in which

$E(s)$ is the transform of the error e
$V(s)$ is the transform of the controller output

The closed-loop transfer function of the feedback controller shown in Figure 21.37 is

$$\mathrm{TF}_{loop} = \frac{G(s)}{1+G(s)H(s)} \tag{21.32}$$

in which

$$G(s) = \mathrm{TF}_{control} \times \mathrm{TF}_{room} \tag{21.33}$$

is the combined transfer function of the controller and process shown in Figure 21.37, where $G(s)$ is sometimes called the *open-loop transfer function*. The corresponding *characteristic equation* is

$$G(s)H(s) + 1 = 0 \tag{21.34}$$

Shortly, we shall see that the characteristic equation contains most of the information needed to assess the performance of linear control systems.

For a thermostat, the following transfer function including both a lag and a first-order time constant is suggested:

$$\mathrm{TF}_{stat} = \frac{e^{-sL_{stat}}}{1+\tau_{stat}s} \tag{21.35}$$

The time lag L_{stat} is on the order of 30–40 s, and the time constant τ_{stat} is on the order of minutes for zone-mounted thermostats.

The first-order approximation to various lag terms, such as that resulting from the distance between a temperature measured at a coil and a controller at a remote chiller or those noted above, is given by the Padé approximation. This is convenient because it preserves a polynomial expression for a system's transfer function—a useful form when one is investigating stability, as we shall see later. To first order in sL, the transform of a lag is

$$e^{-sL} \approx \frac{2-sL}{2+sL} \tag{21.36}$$

Little is known of the transfer functions of other HVAC equipment and systems. It is therefore necessary to determine them experimentally, as we shall discuss shortly.

21.7.3 Control System Stability

One of the key results that Laplace transform analysis of building systems and controls can provide is a judgment as to whether the control system is stable when a step forcing function is imposed. As described in standard controls texts, a system is stable if all roots of the characteristic equation (Equation 21.34) have no positive real part (recall that s is complex and that roots of the characteristic equation are complex in general). For complicated systems, it is laborious to find roots explicitly, but certain global characteristics of the equation can indicate stability. Two *necessary* conditions for stability are as follows:

1. All powers of s must be present in the characteristic equation from zero to the highest order.
2. All coefficients in the characteristic equation must have the same sign.

These are necessary but not sufficient conditions. Routh's criterion is an algebraic method for further assessing stability. A simple algorithm is used to construct a matrix from the characteristic-equation coefficients. Then, the first column of this matrix is examined for sign changes. The number of sign changes in the first column is equal to the number of positive roots. One positive (real-part) root or more indicates instability (Letherman, 1981). Examples 21.4 and 21.5 illustrate two methods for analyzing the stability of a simple proportional controller.

Example 21.4: Stability of a Second-Order Controller

Consider the stability of the heating coil controller shown in Figure 21.2b with proportional-only control (i.e., K_i = 0.0). Examine the behavior of the system as the proportional gain is varied. The transfer function of each component of the system is shown in parentheses in the figure. Note that the dynamics of the controlled device (i.e., valve and actuator) are not considered.

Given: Component transfer functions

Find: Stability conditions

Figure: See Figures 21.2b and 21.37.

Assumptions: The coil system components and control systems are linear.

Solution

The open-loop transfer function $G(s)$ is the product of the controller, device, and process transfer functions:

$$G(s) = K_p K_d \frac{\varepsilon}{1+\tau_{coil}s}$$

where
ε is the coil effectiveness
τ_{coil} is the coil time constant*

The sensor transfer function $H(s)$ is

$$H(s) = 1/(1+\tau_s \cdot S)$$

These two expressions can be combined to find the characteristic equation, Equation 21.34, for this system:

$$G(s)H(s)+1 = \frac{\tau_{coil}\tau_s s^2 + (\tau_{coil}+\tau_s)s + (K_p K_d \varepsilon + 1)}{(1+\tau_s s)(1+\tau_{coil}s)} = 0$$

The denominator is nonzero; therefore, if any real root of the numerator has a positive part, the system is unstable. In this simple example, the quadratic formula can be used to find the roots r. We will then examine the result for stability. The roots of the characteristic equation are

$$r_1, r_2 = \frac{-(\tau_{coil}+\tau_s) \pm \sqrt{(\tau_{coil}+\tau_s)^2 - 4\tau_s\tau_{coil}(K_p K_d \varepsilon + 1)}}{2\tau_{coil}\tau_s}$$

The real part of these roots cannot be positive; therefore, the system is stable. That is, for any physically realistic values of the five system parameters (of course, ε, the gains K, and the time constants τ must all be positive), the system is unconditionally stable if we include damped sinusoidal functions within the definition of stability. However, as the gain K_p increases, the frequency of system oscillation increases as the square root of the gain.

Comments

Routh's criterion described in Example 21.5 could have been used to examine the stability of this system with the same result. It is an easier method of checking for instability than the approach used earlier. In addition, if the characteristic equation is of higher order, finding the roots is a much less convenient method than Routh's criterion.

Example 21.5: Proportional Controller Stability—Routh's Criterion

The transfer function of a heating coil (the process) TF_{proc} is determined experimentally to be†

$$TF_{proc} = \frac{1}{s(s+4)(s+6)}$$

and

$$TF_{control} = K_p$$

For simplicity, we ignore the dynamics of the duct temperature sensor. The coil is controlled by a proportional controller with gain K_p. For what values of gain is the system stable?

Given: TF_{proc}, $K_p > 0$

Find: Range of K_p for stability

Figure: See Figure 21.25 (ignore ambient temperature reset feature) and Figure 21.37.

Assumption: The system is linear.

Solution

The characteristic equation, according to the Equation 21.34,‡ is

$$\frac{K_p}{s(s+4)(s+6)} + 1 = 0 \quad (21.37)$$

Clearing fractions gives

$$s^3 + 10s^2 + 24s + K_p = 0$$

* In this example, the coil is treated as a lumped capacitance.

† Each term in the denominator relates to the dynamics of a piece of equipment in the process, e.g., the valve actuator, the valve, and the coil.

‡ Since we are ignoring duct sensor dynamics, $H(s) = 1$.

It is seen that this equation satisfies the two conditions necessary for stability since all powers are present and since all coefficients have the same sign. Routh's criterion can be used to examine stability further. The Routh table for this equation is

$$\begin{array}{|c|c|} 1 & 24 \\ 10 & K_p \\ \dfrac{240-K_p}{10} & 0 \\ K_p & 0 \end{array}$$

The first two rows of the table are constructed by entering the coefficients in a "zigzag" fashion, starting at the upper left-hand corner, as is evident. The first term in the third row is the determinant of the four elements immediately above ($24 \times 10 - K_p \times 1$) divided by the coefficient immediately above (10). The element in the fourth row is found in the same way from the determinants of rows 2 and 3:

$$\frac{K_p(240-K_p)/10 - 10(0)}{(240-K_p)/10} = K_p$$

All elements of the first column except the third are uniformly positive, as required by Routh's criterion for stability. For this term to be positive, the gain must be less than 240:

$$K_p < 240 \tag{21.38}$$

Comments

Integral and derivative control can also be analyzed in this manner. In Problem 21.29, we will see that integral control reduces stability since it increases the order of the characteristic equation by 1. On the other hand, derivative control can stabilize an unstable proportional controller since it reduces the order of the characteristic equation by 1. The control constants K_p, K_i, and K_d must therefore be selected with care. This is the final topic in this section.

21.7.4 Selection of Control Constants: Control System Simulation

There are two methods to empirically determine the proper control constants for an existing system that can be exercised and measured in the field. Either method can be used to finalize controller settings in the field from estimates by the HVAC designer. Figure 21.38a shows the temporary modifications to the controller needed to use the *reaction curve* method. A step change of magnitude M is applied to the process. The response O at the output of the summing junction is measured and plotted, as shown in Figure 21.38b. From the data, three parameters are determined: a lag L, a time constant T, and the steady-state gain O_∞/M. These are used to construct an empirical first-order process transfer function:

$$\mathrm{TF}_{proc} = \frac{(O_\infty/M)e^{-sL}}{1+Ts} \tag{21.39}$$

where O_∞/M is called the *system gain*. From this transfer function, acceptable values of the control constants can be found by using either a method due to Ziegler and Nichols (1942) or a more recent method due to Cohen and Coon (1953). Either method produces constants for P, PI, or PID controllers based on combinations of the three parameters shown in Figure 21.38b. The constants so calculated give stable control with only a few oscillations whose amplitudes decrease by 75%/cycle. See the tables in the online HCB software for details.

If this modest overshoot is undesirable, the *pole-zero cancellation* method of MacArthur et al. (1989) can be used to find an approach to control the setpoint that does not involve any overshoot. Briefly, this method applies the controller transfer function, Equation 21.31 (without the derivative term), to the system transfer function, Equation 21.39, to find the characteristic equation. The result involves a quotient of polynomials. By selecting the proportional and integral gains appropriately, terms in the numerator and denominator of the characteristic equation can be made to cancel, thereby lowering the order of the equation. Using this approach, one can find proportional and integral gains as a function of system time delay and system gain that result in the most rapid approach possible to the setpoint without any overshoot.

The *ultimate frequency* method uses the undisturbed control shown in Figure 21.37 but disables the integral and derivative terms by setting the integral and derivative gains to very small values. The proportional gain is then adjusted so that steady oscillations with neither increasing nor decreasing amplitude result. This specific value of K_p and the period of the ultimate steady-state frequency are used to find the recommended proportional, integral, and derivative gains by using rules in Letherman (1981). See the tables in the online HCB software for details.

Both methods can be used to establish "good" control constants for the conditions under which the tests were conducted. However, most HVAC system controls have significant nonlinearities. As a result, settings at one condition will not necessarily be optimal at another. A recent solution to this problem uses the concept of *adaptive* control, whereby the control constants are adjusted in near real time as operating conditions change. The calculations needed for this method can be performed by a small computer in DDC systems. Underwood (1989) and Curtiss et al. (1993) report on HVAC applications of adaptive (self-tuning) controls.

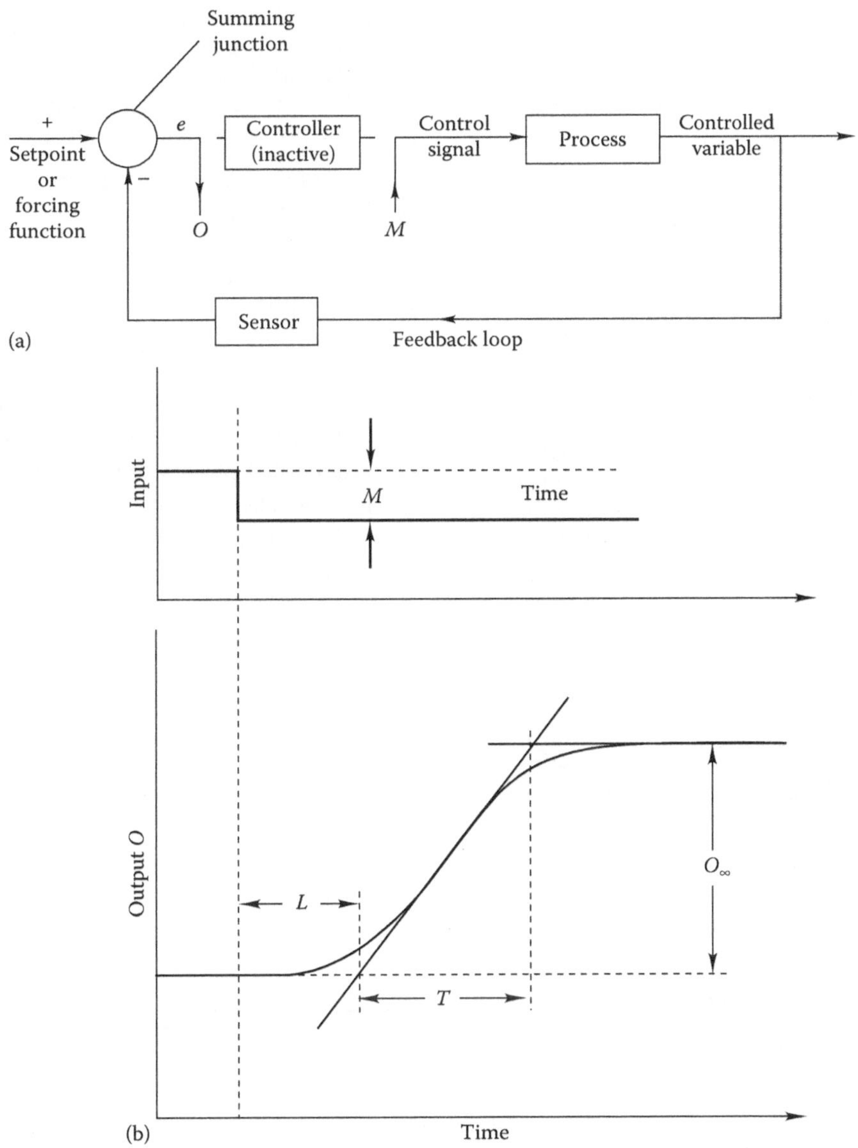

FIGURE 21.38
(a) Feedback control loop modified temporarily for reaction curve test. A control signal step change M is introduced into the process, and the system response (error) O is measured. (b) Typical reaction curve data showing steady output O_∞, lag L, and time constant T that result from an input step of M.

Another method of establishing control system constants is to simulate both the HVAC system and its controls. Details of this method are beyond the scope of this book, but Example 21.6 indicates the approach.

Example 21.6: Heating Coil Control Simulation

Use the online HCB software to study the response of a heating coil to a step change in the temperature setpoint with the following system characteristics. Is the control system stable?

Given: Coil inlet temperature: 30°F
Original coil air outlet setpoint: 55°F
New coil air outlet setpoint: 70°F
Coil throttling range: 50°F
Coil-to-sensor lag time: 0 s
Coil time constant: 120 s
Sensor time constant: 15 s
Proportional gain K_p: 1.0
Integral gain K_i: 0.038
Derivative gain K_d: 0.000

Find: Coil outlet air temperature, controller error, and valve position as a function of time by using the HCB software

Figure: See Figure 21.2b.

Assumptions: The control system is linear, and the valve is quick-acting; i.e., there is no significant lag in valve response to the controller signal.

Solution

Figure 21.39 shows the extended output of the online HCB software for the conditions listed for a simulation period of 10 min (five times the longest time constant in the system).

Comments

The coil outlet temperature overshoots the 70°F setpoint by about 7°F. This occurs because the integral term gets "wound up" immediately after the setpoint change (Figure 21.39b) and dominates the controller output signal, forcing the valve to stay open. Once the overshoot is experienced for enough time, the integral term and the proportional term have relatively equal effect on the process, creating the characteristic "ringing" response. After several cycles, the setpoint is achieved, and the error is essentially zero. The valve position shown in Figure 21.39c is as expected: Immediately after the setpoint change, the valve opens fully where it remains for about 5 min. After 10 min has elapsed, the valve is near its final 80% position (the new setpoint is 40°F, or 80% of the valve throttling range of 50°F, above the coil air inlet temperature).

The reader is encouraged to try the following variations on Example 21.6:

1. Delete the integral control term. Note that the operation is stable but that an offset to the setpoint occurs after steady state is reached.
2. Add a time delay between the sensor and the controller. You will find that just a few seconds of delay here will cause control instability; therefore, one must place a sensor near the outlet of a coil.
3. Add a derivative term and study a range of derivative values. Observe the nature of stability effects.

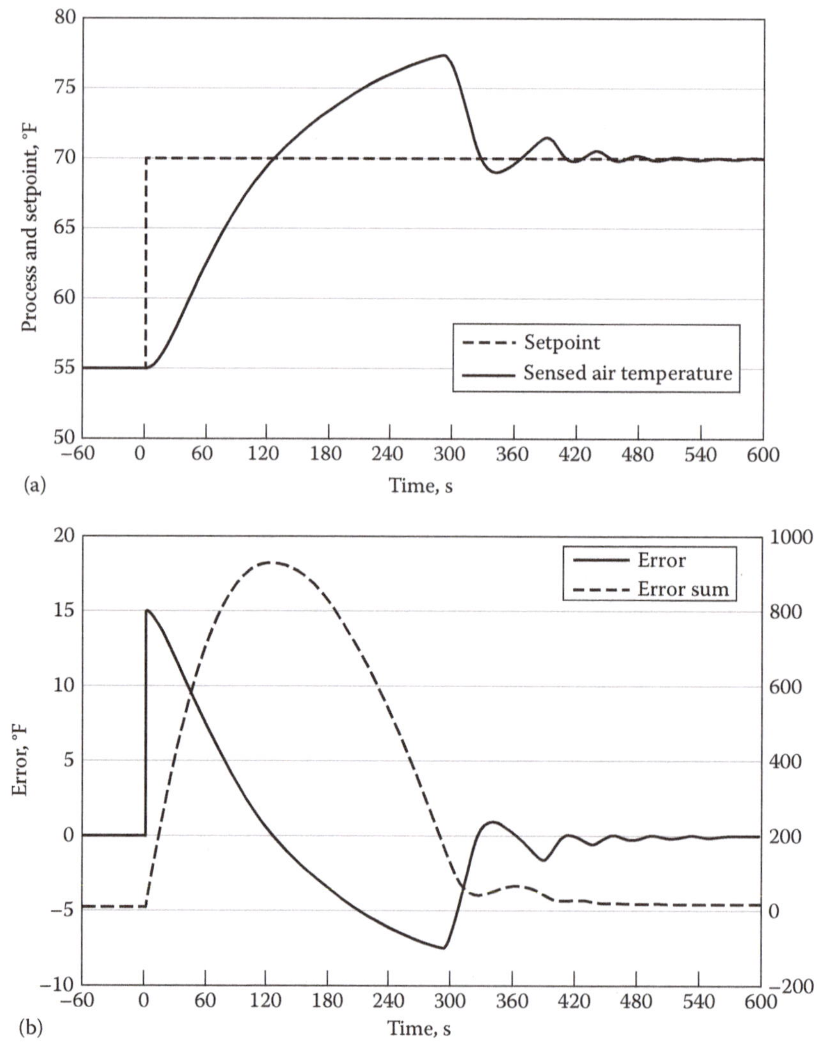

FIGURE 21.39
(a) Dynamic coil outlet temperature plot, (b) corresponding controller error and error sum signals. *(Continued)*

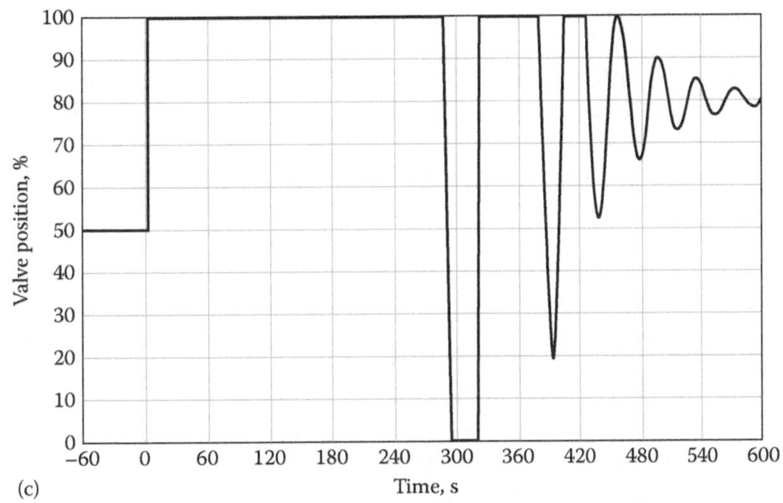

FIGURE 21.39 (*Continued*)
(c) Valve position time history.

21.8 Summary

This chapter introduced the important features of properly designed control systems for HVAC applications. Sensors, actuators, and control methods have been described. The method for determining control system characteristics either analytically or empirically has been discussed. The following rules (ASHRAE, 2015) should be followed to ensure that the control system is as energy-efficient as possible:

1. Operate HVAC equipment only when the building is occupied or when heat is needed to prevent freezing. (In the morning, begin operation 30 min before occupancy to remove stale air.)
2. Consider the efficacy of night setback vis-à-vis building mass. Massive buildings may not benefit from night setback due to overcapacity needed for the morning pickup load.
3. Do not supply heating and cooling simultaneously. Do not supply humidification and dehumidification simultaneously.
4. Reset the heating and cooling air or water temperature to provide only the heating or cooling needed.
5. Use the most economical source of energy first, the most costly, last.
6. Minimize the use of outdoor air while complying with standards during the deep heating and cooling seasons.
7. Consider the use of "dead-band" or "zero-energy" thermostats.
8. Establish control settings for stable operation to avoid system wear and to achieve proper comfort.

Problems

Numbers 1–4 given in parenthesis denote the degree of difficulty.

21.1 The thermostat inserted in the cooling system of an automotive engine controls the engine coolant temperature. Identify the sensor, controller, actuator, process, and controlled variable. *(1)*

21.2 Residential natural gas–based heating system controls typically include a wall thermostat, a gas valve, and the furnace. Identify the sensor, controller, actuator, process, and controlled variable for the control system. *(1)*

21.3 An electric duct heater has a capacity ranging from 0 to 10 kW. The proportional controller signal that controls heater power input ranges from 0 to 5 V dc. What are the throttling range and the proportional gain? *(1)*

21.4 Write the equation relating heater output to sensed temperature for the controller described in Problem 21.3 if the thermostat output voltage decreases linearly with temperature between 90°F (32°C) and 60°F (16°C) for a nominal thermostat setpoint of 75°F (24°C). *(2)*

21.5 What is the steady-state error (in temperature) for the controller in Problem 21.3 if the heat loss from the space heated by the electric heater is (a) 7 kW (b) 10 kW? *(2)*

21.6 What is the resistance of a platinum RTD at −40°C and 120°C? Its nominal resistance at 0°C is 100 Ω. *(1)*

21.7 What is the resistance-temperature equation for a thermistor with the following measured data? *(2)*

Temperature, K	283	288	293	298	303
Resistance, Ω	112,000	72,000	50,000	34,000	22,000

21.8 A thermometer inserted in a 55°C airstream has a time constant of 5.0 s. If the air temperature is rapidly increased to 75°C, when does the thermometer read 65°C? What is the thermometer reading after three time constants have passed? *(2)*

21.9 What is the C_v of a linear water control valve that flows at 100 gal/min (6.3 L/s) with a pressure drop of 5 psi (35 kPa)? What is the gain of the valve, expressed in two ways: on the basis of its total stem travel [1.0 in. (2.5 cm)] and on the basis of the control actuator voltage (0–10 V) that moves the stem over its full travel? *(2)*

21.10 The flow through a coil and linear valve subsystem (Figure 21.19a) is determined by the distribution of the available constant pressure difference (85 kPa) between the valve and the coil. Assume that the coil pressure drop is given by $\Delta P_{coil} = 2.5\dot{V}^2$ kPa, where $\dot{V}$ L/s is the volumetric flow rate. Even though the valve flow at constant imposed pressure across the valve would vary linearly with position, the loop flow does not because of the nonlinear nature of the coil. Plot the loop flow versus the percentage of valve travel for two values: $C_v = 0.5$ and $C_v = 1.3$. (The valve equation is $\dot{V}\,(\text{L/s}) = C_v\sqrt{\Delta p}$ (kPa). Which value of C_v results in the most nearly linear valve travel compared to the loop flow characteristic? *(2)*

21.11 What is the gain of the valve *as installed in the loop* for the configuration described in Problem 21.10 for both C_v values? Note that the gain is not constant. If you had to pick one value of proportional gain for operation in the control midrange, what would it be? *(2)*

21.12 If the coil described in Problem 21.10 is connected in a local loop with a linear three-way valve ($C_v = 0.5$), as shown in Figure 21.19b, what pressure drop should be set across the balancing valve? What will the total loop flow be at valve midtravel? *(3)*

21.13 The flow characteristic of an equal-percentage valve is given by Equation 21.12. A valve of this type with $K = 20$ and $k = 1.0$ is connected in the coil loop described in Problem 12.10 with the valve having $C_v = 1.3$. Plot the valve flow versus position characteristic for the loop, and comment on its linearity. *(2)*

21.14 What is the C_v value of a linear water control valve that flows at 20 gal/min (1.2 L/s) with a pressure drop of 4 psi (28 kPa)? What is the gain of the valve expressed on the basis of its total stem travel [2.0 in. (5 cm)] and on the basis of the control actuator voltage (0–5 V) that moves the stem over its full travel? *(2)*

21.15 Derive the steady-state transfer function of an air heat transfer coil that has two independent inputs—the inlet water and air temperatures. Draw a block diagram with the two inputs and the single output (the outlet air temperature) shown. *(3)*

21.16 Modify the model developed in Problem 21.15 to include coil dynamics. Consider it to be a lumped-capacitance time constant τ_{hx} with the effectiveness ε. Modify the block diagram from Problem 21.15 as appropriate. *(3)*

21.17 Select a steam control valve using saturated 80 psig steam to heat atmospheric pressure water for the following conditions. Water is to be heated to 20°F (11°C) at 82.5 gal/min (5.2 L/s). What is the value of C_v? Find both the critical flow pressure drop and the difference between supply and return steam pressure (atmospheric pressure) to find the proper entry in Equation 21.19. *(3)*

21.18 The pressure drop through a water coil flowing at 15 gal/min (1 L/s) is 6 psi (42 kPa). What is the required value of C_v for good flow control as ensured by selecting a valve authority of 0.50? *(2)*

21.19 Determine the C_v value of a three-way valve used to control water flow through a cooling coil. The total coil and piping pressure drop under full-flow conditions in the coil loop is 3 ftWG. If the peak coil flow is 70 gal/min (4.2 L/s), what is the recommended C_v value and what type of valve should be selected? *(2)*

21.20 What is the inverse Laplace transform for the following expression? *(1)*

$$\frac{s+3}{(s+4)(s-2)}$$

21.21 What is the Laplace transform of the following differential equation? Solve the equation if the initial conditions are $Y(t=0) = 3$ and $Y'(t=0)$. *(2)*

$$\frac{d^2T}{dt^2} + 2\frac{dT}{dt} + 4 = 0$$

21.22 What is the transfer function for a temperature sensor (in a water stream) with mass m, convection heat transfer coefficient h, specific heat c, and surface area A? The transfer function expressed in the Laplace transform domain is the ratio of the transform of the sensed temperature to the transform of the fluid temperature. *(2)*

21.23 A water heating tank is heated electrically with a 3 kW heater. The heat loss coefficient from the tank UA = 40 W/K, and the volume of water in the tank is 80 L. What is the open-loop transfer function relating the tank temperature to heat input (the tank-to-air temperature difference is a convenient variable here)? *(2)*

21.24 A proportional controller with gain k is connected to the tank described in Problem 21.23. If the open-loop transfer function is expressed as

$$\frac{K/(\text{UA})}{1+\tau s}$$

what is the time constant τ? At time zero, the setpoint is rapidly changed from a steady value of 15°C (also the environmental temperature) to 25°C. If the steady-state offset is found to be 1°C, what is the value of K? If the gain is double, what is the resulting offset? *(2)*

21.25 A proportional controller is operated so that its setpoint is increased by 5%/min. After an initial transient, the output increases by 1%/min. What is the gain? *(1)*

21.26 Write the transfer function for a baseboard room heating system including the dynamic effects of the room itself, the baseboard heating units, the thermostat, and the PI controller. The controlled variable is the room temperature. *(2)*

21.27 What are the stability criteria for K, given these characteristic equations? *(2)*

$$as^2+bs+K=0$$

$$as^3+bs^2+cs+K=0$$

21.28 For what values of K is the following characteristic equation stable? *(2)*

$$s^4+7s^3+11s^2+7s+(K+1)=0$$

21.29 A coil temperature PI controller with a fast-acting valve has a sensor time constant τ_s and a heating coil time constant τ_{coil}. Draw the system block diagram and write the characteristic equation if the coil is controlled by a PI controller. Determine the stability criteria on the proportional and integral gains for stability. *(3)*

21.30 Is the system described in Problem 21.29 with a proportional gain of 3 and integral gain of 0 stable if the sensor time constant is 3 s and the heating coil time constant is 60 s? *(3)*

21.31 For the system described in Problem 21.29, is a system with a proportional gain of 3 and air integral gain of 0.05 stable if the sensor time constant is 3 s and the heating coil time constant is 60 s? *(3)*

21.32 The outlet temperature of a heating coil subjected to a step increase in inlet air temperature is shown in Figure 21.2a. What is the process transfer function in the form of Equation 21.39? *(2)*

21.33 Write the characteristic equation for a controller of the coil shown in Figure 21.2a if the outlet temperature sensor has a time constant τ_s. Use the Padé approximation for the delay term, and ignore the dynamics of the control valve. If the sensor time constant is 10 s and a proportional-only control is used, what is the upper limit of the proportional gain for stability? *(4)*

21.34 Pneumatic controls can take advantage of small air pressures by varying the size of the diaphragm used in the actuator. What actuator force can a 15 psi control signal exert if the round diaphragm has a 3 in. diameter? What if the diaphragm has a 6 in. diameter? What if the diaphragm has a 12 in. diameter? *(1)*

21.35 In the system schematic of Figure 21.2a, what will be the effect on the controlled variable as the temperature sensor is moved farther away from the coil? *(1)*

21.36 The resistance of a platinum RTD is measured using a constant current of 1 mA. The nominal resistance of the RTD is 100 Ω at 0°C. The air around the RTD is at 20°C, but the current running through the RTD causes it to heat up.

(a) What is the theoretical resistance of the RTD at 20°C?

(b) What is the power dissipated by the RTD in the form of heat?

(c) If the heat loss from the RTD to the environment is 0.05 W/°C, what is the actual temperature of the RTD, including the effect of self-heating? *(2)*

21.37 The temperature of an airstream is measured by connecting a platinum RTD to a building energy management system using 200 ft of 20-gauge wire. The wire has a resistance of 10 Ω/1000 ft. How much error does the wire resistance introduce into the measurement if the nominal resistance of the RTD is 100 Ω at 32°F? How much error would the wire resistance add if the sensor were a thermistor with a nominal resistance of 25 kΩ at 32°F? *(2)*

21.38 An *analog-to-digital* converter is used to generate a computer-readable value from a measured voltage. Suppose a temperature sensor produces a 0–10 V dc signal over the range of −20°F to 120°F. This signal is converted to an 8-bit value, which means that the total voltage range is divided into $2^8 = 256$ discrete levels. What is the minimum temperature change that can then be recorded using this temperature sensor? *(1)*

21.39 Air at 55°F enters a heating coil and comes out at 75°F. You measure the relative humidity with a handheld probe and find an inlet relative humidity of 59% and an outlet relative humidity of 31%. If there is no moisture added to the airstream, do these numbers make sense? *(1)*

21.40 An economizer mode attempts to mix outside air and building return air to minimize the amount of energy needed to condition the resulting mixed-air stream to match the desired supply air conditions. Suppose the conditions are outside air at 90°F and 40% relative humidity, the return air is at 80°F and 70% relative humidity, and the supply air setpoint is 55°F at 80% relative humidity. Should the economizer control use mostly outside air or mostly building return air? *(2)*

21.41 A heating coil with a throttling range of 20°F is controlled using proportional-only control. If the inlet temperature is 40°F, the setpoint is 55°F, and the controller gain is 0.1, what is the steady-state error? What if the gain is 1.0? What if the gain is 10.0? *(2)*

21.42 The hot water coil on an air handling unit has 120°F inlet water temperature, 110°F outlet water temperature, and a water flow rate of 30 gal/min. Also 10,000 ft³/min of air enters this coil at 57°F and leaves at 73°F. If the temperature sensors are accurate to within ±1°F, does it appear that energy is conserved in this coil? Assume that the flow rates of air and water are measured without error. *(2)*

21.43 A damper controller controls three sets of dampers as shown in Figure P21.43. The three dampers are modulated to specify the amount of outdoor air entering the building. How can you calculate the fraction of outdoor air entering the air handling unit if you only know the return air temperature, the outdoor air temperature, and the mixed-air temperature? At what point does this relationship not make sense? *(3)*

21.44 Airflow measurements are often made at several points across the face of an airstream so that the average airflow rate can be determined. A Pitot tube (see Section 16.7) is often used for these kinds of measurements. If the dynamic pressure measurements at five points are 0.23, 0.20, 0.18, 0.25, and 0.20 (all in equivalent inches of water), what is the average velocity of the airstream? *(2)*

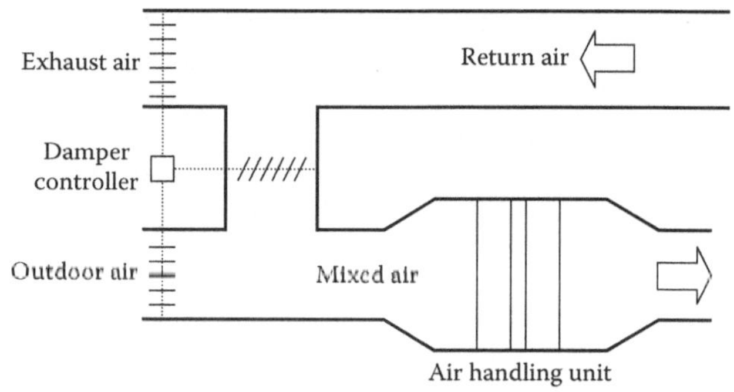

FIGURE P21.43

21.45 Some standard automobiles will not start unless you depress the clutch. Create a simple ladder diagram that shows the start-up sequence for a car engine with this feature. Also, include a rung on this ladder for the warning bell that sounds when you start the car without wearing a seat belt. *(2)*

21.46 Suppose you run a reaction curve test on a heating coil. The valve of the coil is changed from 20% to 80%, resulting in a change of the coil outlet temperature of 48°F–63°F. The outlet temperature does not start increasing until about 10 s after the valve change and does not reach the final temperature until about 70 s after the valve change. What are the recommended PID control constants for this coil? *(3)*

21.47 Most home thermostats include an *anticipator* that is used to shut off the heating system before the room actually reaches the setpoint value. The anticipator is a small electronic heating coil that fools the thermostat into thinking the room is warmer than it actually is. Describe why the thermostat would need such a feature. Most anticipators are adjustable. Why would you want to adjust the amount of anticipation? *(2)*

21.48 Many Scandinavian residences are heated using radiant floor panels. A network of pipes is embedded in a concrete floor, and a pump circulates warm water through the pipes. The pipes warm the slab and the slab then warms the room air by convection. Assuming a constant hot water temperature and a constant water flow rate when the pump is running, what is the transfer function that relates the hot water pump control and the temperature of the room? *(4)*

21.49 In some air systems, airflow to a zone is measured using an array of Pitot tubes that measure the dynamic pressure of the moving air at many points. Most Pitot tube arrays use the *average* dynamic pressure to calculate the velocity of the air when in fact the proper way to make such a measurement is to use the average of all the velocities calculated at each point. Imagine a Pitot tube array in which one-half of the Pitot tubes experience an actual velocity of 500 ft/min while the other half experience a velocity of 350 ft/min. What is the measurement error if the velocity is calculated using the average dynamic pressure? *(4)*

References

ANSI/ASHRAE Standard 135 (2012). *Standard 135-2012-BACnet: A Data Communication Protocol for Building Automation and Control Networks.* American Society of Heating, Refrigerating and Air-Conditioning Engineers, Atlanta, GA.

ASHRAE Applications (2015). *Handbook of HVAC Applications.* American Society of Heating, Refrigerating and Air-Conditioning Engineers, Atlanta, GA.

ASHRAE Fundamentals (2013). *Fundamentals Handbook.* American Society of Heating, Refrigerating and Air-Conditioning Engineers, Atlanta, GA.

Braun, J.E., J.W. Mitchell, S.A. Klein, and W.A. Beckman (1989). Applications of optimal control to chilled water systems without storage. *ASHRAE Trans.*, 95(1), 663–675.

Bushby, S.T. and H.M. Newman (October 2002). BACnet Today: Significant new features and future enhancements. *ASHRAE J.*, 44(10), 10–18.

Coad, W.J. (1990). Temperature controls: The industry's dilemma. *Heat. Piping Air Cond.*, 62(October), 103–104.

Cohen, G.H. and G.A. Coon (1953). Theoretical considerations for retarded control. *Trans. ASME*, 75, 827–834.

Curtiss, P.S., J.F. Kreider, and M.J. Brandemuehl (1993). Adaptive control of HVAC processes using predictive neural networks. *ASHRAE Trans.*, 99(1), 496.

Grimm, N.R. and R.C. Rosaler (1990). *Handbook of HVAC Design.* McGraw-Hill, New York.

Haines, R.W. (1987). *Control Systems for Heating, Ventilating and Air Conditioning*, 4th ed. Van Nostrand Reinhold, New York.

Honeywell (1988). *Engineering Manual of Automatic Control.* Honeywell, Inc., Minneapolis, MN.

Horan, T. (1997). *Control Systems and Applications for HVAC/R.* Prentice Hall, Upper Saddle River, NJ.

Huang, P.H. (1991). Humidity measurements and calibration standards. *ASHRAE Trans.*, 97(2), 298–304.

Letherman, K.M. (1981). *Automatic Controls for Heating and Air Conditioning.* Pergamon, New York.

Levenhagen, J.I. and D.H. Spethmann (1993). *HVAC Controls and Systems.* McGraw-Hill, New York.

MacArthur, J.W., E.W. Gerald, and A.F. Konar (1989). An effective approach for dynamically compensated adaptive control. *ASHRAE Trans.*, 95(2), 411–423.

Madsen, T.L. (1989). How important is the location of the room thermostat? *ASHRAE Trans.*, 95(2), 1041–1049.

McGowan, J.J. (1991). *Networking for Building Automation & Control Systems.* Fairmont Press/Prentice Hall, Lilburn, GA/Englewood Cliffs, NJ.

Price (2011). *Engineer's HVAC Handbook: A Comprehensive Guide to HVAC Fundamentals.* Price Industries Limited, Winnipeg, Manitoba, Canada.

Shadpour, F. (2001). *The Fundamentals of HVAC Direct Digital Control*, 2nd ed. Hacienda Blue Publising, Escondido, CA.

Swan, W. (July 1996). The language of BACnet. *Eng. Syst.*, 13(7), 24–32. http://www.bacnet.org/Bibliography/ES-7-96/ES-7-96.htm.

Underwood, D.M. (1989). Response of self-tuning single loop digital controllers to a computer simulated heating coil. *ASHRAE Trans.*, 95(2), 424–430.

Wylie, C.R. and L.C. Barrett (1982). *Advanced Engineering Mathematics*, 5th ed. McGraw-Hill, New York.

Ziegler, J.G. and N.B. Nichols (1942). Optimum settings for automatic controllers. *Trans. ASME*, 64, 759–763.

22

Lighting and Daylighting

ABSTRACT We start by reviewing some basic principles and definitions related to lighting. The various sources of electric light and the effect of luminaires that house the lamp are discussed. We then treat daylighting (or natural lighting) covering some basic notions and a description of how to perform a daylighting design calculation using a simple method known as the lumen method. How to apply the method to both toplighting and sidelighting is discussed along with illustrative examples. In the last section, we examine the electric energy savings achievable in commercial buildings with proper daylighting design. Several case studies are described in order to illustrate this important energy efficiency option.

Nomenclature

A_{eff}	Effective aperture areas (Equation 22.14), m^2 (ft^2)
A_h	Projected horizontal aperture area of skylights, m^2 (ft^2)
A_w	Area of work plane, m^2 (ft^2)
C_u	Coefficient of utilization
COP	Coefficient of Performance
D	Dimming controls
E_{xh}	Exterior horizontal illuminance, lx (fc)
$E_{xh,s/2}$	Exterior half-sky illuminance on horizontal surface in front of window (Figure 22.6), lx (fc)
E_{xv}	Exterior vertical illuminance (Figure 22.4), lx (fc)
$E_{xv,g/2}$	Vertical illuminance reflected from ground in front of window, lx (fc)
$E_{xv,s/2}$	Exterior vertical illuminance incident on window from half sky in front of window (Figure 22.5), lx (fc)
E_w	Average illuminance of work plane, fc (lx)
F	Total flux produced by lamps (not counting absorption by luminaires), lm
K_e	Ratio of visible to thermal transmittance (Equation 22.15)
L_{dirt}	Loss factor due to dirt (Table 22.5)
l	Length of room, m (ft)
ND	No dimming controls
$\dot{Q}$	Electric power consumed by lights, W
R_A	Ratio of net to gross skylight area, to account for blocking by frame or mullions
SC	Shading coefficient
T_{dif}, T_{dir}	Transmissivities of glazing for diffuse and direct radiation
T_{well}	Transmissivity of light well, if present (well efficiency of Figure 22.7)
U	Overall heat transfer coefficient W/($m^2 \cdot K$) [Btu/($h \cdot ft^2 \cdot °F$)]
w	Width, m (ft)

Greek

Δh	Room height—work plane height, m (ft)
θ_p	Zenith angle of plane (tilt from horizontal, up > 0)
θ_s	Zenith angle of sun
ρ	Reflectivity
λ	Latitude
τ	Transmissivity
ϕ_p	Azimuth of plane (positive for orientations west of south)
ϕ_s	Azimuth of sun

Subscripts

apert	Aperture
control	Control device
dif	Diffuse
dir	Direct
eff	Effective
g	Ground
s	Sky
tot	Total
v	Visible

22.1 Principles of Lighting

Light is the portion of the electromagnetic spectrum visible to the human eye. It ranges from the red end to the violet end of the spectrum, with wavelengths from 700 to 400 nm and frequencies from 4.3×10^{14} to 7.5×10^{14} Hz. Like all electromagnetic radiation, it travels through empty space at a speed of about 186,000 mi/s (300,000 km/s). In the mid-nineteenth century, light was described by James Clerk Maxwell in terms of electromagnetic waves, but twentieth-century physicists showed that it exhibits properties of particles as well; its carrier particle is the photon. Light is the basis for the sense of sight and for the perception of color.

The quantity of light emitted, per unit time, by a source of radiation is called the *luminous flux*, and it is measured in units of *lumen* (abbreviated lm). The lumen is defined in terms of radiative power weighted by the spectral sensitivity of the human eye. For energy-conscious design, a quantity of special importance is the *luminous efficacy* of a light source, defined as the ratio of its light output (in lumens) to the necessary power input (in watts). The theoretical upper limit is 683 lm/W corresponding to lossless conversion of the input to monochromatic light at the wavelength of 555 nm, where the eye is most sensitive. For other spectral distributions, the efficacy can only be lower. For instance, much of the solar spectrum is outside the visible range, and the luminous efficacy of daylight is only around 100–120 lm/W (relative to its radiative power). The luminous efficacy of incandescent lamps is around 10–20 lm/W, and that of fluorescent lamps is around 50–90 lm/W. The relatively high efficacy of daylight explains its interest for energy efficiency.

Of particular interest to the designer is the *illuminance* of the surface to be lighted. The illuminance is the luminous flux divided by the area on which it is incident; its unit is the *lux* (abbreviated lx):

$$1\,\text{lx} = 1\,\text{lm/m}^2 \qquad (22.1\ \text{SI})$$

In IP units, the analogous quantity is the *footcandle* (abbreviated fc):

$$1\,\text{fc} = 1\,\text{lm/ft}^2 \qquad (22.1\ \text{IP})$$

In addition to the flux, one may be interested in the angular distribution. The luminous flux emitted by a source into an angular region of unit solid angle (measured in steradians [sr]) is the *luminous intensity*; its unit is the *candela* (abbreviated cd):

$$1\,\text{cd} = 1\,\text{lm/sr} \qquad (22.2)$$

The apparent brightness of a surface depends on the flux received by the eye. This can be characterized as *luminance*, defined as the luminous flux emitted per unit area and per solid angle in a given direction.

Apart from energy consumption, the principal considerations for the design of lighting systems are

- Luminous flux
- Spectral composition (color rendition)
- Spatial and angular distribution of the radiation

As for the last consideration, a major concern is the possibility of extremely bright sources of light within the normal field of vision, something called *glare*. To understand this concept, consider, e.g., light bulbs that are visible rather than hidden inside a luminaire. They cause less discomfort if they have a white coating rather than clear glass; instead of coming from a small bright filament, the same amount of light is seen to emerge from the larger and less bright surface of the bulb, thus greatly reducing the contrast with the background. Quite generally, glare should be avoided.

But even if spectral and angular distributions can be quantified, lighting design choices are often based on criteria that defy easy analysis, such as prestige and aesthetics.

22.2 Electric Lighting

There are a wide variety of electric light sources to choose from; some characteristics are summarized in Table 22.1. The data for efficacy and lifetime are approximate. Actual values can vary quite a bit from one model to another. The light output of a lamp decreases during its life, and in many applications, one will replace a lamp long before it would burn out.

Incandescent lamps produce light by simple thermal radiation from a heated filament. Their efficacy is low because the temperature limitations of even the best filament material (tungsten) imply that only a small portion of the emitted power lies in the visible part of the electromagnetic spectrum; most of the power is given off as infrared radiation. In addition to the ordinary incandescent lamps that have been in use since Edison, a variant has become popular in recent years: the halogen lamp. It utilizes the tungsten–halogen cycle to counteract the principal cause of gradual degradation in ordinary incandescent lamps. This cycle makes the tungsten molecules that have evaporated return to the filament instead of condensing on the glass envelope. Furthermore, the filament can burn at higher temperature, thus improving both efficacy (by about 30%–50% relative to ordinary incandescent) and color rendition.

Gaseous discharge lamps are based on quantum transitions between discrete energy levels, and they reach much higher efficacy. The wavelengths corresponding to these transitions do not, in general, lie in the desired region of the visible spectrum, but they can be shifted to longer wavelengths by means of special coatings. For instance, in fluorescent lamps, a gaseous discharge in a tube filled with Ar, Ar–Ne, or Kr produces ultraviolet radiation that is converted to light by phosphor coatings ("fluorescent") on the inside of the tube. Special coatings

TABLE 22.1

Comparison of Light Sources

Type	Efficacy, lm/W	Life, h	Comments
Incandescent			Low first cost, low efficiency, low life
Ordinary	12–18	750–1,500	Best for seldom used spaces
Halogen	16–29	2–4,000	Excellent color rendition; output fairly constant over entire lifetime; good for task lights and special effects
Gaseous discharge			Medium first cost, high life
Fluorescent	50–90	10,000	Offices and other commercial uses
Mercury vapor	50–60	10,000	Indoor commercial and outdoor uses
High-pressure sodium	100–150	15,000	Fairly good color rendition; indoor commercial and outdoor uses
Low-pressure sodium	200		Yellow; outdoor, especially roads
Compact fluorescent	60–70	6–14,000	Low to medium first cost, high energy efficiency, long life, medium color rendition
Light-emitting diodes	20–50	Up to 100,000	Medium to high cost, high energy efficiency, longest life, medium color rendition, comes in several colors

Source: From EPA, Lighting technologies: A guide to energy-efficient illumination, Energy STAR Program, U.S. Environmental Protection Agency, 2005, www.energystar.gov.

Note: For a bit of historical perspective, note that artificial light used to be very expensive when the best source was a candle with 0.15 lm/W, lasting less than a day (Competitek, 1988).

TABLE 22.2

Lamp Wall Temperature (MLWT), Light Output (lm Index), and Luminous Efficacy (lm/W Index) for Several Fluorescent System Configurations

Luminaire Configuration	MLWT, °C	m Index[a]	lm/W Index[a]
Nonairflow four-lamp lensed troffer			
No airflow	56.6	78.3	89.4
Airflow four-lamp lensed troffer			
No airflow	55.8	79.2	90.0
20 ft^3/min	36.7	98.3	*99.3*
50 ft^3/min	31.5	99.4	98.0
Nonairflow four-lamp louvered parabolic troffer			
No airflow	53.1	82.2	91.9
Airflow four-lamp parabolic troffer			
No airflow	51.8	83.8	93.1
20 ft^3/min	40.9	95.6	98.8
50 ft^3/min	35.7	99.0	*99.8*

Source: Competitek, Inc., Lighting Report, Rocky Mountain Institute, Snowmass, CO, 1988.

[a] As percent of output and efficacy under ANSI test conditions (open-air strip fixture at 25°C ambient).

are available for improving the color rendition, although generally at increased cost and energy consumption. In recent years, compact fluorescent tubes have been developed that offer high efficacy in a size comparable to incandescent bulbs.

The output of a fluorescent lamp varies with the temperature of its environment. The temperature dependence for a typical tube is shown in Table 22.2; details can vary from one model to another. It is advisable to pay attention to the expected temperature of the lamp environment when one is selecting a fluorescent tube.

Gaseous discharge lamps must be connected to the mains via a special device, called the *ballast*. It prevents the current from exceeding the design value, it provides the starting voltage kick, and it corrects the power factor. The power consumption of the ballast can be a significant item, on the order of 10% of the total. More efficient, high-frequency ballasts have been developed in recent years.

Compact florescent lamps (CFLs) have been introduced about 40 years back, but only in the last 15 years or so have they become popular. They are the most energy-efficient choice readily available in the market. CFLs use gases and phosphor inside the lamps to create light. They come in screw-in and pin-based configurations. Most CFLs do not operate well with remote or dimmer switches, and specialty switches have to be used. They also contain a small amount of mercury.

Light-emitting diodes are becoming very popular in recent years due to the great energy efficiency strides made (with promise of more improvements to come). They use solid-state lighting technology and are essentially solar PV cells acting backward. They are used increasingly in residences and commercial buildings and are very resistant to physical abuse. They are in essence point source devices and are ideal for spot or narrow-beam applications.

Among commercially available lamps, the record for luminous efficacy, around 200 lm/W, is held by the low-pressure sodium lamp. Unfortunately, it produces an almost monochromatic yellow light, suitable only for outdoor applications such as roads. High-pressure sodium lamps yield white light with fairly good color rendition, but the efficacy is lower.

Only part of the light produced by a lamp reaches the work plane, with the rest being absorbed in the luminaire or by intervening objects. The *luminaire* is the housing of the lamp; it contains various reflective or refractive elements, designed to redistribute the light in the desired fashion. The *luminaire efficiency*, defined as the ratio of the light leaving the luminaire to the light produced by the lamp, varies a great deal from one model to another. For instance, in a survey of luminaires for fluorescent tubes, efficiencies were found to range from 30% to 70% for devices that serve essentially the same purpose (direct illumination of a workspace below, with restricted exit angles to reduce glare) (Rabl, 1990). The low efficiencies were due to diffuse reflectors that trapped much of the radiation instead of sending it to the target area. Specular surfaces of high reflectivity perform far better, e.g., anodized or vacuum-deposited aluminum. Silver has the highest reflectivity, around 0.95, but it requires careful protection or it will tarnish. In recent years, silvered plastic films with durable coatings have been developed, and they are being used increasingly for luminaires (Competitek, 1988).

Most luminaires emit light not only straight toward the work plane but in other directions as well. Exact calculation of the amount of light that reaches the work plane would involve all the interreflections from walls, ceiling, etc., a daunting task that is usually simplified by various approximations. For hand calculations, the following method, known as the *lumen method*, is convenient and widely used. It determines the average illuminance E_w (fc or lx) of the work plane by the formula

$$E_w = \frac{C_u F}{A_w} \tag{22.3}$$

where

F is the total light output of lamps (not counting absorption by luminaires), lm

A_w is the area of work plane

C_u is the coefficient of utilization

If all the light from the lamp reached the work plane, the average illuminance would be F/A_w. Thus, the coefficient of utilization represents the efficiency with which the luminaire–room combination transfers the light to the work plane. It depends on the geometry and the reflectivities of the room, and on the luminaire; the geometry enters only via the room ratio (= 2 × ratio of floor area and wall area above work plane):

$$\text{Room ratio} = \frac{wl}{\Delta h(w + l)} \tag{22.4}$$

where

Δh is the room height − work plane height

w is the width

l is the length of room

The coefficient of utilization of a luminaire is provided in tabular form by the manufacturer; an example is shown in Table 22.3. The method determines only the average illuminance over the work plane. The designer should verify that the room geometry and spacing of luminaires are such as to make the illuminance sufficiently uniform.

TABLE 22.3

Typical Coefficient of Utilization Table for Particular Ceiling-Mounted Luminaire with Two Fluorescent Tubes, as a Function of Room Ratio (Equation 22.4) and of Reflectivities (Work Plane, Ceiling, and Wall)

Reflectivities of Work Plane	0.3					0.1								0.0
Of Ceiling	0.7			0.5		0.7			0.5			0.3		0.0
Of Wall	0.5	0.3	0.1	0.5	0.3	0.5	0.3	0.1	0.5	0.3	0.1	0.3	0.1	0.0
Room ratio														
0.60	0.26	0.22	0.19	0.25	0.21	0.25	0.21	0.19	0.24	0.21	0.19	0.21	0.19	0.18
0.80	0.31	0.27	0.24	0.30	0.26	0.29	0.26	0.23	0.29	0.26	0.23	0.25	0.23	0.22
1.00	0.35	0.31	0.28	0.34	0.30	0.33	0.29	0.27	0.32	0.29	0.27	0.29	0.27	0.26
1.25	0.39	0.35	0.32	0.37	0.34	0.36	0.33	0.31	0.35	0.33	0.30	0.32	0.30	0.29
1.50	0.42	0.38	0.35	0.40	0.37	0.39	0.36	0.33	0.38	0.35	0.33	0.35	0.33	0.32
2.00	0.47	0.43	0.40	0.44	0.41	0.42	0.40	0.38	0.41	0.39	0.37	0.38	0.37	0.36
2.50	0.49	0.46	0.44	0.47	0.44	0.44	0.42	0.40	0.43	0.41	0.40	0.41	0.39	0.38
3.00	0.51	0.49	0.46	0.49	0.46	0.46	0.44	0.42	0.45	0.43	0.42	0.43	0.41	0.40
4.00	0.54	0.52	0.50	0.51	0.49	0.48	0.46	0.45	0.47	0.45	0.44	0.45	0.44	0.42
5.00	0.56	0.54	0.52	0.52	0.51	0.49	0.47	0.46	0.48	0.47	0.46	0.46	0.45	0.44

Source: Data in Philips, Philirama 89/90, Catalog of products, Companie Philips Eclairage, Boulogne-Billancourt, France, 1989.

Example 22.1: Lighting Systems Design

A room is to be illuminated with fluorescent tubes, using the luminaire of Table 22.3. The room characteristics are

$$\text{Height } \Delta h = 6.3 \text{ ft } (1.9 \text{ m})$$

between ceiling and work plane

$$\text{Width} = 40 \text{ ft } (12.2 \text{ m})$$

$$\text{Length} = 30 \text{ ft } (9.1 \text{ m})$$

and the reflectivities are 0.5 for the ceiling and walls and 0.2 for the work plane.

How many tubes and how many luminaires are needed to ensure a minimum illuminance of 50 fc (543 lx), averaged over the work plane, if each tube produces 3000 lm? What is the electric power if each tube, together with its ballast, requires 40 W?

Given: Luminaire of Table 22.3.
$A_w = 1200$ ft^2 (111 m^2).
$E_w > 50$ fc (543 lx).

Find: Number of tubes, number of luminaires, and total electric power.

Intermediate quantities: Room ratio = 2.45, from Equation 22.4. For C_u, one must interpolate between room ratios of 2.5 and 3.0 and between work plane reflectivities of 0.1 and 0.3. One finds $C_u = 0.46$, interpolated from Table 22.3 for wall and ceiling reflectivity of 0.5.

Solution

$$\text{Required flux } F = \frac{E_w A_w}{C_u} \quad \text{from Equation 22.3}$$

$$= (50 \text{ lm/ft}^2)\left(\frac{1{,}200 \text{ ft}^2}{0.46}\right)$$

$$= 130{,}435 \text{ lm}$$

$$\text{No. tubes} = 130{,}435 \text{ lm}/3{,}000 \text{ lm/tube}$$

$$= 43.5 \text{ rounded up to } 44$$

$$\text{No. luminaries} = \frac{44 \text{ tubes}}{2 \text{ tubes/luminaire}} = 22 \text{ luminaires}$$

$$\text{Electric power} = 44 \text{ tubes} \times 40 \text{ W/tube} = 1{,}760 \text{ W}$$

Comments

The luminous efficacy of the tubes is 3000 lm/40 W = 75 lm/W. Only 45% of the light produced by the tubes actually reaches the work plane. The electric power per floor area is 1760 W/1200 ft^2 = 1.47 W/ft^2 (15.82 W/m^2), and if it all stays in the room, it imposes a cooling load of 1760 W_t = 1760 W/(3.517 kW/ton) = 0.50 ton for this 1200 ft^2 (111 m^2) area. It will be interesting to compare this with the performance of the skylight in Example 22.2.

22.3 Daylighting

22.3.1 Importance

Some 5% of the total U.S. energy consumption is used for lighting. In commercial buildings, lighting is the major consumer of electricity; in addition, it induces energy consumption for removing the heat of the lights. Clearly, lighting and the interaction between lighting and thermal loads deserve some attention by HVAC design engineers, even if in the past it may not have been considered as part of their domain. By the first law of thermodynamics, if a lamp consumes $\dot{Q}_{light}$ of electric power, it decreases the heating load and increases the cooling load by $(1 - f)\,\dot{Q}_{light}$, where f is the fraction of $\dot{Q}_{light}$ that escapes to the outside. While the energy that escapes directly as light through the windows is usually negligible, a significant fraction of the heat can be carried off by the ventilation air and extracted to the outside. This is the case in commercial buildings where the lights are attached to the plenum and the air from a room is exhausted through the plenum. In such systems, f is approximately the product of the fraction of heat given off by convection and the fraction of the return airflow that is exhausted to the outside.

The HVAC design engineer is implicated even more if one wants to reduce the energy consumption by using daylight. Optimization of daylighting design is complicated because one has to trade off the solar heat gains due to increased daylight against the decreased electricity consumption for lights. The question of design optimization is addressed in Section 23.5.

22.3.2 Principles

Daylighting (also known as *natural lighting*) is, of course, the oldest means of illumination, but it has gained new popularity since the oil crises. It has psychological appeal in addition to its promise of energy savings. The potential for savings is greatest in commercial buildings for a number of reasons: they are occupied during the day and use much electricity for lighting, the electricity during the day (peak period) is expensive, and the heat of the lights imposes a cooling load during much of the year. Having higher luminous efficacy, daylight is cooler than electric light. More than one-half of commercial space is located directly under a roof or adjacent to a window and thus is accessible to daylight. In residential buildings, by contrast, most of the light is needed when there is little daylight, and cooling loads are less important.

There are a wide variety of architectural designs for daylighting. The most conventional types are shown in

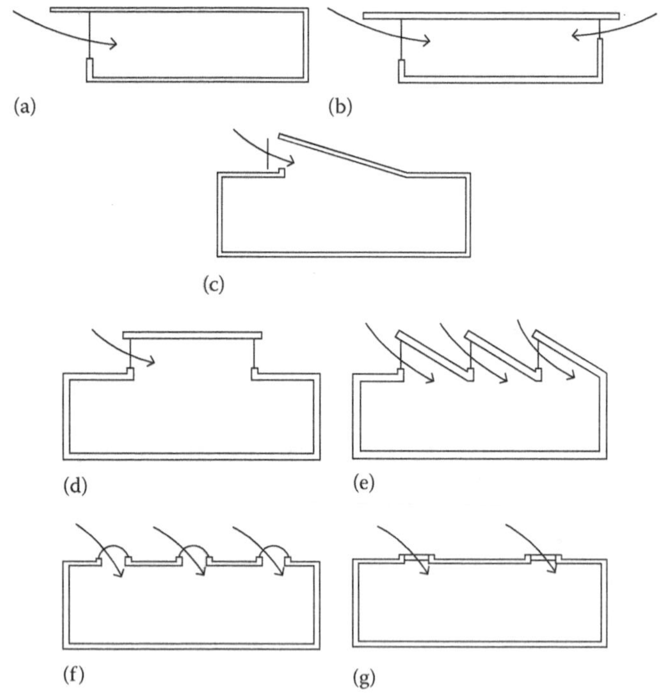

FIGURE 22.1
Design types for daylighting. (a) Unilateral lighting section, (b) bilateral lighting section, (c) clerestory lighting section, (d) roof monitor lighting section, (e) sawtooth lighting section, (f) domed skylight section, and (g) flat skylight section. (From *IES Lighting Handbook: Reference Volume*, Kaufman, J.E., ed. and Christensen, J.F., assoc ed., Illuminating Engineering Society of North America, New York, 1984a.)

Figure 22.1, including ordinary windows, clerestories, and skylights. Providing adequate daylight is a challenge because inside a building the transmitted light decreases rapidly with the distance from the opening. As a rule of thumb for ordinary windows, the maximum distance to which useful light can be brought into a building is 1.5–2 times the height of the upper window edge. The great advantage of skylights is their ability to spread the light quite evenly over a large work plane; their effectiveness for saving energy will be confirmed by the results in Section 22.5. However, skylights tend to be more expensive per unit area than vertical windows, because skylights require better waterproofing.

A major concern with daylighting is the avoidance of glare, i.e., excessive contrast, during periods of direct sunlight. A simple and effective countermeasure is to employ diffuse instead of clear glazing. Of course, that may not be an attractive solution if it interferes with a view to the outside. But diffuse glazing can be quite acceptable in skylights, clerestories, and other windows not used for vision. In any case, the simple measure of illuminance fails to capture all that is important about daylighting in practice. Aesthetics is an essential quality of daylighting, yet it eludes quantification. For example, if a horizontal skylight with diffuse glazing looks just like a fluorescent fixture, an architectural opportunity has been missed. As an alternative to diffuse glazing, one can avoid glare by interposing blinds or diffuse reflecting surfaces. Or one can tilt the aperture toward the north. The energy consequences are examined in Section 22.5.

Figure 22.2 shows how skylights can be used to create an environment more pleasant and open than a conventional office space, and Figure 22.3 shows a library lit by skylights. In recent years, there has been some experimentation with innovative schemes based on light shelves or on tracking reflectors with light guides, with the goal of transporting daylight deeper into the building.

FIGURE 22.2
Roof structure with linear skylights. (From Place, J.W. et al., *ASHRAE Trans.*, 93(1), 1987.)

FIGURE 22.3
Boulder, Colorado, Public Library lighted by clerestories and skylights. (Courtesy of Lightforms, Inc., New York.)

22.4 Analysis of Daylighting

For the analysis of daylighting, the following quantities need to be determined successively:

1. The daylight incident on the aperture, keeping track of three separate components: direct radiation from the solar disk, radiation from the sky, and radiation reflected by the ground
2. The fraction transmitted through the aperture, including losses due to dirt
3. The fraction transmitted to the work plane, taking into account multiple reflections at walls and other intervening surfaces (including the attenuation by a light well, if present)
4. The electricity that can be saved

Unlike the design conditions for thermal loads, the choice of design conditions for daylighting is not unique because daylighting is rarely designed as a standalone system for extreme conditions. In regions far from the equator, the seasonal change of day length makes it difficult or impossible to rely entirely on daylight for the duration of normal working hours. Complete backup with electric lamps is imperative in almost all situations. This reduces the role of design conditions from being an absolute minimum requirement to being merely a guide. One may want to require, somewhat arbitrarily, that no electric backup be necessary during certain periods, e.g., clear summer days, moderately overcast summer days, or winter days at noon.

The performance depends on the geometry of the daylighting system, on site and climate, and on the control of the backup. Calculations are difficult and tedious, if accuracy is desired. Fortunately, the chore can be performed by computer programs, such as EnergyPlus (2009) or Windows (LBNL, 2011). In this book, we present only a simple method for design calculations, the *lumen method* of the IES (1989).

22.4.1 Daylight Data

The choice of data depends on what is obtainable and what accuracy is desired. At the crudest level, one uses just the design conditions, as presented in Section 22.4.1. Estimation of electricity savings is more difficult because it depends on the variability of the available radiation. At the level of hourly simulations, the simplest approach is to assume proportionality between daylight (in lumens) and solar radiation (in watts) and to treat the diffuse component as isotropic. But in reality the lumen distribution is not isotropic, even under a uniform clear or a uniform overcast sky, to say nothing of the effect of clouds. For instance, for an overcast sky, the IES model assigns a distribution proportional to

$$1+2\cos\theta_s$$

with zenith angle θ_s. Furthermore, the luminous efficacy of solar radiation varies as a function of meteorological conditions. These details are treated in more sophisticated simulation models. The key parameters needed for a detailed modeling of daylight availability are the astronomical coordinates of the sun (zenith, azimuth, solar time), the geographic location of the site (longitude, latitude, elevation above sea level), and the atmospheric conditions (turbidity, cloudiness).

The necessary solar geometric relationships were discussed already in Chapter 4. While the calculation of daylight availability is basically like the calculation of solar heat gain factors, with the replacement of watts by lumens, there is a difference in the detail to which angular distributions should be taken into account. For heat gains, it usually does not matter very much whether solar radiation is absorbed by a wall or by the floor, 1 or 3 m from a window; once a ray has entered, it can be counted as heat gain. But the daylighting effect is quite sensitive to the direction of the rays: do they strike a wall or ceiling, or do they reach the target directly? Daylighting designers have developed their own solar radiation models, whose results may differ from models for thermal applications.

Even if a detailed modeling of the angular distribution of incident daylight is not feasible, it is recommended to carry out separate calculations for direct radiation (i.e., from the solar disk), for sky radiation, and for radiation reflected from the ground (the last two components are diffuse). Accordingly, the design data of the IES Calculation Procedures Committee (IES, 1984b) are presented in a way that permits the separation of these three components. Direct illuminance is shown in Figure 22.4, and sky illuminance in Figure 22.5, all for the horizontal plane and for several vertical planes. Finally, to help determine the illuminance received from the ground, Figure 22.6 shows the diffuse illuminance on the horizontal, indicated by the label *horizontal (full sky)*. (In addition, Figure 22.6 shows a set of curves labeled *horizontal [half sky]*; they will be needed as a supplementary variable to select a coefficient of utilization in Table 22.6.)

Three sky conditions are considered: clear sky (part a), partly cloudy sky (part b), and overcast sky (part c). Since there is no direct radiation under overcast skies, there is no Figure 22.4c. All graphs are plotted as function of the solar altitude angle (the complement 90° − θ_s of the zenith angle θ_s). The curves for vertical surfaces depend on the absolute value of the difference $\phi_s - \phi_p$ between

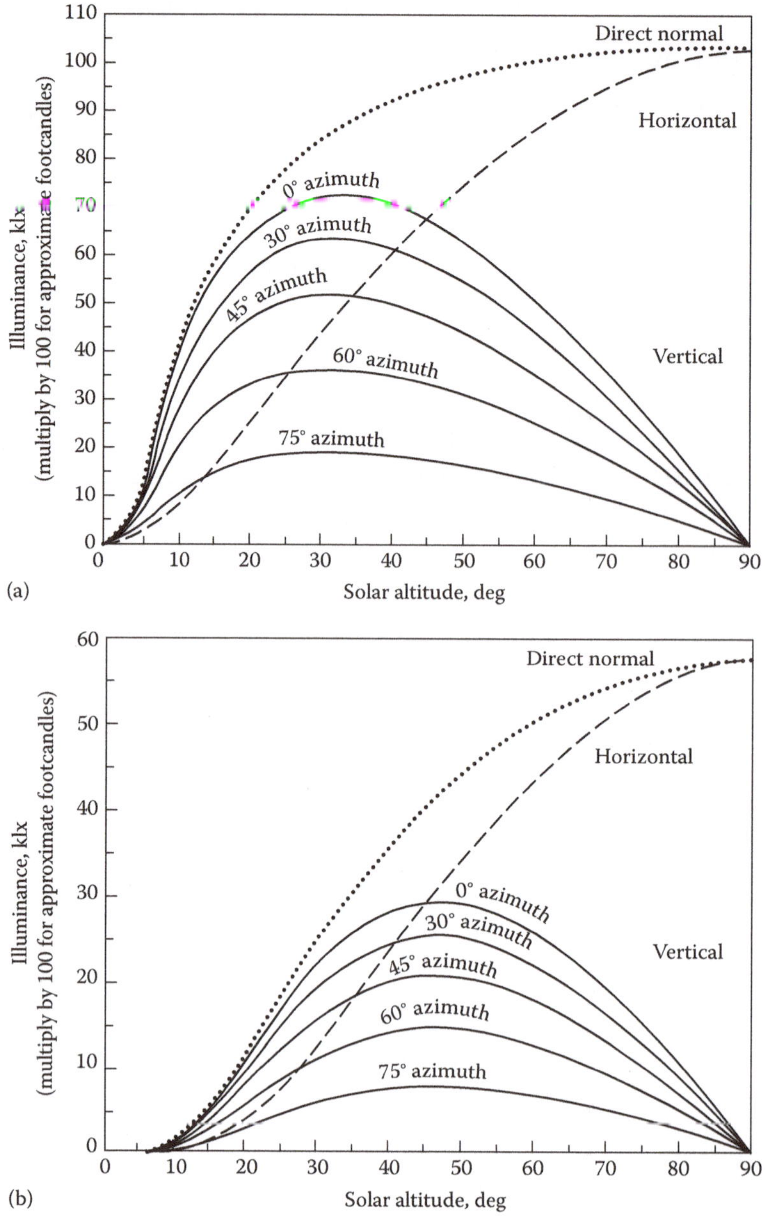

FIGURE 22.4
Direct component of design illuminance as a function of solar altitude (90° − θ_s). Azimuth label indicates $|\phi_s - \phi_p|$ of azimuth angles of sun ϕ_s and of surface ϕ_p, respectively. Horizontal, $E_{xh,dir}$; vertical, E_{xvdir}. (a) Clear sky, (b) partly cloudy sky. (From *IES Lighting Handbook: Reference Volume*, Kaufman, J.E., ed and Christensen, J.F., assoc ed., Illuminating Engineering Society of North America, New York, 1984a.)

the azimuth angles of the sun ϕ_s and of the surface ϕ_p; in the graphs, it is indicated by the label *azimuth*. For the overcast sky, the illuminance is independent of the azimuth. In using these data, one should not forget the enormous variability of the atmosphere; these design conditions are merely a representation of what might be typical.

A quick look at the scales of these figures suggests that when direct illuminance can reach a surface, it is likely to dominate the other terms by far. For instance, when the altitude is 30° and the azimuth is 45° (for a vertical surface), Figure 22.4a shows a direct illuminance of 50 klx under clear sky conditions. The corresponding diffuse sky illuminance is only 11.5 klx, from Figure 22.5a. For an overcast sky, Figure 22.5c shows that the surface receives only 4.5 klx.

22.4.2 Lumen Method for Toplighting

Skylights provide lighting from the top. So it comes as no surprise that the calculation is similar to the one described in Section 22.2 for electric lighting.

For simplicity, we consider only horizontal apertures. Thus, we need not worry about radiation reflected from the ground. Disregarding for the moment the directional dependence of the transmissivity of the glazing, we can state the basic formula for the average illuminance E_w on the indoor work plane in the form

$$E_w = E_{xh}\tau C_u \frac{A_h}{A_w} \tag{22.5}$$

where
- A_h is the projected horizontal aperture area of skylights
- A_w is the area of work plane (horizontal)
- τ is the transmissivity of skylight (net, i.e., including losses due to dirt and due to frames, mullions and solar control devices, and, if appropriate, losses in light well)
- C_u is the room coefficient of utilization (Table 22.4)
- E_{xh} is the exterior illuminance on horizontal surface

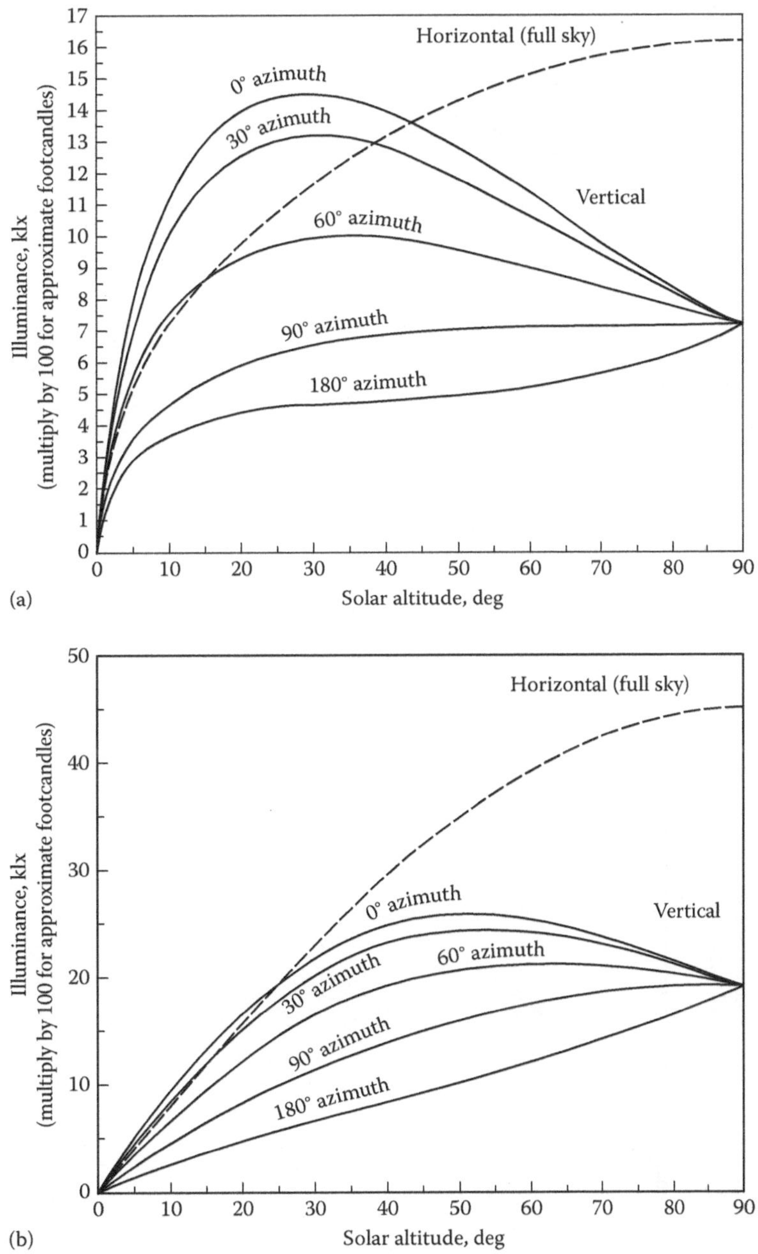

FIGURE 22.5
Diffuse sky component of design illuminance as a function of solar altitude (90° − θ_s). Azimuth label indicates $|\phi_s - \phi_p|$ of azimuth angles of sun ϕ_s and of surface ϕ_p, respectively. Horizontal, $E_{xh,s}$; vertical, $E_{xv,s/2}$. (a) Clear sky, (b) partly cloudy sky. (From *IES Lighting Handbook: Reference Volume*, Kaufman, J.E., ed. and Christensen, J.F., assoc ed., Illuminating Engineering Society of North America, New York, 1984a.)
(Continued)

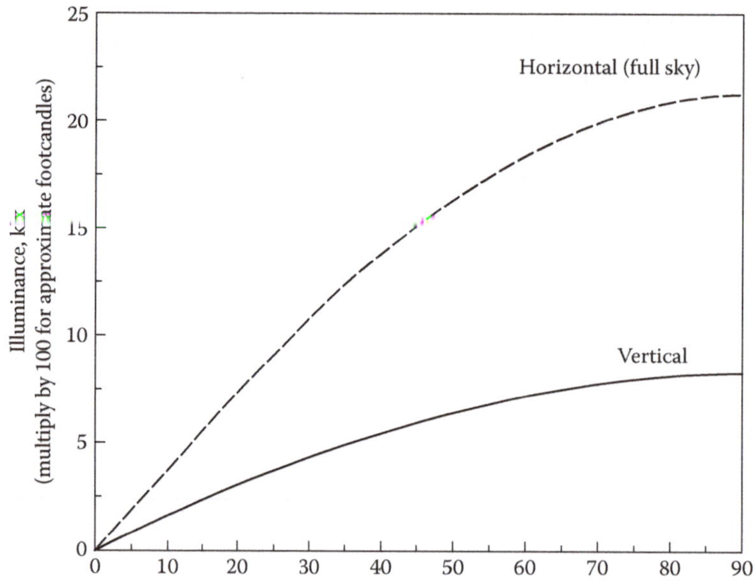

(c)

FIGURE 22.5 ***(Continued)***
Diffuse sky component of design illuminance as a function of solar altitude (90° − θ_s). Azimuth label indicates $|\phi_s - \phi_p|$ of azimuth angles of sun ϕ_s and of surface ϕ_p, respectively. Horizontal, $E_{xh,s}$; vertical, $E_{xv,s/2}$. (c) overcast sky. (From *IES Lighting Handbook: Reference Volume*, Kaufman, J.E., ed. and Christensen, J.F., assoc ed., Illuminating Engineering Society of North America, New York, 1984a.)

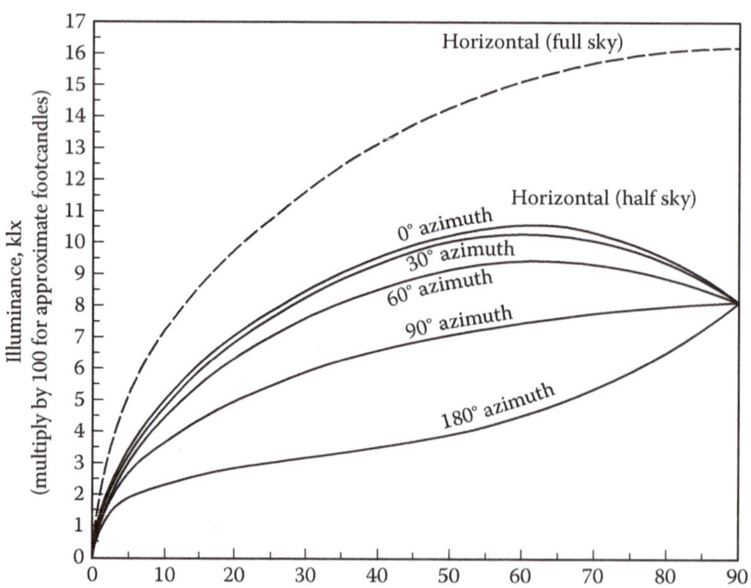

(a)

FIGURE 22.6
Diffuse horizontal half-sky component (= $E_{xh,s/2}$ = solid lines) of design illuminance on ground in front of a vertical surface as a function of solar altitude (90°−θ_s). Azimuth label indicates $|\phi_s - \phi_p|$ of azimuth angles of sun ϕ_s and of surface ϕ_p, respectively. (a) Clear sky. (From IES, *J. Illum. Eng. Soc.* (July), 381, 1984b.)

(Continued)

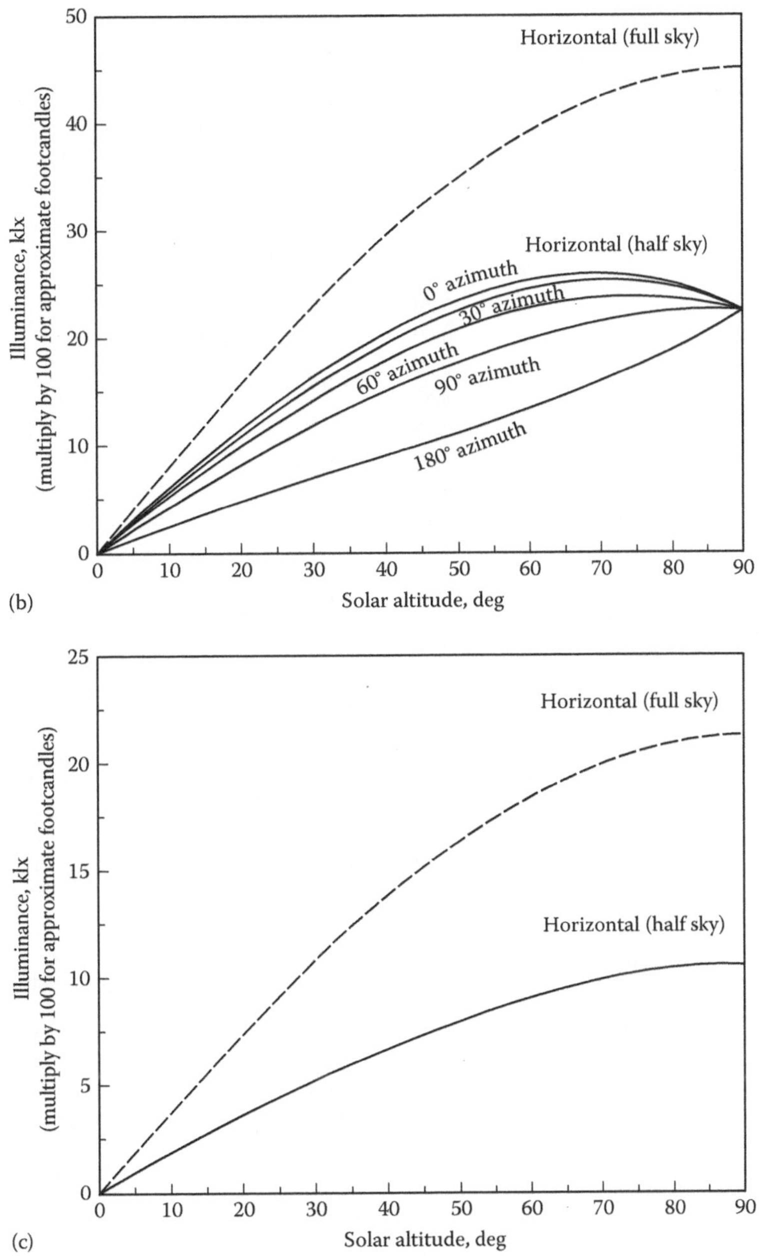

FIGURE 22.6 ***(Continued)***
Diffuse horizontal half-sky component (= $E_{xh,s/2}$ = solid lines) of design illuminance on ground in front of a vertical surface as a function of solar altitude (90°–θ_s). Azimuth label indicates $|\phi_s - \phi_p|$ of azimuth angles of sun ϕ_s and of surface ϕ_p, respectively. (b) Partly cloudy sky, and (c) overcast sky. (From IES, *J. Illum. Eng. Soc.*, July, 381, 1984b.)

The transmissivity τ is different for direct and for diffuse radiation. Adding appropriate subscripts, one can combine the contributions to the net transmissivity in the form

$$\tau_{dif} = T_{dif} T_{well} T_{control} R_A L_{dirt} \quad \text{for diffuse radiation} \tag{22.6}$$

and

$$\tau_{dir} = T_{dir} T_{well} T_{control} R_A L_{dirt} \quad \text{for direct radiation} \tag{22.7}$$

where

T_{dif}, T_{dir} are the transmissivity of glazing for diffuse and direct radiation (from manufacturer's brochure)

T_{well} is the transmissivity of light well, if present (= well efficiency of Figure 22.7)

$T_{control}$ is the transmissivity of control devices such as louvers or diffusers, if present

R_A is the ratio of net to gross skylight area, to account for blocking by frame or mullions

L_{dirt} is the loss factor due to dirt (Table 22.5)

TABLE 22.4

Room Coefficient of Utilization C_u for Skylights, Based on Floor Reflectivity = 20%

		Wall Reflectivity, %		
Ceiling Reflectivity, %	**RCR**[a]	**50**	**30**	**10**
80	0	1.19	1.19	1.19
	1	1.05	1.00	0.97
	2	0.93	0.86	0.81
	3	0.83	0.76	0.70
	4	0.75	0.67	0.60
	5	0.67	0.59	0.53
	6	0.62	0.53	0.47
	7	0.57	0.49	0.43
	8	0.54	0.47	0.41
	9	0.53	0.46	0.41
	10	0.52	0.45	0.40
50	0	1.11	1.11	1.11
	1	0.98	0.95	0.92
	2	0.87	0.83	0.78
	3	0.79	0.73	0.68
	4	0.71	0.64	0.59
	5	0.64	0.57	0.52
	6	0.59	0.52	0.47
	7	0.55	0.48	0.43
	8	0.52	0.46	0.41
	9	0.51	0.45	0.40
	10	0.50	0.44	0.40
20	0	1.04	1.04	1.04
	1	0.92	0.90	0.88
	2	0.83	0.79	0.76
	3	0.75	0.70	0.66
	4	0.68	0.62	0.58
	5	0.61	0.56	0.51
	6	0.57	0.51	0.46
	7	0.53	0.47	0.43
	8	0.51	0.45	0.41
	9	0.50	0.44	0.40
	10	0.49	0.44	0.40

Source: IES recommended practice for the lumen method of daylight calculations, IES Report RP-23-1989, Illuminating Engineering Society of North America, New York, 1989.

[a] RCR, room cavity ratio of Equation 22.9.

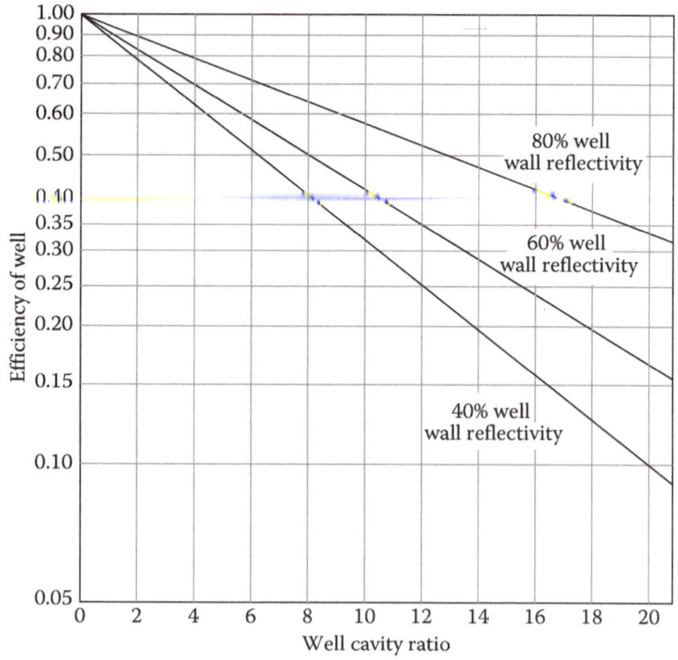

FIGURE 22.7
Well efficiency (transmissivity) of light well, for three values of wall reflectivity, as a function of height, length, and width of well, via the well cavity ratio (5 × height × [width + length])/(length × width). (From IES recommended practice for the lumen method of daylight calculations, IES Report RP-23-1989, Illuminating Engineering Society of North America, New York, 1989.)

TABLE 22.5

Loss Factor due to Dirt L_{dirt} as Function of Orientation of Glazing

	Orientation		
Location	**Vertical**	**Sloped**	**Horizontal**
Clean areas	0.9	0.8	0.7
Industrial areas	0.8	0.7	0.6
Very dirty areas	0.7	0.6	0.5

Source: IES recommended practice for the lumen method of daylight calculations, IES Report RP-23-1989, Illuminating Engineering Society of North America, New York, 1989.

The dependence of the transmissivity of the glazing on the incidence angle necessitates a slight complication of Equation 22.5. Under an overcast sky, there is only diffuse light, and one can write

$$E_w = E_{xh,s}\tau_{dif}C_u\frac{A_h}{A_w} \quad \text{for overcast sky} \tag{22.8}$$

where the exterior horizontal illuminance $E_{xh,s}$ from the sky can be found in Figure 22.5c, and C_u comes from Table 22.4.

If a light well is present, the transmissivity is multiplied by an additional factor, the well efficiency, obtained from the ordinate of Figure 22.7 as a function of well wall reflectivity and geometry. Of the light that has entered the room, the fraction that reaches the work plane is given by the room coefficient of utilization C_u, listed in Table 22.4 for several reflectivities and as a function of geometry, represented by the room cavity ratio:

$$\text{Room cavity ratio} = \frac{5\times\Delta h(w+l)}{wl} \tag{22.9}$$

where

Δh is the room height – work plane height
w is the width
l is the length of room (the room cavity ratio is five times the inverse of the room ratio of Equation 22.4)

The result is the average illuminance over the entire work plane. The illuminance will approach uniformity to the extent that the distance between individual skylight openings is small compared to Δh.

For clear and for partly cloudy days, Equation 22.5 is generalized to contain two terms, one for diffuse illuminance from the sky and one for direct illuminance from the sun:

$$E_w = (E_{xh,s}\tau_{dif} + E_{xh,dir}\tau_{dir})C_u \frac{A_h}{A_w} \text{ for clear or partly cloudy skies} \quad (22.10)$$

where

$E_{xh,s}$ is the diffuse illuminance from sky (Figure 22.5a or b)
$E_{xh,dir}$ is the direct illuminance from sun (Figure 22.4a or b)
C_u is the room coefficient of utilization (Table 22.4)

The skylight transmissivity τ_{dir} for direct radiation is different from τ_{dif} and varies with the angle of incidence. But the room coefficient of utilization is assumed independent of the angle of incidence, within the approximation of the lumen method.

Example 22.2: Illuminance Calculations with Horizontal Skylights

Find the illuminance for overcast and for clear days at noon, equinox, on the work plane of a room with horizontal skylights, without light well or control devices.

Given: Six evenly distributed horizontal skylights of total net area.

$A_h = 6 \times 3$ ft $\times 3$ ft (6×0.9144 m $\times 0.9144$ m).

Floor area (= work plane of area) $A_w = 30$ ft $\times$ 40 ft (9.144 m $\times$ 12.192 m).

Height difference between skylight and work plane $\Delta h = 6.3$ ft (1.92 m).

Latitude $\lambda = 40°$.

$T_{dif} = 0.79$, $T_{dir} = 0.92$.

Wall and ceiling reflectivity = 0.50.

Work plane reflectivity = 0.20.

Clean location.

Find: E_w for overcast and for clear days.

Assumption: Use lumen method for design conditions.

Lookup values: Declination at equinox $\delta = 0°$ from Equation 4.3.

Zenith angle $\theta_s = \lambda - \delta = 40°$ from Equation 4.9.

Solar altitude = $90° - \theta_s = 50°$.

Azimuth = 0 at noon from Equation 4.10.

$E_{xh,s}$ = 1600 fc (≈16 klx) from Figure 22.5c (overcast, horizontal).

$E_{xh,s}$ = 1400 fc (≈14 klx) from Figure 22.5a (clear sky, horizontal).

$E_{xh,dir}$ = 7500 fc (≈75 klx) from Figure 22.4a (clear sky, horizontal).

L_{dirt} = 0.7 from Table 22.5 for clean areas, horizontal.

$\tau_{dif} = 0.79 \times 0.7 = 0.553$ (no area reduction because A_h is already net area).

$\tau_{dir} = 0.92 \times 0.7 = 0.644$.

Room cavity ratio = $(5 \times 6.3 \times [40 + 30])/(40 \times 30)$ = 1.84 from Equation 22.9.

C_u = 0.89 interpolated from Table 22.4 for wall and ceiling reflectivity of 0.5.

Area ratio $A_h/A_w = 0.045$.

Solution

For overcast days, the illuminance is, from Equation 22.8,

$$E_w = 1600 \text{ fc} \times 0.553 \times 0.89 \times 0.045 = 36 \text{ fc } (\approx 360 \text{ lx})$$

For sunny days, the illuminance is, from Equation 22.10,

$$E_w = (1400 \text{ fc} \times 0.553 + 7500 \text{ fc} \times 0.644) \times 0.89 \times 0.045$$

$$= 31 + 192 \text{ fc} = 223 \text{ fc } (\approx 2230 \text{ lx})$$

Comments

This is the same room as in Example 22.1. For the sunny design day, the skylight, with a mere 4.5% of the floor area, provides more than four times the illuminance level of 50 fc (≈500 lx) that is typically required in offices. The total flux for 1,200 ft² is 223 fc × 1,200 ft² = 268,000 lm. Since sunlight has a luminous efficacy around 110 lm/W, the corresponding cooling load is (268,000 lm)/(110 lm/W) = 2,440 W_t = 2,440 W/(3.517 kW/ton) = 0.69 ton; this is not much larger than the cooling load of 1,760 W_t for the electric system in Example 22.1, while providing more than four times as much light. The cooling load would be less than one-third of that imposed by the electric system, if the transmissivity of the skylight was controlled with an exterior shading device to give the same illuminance (an interior shading device would absorb much of the radiation—not good for reducing the cooling load). A selective coating could also help if it has high visible transmissivity and low shading coefficient.

This example illustrates why the optimization of daylighting is challenging. Under the right conditions or with the right control, daylight can indeed have a much higher efficacy than electricity, but for the average performance, this advantage is less obvious. Without adjustable transmissivity, the illuminance will be either too high or too low most of the time, and finding the optimal design requires careful analysis of the annual performance. In Example 22.2, should the skylight be scaled up to yield 50 fc instead of 33 fc on the overcast day? Or should it be reduced to limit overheating during sunny days? The design conditions give no answer.

We will take up this question again in Section 22.5, where we will present some optimization results obtained by means of hourly computer simulations. They will show that the skylight area ratio $A_w/A_h = 0.045$ of Example 22.2 is not a bad guess, for a wide range of applications. And like most optimum values, there is fair tolerance to deviations from the optimum.

22.4.3 Lumen Method for Sidelighting

The lumen method for vertical apertures uses an equation similar to the one for toplighting. But there is a complication. Whereas with toplighting one can assume that all points in the work plane receive about the same amount of light, with sidelighting the illuminance varies strongly between different points in the room that are at different distances from the aperture. Therefore, the calculation should be done for several reference points located on a line perpendicular to the aperture. The procedure of IES (1989) recommends five reference points at 10%, 30%, 50%, 70%, and 90% of room depth, as indicated in Figure 22.8.

The lumen method gives no credit for direct sunlight in the room; one assumes that it is kept out by overhangs, blinds, and shades in order to avoid problems with glare. Glare from direct sunlight can also be eliminated by using diffuse glazing, a technique that is convenient for transom windows above ordinary vision windows. Here, we present only the simplest case: a window unshaded by plants, overhangs, or blinds.

Disregarding certain complications of angular distributions, the basic formula for the illuminance E_w on the work plane can be stated in the form

$$E_w = E_{xv} \tau C_u \qquad (22.11)$$

where

E_{xv} is the exterior vertical illuminance on the window

$\tau = \tau_{dif}$ is the diffuse transmissivity, calculated by Equation 22.6 as for skylights

No area ratio appears in the equation; it is implicit in the coefficient of utilization C_u. But for greater accuracy, the contributions from the sky and from the ground are calculated separately, with separate values for C_u, and then added:

$$E_w = \tau(E_{xv,s/2} C_{u,s} + E_{xv,g/2} C_{u,g}) \qquad (22.12)$$

where

$E_{xv,s/2}$ is the vertical illuminance incident on window from the half sky in front of the window (Figure 22.5)

$E_{xv,g/2}$ is the vertical illuminance reflected from the ground in front of the window, obtained from the horizontal illuminance E_{xh} by

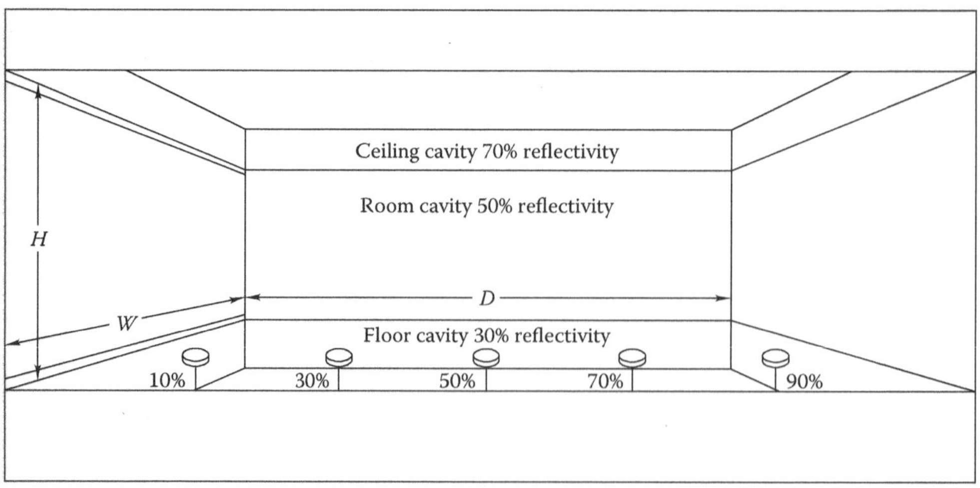

FIGURE 22.8
Standard conditions for the calculation of sidelighting. (From IES recommended practice for the lumen method of daylight calculations, IES Report RP-23-1989, Illuminating Engineering Society of North America, New York, 1989.)

$$E_{xv,g/2} = \frac{\rho_g}{2} E_{xh} \qquad (22.13)$$

analogous to Equation 4.33; ρ_g is the reflectivity of the ground. For overcast days, E_{xh} is the full-sky value from Figure 22.5c. For clear or for partly cloudy days, one takes E_{xh} as the sum of the direct horizontal component (Figure 22.4a or b) and the horizontal diffuse sky component (Figure 22.5a or b).

The coefficient of utilization C_u for sidelighting depends on the angular distribution of the light incident on the aperture. As a simple approximation for the values in Table 22.6, standard distributions have been assumed for sky radiation and for ground radiation, and separate tables are provided for different ratios of

$$E_{xv,s/2} = \text{Vertical half-sky illuminance (Figure 22.5)}$$

$$E_{xh,s/2} = \text{Horizontal half-sky illuminance (Figure 22.6)}$$

Here, we cite only a few excerpts to illustrate the method; for a full set of tables, including the effects of venetian blinds, we refer to IES (1989). Table 22.6 is arranged according to the ratio of room depth to window height (reading from top to bottom) and the ratio of window width to window height (reading from left to right). For each combination of these ratios, a set of five coefficients is listed, one for each of the five reference points.

Example 22.3: Illuminance Calculation with Vertical Windows

Suppose the room of Example 22.2 has a south-facing vertical window instead of the skylights. Find the illuminance in the center of the room for an overcast equinox day at noon.

Given: Floor area (= work plane area) A_w = 30 ft (9.144 m) (perpendicular to window) × 40 ft (12.192 m) (along window).

Room depth = 30 ft (9.144 m) (*D* in Figure 22.8).

Window width = 20 ft (6.096 m) (*W* in Figure 22.8).

Window height = 5 ft (1.524 m) (*H* in Figure 22.8).

Height difference between ceiling and work plane Δh = 6.3 ft (1.92 m).

Wall and ceiling reflectivity = 0.50.

Work plane reflectivity = 0.20.

Ground reflectivity = 0.20.

Find: E_w at depth of 15 ft (= 50% point in Figure 22.8).

Intermediate quantities: Solar altitude = 90° − θ_s = 50° (from Example 22.2).

Azimuth of sun ϕ_s = 0° (from Equation 4.10).

Azimuth of window ϕ_p = 0° (south facing).

Azimuth difference ("azimuth" of Figures 22.4 through 22.6) $|\phi_s - \phi_p|$ = 0°.

$E_{xv,s/2} \approx$ 590 fc exterior vertical illuminance on window, from Figure 22.5c.

$E_{xh} \approx$ 1500 fc exterior horizontal illuminance, from Figure 22.5c.

$E_{xh,s/2} \approx$ 750 fc horizontal half-sky exterior illuminance, from Figure 22.6c.

L_{dirt} = 0.9 from Table 22.5 for clean areas, vertical.

τ_{dif} = 0.79 × 0.9 = 0.711 (no area reduction because A_h is already net area).

Solution

The vertical illuminance reflected from the ground is, from Equation 22.13,

$$E_{xv,g/2} = \frac{E_{xh}\rho_g}{2} = 1500 \text{ fc} \times \frac{0.2}{2} = 150 \text{ fc}$$

To determine the coefficients of utilization, we need several ratios:

$$\frac{\text{Window width}}{\text{Window height}} = \frac{20 \text{ ft}}{5 \text{ ft}} = 4.0$$

$$\frac{\text{Room depth}}{\text{Window height}} = \frac{30 \text{ ft}}{5 \text{ ft}} = 6.0$$

Horizontal half-sky exterior illuminance is $E_{xh,s/2}$ = 750 fc.

Vertical/horizontal half-sky exterior illuminance ratio is

$$\frac{E_{xv,s/2}}{E_{xh,s/2}} = \frac{590}{750} = 0.79$$

$$C_{u,s} = \begin{cases} 0.049 & \text{at } E_{xv,s/2}/E_{xh,s/2} = 0.75, \text{ from Table 22.6a} \\ 0.064 & \text{at } E_{xv,s/2}/E_{xh,s/2} = 1.00, \text{ from Table 22.6b} \end{cases}$$

By interpolation,

$$C_{u,s} = 0.049 + (0.064 - 0.049)\left(\frac{0.79 - 0.75}{1.00 - 0.75}\right) = 0.051$$

For the ground component, $C_{u,g}$ = 0.054, from Table 22.6d.

Finally, the illuminance at the work plane, from Equation 22.12, is

$$E_w = 590 \times 0.711 \times 0.051 + 150 \times 0.711 \times 0.054 = 27 \text{ fc}$$

Repeating the calculation for other values of percentage of room depth, we obtain the results plotted in Figure 22.9.

Comments

The illuminance decreases rapidly over the first 10 ft (≈3 m).

TABLE 22.6

Coefficient of Utilization C_u for Window without Blinds, for Room of Figure 22.8

(a) Vertical/horizontal half-sky exterior illuminance ratio = $E_{xv,s/2}/E_{xh,s/2} = 0.75$

Room Depth/Window Height	Percent D	Window Width/Window Height: 0.5	1	2	3	4	6	8	Infinite
1	10	0.824	0.864	0.870	0.873	0.875	0.879	0.880	0.883
	30	0.547	0.711	0.777	0.789	0.793	0.798	0.799	0.801
	50	0.355	0.526	0.635	0.659	0.666	0.669	0.670	0.672
	70	0.243	0.386	0.505	0.538	0.548	0.544	0.545	0.547
	90	0.185	0.304	0.418	0.451	0.464	0.444	0.446	0.447
2	10	0.667	0.781	0.809	0.812	0.813	0.815	0.816	0.824
	30	0.269	0.416	0.519	0.544	0.551	0.556	0.557	0.563
	50	0.122	0.204	0.287	0.319	0.351	0.339	0.341	0.345
	70	0.068	0.116	0.173	0.201	0.214	0.223	0.226	0.229
	90	0.050	0.084	0.127	0.151	0.164	0.167	0.171	0.172
3	10	0.522	0.681	0.739	0.746	0.747	0.749	0.747	0.766
	30	0.139	0.232	0.320	0.350	0.360	0.366	0.364	0.373
	50	0.053	0.092	0.139	0.163	0.174	0.183	0.182	0.187
	70	0.031	0.053	0.081	0.097	0.106	0.116	0.116	0.119
	90	0.025	0.041	0.061	0.074	0.082	0.089	0.090	0.092
4	10	0.405	0.576	0.658	0.670	0.673	0.675	0.674	0.707
	30	0.075	0.134	0.197	0.224	0.235	0.243	0.243	0.255
	50	0.028	0.050	0.078	0.094	0.104	0.112	0.114	0.119
	70	0.018	0.031	0.048	0.059	0.065	0.073	0.074	0.078
	90	0.016	0.026	0.040	0.048	0.053	0.059	0.061	0.064
6	10	0.242	0.392	0.494	0.516	0.521	0.524	0.523	0.588
	30	0.027	0.054	0.086	0.102	0.111	0.119	0.120	0.135
	50	0.011	0.023	0.036	0.044	0.049	0.055	0.056	0.063
	70	0.009	0.018	0.027	0.032	0.035	0.040	0.041	0.046
	90	0.008	0.016	0.023	0.028	0.031	0.034	0.035	0.040
8	10	0.147	0.257	0.352	0.380	0.387	0.391	0.392	0.482
	30	0.012	0.026	0.043	0.054	0.060	0.067	0.070	0.086
	50	0.006	0.013	0.021	0.026	0.029	0.333	0.035	0.043
	70	0.005	0.011	0.017	0.021	0.023	0.026	0.027	0.034
	90	0.004	0.010	0.015	0.019	0.021	0.023	0.025	0.030
10	10	0.092	0.168	0.248	0.275	0.284	0.290	0.291	0.395
	30	0.006	0.014	0.026	0.032	0.036	0.041	0.044	0.059
	50	0.003	0.008	0.014	0.017	0.019	0.022	0.024	0.032
	70	0.003	0.007	0.012	0.014	0.016	0.018	0.019	0.026
	90	0.003	0.006	0.011	0.013	0.015	0.016	0.017	0.024

(b) Vertical/horizontal half-sky exterior illuminance ratio = $E_{xv,s/2}/E_{xh,s/2} = 1.00$

Room Depth/Window Height	Percent D	Window Width/Window Height: 0.5	1	2	3	4	6	8	Infinite
1	10	0.671	0.704	0.711	0.715	0.717	0.726	0.726	0.728
	30	0.458	0.595	0.654	0.668	0.672	0.682	0.683	0.685
	50	0.313	0.462	0.563	0.589	0.598	0.607	0.608	0.610
	70	0.227	0.362	0.478	0.515	0.527	0.530	0.532	0.534
	90	0.186	0.306	0.424	0.465	0.481	0.468	0.471	0.472
2	10	0.545	0.636	0.658	0.660	0.661	0.665	0.666	0.672
	30	0.239	0.367	0.459	0.484	0.491	0.499	0.501	0.506
	50	0.121	0.203	0.286	0.320	0.335	0.348	0.351	0.355
	70	0.074	0.128	0.192	0.226	0.243	0.259	0.264	0.267
	90	0.058	0.101	0.156	0.188	0.207	0.215	0.221	0.223
3	10	0.431	0.561	0.607	0.613	0.614	0.616	0.615	0.631
	30	0.133	0.223	0.306	0.337	0.348	0.357	0.357	0.366
	50	0.058	0.103	0.155	0.183	0.197	0.211	0.213	0.218
	70	0.037	0.064	0.098	0.119	0.132	0.147	0.150	0.154
	90	0.030	0.051	0.079	0.098	0.110	0.122	0.126	0.129
4	10	0.339	0.482	0.549	0.560	0.563	0.566	0.565	0.593
	30	0.078	0.139	0.204	0.234	0.247	0.258	0.260	0.272
	50	0.033	0.060	0.094	0.114	0.126	0.139	0.143	0.150
	70	0.022	0.039	0.061	0.074	0.083	0.095	0.099	0.104
	90	0.019	0.032	0.050	0.061	0.070	0.080	0.084	0.089
6	10	0.211	0.343	0.433	0.453	0.458	0.461	0.461	0.518
	30	0.033	0.065	0.103	0.123	0.135	0.145	0.148	0.167
	50	0.015	0.029	0.047	0.057	0.064	0.073	0.077	0.086
	70	0.011	0.021	0.033	0.040	0.045	0.051	0.054	0.060
	90	0.010	0.019	0.028	0.034	0.038	0.044	0.046	0.052
8	10	0.135	0.238	0.326	0.353	0.362	0.366	0.367	0.452
	30	0.016	0.034	0.058	0.072	0.080	0.090	0.094	0.116
	50	0.008	0.017	0.027	0.034	0.039	0.045	0.048	0.059
	70	0.006	0.013	0.021	0.026	0.028	0.032	0.035	0.043
	90	0.005	0.012	0.019	0.023	0.025	0.029	0.031	0.038
10	10	0.090	0.165	0.244	0.272	0.283	0.290	0.291	0.395
	30	0.009	0.020	0.036	0.045	0.052	0.060	0.064	0.087
	50	0.005	0.010	0.019	0.023	0.026	0.030	0.033	0.044
	70	0.004	0.009	0.015	0.018	0.020	0.023	0.025	0.033
	90	0.003	0.008	0.014	0.016	0.018	0.020	0.022	0.030

(Continued)

TABLE 22.6 (*Continued*)

Coefficient of Utilization C_u for Window without Blinds, for Room of Figure 22.8

Room Depth/ Window Height	Percent *D*	Window Width/Window Height 0.5	1	2	3	4	6	8	Infinite	Room Depth/ Window Height	Percent *D*	Window Width/Window Height 0.5	1	2	3	4	6	8	Infinite
(c) Vertical/horizontal half-sky exterior illuminance ratio = $E_{xv,s/2}/E_{xh,s/2} = 1.50$										(d) Ground component									
1	10	0.503	0.528	0.536	0.541	0.544	0.557	0.558	0.559	1	10	0.105	0.137	0.177	0.197	0.207	0.208	0.210	0.211
	30	0.359	0.464	0.514	0.528	0.534	0.549	0.550	0.552		30	0.116	0.157	0.203	0.225	0.235	0.241	0.243	0.244
	50	0.261	0.384	0.471	0.499	0.508	0.524	0.526	0.527		50	0.110	0.165	0.217	0.241	0.252	0.267	0.269	0.270
	70	0.204	0.325	0.432	0.470	0.485	0.497	0.499	0.500		70	0.101	0.162	0.217	0.243	0.253	0.283	0.285	0.286
	90	0.179	0.295	0.412	0.456	0.475	0.474	0.477	0.478		90	0.091	0.146	0.199	0.230	0.239	0.290	0.292	0.293
2	10	0.412	0.477	0.490	0.492	0.493	0.498	0.499	0.505	2	10	0.095	0.124	0.160	0.178	0.186	0.186	0.189	0.191
	30	0.201	0.304	0.379	0.402	0.410	0.422	0.424	0.429		30	0.082	0.132	0.179	0.201	0.212	0.219	0.222	0.225
	50	0.115	0.192	0.269	0.304	0.320	0.339	0.343	0.347		50	0.062	0.113	0.165	0.189	0.202	0.214	0.218	0.220
	70	0.078	0.136	0.204	0.241	0.261	0.286	0.292	0.295		70	0.051	0.093	0.141	0.165	0.179	0.194	0.198	0.200
	90	0.066	0.117	0.183	0.221	0.246	0.262	0.271	0.273		90	0.045	0.079	0.118	0.140	0.153	0.179	0.183	0.185
3	10	0.331	0.426	0.458	0.461	0.462	0.465	0.465	0.477	3	10	0.088	0.120	0.157	0.175	0.183	0.185	0.163	0.167
	30	0.121	0.202	0.275	0.304	0.316	0.327	0.329	0.337		30	0.059	0.107	0.154	0.176	0.187	0.198	0.193	0.198
	50	0.062	0.109	0.164	0.193	0.209	0.228	0.232	0.238		50	0.039	0.074	0.114	0.134	0.146	0.157	0.166	0.170
	70	0.041	0.073	0.114	0.138	0.154	0.176	0.183	0.188		70	0.031	0.055	0.085	0.101	0.111	0.122	0.127	0.130
	90	0.035	0.062	0.099	0.123	0.141	0.159	0.169	0.173		90	0.028	0.047	0.070	0.083	0.092	0.107	0.113	0.115
4	10	0.265	0.372	0.422	0.430	0.433	0.435	0.435	0.456	4	10	0.073	0.113	0.154	0.174	0.183	0.187	0.176	0.184
	30	0.077	0.137	0.199	0.229	0.243	0.256	0.259	0.272		30	0.040	0.082	0.127	0.148	0.159	0.170	0.177	0.185
	50	0.037	0.069	0.107	0.130	0.144	0.161	0.167	0.175		50	0.025	0.049	0.078	0.094	0.103	0.113	0.117	0.123
	70	0.026	0.046	0.073	0.089	0.101	0.119	0.126	0.132		70	0.020	0.036	0.054	0.065	0.071	0.079	0.083	0.087
	90	0.022	0.039	0.063	0.078	0.090	0.106	0.114	0.120		90	0.019	0.032	0.046	0.054	0.060	0.069	0.073	0.076
6	10	0.173	0.281	0.351	0.368	0.373	0.375	0.375	0.422	6	10	0.056	0.106	0.143	0.164	0.175	0.184	0.173	0.194
	30	0.037	0.073	0.115	0.137	0.151	0.164	0.168	0.189		30	0.021	0.050	0.081	0.098	0.107	0.117	0.123	0.138
	50	0.018	0.036	0.058	0.071	0.080	0.092	0.098	0.110		50	0.013	0.027	0.041	0.049	0.054	0.060	0.064	0.072
	70	0.013	0.026	0.040	0.049	0.056	0.064	0.069	0.078		70	0.011	0.021	0.029	0.033	0.035	0.039	0.041	0.046
	90	0.012	0.023	0.035	0.043	0.048	0.057	0.062	0.070		90	0.011	0.020	0.026	0.030	0.032	0.035	0.037	0.042
8	10	0.117	0.207	0.282	0.305	0.314	0.319	0.320	0.393	8	10	0.036	0.082	0.122	0.143	0.156	0.166	0.170	0.208
	30	0.020	0.042	0.071	0.087	0.098	0.111	0.116	0.143		30	0.011	0.029	0.050	0.062	0.070	0.078	0.082	0.101
	50	0.010	0.021	0.035	0.044	0.050	0.058	0.063	0.078		50	0.007	0.016	0.024	0.028	0.031	0.035	0.038	0.046
	70	0.007	0.016	0.026	0.032	0.036	0.041	0.045	0.055		70	0.006	0.013	0.018	0.020	0.021	0.023	0.025	0.030
	90	0.076	0.014	0.023	0.028	0.031	0.036	0.040	0.049		90	0.006	0.013	0.017	0.019	0.020	0.022	0.023	0.028
10	10	0.082	0.153	0.224	0.250	0.262	0.269	0.271	0.368	10	10	0.024	0.061	0.109	0.120	0.131	0.144	0.147	0.200
	30	0.012	0.026	0.047	0.059	0.068	0.078	0.084	0.114		30	0.006	0.017	0.034	0.040	0.046	0.053	0.056	0.076
	50	0.006	0.014	0.024	0.030	0.034	0.040	0.044	0.060		50	0.004	0.010	0.016	0.018	0.020	0.023	0.024	0.033
	70	0.005	0.011	0.019	0.022	0.025	0.029	0.032	0.043		70	0.004	0.009	0.013	0.014	0.015	0.016	0.016	0.022
	90	0.004	0.010	0.017	0.020	0.023	0.026	0.028	0.038		90	0.004	0.009	0.013	0.013	0.014	0.015	0.016	0.021

Source: IES recommended practice for the lumen method of daylight calculations, IES Report RP-23-1989, Illuminating Engineering Society of North America, New York, 1989.

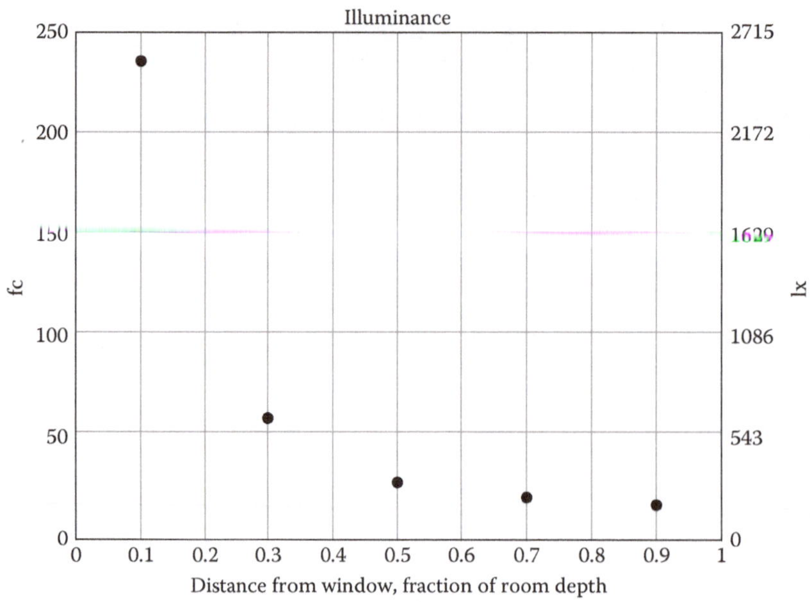

FIGURE 22.9
Illuminance on work plane of Example 13.3 as a function of distance from the window.

22.5 Design of Buildings for Daylighting

In commercial buildings, the use of daylight is potentially the most important envelope design option for efficiency. There are numerous reasons. Energy consumption for lighting is large, and lights are used during daytime; thus daylight can displace expensive electricity during peak periods. Much of commercial floor space is sufficiently close to the outside, at the perimeter, or under the roof to enjoy easy access to daylight. The fraction directly under the roof is one-half by itself, according to Place et al. (1987). Furthermore, sunlight has high luminous efficacy, around 110 lm/W, about twice as high as conventional fluorescent lamps. Being cooler, sunlight offers the possibility of reducing the cooling loads of a building. And then there are, of course, important psychological benefits because daylighting can provide a view of the outside, a sense of openness, and a good spectral composition of the light.

By contrast, in residential buildings, the demand for lighting is usually quite limited during daytime; that is the reason why we discuss daylighting only in the commercial context. In residential buildings, it is more appropriate to consider the heat collection potential of windows, because residential heating loads tend to be relatively larger and better matched to the sun than those of commercial buildings.

Ordinary (vertical) windows are, of course, the simplest daylighting scheme; they do not call for any special ingredient beyond the installation of appropriate light controllers, but they are limited to the perimeter of a building. Skylights can provide daylight to any zone directly under the roof, but they imply higher construction cost than for a simple opaque roof. Finally, there are special schemes, based on light shelves or light guides. In this section, we discuss only the role of windows and of skylights, citing results from some important studies carried out at Lawrence Berkeley National Laboratory (LBNL). They are based on simulations of standard building designs. The results are indicative of general trends and provide valuable guidelines for the design of daylit buildings, all the more so since many features of daylighting have been found to be fairly insensitive to climate. Thus, when carrying out simulations for a building under design, the designer knows in what range of parameters to look for the optimum.

The design variables to optimize are the aperture size, glazing type, and choice of shading device. The glazing is characterized by three variables: the shading coefficient SC,* the transmissivity for visible radiation τ_v, and the conductance U (see Chapter 5). For the performance analysis of daylighting, it is convenient to group certain variables. One group is the *effective aperture* A_{eff}, normalized by the total area A_{tot} of the wall or roof:

$$A_{eff} = \frac{\tau_v A_{apert}}{A_{tot}} \tag{22.14}$$

* The SC index has been supplanted by the SHGC index (see Equations 5.16 and 5.17). The SC of a glazing is equal to its SHGC divided by that of standard double strength glass (=0.87).

The other is the ratio of visible to thermal transmittance:

$$K_e = \frac{\tau_v}{\text{SC}} \tag{22.15}$$

It is proportional to the instantaneous luminous efficacy of the daylight transmitted into the building.

22.5.1 Windows

For the daylighting performance of ordinary (vertical) windows, we cite a study by Arasteh et al. (1986) and by Sweitzer et al. (1987). These authors have considered the perimeter zones of an intermediate floor of a standard office building, as shown in Figure 22.10. They assume continuous dimmers for the control of the lights and, in the interest of thermal and visual comfort, the use of shades whenever the direct solar radiation transmitted through a window exceeds 63 W/m². Three levels of installed lighting power are evaluated (7.5, 18.3, and 29.1 W/m²), all for the same design illumination of 50 fc (540 lx). The HVAC system consists of a separate constant-volume system with fan coil and economizer for each zone, and the average COP of the cooling plant is 3.0.

The resulting energy consumption, as a function of the effective aperture, is shown in Figure 22.11 for a hot and humid climate: Lake Charles, Louisiana. Total

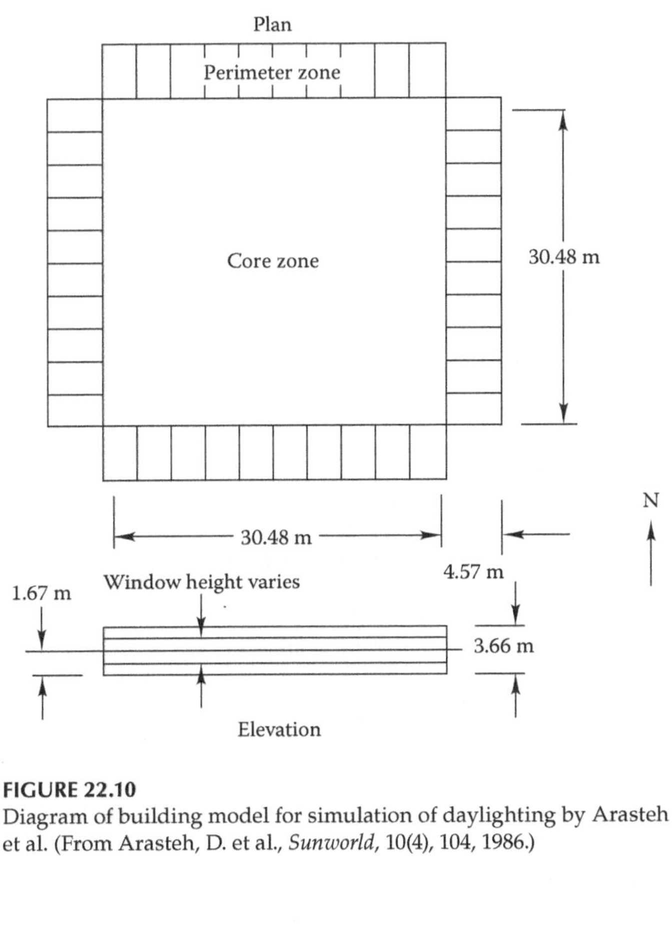

FIGURE 22.10
Diagram of building model for simulation of daylighting by Arasteh et al. (From Arasteh, D. et al., *Sunworld*, 10(4), 104, 1986.)

consumption is shown, as well as the components of lighting, cooling, office equipment, and fans. The dashed lines correspond to continuous dimming (D) controls, the solid lines to the absence of daylighting. With daylighting, the total consumption drops appreciably as the aperture increases from 0 to about 0.1. Beyond that value, further gains in lighting are small, while the cooling load continues to increase; thus the total consumption grows again.

Without dimming controls (ND), there is no initial drop; the consumption increases monotonically with aperture. This is most plausible: In a hot climate, the only energy benefit of windows could come from daylight—and without dimming controls, there are no savings.

The magnitude of the savings depends on the desired illumination level, efficacy of the electric lighting system, lighting control, HVAC system, etc. The higher the installed lighting capacity, the higher the savings from daylight and the higher the optimal effective aperture, as shown in Figure 22.12. The optimal aperture is typically in the range of 0.1–0.2.

Of course, the consumption is very sensitive to the choice of the glazing, especially the ratio $K_e = \tau_v/\text{SC}$. The higher, the better. The theoretical maximum is about 2.8. Most glazing for commercial buildings is in the range of 0.5–1.0, depending on its tint. The total electricity consumption (excluding space heat) for a south zone in Madison, Wisconsin, and in Lake Charles, Louisiana, is plotted versus aperture in Figure 22.13 for several values of K_e. The performance of specific glazings has been evaluated at a window/wall ratio WWR of 0.75. The figures also apply to WWR of 0.25 and 0.50 provided the visible transmittance value τ_v is modified as per the their graded scales provided.

The differences between different glazings are large, often on the order of 10% in total electricity. As for the benefit of increased K_e values, it is most pronounced when going from 0.5 to 1.0; beyond $K_e = 1.5$, further gains are negligible. As in the previous figures, the saturation of daylighting causes the consumption to increase at large apertures (beyond $A_e \approx 0.3$). No influence of U value is observable in this figure, because space heat is not included; the annual cooling energy is not very sensitive to the U value, the average temperature difference during the cooling season being small. It is interesting to note, both here and for the skylights in Figure 22.14 that the optimal aperture seems to be quite insensitive to climate.

The savings due to daylight are worth on the order of \$6 per square meter of floor area per year (Arasteh et al., 1986). This is to be compared with the cost of dimming controls, estimated around \$10–\$16 per square meter in new construction, \$43–\$54 per square meter in retrofits. In new construction, the economics are more favorable because this cost is partially offset by savings

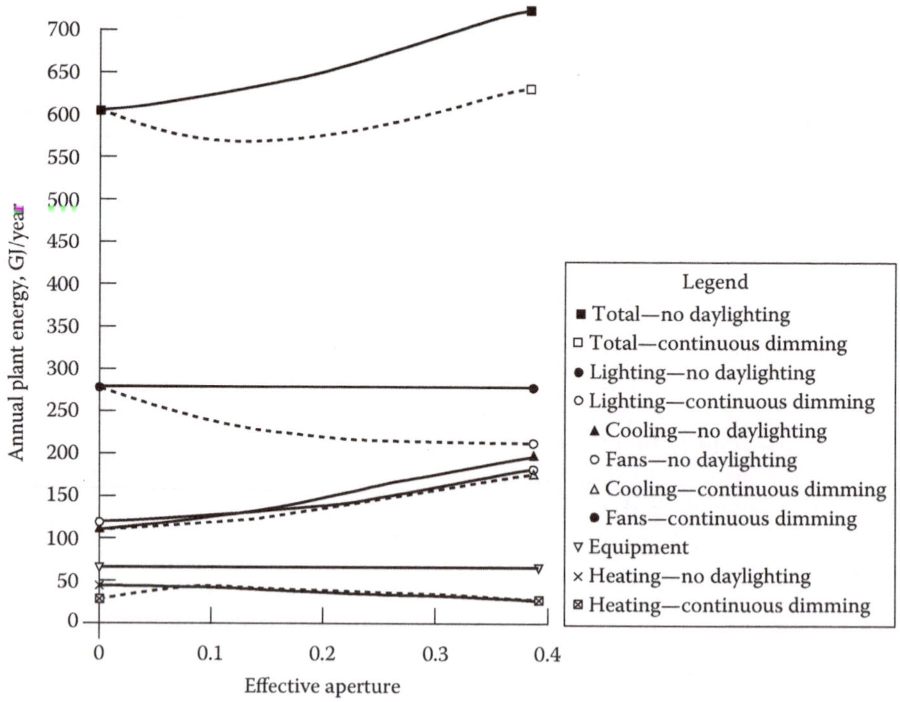

FIGURE 22.11
Annual plant energy requirements with daylighting (dashed lines) and without daylighting (solid lines) for the total building (five zones) of Figure 22.10 as a function of the effective aperture, for Lake Charles, Louisiana. (From Arasteh, D. et al., *Sunworld*, 10(4), 104, 1986.)

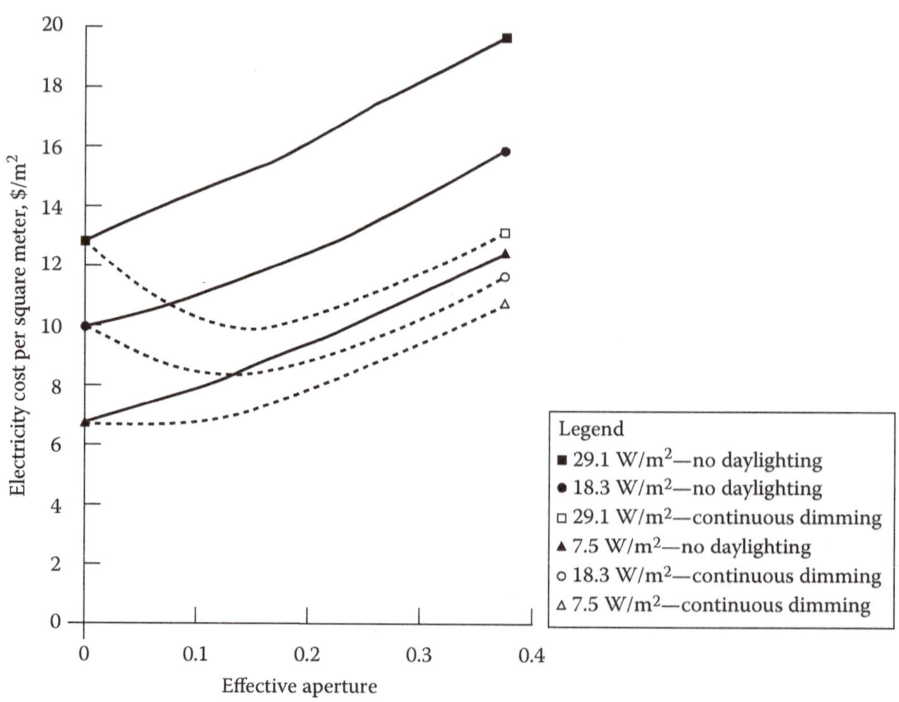

FIGURE 22.12
Annual total electricity cost versus A_{eff} for three levels of installed lighting capacity for the building of Figure 22.10 as a function of the effective aperture, for Lake Charles, Louisiana. (From Arasteh, D. et al., *Sunworld*, 10(4), 104, 1986.)

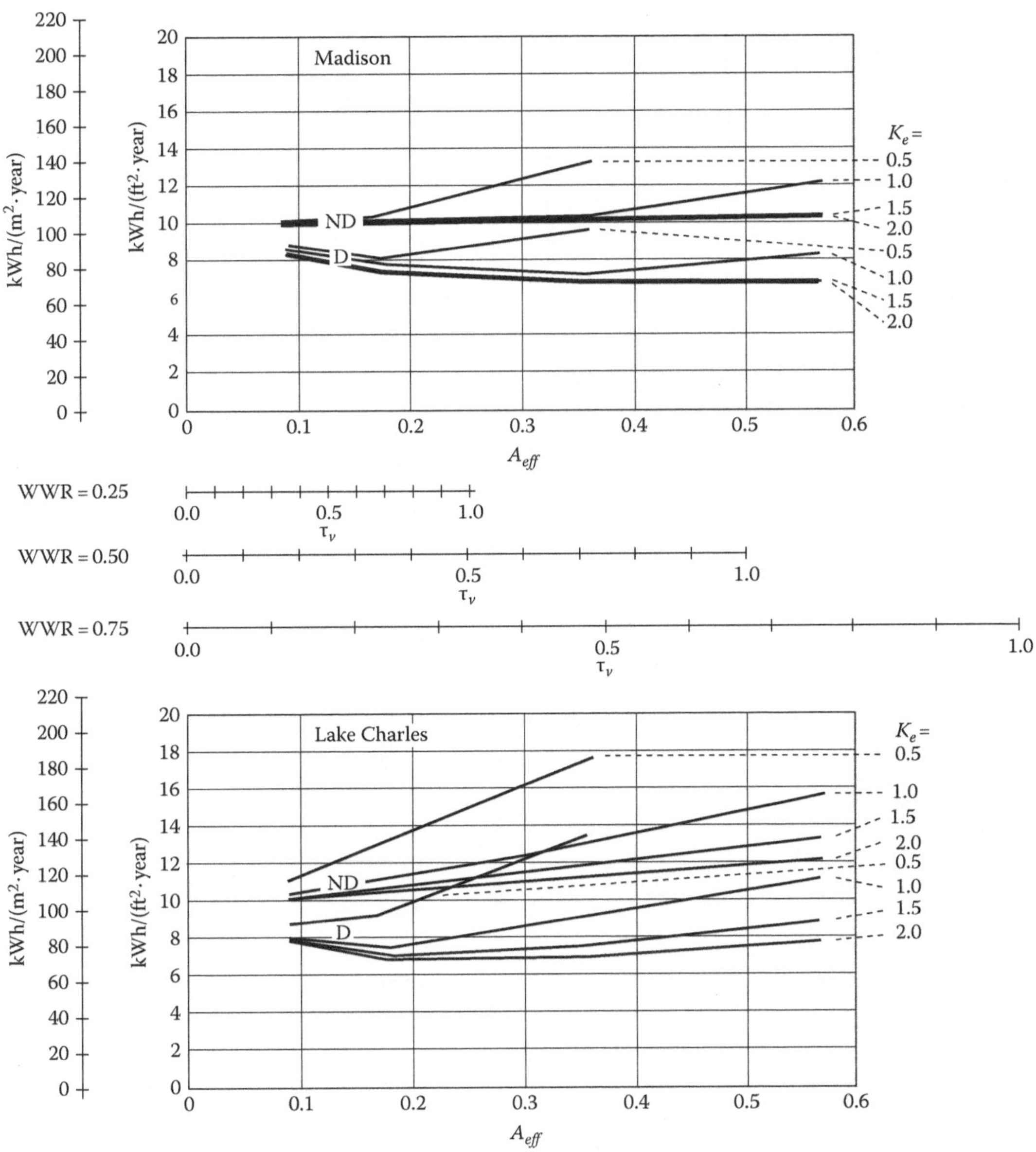

FIGURE 22.13
Total electricity use (lights, cooling, and fans, but not space heat) versus effective aperture A_{eff} with (= D) and without (= ND) dimming controls, for south zone of building of Figure 22.10, for four values of $K_e = \tau_v/SC$: 0.5, 1.0, 1.5, and 2.0 (some of the curves are almost on top of each other). *Top*, Madison, Wisconsin; *bottom*, Lake Charles, Louisiana. (From Sweitzer, G. et al., *ASHRAE Trans.*, 93, 1, 1987.)

in cooling equipment. Thus, the payback times for daylight controls with vertical windows appear to be on the order of a few years in new construction and around 8 years in retrofits.

22.5.2 Roof Apertures (Skylights)

The potential of skylights can be appreciated from a study by Place et al. (1987). Using a simulation program, these authors evaluated a variety of roof apertures for the top floor of a standard office building of 100 ft × 100 ft (30 m × 30 m) base area.

They considered both linear apertures, as in Figure 22.2, and localized apertures, with three tilts: horizontal, 60°, and vertical. Diffuse glazing or diffusers were assumed to prevent problems with glare. The design illumination level is 50 fc (540 lx) on the work plane. Power to the electric lights is reduced linearly in response to daylight while maintaining constant total illumination (except that the electricity cannot be reduced below a minimum of 20% of full power, typical of controllers available at the time of the study). The peak electric power for lighting is 2.5 W/ft², corresponding to a system luminous efficacy of 20 lm/W (i.e., relative to

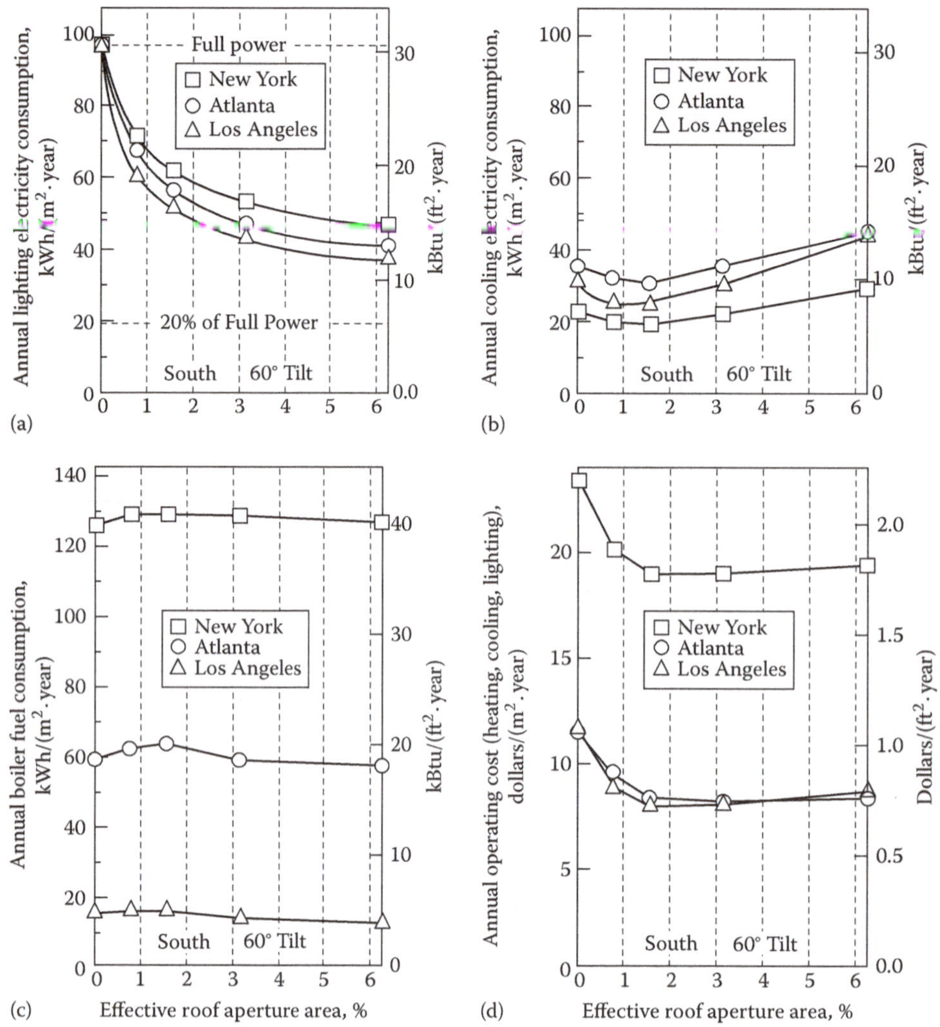

FIGURE 22.14
Annual consumption for office building with daylighting by skylights as a function of effective aperture area. (a) Electricity for lighting, (b) electricity for cooling, (c) boiler fuel, and (d) operating costs for cooling, heating, and lighting. (From Place, J.W. et al., *ASHRAE Trans.*, 93(1), 1987.)

the illumination delivered on the work plane). The *system* luminous efficacy of daylight is taken as 72 lm/W (as opposed to the outdoor value of 110 lm/W).

The resulting annual consumption as a function of effective aperture is shown in Figure 22.14, for lighting, cooling, and heating, together with the total annual operating cost for these terms, at three sites: New York City, Los Angeles, and Atlanta. The basic pattern of Figure 22.14 is similar to that in the preceding figures for vertical apertures. There is a rapid drop in total consumption as the aperture is increased from zero. Diminishing returns are reached with saturation of daylighting, and at large apertures, the consumption goes up again. The main difference between vertical and horizontal apertures lies in the position of the optimum. While the optimal aperture was found to occur around 0.1–0.3 for vertical windows, it is an order of magnitude smaller, around 0.01–0.06, for skylights. Most of the savings are due to reduced consumption for lighting, but reductions in cooling loads are not to be overlooked. The curves are quite similar in shape for the three locations, even though the magnitudes are different. Likewise, the utility rates have a strong effect on the magnitude of the savings, but not on the choice of the design.

As for the performance of different skylight designs, horizontal apertures yield higher savings per aperture area, because they are the most effective for beam daylighting, but vertical glazings (with orientations S, S + N, S + W + N + E, Sw + Se, or Nw + Ne + Sw + Se) produce even lower total annual cost because they perform better at low sun angles; see Figure 22.15. The optimal aperture for vertical skylights is larger, around 0.03–0.06, than that for horizontal skylights. A north aperture alone is not as good because it cannot utilize beam radiation. Psychological considerations also play a role: the

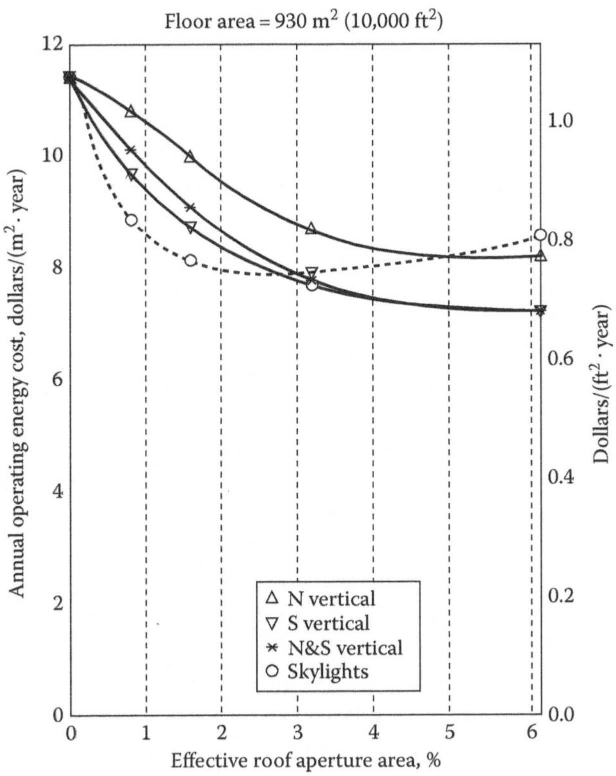

FIGURE 22.15
Annual operating energy cost for office building in Atlanta, Georgia, with daylighting by several types of skylights as a function of effective aperture area. (From Place, J.W. et al., *ASHRAE Trans.*, 93(1), 1987.)

desire for sunlight is stronger in winter than in summer; hence, skylights that are oriented toward the south may be preferable. All these designs presuppose that diffusing elements are used to prevent glare.*

The annual savings achievable with skylights can be as large as \$4 per square meter of floor area for A_{eff} around 0.04–0.06. That is to be compared with the incremental first costs for dimmers and for skylights minus the savings from reduced cooling plant capacity. These costs are very much dependent on the details of construction and are difficult to estimate in general. To translate the \$4 per square meter of floor area into permissible cost per skylight area, take the example of $A_{eff} = 0.04$ with a solar transmittance of $\tau_v = 0.50$, which implies that $A_{sky}/A_{floor} = 0.04/\tau_v = 0.08$; hence, the annual savings are \$4/0.08 m² = \$50 per square meter of A_{sky}. While the savings per area of glazing are highest for horizontal skylights, this is not necessarily the appropriate criterion because of fixed costs associated with the installation of any skylight.

To conclude this chapter, note that *windows can be more efficient than opaque walls.* This is reassuring, since people will demand windows anyway (psychological benefits of daylight are an important bonus, even if an unquantifiable one). The challenge is laid down for the designer: find energy-efficient solutions. Skylight designs have been tested in scale models and in real buildings; the results confirm that good illumination and high luminous system efficacy can indeed be achieved.

* And, of course, that attention will be paid to avoid practical problems such as leaks of rain and excessive snow accumulation.

Problems

Numbers 1–4 given in parenthesis denote the degree of difficulty.

22.1 Define luminous efficacy. What are typical values for the most important sources of light? *(1)*

22.2 Consider a fluorescent tube that consumes 40 W to produce 3000 lm.

(a) What is the luminous efficacy of the tube?

(b) What is the system luminous efficacy if the tube is used in a luminaire with a coefficient of utilization of 0.45? *(1)*

22.3 Consider two options for the luminaires in a one-story office building with floor area 100 ft × 100 ft (30.48 m × 30.48 m). Each luminaire contains a fluorescent tube, consuming 40 W (including the ballast) and producing 3000 lm of light. For the illumination in the work plane, one luminaire has a coefficient of utilization of 0.60, while the other has 0.45; they are similar in other characteristics. The average illumination in the work plane is to be 50 fc (543 lx), after allowing for 10% degradation due to dirt and aging.

(a) How many luminaires are needed, and what is the electric power per floor area, for each of these options?

(b) If the utilization is 2500 h/year and the electricity cost is \$0.10/kWh, what is the difference in annual electricity cost?

(c) If the system life is 20 years and the highest permissible payback time is 10 years, how much more can one pay for the most efficient luminaire? Express the answer in dollars for the building and in dollars per luminaire. *(3)*

22.4 Design the lighting for the corridors of a commercial building, using fluorescent tubes in a luminaire with the coefficients of utilization of Table 22.2. Make the following assumptions:

- Average illumination on floor 50 fc (543 lx)
- Luminous efficacy of tubes 80 lm/W

- One 40 W tube per luminaire (including ballast)
- Corridor height of 8 ft (2.44 m)
- Corridor width of 8 ft (2.44 m)
- Corridor length large compared to height and width
- Ceiling reflectivity = 0.7
- Wall reflectivity = 0.5
- Floor reflectivity = 0.2

(a) What is the spacing between luminaires?

(b) What is the installed power in watts per square foot (meter)?

(c) What is the annual electricity consumption if the lights are on 10 h/day for 6 days/week? *(3)*

22.5 As an alternative to the electric lighting of Problem 22.4, consider horizontal skylights for the corridors that are on the top floor. Corridors are a good place for daylighting because wide variability of the illumination is usually acceptable. In this case, suppose that the system is considered adequate if it provides 50 fc (543 lx) under an overcast sky at noon of winter solstice. Assume the following values: latitude = 40°N, T_{dif} = 0.79, L_{dirt} = 0.9, and R_A = 0.85.

(a) Estimate the area of the necessary skylights.

(b) How large are the electricity savings per square foot (meter) of corridor, relative to Problem 22.4, if the electric backup is needed 10% of the time (full-power equivalent)?

(c) Estimate the concomitant savings in cooling energy if the cooling season lasts 1000 h and if the coefficient of performance (COP) of the cooling system is 3.0. Assume that 80% of the heat produced by the electric lights must be removed during the cooling season and that the luminous efficacy of daylight in this application is 90 lm/W, averaged over the cooling season. *(4)*

22.6 Suppose the room of Example 22.1 has a window of height 4 ft (1.22 m) running along its entire width. Assume the following values: latitude = 40°N, T_{dif} = 0.79, L_{dirt} = 0.9, R_A = 0.85.

(a) Evaluate the illumination due to daylight for a clear equinox sky, assuming no direct radiation on the window and reflectivity of ground = 0.25.

(b) If one does not take credit for any direct solar illumination in the room, to what extent does the result depend on the orientation of the room?

(c) Suppose that 60% of the electricity for the outer 15 ft (4.57 m) of the room can be saved, averaged over the year, if daylight sensors are used. How large are the savings? *(4)*

22.7 A compact fluorescent bulb can be used to replace a standard incandescent bulb. Suppose a particular task requires a total 200 lm. If a 50 W fluorescent bulb has a life of 7500 h and an efficacy of 70 lm/W, while a 50 W incandescent bulb has a life of 1000 h and an efficacy of 15 lm/W, does it make more sense over the long run to use the compact fluorescent? Assume the compact fluorescent bulb costs \$20, the incandescent bulb costs \$1, and electricity costs 8¢/kWh. *(2)*

22.8 A 25 ft × 40 ft room at 35° latitude is to have an illumination level of 50 lm/ft² at the work plane. Two design options are to be evaluated: (1) 100% fluorescent lighting at the ceiling using the luminaire of Table 22.3 and lamps with 70 lm/W and (2) horizontal skylights with ordinary single glazing and an aperture that provides 50 lm/ft² on June 21 at noon if the sky is clear. The ceiling is 6 ft above the work plane and the reflectivities of ceiling, wall, and work plane are 0.5, 0.3, and 0.2, respectively.

(a) What is the electric power of the fluorescent option?

(b) Using the CLFs of Chapter 9, estimate the corresponding cooling load at noon if the lights are turned on from 8 AM to 6 PM, and the "*a*" and "*b*" coefficients are 0.55 and *D*.

(c) How large are the skylights for the daylighting option?

(d) Using the CLFs of Chapter 9, estimate the corresponding cooling load at noon. Ignore any changes in conductive loads due to the replacement of ceiling by skylights. *(4)*

22.9 What is the instantaneous sensible cooling load for the conditions of Problem 22.8 if the lighting level of 50 lm/ft² is to be achieved on an overcast day? Assume noon on June 21 and ignore any changes in conductive loads due to the replacement of ceiling by skylight. *(2)*

22.10 Suppose you have to design a daylighting system using skylights for a one-story office building with floor area 200 ft × 100 ft (60.96 m × 30.48 m) in Atlanta. How large would you choose the aperture area, and what type of glazing would you use? Approximately how large are the annual energy savings? *(3)*

References*

Arasteh, D., R. Johnson, S. Selkowitz, and D. Connell (1986). Cooling energy and cost savings with daylighting in a hot and humid climate. *Sunworld*, 10(4), 104.

Competitek, Inc. (1988). Lighting report. Rocky Mountain Institute, Snowmass, CO.

Energy Plus (2009). Energy plus building energy simulation software, developed by the National Renewable Energy Laboratory (NREL) for the U.S. Department of Energy, under the Building Technologies program, Washington, DC. http://www.nrel.gov/buildings/energy_analysis.html#energyplus. Accessed on September 2014.

EPA (2005). Lighting technologies: A guide to energy-efficient illumination. Energy STAR Program, U.S. Environmental Protection Agency, www.energystar.gov. Accessed on September 2014.

IES (1984a). *IES Lighting Handbook: Reference Volume* (J.E. Kaufman, ed. and J. F. Christensen, assoc. ed.). Illuminating Engineering Society of North America, New York.

IES (1984b). Recommended practice for the calculation of daylight availability. *J. Illum. Eng. Soc.*, July, 381–391.

IES (1987). *IES Lighting Handbook: Application Volume* (J.E. Kaufman, ed. and J.F. Christensen, assoc. ed.). Illuminating Engineering Society of North America, New York.

IES (1989). IES recommended practice for the lumen method of daylight calculations. IES Report RP-23-1989. Illuminating Engineering Society of North America, New York.

LBNL (2011). Windows 6.3. Lawrence Berkeley National laboratory, Berkeley, CA. www.windows.lbl.gov/software/window/6/6.3.36/window_releasenotes.htm. Accessed on September 2014.

Philips (1989). Philirama 89/90. Catalog of products, Companie Philips Eclairage, Boulogne-Billancourt, France.

Place, J.W., J.P. Coutier, M.R. Fontoynont, R.C. Kammerud, B. Andersson, W.L. Carroll, M.A. Wahlig, F.S. Bauman, and T.L. Webster (1987). The impact of glazing orientation, tilt, and area on the energy performance of room apertures. *ASHRAE Trans.*, 93, 1.

Rabl, A. (October 1990). Technologies d'éclairage: Luminaires performants (Lighting technologies: Efficient luminaires). *Revue de l'énergie*, 424, 482–488.

Sweitzer, G., D. Arasteh, and S. Selkowitz (1987). Effects of low-emissivity glazings on energy use patterns in nonresidential daylighted buildings. *ASHRAE Trans.*, 93, 1–14.

* The standard references on lighting are the two parts of the *IES Lighting Handbook* (IES, 1984a, 1987).

23

Costing and Economic Analysis

ABSTRACT We start by discussing basic notions such as time value of money and inflation, which are relevant to analyzing future costs. We then present historic trends of the consumer price index and those for different sources of energy. Next, concepts that influence cash flow analysis are covered followed by an in-depth treatment of the various terms that need to be considered while performing a life cycle assessment of design alternatives. We then discuss the various issues relevant to electricity costs in buildings, such as energy and demand costs, and on-peak and off-peak electric rate structures. Subsequently, we address how to perform an economic analysis in terms of payback period and internal rate of return methods. This is followed by a discussion of the role of risk and uncertainty associated with different aspects of the economic analysis. Next, various sources of reference material that provide service life and maintenance costs of various HVAC components are identified. We then describe and illustrate a simple but elegant model of annual maintenance costs of HVAC systems in office buildings. Finally, pertinent issues related to optimization in a design situation are addressed.

Nomenclature

A	Annual payment
A_{floor}	Area of the building, m^2 (ft^2)
A_{life}	Levelized annual cost
A_M	Annual cost for maintenance, first-year \$
$(A/P, r, N)$	Capital recovery factor
C	Cost at size S
C_{cap}	Capital cost, first-year \$
C_{life}	Life cycle cost
C_n	Single expense in nth year
COP	Coefficient of performance
CPI	Consumer price index
C_{rcf}	Cost at reference size S_r
C_{salv}	Salvage value, first-year \$
D	Annual degree-seconds, K·s
F_n	Future cash amount in nth year
f_{dep}	Present value of total depreciation as fraction of C_{cap}
$f_{dep,n}$	Depreciation during year n as fraction of C_{cap}
f_l	Fraction of investment paid by loan
I_n	Interest payment during nth year
k	Conductivity, W/(m·K)
L	Loan amount
m	Number of compounding intervals per year
m	Exponent in Equation 23.62
N	System life, number of years
N_2	Doubling time, number of years
N_{dep}	Depreciation period, year
N_l	Loan period, year
N_p	Payback time, year
n	Year
P	Present cash amount
$(P/F, r, n)$	Present worth factor
P_{int}	Present value of interest payments
P_{max}	Peak demand, kW
P_n	Principal during nth payment period
p_{dem}	Demand charge, \$/(kW·month)
p_e	Energy price
$\bar{p}_e$	Levelized energy price
p_{ins}	Price of insulation, \$/m^3
Q	Energy consumption, J (Btu)
r	Nominal growth rate
r_o	Real growth rate given by Equation 23.5
$r_{d,e}$	Real energy growth rate given by Equation 23.20
$r_{d,l}$	Growth rate defined by Equation 23.39
$r_{d,M}$	$(r_d - r_M)/(1 + r_M)$
r_d	Market discount rate
r_e	Market energy price escalation rate
r	Interest rate
r_{int}	Interest rate
r_{inf}	General inflation rate
r_l	Market loan interest rate
r_M	Market escalation rate for maintenance costs
r_r	Internal rate of return
S	$=-C_{life} + C_{life,ref}$, life cycle savings
S	Size of equipment
t	Thickness of insulation, m (ft)

Greek

τ	Incremental tax rate
τ_{cred}	Tax credit

Subscripts

ann Annual growth rate
cont Continuous growth rate
cred Tax credit
dem Demand
dep Depreciation
e Energy
l Loan
0 Real rate
m Maintenance, monthly
n Index for year or payment interval
ref Reference

23.1 Comparing Present and Future Costs

We do not live in paradise, and our resources are limited. Therefore, it behooves us to try to reduce the cost of heating and cooling to a minimum—subject, of course, to the constraint of providing the desired indoor environment and services. While capital costs and operating costs are readily stated in financial terms, other factors, such as comfort, convenience, and aesthetics, may be difficult or impossible to quantify. Furthermore, there is uncertainty: future energy prices, future rental values, future equipment performance, and future uses of a building are all uncertain.

As a way around these difficulties, it is best to approach the design optimization process in the following manner. First, one evaluates the total cost for each proposed design or design variation by properly combining all capital and operating costs. Then, knowing the cost of each design, one can select the "best," much like selecting the best product in a store where each product carries a price tag. Proceeding in this way, one separates the factors that can be quantified unambiguously (i.e., the calculation of the price tag) from those that are less tangible (e.g., aesthetics). The calculation of the price tag is the essence of engineering economics; it forms the main part of this chapter. Optimization and some effects of uncertainty are addressed at the end.

23.1.1 Effect of Time on the Value of Money

Before one can compare first costs (i.e., capital costs) and operating costs, one must apply a correction because a dollar (or any other currency unit) to be paid in the future does not have the same value as a dollar today. This time dependence of money is due to two, quite different causes. The first is *inflation*, the well-known and ever-present erosion of the value of our currency. The second is the *time value* of money and reflects the fact that a dollar today can buy goods to be enjoyed immediately, or it can be invested to increase its value by profit or interest. Thus, a dollar that becomes available in the future is less desirable than a dollar today; its value must be discounted. This is true even if there is no inflation. Both inflation and discounting are characterized in terms of annual rates.

Let us begin with inflation. To avoid confusion, it is advisable to add subscripts to the currency signs, indicating the year in which the currency is specified. For example, during the mid-1980s, the inflation rate r_{inf} in Western industrial countries was around $r_{inf} = 4\%$. Thus, a dollar bill in 1986 is worth only 1/(1 + 0.04) as much as the same dollar bill 1 year before:

$$\$_{1986}1.00 = \$_{1985}\frac{1}{1+r_{inf}} = \$_{1985}\frac{1}{1+0.04} = \$_{1985}0.96$$

Actually, the definition and measure of the inflation rate are not without ambiguities since different prices escalate at different rates and the inflation rate depends on the mix of goods assumed. Probably, the most common measure is the *consumer price index* (CPI), an index that has arbitrarily been set at 100 during 1983–1984. Its evolution is shown in Figure 23.1a, along with another index of interest to the HVAC designer, namely, the construction cost index provided by the *Engineering News Record* (Figure 23.1b). In terms of the CPI, the average inflation rate from year *ref* to year *ref* + *n* is given by*

$$(1+r_{inf})^n = \frac{CPI_{ref+n}}{CPI_{ref}} \tag{23.1}$$

Suppose $\$_{1985}1.00$ has been invested at an interest rate $r_{int} = 10\%$, the *nominal* or *market* rate, as usually quoted by financial institutions. Then after 1 year, this dollar has grown to $\$_{1986}1.10$, but it is worth only $\$_{1985}1.10/1.04 = \$_{1985}1.06$. To show the increase in the real value, it is convenient to define the real interest rate r_{int0} by the relation

$$1+r_{int0} = \frac{1+r_{int}}{1+r_{inf}} \tag{23.2}$$

or

$$r_{int0} = \frac{r_{int}-r_{inf}}{1+r_{inf}} \tag{23.3}$$

The simplest way of dealing with inflation is to eliminate it from the analysis right at the start by using *constant currency* and expressing all growth rates (interest,

* For simplicity we write the equations as if all growth rates were constant. Otherwise, the factor $(1 + r)^n$ would have to be replaced by the product of factors for each year $(1+r_1)(1+r_2)\cdots(1+r_n)$. Such a generalization is straightforward but tedious, and of dubious value in practice; it is uncertain enough to predict average trends without trying to guess a detailed scenario.

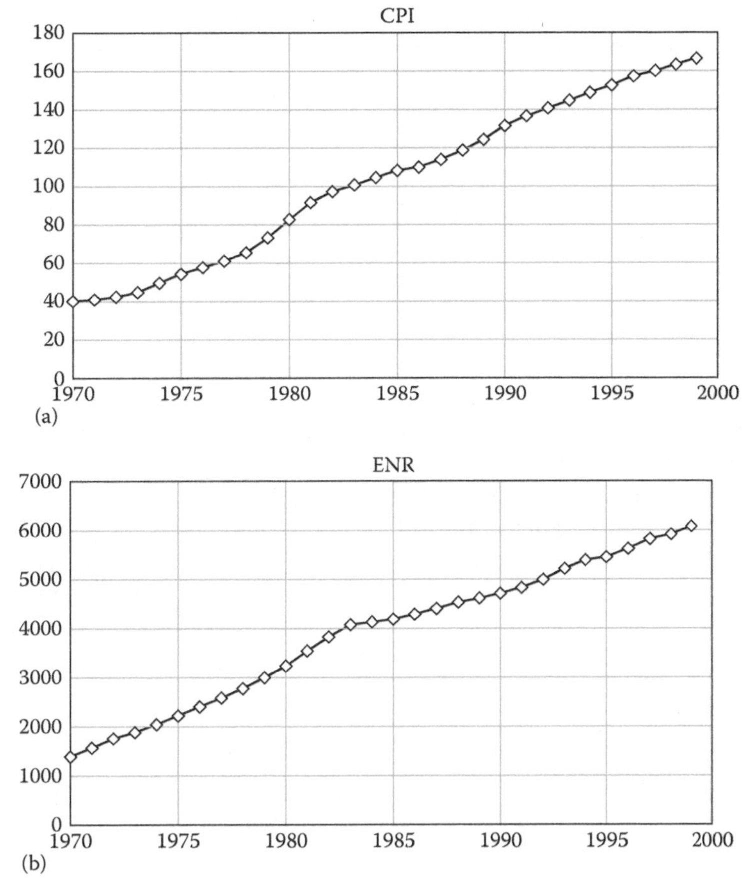

FIGURE 23.1
History of various cost indices: (a) CPI = consumer price index. (From ftp://ftp.bls.gov/pub/special.requests/cpi/cpiai.txt.) (b) ENR = *Engineering News Record* construction cost index.

energy price escalation, etc.) as real rates, relative to constant currency. After all, one is concerned about the real value of cash flows, not about their nominal values in a currency eroded by inflation. Constant currency is obtained by expressing the *current* or *inflating* currency of each year (i.e., the nominal value of the currency) in terms of equivalent currency of an arbitrarily chosen reference year *ref*. Thus, the current dollar of year *ref* + *n* has a constant-dollar value of

$$\$_{ref} = \frac{\$_{ref+n}}{(1+r_{inf})^n} \tag{23.4}$$

A *real growth rate* r_0 is related to the *nominal growth rate r* in a way analogous to Equation 23.2:

$$r_0 = \frac{r - r_{inf}}{1 + r_{inf}} \tag{23.5}$$

For low inflation rates, one can use the approximation

$$r_0 \approx r - r_{inf} \quad \text{if } r_{inf} \text{ small} \tag{23.6}$$

As proven in Section 23.2.6, an analysis in terms of constant currency and real rates is exactly equivalent to one with inflating currency and nominal rates, if the investment is paid out of equity (i.e., without loan) and without tax deduction for depreciation or interest. Slight real differences between the two approaches can arise from the formulas for depreciation and for loan payments (in the United States, loan payments are usually arranged to have fixed amounts in current currency, and the real value of annual loan payments differs between the two approaches). Therefore, the inflating-dollar approach is commonly chosen in the U.S. business world.

However, when the constant-dollar approach is correct, it offers several advantages. Having one variable less, it is simpler and clearer. What is more important is that the long-term trends of real growth rates are fairly well known even if the inflation rate turns out to be erratic. For example, from 1955 to 1980 the real interest rate on high-quality corporate bonds consistently hovered about 2.2% despite large fluctuations of inflation (Jones, 1982), while the high real interest rates of the 1980s have probably been a short-term anomaly.

Riskier investments, such as the stock market, may promise higher returns, but they, too, tend to be more constant in constant currency.

The historic evolution of energy prices in terms of real currency (or adjusted for inflation against constant 2014 dollars) and market or nominal price (which is subject to inflation) is depicted in Figure 23.2. We note that the two prices track each other fairly well for imported crude oil, while those for electricity price suggest a significant drop in real price over the years (Figure 23.2b). Another important observation is that real prices are much less volatile than market prices. For example, the market price (= price in inflating currency) of crude oil reached a peak of \$36/barrel in 1981, which is 10 times higher than the market price during the 1960s, while in terms of constant currency the price increase over the same period was only a factor of 5. Crude oil during the oil crises is, of course, an example of extreme price fluctuations. For other goods, the price in constant currency is far more stable (it would be exactly constant in the absence of relative price shifts among different goods). Therefore, it is instructive to think in terms of real rates and real currency.

Example 23.1: Escalation Rates for Residential Electricity Prices

Find the nominal and real escalation rates for residential electricity prices between 1985 and 2010.

Given: Data in Figure 23.2b, the real prices are in 2014 dollars.

Find: Real growth rate r_0 and nominal growth rate r

Solution

In 1985, the nominal (or market) and real (based on 2014) prices were $p_{1985} = 7.79$ and 17.21 ¢/kWh, respectively. In 2010, the prices were $p_{2010} = 11.54$ and 12.57¢/kWh, respectively. The number of

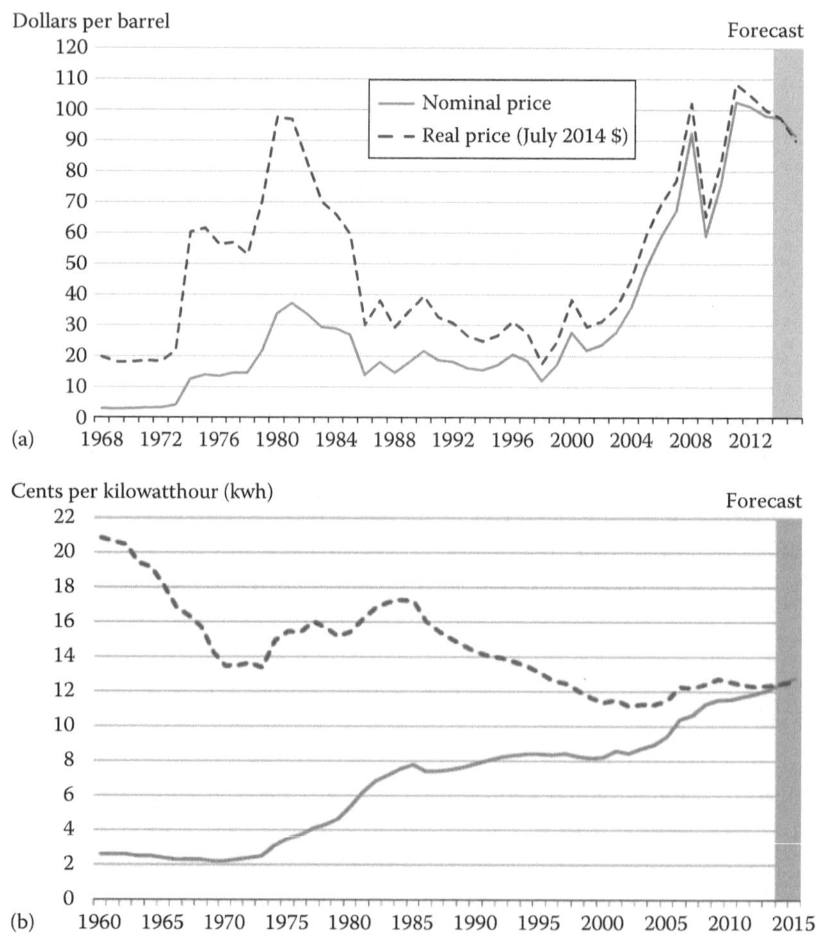

FIGURE 23.2
Annual energy prices, in real or constant July 2014 dollars (dotted line) and in nominal market prices (solid line). (a) Imported crude oil. (b) Residential electricity price. (From EIA, Short-term energy outlook, U.S. Energy Information Administration, Washington, DC, July 2014, http://www.eia.gov/forecasts/steo/realprices/.)

years is $n = 2010 - 1985 = 25$. From Equation 23.1, the real growth rate is given by

$$r_0 = -1 + \sqrt[n]{\frac{p_{2010}}{p_{1985}}}$$

$$= -1 + \sqrt[25]{\frac{12.57}{17.21}} = -1 + (0.7304)^{1/25} = -0.0125$$

The nominal or market growth rate is

$$r = -1 + \sqrt[25]{\frac{11.54}{7.79}} = -1 + (1.4814)^{1/25} = 0.0158$$

Comments

This highlights the importance of distinguishing between real and nominal growth rates. While the apparent market price has grown by 1.58%/year, the real price has actually decreased by 1.25%/year; the inflation rate averaged about 2.9% during this period.

23.1.2 Discounting of Future Cash Flows

As mentioned earlier, even if there were no inflation, a future cash amount F is not equal to its *present value* P; it must be discounted. The relation between P and its future value F_n in n years from now is given by the *discount rate* r_d, defined as

$$P = \frac{F_n}{(1+r_d)^n} \tag{23.7}$$

The higher the discount rate, the lower the present value of future transactions.

To determine the appropriate value of the discount rate, one has to ask: At what value of r_d is one indifferent between an amount P today and an amount $F_1 = P/(1 + r_d)$ a year from now? That depends on the circumstances and on individual preferences. Consider a consumer who would put his money in a savings account with 5% interest. His discount rate is 5%, because by putting the $1000 into this account, he, in fact, accepts the alternative of (1 + 5%) × $1000 a year from now. If instead he used it to pay off a car loan at 10%, then his discount rate would be 10%; paying off the loan is like putting the money into a savings account that pays at the loan interest rate. If the money allowed him to avoid an emergency loan at 20%, then his discount rate would be 20%. At the other extreme, if he hid the money in his mattress, his discount rate would be zero.

The situation becomes more complex when there are several different investment possibilities offering different returns at different risks, such as savings accounts, stocks, real estate, or a new business venture. By and large, if one wants the prospect of a higher rate of return, one has to accept a higher risk. Thus, as a more general rule, we can say that the appropriate discount rate for the analysis of an investment is the rate of return on alternative investments of comparable risk. In practice, that is sometimes quite difficult to determine, and it may be desirable to have an evaluation criterion that bypasses the need to choose a discount rate. Such a criterion is obtained by calculating the profitability of an investment in terms of an unspecified discount rate and then solving for the value of the rate at which the profitability goes to zero. That method, called the "internal rate of return method," will be explained in Section 23.3.2.

Just as with other growth rates, one can specify the discount rate with or without inflation. If F_n is given in terms of constant currency, designated as F_{n0}, then it must be discounted with the real discount rate r_{d0}. The latter is, of course, related to the market discount rate r_d by

$$r_{d0} = \frac{r_d - r_{inf}}{1 + r_{inf}} \tag{23.8}$$

according to Equation 23.5. Present values can be calculated with real rates and real currency or with market rates and inflating currency; the result is readily seen to be the same because multiplying numerator and denominator of Equation 23.7 by $(1 + r_{inf})^n$, one obtains

$$P = \frac{F_n}{(1+r_d)^n} = \frac{F_n(1+r_{inf})^n}{(1+r_{inf})^n(1+r_d)^n}$$

which is equal to

$$P = \frac{F_{n0}}{(1+r_{d0})^n}$$

since from Equation 23.4

$$F_{n0} = \frac{F_n}{(1+r_{inf})^n} \tag{23.9}$$

The ratio P/F_n of present and future value is called the "present worth factor." We shall designate it with the mnemonic notation

$$(P/F, r, n) = \frac{P}{F_n} = (1+r)^{-n} \tag{23.10}$$

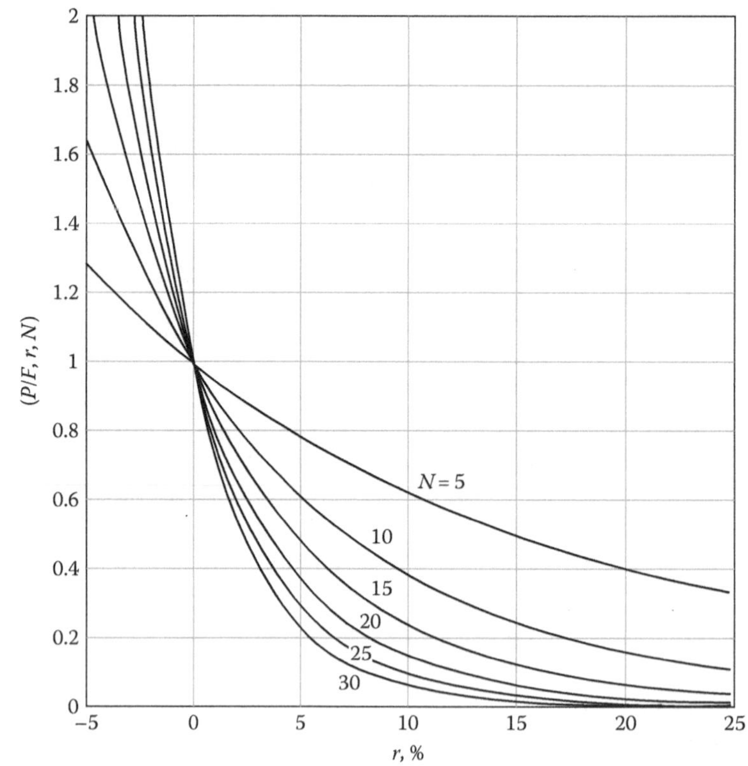

FIGURE 23.3
The present worth factor $(P/F, r, N)$ as function of rate r and number of years N.

It is plotted in Figure 23.3. Its inverse

$$(F/P, r, n) = \frac{1}{(P/F, r, n)} \tag{23.11}$$

is called the "compound amount factor." These factors are the basic tools for comparing cash flows at different times. Note that we have chosen the so-called end-of-year convention by designating F_n as the value at the end of the nth year. Also, we have assumed annual intervals, which are generally an adequate time step for engineering economic analysis; accountants, by contrast, tend to work with monthly intervals, corresponding to the way most regular bills are paid. The basic formulas are the same, but the numerical results differ because of differences in the compounding of interest; this point will be explained more fully in Section 23.1.4 when we pass to the continuous limit by letting the time step approach zero.

Example 23.2: Real Discount Rate of Energy Savings

What might be an appropriate discount rate for analyzing the energy savings from a proposed new cogeneration plant for a university campus? Consider the fact that from 1990 to 2010 the endowment of the university has grown by a factor 8 (current dollars) due to profits from investments.

Given:

Growth factor in current dollars = 8.0

Increase in CPI $= \frac{218.1}{130.7} = 1.67$, from Figure 23.1a over $N = 20$ years

Find: Real discount rate r_{d0}

Solution

We take the real growth factor, 8.0/1.67, and set it equal to $(1 + r_{d0})^N$. The result is $r_{d0} = 21.3\%$.

Comments

Choosing a discount rate is not without pitfalls. For this example, the comparison with the real growth of other long-term investments seems appropriate; of course, there is no guarantee that the endowment will continue growing at the same real rate in the future.

23.1.3 Equivalent Cash Flows and Levelizing

It will be convenient to express payments that are irregular or variable as equivalent equal payments in regular intervals; in other words, one replaces nonuniform series by equivalent uniform series. We shall refer to this technique as *levelizing*. It is useful because regularity facilitates understanding and planning. To develop the formulas, let us calculate the present value P of a series of N equal annual payments A. If the first payment occurs at the end of the first year, its present value is $A/(1 + r_d)$. For the second year, it is $A/(1 + r_d)^2$. Adding all the present values from years 1 to N, we find the total present value

$$P = \frac{A}{1+r_d} + \frac{A}{(1+r_d)^2} + \cdots + \frac{A}{(1+r_d)^N} \tag{23.12}$$

This is a simple geometric series, and the result is readily summed to

$$P = A\frac{1-(1+r_d)^{-N}}{r_d} \quad \text{for } r_d \neq 0 \tag{23.13}$$

For zero discount rate, this equation is indeterminate, but its limit $r_d \to 0$ is A/N, reflecting the fact that the N present values all become equal to A in that case. Analogous to the notation for the present worth factor, we designate the ratio of A and P by

$$(A/P, r_d, N) = \begin{cases} \dfrac{r_d}{1-(1+r_d)^{-N}} & \text{for } r_d \neq 0 \\ \dfrac{1}{N} & \text{for } r_d = 0 \end{cases} \tag{23.14}$$

It is called the "capital recovery factor," and we have plotted it in Figure 23.4. For the limit of long life $N \to \infty$, it is worth noting that $(A/P, r_d, N) \to r_d$ if $r_d > 0$.

The inverse is known as the *series present worth factor* since P is the present value of a series of equal payments A.

With the help of the present worth factor and capital recovery factor, any single expense C_n that occurs in year n, for instance, a major repair, can be expressed as an equivalent annual expense A that is constant during each of the N

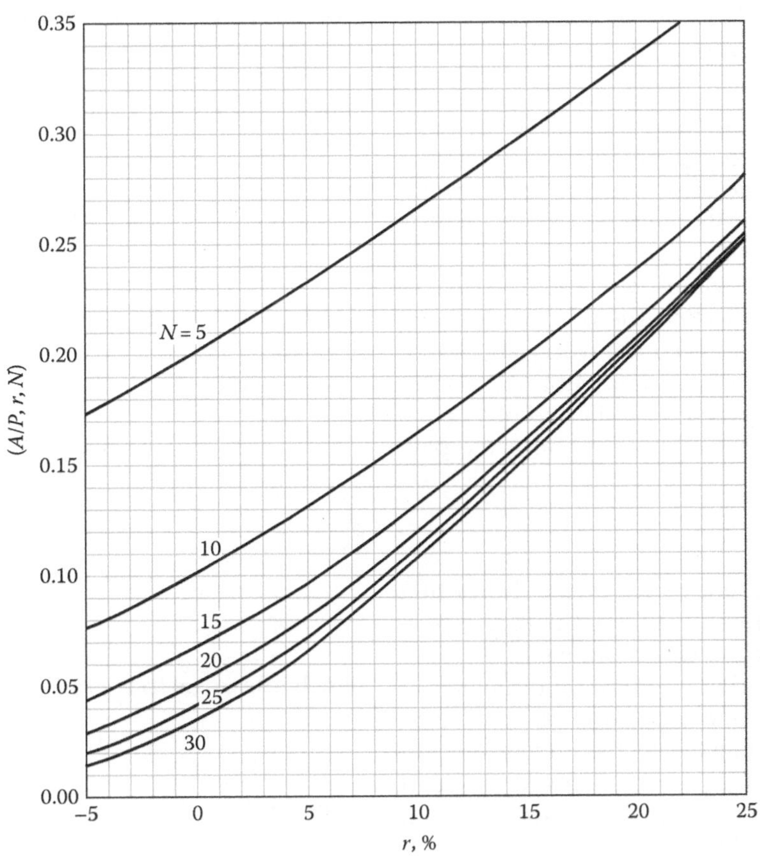

FIGURE 23.4
The capital recovery factor $(A/P, r, N)$ as function of rate r and number of years N.

years of the life of the system. The present value of C_n is $P = (P/F, r_d, n)C_n$, and the corresponding annual cost is

$$A = (A/P, r_d, N)(P/F, r_d, n)C_n$$

$$= \frac{r_d}{1-(1+r_d)^{-N}}(1+r_d)^{-n}C_n \tag{23.15}$$

Example 23.3: Annual Levelized Cost of Salvage

A system has salvage value of $1000 at the end of its useful life of $N = 20$ years. What is the equivalent levelized annual value if the discount rate is 8%?

Given: C_{20} = $1000, $N = n = 20$, and $r_d = 0.08$

Find: A

Lookup values:
$(A/P, r_d, N) = 0.1019$ for Figure 23.4 or Equation 23.14
$(P/F, r_d, n) = 0.2145$ from Figure 23.3 or Equation 23.10

Solution

Insert into Equation 23.15

$$A = (A/P, r_d, N)(P/F, r_d, n)C_{20}$$
$$= 0.1019 \times 0.2145 \times 1000\$/\text{year} = 21.86\$/\text{year}$$

A very important application of the capital recovery factor is the calculation of loan payments. In principle, a loan could be repaid according to any arbitrary schedule, but in practice the most common arrangement is based on constant payments in regular intervals. The portion of A due to interest varies, in a way to be calculated in Section 23.2.2; but to find the relation between A and the loan amount L, we need not worry about that issue. Let us first consider a loan of amount L_n that is to be repaid with a single payment F_n at the end of n years. With n years of interest, at loan interest rate r_l, the payment must be

$$F_n = L_n(1+r_l)^n$$

Comparison with the present worth factor shows that the loan amount is the present value of the future payment F_n, discounted at the loan interest rate.

A loan that is to be repaid in N equal installments can be considered as the sum of N loans, the nth loan to be repaid in a single installment A at the end of the nth year. Discounting each of these payments at the loan interest rate and adding them, we find the total present value; it is equal to the total loan amount

$$L = P = \frac{A}{1+r_l} + \frac{A}{(1+r_l)^2} + \cdots + \frac{A}{(1+r_l)^N} \tag{23.16}$$

This is just the series of the capital recovery factor. Hence, the relation between annual loan payment A and loan amount L is

$$A = (A/P, r_l, N)L \tag{23.17}$$

Now, the reason for the term "capital recovery factor" becomes clear: It is the rate at which a bank recovers its investment in a loan.

Example 23.4: Annual Payments of a Mortgage

A home buyer obtains a mortgage of $100,000 at an interest rate of 8% over 20 years. What are the annual payments?

Given: L = $100,000, r_l = 8%, and N = 20 year

Find: A

Solution

From Figure 23.4, the capital recovery factor is 0.1019, and the annual payments are $10,190, approximately one-tenth of the loan amount.

Some payments increase or decrease at a constant annual rate. It is convenient to replace a growing or diminishing cost by an equivalent constant or *levelized* cost. Suppose the price of energy is p_e at the start of the first year, escalating at an annual rate r_e while the discount rate is r_d. If the annual energy consumption Q is constant, then the present value of all the energy bills during the N years of system life is

$$P_e = Qp_e\left[\left(\frac{1+r_e}{1+r_d}\right)^1 + \left(\frac{1+r_e}{1+r_d}\right)^2 + \cdots + \left(\frac{1+r_e}{1+r_d}\right)^N\right] \tag{23.18}$$

(assuming the end-of-year convention described earlier). As in Equation 23.4, we introduce a new variable $r_{d,e}$, defined by

$$1 + r_{d,e} = \frac{1+r_d}{1+r_e} \tag{23.19}$$

or

$$r_{d,e} = \frac{r_d - r_e}{1+r_e} \quad (\approx r_d - r_e \quad \text{if } r_e \ll 1) \tag{23.20}$$

which allows us to write P_e as

$$P_e = (P/A, r_{d,e}, N)Qp_e \tag{23.21}$$

Since $(A/P, r_d, N)$ is the inverse of $(A/P, r_d, N)$, we can write this as

$$P_e = (P/A, r_d, N)Q\left[\frac{(A/P, r_d, N)}{(A/P, r_{d,e}, N)}p_e\right]$$

If the quantity in brackets were the price, this would be just the formula without escalation. Let us call this quantity the *levelized* energy price $\bar{p}_e$, where

$$\bar{p}_e = \frac{(A/P, r_d, N)}{(A/P, r_{d,e}, N)} p_e \tag{23.22}$$

It allows us to calculate the costs as if there were no escalation. Levelized quantities can fill a gap in our intuition that is ill prepared to gauge the effects of exponential growth over an extended period. The levelizing factor

$$\text{Levelizing factor} = \frac{(A/P, r_d, N)}{(A/P, r_{d,e}, N)} \tag{23.23}$$

tells us in effect the average of a quantity that changes exponentially at a rate r_e while being discounted at a rate r_d, over a lifetime of N years. It is plotted in Figure 23.5 for a wide range of the parameters.

Example 23.5: Levelized Fuel Price

The price of fuel is p_e = \$5/GJ at the start of the first year, growing at a rate r_e = 4%, while the discount rate is r_d = 6%. What is the equivalent levelized price over N = 20 years?

Given: p_e = \$5/GJ, r_e = 4%, r_d = 6%, and N = 20 year

Find: $\bar{P}_e$

Solution

From Figure 23.5, the levelizing factor is 1.44. Hence, the levelized fuel price is $\bar{p}_e = 1.44 \times \$5/\text{GJ} = \$7.20\ \text{GJ}$.

Several features may be noted in Figure 23.5. First, the levelizing factor increases with cost escalation r_e, being unity if r_e = 0. Second, for a given escalation rate, the levelizing factor decreases as the discount rate increases, reflecting the fact that a high discount rate deemphasizes the influence of high costs in the future.

23.1.4 Discrete and Continuous Cash Flows

The aforementioned formulas suppose that all costs and revenues occur in discrete intervals. That is common engineering practice, consistent with the fact that bills are paid in discrete installments. Thus, growth rates are quoted as annual changes even if growth is continuous. It is instructive to consider the continuous case.

Let us establish the connection between continuous and discrete growth by way of an apocryphal story about the discovery of *e*, the basis of natural logarithms. Before the days of compound interest, a mathematician who was an inveterate penny pincher thought about the possibilities of increasing the interest he earned on his money. He realized that if the bank gives interest at a rate of r per year, he could get even more by taking the money out after half a year and reinvesting it to earn interest on the interest as well. With m such compounding intervals per year, the money would grow by a factor

$$\left(1 + \frac{r}{m}\right)^m$$

and the larger m, the larger this factor. Of course, he looked at the limit $m \to \infty$ and found the result

$$\lim_{m\to\infty}\left(1 + \frac{r}{m}\right)^m = e^r \quad \text{with } e = 2.71828\ldots \tag{23.24}$$

At the end of 1 year, the growth factor is $1 + r_{ann}$ with annual compounding at a rate r_{ann}, while with continuous compounding at a rate r_{cont}, the growth factor is exp r_{cont}. If the two growth factors are to be the same, the growth rates must be related by

$$1 + r_{ann} = \exp(r_{cont}) \tag{23.25}$$

With this replacement of rates, the continuous formulas in Table 23.1 yield the same results as the discrete ones. Similarly, with m compounding intervals at rate r_m continuous compounding is equivalent to annual compounding if one takes

$$1 + r_{ann} = \left(1 + \frac{r_m}{m}\right)^m \tag{23.26}$$

Example 23.6: Growth Rates from Compounding

A bank quotes a nominal interest rate of 10% (i.e., annual growth without compounding). What is the equivalent annual growth rate with monthly, daily, and continuous compounding?

Solution

For monthly compounding take m = 12 in Equation 23.26, with the result

$$r_{12} = \left(1 + \frac{0.1}{12}\right)^{12} - 1 = 0.104713$$

With daily compounding we have

$$r_{365} = \left(1 + \frac{0.1}{365}\right)^{365} - 1 = 0.105155$$

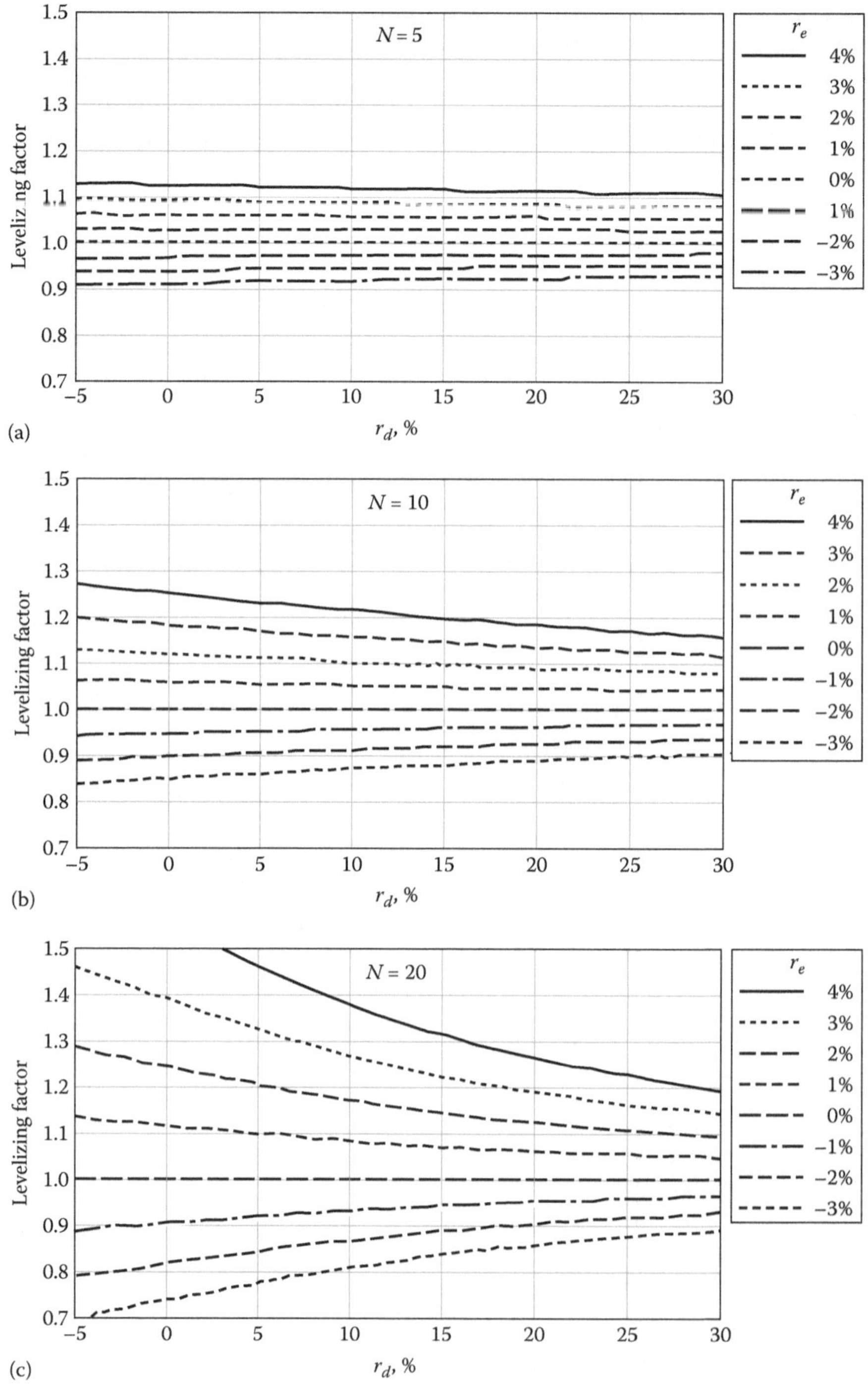

FIGURE 23.5
Levelizing factor $(A/P, r_d, N)/(A/P, r_{d,e}, N)$/as function of r_d and r_e for (a) $N = 5$ years, (b) $N = 10$ years, and (c) $N = 20$ years.

and with continuous compounding

$$r_{cont} = \exp(0.1) - 1 = 0.105171$$

Comment

Beyond monthly compounding the differences are very small.

For small rates, the first three terms in the series expansion of the exponential give an approximation

$$r_{ann} \approx r_{cont}\left(1 + \frac{r_{cont}}{2}\right)$$

which is convenient if one does not have a calculator at hand.

TABLE 23.1

Discrete and Continuous Formulas for Economic Analysis, with Growth Rate r and Time Horizon N

Quantity Known	Quantity to Be Found	Factor	Expression for Discrete Analysis	Expression for Continuous Analysis
P	F	$(F/P, r, N)$	$(1+r)^N$	$\exp(rN)$
F	P	$(P/F, r, N)$	$(1+r)^{-N}$	$\exp(-rN)$
P	A	$(A/P, r, N)$	$\frac{r}{1-(1+r)^{-N}}$	$\frac{\exp(r)-1}{1-\exp(-rN)}$
A	P	$(P/A, r, N)$	$\frac{1-(1+r)^{-N}}{r}$	$\frac{1-\exp(-rN)}{\exp(r)-1}$

The rates for the discrete and continuous formulas are related by Equation 23.25.

23.1.5 Rule of 70 for Doubling Times

Most of us do not have a good intuition for exponential growth. As a helpful tool, we present therefore the rule of 70 for doubling times. The doubling time N_2 is related to the continuous growth rate r_{cont} by

$$2 = \exp(N_2\, r_{cont}) \tag{23.27}$$

Solving the exponential relation for N_2, we obtain

$$N_2 = \frac{\ln(2)}{r_{cont}} = \frac{0.693\ldots}{r_{cont}}$$

The product of doubling time and growth rate in units of percent is very close to 70 years:

$$N_2 r_{cont} \times 100 = 69.3\ldots \text{ years} \approx 70 \text{ years} \tag{23.28}$$

In terms of annual rates the relation would be

$$N_2 = \frac{\ln 2}{\ln(1+r_{ann})}$$

numerically close to Equation 23.28 for small rates, but less convenient.

Example 23.7: Doubling Time

Population growth rates average around 2% for the world as a whole and reach 4% in certain countries. What are the corresponding doubling times?

Solution

$$\text{Doubling time} = \frac{70}{2} = 35 \text{ years for the world}$$

$$\frac{70}{4} = 17.5 \text{ years for countries with 4\% growth}$$

Example 23.8: Assessing Validity of an Economic Claim

A consultant presents an economic analysis of an energy investment with $N = 20$ years, assuming a 10% escalation rate for energy prices without stating the inflation rate. Is that reasonable?

Solution

A growth rate of 10% implies a doubling time of 7 years. There would be almost three doublings in 20 years, with a final energy price almost eight times the original. In constant dollars that would clearly be an absurd hypothesis. A totally different conclusion emerges if the inflation were 6%: then the real growth rate would be only 10% − 6% = 4%, and the doubling time 17.5 years in constant currency; although extreme, that is not inconceivable given our experience since the 1973 oil shock.

Comment

There are two lessons:

1. Never state the growth rate or discount rate without indicating the corresponding inflation!
2. Be careful about assuming large growth rates over long time horizons! Use the rule of 70 to check whether the implications for the end of the time horizon make sense.

23.2 Life Cycle Cost

23.2.1 Cost Components

A rational decision is based on the true total cost. That is the sum of the present values of all cost components, and it is called "life cycle cost." The cost components relevant for the HVAC engineer are as follows: (1) capital cost (= total initial investment), (2) energy costs, (3) costs for maintenance including major repairs, (4) resale value, (5) insurance, and (6) taxes.

There is some arbitrariness in this assignment of categories. One could make a separate category for repairs, or one could include energy among operation and maintenance (O&M) cost, as it is done in some industries. There is, however, a good reason for keeping energy apart. In buildings, energy costs dominate the other O&M costs, and they can grow at a different rate. Electric rates usually contain charges for peak demand in addition to charges for energy (see Section 23.2.4). As a general rule, if an item is important, it merits separate treatment.

Rental income needs to be included if one wants to evaluate the profitability of the building or if one wants to compare design options that would affect the rent. It can be left out of the picture if one is only concerned with comparing design options that do not differ in their effect on rental income. Likewise for cleaning, security, and fire protection. *Quite generally, when one is comparing two options, there is no need to include terms that would be the same for each.* For instance, when choosing between two chillers, one can focus on the costs associated directly with the chillers (capital cost, energy, maintenance), without worrying about the heating system if that is not affected. In some cases, it becomes necessary to account for the effects of taxes, due to tax deductions for interest payments and depreciation; hence, we discuss these items first, before presenting the equation for the complete system cost.

23.2.2 Principal and Interest

In the United States, many types of interest payments are deductible from the income tax, while payments for the reimbursement of the loan are not. A tax-paying investor therefore needs to know what fraction of a loan payment is due to interest. As explained in Section 23.1.3, we assume that a loan of duration N_l is repaid in N_l regular and equal payments A. (In this chapter, we take N and N_l in years, but the formulas are valid for any choice of units, and for billing purposes the month is frequently chosen as payment period; slight numerical differences in the payments are due to compound interest.) Consider the nth payment, and let I_n = interest and P_n = principal (= loan reimbursement); their sum A is constant

$$I_n + P_n = A \tag{23.29}$$

Up to this point $n - 1$ payments have been made, and so the debt remaining (on a loan of amount L) is

$$\text{Remaining debt} = L - P_1 - P_2 - \cdots - P_{n-1} \tag{23.30}$$

At a loan interest rate r_l the interest for the nth period is

$$I_n = r_l(L - P_1 - P_2 - \cdots - P_{n-1}) \tag{23.31}$$

Comparing I_{n+1} with I_n, one finds

$$I_{n+1} - I_n = -r_l P_n \tag{23.32}$$

By means of Equation 23.29 one can eliminate P_n with the result

$$I_{n+1} = (1 + r_l)I_n - r_l A \tag{23.33}$$

This recursion relation has the solution

$$I_n = (1 + r_l)^{n-1} r_l L + [1 - (1 + r_l)^{n-1}]A \tag{23.34}$$

as can readily be proved by mathematical induction. Since A and L are related by $A = (A/P, r_l, N_l)L$, where N_l is the duration of the loan and $(A/P, r_l, N_l)$ is the capital recovery factor of Equation 23.14, this can be rewritten in the form

$$\frac{I_n}{A} = 1 - (1 + r_l)^{n-1-N_l} \tag{23.35}$$

It is worth noting that the period n enters only in the combination $(n - N_l)$. This implies that the fractional allocation to principal and interest depends only on the number of periods $(n - N_l)$ left in the loan, not on the original life of the loan. A loan has no memory, so to speak.

In general, the loan interest rate r_l differs from the discount rate r used for the economic analysis, and the loan life N_l may be different from the system life N. Inserting $A = (A/P, r_l, N_l)$ into Equation 23.35, we find that the interest payment I is related to loan amount L by

$$I_n = [1 - (1 + r_l)^{n-1-N_l}](A/P, r_l, N_l)L \tag{23.36}$$

The present value P_{int} of the total interest payments is found by discounting each I with the discount rate r and summing over n:

$$P_{int} = \sum_{n=1}^{N_l} \frac{1 - (1 + r_l)^{n-1-N_l}}{(1 + r_d)^n} (A/P, r_l, N_l)L \tag{23.37}$$

By using the formula for geometric series, this can be transformed to

$$P_{int} = \left\{ \frac{(A/P, r_l, N_l)}{(A/P, r_d, N_l)} - \frac{(A/P, r_l, N_l) - r_l}{(1 + r_l)(A/P, r_{dl}, N_l)} \right\} L \tag{23.38}$$

with

$$r_{dl} = \frac{r_d - r_l}{1+r_l} \tag{23.39}$$

If the incremental tax rate is τ, the total tax payments are reduced by τP_{int} (assuming a constant tax rate; otherwise, the tax rate would have to be included in the summation).

Example 23.9: Tax-Deductible Interest Payments

A solar water heating system costing \$2000 is financed with a 5-year loan at r_l = 8%. The tax rate is τ = 40%. How much is the tax deduction for interest worth if the discount rate is r_d = 8%?

Given: L = \$2000, r_l = 8%, r_d = 8%, and τ = 40%

Find: P_{int}

Lookup values: $(A/P, r_{dl}, N_l) = 0.20$ and $(A/P, r_l, N_l) = 0.2505$ from Equation 23.14

Solution

We have $r_{dl} = 0$, since $r_l = r_d$. Also $(A/P, r_{dl}, N_l) = 0.20$ and $(A/P, r_l, N_l) = 0.2505 = (A/P, r_d, N_l)$. Thus, the present value of the interest payments is, from Equation 23.38,

$$P_{int} = \left(1 - \frac{0.2505 - 0.08}{1.08 \times 0.20}\right)(\$2000) = \$421$$

At the stated tax rate that is worth 0.40 × \$421 = \$168.

23.2.3 Depreciation and Tax Credit

U.S. tax law allows business property to be depreciated. This means that for tax purposes the value of the property is assumed to decrease by a certain amount each year, and this decrease is treated as a tax-deductible loss. For the economic analysis, one needs to express the depreciation as an equivalent present value. The details of the depreciation schedule have been changing with the tax reform of the 1980s. Instead of trying to present the full details, which can be found in the publications of the Internal Revenue Service, we merely note the general features. In any year n, a certain fraction $f_{dep,n}$ of the capital cost (minus salvage value) can be depreciated. For example, in the simple case of straight-line depreciation over N_{dep} years

$$f_{dep,n} = \frac{1}{N_{dep}} \tag{23.40}$$

To obtain the total present value, one multiplies by the present worth factor and sums over all years from 1 to N:

$$f_{dep} = \sum_{n=1}^{N_{dep}} f_{dep,n}(P/F, r_d, n) \tag{23.41}$$

For straight-line depreciation, the sum is

$$f_{dep} = \frac{(P/A, r_d, N_{dep})}{N_{dep}} \tag{23.42}$$

A further feature of some tax laws is the tax credit. For instance, in the United States for several years around 1980 and again from 2005 or so onward, tax credits were granted for certain renewable energy systems. If the tax credit rate is τ_{cred} for an investment C_{cap}, the tax liability is reduced by $\tau_{cred} C_{cap}$.

Example 23.10: Present Value of Depreciation

A machine costs \$10,000 and is depreciated with straight-line depreciation over 5 years, the salvage value after 5 years being \$1,000. Find the present value of the tax deduction for depreciation if the incremental tax rate is τ = 40% and the discount rate r_d = 15%.

Given: C_{cap} = 10 k\$
C_{salv} = 1 k\$
N = 5 years
τ = 0.4
r_d = 0.15

Find: $\tau \times f_{dep}(C_{cap} - C_{salv})$

Lookup values:

$$f_{dep} = \frac{(P/A, 0.15, 5\text{ years})}{5\text{ years}} = \frac{1/0.2983}{5} = \frac{3.3523}{5}$$
$$= 0.6705 \quad \text{from Equation 23.42}$$

Solution

For tax purposes, the net amount to be depreciated is the difference

$$C_{cap} - C_{salv} = (10 - 1)\text{ k\$} = 9\text{ k\$}$$

and with straight-line depreciation $1/N_{dep} = \frac{1}{5}$ of this can be deducted from the tax each year. Thus, the annual tax is reduced by $\tau \times \frac{1}{5} \times 9$ k\$ = 0.40×1.8 k\$ = 0.72 k\$ for each of the 5 years.

The present value of this tax reduction is

$$\tau \times f_{dep}(C_{cap} - C_{salv}) = 0.40 \times 0.6705 \times 9 \text{ k\$} = 2.41 \text{ k\$}$$

Comment

The present value of the reduction would be equal to 5 × 0.72 k\$ – 3.6 k\$ if r_d were zero. The discount rate of 15% reduces the present value by almost a one-third to $0.6705 = f_{dep}$.

23.2.4 Energy and Demand Charges

The cost of producing electricity has two major components: fuel and capital (for the power plant and distribution system). To the extent that it is practical, utility companies try to base the rate schedule on their production cost. Thus, the rates for large customers contain four parts: the energy cost, a charge proportional to the peak demand, adjustments and surcharges, and taxes. Residential rates often omit the demand component because the cost of separate meters used to be considered too high. With the advent of "smart" meters, the metering landscape has undergone a dramatic shift, and several utilities now offer electric rate structures to residential customers that include a demand component as well.

For the energy charge, a number of versions are commonly used. The simplest is a *flat rate*, with a single cost per kilowatt hour consumed. In commercial and industrial rates, however, this is rarely the case. *Time-of-use rates* vary depending on the hour and type of day. Typically, the *on-peak* period occurs between 8 a.m. and 8 p.m. during weekdays, and the *off-peak* period is all other times. There may also be one or more *shoulder periods* between the on-peak and off-peak periods. The on-peak prices are usually 1.2–3 times higher than off-peak prices. *Block rates* vary the energy cost by the total amount of electricity used. For example, the first 1000 kWh may cost 6¢/kWh, the next 1000 kWh may cost 4¢/kWh, and any additional consumption would be 2¢/kWh. Note that there may be up to a half dozen blocks and that the rates do not necessarily decrease as the blocks "fill up." Furthermore, the amount paid for energy use in any given block may be adjusted by the total peak demand incurred while that block was in effect.

Time-of-use and block rates are also frequently applied to the demand component of commercial and industrial rates. For example, the first 100 kW of demand between 10 a.m. and 3 p.m. may be charged at one rate, while the next 100 kW of demand during this time interval is charged at another rate, and so on. The demand charge may also be *ratcheted;* i.e., the highest demand incurred during a given month is used to calculate the demand charge for successive months. The ratchet period is typically 6–12 months and may use some percentage of the ratchet period peak demand or the current month's peak demand, whichever is higher.

The energy and demand rates can change seasonally. Most rates change between summer and winter, but a four-season split is not uncommon. Both electricity and gas bills are also adjusted, respectively, by the *energy cost adjustment* or the *purchased gas adjustment* and by any *surcharges.* They are essentially adders to the bill to account for monthly changes in the cost of fuel or energy transport and are applied either as a percentage of the total bill or as a function of the consumption component. Finally, state and local taxes are usually applied to the entire bill. There may also be a number of riders that dictate special fees and charges based on consumption and demand characteristics.

Several utilities now offer *real-time pricing* electricity rates where the marginal costs of the utility are passed directly to the consumer. In these rates, the cost of electricity changes every hour of the day and can vary by a factor of 50 between the lowest and highest hourly costs. The highest costs occur only during 2% or 3% of the year.

As a simple example, consider a rate schedule with a monthly demand charge p_{dem} = \$10/kW and the energy charge p_e = \$0.07/kWh. A customer with monthly energy consumption Q_m and peak demand P_{max} will receive a total bill of

$$\text{Monthly bill} = Q_m p_e + P_{max} p_{dem} \tag{23.43}$$

In most cases p_e and p_{dem} depend on time of day and time of year, being higher during the system peak than off peak. In regions with extensive air conditioning, the system peak occurs in the afternoon of the hottest days. In regions with much electric heating, the peak is correlated with outdoor temperature.

Example 23.11: Electric Bill Calculation

A 100 ton electric chiller with COP = 3 is used for 8 months of the year (running at 100% capacity at least once per month during 4 months and at 50% capacity at least once per month during 4 months), and the total load is equivalent to 1000 h at peak capacity (a typical value around the belt from New York to Denver). What is the annual electricity bill if p_e = \$0.10/kWh$_e$ and p_{dem} = \$10/kW$_e$ per month?

Given: $P_{max,t}$ = 100 tons × 3.516 kW$_t$/ton, with COP = 3 kW$_t$/kW$_e$

Annual energy = P_{max} × 1000 h

Demand P_{max} for 4 months and 0.5 × P_{max} for 4 months

p_e = \$0.10/kWh$_e$

p_{dem} = \$10/kW$_e$ per month

Find: Annual bill

Solution

$$\text{Peak demand } P_{\max} = \frac{100 \text{ tons} \times 3.516 \text{ kW}_t/\text{ton}}{3 \text{ kW}_t/\text{kW}_e}$$

$$= 117.2 \text{ kW}_e$$

$$\text{Annual energy } Q = P_{\max,e} \times 1{,}000 \text{ h} = 117{,}200 \text{ kWh}_e$$

$$\begin{aligned} \text{Annual bill} &= Qp_e + P_{\max} p_{dem} \times (4 \times 1 + 4 \times 0.5) \\ &= 117{,}200 \text{ kWh}_e \times \$0.10/\text{kWh}_e \\ &\quad + 117.2 \text{ kW}_e \times \$10/\text{kW}_e \times 6 \\ &= \$11{,}720 + \$7{,}032 \\ &= \$18{,}752 \end{aligned}$$

Comments

In a real building, the precise value of the peak demand may be difficult to predict because it depends on the coincidence of the demands of individual pieces of equipment.

The total cost per kilowatthour depends on the load profile. The more uneven the profile, the higher the cost. To take an extreme example, suppose the chiller were used only 1 h/year, at full capacity. Then, with the rate structure of this example, the demand charge would be \$1172, while the energy charge would be only \$11.72. The total cost per kilowatthour would be \$1172 + 11.72 = \$1183.72 for 117.2 kWh, an effective electricity price of \$10.10/kWh. This illustrates the interest of load leveling devices, such as cool storage for electric chillers (see Section 18.9).

23.2.5 Complete Formula

The equations for a business investment can be stated in terms of before-tax cost or after-tax cost. Consider the purchase of fuel, with a market price of \$5/GJ, by a business that is subject to an income tax rate $\tau = 40\%$. Fuel, like all business expenditures, is tax deductible. And, ultimately, it is paid by profits. To purchase 1 GJ, one takes \$5 of profits, before taxes; this reduces the tax liability by \$5 × 40% = \$2, resulting in a net cost of only \$3 after taxes.

We could do the accounting before or after taxes: the former counts the cash payments and the latter their net (after-tax) values. The two modes differ by a factor $1 - \tau$, where τ is the income tax rate. For example, if the market price of fuel is \$5/GJ and the tax rate $\tau = 40\%$, then the before-tax cost of fuel is \$5/GJ and the after-tax cost is $(1 - \tau)$ (\$5/GJ) = \$3/GJ. Stated in terms of *after-tax cost*, the complete equation for the life cycle cost of an energy investment can be written in the form

$$\begin{array}{ll} C_{life} = & \\ C_{cap}\Big\{1 - f_l & \text{down payment} \\ + f_l \dfrac{(A/P, r_l, N_l)}{(A/P, r_d, N_l)} & \text{cost of loan} \\ - \tau f_l \left[\dfrac{(A/P, r_l, N_l)}{(A/P, r_d, N_l)} - \dfrac{(A/P, r_l, N_l) - r_l}{(1 + r_l)(A/P, r_{d,l}, N_l)}\right] & \text{tax deduction for interest} \\ - \tau_{cred} & \text{tax credit} \\ - \tau f_{dep}\Big\} & \text{depreciation} \\ - C_{salv}\left(\dfrac{1 + r_{inf}}{1 + r_d}\right)^N (1 - \tau) & \text{salvage} \\ + Qp_e \dfrac{1 - \tau}{(A/P, r_{d,e}, N)} & \text{cost of energy} \\ + P_{\max} p_{dem} \dfrac{1 - \tau}{(A/P, r_{d,dem}, N)} & \text{cost of demand} \\ + A_M \dfrac{1 - \tau}{(A/P, r_{d,M}, N)} & \text{cost of maintenance} \end{array} \tag{23.44}$$

where

A_M is the annual cost for maintenance, first-year \$
C_{cap} is the capital cost, first-year \$
C_{salv} is the salvage value, first-year \$
f_{dep} is the present value of depreciation, as fraction of C_{cap}
f_l is the fraction of investment paid by loan
N is the system life, year
N_l is the loan period, year
p_e is the energy price, first-year \$/GJ
Q is the annual energy consumption, GJ
r_d is the market discount rate
r_e is the market energy price escalation rate
$r_{d,e} = (r_d - r_e)/(1 + r_e)$
r_{dem} is the market demand charge escalation rate
$r_{d,dem} = (r_d - r_{dem})/(1 + r_{dem})$
r_{inf} is the general inflation rate
r_l is the market loan interest rate
$r_{d,l} = (r_d - r_l)/(1 + r_l)$
r_M is the market escalation rate for maintenance costs
$r_{d,M} = (r_d - r_M)/(1 + r_M)$
τ is the incremental tax rate
τ_{cred} is the tax credit

If there are several forms of energy, e.g., gas and electricity, the term Qp_e is to be replaced by a sum over the individual energy terms. Many other variations and complications are possible; for instance, the salvage tax rate could be different from τ.

Example 23.12: Chiller Life Cycle Costing

Find the life cycle cost of the chiller of Example 23.11 under the following conditions:

Given:
System life $N = 20$ years
Loan life $N_l = 10$ years
Depreciation period $N_{dep} = 10$ years, straight line depreciation
Discount rate $r_d = 0.15$
Loan interest rate $r_l = 0.15$
Energy escalation rate $r_e = 0.01$
Demand charge escalation rate $r_{dem} = 0.01$
Maintenance cost escalation rate $r_M = 0.01$
Inflation $r_{inf} = 0.04$
Loan fraction $f_l = 0.7$
Tax rate $\tau = 0.5$
Tax credit rate $\tau_{cred} = 0$
Capital cost (at \$400/ton) $C_{cap} = 40$ k\$
Salvage value $C_{salv} = 0$
Annual cost of maintenance $A_M = 0.8$ k\$/year (= 2% of C_{cap})
Capacity 100 tons = 351.6 kW_t
Peak electric demand 351.6 kW_t/COP = 117.2 kW_e
Annual energy consumption $Q = 100$ kton·h = 351.6 MWh_t
Electric energy price p_e = 10¢/kWh_e = 100 \$/MWh_e
Demand charge p_{dem} = 10 \$/kW_e·month, effective during 6 months of year
The rates are market rates.

Find: C_{life}

Lookup values:
$r_{d,l} = 0.0000$, $(A/P, r_l, N_l) = 0.1993$
$r_{d,e} = 0.1386$, $(A/P, r_d, N_l) = 0.1993$
$r_{d,dem} = 0.1386$, $(A/P, r_{d,l}, N_l) = 0.1000$
$r_{d,M} = 0.1386$, $(A/P, r_d, N) = 0.1598$
$(1 + r_{inf})/(1 + r_d) = 0.9043$, $(A/P, r_{d,e}, N) = 0.1498$
$f_{dep} = 0.502$ from Equation 23.42, $(A/P, r_{d,dem}, N) = 0.1498$, $(A/P, r_{d,M}, N) = 0.1498$

Solution

The following are the components of C_{life} (all in k\$) per Equation 23.44:

Down payment	12.0
Cost of loan	28.0
Tax deduction for interest	−8.0
Tax credit	0.0
Depreciation	−10.0
Salvage value	0.0
Cost of energy	39.1
Cost of demand charge	23.5
Cost of maintenance	2.7
Total = C_{life}	87.3

Comments

A spreadsheet is recommended for this kind of calculation. The software provided with the book can also be used. The cost of energy and demand is higher than the capital cost.

23.2.6 Cost per Unit of Delivered Service

Sometimes it is required to determine the cost per unit of delivered service (for instance, cost per ton-hour of cooling), analogous to the cost per driven mile for cars. This can be calculated as the ratio of levelized annual cost to annual delivered service. The levelized annual cost is obtained by multiplying the life cycle cost by the capital recovery factor for discount rate and system life. There appear two possibilities: the real discount rate r_{d0} and the market discount rate r_d. The quantity $(A/P, r_{d0}, N)C_{life}$ is the annual cost in constant dollars (of the initial year), whereas $(A/P, r_d, N)C_{life}$ is the annual cost in inflating dollars. The latter is difficult to interpret because it is an average over dollars of different real value. Therefore, we levelize with the real discount rate because it expresses everything in first-year dollars, consistent with the currency of C_{life}. Thus, we write the annual cost in initial dollars as

$$A_{life} = (A/P, r_{d0}, N)C_{life} \tag{23.45}$$

The effective total cost per delivered service is therefore

$$\text{Effective cost per energy} = \frac{A_{life}}{Q} \tag{23.46}$$

where Q is the annual delivered service (assumed constant, for simplicity).

The reader may wonder why we do not simply divide C_{life} by the service NQ delivered by the system over its lifetime. That would not be consistent because C_{life} is the present value, while NQ contains service flows (and thus monetary values) that are associated with future times. Strictly speaking, one must allocate service flows and costs within the same time frame. That is accomplished by dividing the levelized annual cost by the levelized annual service—and the latter is equal to Q because we have assumed that the consumption is constant from year to year.

Example 23.13: Chiller Energy Cost

What is the cost per ton-hour for the chiller of Example 23.12?

Given: $C_{life} = 96.9$ k\$ and $Q = 100$ kton·h

Find: A_{life}/Q

Lookup values:
$r_{d0} = 0.1058$ from Equation 23.15
$(A/P, r_{d0}, N) = 0.1221$ from Equation 23.14

Solution

Levelized annual cost in first-year dollars is

$$A_{life} = (A/P, r_{d0}, N)C_{life} = 0.1221 \times 87.3\ \text{k}$$
$$= \$10{,}659/\text{year}$$

$$\text{Cost per ton-hour} = \frac{A_{life}}{Q} = \$0.107/\text{ton}\cdot\text{h}$$

23.2.7 Constant Currency versus Inflating Currency

In the life cycle cost equation, all cost components have been converted to equivalent present values (i.e., first-year costs). Let us see to what extent the result is the same whether one uses constant currency and real rates or inflating currency and market rates. In the term for energy cost, only the variable $r_{d,e} = (r_d - r_e)/(1 + r_e)$ depends on this choice. Inserting real rates according to

$$1 + r_{d0} = \frac{1 + r_d}{1 + r_{inf}} \tag{23.47}$$

and

$$1 + r_{e0} = \frac{1 + r_e}{1 + r_{inf}} \tag{23.48}$$

one finds that

$$r_{d,e} = \frac{(1 + r_{d0})(1 + r_{inf}) - (1 + r_{e0})(1 + r_{inf})}{(1 + r_{e0})(1 + r_{inf})} \tag{23.49}$$

and after canceling the factor $1 + r_{inf}$, one sees that this is equal to

$$r_{d,e} = \frac{1 + r_{d0} - (1 + r_{e0})}{1 + r_{e0}} = r_{d,e0} \tag{23.50}$$

The energy cost is the same, whether one uses real rates or market rates. The same holds for the maintenance cost term. The salvage term is also independent of this choice because

$$\frac{1 + r_{\text{inf}}}{1 + r_d} = \frac{1}{1 + r_{d0}}$$

By contrast, the ratio of capital recovery factors in the loan terms is not invariant, as one can see by inserting numerical values. For example, with $r_{d0} = 0.08$, $r_{l0} = 0.12$, and $r_{inf} = 0.05$, one finds, with $N = 20$ years,

$$\frac{(A/P, r_{l0}, N)}{(A/P, r_{d0}, N)} = 1.31$$

The corresponding market rates are $r_d = 0.134$ and $r_l = 0.176$, and the ratio becomes

$$\frac{(A/P, r_l, N)}{(A/P, r_d, N)} = 1.26$$

The difference arises from the fact that the cash flows are different. For a loan that is based on real rates, the annual payments are constant in constant currency, whereas for one based on market rates, the payments are constant in inflating currency. Likewise, the depreciation terms can depend on inflation.

It follows that the two approaches, constant currency and inflating currency, yield identical results for equity investments ($f_l = 0$) without depreciation. But if f_l or f_{dep} is not zero, there can be differences. Numerically, the effect is not large, at most on the order of 10 percent, for inflation rates below 10% (Dickinson and Brown, 1979). The effect has opposite signs for the loan term and the depreciation term, leading to partial cancellation of the error.

23.3 Economic Evaluation Criteria

23.3.1 Life Cycle Savings

Having determined the life cycle cost of each relevant design alternative, one can select the "best," i.e., the one that offers all desirable features at the lowest life cycle cost. Frequently, one takes one design as reference and considers the difference between it and each alternative design. The difference is called "life cycle savings" relative to the reference case

$$S = -\Delta C_{life} \quad \text{with } \Delta C_{life} = C_{life} - C_{life,ref} \tag{23.51}$$

Often, the comparison can be quite simple because only those terms that are different between the designs under consideration need to be considered. For simplicity, we write the equations of this section only for an equity investment without tax. Then, the loan fraction f_l in Equation 23.41 is zero, and most of the complications of that equation drop out. Of course, the concepts of life cycle savings, internal rate of return, and payback time are perfectly general, and tax and loan can be readily included.

A particularly important case is the comparison of two designs that differ only in capital cost and operating cost: the one that saves operating costs has higher initial cost (otherwise, the choice would be obvious, without any need for an economic analysis). Setting

$f_l = 0$ and $\tau = 0$ in Equation 23.40 and taking the difference between the two designs, one obtains the life cycle savings as

$$S = \frac{-\Delta Q \times p_e}{(A/P, r_{d,e}, N)} - \Delta C_{cap} \tag{23.52}$$

where

$\Delta Q = Q - Q_{ref}$

$=$ difference in annual energy consumption

$\Delta C_{cap} = C_{cap} - C_{cap,ref} =$ difference in capital cost

$r_{d,e} = (r_d - r_e)/(1 + r_e)$

(If the reference design has higher consumption and lower capital cost, ΔQ is negative and ΔC_{cap} is positive with this choice of signs.)

Example 23.14: Life Cycle Savings of Chiller

Compared to a one-stage model, a two-stage absorption chiller is more efficient, but its first cost is higher. Find the life cycle savings of a two-stage model for the followings situation.

Given:

The required chiller capacity is 1000 kW_t operating at 1000 h/year full-load equivalent.

A single-stage absorption chiller has COP = 0.7 and costs \$100/kW_t (= reference system), while a two-stage absorption chiller has COP = 1.1 and costs \$130/kW_t.

Gas price p_e = \$4/GJ at the start.

Escalating at r_e = 0% (real).

Discount rate r_d = 8% (real).

Find: Life cycle savings for the two-stage chiller

Lookup value: $(A/P, r_d, N) = 0.1019$

Solution

The annual energy consumption is

$$1000\ \text{kW}_t \times 1000\ \text{h/COP} = 1.0\ \text{MWh}_t/\text{COP}$$

$$= \begin{cases} 5.143 \times 10^3\ \text{GJ}_{gas} & \text{for COP} = 0.7 \\ 3.273 \times 10^3\ \text{GJ}_{gas} & \text{for COP} = 1.1 \end{cases}$$

Thus, the difference in energy cost is

$$\Delta Q \times p_e = (3.273 - 5.143) \times 10^3\ \text{GJ} \times \$4/\text{GJ} = -\$7,481/\text{year}$$

and the difference in capital cost is

$$\Delta C_{cap} = (130 - 100)(1,000) = \$30,000$$

From Equation 23.52, we find the life cycle savings

$$S = \frac{-\Delta Q \times p_e}{(A/P, r_{d,e}, N) - \Delta C_{cap}}$$

$$= \frac{\$7,481}{0.1019} - 30,000 = \$73,415 - 30,000 = \$43,415$$

Comment

Even though the discount rate in this example is rather high (5% might be more appropriate), the life cycle savings are large. The investment certainly pays off.

23.3.2 Internal Rate of Return

The life cycle savings are the true savings if all the input is known correctly and without doubt. But future energy prices or system performance is uncertain, and the choice of the discount rate is not clear-cut. An investment in a building or its equipment is uncertain, and it must be compared with competing investments that have their own uncertainties. The limitation of the life cycle savings approach can be circumvented if one evaluates the profitability of an investment by itself, expressed as a dimensionless rate. Then, one can rank different investments in terms of profitability and in terms of risk. General business experience can serve as a guide for expected profitability as function of risk level. Among investments of comparable risk, the choice can then be based on profitability.

More precisely, the profitability is measured as the *internal rate of return* r_r, defined as that value of the discount rate r_d at which the life cycle savings S are zero

$$S(r_d) = 0 \quad \text{at } r_d = r_r \tag{23.53}$$

For an illustration, take the case of Equation 23.52 with energy escalation rate $r_e = 0$ (so that $r_{d,e} = r_d$), and suppose an extra investment ΔC_{cap} is made to provide annual energy savings ($-\Delta Q$). The initial investment ΔC_{cap} provides an annual income from energy savings of

$$\text{Annual income} = (-\Delta Q) \times p_e \tag{23.54}$$

If ΔC_{cap} were placed in a savings account instead, bearing interest at a rate r_r, the annual income would be

$$\text{Annual income} = (A/P, r_r, N)\ \Delta C_{cap} \tag{23.55}$$

The investment behaves as a savings account whose interest rate r_r is determined by the equation

$$(A/P, r_r, N)\Delta C_{cap} = (-\Delta Q) \times p_e \tag{23.56}$$

Dividing by $(A/P, r_r, N)$, we see that right and left sides correspond to the two terms in Equation 23.52 for the life cycle savings

$$S = \frac{-\Delta Q \times p_e}{(A/P, r_d, N)} - \Delta C_{cap} \tag{23.57}$$

and that r_r is indeed the discount rate r_d for which the life cycle savings are zero; it is the internal rate of return. Now the reason for the name is clear: it is the profitability of the project by itself, without reference to an externally imposed discount rate. When the explicit form of the capital recovery factor is inserted, one obtains an equation of Nth degree, generally not solvable in closed form. Instead one resorts to iterative or graphical solution. (There could be up to N different real solutions, and multiple solutions can indeed occur if there are sign changes in the stream of annual cash flows. But, not to worry, the solution is unique for the case of interest here: an initial investment that brings a stream of annual savings.)

Example 23.15: Internal Rate of Return

What is the rate of return for Example 23.14?

Given: $S = \frac{-\Delta Q p_e}{(A/P, r_{d,e}, N)} - \Delta C_{cap}$ with $r_{d,e} = r_d$ (because $r_e = 0$)

$$(-\Delta Q)p_e = \$7{,}481$$

$$\Delta C = \$30{,}000$$

Find: r_r

Solution

$S = 0$ for

$$(A/P, r_r, N) = \frac{-\Delta Q p_e}{\Delta C_{cap}}$$

$$= \frac{7{,}481}{30{,}000} = 0.2494 \quad \text{with } N = 20$$

By iteration one finds $r_r = 0.246 = 24.6\%$.

23.3.3 Payback Time

The *payback time* N_p is defined as the ratio of extra capital cost ΔC_{cap} to first-year savings

$$N_p = \frac{\Delta C_{cap}}{\text{First-year savings}} \tag{23.58}$$

(The inverse of N_p is sometimes called the "return on investment.") If one neglects discounting, one can say that after N_p years the investment has paid for itself and any revenue thereafter is pure gain. The shorter N_p, the higher the profitability. As a selection criterion, the payback time is simple, intuitive, and obviously wrong because it neglects some of the relevant variables. Attempts have been made to correct for that by constructing variants such as a discounted payback time (by contrast to which Equation 23.58 is sometimes called simple payback time), but the resulting expressions become so complicated that one might as well work directly with life cycle savings or internal rate of return.

The simplicity of the simple payback time is, however, irresistible. When investments are comparable to each other in terms of duration and function, the payback time can give an approximate ranking that is sometimes clear enough to be able to discard certain alternatives right from the start, thus avoiding the effort of detailed evaluation.

To justify the use of the payback time, let us recall Equation 23.56 for the internal rate of return and note that it can be written in the form

$$(A/P, r_r, N) = \frac{1}{N_p} \quad \text{or} \quad (P/A, r_r, N) = N_p \tag{23.59}$$

The rate of return is uniquely determined by the payback time N_p and the system life N. This equation implies a simple graphical solution for finding the rate of return if one plots $(P/A, r_r, N)$ on the x-axis versus r_r on the y-axis, as in Figure 23.6. Given N and N_p, one simply looks for the intersection of the line $x = N_p$ (i.e., the vertical line through $x = N_p$) with the curve labeled by N; the ordinate (y-axis) of the intersection is the rate of return r_r.

This graphical method can be generalized to the case where the annual savings change at a constant rate r_e. In that case, Equation 23.21 implies that the rate of return is replaced by $r_{r,e} = (r_r - r_e)/(1 + r_e)$, and Figure 23.6 yields $r_{r,e}$ rather than r_r. In other words, Equation 23.59 becomes

$$(P/A, r_{r,e}, N) = N_p \quad \text{with } r_{r,e} = \frac{r_r - r_e}{1 + r_e} \tag{23.60}$$

The graph yields $r_{r,e}$, which is readily solved for

$$r_r = r_{r,e}(1 + r_e) + r_e \tag{23.61}$$

In particular, if r_e is equal to the general inflation rate r_{inf}, then $r_{r,e}$ is the real rate of return $r_{r,0}$.

Example 23.16: Payback Period

Find payback time for Example 23.15 and check the rate of return graphically.

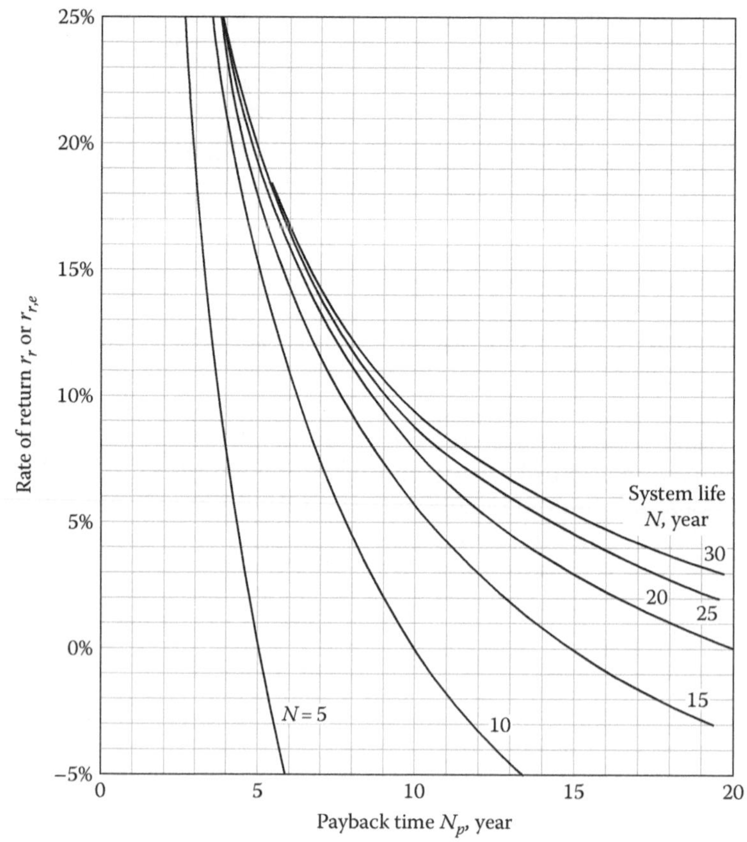

FIGURE 23.6
Relation between rate of return r_r, system life N, and payback time N_p. If r_e = escalation rate of annual savings $\neq 0$, the vertical axis is the variable $r_{r,e}$ from which r_r is obtained as $r_r = r_{r,e}(1 + r_e) + r_e$.

Given: First-year savings $(-\Delta Q)p_e = \$7{,}481$

Extra investment $\Delta C_{cap} = \$30{,}000$

Find: N_p and r_r for $r_e = 0\%$ and 2%

Solution

$$N_p = \frac{30{,}000}{7{,}481} = 4.01 \text{ years}$$

It is independent of r_e. Then $r_r = 0.246$ for $r_e = 0$, from Figure 23.6, and $r_r = 0.271$ for $r_e = 2\%$, from Equation 23.60.

Generally, a real (i.e., corrected for inflation) rate of return above 10% can be considered excellent if there is low risk—a look at savings accounts, bonds, and stocks shows that it is difficult to find better. From the graph we see immediately that $r_{r,e}$ is above 10% if the payback time is shorter than 8.5 years (6 years), for a system life of 20 years (10 years). And $r_{r,e}$ is close to the real rate of return if the annual savings growth is close to the general inflation rate.

23.4 Complications of the Decision Process

In practice, the decision process is likely to bump into some obstacles. Suppose, e.g., that the annual operating cost of a proposed office building can be reduced by $1000 if one installs daylight sensors and dimmers for the lights, at an extra cost of $2000. The payback time is only 2 years. It looks like an irresistible investment opportunity, with a rate of return well above 25%, as shown by Figure 23.6 (the exact value depends somewhat on lifetime and taxes, but that is beside the point). However, quite a few hurdles stand in the way.

First, to find out about this opportunity, the design engineer has to obtain the necessary information. Requesting catalogues, reading technical reviews of the equipment, and carrying out the calculations of cost and performance all take time and effort. Under the pressures of the job, the engineer may not be willing to spend the extra time or to neglect other items that compete for her or his attention.

Suppose our engineer has done a good analysis and tries to convince the builder to spend the extra money. In the case of a speculative office building, the builder is likely to say, "Why should I pay a penny more, if only the future tenant will reap the benefit?" So the design engineer is forced to aim for lowest first cost.

Even if the builder is willing to spend a bit more for efficiency, with hopes that the prospect of reduced energy bills will make it easier to find tenants, the decision is not obvious. Can the builder trust the claims of the sales brochure or the calculations of the engineer? Daylight controls are relatively new, and perhaps the builder has heard that some of the first models did not live up to expectations. If malfunctions reduce the productivity of the workers, the hassle and the costs could more than nullify the expected savings. So the builder may refuse to take what he or she perceives as an excessive risk. The threat of a liability suit is a potent inhibiter; no wonder the building industry has a reputation for extreme conservatism.

This example illustrates the basic mechanisms that so frequently prevent the adoption of efficient technologies:

- Lack of information or excessive cost of obtaining the information
- Purchase decision made by someone who does not have to pay the operating costs
- Uncertainty (about future costs, reliability, etc.)

Any one of the hurdles can be sufficient cause to reject an investment. In the aforementioned example, the decision to reject the lighting controller looks as if the discount rate were higher than 25%. Quite generally, these mechanisms have the effect of raising the apparent discount rate or foreshortening the time horizon. The resulting decisions appear irrational: people do not spend as much for energy efficiency as would be optimal according to a life cycle cost analysis with the correct discount rate. In reality, this irrationality is but a reflection of other problems.

In the world of business, risk and uncertainty are all-pervasive, so much so that most decision makers insist on very short payback times—almost always less than 5 years and frequently less than 2 years. But it depends on the nature of the business and on the circumstances. There are industries such as electric power plants where profits are sure (albeit moderate): once a power plant has been built, it is expected to run smoothly for at least 30 years. Here, the discount rates are low, and payback times are longer than 10 years. Governments, being charged with the long-term welfare of the citizens, also tend to have a long time horizon.

What does all this mean for the HVAC engineer? *The more a design choice involves unproven technology or is dependent on occupant behavior for proper functioning, the more risky it is.* For instance, a daylighting strategy that relies on manual control of shading devices by the occupants may not bring the intended savings because the occupants may not follow the intentions of the designer. Likewise, when one is considering a new design or a new piece of equipment without a track record, it is not irrational to demand short payback times.

By contrast, paying extra for an efficient boiler or chiller is a safe investment (assuming the equipment has a good reputation) because the occupant does not care how the heating or cooling is produced, as long as the environment is comfortable, pleasant, and healthy. Also, the building will certainly be heated and cooled over its entire life. Here a life cycle cost analysis with the correct discount rate is certainly in order, and it would be shortsighted to insist on payback times of less than 2 years.

Finally, what about the problem of the builder or landlord who refuses to pay for measures that would only reduce the energy bill of the tenant or of a future owner? This is a serious difficulty, indeed. In an ideal market, the information about reduced energy cost would translate as higher rent or resale value, but in practice, this process is slow and inefficient (there is a market failure, in the language of the economists). This justifies energy efficiency standards and their enforcement by government regulations.

23.5 Cost Estimation

23.5.1 Capital Costs

For mass-produced consumer products, such as cars or cameras, the capital cost (i.e., the purchase price) is easy to determine; look at catalogues or newspaper ads or call the store. Even then there may be uncertainties: When you actually go the store, a discount may be offered on the spot to beat a competitor. Different prices can be found in different stores for identical products, not only because of differences in service or transportation but also because of the sheer difficulty of obtaining the price information.

And, of course, price is not the only criterion. Even more important, and more difficult to ascertain and compare, are the various characteristics of a product: the features it offers, the quality, the operating costs, and so on. Economists have even coined a special term, "cost of information," so universal is the difficulty of finding the pertinent information.

For HVAC equipment, the problems tend to be more complicated than for consumer goods. Transport and installation costs are important items in addition to the cost of the equipment at the factory. The determination of the cost can become a major undertaking, especially for complex or custom-made systems. The capital cost of a system or component is known with certainty only when one has a firm contract from a vendor. But asking for bids on each design variation is simply not feasible—the cost of information would become prohibitive.

The more a design engineer wants to be sure of coming close to the optimal design, the more she or he needs to learn about the details of the cost calculation. Information on costs is available from a number of sources, e.g., Boehm (1987) and Konkel (1987). An important feature is the variation of the cost with size. Because of fixed costs and economies of scale, simple proportionality between cost and size is not the rule. But usually one can assume the following functional form over a limited range of sizes:

$$C = C_{ref}\left(\frac{S}{S_{ref}}\right)^m \qquad \text{for } S_{\min} < S < S_{\max} \qquad (23.62)$$

where

C is the cost at size S
C_{ref} is the cost at a reference size S_{ref}
m is the exponent

Typically, m is in the range of 0.5–1.0; exponents less than unity are a reflection of economies of scale. On a logarithmic plot, m is the slope of ln C versus ln S. If m is not known, a value of 0.6 can be recommended as default. Tables on the online HCB software summarize cost data for HVAC equipment in this form.

When interpreting such cost figures, one has to be careful about what is included and what is not. Is it the cost at the factory free on board (i.e., excluding transportation), the cost delivered to the site, or the cost installed? For items such as cool storage the space requirements may impose additional costs. And finally, what are the specific features and how is the quality?

Costs change not only with general inflation but also with the evolution of technology. The first models of a novel product tend to be expensive. Gradually, mass production, technological advances, and competition combine to drive the prices down. General inflation or increases in the cost of some input, for instance, energy, will push in the opposite direction. The resulting evolution of the price of the product may be difficult to predict. Cost reductions due to technological advances are more likely with products of high technology (e.g., energy management systems) than with mature products that cannot be miniaturized (e.g., fans and motors). In some cases, there is an improvement in a product rather than a reduction of its cost; variable-speed motors, for instance, are more expensive than constant-speed motors but allow better control or higher system efficiency.

Cost tabulations are based on sales or projects of the past, and they must be updated to the present by means of correction factors. For that purpose one could use general inflation (i.e., the CPI discussed in Section 23.1.1), but that is less reliable than specific cost indices for that class of equipment or that sector of the economy. The following two indices are particularly pertinent for buildings and HVAC equipment. One is the Marshall and Swift equipment cost index, the values of which are published regularly in *Chemical Engineering*. Another one is the construction cost index as plotted in Figure 23.1(b).

It is important during the design process to have a realistic understanding of all the relevant costs, yet the effort of obtaining these costs should not be prohibitive. Konkel (1987) describes a method that seems to be a good compromise between these conflicting requirements. The basic idea is to group certain portions of a project into what is called "unit operations." The components of the unit operations, called "unit assemblies," are itemized, priced, and plotted by size of unit operation. A boiler is an example of a unit operation; its unit assemblies include burner, air intake, flue, shutoff valves, piping, fuel supply, expansion tank, water makeup valves, and deaerator. The sizes and costs vary with the size of the boiler. Once the size–price relations have been found for each component, the size–price relation for the boiler as a whole is readily derived. Knowing the size–price relation for the unit operations, the designer can estimate the total cost of a project and its design variations without too much effort.

23.5.2 Maintenance and Energy

Maintenance cost and energy prices may evolve differently from general inflation and from each other. It is instructive to correct energy prices for general inflation, as in Figure 23.2 where the prices of oil, gas, and electricity are shown in both current and constant dollars. One can see that some adjustments have occurred since the oil shocks of the 1970s. In fact, the real price of crude oil was almost the same in 1998 as during the 1960s.

What should we assume for the future? Projections of energy prices are published periodically by several organizations, for instance, the American Gas Association and the National Institute of Standards and Technology (Lippiat and Ruegg, 1990; Fuller and Petersen, 1996). Most analysts predict real escalation rates in the range of 0%–3%, averaged over the next two decades. This is based on the gradual exhaustion of cheap oil and gas

TABLE 23.2

Median Service Life of a Few Typical HVAC Equipment

Equipment Type	Median Service Life, year	Equipment Type	Median Service Life, year
DX air distribution equipment	>24	Gas hot water boiler, steel	>22
Chillers, centrifugal	>25	Pneumatic electronic controls	>7
Cooling towers, metal	>22	Portable electric hot water heaters	>21

reserves and the fact that alternatives, i.e., wind and solar, are gradually getting cheaper. However, such predictions are fraught with uncertainty. For example, gas fracking has resulted in a major rethinking of our energy futures. How will public acceptance of nuclear power evolve? How much can be saved by improved energy efficiency, and at what cost? What constraints will be imposed by environmental and climate change concerns?

Data on service life and maintenance costs of buildings and HVAC systems can be obtained from sources such as BOMA published annually (BOMA, 2014). The median service life of a few typical HVAC systems is shown in Table 23.2. For greater detail, the interested reader can refer to Akalin (1978) and to the chapter on owning and operating costs from the ASHRAE Applications Handbook (2011). ASHRAE has created two free interactive web-based databases (www.ashrae.org/database) for all major HVAC equipment in terms of service life and maintenance costs. These databases can be custom queried so that the search matches specific criteria such as location, building function, size and age of building, height, and BOMA class.

For simpler types of analysis relevant to maintenance costs of HVAC equipment in office buildings, a succinct equation is proposed (ASHRAE TC1.8, 1985), which is based on an analysis of survey data on HVAC maintenance costs. It frames the annual cost A_M for maintenance, in dollars per floor area A_{floor} as

$$\frac{A_M}{A_{floor}} = C_{base} + an + h + c + d \tag{23.63}$$

where

C_{base} is the value for base system (fire-tube boilers for heating, centrifugal chillers for cooling, and VAV for distribution, during the first year)

n is the age of equipment, years

a is the coefficient for age of equipment

h, c, and d are the coefficients that allow the adjustment to other systems

Numerical values for C_{base}, a, h, c, and d are listed in Table 23.3. These values are in 1983 dollars. They still need to be adjusted to the year of interest by multiplication by the corresponding ratio of CPI values, as explained in Section 23.1. In using this equation one should keep in mind that it is based on a survey of office buildings originally published in 1986. Extrapolation to other building types or newer technologies may introduce large and unknown uncertainties.

TABLE 23.3

HVAC Maintenance Costs of Equation 23.61: Cost C_{base} of Base System and Coefficients for Adjustment, 1983 U.S. Dollars

	$/ft²	$/m²
C_{base}	0.3335	3.590
Coefficient a for age per year	0.0018	0.019
Heating equipment, *coefficient h*		
Water-tube boiler	0.0077	0.083
Cast-iron boiler	0.0094	0.101
Electric boiler	−0.0267	−0.287
Heat pump	−0.0969	−1.043
Electric resistance	−0.1330	−1.432
Cooling equipment, *coefficient c*		
Reciprocating chiller	−0.0400	−0.431
Absorption chiller (single stage)	0.1925	2.072
Water source heat pump	−0.0472	−0.508
Distribution system, *coefficient d*		
Single zone	0.0829	0.892
Multizone	−0.0466	−0.502
Dual duct	−0.0029	−0.031
Constant volume	0.0881	0.948
Two-pipe fan coil	−0.0277	−0.298
Four-pipe fan coil	0.0580	0.624
Induction	0.0682	0.734

Source: ASHRAE, *Design/Data Manual for Closed-Loop Ground Coupled Heat Pump Systems*, ASHRAE, Atlanta, GA, 1985. Copyright ASHRAE, www.ashrae.org.

Example 23.17: Maintenance Cost of a HVAC System

Estimate the annual HVAC maintenance cost for an office building that has a floor area of 1000 m² and is n = 10 years old in the year 2003. The system consists of an electric boiler, a reciprocating chiller, and a constant-volume distribution system. Suppose the CPI is 180 in 2003.

Given: $A_{floor} = 1000\ \text{m}^2$
$n = 10$ year

$$\frac{\text{CPC}_{2003}}{\text{CPI}_{1983}} = \frac{180}{100} = 1.80$$

Lookup values: From Table 23.3

$$C_{base} = 3.59, \quad a = 0.019, \quad h = -0.287, \quad c = -0.431,$$

$$d = 0.948\ (\text{in } \$_{1983}/\text{m}^2)$$

Find: A_M

Solution

Using Equation 23.63, one finds

$$A_M = 1000\ \text{m}^2 \times [(3.59 + 0.019 \times 10 - 0.287 - 0.431 + 0.948)\ \$_{1983}/\text{m}^2] = \$_{1983}4010/\text{m}^2$$

To convert to 2003 dollars, multiply by the CPI ratio

$$A_M = \$_{1983}4010/\text{m}^2 \times 1.80 = \$_{2003}7218/\text{m}^2$$

23.6 Optimization

In principle, the process of optimizing the design of a building is simple: Evaluate all possible design variations and select the one with the lowest life cycle cost. Who would not want to choose the optimum? In practice, it would be a daunting task to find the true optimum, among all conceivable designs. The difficulties, some of which have already been discussed, are

- The enormous number of possible design variations (building configuration and materials, HVAC systems, types and models of equipment, control modes)
- Uncertainties (costs, energy prices, reliability, occupant behavior, future uses of building)
- Imponderables (comfort, convenience, aesthetics)

Fortunately, there is a certain tolerance for moderate errors, as we will show in this section. That facilitates the job greatly, because one can reduce the number of steps in the search for the optimum. Also, within narrow ranges some variables can be suboptimized without worrying about their effect on others.

Some quantities are easier to optimize than others. Optimizing the heating and cooling equipment, for a given building envelope, is less problematic than trying to optimize the envelope—the latter touches on the imponderables of aesthetics and image.

It is instructive to illustrate the optimization process with a very simple example: the thickness of insulation (Rabl, 1985). The annual heat flow Q across the insulation is

$$Q = \frac{AkD}{t} \tag{23.64}$$

where
A is the area, m^2
k is the conductivity, $\text{W/(m}\cdot\text{K)}$
D is the annual degree-seconds, $\text{K}\cdot\text{s}$
t is the thickness of insulation, m

The capital cost of the insulation is

$$C_{cap} = Atp_{ins} \tag{23.65}$$

where p_{ins} is the price of insulation ($/m³). The life cycle cost is

$$C_{life} = C_{cap} + Q\frac{p_e}{(A/P, r_{d,e}, N)} \tag{23.66}$$

where
p_e is the first-year energy price
$r_{d,e}$ is related to discount rate and energy escalation rate as in Equation 23.20

We want to vary the thickness t to minimize the life cycle cost, keeping all the other quantities constant. (This model is a simplification that neglects the fixed cost of insulation as well as possible feedback of t on D.) Eliminating t in favor of C_{cap}, one can rewrite Q as

$$Q = \frac{K}{C_{cap}} \tag{23.67}$$

with a constant

$$K = A^2kDp_{ins} \tag{23.68}$$

Then, the life cycle cost can be written in the form

$$C_{life} = C_{cap} + \frac{PK}{C_{cap}} \tag{23.69}$$

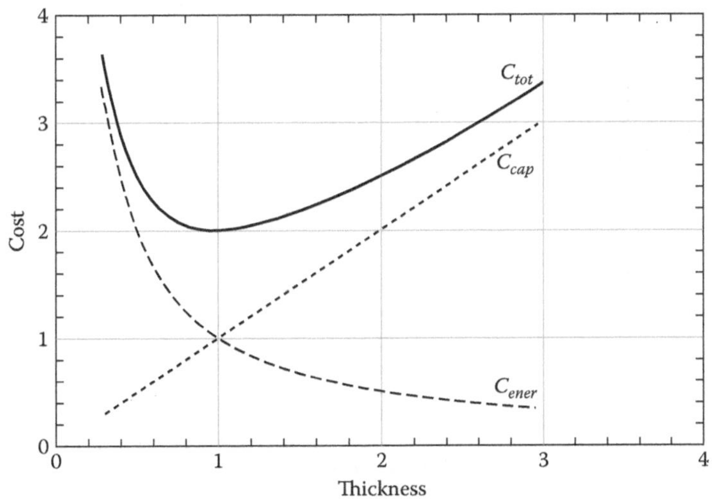

FIGURE 23.7
Optimization of insulation thickness. Insulation cost = C_{cap}, energy cost = C_{ener}, and life cycle cost = C_{tot}.

where the variable

$$P = \frac{p_e}{(A/P, r_{d,e}, N)} \tag{23.70}$$

contains all the information about energy price and discount rate. Here K is fixed, and the insulation investment C_{cap} is to be varied to find the optimum. Variable C_{life} and its components are plotted in Figure 23.7. As t is increased, capital cost increases and energy cost decreases; C_{life} has a minimum at some intermediate value. Setting the derivative of C_{life} with respect to C_{cap} equal to zero yields the optimal value $C_{cap,o}$

$$C_{cap,o} = \sqrt{KP} \tag{23.71}$$

Now here is an interesting question: What is the penalty for not optimizing correctly? In general, the following causes could prevent correct optimization:

- Insufficient accuracy of the algorithm or program for calculating the performance
- Incorrect information on economic data (e.g., the factor P in Equation 23.71)
- Incorrect information on technical data (e.g., the factor K in Equation 23.71)
- Unanticipated changes in the use of the building

Misoptimization would produce a design at a value C_{cap} different from the true optimum $C_{cap,o}$. For the example of insulation thickness, the effect on the life cycle cost can be seen directly with the solid curve in Figure 23.7. For instance, a +10% or −10% error in $C_{cap,o}$ would increase C_{life} by only +1%. Thus, the penalty is not excessive for small errors.

This relatively large insensitivity to misoptimization is a feature much more general than the insulation model. As shown by Rabl (1985), the greatest sensitivity likely to be encountered in practice corresponds to the curve

$$\frac{C_{life,true}(C_{cap,o,guess})}{C_{life,true}(C_{cap,o,true})} = \frac{x}{1+\log x} (upper\ bound) \tag{23.72}$$

which is also shown in Figure 23.8, with the label *upper bound*. Even here the minimum is broad; if the true energy price differs by +10% (−10%) from the guessed price, the life cycle cost increases only 0.4% (0.6%) over the minimum. Even when the difference in prices is 30%, the life cycle cost penalty is less than 8%.

Errors in the factor K (due to wrong information about price or conductivity of the insulation material) can be treated in the same way, because K and P play an entirely symmetric role in the aforementioned equations. Therefore, curves in Figure 23.8 also apply to uncertainties in other input variables.

The basic phenomenon is universal: any smooth function is flat at an extremum. The only question is, "How flat?" For energy investments that question has been answered with the curves of Figures 23.7 and 23.8. We can conclude that misoptimization penalties are definitely less than 1% (10%) when the uncertainties of the input variables are less than 10% (30%).

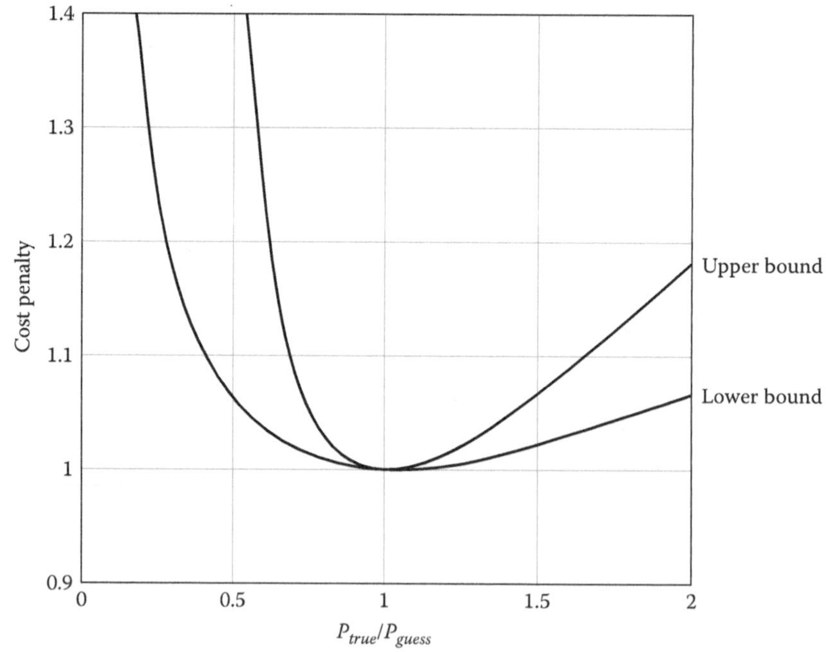

FIGURE 23.8
Life cycle cost penalty versus energy price ratio.

Problems

Numbers 1–4 given in parenthesis denote the degree of difficulty.

23.1 Estimate your personal discount rate, in both constant and inflating dollars. *(1)*

23.2 Suppose you are offered the choice between \$1000 today and \$5000 in 10 years from now, and the average inflation rate is 4%. What are the real and the market discount rates at which you are indifferent between these two choices? *(2)*

23.3 A bank advertises an interest rate of 10% (annual rate, no compounding). Use the rule of 70 (Section 23.1.5) to estimate how long will it take to double your money (a) in inflating currency and (b) in constant currency, if the inflation rate is 5%. *(2)*

23.4 Suppose that market fuel prices escalate at a rate of $r_e = 8\%$, while the general inflation rate is $r_{inf} = 6\%$.

(a) If fuel costs \$5/MBtu (\$5.275/GJ) today, what will be the price after 20 years, in both constant and inflating dollars?

(b) What is the corresponding levelized fuel price if the discount rate is $r_{d,0} = 5\%$ (constant dollars)? *(2)*

23.5 A building is to be heated with a furnace that costs \$2500 and has a seasonal average efficiency of 90%. Suppose the annual heating load is 100 MBtu (105.5 GJ) and the fuel price is \$6.00/MBtu (\$6.33/GJ). Assume a lifetime of 20 years and a real discount rate of 5%.

(a) Calculate the life cycle cost of heating this building.

(b) Calculate the levelized total cost of delivered heat (in \$/MBtu (\$/GJ)). What fraction is due to capital and what fraction is due to energy? *(3)*

23.6 Suppose a system of heat recovery from exhaust air costs \$10,000 and reduces the annual heating bill by \$2,000. Assume a real discount rate of 5%.

(a) What is the payback time?

(b) Plot the life cycle savings versus the lifetime of the system for values from 5 to 20 years.

(c) Plot the rate of return versus the lifetime of the system for values from 5 to 20 years. Does the rate of return depend on the discount rate assumed for part b? *(3)*

23.7 Compare two chillers, both rated for an output of 5 kW_t (1.58 ton refrigeration):

- Model A has a COP of 2.0 and costs \$1500.
- Model B has a COP of 2.5 and costs \$2000.

They are used for 500 h/year. Assume an electricity price of 10¢/kWh$_e$, a lifetime of 15 years, and a real discount rate of 5%.

(a) Calculate the life cycle costs.

(b) Find the payback time and rate of return for using model B instead of A.

(c) What is the highest price at which model B is competitive with A? *(3)*

23.8 Consider two choices for the furnace to heat a building:

- An ordinary gas furnace with efficiency of 75% (seasonal average), costing $5,000
- A condensing furnace with efficiency of 95% (seasonal average), costing $10,000

Suppose the annual heating load is 1000 MBtu (1055 GJ) and the fuel price is $5.00/MBtu ($5.275/GJ). Assume that both furnaces can be expected to last 15 years, and use a real discount rate of 5%.

(a) Calculate the life cycle cost for each.

(b) Calculate the levelized total cost of delivered heat for each. What fraction is due to capital, and what fraction is due to energy?

(c) Find the payback time and rate of return for using model B instead of A. *(3)*

23.9 A solar domestic water heater costs C = $2000 and supplies Q = 20 GJ (18.96 MBtu) per year. The backup uses electric resistance heating with an efficiency of 100%. The price of electricity is p_e = 6¢/kWh, and the discount rate is 5% (real). Find the payback time, life cycle savings, and rate of return for the following cases:

(a) p_e and Q constant, life N = 20 years

(b) p_e and Q constant, life N = 10 years

(c) p_e and Q constant, life N = 30 years

(d) p_e escalates at 4%/year, Q constant, N = 20 years

(e) p_e constant, Q degrades at 1%/year, N = 20 years *(4)*

23.10 Evaluate and compare the cost of generating electric power with the following three types of power plants, as a function of capacity factor (= annual average power produced per peak power rating). Assume a system life of 30 years and a real discount rate of $r_{d,0}$ = 6%; neglect O&M. Calculate the cost of electricity (¢/kWh$_e$) for each of these two power plants and for each of these values of the capacity factor: 0.70 (= baseload) and 0.10 (= peak load).

(a) *Coal power plant*: Capital cost C_{cap} = $1500/kW$_e$, efficiency (fuel to electric)

η = 33.3%, fuel price p_f = $2.50/GJ.

(b) *Simple gas turbine*: Capital cost C_{cap} = $250/kW$_e$, efficiency (fuel to electric)

η = 20%, fuel price p_f = $5.00/GJ.

(c) *Solar power plant*: Capital cost C_{cap} = $3000/kW$_e$, annual production corresponds to a capacity factor of 0.15 (this number depends on the amount of sunshine at the site).

(d) At what capital cost does the solar plant become competitive with the others?

(e) The total demand of an electric grid contains components that are relatively constant (typical of many industrial loads) and components that are more or less variable (typical of building loads, especially cooling). Which plant type would you use for demand peaks and which for constant demand? *(4)*

References

Akalin, M.T. (1978). Equipment life and maintenance cost survey (RP-186). *ASHRAE Trans.*, 84(2), 94–106. American Society of Heating, Refrigerating and Airconditioning Engineers, Atlanta, GA.

ASHRAE Applications (2011). Chapter 37: Owning and operating costs, in *HVAC Applications Handbook*. American Society of Heating, Refrigerating and Airconditioning Engineers, Atlanta, GA.

ASHRAE TC1.8 (1985). Analysis of survey data on HVAC maintenance costs. ASHRAE Technical Committee 1.8 Research Project 382, ASHRAE, Atlanta, GA.

Boehm, R.F. (1987). *Design, Analysis of Thermal Systems*. Wiley, New York.

BOMA (2014). BOMA experience exchange report: Income/expense analysis for office buildings. Building Owners and Managers Association International, Washington, DC.

Dickinson, W.C. and K.C. Brown (1979). Economic analysis for solar industrial process heat. Report UCRL-52814. Lawrence Livermore Laboratory, Livermore, CA.

EIA (2014). Short-term energy outlook. U.S. Energy Information Administration, Washington, DC. http://www.eia.gov/forecasts/steo/realprices/. Accessed on January 2015.

Fuller, S.K. and S.R. Petersen (1996). *Life Cycle Costing Manual for the Federal Energy Management Program*, NIST Handbook 135. National Institute of Standards and Technology, Building and Fire Research Laboratory, Gaithersburg, MD.

Jones, B.W. (1982). *Inflation in Engineering Economic Analysis*. Wiley Interscience, New York.

Konkel, J.H. (1987). *Rule of Thumb Cost Estimating for Building Mechanical Systems: Accurate Estimating and Budgeting Using Unit Assembly Costs*. McGraw-Hill, New York.

Lippiat, B.C. and R.T. Ruegg (1990). Energy prices and discount factors for life-cycle cost analysis 1990. Report NISTIR 85-3273-4, Annual Supplement to NBS Handbook 135 and NBS Special Publication 709. National Institute of Standards and Technology, Applied Economics Group, Gaithersburg, MD.

Rabl, A. (October 1985). Optimizing investment levels for energy conservation: Individual versus social perspective and the role of uncertainty. *Energy Econ.*, 7(4), 259.

24

Design for Energy Efficiency

ABSTRACT We begin by stating the design goal and describing the process that can lead to efficient building designs. It is crucial that the designer have a good intuitive understanding of the important elements of a building and their influence on energy consumption. For this purpose, we discuss the various broad categories of design elements. We then proceed to specific design procedures and energy-efficient recommendations for residential buildings including considerations of glazing, thermal mass and the use of passive solar technologies. Recommendations specific to HVAC systems for commercial buildings are then presented based on the results of previous studies. We then cover in brief several important solar technologies, including water heating, preheating of ventilation air, and photovoltaic (PV) systems. A brief overview of passive and hybrid cooling methods is also provided. Next, we address the issue of uncertainties in simulated building energy predictions and discuss the role of energy benchmarks and rating methods. We then discuss the historical and ongoing development of building energy codes and standards and describe the various compliance pathways. Finally, we provide a broad overview of green or high-performance buildings and describe current efforts to designing, constructing, and operating such buildings.

Nomenclature

A	Area, m² (ft²)
ACH	Air changes per hour, h^{-1}
C	Heat capacity, kJ/K (Btu/°F)
c_p	Specific heat at constant pressure, kJ/(kg·K) [Btu/(lb_m·°F)]
COP	Coefficient of performance of a vapor cooling system
$D_h(T_{bal})$	Heating degree-days for base T_{bal}, K·days (°F·days)
K_{tot}	Total heat transmission coefficient of building, W/K [Btu/(h·°F)]
LCR	Load collector ratio defined by Equation 24.4
M	Mass, kg (lbm)
NLC	Net load coefficient defined by Equation 24.2
p	Pressure, Pa (lb_f/in.²)
Q	Heat gains, MJ/h (Btu/h)
$\dot{Q}$	Heat rate or heat flow per unit time, W (Btu/h)
R	Thermal resistance, K/W [(°F·h)/Btu] [$\equiv \Delta T/\dot{Q}$]
S	Monthly solar radiation on aperture MJ/m² (Btu/ft²)
SC	Shading coefficient of window
SSF	Solar savings fraction
T	Temperature, K or °C (°R or °F)
$\dot{V}$	Ventilation rate, m³/s or L/s (ft³/min)
V	Volume, m³(ft³)
U	Overall heat transfer coefficient, W/(m²·K) [Btu/(h·ft²·°F)]

Greek

ρ	Density, kg/m³ (lb_m/ft³)
τ	Transmissivity
η	Efficiency

Subscripts

av	Average
aux	Auxiliary
bal	Balance point
cond	Conductive
eff	Effective
f	Frame of window
g	Glass
i	Indoor, internal
inf	Infiltration
max	Maximum
min	Minimum
o	Outdoor, external
obs	Observed or measured
proj	Projected
sol	Solar
sup	Supply
tot	Total
the	Theoretical or predicted
vent	Ventilation

24.1 Road to Efficiency

24.1.1 Introduction

The search for the "optimal" building design involves two kinds of analyses: calculation of performance and calculation of costs. Most of the building science and HVAC equipment-related concepts and performance calculational procedures have already been presented in previous chapters, and the calculation of costs has been discussed in Chapter 23. This chapter focuses on the design process, presents design recommendations, and discusses relevant codes and standards. The recommendations attempt to distill the experience gained as a result of the intensive research efforts on energy efficiency that were stimulated by the oil crisis of the 1970s. This chapter also presents concepts related to passive building elements, an aspect that has waned in importance in the last few decades, and also discusses green or high-performance buildings that are acquiring increasing relevance.

The traditional academic view has been that, in principle, the design process is an optimization problem that could be solved by brute-force enumeration of all possible designs, evaluating the cost and performance of each and then selecting the "best" one. There are two problems with this line of reasoning. One, the number of design options is so enormous that the job would be hopeless without intelligent guidelines based on past experience. Two, building design has a strong "art" element that escapes a crisp mathematical objective function formulation. It is more pertinent to view the design process as a "satisficing" search (satisfy + suffice)—a term coined by H. A. Simon in the context of economic theory (Addison, 1988). Choosing from the wide variety of innovative technologies and energy efficiency measures available today, a designer has to compensate environmental, energy, financial, and social factors in order to reach the best possible solution that will ensure the maximization of the energy efficiency of a building while satisfying the final user/owner desires and requirements. Thus, the need to address multicriteria considerations makes it more valuable for a designer to know the "latitude" or "degrees of freedom" he or she has in changing certain design variables while achieving certain preset thresholds in energy performance, life cycle cost, environmental impacts, etc. What is required then is a framework that will allow designers to *explore the consequences of decisions* relating to these variables. This necessitates a synergy between automated performance prediction/visualization and the human capabilities to perceive, relate, and ultimately select a satisficing solution. One such framework has been proposed and illustrated by Dutta et al. (2016). It involves the necessary set of simulations, data mining analysis of building energy, and a regression analysis the results of which are embedded in a computational engine, which dynamically powers a real-time visualization plot of the effect of different design variables. We envision similar sophisticated tools to evolve in the future and be used routinely by building designers.

Another, often lamented, stumbling block has been the lack of communication between architects and engineers. Some engineers may even believe that it would be futile for them to worry about the design of the envelope of a building because those decisions are usually made by the architect or the builder. That need not be the case: many thermal characteristics of the envelope, e.g., insulation thickness, have little or no conflict with any architectural aspects. Actually, an even greater obstacle on the road to efficiency has been the difficulty of carrying out sufficient analysis early enough in the design process. When accurate load calculations are not available until the design development phase (see the design phases in Table 1.2), it is too late for significant changes. This is currently changing especially in the larger design firms where advanced computer-aided design methods are being increasingly used to facilitate the design process. If load calculations and cost estimation procedures are integrated into the drafting software used by the architect, the cost implications of each design element and of each design change can be observed immediately and interactively, beginning with the initial sketches. Thus, the architect can directly benefit from the engineer's understanding of the energy aspects of the envelope. For these reasons, we do not hesitate to discuss envelope design options in a book primarily meant for engineers.

There are several market barriers to energy efficiency in buildings other than technical:

1. *Occupant behavioral* patterns (ignorance or disregard to conservation) can greatly impact energy use especially in residences.
2. *Psychological* factors (which include inertia, fear of new technology, and avoidance of risk) also plague both residential and commercial building customers, designers, and engineers.
3. *Economic* barriers generally include high initial costs and small monthly savings (savings come in small dribbles).

With traditional design practice, it has not been easy to realize energy-efficient buildings. One obstacle is the still too frequent tendency to minimize initial cost rather than life cycle cost, especially in cases where the builder does not expect to be the occupant. Such an outlook is gradually changing and life cycle costing (LCC) is increasingly becoming common practice.

This chapter is primarily concerned with the technical aspects related to the design of energy-efficient buildings. Codes and standards are important drivers, while societal aspirational goals are often reflected in policy making, road-mapping and setting of targets, financial rebates, and grassroot movements. These aspects are discussed in Section 24.8.

24.1.2 Design Goal

Quite generally, efficiency is easy to define for processes with a single input and a single output, both of which are quantifiable. Efficiency is a valuable concept, and we have already encountered it in the analysis of HVAC equipment. For a building as a whole, however, a precise definition of efficiency remains elusive. Already on the input side, there is the complication of different forms of energy, even if one considers energy as the only input—quite apart from the question of whether the relevant input should be taken as money rather than energy. The output defies quantification since it involves all the services that a building is supposed to provide: protection from the elements, a comfortable and quiet indoor environment, and good air quality, not to mention aspects such as security, prestige and aesthetics.

Therefore, at the level of the building as a whole, we attempt only a qualitative formulation of efficiency: *an efficient building is one that provides the required conditions of comfort, convenience, etc., under the specified conditions of utilization for the lowest life cycle cost.** It is in this sense that we describe the design goal as the search for efficiency.

We also note a further complication: the difficulty of foreseeing the future utilization of the building. For example, suppose that a building has been designed for internal gains of 15 W/m^2 but the occupants bring extra equipment that raises the gains to 30 W/m^2 (a common occurrence during the onset of the computer revolution about 30 years ago); surely, the design would have been different, had one known this in advance. Therefore, it is advisable to plan for flexibility and to avoid designs that are too vulnerable to changes in occupancy or utilization. A few additional sensitivity analyses during the design phase can provide the needed insights.

24.1.3 Design Process

How can designers know how close they have come to the optimal design or whether and how they could reduce the energy consumption even further? As stated previously, the search for the optimal design is not a straightforward optimization problem. The number of design options and variables is so large, especially for commercial buildings, that optimization by exhaustive search is simply not feasible, at least with the computers of the foreseeable future. Rather the search must be guided by intelligence involving a combination of search (by parametric studies) and guidance (by performance indices). Experience and intuition are essential. To help develop an intuitive understanding of the influence of various building elements on energy consumption, we present general rules and recommendations in Section 24.2.

In *parametric studies*, one varies one parameter at a time and displays the effect on loads (peak and average) and on costs. Good graphic display of the results is important so the designer can see in which direction to pursue the search. It is instructive to try to identify, as far as possible, the contributions of individual elements of the design, e.g., the heat loss due to air exchange. This was illustrated in Example 10.5 using the degree-day method.

For judging the performance of a design, there are two types of criteria. One is to *compare the observed annual energy consumption per unit of floor area with that of other buildings* that serve the same function. Such data are presented in Section 24.7. Preferably, the comparisons should be broken down by end uses (heating, cooling, lighting, ventilation, as well as peak loads), assuming, of course, that one has the necessary data.

But relying on observed data alone is a bit like the blind leading the blind: there is no guarantee that others are efficient. To go beyond that, one needs *theoretical guidelines*. There are two types of guidelines: systematic simulation studies and fundamental limits. A wide range of parametric studies of hypothetical buildings have been carried out, especially with the goal of formulating performance standards for buildings. The collective experience gained by such studies has been incorporated in standards such as ASHRAE 90.1 (2013) or ASHRAE 189.1 (2014). Examples of fundamental limits for heat pumps (Section 14.10) and of secondary air delivery systems (Section 19.7) have been previously provided.

24.1.4 Role of Building Type and Utilization

The design process depends on the type and utilization of the building in question. For instance, if skin loads dominate, one should concentrate one's efforts on improving the thermal performance of the envelope. High internal gains suggest that one should examine the efficiency of the lighting and pay special attention to the efficiency of the HVAC system. If fresh air requirements are large, a look at heat recovery is advisable. These matters depend on the type and utilization of the building. Therefore, we

* Many people, including the authors, have used the term "energy-efficient building" to express this idea. But in a literal sense, that term is not meaningful, for if one tried to define energy efficiency by substituting *energy* for *cost* in the statement of the design goal, the conclusion would be absurd: without cost constraints one could add insulation, heat recovery, and the like, ad infinitum, reducing the energy consumption of a building to zero.

TABLE 24.1

Some Typical Differences between Commercial and Residential Buildings

	Commercial Buildings	Residential Buildings
Occupancy	Mainly daytime, rarely nights/weekends	Almost always at night, often reduced or variable/irregular during day
Density (occupants per floor area)	Depending on building type; in office buildings relatively high density (around 0.05–0.1 person/m²)	Usually low density (around 0.02 person/m² in wealthy countries)
Internal gains	Often relatively large	Relatively small
Ventilation	Usually central; outdoor air by central ventilation (in the United States)	Outdoor air by infiltration (in the United States)
HVAC system	Usually (in the United States) large, central, and relatively complex systems	Simple; usually single zone; air not shared between residential multifamily units
Size	Often quite large	Small (in the case of houses)
Surface/volume ratio	Often fairly small	Usually quite large, even in apartment buildings
Important loads	Internal gains tending to be more important than envelope loads (lighting and cooling often dominant)	Envelope loads tending to be more important than internal gains (heating dominant in temperate and cold climates)
Design	Usually custom design	More or less mass produced

have listed in Table 24.1 some typical differences between the two major categories of buildings: commercial and residential. Of course, it should be understood that such a table is a gross simplification in view of the enormous variety of commercial buildings.

24.2 Design Elements and Recommendations

24.2.1 Categories

The process of designing a building involves many choices, from the conception of the envelope to the specification of the HVAC equipment. As a general guide, we have listed in Table 24.2 the major elements, grouped in four categories: (1) environment, (2) structure, (3) equipment, and (4) system and controls.

As far as is practical, we have suggested simple rules to indicate how to choose each element for maximal energy efficiency *if cost were no object*. The rules are discussed in the four sub-sections below. Between some of these elements, there is strong coupling. In particular, the interaction between the envelope and the HVAC system makes coordination between architect and engineer critical.

Obviously such a simple table cannot capture all the subtleties of building design (Burt Hill Kosar Rittelmann Associates, 1985; Crabb, 1988). There are an enormous variety of building types and utilization patterns, and in particular, there are great differences between residential and commercial buildings (see Table 24.1). Nonetheless, we believe that a simple summary is a good starting point for honing one's intuition. The remainder of this section presents more specific comments on the elements of Table 24.2.

24.2.2 Environment

Here, we have factors such as site and vegetation that are somewhat difficult to quantify. The site affects wind-induced heat transfer as well as solar gains due to shading by adjacent buildings or trees or due to reflected radiation. But in many situations, external constraints leave little freedom to influence these factors.

Shading should be carefully taken into account; it affects not only heating and cooling loads but also daylighting. A particularly noticeable (and notorious) site effect occurs when a new structure is put up next to an existing building with large windows, casting either a shadow or a large amount of reflected radiation on it.

Wind patterns can affect infiltration and the surface heat transfer coefficient. Wind breaks can save heating energy, but they reduce the ventilation that is desirable for summer comfort in buildings without air conditioning. The tighter and better insulated a building, the lower its sensitivity to wind. Mechanical ventilation can also reduce the sensitivity to wind, as shown in Section 6.4.3. In current conventional construction with infiltration rates below 0.5 (air change per hour) and a reasonable amount of double glazing, the sensitivity of the total heat loss coefficient K_{tot} to wind does not appear very significant. For instance, K_{tot} of the house in Example 7.6 increases only 10% between the wind speeds of 7.5 mi/h (3.4 m/s) and 15 mi/h (6.7 m/s) of the value recommended by ASHRAE Fundamentals (2013) for summer and winter design conditions, and most of that variation is due to infiltration. Future construction is likely to be even less sensitive to wind. Wind patterns are the most relevant in buildings that are the most difficult to change—around old buildings.

Plants, especially large trees, can provide multiple benefits, breaking wind currents, attenuating noise, and offering shade and evaporative cooling, to say nothing about

TABLE 24.2

Design Elements and Their Effect on Energy Consumption

Element	Variable	Recommendation (If Cost Were No Object)
Environment		
Site and orientation		Pay attention to climate and to shading by surrounding objects (and to wind if natural ventilation is important).
Vegetation		Plants offer wind break, sound attenuation, shade, and evaporative cooling, plus beauty.
Structure		
Shape	S/V	If all else is fixed, heat loss is minimized by minimizing surface per volume.
Internal walls	Mc_p	Benefit of thermal mass depends on thermostat control (can store heat gains, but inhibits savings from thermostat setback).
Foundation		Do not overlook heat loss to ground.
		Ground coupling may be beneficial for cooling.
External walls, roof	U, K_{tot}	This should be as low as possible.
Doors		Use insulated doors with seals.
		If many door openings per day, install vestibule or revolving door.
	$\dot{V}_{inf}$	The doors should be no larger than needed for comfort. Use tight construction. Mechanical ventilation allows better control of air exchange than natural infiltration.
Windows		Use tight, insulating frames for windows.
		Movable night insulation for windows can be very effective.
Shading devices	$\dot{Q}_{sol}$	Use shading device as needed to control solar heat gains.
		Exterior shades are more effective than interior ones (especially if glass has high absorptivity).
		Solar gains are beneficial for heating, but undesirable for cooling. Beware of large east or west glazing without solar control.
Skylights	Daylight	This can save energy if lights are controlled by daylight sensor (but be careful with heat gains, especially from large, unshaded south-facing skylights).
Sunspaces		Sunspaces can save heating energy if temperature is allowed to float.
Equipment		
Service hot water		
Heating plant	η (COP)	This should be as high as possible.
		Heat pumps are more efficient than resistance heat, especially if source is not too cold.
Cooling plant		
Fans, pumps		They should have variable speed/capacity for variable load.
Air outlets	Capacity, Throw	Avoid stagnation of air.
Lights	Luminous efficacy	Aim for high luminous efficacy (lm/W).
System and control		
System		Minimize simultaneous heating and cooling.
Zoning		Match utilization, heat gains, and losses.
Ducts, pipes	ΔP	The pressure drop should be as low as possible.
Airflow	$\dot{V}_{sup}$	Use variable rather than fixed amount.
Air supply temperature	T_{sup}	Use variable rather than fixed amount.
Thermostat	T_{min}, T_{max}	Set limits as wide as allowed by comfort.
Building management system		Provides diagnostics (compare actual and predicted performances).
		Detect degradation of equipment by comparison with historical records.
Sensors (daylight, occupancy)		Turn equipment off when not needed.
Economizer	$\dot{V}_{vent}$	Use this mode of operation instead of chiller whenever possible (enthalpy control is more efficient than temperature control).

(Continued)

TABLE 24.2 (*Continued*)

Design Elements and Their Effect on Energy Consumption

Element	Variable	Recommendation (If Cost Were No Object)
Night ventilation		Consider if sensible cooling outweighs fan energy and latent gains; this is especially effective in massive buildings.
Cool storage	Q, Q, η	Look at possible savings from lower electricity cost at night (energy and demand charge) and from reduced chiller capacity.
Hot storage		Look at possible savings from lower electricity cost at night (energy and demand charge) and from reduced heating equipment capacity. Also consider if free heat is available (e.g., from heat recovery) but mismatched in time from heating loads.
Heat recovery		Consider if free heat is available (e.g., from condenser of chiller or from exhaust ventilation).

Braces and arrows indicate connection between elements, variables, and recommendations. In practice, the recommendations must, of course, be tempered by cost and other constraints.

their beauty. A number of studies have emphasized the cooling benefit of plants (LBL, 1988). For example, by comparing air temperatures in neighboring sites with and without forest cover, differences around 4°F (2°C) have been observed by Akbari et al. (as reported in LBL, 1988). Meier (1990) has reviewed all the publications till then that report measured data for the effect of vegetation on cooling, and his findings are summarized in Table 24.3. Part (a) shows reductions in surface temperature; part (b) shows reductions in energy consumption that can be attributed to the vegetation. Despite the enormous variety of vegetative landscaping, the surface temperature reductions in Table 24.3a are quite similar, mostly in the range of 20°F–40°F (10–20 K) for peak conditions. The savings in cooling energy range from about 25% to 90%. The cooling benefits of plants can be very important!

On the other hand, compared to conventional equipment, the engineer may find the use of plants less predictable. Mechanical equipment, such as a chiller or a shading device, can usually be counted on to be ready in time for occupancy, and there is much less uncertainty about its performance. As for the landscaping, the engineer may have little influence, and then plants may take many years to grow to full size. And, of course, plants impose certain maintenance costs, for watering, fertilizing, removal of weeds and dead leaves, etc. These obstacles may be more serious in practice than the fact that the heat transfer mechanisms are complex and difficult to analyze with precision. Again, factors beyond energy alone may dominate the decision. In any case, it is important for the decision makers to be aware of the cooling benefits of plants.

24.2.3 Structure and Envelope

For a given floor space, the environmental driving terms are minimized by minimizing the *surface-to-volume ratio*. Since spherical shapes are rarely suitable for buildings, the cube is the best according to this criterion. But in practice the shape is likely to be determined by other considerations. In particular, people like windows, and if one tries to maximize the number of rooms with windows, one ends up with a large surface-to-volume ratio.

The main effects of the structure on the thermal behavior of a building can be characterized by the following five variables, shown in Table 24.2:

1. Total heat transmission coefficient K_{tot}
2. Air exchange rate $\dot{V}$ due to infiltration and/or ventilation
3. Thermal mass (= total effective heat capacity C_{eff})
4. Solar heat gains $\dot{Q}_{sol}$
5. Utilizable daylight

The last is relevant for energy efficiency only to the extent that daylight has a value, e.g., by virtue of appropriate control of the electric lighting. The light entering through windows, skylights, and other solar apertures can help save electricity for lighting—but only if lights are turned off in response to daylight. If properly designed and functioning correctly, a building with *daylighting* can save much energy, as discussed in Section 22.5.

The lower the *U values* of the building envelope, the lower the heating load and the lower the temperature-dependent component of the cooling load. The recommendation for low *U* values bears on comfort, too. The better the envelope, the lower the risk of discomfort caused by temperature gradients inside the building, by drafts from cold leaky windows, or by low radiant temperatures in perimeter zones. This is an instance where comfort and energy conservation go hand in hand.

A sizable part of the heating and cooling loads is proportional to the air exchange rate. Of course, a minimum of fresh outdoor air is required for comfort and health. But uncontrolled *infiltration* will rarely provide just what is needed. Infiltration depends on the details of the construction that are impossible to predict with precision, and it will vary with wind speed and temperature difference. The resulting air exchange is likely to be much higher than the optimum in winter, and in summer, it

TABLE 24.3

Measured Data for Effect of Vegetation on Cooling

Author and Year	Location and Climate	Type of Planting	Wall–Vegetation Distance	Difference Measured	ΔT, °C	Notes
(a) Surface temperature reductions						
Parker (1981)	Miami, FL, hot/humid	Shrubs and trees	Shrubs < 1 m, trees < 10 m	Wall with and without plants	16	Westwall, 5 p.m., maximum value about 1 month apart
Hoyano (1988)	Tokyo, Japan, hot/humid	Ivy covering	Touching	Wall with and without ivy	18	Westwall, 3 p.m., maximum value; 1 year apart
Hoyano (1988)	Tokyo, Japan, hot/humid	Dense canopy ever-greens (Kaizuka hort)	0.2–0.6 m	Wall and inside plant surface	5–20	Westwall, 3 p.m.; parallel measurements
McPherson (1989)	Tucson, AZ, hot/dry, desert	18 shrubs and 5 cm decomposed granite	0.5 m	Wall with shrubs and no shrubs	17	Westwall, 3 p.m.; different buildings
McPherson (1989)	Tucson, AZ, hot/dry, desert	Turf, extending about 5 m from structure	Surrounding building	Wall with turf versus decomposed granite	6	Westwall, 3 p.m.; different buildings
Makzoumi and Jaff (1987)	Baghdad, Iraq, hot/dry desert	Vine (*Luffa cylindrica*) on trellis	0.1–0.4 m	Wall with and without vines	17	Southwest wall, 3 p.m.; maximum value; different building
Harazono (1989)	Osaka, Japan, hot/humid	Rooftop hydroponic using lightweight planting substrates and mixed plants	0.1 m	One-half of roof with, one-half without	21	Average for 10 a.m. to 6 p.m. on clear August day
Halvorson (1984)	Pullman, WA, temperate	Vertical vine canopy	n.a.	Wall with and without vine	20	
(b) Energy savings						
Air conditioning			**Energy Measurement**	**ΔE**	**ΔE%**	
Dewalle et al. (1983)	Central Pennsylvania, temperate	Forest site versus clear site	AC electricity for identical mobile homes	230 W	80	37-day test period
Parker (1983)	Miami, Florida, hot/humid	Shrubs and trees	AC electricity with and without landscaping	5000 W	58 24	6 h (afternoon) Test period 10-day periods
McPherson et al. (1989)	Tucson, AZ, hot/dry desert	Shrubs surrounding model house	AC electricity with and without shrubs	104 W	27	2-week period
McPherson et al. (1989)	Tucson, AZ, hot/dry desert	Turf surrounding model house	AC electricity with and without turf	100 W	25	2-week period
Parker (1990)	Palm Beach, FL, hot/humid	Miscellaneous trees and shrubs	Annual electricity for whole houses	1.8 W/m²	34	Inferred from regression of 25 houses, from landscape class 0→2,3
Heat gain						
Hoyano (1988)	Tokyo, Japan, hot/humid	Vine-covered wall	Heat gain with and without vines	175 W/m²	75	Peak value at 4 p.m. on west wall
Hoyano (1988)	Tokyo, Japan, hot/humid	Row of evergreens next to wall	Heat gain through wall	>60 W/m²	>50	Peak value at 4 p.m. on west wall for widely spaced trees
Harazono (1989)	Osaka, Japan, hot/humid	Rooftop vegetation	One-half of roof with, one-half without	130 W/m²	90	Average from 10 a.m. to 4 p.m.

Source: Meier, A.K., Measured cooling savings from vegetative landscaping, in *ACEEE 1990 Summer Study on Energy Efficiency in Buildings*, vol. 4, American Council for an Energy Efficient Economy, Asilomar, CA, 1990, p. 133.

may not be enough. In climates with large heating or cooling loads, airtight construction with mechanical ventilation is more conducive to energy efficiency than uncontrolled infiltration. If one opts for mechanical ventilation sufficient to supply the required flow of fresh air, then the envelope should be very tight. Swedish houses set a good example with rates around or below 0.1 ach. Mechanical ventilation opens up a further opportunity for energy saving, namely, heat recovery.

Windows demand the greatest attention. It is through the windows that the environmental driving terms affect the building most strongly. With conventional construction, conductive heat transfer and solar heat gains are an order of magnitude larger per unit area through glazing than through opaque walls. *Windows can make the difference between efficiency and waste.* One very effective way of reducing heat losses through windows is to cover them with movable insulation at night. Although this is not common practice in the United States,* automated shutters could be sufficiently convenient to be accepted.

Architectural considerations tend to override U values when it comes to windows. Fortunately, the technology of windows is advancing. Windows with low heat loss coefficients are commercially available thanks to advances in low-emissivity coatings, aerogels, and evacuated panels. Combining low U values with control of solar transmissivity, via adjustable shading devices or electrochromic coatings, one can produce windows that are more energy efficient than opaque walls. The recommendation, in Table 24.2, against large windows on east or west facades applies most strongly to conventional glazing. The better the control of solar heat gains, the more freedom one has in placing windows anywhere. But in any case, large, tilted, south-facing skylights without shading should be avoided for they would suffer from high solar heat gains.

Solar heat gains are beneficial only when they reduce the heating loads; in summer, they can impose a severe penalty. It is crucial to pay attention to this term and to reduce it in summer, by using shading devices or glazing with a low shading coefficient. Solar heat gains tend to be less beneficial in commercial than in residential buildings because the gains coincide with the rather substantial heat gains from lights, office equipment, and occupants. Indeed, for many commercial buildings, cooling loads are more important than heating loads, even in cold climates.

As for *shading devices*, a basic choice is between interior and exterior shades. Interior devices can be lightweight and inexpensive, but they reduce heat gains only to the extent that they reflect radiation back out of the building. In practice, a sizable fraction of the solar radiation is absorbed by the interior shades or by the glazing (on the way in or on the way out). The ability to withstand wind loading makes exterior shades more costly, especially if they are adjustable or movable, but their thermal performance is better.

Since the 1970s, *sunspaces*, such as greenhouses and atria, have been in vogue. Even though they are frequently advertized as energy savers, the real savings depend on the details of design and control, in particular on occupant behavior. From a thermal point of view, a sunspace is very much like the airspace inside a double-glazed window, with extra surface for absorption of solar radiation and extra mass for its storage. *If* the temperature of the sunspace is allowed to float, then the air can act as a layer of insulation and solar heat gains can be stored by the thermal mass of the sunspace. However, if the sunspace is conditioned, the thermal boundary of the building is, in fact, moved to the outer glazing; the resulting load is larger than that in a building without this sunspace because the glazing provides less insulation than the opaque wall that would have been the thermal boundary.

Therefore, *to save energy, a sunspace must remain unconditioned.* Like adding insulation, adding a sunspace can save energy; but unlike savings from insulation, savings from sunspaces are highly dependent on occupant behavior. Furthermore, in summer, a sunspace may become unbearable unless there is effective control of solar gains and ventilation. In winter, there is the risk of drafts and frost in the sunspace. In practice, the value of energy savings, if any, is often secondary to other considerations: the cost of the structure and its aesthetic appeal.

Thermal mass† is a more complicated matter. The effects of interior mass and mass in the envelope differ somewhat, even though we have sometimes used a single variable for simplicity: the effective heat capacity C_{eff}. The effect of thermal mass depends on the control of the thermostat. If the thermostat is always kept at the same setting, the heat capacity in the interior of the building has almost no effect: heat exchange with the mass is negligible. If one tries to save energy by setback during unoccupied periods, thermal mass is an obstacle because it slows the cooldown and necessitates greater capacity of the heating equipment; the setback recovery occurs during the early morning hours just at the time of the peak heating load. During the cooling season, this requirement for excess capacity is less of a problem because the peak cooling load is not coincident with setback recovery.

But thermal mass can also be desirable, especially during the cooling season or in buildings with passive solar heating. Peak cooling loads, as shown in Chapter 9, are significantly reduced by thermal mass. Even if the thermostat set point is constant, the temperature of the mass can differ somewhat from that of the air in the building, and

* By contrast to Europe, where external or internal shutters are a common feature in residential buildings and where people are well accustomed to the routine of closing them at night—often more for security than for comfort or energy conservation.

† Also known as *thermal inertia or heat capacity*. See Section 8.1 for a discussion of the effective heat capacity.

the mass can soak up radiative energy from lights and solar radiation. The building shell can buffer the exterior heat pulse. In a climate where days are hot and nights are cool, while the 24 h average stays within comfort limits, the interior of a massive building can remain very comfortable without any space conditioning, whereas a lightweight structure would be intolerable. This principle can be applied in any building where there are substantial heat gains during part of the day, which can be stored by the thermal mass, to be released later.

Thus, the recommendation for thermal mass involves a trade-off between two opposing strategies to reduce heating energy: reduction in the indoor–outdoor temperature difference by thermostat setback, on one hand, and utilization of free heat gains, on the other hand. In addition, *thermal mass, especially thermal mass in the envelope, reduces peak cooling loads*; thus, it can save equipment capacity. The bottom line depends on the climate, temporal distribution of heat gains, and occupancy pattern. Both heating and cooling should be taken into account. We present further analysis of this issue in Section 24.3.2, showing that thermal mass is beneficial for energy conservation in typical residential applications in the United States.

In practice, the potential savings may not be sufficient to justify a modification of customary construction methods for the sake of increasing or decreasing the thermal mass. However, in some cases, it may be fairly easy to modify the effective heat capacity by modifying the heat transfer between mass and air. Mass in external walls or the roof can be coupled (decoupled) permanently by placing the insulation on the exterior (interior) surface. In commercial buildings with ceiling plenum and concrete floors between different stories, the effective mass can be increased by circulating ventilation air through the ceiling plenum, in direct contact with the concrete.

24.2.4 Equipment

In the third part of Table 24.2, the basic recommendation is very simple: *choose equipment with high efficiency* (high coefficient of performance [COP] for the case of heat pumps and air conditioners). Heat pumps are more efficient than electric resistance heating. Condensing furnaces are more efficient than noncondensing ones. Two-stage absorption chillers are more efficient than those with a single stage. Fluorescent lights and, more recently, light-emitting diodes are far more efficient than incandescent lights, and compact fluorescents. Always look for the most efficient equipment in the catalog and do at least a quick evaluation; chances are it will pay for itself. Even if it does not, one gets an idea of the relative importance of efficiency in a given situation. Information on efficient equipment can be found in several special publications, e.g., ACEEE (1989) for the residential sector and Usibelli et al. (1985) for the commercial sector. For electric equipment, one can find excellent guidelines in the publications of the Electric Power Research Institute.

The efficiency of *heat pumps* and *chillers* depends not only on the machine itself but also on the method of extracting or rejecting the heat. Chiller COPs can range from about two for air-cooled window air conditioners to six for large centrifugal chillers with cooling towers. The smaller the temperature difference across which heat is to be pumped, the higher the COP. Furthermore, the COP increases somewhat with larger capacities. The COP varies with operating conditions; nominal or design values are usually quoted for full capacity and standard temperatures. When interpreting quoted performance values, one has to be careful of the operating conditions and how much of the ancillary equipment is included, such as fans and the cooling tower. Part-load characteristics must be considered for the calculation of annual energy consumption.

While air source heat pumps for space heating are very efficient in mild weather, they suffer a double handicap in cold weather. Both their COP and their capacity decrease, reaching the minimum when one needs the most heat. For that reason, air source heat pumps in the United States have been installed mostly to the south of the 40° latitude. Typical COPs have been around 1.5, averaged over a heating season. Much better COPs can be achieved during cold weather if groundwater or the earth can be used as a heat source, in places where this option is practical. The economics of heat pumps may be improved if they can do double duty as chillers in summer. For ground coupling that is also the best arrangement because it maintains, at least approximately, the long-term thermal balance of the ground.

HVAC loads in most buildings are variable. For variable loads, there is a further rule: avoid penalties from poor part-load system efficiency. Thermal storage can be very effective for avoiding part-load operation of heating and cooling plants. For slowing down pumps and fans, *variable-speed* electric motor drives are a more efficient means than throttling or recirculating the flow.

The design and placement of *air diffusers* must not be neglected. Avoid short-circuiting or zones of stagnant air. The risk of stagnation is increased if cold air is blown into a room from below or hot air from above (be especially careful when heating tall rooms from above). The occupants may try to compensate for the lack of comfort by adjusting the thermostat and placing a greater load on the heating or cooling plant, but that does not solve the air quality problems. For the best performance and comfort, perimeter zones should have two sets of diffusers, one for heating and one for cooling.

24.2.5 System and Controls

Here, HVAC designers can test their mettle. The number of design options and variables is staggering, as one can see, e.g., from a look at the manual of the DOE2.1

computer code (Birdsall et al., 1990) where a rather terse summary of system options sprawls over 300 pages, while the technical manual for Energy Plus (2009) is over 1000 pages. The choices begin with some basic questions: Does one want heating, does one want cooling, does one want mechanical ventilation, or does one want humidity control? Sometimes, the answer hinges more on custom than on climate; e.g., most older office buildings in Europe do not employ air conditioning or mechanical ventilation even where the cooling degree-days are comparable to those in the United States.

Another fundamental decision is the choice between *local* and *central* HVAC systems. Actually, there is a spectrum of centralization. Air conditioning in a multifamily building may be supplied by window air conditioners, by air conditioners in each apartment, or by a central chiller for a building or group of buildings. *Centralization is cost-effective if economies and efficiencies of scale outweigh the cost and losses of the distribution system.* As an example of the cost crossover between the local and the central systems, Figure 24.1 compares the life cycle cost of a central HVAC system with the local alternative, as a function of the number of apartments; in this particular case, the central system offers a lower life cycle cost (at 7% discount rate) if the number of apartments exceeds six. In office buildings, one usually opts for central systems. There are also considerations besides energy. Aesthetics, comfort, and noise can have a bearing, e.g., in the preference for central over window air conditioners. If several independent users share a system, the apportioning of costs may be troublesome because thermal energy is relatively difficult to meter.

The *distribution* of thermal energy is usually accomplished by *air* or *water*. In the past, steam was a common transport fluid for heating within buildings. But steam pipes undergo large temperature excursions under part-load conditions, and inside buildings the cacophony of contracting and expanding pipes can be quite a nuisance. Another problem arises from the frequent and notorious failures of steam traps and the resulting inefficiencies.

With central cooling, air is widely used as a distribution fluid, especially in the United States. Hydronic systems are also possible, distributing chilled water by pipes to local fan coils or induction coils; such systems are, in fact, quite common in Europe. In hydronic systems, one can avoid the need for an extra drainpipe for the condensate from the local cooling coils, if the outdoor air is supplied by a central ventilation system with central dehumidification and if the latent loads inside the building are not too large (this should be checked by calculation). One must pay attention to the pressure drop in ducts and pipes, because energy consumption increases with the pressure drop.

The system and control part of Table 24.2 states some apparently simple rules, but their implementation is not always obvious. *Minimization of simultaneous heating and cooling* is just plain common sense from the point of view of thermodynamics, although it is not always the solution with the lowest capital cost. The occurrence of simultaneous heating and cooling is especially likely in systems with *electric reheat*. If such a system is poorly adjusted, the electric reheat can easily compensate to maintain comfort, but the overconsumption of energy is costly and the problem may be difficult to detect.

For *zoning*, the general recommendation is to *match the distribution of heat gains and the utilization patterns of the building*. Along the perimeter, one should plan separate zones for facades that face in different directions, to accommodate different solar gains. In large buildings,

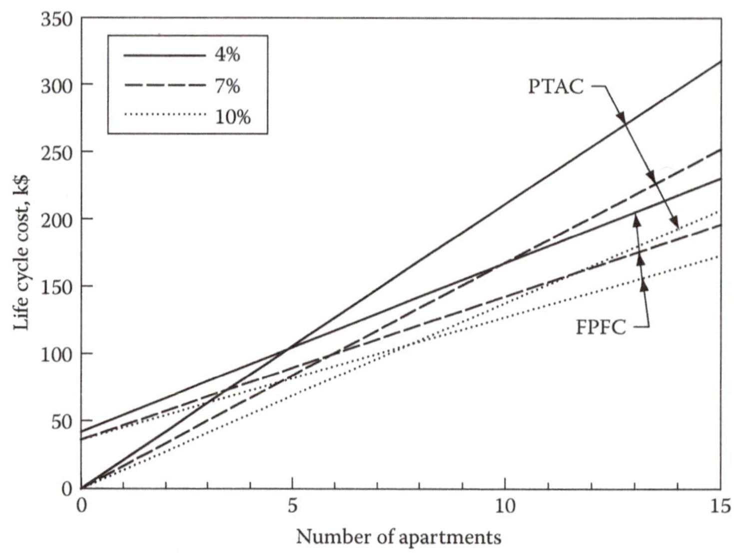

FIGURE 24.1
Life cycle cost versus number of apartments for local system and for central system (PTAC, packaged terminal air conditioners; FPFC, four-pipe fan coils) and three discount rates (4%, 7%, and 10%) in Chicago. (From Byrne, S.J. and Fay, J.M., *ASHRAE Trans.*, 95, 2, 1989.)

at least one separate zone for the interior is advisable. Computer centers, used around the clock, and conference rooms, used occasionally, should be zoned apart from offices with normal working hours. And, of course, one invites trouble when one tries to control rooms with different loads by the same thermostat, e.g., a windowless laboratory and an office with windows.

The cooling energy can be reduced by resetting the supply air temperature in response to the loads. With regard to *thermostat set points*, the recommendation is to heat and cool no more than is necessary for comfort and safety. Certainly, there should be a dead band between the onset of heating and the onset of cooling.

This brings us to a fundamental recommendation: *turn equipment down or off when there is no need for it.* This applies to all equipment, from chillers to lights and office equipment. For instance, during the morning warm-up of an office building, do not open the outdoor-air intake until occupancy begins. One of the advantages of a computerized building management system is its ability to ensure that the entire HVAC system operates just in time for each need and no more than necessary. Of course, there are items such as personal computers, printers, and copiers that one may not want to turn off each time one stops using them, simply because the start-up takes too long (with current designs and operating systems). Here, one needs a controller that senses when equipment is not in use and turns it off automatically, followed by automatic and instantaneous restart the moment it is needed again (and, in the case of computers, without any loss of information).

For lights, the solution is automatic control by daylight sensors and occupancy sensors. The latter are particularly important for places such as conference rooms that are used intermittently by a large number of people. The lights are frequently left on in such places because nobody seems to feel an obligation to turn off the switch. Daylight sensors can save much electricity in offices with windows. Note that occupancy sensors can also bring important benefits as part of the security system of a building.

The *economizer,* rather than the chiller, should be used whenever the enthalpy of outdoor air is below the specified supply air enthalpy. Enthalpy control is somewhat more efficient than control based on the dry-bulb temperature of outdoor air (provided the enthalpy sensor works reliably).

A few further design options are mentioned in Table 24.2. *Heat recovery* in all-air systems deserve special attention, and this topic has been discussed previously in Section 19.6.3. Cool storage is another attractive design option (Section 18.9) can yield considerable savings if the total cost of electricity (energy charge plus demand charge, see Section 23.2.4) at night is lower than that during the day, a common situation for large consumers. Just as important are capacity savings in buildings with daytime-only cooling loads; with storage, the same daily cooling energy can be produced by a smaller chiller running continuously. Compared to water tanks, cool storage with ice is more compact and can be bought as a prefabricated unit. For the analysis of storage, the most important variables are the energy capacity, charging and discharging rates, operating temperature range, and storage efficiency (fraction of stored energy that is recovered). Many publications on design and performance of thermal storage are available (See Section 18.9).

24.2.6 Solar Energy

There are two broad conversion areas of solar energy utilization: solar thermal and PVs. The former is concerned with the utilization of solar radiation to provide useful heating (from low-temperature domestic hot water (DHW) applications to high temperature for power production). PVs, on the other hand, apply to the direct conversion of solar energy to electricity, mainly through the use of solar cells. In the last decade or so, there has been an exponential growth in solar PV systems on buildings, especially residences.

Solar thermal energy for buildings has had its fashions, hopes, and disappointments. The oldest and so far the most successful application has been service hot water in climates with little or no frost, while active systems for space heating and cooling have turned out to be too costly, in most cases (Kreider and Kreith, 1990). But passive solar heating, the approach by which the building itself is utilized as a collector, can be cost-effective. This topic is addressed in Section 24.3.3. Another successful application is solar preheating of ventilation air (see Section 24.5.2). Yet another form of solar energy is daylight. In commercial buildings, daylighting holds much promise, especially in perimeter zones where people are sitting near the window anyway and only a light sensor is needed to reap the electricity savings. The savings achievable by daylighting have been discussed in Section 22.5.

Designing for solar energy requires careful attention to the interrelation among solar radiation, outdoor temperature, and utilization of the building. For example, the heating load of a commercial building with high internal gains and a 40 h/week occupancy is likely to be so small that passive solar heating would not be appropriate. Solar energy utilization in buildings is more difficult than energy-efficient measures such as the use of insulation or the choice of equipment with high efficiency. If one adds too much insulation, the cost of the building will be higher than the optimum, but there is little risk of poor performance or discomfort. But if one makes the collector area of a building with passive solar heating too large, the consequences for comfort may be severe.

Much experience about passive solar design for commercial buildings has been gained from a series of 19 demonstration projects that were built around 1980 and followed by extensive performance evaluation. Detailed information about these buildings and their evaluation can be found in a book by Burt Hill Kosar Rittelmann Associates and Min Kantrowitz Associates (1987). These designs involved a variety of strategies, often in combination, especially direct gain or massive storage walls for passive heating, daylighting, shading devices, and night ventilation for cooling. While it is difficult to compress the conclusions to a few lines, one can say that these design approaches can be very successful. However, the design process is more difficult (and thus more costly), and the risk of unanticipated problems is greater than with conventional design. There are risks when one uses unconventional technologies or when a design relies too much on specific behavior patterns of the occupants (e.g., manual adjustment of shading devices may not be followed as intended or occupancy schedules may change). Nevertheless, these 19 buildings were found to cost little, if any, more than conventional construction.

24.3 Residential Buildings

Most residential HVAC systems are relatively simple because they are designed to serve a single zone. This tends to be true even for multifamily buildings, with each residence having its own single-zone HVAC system. The loads are dominated by the building envelope. Thus, the basic recommendations for efficiency are insulation, airtightness, and efficient equipment, as we discussed at the start of this chapter. The prime example is Swedish housing. Their infiltration rate is so low that mechanical ventilation is used, with heat recovery from the exhaust airstream.

For residential buildings, the only major design complication arises from windows and the interaction with thermal mass. As the window area is increased, one gains free solar heat but at the price of increased conductive loss and cooling load. The trade-offs are not obvious. In particular, they depend on the thermal mass of the building. The role of the relevant window variables (area, orientation, glazing type, and shading device) for lightweight houses is discussed next. Then, we consider, in Section 24.3.2, the effect of varying the thermal mass of the walls. This will be followed by discussion of passive solar buildings where the heat capacity is greatly increased to improve the utilization of solar heat.

24.3.1 Lightweight Houses: Energy and Window *U* Values

Here, we cite an interesting series of papers by Sullivan and Selkowitz (1985, 1987). Using the DOE2.1 simulation program (Birdsall et al., 1990) with hourly weather data, these authors have analyzed the effects of varying the window size, glazing type, and building orientation for a single-family house of typical U.S. construction. The dimensions of this ranch-style house are shown in Figure 24.2; the height is 8.0 ft (2.44 m), with an additional 5.35 ft (1.63 m) for the roof structure. The wall and roof are of wood-frame construction with insulation levels of $R = 11$ and 30 (h·ft²·°F)/Btu, respectively [$U = 0.52$ and 0.19 W/(m²·K)]. The major portion of the heat capacity lies in the slab-on-grade floor: a carpet-covered 4 in. (0.1 m) thick slab of concrete on top of insulation and gravel.

The total floor area is 1539 ft² (143 m²). The window sizes are fixed on three sides at a total of 8.65% of the floor area, while the size of a fourth window, called the "primary window," is varied, such that the grand-total window area varies from 8.65% to 25.79% of the floor area. The glazing type is also varied, on all windows at the same time. To test the effect of orientation, the house is rotated in 45° increments, the primary window facing

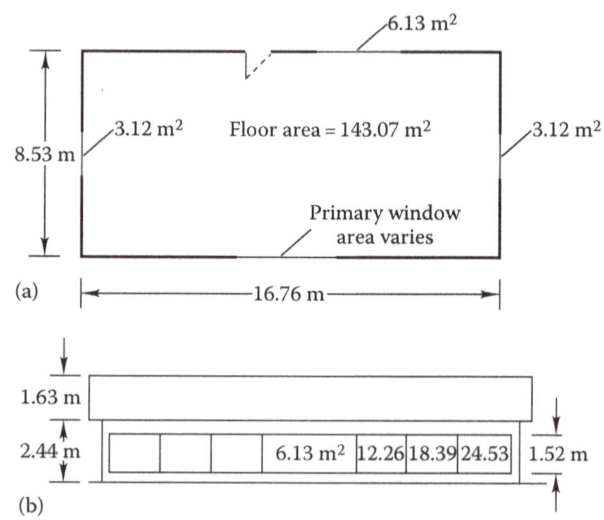

U of wall ($R = 11$)	0.091 Btu/(h·ft²·°F) [0.52 W/(m²·K)]
U of roof ($R = 30$)	0.033 Btu/(h·ft²·°F) [0.19 W/(m²·K)]
Outdoor air	0.7 ach, winter average 10 ach when natural ventilation can obviate mechanical cooling
T_i heating	70°F (21.1°C) setback to 60°F (15.6°C) 24:00 to 6:00
T_i cooling	78°F (25.6°C) all hours
Nonsolar heat gains, sensible	2248 Btu/h (659 W), all day average
Heat gains, latent	508 Btu/h (149 W), all day average
Heating system efficiency	0.74
Cooling system COP	2.174

Area of primary window is varied, while that of the other three windows is held fixed (at 3.12 + 6.13 + 3.12 m², which is 8.65% of the floor area).

(c)

FIGURE 24.2
The house analyzed by Sullivan and Selkowitz (1985, 1987): (a) floor plan, (b) facade with primary window (placed symmetrically; numbers indicate sizes considered in this study), and (c) principal assumptions.

south, southwest, etc. The simulations were carried out for three locations: Madison, Wisconsin (cold continental climate); Lake Charles, Louisiana (hot and humid climate); and Phoenix, Arizona (hot and dry climate). Here, we present an extract of their results.

Sullivan and Selkowitz employed an interesting technique to enhance the flexibility of presenting and using the results. Normally, each simulation yields results only for one set of input parameters, but here one would like the possibility of varying area A, conductance U, and shading coefficient SC continuously. The authors used linear least-squares regression analysis of simulation results at a few discrete values to obtain simple functional forms that summarize the essential trends. Thus, they obtained polynomials of first and second order for the annual heating and for cooling energy as a function of AU and ($A \times$ SC) of the windows, as well as internal gains, air change, and conductance of opaque surfaces. The functional form is linear in all these variables, except for a quadratic term in ($A \times$ SC) for the primary window. The seasonal energy consumption includes the performance of the HVAC system: an efficiency of 0.74 for heating and a COP of 2.17 for cooling are assumed.

The resulting equation makes it easy to assess the role of the windows. In Figure 24.3a through d, the annual energy for heating and cooling is plotted versus primary window area, with separate curves for different values of U and SC of the glazing. The curves shown here assume that the primary window faces (a) south for a cold climate (Madison) and (b) north for a hot climate (Lake Charles). The U values shown—1.006 Btu/(h·ft^2·°F) [5.713 W/(m^2·K)], 0.471 Btu/(h ft^2·°F) [2.675 W/(m^2·K)], and 0.302 Btu/(h·ft^2·°F) [1.715 W/(m^2·K)]—correspond roughly to single, double, and triple glazings, while 0.094 Btu/(h·ft^2·°F) [0.534 W/(m^2·K)] could be achieved with aerogel or evacuated windows. Interpolating between the curves, one can determine the consumption for any other window type. We note the following:

1. The figures show that the *heating load is very sensitive to the U value*: The smaller U, the lower the heating load (even in Lake Charles, although here the magnitude is fairly small). The *cooling load, by contrast, varies little with U*; the slight decrease with increasing U is due to the fact that the average of T_o is slightly below that of T_i [= 78°F (25.6°C)] during the cooling season, even in Lake Charles. To understand these trends, note that heating loads are dominated by the indoor–outdoor temperature difference, whereas cooling loads arise largely from heat gains, with the average indoor–outdoor temperature difference during the cooling season being rather small.
2. The variation with *shading coefficient* is as expected: a high SC reduces heating loads and increases cooling loads, which carries more weight in the total energy bill depends on the climate and on energy prices. Obviously, in hot climates, a low SC is needed for low consumption.
3. Whether *solar gains or conductive losses* are more important can be seen from the variation of the heating load with the window area. With a single pane, the heating load increases with area; with a double pane, it is roughly constant; and when U is even lower, the windows bring a definite gain during the heating season (for the orientations and climates considered here). But note that the cooling load will invariably increase with the window area (unless it is compensated by a reduced lighting load, which is not considered here because it is unlikely in residential buildings). *Therefore, if windows are proposed as energy savers, evaluation of their consequences for cooling is imperative.*
4. To show the *effect of orientation*, Figure 24.4 plots the incremental annual energy as a function of the orientation of the primary window, with the incremental annual energy being the change in annual heating and cooling energy as a result of increasing the primary window area from 0 to 198 ft^2 (18.39 m^2). In this figure, specific glazing types are assumed, as listed in Table 24.4. For cooling in Madison, only two glazings are shown because there is little difference between different types. For Lake Charles, only glazings of interest for hot climates are included. The low sensitivity of cooling loads to orientation may appear surprising, but it reflects the shade management assumed for these figures (solar gains reduced by 40% whenever direct solar gain on a window exceeded 63 W/m^2 during the cooling season). The heating loads do change significantly with orientation. For the glazings considered here, windows with northerly orientations increase the heating load, by contrast to Figure 24.4a and b where southern exposure was assumed.

Example 24.1: Annual Savings from Window Replacement

Find the annual savings for the house of Figure 24.2 in Madison at a fuel price of \$7/GJ if single-pane glazing with U = 5.713 W/(m^2·K) is replaced by double-pane glazing with U = 2.675 W/(m^2·K), the shading coefficient remaining the same at SC = 1.0. The area of the house is 143.06 m^2 and the total window is 8.65% of this value. Then total window area of 12.37 m^2 (= 0.0865 × 143.06 m^2), i.e., primary window area = 0. Express the result as savings per window area.

Given: Figure 24.3b at A = 0, SC = 1.0

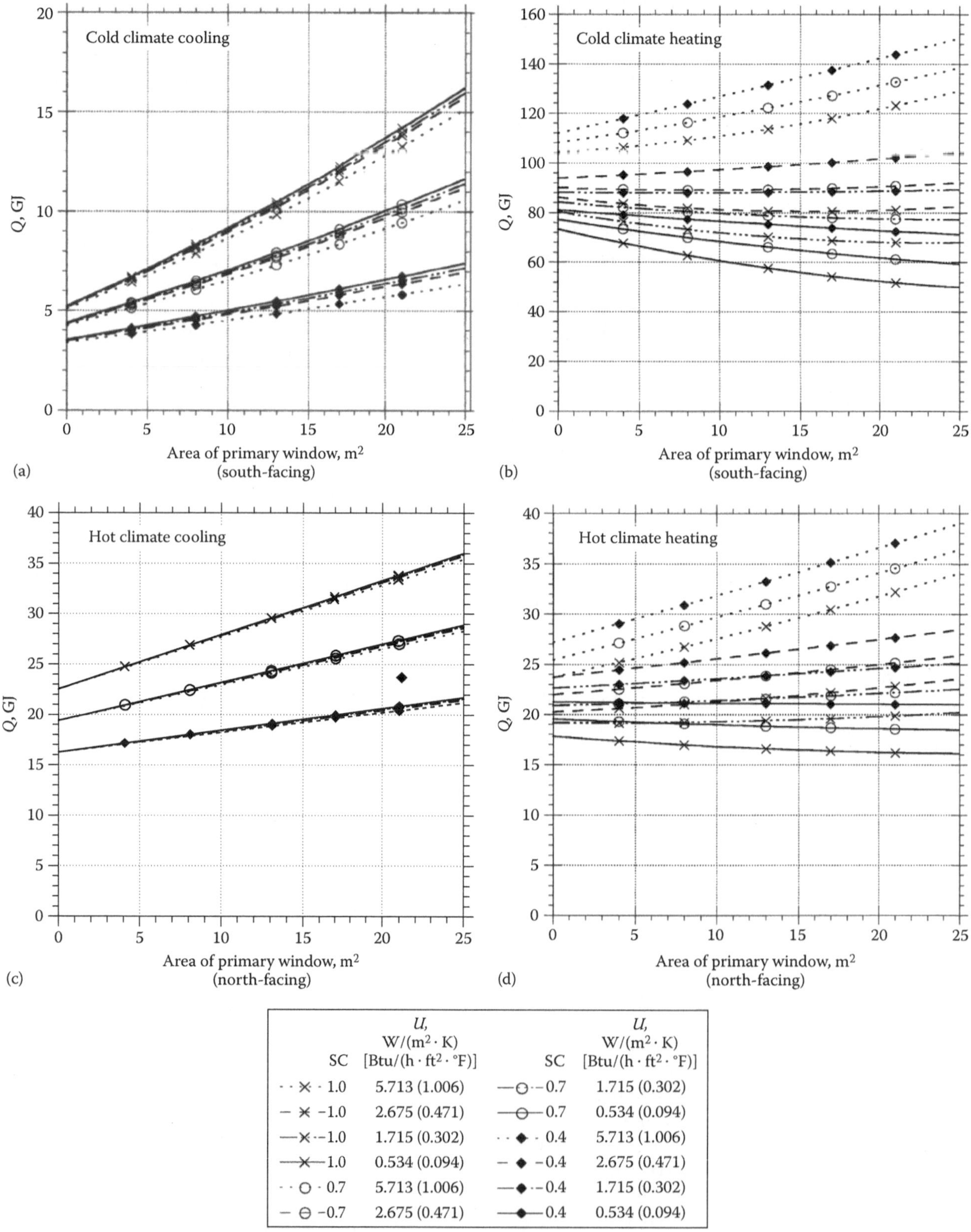

FIGURE 24.3
Annual energy consumption for heating and cooling of the house analyzed by Sullivan and Selkowitz (1985) as a function of area of primary glazing, for several values of shading coefficient SC and of U value of glazing (same glazing on primary and nonprimary windows). Note different scales in different graphs. (a) Cold climate (Madison, Wisconsin), cooling. (b) Cold climate (Madison, Wisconsin), heating. (c) Hot climate (Lake Charles, Louisiana), cooling. (d) Hot climate (Lake Charles, Louisiana), heating.

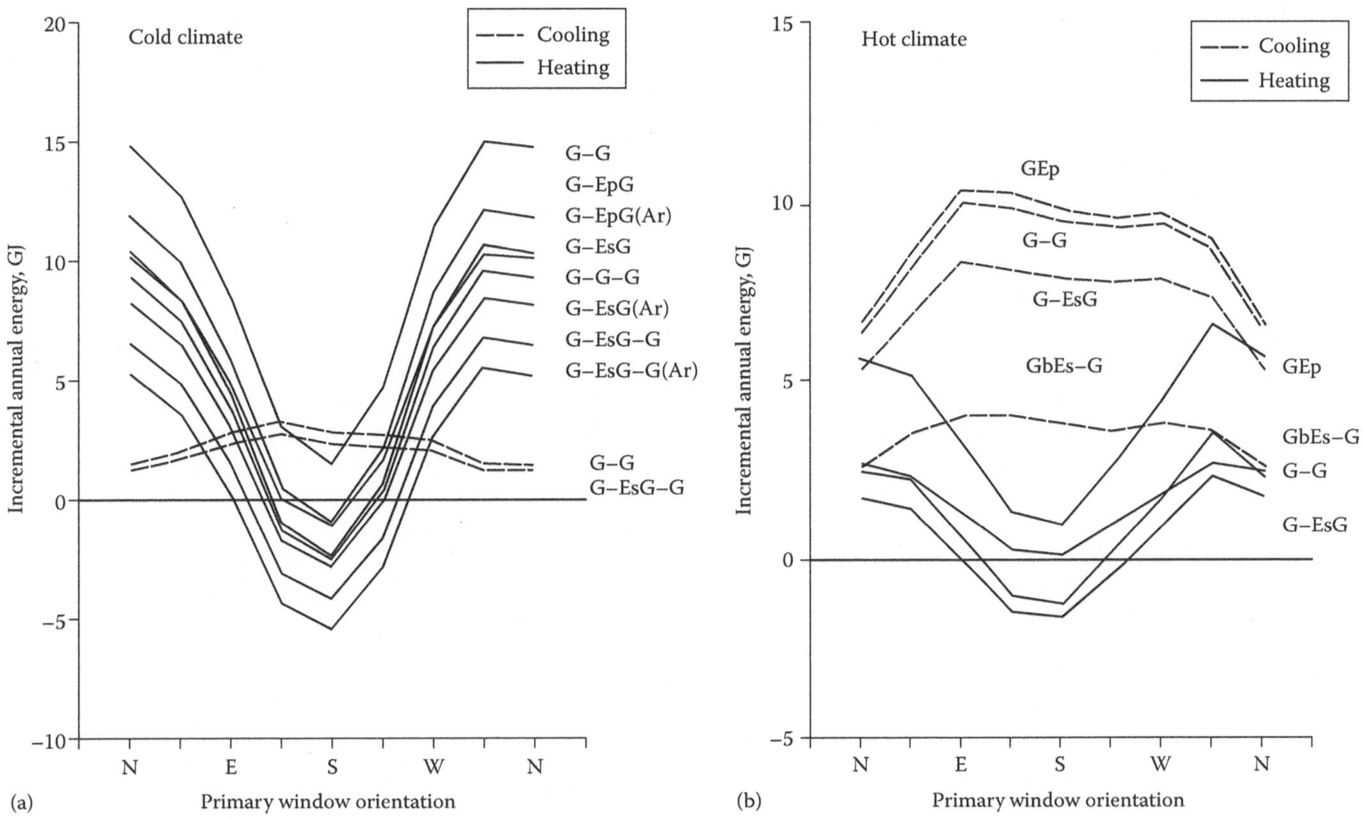

FIGURE 24.4
Change in annual heating and cooling energy for the house analyzed by Sullivan and Selkowitz (1985) as a function of orientation of primary glazing, for glazing types of Table 24.4. (a) Cold climate (Madison, Wisconsin) and (b) hot climate (Lake Charles, Louisiana).

TABLE 24.4
Glazing Properties Assumed for Figure 24.4

Window Design	Gas Fill	Winter *U* Value	Summer *U* Value	Shading Coefficient	Solar Transmittance	Visible Transmittance
G–G[a]	Air	2.85 (0.50)	3.16 (0.56)	0.88	0.71	0.82
G–G–G	Air	1.86 (0.33)	2.20 (0.39)	0.79	0.61	0.74
G–EpG	Air	2.34 (0.41)	2.63 (0.46)	0.86	0.64	0.73
G–EsG[a]	Air	1.94 (0.34)	2.00 (0.35)	0.73	0.58	0.74
G–EpG	Argon	2.09 (0.37)	2.38 (0.42)	0.86	0.64	0.73
G–EsG	Argon	1.62 (0.28)	1.68 (0.30)	0.73	0.58	0.74
G–EsG–G	Air	1.32 (0.23)	1.53 (0.27)	0.71	0.52	0.71
G–EsG–G	Argon	1.11 (0.19)	1.30 (0.23)	0.72	0.52	0.71
GEp[a]	—	5.05 (0.89)	4.54 (0.80)	0.92	0.75	0.80
GbEs–G[a]	Air	1.94 (0.34)	2.05 (0.36)	0.37	0.26	0.46

Source: Sullivan, R. and Selkowitz, S., *ASHRAE Trans.*, 91(2A), 320, 1985.
Notes:
1. The units of the *U* value are $W/(m^2 \cdot °C)$ $[Btu/(h \cdot ft^2 \cdot °F)]$.
2. G denotes glazing layer; Ep, a pyrolytic low-*e* coating (e = 0.35); and Es, a sputtered low-*e* coating (e = 0.15) on one side of the glazing; Gb denotes a bronze reflective coating. For the G–EsG–G and G–EpG units, the middle layer can be low-*e* coated glass or a low-e coated polyester film. Also, several other coatings on polyester film offer lower SCs with equivalent conductances to those shown.
3. The gap width between glazing layers is 12.7 mm (0.5 in.).

[a] Windows examined in cooling-dominated locations.

Find: Difference between curves for $U = 5.713$ and $2.675\ \text{W/(m}^2\cdot\text{K)}$

Solution

Reading the curves, we find $Q = 104$ GJ with single glazing and $Q = 86$ GJ with double glazing, a difference of 18 GJ, or roughly one-fifth of the load. This is worth 18 GJ × \$7/GJ = \$126/year.

Dividing by 12.37 m², we obtain \$10.19/m² of window per year.

Comments

Suppose the incremental cost of double glazing is \$50/m². Then, the payback time is 50/10.19 = 4.9 years. *Savings due to good glazing are very important!*

When going to high-performance windows, one should not overlook the heat losses through window frames. Sullivan and Selkowitz have found that *the U value of the window frame is crucial in cold climates.* The aforementioned results assume good insulated frames with $U = 2.0\ \text{W/(m}^2\cdot\text{K)}$ and the frame area being 20% of the window area. The effect of the frame on the average U value, denoted U_{av}, can be seen from the equation (a simplified version of Equation 5.5)

$$U_{av} = \frac{U_g A_g + U_f A_f}{A_g + A_f} \tag{24.1}$$

where the subscripts g and f refer to glass and frame, respectively. If the frame area represents 20% of the window area, we have

$$U_{av} = 0.8U_g + 0.2U_f$$

For example, with $U_g = 2.0$ and $U_f = 5.2\ \text{W/(m}^2\cdot\text{K)}$ (conventional aluminum frame *with* thermal barrier), we would find

$$U_{av} = 0.8 \times 2.0 + 0.2 \times 5.2 = 2.64\ \text{W/(m}^2\cdot\text{K)}$$

a substantial increase over the value of the glazing itself. A conventional frame *without* thermal barrier has 12 W/(m²·K), and the penalty would be even larger.

24.3.2 Effects of Thermal Mass

The previous results presented are based on a fixed value of the heat capacity of the building, corresponding to the typical lightweight construction of U.S. houses. Now, we consider how loads vary with the heat capacity. A large number of studies have been carried out to investigate this question (e.g., Byrne and Ritschard, 1985; Christian, 1991; and references cited in these articles). The general trends are clear; but due to the complex interplay of many different variables, it would be difficult to summarize the results in a form that is simple, accurate, and applicable in all situations. Therefore, we indicate only the general trends to guide the designer; for a precise analysis in a specific situation, a detailed dynamic calculation is recommended.

To begin, it is instructive to consider two extreme limits: zero heat capacity and infinite heat capacity. As pointed out by Mitchell and Beckman (1989), both cases can be analyzed with a simple static calculation, in terms of the balance-point temperature T_{bal} and heat loss coefficient K_{tot}. If the heat capacity is zero, there are no dynamic effects, and one can use the variable-base degree-day method of Section 10.2. In the limit of infinite heat capacity, the indoor temperature and the balance-point temperature approach constant values, and once again a steady-state method can be used. The difference in energy consumption between the two cases depends on the fluctuations of outdoor temperature T_o. There is no difference if T_o is always below T_{bal}. But if T_o fluctuates around T_{bal}, the zero-heat-capacity building needs heating whenever the *instantaneous* T_o is below T_{bal}, while the infinite-heat-capacity building needs heating only when the *average* T_o is below T_{bal}.

Mitchell and Beckman (1989) have evaluated these limits on a monthly time scale, i.e., interpreting *infinite capacity* to mean that heat gains are stored and reused during the course of each month. Some of their results are shown in Figure 24.5, as daily heating requirement versus monthly average outdoor temperature, for three values of the heat capacity, keeping K_{tot} and T_{bal} fixed at values representative of conventional construction in Madison. The line labeled "large capacity" is a straight line, for $T_o < T_{bal}$, whose slope is the heat loss coefficient K_{tot} of the building; for $T_o > T_{bal}$, the heating requirement is zero. The zero-capacity line approaches the large-capacity line at low values of T_o, but lies significantly above it when T_o is comparable to T_{bal}. The behavior of a conventional house is indicated by the line labeled "typical capacity." When the monthly average T_o is equal to T_{bal}, the mass effect is quite pronounced: the zero-capacity house needs 120 MJ/day, the typical-capacity house needs 60 MJ/day, and the large-capacity house needs nothing. At very low T_o, the three curves are almost the same.

This pattern remains true even when one accounts for more realistic conditions. For example, the DOE2.1 simulations of Byrne and Ritschard (1985) include nighttime thermostat setback during the heating season (from 21°C to 16°C, midnight through 6 a.m.). Their results imply that as the heat capacity is increased, the energy savings from storage of free heat in the building mass more than compensate for the reduction in savings from (moderate) thermostat setback.

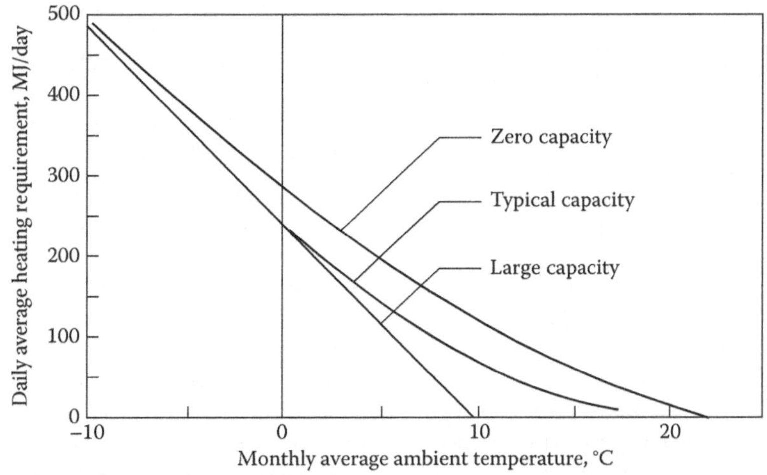

FIGURE 24.5
Daily average heating requirements as a function of monthly average outdoor temperature for a house in Madison, Wisconsin, for three values of the heat capacity. (From Mitchell, J.W. and Beckman, W.A., *Solar Energy*, 42, 113, 1989.)

Thermal mass brings significant energy savings in any case, for heating and for cooling. The mass benefits for cooling are easy to understand because during much or most of the cooling season, the daily average of T_o is below the thermostat set point for T_i. The mass averages in effect over the diurnal swings and causes the cooling load to follow the average rather than the instantaneous values of T_o.

For the practicing designer, it is helpful to show the benefit of thermal mass in terms of equivalent insulation. A table for this purpose has been prepared by the Council of American Building Officials (CABO, 1987), and we present here, in Table 24.5, an extract of a more recent and refined version that has been recommended by Christian (1991). This table shows the U value of a massive wall that is equivalent to that of a lightweight wall, in the sense of yielding approximately the same annual energy use for typical residential applications. The massive wall has a heat capacity of 6 Btu/(°F·ft²) [123 kJ/(K·m²)] that corresponds to 2.5 in. (6.5 cm) of heavyweight concrete, and three cases are considered for the placement of the insulation: external insulation, internal insulation, and insulation integral with the mass (e.g., insulation between two layers of brick or concrete). Since the savings depend on climate, separate entries are shown for different ranges of heating degree-days.

For example, in a climate with less than 2000°F·days, a lightweight wall with $U = 0.10$ Btu/(h·°F·ft²) is equivalent to a massive wall with $U = 0.16$ Btu/ (h·°F·ft²) and insulation on the outside; in other words, it needs $\Delta R = 1/0.10 - 1/0.16 = 10 - 6.25 = 3.75$ (h·°F·ft²)/Btu less of insulation. The difference between massive and lightweight walls is less important in colder climates, and above 8000°F.days the table shows no difference in U value. This is consistent with the explanation at the

TABLE 24.5
U Value of a Massive Wall That Is Equivalent in Terms of Annual Thermal Performance to a Lightweight Wall

	Equivalent *U* of Massive Wall									
°F Days	***U* of Lightweight Wall**	**0.20**			**0.10**			**0.04**		
(Base 65°F)	**Placement of Insulation**	**Exterior**	**Integral**	**Interior**	**Exterior**	**Integral**	**Interior**	**Exterior**	**Integral**	**Interior**
<2000		0.28	0.28	0.25	0.16	0.15	0.12	0.08	0.07	0.04
2001–4000		0.27	0.27	0.24	0.15	0.14	0.12	0.08	0.06	0.04
4001–5501		0.25	0.26	0.23	0.14	0.13	0.11	0.07	0.06	0.04
5501–6501		0.23	0.24	0.22	0.12	0.12	0.11	0.06	0.05	0.04
6501–8000		0.22	0.22	0.21	0.11	0.11	0.10	0.05	0.05	0.04
>8000		0.20	0.20	0.20	0.10	0.10	0.10	0.04	0.04	0.04

Source: Christian, J.E., *ASHRAE Trans.*, 97, 2, 1991.
Notes: U is in Btu/(h·°F·ft²) [1 Btu/(h·°F·ft²) = 5.678 W/(K·m²)], the heat capacity of massive wall is 6 Btu/(°F·ft²) [123 kJ/(K·m²)], and the three placements of insulation relative to mass are exterior, integral, and interior; valid for residential applications.

beginning of this section (see Figure 24.5). For interior insulation, the benefit is less pronounced than for exterior insulation, and integral insulation is intermediate between the interior and exterior cases.

To give a brief summary of thermal mass effects that have been discussed here or elsewhere in this book, we can say that thermal mass

1. Is beneficial for cooling (it reduces both peak load and energy consumption)
2. Reduces energy savings achievable by thermostat setback during the heating season
3. Improves utilization of free heat gains when $(T_{bal} - T_o)$ changes sign during the day (especially important in sunny climates)

The result is that on balance, in typical U.S. residential building designs, the addition of thermal mass can yield significant net energy savings, even during the heating season. Also, in commercial buildings, the heating energy savings from a low-mass design with long and deep thermostat setback may outweigh the other benefits from thermal mass, and a detailed analysis is required. The extra capacity required for thermostat setback recovery in winter (the "pickup load") increases with thermal mass, but this can be avoided by reducing the depth or duration of the setback during the coldest weather.

24.3.3 Passive Solar Heating

Passive solar heating relies on the capacity of the building to collect a significant amount of solar heat during the day and to store some of it to offset heat losses during the night. Storage by phase-change materials would be interesting because they require less volume (which is at a premium in buildings). But given the difficulties of finding suitable phase-change materials, one usually is limited to sensible heat storage; brick or concrete is the most common choice, although water tanks have also been used. Thus, a substantial temperature excursion, at least several degrees Celsius, is necessary if one wants to achieve a substantial solar contribution while keeping the amount of mass (and its cost) within reason.

The principal approaches to passive heating are shown in Figure 24.6a through d. The *direct gain* system (Figure 24.6a) is simply a sufficiently heavy floor (or other storage element) behind a large south-facing window. Flexibility can be gained by adding clerestory and/or circulating fan (Figure 24.6b). The window should have a low U value, and movable night insulation greatly enhances the performance. *Sunspaces* such as attached greenhouses or atria are quite popular (Figure 24.6c), and the energy savings, if any, depend on whether the sunspace is conditioned.

The *storage wall* (also known as the "Trombe wall," after its inventor), shown in Figure 24.6d, is basically a vertical south-facing flat-plate collector with a heavy

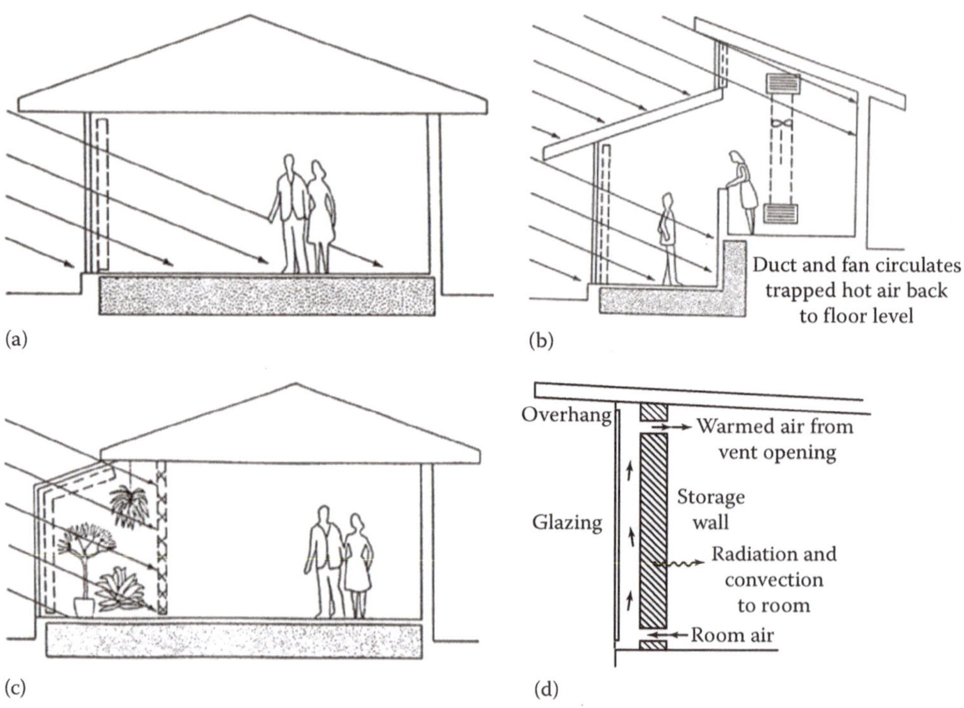

FIGURE 24.6
Schematic diagrams of passive solar heating methods. Additional option of movable insulation is shown as dotted line: (a) direct gain, (b) direct gain with additional clerestory and fan, (c) sunspace (atrium, greenhouse), and (d) Trombe wall.

brick or concrete wall as absorber and storage. The wall thickness (on the order of 0.3 m) is sufficient to store the diurnal heat input and to release it to the room over the course of the night. To improve the control of the heat transfer, vents can be added at bottom and top, as indicated. Selective coatings for the absorber boost the performance.

Passive solar heating can be very effective at reducing energy bills, but it imposes certain constraints not compatible with everybody's preferences. The building must have a large window facing a southerly direction, with a shading device to avoid overheating in summer. A Trombe wall blocks at least part of the view to the south. With direct gain, one can enjoy a full view, but furnitures and carpets should not interfere with the absorption of radiation in the floor. The intense light may bleach textiles and artworks in the room. And then there are the aforementioned temperature fluctuations.* Residents who want a steady 21°C might not be satisfied. Above all, the designer should pay close attention to summer performance. *Never recommend a passive solar design unless summer temperatures and cooling loads have been evaluated.* Otherwise, the building may be uncomfortable in summer, or the problems will be covered up by using the air conditioner all the more.

To design a passive solar heating system, one has to choose appropriate values for the solar aperture (which depends on the size, orientation, and transmissivity of the glazing) and for the storage capacity (which depends on the amount, specific heat, and distribution of the mass). The choice is made on the basis of annual energy consumption—unlike conventional heating systems that are designed on the basis of peak loads. The reason lies in the fact that stand-alone solar energy systems would be far too expensive; a backup is almost always required for peak conditions. In most climates, the coldest weather may occur during overcast periods of such duration that it is advisable to choose the capacity of the backup with a conventional peak-load calculation.

Since temperature fluctuations in a passive solar building are important, steady-state methods such as the degree-day or bin methods (Chapter 10) are not appropriate for the prediction of annual performance. There are several dynamic simulation programs (requiring hourly weather data) that are well suited to the analysis of passive solar systems, for example, BLAST (1986), SERIRES, or CALPAS3 (Berkeley Solar Group, 1981). Such programs have the advantage of providing, in addition to the calculation of annual energy, an evaluation of summer conditions. As an alternative, a variety of shorthand methods have been developed. The principal limitation of the existing shorthand methods is their inability to assess summer performance.

Perhaps best known among shorthand methods is the *solar load ratio (SLR) method*, developed as correlation of a large number of hour-by-hour simulations (ASHRAE, 1984; Jones, 1983). The correlations are simple equations with a couple of coefficients that depend on the design and construction of the building. One calculation is needed for each month of the heating season; it can be programmed with a spreadsheet. Different designs, e.g., differing in thickness of storage floor, require different coefficients. Correlations have been developed for all the standard configurations and can be found in the cited references. Here, we present only one particular example, designated sunspace D1; see Figure 24.7. The key parameters are listed in Example 24.2, and the correlation is shown as a graph in Figure 24.8.

To explain this method, one first has to address the interpretation of the performance of a passive solar element, such as a sunspace. By contrast to conventional heating systems, passive solar heating elements are an intrinsic part of the structure of a building. This is awkward if one wants to compare a building with passive solar heating to a building without passive solar heating. To resolve this difficulty, the SLR method uses a heat loss coefficient that is based on the nonsolar parts of the envelope rather than on the total envelope of the building. This coefficient is called the "net load coefficient" (NLC) and has units of Btu per degree Fahrenheit per day:

$$\text{NLC} = \frac{24\text{ h}}{1\text{ day}}\left(\rho c_p \dot{V} + \sum_{nonsolar} UA\right) \tag{24.2}$$

It differs from K_{tot} of Equation 7.21 by the exclusion of the solar elements of the envelope and by the factor 24 h/1 day. When comparing a solar building and a nonsolar building, one assumes that the solar parts of the

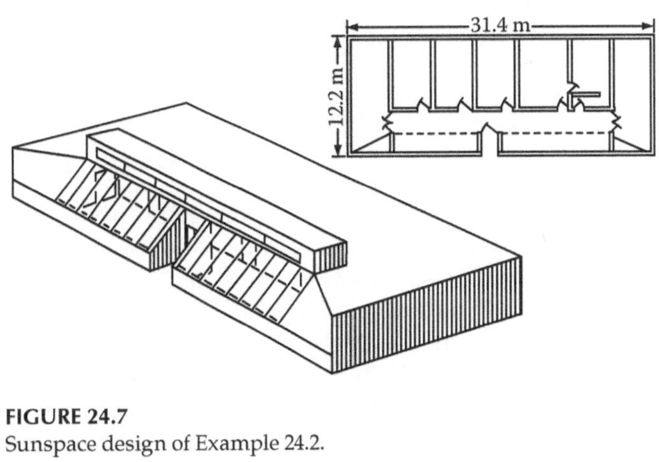

FIGURE 24.7
Sunspace design of Example 24.2.

* For instance, the *solar load ratio* (SLR) design method, discussed in the following, allows a range from 18.3°C (65°F) to 23.9°C (75°F).

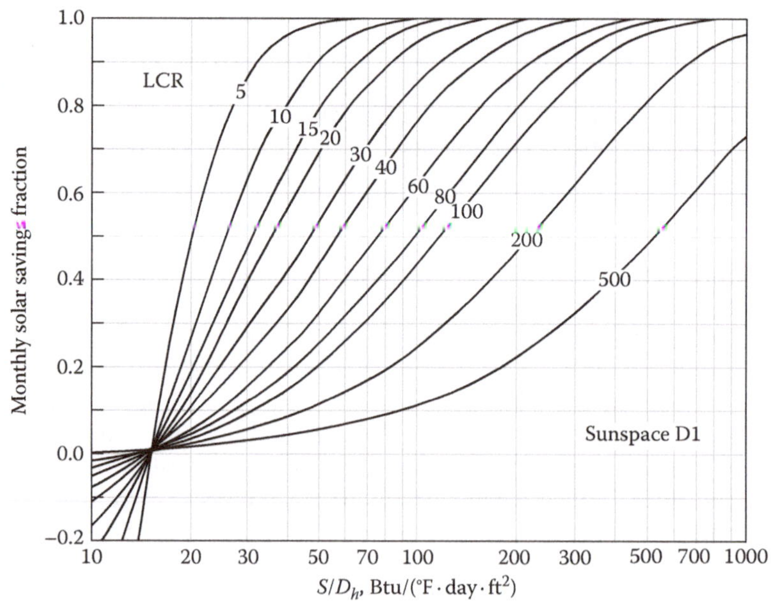

FIGURE 24.8
Correlation for monthly solar savings fraction (SSF) for the sunspace design (D1) whose key parameters are listed in Example 24.2 (From Jones, R.W., ed., *Passive Solar Design Handbook*, vol. 3, American Solar Energy Society, Boulder, CO, 1983.)

envelope are replaced by an energy-neutral wall (i.e., an opaque wall with zero conductivity).

Another key parameter of this method is the *projected collector area* A_{proj}; it is the projection of the aperture area of the solar elements on the vertical south-facing plane. The area is counted net, i.e., excluding mullions and shaded portions. For example, the building in Figure 24.7 has a sunspace with south-facing glazing of net area 51 m², its normal being tilted at 50° from the vertical; its projected area is $A_{proj} = \sin 50° \times 51\ \text{m}^2 = 39\ \text{m}^2$.

As for weather data, the method requires the *monthly solar radiation S* on the collector area and the *monthly heating degree-days* $D_h(T_{bal})$ based on the balance-point temperature T_{bal}. The latter is calculated as in Equation 10.4, but excluding the solar gains of the aperture. In terms of the NLC and daily solar gain $\dot{Q}_{gain,day}$ (in Btu per day), we have

$$T_{bal} = T_i - \frac{\dot{Q}_{gain,day}}{\text{NLC}} \tag{24.3}$$

For example, in Figure 24.7, the solar gains from windows on the east, west, and north facades are included, but those from the sunspace are excluded from the calculation of T_{bal}. The solar radiation S on the aperture can be calculated from the common horizontal insolation data by using the methods described in Section 4.9.

Having determined these quantities for the design in question, one obtains the *solar savings fraction* (SSF) from an empirical linear correlation with two ratios: the ratio S/D_h and the load collector ratio (LCR):

$$\text{LCR} = \frac{\text{NLC}}{A_{proj}} \tag{24.4 IP}$$

Note that the correlation is dimensional and LCR must be stated with the dimensions of Btu per degree Fahrenheit per day per square foot. Finally, the auxiliary heating energy to be supplied by the backup is

$$Q_{aux} = \text{NLC} \times D_h \times (1 - \text{SSF}) \tag{24.5}$$

where SSF is the fraction of the load NLC × D_h that is supplied by the solar elements. This procedure is repeated for all months of the heating season to obtain the annual total.

Example 24.2: Use of the Solar Load Ratio Method

Find the auxiliary heating energy for the building of Figure 24.7 in Denver, Colorado, in January.

Given: T_i = 68°F (20°C)

$$\dot{Q}_{gain} = 4200\ \text{Btu/h}\ (1231\ \text{W})$$

Sunspace design type D1 (12 in. [0.3 m] masonry wall between sunspace and conditioned space)

Sunspace being unconditioned and Semi-enclosed by conditioned space

Aperture of sunspace double glazed with an area of 550 ft² (51.1 m²), its normal being tilted at 50° from vertical

South facing, without night insulation

The envelope has the following characteristics:

	A or $\dot{V}$, ft² or ft³/min	U, Btu/(h·°F·ft²)	UA or $\rho c_p \dot{V}$, Btu/(h·°F)
Opaque walls	2,000	0.040	80.0
Roof	3,000	0.030	90.0
Floor (over crawl space)	3,000	0.040	120.0
Windows (east, west, north)	100	0.550	55.0
Subtotal			345.0
Infiltration	200 ft³/min (at 0.5 ach with V = 24,000 ft³)	$\rho c_p = 0.015$ Btu/(ft³·°F) in Denver	180.0
Subtotal			*525.0*
Sunspace	550	0.55	302.2
Total			827.2

The heat loss coefficient without the sunspace is indicated in italic as 525.0 Btu/(h·°F) (277 W/K). Hence, the NLC is

$$\text{NLC} = 24\ \text{h/day} \times 525.0\ \text{Btu/(h}\cdot\text{°F)}$$
$$= 12{,}600\ \text{Btu / (°F}\cdot\text{day)}$$

Find: Q_{aux}

Lookup values: We need to look up solar radiation S and degree-days D_h for January in Denver. (For simplicity, we take these values as known; they could be calculated by using the methods of Sections 4.9 and 10.3.)

$$S = 43{,}600\ \text{Btu/ft}^2$$
$$= \text{Monthly solar radiation on aperture}$$

For the degree-days, we first determine the balance-point temperature from Equation 24.3:

$$T_{bal} = 68\text{°F} - \frac{24\ \text{h} \times 4{,}200\ \text{Btu/h}}{12{,}600\ \text{Btu/(°F}\cdot\text{day)}} = 60\text{°F}$$

The corresponding degree-days are

$$D_h = 930\text{°F}\cdot\text{days}$$
$$= \text{Monthly degree-days } D_h(T_{bal}) \text{ for base } T_{bal}$$

Solution

The projected area is

$$A_{proj} = 550\ \text{ft}^2 \times \sin 50° = 421\ \text{ft}^2$$

From Equation 24.4,

$$\text{LCR} = \frac{12{,}600\ \text{Btu/(°F}\cdot\text{day}\cdot\text{ft}^2)}{421\ \text{ft}^2} = 29.9\ \text{Btu/(°F}\cdot\text{day)}$$

For the ratio S/D_h, we have

$$\frac{S}{D_h} = \frac{43.600\ \text{Btu/ft}^2}{930\text{°F}\cdot\text{day}} = 46.9\ \text{Btu/(°F}\cdot\text{day}\cdot\text{ft}^2)$$

Reading Figure 24.8 at these values, we find

$$\text{SSF} = 0.51$$

This means that 51% of the heating requirement of the building is provided by the sunspace in January. Equation 24.5 yields the auxiliary heating requirement for the month:

$$\dot{Q}_{aux} = \text{NLC} \times D_h \times (1 - \text{SSF})$$
$$= 12{,}600\ \text{Btu/(°F}\cdot\text{day)} \times 930\text{°F}\cdot\text{days} \times (1 - 0.51)$$
$$= 5.74\ \text{MBtu/month}$$

Comments

In accordance with the assumptions of the method, this result presupposes that the thermostat dead band is 65°F–75°F for the conditioned part of the building and 45°F–95°F for the sunspace. The designer should check whether this is agreeable with the future occupants of the building.

As a general rule, annual SSF values in the range of 0.4–0.7 are achievable with reasonable designs, except in the cloudiest regions of the United States; in other words, the passive solar approach can reduce the heating bill by about 40%–70%.

A more general method than the SLR method has been proposed by Monsen et al. (1982) called the unutilizability method meant for buildings with auxiliary heating backup. It has been demonstrated for direct gain systems and storage walls designs. As before, for each month, the fraction of the solar radiation that is not utilizable can be determined from empirical correlations determined from computer simulations using two parameters that represent certain climatic data and building parameters. It is then a simple step to deduce the monthly SSF.

Gordon and Zarmi (1981a,b) have developed a shorthand analytical method based on combining a thermal network model for the building response with location-specific climatic patterns that allow convenient prediction of the long-term thermal performance of passively heated solar houses. Assorted building modifications such as glazing area, thermal mass, storage absorptance, and night insulation can be analyzed either on monthly time scales or for the entire season. This approach has been developed by and further elaborated by Cowing and Kreider (1987). For greatest accuracy, the data for solar radiation and heating degree-days must be preprocessed in the form of a special frequency distribution. For the 26 SOLMET (1978) stations in the United States, these distributions have already been prepared by Cowing and Kreider.

24.4 Commercial Buildings: HVAC Systems

Commercial buildings consume much more electricity than homes, mostly because of higher lighting loads. That they also tend to require more fuel may appear perverse: With higher internal heat gains and a lower surface-to-volume ratio, their heating loads should be much lower. That is a manifestation of simultaneous heating and cooling loads in different zones of commercial buildings, a problem we discussed in Section 19.7.

Section 22.5 discussed the issue of daylighting in commercial buildings and how proper design can greatly reduce lighting electricity use. A summary was provided for the results of the published studies that investigated the impact of different window apertures of vertical windows and skylights along with different types of lighting controls on the heating and cooling energy use in commercial buildings. Estimates of payback periods for such measures were also discussed.

HVAC systems consume more primary energy than what is estimated purely from the building loads. This energy penalty is caused (1) by the manner in which the HVAC systems are operated, (2) by improper sizing of equipment and systems, (3) due to their part-load inefficiencies, and (4) by energy bucking or simultaneous heating and cooling. This excess energy consumption is quite significant both in terms of the building owner perspective (site energy) and to society as a whole (source energy).

To what extent the energy consumption of commercial buildings depends on the HVAC system can be deduced from an interesting series of studies by Kao (1982, 1983, 1985). Using the program BLAST, Kao simulated four types of commercial buildings: a small office, a large office, a school, and a retail store. Building structure and utilization were kept fixed, while varying the HVAC system. A wide variety of HVAC systems were evaluated, about 20 of the most common types for each building. Here, we present some selected results, as heating and cooling loads, excluding the efficiency of the primary system (or, to use the language of DOE2.1, only the performance of the "system" is considered, without "plant").

Six sites were considered, covering a broad range of climates; the heating degree-days ranged from 832 K·days (1498°F·days) in Lake Charles, Louisiana, to 4294 K·days (7730°F·days) in Madison, Wisconsin, and the cooling degree-days from 47 K·days (84°F·days) in Santa Maria, California, to 1522 K·days (2739°F·days) in Lake Charles. The same envelope is assumed for all climates, even though different envelope designs might be optimal or required by building codes in different climates. The simulation results have been correlated with degree-days to highlight trends and to permit interpolation to other sites.

A summary description of buildings and operating conditions is presented in Table 24.6. We have only presented results for two building types (small office and large office), but other building types were also investigated. The comfort conditions (outdoor airflow and thermostat set points for occupied periods) may appear a bit extreme today; they reflect standards imposed by the government during the energy crisis. Obeyed in public buildings but not in all private

TABLE 24.6

Summary of Building Characteristics for Figures 24.9 and 24.10

	Units	Small Office	Large Office
Zones		2*(ESWN + Int)	ESWN + Int
Stories		3	12
A_{floor}	m² (ft²)	2,787 (30,000)	22,297 (240,000)
A_{glass}/A_{floor}		0.15	0.14
Glass type		Double	Single, $\tau = 0.24$
A_{floor}/occupant (peak)	m² (ft²)	11.6 (125)	9.3 (100)
Min. outdoor air/occupant	L/s (ft³/min) per occupant	2.8 (6)	3.8 (8)
Min. outdoor air/ventil. air		0.1	0.1
K_{cond}/A	W/(K·m²) [Btu/(h·ft²)]	0.676 (0.119)	1.040 (0.183)
$T_{i,\min}$	°C (°F)	18.9 (66.0)	20.0 (68.0)
$T_{i,\max}$	°C (°F)	25.6 (78.0)	25.6 (78.0)
$T_{setback}$	°C (°F)	13.3 (56.0)	11.1 (52.0)
Heat gains (average during occupancy)	W/m² [Btu/(h·ft²)]	34.7 (11.0)	49.2 (15.6)

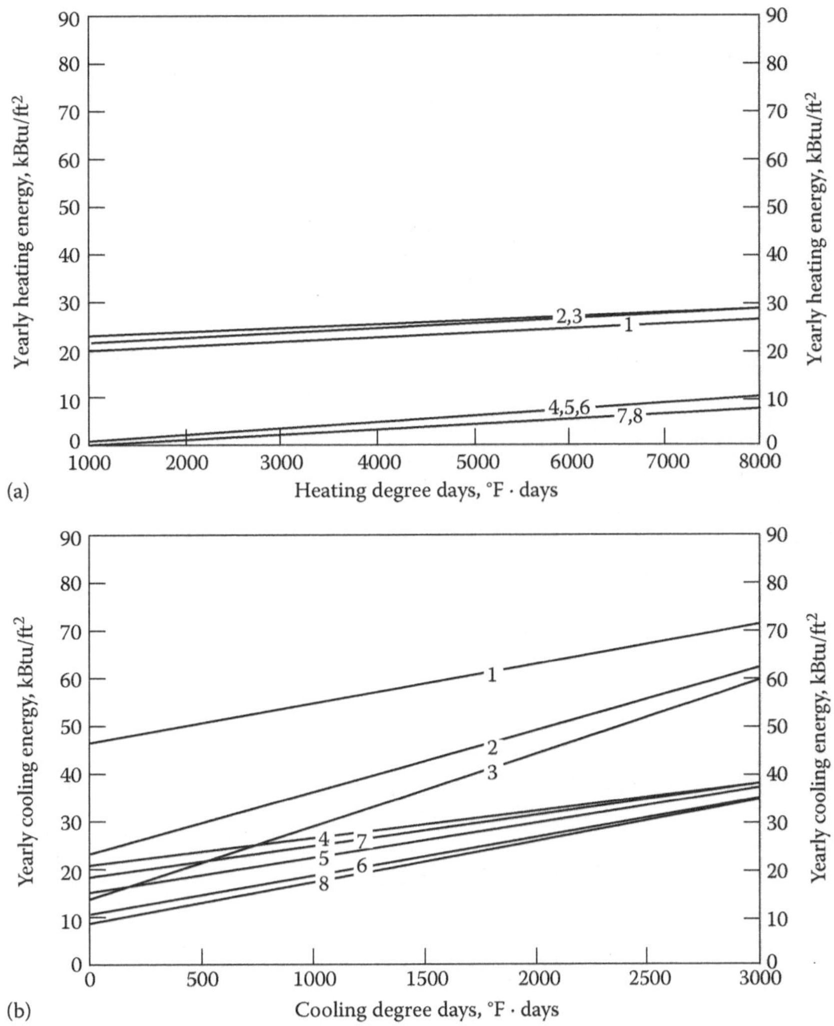

FIGURE 24.9
(a and b) Heating and cooling loads for small office building with different HVAC systems as a function of degree-days. 1—RH, base (one central air handler with preheat coil and cooling coil distributing conditioned air to each zone; heating and cooling control in each zone accomplished by reheat coils in supply air ducts of each zone; no reheat coil in zone one [interior first two floors, with no surfaces to outside]; T_{sup} set at 16.7°C [62°F] during working hours); 2—RH, temperature economy (like 1 but with economizer controlled by temperature); 3—RH, enthalpy economy (like 1 but with economizer controlled by enthalpy); 4—VAV, base (central variable-volume fan [with variable-pitch inlet vanes] supplying air to all 10 zones, each zone having VAV dampers, with minimal airflow below 24.4°C [76°F] and airflow increasing above this temperature to a maximum at 25.6°C [78°F]; T_{sup} set at 12.8°C [55°F] during working hours); 5—VAV, temperature economy (like 4 but with economizer controlled by temperature); 6—VAV, enthalpy economy (like 4 but with economizer controlled by enthalpy); 7—VAV, zone reset (like 4 but with T_{sup} variable between 12.8°C [55°F] and 16.7°C [62°F] during working hours according to the zones with largest instantaneous cooling load); 8—VAV, enthalpy economy, zone reset (like 7 but with economizer controlled by enthalpy). (From Kao, J.Y., *ASHRAE Trans.*, 91(2B), 810, 1985.)

buildings, they have gradually been abandoned with the new energy affluence.

The results are shown as a function of degree-days in Figures 24.9 and 24.10 for the small and large office buildings, respectively with a brief description of the HVAC systems in each caption and with the numerical order corresponding roughly to the ranking in energy performance. *The highest and the lowest can easily differ by more than a factor of 2.* Generally, the systems with the highest consumption are the ones that use constant volume, without economizer and with fixed supply air temperature. These conclusions for the economizer are in broad agreement with the simulations by Spitler et al. (1987), although the latter imply somewhat lower benefits for enthalpy control. The lowest consumption is achieved by VAV systems with enthalpy economizer and with variable T_{sup} reset according to the zone load or outdoor temperature. Such reset lowers the amount of reheat.

VAV systems use far less than constant-volume systems for heating and almost always less for cooling (possibly with slight exceptions at low cooling loads). The economizer will always reduce the cooling load, although it may

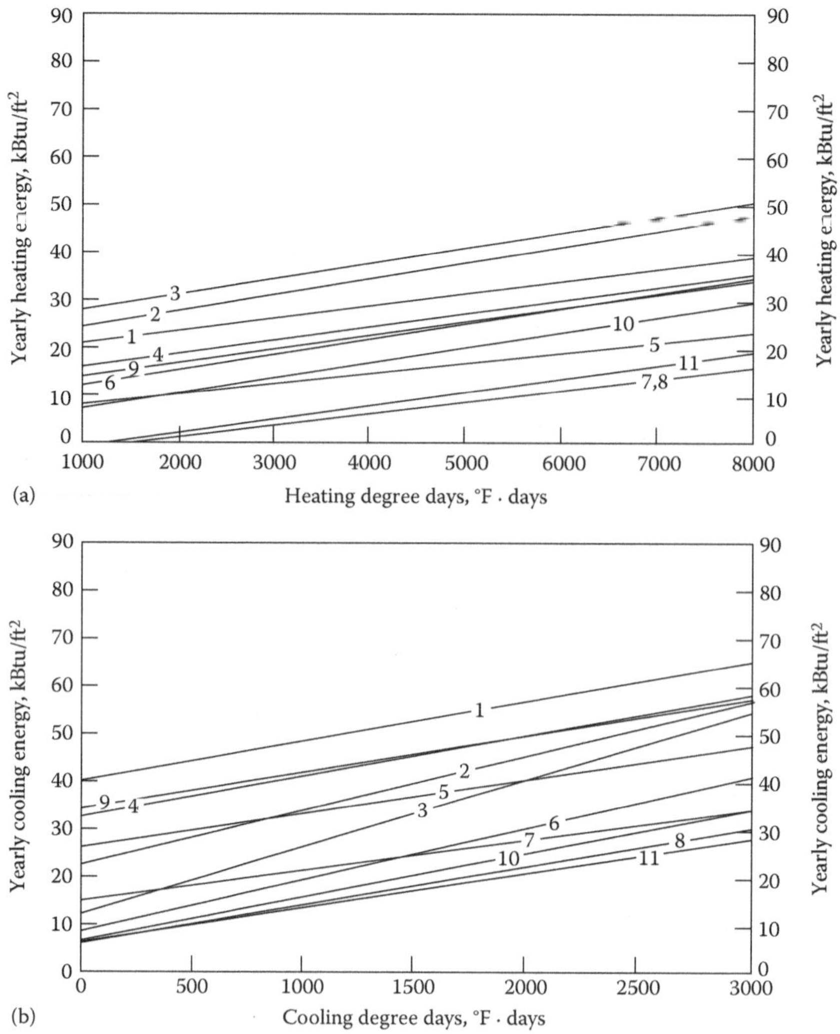

FIGURE 24.10
(a and b) Heating and cooling loads for large office building with different HVAC systems as a function of degree-days. 1—RH, base (constant-volume system with terminal reheat; T_{sup} = 15.0°C [59°F] including fan heat gain; minimum outdoor air, during occupied hours only, at 10% of supply air); 2—RH, temperature economy (like 1 but with economizer controlled by temperature); 3—RH, enthalpy economy (like 1 but with economizer controlled by enthalpy); 4—RH, OA reset (like 1 but with T_{sup} of perimeter zones varied in linear fashion according to the outdoor air T_o, between 15°C [at T_o = 32.2°C] and 17.2°C [at T_o = 21.2°C]); 5—RH, zone reset (like 1 but with T_{sup} of perimeter and interior zones varied according to the load of each zone; no limits on T_{sup} imposed); 6—RH, enthalpy economy, zone reset (like 1 but with economizer controlled by enthalpy and with T_{sup} of perimeter and interior zones varied according to the load of each zone); 7—VAV, base (VAV for interior and perimeter, flow variable from 20 to 100 percent of full flow; constant flow of outdoor air during occupied hours; constant T_{sup} = 15°C [59°F]; reheat coils in perimeter zones only); 8—VAV, enthalpy economy (like 7 but with economizer controlled by enthalpy); 9—Int-VAV enthalpy economy, Ext-DD base (VAV base [like 7] for interior zones and DD [dual-duct] system for perimeter zones; outdoor airflow constant during occupied hours; cold duct T_{sup} = 15°C (59°F) and T_{hot} = 60°C (140°F) at all times); 10—Int-VAV enthalpy economy, Ext-DD enthalpy economy, zone reset (like 9 but with economizer controlled by enthalpy for entire building and cold and hot T_{sup} of dual-duct system reset according to the zone loads); 11—Int-VAV enthalpy economy, Ext-FC (like 8 for interior [VAV with enthalpy economizer] and four-pipe fan coil system for perimeter [with hot and cold water available year-round]). (From Kao, J.Y., *ASHRAE Trans.*, 91(2B), 810, 1985.)

cause slightly increased heating. To minimize this heating penalty, *reset of T_{sup} is strongly advisable.* Enthalpy is more efficient than temperature control for the economizer, attaining appreciably lower cooling load, with little effect on heating. Both absolute and relative savings are larger for low cooling degree-day climates (about 25% relative savings) than for high cooling degree-day climates (about 5% relative savings).

A number of other issues have not been addressed, in particular, the risk of discomfort and the flexibility in adjusting to changes in operating conditions or internal loads. Installation of reheat is the easiest protection against such eventualities, albeit at the price of increased energy consumption. Also, reheat systems render equipment failures more difficult to detect.

A more up-to-date study than the Kao (1985) study was undertaken by Huang and Franconi, 1999), which summarized the results of simulating 120 commercial building prototypes (12 building types, 5 climatic zones, and 2 vintages) to determine the relative contributions of roofs, walls, windows, infiltration, ventilation, lighting, equipment, and people. Hourly component loads were extracted and plotted as pie charts for all the prototypes. The largest contributors to heating loads were found to be windows, walls, and infiltration, while those for cooling were lighting, solar gain, and equipment. The study also introduced the concept of *system and plant factors* to quantify net efficiencies of air-handling systems in meeting building loads and of central plants in providing heating and cooling energy needs. The overall source efficiency for the air-conditioning system including both system and plant factors was 0.33 for heating and 0.57 for cooling.

24.5 Alternative Energy Technologies

Section 24.3.3 discussed passive solar space heating applicable to residences. Solar energy is a source of low-temperature heat that has selected applications to buildings. For example, it can be used for hot water heating and for preheating air. A brief overview of traditional passive cooling techniques and how they are being integrated with mechanical HVAC systems is also provided. The adoption of PV systems in buildings has seen an explosion in the first few years, and this passive solar energy application is also treated in this chapter.

24.5.1 Solar Water Heating

Solar water heating is a particularly effective method of using this renewable resource since low to moderate temperature water (up to 140°F [60°C]) can be produced by readily available flat-plate collectors (Goswami et al., 2000). Figure 24.11 shows one system for heating service water for residential or commercial needs by using solar collectors. The system consist of three loops; it is instructive to describe the system's operation based on these three. First, the collector loop (filled with a nonfreezing solution, if needed) operates whenever the DHW controller determines that the collector is warmer, by a few degrees, than the storage tank. Heat is transferred from the solar-heated fluid by a counterflow or plate heat exchanger to the storage tank in the second loop of the system. Storage is needed since the availability of solar heat rarely matches the instantaneous water-heating load. The check valve in the collector loop is needed to prevent reverse flow at night in systems where the collectors (which are cold at night) are mounted above the storage tank.

The third fluid loop is the hot water delivery loop. Hot water drawn off to the load is replaced by cold water supplied to the solar preheat tank, where it is heated as much as possible by solar heat. If solar energy is insufficient to

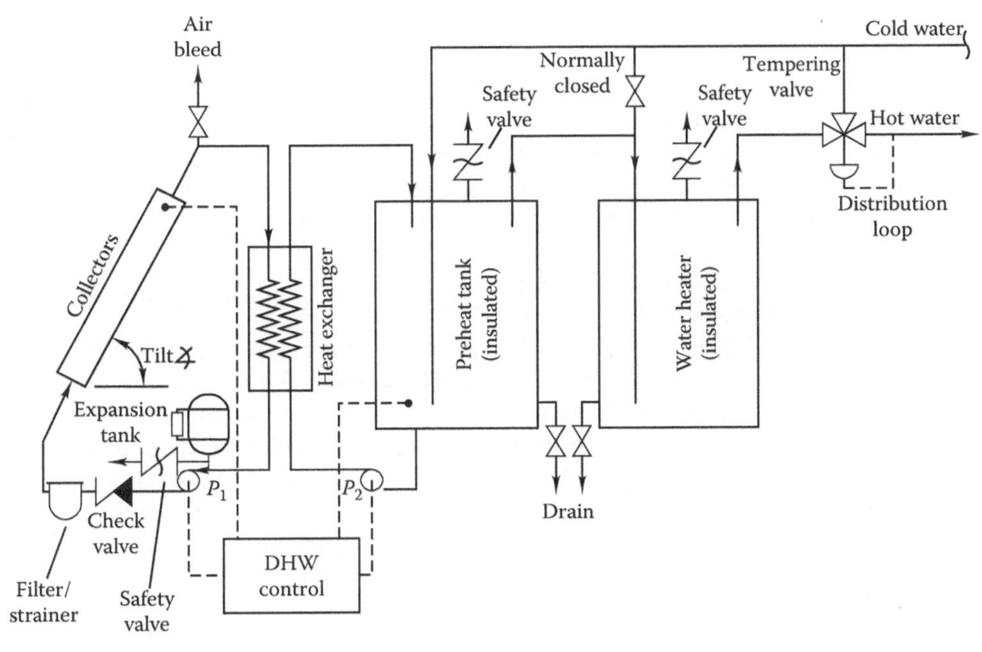

FIGURE 24.11
Solar water-heating system including collectors, pumps, heat exchanger, and storage tanks along with piping and ancillary fittings. Collectors are tilted up from the horizontal at a fixed angle roughly equal to the local latitude.

heat the water to its set point, conventional fuels can finish the heating in the water heater tank, shown on the right of the figure. The tempering valve in the distribution loop is used to limit the temperature of water dispatched to the building if the solar tank should be above the water heater set point in summer.

The energy delivery of DHW systems can be found by using the *f*-chart method described in Goswami et al. (2000). As a rough rule of thumb, 1 ft^2 of collector can provide 1 gal of hot water per day (45 L/m^2) on average in sunny climates. Design pump flows are to be 0.02 gal/min/ft^2 of collector [0.01 L/(s·m^2)], and a heat exchanger effectiveness of at least 0.75 can be justified economically. Tank should be insulated so that no more than 2% of the stored heat is lost overnight (Goswami et al., 2000).

Solar heating should be assessed on an economic basis. If the cost of delivered solar heat, including the amortized cost of the delivery system and its operation, is less than that of competing energy sources, then an incentive exists for using the solar resource. The collector area needed on commercial buildings can be large; if possible, unused roof space can be used to hold the collector arrays. Of course, large high-rise buildings in urban locations are not likely candidates for solar heating, since very little roof or nearby ground area is likely to be available for collector mounting.

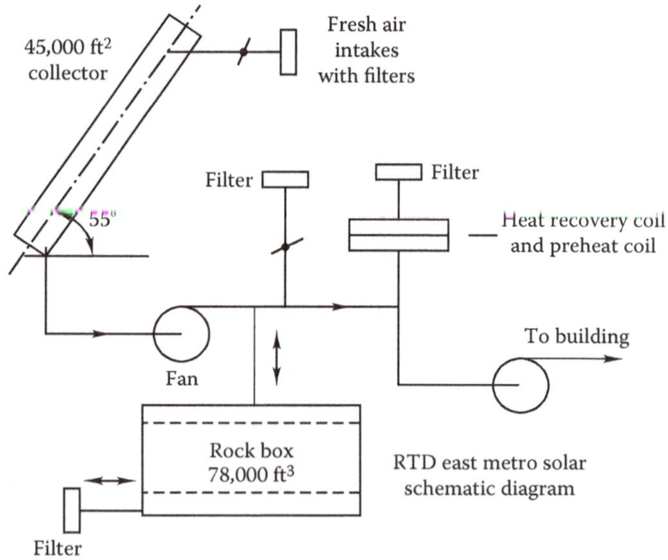

FIGURE 24.12
Solar air preheating system for ventilation air with optional storage. The collector and storage sizes shown in the figure represent those used on the first system of this type built in Denver, Colorado. (Courtesy of Jan F. Kreider and Associates, Inc., Boulder, CO.)

24.5.2 Solar Preheated Air

The second practical solar system for HVAC applications, namely, ventilation air preheating, is shown in Figure 24.12. In this application, ventilation air is passed through a filter and through an air-type solar collector prior to being introduced to the HVAC system. In this way, solar heat can provide some preheating of all outdoor air in winter, regardless of the solar flux available. With a typical flat-plate collector, about 50% of the incident sunlight can be converted to useful heat. Flow rates of these systems range between 1 and 2 ft^3/(min·ft^2) [5 and 10 L/(s·m^2)].

No special controls are needed; whenever ventilation air is needed, it is drawn through the collectors, and is heated. The storage shown in the figure is optional. If solar availability exceeds the load, heat can be stored for use at night. If insufficient solar heat is available, steam is used to heat the ventilation air, as shown. Air preheat is also accomplished by recovering heat from exhaust air, as described in Section 19.6.3.

From the 1990s, there has been great interest in integrating the solar collectors with the building envelope. One such application is the use of *transpired air collectors* in cold climates. As shown in Figure 24.13, the outdoor air heats up as it is drawn through holes in the vertical south-facing unglazed solar collector, collected in a plenum and then ducted to interior spaces by a circulating fan. Part of the heat being conducted outdoors through the wall is also collected by the air. The pores generally cover 0.5%–2% of the collector area, and thermal efficiencies around 60% have been reported. A more recent technological development is to combine such thermal designs with building-integrated PV systems, is referred to as BIPV/T (Athienitis et al., 2011). The PV subsystem efficiency is said to be typically around 15%, while that of the thermal system is about 30%. The BIPV/T concept has been applied to a full-scale office building demonstration project in Montreal, Canada. The system covers 3100 ft^2 (288 m^2); the solar installation can produce 25 kW of electricity for use within the building and approximately 260 MBtu/h (75 kW) of preheated fresh air at a rate of 10,000 cfm (4.7 m^3/s).

24.5.3 Passive and Hybrid Cooling

Passive systems are those that use on-site energy flows available from the natural environment in combination with architectural features of the building (shape, orientation, mass, envelope insulation, air passage through building, etc.), to heat or cool the building. They are much more sensitive to natural ambient fluctuations, and so may not provide the type of rigid indoor air temperature control that mechanical HVAC systems do. Though passive buildings have had a long evolutionary growth and adaptation over the years, many of the design practices have been discarded in favor of the much easier to design modern buildings using mechanical HVAC systems. This has been fuelled by the increase in occupant

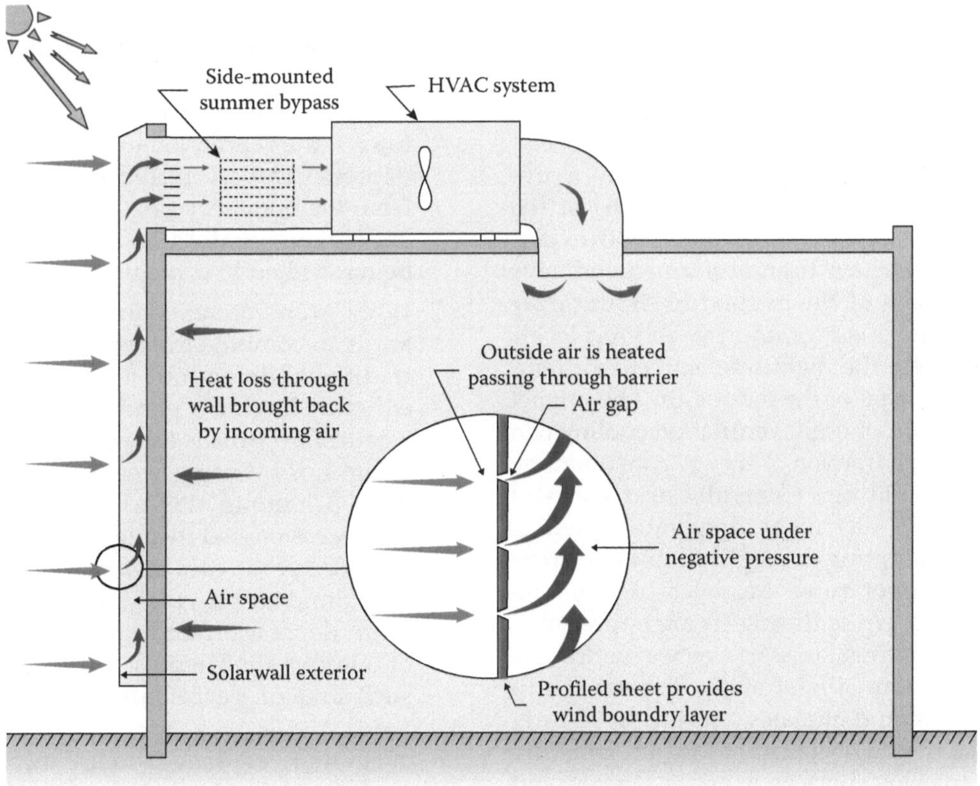

FIGURE 24.13
Conceptual sketch of a transpired solar air heater with ducting. (From Conserval Engineering, Inc., 2012, www. monstercommercial.com, accessed May 2015.)

demand for tight indoor temperature control. The push for low-energy buildings has reawakened interest in these passive and climate-responsive technologies leading to hybrid and mixed-mode (MM) buildings. A very short overview of some of the low-energy cooling methods is covered here with passive heating already covered previously in Section 24.3.3.

The interested reader can refer to the series of technical reports freely available online containing description, early selection criteria, design guidelines, and monitored results of actual demonstration case studies of a number of low-energy cooling technologies suitable for European countries (IEA, 1995). The intent was to promote the use of such low-energy cooling strategies and technologies by showing that they could be integrated with conventional HVAC systems and have lower life cycle costs while meeting indoor thermal comfort standards. A good textbook that deals with both scientific fundamentals and practical and architectural design issues of passive and low-energy cooling is that by Givoni (1994). Other good books on passive/solar low-energy buildings are those of Mazria (1979), Sodha et al. (1986), Cook (1989), Santamouris and Asimakopoulos (1996), Athienitis and Santamouris (2002), and Nayak and Prajapathi (2006).

Passive cooling is much harder to achieve than passive heating. Low-energy cooling technologies are alternatives to the purely mechanical type of cooling systems discussed in previous chapters (good references are those of Eicker in 2003 and 2009). The terminology is not precise but generally refers to various methods such as passive and hybrid systems, which utilize natural heat sinks (air, water, wind, sky, and ground) to deliver cooling to the space. The cooling capacity of such natural heat sinks varies greatly with ambient climatic conditions and is not matched with the building loads. The cooling capacity is high when ambient conditions are milds and often nonexistent during peak cooling conditions. Hence, designers in the future may evaluate/adopt one or more of the following strategies:

1. *Precooling strategies* making use of the building thermal mass to maximize potential cooling potentials of cool night air
2. *Induced ventilation* with or without evaporative cooling, with outdoor airflow enhanced by updraft or downdraft towers (also called wind catchers)
3. *Integration* with conventional mechanical systems, i.e., use of hybrid systems

Nighttime ventilation is one of the most widely used precooling strategy for passive cooling of residences and small commercial buildings in climates with cool nights. The building's thermal mass can be cooled at night by ventilating the inside of the space with the relatively lower outdoor-air temperatures, thereby maintaining lower indoor temperatures during part of the warmer daytime period. Numerous studies, both experimental and theoretical, have been performed and have shown the effectiveness of the method to significantly reduce air-conditioning loads or improve comfort levels in those climates where the nighttime ambient air temperature drops below that of the indoor air. The impact of widespread adoption of night ventilation cooling can be large, given the large fraction of energy consumed by air conditioning of buildings. Generally, arid zones of modest elevations have day/night temperature swings that are suitable for adopting the night ventilation strategy. The daytime temperatures are much higher than human comfort levels, while the nights are appreciably cooler. There are numerous large arid zones worldwide that would benefit from this strategy. For air-conditioned buildings in humid climates, one has to be careful not to increase the daytime latent load by moisture absorbed at the interior surfaces at night (Fairey and Kerestecioglu, 1985). Givoni (1994) discusses other such precooling strategies, namely, radiant cooling to the night sky, slab cooling using earth tubes, and evaporative cooling using roof ponds.

The second type of strategy involves inducing and enhancing natural ventilation, which may provide relief depending on the outdoor-air conditions. There are in essence two variants:

1. Without evaporative cooling. This category, includes two subvariants: the updraft tower, which operates on the stack effect, and the wind catcher or downdraft wind towers (Figure 24.14), which are traditional techniques common in the middle east. The driving force for induced ventilation is provided by the positive air pressure created at the windward entrance of the wind tower. In case of the two-sided towers or double towers, the ventilation inflow is enhanced by the depressurization created at the leeward side and from the buoyancy or stack effect of the indoor air. Hughes et al. (2011) describe enhancements being studied in contemporary tower designs.
2. With evaporative cooling. The second variant is to combine evaporative cooling with the traditional downdraft tower. This can be done either by having a pond inside the house or in a courtyard (this is the traditional architectural design) or by using wet pads inside the tower itself (Chalfoun, 1997; Givoni, 1994). The outdoor air is passed through wet pads close to the entrance of the cool tower, and the corresponding humid air provides both a dry-bulb temperature depression and enhances the downflow of outdoor air. There is still ongoing interest in such strategies especially in the middle east.

FIGURE 24.14
Photograph of downdraft wind towers.

The integration of conventional mechanical systems with traditional low-energy techniques is an area of active research interest. Mixed-mode (*MM*) ventilation refers to a hybrid approach to space conditioning that uses a combination of natural ventilation from operable windows (either manually or automatically controlled) or other passive inlet vents and mechanical systems that provide air distribution and some form of cooling (Brager, 2006). A well-designed MM building allows spaces to be naturally ventilated during periods of the day or year when it is feasible or desirable and uses air conditioning for supplemental cooling when natural ventilation is not sufficient. The goal is to provide acceptable comfort while minimizing the significant energy use and operating costs of air conditioning. MM buildings are often classified in terms of their operation strategies, such as

1. Concurrent strategies (mechanical cooling and natural ventilation can operate in the same space at the same time)
2. Changeover strategies (the building switches between mechanical cooling and natural ventilation on a seasonal or daily basis)
3. Zoned strategies (mechanical cooling and natural ventilation operate in different areas of the building)

Some work has been done in MM buildings that employ a combination of natural ventilation and mechanical systems and intelligently switching between the two to minimize energy use, while preserving occupant comfort. A major consideration is the manner in which such switching is done given that there are several zones in a building. Reductions in cooling- and ventilation-related energy use from 20% to 50% over code-designed buildings have been reported, and such MM buildings are said to consistently outperform conventional buildings on thermal comfort and occupant satisfaction (Henze and May-Ostendorp, 2012).

24.5.4 Solar Photovoltaics

Photovoltaic (PV) technology directly converts solar energy into electric power. Its applications are diverse: from small-scale residential systems of a few kW to large centralized installations of tens of MW. There are several books on this topic ranging from cell solid state theory to PV system applications, modeling, and sizing, (for example, Masters, 2004; Messenger and Ventre, 2010) and to building PV systems (Eicker, 2003). The application of solar PV in buildings has seen an exponential increase in the few decades due to a number of reasons: rapid reduction in PV costs, low system maintenance, state and federal financial incentives, rising costs of traditional electric power, mature financing mechanisms, etc. No less important is the element of its social appeal given that PV systems are environmentally benign and are an attractive alternative to fossil-fuel-based power generation. The use of PV systems is an essential element in achieving net-zero buildings that, on an annual basis, consume as much energy as they generate on-site. This is a widely sought-after goal in buildings worldwide.

PV systems can be classified based on different criteria, with the most common ones relating to the cell material and whether solar concentration is used or not. Cell material criteria differentiate systems based on the semiconductor material used to form the solar cells, which may also be divided into two main groups: crystalline silicon based and thin film. The former can be further subdivided into single crystalline and polycrystalline. Thin films include a diverse group of materials such as amorphous Si, gallium arsenide, indium phosphide, and cadmium telluride. These materials may differ widely in their conversion efficiency as well as in their production costs. Building PV systems tend to be predominantly crystalline silicon or amorphous silicon under fixed stationary mount (i.e., nonconcentrating).

Building PV systems can be either roof mounted or building integrated. BIPVs can either replace or be integrated seamlessly with building elements (facades, roofs, windows, or walls) and can be made to look aesthetically pleasing with different colors and shapes. BIPV tile products are available commercially, which can substitute traditional tiles in size, shape, and curvature. Because of cost, residential solar systems are generally roof mounted, while commercial buildings may include both types of systems, with the extra cost of BIPV offset by aesthetic advantages. For commercial buildings with land around them, the PV systems can be mounted on external structures such as car parks with the cost of the roof structure partially offset by the solar panels.

The individual cells are connected in series and in parallel to form modules that are then mounted on the roof of buildings. The efficiencies of single-crystalline modules are about 18%, those for polycrystalline about 15%, and those for the different types of thin film modules ranging from 8% to 12% . The efficiency of solar panels does degrade a little bit over time. The general rule of thumb is that the power output drops by about 0.5% every year. Solar panel manufacturers often offer a warranty that guarantees the power output stays above 80% after 25 years.

Figure 24.15 shows one of the several configurations of a grid-connected PV system. The design shown has solar panels (which are made of several individual solar cells connected in series and in parallel), which are in turn connected in series. In the figure, individual panels generate 24 V DC, which is then converted to 120 V AC by individual inverters as shown. A more common and cheaper design, especially for smaller PV systems, is

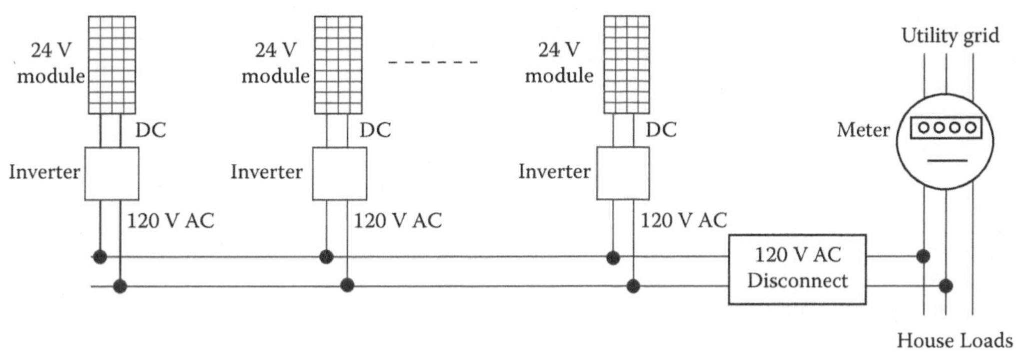

FIGURE 24.15
Sketch of a grid-intertie PV system design with an inverted for each module. Other types of system configurations are also used.

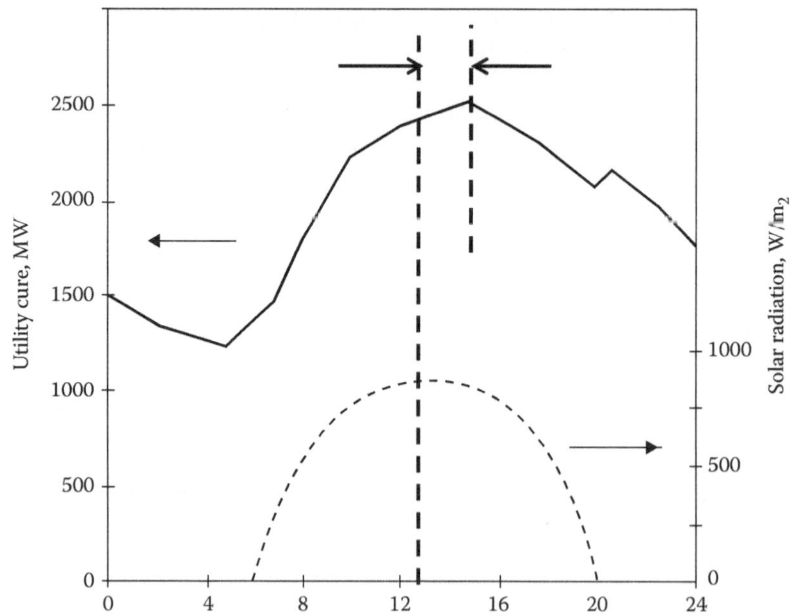

FIGURE 24.16
Figure illustrating the time shift in diurnal profiles of solar radiation availability and the electric utility load.

to have a single inverter just before the breaker shown. Other types of system configurations are also used. The advantage of the design in Figure 24.15 is that it allows convenient system capacity increases in the future.

Having a small electric battery coupled to the PV system has tangible benefits: this dampens short-term intermittent fluctuations in solar radiation and, more importantly, can compensate for the small diurnal mismatch between solar radiation availability (which is proportional to the electric power generated on-site) and the electric load of the building (as shown in Figure 24.16).* This shift (without storage) results in higher demand costs to the customer because the electric utility has to operate costly peaking turbines during late afternoons. The solar electric storage would allow better matching and result in some load leveling, which is of value to both utility and to the customer. However, current building grid-connected PV systems usually do not have battery backup because of cost considerations. This system option is likely to become attractive in the future as battery costs are bound to reduce given the great R&D efforts currently underway in this regard.

There are two common schemes for pricing the electric generation for the PV system:

1. *Net metering* where the utility meter runs backward when solar PVs generate power in excess of house needs, thereby building up a credit with the utility. This metering method is prevalent in the United States where the customer receives upfront federal and state rebates while purchasing the PV system.
2. *Feed-in tariff* wherein the utility installs an additional meter to monitor PV system generation. This scheme allows for differential electricity rates of production (from the PV system) and consumption (from the electric utility meter). The electricity rate sold back to the electric utility is several times greater than that which the customer would be charged. This scheme is more common in Europe and certain parts of Canada where financial incentives do not include upfront rebates.

Costs of solar systems include the solar module and the balance of system, which include inverters, mounting hardware, labor, permitting fees, shipping, taxes, and profits. The installed costs of building PV systems have declined greatly in the last two decades and are still declining. According to the latest edition of "Tracking the Sun," an annual PV cost tracking report produced by Lawrence Berkeley National Laboratory of the U.S. Department of Energy (DOE), the median installed price (excluding rebates) of PV systems completed in 2012 was \$5.30/W for residential and small commercial systems smaller than 10 kilowatts (kW) and was \$4.60/W for commercial systems of 100 kW or more in size. The report is based on data from more than 200,000 residential-, commercial-, and utility-scale PV systems

* Note that the solar radiation is shown to peak and be symmetric about 1 pm and not noon; this is due to the 1 h shift from daylight savings.

installed between 1998 and 2012 across 29 states, representing roughly 72% of all grid-connected PV capacity in the United States.

24.6 Uncertainties in Simulations

24.6.1 Sources

The ultimate test of any theory is, of course, the confrontation with observation. Another dimension is: "How close are the load and energy predictions from different simulation programs?" Much work has been done to ensure confidence in the validity of building models. However, there are large differences between simulated and measured building energy performance (even with no uncertainty in the measurements themselves), and here, we present some of the causes. The differences (which can be interpreted as uncertainties in the simulated results) can be classified as follows (De Witt, 2003):

1. *Specification uncertainty* is due to differences between the way the building (physical construction as well as its materials and systems) is specified in the simulation model and the way it is actually built.
2. *Modeling uncertainty* is due to the simplifying assumptions in the models of the various energy flows in the building and of HVAC systems. To begin, there is no uncertainty about the principles; the relevant laws of physics are known beyond a shadow of a doubt. How they have been applied to the engineering situation is another matter. It is not obvious which phenomena are important in real buildings and in how much detail they should be modeled. Is hourly time resolution sufficient? Some of the processes are extremely complicated and difficult to calculate; approximations made during model formulation are likely to lead to some error, the impact of which is building specific.
3. *Numerical uncertainty* arises because of the manner of coding and solving the various equations (discretization, numerical convergence, etc.). A methodology to identify and diagnose differences in simulation predictions that may be caused by algorithmic differences, modeling limitations, coding errors, or input errors has been developed by Judkoff and Neymark (2006), which has been made into a standard (ANSI/ASHRAE Standard 140, 2011).
4. *Scenario uncertainty* is due to improper specification of the forcing functions. For example, the outdoor-air conditions used for the building simulation may be from a location several miles away from the building. Further, there is stochasticity in the diurnal scheduling of the building over the year (after all, buildings are occupied by people), which current-day simulations fail to capture satisfactorily. The book chapter by De Witt (2003) addresses this issue in some detail.

24.6.2 Validation

For the validation, there are basically two approaches: comparison of models with each other and comparison with measured data. Such comparisons can be carried out at different levels, from tests of components (e.g., a wall segment or a fan) to entire buildings. Generally the difficulties grow with the size of the system. While the validation of component models by laboratory tests is relatively straightforward, the instrumentation and evaluation of a large occupied building are a formidable undertaking. In recent years, the collection of data has become easier thanks to computerized energy management systems, but the analysis still requires a great deal of effort.

Test cells and small laboratory buildings are an intermediate solution, and they have been used for testing the simulation of dynamic behavior. The main challenge is the modeling of rapid changes, e.g., during setback recovery. The tests have shown that models such as DOE2.1 (Birdsall et al., 1990) and Energy Plus (2009) are acceptable for modeling phenomena with hourly time resolution.

One reason why validation by real buildings is so difficult lies in the near impossibility of performing observations under strict control of all variables. Therefore, one usually resorts to the alternative of validating different simulation programs for buildings against each other. An interesting exercise of that type has been undertaken by Kusuda (1981). In an attempt to evaluate the accuracy of a steady-state method (the TC4.7 method described in Knebel, 1983) for calculating annual energy consumption, the same building was analyzed by seven different analysts; each analyst used a different dynamic simulation program and compared his results with the TC4.7 method. As it turned out, the discrepancies between the results of the seven analysts were generally greater than the differences that the analysts found between their own simulation and their own TC4.7 result. The explanation can be found in the complexity of the input needed to give a reasonably realistic description of a typical building—and in the fact that different simulations use different approaches and require different information. So it is not surprising that different analysts would come up with different interpretations, even though the ground rules are the same.

24.6.3 Comparison with Measured Data

Measured performance data are crucial. They are the ultimate test of predictions, and they provide reference values to guide the design of new buildings. There are a wide range of data types and data sources. At one end of the spectrum are the meter readings of the utility companies. They are very reliable and accurate but contain little information beyond the total consumption of a building during the billing interval (monthly in the United States, even longer time intervals in many other countries). For large consumers, the electricity consumption is metered separately during peak (daytime weekdays), intermediate, and off-peak periods (nights and weekends), and the peak instantaneous power may be monitored as well.

How close are the energy use between predicted (or modeled during the design phase) and measured performance once the building is constructed? A good illustration is provided by Figure 24.17 taken from a study by Turner and Frankel (2008) who studied 121 North American new construction buildings. The Leadership in Energy and Environmental Design (LEED) program (described briefly in Section 24.8.3) awards energy performance points for a new building based on predicted energy cost savings compared to a modeled code baseline building. Since the design documents are to be evaluated by a third party, the designers tend to take extra care while performing the simulations. Even then, it is seen that there are large discrepancies between individual buildings, but that on average the agreement was quite good. The report pointed out that there are several buildings whose measured performance is very much worse than predicted, and, in fact, even worse than stipulated by code (which is about 30% lower).

For comparisons with measured data from residences, we show results for a group of 82 new single-family houses in France (Marchio and Rabl, 1991). On average, the agreement is quite good, but again there is enormous scatter. Many detailed investigations have found that behavioral differences play a large role in residential energy consumption. Figure 24.18 contains an interesting illustration of this effect: The six houses at the predicted consumption C_{the} = 10,300 kWh have the same construction, yet actual consumption varies by almost a factor of 2. This appears to be quite typical. In apartment buildings, the scatter can be even larger because of "heat theft" between neighboring units with different thermostat settings.

Multiterabyte databases of measured hourly and subhourly data on millions of buildings in the foreseeable future will be available. In the last 10–15 years, a

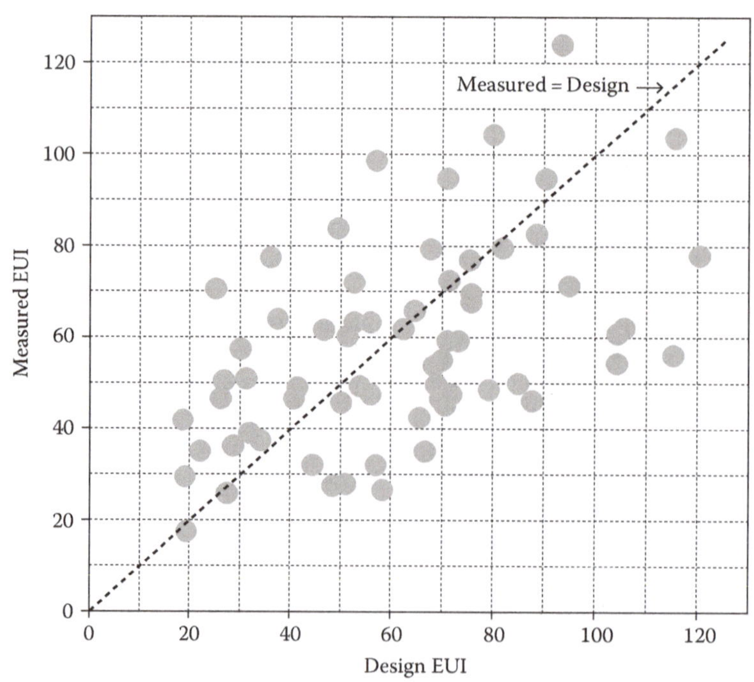

FIGURE 24.17
Comparison of modeled and measured energy use in several buildings that applied for LEED certification. EUI is the energy use index in units of kBtu/(ft^2-yr). (From Turner, C. and Frankel, M., Energy performance of LEED for new construction buildings, Report prepared by New Buildings Institute, Vancouver, WA for U.S. Green Building Council, Washington, DC, May 2008.)

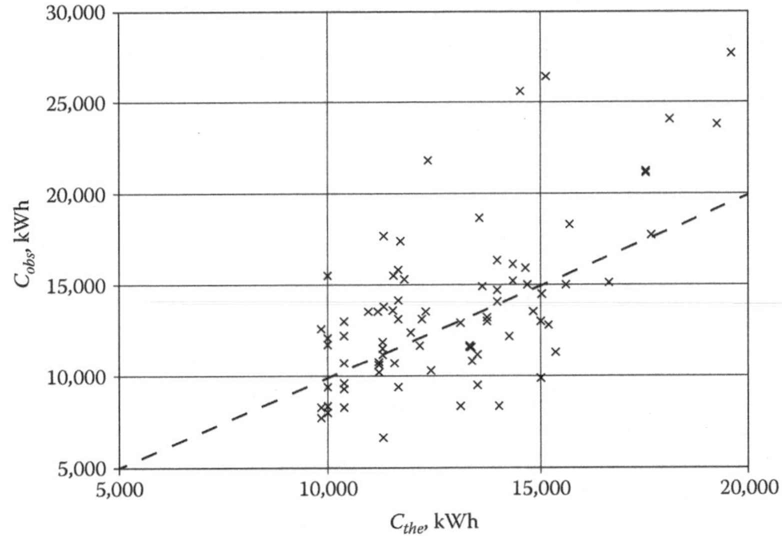

FIGURE 24.18
Comparison of predicted and observed energy consumption for 82 new single-family houses in France. C_{the} is the predicted and C_{obs} is the observed consumption. The dashed line corresponds to perfect agreement (note suppressed zero and different scales). 1 kWh = 3.6 and MJ = 3.412 kBtu. (From Marchio, D. and A. Rabl., *Energy Build.*, 17, 131, 1991.)

substantial number of U.S. electric utilities are switching to advanced metering infrastructure (AMI) to collect, store, and allow data mining of these data. The data collected will be whole house hourly data for residential and small commercial buildings and 15 min data for industrial and large commercial. Further, state legislatures such as the "Green Button Initiative" in California are passing laws, whereby customers can transfer the rights of data to commercial companies. Such laws will overcome the issues of data confidentiality, which currently restricts the free distribution of energy use data. Companies can then, in return, provide valuable services such as insights into occupant behavior patterns in residences or improved operational efficiency in commercial buildings. Note that such databases provide only aggregated information. Submetering is rare, even though it is most desirable to identify heating, cooling, lights, equipment, etc., and to correct for variations in weather or operating schedule. The separation of heating and cooling consumption is usually not provided in commercial buildings unless different fuels are used.

24.7 Energy Benchmarking and Rating

Energy benchmarking is the process of comparing and ranking the measured energy performance of a particular building against a distribution of buildings (along with their energy systems) with similar features. This is akin to *Energy Guide* labels on appliances that indicate the relative performance of a particular appliance within the range from least efficient to the most efficient of similar appliances. An important issue in building energy benchmarking is the use of performance indices to characterize the building. The performance indices can sometimes serve as benchmarks by themselves. Energy use indices (EUIs), in units of kWh/sqft/year or Btu/sqft/year, are widely used as energy benchmarks in building energy analysis since it is an attempt to normalize the energy use corresponding to a strong determinant, namely, the square footage of the building.* By normalizing out strong determinants, wide differences between building EUIs would be indicators of inefficient buildings or systems where improvements can be made (Sharp, 1996).

EUI is a standard unit of measurement for building energy analysis and has been studied for use as a whole buildings energy target. Despite being normalized for area, EUIs vary considerably and are, thus, ambiguous energy benchmarks as indicators of energy performance of an individual building. To overcome this ambiguity, simple statistical regression models have been developed to correct for variations in building characteristics, thereby providing more accurate benchmarks or estimators of energy use in a specific building. Benchmarking tools that use this approach include the U.S. Environmental Protection Agency's ENERGY STAR® Portfolio Manager (EPA, 2011) and ASHRAE'S Building Energy Quotient (BEQ) (ASHRAE BEQ, 2014).

* A rather troubling ambiguity in all these performance data is the definition of the floor area itself. Is it gross area, net area, or conditioned area, and does it include or exclude the parking area (which may be partially conditioned)?

The ASHRAE BEQ rating is the ratio of the source EUI use of the rated building to the median energy use of its building type. A net-zero building will have a rating of zero. The median of building performance for that particular building type is set at 100. There is no theoretical upper end to the building EQ rating scale to track poor energy performance; however, for pragmatic reasons, a building with a score of 125 is assigned a rating of "poor." Net-zero energy buildings that also produce an energy surplus can have a rating of less than zero. The rating system has two components, an *operational rating* based on measured data during actual performance and an *asset rating* that characterizes the building's simulated energy use under predetermined conditions that would be useful for real-estate transactions (Jarnagin, 2009).

The database widely used to obtain energy information on the entire gamut of commercial buildings in the United States is the Commercial Buildings Energy Consumption Survey (CBECS, 2003) database. This database contains building characteristics, system descriptions, energy expenditure, and energy consumption for 6380 commercial buildings across the U.S. commercial buildings including several building types such as office buildings, K-12 schools, hotels, retail stores, supermarkets, data centers, correctional institutions, and buildings used for religious worship, to name a few. There are two types of building and system features in the CBECS database, categorical and continuous. The method used for data collection was the *computer-assisted personal interviewing* with building owners, tenants, and managers. Consequently, the accuracy of the self-reported data depends on how well the respondents knew about their buildings. Recently, a few studies (for example, Kaskhedikar et al., 2015) cast doubts on the accuracy and completeness of the data contained in CBECS database. The paper concluded with the cautious statement urging more careful appraisal given that this database is (1) widely used for ongoing building energy conservation studies and (2) the basis of energy reporting and disclosure legislation being enforced by numerous U.S. cities for new buildings.

Li and Hong (2014) conducted an analysis of monitored energy use in 51 so-called high-performance buildings (21 in the United States, 11 in Europe, 5 in Australia, 7 in Asia, and 7 in China) using a complete year of data. They report large EUI variation, greater than a factor of 10 among buildings; smaller buildings tended to have lower EUI values. They found that the weather, though important, was not a decisive factor on EUI variability. The statistical analysis was unable to find a direct correlation between specific building technologies and low EUI values. This result, as well as other similar studies, seem to suggest that EUI may not be a properly normalized performance index in the first place. More research is warranted in this general area.

24.8 Drivers for Efficiency

The U.S. Congress has set ambitious goals for buildings of the future through the *Energy Independence and Security Act of 2000*. It passed the *Net-Zero Energy Commercial Building Initiative* to support the goal of net-zero energy for all new commercial buildings by 2030 and specifies a net-zero target for 50% of the U.S. commercial building sector by 2040 and net zero for all U.S. commercial buildings by 2050. How are we to get there? From a technological perspective, this requires, at the least, that building energy codes be gradually tightened, that building designers have appropriate design guides to accomplish the objectives, and that voluntary initiatives be actively promoted by proper coordination of appropriate governmental agencies and professional societies, and that the society at large embrace the "green" movement. This section describes these three broad pathways.

24.8.1 Codes and Standards

The design and construction of buildings is subject to codes, i.e., laws, ordinances, and other types of regulations. Similarly, there are building energy codes and standards that set minimum requirements for energy-efficient design and construction for new and renovated buildings that impact energy use and emissions for the life of the building. It was in 1975 that the first building energy standard, namely, ASHRAE Standard 90.1, Energy Conservation in New Building Design, was published. Since no U.S. national standard existed previously, this was the first time many architects and engineers were alerted to the energy consequences of their design. Standard 90.1 has been widely adopted as a benchmark for building efficiency in commercial buildings (another standard was developed for residential buildings). These standards are incrementally upgraded in terms of energy efficiency and revised every few year through voluntary consensus and public hearing processes that are critical to widespread support for their adoption. ASHRAE 90.1 (2013) provides minimum energy-efficient requirements for the design, construction, and plan for operation and maintenance for new buildings, new portions, or new systems in existing buildings and in some cases for industrial and manufacturing processes. The standard also provides criteria to determine compliance with these requirements. The standard covers the following main areas: (1) building envelope, (2) heating, ventilating and air conditioning, (3) service hot water, (4) power, and (5) lighting.

ASHRAE 90.1 provides some flexibility in the sense that two alternative paths are advocated in the standard (Figure 24.19): (1) prescriptive path and (2) the energy cost budget (ECB) method. Both these methods must

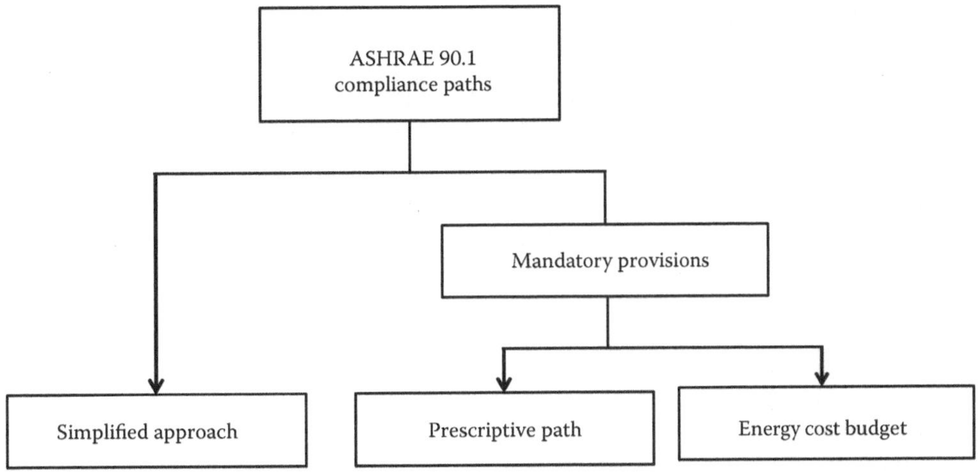

FIGURE 24.19
Different compliance paths for ASHRAE Standard 90.1.

satisfy the *general and mandatory provisions* primarily for the building envelope (walls, windows, doors, and air leakage)—for example, minimum material thermal resistances are specified, as we have minimum HVAC equipment efficiency such as chiller efficiency—and similarly for lighting, power, etc. For simple cases, the standard allows a simplified approach, for example, smaller buildings (less than 25,000 ft^2 or less than two-story building). The standard clearly defines when this option is applicable.

As stated previously, the standard relates to several individual building subsystem areas, each of which consist of two alternative options:

1. *Prescriptive path* wherein preset criteria for components and systems of the building in each of the areas listed earlier must be met. No trade-offs are allowed.
2. *ECB method* whereby certain subsystem performance targets must be met. With this option, the designer has the option to trade off between different components such that the subsystem as a whole meets some preset target.

The ECB methods looks at the performance of the building as a whole in terms of annual energy cost by utilizing any combination of material, systems, and operating procedures included in the prescriptive requirements sections. The ECB method is used to demonstrate compliance for cases where the architect or engineer needs to deviate from the prescriptive path requirements; this allows some flexibility in the building design. Another section of the standard (Appendix G of the standard) is used to quantify performance that substantially exceeds the requirements of Standard 90.1. Appendix G (also known as "performance rating method") is a modification of the ECB method. Entities such as the U.S. Green Building Council (USGBC) and others have been using this method. This approach allows greater flexibility for innovation for building owners seeking LEED certification and other types of recognition.

The *ECB method* utilizes an approved building energy simulation program to compare the energy cost of proposed (actual) building design (also known as "design energy cost" [DEC]) to the energy cost of budget building design (also known as ECB). DEC energy use must be lower or equal to that of ECB to demonstrate compliance. The ECB section (section 11 of the standard) provides detailed information and modeling rules on how to develop the building energy models.

Standard 90.1 is a code-intended standard. As such, it is written in unambiguous language intended to allow a code official to determine whether the building either complies with the standard or not. All states in the United States and most of the larger municipalities have promulgated local building codes that are strictly enforced. Such measures dovetail into larger efforts. The *Energy Policy Act* (or EPACT92) was passed by the U.S. Congress in 1992, which set goals, created mandates, and amended utility laws to increase clean energy use and improve overall energy efficiency in the United States. The scope of EPACT92 is much broader than buildings, but the "title" related to buildings requires all states to adopt building energy codes for commercial and high-rise multifamily residential buildings that at least meet Standard 90.1.

The gradual improvements in the ASHRAE Standard 90.1 is being actively supported by the DOE through the development DOE's commercial reference building models (Deru et al., 2011). The reference building models are meant to provide consistency in modeling

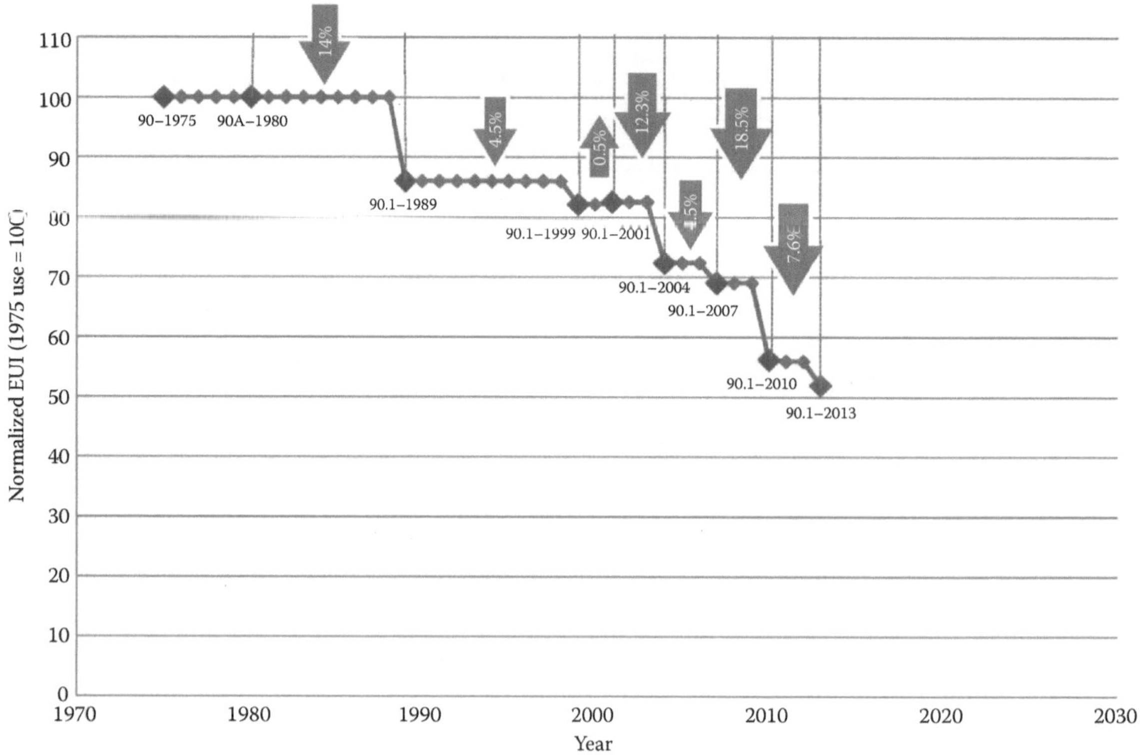

FIGURE 24.20
Energy efficiency improvements in ASHRAE Standard 90.1 for commercial buildings in successive versions from 1975 till 2013. (Courtesy of Bing Liu, Pacific Northwest National Laboratory, Richland, WA, 2015.)

approaches and implementation across commercial buildings but are not intended to represent energy use in any particular building. Rather, they are hypothetical models with ideal operations that meet certain minimum requirements.

The interested reader can refer to Thornton et al. (2011) for a description of the simulation and analysis framework driving the energy efficiency in Standard 90.1. The purpose of these building models is to represent new and existing buildings so as to "assess new technologies; optimize designs; analyze advanced controls; develop energy codes and standards; and to conduct lighting, daylighting, ventilation, and indoor air quality studies." Approximately 70% of the U.S. building stock are said to be represented by 15 commercial building types* and one multifamily residential building in 17 U.S. locations. All these 16 building types are simulated for different energy efficiency measures and evaluated in the 17 locations representative of the climatic variability of the United States. The energy results are then weighted by a process described by Thornton et al. to determine the energy reduction impact at the national level. Such an approach is said to provide a common starting point to measure the progress of DOE energy efficiency goals for commercial buildings through modifications to the building shell, ventilation needs, and its equipment (HVAC system, lighting, and plug loads).

Historically, the sets of building models have undergone small changes over the years. Figure 24.20 shows the history of ASHRAE Standard 90.1 in terms of how commercial building sector EUI has gradually decreased over the years with each successive version. We note that compared to the 1975 version of the standard, the 2004 version has achieved about a 30% reduction in EUI. The 2013 version was meant to reduce EUI by another 30% from the 2004 level, which is approximately a reduction of 50% from the 1975 standard. The ultimate goal is to achieve net-zero target by 2030. Whether this target will be met or not remains to be seen.

24.8.2 Design Guides

In an effort to provide designers with specific design guidelines, several organizations have developed design guides on how to achieve energy reductions in buildings. Perhaps the most technically complete are the ASHRAE Advanced Energy Design Guides, which are a series of publications designed to provide recommendations for achieving prestipulated

* See Figure 1.6 that gives the 2003 USDOE statistics by building type and energy intensity.

energy savings compared to Standard 90.1 (1999). Two sets of guides, which are specific to a building type, have been published: (1) the 30% reduction guides, which include six building types, and (2) the 50% energy reduction guides, which include five building types. Note that these reductions are consistent with the EUI reduction targets set by successive versions of the 90.1 standard, as described earlier. These guides can be freely downloaded from https://www.ashrae.org/standards-research–technology/advanced-energy-design-guides

Additional guides are also being developed. It must be cautioned that these documents are meant to be guidelines only and should be treated as such. They have been developed based on detailed simulation studies and have not been fully vetted against actual monitoring and measured data analysis.

24.8.3 Green Building Movement

In recent years, the *sustainability* movement has acquired such an impetus that it touches on all facets of human endeavor. In essence, this is a holistic movement that includes three elements relevant to current and future generations: minimizing the adverse environmental impacts of human activity while simultaneously maintaining economic progress and assuring social equity to people worldwide. Given the scope of this book, we shall limit our discussion to the first element, which is often referred to as the *green* movement. The green movement, in turn, may be viewed as consisting of three solution pathways:

1. Technoeconomic
2. Policy, which includes codes and financial incentives
3. Social, which includes behavioral change, rethinking of human values, awareness of the implications of current choices, and making voluntary choices

Most of this book has dealt with the technoeconomic considerations in designing energy-efficient buildings. While energy efficiency (and indoor air quality) are certainly critical issues, green buildings (a term used synonymously with *high-performance* buildings) include other aspects as well. There are numerous books on this subject (see, for example, Randolph and Masters, 2008) and professional societies such as ASHRAE, AIA, and USGBC also have several pertinent publications.

In general, the technoeconomic considerations of green buildings build on the current and ongoing thrusts of enhancing energy efficiency measures and integrating renewable energy systems. Green building design goes a step further by including requirements for site sustainability, water use efficiency, indoor environmental quality, and the building's impact on the atmosphere, materials, and resources. Hence, rather than the traditional LCC, building designs tend to be based on methods such as life cycle analysis, which consider embodied energy in the different materials that constitute the building, and the carbon footprint of constructing and operating the building. ASHRAE 189.1 (2014) is a green building standard that defines *minimum requirements* for designing, constructing, commissioning, maintaining, and operating green buildings (it covers the same group of building spaces as does Standard 90.1). It is neither a design guide nor a rating system but establishes mandatory criteria similar to ASHRAE 90.1. Useful design guidance and bulleted tips to address these various issues are provided in another book (ASHRAE Green Guide, 2010).

Extensions to the energy rating methods (see Section 24.7) to be applicable to green buildings have also been developed by several organizations. These are voluntary programs that help motivate customers to recognize the value of green building designs. The leading rating method in the United States is the LEED developed by the USGBC. It started out with a basic LEED rating method for new construction, but over the years several rating documents have been released for a variety of building projects (existing buildings, schools, retail stores, healthcare centers, homes, etc.). All LEED programs award certification on a numerical scale extending up to 110. Points are awarded depending on certain specific design criteria grouped in seven categories as follows (with maximum points allocated shown in brackets):

1. Sustainable site (26 points)
2. Water efficiency (10 points)
3. Energy and atmosphere (35 points)
4. Materials and resources (14 points)
5. Indoor environmental quality (15 points)
6. Innovation in design (6 points)
7. Regional priority credits (4 points)

LEED certification includes the following categories: certification (40–49 points), silver (50–59 points), gold (60–79 points), and platinum (80+). Other countries are also developing green building rating methods better suited to their unique circumstances. Another widely used voluntary, consensus-based rating method, originally developed in the United Kingdom for office buildings, is the Building Research Establishment Environmental Assessment Method (BREEAM).

Problems

Numbers 1–4 given in parenthesis denote the degree of difficulty.

24.1 A central station converts fuel to electricity with a typical efficiency of 33%. How high must the COP of a heat pump be so that the overall fuel utilization is better with electric heating than with burning of fuel in local furnaces of 80% efficiency? Briefly state one or two reasons why this simple argument is not sufficient for a decision, quite apart from the cost considerations. *(1)*

24.2 Consider the comparison of HVAC systems for the small office building in Figure 24.9, and try to estimate qualitatively how the comparison would change if the window/wall ratio (WWR) were changed? *Hint:* Consider the variability of zone loads for two extreme values of the WWR—0 and 1. *(2)*

24.3 (This problem may take a fair amount of effort.) Consider a VAV system that supplies two zones: an interior zone and a perimeter zone. Both are maintained at T_{int} = 23°C (73.4°F). Suppose the cooling load of the interior zone is constant at $\dot{Q}_1 = 1\,\text{kW}_t\,(3412\,\text{Btu/h})$, while solar gains cause the load of the perimeter zone to vary from $\dot{Q}_1$ to $4\dot{Q}_1$ at a time when the outdoor temperature is at 33°C (91.4°F) and 50% relative humidity. These loads exclude the contribution of the outdoor air. The minimum outdoor-air requirement is $\dot{V}_o = 34.7\,\text{L/s}\,(73.6\,\text{ft}^3/\text{min})$, in each zone. Suppose that the supply air temperature is constant at T_{sup} = 14°C (57.2°F) and 100% relative humidity. (a) Draw a schematic of the system and indicate how the airflow rates are controlled. (b) Find the maximum and minimum flow rates for the central supply airflow. (c) Find the maximum and minimum of the total cooling load, including outdoor air. (d) How much could the total cooling load be reduced if the zones were supplied by separate systems? (e) How would the results change if T_{sup} were variable? (f) Evaluate the multizone efficiency of Equation 19.26. *(4)*

24.4 (This problem may take a fair amount of effort.) Consider a VAV system that supplies two zones, an interior zone and a perimeter zone, each with 50 m² (538 ft²) floor area and height of 2.5 m (8.2 ft). Both are maintained at T_i = 23°C (73.4°F). Suppose that the cooling load of the interior zone is constant at $\dot{Q}_1 = 1\,\text{kW}_t\,(3412\,\text{Btu/h})$, while the perimeter zone has a heating load of $\dot{Q}_2 = (UA)_{per}(T_i - T_o)$. These loads exclude the contribution of the outdoor air. The minimum outdoor-air requirement is $\dot{V}_o = 34.7\,\text{L/s}\,(73.6\,\text{ft}^3/\text{min})(2.5\,\text{m} \times 50\,\text{m}^2 \times 1.0\,\text{ach})$ in each zone. The supply air temperature is 14°C (57.2°F). Neglect the effects of humidity. (a) Draw a schematic of the system, and indicate how the system is controlled. (b) Find the cutoff temperature of the economizer. (c) Find the supply airflow rate. (d) How much reheat is needed as a function of T_o? (e) Calculate and plot the total thermal power as a function of T_o. (f) Evaluate the multizone efficiency of Equation 19.26. *(4)*

24.5 Consider a building whose cooling load is proportional to the temperature difference between T_o and 65°F (18.3°C), reaching a peak of 200 tons (703 kW) at 100°F (37.8°C). Set up a spreadsheet with bin data for Albuquerque, New Mexico, to evaluate the annual energy consumption for these two options: a single 200 tons (703 kW_t) chiller with COP = 4.5 and two chillers with 100 tons (351.5 kW_t) each and COP = 4.0. Both have the same part-load efficiency curve, given by Equation 14.32 with a = 0.160, b = 0.316, and c = 0.519 (the coefficients of Example 14.9). For the two-chiller system, you need to decide (you can use trial and error) at each temperature bin how to split the load among the two chillers to maximize the efficiency. *(4)*

24.6 As a simple approximation of a house with passive solar heating using a water Trombe wall (see Figure 24.6d), assume that the only solar gains come from a south-facing window and that they are entirely absorbed in a perfectly mixed water tank. The house is characterized by the thermal network of Example 8.7 and Figure 8.17, and the storage tank is included in the heat capacity C_i; the other parameters are the same as in Example 8.7:

$$C_i = 1.0\,\text{kWh/K} + \rho V c_p \text{ of storage tank}$$

$$C_e = 5.0\,\text{kWh/K}\,(9.48\,\text{kBtu/°F})$$

$$R_i = 2.5\,\text{K/kW}\,[1.32\,(\text{h}\cdot\text{°F})/\text{Btu}] = R_o$$

Assume that the transmitted solar flux per area A of the window is given by

$$I(t) = \max\left\{0 - 0.2 + \cos\left[(t - 12\,\text{h})\frac{\pi}{12\,\text{h}}\right]\right\} \times 800\,\text{W/m}^2$$

as function of time of day t. The outdoor temperature is

$$T_o(t) = \cos\left[(t - 16\,\text{h})\frac{\pi}{12\,\text{h}}\right] \times 5\text{°C}$$

(a) Write down the differential equations and specify all the input.

(b) Set up a spreadsheet or program to solve the equation for T_i numerically for the case where the only other energy input is a constant base level of 1.0 kW (you can use, e.g., the method illustrated in Example 8.8).

(c) Vary window area A and storage volume V until T_i will always be within the range of 20°C–24°C (68°F–75.2°F). *(4)*

24.7 Suppose you have to design a daylighting system using skylights for a one-story office building with floor area 200 ft × 100 ft (60.96 m × 30.48 m) in Atlanta. How large would you choose the aperture area, and what type of glazing would you use? Approximately how large are the annual energy savings? *(4)*

24.8 A rooftop solar hot water heater with collector areas of 8 m^2 and storage of 640 L is installed in Phoenix, Arizona, on a house that uses 160 L/day of hot water at 60°C. Seventy percent of the total energy required is supplied by solar energy (this is often referred to as a solar fraction of 70%) and 30% is supplemented with an electric heater. The solar system costs $3000 for installation, whereas an electric water heater and tank for the same supply costs $800 for installation. The temperature of the water in the main from which it is to be heated is 10°C.

(a) Assuming that the electricity cost is $0.10/kWh and does not change over time, estimate the simple payback time for the solar hot water heater.

(b) Repeat the aforementioned problem, but assume that the cost of installation is to be repaid at 7% discount over 15 years and that the electricity costs escalate at 10%/year. You will solve this by preparing a spreadsheet showing the net present value for each year. *(3)*

24.9 Write an essay on the Energy Star program started by USEPA—their description, origin, scope, ways of implementation, and impact. *(2)*

24.10 An environmentally friendly *green* house costs about 10% more to construct than a conventional home. The green home is assumed to save 50%/year on energy expenses to heat and cool the dwelling. Assume the following:

- It is an all-electric home needing 8 MWh/year for heating and cooling and 6 MWh/year for equipment in the house (ordinary home).
- The conventional home costs $250,000 and its life span is 30 years.
- The cost of electricity is $15 cents assumed constant.
- The *green* home has the same life span and no additional value at the end or 30 years.
- The discount rate is 6% per year.
  - Is the *green* home a worthwhile investment purely from an economic point of view?
  - If the government imposes a carbon tax of $50/ton, how would you go about reevaluating the added benefit of the *green* home. Assume 170 lbm of carbon are released per MWh of electricity generation (corresponds to the rather progressive state of California).
  - If an environmental assessment were to be undertaken, what other issues would you consider? *(3)*

24.11 Research in more detail the LEED rating system for new construction and write a two-page essay describing it in more detail. You will refer to Turner and Frankel (2008) and discuss differences between measured and modeled building performances. This report can be located on the Internet. *(2)*

24.12 Solar PV systems can be mounted on the roofs of buildings. You will use the PVWATTS software program developed by the U.S. DOE's National Renewable Energy Laboratory to allow nontechnical experts the ability to estimate the electric output of grid-connected PV systems. It is a free web-based program with an in-built database of relevant climatic conditions in hundreds of locations worldwide. It has a number of system options that the user must select appropriately.

(a) Read the online tutorial and state the limitations and assumptions of the software program.

(b) You will use this program (freely available online) to determine the approximate range of electric power produced per year per square meter of PV module area at a few locations in the United States that are representative of extreme solar radiation availability. Assume the solar panels to be fixed, nonconcentrating, south facing, and tilted at an angle equal to the latitude of the locations. Assume a ground-cover ratio of 50%.

(c) How does this electric output compare with the electric requirements of a typical 200 m^2 all-electric single-story house. *(4)*

References

ACEEE (1989). *The Most Energy-Efficient Appliances*. American Council for an Energy-Efficient Economy (ACEEE), Washington, DC.

Addison, M.S. (1988). A multiple criteria satisficing methodology for the design of energy-efficient buildings. Unpublished Master of Environmental Planning thesis, Arizona State University, Tempe, AZ.

ANSI/ASHRAE Standard 140 (2011). *Standard Method of Test for the Evaluation of Building Energy Analysis Computer Programs*. American Society of Heating, Refrigerating and Air-Conditioning Engineers, Atlanta, GA.

ASHRAE (1984). *Passive Solar Heating Analysis: A Design Manual*. American Society of Heating, Refrigerating and Air-Conditioning Engineers, Atlanta, GA.

ASHRAE 189.1 (2014). *ASHRAE Standard 189.1-2014, Standard for the Design of High-Performance Green Buildings*. American Society of Heating, Refrigerating and Air-Conditioning Engineers, Atlanta, GA.

ASHRAE 90.1 (1999, 2004, 2007, 2010, 2013). *ANSI/ASHRAE/IESNA 90.1-2007, Energy Standard for Buildings Except Low-Rise Residential Buildings*. American Society of Heating, Refrigerating and Air-Conditioning Engineers, Atlanta, GA.

ASHRAE BEQ (2014). *Building Energy Quotient*. American Society of Heating, Refrigerating and Air-conditioning Engineers, Atlanta, GA. http://buildingenergyquotient.org/what-is-bEQ.html. Accessed on January 2014.

ASHRAE Fundamentals (2013). *Handbook of Fundamentals*. American Society of Heating, Refrigerating and Air-Conditioning Engineers, Atlanta, GA.

ASHRAE Green Guide (2010). *ASHRAE Green Guide: The Design, Construction and Operation of Sustainable Buildings*. American Society of Heating, Refrigerating and Air-Conditioning Engineers, Atlanta, GA.

Athienitis, A.K. and M. Santamouris (2002). *Thermal Analysis and Design of Passive Solar Buildings*. James and James, London, U.K.

Athienitis, A.K., J. Bambara, B. O'Neill, and J. Faille (2011). A prototype photovoltaic/thermal system integrated with transpired collector. *Solar Energy*, 85, 139–153.

Berkeley Solar Group (1981). *CALPAS3 User's Manual*. Berkeley Solar Group, Berkeley, CA.

Birdsall, B., W.F. Buhl, K.L. Ellington, A.E. Erdem, and F.C. Winkelmann (1990). Overview of the DOE 2.1 building energy analysis program. Report LBL-19735, Revised 1. Lawrence Berkeley Laboratory, Berkeley, CA.

BLAST (1986). The building load analysis and system thermodynamics program, Version 3. User's Manual. BLAST Support Office, University of Illinois, Champaign-Urbana, IL.

Brager, G.S. (August 2006). Mixed-mode cooling. *ASHRAE J.*, 48, 30–37.

Burt Hill Kosar Rittelmann Associates (1985). *Small Office Building Design Handbook: Design for Reducing First Costs and Utility Costs*. Van Nostrand Reinhold, New York.

Burt Hill Kosar Rittelmann Associates and Min Kantrowitz Associates (1987). *Commercial Building Design: Integrating Climate, Comfort and Cost*. Van Nostrand Reinhold, New York.

Byrne, S.J. and J.M. Fay (1989). A comparison of central and individual systems for space conditioning and domestic hot water in new multifamily buildings. *ASHRAE Trans.*, 95, 2.

Byrne, S.J. and R.L. Ritschard (1985). A parametric analysis of thermal mass in residential buildings. Report LBL-20288. Lawrence Berkeley Laboratory, Berkeley, CA.

CABO (1987). *Model Energy Code*. Council of American Building Officials, Falls Church, VA.

CBECS (2003). Commercial building energy consumption survey (CBECS) survey. Retrieved December 2012, http://www.eia.gov/consumption/commercial/data/2003/.

Chalfoun, N.V. (1997). Design and application of natural down-draft evaporative cooling devices, in *26th American Solar Energy Society (ASES) and the 22nd National Passive Solar Annual Conference*, Washington, DC. Retrieved from http://hed.arizona.edu/hed/docs/ASES-97.pdf.

Christian, J.E. (1991). Thermal mass credits relating to building envelope energy standards. *ASHRAE Trans.*, 97, 2.

Cook, J. (ed.) (1989). *Passive Cooling*. The MIT Press, Cambridge, MA.

Cowing, T. and J.F. Kreider (1987). Solar load ratio statistics for the United States and their use for analytical solar performance prediction. *ASME J. Solar Energy Eng.*, 109, 281–288.

Crabb, J.A. (1988). An approach to the design of energy efficient heated buildings. SWEG Report 41. Energy Studies Unit, Department of Physics, University of Exeter, Devon, U.K.

Deru, M.K., D. Field, K. Studer, B. Benne, P. Griffith, B. Torcellini, M. Liu et al. (February 2011). U.S Department of Energy Commercial Reference Building Models of the National Building Stock. Technical Report NREL/TP-5500-46861. National Renewable Energy Laboratory, Golden, CO.

De Witt, S. (2003). Uncertainty in building simulation, Chapter 2, in *Advanced Building Simulation* (A.M. Malkawi and G. Augenbroe, eds.). Spon Press, New York.

Dutta, R., T.A. Reddy, and G. Runger (January 2016). A visual analytics based methodology for multi-criteria evaluation of building design alternatives. *ASHRAE Conference Paper OR-16-6051*, 6pp., Orlando, FL.

Eicker, U. (2003). *Solar Technologies for Buildings*. John Wiley & Sons, Chichester, U.K.

Eicker, U. (2009). *Low Energy Cooling for Sustainable Buildings*. John Wiley & Sons, Chichester, U.K.

Energy Plus (2009). Energy building energy simulation software. Developed by the National Renewable Energy Laboratory (NREL) for the U.S. Department of Energy, under the Building Technologies program, Washington, DC. http://www.nrel.gov/buildings/energy_analysis.html#energyplus, accessed May 2014.

EPA (2011). Energy star performance ratings: Technology methodology. http://www.energystar.gov/ia/business/evaluate_performance/General_Overview_tech_methodology.pdf. Accessed on January 2014.

Fairey, P.W. and A.A. Kerestecioglu (1985). Dynamic modelling of combined thermal and moisture transport in buildings: Effects on cooling loads and space conditions. *ASHRAE Trans.*, 91(2A), 461.

Givoni, B. (1994). *Passive and Low Energy Cooling of Buildings.* Van Nostrand, New York.

Gordon, J.M. and Y. Zarmi (1981a). Analytic model for passively heated solar houses: 1. Theory. *Solar Energy*, 27, 331.

Gordon, J.M. and Y. Zarmi (1981b). Analytic model for passively heated solar houses: 2. User's guide. *Solar Energy*, 27, 343.

Goswami, D.Y., F. Kreith, and J.F. Kreider (2000). *Principles of Solar Engineering*, 2nd ed. Taylor & Francis Group, Philadelphia, PA.

Henze, G. and P. May-Ostendorp (May 2012). HVAC control algorithms for mixed mode buildings. Final Research Report submitted to United States Green Building Council by University of Colorado, Boulder, CO.

Huang, J. and E. Franconi (1999). Commercial heating and cooling loads component analysis. LBL-37208, Report prepared by Lawrence Berkeley National Laboratory for the U.S. Department of Energy, Washington, DC.

Hughes, B.R., H.N. Chaudhry, and A.A. Ghani (2011). A Review of sustainable cooling technologies in buildings. *Renew. Sust. Energ. Rev.*, 15, 3112–3120.

IEA (1995). Annex 28—Review of low energy cooling technologies. International Energy Agency, Building Research Establishment, Britain. http://www.ecbcs.org/docs/annex_28_review.pdf.

Jarnagin, R.E. (December 2009). ASHRAE building EQ. *ASHRAE J.*, 51(12), 18–19.

Jones, R.W., ed. (1983). *Passive Solar Design Handbook*, vol. 3. American Solar Energy Society, Boulder, CO.

Judkoff, R. and J. Neymark (January 2006). Model validation and testing: The methodological foundation of ASHRAE standard 140. *ASHRAE Trans.*, 112 (2), 367–376.

Kao, J.Y. (1982). Strategies for energy conservation in small office buildings. Center for Building Technology Report NBSIR 84-2489 (June), National Bureau of Standards, Gaithersburg, MD.

Kao, J.Y. (1983). Strategies for energy conservation for a large office building. Center for Building Technology Report NBSIR 84-2746, National Bureau of Standards, Gaithersburg, MD.

Kao, J.Y. (1985). Control strategies and building energy consumption. *ASHRAE Trans.*, 91(2B), 810–817.

Kaskhedikar, A., T.A. Reddy, and G. Runger (January 2015). Use of random forest algorithm to evaluate model based EUI benchmarks. *ASHRAE Trans.*, 121(1), 17–28.

Knebel, D.E. (1983). *Simplified Energy Analysis Using the Modified Bin Method.* American Society of Heating, Refrigerating and Air-Conditioning Engineers, Atlanta, GA. See also Knebel, D. and S. Silver (1985). Upgraded documentation of the TC4.7 simplified energy analysis procedure. *ASHRAE Trans.*, 91(2A), 1985.

Kreider, J.F. and F. Kreith (1990). *Solar Design: Components, Systems and Economics.* Hemisphere, New York.

Kusuda, T. (1981). A comparison of energy calculation procedures. *ASHRAE J.* (August), 21.

LBL (1988). Energy and environment division: Annual report. Report LBL-26585, Energy Efficient Buildings Program, Lawrence Berkeley Laboratory, Berkeley, CA.

Li, C. and T. Hong (January 2014). Revisit of energy use and technologies of high performance buildings, in *ASHRAE Conference Paper SE-14-C033*, 120(2).

Marchio, D. and A. Rabl (1991). Energy-efficient gas-heated housing in France: Predicted and observed performance. *Energy Build.*, 17, 131–139.

Masters, G.M. (2004). *Renewable and Efficient Electric Power Systems.* Wiley Interscience, Hoboken, NJ.

Mazria, E. (1979). *The Passive Solar Energy Book.* Rodale Press, Emmanus, PA.

Meier, A.K. (1990). Measured cooling savings from vegetative landscaping, in *ACEEE 1990 Summer Study on Energy Efficiency in Buildings*, vol. 4, American Council for an Energy-Efficient Economy, Asilomar, CA, p. 133.

Messenger, R.A. and J. Ventre (2010). *Photovoltaic Systems Engineering*, 3rd ed. CRC Press, Boca Raton, FL.

Mitchell, J.W. and W.A. Beckman (1989). Theoretical limits for storage of energy in buildings. *Solar Energy*, 42, 113–120.

Monsen, W.A., S.A. Klein and W.A. Beckman (1982). The unutilizability design method for collector storage walls. *Solar Energy*, 29(5), 421–429.

Nayak, J.K. and J.A. Prajapati (2006). *Handbook on Energy Conscious Buildings.* Indian Institute of Technology, Bombay for the Solar Energy Centre, Ministry of Non-Conventional Energy Sources, New Delhi, India. Free download: http://mnre.gov.in/centers/about-sec-2/hand-book-on-energy-conscious-buildings/.

Randolph, J. and G.M. Masters (2008). *Energy for Sustainability: Technology, Planning and Policy.* Island Press, Washington, DC.

Santamouris, M. and D. Asimakopoulos (1996). *Passive Cooling of Buildings.* James and James, London, U.K.

Sharp, T. (1996). Energy benchmarking in commercial office buildings, in *ACEEE 1996 Summer Study on Energy Efficiency in Buildings*, vol. 4, pp. 321–329, Asilomar, CA.

Sodha, M.S., N.K. Bansal, A. Kumar and M.A.S. Malik (1986). *Solar Passive Building: Science and Design.* Pergamon Press, Oxford, U.K.

SOLMET (1978). Volume 1: User's manual, vol. 2: Final report, hourly solar radiation surface meteorological observations. Report TD-9724. National Climatic Center, Asheville, NC.

Spitler, J.D., D.C. Hittle, D.L. Johnson, and C.O. Pederson (1987). A comparative study of the performance of temperature-based and enthalpy-based economy cycles. *ASHRAE Trans.*, 93(2), 13–22.

Sullivan, R. and S. Selkowitz (1985). Energy performance analysis of fenestration in a single-family residence. *ASHRAE Trans.*, 91(2A), 320–336.

Sullivan, R. and S. Selkowitz (1987). Residential heating and cooling energy cost implications associated with window type. *ASHRAE Trans.*, 93, 1.

Thornton, B.A., M.I. Rosenberg, E.E. Richman, W. Wang, Y. Xie, J. Zhang, H. Cho, V.V. Mendon, R.A. Athalye, and B. Liu (May 2011). Achieving the 30% energy and cost savings analysis of ASHRAE Standard 90.1-2010, PNNL-20405. Pacific Northwest National Laboratory, Prepared for the U.S. Department of Energy, Washington, DC.

Turner, C. and M. Frankel (May 2008). Energy performance of LEED for new construction buildings. Report prepared by New Buildings Institute, Vancouver, WA for U.S. Green Building Council, Washington, DC.

Usibelli, A., S. Greenberg, M. Meal, A. Mitchell, R. Johnson, G. Sweitzer, F. Rubinstein, and D. Avasteh (February 1985). Commercial sector conservation technologies. Report LBL 18543. Lawrence Berkeley Laboratory, Berkeley, CA.

Appendix

TABLE A.1

(IP Units) Transport Properties of Standard, Dry Air (14.7 psia)

°R	°F	ρ (lb_m/ft^3)	c_p (Btu/lb·°R)	k (c_p/c_v)	μ x 10^4 (lb_m/ft·s)
180	−280	0.2247	0.2456		0.0466
198	−262	0.2033	0.2440	1.4202	0.0513
216	−244	0.1858	0.2430	1.4166	0.0559
234	−226	0.1711	0.2423	1.4139	0.0604
252	−208	0.1586	0.2418	1.4119	0.0648
270	−190	0.1478	0.2414	1.4102	0.0691
288	−172	0.1384	0.2411	1.4089	0.0733
306	−154	0.1301	0.2408	1.4079	0.0774
324	−136	0.1228	0.2406	1.4071	0.0815
342	−118	0.1163	0.2405	1.4064	0.0854
360	−100	0.1104	0.2404	1.4057	0.0892
369	−91	0.1078	0.2403	1.4055	0.0911
378	−82	0.1051	0.2403	1.4053	0.0930
387	−73	0.1027	0.2403	1.4050	0.0949
396	−64	0.1003	0.2402	1.4048	0.0967
405	−55	0.0981	0.2402	1.4046	0.0986
414	−46	0.0959	0.2402	1.4044	0.1004
423	−37	0.0939	0.2402	1.4042	0.1022
432	−28	0.0919	0.2401	1.4040	0.1039
441	−19	0.0901	0.2401	1.4038	0.1057
450	−10	0.0882	0.2401	1.4036	0.1074
459.7	0	0.0865	0.2401	1.4034	0.1092
468	8	0.0848	0.2401	1.4032	0.1109
486	26	0.0817	0.2492	1.4029	0.1143
504	44	0.0787	0.2402	1.4024	0.1176
522	62	0.0760	0.2403	1.4020	0.1208
540	80	0.0735	0.2404	1.4017	0.1241
558	98	0.0711	0.2405	1.4013	0.1272
576	116	0.0689	0.2406	1.4008	0.1303
594	134	0.0668	0.2407	1.4004	0.1334
612	152	0.0648	0.2409	1.3999	0.1364
630	170	0.0630	0.2411	1.3993	0.1394
648	188	0.0612	0.2412	1.3987	0.1423
666	206	0.0595	0.2415	1.3981	0.1452
684	224	0.0580	0.2417	1.3975	0.1479
702	242	0.0565	0.2420	1.3968	0.1508
720	260	0.0551	0.2422	1.3961	0.1536
738	278	0.0537	0.2425	1.3953	0.1563
756	296	0.0525	0.2428	1.3946	0.1590
774	314	0.0512	0.2432	1.3938	0.1617
792	332	0.0501	0.2435	1.3929	0.1643
810	350	0.0490	0.2439	1.3920	0.1670

Source: CRC Press, *Handbook of Tables for Applied Engineering Science*, CRC Press, Inc., Boca Raton, FL, 1973.

TABLE A.1 (*Continued*)

(SI Units) Transport Properties of Standard, Dry Air (101.3 kPa)

K	°C	ρ (kg/m³)	c_p (kJ/kg·K)	k (c_p/c_v)	μ x 10⁶ (Pa-s)
100	−173.15	3.598	1.028		6.929
110	−163.15	3.256	1.022	1.420	7.633
120	−153.15	2.975	1.017	1.416	8.319
130	−143.15	2.740	1.014	1.413	8.990
140	−133.15	2.540	1.012	1.411	9.646
150	−123.15	2.367	1.010	1.410	10.28
160	−113.15	2.217	1.009	1.408	10.91
170	−103.15	2.085	1.008	1.407	11.52
180	−93.15	1.968	1.007	1.407	12.12
190	−83.15	1.863	1.007	1.406	12.71
200	−73.15	1.769	1.006	1.405	13.28
210	−63.15	1.684	1.006	1.405	13.85
220	−53.15	1.607	1.006	1.404	14.40
230	−43.15	1.537	1.006	1.404	14.94
240	−33.15	1.473	1.005	1.404	15.47
250	−23.15	1.413	1.005	1.403	15.99
260	−13.15	1.359	1.005	1.403	16.50
270	−3.15	1.308	1.006	1.402	17.00
280	6.85	1.261	1.006	1.402	17.50
290	16.85	1.218	1.006	1.402	17.98
300	26.85	1.177	1.006	1.401	18.46
310	36.85	1.139	1.007	1.401	18.93
320	46.85	1.103	1.007	1.400	19.39
330	56.85	1.070	1.008	1.400	19.85
340	66.85	1.038	1.008	1.399	20.30
350	76.85	1.008	1.009	1.399	20.75
360	86.85	0.980	1.010	1.398	21.18
370	96.85	0.953	1.011	1.398	21.60
380	106.85	0.928	1.012	1.397	22.02
390	116.85	0.905	1.013	1.396	22.44
400	126.85	0.882	1.014	1.396	22.86
410	136.85	0.860	1.015	1.395	23.27
420	146.85	0.840	1.017	1.394	23.66
430	156.85	0.820	1.018	1.393	24.06
440	166.85	0.802	1.020	1.392	24.85
450	176.85	0.784	1.021	1.392	24.85

Source: CRC Press, *Handbook of Tables for Applied Engineering Science*, CRC Press, Inc., Boca Raton, FL, 1973.

TABLE A.2
(IP Units) Properties of Liquid Water

	At 1 atm or 14.7 psia				At 1000 psia (68 atm)			
Temp. °F	Density (lb_m/ft^3)	Specific Heat ($Btu/lb_m \cdot °R$)	Viscosity ($lb_f \cdot s/ft^2 \times 10^{-7}$)	Thermal Conductivity ($Btu/h \cdot ft \cdot °R$)	Density (lb_m/ft^3)	Specific Heat ($Btu/lb_m \cdot °R$)	Viscosity ($lb_f \cdot s/ft^2 \times 10^{-7}$)	Thermal Conductivity ($Btu/h \cdot ft \cdot °R$)
32	62.42	1.007	366	0.3286	62.62	0.999	365	0.3319
40	62.42	1.004	323	0.3340	62.62	0.997	323	0.3370
50	62.42	1.002	272	0.3392	62.62	0.995	272	0.3425
60	62.38	1.000	235	0.3450	62.58	0.994	235	0.3480
70	62.31	0.999	204	0.3500	62.50	0.994	204	0.3530
80	62.23	0.998	177	0.3540	62.42	0.994	177	0.3580
90	62.11	0.998	160	0.3590	62.31	0.994	160	0.3620
100	62.00	0.998	142	0.3633	62.19	0.994	142	0.3666
110	61.88	0.999	126	0.3670	62.03	0.994	126	0.3710
120	61.73	0.999	114	0.3710	61.88	0.995	114	0.3740
130	61.54	0.999	105	0.3740	61.73	0.995	105	0.3780
140	61.39	0.999	96	0.3780	61.58	0.996	96	0.3810
150	61.20	1.000	89	0.3806	61.39	0.996	89	0.3837
160	61.01	1.001	83	0.3830	61.20	0.997	83	0.3860
170	60.79	1.002	77	0.3860	60.98	0.998	77	0.3890
180	60.57	1.003	72	0.3880	60.75	0.999	72	0.3910
190	60.35	1.004	68	0.3900	60.53	1.001	68	0.3930
200	60.10	1.005	62.5	0.3916	60.31	1.002	62.9	0.3944
250	Boiling point 212°F				59.03	1.001	47.8	0.3994
300					57.54	1.024	38.4	0.3993
350					55.83	1.044	32.1	0.3944
400					53.91	1.072	27.6	0.3849
500					49.11	1.181	21.6	0.3508
600					Boiling point 544.58°F			

Source: CRC Press, *Handbook of Tables for Applied Engineering Science*, CRC Press, Inc., Boca Raton, FL, 1973.

TABLE A.3

(IP Units) Properties of Saturated Steam

		Specific Volume (ft^3/lb_m)		Internal Energy (Btu/lb_m)			Enthalpy (Btu/lb_m)			Entropy ($Btu/(lb_m \cdot °R)$)		
Pressure (psia)	Saturation Temp. (°F)	Sat. Liquid	Sat. Vapor	Sat. Liquid	Evap.	Sat. Vapor	Sat. Liquid	Evap.	Sat. Vapor	Sat. Liquid	Evap.	Sat. Vapor
p	T_{sat}	v_f	v_g	u_f	u_{fg}	u_g	h_f	h_{fg}	h_g	s_f	s_{fg}	s_g
0.08866	32.018	0.016022	3302	0.00	1021.2	1021.2	0.01	1075.4	1075.4	0	2.1869	2.1869
0.09992	35	0.016021	2948	2.99	1019.2	1022.2	3.00	1073.7	1076.7	0.00607	2.1704	2.1764
0.12166	40	0.016020	2445	8.02	1015.8	1023.9	8.02	1070.9	1078.9	0.01617	2.1430	2.1592
0.14748	45	0.016021	2037	13.04	1012.5	1025.5	13.04	1068.1	1081.1	0.02618	2.1162	2.1423
0.17803	50	0.016024	1704.2	18.06	1009.1	1027.2	18.06	1065.2	1083.3	0.03607	2.0899	2.1259
0.2563	60	0.016035	1206.9	28.08	1002.4	1030.4	28.08	1059.6	1087.7	0.05555	2.0388	2.0943
0.3632	70	0.016051	867.7	38.09	995.6	1033.7	38.09	1054.0	1092.0	0.07463	1.9896	2.0642
0.5073	80	0.016073	632.8	48.08	988.9	1037.0	48.09	1048.3	1096.4	0.09332	1.9423	2.0356
0.6988	90	0.016099	467.7	58.07	982.2	1040.2	58.07	1042.7	1100.7	0.11165	1.8966	2.0083
0.9503	100	0.016130	350.0	68.04	975.4	1043.5	68.05	1037.0	1105.0	0.12963	1.8526	1.9822
1	101.70	0.016136	333.6	69.74	974.3	1044.0	69.74	1036.0	1105.8	0.13266	1.8453	1.9779
2	126.04	0.016230	173.75	94.02	957.8	1051.8	94.02	1022.1	1116.1	0.17499	1.7448	1.9198
3	141.43	0.016300	118.72	109.38	947.2	1056.6	109.39	1013.1	1122.5	0.20089	1.6852	1.8861
4	152.93	0.016358	90.64	120.88	939.3	1060.2	120.89	1006.4	1127.3	0.21983	1.6426	1.8624
5	162.21	0.016407	73.53	130.15	932.9	1063.0	130.17	1000.9	1131.0	0.23486	1.6093	1.8441
6	170.03	0.016451	61.98	137.98	927.4	1065.4	138.00	996.2	1134.2	0.24736	1.5819	1.8292
8	192.84	0.016526	47.35	150.81	918.4	1069.2	150.84	988.4	1139.3	0.26754	1.5383	1.8058
10	193.19	0.016590	38.42	161.20	911.0	1072.2	161.23	982.1	1143.3	0.28358	1.5041	1.7877
14.696	211.99	0.016715	26.80	180.10	897.5	1077.6	180.15	970.4	1150.5	0.31212	1.4446	1.7567
15	213.03	0.016723	26.29	181.14	896.8	1077.9	181.19	969.7	1150.9	0.31367	1.4414	1.7551
20	227.96	0.016830	20.09	196.19	885.8	1082.0	196.26	960.1	1156.4	0.33580	1.3962	1.7320
25	240.08	0.016922	16.306	208.44	876.9	1085.3	208.52	952.2	1160.7	0.35345	1.3607	1.7142
30	250.34	0.017004	13.748	218.84	869.2	1088.0	218.93	945.4	1164.3	0.36821	1.3314	1.6996
35	259.30	0.017073	11.900	227.93	862.4	1090.3	228.04	939.3	1167.4	0.38093	1.3064	1.6873
40	267.26	0.017146	10.501	236.03	856.2	1092.3	236.16	933.8	1170.0	0.39214	1.2845	1.6767
45	274.46	0.017209	9.403	243.37	850.7	1094.0	243.51	928.8	1172.3	0.40218	1.2651	1.6673
50	281.03	0.017269	8.518	250.08	845.5	1095.6	250.24	924.2	1174.4	0.41129	1.2476	1.6589
55	287.10	0.017325	7.789	256.28	840.8	1097.0	256.46	919.9	1176.3	0.41963	1.2317	1.6513
60	292.73	0.017378	7.177	262.06	836.3	1098.3	262.25	915.8	1178.0	0.42733	1.2170	1.6444
65	298.00	0.017429	6.657	267.46	832.1	1099.5	267.67	911.9	1179.6	0.43450	1.2035	1.6380
70	302.96	0.017478	6.209	272.56	828.1	1100.6	272.79	908.3	1181.0	0.44120	1.1909	1.6321
75	307.63	0.017524	5.818	277.37	824.3	1101.6	277.61	904.8	1182.4	0.44749	1.1790	1.6265
80	312.07	0.017570	5.474	281.95	820.6	1102.6	282.21	901.4	1183.6	0.45344	1.1679	1.6214
85	316.29	0.017613	5.170	286.30	817.1	1103.5	286.58	898.2	1184.8	0.45907	1.1574	1.6165
90	320.31	0.017655	4.898	290.46	813.8	1104.3	290.76	895.1	1185.9	0.46442	1.1475	1.6119
95	324.16	0.017696	4.654	294.45	810.6	1105.0	294.76	892.1	1186.9	0.46952	1.1380	1.6076
100	327.86	0.017736	4.434	298.28	807.5	1105.8	298.61	889.2	1187.8	0.47439	1.1290	1.6034
110	334.82	0.017813	4.051	305.52	801.6	1107.1	305.88	883.7	1189.6	0.48355	1.1122	1.5957
120	341.30	0.017886	3.730	312.27	796.0	1108.3	312.67	878.5	1191.1	0.49201	1.0966	1.5886
130	347.37	0.017957	3.457	318.61	790.7	1109.4	319.04	873.5	1192.5	0.49989	1.0822	1.5821
140	353.08	0.018024	3.221	324.58	785.7	1110.3	325.05	868.7	1193.8	0.50727	1.0688	1.5761
150	358.48	0.018089	3.016	330.24	781.0	1111.2	330.75	864.2	1194.9	0.51422	1.0562	1.5704

Source: Courtesy of ASHRAE, *Handbook of Fundamentals*, American Society of Heating, Refrigerating and Air-Conditioning Engineers, Atlanta, GA, 1997. With permission.

TABLE A.3
(SI Units) Properties of Saturated Steam

		Specific Volume (m^3/kg)		Internal Energy (kJ/kg)			Enthalpy (kJ/kg)			Entropy (kJ/(kg·K))		
Pressure (kPa)	Saturation Temp. (°C)	Sat. Liquid	Sat. Vapor	Sat. Liquid	Evap.	Sat. Vapor	Sat. Liquid	Evap.	Sat. Vapor	Sat. Liquid	Evap.	Sat. Vapor
p	T_{sat}	v_f	v_g	u_f	u_{fg}	u_g	h_f	h_{fg}	h_g	s_f	s_{fg}	s_g
0.6113	0.01	0.001	206.14	0	2375.3	2375.3	0.01	2501.3	2501.4	0	9.1562	9.1562
1.0	6.98	0.001	129.21	29.30	2355.7	2385.0	29.30	2484.9	2514.2	0.1059	8.8697	8.9756
1.5	13.03	0.001001	87.98	54.71	2338.6	2393.3	54.71	2470.6	2525.3	0.1957	8.6322	8.8279
2.0	17.50	0.001001	67.00	73.48	2326.0	2399.5	73.48	2460.0	2533.5	0.2607	8.4629	8.7237
2.5	21.08	0.001002	54.25	88.48	2315.9	2404.4	88.49	2451.6	2540.4	0.3120	8.3311	8.6432
3.0	24.08	0.001003	45.67	101.04	2307.5	2408.5	101.05	2444.5	2545.5	0.3545	8.2231	8.5776
4.0	28.96	0.001004	34.80	121.45	2293.7	2415.2	121.46	2432.9	2554.4	0.4226	8.0520	8.4746
5.0	32.88	0.001005	28.19	137.81	2282.7	2420.5	137.82	2423.7	2561.5	0.4764	7.9187	8.3951
7.5	40.29	0.001008	19.24	168.78	2261.7	2430.5	168.79	2406.0	2574.8	0.5764	7.6750	8.2515
10	45.81	0.001010	14.67	191.82	2246.1	2437.9	191.83	2392.8	2584.7	0.6493	7.5009	8.1502
15	53.97	0.001014	10.02	225.92	2222.8	2448.7	225.94	2373.1	2599.1	0.7549	7.2536	8.0085
20	60.06	0.001017	7.649	251.38	2205.4	2456.7	251.40	2358.3	2609.7	0.8320	7.0766	7.9085
25	64.97	0.001020	6.204	271.90	2191.2	2463.1	271.93	2346.3	2618.2	0.8931	6.9383	7.8314
30	69.10	0.001022	5.229	289.20	2179.2	2468.4	289.23	2336.1	2625.3	0.9439	6.8247	7.7686
40	75.87	0.001027	3.993	317.53	2159.5	2477.0	317.58	2319.2	2636.8	1.0259	6.6441	7.6700
50	81.33	0.001030	3.240	340.44	2143.4	2483.9	340.49	2305.4	2645.9	1.0910	6.5029	7.5939
75	91.78	0.001037	2.217	384.31	2112.4	2496.7	384.39	2278.6	2663.0	1.2130	6.2434	7.4564
100	99.63	0.001043	1.6940	417.36	2088.7	2506.1	417.46	2258.0	2675.5	1.3026	6.0568	7.3594
125	105.99	0.001048	1.3749	444.19	2069.3	2513.5	444.32	2241.0	2685.4	1.3740	5.9104	7.2844
150	111.37	0.001053	1.1593	466.94	2052.7	2519.7	467.11	2226.5	2693.6	1.4336	5.7897	7.2233
175	116.06	0.001057	1.0036	486.80	2038.1	2524.9	486.99	2213.6	2700.6	1.4849	5.6868	7.1717
200	120.23	0.001061	0.8857	504.49	2025.0	2529.5	504.70	2201.9	2706.7	1.5301	5.5970	7.1271
225	124	0.001064	0.7933	520.47	2013.1	2533.6	520.72	2191.3	2712.1	1.5706	5.5173	7.0878
250	127.44	0.001067	0.7187	535.10	2002.1	2537.2	535.37	2181.5	2716.9	1.6072	5.4455	7.0527
275	130.6	0.001070	0.6573	548.59	1991.9	2540.5	548.89	2172.4	2721.3	1.6408	5.3801	7.0209
300	133.55	0.001073	0.6058	561.15	1982.4	2543.6	561.47	2163.8	2725.3	1.6718	5.3201	6.9919
325	136.3	0.001076	0.5620	572.90	1973.5	2546.4	573.25	2155.8	2729.0	1.7006	5.2646	6.9652
350	138.88	0.001079	0.5243	583.95	1965.0	2548.9	584.33	2148.1	2732.4	1.7275	5.2130	6.9405
375	141.32	0.001081	0.4914	594.40	1956.9	2551.3	594.81	2140.8	2735.6	1.7528	5.1647	6.9175
400	143.63	0.001084	0.4625	604.31	1949.3	2553.6	604.74	2133.8	2738.6	1.7766	5.1193	6.8959
450	147.93	0.001088	0.4140	622.77	1934.9	2557.6	623.25	2120.7	2743.9	1.8207	5.0359	6.8565
500	151.86	0.001093	0.3749	639.68	1921.6	2561.2	640.23	2108.5	2748.7	1.8607	4.9606	6.8213
550	155.48	0.001097	0.3427	655.32	1909.2	2564.5	655.93	2097.0	2753.0	1.8973	4.8920	6.7893
600	158.85	0.001101	0.3157	669.90	1897.5	2567.4	670.56	2086.3	2756.8	1.9312	4.8288	6.7600
650	162.01	0.001104	0.2927	683.56	1886.5	2570.1	684.28	2076.0	2760.3	1.9627	4.7703	6.7331
700	164.97	0.001108	0.2729	696.44	1876.1	2572.5	697.22	2066.3	2763.5	1.9922	4.7158	6.7080
750	167.78	0.001112	0.2556	708.64	1866.1	2574.7	709.47	2057.0	2766.4	2.0200	4.6647	6.6847
800	170.43	0.001115	0.2404	720.22	1856.6	2576.8	721.11	2048.0	2769.1	2.0462	4.6166	6.6628
850	172.96	0.001118	0.2270	731.27	1847.4	2578.7	732.22	2039.4	2771.6	2.0710	4.5711	6.6421
900	175.38	0.001121	0.2150	741.83	1838.6	2580.5	742.83	2031.1	2773.9	2.0946	4.5280	6.6226
950	177.69	0.001124	0.2042	751.95	1830.2	2582.1	753.02	2023.1	2776.1	2.1172	4.4869	6.6041
1000	179.91	0.001127	0.19444	761.68	1822.0	2583.6	762.81	2015.3	2778.1	2.1387	4.4478	6.5865
1100	184.09	0.001133	0.17753	780.09	1806.3	2586.4	781.34	2000.4	2781.7	2.1792	4.3744	6.5536
1200	187.99	0.001139	0.16333	797.29	1791.5	2588.8	798.65	1986.2	2784.8	2.2166	4.3067	6.5233
1300	191.64	0.001144	0.15125	813.44	1777.5	2591.0	814.93	1972.7	2787.6	2.2515	4.2438	6.4953

Source: Courtesy of ASHRAE, *Handbook of Fundamentals*, American Society of Heating, Refrigerating and Air-Conditioning Engineers, Atlanta, GA, 1997. With permission.

TABLE A.4
(IP Units) Properties of Superheated Steam

Pressure (psia)	Temperature (°F)	v (ft^3/lb_m)	u (Btu/lb_m)	h (Btu/lb_m)	s ($Btu/lb_m \cdot °R$)
1.0	Sat.	333.6	1044.0	1105.8	1.9779
(101.70°F)	200	392.5	1077.5	1150.1	2.0508
	240	416.4	1091.2	1168.3	2.0775
	280	440.3	1105.0	1186.5	2.1028
	320	464.2	1118.9	1204.8	2.1269
	360	488.1	1132.9	1223.2	2.1500
	400	511.9	1147.0	1241.8	2.1720
	440	535.8	1161.2	1260.4	2.1932
	500	571.5	1182.8	1288.5	2.2235
	600	631.1	1219.3	1336.1	2.2706
	700	690.7	1256.7	1384.5	2.3142
	800	750.3	1294.9	1433.7	2.3550
	1000	869.5	1373.9	1534.8	2.4294
	1200	988.6	1456.7	1639.6	2.4967
	1400	1107.7	1543.1	1748.1	2.5584
5	Sat.	73.53	1063.0	1131.0	1.8441
(162.21°F)	200	78.15	1076.3	1148.6	1.8715
	240	83.00	1090.3	1167.1	1.8987
	280	81.83	1104.3	1185.5	1.9244
	320	92.64	1118.3	1204.0	1.9487
	360	97.45	1132.4	1222.6	1.9719
	400	102.24	1146.6	1241.2	1.9941
	440	107.03	1160.9	1259.9	2.0154
	500	114.20	1182.5	1288.2	2.0458
	600	126.15	1219.1	1335.8	2.0930
	700	138.08	1256.5	1384.3	2.1367
	800	150.01	1294.7	1433.5	2.1775
	1000	173.86	1373.9	1534.7	2.2520
	1200	197.70	1456.6	1639.5	2.3192
	1400	221.54	1543.1	1748.1	2.3810
14.696	Sat.	26.80	1077.6	1150.5	1.7567
(211.99°F)	240	28.00	1087.9	1164.0	1.7764
	280	29.69	1102.4	1183.1	1.8030
	320	31.36	1116.8	1202.1	1.8280
	360	33.02	1131.2	1221.0	1.8516
	400	34.67	1145.6	1239.9	1.8741
	440	36.31	1160.1	1258.8	1.8956
	500	38.77	1181.8	1287.3	1.9263
	600	42.86	1218.6	1335.2	1.9737
	700	46.93	1256.1	1383.8	2.0175
	800	51.00	1294.4	1433.1	2.0584
	1000	59.13	1373.7	1534.5	2.1330
	1200	67.25	1456.5	1639.3	2.2003
	1400	75.36	1543.0	1747.9	2.2621
	1600	83.47	1633.2	1860.2	2.3194

(Continued)

TABLE A.4 (*Continued*)

(IP Units) Properties of Superheated Steam

Pressure (psia)	Temperature (°F)	v (ft³/lb$_m$)	u (Btu/lb$_m$)	h (Btu/lb$_m$)	s (Btu/lb$_m$·°R)
20	Sat.	20.09	1082.0	1156.4	1.7320
(227.96°F)	240	20.47	1086.5	1162.3	1.7405
	280	21.73	1101.4	1181.8	1.7676
	320	22.98	1116.0	1201.0	1.7930
	360	24.21	1130.6	1220.1	1.8168
	400	25.43	1145.1	1239.2	1.8395
	440	26.64	1159.6	1258.2	1.8611
	500	28.46	1181.5	1286.8	1.8919
	600	31.47	1218.4	1334.8	1.9395
	700	34.47	1255.9	1383.5	1.9834
	800	37.46	1294.3	1432.9	2.0243
	1000	43.44	1373.5	1534.3	2.0989
	1200	49.41	1456.4	1639.2	2.1663
	1400	55.37	1542.9	1747.9	2.2281
	1600	61.33	1633.2	1860.1	2.2854

Source: Courtesy of ASHRAE, *Handbook of Fundamentals*, American Society of Heating, Refrigerating and Air-Conditioning Engineers, Atlanta, GA, 1997. With permission.

TABLE A.4

(SI Units) Properties of Superheated Steam

Pressure (MPa)	Temperature (°C)	v (m³/kg)	u (kJ/kg)	h (kJ/kg)	s (kJ/kg·K)
0.01	Sat.	14.674	2437.9	2584.7	8.1502
	50	14.869	2443.9	2592.6	8.1749
	100	17.196	2515.5	2687.5	8.4479
	150	19.512	2587.9	2783.0	8.6882
	200	21.825	2661.3	2879.5	8.9038
	250	24.136	2736.0	2977.3	9.1002
	300	26.445	2812.1	3076.5	9.2813
	400	31.063	2968.9	3279.6	9.6077
	500	35.679	3132.3	3489.1	9.8978
	600	40.295	3302.5	3705.4	10.1608
	700	44.911	3479.6	3928.7	10.4028
	800	49.526	3663.8	4159.0	10.6281
	900	54.141	3855.0	4396.4	10.8396
	1000	58.757	4053.0	4640.6	11.0393
	1100	63.372	4257.5	4891.2	11.2287
	1200	67.987	4467.9	5147.8	11.4091
	1300	72.602	4683.7	5409.7	11.5811
0.05	Sat.	3.240	2483.9	2645.9	7.5939
	100	3.418	2511.6	2682.5	7.6947
	150	3.889	2585.6	2780.1	7.9401
	200	4.356	2659.9	2877.7	8.1580
	250	4.820	2735.0	2976.0	8.3556
	300	5.284	2811.3	3075.5	8.5373
	400	6.209	2968.5	3278.9	8.8642
	500	7.134	3132.0	3488.7	9.1546
	600	8.057	3302.2	3705.1	9.4178
	700	8.981	3479.4	3928.5	9.6599
	800	9.904	3663.6	4158.9	9.8852
	900	10.828	3854.9	4396.3	10.0967
	1000	11.751	4052.9	4640.5	10.2964
	1100	12.674	4257.4	4891.1	10.4859
	1200	13.597	4467.8	5147.7	10.6662
	1300	14.521	4683.6	5409.6	10.8382
0.2	Sat.	0.8857	2529.5	2706.7	7.1272
	150	0.9596	2576.9	2768.8	7.2795
	200	1.0803	2654.4	2870.5	7.5066
	250	1.1988	2731.2	2971.0	7.7086
	300	1.3162	2808.6	3071.8	7.8926
	400	1.5493	2966.7	3276.6	8.2218
	500	1.7814	3130.8	3487.1	8.5133
	600	2.013	3301.4	3704.0	8.7770
	700	2.244	3478.8	3927.6	9.0194
	800	2.475	3663.1	4158.2	9.2449
	900	2.705	3854.5	4395.8	9.4566
	1000	2.937	4052.5	4640.0	9.6563
	1100	3.168	4257.0	4890.7	9.8458
	1200	3.399	4467.5	5147.5	10.0262
	1300	3.630	4683.2	5409.3	10.1982

(Continued)

TABLE A.4 (*Continued*)

(SI Units) Properties of Superheated Steam

Pressure (MPa)	Temperature (°C)	v (m³/kg)	u (kJ/kg)	h (kJ/kg)	s (kJ/kg·K)
0.3	Sat.	0.6058	2543.6	2725.3	6.9919
	150	0.6339	2570.8	2761.0	7.0778
	200	0.7163	2650.7	2865.6	7.3115
	250	0.7964	2728.7	2967.6	7.5166
	300	0.8753	2806.7	3069.3	7.7022
	400	1.0315	2965.6	3275.0	8.0330
	500	1.1867	3130.0	3486.0	8.3251
	600	1.3414	3300.8	3703.2	8.5892
	700	1.4957	3478.4	3927.1	8.8319
	800	1.6499	3662.9	4157.8	9.0576
	900	1.8041	3854.2	4395.4	9.2692
	1000	1.9581	4052.3	4639.7	9.4690
	1100	2.1121	4256.8	4890.4	9.6585
	1200	2.2661	4467.2	5147.1	9.8389
	1300	2.4201	4683.0	5409.0	10.0110

Source: Courtesy of ASHRAE, *Handbook of Fundamentals*, American Society of Heating, Refrigerating and Air-Conditioning Engineers, Atlanta, GA, 1997. With permission.

TABLE A.5

(IP Units) Properties of Saturated Liquid and Saturated Vapor of R-22

Temp. (°F)	Pressure (psia)	Density (lb/ft³)	Volume (ft³/lb)	Enthalpy (Btu/lb)		Entropy (Btu/lb·°F)	
		Liquid	*Vapor*	*Liquid*	*Vapor*	*Liquid*	*Vapor*
−250	—	107.37	—	−63.169	76.604	−0.21914	0.44952
−240	—	106.41	—	−56.462	77.629	−0.18786	0.42332
−230	—	105.48	—	−51.569	78.669	−0.16605	0.40101
−220	0.002	104.58	16805	−47.705	79.724	−0.14958	0.38211
−210	0.004	103.7	6982.6	−44.426	80.796	−0.13616	0.36538
−200	0.01	102.81	3151.5	−41.474	81.882	−0.12457	0.35048
−190	0.022	101.92	1527.4	−38.706	82.984	−0.11411	0.33715
−180	0.044	101.03	787.79	−36.038	84.1	−0.10439	0.32518
−170	0.084	100.12	429.22	−33.424	85.23	−0.09521	0.31441
−160	0.151	99.22	245.51	−30.839	86.373	−0.08644	0.3047
−150	0.262	98.3	146.65	−28.269	87.528	−0.078	0.29594
−140	0.435	97.38	91.059	−25.708	88.692	−0.06986	0.28801
−130	0.696	96.46	58.544	−23.15	89.864	−0.06198	0.28082
−120	1.08	95.53	38.833	−20.594	91.04	−0.05435	0.2743
−110	1.626	94.6	26.494	−18.038	92.218	−0.04694	0.26838
−100	2.384	93.66	18.54	−15.481	93.397	−0.03973	0.26298
−90	3.413	92.71	13.275	−12.921	94.572	−0.03271	0.25807
−80	4.778	91.75	9.7044	−10.355	95.741	−0.02587	0.25357
−70	6.555	90.79	7.2285	−7.783	96.901	−0.01919	0.24945
−60	8.83	89.81	5.4766	−5.201	98.049	−0.01266	0.24567
−50	11.696	88.83	4.2138	−2.608	99.182	−0.00627	0.2422
−45	13.383	88.33	3.716	−1.306	99.742	−0.00312	0.24056
−41.44	*14.696*	*87.97*	*3.4048*	*−0.377*	*100.138*	*−0.0009*	*0.23944*
−40	15.255	87.82	3.288	0	100.296	0	0.23899
−35	17.329	87.32	2.9185	1.31	100.847	0.00309	0.23748
−30	19.617	86.81	2.5984	2.624	101.391	0.00616	0.23602
−25	22.136	86.29	2.3202	3.944	101.928	0.0092	0.23462
−20	24.899	85.77	2.0774	5.268	102.461	0.01222	0.23327
−15	27.924	85.25	1.865	6.598	102.986	0.01521	0.23197
−10	31.226	84.72	1.6784	7.934	103.503	0.01818	0.23071
−5	34.821	84.18	1.5142	9.276	104.013	0.02113	0.22949
0	38.726	83.64	1.3691	10.624	104.515	0.02406	0.22832
5	42.96	83.09	1.2406	11.979	105.009	0.02697	0.22718
10	47.538	82.54	1.1265	13.342	105.493	0.02987	0.22607
15	52.48	81.98	1.025	14.712	105.968	0.03275	0.225
20	57.803	81.41	0.9343	16.09	106.434	0.03561	0.22395
25	63.526	80.84	0.8532	17.476	106.891	0.03846	0.22294
30	69.667	80.26	0.7804	18.871	107.336	0.04129	0.22195
35	76.245	79.67	0.715	20.275	107.769	0.04411	0.22098
40	83.28	79.07	0.6561	21.688	108.191	0.04692	0.22004
45	90.791	78.46	0.6029	23.111	108.6	0.04972	0.21912
50	98.799	77.84	0.5548	24.544	108.997	0.05251	0.21821
55	107.32	77.22	0.5111	25.988	109.379	0.05529	0.21732
60	116.38	76.58	0.4715	27.443	109.748	0.05806	0.21644
65	126	75.93	0.4355	28.909	110.103	0.06082	0.21557
70	136.19	75.27	0.4026	30.387	110.441	0.06358	0.21472
75	146.98	74.6	0.3726	31.877	110.761	0.06633	0.21387
80	158.4	73.92	0.3451	33.381	111.066	0.06907	0.21302
85	170.45	73.22	0.3199	34.898	111.35	0.07182	0.21218

(Continued)

TABLE A.5 (*Continued*)

(IP Units) Properties of Saturated Liquid and Saturated Vapor of R-22

		Density (lb/ft³)	Volume (ft³/lb)	Enthalpy (Btu/lb)		Entropy (Btu/lb·°F)	
Temp. (°F)	**Pressure (psia)**	***Liquid***	***Vapor***	***Liquid***	***Vapor***	***Liquid***	***Vapor***
90	183.17	72.51	0.2968	36.43	111.616	0.07456	0.21134
95	196.57	71.79	0.2756	37.977	111.859	0.0773	0.2105
100	210.69	71.05	0.256	39.538	112.081	0.08003	0.20965
105	225.53	70.29	0.2379	41.119	112.278	0.08277	0.20879
110	241.14	69.51	0.2212	42.717	112.448	0.08552	0.20793
115	257.52	68.71	0.2058	44.334	112.591	0.08827	0.20705
120	274.71	67.89	0.1914	45.972	112.704	0.09103	0.20615
125	292.73	67.05	0.1781	47.633	112.783	0.09379	0.20522
130	311.61	66.17	0.1657	49.319	112.825	0.09657	0.20427
135	331.38	65.27	0.1542	51.032	112.826	0.09937	0.20329
140	352.07	64.33	0.1434	52.775	112.784	0.1022	0.20227
145	373.71	63.35	0.1332	54.553	112.692	0.10504	0.20119
150	396.32	62.33	0.1237	56.37	112.541	0.10793	0.20006
160	444.65	60.12	0.1063	60.145	112.035	0.11383	0.19757
170	497.35	57.59	0.0907	64.175	111.165	0.12001	0.19464
180	554.82	54.57	0.0763	68.597	109.753	0.12668	0.19102
190	617.53	50.62	0.0625	73.742	107.398	0.13432	0.18613
200	686.11	44.44	0.0478	80.558	102.809	0.14432	0.17805
205.06	**723.74**	**32.7**	**0.0306**	**91.052**	**91.052**	**0.15989**	**0.15989**

Source: Courtesy of ASHRAE, *Handbook of Fundamentals,* American Society of Heating, Refrigerating and Air-Conditioning Engineers, Atlanta, GA, 1997. With permission.

Notes: Values shown in italics are at the boiling point. Values shown in bold are at the critical point.

TABLE A.5

(SI Units) Properties of Saturated Liquid and Saturated Vapor of R-22

Temp. (°C)	Pressure (MPa)	Density (kg/m^3)	Volume (m^3/kg)	Enthalpy (kJ/kg)		Entropy (kJ/kg·K)	
		Liquid	*Vapor*	*Liquid*	*Vapor*	*Liquid*	*Vapor*
−150.00	—	1701.5	—	26.01	335.85	0.0566	2.5752
−140.00	—	1675.3	—	43.84	340.24	0.1961	2.4222
−130.00	0.00006	1649.7	229.29	57.00	344.75	0.2916	2.3017
−120.00	0.00023	1624.0	63.648	68.51	349.38	0.3694	2.2033
−110.00	0.00074	1598.0	21.311	79.47	354.11	0.4386	2.1220
−100.00	0.00200	1571.7	8.2980	90.24	358.93	0.5027	2.0545
−90.00	0.00480	1545.1	3.6548	100.95	363.82	0.5629	1.9982
−80.00	0.01035	1518.3	1.7816	111.66	368.75	0.6197	1.9508
−70.00	0.02044	1491.1	0.94476	122.36	373.68	0.6738	1.9109
−60.00	0.03747	1463.6	0.53734	133.11	378.58	0.7253	1.8770
−50.00	0.06449	1435.5	0.32405	143.91	383.39	0.7748	1.8480
−48.00	0.07140	1429.8	0.29469	146.08	384.35	0.7844	1.8427
−46.00	0.07890	1424.1	0.26849	148.25	385.29	0.7940	1.8376
−44.00	0.08700	1418.4	0.24507	150.43	386.23	0.8035	1.8326
−42.00	0.09575	1412.6	0.22410	152.61	387.17	0.8130	1.8277
−40.80	*0.10132*	*1409.1*	*0.21256*	*153.93*	*387.72*	*0.8186*	*1.8249*
−40.00	0.10518	1406.8	0.20526	154.80	388.09	0.8224	1.8230
−38.00	0.11533	1401.0	0.18832	156.99	389.01	0.8317	1.8184
−36.00	0.12623	1395.1	0.17306	159.19	389.93	0.8410	1.8140
−34.00	0.13793	1389.2	0.15927	161.40	390.84	0.8502	1.8096
−32.00	0.15045	1383.3	0.14680	163.61	391.74	0.8594	1.8054
−30.00	0.16384	1377.3	0.13551	165.82	392.63	0.8685	1.8013
−28.00	0.17815	1371.3	0.12525	168.04	393.52	0.8776	1.7973
−26.00	0.19340	1365.2	0.11593	170.27	394.39	0.8866	1.7934
−24.00	0.20965	1359.1	0.10744	172.51	395.26	0.8955	1.7896
−22.00	0.22693	1352.9	0.09970	174.75	396.12	0.9044	1.7859
−20.00	0.24529	1346.8	0.09262	177.00	396.97	0.9133	1.7822
−18.00	0.26477	1340.5	0.08615	179.26	397.81	0.9222	1.7787
−16.00	0.28542	1334.2	0.08023	181.53	398.64	0.9309	1.7752
−14.00	0.30728	1327.9	0.07479	183.81	399.46	0.9397	1.7719
−12.00	0.33040	1321.5	0.06979	186.09	400.27	0.9484	1.7686
−10.00	0.35482	1315.0	0.06520	188.38	401.07	0.9571	1.7653
−8.00	0.38059	1308.5	0.06096	190.69	401.85	0.9657	1.7621
−6.00	0.40775	1301.9	0.05706	193.00	402.63	0.9743	1.7590
−4.00	0.43636	1295.3	0.05345	195.32	403.39	0.9829	1.7560
−2.00	0.46646	1288.6	0.05012	197.66	404.14	0.9915	1.7530
0.00	0.49811	1281.8	0.04703	200.00	404.87	1.0000	1.7500
2.00	0.53134	1275.0	0.04417	202.35	405.59	1.0085	1.7471
4.00	0.56622	1268.1	0.04152	204.72	406.30	1.0170	1.7443
6.00	0.6028	1261.1	0.03906	207.10	406.99	1.0254	1.7415
8.00	0.6411	1254.0	0.03676	209.49	407.67	1.0338	1.7387
10.00	0.6812	1246.9	0.03463	211.89	408.33	1.0422	1.7360
12.00	0.7231	1239.7	0.03265	214.31	408.97	1.0506	1.7333
14.00	0.7670	1232.4	0.03079	216.74	409.60	1.0590	1.7306
16.00	0.8128	1225.0	0.02906	219.18	410.21	1.0673	1.7280
18.00	0.8606	1217.6	0.02744	221.63	410.80	1.0756	1.7254
12.00	0.9104	1210.0	0.02593	224.10	411.38	1.0840	1.7228
22.00	0.9624	1202.4	0.02451	226.59	411.93	1.0923	1.7202
24.00	1.0165	1194.6	0.02319	229.09	412.46	1.1006	1.7177

(Continued)

TABLE A.5 (*Continued*)

(SI Units) Properties of Saturated Liquid and Saturated Vapor of R-22

Temp. (°C)	Pressure (MPa)	Density (kg/m^3)	Volume (m^3/kg)	Enthalpy (kJ/kg)		Entropy (kJ/kg·K)	
		Liquid	*Vapor*	*Liquid*	*Vapor*	*Liquid*	*Vapor*
26.00	1.0728	1186.8	0.02194	231.60	412.98	1.1088	1.7151
28.00	1.1314	1178.8	0.02077	234.14	413.46	1.1171	1.7126
30.00	1.1924	1170.7	0.01968	236.69	413.93	1.1254	1.7101
32.00	1.2557	1162.5	0.01864	239.25	414.37	1.1336	1.7075
34.00	1.3215	1154.2	0.01767	241.84	414.79	1.1419	1.7050
36.00	1.3898	1145.7	0.01675	244.44	415.18	1.1501	1.7024
38.00	1.4606	1137.1	0.01589	247.06	415.54	1.1584	1.6999
40.00	1.5341	1128.4	0.01507	249.71	415.87	1.1667	1.6973
42.00	1.6103	1119.5	0.01430	252.37	416.17	1.1749	1.6947
44.00	1.6892	1110.4	0.01357	255.06	416.44	1.1832	1.6921
46.00	1.7709	1101.2	0.01288	257.77	416.68	1.1915	1.6894
48.00	1.8555	1091.8	0.01223	260.51	416.87	1.1998	1.6867
50.00	1.9431	1082.1	0.01161	263.27	417.03	1.2081	1.6840
55.00	2.1753	1057.1	0.01020	270.31	417.24	1.2291	1.6768
60.00	2.4274	1030.5	0.00895	277.56	417.14	1.2503	1.6692
65.00	2.7008	1001.8	0.00784	285.06	416.65	1.2718	1.6610
70.00	2.9967	970.4	0.00684	292.90	415.69	1.2940	1.6518
75.00	3.3168	935.3	0.00594	301.18	414.09	1.3169	1.6413
80.00	3.6627	894.8	0.00511	310.10	411.60	1.3413	1.6287
85.00	4.0368	845.1	0.00433	320.05	407.72	1.3680	1.6128
90.00	4.4416	777.5	0.00355	331.98	401.33	1.3998	1.5907
95.00	4.8820	665.4	0.00264	348.86	387.46	1.4442	1.5491
96.14	**4.9900**	**523.8**	**0.00191**	**366.59**	**366.59**	**1.4918**	**1.4918**

Source: Courtesy of ASHRAE, *Handbook of Fundamentals*, American Society of Heating, Refrigerating and Air-Conditioning Engineers, Atlanta, GA, 1997. With permission.

Notes: Values shown in italics are at the boiling point. Values shown in bold are at the critical point.

TABLE A.6

(IP Units) Properties of Saturated Liquid and Saturated Vapor of R-134a

		Density (lb/ft³)	Volume (ft³/lb)	Enthalpy (Btu/lb)		Entropy (Btu/lb·°F)	
Temp. (°F)	Pressure (psia)	*Liquid*	*Vapor*	*Liquid*	*Vapor*	*Liquid*	*Vapor*
−153.94	0.057	99.34	564.85	32.989	80.235	0.09154	0.27880
−150	0.072	98.95	449.29	31.902	80.783	0.08801	0.27588
−140	0.130	97.98	259.15	29.093	82.190	0.07908	0.26903
−130	0.222	97.01	155.69	26.238	83.618	0.07029	0.26294
−120	0.367	96.05	97.027	23.359	85.066	0.06169	0.25752
−110	0.586	95.09	62.509	20.467	86.531	0.05330	0.25270
−100	0.906	94.13	41.496	17.569	88.011	0.04513	0.24842
−90	1.363	93.17	28.303	14.665	89.504	0.03717	0.24462
−80	1.997	92.21	19.783	11.755	91.005	0.02940	0.24125
−75	2.396	91.73	16.680	10.297	91.759	0.02559	0.23972
−70	2.859	91.25	14.138	8.837	92.514	0.02182	0.23827
−65	3.393	90.77	12.045	7.374	93.270	0.01809	0.23691
−60	4.006	90.28	10.310	5.907	94.026	0.01440	0.23563
−55	4.707	89.80	8.8656	4.437	94.783	0.01075	0.23443
−50	5.505	89.31	7.6569	2.963	95.539	0.00713	0.23331
−45	6.409	88.82	6.6405	1.484	96.295	0.00355	0.23225
−40	7.429	88.32	5.7819	0.000	97.050	0.00000	0.23125
−35	8.577	87.83	5.0533	1.489	97.804	0.00352	0.23032
−30	9.862	87.33	4.4325	2.984	98.556	0.00701	0.22945
−25	11.297	86.82	3.9014	4.484	99.306	0.01048	0.22863
−20	12.895	86.32	3.4452	5.991	100.054	0.01392	0.22786
−15	14.667	85.81	3.0519	7.505	100.799	0.01733	0.22714
−14.92	*14.696*	*85.80*	*3.0462*	*7.529*	*100.811*	*0.01739*	*0.22713*
−10	16.626	85.29	2.7116	9.026	101.542	0.02073	0.22647
−5	18.787	84.77	2.4161	10.554	102.280	0.02409	0.22584
0	21.162	84.25	2.1587	12.090	103.015	0.02744	0.22525
5	23.767	83.72	1.9337	13.634	103.745	0.03077	0.22470
10	26.617	83.18	1.7365	15.187	104.471	0.03408	0.22418
15	29.726	82.64	1.5630	16.748	105.192	0.03737	0.22370
20	33.110	82.10	1.4101	18.318	105.907	0.04065	0.22325
25	36.785	81.55	1.2749	19.897	106.617	0.04391	0.22283
30	40.768	80.99	1.1550	21.486	107.320	0.04715	0.22244
35	45.075	80.42	1.0484	23.085	108.016	0.05038	0.22207
40	49.724	79.85	0.9534	24.694	108.705	0.05359	0.22172
45	54.732	79.26	0.8685	26.314	109.386	0.05679	0.22140
50	60.116	78.67	0.7925	27.944	110.058	0.05998	0.22110
55	65.895	78.07	0.7243	29.586	110.722	0.06316	0.22081
60	72.087	77.46	0.6630	31.239	111.376	0.06633	0.22054
65	78.712	76.84	0.6077	32.905	112.019	0.06949	0.22028
70	85.787	76.21	0.5577	34.583	112.652	0.07264	0.22003
75	93.333	75.57	0.5125	36.274	113.272	0.07578	0.21979
80	101.370	74.91	0.4715	37.978	113.880	0.07892	0.21957
85	109.920	74.25	0.4343	39.697	114.475	0.08205	0.21934
90	119.000	73.57	0.4004	41.430	115.055	0.08518	0.21912
95	128.630	72.87	0.3694	43.179	115.619	0.08830	0.21890
100	138.830	72.16	0.3411	44.943	116.166	0.09142	0.21868
105	149.630	71.43	0.3153	46.725	116.694	0.09454	0.21845
110	161.050	70.68	0.2915	48.524	117.203	0.09766	0.21822
115	173.110	69.91	0.2697	50.343	117.690	0.10078	0.21797

(Continued)

TABLE A.6 (*Continued*)

(IP Units) Properties of Saturated Liquid and Saturated Vapor of R-134a

		Density (lb/ft³)	Volume (ft³/lb)	Enthalpy (Btu/lb)		Entropy (Btu/lb·°F)	
Temp. (°F)	**Pressure (psia)**	***Liquid***	***Vapor***	***Liquid***	***Vapor***	***Liquid***	***Vapor***
120	185.840	69.12	0.2497	52.181	118.153	0.10391	0.21772
125	199.250	68.31	0.2312	54.040	118.591	0.10704	0.21744
130	213.380	67.47	0.2141	55.923	119.000	0.11018	0.21715
135	228.250	66.60	0.1983	57.830	119.377	0.11333	0.21683
140	243.880	65.70	0.1836	59.764	119.720	0.11650	0.21648
145	260.310	64.77	0.1700	61.727	120.024	0.11968	0.21609
150	277.570	63.80	0.1574	63.722	120.284	0.12288	0.21566
155	295.690	62.78	0.1455	65.752	120.495	0.12611	0.21517
160	314.690	61.72	0.1345	67.823	120.650	0.12938	0.21463
165	334.620	60.60	0.1241	69.939	120.739	0.13268	0.21400
170	355.510	59.42	0.1144	72.106	120.753	0.13603	0.21329
175	377.400	58.16	0.1052	74.335	120.677	0.13945	0.21247
180	400.340	56.80	0.0965	76.636	120.493	0.14295	0.21151
185	424.370	55.33	0.0881	79.027	120.175	0.14655	0.21037
190	449.550	53.70	0.0801	81.534	119.684	0.15029	0.20901
195	475.950	51.86	0.0723	84.196	118.963	0.15423	0.20733
200	503.640	49.70	0.0646	87.088	117.906	0.15847	0.20519
205	532.720	47.00	0.0566	90.368	116.289	0.16326	0.20226
210	563.340	43.03	0.0474	94.548	113.411	0.16933	0.19750
213.85	**588.270**	**32.04**	**0.0312**	**103.775**	**103.775**	**0.18128**	**0.18128**

Source: Courtesy of ASHRAE, *Handbook of Fundamentals*, American Society of Heating, Refrigerating and Air-Conditioning Engineers, Atlanta, GA, 1997. With permission.

Notes: Values shown in italics are at the boiling point. Values shown in bold are at the critical point.

TABLE A.6

(SI Units) Properties of Saturated Liquid and Saturated Vapor of R-134a

		Density (kg/m³)	Volume (m³/kg)	Enthalpy (kJ/kg)		Entropy (kJ/kg·K)	
Temp. (°C)	Pressure (MPa)	*Liquid*	*Vapor*	*Liquid*	*Vapor*	*Liquid*	*Vapor*
−103.30	***0.00039***	***1591.2***	***35.263***	***71.89***	***335.07***	***0.4143***	***1.9638***
−100	0.00056	1581.9	25.039	75.71	337.00	0.4366	1.9456
−90	0.00153	1553.9	9.7191	87.59	342.94	0.5032	1.8975
−80	0.00369	1526.2	4.2504	99.65	349.03	0.5674	1.8585
−70	0.00801	1498.6	2.0528	111.78	355.23	0.6286	1.8269
−60	0.01594	1471.0	1.0770	123.96	361.51	0.6871	1.8016
−50	0.02945	1443.1	0.60560	136.21	367.83	0.7432	1.7812
−40	0.05122	1414.8	0.36095	148.57	374.16	0.7973	1.7649
−30	0.08436	1385.9	0.22596	161.10	380.45	0.8498	1.7519
−28	0.09268	1380.0	0.20682	163.62	381.70	0.8601	1.7497
−26.07	*0.10132*	*1374.3*	*0.19016*	*166.07*	*382.90*	*0.8701*	*1.7476*
−26	0.10164	1374.1	0.18961	166.16	382.94	0.8704	1.7476
−24	0.11127	1368.2	0.17410	168.70	384.19	0.8806	1.7455
−22	0.12160	1362.2	0.16010	171.26	385.43	0.8908	1.7436
−20	0.13268	1356.2	0.14744	173.82	386.66	0.9009	1.7417
−18	0.14454	1350.2	0.13597	176.39	387.89	0.9110	1.7399
−16	0.15721	1344.1	0.12556	178.97	389.11	0.9211	1.7383
−14	0.17074	1338.0	0.11610	181.56	390.33	0.9311	1.7367
−12	0.18516	1331.8	0.10749	184.16	391.55	0.9410	1.7351
−10	0.20052	1325.6	0.09963	186.78	392.75	0.9509	1.7337
−8	0.21684	1319.3	0.09246	189.40	393.95	0.9608	1.7323
−6	0.23418	1313.0	0.08591	192.03	395.15	0.9707	1.7310
−4	0.25257	1306.6	0.07991	194.68	396.33	0.9805	1.7297
−2	0.27206	1300.2	0.07440	197.33	397.51	0.9903	1.7285
0	0.29269	1293.7	0.06935	200.00	398.68	1.0000	1.7274
2	0.31450	1287.1	0.06470	202.68	399.84	1.0097	1.7263
4	0.33755	1280.5	0.06042	205.37	401.00	1.0194	1.7252
6	0.36186	1273.8	0.05648	208.08	402.14	1.0291	1.7242
8	0.38749	1267.0	0.05284	210.80	403.27	1.0387	1.7233
10	0.41449	1260.2	0.04948	213.53	404.40	1.0483	1.7224
12	0.44289	1253.3	0.04636	216.27	405.51	1.0579	1.7215
14	0.47276	1246.3	0.04348	219.03	406.61	1.0674	1.7207
16	0.50413	1239.3	0.04081	221.80	407.70	1.0770	1.7199
18	0.53706	1232.1	0.03833	224.59	408.78	1.0865	1.7191
20	0.57159	1224.9	0.03603	227.40	409.84	1.0960	1.7183
22	0.60777	1217.5	0.03388	230.21	410.89	1.1055	1.7176
24	0.64566	1210.1	0.03189	233.05	411.93	1.1149	1.7169
26	0.68531	1202.6	0.03003	235.90	412.95	1.1244	1.7162
28	0.72676	1194.9	0.02829	238.77	413.95	1.1338	1.7155
30	0.77008	1187.2	0.02667	241.65	414.94	1.1432	1.7149
32	0.81530	1179.3	0.02516	244.55	415.90	1.1527	1.7142
34	0.86250	1171.3	0.02374	247.47	416.85	1.1621	1.7135
36	0.91172	1163.2	0.02241	250.41	417.78	1.1715	1.7129
38	0.96301	1154.9	0.02116	253.37	418.69	1.1809	1.7122
40	1.0165	1146.5	0.01999	256.35	419.58	1.1903	1.7115
42	1.0721	1137.9	0.01890	259.35	420.44	1.1997	1.7108
44	1.1300	1129.2	0.01786	262.38	421.28	1.2091	1.7101
46	1.1901	1120.3	0.01689	265.42	422.09	1.2185	1.7094
45	1.2527	1111.3	0.01598	268.49	422.88	1.2279	1.7086

(Continued)

TABLE A.6 (*Continued*)

(SI Units) Properties of Saturated Liquid and Saturated Vapor of R-134a

		Density (kg/m³)	Volume (m³/kg)	Enthalpy (kJ/kg)		Entropy (kJ/kg·K)	
Temp. (°C)	Pressure (MPa)	*Liquid*	*Vapor*	*Liquid*	*Vapor*	*Liquid*	*Vapor*
50	1.3177	1102.0	0.01511	271.59	423.63	1.2373	1.7078
52	1.3852	1092.6	0.01430	274.71	424.35	1.2468	1.7070
54	1.4553	1082.9	0.01353	277.86	425.03	1.2562	1.7061
56	1.5280	1073.0	0.01280	281.04	425.68	1.2657	1.7051
58	1.6033	1062.8	0.01212	284.25	426.29	1.2752	1.7041
60	1.6815	1052.4	0.01146	287.49	426.86	1.2847	1.7031
62	1.7625	1041.7	0.01085	290.77	427.37	1.2943	1.7019
64	1.8464	1030.7	0.01026	294.08	427.84	1.3039	1.7007
66	1.9334	1019.4	0.00970	297.44	428.25	1.3136	1.6993
68	2.0234	1007.7	0.00917	300.84	428.61	1.3234	1.6979
70	2.1165	995.6	0.00867	304.29	428.89	1.3332	1.6963
72	2.2130	983.1	0.00818	307.79	429.10	1.3430	1.6945
74	2.3127	970.0	0.00772	311.34	429.23	1.3530	1.6926
76	2.4159	956.5	0.00728	314.96	429.27	1.3631	1.6905
78	2.5227	942.3	0.00686	318.65	429.20	1.3733	1.6881
80	2.6331	927.4	0.00646	322.41	429.02	1.3837	1.6855
85	2.9259	886.2	0.00550	332.27	427.91	1.4105	1.6775
90	3.2445	836.9	0.00461	343.01	425.48	1.4392	1.6663
95	3.5916	771.6	0.00374	355.43	420.60	1.4720	1.6490
100	3.9721	646.7	0.00265	374.02	407.08	1.5207	1.6093
101.03	**4.056**	**513.3**	**0.002**	**389.79**	**389.79**	**1.5593**	**1.5593**

Source: Courtesy of ASHRAE, *Handbook of Fundamentals*, American Society of Heating, Refrigerating and Air-Conditioning Engineers, Atlanta, GA, 1997. With permission.

Notes: Values shown in bold italics are at the triple point. Values shown in italics are at the boiling point. Values shown in bold are at the critical point.

TABLE A.7

(IP Units) Properties of Superheated Vapor of R-22

Pressure (psia)	Properties	Temperature −22°F	−4°F	14°F	32°F	50°F	68°F	86°F	104°F	122°F	140°F	158°F
20.31	Volume (ft³/lb)	2.566	2.687	2.807	2.926	3.044	3.16	3.276	3.392	3.506	3.621	3.734
	Enthalpy (Btu/lb)	102.4	105.1	107.8	110.6	113.3	116.1	119	121.9	124.8	127.8	130.8
	Entropy (Btu/lb·R)	0.237	0.243	0.249	0.255	0.26	0.266	0.271	0.276	0.281	0.286	0.291
26.11	Volume (ft³/lb)	—	2.066	2.162	2.256	2.35	2.442	2.534	2.624	2.715	2.805	2.894
	Enthalpy (Btu/lb)	—	104.7	107.5	110.2	113	115.9	118.8	121.7	124.6	127.6	130.6
	Entropy (Btu/lb·R)	—	0.237	0.243	0.249	0.254	0.26	0.265	0.27	0.275	0.28	0.285
29.01	Volume (ft³/lb)	—	1.849	1.936	2.022	2.107	2.191	2.274	2.356	2.438	2.519	2.6
	Enthalpy (Btu/lb)	—	104.5	107.3	110.1	112.9	115.7	118.6	121.6	124.5	127.5	130.6
	Entropy (Btu/lb·R)	—	0.234	0.24	0.246	0.252	0.257	0.262	0.268	0.273	0.278	0.283
36.26	Volume (ft³/lb)	—	—	1.529	1.6	1.669	1.738	1.805	1.872	1.939	2.005	2.07
	Enthalpy (Btu/lb)	—	—	106.8	109.6	112.5	115.4	118.3	121.3	124.3	127.3	130.3
	Entropy (Btu/lb·R)	—	—	0.234	0.24	0.246	0.251	0.257	0.262	0.267	0.273	0.278
43.51	Volume (ft³/lb)	—	—	1.257	1.318	1.377	1.436	1.493	1.55	1.606	1.662	1.717
	Enthalpy (Btu/lb)	—	—	106.3	109.2	112.1	115.1	118	121	124	127	130.1
	Entropy (Btu/lb·R)	—	—	0.229	0.235	0.241	0.247	0.252	0.258	0.263	0.268	0.273
50.76	Volume (ft³/lb)	—	—	1.063	1.116	1.169	1.22	1.27	1.32	1.368	1.417	1.465
	Enthalpy (Btu/lb)	—	—	105.8	108.8	111.7	114.7	117.7	120.7	123.7	126.8	129.9
	Entropy (Btu/lb·R)	—	—	0.225	0.231	0.237	0.243	0.248	0.254	0.259	0.264	0.269
58.02	Volume (ft³/lb)	—	—	—	0.965	1.012	1.058	1.103	1.147	1.19	1.233	1.275
	Enthalpy (Btu/lb)	—	—	—	108.3	111.3	114.3	117.4	120.4	123.5	126.5	129.7
	Entropy (Btu/lb·R)	—	—	—	0.227	0.233	0.239	0.245	0.25	0.256	0.261	0.266
65.27	Volume (ft³/lb)	—	—	—	0.847	0.89	0.932	0.972	1.012	1.051	1.09	1.128
	Enthalpy (Btu/lb)	—	—	—	107.9	110.9	114	117	120.1	123.2	126.3	129.4
	Entropy (Btu/lb·R)	—	—	—	0.224	0.23	0.236	0.242	0.247	0.253	0.258	0.263
72.52	Volume (ft³/lb)	—	—	—	—	0.792	0.83	0.868	0.904	0.94	0.975	1.01
	Enthalpy (Btu/lb)	—	—	—	—	110.5	113.6	116.7	119.8	122.9	126	129.2
	Entropy (Btu/lb·R)	—	—	—	—	0.227	0.233	0.239	0.244	0.25	0.255	0.26
87.02	Volume (ft³/lb)	—	—	—	—	0.645	0.678	0.711	0.743	0.773	0.804	0.833
	Enthalpy (Btu/lb)	—	—	—	—	109.7	112.8	116	119.2	122.3	125.5	128.7
	Entropy (Btu/lb·R)	—	—	—	—	0.222	0.228	0.234	0.239	0.245	0.25	0.256
101.53	Volume (ft³/lb)	—	—	—	—	—	0.569	0.599	0.627	0.654	0.681	0.707
	Enthalpy (Btu/lb)	—	—	—	—	—	112	115.3	118.5	121.8	125	128.2
	Entropy (Btu/lb·R)	—	—	—	—	—	0.223	0.229	0.235	0.241	0.246	0.251
116.03	Volume (ft³/lb)	—	—	—	—	—	0.487	0.514	0.54	0.564	0.589	0.612
	Enthalpy (Btu/lb)	—	—	—	—	—	111.2	114.6	117.9	121.2	124.5	127.8
	Entropy (Btu/lb·R)	—	—	—	—	—	0.219	0.225	0.231	0.237	0.242	0.248
130.54	Volume (ft³/lb)	—	—	—	—	—	0.423	0.448	0.472	0.495	0.517	0.538
	Enthalpy (Btu/lb)	—	—	—	—	—	110.3	113.8	117.2	120.6	123.9	127.3
	Entropy (Btu/lb·R)	—	—	—	—	—	0.215	0.221	0.227	0.233	0.239	0.245
145.04	Volume (ft³/lb)	—	—	—	—	—	—	0.395	0.417	0.438	0.459	0.479
	Enthalpy (Btu/lb)	—	—	—	—	—	—	113	116.5	119.9	123.4	126.7
	Entropy (Btu/lb·R)	—	—	—	—	—	—	0.218	0.224	0.23	0.236	0.241
159.54	Volume (ft³/lb)	—	—	—	—	—	—	0.351	0.372	0.392	0.412	0.43
	Enthalpy (Btu/lb)	—	—	—	—	—	—	112.2	115.8	119.3	122.8	126.2
	Entropy (Btu/lb·R)	—	—	—	—	—	—	0.214	0.221	0.227	0.233	0.239
174.05	Volume (ft³/lb)	—	—	—	—	—	—	—	0.335	0.354	0.372	0.39
	Enthalpy (Btu/lb)	—	—	—	—	—	—	—	115	118.6	122.2	125.7
	Entropy (Btu/lb·R)	—	—	—	—	—	—	—	0.218	0.224	0.23	0.236

(*Continued*)

TABLE A.7 (*Continued*)

(IP Units) Properties of Superheated Vapor of R-22

Pressure (psia)	Properties	Temperature −22°F	−4°F	14°F	32°F	50°F	68°F	86°F	104°F	122°F	140°F	158°F
188.55	Volume (ft³/lb)	—	—	—	—	—	—	—	0.302	0.321	0.339	0.355
	Enthalpy (Btu/lb)	—	—	—	—	—	—	—	114.2	118	121.6	125.1
	Entropy (Btu/lb·R)	—	—	—	—	—	—	—	0.215	0.222	0.228	0.234
203.06	Volume (ft³/lb)	—	—	—	—	—	—	—	0.275	0.293	0.31	0.326
	Enthalpy (Btu/lb)	—	—	—	—	—	—	—	113.4	117.2	121	124.6
	Entropy (Btu/lb·R)	—	—	—	—	—	—	—	0.212	0.219	0.225	0.231
217.56	Volume (ft³/lb)	—	—	—	—	—	—	—	0.25	0.268	0.285	0.3
	Enthalpy (Btu/lb)	—	—	—	—	—	—	—	112.5	116.5	120.3	124
	Entropy (Btu/lb·R)	—	—	—	—	—	—	—	0.209	0.216	0.223	0.229
253.82	Volume (ft³/lb)	—	—	—	—	—	—	—	—	0.218	0.234	0.248
	Enthalpy (Btu/lb)	—	—	—	—	—	—	—	—	114.5	118.6	122.5
	Entropy (Btu/lb·R)	—	—	—	—	—	—	—	—	0.21	0.217	0.224
290.08	Volume (ft³/lb)	—	—	—	—	—	—	—	—	—	0.195	0.209
	Enthalpy (Btu/lb)	—	—	—	—	—	—	—	—	—	116.7	120.9
	Entropy (Btu/lb·R)	—	—	—	—	—	—	—	—	—	0.212	0.218
326.34	Volume (ft³/lb)	—	—	—	—	—	—	—	—	—	0.163	0.177
	Enthalpy (Btu/lb)	—	—	—	—	—	—	—	—	—	114.6	119.2
	Entropy (Btu/lb·R)	—	—	—	—	—	—	—	—	—	0.206	0.214
362.60	Volume (ft³/lb)	—	—	—	—	—	—	—	—	—	—	0.152
	Enthalpy (Btu/lb)	—	—	—	—	—	—	—	—	—	—	117.2
	Entropy (Btu/lb·R)	—	—	—	—	—	—	—	—	—	—	0.209
398.86	Volume (ft³/lb)	—	—	—	—	—	—	—	—	—	—	0.13
	Enthalpy (Btu/lb)	—	—	—	—	—	—	—	—	—	—	115
	Entropy (Btu/lb·R)	—	—	—	—	—	—	—	—	—	—	0.203

Source: Courtesy of ASHRAE, *Handbook of Fundamentals*, American Society of Heating, Refrigerating and Air-Conditioning Engineers, Atlanta, GA, 1997. With permission.

TABLE A.7

(SI Units) Properties of Superheated Vapor of R-22

Pressure (kPa)	Properties	Temperature										
		−30°C	−20°C	−10°C	0°C	10°C	20°C	30°C	40°C	50°C	60°C	70°C
120	Volume (m³/kg)	0.188	0.196	0.205	0.214	0.222	0.231	0.239	0.247	0.255	0.264	0.272
	Enthalpy (kJ/kg)	393.5	399.7	405.9	412.3	418.7	425.2	431.8	438.5	445.3	452.2	459.2
	Entropy (kJ/kg·K)	1.832	1.857	1.881	1.905	1.928	1.951	1.973	1.995	2.016	2.037	2.058
140	Volume (m³/kg)	0.16	0.167	0.175	0.182	0.19	0.197	0.204	0.211	0.218	0.226	0.233
	Enthalpy (kJ/kg)	393	399.2	405.5	411.9	418.4	424.9	431.5	438.3	445.1	452	459
	Entropy (kJ/kg·K)	1.816	1.841	1.865	1.889	1.912	1.935	1.957	1.979	2.001	2.022	2.043
180	Volume (m³/kg)	—	0.129	0.135	0.141	0.146	0.152	0.158	0.164	0.169	0.175	0.18
	Enthalpy (kJ/kg)	—	398.3	404.7	411.1	417.7	424.3	431	437.8	444.6	451.6	458.6
	Entropy (kJ/kg·K)	—	1.814	1.839	1.863	1.886	1.909	1.932	1.954	1.975	1.997	2.017
200	Volume (m³/kg)	—	0.115	0.121	0.126	0.131	0.137	0.142	0.147	0.152	0.157	0.162
	Enthalpy (kJ/kg)	—	397.8	404.3	410.8	417.3	424	430.7	437.5	444.4	451.4	458.4
	Entropy (kJ/kg·K)	—	1.803	1.828	1.852	1.875	1.898	1.921	1.943	1.965	1.986	2.007
250	Volume (m³/kg)	—	—	0.095	0.1	0.104	0.108	0.113	0.117	0.121	0.125	0.129
	Enthalpy (kJ/kg)	—	—	403.2	409.8	416.5	423.2	430	436.8	443.8	450.8	457.9
	Entropy (kJ/kg·K)	—	—	1.803	1.828	1.852	1.875	1.898	1.92	1.942	1.963	1.984
300	Volume (m³/kg)	—	—	0.078	0.082	0.086	0.089	0.093	0.097	0.1	0.104	0.107
	Enthalpy (kJ/kg)	—	—	402.1	408.8	415.6	422.4	429.2	436.2	443.2	450.2	457.4
	Entropy (kJ/kg·K)	—	—	1.783	1.808	1.832	1.856	1.879	1.901	1.923	1.945	1.966
350	Volume (m³/kg)	—	—	0.066	0.07	0.073	0.076	0.079	0.082	0.085	0.088	0.091
	Enthalpy (kJ/kg)	—	—	400.9	407.8	414.6	421.5	428.5	435.5	442.5	449.7	456.9
	Entropy (kJ/kg·K)	—	—	1.765	1.79	1.815	1.839	1.862	1.885	1.907	1.929	1.95
400	Volume (m³/kg)	—	—	—	0.06	0.063	0.066	0.069	0.071	0.074	0.077	0.079
	Enthalpy (kJ/kg)	—	—	—	406.7	413.7	420.7	427.7	434.8	441.9	449.1	456.3
	Entropy (kJ/kg·K)	—	—	—	1.774	1.8	1.824	1.847	1.87	1.893	1.914	1.936
450	Volume (m³/kg)	—	—	—	0.053	0.055	0.058	0.061	0.063	0.066	0.068	0.07
	Enthalpy (kJ/kg)	—	—	—	405.7	412.8	419.9	427	434.1	441.3	448.5	455.8
	Entropy (kJ/kg·K)	—	—	—	1.76	1.786	1.81	1.834	1.857	1.88	1.902	1.923
500	Volume (m³/kg)	—	—	—	—	0.049	0.052	0.054	0.056	0.059	0.061	0.063
	Enthalpy (kJ/kg)	—	—	—	—	411.8	419	426.2	433.4	440.6	447.9	455.2
	Entropy (kJ/kg·K)	—	—	—	—	1.773	1.798	1.822	1.846	1.868	1.89	1.912
600	Volume (m³/kg)	—	—	—	—	0.04	0.042	0.044	0.046	0.048	0.05	0.052
	Enthalpy (kJ/kg)	—	—	—	—	409.8	417.2	424.6	431.9	439.3	446.7	454.1
	Entropy (kJ/kg·K)	—	—	—	—	1.75	1.776	1.801	1.825	1.848	1.87	1.892
700	Volume (m³/kg)	—	—	—	—	—	0.035	0.037	0.039	0.041	0.042	0.044
	Enthalpy (kJ/kg)	—	—	—	—	—	415.4	422.9	430.5	438	445.5	453
	Entropy (kJ/kg·K)	—	—	—	—	—	1.757	1.782	1.806	1.83	1.853	1.875
800	Volume (m³/kg)	—	—	—	—	—	0.03	0.032	0.034	0.035	0.037	0.038
	Enthalpy (kJ/kg)	—	—	—	—	—	413.4	421.2	428.9	436.6	444.2	451.9
	Entropy (kJ/kg·K)	—	—	—	—	—	1.739	1.765	1.79	1.814	1.837	1.86
900	Volume (m³/kg)	—	—	—	—	—	0.026	0.028	0.029	0.031	0.032	0.034
	Enthalpy (kJ/kg)	—	—	—	—	—	411.4	419.5	427.4	435.2	443	450.7
	Entropy (kJ/kg·K)	—	—	—	—	—	1.722	1.749	1.775	1.799	1.823	1.846
1000	Volume (m³/kg)	—	—	—	—	—	—	0.025	0.026	0.027	0.029	0.03
	Enthalpy (kJ/kg)	—	—	—	—	—	—	417.6	425.7	433.7	441.7	449.6
	Entropy (kJ/kg·K)	—	—	—	—	—	—	1.734	1.761	1.786	1.81	1.833
1100	Volume (m³/kg)	—	—	—	—	—	—	0.022	0.023	0.024	0.026	0.027
	Enthalpy (kJ/kg)	—	—	—	—	—	—	415.7	424	432.2	440.3	448.3
	Entropy (kJ/kg·K)	—	—	—	—	—	—	1.72	1.748	1.773	1.798	1.822

(Continued)

TABLE A.7 (*Continued*)

(SI Units) Properties of Superheated Vapor of R-22

Pressure (kPa)	Properties	Temperature −30°C	−20°C	−10°C	0°C	10°C	20°C	30°C	40°C	50°C	60°C	70°C
1200	Volume (m³/kg)	—	—	—	—	—	—	—	0.021	0.022	0.023	0.024
	Enthalpy (kJ/kg)	—	—	—	—	—	—	—	422.3	430.7	439	447.1
	Entropy (kJ/kg·K)	—	—	—	—	—	—	—	1.735	1.761	1.787	1.811
1300	Volume (m³/kg)	—	—	—	—	—	—	—	0.019	0.02	0.021	0.022
	Enthalpy (kJ/kg)	—	—	—	—	—	—	—	420.5	429.1	437.5	445.8
	Entropy (kJ/kg·K)	—	—	—	—	—	—	—	1.723	1.75	1.776	1.8
1400	Volume (m³/kg)	—	—	—	—	—	—	—	0.017	0.018	0.019	0.02
	Enthalpy (kJ/kg)	—	—	—	—	—	—	—	418.6	427.5	436.1	444.6
	Entropy (kJ/kg·K)	—	—	—	—	—	—	—	1.711	1.739	1.765	1.79
1500	Volume (m³/kg)	—	—	—	—	—	—	—	0.016	0.017	0.018	0.019
	Enthalpy (kJ/kg)	—	—	—	—	—	—	—	416.5	425.7	434.6	443.2
	Entropy (kJ/kg·K)	—	—	—	—	—	—	—	1.699	1.728	1.755	1.781
1750	Volume (m³/kg)	—	—	—	—	—	—	—	—	0.014	0.015	0.015
	Enthalpy (kJ/kg)	—	—	—	—	—	—	—	—	421.1	430.6	439.7
	Entropy (kJ/kg·K)	—	—	—	—	—	—	—	—	1.702	1.731	1.758
2000	Volume (m³/kg)	—	—	—	—	—	—	—	—	—	0.012	0.013
	Enthalpy (kJ/kg)	—	—	—	—	—	—	—	—	—	426.3	436
	Entropy (kJ/kg·K)	—	—	—	—	—	—	—	—	—	1.708	1.737
2250	Volume (m³/kg)	—	—	—	—	—	—	—	—	—	0.01	0.011
	Enthalpy (kJ/kg)	—	—	—	—	—	—	—	—	—	421.3	431.9
	Entropy (kJ/kg·K)	—	—	—	—	—	—	—	—	—	1.685	1.716
2500	Volume (m³/kg)	—	—	—	—	—	—	—	—	—	—	0.009
	Enthalpy (kJ/kg)	—	—	—	—	—	—	—	—	—	—	427.4
	Entropy (kJ/kg·K)	—	—	—	—	—	—	—	—	—	—	1.696
2750	Volume (m³/kg)	—	—	—	—	—	—	—	—	—	—	0.008
	Enthalpy (kJ/kg)	—	—	—	—	—	—	—	—	—	—	422.2
	Entropy (kJ/kg·K)	—	—	—	—	—	—	—	—	—	—	1.674

Source: Courtesy of ASHRAE, *Handbook of Fundamentals*, American Society of Heating, Refrigerating and Air-Conditioning Engineers, Atlanta, GA, 1997. With permission.

TABLE A.8

(IP Units) Properties of Superheated Vapor of R-134a

Pressure (psia)	Properties	Temperature −22°F	−4°F	14°F	32°F	50°F	68°F	86°F	104°F	122°F	140°F	158°F
11.60	Volume (ft³/lb)	3.819	4.001	4.181	4.358	4.533	4.706	4.878	5.048	5.218	5.387	5.555
	Enthalpy (Btu/lb)	99.9	103.3	106.7	110.3	113.9	117.5	121.3	125.1	129	132.9	136.9
	Entropy (Btu/lb·R)	0.229	0.236	0.244	0.251	0.258	0.265	0.272	0.279	0.286	0.293	0.299
13.05	Volume (ft³/lb)	—	3.541	3.703	3.861	4.018	4.173	4.327	4.479	4.631	4.782	4.932
	Enthalpy (Btu/lb)	—	103.1	106.6	110.1	113.8	117.4	121.2	125	128.9	132.9	136.9
	Entropy (Btu/lb·R)	—	0.234	0.241	0.249	0.256	0.263	0.27	0.277	0.284	0.29	0.297
14.50	Volume (ft³/lb)	—	3.173	3.32	3.464	3.607	3.747	3.886	4.024	4.161	4.297	4.433
	Enthalpy (Btu/lb)	—	103	106.5	110	113.7	117.3	121.1	124.9	128.8	132.8	136.8
	Entropy (Btu/lb·R)	—	0.231	0.239	0.246	0.254	0.261	0.268	0.275	0.281	0.288	0.295
14.70	Volume (ft³/lb)	—	3.129	3.275	3.418	3.558	3.697	3.834	3.971	4.106	4.241	4.374
	Enthalpy (Btu/lb)	—	102.9	106.4	110	113.6	117.3	121.1	124.9	128.8	132.8	136.8
	Entropy (Btu/lb·R)	—	0.231	0.239	0.246	0.253	0.26	0.267	0.274	0.281	0.288	0.294
17.40	Volume (ft³/lb)	—	2.62	2.745	2.868	2.989	3.108	3.225	3.342	3.457	3.571	3.685
	Enthalpy (Btu/lb)	—	102.6	106.2	109.8	113.4	117.2	120.9	124.8	128.7	132.7	136.7
	Entropy (Btu/lb·R)	—	0.227	0.235	0.242	0.25	0.257	0.264	0.271	0.278	0.284	0.291
20.31	Volume (ft³/lb)	—	—	2.335	2.442	2.548	2.651	2.753	2.854	2.954	3.052	3.151
	Enthalpy (Btu/lb)	—	—	105.9	109.5	113.2	117	120.8	124.6	128.6	132.5	136.6
	Entropy (Btu/lb·R)	—	—	0.231	0.239	0.246	0.254	0.261	0.268	0.274	0.281	0.288
26.11	Volume (ft³/lb)	—	—	1.787	1.874	1.959	2.042	2.123	2.203	2.282	2.361	2.438
	Enthalpy (Btu/lb)	—	—	105.3	109	112.8	116.6	120.4	124.3	128.3	132.3	136.4
	Entropy (Btu/lb·R)	—	—	0.226	0.233	0.241	0.248	0.255	0.262	0.269	0.276	0.283
29.01	Volume (ft³/lb)	—	—	1.595	1.675	1.753	1.829	1.903	1.976	2.047	2.118	2.189
	Enthalpy (Btu/lb)	—	—	105	108.8	112.6	116.4	120.3	124.2	128.1	132.2	136.3
	Entropy (Btu/lb·R)	—	—	0.223	0.231	0.238	0.246	0.253	0.26	0.267	0.274	0.281
36.26	Volume (ft³/lb)	—	—	—	1.316	1.381	1.444	1.506	1.566	1.624	1.682	1.74
	Enthalpy (Btu/lb)	—	—	—	108.1	112	115.9	119.8	123.8	127.8	131.9	136
	Entropy (Btu/lb·R)	—	—	—	0.226	0.233	0.241	0.248	0.255	0.262	0.269	0.276
43.51	Volume (ft³/lb)	—	—	—	—	1.133	1.188	1.241	1.292	1.342	1.392	1.44
	Enthalpy (Btu/lb)	—	—	—	—	111.4	115.4	119.4	123.4	127.4	131.5	135.7
	Entropy (Btu/lb·R)	—	—	—	—	0.229	0.236	0.244	0.251	0.258	0.265	0.272
50.76	Volume (ft³/lb)	—	—	—	—	0.955	1.004	1.051	1.096	1.141	1.184	1.226
	Enthalpy (Btu/lb)	—	—	—	—	110.8	114.9	118.9	123	127.1	131.2	135.4
	Entropy (Btu/lb·R)	—	—	—	—	0.225	0.233	0.24	0.248	0.255	0.262	0.269
58.02	Volume (ft³/lb)	—	—	—	—	0.822	0.866	0.909	0.95	0.989	1.028	1.066
	Enthalpy (Btu/lb)	—	—	—	—	110.2	114.3	118.4	122.6	126.7	130.9	135.1
	Entropy (Btu/lb·R)	—	—	—	—	0.221	0.229	0.237	0.244	0.252	0.259	0.266
65.27	Volume (ft³/lb)	—	—	—	—	—	0.758	0.798	0.835	0.871	0.907	0.941
	Enthalpy (Btu/lb)	—	—	—	—	—	113.8	118	122.1	126.3	130.6	134.8
	Entropy (Btu/lb·R)	—	—	—	—	—	0.226	0.234	0.242	0.249	0.256	0.263
72.52	Volume (ft³/lb)	—	—	—	—	—	0.672	0.709	0.744	0.777	0.809	0.841
	Enthalpy (Btu/lb)	—	—	—	—	—	113.2	117.5	121.7	126	130.2	134.5
	Entropy (Btu/lb·R)	—	—	—	—	—	0.223	0.231	0.239	0.246	0.254	0.261
87.02	Volume (ft³/lb)	—	—	—	—	—	—	0.575	0.606	0.635	0.663	0.691
	Enthalpy (Btu/lb)	—	—	—	—	—	—	116.4	120.8	125.2	129.5	133.9
	Entropy (Btu/lb·R)	—	—	—	—	—	—	0.226	0.234	0.242	0.249	0.256
101.53	Volume (ft³/lb)	—	—	—	—	—	—	0.478	0.507	0.533	0.559	0.583
	Enthalpy (Btu/lb)	—	—	—	—	—	—	115.3	119.9	124.4	128.8	133.3
	Entropy (Btu/lb·R)	—	—	—	—	—	—	0.222	0.23	0.238	0.245	0.253

(*Continued*)

TABLE A.8 (*Continued*)

(IP Units) Properties of Superheated Vapor of R-134a

Pressure (psia)	Properties	Temperature										
		−22°F	−4°F	14°F	32°F	50°F	68°F	86°F	104°F	122°F	140°F	158°F
116.03	Volume (ft³/lb)	—	—	—	—	—	—	—	0.432	0.457	0.48	0.502
	Enthalpy (Btu/lb)	—	—	—	—	—	—	—	118.9	123.5	128.1	132.6
	Entropy (Btu/lb·R)	—	—	—	—	—	—	—	0.226	0.234	0.242	0.249
130.54	Volume (ft³/lb)	—	—	—	—	—	—	—	0.373	0.397	0.419	0.439
	Enthalpy (Btu/lb)	—	—	—	—	—	—	—	117.9	122.6	127.3	131.9
	Entropy (Btu/lb·R)	—	—	—	—	—	—	—	0.222	0.231	0.238	0.246
145.04	Volume (ft³/lb)	—	—	—	—	—	—	—	0.326	0.348	0.369	0.389
	Enthalpy (Btu/lb)	—	—	—	—	—	—	—	116.8	121.7	126.5	131.2
	Entropy (Btu/lb·R)	—	—	—	—	—	—	—	0.219	0.227	0.235	0.243
159.54	Volume (ft³/lb)	—	—	—	—	—	—	—	—	0.308	0.329	0.347
	Enthalpy (Btu/lb)	—	—	—	—	—	—	—	—	120.7	125.7	130.5
	Entropy (Btu/lb·R)	—	—	—	—	—	—	—	—	0.224	0.232	0.24
174.05	Volume (ft³/lb)	—	—	—	—	—	—	—	—	0.275	0.294	0.312
	Enthalpy (Btu/lb)	—	—	—	—	—	—	—	—	119.7	124.8	129.8
	Entropy (Btu/lb·R)	—	—	—	—	—	—	—	—	0.221	0.23	0.238
188.55	Volume (ft³/lb)	—	—	—	—	—	—	—	—	0.246	0.265	0.283
	Enthalpy (Btu/lb)	—	—	—	—	—	—	—	—	118.6	123.9	129
	Entropy (Btu/lb·R)	—	—	—	—	—	—	—	—	0.218	0.227	0.235
203.06	Volume (ft³/lb)	—	—	—	—	—	—	—	—	—	0.24	0.257
	Enthalpy (Btu/lb)	—	—	—	—	—	—	—	—	—	123	128.2
	Entropy (Btu/lb·R)	—	—	—	—	—	—	—	—	—	0.224	0.233
217.56	Volume (ft³/lb)	—	—	—	—	—	—	—	—	—	0.218	0.235
	Enthalpy (Btu/lb)	—	—	—	—	—	—	—	—	—	122	127.4
	Entropy (Btu/lb·R)	—	—	—	—	—	—	—	—	—	0.221	0.23
253.82	Volume (ft³/lb)	—	—	—	—	—	—	—	—	—	—	0.189
	Enthalpy (Btu/lb)	—	—	—	—	—	—	—	—	—	—	125.1
	Entropy (Btu/lb·R)	—	—	—	—	—	—	—	—	—	—	0.224

Source: Courtesy of ASHRAE, *Handbook of Fundamentals*, American Society of Heating, Refrigerating and Air-Conditioning Engineers, Atlanta, GA, 1997. With permission.

TABLE A.8

(SI Units) Properties of Superheated Vapor of R-134a

Pressure (kPa)	Properties	Temperature										
		−30°C	−20°C	−10°C	0°C	10°C	20°C	30°C	40°C	50°C	60°C	70°C
80	Volume (m³/kg)	0.238	0.249	0.261	0.272	0.282	0.293	0.304	0.315	0.325	0.336	0.346
	Enthalpy (kJ/kg)	380.2	388.1	396.2	404.4	412.8	421.3	430	438.9	447.9	457.1	466.5
	Entropy (kJ/kg·K)	1.755	1.787	1.818	1.849	1.879	1.908	1.937	1.966	1.995	2.023	2.05
90	Volume (m³/kg)	—	0.221	0.231	0.241	0.25	0.26	0.27	0.279	0.289	0.298	0.307
	Enthalpy (kJ/kg)	—	387.8	395.9	404.2	412.6	421.1	429.8	438.7	447.8	457	466.3
	Entropy (kJ/kg·K)	—	1.776	1.807	1.838	1.868	1.898	1.927	1.956	1.985	2.013	2.04
100	Volume (m³/kg)	—	0.198	0.207	0.216	0.225	0.233	0.242	0.251	0.259	0.268	0.276
	Enthalpy (kJ/kg)	—	387.4	395.6	403.9	412.3	420.9	429.6	438.5	447.6	456.8	466.2
	Entropy (kJ/kg·K)	—	1.766	1.798	1.829	1.859	1.889	1.918	1.947	1.976	2.004	2.031
101	Volume (m³/kg)	—	0.195	0.204	0.213	0.222	0.23	0.239	0.247	0.256	0.264	0.273
	Enthalpy (kJ/kg)	—	387.4	395.5	403.8	412.3	420.9	429.6	438.5	447.6	456.8	466.2
	Entropy (kJ/kg·K)	—	1.765	1.797	1.828	1.858	1.888	1.917	1.946	1.975	2.003	2.03
120	Volume (m³/kg)	—	0.163	0.171	0.179	0.186	0.194	0.201	0.208	0.215	0.223	0.23
	Enthalpy (kJ/kg)	—	386.7	394.9	403.3	411.8	420.5	429.2	438.2	447.3	456.5	465.9
	Entropy (kJ/kg·K)	—	1.749	1.781	1.812	1.843	1.873	1.902	1.931	1.96	1.988	2.016
140	Volume (m³/kg)	—	—	0.145	0.152	0.159	0.165	0.172	0.178	0.184	0.19	0.196
	Enthalpy (kJ/kg)	—	—	394.3	402.7	411.3	420	428.9	437.8	447	456.2	465.7
	Entropy (kJ/kg·K)	—	—	1.767	1.798	1.829	1.859	1.889	1.918	1.947	1.975	2.003
180	Volume (m³/kg)	—	—	0.111	0.117	0.122	0.127	0.132	0.137	0.142	0.147	0.152
	Enthalpy (kJ/kg)	—	—	392.9	401.6	410.3	419.1	428.1	437.1	446.3	455.7	465.2
	Entropy (kJ/kg·K)	—	—	1.742	1.774	1.806	1.836	1.866	1.896	1.925	1.953	1.981
200	Volume (m³/kg)	—	—	0.099	0.104	0.109	0.114	0.119	0.123	0.128	0.132	0.136
	Enthalpy (kJ/kg)	—	—	392.3	401	409.8	418.7	427.7	436.8	446	455.4	464.9
	Entropy (kJ/kg·K)	—	—	1.732	1.764	1.796	1.827	1.857	1.886	1.915	1.944	1.972
250	Volume (m³/kg)	—	—	—	0.082	0.086	0.09	0.094	0.098	0.101	0.105	0.108
	Enthalpy (kJ/kg)	—	—	—	399.5	408.5	417.5	426.6	435.9	445.2	454.6	464.2
	Entropy (kJ/kg·K)	—	—	—	1.742	1.774	1.805	1.836	1.866	1.895	1.924	1.953
300	Volume (m³/kg)	—	—	—	—	0.071	0.074	0.077	0.081	0.084	0.087	0.09
	Enthalpy (kJ/kg)	—	—	—	—	407.1	416.3	425.6	434.9	444.4	453.9	463.6
	Entropy (kJ/kg·K)	—	—	—	—	1.756	1.788	1.819	1.849	1.879	1.908	1.936
350	Volume (m³/kg)	—	—	—	—	0.06	0.063	0.065	0.068	0.071	0.074	0.076
	Enthalpy (kJ/kg)	—	—	—	—	405.7	415.1	424.5	434	443.5	453.2	462.9
	Entropy (kJ/kg·K)	—	—	—	—	1.739	1.772	1.803	1.834	1.864	1.893	1.922
400	Volume (m³/kg)	—	—	—	—	0.051	0.054	0.057	0.059	0.062	0.064	0.066
	Enthalpy (kJ/kg)	—	—	—	—	404.2	413.8	423.4	433	442.7	452.4	462.2
	Entropy (kJ/kg·K)	—	—	—	—	1.724	1.758	1.79	1.821	1.851	1.881	1.91
450	Volume (m³/kg)	—	—	—	—	—	0.047	0.05	0.052	0.054	0.056	0.059
	Enthalpy (kJ/kg)	—	—	—	—	—	412.6	422.3	432	441.8	451.6	461.5
	Entropy (kJ/kg·K)	—	—	—	—	—	1.745	1.777	1.809	1.84	1.869	1.899
500	Volume (m³/kg)	—	—	—	—	—	0.042	0.044	0.046	0.048	0.05	0.052
	Enthalpy (kJ/kg)	—	—	—	—	—	411.2	421.2	431	440.9	450.8	460.8
	Entropy (kJ/kg·K)	—	—	—	—	—	1.732	1.766	1.798	1.829	1.859	1.889
600	Volume (m³/kg)	—	—	—	—	—	—	0.036	0.038	0.04	0.041	0.043
	Enthalpy (kJ/kg)	—	—	—	—	—	—	418.8	429	439.1	449.2	459.4
	Entropy (kJ/kg·K)	—	—	—	—	—	—	1.745	1.778	1.81	1.841	1.871
700	Volume (m³/kg)	—	—	—	—	—	—	0.03	0.032	0.033	0.035	0.036
	Enthalpy (kJ/kg)	—	—	—	—	—	—	416.2	426.8	437.2	447.6	457.9
	Entropy (kJ/kg·K)	—	—	—	—	—	—	1.726	1.76	1.793	1.824	1.855

(Continued)

TABLE A.8 (*Continued*)

(SI Units) Properties of Superheated Vapor of R-134a

Pressure (kPa)	Properties	Temperature −30°C	−20°C	−10°C	0°C	10°C	20°C	30°C	40°C	50°C	60°C	70°C
800	Volume (m³/kg)	—	—	—	—	—	—	—	0.027	0.028	0.03	0.031
	Enthalpy (kJ/kg)	—	—	—	—	—	—	—	424.5	435.3	445.9	456.4
	Entropy (kJ/kg·K)	—	—	—	—	—	—	—	1.743	1.777	1.809	1.841
900	Volume (m³/kg)	—	—	—	—	—	—	—	0.023	0.025	0.026	0.027
	Enthalpy (kJ/kg)	—	—	—	—	—	—	—	422.1	433.2	444.1	454.8
	Entropy (kJ/kg·K)	—	—	—	—	—	—	—	1.728	1.763	1.796	1.828
1000	Volume (m³/kg)	—	—	—	—	—	—	—	0.02	0.022	0.023	0.024
	Enthalpy (kJ/kg)	—	—	—	—	—	—	—	419.6	431.1	442.2	453.2
	Entropy (kJ/kg·K)	—	—	—	—	—	—	—	1.713	1.749	1.783	1.815
1100	Volume (m³/kg)	—	—	—	—	—	—	—	—	0.019	0.02	0.022
	Enthalpy (kJ/kg)	—	—	—	—	—	—	—	—	428.8	440.3	451.6
	Entropy (kJ/kg·K)	—	—	—	—	—	—	—	—	1.735	1.771	1.804
1200	Volume (m³/kg)	—	—	—	—	—	—	—	—	0.017	0.018	0.019
	Enthalpy (kJ/kg)	—	—	—	—	—	—	—	—	426.4	438.3	449.8
	Entropy (kJ/kg·K)	—	—	—	—	—	—	—	—	1.722	1.759	1.793
1300	Volume (m³/kg)	—	—	—	—	—	—	—	—	0.015	0.017	0.018
	Enthalpy (kJ/kg)	—	—	—	—	—	—	—	—	423.8	436.2	448.1
	Entropy (kJ/kg·K)	—	—	—	—	—	—	—	—	1.709	1.747	1.782
1400	Volume (m³/kg)	—	—	—	—	—	—	—	—	—	0.015	0.016
	Enthalpy (kJ/kg)	—	—	—	—	—	—	—	—	—	434	446.2
	Entropy (kJ/kg·K)	—	—	—	—	—	—	—	—	—	1.736	1.772
1500	Volume (m³/kg)	—	—	—	—	—	—	—	—	—	0.014	0.015
	Enthalpy (kJ/kg)	—	—	—	—	—	—	—	—	—	431.6	444.2
	Entropy (kJ/kg·K)	—	—	—	—	—	—	—	—	—	1.724	1.762
1750	Volume (m³/kg)	—	—	—	—	—	—	—	—	—	—	0.012
	Enthalpy (kJ/kg)	—	—	—	—	—	—	—	—	—	—	438.9
	Entropy (kJ/kg·K)	—	—	—	—	—	—	—	—	—	—	1.737

Source: Courtesy of ASHRAE, *Handbook of Fundamentals*, American Society of Heating, Refrigerating and Air-Conditioning Engineers, Atlanta, 1997. With permission.

Index

I

R

S

T